ATLAS OF MICROSURGERY OF THE LATERAL SKULL BASE

Editors

Mario Sanna
Wenlong Tang
Gianluca Piras

With the collaboration of

Abdelkader Taibah
Alessandra Russo
Fernando Mancini
Enrico Piccirillo
Antonio Caruso
Lorenzo Lauda
Annalisa Giannuzzi
Diana Vlad

ATLAS OF MICROSURGERY OF THE LATERAL SKULL BASE

人民卫生出版社
PMPH PEOPLE'S MEDICAL PUBLISHING HOUSE

ACADEMIC PRESS
An imprint of Elsevier

Academic Press is an imprint of Elsevier
125 London Wall, London EC2Y 5AS, United Kingdom
525 B Street, Suite 1650, San Diego, CA 92101, United States
50 Hampshire Street, 5th Floor, Cambridge, MA 02139, United States

ISBN 978-0-443-19083-4

For information on all Academic Press publications
visit our website at https://www.elsevier.com/books-and-journals

Publisher: Megan Ball
Acquisitions Editor: Megan McManus
Editorial Project Manager: Himanshi Yadav
Production Project Manager: Punithavathy Govindaradjane
Cover Designer: Mark Rogers

Typeset by STRAIVE, India

Printed in Great Britain
Last digit is the print number: 10 9 8 7 6 5 4 3 2

CONTENTS

AUTHOR BIOGRAPHIES

Mario Sanna, MD
Professor of Otolaryngology
Director, Gruppo Otologico
Founder, Mario Sanna Foundation
Piacenza and Rome, Italy

Mario Sanna is the Founder and Director of the Gruppo Otologico (Piacenza–Rome, Italy), an internationally acclaimed center for otology, hearing implantology, and skull base surgeries. Prof. Sanna is a master in his field, known across the world for his pioneering work in otology and skull base surgery. He is a founding member of the European Skull Base Society and the Italian Skull Base Society, and an honorary member of the French and Spanish Otolaryngological Societies. Prof. Sanna has authored 15 books and over 375 articles in peer-reviewed journals. His illustrated textbooks are extremely popular, and some of them have won international awards. He has been an invited faculty at over 400 conferences and academic meetings. Over 200 fellows and 1000 observers have been trained at the Gruppo Otologico. Prof. Sanna has contributed tremendously to the knowledge of otology and skull base surgery and is considered by his peers to be one of the pioneers of the modern skull base surgery.

Wenlong Tang, MD
Otologist and Skull Base Surgeon
Shenzhen Longgang Otolaryngology Hospital
Shenzhen Otolaryngology Research Institute
Shenzhen, China

Wenlong Tang is a master supervisor, neurosurgeon, otologist, and skull base surgeon at Shenzhen Longgang Otolaryngology Hospital. Director of the Anatomical Research Department at Shenzhen Otolaryngology Research Institute. Director of Otolaryngology, Head and Neck Surgery Anatomical Training Center, Vice Director of the Institute of Skull Base Surgery at the Affiliated Heping Hospital of Changzhi Medical College. Dr. Tang earned his B.S. degree at Changzhi Medical College in 2012 and his M.S. degree at Zunyi Medical College in 2016, supervised by Prof. Shuqi Qiu, continued with his clinical fellowship at the Gruppo Otologico, Piacenza–Rome, Italy, from 2016 to 2017, instructed by Prof. Mario Sanna. He authored *Atlas of Lateral Skull Base Microsurgical Anatomy* (PMPH, 2015), *The Temporal Bone: Anatomical Dissection and Surgical Approaches* (Thieme, 2018), and *Atlas of Temporal Bone and Lateral Skull Base Surgical Approaches* (PMPH, 2020). As an Executive Committee Member of the ENT, Head and Neck Surgery Branch of the Chinese Society of Anatomy, Executive Committee Member of the Neurosurgery Branch of the Chinese Society of Anatomy, and Executive Committee Member of the Clinical Neuroanatomy Branch of the Chinese Society of Anatomy. Dr. Tang has been the leading investigator in multiple fundamental research projects at the level of both municipalities and Guangdong–Hong Kong–Macao Greater Bay

Area, and participated in one joint research project of the National Natural Science Foundation of China and has completed the translation of seven books and the publication of seven articles in core journals. meanwhile, hosted six national temporal bone dissection training courses. Dr. Tang specializes in the clinical applications of cerebral and skull base anatomic techniques, with his research interests lying in lateral skull base microsurgical anatomy and endoscopic skull base anatomy. Engaged in clinical anatomy and application research of the skull and base for more than 15 years, specializing in the diagnosis and treatment of diseases in the middle ear and skull base region.

Gianluca Piras, MD
Otologist and Skull Base Surgeon
Gruppo Otologico
Piacenza and Rome, Italy

Gianluca Piras graduated in medicine, specializing in Otolaryngology-Head and Neck Surgery, from the University of Bologna. He also worked as a researcher and tutor in microsurgical techniques at the Centre for Applied Biomedical Research (CRBA), University of Bologna. He was certified as an Associate Professor of Otolaryngology and Audiology by the Italian National Scientific Board. He has been working in the Gruppo Otologico since 2015 as an otologist and lateral skull base surgeon. His focus is on posterior fossa and skull base tumors, skull base paragangliomas and hearing implantology. He has been a speaker and faculty member at the most important international conferences on lateral skull base surgery. He is the author of more than 30 publications in scientific journals, six book chapters, and co-author of three books published by Thieme (Color Atlas of Endo-Otoscopy. Examination-Diagnosis-Treatment; The Temporal Bone: Anatomical Dissection and Surgical Approaches; Microsurgical Management of Middle Ear and Petrous Bone Cholesteatoma), translated into Chinese.

CONTRIBUTING AUTHORS

MAIN AUTHORS

Mario Sanna, MD
Professor of Otolaryngology
Director, Gruppo Otologico
Founder, Mario Sanna Foundation
Piacenza and Rome, Italy

Wenlong Tang, MD
Neurosurgeon, Otologist and Skull Base Surgeon
Shenzhen Longgang Otolaryngology Hospital
Shenzhen Otolaryngology Research Institute
Shenzhen, China

Gianluca Piras, MD
Otologist and Skull Base Surgeon
Gruppo Otologico
Piacenza and Rome, Italy

WITH THE COOPERATION OF

Abdelkader Taibah, MD
Neurosurgeon, Otologist, and Skull Base Surgeon
Gruppo Otologico
Piacenza and Rome, Italy

Alessandra Russo, MD
Otologist and Skull Base Surgeon
Gruppo Otologico
Piacenza and Rome, Italy

Fernando Mancini, MD
Otologist and Skull Base Surgeon
Gruppo Otologico
Piacenza and Turin, Italy

Enrico Piccirillo, MD
Otologist and Skull Base Surgeon
Gruppo Otologico
Piacenza and Rome, Italy

Antonio Caruso, MD
Otologist and Skull Base Surgeon
Gruppo Otologico
Piacenza and Rome, Italy

Lorenzo Lauda, MD
Otologist and Skull Base Surgeon
Gruppo Otologico
Piacenza and Rome, Italy

Annalisa Giannuzzi, MD
Otologist and Skull Base Surgeon
Gruppo Otologico
Piacenza and Rome, Italy

Diana Vlad, MD
ENT, Otologist
Military Hospital “Dr. Constantin Papilian”
Cluj Napoca, Romania

PREFACE

This volume represents the third edition of our "long journey" within lateral skull base surgery. A long journey because the Gruppo Otologico, founded in 1983, has now blown out 40 candles. Much has changed in these 40 years. Skull base surgery has gone from being a pioneering and dangerous surgery, complicated and reserved for a few centers, to a global expansion. This expansion is also due to the work we have done all these years and has led to the welcoming of thousands of fellows and observers from all over the world. It is a great satisfaction to see our surgical principles applied from India to China, from Japan to Australia, and from Europe to the Arab countries. In this context, the fruitful collaboration we have established with Dr. Wenlong Tang, our Fellow for about 2 years, and with the Shenzhen Center, will see the rise of a new "rib" of the Gruppo Otologico. Therefore we would also like to thank People's Medical Publishing House, which, together with Elsevier, has made the publication of this book possible, taking up the baton of the fruitful past collaboration with Thieme.

None of this would have been possible without the masters before us somehow leading the way.

William House was the pioneer of skull base surgery and the first to use transtemporal approaches to the cerebellopontine angle. Ugo Fisch, the father of the modern skull base surgery, for systematizing approaches to the infratemporal fossa and improving surgical control of the facial nerve and the internal carotid artery.

This volume complements what has been published previously. The various surgical approaches to the lateral skull base are described in detail with excellent step-by-step dissections accompanied by illustrative surgical cases. All this is complemented by clinical and radiological diagnosis of the various pathologies of the lateral skull base. It is hoped that, thanks to this volume, surgery on the lateral skull base will become even more routine and standardized, but performed in organized centers that can guarantee an adequate surgical volume and trained doctors. Centralization of pathology will become a fundamental requirement for the proper performance of this complex surgery. This will require multidisciplinarity, especially between otosurgeons and neurosurgeons, making these two figures interchangeable but indispensable. It is also worth mentioning the crucial role of the neuroradiologist, especially the interventionalist, who has made it possible to operate on lesions that were previously considered inoperable.

The introduction of the internal carotid artery stenting in paraganglioma surgery has been a real breakthrough in the management of this complex anatomy. I would therefore like to thank the late Dr. Paolo Piazza for all the pioneering work we have done together. The importance of interventional neuroradiology is reiterated in the final chapters of the book.

This volume is the fruit of the invaluable work of all the members of the Gruppo Otologico, which has allowed the center to become a world reference in otology and lateral skull base surgery.

Finally, special thanks are due to our families for their support and patience in carrying out such a complex and important work.

Mario Sanna
Wenlong Tang
Gianluca Piras

CHAPTER 1

Historical Overview on Lateral Skull Base Surgery

Lateral skull base (LSB) surgery has had an interesting history since its inception in the late nineteenth century, with major advances in anesthesiology, diagnostic and interventional radiology, and the introduction of multidisciplinary treatment teams. Initially considered inoperable, LSB lesions, when removed surgically in the early days, were associated with a mortality rate of 80%, which has gradually decreased to 0.5% and less in current practice, thanks to the efforts of several pioneering surgeons [1]. In modern medicine, the philosophy has shifted significantly toward cranial nerve preservation and functional rehabilitation.

Descriptions of LSB tumors date back to the mid-1800s when Eduard Sandifort described what was thought to be a vestibular schwannoma (VS) during an autopsy, while Valentine described the first tympanojugular paraganglioma (TJP) in 1840 [2]. With the introduction of general anesthesia by Morton (1946), antiseptic operating conditions and X-rays by Roentgen, brain surgery took a completely different course. From its birth, skull base surgery was naturally the domain of neurosurgeons, starting in 1893 when Sir William Macewen published his book "Pyogenic Infective Diseases of the Brain and Spinal Cord" [3] and is considered the first true skull base surgeon. The following year, Charles Balance performed the first successful excision of an LSB tumor, thought to be a neurinoma or possibly a meningioma, using a suboccipital cranial flap [4]. The patient survived with permanent facial and trigeminal nerve palsy. In the following years, Hartley and Feodor Krause (1857–1937) introduced the middle cranial fossa approach (known as the Hartley-Krause procedure) for the resection of trigeminal nerve rootles in patients with trigeminal neuralgia [5]. In the following years, the procedure became the treatment of choice for this condition, with a cure rate of up to 99%. Krause himself also performed the suboccipital approach for cerebellopontine angle tumors and facial nerve resection for intractable tinnitus.

The most important figure in the development of LSB surgery is Harvey Cushing (1869–1939) (Fig. 1.1), considered the founder of modern neurosurgery. A brilliant mind, he introduced the use of surgical clips, intracapsular dissection of acoustic tumors, and Bovie's monopolar diathermy to neurosurgery. With the use of electrocautery, he was able to perform more complex neurosurgical procedures that had previously been discouraged because of the high risk of hemorrhage. In 1917 he published his work "Tumors of the Nervus Acusticus and the Syndrome of the Cerebello-Pontine Angle." Like Krause, he was a strong advocate of the use of radiography to diagnose intracranial tumors and plan their removal [6]. He postulated the practice of a scrupulous surgical technique using an extended bilateral suboccipital approach and subtotal intracapsular tumor resection to preserve the nervous structures and brainstem vasculature intact [1]. He also emphasized the importance of gentle cerebellar retraction and the use of supratentorial ventricular puncture to reduce CSF pressure and increase the surgical space. During his career, he reduced the mortality rate for VS surgery from 80% to 20%.

Atlas of Microsurgery of the Lateral Skull Base. https://doi.org/10.1016/B978-0-443-19083-4.00001-5

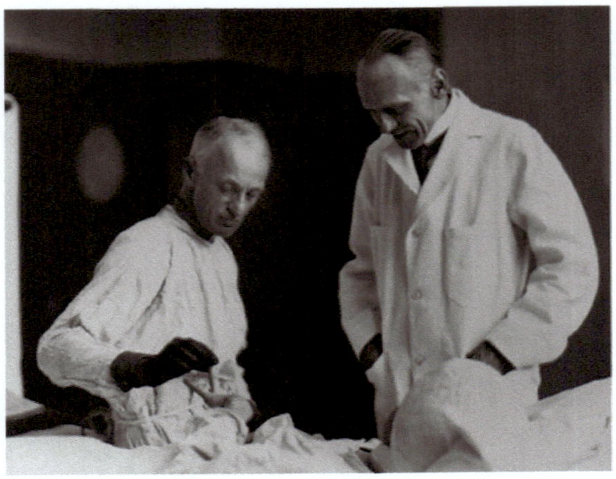

Fig. 1.1 Picture of Cushing performing a postsurgical examination with Otfrid Forrester (a visiting surgeon) [6].

Continuing Cushing's work, his student Walter Dandy promoted the practice of the unilateral suboccipital technique and total capsular tumor dissection. Improving his masters' techniques and aided by advances in diagnostic equipment, he achieved a mortality rate of 10.87% after the complete removal of 46 VS [7]. Nevertheless, the cerebellopontine angle was a major challenge for neurosurgeons at the time (Cushing called it the "bloody angle"), and total facial paralysis was an inherent postoperative complication because the outer cap of the cerebellum was usually sacrificed to achieve tumor excision. Olivercrona was perhaps the first surgeon to emphasize the preservation of the facial nerve, achieving a 65% preservation rate during his practice, considered an extraordinary result for the time. In the early and mid-twentieth century, there were some attempts by a few otologists, such as Rudolf Panse, Franciscus Hubertus Quix, and Schmiegelow, to introduce and promote the transtemporal translabirynthine technique for VS removal, but they were categorically overshadowed and rejected by the neurosurgeons of the time, who considered the approach dangerous and lacking in exposure. Gradually, advances in audiology, otology, and radiology made it possible to diagnose early tumors, but neurosurgeons were reluctant to operate on these lesions because the associated morbidity was too high. Over time, the removal of low-grade VS became the goal of otologists due to early diagnosis, the use of the microscope, and the introduction of the motor drill into surgical practice.

The microscope was first used in ear surgery in 1921 by the Swedish otologist Carl Olof Nylen [8]. Gunnar Holmgren, recognizing the great advantages of magnification, quickly replaced the monocular model with the binocular microscope (Fig. 1.2), which he developed in 1922 [9]. This was later improved and perfected by Littmann and Zeiss in the 1950s. Holmgren was also perhaps the first surgeon to use a dental drill in otological and skull base surgery (1923).

The transition from neurosurgery to microscopic neurotology was significantly influenced by William House (1923–2012) (Fig. 1.3), who was deeply concerned by the mortality and morbidity associated with VS surgery and

Fig. 1.2 Holmgren's binocular microscope [8].

Fig. 1.3 Dr. William F. House at an operating microscope.

believed that early intervention would lead to better outcomes. He was the first to introduce the microscope into the management of VS [10], allowing better visualization of the neurovascular elements within the CPA and complete tumor resection with reduced morbidity [11,12]. Facial nerve preservation after tumor resection became possible for the first time. His harmonious collaboration with neurosurgeon Bill Hitselberger, unique at the time, led to the resurrection of the translabyrinthine procedure. In 1968, they reported a series of 200 cases treated with the TLA technique over 7 years, with a mortality rate of 7% and a facial nerve preservation rate of 95%. These were more than remarkable results at the time [13]. With experience and surgical refinement, outcomes have gradually improved [14]. During his long and distinguished career, House introduced other approaches to the removal of skull base tumors, including the transcochlear approach, which allowed access to the clivus, the extended middle fossa (1976), and pioneering work on cochlear implantation [15,16]. An engineer with whom he worked closely, Jack Urban, developed a device that increased the speed of the Jordan-Day otological drill, greatly improving its efficiency. The use of the microscope also enabled the development of hearing preservation surgery in the form of the middle fossa and the modern retrosigmoid approach.

Another major figure in LSB, Ugo Fisch, introduced and promoted the transotic approach to CPA surgery. By drilling the cochlea, he gained access to excise tumors with anterior extension, but gradually he applied the technique to all cases of VS [17,18].

In the years that followed, many centers around the world adopted and refined House's techniques, achieving excellent results. The senior author of this book is credited with developing the transapical extension of the TLA approach and many other techniques described in this book, having performed over 3500 procedures for VS to date [19–21]. The mortality rate recorded by the center is less than 0.1% with a facial nerve preservation rate of 96%. Meanwhile, other surgical groups were working on another important pathological entity, the pathology of glomus tumors. In 1945, Rosenwasser was the first to attempt radical removal of a glomus tumor by performing a radical mastoidectomy, describing a "carotid-like" tumor of the middle ear [22]. Gradually, together with other surgeons, he became one of the first to offer a systemic surgical management of these tumors during his 25 years of experience in the field [23]. Middle ear tumors were removed by a transtympanic approach, tumors with extension to the mastoid by mastoidectomy, and tumors with jugular foramen or intracranial extension by wide resection with control of the neck vessels and postoperative radiation in case of incomplete removal. Capps (1952) was the first to report anterior transposition of the facial nerve with the closure of the sigmoid sinus and ligation of the internal jugular vein (IJV) for removal of TJP. However, due to significant complications, he did not advocate this approach, and radiotherapy became the mainstay of therapy for glomus tumors for many years [24]. In 1953, Guild published findings on 248 glomus bodies, 135 along the Jacobson nerve (tympanic branch of the glossopharyngeal nerve) and 113 along the Arnold nerve (auricular branch of the vagus nerve) in 88 temporal bones. Each body was ~0.5–0.25 mm in size, with the greatest number appearing in middle age, with a marked decrease in number after the age of 60 [25].

Having gained experience with this challenging lesion, in 1964 Shapiro and Neues described the combination of mastoidectomy, anterior facial nerve transposition, and neck exposure with ligation of the IJV and arterial branches to the tumor [26]. These authors pioneered the use of severe hypotension and hypothermia as critical factors in avoiding death from severe hemorrhage. Glasscock et al. reported their approach using both a combination of Shapiro's wide exposure of the skull base and House's extended facial recess procedure [27]. He was also the first to distinguish a glomus tympanicum from a glomus jugulare. A notable historical contribution to this field was made by Ugo Fisch (Fig. 1.4), who in the 1970s published a systematic classification system for TPJ and established a management scheme for class C tumors. Together with his associates, he described the infratemporal fossa (IFT) approaches A, B, and C for the removal of type C and D TPJ and other lesions involving the jugular foramen, internal carotid artery (ICA), and petrous apex, many of which were previously considered unresectable [28,29]. The hallmarks of ITFA are anterior facial nerve rerouting, blind sac closure of the external auditory canal, and double sigmoid sinus and jugular vein ligation. The key feature was to achieve wide, uninterrupted access to the jugular fossa and proximal and distal control of the ICA. He emphasized that the ICA was the key to LSB surgery. Later refinements to his classification and technique improved the management of TJP. Kinney was the first to report the rationale for single and staged resections of intracranial extension due to the risk of CSF leak, which was further implemented into current practice [30]. In 2006, Sanna et al. reported the first series with the use of intra-arterial ICA stenting for the management of TPJ [31,32], which has been instrumental in achieving safe tumor dissection involving the ICA.

None of these surgical achievements would have been possible without the support of developments in radiology, together with a range of technical tools, including balloon occlusion testing introduced in the mid-1970s, preoperative tumor embolization, the use of electrophysiological monitoring of the facial nerve introduced by Delgado, auditory brainstem responses, continued advances in microscope technology and improved instrumentation, all of which have greatly aided tumor resection, improving safety and surgical proficiency [1]. High-resolution computed tomography and magnetic resonance imaging (MRI) techniques have revolutionized the diagnosis and management of skull base lesions. Early tumor detection, together with improved MRI resolution, has allowed clinicians to better understand the natural history of these lesions and has modified the treatment algorithm. Over time, radiotherapy became an important arm of treatment for skull base tumors with the introduction and advancement of Gamma Knife, although some controversies remain. Ongoing advances in neurointerventional radiology, surgical techniques, and even radiotherapy will continue to improve the management of these challenging lesions.

Fig. 1.4 Ugo Fisch.

REFERENCES

[1] Mcrackan TR, Brackmann DE. Historical perspective on evolution in Management of Lateral Skull Base Tumors. Otolaryngol Clin North Am 2015;48(3):397–405. https://doi.org/10.1016/j.otc.2015.02.002.

[2] Valentin G. Über eine gangliose Anschwellung in der Jacobsenschen Anastomose des Menschen. Arch Anat Physiol 1840;89:287–90.

[3] Lustig LR. Anesthesia, antisepsis, microscope: the confluence of neurotology. Otolaryngol Clin North Am 2007;40(3):415–37 [vii].

[4] Glasscock 3rd ME. History of the diagnosis and treatment of acoustic neurinoma. Arch Otolaryngol 1968;88:578–85.

[5] Rosegay H. The Krause operations. J Neurosurg 2005;102:745–52.

[6] Malekpour M, Cohen-Gadol AA. Making the "inoperable" tumors "operable": Harvey Cushing's contributions to the surgery of posterior fossa tumors. Neurosurg Focus 2014;36(4):E15. https://doi.org/10.3171/2014.2.FOCUS13580.

[7] Dandy W. Results of removal of acoustic tumors by the unilateral approach. Arch Surg 1941;42(6):1026–33. https://doi.org/10.1001/archsurg.1941.01210120061006.

[8] Mudry A. The history of the microscope for use in ear surgery. Am J Otol 2000;21(6):877–86 [PMID: 11078079].

[9] Glasscock ME, Gulya AJ. Glasscock-Shambaugh surgery of the ear. Hamilton, Ont: BC Decker; 2003.

[10] Monfared A, Mudry A, Jackler R. The history of middle cranial fossa approach to the cerebellopontine angle. Otol Neurotol 2010;31(4):691–6. https://doi.org/10.1097/MAO.0b013e3181c0e98e.

[11] House WF. Surgical exposure of internal auditory canal and its contents through the middle cranial fossa. Arch Laryngoscope 1961;71:1363–85.

[12] House WF. Middle cranial fossa approach to the petrous pyramid. Report of 50 cases. Arch Otolaryngol 1963;78:460–9.

[13] House WF, Hitselberger WW. Preservation of the facial nerve in acoustic tumor surgery. Arch Otolaryngol 1968;88:655–8.

[14] Brackmann DE, Green D. Translabyrinthine approach for acoustic tumor removal. Otolaryngol Clin North Am 1992;25:311–29.

[15] House WF, Hitselberger WE. The transcochlear approach to the skull base. Arch Otolaryngol 1976;102:334–42.

[16] House WF. Transcochlear approach to the petrous apex and clivus. Trans Sect Otolaryngol Am Acad Ophthalmol Otolaryngol 1977;84. ORL 927–931.

[17] Browne JD, Fisch U. Transotic approach to the cerebellopontine angle. Otolaryngol Clin North Am 1992;25:331–46.

[18] Jenkins HA, Fisch U. The transotic approach to resection of difficult acoustic tumors of the cerebellopontine angle. Am J Otol 1980;2:70–6.

[19] Angeli RD, Piccirillo E, Di Trapani G, Sequino G, Taibah A, Sanna M. Enlarged translabyrinthine approach with transapical extension in the management of giant vestibular schwannomas: personal experience and review of literature. Otol Neurotol 2011;32:125–31.

[20] Sanna M, Agarwal M, Mancini F, Taibah A. Transapical extension in difficult cerebellopontine angle tumors. Ann Otol Rhinol Laryngol 2004;113:676–82.

[21] Sanna M, Agarwal M, Jain Y, Russo A, Taibah AK. Transapical extension in difficult cerebellopontine angle tumours: preliminary report. J Laryngol Otol 2003;117:788–92.

[22] Rosenwasser H. Carotid body tumor of the middle ear and mastoid. Arch Otolaryngol 1945;41:64–7.

[23] Rosenwasser H. Glomus jugulare tumors. Long-termtumors. Arch Otolaryngol 1969;89(1):160–6.

[24] Capps FC. Glomus jugulare tumours of the middle ear. J Laryngol Otol 1952;66(7):302–14.

[25] Guild SR. Theglomus jugulare, an on chromaffin paraganglion in man. Ann Otol Rhinol Laryngol 1953;62(4):1045–71.

[26] Shapiro MJ, Neues DK. Technique for removal of glomus jugulare tumors. Arch Otolaryngol 1964;79:219–24.

[27] Glasscock MEIII, Harris PF, Newsome G. Glomus tumors: diagnosis and treatment. Laryngoscope 1974;84(11):2006–32.

[28] Fisch U. Infratemporal fossa approach to tumors of the temporal bone and base of the skull. J Laryngol Otol 1978;92(11):949–67.

[29] Fisch U, Pillsbury HC. Infratemporal fossa approach to lesions in the temporal bone and base of the skull. Arch Otolaryngol 1979;105:99–107.

[30] Kinney SE. Glomus jugulare tumors with intracranial extension. Am J Otol 1979;1(2):67–71.

[31] Sanna M, Khrais T, Menozi R, Piaza P. Surgical removal of jugular paragangliomas after stenting of the intratemporal internal carotid artery: a preliminary report. Laryngoscope 2006;116(5):742–6.

[32] Piazza P, DiLella F, Menozzi R, Bacciu A, Sanna M. Absence of the contralateral internal carotid artery: a challenge for management of ipsilateral glomus jugulare and glomus vagale tumors. Laryngoscope 2007;117(8):1333–7.

CHAPTER 2

Operating Room—Special Considerations in Skull Base Surgery

Operating Room Setup

Operating Room Arrangement

The operating room setup is seen in Figs. 2.1–2.4. The surgeon sits at the side of the table. The instrument table is on the opposite side. The scrub nurse sits at this table facing the head of the operating table. From this position, she can hand the instruments to the surgeon without interfering with the surgical field. The scrub nurse controls the foot pedals of the drill, the monopolar and bipolar coagulation, and the water irrigation set. The microscope is placed at the head of the operating table, whereas the anesthetist is placed at the leg of the table. A television monitor is placed at the head of the table where the scrub nurse can comfortably watch the microsurgical surgical procedures (Fig. 2.4). This enables the proper assistance, particularly regarding the size of the burr ends, the choice of the proper instruments, and the rate of supply of the irrigating water. In the middle fossa and the subtemporal approaches, the surgeon sits at the head of the operative table, whereas the microscope is placed on the side of the table; the position of the scrub nurse remains the same (Figs. 2.2 and 2.4).

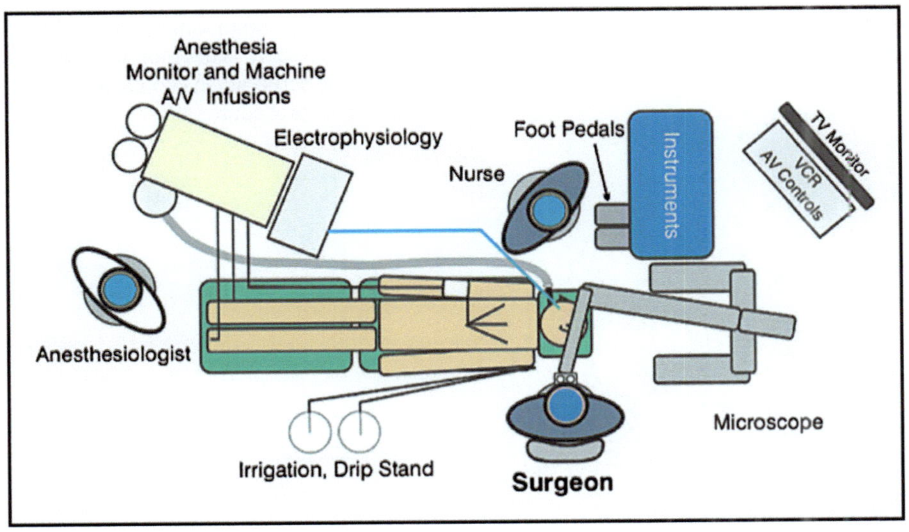

Fig. 2.1 Schematic drawing of the operating room arrangement during most of the skull base approaches.

Atlas of Microsurgery of the Lateral Skull Base. https://doi.org/10.1016/B978-0-443-19083-4.00002-7

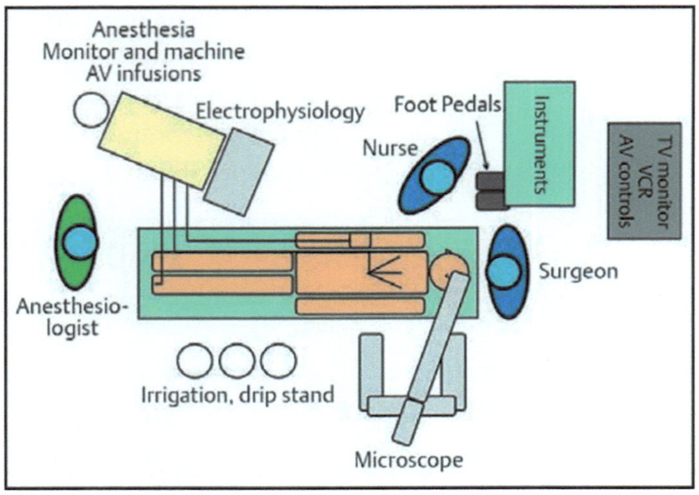

Fig. 2.2 Schematic drawing of the operating room arrangement during the middle fossa transpetrous and the subtemporal approaches.

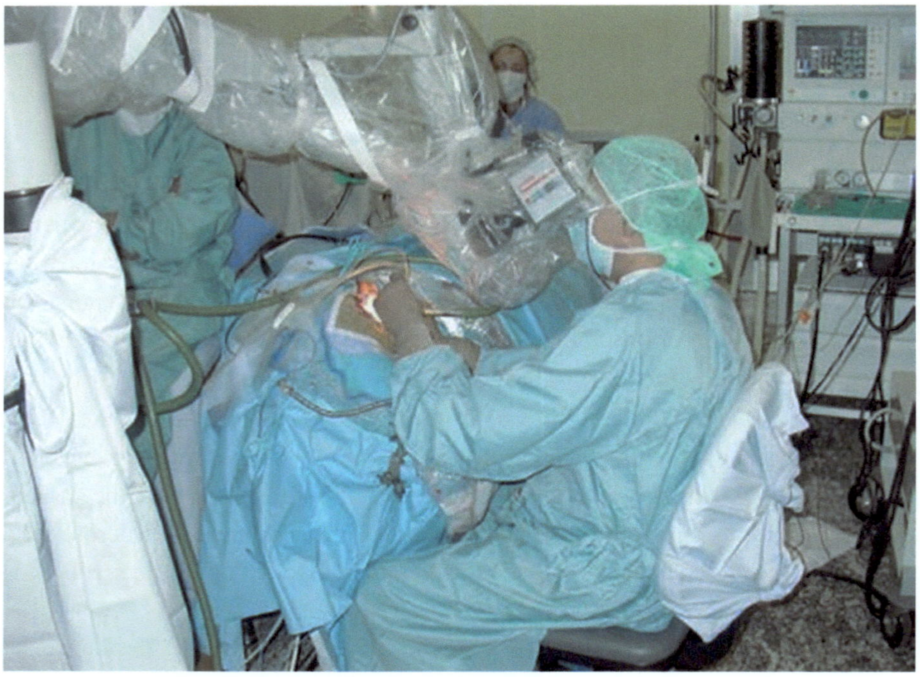

Fig. 2.3 Operating room during skull base surgery.

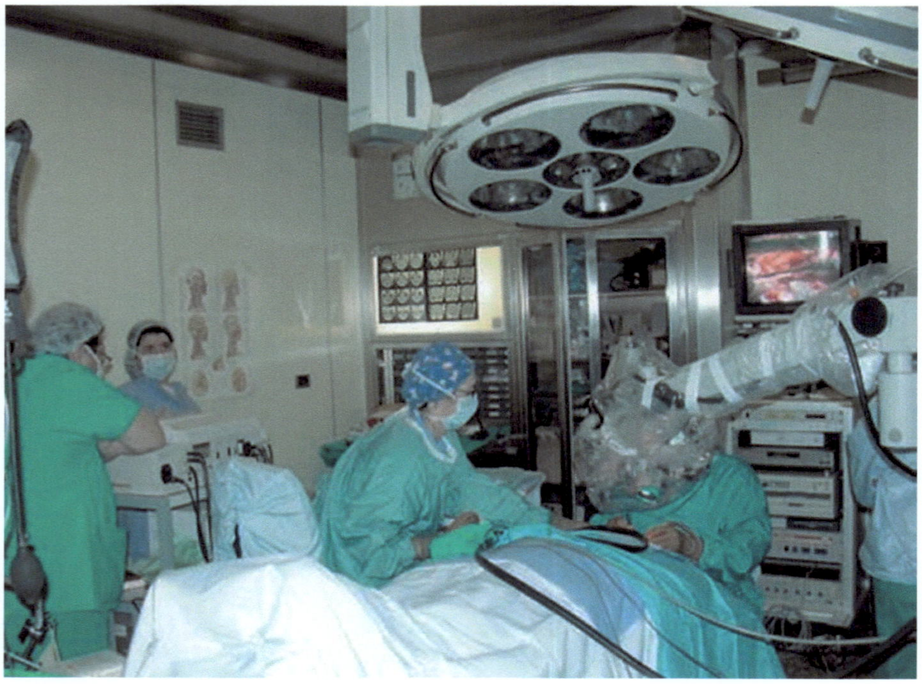

Fig. 2.4 Operating room during an enlarged middle fossa approach. The monitor stand is placed so that the scrub nurse can clearly follow the microsurgical procedure.

The Patient Position and the Operating Room

In most approaches, the patient lies supine, with the head rotated to the contralateral side. The patient is usually placed so that the head is on the foot end of the table. This provides maximum room under the patient's head for the surgeon's legs and the chair. The patient is secured to the bed at the wrists and the thighs (Fig. 2.5). During surgery, the bed position can be changed by whole-body up and down and whole-body rotation to the right or left (Fig. 2.6).

Monitoring

A Nerve Integrity Monitor-2 (NIM-2) (Xomed-Treace) is used for facial nerve monitoring. Pairs of electrodes are placed in the orbicularis oculi and orbicularis oris muscles to allow continuous electromyographic monitoring of the facial nerve. When the aim is preservation of hearing, an ipsilateral ear stimulator is applied, and electrodes are placed for auditory brainstem responses and direct eighth nerve stimulation (Fig. 2.7).

Preparation of the Operative Site

The operative site is shaved. The area is then prepared with ether 10 ppm, followed by Citrosil (Glaxo, Italy). The abdomen is also prepared in the same manner. The patient is then draped in layers (Fig. 2.8) using sterile adhesive drapes incorporating a clear window for the operative site and a fluid collecting system.

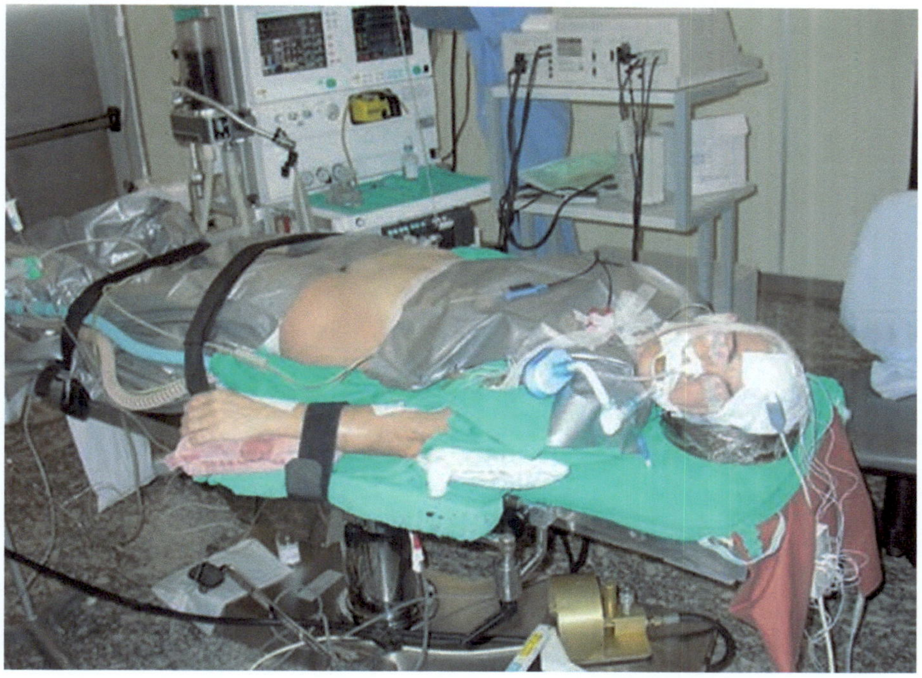

Fig. 2.5 The patient is secured to the table at the wrist and thigh. The anesthesia machine is placed at the end of the table.

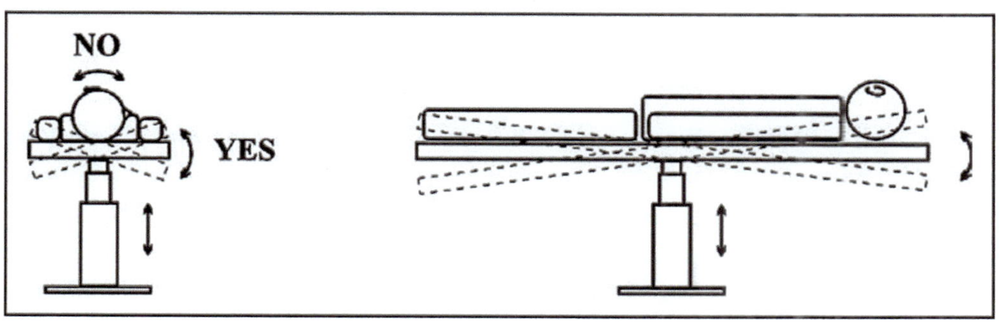

Fig. 2.6 Adjustment of the bed position.

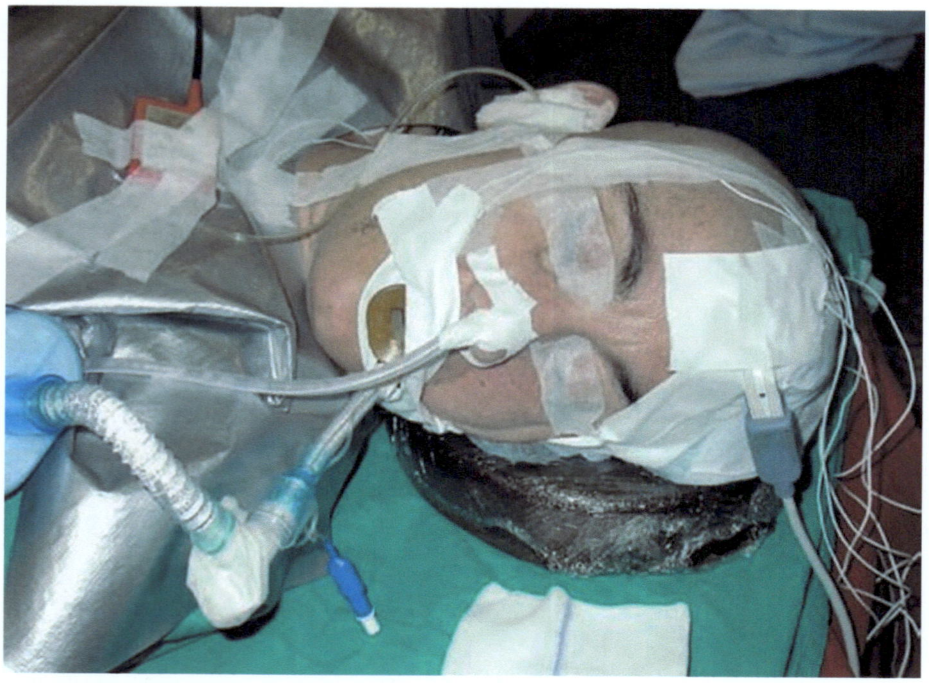

Fig. 2.7 A patient prepared for surgery in a hearing preservation procedure. The electrodes for facial nerve monitoring can be seen. Notice the additional electrodes and the microphone placed in the ear for eighth nerve monitoring.

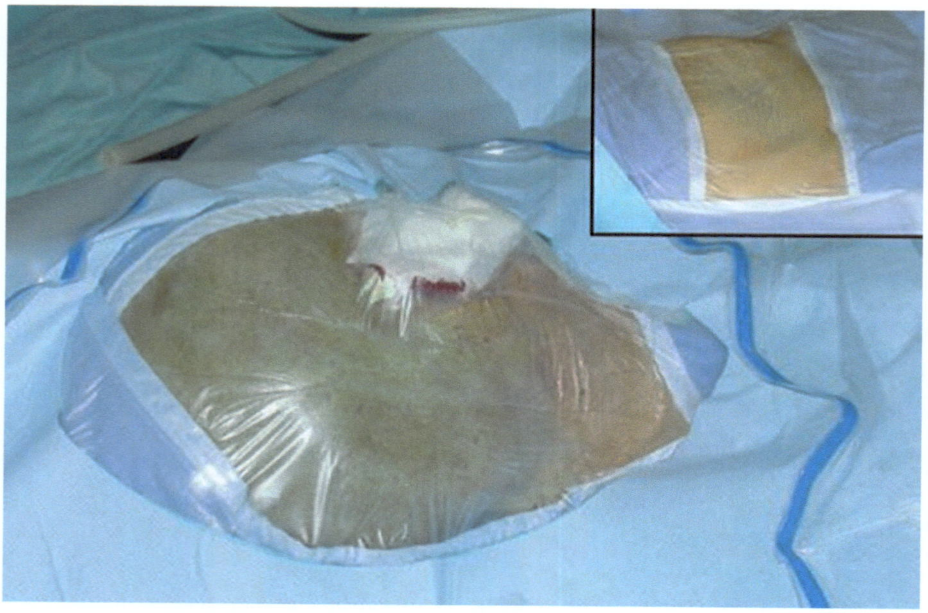

Fig. 2.8 The patient is draped with adhesive drapes. Inset (*upper right*) shows preparation of the abdomen for obtaining the fat used for closure at the end of the procedure.

Surgeon's Position

The surgeon should sit comfortably on a chair with a draped backrest (Figs. 2.9 and 2.10). The chair should have wheels to allow free movement. Most instruments are held like a pen. During fine manipulations, the surgeon's hand should be supported on the patient's head. This enables precise manipulations without tremors. It also lessens the tension on the arms and decreases fatigue.

Microscope

We use a Zeiss S-21 operating microscope with a floor stand and an objective lens of 250mm focal length. The microscope is equipped with straight eyepieces. A still camera and a 3CCD video camera are mounted on the microscope (Fig. 2.11). The 3CCD camera is attached to a monitor, a video recorder, and a digital video printer. The microscope is covered with a sterile drape.

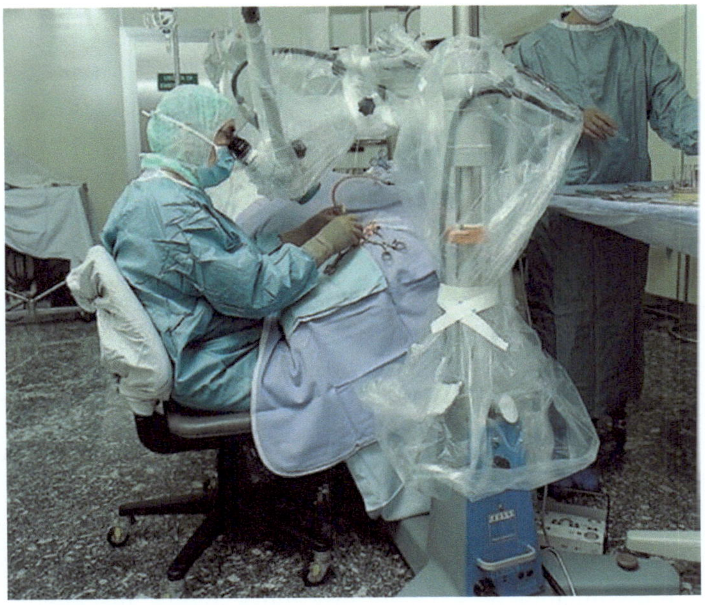

Fig. 2.9 Correct position of the surgeon.

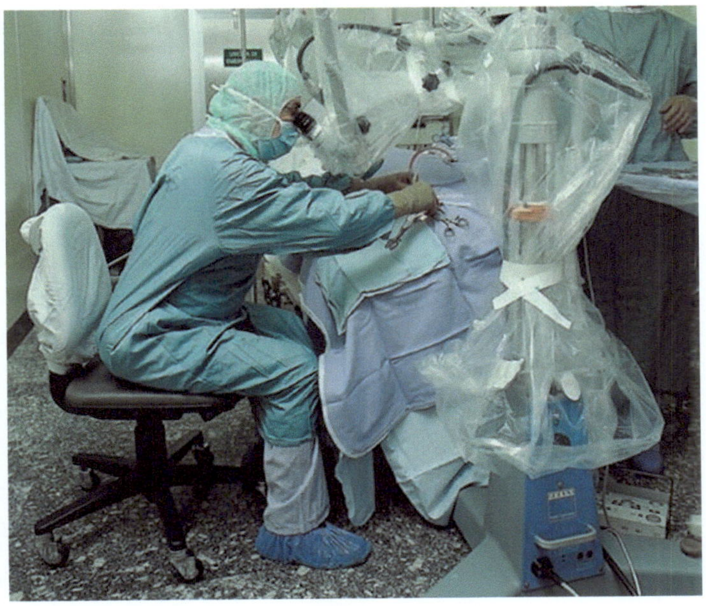

Fig. 2.10 Incorrect position of the surgeon.

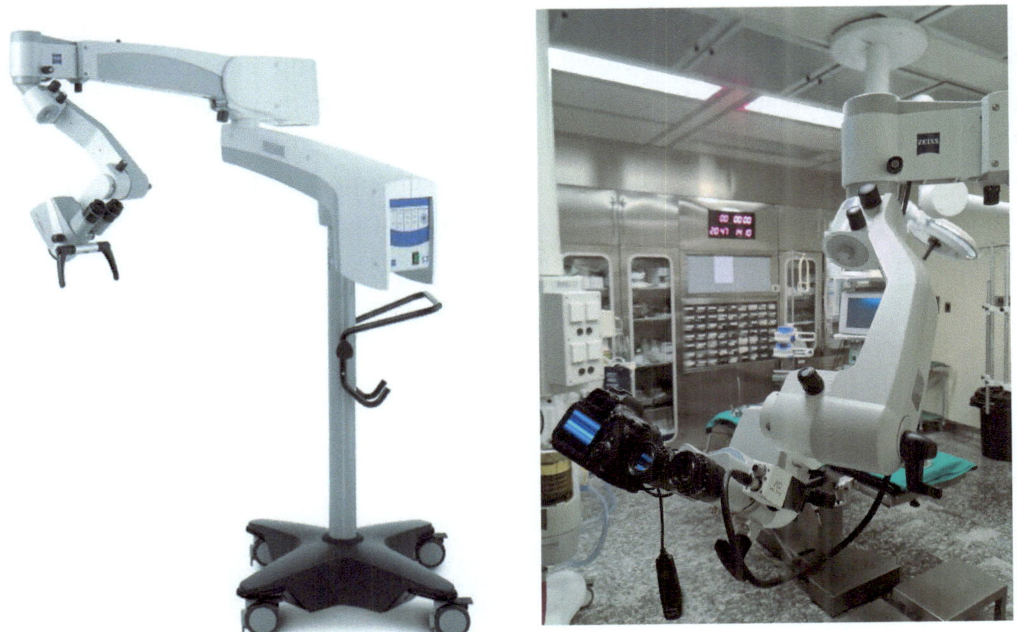

(A) (B)

Fig. 2.11 The Zeiss Movena S7 microscope (A). The microscope is equipped with an HD camera and a digital photographic machine (B).

Instruments

SUCTION IRRIGATION

Suction irrigators of different sizes with finger control are used throughout surgery (Fig. 2.12A). The channel for irrigation is connected to an infusion set, and the rate of infusion is controlled by the scrub nurse. The Brackmann suction irrigation tips are special types equipped with side holes near the blunt tip and are used for intradural work or during manipulations of fine structures (e.g., during facial nerve transposition) (Fig. 2.12B). Traditional suction tips can be very dangerous when used near delicate neurovascular structures. The side holes in the Brackmann type enable the distribution of the suction power and hence provide effective but indirect suctioning without endangering the fine structures lying near the tip.

DRILLS

We use the Bien Air drill system for the initial bone work. This system includes a multifunctional control unit in the form of a foot pedal. A forward/reverse switch determines the direction of burr rotation. Both straight and angled handpieces are used (Fig. 2.13).

BIPOLAR COAGULATION

Bipolar coagulation is an essential tool in skull base surgery. Different tip sizes should be available. Fine tips (0.3 mm) are used in the most delicate work for coagulating fine vessels intracranially. Larger tips (1 and 1.3 mm) are used for debulking and coagulation of the tumor and tumor capsule (Fig. 2.14A and B).

In our practice, we prefer the use of the Vesalius machine (molecular resonance generator) (Fig. 2.15). The scalpel produces an alternating current characterized by high-frequency waves combined in a particular way (CSS method); the fundamental wave is at 4 MHz, and the subsequent ones are at 8, 12, and 16 MHz, their amplitude decreasing with increasing frequency. Energy is delivered in quanta that are calibrated for human tissue and thus equal to the bond energy of the "human cell bonds" of the target particle. All the energy will be used to break these bonds, avoiding an increase in the kinetic energy and hence avoiding a rise in temperature.

The Vesalius machine can be used in a number of modes:

- Cutting only
- Combined cutting coagulation
- Coagulation only.

Cutting is caused by the breaking of the cellular structure through the energy supplied from the tip of the electronic scalpel and is cold, precise, and with a variable thickness. From a clinical point of view, this means that the amount of secondary necrosis is minimal and that the recovery will be by primary intention with reduction in edema and postoperative pain. Coagulation is obtained through denaturation of protein. The quaternary structure of the protein is altered, triggering the physiological cascade of coagulation.

The coagulation reaction is obtained by means of a particular electrical wave and occurs at a temperature of approximately 63°C. This procedure avoids the creation of a necrosis plug, which requires thousands of volts and exposes the patient to dangerous voltages and large areas of secondary necrosis.

Advantages that we have noted using this machine include:

- Inflammatory reaction and postoperative pain are reduced.
- The biological tissues around the area operated on suffer less thermal damage, resulting in fewer complications and greater preservation of nervous tissues both anatomically and functionally and with faster recovery.
- Cutting is clear and precise; thus, there is no damage or alteration of tissue taken for biopsy.
- The effective and reliable coagulation of vascular structures helps immensely in achieving a bloodless operative field.

MICROINSTRUMENTS

The minimum number of instruments should be used. An overcrowded instrument table with numerous useless items is a disturbing nightmare for the scrub nurse in these lengthy procedures. The instrument table should always be organized in the same order to facilitate handling and proper and prompt management by the scrub nurse (Figs. 2.16–2.22).

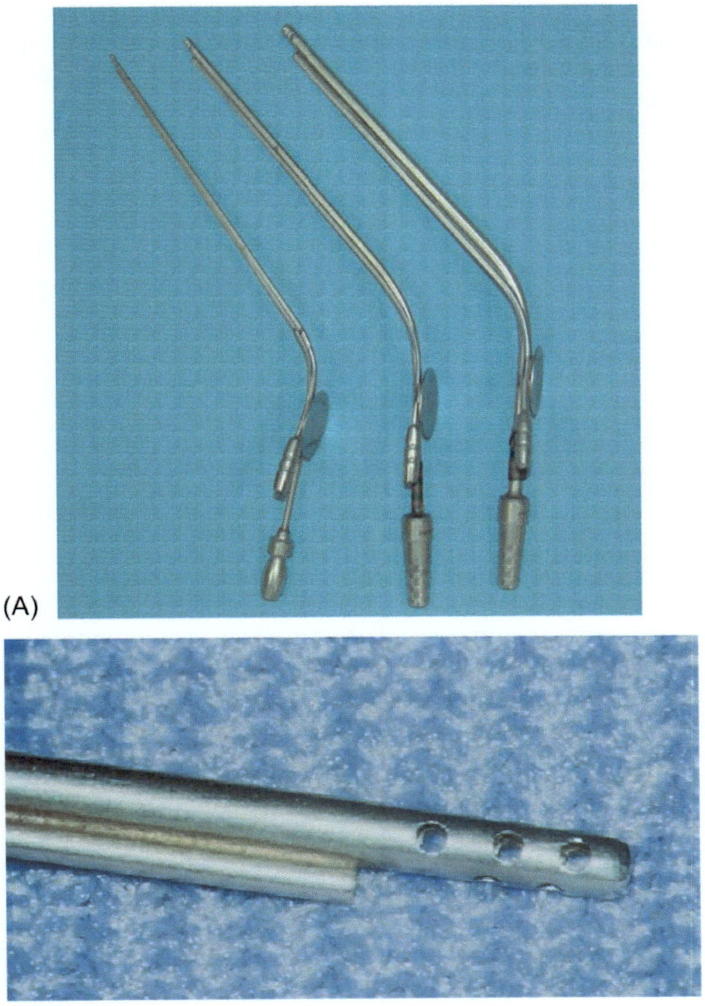

(A)

(B)

Fig. 2.12 (A) Different sizes of the suction irrigators. (B) View of the Brackmann-type suction irrigator. Side holes are clearly visible.

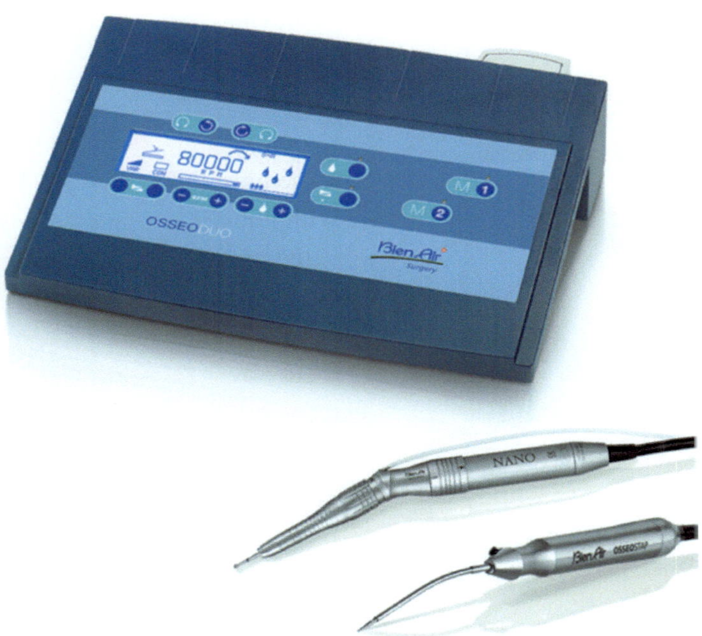

Fig. 2.13 The Bien Air drill system.

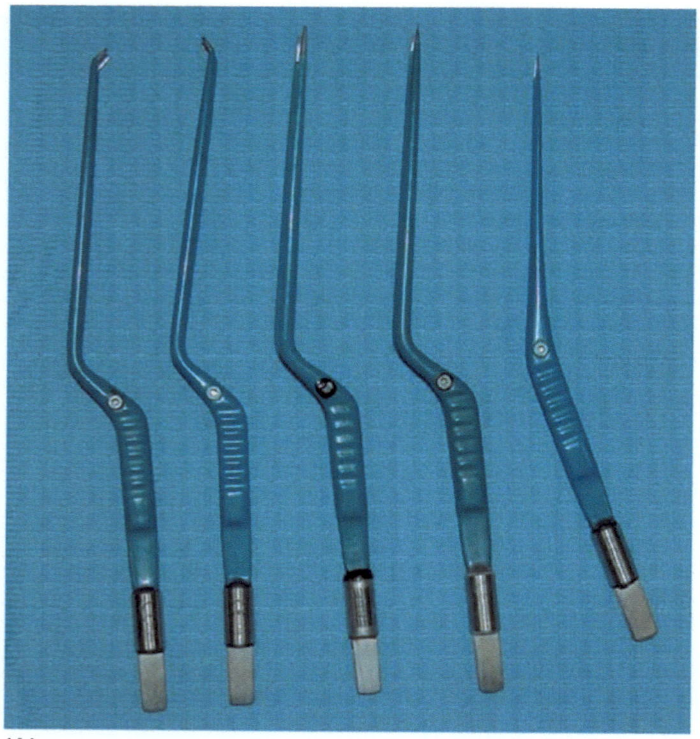

(A)

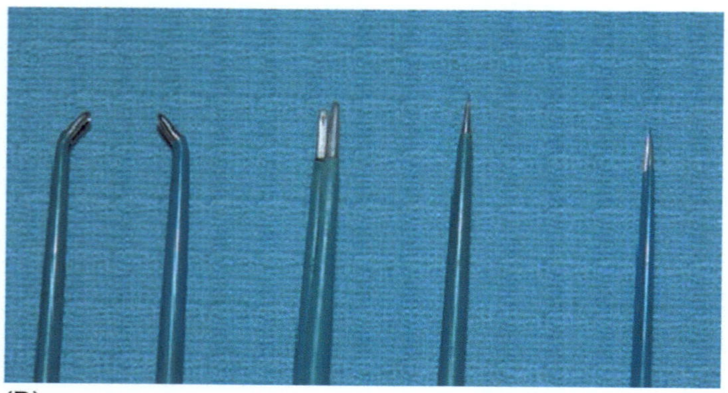

(B)

Fig. 2.14 (A) Different bipolar forceps. These include angled forceps, bayonet forceps with straight tips, and bayonet forceps with angled tips. (B) A higher magnification showing the different tip sizes of the bipolar forceps.

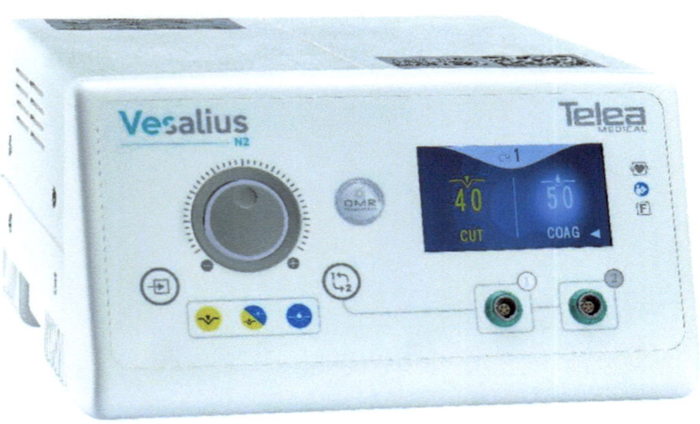

Fig. 2.15 The Vesalius molecular resonance generator.

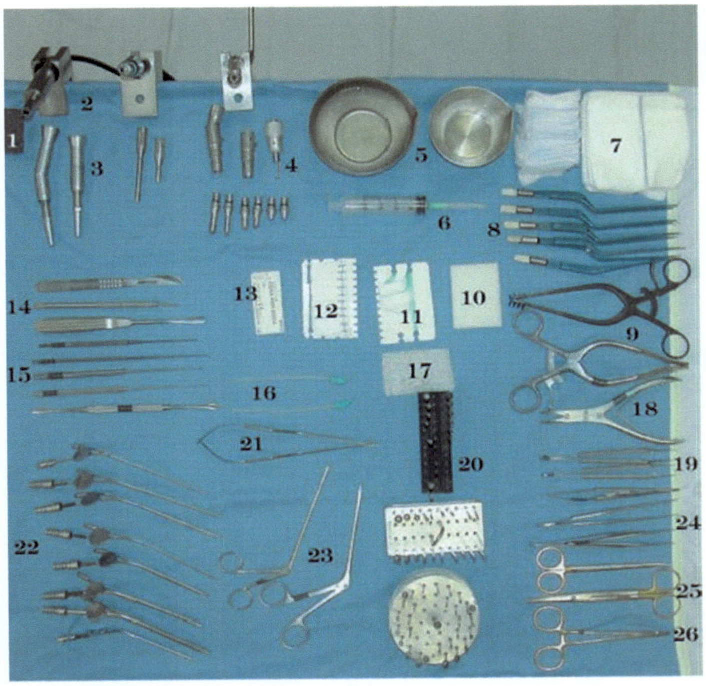

Fig. 2.16 Instrument table setup for skull base surgery. (1) Scratchpad, (2) Micromotors, (3) Handpieces, (4) Midas Rex craniotome handpiece, (5) Metal cups for saline and placement of the abdominal fat, (6) 20 mL syringe, (7) Gauze, (8) Various bipolar forceps, (9) Self-retaining retractors, (10) Absorbable gelatin sponge (Spongostan, Ferrosan), (11) Surgical patties (Merocel), 12 American surgical sponges, strung (American Silk Sutures Inc.), (13) Bone wax, (14) Surgical knives, (15) Dissectors and elevators, including Lempert periosteal elevator and Freer elevator, (16) Hooks for flap retraction, (17) Oxidized cellulose (Oxycel, Parke-Davis), (18) Bone rongeur, (19) Microtissue forceps and needle holder, (20) Burrs arranged on their racks, (21) Microscissors, (22) Suction irrigation tubes; both Brackmann and regular tips, (23) Microforceps, (24) Toothed and non-toothed forceps, (25) Curved scissors, (26) Needle holder.

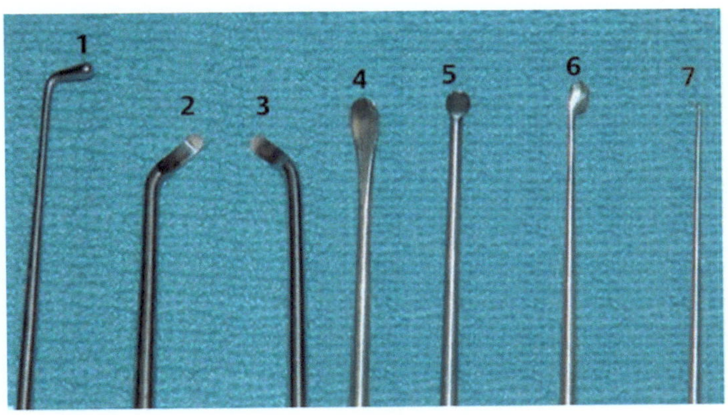

Fig. 2.17 Enlarged view of the microinstruments. (1) Blunt tip right-angled hook, (2) Double curved raspatory (*right*-sided), (3) Double curved raspatory (*left*-sided), (4) Straight microdissector, (5) Round dissector 45°, (6) Vertical dissector. (7) 90°-angled sharp hook.

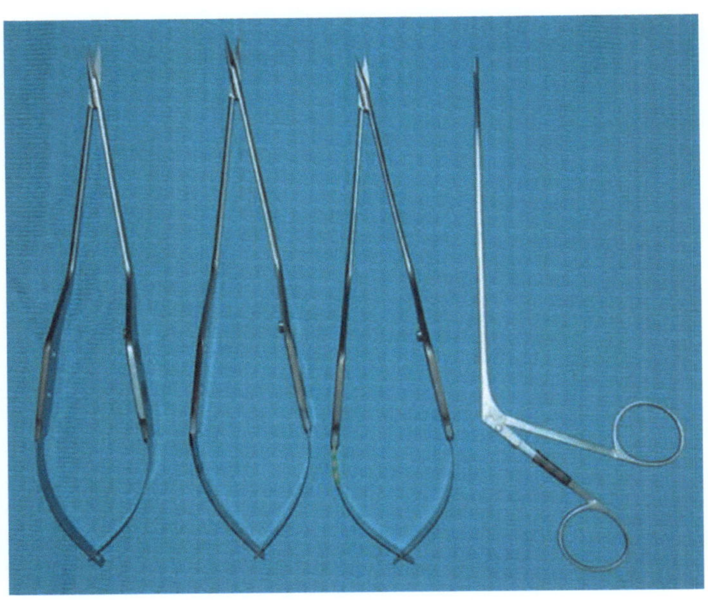

Fig. 2.18 Different types of microscissors.

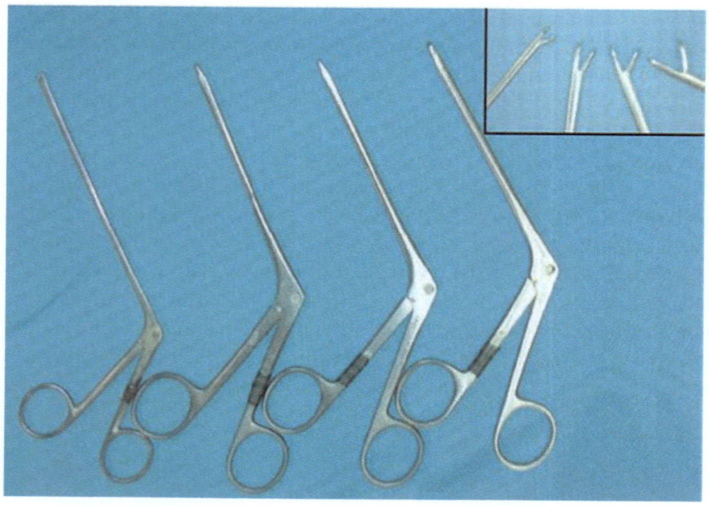

Fig. 2.19 Different types of micro forceps. Inset at *upper right* shows an enlarged view of the tips, demonstrating the different types and sizes.

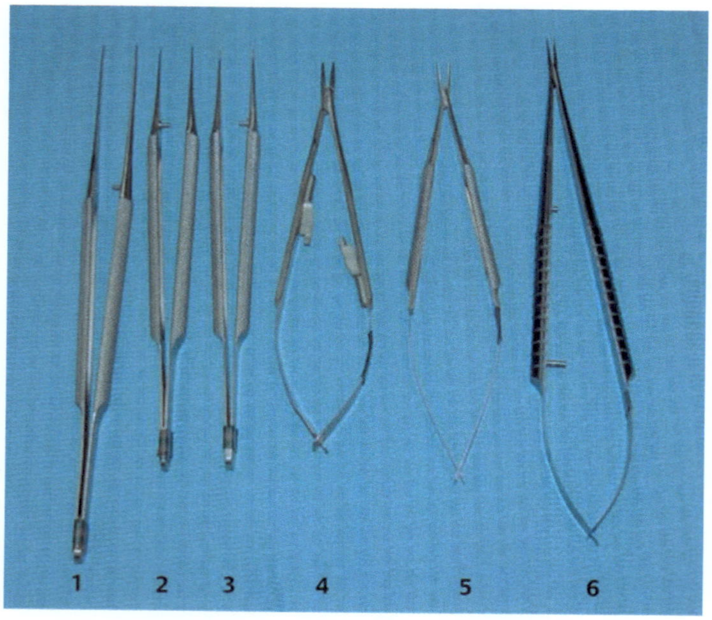

Fig. 2.20 Microsurgical instruments for facial and other cranial nerve anastomosis. (1) and (2) Straight microtissue forceps of different lengths, (3) Microtissue forceps with angled tip, (4) Microsurgery needle holder with latch, (5) Microscissors, (6) Microsuturing forceps.

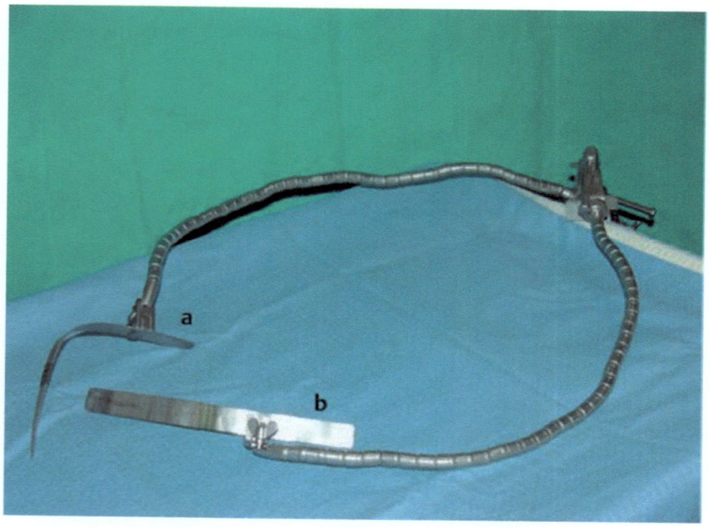

Fig. 2.21 Leyla retractor with Fukushima blade (A) and malleable blade (B).

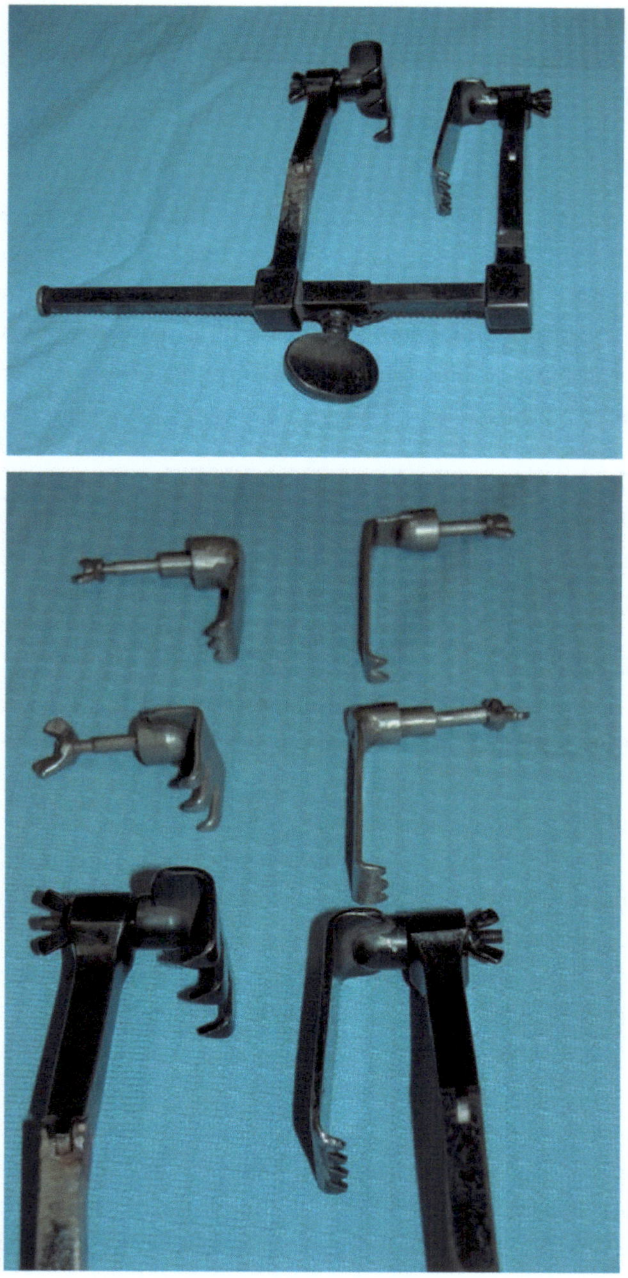

(A) (B)

Fig. 2.22 (A) Fisch infratemporal fossa retractor. (B) Different blades that we use with the infratemporal fossa retractor in different situations.

Postoperative Care

At the end of the procedure, the wound is dressed in the usual manner and a crepe band is applied. In intradural procedures, no drains are inserted. However, a drain is placed in infratemporal fossa approaches for extradural tumors.

Extubation is carried out on the operating table. We do not recommend prolonged intubation as a routine postoperative procedure since it can interfere with monitoring the patient for complications.

The patient is then transferred to the intensive care unit and is monitored for pulse rate, respiratory rate, arterial oxygen saturation, and electrocardiogram tracing. The blood pressure monitor takes recordings at intervals. In addition, the level of consciousness of the patient, pupillary reflexes, and motor responses are also assessed. All these parameters are recorded every 15 min for the first six postoperative hours and then half-hourly for 12–18 h.

After 24 h, the patient is transferred to the ward. Observation of the vital signs and level of consciousness continues every 2 h during the first day in the ward and every 4 h thereafter until discharge. The indwelling urinary catheter and the nasogastric tube are removed after transfer to the ward. Oral fluids are usually started at that time. Early ambulation is encouraged after the first 24 h to minimize the risk of pulmonary embolism.

Special Considerations in Skull Base Surgery

Introduction

Skull base surgery has certain characteristics that distinguish it from traditional surgery. Besides having the basic surgical knowledge and skills, the lateral skull base surgeon should be well aware of particular techniques that are combinations of skills involved in microsurgery, otology, neurosurgery, and head and neck surgery. The skull base surgeon should be well cognizant of the three-dimensional anatomy of the temporal bone and the base of the skull before embarking on surgery and should master drilling of the temporal bone as well as the refined neurosurgical skills of dealing with brain tissue and effecting hemostasis in both situations. This section will emphasize only those particular techniques that are peculiar to this type of surgery. These techniques can be grouped under the following headings:

- General guidelines for drilling
- Management of bleeding
- How to deal with the brain and other neurovascular structures
- Techniques of tumor dissection and removal
- How to widen the access to the tumor.

General Guidelines for Drilling

The motor can be either air-driven or electrically driven. We use both systems in our center. The electrically driven system (Bien Air ORL-R-92 System) is mainly used in the gross bone work in the initial phase of surgery, whereas the air-driven Midas Rex system and the Anspach are utilized during finer bone work. These are high-speed drills that need almost no pressure while performing the bony work, thereby offering maximum safety and speed.

- The handpiece should be held as a pen. The little and ring fingers are placed on the patient's head to give support to the surgeon's hand. Drilling should be performed with the side rather than with the tip of the burr end. Only minimal pressure should be applied, particularly when coming near to important structures.
- The largest possible burr should be used. In the initial phases of mastoidectomy, a large conical burr is preferred as the working area of this burr is much greater than that of a Rosen head burr. The majority of bone work is performed by cutting burrs. Diamond burrs are reserved for working near delicate structures such as the facial nerve, sigmoid sinus, or dura. Diamond burrs are also used without irrigation to stop bleeding originating from the bone.
- Drilling should always be parallel to important structures and should start from the most dangerous area to the safest area.
- Continuous suction irrigation is a must while drilling. This helps in clearing the bone dust that impedes visualization and prevents clogging of the flukes of the burr end. Irrigation also helps to cool the drilled bone and prevents thermal injury.
- The suction irrigator should wander about the burr. The suction irrigator can be placed between important structures (e.g., an exposed facial nerve) and the burr in such a way that if the burr slips, it hits against the suction irrigator rather than the structure itself.

Management of Bleeding

Bleeding during neuro-otological and skull base surgery is quite different from that in head and neck surgery since it cannot be controlled by ligation of the vessels. Different instruments and hemostatic materials have been devised for control of bleeding in such surgery. In our center, we use the following.

MONOPOLAR COAGULATION

Monopolar coagulation is used to achieve hemostasis while making the initial incisions of the skin and musculoperiosteal tissue.

BIPOLAR COAGULATION

Bipolar coagulation is the most important tool in neuro-otological and skull base procedures. As mentioned in Chapter 1, in our practice we prefer to use the Vesalius machine (molecular resonance generator).

Both ordinary and fine bipolar forceps tips are necessary. Ordinary tips are used for extradural structures and surface coagulation of the tumor, while fine tips (tip size 0.3 mm) are used for coagulating tumor vessels branching from major cerebellopontine arteries or bleeding vessels over the brainstem.

Copious irrigation during bipolar coagulation is important to prevent the tips from sticking to the structure being coagulated. Sticking may cause rebleeding or tearing and further injury. Black deposits forming on the tips make the bipolar device less efficient and may cause sticking. Such deposits should be cleaned by the nurse using a scratch pad. The use of a scalpel for such a task should be discouraged as it causes scratching and irregularity of the tip surface.

Small tears of the sigmoid sinus can be managed by the judicious use of bipolar coagulation. For such cases, the bipolar device is adjusted to a low setting, continuous irrigation is used, and the forceps tips are used to approximate the tear edges while performing coagulation. On the other hand, the bipolar device should never be used to manage bleeding from small tears of the jugular bulb, as more bleeding usually occurs.

BLEEDING FROM BONY SURFACES

Bleeding from bony surfaces is controlled by using diamond burrs without irrigation. The heat generated by the diamond burr is usually sufficient to stop such minor bleeding.

Significant bleeding from bony surfaces can be stopped using bone wax. A small piece of bone wax is applied and pressed firmly with the thumb. When the bleeding surface is deeply located in the operative field, a small piece of bone wax shaped like a sphere is applied using a septal dissector and is compressed using cottonoids.

GELATIN SPONGE

This agent, e.g., Spongostan (Ferrosan) or Gelfoam (Johnson & Johnson), has an enormous capacity to absorb blood. The pressure exerted by the weight of the blood-soaked Gelfoam in contact with an oozing surface assists in hemostasis. The advantage of this material is that it can be left in place to be absorbed in 20–45 days. It is used for the control of capillary ooze and bleeding from small arterioles, particularly in those locations where the bipolar coagulator can be injurious, such as bleeding from vessels over the facial nerve. It is also used to stop bleeding from venous sinuses. In such cases, a cottonoid is placed over the gelatin sponge to prevent its dislodgment.

OXIDIZED CELLULOSE AND OXIDIZED REGENERATED CELLULOSE

Oxidized cellulose, e.g., Oxycel (Parke-Davis), and oxidized regenerated cellulose, e.g., Surgicel (Johnson & Johnson), act as an organic acid and hence behave like a caustic agent that reacts with blood to produce a reddish-brown mass that functions as an artificial clot. These agents have also proved to have potent bactericidal action.

These forms of cellulose are mainly used to stop capillary ooze. They can also be used for intraluminal packing of venous sinuses in case of inadvertent injury or intentional transection as a part of the approach.

Because of its mild tissue reaction, excessive Surgicel left in the cerebellopontine angle may lead to postoperative fever. As these agents swell when in contact with blood, they should be used carefully in confined spaces since such swelling might cause pressure neuropathy of adjacent nerves.

How to Deal With Brain Tissue and Other Neurovascular Structures

The following are general rules adopted in our center:

- Brain retraction should be avoided whenever possible. The hazards of brain retraction depend mainly on the force used and the length of time the brain is retracted. The use of prolonged static brain retraction is hazardous and should be avoided. However, in certain approaches, e.g., the middle fossa transpetrous approach, a static retractor is applied in the initial phase of the approach and if possible is removed during the stage of tumor removal. Intermittent retraction is practically achieved with the suction irrigator or the bipolar forceps during the subsequent phases of tumor resection.
- Whenever needed, gentle elevation and support of the brain are performed, thus exposing the cisterns. The arachnoid of the cistern is opened next, allowing cerebrospinal fluid (CSF) to leak. This results in spontaneous shrinkage and retraction of the brain, providing the necessary space for tumor removal. Every attempt should be made to perform this maneuver in the initial phase of tumor removal since it practically obviates brain retraction.
- Direct contact between the instruments and the brain surface should be avoided. Once the dura is opened, Merocel is applied over the brain surface to afford protection. When the tumor is dissected free from attached neurovascular structures, Merocel is insinuated between the tumor and these structures to protect them during the subsequent phases of tumor resection. Merocel can also act as a dissector. It can be propelled into the interface between the tumor and normal tissue, thereby gradually separating them (Fig. 2.23). We prefer Merocel over neurosurgical cottonoids as the latter adhere to the brain surface and might be injurious on attempting to remove it.
- Once the dura is opened, the Brackmann suction irrigator is utilized (see Fig. 2.23). This is a particular type of suction tip that has a blunt tip and side holes, which avoids direct suction (with resultant trauma) over the sensitive neurovascular tissues.

Techniques of Tumor Dissection and Removal

Tumor removal in skull base and neuro-otological surgery differs greatly from tumor removal elsewhere. The reason for this is that skull base tumors lie adjacent to vital structures. Tumor removal should therefore be performed with little or no compromise of these structures. Most of these tumors are benign, so tumor debulking and piecemeal removal are the techniques used for their removal. Many of these tumors have an aggressive nature with invasion of the dura and bone leading to a high tendency to recurrence (e.g., meningioma, chordoma, and chemodectoma). Total tumor removal should therefore entail the removal of the dura and complete drilling of the involved bone.

The following are general rules for tumor removal:

- Particularly in vascular tumors, bipolar coagulation of the tumor surface under continuous suction irrigation is performed initially. This coagulates tumor surface vessels and reduces bleeding to a minimum when the tumor capsule is opened.

In intradural tumors:

- The tumor is usually surrounded by a double layer of arachnoid. Finding this arachnoid plane is the most important initial step before attempting tumor dissection. Staying in the arachnoid plane ensures protection of the intradural neurovascular structures and obviates troublesome bleeding that may ensue if this plane is mistaken.
- The most important instruments we use for tumor dissection include Freer septal dissector, bipolar forceps, suction irrigation, and, in delicate areas, neurosurgical hooks.
- Continuous suction irrigation with warmed (37°C) Ringer's solution is a must during tumor removal. Suction irrigation clears the field of blood and blood clots and hence enhances visualization. Irrigation also maintains the brain tissues in a more physiological environment.
- Before arachnoid webs are severed or tumor surfaces are cut, bipolar coagulation is affected first. This avoids bleeding that may be difficult to control because the cut edges usually retract deep into the subarachnoid space and become relatively inaccessible.
- Small tumor vessels stemming from significant arteries (e.g., anterior inferior cerebellar artery) should be coagulated and severed away from the main artery. Bipolar coagulation is performed over a long segment of the tumor vessel so that one can be sure of its complete closure before cutting it.
- Except for small tumors, intracapsular debulking of the tumor core is performed first. For this step, we use simple instruments, namely, Freer septal dissector, bayonet neurosurgical scissors, and tumor-cupped forceps. Bleeding from the tumor core is controlled by bipolar coagulation and/or temporary Surgicel packing.

In the majority of lateral transtemporal approaches to the skull base, the tumor lies between the surgeon and the neurovascular structures; that is, the tumor hides these structures from the surgeon. Nevertheless, blind separation of the tumor from these structures should **never** be performed. In large tumors, it is a difficult task to achieve the goal of tumor removal with the preservation of the related neurovascular structures. The following strategy is therefore performed: First, intracapsular debulking is done to reduce the tumor bulk. This is followed by separation of the tumor from the surrounding structures. Once a piece of tumor has been dissected, it is excised with its capsule. Thus, tumor size gradually decreases. At the end, a significantly smaller tumor mass remains attached to neurovascular tissues that lie anterior to it. By pushing the tumor capsule to one side, the plane of cleavage between the tumor and these structures can be clearly seen and tumor removal can thus be completed (Fig. 2.24A–C).

How to Widen the Access to the Tumor

Skull base tumors have the disadvantage of being deeply located in an area of complex neurovascular structures. Adequate exposure has always been a major concern in this type of surgery. Traditional neurosurgical techniques entailed a relatively small craniotomy through which access is mainly gained by brain retraction. The concept of modern skull base surgery is to remove the bone and leave the brain alone. Bone removal provides the necessary access while obviating the need for excessive brain retraction. In addition, various surgical tricks are used to enhance exposure and improve visualization. The following are some of these tricks:

- We do not use self-retaining retractors for the skin and subcutaneous tissue. Retractors impede surgical access and make the working angle narrower and the surgical field deeper (Fig. 2.25).
- Extensive bone removal greatly enhances the approach: for example, extensive bone removal uncovering the posterior fossa posterior to the sigmoid sinus, as well as complete uncovering of the sigmoid sinus, allows retraction of the sigmoid sinus and widens the angle of vision greatly in the translabyrinthine and transcochlear approaches (Fig. 2.26).
- Temporary removal of bony boundaries that are reattached at the end of the approach: for example, the removal of the zygomatic arch and the orbital rim and the wide craniotomy in the orbitozygomatic approach, which markedly enhances the exposure gained by this approach.
- A useful trick is to coagulate the dura using bipolar coagulation in vertical strokes and under suction irrigation. With this technique, the dura retracts and provides more space for work (Fig. 2.27).
- Changing the bed position is an important step during different phases of surgery as it affects the angle of vision and provides a better exposure to the structures involved by the tumor. The same also applies to changing the angulation of the microscope and sometimes the surgeon's position (Fig. 2.28).
- Overcoming anatomical obstacles. This is an important part of skull base surgery and can be divided into two categories:
    1. Overcoming normal anatomical structures, as in transposing the facial nerve either anteriorly (infratemporal fossa approach type A) or posteriorly (modified transcochlear approach). It also involves mobilization of the transpetrous internal carotid artery, as in the infratemporal fossa type B approach to access the petrous apex.
    2. Overcoming anatomical variants that hinder the approach, e.g., a high jugular bulb or a far-advanced sigmoid sinus.

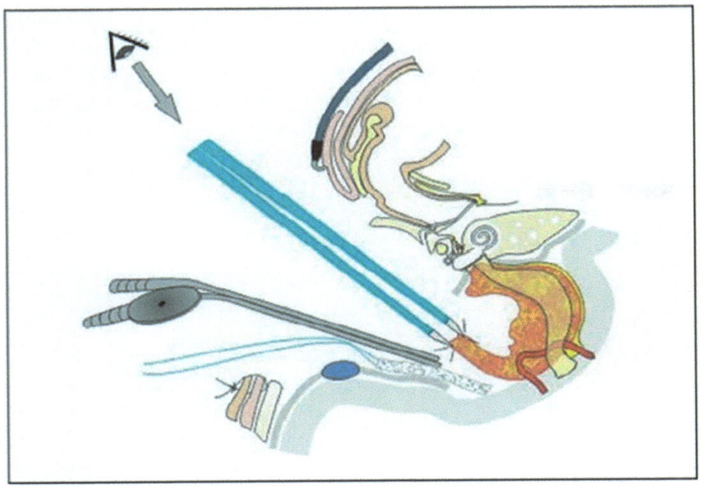

Fig. 2.23 Merocel is propelled between the tumor and the cerebellum, giving protection for the brain and also acting as a dissector.

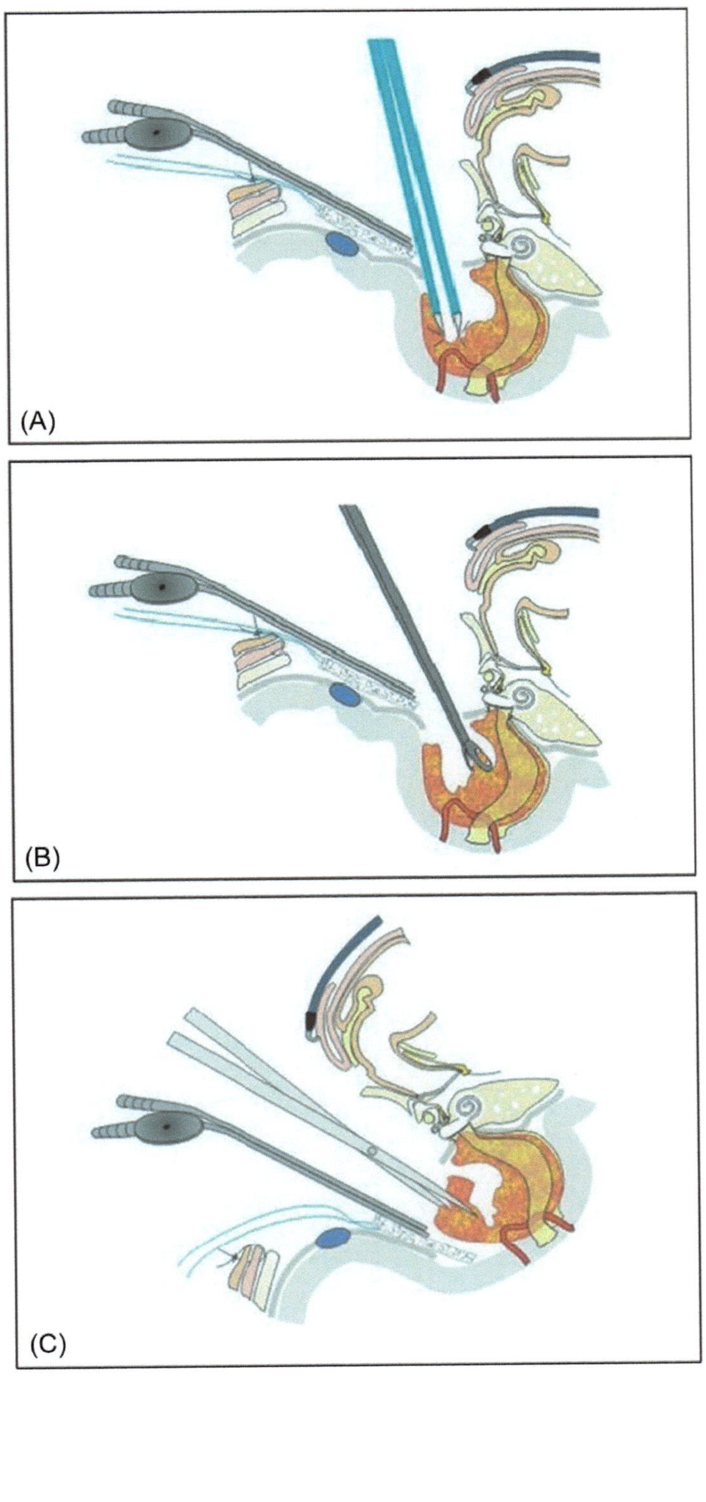

Fig. 2.24 (A) Tumor debulking. Intracapsular bipolar coagulation. (B) Intracapsular tumor debulking with micro forceps. (C) Tumor reduction. The completely dissected part is excised using microscissors.

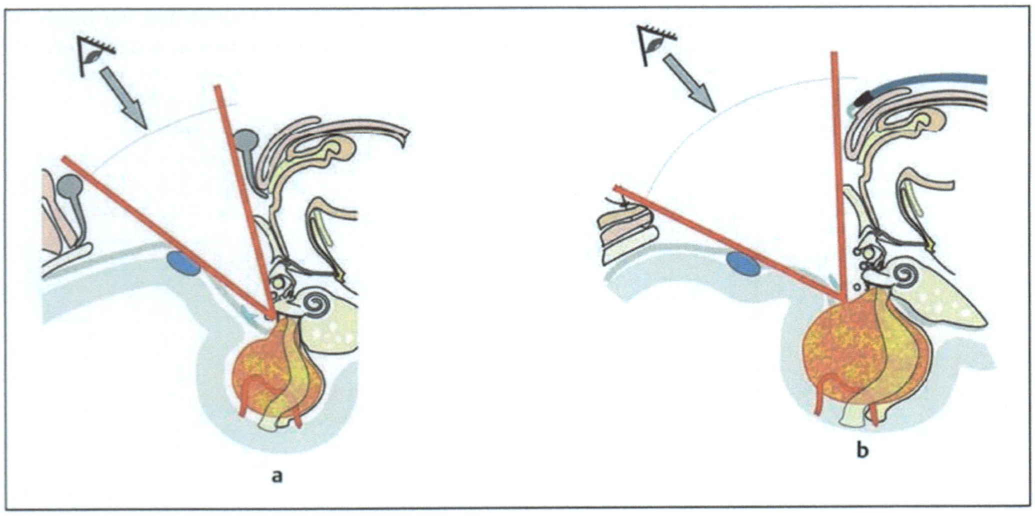

Fig. 2.25 (A) Angle of vision using self-retaining retractors. (B) The angle is markedly enhanced after removing the retractors and using sutures.

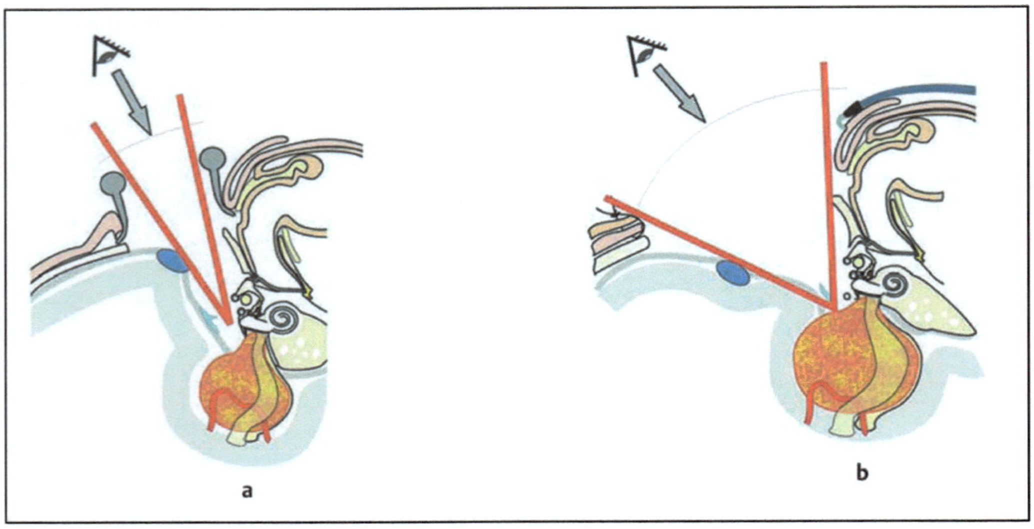

Fig. 2.26 Difference between classical approach (A) and enlarged approach (B). In the enlarged approach, the removal of retractors and, more importantly, the wide bone removal allows retraction of the sigmoid sinus and provides a much wider access.

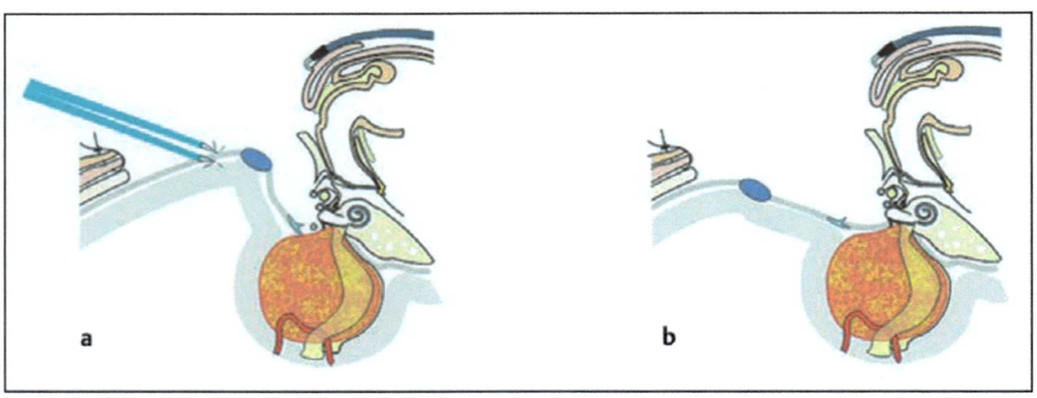

Fig. 2.27 The advanced dura in (A) limits the approach. Bipolar coagulation (B) leads the dura to retract, giving more space.

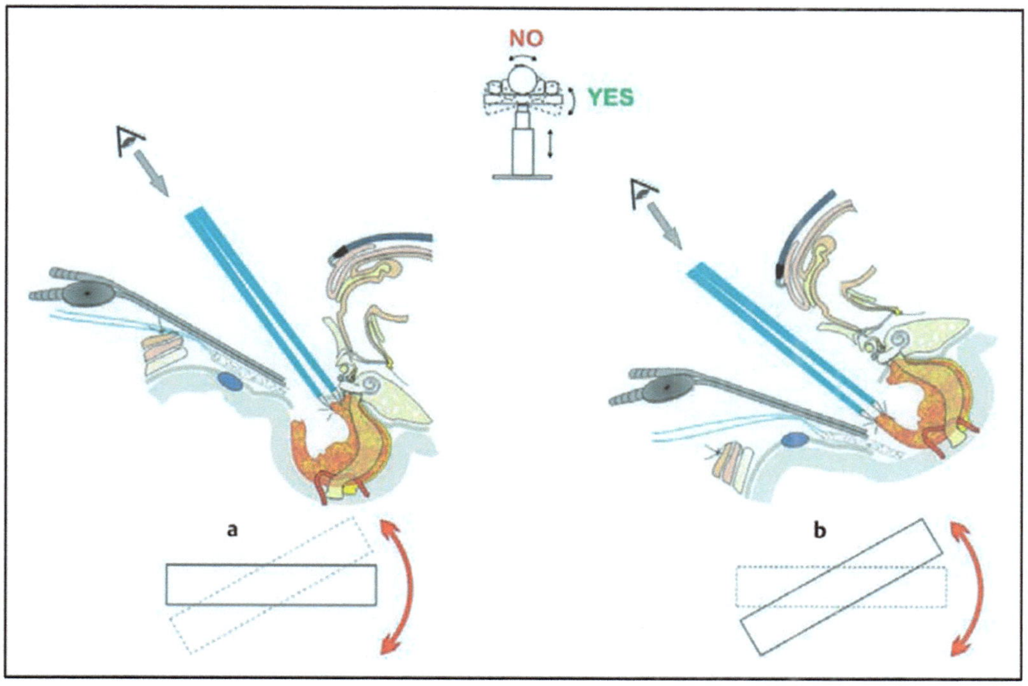

Fig. 2.28 Example of changing the bed position in the translabyrinthine approach. (A) Bed away from the surgeon allows better control of the anterior pole of the tumor. (B) Bed toward the surgeon allows proper exposure of the posterior pole of the tumor and separation of it from the cerebellum.

CHAPTER 3

Introduction to Lateral Skull Base Surgery

Classification of the Lateral Skull Base Approaches

Lateral skull base approaches can be classified according to the relation of the target area and the approach to the otic capsule. Fig. 3.1 shows the range of skull base that can be covered by current lateral skull base surgical approaches. Thus, approaches can be divided into those passing through part or all of the otic capsule and those conserving the otic capsule and passing either posteriorly, superiorly, anteriorly, or inferiorly to it.

Approaches passing through the otic capsule:

- The translabyrinthine approach.
- The transcochlear approaches: these include the original transcochlear approach of House, the transotic approach of Fisch, and the system of the modified transcochlear approach.

Approaches conserving the otic capsule (extralabyrinthine approaches):

- Passing above the otic capsule: the middle cranial fossa, enlarged middle cranial fossa, and middle fossa transpetrous approaches.
- Passing behind the otic capsule: the retrosigmoid, the retrolabyrinthine, and the retrolabyrinthine transtentorial approaches.
- Passing inferior to the otic capsule:

Approaches to the jugular foramen area: infratemporal type A; petro-occipital transsigmoid (POTS).

– Approach dealing mainly with the craniovertebral junction: extreme lateral approach.

- Passing anterior to the otic capsule: the preauricular infratemporal subtemporal fossa approach and the infratemporal fossa approaches types B, C, and D.

Combinations of any of the above are termed combined approaches and are used in certain skull base lesions (see Chapter 16).

Fig. 3.2 shows schematic drawings of a series of intradural lateral approaches to the skull base. The suboccipital and the retrolabyrinthine give a limited view of the anterior cerebellopontine angle. This view is successively increased by further anterolateral extension (translabyrinthine, transotic, and modified transcochlear). Note that the preauricular infratemporal subtemporal approach is mainly centered anteriorly.

Atlas of Microsurgery of the Lateral Skull Base. https://doi.org/10.1016/B978-0-443-19083-4.00003-9

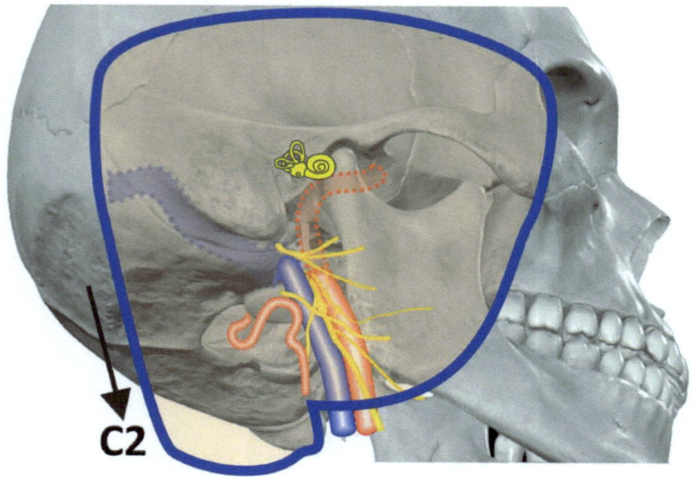

Fig. 3.1 Schematic drawing showing the range of skull base that can be covered by current lateral skull base surgical approaches.

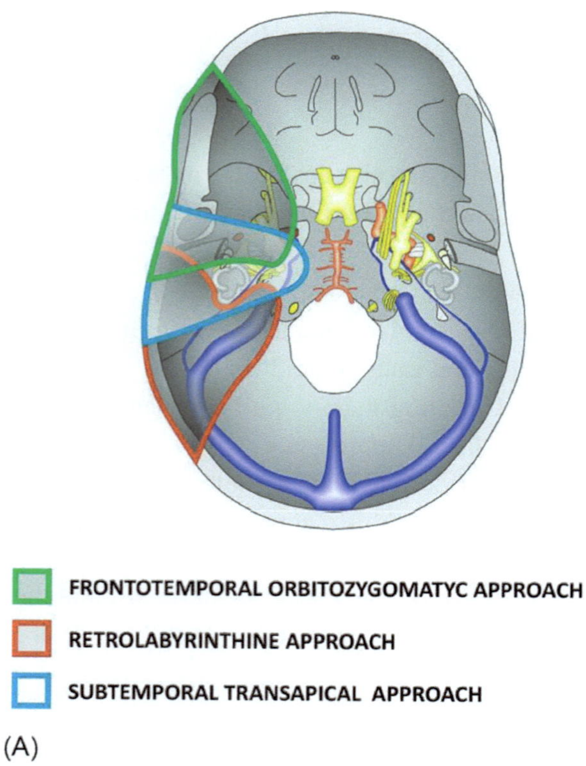

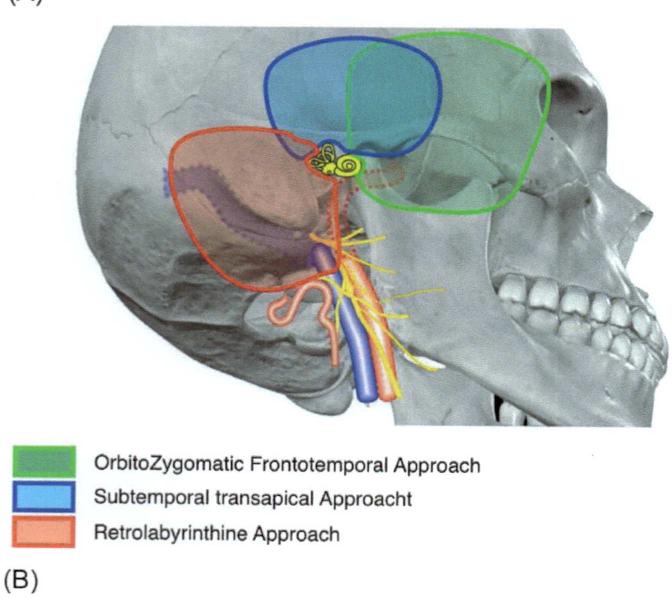

Fig. 3.2 (A and B) Schematic drawings of a series of intradural lateral approaches to the skull base.

CHAPTER 4

Anatomy of the Temporal Bone and Related Structures

Osseous Relationships of the Temporal Bone

The temporal bone actually consists of four fused parts: the squamous, tympanic, mastoid, and petrous bones (Figs. 4.1–4.8).

Squamous Bone

The squamous part of the bone represents the major part of the lateral surface of the bone. Above the level of the zygomatic process, the vertical portion of the squamous bone extends upward to cover part of the temporal lobe of the brain. The zygomatic process is actually part of the squamous portion of the bone. It originates anterior to the external auditory canal at the level of the junction of the vertical and horizontal parts of the squamous bone. The root of the zygomatic process shows an initial swelling known as the posterior zygomatic tubercle. Traced anteriorly, the root thins out to form the glenoid fossa for the articulation of the head of the mandible and then thickens again to form the anterior zygomatic tubercle. The zygomatic process then thins out and flattens as it separates from the squamous bone and ends by articulation with the zygomatic bone. Posterior to the external auditory canal, the zygomatic process can be traced as a somewhat faint line, the temporal line (supramastoid crest), indicating the level of the middle cranial fossa. The squamous bone then extends inferiorly in its retromeatal portion, forming the flattened lateral part of the mastoid process. The squamous part of the temporal bone also forms the superior parts of both the anterior and posterior walls of the bony external auditory canal. On the posterosuperior border of the canal, the spine of Henle can be seen.

Tympanic Bone

The gutter-shaped tympanic bone forms the inferior wall and major parts of the anterior and posterior walls of the bony external auditory canal. Two sutures between the elementary structures that form the temporal bone appear in the canal. The tympanosquamous suture is located anterosuperiorly, and the tympanomastoid suture posteroinferiorly. Connective tissue enters these suture lines, and sharp dissection may be required during meatal skin elevation. The temporomandibular joint is located just anterior to the canal and is separated from the canal only by a thin bony shell. The lateral border of the tympanic bone is roughened for the attachment of the cartilaginous part of the external auditory canal, which forms the outer two-thirds of the canal. The inferior edge of the tympanic bone expands to form the vaginal process where the styloid process lies. The styloid process is ensheathed at its base by the tympanic part and projects downward, serving as the attachment of three styloid muscles.

Mastoid Process

The mastoid process can be seen at the posterior and inferior border of the temporal bone, protruding anteroinferiorly to variable levels, depending on the pneumatization of the mastoid. The process serves as the anterior part of the attachment of the sternocleidomastoid muscle. On its medial surface lies the digastric groove, from which the posterior belly of the digastric muscle originates. On the posteromedial end of

Atlas of Microsurgery of the Lateral Skull Base. https://doi.org/10.1016/B978-0-443-19083-4.00004-0

the groove, an impression of the occipital artery can be seen. The stylomastoid foramen, from which the main trunk of the facial nerve exits the temporal bone, can be seen at the anterior border of the digastric ridge posterior to the styloid process.

Petrous Bone

The most prominent feature of the medial aspect of the temporal bone is the petrous part. Shaped like a pyramid, this part protrudes in an anteromedial direction, with the base located laterally and formed by the semicircular canals, vestibule, cochlea, and carotid artery. The apex of this bone forms part of the anterior foramen lacerum. Through the apex, the internal carotid artery exits the petrous bone to the anterior foramen lacerum, where it curves superiorly on its way to the cavernous sinus. The end of the bony part of the eustachian tube, the isthmus, is also located in the apex anterior to the carotid opening and just medial to the spine of the sphenoid. The superior surface of the petrous bone forms part of the middle cranial fossa. It begins from the arcuate eminence and ends at the foramen lacerum. The groove of the greater superficial petrosal nerve can be seen coursing close to the bone near the anterior border of this surface; in 10% of cases, the nerve can be traced posteriorly into a dehiscent geniculate ganglion. The bisection of the angle is formed by this groove and the arcuate eminence marks the position of the internal auditory canal. Near the foramen lacerum, the impression of Meckel's cavity can be seen. The posterior border of this surface is marked by the groove for the superior petrosal sinus, which separates the superior and posterior surfaces. The posterior surface of the petrous bone forms part of the posterior cranial fossa. The opening for endolymphatic duct and sac can be seen at the lateral end of this surface. This opening represents an important landmark for the posterior semicircular canal in procedures using the retrosigmoid approach. The most important feature of the posterior surface is the internal auditory meatus.

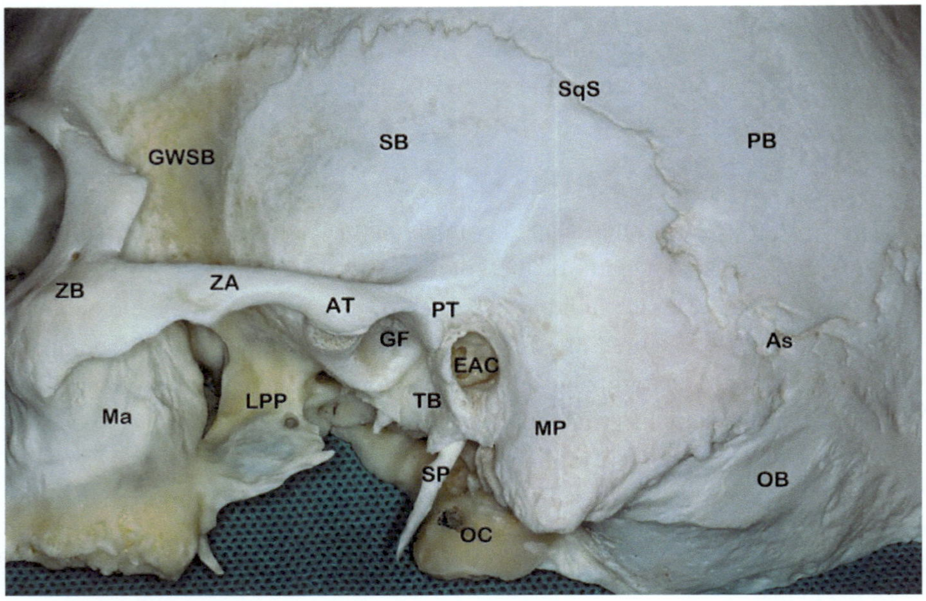

Fig. 4.1 Lateral view of the articulated temporal bone. *GWSB*, greater wing of the sphenoid bone; *SB*, squamosal part of the temporal bone; *PB*, parietal bone; *ZB*, zygomatic bone; *ZA*, zygomatic arch; *AT*, anterior zygomatic tubercle; *PT*, posterior zygomatic tubercle; *GF*, glenoid fossa; *EAC*, external auditory canal; *As*, asterion; *OB*, occipital bone; *MP*, mastoid process; *TB*, tympanic bone; *SP*, styloid process; *LPP*, lateral pterygoid plate; *OC*, occipital condyle; *Ma*, maxilla.

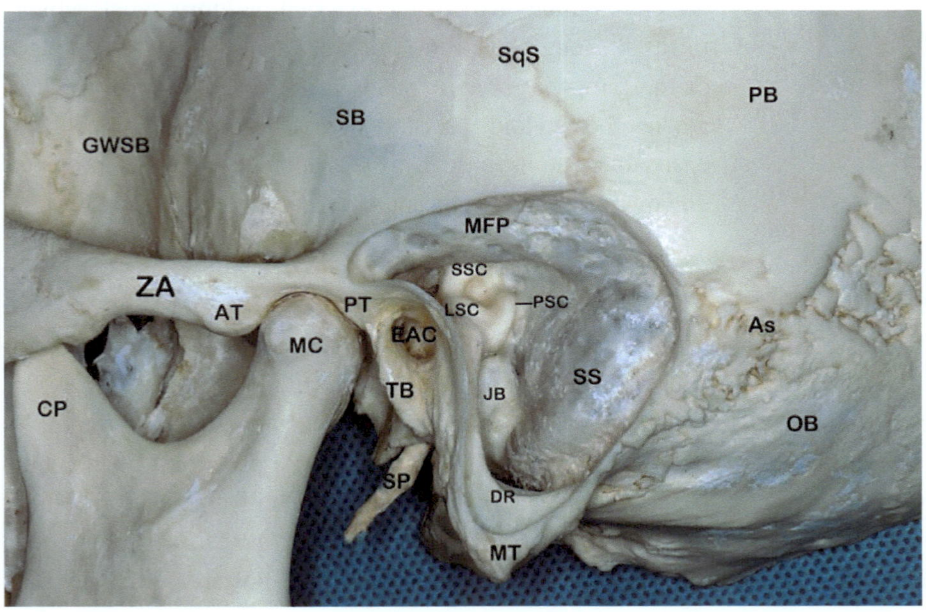

Fig. 4.2 Lateral view of the articulated temporal bone. The cortical mastoidectomy has been performed. *GWSB*, greater wing of the sphenoid bone; *SB*, squamosal part of the temporal bone; *SqS*, squamous suture; *PB*, parietal bone; *MC*, mandibular condyle; *AT*, anterior zygomatic tubercle; *PT*, posterior zygomatic tubercle; *EAC*, external auditory canal; *SSC*, superior semicircular canal; *LSC*, lateral semicircular canal; *PSC*, posterior semicircular canal; *MFP*, middle fossa plate; *JB*, jugular bulb; *SS*, sigmoid sinus; *TB*, tympanic bone; *SP*, styloid process; *DR*, digastric ridge; *CP*, coronoid process; *ZA*, zygomatic arch; *As*, asterion; *OB*, occipital bone; *MT*, mastoid tip.

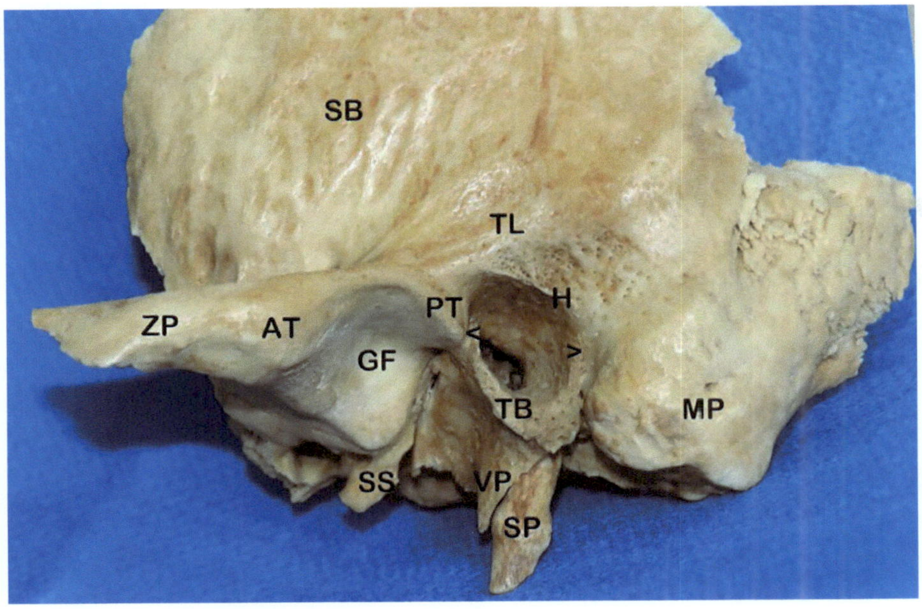

Fig. 4.3 The lateral surface of a left temporal bone. *SB*, squamosal part of the temporal bone; *TL*, temporal line; *ZP*, zygomatic process; *AT*, anterior zygomatic tubercle; *PT*, posterior zygomatic tubercle; *GF*, glenoid fossa; *H*, Henle's spine; *MP*, mastoid process; *TB*, tympanic bone; *SP*, styloid process; *VP*, vaginal process; *SS*, spine of the sphenoid bone; <, tympanosquamous suture; >, tympanomastoid suture.

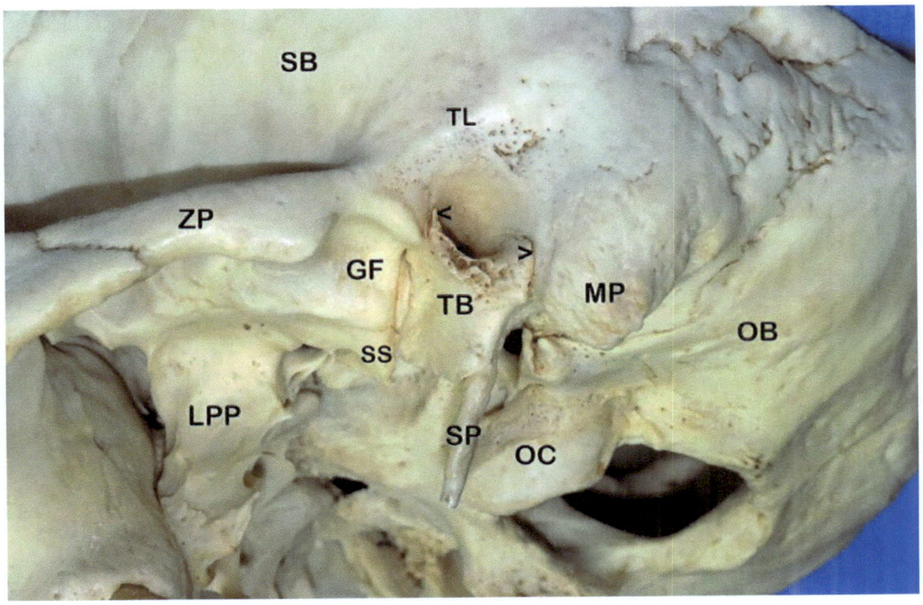

Fig. 4.4 Oblique enlarged inferior view of a left articulated temporal bone. *SB*, squamosal part of the temporal bone; *TL*, temporal line; *ZP*, zygomatic process; *GF*, glenoid fossa; *TB*, tympanic bone; *MP*, mastoid process; *SS*, spine of the sphenoid bone; *SP*, styloid process; *OB*, occipital bone; *OC*, occipital condyle; *LPP*, lateral pterygoid plate; <, tympanosquamous suture; >, tympanomastoid suture.

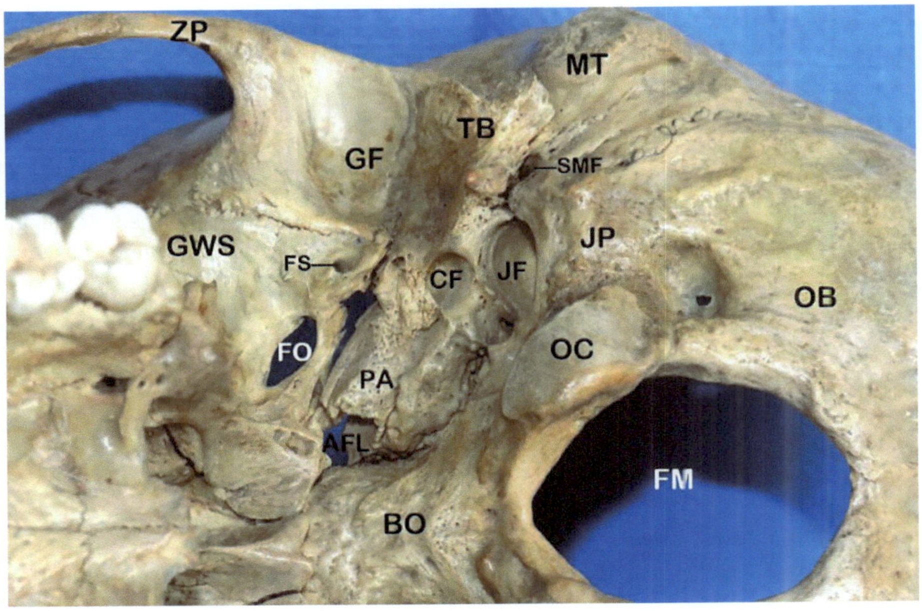

Fig. 4.5 An inferior view of a right articulated temporal bone. *ZP*, zygomatic process; *MT*, mastoid tip; *GF*, glenoid fossa; *TB*, tympanic bone; *SMF*, stylomastoid foramen; *GWS*, greater wing of the sphenoid bone; *FS*, foramen spinosum; *CF*, carotid foramen; *JF*, jugular foramen; *JP*, jugular process of the occipital bone; *FO*, foramen ovale; *PA*, petrous apex; *AFL*, anterior foramen lacerum; *OC*, occipital condyle; *OB*, occipital bone; *BO*, basilar part of the occipital bone; *FM*, foramen magnum.

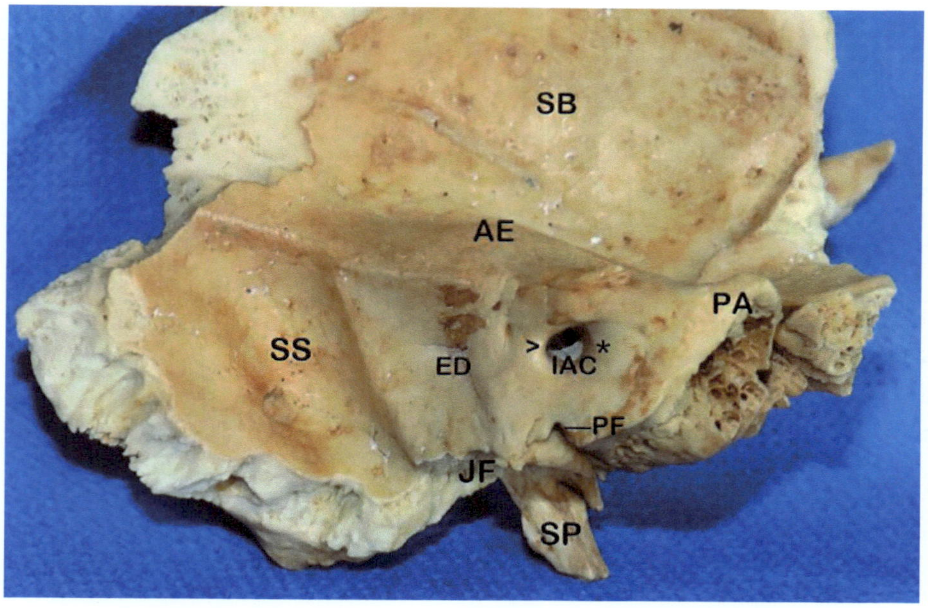

Fig. 4.6 Oblique posterior view of another left temporal bone. *SB*, squamosal part of the temporal bone; *AE*, arcuate eminence; *PA*, petrous apex; *SS*, sigmoid sinus groove; *ED*, endolymphatic duct; *IAC*, internal auditory canal; *PF*, pyramidal fossa; *JF*, jugular foramen; *SP*, styloid process; >, posterior lip; *, anterior lip.

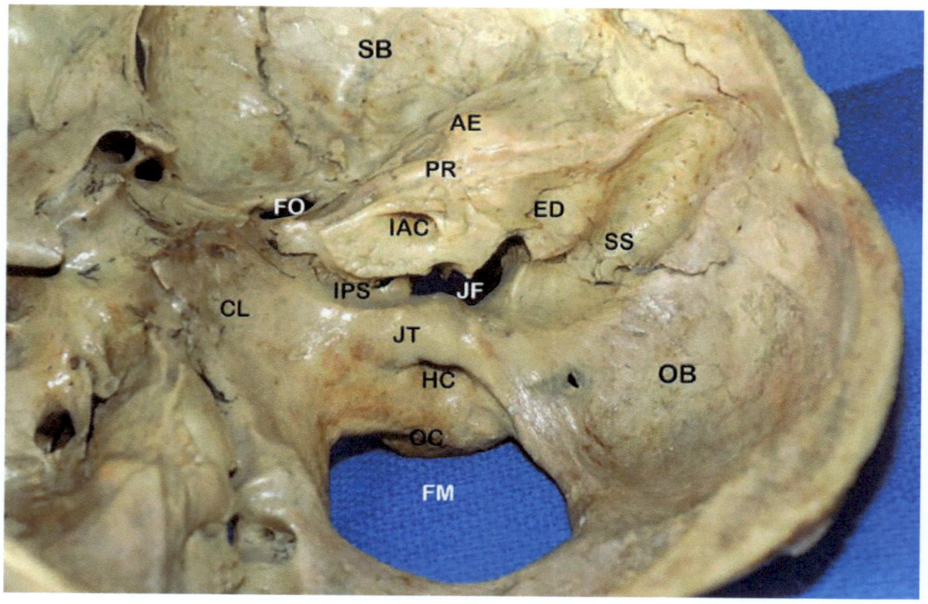

Fig. 4.7 A view of the posterior surface of a right articulated temporal bone. *SB*, squamosal part of the temporal bone; *AE*, arcuate eminence; *PR*, petrous ridge; *FO*, foramen ovale; *IAC*, internal auditory canal; *ED*, endolymphatic duct; *SS*, sigmoid sinus groove; *IPS*, groove for inferior petrosal sinus; *JF*, jugular foramen; *CL*, clivus; *JT*, jugular tubercle; *HC*, hypoglossal canal; *OC*, occipital condyle; *FM*, foramen magnum.

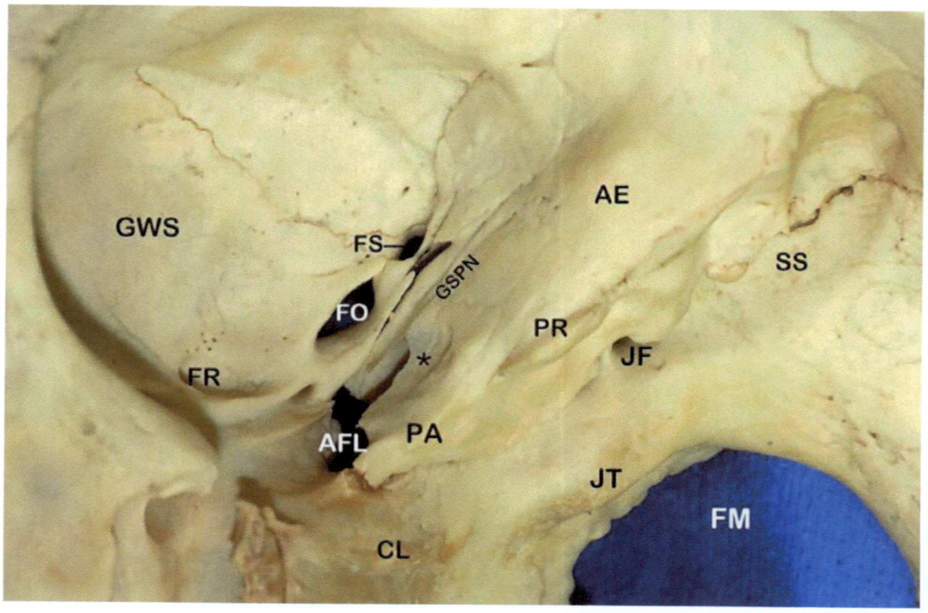

Fig. 4.8 A superior view of a right articulated temporal bone. *GWS*, greater wing of the sphenoid bone; *FS*, foramen spinosum; *FO*, foramen ovale; *FR*, foramen rotundum; *GSPN*, groove for the greater superficial petrosal nerve; *AE*, arcuate eminence; *PR*, petrous ridge; *JF*, jugular foramen; *SS*, sigmoid sinus groove; *AFL*, anterior foramen lacerum; *PA*, petrous apex; *JT*, jugular tubercle; *CL*, clivus; *FM*, foramen magnum; *, Meckel's cave impression.

The Bony Labyrinth

The bony labyrinth consists of three parts: the semicircular canals, the vestibule, and the cochlea.

SEMICIRCULAR CANALS

The prominence of the lateral semicircular canal in the medial wall of the antrum slopes at about 30 degrees, running from anterosuperior to posteroinferior. The bony capsule of the labyrinth is compact and hard, and more resistant to bone erosion. However, due to its proximity to the antrum, the lateral semicircular canal is the most vulnerable labyrinth to pathologies that erode the medial wall of the antrum, such as cholesteatoma. At the anterior end of the lateral semicircular canal lies the ampulla, which accommodates sensory cells and opens to the utricles. The ampulla is located in the medial wall of the posterior attic. The other two semicircular canals are sited nearly perpendicularly to the lateral semicircular canal (Fig. 4.9).

The posterior semicircular canal lies just posterior to the lateral semicircular canal, and the posterior edge of the lateral semicircular canal points almost to the center of the posterior semicircular canal. The posterior semicircular canal courses nearly in parallel with the posterior fossa dura. Its ampulla is located at its inferior end, just medial to the mastoid segment of the facial nerve. The superior end of the posterior semicircular canal joins the superior semicircular canal, forming the common crus (Fig. 4.10).

The superior semicircular canal is located just beneath the middle cranial fossa plate. Its ampulla is at the anterior end, just superomedial to the ampulla of the lateral semicircular canal. It courses nearly perpendicular to the long axis of the pyramid, which places the canal deeper and farther from the antrum posteriorly. The canal is therefore not seen in the majority of cases, and the entire canal may be exposed in limited cases in which intensive removal of perilabyrinthine cells is required, such as petrous bone cholesteatoma. In rare cases, the canal is dehiscent and is directly in contact with the middle fossa dura. The ampullae of the superior and lateral semicircular canals are located in the medial wall of the posterior attic. If it is necessary to drill deeply into the medial wall of the attic, great care should therefore be taken not to open these two ampullae. The labyrinth is less resistant to an insult to the ampulla than to the semicircular canals.

VESTIBULE

The vestibule is a hollow space within the petrous bone that contains the utricle and the saccule. This space lies anterior to the semicircular canals, medial to the oval window, lateral to the fundus of the internal auditory canal, and posterior to the cochlea (Figs. 4.11 and 4.12). Its posterior surface receives the five openings of the semicircular canals. Anteriorly, the ductus reuniens connects the vestibule to the scala vestibuli of the cochlea.

COCHLEA

The cochlea, a spirally curved tube that forms two and a half turns, lies in front of the vestibule (Fig. 4.12). It has a wide base and a narrow apex. The basal turn of the cochlea projects into the middle ear, forming the promontory bulge. The coils of the cochlea curve around a central cone of bone known as the modiolus, which arises from the fundus of the internal auditory canal and carries within it the fibers of the cochlear nerve (Fig. 4.13). From the modiolus, a shelf-like bony projection known as the bony spiral lamina extends halfway into the cavity of the cochlear turns. The membranous spiral lamina or basilar membrane connects the edge of the bony spiral lamina to the lateral surface of the cochlear turns, thus dividing the space into two compartments—the scala vestibule superiorly and the scala tympani inferiorly.

The cochlea has an intimate relationship with the carotid artery at its transition from the vertical to the horizontal segment. The basal turn is anterosuperior to the jugular bulb (Fig. 4.14), with the infracochlear space extending among these two structures and the carotid artery (Fig. 4.15).

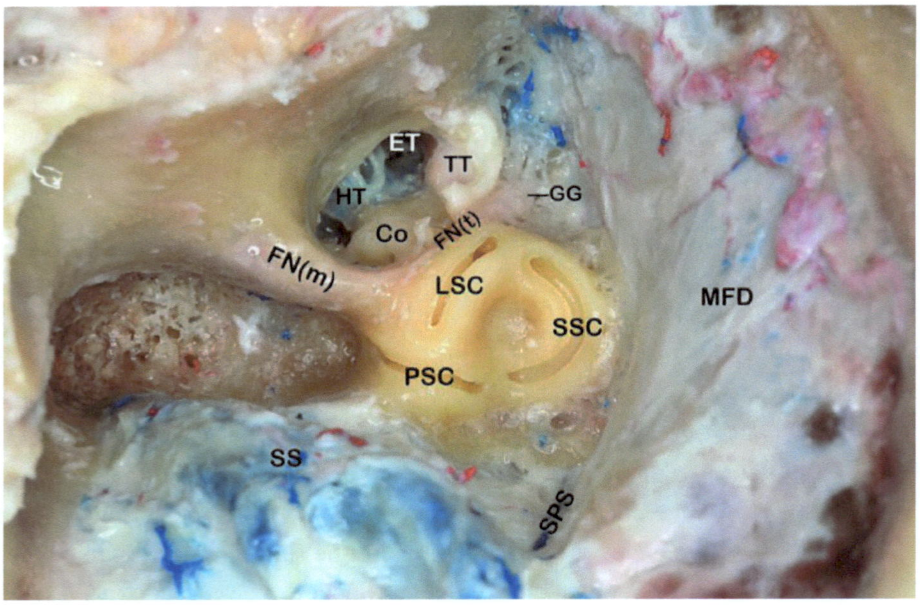

Fig. 4.9 In this left temporal bone, the superior semicircular canal, lateral semicircular canal, and posterior semicircular canal have been identified to show their anatomy. *ET*, eustachian tube; *TT*, tensor tympani muscle; *HT*, hypotympanum; *GG*, geniculate ganglion; *Co*, cochlea; *FN(t)*, tympanic segment of facial nerve; *FN(m)*, mastoid segment of facial nerve; *SSC*, superior semicircular canal; *LSC*, lateral semicircular canal; *PSC*, posterior semicircular canal; *MFD*, middle fossa dura; *SPS*, superior petrosal sinus; *SS*, sigmoid sinus.

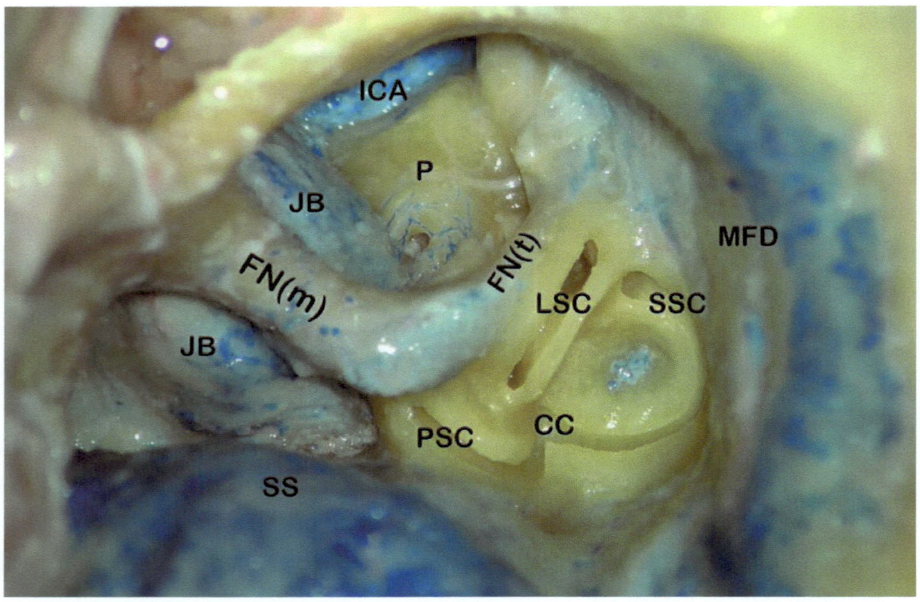

Fig. 4.10 In this left temporal bone, the posterior end of the superior canal and the upper end of the posterior canal join deep to form the common crus. *P*, promontory; *ICA*, internal carotid artery; *SSC*, superior semicircular canal; *LSC*, lateral semicircular canal; *PSC*, posterior semicircular canal; *CC*, common crus; *FN(m)*, mastoid segment of facial nerve; *FN(t)*, tympanic segment of the facial nerve; *MFD*, middle fossa dura; *SS*, sigmoid sinus; *JB*, jugular bulb.

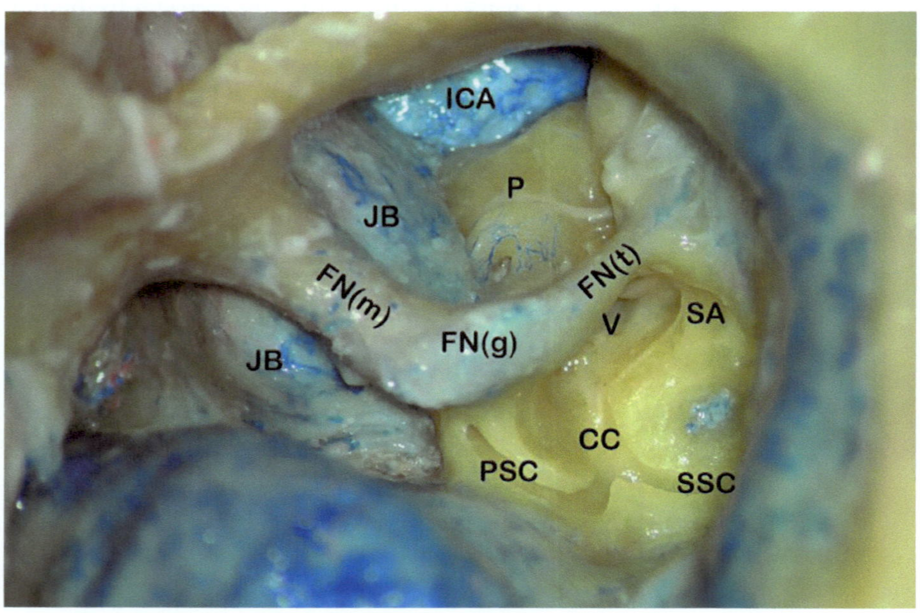

Fig. 4.11 In this left temporal bone, the vestibule has been opened. Note the close relationship between the high jugular bulb and the mastoid segment of the facial nerve. *ICA*, internal carotid artery; *P*, promontory; *SA*, ampullated end of superior semicircular canal; *SSC*, superior semicircular canal; *V*, vestibule; *CC*, common crus; *PSC*, posterior semicircular canal; *FN(m)*, mastoid segment of the facial nerve; *FN(g)*, second genu of facial nerve; *FN(t)*, tympanic segment of the facial nerve; *JB*, jugular bulb.

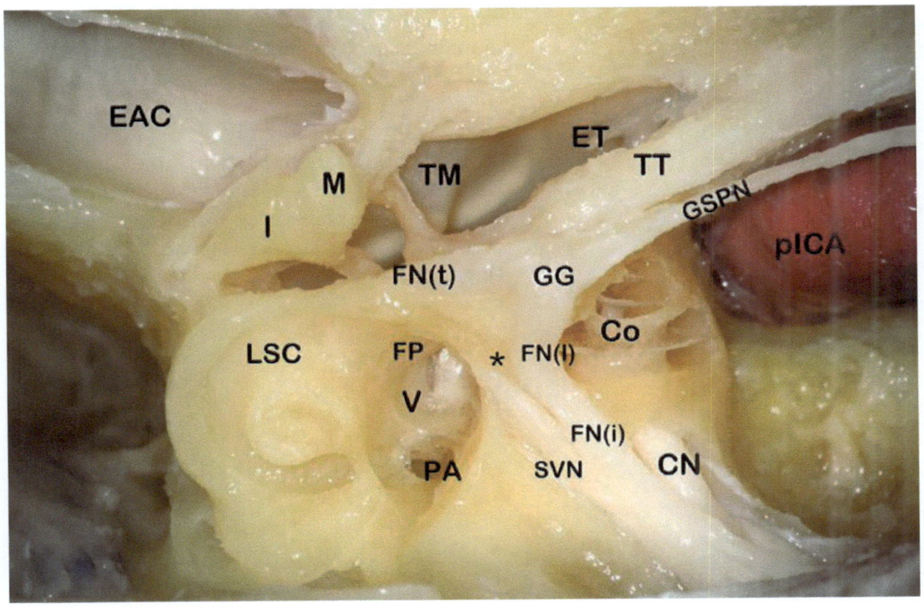

Fig. 4.12 A superior view of the inner ear components as seen from the middle cranial fossa of a left temporal bone. *EAC*, external auditory canal; *I*, incus; *M*, malleus; *TM*, tympanic membrane; *ET*, eustachian tube; *TT*, tensor tympani muscle; *pICA*, petrous segment of internal carotid artery; *GSPN*, greater superficial petrosal nerve; *GG*, geniculate ganglion; *FN(t)*, tympanic segment of the facial nerve; *FN(l)*, labyrinthine segment of the facial nerve; *FN(i)*, internal auditory canal segment of the facial nerve; *SVN*, superior vestibular nerve; *CN*, cochlear nerve; *Co*, cochlea; *LSC*, lateral semicircular canal; *FP*, footplate of the stapes; *V*, vestibule; *PA*, ampullated end of posterior semicircular canal; *, Bill's bar.

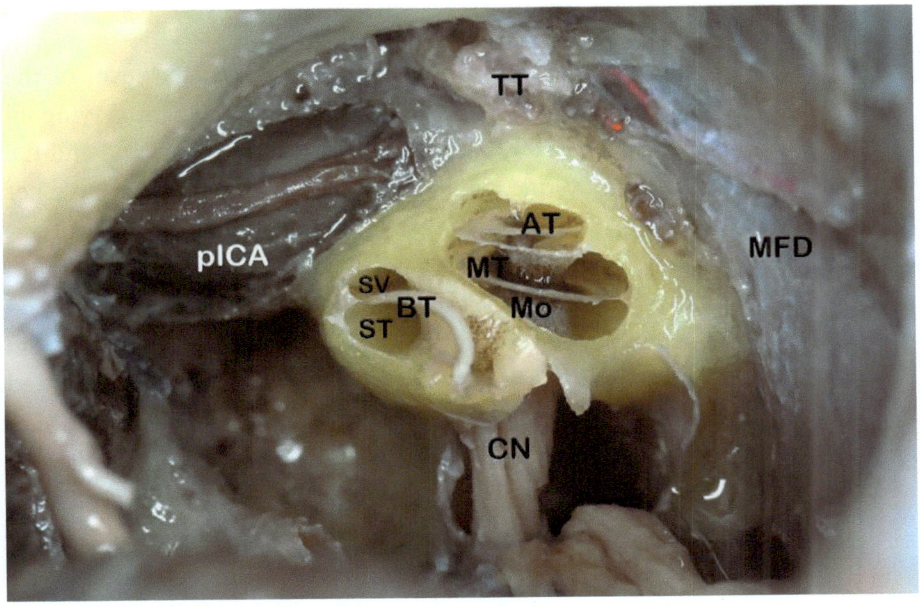

Fig. 4.13 The cochlea has been drilled open, and the basal turn, middle turn, and apical turn can be seen. The cochlear nerve can be seen here passing from the internal auditory canal to the modiolus. *TT*, tensor tympani muscle; *pICA*, petrous segment of internal carotid artery; *MFD*, middle fossa dura; *AT*, apical turn of cochlea; *MT*, middle turn of cochlea; *BT*, basal turn of cochlea; *Mo*, modiolus; *SV*, scala vestibuli; *ST*, scala tympani; *CN*, cochlear nerve.

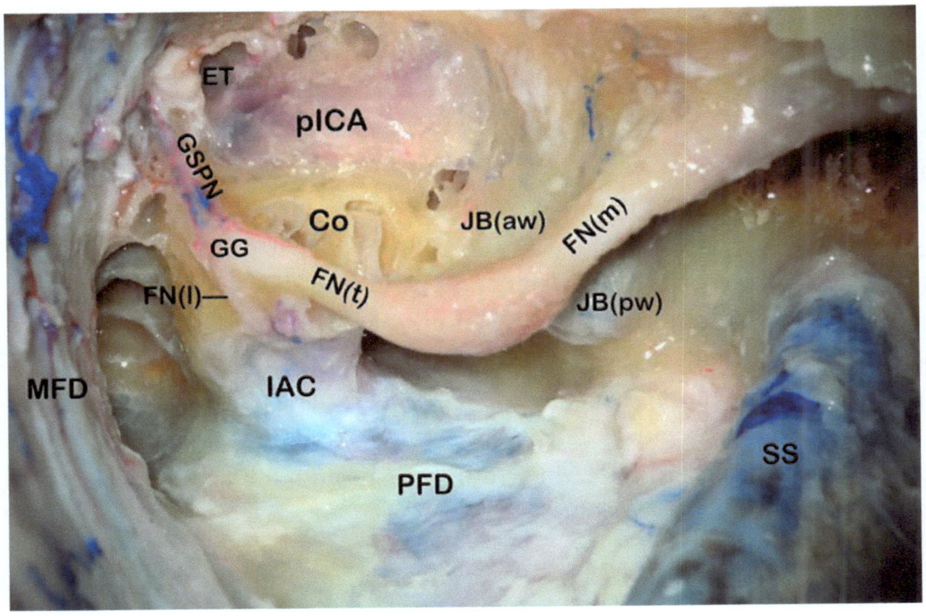

Fig. 4.14 The cochlea has an intimate relationship with the petrous carotid artery. The basal turn of the cochlea is anterosuperior to the jugular bulb. *ET*, eustachian tube; *pICA*, petrous segment of internal carotid artery; *GSPN*, greater superficial petrosal nerve; *GG*, geniculate ganglion; *Co*, cochlea; *FN(t)*, tympanic segment of the facial nerve; *FN(l)*, labyrinthine segment of the facial nerve; *FN(m)*, mastoid segment of the facial nerve; *JB(aw)*, anterior wall of the jugular bulb; *JB(pw)*, posterior wall of the jugular bulb; *IAC*, internal auditory canal; *MFD*, middle fossa dura; *PFD*, posterior fossa dura; *SS*, sigmoid sinus.

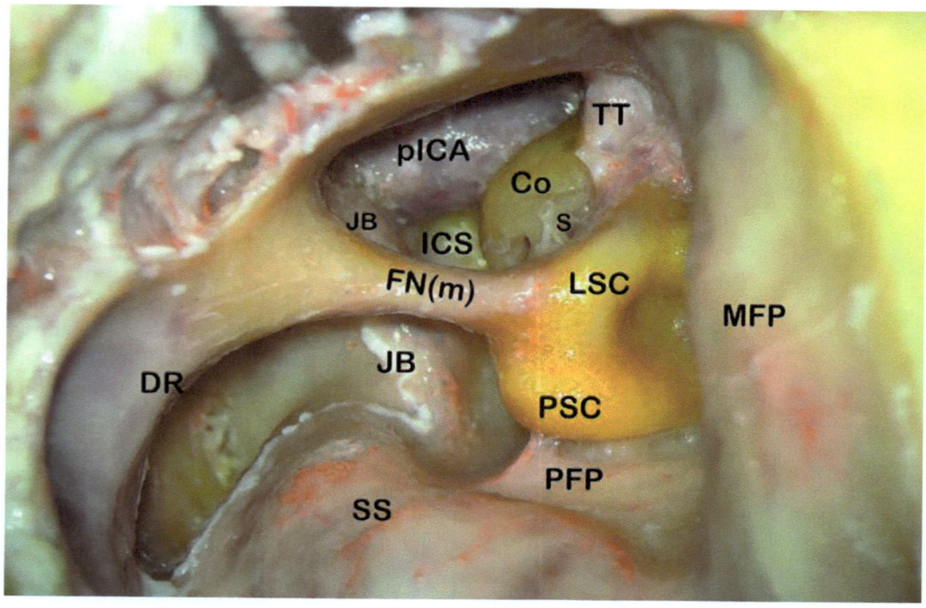

Fig. 4.15 The infracochlear space extending among the cochlea, jugular bulb, and the carotid artery. *TT*, tensor tympani muscle; *pICA*, petrous segment of internal carotid artery; *Co*, cochlea; *S*, stapes; *JB*, jugular bulb; *ICS*, infracochlear space; *FN(m)*, mastoid segment of the facial nerve; *LSC*, lateral semicircular canal; *PSC*, posterior semicircular canal; *MFP*, middle fossa plate; *PFP*, posterior fossa plate; *DR*, digastric ridge; *SS*, sigmoid sinus.

Tympanic Cavity

The tympanic cavity is a narrow air-filled space between the tympanic membrane laterally and the promontory containing the auditory and vestibular labyrinth medially (Fig. 4.16). It communicates posteriorly with the mastoid antrum and anteriorly through the eustachian tube with the nasopharynx. It contains the malleus, incus, and stapes (Fig. 4.17). The tympanic cavity opens upward into the epitympanic recess, which contains the head of the malleus and the body of the incus. The roof of the tympanic cavity is formed by a thin plate, the tegmen tympani, which separates the middle fossa and tympanic cavities, and also roofs the mastoid antrum and the tensor tympani. The thin floor of the tympanic cavity separates the cavity from the jugular bulb. The medial part of the floor is perforated by an opening for the tympanic branch of the glossopharyngeal nerve. The tympanic plexus lies on the promontory, the parasympathetic supply of which is provided by the inferior tympanic branch of the glossopharyngeal nerve (Fig. 4.18). The inferior tympanic artery, a branch of the ascending pharyngeal artery accompanies the nerve and usually occupies a groove in the promontory. The lateral wall is formed by the tympanic membrane and the osseous ring to which the membrane attaches. The ring is deficient above near the openings of the anterior and posterior canaliculi for the chorda tympani. The posterior canaliculus for the chorda tympani arises from the facial canal a few millimeters above the stylomastoid foramen and ascends in front of the facial canal to open into the tympanic cavity at the level of the upper part of the handle of the malleus. The chorda tympani passes in close relation to the tympanic membrane and the upper part of the handle of the malleus and forward to enter its anterior canaliculus at the medial aspect of the petrotympanic fissure (Fig. 4.19), and descends vertically medial to the sphenoid spine and lateral pterygoid muscle to join the lingual nerve. The nerve contains sensory fibers for taste and secretory fibers innervating the submandibular and sublingual glands.

The anterior or carotid wall of the middle ear is separated from the ICA by a thin layer of cortical bone. This is perforated usually by two caroticotympanic arteries. The eustachian tube and tensor tympani open into the anterosuperior aspect of the middle ear and lie lateral to the ICA (Fig. 4.12). In 2% of cases, the bony shell separating the carotid artery from the eustachian tube is absent.

Middle ear landmarks for the facial nerve are also of critical importance (Fig. 4.20). The nerve runs immediately superior to the oval window, and the cochleariform process lies immediately inferior to the nerve before it enters the geniculate ganglion. The stapes superstructure lies in close association with the tympanic segment of the nerve. The cochleariform process provides an attachment for the tendon of the tensor tympani prior to its attachment to the neck of the malleus. The cog is a bony process situated on the roof of the epitympanum and lies in the same plane as the cochleariform process; it can provide an additional landmark when there has been significant erosion by tumor.

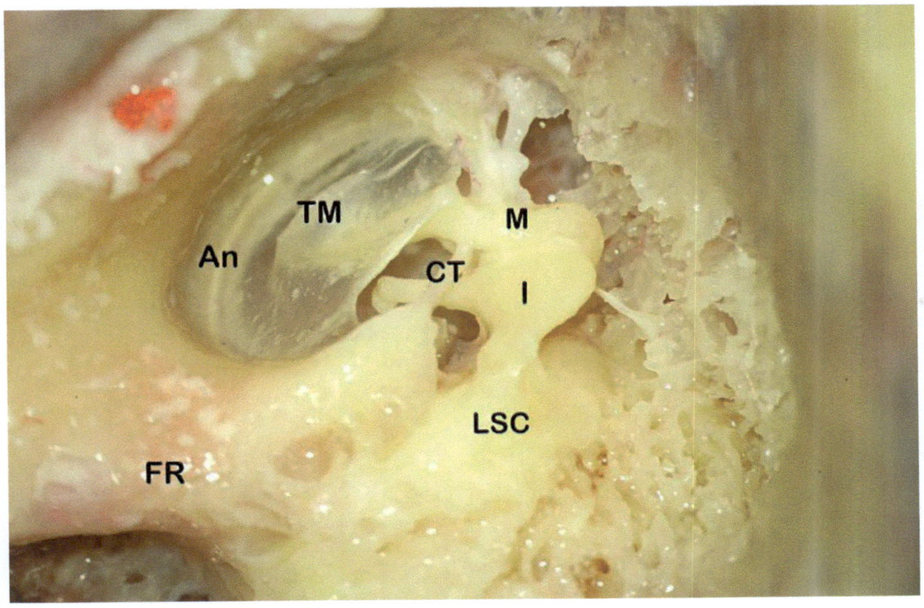

Fig. 4.16 A microscopic view of the left tympanic membrane and articulated ossicular chain. *An*, annulus; *TM*, tympanic membrane; *CT*, chorda tympani; *M*, malleus; *I*, incus; *LSC*, lateral semicircular canal; *FR*, facial ridge.

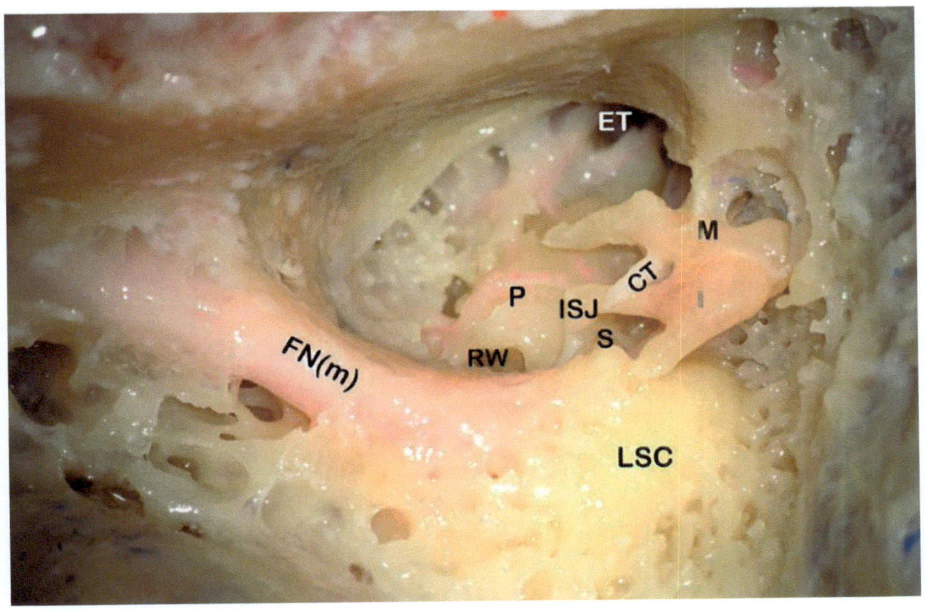

Fig. 4.17 A view of the articulated ossicular chain of the left ear. *ET*, eustachian tube; *CT*, chorda tympani; *M*, malleus; *I*, incus; *S*, stapes; *ISJ*, incudostapedial joint; *P*, promontory; *RW*, round window; *FN(m)*, mastoid segment of the facial nerve; *LSC*, lateral semicircular canal.

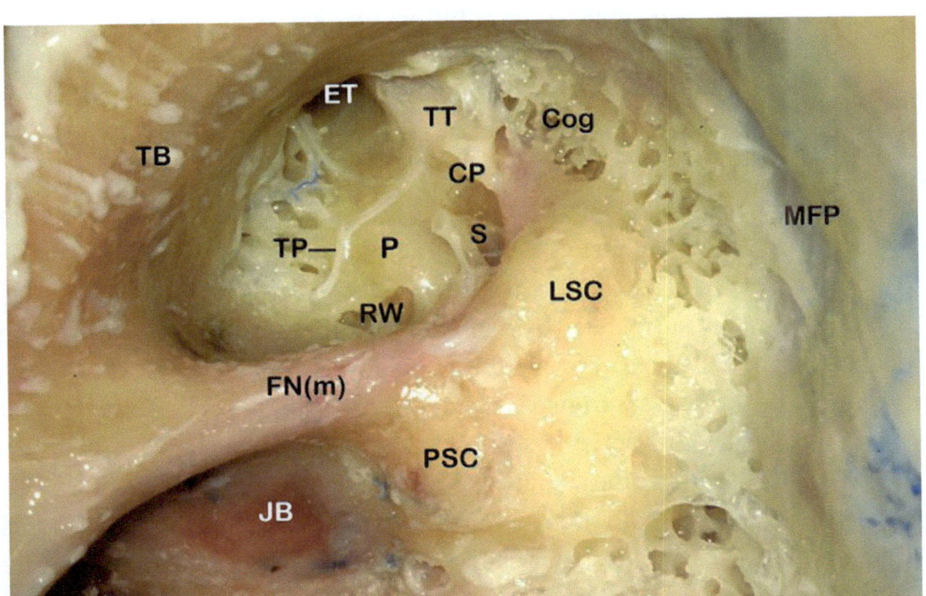

Fig. 4.18 The tympanic plexus lies on the promontory. *ET*, eustachian tube; *TT*, tensor tympani muscle; *TB*, tympanic bone; *CP*, cochleariform process; *TP*, tympanic plexus; *P*, promontory; *S*, stapes; *RW*, round window; *LSC*, lateral semicircular canal; *FN(m)*, mastoid segment of the facial nerve; *PSC*, posterior semicircular canal; *JB*, jugular bulb; *MFP*, middle fossa plate.

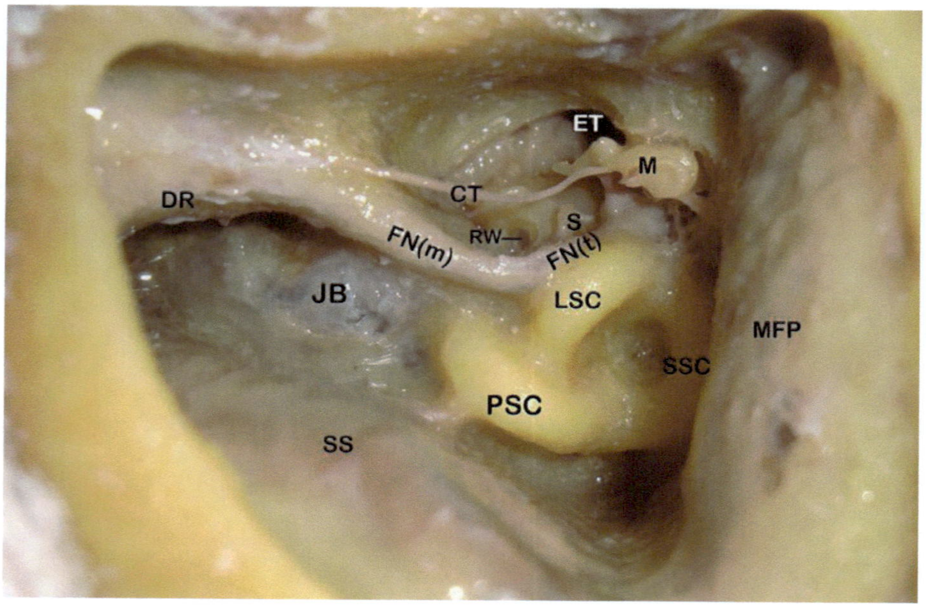

Fig. 4.19 The chorda tympani passes to the upper part of the handle of the malleus and forward to enter its anterior canaliculus at the medial aspect of the petrotympanic fissure. *ET*, eustachian tube; *M*, malleus; *CT*, chorda tympani; *S*, stapes; *RW*, round window; *FN(m)*, mastoid segment of the facial nerve; *FN(t)*, tympanic segment of the facial nerve; *JB*, jugular bulb; *DR*, digastric ridge; *SSC*, superior semicircular canal; *LSC*, lateral semicircular canal; *PSC*, posterior semicircular canal; *SS*, sigmoid sinus; *MFP*, middle fossa plate.

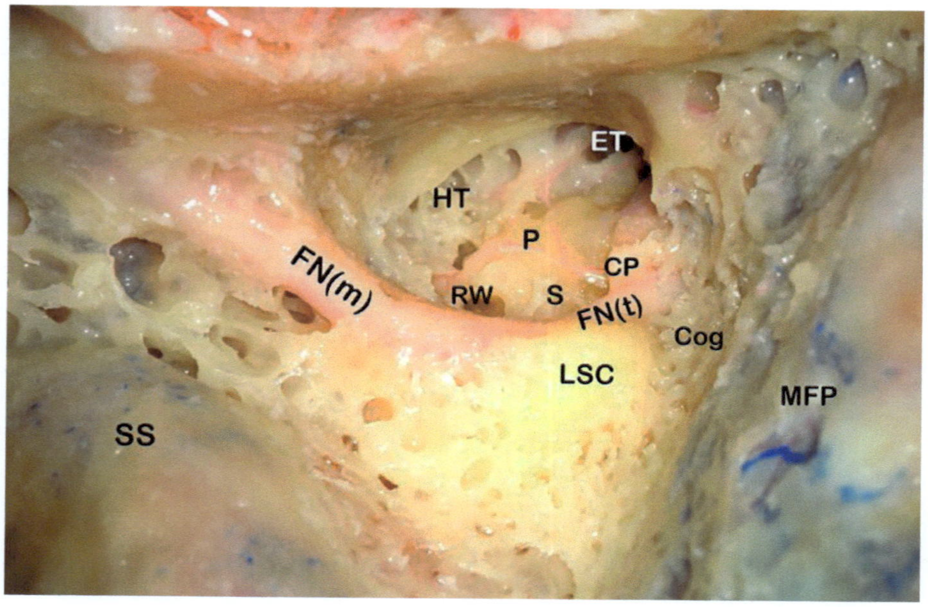

Fig. 4.20 A view of the tympanic cavity of a left ear after removal of the superior and posterior canal walls, the malleus, and incus. *ET*, eustachian tube; *HT*, hypotympanum; *P*, promontory; *CP*, cochleariform process; *RW*, round window; *S*, stapes; *FN(m)*, mastoid segment of the facial nerve; *FN(t)*, tympanic segment of the facial nerve; *LSC*, lateral semicircular canal; *SS*, sigmoid sinus; *MFP*, middle fossa plate.

Internal Auditory Canal

The internal auditory canal is a canal almost 1 cm in length that runs in a lateral direction from the cerebellopontine angle through the petrous bone (Figs. 4.21 and 4.22). The porus of the internal auditory canal can be seen on the posterior surface of the petrous bone (Fig. 4.6). Its posterior border is formed by an acute angle, while the anterior border is flatter. The dura of the posterior cranial fossa continues into the internal auditory canal (Fig. 4.23), lining the whole length, and ends by merging with the contained nerves as these enter their corresponding foramina. The long axis of the canal lies in line with the long axis of the external auditory canal.

At the lateral end of the internal auditory canal lies the fundus (Figs. 4.24 and 4.25). The horizontal crest divides this area into a smaller upper zone and a larger lower zone. The upper zone is further divided by the vertical bar (Bill's bar) into an anterior foramen for the passage of the facial nerve and a posterior one for the passage of the superior vestibular nerve. The superior ampulla is immediately lateral to the superior ampullary nerve (continuation of the superior vestibular nerve) (Fig. 4.26), which itself is lateral to the labyrinthine segment of the facial nerve (Fig. 4.27). The cochlear nerve passes to the cochlea through a central canal surrounded by multiple small foramina located anteriorly in the lower zone of the fundus. The inferior vestibular nerve passes from the posterior part of this zone. Posteroinferior to this nerve lies the canal for the singular nerve, which supplies the ampulla of the posterior semicircular canal. In addition to this nerve, the canal also contains the internal auditory artery and vein, and a loop of the anterior inferior cerebellar artery may also be present.

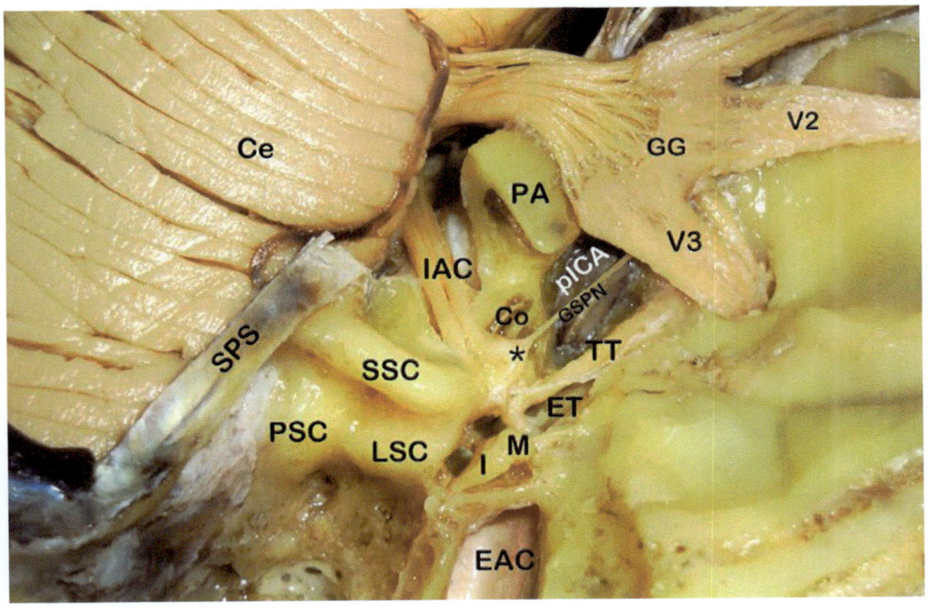

Fig. 4.21 The anatomic view of the right middle fossa. *Ce*, cerebellum; *GG*, gasserian ganglion; *V2*, maxillary nerve; *V3*, mandibular nerve; *PA*, petrous apex; *IAC*, internal auditory canal; *SPS*, superior petrosal sinus; *SSC*, superior semicircular canal; *LSC*, lateral semicircular canal; *PSC*, posterior semicircular canal; *Co*, cochlea; *pICA*, petrous segment of internal carotid artery; *GSPN*, greater superficial petrosal nerve; *ET*, eustachian tube; *TT*, tensor tympani muscle; *M*, malleus; *I*, incus; *EAC*, external auditory canal; *, geniculate ganglion.

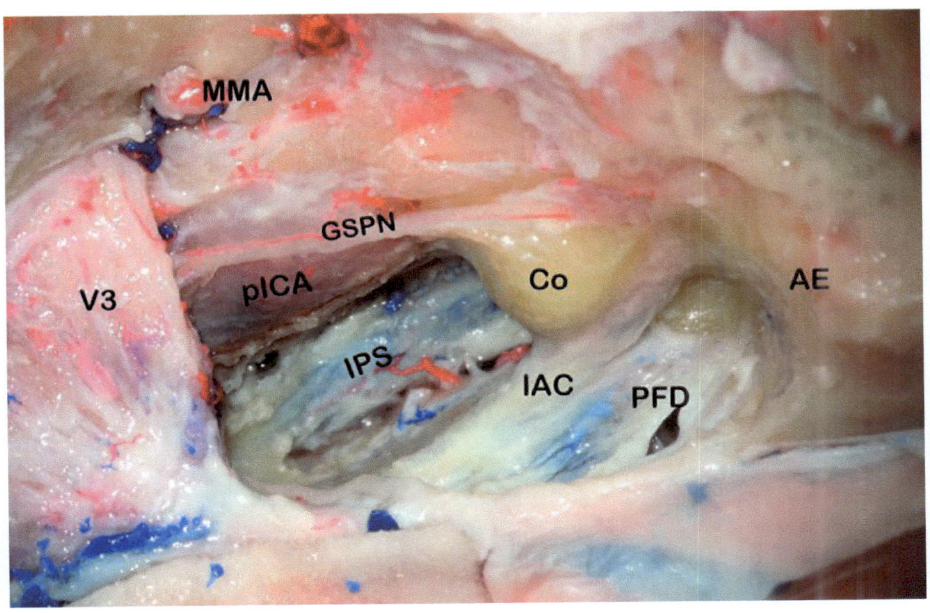

Fig. 4.22 The view through the middle fossa approach of a right temporal bone illustrates the relationship between the internal auditory canal and various structures in this area. *MMA*, middle meningeal artery; *V3*, mandibular nerve; *GSPN*, greater superficial petrosal nerve; *pICA*, petrous segment of internal carotid artery; *Co*, cochlea; *AE*, arcuate eminence; *IPS*, inferior petrosal sinus; *IAC*, internal auditory canal; *PFD*, posterior fossa dura.

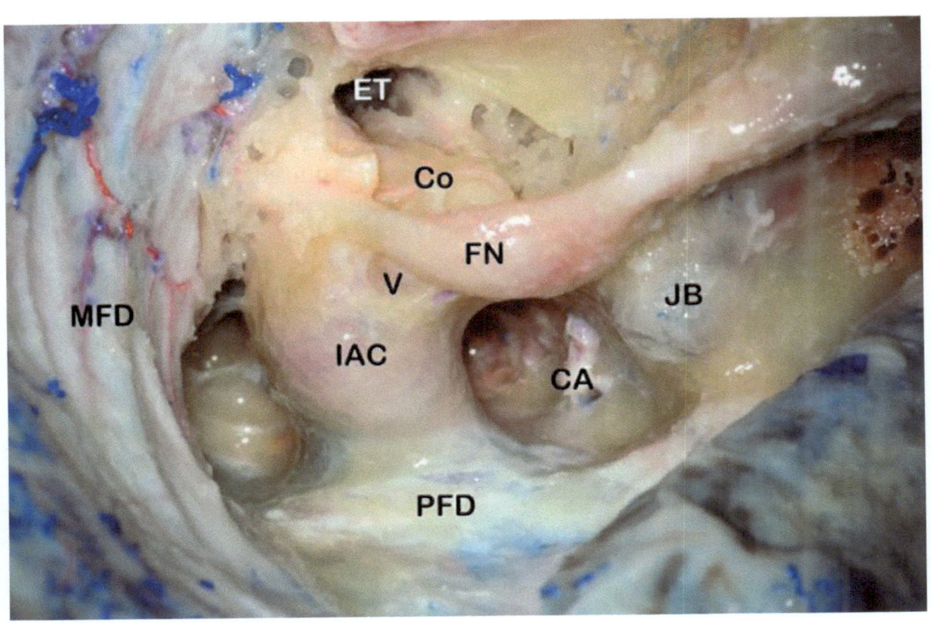

Fig. 4.23 Skeletonization of the internal auditory canal of a right temporal bone has been done. The bone overlying the internal auditory canal has been completely removed. *ET*, eustachian tube; *Co*, cochlea; *FN*, facial nerve; *V*, vestibule; *IAC*, internal auditory canal; *PFD*, posterior fossa dura; *CA*, cochlear aqueduct; *JB*, jugular bulb; *MFD*, middle fossa dura.

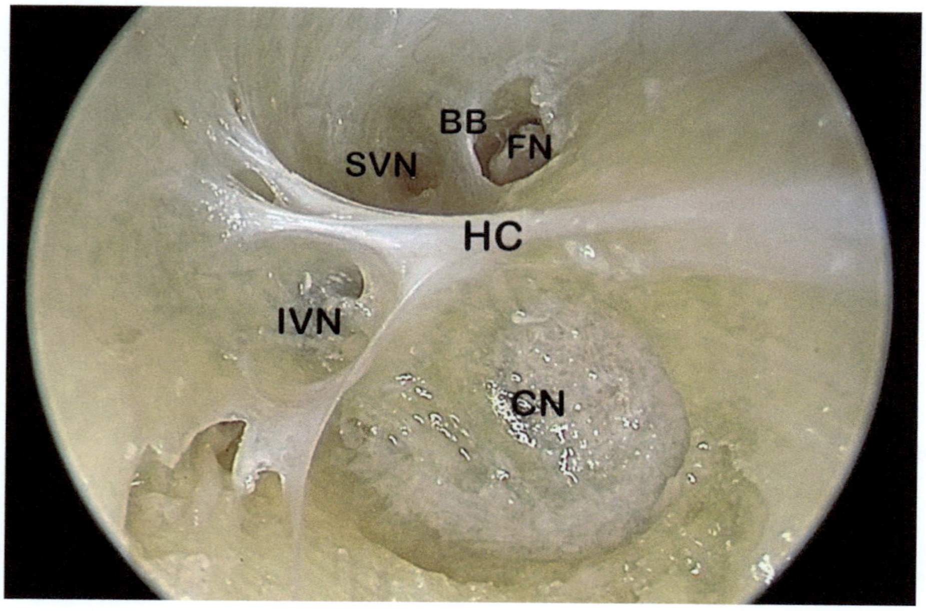

Fig. 4.24 An endoscopic view of the fundus of a left internal auditory canal. The opening of the fallopian canal for the passage of the facial nerve can be seen lying anterior to Bill's bar. *BB*, Bill's bar; *HC*, horizontal crest; *SVN*, superior vestibular nerve; *FN*, facial nerve; *IVN*, inferior vestibular nerve; *CN*, cochlear nerve.

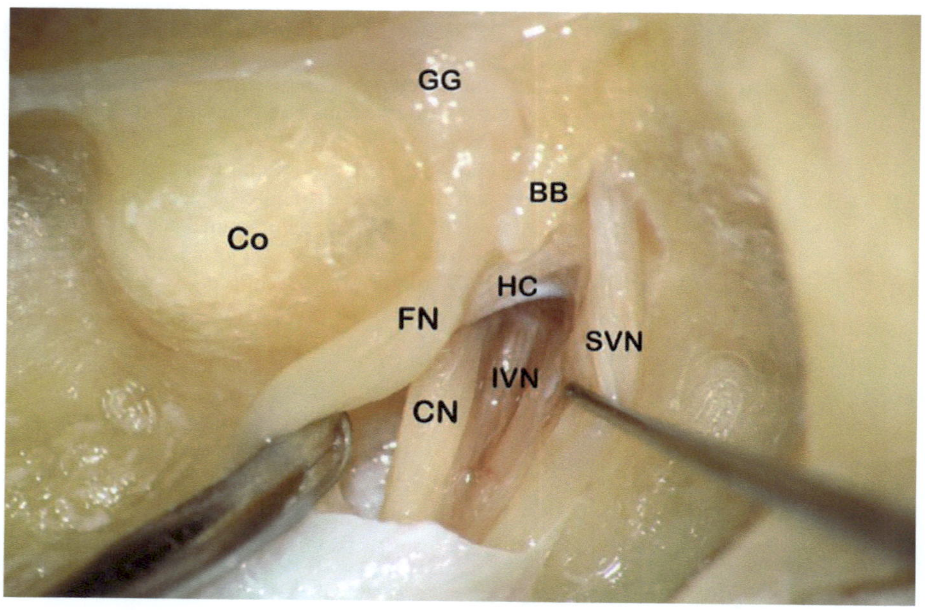

Fig. 4.25 The view through the middle fossa approach of a right temporal bone. The transverse crest divides the fundus of the canal into an upper and a lower compartment. The facial and superior vestibular nerve course above the transverse crest and the cochlear and inferior vestibular nerves pass below the crest. The vertical crest (Bill's bar) separates the facial and superior vestibular nerves. *GG*, geniculate ganglion; *Co*, cochlea; *BB*, Bill's bar; *HC*, horizontal crest; *SVN*, superior vestibular nerve; *FN*, facial nerve; *IVN*, inferior vestibular nerve; *CN*, cochlear nerve.

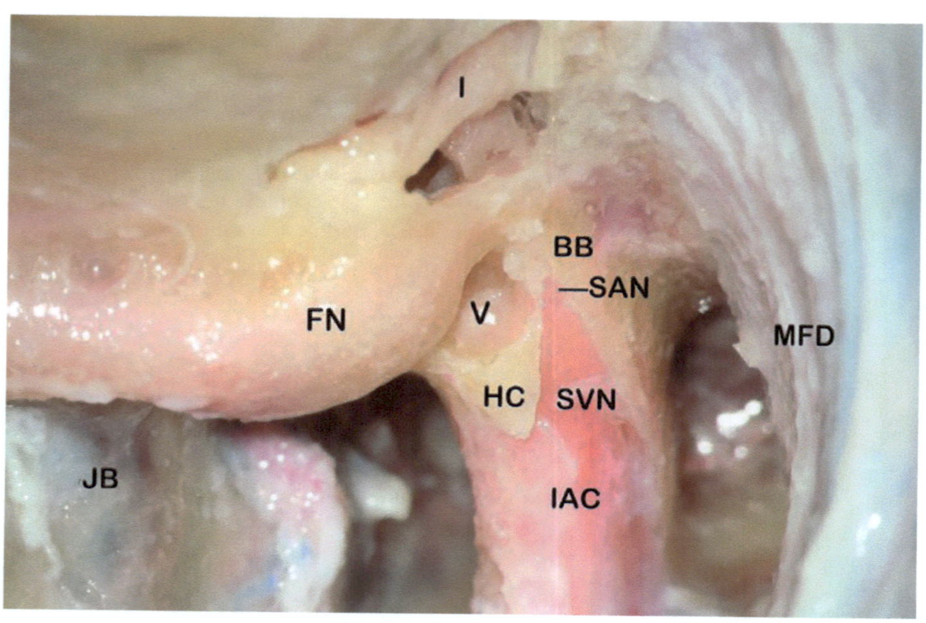

Fig. 4.26 The horizontal crest separates the inferior vestibular nerve from the superior vestibular nerve. The superior vestibular nerve is followed laterally where it leaves the fundus of the internal auditory canal. The nerve then lies in a tiny canal that enters the ampulla of the superior semicircular canal as the superior ampullary nerve. *I*, incus; *BB*, Bill's bar; *SAN*, superior ampullary nerve; *SVN*, superior vestibular nerve; *V*, vestibule; *FN*, facial nerve; *HC*, horizontal crest; *MFD*, middle fossa dura; *IAC*, internal auditory canal; *JB*, jugular bulb.

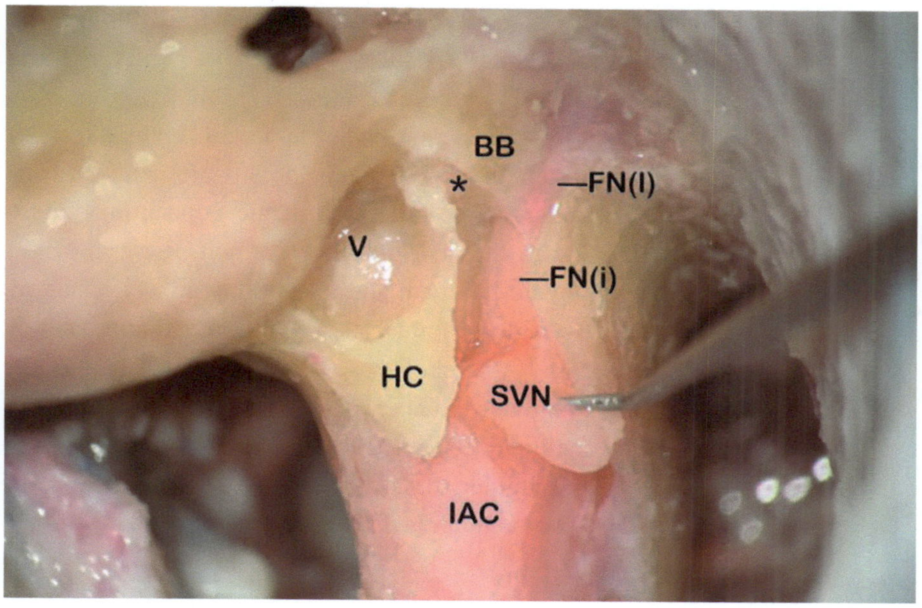

Fig. 4.27 The hook is moved medially toward the fundus, with subsequent separation and establishment of a good plane of cleavage between the superior vestibular nerve and the anteriorly lying facial nerve. Once the superior ampullary nerve is detached and reflected medially and posteriorly, the facial nerve can be clearly observed. *BB*, Bill's bar; *FN(l)*, labyrinthine segment of the facial nerve; *FN(i)*, internal auditory canal segment of the facial nerve; *SVN*, superior vestibular nerve; *HC*, horizontal crest; *IAC*, internal auditory canal; *V*, vestibule; *, canal for superior vestibular nerve.

Jugular Foramen

Overview

The anatomy of the jugular foramen itself is highly complex and is surrounded by several compartments and structures that limit access to this area. The initial exposition of the jugular foramen anatomy provides an ideal core to expand upon and describe the adjacent compartments. These include the middle ear and mastoid, the upper neck, and the posterior fossa compartments.

The anatomy of the jugular foramen area is one of the most variable regions at the skull base. There is some divergence in the literature regarding the nomenclature of this area, with the terms jugular foramen and jugular fossa often used interchangeably. Jugular foramen is a three-dimensional space containing the inferior petrosal sinus, lower cranial nerves, and jugular bulb. From an anatomical point of view, it is a very complex area, representing the connection between the intracranial and the extracranial compartments.

Bony Margins of the Jugular Foramen

The jugular foramen is located immediately behind the entrance of the carotid canal, lateral to the anterior half of the occipital condyle, medial to the stylomastoid foramen, and posteromedial to the styloid process and posteroinferior to the middle ear cleft (Fig. 4.4). It shows a tremendous variability in all three spatial dimensions. Asymmetry of the bulbs is frequent, with the right side larger in ~70% of cases.

The jugular foramen is bounded by the temporal and occipital bones (Fig. 4.7). The anterolateral wall of the foramen is formed by the petrous portion of the temporal bone, and the posteromedial wall is formed by the jugular process of the condylar part of the occipital bone.

The petroclival fissure between the lateral edge of the clival part of the occipital bone and the petrous part of the temporal bone intersects the anteromedial edge of the foramen, and the occipitomastoid suture between the mastoid portion of the temporal bone and the condylar part of the occipital bone intersects its posterolateral edge.

The temporal component of the jugular foramen can be seen anteromedial to the stylomastoid foramen and medial to both the tympanic bone and the styloid process. From the lateral border of the foramen, the jugular spine of the temporal bone can be seen extending into the foramen toward its occipital counterpart and separating the foramen into what are known as the vascular and nervous compartments. Through the fossa and at a more superior level, the dome of the jugular bulb can be seen. Posteriorly lies the small canal for the passage of Arnold's nerve (the auricular branch of the vagus nerve), while anteriorly the end of the groove of the inferior petrosal sinus can be seen lateral and anterior to the opening of the cochlear aqueduct. The foramen of the internal carotid artery is separated from the anterior border of the jugular foramen by a thin wedge of bone called the jugulocarotid spine, through which a canal for the passage of Jacobson's nerve (tympanic nerve) to the tympanic cavity lies (Fig. 4.28).

Internal Anatomy

The jugular foramen can be divided into an anteromedial compartment (petrosal part of jugular foramen), usually containing the glossopharyngeal nerve, inferior petrosal sinus, and meningeal branch of the ascending pharyngeal artery, and a posterolateral compartment (sigmoid part of jugular foramen) containing the vagus and spinal accessory nerves with the jugular bulb. This arrangement shows a high degree of variability.

The two compartments are divided by the jugular spine of the temporal and occipital bones, which are connected by a fibrous band that can occasionally ossify (Figs. 4.29 and 4.30). The endocranial opening is divided by a dural fold with the glossopharyngeal nerve usually running anterosuperomedially and the vagus and spinal accessory nerves posteroinferolaterally.

The sigmoid sinus joins the jugular foramen posteriorly. At this level it usually makes an almost 180 degrees turn at the jugular bulb, descending from there into the neck as the jugular vein (Fig. 4.31). During the transition between the sigmoid sinus and the bulb, the vessel progressively loses its dural envelopment, becoming more fragile.

At the level of the junction between the sigmoid sinus and the bulb, the inferomedial wall presents the opening of the posterior condylar vein (Figs. 4.32 and 4.33); the opening of the anterior condylar vein is usually located close to the orifice of the inferior petrosal sinus if not in the inferior petrosal sinus itself. The two condylar veins then run respectively into the posterior condylar and the hypoglossal canal, establishing a connection with the perivertebral venous plexus.

The opening of the inferior petrosal sinus is usually located in the anteromedial wall of the bulb (Fig. 4.34). More openings may be present.

The lower cranial nerves make a 90 degrees rotation after entering the foramen, with the glossopharyngeal nerve in the upper position, in close association with the opening of the cochlear aqueduct at the site (Figs. 4.35 and 4.36). This is represented by the pyramidal fossa or glossopharyngeal fossa at the anterolateral margin of the bony jugular foramen (Fig. 4.37).

The exocranial end of the jugular fossa is the portal of exit of the glossopharyngeal, vagus, and spinal accessory nerves, and of the jugular vein at the skull base into the post-styloid parapharyngeal space.

LATERAL RELATIONSHIPS

The vertical or mastoid segment of the facial nerve is lateral to the jugular bulb, with the larger component of the bulb lying posterior to the vertical plane of the nerve in the majority of cases (Fig. 4.14). From the stylomastoid foramen, the nerve turns anteriorly, entering the parotid gland.

The styloid process of the temporal bone averages ~2.4 cm in length and projects downward and forward from the undersurface of the tympanic bone immediately anteromedial to the stylomastoid foramen. It gives origin to the stylohyoid ligament and the stylohyoid, stylopharyngeus, and styloglossus muscles. From a lateral approach, it blocks access to the skull base entry points of the carotid artery and jugular vein (Fig. 4.38). The mastoid process is variably pneumatized, and further obscures posterolateral access to the jugular fossa area, with the sternocleidomastoid muscle attaching laterally and the digastric muscle attaching medially.

SUPERIOR RELATIONSHIPS

The jugular plate separates the jugular fossa from the hypotympanic area of the middle ear and is usually a thin plate of cortical bone. The jugular bulb may protrude into the middle ear in a small percentage of cases (Fig. 4.39), and sometimes the jugular plate may be missing. Superolateral to the jugular bulb is the inferior portion of the tympanic bone; this may be infiltrated by the lesion, which in this way is able to reach the external auditory canal to produce the classic otoscopic appearance of a "rising sun."

Progressing in a lateral to medial direction, the superior relationships of the jugular bulb are represented by the posterior semicircular canal (Fig. 4.40) and the internal auditory canal (Fig. 4.41). The relationship with both these structures may be highly variable, depending on the size, shape, and position of the jugular bulb. The internal auditory canal is involved only in the advanced stages of tumor. Several foramina and canaliculi of importance to the jugular fossa area are present on the inferior aspect of the skull base.

The tympanic canaliculus is seen in the jugulotympanic spine, transmitting the inferior tympanic artery and tympanic nerve (Jacobson's nerve) from the glossopharyngeal nerve into the middle ear. The nerve then runs on the surface of the promontory accompanied by the inferior tympanic artery, reaching the area of the cochleariform process.

The mastoid canaliculus at the medial aspect of the jugular fossa transmits the auricular branch of the vagus nerve (Arnold's nerve) arising from the superior vagal ganglion. The importance of these canaliculi is due to the distribution of paraganglial tissue along the course of these nerves as well as within the adventitia of the jugular bulb itself.

The cochlear aqueduct is a bony channel with a pyramidal shape connecting the perilymphatic space of the scala tympani in close proximity to the round window with the subarachnoid space at the level of the jugular foramen (Fig. 4.42).

ANTERIOR RELATIONSHIPS

The vertical portion of the carotid canal lies immediately anterior to the jugular fossa, separated by the jugulocarotid spine, or carotid crest (Figs. 4.28 and 4.43). The close relationship explains the frequent involvement of the canal as well as the artery during the growth of tumor.

MEDIAL RELATIONSHIPS

The intracranial opening of the jugular fossa provides the entry point for the glossopharyngeal, vagus, and spinal accessory nerves as well as the sigmoid sinus and inferior petrosal sinus. At the point of entry into the fossa, the nerves rotate 90 degrees. The glossopharyngeal, vagus, and accessory nerves pierce the dural roof of the jugular foramen. The glossopharyngeal nerve is separated from the vagus nerve and spinal accessory nerves by a narrow dural septum (Fig. 4.44). The glossopharyngeal nerve enters the glossopharyngeal meatus and the vagus nerve enters the vagal meatus with the branches of the accessory nerve. The jugular dural fold projects downward and medially from the lateral and upper margin of the jugular foramen over the site at which the nerves enter the dura roof of the foramen. A bridging vein from the medulla joins the inferior petrosal sinus on the medial side of the jugular bulb (Fig. 4.44).

The jugular tubercle is a rounded prominence located at the junction of the basilar and condylar portions of the occipital bone. It is situated above the hypoglossal canal and medial to the lower half of the intracranial end of the jugular foramen (Fig. 4.45). The glossopharyngeal, vagus, and accessory nerves cross the posterior portion of the jugular tubercle in passing from the brainstem to the jugular foramen (Fig. 4.46), sometimes coursing in a shallow groove on the surface of the tubercle.

INFERIOR RELATIONSHIPS

The occipital condyles are part of the occipital bone (Figs. 4.4 and 4.5). They are oval in shape, face downward and laterally, and articulate with the atlas. The posterior half of each occipital condyle forms part of the inferomedial wall of the jugular fossa. The occipital condyles are composed of cancellous bone surrounded by a thin layer of cortical bone. The presence of cancellous bone can be helpful from an imaging perspective, but it can make a clear delineation from infiltrated bone challenging. The condyles are crossed by two canals: the anterior condylar canal, also known as hypoglossal canal, and the posterior condylar canal (Fig. 4.47). The hypoglossal canal is directed forward and laterally at a 44 degrees angle with the sagittal plane. The position of the hypoglossal canal is consistently located just anterior, inferior, and slightly medial to the anterior inferior edge of the jugular bulb.

The extracranial end of the hypoglossal canal is located immediately above the junction of the anterior and middle one-third of the occipital condyle. The hypoglossal canal contains a venous plexus (anterior condylar vein) and the hypoglossal nerve. Very often the nerve originates via two distinct routes that fuse in the canal into a single nerve (Figs. 4.48 and 4.49).

As the hypoglossal nerve exits the canal it is usually anteromedial to the vagus nerve and then observes a 270 degrees turn to pass anterior to it.

The posterior condylar canal contains the posterior condylar vein, establishing a connection between the bulb and the perivertebral venous plexus.

The atlas is located inferiorly to the skull and its articular facets articulate with the occipital condyles (Fig. 4.50). The atlas and the occipital bone are united by the articular capsules surrounding the atlanto-occipital joints and by the anterior and posterior atlanto-occipital membranes. The superior surface of the atlas contains a groove to accommodate the V3 portion of the vertebral artery.

The transverse process extends from the lateral mass of the atlas and provides a high cervical landmark for the internal jugular vein (Figs. 4.51 and 4.52). The rectus capitis lateralis, levator scapulae, and superior and inferior oblique muscles attach to it, and it is located below and behind the jugular foramen. The rectus capitis lateralis muscle attaches to the jugular process of the occipital bone immediately behind the jugular foramen.

POSTERIOR RELATIONSHIPS

Posteriorly the jugular bulb is connected with the sigmoid sinus (Fig. 4.31) contained in a groove in the posterior surface of the temporal bone (Fig. 4.7). Like all the other sinuses, the sigmoid sinus is enveloped in a double dural layer.

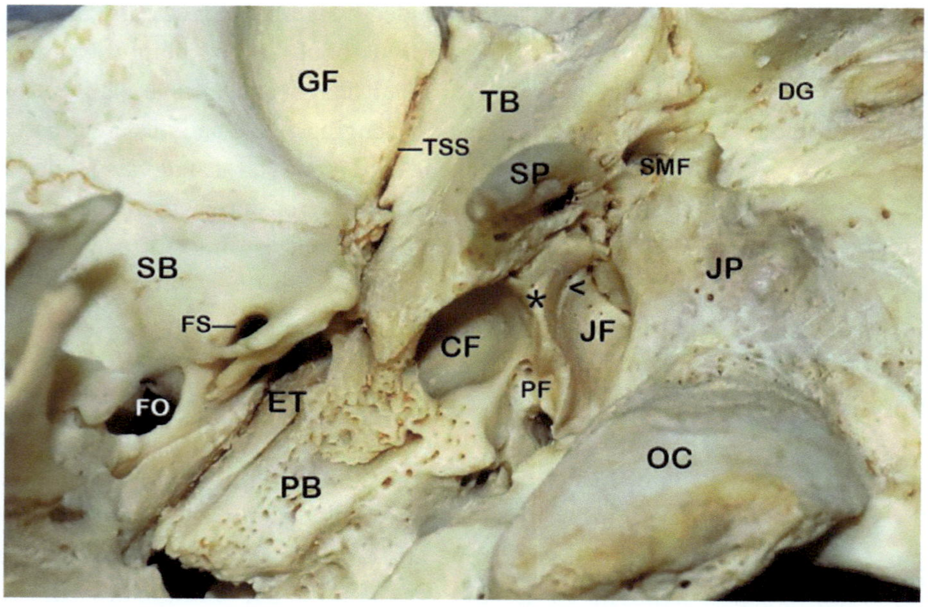

Fig. 4.28 An inferior view of bony jugular foramen and related structures. *GF*, glenoid fossa; *TSS*, tympanosquamous, suture; *TB*, tympanic bone; *DG*, digastric groove; *SP*, styloid process; *SMF*, stylomastoid foramen; *JP*, jugular process of the occipital bone; *SB*, sphenoid bone; *FS*, foramen spinosum; *FO*, foramen ovale; *ET*, sulcus of eustachian tube; *CF*, carotid foramen; *JF*, jugular foramen; *PF*, pyramidal fossa; *PB*, petrous bone; *OC*, occipital condyle; *, jugulo-carotid spine (carotid crest); <, Jacobson's nerve canal.

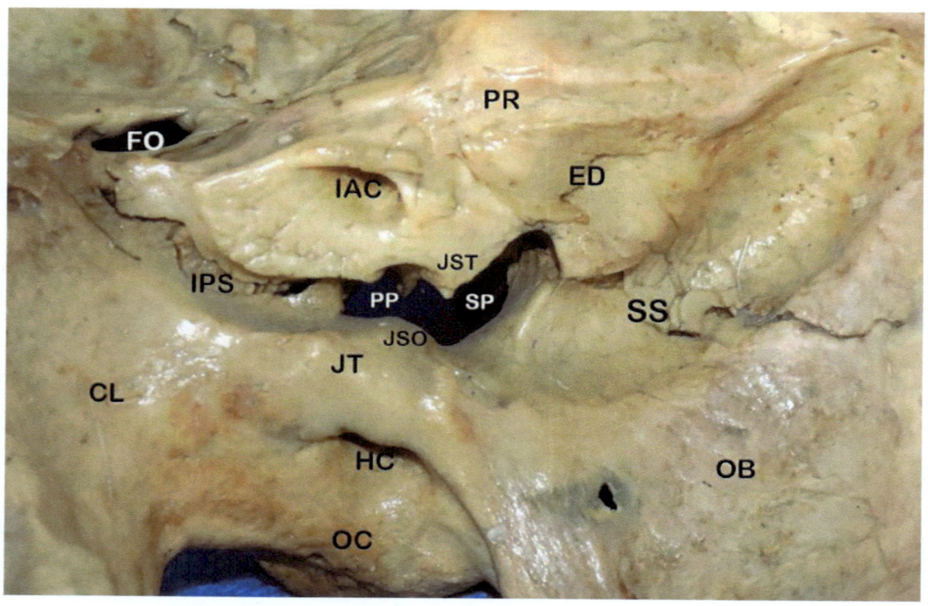

Fig. 4.29 Anatomy of the bony jugular foramen and related structures. The jugular tubercle is located at the junction of the basilar and condylar portions of the occipital bone. *FO*, foramen ovale; *PR*, petrous ridge; *IAC*, internal auditory canal; *ED*, endolymphatic duct; *IPS*, groove for inferior petrosal sinus; *PP*, petrosal part of jugular foramen; *SP*, sigmoid part of jugular foramen; *JST*, jugular spine of the temporal bone; *JSO*, jugular spine of the occipital bone; *SS*, sigmoid sinus groove; *JT*, jugular tubercle; *CL*, clivus; *HC*, hypoglossal canal; *OC*, occipital condyle; *OB*, occipital bone.

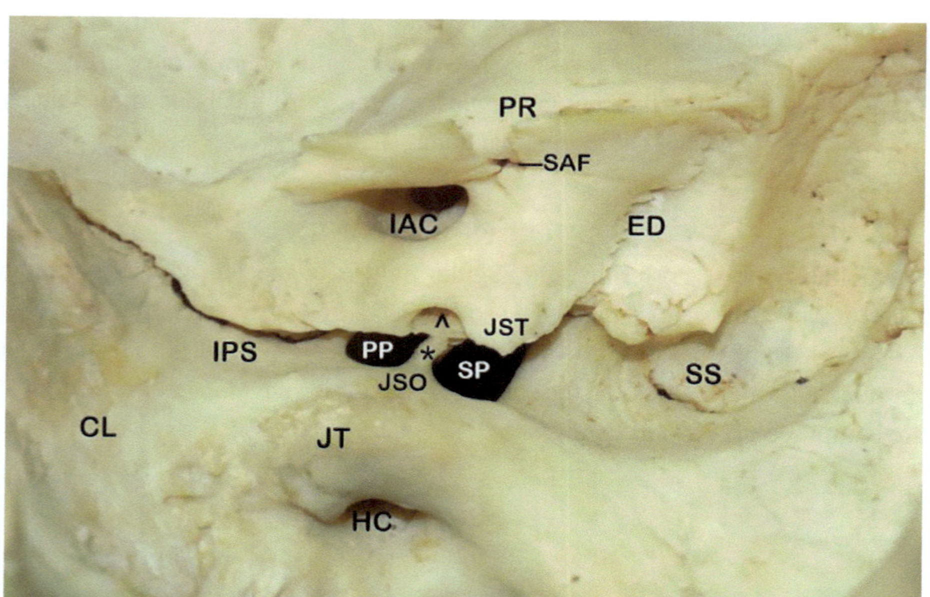

Fig. 4.30 The two compartments are divided by the jugular spine of the temporal and occipital bones, which are connected by a fibrous band that can occasionally ossify (*). *PR*, petrous ridge; *IAC*, internal auditory canal; *SAF*, subarcuate fossa; *ED*, endolymphatic duct; *IPS*, groove for inferior petrosal sinus; *SS*, sigmoid sinus groove; *JT*, jugular tubercle; *CL*, clivus; *HC*, hypoglossal canal; *PP*, petrosal part of jugular foramen; *SP*, sigmoid part of jugular foramen; *JST*, jugular spine of the temporal bone; *JSO*, jugular spine of the occipital bone; ^, pyramidal fossa.

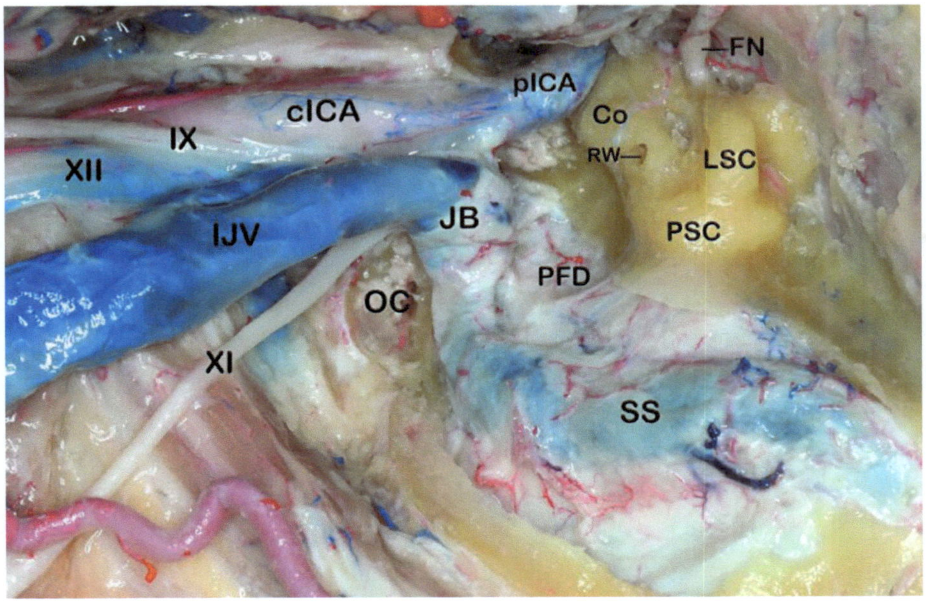

Fig. 4.31 The sigmoid sinus joins the jugular bulb posteriorly. At this level it usually makes an almost 180 degrees turn at the jugular bulb, descending from there into the neck as the internal jugular vein. *FN*, rerouted part of the facial nerve; *Co*, cochlea; *RW*, round window; *LSC*, lateral semicircular canal; *PSC*, posterior semicircular canal; *pICA*, petrous segment of internal carotid artery; *cICA*, cervical segment of internal carotid artery; *PFD*, posterior fossa dura; *JB*, jugular bulb; *SS*, sigmoid sinus; *OC*, occipital condyle; *IJV*, internal jugular vein; *IX*, glossopharyngeal nerve; *XI*, accessory nerve; *XII*, hypoglossal nerve.

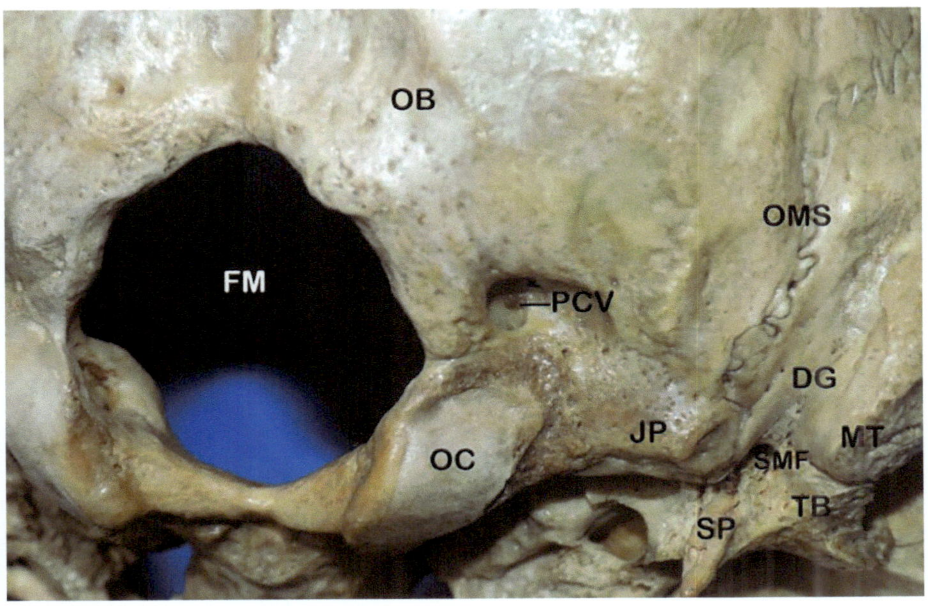

Fig. 4.32 The posterior view of occipital bone. *OB*, occipital bone; *FM*, foramen magnum; *PCV*, posterior condylar vein; *OMS*, occipitomastoid suture; *DG*, digastric groove; *OC*, occipital condyle; *JP*, jugular process of the occipital bone; *SMF*, stylomastoid foramen; *TB*, tympanic bone; *MT*, mastoid tip; *SP*, styloid process.

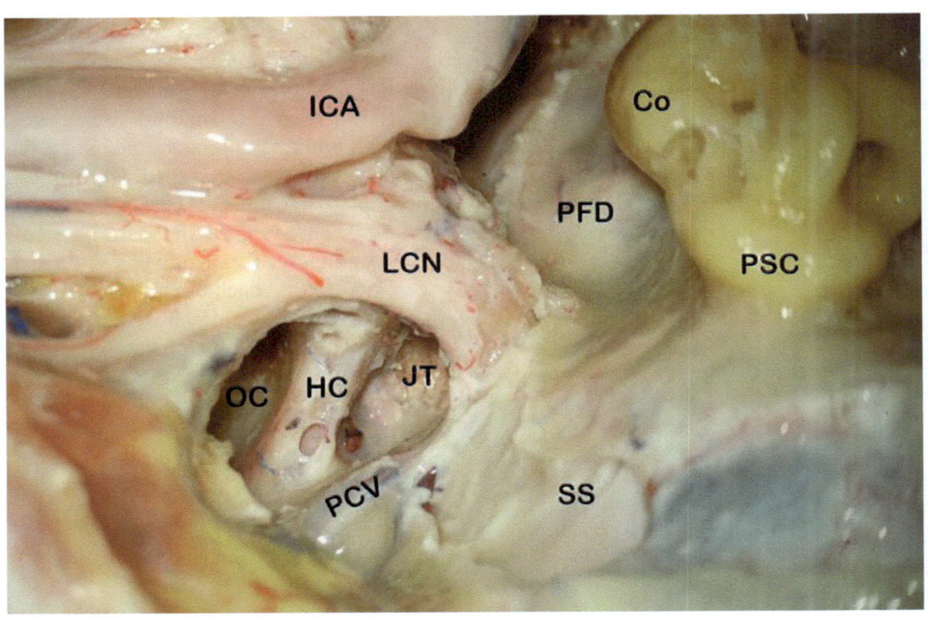

Fig. 4.33 At the level of the junction between the sigmoid sinus and the bulb, the inferomedial wall presents the opening of the posterior condylar vein. *ICA*, internal carotid artery; *PFD*, posterior fossa dura; *Co*, cochlea; *PSC*, posterior semicircular canal; *LCN*, lower cranial nerves; *JT*, jugular tubercle; *HC*, hypoglossal canal; *OC*, occipital condyle; *PCV*, posterior condylar vein; *SS*, sigmoid sinus.

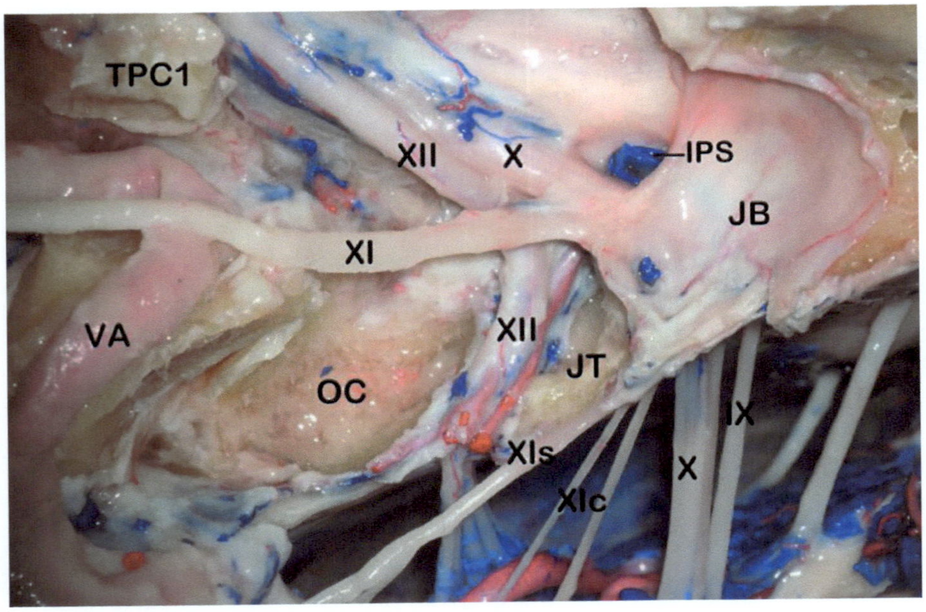

Fig. 4.34 The opening of the inferior petrosal sinus is usually located in the anteromedial wall of the jugular bulb. *TPC1*, transverse process of atlas; *IX*, glossopharyngeal nerve; *X*, vagus nerve; *XI*, accessory nerve; *XIc*, cranial rootlets of accessory nerve; *XIs*, spinal root of accessory nerve; *XII*, hypoglossal nerve; *VA*, vertebral artery; *JT*, jugular tubercle; *OC*, occipital condyle; *JB*, jugular bulb; *IPS*, opening of inferior petrosal sinus.

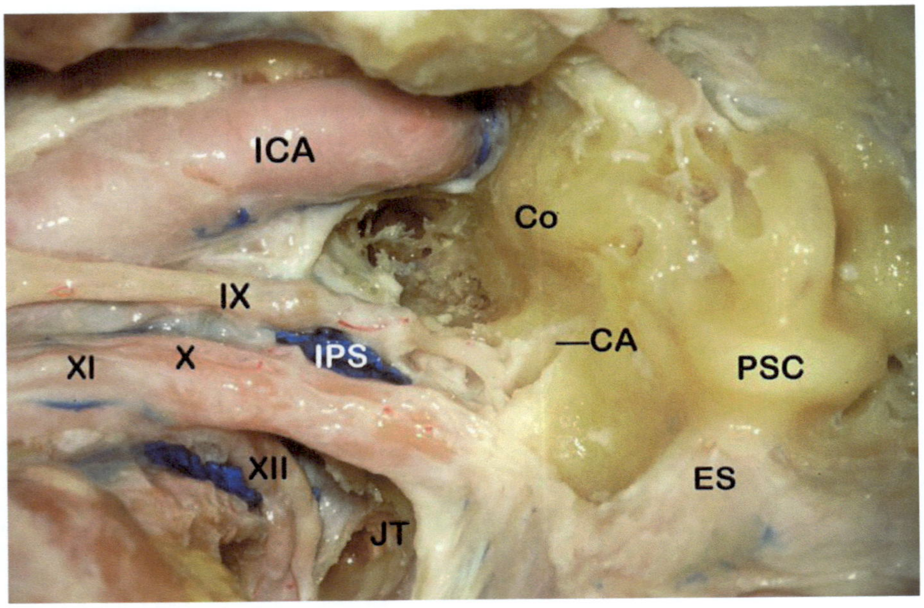

Fig. 4.35 The lower cranial nerves make a 90 degrees rotation after entering the jugular foramen, with the glossopharyngeal nerve in the upper position, in close association with the opening of the cochlear aqueduct at the site. *ICA*, internal carotid artery; *Co*, cochlea; *PSC*, posterior semicircular canal; *CA*, cochlear aqueduct; *ES*, endolymphatic sac; *IPS*, opening of inferior petrosal sinus; *IX*, glossopharyngeal nerve; *X*, vagus nerve; *XI*, accessory nerve; *XII*, hypoglossal nerve; *JT*, jugular tubercle.

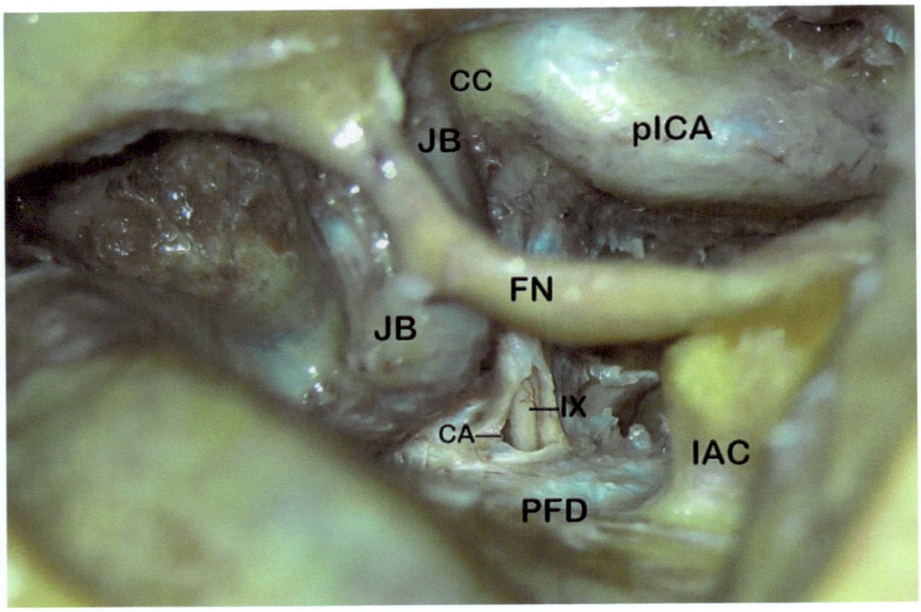

Fig. 4.36 Relationship between the cochlear aqueduct and the glossopharyngeal nerve. Since the nerve lies just inferior to the cochlear aqueduct, the latter is used as a landmark to the nerve in the translabyrinthine approach, indicating the lower limit of drilling in order to avoid injury to the glossopharyngeal nerve. *CC*, carotid crest; *pICA*, petrous segment of internal carotid artery; *JB*, jugular bulb; *FN*, facial nerve; *IX*, glossopharyngeal nerve; *CA*, cochlear aqueduct; *PFD*, posterior fossa dura; *IAC*, internal auditory canal.

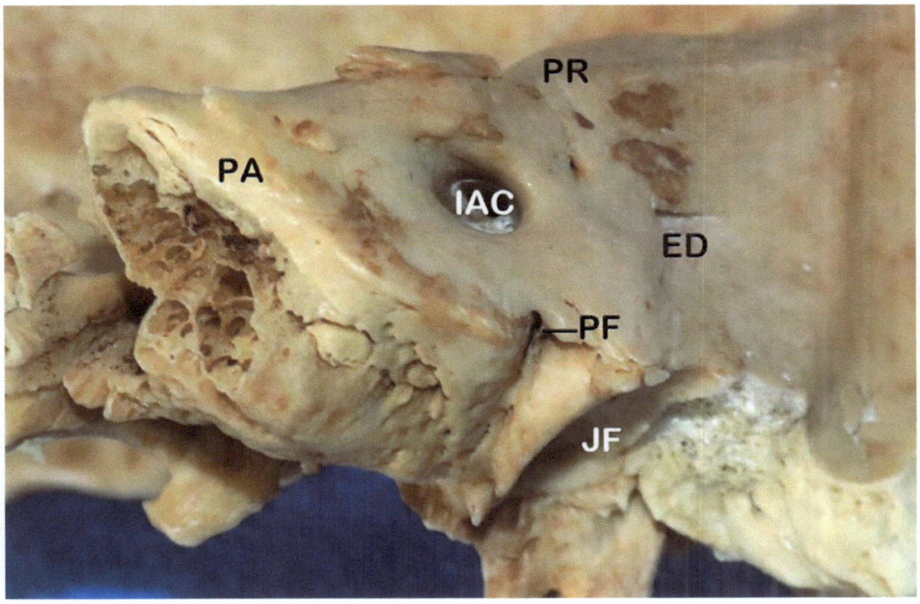

Fig. 4.37 The pyramidal fossa located at the anteromedial margin of the bony jugular foramen. *PR*, petrous ridge; *PA*, petrous apex; *IAC*, internal auditory canal; *ED*, endolymphatic duct; *PF*, pyramidal fossa; *JF*, jugular foramen.

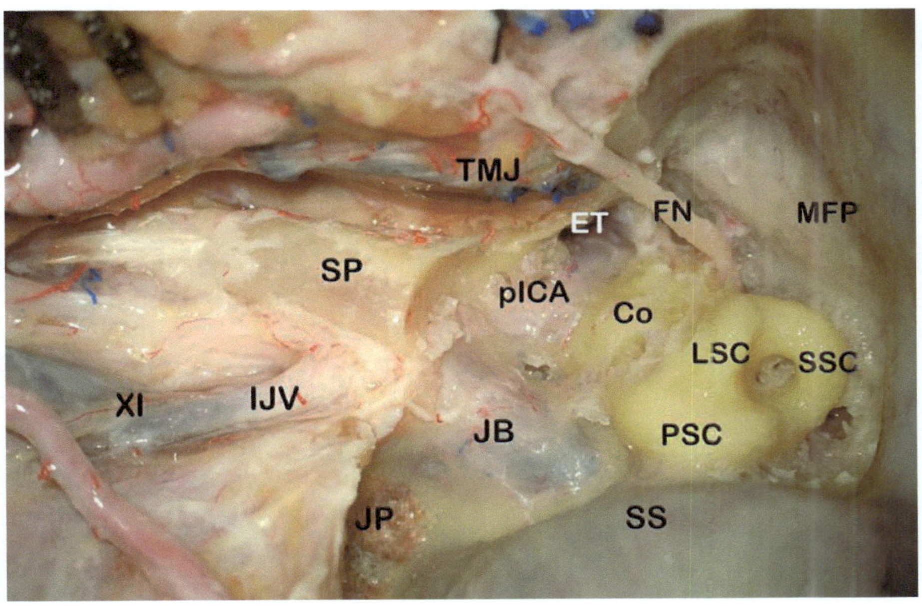

Fig. 4.38 From a lateral approach the styloid process blocks access to the skull base entry points of the carotid artery and jugular vein. *FN*, rerouted part of the facial nerve; *MFP*, middle fossa plate; *TMJ*, temporamandibular joint; *ET*, eustachian tube; *pICA*, petrous segment of internal carotid artery; *SP*, styloid process; *Co*, cochlea; *SSC*, superior semicircular canal; *LSC*, lateral semicircular canal; *PSC*, posterior semicircular canal; *JB*, jugular bulb; *SS*, sigmoid sinus; *IJV*, internal jugular vein; *XI*, accessory nerve; *JP*, jugular process of the occipital bone.

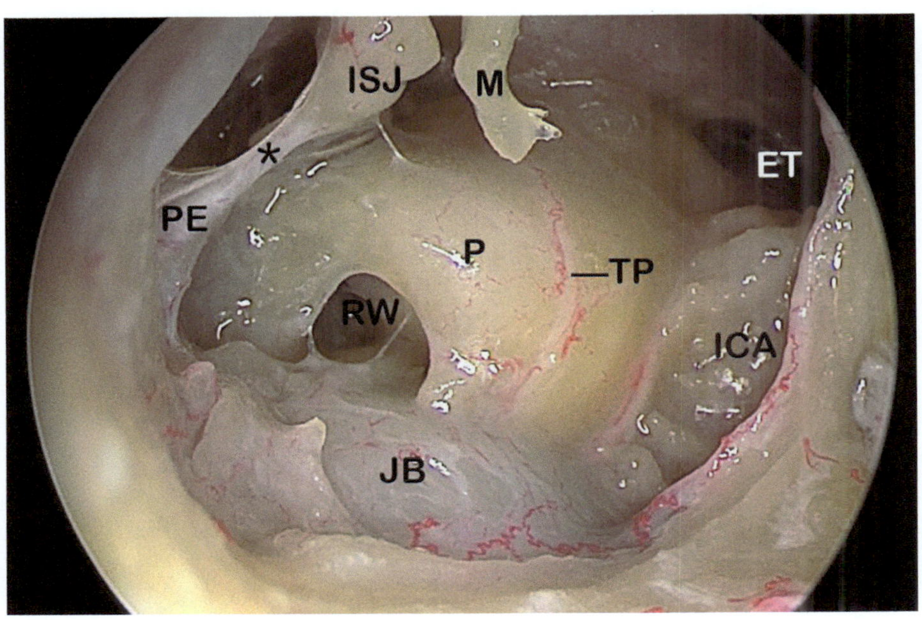

Fig. 4.39 The jugular bulb may protrude into the middle ear, and sometimes the jugular plate may be missing. *ISJ*, incudostapedial joint; *M*, manubrium of the malleus; *PE*, pyramidal eminence; *P*, promontory; *RW*, round window; *TP*, tympanic plexus; *ET*, eustachian tube; *ICA*, internal carotid artery; *JB*, jugular bulb; *, tendon of stapedius muscle.

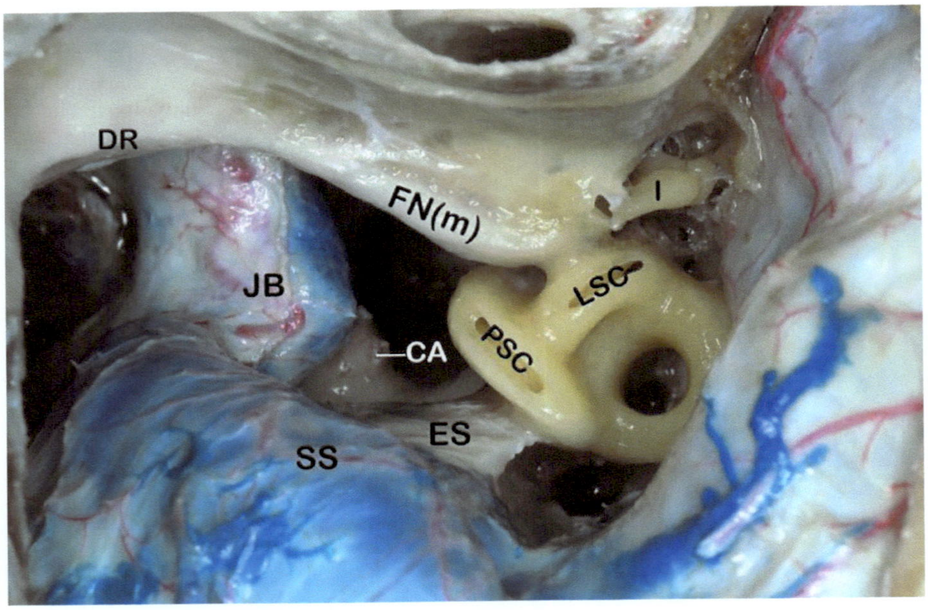

Fig. 4.40 Relationship between the jugular bulb and the posterior semicircular canal can be appreciated. *I*, incus; *FN(m)*, mastoid segment of facial nerve; *DR*, digastric ridge; *JB*, jugular bulb; *SS*, sigmoid sinus; *CA*, cochlear aqueduct; *ES*, endolymphatic sac; *LSC*, lateral semicircular canal; *PSC*, posterior semicircular canal.

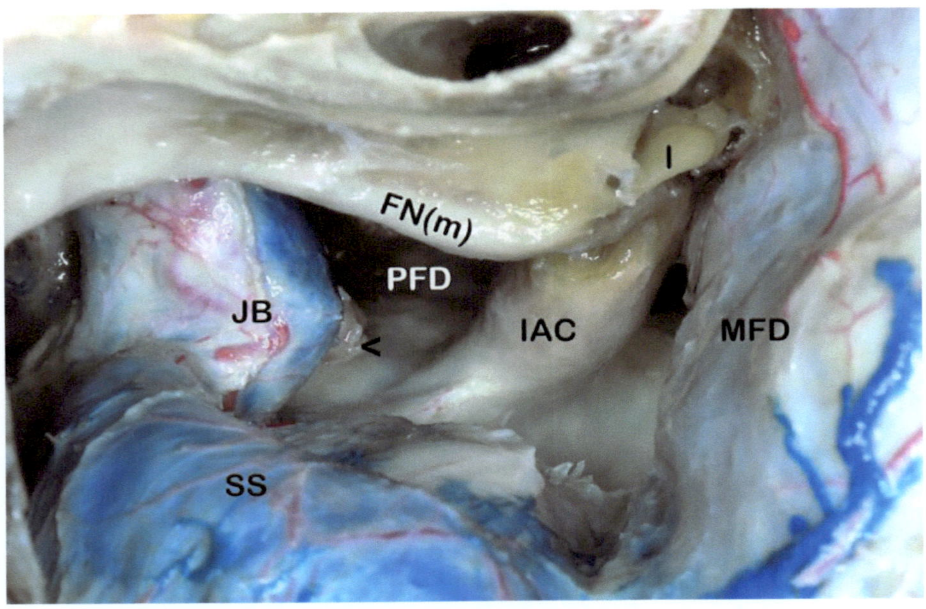

Fig. 4.41 Relationship between the jugular bulb and the internal auditory canal can be appreciated. *I*, incus; *FN(m)*, mastoid segment of facial nerve; *JB*, jugular bulb; *SS*, sigmoid sinus; *PFD*, posterior fossa dura; *IAC*, internal auditory canal; *MFD*, middle fossa dura; <, cochlear aqueduct.

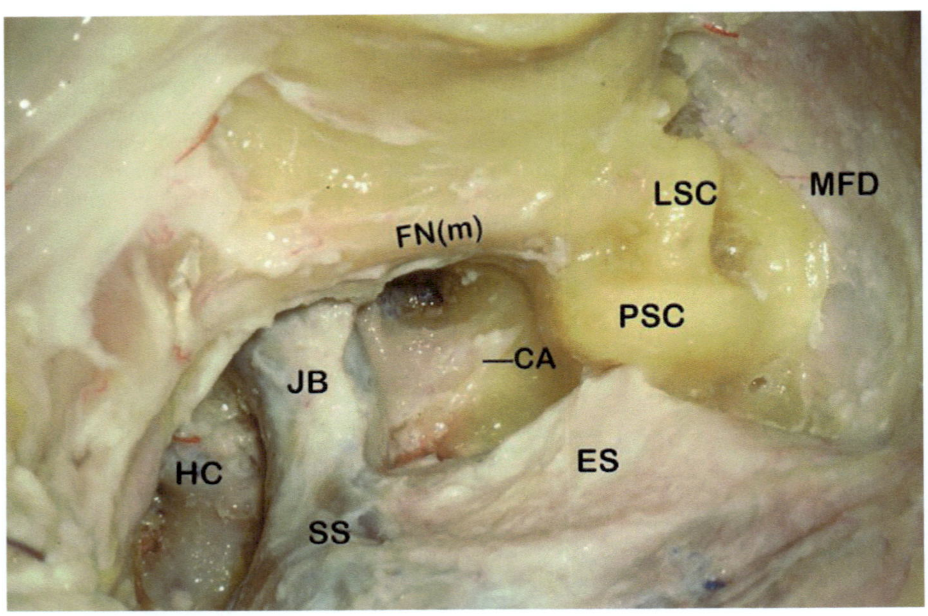

Fig. 4.42 The cochlear aqueduct is a bony channel with a pyramidal shape connecting the perilymphatic space of the scala tympani with the subarachnoid space at the level of the jugular foramen. *FN(m)*, mastoid segment of facial nerve; *LSC*, lateral semicircular canal; *PSC*, posterior semicircular canal; *MFD*, middle fossa dura; *JB*, jugular bulb; *SS*, sigmoid sinus; *CA*, cochlear aqueduct; *ES*, endolymphatic sac; *HC*, hypoglossal canal.

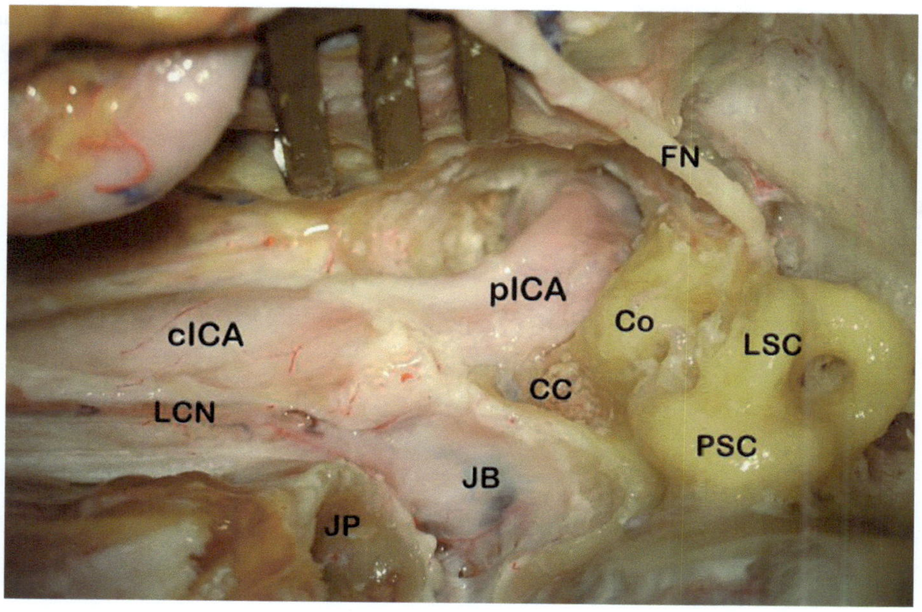

Fig. 4.43 The vertical portion of the petrous carotid artery lies immediately anterior to the jugular bulb, separated by carotid crest. *FN*, rerouted part of the facial nerve; *pICA*, petrous segment of internal carotid artery; *cICA*, cervical segment of internal carotid artery; *LCN*, lower cranial nerves; *JB*, jugular bulb; *JP*, jugular process of the occipital bone; *Co*, cochlea; *LSC*, lateral semicircular canal; *PSC*, posterior semicircular canal.

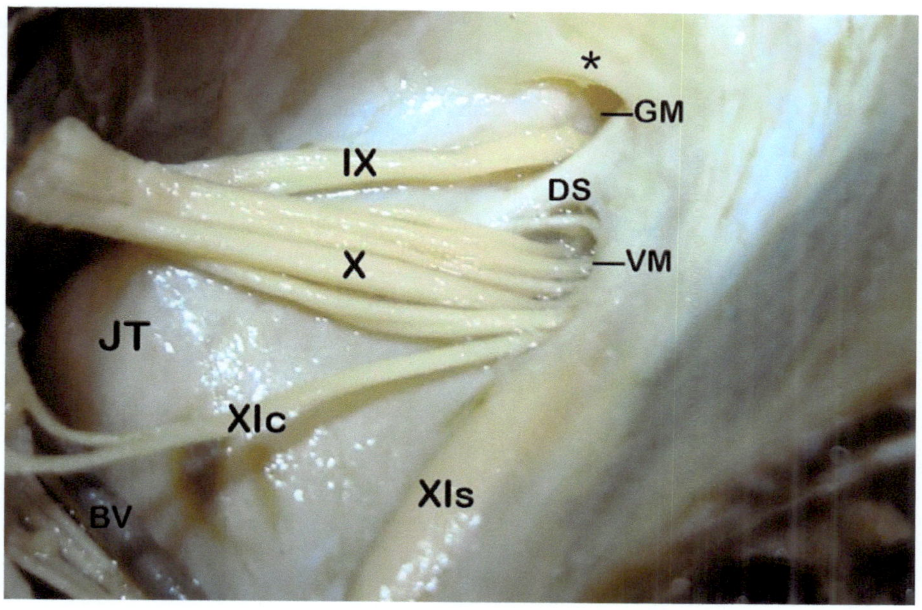

Fig. 4.44 Posterior view of the intracranial aspect of the left jugular foramen. The glossopharyngeal nerve is separated from the vagus nerve by a narrow dural septum. *GM*, glossopharyngeal meatus; *VM*, vagal meatus; *DS*, dural septum; *IX*, glossopharyngeal nerve; *X*, vagus nerve; *XI*, accessory nerve; *XIc*, cranial rootlets of accessory nerve; *XIs*, spinal root of accessory nerve; *JT*, jugular tubercle; *BV*, bridging vein; *, jugular dural fold.

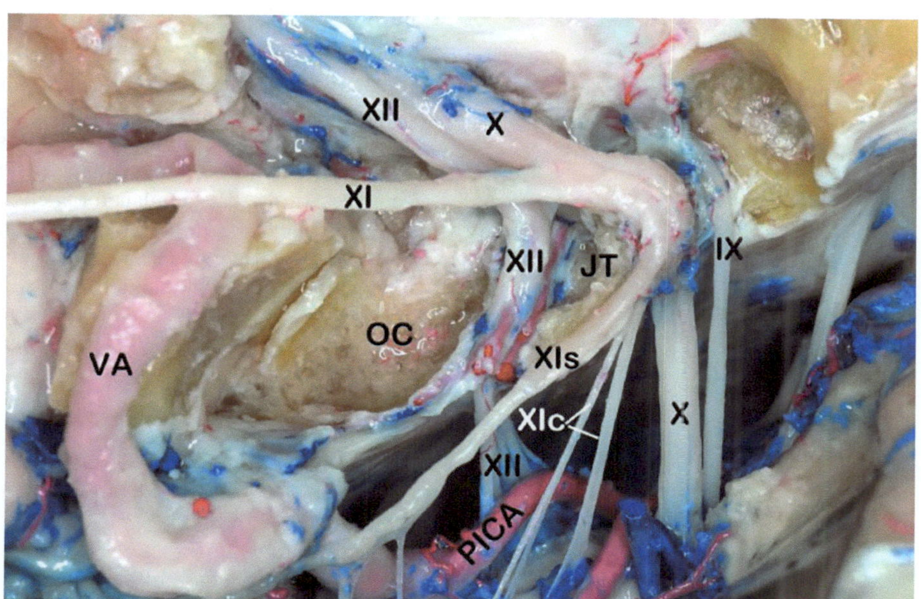

Fig. 4.45 The jugular tubercle is situated above the hypoglossal canal. The glossopharyngeal, vagus, and accessory nerves cross the posterior portion of the jugular tubercle in passing from the brainstem to the jugular foramen. *IX*, glossopharyngeal nerve; *X*, vagus nerve; *XI*, accessory nerve; *XIc*, cranial rootlets of accessory nerve; *XIs*, spinal root of accessory nerve; *XII*, hypoglossal nerve; *VA*, vertebral artery; *JT*, jugular tubercle; *OC*, occipital condyle; *PICA*, posteroinferior cerebellar artery.

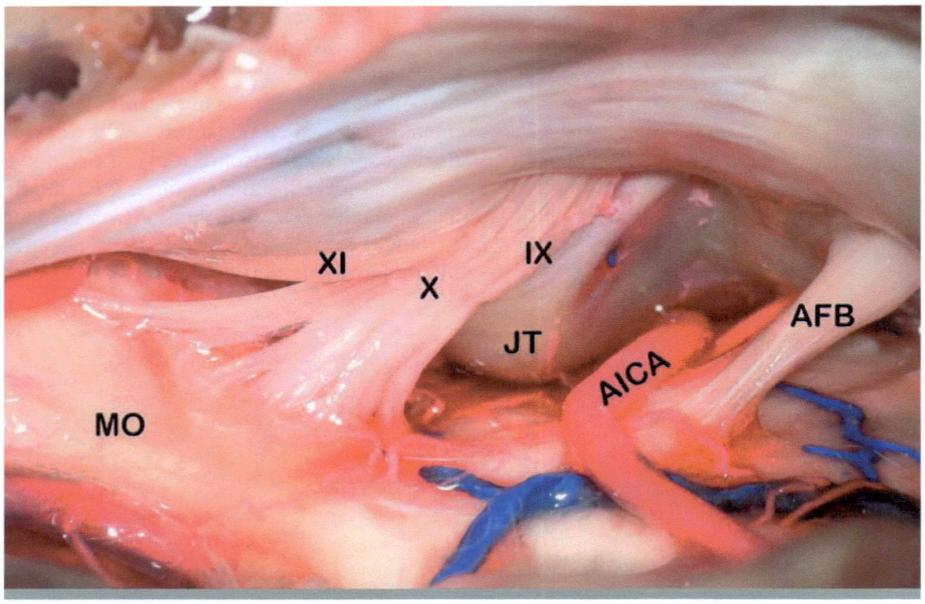

Fig. 4.46 The lower cranial nerves entering the jugular foramen have been exposed. *IX*, glossopharyngeal nerve; *X*, vagus nerve; *XI*, accessory nerve; *JT*, jugular tubercle; *AFB*, acousticofacial bundle; *AICA*, anteroinferior cerebellar artery; *MO*, medulla oblongata.

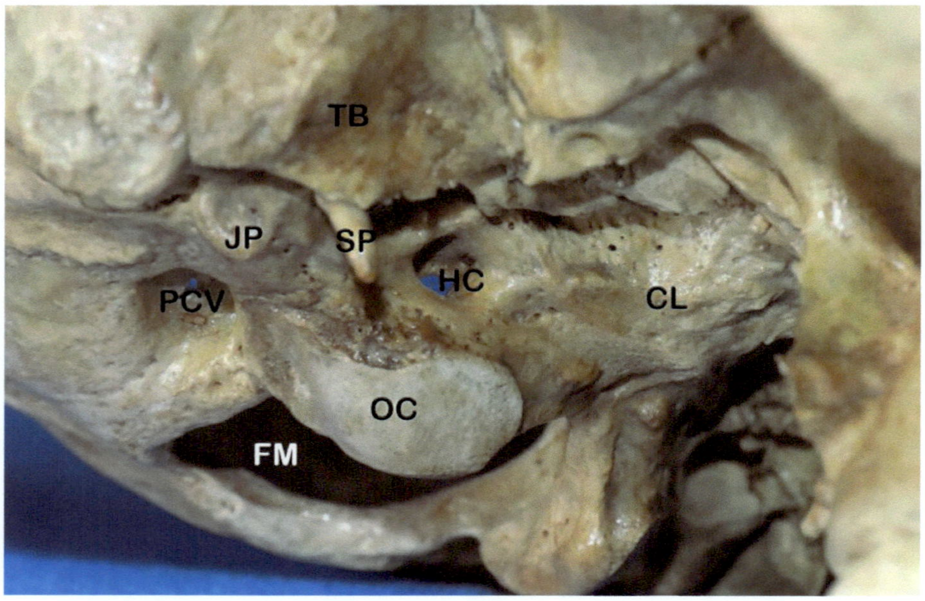

Fig. 4.47 Relationship of the hypoglossal canal and the occipital condyle. The extracranial end of the hypoglossal canal is located immediately above the junction of the anterior and middle one-third of the occipital condyle. *TB*, tympanic bone; *JP*, jugular process of the occipital bone; *SP*, styloid process; *HC*, hypoglossal canal; *FM*, foramen magnum; *PCV*, posterior condylar vein; *OC*, occipital condyle; *CL*, clivus.

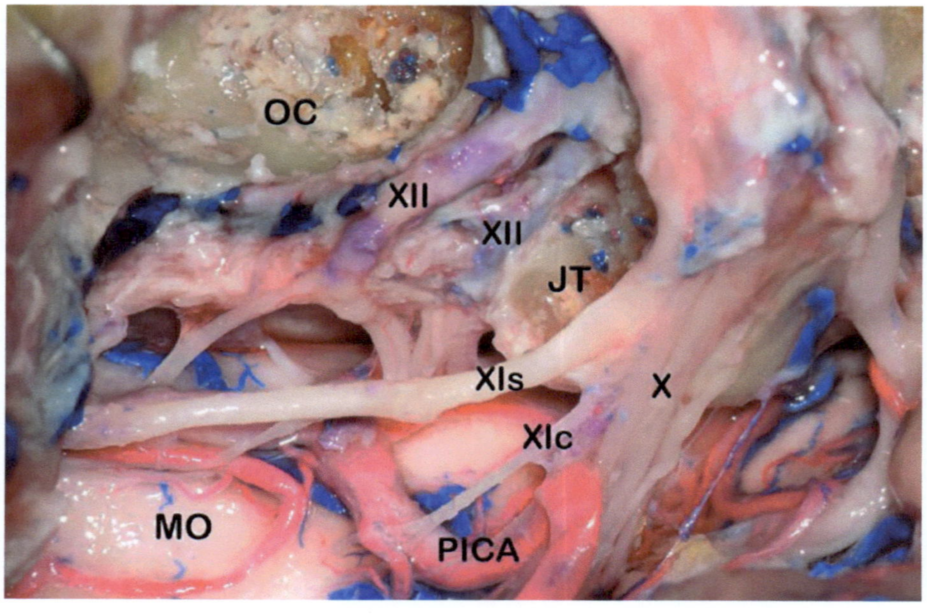

Fig. 4.48 The jugular bulb and adjoining segment of the internal jugular vein have been removed to expose the glossopharyngeal, vagus, and accessory nerves passing through the jugular foramen. The cortical bone lining the hypoglossal canal has been removed to expose the hypoglossal nerve exiting the jugular foramen to descend in the carotid sheath. *OC*, occipital condyle; *XII*, hypoglossal nerve; *JT*, jugular tubercle; *X*, vagus nerve; *XIc*, cranial rootlets of accessory nerve; *XIs*, spinal root of accessory nerve; *MO*, medulla oblongata; *PICA*, posteroinferior cerebellar artery.

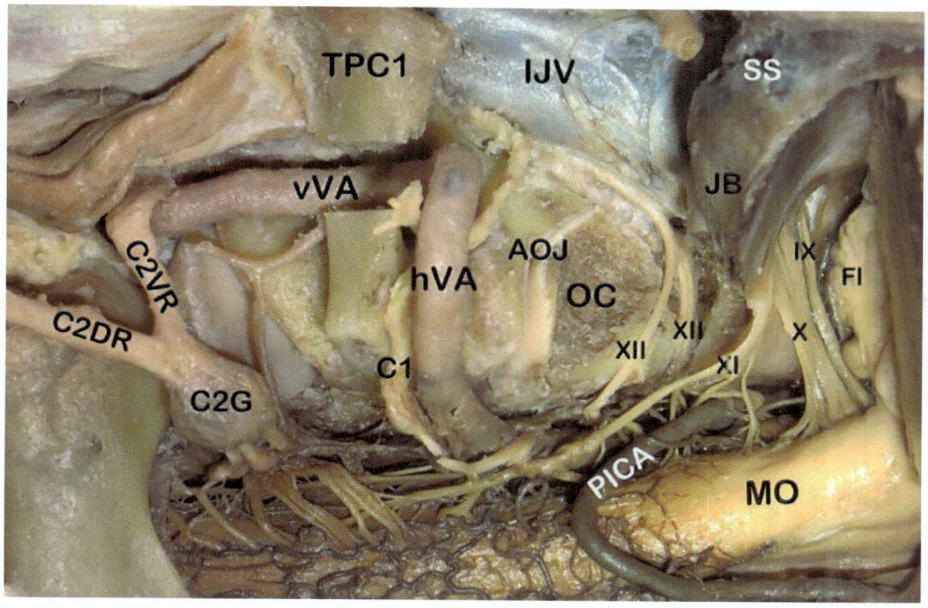

Fig. 4.49 Very often the hypoglossal nerve originates via two distinct routes that fuse in the canal into a single nerve. *TPC1*, transverse process of atlas; *IJV*, internal jugular vein; *JB*, jugular bulb; *SS*, sigmoid sinus; *Fl*, flocculus; *IX*, glossopharyngeal nerve; *X*, vagus nerve; *XI*, accessory nerve; *XII*, hypoglossal nerve; *MO*, medulla oblongata; *PICA*, posteroinferior cerebellar artery; *OC*, occipital condyle; *AOJ*, atlanto-occipital joints; *vVA*, vertical segment of vertebral artery; *hVA*, horizontal segment of vertebral artery; *C2VR*, ventral ramus of C2; *C2DR*, dorsum ramus of C2; *C2G*, ganglion of C2.

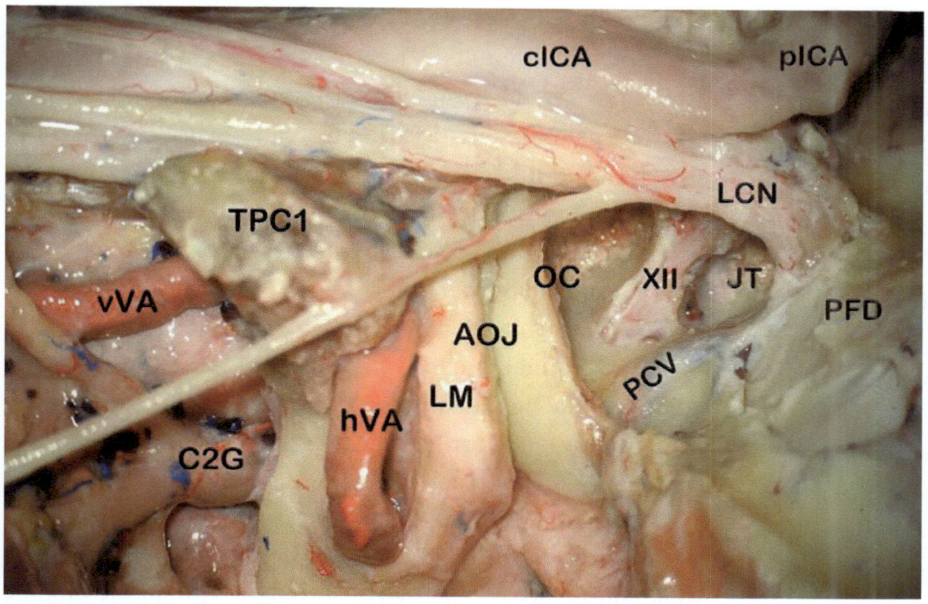

Fig. 4.50 The atlas is located inferiorly to the skull and its articular facets articulate with the occipital condyle. *pICA*, petrous segment of internal carotid artery; *cICA*, cervical segment of internal carotid artery; *LCN*, lower cranial nerves; *PFD*, posterior fossa dura; *JT*, jugular tubercle; *XII*, hypoglossal nerve; *OC*, occipital condyle; *AOJ*, atlanto-occipital joints; *PCV*, posterior condylar vein; *LM*, lateral mass; *TPC1*, transverse process of atlas; *vVA*, vertical segment of vertebral artery; *hVA*, horizontal segment of vertebral artery; *C2G*, ganglion of C2.

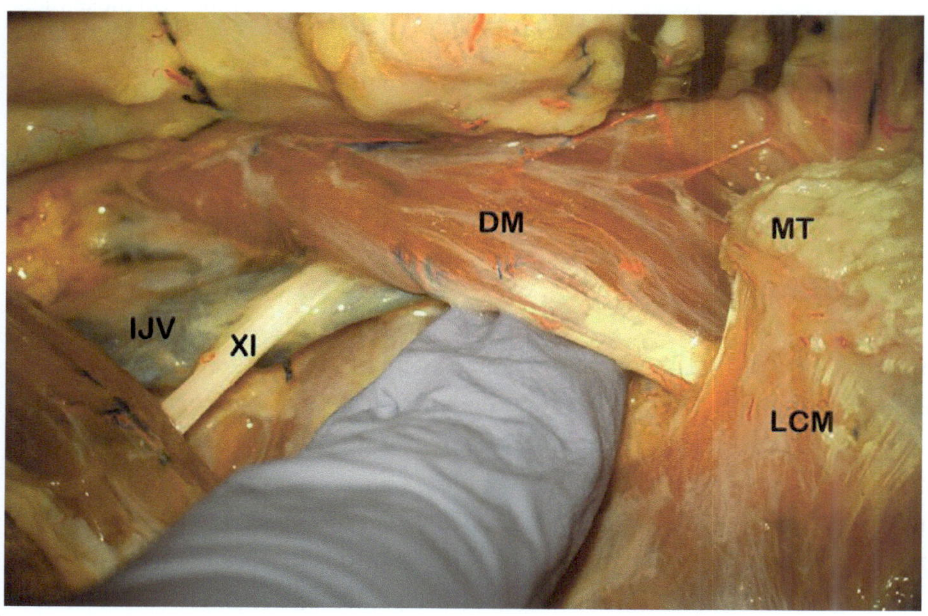

Fig. 4.51 The lateral process of C1 is palpated as a landmark to identify the internal jugular vein. *DM*, digastric muscle; *MT*, mastoid tip; *LCM*, longissimus capitis muscle; *XI*, accessory nerve; *IJV*, internal jugular vein.

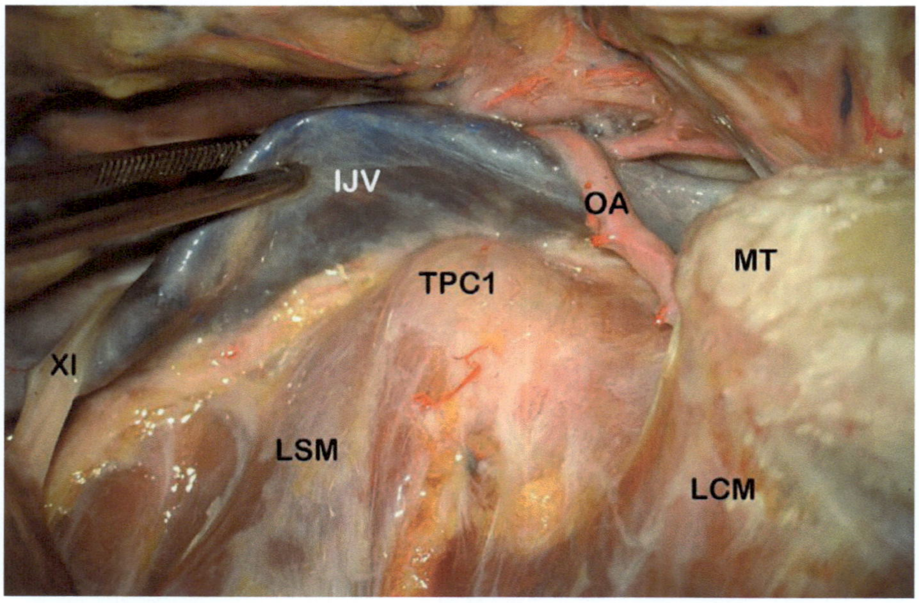

Fig. 4.52 The transverse process extends from the lateral mass of the atlas and provides a high cervical landmark for the internal jugular vein. *IJV*, internal jugular vein; *OA*, occipital artery; *MT*, mastoid tip; *TPC1*, transverse process of atlas; *XI*, accessory nerve; *LCM*, longissimus capitis muscle; *LSM*, levator scapular muscle.

Facial Nerve

The facial nerve (CN VII) is a mixed nerve containing motor, sensory, and parasympathetic fibers. The motor component forms the major part of the nerve and receives contributions from three sources.

Fibers originating in the lower part of the precentral gyrus of the cortex pass down to reach the posterior portion of the internal capsule, where they are tightly packed together. From there, they pass in proximity to the pyramidal tract in the basal part of the pons, where the majority of fibers cross to the contralateral facial nucleus. Only a small proportion of the fibers innervating the occipitofrontalis and the orbicularis oculi muscles contribute to the ipsilateral facial nucleus. From the thalamus and globus pallidus, fibers concerned with emotional facial expressions reach the nerve via the reticular formation. The last group of neurons contributing to the motor component of the facial nerve are derived from the central brainstem nuclei. The neurons of this group are concerned with the complex facial nerve reflexes and reach the facial nerve nucleus through the connections with the corresponding nucleus. The most important are the corneal reflex through the trigeminal nuclei and the stapedius reflex through the acoustic nuclei.

The second group of efferent fibers carried by the facial nerve are the visceral efferents. These fibers are concerned with the secretomotor function of the facial nerve and supply the submandibular, sublingual, and lacrimal glands, and the secretory glands of the pharynx, palate, and nose. They originate in the superior salivatory nucleus near the facial nucleus and reach the facial nerve after passing through the nervus intermedius.

In addition to efferent fibers, the facial nerve embraces two types of afferents. Visceral afferents carry taste sensation from the anterior two-thirds of the tongue and from the soft palate. Somatic afferents carry sensation from the concha from a small part of the external auditory meatus, and from a small spot of skin posterior to the auricle. Both afferents enter the brainstem after passing through the nervus intermedius. Thereafter, visceral afferents pass to the nucleus of the tractus solitarius, while somatic afferents pass to the trigeminal nuclei.

After leaving the facial motor nucleus, neurons pass posterolaterally and turn around the nucleus of the abducent nerve (CN VI) to reach their exit from the brainstem.

The course of the facial nerve after emerging from the brainstem is traditionally divided into three parts: intracranial, intratemporal, and extratemporal. Numerous studies have reported the measurements of each portion of the facial nerve. In our practice we do not depend much on these measurements because of the relatively wide differences between individuals, preventing any significant predictive value, especially in delicate surgery such as that of the skull base.

Anatomy of the Intracranial Facial Nerve

The intracranial part includes both the cerebellopontine angle (CPA) and the intracanalicular segments of the facial nerve. The CPA segment starts after the facial nerve leaves the pons ventrolaterally near its posterior border above the olive (Fig. 4.53). The nervus intermedius (Fig. 4.54) emerges at the lower border of the pons between the pons and the cerebellar peduncle lying closer to the origin of the vestibulocochlear nerve (CN VI). The three nerves travel together in the pontine cistern in a lateral direction to reach the porus of the internal auditory canal (IAC) with the nervus intermedius lying in the middle all the way to the fundus. Near the brainstem the facial nerve lies anterior and inferior to CN VIII. As they move laterally, the facial nerve gently curves superiorly to form an anterosuperior relation to CN VIII at the level of the porus. This relation is also maintained in the IAC. A sleeve of meninges invests and is slightly adherent to the inner surface of the IAC. Near the fundus of the IAC the dura merges with the facial nerve and nervus intermedius, forming a single trunk.

At the lateral end of the internal auditory canal, the fundus forms the line of demarcation between the intradural and intratemporal parts of the nerve. The fundus is divided into three compartments by two bony bars (Figs. 4.24 and 4.25). The horizontal crest is a bar of bone that extends along the horizontal meridian of the fundus. From its center, the vertical crest (Bill's bar) extends to the superior border of the canal. The joint facial nerve passes through the anterosuperior compartment of the fundus to reach the intralabyrinthine canal. The superior vestibular nerve passes through the posterosuperior compartment to continue thereafter as the superior ampullary nerve. This nerve runs in a small canal that connects the posterosuperior compartment and the ampullae of the lateral and superior semicircular canals. Identification of the canal in translabyrinthine surgery acts as the best landmark to identify the facial nerve in the internal auditory canal. At this level, Bill's bar protects the facial nerve lying anterior to it from being injured during the dislodging of the ampullary nerve from its canal. Once the ampullary nerve is dislodged, it can be followed to its origin from the superior vestibular nerve. On dislocating the latter posteriorly from the internal auditory canal, the facial nerve can be seen lying anterior to it.

Anatomy of the Intratemporal Facial Nerve

This part of the facial nerve is encased within the fallopian canal. It is divided into three segments that are separated from each other by two genuses (Fig. 4.55).

The labyrinthine segment is the thinnest and shortest segment of the facial nerve. It passes from medial to lateral from the fundus to the geniculate ganglion. The borders of the narrow canal containing this segment are formed by the cochlea anteriorly, the superior semicircular canal posteriorly, the vestibule inferiorly, and a thin plate of bone separating the nerve from the middle cranial fossa dura superiorly (Fig. 4.12).

The geniculate ganglion is a swelling that marks the first genu of the facial nerve and contains the cell bodies of all the sensory fibers of the facial nerve. The thin plate of bone separating the geniculate ganglion from the middle cranial fossa dura may be absent in 10%–14% of cases, putting the facial nerve at risk during middle cranial fossa surgery. The first branch of the facial nerve is given out at this level. The greater superficial petrosal nerve leaves the anterior surface of the geniculate ganglion (Fig. 4.56). Lying extradurally in a groove that follows the orientation of the petrous bone, anteromedially this branch continues beneath the trigeminal ganglion to reach the foramen lacerum. There it is joined by the deep petrosal nerve to form the vidian nerve. The greater superficial petrosal nerve carries the secretomotor fibers to the lacrimal gland and to the secretory glands of the pharynx, palate, and nose. It also carries taste fibers from the soft palate. Another array of smaller sensory nerves leaves the geniculate ganglion and joins the tympanic plexus.

Curving posteriorly at an angle ranging between 60 and 90 degrees, the facial nerve continues as the tympanic segment. This segment lies on the medial wall of the tympanum and protrudes into the cavity covered by a thin shell of bone (Fig. 4.57). The beginning of this segment is marked by the cog superiorly and by the cochleariform process inferiorly. As the nerve passes posteriorly, it can be seen slanting inferiorly to assume an inferior position under the protrusion of the lateral semicircular canal on the medial wall of the middle ear. Underneath the nerve at this level lie the bulge of the promontory and the oval window. The incidence of dehiscence of the fallopian canal covering this segment is high and in some series is reported to be up to 40%. When the nerve reaches the level of the oval window, it starts to curve inferiorly, forming the second genu, which lies within the curvature of the lateral semicircular canal. Just before the nerve reaches the genu, the ampullae of the superior and lateral semicircular canal lie medial to the facial nerve separated from it by a thin plate of bone.

Medial to the short process of the incus, the nerve assumes a descending course starting at the mastoid segment of the nerve. The pyramidal eminence, a bony protrusion that houses the stapedius muscle, lies a few millimeters inferior to the short process of the incus in contact with the anterior surface of the nerve. At this level (the pyramidal turn) the facial nerve protrudes posterolateral to the lateral semicircular canal, putting the nerve at risk of injury during mastoid surgery. At this level, the vestibule lies medial to the second genu of the nerve.

The descending or mastoid segment of the facial nerve can be traced from the second genu to the anterior edge of the digastric ridge where the stylomastoid foramen lies, marking the end of the intratemporal part of the facial nerve (Fig. 4.58).

The relationship of this segment to the tympanic annulus is important in the surgery of the external auditory canal. The posteromedial relationship of the facial nerve to the annulus is more constant near the posterosuperior quadrant of the annulus, whereas reaching the posteroinferior quadrant there is a possibility that the nerve crosses lateral to the plane of the annulus, making it susceptible to injury there.

The ampulla of the posterior semicircular canal lies medial to the midportion of this segment, forming a useful landmark in translabyrinthine surgery (Fig. 4.59). The lower part of this segment is usually centered on the lateral surface of the jugular bulb. The superiorlateral extension of which determines the space available for an infrafacial tympanotomy between the facial nerve, jugular bulb, and inferior semicircular canal.

The first muscular branch is the nerve to the stapedius. This nerve is given out from the anterior surface of the facial nerve at the level of the pyramidal eminence. After branching out from the facial nerve, it passes in a small canal within the pyramid to reach the stapedius muscle, which originates from within the pyramidal eminence.

The chorda tympani (Fig. 4.19) is the branch that carries the secretomotor fibers to the sublingual and submandibular glands and the taste fibers from the anterior two-thirds of the tongue. The point of origin of this branch from the facial nerve can lie anywhere between the lateral semicircular canal and the stylomastoid foramen. An origin from the extratemporal part of the facial nerve is also reported. In such instances, the chorda tympani travels back to enter the stylomastoid foramen. After originating from the facial nerve, the chorda tympani travels for a short distance within the posterior wall of the tympanic cavity. From there, the chorda tympani enters the tympanic cavity, passes in an anterior direction medial to the neck of the malleus, and exits the tympanic membrane at its anterior border through the gasserian fissure (tympanosquamous fissure).

The blood supply of the mastoid segment of the facial nerve traverses the medial surface of the bone to reach the nerve (Fig. 4.60). In this portion the nerve is adherent to the canal, a point to be taken into consideration when dissecting the nerve from the canal in facial nerve rerouting. To avoid injury, these adherences should be released using a sharp instrument.

Anatomy of the Extratemporal Facial Nerve

The definite arrangement of the bundles of facial nerve motor fibers and sometimes the splitting up of the facial nerve into its main divisions start in this segment.

The main trunk of the facial nerve emerges from the stylomastoid foramen at the anterior border of the digastric ridge (Fig. 4.61). The mastoid process lies lateral to the facial nerve at its exit from the stylomastoid foramen, protecting the nerve from possible injury during the performance of the lower part of a retroauricular incision. In young children, however, the mastoid process is not well developed, rendering the superficially lying facial nerve vulnerable to surgical injury. This is why the lower part of a retroauricular incision should be directed backward in young patients.

Following the posterior belly of the digastric muscle backward to its attachment to the digastric groove medial to the mastoid process leads to the main trunk of the facial nerve at its exit, thus forming an important landmark for the identification of the facial nerve during surgery in that region (Fig. 4.62). Another commonly described landmark is the tragal pointer, where the main trunk of the facial nerve lies perpendicular to the bisector of the line joining the mastoid process to the tragal pointer. Bleeding from the branches of the posterior auricular artery lying just lateral to the facial nerve serves to inform the surgeons that they are approaching the nerve (Fig. 4.63).

After leaving the stylomastoid foramen, the nerve gives off two branches. The posterior auricular nerve runs superiorly between the external auditory canal and the anterior surface of the mastoid to supply the intrinsic muscles of the ear and the occipital belly of the occipitofrontal muscle. A second branch runs inferiorly to supply the posterior belly of the digastric and the stylohyoid muscle. Occasionally each of these two muscles receives a separate branch from the main trunk. After giving off these branches, the main trunk of the facial nerve continues anterolaterally in the retromandibular fossa and crosses the lateral surface of the styloid process to enter the posterior surface of the parotid gland.

The division of the facial nerve into the branches supplying the muscles of facial expression is very variable. As mentioned earlier, the division of the nerve sometimes starts within the mastoid canal, thus putting the nerve at risk of injury if the surgeon is not aware of this possibility. In the commonest pattern, however, the division into the temporofacial and cervicofacial trunk starts within the substance of the parotid gland (Fig. 4.64). Following, these two lead to five major branches in the vast majority of cases.

The temporal branch crosses the zygomatic process lateral to its periosteum to supply the anterior and superior auricular muscles and the frontal belly of the occipitofrontalis. The zygomatic branch after crossing the zygomatic process lateral to the covering periosteum, splits to supply the upper and lower parts of the orbicularis oculi. To maintain its integrity in the infratemporal approaches, the periosteum of the zygomatic arch should be elevated and secured before cutting the arch. Passing more or less horizontally, the buccal branch runs forward to supply the buccinator and the muscles of the upper part of the lip. The mandibular branch, after emerging from the lower part of the gland, runs below the level of the mandible deep to the platysma, then turns superiorly to reach the muscles of the lower part of the lip, which it supplies. During surgery on the submandibular gland, the part of the nerve passing beneath the mandible is at risk: in order to avoid injury to it, the skin incision should be made at least 4 cm from the edge of the mandible and the flap elevated should include the deep cervical fascia which lies deep to the nerve, thus covering it. The last branch is the cervical branch, which passes behind the mandible after emerging from the posterior surface of the gland and continues to supply the platysma.

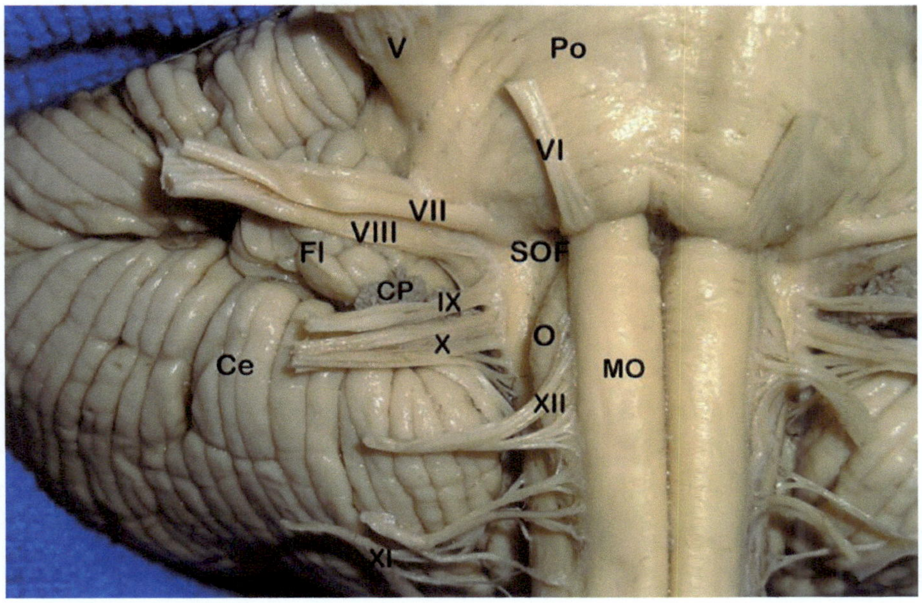

Fig. 4.53 Brainstem and petrosal surface of the cerebellum. *V*, trigeminal nerve; *Po*, pons; *VI*, abducent nerve; *VII*, facial nerve; *VIII*, vestibulocochlear nerve; *Fl*, flocculus,; *CP*, choroid plexus; *SOF*, supraolivary fossette; *O*, olive; *MO*, medulla oblongata; *Ce*, cerebellum; *IX*, glossopharyngeal nerve; *X*, vagus nerve; *XI*, accessory nerve; *XII*, hypoglossal nerve.

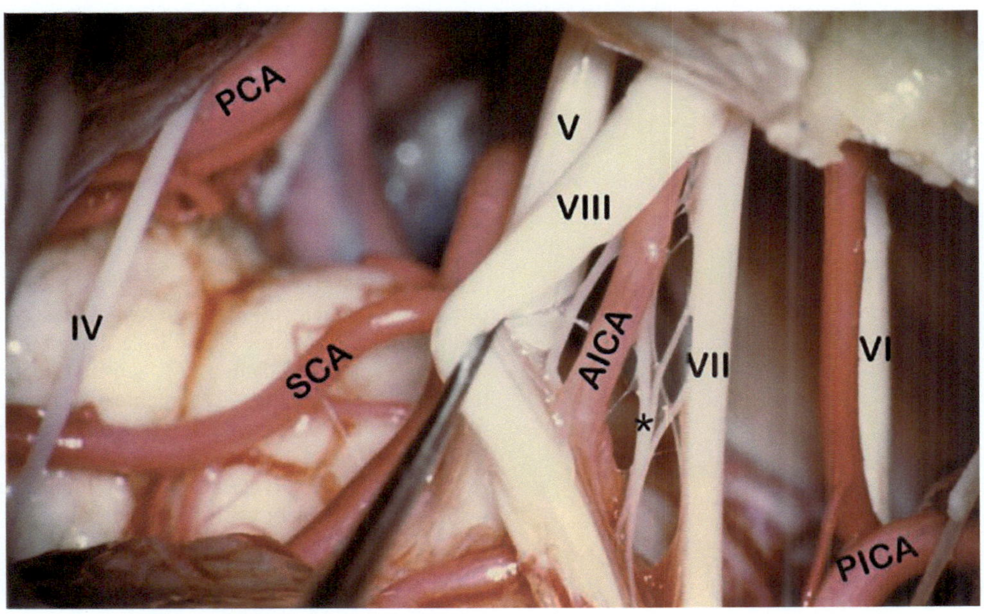

Fig. 4.54 The nervus intermedius emerges at the lower border of the pons between the pons and the cerebellar peduncle lying closer to the origin of the vestibulocochlear nerve. *PCA*, posterior cerebral artery; *SCA*, superior cerebellar artery; *IV*, trochlear nerve; *V*, trigeminal nerve; *VI*, abducent nerve; *VII*, facial nerve; *VIII*, vestibulocochlear nerve; *AICA*, anteroinferior cerebellar artery; *PICA*, posteroinferior cerebellar artery; *, nervus intermedius.

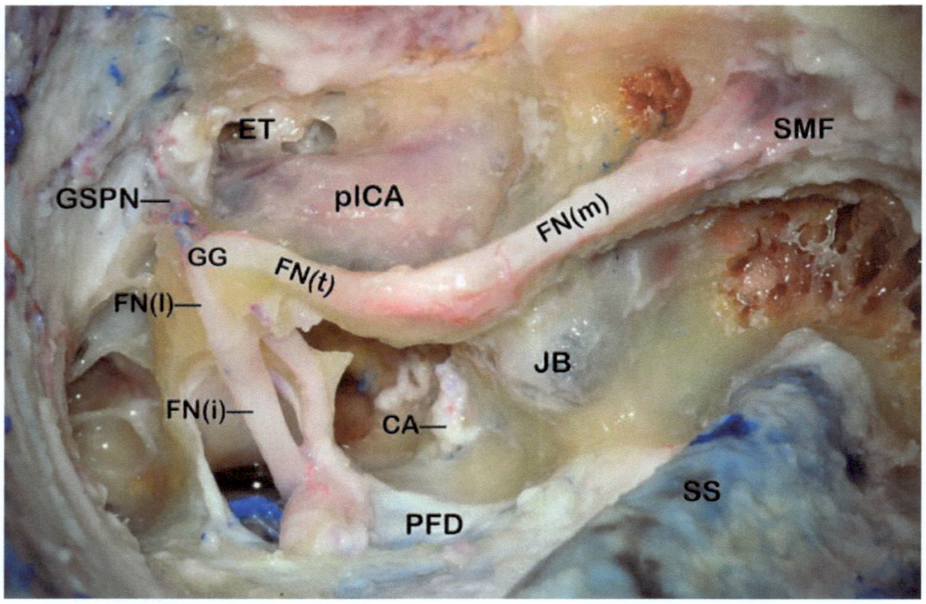

Fig. 4.55 The intratemporal facial nerve divided into three segments that are separated from each other by two genuses. *ET*, eustachian tube; *GSPN*, greater superficial petrosal nerve; *pICA*, petrous segment of internal carotid artery; *FN(m)*, mastoid segment of facial nerve; *FN(t)*, tympanic segment of facial nerve; *GG*, geniculate ganglion; *FN(l)*, labyrinthine segment of facial nerve; *FN(i)*, internal auditory canal segment of facial nerve; *PFD*, posterior fossa dura; *CA*, cochlear aqueduct; *JB*, jugular bulb; *SMF*, stylomastoid foramen; *SS*, sigmoid sinus.

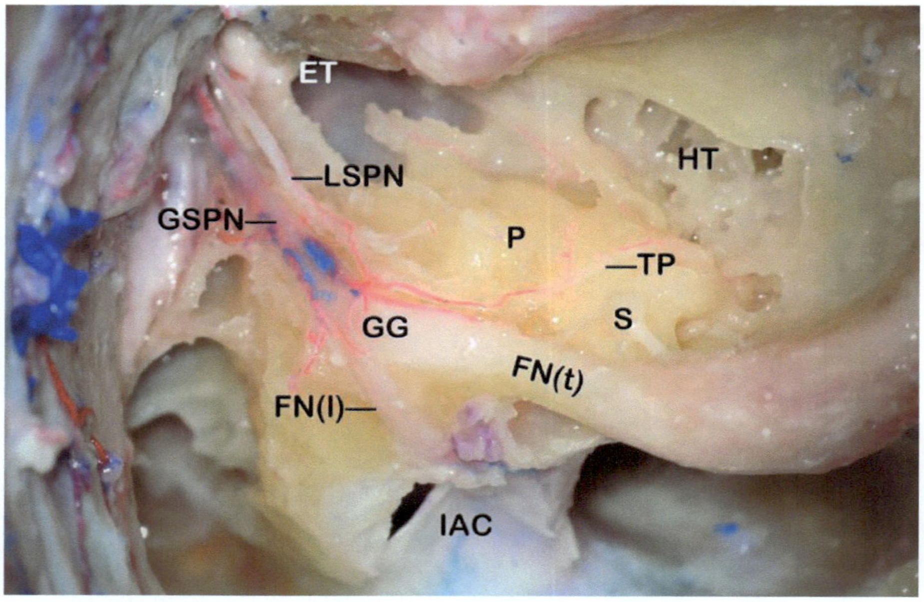

Fig. 4.56 The greater superficial petrosal nerve leaves the anterior surface of the geniculate ganglion. *ET*, eustachian tube; *GSPN*, greater superficial petrosal nerve; *LSPN*, lesser superficial petrosal nerve; *P*, promontory; *TP*, tympanic plexus; *S*, stapes; *HT*, hypotympanum; *FN(l)*, labyrinthine segment of facial nerve; *FN(t)*, tympanic segment of facial nerve; *GG*, geniculate ganglion; *IAC*, internal auditory canal.

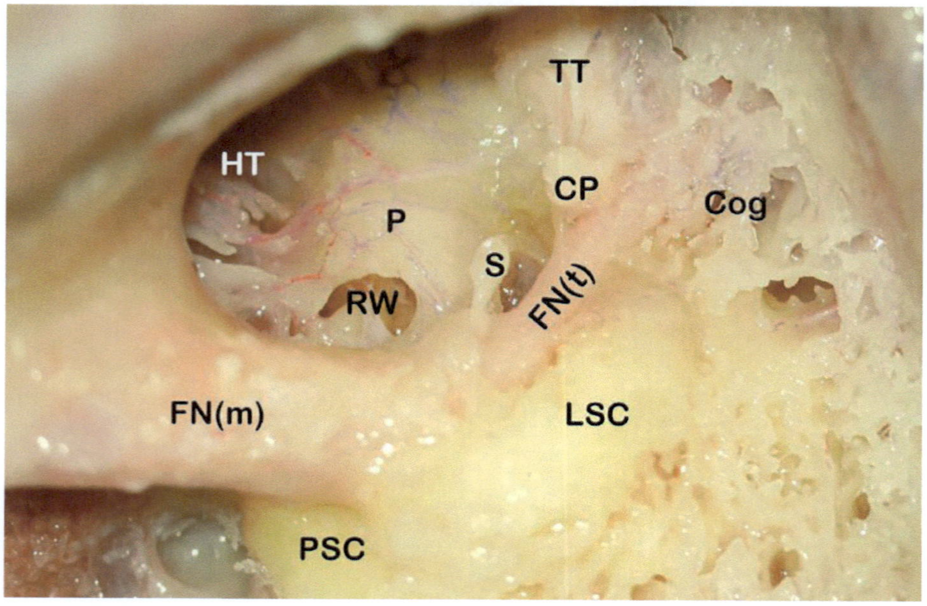

Fig. 4.57 The tympanic segment of facial nerve lies on the medial wall of the tympanum and protrudes into the cavity covered by a thin shell of bone. *HT*, hypotympanum; *TT*, tensor tympani muscle; *P*, promontory; *CP*, cochleariform process; *S*, stapes; *RW*, round window; *FN(t)*, tympanic segment of facial nerve; *FN(m)*, mastoid segment of facial nerve; *LSC*, lateral semicircular canal; *PSC*, posterior semicircular canal.

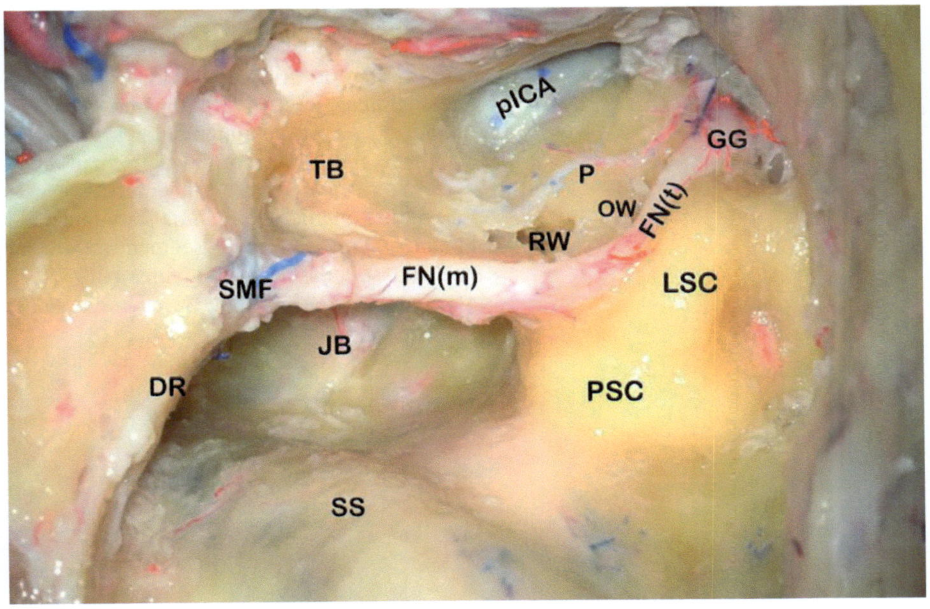

Fig. 4.58 The mastoid segment of the facial nerve can be traced from the second genu to the anterior edge of the digastric ridge where the stylomastoid foramen lies, marking the end of the intratemporal part of the facial nerve. *pICA*, petrous segment of internal carotid artery; *TB*, tympanic bone; *P*, promontory; *OW*, oval window; *RW*, round window; *SMF*, stylomastoid foramen; *FN(t)*, tympanic segment of facial nerve; *FN(m)*, mastoid segment of facial nerve; *GG*, geniculate ganglion; *LSC*, lateral semicircular canal; *PSC*, posterior semicircular canal; *JB*, jugular bulb; *DR*, digastric ridge; *SS*, sigmoid sinus.

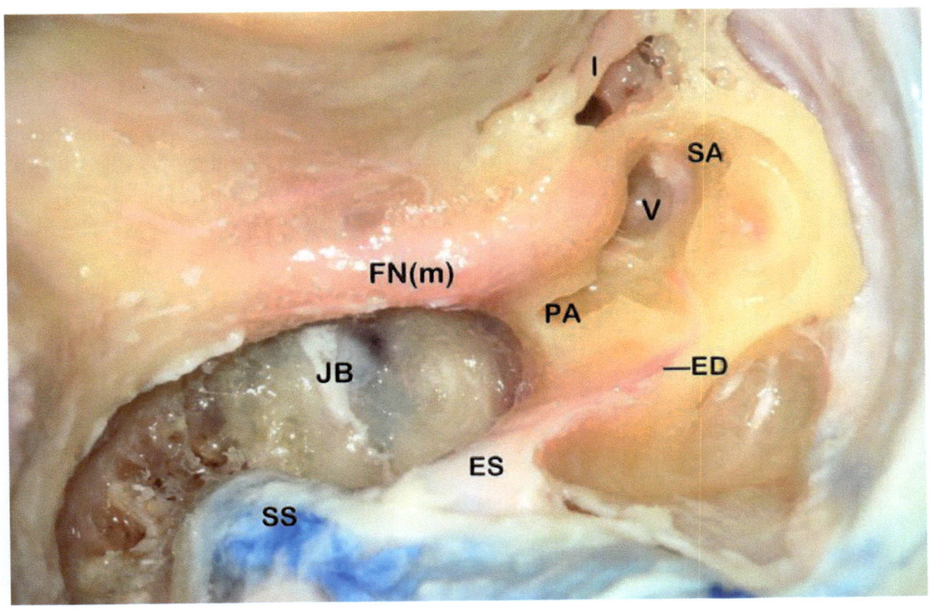

Fig. 4.59 The ampulla of the posterior semicircular canal lies medial to the midportion of this segment, forming a useful landmark in translabyrinthine surgery. *I*, incus; *SA*, ampullated end of superior semicircular canal; *PA*, ampullated end of posterior semicircular canal; *V*, vestibule; *FN(m)*, mastoid segment of facial nerve; *JB*, jugular bulb; *ED*, endolymphatic duct; *ES*, endolymphatic sac; *SS*, sigmoid sinus.

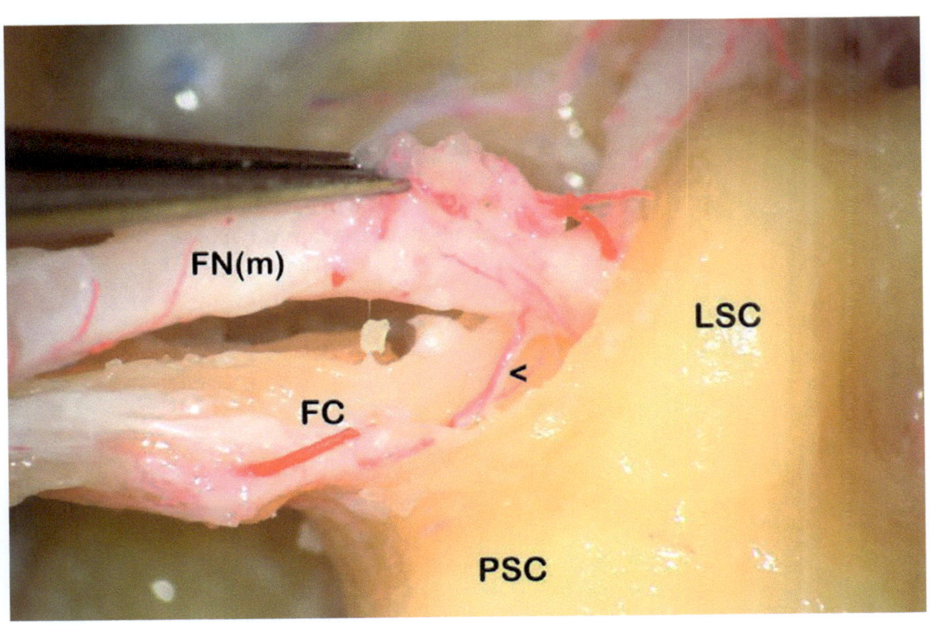

Fig. 4.60 The blood supply of the mastoid segment of the facial nerve traverses the medial surface of the bone to reach the nerve (<). *FN(m)*, mastoid segment of facial nerve; *FC*, fallopian canal; *LSC*, lateral semicircular canal; *PSC*, posterior semicircular canal.

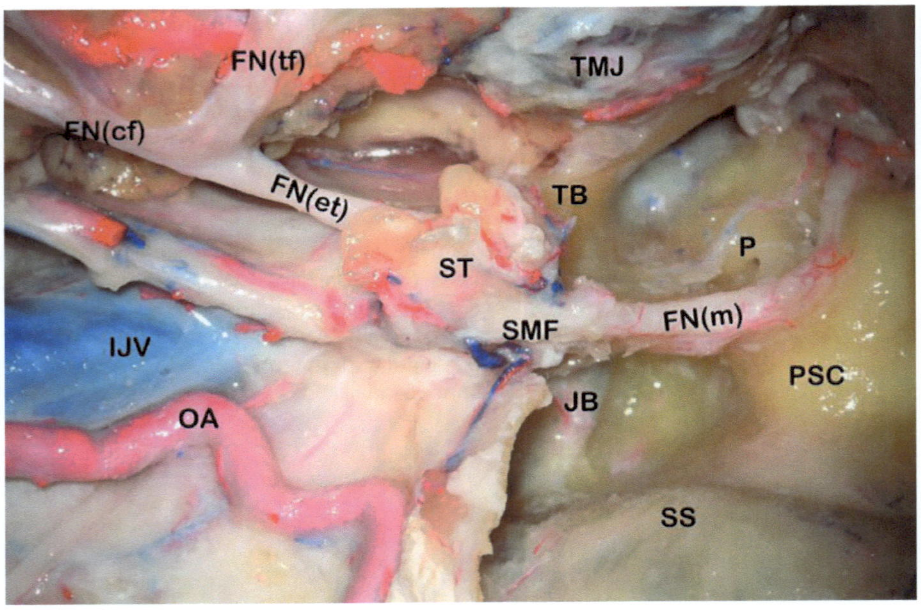

Fig. 4.61 The main trunk of the facial nerve emerges from the stylomastoid foramen at the anterior border of the digastric ridge. *TMJ*, temporamandibular joint; *TB*, tympanic bone; *P*, promontory; *FN(et)*, extratemporal segment of facial nerve; *FN(tf)*, temporofacial trunk of facial nerve; *FN(cf)*, cervicofacial trunk of facial nerve; *IJV*, internal jugular vein; *SMF*, stylomastoid foramen; *FN(m)*, mastoid segment of facial nerve; *ST*, soft tissues; *JB*, jugular bulb; *PSC*, posterior semicircular canal; *SS*, sigmoid sinus; *OA*, occipital artery.

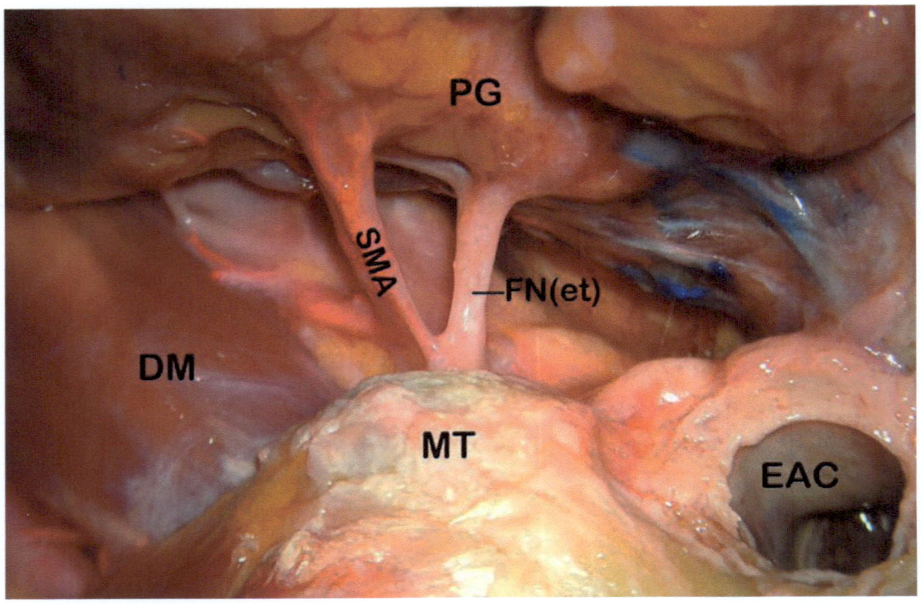

Fig. 4.62 The posterior belly of digastric muscle is an important landmark for the identification of the main trunk of the facial nerve during surgery. *PG*, parotid gland; *FN(et)*, extratemporal segment of facial nerve; *SMA*, stylomastoid artery; *DM*, digastric muscle; *MT*, mastoid tip; *EAC*, external auditory canal.

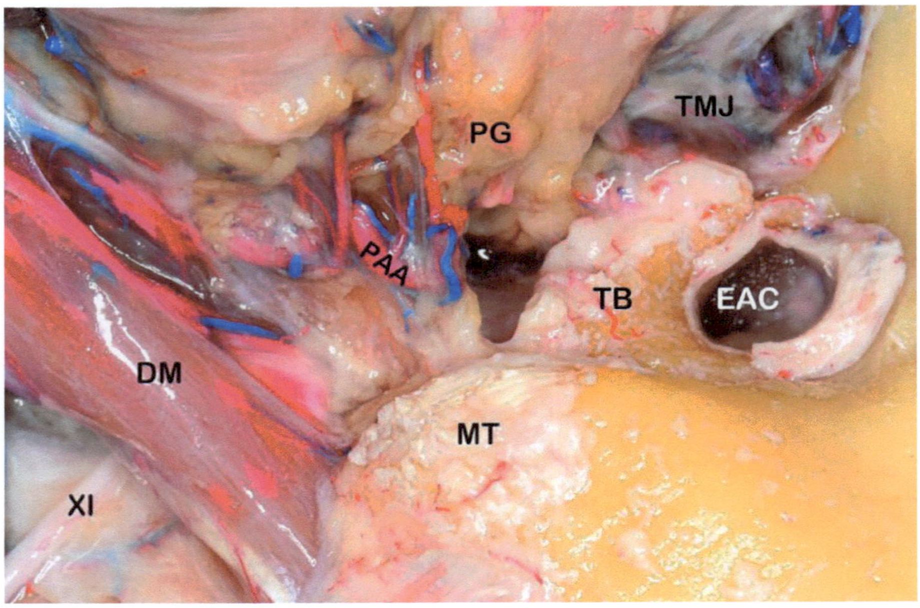

Fig. 4.63 Bleeding from the branches of the posterior auricular artery lying just lateral to the facial nerve serves to inform the surgeons that they are approaching the facial nerve. *TMJ*, temporamandibular joint; *PG*, parotid gland; *PAA*, posterior auricular artery; *DM*, digastric muscle; *XI*, accessory nerve; *MT*, mastoid tip; *EAC*, external auditory canal; *TB*, tympanic bone.

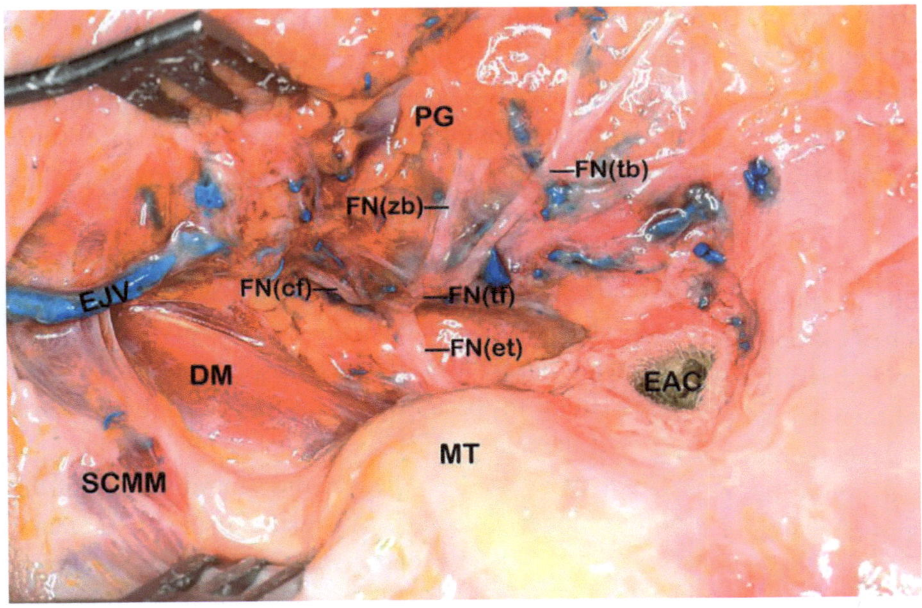

Fig. 4.64 The division into the temporofacial and cervicofacial trunk starts within the substance of the parotid gland. Following these two trunks divided into five major branches in the vast majority of cases. *FN(tf)*, temporofacial trunk of facial nerve; *FN(cf)*, cervicofacial trunk of facial nerve; *FN(et)*, extratemporal segment of facial nerve; *FN(tb)*, temporal branches of facial nerve; *FN(zb)*, zygomatic branches of facial nerve; *PG*, parotid gland; *EJV*, external jugular vein; *MT*, mastoid tip; *DM*, digastric muscle; *SCMM*, sternocleidomastoid muscle; *EAC*, external auditory canal.

Intradural Neurovascular Anatomy

Brainstem

The brainstem in the posterior fossa is composed of the mesencephalon, pons, and medulla oblongata (Fig. 4.65). The mesencephalon consists of the cerebral peduncles, the tegmentum, and the tectum. It is demarcated superiorly from the diencephalon by the sulcus between the optic tracts and the cerebral peduncles, and inferiorly from the pons by the pontomesencephalic sulcus. The pontomesencephalic sulcus runs from the superior foramen cecum around the cerebral peduncles to join the lateral mesencephalic sulcus, a vertical sulcus between the tegmentum and the cerebral peduncle.

The belly of the pons is convex from side to side, as well as from top to bottom, and is continuous on each side with the middle cerebellar peduncles. It has a shallow midline groove, the basilar sulcus, which extends from its superior to its inferior border. The posterior root of the trigeminal nerve emerges from the upper portion of the middle cerebellar peduncle just below the anterior angle of the cerebellum. The pons is demarcated inferiorly from the medulla by the pontomedullary sulcus, which extends laterally from the inferior foramen cecum (a midline dimple) to the supraolivary fossette (a depression located rostral to the olive). The rootlets of the facial and the vestibulocochlear nerves arise superior to this fossette and the rootlets of the glossopharyngeal and the vagal nerves originate dorsal to it (Fig. 4.53). The abducens nerves emerge from the pontomedullary sulcus rostral to the pyramids.

The anterior surface of the medulla oblongata is formed by the medullary pyramids, which face the clivus, the anterior edge of the foramen magnum, and the rostral part of the odontoid process (Fig. 4.66). The anteromedian sulcus divides the upper medulla oblongata in the anterior midline between the pyramids and disappears on the lower medulla at the level of the decussation of the pyramids, but it reappears below the decussation and is continuous caudally with the anteromedian fissure of the spinal cord. The lateral surface of the medulla oblongata is formed predominantly by the inferior olives, which are situated lateral to and separated from the pyramids by the anterolateral (preolivary) sulcus. The rootlets of the hypoglossal nerves arise in the anterolateral sulcus (Fig. 4.53). The lateral surface is demarcated posteriorly by the exits of the rootlets of the glossopharyngeal, vagus, and accessory nerves just dorsal to the posterolateral (postolivary) sulcus, which courses along the dorsal margin of the olive and is continuous below with the posterolateral sulcus of the spinal cord. The lower medulla oblongata blends indistinguishably into the upper spinal cord at the level of the C1 nerve roots.

Foramen of Luschka

The structures related to the lateral recess of the fourth ventricle that have a consistent relationship to the facial and vestibulocochlear nerves are the foramen of Luschka and its choroid plexus, and the flocculus (Fig. 4.67). The foramen of Luschka is situated at the lateral margin of the pontomedullary sulcus, just behind the junction of the glossopharyngeal nerve with the brainstem, and immediately posteroinferior to the junction of the facial and vestibulocochlear nerves with the brainstem. The foramen of Luschka is infrequently well visualized. A consistently identifiable choroid plexus, however, hangs out of the foramen of Luschka and sits on the posterior surface of the glossopharyngeal and vagus nerves just inferior to the junction of the facial and vestibulocochlear nerves with the brainstem (Fig. 4.68). Another structure related to the lateral recess, the flocculus, projects from the margin of the lateral recess and foramen of Luschka into the cerebellopontine angle, just posterior to where the facial and vestibulocochlear nerves join the pontomedullary sulcus.

Vestibulocochlear Nerve

The complex formed by the facial and vestibulocochlear nerve originates at the level of the lateral portion of the pontomedullary junction and both nerves course anterolaterally through the CPA cistern and enter the porus of the internal auditory canal as two separate structures: the facial nerve and vestibulocochlear bundle (Fig. 4.69). At all times the vestibulocochlear nerve maintains a lateral position with respect to the facial nerve.

The vestibular and cochlear components of the eighth nerve enter the meatus together and divide into separately identifiable nerves within the internal auditory canal in the most lateral 3–4 mm of the canal.

Lower Cranial Nerves

The lower cranial nerves are positioned in the inferior area of the posterior cranial fossa. All the nerves originate from the medulla oblongata and run superolaterally to enter the jugular foramen (Fig. 4.70).

The glossopharyngeal, vagus, and spinal accessory nerves arise in that order from superior to inferior along the postolivary sulcus (Fig. 4.53). The glossopharyngeal nerve is constituted by a single root and is positioned superiorly; the multiple roots of the vagus nerve are positioned immediately inferior to the glossopharyngeal nerve; at the most inferior extremity are located the two distinct roots, the cranial and the spinal root of accessory nerves. The cranial root is smaller and is located inferiorly to the fibers of the vagus nerve. The larger spinal root originates as multiple roots from the spinal cord; these roots fuse into a single trunk that ascends the upper portion of the cervical canal and enters the posterior fossa through the foramen magnum.

The glossopharyngeal nerve and the vagus nerve with the spinal accessory nerve invariably enter the medial wall of the jugular fossa in separate dural sheaths. The opening of the inferior petrosal sinus or multiple openings also separates the glossopharyngeal nerve from the vagus and spinal accessory nerves (Fig. 4.71).

The nerves then undergo a 90 degrees rotation within the jugular fossa. These nerves, especially the vagus nerve, maintain a multifascioular architecture through the jugular fossa. The superior ganglion of the vagus nerve ends at the extracranial opening of the jugular fossa.

There is a high variability in the course of the nerves during their course through the fossa, which is significant for the relationship to the inferior petrosal sinus openings (Fig. 4.34). They run through a septum of connective tissue in continuity with the dura mater and the pericranium. The tympanic branch of the glossopharyngeal nerve (Jacobson's nerve) and the auricular branch of the vagus nerve (Arnold's nerve) also cross the jugular fossa.

At the exit point, the glossopharyngeal nerve remains anterolateral, coursing lateral to the ICA within the carotid sheath. The vagus nerve emerges more deeply between the ICA and IJV, and in 80% of cases the spinal accessory nerve courses lateral to the upper IJV at the level of the lateral process of the atlas.

The hypoglossal nerve leaves the ventral surface of the medulla oblongata medially with respect to the lower cranial nerves (Fig. 4.53). The roots run laterally toward the hypoglossal canal. In the majority of cases, the nerve fibers are grouped into two main trunks that usually fuse once inside the hypoglossal canal.

The two roots of the hypoglossal nerve are in close relationship with the posterior inferior cerebellar artery that runs between them (Fig. 4.72). Inferomedially to the lower cranial nerves, the vertebral artery enters the skull base through the foramen magnum.

Involvement of the hypoglossal nerve by a skull base lesion results in dysarthria and oral-phase dysphagia. As an isolated deficit, hypoglossal loss may be compensated by swallowing and articulatory therapy. In combination with other nerve deficits, however, hypoglossal nerve injury increases functional loss and affects rehabilitation.

Basilar Artery

The basilar artery is a large midline or paramedian vessel that is formed by the confluence of the two vertebral arteries (Fig. 4.72). It extends superiorly in the prepontine cistern, from its origin near the pontomedullary junction to its terminal bifurcation into the two posterior cerebral arteries, in the interpeduncular cistern adjacent to the dorsum sellae or in the suprasellar cistern below the level of the third ventricular floor.

The basilar artery has labyrinthine, perforating, cerebellar, and cerebral hemispheric branches:

- The labyrinthine arteries are long arteries that arise either directly from the basilar artery or from the superior or anterior inferior cerebellar arteries; they accompany the facial and vestibulocochlear nerves into the internal auditory meatus.
- Along its course in the prepontine cistern the basilar artery gives off numerous pontine perforating arteries. The median and paramedian pontine perforating arteries arise at right angles from the posterior aspect of the basilar artery, penetrate the pons, and extend to the floor of the fourth ventricle. The lateral pontine arteries arise from the

posterolateral surface of the basilar artery, encircle the brainstem and, along their course, give rise to many small perforating branches that penetrate the pons at right angles to their parent vessels.

Superior Cerebellar Artery

The superior cerebellar artery arises at or near the bifurcation of the basilar artery; as it leaves the basilar artery, it turns around the anterolateral aspect of the upper pons (Figs. 4.73 and 4.74). After passing one-half of the distance around the pons, the superior cerebellar artery turns dorsal and gives off a branch termed the anterior lateral marginal artery, which is a cortical branch. As it continues its course around the brainstem, the superior cerebellar artery gives off one or two hemispheric branches, which ramify over the superior surface of the cerebellar hemisphere: these are termed the medial hemispheric and lateral hemispheric branches. The main stem continues as the superior vermian branch; there it may anastomose end to end with the vermian branch of the PICA.

Anterior Inferior Cerebellar Artery

The anterior inferior cerebellar artery (AICA) originates from the basilar artery (Fig. 4.72) in the majority of the cases and runs laterally toward the VII–VIII nerve complex (Fig. 4.75).

The AICA normally arises above the junction of the two vertebral arteries and at a variable distance from it. The size and course of this artery are variable and are dependent in part upon the size and distribution of the PICA; the variation in size of the AICA is in inverse proportion to the size of the PICA.

After its origin, the AICA extends laterally and downward across the pons, passing behind the internal auditory canal, into which it frequently makes a tight loop; at this level, it gives origin to the labyrinthine artery.

In the classical pattern, at the level of the cerebellopontine angle the artery divides into two major branches that supply the petrosal surface of the cerebellum: a lateral-rostral branch for the flocculus and semilunar lobules, and a medial-caudal branch for the biventral lobule and middle cerebellar peduncle. In case of a small or absent posterior inferior cerebellar artery the cortical territory of the anterior inferior cerebellar artery is more extended.

Posterior Inferior Cerebellar Artery

The posterior inferior cerebellar artery (PICA) arises from the vertebral artery near the inferior olive and passes posteriorly around the medulla (Figs. 4.72 and 4.76). It usually originates within the dura mater, but it may infrequently originate from the terminal extradural part of the vertebral artery. At the anterolateral margin of the medulla, it passes rostral or caudal to or between the rootlets of the hypoglossal nerve, and at the posterolateral margin of the medulla, it courses rostral to or between the fila of the glossopharyngeal, vagus, and accessory nerves. After passing the latter nerves, it courses around the cerebellar tonsil, enters the cerebellomedullary fissure, and passes posterior to the lower half of the roof of the fourth ventricle. On exiting the cerebellomedullary fissure, its branches are distributed to the vermis and hemisphere of the suboccipital surface. The PICA gives off perforating, choroidal, and cortical arteries. The cortical arteries are divided into vermian, tonsillar, and hemispheric groups.

PICA is the most complex and variable of the cerebellar arteries. It supplies the lower medulla and inferior aspect of the fourth ventricle, the tonsils, the vermis, and the inferolateral cerebellar hemispheres.

The size of the territory of the PICA is in inverse proportion to that of the ipsilateral anterior inferior cerebellar artery (AICA), the larger of the two being referred to as dominant. It is common for a dominant PICA to arise from the larger of the vertebral arteries, with a small ipsilateral and a dominant contralateral AICA. An even more extensive distribution of the PICA is seen in 14% of brains when the vermian branches of the PICA cross the midline to establish a bilateral supply.

After arising from the vertebral artery, the PICA turns around the medulla (medullary segment). At this point the medullary segment may loop inferiorly or superiorly or may form a complex curve; it provides direct perforator and circumflex medullary branches.

On the posterior surface of the medulla, the PICA is confronted with the tonsillar projection of the cerebellar hemisphere (tonsillar segment); the course that the PICA takes to loop over the tonsil, close to the midline. The ascending limb of this loop lies immediately behind the lower roof of the lower half of the fourth ventricle.

The tip of the supratonsillar arch is known as the choroidal point; when occlusion or sacrifice of a PICA becomes necessary, subsequent morbidity may be significantly less if the occlusion is performed distal to this point. At or soon after the zenith of the supratonsillar curve, the PICA takes a posteroinferior course and bifurcates into two main trunks: the smaller (vermian) trunk stays close to the midline and supplies the median structures along the inferior surface of the vermis; the larger (tonsillohemispheric) trunk serves the lateral inferior cerebellar surface. The vermian and hemispheric trunks have prominent anastomoses with the superior cerebellar artery.

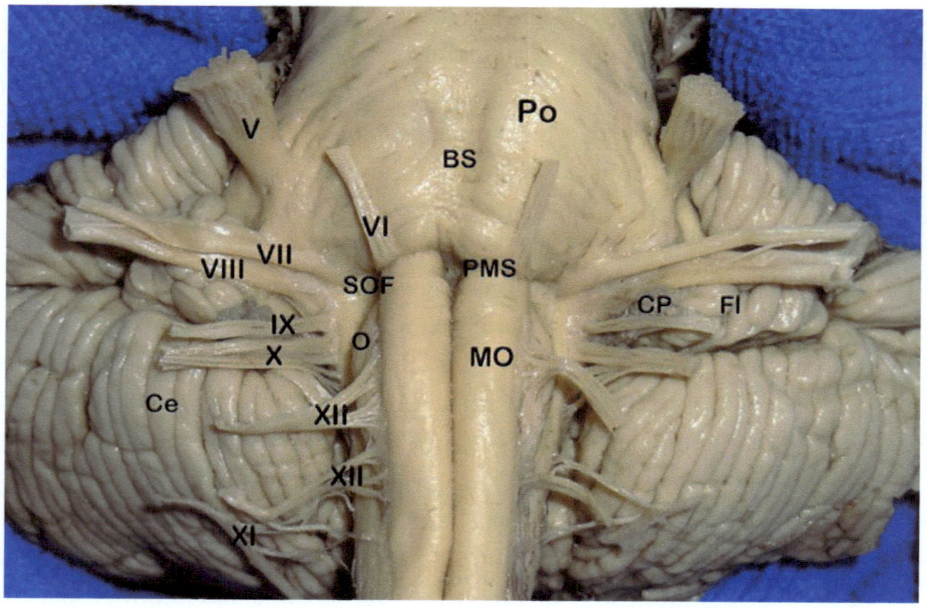

Fig. 4.65 The ventral surface of the brainstem. *V*, trigeminal nerve; *Po*, pons; *BS*, basilar sulcus; *VI*, abducent nerve; *VII*, facial nerve; *VIII*, vestibulocochlear nerve; *Fl*, flocculus; *CP*, choroid plexus; *PMS*, pontomedullary sulcus; *SOF*, supraolivary fossette; *O*, olive; *MO*, medulla oblongata; *Ce*, cerebellum; *IX*, glossopharyngeal nerve; *X*, vagus nerve; *XI*, accessory nerve; *XII*, hypoglossal nerve.

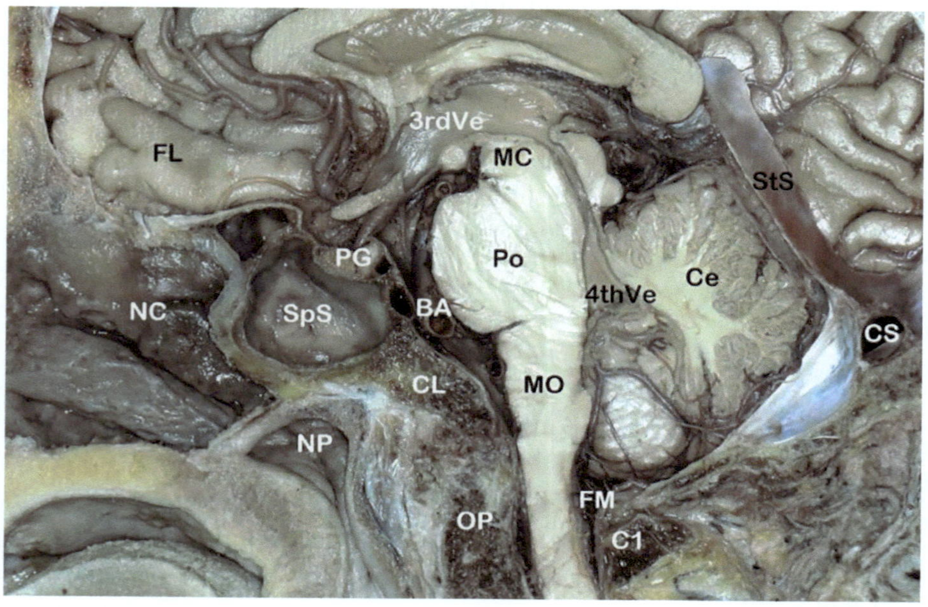

Fig. 4.66 Midsagittal views of the brainstem. *FL*, frontal lobe; *3rdVe*, third ventricle; *MC*, mesencephalon; *Po*, pons; *MO*, medulla oblongata; *4thVe*, fourth ventricle; *Ce*, cerebellum; *PG*, pituitary gland; *SpS*, sphenoid sinus; *NC*, nasal cavity; *BA*, basilar artery; *CL*, clivus; *NP*, nasopharynx; *OP*, odontoid process; *FM*, foramen magnum; *C1*, atlas; *StS*, straight sinus; *CS*, confluence of sinuses.

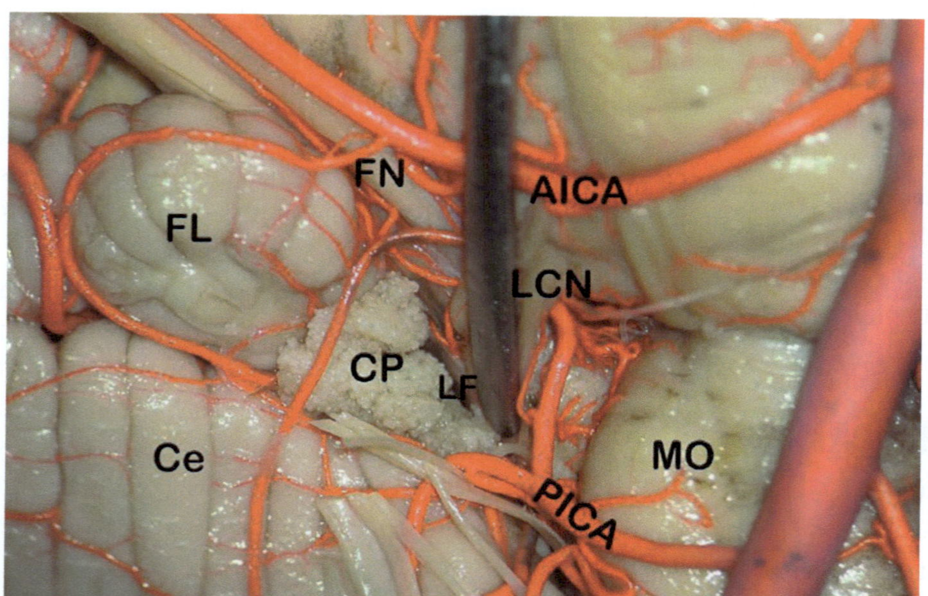

Fig. 4.67 Both the choroid plexus and foramen of Luschka are important landmarks for access to the lateral recess of the fourth ventricle. The choroid plexus and foramen of Luschka can be found at the base of the vestibulocochlear nerve and glossopharyngeal nerves, located just posteriorly. *AICA*, anteroinferior cerebellar artery; *FN*, facial nerve; *FL*, flocculus; *CP*, choroid plexus; *LF*, foramen of Luschka; *PICA*, posteroinferior cerebellar artery; *MO*, medulla oblongata; *Ce*, cerebellum; *LCN*, lower cranial nerves.

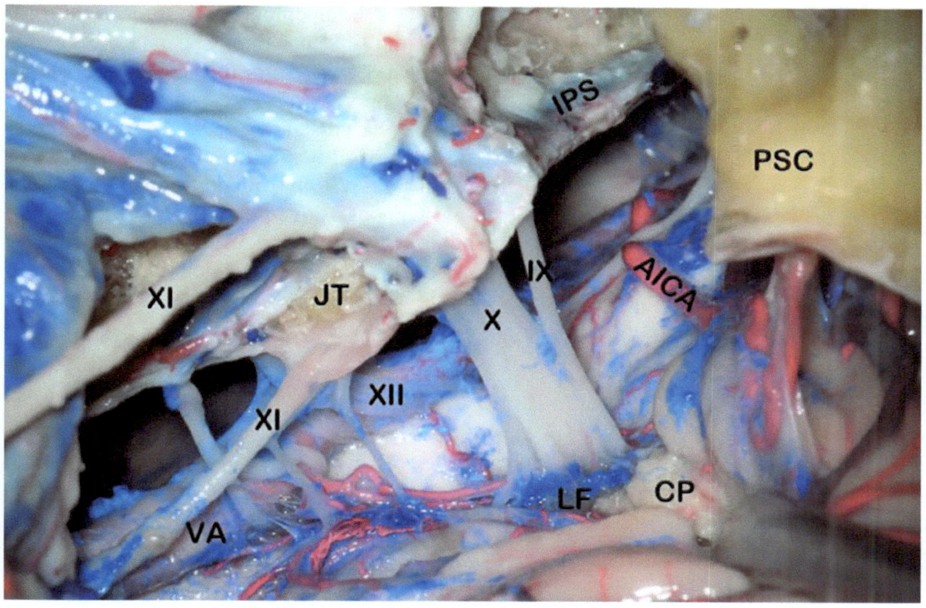

Fig. 4.68 Progressive retraction of the flocculus is required to have access to the Luschka foramen. *IPS*, inferior petrosal sinus; *PSC*, posterior semicircular canal; *AICA*, anteroinferior cerebellar artery; *CP*, choroid plexus; *LF*, foramen of Luschka; *IX*, glossopharyngeal nerve; *X*, vagus nerve; *XI*, accessory nerve; *XII*, hypoglossal nerve; *JT*, jugular tubercle; *VA*, vertebral artery.

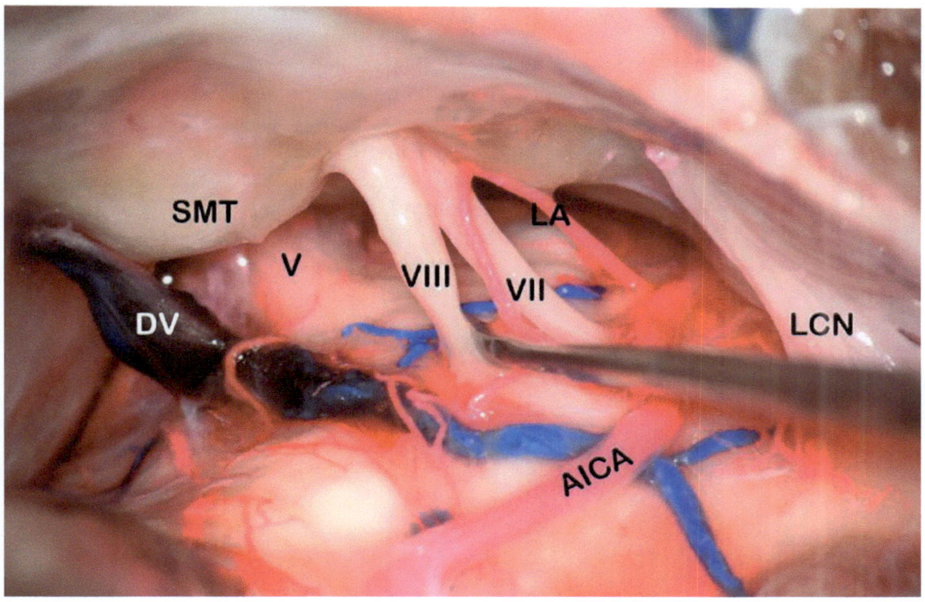

Fig. 4.69 The cleavage plane between the vestibulocochlear nerve and the facial nerve can be identified using a small probe. The labyrinthine artery runs between the nerves. *DV*, Dandy's vein (superior petrosal vein); *SMT*, suprameatal tubercle; *V*, trigeminal nerve; *VII*, facial nerve; *VIII*, vestibulocochlear nerve; *LA*, labyrinthine artery; *AICA*, anteroinferior cerebellar artery; *LCN*, lower cranial nerves.

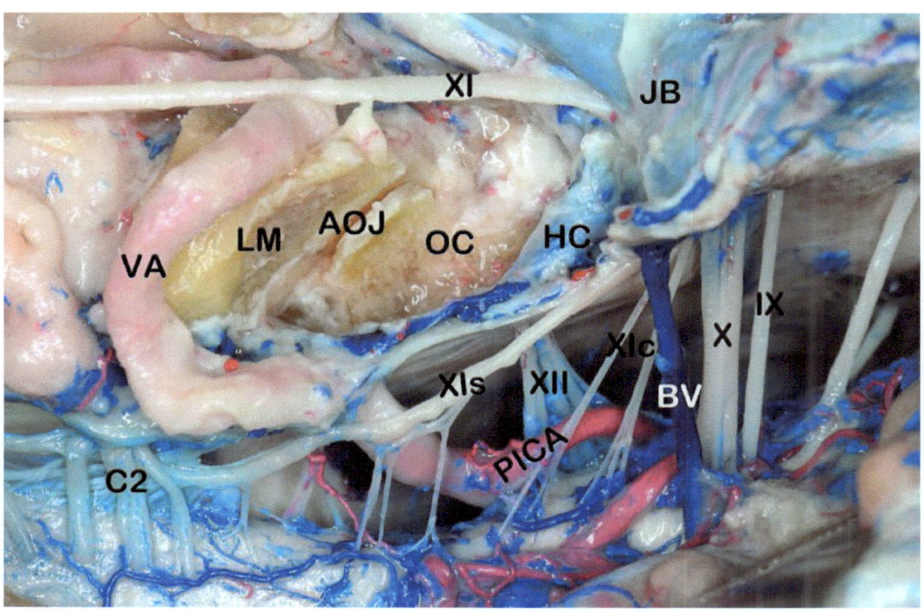

Fig. 4.70 The lower cranial nerves are positioned in the inferior area of the posterior cranial fossa. All the nerves originate from the medulla oblongata and run superolaterally to enter the jugular foramen. *JB*, jugular bulb; *HC*, hypoglossal canal; c*XI*, cranial rootlets of accessory nerve; s*XI*, spinal root of accessory nerve; *XI*, accessory nerve; *XII*, hypoglossal nerve; *VA*, vertebral artery; *PICA*, posteroinferior cerebellar artery; *BV*, bridge vein; *IX*, glossopharyngeal nerve; *X*, vagus nerve; *OC*, occipital condyle; *AOJ*, atlanto-occipital joints; *LM*, lateral mass; *C2*, rootlets of C2.

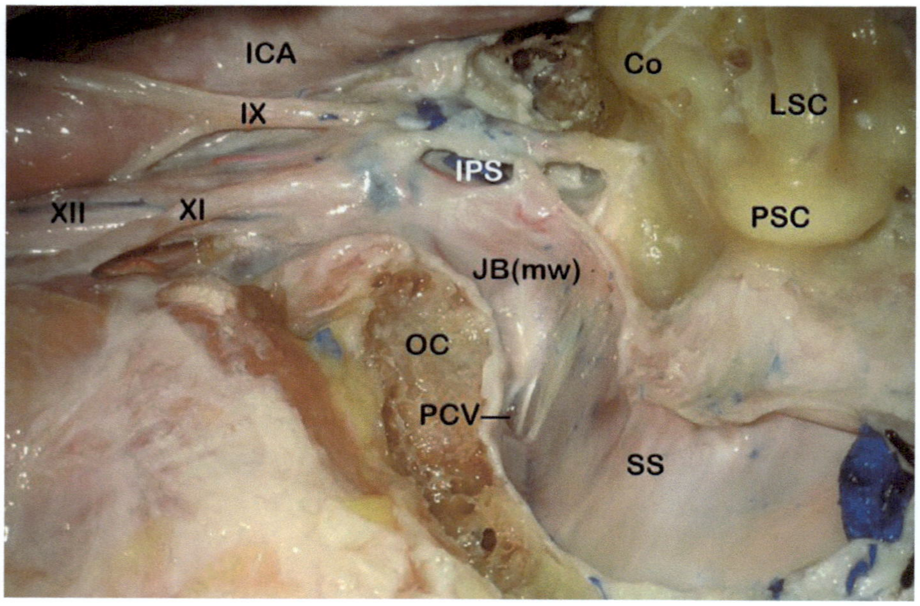

Fig. 4.71 The opening of the inferior petrosal sinus or multiple openings separate the glossopharyngeal nerve from the vagus and spinal accessory nerves. *ICA*, internal carotid artery; *IX*, glossopharyngeal nerve; *XI*, accessory nerve; *XII*, hypoglossal nerve; *IPS*, opening of inferior petrosal sinus; *JB(mw)*, medial wall of the jugular bulb; *PCV*, posterior condylar vein; *OC*, occipital condyle; *Co*, cochlea; *LSC*, lateral semicircular canal; *PSC*, posterior semicircular canal; *SS*, sigmoid sinus.

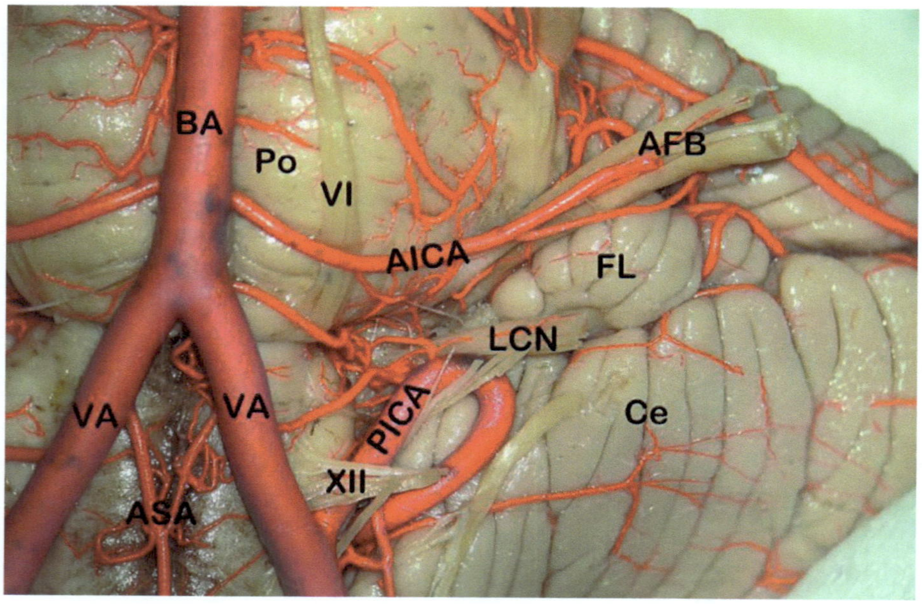

Fig. 4.72 The posterior inferior cerebellar artery arises from the vertebral artery near the inferior olive and passes posteriorly around the medulla. *BA*, basilar artery; *AICA*, anteroinferior cerebellar artery; *VA*, vertebral artery; *PICA*, posteroinferior cerebellar artery; *ASA*, anterior spinal artery; *Po*, pons; *FL*, flocculus; *Ce*, cerebellum; *VI*, abducent nerve; *AFB*, acousticofacial bundle; *XII*, hypoglossal nerve; *LCN*, lower cranial nerves.

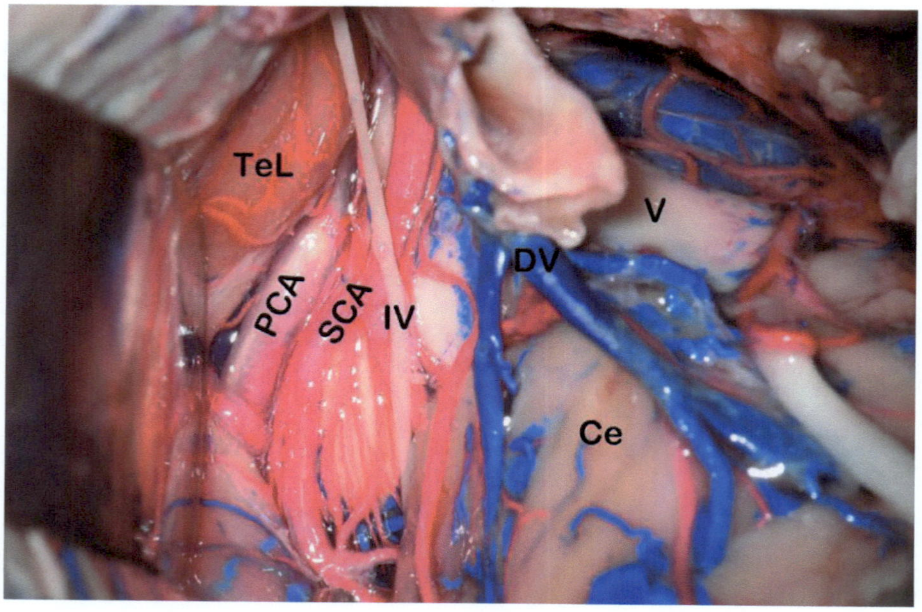

Fig. 4.73 The superior cerebellar artery turns around the anterolateral aspect of the upper pons in ambient cistern. *TeL*, temporal lobe; *V*, trigeminal nerve; *Ce*, cerebellum; *DV*, Dandy's vein (superior petrosal vein); *IV*, trochlear nerve; *SCA*, superior cerebellar artery; *PCA*, posterior cerebral artery.

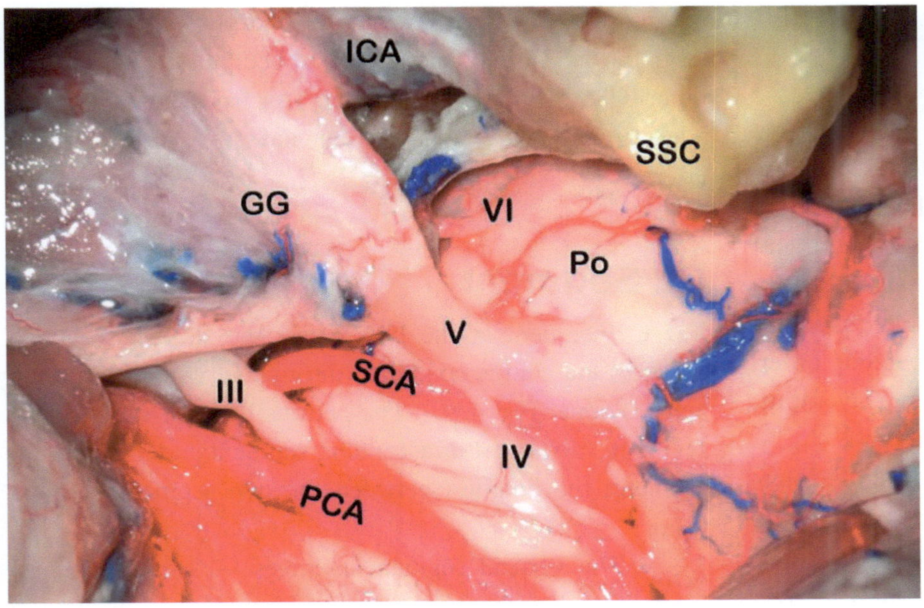

Fig. 4.74 The superior cerebellar artery arises at or near the bifurcation of the basilar artery; as it leaves the basilar artery, it turns around the anterolateral aspect of the upper pons. *ICA*, internal carotid artery; *GG*, gasserian ganglion; *VI*, abducent nerve; *SSC*, superior semicircular canal; *Po*, pons; *V*, trigeminal nerve; *IV*, trochlear nerve; *SCA*, superior cerebellar artery; *PCA*, posterior cerebral artery; *III*, oculomotor nerve.

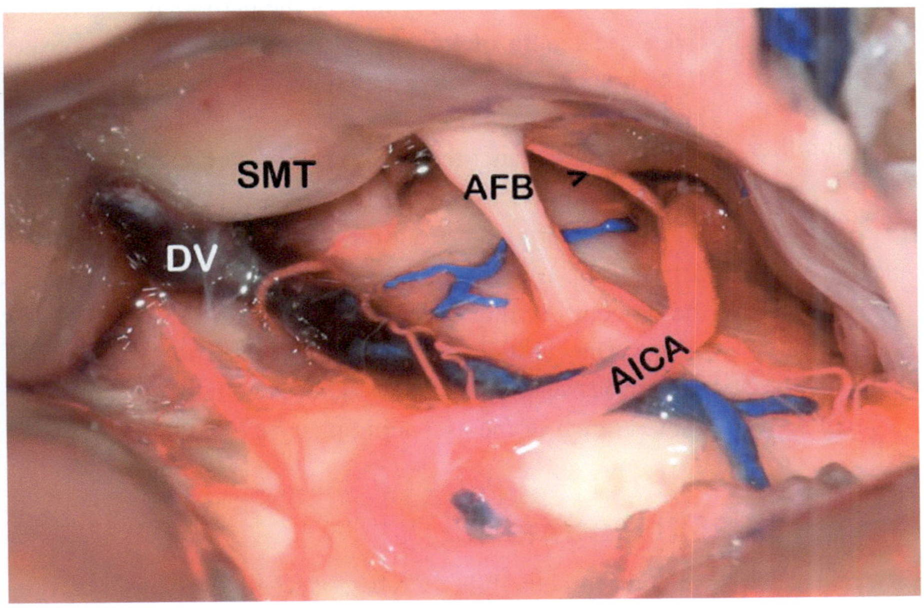

Fig. 4.75 The anterior inferior cerebellar artery runs laterally toward the acousticofacial bundle. *DV*, Dandy's vein (superior petrosal vein); *SMT*, suprameatal tubercle; *AFB*, acousticofacial bundle; *AICA*, anteroinferior cerebellar artery; >, labyrinthine artery.

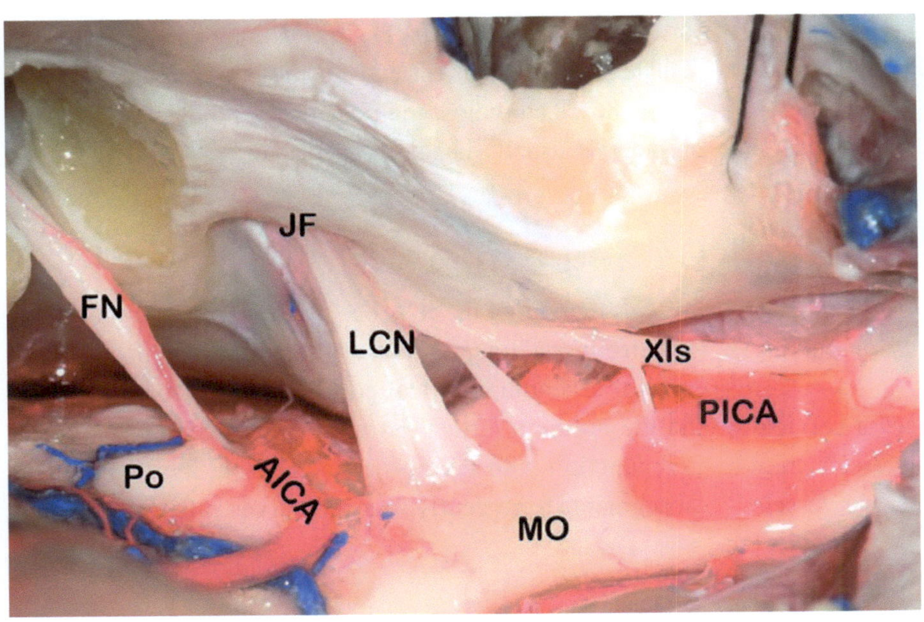

Fig. 4.76 The posterior inferior cerebellar artery arises from the vertebral artery near the inferior olive and passes posteriorly around the medulla. *FN*, facial nerve; *Po*, pons; *AICA*, anteroinferior cerebellar artery; *LCN*, lower cranial nerves; *PICA*, posteroinferior cerebellar artery; *MO*, medulla oblongata; *XIs*, spinal root of accessory nerve; *JF*, jugular foramen.

Anatomy of the Internal Carotid Artery

The common carotid artery usually bifurcates at the third and fourth cervical vertebrae into an external carotid artery trunk and an internal carotid artery (ICA). The anatomy of the internal carotid artery is named in segments between the common carotid bifurcation in the neck and the supraclinoid internal carotid artery bifurcation. In ascending order, these consist broadly of the cervical, petrous, precavernous, cavernous, paraclinoid, and supraclinoid segments.

Various numbering and descriptive schemes for the internal carotid artery have been introduced: some schemes number only the intracranial segments of the internal carotid artery, some the entire course. Some schemes number with the direction of flow, others number against the direction of flow.

To encompass the babel of nomenclature and numbers for the segments of the ICA, Bouthillier et al. described seven segments that will be used as an outline for the following description (Fig. 4.77). The classification of Bouthillier et al. identifies ICA segments according to their adjacent structures and the compartments they traverse. Seven anatomically distinct segments are described, numbered in the direction of normal blood flow:

- Cervical segment (C1)
- Petrous segment (C2)
- Lacerum segment (C3)
- Cavernous segment (C4)
- Clinoid segment (C4)
- Ophthalmic segment (C6)
- Communicating segment (C7)

C1 Segment

The C1 segment of the ICA has two parts represented by the carotid sinus and the ascending cervical segment.

- The carotid sinus is the most proximal aspect of the cervical internal carotid artery and forms a significant focal dilatation; the diameter of the normal sinus is 8–9 mm, compared with diameters of 7 mm for the common carotid artery and 4 mm for the ICA distal to the sinus.
- The ascending cervical segment courses cephalad within the carotid space (Fig. 4.78); tortuosity or kinking is relatively common (Fig. 4.79).

No named branches arise from the carotid sinus or cervical ICA segment; however, several anomalous vessels may originate from the extracranial ICA. These include vessels that normally originate from the external carotid artery (e.g., the ascending pharyngeal and occipital arteries), from other ICA segments (vidian artery), or from the vertebrobasilar circulation (e.g., cerebellar and posterior meningeal arteries).

C2 Segment

The C2 segment of the ICA enters the skull base at the carotid canal and is contained throughout its course by the petrous bone (Fig. 4.80). The petrous ICA has two distinct segments: a vertical or ascending segment and a horizontal segment; the junction between the two segments form a genu.

- The vertical segment averages −10 mm in length; from the genu, the ICA turns anteromedially toward the apex of the petrous bone (Fig. 4.81).
- The horizontal segment of the petrous ICA is nearly twice the length of its vertical segment. The intrapetrous ICA branches are represented by the caroticotympanic and vidian arteries (Figs. 4.82 and 4.83).
- The vidian artery, also termed the artery of the pterygoid canal and usually arising from the external carotid artery, may originate from the horizontal petrous segment; it passes from the foramen lacerum and anastomoses with the branches of the external carotid artery.
- The caroticotympanic artery arises near the genu of the petrous segment and anastomoses with the inferior tympanic branch of the ascending pharyngeal artery.

C3 Segment

The C3 segment of the ICA begins where the petrous carotid canal ends (Figs. 4.84 and 4.85); here the ICA runs above the foramen lacerum and ascends in a posterior extension of the carotid sulcus, part of the basisphenoid.

The lacerum segment ends at the petrolingual ligament (Fig. 4.86), 1 cm below the posterior genu of the cavernous segment; it usually has no branches.

C4 Segment

The C4 segment of the ICA begins at the superior margin of the petrolingual ligament. It has three subsegments: the posterior or ascending vertical portion; the horizontal portion; and the anterior vertical portion (Figs. 4.85 and 4.86). At their junction with the horizontal portion, the two vertical subsegments form gently rounded curves called the anterior genu and the posterior genu. The C4 segment exits the cavernous sinus through a dural ring in the superior wall.

Several branches arise from the C4 segment: the posterior trunk, also called the meningohypophyseal artery, arises from the posterior genu; two major groups of arteries, the capsular arteries and the inferolateral trunk, arise from the horizontal segment. The posterior trunk has hypophyseal, dural tentorial, and clival branches.

- The posteroinferior hypophyseal artery is directed medially and supplies the neurohypophysis and peripheral adenohypophysis of the pituitary gland. It also gives a medial clival branch, which runs near the midline along the clivus, to meet the ascending dival branch of the hypoglossal artery, deriving from the ascending pharyngeal artery.
- The tentorial branches are represented by the marginal tentorial artery and the basal tentorial artery.
  - The basal tentorial artery extends laterally and horizontally along the petrous ridge.
  - The marginal tentorial artery follows the tentorial incisura posteriorly and superiorly, progressively approaching the midline.
- The lateral dival artery gives lateral and medial branches along the course of the superior and inferior petrosal sinuses, respectively. It anastomoses with the middle meningeal artery and with the jugular branch of the ascending pharyngeal artery.

The inferolateral trunk arises in the midportion of the horizontal cavernous carotid artery and is directed inferiorly; a pattern of branching into three or four divisions is commonly seen. The most proximal branch is directed posteriorly and supplies the free margin of the tentorium. A branch to the superior orbital fissure is directed anteriorly, where it meets the deep recurrent ophthalmic branch from the ophthalmic artery.

The inferolateral trunk is then directed laterally and provides branches to the region of the foramen rotundum, which anastomoses with the internal maxillary artery, and the recurrent artery of the foramen lacerum, which anastomoses with the carotid branch of the superior pharyngeal branch of the ascending pharyngeal artery.

C5 Segment

The clinoid segment begins at the proximal dural ring and ends at the distal dural ring where the ICA becomes intradural. The distal dural ring is the only complete ring that entirely surrounds the ICA and fuses laterally with the adventitia of the ICA. This ring is continuous with the adjacent dura of the falciform ligament, the anterior clinoid process, and the cavernous sinus roof. The clinoid segment is part of the anterior loop of the ICA.

The clinoid segment is short, coursing obliquely between the anterior clinoid process laterally and the carotid sulcus of the basisphenoid medially. The periosteum of the anterior clinoid and carotid sulcus covers much of this segment. Posterosuperiorly, the clinoid segment is covered by dura continuous with the roof of the cavernous sinus. If the anterior and middle clinoid processes fuse, the clinoid segment of the ICA will be entirely surrounded by bone. This segment of the artery is wedge-shaped, as illustrated by Fischer, because the distal and proximal dural rings fuse posteriorly at the cavernous sinus roof. Although this segment is not intracavernous, veins of the anterior cavernous sinus often extend along this segment through an incomplete or incompetent proximal dural ring.

C6 Segment

The ophthalmic segment begins at the distal dural ring and ends just proximal to the origin of the posterior communicating artery (PComA). The proximal ophthalmic segment is the intradural portion of the anterior loop of the ICA. Two major arterial branches commonly arise from this segment; they are the ophthalmic and superior hypophyseal arteries. The ophthalmic artery may arise as a branch of the inferolateral trunk of the cavernous segment. The superior hypophyseal artery may arise as single or multiple branches. The lateral surface of the proximal ophthalmic segment is obscured by the anterior clinoid process. Occasionally, a dural evagination along the mesial aspect of the ophthalmic segment at the level of the distal dural ring results in an extension of the intradural space called the carotid cave.

C7 Segment

The communicating segment begins just proximal to the origin of the PComA and ends at the ICA bifurcates into its two terminal branches, the anterior and middle cerebral arteries. Two major branches, the posterior communicating artery and the anterior choroidal artery, arise from the communicating segment.

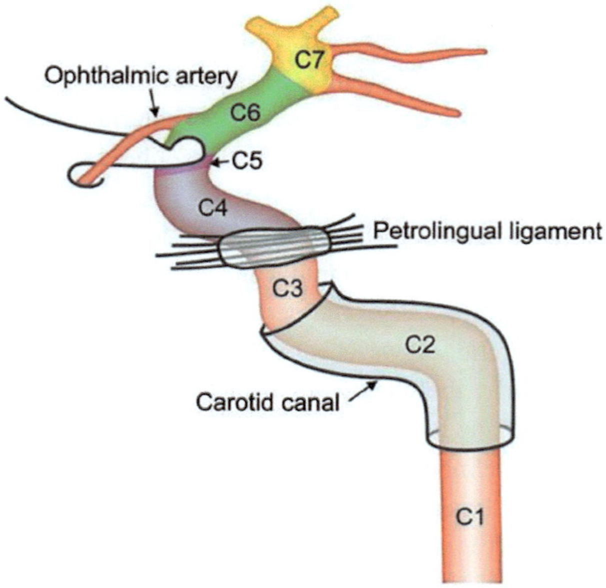

Fig. 4.77 Schematic drawing of the internal carotid artery and its subdivision into seven segments, according to Bouthillier; see text for explanation.

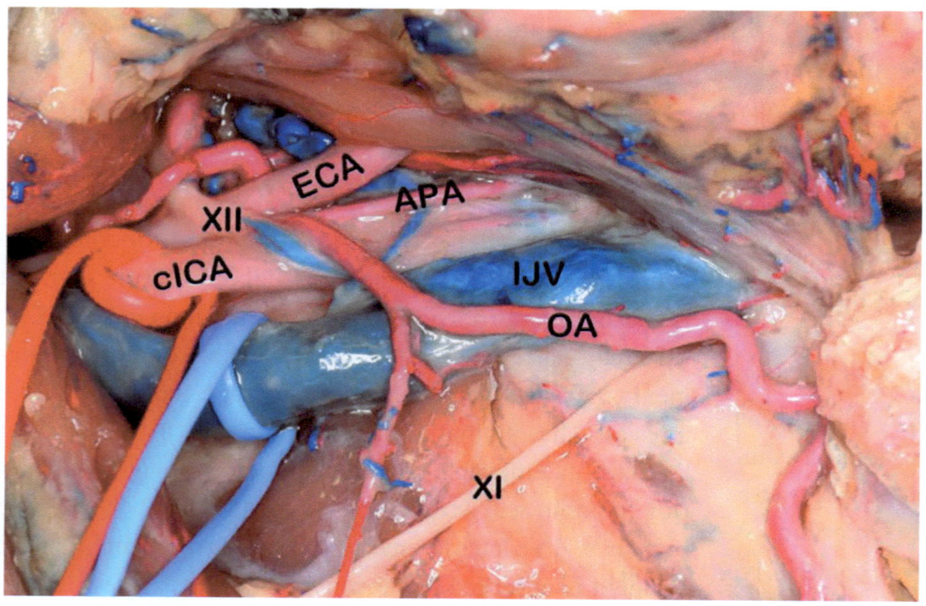

Fig. 4.78 The ascending cervical segment of internal carotid artery courses cephalad within the carotid space. *ECA*, external carotid artery; *APA*, ascending pharyngeal artery; *XII*, hypoglossal nerve; *cICA*, cervical segment of internal carotid artery; *IJV*, internal jugular vein; *OA*, occipital artery; *XI*, accessory nerve.

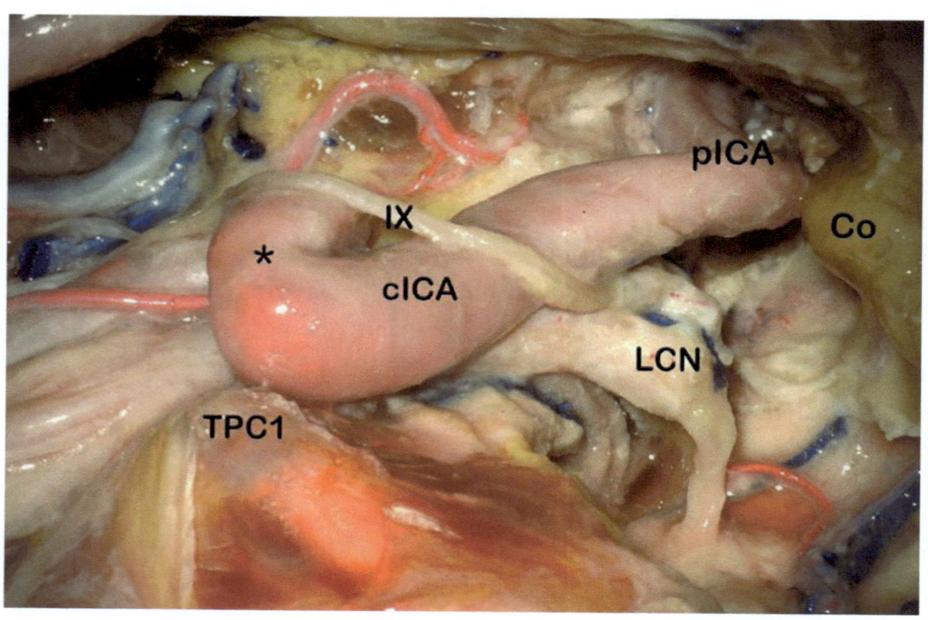

Fig. 4.79 Infratemporal fossa type A approach with transcondylar transtubercolar extension. The relationship between the cochlea and the internal carotid artery can be appreciated. The cervical segment of internal carotid artery shows a kinking on its cervical portion (*), which precludes arterial stenting. *IX*, glossopharyngeal nerve; *pICA*, petrous segment of internal carotid artery; *cICA*, cervical segment of internal carotid artery; *Co*, cochlea; *LCN*, lower cranial nerves; *TPC1*, transverse process of C1.

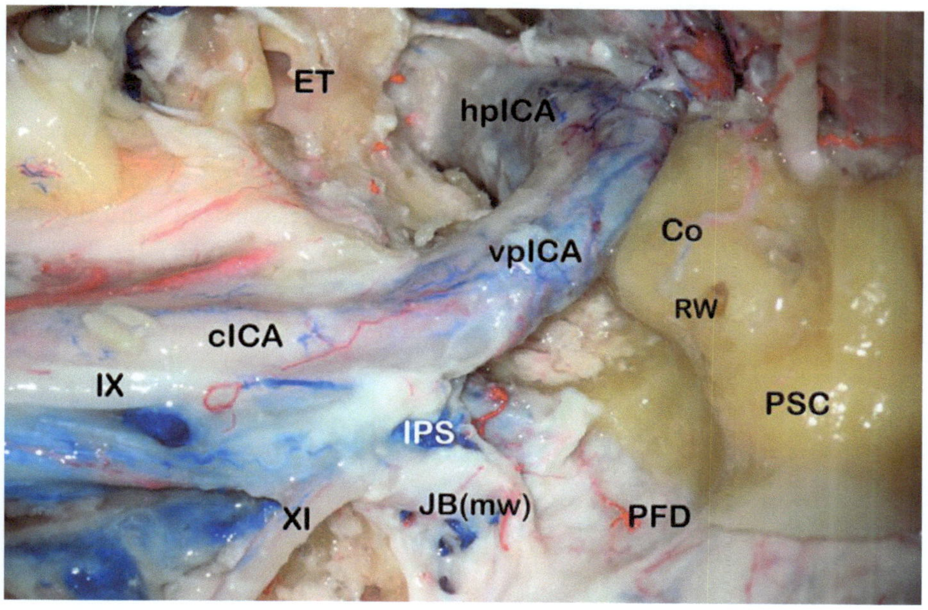

Fig. 4.80 The horizontal portion of the petrous carotid artery is under control after drilling the petrous bone lateral to the petrous carotid artery. *ET*, eustachian tube; *hpICA*, horizontal portion of petrous segment of internal carotid artery; *vpICA*, vertical portion of petrous segment of internal carotid artery; *cICA*, cervical segment of internal carotid artery; *Co*, cochlea; *RW*, round window; *PSC*, posterior semicircular canal; *IX*, glossopharyngeal nerve; *XI*, accessory nerve; *IPS*, opening of inferior petrosal sinus; *JB(mw)*, medial wall of the jugular bulb; *PFD*, posterior fossa dura.

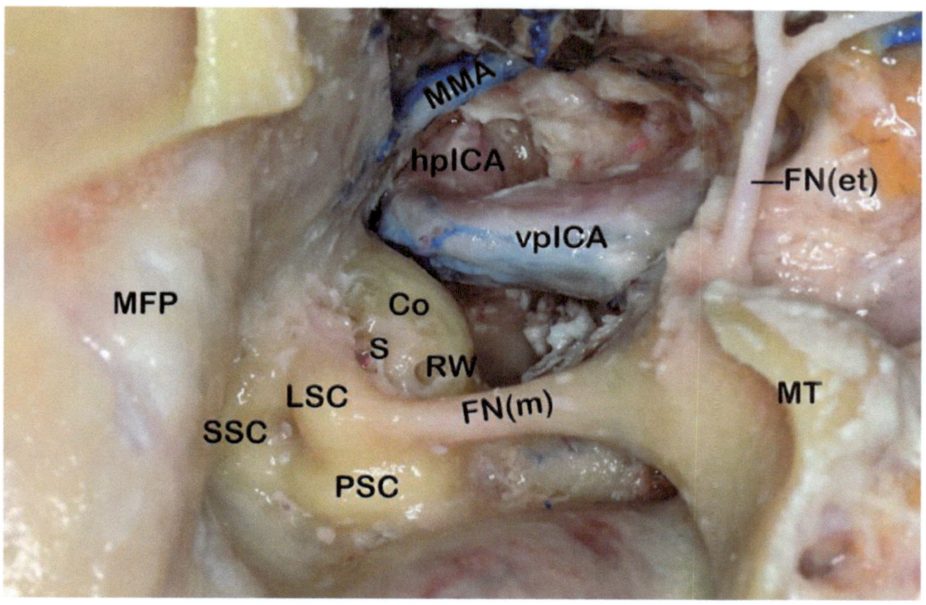

Fig. 4.81 Infratemporal fossa type B approach. The vertical portion of petrous segment of internal carotid artery from the genu of the ICA turns anteromedially toward the apex of the petrous bone. *MMA*, middle meningeal artery; *FN(et)*, extratemporal segment of facial nerve; *hpICA*, horizontal portion of petrous segment of internal carotid artery; *vpICA*, vertical portion of petrous segment of internal carotid artery; *Co*, cochlea; *RW*, round window; *S*, stapes; *FN(m)*, mastoid segment of facial nerve; *MT*, mastoid tip; *LSC*, lateral semicircular canal; *PSC*, posterior semicircular canal; *SSC*, superior semicircular canal; *MFP*, middle fossa plate.

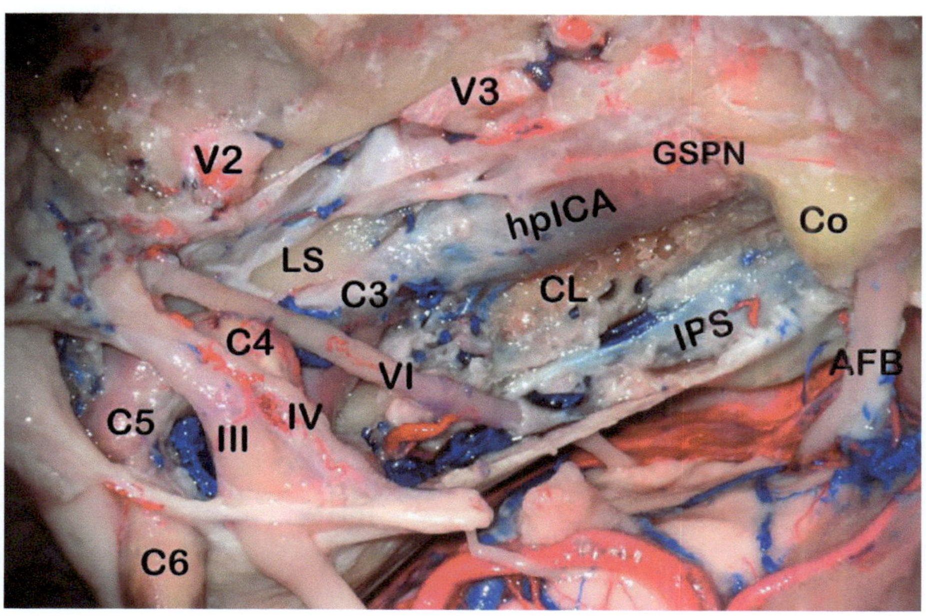

Fig. 4.82 The horizontal portion of petrous segment of internal carotid artery is nearly twice the length of its vertical segment. *V2*, maxillary nerve; *V3*, mandibular nerve; *GSPN*, greater superficial petrosal nerve; *hpICA*, horizontal portion of petrous segment of internal carotid artery; *LS*, Lingula of sphenoid bone; *CL*, clivus; *Co*, cochlea; *AFB*, acousticofacial bundle; *IPS*, inferior petrosal sinus; *III*, oculomotor nerve; *IV*, trochlear nerve; *VI*, abducent nerve; *C3*, lacerum segment of ICA; *C4*, cavernous segment of ICA; *C5*, clinoid segment of ICA; *C6*, ophthalmic segment of ICA.

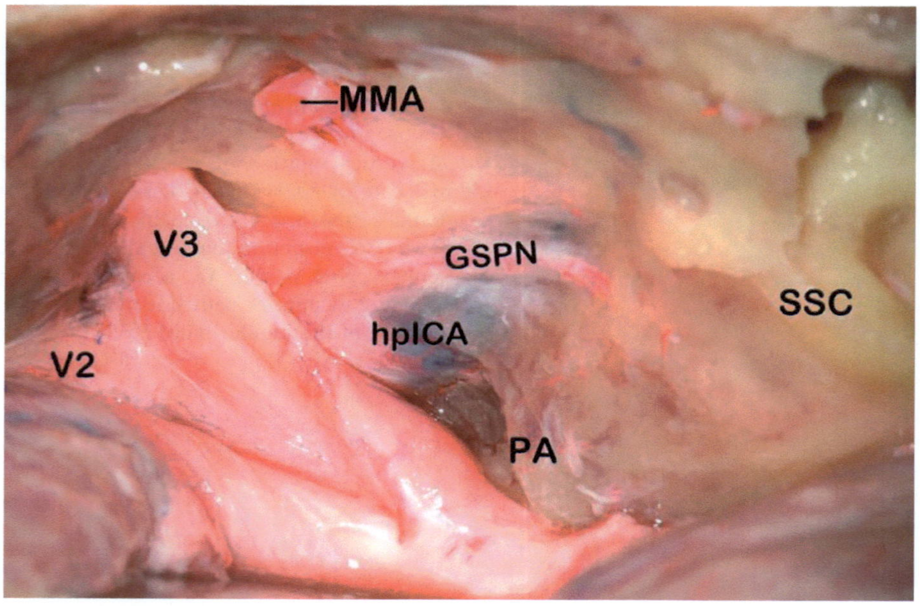

Fig. 4.83 The view through the middle fossa approach of a right temporal bone. In this case, the terminal part of the petrous carotid artery is exposed under the dura and below the greater superficial petrosal nerve, as occurs in approximately 14% of temporal bones. The petrous carotid artery is usually covered by bone up to the lateral side of the third trigeminal division. *MMA*, middle meningeal artery; *V2*, maxillary nerve; *V3*, mandibular nerve; *GSPN*, greater superficial petrosal nerve; *hpICA*, horizontal portion of petrous segment of internal carotid artery; *SSC*, superior semicircular canal; *PA*, petrous apex.

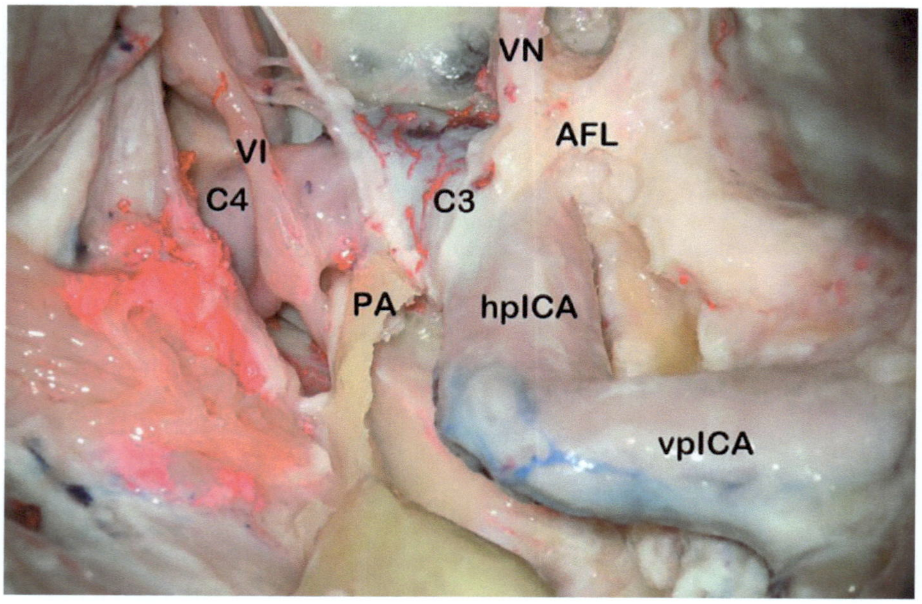

Fig. 4.84 The infratemporal fossa approach type C. The C3 segment of the ICA begins where the petrous carotid canal ends; here the ICA runs above the foramen lacerum and ascends into the cavernous sinus. *C3*, lacerum segment of ICA; *C4*, cavernous segment of *ICA*; *hpICA*, horizontal portion of petrous segment of internal carotid artery; *vpICA*, vertical portion of petrous segment of internal carotid artery; *PA*, petrous apex; *VI*, abducent nerve; *VN*, vidian nerve; *AFL*, anterior foramen lacerum.

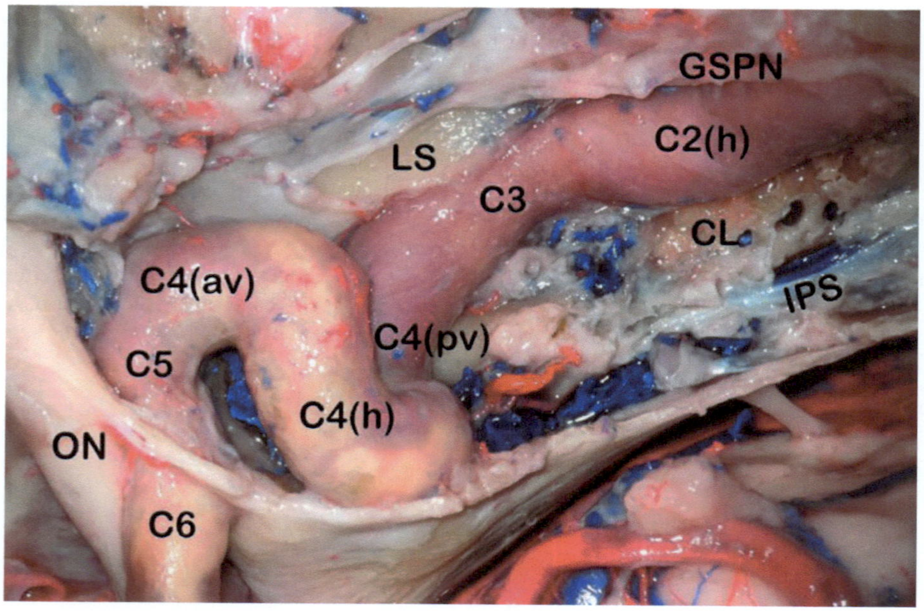

Fig. 4.85 The C4 segment of the ICA begins at the superior margin of the petrolingual ligament. It has three subsegments: the posterior or ascending vertical portion, the horizontal portion, and the anterior vertical portion. *C2(h)*, horizontal portion of petrous segment of ICA; *C3*, lacerum segment of ICA; *C4(pv)*, posterior vertical portion of cavernous segment of ICA; *C4(h)*, horizontal portion of cavernous segment of ICA; *C4(av)*, anterior vertical portion of cavernous segment of ICA; *C5*, clinoid segment of ICA; *C6*, ophthalmic segment of ICA; *GSPN*, greater superficial petrosal nerve; *LS*, Lingula of sphenoid bone; *CL*, clivus; *IPS*, inferior petrosal sinus; *ON*, optic nerve.

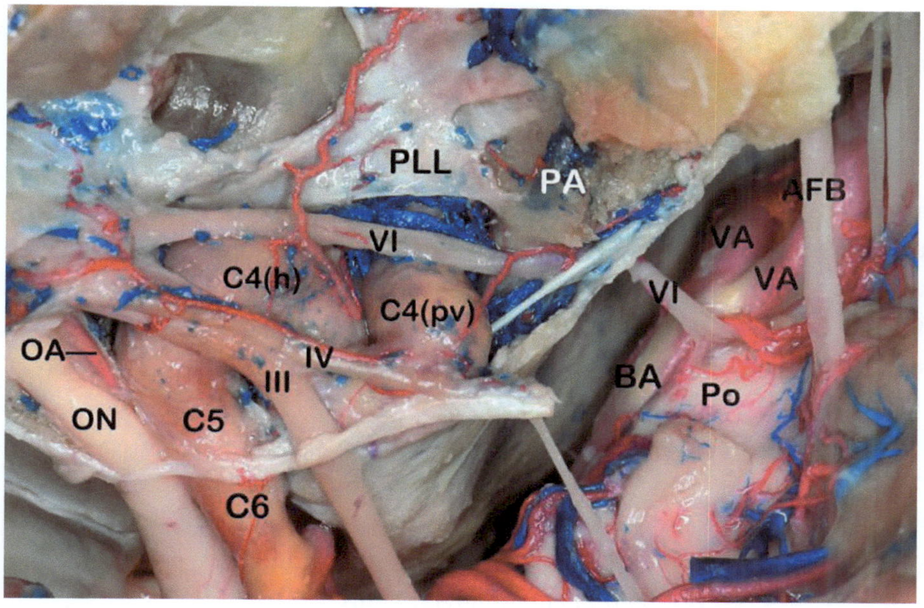

Fig. 4.86 The trigeminal nerve has been removed to expose the petrolingual ligament, which extends above the internal carotid artery just proximal to the artery's entry into the cavernous sinus. *PLL*, petrolingual ligament; *PA*, petrous apex; *III*, oculomotor nerve; *IV*, trochlear nerve; *VI*, abducent nerve; *AFB*, acousticofacial bundle; *BA*, basilar artery; *VA*, vertebral artery; *Po*, pons; *ON*, optic nerve; *OA*, ophthalmic artery; *C4(pv)*, posterior vertical portion of cavernous segment of ICA; *C4(h)*, horizontal portion of cavernous segment of ICA; *C5*, clinoid segment of ICA; *C6*, ophthalmic segment of ICA.

Anatomy of the Vertebral Artery

The paired vertebral arteries are the first and largest branches of the subclavian arteries. They arise from the subclavian arteries, usually ascend through the transverse processes of the upper six cervical vertebrae, pass behind the lateral masses of the atlas, enter the dura mater behind the occipital condyles, ascend through the foramen magnum to the front of the medulla, and join the contralateral vertebral artery to form the basilar artery at the pontomedullary junction. Each artery is divided into intradural and extradural parts.

Segments of Vertebral Artery

The vertebral artery (VA) is divided into four segments (Fig. 4.87).

V1 SEGMENT

The V1 segment arises from the subclavian artery and courses posterosuperiorly to enter the foramen of the sixth cervical vertebra.

V2 SEGMENT

The V2 segment ascends vertically, passing through the foramina of C2–C6 the V3 segment, makes a first bend at the level of the C2 foramen and a second bend before running superiorly through the C1 transverse foramen. It gives off multiple small, unnamed branches that supply the deep cervical musculature, and segmental spinal branches that supply the spinal cord and its coverings. The anterior meningeal artery arises from the distal portion of the V2 segment and supplies the dura around the foramen magnum.

V3 SEGMENT

The V3 segment extends from the transverse foramen of the axis to the dural penetration by the vertebral artery; it is further subdivided into two parts: a proximal vertical part (V3v) and a distal horizontal part (V3h) (Fig. 4.88). The V3 segment possesses four vascular loops:

- The inferomedial loop appears at the transverse foramen of the axis and directs the artery laterally and slightly posteriorly and upward.
- The inferolateral loop directs the artery distinctly upward and slightly anterior, toward the transverse foramen of the atlas.
- The superolateral loop is located at the point where the V3v turns into a horizontal position in the sulcus of the posterior arch of the atlas and becomes V3h. The V3v and V3h parts each have two constant branches:
  - The muscular artery of V3v arises at the inferolateral arterial loop; it communicates with the branches of the ascending pharyngeal artery.
  - The radiculomuscular artery of V3v arises below the transverse foramen of the atlas, and gives rise to the medial branch (a radiculomedullary artery) vascularizing the C2 ganglion, the C2 nerve, and the spinal cord; and the lateral branch, a muscular branch for the suboccipital muscles.
  - The muscular artery of V3h vascularizes the muscles of the deep muscular layer and communicates with the branches of the occipital artery.
  - The posterior meningeal artery of V3h arises at the superomedial loop and vascularizes the neighboring portion of the posterior fossa dura, the falx cerebelli, and the posterior portion of the tentorium (Fig. 4.89).

• The superomedial loop surrounds the lateral mass of the atlas and brings the V3 to its dural foramen.

V4 SEGMENT

The V4 segment enters the skull through the foramen magnum: after traversing the atlanto-occipital membrane, the artery pierces the dura and enters the subarachnoid space, usually in the lateral medullary cistern, and then runs superomedially in the premedullary cistern behind the clivus and unites with the contralateral vertebral artery at the pontomedullary junction to form the basilar artery (Fig. 4.72).

The branches of the intracranial segment of the vertebral artery have been classified into medial and lateral branches:

• The medial branches arise from the posterior and medial aspect of the vertebral artery, particularly the distal half, and supply the anterior medulla and pyramid; they comprise the anterior spinal artery and some small perforator vessels for the foramen cecum.

• The lateral branches of the intracranial segment of the vertebral artery include the posterior inferior cerebellar artery and circumferential branches to the inferior cerebellar peduncle, the lateral medulla, and the olivary structures.

Three Layers of Muscles Cover the Extraosseous Portion of the Vertebral Artery

• The superficial layer is represented by the sternocleidomastoid (Fig. 4.90).

• The middle layer is formed by the splenius capitis, which lies perpendicular to the sternocleidomastoid muscle, and the longissimus capitis, attached to its medial surface (Figs. 4.90 and 4.91).

• The deeper layer consists of the superior oblique and inferior oblique muscles, both inserted into the transverse process of the atlas (Figs. 4.92–4.94).

The superior and inferior oblique muscles, together with the first cervical vertebra, are constant landmarks for the vertebral artery. These muscles form the suboccipital triangle that houses the terminal extradural segment of the vertebral artery (V3) and the first cervical nerve.

Inferior reflection of the inferior oblique muscle allows the exposure of the vertebral artery between the transverse process of the first and second cervical vertebrae as it courses laterally and superiorly to reach the foramen transversarium of the atlas. At this level, the anterior branch of the second cervical nerve crosses the lateral surface of the artery (Figs. 4.88 and 4.89).

Reflection of the superior oblique muscle exposes the last extradural segment of the artery between the transverse process of the atlas and its entrance into the atlanto-occipital dura (Fig. 4.89). The artery runs initially in an anteroposterior direction into a groove of the superior surface of the lateral arch of the atlas. After passing back to the atlanto-occipital joint it makes a sharp anterosuperior bend to enter the foramen magnum anteriorly to the occipital condyle. At this level, the artery pierces the dura and enters the intracranial space. At the entrance point, the dura is tightly adherent to the artery.

In the extracranial segment, the vertebral artery is surrounded by a rich venous plexus that becomes a true cavernous sinus at the level of V4 (Fig. 4.95). This plexus is connected with the jugular bulb through the posterior, anterior, and lateral condylar veins.

Vertebral Artery Branches

The branches arising from the vertebral artery can be divided into two groups: a cervicospinal group that includes the lateral spinal and muscular arteries and a cranial group located in the region of the foramen magnum and the craniovertebral junction that includes the posterior and anterior spinal, and anterior and posterior meningeal arteries, and the PICA.

CERVICOSPINAL GROUP

The lateral spinal branches enter the spinal canal through the intervertebral foramina and divide into branches that pass along the dorsal and ventral nerve roots to reach and supply the surrounding part of the spinal cord and its membranes and anastomose with arteries from adjacent levels. These branches also divide into ascending and descending branches that anastomose with ascending and descending branches from adjacent levels. The lateral spinal branches also give off small branches near the attachment of the pedicles that supply the vertebral body and periosteum and join similar branches from the contralateral side so that a central anastomotic chain is found on the posterior surface of the vertebral body.

MUSCULAR BRANCHES

These branches supply the deep muscles of the neck, near the lateral mass of the atlas, and anastomose with the occipital, ascending pharyngeal, and cervical arteries.

POSTERIOR INFERIOR CEREBELLAR ARTERY

The detail see "Posterior inferior cerebellar artery" section.

POSTERIOR SPINAL ARTERY

The posterior spinal artery usually arises from the posteromedial surface of the vertebral artery, just outside the dura mater (Fig. 4.96), but it may also arise from the initial intradural part of the vertebral artery, or the PICA. Care should be taken to preserve the posterior spinal artery during dural opening because it may be incorporated into the dural cuff around the vertebral artery. As the posterior spinal artery passes through the dura mater, it is surrounded by the same fibrous tunnel as the vertebral artery and the first cervical nerve root. In the subarachnoid space, it courses medially behind the rostral attachments of the dentate ligament.

On reaching the lower medulla, the posterior spinal artery divides into ascending and descending branches (Fig. 4.96). The ascending branch courses through the foramen magnum and supplies the restiform body, gracile and cuneate tubercles, rootlets of the accessory nerve, and the choroid plexus. The descending branch passes downward between the dorsal rootlets and the dentate ligament on the posterolateral surface of the spinal cord (Fig. 4.96) and supplies the superficial part of the dorsal half of the cervical spinal cord. It anastomoses with the posterior branches of the radicular arteries entering the vertebral foramen at lower levels.

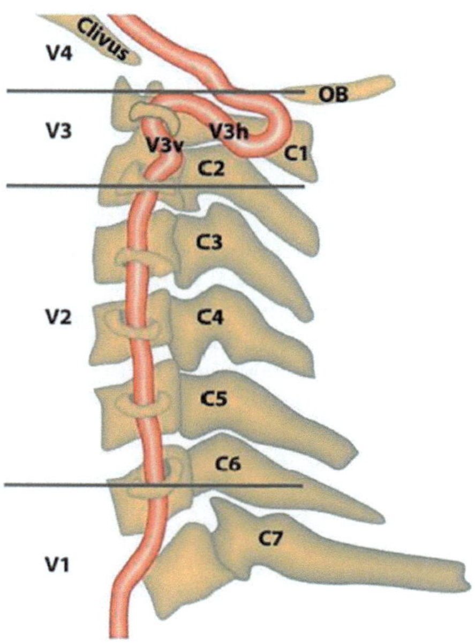

Fig. 4.87 Schematic drawing and selective injection of the vertebral artery in lateral projection.

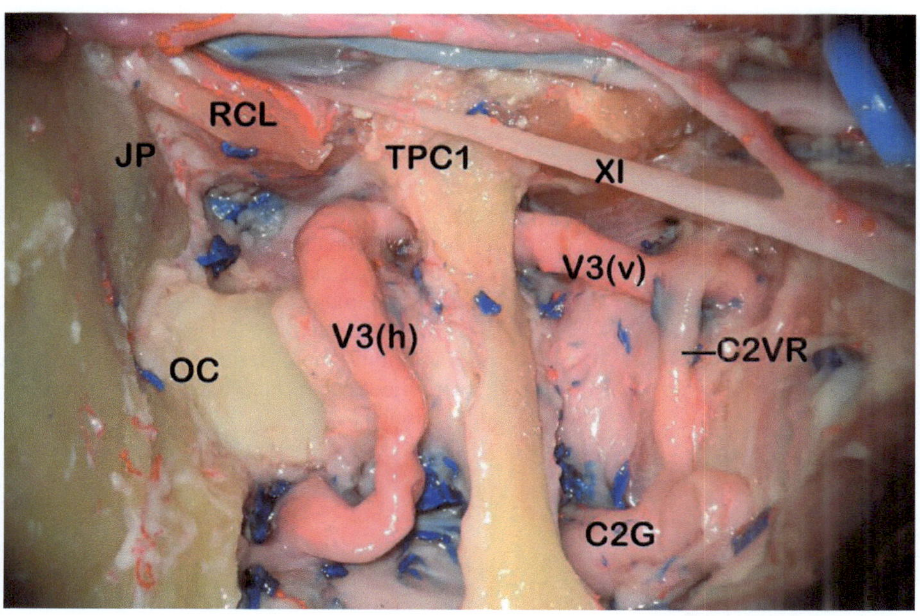

Fig. 4.88 The V3 segment extends from the transverse foramen of the axis to the dural penetration by the vertebral artery; it is further subdivided into two parts: a proximal vertical part (V3v) and a distal horizontal part (V3h). Note the rectus capitis lateralis extends from the transverse process of C1 to the occipital bone behind the jugular foramen. *JP*, jugular process of the occipital bone; *RCL*, rectus capitis lateralis; *XI*, accessory nerve; *OC*, occipital condyle; *TPC1*, transverse process of atlas; *V3(v)*, vertical segment of vertebral artery; *V3(h)*, horizontal segment of vertebral artery; *C2G*, ganglion of C2; *C2VR*, ventral ramus of C2.

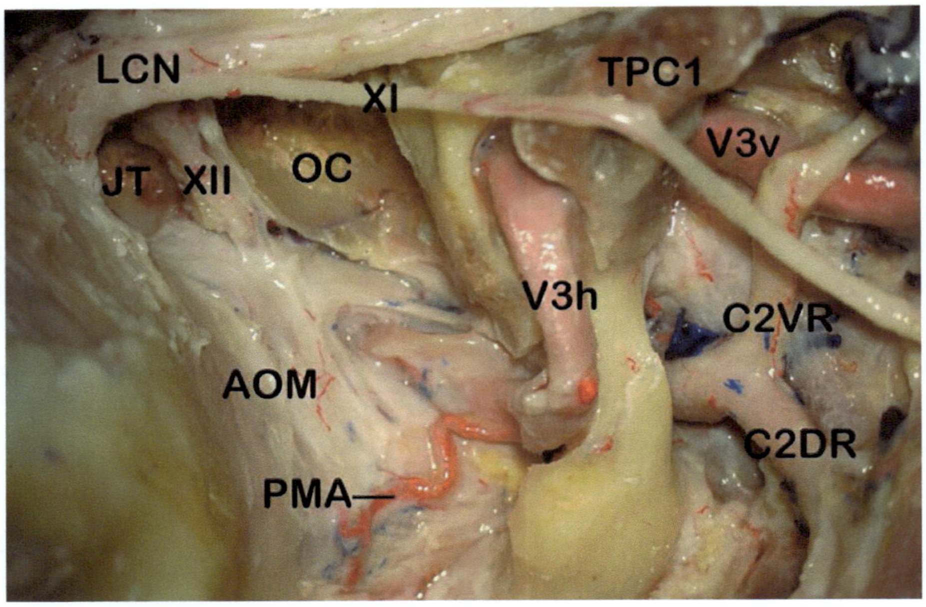

Fig. 4.89 The superior and inferior oblique muscles have been removed. The V3 passes behind the atlantal condyle, gives rise to a posterior meningeal branch, and penetrates the posterior atlanto-occipital membrane to enter the dura. *LCN*, lower cranial nerves; *XI*, accessory nerve; *TPC1*, transverse process of atlas; *JT*, jugular tubercle; *OC*, occipital condyle; *XII*, hypoglossal nerve; *V3(v)*, vertical segment of vertebral artery; *V3(h)*, horizontal segment of vertebral artery; *AOM*, atlanto-occipital membrane; *C2VR*, ventral ramus of C2; *C2DR*, dorsum ramus of C2; *PMA*, posterior meningeal artery.

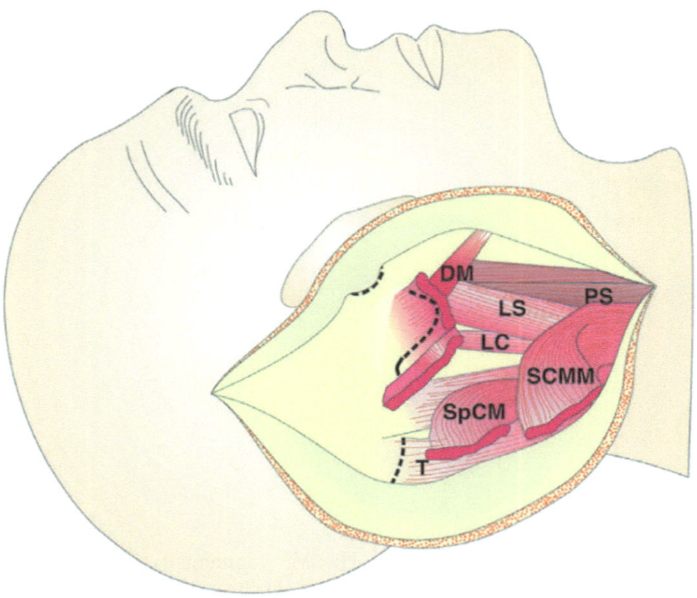

Fig. 4.90 Posterior neck muscles. The muscles in this region can be grouped into three levels. The sternocleidomastoid is the most superficial. More medially, and perpendicular to the sternocleidomastoid, lies the splenius capitis, and attached to its medial surface lies the slender longissimus capitis. *DM*, posterior belly of the digastric muscle; *LC*, longissimus capitis muscle; *LS*, levator scapulae muscle; *PS*, posterior scalene muscle; *SCMM*, sternocleidomastoid muscle; *SpCM*, splenius capitis muscle; *T*, trapezius muscle.

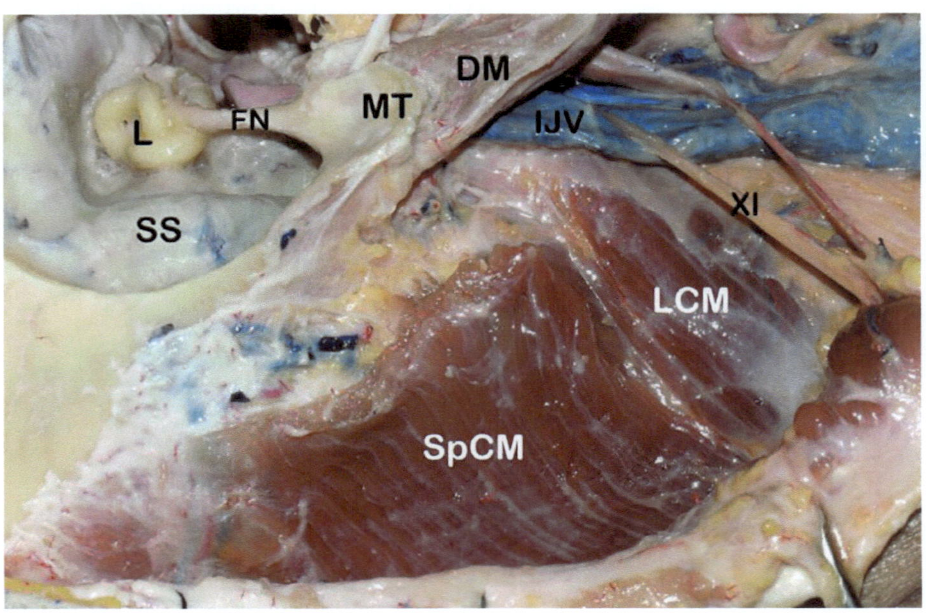

Fig. 4.91 The middle layer of the posterior neck muscles is formed by the splenius capitis, which lies perpendicular to the sternocleidomastoid muscle, and the longissimus capitis, attached to its medial surface. *L*, labyrinth; *FN*, facial nerve; *MT*, mastoid tip; *DM*, posterior belly of the digastric muscle; *IJV*, internal jugular vein; *SS*, sigmoid sinus; *XI*, accessory nerve; *LCM*, longissimus capitis muscle; *SpCM*, splenius capitis muscle.

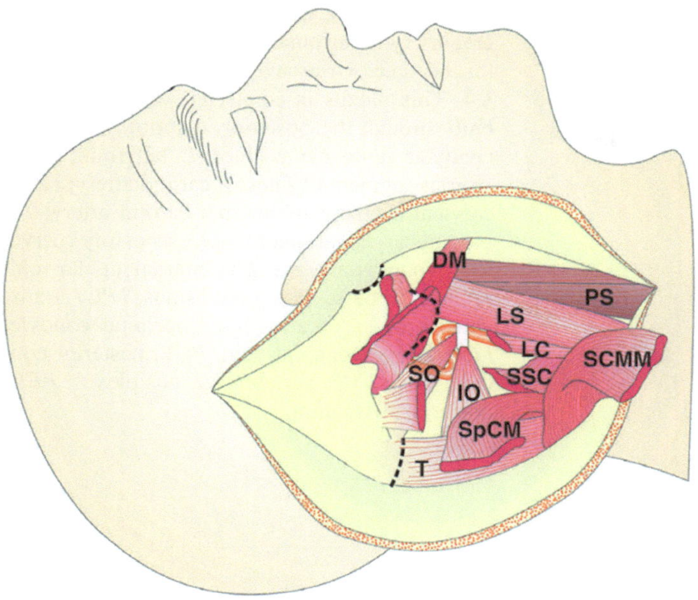

Fig. 4.92 Illustration of deeper posterior neck muscles. *DM*, posterior belly of the digastric muscle; *IO*, inferior oblique muscle; *LC*, longissimus capitis muscle; *LS*, levator scapulae muscle; *PS*, posterior scalene muscle; *SCMM*, sternocleidomastoid muscle; *SO*, superior oblique muscle; *SpCM*, splenius capitis muscle; *SSC*, semispinalis capitis muscle; *T*, trapezius muscle.

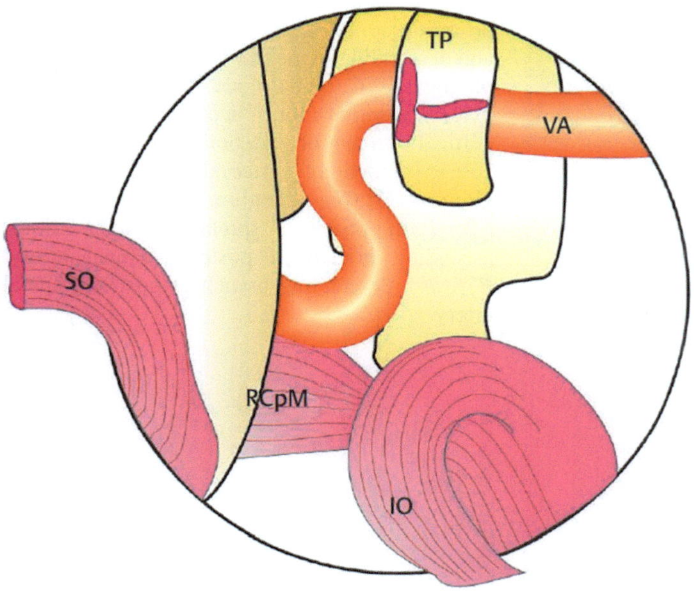

Fig. 4.93 Magnified view during an extreme lateral extension of the approach. The transverse process of the atlas (TP) is an important landmark. It provides attachment to the levator scapulae as well as the superior (SO) and inferior oblique (IO) muscles. The vertebral artery (VA) bends sideways after leaving the transverse process of the axis, ascends to enter the transverse foramen of the atlas, and then curves backward and medially to cross over the posterior arch of atlas. *RCpM*, rectus capitis posterior muscle.

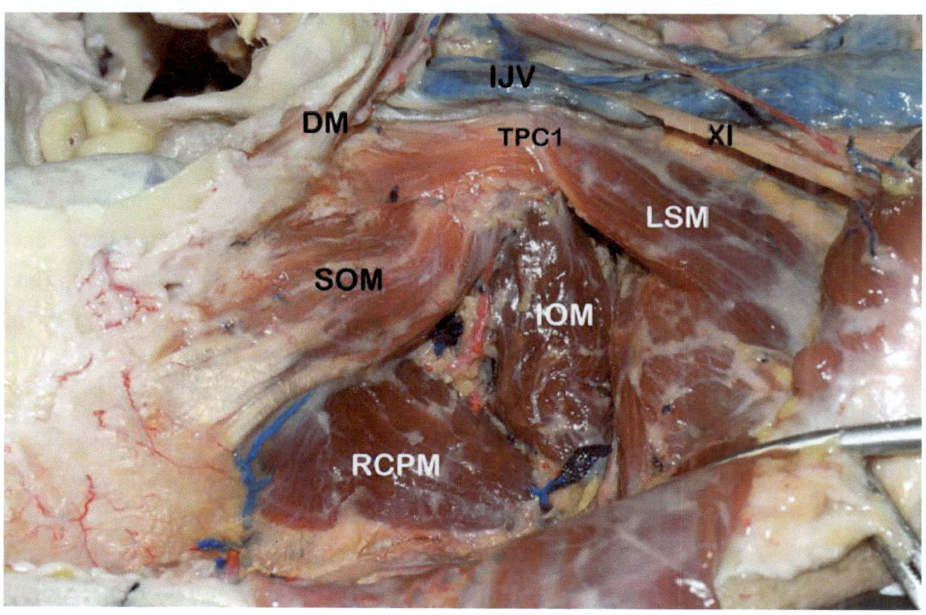

Fig. 4.94 The deeper layer of posterior neck muscles consists of the superior oblique and inferior oblique muscles, both inserted into the transverse process of the atlas. *DM*, posterior belly of the digastric muscle; *IJV*, internal jugular vein; *XI*, accessory nerve; *TPC1*, transverse process of atlas; *LSM*, levator scapulae muscle; *SOM*, superior oblique muscle; *IOM*, inferior oblique muscle; *RCPM*, rectus capitis posterior muscle.

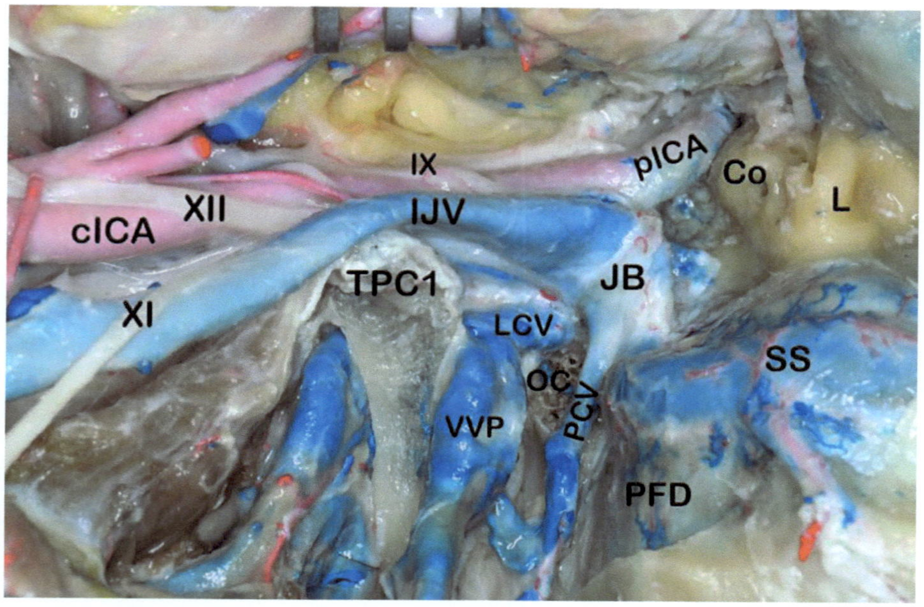

Fig. 4.95 In the extracranial segment the vertebral artery is surrounded by a rich venous plexus that becomes a true cavernous sinus at the level of V4. This plexus is connected with the jugular bulb through the posterior, anterior, and lateral condylar veins. *Co*, cochlea; *L*, labyrinth; *pICA*, petrous segment of internal carotid artery; *cICA*, cervical segment of internal carotid artery; *IX*, glossopharyngeal nerve; *XI*, accessory nerve; *XII*, hypoglossal nerve; *IJV*, internal jugular vein; *JB*, jugular bulb; *SS*, sigmoid sinus; *TPC1*, transverse process of atlas; *OC*, occipital condyle; *LCV*, lateral condylar vein; *PCV*, posterior condylar vein; *VVP*, vertebral venous plexus; *PFD*, posterior fossa dura.

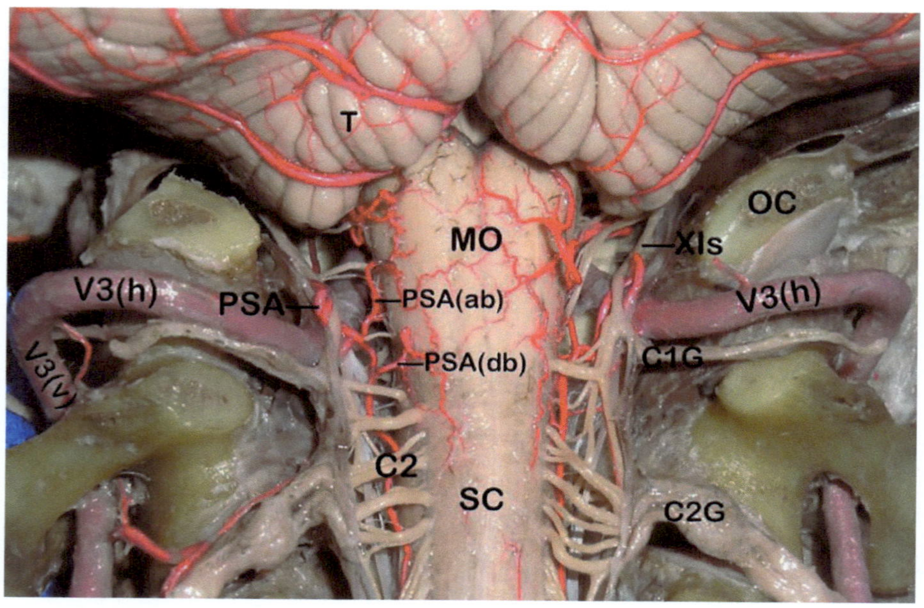

Fig. 4.96 On reaching the lower medulla, the posterior spinal artery divides into ascending and descending branches. *T*, tonsil; *MO*, medulla oblongata; *SC*, spinal code; *OC*, occipital condyle; *XIs*, spinal root of accessory nerve; *V3(v)*, vertical segment of vertebral artery; *V3(h)*, horizontal segment of vertebral artery; *C1G*, ganglion of C1; *C2G*, ganglion of C2; *C2*, roots of C2; *PSA*, posterior spinal artery; *PSA(ab)*, ascending branch of posterior spinal artery; *PSA(db)*, descending branch of posterior spinal artery.

The Craniocervical Venous System

Most of the cerebral venous drainage collects ultimately into the transverse and sigmoid sinuses of the skull base. Two descending cerebral venous outflow tracks can be distinguished: the internal jugular vein anteriorly and the vertebral venous system posteriorly. In the supine position, cerebral venous drainage occurs primarily through the internal jugular veins, as seen on venousphase cerebral angiograms. However, several authors have shown that the internal and external vertebral venous systems represent the major outflow track of cerebral venous drainage in the erect position and that outflow through the internal jugular veins is absent or negligible in this position.

Whereas the pathway from the dural sinuses of the posterior fossa into the internal jugular veins is anatomically obvious through the jugular foramen, connections with the vertebral venous plexus are numerous and complex. Several authors have stressed the importance of emissary veins of the posterior and middle cranial fossae; emissary veins must no longer be considered to be of secondary physiological importance, acting mainly as safety venous outlet routes, but as obligatory outflow pathways for encephalic drainage in the upright position.

For the intracranial venous system, we will give a detailed description of the venous sinuses related to the base of the skull; a description of the superior and inferior sagittal sinuses is outside the scope of this work. We will then describe the emissary veins, which represent an important connection between the intra- and extracranial venous systems. We will end our presentation by considering the main components of the extracranial venous system, represented by the pterygoid venous plexuses, the vertebral venous system in the suboccipital region, and veins of the neck, namely, the internal and external jugular veins.

Skull Base Dural Venous Sinuses

The dural venous sinuses are endothelium-lined channels located between the periosteal and meningeal layers of the dura that act as the walls of the sinuses. Although many texts describe the dural sinuses as simple channels, most are complex trabeculated structures. The dural venous sinuses form the major drainage pathway for the cranial cavity and its contents; via a network of emissary veins that pass directly through the cranial vault and basilar foramina, the dural sinuses communicate with the extracranial venous system; these intercommunications may provide an important potential pathway for collateral venous drainage in case of venous occlusion.

The skull base dural venous sinuses are the cavernous and intercavernous sinuses, the superior and inferior petrosal sinuses, the occipital sinus, and the lateral sinuses. A brief description of the straight sinus, which is not properly a skull base structure, will be given because of its relationship to the confluence of sinuses.

STRAIGHT SINUS

The straight sinus is formed by the confluence of the inferior sagittal sinus and the vein of Galen; it runs posteroinferiorly within the confluence of the falx and tentorium cerebelli; it receives cerebellar vermis, hemispheric, and tentorial tributaries; it terminates in the confluence of sinuses, at the internal occipital protuberance. In 84% of cases, it is represented by a single midline channel, 4 cm long.

CONFLUENCE OF SINUSES

The confluence of sinuses is formed by the union of the superior sagittal sinus, the straight sinus, and the right and left transverse sinuses. The typical arrangement, in which both transverse sinuses are equal in size, is present in 50% of cases. Nevertheless, great anatomical variability is seen and in 25% of cases the left transverse sinus is relatively small and the superior sagittal sinus continues as the dominant right transverse sinus. In 10% of cases, the right transverse sinus is relatively small and the superior sagittal sinus continues as the dominant left transverse sinus.

Less common patterns are represented by segmental hypoplasia or aplasia of the proximal segment of a small transverse sinus, which is reconstituted distally by a vein of Labbe, or absence of sinus confluence with the superior sagittal sinus becoming the right transverse sinus and the straight sinus becoming the left transverse sinus.

TRANSVERSE SINUSES

The transverse sinuses, also known as the lateral sinuses, are contained within the tentorium and the petrous bone (Fig. 4.97). The main venous structures draining into the transverse sinuses are the inferior cerebellar veins, the occipital veins, the temporal veins, the superior petrosal sinus, the straight sinus, and the superior sagittal sinus.

An equal division of blood from the superior sagittal sinus into the right and left transverse sinuses is seen in only ~20% of patients; the right transverse sinus is more often a continuation of the superior sagittal sinus, whereas the left transverse sinus receives most of its supply from the straight sinus.

In approximately half of all cases both the transverse sinuses are well formed but asymmetric in size; in 4%–20% of anatomical specimens a narrowed or atretic segment can be identified in at least one of the transverse sinuses.

SIGMOID SINUS

The sigmoid sinus represents the anteroinferior continuation of the transverse sinus (Fig. 4.97); it begins at the base of the petrous bone, where the transverse sinus leaves the tentorial margin; it courses along the groove of the squamous portion of the occipital bone, passing inferomedially in an S-shaped curve and terminates at the level of the jugular foramen. It receives inconstant veins from the pons and medulla, the condylar emissary vein, and the mastoid emissary vein.

SUPERIOR PETROSAL SINUS

The superior petrosal sinus is a channel that extends from the cavernous sinus to the sigmoid sinus (Fig. 4.98); it begins at the end of the transverse sinus, runs along the attachment of the tentorium cerebelli to the dorsal ridge of the petrous bone, and ends in the posterior portion of the cavernous sinus. The direction of flow is probably from posterior to anterior. Its main tributaries include the superior petrosal vein (Dandy's vein) (Fig. 4.99), the lateral mesencephalic vein, anterior cerebellar veins, inferior cerebral veins, and veins draining the middle and inner ear.

INFERIOR PETROSAL SINUS

The inferior petrosal sinus is a dural venous sinus that lies in a groove between the petrous apex and the clivus, extending laterally along the petro-occipital fissure and connecting the posterior aspect of the ipsilateral cavernous sinus usually with the internal jugular vein (Figs. 4.100 and 4.101). An anatomical classification according to the level of the inferior petrosal sinus-internal jugular vein junction has been developed; the level of the junction between the sinus and the vein is variable.

CAVERNOUS SINUS

The cavernous sinus is defined as the dural envelope within which the cavernous segment of the internal carotid artery courses. It is the site of a venous confluence that receives the terminal ends of veins draining the orbit, the Sylvian fissure, and the middle and anterior cranial fossae, and has free communication with the basilar plexus and the superior and inferior petrosal and intercavernous sinuses. It narrows anteriorly adjacent to the superior orbital fissure and is widest posteriorly, lateral to the dorsum sellae, where it opens into the venous confluence, formed by the junction of the basilar, cavernous, superior, and inferior petrosal sinuses.

The cavernous sinus is connected to the orbit by the superior and inferior ophthalmic veins; to the cerebral hemisphere through the middle and inferior cerebral veins; to the retina by the central retinal vein; to the dura by the tributaries of the middle meningeal vein; to the transverse sinus via the superior petrosal sinus; to the jugular bulb by way of the inferior petrosal sinus; to the pterygoid plexus by emissary veins passing through the cranial foramina; and to the facial veins through the ophthalmic veins.

There are three main venous spaces within the cavernous sinus, identified by their relationship to the carotid artery: these are the medial, the anteroinferior, and the posterosuperior compartments.

- The medial compartment is situated between the pituitary gland and the internal carotid artery; this medial space may be as wide as 7 mm but may be obliterated by a tortuous carotid artery that indents the pituitary gland.
- The anteroinferior space is located in the concavity below the first curve of the intracavernous carotid artery, where the superior and inferior ophthalmic veins commonly open into the sinus.
- The posterosuperior space is located between the carotid artery and the posterior half of the roof of the sinus; the basilar sinus opens in this space and the meningohypophyseal artery arises in this space.

BASILAR SINUS

The basilar sinus is the largest and most constant intercavernous connection across the midline; it passes posteriorly to the dorsum sellae and upper clivus, and connects the posterior aspect of both cavernous sinuses; the superior and inferior petrosal sinuses open in the lateral part of the basilar sinus.

Emissary Veins

Generally speaking, an emissary vein represents a connection between extracranial veins and the cranial dural venous sinuses.

- The mastoid emissary vein receives the deep cervical veins and occipital and posterior auricular veins, which are scalp veins; passing through the mastoid foramen, together with the meningeal branch of the occipital artery, it drains into the sigmoid sinus.
- The posterior condylar emissary vein originates from the jugular bulb, passes through the posterior condylar canal, and drains into the suboccipital cavernous sinus and the deep posterior cervical veins (Fig. 4.95).
- The anterior condylar emissary vein originates from the inferior petrosal sinus, passes through the anterior condylar canal, and joins the anterior internal vertebral venous plexus, the basilar plexus, and the marginal sinus.
- The sphenobasal emissary vein receives the pterygoid plexus and, passing through the foramina ovale, spinosum, and lacerum, enters the cavernous sinus.

Veins of the Neck

With respect to the superficial cervical fascia, the veins of the neck can be subdivided into a superficial group and a deep group:

- The superficial group drains the subcutaneous structures and are tributaries of the anterior jugular vein, the external jugular vein, and the posterior jugular vein.
- The deep group drains all structures located deep to the superficial cervical fascia and are tributaries of the internal jugular vein and the subclavian vein.

EXTERNAL JUGULAR VEIN

The external jugular vein begins at the level of the mandible and is formed mainly by the posterior division of the retromandibular vein and the posterior auricular vein (Fig. 4.102). It receives tributaries from the lateral scalp, the pinna, and deep facial structures, and is directed downward, toward the midpoint of the clavicle; it terminates in the subclavian vein.

- The posterior jugular vein is an inconstant branch that drains the posterior part of the scalp and empties in the midportion of the external jugular vein.
- The anterior jugular vein is located in the anterior part of the neck.

INTERNAL JUGULAR VEIN

The internal jugular vein begins in the jugular fossa with a slight dilatation called the jugular bulb. It courses inferiorly within the carotid space and terminates by uniting with its respective subclavian vein to form the brachiocephalic vein (Fig. 4.95). The main tributaries of the bulb of the internal jugular vein are represented by the sigmoid sinus, the inferior petrosal sinus, and the anterior condylar confluent; the main tributaries of the lower internal jugular vein are the common facial vein and the lingual, pharyngeal, and thyroid veins. Sometimes an inferior petrosal sinus-internal jugular vein junction is observed in the lower portion of the internal jugular vein.

Pterygoid Venous Plexus

The pterygoid venous plexus is an extensive network of small vascular channels that lies in the masticator space. It receives numerous palatine, deep temporal, masseteric, buccal, dental, and orbital tributaries, communicates freely with the facial veins, and has extensive anastomoses with the cavernous sinus via sphenobasal emissary veins that pass through the foramina ovale, spinosum, and lacerum.

Vertebral Venous System in the Suboccipital Region

The venous structures in the suboccipital region are represented by:

- The suboccipital venous plexus.
- The vertebral venous plexus, related to the spine.
- The vertebral artery venous plexus at the atlantoaxial interspace, around the vertical portion of the V3 segment of the vertebral artery.
- The vertebral artery venous plexus at the occipitoatlantal interspace around the horizontal portion of the V3 segment of the vertebral artery (the so-called suboccipital cavernous sinus).

SUBOCCIPITAL VENOUS PLEXUS

The suboccipital venous plexus is located between the muscular layers and continues inferiorly into the deep cervical veins; it communicates with the vertebral venous plexus via an anastomotic vein at the atlanto-occipital interspace, with the transverse sigmoid sinus via the mastoid emissary vein, and with the vertebral artery venous plexus, which is around the vertical and horizontal portions of the V3 segment of the vertebral artery.

VERTEBRAL VENOUS PLEXUS

The vertebral venous plexus is a rich network that is found around the vertebral column; it is an inferior continuation of the occipital and marginal sinuses and the basilar venous plexus. The vertebral venous plexus is an accessory route of venous return from the intracranium during the flexion and extension, as well as the rise in intra-abdominal pressure, when the body is in an upright position.

At the occipitoatlantal and atlantoaxial interspaces the vertebral venous plexus communicates with venous plexus around the V3 segment of the vertebral artery of both sides. After bilateral gradual ligation of internal jugular veins, the vertebral venous plexus assumes the role of venous drainage from the intracranial compartment, during which contrast medium has been detected exiting the cavernous sinus through the ophthalmic veins and pterygoid plexuses.

VERTEBRAL ARTERY VENOUS PLEXUS (VERTEBRAL VEIN)

The vertebral artery venous plexus is a rich network that is found around the V3 segment of the vertebral artery (Fig. 4.95); it is formed in the suboccipital triangle by tributaries from the suboccipital venous plexus and by small muscular veins.

In its proximal portion, it surrounds the horizontal portion of the V3 segment of the vertebral artery, enters the foramen in the transverse process of the atlas, and descends surrounding the vertical portion of the V3 segment; the venous plexus then follows the V2 segment of the vertebral artery and ends as single trunk emerging from the foramen transversarium of the sixth cervical vertebra to empty into the brachiocephalic vein.

Because of the anatomical similarities with the cavernous sinus, the portion of the vertebral artery venous plexus that surrounds the horizontal portion of the V3 segment is also called the "suboccipital cavernous sinus."

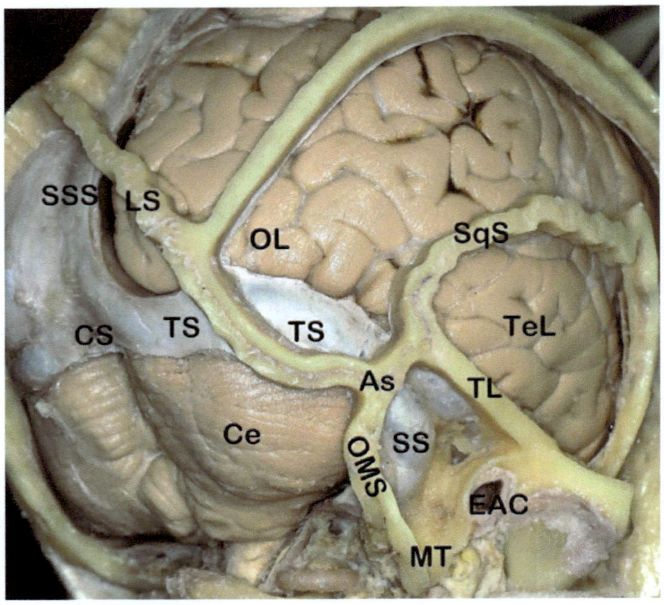

Fig. 4.97 The main venous structures of the head from a lateral perspective showing the superior sagittal sinus, confluence of sinuses, transverse sinus, and sigmoid sinus. The sigmoid sinus represents the anteroinferior continuation of the transverse sinus at asterion level. *SSS*, superior sagittal sinus; *CS*, confluence of sinuses; *TS*, transverse sinuses; *SS*, sigmoid sinus; *LS*, lambdoid suture; *SqS*, squamous suture; *OMS*, occipitomastoid suture; *TL*, temporal line; *EAC*, external auditory canal; *MT*, mastoid tip; *TeL*, temporal lobe; *OL*, occipital lobe; *Ce*, cerebellum; *As*, asterion.

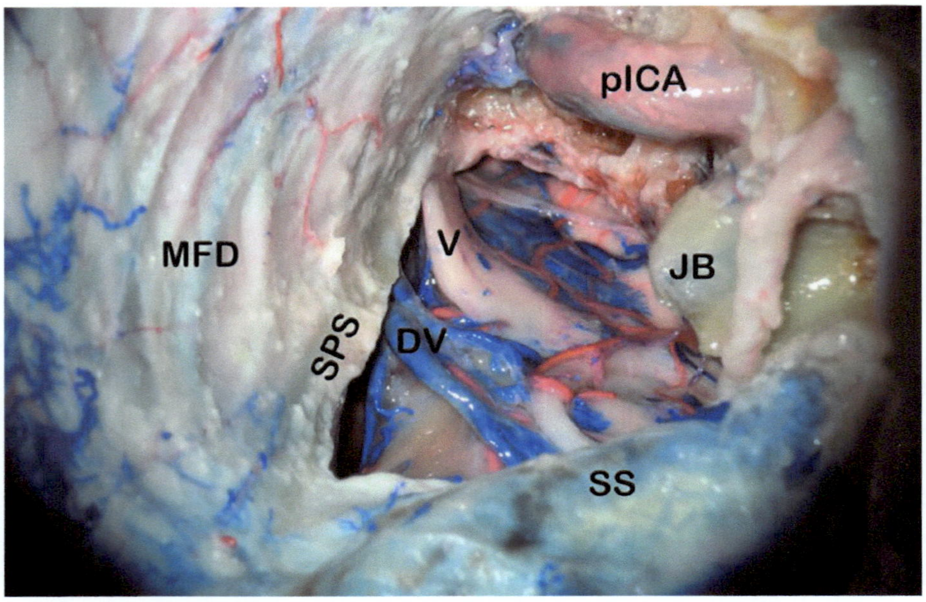

Fig. 4.98 The superior petrosal sinus is a channel that extends from the cavernous sinus to the sigmoid sinus; it begins at the end of the transverse sinus, runs along the attachment of the tentorium cerebelli to the dorsal ridge of the petrous bone, and ends in the posterior portion of the cavernous sinus. *pICA*, petrous segment of internal carotid artery; *V*, trigeminal nerve; *MFD*, middle fossa dura; *SPS*, superior petrosal sinus; *DV*, Dandy's vein (superior petrosal vein); *JB*, jugular bulb; *SS*, sigmoid sinus.

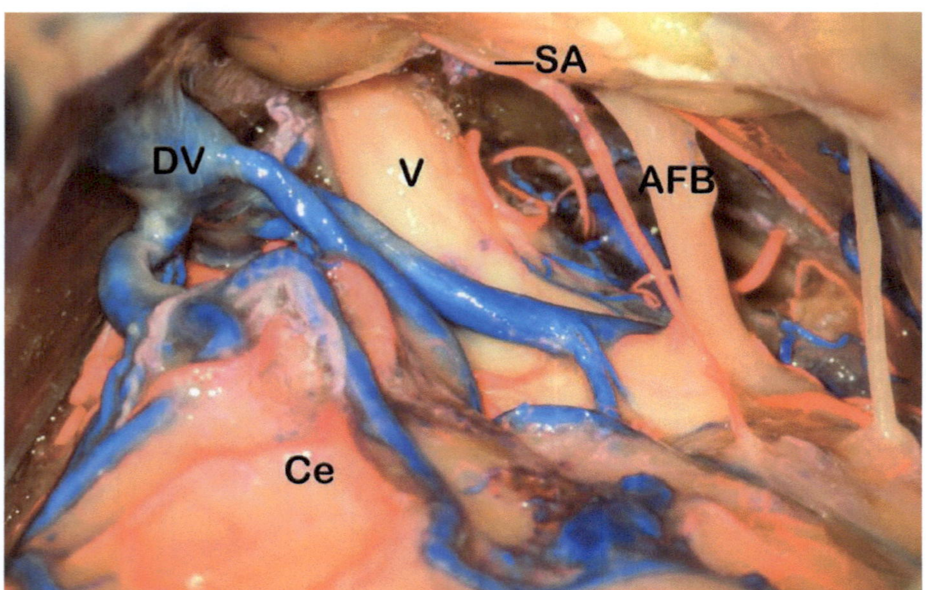

Fig. 4.99 The Dandy's vein (superior petrosal vein) are among the largest and most frequently encountered veins in the posterior fossa and drains into the superior petrosal sinus. *DV*, Dandy's vein (superior petrosal vein); *V*, trigeminal nerve; *SA*, subarcute artery; *AFB*, acousticofacial bundle; *Ce*, cerebellum.

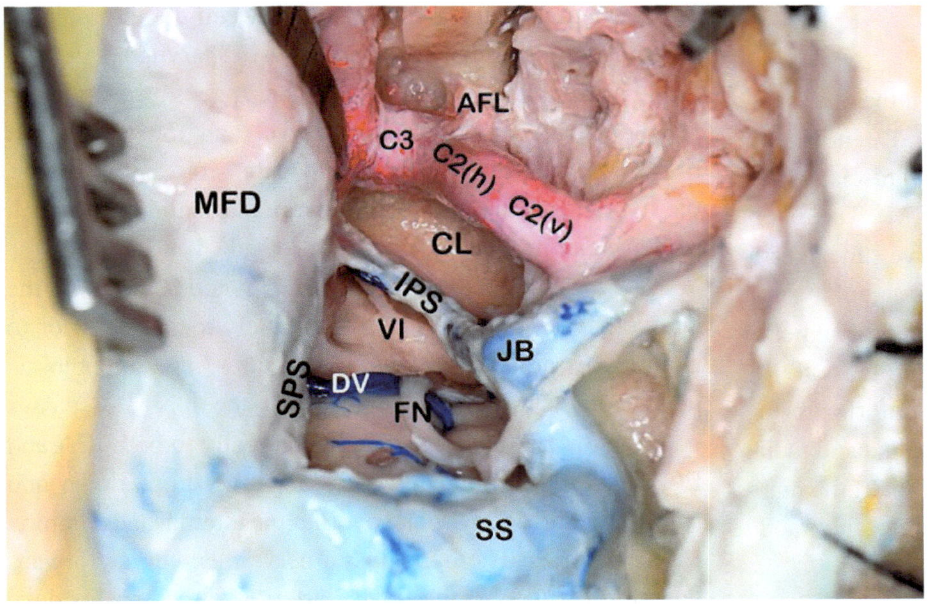

Fig. 4.100 The inferior petrosal sinus, which courses on the intracranial surface of the petroclival fissure, communicates the cavernous sinus and basilar venous plexus at its upper end and with the jugular bulb at its lower end. *MFD*, middle fossa dura; *CL*, clivus; *SPS*, superior petrosal sinus; *IPS*, inferior petrosal sinus; *VI*, abducent nerve; *FN*, facial nerve; *DV*, Dandy's vein (superior petrosal vein); *JB*, jugular bulb; *SS*, sigmoid sinus; *C3*, lacerum segment of internal carotid artery; *C2(h)*, horizontal portion of petrous segment of internal carotid artery; *C2(v)*, vertical portion of petrous segment of internal carotid artery; *AFL*, anterior foramen lacerum.

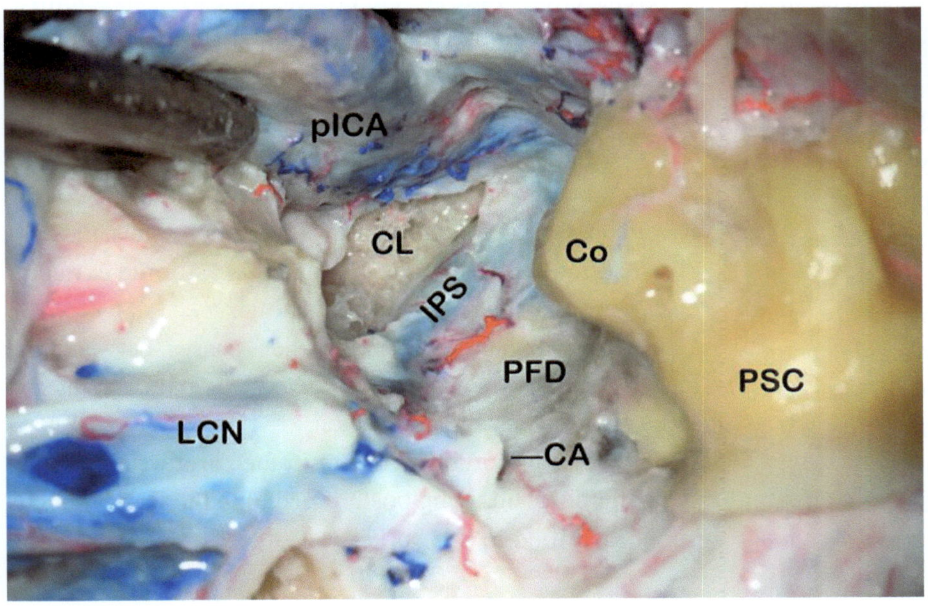

Fig. 4.101 The tip of the suction is used to displace the petrous carotid artery laterally while the petrous bone medial to the petrous carotid artery has been drilled to expose the inferior petrosal sinus and clivus. *pICA*, petrous segment of internal carotid artery; *CL*, clivus; *IPS*, inferior petrosal sinus; *LCN*, lower cranial nerves; *PFD*, posterior fossa dura; *CA*, cochlear aqueduct; *Co*, cochlea; *PSC*, posterior semicircular canal.

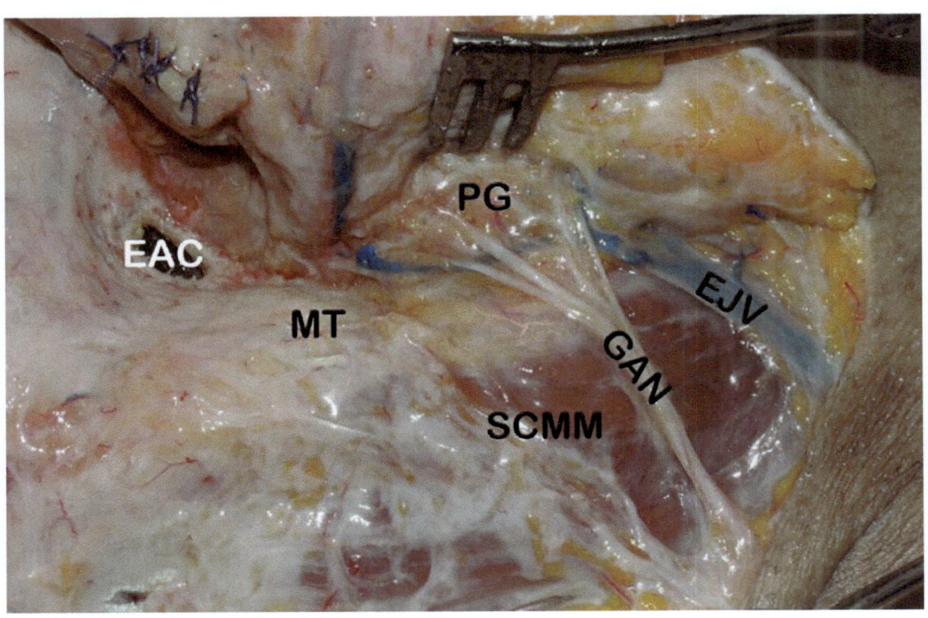

Fig. 4.102 The external jugular vein and the greater auricular nerve run superficial to the superficial layer of the deep cervical fascia. *EAC*, external auditory canal; *MT*, mastoid tip; *PG*, parotid gland; *SCMM*, sternocleidomastoid muscle; *GAN*, greater auricular nerve; *EJV*, external jugular vein.

The Anterolateral Neck

Deep to the skin, the next robust layer is that of the platysma, which thins toward the midline and laterally. Deep in turn, this is the superficial layer of the deep cervical fascia, which envelops the sternocleidomastoid muscle (SCM). The external jugular vein and the greater auricular nerve run superficial to this layer (Fig. 4.102). When raising subplatysmal flaps, the position of the marginal branch of the facial nerve must be identified. This exits the parotid gland close to the angle of the mandible and invariably loops inferior to the lower border of the mandible deep to the platysma and facial vein. It then loops superiorly to cross the margin of the mandible approximately at its midpoint, close to the facial artery.

In either posterior retraction or transection of the SCM, the digastric muscle is revealed, with only the facial vein running laterally to this structure. Either superolateral retraction or transection reveals the carotid sheath, allowing identification of its contents.

The level II lymph nodes are here divided into IIa and IIb by the course of the spinal accessory nerve, which passes deep the SCM at the junction of its upper and middle thirds anteriorly. The spinal accessory nerve crosses laterally to the IJV in 80% of cases (Fig. 4.103), at the level of the lateral process of the atlas, which is easily palpated (Fig. 4.41). The remaining lower cranial nerves (IX and X) (Fig. 4.104) and the hypoglossal nerve (Fig. 4.105) are situated on the medial side of the IJV at this level. The junction of level II and III nodes is marked by the level at by which the omohyoid crosses the internal jugular vein.

The neural structures that pass between the internal and external carotid artery include the glossopharyngeal nerve in close association with the stylopharyngeus, the superior laryngeal nerve of the vagus nerve, and the pharyngeal branch of the vagus nerve. With extension of the tumor, the hypoglossal nerve and the marginal mandibular branch of the facial nerve need to be addressed.

The vagus nerve lies deep to the IJV and ICA within the carotid sheath and should be definitively identified prior to placing ties around the arterial structures. The hypoglossal nerve from its initial medial position to the upper IJV, passes between the IJV and ICA and then laterally to both ICA and ECA. The descending hypoglossal nerve drops down and travels medial to the IJV and overlies the internal and common carotid arteries. The hypoglossal nerve then curves forward lateral to the loop of the lingual artery and then passes deep to the mylohyoid and superficial to the hyoglossus muscle.

Transection of the styloid process following skeletonization of its musculotendinous attachments allows improved access to the upper carotid sheath, allowing identification of the lower cranial nerves as they exit the jugular foramen. The exit point of the glossopharyngeal nerve from the jugular foramen is located just on the posteromedial side of the styloid process and the stylomastoid foramen. Thus, the base of the styloid process is the first landmark for the extracranial portion of the glossopharyngeal nerve.

The glossopharyngeal nerve then passes between the internal and external carotid arteries lying on the lateral surface of stylopharyngeus muscle.

The other important structure in the carotid space is the sympathetic chain, lying posteromedial to the ICA, which is rarely identified in this surgery.

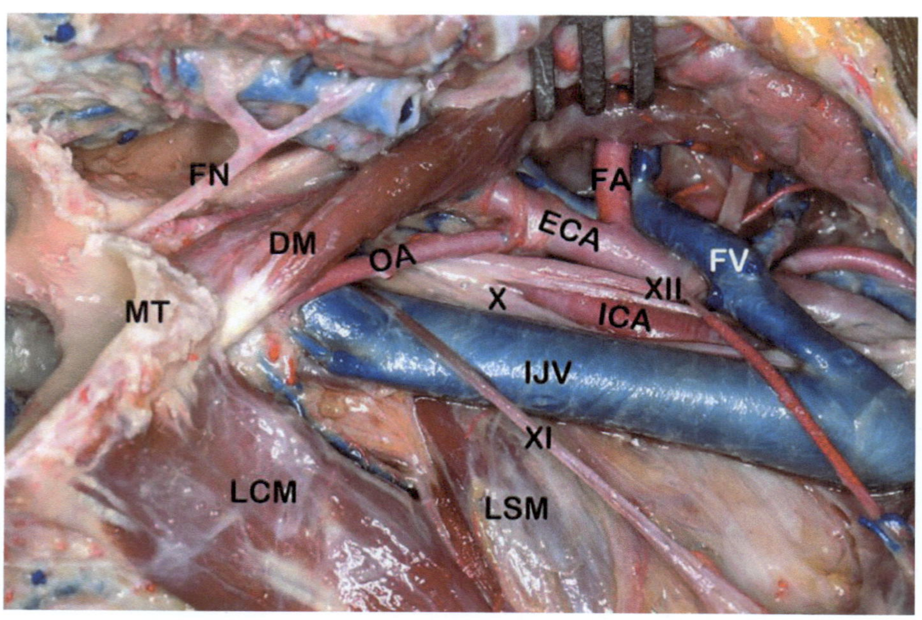

Fig. 4.103 The spinal accessory nerve crosses laterally to the internal jugular vein in 80% of cases. *FN*, facial nerve; *MT*, mastoid tip; *DM*, digastric muscle; *ECA*, external carotid artery; *FA*, facial artery; *FV*, facial vein; *OA*, occipital artery; *IJV*, internal jugular vein; *ICA*, internal carotid artery; *X*, vagus nerve; *XII*, hypoglossal nerve; *XI*, accessory nerve; *LCM*, longissimus capitis muscle; *LSM*, levator scapular muscle.

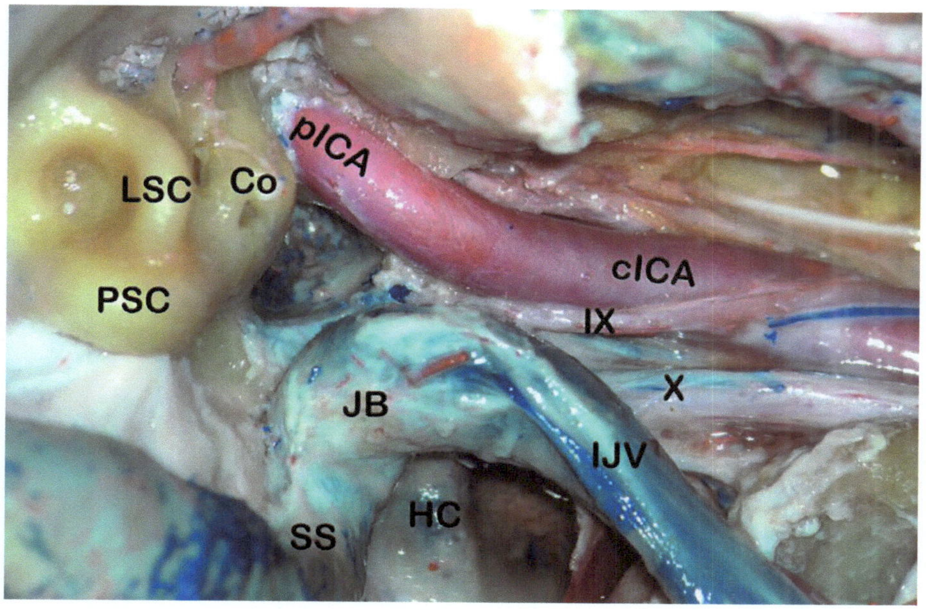

Fig. 4.104 The glossopharyngeal nerve and vagus nerve are situated on the medial side of the internal jugular vein at the level of the lateral process of the atlas. *SSC*, superior semicircular canal; *PSC*, posterior semicircular canal; *Co*, cochlea; *pICA*, petrous segment of internal carotid artery; *cICA*, cervical segment of internal carotid artery; *IX*, glossopharyngeal nerve; *X*, vagus nerve; *SS*, sigmoid sinus; *IJV*, internal jugular vein; *JB*, jugular bulb; *HC*, hypoglossal canal.

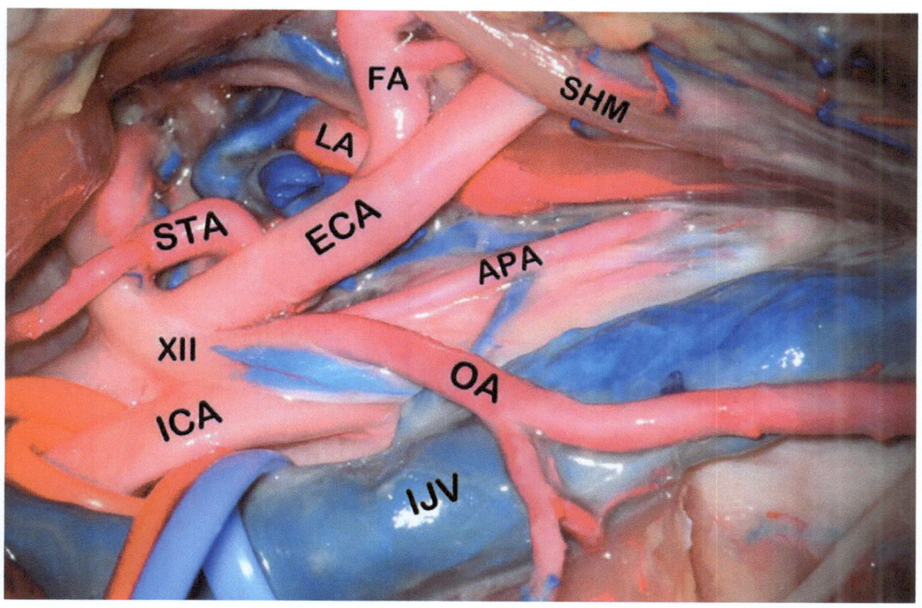

Fig. 4.105 The hypoglossal nerve from its initial medial position to the upper *IJV*, passes between the *IJV* and *ICA* and then laterally to both *ICA* and *ECA*. *FA*, facial artery; *LA*, lingual artery; *ECA*, external carotid artery; *STA*, superior thyroid artery; *APA*, ascending pharyngeal artery; *OA*, occipital artery; *IJV*, internal jugular vein; *XII*, hypoglossal nerve; *ICA*, internal carotid artery; *SHM*, stylohyoid muscle.

CHAPTER 5

Radiology of Lateral Skull Base

The lateral skull base (LSB) presents a highly intricate anatomy, including numerous fissures and foramina that communicate with facial structures and deep fascial spaces in the upper neck. Tumors involving this area are often challenging due to their profound location, diversity, and proximity to vital neurovascular structures. Fundamentally, imaging plays an essential role in the management of such lesions, representing the key instrument for:

- preoperative differential diagnosis
- assessing tumor extent and respectability
- optimal treatment planning
- follow-up and recurrence

A thorough knowledge of this field as well as a good collaboration with the radiologist are mandatory for any neurootologist. To avoid misdiagnosis, the surgeon himself should be prepared to interpret the imaging and to discuss the challenging aspects with the radiologist pre- and postoperatively. This will allow an improved correlation between radiology, surgical anatomy, and associated pathology leading to proficient surgical practice.

Computed tomography (CT) and magnetic resonance imaging (MRI) are the main techniques used for cross-sectional imaging of skull base pathology. They play complementary roles in a comprehensive description of lesions; CT is ideal for demonstrating bone structure while MR is the modality of choice for evaluating the soft tissues. In this chapter, we highlight the most important radiological data that the surgeon seeks for proper preoperative differential diagnosis and surgical planning.

With densely packed neural structures and vessels traversing multiple tissue layers, there is limited tolerance to manipulation or mass effect.

Imaging of Skull Base Anatomy

Adequate intraoperative exposure is an essential step for optimal tumor removal that involves the temporal bone, middle and posterior cranial fossae, and the infratemporal fossa while assuring minimal morbidity and mortality. For proper preoperative planning and for understanding pathology one must possess a good radiological knowledge of the anatomy. We will focus here on the identification and description of the various landmarks needed to interpret radiological images. Axial and coronal CT images, axial T2MRI images, and gadolinium (Gd)-enhanced T1 MRI and coronal T1 MRI images will be illustrated.

The skull base is composed of five bones: ethmoid, sphenoid, occipital, paired temporal, and paired frontal, these bones form the floors of the anterior, middle, and posterior cranial fossa (Fig. 5.1).

Atlas of Microsurgery of the Lateral Skull Base. https://doi.org/10.1016/B978-0-443-19083-4.00005-2

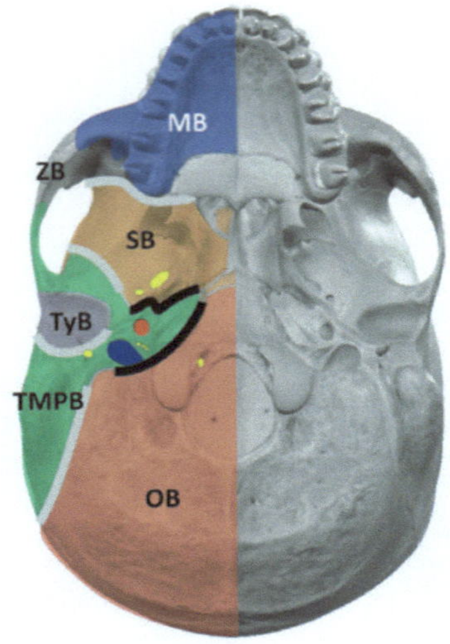

Fig. 5.1 Drawing of the lateral skull base. Please note the delimitation of the bones that compose the skull base and the in-between sutures. *MB*, maxillary bone: *OB*, occipital bone; *TMPB*, temporal-mastoid-petrous bone; *TyB*, tympanic bone; *SB*, sphenoid bone; *ZR*, zygomatic bone.

The middle cranial fossa can be divided for didactic reasons into two compartments: central and lateral. The central part comprises the sella turcica and parasellar region flanked by the cavernous sinuses, skull base foramina and cranial nerves II–VI, and the internal carotid artery (ICA). The lateral compartment is represented by the grater wing of the sphenoid, the squamous part of temporal bone, and temporomandibular joint. The posterior skull base is formed by the posterior temporal and occipital bones. It contains the internal acoustic canal (IAC) transmitting the facial (VII) and vestibulocochlear (VIII) cranial nerves, the jugular foramen traversed by the jugular vein and lower cranial nerves IX–XI, the hypoglossal canal, and the foramen magnum.

The infratemporal fossa represents the anatomical area limited superiorly by the squamous temporal bone and greater wing of the sphenoid, laterally by the ramus of mandible and medially by the pterygoid process and the parapharyngeal space [1]. Tumors or inflammatory processes that arise in this location can often extend up involving skull base. For a more detailed anatomical description of the skull base, see Chapter 4.

Axial CT, Bone Windows

See Figs. 5.2–5.5.

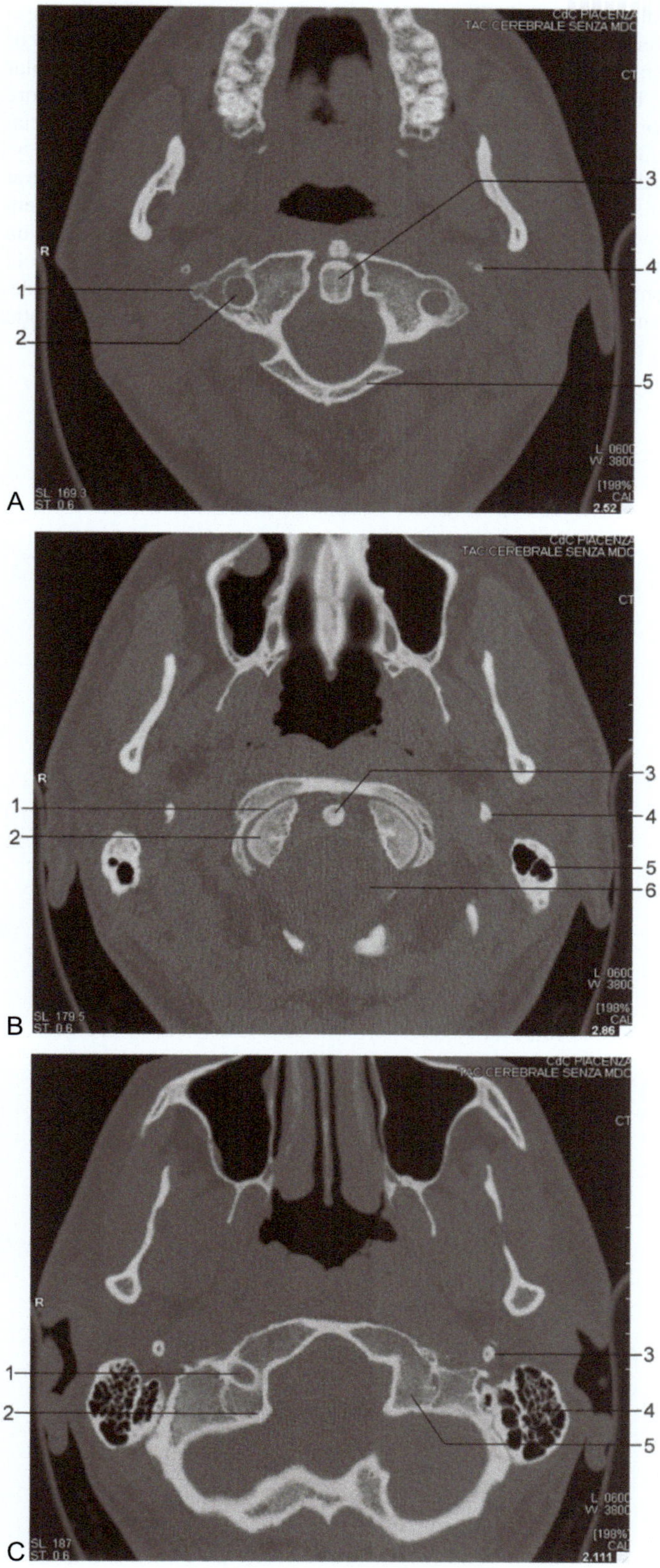

Fig. 5.2 (A) CT image through the upper cervical spine. The transverse processes of C1 and the odontoid process are seen, as are the ascending ramus of the mandible, the dentition of the maxilla, and the tip of the styloid process. 1. transverse process of the atlas, 2. transverse foramen of the atlas/vertebral artery, 3. odontoid, 4. styloid process, 5. posterior arch of the atlas. (B) Inferiormost aspect of the skull base with the atlanto-occipital joint visible. This is in the same plane as the inferior nasopharynx and a well-pneumatized mastoid tip. 1. atlanto-occipital joint, 2. occipital condyle, 3. odontoid tip, 4. styloid process, 5. mastoid tip, 6. foramen magnum. (C) The posterior cranial fossa is now seen, with the lower stylomastoid foramen and upper extent of the nasopharynx and fossa of Rosenmüller identified. 1. posterior condylar vein, 2. jugular tubercle, 3. styloid process, 4. mastoid tip, 5. occipital bone.

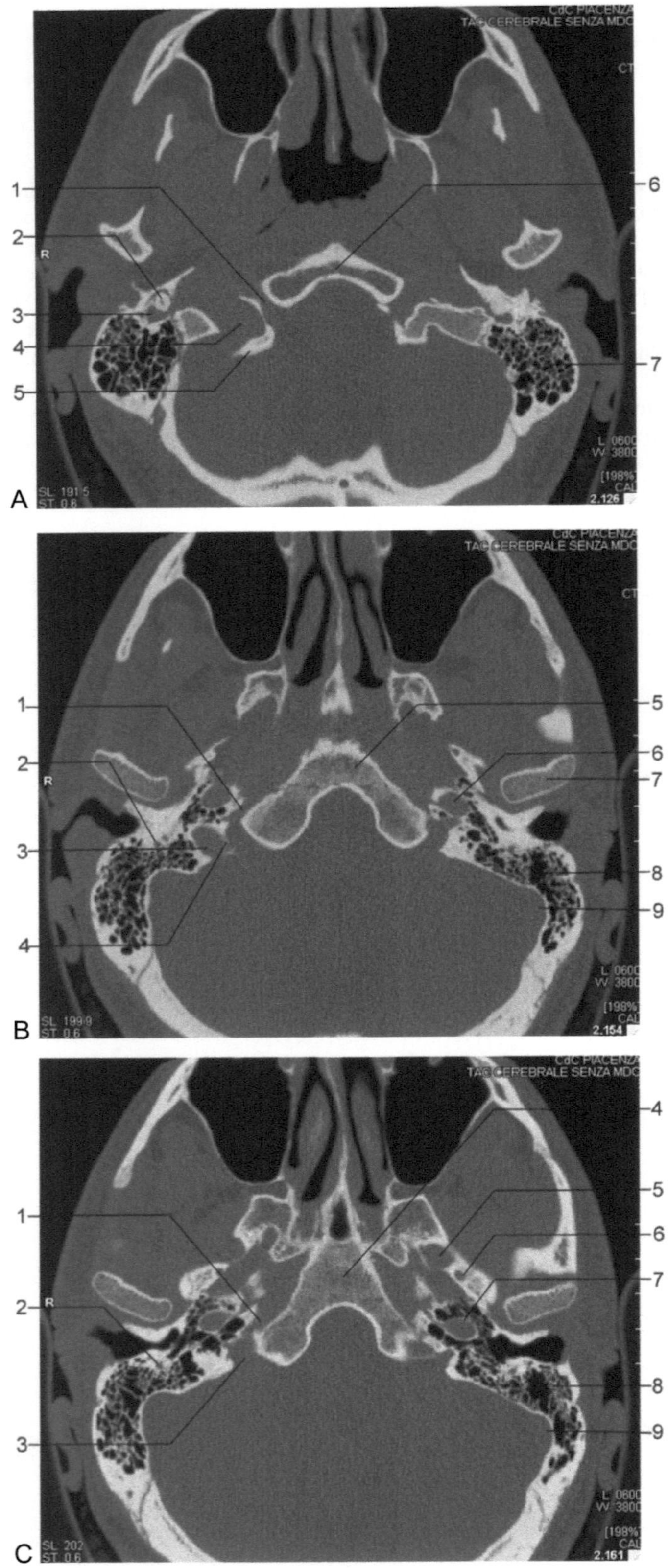

Fig. 5.3 (A) The lower aspect of the jugular bulb and its relationship to the jugular tubercle are seen. Note the angle of the hypoglossal canal, and the conelike narrowing of the stylomastoid foramen as the facial nerve enters the mastoid. The occipitomastoid suture is seen. 1. hypoglossal canal, 2. styloid process, 3. stylomastoid foramen, 4. jugular bulb, 5. jugular tubercle, 6. clivus, 7. mastoid air cells. (B) This image shows the inferior petrosal sinus entering the jugular bulb. The inferior aspect of the intrapetrous carotid is separated from the jugular bulb by the jugular spine. The nasolacrimal duct and pterigopalatine fossa are seen more anteriorly. 1. inferior petrosal sinus, 2. mastoid segment of the facial nerve, 3. jugular bulb, 4. jugular spine, 5. clivus, 6. internal carotid artery, 7. mandibular condyle, 8. mastoid air cells, 9. sigmoid sinus. (C) Image at the inferior level of the external auditory canal. The vertical segment (mastoid) of the facial nerve is seen. Note the degree of pneumatization around the internal carotid artery. 1. inferior petrosal sinus, 2. mastoid segment of the facial nerve, 3. jugular foramen, 4. clivus, 5. foramen ovale, 6. foramen spinosum, 7. vertical segment of the petrous internal carotid artery, 8. mastoid air cells, 9. sigmoid sinus.

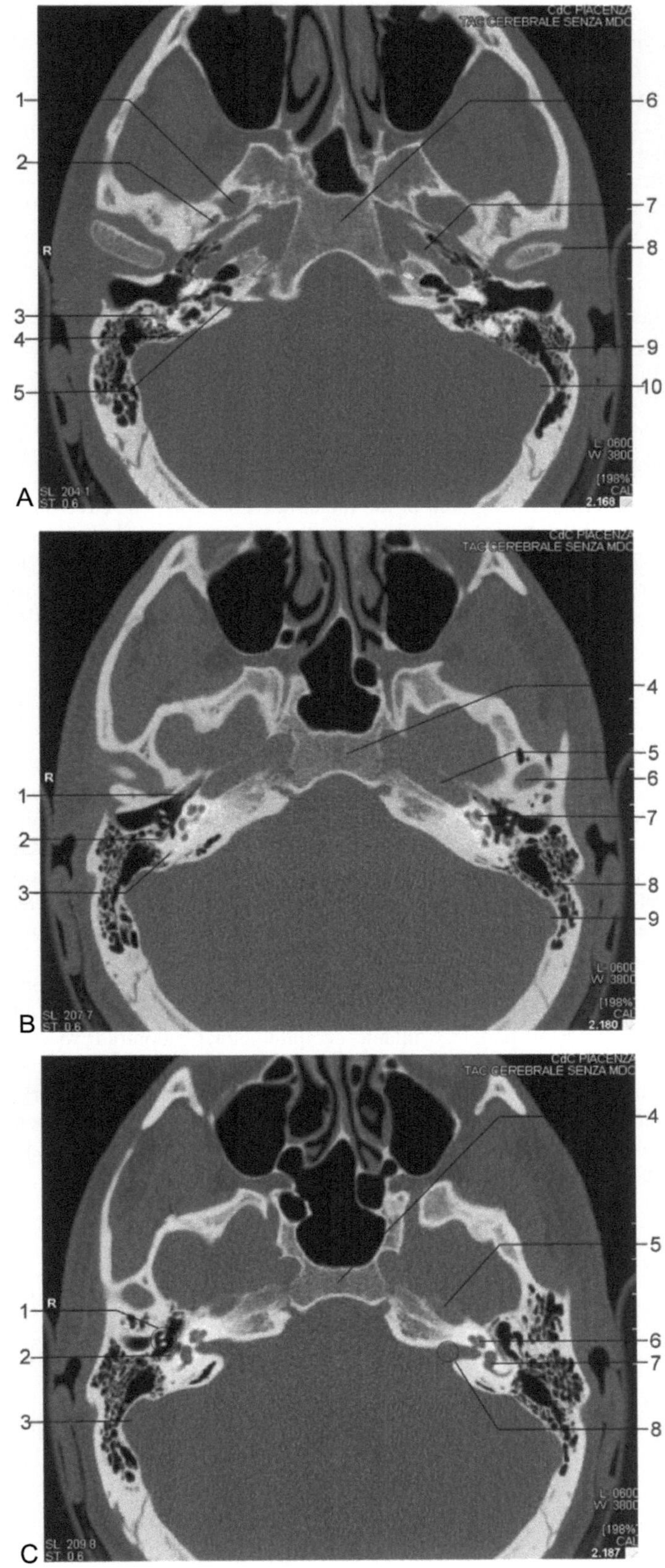

Fig. 5.4 (A) Image showing the inferior aspect of the sphenoid sinus and the genu of the intrapetrous carotid. Note the position of the bony eustachian tube lateral to the carotid and the anteromedial plane of the horizontal segment. Note also the cochlear and vestibular aqueducts, structures of importance to the skull base surgeon. 1. foramen ovale, 2. foramen spinosum, 3. mastoid segment of the facial nerve, 4. vestibular aqueduct, 5. cochlear aqueduct, 6. clivus, 7. horizontal segment of the petrous internal carotid artery, 8. mandibular condyle, 9. mastoid air cells, 10. sigmoid sinus. (B) Image showing the inferior aspect of the middle cranial fossa. The tensor tympani muscle is seen, superior to the plane of the eustachian tube. In this image the facial nerve is nearing the second genu and the sinus tympani is seen. Note the poorly pneumatized petrous apex and the inferior aspect of the cavernous carotid. Note the pterygoid canal extending anteriorly to enter the pterygopalatine foramen. Koerner's septum is also well seen in the mastoid air cells. 1. tensor tympani muscle, 2. mastoid segment of the facial nerve, 3. posterior semicircular canal, 4. clivus, 5. horizontal segment of the petrous internal carotid artery, 6. mandibular fossa, 7. cochlea, 8. mastoid antrum, 9. sigmoid sinus. (C) Image in the plane of the internal auditory canal. The malleus and incus are seen, as is the tympanic or horizontal segment of the facial nerve. 1. malleus/incus, 2. tympanic segment of the facial nerve, 3. sigmoid sinus, 4. clivus, 5. horizontal segment of the petrous internal carotid artery, 6. cochlea, 7. vestibule/lateral semicircular canal, 8. internal auditory canal.

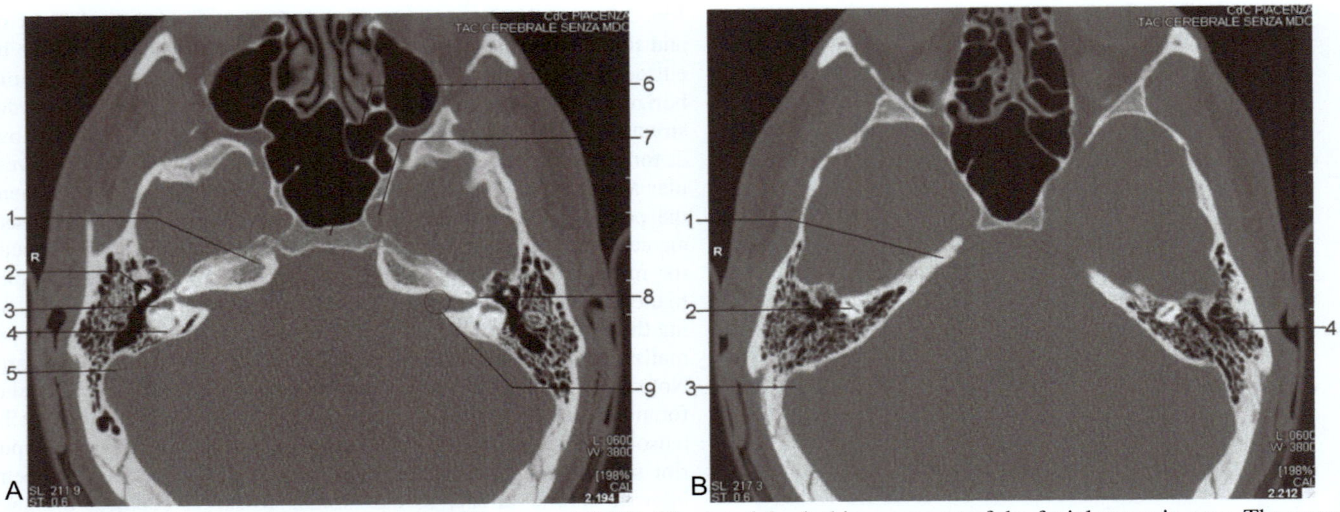

Fig. 5.5 (A) Plane the through upper internal auditory canal (IAC). The intralabyrinthine segment of the facial nerve is seen. The cavernous segment of the internal carotid artery is abutting the lateral wall of the sphenoid sinus. 1. petrous apex, 2. malleus/incus, 3. vestibule/lateral semicircular canal, 4. posterior semicircular canal, 5. sigmoid sinus, 6. clivus, 7. cavernous segment of the internal carotid artery, 8. intralabyrinthine segment of the facial nerve, 9. internal auditory canal. (B) The superior semicircular canals are seen running in a plane perpendicular to the plane of the posterior face of the petrous temporal bones. 1. petrous apex, 2. superior semicircular canal, 3. sigmoid sinus, 4. mastoid air cells.

Coronal CT, Bone Windows

See Figs. 5.6.–5.8.

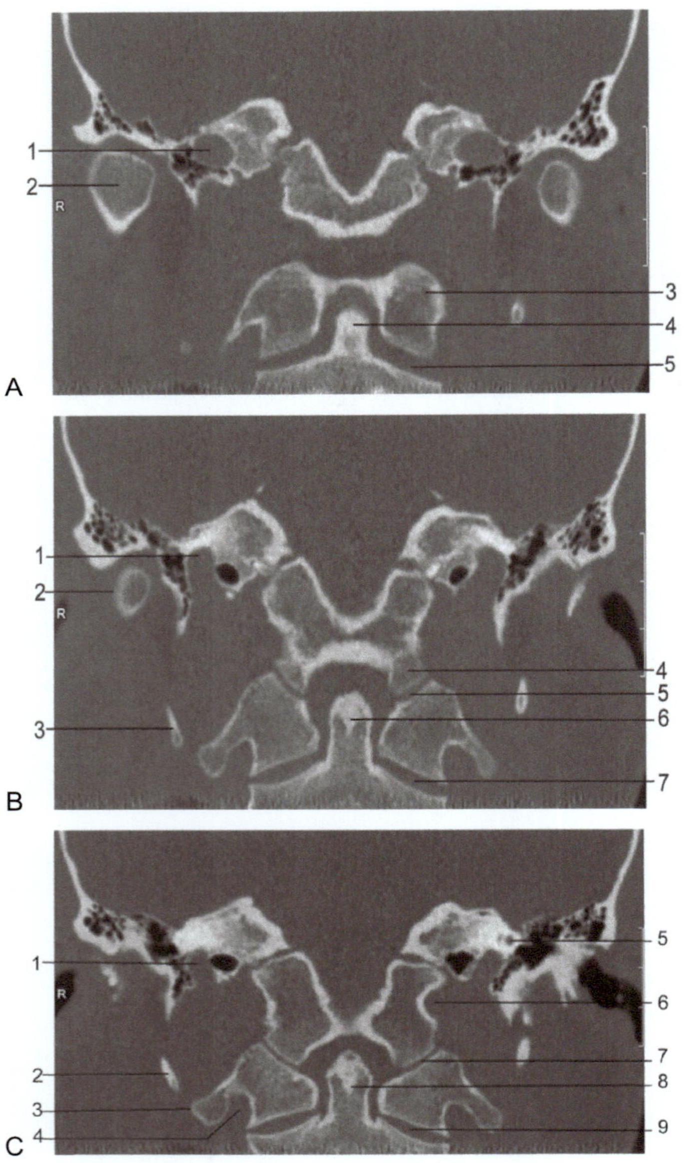

Fig. 5.6 (A) Plane through the petrous apex showing the horizontal internal carotid artery (ICA). Note the relationship of the ICA to the floor of the middle cranial fossa. This plane also passes though the mandibular condyle, ondontoid process, and lateral mass of the atlas. 1. horizontal segment of the petrous internal carotid artery, 2. mandibular condyle, 3. lateral mass of the atlas, 4. odontoid process, 5. atlantoaxial joint. (B) The atlanto-occipital joint and atlantoaxial joints are seen, as is the vertical segment of the internal carotid artery. 1. vertical segment of the petrous internal carotid artery, 2. mandibular condyle, 3. styloid process, 4. occipital condyle, 5. atlanto-occipital joint, 6. odontoid process, 7. atlantoaxial joint, (C) Note the relationship between the carotid and the cochlea. 1. vertical segment of the petrous internal carotid artery, 2. styloid process, 3. transverse process of the atlas, 4. transverse foramen of the atlas, 5. cochlea, 6. hypoglossal canal, 7. atlanto-occipital joint, 8. odontoid process, 9. atlantoaxial joint.

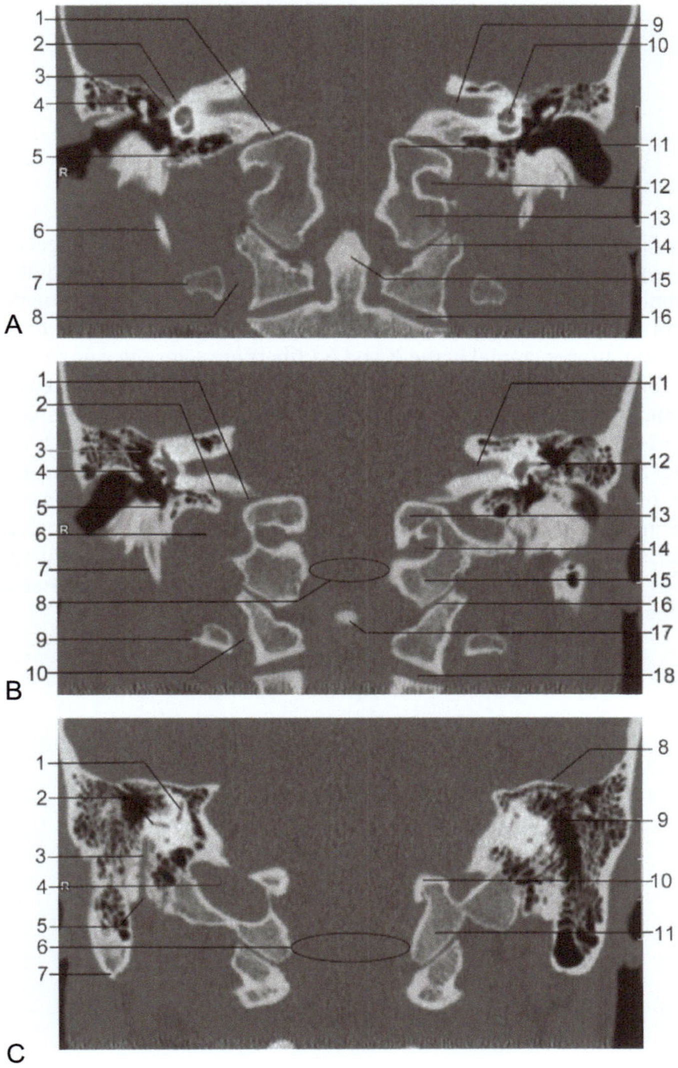

Fig. 5.7 (A) Plane through the internal auditory canal showing the head of the malleus. Note the angle of the atlanto-occipital joint and the petro-occipital suture. Note also the bird's-beak appearance of the jugular tubercle and the hypoglossal canal. The extent of the tympanic bone is also well seen. 1. petro-occipital fissure, 2. intralabyrinthine segment of the facial nerve, 3. tympanic segment of the facial nerve, 4. malleus head, 5. hypotympanum, 6. styloid process, 7. transverse process of the atlas, 8. transverse foramen of the atlas, 9. internal auditory canal, 10. cochlea, 11. jugular tubercle, 12. hypoglossal canal, 13. occipital condyle, 14. atlanto-occipital joint, 15. odontoid process, 16. atlantoaxial joint, (B) This plane shows the jugular foramen and jugular bulb, with the relationship of the cochlear aqueduct visible. 1. petro-occipital fissure, 2. cochlear aqueduct, 3. mastoid antrum, 4. tympanic segment of the facial nerve, 5. hypotympanum, 6. jugular bulb, 7. styloid process, 8. foramen magnum, 9. transverse process of the atlas, 10. transverse foramen of the atlas, 11. internal auditory canal, 12. lateral semicircular canal, 13. jugular tubercle, 14. hypoglossal canal, 15. occipital condyle, 16. atlanto-occipital joint, 17. odontoid process, 18. atlantoaxial joint. (C) The vertical segment of the facial nerve is seen, lateral to the jugular bulb, in this case well separated by pneumatized bone. The superior semicircular canal is seen, well covered by bone. 1. superior semicircular canal, 2. lateral semicircular canal, 3. mastoid segment of the facial nerve, 4. jugular bulb, 5. stylomastoid foramen, 6. foramen magnum, 7. mastoid tip, 8. tegmen, 9. mastoid antrum, 10. jugular tubercle, 11. occipital condyle.

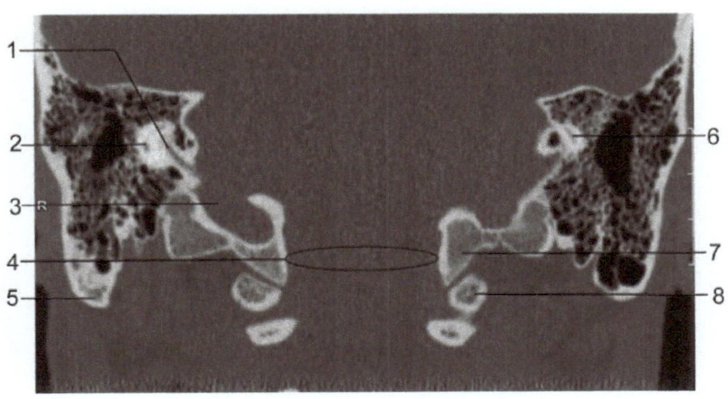

Fig. 5.8 Note the relationship between the posterior semicircular canal and the vestibular aqueduct. 1. vestibular aqueduct, 2. posterior semicircular canal, 3. jugular bulb, 4. foramen magnum, 5. mastoid tip, 6. posterior semicircular canal, 7. mastoid antrum, 8. occipital condyle.

Axial Heavily Weighted T2 MRI (High-Resolution T2-CISS Sequence)

See Figs. 5.9–5.11.

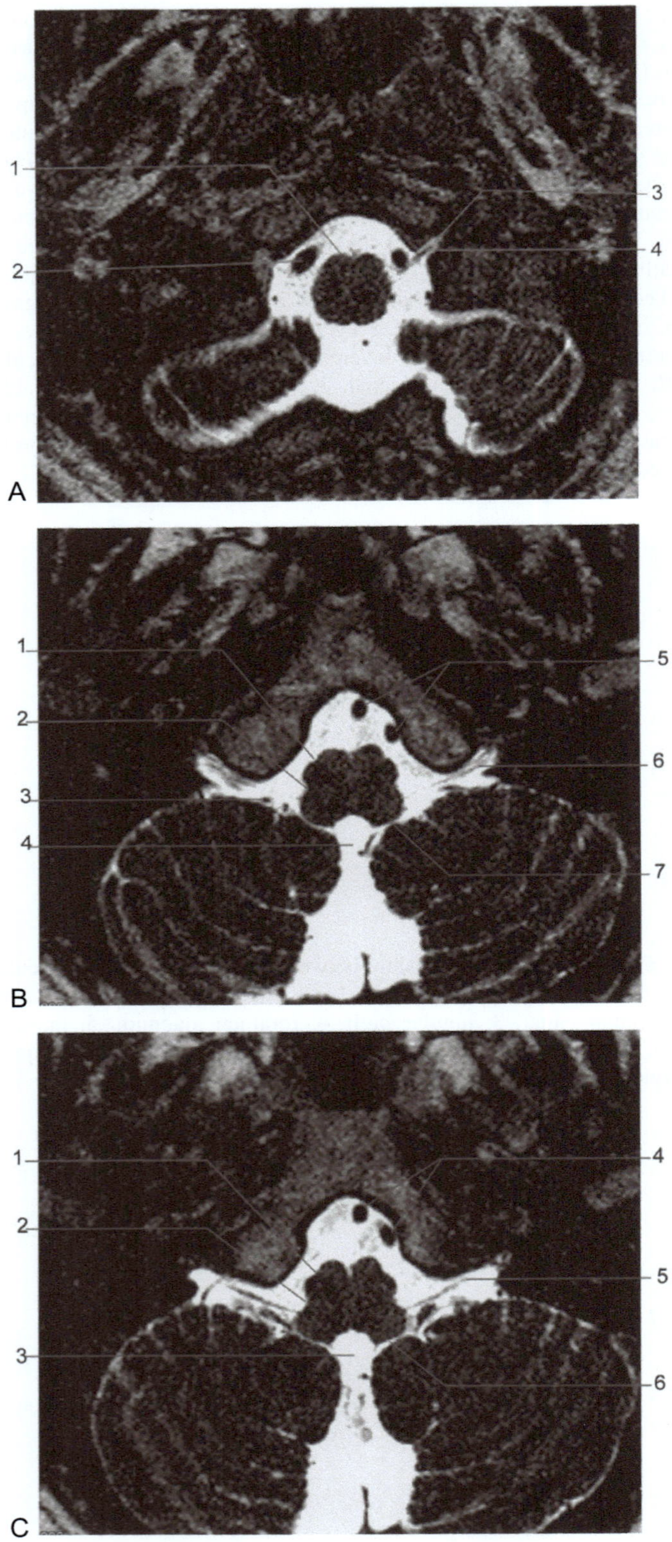

Fig. 5.9 (A) Cerebrospinal fluid (CSF) within the medial aspect of the hypoglossal canal highlights the hypoglossal nerve. 1. medullary pyramid, 2. vertebral artery, 3. hypoglossal canal, 4. hypoglossal nerve. (B) In this plane the foramen of Luschka is seen opening into the fourth ventricle. Lower cranial nerve X and XI fibers are seen entering the jugular foramen surrounded by CSF. The vertebral arteries are seen coming together to form the basilar artery farther superiorly. 1. preolivary sulcus, 2. postolivary sulcus, 3. posterior inferior cerebellar artery, 4. fourth ventricle, 5. vertebral arteries, 6. vagus and spinal accessory nerves, 7. foramen of Luschka. (C) Cranial nerve IX is seen entering the anterior margin of the jugular foramen, with the posterior inferior cerebellar artery (PICA) looping toward the jugular foramen. 1. preolivary sulcus, 2. postolivary sulcus, 3. fourth ventricle, 4. vertebral arteries, 5. glossopharyngeal nerve, 6. foramen of Luschka.

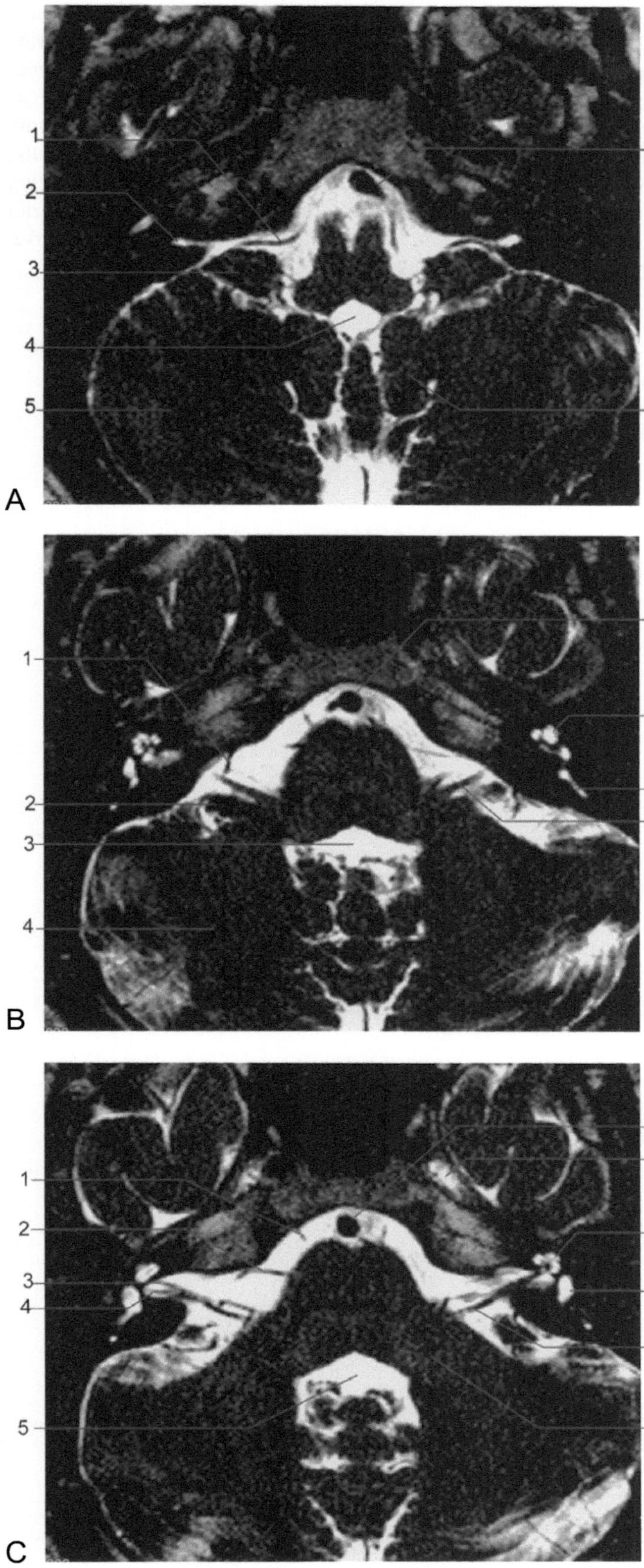

Fig. 5.10 (A) The cochlear aqueduct is seen running from the inferior aspect of the cochlea to the jugular foramen at the inferior aspect of the cerebellopontine angle. The flocculus of the cerebellum and the junction of the vertebral arteries are seen. 1. anterior inferior cerebellar artery, 2. cochlear aqueduct, 3. flocculus, 4. fourth ventricle, 5. posterior lobe, 6. vertebrobasilar junction, 7. cerebellar tonsil. (B) The cochlear nerve is seen leaving the brainstem. 1. anterior inferior cerebellar artery, 2. flocculus, 3. fourth ventricle, 4. inferior cerebellar hemisphere, 5. basilar artery, 6. cochlea, 7. posterior semicircular canal, 8. cochlear nerve. (C) The inferior aspect of the internal auditory canal showing the cochlear and vestibular nerves. 1. abducent nerve, 2. anterior inferior cerebellar artery, 3. cochlear nerve, 4. vestibular nerve, 5. fourth ventricle, 6. basilar artery, 7. Meckel's cave, 8. cochlea, 9. vestibule, 10. cochleovestibular nerve, 11. middle cerebellar peduncle.

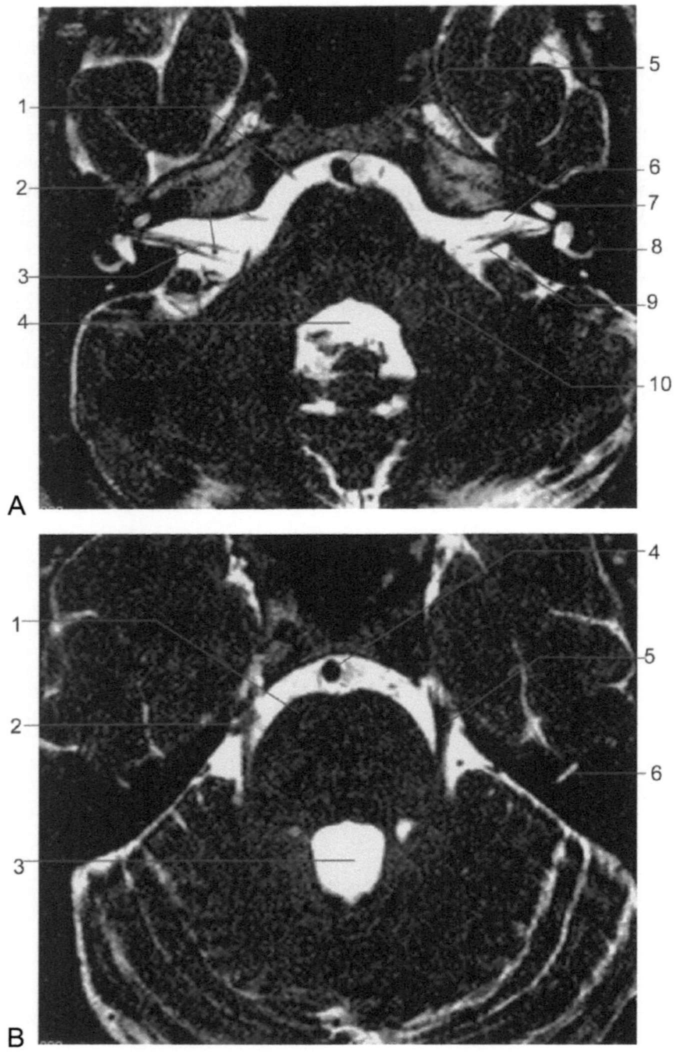

Fig. 5.11 (A) More superior aspect of the internal auditory canal showing a loop of the anterior inferior cerebellar artery (AICA) entering the canal and the artery passing between the facial and cochlear nerves. 1. abducent nerve, 2. anterior inferior cerebellar artery, 3. facial nerve, 4. fourth ventricle, 5. basilar artery, 6. AICA loop in the internal auditory canal, 7. cochlea, 8. lateral semicircular canal, 9. cochleovestibular nerve, 10. middle cerebellar peduncle. (B) The superior cerebellar artery is seen in close association with the trigeminal nerve. 1. pons, 2. superior cerebellar artery, 3. fourth ventricle, 4. basilar artery, 5. trigeminal nerve, 6. superior semicircular canal.

Axial T1 Gadolinium-Enhanced MRI

See Figs. 5.12, 5.13.

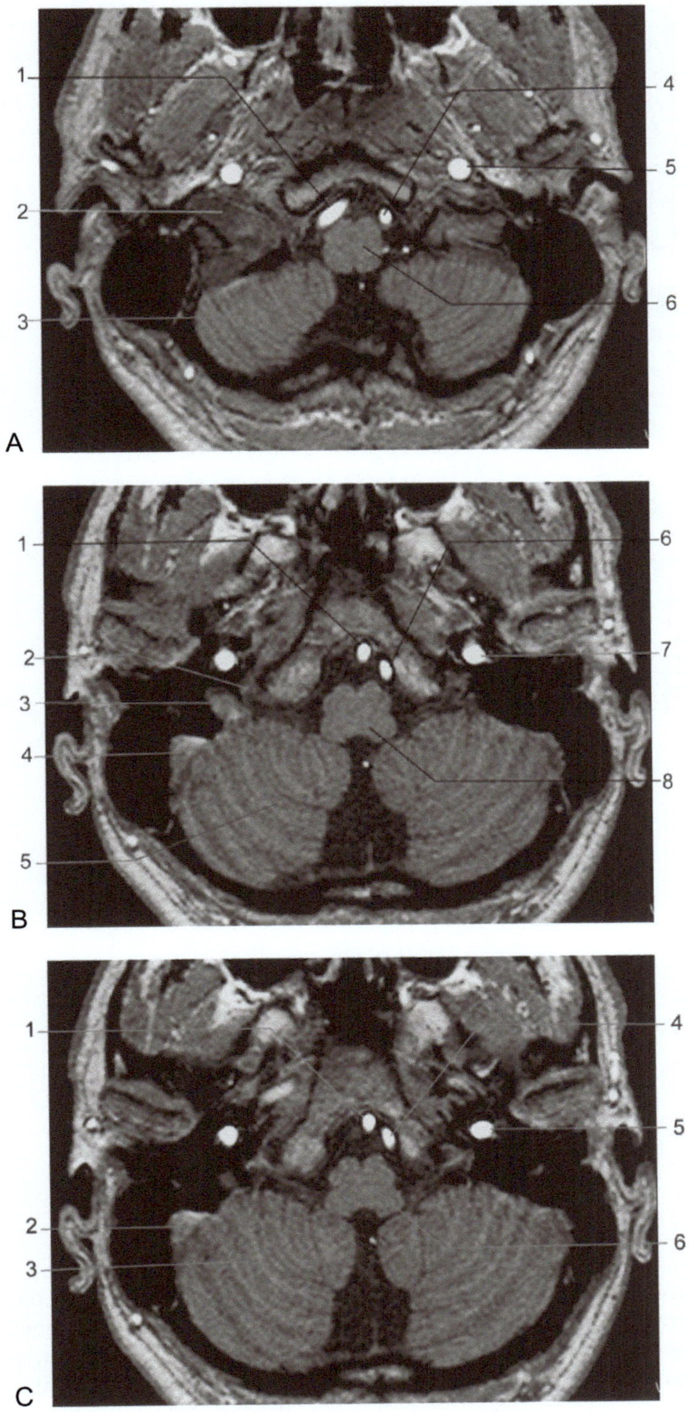

Fig. 5.12 (A) The jugular bulb is more prominent on the right side, as is usually the case. The hypoglossal canal is also seen in this plane. 1. vertebral artery, 2. jugular bulb, 3. cerebellar hemisphere, 4. vertebral artery, 5. internal carotid artery, 6. medulla oblongata. (B) The divisions of the jugular foramen are seen, with the neural bundle traveling anteriorly to the jugular bulb. 1. vertebral artery, 2. pars nervosa, 3. pars vascularis, 4. sigmoid sinus, 5. cerebellar hemisphere, 6. vertebral artery, 7. internal carotid artery, 8. medulla oblongata. (C) The vertebral and internal carotid arteries are seen. 1. vertebral artery, 2. sigmoid sinus, 3. posterior lobe, 4. vertebral artery, 5. internal carotid artery, 6. medulla oblongata.

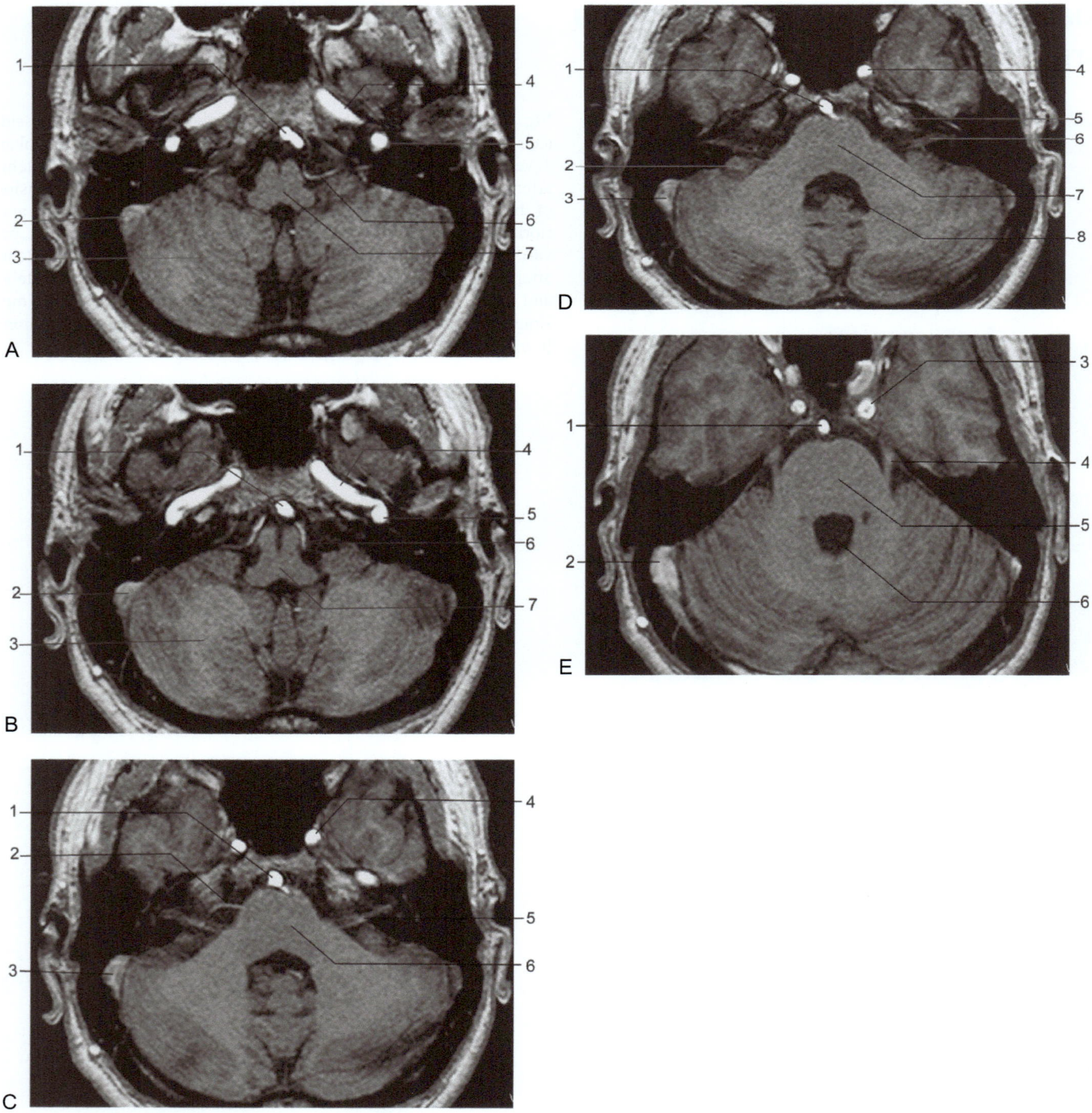

Fig. 5.13 This series of images follows the internal carotid artery superiorly into its cavernous segment. The PICA is seen arising from the vertebral arteries and the AICA from the basilar arteries. Note the fatty marrow within the clivus. (A) 1. vertebrobasilar junction, 2. sigmoid sinus, 3. posterior lobe, 4. horizontal segment of the petrous internal carotid artery, 5. vertical segment of the internal carotid artery, 6. posterior inferior cerebellar artery, 7. medulla oblongata. (B) 1. basilar artery, 2. sigmoid sinus, 3. posterior lobe, 4. horizontal segment of the petrous internal carotid artery, 5. vertical segment of the internal carotid artery, 6. posterior inferior cerebellar artery, 7. medulla oblongata. (C) 1. basilar artery, 2. anterior inferior cerebellar artery, 3. sigmoid sinus, 4. cavernous segment of the internal carotid artery, 5. facial and cochleovestibular nerves, 6. pons. (D) 1. basilar artery, 2. flocculus, 3. sigmoid sinus, 4. cavernous segment of the internal carotid artery, 5. bone marrow in the petrous apex, 6. facial and cochleovestibular nerves, 7. pons, 8. fourth ventricle. (E) 1. basilar artery, 2. sigmoid sinus, 3. cavernous segment of the internal carotid artery, 4. trigeminal nerve, 5. pons, 6. fourth ventricle.

T1 MRI Coronal View

See Fig. 5.14.

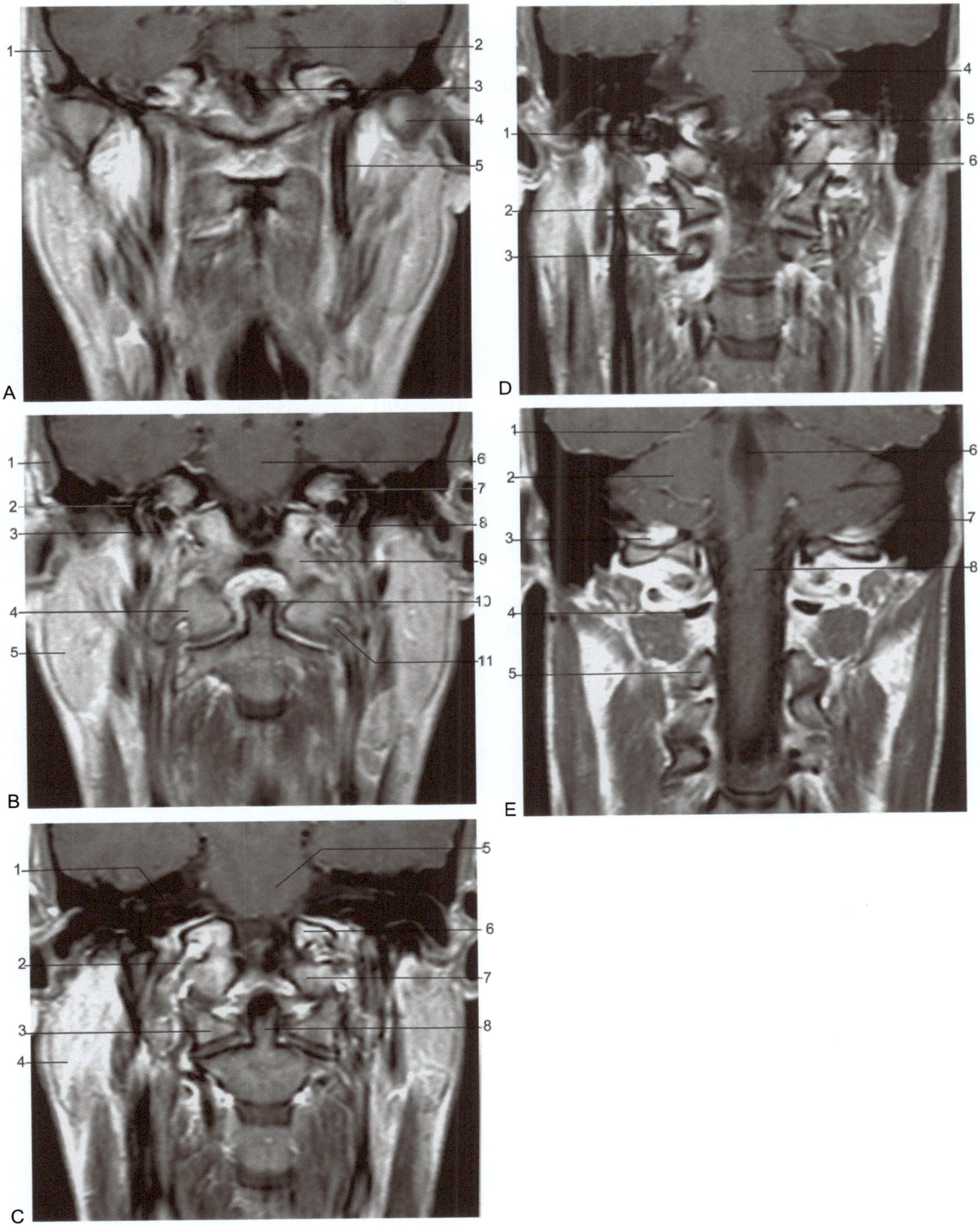

Fig. 5.14 (A) The internal carotid artery is seen as a flow void on T1-weighted MRI. 1. temporalis muscle, 2. midbrain, 3. basilar artery, 4. mandibular condyle, 5. internal carotid artery. (B) The fatty marrow is seen within the petrous apex, occipital condyles, and jugular tubercles. 1. temporalis muscle, 2. genu of the petrous internal carotid artery, 3. hypoglossal canal, 4. body of the atlas, 5. parotid gland, 6. midbrain, 7. bone marrow of the petrosal apex, 8. jugular tubercle, 9. occipital condyle, 10. odontoid process, 11. transverse foramen of the atlas/vertebral artery. (C) The venous plexus within the hypoglossal canal is hyperintense on T1-weighted MRI as it represents slow-moving blood. 1. internal auditory canal, 2. hypoglossal canal, 3. body of the atlas, 4. parotid gland, 5. midbrain, 6. jugular tubercle, 7. occipital condyle, 8. odontoid process. (d) 1. jugular bulb, 2. atlas, 3. axis, 4. pons, 5. jugular tubercle, 6. foramen magnum. (E) 1. tentorium cerebelli, 2. cerebellum, 3. agular bulb, 4. vertebral artery, 5. axis, 6. fourth ventricle, 7. sigmoid sinus.

General Considerations in Skull Base Pathology Imaging

The LSB can be affected by a large variety of tumors with individual behavior. Preoperative imaging provides important information regarding the lesion site and extent, the morphology (consistency, vascularity, differentiate benign from malignant tumors), and the involvement of nearby structures. It facilitates not only the differential diagnosis but also the patents' prognosis (staging, morbidity, mortality) [2].

The main imaging techniques used under different protocols in current practice are:

- CT
- MRI
- Positron emission tomography/computed tomography (PET-CT)
- Digital subtraction angiography (DSA)

High-Resolution CT (HRCT)

High-resolution CT is the best modality for defining the bone anatomy and pathology, the latter including the presence of erosion, reactive bone changes, dysplasia, infection, tumoral invasion, fracture, and CSF leak [2–4]. Subtle modifications within the cortical margins and neurovascular foramina are also obvious in HRCT in cases of tumoral invasion or perineural spread [3]. The presence and pattern of intratumor calcification identified on CT (Fig. 5.15), can also help in orienting the diagnosis.

Present generation multidetector scanners have the unique performance of acquiring in a short time, isotropic volumetric data instead of individual slice data. Reconstruction of high-resolution images can be done in any plane with section thicknesses of up to 0.5 mm, without compromising the spatial resolution of the original axial images. The advantage is the reduced time needed to perform the scanning and enhanced details of the images [5]. Both bone window and soft tissue algorithms are usually used for skull base pathology. Intravenous contrast is not usually necessary as the MRI will provide soft tissue detail [6]; however, it must be performed in cases where MRI is contraindicated.

Tumoral behavior can be predicted on CT scans through the aspect of peritumoral bone involvement. Aggressive infections and malignant lesions will show a permeative/destructive pattern (Fig. 5.16B,D) while smooth cortical remodeling and cortical thinning are associated with benign slow-growing processes [3] (Fig. 5.16A,C). Erosion of the internal carotid artery canal is another important element to evaluate for proper tumor staging (Fig. 5.16B), surgical planning, and preoperative ICA management (stenting, occlusion).

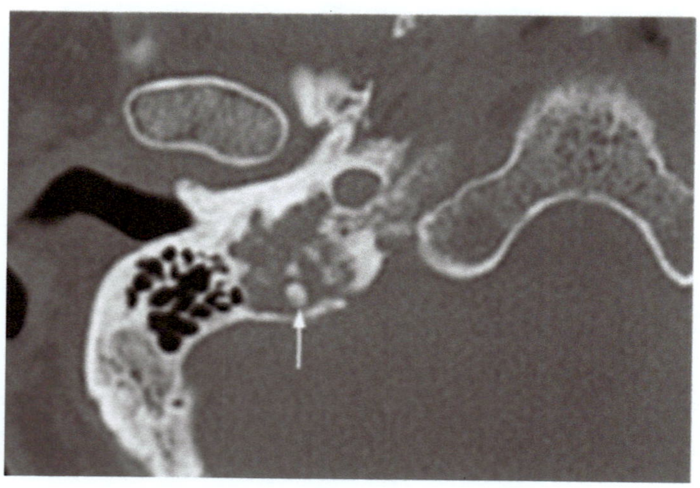

Fig. 5.15 CT scan of a petrous apex lesion with the presence of bone erosion and intratumor calcifications (*white arrow*).

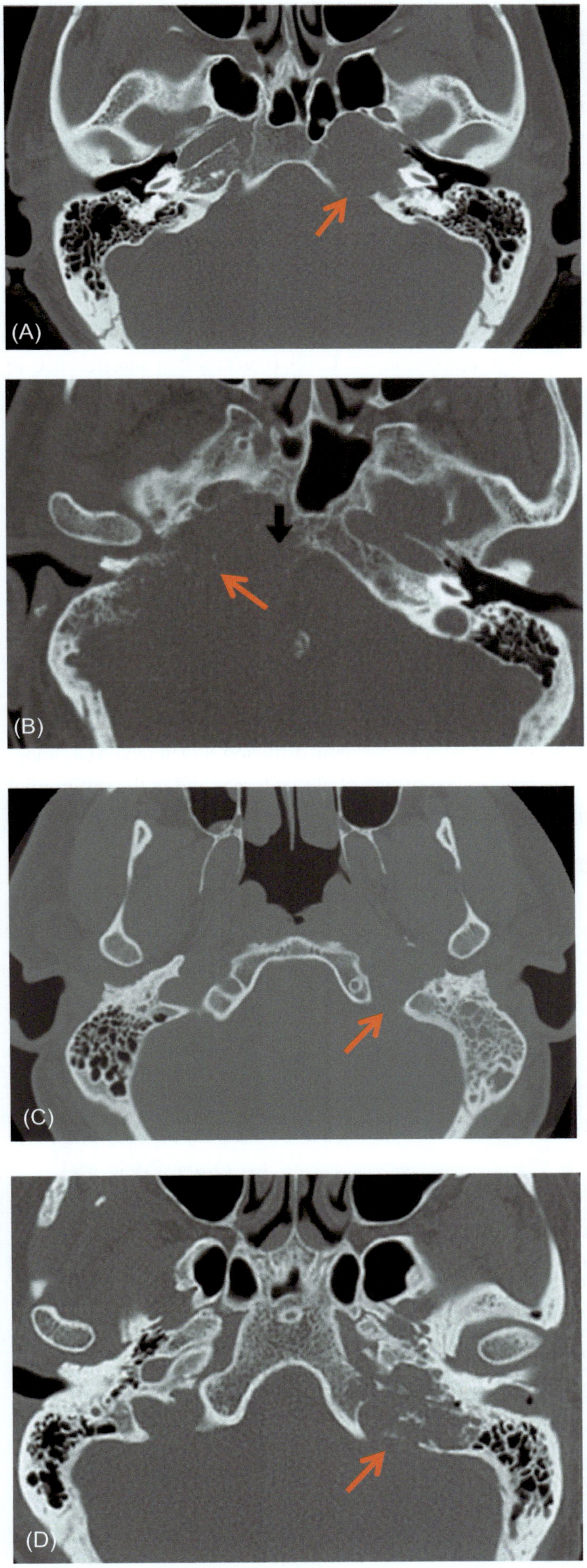

Fig. 5.16 Different patterns of peritumoral bone remodeling: (A,C): smooth bone remodeling produced by tumor growth at the level of the petrous apex (A) and jugular foramen (C); (B,D): permeative pattern of bone invasion of the petrous apex (B) and jugular foramen (D), note the erosion of the horizontal portion of the ICA bony canal.

MRI

MRI is superior to CT for soft tissue characterization and has become the primary imaging modality for evaluation of non-osseous skull base lesions. It can show in great detail tumoral invasion of nearby structures, such as dural, perineural, perivascular, leptomeningeal or brain tissue spread; the same goes for extension into the orbit, the cavernous sinus or skull base fissures and foramina [2,3,7–9].

A routine MRI study comprises axial and coronal images using fast spin echo T1 and T2 weighted images and post-contrast T1 with iv administration of gadolinium. The introduction of 3-T (Tesla) field strength, which allows improved image resolution, and developments in pulse sequence that permits submillimeter scanning have significantly improved characterization of the structures within the skull base [10]. Unenhanced Tl-weighted (T1W) images demonstrate high signals from structures containing fat while T2W are sensitive to the presence of increased water content. In general, inflammatory and benign lesions have less cellularity and higher water composition appearing bright on T2W sequence when compared to gray matter, while malignancies present an intermediate intensity due to the dense cellular structure.

Fluid-Attenuated Inversion Recovery (FLAIR) is a T2-weighted sequence obtained by using an inversion pulse to suppress the signal of the cerebrospinal fluid (CSF), so it will appear dark. It is convenient to accentuate lesions that develop near the ventricles or brain sulci. The lack of CSF attenuation of the sulcal signal on FLAIR is indicative of leptomeningeal involvement with the replacement of normal CSF by blood or tumor [11].

The use of gadolinium contrast in T1-weighted images will enhance active lesions such as inflammation and neoplasms. Fat-suppression techniques are suitable for increasing contrast resolution resulting in a better definition of the lesion, but its' use in skull base pathology is debatable because incomplete fat suppression can be mistaken for tumoral enhancement [12]. High-resolution T2 CISS (constructive interference in steady state) provides excellent contrast between cerebrospinal fluid (CSF) and other structures in the brain being very useful for evaluating structures surrounded by CSF (e.g., internal acoustic canal) (Fig. 5.10,5.11).

Advanced MRI techniques like diffusion-weighted imaging (DWI), susceptibility-weighted imaging (SWI), perfusion-weighted imaging (PWI), and MR spectroscopy (MRS) give much more information regarding tissue characteristics of the lesion and can lead to a more specific differential diagnosis especially in difficult to interpret cases [13]. DWI sequences demonstrate the molecular motion or diffusion of water protons within tissue using an apparent diffusion coefficient(ADC)map. It is useful in identifying cholesteatomas and in the differential diagnosis of benign vs malignant tumors [11,14] (Fig. 5.17).

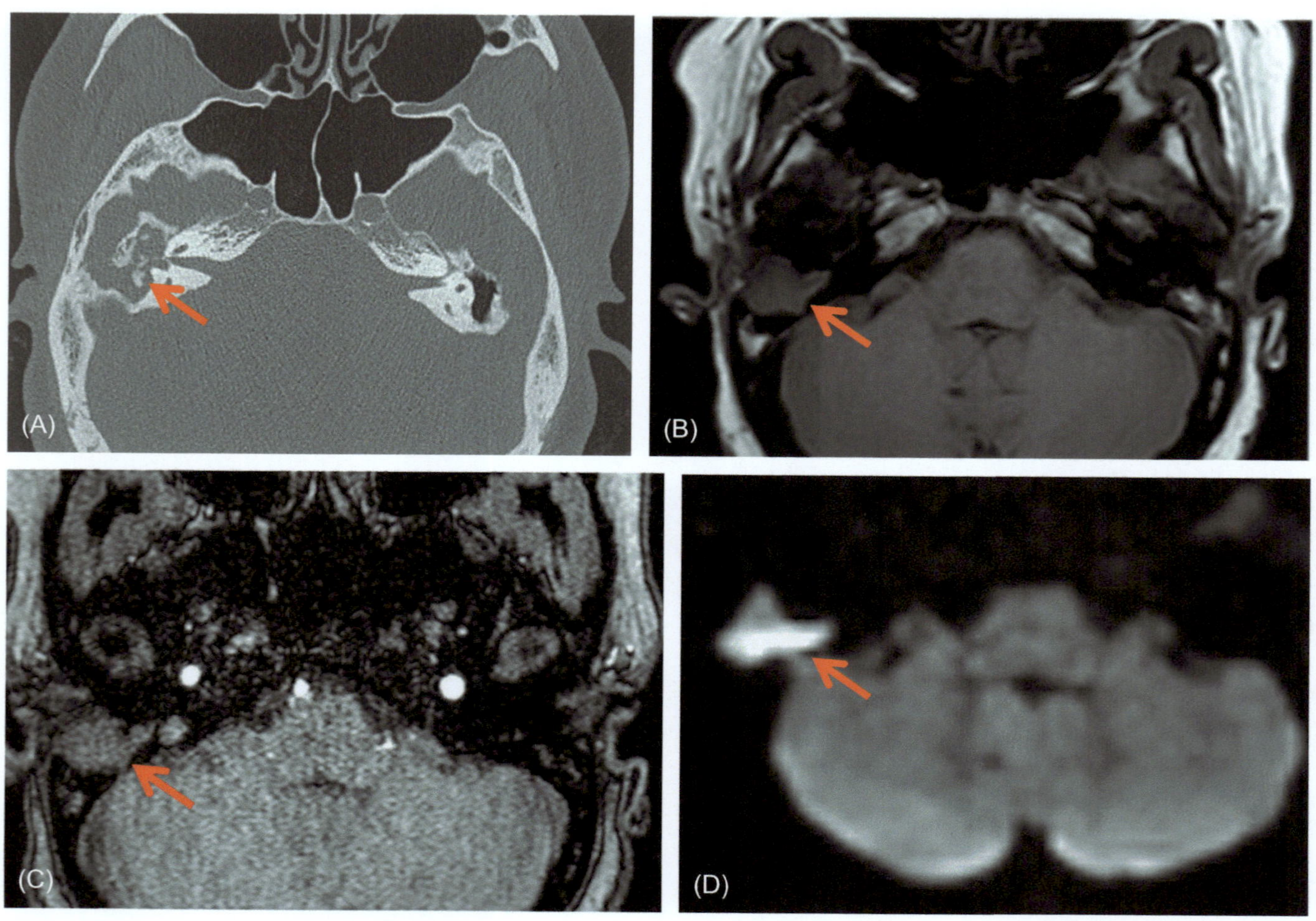

Fig. 5.17 (A) CT scan axial view of a petrous bone cholesteatoma (PBC) with erosion of the bony labyrinth (*red arrow*). (B). T1W nonenhancing MRI axial image showing the same PBC which appears isodense and does not pick up contrast after Gd administration (C). (D) Non-epi DWI axial MRI view of the PBC as a highly hyperintense mass in the right TB.

As mentioned above, tumoral involvement of nearby structures can best be evaluated on MRI. Dural, leptomeningeal, and parenchymal invasion are best appreciated on T1WI (postcontrast) and T2WI/FLAIR. Dural invasion is one of the most important factors in conditioning surgical planning and prognosis. Thick linear enhancement >5mm or nodular enhancement of the dura is considered pathologic [15]. The absence of the hypointense layer between the tumor and the overlying enhancing dura also suggests that the dura is affected [3]. Brain involvement is characterized by enhancement or edema of the nervous tissue adjacent to tumor.

Invasion of the cavernous sinus currently represents a relative contraindication for complete tumor resection, in any case, it warrants much caution during preoperative planning [16]. It can be appreciated by the presence of compression of the cavernous carotid artery; loss of contrast enhancement of the cavernous sinus, and bulging of the lateral sinus wall [3] (Fig. 5.18).

Perineural spread is another well-known negative prognostic factor that may impede cure if unrecognized [17]. Postcontrast fat-saturation T1W thin section is the most sensitive technique for preoperative treatment mapping for either surgery or radiation therapy in patients with skull base tumors with perineural spread [13]. Tumoral invasion will disrupt the blood-nerve barrier increasing capillary permeability and resulting in higher uptake of the contrast medium [18]. Segmental/diffuse enhancement and enlargement of the cranial nerve, enlargement of the bony foramen, and replacement of the fat within the foramen are signs associated with nerve involvement.

The use of intraoperative MRI has been proven beneficial in anterior skull base surgery [19,20], however in lateral skull base this is uncharted territory. Recent studies have noted that intraoperative MRI may help predict postoperative facial nerve outcomes after vestibular schwannoma removal based on the consistency of the tumor appreciated with the help of MRI, which may affect treatment and outcomes [21].

Magnetic Resonance Angiography and Venography

Magnetic Resonance Angiography (MRA) and Venography (MRV) are non-invasive techniques used for three-dimensional representation of the cervico-cerebral vasculature (Fig. 5.19). It is helpful for the preoperative evaluation of the neoplastic vascular supply, vascular

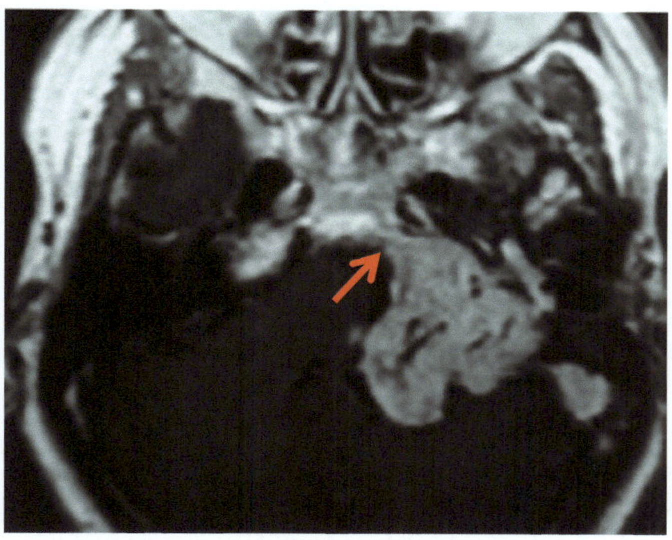

Fig. 5.18 T1W with Gd-enhancement MRI axial view of a left-sided tympanojugular paraganglioma (Class C3 Di2) with intradural extension and invasion of the cavernous sinus (*arrow*).

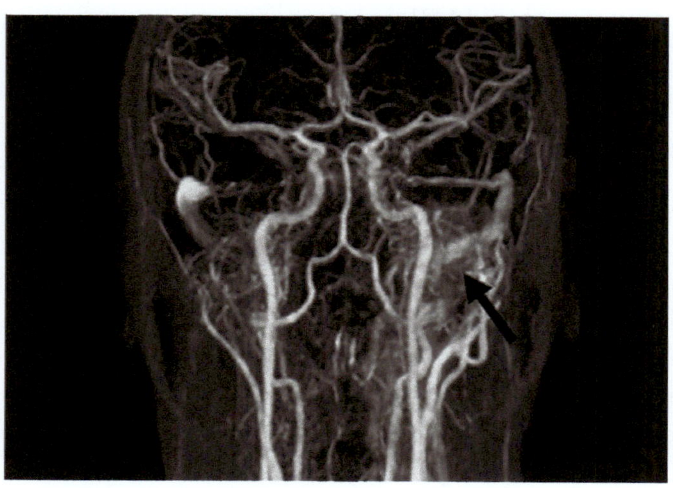

Fig. 5.19 Head and neck magnetic resonance angiography showing a class C2 tympanojugular paraganglioma seen as the hypervascular area (*black arrow*) lateral to the vertical portion of the carotid artery.

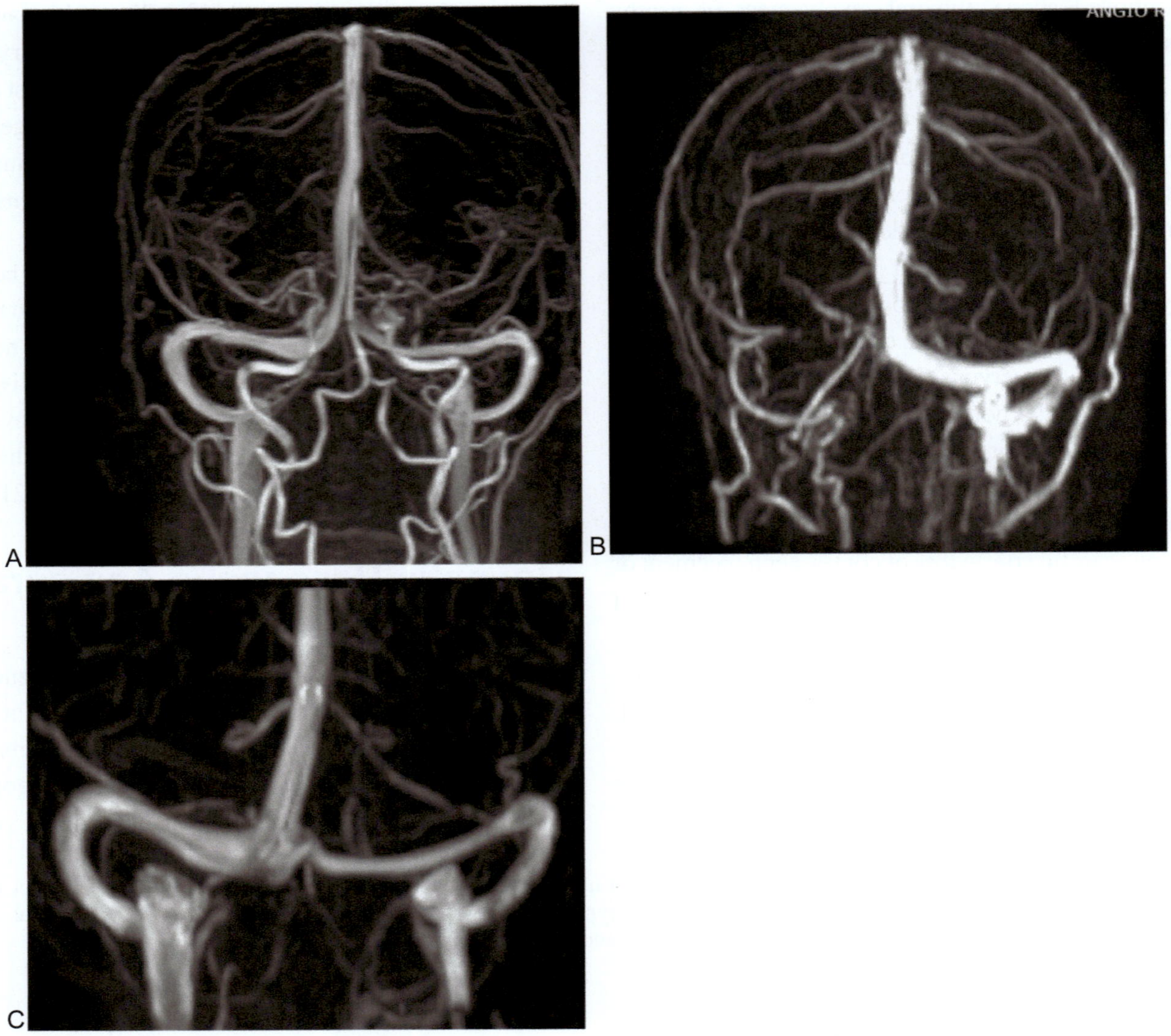

Fig. 5.20 MR venography of the craniocervical venous system. (A) Symmetric sigmoid sinuses. (B) Unique left-sided sigmoid sinus. (C) Dominant right-sided sigmoid sinus with high jugular bulb.

malformation, or aneurysms, to evaluate known or suspected venous thrombosis and for the differential diagnosis of jugular foramen tumors (Fig. 5.19). MRI venography is essential to evaluate the cerebral venous anatomy, the dominance of sigmoid sinuses and connection to the transverse/sagittal sinus, temporal lobe draining patterns, the vein of Labbé, and jugular bulb size. They can influence the preoperative decision process regarding approach and surgical technique in cases of unique/dominant sigmoid sinus or very high jugular bulb on the ipsilateral side as the tumor (Fig. 5.20). Gadolinium enhancement often allows visualization of the difference in the size of the sigmoid sinus on each side.

Positron Emission Tomography/ Computed Tomography (PET-CT)

PET-CT using 18 F-fluorodeoxyglucose (FDG) has been used to characterize various malignancies, as lesions with intense metabolic activity uptake FDG in higher concentration. PET-CT can be an important tool for initial diagnostic and following surveillance of skull base primary cancers or metastasis [22]. Non the less interpretation can be challenging as nonmalignant lesions can also intensify, such as inflammation, infection, and physiologic uptake of brain, resulting in a low positive predictive value (PPV). In case of doubt, calculating the standardized uptake value (SUV) may be useful but experience is needed [23].

Digital Subtraction Angiography (DSA)

Angiography has superior resolution to MRA, but it is an invasive procedure associated with a higher risk of complications. It plays a critical role in the management of tympanojugular paragangliomas but is rarely necessary for diagnosis. Paragangliomas display a characteristic intense tumor blush and rapid venous diffusion (Fig. 5.21). This allows differentiation in cases where CT and MRI remain equivocal. Just as importantly, it allows detailed analysis

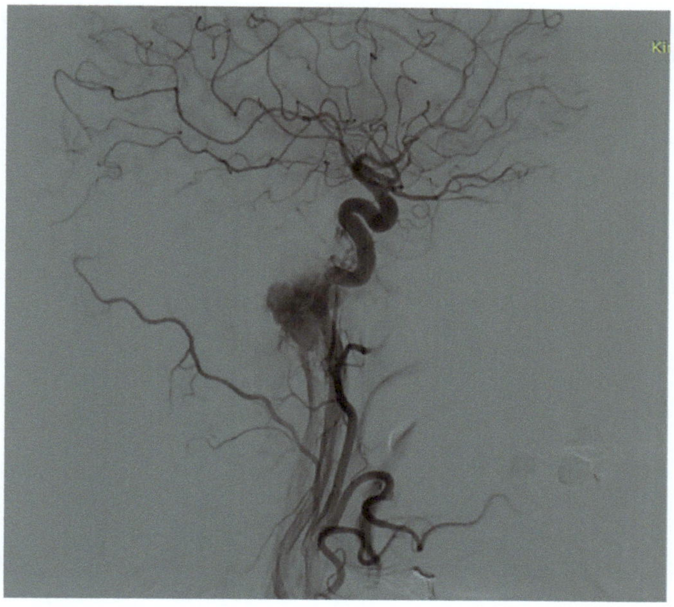

Fig. 5.21 ICA injection showing the tumoral blush of a grade C2 paraganglioma, that invades the vertical segment of the ICA.

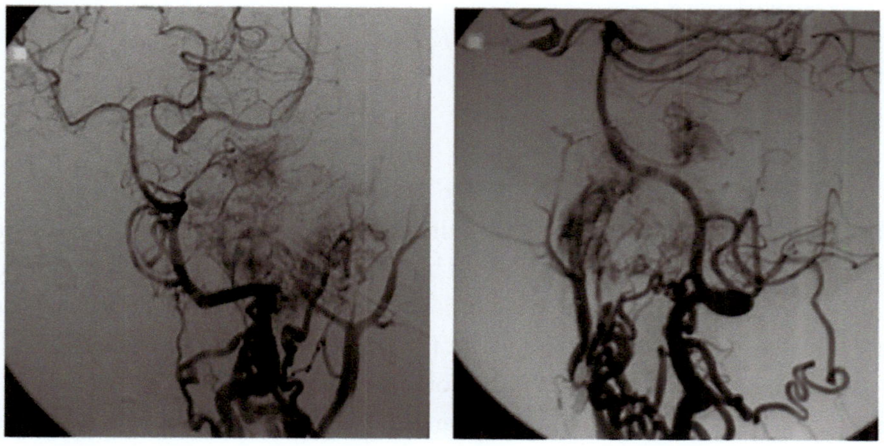

Fig. 5.22 *Left* vertebral artery injection in both views, after embolization with glue; there is great reduction of the tumoral blush with patency of the arterial branches.

of tumor vascular supply, and embolization if operative management is planned (Fig. 5.22), as well as assessment of ICA involvement, contralateral cerebral blood flow, and venous drainage.

Other radiological tools such as CT angiography, CT venography, cisternography, or plain RX have a limited contribution to skull base pathology. Scintigraphy is recommended for cases of suspected malignant otitis externa. CT angiography and venography seem to be valuable in complex cases of skull base tumors [24] but further studies are needed in this field.

Imaging of Skull Base Pathology

Even though a large variety of lesions can involve the skull base, based on location and imaging features the differential diagnosis can be frequently narrowed down to a few possibilities [25]. Non the less, it is difficult to establish a reliable classification system for these tumors due to overlaps that ensue in regard to site and imagistic features. Depending on their origin lesions can be intrinsic to the skull base, arising from the intracranial or the extracranial compartment. According to the site, they can be grouped into cerebellopontine angle, petrous apex, petroclival and clival, jugular foramen, and sellar and parasellar. Lesions can also be classified as intradural, transdural, or extradural. Based on etiology LSB pathology can be congenital, inflammatory (infective and noninfective), traumatic, and neoplastic (benign and malignant). Lesions may be also classified according to morphology and behavior as cystic or solid, and aggressive or nonaggressive.

In general, location is more important than factor as impact on patient symptomatology, small lesions can have a greater clinical effect than large ones that can be asymptomatic for a long time. Tumor site and extent also dictate the surgical approach to be used. As such we have chosen a subdivision of lesions according to their anatomical location as intra-axial lesions: sellar/parasellar, clival/paraclival, and occipital foramen and extra-axial lesions: petrous apex, cerebellopontine angle, and jugular fossa (Fig. 5.23).

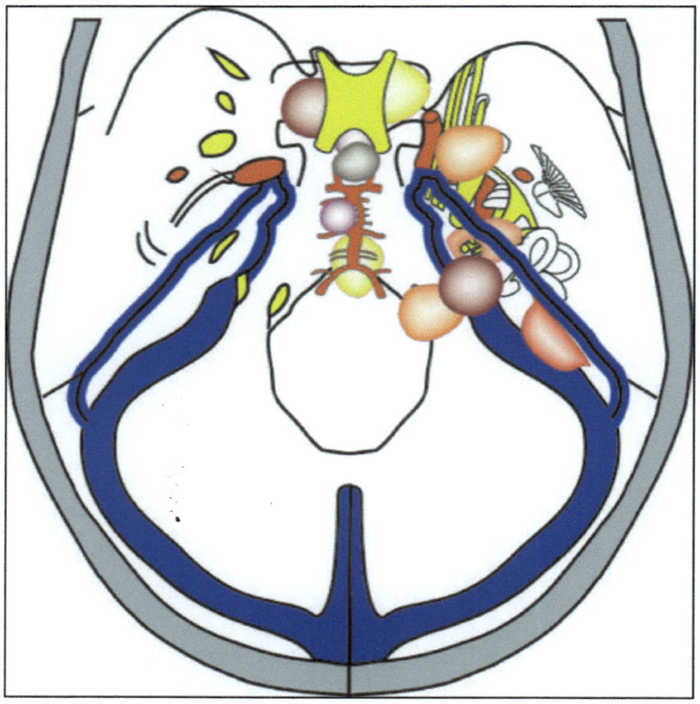

Fig. 5.23 Image of the skull base and possible development of tumors at the different anatomical sites.

Intra-Axial Lesions

SELLAR AND PARASELLAR TUMORS

Sellar and parasellar tumors are currently addressed by using transsphenoidal endoscopic surgery so they are not in focus of this book. Lateral approaches with skull base techniques are rarely necessary and are usually used in combination with other techniques. Pituitary adenomas comprise up to 95% of tumors located in the region followed by meningiomas, craniopharyngiomas, Rathke' s cleft cysts, and aneurysms [26]. Indications for transcranial approaches are optic apparatus decompression, tumor extension beyond a transsphenoidal reach, evidence of a dural tail, and a firm consistency with intimate adherence to the neurovascular structures [27]. The two classic lateral surgical approaches to this region are the pterional and orbitozygomatic craniotomies.

CLIVAL AND PETROCLIVAL LESIONS

Lesion that arise in the clival/paraclival area are specific, as they are conditioned by the location in the skull base and the local embryologic development. They can be either intradural or extradural. Transdural cases are those that invade the dura and extend on both sides. Intradural lesions include meningioma, epidermoids, trigeminal neurinoma, and basilar artery aneurysm (Table 5.1). Extradural lesions include chordoma, chondroma, chondrosarcoma, cholesterol granuloma, extensive glomus tumors, pituitary adenoma, and nasopharyngeal carcinoma. Chordomas and metastasis are the most frequent destructive lesions of the site. Imaging is helpful in differentiating tumor from inflammation and benign vs malignant neoplasms but the definitive diagnosis necessitates biopsy or excision with histologic examination. Studies have documented that the endoscopic endonasal

TABLE 5.1 Clival and petroclival lesions.

Neoplastic lesion	Pseudoneoplastic lesions	Extraclival invasive tumors
Chordoma	Fibrous dysplasia	Nasopharyngeal carcinoma
Chondroma	Paget disease	Nasopharyngeal rhabdomyosarcoma
Chondrosarcoma	Langerhans histiocytosis	Juvenile angiofibroma
Meningioma	Radionecrosis	Mucocele
Trigeminal neurinoma	Osteomyelitis	Pituitary adenoma
Epidermoids	Arachnoid granulations	Extensive paraganglioma
Lymphoma		
Multiple myeloma		
Metastatic tumors		

technique is safe and effective for the resection of clival lesions and is recommended whenever reasonable.

CHORDOMAS

Chordomas are most the common tumors of the clival region. They are low-grade malignant tumors that derive from remnants of the notochord. They onset in adulthood inducing headache and cranial nerve palsies. On CT, chordomas present as central bone-erosive, lobulated soft-tissue masses with fine calcifications resulting from bone destruction. On MRI, chordomas are isointense or hypointense on T1-weighted and markedly hyperintense on T2-weighted images but are usually heterogeneous (Fig. 5.24). They enhance after gadolinium administration. Clival chordomas show four patterns of extension: posterior into the posterior fossa and prepontine cistern; anterior to involve the nasopharynx, sphenoid sinus, and less commonly the orbit and the maxillary sinus; lateral to the parasellar region and petrous apex; less frequently, caudal, particularly in children. MRI is useful for studying the relation between the tumor the surrounding structures and the major cerebral arteries, which are more commonly displaced by the tumor. Less frequently the tumor encases the artery or rarely stenoses the lumen. Angiography is indicated in such cases, particularly for the proper evaluation of the carotid siphon. The high local recurrence rate is common for this pathology despite radical surgical treatment resulting is a poor prognosis [28]; distant metastases are unusual.

CHONDROMAS AND CHONDROSARCOMAS

Chondromas and chondrosarcomas originate from the cartilaginous remnants within the cranial base synchondroses, most frequently the petro-occipital junction. They are usually singular paramedial lesions that can extend from the middle to the posterior cranial fossa, particularly they develop in the prasellar space. These lesions typically present bone involvement and intratumoral calcifications, often with the pattern of rings and arcs [29], that are best identified on CT scans with triplanar bone algorithm. MRI will illustrate the high-water content of these lesions as high signal intensity on T2W, and hypointense on T1W, very heterogenous [28–30], and will also reveal the associations between lesions and surrounding anatomical structures (Fig. 5.25).

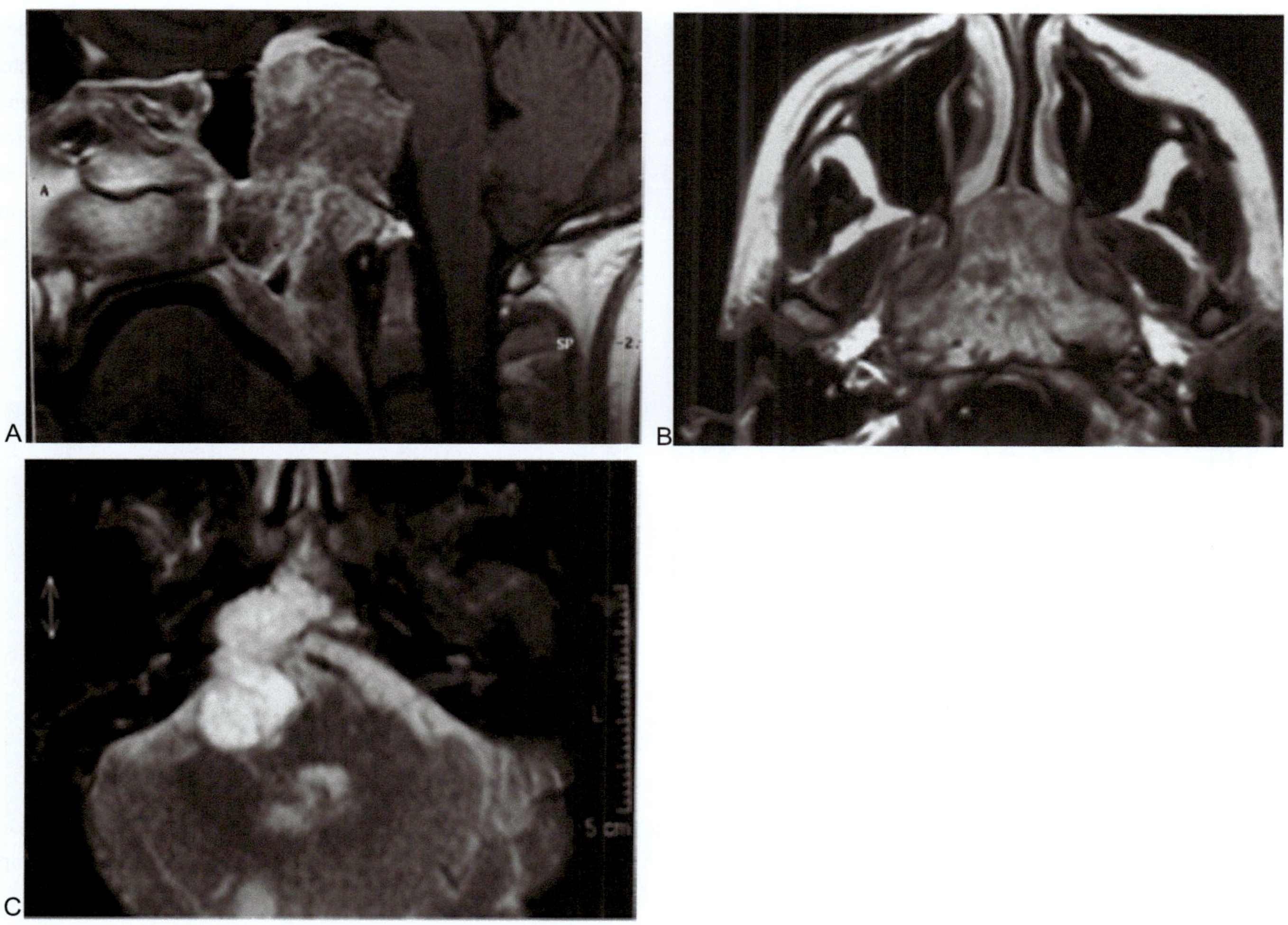

Fig. 5.24 T1W with contrast sagittal (A) and axial (B) MRI image of a clival chordoma showing extension into the nasopharynx and sphenoid sinus, with caudal extension. c-T2W weighted axial MRI image of a clival chordoma showing as a hyperintense tumor with posterior extension.

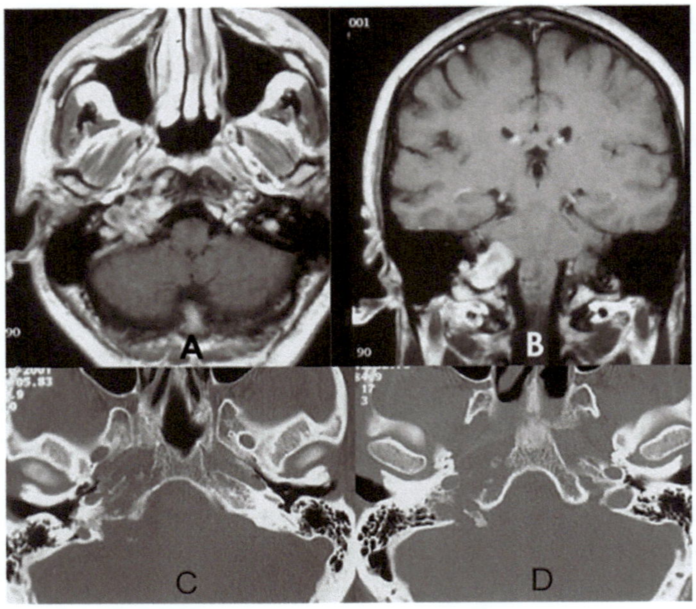

Fig. 5.25 Preoperative imaging of the case. (A) Axial MRI with gadolinium enhancement and (B) coronal MRI with gadolinium enhancement showing an extradural petrous apex tumor. (C) CT scan with bone window showing the tumor intimately related to the horizontal internal carotid artery. (D) CT scan with bone window at a lower level. The tumor also involves the vertical internal carotid artery.

Differentiating between chordoma and chondrosarcoma is challenging due to their similar radiological and histopathological aspects [31]. Chordomas develop centrally whereas chondrosarcomas are usually paramedian lesions that affect patients towards the 6th decade of life [17]. Recent studies have shown that they demonstrate differences on DWI MRI that may help distinguish the two entities, with chondrosarcoma having a higher average apparent diffusion coefficient than chordoma [32,33]. Separating the two entities is essential since chondrosarcoma tends to have a better prognosis than chordoma [34].

TRIGEMINAL NEURINOMAS

These show the same radiological features of neurinomas elsewhere, namely, isointense to hypointense on T1weighted MRI, hyperintense on T2-weighted, and uniform enhancement with gadolinium (Figs. 5.26, 5.27). Trigeminal neurinoma can be classified into four types: type I, predominantly middle fossa; type II, predominantly posterior fossa; type III, present in both fossae; and type IV, which is entirely extradural and involves the infratemporal fossa. Cases in the middle fossa usually cause erosion of the floor of the middle fossa involving the foramen ovale of the anteromedial portion of the petrous apex. Type III lesions show

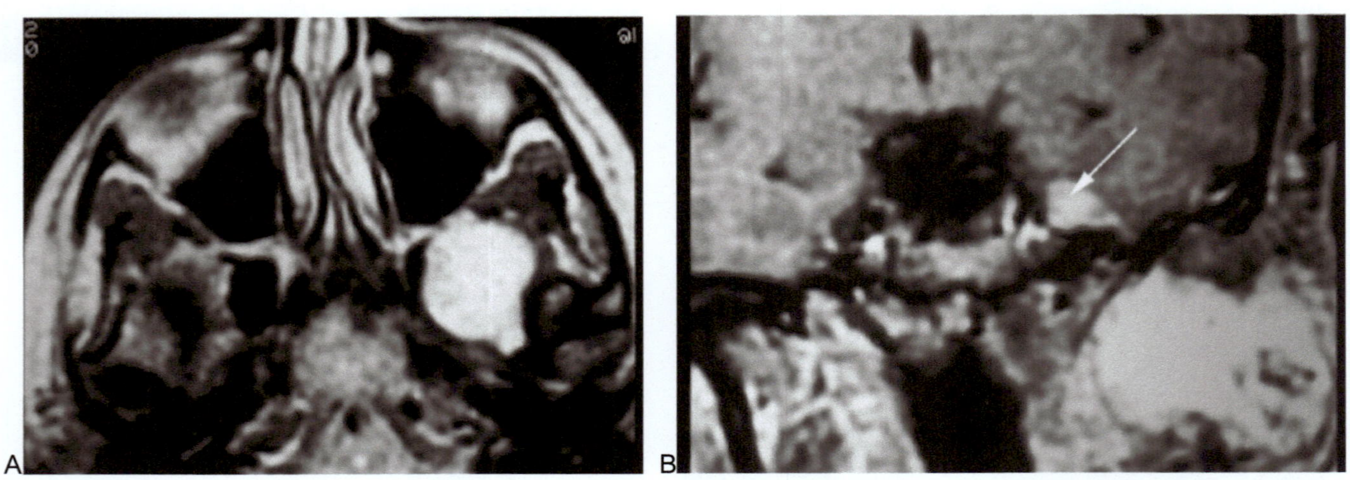

Fig. 5.26 (A,B) Preoperative MRI with gadolinium enhancement. (A) Axial view showing the tumor massively involving the infratemporal fossa. (B) Coronal view. The tumor involves Meckel's cave (*arrow*) and extends into the infratemporal fossa in a very low position.

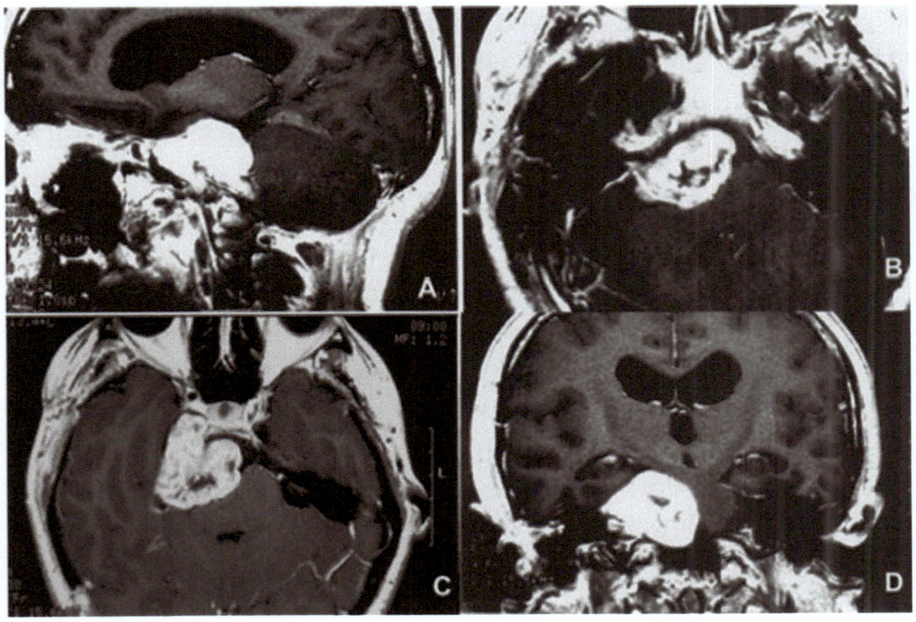

Fig. 5.27 MRI with gadolinium enhancement showing the tumor. (A) Sagittal view showing the tumor involving both the posterior and middle cranial fossa. (B) Axial view. The tumor is seen in the posterior fossa lying anterior to the internal auditory canal and involving the prepontine cistern. (C) Axial view at a higher level. The parasellar extension of the tumor is seen. (D) Coronal view. Note the marked displacement of the brainstem. Involvement of Meckel's cave is appreciated.

a dumbbell-shaped appearance. Orbital extension can occur in some cases. Extensive lesions rarely engulf the internal carotid artery.

NASOPHARYNGEAL CARCINOMA

This is the most common malignant tumor involving the nasopharynx. Up to 80% origin in the posterolateral recess of the pharyngeal wall (Rosenmüller fossa). Extensive lesions usually cause bone destruction of the central clivus. Tumors can extend into the pterygopalatine fossa (Fig. 5.28) and further superiorly through the inferior orbital fissure to the orbit or intracranially through the superior orbital fissure or by perineural spread. MRI is the gold standard for diagnosing the disease, being able to detect even subclinical lesions that are missed on rhinopharyngeal endoscopy. The tumor is hypo- to isointense on T1W images and hypointense on T2W (relative to muscles), it shows moderate-to-intense contrast enhancement [35,36].

METASTASES

Primary neoplasms that usually display hematogenous spread to the skull base are breast, prostate, lung, and kidney cancers. They may occur anywhere but the clivus, petrous apex, and sphenoid triangle are most frequently affected because they present high marrow content [17]. Nonetheless, clival involvement is relatively rare, but its' incidence seems to be increasing in the reported literature [37]. They are usually lithic lesions of variable intensity on both T1 and T2 sequences depending on the tumor of origin, therefore PET-CT and knowledge of the primary tumor are important factors in guiding the diagnosis.

FIBROUS DYSPLASIA

Fibrous dysplasia is a pseudoneoplastic congenital process, characterized by the replacement of normal bone with fibrous stroma and isles of immature bone. On CT,

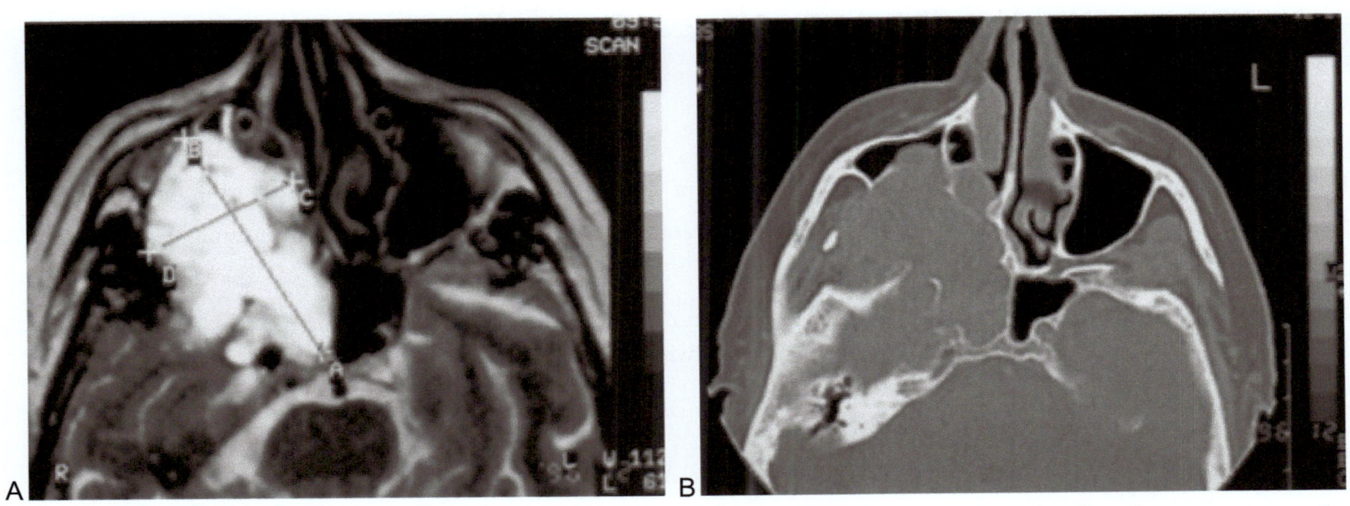

Fig. 5.28 (A) MRI, T2-weighted image. Note the hyperintense tumor. (B) CT scan, axial view. The tumor involves the petrous apex, sphenoid sinus, pterygopalatine fossa, infratemporal fossa, and maxillary sinus.

it exhibits a typical ground-glass-like appearance, while on MRI, it presents low signal on both T1 and T2 images. If asymptomatic, they do not require treatment, but follow-up is necessary due to the possibility of secondary osteosarcoma growth.

FORAMEN MAGNUM TUMORS

Foramen magnum (FM) lesions account for 1.1%–3.8% of all spinal cord and brain tumors [38]. The most common lesions arising in this space are meningiomas that have the same radiological features as anywhere along the skull base. Neurinomas, chordomas, and bone tumors can also arise in this region. These are usually under the care of neurosurgeons and are resected through a posterior approach. Lateral approaches with or without partial condyle resection, can also be employed.

Extra-Axial Lesions

PETROUS APEX LESIONS

The petrous apex can be afflicted by a series of lesions that cause mass effect or direct invasion of the neurovascular structures found here. These lesions can be primary or secondary, as metastases or as direct extensions from adjacent tumors. Primary pathologic processes include cholesterol granuloma, petrous bone cholesteatoma, petrous apex effusion, internal carotid artery aneurysm, and chondrosarcoma. Secondary lesions include chordoma, glomus tumors, neuromas, meningiomas, osteomyelitis of the temporal bone, and metastases (Table 5.2).

CHOLESTEROL GRANULOMA

Cholesterol granuloma is by far the commonest petrous apex lesion. These are expansile, slow-growing cystic formations containing cholesterol crystals surrounded by chronic inflammatory tissue. They can extend in multiple directions most often they evolve posterolaterally into the mastoid segment, the internal auditory canal, or the middle ear, causing auditory symptoms and/or headache. On otoscopy, it may appear as a blue tympanum. On CT it appears as a smoothly marginated expansile lesions isodense with the brain tissue and nonenhancing (Fig. 5.29). It is usually difficult to differentiate these lesions from petrous bone cholesteatoma on CT scan. However, on MRI, distinction is usually possible as cholesterol granulomas show a typical hyperintense signal on both T1 an T2 due to the content of blood-by products and proteinaceous debris. Small lesions are relatively uniform, but as they grow, they can be more heterogenous. Usually, they present a subtle peripheral enhancement on postcontrast images secondary to the marginal inflammatory tissue [39], but rarely they may show central enhancement if they have a more solid component. In cases of acquired petrous bone cholesteatoma, careful examination of the CT scan is needed since labyrinthine invasion may be evident even without preoperative sensorineural hearing loss and this might have some implications for the choice of the surgical approach. Cholesterol granulomas can be managed conservatively, but in case of large lesions involving essential neurovascular structures or reluctant symptomatology surgical drainage or resection is indicated [40].

TABLE 5.2 Differential diagnosis of petrous apex lesions.

	MRI			
Lesion	**T1**	**T2**	**Gd**	**Special features**
Cholesterol granuloma	↑	↑	No	Smooth expansile lesion
Petrous bone cholesteatoma	↓	↑	No	Smooth bone erosion
Chordoma	↓	↑	Yes	Heterogeneous
Petrous apicitis	↓	↑	Rim enhancement	Irregular bone erosion, history of otitis
Effusion	↓↑	↑	Mucosal enhancement	No bone erosion, preserved partitions between air cells. Pneumatized apex
Carotid aneurysm	↓	↑↓	Rim enhancement	Smooth bone erosion

T1, T1-weighted; *T2*, T2-weighted; *Gd*, gadolinium-enhanced; ↑, hyperintense; ↓, hypointense.

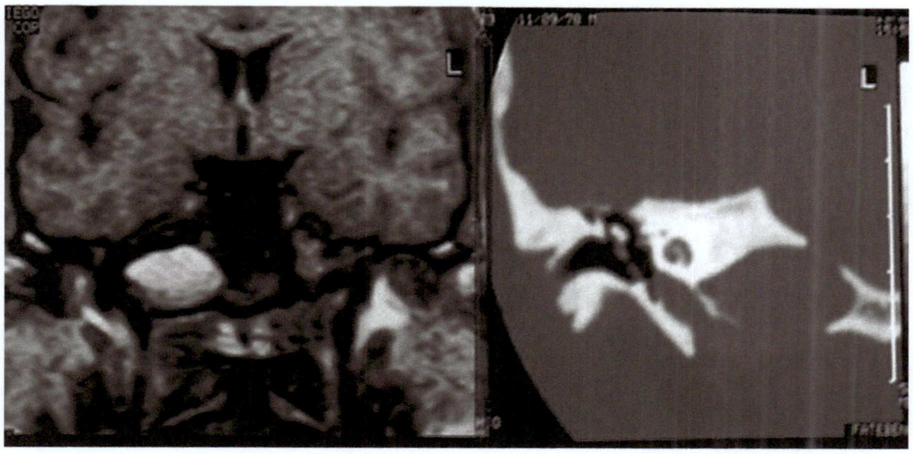

Fig. 5.29 A case of cholesterol granuloma of the right petrous apex. Preoperative imaging: *left*, T1-weighted MRI without gadolinium enhancement, coronal view, showing a large hyperintense mass; *right*, high-resolution CT scan, coronal view. Note the smooth bone erosion of the petrous apex caused by the lesion.

EFFUSION

Effusion within a pneumatized petrous apex is probably due to trapped fluid from previous otitis media. Opacified air cells show a high T2-weighted signal (Fig. 5.30A,B). On T1-weighted images, the signal is usually hypointense, with a rim of gadolinium enhancement reflecting surrounding mucosal hyperemia. It appears on CT scan as a soft-tissue density filling an air cell tract without causing bone destruction (Fig. 5.30). Some cases with high protein content will present with an intermediate or even high signal in T1-weighting, mimicking cholesterol granuloma. In these latter cases, the absence of bone destruction and the lack of an expansile lesion confirm the diagnosis.

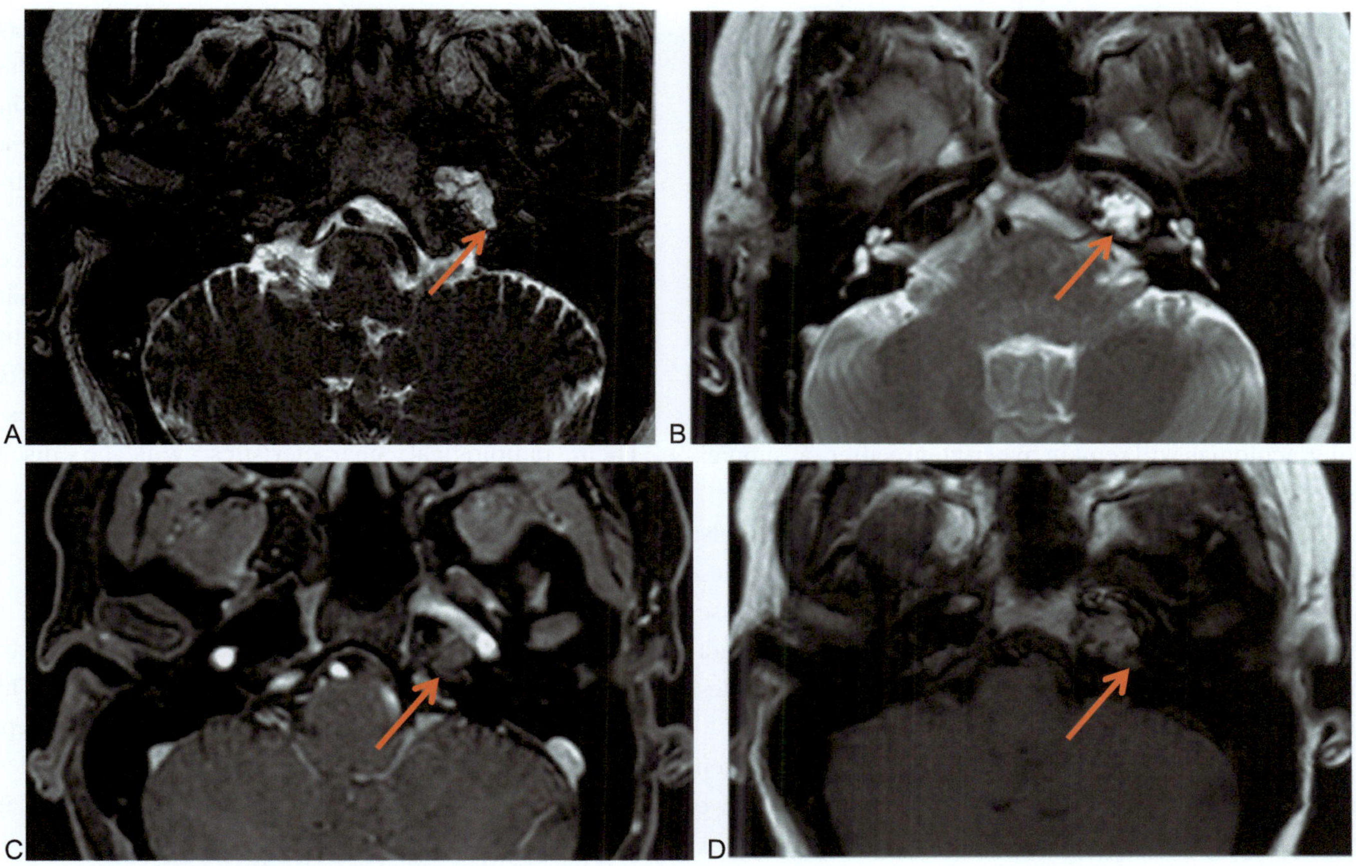

Fig. 5.30 MRI axial sequence showing an expanded left petrous apex lesion (*red arrow*) which on axial T2 weighted images (A,B) shows a high signal intensity. On T1-weighted sequence, this lesion was hypointense (C) and did not show any enhancement with contrast administration (D).

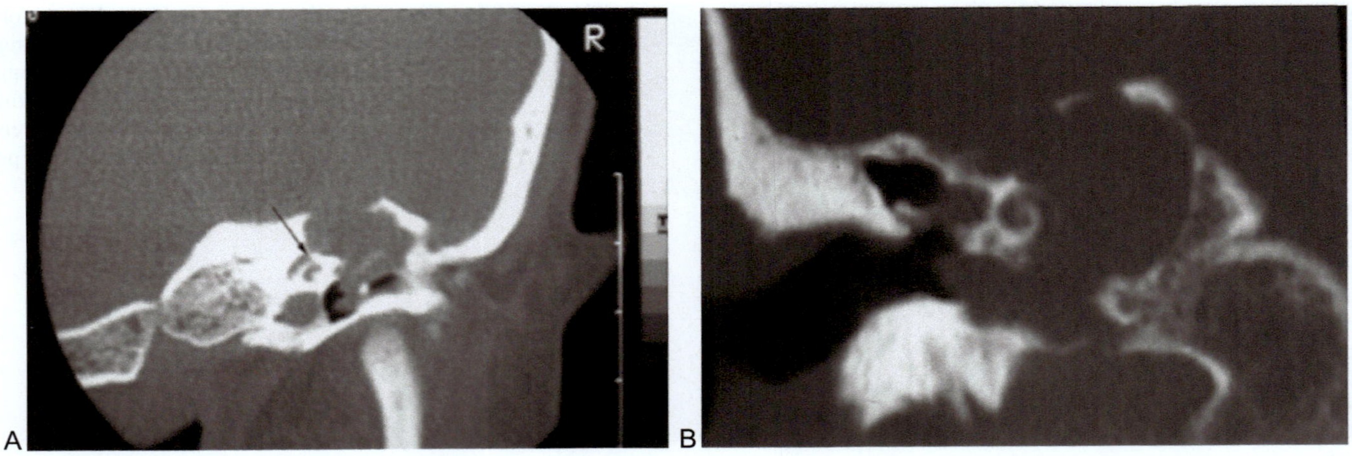

Fig. 5.31 CT scan coronal view of a supralabyrinthine petrous bone cholesteatoma (A) and a massive petrous bone cholesteatoma (B).

PETROUS BONE CHOLESTEATOMA (EPIDERMOID)

Petrous bone cholesteatomas are more often congenital than acquired, affecting obviously children and young adults. They arise from aberrant ectoderm that is trapped during embryogenesis [40]. On CT scan they are isodense expansile lesions. Typically, on MRI, cholesteatoma is hypointense on T1, hyperintense on T2, and does not enhance after gadolinium administration. They have intermediate signal on FLAIR images, but characteristic of these lesions are DWI sequences as they show restricted diffusion and appear bright. This feature is most important during postoperative surveillance, in the detection of recurrent cholesteatomas (Fig. 5.31).

Asymmetric petrous apex pneumatization can cause diagnosis problems due to the fact that the fatty bone marrow contained may appear hyperintense on T1-weighted images and may be mistaken for a lesion. Fat-suppressed imaging allows differentiation of the two entities, as cholesterol granulomas will remain hyperintense after fat suppression [13,17].

Clival chordomas and condrosarcomas can also extend laterally to involve the petrous apex. On CT, these lesions appear as destructive lobulated soft-tissue masses that typically enhance. Although much overlap exists, chordomas are usually located near the midline and enhance less with gadolinium when compared to chondrosarcomas.

Neuromas involving the petrous apex are usually extensions from other areas, e.g., trigeminal neuromas and lower cranial nerve neuromas. Infrequently, facial nerve neuromas can be extensive and involve the petrous apex. They typically show contrast enhancement, a fact that distinguishes them from petrous bone cholesteatomas.

Paragangliomas can also involve the petrous apex by secondary extension and will be considered with tumors of the jugular foramen.

Meningiomas of the petrous bone usually represent an extension from an intradural lesion. More rarely, isolated meningioma of the temporal bone can be found and the preoperative diagnosis cannot be usually obtained. Once a meningioma of the temporal bone has been diagnosed, MRI is usually needed to verify the presence or absence of any intradural location of the lesion.

Hematogenous metastasis—As part of the skeletal system, the skull base is susceptible to hematogenously spread metastatic disease. Within the temporal bone, the petrous apex is the most commonly affected site and is involved in 80% of reported cases [5]. Their radiological characteristics will be discussed further on.

CEREBELLOPONTINE ANGLE TUMORS

Vestibular schwannoma (acoustic neurinoma) constitutes 75%–90% of cerebellopontine angle tumors. This is followed in order of frequency by meningioma, epidermoids, and nonacoustic neurinomas. The main lesions occurring in the cerebellopontine angle are summarized in Table 5.3.

TABLE 5.3 Radiological differential diagnosis of extra-axial cerebellopontine angle lesions.

Lesion	MRI				
	T1	T2	Gd	Contents	Special features
Acoustic neurinoma	↓→	→↑	Yes	Many contain microcysts and macrocysts	IAC involved, acute bone-tumor angle
Meningioma	↓→	↑↓	Yes		Dural tail. Obtuse bone-tumor angle
Epidermoid cyst	↑	↓	No	Heterogeneous	Encases neurovascular structures. Irregular surface
Arachnoid cyst	↑	↓	No	Homogeneous	Isointense to CSF. Displaces neurovascular structures. Smooth surface
Lipoma	↑	↑	No	Homogeneous	Decreased intensity with fat suppression
Nonacoustic neurinomas	↓→	→↑	Yes		Follows the anatomical course of the nerve
Metastasis	↓	↑	Possible		Brain edema
Aneurysm	↓	↓	Possible (in thrombosed aneurysm)		Well circumscribed

T1, T1-weighted; *T2*, T2-weighted; *Gd*, gadolinium-enhanced; ↑, hyperintense; ↓, hypointense; →, isointense; *IAC*, internal auditory canal; *CSF*, cerebrospinal fluid.

ACOUSTIC NEURINOMAS

Acoustic neurinomas (AN) or vestibular schwannomas (VS) are benign encapsulated tumors arising from the nerve sheaths of the vestibular nerve. They occur as proliferations of Schwann cells. The inferior vestibular nerve has been reported to be the most frequent site of origin (up to 91%), and the superior vestibular nerve accounts for the rest. The mean tumoral growth rate of acoustic neurinomas is thought to be between 0.9 and 1.2 mm/year. Further growth can occur only by compressing or displacing the brainstem. When the tumor reaches about 4 cm, it often extends far anteriorly and compresses the trigeminal and abducent nerves. It can also extend inferiorly to the jugular foramen thereby deforming the lower cranial nerves.

Acoustic neurinomas appear isodense to hypodense on non-enhanced computed tomography (CT) and enhance homogeneously after injection of contrast agent (Fig. 5.32). Sometimes the tumor growth produces a progressive enlargement of the IAC, detected by the bone algorithm CT (Fig. 5.33). However, CT is not ordered routinely because it is inferior to MRI in its resolution of soft tissue and can miss tumors up to 1 cm into the CPA. At present, it is only indicated for patients who cannot undergo an MRI scan.

MRI is the imaging study of choice to diagnose an acoustic neurinoma, on CT scan they are isodense with brain tissue but can be identified after contrast uptake. On MRI, acoustic neurinomas, like all neurinomas, are shown as

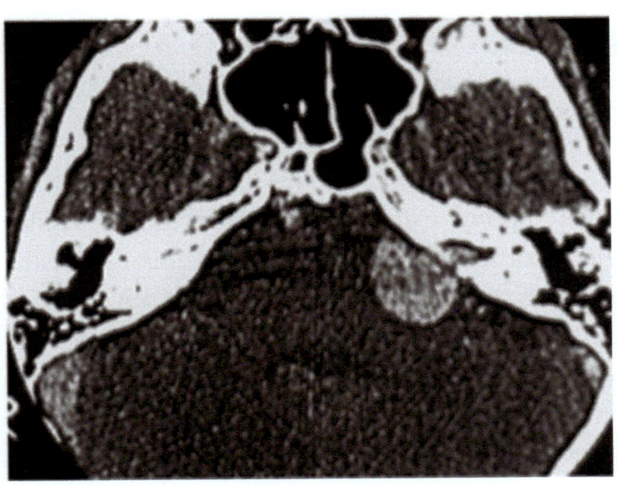

Fig. 5.32 Soft-tissue algorithm CT after contrast medium infusion, showing a 2-cm vestibular schwannoma on the left side.

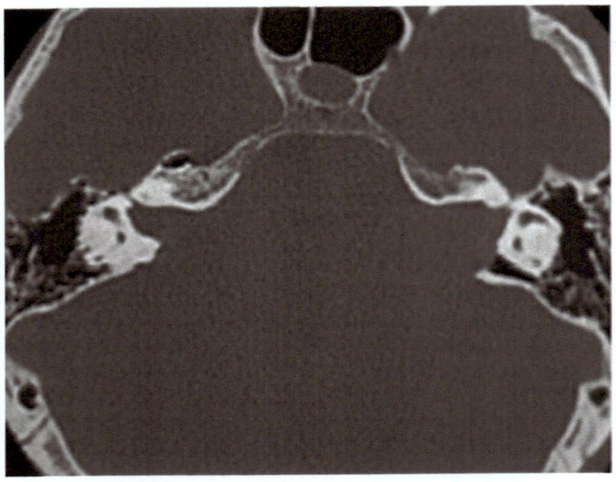

Fig. 5.33 Bone algorithm CT scan showing a typical erosion of the *left* internal auditory canal. The size of the internal auditory canal can be compared with the normal side.

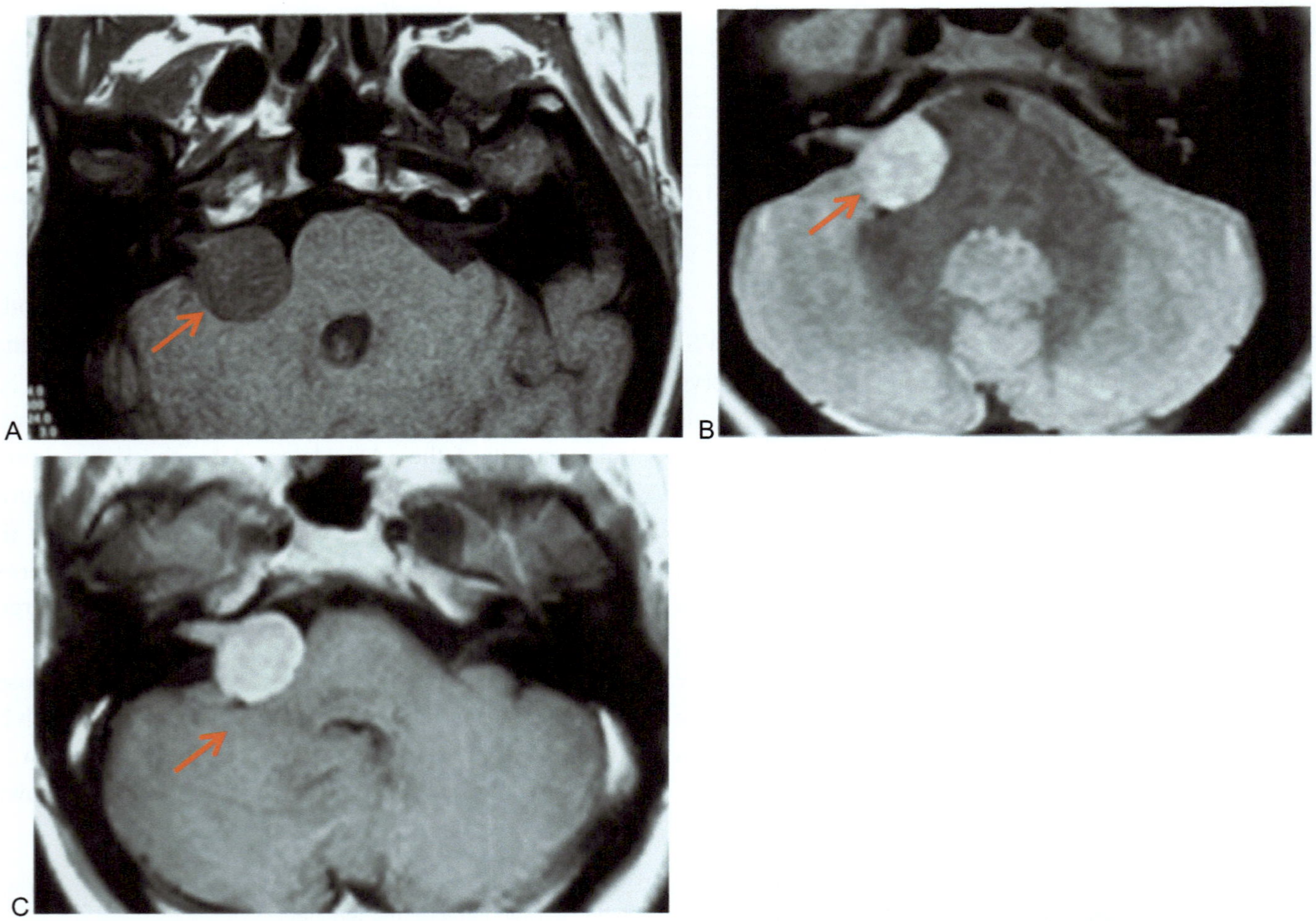

Fig. 5.34 MRI axial image of a right-sided acoustic neurinoma showing as a lesion occupying the right CPA and IAC (*red arrow*) which is hypointense in T1-weighted image (A), hyperintense in T2W (B) and enhancing after Gd administration in T1W (C).

isointense to hypointense lesion on T1W (Fig. 5.34A) and as hyperintense on T2W imaging (Fig. 5.34B). The tumor enhances intensely after infusion of gadolinium contrast (Fig. 5.34C) making possible even the visualization on lesions up to 2–3mm in size (Fig. 5.35) that could be missed otherwise. Axial and coronal (Fig. 5.36) sequences are used in current practice, while sagittal sequences are not routinely of any additional help. They usually display the shape of an ice cream cone, with the intracanalicular part appearing as the cone and the round CPA piece as the cream. As mentioned above, high-resolution T2W CISS sequences have added an additional tool to acoustic neurinoma diagnosis and characterization, as the high contrast between the cerebrospinal fluid and the other

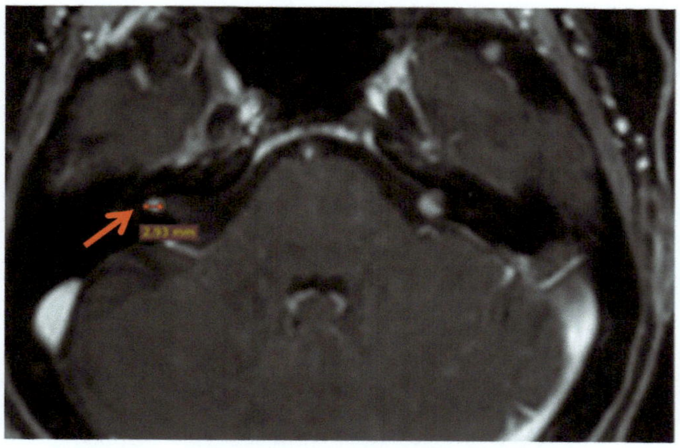

Fig. 5.35 T1W with Gd-enhancement axial MRI image of a right-sided intracanalicular acoustic neurinoma of 2.93 mm diameter (*red arrow*).

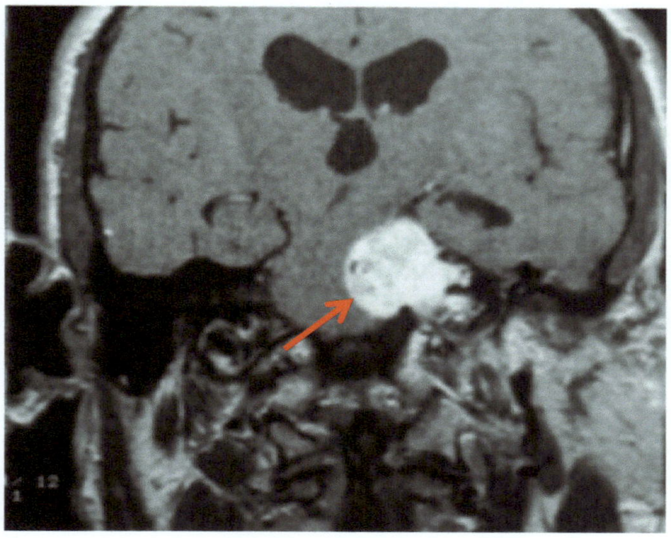

Fig. 5.36 T1W with Gd-enhancement coronal MRI image of a large left-sided acoustic neurinoma (*red arrow*) occupying the CPA. Note the displacement of the brainstem.

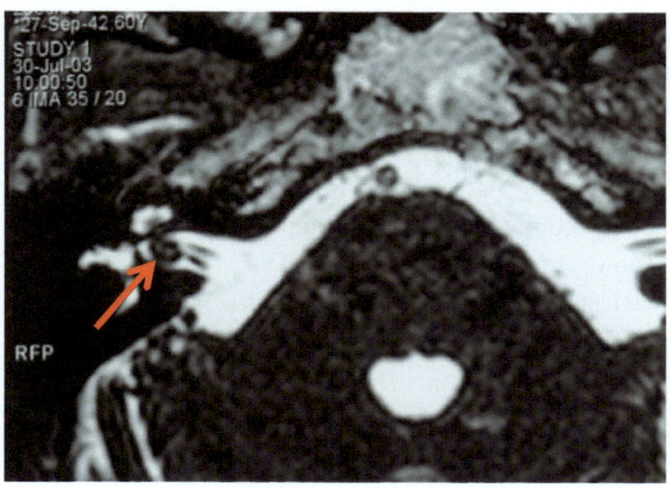

Fig. 5.37 High-resolutionT2-weighted sequence showing a small left-sided acoustic neuroma (*red arrow*) arising toward the fundus of the IAC.

structures provides an excellent view of the nerves and the other structures of the cerebellopontine angle (CPA) (Fig. 5.37). Intralesional hemorrhage (Fig. 5.38A) and cysts (Fig. 5.38B–D) determine inhomogeneous enhancement of the schwannoma and are associated with a more rapid evolution and higher degree of hearing loss [41]. They have hyperintense T2 and variable T1 signal. Cystic lesions comprise around 10% of VSs and respond poorly to radiotherapy [42]. Cysts may be particularly important when deciding on total or subtotal tumor removal, particularly if they are peripherally located and present thin walls (Fig. 5.38c,d)

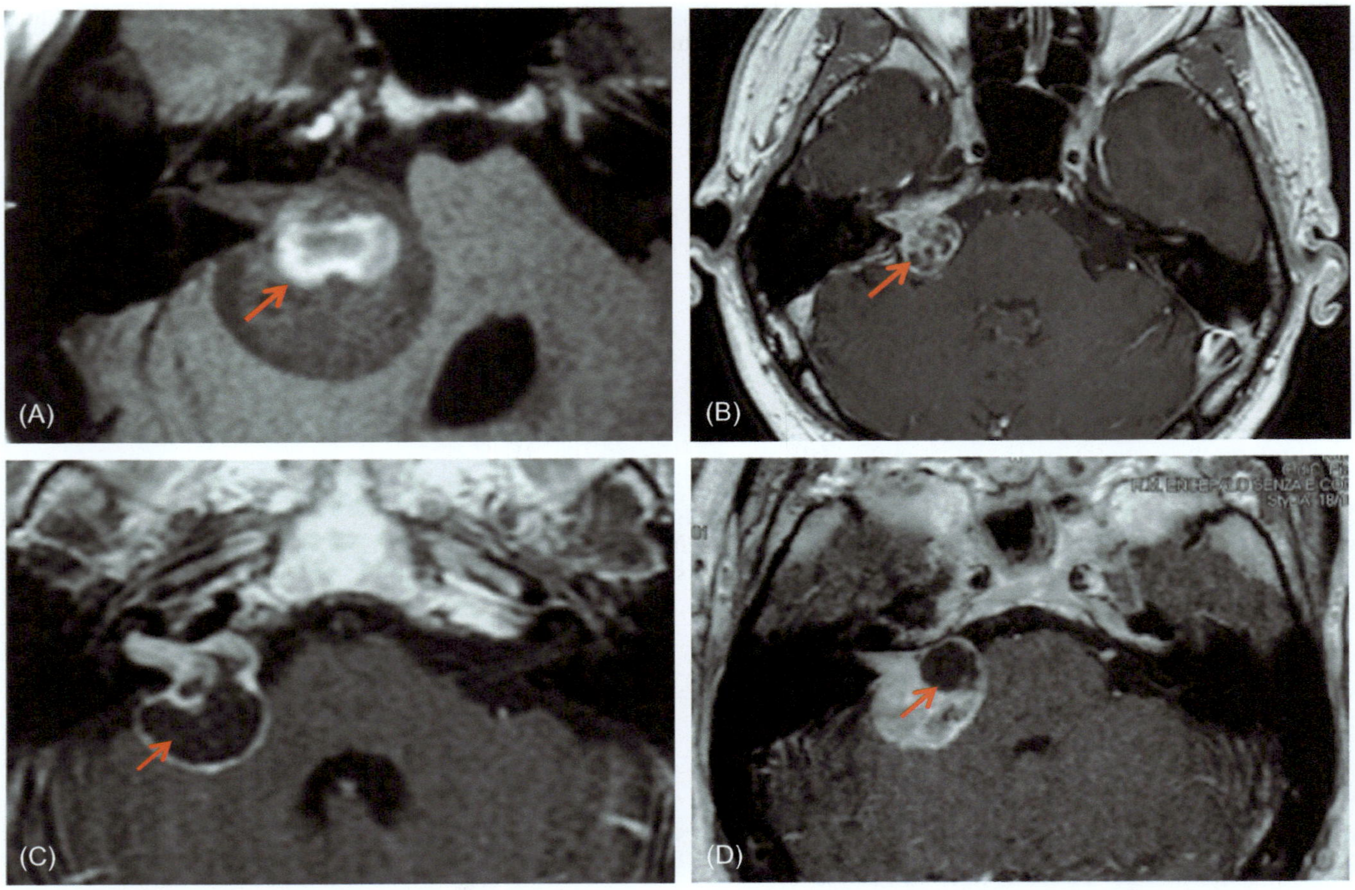

Fig. 5.38 (A) T1W axial MRI image of a right-sided AN with intratumoral hemorrhage (*red arrow*). (B) T1W axial image with contrast of a right-sided polycystic AN (*red arrow*). (C and D) T1W axial image with contrast of a right-sided AN with a small anterior peripheric cyst (*red arrow*) (C) and a mainly cystic acoustic neurinoma (*red arrow*) (D).

Although new imaging techniques allow the volume of the tumor to be assessed, the current standard is to grade tumors according to the largest extrameatal diameter. Intrameatal tumors are designated as size 0 and are classified into:

- Grade 0: intracanalicular (Fig. 5.39A).
- Grade 1: less than 1 cm in greatest diameter (Fig. 5.39B).
- Grade 2: between 1 and 2 cm in greatest diameter (Fig. 5.39C).
- Grade 3: between 2 and 3 cm in greatest diameter (Fig. 5.39D).
- Grade 4: between 3 and 4 cm in greatest diameter (Fig. 5.39E).
- Grade 5: more than 4 cm in greatest diameter (Fig. 5.39F).

For small VS it is important to evaluate the invasion of the fundus of the IAC especially when planning a hearing conservation procedure, because the rate of success seems to be much higher when the fundus is free (Fig. 5.39A). In rare cases tumors may extend into the cochlea or vestibule making hearing preservation impossible (Fig. 5.40). This situation requires a transotic approach for total tumor resection.

MRI also provides important additional information in regard to tumor surroundings and anatomy, brain stem or cerebellar compression or identation, peritumoral edema (Fig. 5.41), hydrocephalus, and tonsillar herniation [43], which may be significant when deciding the most appropriate treatment modality.

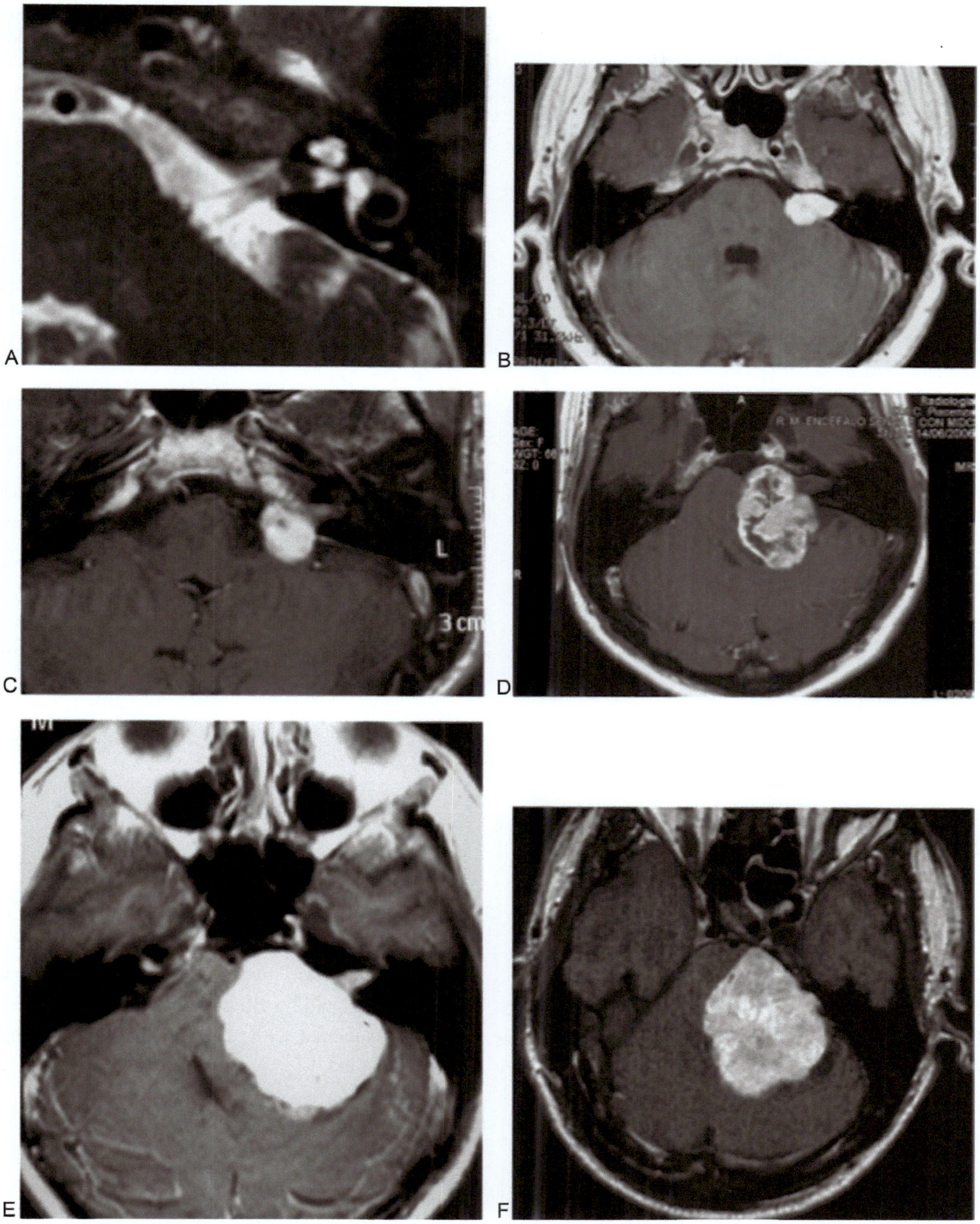

Fig. 5.39 (A–F) Grading of acoustic neurinoma. (A) Grade 0 (intracanalicular neurinoma). (B) Grade 1 (<1 cm). (C) Grade 2 (between 1 and 2 cm). (D) Grade 3 (between 2 and 3 cm). (E) Grade 4 (between 3 and 4 cm). (F) Grade 5 (>4 cm).

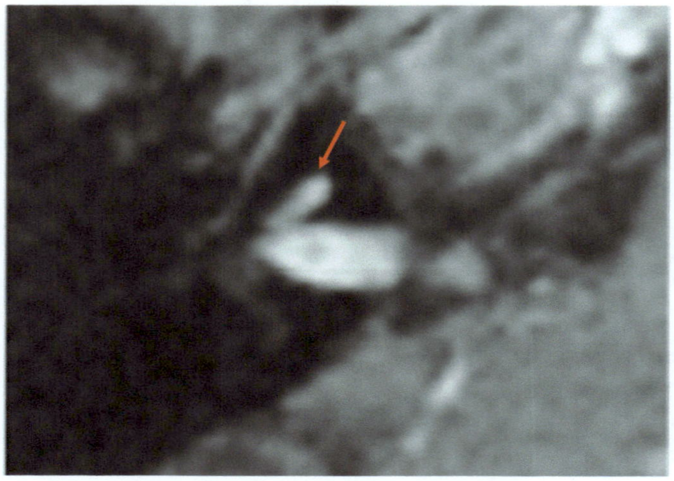

Fig. 5.40 Axial T1W MRI sequence with Gd-enhancement showing a right-sided VS with invasion of the inner ear limited to the basal turn of the cochlea (*red arrow*).

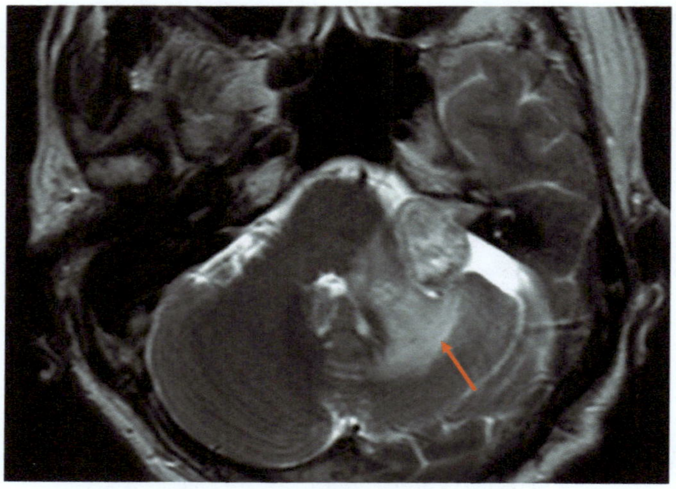

Fig. 5.41 Axial T2W MRI sequence showing a left-sided VS with peritumoral edema (*red arrow*).

NEUROFIBROMATOSIS TYPE 2

Neurofibromatosis type 2 is a genetic disease that is characterized mainly by the presence of bilateral vestibular schwannomas. The tumor's size may include every combination, from two very small lesions to two giant tumors (Fig. 5.42). Bilateral schwannomas may or may not be associated with additional lesions both intracranial or in the spine.

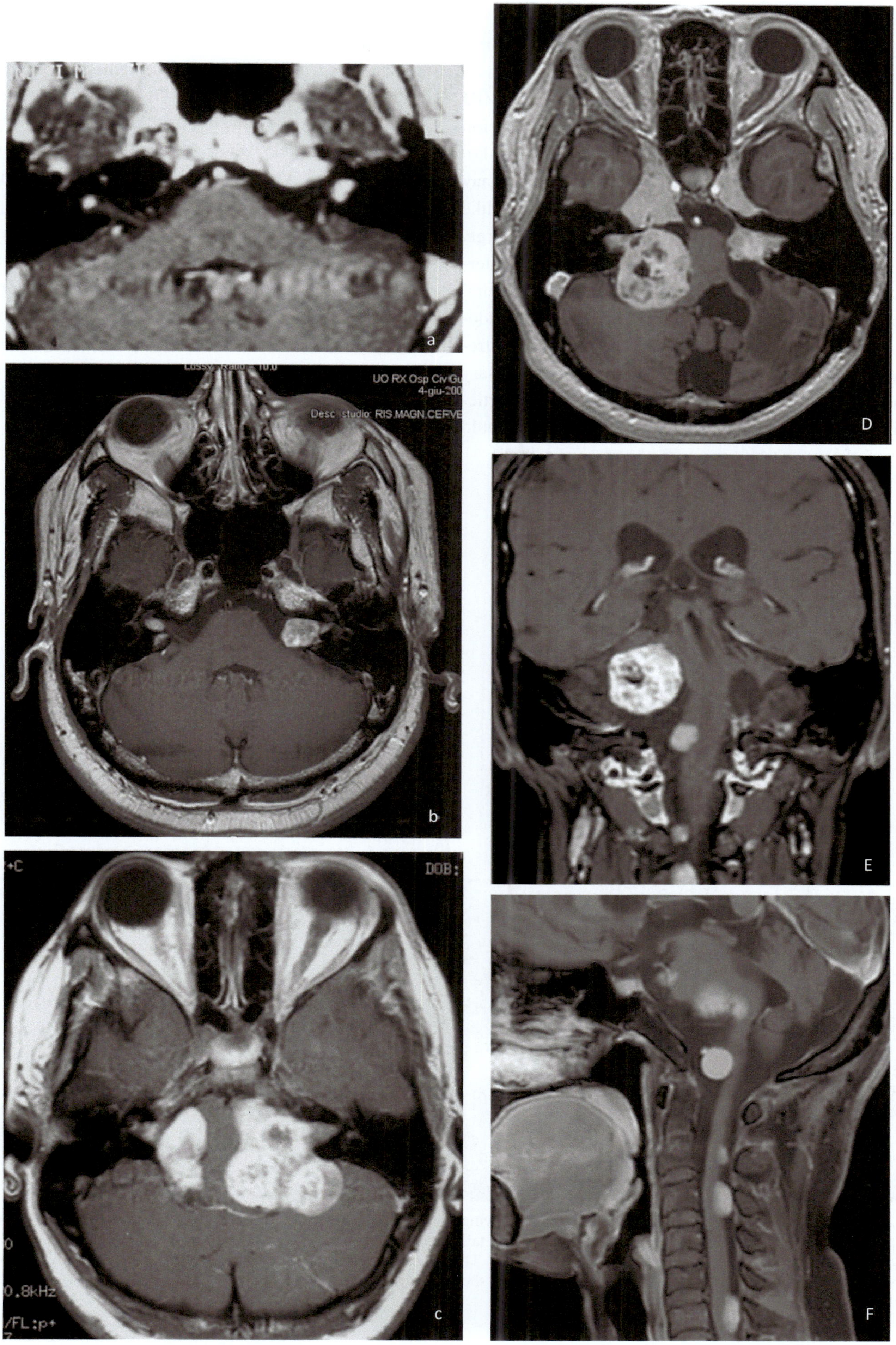

Fig. 5.42 Neurofibromatosis type 2. (A) Bilateral intracanalicular vestibular schwannomas. (B) Intracanalicular right-sided vestibular schwannoma associated with a 1.5-cm tumor on the contralateral side. (C) Medium right-sided vestibular schwannoma associated with a huge tumor on the contralateral side. (D) Bilateral vestibular schwannomas associated with sphenoclival meningiomas involving both internal carotid arteries. (E) The posterior fossa tumors are associated with multiple intramedullary lesions (coronal view). (F) The posterior fossa tumors are associated with multiple intramedullary lesions (sagittal view).

MENINGIOMAS

Meningiomas represent the main pathology to be discriminated from VS in the CPA, this is achievable in most cases based on the features present on MR images. Classic MRI appearances are those of a T1 and T2 isointense lesion with avid uniform contrast enhancement (Fig. 5.43A–D). Depending on their size and location, these lesions may show for years an unexpected lack of growth [17]. Unlike acoustic neurinomas, the majority of meningiomas are large, sessile with a wide dural base, and lie eccentric to the internal auditory canal (Fig. 5.43). An enhancement of a dural tail adjacent to the tumor bulk after administration of gadolinium contrast agent is a characteristic sign that is frequently found in meningiomas, but this is not a necessary or pathognomonic feature [17]. Dural enhancement, particularly if irregular, might indicate tumor invasion and should be resected. In our experience, a broad dural attachment has been the most constant distinguishing finding in meningiomas. Very infrequently, in large en-globe tumors, this finding is less evident (Fig. 5.44).

Extension of the tumor into the internal auditory canal does not exclude the presence of a meningioma (Fig. 5.43A,C). Very rarely, intracanalicular meningiomas can be found (Fig. 5.43D) and they cannot be distinguished from a VS prior surgery or histopathologic examination. Posterior petrous surface meningiomas can be classified according to their position in relation to the internal auditory canal using the Desgeorges classification (A anterior to the acoustic meatus—Fig. 5.46, M centered on the auditory meatus, P posterior to the internal auditory canal-Fig. 5.45, AM both anterior and centered on the meatus, MP centered on the meatus and posterior, AMP A, M, and P combined) [56].

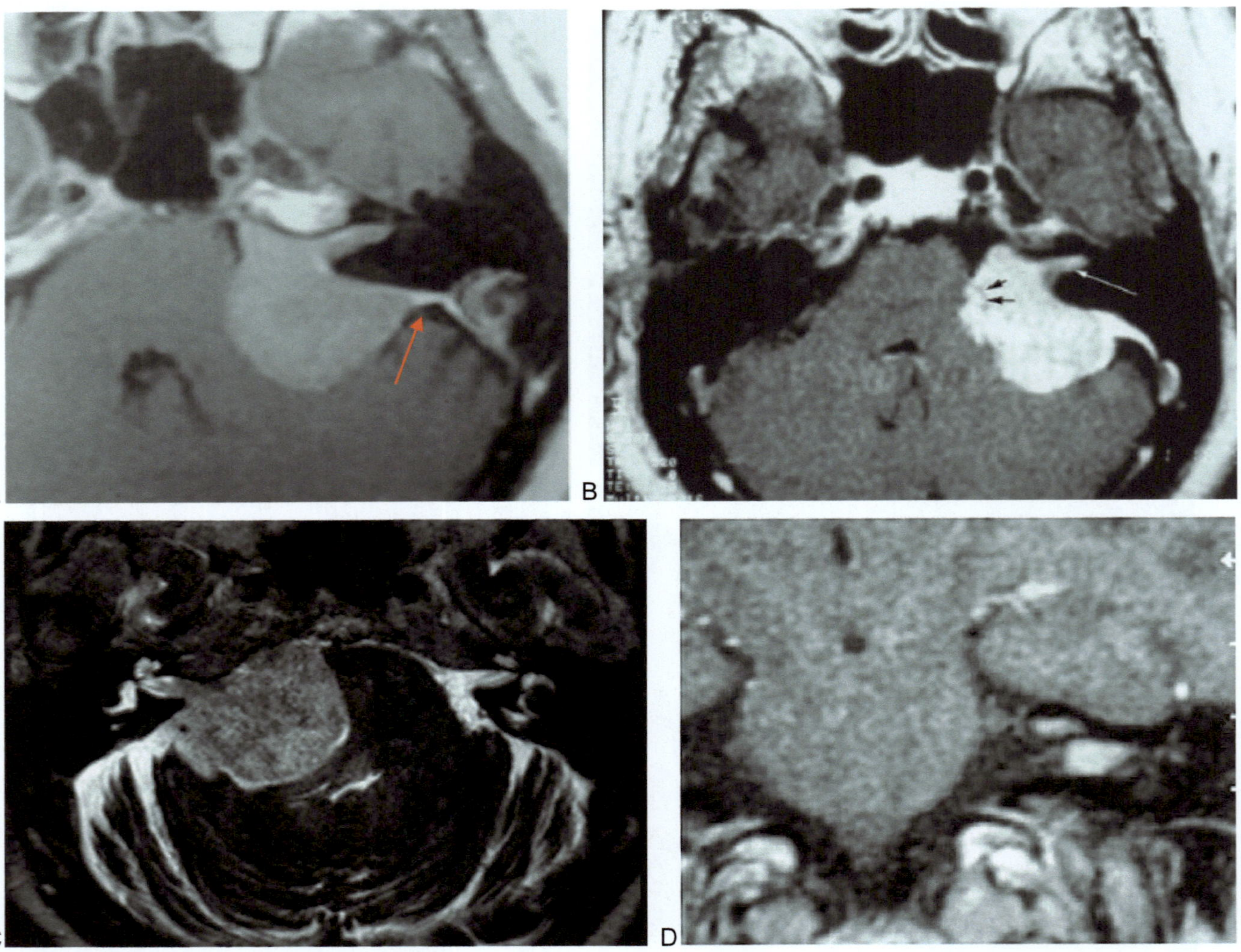

A B C D

Fig. 5.43 (A–D) Meningiomas. (A) T1W axial MRI image showing a left posterior fossa meningioma involving the entire posterior surface of the temporal bone (class AMP); note the broad base and "dural tail" (*red arrow*). (B) T1W MRI with gadolinium enhancement; axial view showing a left posterior petrous surface meningioma, class AMP. The tumor extends into the internal auditory canal (*white arrow*) and causes compression of the brainstem (*black arrows*). (C) T2W axial MRI showing a right-sided meningioma (D) T1W axial MRI with contrast showing a left intracanalicular meningioma.

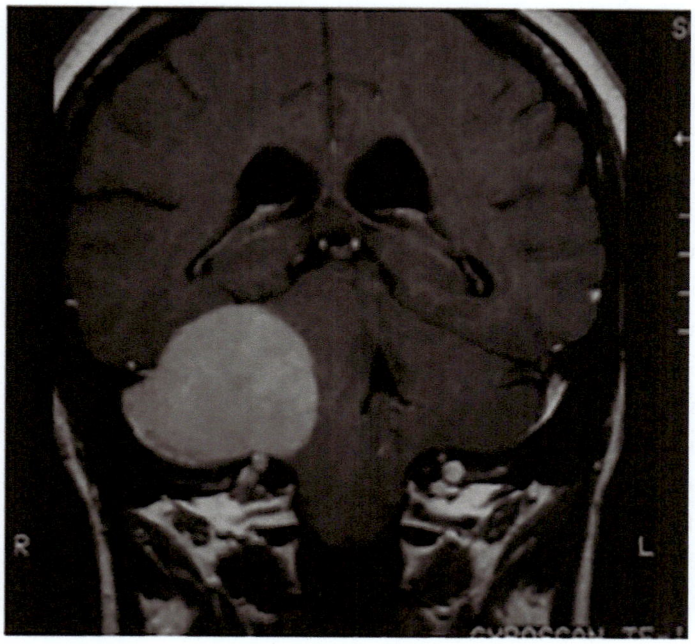

Fig. 5.44 T1W coronal MRI with contrast showing a large meningioma compressing the cerebellum.

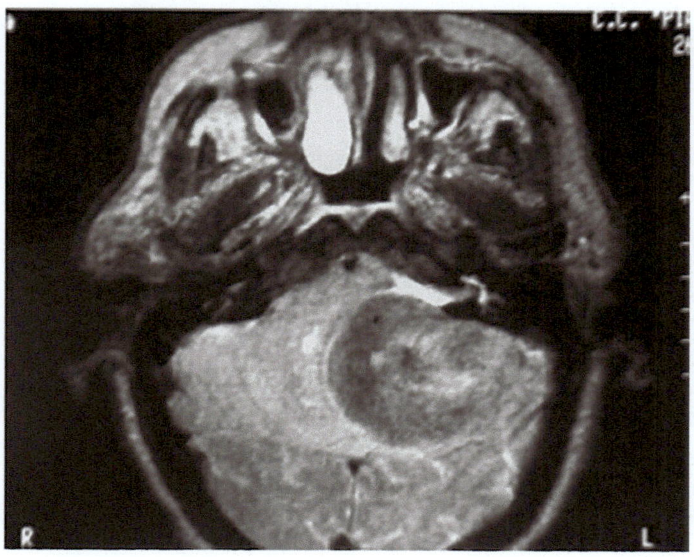

Fig. 5.45 T2-weighted MRI, axial view, showing a huge left meningioma lying posterior to the internal auditory canal.

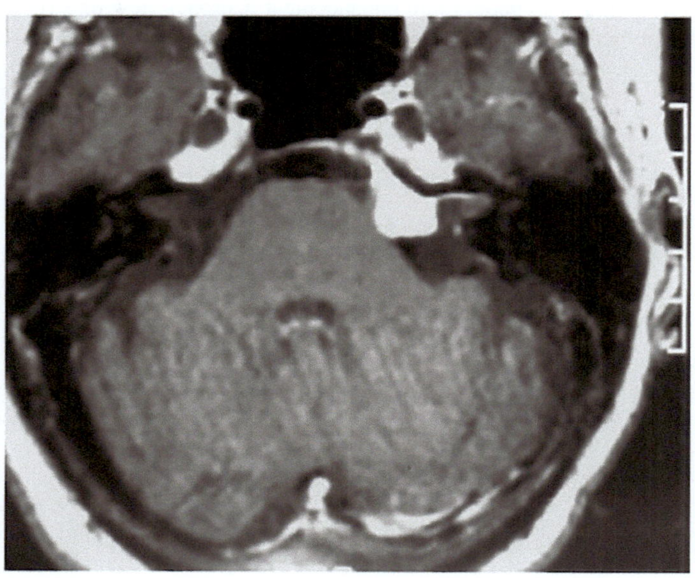

Fig. 5.46 MRI with gadolinium enhancement; axial view. A small meningioma lying anterior to the internal auditory canal can be seen.

Petroclival meningiomas are those whose dural attachment is anterior to the brainstem. These lesions frequently have supratentorial extension, involve Meckel's cave, and/or involve the cavernous sinus (Figs. 5.47-5.49).

Invasion of the bone of the skull base by the tumor is a frequent finding and is better visualized on CT scan (Figs. 5.50,5.51). Proper evaluation of the CT scan is therefore of paramount importance as total tumor removal necessitates removal of the involved bone. Hyperostosis or reactional bone thickening is also common and is frequently seen in en plaque meningiomas (Fig. 5.52).

Skull base meningiomas tend to adhere to the surrounding neurovascular structures with no intervening arachnoid planes. While the adherence to the nerves cannot be visualized radiologically, vessel involvement can be better visualized on MRI. Encasement of large arteries (internal carotid artery, basilar, vertebral) and or medium-sized arteries (superior cerebellar artery, posterior inferior cerebellar artery) is well demonstrated on MRI (Figs. 5.53, 5.54). However, the involvement of the anterior inferior cerebellar artery is more difficult to visualize. An important aspect to study is the arachnoid plane between the tumor and the related brainstem and cerebellum. A good plane of cleavage is demonstrated by thin line of CSF intensity between the tumor and the brain tissue (Fig. 5.55). In other cases, this line can be hypointense both in T1- and T2-weighted images. Meningiomas that adhere to the surrounding brain tissue usually present perilesional edema; which is hyperintense on T2-weighted images and hypodense on CT (Figs. 5.56). These cases usually show an altered plane of cleavage and are very difficult to remove totally without causing damage to the adherent brain tissue.

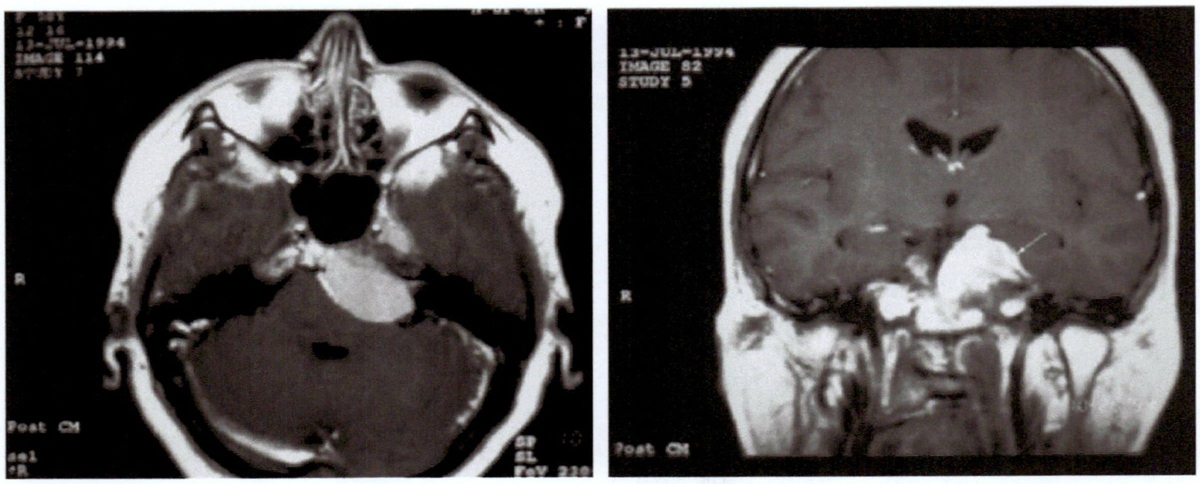

Fig. 5.47 (A) MRI with gadolinium enhancement; axial view. A large left petroclival meningioma can be seen. Note the extension into the prepontine cistern. The tumor lies in front of the brainstem and crosses the midline. (B) MRI with gadolinium enhancement; coronal view of the same patient. The tumor extension into Meckel's cave can be seen (*arrow*).

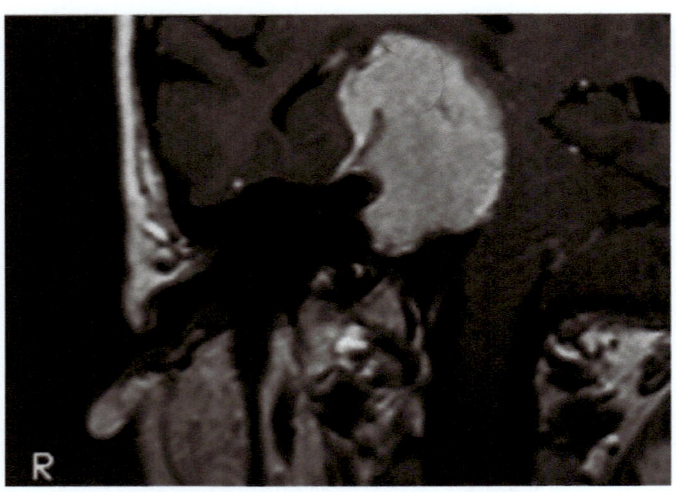

Fig. 5.48 MRI with gadolinium enhancement. Coronal view of a right petroclival meningioma showing supratentorial extension.

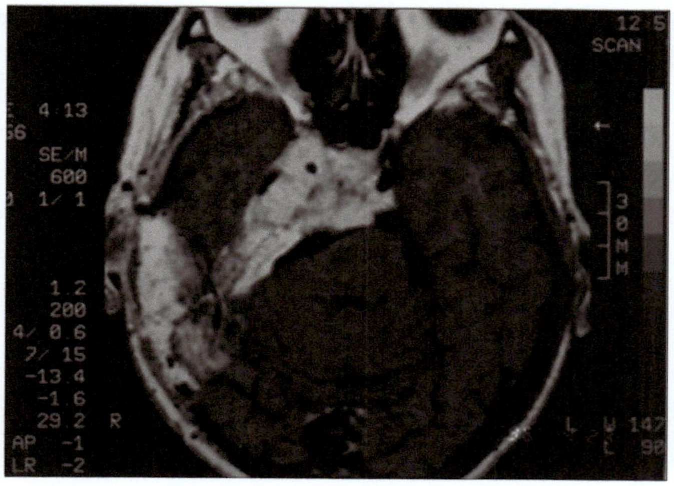

Fig. 5.49 MRI with gadolinium enhancement in a right petroclival meningioma with supratentorial extension. Note the involvement of the cavernous sinus and the encasement of the internal carotid artery.

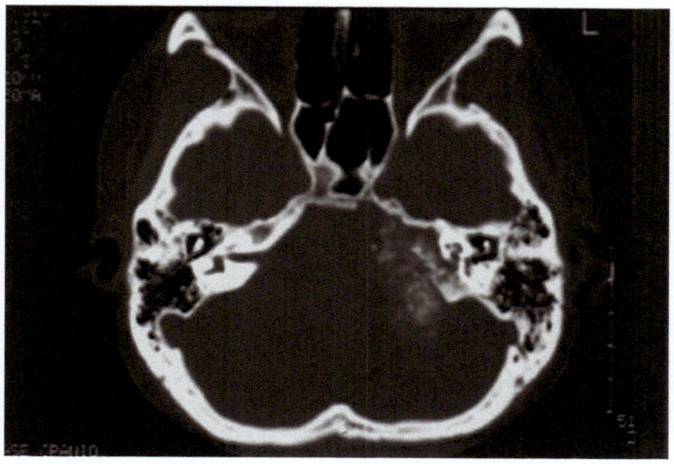

Fig. 5.50 CT scan of a meningioma involving the left posterior surface of the temporal bone. Note the massive erosion and involvement of the temporal bone. Intratumoral calcification can be seen.

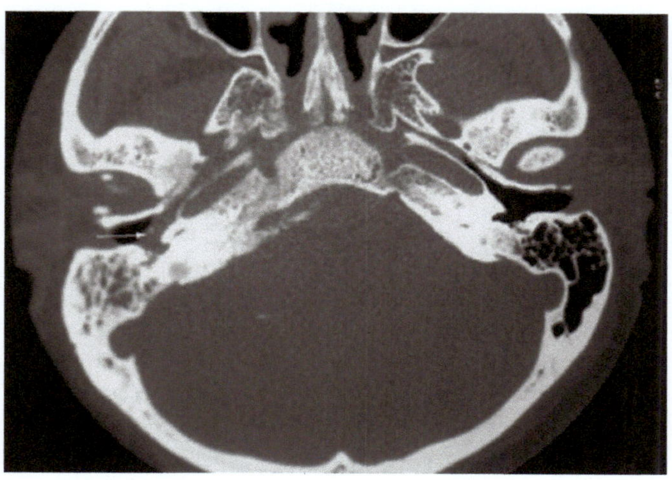

Fig. 5.51 CT scan of a right-sided meningioma of the posterior surface of the temporal bone. Bone invasion (*arrow*) denotes middle ear involvement.

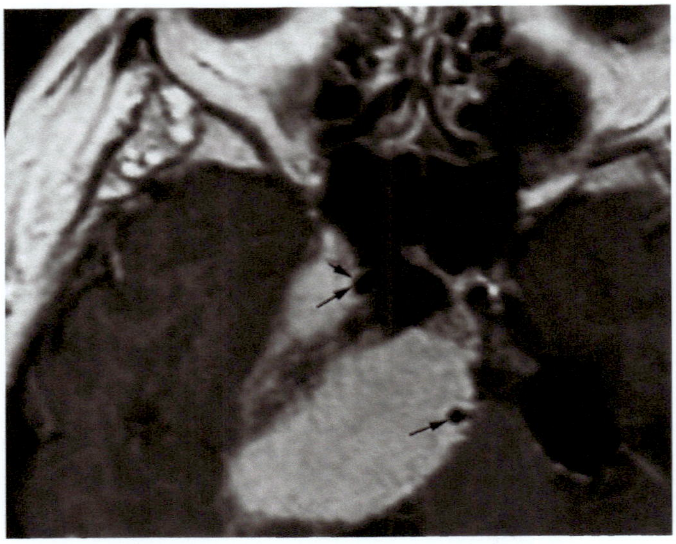

Fig. 5.52 A petroclival meningioma involving the basilar and internal carotid arteries (*arrows*).

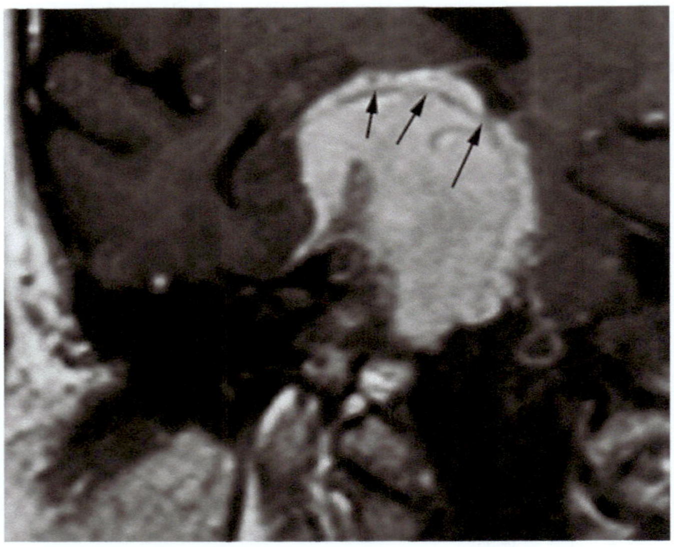

Fig. 5.53 The superior cerebellar artery (*arrows*) can be seen within the meningioma.

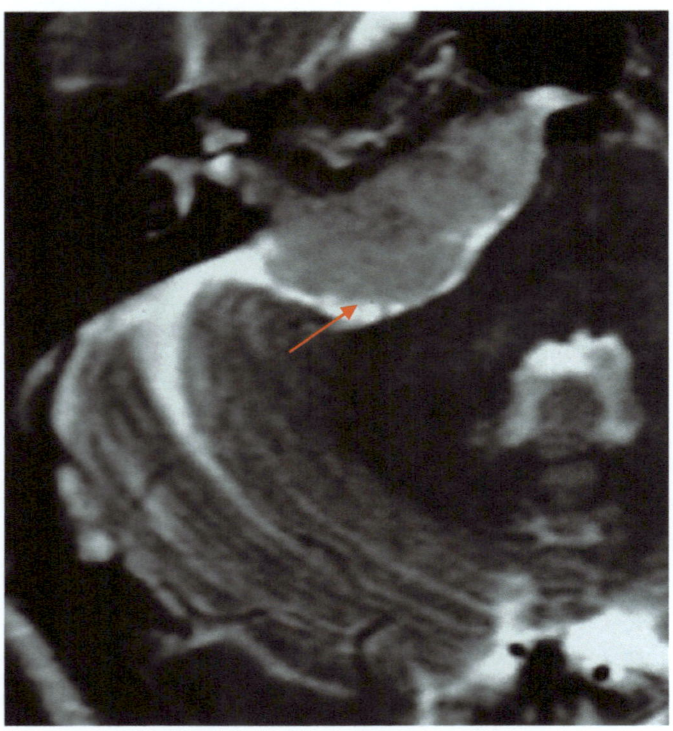

Fig. 5.54 A perfect dissection plane, as there is a rim of cerebrospinal fluid between the tumor and the brainstem (*arrow*).

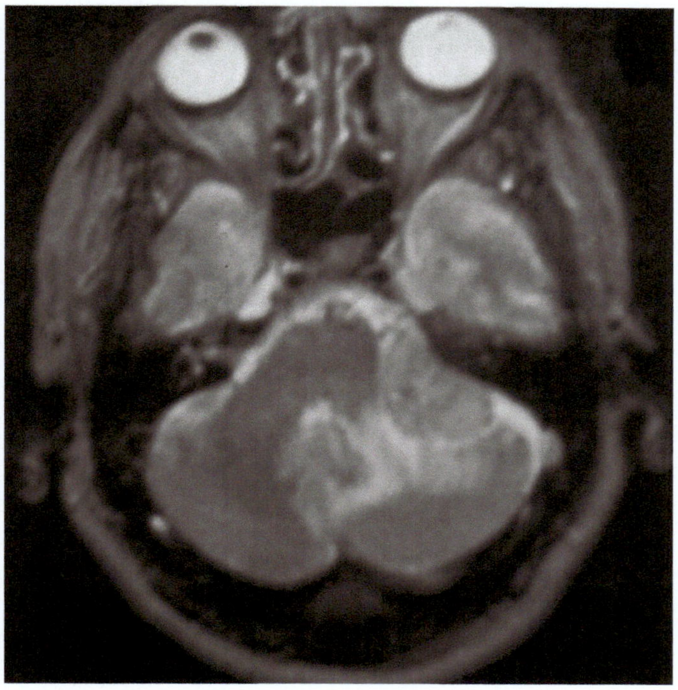

Fig. 5.55 Absence of a dissection plane between the meningioma and the cerebellum, in addition to the edema of the cerebellum, predicts a poor dissection plane between the tumor and brain tissue.

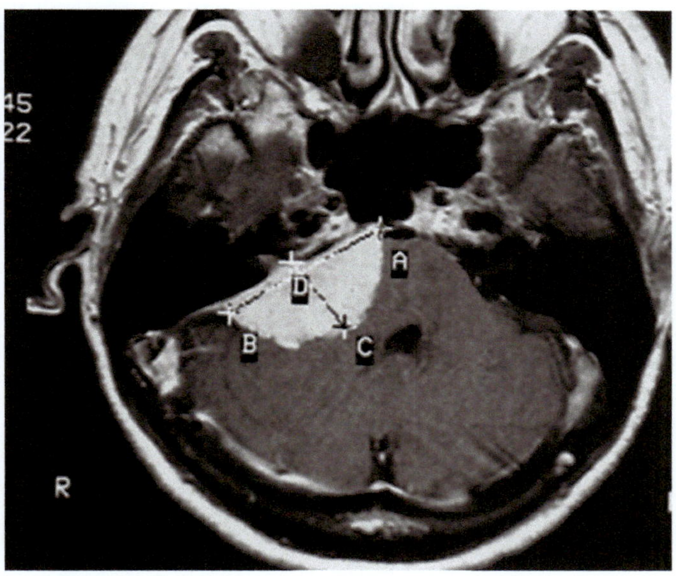

Fig. 5.56 No dissection plane between the meningioma (class AMP) and the brainstem.

EPIDERMOIDS

Epidermoids are characterized by slow growth. They tend to expand and spread along normal cleavage planes and thus they usually occupy more than one intracranial compartment. They have irregular "scalloped" margins. Epidermoids also tend to envelop vital neurovascular structures, making total excision a difficult task (Fig. 5.57). Epidermoids may be mistaken for arachnoid cysts. Both lesions are nonenhancing and are hypointense on T1-weighted and hyperintense on T2-weighted images. While arachnoid cysts show signal intensity similar to CSF, epidermoids have a slightly hyperintense segment when compared to CSF.

Unlike epidermoids, **arachnoid cysts** are homogeneous and have smooth surfaces. They also tend to displace the neurovascular structures rather than surround them (Fig. 5.58). Suppression of CSF signal allows for distinction between epidermoid cysts (*bright*) and arachnoid cysts (*dark*).

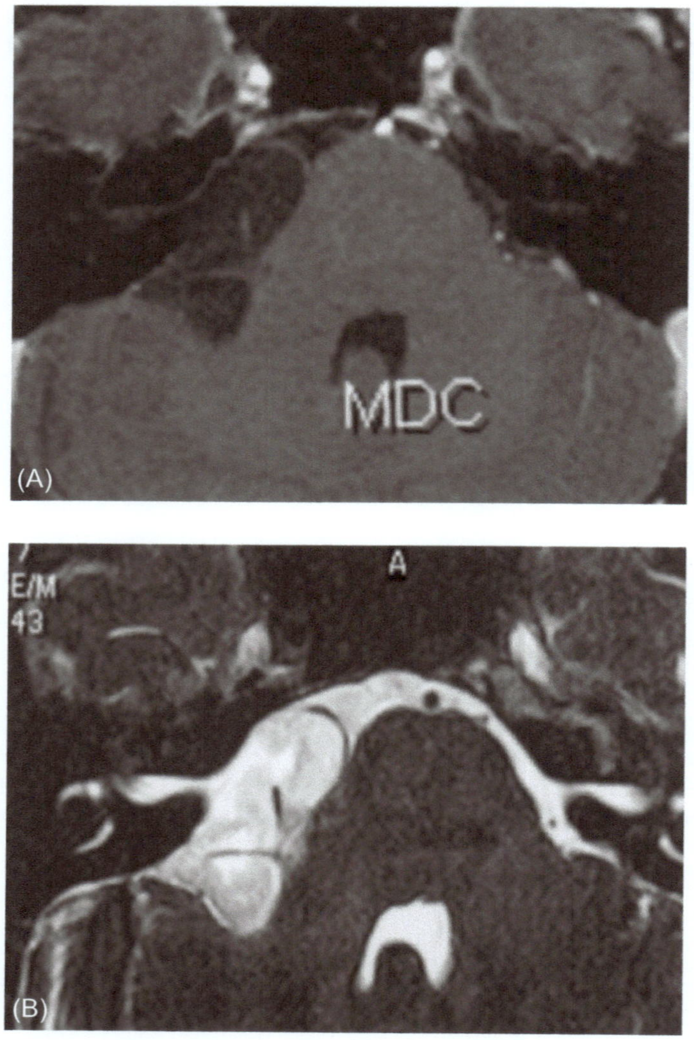

Fig. 5.57 Axial MRI sequence of right-sided epidermoid which appears hypointense in T1 (A) and hyperintense in T2 (B).

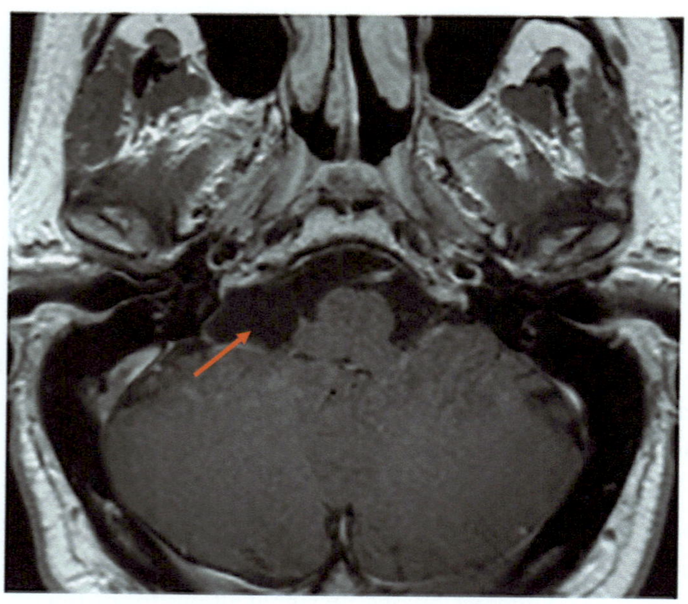

Fig. 5.58 Axial T1 fat-suppressed postcontrast image demonstrates a left CPA arachnoid cyst (*red arrow*) extending into the prepontine cistern, note there is no enhancement of the CSF-signal-intensity mass.

Facial nerve schwannomas: Facial nerve schwannomas entirely located in the CPA and IAC are impossible to differentiate from VSs. In most cases, facial nerve schwannomas show an involvement of the labyrinthine portion as well as the geniculate ganglion (Figs. 5.59, 5.60). Enlargement of the labyrinthine portion of the fallopian canal is confirmed by CT scan.

Trigeminal neurinomas are quite uncommon. Rarely, an isolated posterior fossa trigeminal neuroma can be seen and can be mistaken for an acoustic neuroma. However, the

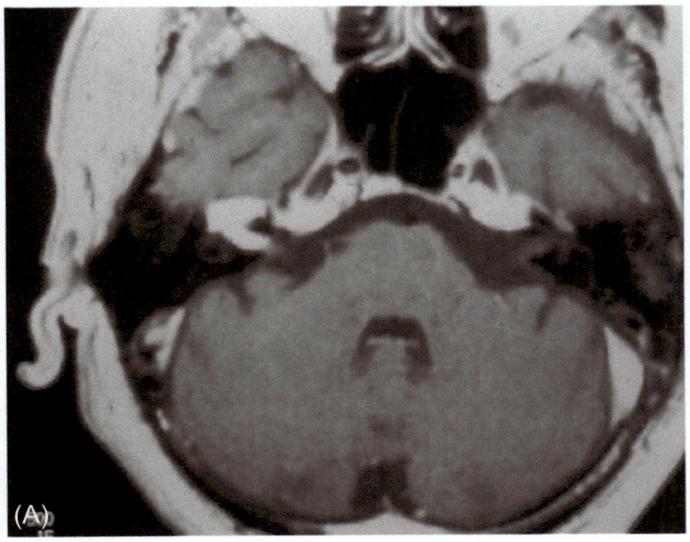

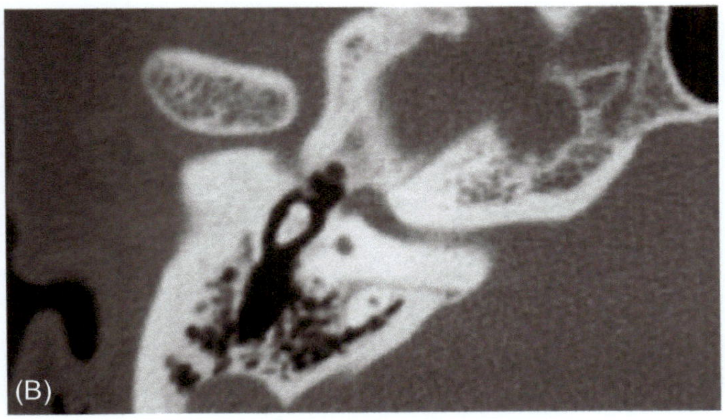

Fig. 5.59 Facial nerve schwannomas. (A) Localization entirely within CPA and IAC with involvement of the labyrinthine portion and the geniculate ganglion. (B) Enlargement of the labyrinthine portion of the fallopian canal.

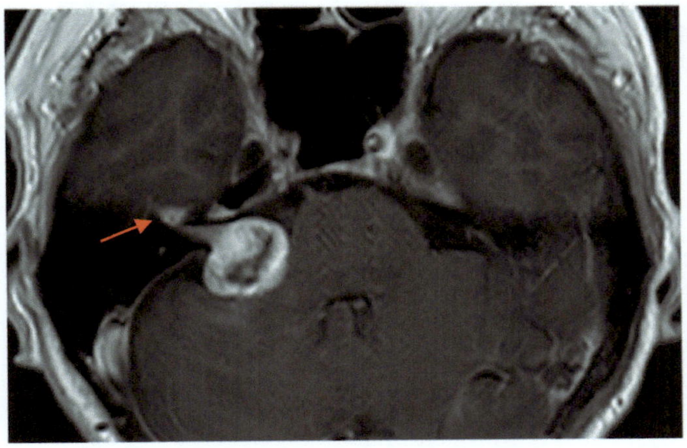

Fig. 5.60 Axial T1W with Gd-enhancement image demonstrating a heterogeneously enhancing mass filling the right IAC/CPA with involvement of the labyrinthine segment of the facial nerve into the geniculate ganglion (*red arrow*).

absence of internal auditory canal involvement as well as the relatively cephalad extension of the lesion my serve as distinctive features.

Lipomas: unlike all other cerebellopontine angle lesions, have a very bright signal on T1-weighted images, which can be suppressed by fat-suppression techniques (Fig. 5.61). These tumors are usually not encapsulated and they envelop the nerves and vessels, making total removal difficult.

Metastases: Intracanalicular metastases (Fig. 5.62) are not distinguishable from intracanalicular VSs. Metastases are often accompanied by multiple intracranial locations With growth, metastases show features of invasiveness and infiltration such as the absence of a clear cleavage plane that allows differentiation from VS.

Lower cranial nerve neurinomas are again uncommon and will be discussed with jugular foramen tumors (Fig. 5.63).

Isolated cerebellopontine angle chordomas with no gross clival or petrous apex involvement are very rare. They can be mistaken for epidermoids. However, they have a slightly higher T1-weighted signal compared with epidermoids and they enhance significantly with gadolinium contrast agent.

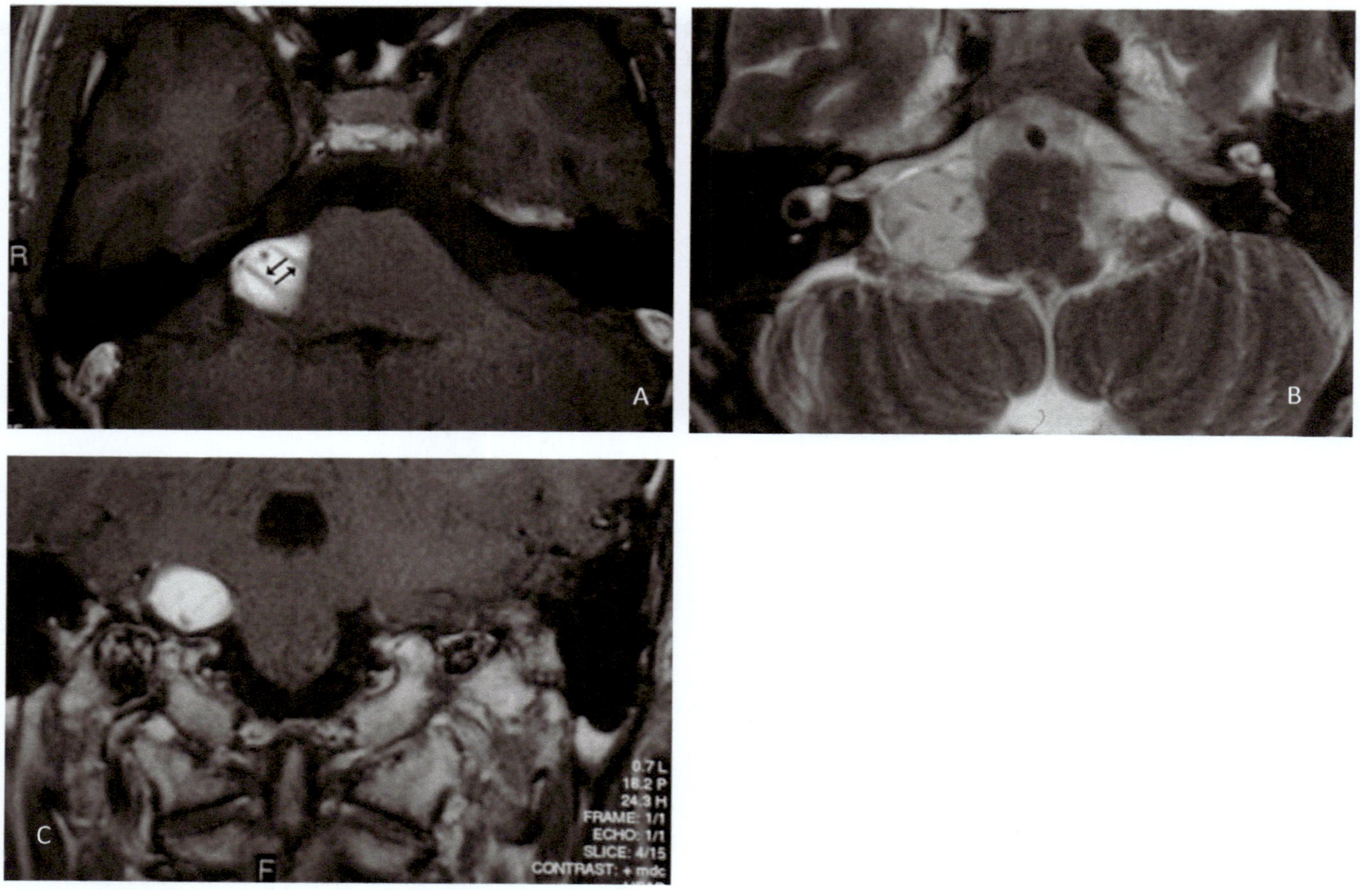

Fig. 5.61 MRI of a case of right cerebellopontine angle lipoma. (A) T1-weighted image without gadolinium enhancement showing a hyperintense lesion. Note the vessels (*arrows*) encased by the tumor. (B) T2-weighted image. The hyperintense mass is seen. (C) T1-weighted image with gadolinium enhancement, coronal view.

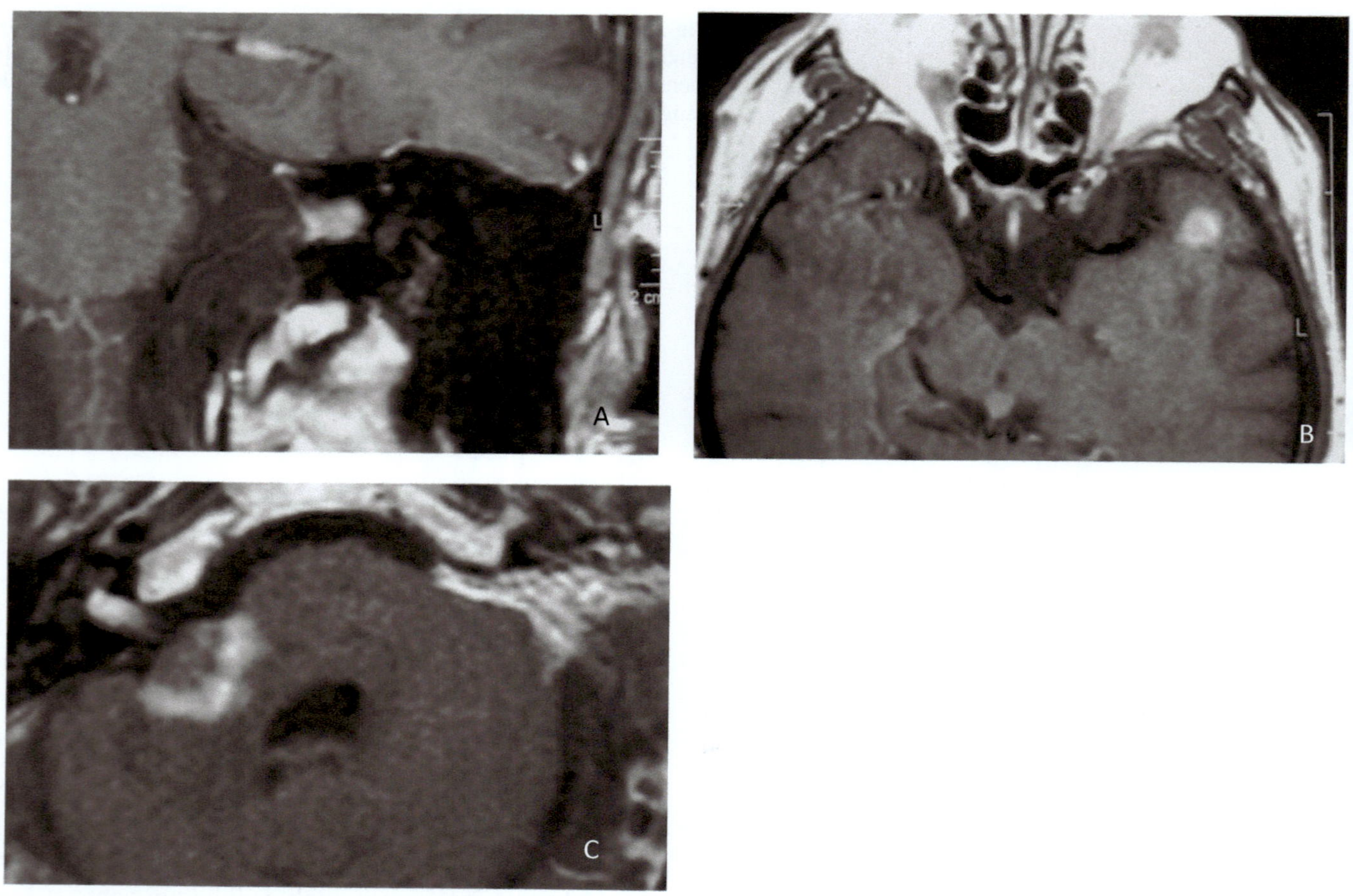

Fig. 5.62 Metastases. (A) Intracanalicular metastases. (B) Metastases in the temporal lobe. (C) Absence of a clear cleavage plane that allows differentiation from vestibular schwannoma.

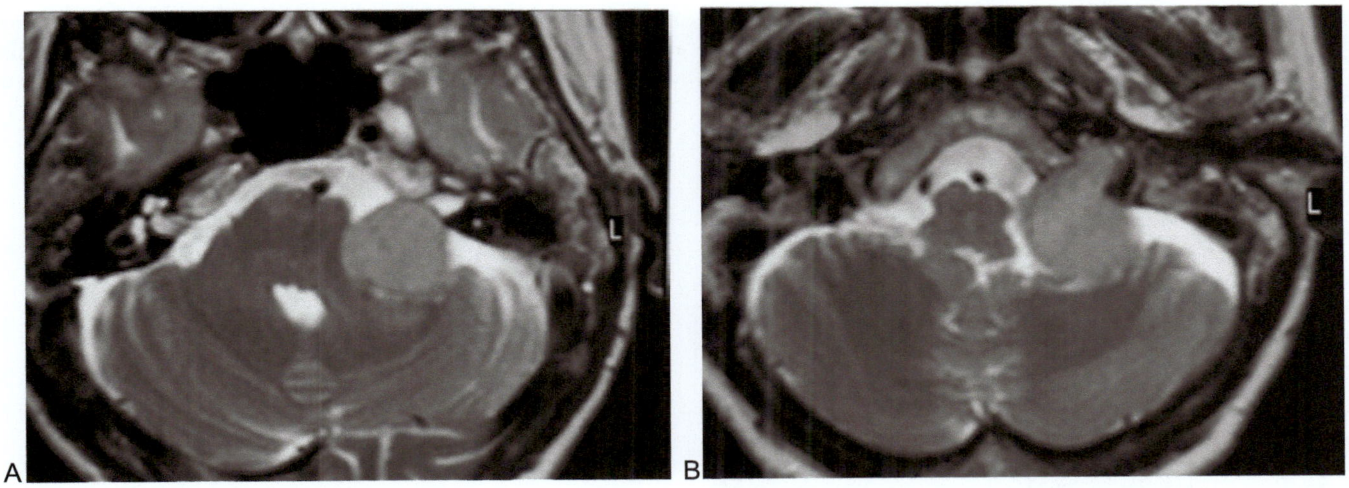

Fig. 5.63 Lower cranial nerve schwannoma. (A) Missing invasion of the IAC. (B) Lower cranial nerve schwannoma inside the jugular foramen.

EXTENDED INTRA-AXIAL LESIONS

Very infrequently, intra-axial lesions grow as exophytic cerebellopontine angle lesions and thus mimic extra-axial lesions. On MRI, certain features help to distinguish these lesions from extra-axial lesions, including irregularity of the tumor-brain interface, peritumoral edema on T2-weighted images, and dilatation of the lateral recess of the fourth ventricle. In addition, the age and clinical picture can help in establishing the diagnosis. These lesions are typically associated with severe neurological manifestations such as hydrocephalus, ataxia, or rapidly evolving multiple cranial nerve palsies. Intra-axial tumors include astrocytoma, medulloblastoma, choroids plexus papilloma, lymphoma, hemangioblastoma, and ependymoma (Figs. 5.64; Table 5.4)

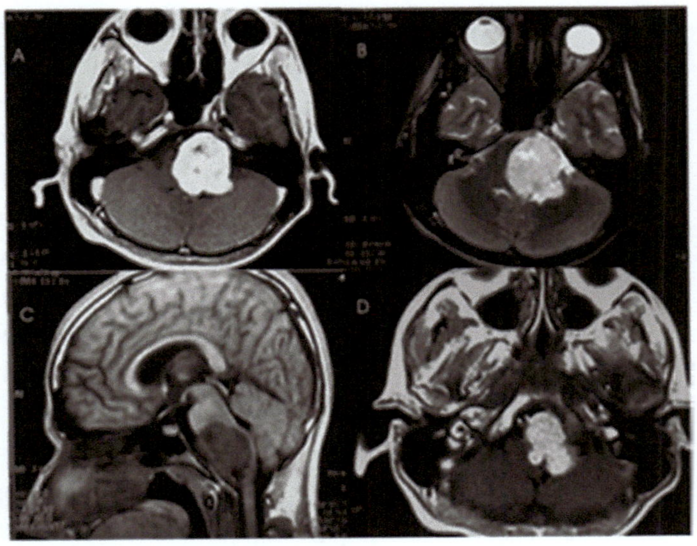

Figs. 5.64 MRI of a case of left posterior fossa ependymoma in a 17-year-old male with a rapid onset of vertigo, vomiting, ataxia, headache, and blurring of vision. (A) Axial view with gadolinium enhancement showing an enhancing large mass markedly displacing the brainstem with no internal auditory canal involvement. (B) T2-weighted image. The heterogeneous appearance of the mass can be appreciated. Some degree of brain edema can noted. (C) Sagittal view, T1-weighted image without gadolinium enhancement. The caudal extent and the multilobular appearance can be appreciated. (D) Axial view with gadolinium enhancement at lower level than (A). Note the lobulated appearance of the tumor.

TABLE 5.4 Radiological differential diagnosis of intra-axial cerebellopontine angle lesions.

	MRI			
Lesion	**T1**	**T2**	**Gd**	**Special features**
Ependymoma	↓	↑	Yes	Heterogeneous
Medulloblastoma	↓	↑	Yes	Arises from cerebellum
Choroid plexus papilloma	↓	↑	Yes	Extends through the foramen of Luschka
Glioma	↓	↑	Variable	Arises from the brainstem

T1, T1-weighted; *T2*, T2-weighted; *Gd*, gadolinium-enhanced; ↑, hyperintense; ↓, hypointense.

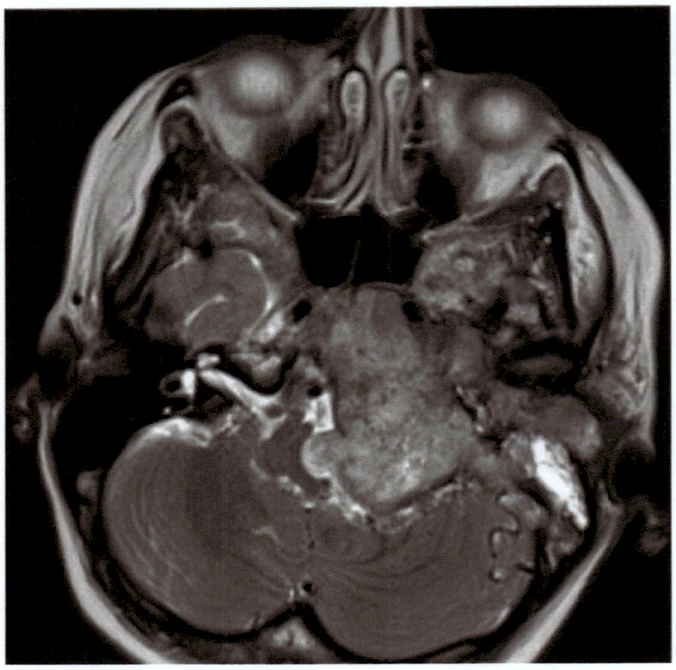

Fig. 5.65 Axial T2-weighted image shows a voluminous heterogenous mass lesion involving the left IAC and CPA cistern, with multiple flow voids within and around the lesion, extending into the prepontine cistern.

SKULL BASE TUMORS EXTENDING INTO THE CEREBELLOPONTINE ANGLE

Lesions arising from the clivus and temporal bone can spread transdurally into the cerebellopontine angle. Bone erosion at the site of origin is generally observed. These lesions include cholesterol granuloma, chordomas and chondrosarcomas, paragangliomas (Fig. 5.65), and giant cell tumors. These lesions will be covered in detail in the discussion of extradural skull base lesions (see below).

JUGULAR FORAMEN TUMORS

The most common tumors involving the jugular foramen are glomus jugulare tumors, followed by lower cranial nerve neurinomas and meningiomas. Less common lesions include chondrosarcoma, chordoma, hemangiopericytoma, and metastases.

PARAGANGLIOMAS (PGG)

Fisch classified paragangliomas into types A–D. Types A and B are tumors limited to the middle ear and mastoid, type C PGG develop in or around the jugular bulband type D represents intracranial extension. We further subdivided middle ear tumors into type A1 (tumor is completely visible on otoscopy) and type A2 (tumor margins are not visible on otoscopy. Tumor may extend anteriorly to the eustachian tube and/or posteriorly to the mesotympanum). Class B tumors can be subdivided into type B1 (tumor confined to the middle ear cleft with extension to the hypotympanum), type B2 (tumor involving the middle ear cleft with extension to the hypotympanum and mastoid), and type B3 (tumor confined to the tympanomastoid compartment with erosion of the carotid canal). [57]. Type C, or tympanojugular (TPJ) tumors, is further sub divided into classes C1–C4 according to the anterior tumor extension and its relation to the internal carotid artery, whereas type D tumors can be either extracranial "De" or intracranial "Di." We no longer consider the De-extension and only pay attention to the Di-component for a possible staged resection. Vertebral artery involvement has also been introduced in our center. Ve means involvement of the vertebral artery in its extradural compartment, Vi means intradural involvement of the vertebral artery [58]. On high-resolution CT, paragangliomas demonstrate an irregular "moth-eaten" erosion of the bone of the jugular foramen (Fig. 5.66). The jugulo-carotid spine (which is the bone separating the jugular bulb from the internal carotid artery) is characteristically eroded (Fig. 5.67). The degree of bony involvement is often difficult to assess, however. The most critical initial step is the differentiation between tympanomastoid paragangliomas and small TJPs, with a coronal image being the most helpful (Fig. 5.68).

If high-resolution CT (HRCT) confirms that the lesion remains free of the jugular fossa, no further imaging is mandatory; as seen in Fig. 5.69, MRl can be very helpful in differentiating tumor from middle ear and mastoid effusions. However, any doubt on any radiological view necessitates further assessment.

PGGs enhance densely and homogeneously with contrast. The degree of bony infiltration can also be difficult to ascertain, and it is often underestimated. This is especially the case in poorly pneumatized temporal bones and in the presence of involvement of the petrous apex (Fig. 5.70), clivus (Fig. 5.71), occipital condyle (Fig. 5.72), and hypoglossal canal (Fig. 5.73), because of the bony marrow usually present in these areas.

MRI provides exquisite detail relating to the soft-tissue relationships in the deep neck spaces as well as intracranially. From 60% to 75% of cases have intracranial extension at presentation, with the rate of intradural involvement being ~30% [44–46]. Several sequences can be employed to delineate both the tumor and also arterial anatomy and venous drainage.

Low to intermediate T1W signal and relatively high T2W signal (Fig. 5.74) are typical for paragangliomas. Small tumors enhance homogeneously but areas of necrosis and hemorrhage are present with increasing size (Fig. 5.75).

A classic "salt and pepper" pattern in lesions larger than 2 cm (in any dimension) is often seen, especially in T2W images (Fig. 5.76). This is due to areas of brightening onT2-weighting due to areas of slow flow within the tumor, and the presence of large intratumoral vasculature appearing as flow voids [47,48]. Axial and coronal images are commonly used to evaluate the lesion, but sagittal images allow appreciation of tumor extension as a whole (Fig. 5.77).

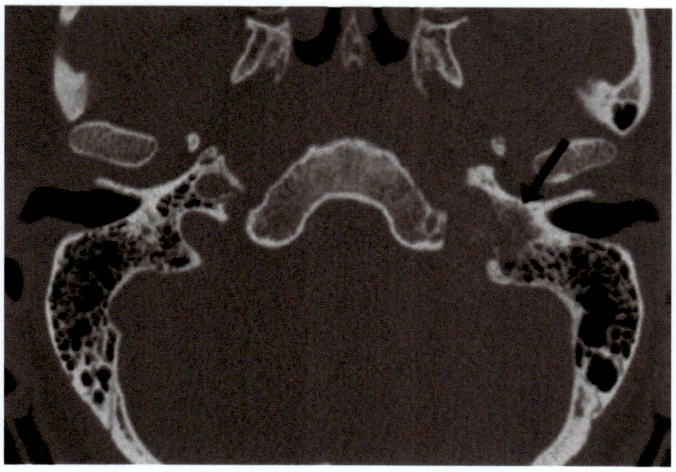

Fig. 5.66 CT. Moth-eaten bone (*arrow*) at the lateral aspect of the jugular fossa.

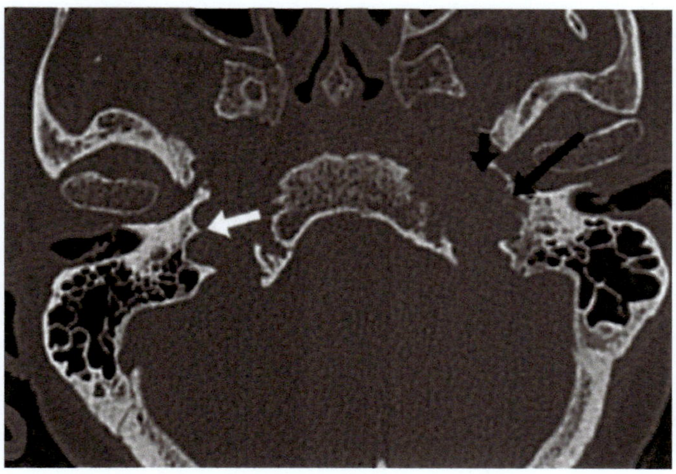

Fig. 5.67 CT. Caroticojugular crest erosion (*two black arrows*). Normal contralateral side (*white arrow*).

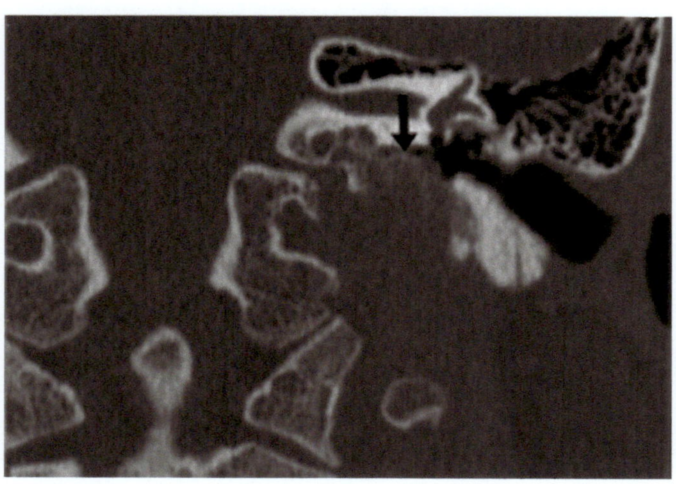

Fig. 5.68 CT. Coronal view with erosion of the jugular plate at the roof of the jugular fossa (*arrow*).

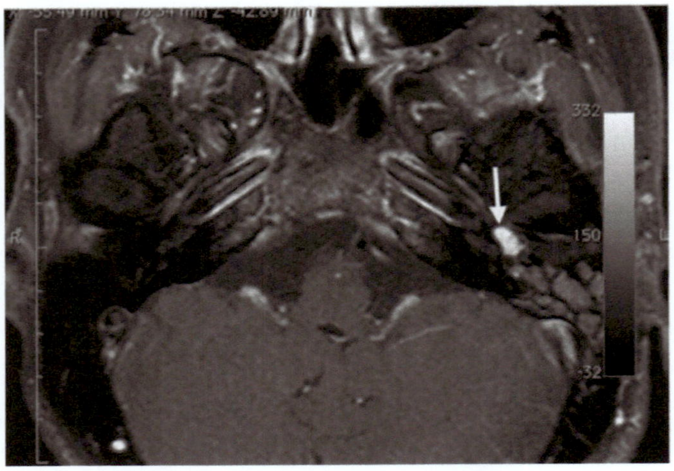

Fig. 5.69 MRI. A small class A2 paraganglioma (*white arrow*) occluding the eustachian tube orifice, producing a mastoid effusion. This shows the utility of MRI scanning in differentiating between tumor and fluid (*black arrow*).

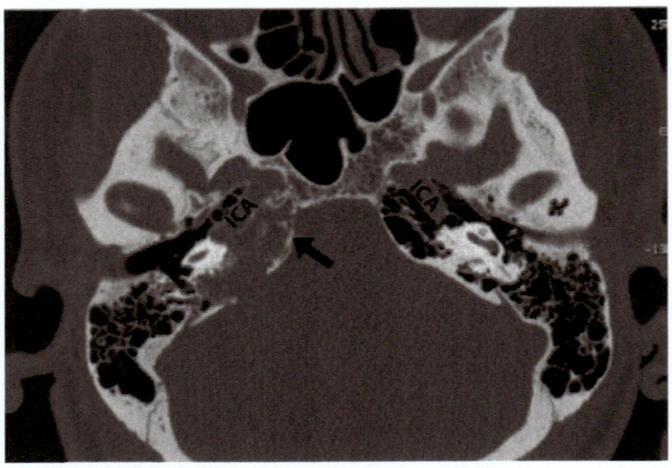

Fig. 5.70 CT. Involvement of petrous apex (*black arrow*). ICA internal carotid artery.

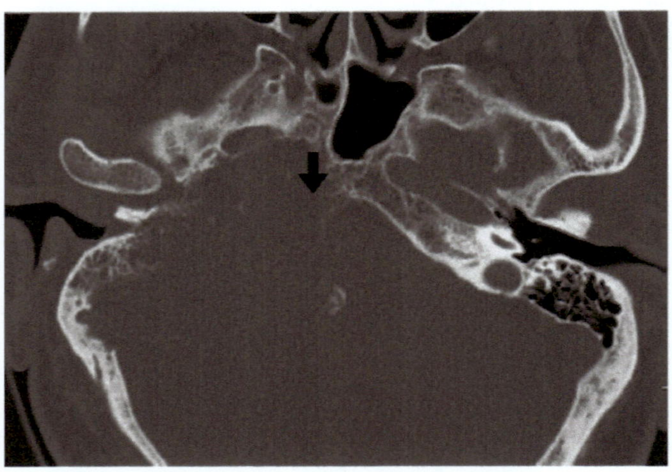

Fig. 5.71 CT. Massive erosion of the petrous bone and clivus (*black arrow*).

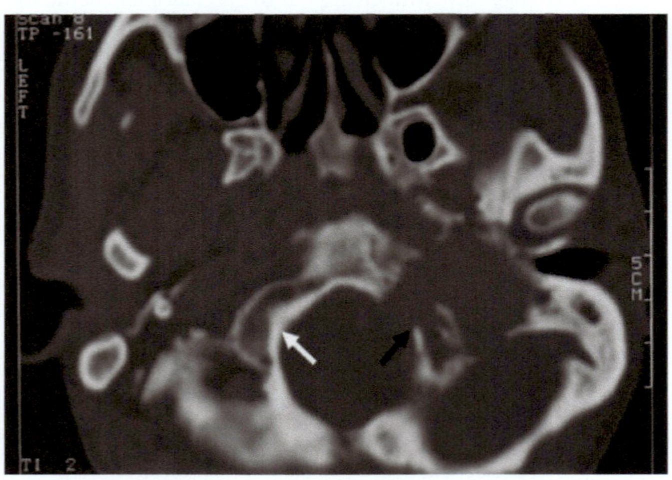

Fig. 5.72 CT. Erosion of the occipital condyle (*black arrow*) *White arrow*: normal occipital condyle on the contralateral side.

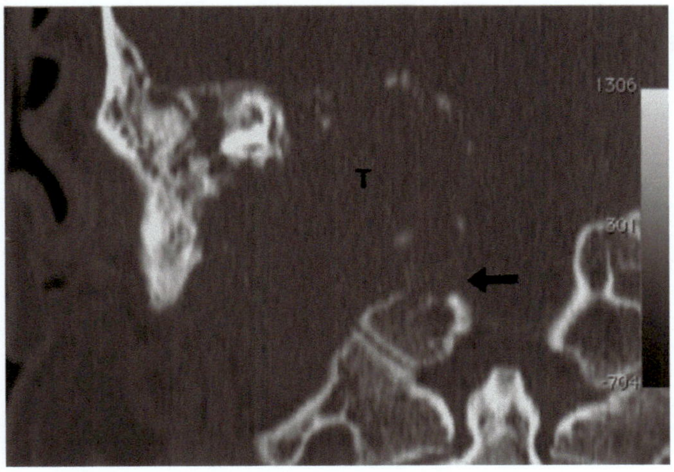

Fig. 5.73 CT. Erosion of the hypoglossal canal (*black arrow*).

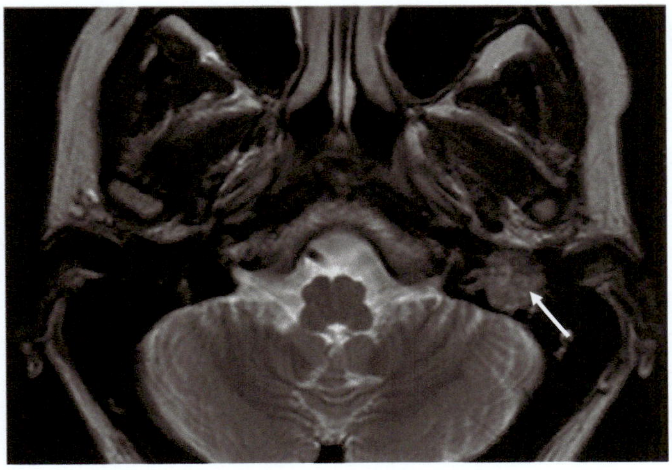

Fig. 5.74 MRI. Tumor (*white arrow*) showing relatively high T2W signal.

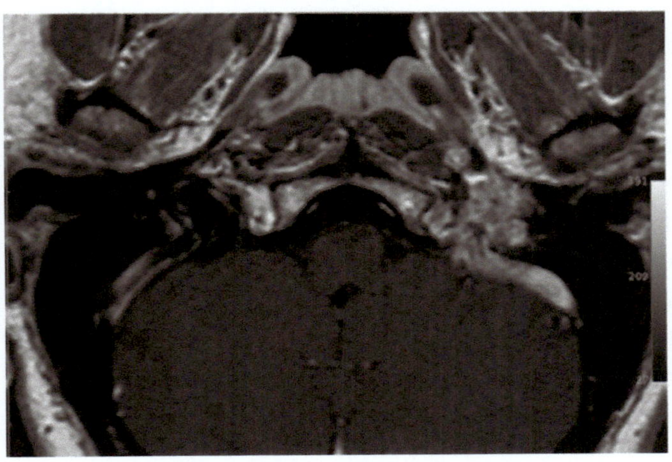

Fig. 5.75 MRI. Homogeneous enhancement of a small tumor.

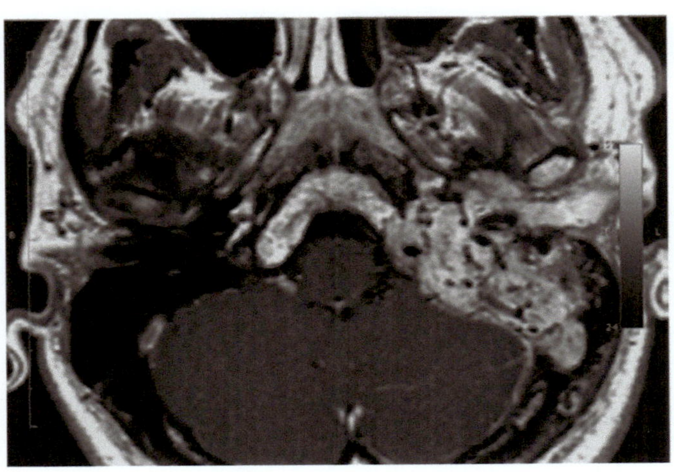

Fig. 5.76 MRI. Areas of necrosis and hemorrhage, as well as intratumoral vessels, are present in large lesions.

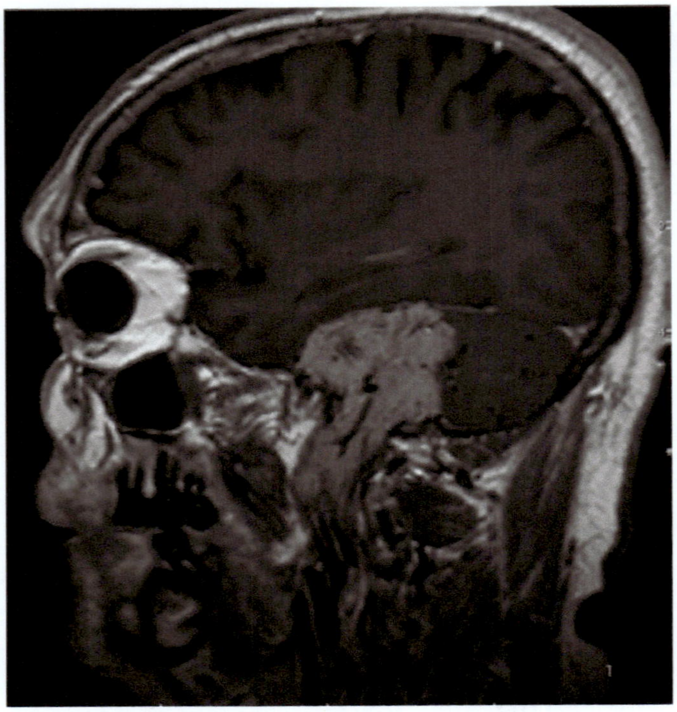

Fig. 5.77 MRI. Sagittal view allows appreciation of the degree of intra- and extracranial extension of the TJP.

The relation to nearby structures is of upmost importance and HRCT scan is the study of choice for highlighting tumoral bone involvement in the carotid canal, the occipital condyle, the hypoglossal canal, the facial canal, the infralabyrinthine compartment, the petrous apex, and the inner ear. The erosion of the vertical segment of the carotid canal (C1–C2) is better evaluated on axial high-resolution CT (Fig. 5.78). Involvement of the horizontal carotid canal and foramen lacerum (C3–C4) are evaluated on both axial and coronal high-resolution CT (Fig. 5.79). While CT better reveals the erosion of the carotid canal, tumor involvement of the internal carotid artery per se is better visualized on MRI. MRI is superior to CT in detecting early intradural extension (Fig. 5.80, 5.81) as well as cavernous sinus involvement (Fig. 5.82). Tumor extension into the venous sinuses and extension into the internal jugular vein are also better seen on MRI. Dual T2 fast spin echo sequences, non-contrast and contrast time-of-flight sequences, and contrast-enhanced MRA and MRV (often with 3D reconstruction) provide further information [49]. Despite these advances, detection of lesions less than 10 mm in size is difficult, except for those arising in the tympanic cavity.

Digital subtraction angiography (DSA) is performed for all cases of glomus jugulare tumors demonstrating the characteristic tumor blush (Fig. 5.83). Angiography shows the type, size, number, and relations of the tumor-feeding vessels. It determines the angioarchitecture of the tumor and its different vascular compartments, and the relation to the vertebral and internal carotid arteries is demonstrated. It also studies the venous drainage and the patency or invasion of the dural sinuses by the tumor. Preoperative embolization is performed next to reduce intraoperative bleeding.

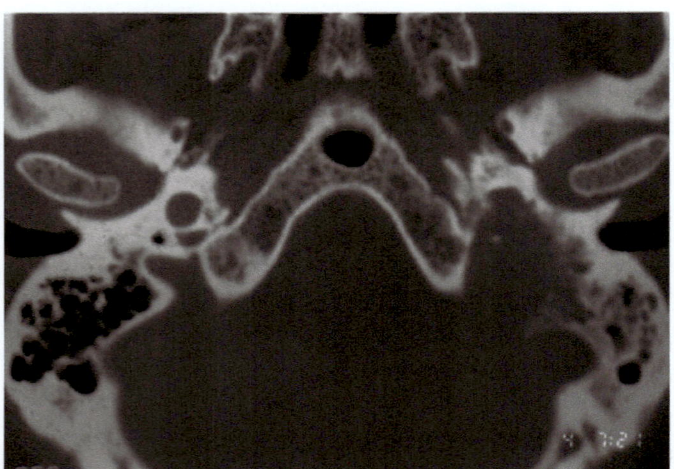

Fig. 5.78 Erosion of the vertical portion of the carotid artery.

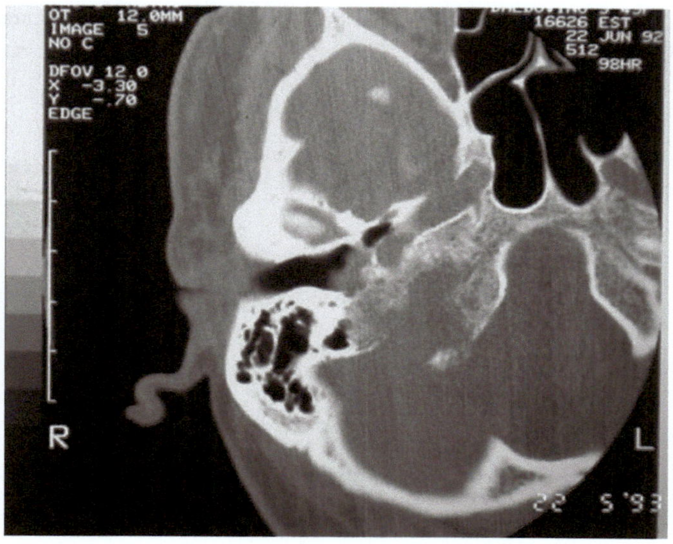

Fig. 5.79 CT scan, axial view, showing a C3 glomus jugulare tumor involving the horizontal segment of the internal carotid artery.

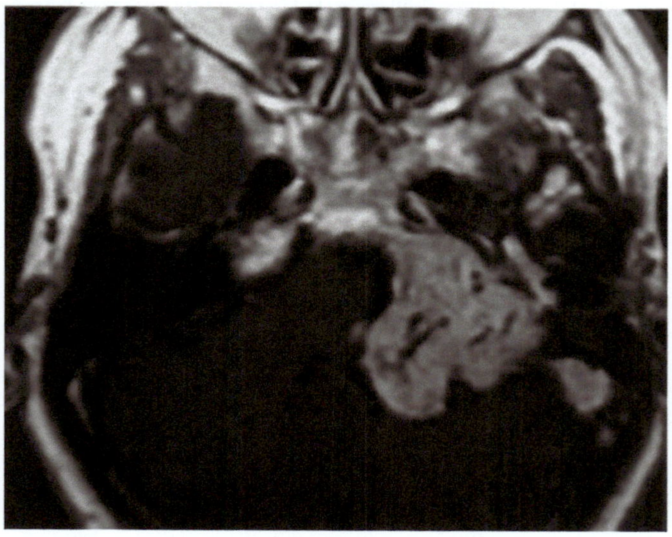

Fig. 5.80 MRI for intradural invasion and cavernous sinus involvement.

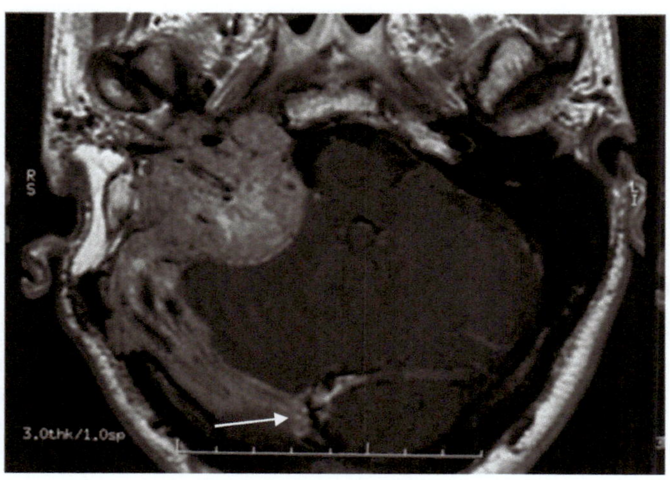

Fig. 5.81 MRI. Intracranial intradural extension of a large paraganglioma (Class Di2). Note the tumor invading the entire sigmoid sinus up to the torcular (*white arrow*).

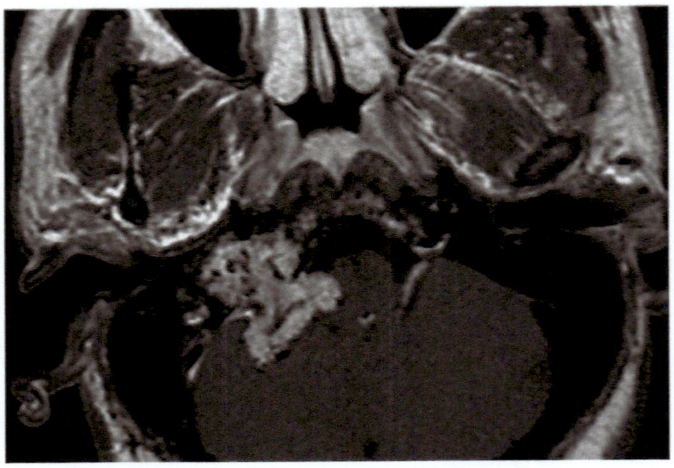

Fig. 5.82 MRl. Large paraganglioma with intradural component (Class Di2).

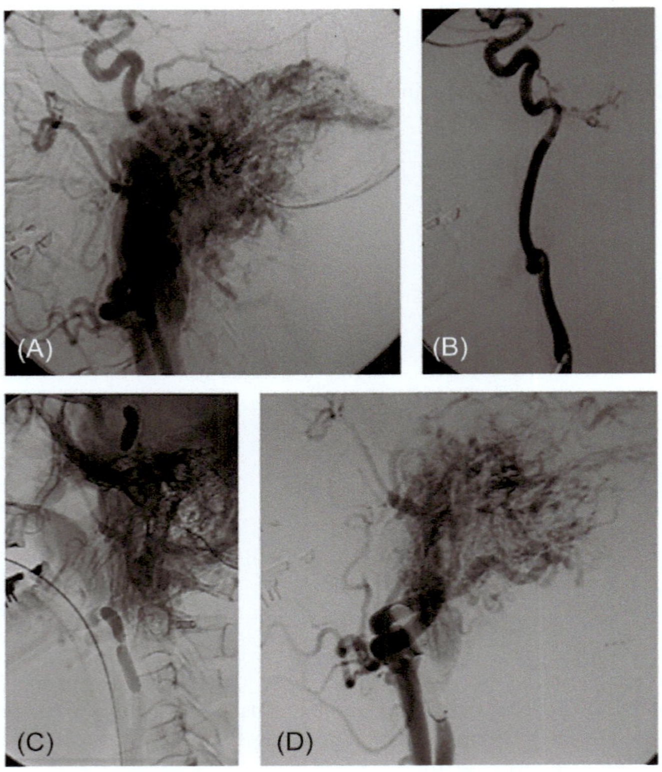

Fig. 5.83 DSA showing a TJP: Injection of the common (A) and internal carotid (B) artery in lateral view, showing high tumoral blush coming mainly from external carotid branches. (C,D) Balloon occlusion was performed because of encasement of the ICA; unsubtracted lateral view, showing three balloons inflated in the ICA (C); common carotid injection (D) at the end of the procedure, confirming ICA occlusion and tumoral blush from external carotid branches.

LOWER CRANIAL NERVE NEURINOMA

Lower cranial nerve neurinomas are characterized by expansion of the jugular foramen with smooth well-defined bony margins (Fig. 5.84). Neurinomas can extend into the basal cistern superiorly or into the carotid space inferiorly, with the characteristic dumbbell-shaped pattern better seen in the coronal or sagittal views. As with all other neurinomas, MRI is isointense or hypointense on T1 weighting and isointense or hyperintense on T2 weighting and enhances with gadolinium (Fig. 5.85). Intratumoral cystic change is commonly seen in LCN schwannomas with temporal bone and clival extension is very rare [47,50]. Large lesions can extend to involve the vertical internal carotid artery and the petrous apex.

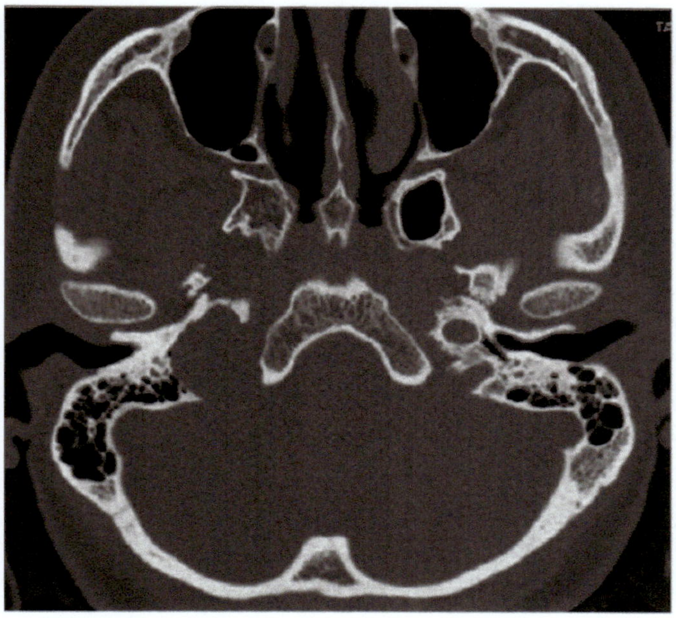

Fig. 5.84 CT. Mass lesion showing expansion centered on the jugular foramen. Note the scalloped, sclerotic margin.

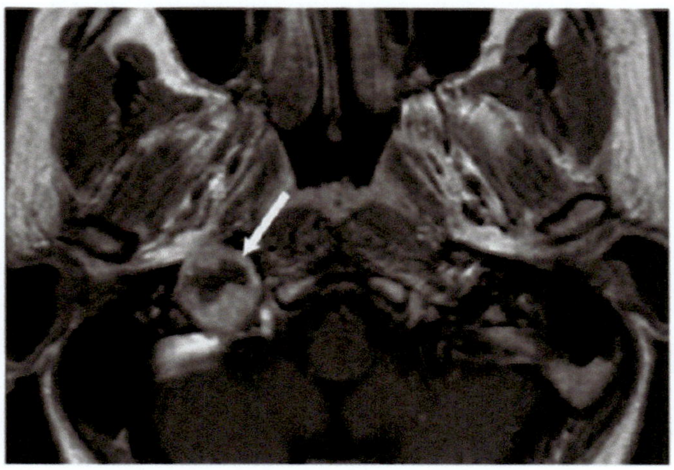

Fig. 5.85 Note the degree of cystic change (*arrow*) in this schwannoma of the lower cranial nerves.

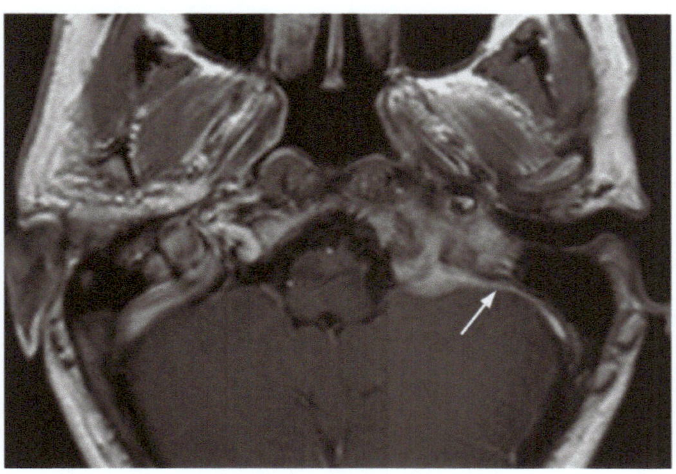

Fig. 5.86 Jugular foramen meningioma with the classic en plaque appearance and with a dural tail (*arrow*).

MENINGIOMA

Meningiomas of the jugular foramen cause irregular enlargement of the jugular foramen and occasionally cause destruction of the temporal bone. The posterior fossa component is usually en-plaque (Fig. 5.86), but can take a globular form, more commonly seen in meningiomas secondarily involving the JF [47]. The presence of a dural tail is common (75%–100% of cases), whereas the classic findings of hyperostosis and intratumoral calcifications, while highly suggestive, are rarely seen at this site. These features can help distinguish meningiomas from neurinomas on CT (Fig. 5.87). Bony margins of the jugular foramen appear irregular, with permeative-sclerotic changes (Fig. 5.88) occurring in a "centrifugal" fashion. In comparison with jugulotympanic paragangliomas, there is relative maintenance of the bony architecture.

On MRI, meningiomas are isointense or hypointense on T1-weighted and hyperintense on T2-weighted images. Meningiomas can be distinguished from neurinoma by the presence of irregular borders, enhancing more intensely with gadolinium and, most often, a lower intensity on T2 weighted images (Fig. 5.89). Uncommonly, meningiomas can extend through the jugular foramen into the carotid space in the neck, thereby mimicking glomus tumors. Unlike with glomus jugulare tumors, bone destruction of the jugular fossa is much less evident. The serpentine flow voids characteristic of glomus tumors are absent in meningioma. On angiography, meningiomas show a much less bright and slower tumor blush compared to glomus tumors. Extensive cases can cause massive temporal bone involvement and hence a proper evaluation of high-resolution CT and MRI is essential to identify the exact tumor extent for appropriate preoperative planning.

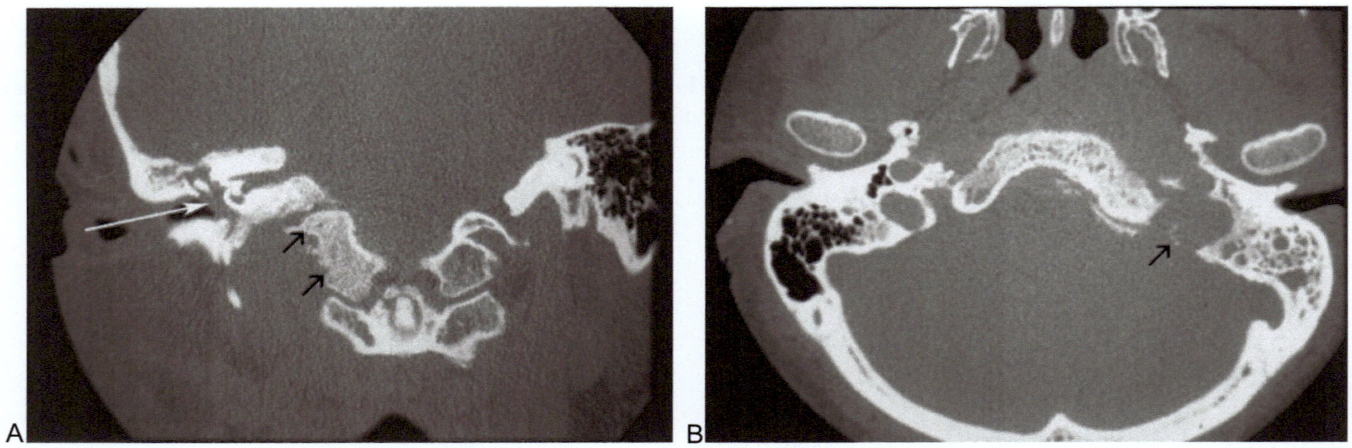

Fig. 5.87 CT scan of a left jugular foramen meningioma. (A) Coronal view. Note the irregular erosion of the borders (black arrows). The middle ear is also invaded by the tumor (*white arrow*). (B) Axial view. The tumor involves the vertical internal carotid artery. Intratumoral calcification is seen (*arrow*).

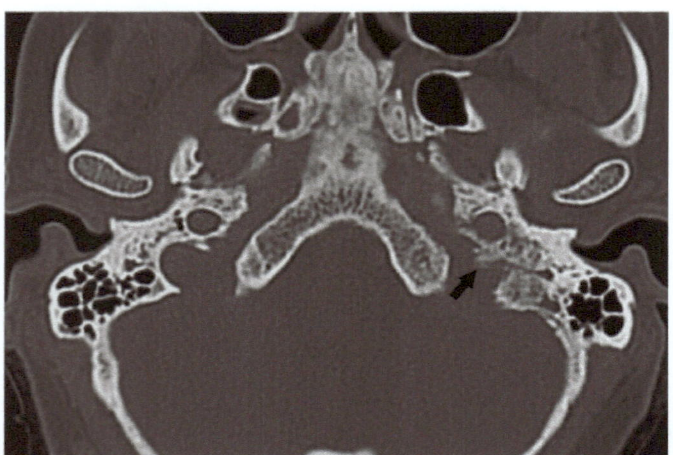

Fig. 5.88 Jugular foramen meningiomas may produce hyperostosis (*arrow*).

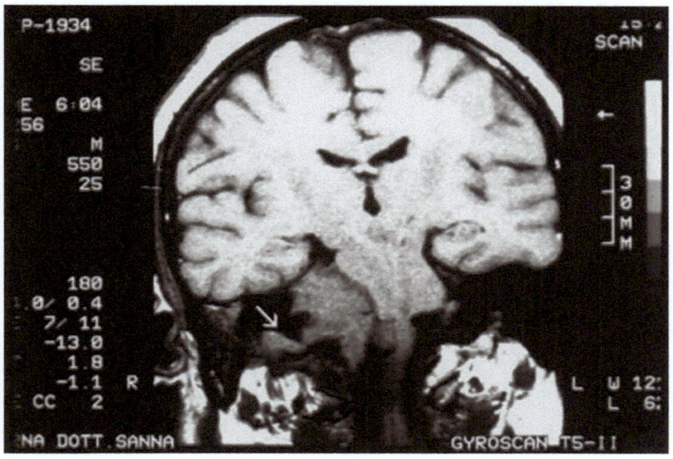

Fig. 5.89 T1-weighted MR image without gadolinium enhancement, coronal view, showing a hypointense right jugular foramen meningioma. Note the extent of the tumor into the jugular foramen (*arrow*).

JUGULAR BULB ANOMALIES

Jugular bulb anomalies can mimic tumor. Sometimes a large asymmetric jugular bulb may be taken for a jugular foramen lesion (Fig. 5.90). The combination of normal bony anatomy on CT and flow-sensitive MRI allows differentiation from true pathology [50]. The right jugular foramen is larger than the left in 75% of the population. When the roof of the jugular bulb is seen above the level of floor of internal auditory canal, it is called a high-riding jugular bulb, which is more common on the right side. This is a dangerous variant and compromises the exposure during translabyrinthine surgery Indian paper.

JUGULAR FORAMEN CHONDROSARCOMA

Chondrosarcoma of the skull base most commonly arises from the petro-occipital fissure, with the JF rarely involved due to postero-inferolateral extension [34,40,51–53]. Much more rarely, chondrosarcoma can arise from the jugular foramen itself. Indeed, in our series, this is the fourth commonest pathology of the jugular fossa. Destructive bony margins are typical, with intratumoral calcification commonly seen (Fig. 5.91). Classic radiological features on MRI reveal intermediateT1W images and high-intensity T2W images. There is usually a moderately heterogeneous pattern of contrast enhancement (Fig. 5.92), but chondrosarcomas display an appreciable level of variability in their imaging characteristics.

ENDOLYMPHATIC SAC TUMOR

Endolymphatic sac tumors are rare vascularized lesions that can mimic a paraganglioma from both otoscopic and radiological points of view. They may involve the superior surface of the jugular bulb due to inferior extension as well as infiltrating the middle ear. MRI appearance is similar to that of paragangliomas, often with intense gadolinium enhancement, but without the presence of large vascular voids (Fig. 5.93). The origin of these tumors is higher, at the level of the endolymphatic sac, centered on the posterior face of the petrous temporal bone, and invasion of the jugular bulb—the rule in paragangliomas—is only rarely present in endolymphatic sac tumors. In doubtful cases, angiography may be of some help as the tumor blush is not as intense as in paragangliomas.

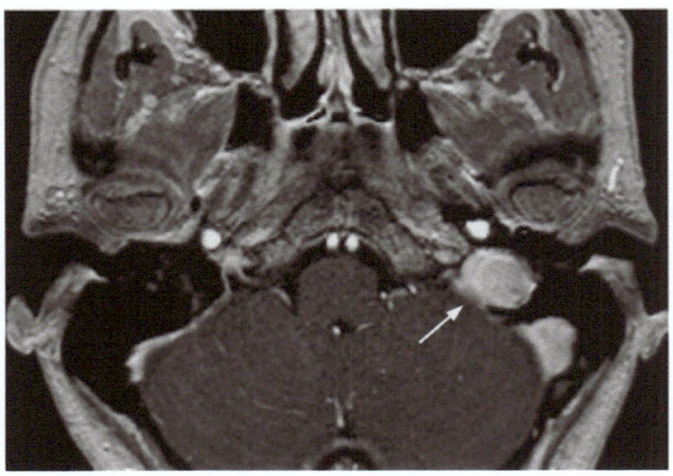

Fig. 5.90 Asymmetry of the jugular bulb (*arrow*).

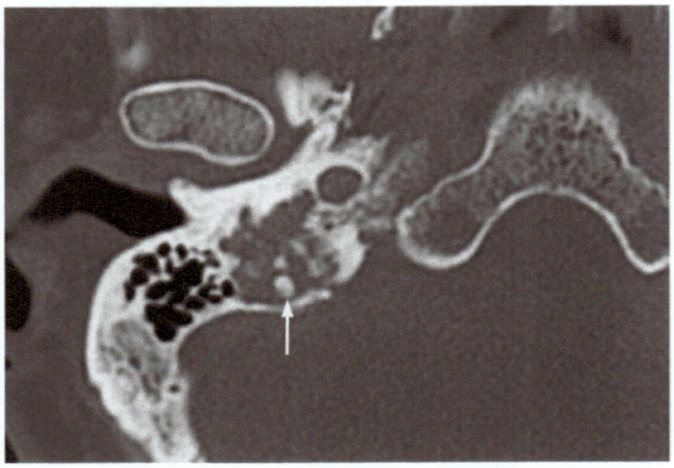

Fig. 5.91 Intratumoral calcification is a feature of jugular foramen chondrosarcomas (*white arrow*) on CT.

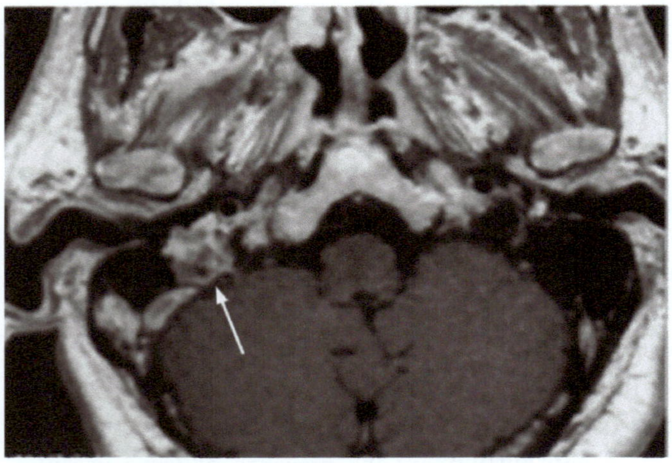

Fig. 5.92 Jugular foramen chondrosarcoma (*arrow*).

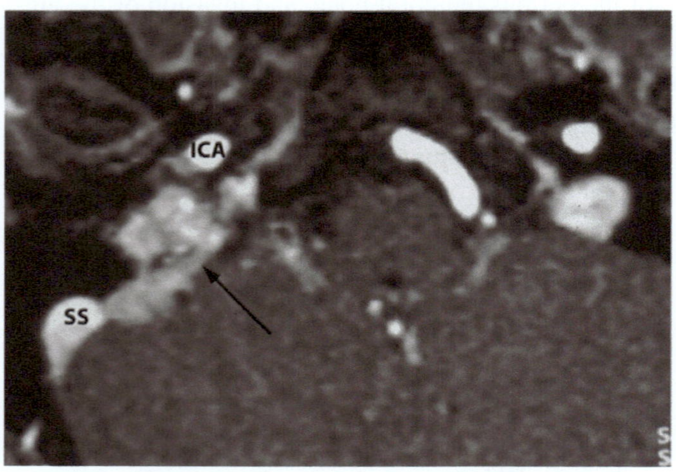

Fig. 5.93 MRI of an endolymphatic sac tumor (*arrow*). *ICA*, internal carotid artery; *SS*, sigmoid sinus.

Follow-Up

WAIT AND SCAN

Many patients affected by small lesions are not candidates for surgery and may initially be followed with a "wait and scan" policy. However, not all these tumors fail to remain stable and in such a case the policy may be switched to surgery or stereotactic radiotherapy (Fig. 5.94A,B).

POSTOPERATIVE FOLLOW-UP

Patients with lateral skull base lesions are usually followed-up with postoperative MRI for some years (5–10) in order to exclude residual tumors with potential for growth. The fat positioned at the end of the surgical procedure that occupies remnant cavity (Fig. 5.95A) usually shrinks to 50% during the first year, allowing re-expansion of the nervous tissue (brain, cerebellum, and the brainstem) (Fig. 5.95B). After partial removal, the residual tumor may be confused with the fat; MRI with fat-suppression technique is indicated for better evaluation of the presence of the residual lesion (Fig. 5.96). Sometimes a tumor component deeply indented into the brainstem is left during surgery in order to avoid neurological complications. With cerebellar and midbrain re-expansion these tumor remnants are usually pushed laterally, making revision surgery easier if required (Fig. 5.97). Because of scars and tissue positioned to pack the canal, especially in the retrosigmoid approach, small enhancements inside the IAC are common in the postoperative period (Fig. 5.98). Progressive growth is the only sure indicator of residual lesion.

RESIDUAL TUMORS

Residual tumors have the same features as the primary ones. Depending on the experience of the surgeon and the difference encountered during surgery, differing amounts of tumor are left in place. Although subtotal removal is adopted as a defined strategy in selected patients (usually elderly patients), sometimes it results in an enlarged biopsy (Fig. 5.99). On PET-CT recurrence and postsurgical fibrosis can be easily differentiated from the primary tumor [3]. Recurrence of cholesteatoma is best appreciated on DWI MRI, avoiding unnecessary surgery (Fig. 5.100) [54,55].

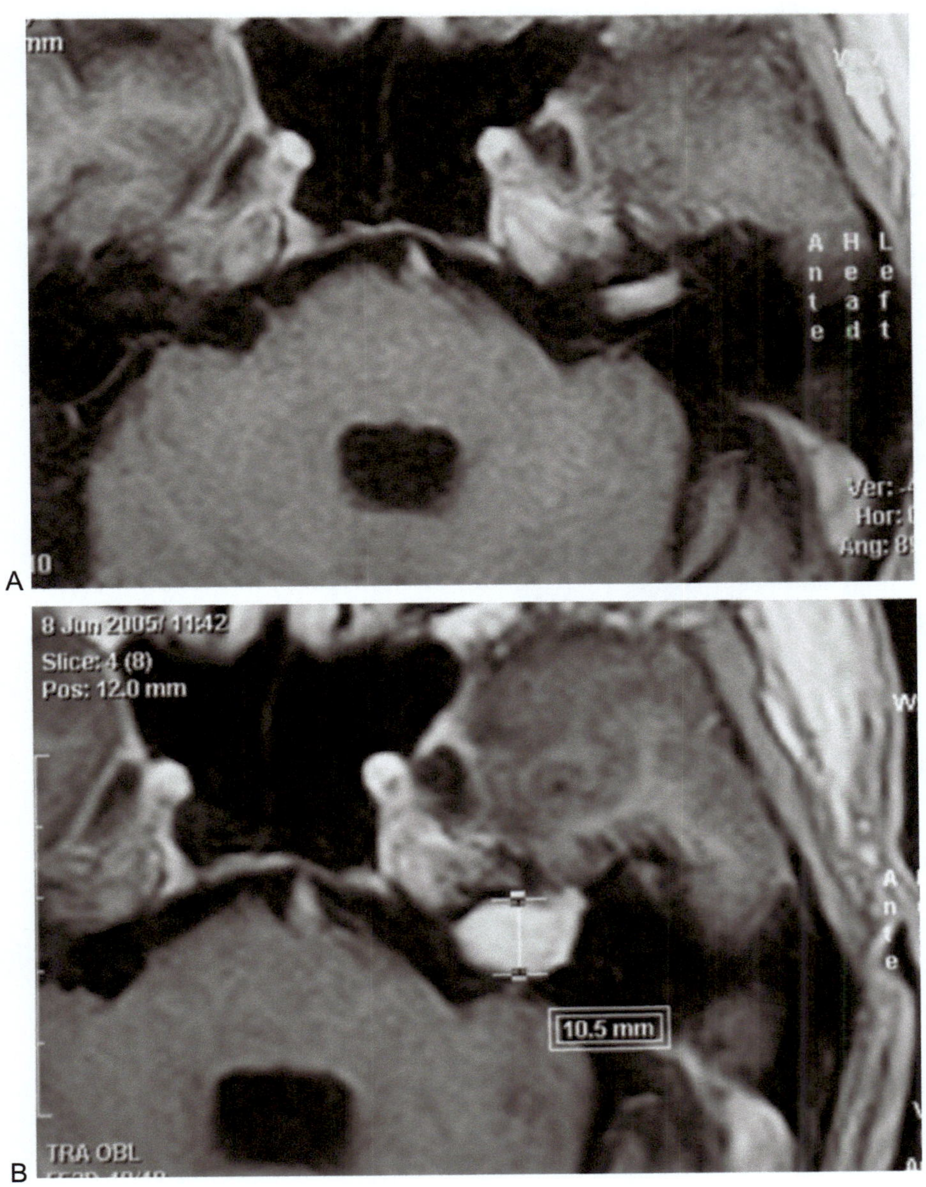

Fig. 5.94 (A) Left-sided intracanalicular vestibular schwannoma diagnosed in 2003. (B) The same case 2 years later with a 1-cm growth into the CPA.

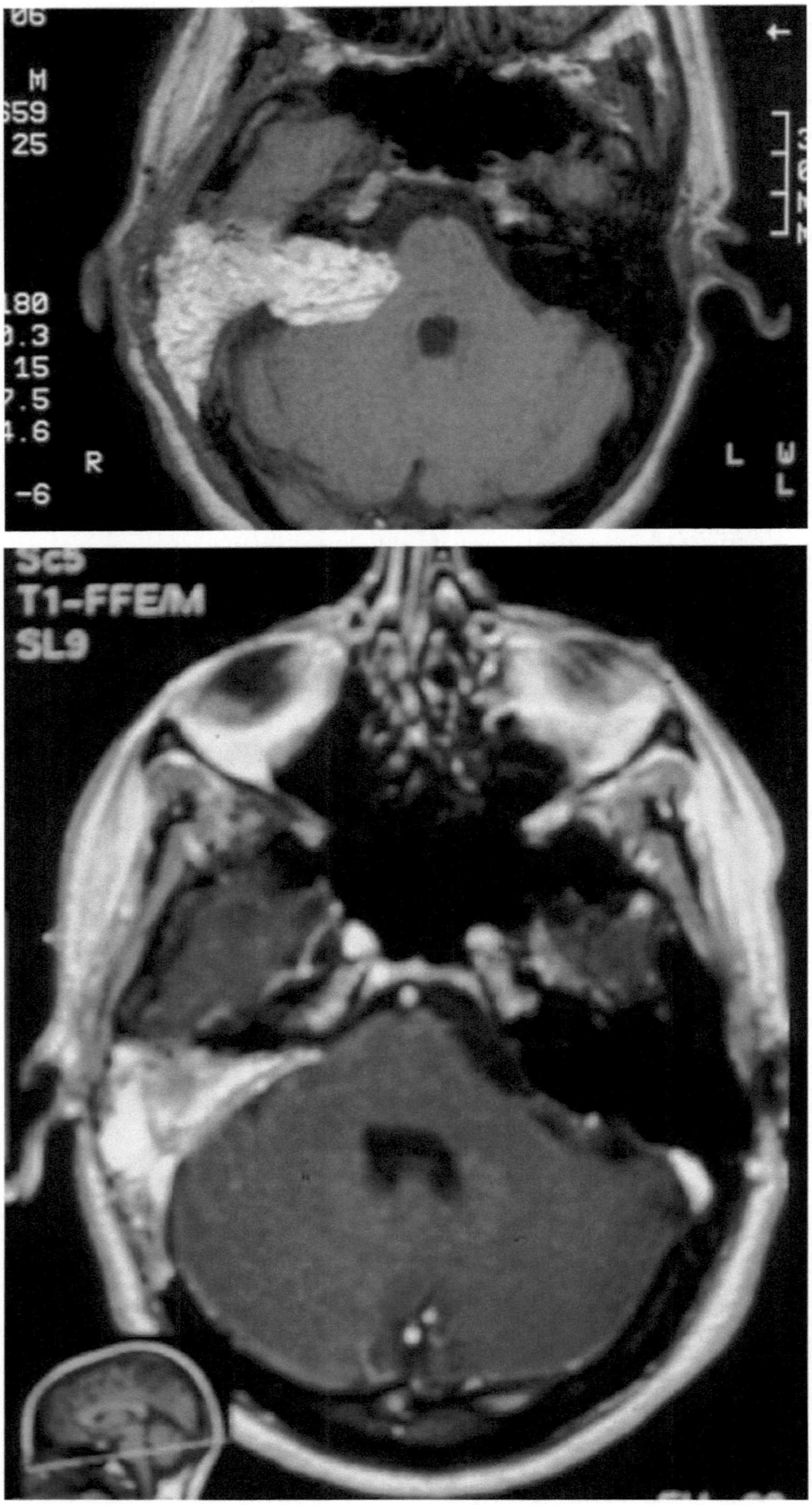

Fig. 5.95 (A) CPA completely filled with fat positioned during the TLA. (B) Shrinking of the fat to 50% during the first year.

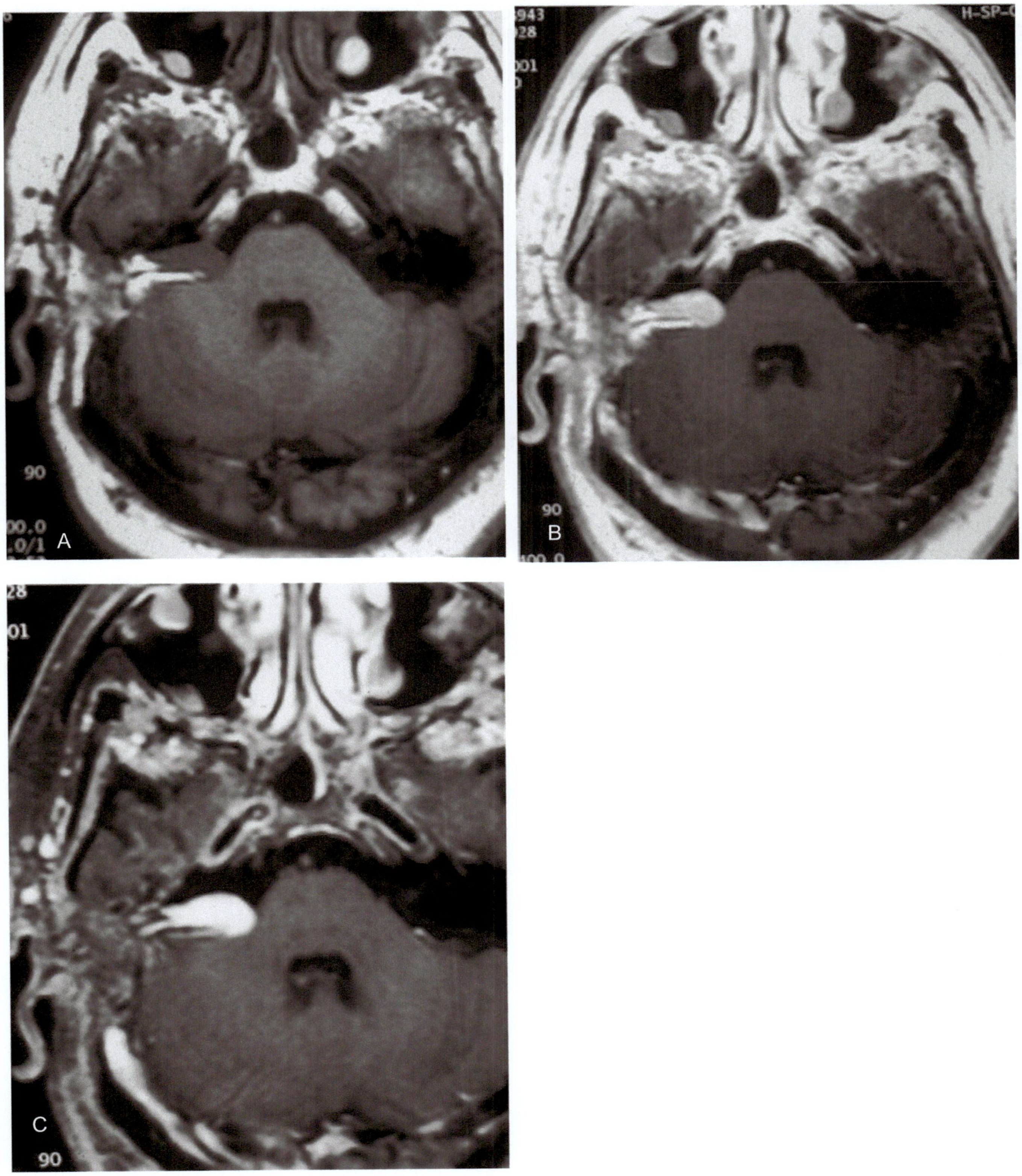

Fig. 5.96 (A) T1-weighting without gadolinium. (B) T1-weighting with gadolinium. (C) T1-weighting with gadolinium and fat-suppression technique.

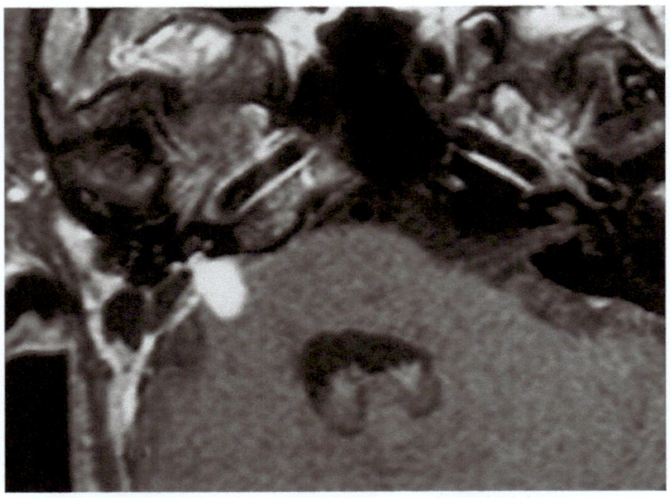

Fig. 5.97 Tumor remnant left inside to avoid neurological complications.

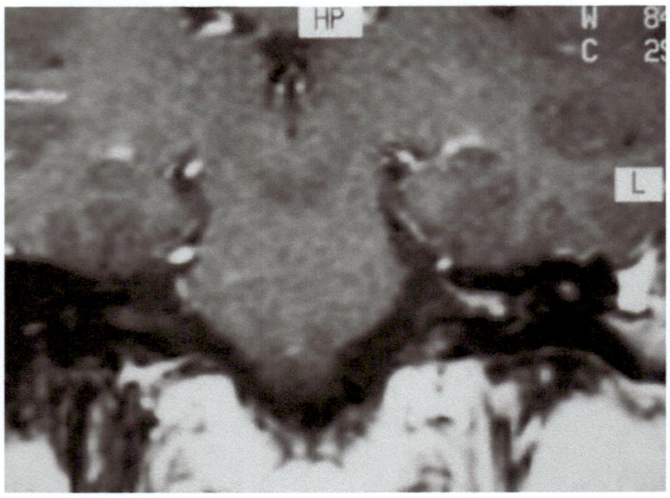

Fig. 5.98 Small enhancement inside the IAC.

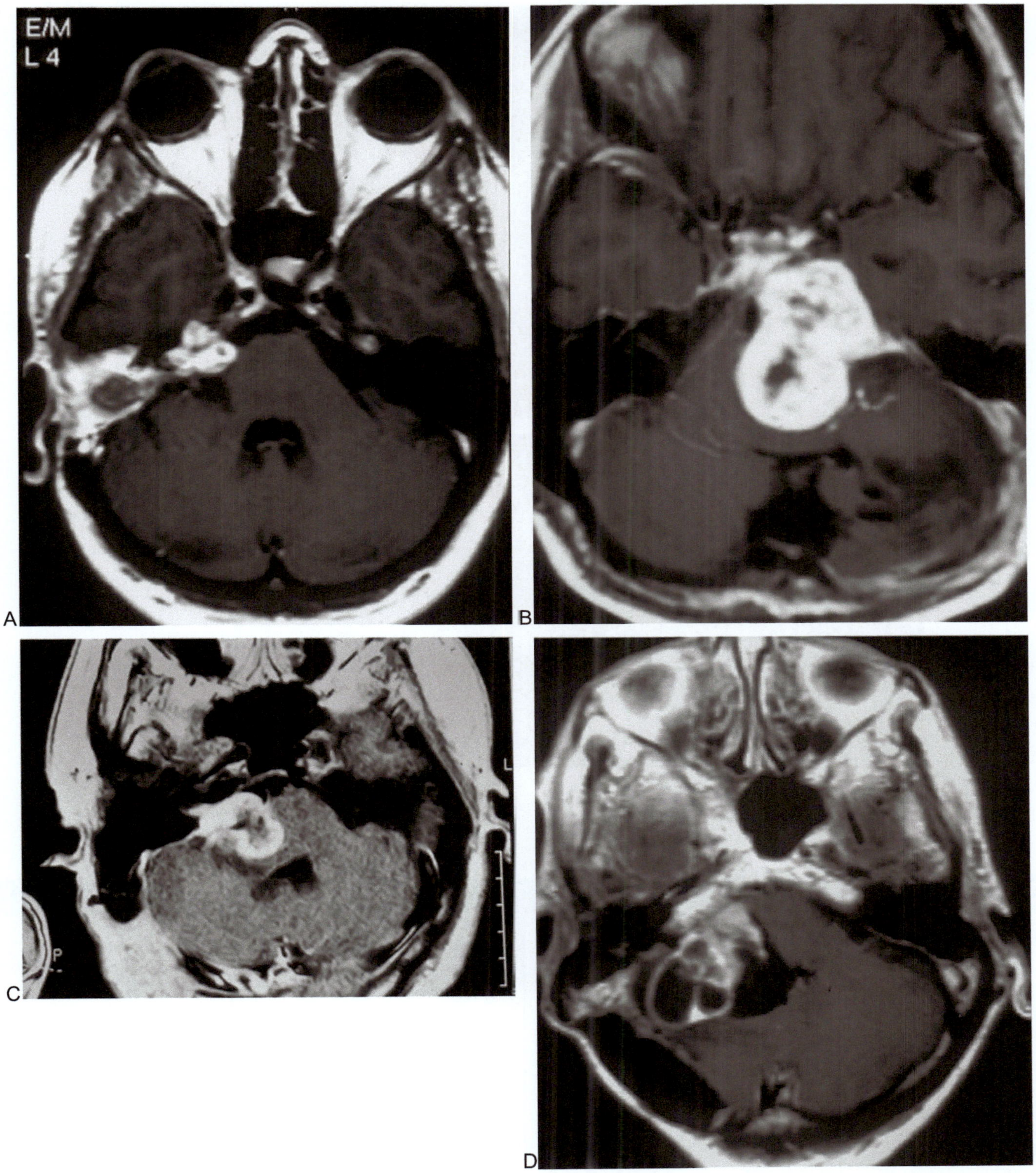

Fig. 5.99 (A) Residual tumor left at the level of the petrous apex. Note aniatrogenic cholesteatoma embedded into the fat used to obliterate the transotic approach. (B) Huge residual lesion occupying the prepontine cistern with significant compression of the brainstem. (C) Residual vestibular schwannoma after a retrosigmoid approach (note the scar adhesions between soft tissues and the dura of the cerebellum). (D) Large residual lesion with intratumoral necrotic changes.

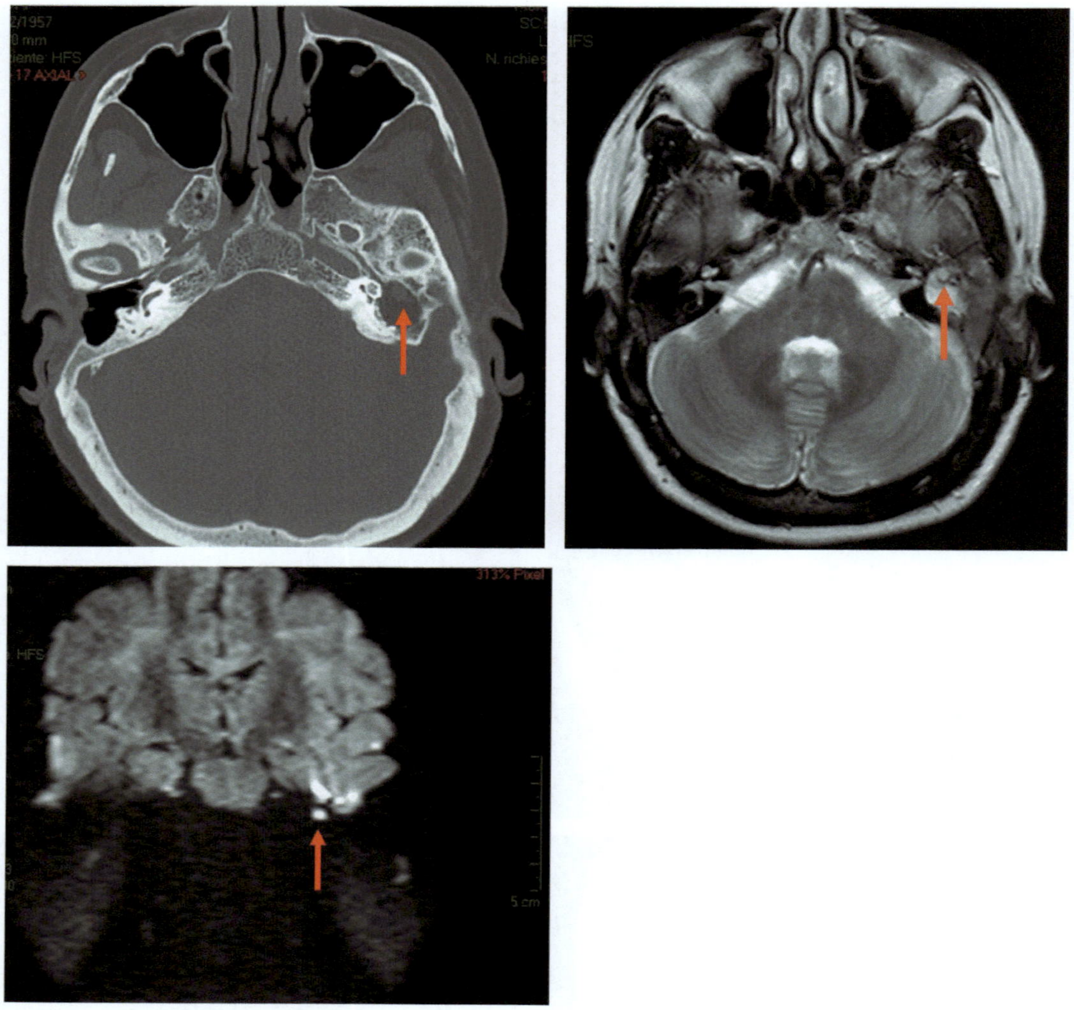

Fig. 5.100 A case of recurrent cholesteatoma.

REFERENCES

[1] Peris-Celda M, Martinez-Soriano F, Rhoton Jr AL. Atlas of head, neck, and brain: 2D and 3D images. New York: Thieme Medical Publishers; 2017. ISBN 978 1 60406 900 6.

[2] Cusimano MD, Di Ieva A, Lee JM. Handbook of skull base surgery. New York: Thieme Medical Publishers; 2016.

[3] Raut AA, Naphade PS, Chawla A. Imaging of skull base: pictorial essay. Indian J Radiol Imaging 2012;22(4):305–16. https://doi.org/10.4103/0971-3026.111485.

[4] Kraus DH, Lanzieri CF, Wanamaker JR, Little JR, Lavertu P. Complementary use of computed tomography and magnetic resonance imaging in assessing skull base lesions. Laryngoscope 1992;102:623–9.

[5] Parizel PM, Carpentier K, Van Marck V, et al. Pneumosinus dilatans in anterior skull base meningiomas. Neuroradiology 2013;55:307–11.

[6] Hudgins PA, Baugnon KL. Head and Neck: Skull Base Imaging. Neurosurgery 2018;82(3):255–67. https://doi.org/10.1093/neuros/nyx492.

[7] Parmar H, Gujar S, Shah G, Mukherji SK. Imaging of the anterior skull base. Neuroimaging Clin N Am 2009;19:427–39.

[8] Ginsberg LE. Perineural tumor spread associated with head and neck malignancies. In: Som PM, Curtin HD, editors. Head and Neck Imaging. 5th ed. St. Louis: Mosby; 2011.

[9] Borges A. Imaging of the central skull base. Neuroimaging Clin N Am 2009;19:669–96.

[10] Kirsch CF. Advances in magnetic resonance imaging of the skull base. Int Arch Otorhinolaryngol 2014;18(Suppl 2):S127–35. https://doi.org/10.1055/s-0034-1390013.

[11] Naidich TP, Castillo M, Cha S, Smirniotopoulos J. Imaging of the brain. Philadelphia: Elsevier/Saunders; 2013.

[12] Ginat D, Mangla R, Yeaney G, Ekholm S. Diffusion weighted imaging of skull lesions. J Neurol Surg B Skull Base 2014;75(3):204–13.

[13] Mathur A, Jain N, Kesavadas C, Thomas B, Kapilamoorthy TR. Imaging of skull base pathologies: Role of advanced magnetic resonance imaging techniques. Neuroradiol J 2015;28(4):426–37. https://doi.org/10.1177/1971400915609341.

[14] Bou-Assaly W, Srinivasan A, Mukherji SK. Head and neck high-field imaging: oncology applications. Neuroimaging Clin N Am 2012;22(2):285–96. xi.

[15] Parmar H, Gujar S, Shah G, Mukherji SK. Imaging of the anterior skull base. Neuroimaging Clin N Am 2009;19 (3):427–39.

[16] Lee TJ, Tsang NM, Huang Y, Hsu CL, Hsin LJ, Lee YH, Chang KP. Cavernous sinus involvement is not a risk factor for the primary tumor site treatment outcome of Sinonasal adenoid cystic carcinoma. J Otolaryngol Head Neck Surg 2018;47(1):12. https://doi.org/10.1186/s40463-018-0257-z.

[17] Thust SC, Yousry T. Imaging of skull base tumours. Rep Pract Oncol Radiother 2016;21(4):304–18. https://doi.org/10.1016/j.rpor.2015.12.008.

[18] Ong CK, Chong VF. Imaging of perineural spread in head and neck tumours. Cancer Imaging 2010;10(A(1A)):S92–8. Published 2010 Oct 4 https://doi.org/10.1102/1470-7330.2010.9033.

[19] Baumann F, Schmid C, Bernays RL. Intraoperative magnetic resonance imaging-guided transsphenoidal surgery for giant pituitary adenomas. Neurosurg Rev 2010;33(1):83–90.

[20] Choudhri O, Mindea SA, Feroze A, Soudry E, Chang SD, Nayak JV. Experience with intraoperative navigation and imaging during endoscopic transnasal spinal approaches to the foramen magnum and odontoid. Neurosurg Focus 2014;36(3):E4.

[21] Copeland WR, Hoover JM, Morris JM, Driscoll CL, Link MJ. Use of preoperative MRI to predict vestibular schwannoma intraoperative consistency and facial nerve outcome. J Neurol Surg B Skull Base 2013;74(6):347–50.

[22] Hines JP, Howard BE, Hoxworth JM, Lal D. Positive and negative predictive value of PET-CT in skull base lesions: case series and systematic literature review. J Neurol Surg Rep 2016;77(1):e39–45. https://doi.org/10.1055/s-0035-1570387.

[23] Tomura N, Mizuno Y, Saginoya T. PET/CT findings for tumors in the base of the skull: comparison of 18 F-FDG with 11 C-methionine. Acta Radiol 2016;57(3):325–32. https://doi.org/10.1177/028418511557 5342. Epub 2015 Mar 19.

[24] Bi W, Brown P, Abolfotoh M, Al-Mefty O, Mukundan S, Dunn IF. Utility of dynamic computed tomography angiography in the preoperative evaluation of skull base tumors. J Neurosurg JNS 2015;123(1):1–8.

[25] Kelly HR, Curtin HD. Imaging of skull base lesions. Handb Clin Neurol 2016;135:637–57. https://doi.org/10.1016/B978-0-444-53485-9.00030-1.

[26] Stacey RJ, Powell MP. Sellar and parasellar tumors. In: Moore AJ, Newell DW, editors. Neurosurgery. Springer, London: Springer Specialist Surgery Series; 2005.

[27] Rangel-Castilla L, Russin JJ, Spetzler RF. Surgical management of skull base tumors. Rep Pract Oncol Radiother 2016;21 (4):325–35. https://doi.org/10.1016/j.rpor.2014.09.002.

[28] Hofmann E, Prescher A. The clivus: anatomy, normal variants and imaging pathology. Clin Neuroradiol 2011;28.

[29] Geng S, Zhang J, Zhang LW, et al. Diagnosis and microsurgical treatment of chondromas and chondrosarcomas of the cranial base. Oncol Lett 2014;8(1):301–4. https://doi.org/10.3892/ol.2014.2072.

[30] Som PM, Curtin HD. Head and Neck Imaging. USA: Elsevier Health Sciences; 2002.

[31] Kristin J, Schipper JH, Jalkhi M, Zahnert T, Wagenmann M. Challenges in histologic diagnosis of nonchordomatous lesions of the clivus. JAMA Otolaryngol Head Neck Surg 2015;141(8):745–50.

[32] Müller U, Kubik-Huch RA, Ares C, Hug EB, Löw R, Valavanis A, Ahlhelm FJ. Is there a role for conventional MRI and MR diffusion-weighted imaging for distinction of skull base chordoma and chondrosarcoma? Acta Radiol 2016;57 (2):225–32.

[33] Santegoeds RGC, Temel Y, Beckervordersandforth JC, Van Overbeeke JJ, Hoeberigs CM. State-of-the-art imaging in human chordoma of the skull base. Curr Radiol Rep 2018;6 (5):16. https://doi.org/10.1007/s40134-018-0275-7.

[34] Almefty K, Pravdenkova S, Colli BO, Al-Mefty O, Gokden M. Chordoma and chondrosarcoma: similar, but quite different, skull base tumors. Cancer 2007;110 (11):2457–67.

[35] Dubrulle F, Souillard R, Hermans R. Extension patterns of nasopharyngeal carcinoma. Eur Radiol 2007;17: 2622–30.

[36] Razek AA, Castillo M. Imaging Lesions of the Cavernous Sinus. Am J Neuroradiol 2009;30(3):444–52. https://doi.org/10.3174/ajnr.A1398.

[37] Deconde AS, Sanaiha Y, Suh JD, Bhuta S, Bergsneider M, Wang MB. Metastatic disease to the clivus mimicking clival chordomas. J Neurol Surg B Skull Base 2013;74(5):292–9. https://doi.org/10.1055/s-0033-1348027.

[38] Das KK, Kumar R, Ashish K, et al. Extramedullary foramen magnum tumors and their surgical management: An experience with 29 cases. Asian J Neurosurg 2014;9(4):223–32. https://doi.org/10.4103/1793-5482.146616.

[39] Royer MC, Pensak ML. Cholesterol granulomas. Curr Opin Otolaryngol Head Neck Surg 2007;15:319–22.

[40] Razek AA, Huang BY. Lesions of the petrous apex: classification and findings at CT and MR imaging. Radiographics. 2012;32(1):151–73. https://doi.org/10.1148/rg.321105758. PMID: 22236899.

[41] Piccirillo E, Wiet MR, Flanagan S, et al. Cystic vestibular schwannoma: classification, management and facial nerve outcomes. Otol Neurotol 2009;30:826–34. https://doi.org/10.1097/MAO.0b013e3181b04e18. pmid:19704364.

[42] Benech F, Perez R, Fontanella MM, et al. Cystic versus solid vestibular schwannomas: a series of 80 grade III-IV patients. Neurosurg Rev 2005;28:209–13. https://doi.org/10.1007/s10143-005-0380-y. pmid:15739069.

[43] Lin EP, Crane BT. The Management and Imaging of Vestibular Schwannomas. Am J Neuroradiol 2017;38(11):2034–43. https://doi.org/10.3174/ajnr.A5213.

[44] Sanna M, Jain Y, De Donato G, Rohit LL, Taibah A. Management of jugular paragangliomas: the Gruppo Otologico experience. Otol Neurotol 2004;25(5):797–804.

[45] Suárez C, Sevilla MA, Llorente JL. Temporal paragangliomas. Eur Arch Otorhinolaryngol 2007;264(7):719–31.

[46] Sanna MK, Mancini T, Russo F, Taibah A. The Facial Nerve in Temporal Bone and Lateral Skull Base Microsurgery. Stuttgart: Thieme; 2006.

[47] Macdonald AJ, Salzman KL, Harnsberger HR, Gilbert E, Shelton C. Primary jugular foramen meningioma: imaging appearance and differentiating features. AJR Am J Roentgenol 2004;182(2):373–7.

[48] Som P, Curtin H. Head and Neck Imaging. St Louis: Mosby; 2003.

[49] Van den Berg R, Verbist BM, Mertens BJ, van der Mey AG, van Buchem MA. Head and neck paragangliomas: improved tumor detection using contrast-enhanced 3D time-of-flight MR angiography as compared with fat-suppressed MR imaging techniques. AJNR Am J Neuroradiol 2004;25(5):863–70.

[50] Eldevik OP, Gabrielsen TO, Jacobsen EA. Imaging findings in schwannomas of the jugular foramen. AJNR Am J Neuroradiol 2000;21(6):1139–44.

[51] Eldevik OP, Gabrielsen TO, Jacobsen EA. Imaging findings in schwannomas of the jugular foramen. AJNR Am J Neuroradiol 2000;21(6):1139–44.

[52] Frank G, Sciarretta V, Calbucci F, Farneti G, Mazzatenta D, Pasquini E. The endoscopic transnasal transsphenoidal approach for the treatment of cranial base chordomas and chondrosarcomas. Neurosurgery 2006;59. ONS50–ONS57.

[53] Wanebo JE, Bristol RE, Porter RR, Coons SW, Spetzler RF. Management of cranial base chondrosarcomas. Neurosurgery 2006;58(2):249–55. discussion 249–255.

[54] Edfeldt L, Strömbäck K, Danckwardt-Lillieström N, Rask-Andersen H, Abdsaleh S, Wikström J. Non-echo planar diffusion-weighted MRI increases follow-up accuracy after one-step step canal wall-down obliteration surgery for cholesteatoma. Acta Otolaryngol 2013;133(6):574–83.

[55] Dremmen MH, Hofman PA, Hof JR, Stokroos RJ, Postma AA. The diagnostic accuracy of non-echo-planar diffusion-weighted imaging in the detection of residual and/or recurrent cholesteatoma of the temporal bone. AJNR Am J Neuroradiol 2012;33(3):439–44.

[56] Sykopetrites V, Taibah A, Piras G, Giannuzzi AL, Mancini F, Sanna M. The otologic approach in the management of posterior petrous surface meningiomas. Eur Arch Otorhinolaryngol 2022;279(12):5655–65. https://doi.org/10.1007/s00405-022-07442-3. Epub 2022 Jun 29. PMID: 35767053.

[57] Sanna M, Fois P, Pasanisi E, Russo A, Bacciu A. Middle ear and mastoid glomus tumors (glomus tympanicum): an algorithm for the surgical management. Auris Nasus Larynx 2010;37(6):661–8. https://doi.org/10.1016/j.anl.2010.03.006. Epub 2010 Apr 18. PMID: 20400250.

[58] Shin SH, Sivalingam S, De Donato G, Falcioni M, Piazza P, Sanna M. Vertebral artery involvement by tympanojugular paragangliomas: management and outcomes with a proposed addition to the fisch classification. Audiol Neurootol 2012;17(2):92–104. https://doi.org/10.1159/000330724. Epub 2011 Aug 18. PMID: 21849774.

CHAPTER 6

Subtotal Petrosectomy

Introduction

Subtotal petrosectomy (STP), with obliteration of the middle ear and mastoid and closure of the external auditory canal (EAC), is an effective solution in mastoid and middle ear diseases that are chronic and recurrent, leaving behind a large surgical cavity, exposing vital structures like the dura, carotid, inner ear fluids, or cerebrospinal fluid (CSF), with no realistic chance of reconstruction of the conductive apparatus [1,2].

Although this technique was introduced in the late 1950s and 1960s [3,4], Ugo Fisch introduced the term "subtotal petrosectomy" in 1965 [2]. He described STP as the complete exenteration of all pneumatic tracts in the temporal bone, including retrofacial, retrosigmoid, antral, retrolabyrinthine, supralabyrinthine, infralabyrinthine, peritubal, and pericarotid cell tracts (Fig. 6.1). Although the initial description involved preservation of the otic capsule, it allowed expanding the procedure, when necessary, to exenterate the cochlea, the labyrinth, or the internal auditory canal.

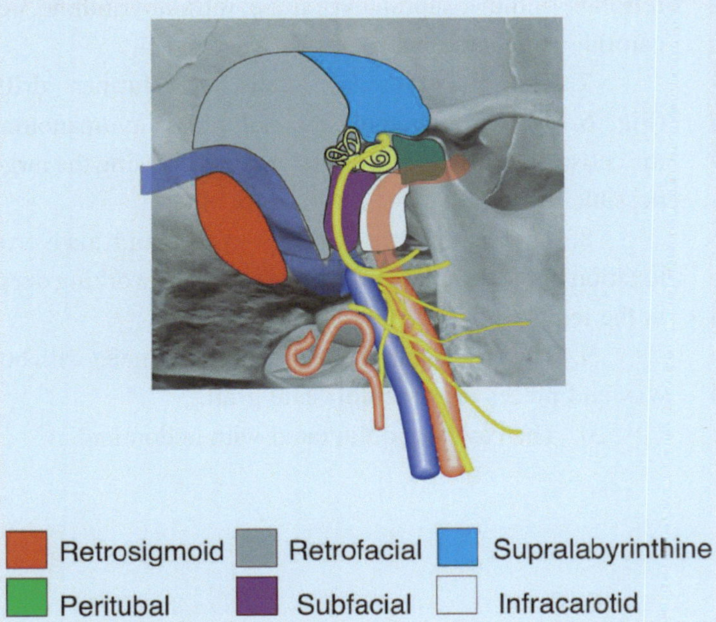

Fig. 6.1 The classical description of subtotal petrosectomy includes complete exenteration of all pneumatic tracts in the temporal bone, including retrofacial, retrosigmoid, antral, retrolabyrinthine, supralabyrinthine, infralabyrinthine, peritubal, and pericarotid cell tracts.

Atlas of Microsurgery of the Lateral Skull Base. https://doi.org/10.1016/B978-0-443-19083-4.00006-4

Indications

– Canal wall down cavity with intractable inflammation with no serviceable hearing.

– Chronic otitis media with no serviceable hearing with multiple previous surgeries.

– Recurrent cholesteatoma with huge mastoid aeration.

– Sustained cerebrospinal fluid (CSF) leak with either spontaneous, posttraumatic, or surgical origin.

– Large meningoencephalic herniation.

– Extensive exposure of middle fossa dura.

– Cochlear implants in special cases.

– Paragangliomas of Class B3.

– Malignancy of the temporal bone.

Surgical Steps (Right Ear)

1. A C-shaped skin incision is made, extending 4–5 cm behind the retroauricular sulcus (Fig. 6.2).

2. The external auditory canal is transected at the bony cartilaginous junction (Fig. 6.3). The skin of the lateral part is dissected free from the cartilage for 1 cm (Fig. 6.4). Two stay sutures are then placed through this skin. Using artery forceps, the sutures are pulled through the external auditory canal, thus everting the free skin edges. These are then closed with 3-0 Vicryl sutures (Fig. 6.5). The cartilage is then folded and sutured in place to reinforce the blind sac closure (Fig. 6.6).

3. A T-shaped musculofascial incision is made with monopolar diathermy. The soft tissues are elevated from the mastoid with the mastoid raspatory. The attachment of the sternocleidomastoid muscle is freed from the lateral portion of the mastoid tip. The soft tissues are held with two articulated self-retaining retractors. The sternocleidomastoid muscle is retracted to expose the entire mastoid tip (Fig. 6.7). A microraspatory is used to elevate the remnants of the external auditory canal skin down to the tympanic annulus.

4. The initial stages of the mastoidectomy are done without magnification (Fig. 6.8). Drilling is extended well above the level of the middle fossa dura and behind the sigmoid sinus to facilitate saucerization of the cavity. All pneumatic cells are removed. A thin sheet of bone is left protecting the dura and sigmoid sinus. The middle cranial fossa dura is identified first, followed by the sigmoid sinus, sinodural angle, mastoid tip cells, and the mastoid antrum. The posterior bony canal wall is lowered to the level of the remaining skin margin. Drilling is done with a cutting burr with continuous suction-irrigation. All overhanging edges of bone are removed to allow the collapse of the overlying tissues when the wound is closed.

5. All air cell tracts associated with the middle ear and mastoid should be totally removed. The facial nerve is identified at the digastric ridge by following the stylomastoid periosteum, which runs medial and anterior to the stylomastoid foramen. The facial bridge is removed with a curette to expose the fallopian canal from the oval window to the stylomastoid foramen. The tympanic membrane is now removed in continuity with the cuff of canal skin to expose the ossicular chain (Fig. 6.9).

6. The incudostapedial joint is separated with an angled hook or joint knife (Fig. 6.10). The incus is then removed with forceps. The tympanic segment of the facial nerve is identified as inferior to the lateral semicircular canal at the oval window (Fig. 6.11). The tensor tympani tendon and anterior malleal ligament are sectioned with microscissors (Fig. 6.12). The anterior epitympanic space is visualized, in particular the cog and its relationship to the tympanic segment of the facial nerve, cochleariform process, and tensor tympani canal (Fig. 6.13). Cog is removed with a diamond burr to expose the supratubal recess (Fig. 6.14). All bone except for a thin covering is removed from the nerve. The pneumatic cells are exenterated in the following order: retrosigmoid, retrofacial, retrolabyrinthine, supralabyrinthine, infralabyrinthine, pericarotid, and supratubal.

7. The hypotympanic cells are further drilled (Fig. 6.15). Through the subfacial recess tympanotomy, the cells under the carotid artery are opened using the largest possible diamond burr (Figs. 6.16–6.18).

8. The stapes arch is removed to avoid inadvertent luxation of the footplate when the surgeon is working deeper in the temporal bone.

9. The eustachian tube orifice is obliterated with bone wax and pieces of musculofascial graft.

10. The cavity is obliterated with abdominal.

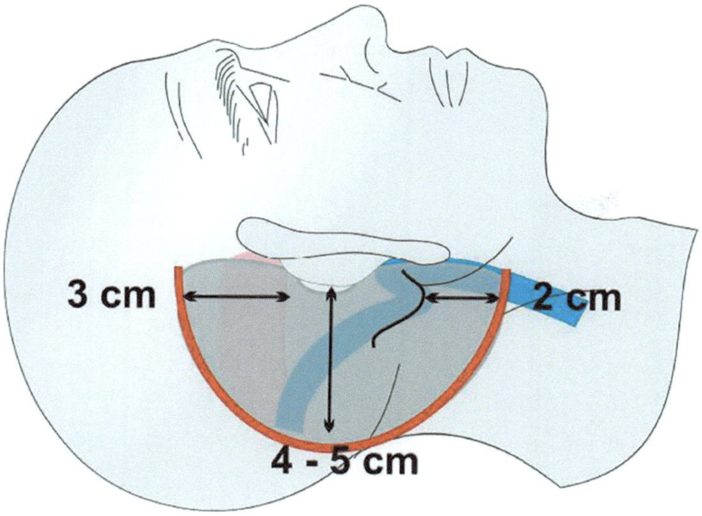

Fig. 6.2 A C-shaped skin incision is made, extending 4–5 cm behind the retroauricular sulcus.

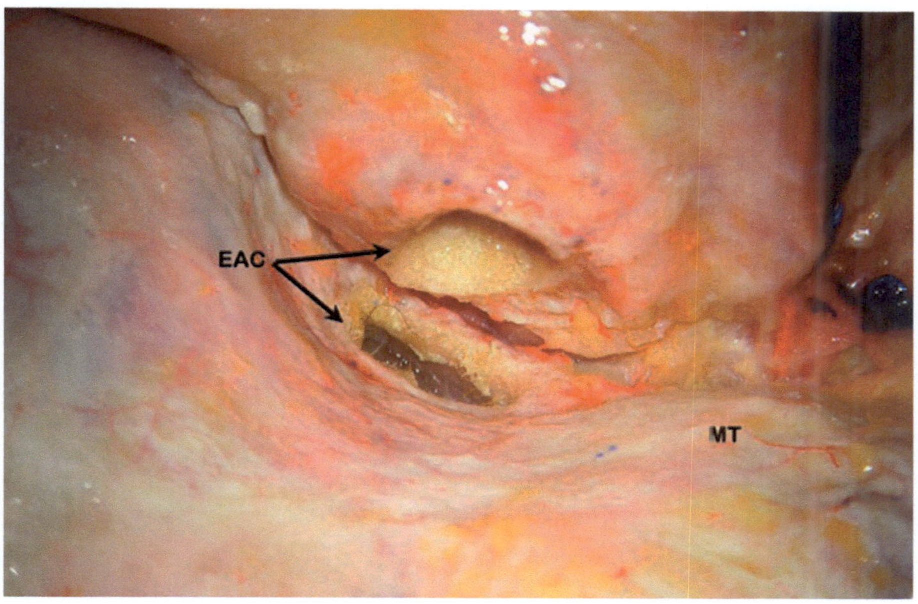

Fig. 6.3 The external auditory canal is transected at the bony cartilaginous junction. *EAC*, external auditory canal; *MT*, mastoid tip.

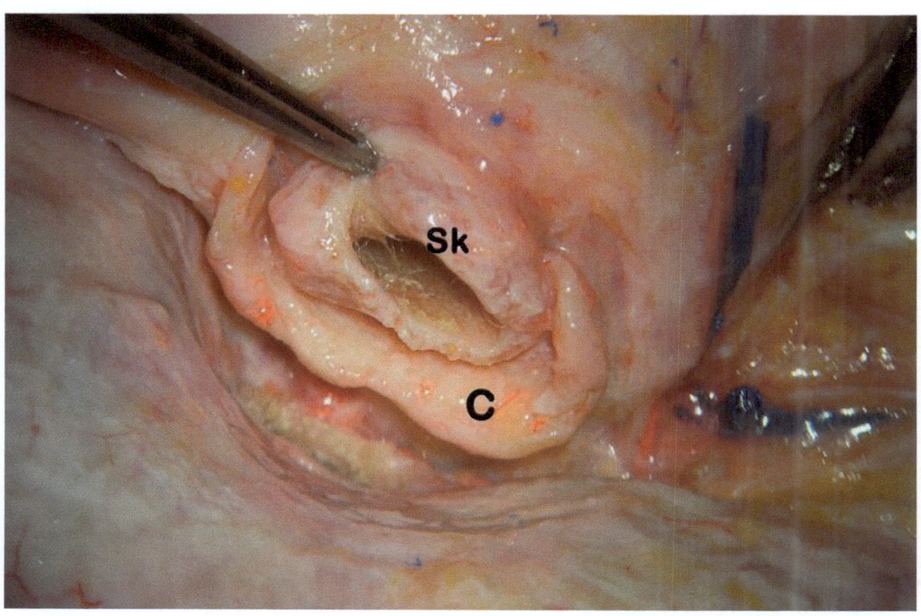

Fig. 6.4 The skin of the lateral part is dissected free from the cartilage. *Sk*, skin; *C*, cartilage.

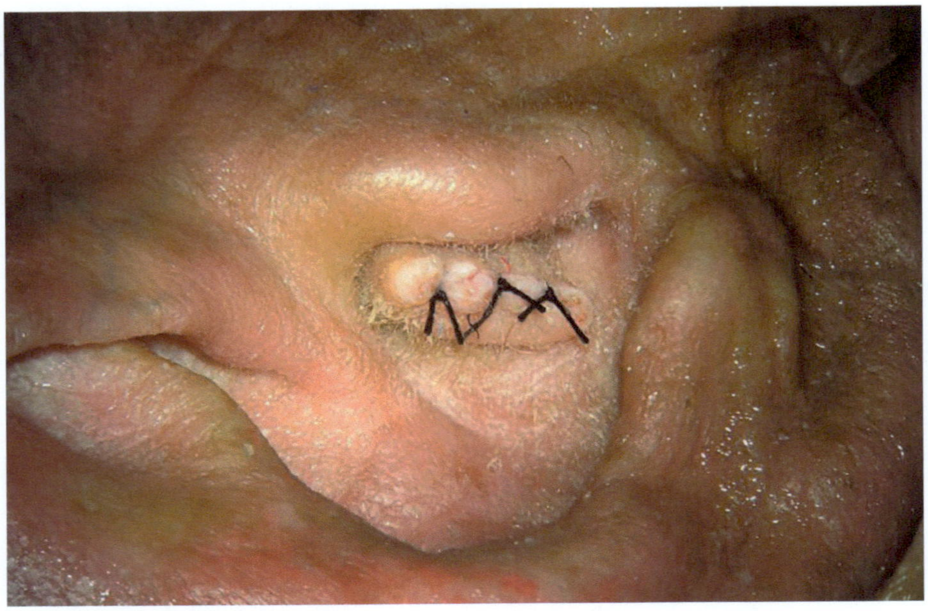

Fig. 6.5 The everted skin is tightly sutured with 3-0 Vicryl sutures.

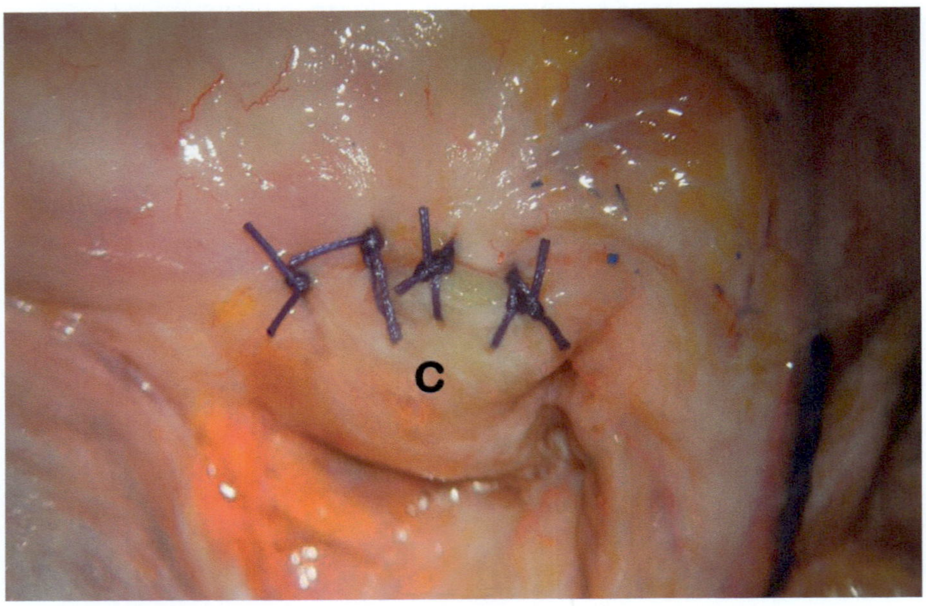

Fig. 6.6 The medial cartilage is then folded and sutured in place to reinforce the blind sac closure. *C*, cartilage.

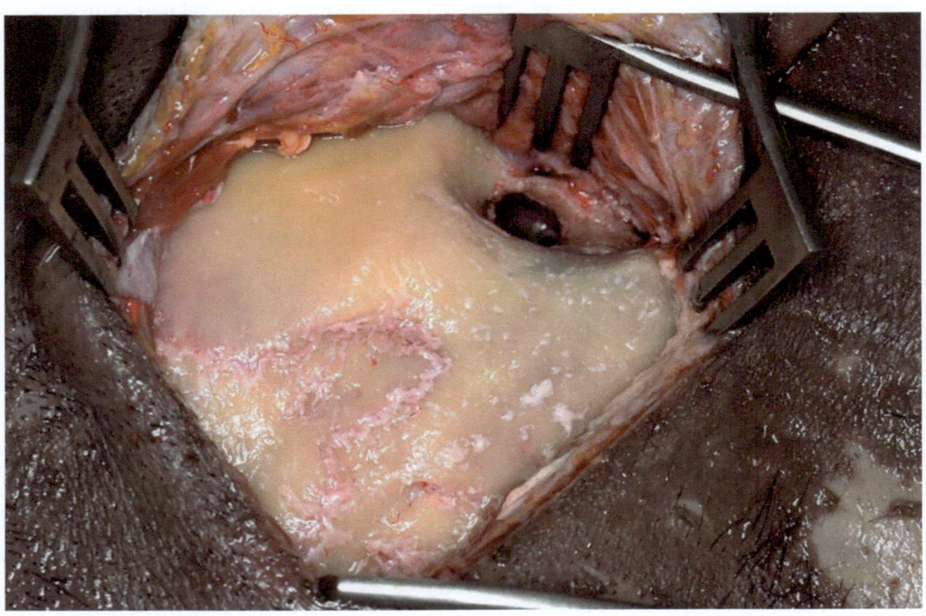

Fig. 6.7 The soft tissues are held with two articulated self-retaining retractors. The sternocleidomastoid muscle is retracted to expose the entire mastoid tip.

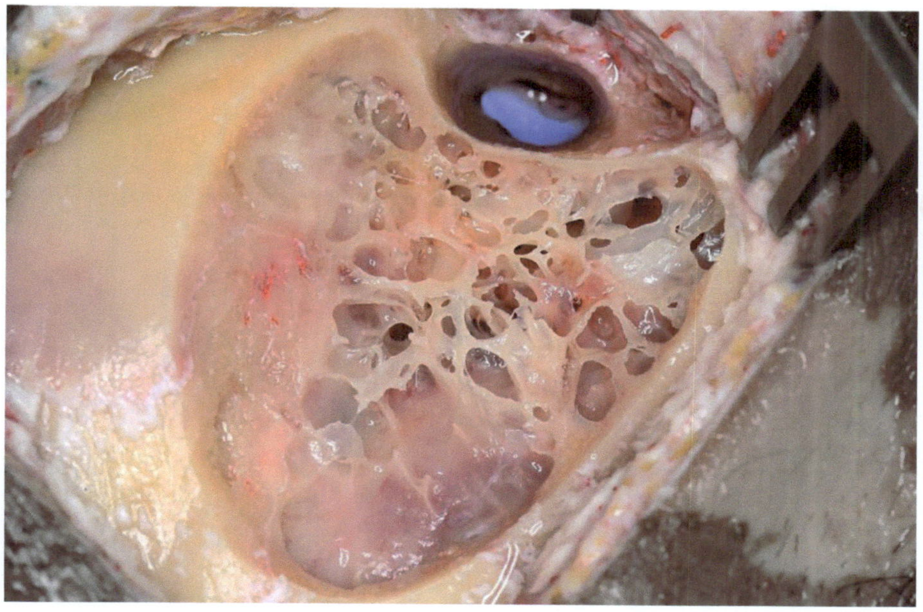

Fig. 6.8 In the early stage of mastoidectomy, drilling is carried out with a large cutting burr. Large cutting burrs allow faster bone removal and easier identification of structures. If drilling is conducted correctly, no important structure in danger is present in the triangle formed by the sigmoid sinus posteriorly, the middle fossa plate superiorly, and the posterior meatal wall anteriorly.

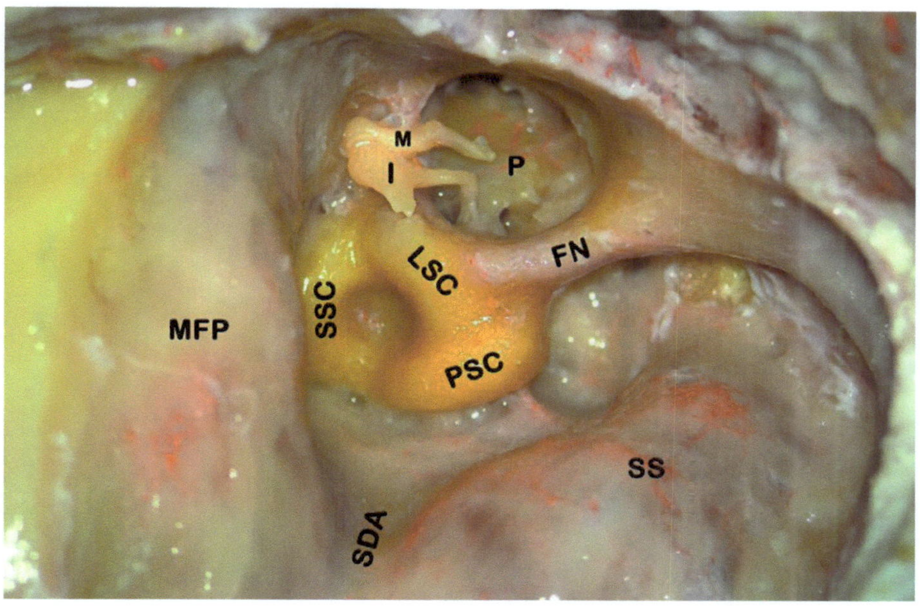

Fig. 6.9 The facial ridge is lowered evenly by drilling the posterior canal wall. All air cell tracts associated with the middle ear and mastoid should be totally removed. The facial bridge is removed with a curette. The tympanic membrane is removed in continuity with the cuff of the canal skin to expose the ossicular chain. *MFP*, middle fossa plate; *SS*, sigmoid sinus; *SDA*, sinodural angle; *LSC*, lateral semicircular canal; *PSC*, posterior semicircular canal; *SSC*, superior semicircular canal; *FN*, facial nerve; *I*, incus; *M*, malleus; *P*, promontory.

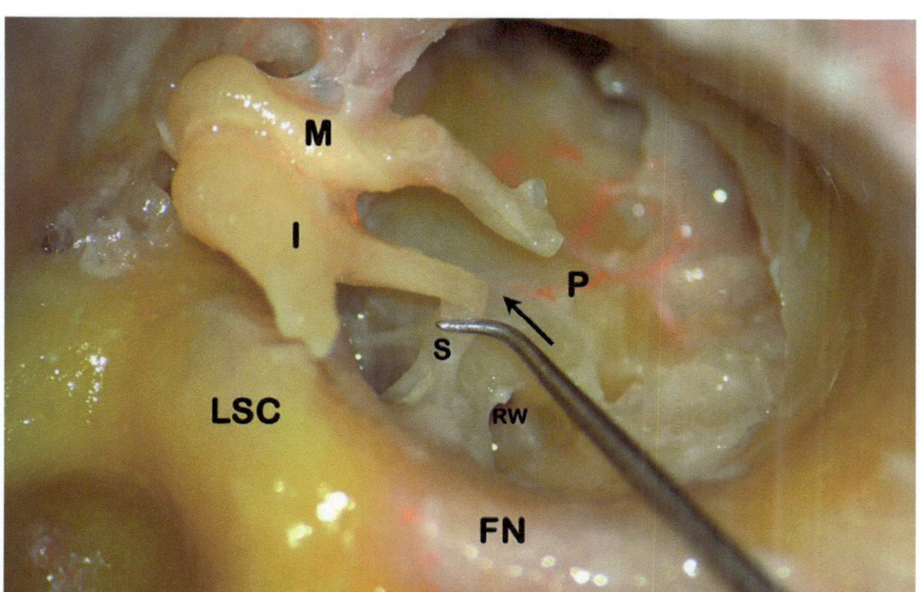

Fig. 6.10 The incudostapedial joint *(arrow)* is disarticulated using a small dissector, with movement from posterior to anterior. *LSC*, lateral semicircular canal; *FN*, facial nerve; *I*, incus; *M*, malleus; *S*, stapes; *P*, promontory; *RW*, round window.

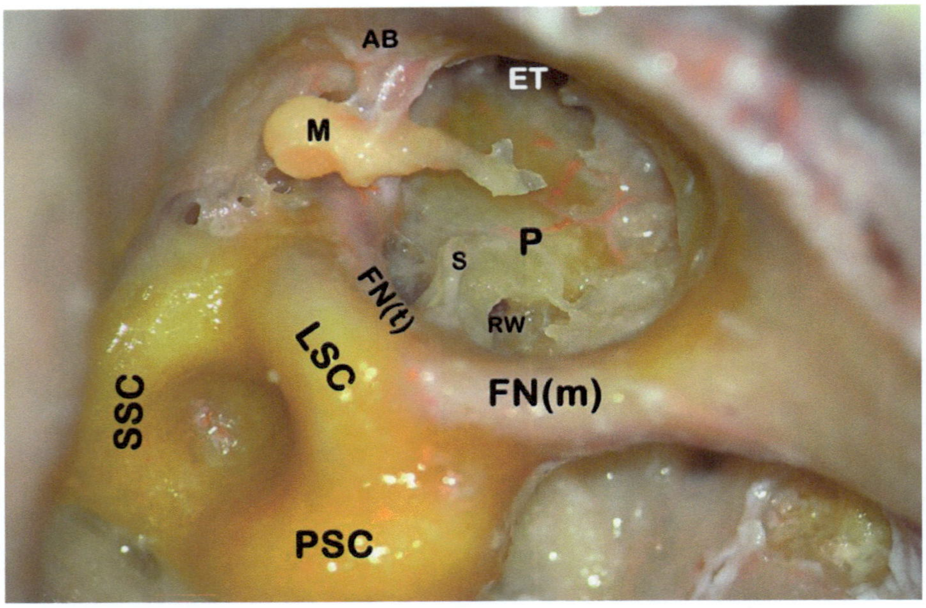

Fig. 6.11 The incus is then removed with forceps to expose the tympanic segment of the facial nerve. *LSC*, lateral semicircular canal; *PSC*, posterior semicircular canal; *SSC*, superior semicircular canal; *FN(t)*, tympanic segment of the facial nerve; *FN(m)*, mastoid segment of the facial nerve; *M*, malleus; *S*, stapes; *P*, promontory; *RW*, round window; *ET*, eustachian tube; *AB*, anterior buttress.

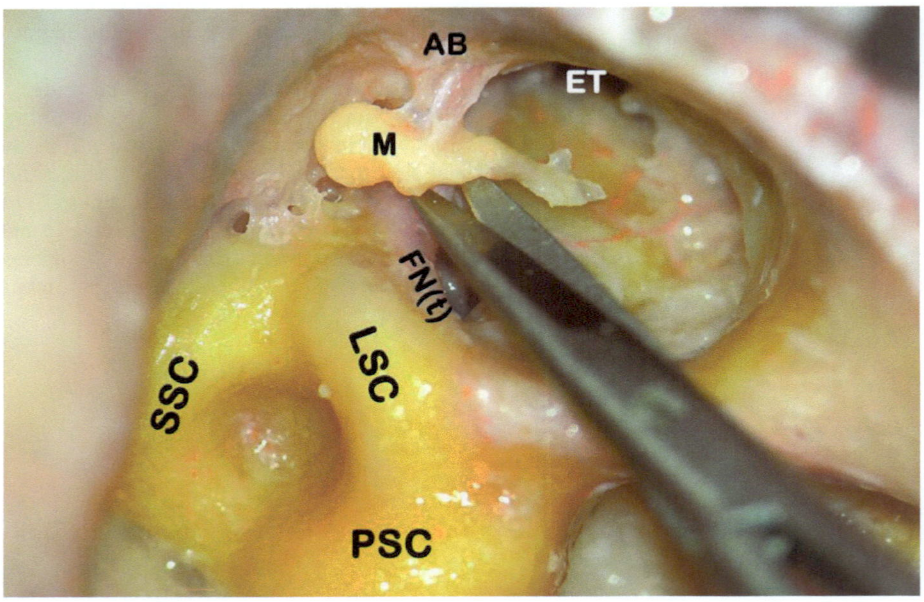

Fig. 6.12 The tensor tympani tendon and anterior malleal ligament are sectioned with microscissors. *LSC*, lateral semicircular canal; *PSC*, posterior semicircular canal; *SSC*, superior semicircular canal; *FN(t)*, tympanic segment of the facial nerve; *M*, malleus; *ET*, eustachian tube; *AB*, anterior buttress.

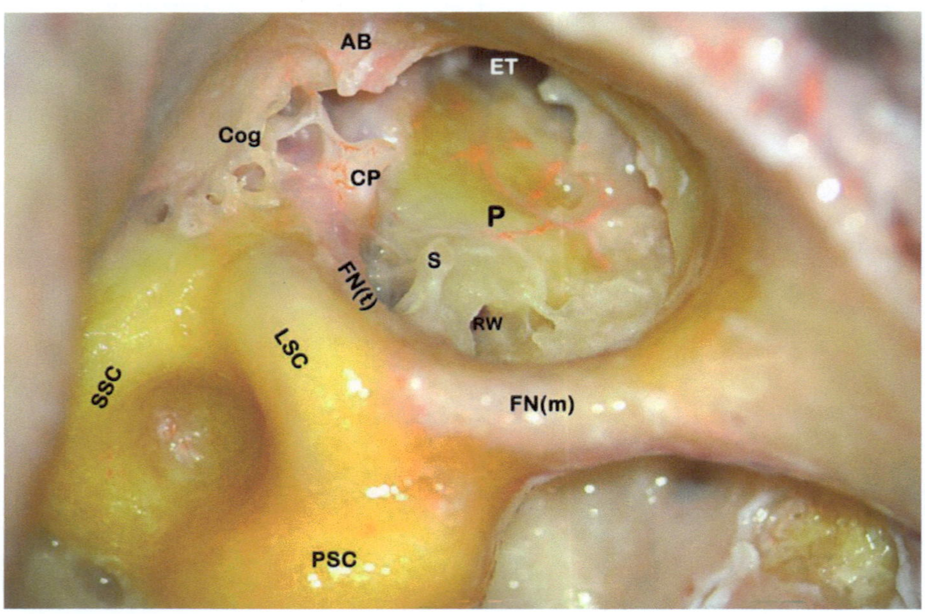

Fig. 6.13 The anterior epitympanic space is visualized, in particular the cog and its relationship to the tympanic segment of the facial nerve, cochleariform process, and tensor tympani canal. The structures around the cochleariform process are shown. *LSC*, lateral semicircular canal; *PSC*, posterior semicircular canal; *SSC*, superior semicircular canal; *FN(t)*, tympanic segment of the facial nerve; *FN(m)*, mastoid segment of the facial nerve; *ET*, eustachian tube; *AB*, anterior buttress; *CP*, cochleariform process; *P*, promontory; *S*, stapes; *RW*, round window.

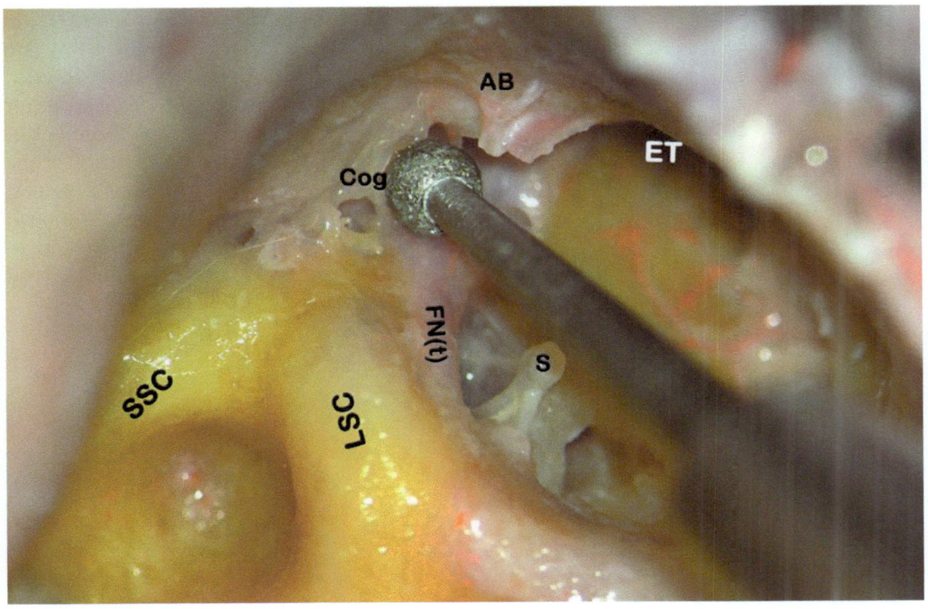

Fig. 6.14 Cog is removed with a diamond burr to expose the supratubal recess. The anterior buttress is removed to make the cavity rounded. *LSC*, lateral semicircular canal; *SSC*, superior semicircular canal; *FN(t)*, tympanic segment of the facial nerve; *ET*, eustachian tube; *AB*, anterior buttress; *S*, stapes.

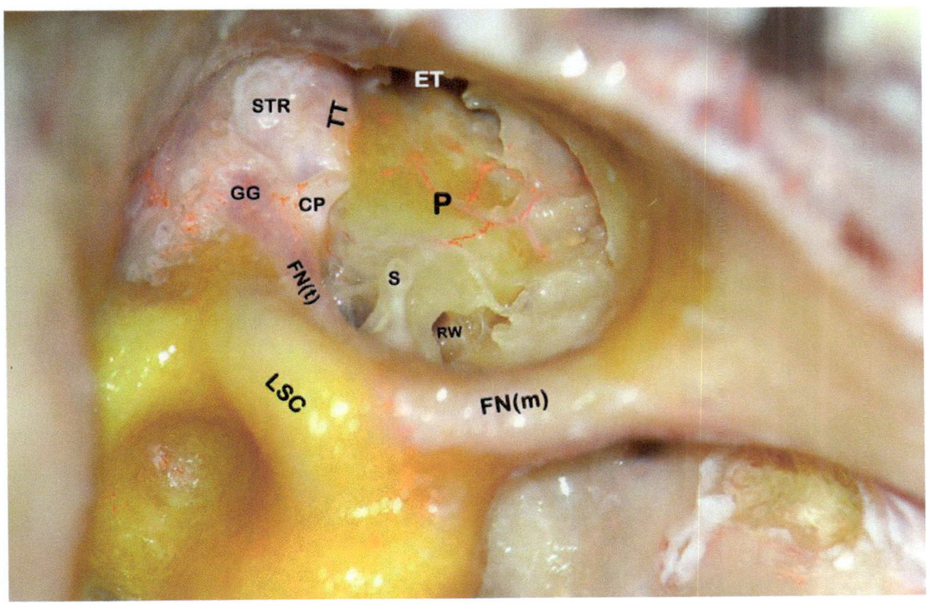

Fig. 6.15 The supratubal recess is completely opened and visualized. *LSC*, lateral semicircular canal; *FN(t)*, tympanic segment of the facial nerve; *FN(m)*, mastoid segment of the facial nerve; *GG*, geniculate ganglion; *ET*, eustachian tube; *S*, stapes; *CP*, cochleariform process; *P*, promontory; *RW*, round window; *TT*, tensor tympani muscle; *STR*, supratubal recess.

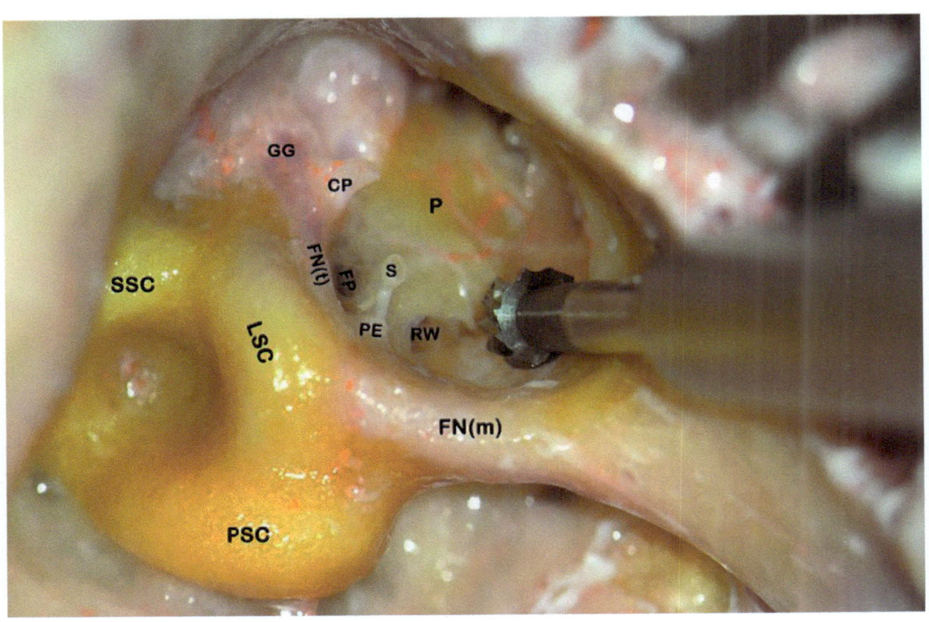

Fig. 6.16 The hypotympanic cells are further drilled. *LSC*, lateral semicircular canal; *PSC*, posterior semicircular canal; *SSC*, superior semicircular canal; *FN(t)*, tympanic segment of the facial nerve; *FN(m)*, mastoid segment of the facial nerve; *GG*, geniculate ganglion; *S*, stapes; *CP*, cochleariform process; *P*, promontory; *RW*, round window; *FP*, footplate of stapes; *PE*, pyramidal eminence.

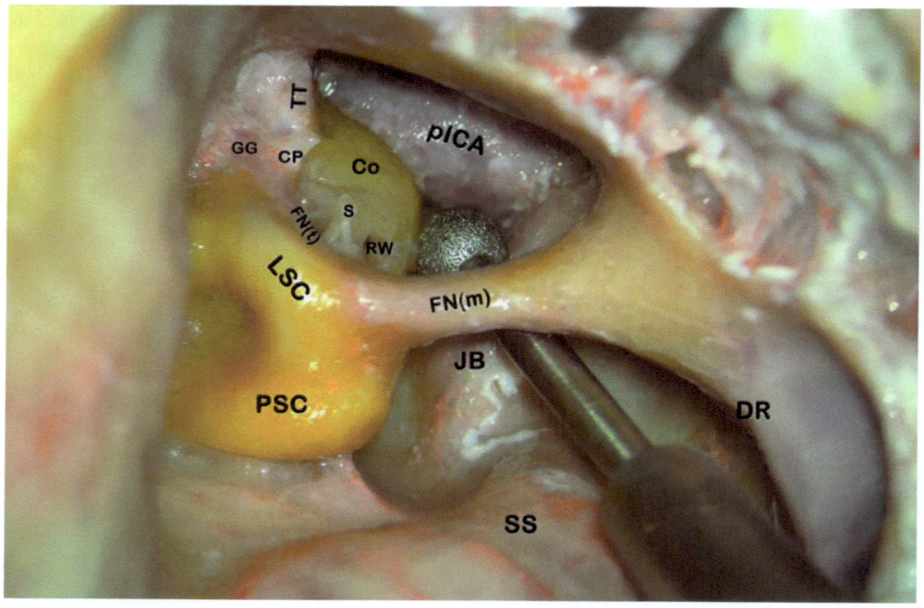

Fig. 6.17 Through the subfacial recess tympanotomy, the cells under the carotid artery are opened using the largest possible diamond burr. *LSC*, lateral semicircular canal; *PSC*, posterior semicircular canal; *FN(t)*, tympanic segment of the facial nerve; *FN(m)*, mastoid segment of the facial nerve; *GG*, geniculate ganglion; *S*, stapes; *CP*, cochleariform process; *RW*, round window; *Co*, cochlea; *TT*, tensor tympani muscle; *JB*, jugular bulb; *pICA*, petrous segment of internal carotid artery; *SS*, sigmoid sinus; *DR*, digastric ridge.

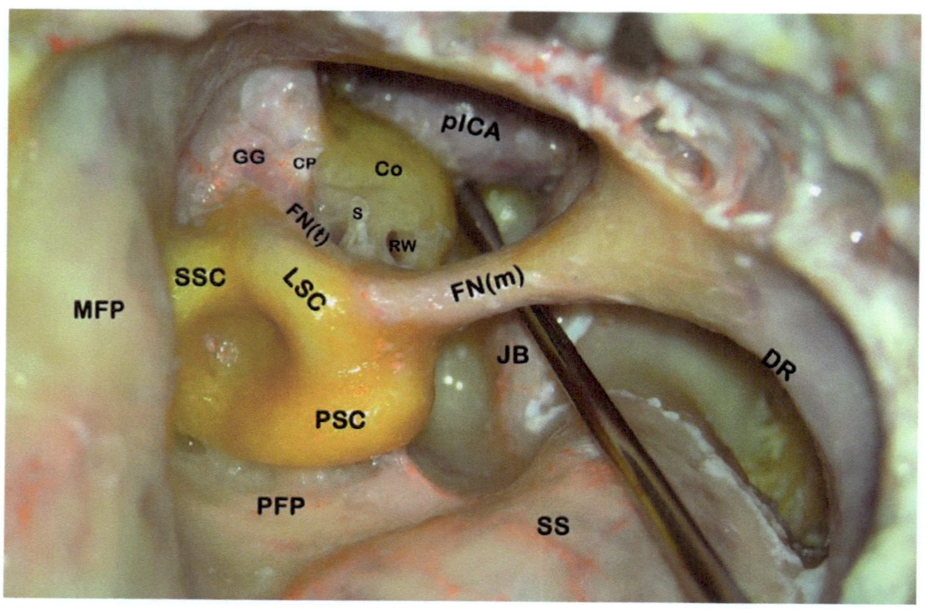

Fig. 6.18 The retrofacial and hypotympanic air cells have been drilled, as shown by the passage of the pointer medial to the mastoid segment of the facial nerve. *LSC*, lateral semicircular canal; *PSC*, posterior semicircular canal; *SSC*, superior semicircular canal; *FN(t)*, tympanic segment of the facial nerve; *FN(m)*, mastoid segment of the facial nerve; *GG*, geniculate ganglion; *S*, stapes; *CP*, cochleariform process; *RW*, round window; *Co*, cochlea; *JB*, jugular bulb; *pICA*, petrous segment of internal carotid artery; *MFP*, middle fossa plate; *SS*, sigmoid sinus; *PFP*, posterior fossa plate; *DR*, digastric ridge.

Clinical Cases

Case 1: Class B3 Tympanomastoid Paraganglioma

See Figs. 6.19–6.56.

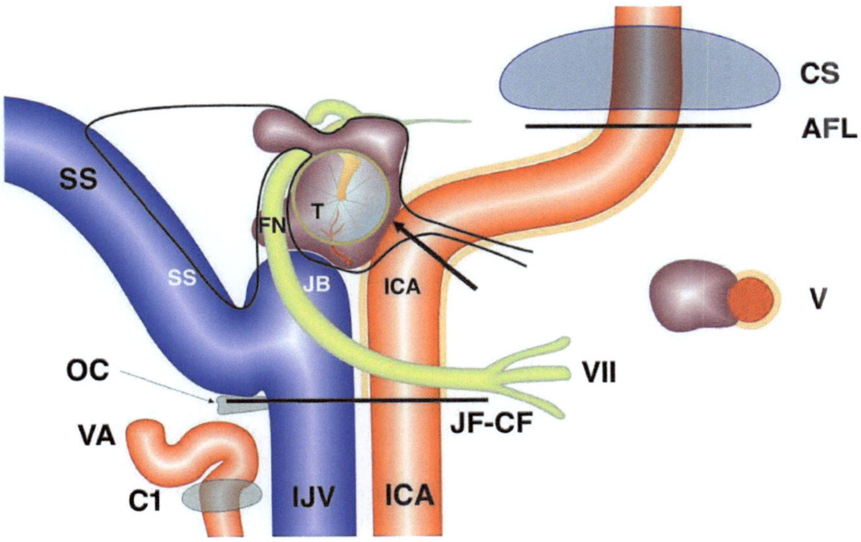

Fig. 6.19 A case of Class B3 glomus tumor in the right ear is shown *(arrow)*. *CS*, cavernous sinus; *AFL*, anterior foramen lacerum; *FN*, facial nerve; *ICA*, internal carotid artery; *V*, vertical portion of petrous carotid artery; *IJV*, internal jugular vein; *OC*, occipital condyle; *VA*, vertebral artery; *JB*, jugular bulb; *C1*, atlas; *JF-CF*, jugular foramen-carotid foramen; *SS*, sigmoid sinus; *T*, tumor.

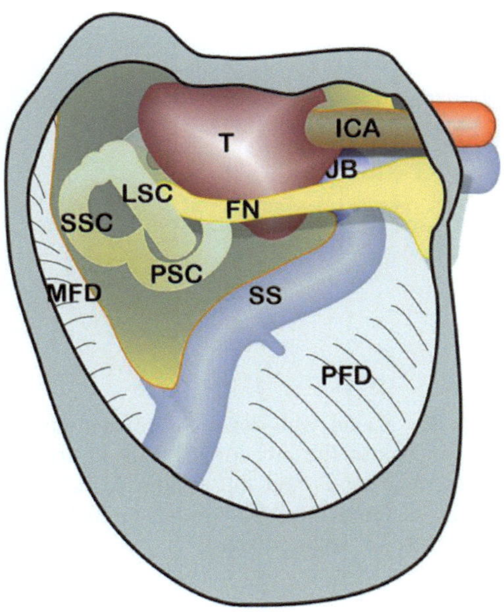

Fig. 6.20 The relationship between the glomus tumor and the internal carotid artery is shown. *ICA*, internal carotid artery; *JB*, jugular bulb; *FN*, facial nerve; *T*, tumor; *SS*, sigmoid sinus; *LSC*, lateral semicircular canal; *PSC*, posterior semicircular canal; *SSC*, superior semicircular canal; *MFP*, middle fossa plate; *PFP*, posterior fossa plate.

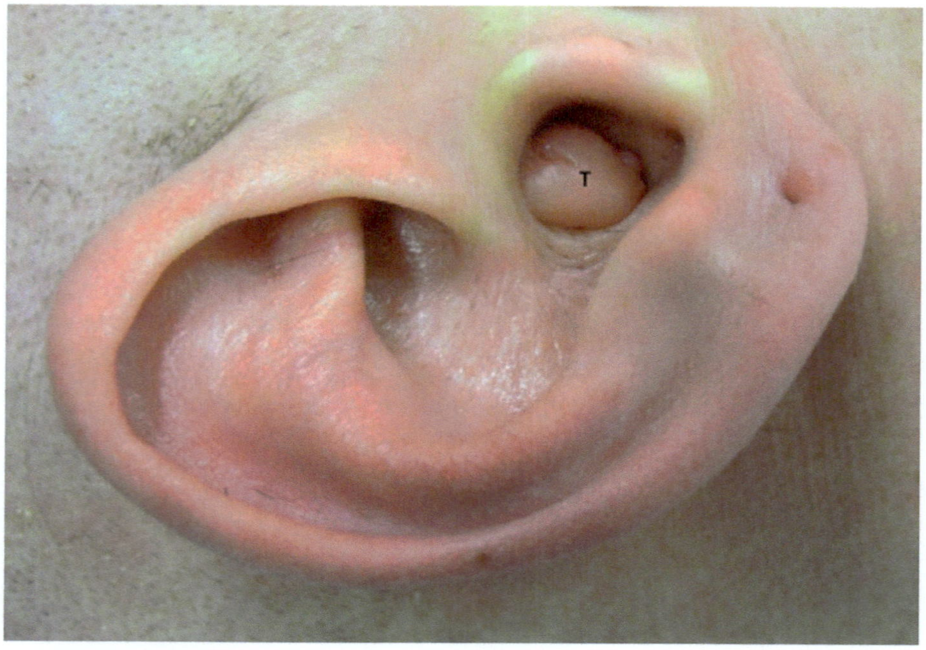

Fig. 6.21 The tumor (T) bulges as a polyp in the external auditory canal.

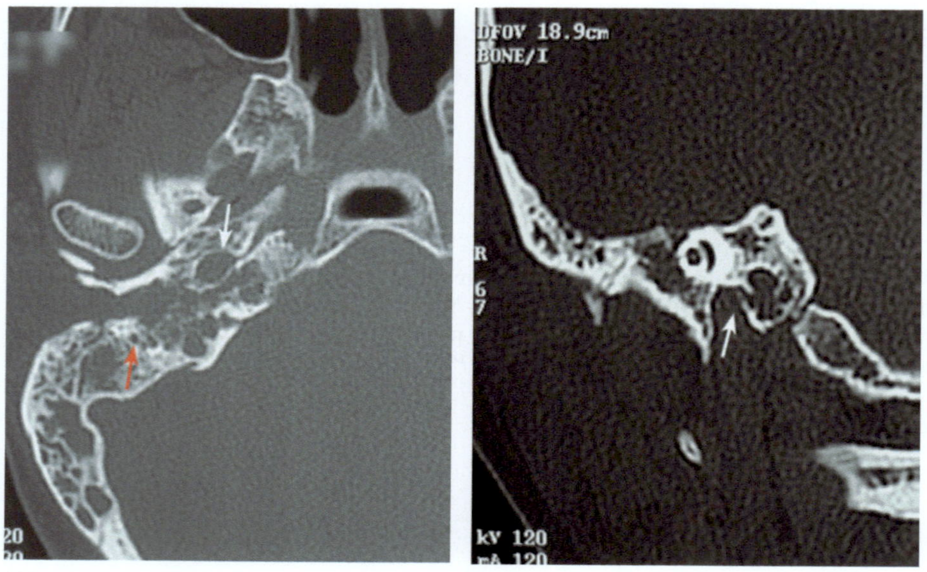

(A) (B)

Fig. 6.22 CT shows the tumor eroding the carotid canal *(white arrows)* and pericarotid cells. (A) The cells located medially to the facial nerve *(red arrow)* are also involved. (B) Erosion in the medial wall of the vertical portion is visualized in the coronal section.

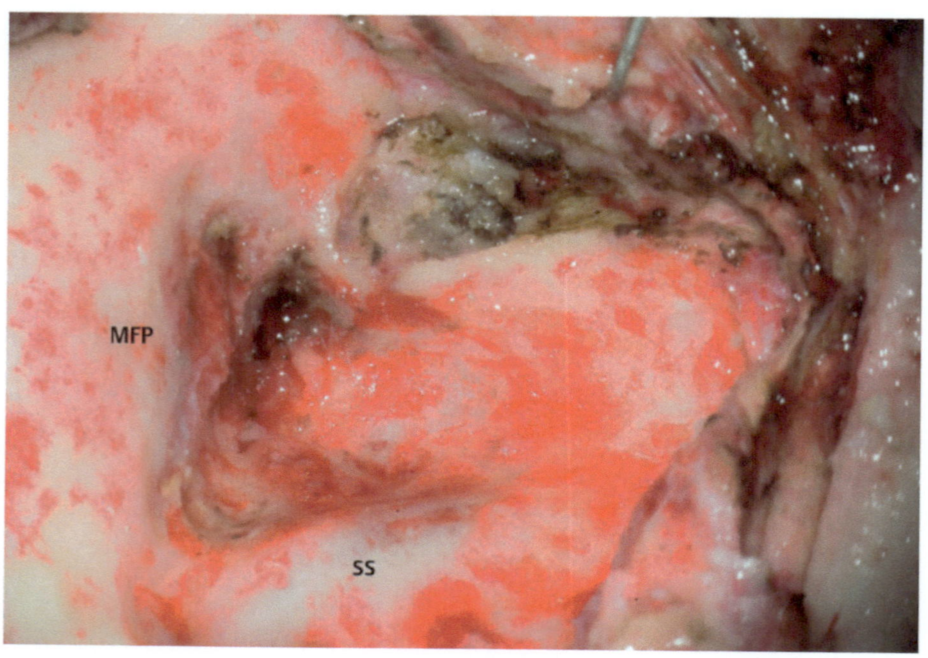

Fig. 6.23 Canal wall-down mastoidectomy is done to expose the area of the carotid artery. The sigmoid sinus and the middle fossa plate are well-skeletonized. *MFP*, middle cranial fossa dural plate; *SS*, sigmoid sinus.

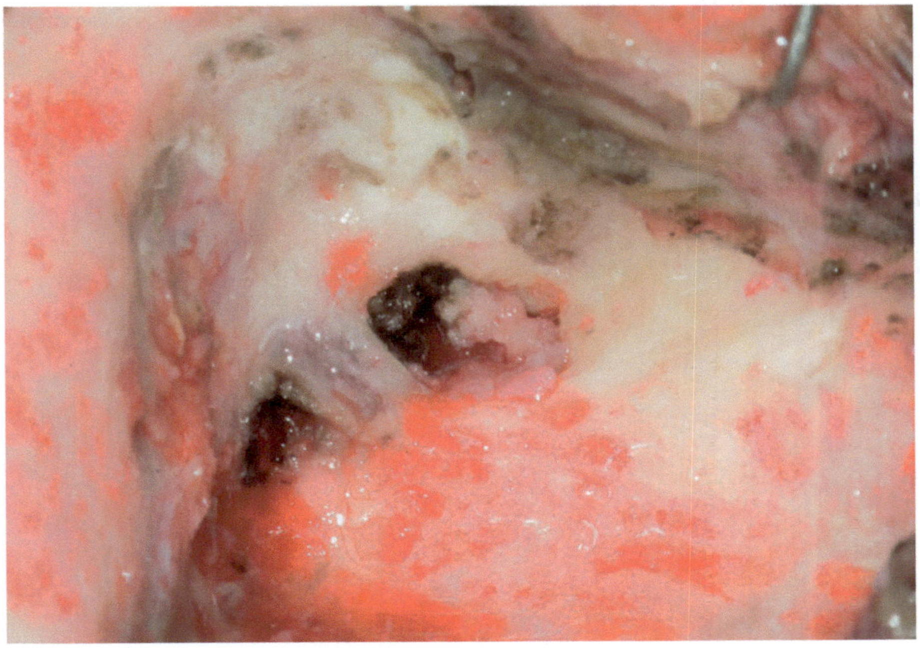

Fig. 6.24 The tumor is filling the attic completely.

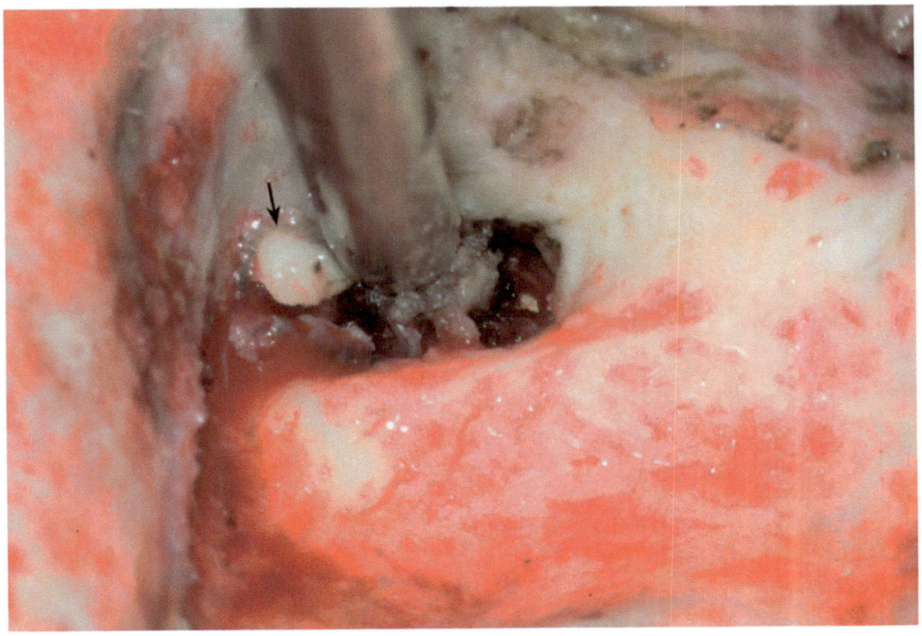

Fig. 6.25 The incus has been removed to expose the tumor and prevent damage to the inner ear. This also allows access to the tumor under the ossicles. The head of the malleus is visible *(arrow)*. The debulking of the tumor should be started from the inferior margin so as to prevent damage to the stapes. The tumor filling the tympanic cavity is removed in a piecemeal manner after sufficient bipolar coagulation.

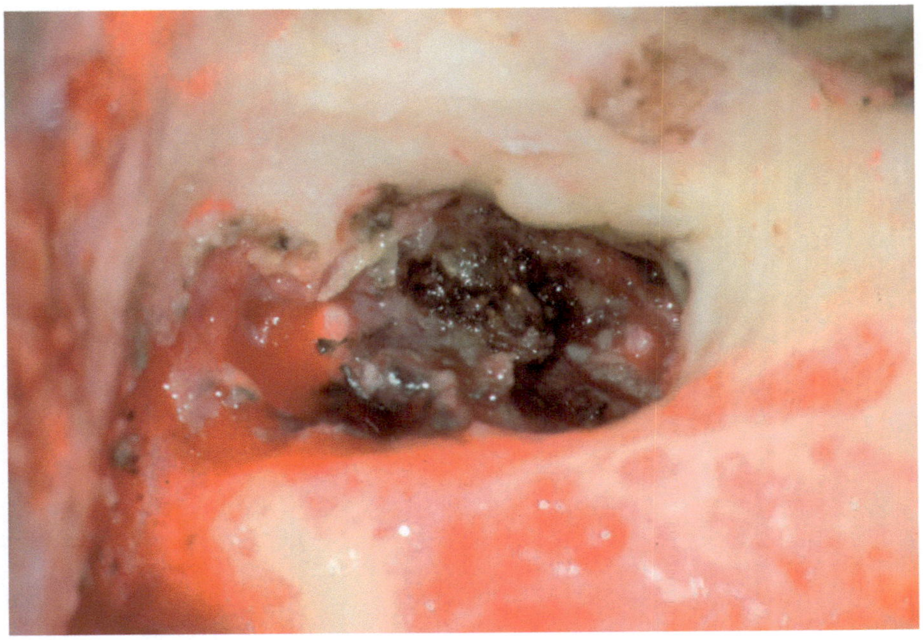

Fig. 6.26 The malleus is removed for further dissection of the tumor from the attic. The superior half of the tumor in the tympanic cavity is left untouched at this stage.

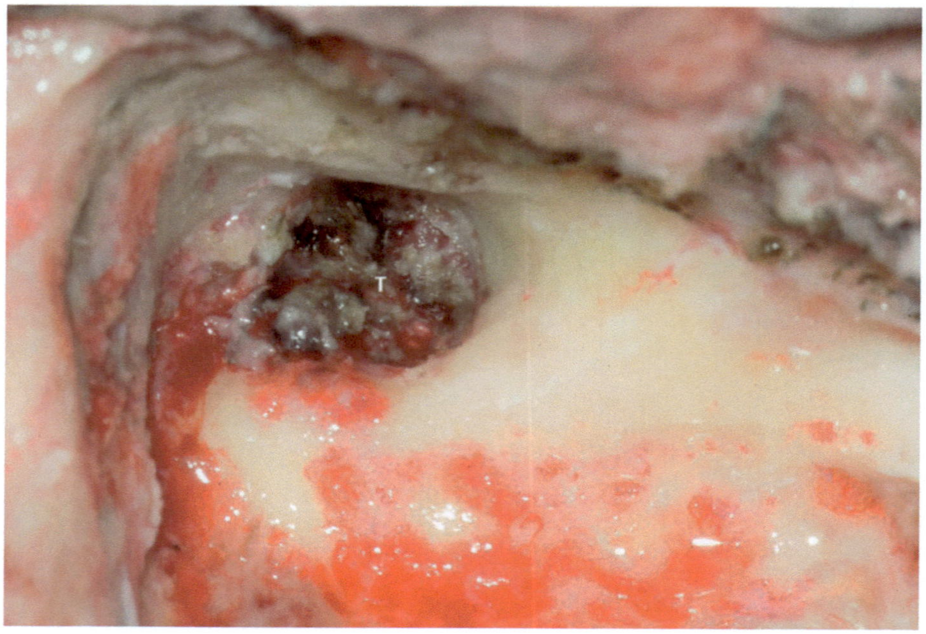

Fig. 6.27 A large tumor (T) protruding from the hypotympanum is visible.

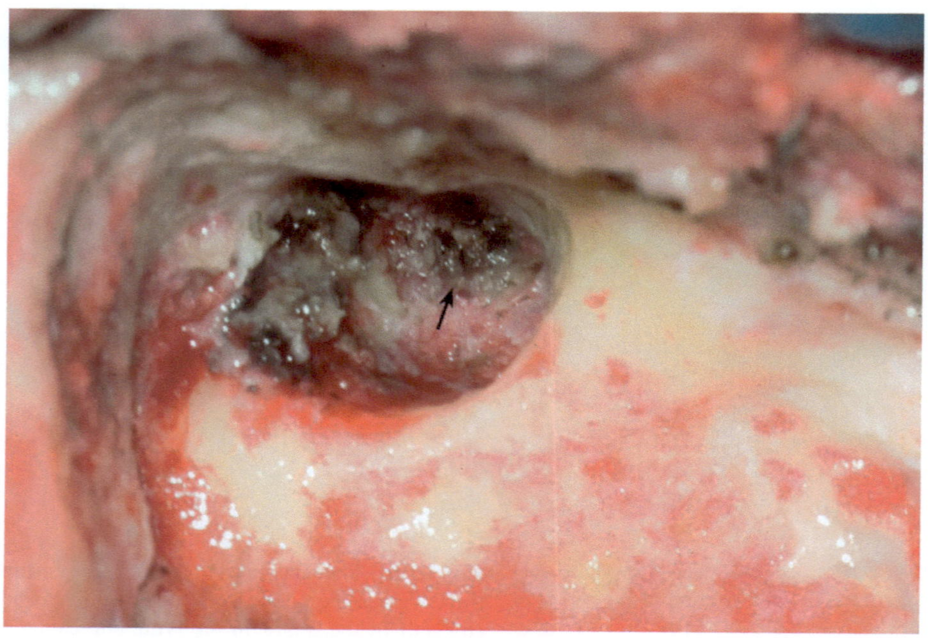

Fig. 6.28 Further debulking is performed to reach the medial wall. The hypotympanic cells are extensively infiltrated by the tumor *(arrow)*.

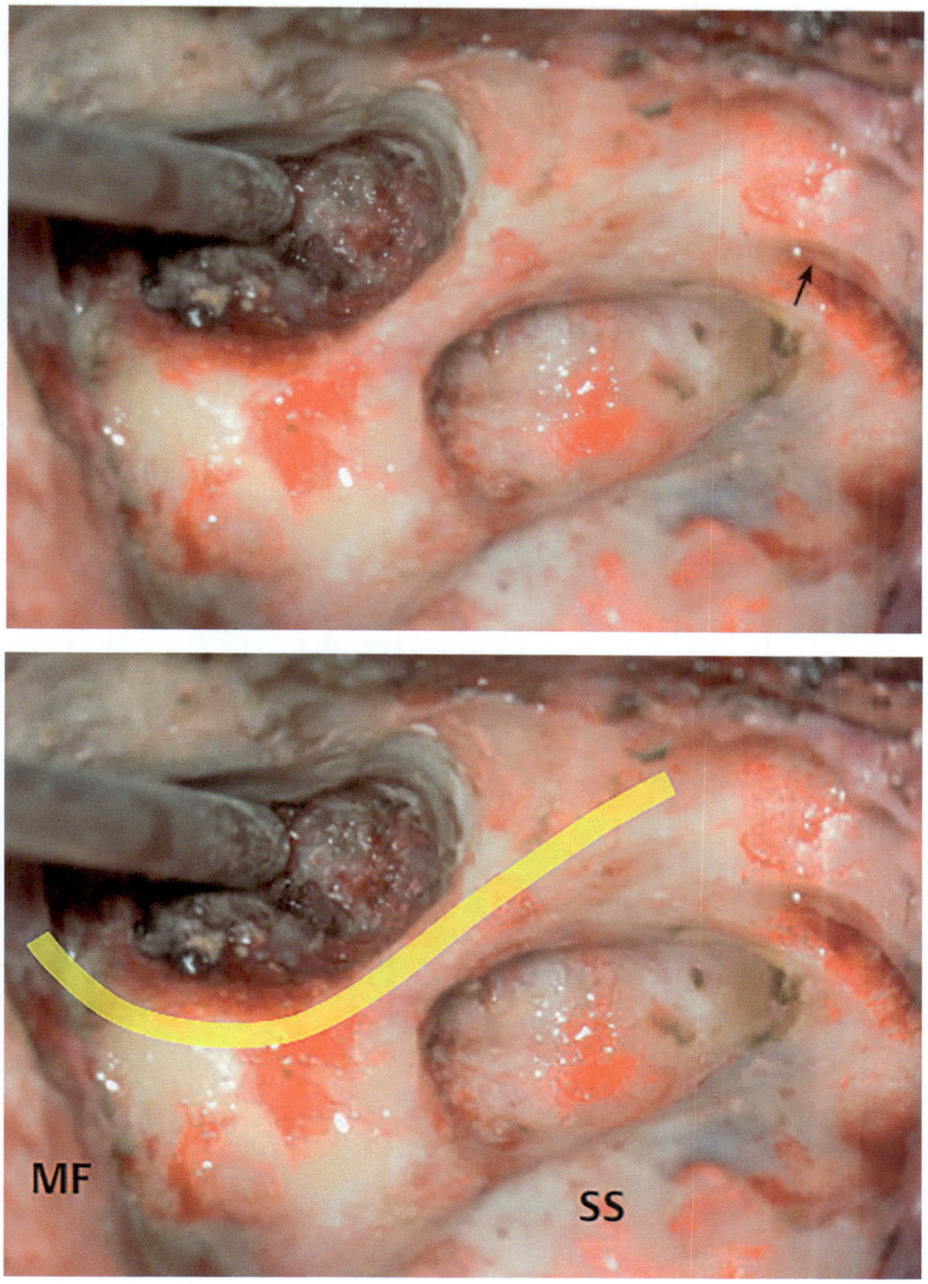

Fig. 6.29 The hypotympanic cells are infiltrated by the tumor. The retrofacial cells are removed to allow access to the tumor extending posteriorly into the cells under the facial nerve. Note that the digastric ridge *(black arrow)* points directly to the course of the facial nerve. *MF*, middle fossa; *SS*, sigmoid sinus.

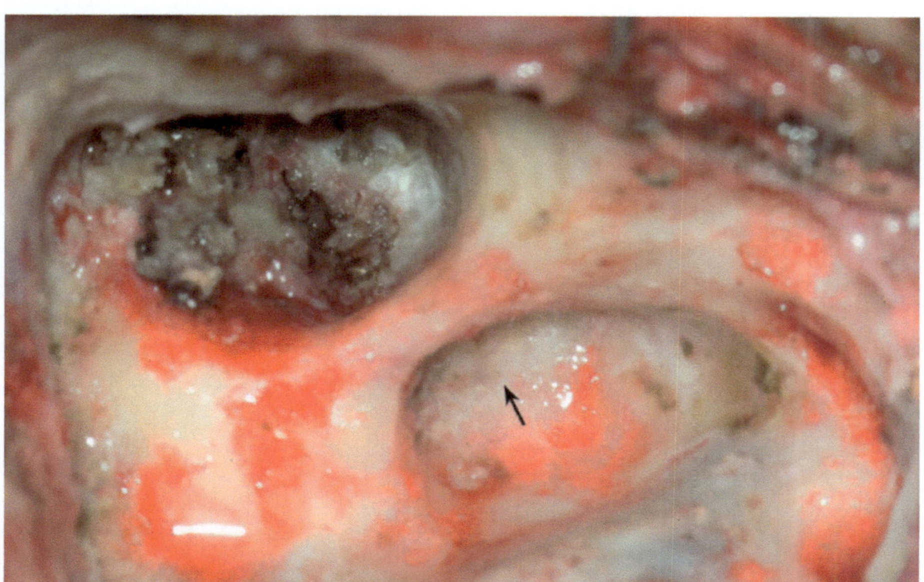

Fig. 6.30 The hypotympanic cells infiltrated by the tumor are further drilled. Hemostasis can be achieved with a bipolar coagulator and Surgicel packing applied from time to time. The bulge of the jugular bulb is visible *(arrow)*.

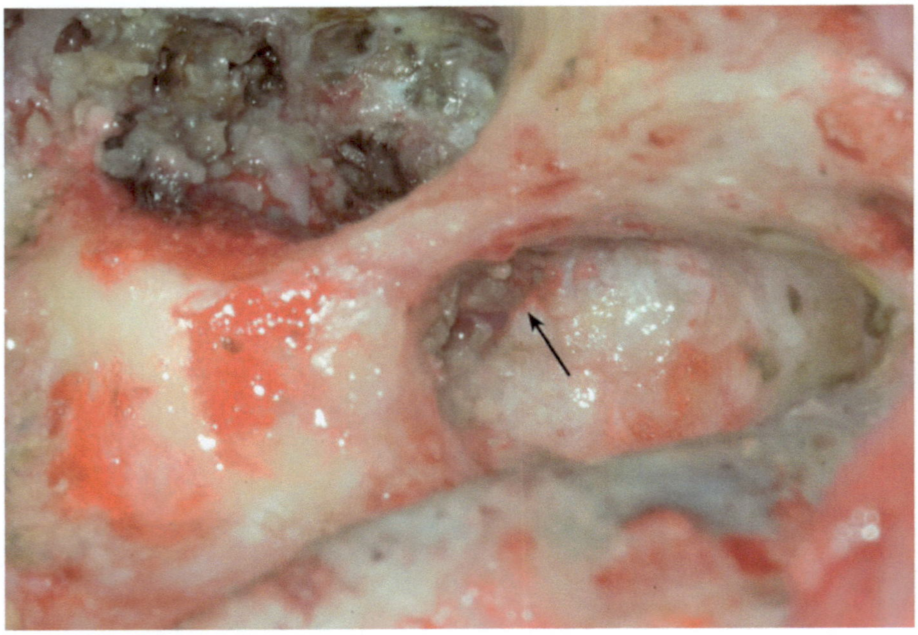

Fig. 6.31 The posterior end of the tumor-infiltrating medial to the facial nerve is exposed through the subfacial recess tympanotomy *(arrow)*.

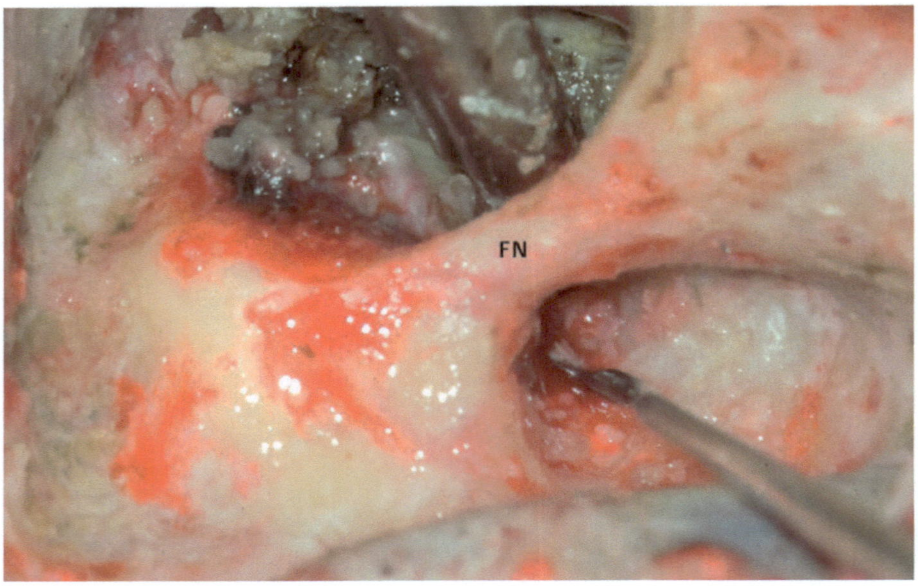

Fig. 6.32 Using the suction tube inserted from the tympanic cavity and the tympanic sinus hook from the subfacial recess tympanotomy, the tumor is dissected from the area above the jugular bulb and under the facial nerve (FN).

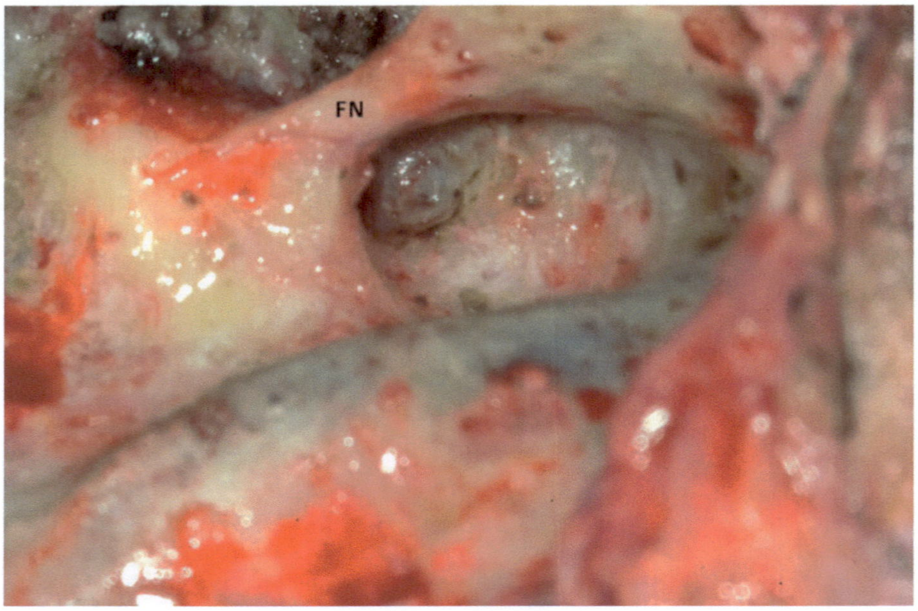

Fig. 6.33 The tumor is eliminated from the area under the facial nerve (FN) with meticulous dissection and drilling.

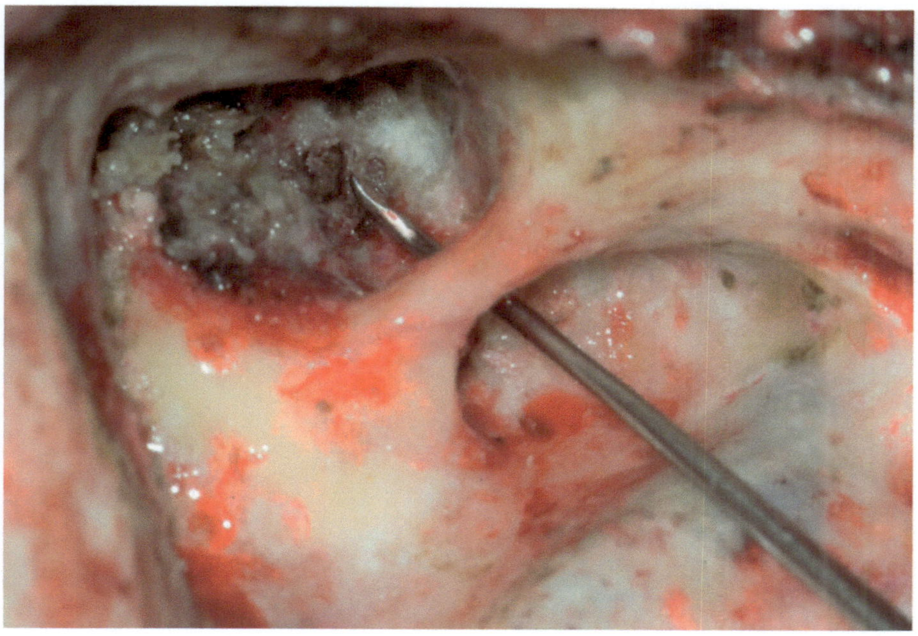

Fig. 6.34 The tumor infiltrating the pericarotid cells is visible, as indicated by the instrument. The tumor is still covering the promontory and the inner ear windows.

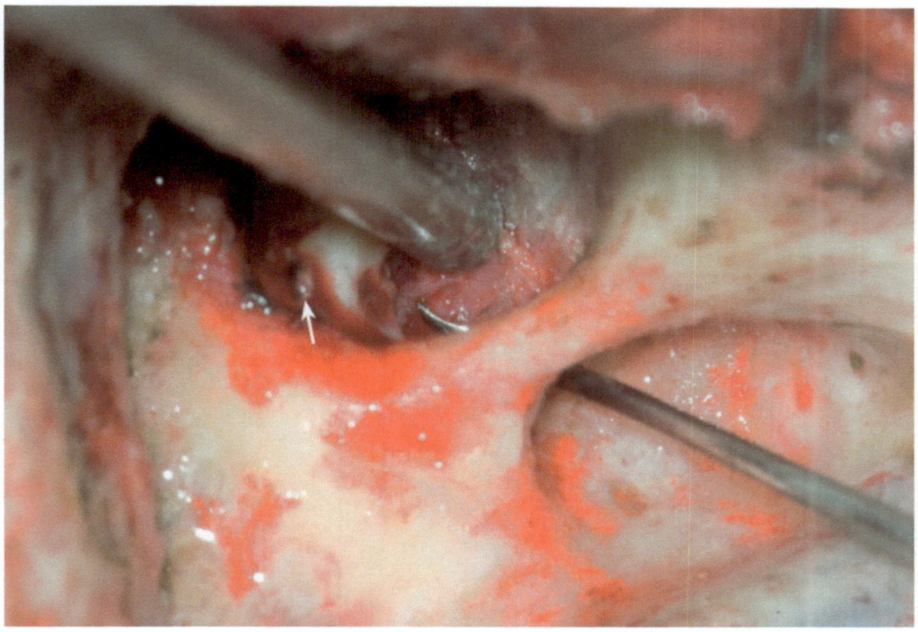

Fig. 6.35 Further dissection of the tumor is performed anteriorly. The head of the stapes is visible *(arrow)*. The round window niche is completely occupied by the tumor. Great care should be taken to avoid damage to the round window membrane.

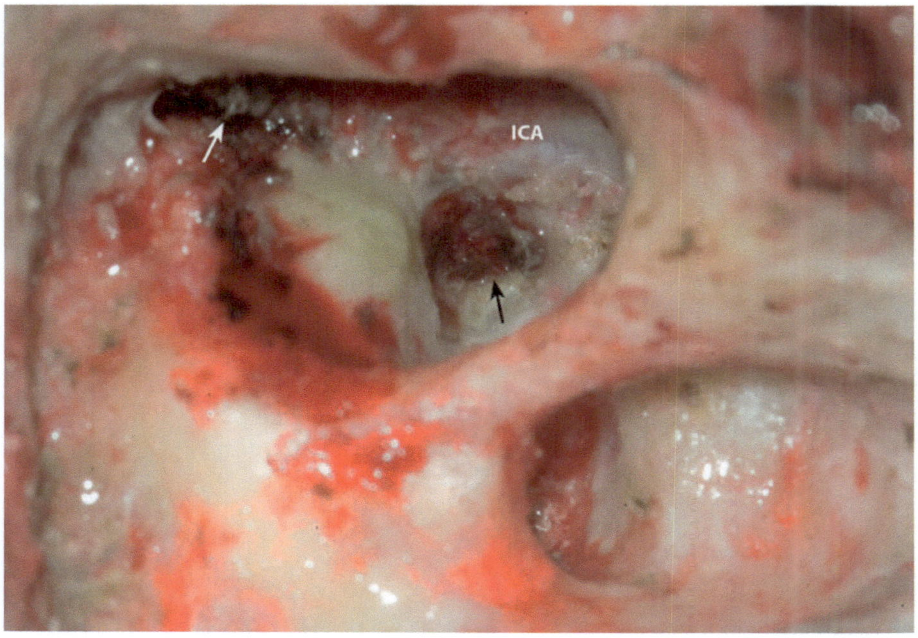

Fig. 6.36 The internal carotid artery is skeletonized (ICA). The tumor infiltrating the cells behind *(black arrow)* and above *(white arrow)* the carotid artery is visible.

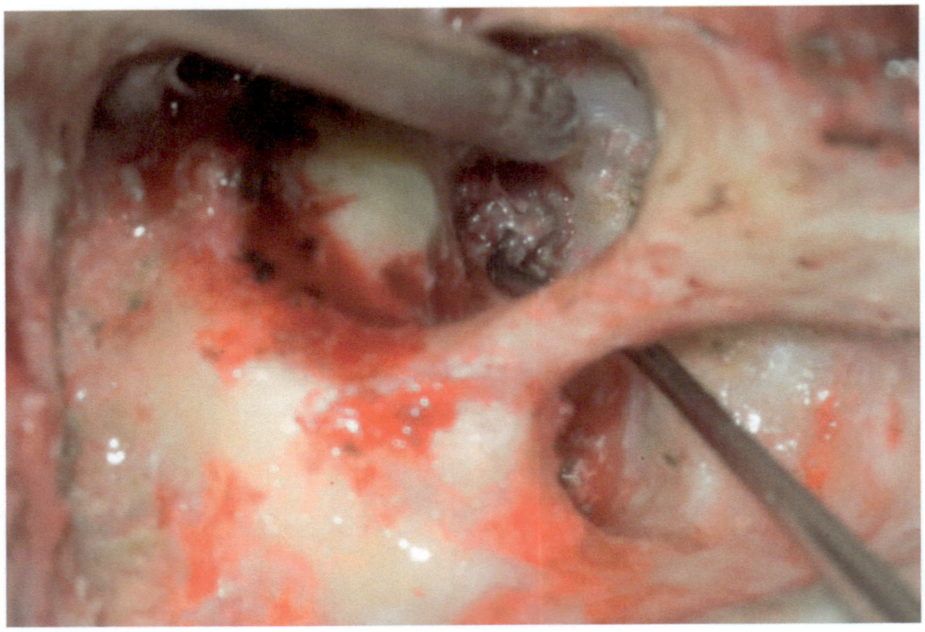

Fig. 6.37 Through the subfacial recess tympanotomy, the cells under the carotid artery are opened, and the tumor is removed using suction and a microdissector.

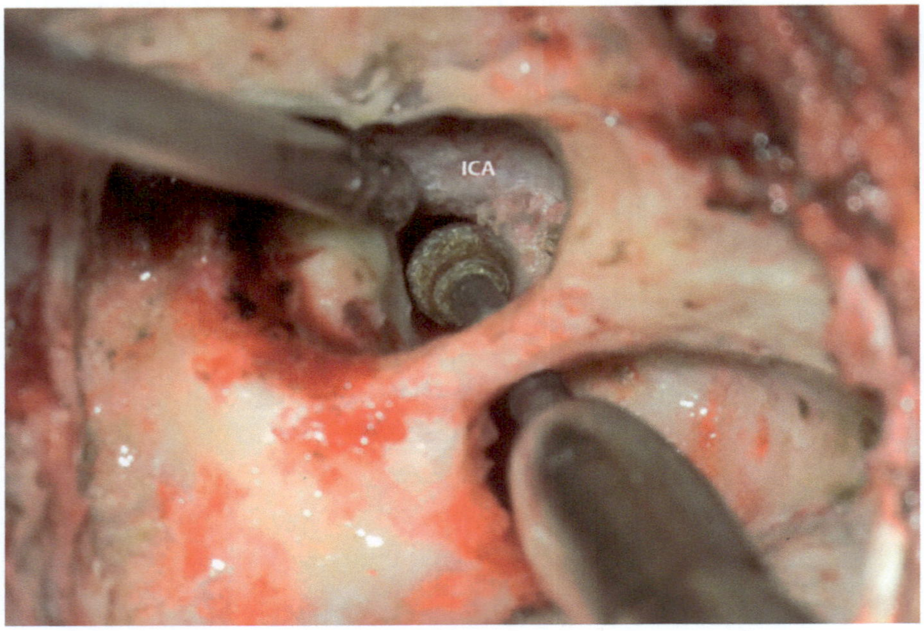

Fig. 6.38 To ensure tumor removal, cells under the internal carotid artery (ICA) are drilled away using the largest possible diamond burr.

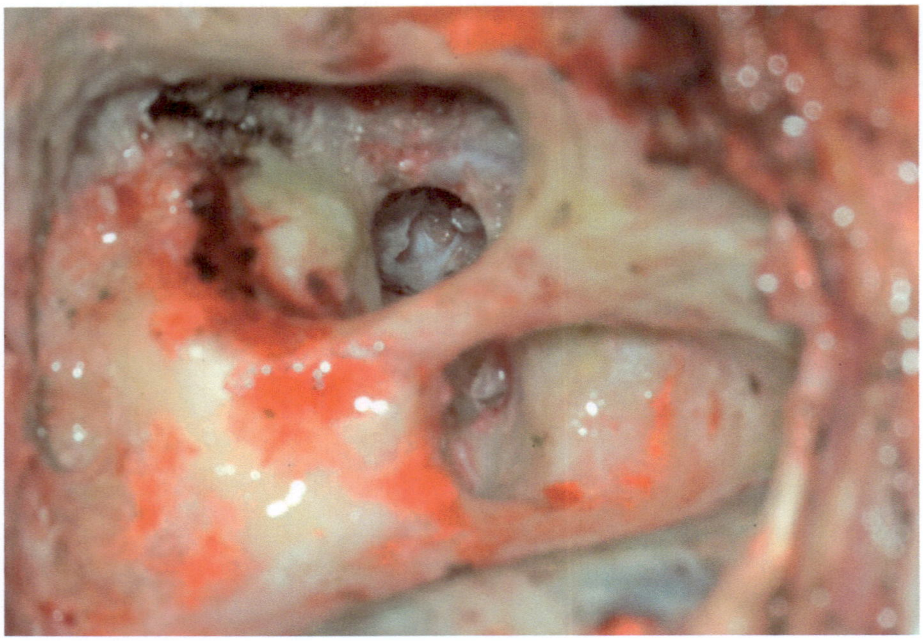

Fig. 6.39 The tumor under the carotid artery has been removed completely.

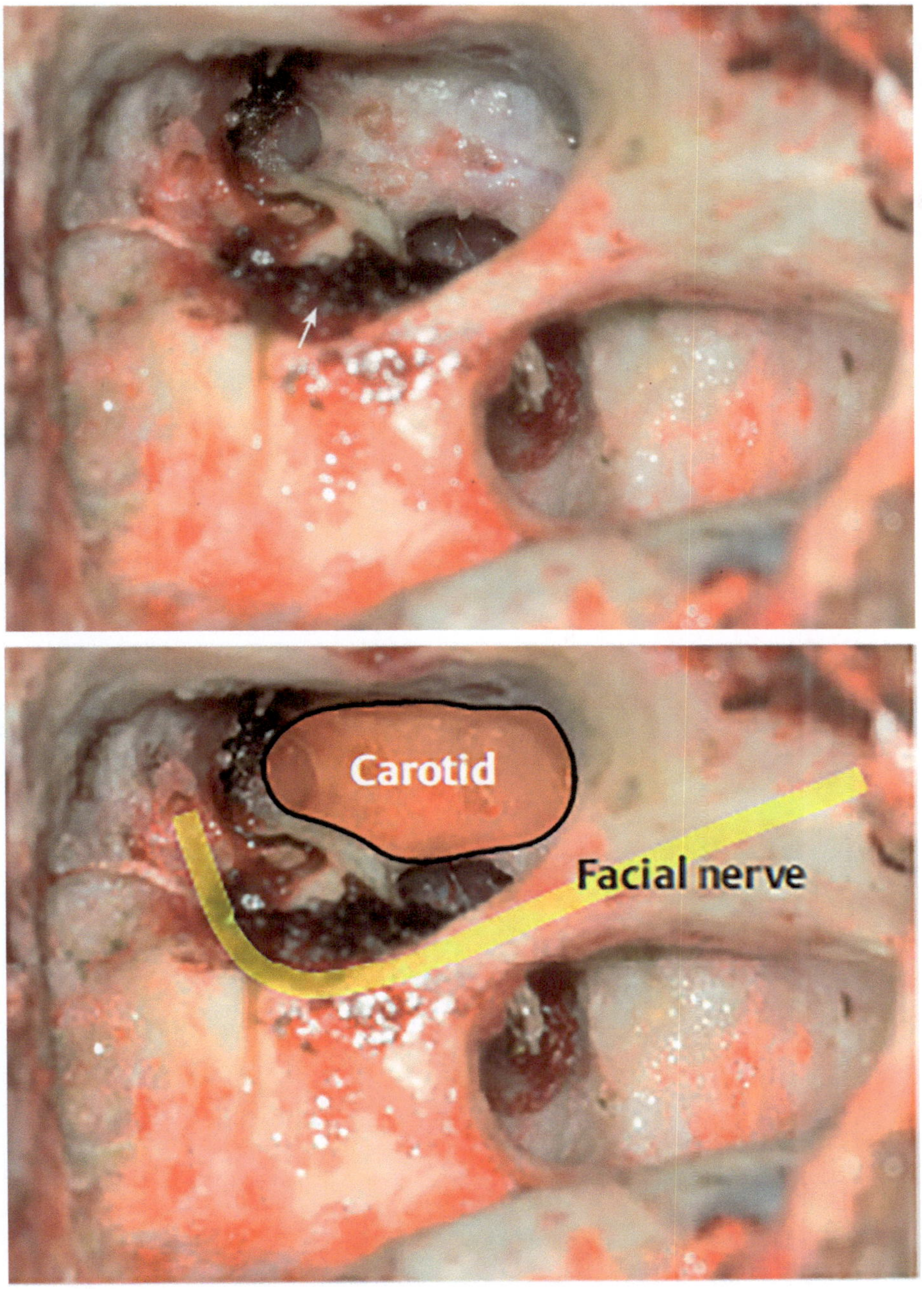

Fig. 6.40 The cells over the carotid artery are opened to remove the tumor-infiltrating into this area. Note that the carotid artery turns nearly at a right angle. Troublesome bleeding from the area of the oval window is managed with a small piece of Surgicel *(arrow)*.

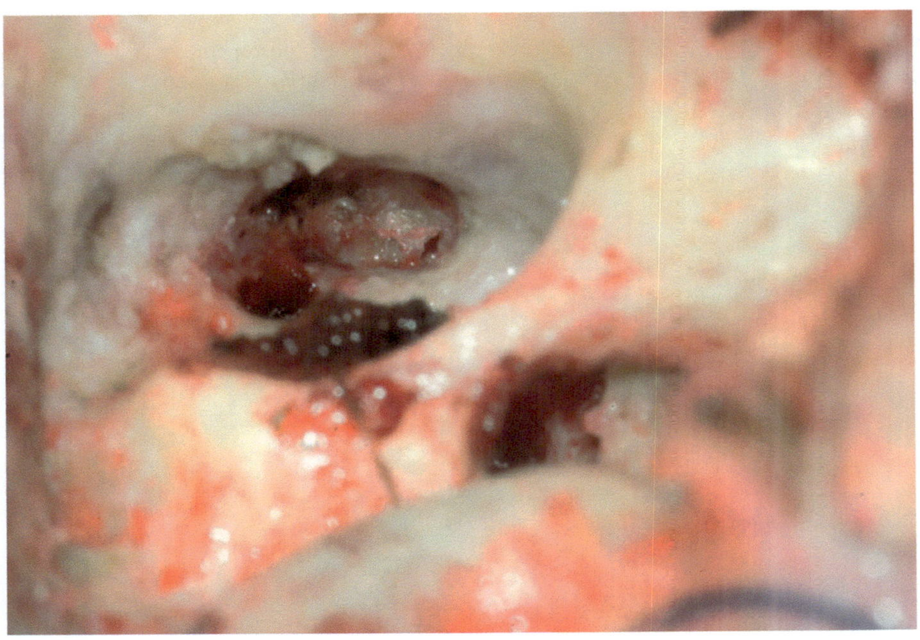

Fig. 6.41 The tumor infiltrating the cells above the carotid artery is completely removed.

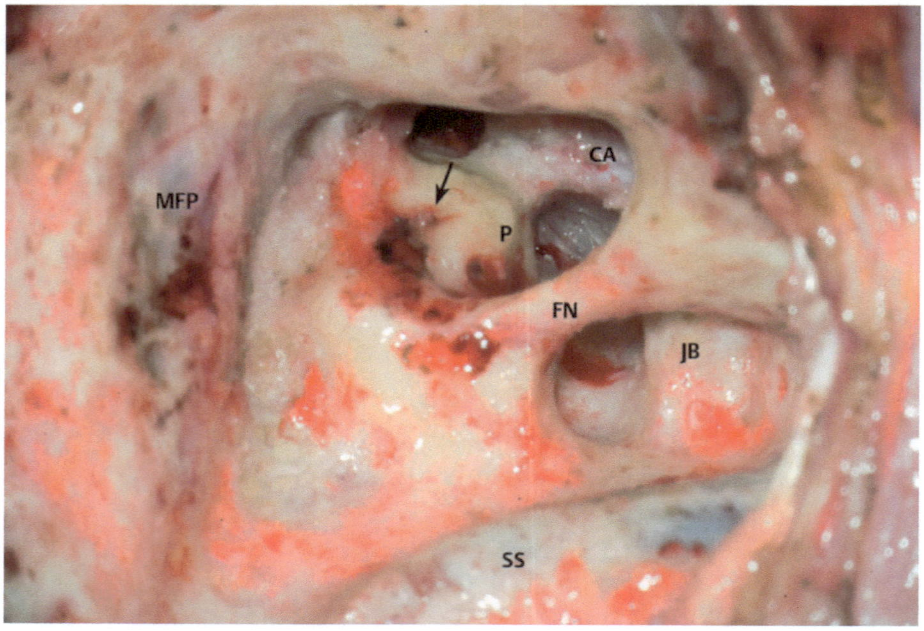

Fig. 6.42 The tumor is completely removed. The suprastructure of the stapes is visible *(arrow)*. *CA*, carotid artery. *FN*, facial nerve; *JB*, jugular bulb; *MFP*, middle cranial fossa dural plate; *P*, promontory; *SS*, sigmoid sinus.

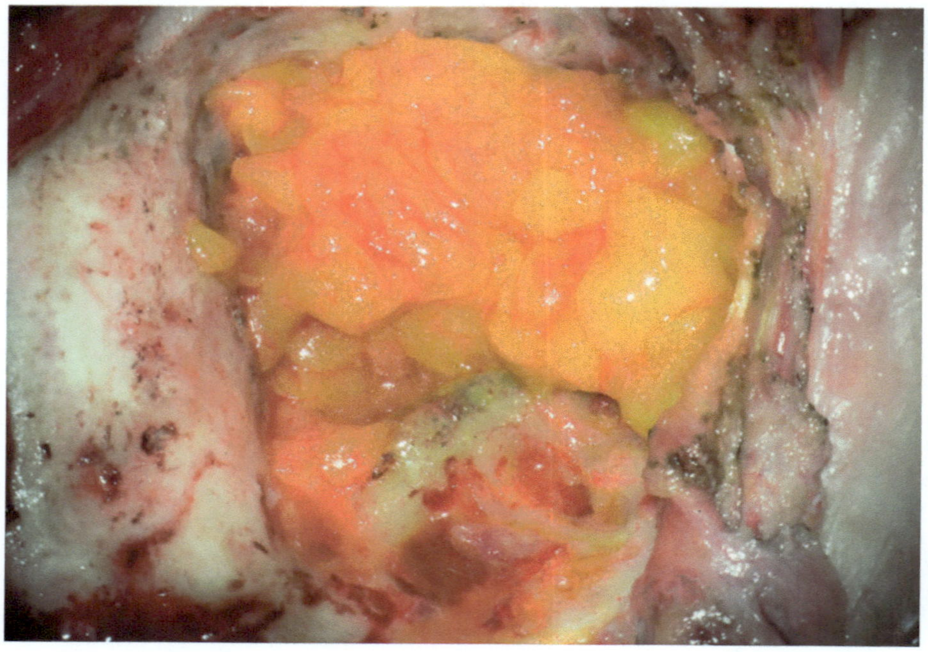

Fig. 6.43 After closure of the eustachian tube with a few pieces of periosteum, the cavity is obliterated with abdominal fat. The retroauricular wound is closed in three layers.

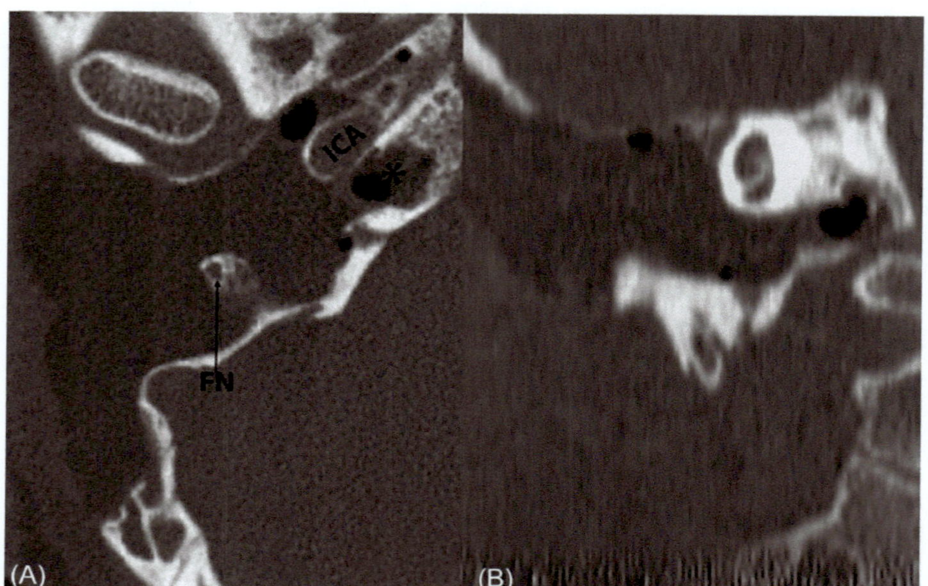

Fig. 6.44 The postoperative CT scan axial (A) and coronal (B) views show the removal of retrofacial cells and bone medial to the carotid artery (asterisk). The cavity is filled with abdominal fat. *FN*, facial nerve; *ICA*, internal carotid artery.

Case 2: Facial Nerve Tumor

See Figs. 6.45–6.67.

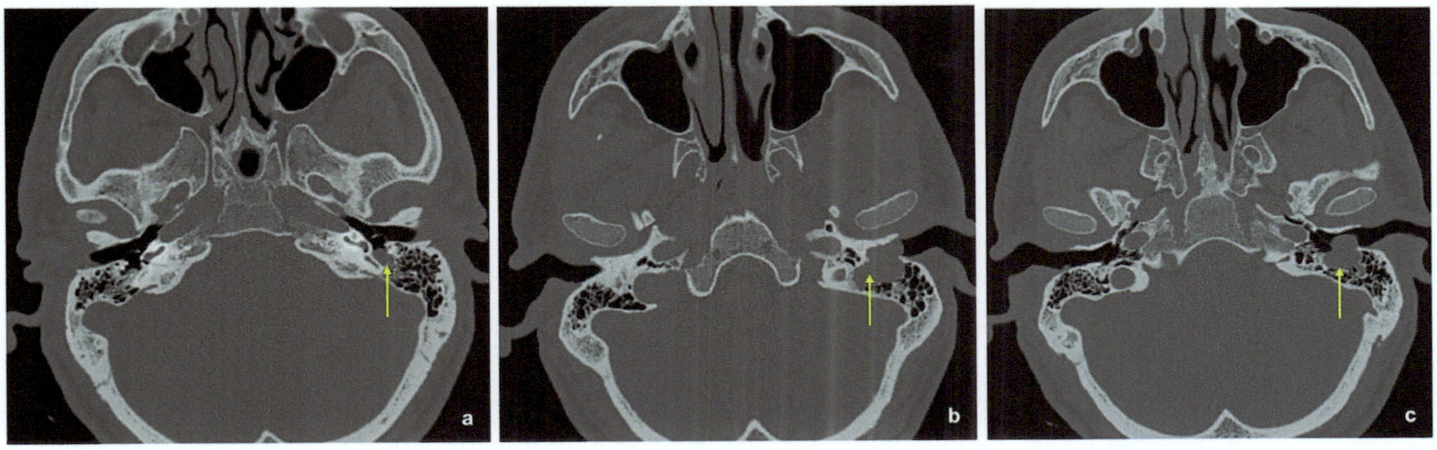

Fig. 6.45 HRCT scan of the temporal bone, axial view. This 42-year-old male patient presented with complete left facial palsy and a lesion extending outside the external auditory canal (EAC). A facial nerve tumor *(yellow arrow)* involving the second genu (A) and the mastoid portion (B) of the facial nerve was seen. The tumor eroded the posterior wall of the EAC (C), extending laterally to the tympanic membrane (C).

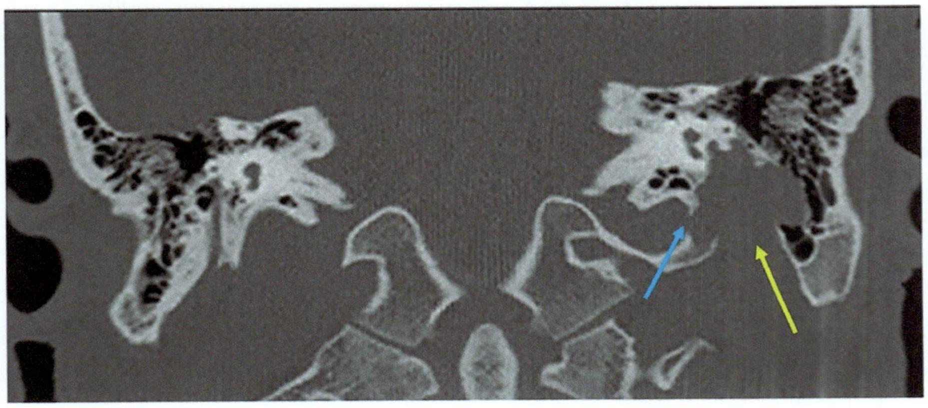

Fig. 6.46 HRCT scan of the temporal bone, coronal view. The tumor extended beyond the stylomastoid foramen *(yellow arrow)* toward the parotid gland. Note the enlargement of the stylomastoid foramen compared to the contralateral side and the erosion of the bone corresponding to the jugular bulb *(blue arrow)*.

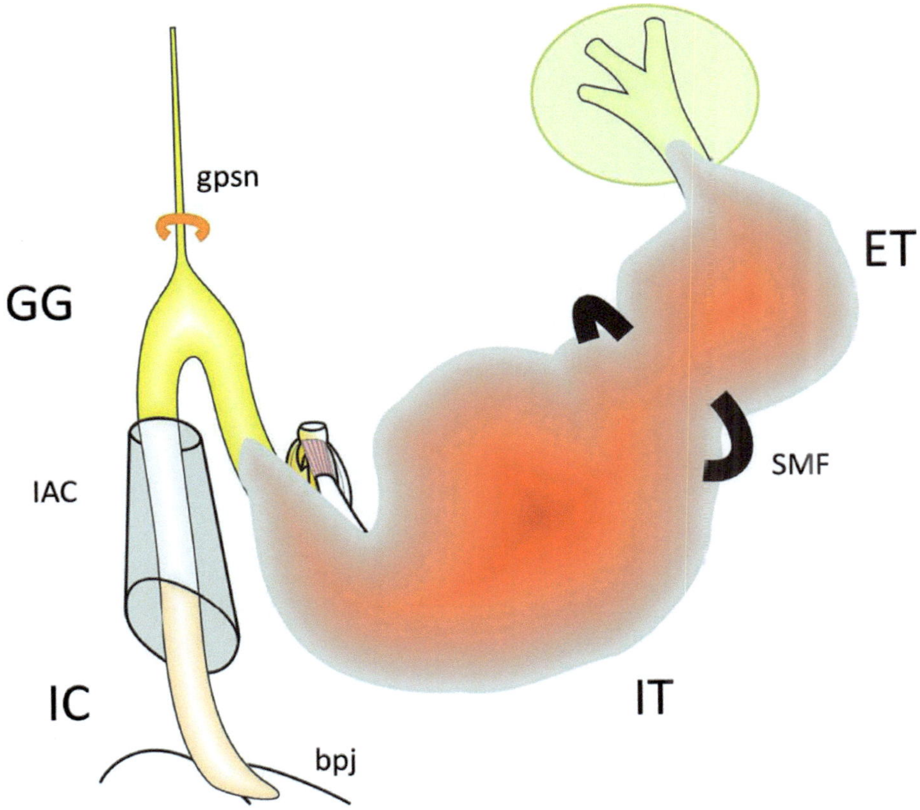

Fig. 6.47 Schematic drawing of the tumor.

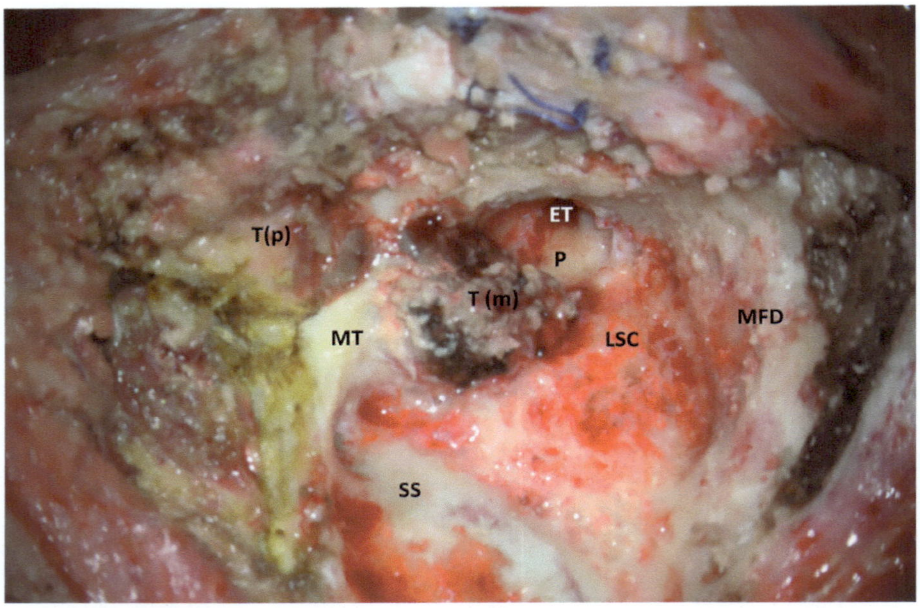

Fig. 6.48 Surgical view. A subtotal petrosectomy has been performed. The tumor involving the facial nerve can be seen both in the mastoid and the parotid gland. *T(p)*, parotid portion of the tumor; *T(m)*, mastoid portion of the tumor; *ET*, eustachian tube; *MT*, mastoid tip; *P*, promontory; *MFD*, middle fossa dura; *SS*, sigmoid sinus; *LSC*, lateral semicircular canal.

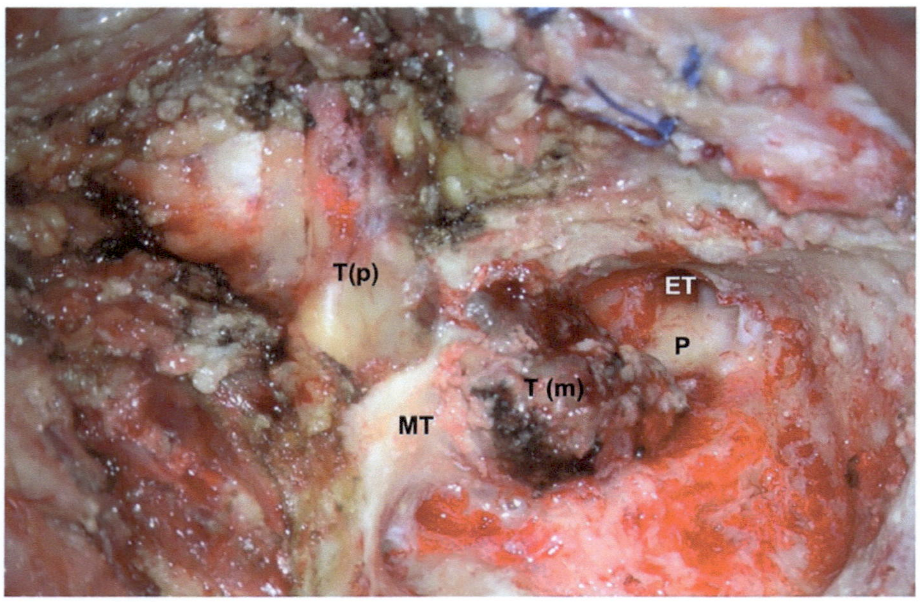

Fig. 6.49 Closer view. *T(p)*, parotid portion of the tumor; *T(m)*, mastoid portion of the tumor; *ET*, eustachian tube; *MT*, mastoid tip; *P*, promontory.

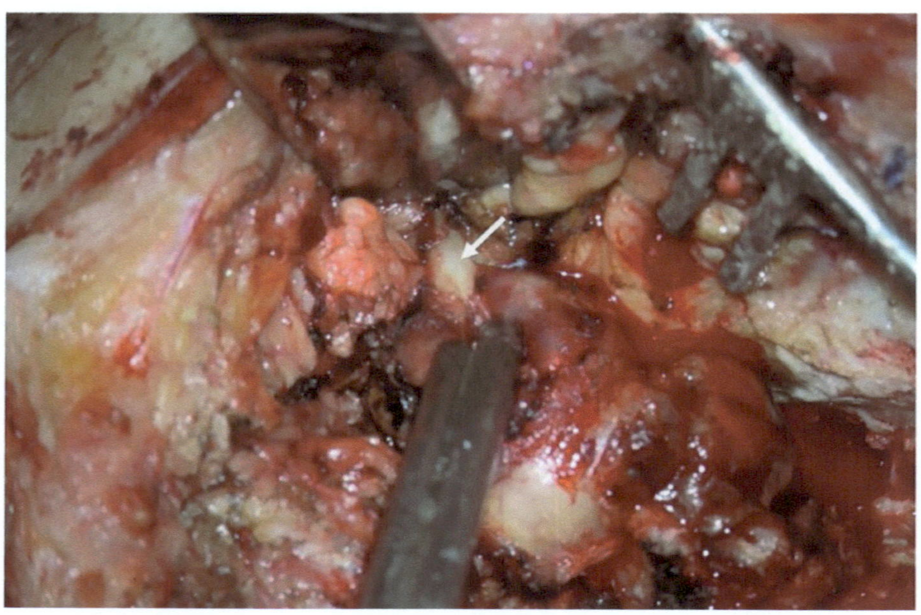

Fig. 6.50 The FN is further dissected in the parotid gland, exposing its nontumoral portion *(white arrow)*.

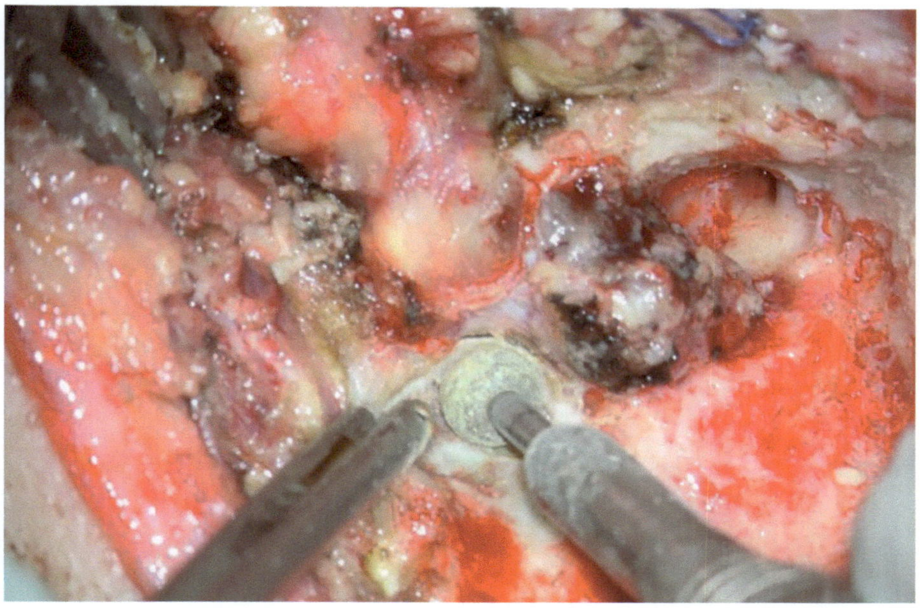

Fig. 6.51 Drilling toward the area of the jugular bulb.

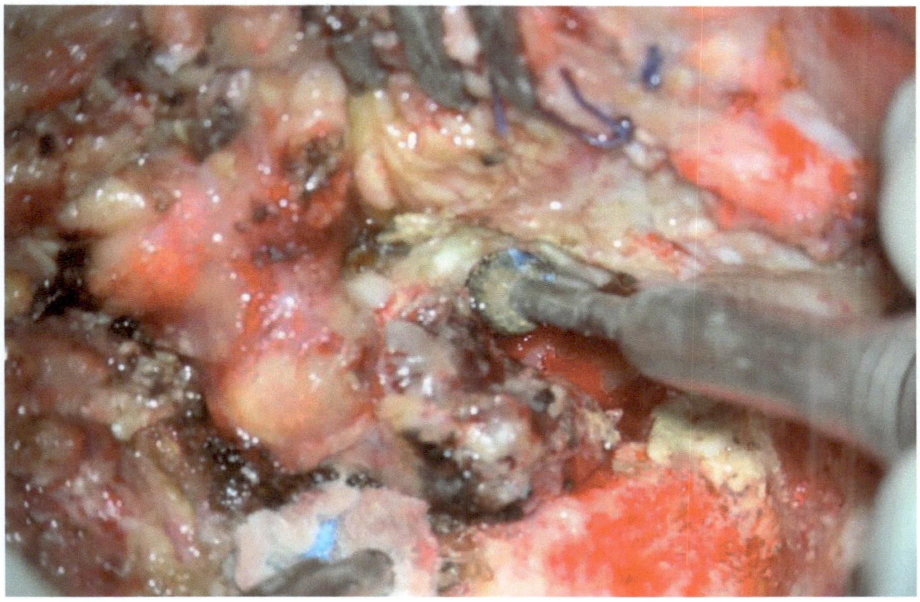

Fig. 6.52 Drilling of the inferior tympanic bone.

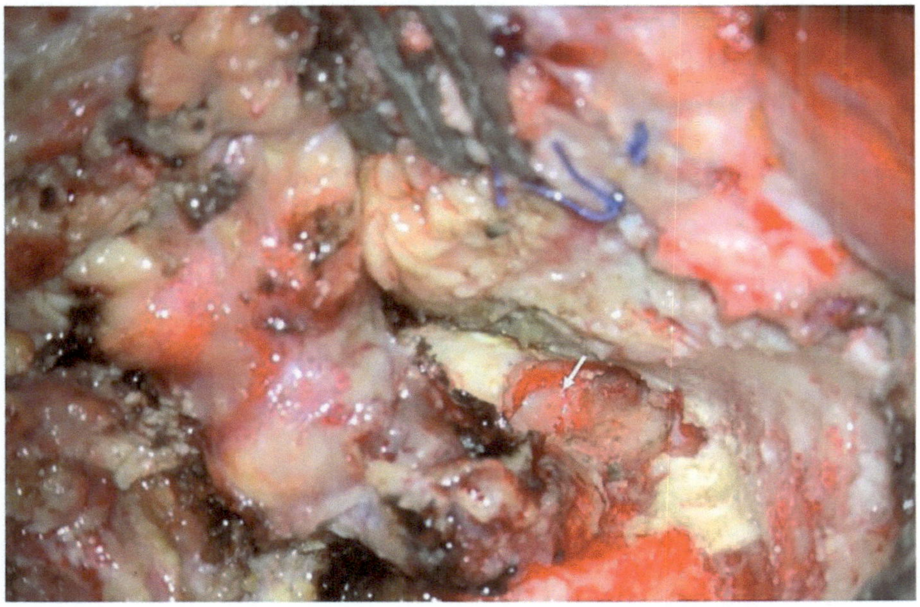

Fig. 6.53 The internal carotid artery *(white arrow)* can be appreciated after removal of the inferior tympanic bone.

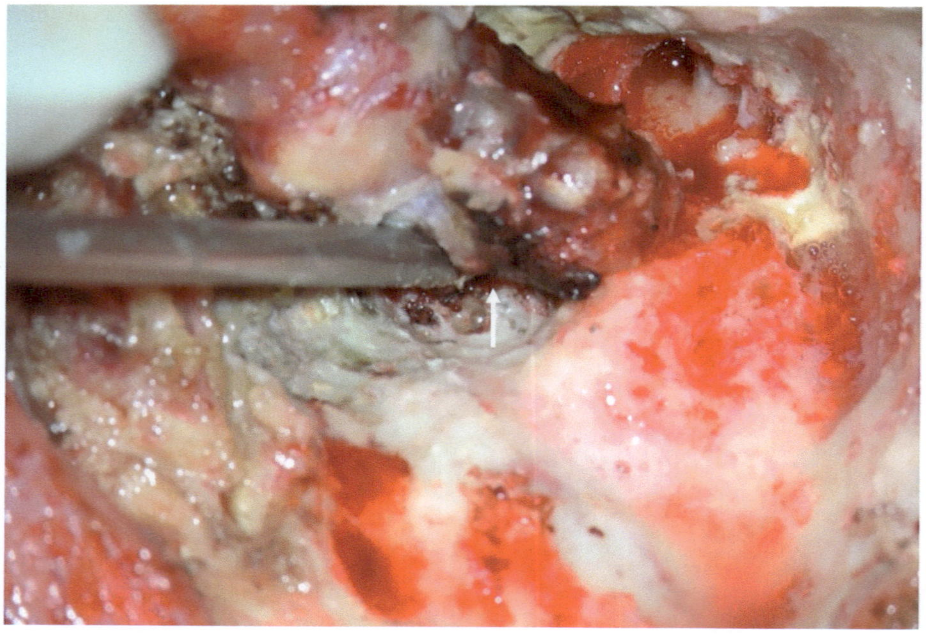

Fig. 6.54 The tumor is dissected from the area of the jugular bulb (suction) and the retrofacial air cells *(white arrow)*.

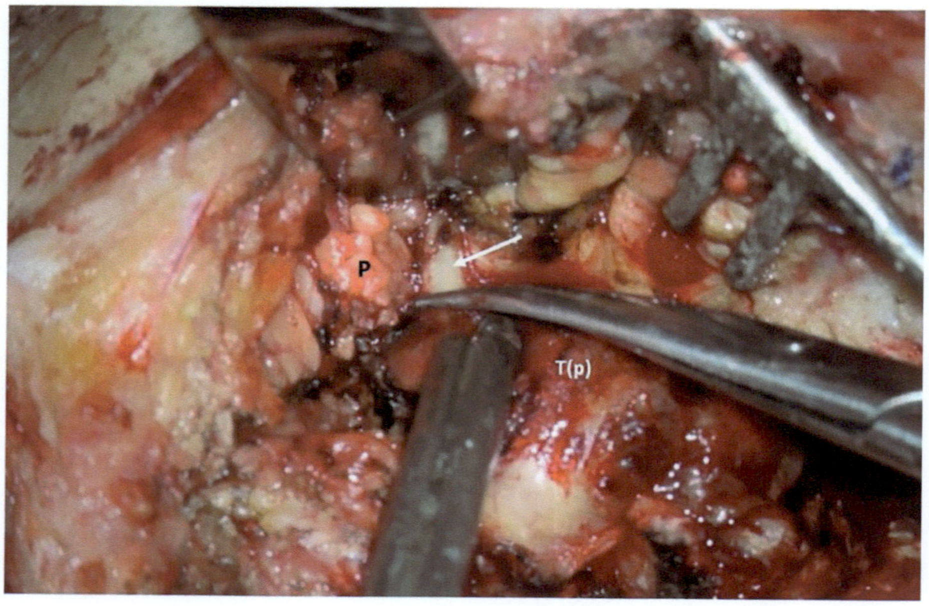

Fig. 6.55 The parotid portion of the tumor is cut. Nearly 1–2 mm of the healthy facial nerve is left on the tumor to have safe margins for further reconstruction. *T(p)*, parotid portion of the tumor, *white arrow*, healthy facial nerve; *P*, parotid.

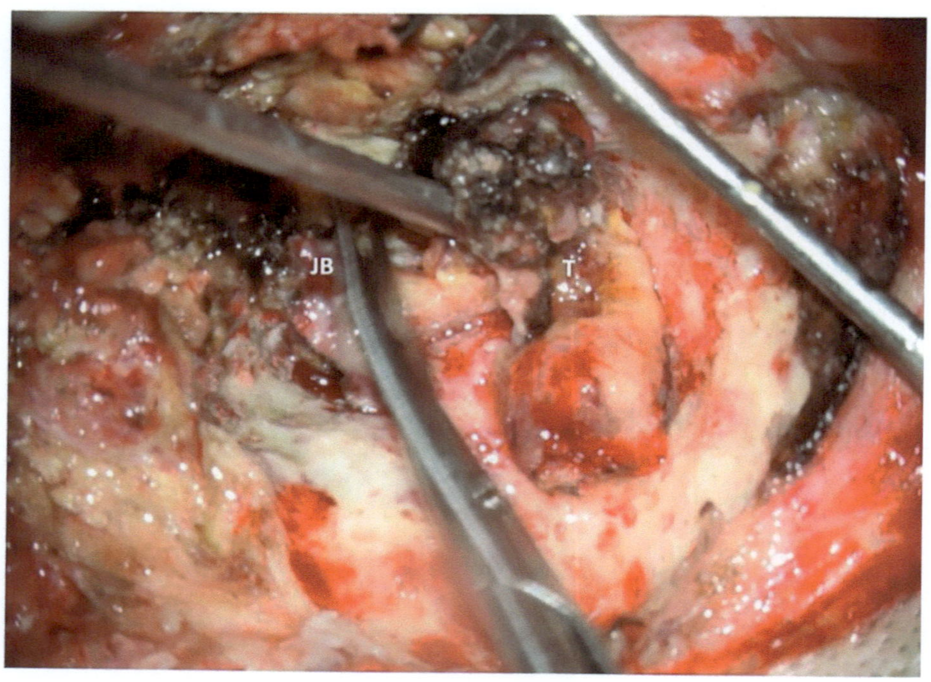

Fig. 6.56 After debulking, the tumor (T) is further dissected into the mastoid cavity. Note the uncovered jugular bulb (JB).

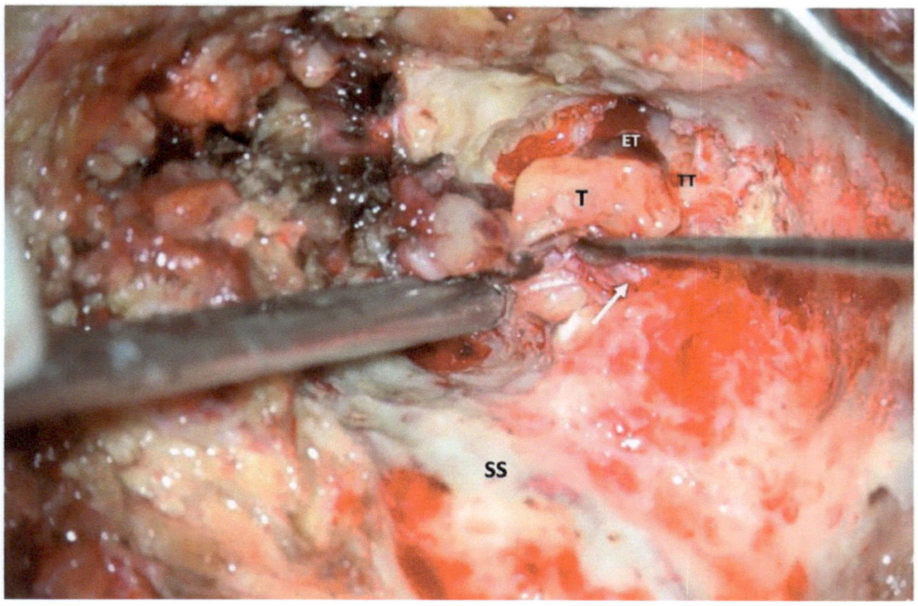

Fig. 6.57 The tumor (T) is reflected superiorly to find the safe proximal stump of the facial nerve (the *white arrow* corresponds to the second genu of the facial nerve). *SS*, sigmoid sinus; *ET*, eustachian tube; *TT*, tensor tympani tendon.

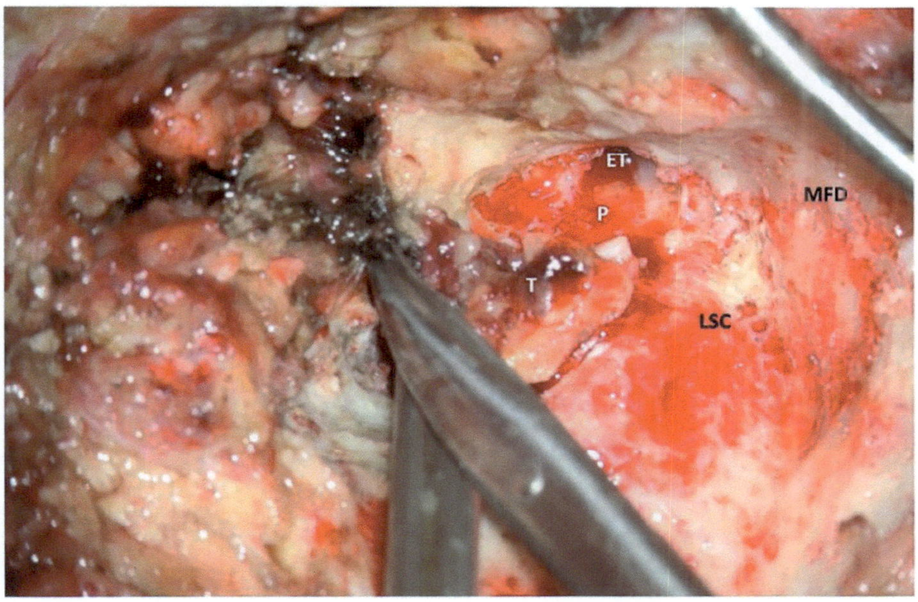

Fig. 6.58 Further tumor dissection at the level of the mastoid tip. *T*, tumor; *ET*, eustachian tube; *P*, promontory; *LSC*, lateral semicircular canal; *MFD*, middle fossa dura.

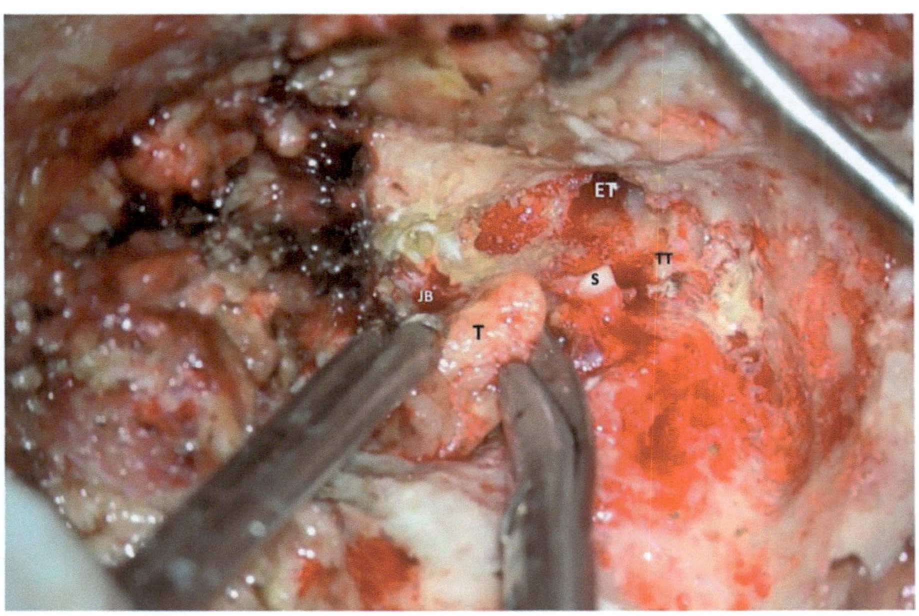

Fig. 6.59 Further tumor debulking. *T*, tumor; *ET*, eustachian tube; *S*, stapes; *TT*, tensor tympani tendon; *JB*, jugular bulb.

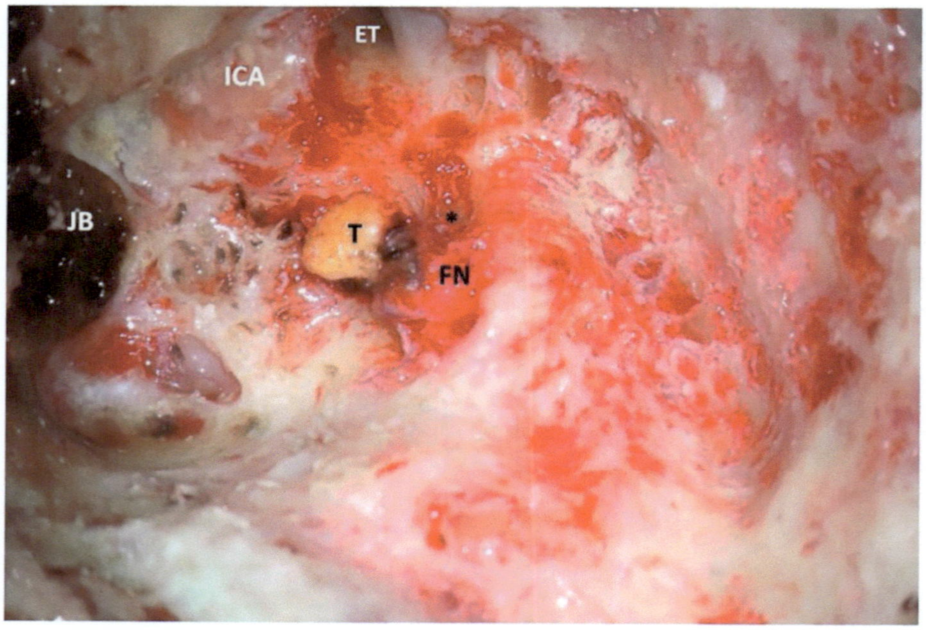

Fig. 6.60 Closer view. The last piece of tumor (T) must be removed together with a 1–2 mm piece of normal facial nerve (FN) to avoid any tumor remnant on the surgical field. *JB*, jugular bulb (protected with Surgicel); *ET*, eustachian tube; *ICA*, internal carotid artery; *, footplate.

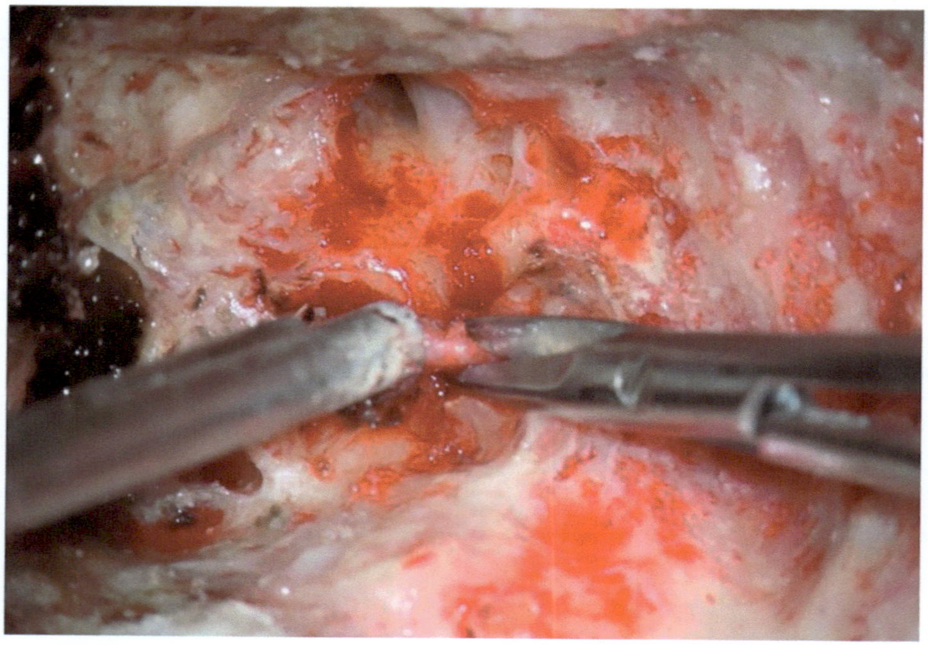

Fig. 6.61 Removal of the last piece of the tumor together with the second genu of the facial nerve.

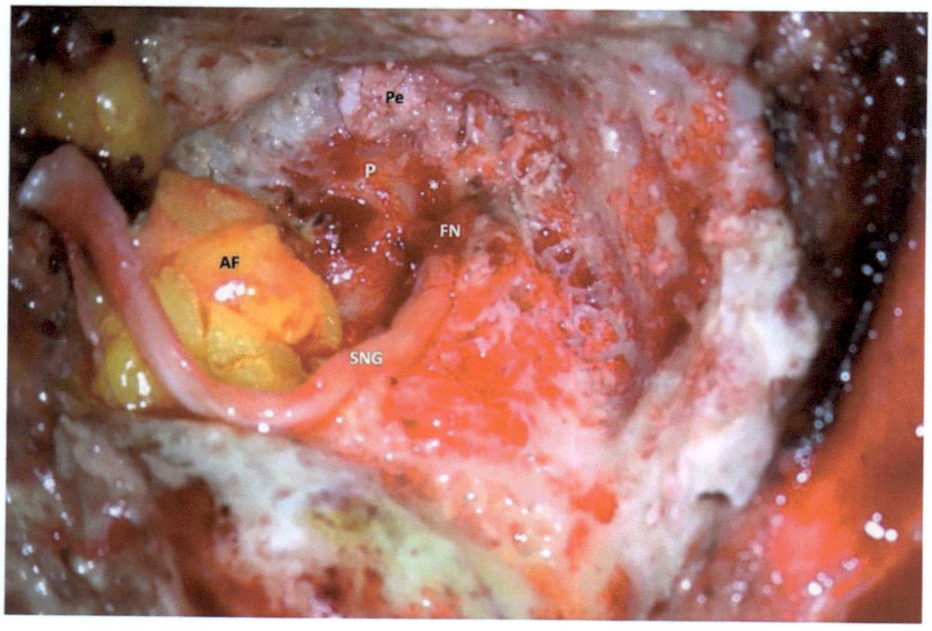

Fig. 6.62 Facial nerve (FN) reconstruction is performed with a sural nerve graft (SNG) anastomosis. The graft is placed on the proximal stump of the FN (tympanic segment). Before the SNG is placed, a piece of abdominal fat is used to both stabilize the reconstruction and seal the surgical cavity. A piece of periosteum (Pe) is used to close the eustachian tube.

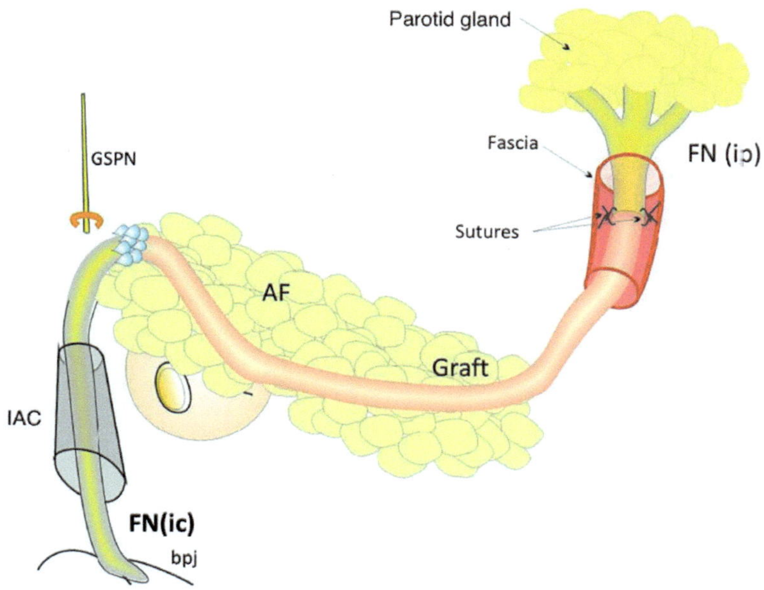

Fig. 6.63 Schematic drawing of facial nerve reconstruction.

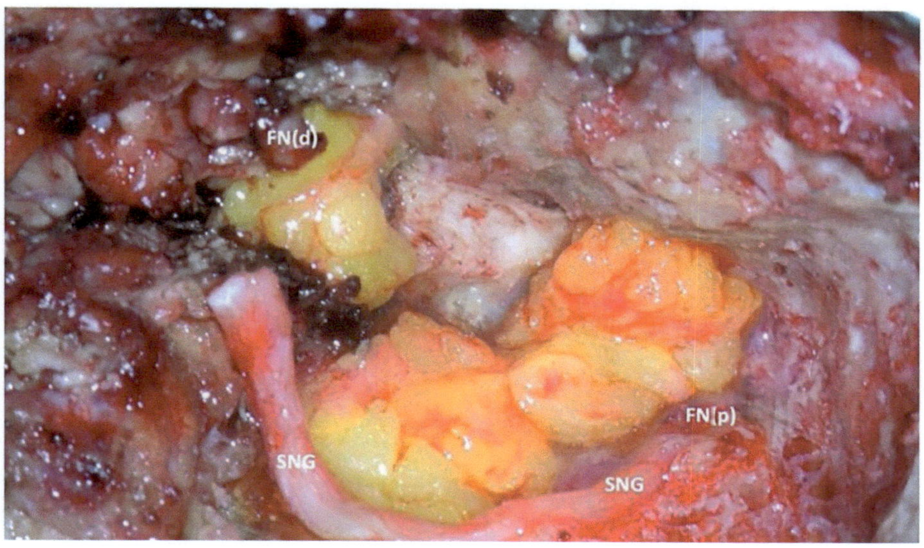

Fig. 6.64 The proximal part of the anastomosis is fixed with fibrin glue. The distal part has to be fixed with sutures. *FN(d)*, distal stump of the facial nerve (parotid); *FN(p)*, proximal stump of the facial nerve (tympanic portion); *SNG*, sural nerve graft.

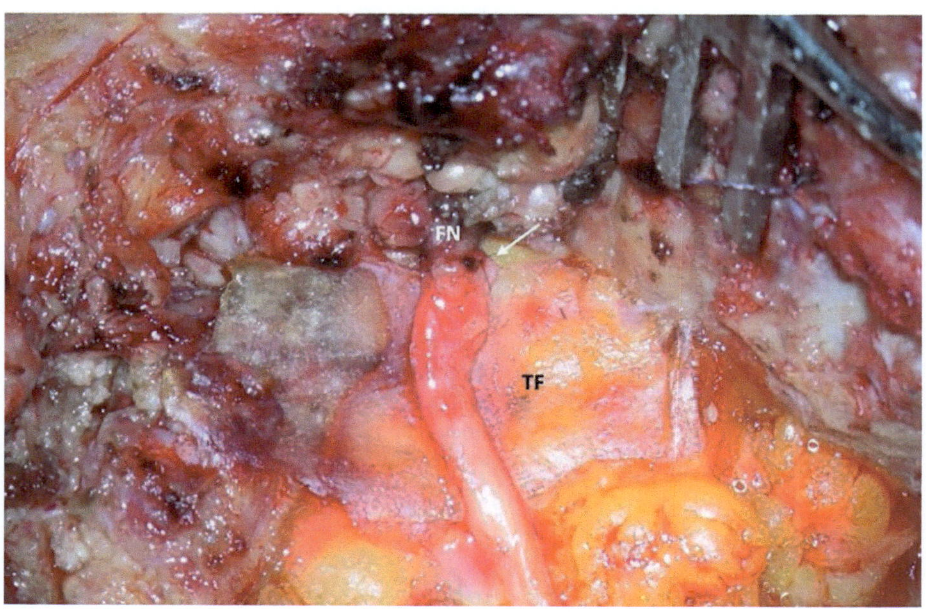

Fig. 6.65 The distal part of the anastomosis is fixed with 10.0 nylon sutures *(white arrow)*. Prior to fixation, more abdominal fat is placed in the cavity together with a temporalis fascia (TF) to protect and stabilize the reconstruction. *FN*, distal stump of the facial nerve (parotid).

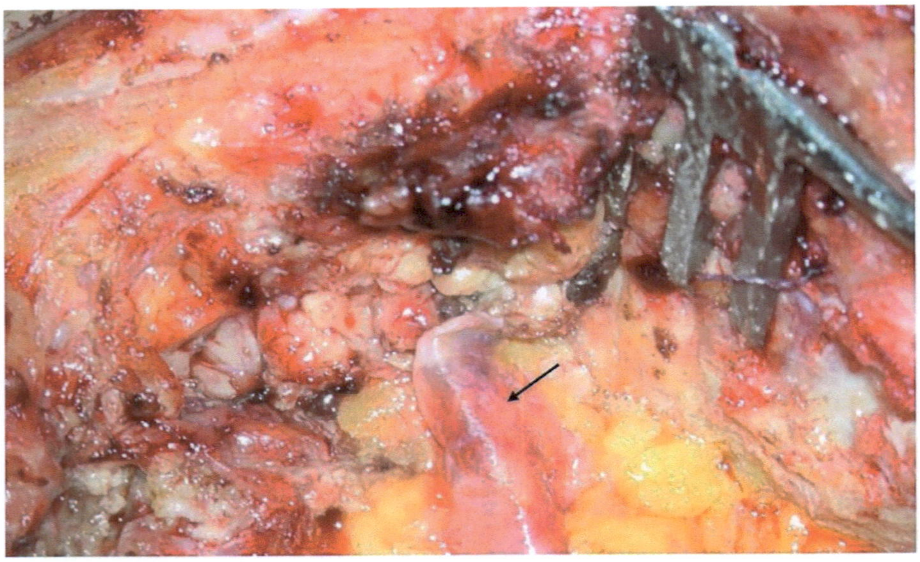

Fig. 6.66 The fascia is wrapped around the reconstruction *(arrow)*.

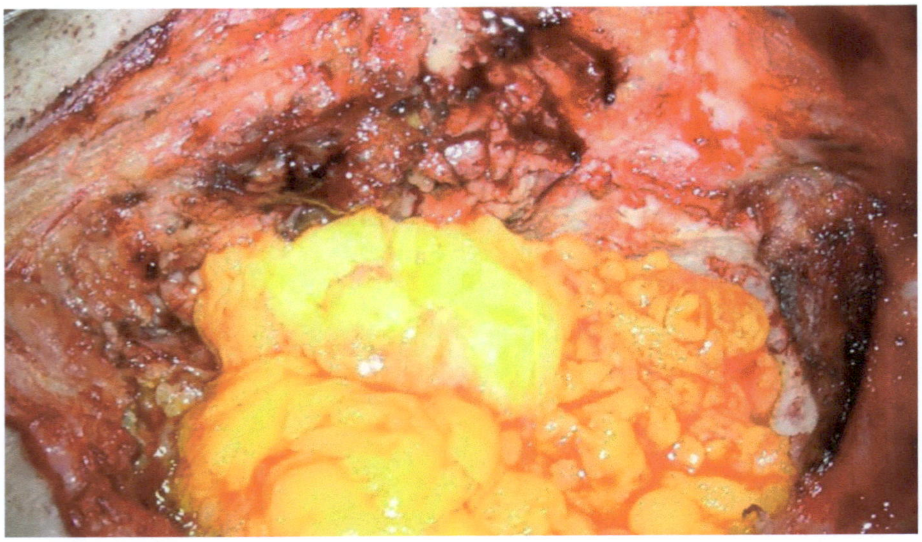

Fig. 6.67 The surgical cavity is completely filled with abdominal fat.

Case 3: Revision Subtotal Petrosectomy and Cochlear Implantation for Recurrent Cholesteatoma

See Figs. 6.68–6.82.

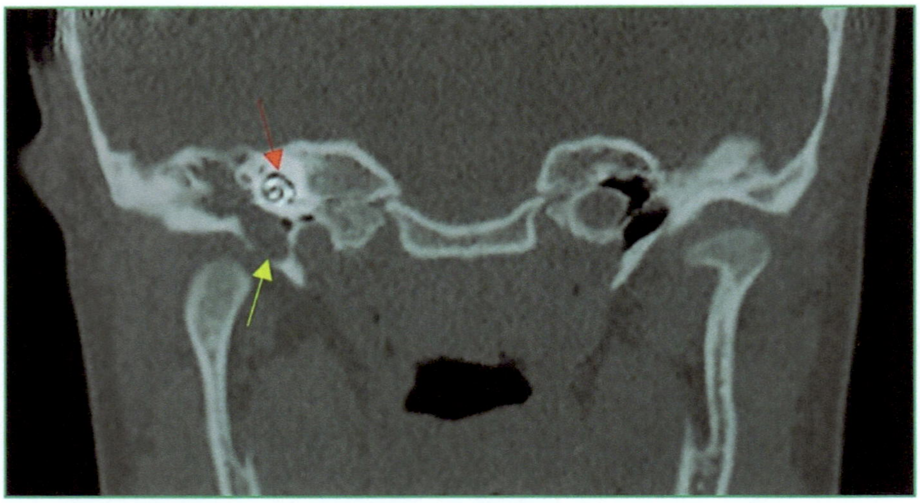

Fig. 6.68 This 13-year-old boy underwent subtotal petrosectomy and cochlear implantation at the age of 3 in another hospital. Due to the reduced performance of the device, he underwent radiological investigations that showed the presence of a suspected cholesteatoma in the surgical cavity. As can be seen on the CT scan of the temporal bone (coronal cut), there is an erosion of the tympanic bone toward the internal carotid artery *(yellow arrow)*, suggesting a residual cholesteatoma. The array is completely inserted into the cochlear turns *(red arrow)*.

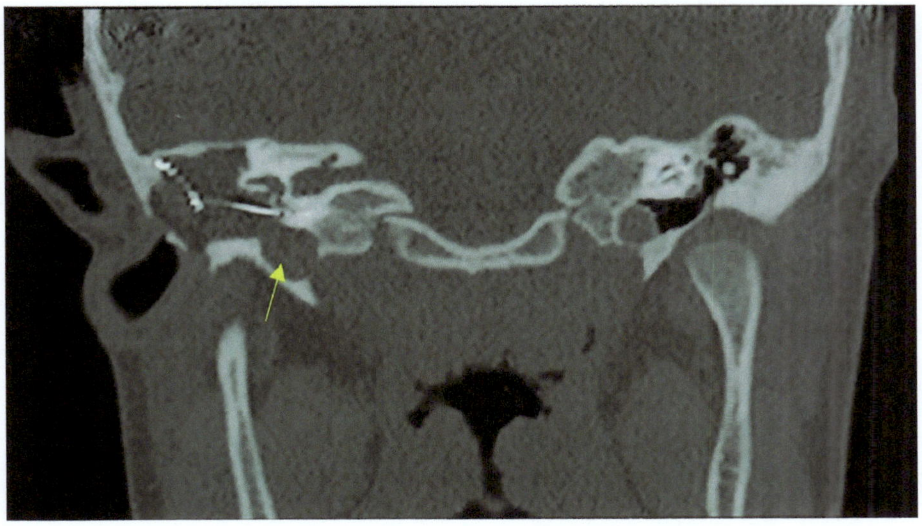

Fig. 6.69 HRCT scan of the temporal bone (coronal cut). Note the erosion of the tympanic bone toward the jugular bulb *(yellow arrow)*.

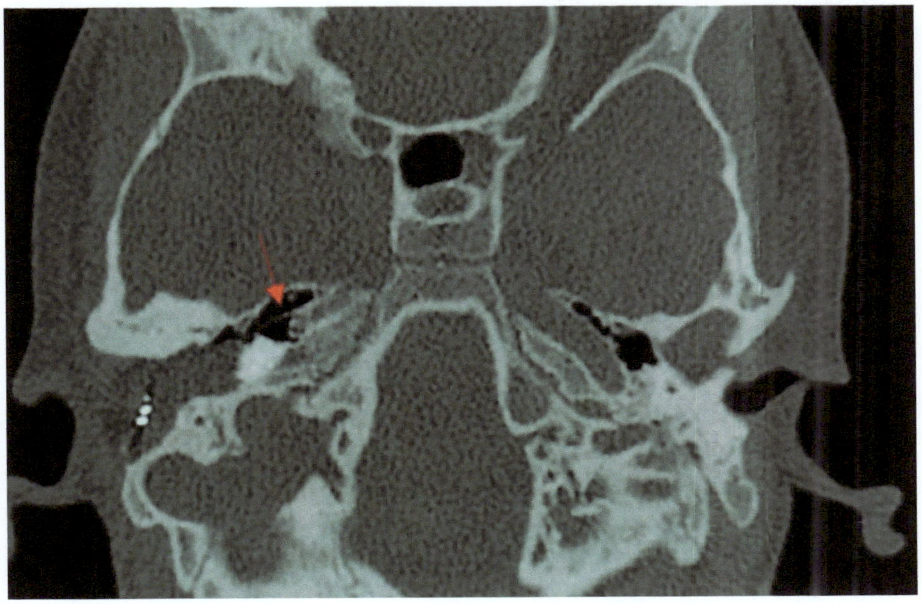

Fig. 6.70 HRCT scan of the temporal bone (axial cut). The tympanic cavity is normally ventilated *(red arrow)* due to reabsorption of the material previously used to obliterate the eustachian tube.

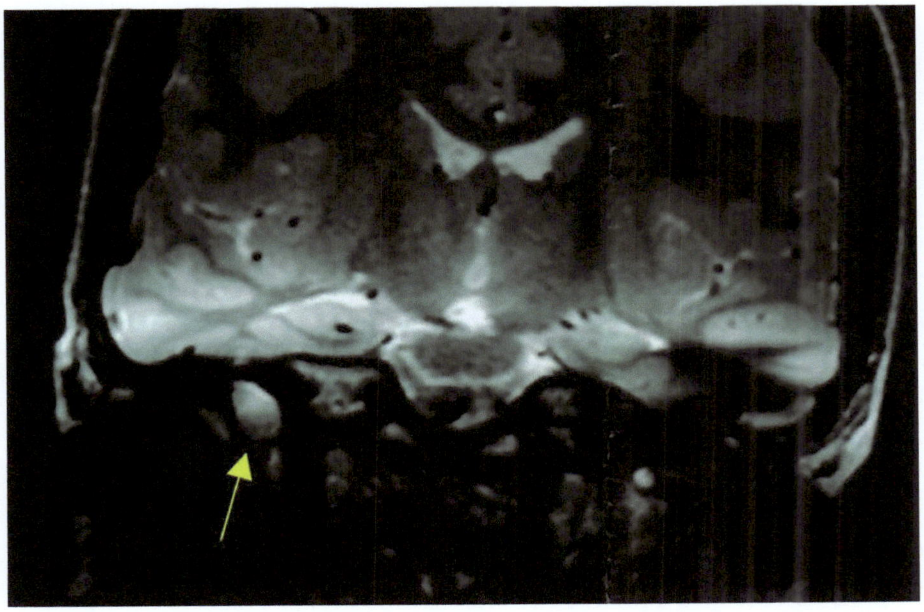

Fig. 6.71 Brain MRI (T2W sequences, coronal cut) after removal of the cochlear implant magnet. There is evidence of hyperintense tissue in the surgical cavity *(yellow arrow)*, consistent with a recurrent cholesteatoma.

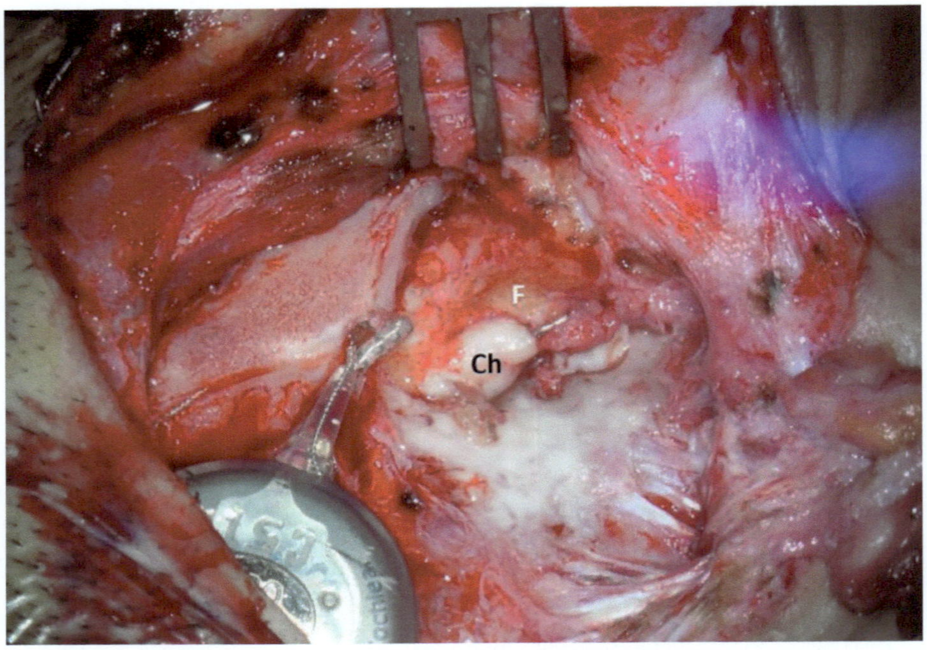

Fig. 6.72 Surgical view. The cavity shows obliterative tissue (abdominal fat—F) mixed with cholesteatoma (Ch).

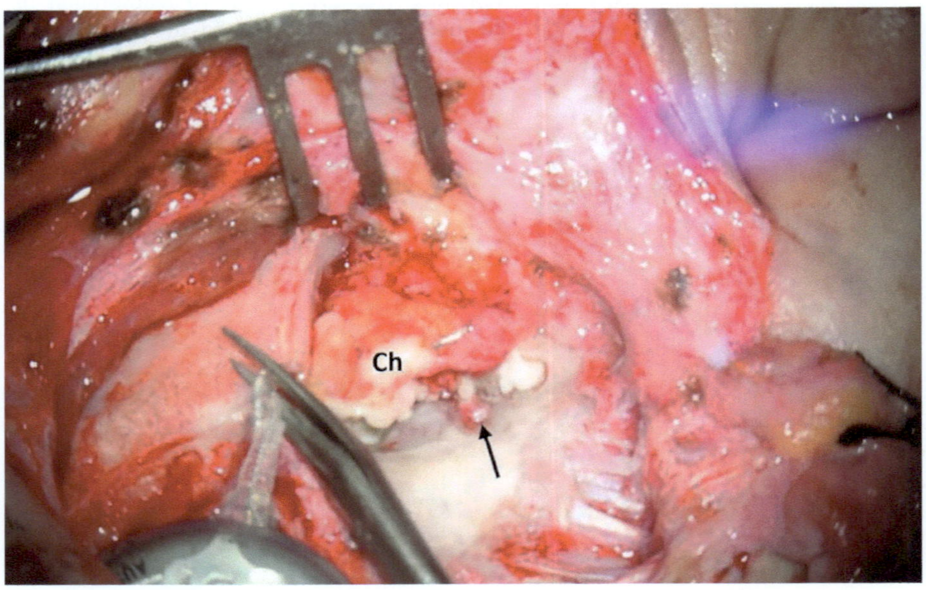

Fig. 6.73 The array is cut. Note the cholesteatoma (Ch) filling the surgical cavity. The sigmoid sinus is partially uncovered *(arrow)*.

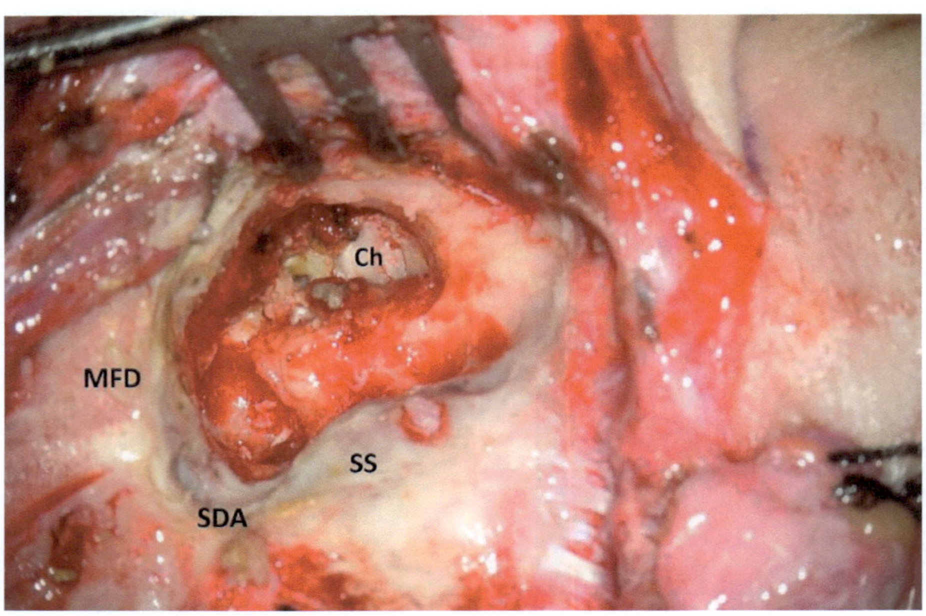

Fig. 6.74 After the removal of the obliterating tissue and part of the cholesteatoma (Ch), the surgical cavity was enlarged to identify the main anatomical landmarks. The cholesteatoma extends into the hypotympanum. *MFD*, middle fossa dura; *SDA*, sinodural angle; *SS*, sigmoid sinus.

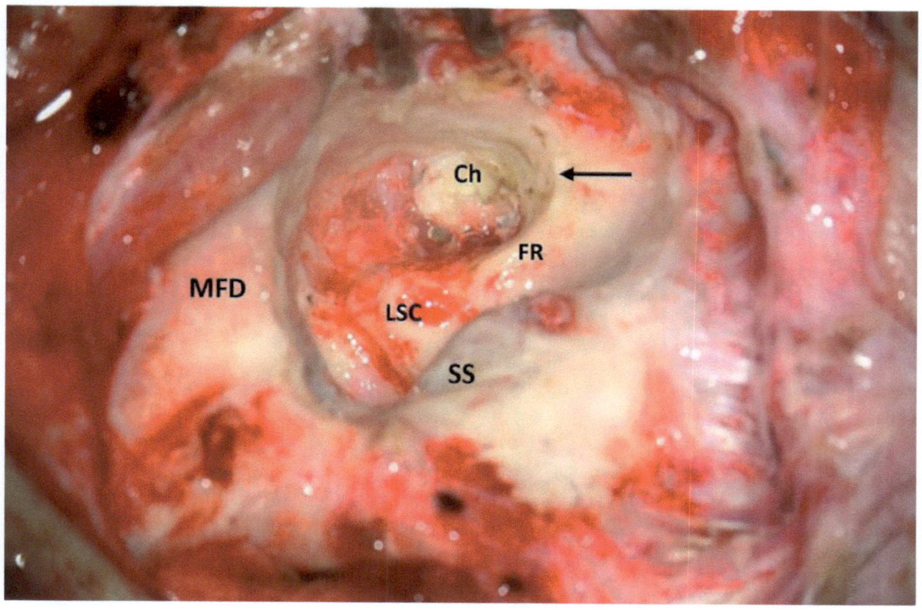

Fig. 6.75 Further removal of the inferior tympanic bone *(black arrow)* allows better isolation of the cholesteatoma (Ch) from the area of the petrous internal carotid artery. *MFD*, middle fossa dura, angle; *SS*, sigmoid sinus; *FR*, facial ridge; *LSC*, lateral semicircular canal.

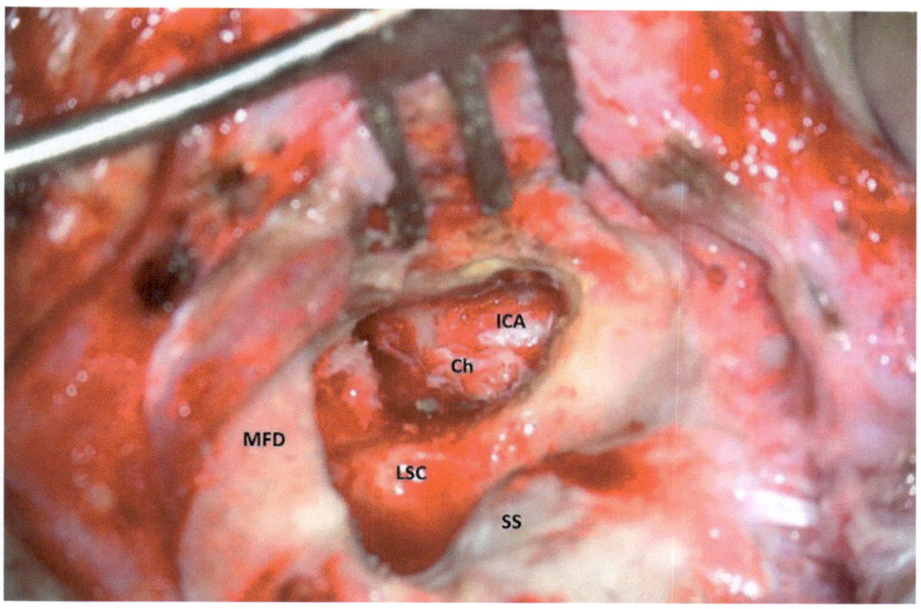

Fig. 6.76 The cholesteatoma (Ch) is progressively removed from the area of the internal carotid artery (ICA). *MFD*, middle fossa dura, angle; *SS*, sigmoid sinus; *LSC*, lateral semicircular canal.

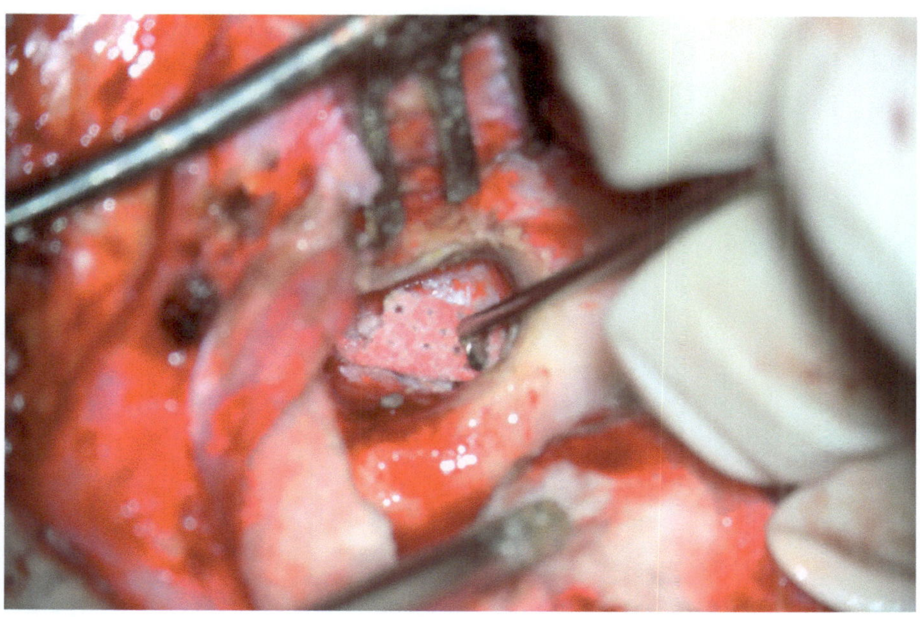

Fig. 6.77 The cholesteatoma matrix is dissected with the help of cottonoid.

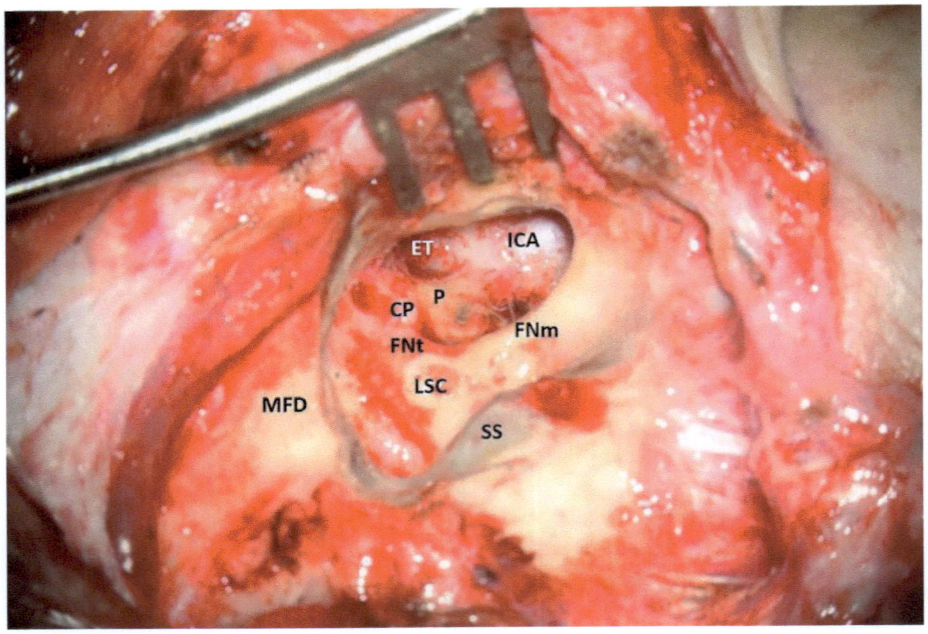

Fig. 6.78 Surgical view after removal of the cholesteatoma. The array is still left in the round window. *MFD*, middle fossa dura; *SS*, sigmoid sinus; *LSC*, lateral semicircular canal; *FNm*, mastoid segment of the facial nerve; *FNt*, tympanic segment of the facial nerve; *CP*, cochleariform process; *P*, promontory; *ET*, eustachian tube; *ICA*, internal carotid artery.

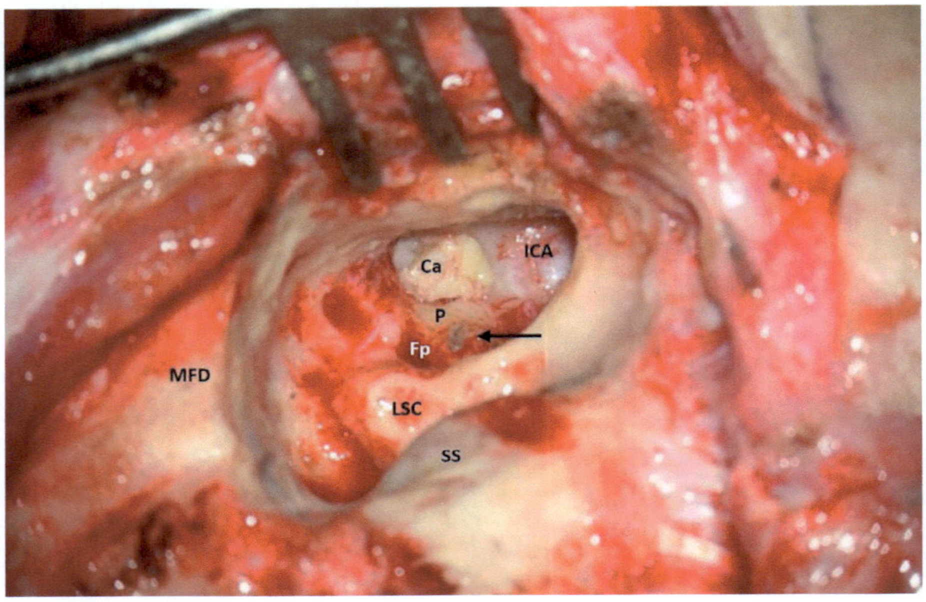

Fig. 6.79 Closer view. Note the array left in the round window *(arrow)*. The eustachian tube is obliterated with periosteum and cartilage (Ca). *MFD*, middle fossa dura; *SS*, sigmoid sinus; *LSC*, lateral semicircular canal; *P*, promontory; *Fp*, footplate; *ICA*, internal carotid artery.

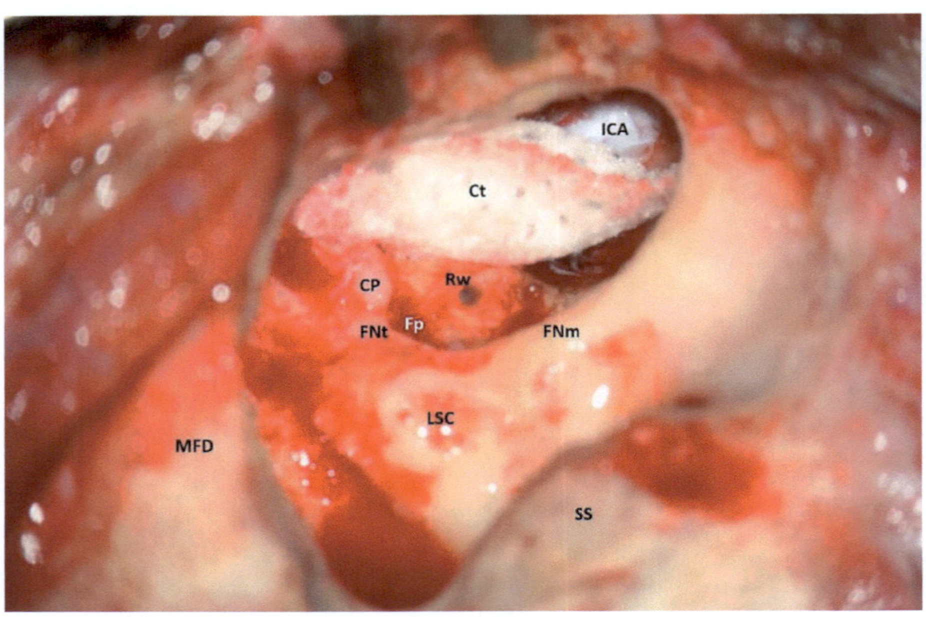

Fig. 6.80 The array is removed to prevent any infection to the subarachnoid space; it will then be replaced with a dummy electrode. *MFD*, middle fossa dura; *SS*, sigmoid sinus; *LSC*, lateral semicircular canal; *Fp*, footplate; *CP*, cochleariform process; *FNm*, mastoid portion of the facial nerve; *FNt*, tympanic portion of the facial nerve; *Rw*, round window; *Ct*, cottonoid; *ICA*, internal carotid artery.

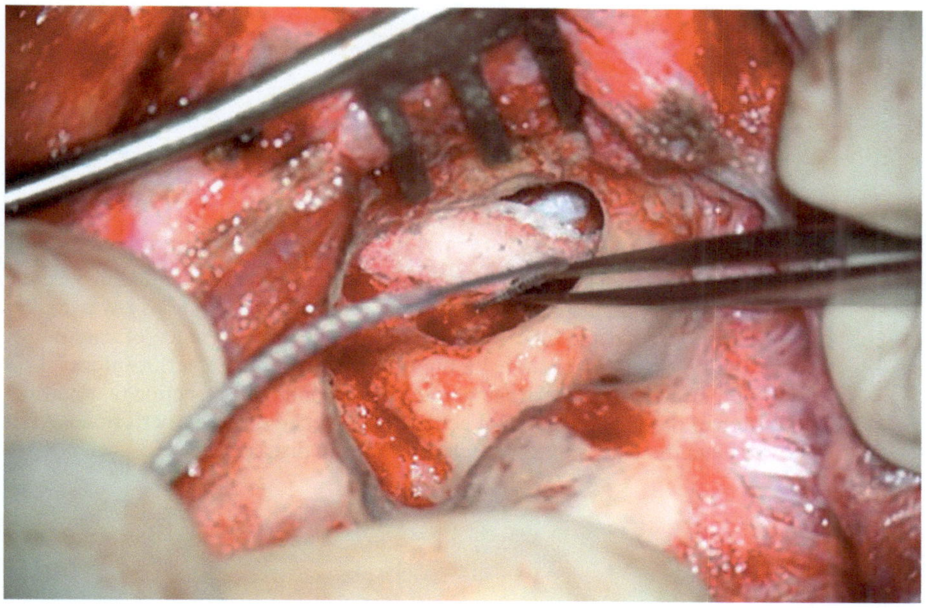

Fig. 6.81 A dummy electrode is inserted into the cochlea to prevent fibrosis and allow further cochlear implantation.

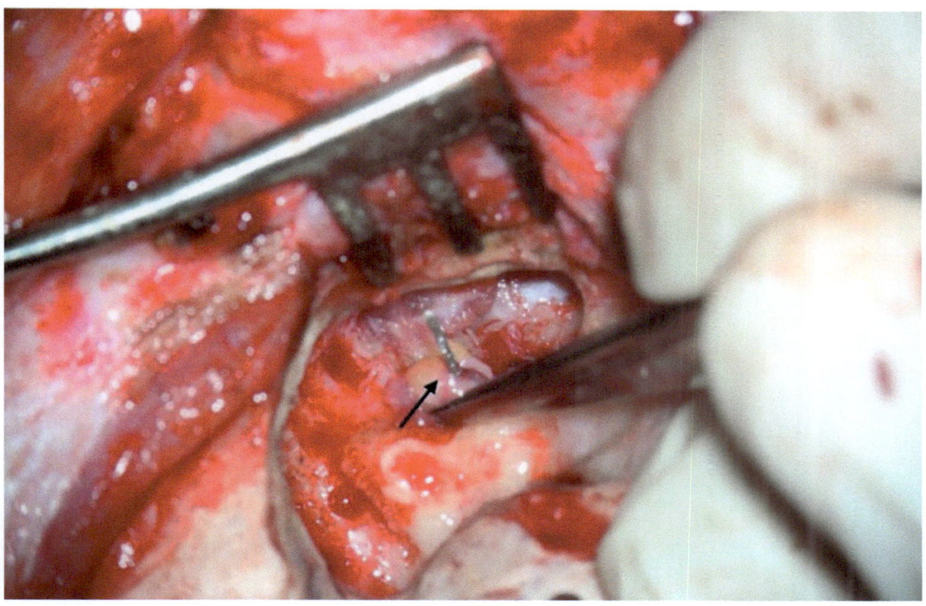

Fig. 6.82 The dummy electrode *(arrow)* is fully inserted into the cochlea and surrounded by muscle and periosteum. The cavity is then filled with abdominal fat.

Case 4: Squamous Cell Carcinoma of the External Auditory Canal (Lateral Temporal Bone Resection)

See Figs. 6.83–6.94.

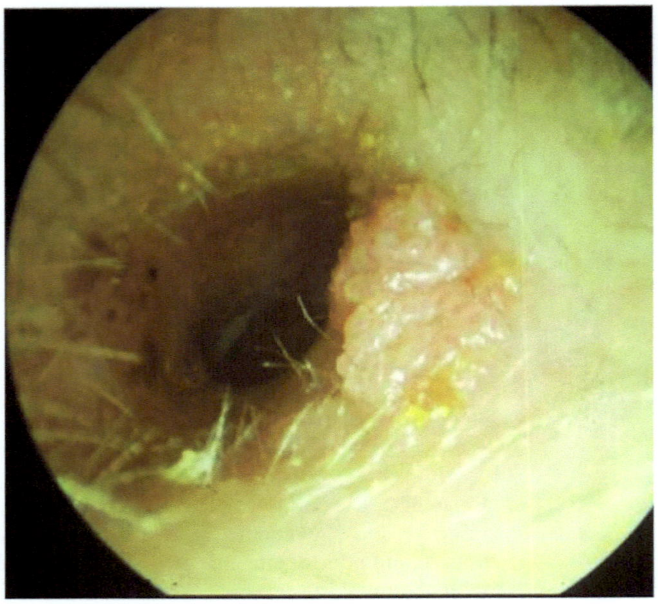

Fig. 6.83 Squamous cell carcinoma of the right external auditory canal. Patient complained of occasional bloody otorrhea. A biopsy was undertaken, and a pathologic examination revealed the presence of carcinoma. The CT scan demonstrated a tumor arising from the anterior-inferior wall of the external auditory canal. Lateral temporal bone resection (a modification of the standard subtotal petrosectomy technique) was performed. After 8 years the patient has no recurrence of the disease.

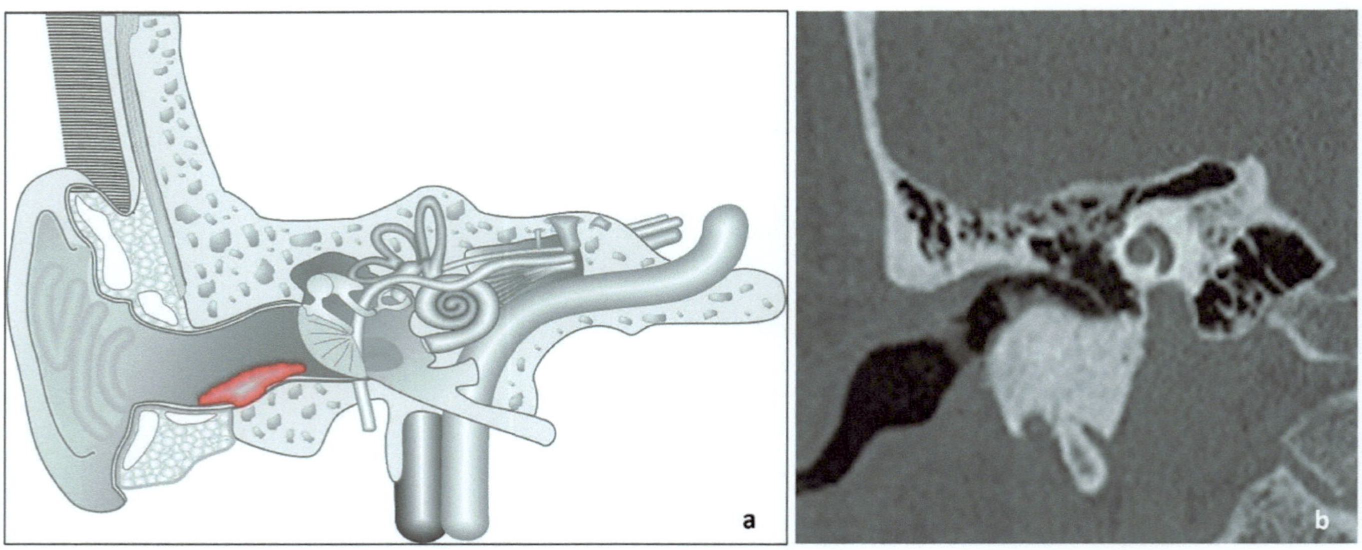

Fig. 6.84 The tumor is confined to the external auditory canal without bony erosion or evidence of soft tissue involvement (stage T1 according to the modified Pittsburgh Classification [cit]). (A) Illustrative picture; (B) CT scan (coronal cut) showing the tumor arising from the inferior tympanic bone.

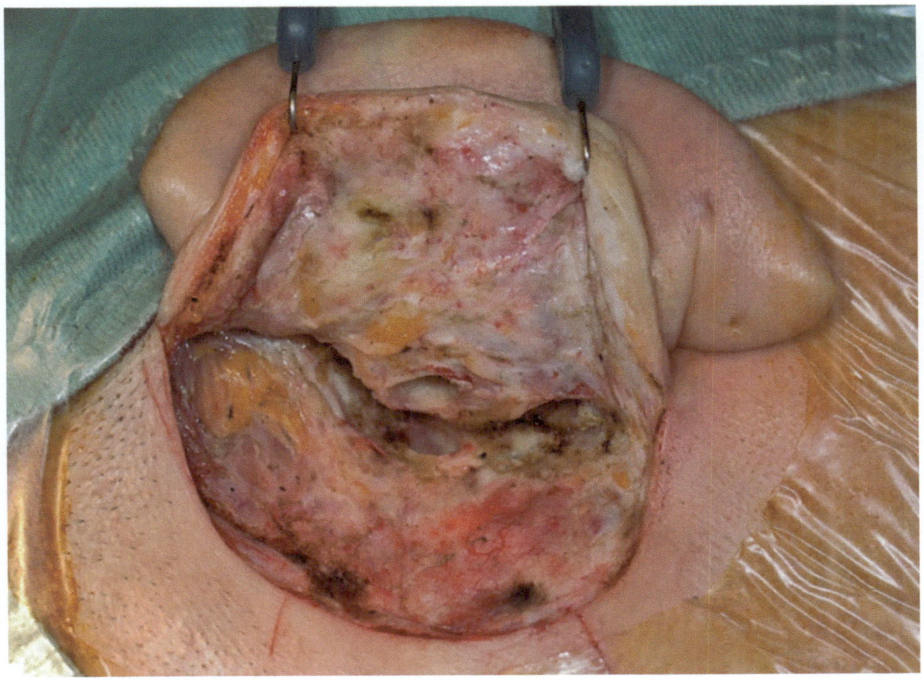

Fig. 6.85 Lateral temporal bone rection (LTBR). This is the surgery of choice for T1 and T2 tumors. The approach is a complete canal wall up mastoidectomy with an extended facial recess opening. The external auditory canal is resected en-bloc along with the tympanic membrane, and the malleus after disarticulation and removal of the incus, with the medial limit defined at the level of the incudostapedial joint. A superficial parotidectomy may be added, especially in T2 tumors. The skin incision starts 3 cm behind the retroauricular sulcus, and the external auditory canal has to be transected laterally to the tumor.

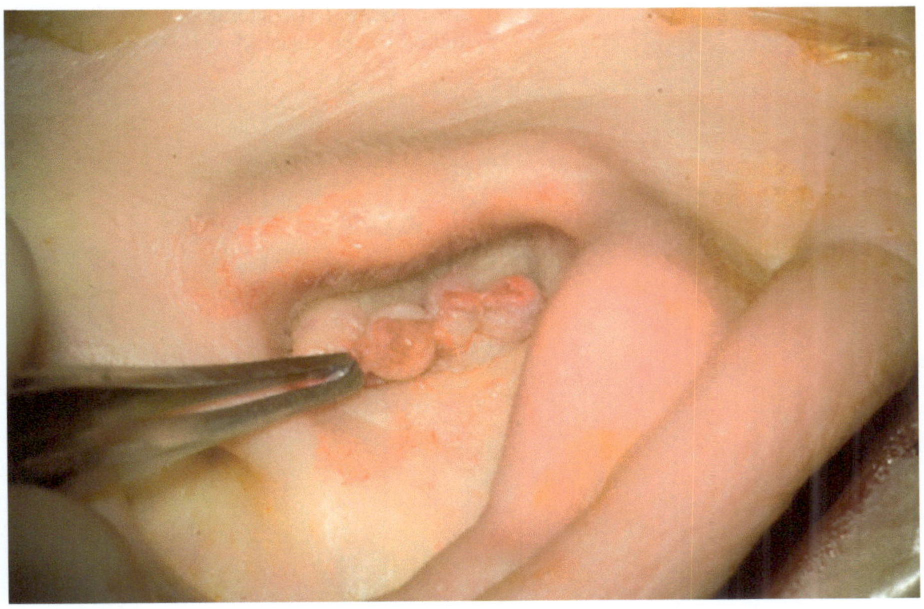

Fig. 6.86 Cul-de-sac closure of the external auditory canal.

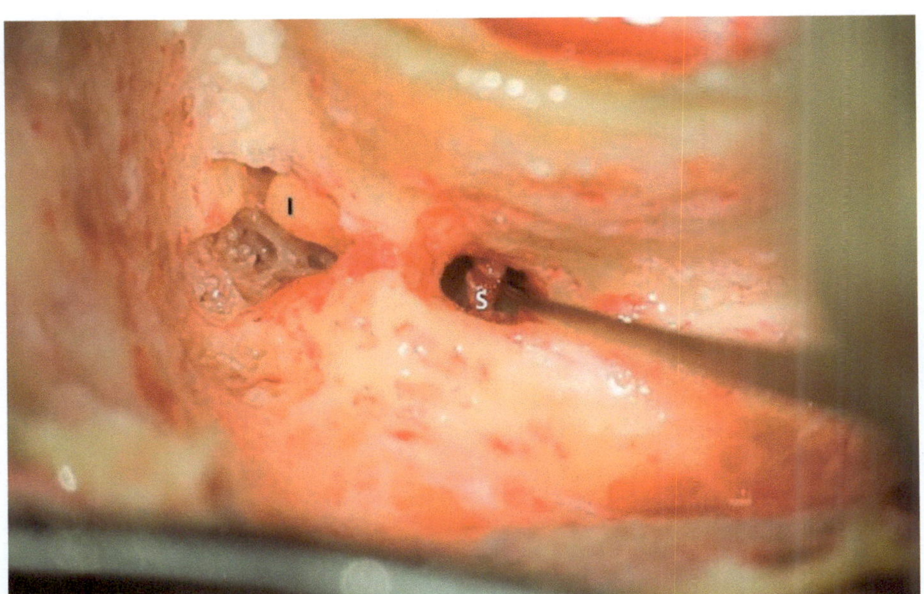

Fig. 6.87 Facial recess tympanotomy. *I*, incus; *S*, stapes.

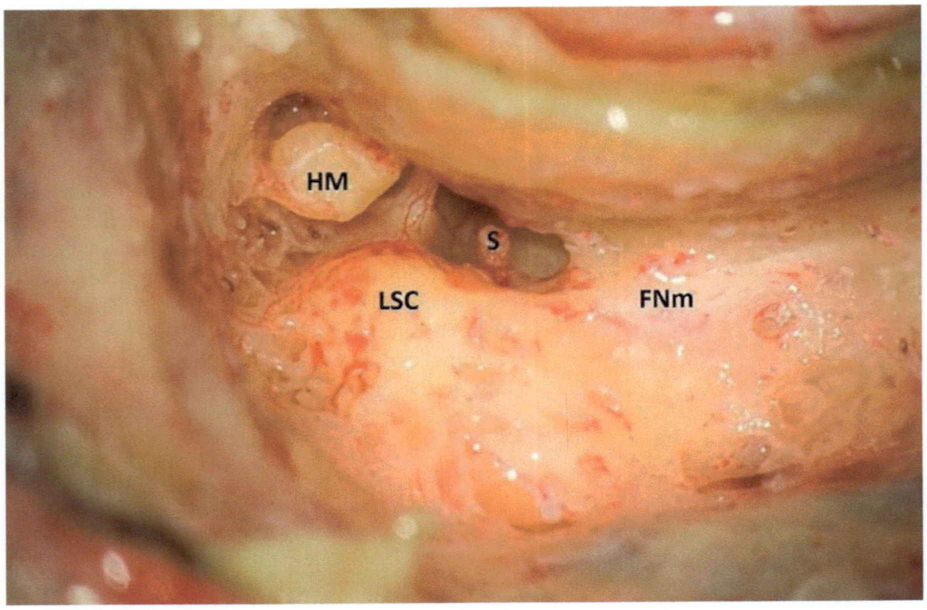

Fig. 6.88 Removal of the incus. *HM*, head of the malleus; *LSC*, lateral semicircular canal; *S*, stapes; *FNm*, mastoid portion of the facial nerve.

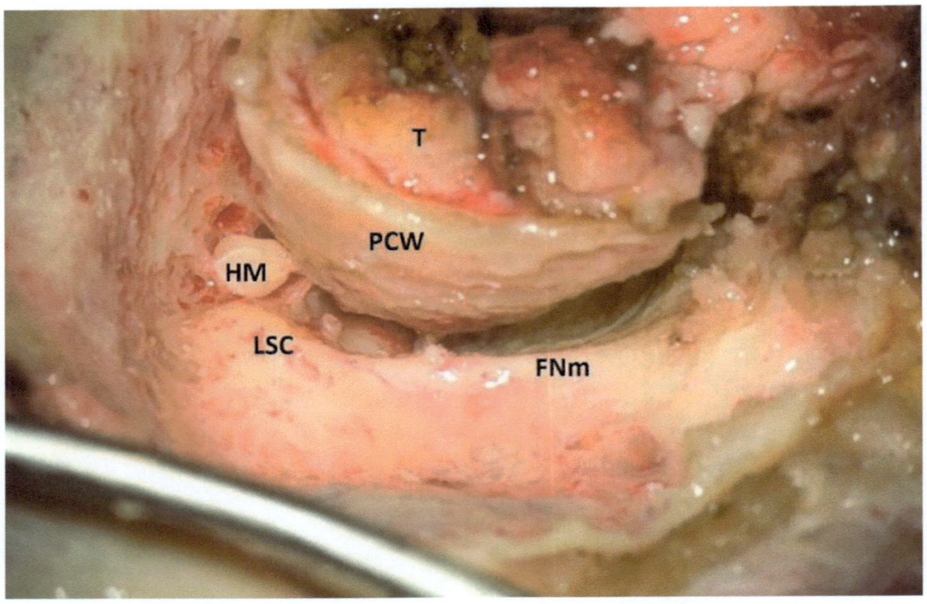

Fig. 6.89 Epi and hypotympanotomy. The chorda tympani has been sacrificed. *HM*, head of the malleus; *LSC*, lateral semicircular canal; *T*, tumor; *FNm*, mastoid portion of the facial nerve.

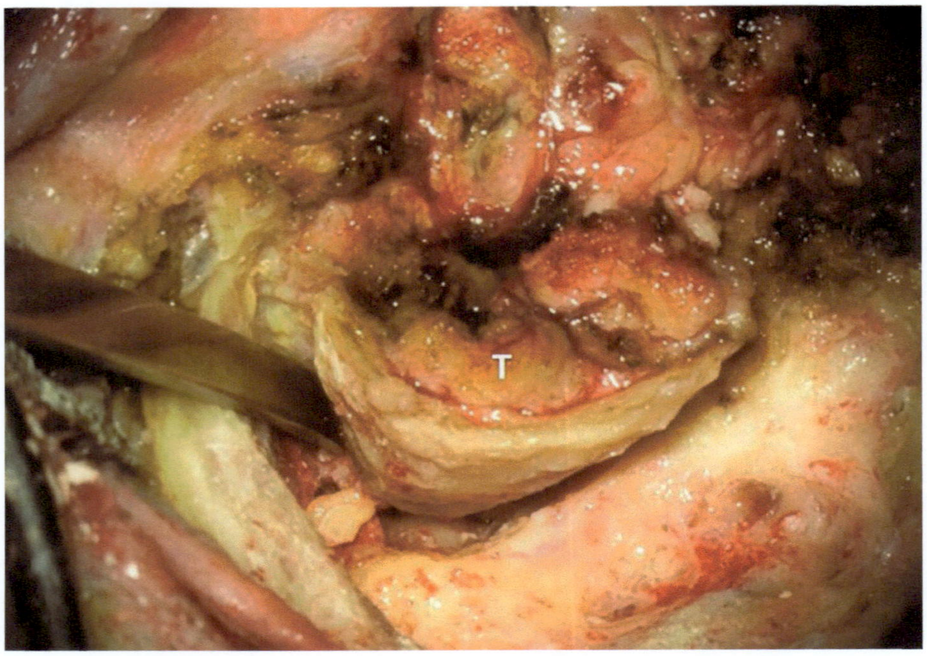

Fig. 6.90 En-bloc resection of the external auditory canal. *T*, tumor.

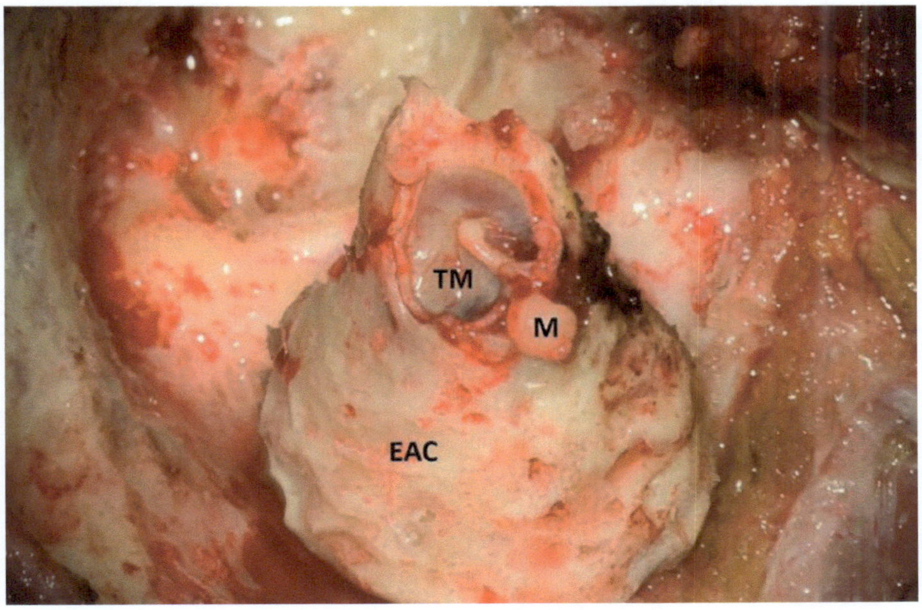

Fig. 6.91 En-bloc specimen including malleus (M), tympanic membrane (TM), and external auditory canal (EAC).

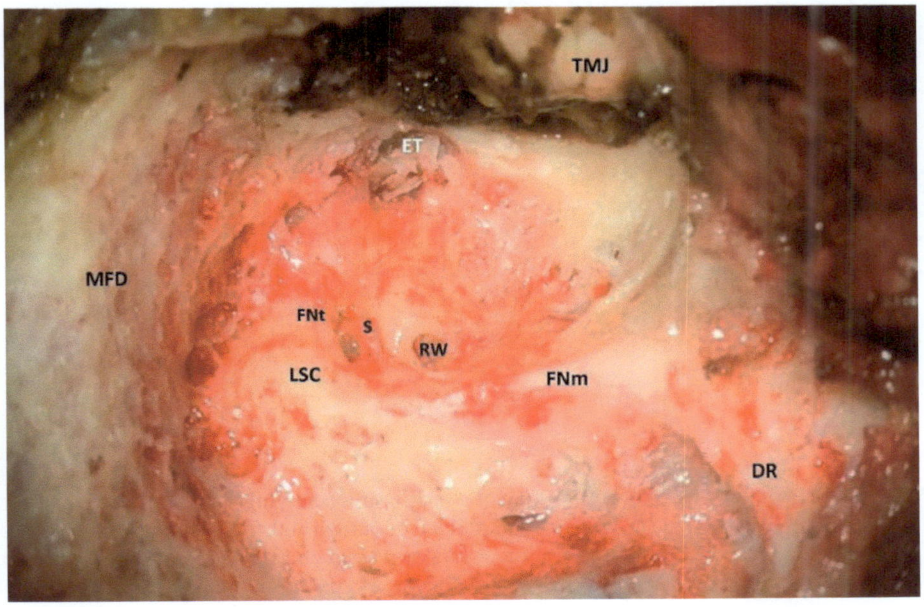

Fig. 6.92 Temporal bone cavity after tumor removal. *MFD*, middle fossa dura plate; *LSC*, lateral semicircular canal; *S*, stapes; *FNm*, mastoid portion of the facial nerve; *DR*, digastric ridge; *FNt*, tympanic portion of the facial nerve; *ET*, eustachian tube (obliterated with cartilage); *TMJ*, temporomandibular joint.

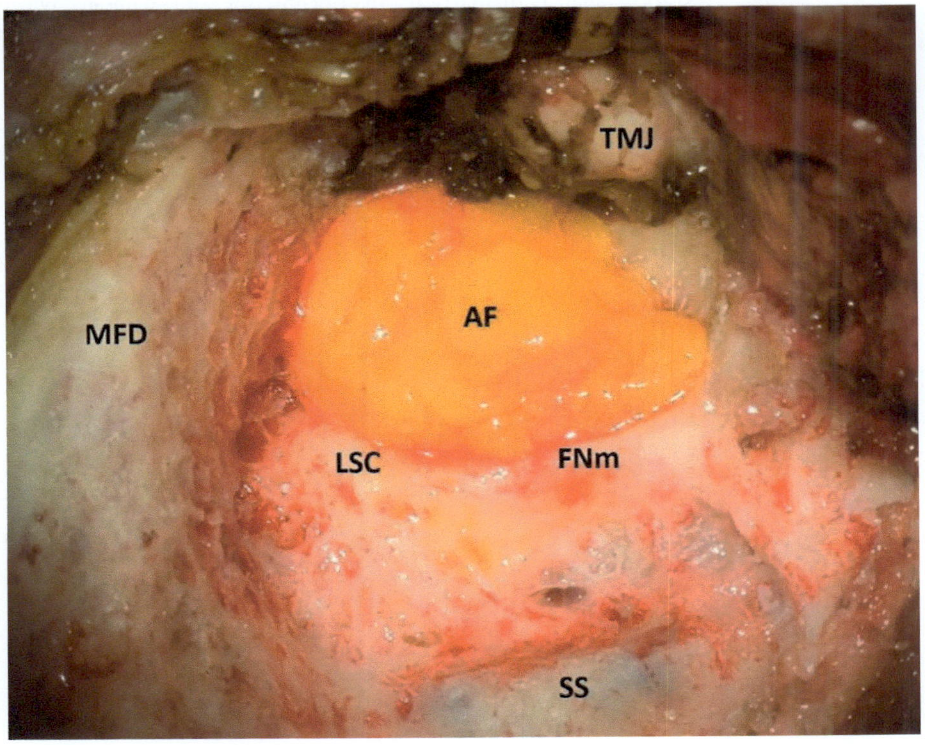

Fig. 6.93 The eustachian tube is further obliterated with periosteum and bone wax. The cavity is then filled with abdominal fat (AF). *TMJ*, temporomandibular joint; *MFD*, middle fossa dura; *LSC*, lateral semicircular canal; *FNm*, mastoid portion of the facial nerve; *SS*, sigmoid sinus.

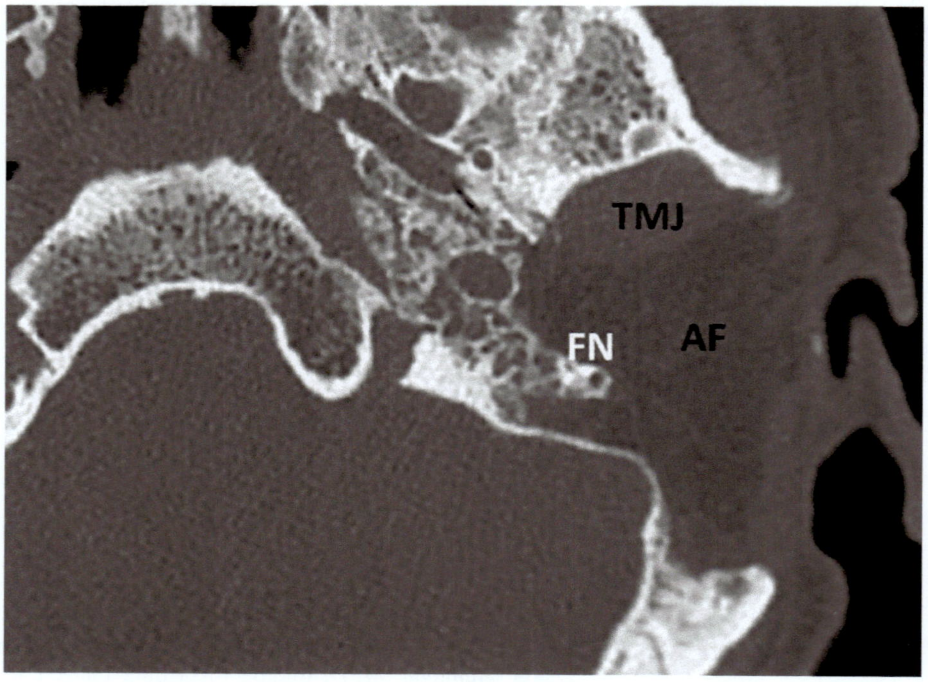

Fig. 6.94 Postoperative HRCT scan of the temporal bone, axial cut. The cavity is completely filled with abdominal fat (AF). The entire anterior tympanic bone has been removed, exposing the temporomandibular joint (TMJ). *FN*, facial nerve.

Case 5: Squamous Cell Carcinoma of the External Auditory Canal (Lateral Temporal Bone Resection With Superficial Parotidectomy)

See Figs. 6.95–6.98.

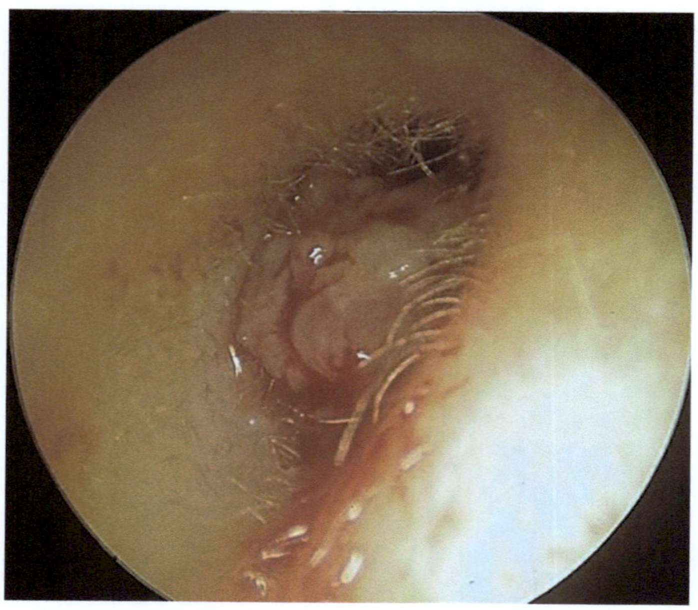

Fig. 6.95 Otoscopy of a neoplasm occupying the left external auditory canal. The patient complained of otalgia and episodes of bloody otorrhea of 1 month's duration. A biopsy was taken, and pathological examination revealed the presence of squamous cell carcinoma of the external auditory canal. The HRCT scan (Fig. 6.96) showed erosion of the external auditory canal, particularly its anterior-inferior wall, without invasion into the glenoid fossa. The tumor was removed en-bloc with superficial parotidectomy. Radiotherapy was administered postoperatively.

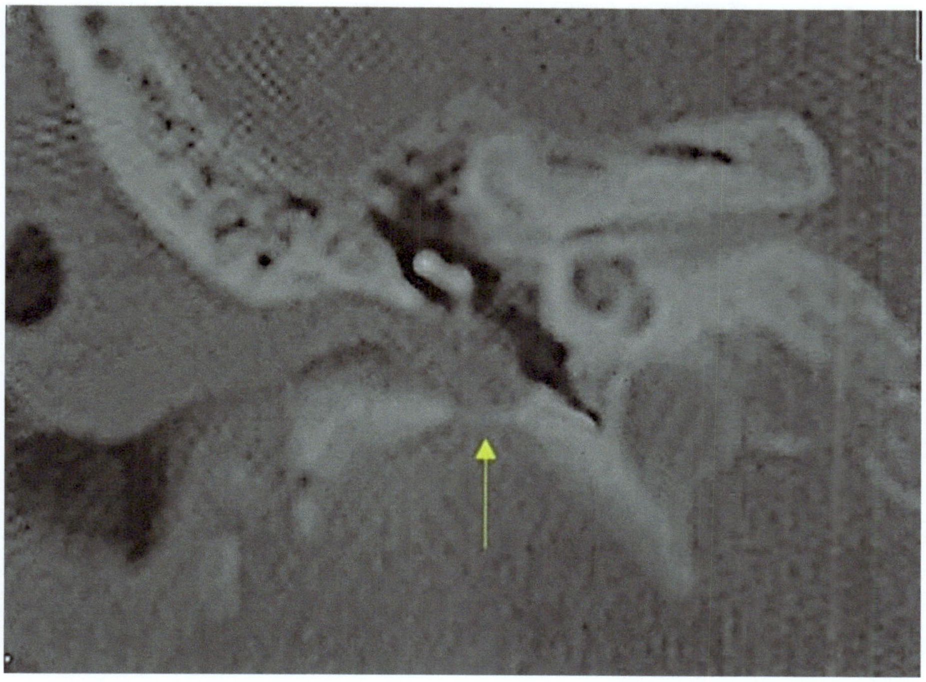

Fig. 6.96 HRCT scan of the temporal bone (coronal cut) shows erosion of the antero-inferior wall of the external auditory canal *(arrow)*. The glenoid fossa is not involved. According to the modified Pittsburg classification, the tumor stage is T2. This means that the tumor is confined to the external auditory canal with bone erosion (not full thickness) or limited soft tissue involvement (less than 5 mm).

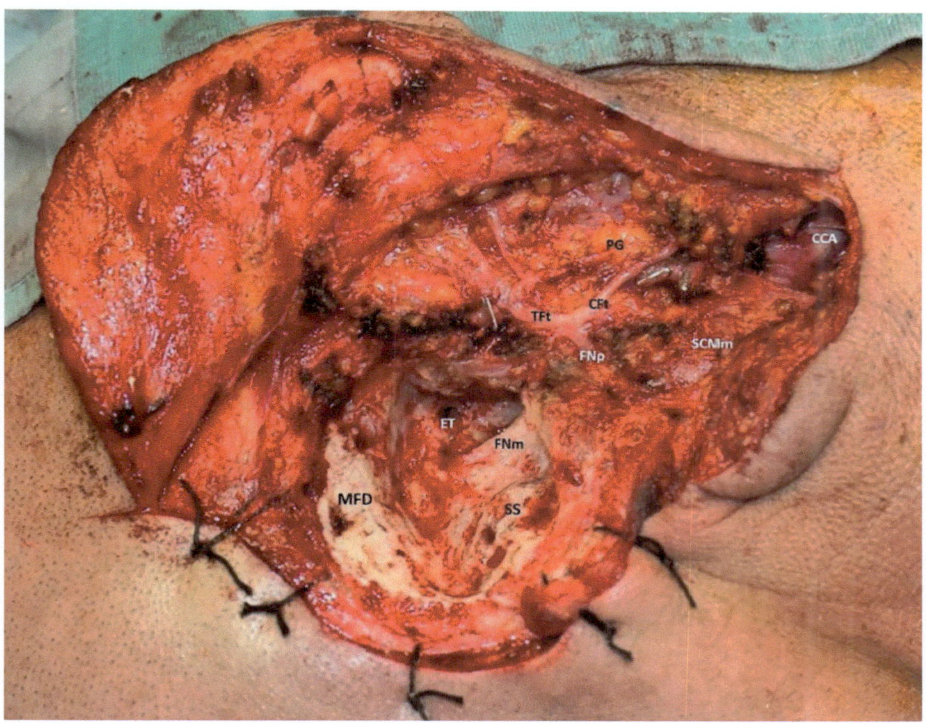

Fig. 6.97 Intraoperative view of the temporal cavity and the neck after tumor resection. *MFD*, middle fossa dura plate; *FNm*, mastoid portion of the facial nerve; *SS*, sigmoid sinus; *ET*, eustachian tube; *FNp*, parotid portion of the facial nerve (main trunk); *TFt*, temporofacial trunk of the facial nerve; *CFt*, cervicofacial trunk of the facial nerve; *SCMm*, sternocleidomastoid muscle; *CCA*, common carotid artery; *PG*, parotid gland (deep lobe).

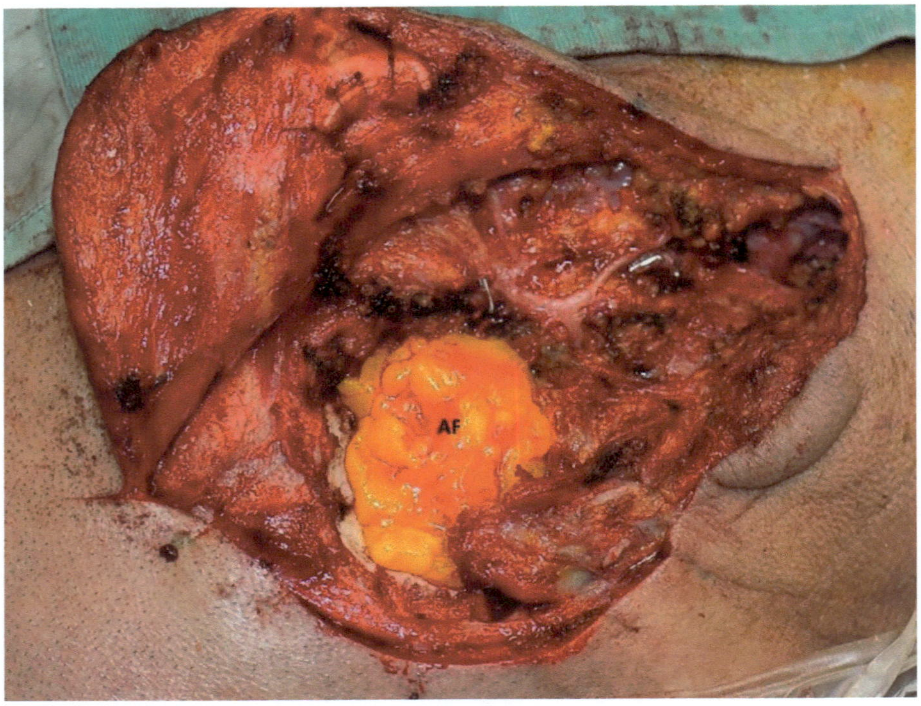

Fig. 6.98 The cavity is filled with abdominal fat (AF).

Case 6: Squamous Cell Carcinoma of the External Auditory Canal With Involvement of the Temporomandibular Joint (Subtotal Temporal Bone Resection)

See Figs. 6.99–6.102.

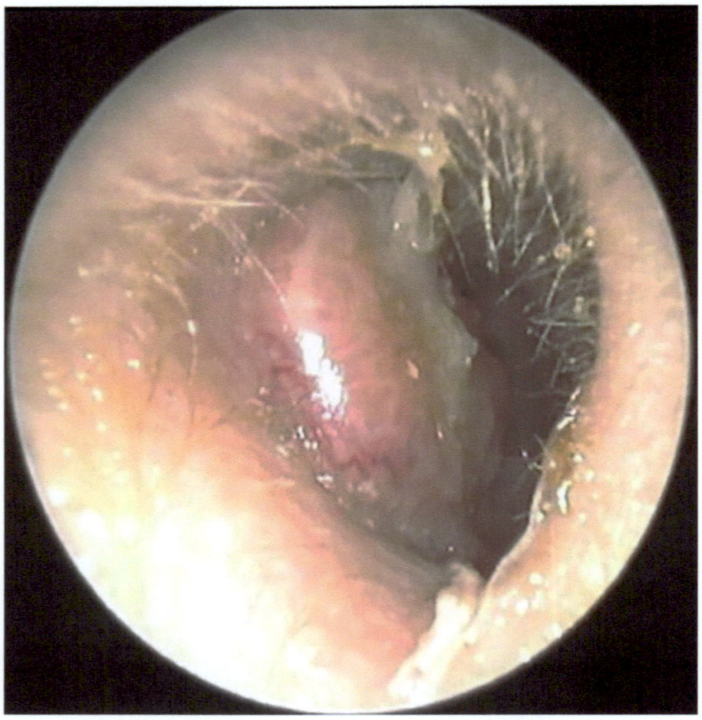

Fig. 6.99 Squamous cell carcinoma of the left external auditory canal. The lesion completely obliterates the meatus, and the tympanic membrane is not visible. The patient complained of hearing loss, persistent otorrhea, otalgia, facial nerve paralysis, and temporomandibular joint pain. A CT and MRI scan showed an extension of the pathology to the temporomandibular joint.

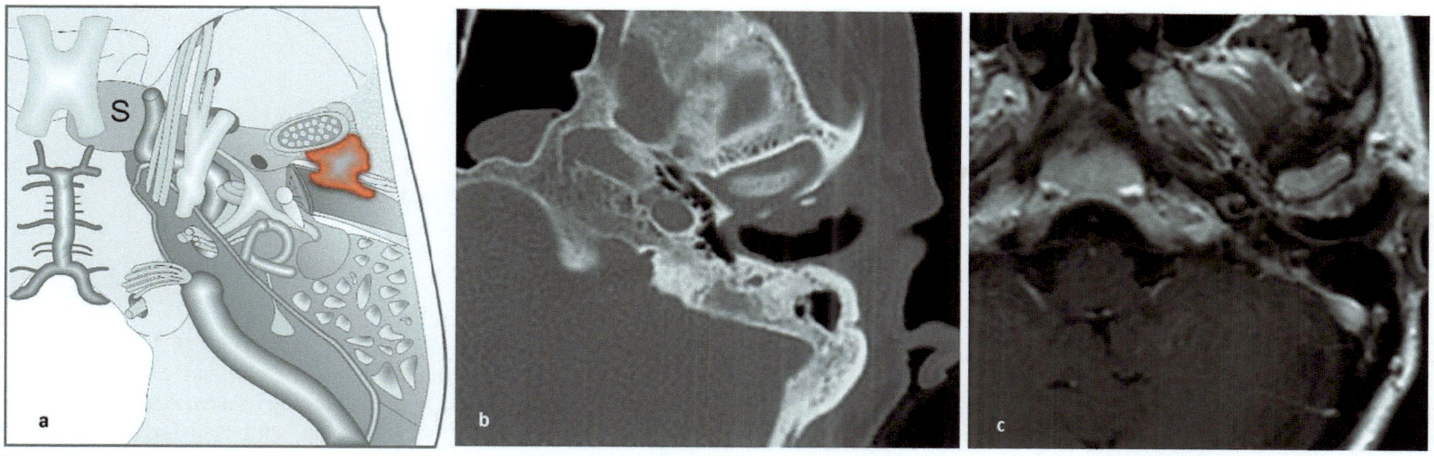

Fig. 6.100 The tumor extends to the temporomandibular joint and is therefore classified as a T4 according to the modified Pittsburg classification. A T4 tumor is defined if there is erosion of the cochlea, petrous apex, medial wall of the middle ear, carotid canal, jugular foramen, or dura; or a tumor with extensive soft tissue involvement (>5 mm), such as involvement of the temporomandibular joint or styloid process involvement; or evidence of facial paresis. (A) Illustrative picture; (B) HRCT scan of the temporal bone, axial cut; (C) T1W MRI scan + Gd enhancement, axial cut.

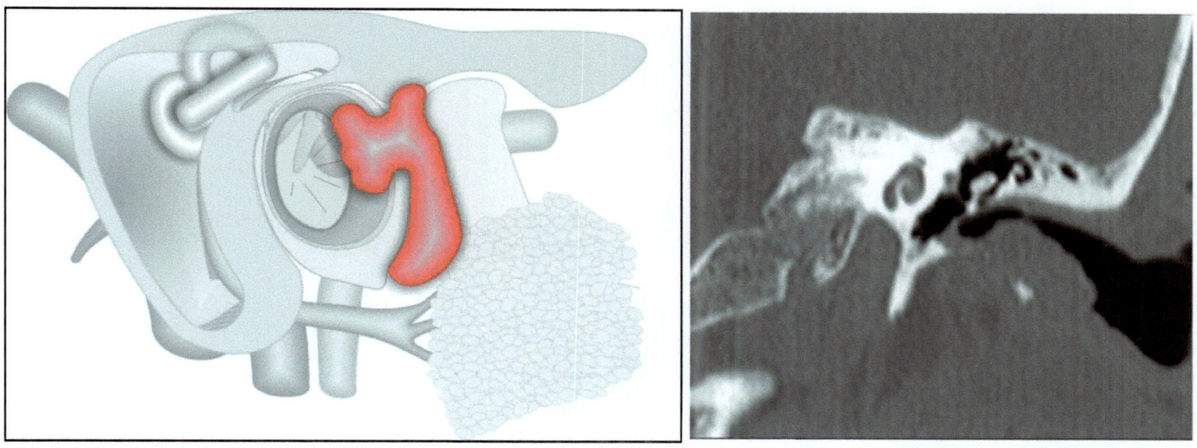

(A) (B)

Fig. 6.101 Coronal view. (A) illustrative picture, (B) HRCT scan of the temporal bone.

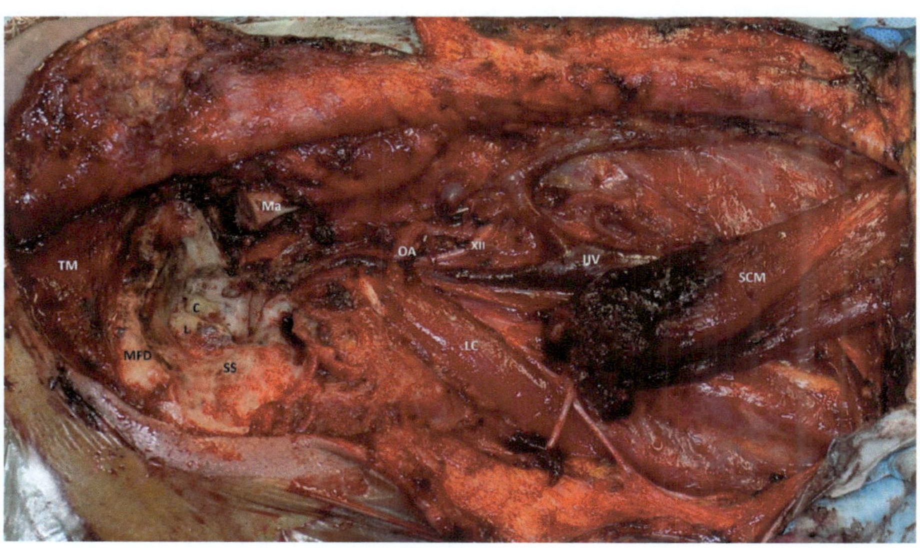

Fig. 6.102 Intraoperative view of the temporal cavity and the neck after tumor resection. A subtotal temporal bone resection (STBR) was performed together with total parotidectomy, modified radical neck dissection, and removal of the mandibular condyle. STBR is a medial extension of lateral temporal bone resection and may include removal of the mastoid tip, labyrinth, cochlea, styloid process, temporomandibular joint, facial nerve, pinna, parotid, dura, sigmoid sinus, and internal carotid artery. *TM*, temporalis muscle; *MFD*, middle fossa dura plate; *SS*, sigmoid sinus; *L*, labyrinth; *C*, cochlea; *Ma*, mandibule (ramus); *OA*, occipital artery; *SCM*, sternocleidomastoid muscle; *XII*, hypoglossal nerve; *IJV*, internal jugular vein; *LC*, longissimus capitis muscle.

Case 7: Squamous Cell Carcinoma of the External Auditory Canal With Involvement of the Pinna (Subtotal Temporal Bone Resection)

See Figs. 6.103–6.106.

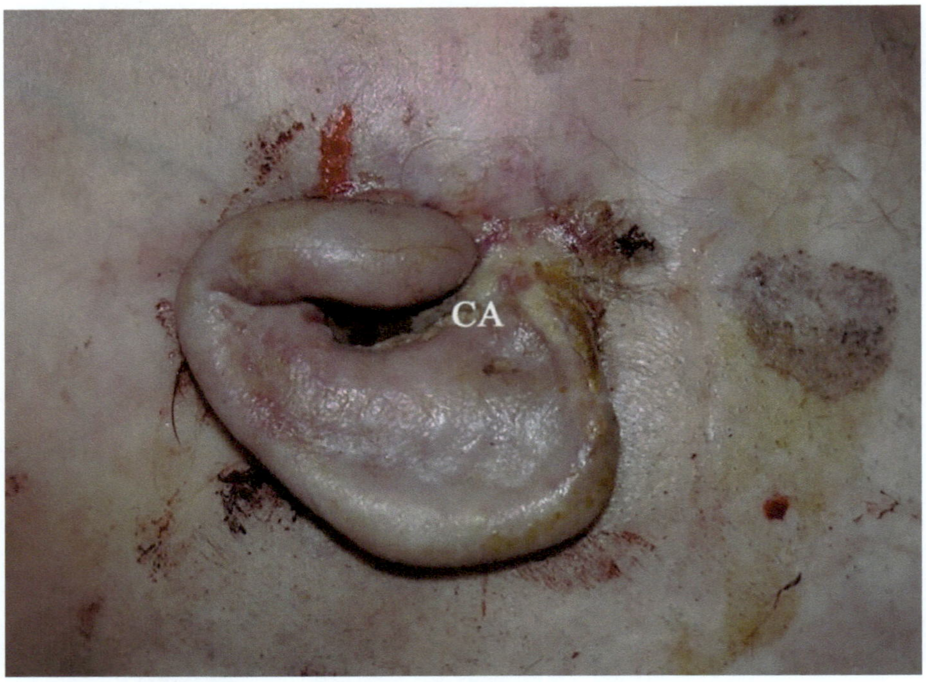

Fig. 6.103 The carcinoma (CA) involves the external auditory meatus and extends into the pinna and surrounding tissues. A subtotal temporal bone resection with excision of the entire pinna and reconstruction with a myocutaneous flap was planned.

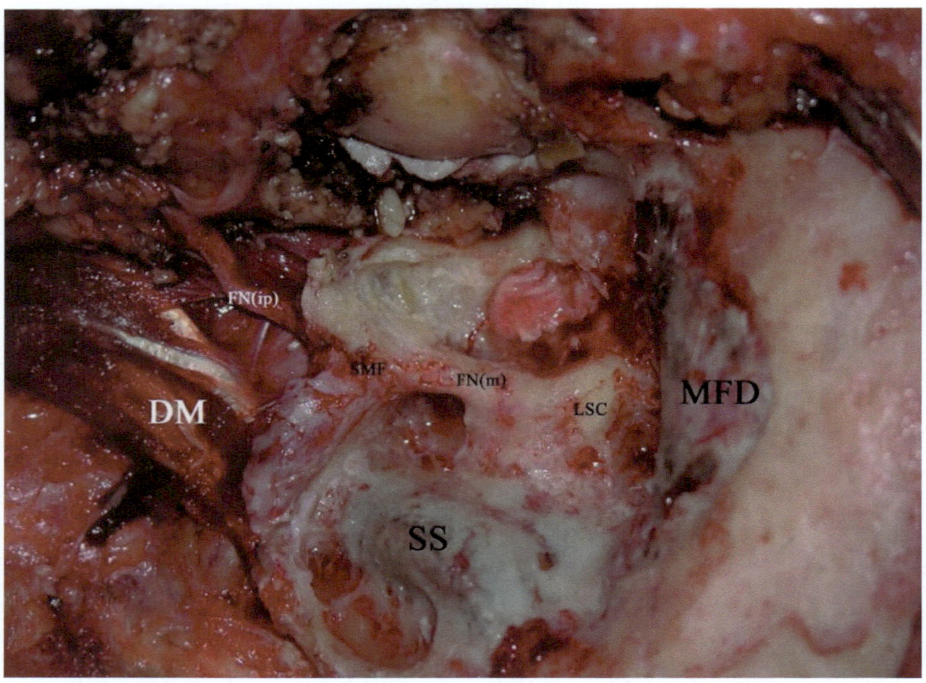

Fig. 6.104 Intraoperative view of the temporal cavity and the neck after tumor resection. A subtotal temporal bone resection (STBR) was performed together with superficial parotidectomy and modified radical neck dissection. *DM*, digastric muscle (posterior belly); *SS*, sigmoid sinus; *FN(m)*, mastoid portion of the facial nerve; *SMF*, stylomastoid foramen; *FN(ip)*, intraparotid portion of the facial nerve; *LSC*, lateral semicircular canal; *MFD*, middle fossa dura.

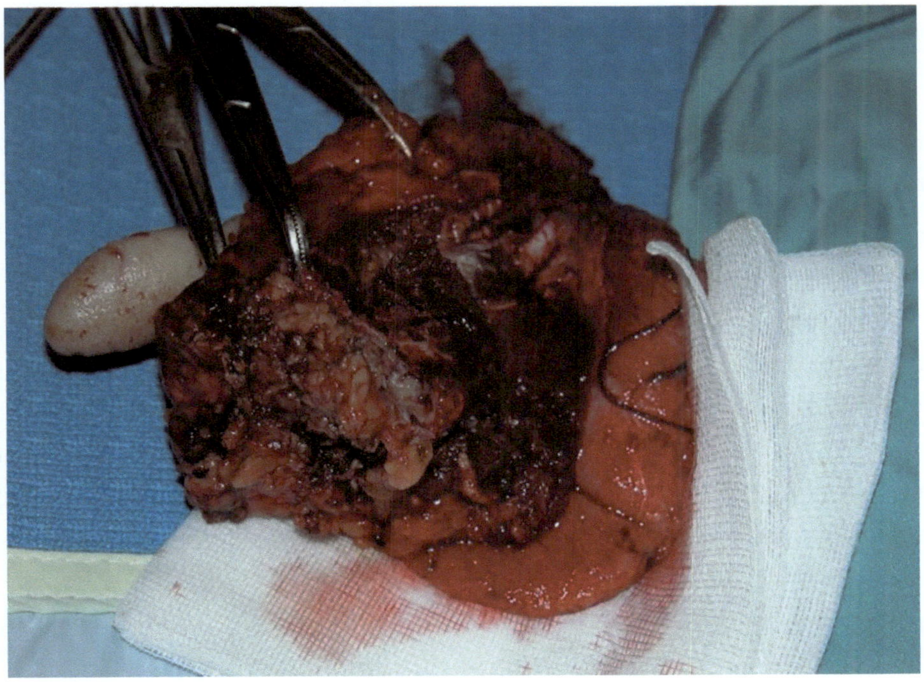

Fig. 6.105 Surgical specimen.

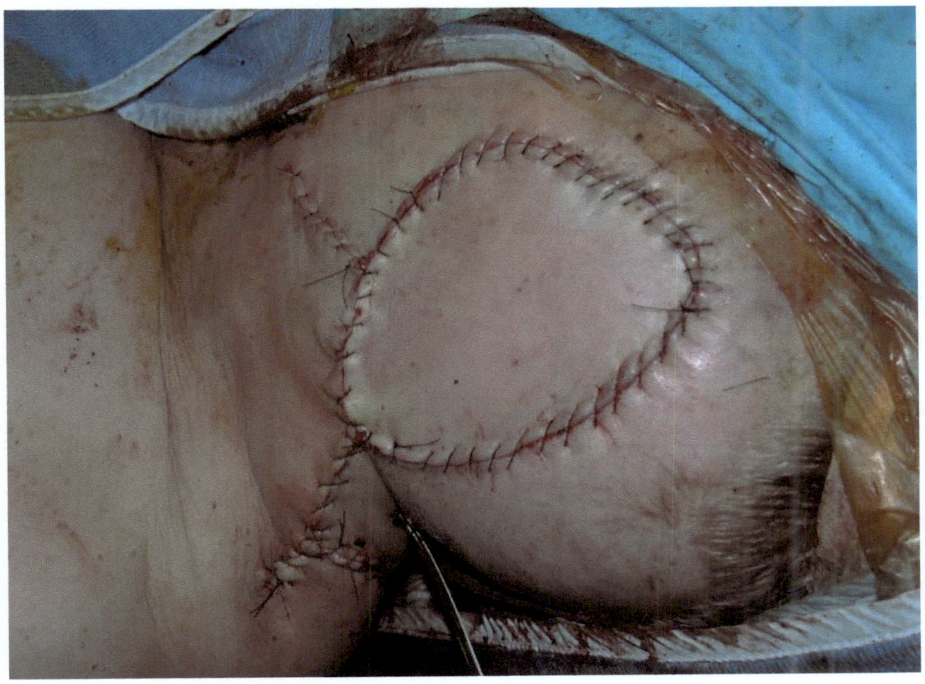

Fig. 6.106 Reconstruction of the cavity with a myocutaneous trapezius flap.

Summary

STP is indicated in any disease primarily in the middle ear and mastoid with limited extensions into deeper parts of the temporal bone that leaves behind a large cavity [1]. In lateral skull base surgery, this procedure is a preliminary step in many approaches, such as transotic, transcochlear, and infratemporal fossa approaches; temporal bone resections; and other skull base procedures for particular cases involving middle and posterior fossa tumors. STP is a very useful and effective surgical tool in the management of a variety of problematic situations; it offers excellent exposure, provides middle ear exclusion, and can be combined safely with hearing implantation procedures [5–17].

HINTS AND PITFALLS

- The postauricular incision must be wide to avoid a fistula that can arise from the friable skin due to previous surgeries.
- In pleuri-operated cases, the tragal cartilage can be brought back posteriorly and sutured to a soft tissue flap as a second layer for external auditory canal closure.
- The skin, annulus, and tympanic membrane with malleus and incus can be removed en bloc to lower the risk of leaving some skin behind.
- While performing the canal wall-down mastoidectomy, it is better to remove as many cell tracts as possible to ensure that there is no residual disease.
- The perilabyrinthine, peritubal, and hypotympanic cell tracts are removed routinely except in cases of cochlear implants in noninfective cases.
- The stapes superstructure has to be removed because leaving it behind could allow fat to exert a piston-like pressure on the stapes or frank subluxation, leading to postoperative vertigo.
- Closure of the eustachian tube must be done meticulously to prevent an opening.
- The wound is closed in two or three layers.
- Follow-up is mandatory with high-resolution computed tomography (HRCT) and magnetic resonance imaging (MRI) scans of the temporal bone and skull base, the latter with diffusion-weighted sequences (MRI-DWI non-EPI).

REFERENCES

[1] Coker NJ, Jenkins HA, Fisch U. Obliteration of the middle ear and mastoid cleft in subtotal petrosectomy: indications, technique, and results. Ann Otol Rhinol Laryngol 1986;95:5–11.

[2] Fisch U. Microsurgery of the skull base. Stuttgart, Germany: Georg Thieme Verlag; 1988.

[3] Rambo JH. Musculoplasty: a new operation for suppurative middle ear deafness. Trans Am Acad Ophthalmol Otolaryngol 1958;62:166–77.

[4] Rambo JH. Primary closure of the radical mastoidectomy wound: a technique to eliminate postoperative care. Laryngoscope 1958;68:1216–27.

[5] Gacek RR. Mastoid and middle ear cavity obliteration for control of otitis media. Ann Otol Rhinol Laryngol 1976;85:305–9.

[6] Nadol Jr JB, Schuknecht HF. Obliteration of the mastoid in the treatment of tumors of the temporal bone. Ann Otol Rhinol Laryngol 1984;93:6–12.

[7] Sanna MM, Free RH, Falcioni M. Subtotal petrosectomy in cochlear implantation surgery for cochlear and other auditory implants. Stuttgart, Germany: Georg Thieme Verlag; 2016.

[8] Sanna MS, Khrais T, Mancini F, Piazza P, Russo A, Taibah A. Atlas of microsurgery of the lateral skull base. Stuttgart, Germany: Georg Thieme Verlag; 2007.

[9] Sanna M, Dispenza F, Flanagan S, De Stefano A, Falcioni M. Management of chronic otitis by middle ear obliteration with blind sac closure of the external auditory canal. Otol Neurotol 2008;29:19–22.

[10] Sanna M, Free R, Merkus P, Falcioni M. Subtotal petrosectomy in cochlear implantation surgery for cochlear and other auditory implants. Stuttgart, Germany: George Thieme Verlag; 2016.

[11] Prasad SC, Roustan V, Piras G, Caruso A, Lauda L, Sanna M. Subtotal petrosectomy: surgical technique, indications, outcomes, and comprehensive review of literature. Laryngoscope 2017;127(12):2833–42.

[12] Moody SA, Hirsch BE, Myers EN. Squamous cell carcinoma of the external auditory canal: an evaluation of a staging system. Am J Otol 2000;21:582–8.

[13] Piras G, Grinblat G, Albertini R, Sykopetrites V, Zhong SX, Lauda L, Sanna M. Management of squamous cell carcinoma of the temporal bone: long-term results and factors influencing outcomes. Eur Arch Otorrinolaringol 2021;278(9):3193–202. https://doi.org/10.1007/s00405-020-06378-w. Epub 2020 Sep 26 PMID: 32979119.

[14] Prasad SC, D'Orazio F, Medina M, Bacciu A, Sanna M. State of the art in temporal bone malignancies. Curr Opin Otolaryngol Head Neck Surg 2014;22(2):154–65. https://doi.org/10.1097/MOO.0000000000000035. PMID: 24573123.

[15] Dandinarasaiah M, Grinblat G, Keshavamurthy VB, Di Rubbo VM, Alessandra R, Lauda L, Sanna M. Characteristics and management of facial nerve schwannomas and hemangiomas. Otol Neurotol 2024;45(1):83–91. https://doi.org/10.1097/MAO.0000000000004033. Epub 2023 Oct 14 PMID: 37853739.

[16] Grinblat G, Vlad D, Caruso A, Sanna M. Evaluation of subtotal petrosectomy technique in difficult cases of cochlear implantation. Audiol Neurootol 2020;25(6):323–35. https://doi.org/10.1159/000507419. Epub 2020 May 29 PMID: 32474562.

[17] Sanna M, Russo A, Caruso A, Taiba A, Piras G. Color atlas of endo otoscopy. Examination-diagnosis-treatment. 1st ed. Thieme; 2017, ISBN:9783132415232.

CHAPTER 7

Translabyrinthine Approaches

Introduction

The translabyrinthine approach (TLA), described by Rudolf Panse in 1904, was first performed for acoustic neuroma excision in 1911 by Franciscus Hubertus Quix of Utrecht University in the Netherlands. Although this first attempt was considered successful, Quix's patient died 6 months later and postmortem examination revealed a substantial residual tumor filling the posterior fossa. After being heavily criticized by two leading pioneers of world neurosurgery, Harvey Cushing and Walter Dandy, TLA faded into obscurity for the next 50 years [1]. It was not until the introduction of the operating microscope that William House popularized TLA again, marking the transition to the era of modern neuro-otology [2]. Since then, the reduction of surgical morbidity has been the main focus of the continuous refinement of the original approach described by House. Over the years, we have introduced some technical modifications to TLA that have allowed us to reduce the mortality rate to <1% (3/4000) with minimal morbidity. Identifying the facial nerve by using the ampullary nerve as a landmark rather than the bill's bar, lowering the high jugular bulb, and extending the bone removal around and anterior to the internal auditory canal (transapical extension) have been the major refinements added to the original technique [3–25].

Enlarged Translabyrinthine Approach

Indications

– Removal of cerebellopontine angle (CPA) lesions, when hearing preservation is not a concern, is as follows:

(a) Acoustic neurinoma:

- With unserviceable preoperative hearing, whatever the size of the tumor is. It should be noted that giant tumors can be safely removed through this approach.
- Acoustic neurinoma more than 1.5 cm in the extrameatal diameter irrespective of the preoperative hearing.
- Cases with neurofibromatosis type 2 (NF2), in which preservation of the cochlear nerve and a concomitant cochlear implant (CI) can be attempted. Furthermore, an auditory brainstem implant (ABI) can be placed through this approach even if the cochlear nerve is not preserved.

(b) Meningiomas, posterior to or centered on the IAC, with unserviceable hearing. Cases that lie anterior to the canal require a transapical extension, whereas large petroclival tumors need a modified transcochlear approach.

(c) Other CPA tumors with unserviceable hearing, such as epidermoids, dermoids, etc.

– Vertigo surgery (labyrinthectomy).

Atlas of Microsurgery of the Lateral Skull Base. https://doi.org/10.1016/B978-0-443-19083-4.00007-5

Contraindications

– This procedure is contraindicated in the only hearing ear. This has become a relative contraindication currently due to the advent of the CI and ABI.

– Ipsilateral chronic otitis media is a further contraindication. In cases of simple tympanic membrane perforation with no active infection, a single-stage enlarged TLA is used, and the external auditory canal is closed as a cul-de-sac. If there is an active infection, a subtotal petrosectomy is first performed, with eradication of the infection, obliteration of the cavity with abdominal fat, and closure of the external auditory canal as a cul-de-sac. The second stage of tumor removal is carried out after 2 or 3 months using the enlarged TLA.

– In some situation, when the ear infection is mild, we may elect to treat it conservatively until the infection is eradicated; then we treat it as a simple tympanic membrane perforation (Fig. 7.1).

SURGICAL STEPS (LEFT EAR)

Surgical Steps to the Mastoidectomy

1. A C-shaped postauricular skin incision is made 4–5 cm behind the retroauricular sulcus, 2–3 cm superior to the attachment of the auricle, and just inferior to the mastoid tip (Fig. 7.2).

2. The skin flap is reflected anteriorly and maintained in position with two hooks. The musculoperiosteal layer is incised in a T-shaped fashion using a cutting diathermy (Fig. 7.3).

3. The musculoperiosteal flaps are elevated with a periosteal elevator. At the level of the mastoid tip, monopolar diathermy is applied to transect the relevant muscles, raising up the flap. Bleeding from the mastoid emissary vein usually occurs at this step, and can be controlled by the use of bone wax. The flaps are sutured to the skin (Fig. 7.4); this helps in reducing the bleeding from the skin incision and at the same time keeps the muscoloperiosteal flaps retracted. We no longer use the conventional retractors, which significantly limit the surgeon's access.

4. An extended mastoidectomy is performed (Fig. 7.5). The middle cranial fossa dura and the sigmoid sinus are identified, leaving a thin shell of bone overlying them. Bone 2–3 cm posterior to the sigmoid sinus is also drilled using a large cutting burr.

Mastoidectomy and Labyrinthectomy; Access to the Internal Auditory Canal and the CPA

5. The mastoid air cells are exenterated, and the antrum is widely opened (Fig. 7.6). The short process of the incus and the lateral semicircular canal are the landmarks for the facial nerve. The digastric ridge is identified, which directly points to the mastoid segment of the facial nerve at the level of the stylomastoid foramen. The facial nerve is only skeletonized, but not uncovered, in order to reveal its precise location. Identification of the facial never by facial nerve stimulator (Fig. 7.7).

6. Using a large diamond burr, the remaining bone covering the middle cranial fossa and the sigmoid sinus as well as the posterior fossa dura posterior to the sinus are further drilled (Figs. 7.8–7.12). The sinodural angle should be clearly exposed to reveal the exact location of the superior petrosal sinus exiting the sigmoid sinus. Three semicircular canals are outlined and the retrofacial air cells are drilled (Fig. 7.13).The sigmoid sinus is followed down to the jugular bulb, and three semicircular canals are outlined (Fig. 7.14). Care should be taken to leave a limited area of bone adjacent to the labyrinth in order to protect the middle and posterior fossa while carrying out the labyrinthectomy.

7. The labyrinthectomy starts by opening of the lateral semicircular canal using a medium-sized cutting burr (Fig. 7.15). The posterior semicircular canal is next opened

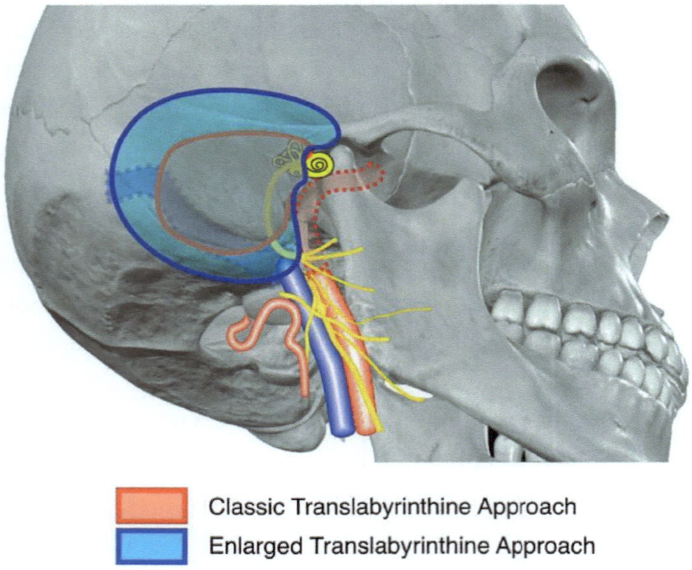

Fig. 7.1 Outlines of the enlarged versus the classic translabyrinthine approach. As can be seen, significantly more bone is removed with the enlarged approach than with the classic approach. This approach can thus create a significant advantage in exposure. Note the lower cranial nerves exit the base of skull anterior to the jugular bulb. As such, it demarcates the anterior limit of bone removal.

(Fig. 7.16), followed by opening of the superior semicircular canal (Fig. 7.17). Care is taken to leave the anterior part of the lateral semicircular canal in order to protect the anteriorly lying facial nerve. The relationship between the three semicircular canals can be well appreciated. Note the five openings into the vestibule (Fig. 7.18).

8. The ampullae of the lateral and superior semicircular canals are drilled (Fig. 7.18). However, the anterior part of the ampullae of these two canals should be left to protect the labyrinthine segment of the facial nerve and to serve as a landmark for the superior vestibular nerve. The vestibule is then opened (Fig. 7.19) widely, but attention should be paid to the following two points: It is important to avoid drilling the floor of the vestibule in order not to enter the internal auditory canal. Drilling the roof of the vestibule may injure the facial nerve, because the nerve runs immediately lateral to the vestibule. Drilling in this area should be performed from superior to inferior or vice versa, but not from medial to lateral, in order to prevent the burr from slipping into the vestibule or onto the facial nerve. Note the endolymphatic duct reaches the vestibule, where it bifurcates into the utricular duct and saccular duct (Fig. 7.20). Identification of the second genu of the facial never by stimulator (Fig. 7.21).

9. The endolymphatic duct extends from the medial surface of the posterior canal to the posterior fossa dura anterior to the sigmoid sinus, and should be transected with the use of a Beaver knife or microscissors (Fig. 7.22). This step facilitates the complete drilling of bone overlying the posterior fossa dura and the subsequent dural retraction (Figs. 7.23–7.24).

10. After completion of the labyrinthectomy and the transection of the endolymphatic duct, the bone left over the posterior and middle fossa is successively reduced. Identification of the lower edge of the internal auditory canal starts by drilling the bone between the jugular bulb and the expected position of the canal (Fig. 7.25). During this step the cochlear aqueduct can be identified by drilling (Fig. 7.26). This is an important landmark for the glossopharyngeal nerve, which lies immediately inferior to it. The cochlear aqueduct is the lower limit of the approach. In small or medium-sized acoustic neurinomas, opening the cochlear aqueduct allows the CSF to drain. This does not occur in large tumors, because the duct is occluded by the tumor. The ampulla of the superior semicircular canal serves as a landmark for the superior border of the internal auditory canal at the fundus.

11. After a wide exposure of the dura is achieved, following the posterior fossa dura allows the porus of the internal auditory canal to be identified. Drilling continues around the internal auditory canal. During this process, drilling should be carried out parallel to the canal in a medial-to-lateral direction. The inferior border of the canal is identified by drilling the retrofacial air cells up to the cochlear aqueduct.

12. The internal auditory canal is outlined with a thin shell of bone still covering (Fig. 7.27). Careful further drilling at the level of the fundus allows identification of the horizontal crest as well as the superior ampullary canal. Bone between the superior border of the canal and the middle fossa dura and between the inferior border of the canal and the jugular bulb can be further drilled, depending on the necessity for tumor removal. We routinely drill the bone 270° around the internal auditory canal. Finally, the remaining thin shell of bone is separated from the underlying dura of the internal auditory canal using a dissector, and is then carefully removed.

Identification of the Facial Nerve at the Fundus

13. The posterior aspect of the internal auditory canal at the level of the fundus is drilled inferiorly. This exposes the inferior vestibular nerve. Further drilling at a more superior level identifies the horizontal crest, which separates the inferior vestibular nerve from the superior vestibular nerve. The superior vestibular nerve is followed laterally where it leaves the fundus of the internal auditory canal. The nerve then lies in a tiny canal that enters the ampulla of the superior semicircular canal as the superior ampullary nerve (Fig. 7.28).

14. The superior ampullary nerve is dissected from its canal using a 90° hook with the tip facing inferiorly (Fig. 7.29). Since the facial nerve diverges anterosuperiorly after leaving the fundus of the internal auditory canal, it is separated from the superior ampullary nerve by a vertical crest of bone, called Bill's bar. This bone protects the facial nerve while this step is being carried out.

15. The hook is moved medially toward the fundus, with subsequent separation and establishment of a good plane of cleavage between the superior vestibular nerve and the anteriorly lying facial nerve. Once the superior ampullary nerve is detached and reflected medially and posteriorly, the facial nerve can be clearly observed (Fig. 7.30). The hook is moved medially toward the fundus, with subsequent separation and establishment of a good plane of cleavage between the inferior vestibular nerve and the anteriorly lying cochlear nerve (Fig. 7.31). Identification of the internal auditory canal segment of the facial never by stimulator (Fig. 7.32). The relationship between the facial and cochlear nerves and the horizontal and vertical crests is well appreciated (Fig. 7.33).

Opening the Dura

16. Bipolar coagulation is performed first, before incision of the dura. The incision starts superiorly 1–2 mm parallel to and inferior to the superior petrosal sinus, running from posterior to anterior using microsurgical scissors (Fig. 7.34). The lower scissors blade should be in close contact with the dura. Once this incision has been performed, Merocel is inserted just under the dura to protect the underlying structures. The utmost care should be taken while

performing the lower dural incision so as not to injure an underlying AICA, which is frequently found at this location, especially in small or medium-sized tumors. A useful technique is to elevate the incised dura, allowing the surgeon to continue the incision under vision. The inferior incision starts just in front of the distal part of the sigmoid sinus, and follows the course of the sinus and jugular bulb to the porus, where it joins the superior incision.

17. Figs. 7.35–7.39 show the different structures seen after opening the dura.

Closure

18. The CPA is temporarily obliterated with cottonoids to prevent blood from trickling in. All the air cells, especially retrofacial cells, should be checked to see whether they have the communication with the tympanic cavity. These cells are obliterated with bone wax.

19. After the incus is disarticulated and removed (Figs. 7.40 and 7.41), the aditus is obliterated with periosteum. A sufficient amount of the periosteum should be inserted through the aditus into the tympanic cavity to close the orifice of the eustachian tube.

20. Sufficient abdominal fat tissue will have been harvested. It is sliced into several long, thin strips, and these strips are placed deeply to reach the CPA.

21. The musculoperiosteal layer is closed using 1–0 Vicryl. This is followed by closure of the subcutaneous tissue and skin

22. A pressure dressing is applied.

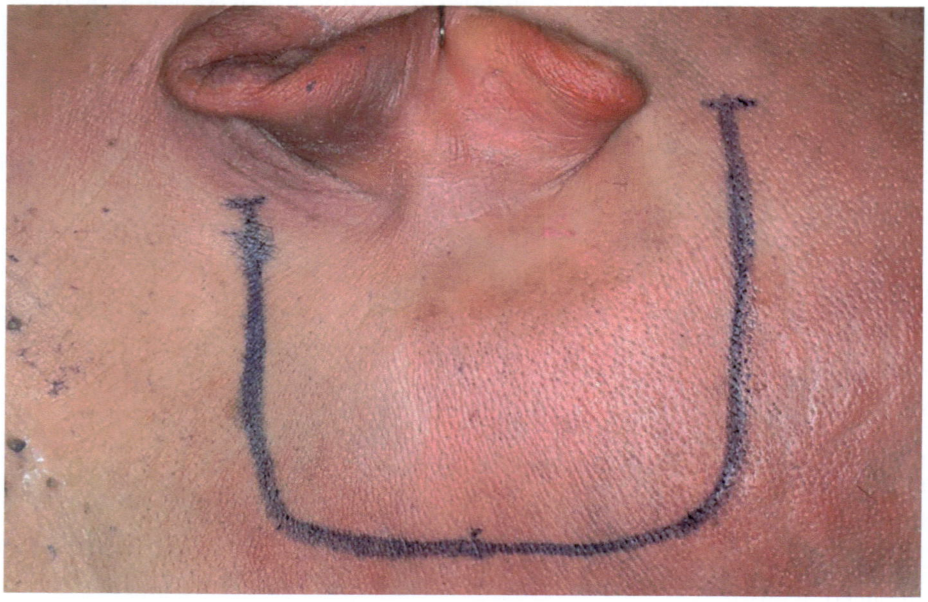

Fig. 7.2 A C-shaped postauricular skin incision is made 4–5 cm behind the retroauricular sulcus, 2–3 cm superior to the attachment of the auricle, and just inferior to the mastoid tip.

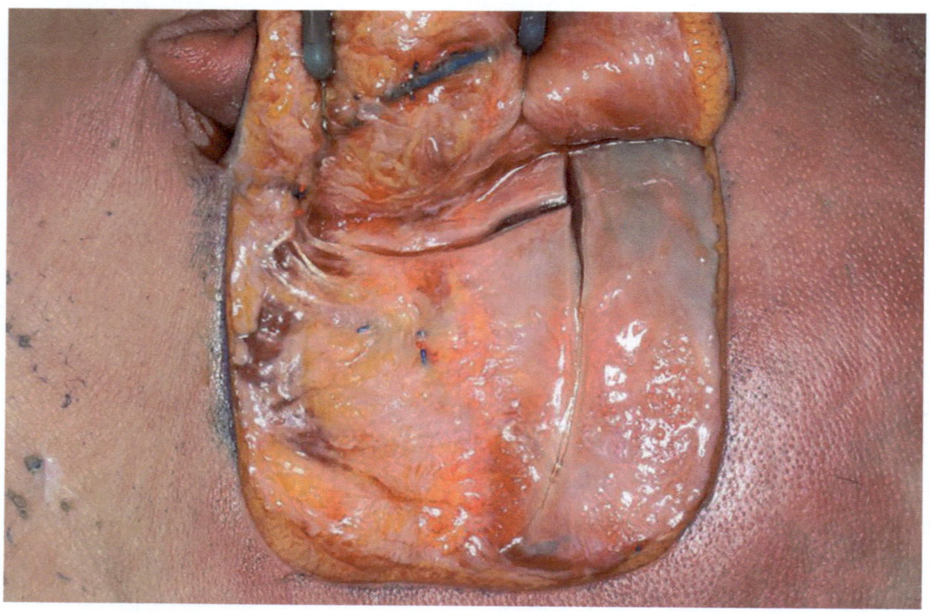

Fig. 7.3 The skin flap is reflected anteriorly and maintained in position with two hooks. The musculoperiosteal layer is incised in a T-shaped fashion using a cutting diathermy.

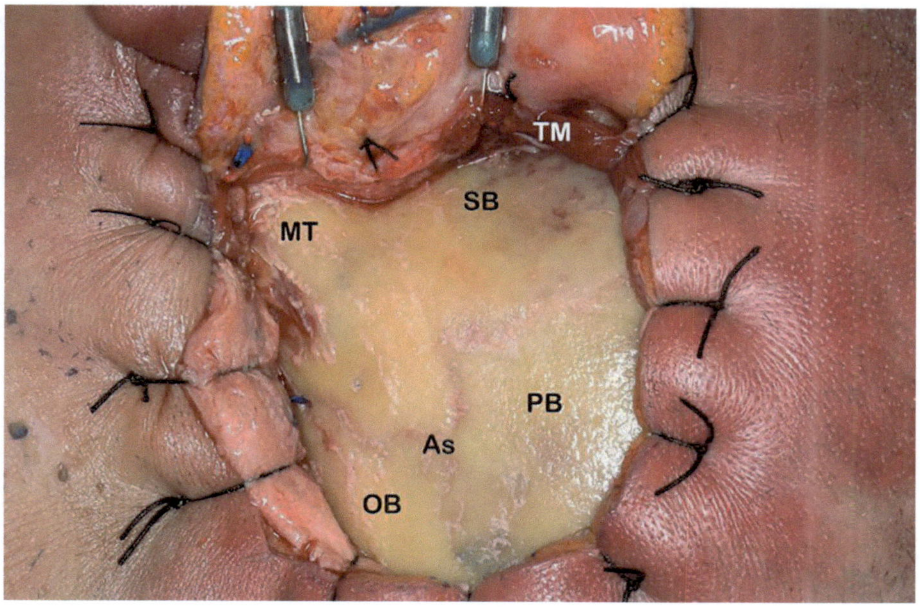

Fig. 7.4 The musculoperiosteal flaps are elevated with a periosteal elevator. The flaps are sutured to the skin; this helps in reducing the bleeding from the skin incision and at the same time keeps the muscoloperiosteal flaps retracted. *As*, asterion; *MT*, mastoid tip; *OB*, occipital bone; *PB*, parietal bone; *SB*, squamosal part of the temporal bone; *TM*, temporalis muscle(retracted).

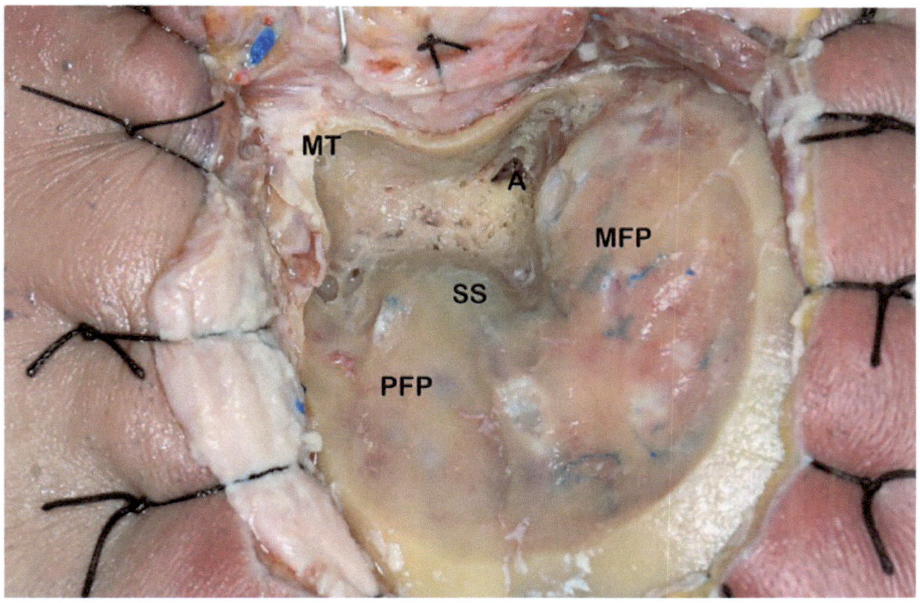

Fig. 7.5 An extended mastoidectomy is performed. The middle cranial fossa dura and the sigmoid sinus are identified, leaving a thin shell of bone overlying them. Bone 2–3 cm posterior to the sigmoid sinus is also drilled using a large cutting burr. *A*, antrum; *MFP*, middle fossa plate; *MT*, mastoid tip; *PFP*, posterior fossa plate; *SS*, sigmoid sinus.

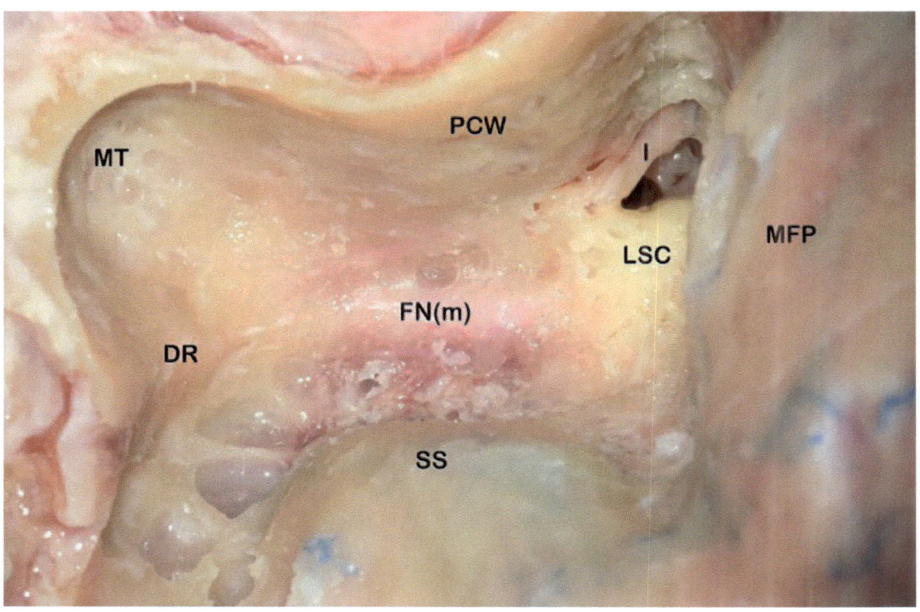

Fig. 7.6 The mastoid air cells are exenterated, and the antrum is widely opened. The short process of the incus and the lateral semicircular canal are the landmarks for the facial nerve. The digastric ridge is identified, which directly points to the mastoid segment of the facial nerve at the level of the stylomastoid foramen. The facial nerve is also skeletonized. *DR*, digastric ridge; *FN(m)*, mastoid segment of facial nerve; *I*, incus; *LSC*, lateral semicircular canal; *MFP*, middle fossa plate; *MT*, mastoid tip; *PCW*, posterior canal wall; *SS*, sigmoid sinus.

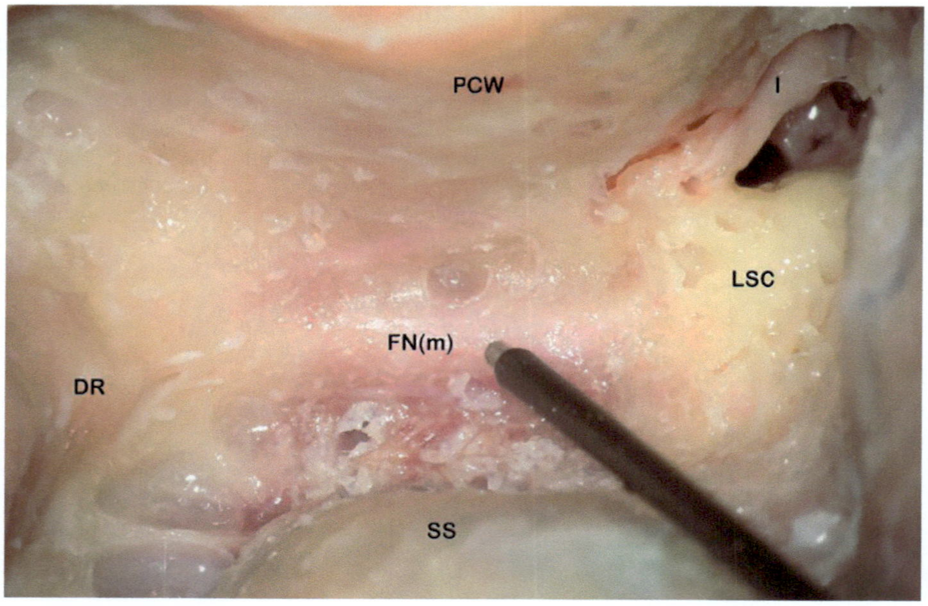

Fig. 7.7 Identification of the facial never by facial nerve stimulator. *DR*, digastric ridge; *FN(m)*, mastoid segment of facial nerve; *I*, incus; *LSC*, lateral semicircular canal; *PCW*, posterior canal wall; *SS*, sigmoid sinus.

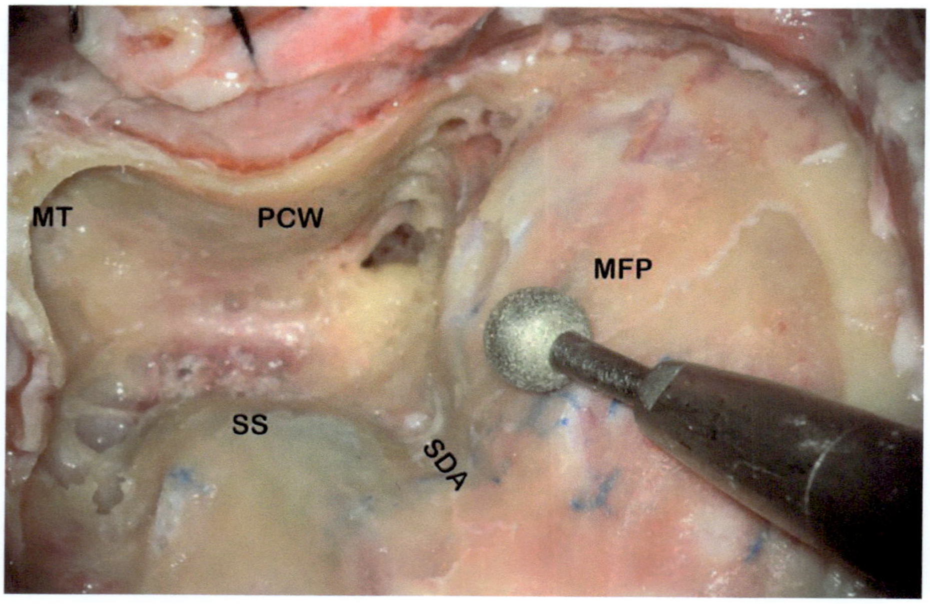

Fig. 7.8 A large diamond burr is used to remove the bone overlying the middle fossa dura. *MFP*, middle fossa plate; *MT*, mastoid tip; *PCW*, posterior canal wall; *SDA*, sinodural angle; *SS*, sigmoid sinus.

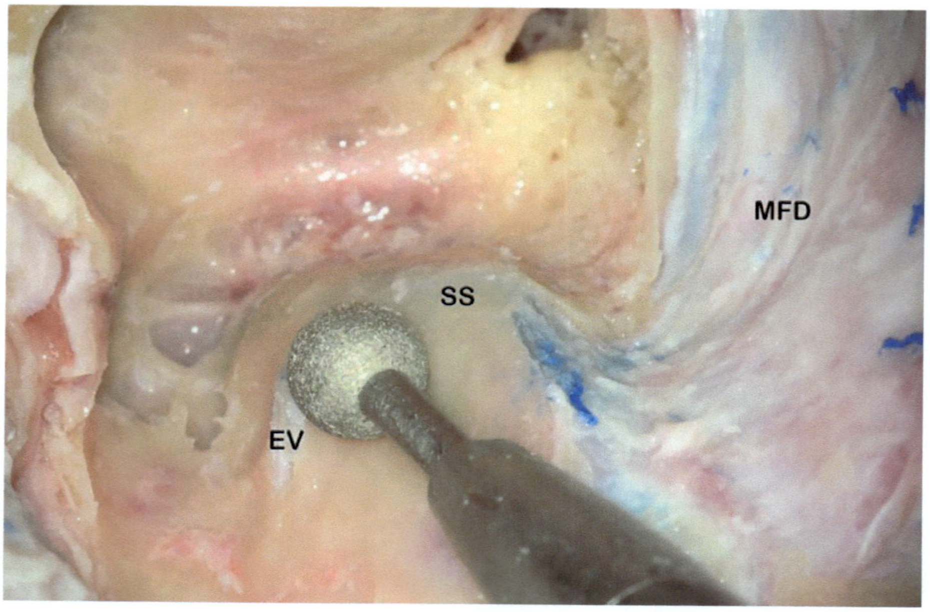

Fig. 7.9 A large diamond burr is used to remove the bone overlying the sigmoid sinus and the posterior fossa dura. *EV*, emissary vein; *MFD*, middle fossa dura; *SS*, sigmoid sinus.

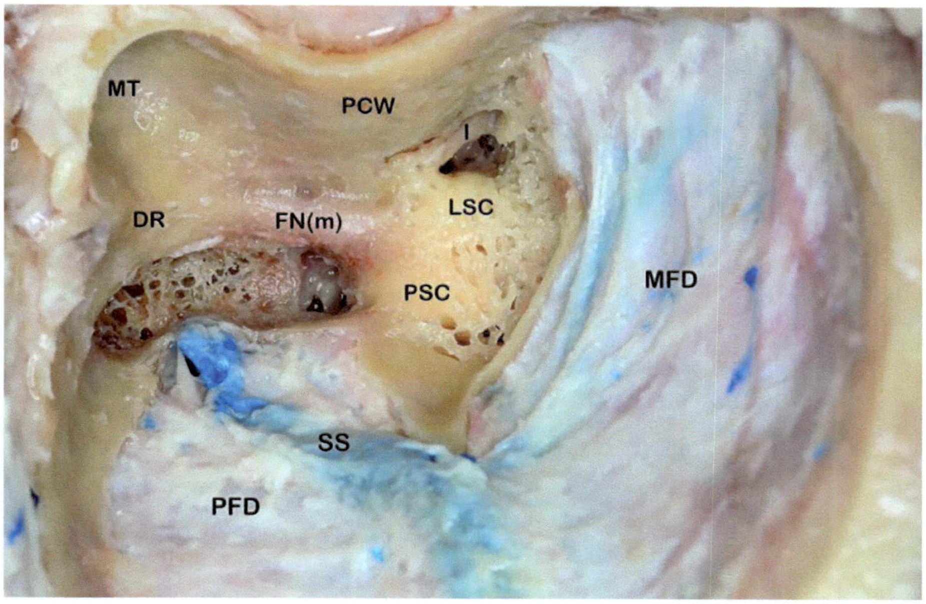

Fig. 7.10 The bone overlying the middle fossa dura, sigmoid sinus and the posterior fossa dura has been adequately removed. *DR*, digastric ridge; *FN(m)*, mastoid segment of facial nerve; *I*, incus; *LSC*, lateral semicircular canal; *MFD*, middle fossa dura; *MT*, mastoid tip; *PCW*, posterior canal wall; *PFD*, posterior fossa dura; *PSC*, posterior semicircular canal; *SS*, sigmoid sinus.

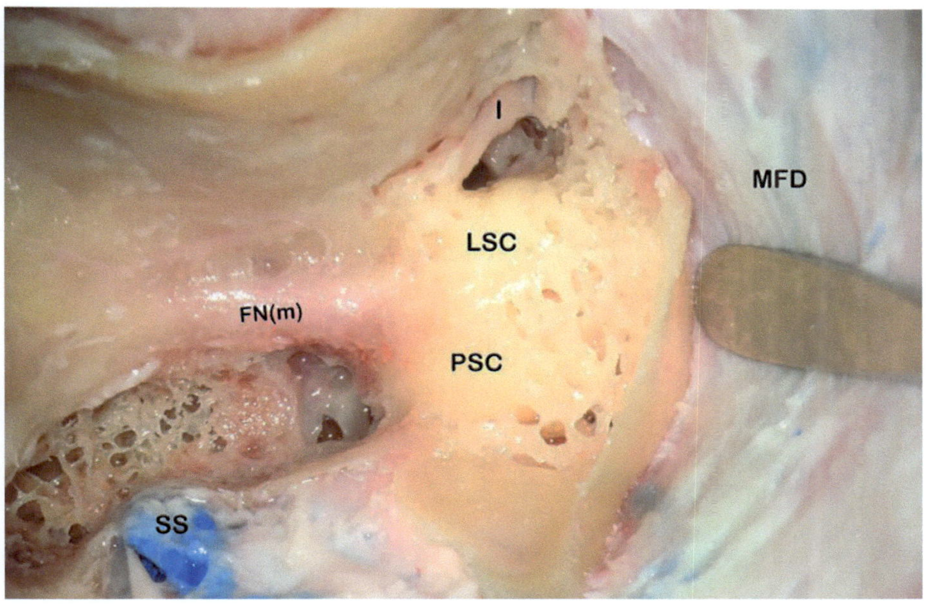

Fig. 7.11 Separating the dura from the drilled bone helps avoid injury to it. *FN(m)*, mastoid segment of facial nerve; *I*, incus; *LSC*, lateral semicircular canal; *MFD*, middle fossa dura; *PSC*, posterior semicircular canal; *SS*, sigmoid sinus.

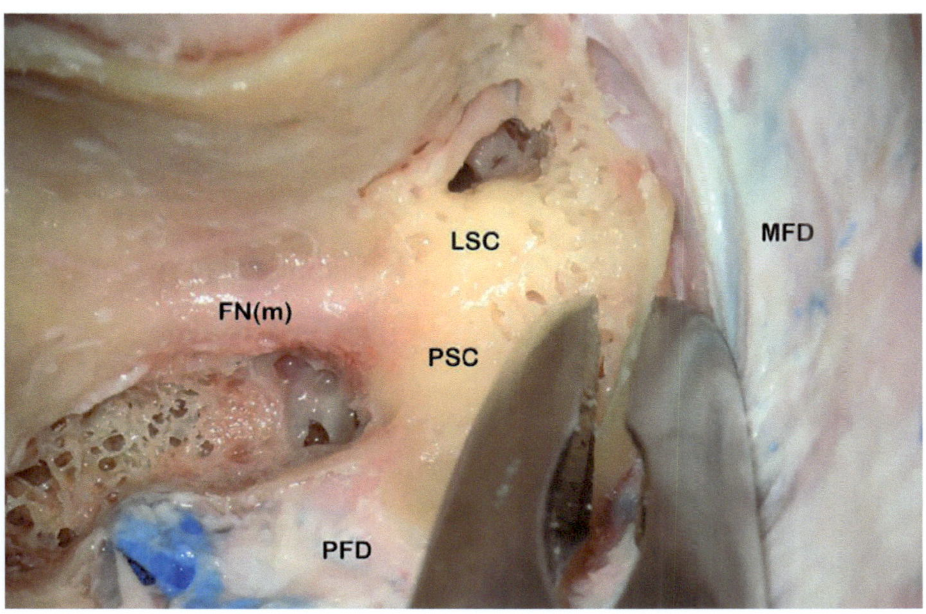

Fig. 7.12 Using a rongeur, the bone over the middle fossa dura and sinodural angle is removed. *FN(m)*, mastoid segment of facial nerve; *LSC*, lateral semicircular canal; *MFD*, middle fossa dura; *PFD*, posterior fossa dura; *PSC*, posterior semicircular canal.

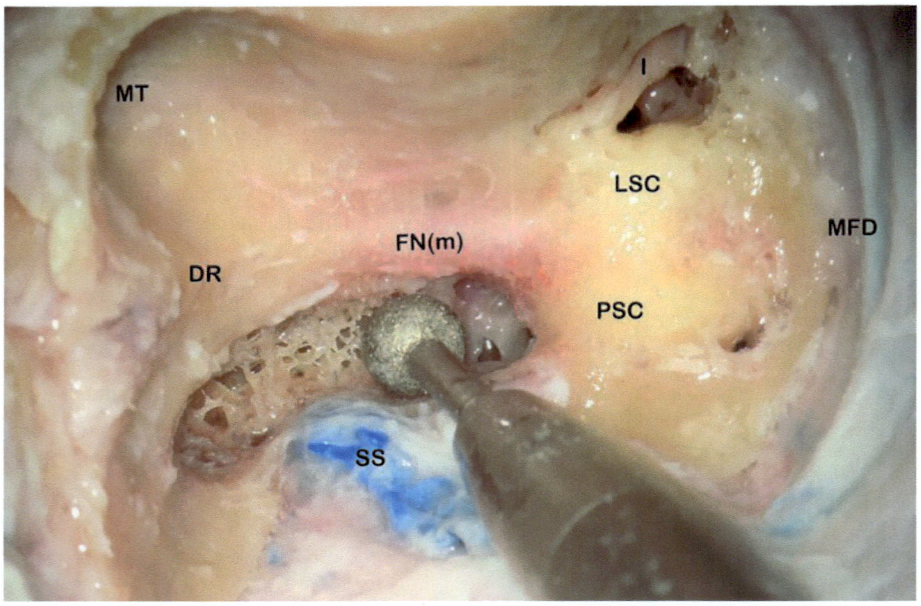

Fig. 7.13 The retrofacial air cells are drilled. *DR*, digastric ridge; *FN(m)*, mastoid segment of facial nerve; *I*, incus; *LSC*, lateral semicircular canal; *MFD*, middle fossa dura; *MT*, mastoid tip; *PSC*, posterior semicircular canal; *SS*, sigmoid sinus.

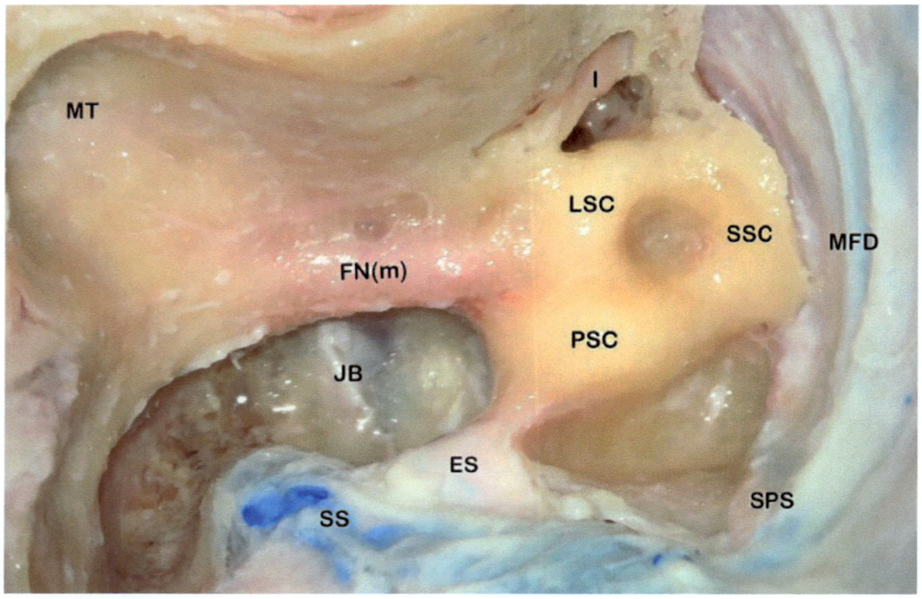

Fig. 7.14 The jugular bulb has been identified medial to the mastoid segment of the facial nerve. *ES*, endolymphatic sac; *FN(m)*, mastoid segment of facial nerve; *I*, incus; *JB*, jugular bulb; *LSC*, lateral semicircular canal; *MFD*, middle fossa dura; *MT*, mastoid tip; *PSC*, posterior semicircular canal; *SPS*, superior petrosal sinus; *SS*, sigmoid sinus; *SSC*, superior semicircular canal.

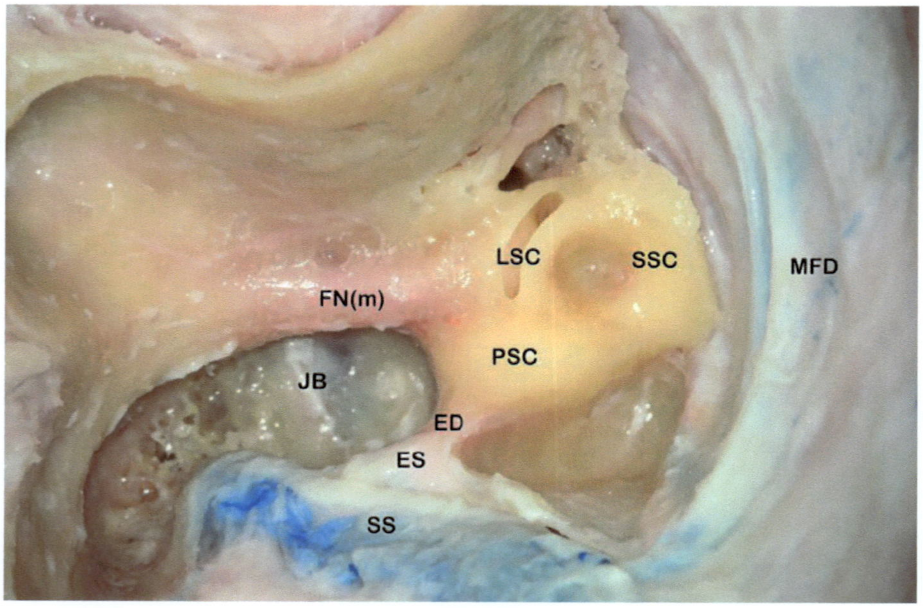

Fig. 7.15 Labirynthectomy starts with opening of the lateral semicircular canal. *ED*, endolymphatic duct; *ES*, endolymphatic sac; *FN(m)*, mastoid segment of facial nerve; *JB*, jugular bulb; *LSC*, lateral semicircular canal; *MFD*, middle fossa dura; *PSC*, posterior semicircular canal; *SS*, sigmoid sinus; *SSC*, superior semicircular canal.

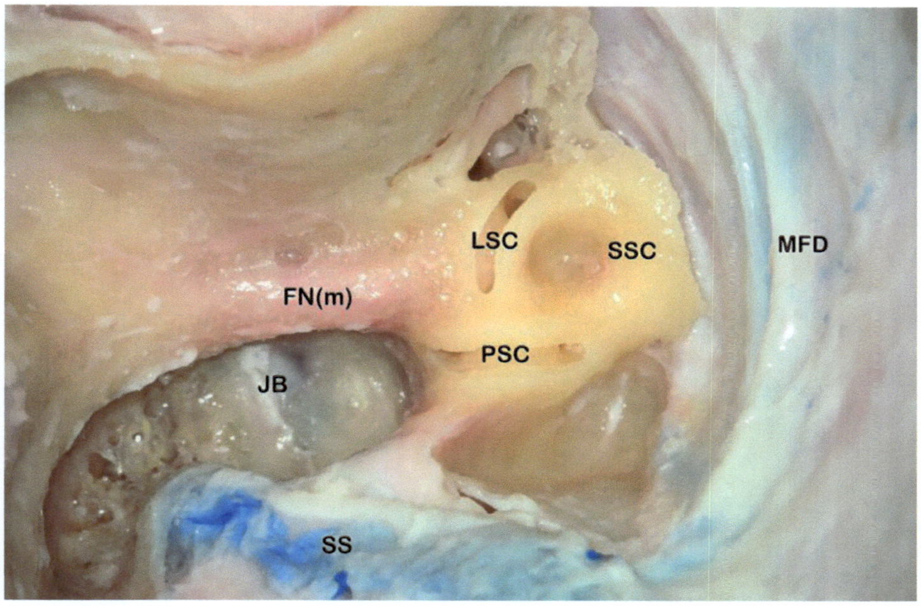

Fig. 7.16 The posterior semicircular canal has been opened. *FN(m)*, mastoid segment of facial nerve; *JB*, jugular bulb; *LSC*, lateral semicircular canal; *MFD*, middle fossa dura; *PSC*, posterior semicircular canal; *SS*, sigmoid sinus; *SSC*, superior semicircular canal.

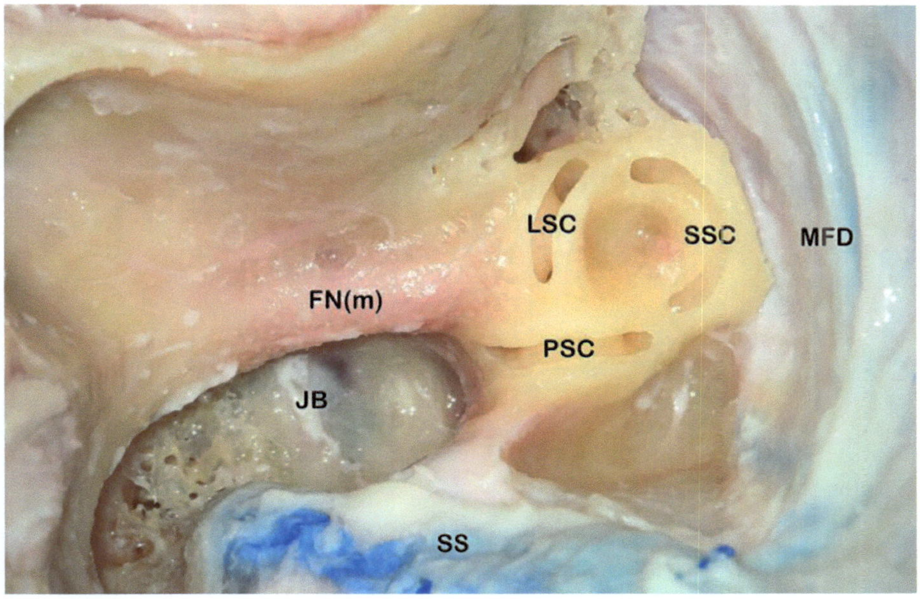

Fig. 7.17 The superior semicircular canal has been opened. *FN(m)*, mastoid segment of facial nerve; *JB*, jugular bulb; *LSC*, lateral semicircular canal; *MFD*, middle fossa dura; *PSC*, posterior semicircular canal; *SS*, sigmoid sinus; *SSC*, superior semicircular canal.

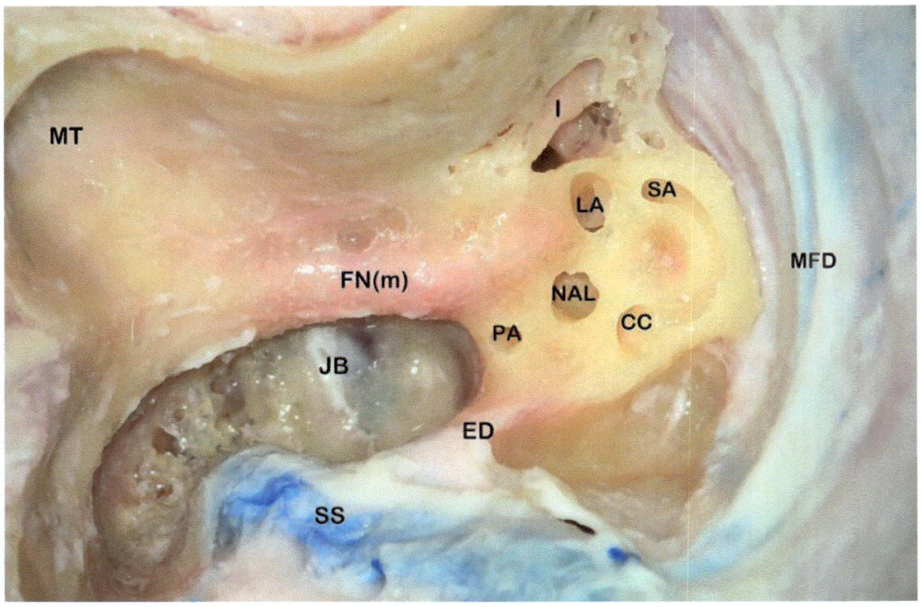

Fig. 7.18 Note the five openings into the vestibule. *CC*, common crus; *ED*, endolymphatic duct; *FN(m)*, mastoid segment of facial nerve; *I*, incus; *JB*, jugular bulb; *LA*, ampullated end of lateral semicircular canal; *MFD*, middle fossa dura; *MT*, mastoid tip; *NAL*, nonampullated end of lateral semicircular canal; *PA*, ampullated end of posterior semicircular canal; *SA*, ampullated end of superior semicircular canal; *SS*, sigmoid sinus.

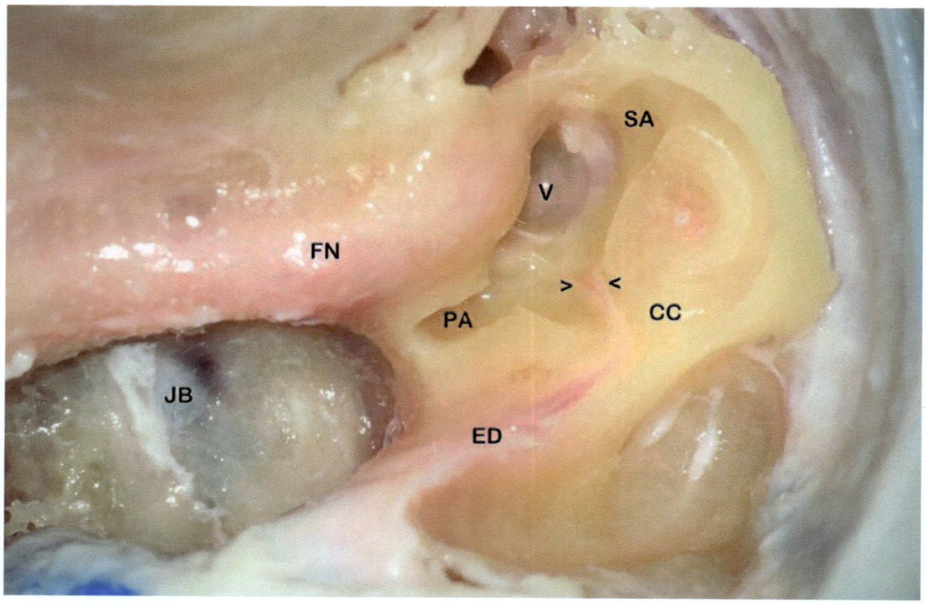

Fig. 7.19 The canals have been drilled away, and the vestibule has been opened. Note the endolymphatic duct reaches the vestibule, where it bifurcates into the utricular duct and saccular duct. <, utricular duct; >, saccular duct; *CC*, common crus; *ED*, endolymphatic duct; *FN*, facial nerve; *JB*, jugular bulb; *PA*, ampullated end of posterior semicircular canal; *SA*, ampullated end of superior semicircular canal; *V*, vestibule.

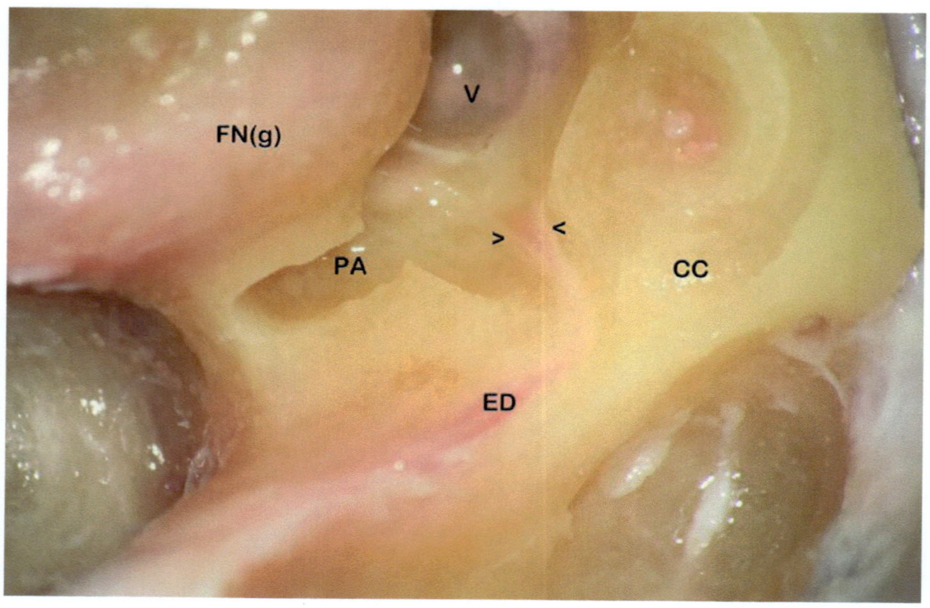

Fig. 7.20 At a higher magnification, the utricular duct and saccular duct can be appreciated. <, utricular duct; >, saccular duct; *CC*, common crus; *ED*, endolymphatic duct; *FN(g)*, second genu of facial nerve; *PA*, ampullated end of posterior semicircular canal; *V*, vestibule.

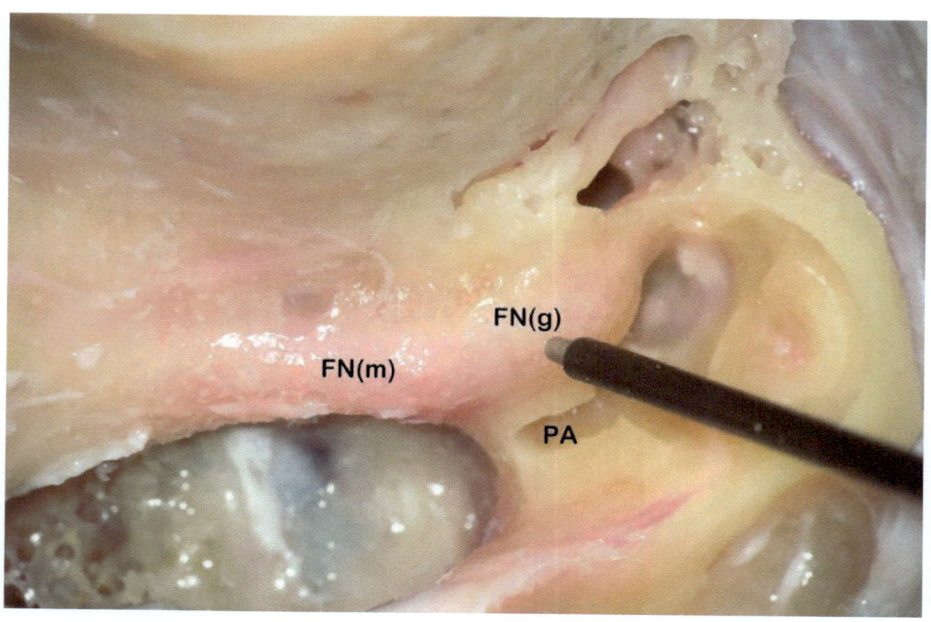

Fig. 7.21 Identification of the second genu of the facial never by stimulator. *FN(g)*, second genu of facial nerve; *FN(m)*, mastoid segment of facial nerve; *PA*, ampullated end of posterior semicircular canal.

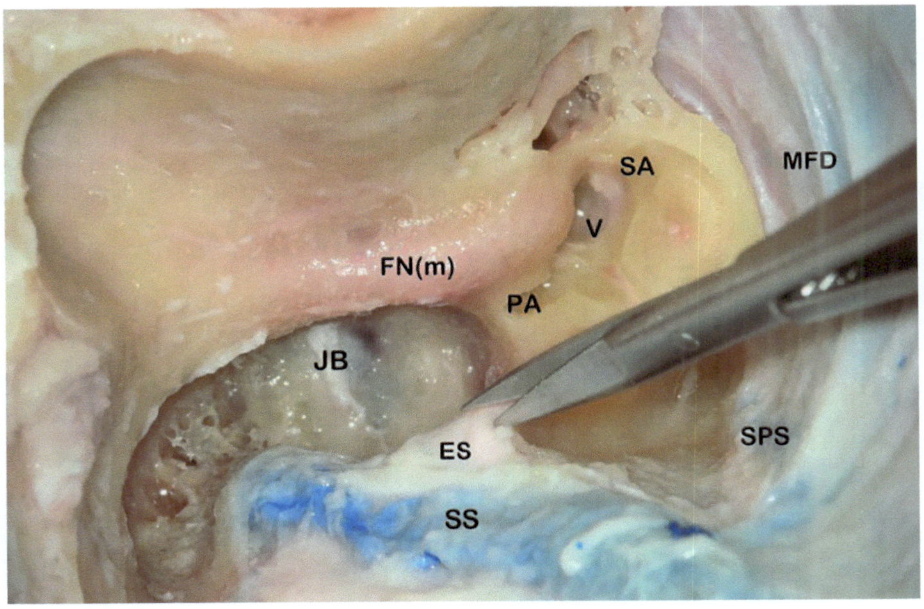

Fig. 7.22 The endolymphatic duct extends from the medial surface of the posterior canal to the posterior fossa dura anterior to the sigmoid sinus, and should be transected with the use of a Beaver knife or microscissors. *ES*, endolymphatic sac; *FN(m)*, mastoid segment of facial nerve; *JB*, jugular bulb; *MFD*, middle fossa dura; *PA*, ampullated end of posterior semicircular canal; *SA*, ampullated end of superior semicircular canal; *SPS*, superior petrosal sinus; *SS*, sigmoid sinus; *V*, vestibule.

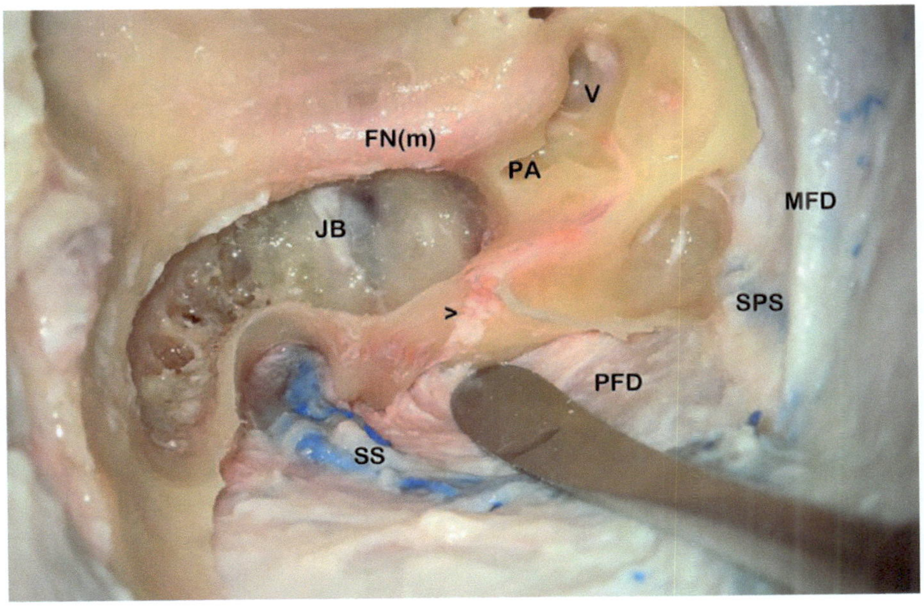

Fig. 7.23 Sharply cut the endolymphatic duct facilitates the complete drilling of bone overlying the posterior fossa dura and the subsequent dural retraction. >, endolymphatic duct(transected); *FN(m)*, mastoid segment of facial nerve; *JB*, jugular bulb; *MFD*, middle fossa dura; *PA*, ampullated end of posterior semicircular canal; *PFD*, posterior fossa dura; *SPS*, superior petrosal sinus; *SS*, sigmoid sinus; *V*, vestibule.

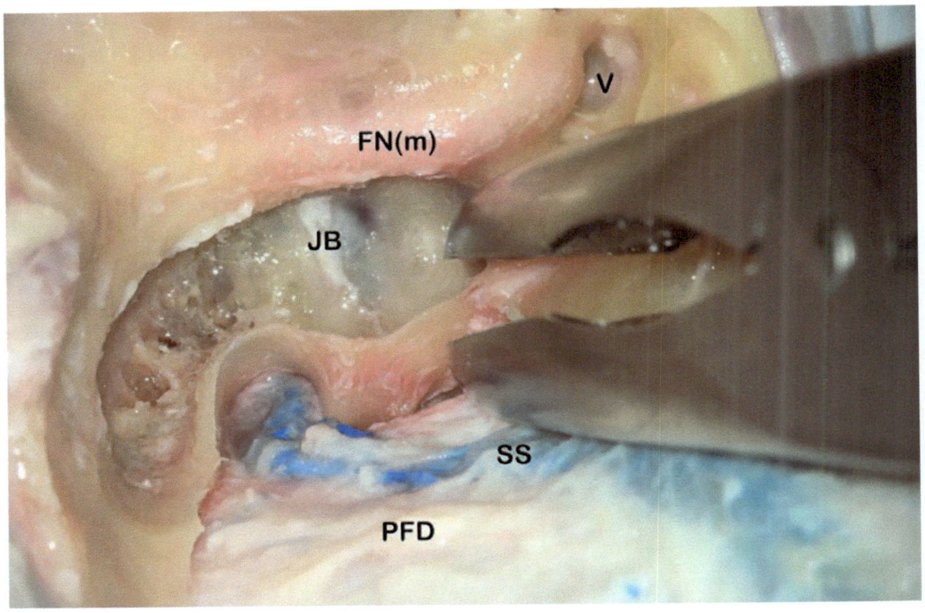

Fig. 7.24 Using a rongeur, the bone over the posterior fossa dura is removed. *FN(m)*, mastoid segment of facial nerve; *JB*, jugular bulb; *PFD*, posterior fossa dura; *SS*, sigmoid sinus; *V*, vestibule.

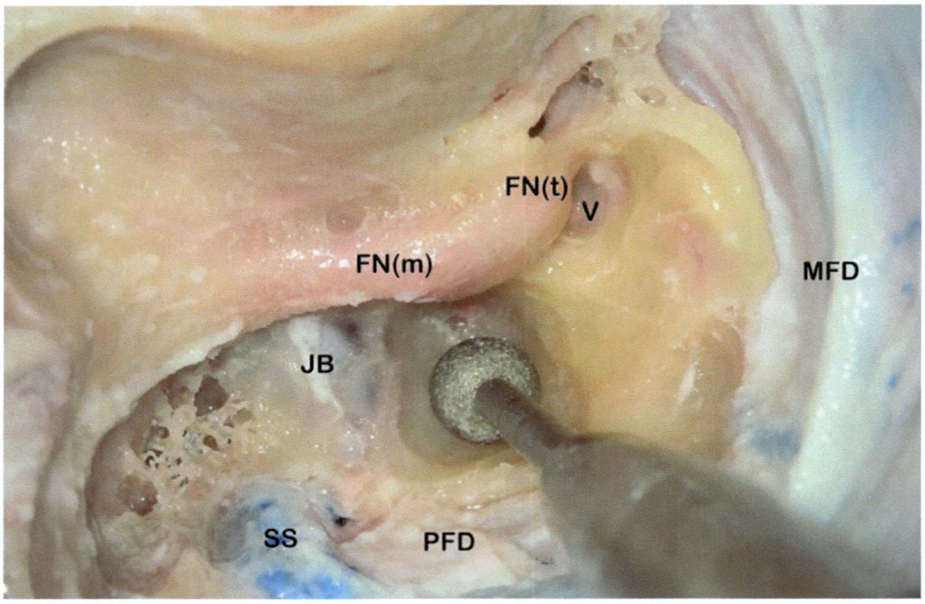

Fig. 7.25 Identification of the lower edge of the internal auditory canal starts by drilling the bone between the jugular bulb and the expected position of the canal. *FN(m)*, mastoid segment of facial nerve; *FN(t)*, tympanic segment of facial nerve; *JB*, jugular bulb; *MFD*, middle fossa dura; *PFD*, posterior fossa dura; *SS*, sigmoid sinus; *V*, vestibule.

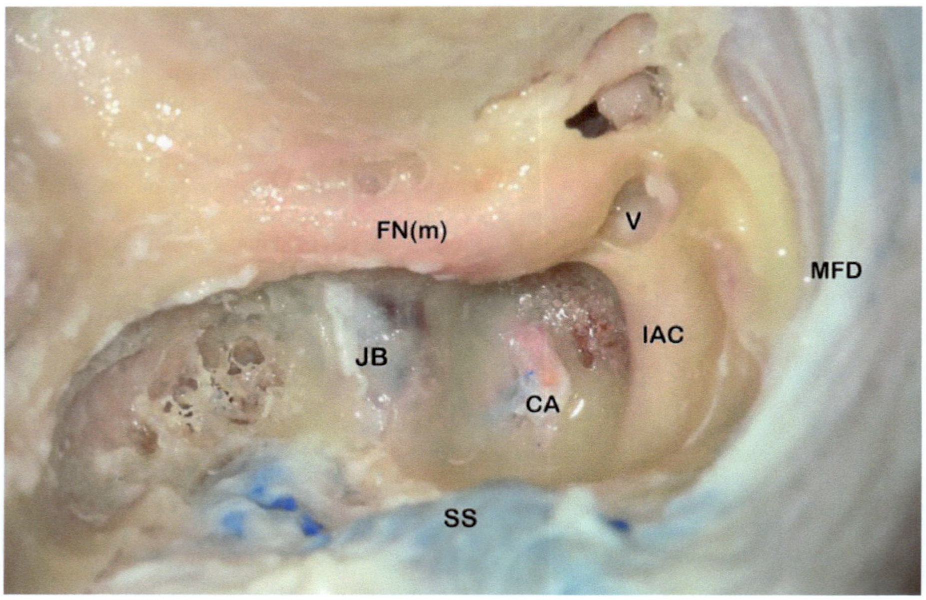

Fig. 7.26 The cochlear aqueduct can be identified by drilling the bone between the jugular bulb and the expected position of the canal. *CA*, cochlear aqueduct; *FN(m)*, mastoid segment of facial nerve; *IAC*, internal auditory canal; *JB*, jugular bulb; *MFD*, middle fossa dura; *SS*, sigmoid sinus; *V*, vestibule.

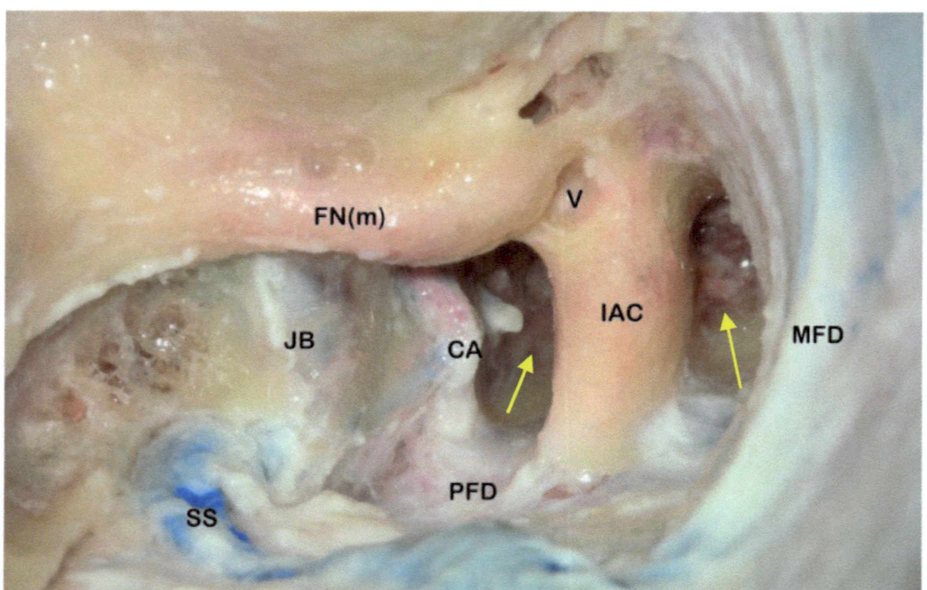

Fig. 7.27 The internal auditory canal is outlined with a thin shell of bone still covering. Note the transapical extension of the approach as indicated by the *arrows*. *CA*, cochlear aqueduct; *FN (m)*, mastoid segment of facial nerve; *IAC*, internal auditory canal; *JB*, jugular bulb; *MFD*, middle fossa dura; *PFD*, posterior fossa dura; *SS*, sigmoid sinus; *V*, vestibule.

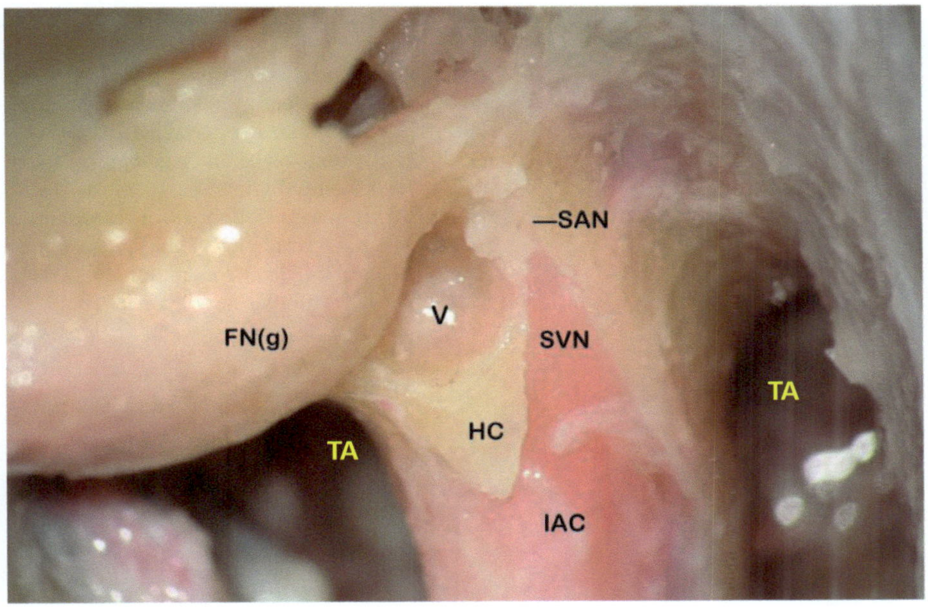

Fig. 7.28 The horizontal crest separates the inferior vestibular nerve from the superior vestibular nerve. The superior vestibular nerve is followed laterally where it leaves the fundus of the internal auditory canal. The nerve then lies in a tiny canal that enters the ampulla of the superior semicircular canal as the superior ampullary nerve. *FN(g)*, second genu of facial nerve; *HC*, horizontal crest; *IAC*, internal auditory canal; *SAN*, superior ampullary nerve; *SVN*, superior vestibular nerve; *TA*, transapical extension of the approach; *V*, vestibule.

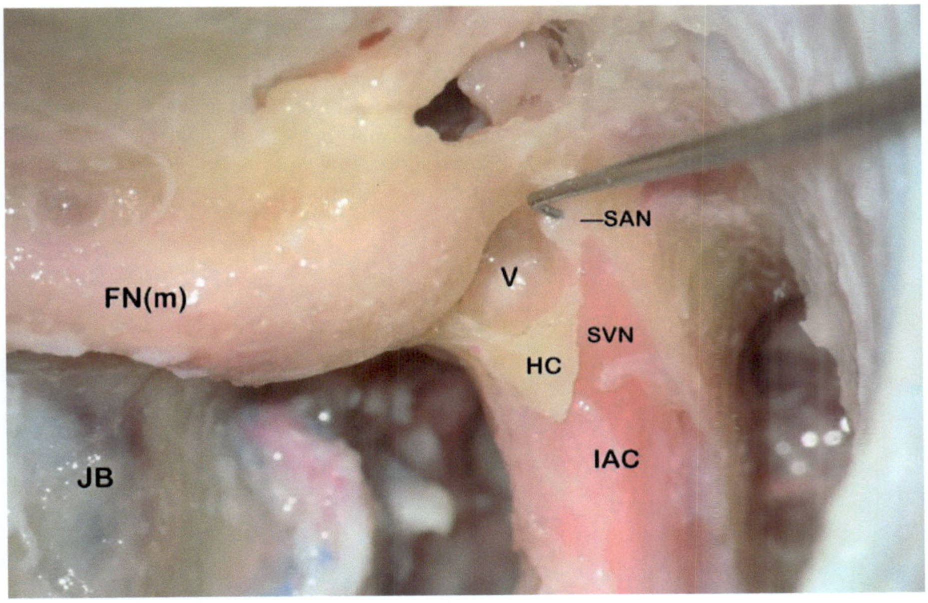

Fig. 7.29 The superior ampullary nerve is dissected from its canal using a 90° hook with the tip facing inferiorly. *FN(m)*, mastoid segment of facial nerve; *HC*, horizontal crest; *IAC*, internal auditory canal; *JB*, jugular bulb; *SAN*, superior ampullary nerve; *SVN*, superior vestibular nerve; *V*, vestibule.

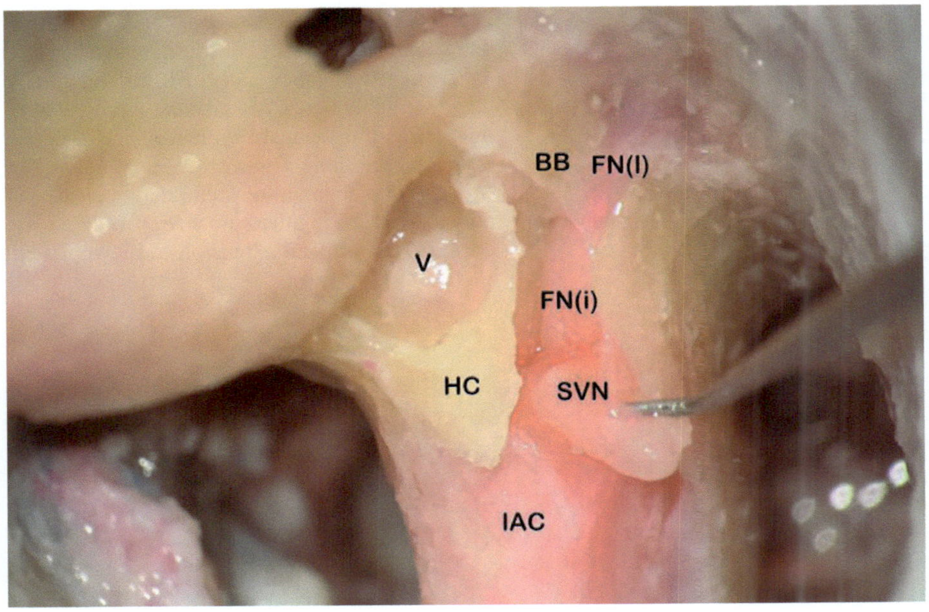

Fig. 7.30 The hook is moved medially toward the fundus, with subsequent separation and establishment of a good plane of cleavage between the superior vestibular nerve and the anteriorly lying facial nerve. Once the superior ampullary nerve is detached and reflected medially and posteriorly, the facial nerve can be clearly observed. *BB*, Bill's bar; *FN(l)*, labyrinthine segment of facial nerve; *FN(i)*, internal auditory canal segment of facial nerve; *HC*, horizontal crest; *IAC*, internal auditory canal; *SVN*, superior vestibular nerve; *V*, vestibule.

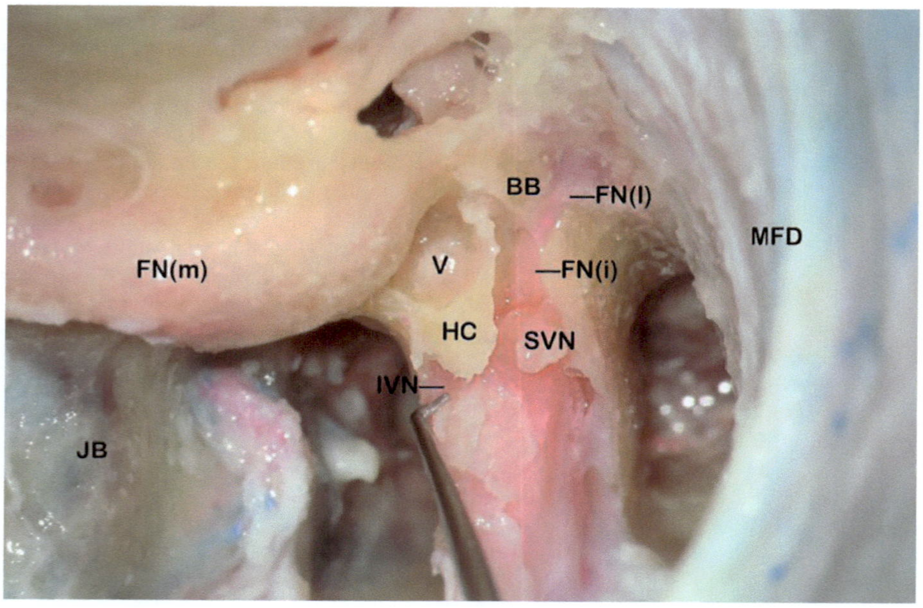

Fig. 7.31 Same image, smaller magnification. The inferior vestibular nerve lies just inferior to the superior vestibular nerve. The horizontal crest divides the superior vestibular nerve from the inferior vestibular nerve. *BB*, Bill's bar; *FN (i)*, internal auditory canal segment of facial nerve; *FN(l)*, labyrinthine segment of facial nerve; *FN(m)*, mastoid segment of facial nerve; *HC*, horizontal crest; *IVN*, inferior vestibular nerve; *JB*, jugular bulb; *MFD*, middle fossa dura; *SVN*, superior vestibular nerve; *V*, vestibule.

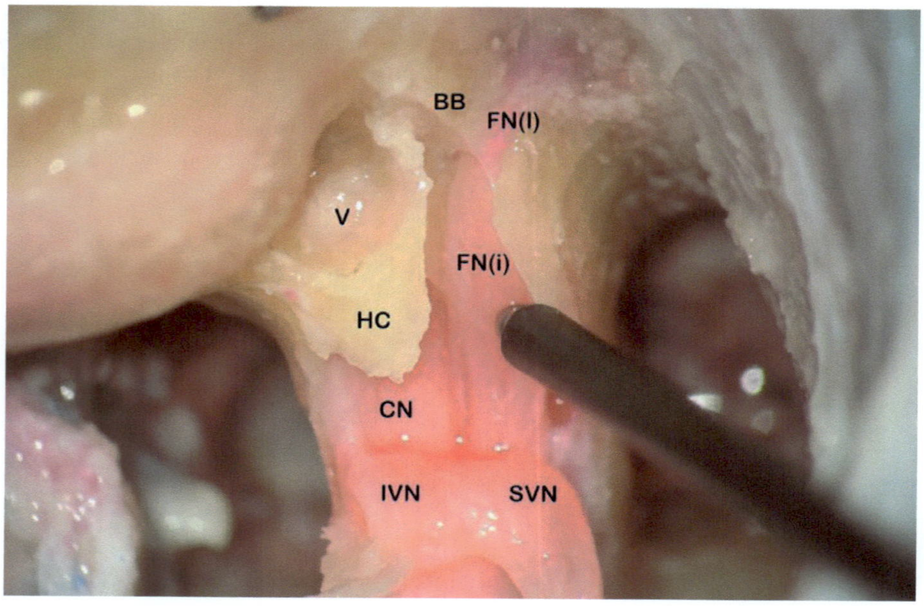

Fig. 7.32 Identification of the internal auditory canal segment of the facial never by stimulator. *BB*, Bill's bar; *CN*, cochlear nerve; *FN(l)*, labyrinthine segment of facial nerve; *FN(i)*, internal auditory canal segment of facial nerve; *HC*, horizontal crest; *IVN*, inferior vestibular nerve; *SVN*, superior vestibular nerve; *V*, vestibule.

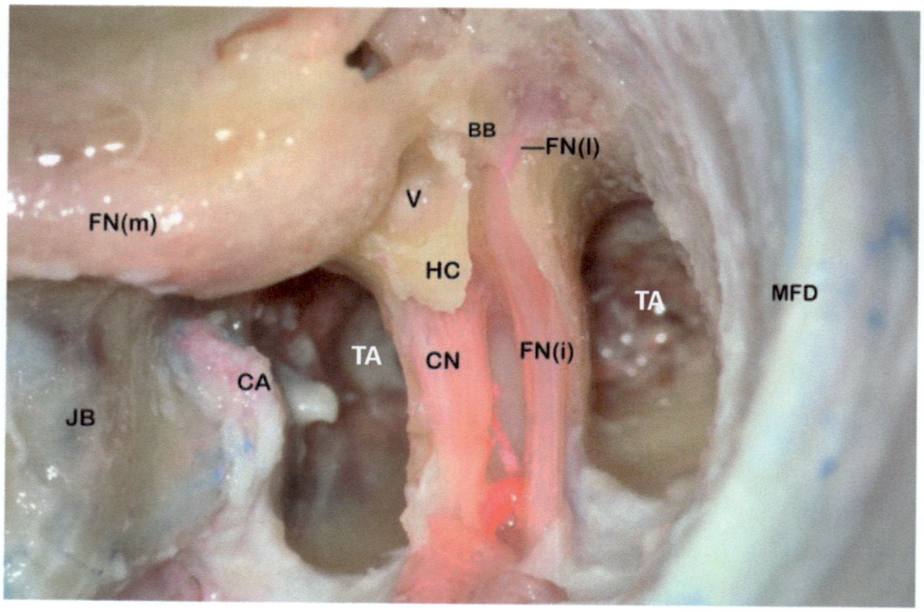

Fig. 7.33 The relationship between the facial and cochlear nerves and the horizontal and vertical crests is well appreciated. *BB*, Bill's bar; *CA*, cochlear aqueduct; *CN*, cochlear nerve; *FN(i)*, internal auditory canal segment of facial nerve; *FN(l)*, labyrinthine segment of facial nerve; *FN (m)*, mastoid segment of facial nerve; *HC*, horizontal crest; *JB*, jugular bulb; *MFD*, middle fossa dura; *TA*, transapical extension of the approach; *V*, vestibule.

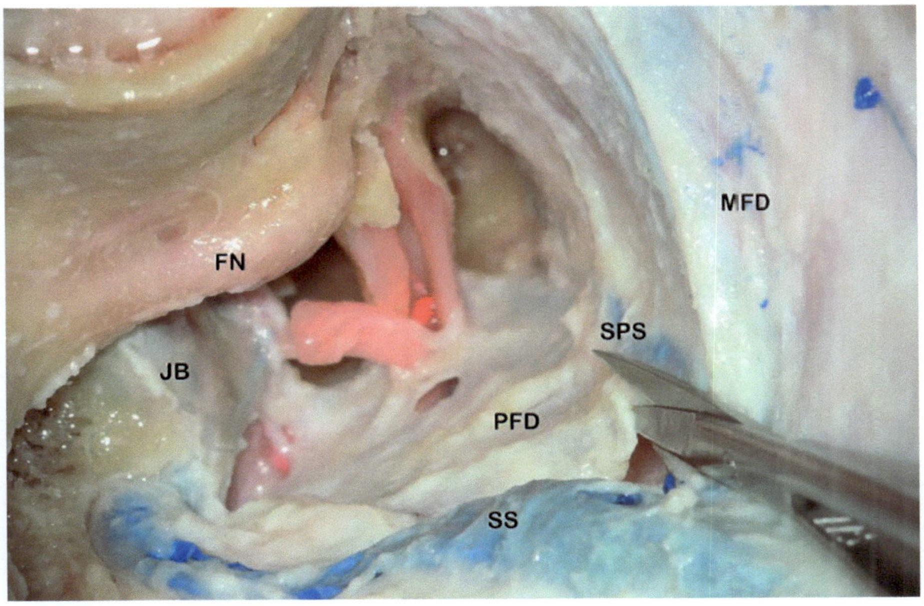

Fig. 7.34 The dural incision starts superiorly 1–2 mm parallel to and inferior to the superior petrosal sinus, running from posterior to anterior using microsurgical scissors. *FN*, facial nerve; *JB*, jugular bulb; *MFD*, middle fossa dura; *PFD*, posterior fossa dura; *SPS*, superior petrosal sinus; *SS*, sigmoid sinus.

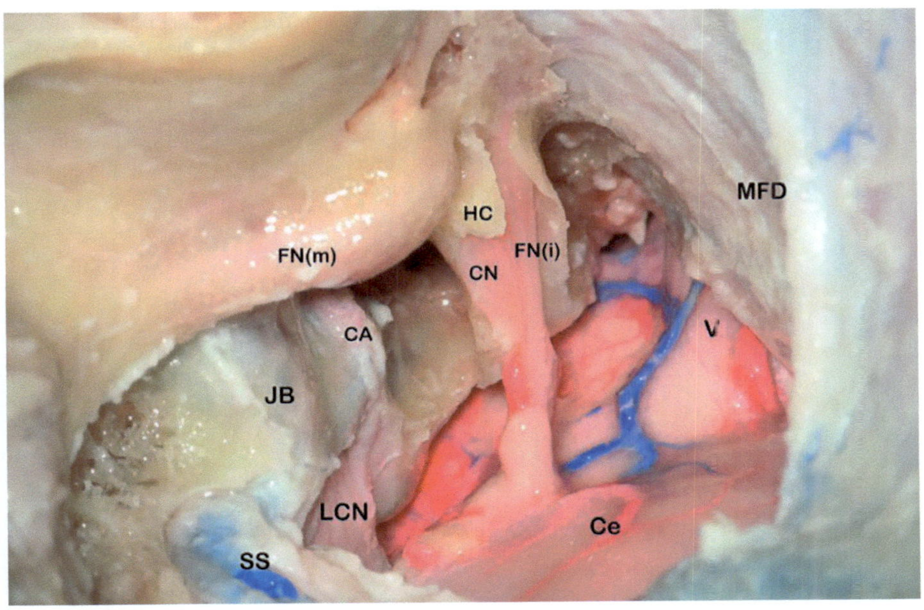

Fig. 7.35 The inferior incision starts just in front of the distal part of the sigmoid sinus, and follows the course of the sinus and jugular bulb to the porus, where it joins the superior incision. After opening the dura, the acousticofacial bundle can be seen. *CA*, cochlear aqueduct; *Ce*, cerebellum; *CN*, cochlear nerve; *FN(i)*, internal auditory canal segment of facial nerve; *FN(m)*, mastoid segment of facial nerve; *HC*, horizontal crest; *JB*, jugular bulb; *LCN*, lower cranial nerves; *MFD*, middle fossa dura; *SS*, sigmoid sinus; *V*, trigeminal nerve.

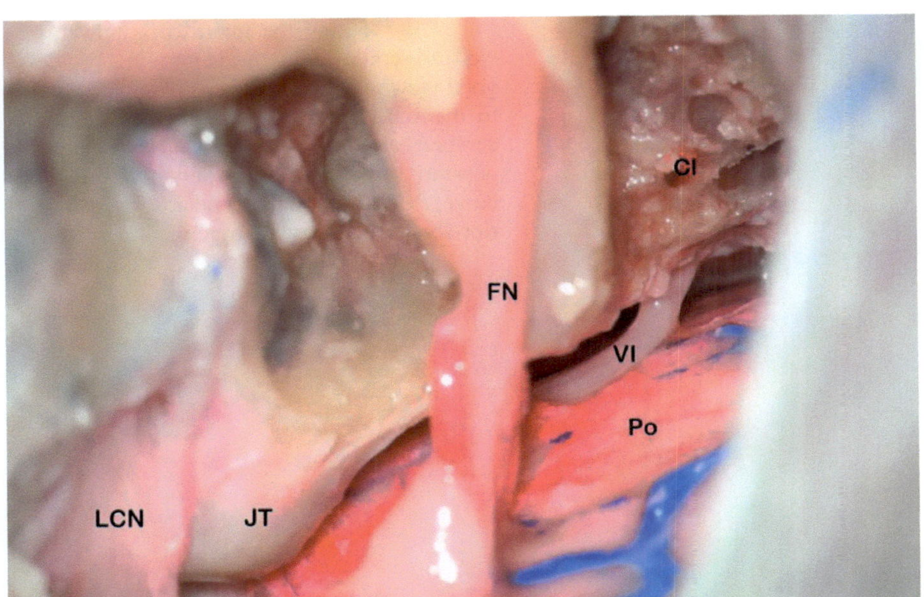

Fig. 7.36 View of the cerebellopontine angle and prepontine cistern as viewed after opening the dura. *Cl*, clivus; *FN*, facial nerve; *JT*, jugular tubercle; *LCN*, lower cranial nerves; *Po*, pons; *VI*, abducent nerve.

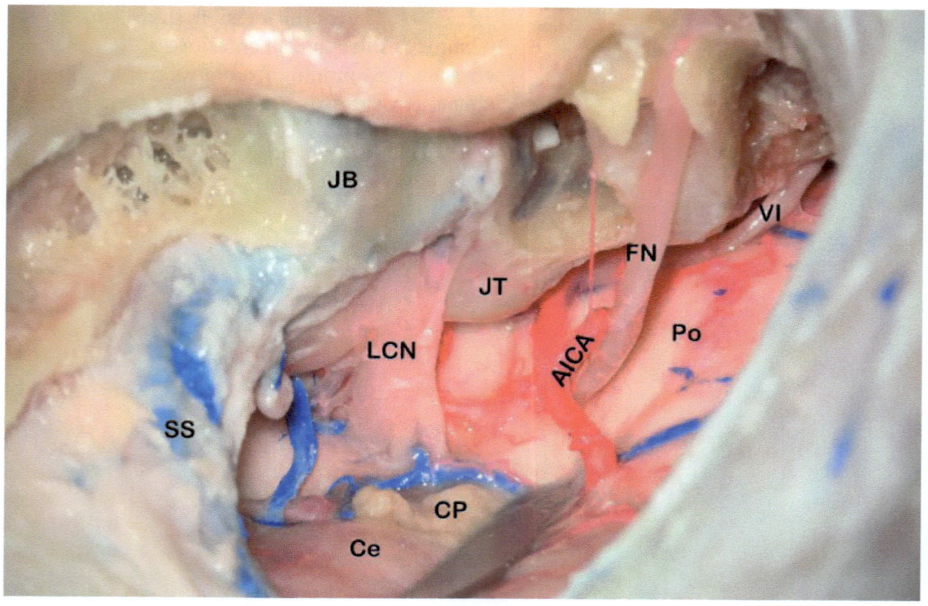

Fig. 7.37 Mild retraction of the cerebellum and flocculus provide a good view of the choroid plexus and lower cranial nerves. *AICA*, anteroinferior cerebellar artery; *Ce*, cerebellum; *CP*, choroid plexus; *FN*, facial nerve; *JB*, jugular bulb; *JT*, jugular tubercle; *LCN*, lower cranial nerves; *Po*, pons; *SS*, sigmoid sinus; *VI*, abducent nerve.

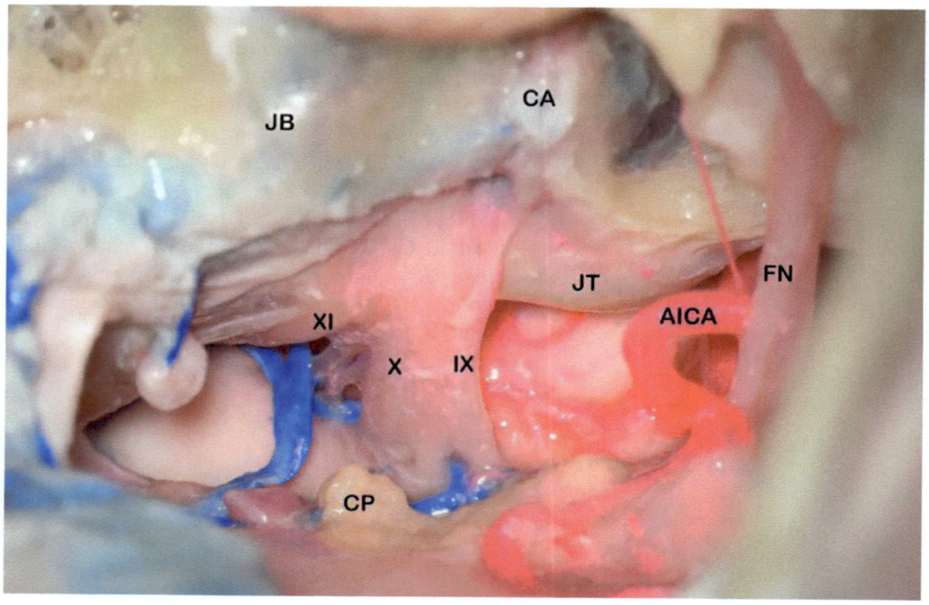

Fig. 7.38 Tilting the microscope downward, the lower cranial nerves are seen. *AICA*, anteroinferior cerebellar artery; *CA*, cochlear aqueduct; *CP*, choroid plexus; *FN*, facial nerve; *IX*, glossopharyngeal nerve; *JB*, jugular bulb; *JT*, jugular tubercle; *X*, vagus nerve; *XI*, accessory nerve.

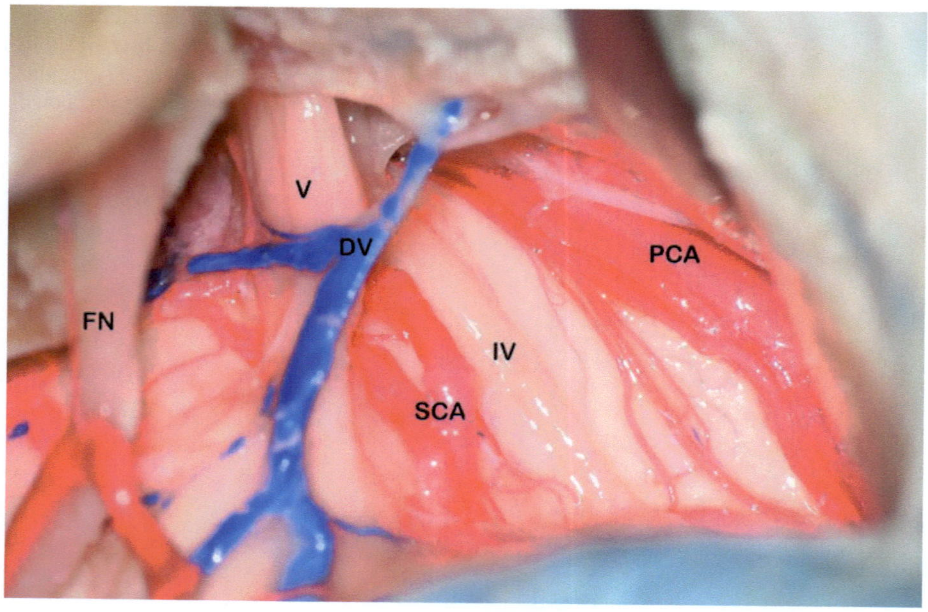

Fig. 7.39 Tilting the microscope upwards as well as retraction of the tentorium provides a good view of the trochlear nerve and its relation to the superior cerebellar artery and posterior cerebral artery. *DV*, Dandy's vein (superior petrosal vein); *FN*, facial nerve; *IV*, trochlear nerve; *PCA*, posterior cerebral artery; *SCA*, superior cerebellar artery; *V*, trigeminal nerve.

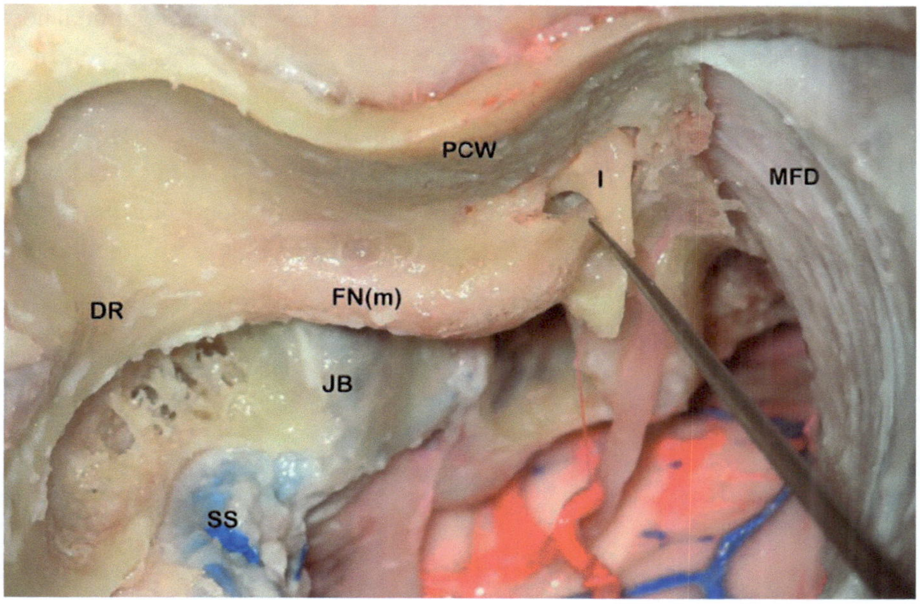

Fig. 7.40 After the incus is disarticulated and removed, the aditus is obliterated with periosteum. *DR*, digastric ridge; *FN(m)*, mastoid segment of facial nerve; *I*, incus; *JB*, jugular bulb; *MFD*, middle fossa dura; *PCW*, posterior canal wall; *SS*, sigmoid sinus.

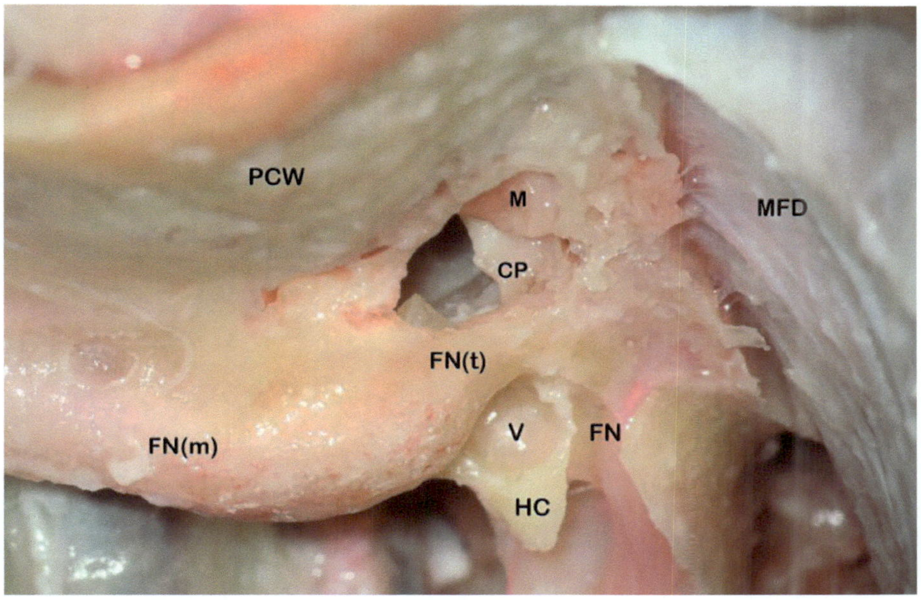

Fig. 7.41 A sufficient amount of the periosteum should be inserted through the aditus into the tympanic cavity to close the orifice of the eustachian tube. *CP*, cochleariform process; *FN (m)*, mastoid segment of facial nerve; *FN(t)*, tympanic segment of facial nerve; *FN*, facial nerve; *HC*, horizontal crest; *M*, malleus; *MFD*, middle fossa dura; *PCW*, posterior canal wall; *V*, vestibule.

Enlarged Translabyrinthine Approach With Closure of the External Auditory Canal

This represents a variation from the classic enlarged TLA. It could be employed in three conditions (Figs. 7.42 and 7.43):

1. High jugular bulb
2. Chronic otitis media (in case of simple tympanic membrane perforation) with no active infection
3. Cases of vestibular schwannoma removal with simultaneous cochlear implantation

SURGICAL STEPS

1. The external auditory canal is transected and closed as cul-de-sac (Figs. 7.44–7.48). The anterior skin flap is retracted anteriorly and maintained in position with two skin hooks.

2. An extended canal wall down mastoidectomy is performed (Fig. 7.49). A large cutting burr is used to drill away the superior and posterior canal wall until the level of the facial nerve (FN) is reached. When the level of the FN is approached, the burr is changed to a diamond burr of the same large size, and drilling is continued parallel to the orientation of the FN with ample suction irrigation.

3. The tensor tympani tendon is cut (Fig. 7.50). The ossicular chain, except for the stapes suprastructure, is removed. The geniculate ganglion and tympanic segment of the facial nerve is further skeletonized (Fig. 7.51). The inferior tympanic bone is also drilled (Fig. 7.52).

4. The anterior wall of the external auditory canal wall is thinned out. The infralabyrinthine and retrofacial air cells are completely drilled (Fig. 7.53). The internal carotid artery (ICA) is identified. Identification of the ICA starts at the

level of the floor of the eustachian tube, or 2 mm anterior to the cochlea. The cochlear aqueduct is the lower limit of the approach.

5. The cochlea and petrous apex bone are drilled until the inferior petrosal sinus is reached (Fig. 7.54). The anteroinferior cerebellar artery (AICA) passes inferior to the abducent nerve, and travels posteriorly to come into close relation to the facial nerve.

6. Tilting the microscope upward, the trigeminal nerve can be well appreciated (Fig. 7.55). Mild retraction of the tentorium provides a good view of the trochlear nerve and its relation to the superior cerebellar artery (SCA) (Fig. 7.56). The inferior petrosal sinus is indicated with a probe (Fig. 7.57). Note the extended control of the prepontine cistern, in comparison with the classic enlarged translabyrinthine approach (see Fig. 7.35).

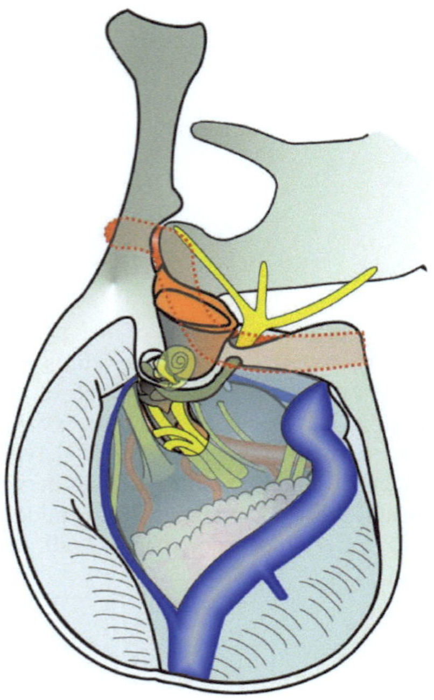

Fig. 7.42 Schematic drawing showing a standard translabyrinthine approach.

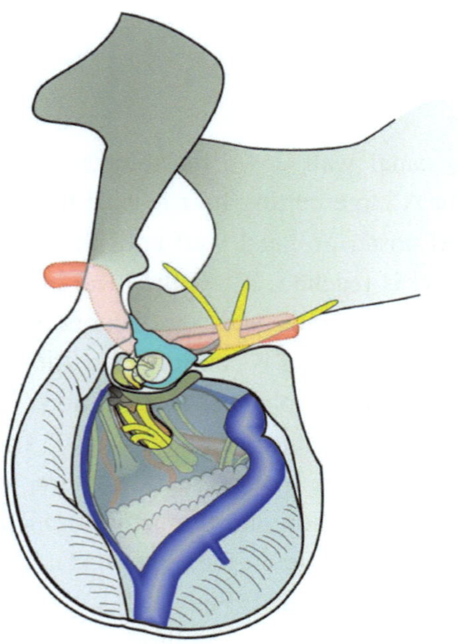

Fig. 7.43 Schematic drawing showing an extended canal wall down mastoidectomy is performed.

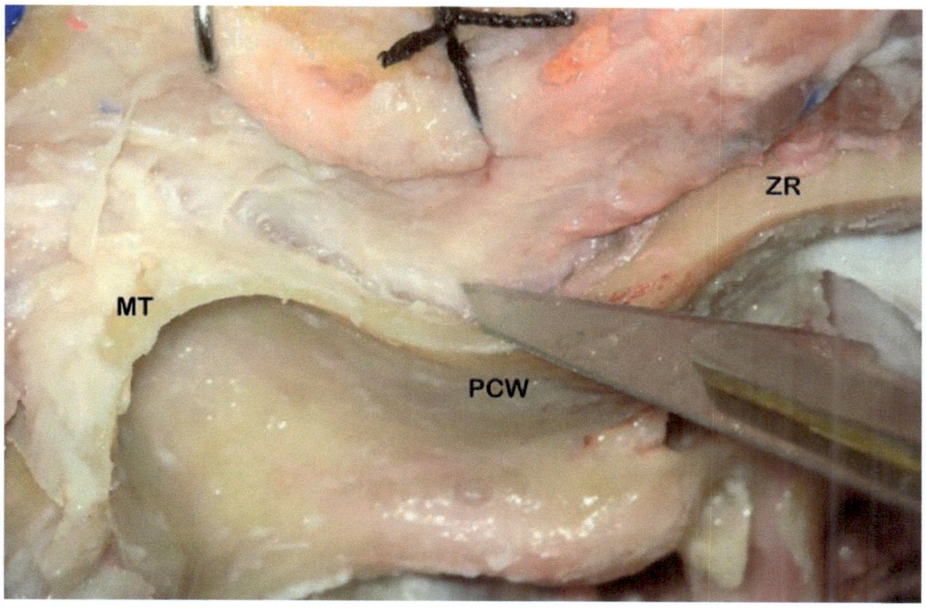

Fig. 7.44 The posterior meatal skin is transected with a Beaver blade. *MT*, mastoid tip; *PCW*, posterior canal wall; *ZR*, zygomatic root.

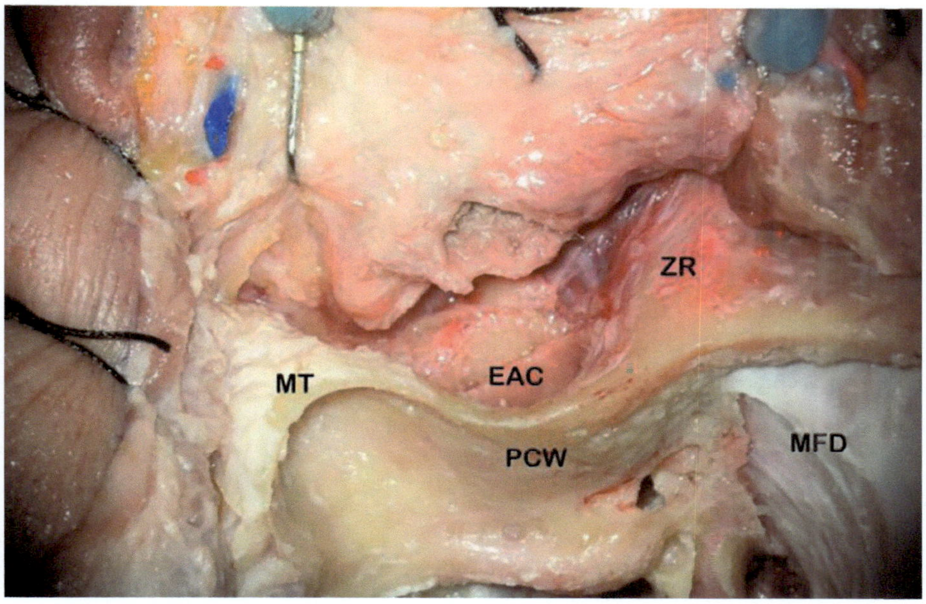

Fig. 7.45 The external auditory canal is transacted. *EAC*, external auditory canal; *MFD*, middle fossa dura; *MT*, mastoid tip; *PCW*, posterior canal wall; *ZR*, zygomatic root.

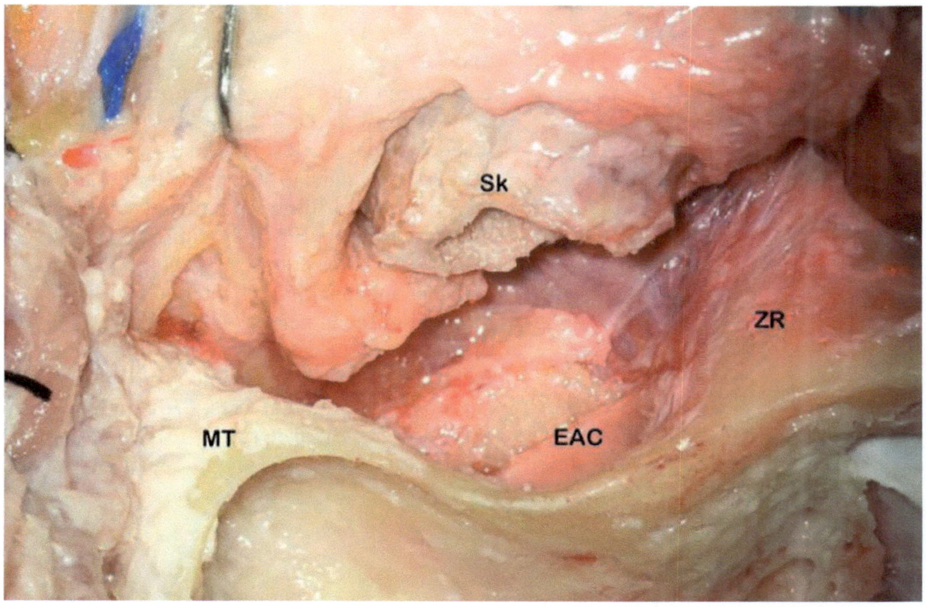

Fig. 7.46 Meticulous dissection of the skin from the anterior cartilage has to be performed. *EAC*, external auditory canal; *MT*, mastoid tip; *Sk*, skin; *ZR*, zygomatic root.

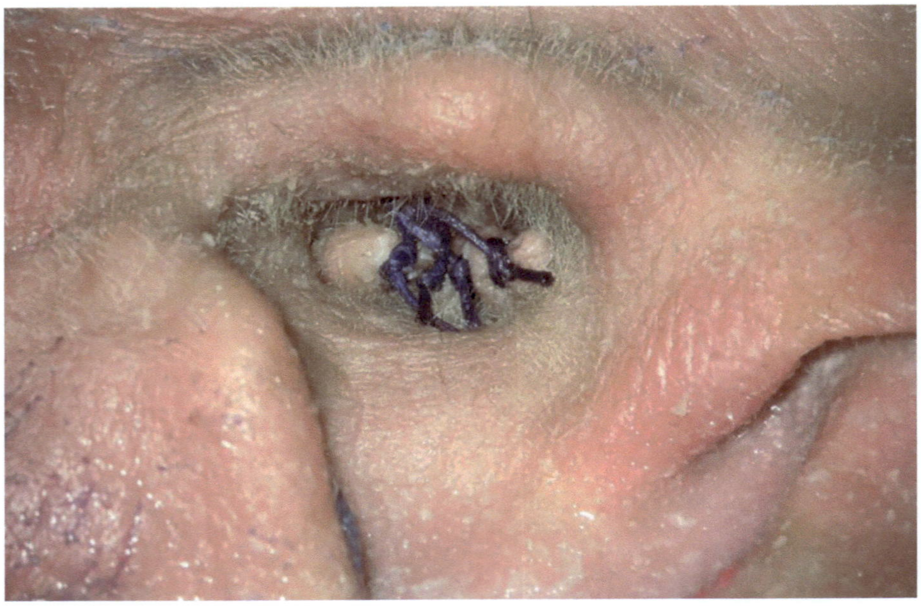

Fig. 7.47 The meatal skin is everted out through the canal and sutured with absorbable sutures.

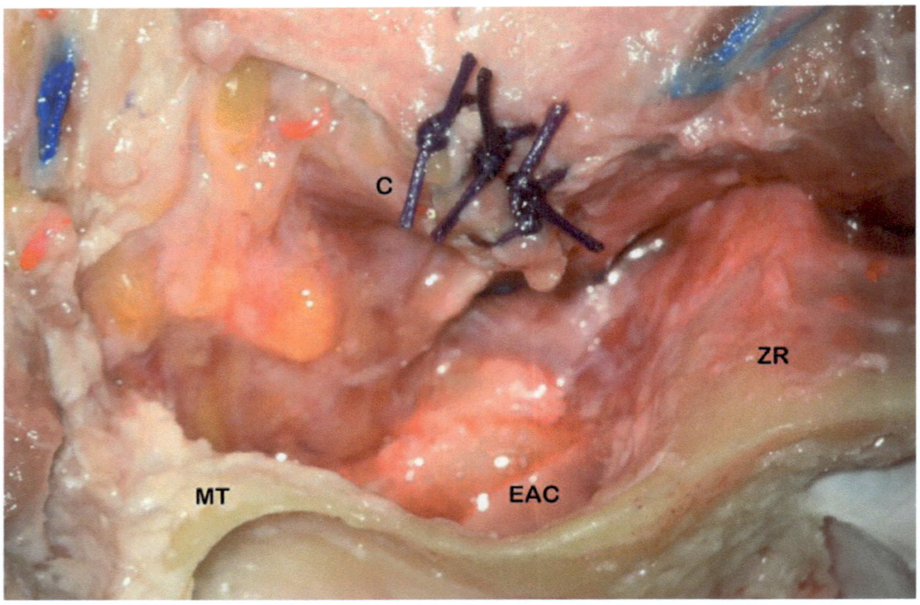

Fig. 7.48 A second layer of closure is achieved by approximating the cartilage of the tragus and the concha. *C*, cartilage; *EAC*, external auditory canal; *MT*, mastoid tip; *ZR*, zygomatic root.

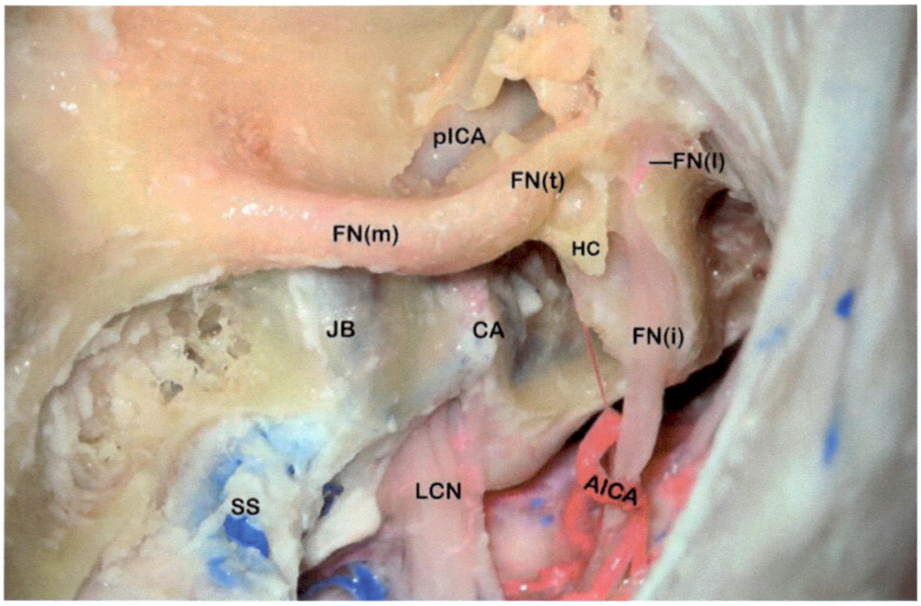

Fig. 7.49 An extended canal wall down mastoidectomy has been performed. A large cutting burr is used to drill away the superior and posterior canal wall until the level of the FN is reached. When the level of the FN is approached, the burr is changed to a diamond burr of the same large size, and drilling is continued parallel to the orientation of the FN with ample suction irrigation. *AICA*, anteroinferior cerebellar artery; *CA*, cochlear aqueduct; *FN(i)*, internal auditory canal segment of facial nerve; *FN(l)*, labyrinthine segment of facial nerve; *FN(m)*, mastoid segment of facial nerve; *FN(t)*, tympanic segment of facial nerve; *HC*, horizontal crest; *JB*, jugular bulb; *LCN*, lower cranial nerves; *pICA*, petrous segment of internal carotid artery; *SS*, sigmoid sinus.

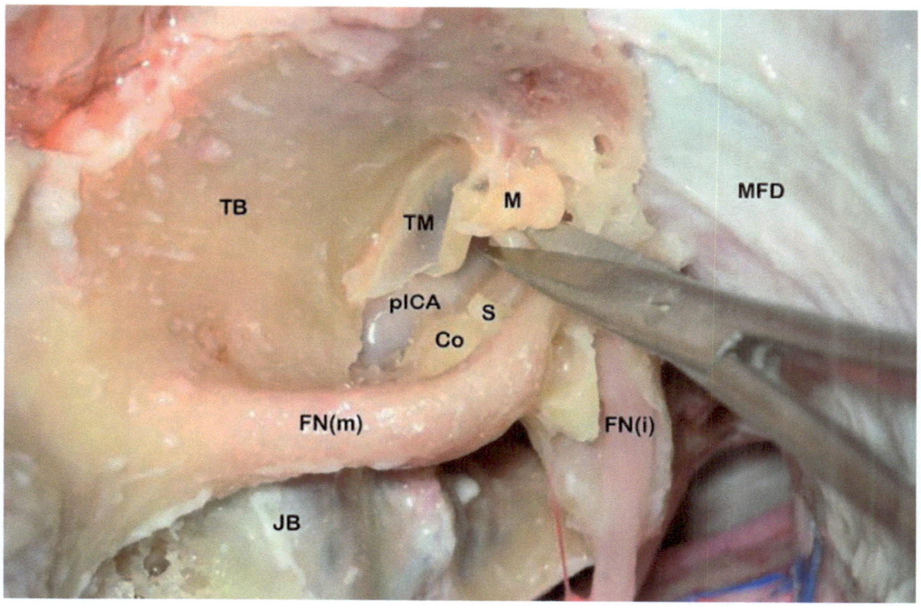

Fig. 7.50 The tensor tympani tendon is being cut. The ossicular chain, except for the stapes infrastructure, is removed. *Co*, cochlea; *FN(i)*, internal auditory canal segment of facial nerve; *FN(m)*, mastoid segment of facial nerve; *JB*, jugular bulb; *M*, malleus; *MFD*, middle fossa dura; *pICA*, petrous segment of internal carotid artery; *S*, stapes; *TB*, tympanic bone; *TM*, tympanic membrane.

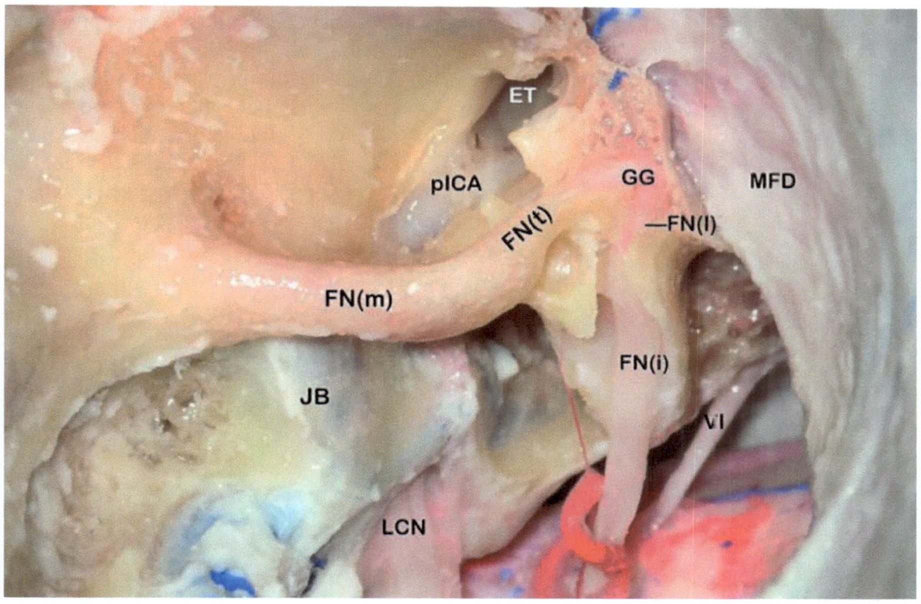

Fig. 7.51 The geniculate ganglion and tympanic segment of the facial nerve is further skeletonized. *ET*, eustachian tube; *FN(i)*, internal auditory canal segment of facial nerve; *FN(l)*, labyrinthine segment of facial nerve; *FN(m)*, mastoid segment of facial nerve; *FN(t)*, tympanic segment of facial nerve; *GG*, geniculate ganglion; *JB*, jugular bulb; *LCN*, lower cranial nerves; *MFD*, middle fossa dura; *pICA*, petrous segment of internal carotid artery; *VI*, abducent nerve.

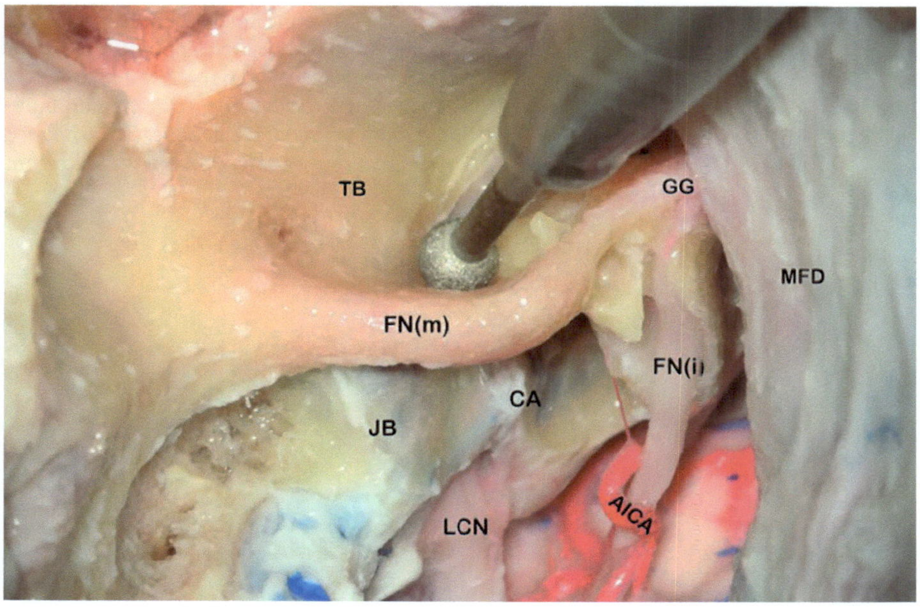

Fig. 7.52 The inferior tympanic bone is drilled. *AICA*, anteroinferior cerebellar artery; *CA*, cochlear aqueduct; *FN(i)*, internal auditory canal segment of facial nerve; *FN(m)*, mastoid segment of facial nerve; *GG*, geniculate ganglion; *JB*, jugular bulb; *LCN*, lower cranial nerves; *MFD*, middle fossa dura; *TB*, tympanic bone.

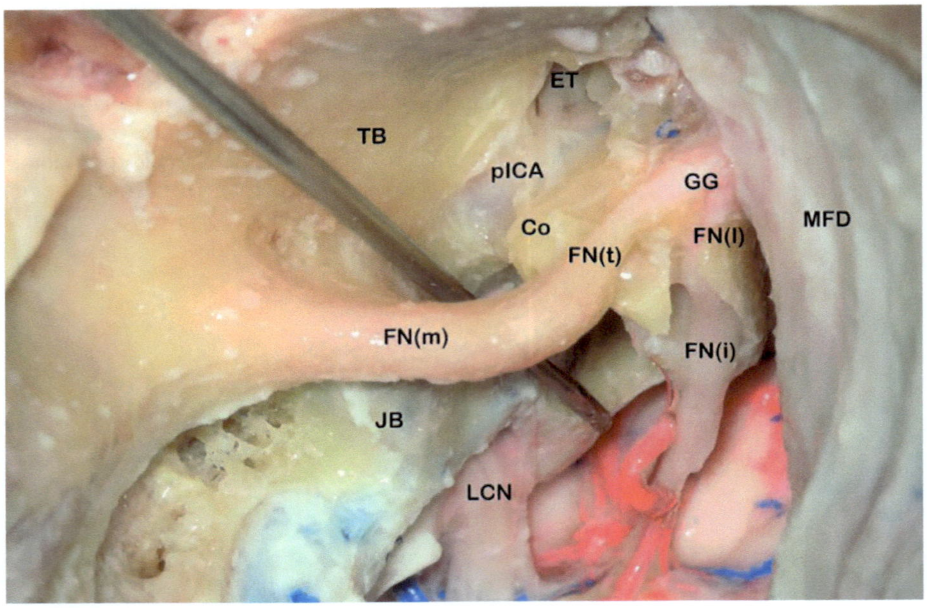

Fig. 7.53 The anterior wall of the external auditory canal wall is thinned out. The infralabyrinthine and retrofacial air cells are completely drilled. The internal carotid artery is identified. The clivus is indicated with a probe. *Co*, cochlea; *ET*, eustachian tube; *FN(i)*, internal auditory canal segment of facial nerve; *FN(l)*, labyrinthine segment of facial nerve; *FN(m)*, mastoid segment of facial nerve; *FN(t)*, tympanic segment of facial nerve; *GG*, geniculate ganglion; *JB*, jugular bulb; *LCN*, lower cranial nerves; *MFD*, middle fossa dura; *pICA*, petrous segment of internal carotid artery; *TB*, tympanic bone.

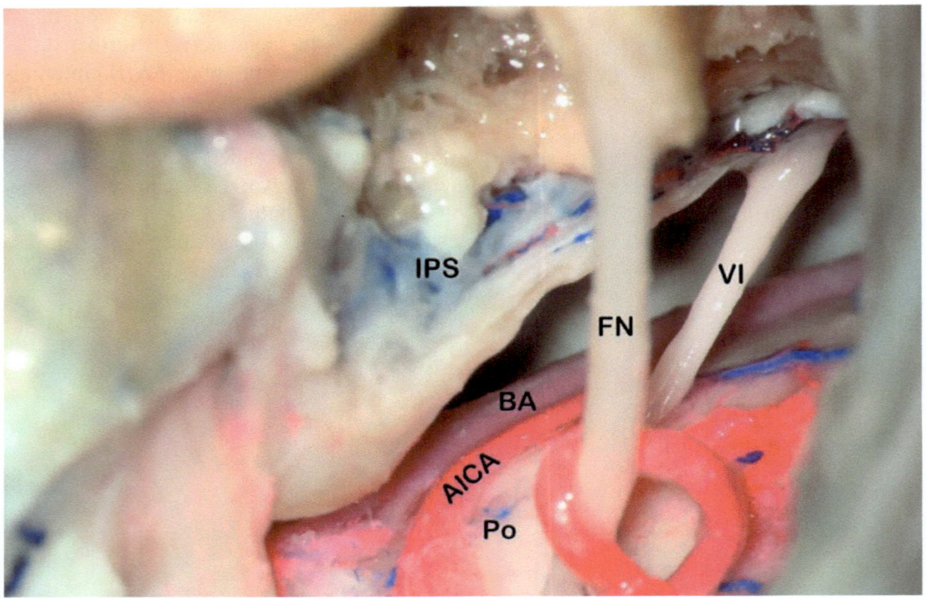

Fig. 7.54 The cochlea and petrous apex bone were drilled until the inferior petrosal sinus is reached. The anteroinferior cerebellar artery passes inferior to the abducent nerve, and travels posteriorly to come into close relation to the facial nerve. *AICA*, anteroinferior cerebellar artery; *BA*, basilar artery; *FN*, facial nerve; *IPS*, inferior petrosal sinus; *Po*, pons; *VI*, abducent nerve.

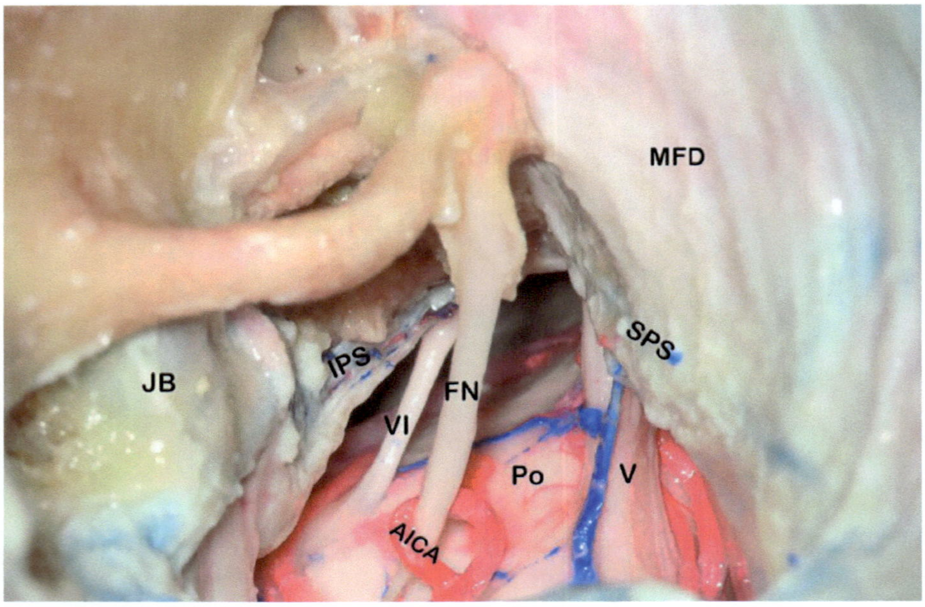

Fig. 7.55 Tilting the microscope upward, the trigeminal nerve can be well appreciated. *AICA*, anteroinferior cerebellar artery; *FN*, facial nerve; *IPS*, inferior petrosal sinus; *JB*, jugular bulb; *MFD*, middle fossa dura; *Po*, pons; *SPS*, superior petrosal sinus; *V*, trigeminal nerve; *VI*, abducent nerve.

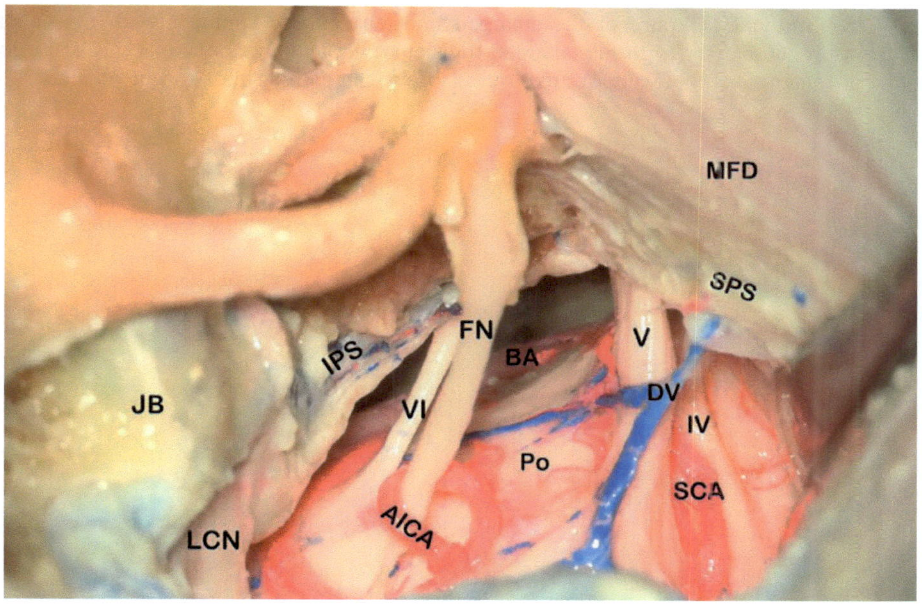

Fig. 7.56 Mild retraction of the tentorium provides a good view of the trochlear nerve and its relation to the superior cerebellar artery. *AICA*, anteroinferior cerebellar artery; *BA*, basilar artery; *DV*, Dandy's vein (superior petrosal vein); *FN*, facial nerve; *IPS*, inferior petrosal sinus; *IV*, trochlear nerve; *JB*, jugular bulb; *LCN*, lower cranial nerves; *MFD*, middle fossa dura; *Po*, pons; *SCA*, superior cerebellar artery; *SPS*, superior petrosal sinus; *V*, trigeminal nerve; *VI*, abducent nerve.

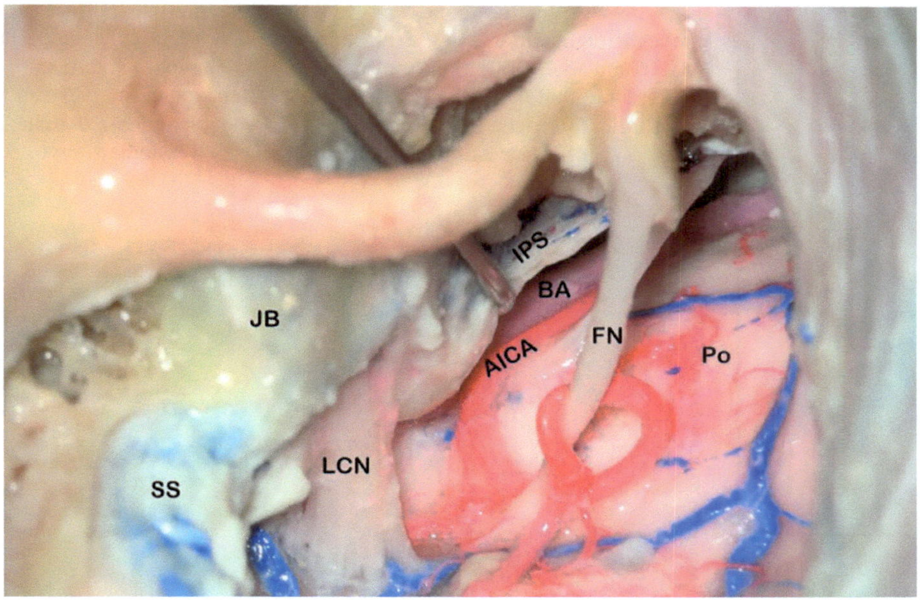

Fig. 7.57 The inferior petrosal sinus is indicated with a probe. Note the extended control of the prepontine cistern here, in comparison with the translabyrinthine approach (see fig. 8.35 in Chapter 8). *AICA*, anteroinferior cerebellar artery; *BA*, basilar artery; *FN*, facial nerve; *IPS*, inferior petrosal sinus; *JB*, jugular bulb; *LCN*, lower cranial nerves; *Po*, pons; *SS*, sigmoid sinus.

Enlarged Translabyrinthine Approach With Transapical Extension

Rationale

This approach is an anterior extension of the enlarged TLA, in which the internal auditory canal is drilled for 320° or 360° of its circumference (types I and II, respectively).

Indications

This approach is indicated for large CPA tumors, in which hearing preservation is not an objective, as follows (Figs. 7.58–7.61):

– Type I approach: Large and giant acoustic neurinomas with anterior extension into the prepontine cisterns.

– Type II approach: Large meningioma of the posterior surface of the temporal bone centered on the internal auditory canal with significant anterior extension.

This approach is also used for even small tumors but with marked anterior extension.

Limitations

This approach has a limitation in lesions that involve the petro-clival region. In such lesions, we use the modified transcochlear approach. However, it is very rare for an acoustic neurinoma to involve the petroclival region, but it is more common for meningiomas.

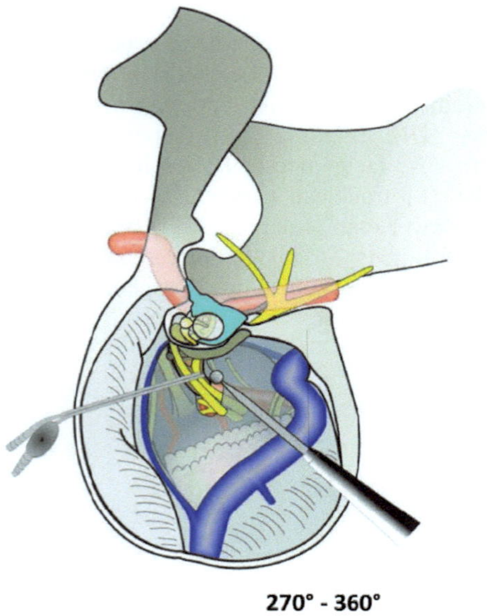

Fig. 7.58 Surgical exposure of the enlarged translabyrinthine approach with transapical extension. In comparison with the enlarged translabyrinthine approach alone, this extension has the advantage that more anterior exposure close to the petrous apex can be obtained.

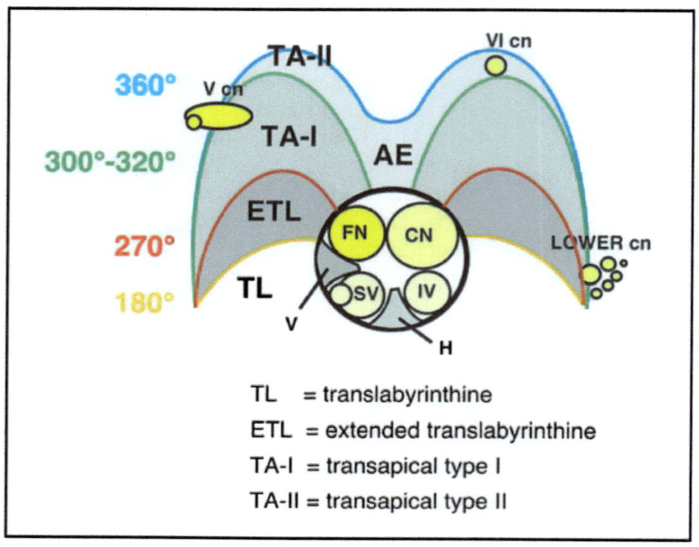

Fig. 7.59 Amount of bone removal around the internal auditory canal in the different variants of the translabyrinthine approach. The exposure is 320° and about 360° in the transapical extension types I and II, respectively. Note that the fifth (Vcn) and sixth (VIcn) cranial nerves can be (un)surgically reached. *cn*, cranial nerve; *CN*, cochlear nerve; *FN*, facial nerve; *IV*, inferior vestibular nerve; *SV*, superior vestibular nerve.

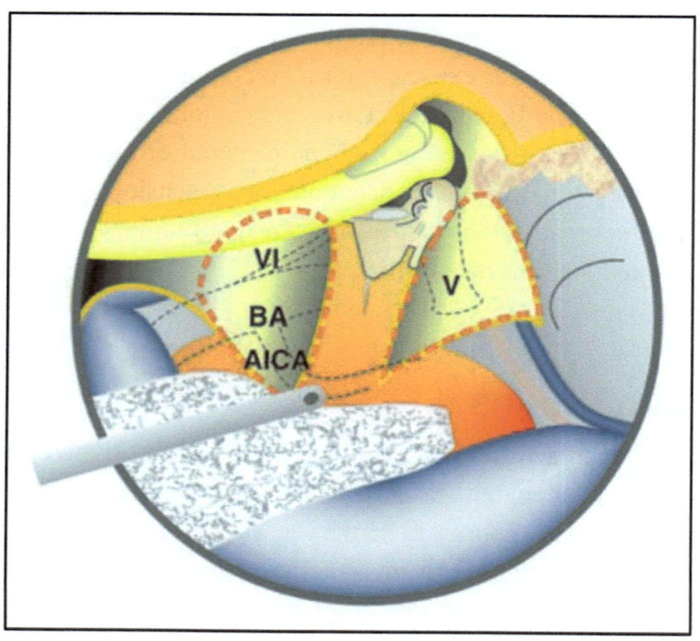

Fig. 7.60 Difference between the conventional 180° translabyrinthine approach and the 320° transapical extension in the surgical view (*left* ear). The dashed line demonstrates the working area for extended bone removal. After this bone work is completed, the important structures can be controlled. *AICA*, anterior inferior cerebellar artery; *BA*, basilar artery; *V*, trigeminal nerve; *VI*, abducent nerve.

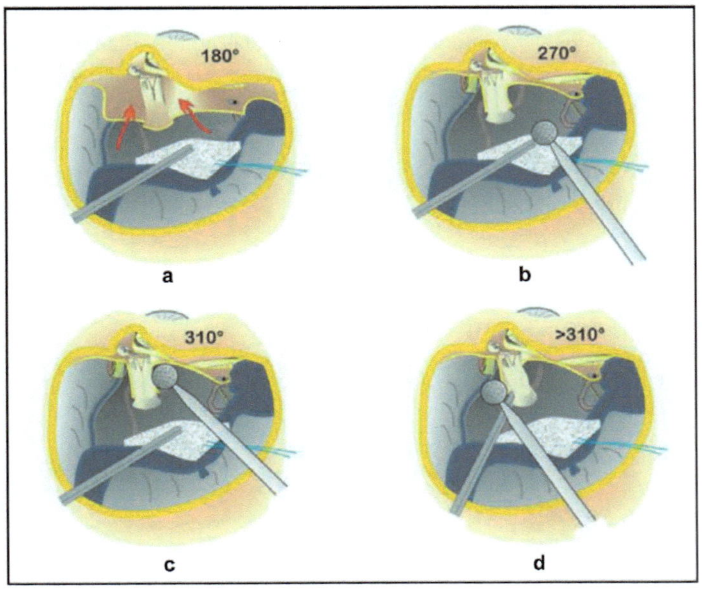

Fig. 7.61 Surgical views of a variety of skeletonizations of the internal auditory canal. (A) Conventional 180°. (B) Our current use for enlarged translabyrinthine 270°. (C) Type I transapical extension 310°. (D) Type II transapical extension by more than 310°.

Surgical Steps (Right Ear, Figs. 7.62–7.70)

1. The enlarged TLA is performed as previously described. Wide bone removal is performed, exposing about 3 cm of the middle fossa dura and the retrosigmoid posterior fossa dura. The sigmoid sinus is completely uncovered.
2. The internal auditory canal is identified and uncovered. Bone is removed for 270° around the canal as previously described.
3. Bone is further removed both inferior to and superior to the internal auditory canal toward the petrous apex as it is required. A 320° exposure of the internal auditory canal is achieved. The contents of the internal auditory canal are pushed inferiorly to allow for drilling of the anterior wall of the canal. Care should be taken not to injure the superior petrosal sinus and facial nerve located far anteriorly.
4. At the end of the procedure, 360° of the internal auditory canal circumference is exposed (type II approach). In the type I approach, it is only necessary to remove 320° of the circumference of the canal.
5. The dura is opened as previously described.

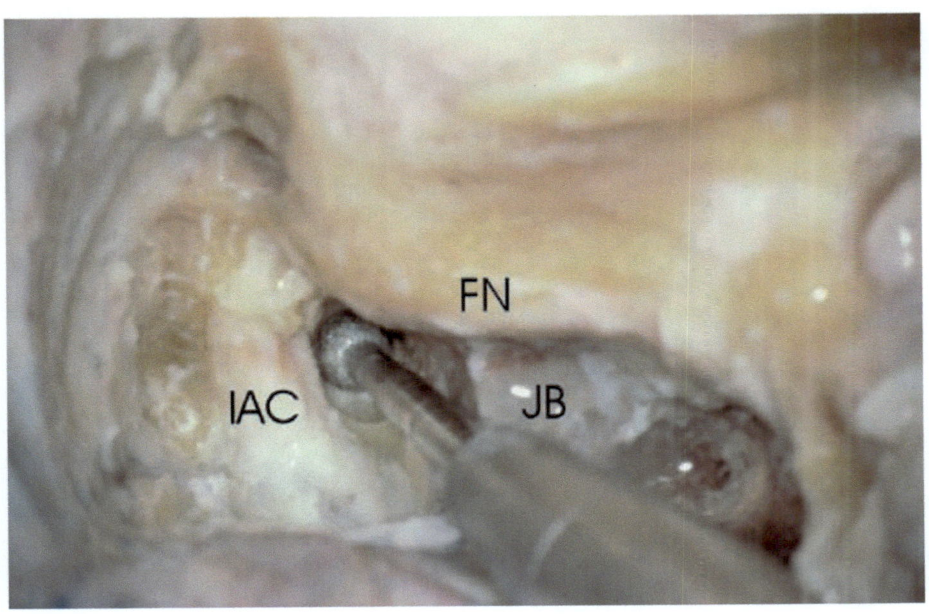

Fig. 7.62 Drilling is being carried out inferior to the internal auditory canal (IAC). *FN*, facial nerve; *JB*, jugular bulb.

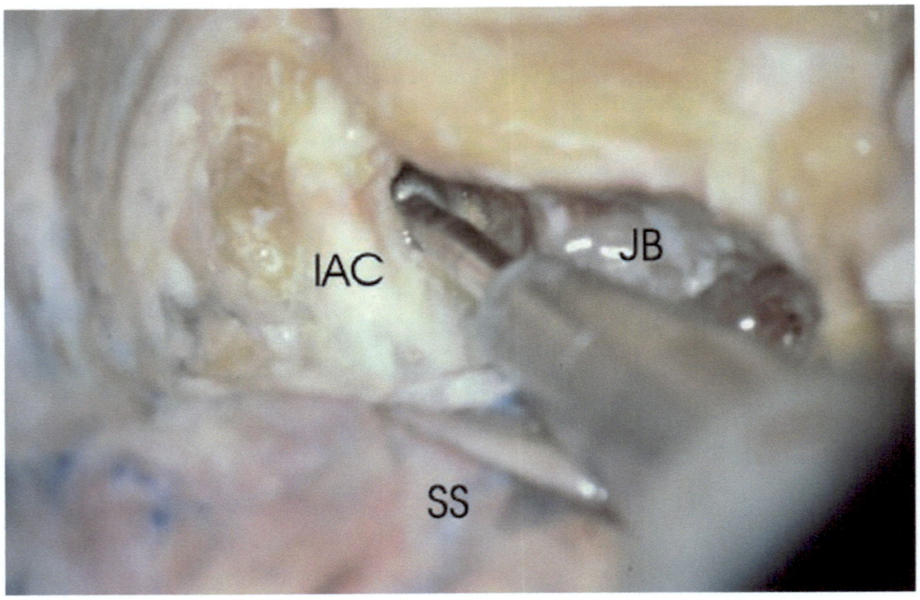

Fig. 7.63 Further extensive drilling is added inferior to the internal auditory canal (IAC) toward the petrous apex. *JB*, jugular bulb; *SS*, sigmoid sinus.

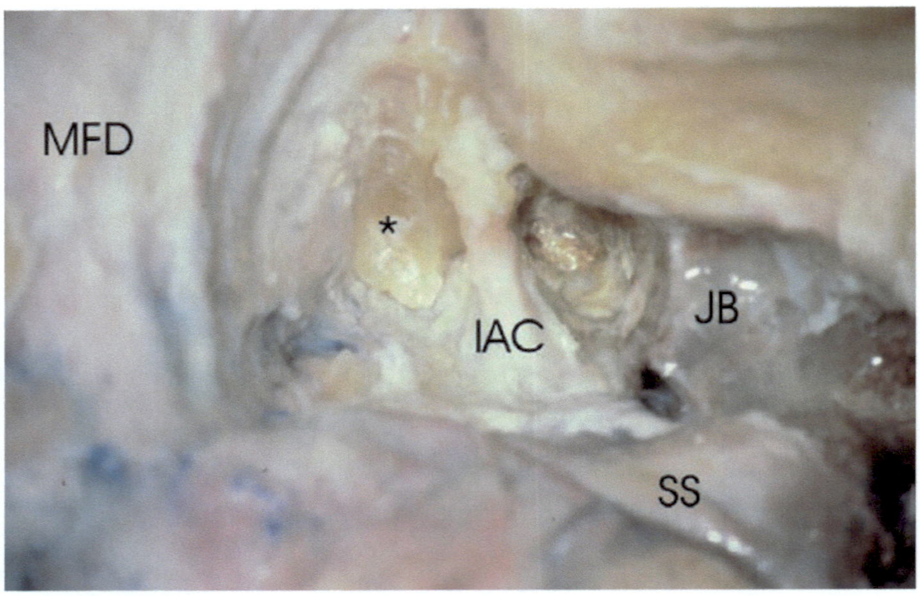

Fig. 7.64 Extensive bone removal inferior and superior to the internal auditory canal (IAC) have been achieved. Bone superior to the canal (*) is still to be removed. *JB*, jugular bulb; *MFD*, middle fossa dura; *SS*, sigmoid sinus.

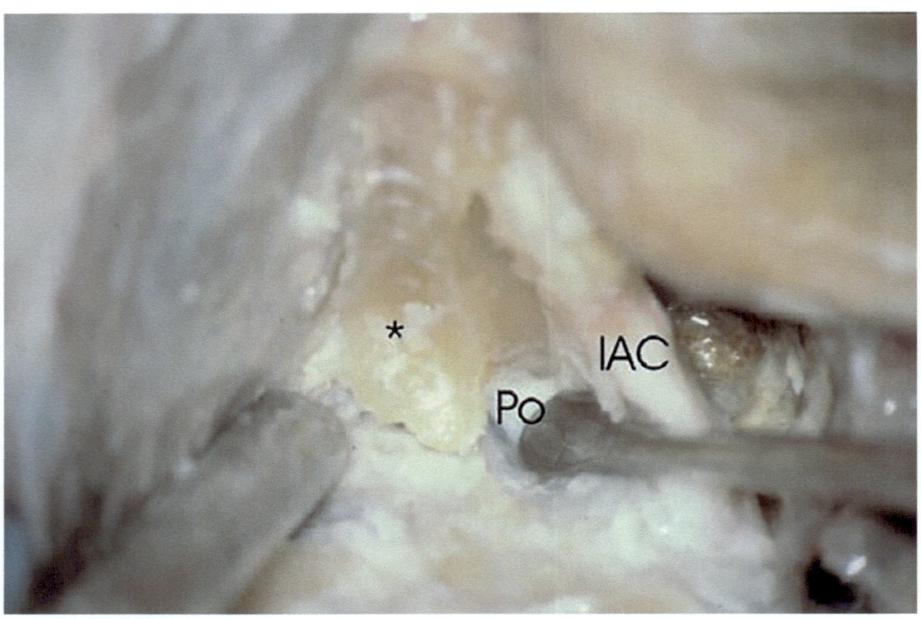

Fig. 7.65 The contents of the internal auditory canal are pushed inferiorly to allow for removal of the remaining bone (*) superior to the internal auditory canal (IAC). *Po*, porus of the internal auditory canal.

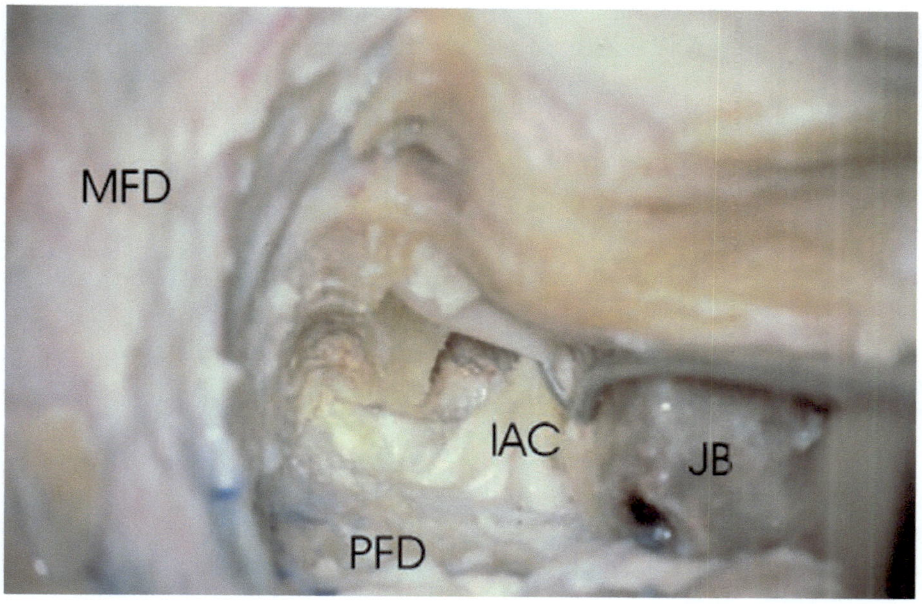

Fig. 7.66 The contents of the canal are displaced inferiorly to show the extent of bone removal. The anterior wall of the canal can be further drilled if needed. *IAC*, internal auditory canal; *JB*, jugular bulb; *MFD*, middle fossa dura; *PFD*, posterior fossa dura.

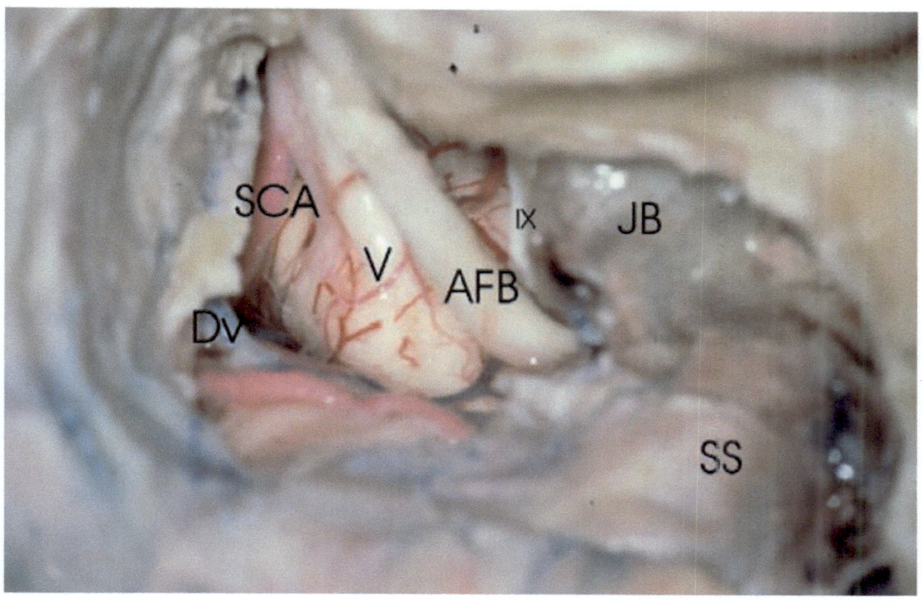

Fig. 7.67 General view of the structures in the right CPA after opening the dura. Note the enhanced exposure of the superior CPA and the excellent exposure of the trigeminal nerve (V) and the superior cerebellar artery (SCA). *AFB*, acousticofacial bundle; *Dv*, Dandy's vein; *IX*, glossopharyngeal nerve; *JB*, jugular bulb; *SS*, sigmoid sinus.

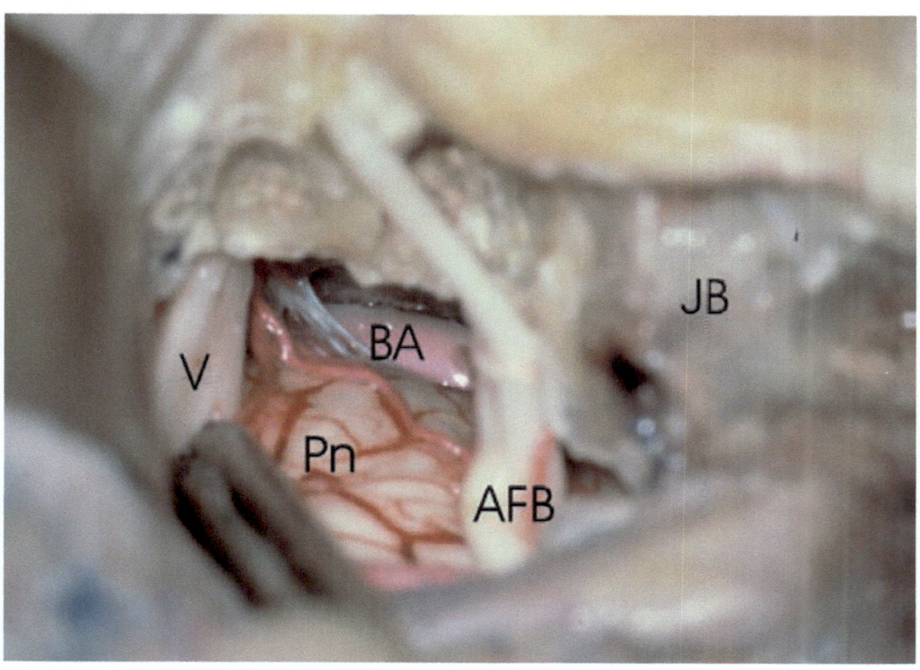

Fig. 7.68 The trigeminal nerve (V) is pushed superiorly. The basilar artery (BA) in the prepontine cistern is clearly visible. *AFB*, acousticofacial bundle; *JB*, jugular bulb; *Pn*, pons.

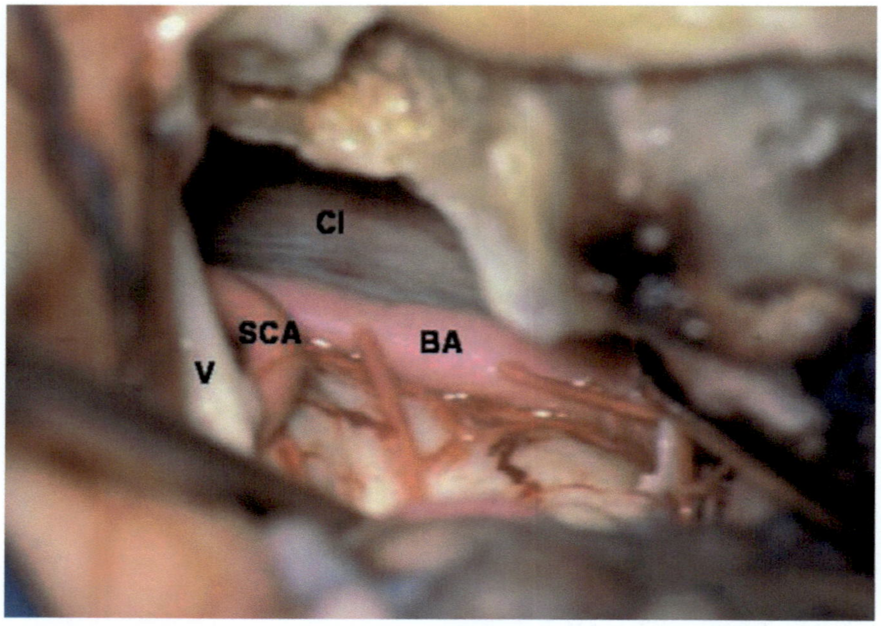

Fig. 7.69 Tilting the microscope anterosuperiorly shows that the superior cerebellar artery (SCA) branches off the basilar artery (BA). The clivus (Cl) can also be visualized. *V*, trigeminal nerve.

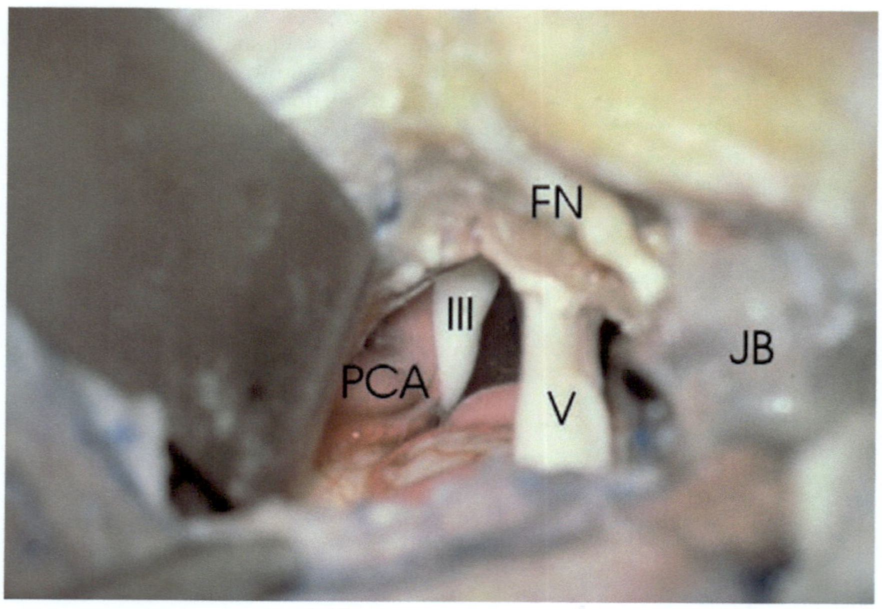

Fig. 7.70 With gentle retraction of the tentorium, the oculomotor nerve (III) is seen between the posterior cerebral artery (PCA) superiorly and the superior cerebellar artery (not labeled) inferiorly. *FN*, facial nerve; *JB*, jugular bulb; *V*, trigeminal nerve.

Case 1: Grade III Vestibular Schwannoma

See (Figs. 7.71–7.96)

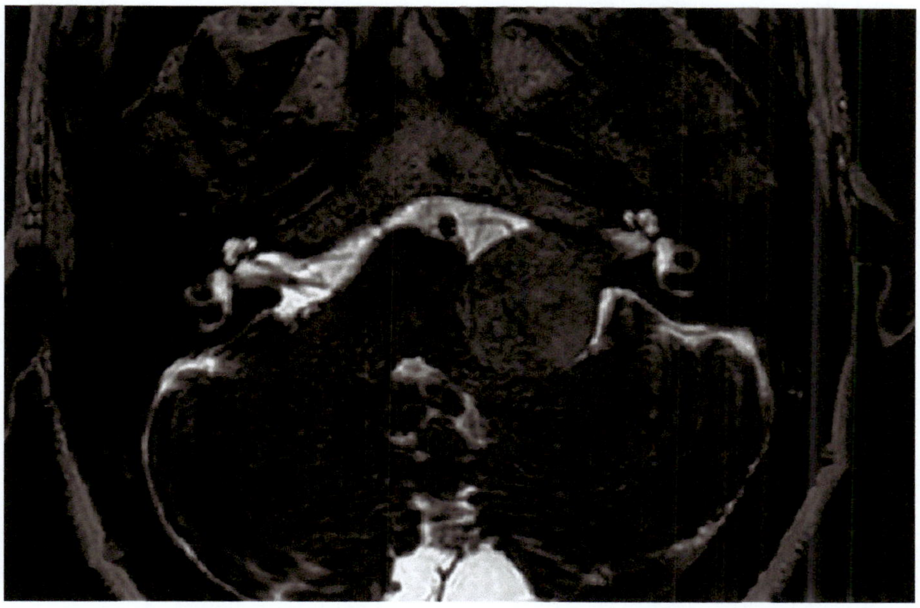

Fig. 7.71 T2W MRI of the brain shows a left vestibular schwannoma in a 53-year-old patient. According to the Tokyo consensus classification, this is a grade III vestibular schwannoma (maximum diameter 21–30 mm in the cerebellopontine angle). An enlarged translabyrinthine approach with transapical extension was planned. *T*, tumor.

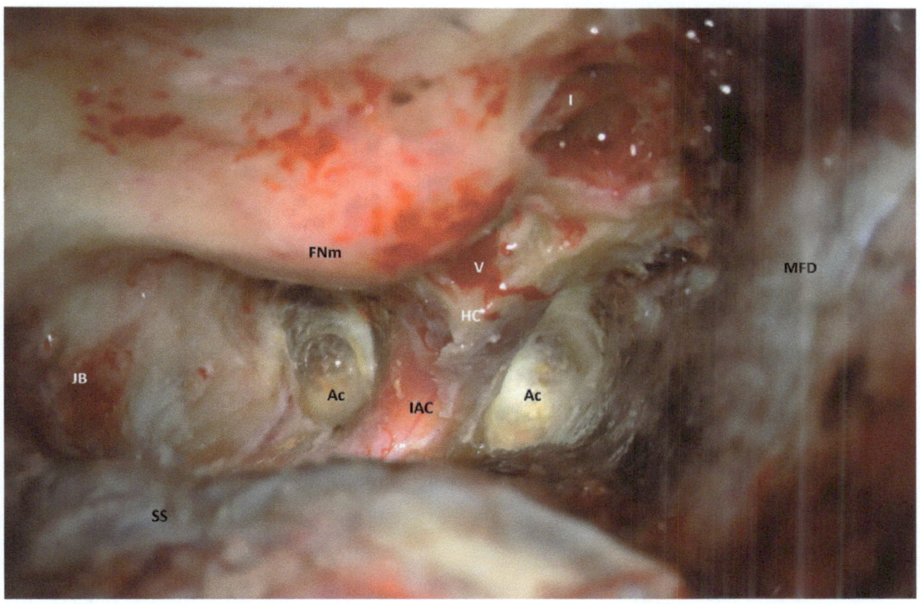

Fig. 7.72 The surgical approach is almost complete and the internal auditory canal (IAC) is fully skeletonised. Note the wide drilling of the apical cells (Ac) to improve anteromedial control of the tumor in the cerebello-pontine angle. *FNm*, mastoid portion of the facial nerve; *HC*, horizontal crest; *I*, incus; *JB*, jugular bulb; *MFD*, middle fossa dura; *SS*, sigmoid sinus; *V*, vestibule.

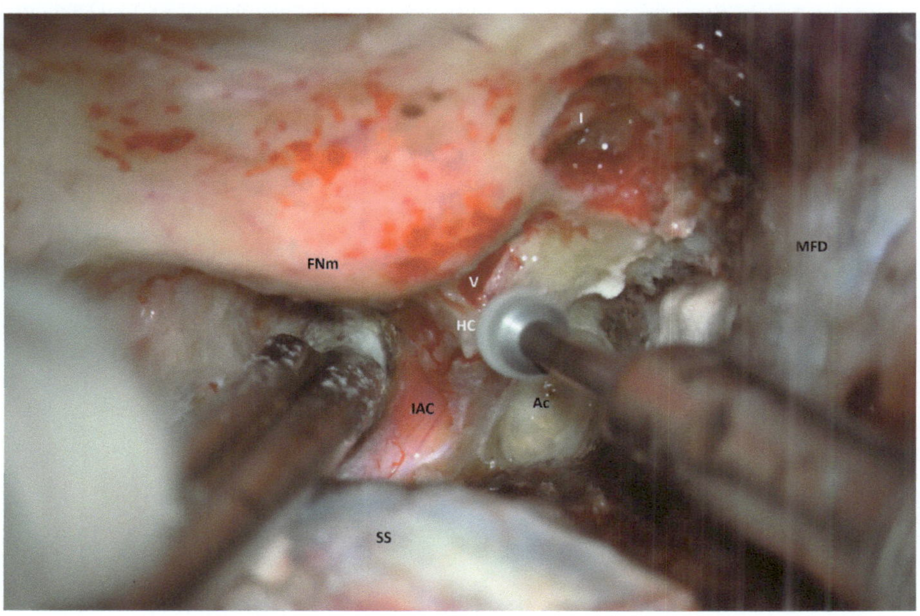

Fig. 7.73 Drilling of the superior border of the vestibule allows visualization of the superior ampullary nerve. *AC*, apical cells; *FNm*, mastoid portion of the facial nerve; *HC*, horizontal crest; *I*, incus; *JB*, jugular bulb; *MFD*, middle fossa dura; *SS*, sigmoid sinus; *V*, vestibule.

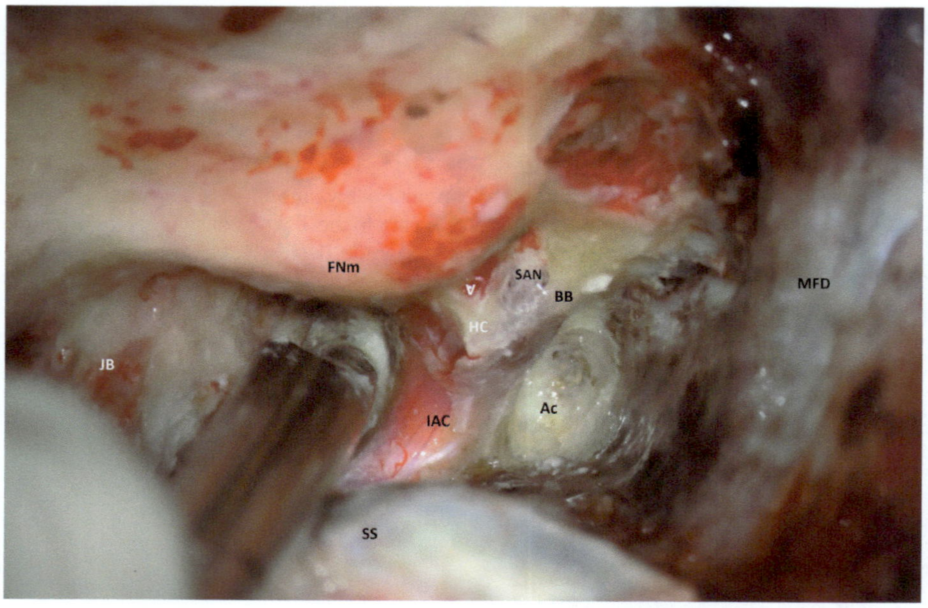

Fig. 7.74 Drilling of the superior border of the vestibule allows visualization of the superior ampullary nerve (SAN). The facial nerve runs just medial to the SAN. Therefore, the horizontal crest (HC), rather than Bill's bar (BB), is the safest landmark for identifying the facial nerve. *AC*, apical cells; *FNm*, mastoid portion of the facial nerve; *I*, incus; *JB*, jugular bulb; *MFD*, middle fossa dura; *SS*, sigmoid sinus; *V*, vestibule.

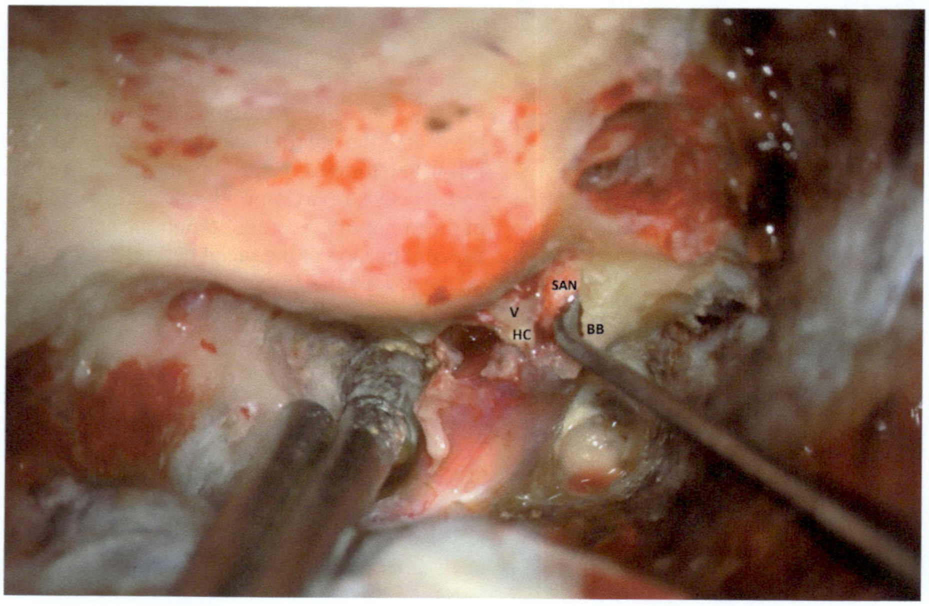

Fig. 7.75 The superior ampullary nerve (SAN) is detached from its canal. *BB*, Bill's bar; *HC*, horizontal crest; *V*, vestibule.

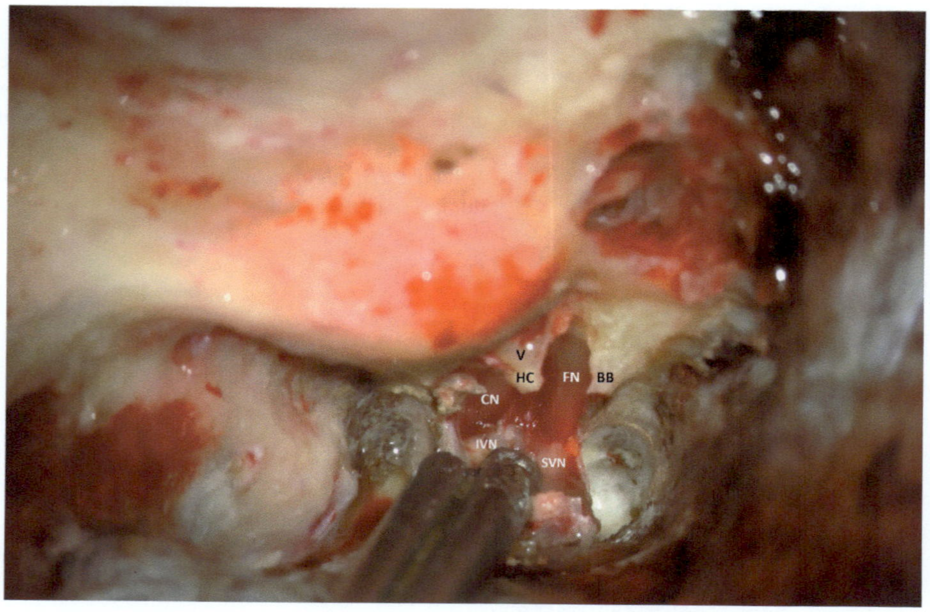

Fig. 7.76 The superior ampullary nerve joins the superior vestibular nerve (SVN). Detachment of both the superior and inferior vestibular nerves exposes the cochlear nerve (CN) and the facial nerve (FN) at the fundus of the internal auditory canal. *BB*, Bill's bar; *HC*, horizontal crest; *IVN*, inferior vestibular nerve; *V*, vestibule.

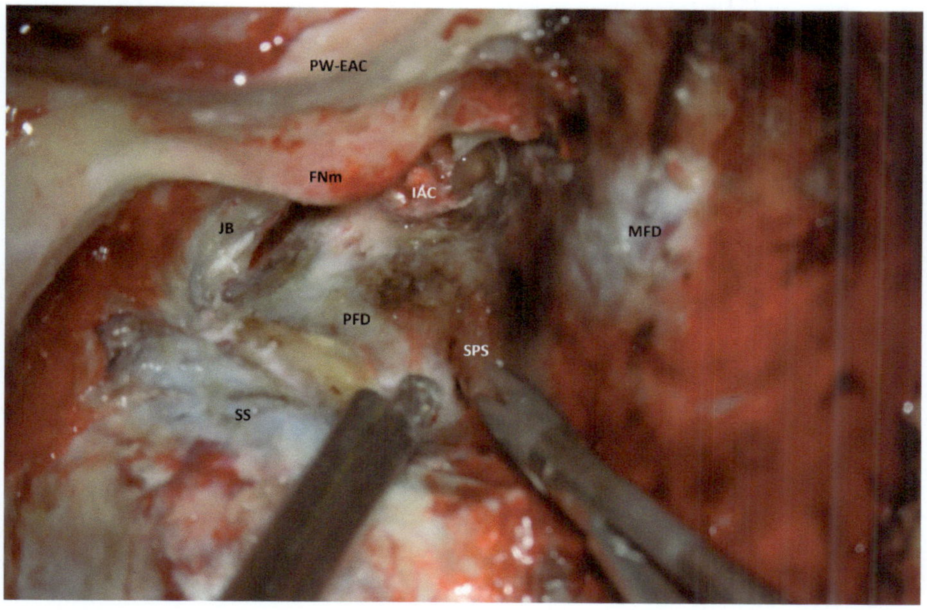

Fig. 7.77 After identifying the facial nerve (FN) at the fundus of the internal auditory canal (IAC), the posterior fossa dura (PFD) is opened. The first incision is made inferiorly and parallel to the superior petrosal sinus (SPS). *FNm*, mastoid portion of the facial nerve; *JB*, jugular bulb; *MFD*, middle fossa dura; *PW-EAC*, posterior wall of the external auditory canal; *SS*, sigmoid sinus.

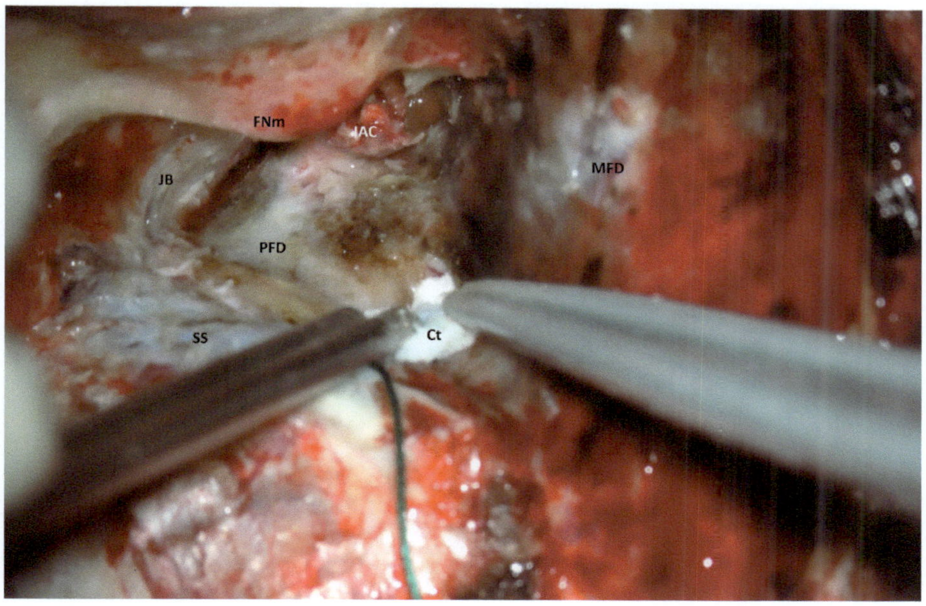

Fig. 7.78 After opening the posterior fossa dura (PFD), a cottonoid (Ct) is used to protect the cerebellum. *FNm*, mastoid portion of the facial nerve; *JB*, jugular bulb; *IAC*, internal auditory canal; *MFD*, middle fossa dura; *SS*, sigmoid sinus.

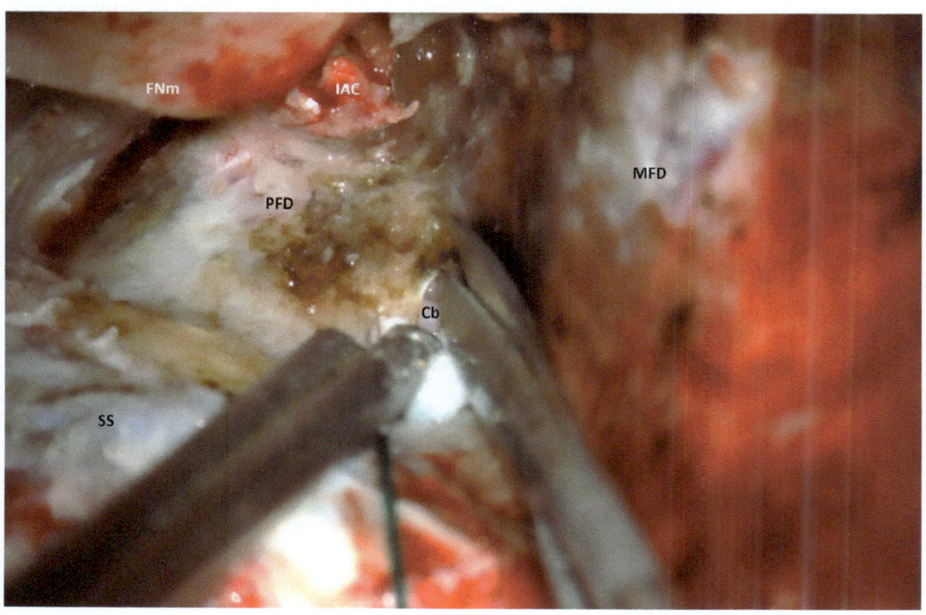

Fig. 7.79 Further anterior opening of the posterior fossa dura (PFD). *Cb*, cerebellum; *FNm*, mastoid portion of the facial nerve; *JB*, jugular bulb; *IAC*, internal auditory canal; *MFD*, middle fossa dura; *SS*, sigmoid sinus.

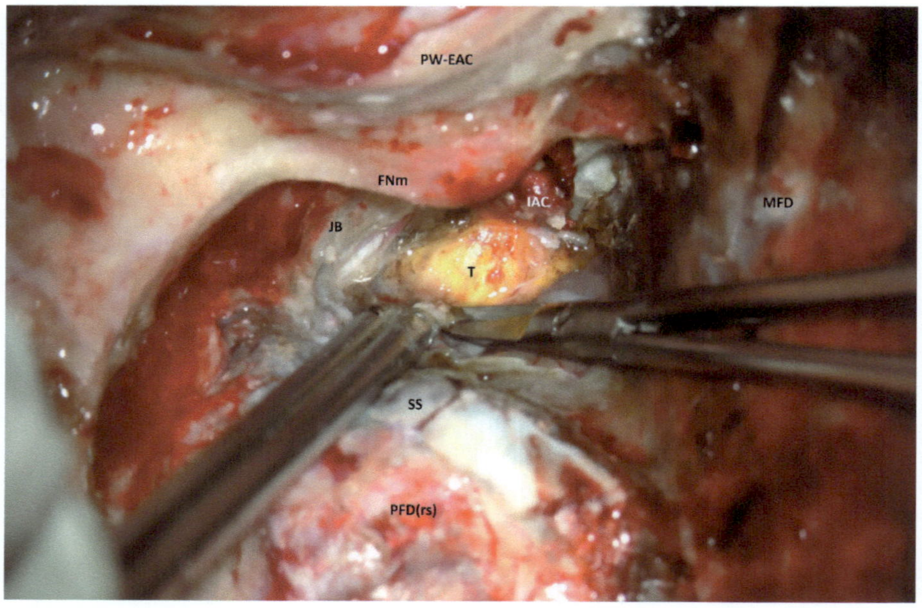

Fig. 7.80 The posterior fossa dura is completely opened and the tumor (T) exposed. The second dural incision is made almost perpendicular to the first one, following the direction of the sigmoid sinus (SS). *FNm*, mastoid portion of the facial nerve; *JB*, jugular bulb; *IAC*, internal auditory canal; *MFD*, middle fossa dura; *PW-EAC*, posterior wall of the external auditory canal.

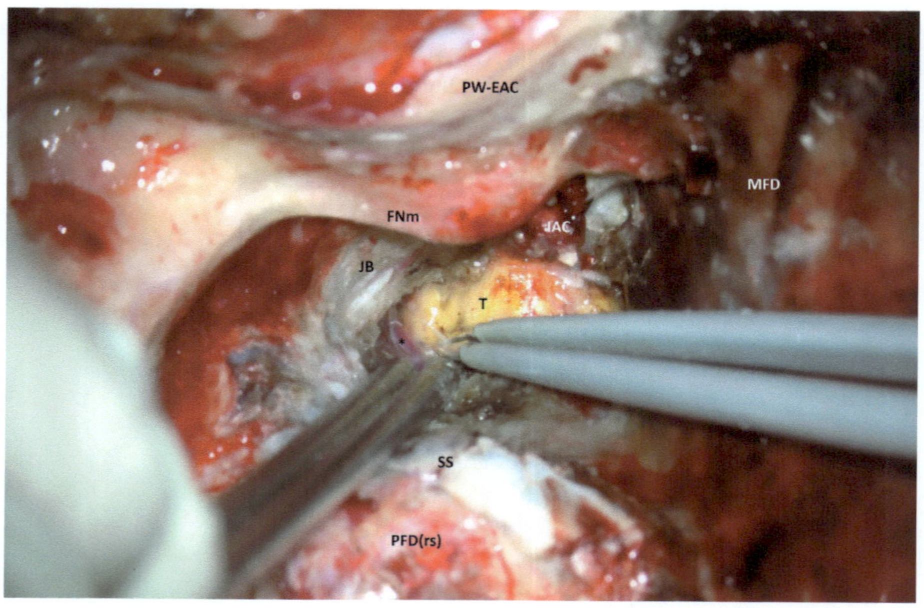

Fig. 7.81 The tumor (T) is dissected from the arachnoid. The tumor is surrounded by a venous plexus at its inferior pole (*). *FNm*, mastoid portion of the facial nerve; *IAC*, internal auditory canal; *JB*, jugular bulb; *MFD*, middle fossa dura; *PFD(rs)*, retrosigmoid posterior fossa dura; *PW-EAC*, posterior wall of the external auditory canal; *SS*, sigmoid sinus.

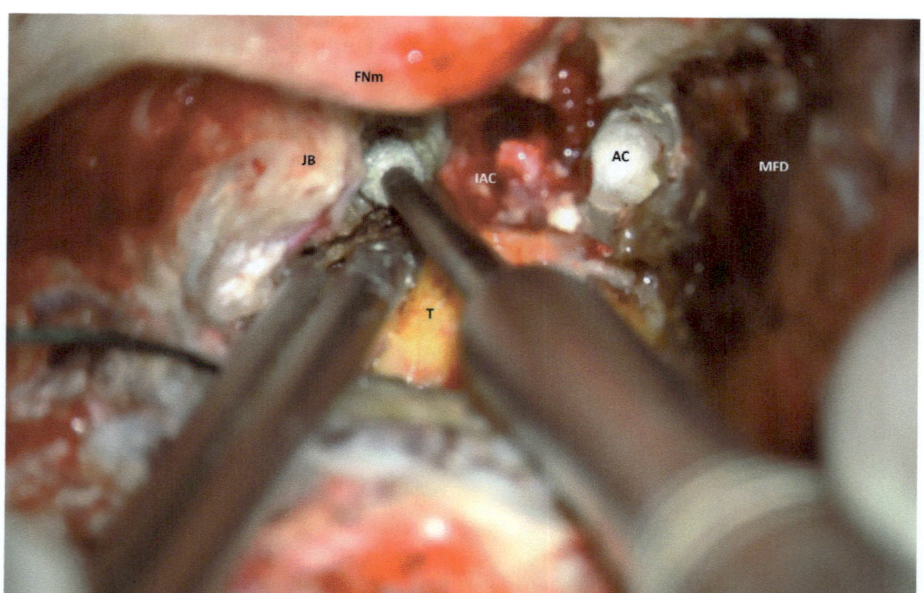

Fig. 7.82 Further transapical drilling is performed to improve control of the infero-medial pole of the tumor (T) within the cerebellopontine angle. *AC*, apical cells; *FNm*, mastoid portion of the facial nerve; *IAC*, internal auditory canal; *JB*, jugular bulb; *MFD*, middle fossa dura.

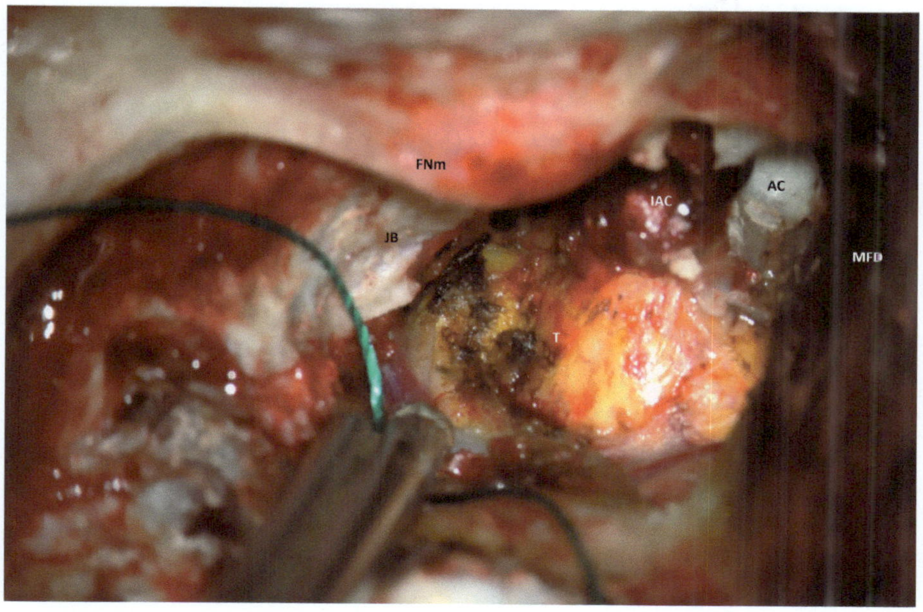

Fig. 7.83 Note the tumor exposure after drilling of the lower apical cells (AC). The tumor (T) is completely dissected from the arachnoid. The venous plexus (*) surrounding the tumor is partially coagulated. *AC*, apical cells; *FNm*, mastoid portion of the facial nerve; *IAC*, internal auditory canal; *JB*, jugular bulb; *MFD*, middle fossa dura.

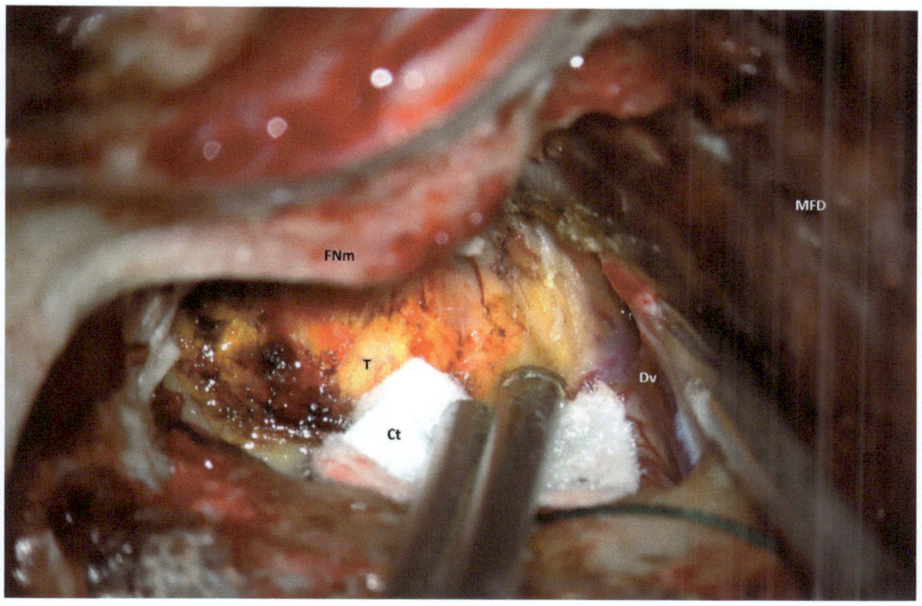

Fig. 7.84 The middle fossa dura (MFD) is gently elevated with visualization of the Dandy's vein (Dv). *Ct*, cottonoid; *FNm*, mastoid portion of the facial nerve; *T*, tumor.

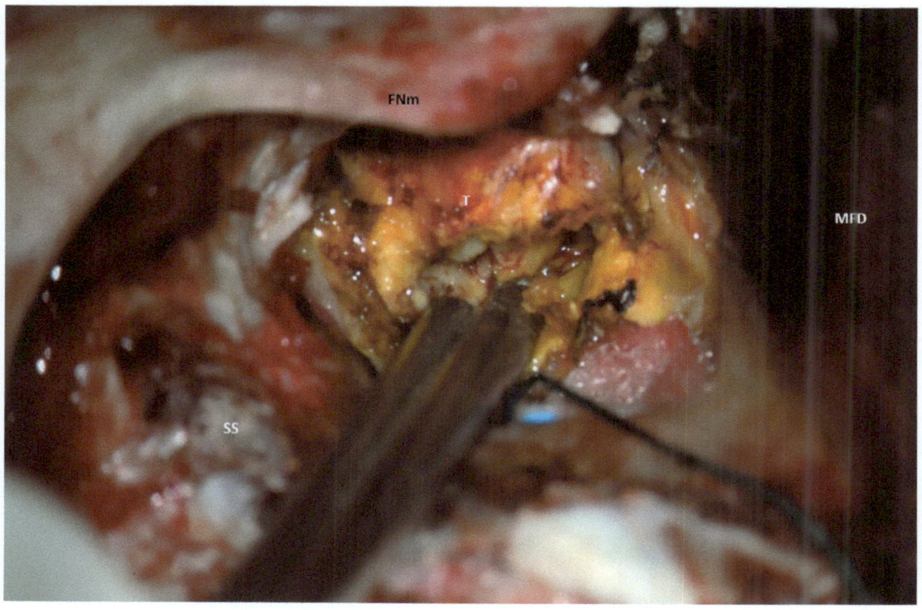

Fig. 7.85 Intratumoral debulking is performed. *FNm*, mastoid portion of the facial nerve; *MFD*, middle fossa dura; *T*, tumor.

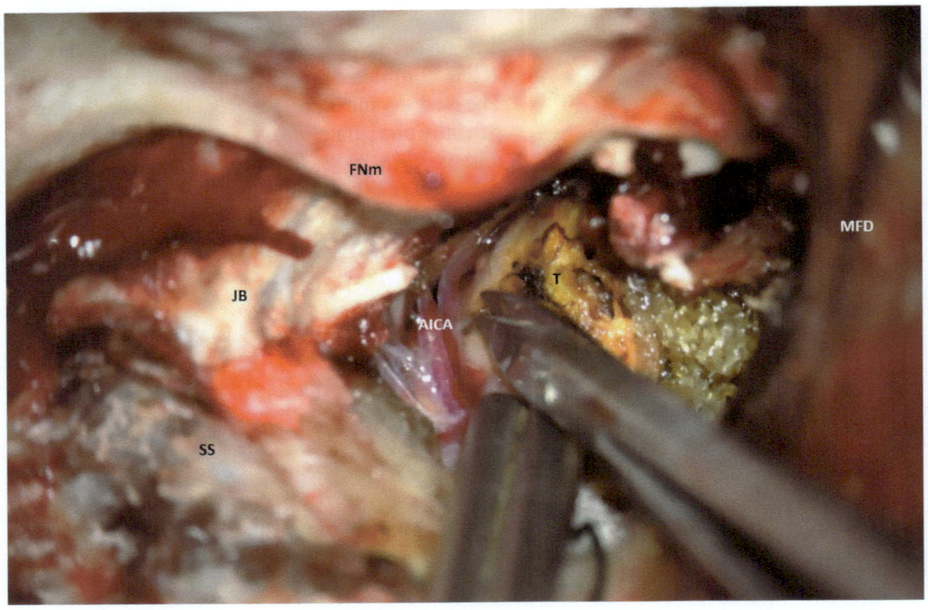

Fig. 7.86 After debulking, the tumor (T) is isolated from the neurovascular structures. *AICA*, anterior-inferior cerebellar artery; *FNm*, mastoid portion of the facial nerve; *JB*, jugular bulb; *MFD*, middle fossa dura; *SS*, sigmoid sinus.

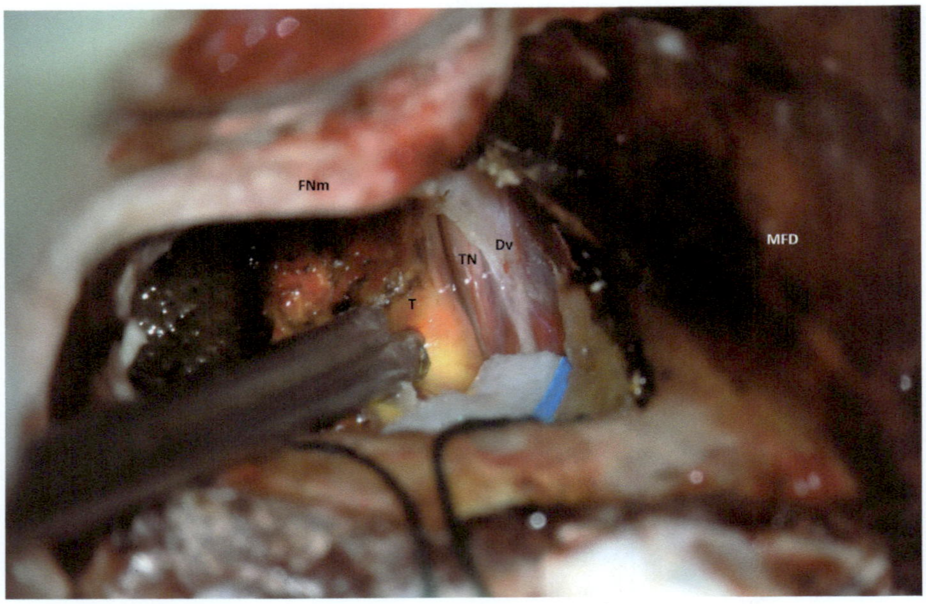

Fig. 7.87 Superiorly, the tumor (T) is dissected from both the Dandy's vein (Dv) and the trigeminal nerve (TN). *MFD*, middle fossa dura; *FNm*, mastoid portion of the facial nerve.

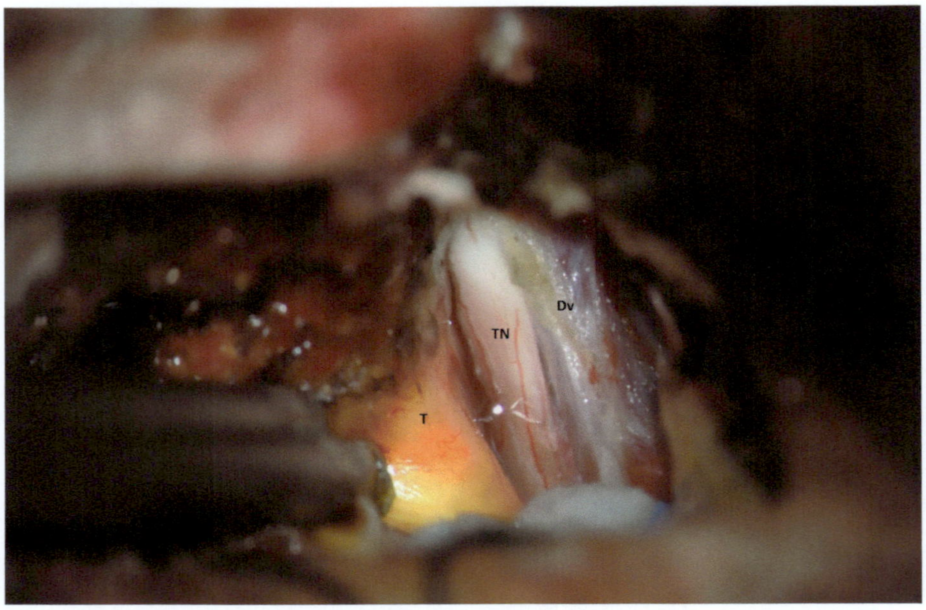

Fig. 7.88 Same view, but with higher magnification. *Dv*, Dandy' s vein; *T*, tumor; *TN*, trigeminal nerve.

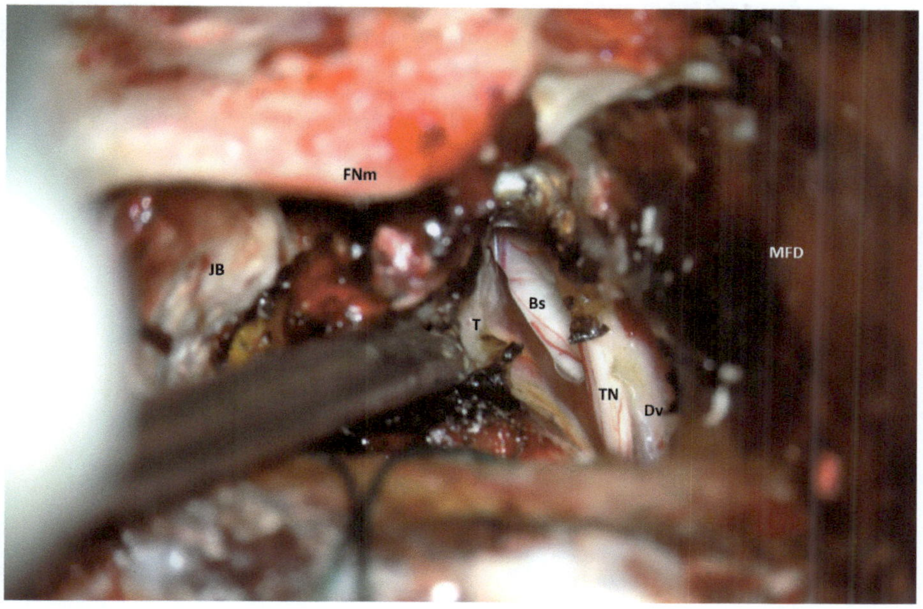

Fig. 7.89 The tumor (T) is further debulked and the superior pole dissected with visualization of the brainstem (Bs). *Dv*, Dandy's vein; *FNm*, mastoid portion of the facial nerve; *MFD*, middle fossa dura; *JB*, jugular bulb; *TN*, trigeminal nerve.

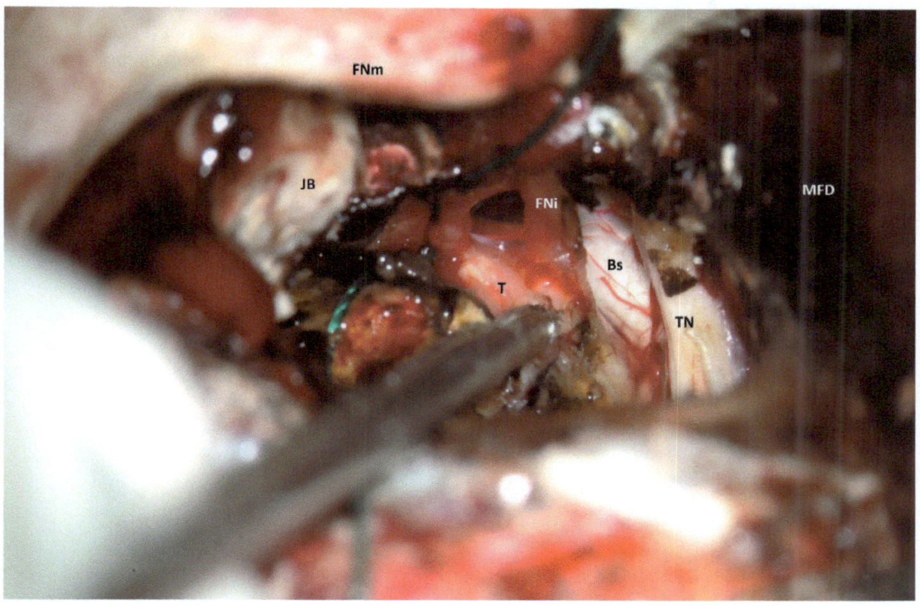

Fig. 7.90 The tumor (T) is dissected from the intradural part of the facial nerve (FNi) at the level of the porus. The Dandy's vein was coagulated and cut to avoid intradural bleeding after tumor retraction. *Bs*, brainstem; *FNm*, mastoid portion of the facial nerve; *JB*, jugular bulb; *MFD*, middle fossa dura; *TN*, trigeminal nerve.

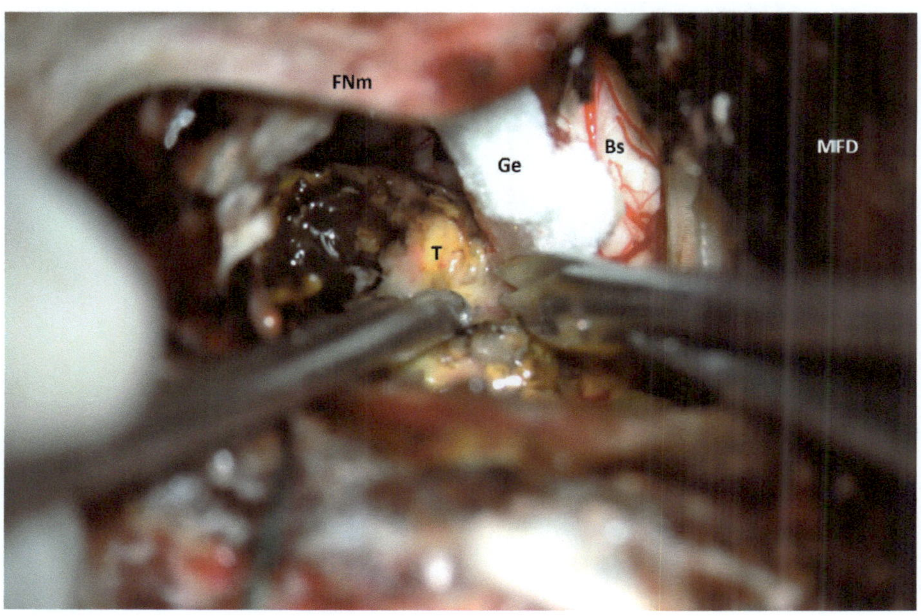

Fig. 7.91 The tumor (T) is further dissected from the intradural part of the facial nerve. Note that the facial nerve is protected with gelfoam (Ge). *Bs*, brainstem; *FNm*, mastoid portion of the facial nerve; *MFD*, middle fossa dura.

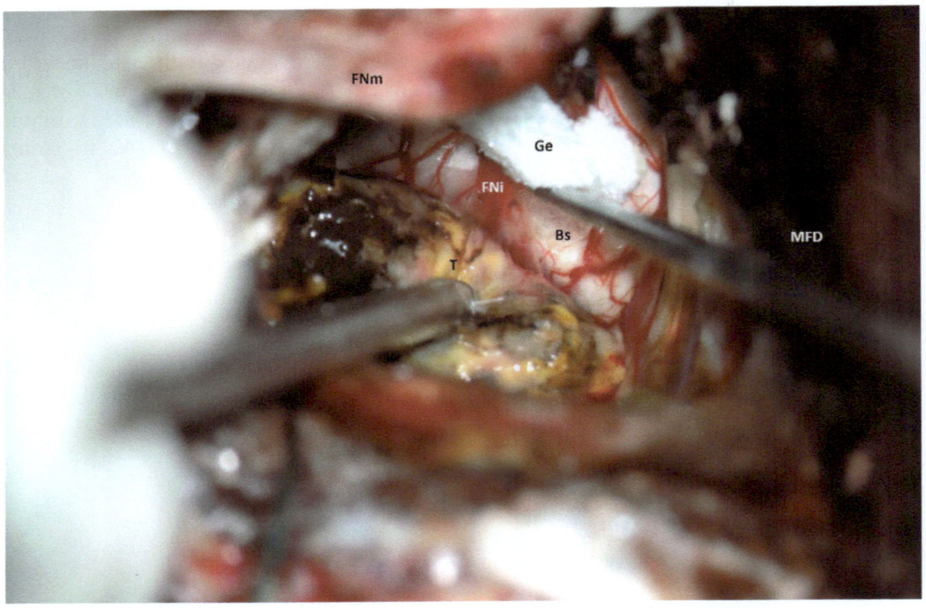

Fig. 7.92 The tumor (T) is dissected from the intradural part of the facial nerve (FNi) at the level of the root entry zone. *Bs*, brainstem; *FNm*, mastoid portion of the facial nerve; *Ge*, gelfoam; *MFD*, middle fossa dura.

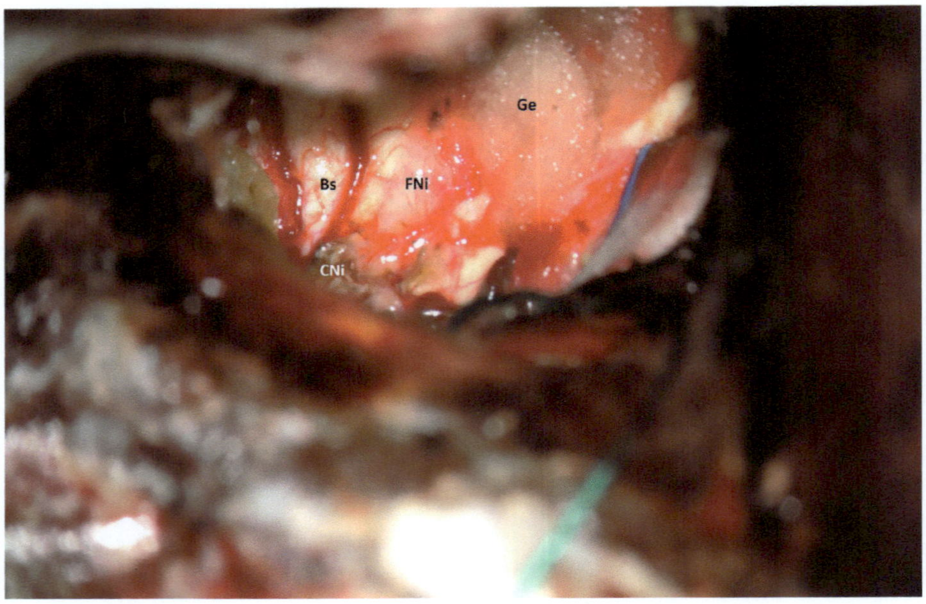

Fig. 7.93 The tumor was completely removed and the facial nerve anatomically and functionally preserved. Note the residual portion of the cochlear nerve (intradural part—CNi). *FNi*, intradural part of the facial nerve; *Bs*, brainstem; *Ge*, gelfoam.

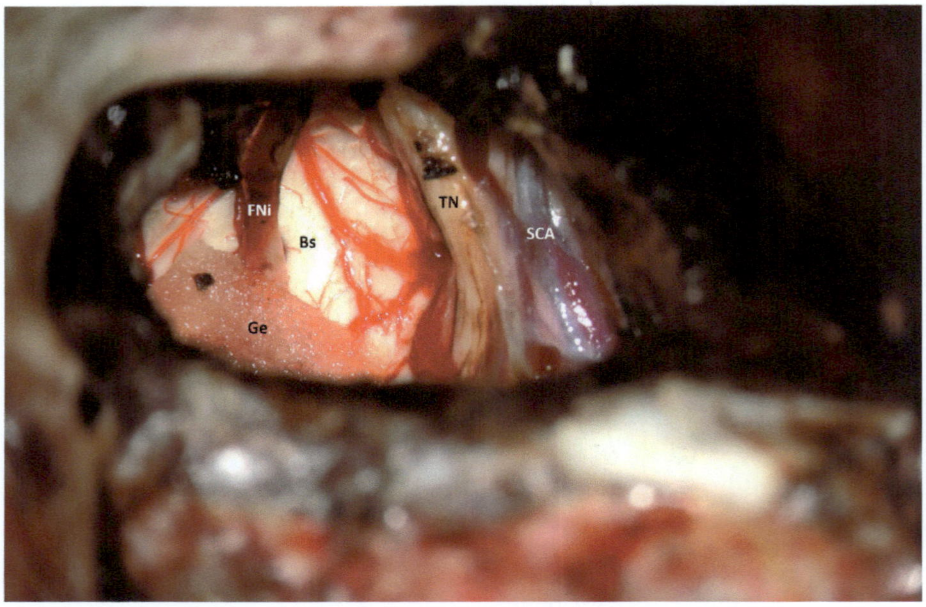

Fig. 7.94 Surgical view after complete removal of the tumor. *Bs*, brainstem; *FNi*, intradural portion of the facial nerve; *Ge*, gelfoam; *TN*, trigeminal nerve; *SCA*, superior cerebellar artery.

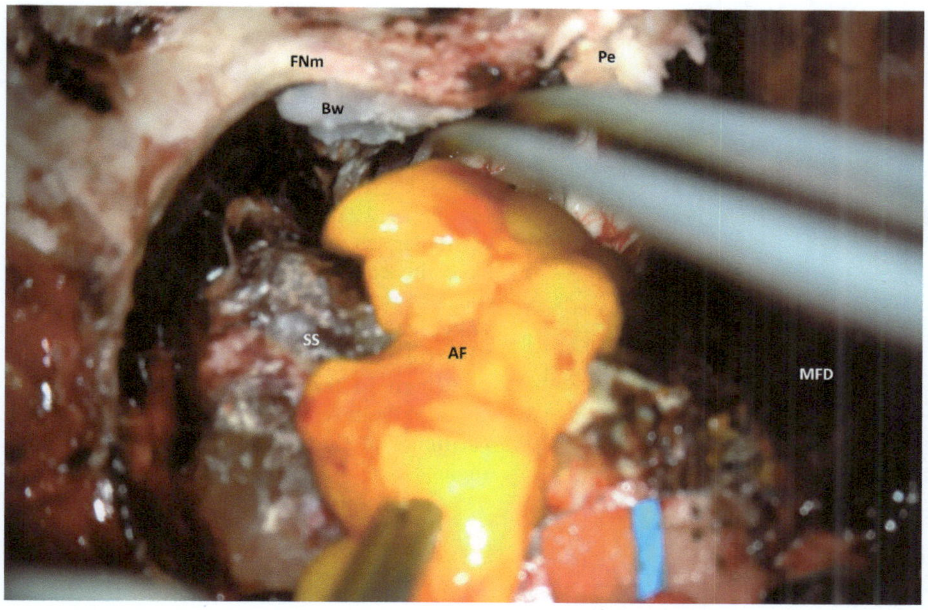

Fig. 7.95 After the incus has been removed, the attic is filled with periosteum (Pe). The cavity is then filled with strips of abdominal fat (AF). In addition, the retrofacial air cells are sealed with bone wax (Bw) to prevent cerebrospinal fluid from leaking into the middle ear. *FNm*, mastoid portion of the facial nerve; *MFD*, middle fossa dura; *SS*, sigmoid sinus.

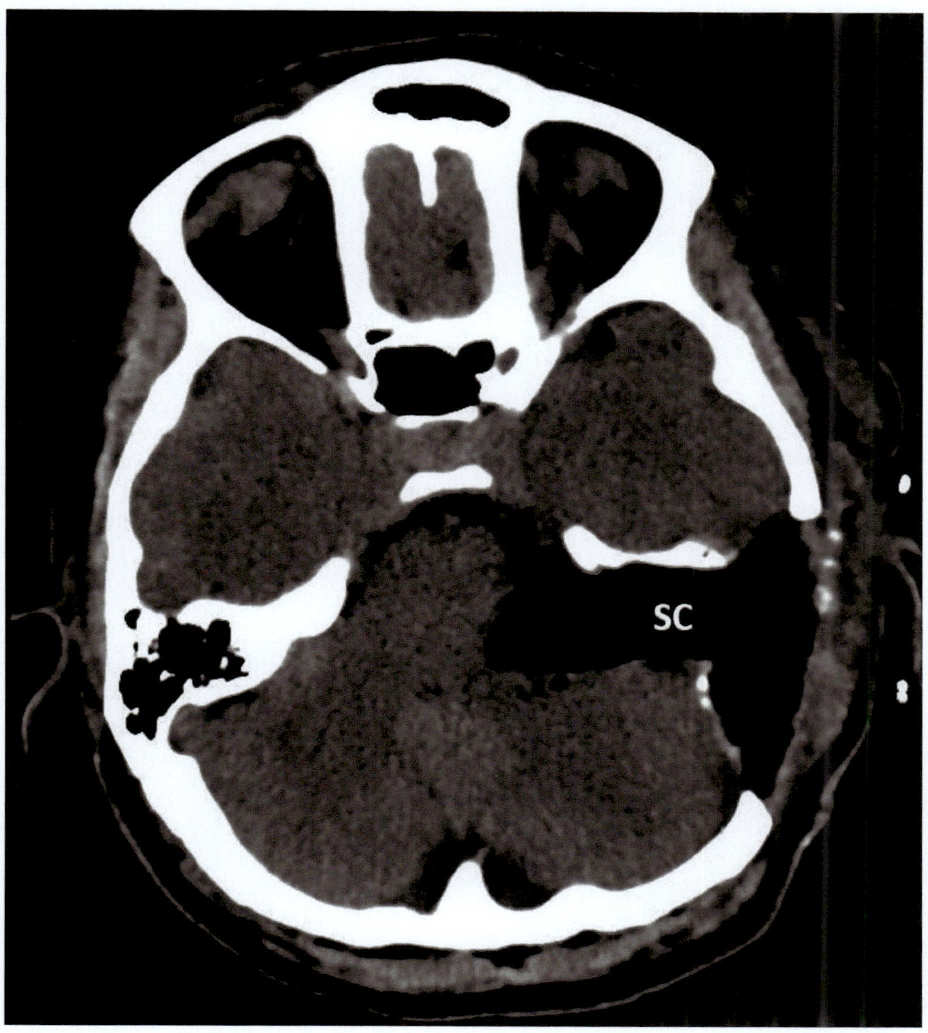

Fig. 7.96 Postoperative CT scan showing the surgical cavity (SC).

Case 2: Grade III Vestibular Schwannoma

See (Figs. 7.97–7.121).

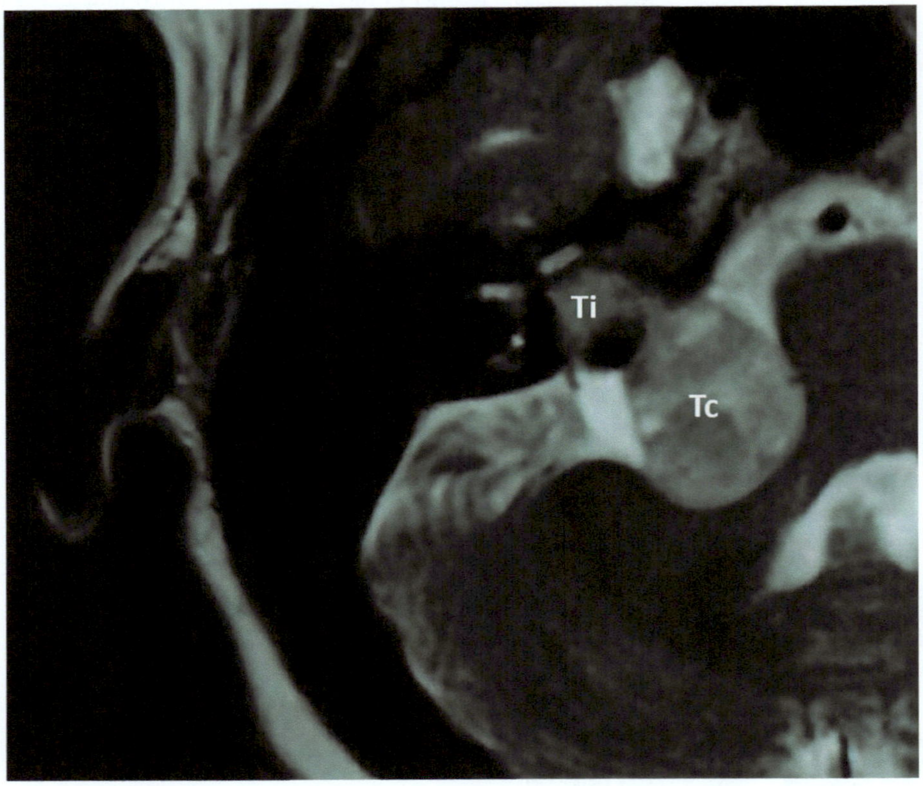

Fig. 7.97 T2W MRI of the brain shows a right vestibular schwannoma in a 40-year-old patient. Note the enlargement of the internal auditory canal caused by the tumor. According to the Tokyo consensus classification, this is a grade III vestibular schwannoma (maximum diameter 21–30 mm in the cerebellopontine angle). *Ti*, tumor at the level of the internal auditory canal; *Tc*, tumor at the level of the cerebellopontine angle.

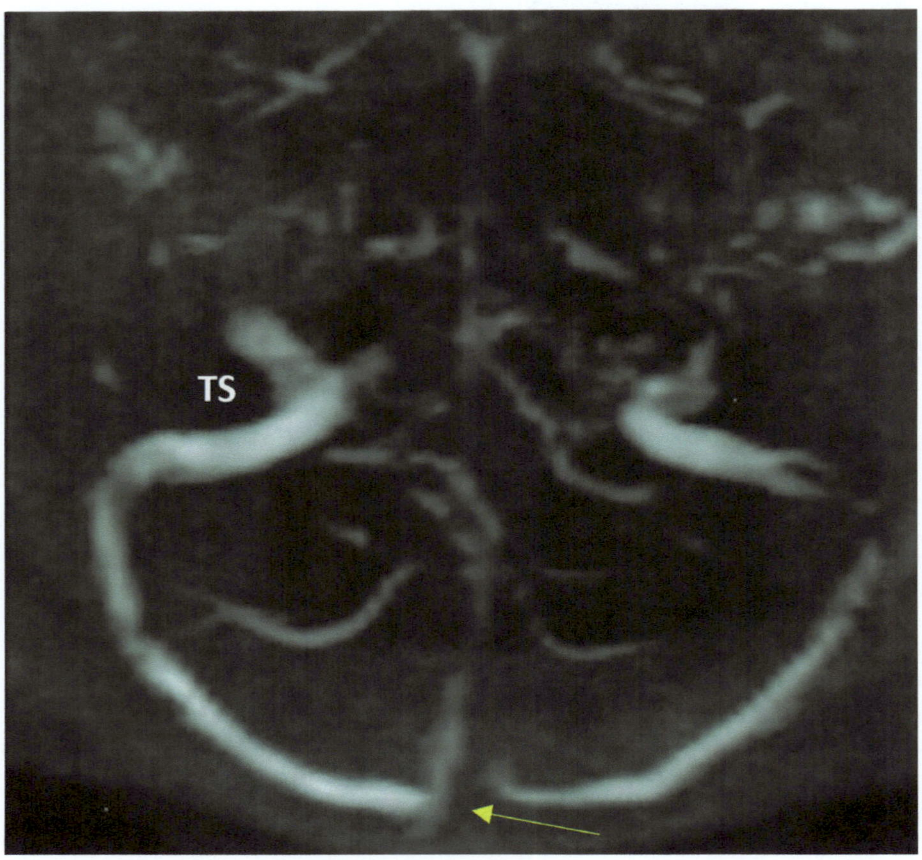

Fig. 7.98 MR-Venography shows a dominant cerebral venous flow on the tumor side (TS), with a weak connection at the level of the torcular (*arrow*). To avoid damage or excessive compression of the dominant venous system (at the level of the transverse sinus, sigmoid sinus and jugular bulb) an enlarged translabyrinthine approach with closure of the external auditory canal was chosen.

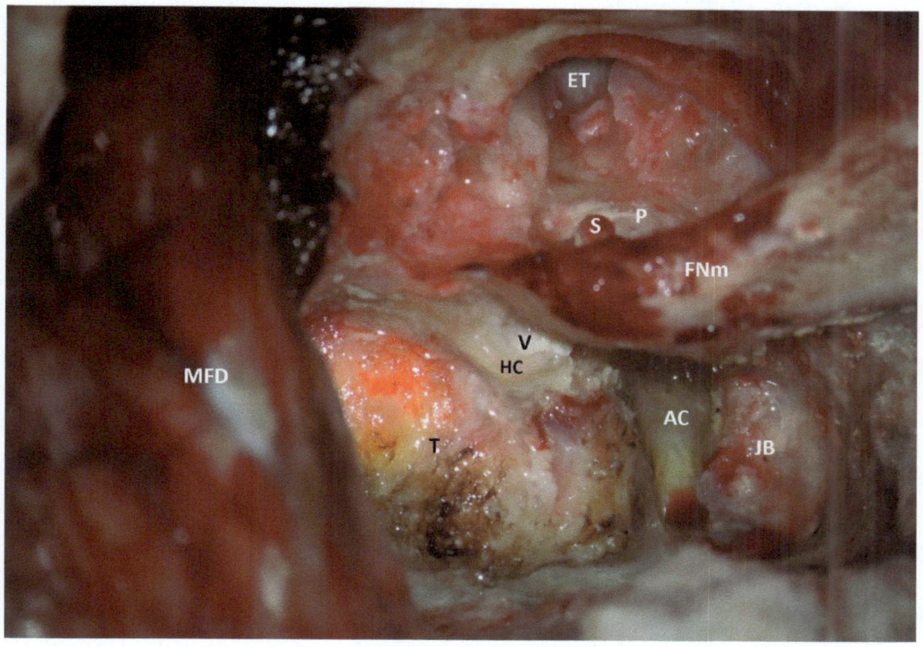

Fig. 7.99 The surgical approach has been completed and the tumor (T) in the internal auditory canal (IAC) fully exposed. Note the enlargement of the IAC, with the horizontal crest (HC) being the only landmark for identification of the facial nerve at the fundus. *AC*, apical cells; *ET*, eustachian tube; *FNm*, mastoid portion of the facial nerve; *JB*, jugular bulb; *MFD*, middle fossa dura; *P*, promontory; *S*, stapes; *V*, vestibule.

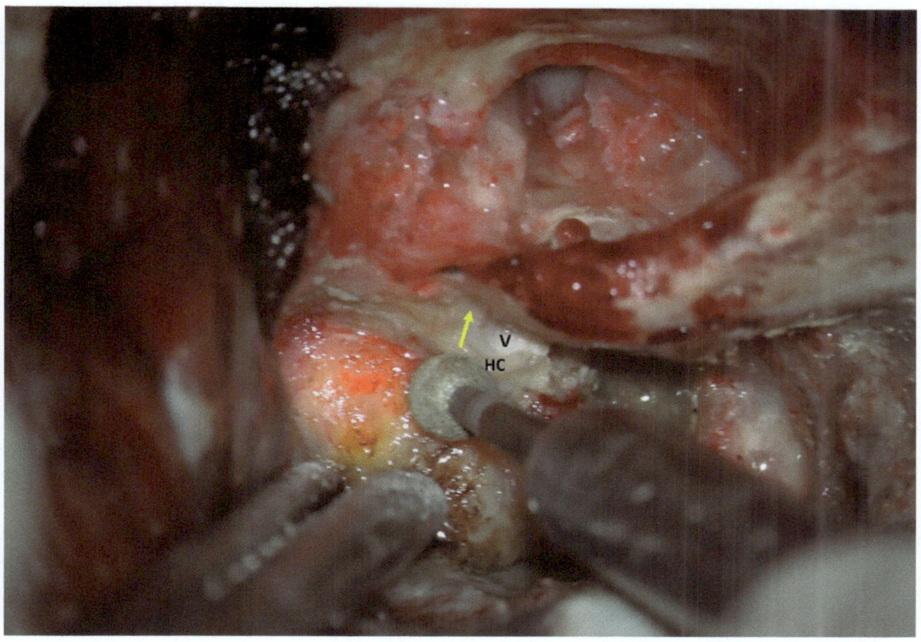

Fig. 7.100 Identification of the superior ampullary nerve starts drilling the superior portion of the vestibule (V). The *arrow* indicates the direction of drilling. *HC*, horizontal crest.

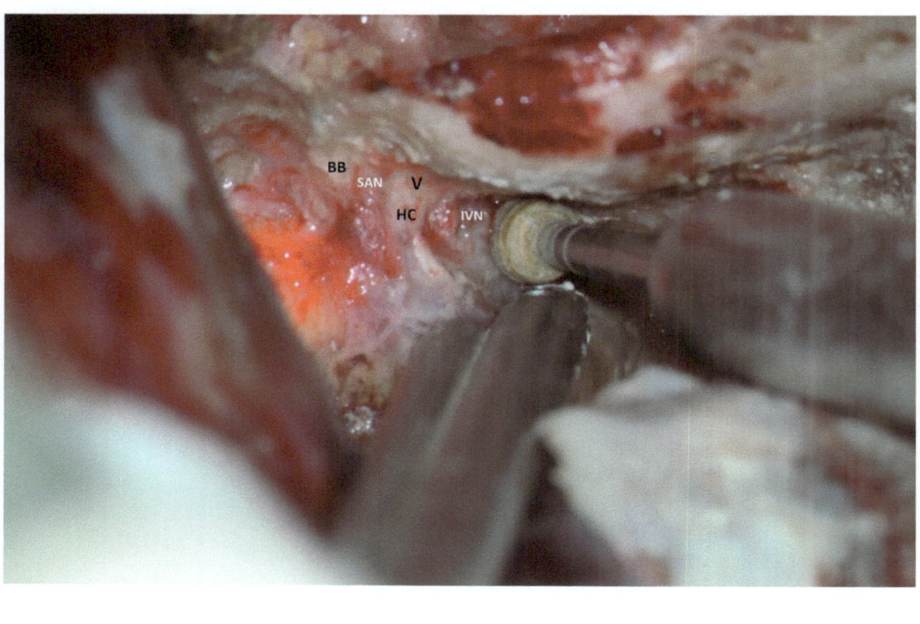

Fig. 7.101 The superior ampullary nerve (SAN) is identified. The inferior vestibular nerve (IVN) is exposed by drilling the inferior part of the horizontal crest (HC). *BB*, Bill's bar; *V*, vestibule.

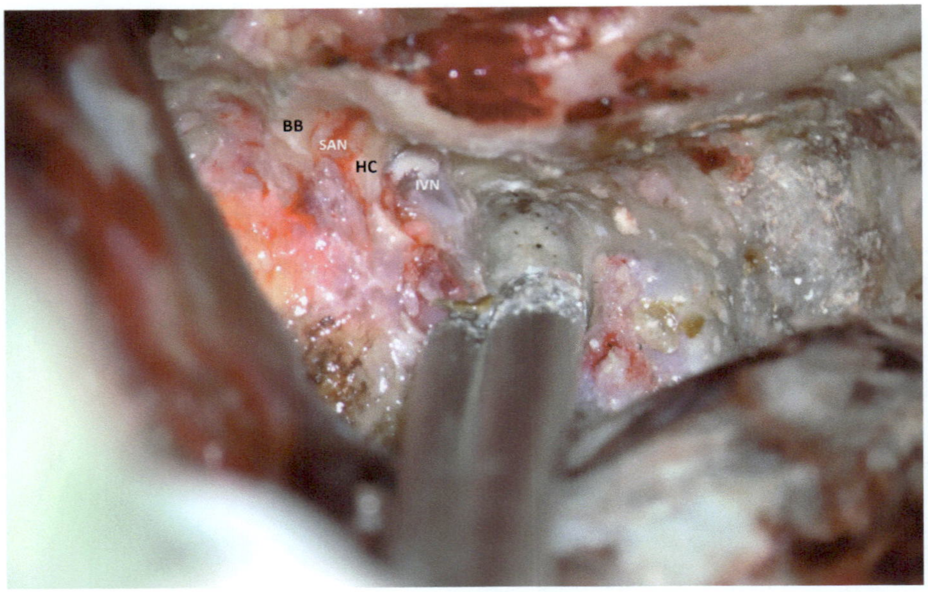

Fig. 7.102 Surgical view after drilling of the lower apical cells. *BB*, bill's bar; *HC*, horizontal crest; *IVN*, inferior vestibular nerve; *SAN*, superior ampullary nerve.

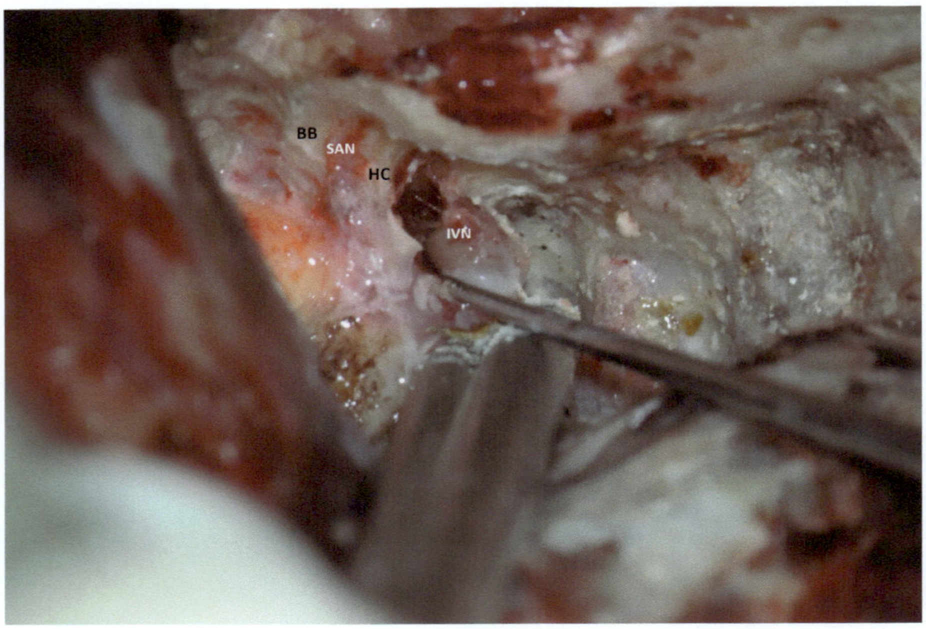

Fig. 7.103 The inferior vestibular nerve (IVN) is detached from its canal. Both the IVN and the cochlear nerve are infiltrated by the tumor. *BB*, bill's bar; *HC*, horizontal crest; *SAN*, superior ampullary nerve.

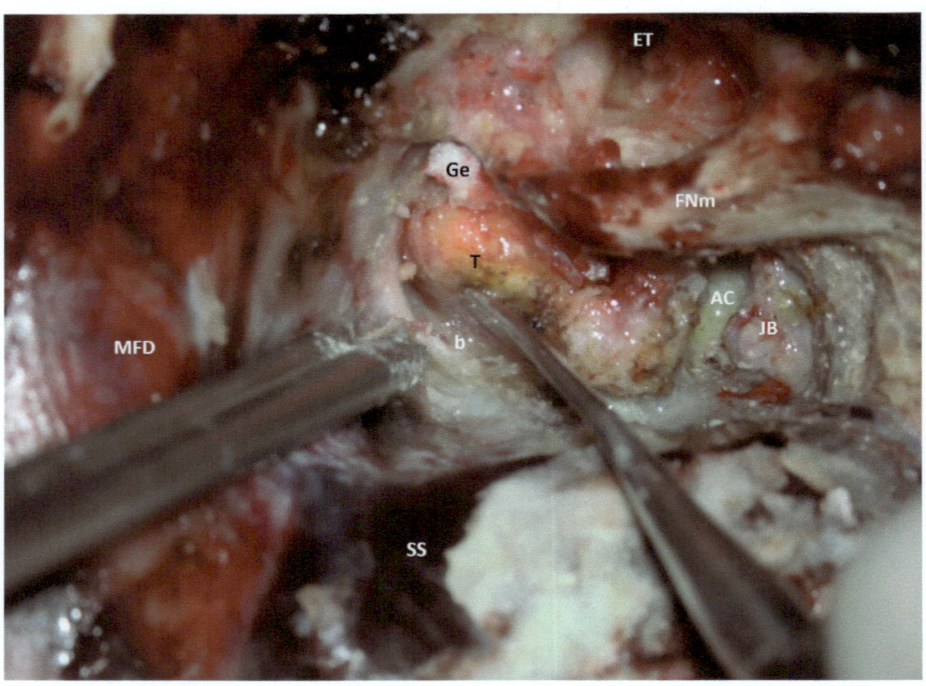

Fig. 7.104 The facial nerve has also been identified at the fundus of the internal auditory canal and protected with Gelfoam (Ge). The tumor (T) still hides the bone (b) at the posterior fossa dura. *AC*, apical cells; *ET*, eustachian tube; *FNm*, mastoid portion of the facial nerve; *JB*, jugular bulb; *MFD*, middle fossa dura; *SS*, sigmoid sinus.

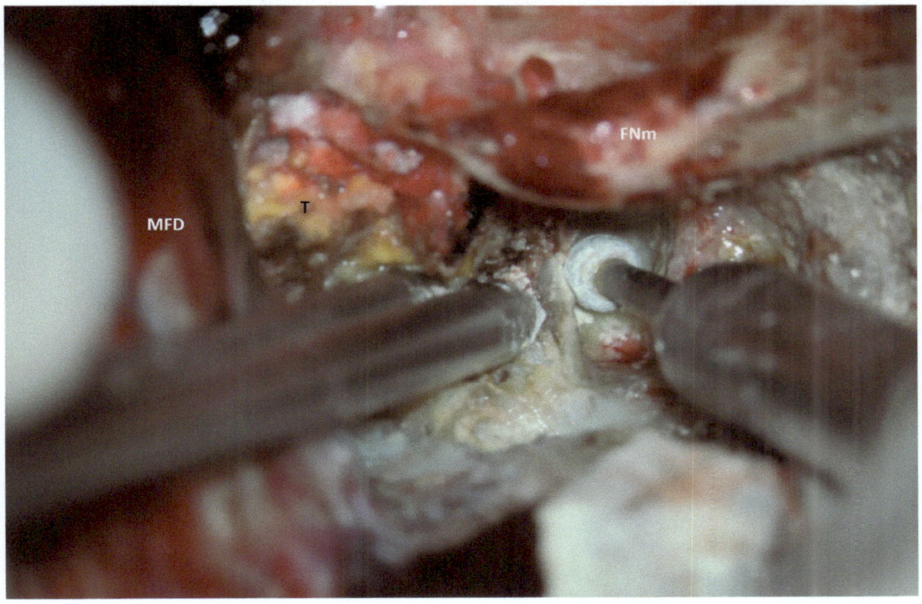

Fig. 7.105 The lower apical cells are further drilled. *FNm*, mastoid portion of the facial nerve; *MFD*, middle fossa dura; *T*, tumor.

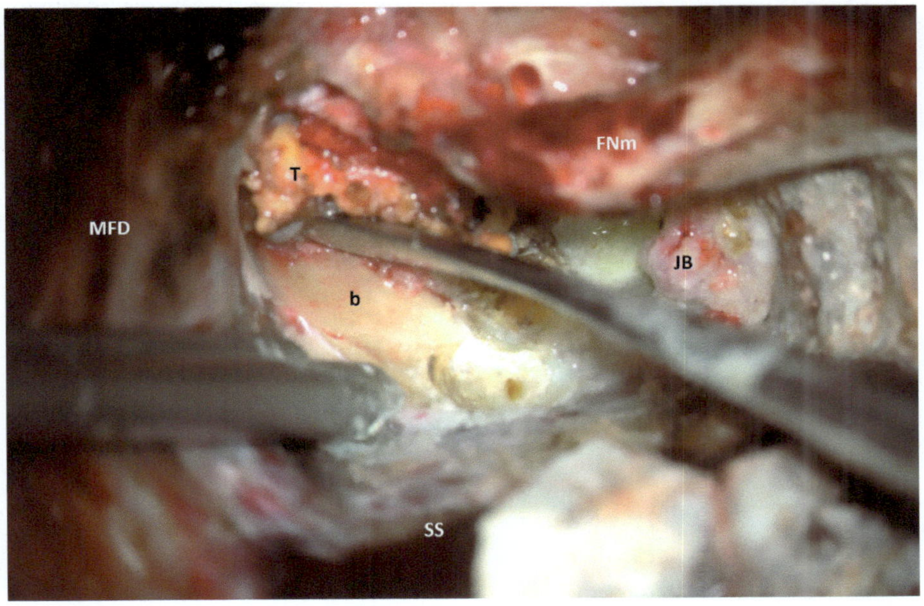

Fig. 7.106 Debulking of the tumor at the internal auditory canal. Note the amount of bone (previously hidden by the tumor) covering the posterior fossa dura. *b*, bone covering the posterior fossa dura; *FNm*, mastoid portion of the facial nerve; *JB*, jugular bulb; *MFD*, middle fossa dura; *SS*, sigmoid sinus; *T*, tumor.

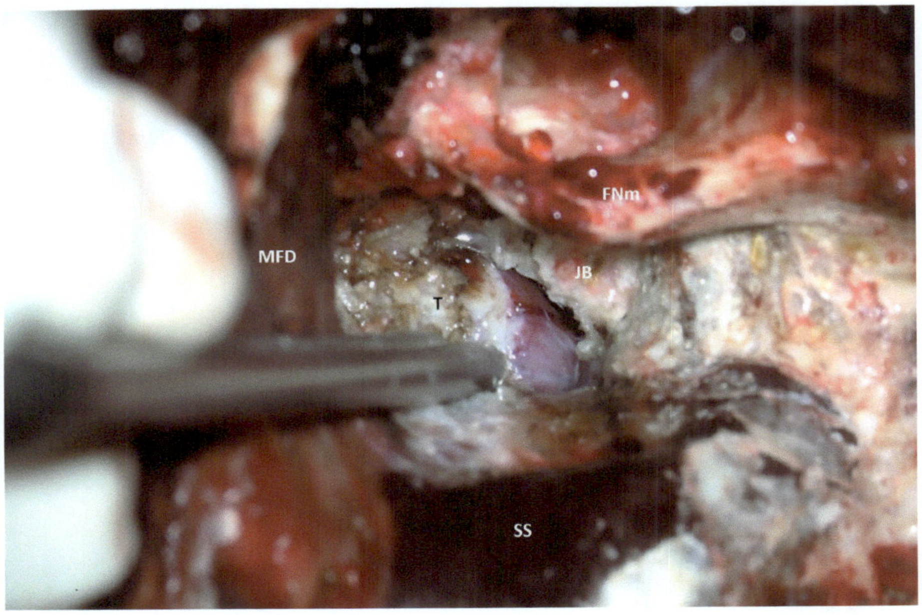

Fig. 7.107 After complete bone removal, the posterior fossa dura is opened and the tumor (T) at the cerebellopontine angle is exposed. *FNm*, mastoid portion of the facial nerve; *JB*, jugular bulb; *MFD*, middle fossa dura; *SS*, sigmoid sinus.

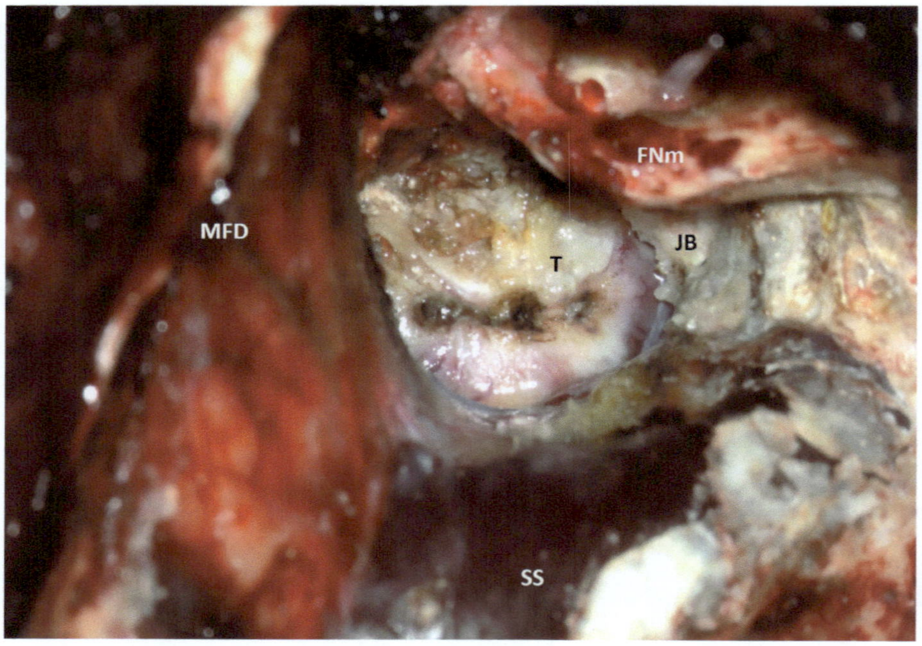

Fig. 7.108 The surface of the tumor (T) is coagulated before debulking. *FNm*, mastoid portion of the facial nerve; *JB*, jugular bulb; *MFD*, middle fossa dura; *SS*, sigmoid sinus.

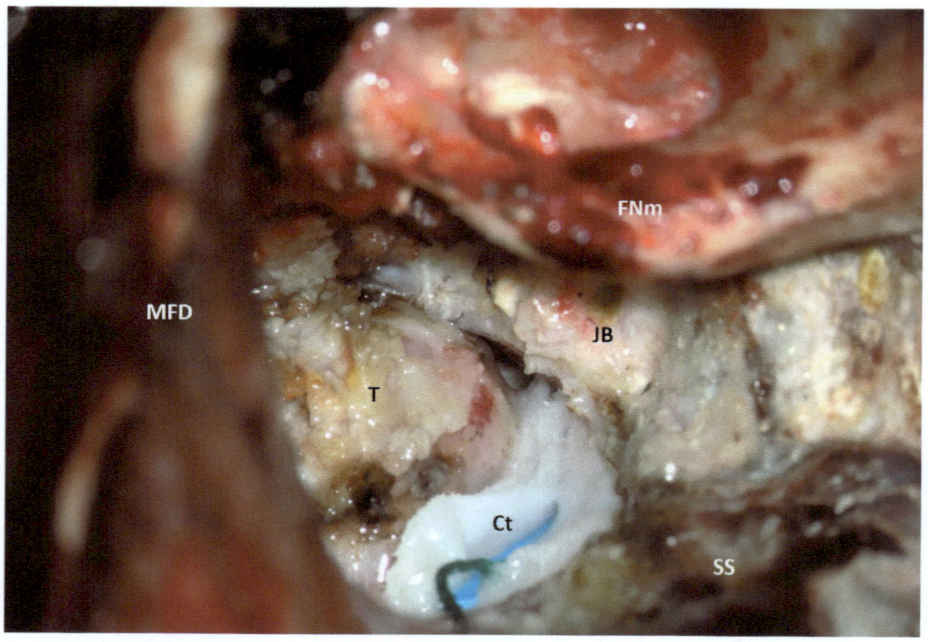

Fig. 7.109 A cottonoid (Ct) is placed inferiorly to protect the area of the lower cranial nerves. *FNm*, mastoid portion of the facial nerve; *MFD*, middle fossa dura; *JB*, jugular bulb; *SS*, sigmoid sinus; *T*, tumor.

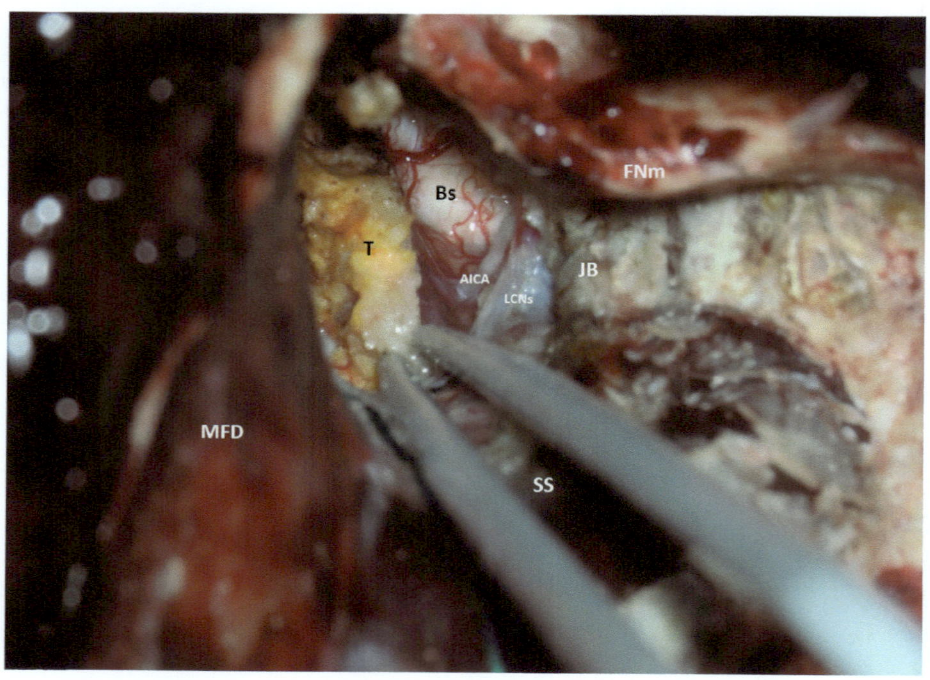

Fig. 7.110 The tumor is reduced and the brainstem (Bs) is visible. Note the anterior-inferior cerebellar artery (AICA) and its branches feeding the tumor. *FNm*, mastoid portion of the facial nerve; *JB*, jugular bulb; *LCNs*, lower cranial nerves; *MFD*, middle fossa dura; *SS*, sigmoid sinus.

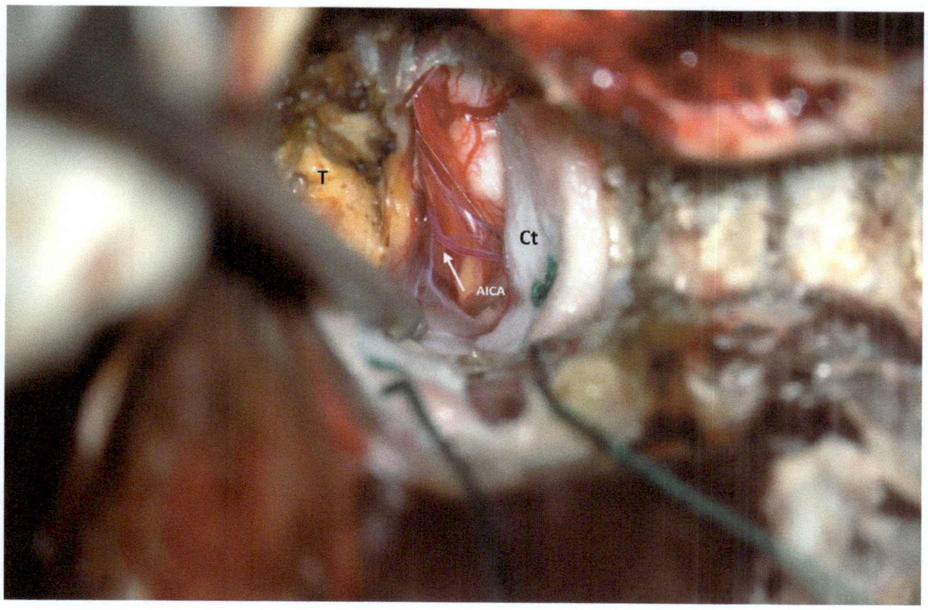

Fig. 7.111 Closer view. Note one of the branches of the anterior-inferior cerebellar artery (AICA) feeding the tumor (T). A cottonoid (Ct) protects the area of the lower cranial nerves.

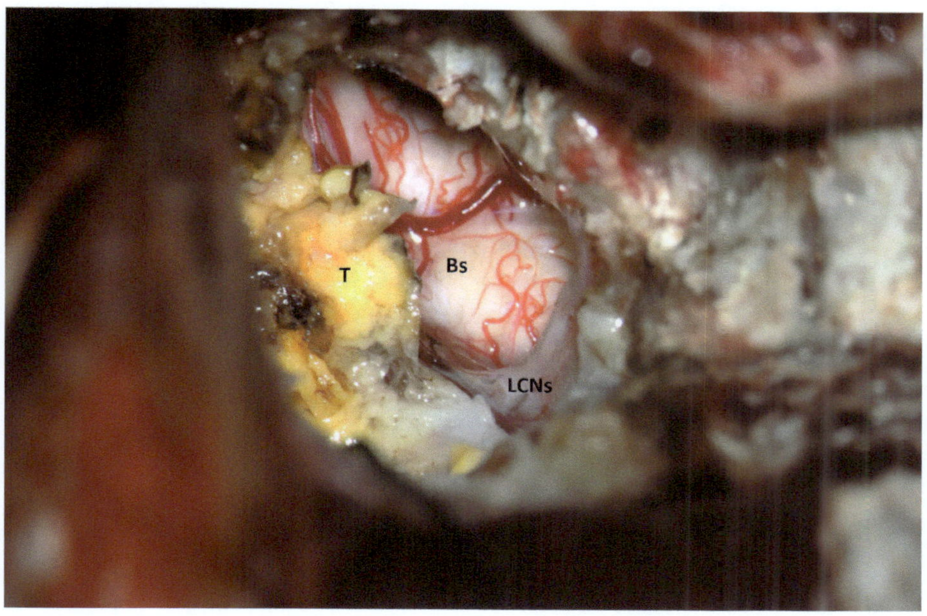

Fig. 7.112 The tumor (T) was devascularised and the anterior-inferior cerebellar artery was displaced. *Bs*, brainstem; *LCNs*, lower cranial nerves.

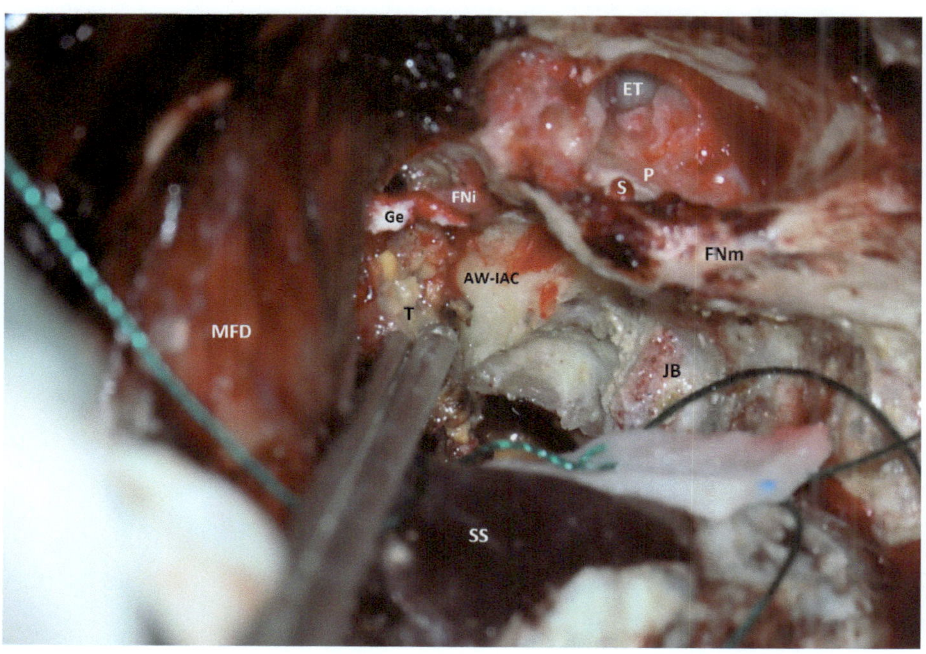

Fig. 7.113 Once the tumor (T) has been reduced in the cerebellopontine angle, the dissection continues at the level of the internal auditory canal to free the facial nerve. *AW-IAC*, anterior wall of the internal auditory canal; *ET*, eustachian tube; *FNi*, internal auditory canal portion of the facial nerve (at the level of the fundus); *FNm*, mastoid portion of the facial nerve; *Ge*, gelfoam; *JB*, jugular bulb; *MFD*, middle fossa dura; *P*, promontory; *S*, stapes; *SS*, sigmoid sinus.

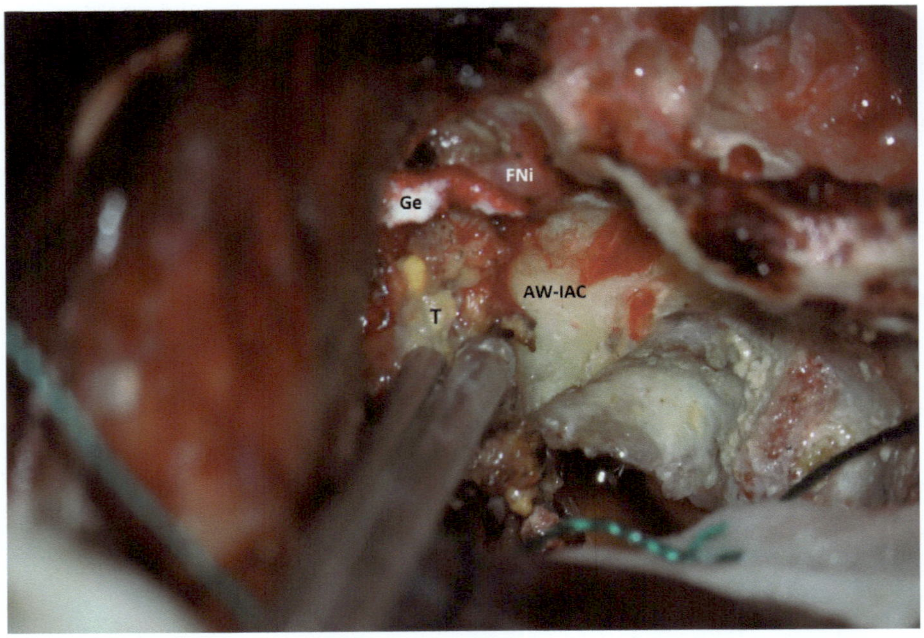

Fig. 7.114 Closer view. *AW-IAC*, anterior wall of the internal auditory canal; *FNi*, internal auditory canal portion of the facial nerve (at the level of the fundus); *Ge*, gelfoam; *T*, tumor.

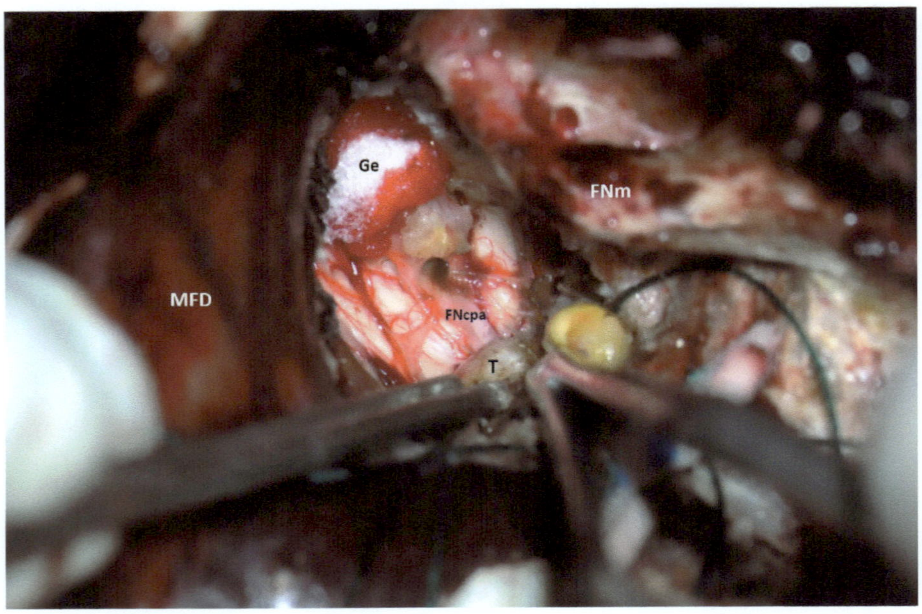

Fig. 7.115 The tumor is further dissected from the facial nerve, proceeding from the internal auditory canal to the cerebellopontine angle. *FNcpa*, cerebellopontine angle portion of the facial nerve; *FNm*, mastoid portion of the facial nerve; *Ge*, gelfoam; *MFD*, middle fossa dura; *T*, tumor.

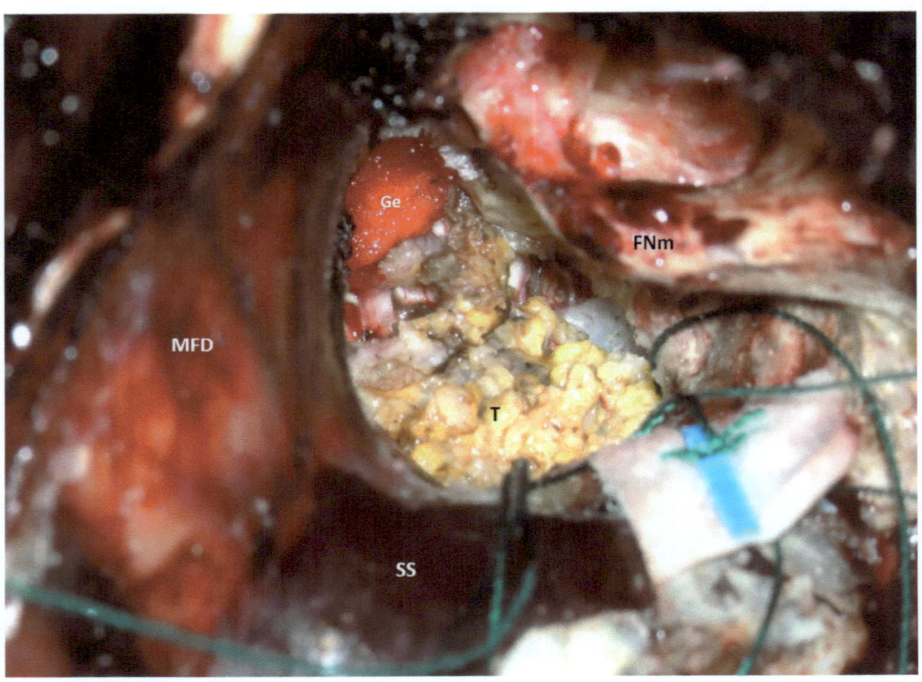

Fig. 7.116 The tumor remnant is reflected superiorly to further check the facial nerve at the level of the root entry zone. *FNm*, mastoid portion of the facial nerve; *Ge*, gelfoam; *MFD*, middle fossa dura; *SS*, sigmoid sinus; *T*, tumor.

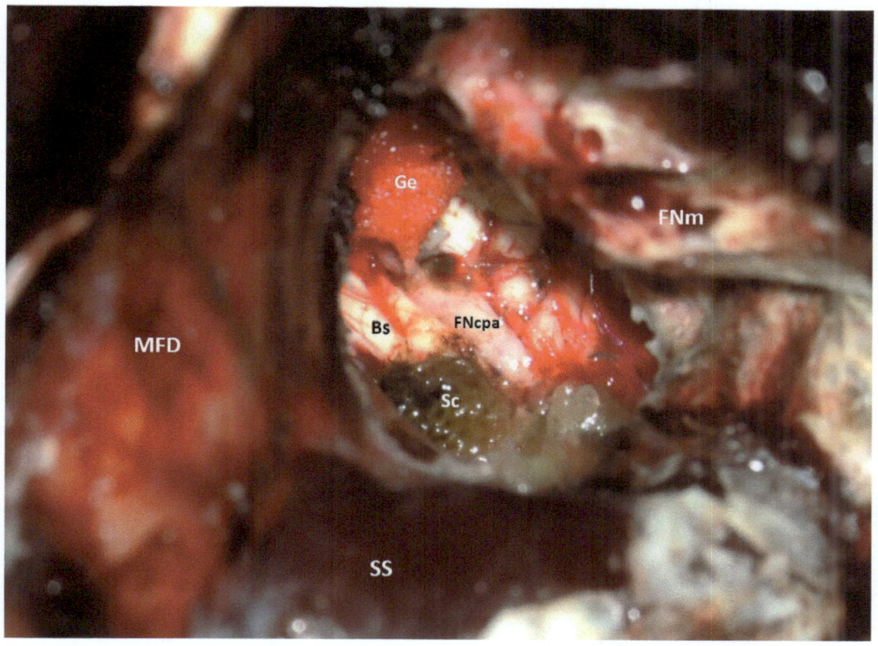

Fig. 7.117 The tumor is removed completely. A piece of surgicel (Sc) is placed on the cerebellum to prevent bleeding from its surface. The facial nerve is anatomically preserved. Gelfoam protects the facial nerve at the level of the internal auditory canal. *Bs*, brainstem; *FNm*, mastoid portion of the facial nerve; *FNcpa*, cerebellopontine angle portion of the facial nerve; *Ge*, gelfoam; *MFD*, middle fossa dura; *SS*, sigmoid sinus.

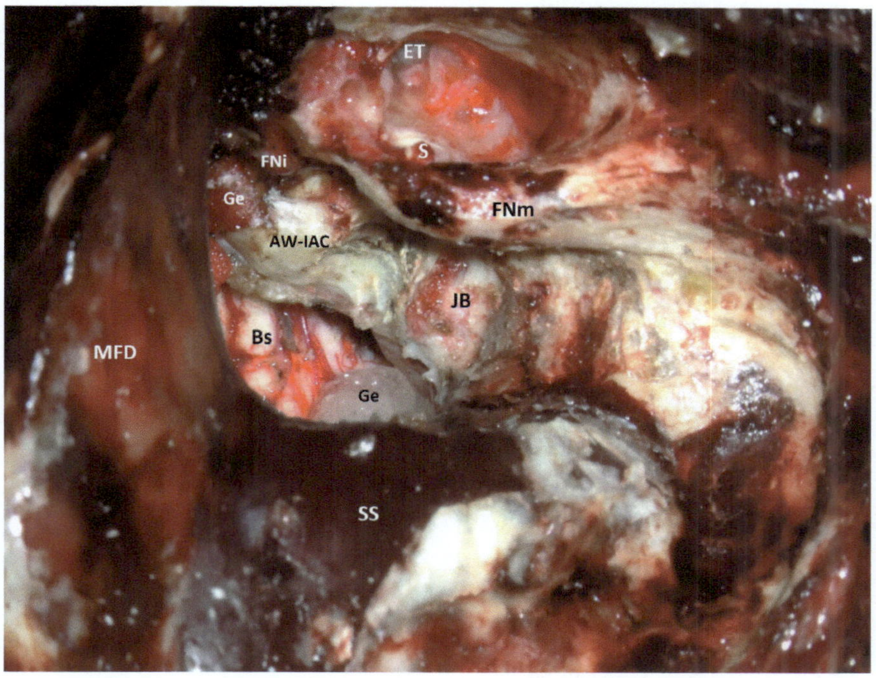

Fig. 7.118 View of the surgical cavity at the end of the procedure. The facial nerve is protected with gelfoam and runs in a superior-medial direction. *AW-IAC*, anterior wall of the internal auditory canal; *Bs*, brainstem; *ET*, eustachian tube; *FNi*, internal auditory canal portion of the facial nerve (at the fundus); *FNm*, mastoid portion of the facial nerve; *JB*, jugular bulb; *MFD*, middle fossa dura; *S*, stapes; *SS*, sigmoid sinus.

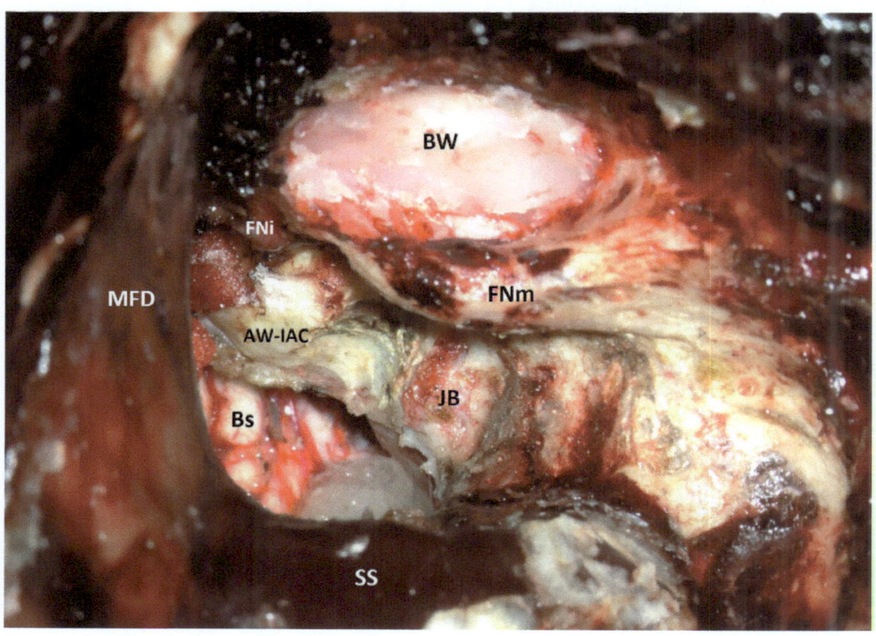

Fig. 7.119 The eustachian tube is closed with bone wax (BW). *AW-IAC*, anterior wall of the internal auditory canal; *Bs*, brainstem; *FNi*, internal auditory canal portion of the facial nerve (at the fundus); *FNm*, mastoid portion of the facial nerve; *JB*, jugular bulb; *MFD*, middle fossa dura; *SS*, sigmoid sinus.

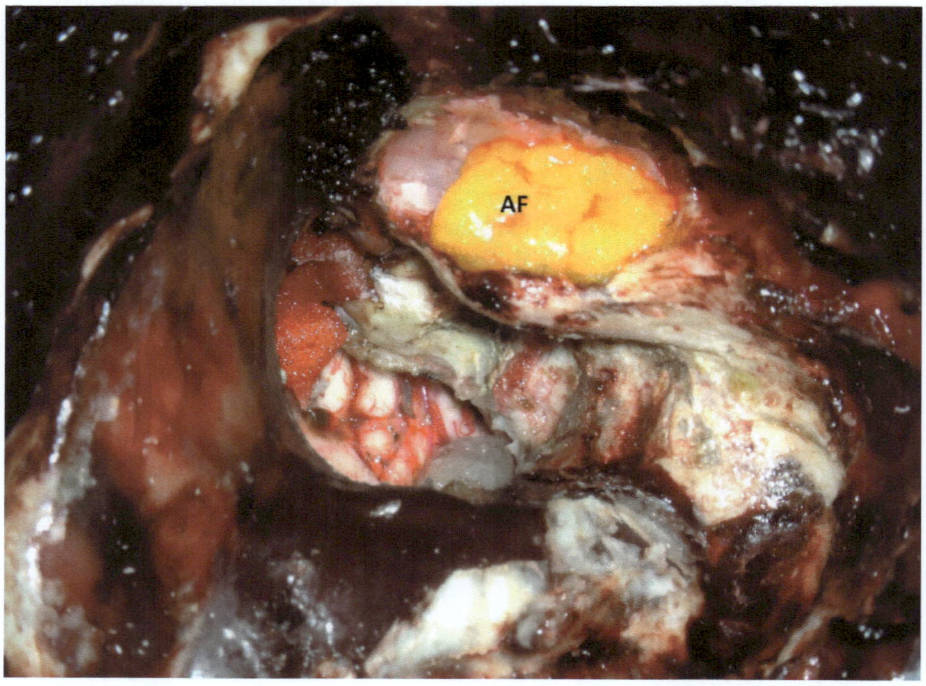

Fig. 7.120 The surgical cavity is filled with abdominal fat (AF). The picture has been cut to remove *dark* part on the middle fossa dura.

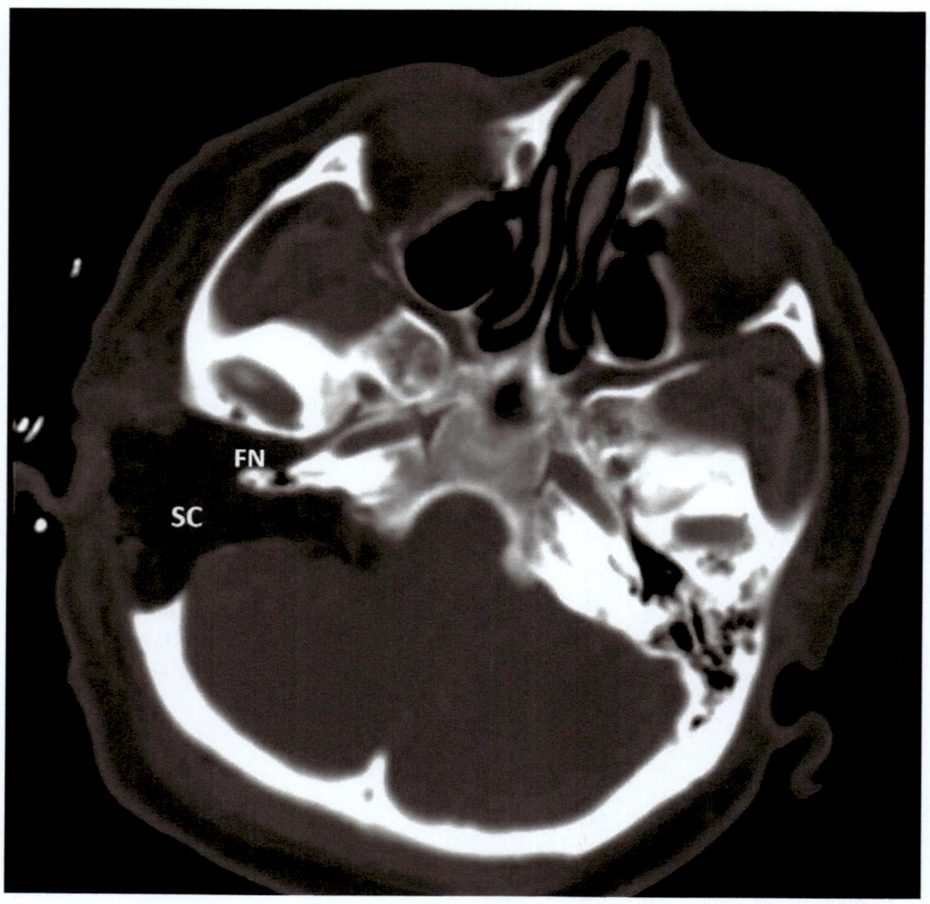

Fig. 7.121 Postoperative CT scan. *FN*, facial nerve; *SC*, surgical cavity.

Case 3: Epidermoid Tumor of the Cerebellopontine Angle

See (Figs. 7.122–7.136).

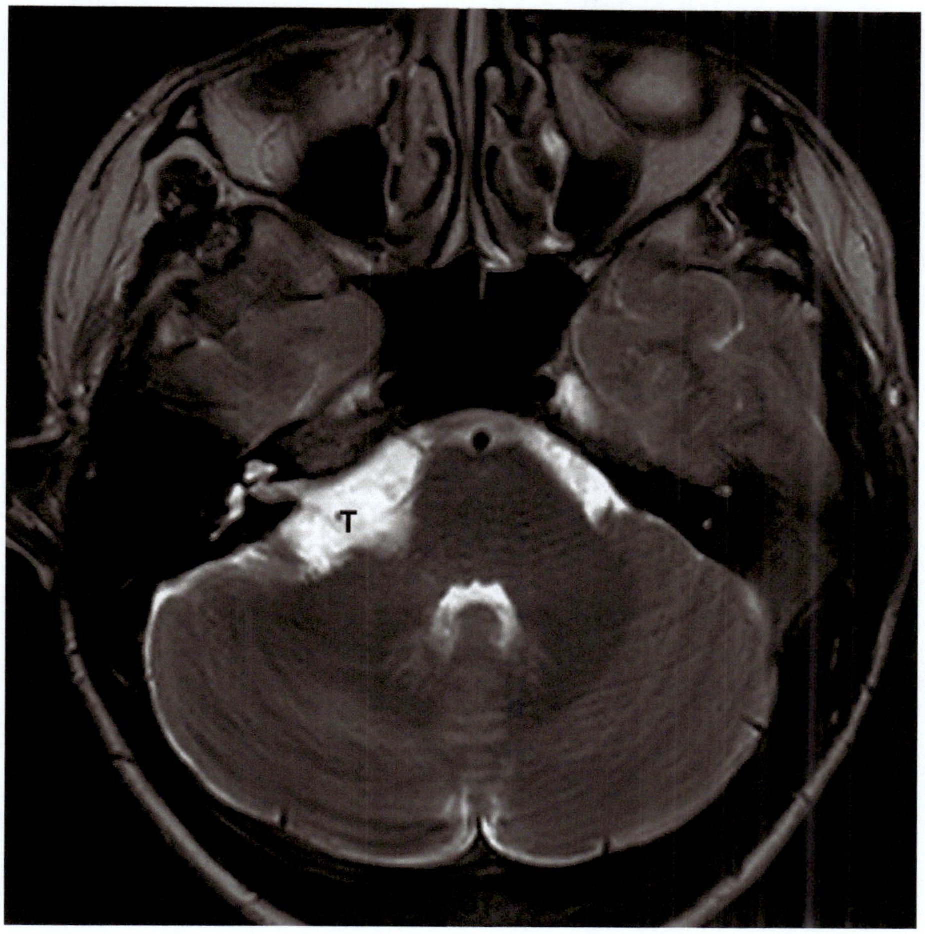

Fig. 7.122 TW2 MRI of the brain showing a right hyperintense lesion consistent with an epidermoid tumor (T) of the cerebellopontine angle. This 34-year-old patient presented with hearing loss and facial numbness.

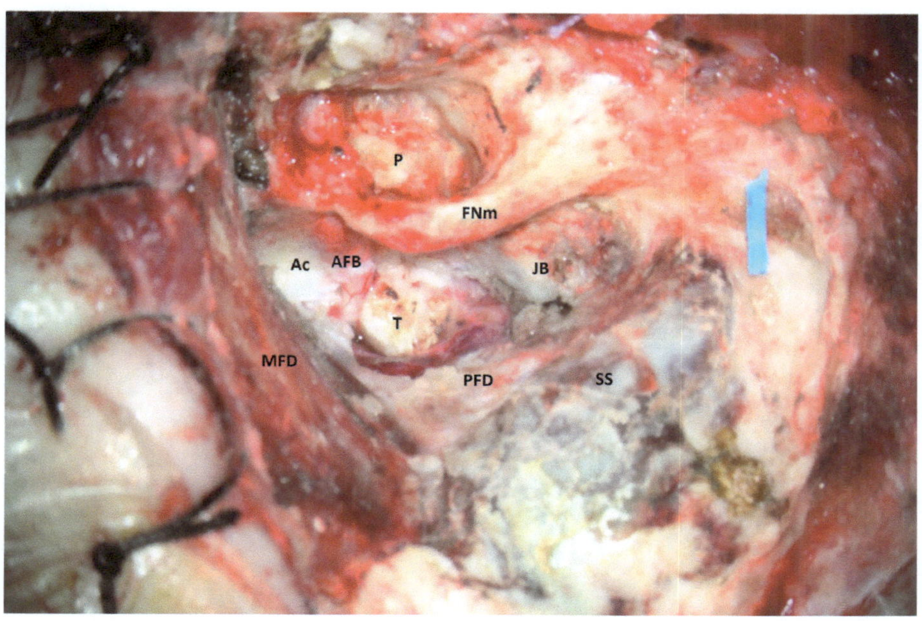

Fig. 7.123 An enlarged translabyrinthine approach with closure of the external auditory canal was chosen to gain control of the anterior part of the cerebello-pontine angle, including the pre-pontine cistern. *Ac*, apical cells; *AFB*, acoustic-facial bundle; *FNm*, mastoid portion of the facial nerve; *JB*, jugular bulb; *MFD*, middle fossa dura; *P*, promontory; *PFD*, posterior fossa dura; *SS*, sigmoid sinus; *T*, tumor.

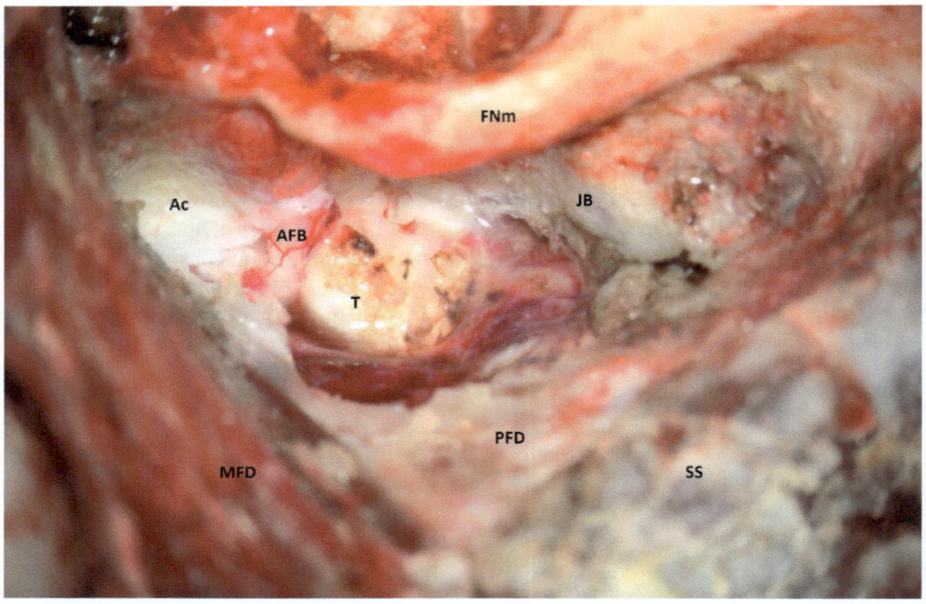

Fig. 7.124 Closer view. The high jugular bulb (JB) was previously lowered with bone wax to gain further control over the area of the lower cranial nerves. *Ac*, apical cells; *AFB*, acoustic-facial bundle; *FNm*, mastoid portion of the facial nerve; *MFD*, middle fossa dura; *PFD*, posterior fossa dura; *SS*, sigmoid sinus; *T*, tumor.

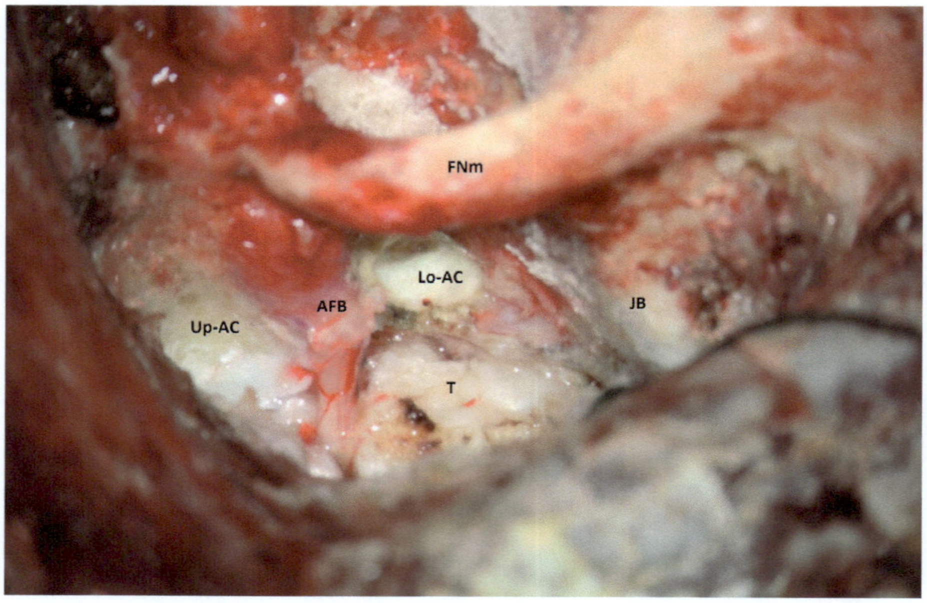

Fig. 7.125 The lower apical cells (Lo-AC) are drilled to gain inferior-medial control of the cerebello-pontine angle. *AFB*, acoustic-facial bundle; *FNm*, mastoid portion of the facial nerve; *Up-AC*, upper apical cells; *JB*, jugular bulb; *T*, tumor.

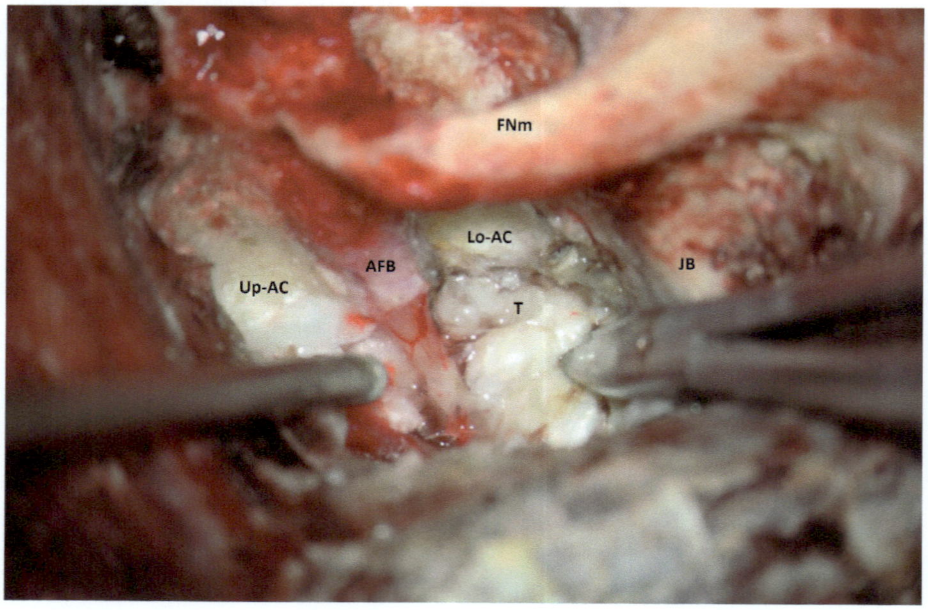

Fig. 7.126 The acoustic-facial bundle (AFB) begins to be isolated from the tumor (T). *FNm*, mastoid portion of the facial nerve; *Up-Ac*, upper apical cells; *Lo-Ac*, lower apical cells; *JB*, jugular bulb.

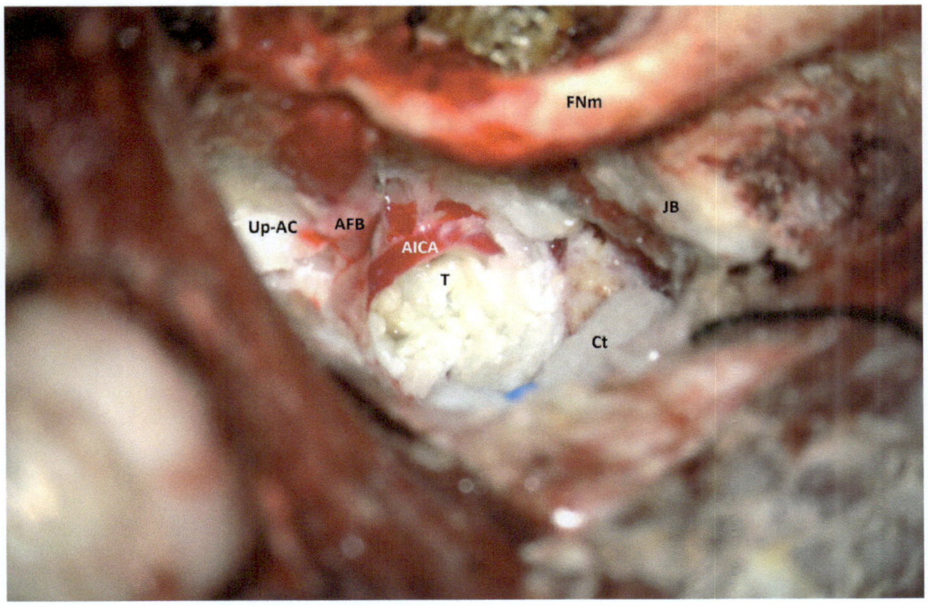

Fig. 7.127 The tumor (T) is debulked, exposing the anterior-inferior cerebellar artery (AICA). *AFB*, acoustic-facial bundle; *Ct*, cottonoid; *FNm*, mastoid portion of the facial nerve; *JB*, jugular bulb; *Up-Ac*, upper apical cells.

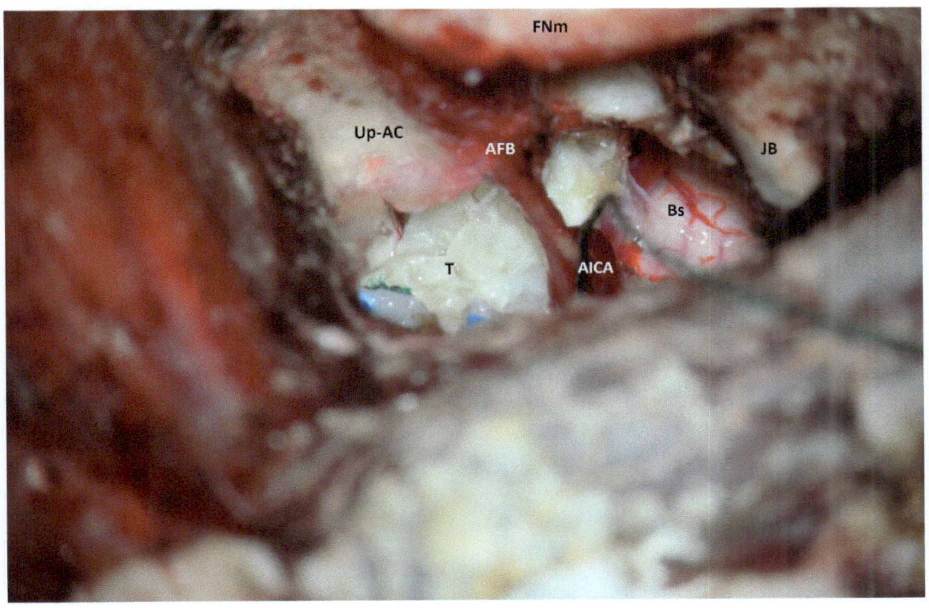

Fig. 7.128 The tumor (T) is further removed and the anterior-inferior cerebellar artery (AICA) is displaced inferiorly. The acoustic-facial bundle (AFB) is located lateral to the tumor. *FNm*, mastoid portion of the facial nerve; *Bs*, brainstem; *JB*, jugular bulb; *Up-AC*, upper apical cells.

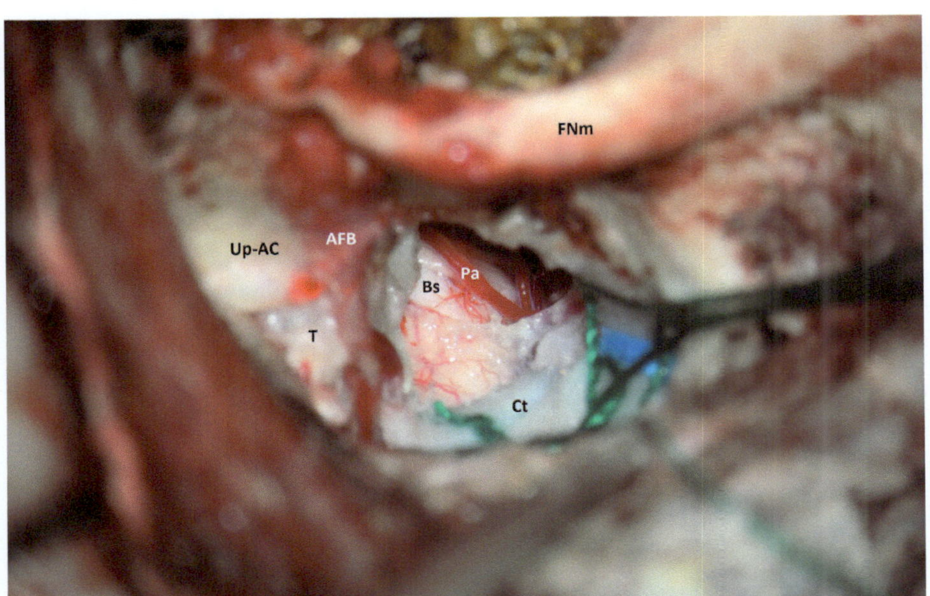

Fig. 7.129 The inferior-medial portion of the cerebello-pontine angle has been freed from the tumor. Note that the lower apical cells have been drilled out. *AFB*, acoustic-facial bundle; *Bs*, brainstem; *Ct*, cottonoid; *FNm*, mastoid portion of the facial nerve; *Pa*, perforating artery; *Up-AC*, upper apical cells.

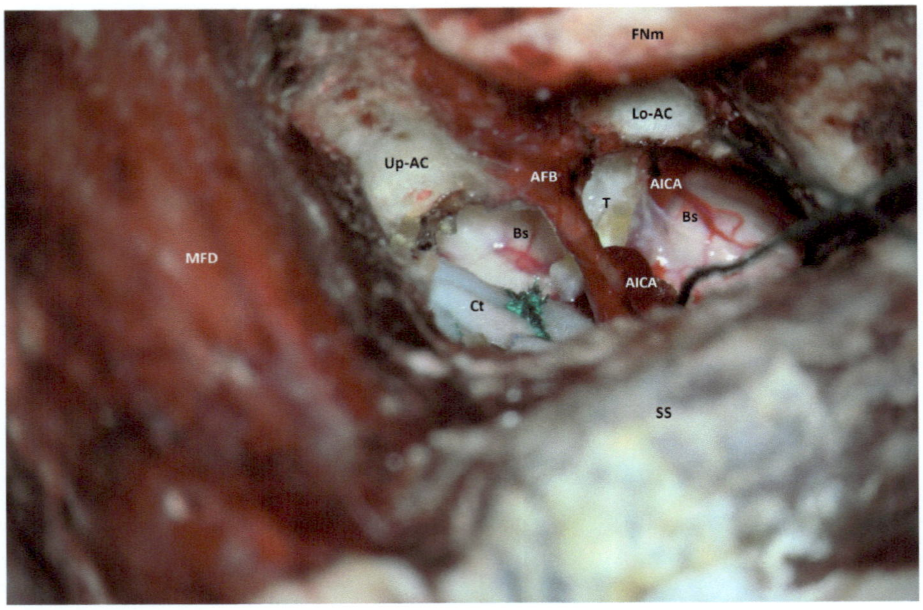

Fig. 7.130 Tumor removal proceeds superiorly. *AFB*, acoustic-facial bundle; *AICA*, anterior-inferior cerebellar artery; *Bs*, brainstem; *CT*, cottonoid; *FNm*, mastoid portion of the facial nerve; *Lo-AC*, lower apical cells; *MFD*, middle fossa dura; *SS*, sigmoid sinus; *Up-AC*, upper apical cells.

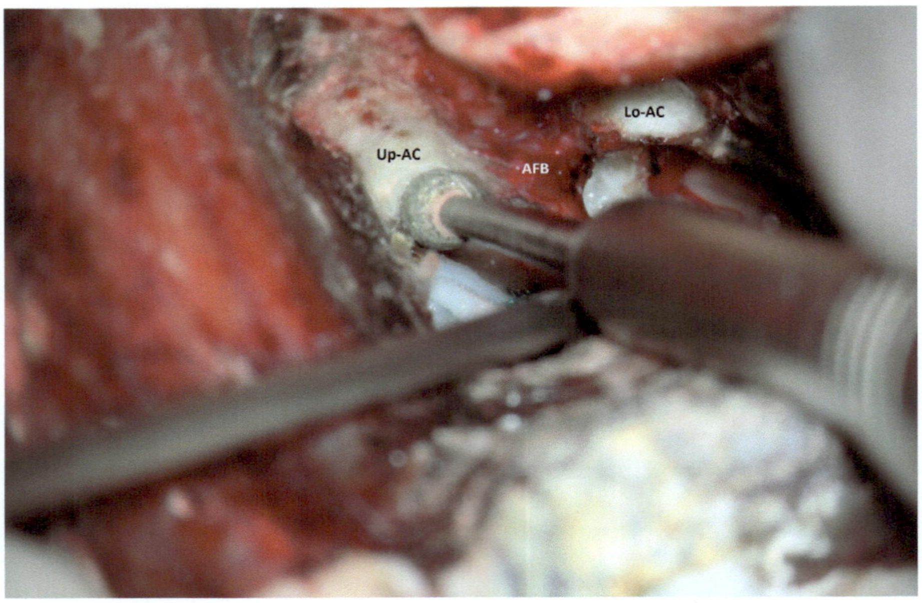

Fig. 7.131 The upper apical cells (Up-AC) are drilled. *AFB*, acoustic-facial bundle; *Lo-AC*, lower apical cells.

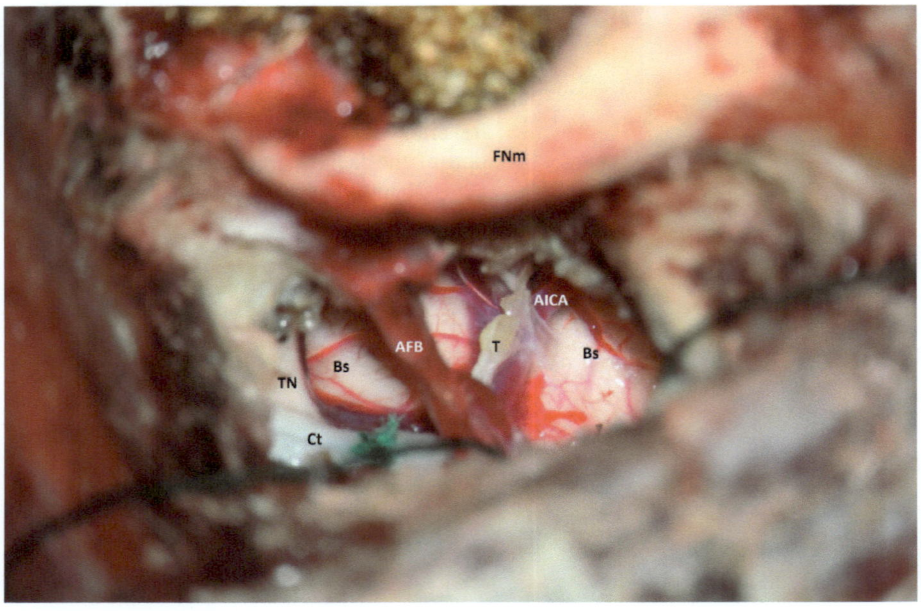

Fig. 7.132 Once the upper apical cells have been drilled out, the trigeminal nerve (TN) is visible. Note the tumor (T) remnant close to the anterior-inferior cerebellar artery (AICA). *AFB*, acoustic-facial bundle; *Bs*, brainstem; *CT*, cottonoid; *FNm*, mastoid portion of the facial nerve.

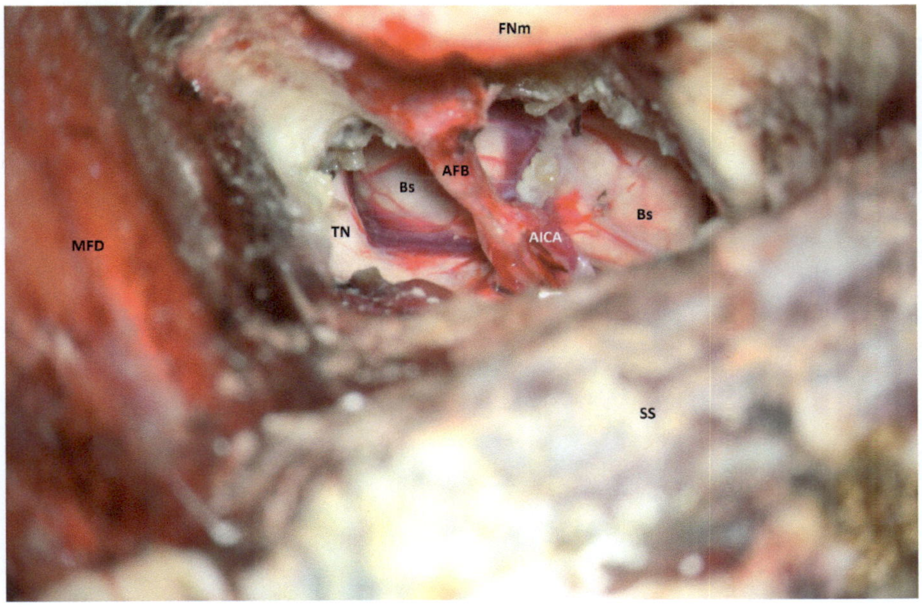

Fig. 7.133 Note the loop of the anterior-inferior cerebellar artery (AICA) at the level of the root-entry zone of the acoustic-facial bundle (AFB). *Bs*, brainstem; *FNm*, mastoid portion of the facial nerve; *MFD*, middle fossa dura; *SS*, sigmoid sinus; *TN*, trigeminal nerve.

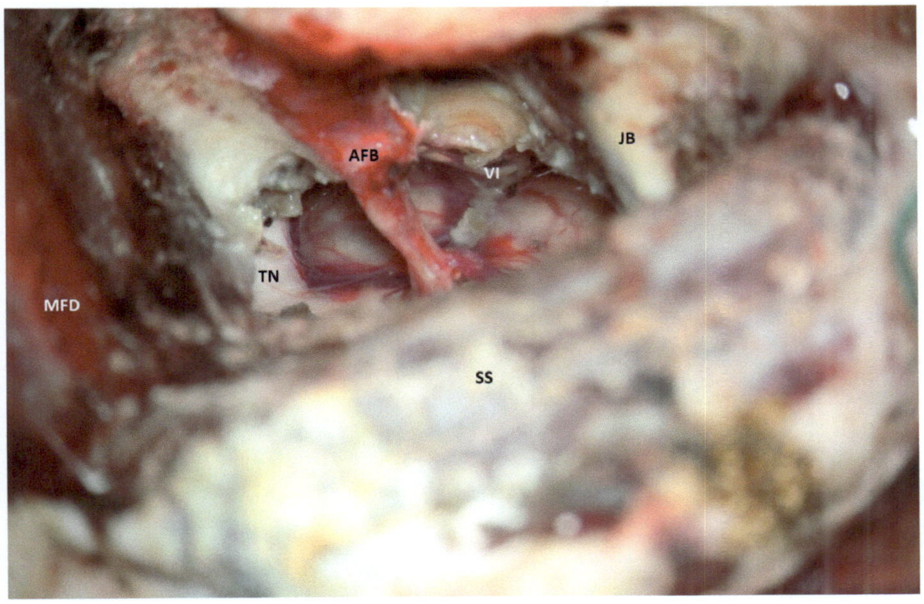

Fig. 7.134 The surgical field is tilted anteriorly, showing the abducens nerve (VI). There is no tumor in the prepontine cistern. *AFB*, acoustic-facial bundle; *JB*, jugular bulb; *MFD*, middle fossa dura; *SS*, sigmoid sinus; *TN*, trigeminal nerve.

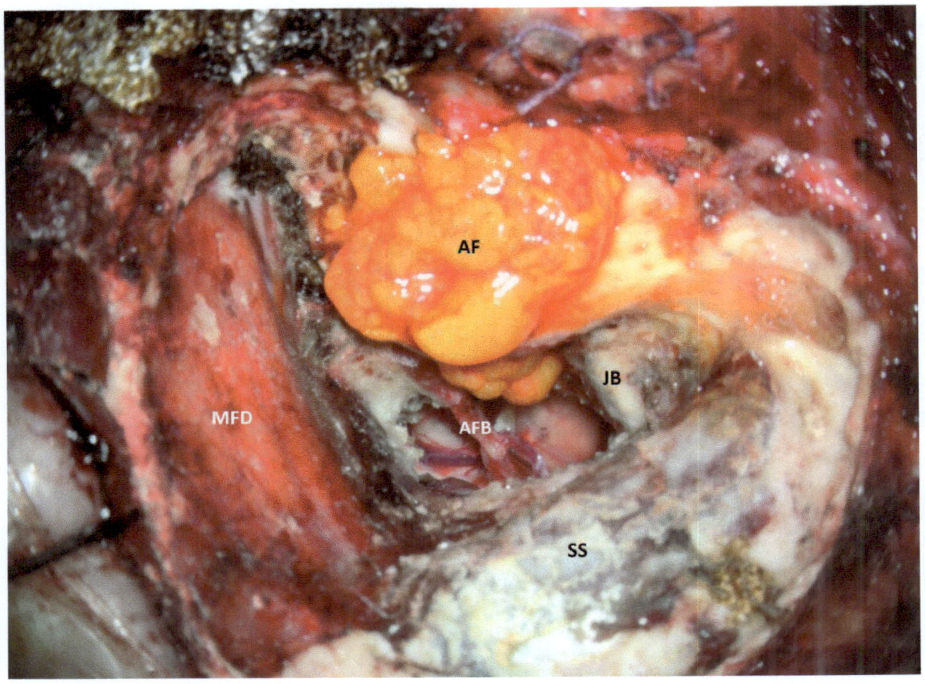

Fig. 7.135 The surgical cavity is filled with abdominal fat (AF). *AFB*, acoustic-facial bundle; *MFD*, middle fossa dura; *JB*, jugular bulb; *SS*, sigmoid sinus.

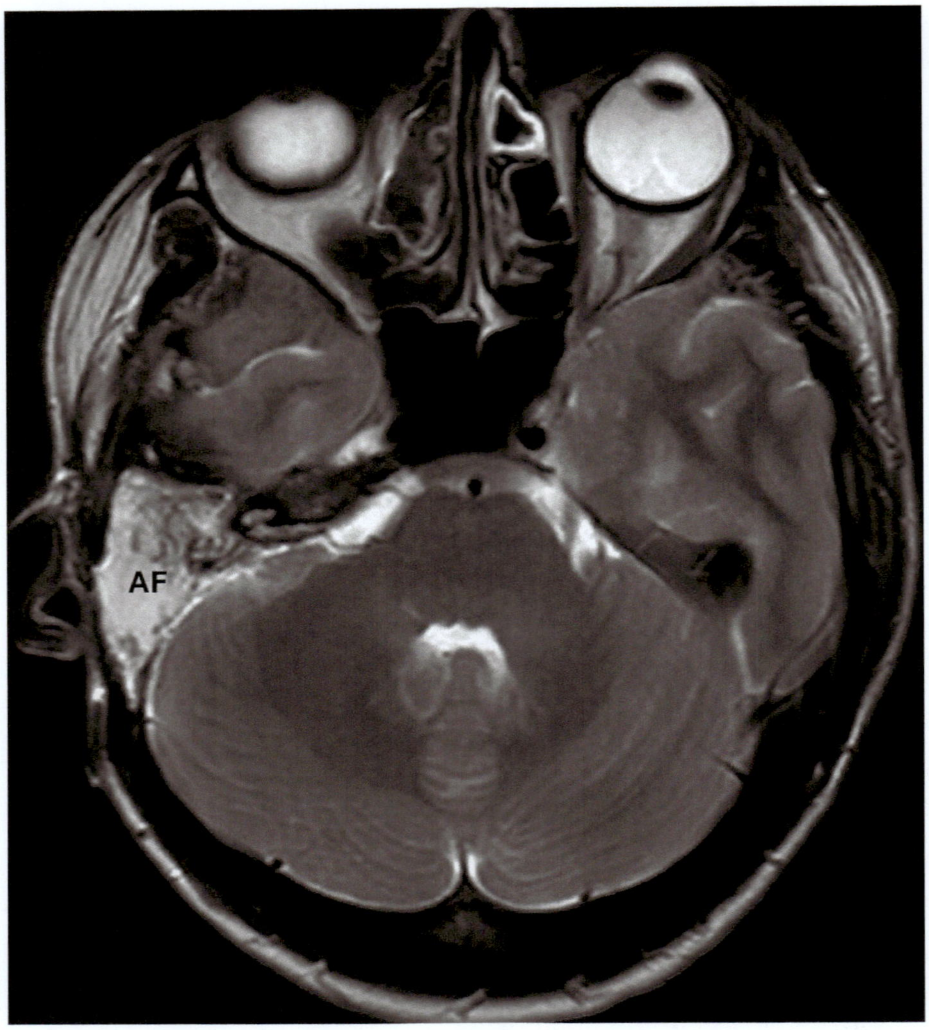

Fig. 7.136 Postoperative MRI. The surgical cavity is filled with abdominal fat (AF).

Case 4: Simultaneous Removal of Vestibular Schwannoma and Cochlear Implantation

See (Figs. 7.137–7.150).

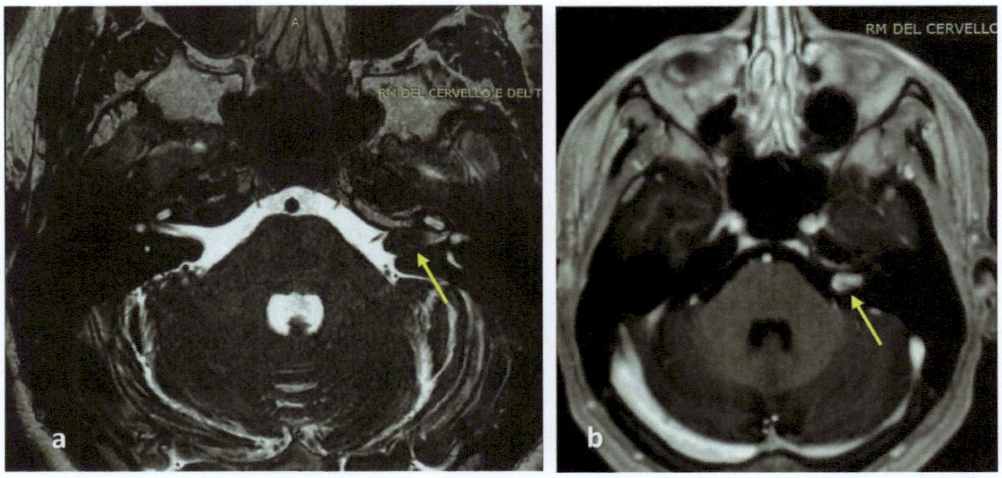

Fig. 7.137 This 37-year-old male patient was evaluated because of left hearing loss and tinnitus. Brain MRI showed a lesion in the internal auditory canal and the left cerebellopontine angle, suggesting a vestibular schwannoma. The patient opted for surgery, which was performed via an enlarged translabyrinthine approach with simultaneous cochlear implantation. (A) T2W-CISS MRI (axial cut): note the tumor occupying the internal auditory canal (arrow); the fundus is free and the tumor extends 3 mm into the cerebellopontine angle. (B) T1W + Gd enhanced MRI (axial cut).

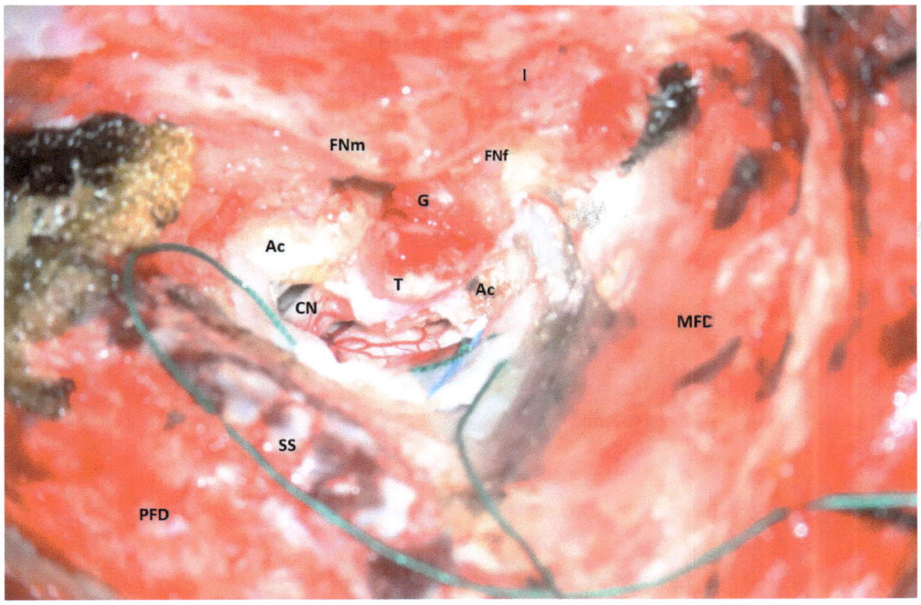

Fig. 7.138 Surgical view after completion of the surgical approach. Note the extent of the craniotomy at the level of the middle and posterior retrosigmoid fossae. The transapical extension allows complete control of the cerebellopontine angle both superiorly (toward the trigeminal nerve) and inferiorly (toward the lower cranial nerves). The vestibular nerves have already been dissected inferiorly along with the most proximal part of the tumor. The facial nerve (medial to the superior vestibular nerve) is protected with Gelfoam ©. *Ac*, apical cells; *CN*, cochlear nerve; *FNf*, facial nerve (at the fundus of the internal auditory canal); *FNm*, facial nerve (mastoid portion); *G*, gelfoam; *I*, incus; *MFD*, middle fossa dura; *PFD*, posterior fossa dura; *SS*, sigmoid sinus; *T*, tumor.

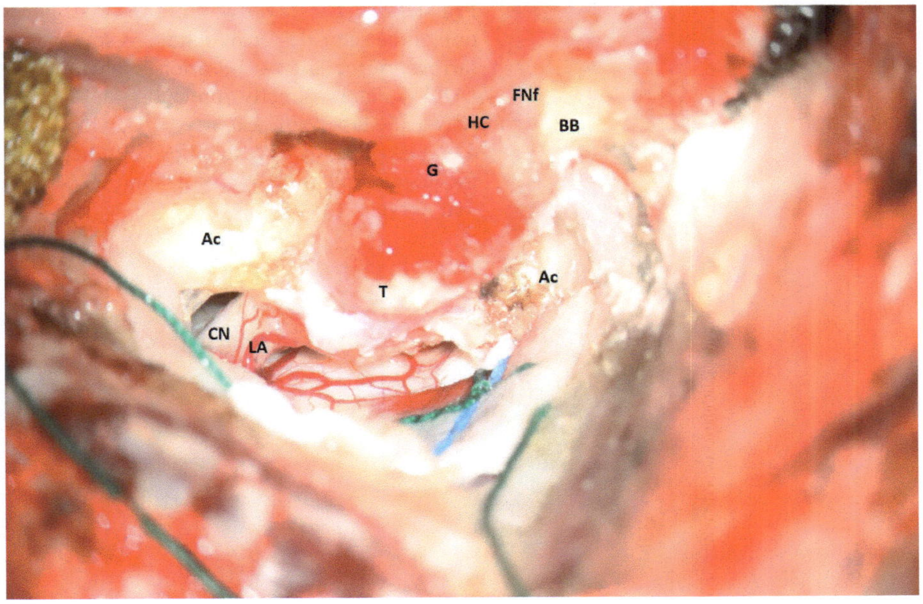

Fig. 7.139 Closer view. Bill's bar (BB) is no longer used as an anatomical landmark to identify the facial nerve because of its proximity to the nerve itself. The horizontal crest (HC) is the most reliable point for locating the facial nerve, which lies just medial to the superior ampullary nerve. *Ac*, apical cells; *CN*, cochlear nerve; *FNf*, facial nerve (at the fundus of the internal auditory canal); *G*, gelfoam; *T*, tumor; *LA*, labyrinthine artery.

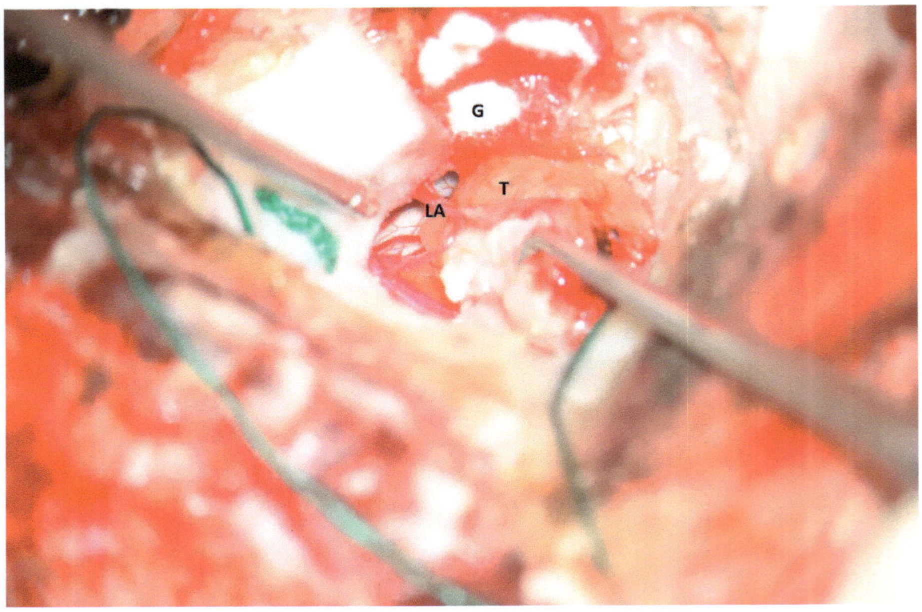

Fig. 7.140 The tumor is carefully dissected from the acoustic-facial bundle. Great care must be taken to preserve the vascularity of the cochlear nerve. *G*, gelfoam; *T*, tumor; *LA*, labyrinthine artery.

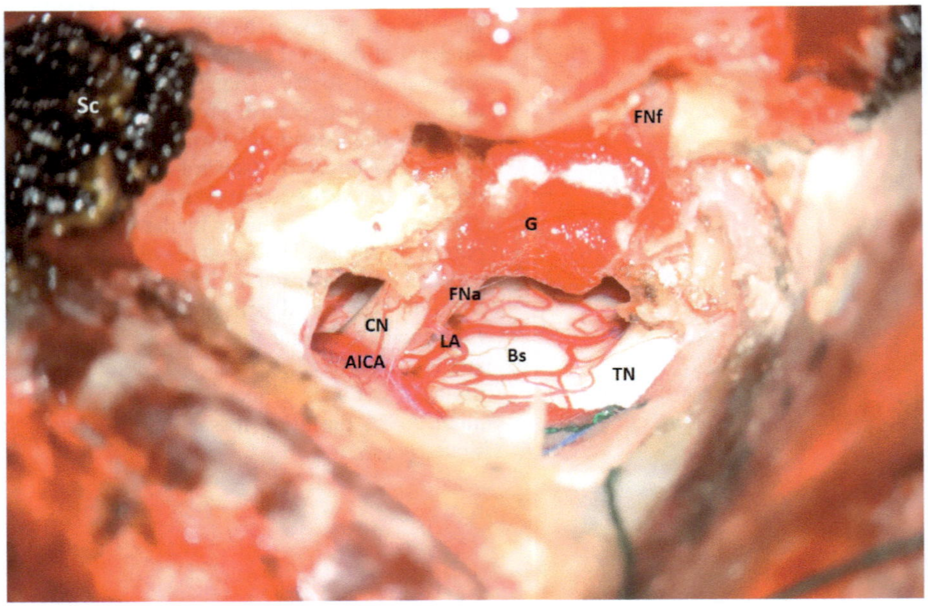

Fig. 7.141 The tumor has been completely removed. Gelfoam © was placed to protect the facial nerve and Surgicel © (Sc) to protect the area of the jugular bulb. *AICA*, anterior inferior cerebellar artery; *Bs*, brainstem; *CN*, cochlear nerve; *FNa*, facial nerve (at the level of the cerebellopontine angle); *FNf*, facial nerve (at the fundus of the internal auditory canal); *LA*, labyrinthine artery; *TN*, trigeminal nerve.

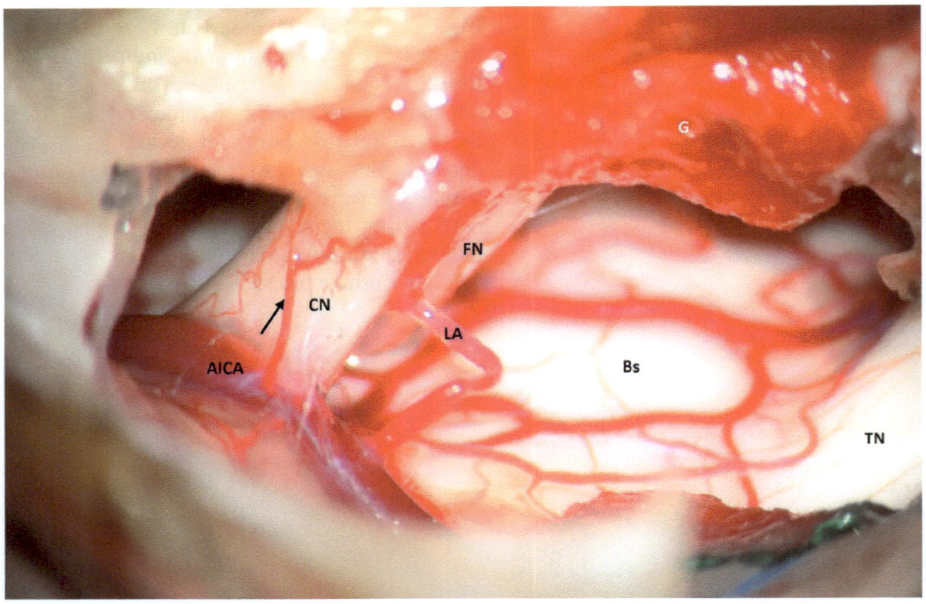

Fig. 7.142 Closer view. Note the vascularization of the acoustic-facial bundle. A small branch (arrow) of the anterior inferior cerebellar artery (AICA) enters directly into the cochlear nerve (CN). The labyrinthine artery (LA), which is also a branch of the AICA, supplies both the CN and the facial nerve (FN). *BS*, brainstem; *TN*, trigeminal nerve.

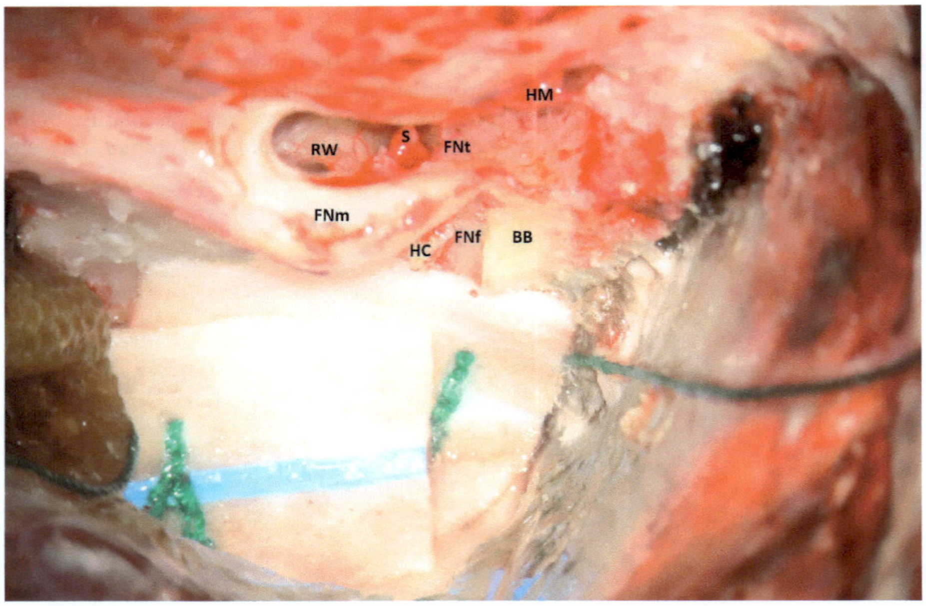

Fig. 7.143 The posterior tympanotomy was carried out with complete visualization of the round window (RW) niche. The incus was previously removed. *BB*, Bill's bar; *FNf*, facial nerve (at the fundus of the internal auditory canal); *FNm*, mastoid portion of the facial nerve; *FNt*, tympanic portion of the facial nerve; *HC*, horizontal crest; *HM*, head of the malleus; *S*, stapes.

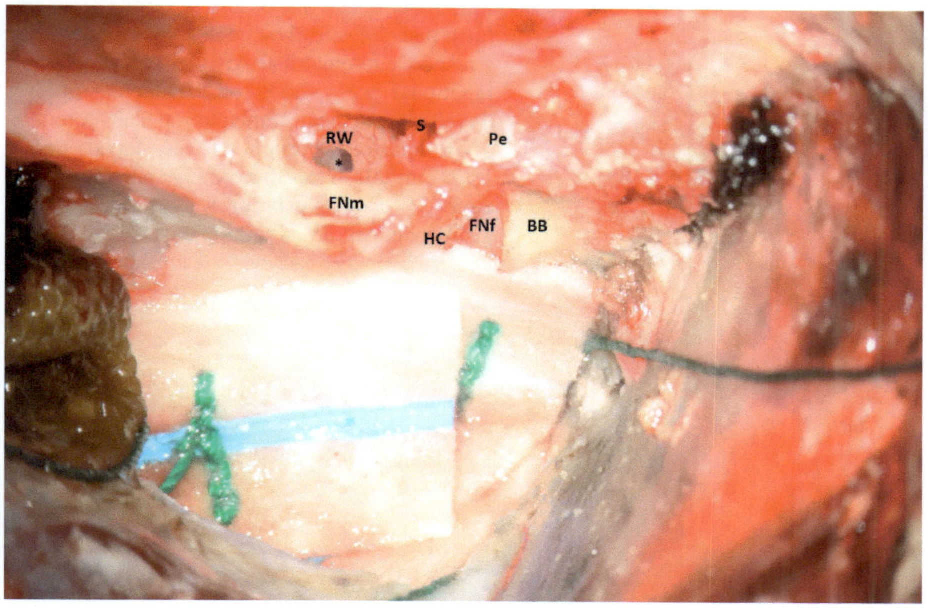

Fig. 7.144 A minimal drill out of the antero-superior portion of the round window niche allows access to the secondary membrane (*). The attic was obliterated with periosteum (Pe). *BB*, Bill's bar; *FNf*, facial nerve (at the fundus of the internal auditory canal); *FNm*, mastoid portion of the facial nerve; *FNt*, tympanic portion of the facial nerve; *HC*, horizontal crest; *RW*, round window niche; *S*, stapes.

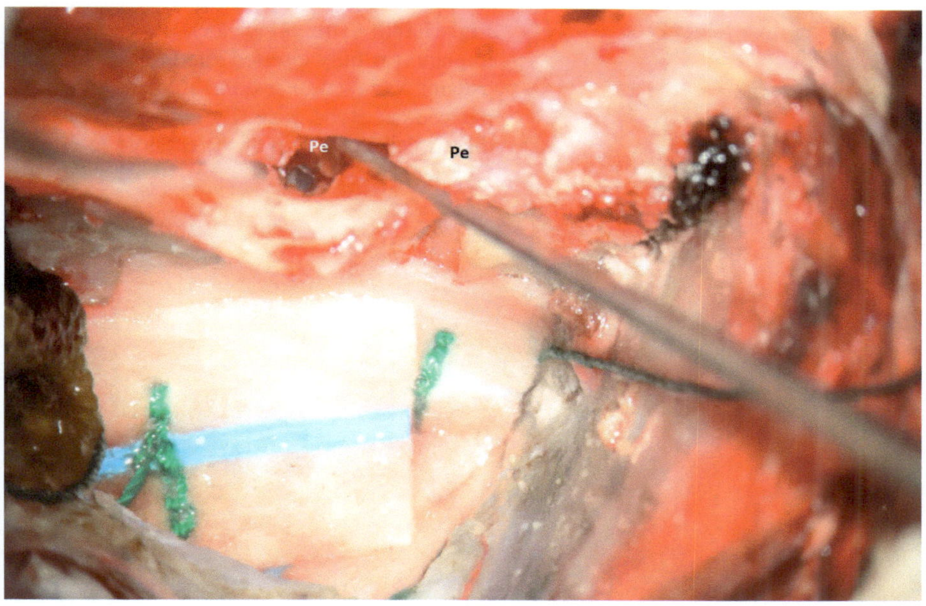

Fig. 7.145 Some more periosteum (Pe) is used to seal the middle ear and prevent cerebrospinal fluid from leaking into the eustachian tube.

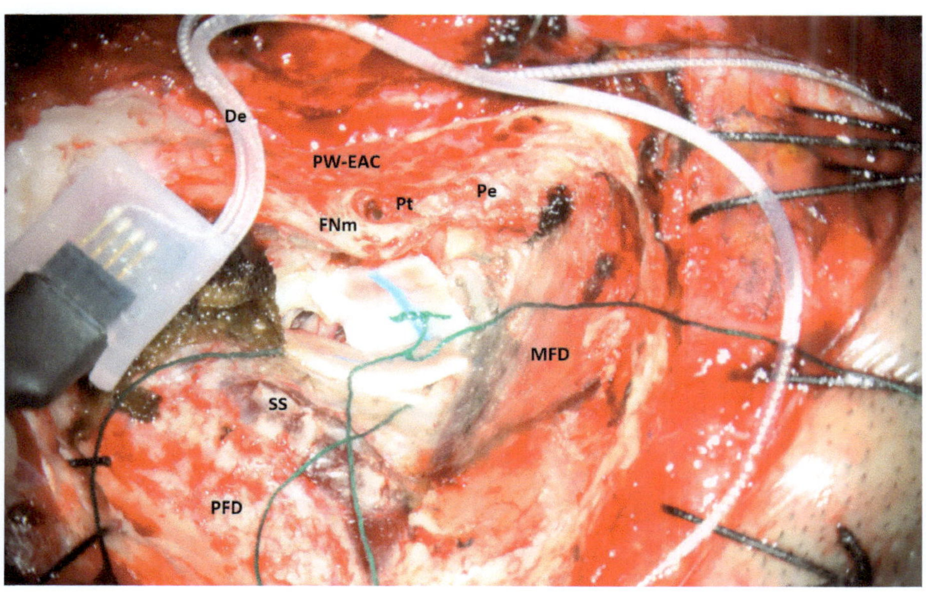

Fig. 7.146 A disposable electrode (De) is used to record the electrical ABR after insertion into the cochlea. *FNm*, mastoid portion of the facial nerve; *MFD*, middle fossa dura; *Pe*, periosteum; *PFD*, posterior fossa dura; *Pt*, posterior tympanotomy area; *PW-EAC*, posterior wall of the external auditory canal; *SS*, sigmoid sinus.

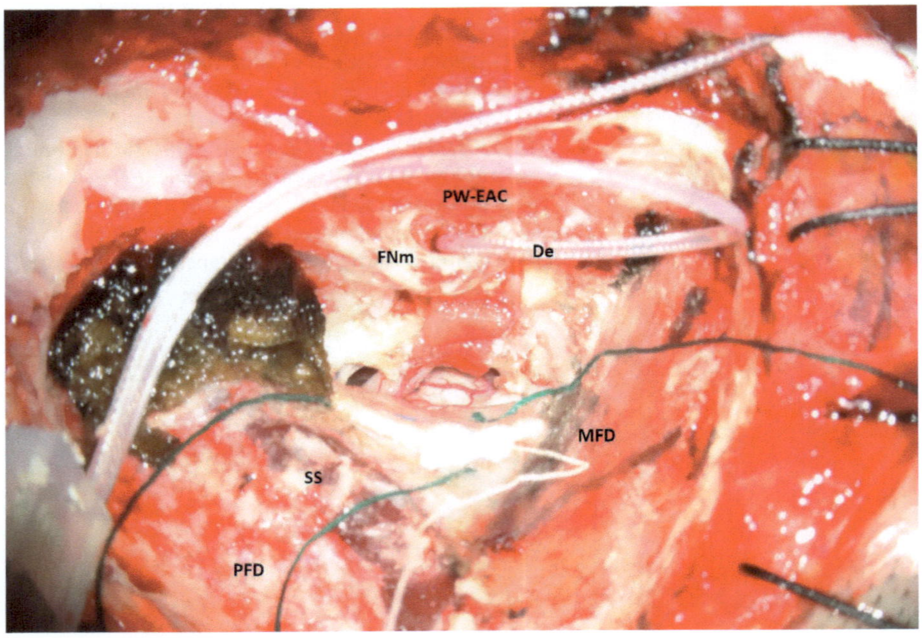

Fig. 7.147 The disposable electrode (De) was fully inserted into the cochlea. The presence of a response to electrical ABR is an indication of functional preservation of the cochlear nerve; therefore, cochlear implantation can continue. *FNm*, mastoid portion of the facial nerve; *MFD*, middle fossa dura; *PFD*, posterior fossa dura; *PW-EAC*, posterior wall of the external auditory canal; *SS*, sigmoid sinus.

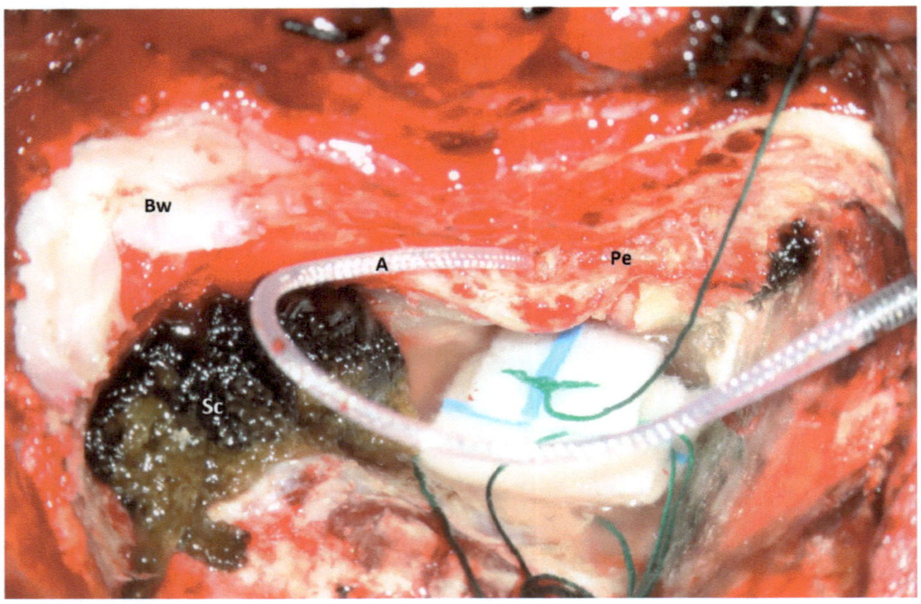

Fig. 7.148 The array (A) of the cochlear implant was fully inserted into the cochlea. Periosteum was used to seal the attic and mesotympanic area. Bone wax (Bw) was used to seal the air cells at the tip of the mastoid to prevent CSF leakage.

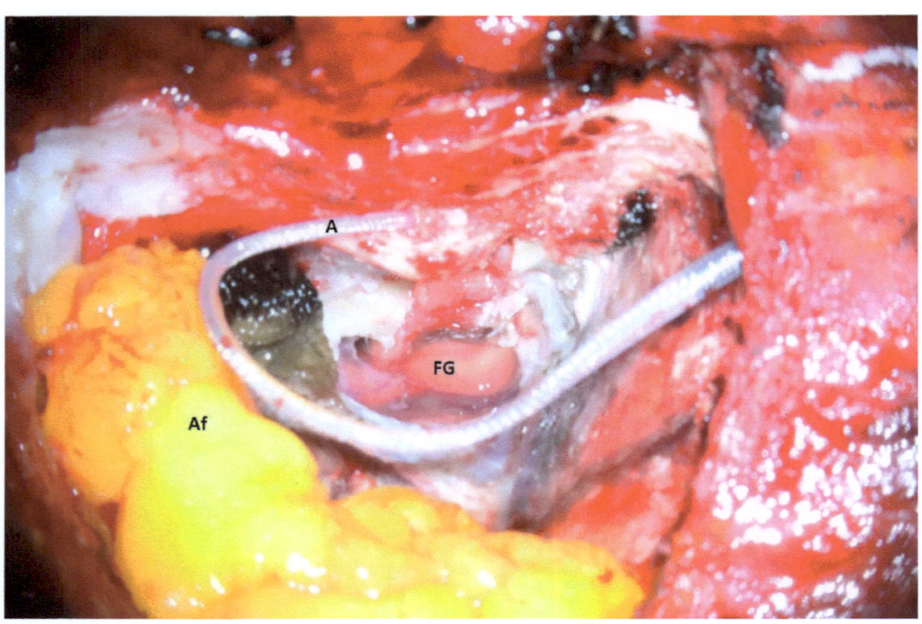

Fig. 7.149 The surgical cavity was then obliterated with abdominal fat (Af) and fibrin glue (FG). A array of the cochlear implant.

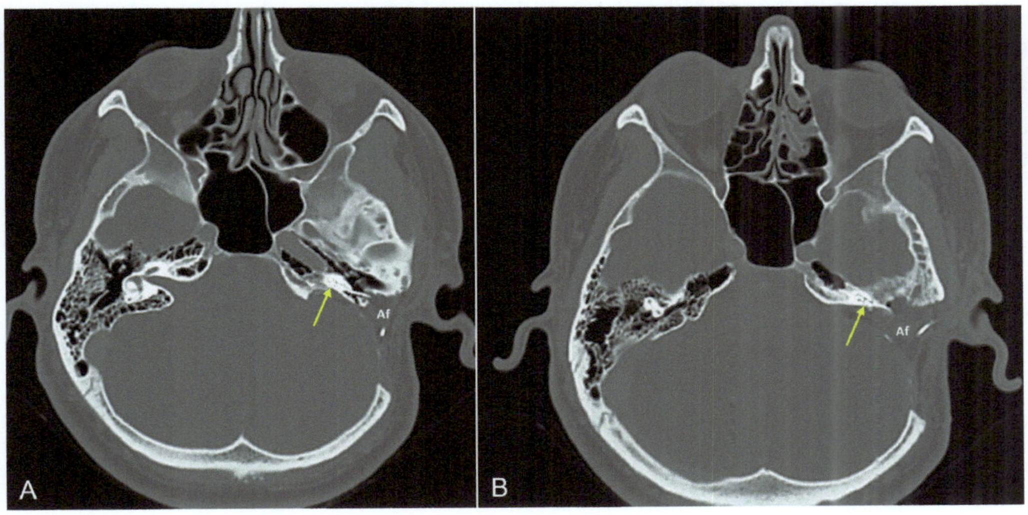

Fig. 7.150 Post-operative HRCT scan, axial cuts (1 year). Note the array (*arrow*) into the basal turn of the cochlea (A) and into the transition point between the middle and the apical turn (B). *Af*, abdominal fat.

Case 5: Simultaneous Removal of Vestibular Schwannoma and Cochlear Implantation (With Closure of the External Auditory Canal)

See (Figs. 7.151–7.163).

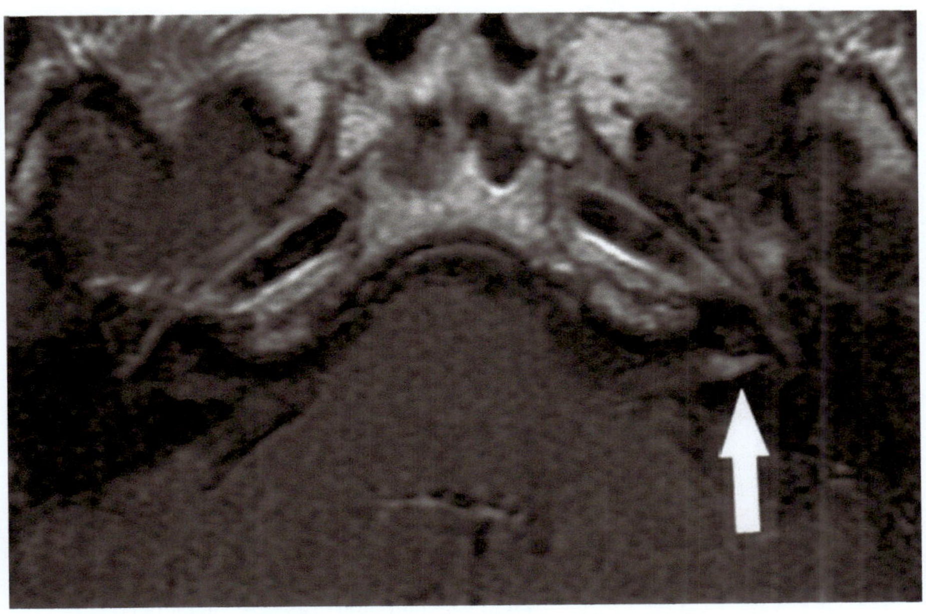

Fig. 7.151 Preoperative MRI showing a intracanalicular vestibular schwannoma on the *left side* (*arrow*).

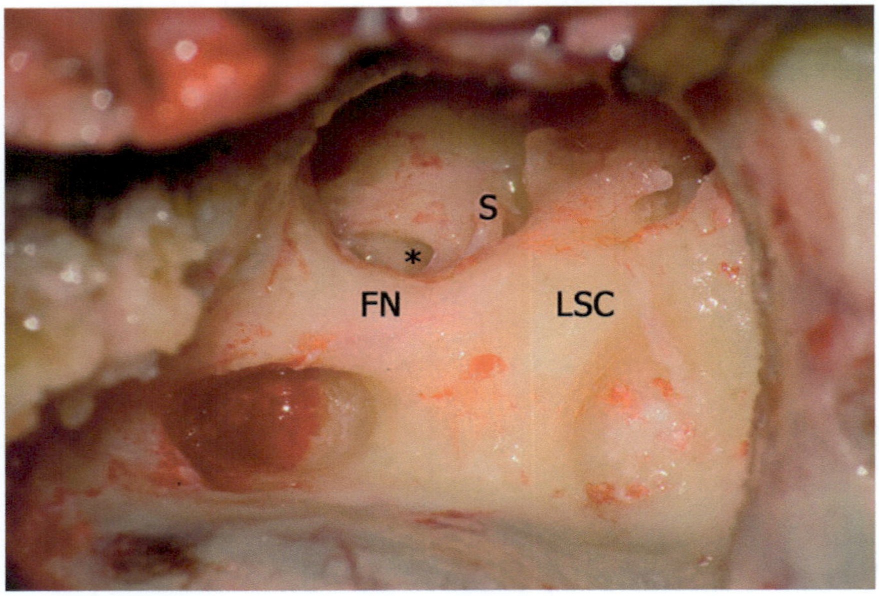

Fig. 7.152 In this case an enlarged translabyrinthine approach with closure of the external auditory canal was adopted to favor the visualization of the round window niche for the cochlear implantation. The canal wall down cavity procedure is performed. The horizontal semicircular canal (LSC) and facial nerve canal (FN) can clearly be seen, as well as the stapes (S) and round window niche (*).

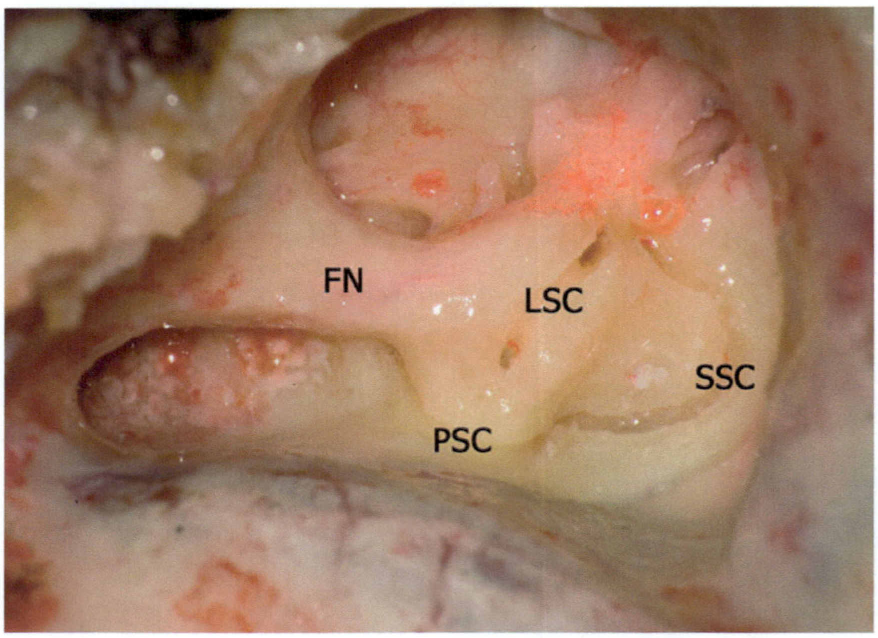

Fig. 7.153 Start of the translabyrinthine procedure with opening of semicircular canals. *FN*, facial nerve; *LSC*, lateral semicircular canal; *PSC*, posterior semicircular canal; *SSC*, superior semicircular canal.

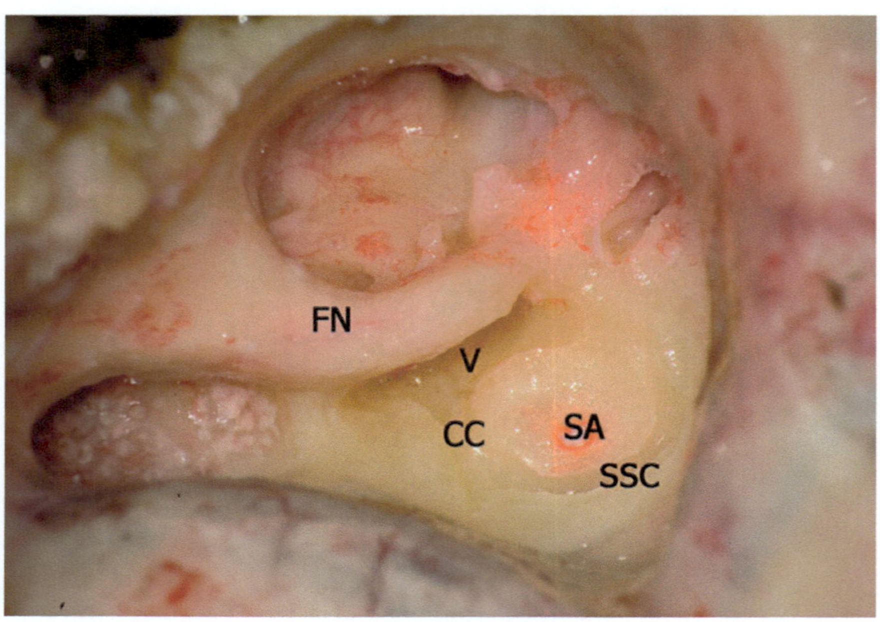

Fig. 7.154 Removal of labyrinth block and opening of the vestibule (V) Note the subarcuate artery (SA) central in the plane of the superior semicircular canal (SSC). *CC*, common crus; *FN*, facial nerve.

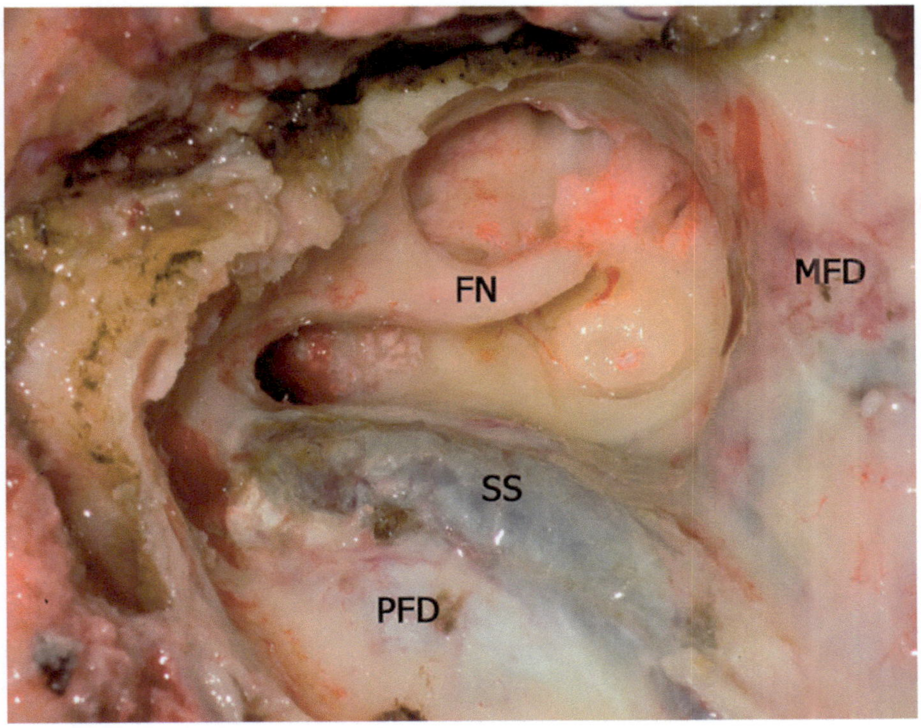

Fig. 7.155 The middle (MFD) and posterior fossa dura (PFD) are exposed for better access. *FN*, facial nerve; *SS*, sigmoid sinus.

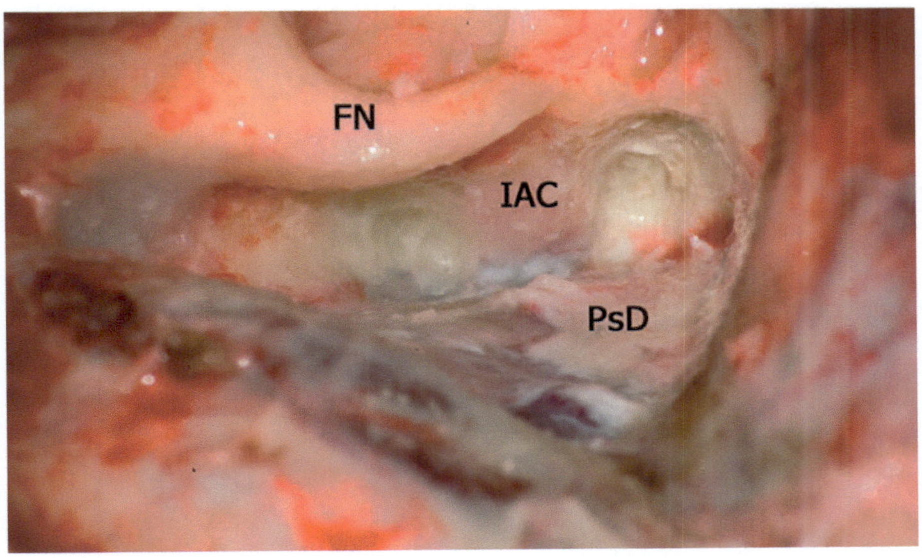

Fig. 7.156 Identification and skeletonization of the internal auditory canal (IAC). *PsD*, presigmoid dura; *FN*, facial nerve.

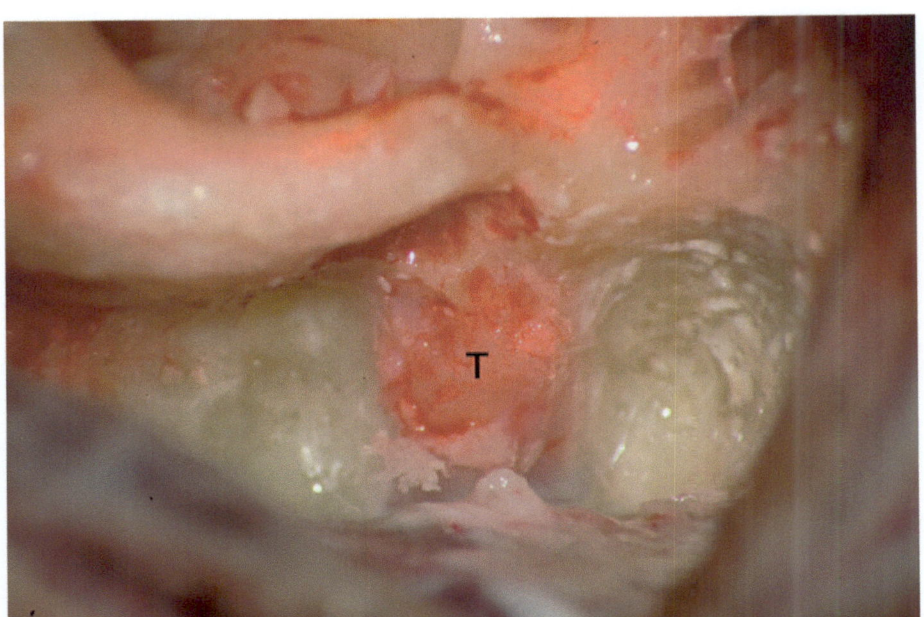

Fig. 7.157 All bone is removed over the internal auditory canal: the intracanalicular schwannoma (T) now becomes visible.

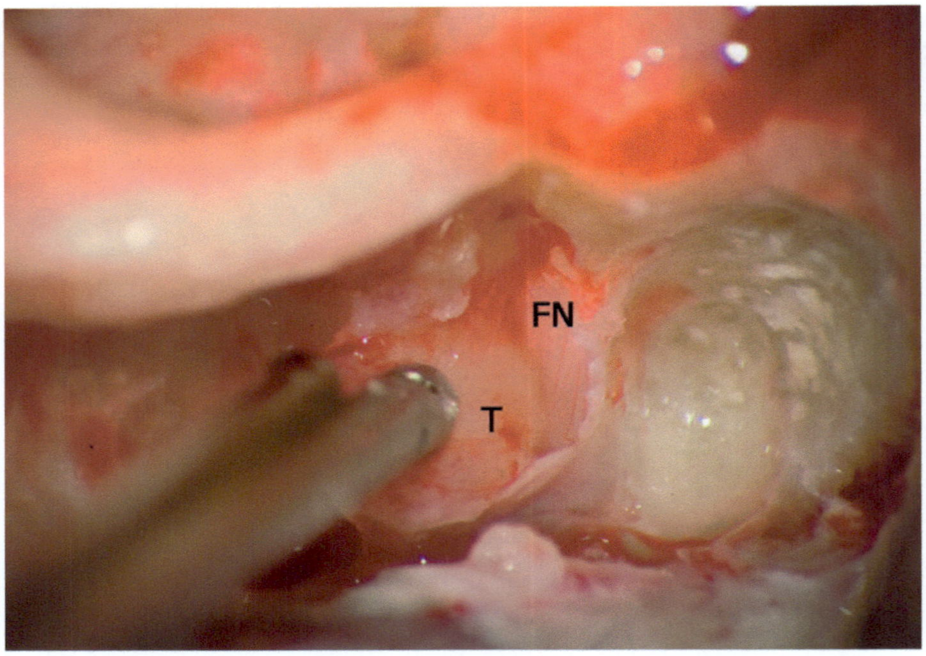

Fig. 7.158 With posterior-inferior displacement of the tumor (T), the facial nerve (FN) is identified at the fundus of the internal auditory canal.

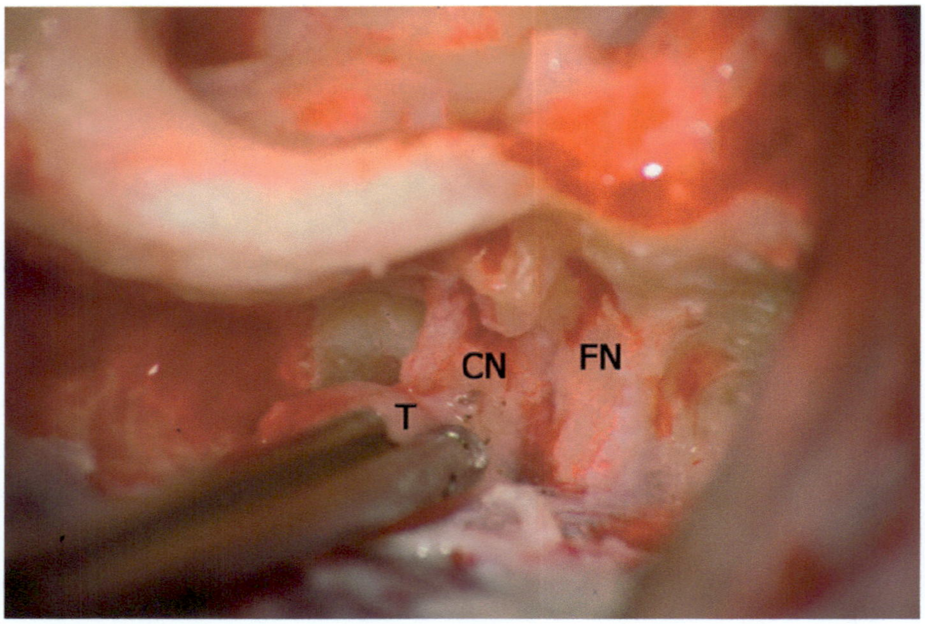

Fig. 7.159 Inferior-anteriorly the cochlear nerve (CN) also becomes visible. *FN*, facial nerve; *T*, tumor.

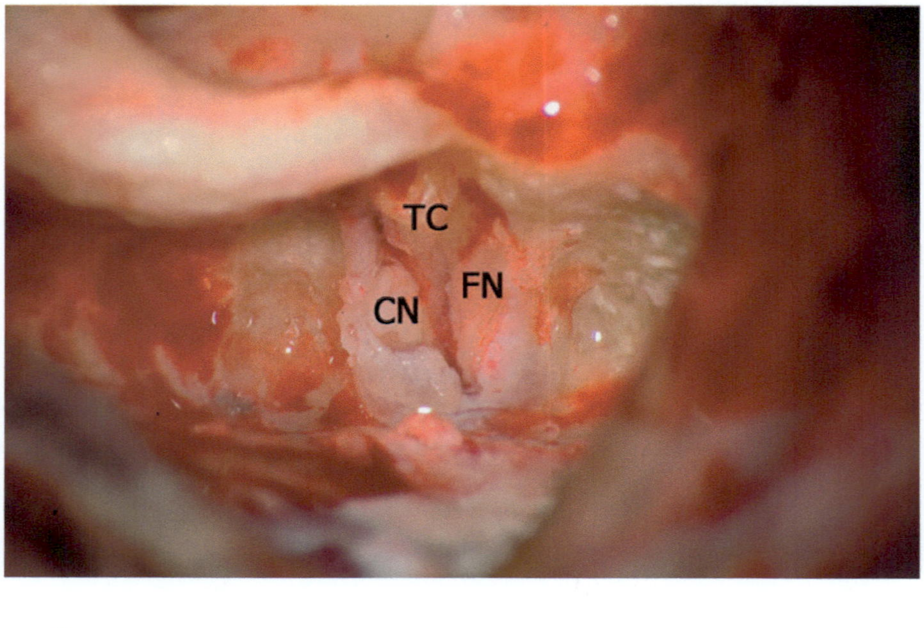

Fig. 7.160 Inferior-anteriorly the cochlear nerve (CN) also becomes visible. *FN*, facial nerve; *T*, tumor.

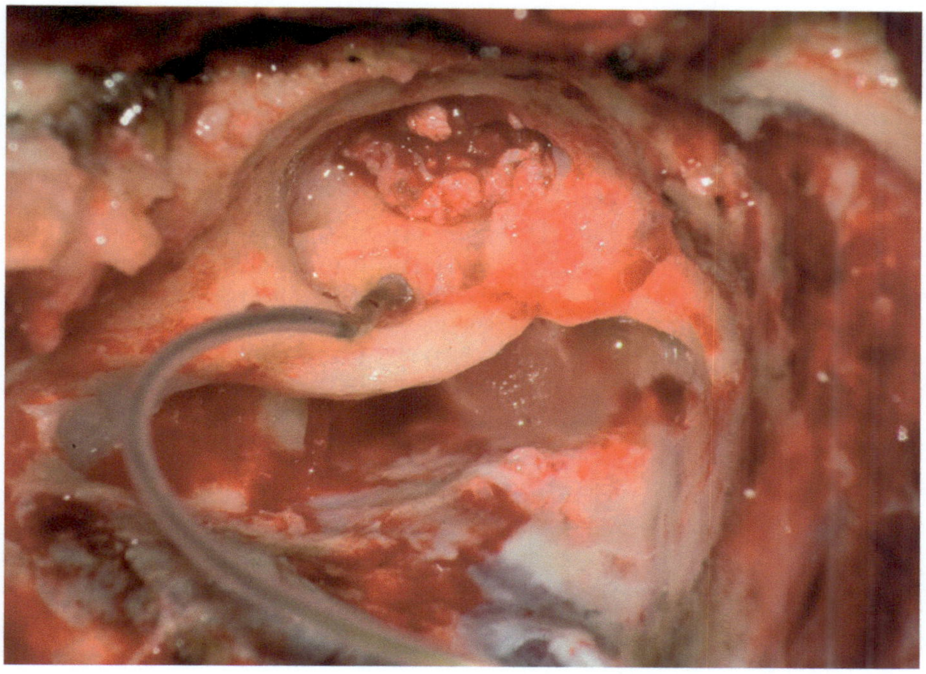

Fig. 7.161 The array of the cochlear implant was fully inserted into the cochlea. This surgical approach provides a direct rehabilitation option incombination with tumor removal. One has to be aware that sometimes the function of the cochlear nerve can be distorted, even though the anatomi-al integrity is intact.

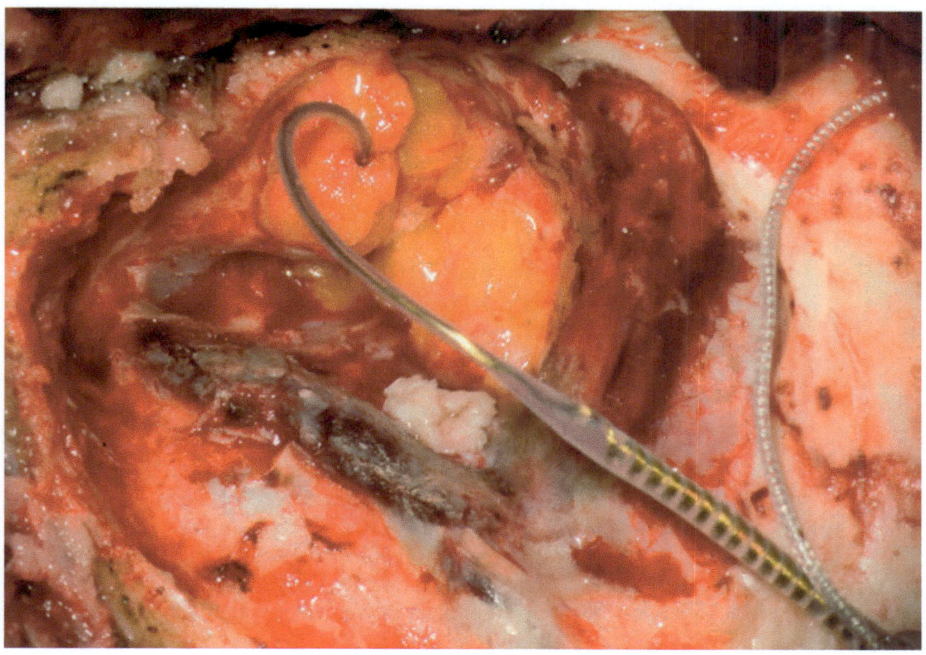

Fig. 7.162 The surgical cavity was then obliterated with abdominal fat and fibrin glue.

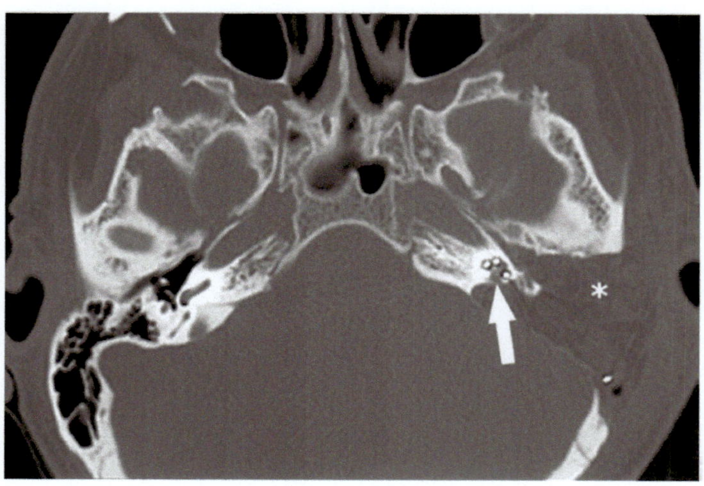

Fig. 7.163 Postoperative CT image, showing the obliterated cavity (*) with the cochlear electrode array inserted in the cochlea (*arrow*).

Auditory Brainstem Implantation—Surgical Anatomy

See (Figs. 7.164–7.169)

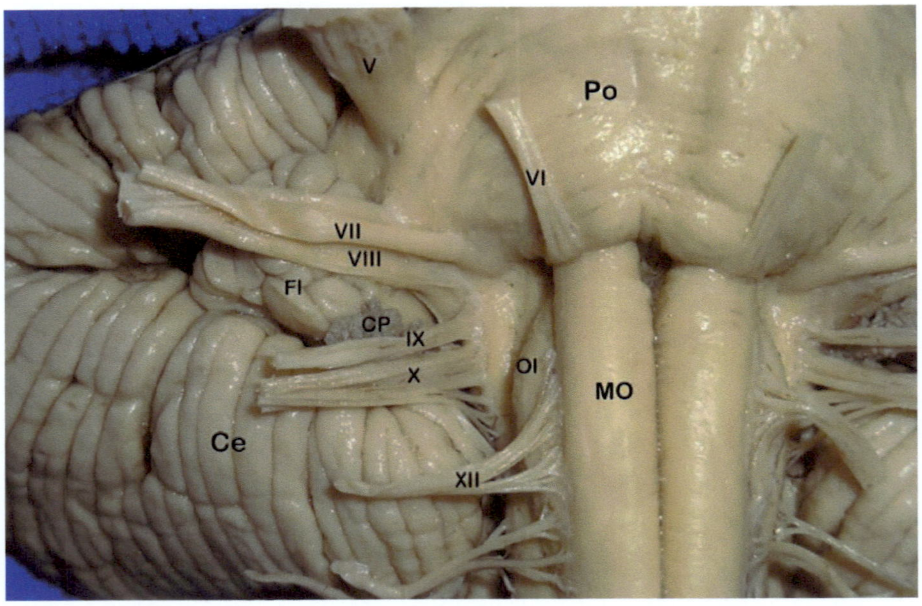

Fig. 7.164 Ventral view of brainstem. The choroid plexus (CP) and Luschka foramen can be found at the base of the vestibulocochlear (VIII) and glossopharyngeal nerves (IX), located just posteriorly. *Ce*, cerebellum; *Fl*, flocculus; *IX*, glossopharyngeal nerve; *MO*, medulla oblongata; *Ol*, olive; *Po*, pons; V, trigeminal nerve; *VI*, abducent nerve; *VII*, facial nerve; *VIII*, vestibulocochlear nerve; *X*, vagus nerve; *XII*, hypoglossal nerve.

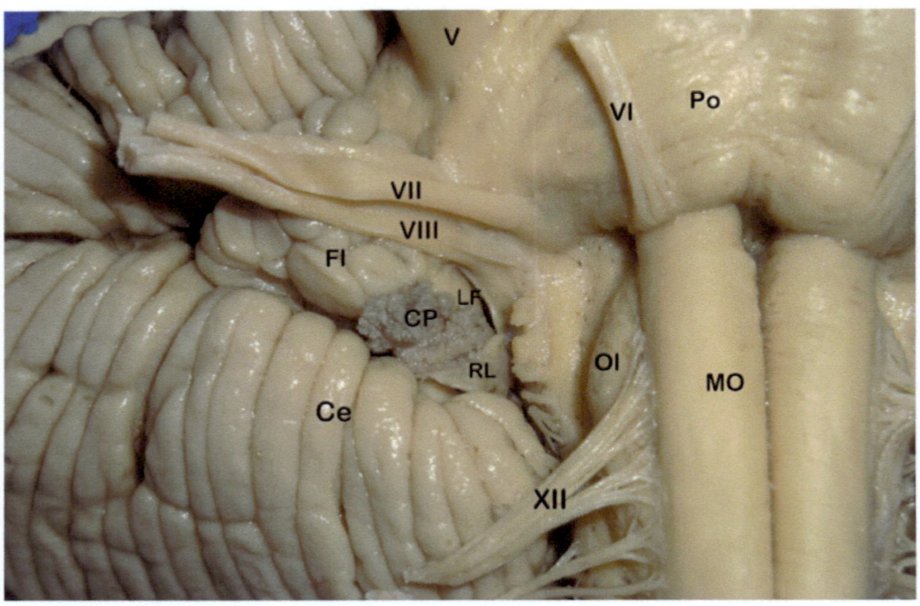

Fig. 7.165 The glossopharyngeal and vagus nerve rootlets has been removed. Ldentification of the flocculus (Fl) of the cerebellum leads to the choroid plexus (CP), which projects from the fourth ventricle and exits through the Luschka foramen (LF). The rhomboid lip (RL), a thin neural membrane in the ventral margin of the lateral recess, extends laterally behind the glossopharyngeal and vagus nerves with the choroid plexus. *Ce*, cerebellum; *MO*, medulla oblongata; *Ol*, olive; *Po*, pons; *V*, trigeminal nerve; *VI*, abducent nerve; *VII*, facial nerve; *VIII*, vestibulocochlear nerve; *XII*, hypoglossal nerve.

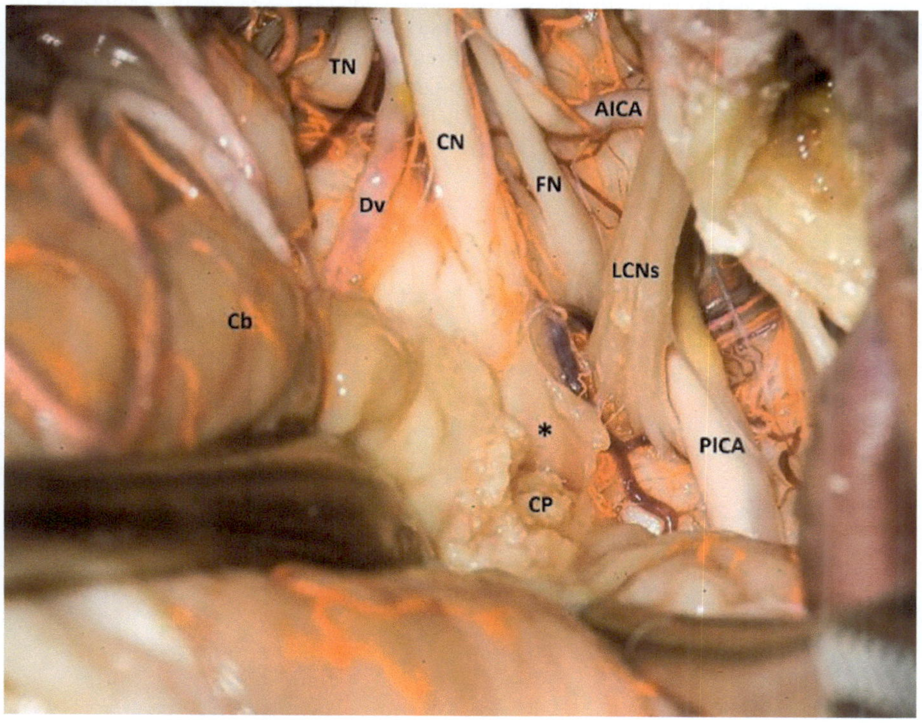

Fig. 7.166 The ABI is inserted through an enlarged translabyrinthine approach (ETLA), which provides the best angle of approach and direct access to the lateral recess of the fourth ventricle (*). A gentle retraction of the flocculus is enough to expose the root of the cochlear nerve (CN) and choroid plexus (CP). The CP provides a constant anatomical landmark for the lateral recess of the fourth ventricle, which lies medial to it. *AICA*, anterior inferior cerebellar artery; *Cb*, cerebellum; *Dv*, Dandy's vein; *FN*, facial nerve; *LCNs*, lower cranial nerves; *PICA*, posterior inferior cerebellar artery; *TN*, trigeminal nerve.

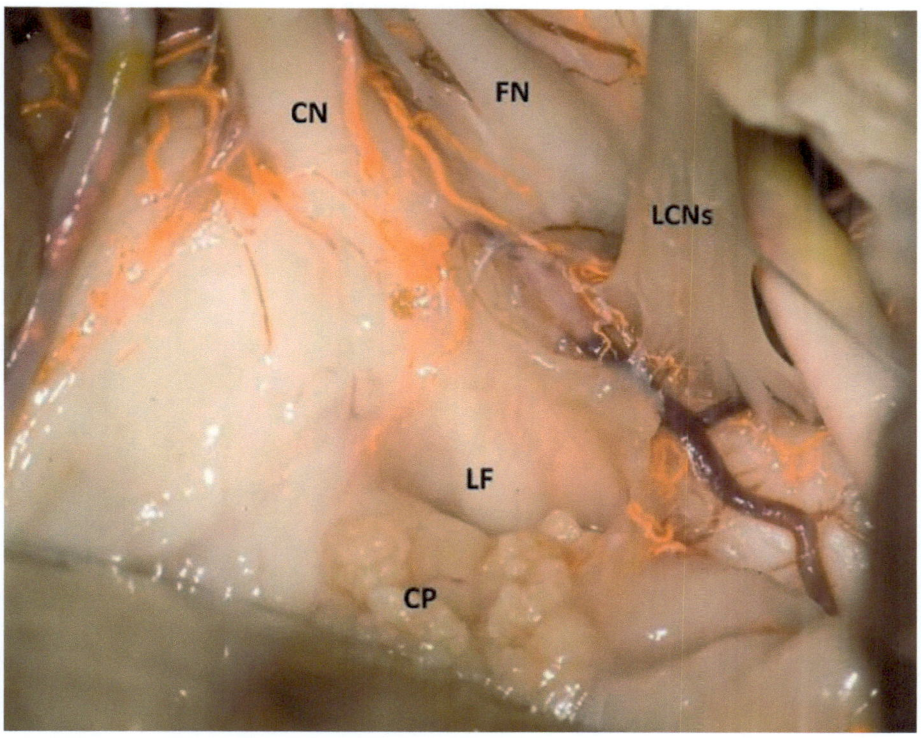

Fig. 7.167 Closer view. *CN*, cochlear nerve; *CP*, choroid plexus; *FN*, facial nerve; *LCNs*, lower cranial nerves; *LF*, foramen of Luschka.

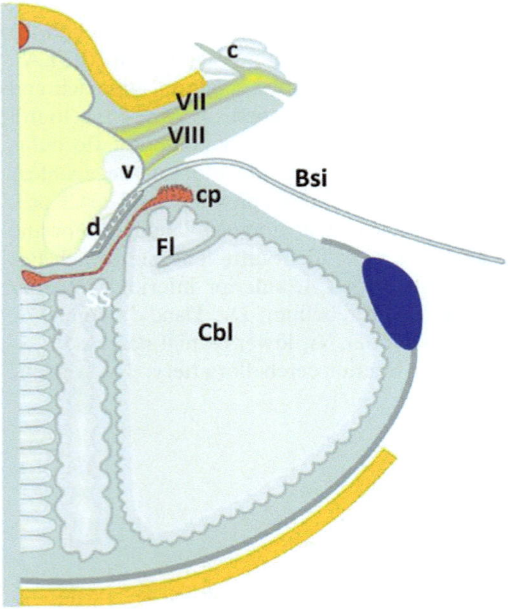

Fig. 7.168 Anatomy of the fourth ventricle and the lateral recess and the location of the cochlear nuclei. This schematic illustration shows an ideal placement of an array of electrodes. The choroid plexus (cp) is shown in red. *c*, cochlea; *Cbl*, cerebellum; *d*, dorsal cochlear nucleus; *Fl*, flocculus; *SS*, sigmoid sinus; *v*, ventral cochlear nucleus; *VII*, facial nerve; *VIII*, stump of the vestibulocochlear nerve.

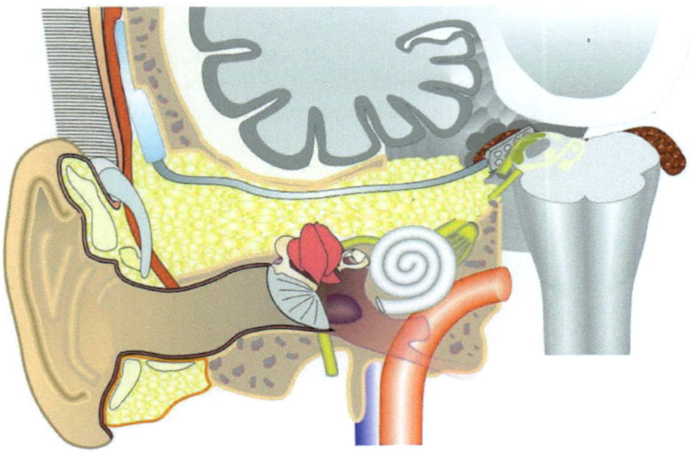

Fig. 7.169 Schematic drawing showing the ABI in place at the end of a standard translabyrinthine approach.

Case 6: Simultaneous Removal of Vestibular Schwannoma and Auditory Brainstem Implantation

See (Figs. 7.170–7.177).

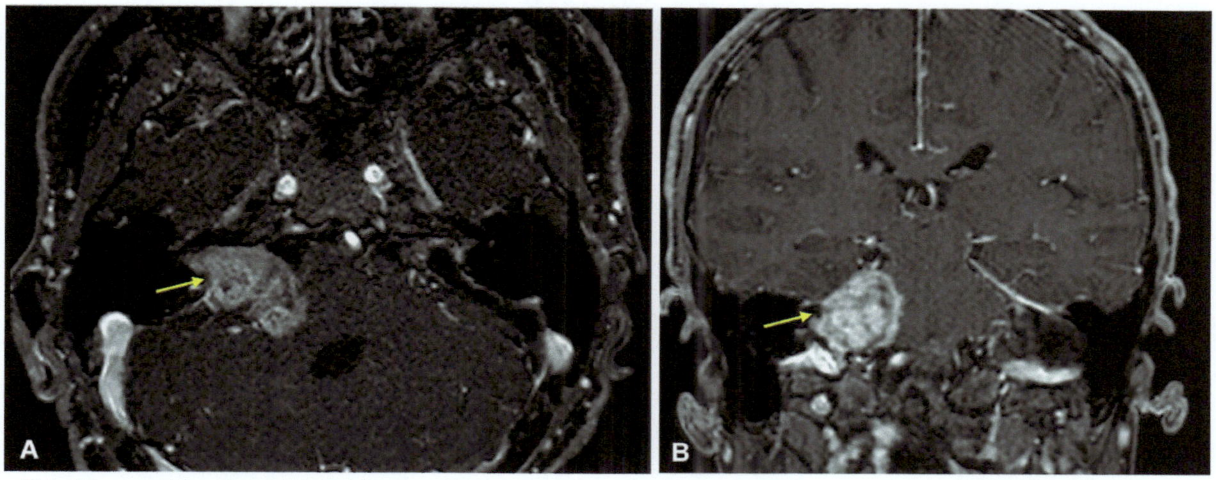

Fig. 7.170 This 26-year-old male patient with neurofibromatosis type 2 (NF2) was evaluated because of previous surgeries performed elsewhere for bilateral vestibular schwannoma. The patient had a residual tumor on the right side (arrow) with bilateral total hearing loss and mild facial nerve palsy on the side of the residual tumor (grade III House-Brackmann scale). He underwent tumor removal with simultaneous auditory brainstem implantation via an enlarged translabyrinthine approach. (A) T1W+Gd enhanced MRI (*axial cut*): note the tumor occupying the internal auditory canal and cerebellopontine angle with compression of the brainstem (*arrow*); (B) T1W+Gd enhanced MRI (*coronal cut*).

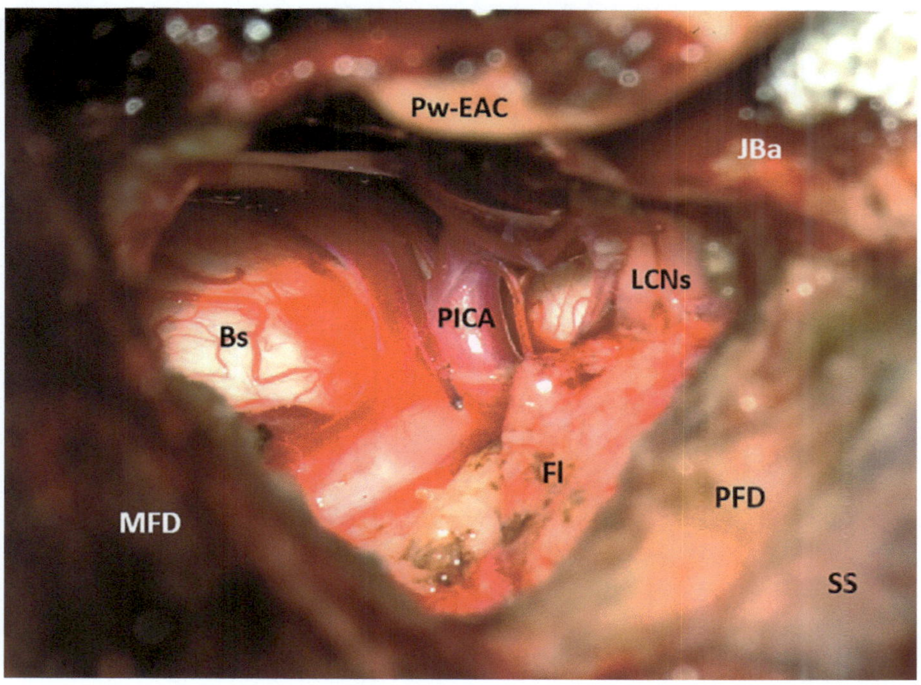

Fig. 7.171 Surgical view after total tumor removal. Both the cochlear and facial nerves were infiltrated by the tumor and sacrificed. *Fl*, flocculus; *JBa*, jugular bulb area; *LCNs*, lower cranial nerves; *MFD*, middle fossa dura; *PFD*, posterior fossa dura; *PICA*, posterior inferior cerebellar artery; *Pw-EAC*, posterior wall of the external auditory canal; *SS*, sigmoid sinus.

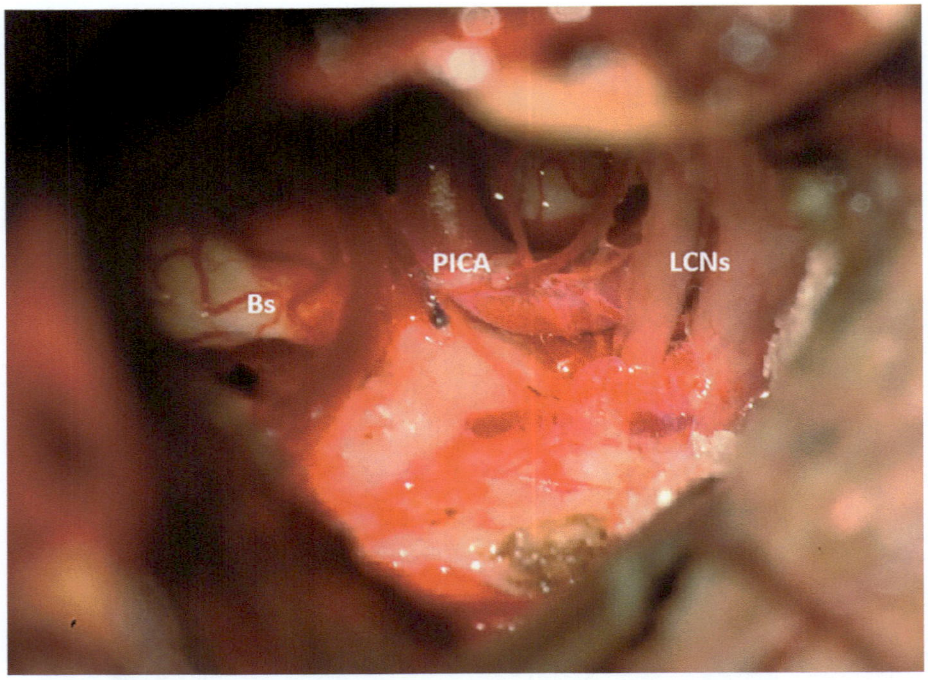

Fig. 7.172 The flocculus is retracted inferiorly to gain access to the foramen of Luschka. *Bs*, brainstem; *LCNs*, lower cranial nerves; *PICA*, posterior inferior cerebellar artery.

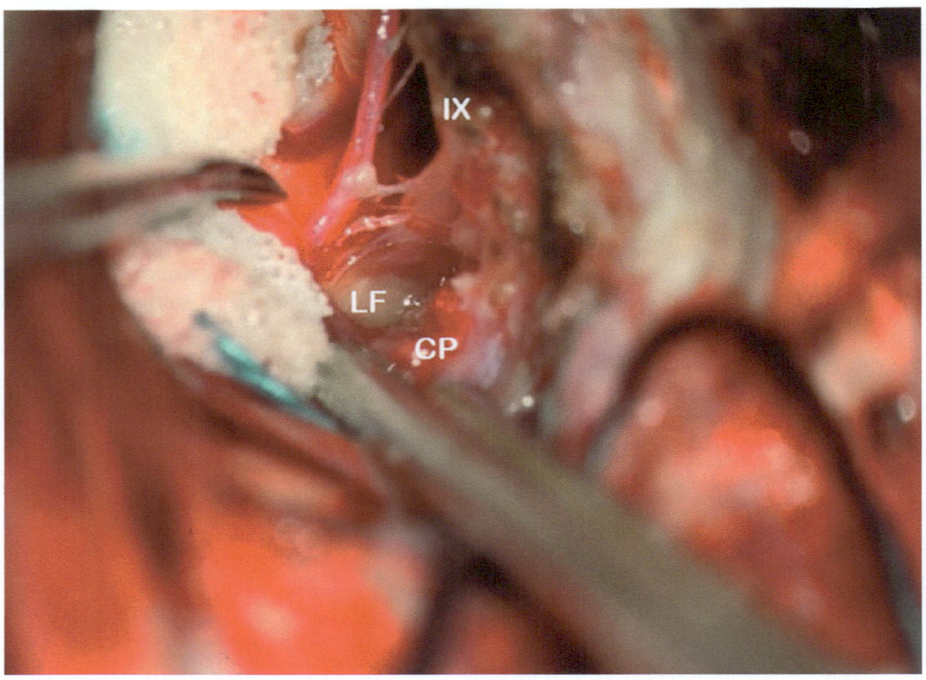

Fig. 7.173 Further retraction of the flocculus exposes both the choroid plexus (CP) and the foramen of Luschka (LF). A Valsalva maneuver performed by the anesthesiologist may help to identify the entrance to the lateral recess of the fourth ventricle. IX glossopharyngeal nerve.

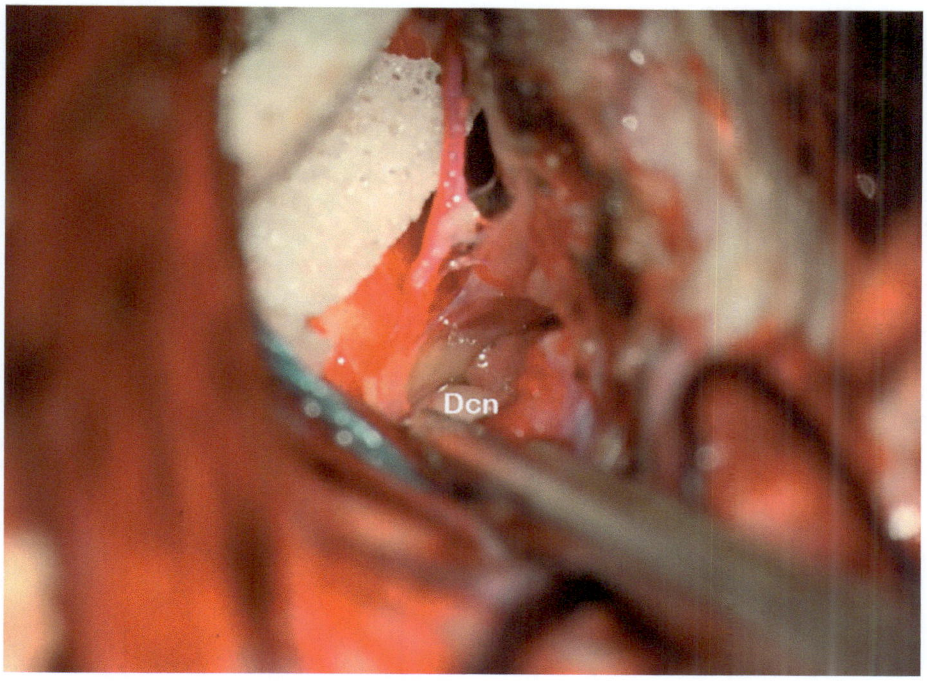

Fig. 7.174 Further retraction exposes the Dorsal cochlear nucleus (Dcn).

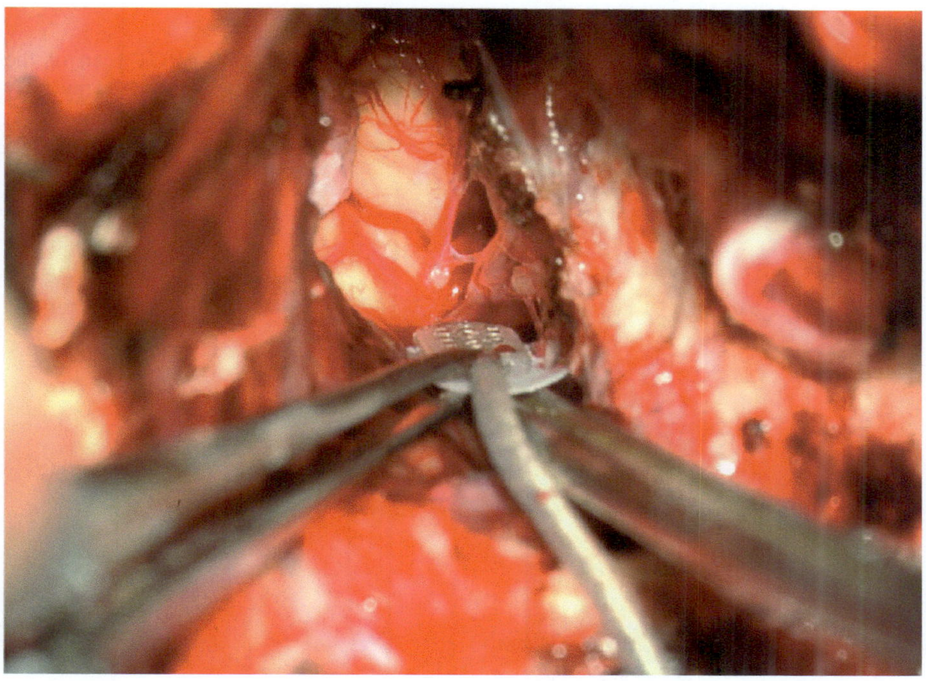

Fig. 7.175 The array of the ABI is inserted into the foramen of Luschka.

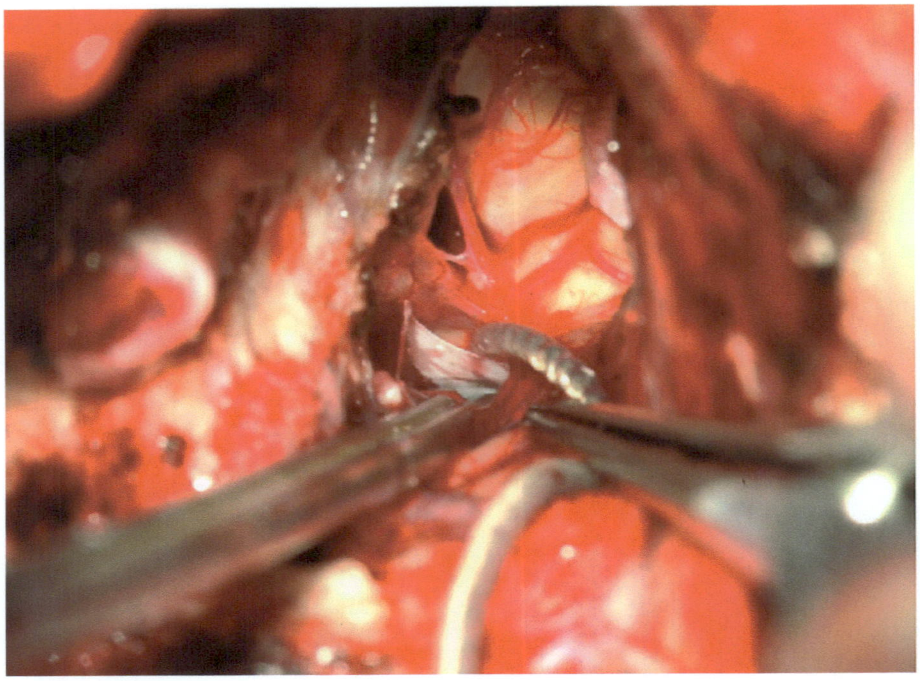

Fig. 7.176 The array is fully inserted into the lateral recess of the fourth ventricle and is in contact with the dorsal and ventral cochlear nuclei. At the end of the procedure the cavity is filled with abdominal fat.

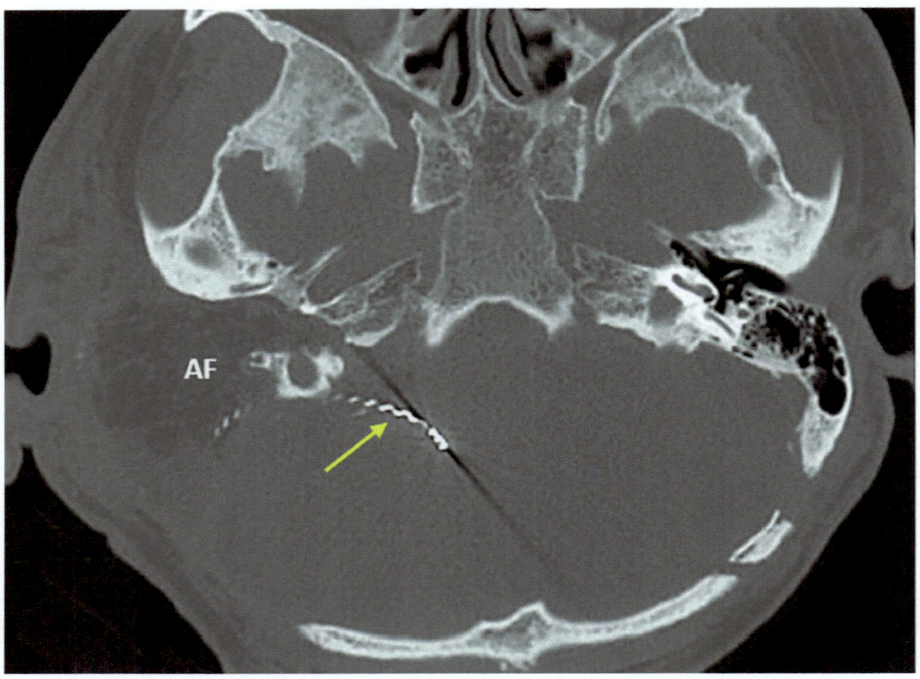

Fig. 7.177 Post-operative HRCT scan of the temporal bone, axial view. Note the array fully inserted into the lateral recess of the fourth ventricle (*arrow*). *AF*, abdominal fat.

HINTS AND PITFALLS

- Extensive removal of bone over the middle fossa dura and posterior to the sigmoid sinus, uncovering the posterior fossa dura, is a fundamental step for obtaining a wide approach adequate for tumor removal, in comparison with the classic translabyrinthine approach.
- The sigmoid sinus has to be completely uncovered. Leaving an island of bone over the sigmoid sinus (Bill's island) limits the degree of sinus retraction. The edges of the bone island may injure the sinus wall during retraction.
- Bone between the terminal part of the sigmoid sinus and the jugular bulb should be removed to allow optimal visualization of the lower portion of the cerebellopontine angle. Removing the bone here is a tedious job, and it should be done with care to avoid injuring the fragile jugular bulb.
- The attachment of the endolymphatic sac to the posterior fossa dura impedes posterior retraction of the latter. For this reason, a blade is used to incise the proximal part of the sac, with the tip directed against the bone.
- The facial nerve lies immediately lateral to the vestibule. The utmost care should be taken to avoid injuring the nerve during drilling to open the vestibule. At this level, drilling should be posterior and never medial to the nerve.
- The IAC should be uncovered over 270 degrees rather than only 180 degrees of its circumference.
- A thin shell of bone left covering the IAC until all the drilling around has been completed helps prevent injury to the contents of the canal from the rotating drill in case it slips.
- The original technique described by House for identifying the facial nerve at the fundus depends on identification of the vertical crest (Bill's bar). This is a constant landmark that separates the facial nerve, located anterosuperiorly, from the superior vestibular nerve, located posteriorly. The only disadvantage of this technique is that it carries a potential risk of injury to the facial nerve, particularly in inexperienced hands. In our practice, we use the horizontal crest and the superior ampullary nerve to identify the facial nerve (as described above), and we no longer use House's method, except in the very rare cases when there is doubt regarding the position of the facial nerve.
- Care should be taken while removing the incus to avoid fracturing the footplate, since this can create a communication between the cerebellopontine angle and the middle ear, with subsequent postoperative CSF leakage. The footplate can be protected by placing a small right-angled pick medial to the long process of the incus and distracting in a posterolateral direction.
- Use periosteum rather than fat for the closure of the attic. Periosteum is easier to manipulate, and provides a better seal.
- Do not displace the jugular bulb inferiorly if it was the only bulb or the predominant sinus. An angio-MRI with evaluation of brain venous drainage should be examined first in order to be sure of the adequacy of the contralateral circulation. If the jugular bulb needs to be lowered to facilitate inferior transapical drilling, decompression is performed and the bulb is then lowered using Surgicel or bone wax (see Figs. 7.178–7.181).
- Transapical drilling should be done with a diamond burr, with extreme care being taken not to injure the facial nerve. A useful trick is to place the suction irrigation tube between the drill and the nerve to provide protection.
- Care should be taken while drilling the petrous apex superior to the IAC so as not to injure the superior petrosal sinus.
- In giant tumors causing high intracranial pressure, the dura usually bulges, hindering access. In such cases, the dura can be opened first to allow drainage of CSF, with a resultant release of pressure and shrinkage of the dura. The drilling can then be continued.
- Particularly in tumors that lie in a far anterior position, the facial nerve is markedly displaced anteriorly. The transapical extension allows much better visualization of the nerve where it starts to curve.
- The transapical extension of the approach also allows better control of the trigeminal nerve up to Meckel's cave, and of the abducent nerve, prepontine cistern, and basilar artery.
- Using this approach, we have abandoned the transotic approach for anteriorly located tumors and limit its use only to cases of concomitant cochlear invasion or vertical IAC involvement.

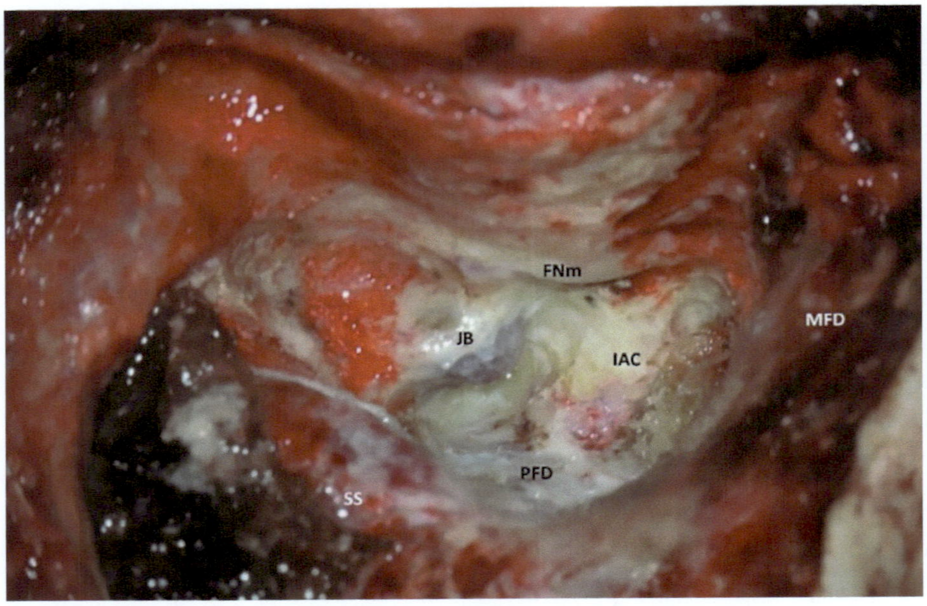

Fig. 7.178 Enlarged translabyrinthine approach for a left-sided vestibular schwannoma. Note the extension of the jugular bulb (JB) obscuring the inferior apical cells. In this case the contralateral venous drainage was normal, symmetrical and connected to the affected side. Therefore, the JB was lowered to improve the inferior extension of the approach to the lower cranial nerves area. *FNm*, mastoid portion of the facial nerve; *IAC*, internal auditory canal; *MFD*, middle fossa dura; *PFD*, posterior fossa dura.

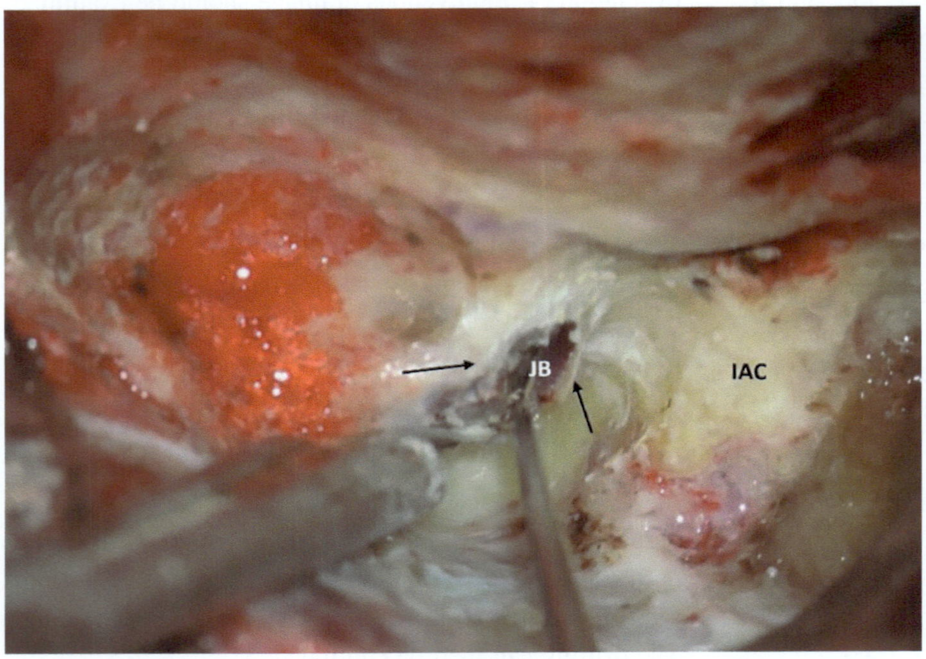

Fig. 7.179 Note the decompression of the jugular bulb (JB). Bone lateral and medial to the JB is preserved to hold the surgicel or bone wax in place during the procedure. *IAC*, internal auditory canal; *MFD*, middle fossa dura.

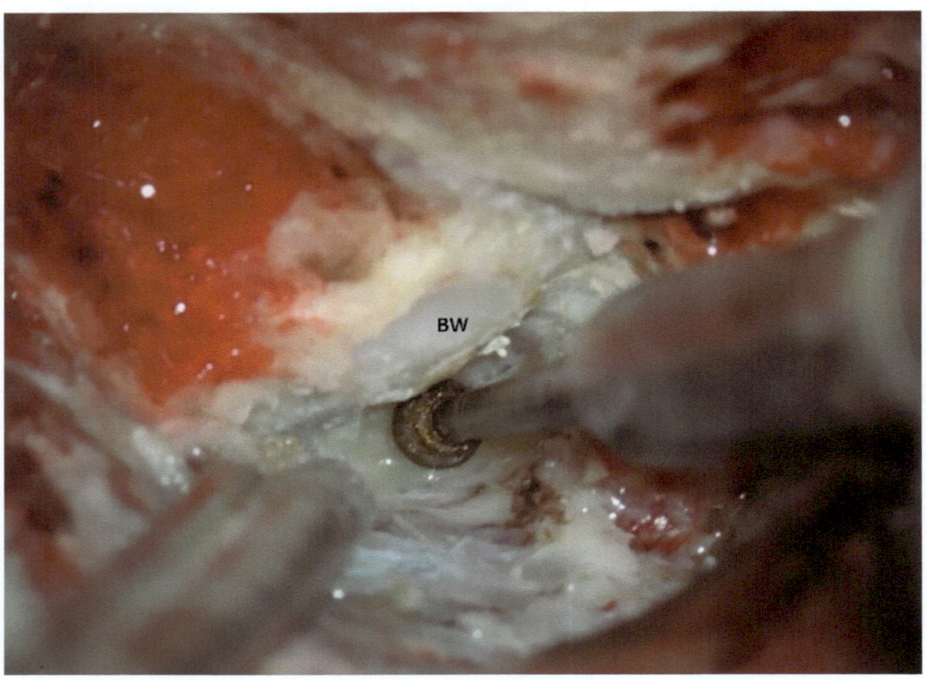

Fig. 7.180 Bone wax (BW) is used to lower the jugular bulb. Drilling of the inferior apical cells is now possible.

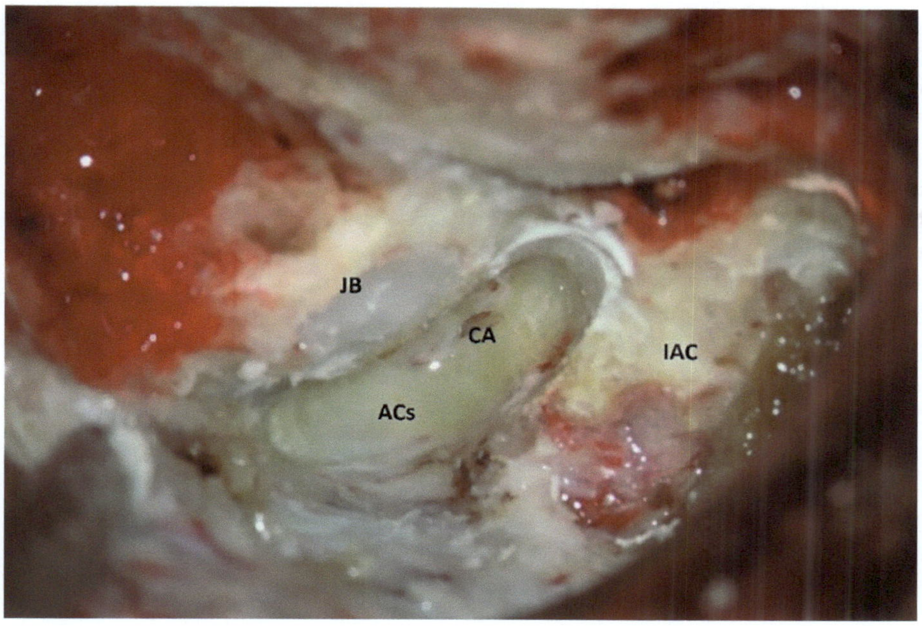

Fig. 7.181 Note the improved visualization of the inferior apical compartment. The cochlear aqueduct (CA) can also be opened for CSF drainage. *ACs*, inferior apical cels; *JB*, jugular bulb; *IAC*, internal auditory canal.

REFERENCES

[1] Nguyen-Huynh AT, Jackler RK, Pfister M, Tseng J. The aborted early history of the translabyrinthine approach: a victim of suppression or technical prematurity? Otol Neurotol 2007;28(2):269–79.

[2] House WF. Transtemporal bone microsurgical removal of acoustic neuromas: report of cases. Arch Otolaryngol 1964;80:617–67.

[3] Sanna M, Mancini F, Russo A, et al. Atlas of acoustic Neurinoma microsurgery. 2nd ed. Stuttgart, Germany: Thieme; 2011.

[4] Sanna M, Saleh E, Russo A, Falcioni M. Identification of the facial nerve in the translabyrinthine approach: an alternative technique. Otolaryngol Head Neck Surg 2001;124 (1):105–6.

[5] Sanna M, Agarwal M, Mancini F, Taibah A. Transapical extension in difficult cerebellopontine angle tumors. Ann Otol Rhinol Laryngol 2004;113(8):676–82.

[6] Saleh EA, Aristegui M, Taibah AK, Mazzoni A, Sanna M. Management of the high jugular bulb in the translabyrinthine approach. Otolaryngol Head Neck Surg 1994;110 (4):397–9.

[7] Sanna M, Rohit SLJ, Jain Y. Technique to prevent postoperative CSF leak in the translabyrinthine excision of vestibular schwannoma. J Laryngol Otol 2003;117(12):965–8.

[8] Falcioni M, Fois P, Taibah A, Sanna M. Facial nerve function after vestibular schwannoma surgery. J Neurosurg 2011;115 (4):820–6.

[9] Ekvall L, Kinnefors A, Bynke O, Rask-Andersen H. Experience and results in 261 consecutive patients with acoustic tumors operated on with the translabyrinthine technique. In: Tos M, Thomsen J, editors. First international conference on acoustic neuroma. Copenhagen, Denmark: Kugler Publications; 1991. p. 347–52.

[10] Thomsen J, Tos M, Borgesen S, Moller H. Surgical results after translabyrinthine removal of 504 acoustic neuromas. In: Tos M, Thomsen J, editors. First international conference on acoustic neuroma. Copenhagen, Denmark: Kugler Publications; 1991. p. 331–5.

[11] Merkus P, Taibah A, Sequino G, Sanna M. Less than 1% cerebrospinal fluid leakage in 1,803 translabyrinthine vestibular schwannoma surgery cases. Otol Neurotol 2010;31(2):276–83.

[12] Khrais TH, Falcioni M, Taibah A, Agarwal M, Sanna M. Cerebrospinal fluid leak prevention after translabyrinthine removal of vestibular schwannoma. Laryngoscope 2004;114 (6):1015–20.

[13] Dandinarasaiah M, Prasad SC, Piccirillo E, Vashishth A, Valentina M, Grinblat G, Codreanu CM, Sanna M. Facial nerve outcomes following Total excision of vestibular schwannoma by the enlarged Translabyrinthine approach. Otol Neurotol 2019;40(2):226–35.

[14] Lassaletta L, Aristegui M, Medina M, Aranguez G, Pérez-Mora RM, Falcioni M, Gavilán J, Piazza P, Sanna M. Ipsilateral cochlear implantation in patients with sporadic vestibular schwannoma in the only or best hearing ear and in patients with NF2. Eur Arch Otorhinolaryngol 2016;273(1): 27–35.

[15] Ben Ammar M, Piccirillo E, Topsakal V, et al. Surgical results and technical refinements in translabyrinthine excision of vestibular schwannomas: the Gruppo Otologico experience. Neurosurgery 2012;70:1481–91. discussion 91.

[16] Brackmann DE, Cullen RD, Fisher LM. Facial nerve function after translabyrinthine vestibular schwannoma surgery. Otolaryngol Head Neck Surg 2007;136:773–7.

[17] Tos M, Thomsen J. The translabyrinthine approach for the removal of large acoustic neuromas. Arch Otorhinolaryngol 1989;246:292–6.

[18] Piccirillo E, Wiet MR, Flanagan S, et al. Cystic vestibular schwannoma: classification, management, and facial nerve outcomes. Otol Neurotol 2009;30:826–34.

[19] Angeli RD, Piccirillo E, Di Trapani G, Sequino G, Taibah A, Sanna M. Enlarged translabyrinthine approach with transapical extension in the management of giant vestibular schwannomas: personal experience and review of literature. Otol Neurotol 2011;32(1):125–31. https://doi.org/10.1097/MAO.0b013e3181ff7562.

[20] Sanna M, Di Lella F, Guida M, Merkus P. Auditory brainstem implants in NF2 patients: results and review of the literature. Otol Neurotol 2012;33(2):154–64. https://doi.org/10.1097/MAO.0b013e318241bc71.

[21] Ben Ammar M, Piccirillo E, Topsakal V, Taibah A, Sanna M. Surgical results and technical refinements in translabyrinthine excision of vestibular schwannomas: the Gruppo Otologico experience. Neurosurgery 2012;70(6):1481–91; discussion 1491. https://doi.org/10.1227/NEU.0b013e31824c010f.

[22] Dandinarasaiah M, Prasad SC, Piccirillo E, Vashishth A, Valentina M, Grinblat G, Codreanu CM, Sanna M. Facial nerve outcomes following total excision of vestibular schwannoma by the enlarged translabyrinthine approach. Otol Neurotol 2019;40(2):226–35. https://doi.org/10.1097/MAO.0000000000002068.

[23] Sanna M, Piccirillo E, Kihlgren C, Cagliero G, Guidi M, Saleh E. Simultaneous Cochlear implantation after translabyrinthine vestibular schwannoma resection: a report of 41 cases. Otol Neurotol 2021;42(9):1414–21. https://doi.org/10.1097/MAO.0000000000003258.

[24] Sykopetrites V, Taibah A, Piras G, Giannuzzi AL, Mancini F, Sanna M. The otologic approach in the management of posterior petrous surface meningiomas. Eur Arch Otorhinolaryngol 2022;279(12):5655–65. https://doi.org/10.1007/s00405-022-07442-3 [Epub 2022 Jun 29].

[25] Wang Y, Piras G, Lauda L, Caruso A, Russo A, Taibah A, Sanna M. Superior petrosal vein sacrifice in translabyrinthine approach for resection of vestibule schwannoma. Eur Arch Otorhinolaryngol 2024;281(3):1195–203. https://doi.org/10.1007/s00405-023-08208-1 [Epub 2023 Sep 4].

CHAPTER 8

Retrosigmoid-Retrolabyrinthine Approach

Introduction

Neurosurgeons introduced the suboccipital approach for cerebellopontine angle (CPA) tumors. Cushing first postulated the practice of a scrupulous surgical technique using an extended bilateral suboccipital approach and subtotal intracapsular tumor resection to preserve the nervous structures and brainstem vasculature intact [1]. Dandy supported the practice of the unilateral suboccipital technique and total capsular tumor dissection, reducing the mortality rate after complete removal of vestibular schwannoma (VS) [2]. Since 1968, there have been cases of VS excision with successful hearing preservation by using the retrosigmoid approach, a technical modification of the suboccipital approach [3–6]. Since then, the debate between otologists and neurosurgeons on the real possibility of auditory preservation after VS removal is still open [7–12]. Similarly, technical modifications such as the use of the endoscope to check the internal auditory canal [9–15] or minimally invasive approaches [16] have been introduced over the years.

We no longer apply the classic retrosigmoid approach for VS removal. A combination of retrosigmoid and retrolabyrinthine craniotomies has been adopted in all of our cases. With this approach, we can significantly decrease the risk of cerebrospinal fluid leak postoperatively due to the exenteration and obliteration of all the retrolabyrinthine air cells, which may potentially communicate with the middle ear space [17]. At the same time, it is possible to obtain a larger access to the CPA without further retraction of the cerebellum by this combination. This has two advantages. We can have an effective spontaneous retraction of the cerebellum by making a small opening on the dura anterior to the sigmoid sinus prior to the retrosigmoid dural incision, which may decrease the intracranial pressure. Additionally, the uncovered sigmoid sinus together with the posterior fossa dura could be displaced anteriorly. This maneuver allows a direct view of the posterior surface of the temporal bone and of the internal auditory canal and its contents.

Indications

The approach is indicated in cerebellopontine angle lesions with good preoperative hearing, and hearing preservation is aimed at:

- Vestibular schwannoma less than 1.5 cm in extrameatal diameter; however, tumors extending to the fundus of the internal auditory canal or enlarging the IAC are considered relative contraindications.
- Vestibular schwannomas even larger than 1.5 cm in extrameatal diameter with serviceable preoperative hearing, in which the translabyrinthine approach is contraindicated, for example, neurofibromatosis type 2. In this special group of patients, we will attempt this approach to give the patient a chance of hearing preservation.
- Posterior fossa meningioma centered on or lying posterior to the internal auditory canal.

Atlas of Microsurgery of the Lateral Skull Base. https://doi.org/10.1016/B978-0-443-19083-4.00008-8

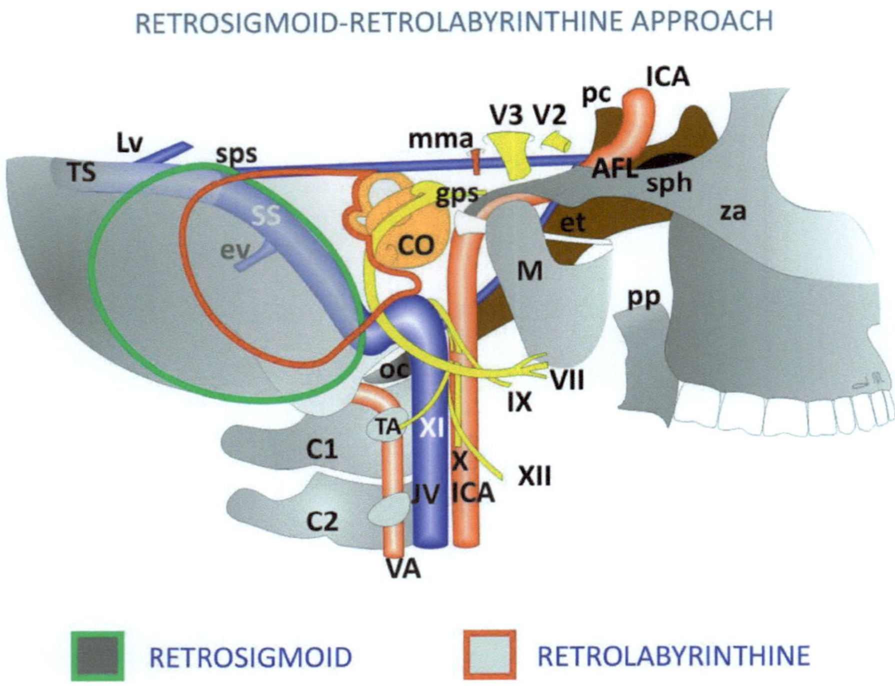

Fig. 8.1 Schematic drawing outlining the retrosigmoid approach and retrosigmoid-retrolabyrinthine approach. *AFL*, anterior foramen lacerum; *C1*, atlas; *C2*, axis; *CO*, cochlea; *et*, eustachian tube; *ev*, emissary vein; *gps*, greater petrosal nerve; *ICA*, internal carotid artery; *JV*, jugular vein; *Lv*, vein of Labbè; *M*, mandibule; *mma*, middle meningeal artery; *oc*, occipital condyle; *pc*, clinoid process; *pp*, pterygoid process; *sph*, sphenoid; *sps*, superior petrosal sinus; *TA*, transverse process of the atlas; *TS*, transverse sinus; *za*, zygomatic arch; *V2*, maxillary nerve; *V3*, mandibular nerve; *VII*, facial nerve; *IX*, glossopharyngeal nerve; *XII*, hypoglossal nerve.

- Other cerebellopontine angle tumors lying posterior to the internal auditory canal, e.g., some cases of epidermoids (Fig. 8.1).

Contraindications

In our experience, this approach is regarded as being contraindicated in the following cases:

- Preoperative unserviceable hearing loss
- Large tumor (i.e., larger than 1.5 cm in extrameatal diameter)
- Very high jugular bulb that extends above the inferior border of the IAC
- Advanced age

Surgical Steps

1. The patient is placed in the supine position with the head turned to the contralateral side. The incision starts about 2 cm above the ear attachment, courses posteriorly (about 6 cm posterior to the postauricular sulcus), and then curves inferiorly 1 cm below the level of the mastoid tip (Fig. 8.2).
2. The skin flap is elevated and fixed anteriorly using skin hooks. A U-shaped, inferiorly based musculoperiosteal flap is outlined (Fig. 8.3). The flap is elevated using a Lempert periosteal elevator (Fig. 8.4). Cutting monopolar diathermy is used against the bone at the level of the mastoid tip and the insertion of the semispinalis muscle. Bleeding from the emissary vein usually occurs during this step and is controlled using bone wax.
3. An enlarged mastoidectomy is carried out (Fig. 8.5). We no longer use the conventional retrosigmoid craniotomy. The sigmoid sinus is identified and completely uncovered between the transverse sinus and the area of the jugular bulb. A retrosigmoid 5 × 5 cm craniotomy is carried out using a diamond burr (Fig. 8.6). The bone is separated from the dura using a septal elevator and is placed in saline for further closure (Fig. 8.7). The posterior fossa dura should be sufficiently exposed approximately 4 cm posterior to the sigmoid sinus. At the same time, a small opening of the dura in the inferior part of the approach is made. This facilitates the CSF leaking during the drilling before a definitive opening of the dura. The dura is held using toothed forceps, and a C-shaped incision is performed using microscissors (Fig. 8.8). The anterior dural flap is reflected together with the sinus using stay sutures (Fig. 8.9).
4. Once the dura is opened, Merocel is applied over the cerebellum to provide protection. The lower surface of the cerebellum is raised gently, and the arachnoid of the cistern is identified. The cistern is opened in a direction parallel to the lower cranial nerve, and CSF is allowed to drain. This allows shrinkage of the cerebellum, providing the necessary space for tumor removal. The arachnoid is removed from the neurovascular structures (Fig. 8.10).

Surgical Anatomy After Opening of the Dura

5. Figs. 8.10–8.14 show the view of the cerebellopontine angle and prepontine cistern as viewed after opening the dura.

6. Figs. 8.15–8.24 show if the tumor involves the internal auditory canal, as in acoustic neurinoma cases, the posterior wall of the internal auditory canal is drilled after elevating an inverted U-shaped dural flap from the bone to be drilled. Drilling the canal continues until the crus commune is blue-lined. This is the most lateral limit of drilling. Pieces of fat and muscle are used to block the internal auditory canal, and its placement is secured by fibrin glue.

7. At the end of the procedure, hemostasis is performed, and the cerebellopontine angle is irrigated with saline for a few minutes. The retrolabyrinthine craniotomy is obliterated with abdominal fat. The dura is closed in a watertight fashion using Vicryl 3/0 sutures. A mixture of bone pate, blood, and fibrin glue is applied over the dura. The craniotomy is returned in place, and the musculoperiosteal tissue is sutured back using Vicryl 1.0.

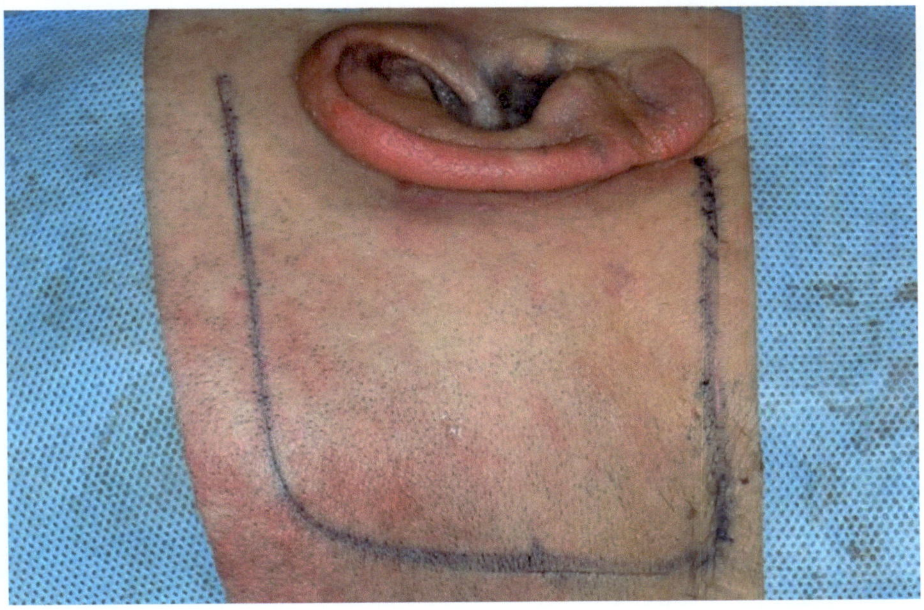

Fig. 8.2 The incision starts about 2 cm above the ear attachment, courses posteriorly (about 6 cm posterior to the postauricular sulcus), and then curves inferiorly 1 cm below the level of the mastoid tip.

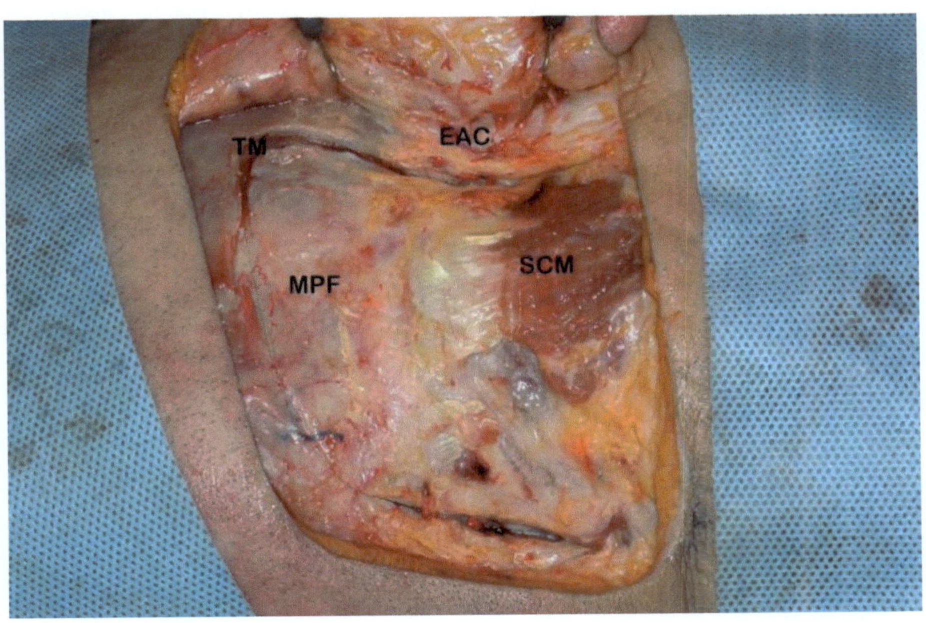

Fig. 8.3 The skin flap has already been elevated anteriorly, and monopolar coagulation was used to cut and delineate an inverted U inferiorly based musculoperiosteal flap. *EAC*, external auditory canal; *SCM*, sternocleidomastoid muscle; *MPF*, musculoperiosteal flap; *TM*, temporalis muscle.

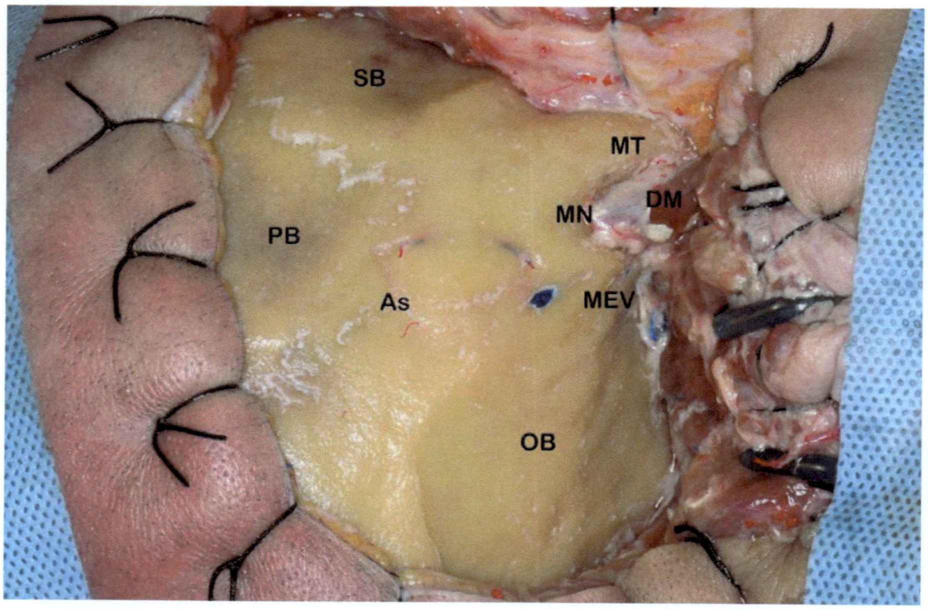

Fig. 8.4 The flap is elevated using a Lempert periosteal elevator. We use 1–0 silk sutures to stitch the soft tissue away from the surgical field, thus avoiding the use of retractors. *SB*, squamosal part of the temporal bone; *MT*, mastoid tip; *DM*, digastric muscle; *MN*, mastoid notch; *MEV*, mastoid emissary vein; *OB*, occipital bone; *As*, asterion; *PB*, parietal bone.

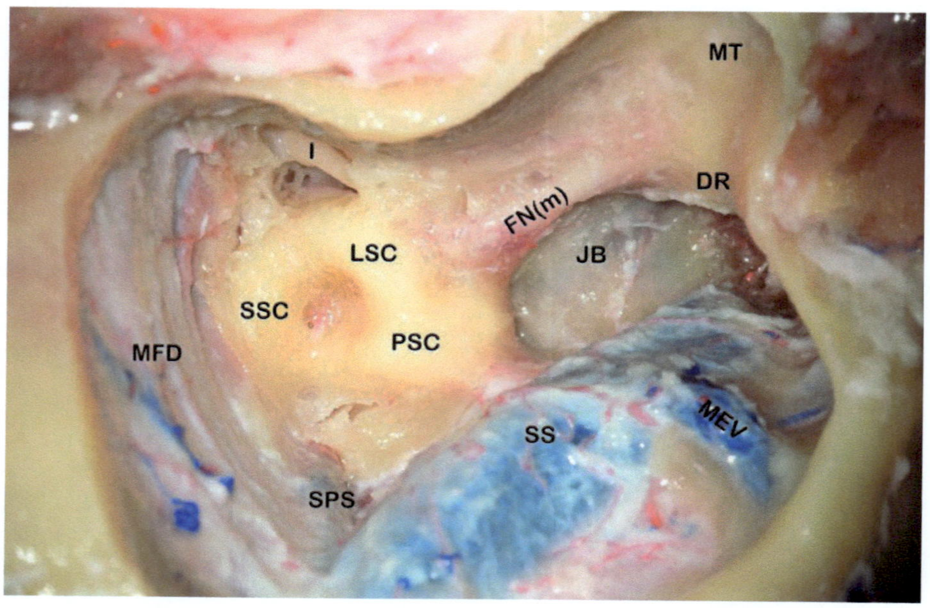

Fig. 8.5 An enlarged mastoidectomy is carried out. After the extended mastoidectomy is accomplished, a wide enough approach has been obtained. The sigmoid sinus is identified and followed to the transverse sinus. *MT*, mastoid tip; *DR*, digastric ridge; *JB*, jugular bulb; *FN(m)*, mastoid segment of facial nerve; *MEV*, mastoid emissary vein; *SS*, sigmoid sinus; *PSC*, posterior semicircular canal; *LSC*, lateral semicircular canal; *I*, incus; *SSC*, superior semicircular canal; *MFD*, middle fossa dura; *SPS*, superior petrosal sinus.

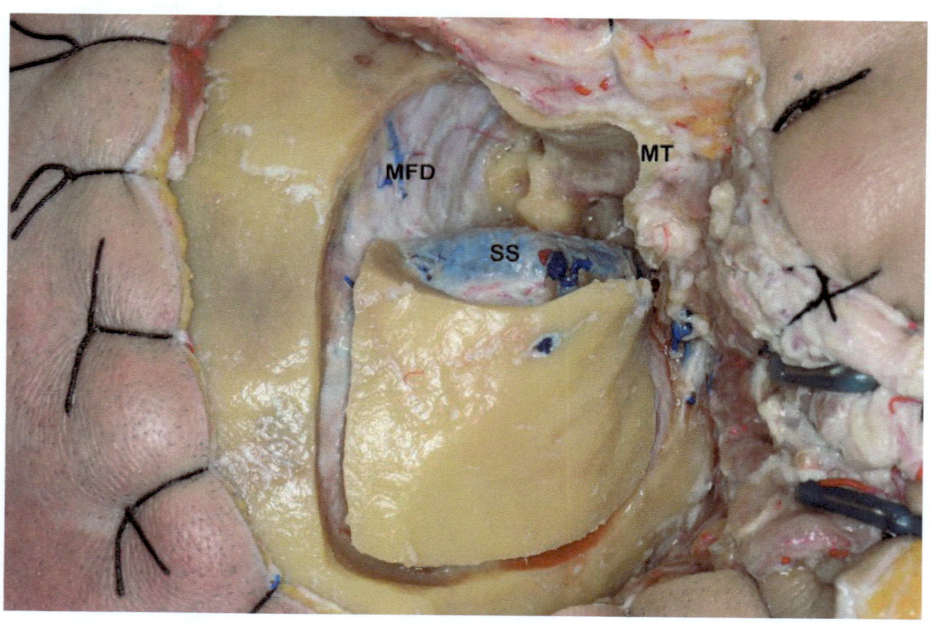

Fig. 8.6 A retrosigmoid 5 × 5 cm craniotomy is carried out using a diamond burr. *MT*, mastoid tip; *MFD*, middle fossa dura; *SS*, sigmoid sinus.

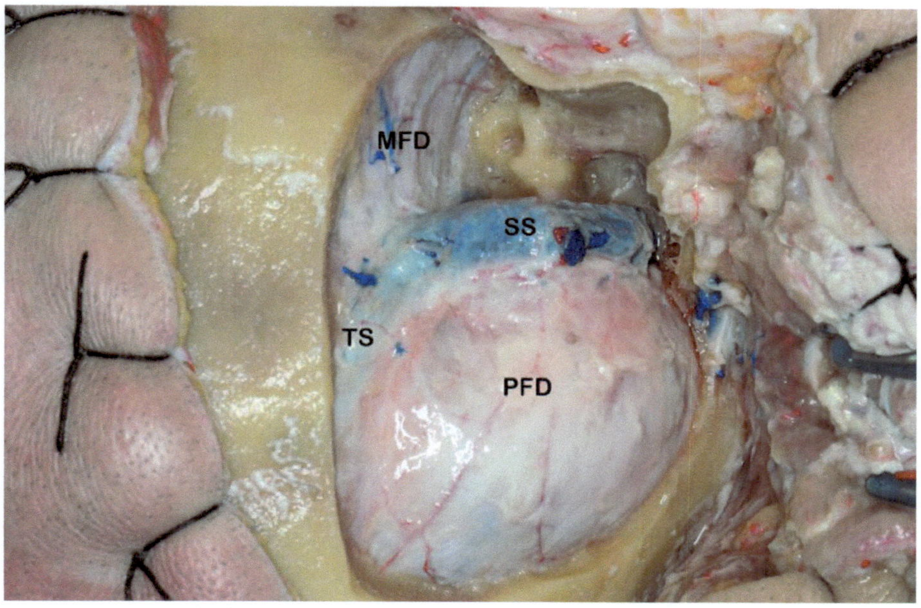

Fig. 8.7 The bone is separated from the dura using a septal elevator. The posterior fossa dura should be sufficiently exposed approximately 4 cm posterior to the sigmoid sinus. *MFD*, middle fossa dura; *SS*, sigmoid sinus; *TS*, transverse sinus; *PFD*, posterior fossa dura.

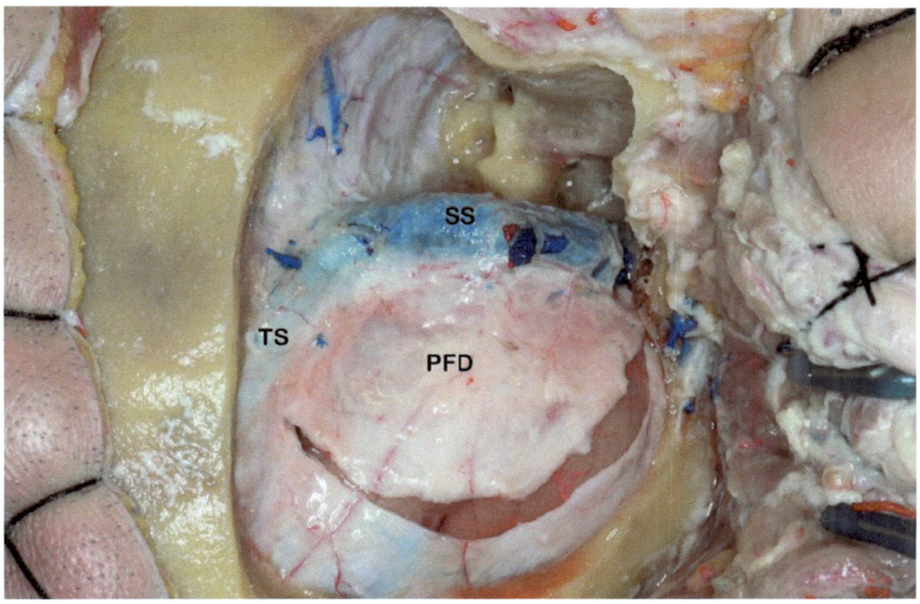

Fig. 8.8 A C-shaped dural flap incision has been made and is performed using microscissors. *SS*, sigmoid sinus; *TS*, transverse sinus; *PFD*, posterior fossa dura.

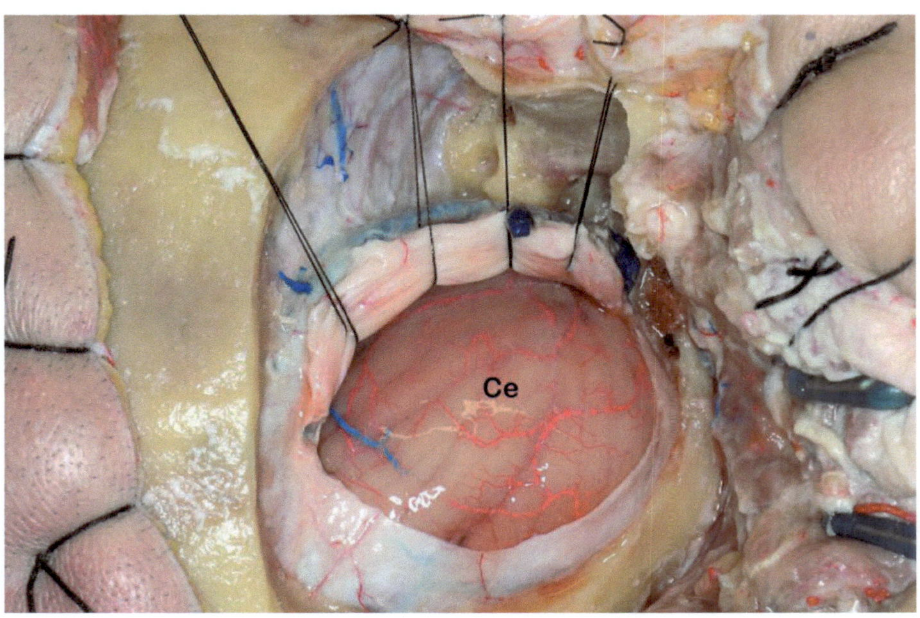

Fig. 8.9 The dural flap has been reflected anteriorly, and the cerebellum (Ce) can be seen.

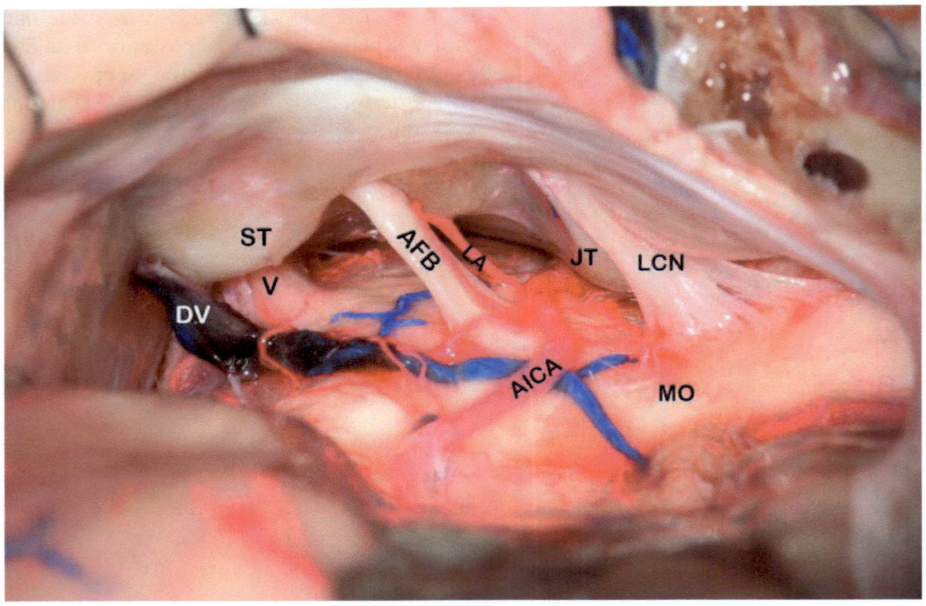

Fig. 8.10 The cerebellum is protected using Merocel, and the retractor is introduced to retract the cerebellum gently. The cerebellum has been elevated to expose the neurovascular structures in the cerebellopontine angle. *MO*, medulla oblongata; *AICA*, anteroinferior cerebellar artery; *LCN*, lower cranial nerves; *JT*, jugular tubercle; *LA*, labyrinthine artery; *AFB*, acousticofacial bundle; *ST*, suprameatal tubercle; *V*, trigeminal nerve; *DV*, Dandy's vein (superior petrosal vein).

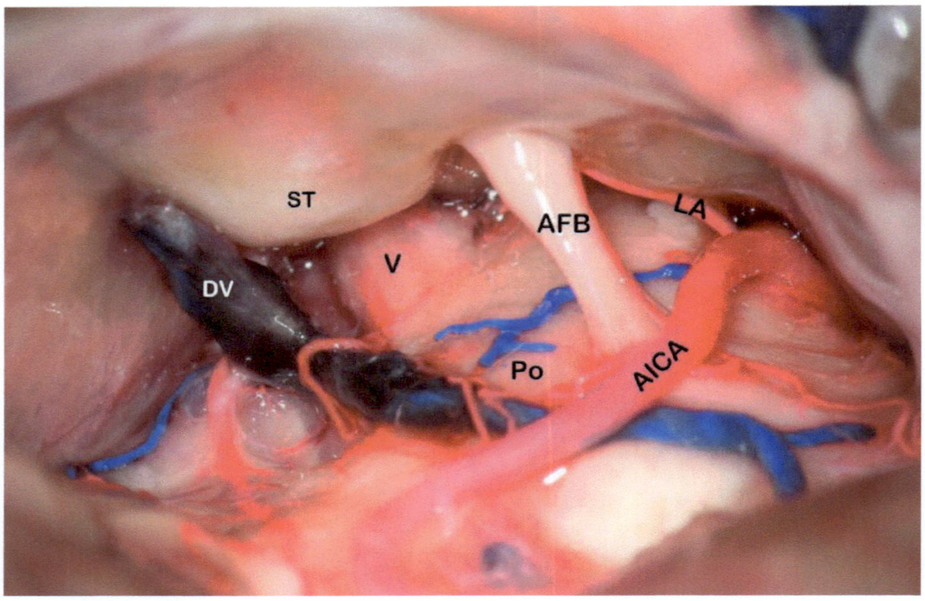

Fig. 8.11 A large Dandy's vein blocks access to the suprameatal area. The suprameatal tubercle is located above the porus of the internal acoustic meatus and lateral to the trigeminal nerve. *LA*, labyrinthine artery; *AFB*, acousticofacial bundle; *AICA*, anteroinferior cerebellar artery; *Po*, pons; *V*, trigeminal nerve; *ST*, suprameatal tubercle; *DV*, Dandy's vein (superior petrosal vein).

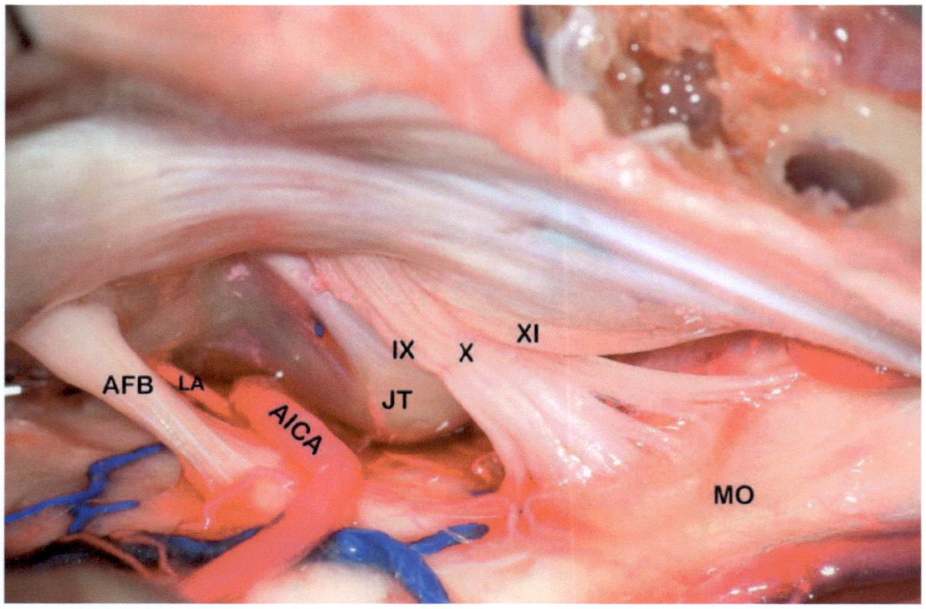

Fig. 8.12 In the lower part of this approach, the lower cranial nerves can be seen. *MO*, medulla oblongata; *XI*, accessory nerve; *X*, vagus nerve; *IX*, glossopharyngeal nerve; *JT*, jugular tubercle; *AICA*, anteroinferior cerebellar artery; *LA*, labyrinthine artery; *AFB*, acousticofacial bundle.

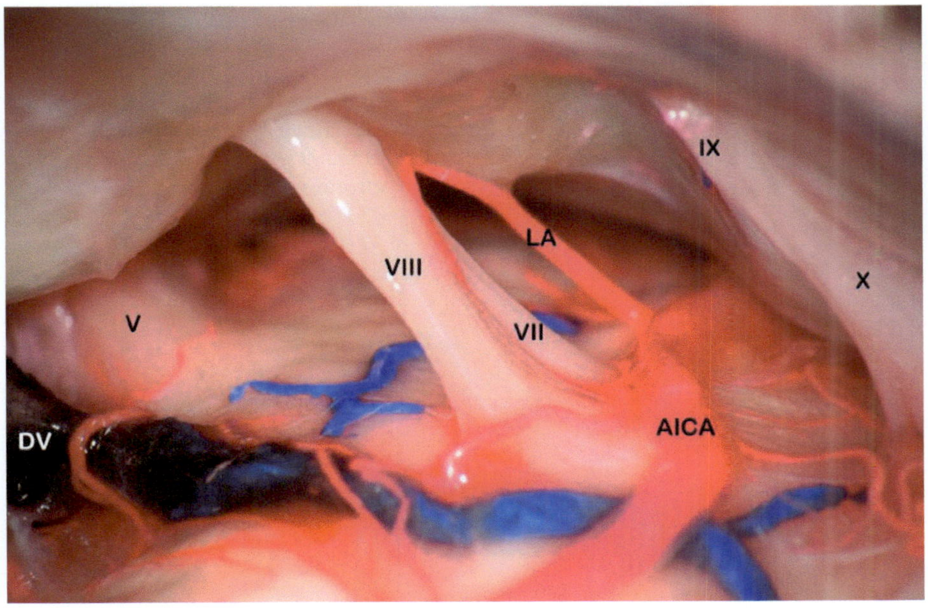

Fig. 8.13 The acousticofacial bundle enters into the internal auditory canal with a labyrinthine branch of the anteroinferior cerebellar artery. *IX*, glossopharyngeal nerve; *X*, vagus nerve; *LA*, labyrinthine artery; *AICA*, anteroinferior cerebellar artery; *VII*, facial nerve; *VIII*, vestibulocochlear nerve; *V*, trigeminal nerve; *DV*, Dandy's vein (superior petrosal vein).

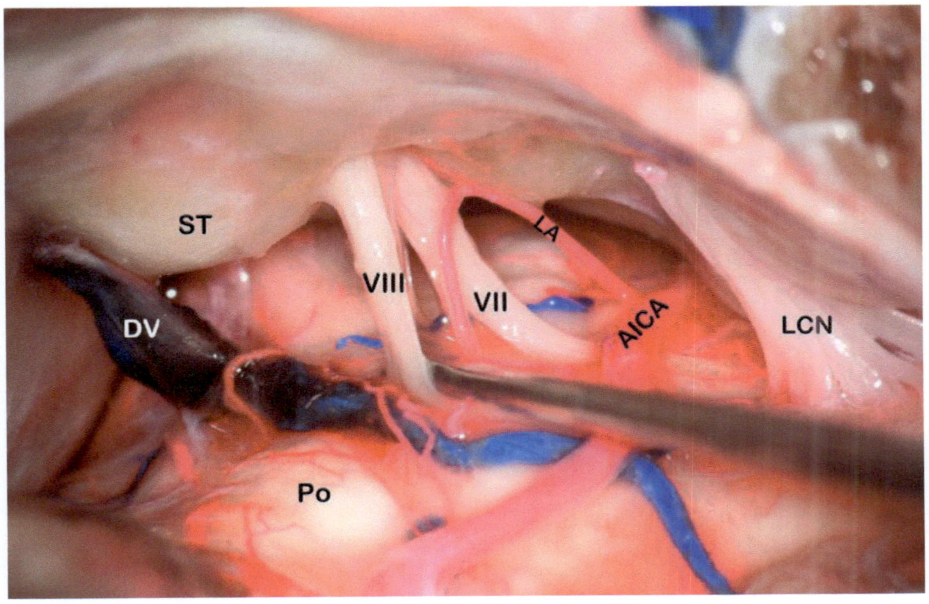

Fig. 8.14 Retracting the vestibulocochlear nerve exposes the anterior-lying facial nerve and labyrinthine artery. *LCN*, lower cranial nerves; *AICA*, anteroinferior cerebellar artery; *LA*, labyrinthine artery; *VII*, facial nerve; *VIII*, vestibulocochlear nerve; *ST*, suprameatal tubercle; *Po*, pons; *DV*, Dandy's vein (superior petrosal vein).

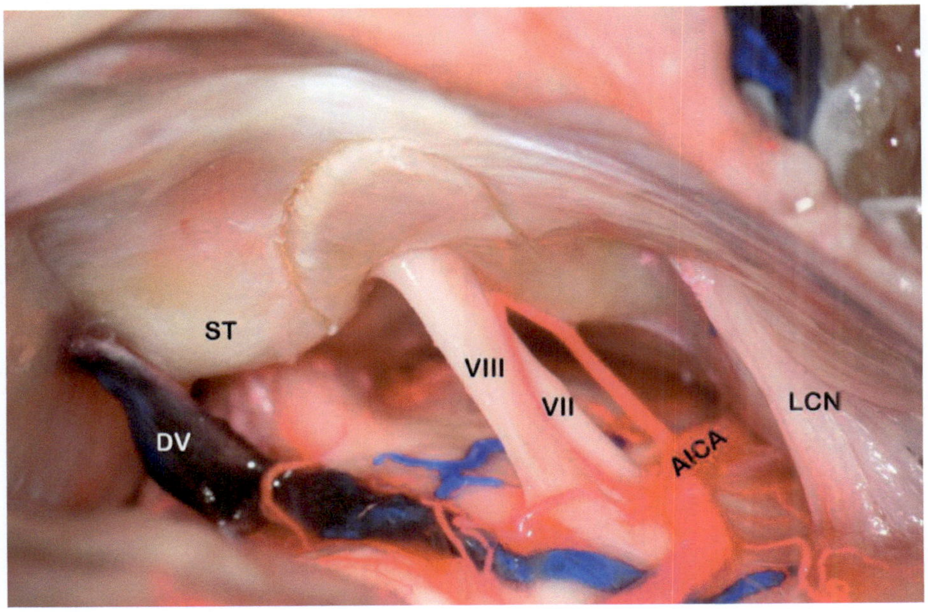

Fig. 8.15 The dura of the posterior lip is cut in an arciform fashion with the base of the arc facing the cranial nerves. *LCN*, lower cranial nerves; *AICA*, anteroinferior cerebellar artery; *VII*, facial nerve; *VIII*, vestibulocochlear nerve; *ST*, suprameatal tubercle; *DV*, Dandy's vein (superior petrosal vein).

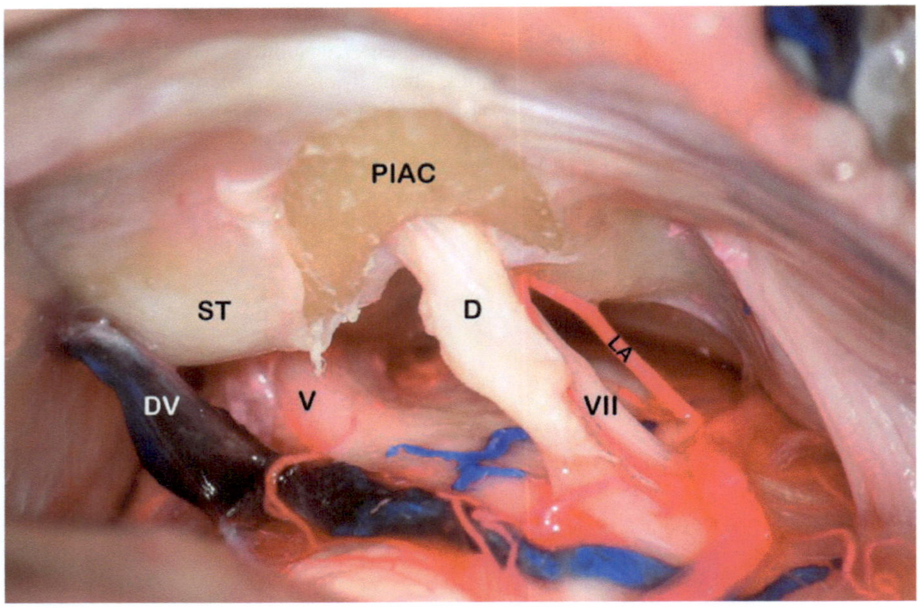

Fig. 8.16 The dura of the posterior lip is reflected. Then, the posterior wall of the IAC could be exposed. *PIAC*, posterior wall of the internal auditory canal; *D*, dura; *LA*, labyrinthine artery; *VII*, facial nerve; *ST*, suprameatal tubercle; *V*, trigeminal nerve; *DV*, Dandy's vein (superior petrosal vein).

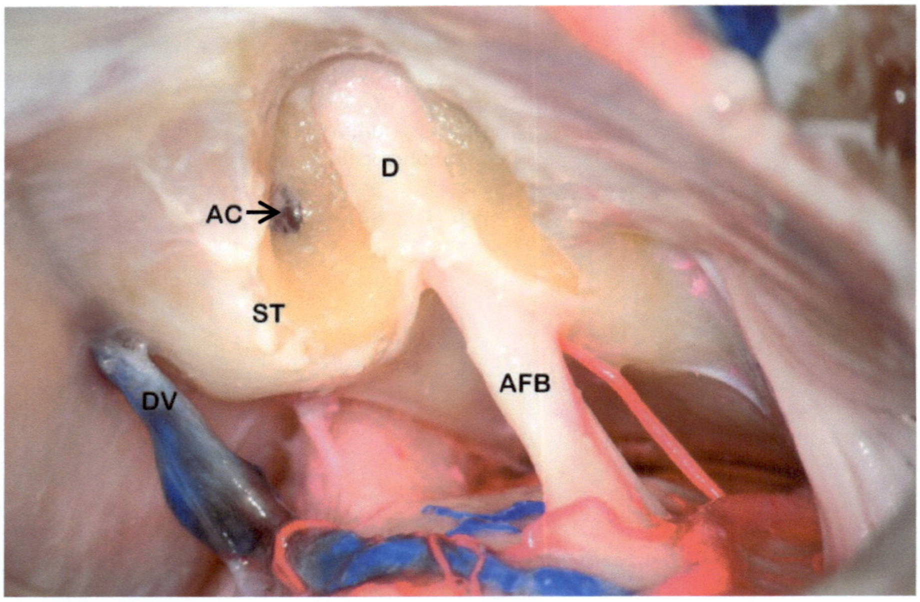

Fig. 8.17 The posterior lip is drilled with a diamond drill with copious irrigation to prevent thermal injury. All exposed air cells are covered with fat after surgery to prevent CSF leakage. *D*, dura; *AFB*, acousticofacial bundle; *ST*, suprameatal tubercle; *AC*, air cells; *DV*, Dandy's vein (superior petrosal vein).

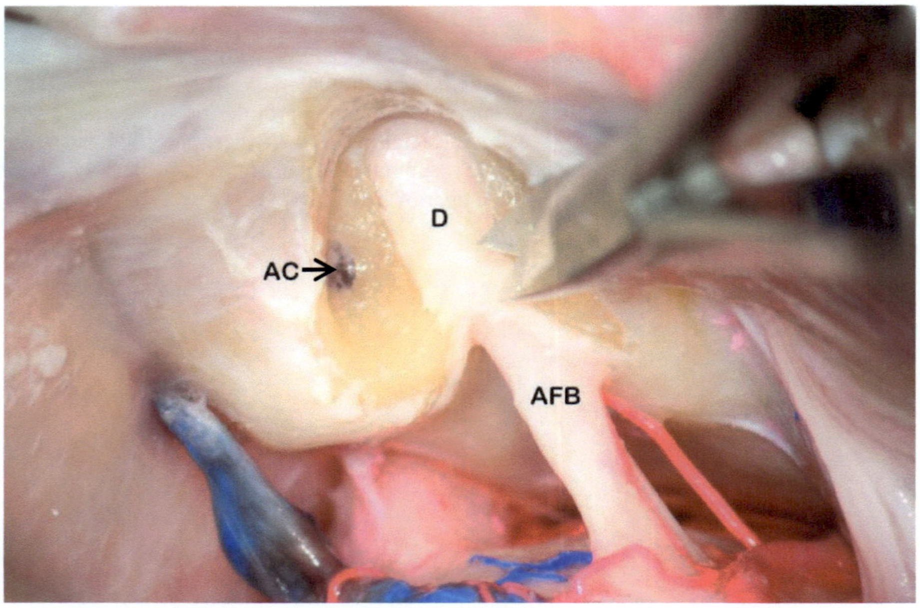

Fig. 8.18 The overlying dura was excised. *D*, dura; *AFB*, acousticofacial bundle; *AC*, air cells.

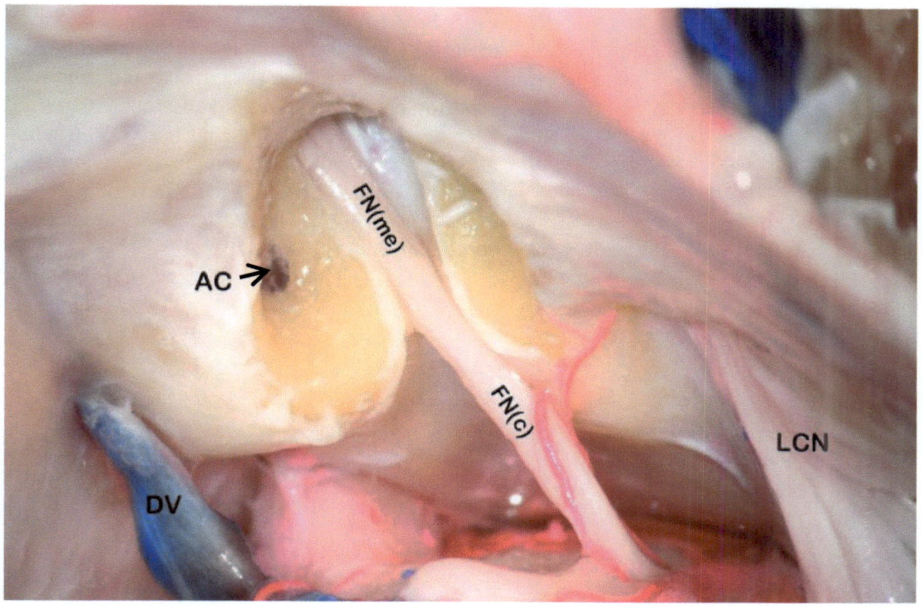

Fig. 8.19 After the incision of the dura of the internal auditory canal, the facial nerve is visualized. Note the air cells were exposed on the surface around the internal auditory canal. *LCN*, lower cranial nerves; *FN(c)*, cisternal segment of facial nerve; *FN(me)*, meatal segment of facial nerve; *AC*, air cells; *DV*, Dandy's vein (superior petrosal vein).

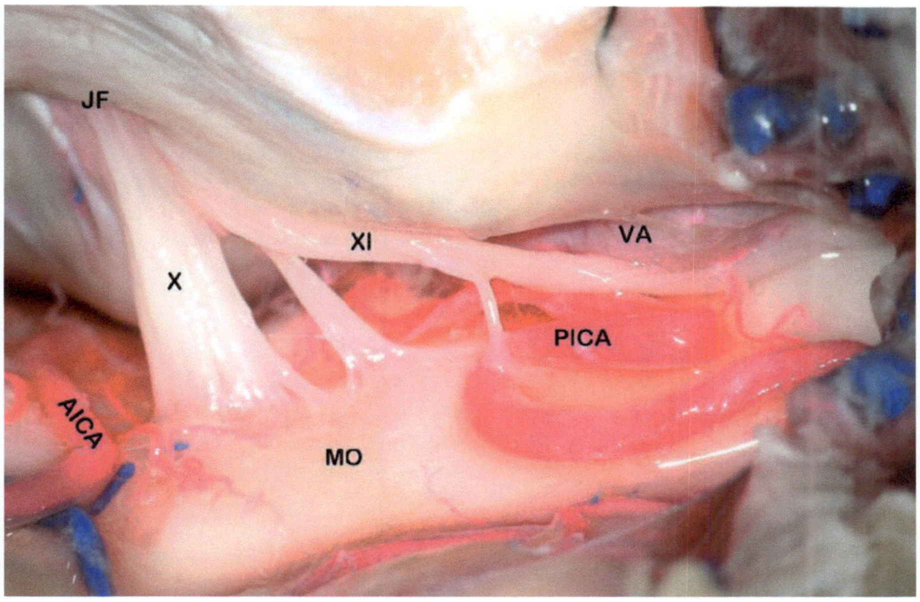

Fig. 8.20 When the microscope is oriented inferiorly, the ipsilateral posteroinferior cerebellar artery and its relationship to the lower cranial nerves are seen. The rootlets of the accessory nerve and vagus nerve are also well demonstrated. *VA*, vertebral artery; *PICA*, posteroinferior cerebellar artery; *XI*, accessory nerve; *X*, vagus nerve; *MO*, medulla oblongata; *JF*, jugular foramen; *AICA*, anteroinferior cerebellar artery.

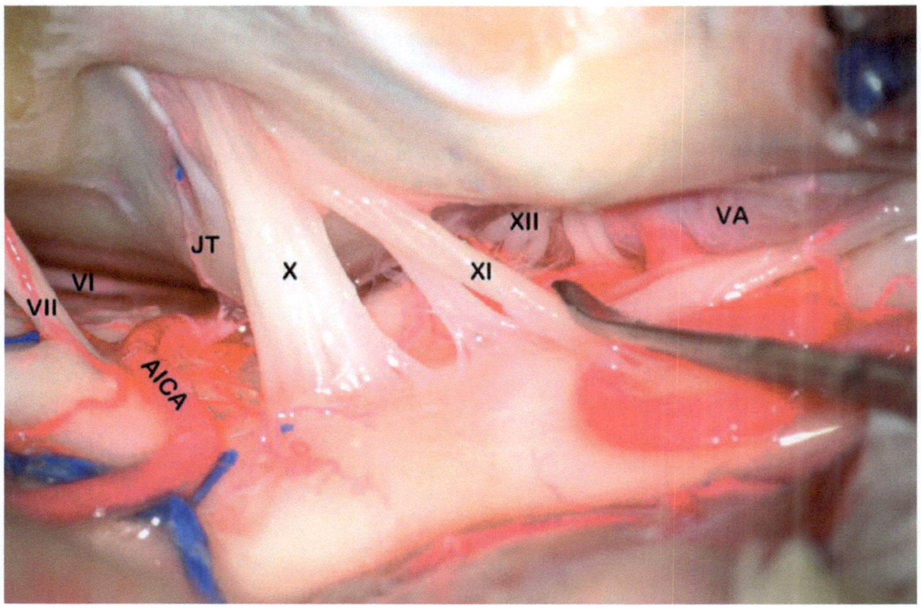

Fig. 8.21 Retracting the accessory nerve exposes rootlets of the hypoglossal nerve. *VA*, vertebral artery; *XII*, hypoglossal nerve; *XI*, accessory nerve; *X*, vagus nerve; *JT*, jugular tubercle; *VI*, abducent nerve; *AICA*, anteroinferior cerebellar artery; *VII*, facial nerve.

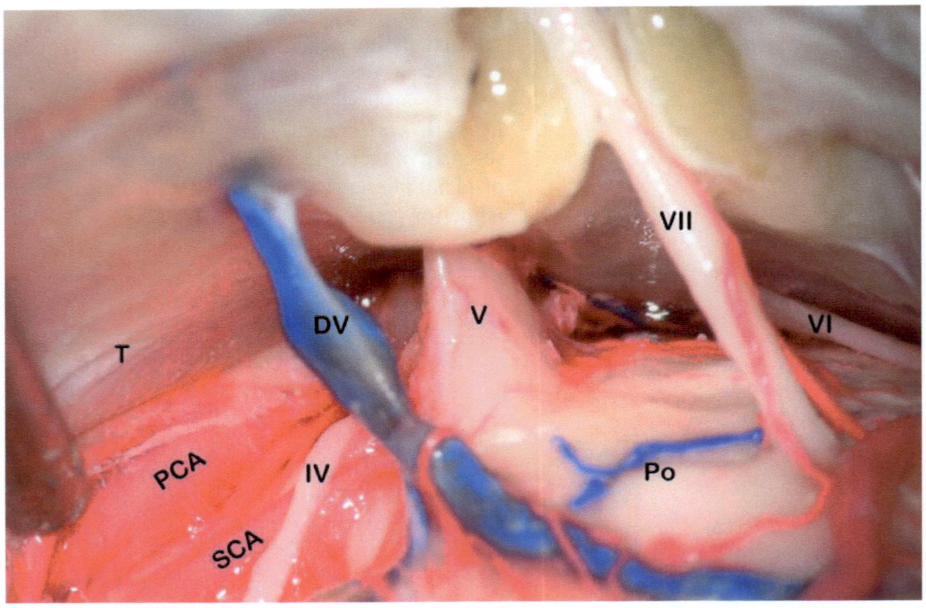

Fig. 8.22 Tilting the microscope upwards identifies the tentorial notch. Retracting the tentorium exposes the posterior cerebral artery, superior cerebellar artery, and trochlear nerve. *VII*, facial nerve; *VI*, abducent nerve; *V*, trigeminal nerve; *Po*, pons; *DV*, Dandy's vein (superior petrosal vein); *IV*, trochlear nerve; *T*, tentorium; *PCA*, posterior cerebral artery; *SCA*, superior cerebellar artery.

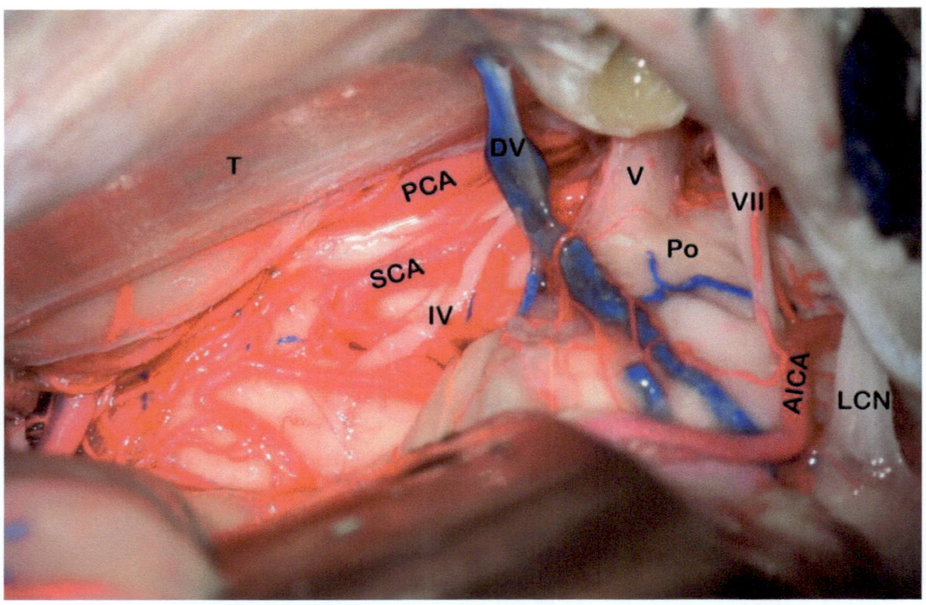

Fig. 8.23 The retractor is introduced to retract the tentorial surface of the cerebellum gently to expose the neurovascular structures in the cerebellomesencephalic fissure and ambient cistern. *LCN*, lower cranial nerves; *AICA*, anteroinferior cerebellar artery; *VII*, facial nerve; *V*, trigeminal nerve; *Po*, pons; *DV*, Dandy's vein (superior petrosal vein); *PCA*, posterior cerebral artery; *SCA*, superior cerebellar artery; *T*, tentorium; *IV*, trochlear nerve.

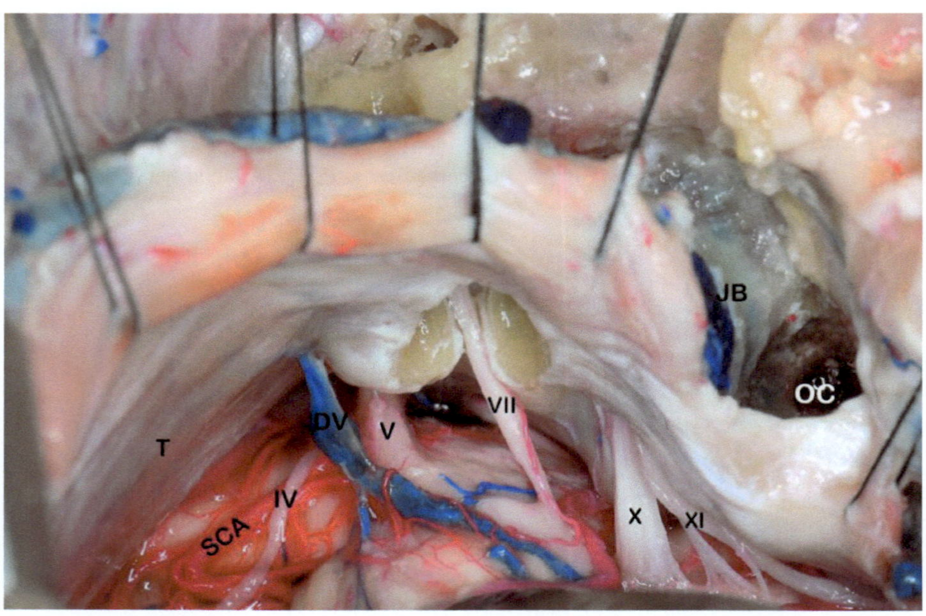

Fig. 8.24 The general view of the cerebellopontine angle after this approach. *OC*, occipital condyle; *JB*, jugular bulb; *VII*, facial nerve; *X*, vagus nerve; *XI*, accessory nerve; *V*, vestibule; *DV*, Dandy's vein (superior petrosal vein); *IV*, trochlear nerve; *SCA*, superior cerebellar artery; *T*, tentorium.

Case 1: Grade I Vestibular Schwannoma

See Figs. 8.25–8.43.

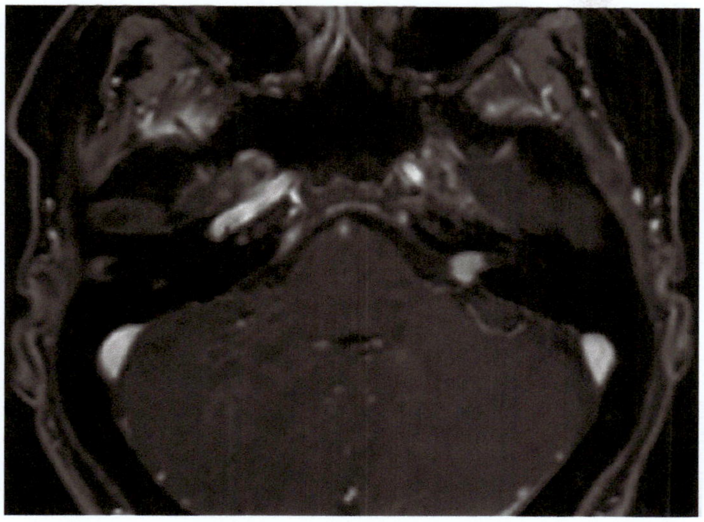

Fig. 8.25 T1W-MRI with Gd enhancement showing a CPA tumor (vestibular schwannoma) on the *left side*. The internal auditory canal is not completely involved by the lesion, which extends 0.8 mm in the CPA (grade I according to the Tokyo consensus classification). This case can be operated through a retrosigmoid-retrolabyrinthine approach with the aim of hearing preservation. A tumor within the IAC is present also on the *right side* (see fig. 12.25 in Chapter 12), leading to an NF2.

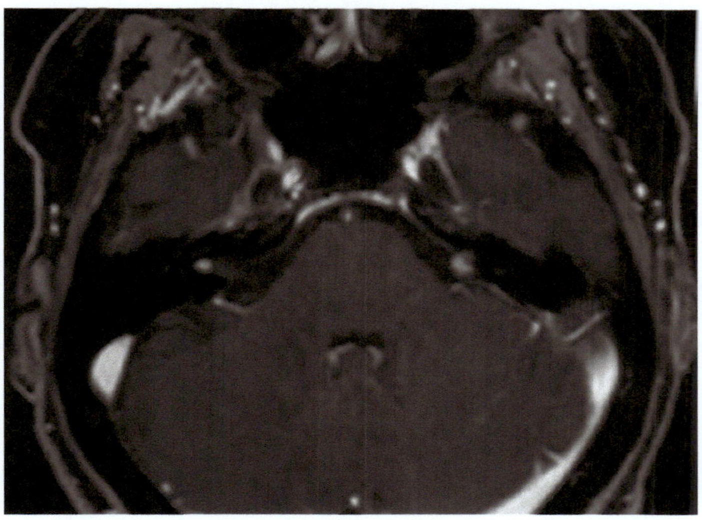

Fig. 8.26 T1W-MRI with Gd enhancement. Note a small VS filling the fundus of the right IAC.

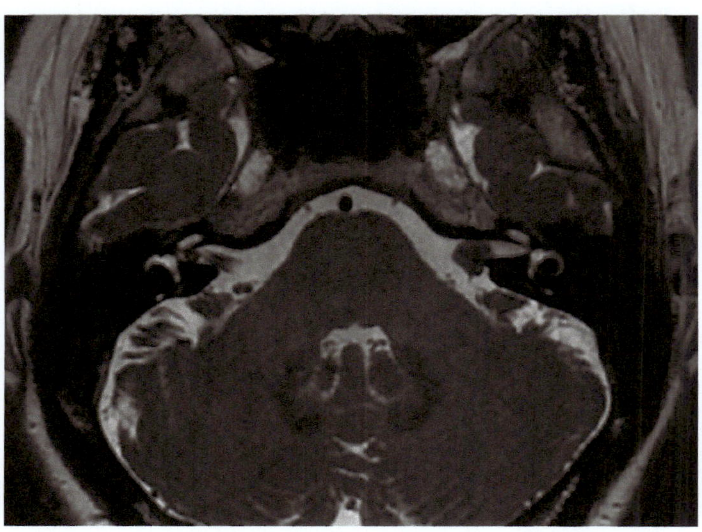

Fig. 8.27 T2W-CISS MRI. The proximal 2/3 of the left IAC is not involved by the tumor.

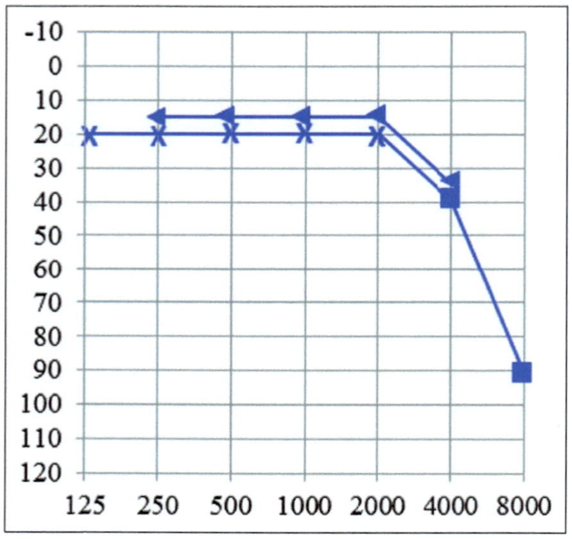

Fig. 8.28 Pure tone audiometry of the *left side*, showing a slight to severe sensorineural hearing loss on high frequencies. The *right side* (not reported) showed normal hearing.

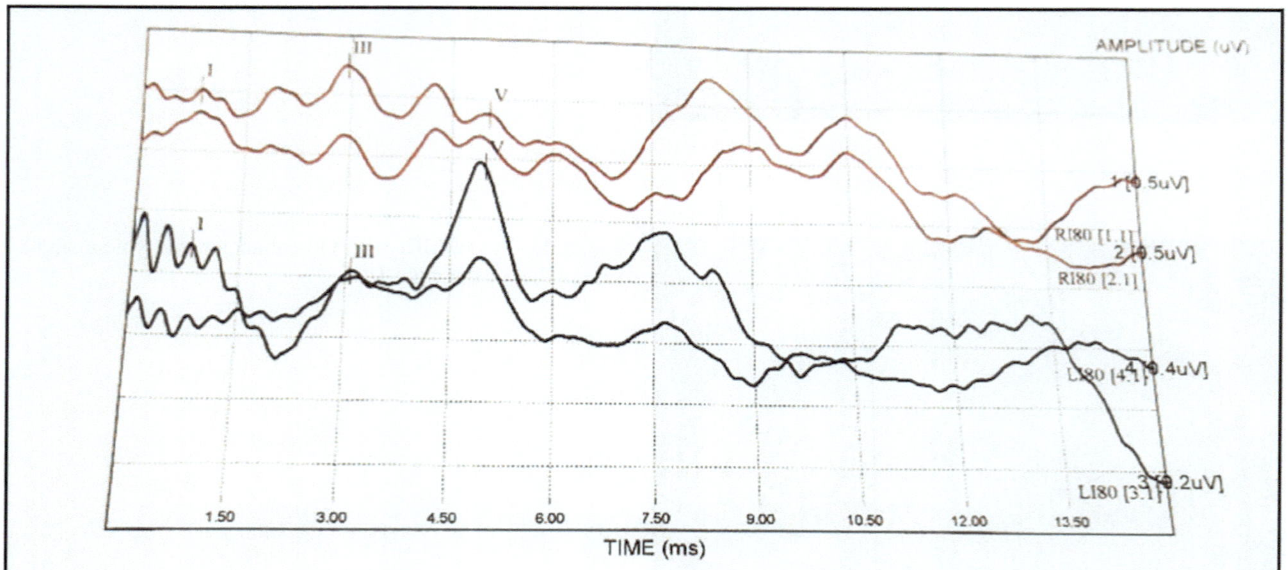

LATENCIES (ms)							
ABR							
Waveform	Ear	I	II	III	IV	V	VI
1	Right	0.83	***	2.95	***	5.05	***
3	Left	0.83	***	3.08	***	5.00	***

INTERPEAK INTERVALS (ms)				
ABR				
Waveform	Ear	I-III	III-V	I-V
1	Right	2.13	2.10	4.22
3	Left	2.25	1.92	4.17

Fig. 8.29 Preoperative ABR showing normal waves, bilaterally.

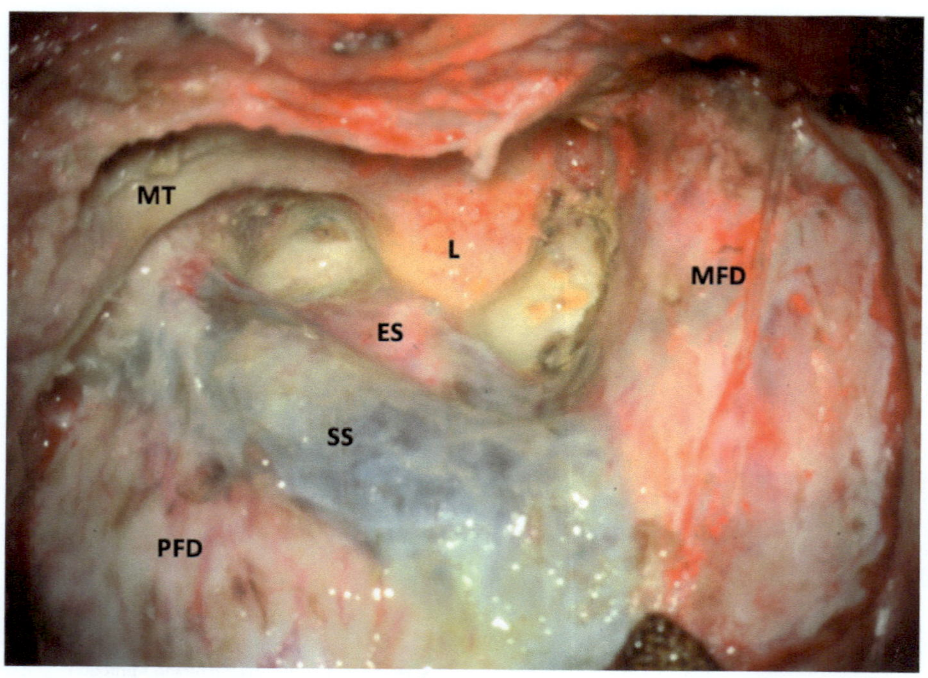

Fig. 8.30 An extended mastoidectomy is performed with sufficient exposure of the retrosigmoid posterior fossa dura. The sigmoid sinus is identified and completely uncovered between the transverse sinus and the area of the jugular bulb. *MT*, mastoid tip; *PFD*, posterior fossa dura; *SS*, sigmoid sinus; *ES*, endolymphatic sac; *MFD*, middle fossa dura; *L*, labyrinth.

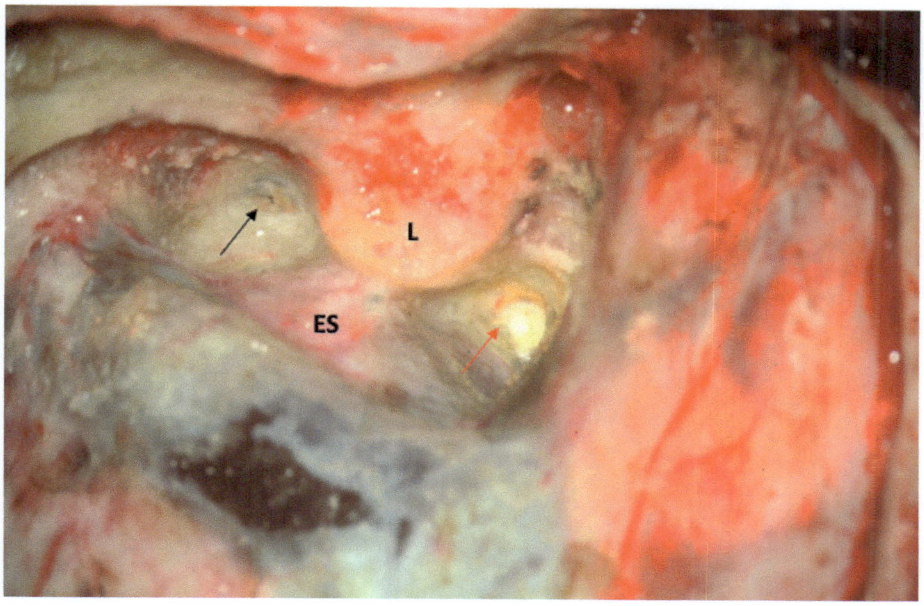

Fig. 8.31 Higher magnification. The retrolabyrinthine part of the approach has been performed. Note the pneumatization of the perilabyrinthine cells (*arrows*) that may lead to postoperative cerebrospinal fluid leak. The superior perilabyrinthine cells (*red arrow*) have been closed with bone wax. The same procedure must be performed with the inferior perilabyrinthine cells (*black arrow*) before opening the posterior fossa dura. *ES*, endolymphatic sac; *L*, labyrinth.

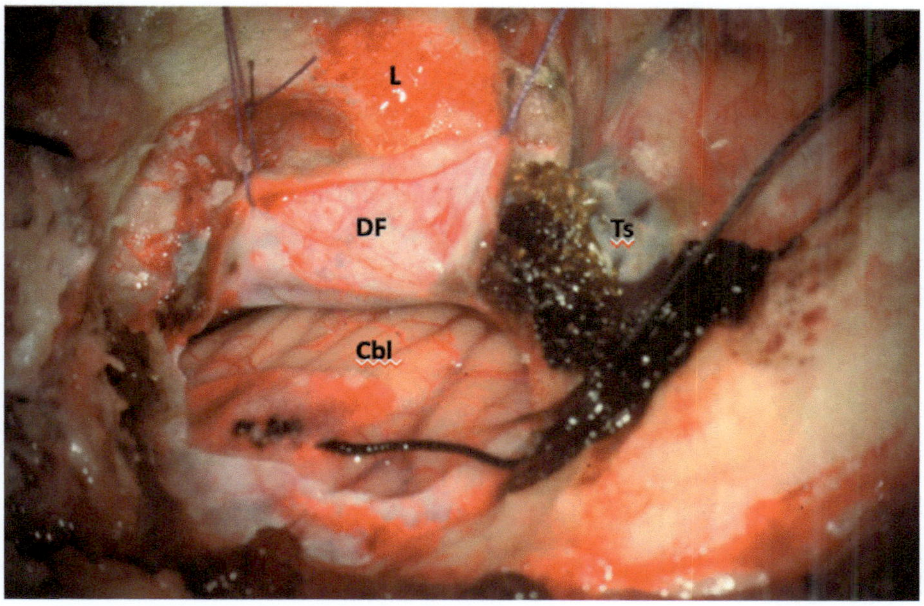

Fig. 8.32 The dura of the posterior fossa has been incised, elevated, and fixed with Vycril sutures. *DF*, dural flap; *Cbl*, cerebellum; *Ts*, transverse sinus; *L*, labyrinth.

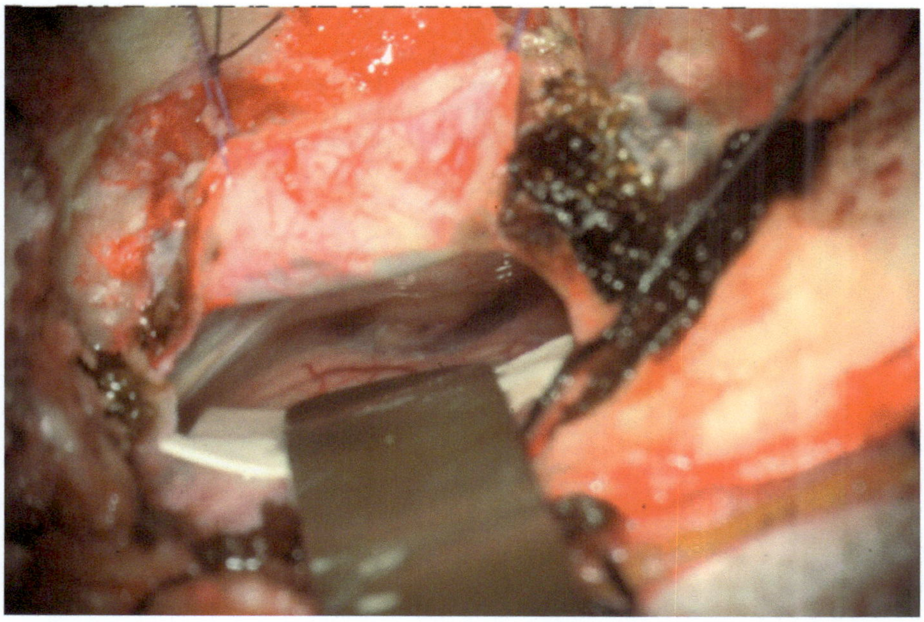

Fig. 8.33 Once the dura is opened, Merocel is applied over the cerebellum to provide protection. The retractor is then applied on the Merocel.

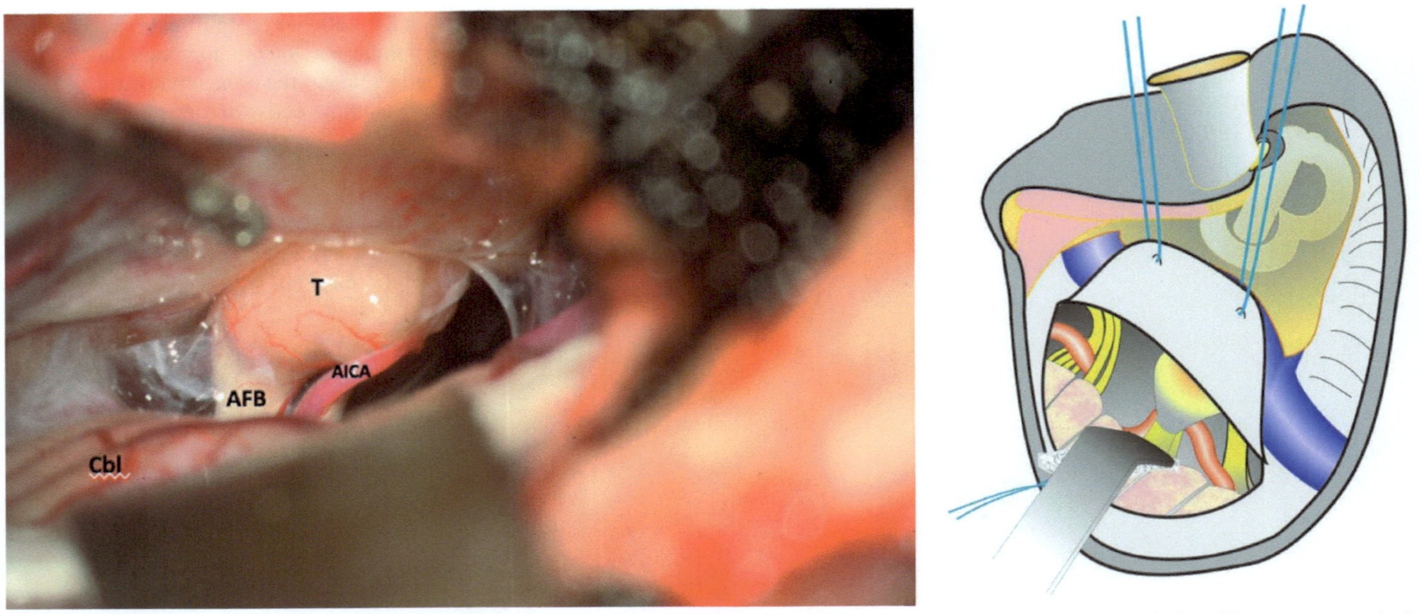

Fig. 8.34 Retraction of the cerebellum exposes the acoustic-facial bundle in the CPA and the tumor. Some arachnoid is still present superiorly and inferiorly. *Cbl*, cerebellum; *T*, tumor; *AICA*, anterior-inferior cerebellar artery; *AFB*, acoustic-facial bundle.

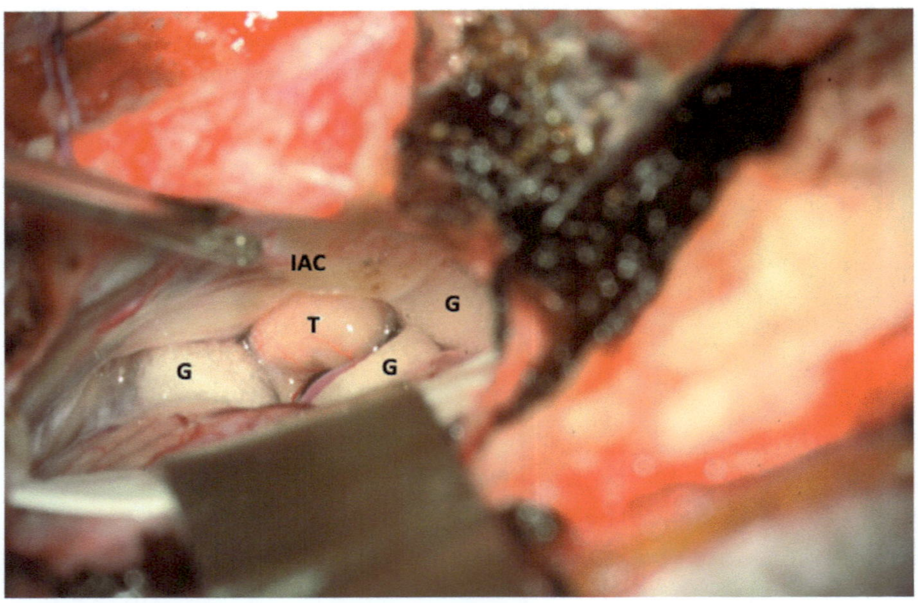

Fig. 8.35 Before drilling the posterior wall of the IAC, Gelfoam is inserted anteriorly into the CPA and around the porus of the IAC to prevent bone dust from falling into the cisterns. The suction indicates the direction of the IAC. *T*, tumor; *IAC*, internal auditory canal; *G*, gelfoam.

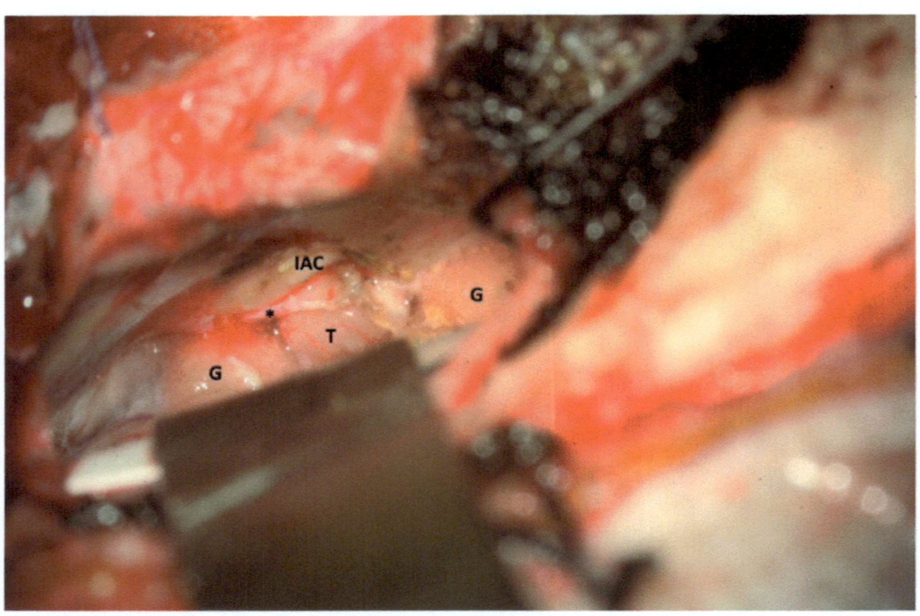

Fig. 8.36 The posterior wall of the IAC is further drilled superiorly to get complete control of the tumor. *T*, tumor; *IAC*, internal auditory canal; *G*, gelfoam.

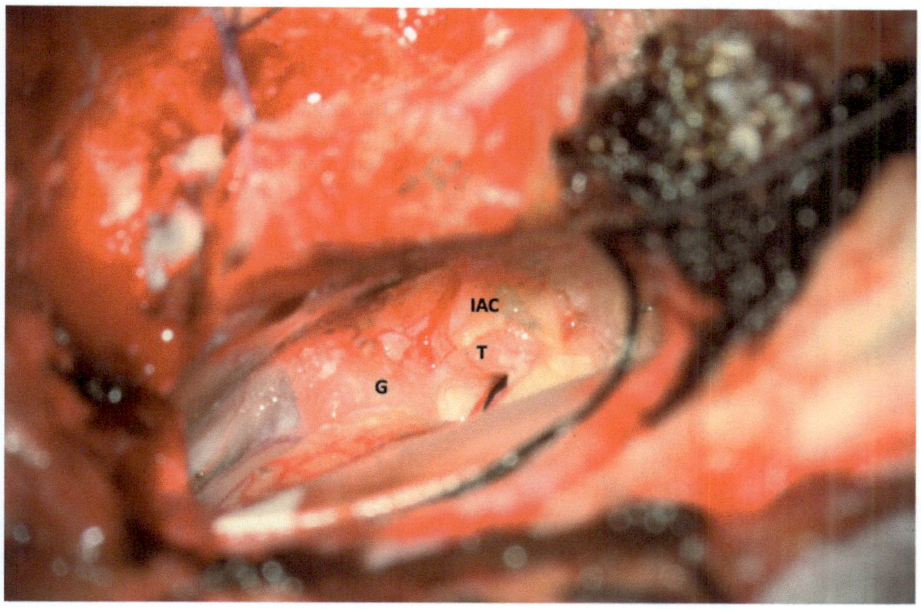

Fig. 8.37 The posterior wall of the IAC has been partially drilled with further exposure of the tumor. *T*, tumor; *IAC*, internal auditory canal; *G*, gelfoam; *, dura of the IAC.

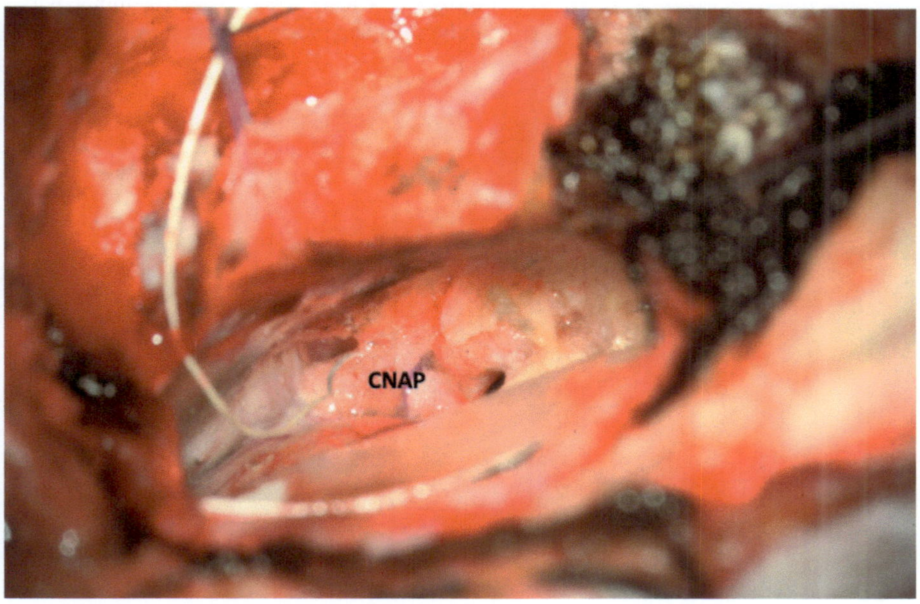

Fig. 8.38 The cochlear nerve action potential can be recorded together with the ABR to assess the functional integrity of the cochlear nerve during and after tumor removal. *CNAP*, cochlear nerve action potential electrode.

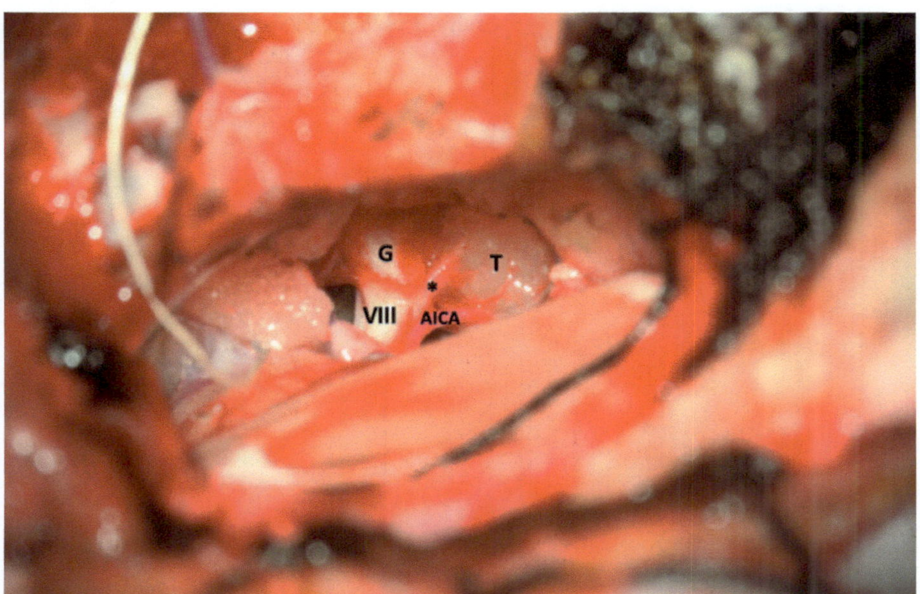

Fig. 8.39 The tumor has been isolated from the cochlear nerve. Note a small branch of the AICA (*) feeding the tumor. Coagulation should be avoided close to the AFB to avoid thermal injury to both the cochlear and facial nerves. A piece of gelfoam is used to protect the proximal portion of the cochlear nerve. *VIII*, cochlear nerve; *G*, gelfoam; *AICA*, anterior-inferior cerebellar artery.

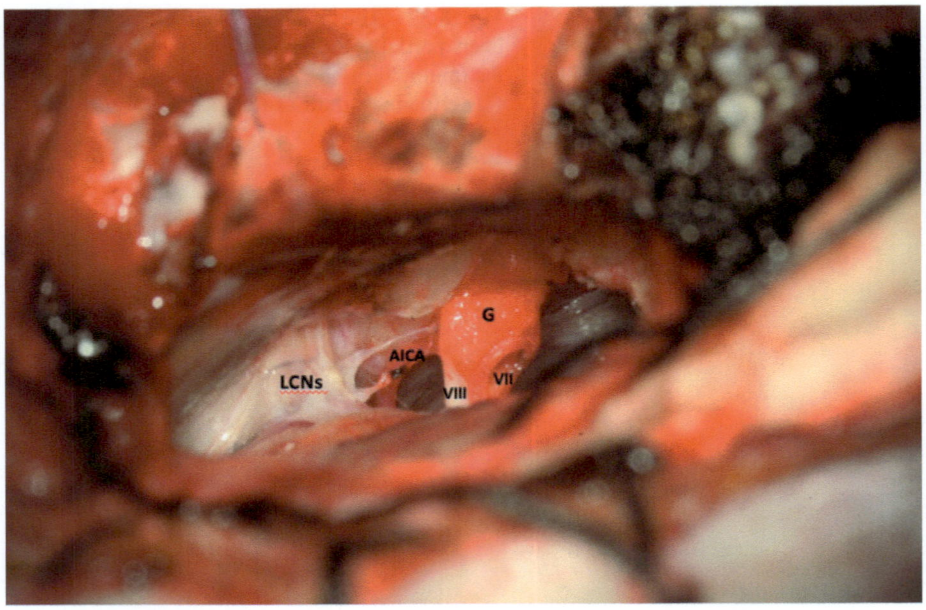

Fig. 8.40 Surgical view after tumor removal. A loop of the AICA can be appreciated around the AFB. A piece of gelfoam has been placed to protect the proximal portion of the AFB at the level of the IAC. *LCNs*, lower cranial nerves area; *VIII*, cochlear nerve; *VII*, facial nerve; *G*, gelfoam; *AICA*, antero-inferior cerebellar artery.

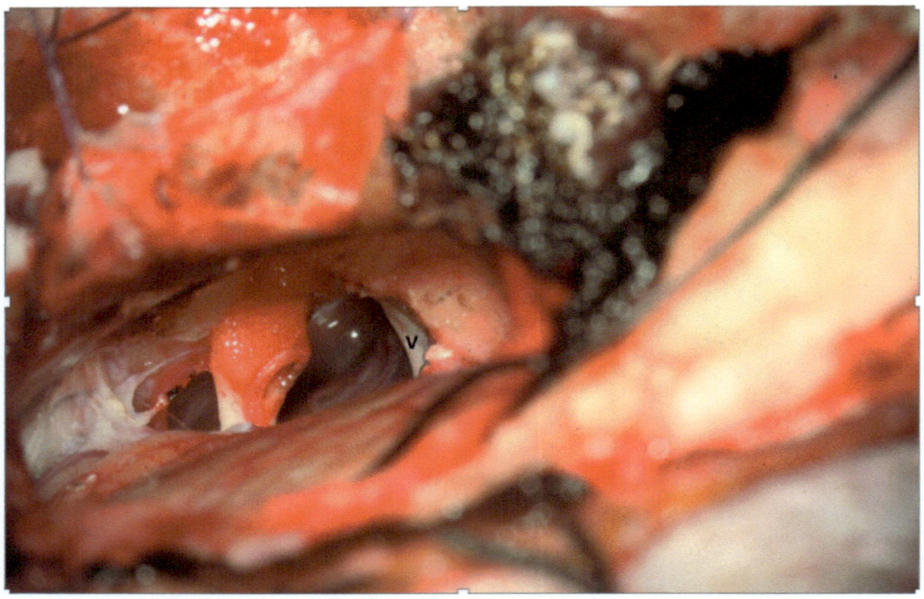

Fig. 8.41 Closer view. The trigeminal nerve can be seen superiorly. *V*, trigeminal nerve.

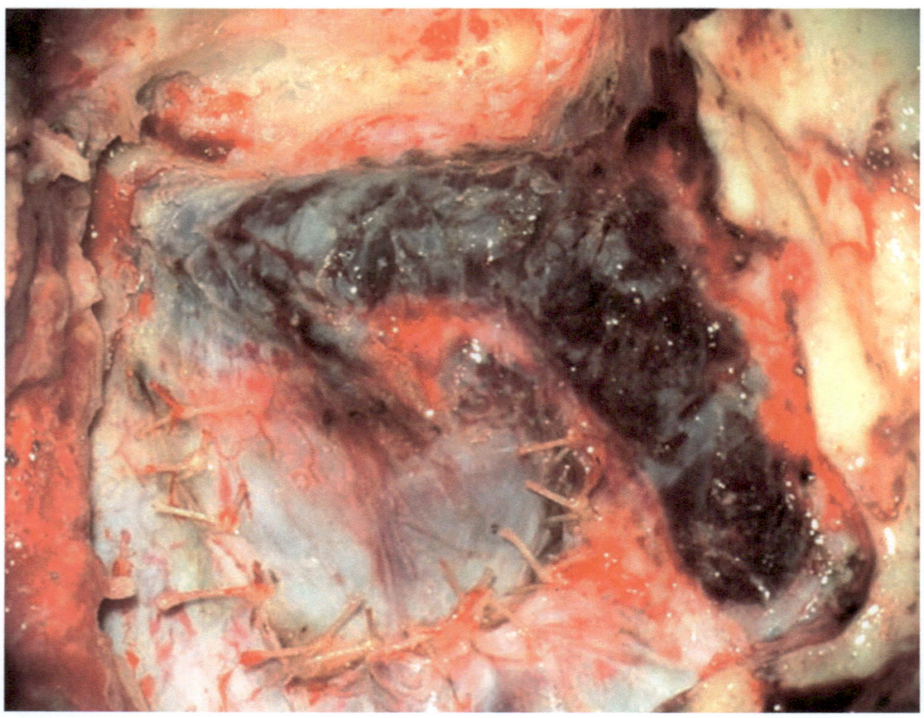

Fig. 8.42 The dura is closed in a watertight fashion using 3.0 Vycril sutures, and then the cavity is obliterated with abdominal fat.

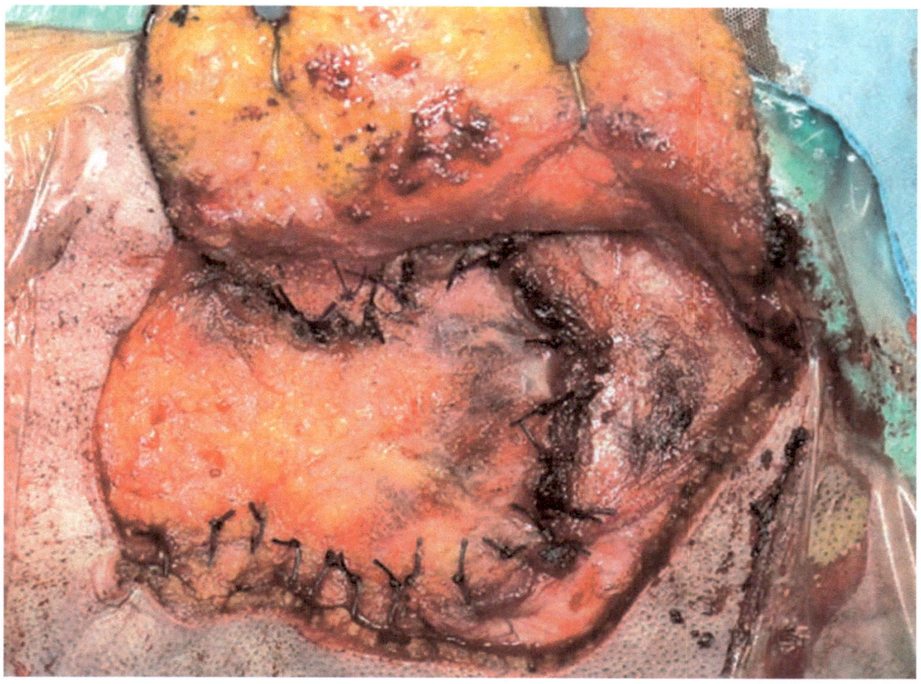

Fig. 8.43 Suture of the musculoperiosteal flap.

Case 2: Intrameatal Vestibular Schwannoma

See Figs. 8.44–8.55.

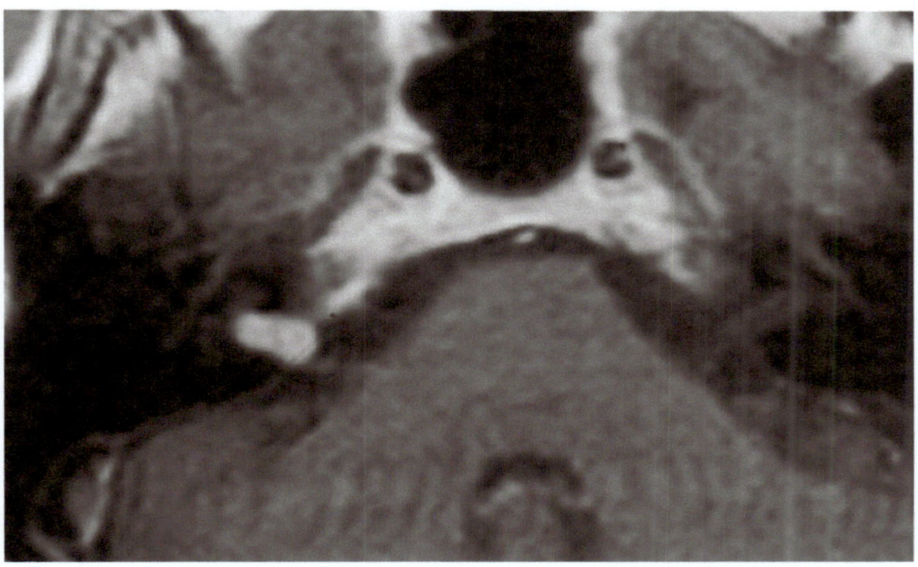

Fig. 8.44 MRI showing an intracanalicular lesion on the *right side*.

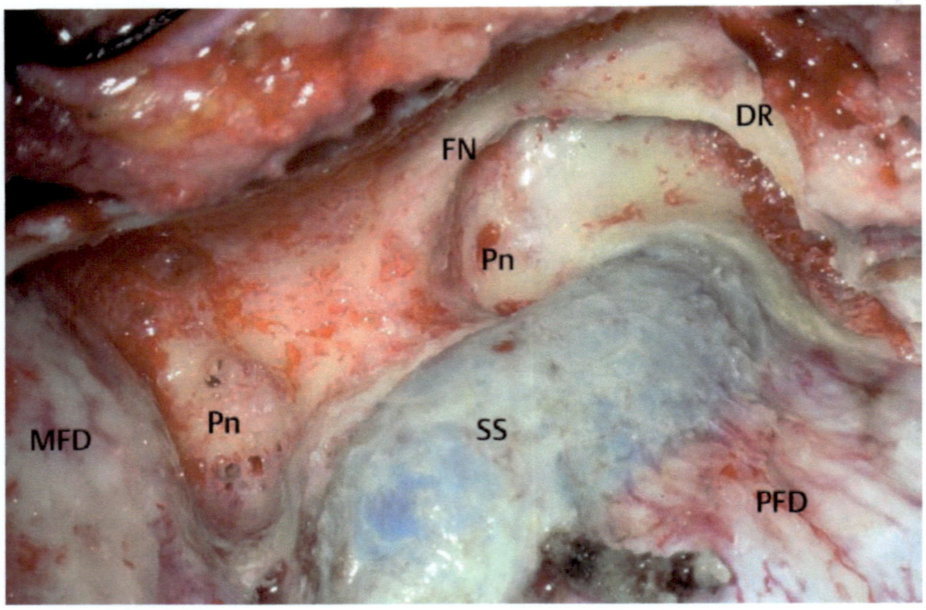

Fig. 8.45 Retrolabyrinthine bone removal; residual pneumatization is visible superiorly and inferiorly to the posterior labyrinth. *FN*, facial nerve; *Pn*, pneumatization; *SS*, sigmoid sinus; *PFD*, posterior fossa dura; *MFD*, middle fossa dura; *DR*, digastric ridge.

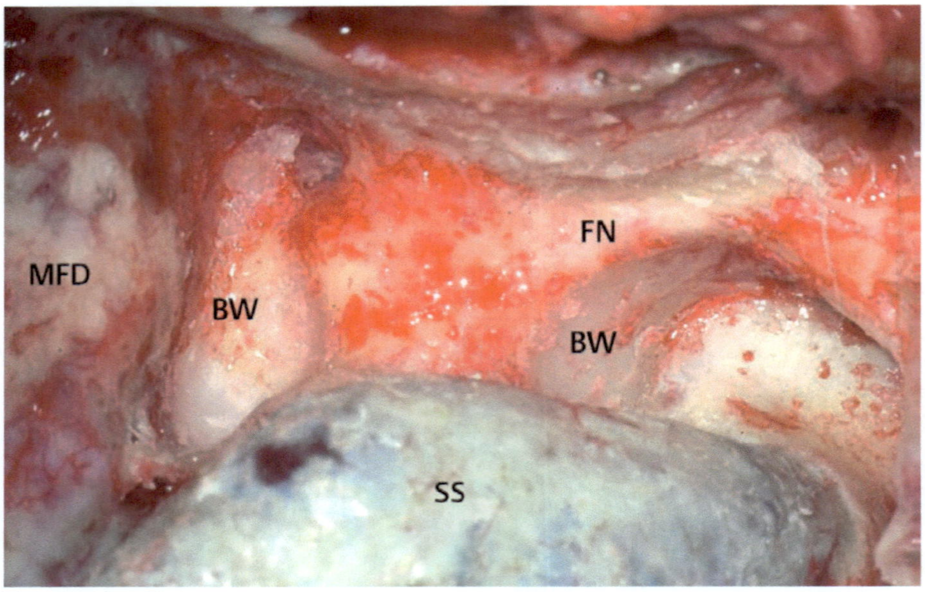

Fig. 8.46 Residual pneumatization has been plugged with bone wax (BW). *FN*, facial nerve; *SS*, sigmoid sinus; *MFD*, middle fossa dura.

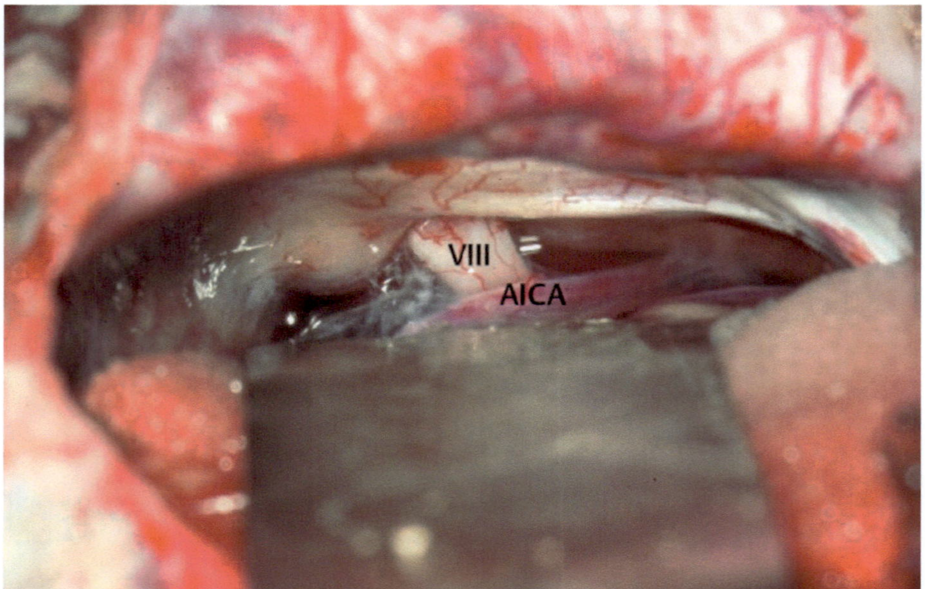

Fig. 8.47 After the opening of the posterior fossa dura, a gentle cerebellar retraction allows visualization of the eighth cranial nerve (VIII). *AICA*, anterior-inferior cerebellar artery.

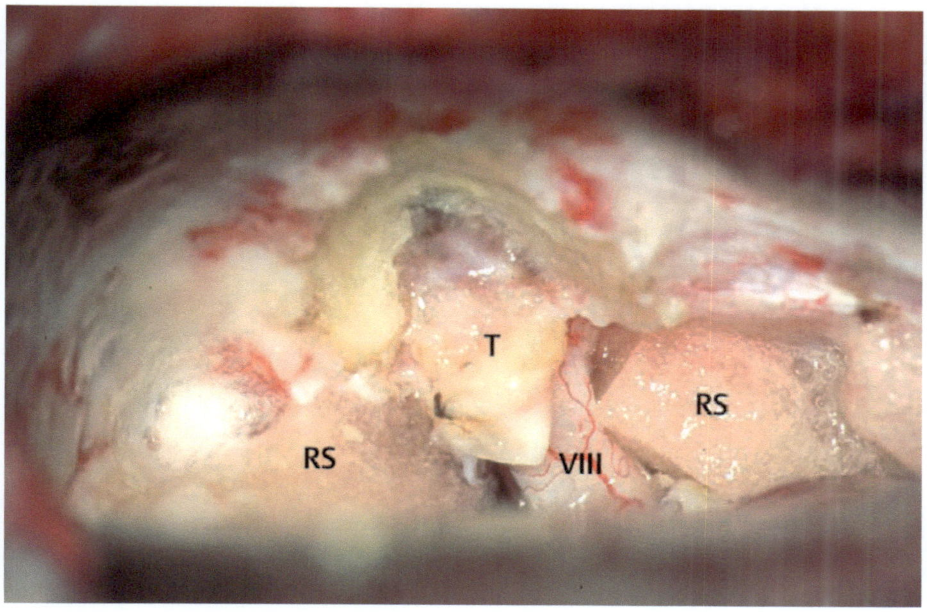

Fig. 8.48 Drilling of the internal auditory canal exposes the intracanalicular tumor (T). Before drilling, the anterior CPA is protected with a resorbable sponge (RS) in order to avoid the dissemination of bone dust. *VIII*, eighth cranial nerve.

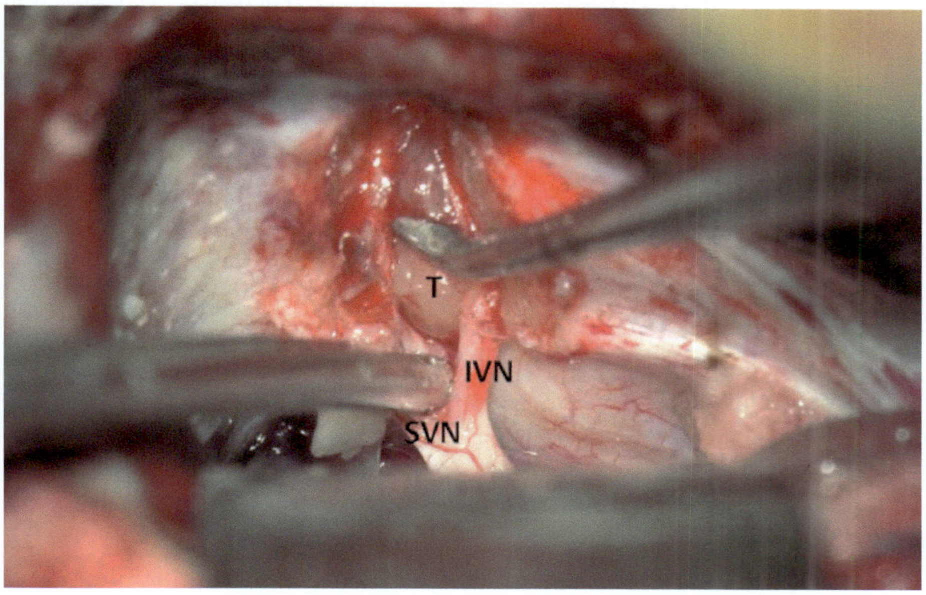

Fig. 8.49 Removal of tumor (T) is performed with blunt dissectors. *SVN*, superior vestibular nerve; *IVN*, inferior vestibular nerve.

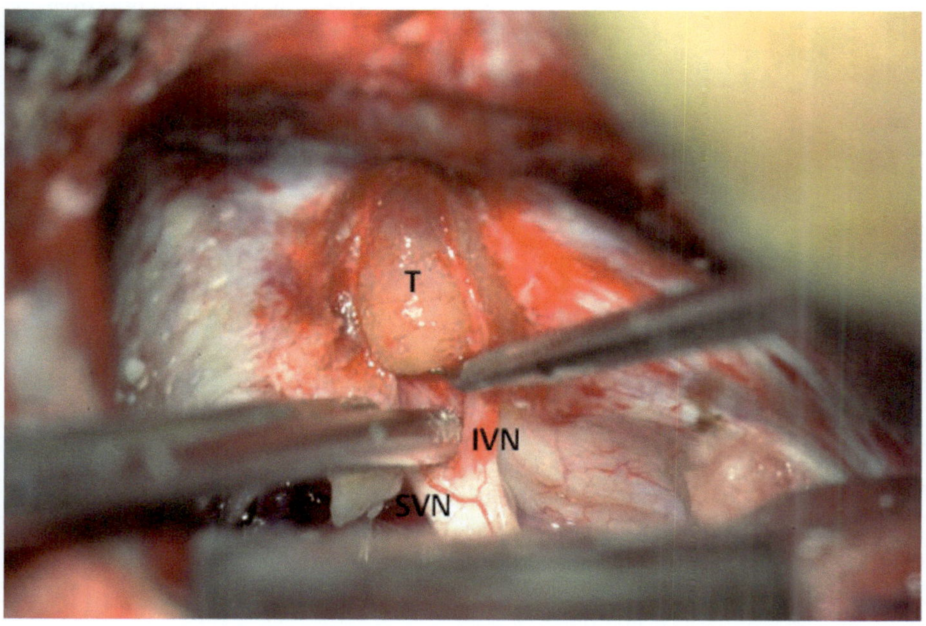

Fig. 8.50 Removal of tumor (T) proceeds with blunt dissectors. *SVN*, superior vestibular nerve; *IVN*, inferior vestibular nerve.

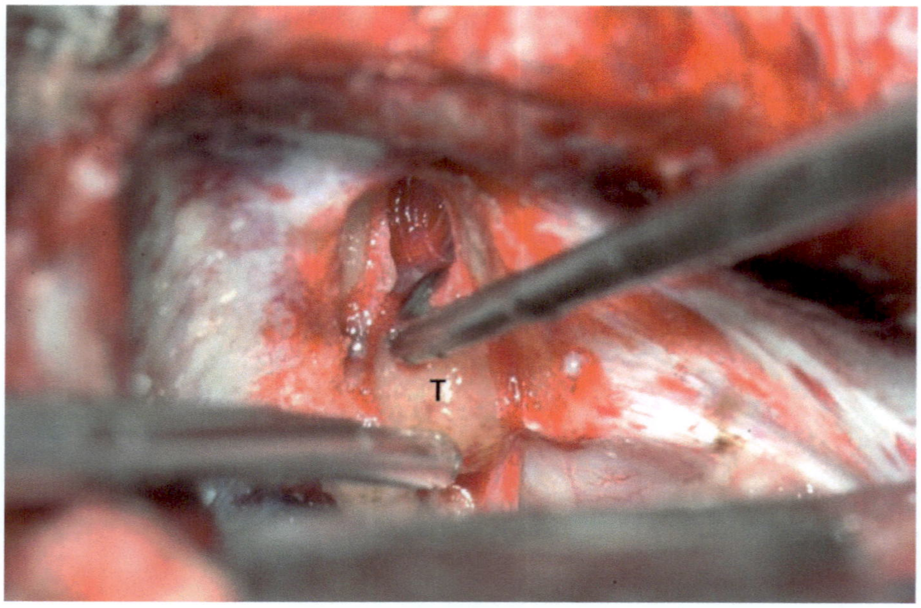

Fig. 8.51 Removal of tumor (T) proceeds with blunt dissectors.

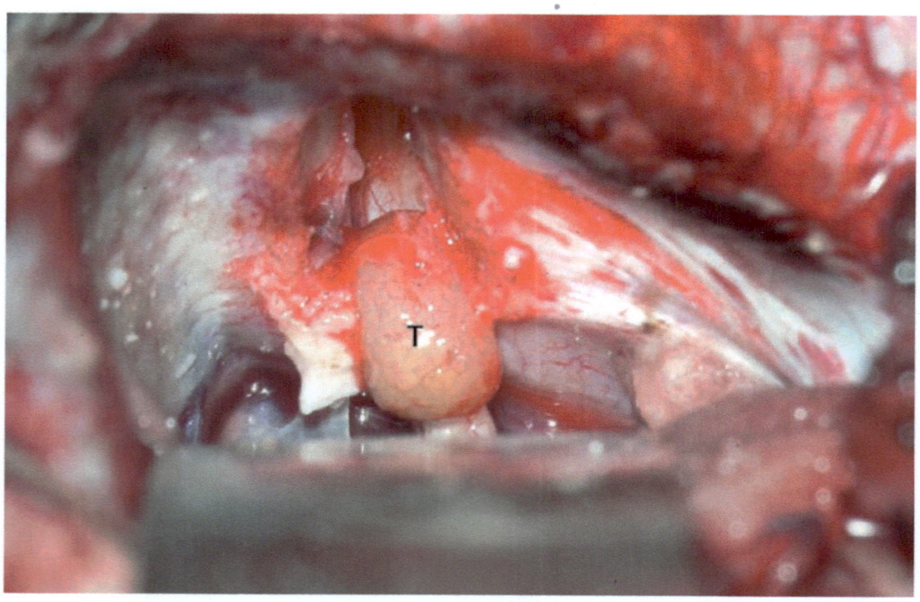

Fig. 8.52 Removal of tumor (T) proceeds with blunt dissectors.

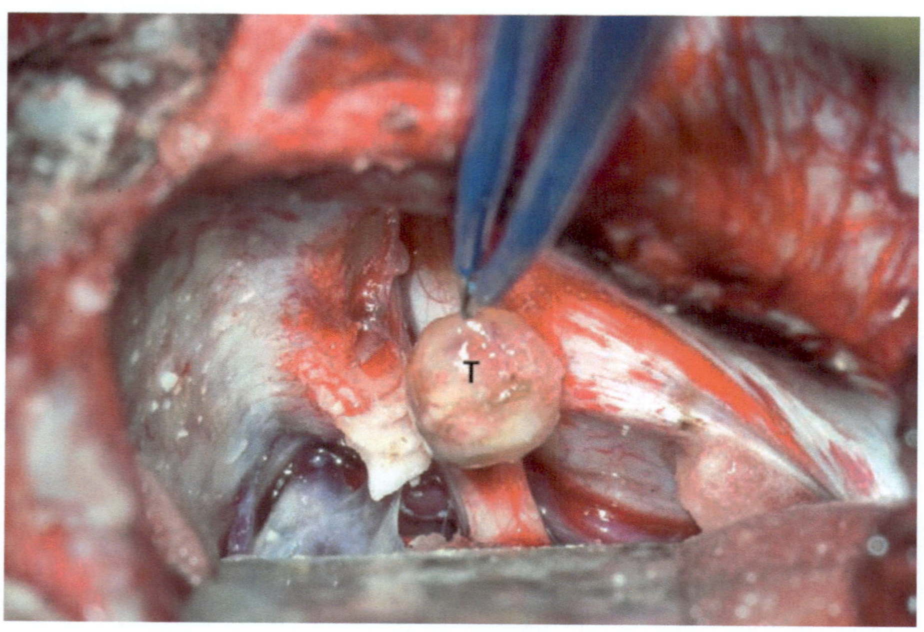

Fig. 8.53 After complete dissection from the eighth cranial nerve, the tumor (T) is removed en bloc.

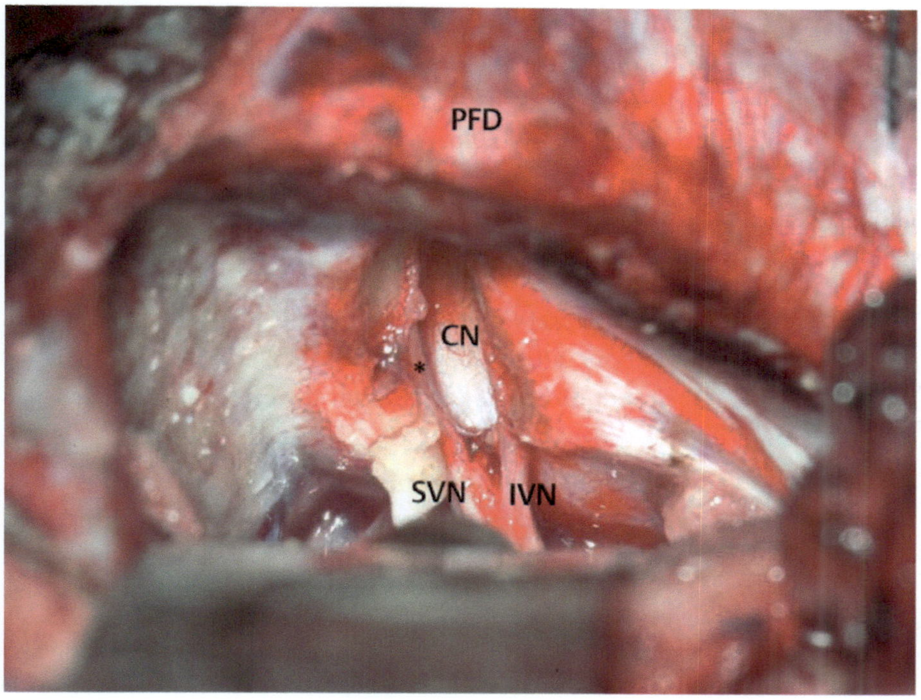

Fig. 8.54 After tumor removal it is possible to visualize the content of the internal auditory canal. *CN*, cochlear nerve; *IVN*, inferior vestibular nerve; *PFD*, posterior fossa dura; *SVN*, superior vestibular nerve; *, facial nerve.

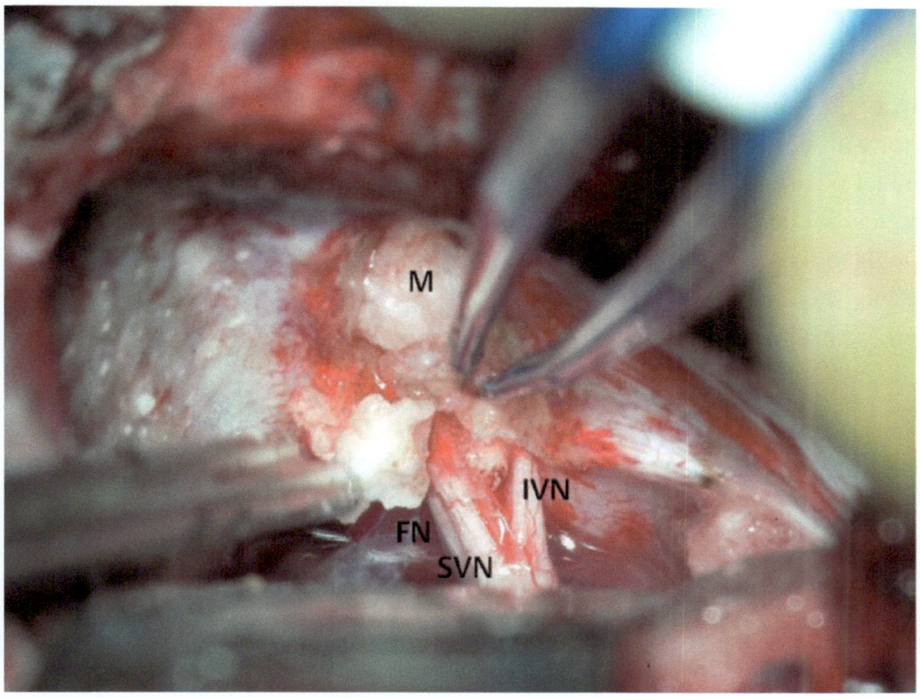

Fig. 8.55 To prevent postoperative CSF leak, the internal auditory canal is packed with muscle (M). *FN*, facial nerve; *SVN*, superior vestibular nerve; *IVN*, inferior vestibular nerve.

HINTS AND PITFALLS

- The supine position with the patient's head rotated to the contralateral side should be preferable to the sitting or semi-sitting position. The latter two positions have a high incidence of air embolism and require special anesthetic care and time-consuming preoperative preparations.
- Most surgeons use a vertical or slightly curved skin incision. The necessary exposure is obtained by the use of self-retaining retractors, which have the disadvantage of making the surgical field deeper and the working angle more acute. We prefer a wide C-shaped incision. Exposure is obtained by the use of self-retaining retractors. Retractors have the disadvantage of making the surgical field deeper and the working angle more acute.
- The addition of the retrolabyrinthine approach allows the sigmoid sinus to be displaced forwards and greatly enhances the exposure. In the case of anteriorly located

tumors, the presigmoid dura can be incised, and the anterior pole of the tumor can be tackled through this opening, thus obviating the need for additional cerebellar retraction. Moreover, by doing the retrolabyrinthine approach, the mastoid air cells are all drilled, and this markedly decreases the incidence of postoperative CSF leak that occurs with the classic retrosigmoid approach.

- Before opening the dura, dehydration measures should be taken by the anesthesiologists. The utmost care should be taken to open the cistern and allow drainage of the CSF as an initial step before tumor removal commences. These measures allow the cerebellum to fall away and reduce the amount needed for cerebellar retraction.
- Before drilling the internal auditory canal, Gelfoam is placed in the angle and around the internal auditory canal, mainly inferior to it, to avoid dispersion of bone dust in the cistern. Drilling is performed using a large diamond burr under continuous suction irrigation. A wide area of bone superior and inferior to the canal is drilled rather than just exposing the posterior wall. This gives more space for safer tumor removal.
- Various methods have been proposed to help avoid injury to the labyrinth during drilling of the posterior wall of the internal auditory canal. The vestibular aqueduct is a constant landmark since it lies immediately medial and inferior to the crus commune, but it is difficult to locate during surgery. Another method is to blue-line the crus commune as shown in the figure. However, there is a risk to opening the labyrinth while doing so.
- A high jugular bulb reaching up to the level of the inferior border of the internal auditory canal is present in about 10% of cases. In these cases, the bulb may be injured while drilling the posterior wall of the canal.
- The fundus of the IAC can be managed using the endoscope.
- Preservation of hearing requires the anatomical and functional preservation of the cochlear nerve. The internal auditory artery, being an end artery, should never be closed. Measures taken to preserve hearing include:
  - Tumor dissection should be from medial to lateral.
  - Bipolar coagulation should be only on the tumor capsule.
  - Bleeding from the facial or cochlear nerves is best managed using Gelfoam.
  - Intraoeprative monitoring of hearing is performed. Maneuvers that alter the ABR waves (e.g., excessive cerebellar retraction) should be abandoned.
- Methods to decrease postoperative CSF leak include:
  - Exenteration of the mastoid air cells (this is achieved by doing the retrolabyrinthine approach)
  - If the internal auditory canal is drilled, air cells surrounding the posterior wall of the canal should be closed by bone wax.
  - Watertight closure of the dura of the posterior fossa. If a small defect is noticed, a small piece of muscle is used to seal the defect.

REFERENCES

[1] Cushing H. Tumors of the nervus acusticus and the syndrome of the cerebellopontine angle. Philadelphia: Saunders; 1917.

[2] Dandy W. Results of removal of acoustic tumors by the unilateral approach. Arch Surg 1941;42(6):1026–33. https://doi.org/10.1001/archsurg.1941.01210120061006.

[3] Rand RW, Kurze T. Preservation of vestibular, cochlear, and facial nerves during microsurgical removal of acoustic tumors: report of two cases. J Neurosurg 1968;28:158–61.

[4] Yang J, Grayeli AB, Barylyak R, Elgarem H, Kalamarides M, Ferrary E, Sterkers O. Functional outcome of retrosigmoid approach in vestibular schwannoma surgery. Acta Otolaryngol 2008;128:881–6.

[5] Samii M, Gerganov V, Samii A. Improved hearing preservation of hearing and facial nerve unction in vestibular schwannoma surgery via the retrosigmoid approach in a series of 200 patients. J Neurosurg 2006;105:527–35.

[6] Sanna M, Mancini F, Russo A, Taibah A, Falcioni M, Di Trapani G. Atlas of acoustic Neurinoma microsurgery. II edition. Thieme; 2010 [ISBN: 9783131102829].

[7] Khrais T, Romano G, Sanna M. Nerve origin of vestibular schwannoma: a prospective study. J Laryngol Otol 2008; 122:128–31.

[8] Khrais T, Sanna M. Hearing preservation surgery in vestibular schwannoma. J Laryngol Otol 2006;120:366–70.

[9] Lassaletta L, Fontes L, Melcon E, Sarria MJ, Gavilan J. Hearing preservation with retrosigmoid approach for vestibular schwannoma: myth or reality? Otolaryngol Head Neck Surg 2003;129:397–401.

[10] Sanna M, Karmarkar S, Landolfi M. Hearing preservation in vestibular schwannoma surgery: fact or fantasy? J Laryngol Otol 1995;109:373–80.

[11] Sanna M, Khrais T, Piccirillo E, Russo A, Augurio A. Hearing preservation surgery in vestibular schwannoma: the hidden truth. Ann Otol Rhinol Laryngol 2004;113:156–63.

[12] Rabelo de Freitas M, Russo A, Sequino G, Piccirillo E, Sanna M. Analysis of hearing preservation and facial nerve function for patients undergoing vestibular schwannoma surgery: the middle cranial fossa approach versus the retrosigmoid

approach- -personal experience and literature review. Audiol Neurootol 2012;17(2):71–81. https://doi.org/10.1159/000329362.

[13] Magnan J, Sanna M. Endoscopy in neuro-otology. Stuttgart, Germany: Georg Thieme Verlag; 1999.

[14] Tatagiba MS, Roser F, Hirt B, Ebner FH. The retrosigmoid endoscopic approach for cerebellopontine-angle tumors and microvascular decompression. World Neurosurg 2014;82 (6 Suppl):S171–6. https://doi.org/10.1016/j.wneu.2014.08.001.

[15] Tatagiba M, Matthies C, Samii M. Microendoscopy of the internal auditory canal in vestibular schwannoma surgery. Neurosurgery 1996;38:737–40.

[16] Magnan J, Barbieri M, Mora R, Murphy S, Meller R, Bruzzo M, Chays A. Retrosigmoid approach for small and medium-sized acoustic neuromas. Otol Neurotol 2002;23(2):141–5.

[17] Falcioni M, Romano G, Aggarwal N, Sanna M. Cerebrospinal fluid leak after retrosigmoid excision of vestibular schwannomas. Otol Neurotol 2008;29:384–6.

CHAPTER 9

The Transotic Approach

Introduction

The transotic (TO) procedure is part of the group of transcochlear (TC) approaches. In 1976, Hitselberger was the first to describe the original TC approach, which includes identification of the internal auditory canal, posterior rerouting of the facial nerve, and removal of the cochlea and petrous apex with preservation of the middle ear and external auditory canal (Fig. 9.1) [10]. In the following years, Fisch (1978) described the TO approach in which he removes the external auditory canal and middle ear but keeps the facial nerve in situ (Fig. 9.2).

The modified TC approach, on the other hand, combines the removal of the external auditory canal and middle ear with the posterior rerouting of the facial nerve, thus removing the major impediment to the anterior extension of the approach. This allows better control of the vertical and horizontal intrapetrous internal carotid artery and facilitates the total removal of the petrous apex. The extensive anterior bone removal provides excellent control of the ventral surface of the brainstem without cerebellar and brainstem retraction. However, currently, the enlarged translabyrinthine approach with the transapical extension enables us to reach the petrous apex and obtain better control of the anteriorly lying lesions. Therefore, the TC approaches are unnecessary in most cases of acoustic neurinoma.

Rationale

The enlarged TO approach is an anterior extension of the translabyrinthine approach through the cochlea without mobilization of the facial nerve (Figs. 9.2 and 9.3), which will allow the preservation of the normal preoperative function of the nerve. The external auditory canal is closed, and the middle ear structures are removed [1–9]. It creates exposure from the superior petrosal sinus to the jugular bulb and from the internal carotid artery to the sigmoid sinus [11,12]. It is indicated in tumors of the middle and posterior cranial fossa with anterior extension. Currently, the TO approach is used for the removal of lesions located anterior to the IAC or inside the vestibule and/or cochlea [13–23]. Others have found the use of the TO approach useful in the treatment of inner ear extension of vestibular schwannomas. The surgical disadvantages of this approach are that the facial nerve is left like a bridge, and the close relationship to the carotid artery limits intraoperative maneuverability (however, the TO approach can be converted to a TC approach if necessary), and complete removal of the petrous apex cannot be achieved. The surgical procedure will of course result in hearing loss, but this is of secondary importance if the hearing in the contralateral ear is functional [12]. The advantages of the TO approach are: wide surgical exposure of the IAC, better exposure of the CPA in cases with challenging anatomy (high jugular bulb

Atlas of Microsurgery of the Lateral Skull Base. https://doi.org/10.1016/B978-0-443-19083-4.00009-X

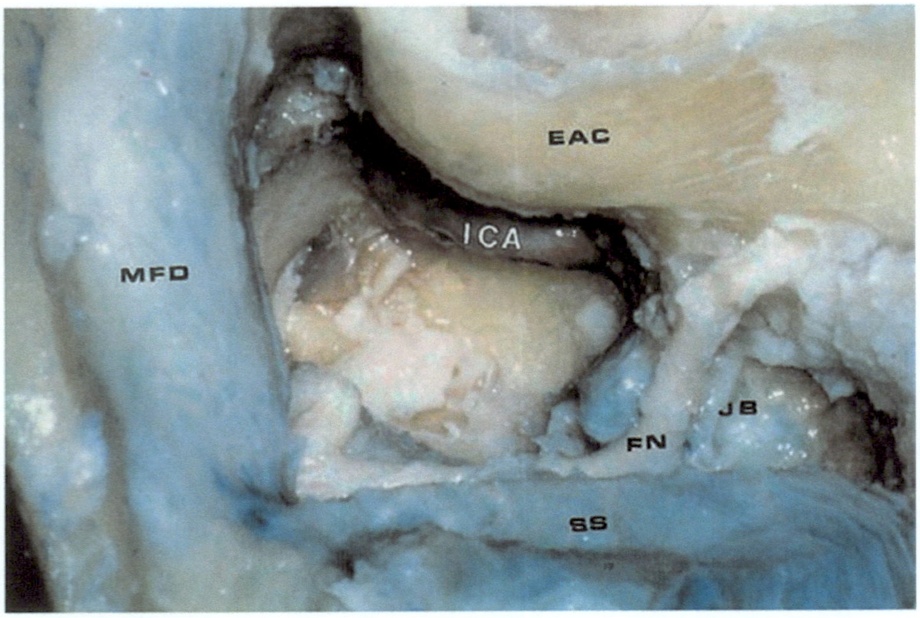

Fig. 9.1 The original transcochlear approach described by House and Hitselberger. *EAC*, external auditory canal; *FN*, facial nerve; *ICA*, internal carotid artery; *JB*, jugular bulb; *MFD*, middle fossa dura; *SS*, sigmoid sinus.

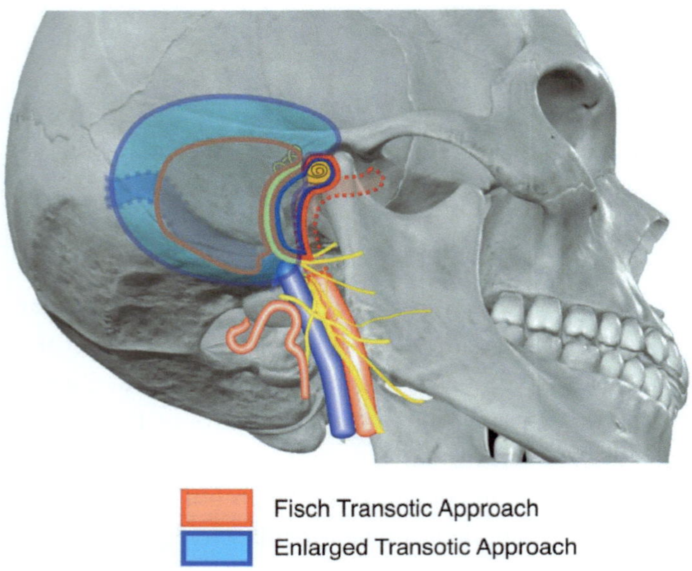

Fig. 9.2 A schematic drawing showing the outlines of the transotic approach (TOA).

and anteriorly positioned sigmoid sinus), increased access to the anterior CPA resulting in better facial nerve preservation rates, closure of the EAC and Eustachian tube, and obliteration of the residual cavity ensuring a low CSF leak.

Indications

- Some cases of cerebellopontine angle tumors with anterior extension and preoperative normal facial nerve function (e.g., epidermoids).
- Some cases of petrous bone cholesteatoma.
- Some cases of petrous bone tumors with preoperative normal facial nerve function.
- Some cases of vestibular schwannoma with invasion of the turns of the cochlea.

Contraindications

- Only hearing ear.
- Active middle ear infection.

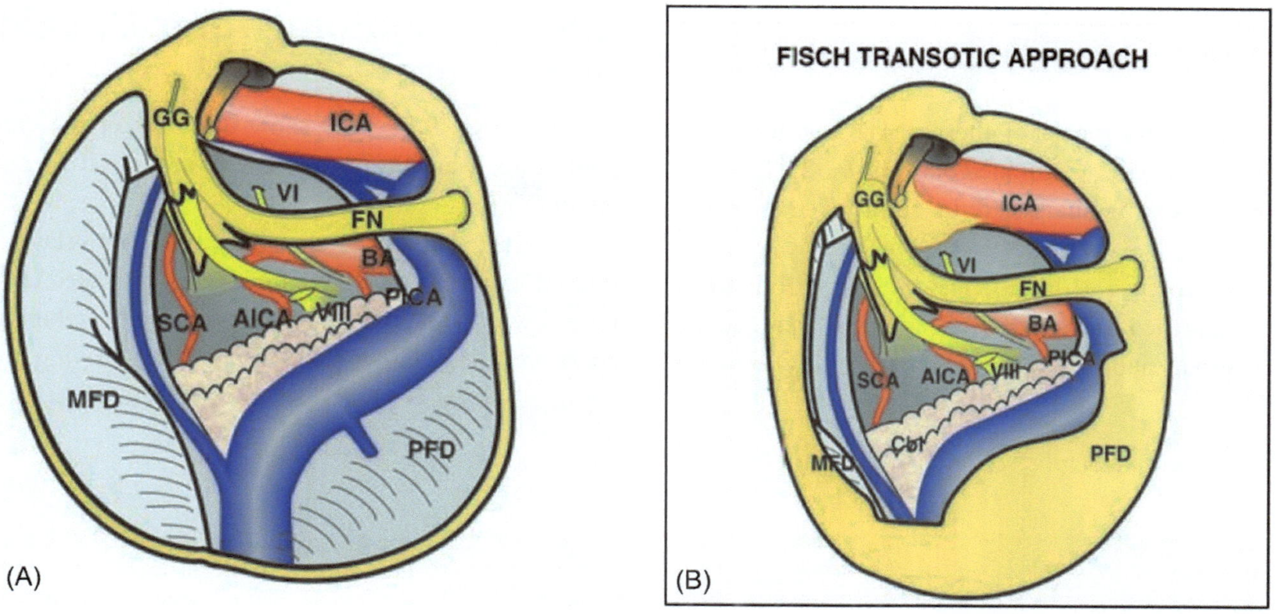

Fig. 9.3 *(Right ear)* (A) Wide exposure obtained with the enlarged transotic approach. (B) Note the differences with the much narrower approach originally described by Fisch. *AICA*, anterior-inferior cerebellar artery; *BA*, basilar artery; *Cbl*, cerebellum; *FN*, facial nerve; *GG*, geniculate ganglion; *ICA*, internal carotid artery; *MFD*, middle fossa dura; *PFD*, posterior fossa dura; *PICA*, posterior inferior cerebellar artery; *SCA*, superior cerebellar artery; *V*, trigeminal nerve; *VI*, abducent nerve; *VIII*, vestibulocochlear nerve.

Surgical Stapes (Left Side) (Figs. 9.4–9.41)

1. A wide C-shaped postauricular incision is made. The incision starts 3 cm above the auricle, curves posteriorly to about 4–5 cm posterior to the postauricular sulcus, and ends inferiorly at the level of the mastoid tip (Fig. 9.4).

2. The skin and subcutaneous tissues are elevated (Fig. 9.5), and the musculoperiosteal layer is incised in a T-shaped pattern (Fig. 9.6).

3. The musculoperiosteal flaps are elevated with a periosteal elevator (Fig. 9.7). Bleeding from the mastoid emissary vein usually occurs at this step (Fig. 9.8) and can be controlled by the use of bone wax. The flaps are sutured to the skin (Fig. 9.9); this helps in reducing the bleeding from the skin incision and at the same time keeps the muscoloperiosteal flaps retracted. We no longer use the conventional retractors, which significantly limit the surgeon's access.

4. The external auditory canal is transected and closed as a cul-de-sac (Figs. 9.10–9.17). The anterior skin flap is retracted anteriorly and maintained in position with two skin hooks (Fig. 9.18).

5. The meatal skin flap is detached from the bony canal wall. And then pushed medially and protected with an aluminum sheet (Fig. 9.19).

6. An extended mastoidectomy is performed, with removal of bone 2–3 cm posterior to the sigmoid sinus and over the middle fossa dura. The posterior and superior canal walls have been lowered (Fig. 9.20). The curette is used to remove the facial bridge (Fig. 9.21) and the ossicular chain can be exposed (Fig. 9.22).

7. The incudostapedial joint is being disarticulated, and the incus has been removed (Fig. 9.23). The tensor tympani tendon is being cut (Fig. 9.24) and the malleus has been removed. The facial nerve is skeletonized. The inferior tympanic bone is also drilled (Fig. 9.25). The cog and anterior buttress have to be removed to get complete control of the anterior attic recess (Fig. 9.26).

8. The decision on whether or not to remove the overlying bone depends on the level of exposure required and is judged in accordance with the pathology being treated. If need be, a large diamond burr is used to remove the bone overlying the middle fossa dura, sigmoid sinus, and posterior fossa dura (Figs. 9.27–9.29).

9. The labyrinthectomy is performed (Fig. 9.30) and the internal auditory canal is identified as previously

described. Bone superior and inferior to the internal auditory canal is drilled, creating two deep troughs around the canal (Fig. 9.31).

10. The anterior wall of the external auditory canal and the tympanic bone are thinned out (Fig. 9.32). The internal carotid artery is identified in front of the cochlea (Fig. 9.33).

11. The cochlea is drilled. The internal carotid artery is better identified. All the bone inferior and superior to the internal auditory canal is completely removed (Fig. 9.34).

Surgical Anatomy After Opening of the Dura

12. Figs. 9.35–9.40 show the view of the cerebellopontine angle and prepontine cistern as viewed after opening the dura.

13. At the end of the procedure, hemostasis is performed, and the cerebellopontine angle is irrigated with saline for a few minutes. The mastoid cavity and cerebellopontine angle are filled with abdominal fat. The overlying musculoperiosteal tissue is sutured tightly with Vicryl 0 (Fig. 9.41).

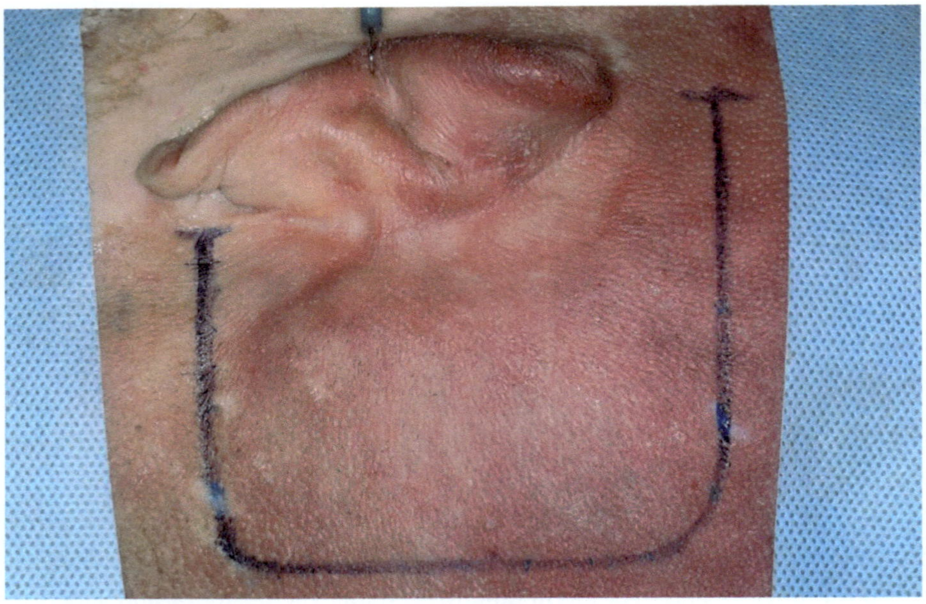

Fig. 9.4 A wide C-shaped postauricular incision is made. The incision starts 3 cm above the auricle, curves posteriorly to about 4–5 cm posterior to the postauricular sulcus, and ends inferiorly at the level of the mastoid tip.

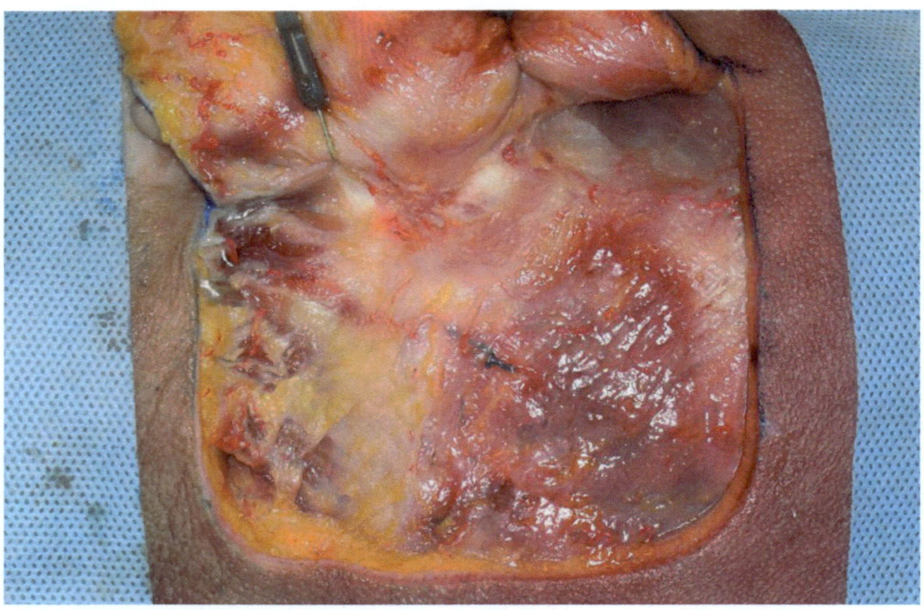

Fig. 9.5 Skin flap is already raised up and held anteriorly by hooks.

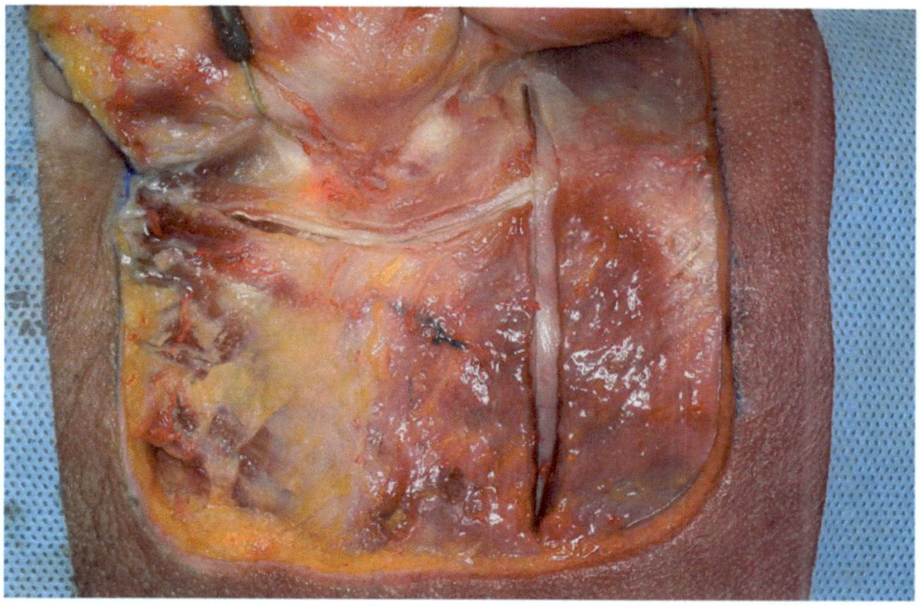

Fig. 9.6 The musculoperiosteal flap is incised using a monopolar diathermy in a "T" shape. The inferior limb of the vertical incision extends up to the mastoid tip.

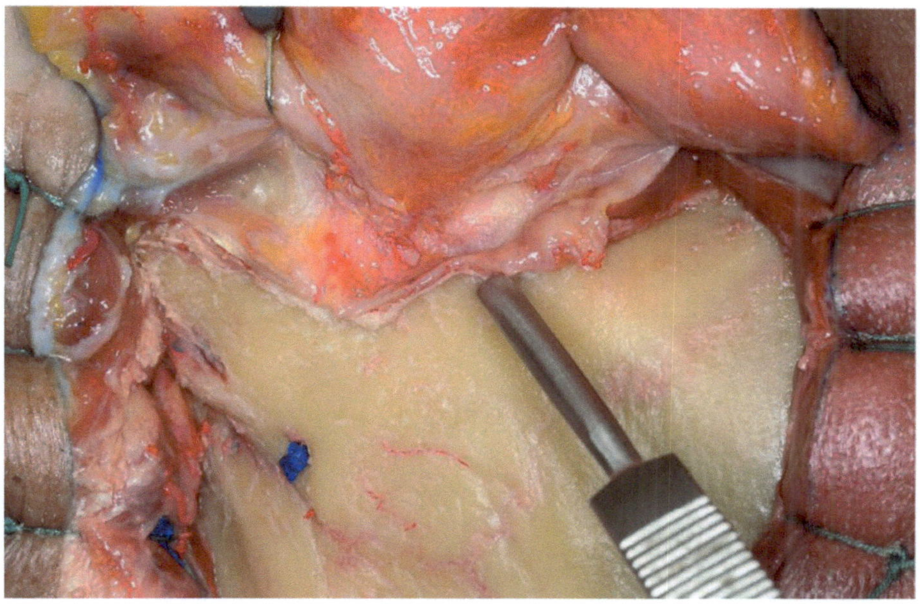

Fig. 9.7 The musculoperiosteal flaps are elevated with a periosteal elevator.

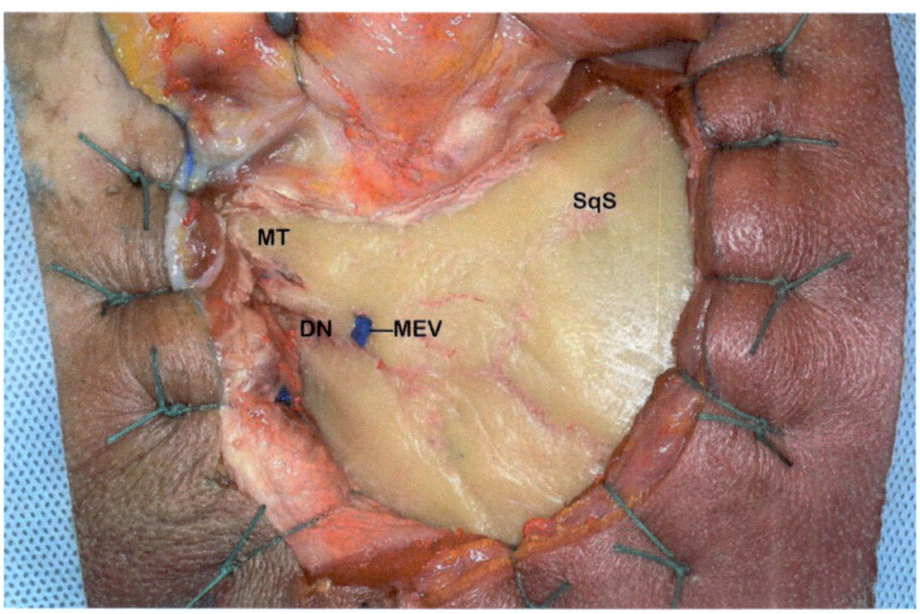

Fig. 9.8 Bleeding from the mastoid emissary vein usually occurs at this step and can be controlled by the use of bone wax. *MEV*, mastoid emissary vein; *SqS*, squamous suture; *MT*, mastoid tip; *DN*, digastric notch.

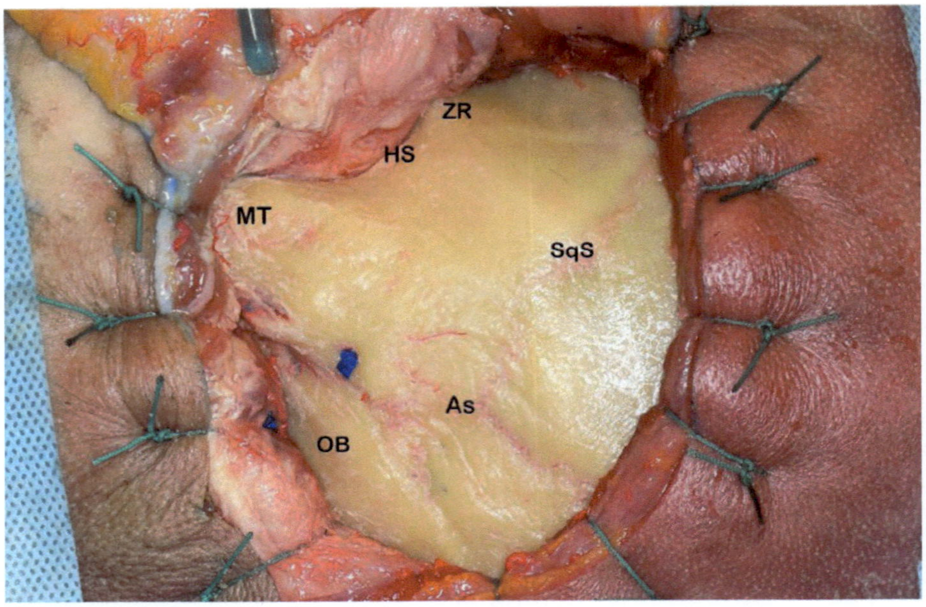

Fig. 9.9 The musculoperiosteal flaps are fixed by suturing. Note that no retractors are used. *ZR*, zygomatic root; *SqS*, squamous suture; *HS*, Henle's spine; *MT*, mastoid tip; *As*, asterion; *OB*, occipital bone.

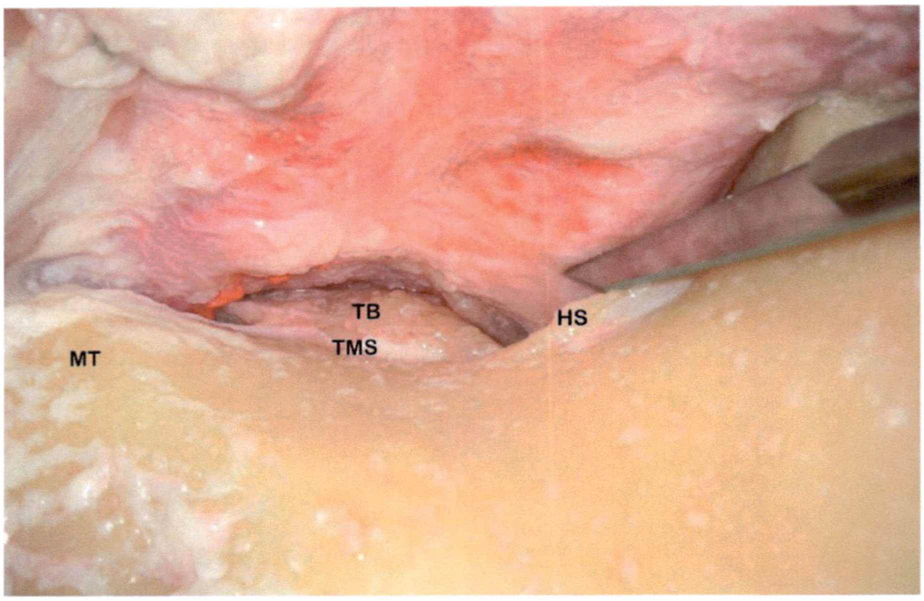

Fig. 9.10 The posterior meatal skin is cut with a Beaver blade. *HS*, Henle's spine; *TB*, tympanic bone; *TMS*, tympanomastoid suture; *MT*, mastoid tip.

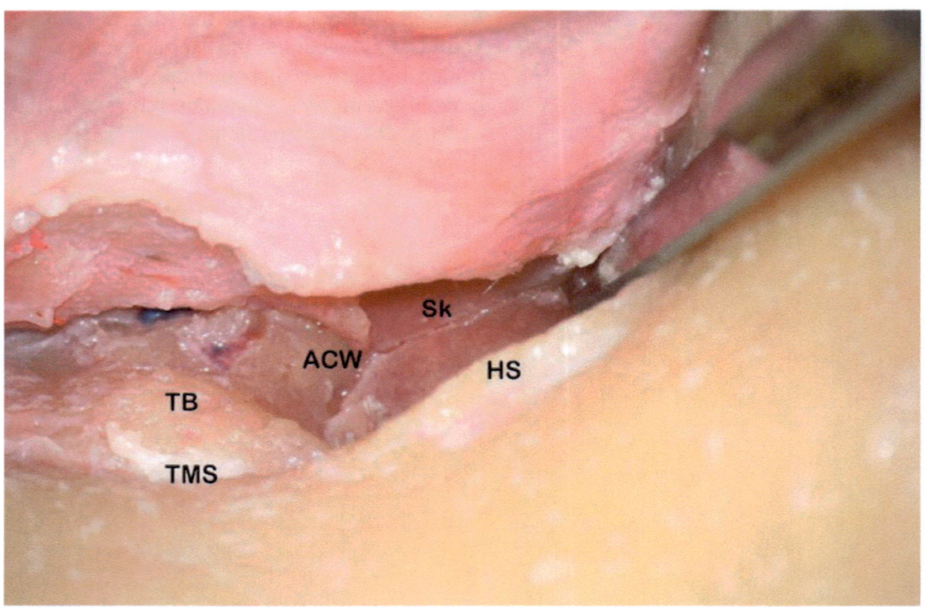

Fig. 9.11 The anterior meatal skin is also cut laterally and partially detached. *HS*, Henle's spine; *Sk*, skin; *ACW*, anterior canal wall; *TB*, tympanic bone; *TMS*, tympanomastoid suture.

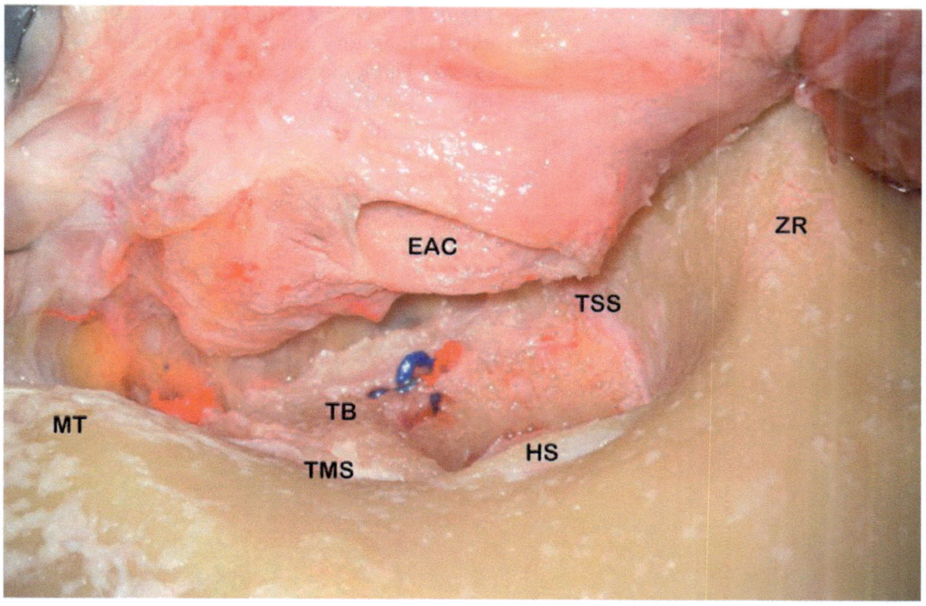

Fig. 9.12 The external auditory canal is transacted. *ZR*, zygomatic root; *TSS*, tympanosquamous suture; *EAC*, external auditory canal; *HS*, Henle's spine; *TB*, tympanic bone; *TMS*, tympanomastoid suture; *MT*, mastoid tip.

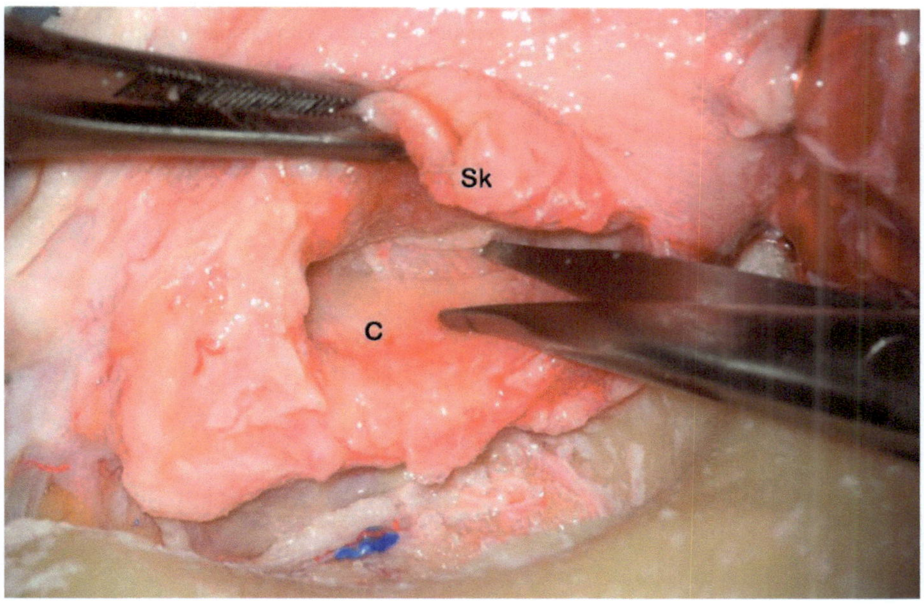

Fig. 9.13 Meticulous dissection of the skin from the anterior cartilage has to be performed. The skin of the cartilaginous canal is elevated for 1 cm from the cut margin toward the external meatus with tympanoplasty scissors to facilitate the eversion of the canal skin. *C*, cartilage; *Sk*, skin.

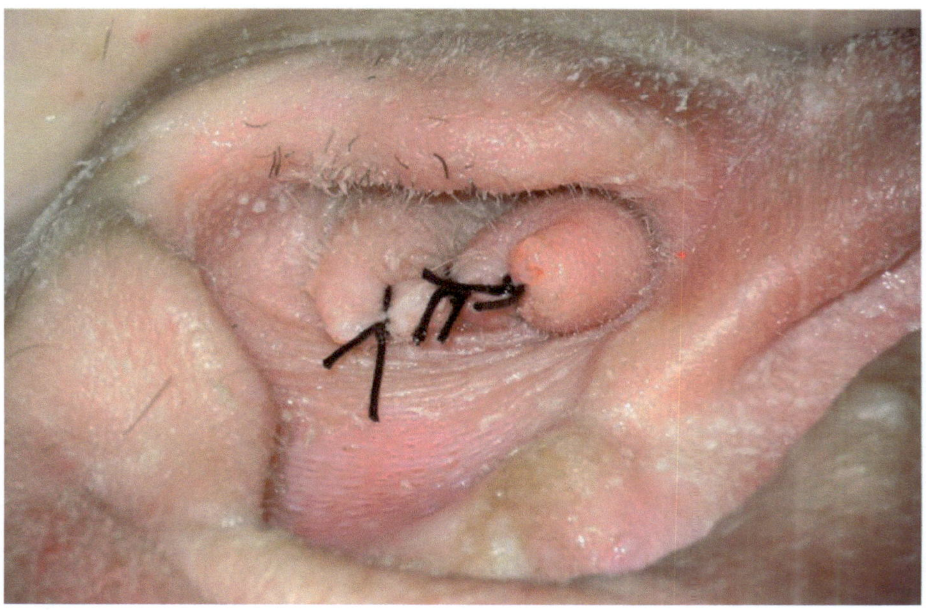

Fig. 9.14 The meatal skin is everted out through the canal and sutured with silk sutures.

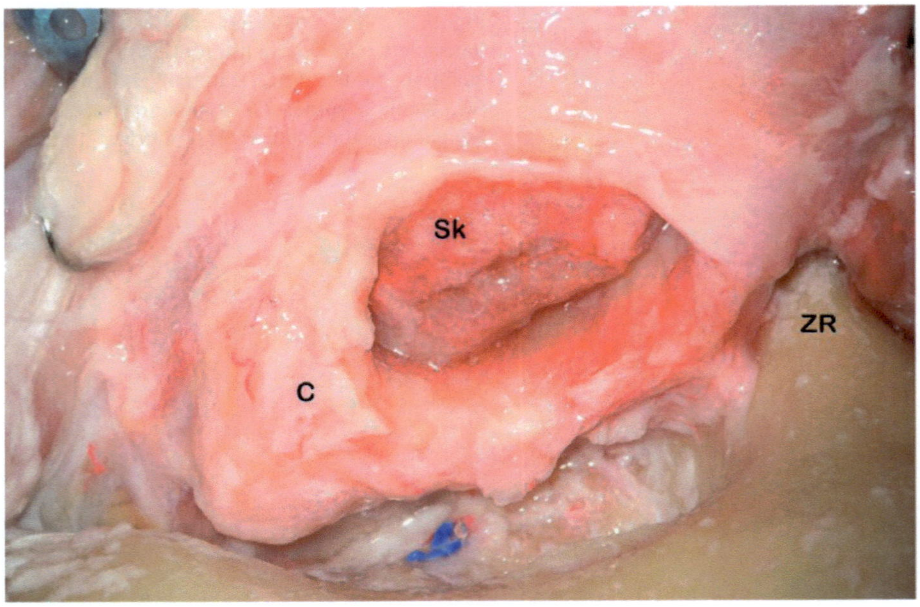

Fig. 9.15 Medially, the anterior cartilage of the external auditory canal is a good material for the second layer closure. *Sk*, skin; *ZR*, zygomatic root; *C*, cartilage.

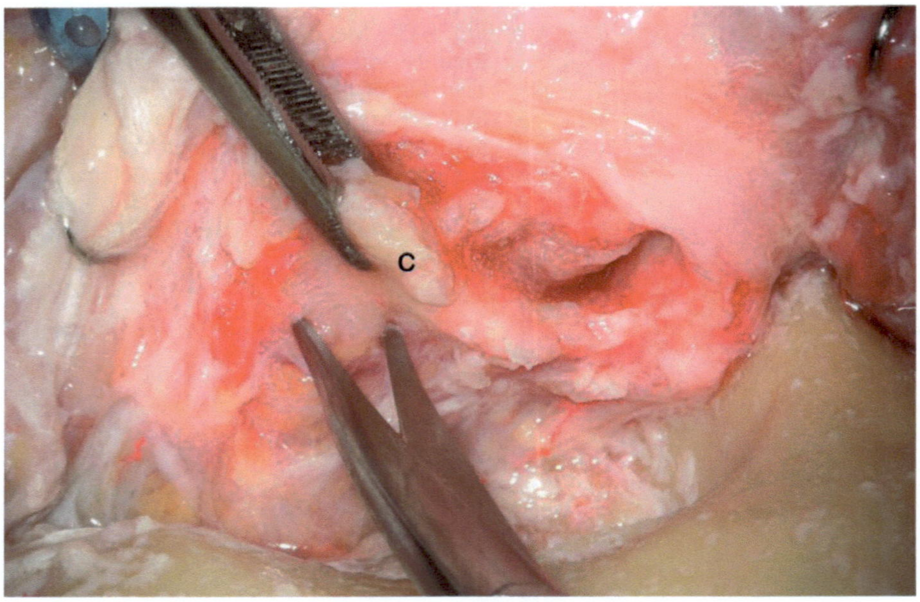

Fig. 9.16 The cartilage of the external auditory canal is elevated with tympanoplasty scissors. *C*, cartilage.

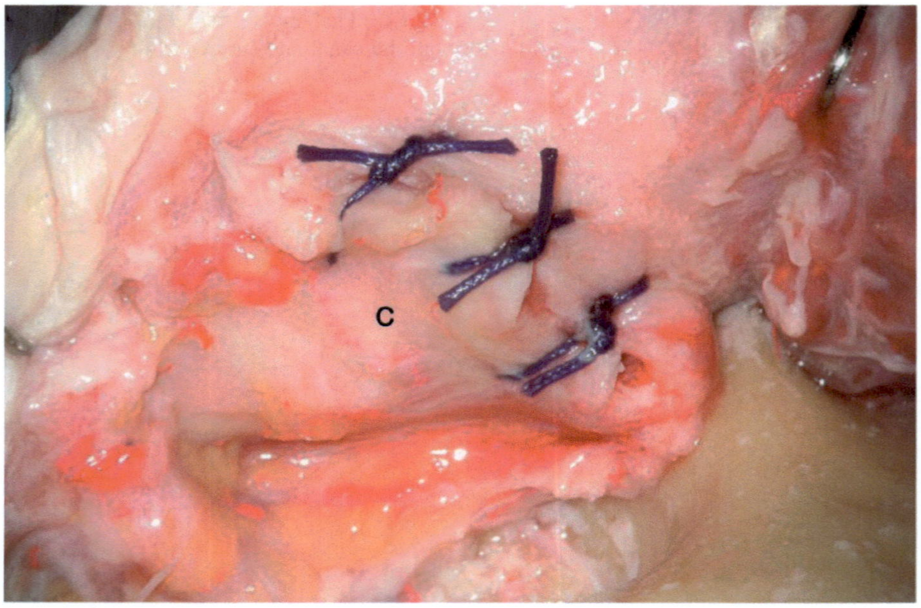

Fig. 9.17 The anterior cartilage of the external auditory canal is sutured with the soft tissues of the concha. This second layer is of utmost importance in case of CSF leakage. *C*, cartilage.

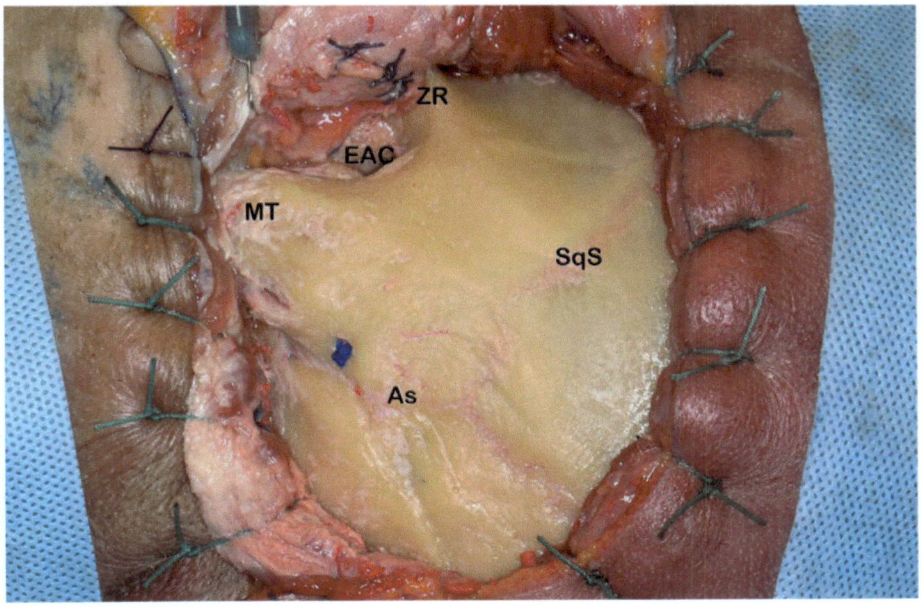

Fig. 9.18 The anterior skin flap is retracted using skin hooks. *ZR*, zygomatic root; *EAC*, external auditory canal; *MT*, mastoid tip; *SqS*, squamous suture; *As*, asterion.

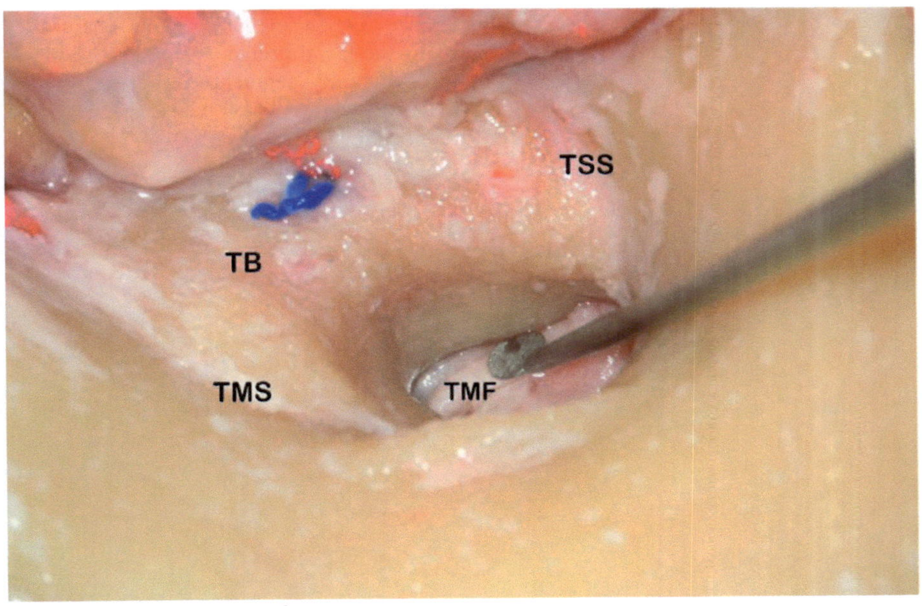

Fig. 9.19 The meatal skin flap is detached from the bony canal wall and then pushed medially. *TSS*, tympanosquamous suture; *TMF*, tympanomeatal flap; *TB*, tympanic bone; *TMS*, tympanomastoid suture.

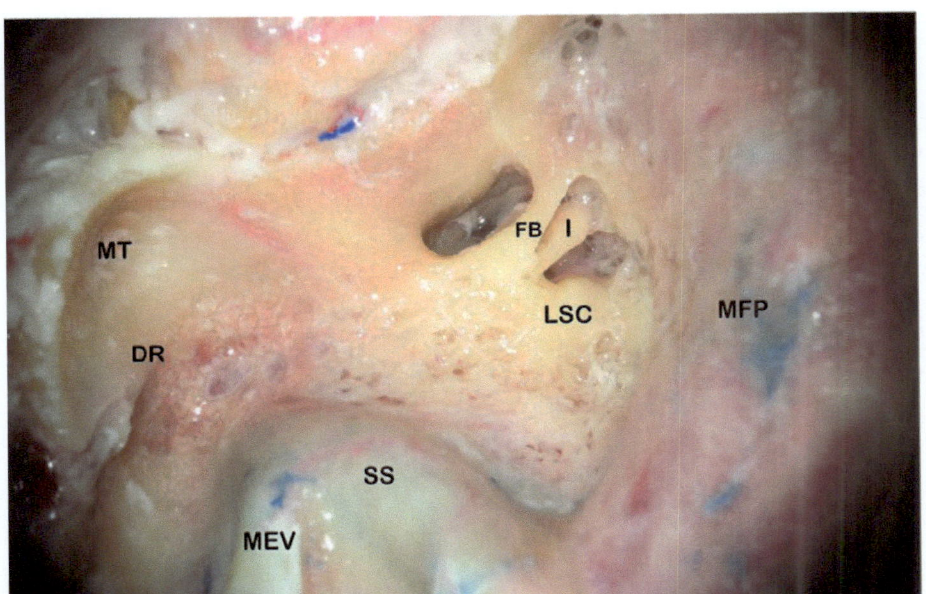

Fig. 9.20 An extended mastoidectomy is performed, with the removal of bone 2–3 cm posterior to the sigmoid sinus and over the middle fossa dura. The posterior and superior canal walls have been lowered. *MFP*, middle fossa plate; *LSC*, lateral semicircular canal; *I*, incus; *FB*, facial bridge; *SS*, sigmoid sinus; *MEV*, mastoid emissary vein; *DR*, digastric ridge; *MT*, mastoid tip.

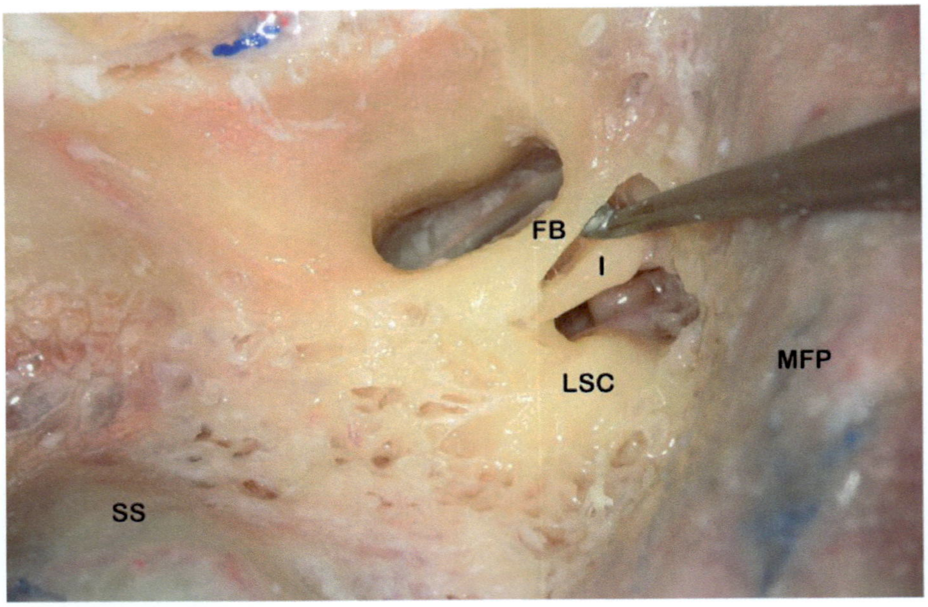

Fig. 9.21 The curette is used to remove the facial bridge to expose the ossicular chain. *MFP*, middle fossa plate; *I*, incus; *LSC*, lateral semicircular canal; *FB*, facial bridge; *SS*, sigmoid sinus.

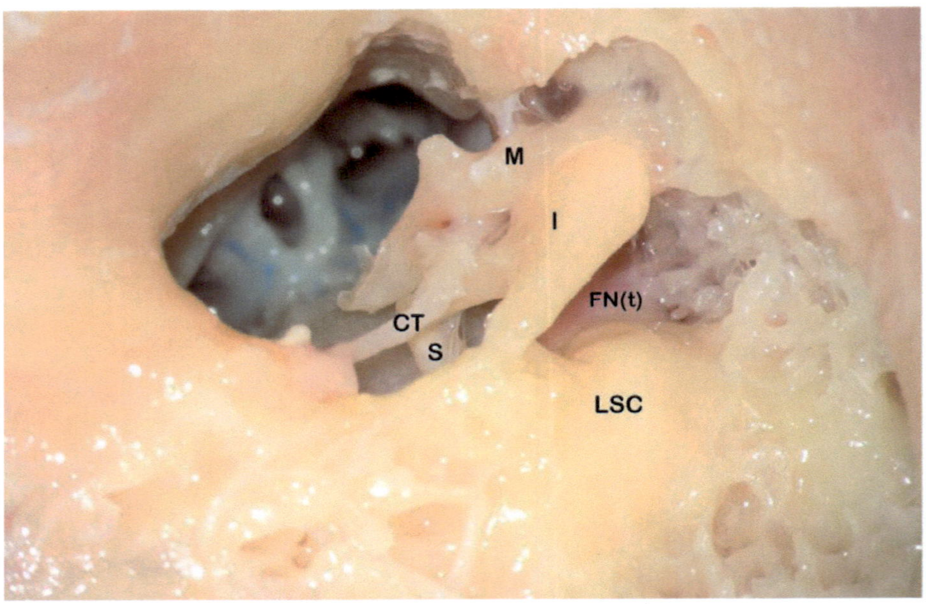

Fig. 9.22 A view of the ossicular chain after removing the facial bridge. *M*, malleus; *I*, incus; *FN(t)*, tympanic segment of facial nerve; *LSC*, lateral semicircular canal; *CT*, chorda tympani; *S*, stapes.

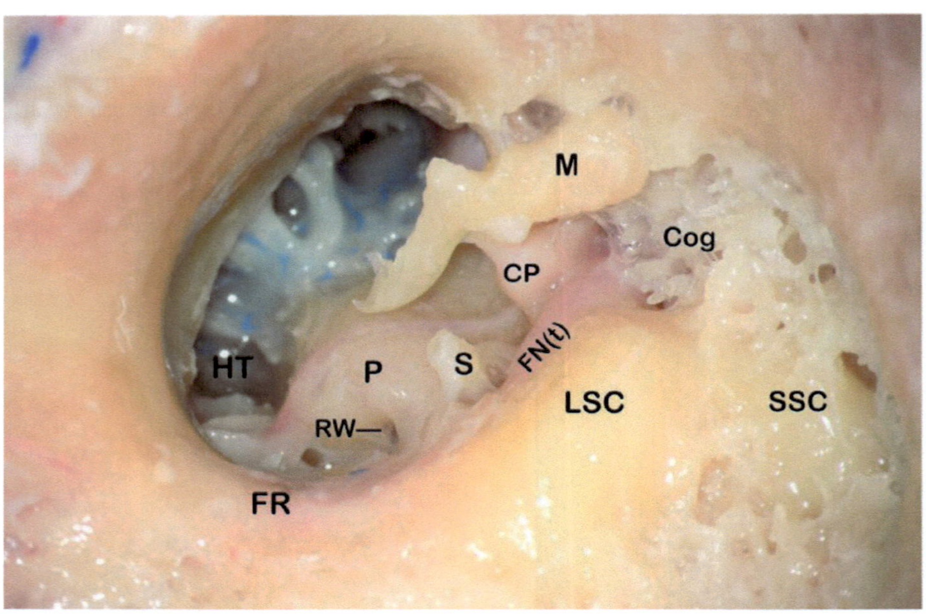

Fig. 9.23 The incudostapedial joint is being disarticulated. The incus has been removed. *M*, malleus; *CP*, cochleariform process; *SSC*, superior semicircular canal; *LSC*, lateral semicircular canal; *FN(t)*, tympanic segment of facial nerve; *S*, stapes; *P*, promontory; *RW*, round window; *HT*, hypotympanum; *FR*, facial ridge.

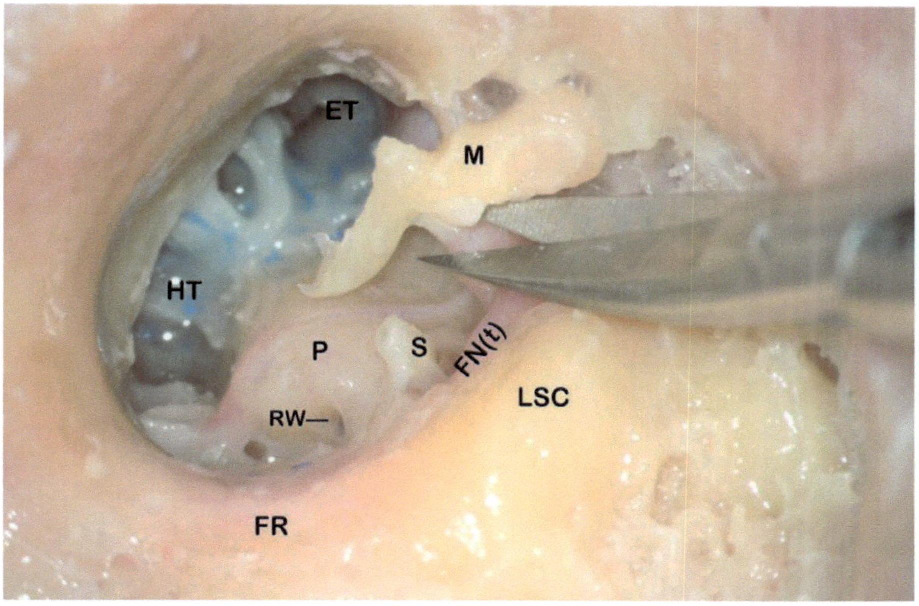

Fig. 9.24 The tensor tympani tendon is being cut. *M*, malleus; *ET*, eustachian tube; *P*, promontory; *S*, stapes; *FN(t)*, tympanic segment of facial nerve; *LSC*, lateral semicircular canal; *RW*, round window; *FR*, facial ridge; *HT*, hypotympanum.

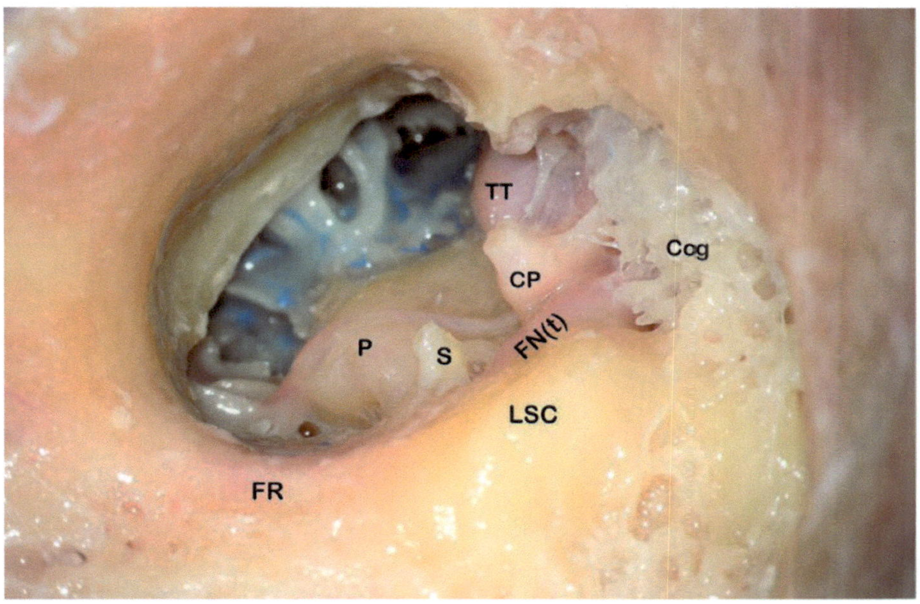

Fig. 9.25 The malleus has been removed. The cog is a bony septum that separates the anterior from the posterior epitympanum. *TT*, tensor tympani muscle; *CP*, cochleariform process; *FN(t)*, tympanic segment of facial nerve; *LSC*, lateral semicircular canal; *S*, stapes; *P*, promontory; *FR*, facial ridge.

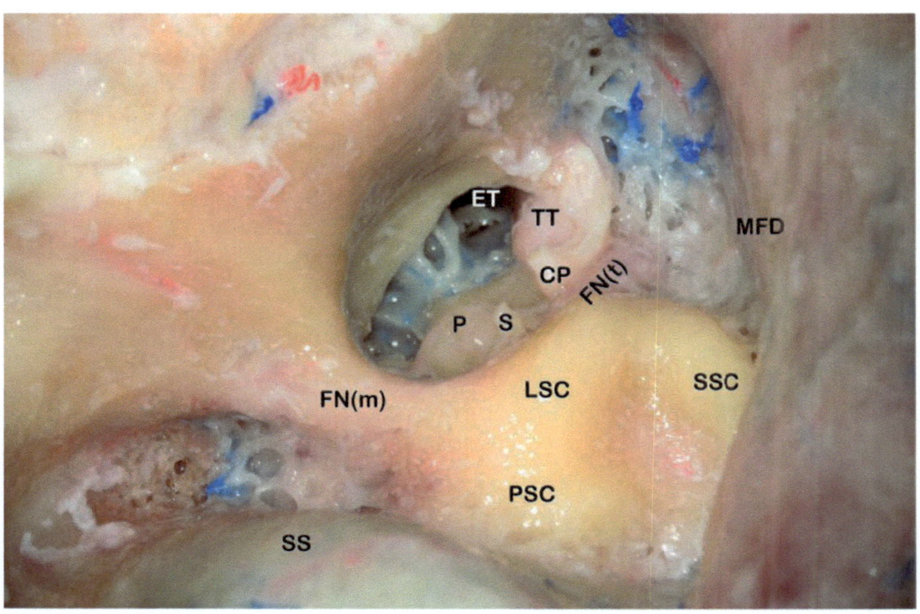

Fig. 9.26 The cog and anterior buttress have to be removed to get complete control of the anterior attic recess. *MFD*, middle fossa dura; *SSC*, superior semicircular canal; *LSC*, lateral semicircular canal; *PSC*, posterior semicircular canal; *FN(t)*, tympanic segment of facial nerve; *FN(m)*, mastoid segment of facial nerve; *CP*, cochleariform process; *TT*, tensor tympani muscle; *ET*, eustachian tube; *S*, stapes; *P*, promontory; *SS*, sigmoid sinus.

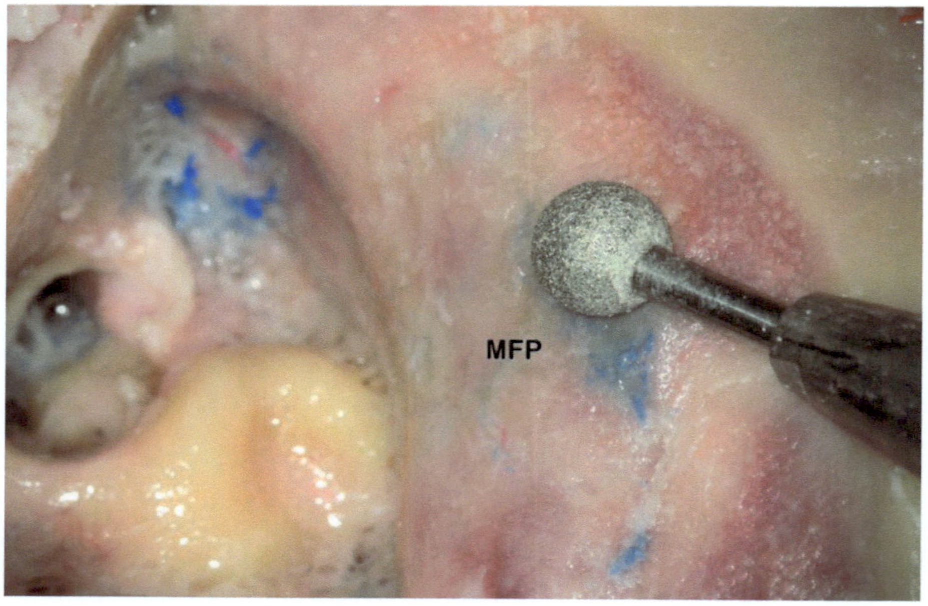

Fig. 9.27 A large diamond burr is used to remove the bone overlying the middle fossa dura. The decision on whether or not to remove the overlying bone depends on the level of exposure required and is judged in accordance with the pathology being treated. *MFP*, middle fossa plate.

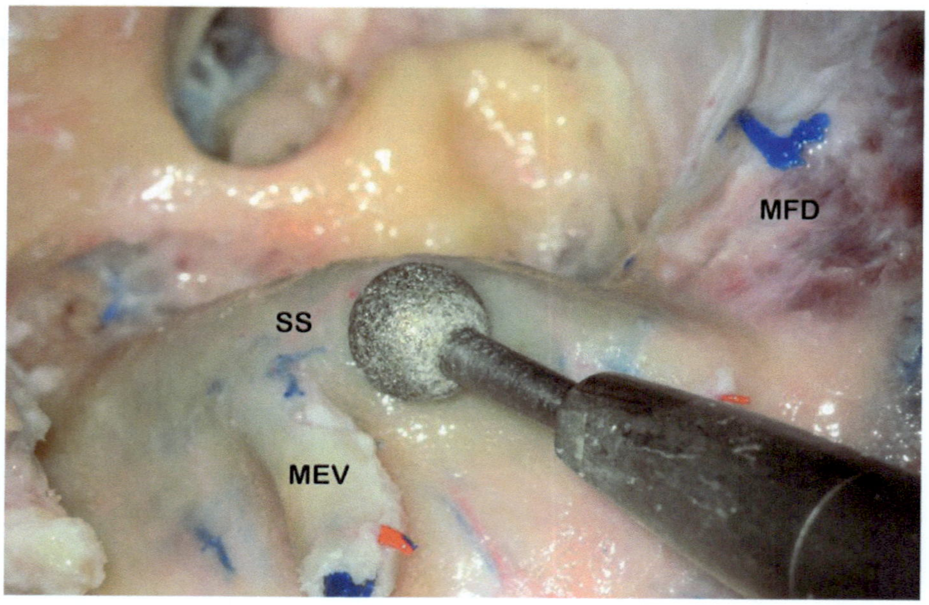

Fig. 9.28 A large diamond burr is used to remove the bone overlying the sigmoid sinus and the posterior fossa dura. *MFD*, middle fossa dura; *SS*, sigmoid sinus; *MEV*, mastoid emissary vein.

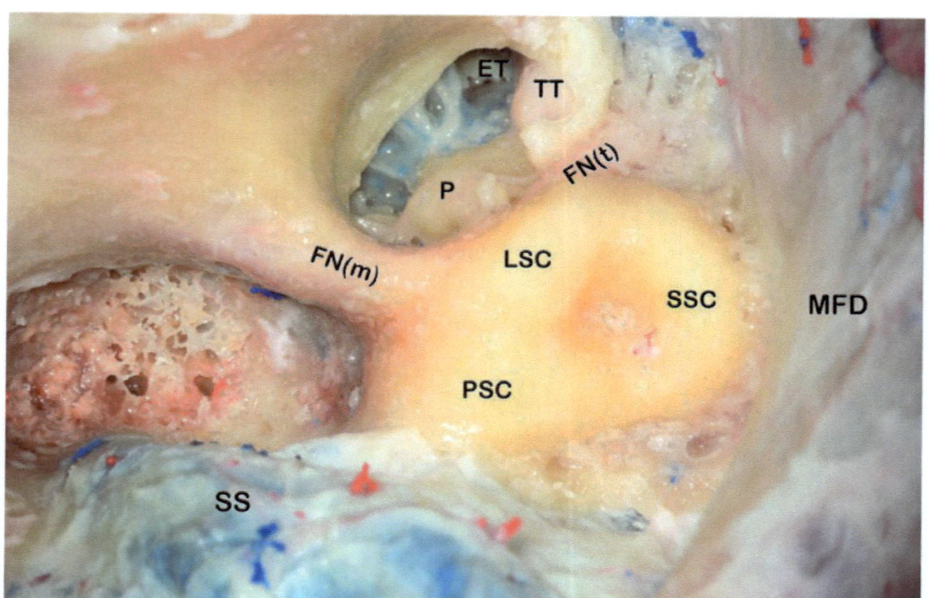

Fig. 9.29 The bone located at the sinodural angle is removed to complete the uncovering of the middle fossa dura and the sigmoid sinus. *MFD*, middle fossa dura; *SSC*, superior semicircular canal; *LSC*, lateral semicircular canal; *PSC*, posterior semicircular canal; *FN(t)*, tympanic segment of facial nerve; *FN(m)*, mastoid segment of facial nerve; *TT*, tensor tympani muscle; *ET*, eustachian tube; *P*, promontory; *SS*, sigmoid sinus.

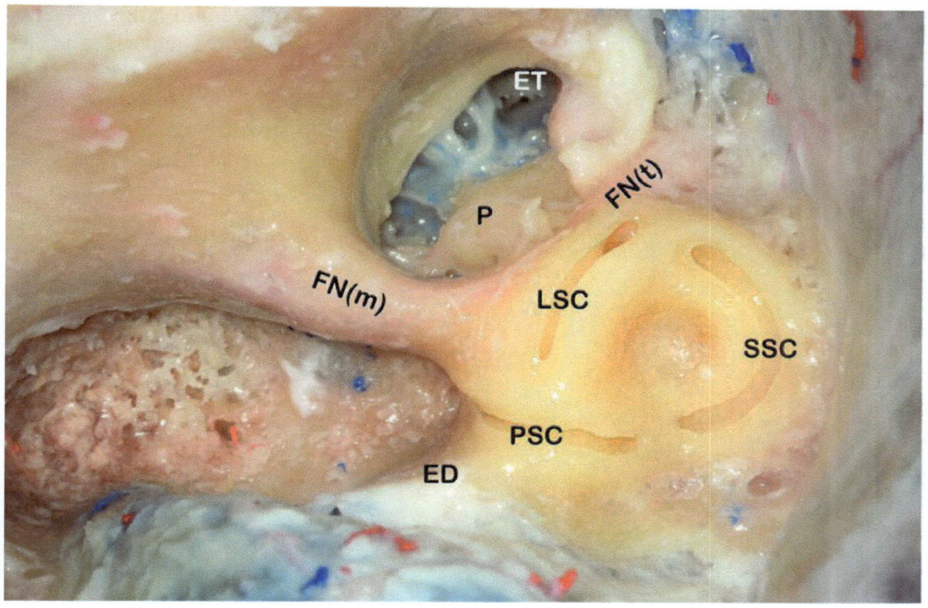

Fig. 9.30 The labyrinthectomy is performed. The lateral semicircular canal, posterior semicircular canal, and superior semicircular canal have been opened. *ET*, eustachian tube; *P*, promontory; *FN(t)*, tympanic segment of facial nerve; *FN(m)*, mastoid segment of facial nerve; *SSC*, superior semicircular canal; *LSC*, lateral semicircular canal; *PSC*, posterior semicircular canal; *ED*, endolymphatic duct.

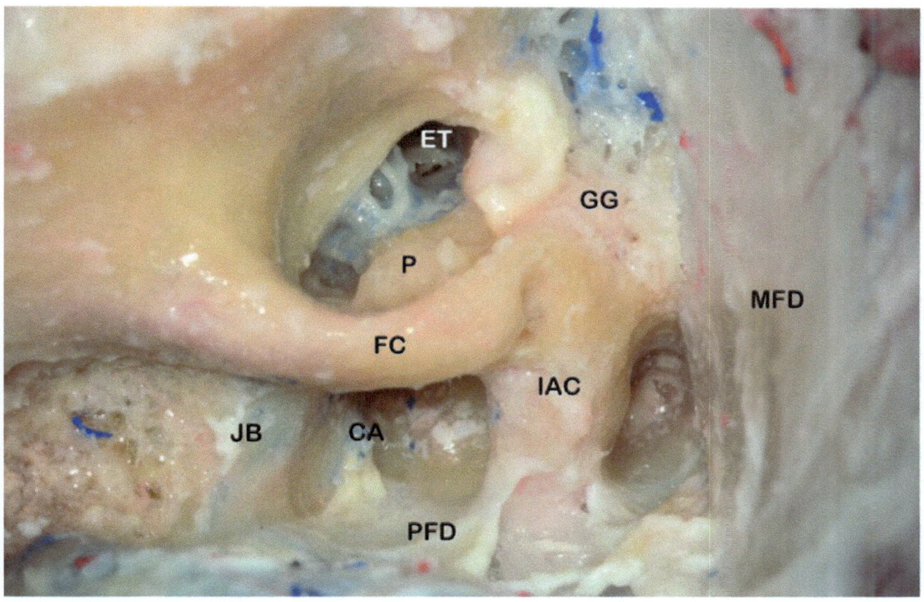

Fig. 9.31 Identification of the internal auditory canal. *MFD*, middle fossa dura; *GG*, geniculate ganglion; *ET*, eustachian tube; *P*, promontory; *FC*, fallopian canal; *CA*, cochlear aqueduct; *IAC*, internal auditory canal; *PFD*, posterior fossa dura; *JB*, jugular bulb.

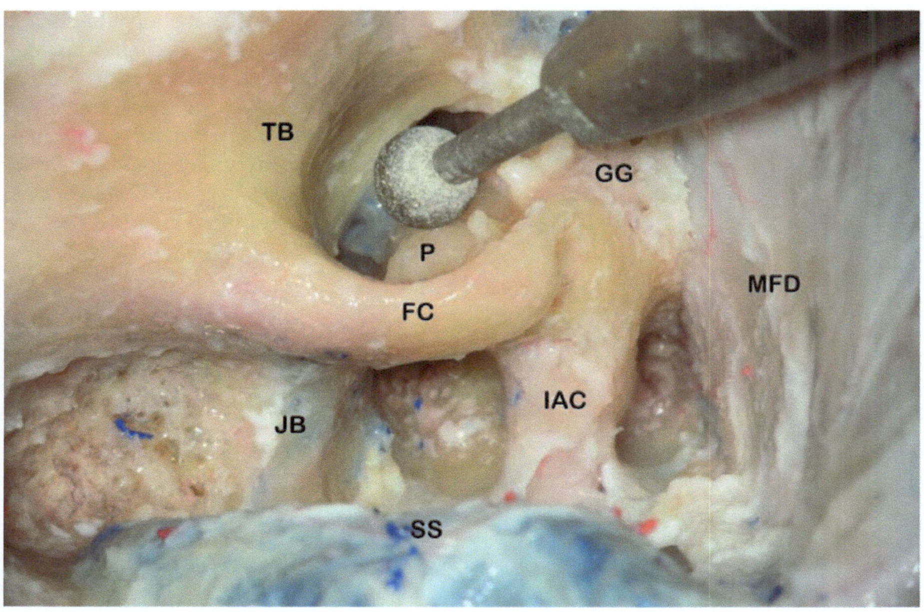

Fig. 9.32 The anterior wall of the external auditory canal and the tympanic bone are thinned out. Identification of the internal carotid artery is either started from the eustachian tube, where the burr is located or at the anteroinferior limit of the promontory. *MFD*, middle fossa dura; *GG*, geniculate ganglion; *P*, promontory; *FC*, fallopian canal; *IAC*, internal auditory canal; *JB*, jugular bulb; *SS*, sigmoid sinus; *TB*, tympanic bone.

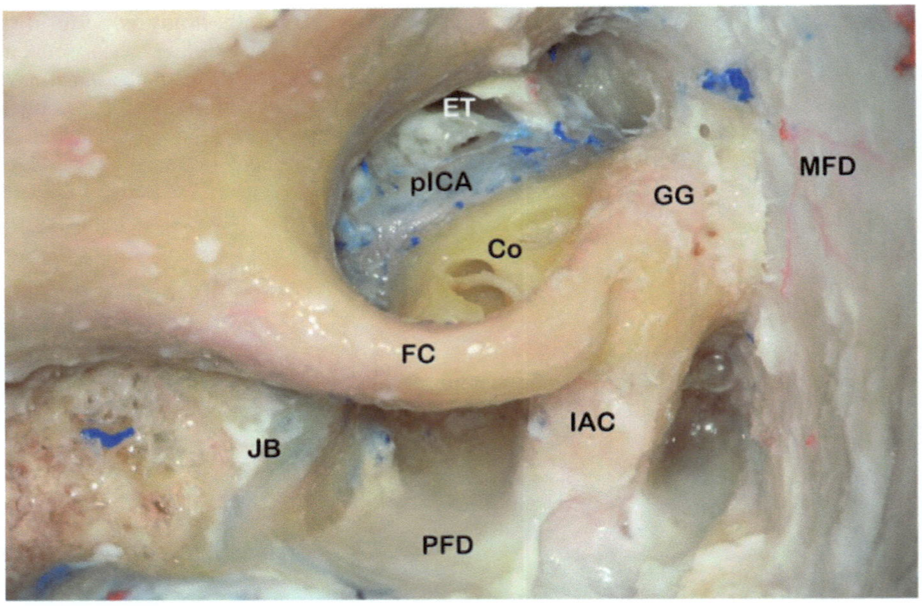

Fig. 9.33 The internal carotid artery is identified in front of the cochlea. The cochlea is partially drilled. *MFD*, middle fossa dura; *GG*, geniculate ganglion; *ET*, eustachian tube; *pICA*, petrous segment of internal carotid artery; *Co*, cochlea; *FC*, fallopian canal; *IAC*, internal auditory canal; *JB*, jugular bulb; *PFD*, posterior fossa dura.

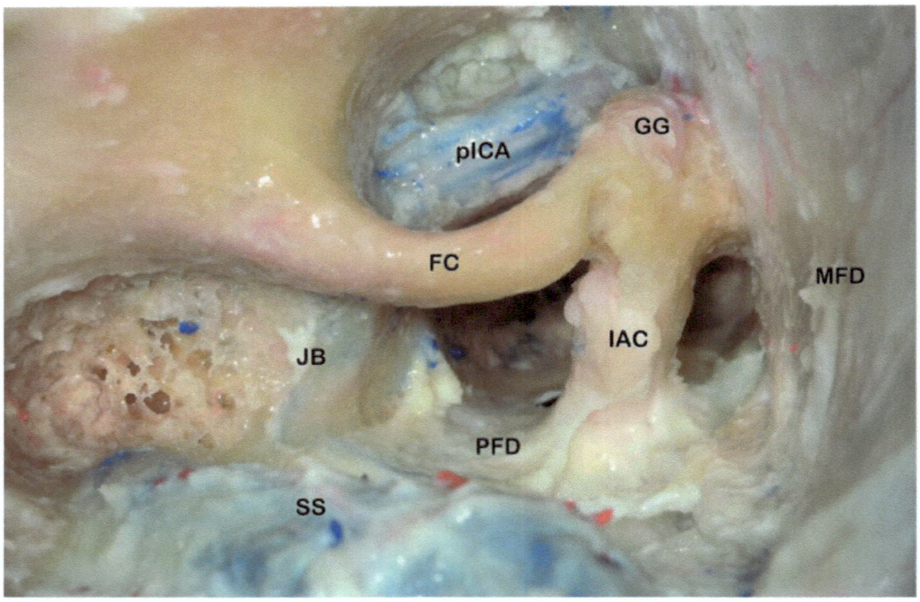

Fig. 9.34 The cochlea is drilled. The internal carotid artery is better identified. All the bone inferior and superior to the internal auditory canal is completely removed. *GG*, geniculate ganglion; *MFD*, middle fossa dura; *IAC*, internal auditory canal; *FC*, fallopian canal; *pICA*, petrous segment of internal carotid artery; *JB*, jugular bulb; *PFD*, posterior fossa dura; *SS*, sigmoid sinus.

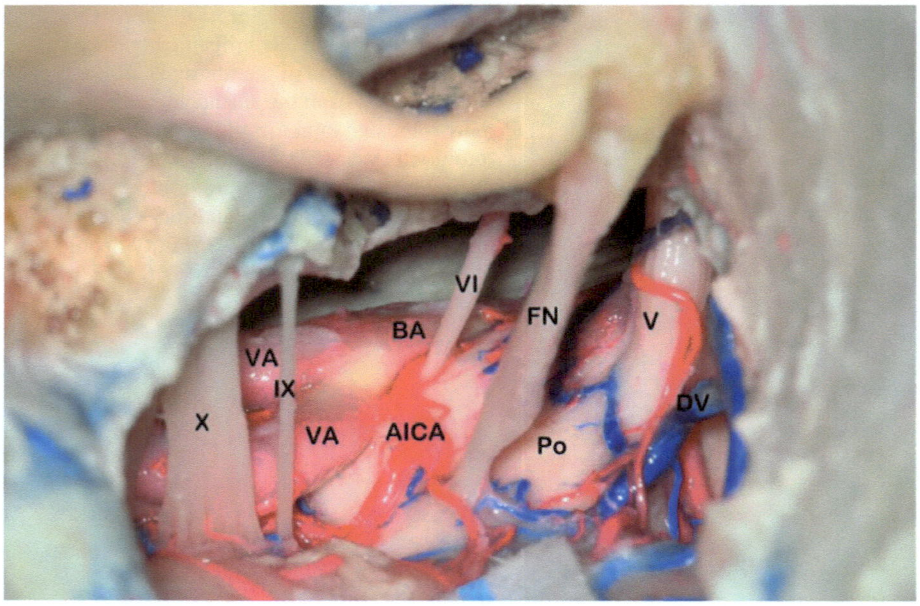

Fig. 9.35 View of the cerebellopontine angle and prepontine cistern as viewed after opening the dura. *DV*, Dandy's vein (superior petrosal vein); *V*, trigeminal nerve; *FN*, facial nerve; *VI*, abducent nerve; *BA*, basilar artery; *AICA*, anteroinferior cerebellar artery; *Po*, pons; *VA*, vertebral artery; *IX*, glossopharyngeal nerve; *VA*, vertebral artery; *X*, vagus nerve.

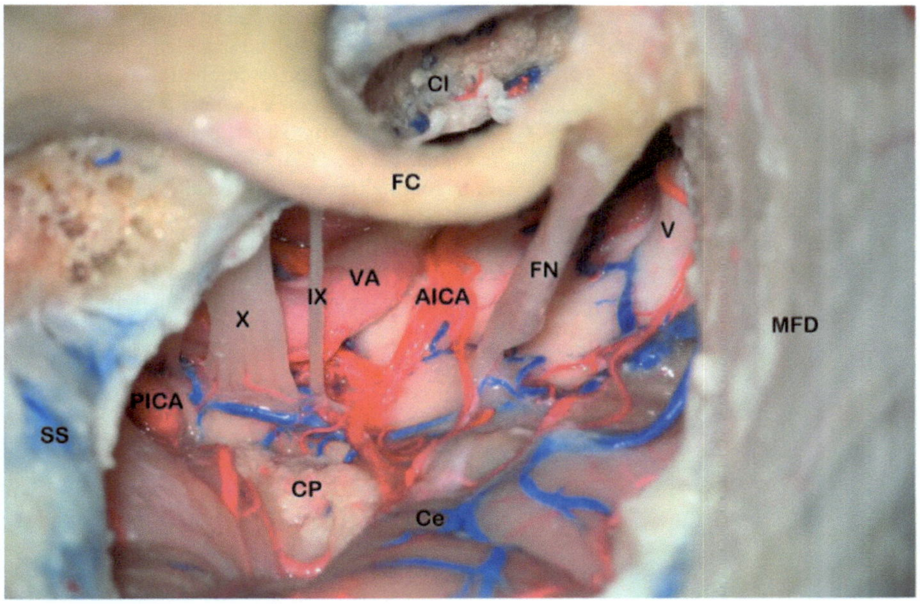

Fig. 9.36 Mild retraction of the flocculus provides a good view of the choroid plexus. *MFD*, middle fossa dura; *V*, trigeminal nerve; *FN*, facial nerve; *Cl*, clivus; *FC*, fallopian canal; *AICA*, anteroinferior cerebellar artery; *VA*, vertebral artery; *IX*, glossopharyngeal nerve; *X*, vagus nerve; *PICA*, posteroinferior cerebellar artery; *SS*, sigmoid sinus; *CP*, choroid plexus; *Ce*, cerebellum.

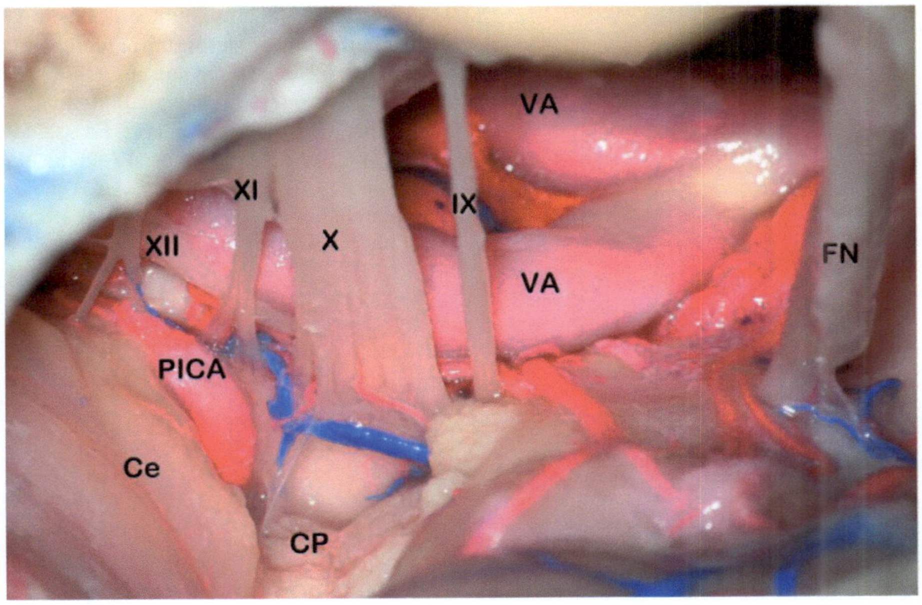

Fig. 9.37 Progressive retraction of the flocculus is required to have access to the Luschka foramen. *FN*, facial nerve; *VA*, vertebral artery; *IX*, glossopharyngeal nerve; *X*, vagus nerve; *XI*, accessory nerve; *XII*, hypoglossal nerve; *PICA*, posteroinferior cerebellar artery; *Ce*, cerebellum; *CP*, choroid plexus.

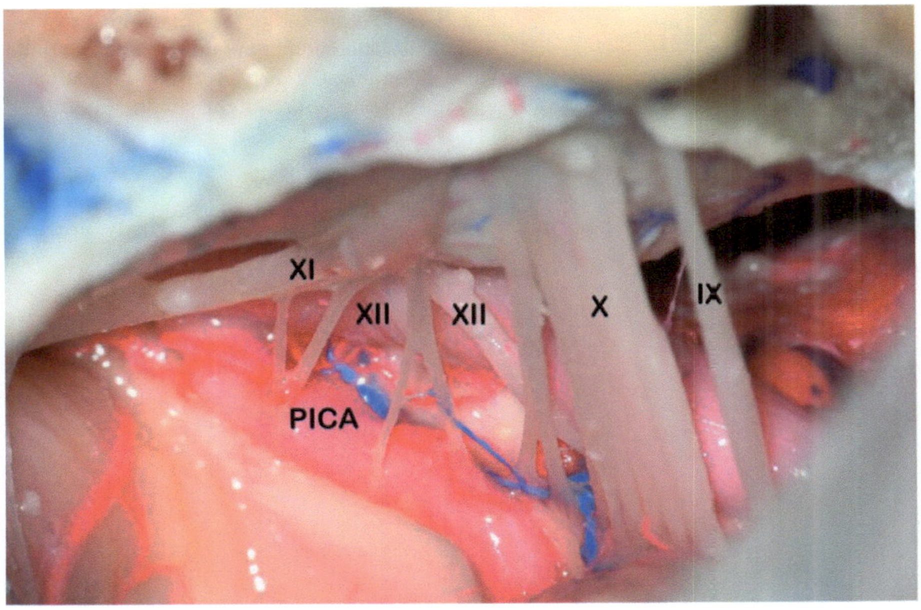

Fig. 9.38 Tilting the microscope downward, the lower cranial nerves are seen. The relation between these nerves and the posterior inferior cerebellar artery is appreciated. *IX*, glossopharyngeal nerve; *X*, vagus nerve; *XI*, accessory nerve; *XII*, hypoglossal nerve; *PICA*, posteroinferior cerebellar artery.

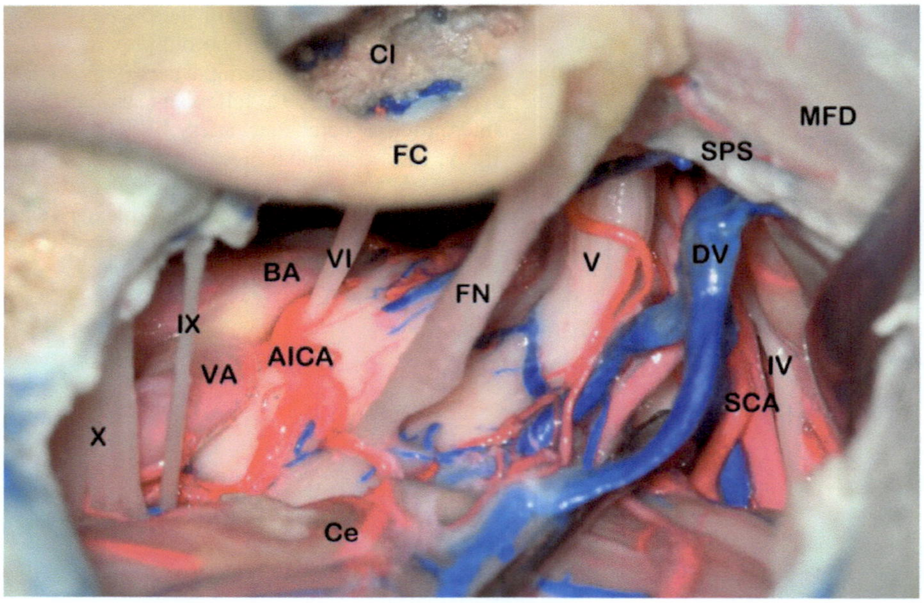

Fig. 9.39 Mild retraction of the tentorium provides a good view of the trochlear nerve and its relation to the superior cerebellar artery. *MFD*, middle fossa dura; *SPS*, superior petrosal sinus; *DV*, Dandy's vein (superior petrosal vein); *IV*, trochlear nerve; *SCA*, superior cerebellar artery; *V*, trigeminal nerve; *FN*, facial nerve; *FC*, fallopian canal; *Cl*, clivus; *VI*, abducent nerve; *BA*, basilar artery; *AICA*, anteroinferior cerebellar artery; *IX*, glossopharyngeal nerve; *VA*, vertebral artery; *X*, vagus nerve; *Ce*, cerebellum.

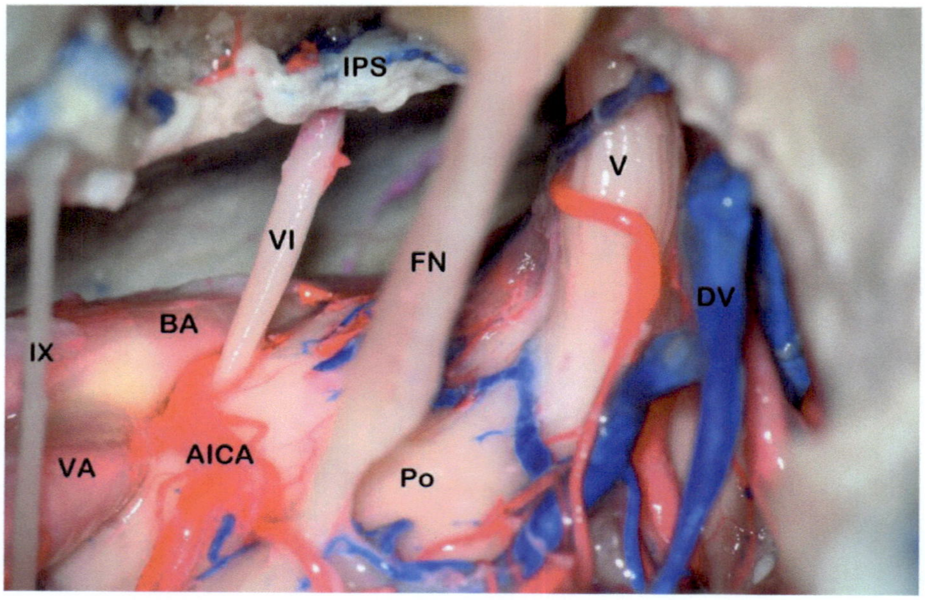

Fig. 9.40 View of the prepontine cistern showing the origin of the anteroinferior cerebellar artery from the basilar artery and its relation to the abducent nerve. *DV*, Dandy's vein (superior petrosal vein); *V*, trigeminal nerve; *FN*, facial nerve; *IPS*, inferior petrosal sinus; *VI*, abducent nerve; *Po*, pons; *BA*, basilar artery; *AICA*, anteroinferior cerebellar artery; *VA*, vertebral artery; *IX*, glossopharyngeal nerve.

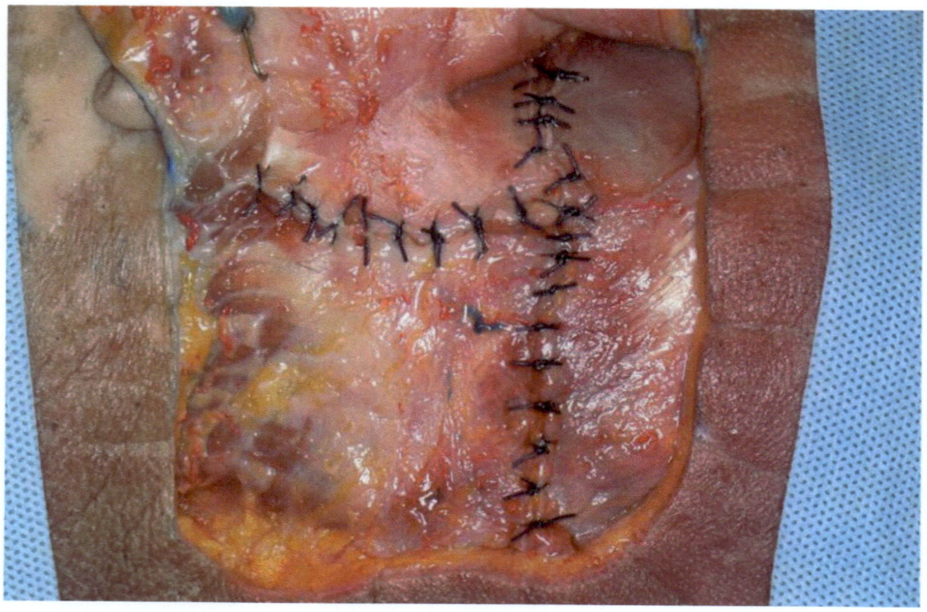

Fig. 9.41 At the end of the procedure, the mastoid cavity and cerebellopontine angle are filled with abdominal fat. The overlying musculoperiosteal tissue is sutured tightly with Vicryl 0.

Case 1: Supralabyrinthine Petrous Bone Cholesteatoma

See Figs. 9.42–9.76.

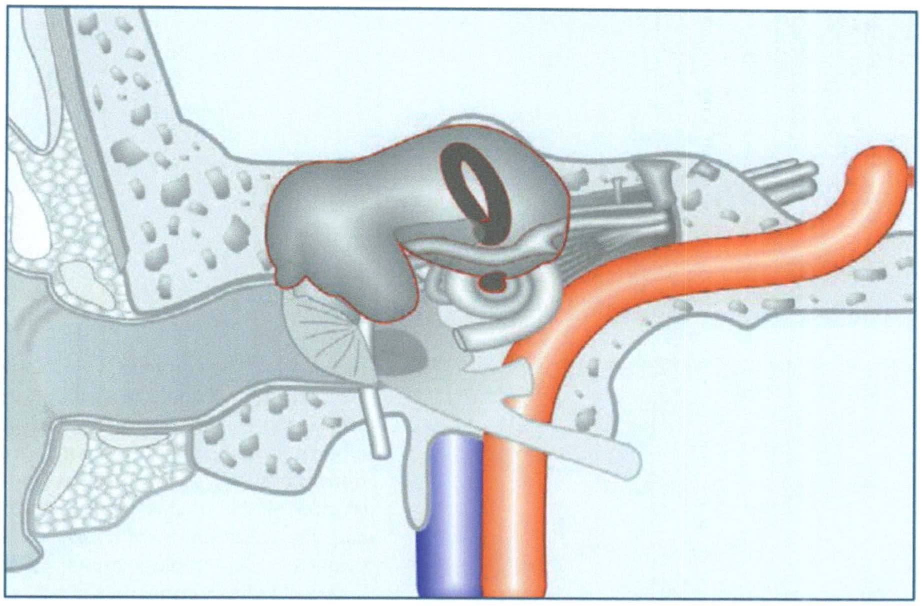

Fig. 9.42 A case of acquired supralabyrinthine petrous bone cholesteatoma with extensive erosion in the posterior labyrinth and a small erosion in the cochlea removed with a transotic approach.

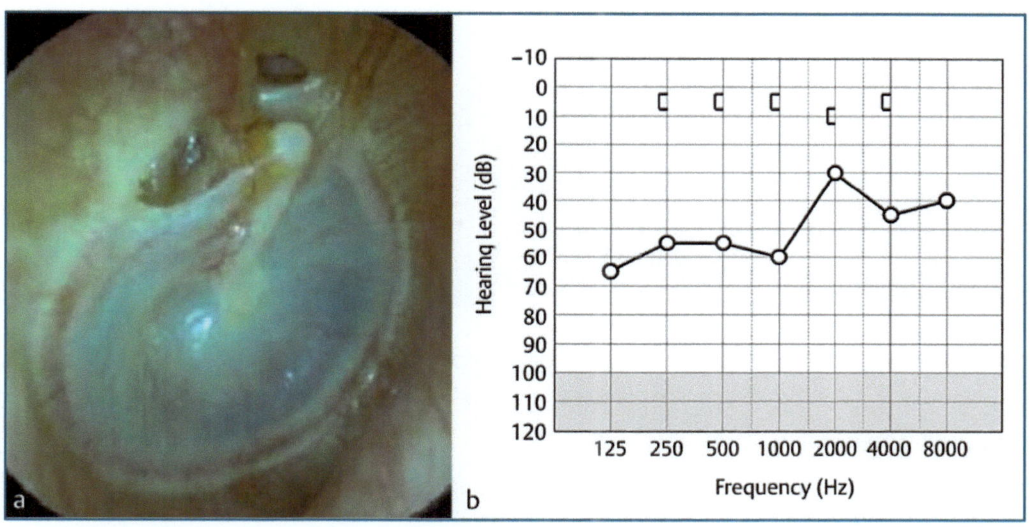

Fig. 9.43 The 57-year-old patient had a history of right-sided hearing loss for years without any sequel until he suffered from sustained complete facial nerve palsy 2 months before. The otoscopy (a) shows a small but deep dimple in the pars flaccida, but no sign of the presence of active cholesteatoma is seen. A preoperative audiogram (b) demonstrates moderate conductive hearing loss in the right with good bone conduction. The hearing in the contralateral ear remains normal.

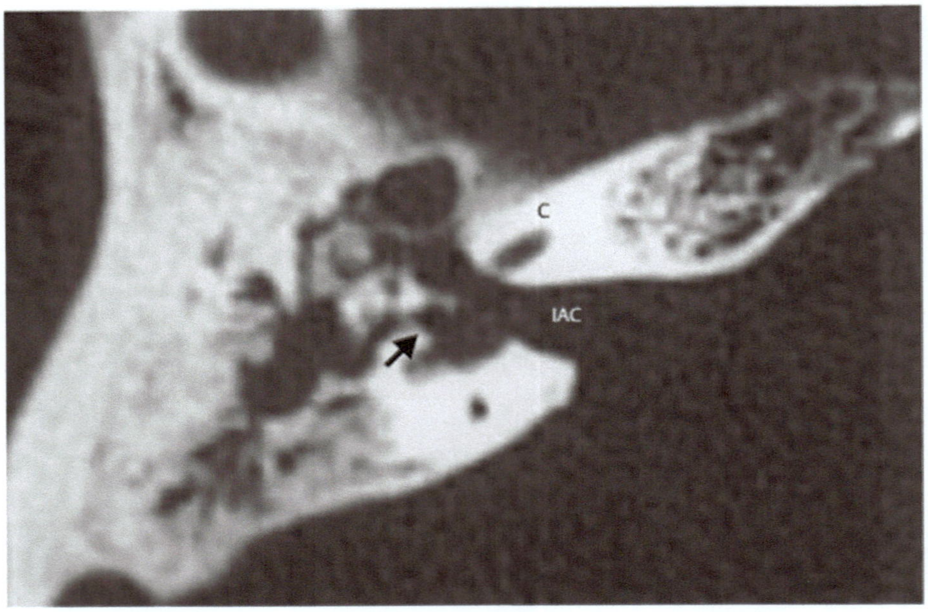

Fig. 9.44 The preoperative axial CT shows the extension of the cholesteatoma matrix toward the fundus of the internal auditory canal, eroding the labyrinthine segment of the fallopian canal and the bone medially to the ampulla of the superior semicircular canal *(arrow)*. *C*, cochlea; *IAC*, internal auditory.

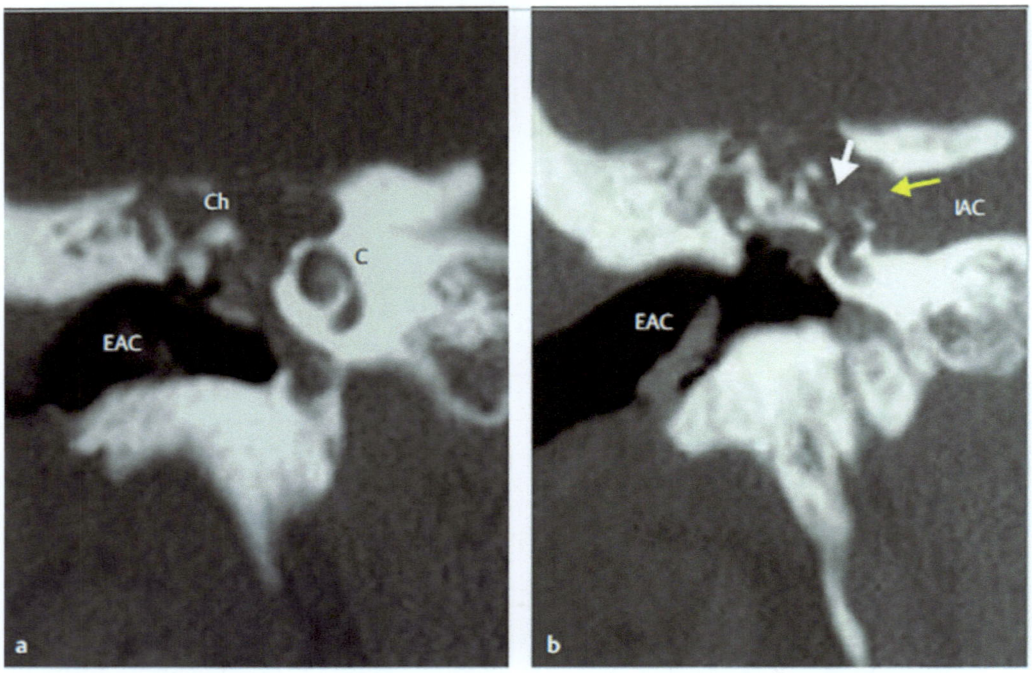

Fig. 9.45 The coronal CT (A) shows the presence of cholesteatoma superiorly to the cochlea, involving the area of the geniculate ganglion. More posterior section (B) shows extensive bone erosion at the fundus *(yellow arrow)* and the vestibule *(white arrow)*. *C*, cochlea; *Ch*, cholesteatoma; *EAC*, external auditory canal.

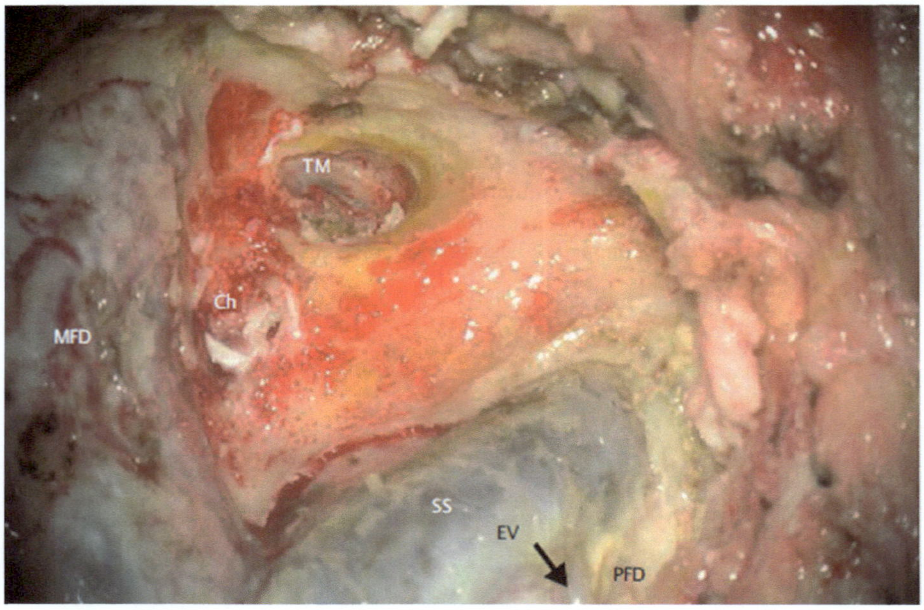

Fig. 9.46 Approach to manage extensive petrous bone cholesteatoma is shown. The middle fossa dura and the sigmoid sinus are widely exposed without leaving any bony coverage. The posterior fossa dura located posteriorly to the sigmoid sinus, is also uncovered for sufficient mobilization of the sigmoid sinus. *Ch*, cholesteatoma; *EV*, emissary vein; *MFD*, middle fossa dura; *PFD*, posterior fossa dura; *SS*, sigmoid sinus; *TM*, tympanic membrane.

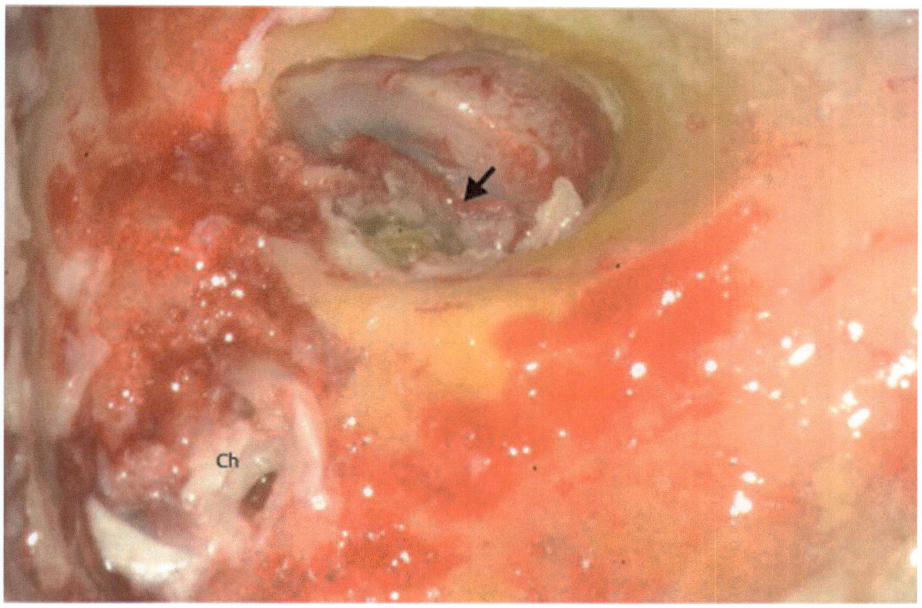

Fig. 9.47 The posterior wall is partially removed, and the skin in the external auditory canal is removed. The anterior and inferior walls of the external auditory canal are also drilled to improve access to the tympanic cavity. The tympanic membrane is left in place at this moment. The cholesteatoma invaginating from the posterosuperior quadrant of the tympanic membrane *(arrow)* to fill the attic is seen. *Ch*, cholesteatoma.

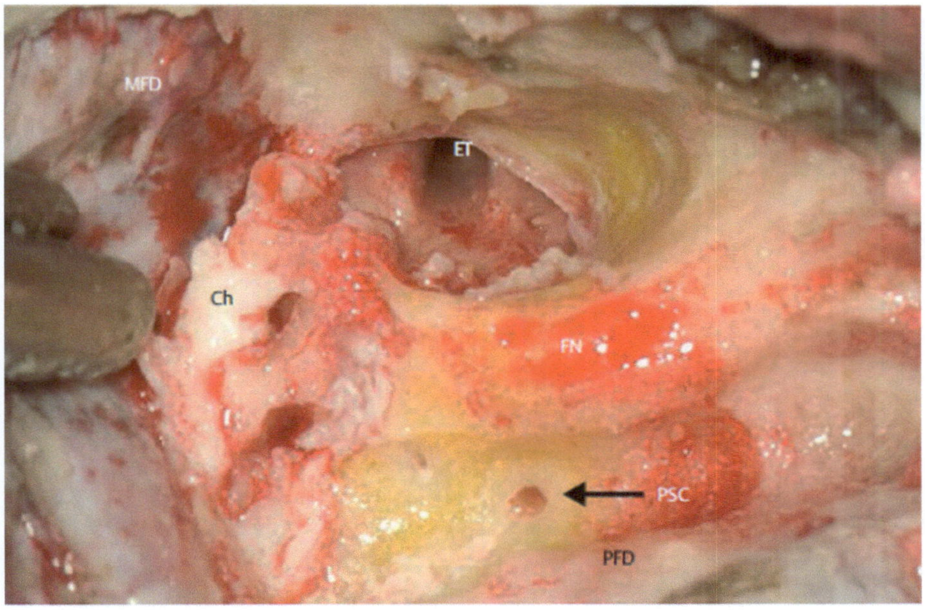

Fig. 9.48 The cholesteatoma advanced posteromedially from the attic to infiltrate the air cells medially to the semicircular canals. The middle fossa dura is pushed superiorly with the suction irrigator to evaluate the extension of cholesteatoma. *Ch*, cholesteatoma; *ET*, eustachian tube; *FN*, facial nerve; *MFD*, middle fossa dura; *PFD*, posterior fossa dura; *PSC*, posterior semicircular canal.

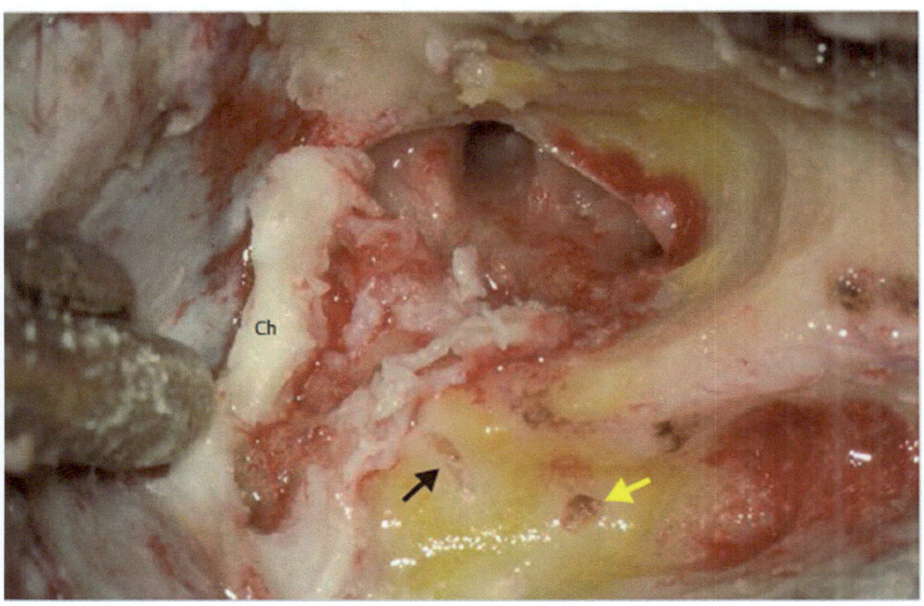

Fig. 9.49 Labyrinthectomy is started to remove cholesteatoma located medially to the labyrinth. The cholesteatoma mainly eroded the anterosuperior part of the labyrinth. The lateral semicircular canal is opened *(black arrow)*, and the posterior semicircular canal *(yellow arrow)* is drilled. *Ch*, cholesteatoma.

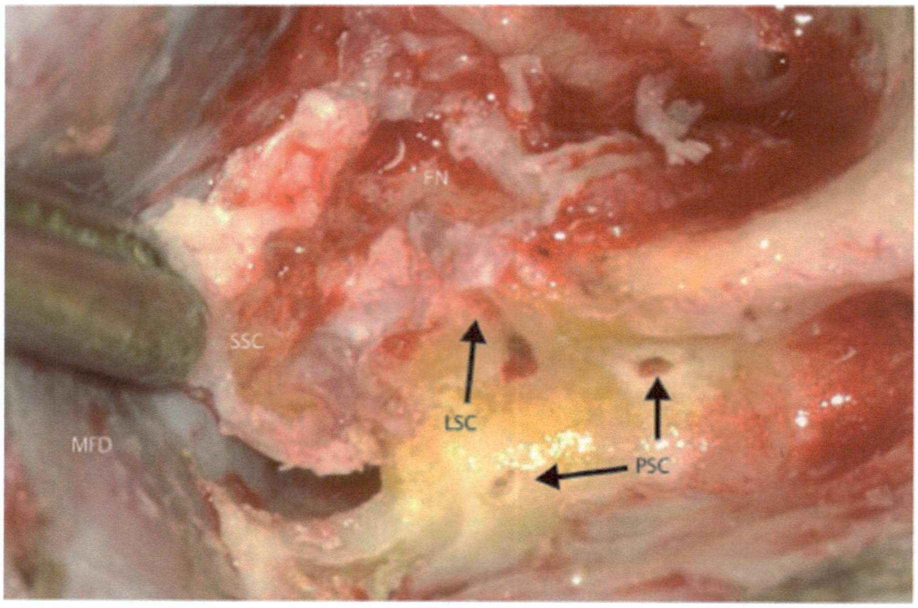

Fig. 9.50 Removal of the posterior labyrinth is further carried out. The anterior half of the lateral semicircular canal is infiltrated by the cholesteatoma. *FN*, facial nerve; *LSC*, lateral semicircular canal; *MFD*, middle fossa dura; *PSC*, posterior semicircular canal; *SSC*, superior semicircular canal.

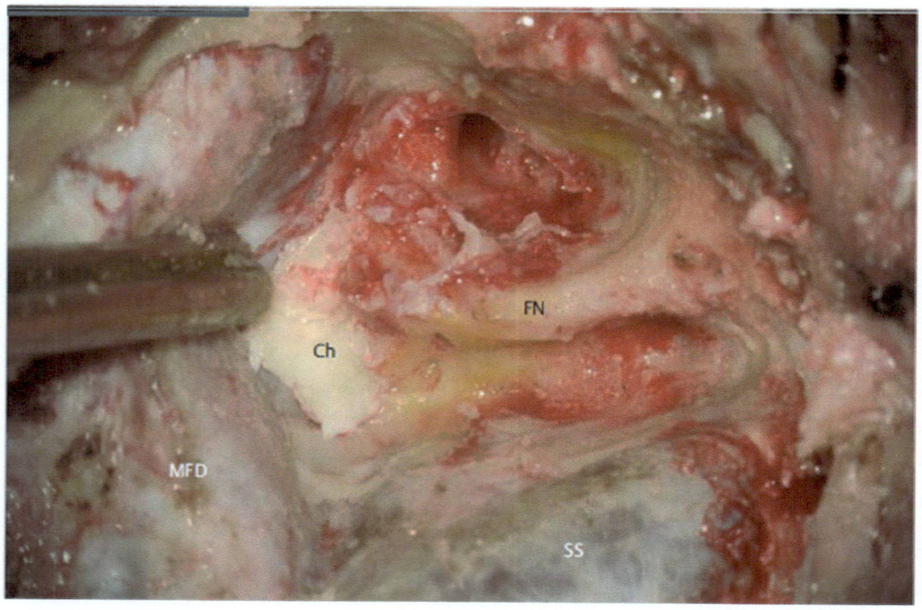

Fig. 9.51 The lateral and the posterior semicircular canals are removed. The cholesteatoma erodes the medial wall of the attic to go deeply toward the internal auditory canal. *Ch*, cholesteatoma; *FN*, facial nerve; *MFD*, middle fossa plate; *SS*, sigmoid sinus.

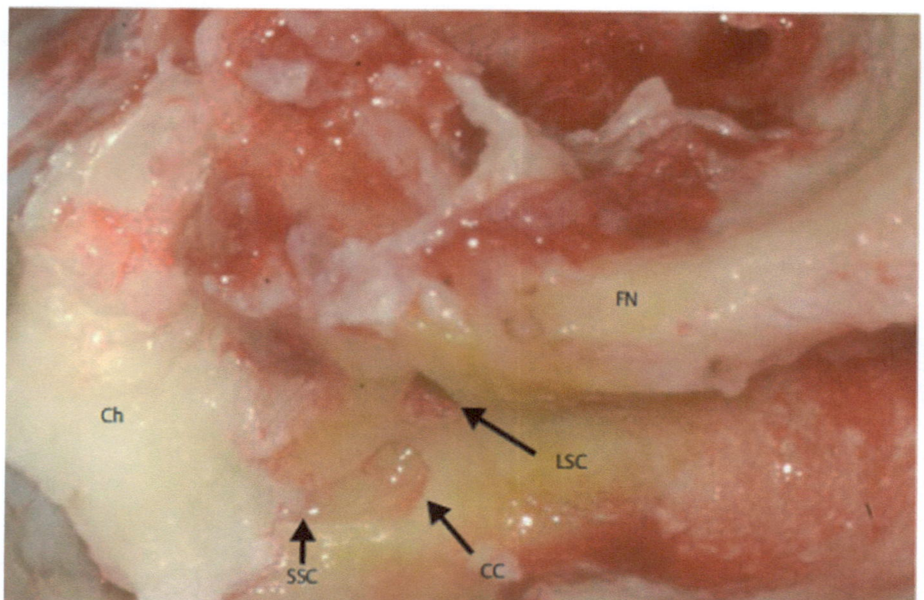

Fig. 9.52 Enlarged view. Removal of the posterior semicircular canal visualized the common crus. Note that the superior semicircular canal is infiltrated by the cholesteatoma. The posterior end of the lateral semicircular canal is seen. *CC*, common crus; *Ch*, cholesteatoma; *FN*, facial nerve; *LSC*, lateral semicircular canal; *SSC*, superior semicircular canal.

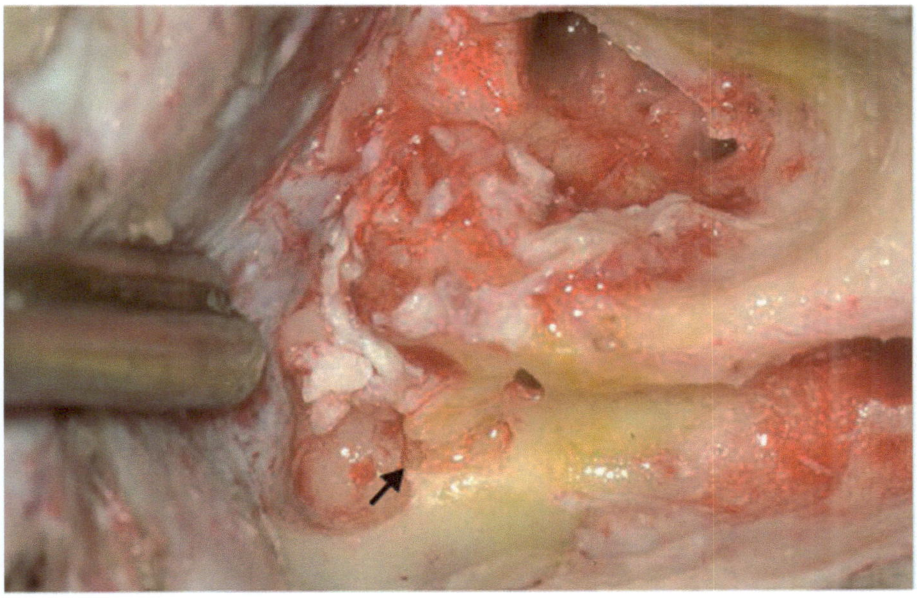

Fig. 9.53 The superior semicircular canal is interrupted by the cholesteatoma *(arrow)*.

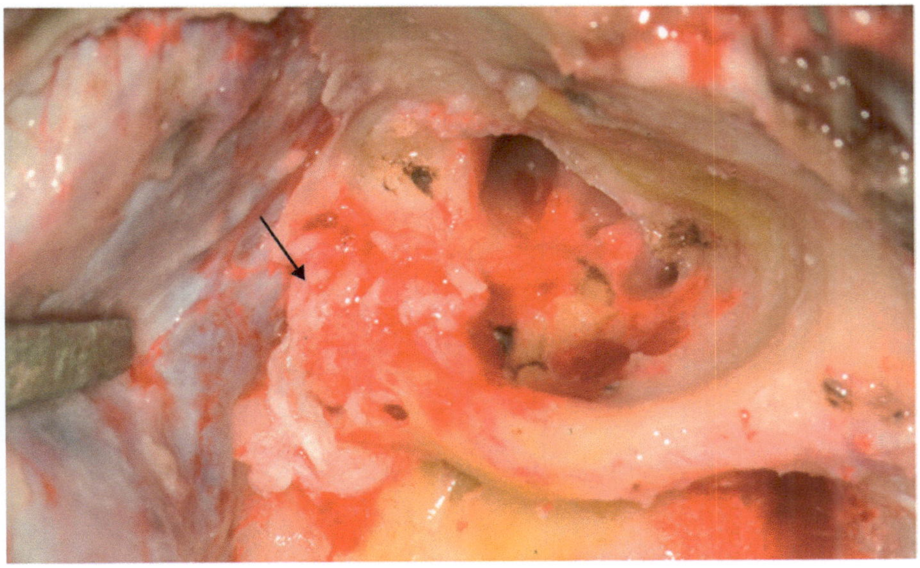

Fig. 9.54 The geniculate ganglion is completely encased in the cholesteatoma *(arrow)*.

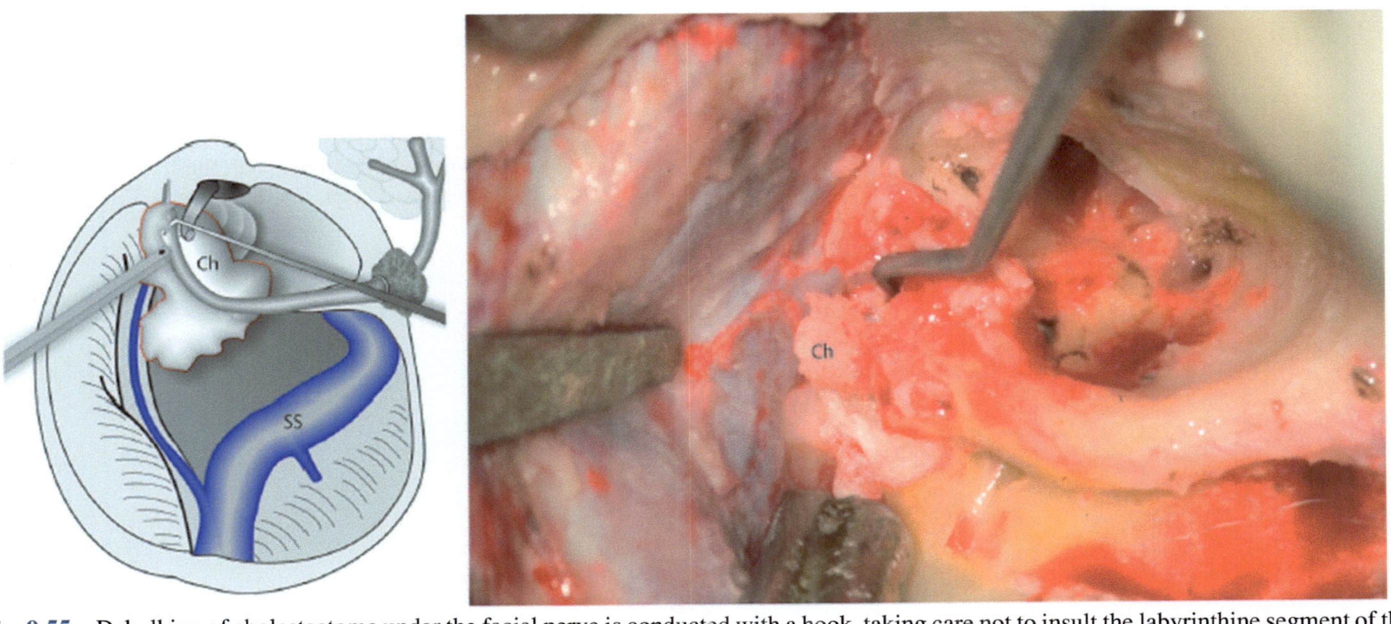

Fig. 9.55 Debulking of cholesteatoma under the facial nerve is conducted with a hook taking care not to insult the labyrinthine segment of the nerve. *Ch*, cholesteatoma.

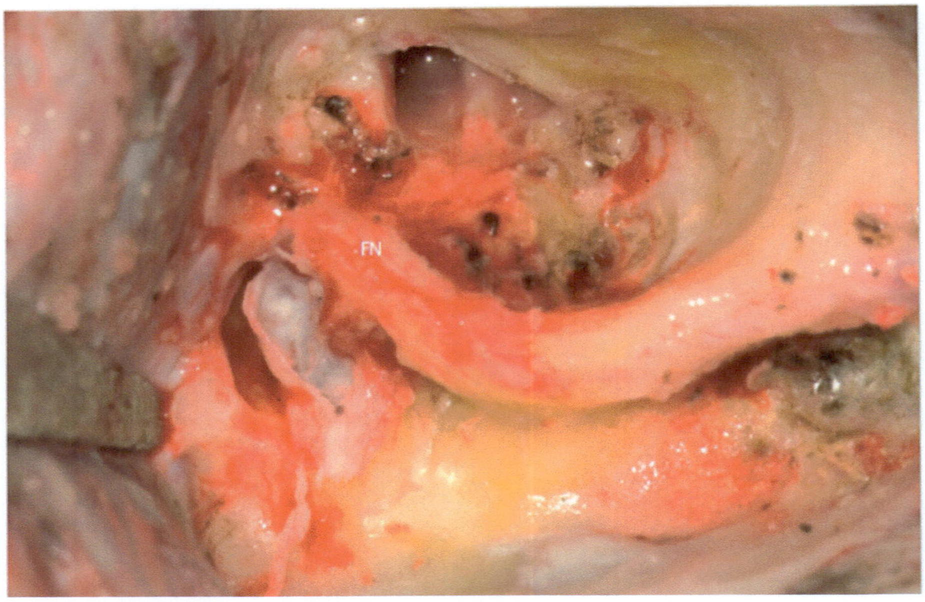

Fig. 9.56 Dissection of the matrix is started from the superior, from the middle fossa dura. The matrix covers the superior face of the internal auditory canal. *FN*, the facial nerve.

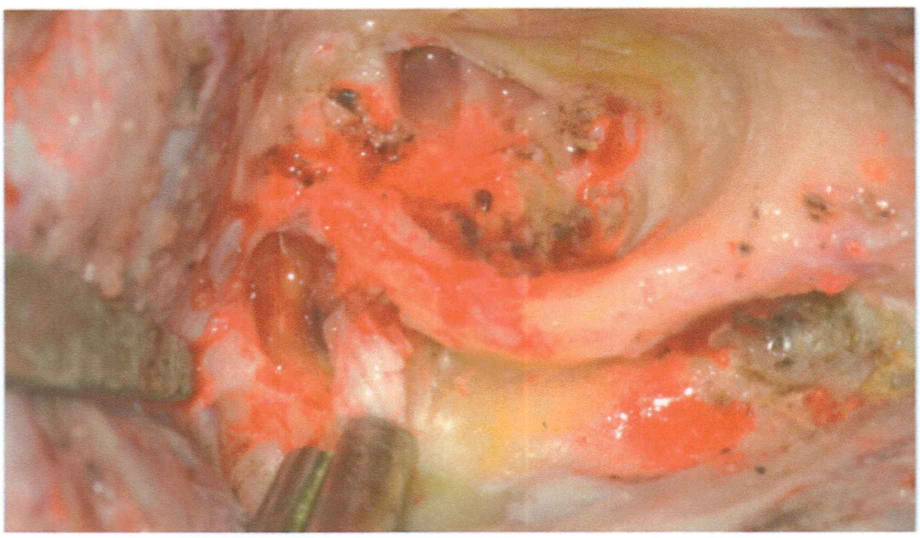

Fig. 9.57 The matrix is detached from the internal auditory canal by pulling it with the suction irrigator.

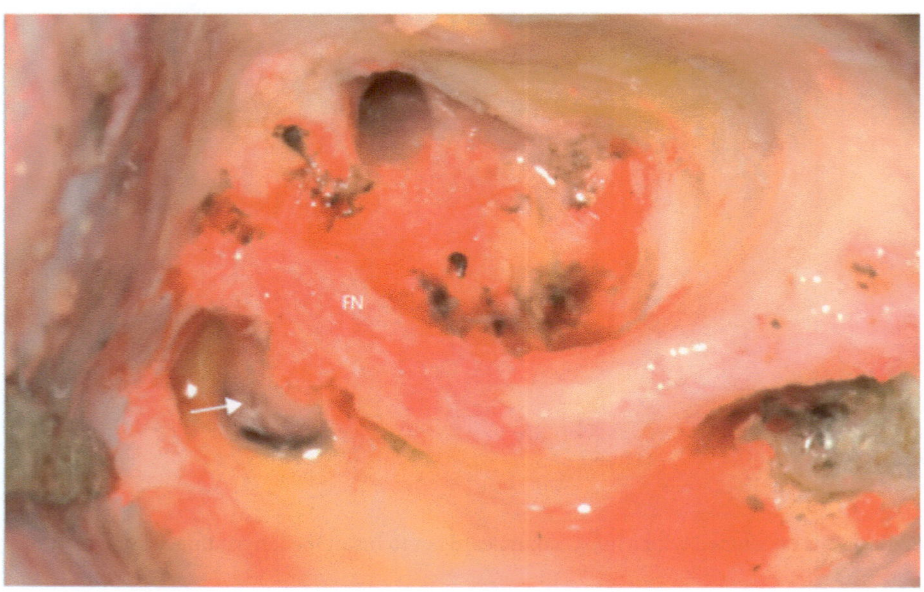

Fig. 9.58 Removal of the matrix exposed the superior face of the internal auditory canal *(arrow)*. Cerebrospinal fluid leaking from the fundus is seen. *FN*, facial nerve.

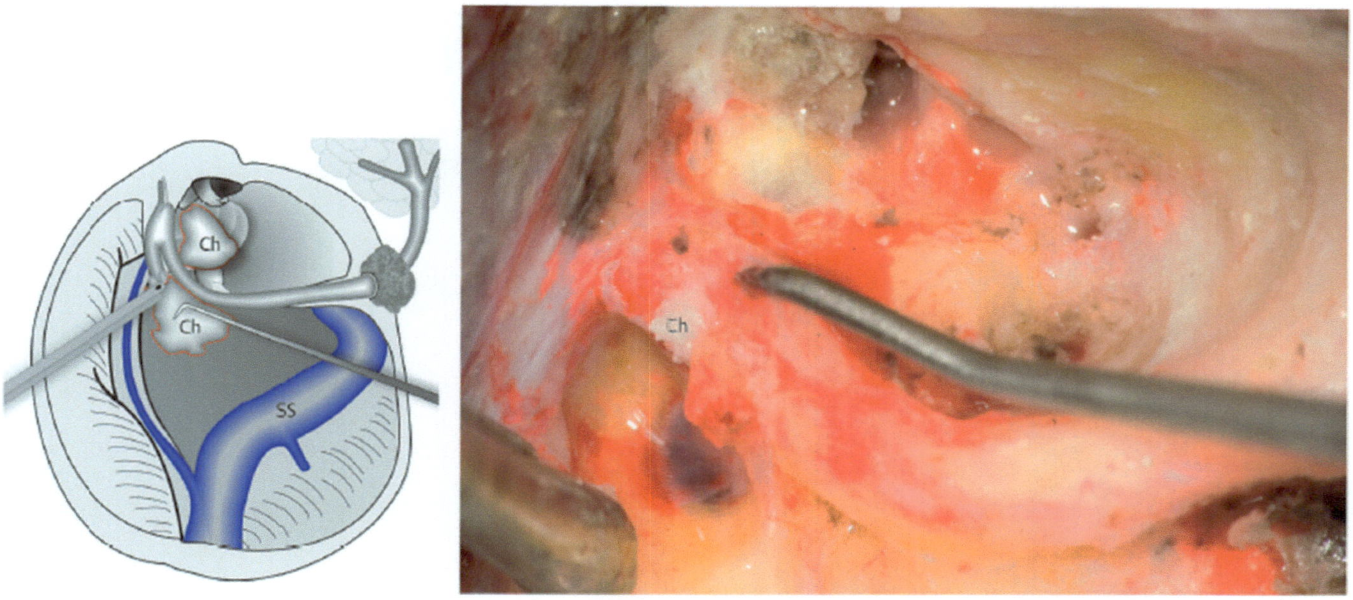

Fig. 9.59 The geniculate ganglion is pushed inferiorly to see better the cholesteatoma and the labyrinthine segment of the facial nerve. *Ch*, cholesteatoma.

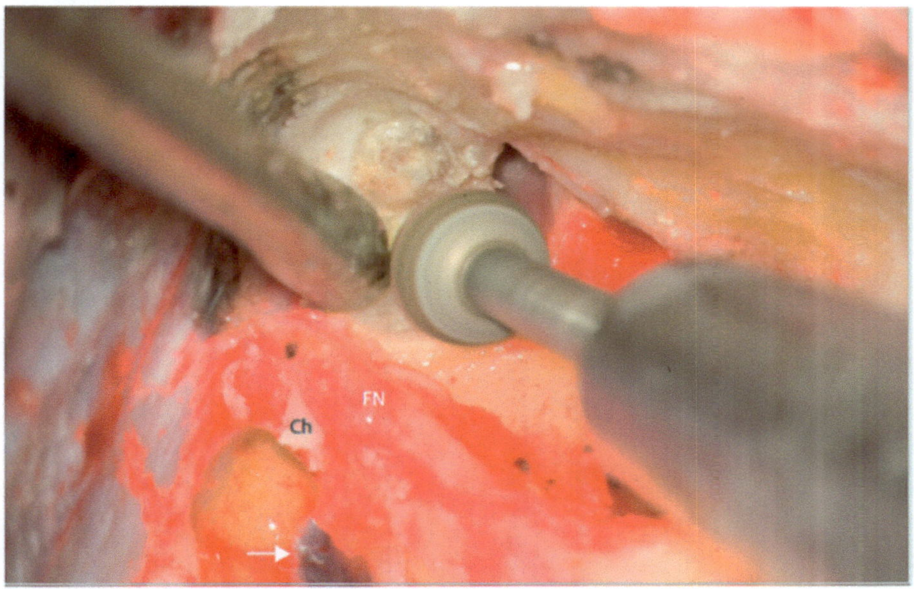

Fig. 9.60 To control the inferior margin of the cholesteatoma, the cochlea is drilled. A very thin dura near the fundus of the internal auditory canal is seen *(arrow)*. *Ch*, cholesteatoma; *FN*, facial nerve.

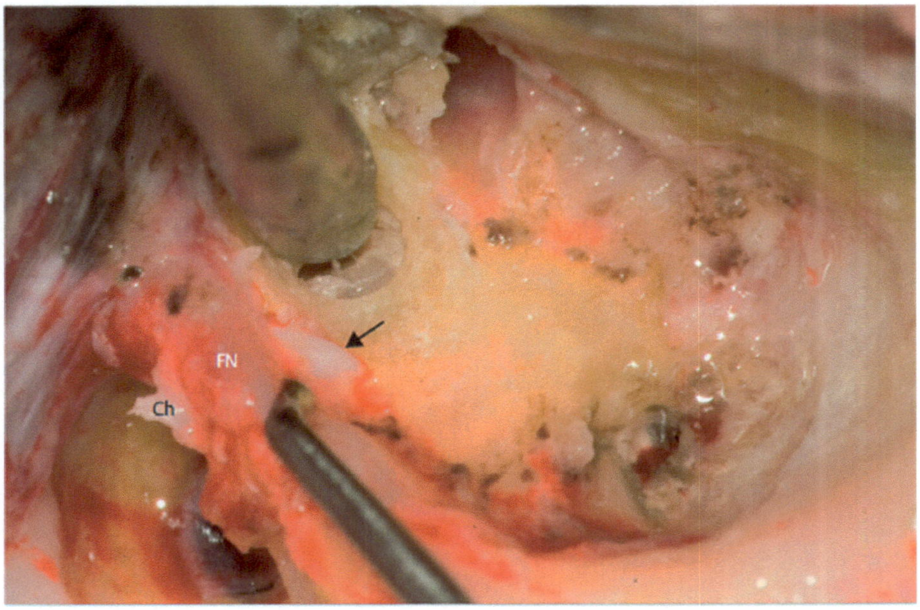

Fig. 9.61 The inferior margin of the cholesteatoma infiltrating the bone under the facial nerve between the cochlea and the vestibule is shown *(arrow)*. An opening of the apical turn of the cochlea is seen under the suction. *Ch*, cholesteatoma; *FN*, facial nerve.

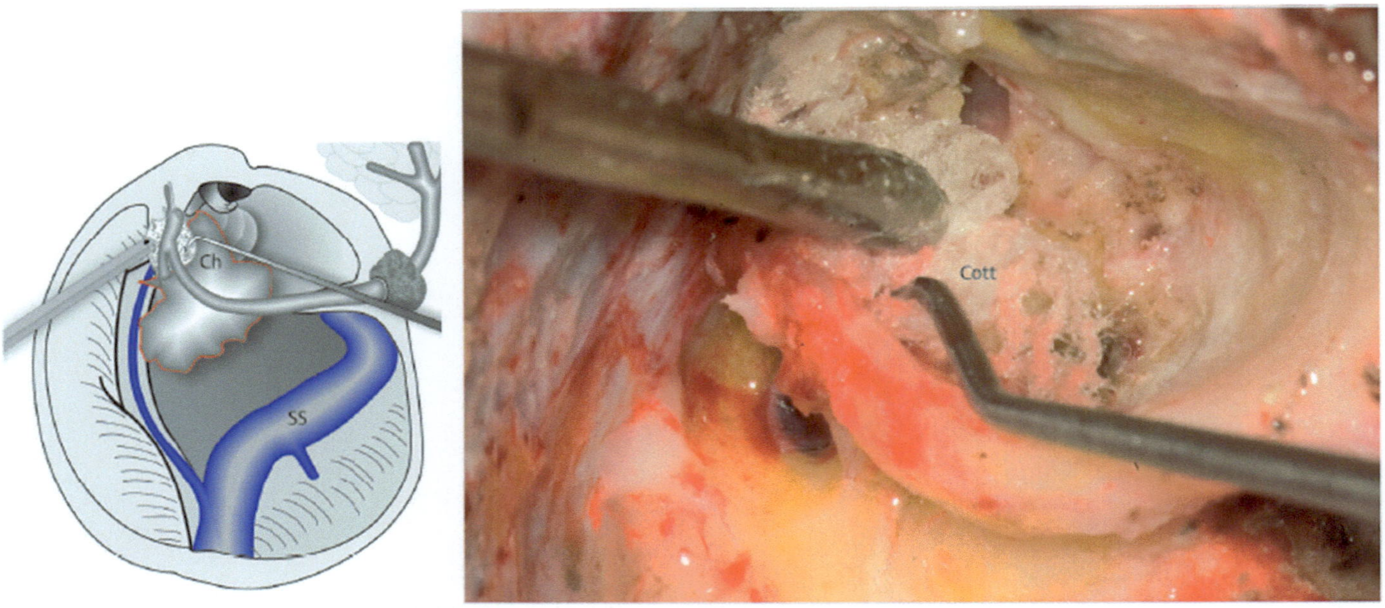

Fig. 9.62 The matrix under the geniculate ganglion is cleaned with cottonoid (Cott).

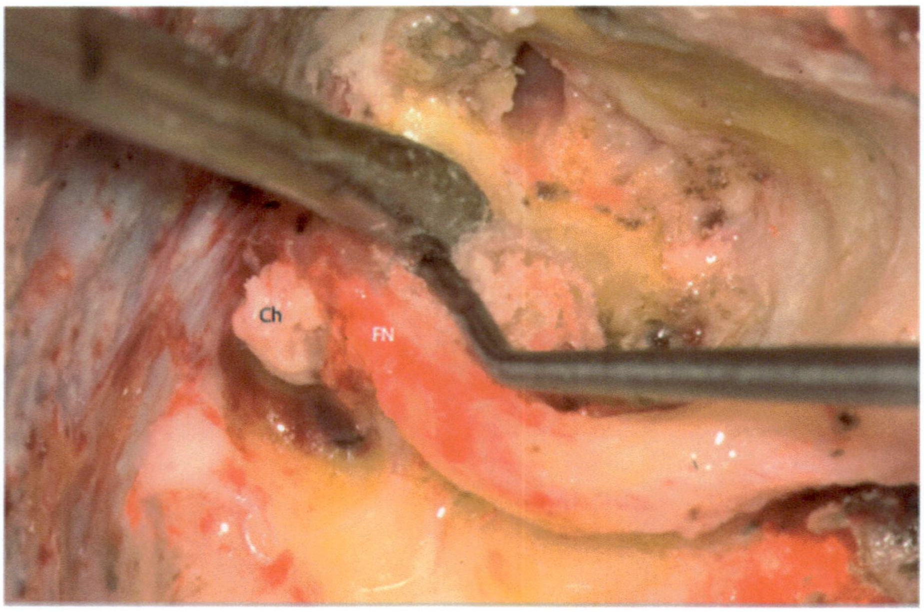

Fig. 9.63 The matrix of the cholesteatoma invaginating under the geniculate ganglion anteriorly to the labyrinthine segment is gently pushed out with the help of a cottonoid. *Ch*, cholesteatoma; *FN*, facial nerve.

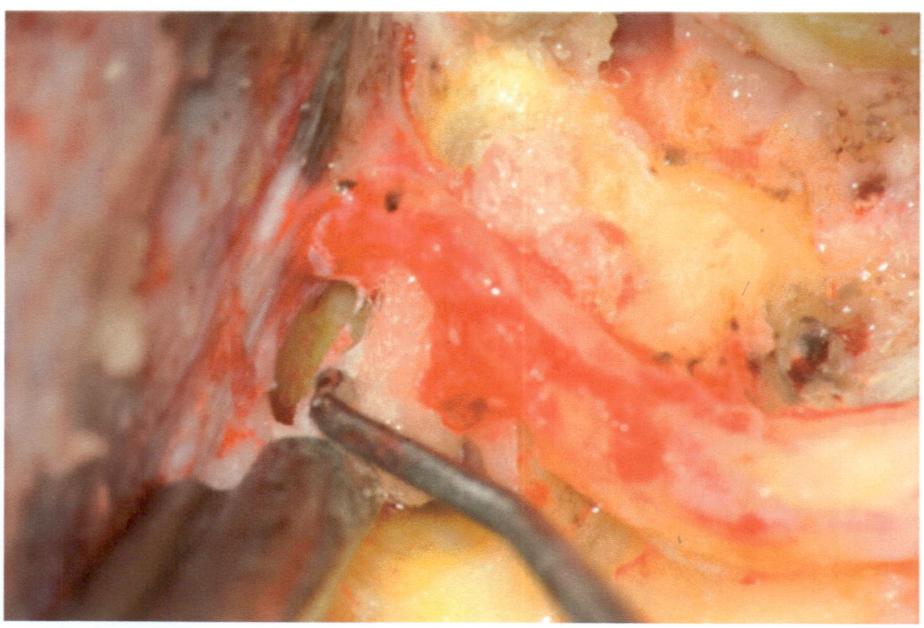

Fig. 9.64 The remaining matrix covering the facial nerve is carefully dissected.

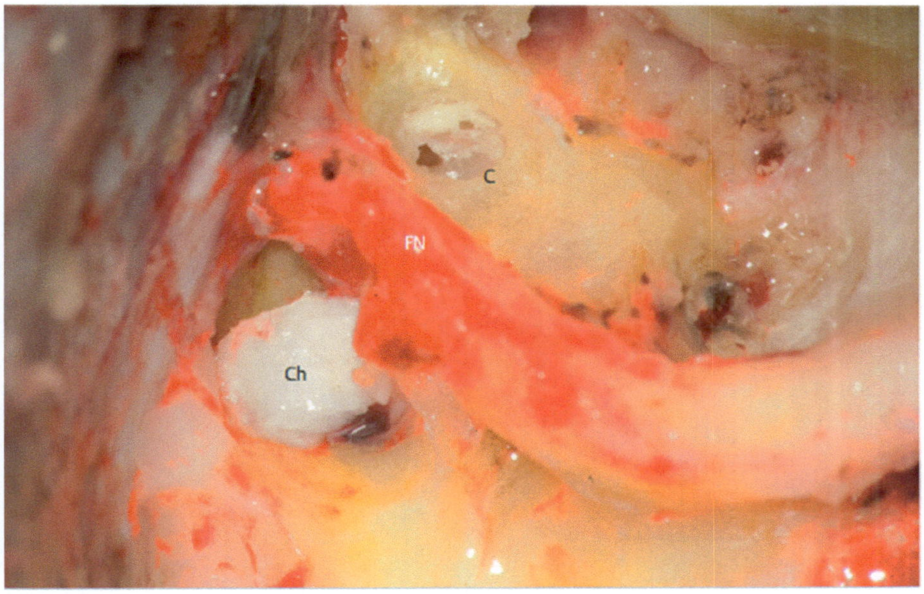

Fig. 9.65 The cholesteatoma posteriorly to the labyrinthine segment is adherent to the facial nerve. *C*, cochlea; *Ch*, cholesteatoma; *FN*, facial nerve.

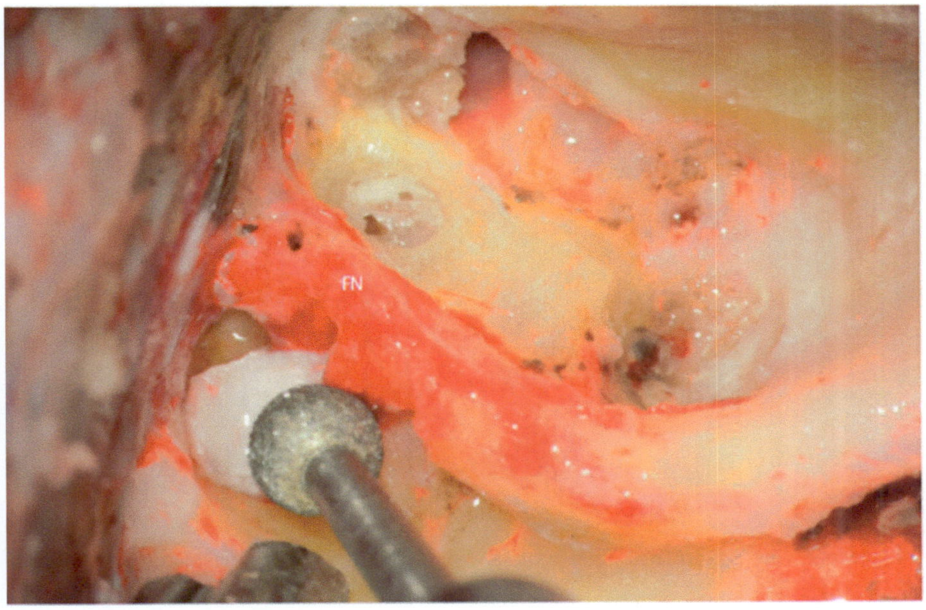

Fig. 9.66 To establish better access to the labyrinthine segment and the fundus of the internal auditory canal, the posterosuperior wall of the internal auditory canal corresponding to the anterosuperior wall of the vestibule is removed with a large diamond burr. *FN*, facial nerve.

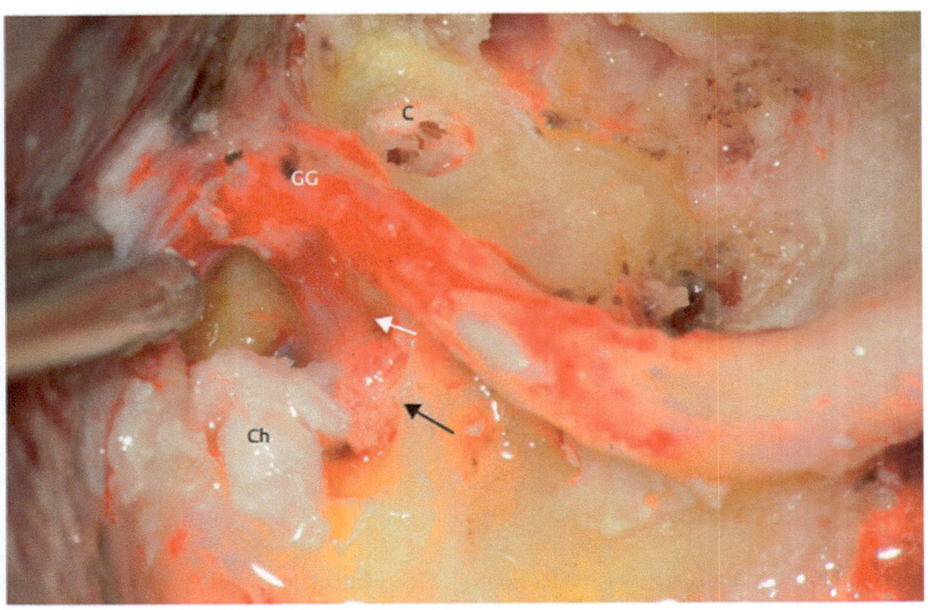

Fig. 9.67 The cholesteatoma is dissected from the posterior face of the labyrinthine segment *(white arrow)*. A transected vestibular nerve *(black arrow)* is seen posteriorly to the facial nerve. *C*, cochlea; *Ch*, cholesteatoma; *GG*, geniculate ganglion.

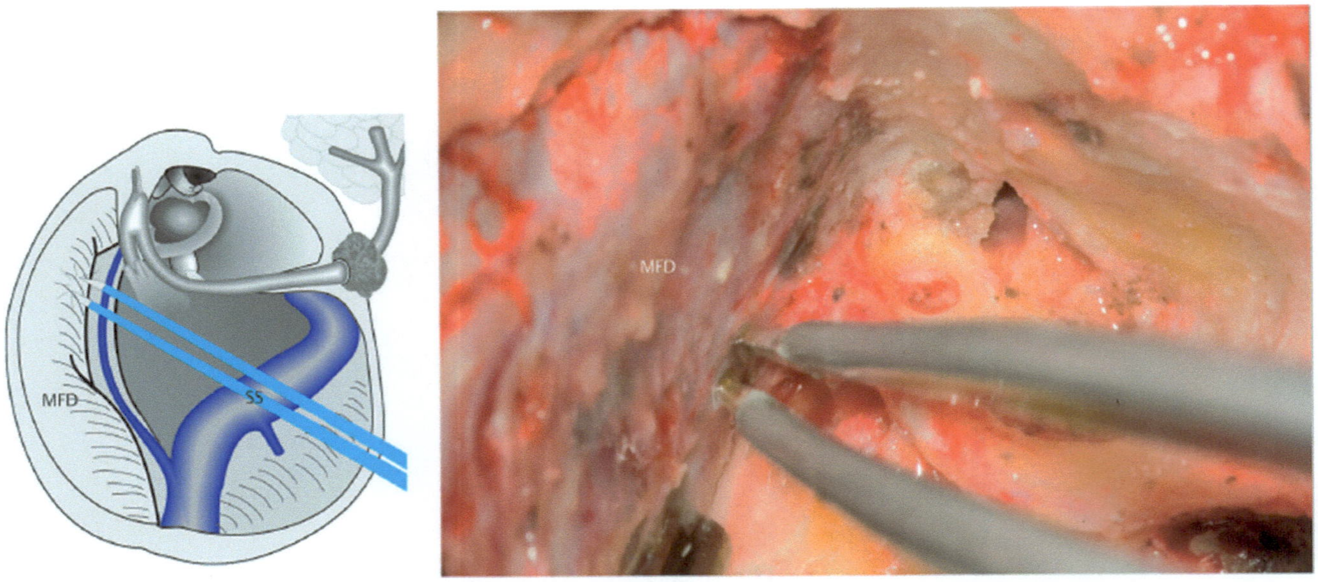

Fig. 9.68 The dura suspicious of the residual matrix should be coagulated with bipolar. *MFD*, middle fossa dura.

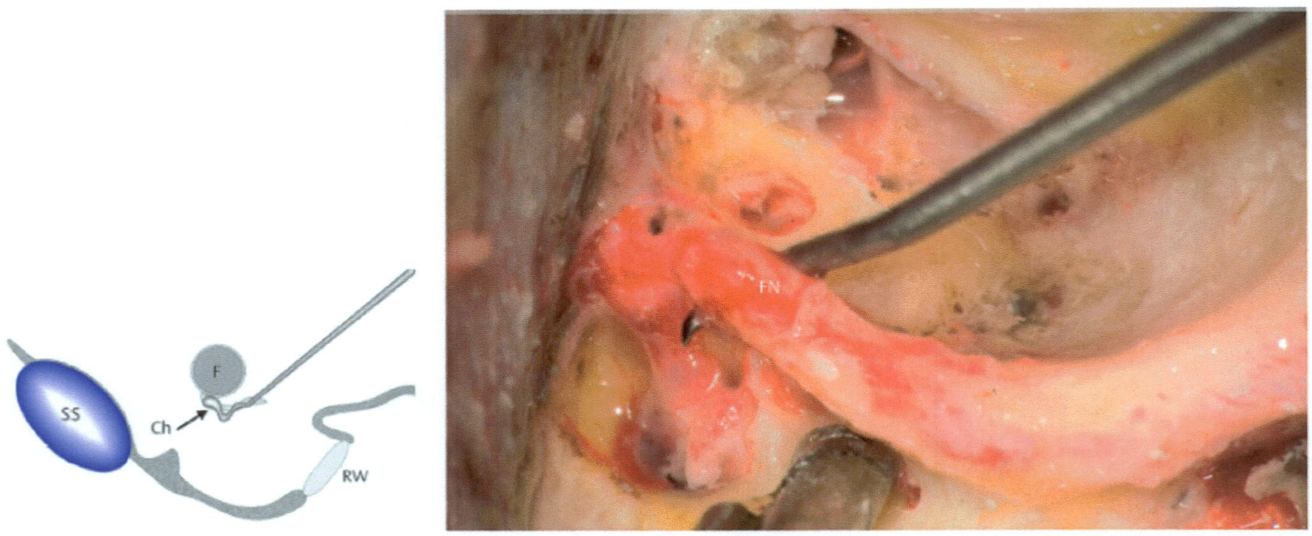

Fig. 9.69 The medial face of the facial nerve (FN) is cleaned with a tympanic sinus hook. *SS*, sigmoid sinus; *Ch*, cholesteatoma; *RA*, round window; *F*, facial nerve.

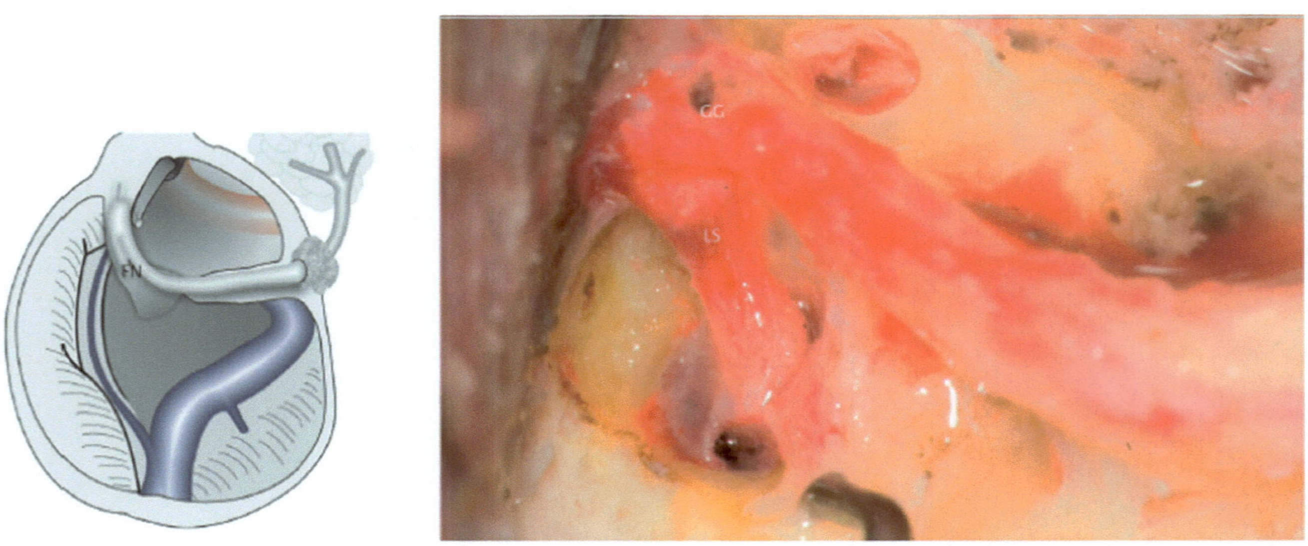

Fig. 9.70 The anterosuperior part of the fundus of the internal auditory canal is under control. *GG*, geniculate ganglion; *LS*, labyrinthine segment of facial nerve.

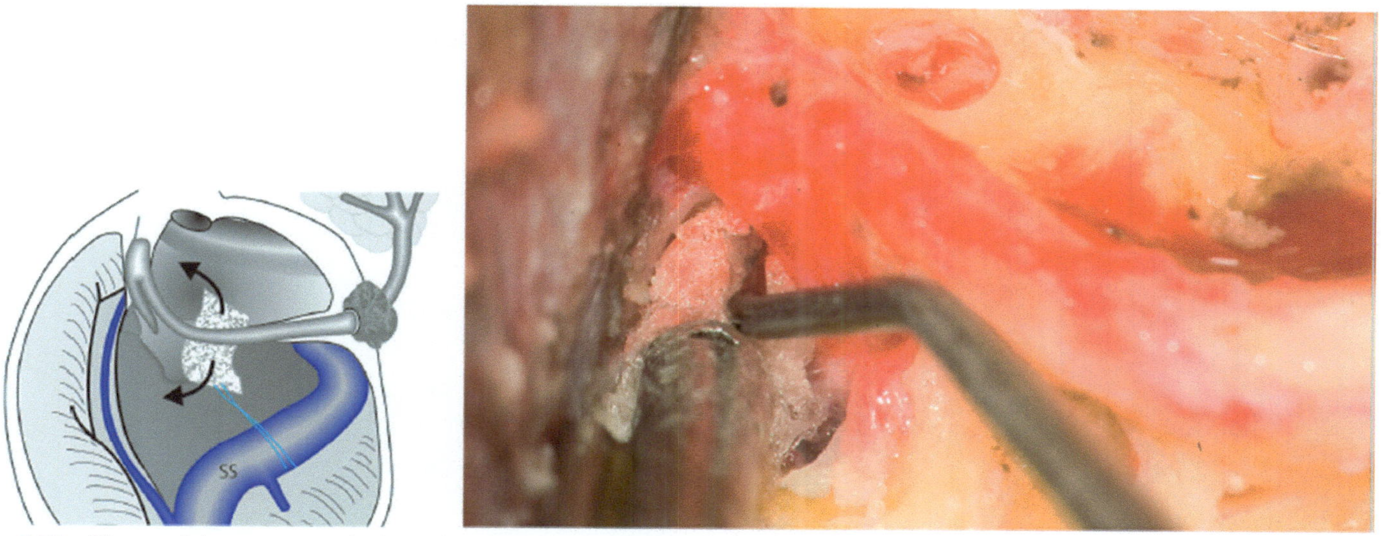

Fig. 9.71 The suspicious area anteriorly to the labyrinthine segment is scrubbed with cottonoid.

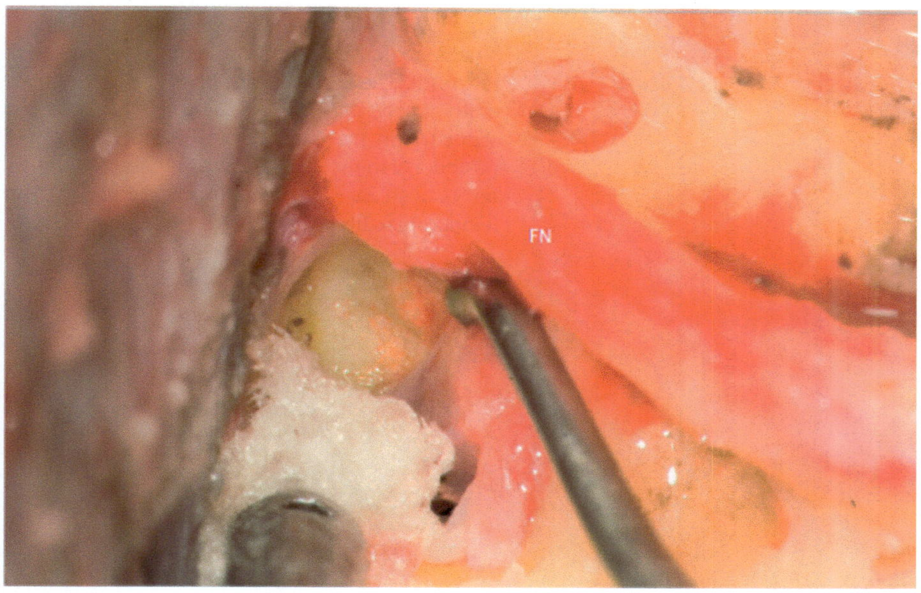

Fig. 9.72 Complete removal of cholesteatoma is checked by pushing the labyrinthine segment of the facial nerve inferiorly. *FN*, facial nerve.

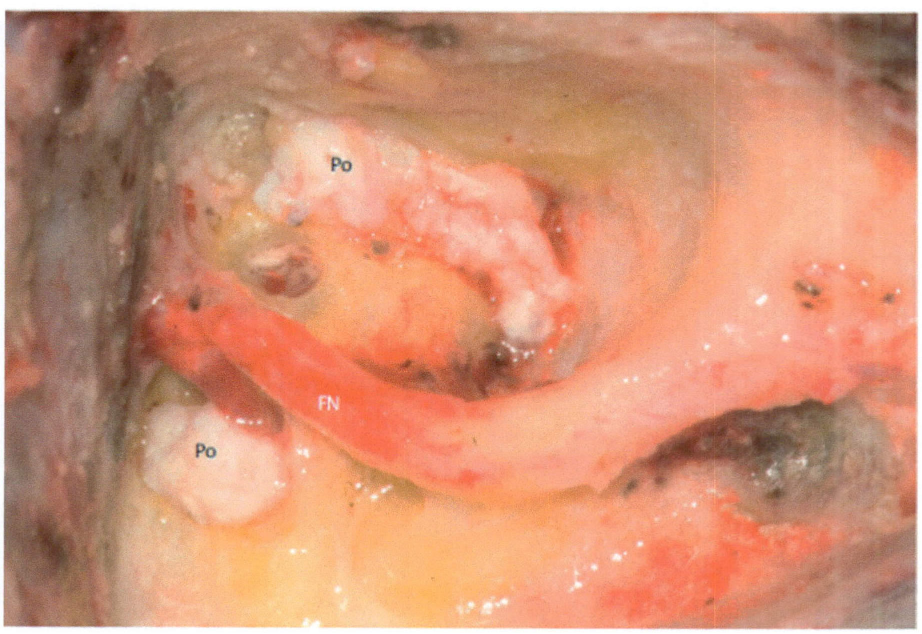

Fig. 9.73 The opening of the internal auditory canal is gently closed, and the eustachian tube is tightly packed with periosteum. *FN*, facial nerve; *Po*, periosteum.

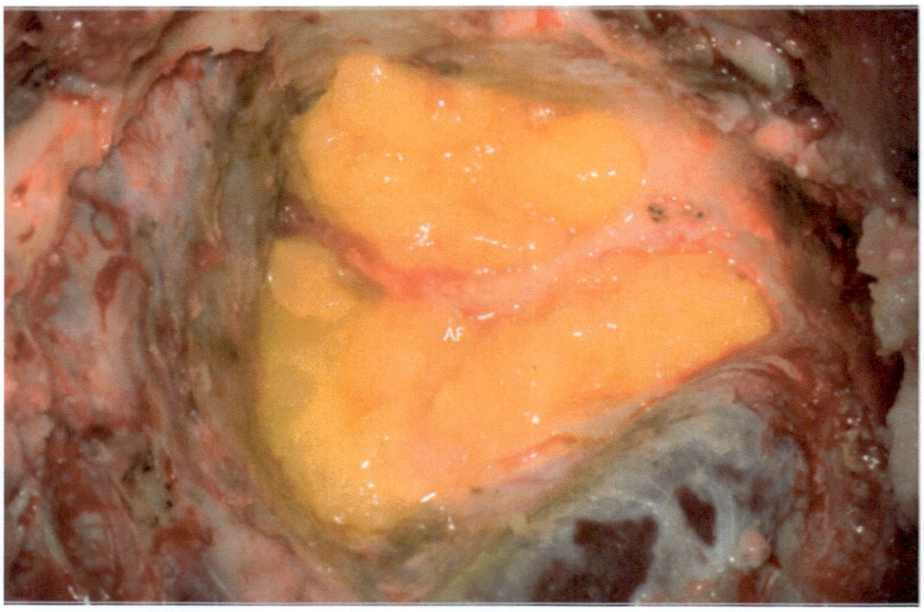

Fig. 9.74 The cavity is obliterated with abdominal fat. *AF*, abdominal fat.

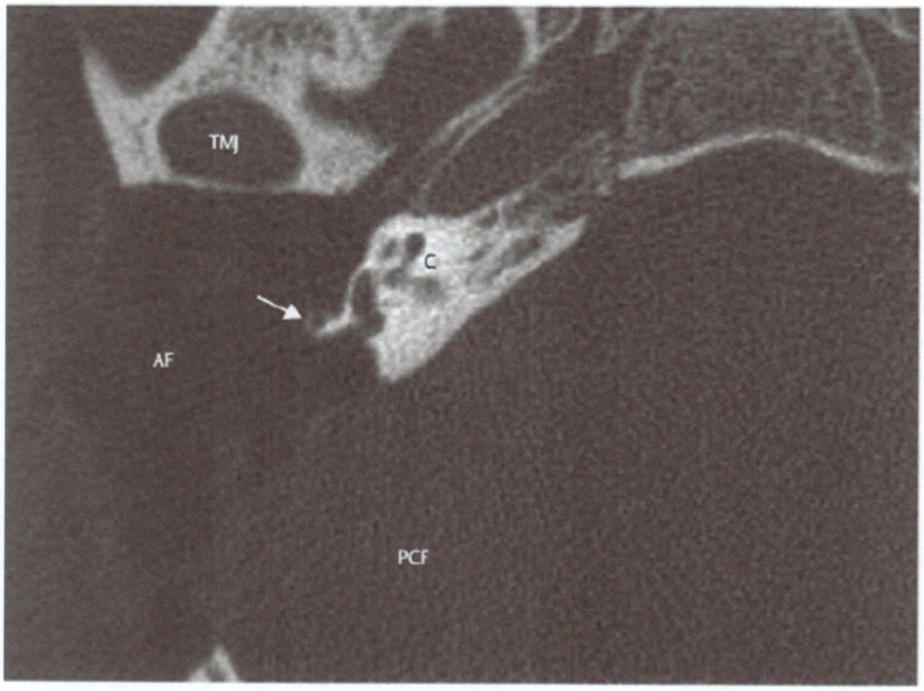

Fig. 9.75 The postoperative axial CT demonstrates extensive removal of the posterior labyrinth. The cochlea, partially drilled posterolaterally, keeps its shape at this level. The mastoid segment of the facial nerve is identified just posteriorly to the cochlea *(arrow)*. *AF*, abdominal fat; *C*, cochlea; *PCF*, posterior cranial fossa; *TMJ*, temporomandibular joint.

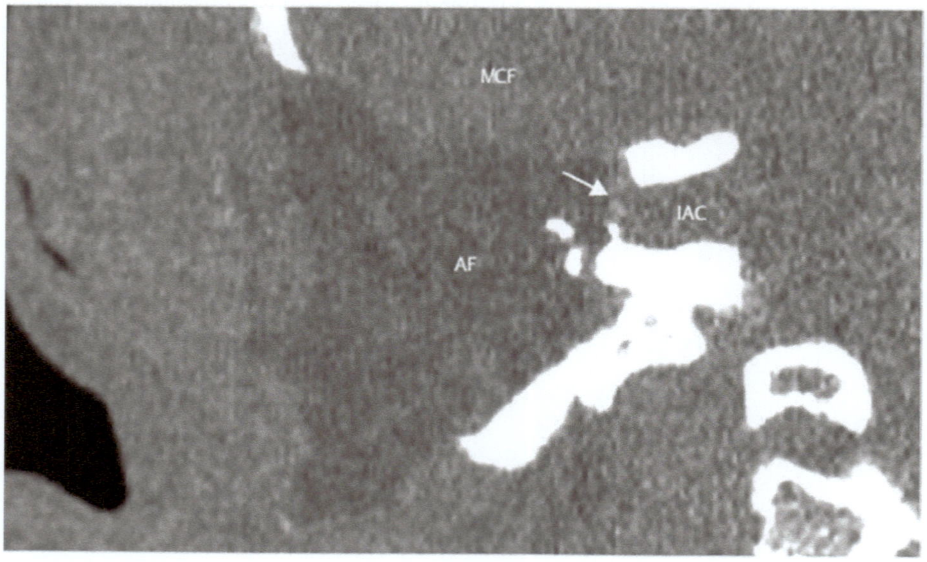

Fig. 9.76 The postoperative coronal CT demonstrates extensive exposure of the middle fossa dura and opening of the internal auditory canal at the fundus *(arrow)*. *AF*, abdominal fat; *IAC*, internal auditory canal; *MCF*, middle cranial fossa.

Case 2: Massive Petrous Bone Cholesteatoma

See Figs. 9.77–9.89.

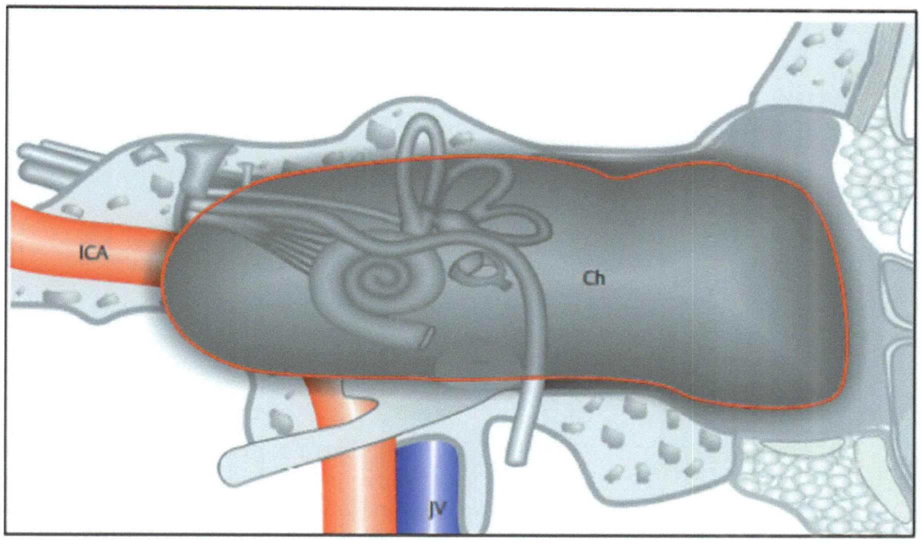

Fig. 9.77 A case of massive petrous bone cholesteatoma formerly treated with posterior fossa approach and reoperated with transotic approach. *Ch*, cholesteatoma; *ICA*, internal carotid artery; *JV*, internal jugular vein.

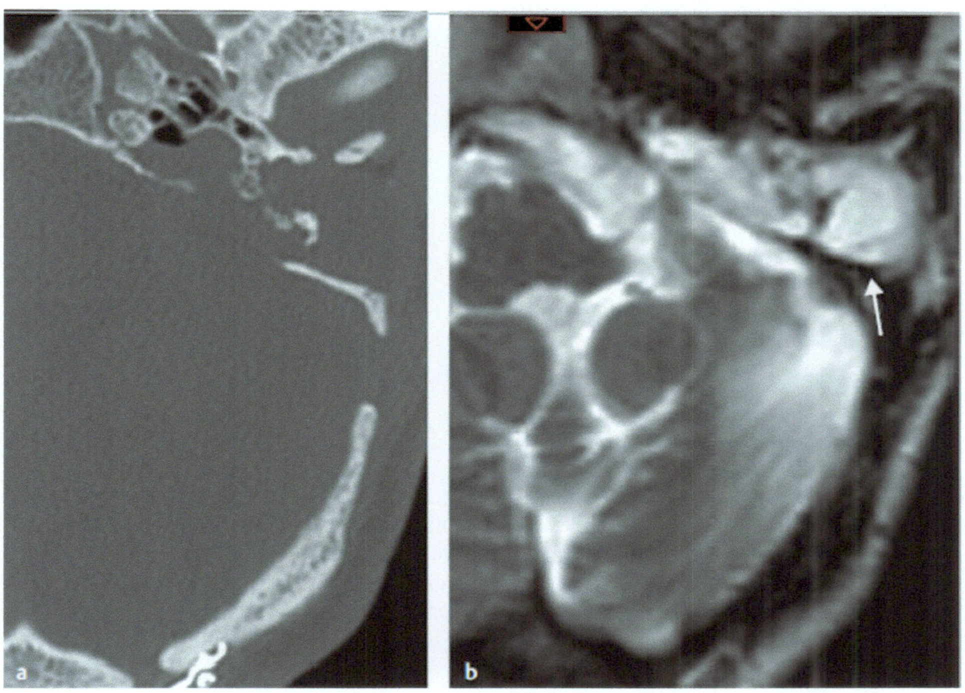

Fig. 9.78 The patient suffered from meningitis with facial weakness with a dead ear 2 years before and underwent two surgeries elsewhere through transmastoid and retrosigmoid approaches. The axial CT (A) shows incomplete access to the petrous apex from both lateral and posterior, and the T2-weighted MRI (B) demonstrated large ischemic change in the left cerebellar hemisphere due to compression during retrosigmoid approach. Recurrent cholesteatoma filling the temporal bone is demonstrated in high signal *(arrow)*.

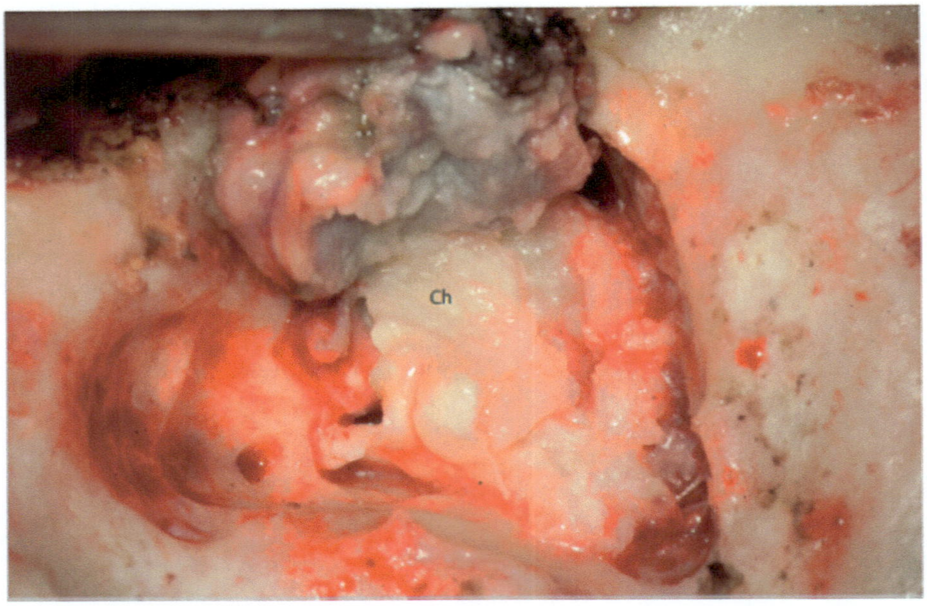

Fig. 9.79 The external auditory canal has already been closed in the former surgery. Scar tissue filling the tympanic and mastoid cavity is reduced, and the cholesteatoma is opened from lateral. *Ch*, cholesteatoma.

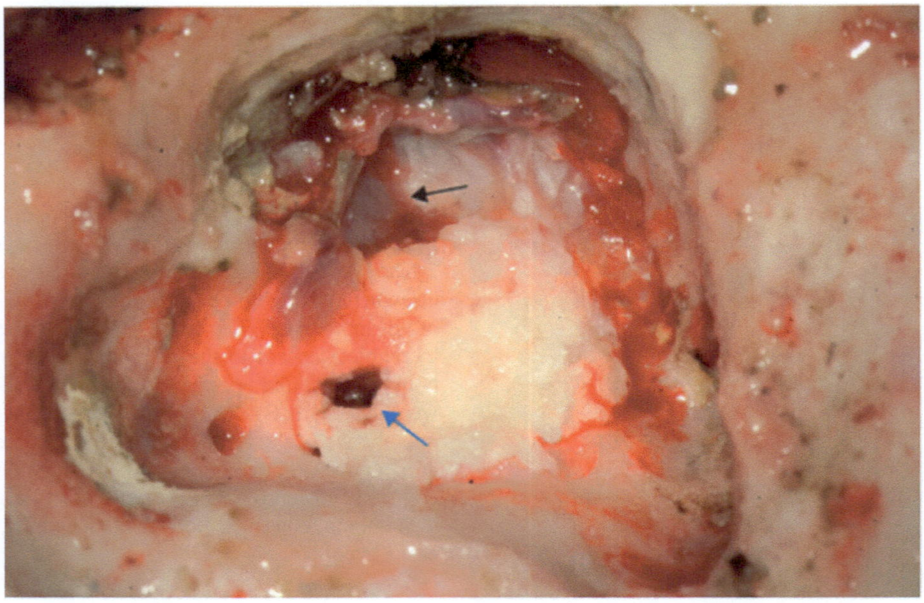

Fig. 9.80 Debris is partially evacuated. A route toward the petrous apex is identified *(blue arrow)*, and granulation tissue covering the hypotympanic area is exposed *(black arrow)*.

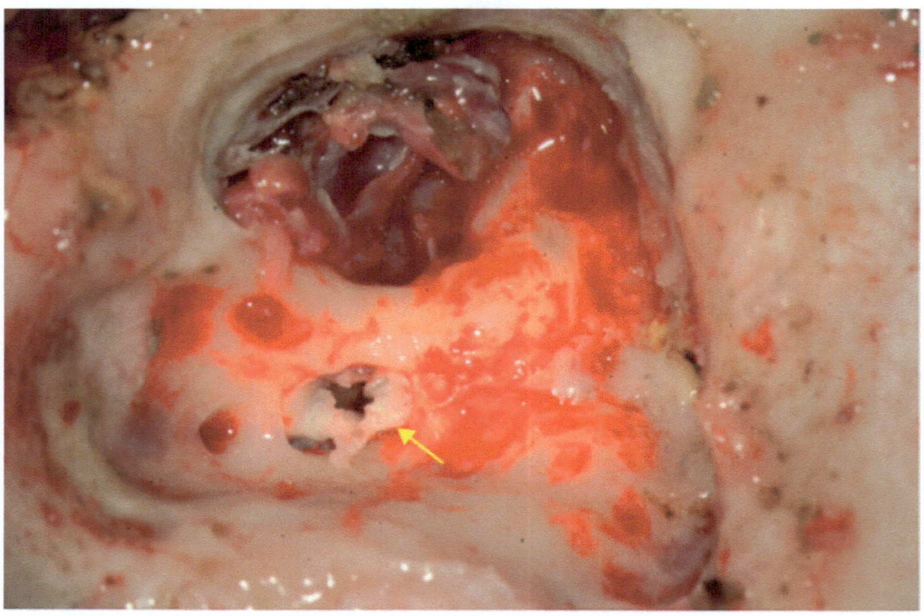

Fig. 9.81 The cholesteatoma in the mastoid has continuity to the petrous apex region through the retrofacial cells *(arrow)*. Scar tissue covering the medial wall of the tympanic cavity is reflected inferiorly.

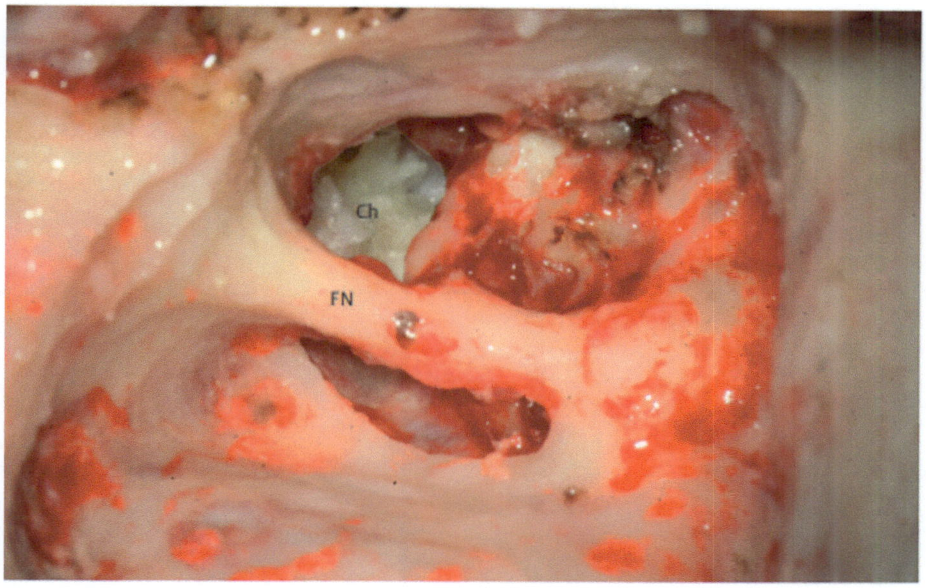

Fig. 9.82 Removal of the granulation tissue covering the hypotympanic area revealed cholesteatoma occupying the petrous apex. The facial nerve is skeletonized, and the retrofacial cells are starting to be removed to enlarge the approach. *Ch*, cholesteatoma; *FN*, facial nerve.

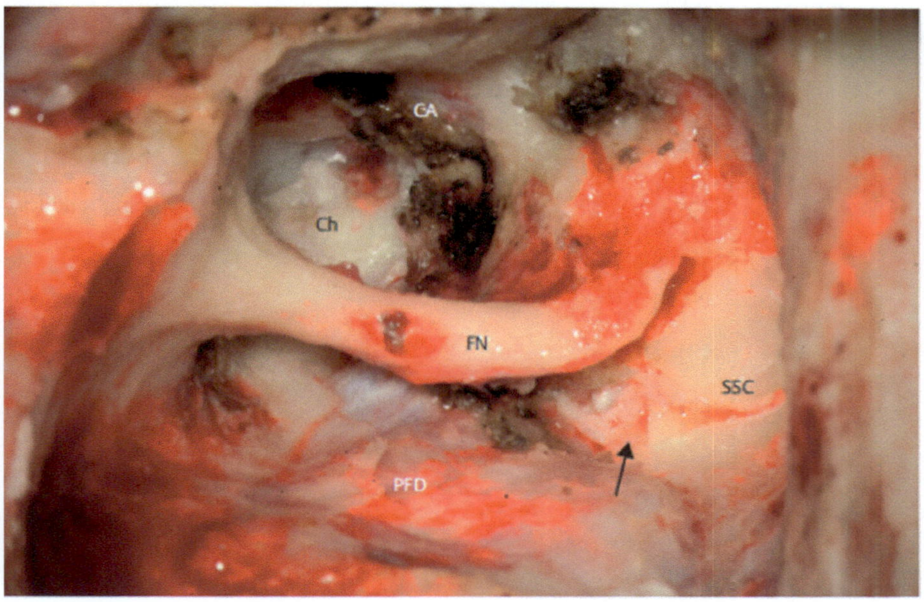

Fig. 9.83 The inferior part of the cochlea is removed, and the carotid artery is skeletonized to enlarge the approach anteriorly to the facial nerve. The posterior fossa dura is exposed and followed medially and superiorly. The posterior and the lateral semicircular canals are removed, and the superior semicircular canal is opened. Cholesteatoma infiltrating medially to the posterior labyrinth is starting to be seen *(arrow)*. *CA*, carotid artery; *Ch*, cholesteatoma; *FN*, facial nerve; *PFD*, posterior fossa dura; *SSC*, superior semicircular canal.

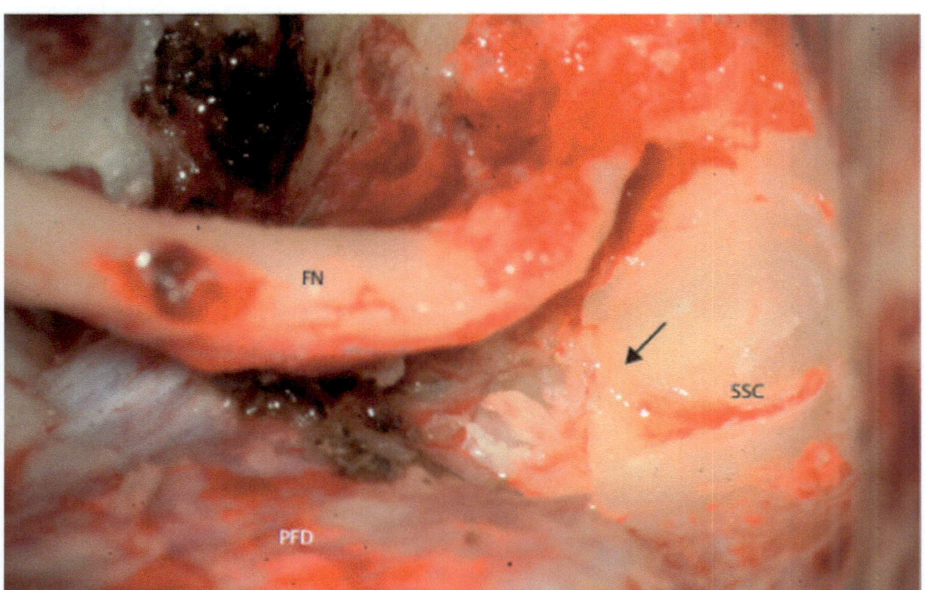

Fig. 9.84 Cholesteatoma located medially to the common crus *(arrow)* is shown. *FN*, facial nerve; *PFD*, posterior fossa dura; *SSC*, superior semicircular canal.

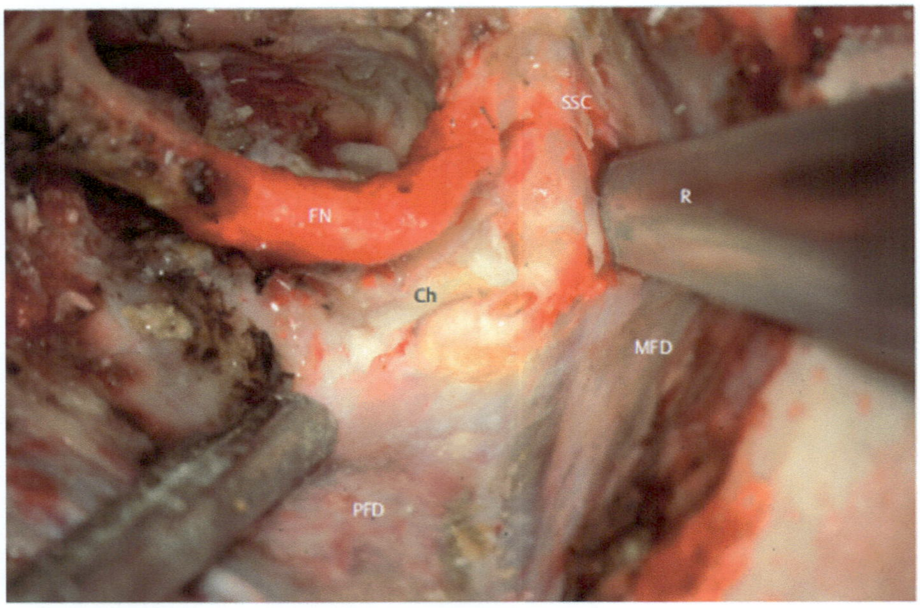

Fig. 9.85 Removal of the posterior half of the superior semicircular canal exposed cholesteatoma infiltrating between the internal auditory canal and the labyrinth. The middle fossa dura is maintained by a self-retaining retractor. *FN*, facial nerve; *MFD*, middle fossa dura; *PFD*, posterior fossa dura; *R*, self-retaining retractor; *SSC*, superior semicircular canal.

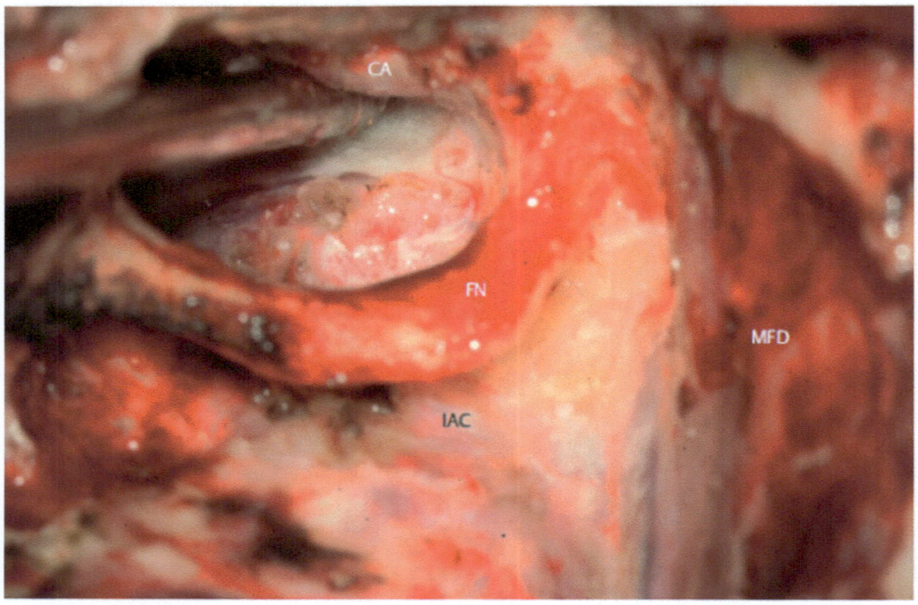

Fig. 9.86 The cholesteatoma infiltrating the area posterosuperior to the internal auditory canal is totally removed, and the matrix covering the carotid artery is removed. The suction tip is introduced to show access to the space medially to the carotid artery. *CA*, carotid artery; *FN*, facial nerve; *IAC*, internal auditory canal; *MFD*, middle fossa dura.

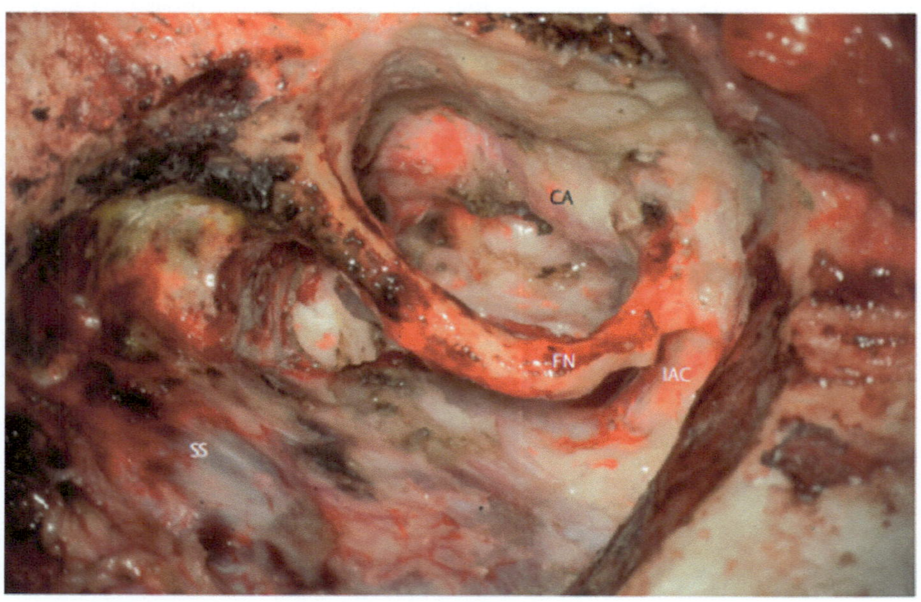

Fig. 9.87 The final shape of the cavity is shown. Sufficient access to this infracochlear-apical lesion is achieved by sufficiently skeletonizing the carotid artery and the facial nerve. *CA*, carotid artery; *FN*, facial nerve; *IAC*, internal auditory canal; *SS*, sigmoid sinus.

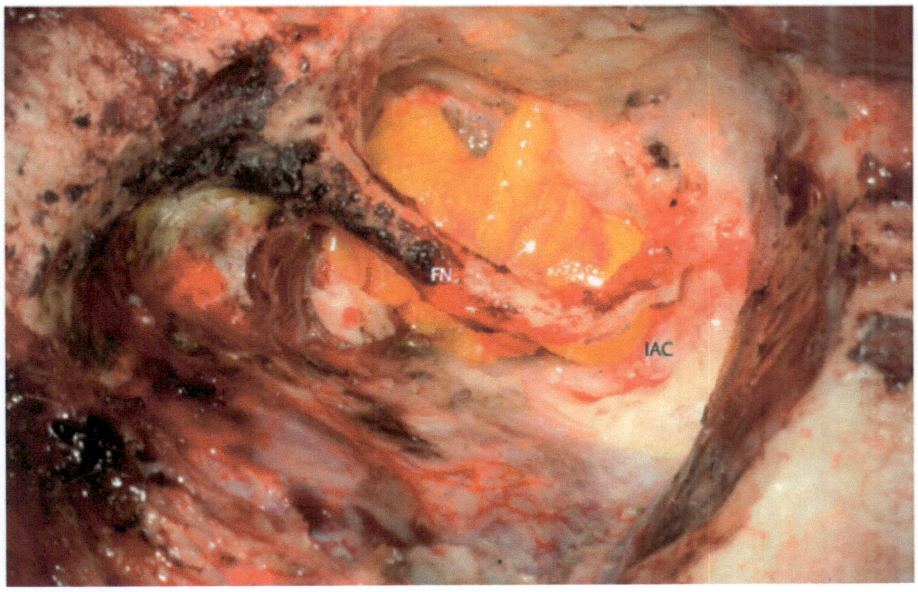

Fig. 9.88 The cavity is obliterated with abdominal fat, and the wound is closed in three layers. *FN*, facial nerve; *IAC*, internal auditory.

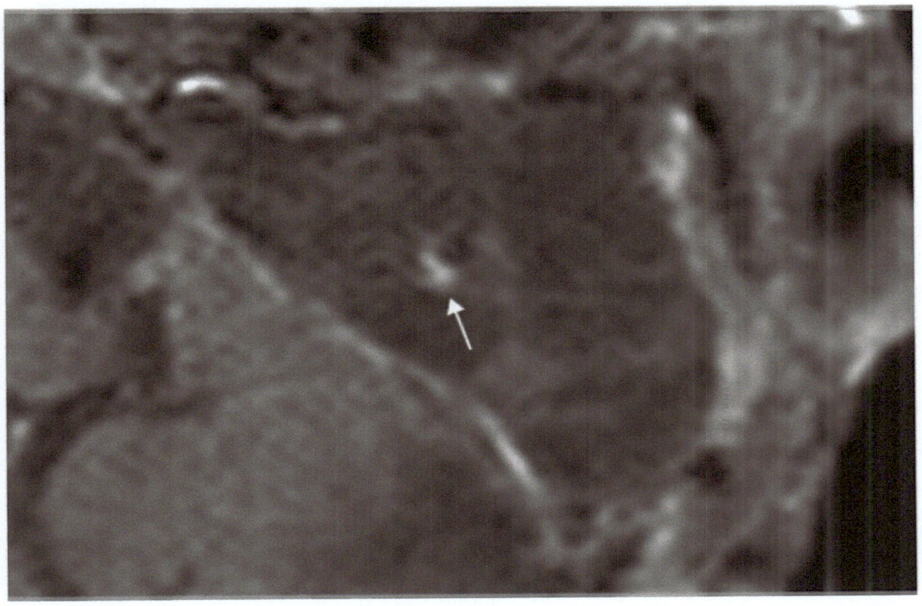

Fig. 9.89 Postoperative MRI 1 week after the surgery is shown. An enhancement of the facial nerve is seen *(arrow)* in the abdominal fat even though the facial nerve function of the patient remained H-B grade I.

Case 3: Grade III Vestibular Schwannoma

See Figs. 9.90–9.122.

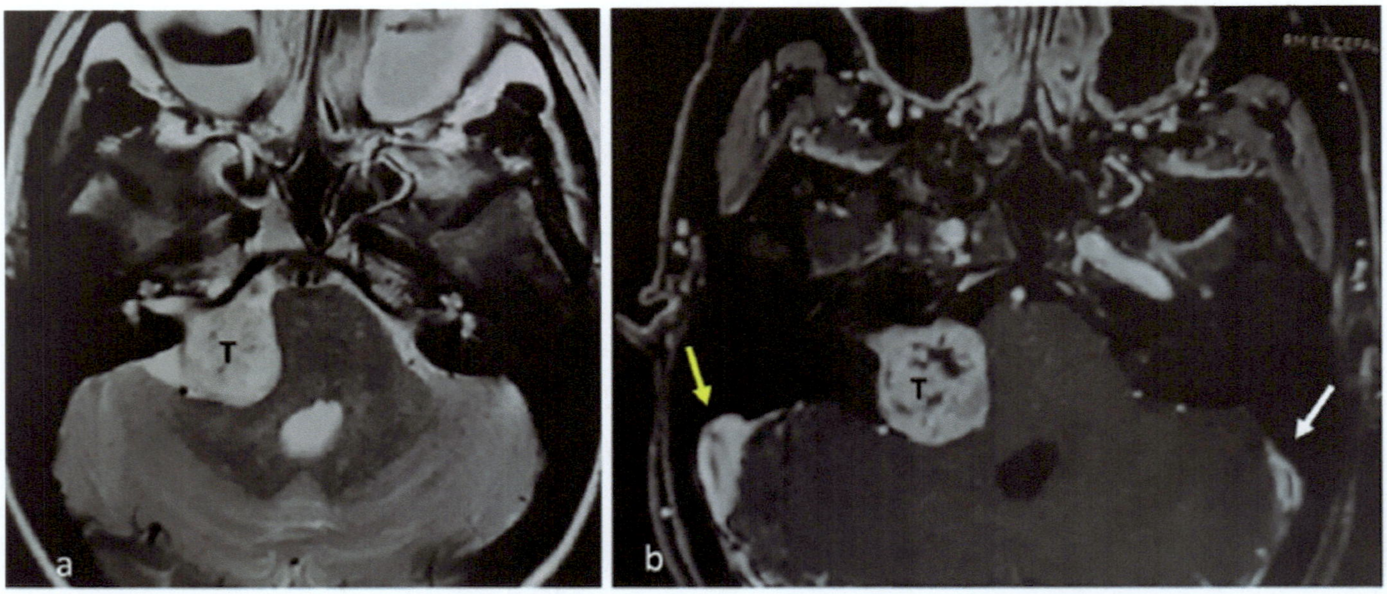

Fig. 9.90 MRI of the brain shows a lesion of the right cerebellopontine angle in a 32-year-old patient. The tumor (T) was histologically a vestibular schwannoma. According to the Tokyo consensus classification, this is a grade 3 vestibular schwannoma (maximum diameter 21–30 mm in the cerebellopontine angle). Note the asymmetry between the right dominant sigmoid sinus *(yellow arrow)* and the left *(white arrow)*. To avoid damage to the dominant sigmoid sinus with possible complications, a transotic approach was chosen to increase the surgical view. (A) T2W sequence, (B) T1W sequence with gadolinium enhancement.

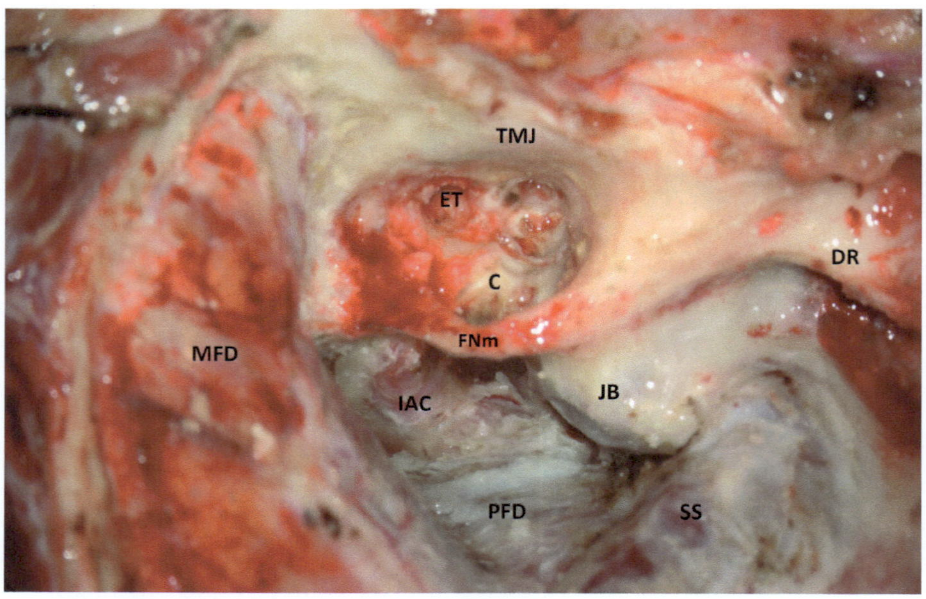

Fig. 9.91 The surgical approach is almost complete. The dominant jugular bulb (JB) was left intact. *MFD*, middle fossa dura; *JB*, jugular bulb; *SS*, sigmoid sinus; *FNm*, mastoid portion of the facial nerve; *PFD*, posterior fossa dura; *DR*, digastric ridge; *C*, cochlea; *ET*, eustachian tube; *TMJ*, temporomandibular joint; *IAC*, internal auditory canal.

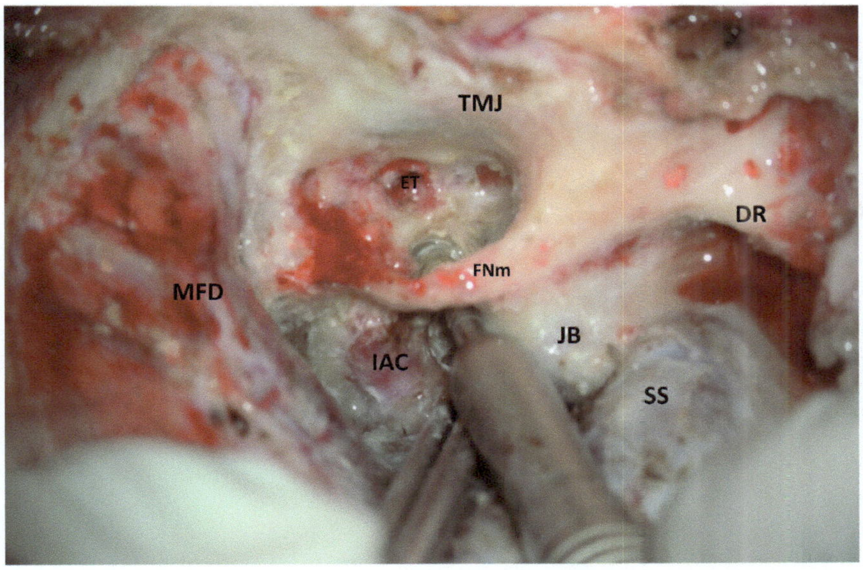

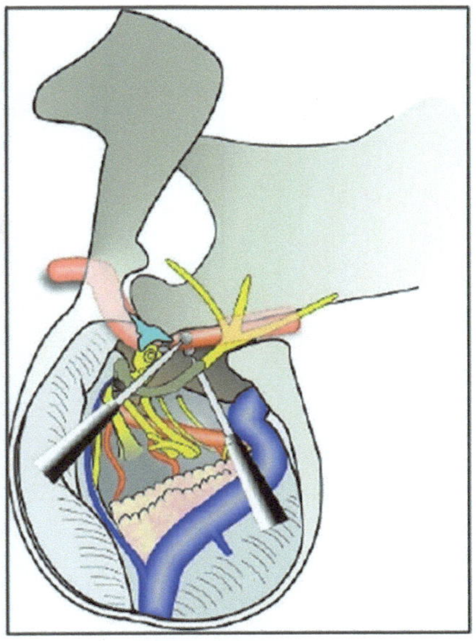

Fig. 9.92 The cochlea and petrous apex are drilled. *MFD*, middle fossa dura; *JB*, jugular bulb; *SS*, sigmoid sinus; *FNm*, mastoid portion of the facial nerve; *PFD*, posterior fossa dura; *DR*, digastric ridge; *C*, cochlea; *ET*, eustachian tube; *TMJ*, temporomandibular joint; *IAC*, internal auditory canal.

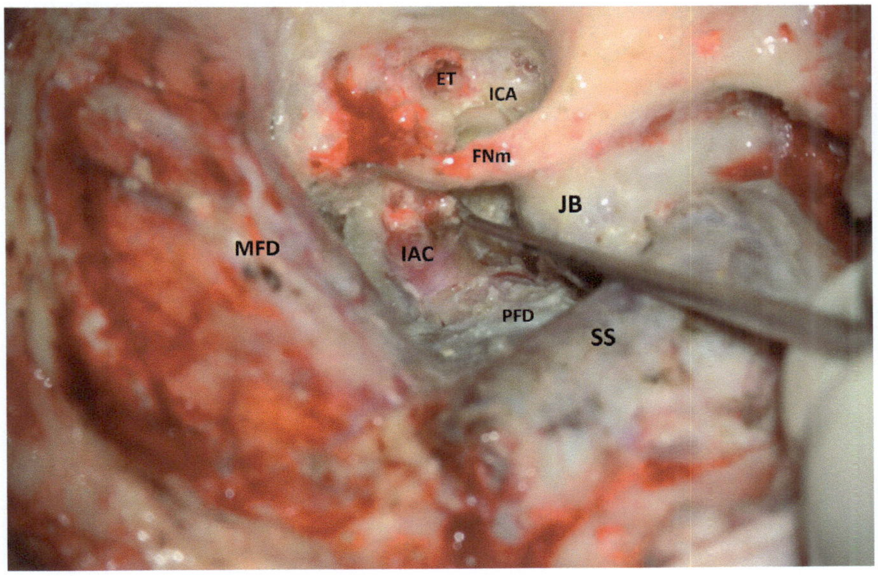

Fig. 9.93 The inferior vestibular nerve is detached from the internal auditory canal (IAC). *MFD*, middle fossa dura; *JB*, jugular bulb; *SS*, sigmoid sinus; *FNm*, mastoid portion of the facial nerve; *PFD*, posterior fossa dura; *DR*, digastric ridge; *C*, cochlea; *ET*, eustachian tube; *ICA*, internal carotid artery.

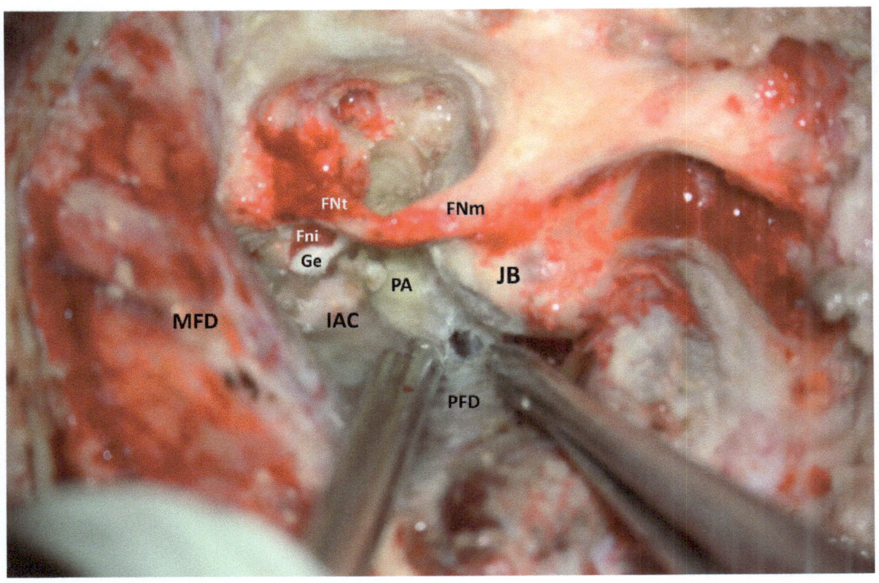

Fig. 9.94 Both the inferior and the superior vestibular nerves were dissected from the internal auditory canal (IAC). The internal auditory canal portion of the facial nerve (Fni) is protected with Gelfoam (Ge). The posterior fossa dura (PFD) is cut with scissors. *MFD*, middle fossa dura; *JB*, jugular bulb; *SS*, sigmoid sinus; *FNm*, mastoid portion of the facial nerve; *FNt*, tympanic portion of the facial nerve.

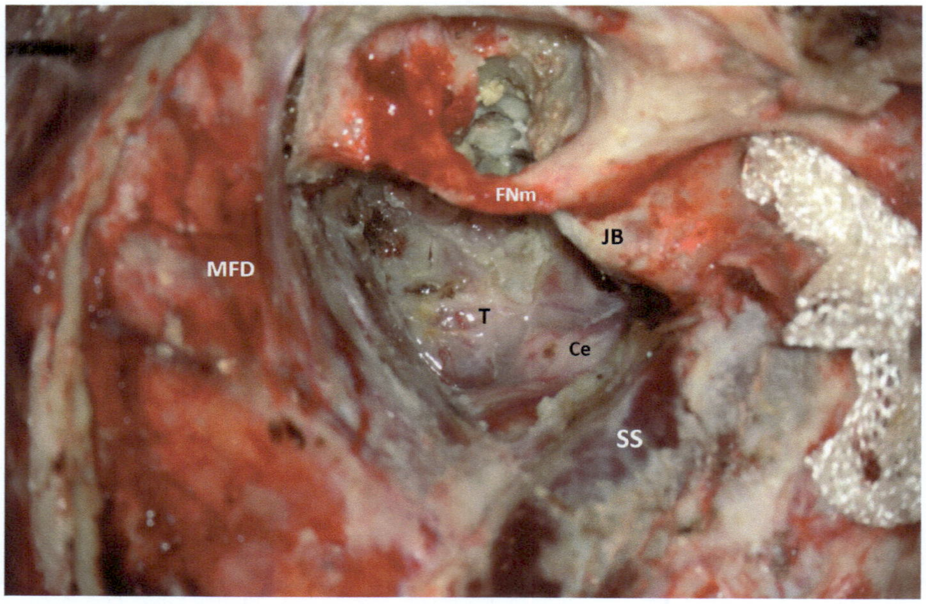

Fig. 9.95 The posterior fossa dura was opened with complete exposure of the tumor (T) in the cerebellopontine angle. *MFD*, middle fossa dura; *JB*, jugular bulb; *SS*, sigmoid sinus; *FNm*, mastoid portion of the facial nerve; *Ce*, cerebellum.

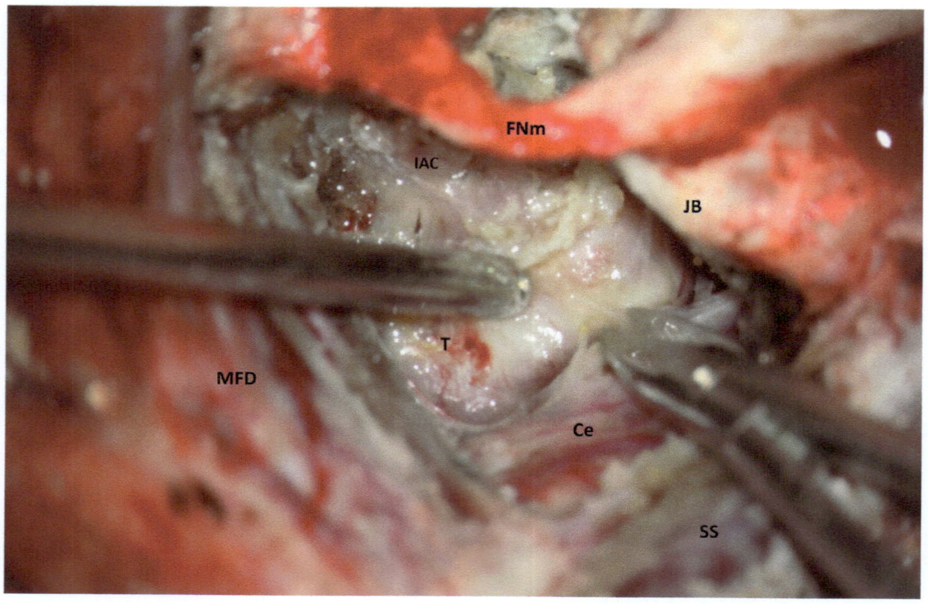

Fig. 9.96 The arachnoid is gently coagulated and cut. *MFD*, middle fossa dura; *JB*, jugular bulb; *SS*, sigmoid sinus; *FNm*, mastoid portion of the facial nerve; *Ce*, cerebellum; *IAC*, internal auditory canal.

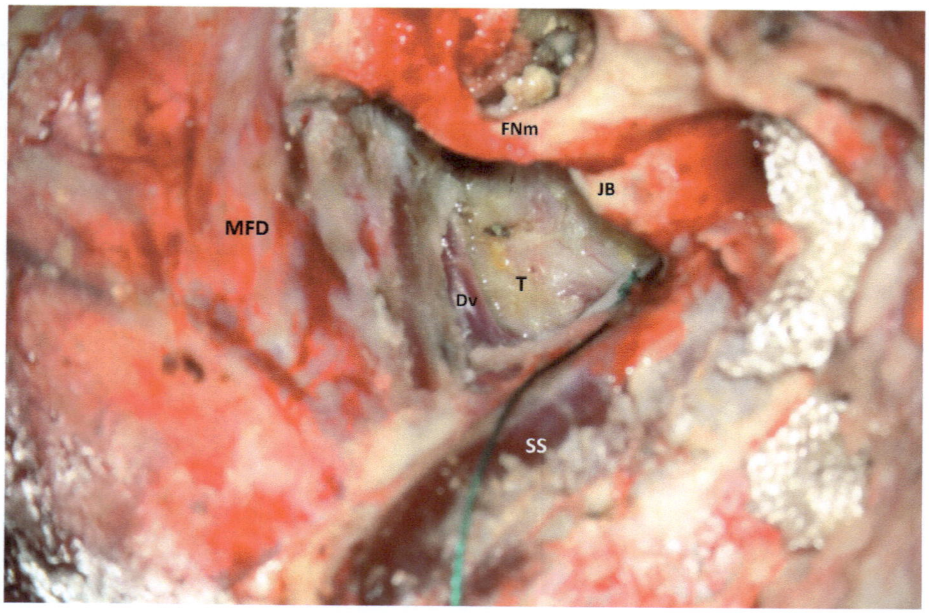

Fig. 9.97 The Dandy's vein (Dv) crosses the upper pole of the tumor (T) and needs to be coagulated to avoid intradural bleeding. *MFD*, middle fossa dura; *JB*, jugular bulb; *SS*, sigmoid sinus; *FNm*, mastoid portion of the facial nerve.

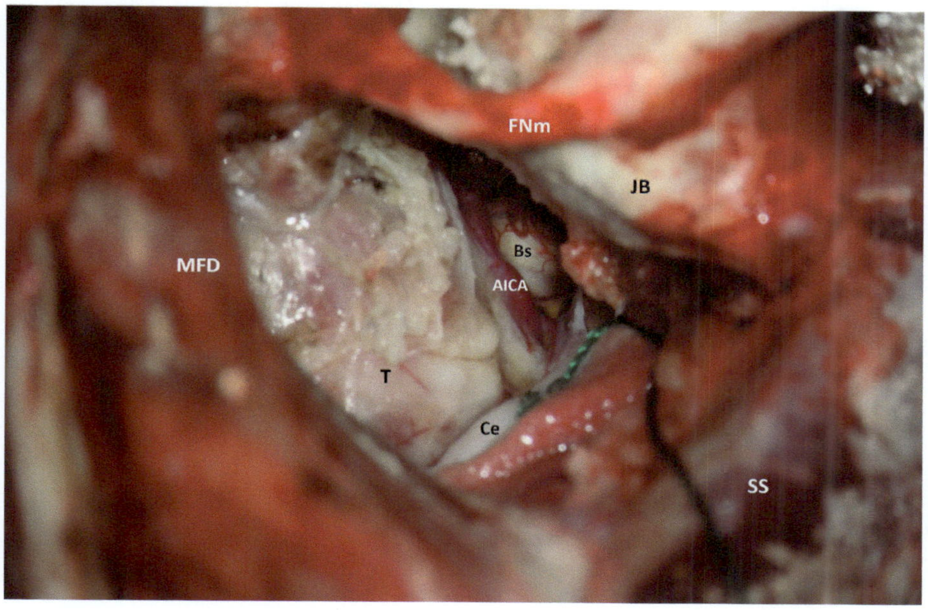

Fig. 9.98 The anterior-inferior cerebellar artery (AICA) is close to the lower pole of the tumor (T). Careful dissection is required to separate the artery from the tumor surface. *MFD*, middle fossa dura; *JB*, jugular bulb; *SS*, sigmoid sinus; *FNm*, mastoid portion of the facial nerve; *Bs*, Brainstem; *Ce*, cerebellum.

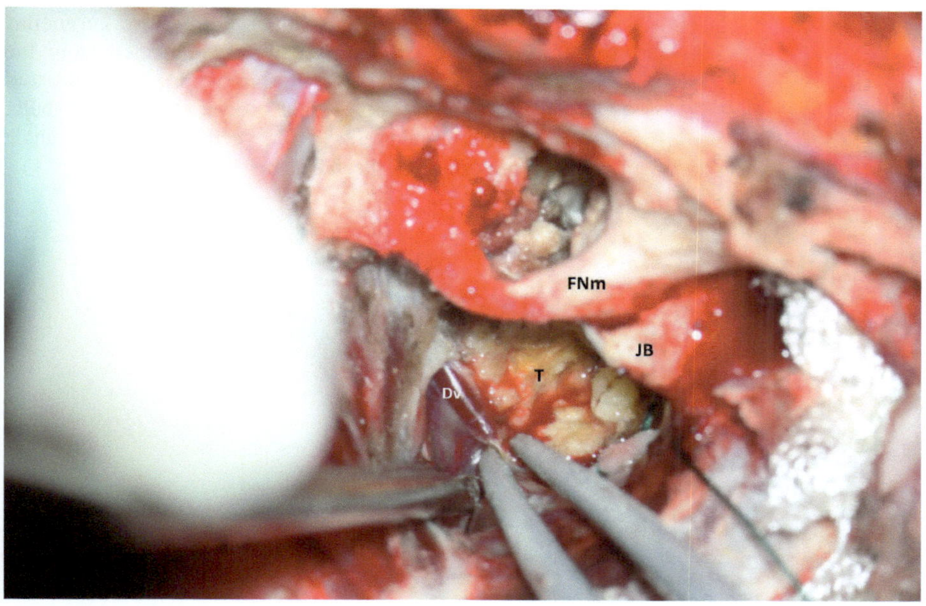

Fig. 9.99 Dandy's vein (Dv) coagulation begins posteriorly at the tumor surface. *FNm*, mastoid portion of the facial nerve; *JB*, jugular bulb; *T*, tumor.

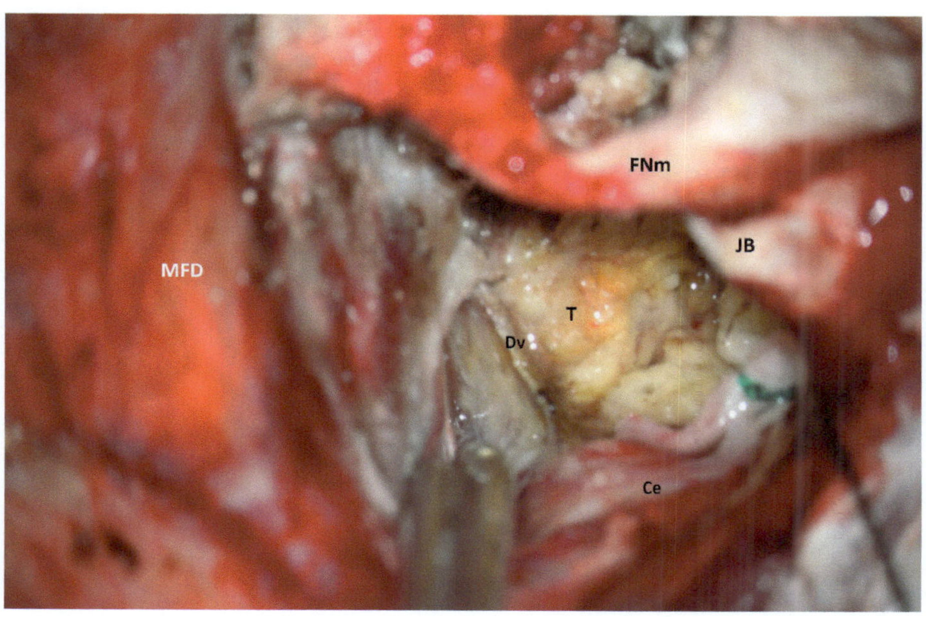

Fig. 9.100 Coagulation of the Dandy's vein (Dv) proceeds anteriorly. *FNm*, mastoid portion of the facial nerve; *JB*, jugular bulb; *T*, tumor; *MFD*, middle fossa dura; *Ce*, cerebellum.

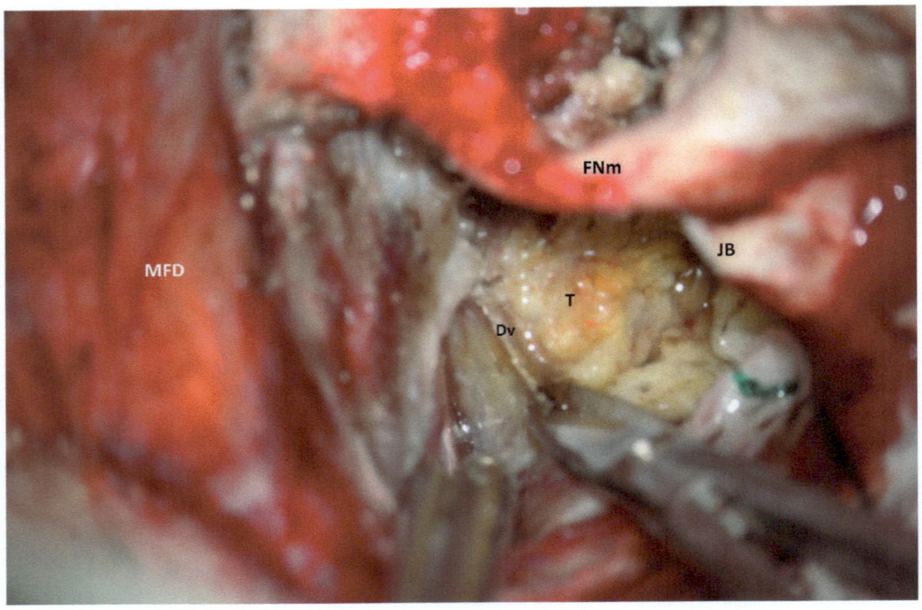

Fig. 9.101 The Dandy's vein (Dv) is cut posteriorly. *FNm*, mastoid portion of the facial nerve; *JB*, jugular bulb; *T*, tumor; *MFD*, middle fossa dura.

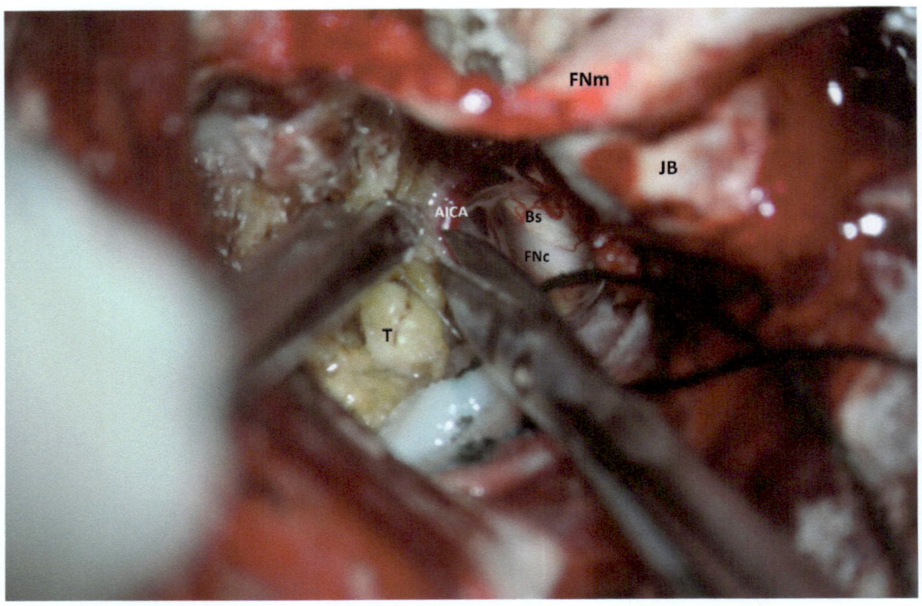

Fig. 9.102 The anterior-inferior cerebellar artery (AICA) is carefully dissected from the lower surface of the tumor (T). *FNm*, mastoid portion of the facial nerve; *JB*, jugular bulb; *Bs*, brainstem; *FNc*, cerebellopontine angle portion of the facial nerve.

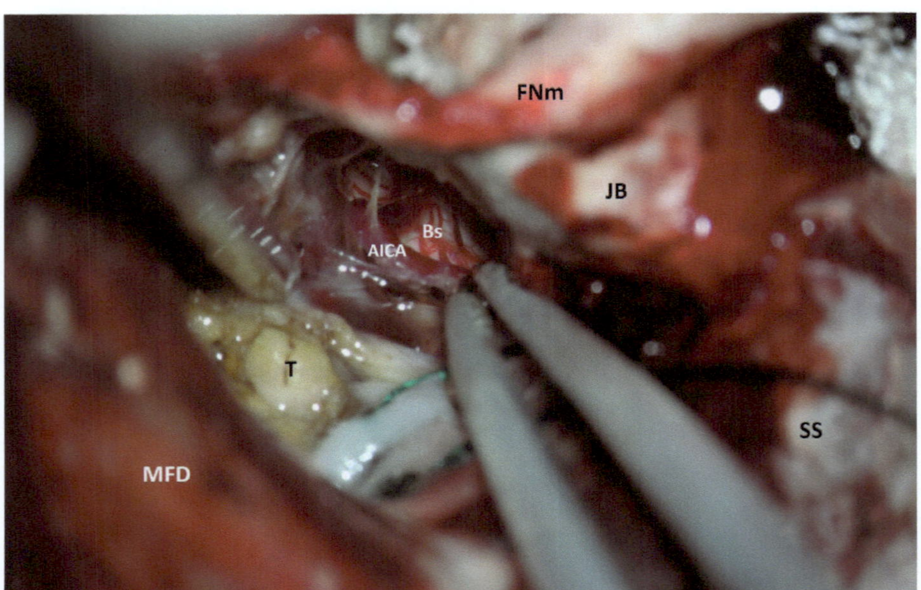

Fig. 9.103 Dissection proceeds inferiorly, coagulating the tumor surface. *FNm*, mastoid portion of the facial nerve; *JB*, jugular bulb; *Bs*, brainstem; *SS*, sigmoid sinus; *MFD*, middle fossa dura; *AICA*, anterior-inferior cerebellar artery.

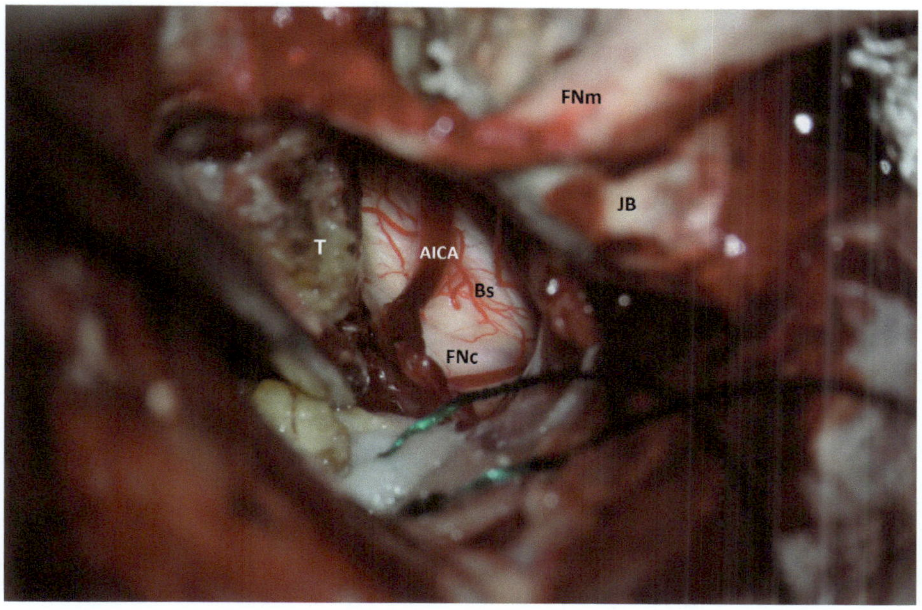

Fig. 9.104 The anterior-inferior cerebellar artery (AICA) has been completely dissected from the tumor (T). *FNm*, mastoid portion of the facial nerve; *JB*, jugular bulb; *Bs*, brainstem; *FNc*, cerebellopontine angle portion of the facial nerve.

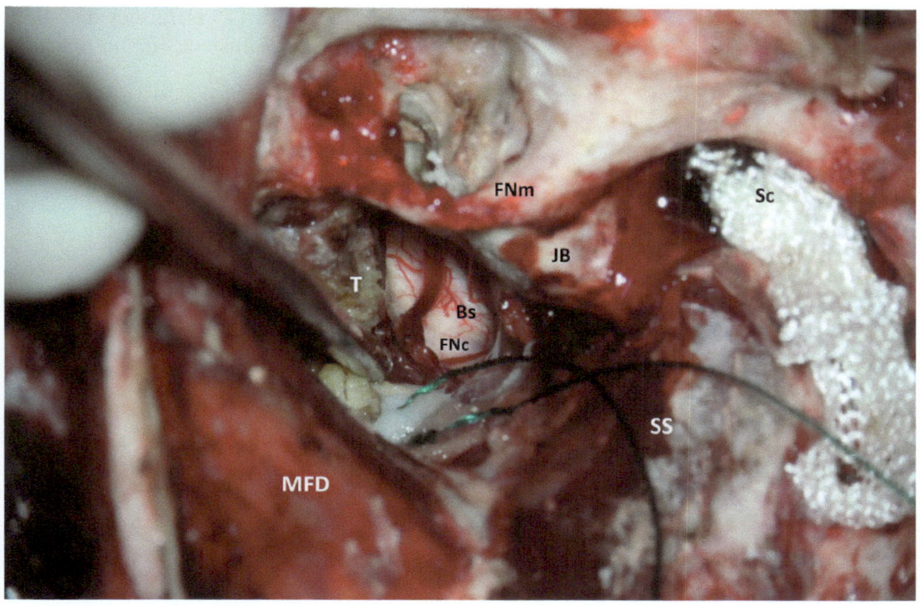

Fig. 9.105 Wider view. *FNm*, mastoid portion of the facial nerve; *JB*, jugular bulb; *Bs*, brainstem; *FNc*, cerebellopontine angle portion of the facial nerve; *FNm*, mastoid portion of the facial nerve; *T*, tumor; *SS*, sigmoid sinus; *Sc*, surgical; *MFD*, middle fossa dura.

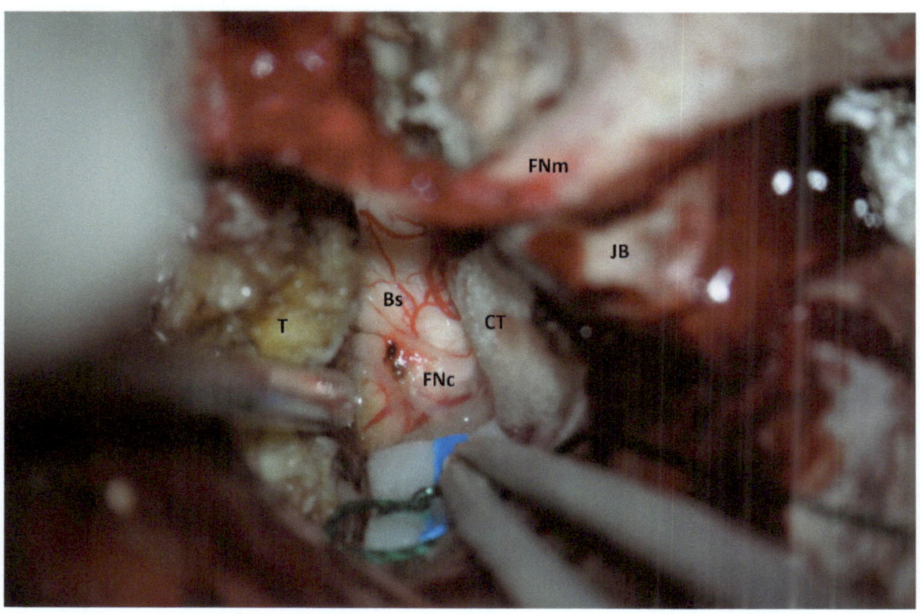

Fig. 9.106 The AICA was displaced inferiorly and protected with a cottonoid (CT). A small arterial branch between the tumor and the cerebellopontine angle portion of the facial nerve (FNc) was coagulated. *JB*, jugular bulb; *Bs*, brainstem; *FNm*, mastoid portion of the facial nerve; *T*, tumor.

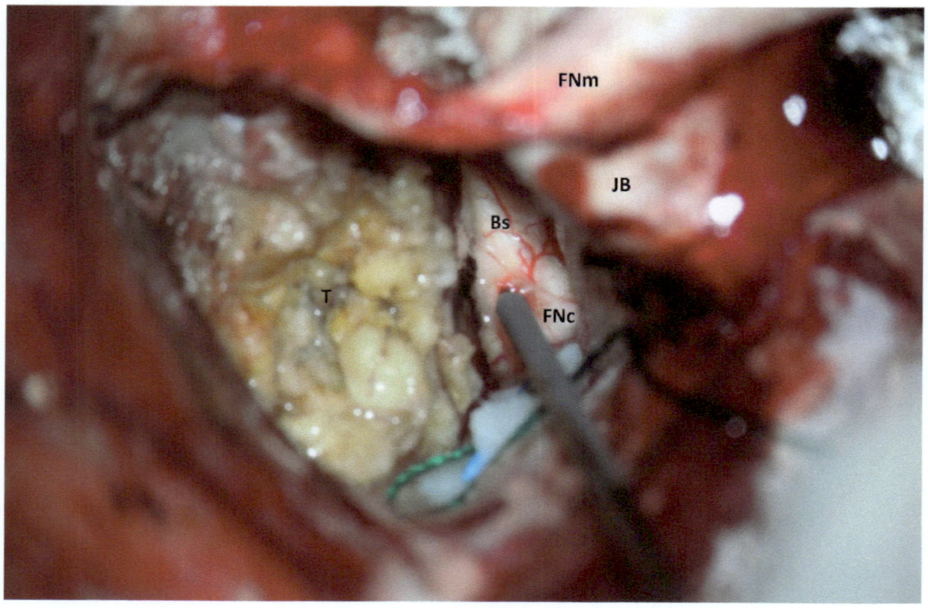

Fig. 9.107 The cerebellopontine angle portion of the facial nerve (FNc) was stimulated. *JB*, jugular bulb; *Bs*, brainstem; *FNm*, mastoid portion of the facial nerve; *T*, tumor.

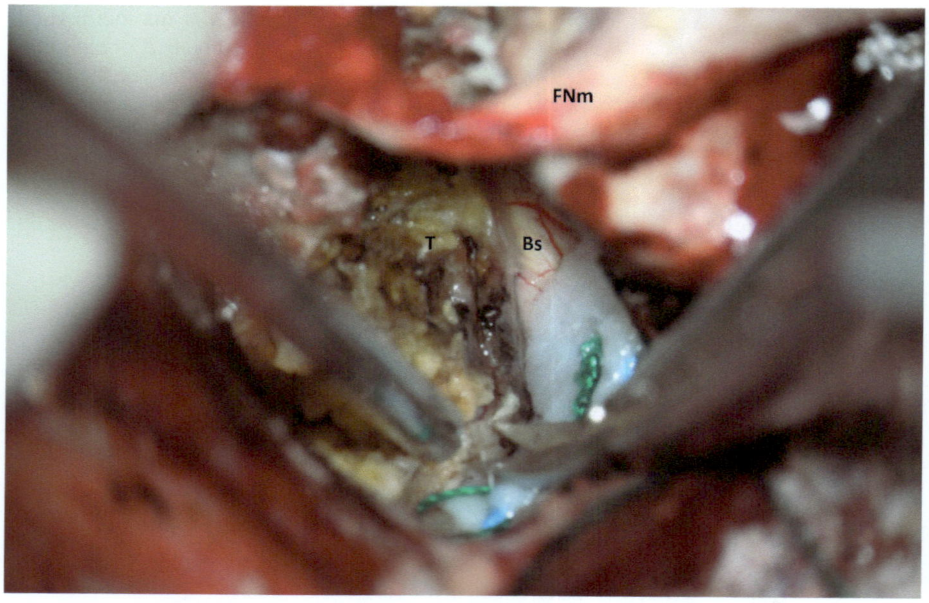

Fig. 9.108 The tumor (T) is further debulked using scissors. *Bs*, brainstem; *FNm*, mastoid portion of the facial nerve.

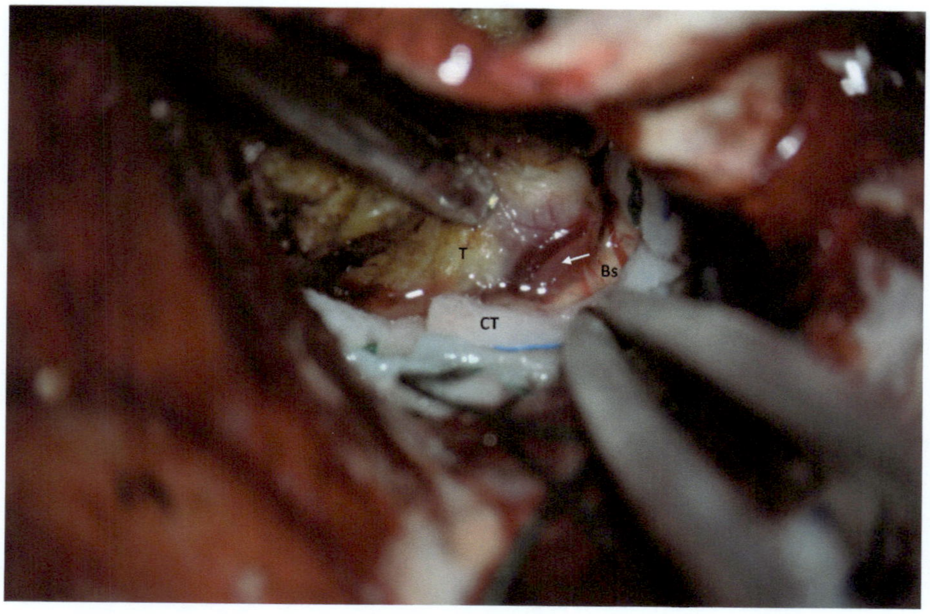

Fig. 9.109 After debulking, the tumor (T) is dissected from the brainstem (Bs). *CT*, cottonoid.

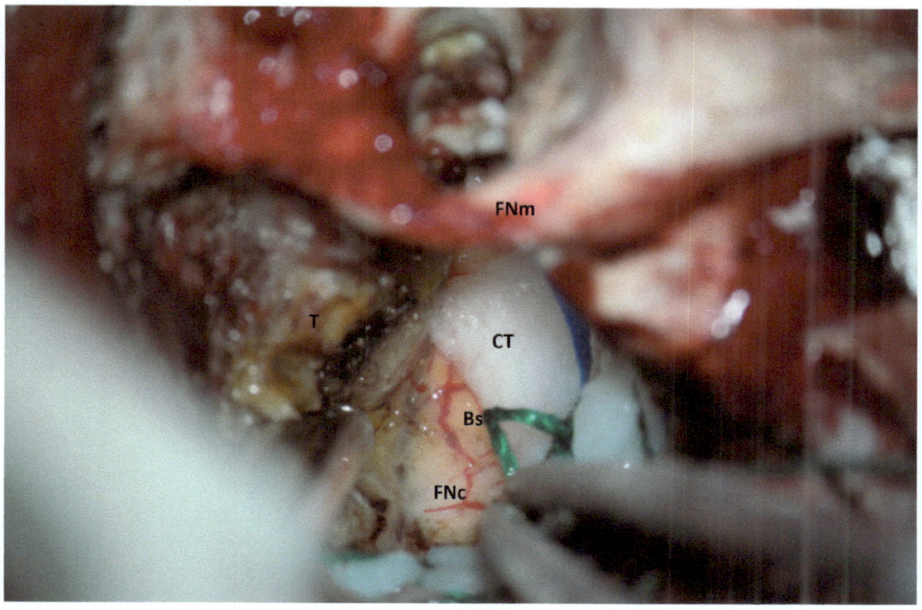

Fig. 9.110 Further retraction of the tumor allows visualization of the proximal cerebellopontine angle portion of the facial nerve (FNc). *Bs*, brainstem; *CT*, cottonoid; *FNm*, mastoid portion of the facial nerve.

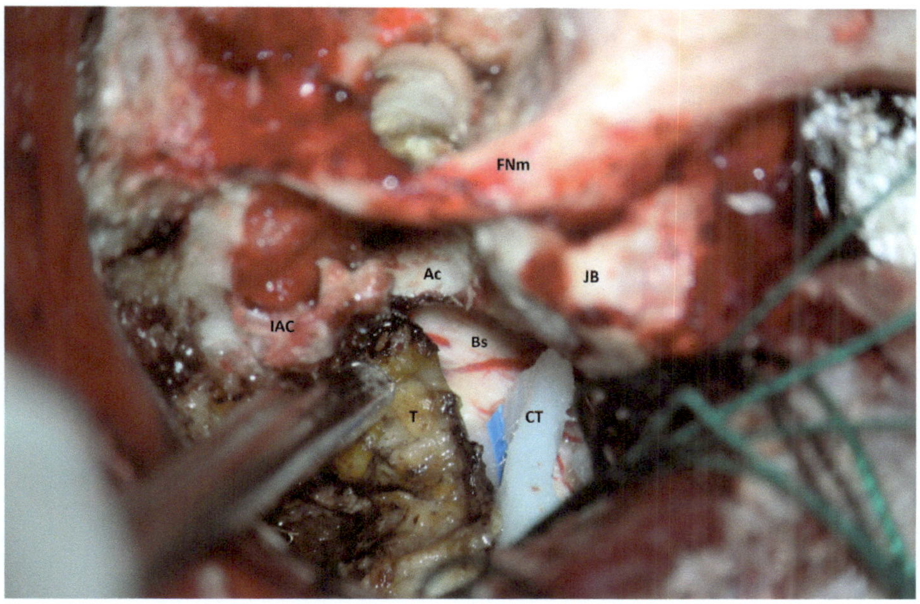

Fig. 9.111 Further drilling of the inferior apical cells (Ac) allows control of tumor extension toward the prepontine cistern. *Bs*, brainstem; *CT*, cottonoid; *FNm*, mastoid portion of the facial nerve; *IAC*, internal auditory canal; *JB*, jugular bulb; *T*, tumor.

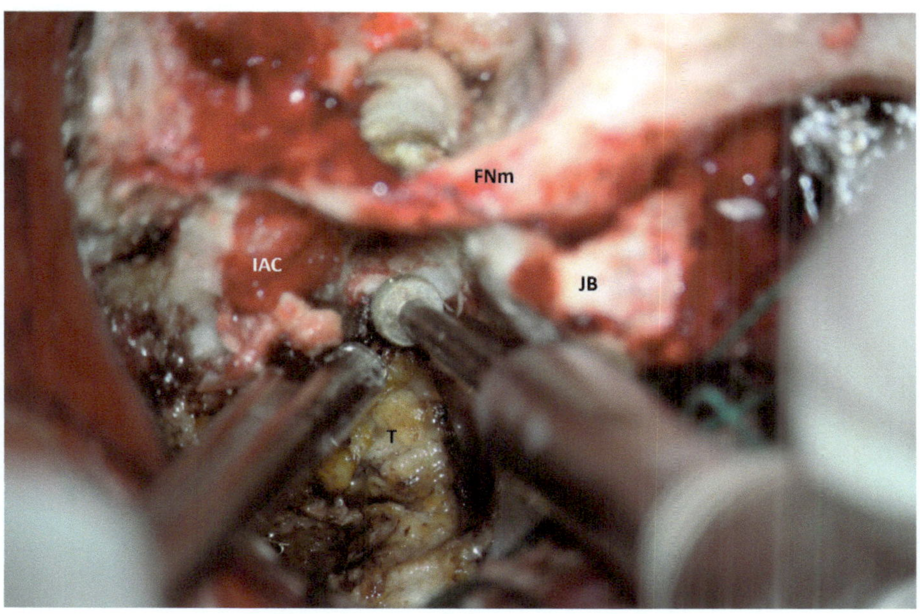

Fig. 9.112 Drilling of the inferior apical cells (Ac) starts. *FNm*, mastoid portion of the facial nerve; *IAC*, internal auditory canal; *JB*, jugular bulb; *T*, tumor.

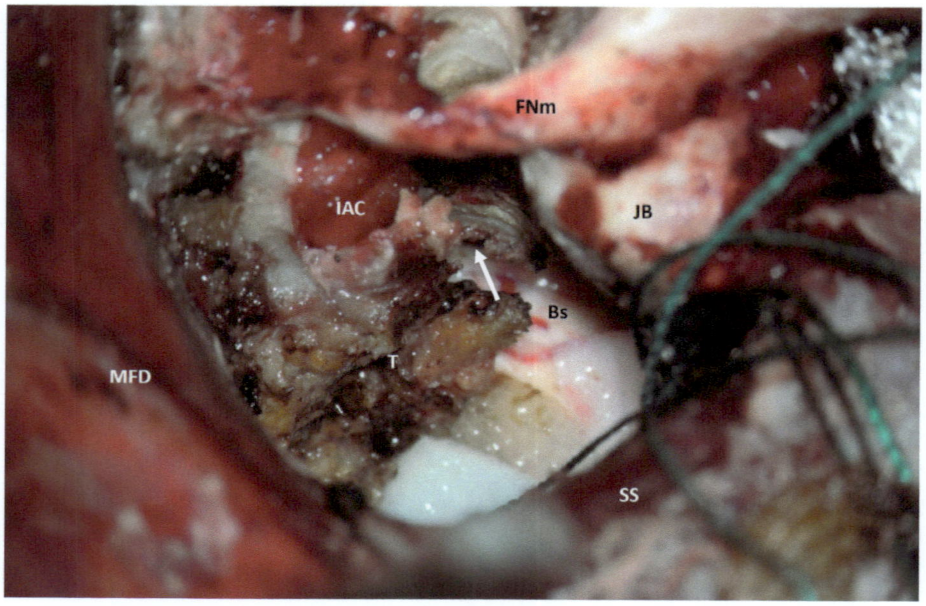

Fig. 9.113 The lower part of the prepontine cistern is now visible *(arrow)*. *FNm*, mastoid portion of the facial nerve; *IAC*, internal auditory canal; *JB*, jugular bulb; *T*, tumor; *Bs*, brainstem; *SS*, sigmoid sinus.

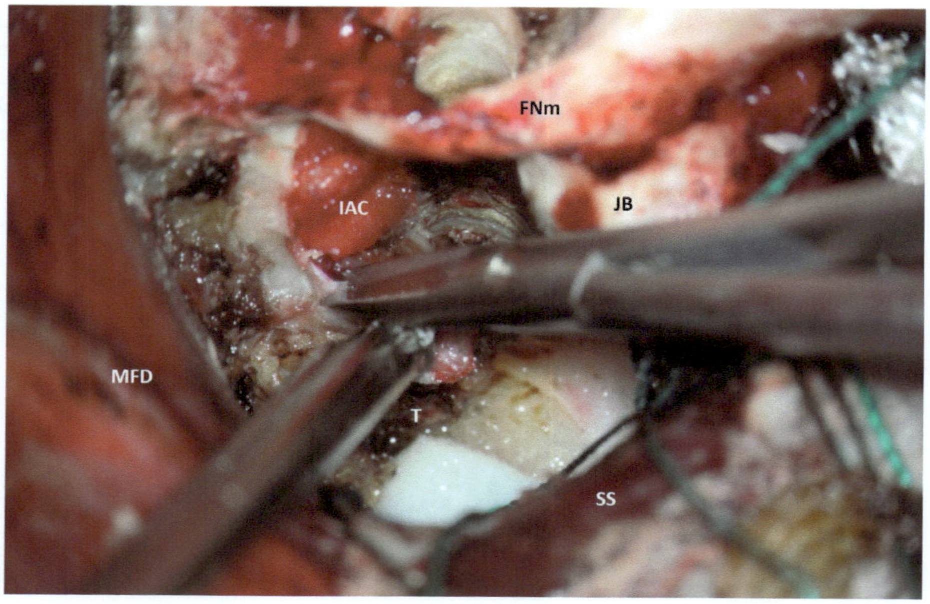

Fig. 9.114 The dura of the internal auditory canal (IAC) is further cut to allow tumor dissection. *FNm*, mastoid portion of the facial nerve; *JB*, jugular bulb; *T*, tumor; *SS*, sigmoid sinus.

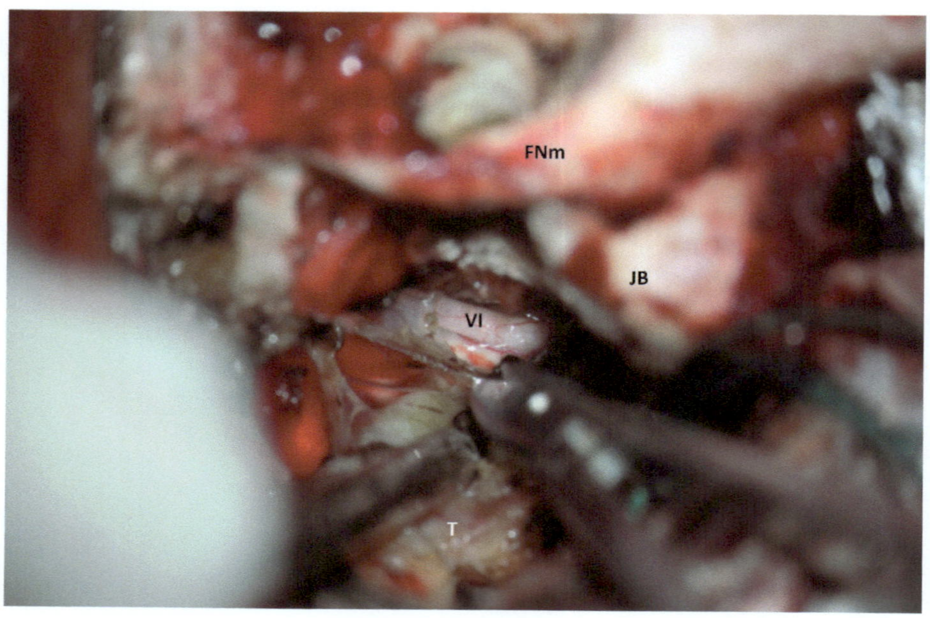

Fig. 9.115 Closer view. After dissection of the arachnoid, the abducens cranial nerve (VI) is visible on the lower part of the prepontine cistern. *FNm*, mastoid portion of the facial nerve; *JB*, jugular bulb.

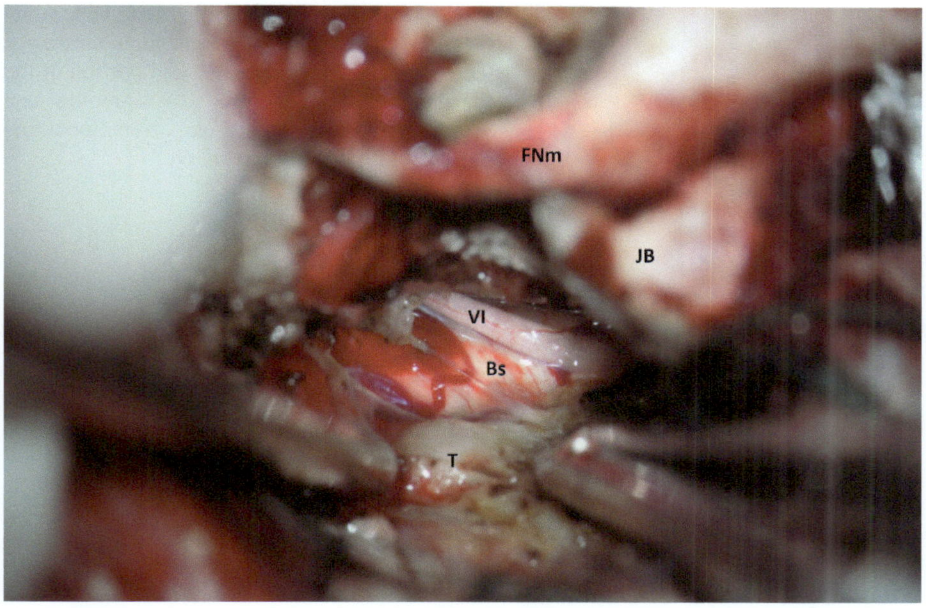

Fig. 9.116 The tumor (T) is now dissected from medial to lateral. *FNm*, mastoid portion of the facial nerve; *JB*, jugular bulb; *VI*, abducens nerve.

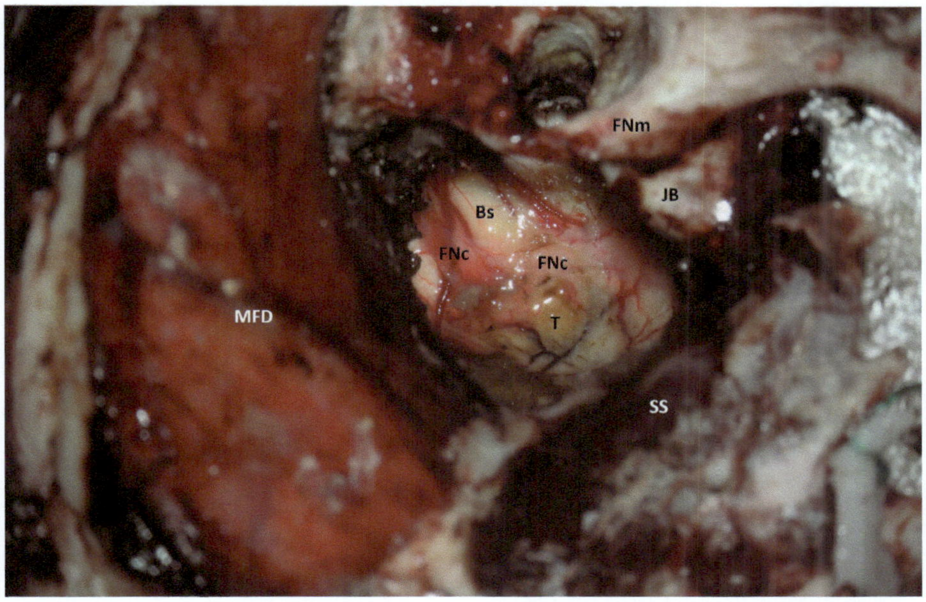

Fig. 9.117 The last piece of tumor is going to be removed after the cerebellopontine angle portion of the facial nerve (FNc) is freed. *FNm*, mastoid portion of the facial nerve; *JB*, jugular bulb; *SS*, sigmoid sinus; *MFD*, middle fossa dura; *Bs*, brainstem.

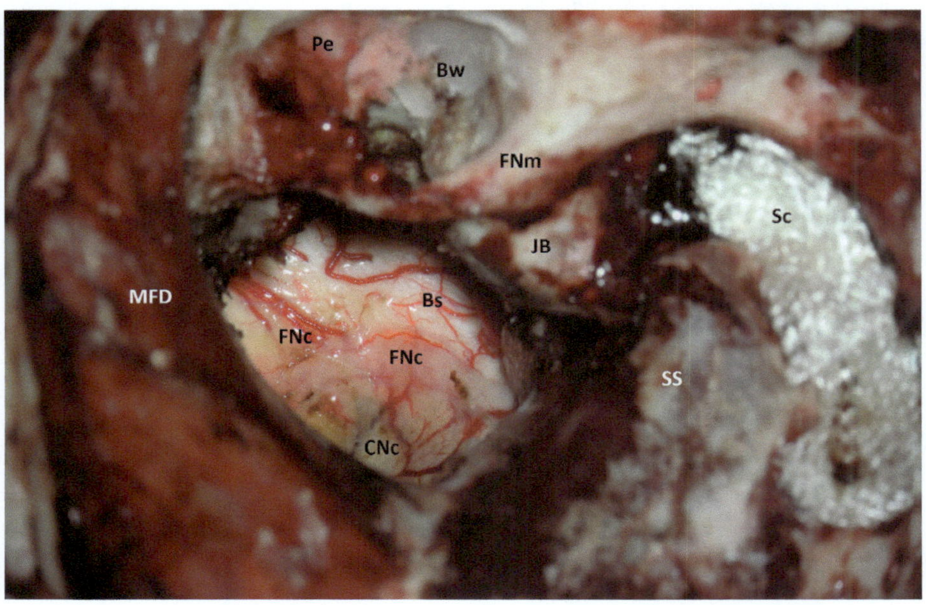

Fig. 9.118 Complete removal of the tumor has been achieved. The facial nerve, even if anatomically preserved, is extremely thinned and distorted by the tumor. Note the remnant of the cerebellopontine angle portion of the cochlear nerve (CNc) at the level of the root-entry zone. The eustachian tube is sealed with periosteum (Pe) and the surrounding air cells with bone wax (BW). *FNm*, mastoid portion of the facial nerve; *JB*, jugular bulb; *SS*, sigmoid sinus; *MFD*, middle fossa dura; *Bs*, brainstem; *FNc*, cerebellopontine angle portion of the facial nerve.

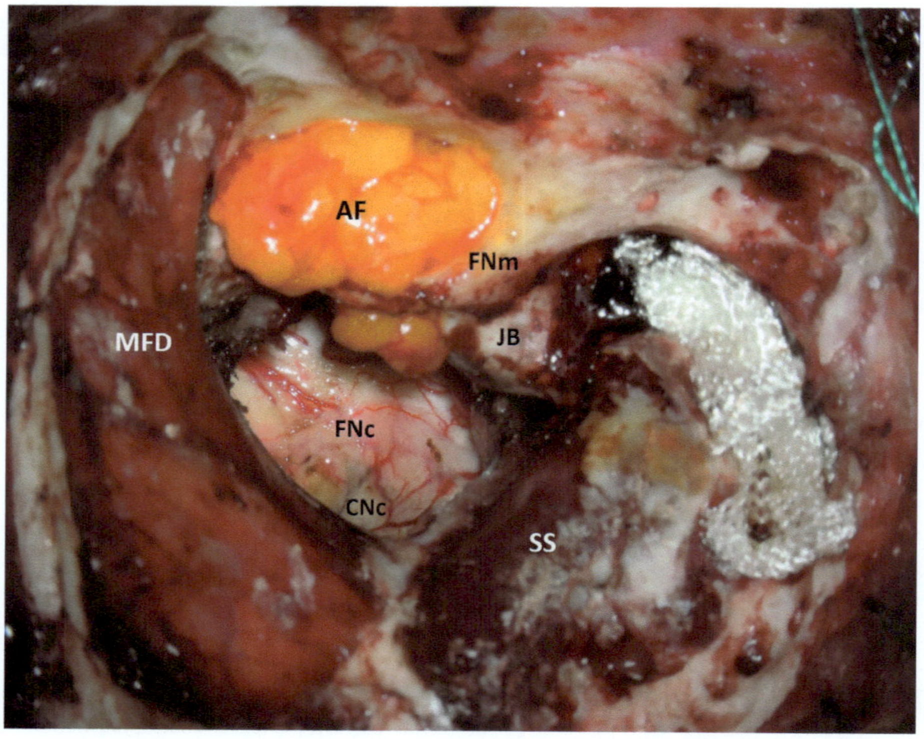

Fig. 9.119 The cavity is filled with abdominal fat (AF). *CNc*, cerebellopontine angle portion of the cochlear nerve; *FNm*, mastoid portion of the facial nerve; *JB*, jugular bulb; *SS*, sigmoid sinus; *MFD*, middle fossa dura; *FNc*, cerebellopontine angle portion of the facial nerve.

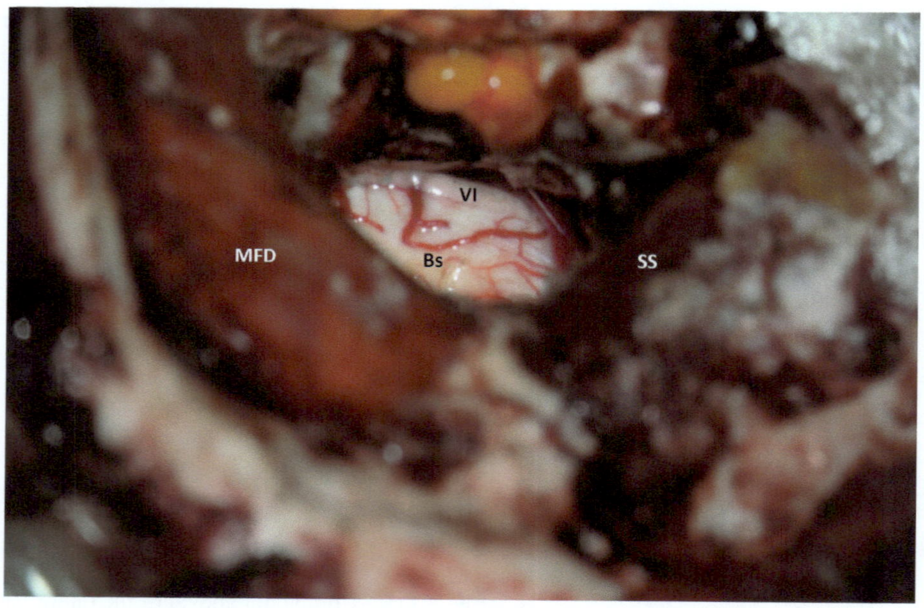

Fig. 9.120 The microscope is tilted forward to show the abducens nerve (VI). *MFD*, middle fossa dura; *SS*, sigmoid sinus; *Bs*, brainstem.

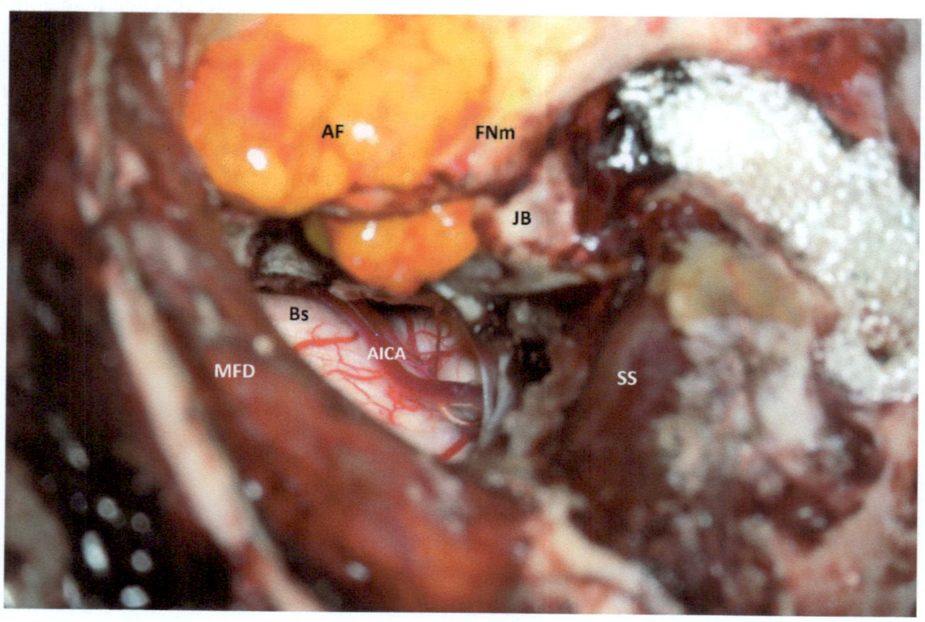

Fig. 9.121 The microscope is tilted inferiorly to show the anterior-inferior cerebellar artery (AICA). *MFD*, middle fossa dura; *SS*, sigmoid sinus; *Bs*, brainstem; *JB*, jugular bulb; *FNm*, mastoid portion of the facial nerve; *AF*, abdominal fat.

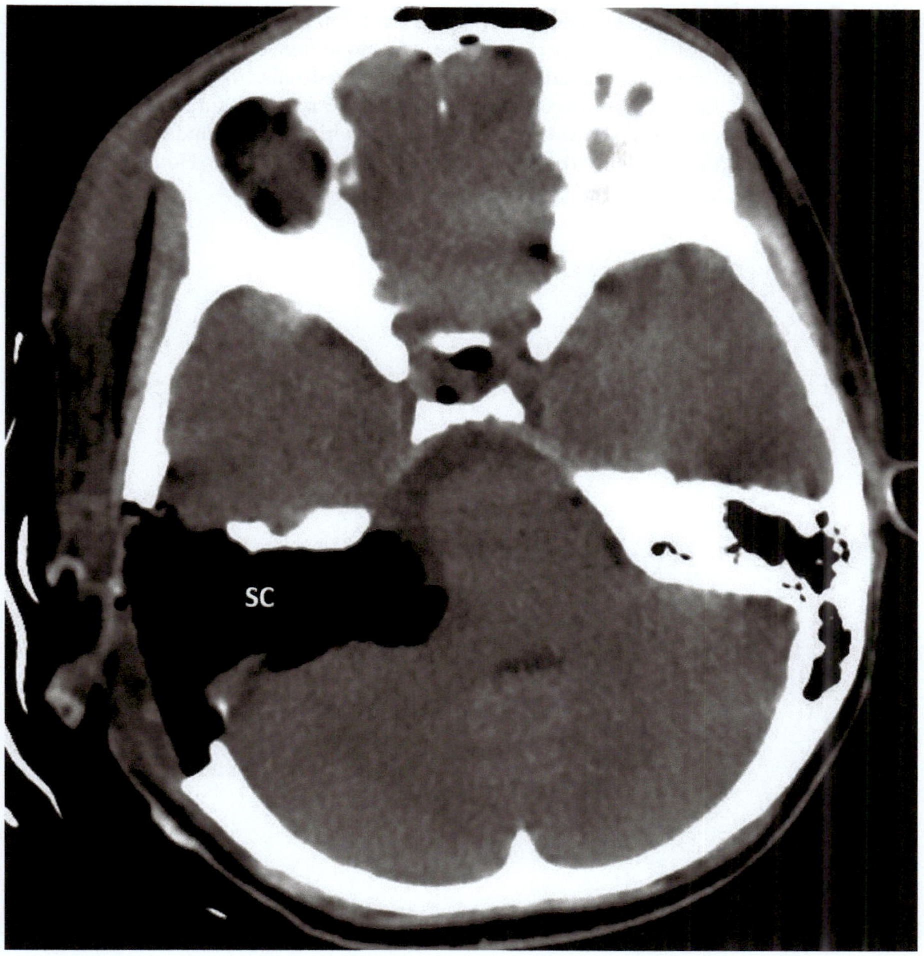

Fig. 9.122 Postoperative CT scan. *SC*, surgical cavity.

Case 4: Supralabyrinthine Petrous Bone Cholesteatoma Extended to the Internal Auditory Canal

See Figs. 9.123–9.177.

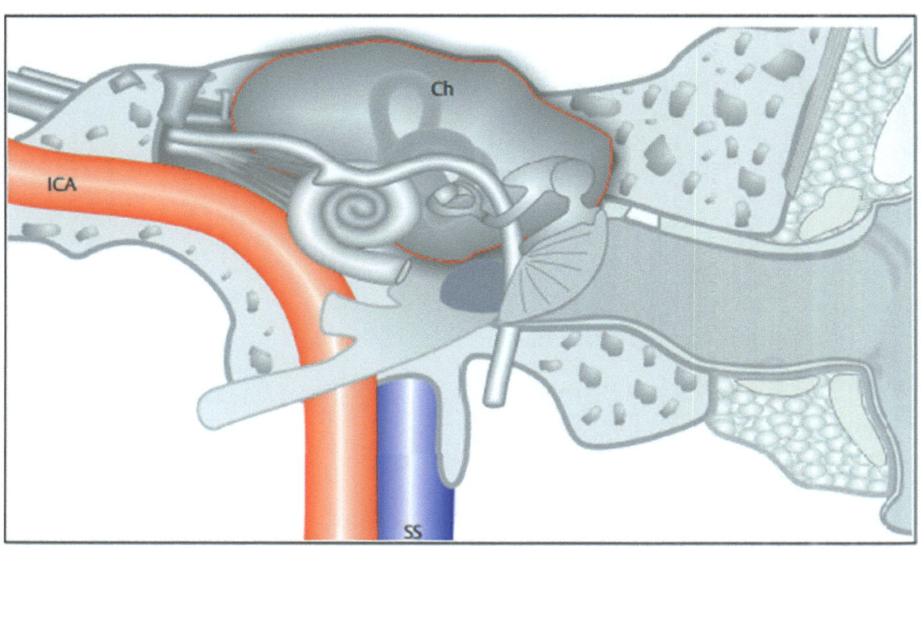

Fig. 9.123 A case of congenital supralabyrinthine petrous bone cholesteatoma extended into the internal auditory canal with erosion in both the cochlea and the posterior labyrinth removed with a transotic approach with facial nerve reconstruction. *Ch*, cholesteatoma; *ICA*, internal carotid artery; *SS*, sigmoid sinus.

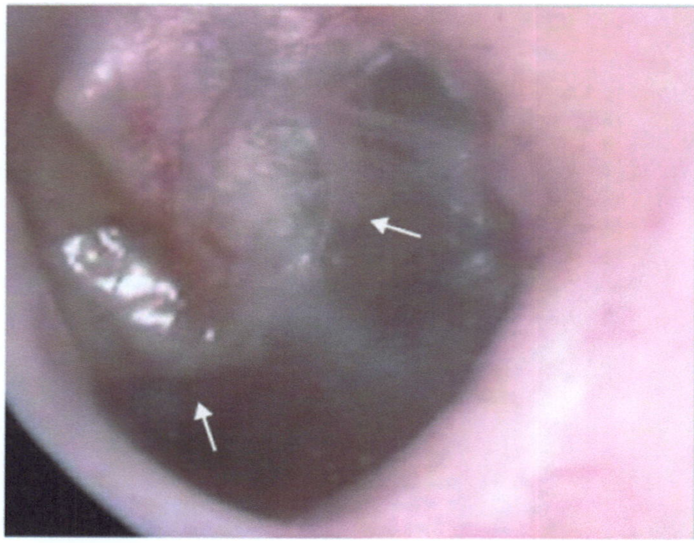

Fig. 9.124 The patient underwent a middle fossa approach by a neurosurgeon for petrous bone cholesteatoma elsewhere after suffering incomplete facial palsy. Relapse of the disease in the form of complete facial palsy forced the patient to have revision surgery. Preoperative otoscopy shows the presence of a whitish mass under the handle of the malleus *(arrows)*. Facial palsy is graded House-Brackmann grade IV.

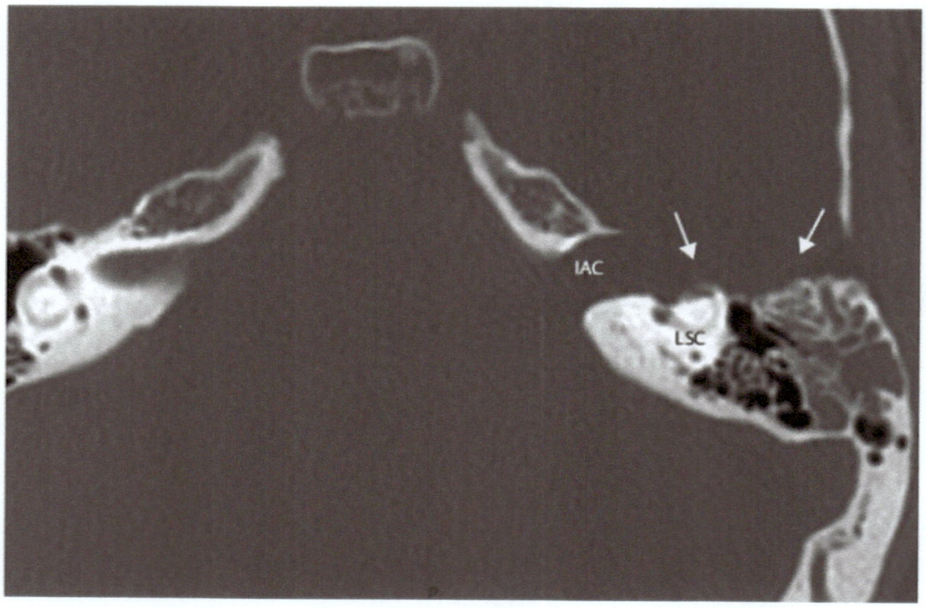

Fig. 9.125 Preoperative CT shows bone erosion in the anterior part of the pyramidal bone *(arrows)*. Erosion of the lateral semicircular canal and enlargement of the internal auditory canal are identified. *IAC*, internal auditory canal; *LSC*, lateral semicircular canal.

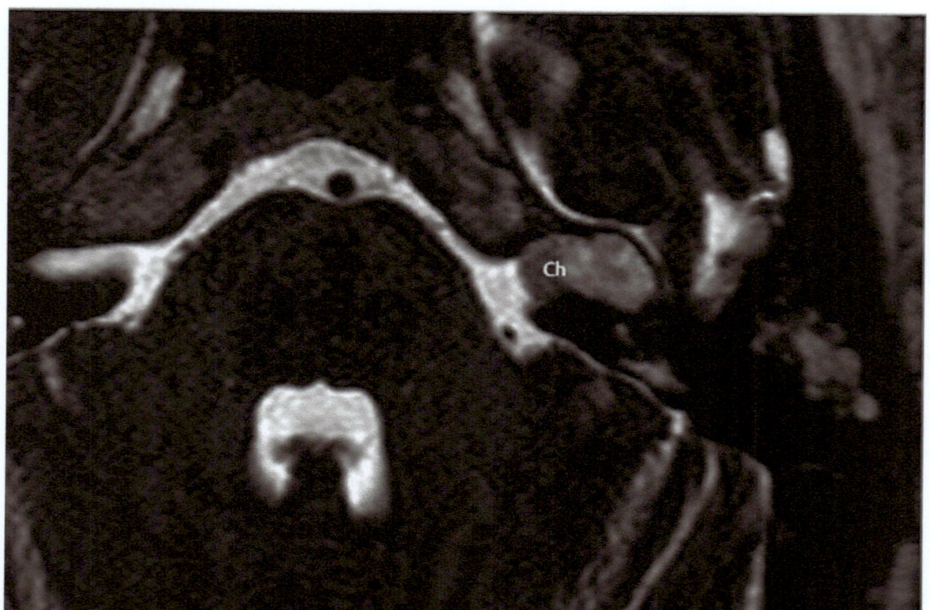

Fig. 9.126 The axial section of the heavy T2-weighted MRI clearly demonstrates the extension of the cholesteatoma into the internal auditory canal. *Ch*, cholesteatoma.

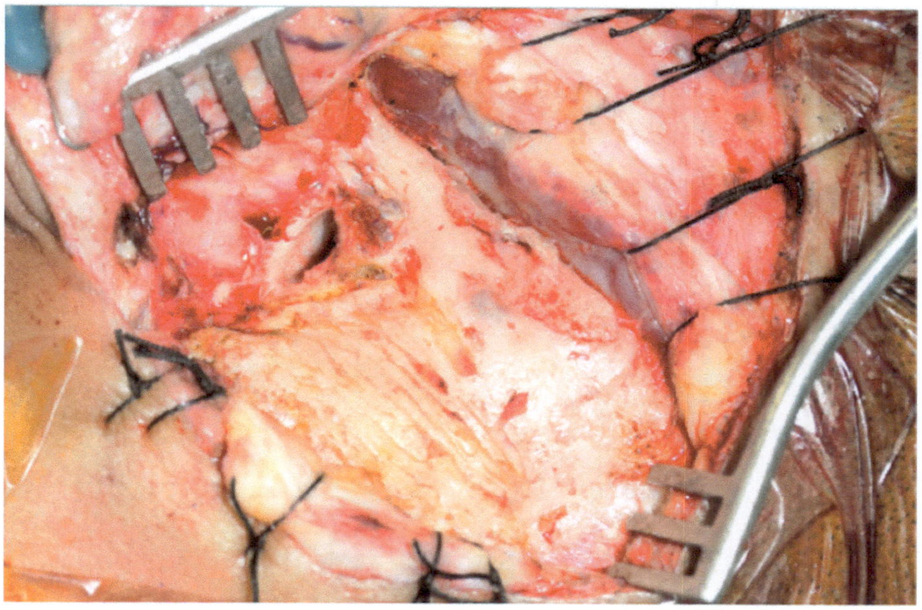

Fig. 9.127 Preparation for petrous bone cholesteatoma surgery is shown. Obtaining sufficient access to the petrous apex requires wide exposure to the lateral surface of the skull, including the anterosuperior part of the occipital bone and the anteroinferior part of the parietal bone.

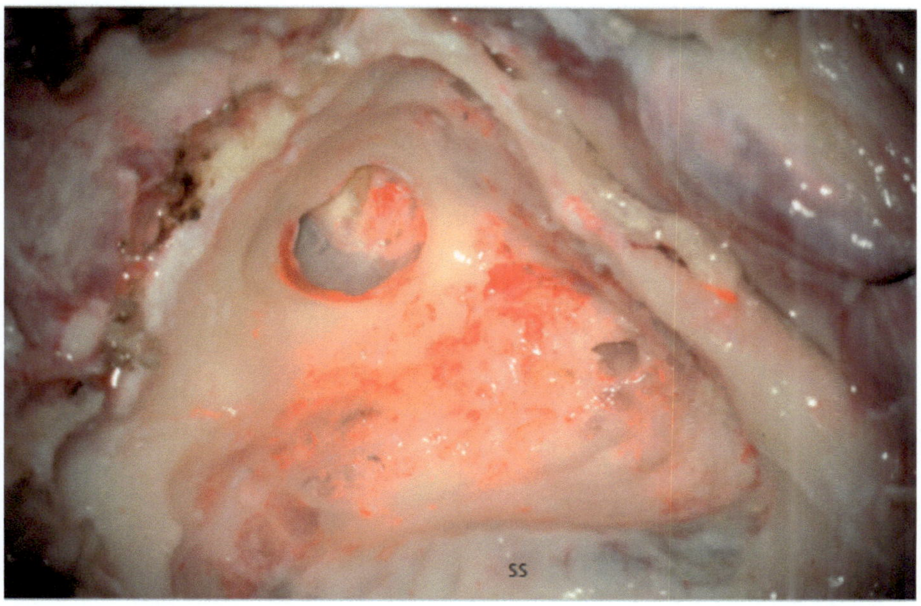

Fig. 9.128 Canal wall down mastoidectomy is carried out. The posterior wall of the bony meatus is partially removed. No cholesteatoma is seen in the mastoid. *SS*, sigmoid sinus.

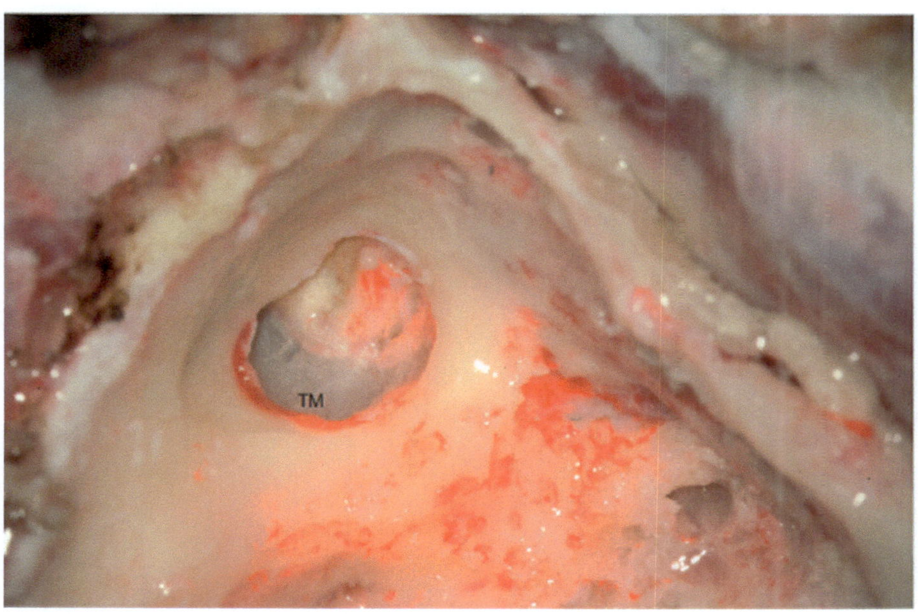

Fig. 9.129 A whitish mass is seen behind the intact tympanic membrane suggesting the congenital nature of the cholesteatoma. *TM*, tympanic membrane.

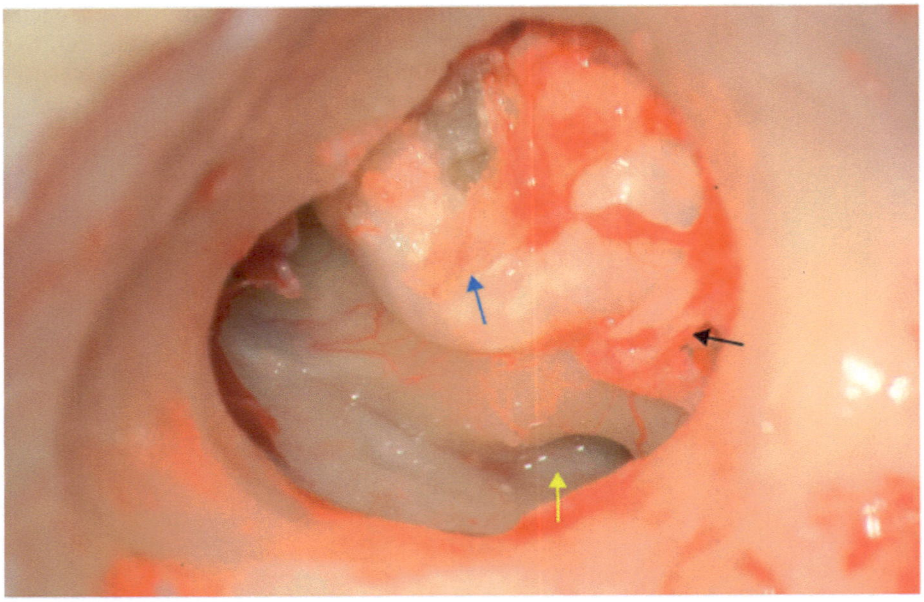

Fig. 9.130 The tympanomeatal flap is totally removed to open the tympanic cavity. The cholesteatoma is seen under the handle of the malleus *(blue arrow)*. The long process of the incus *(black arrow)* and the round window niche *(yellow arrow)* are visualized.

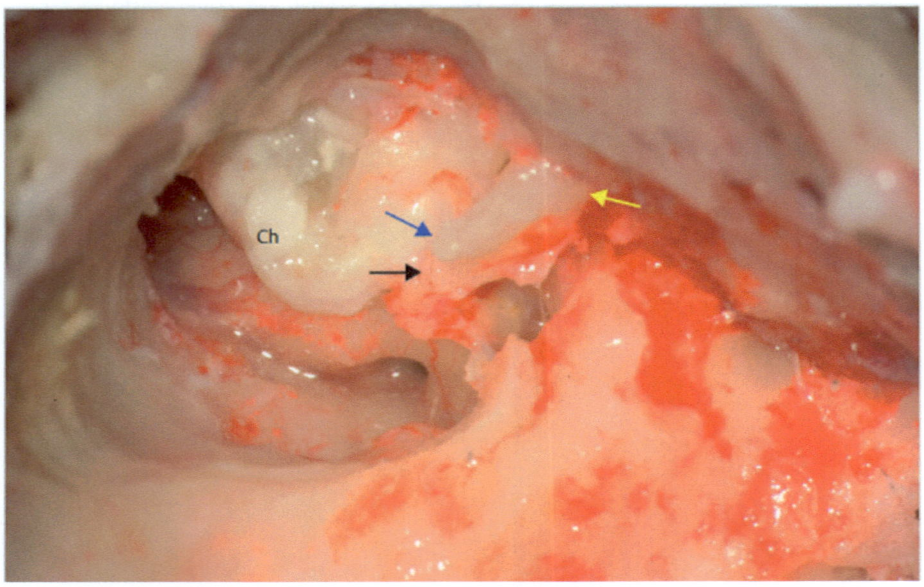

Fig. 9.131 The posterior wall of the meatus is removed, and the attic is partially opened. The incus (*blue arrow*; short process, *yellow arrow*; body, *black arrow*; long process) is totally dislocated as a result of a mistake in the former surgery. *Ch*, cholesteatoma.

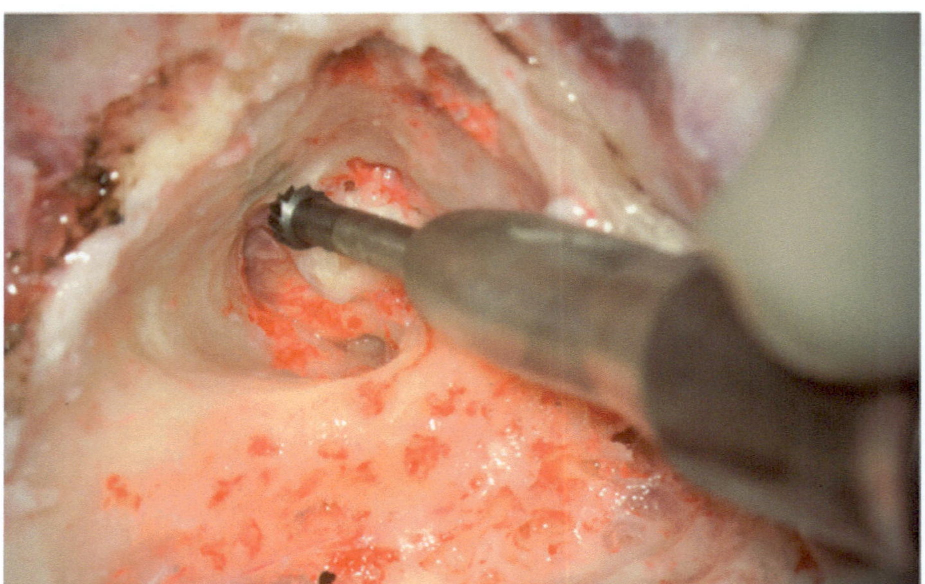

Fig. 9.132 The drilling of the attic is advanced anteriorly to open the supratubal recess.

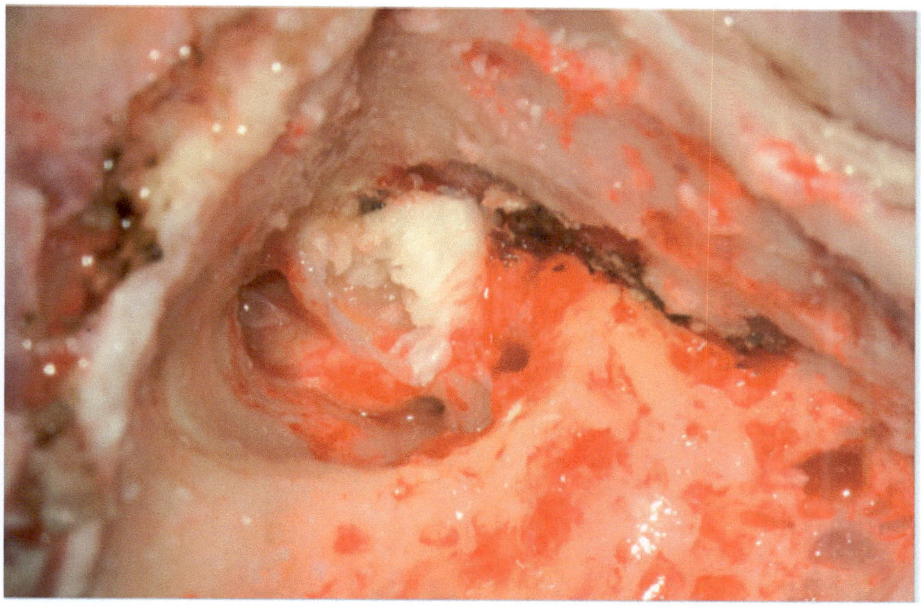

Fig. 9.133 The cholesteatoma eroding the medial wall of the mesotympanum and the attic is shown.

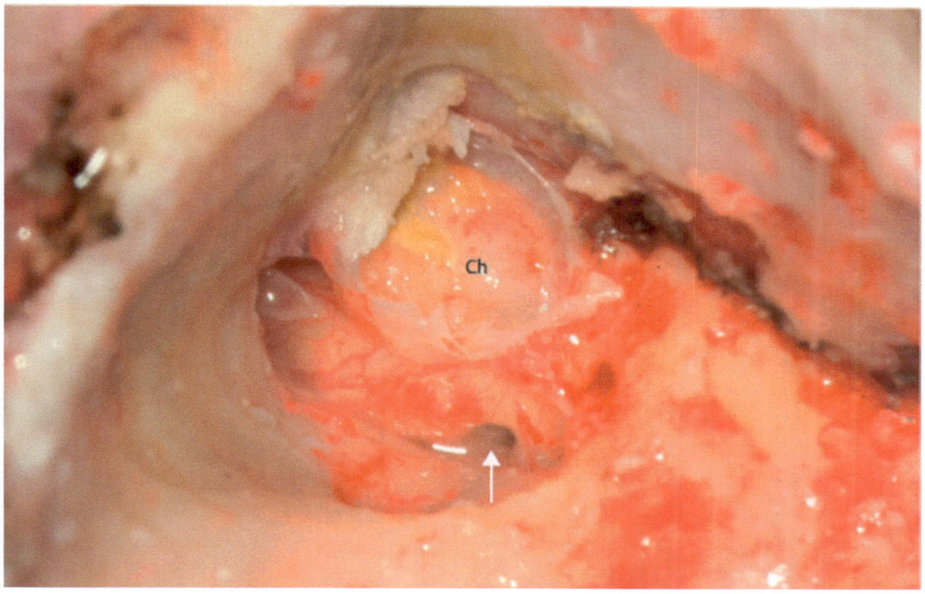

Fig. 9.134 Debris filling the cholesteatoma is partially removed. A thin matrix covering the medial wall eroding the promontory anterior to the round window *(arrow)* is seen. *Ch*, cholesteatoma.

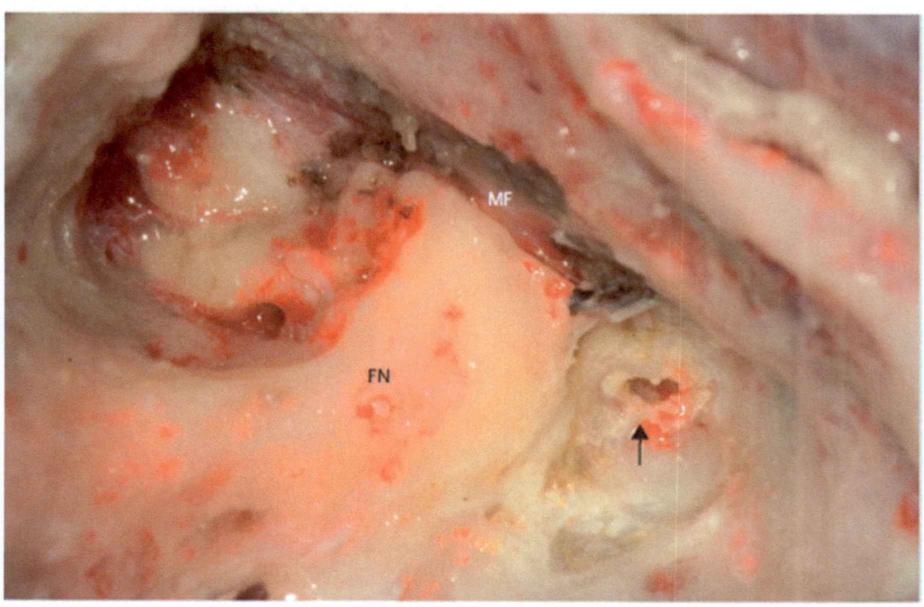

Fig. 9.135 The matrix is removed from the medial wall of the tympanic cavity. Drilling of the semicircular canals revealed extensive ossification of the labyrinth. Cholesteatoma is starting to be exposed *(arrow)* in the perilabyrinthine cells within the superior semicircular canal. As seen, the middle fossa plate in the tegmen of the attic has been removed in the former surgery. *FN*, facial nerve; *MF*, middle fossa dura.

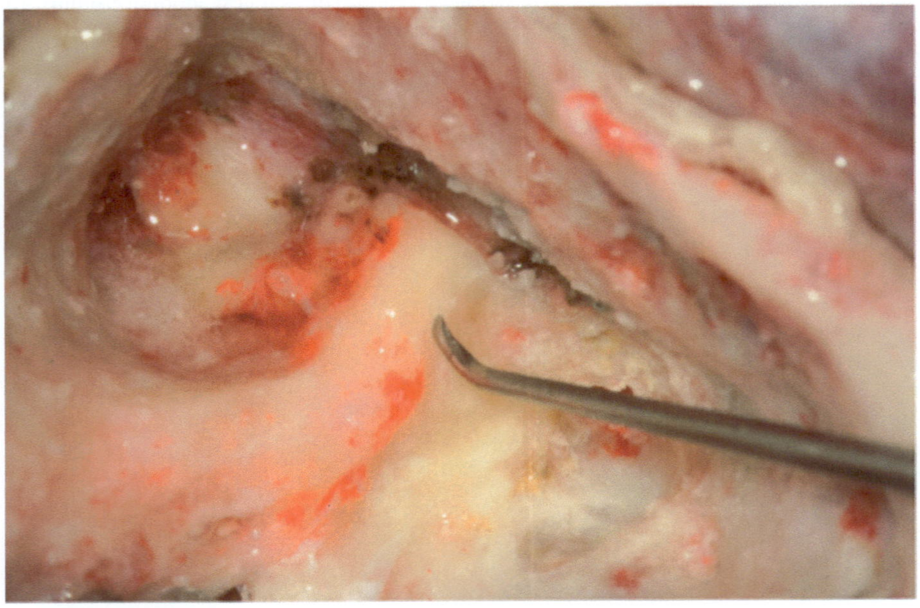

Fig. 9.136 The completely ossified lateral semicircular canal is pointed with the instrument.

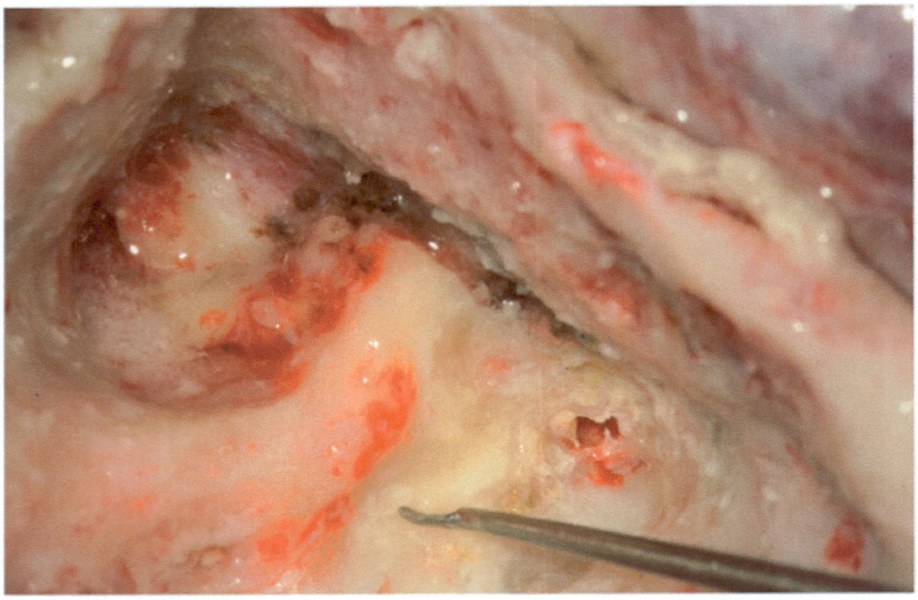

Fig. 9.137 The instrument points to the completely ossified posterior semicircular canal.

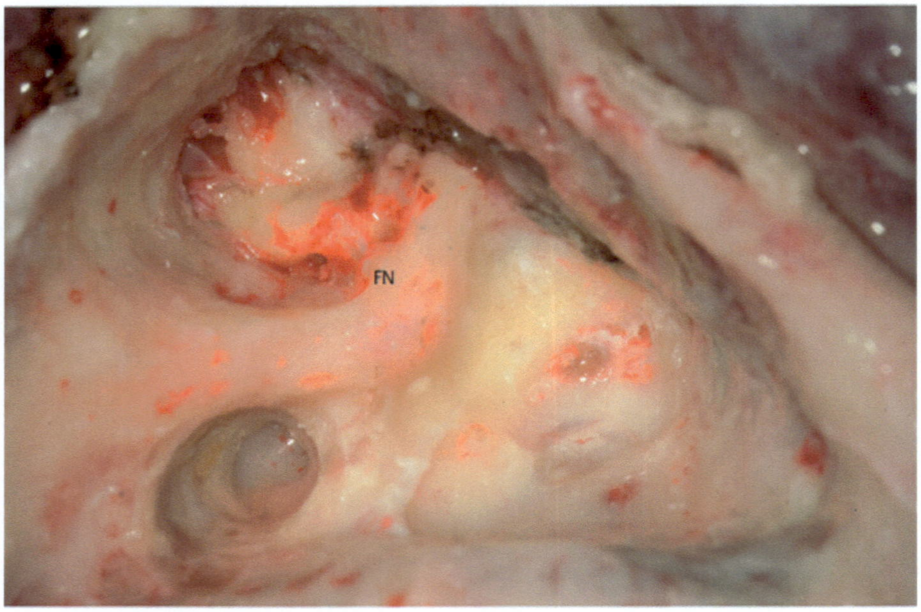

Fig. 9.138 Subfacial tympanotomy to remove the cholesteatoma in hypotympanic cells is carried out, and the cholesteatoma is completely removed. Note the wide opening of the subfacial area connected to the hypotympanum. *FN*, facial nerve.

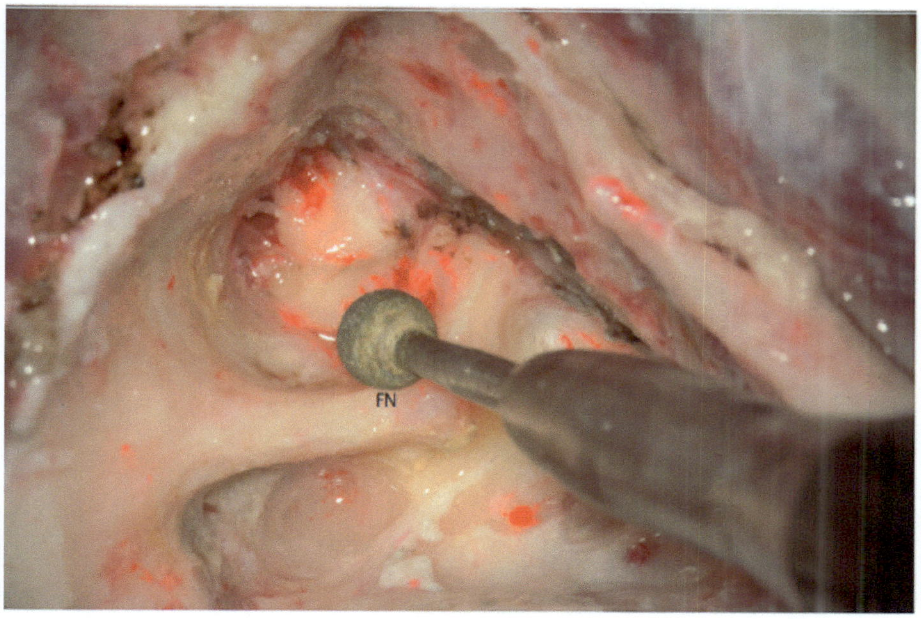

Fig. 9.139 The bone covering the facial nerve is thinned with a large diamond bur.

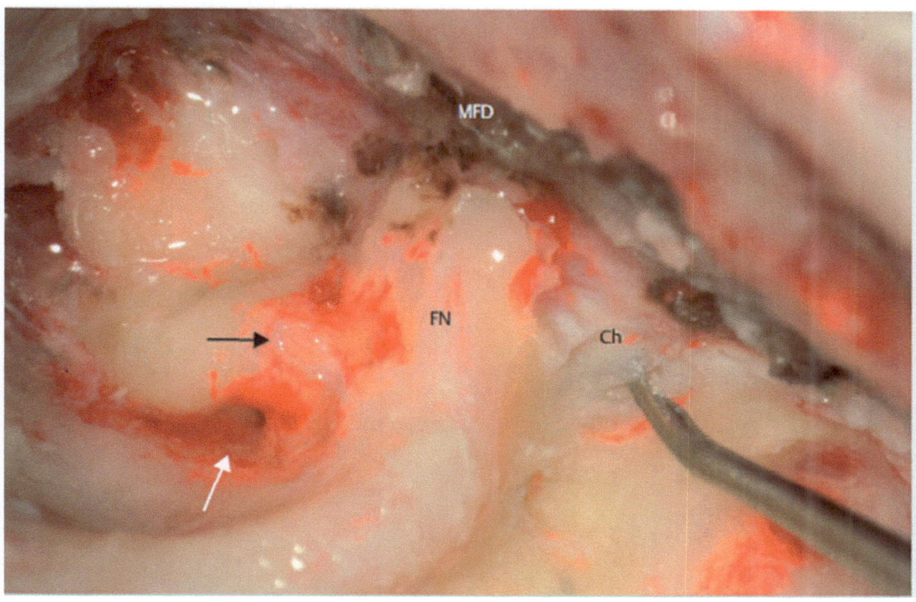

Fig. 9.140 The drilling is advanced anteriorly to remove the medial wall of the attic. The matrix of the cholesteatoma goes deeply along the middle fossa dura. The stapes *(black arrow)* are seen between the facial nerve and the round window niche *(white arrow)*. *Ch*, cholesteatoma; *FN*, facial nerve; *MFD*, middle fossa dura.

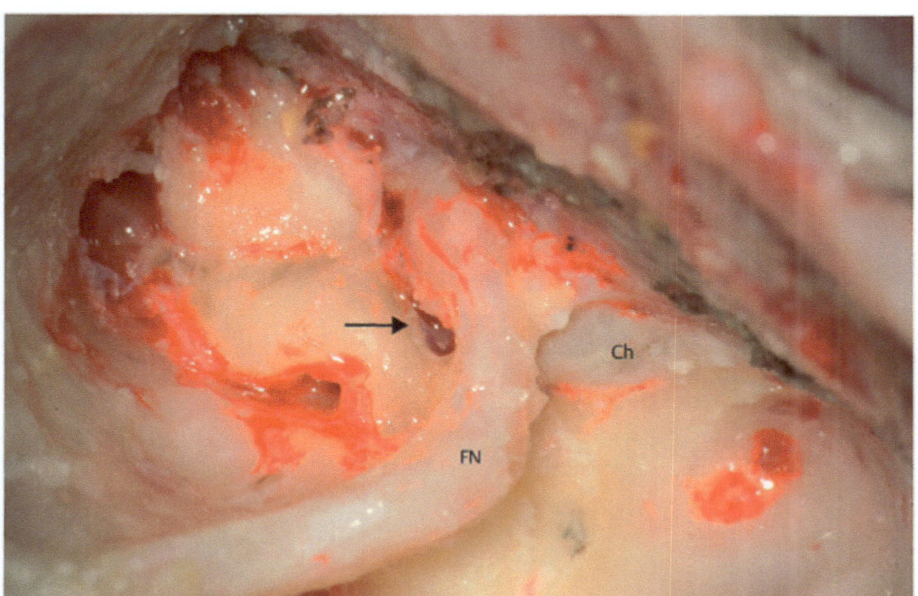

Fig. 9.141 The tympanic segment of the facial nerve is exposed. Note the swelling of the nerve superiorly to the oval window *(arrow)*. *Ch*, cholesteatoma; *FN*, facial nerve.

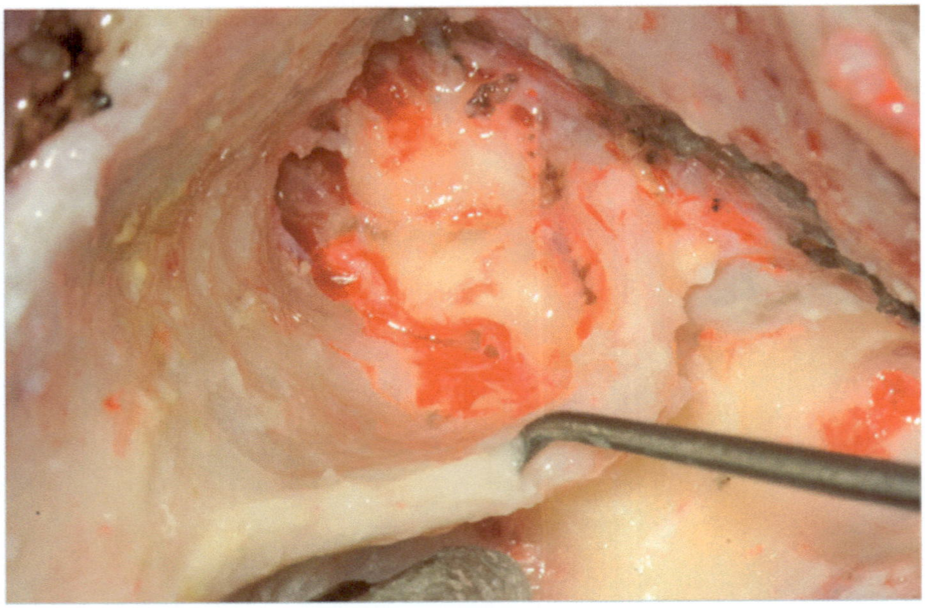

Fig. 9.142 Exposure of the mastoid segment of the facial nerve by removing the thin bony shell is carried out.

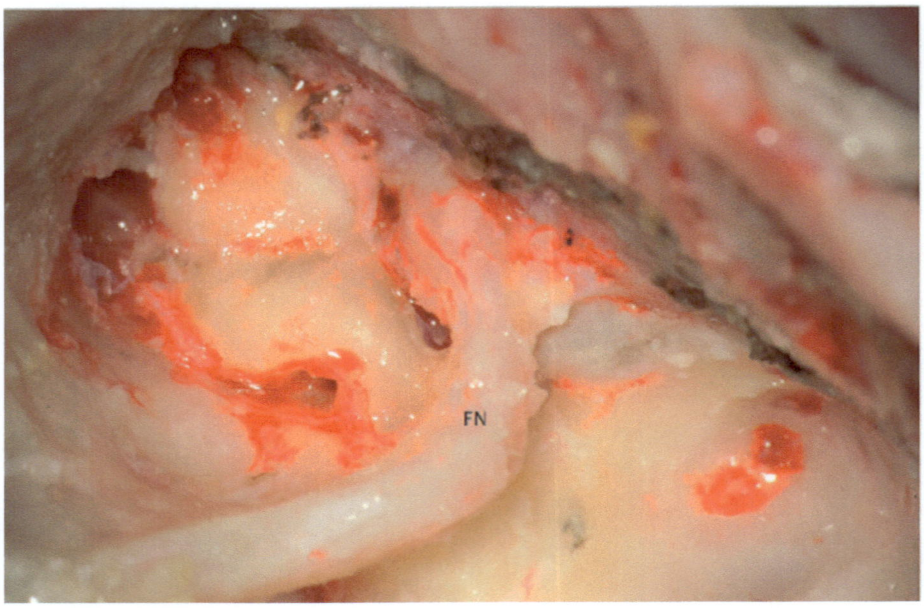

Fig. 9.143 The facial nerve is exposed from the great superficial petrosal nerve to the mastoid segment. *FN*, facial nerve.

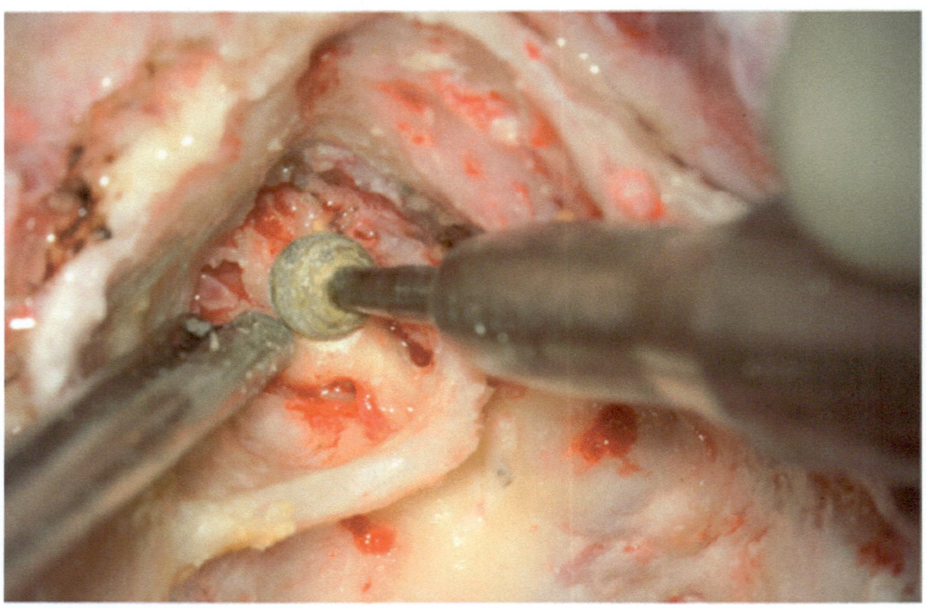

Fig. 9.144 The cochlea is drilled with a large diamond burr.

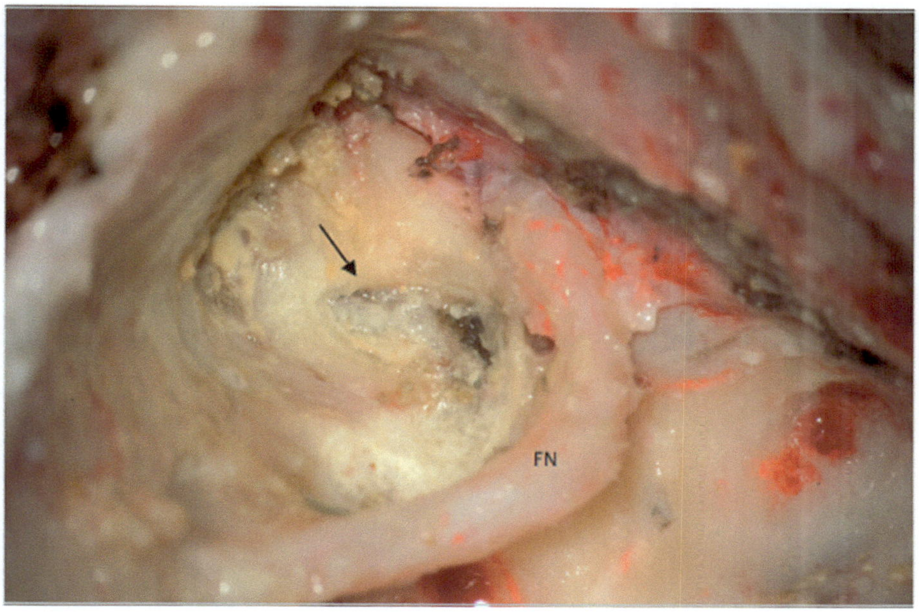

Fig. 9.145 The basal turn of the cochlea filled with fibrous tissue is opened *(arrow)*. *FN*, facial nerve.

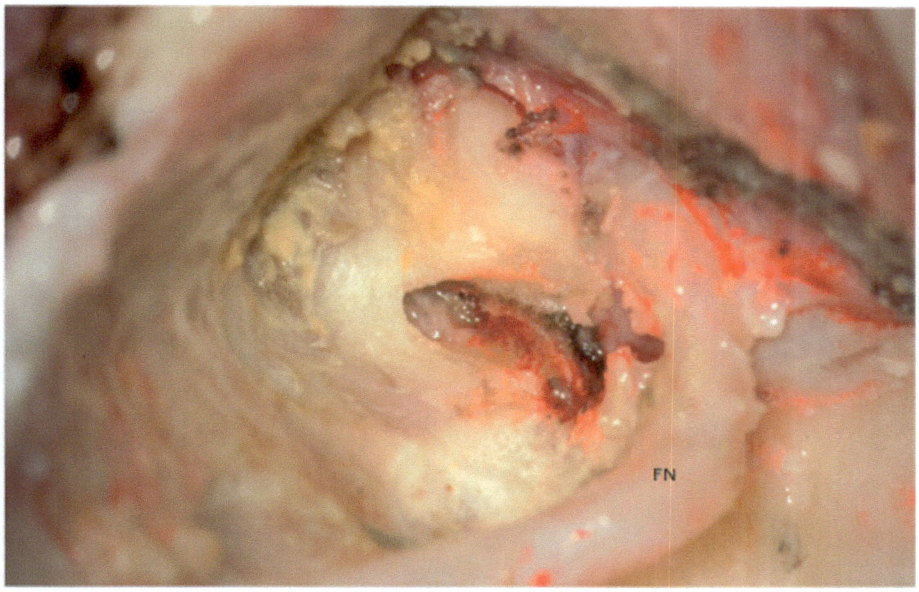

Fig. 9.146 The fibrous tissue filling the basal turn is removed to show the lumen of the turn. *FN*, facial nerve.

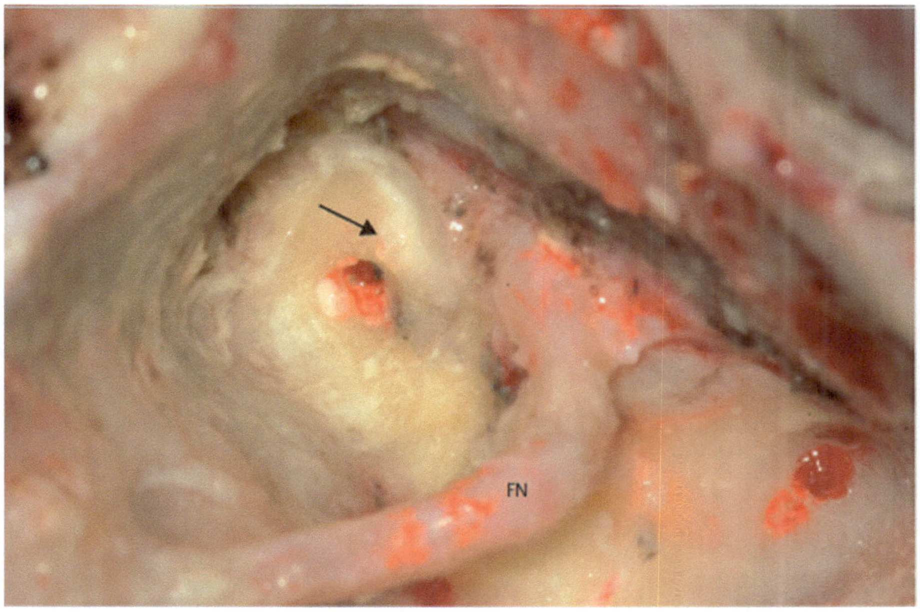

Fig. 9.147 Drilling of the cochlea is advanced superiorly. Complete occlusion of apical two turns with ossification is shown *(arrow)*. *FN*, facial nerve.

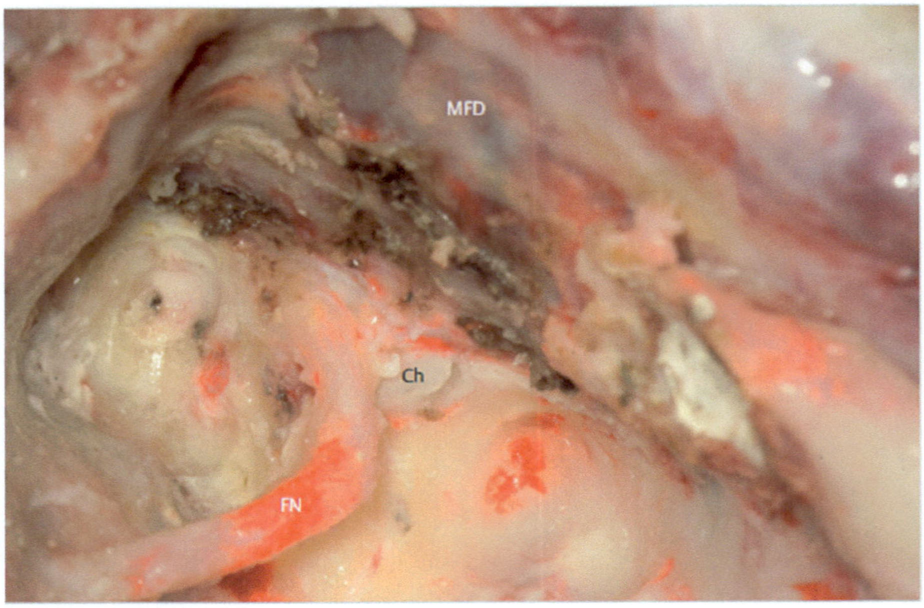

Fig. 9.148 The bone lateral to the formerly exposed middle fossa dura is removed to gain access to the petrous apex. The dura is suspicious that the residual matrix should be carefully coagulated with bipolar. *Ch*, cholesteatoma; *FN*, facial nerve; *MFD*, middle fossa dura.

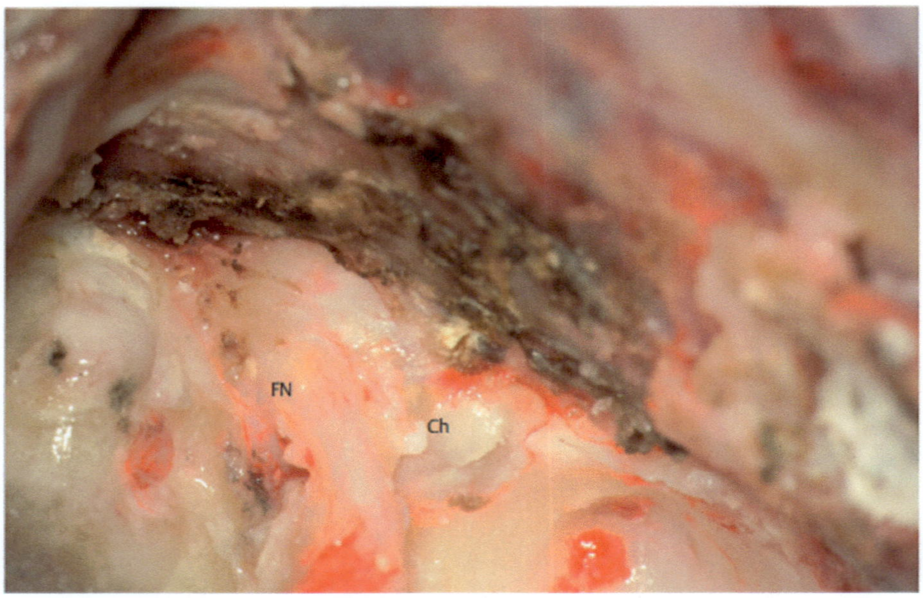

Fig. 9.149 A massive cholesteatoma occupying the area superomedially to the geniculate ganglion is shown. *Ch*, cholesteatoma; *FN*, facial nerve.

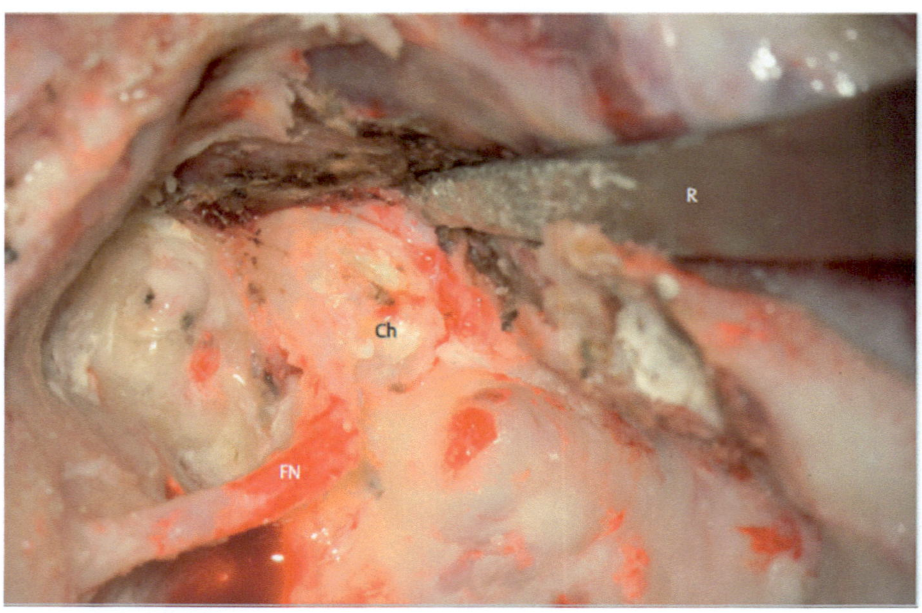

Fig. 9.150 A retractor for the middle fossa approach is applied to maintain good access to the cholesteatoma occupying the supralabyrinthine area. *Ch*, cholesteatoma; *R*, retractor.

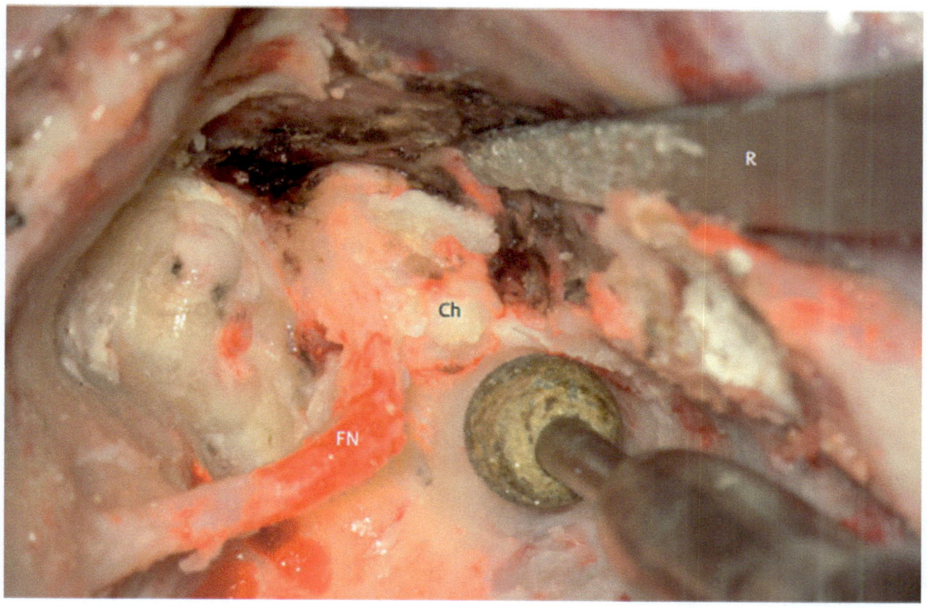

Fig. 9.151 Removal of the bone from the area superiorly to the internal auditory canal is conducted with a large diamond bur. *Ch*, cholesteatoma; *FN*, facial nerve; *R*, middle fossa retractor.

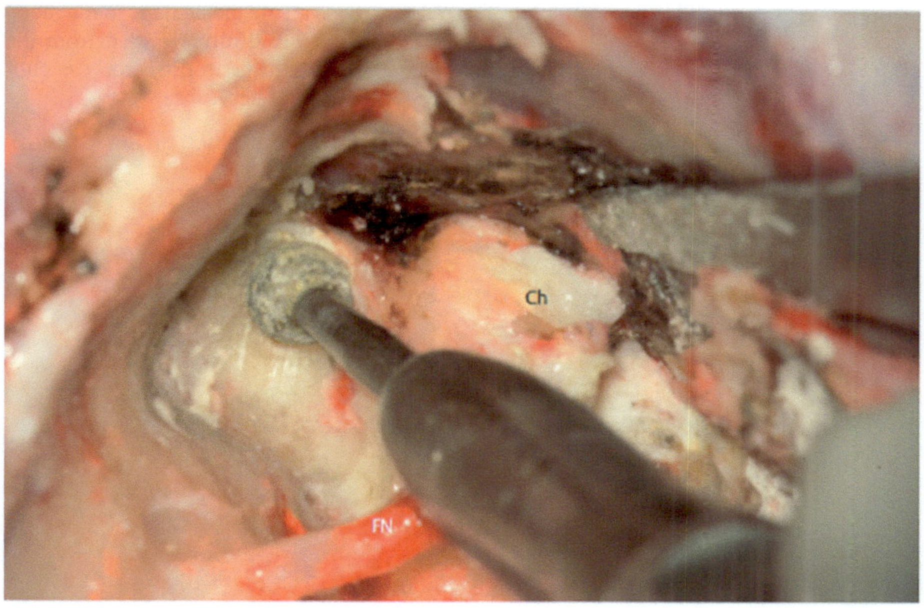

Fig. 9.152 Drilling of the ossified cochlea is carried out. *Ch*, cholesteatoma; *FN*, facial nerve.

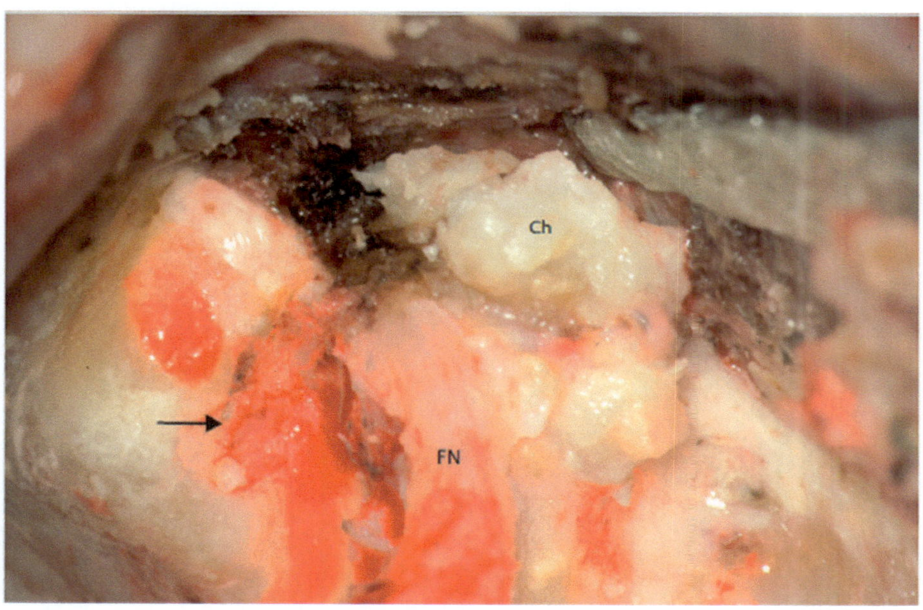

Fig. 9.153 The tissue covering the cholesteatoma filling the area of the cochlea near the fundus is exposed *(arrow)* inferiorly to the facial nerve. *Ch*, cholesteatoma; *FN*, facial nerve.

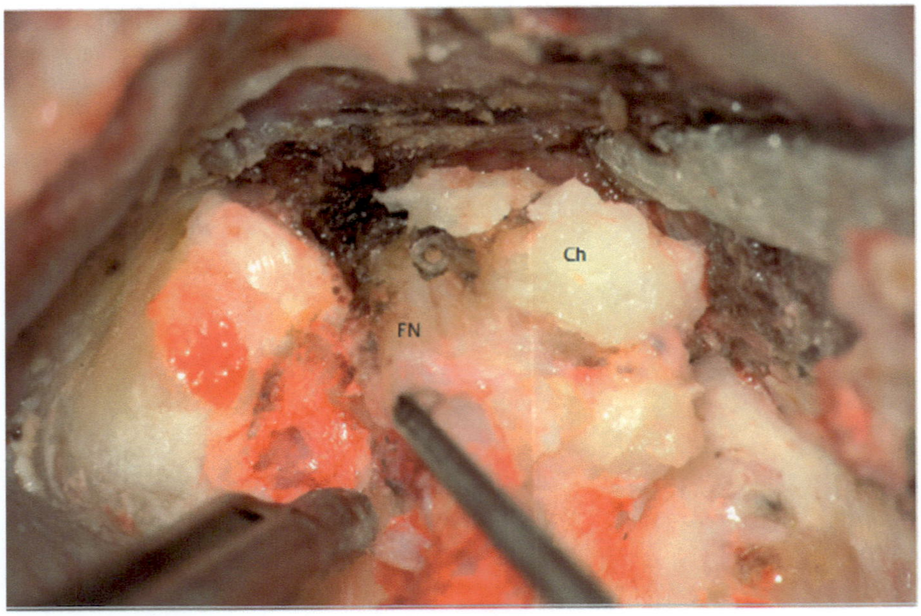

Fig. 9.154 The massive cholesteatoma passing under the facial nerve is shown. *Ch*, cholesteatoma; *FN*, facial nerve.

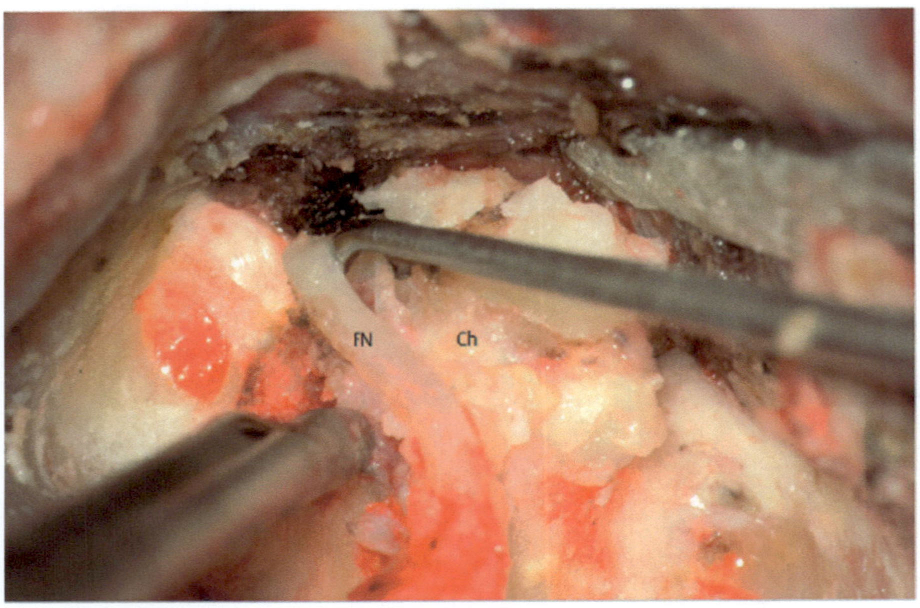

Fig. 9.155 To evaluate the damage to the facial nerve, the nerve is carefully elevated from the fallopian canal with the microdissector. *Ch*, cholesteatoma; *FN*, facial nerve.

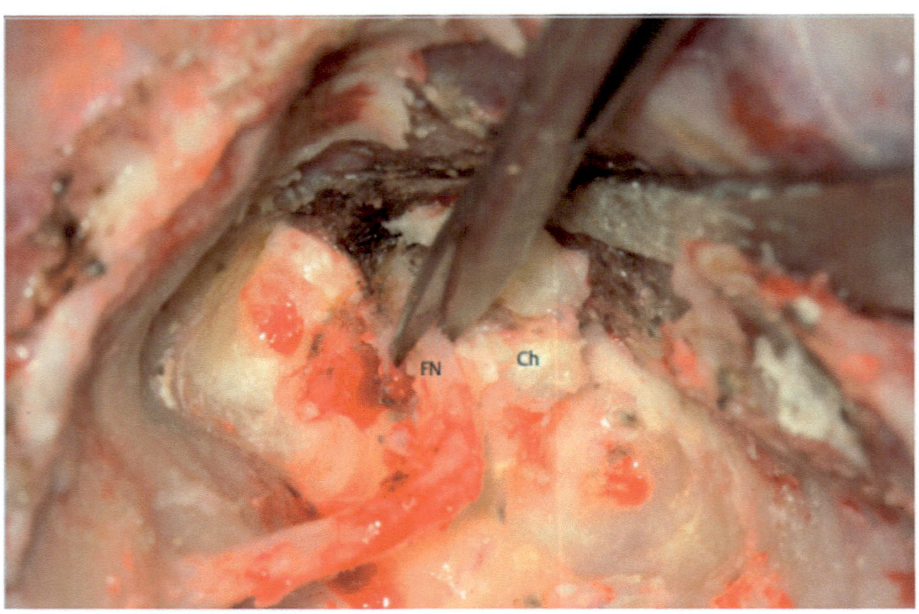

Fig. 9.156 Since the facial nerve is scarred around the geniculate ganglion, a repair with nerve grafting is indicated. The facial nerve is transected at the level of the tympanic segment just superior to the stapes. *Ch*, cholesteatoma; *FN*, facial nerve.

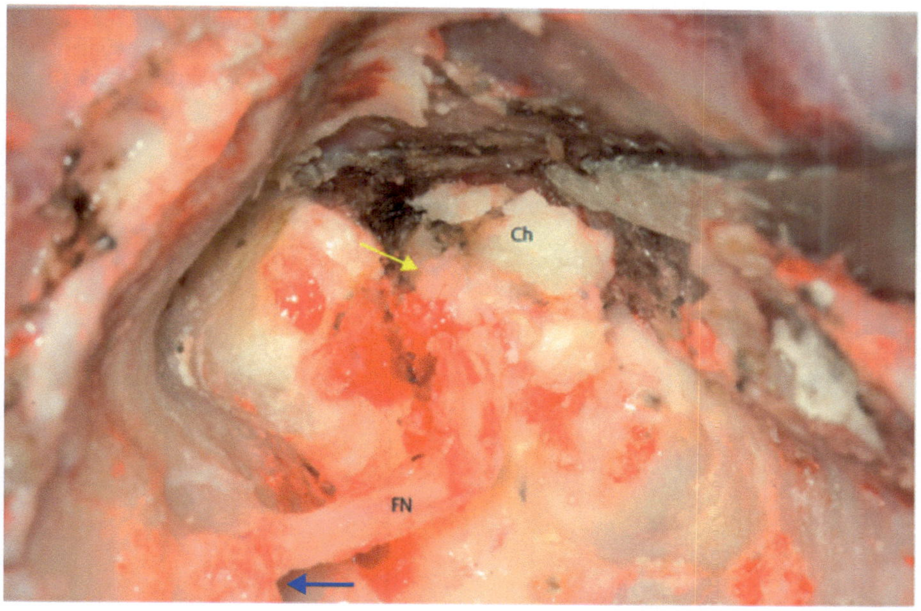

Fig. 9.157 The distal portion of the nerve *(yellow arrow)* is detached from the fallopian canal and reflected posteriorly. The proximal end of the nerve *(blue arrow)* is still in place. *Ch*, cholesteatoma; *FN*, facial nerve.

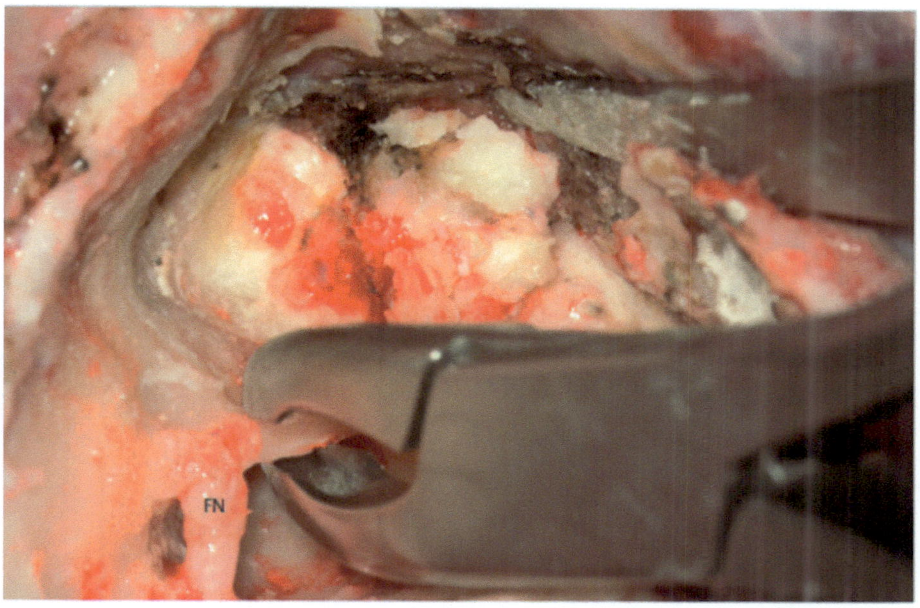

Fig. 9.158 The bone under the fallopian canal is removed with a rongeur. *FN*, facial nerve.

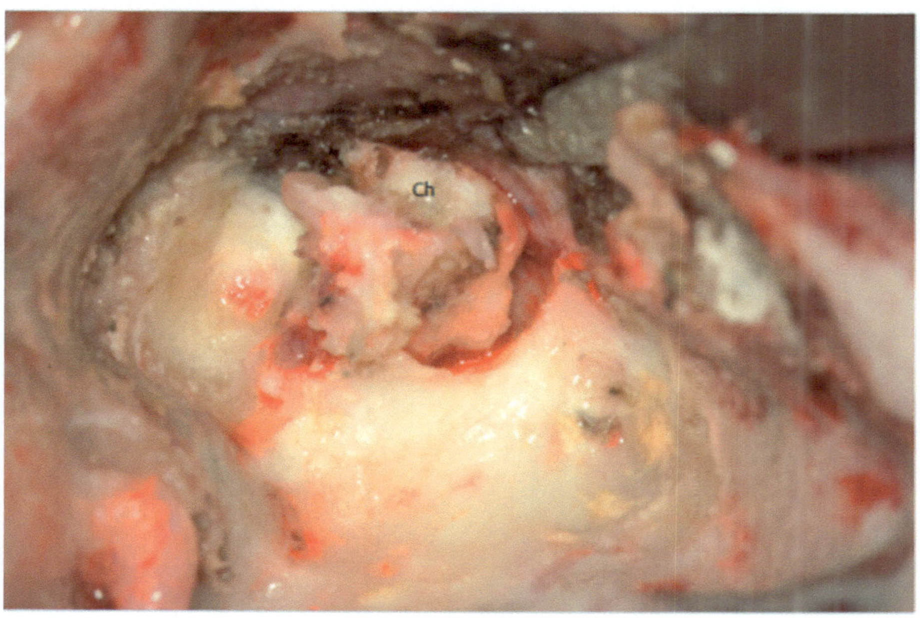

Fig. 9.159 Transection of the facial nerve and removal of the bone under the nerve create wide access to the cholesteatoma. Dissection of cholesteatoma from the middle fossa dura is advanced medially. *Ch*, cholesteatoma.

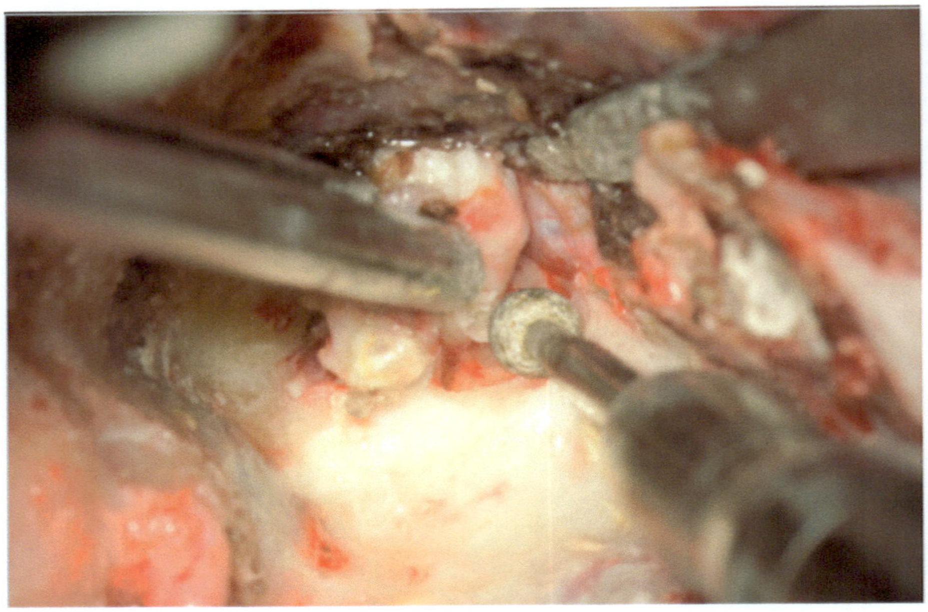

Fig. 9.160 The bone impeding access to the cholesteatoma is removed, and the contents of the internal auditory canal are started to be exposed.

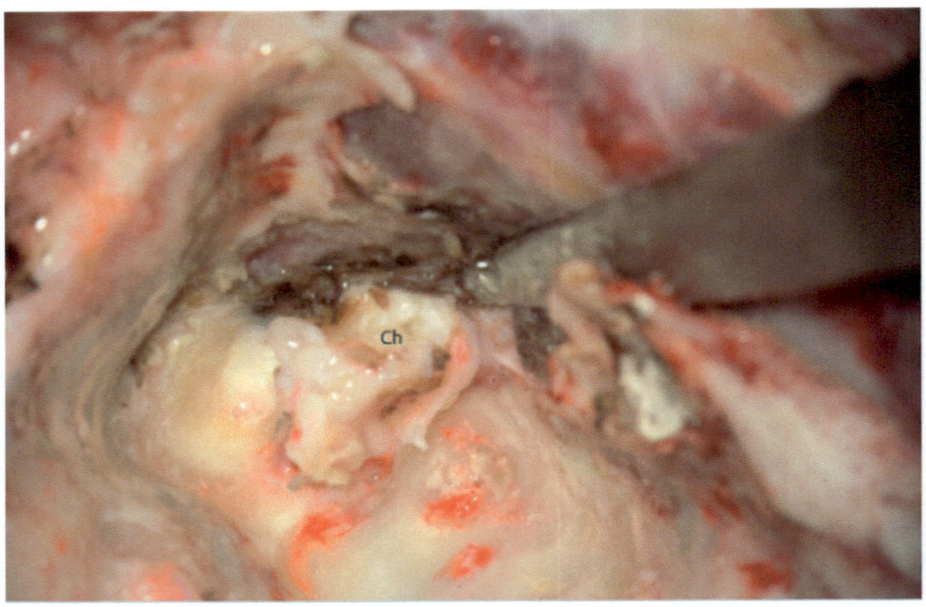

Fig. 9.161 The matrix and granulation tissue covering the internal auditory canal is shown. *Ch*, cholesteatoma.

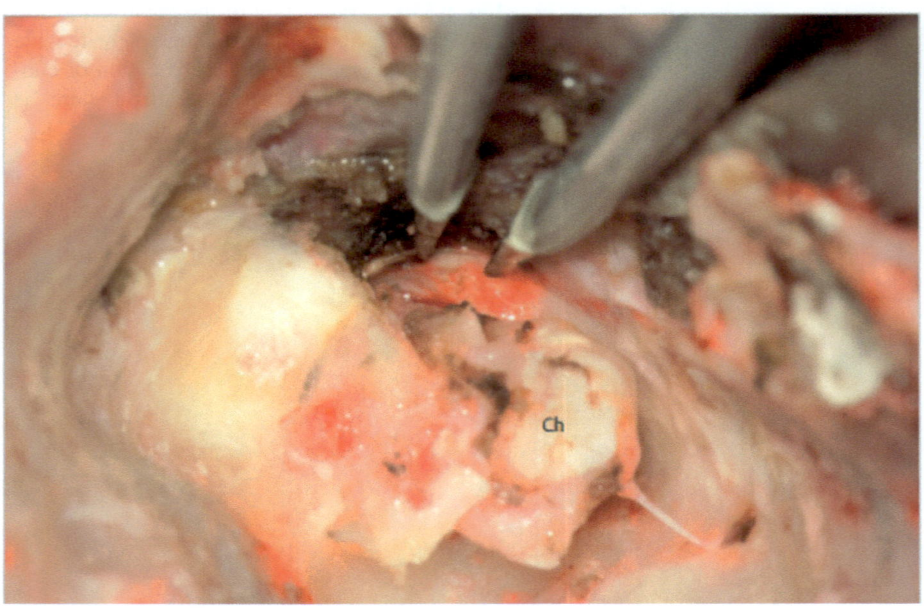

Fig. 9.162 The middle fossa dura needs to be coagulated intensively due to the intimate attachment of the cholesteatoma matrix. *Ch*, cholesteatoma.

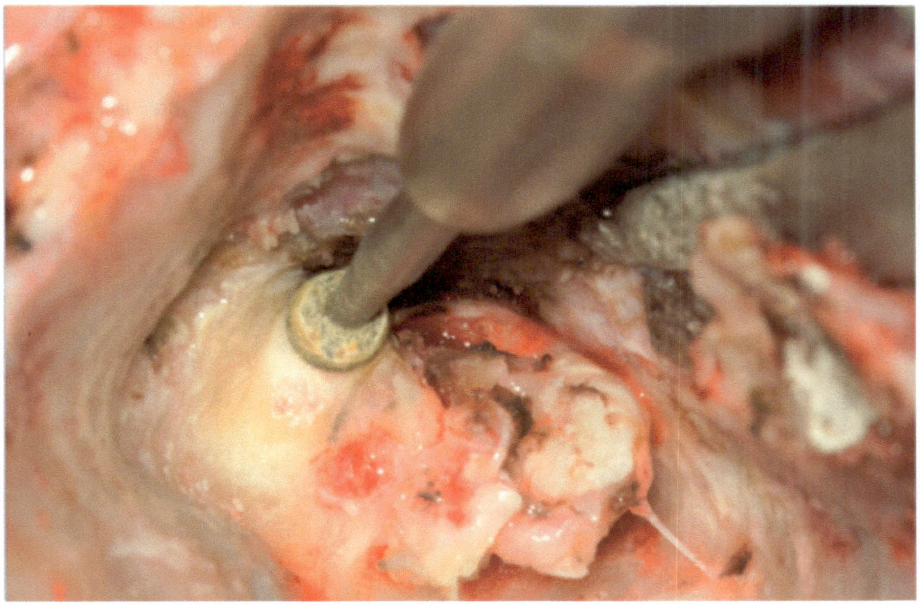

Fig. 9.163 The bone anterior to the internal auditory canal is removed to expose the most medial part of the cholesteatoma.

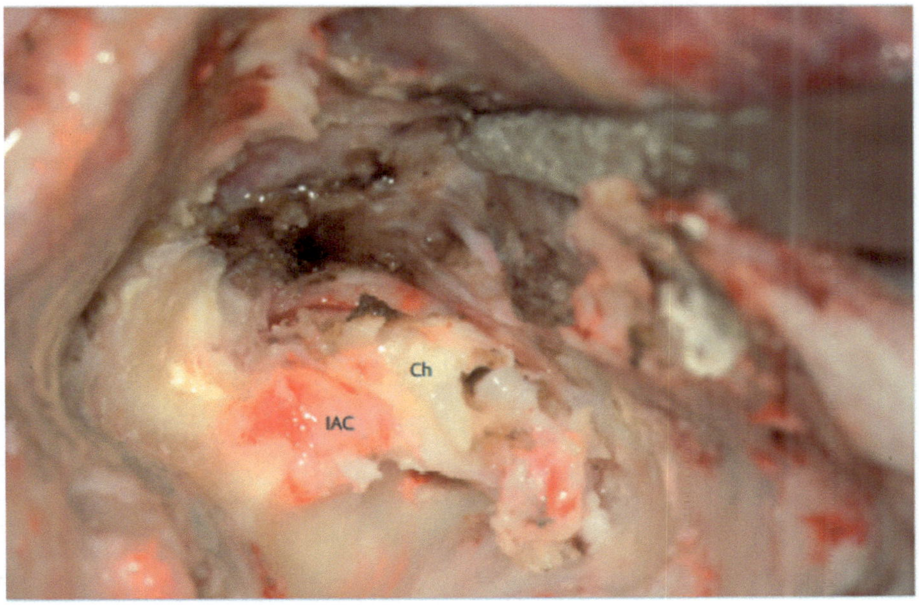

Fig. 9.164 The drilling exposed a cholesteatoma eroding the superior face of the internal auditory canal and invaginating into it is shown. *Ch*, cholesteatoma; *IAC*, internal auditory canal with nerves.

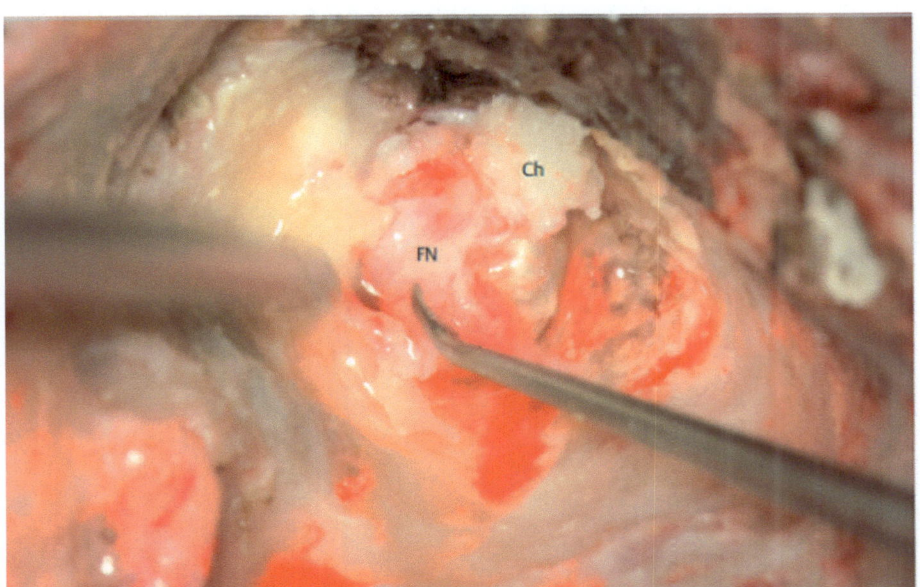

Fig. 9.165 The inferior aspect of the internal auditory canal, covered with periosteum, is detached from the bony wall. Note the cholesteatoma inside the internal auditory canal compressing the contents of it. *Ch*, cholesteatoma; *FN*, facial nerve.

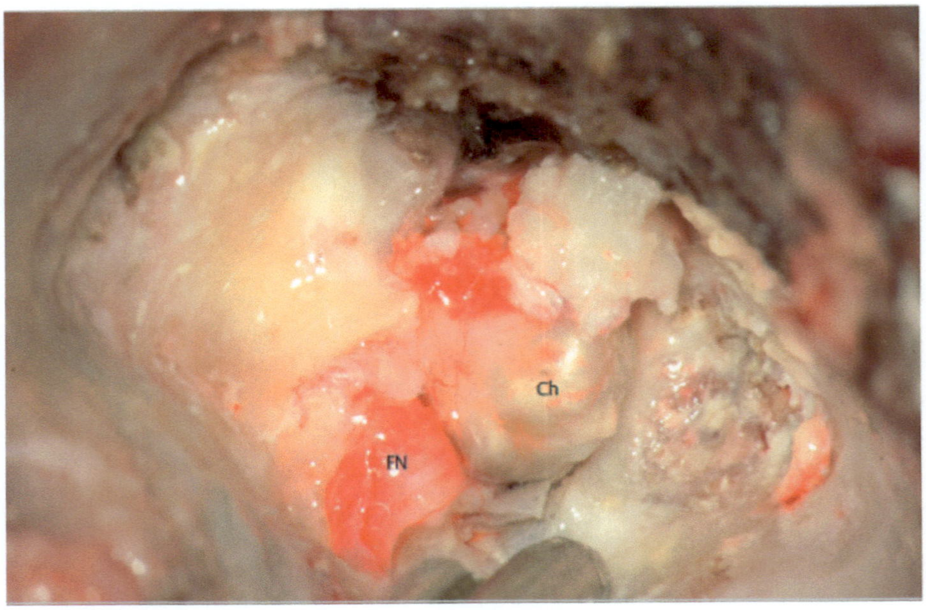

Fig. 9.166 The most medial part of the cholesteatoma invaginating into the internal auditory canal is exposed. The lumpy glazing and whitish appearance are characteristics of the intradural component of the cholesteatoma or epidermoid cyst of the cerebellopontine angle. *Ch*, cholesteatoma; *FN*, facial nerve.

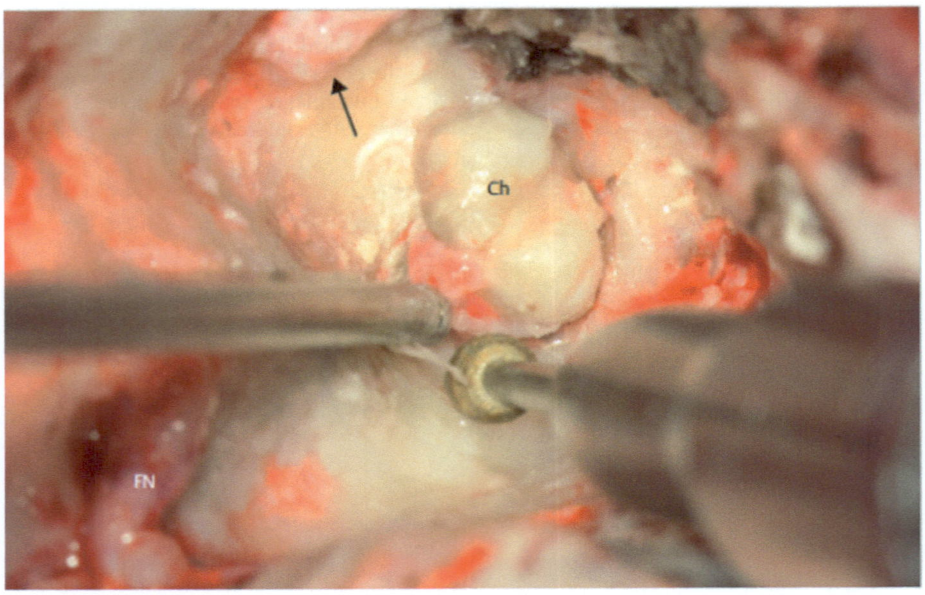

Fig. 9.167 After closing the eustachian tube with pieces of connective tissue *(arrow)*, a groove for a new fallopian canal to accommodate the reanimated facial nerve is created with a diamond bur. The most medial part of the cholesteatoma is left in place to avoid bone dust falling down to the cerebellopontine angle. *Ch*, cholesteatoma; *FN*, facial nerve, distal portion.

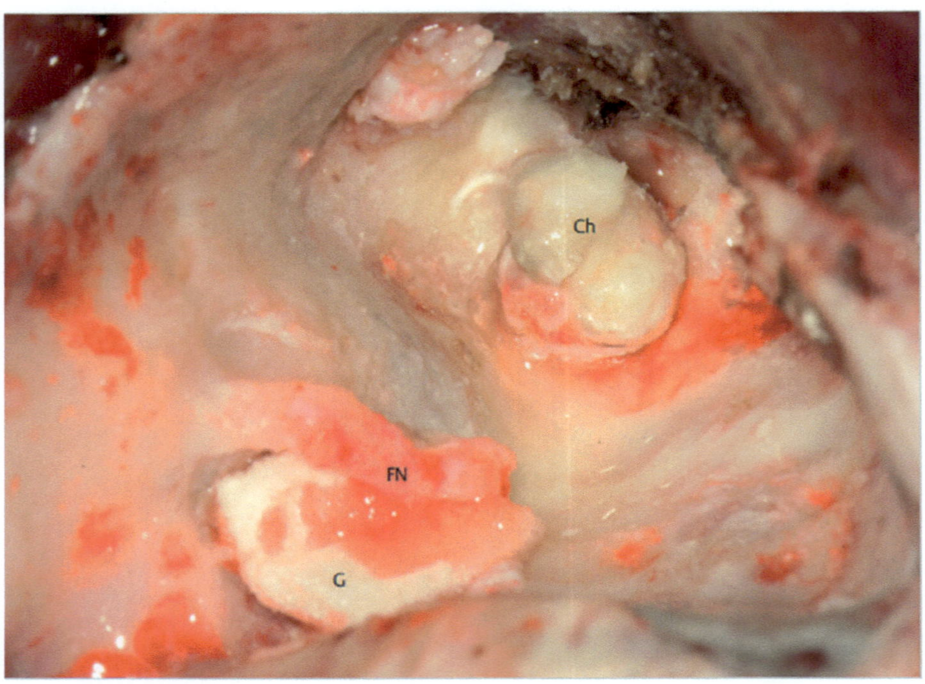

Fig. 9.168 The distal portion of the facial nerve is secured with a large piece of Gelfoam. *Ch*, cholesteatoma; *FN*, facial nerve; *G*, Gelfoam.

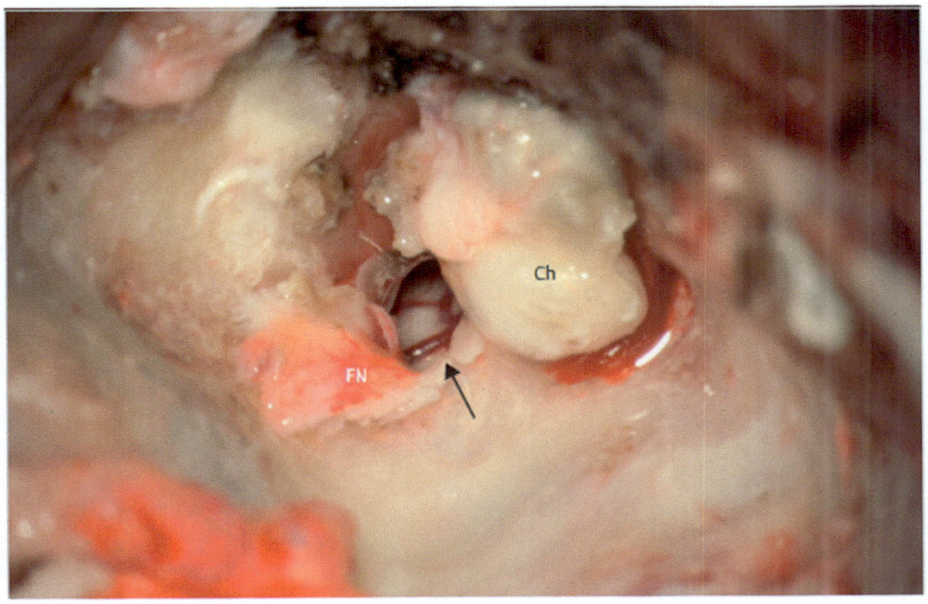

Fig. 9.169 The last part of the cholesteatoma is dissected from the internal auditory canal. The cerebellopontine angle is seen *(arrow)* through the porus of the internal auditory canal. *Ch*, cholesteatoma; *FN*, facial nerve.

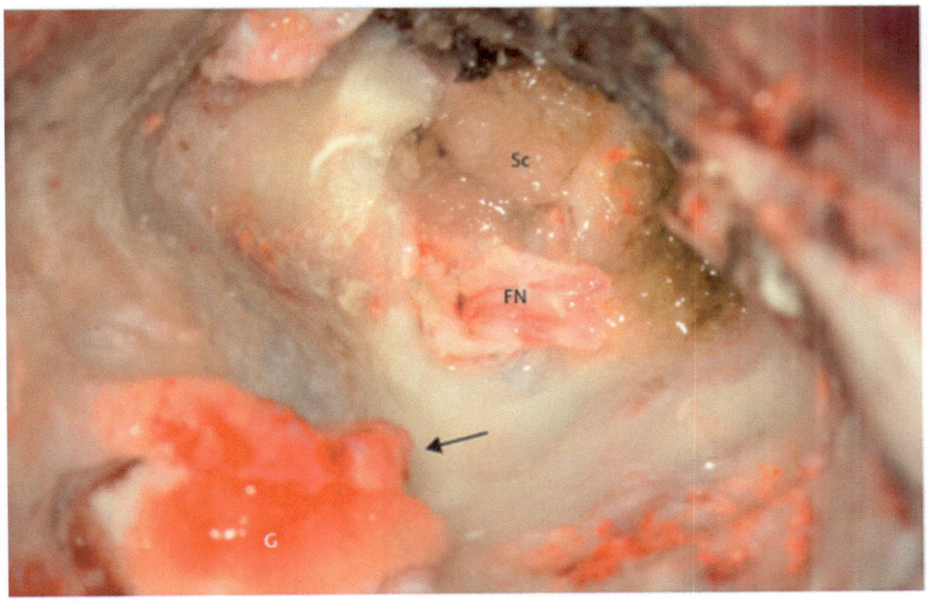

Fig. 9.170 The cholesteatoma is totally removed. The position of the facial nerve is secured by gently packing the porus of the internal auditory canal with periosteum to secure the position of the proximal end of the facial nerve *(arrow)*, so as not to fall down into the cerebellopontine (CP) angle. The distal stump of the facial nerve is refreshed and ready for reconstruction *(arrow)*. *FN*, facial nerve; *G*, Gelfoam; *Sc*, Surgicel.

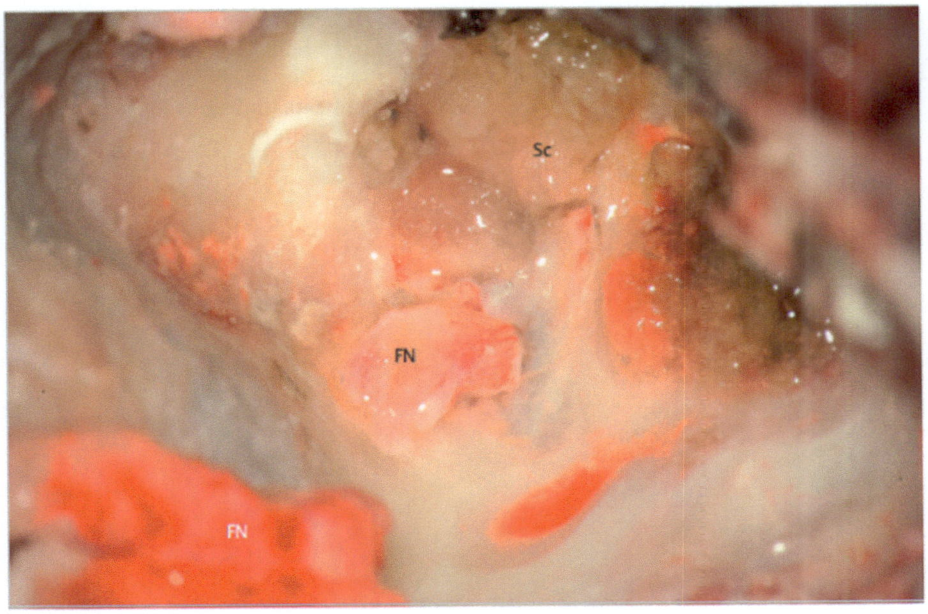

Fig. 9.171 Enlarged view. The proximal stump of the facial nerve is shown. *FN*, facial nerve; *Sc*, Surgicel.

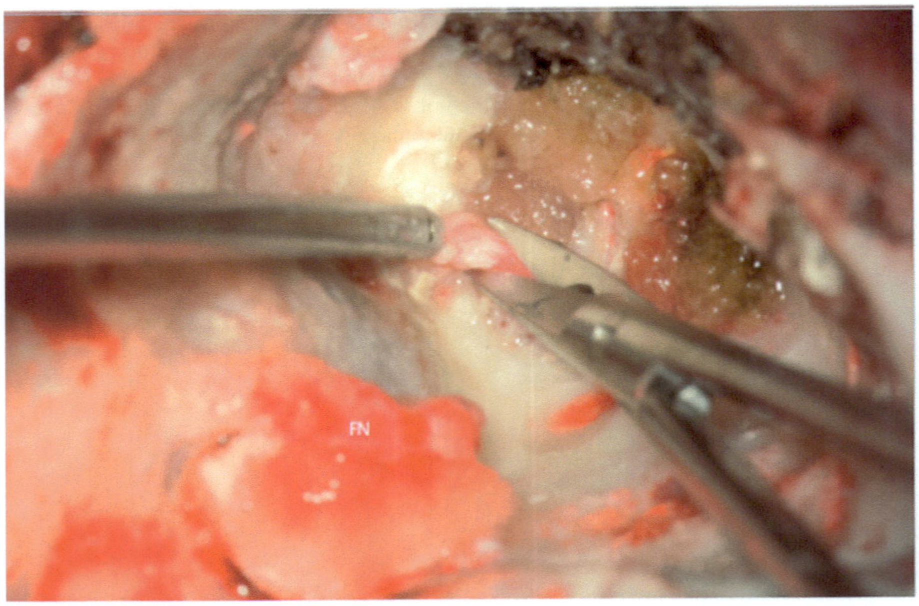

Fig. 9.172 The proximal stump of the facial nerve is refreshed with sharp scissors. *FN*, facial nerve.

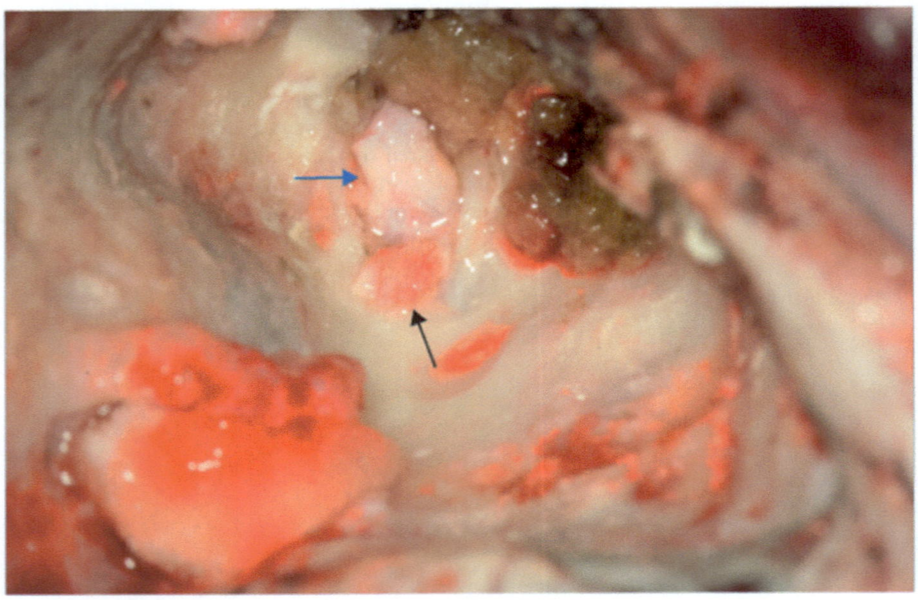

Fig. 9.173 To fix the proximal end of the facial nerve *(black arrow)*, a piece of connective tissue *(blue arrow)* is applied over the nerve.

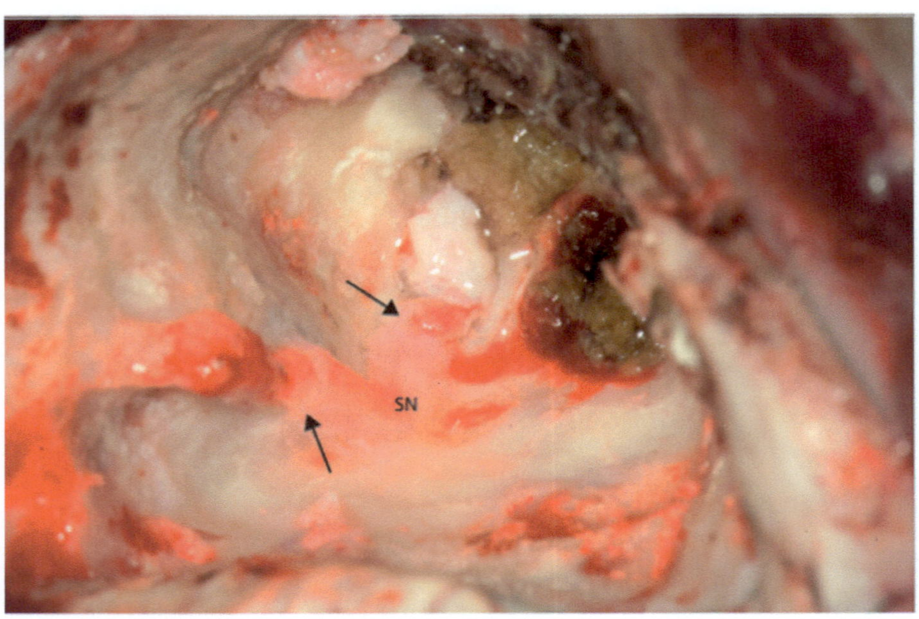

Fig. 9.174 A sural nerve is used as a grafting material. After intimately approximating the graft to both ends of the facial nerve *(arrows)*, the graft is fixed with fibrin glue to the bony wall. *SN*, sural nerve graft.

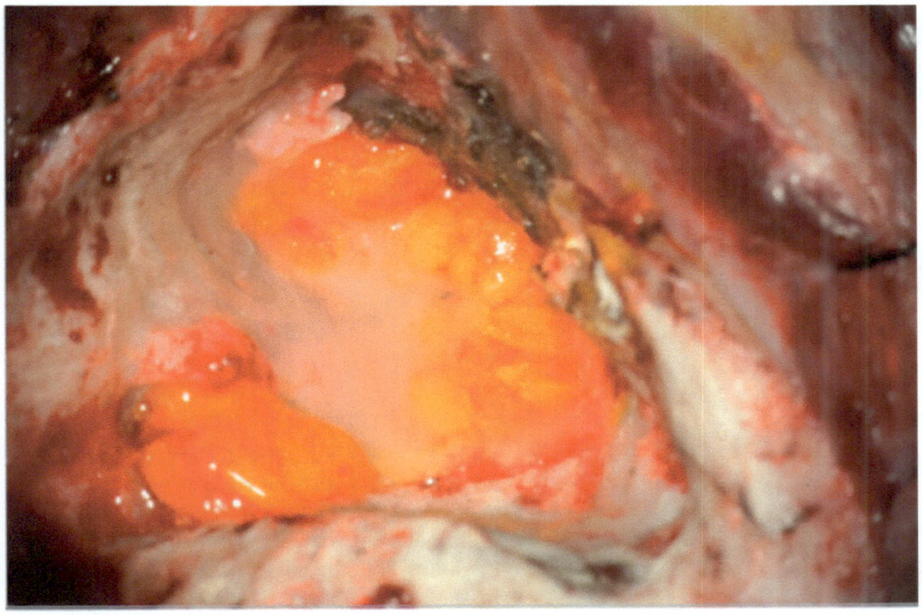

Fig. 9.175 The cavity is packed with abdominal fat.

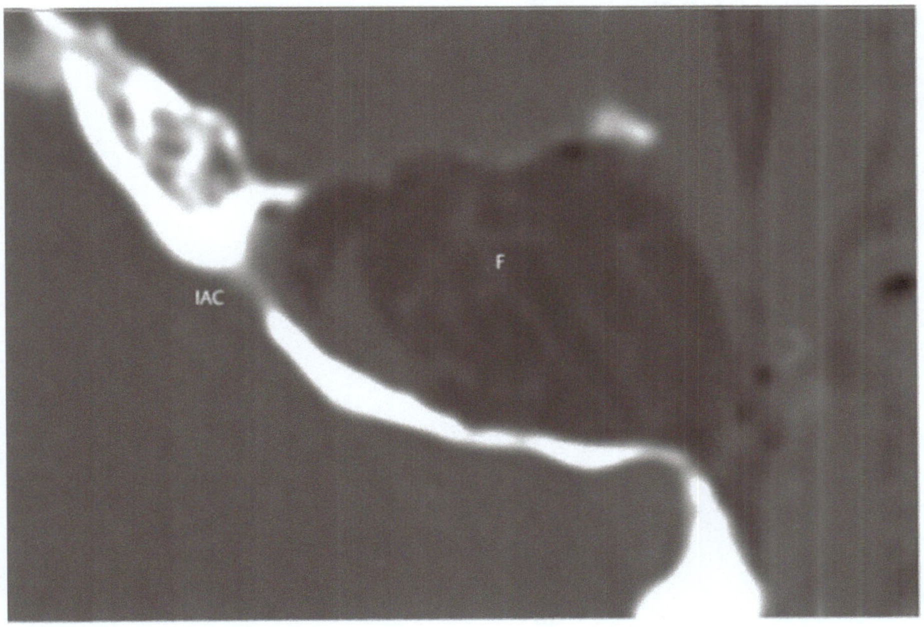

Fig. 9.176 The postoperative CT demonstrates complete removal of the labyrinth. Posterior fossa dura is not uncovered in this case due to anterosuperior localization of the pathology. *F*, abdominal fat; *IAC*, internal auditory canal.

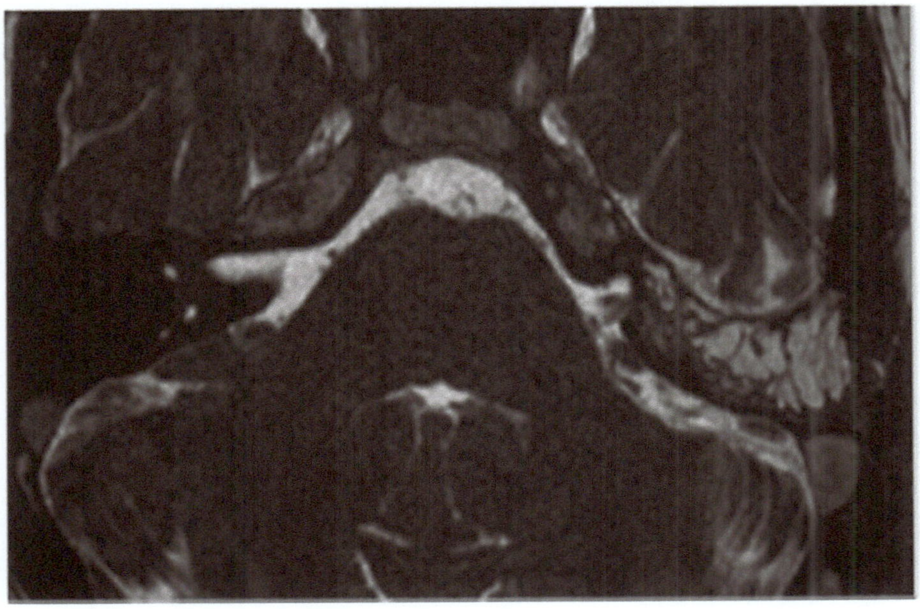

Fig. 9.177 Postoperative MRI shows extensive removal of bone from the temporal bone.

Case 5: Facial Nerve Neuroma Removed Through a Combined TO and Subtemporal Craniotomy Approach

See Figs. 9.178–9.192.

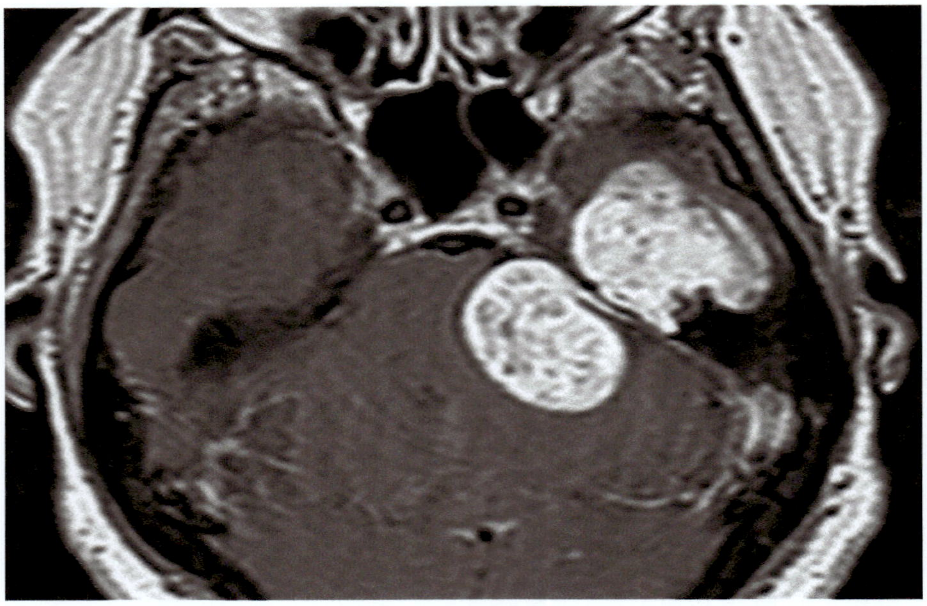

Fig. 9.178 Brain MRI (T1W+Gd enhancement) in a 50-year-old patient with a 6-month history of complete left facial paralysis. The examination revealed a dumbbell-shaped tumor involving both the posterior fossa and the middle fossa.

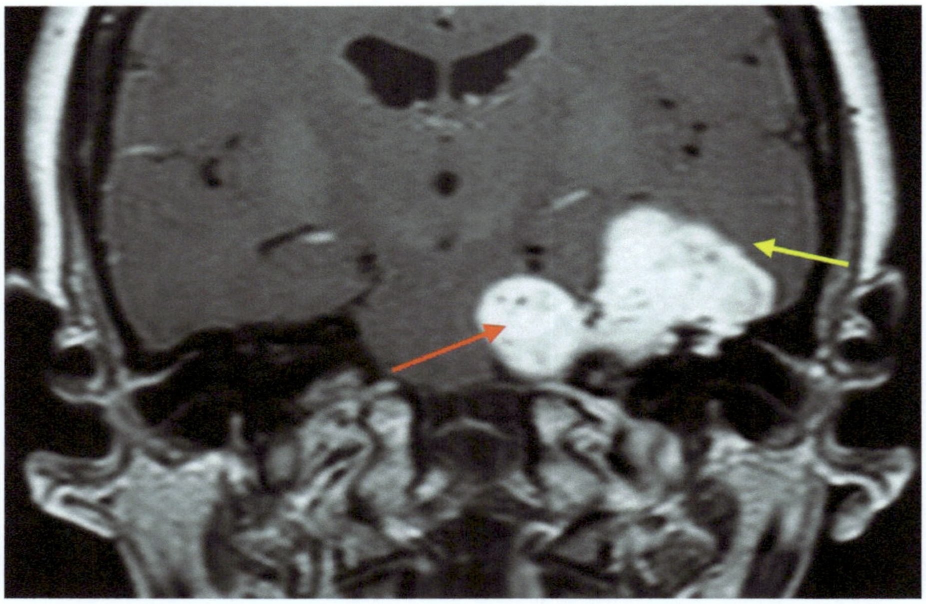

Fig. 9.179 Same patient, coronal view. Note the tumor compression of the temporal lobe *(yellow arrow)* due to the supratentorial extension and the brainstem *(red arrow)*.

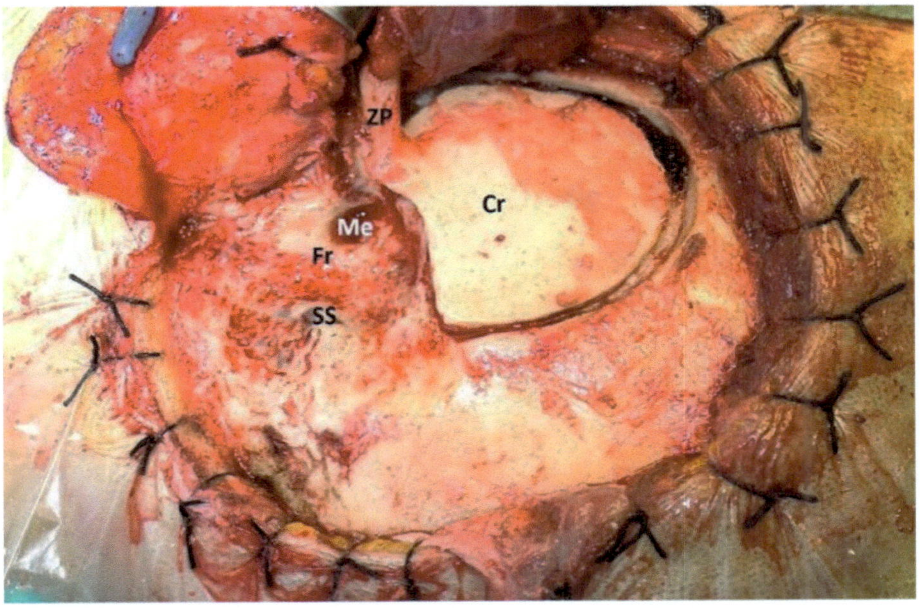

Fig. 9.180 Note the extent of the surgical approach to allow control of both the middle and posterior fossa. Mastoidectomy and temporal craniotomy (Cr) have been performed. *ZP*, zygomatic process; *SS*, sigmoid sinus; *Fr*, facial ridge; *Me*, middle ear.

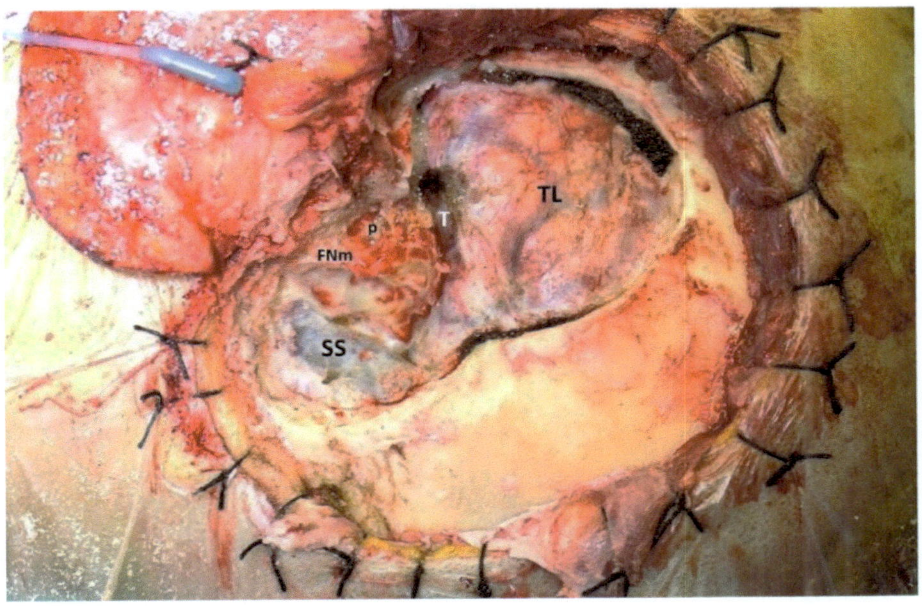

Fig. 9.181 The craniotomy plug has been removed, exposing the temporal lobe (TL). Note the tumor (T) extending to the middle fossa. *SS*, sigmoid sinus; *FNm*, mastoid portion of the facial nerve; *p*, promontory.

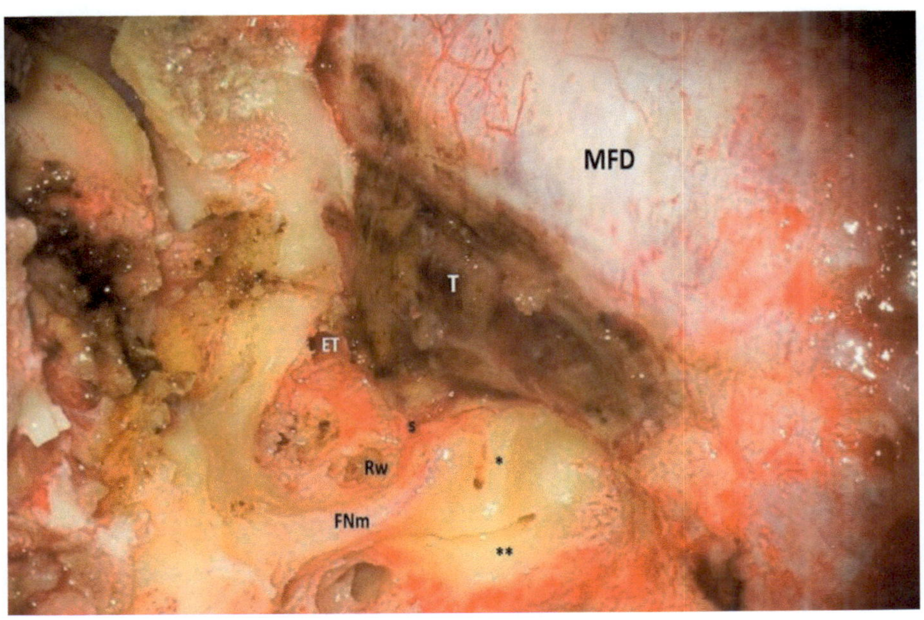

Fig. 9.182 Closer view. The labyrinthectomy was started by opening the lateral semicircular canal (*) and the posterior semicircular canal (**). *MFD*, middle fossa dura; *T*, tumor; *ET*, eustachian tube; *s*, stapes; *Rw*, round window.

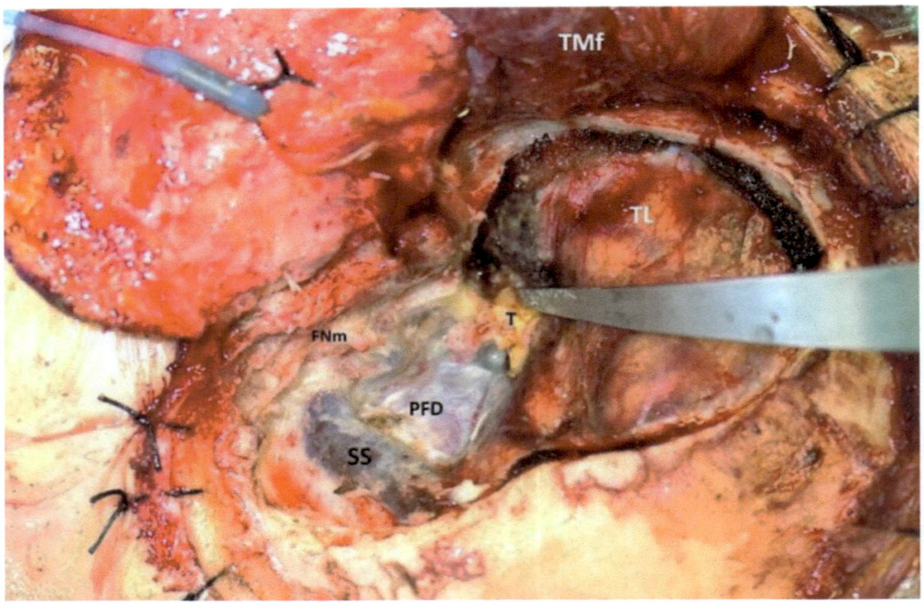

Fig. 9.183 The temporal lobe (TL) has been elevated with a middle fossa retractor, exposing the tumor (T). Note the decompression of both the sigmoid sinus (SS) and the posterior fossa dura (PFD). The facial nerve has been cut at the mastoid segment for subsequent anastomosis. *FNm*, mastoid segment of the facial nerve; *TMf*, temporalis muscle flap.

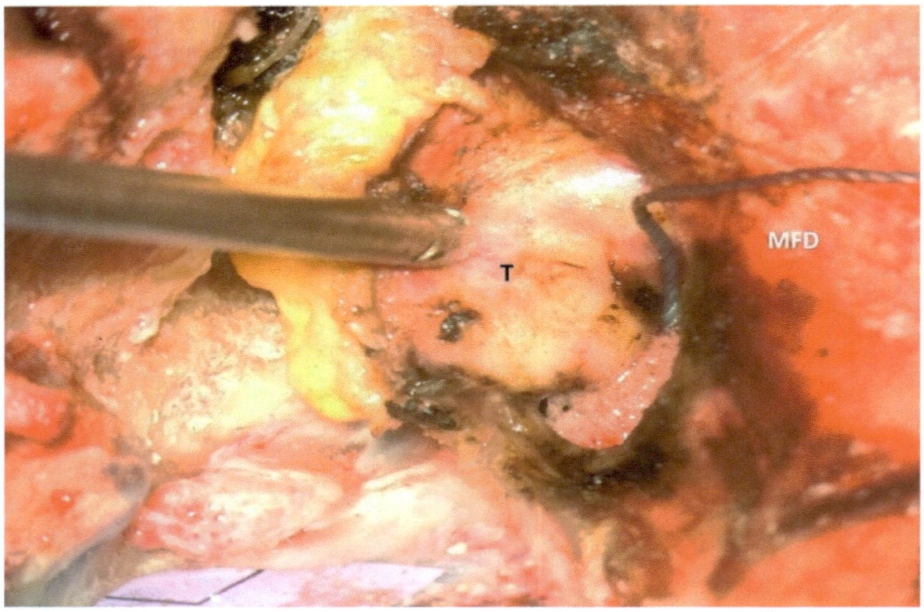

Fig. 9.184 Closer view. The tumor is dissected from the temporal lobe. There is a supratentorial extension of the tumor. *T*, tumor; *MFD*, middle fossa dura.

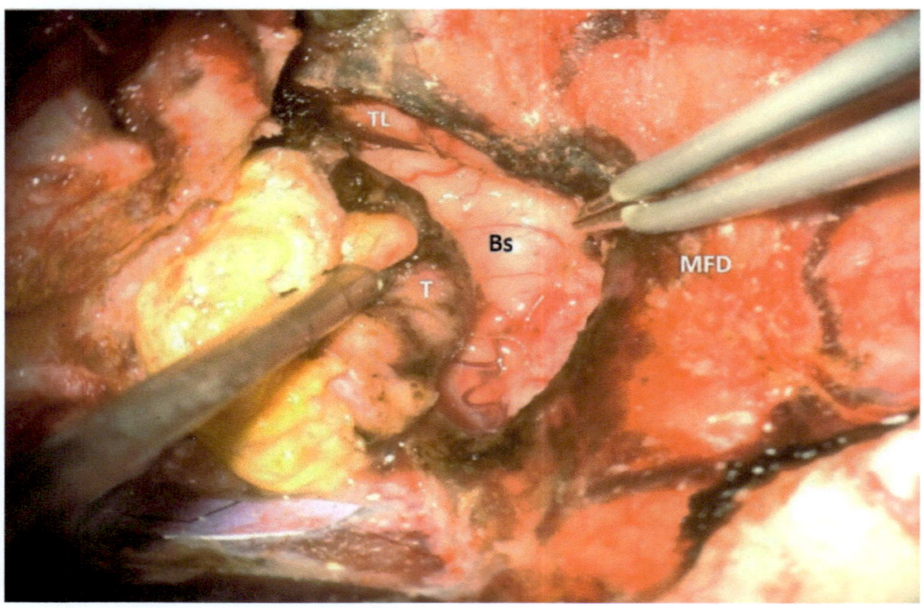

Fig. 9.185 The tumor (T) is dissected from the middle fossa dura (MFD) and the temporal lobe (TL), opening the tentorium. The tumor is then dissected from the posterior fossa. *Bs*, brainstem.

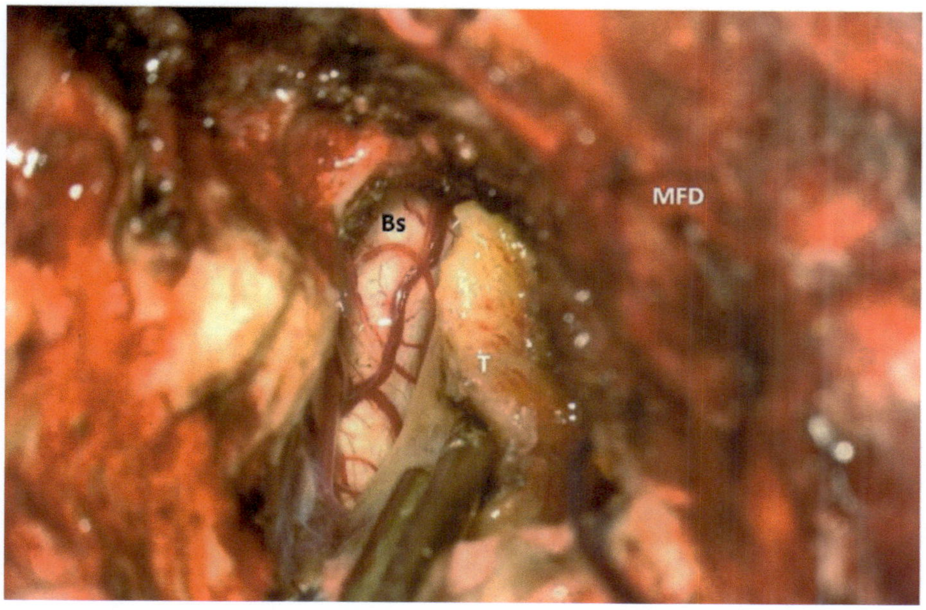

Fig. 9.186 The tumor (T) is further debulked. *MFD*, middle fossa dura; *Bs*, brainstem.

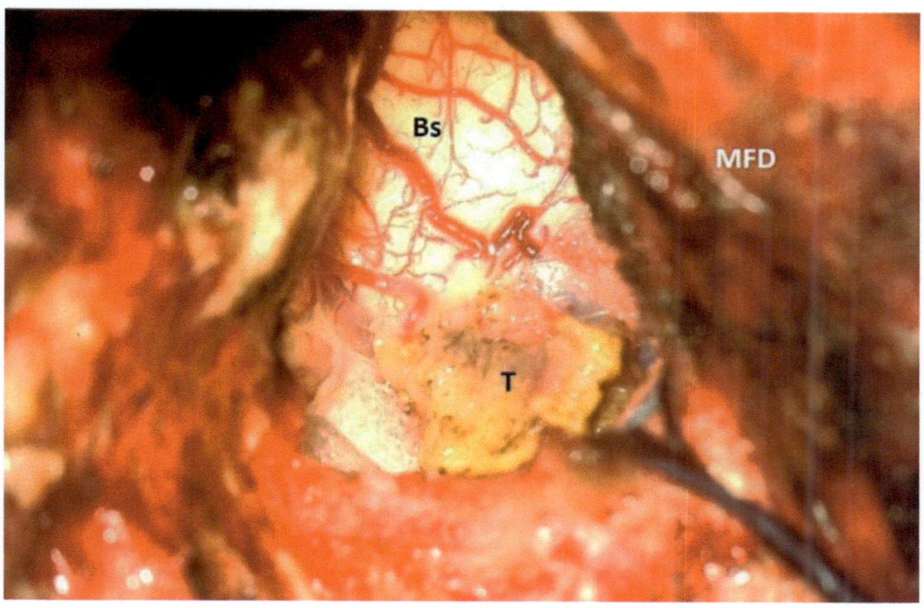

Fig. 9.187 The last piece of tumor (T) is still to be removed. *MFD*, middle fossa dura; *Bs*, brainstem.

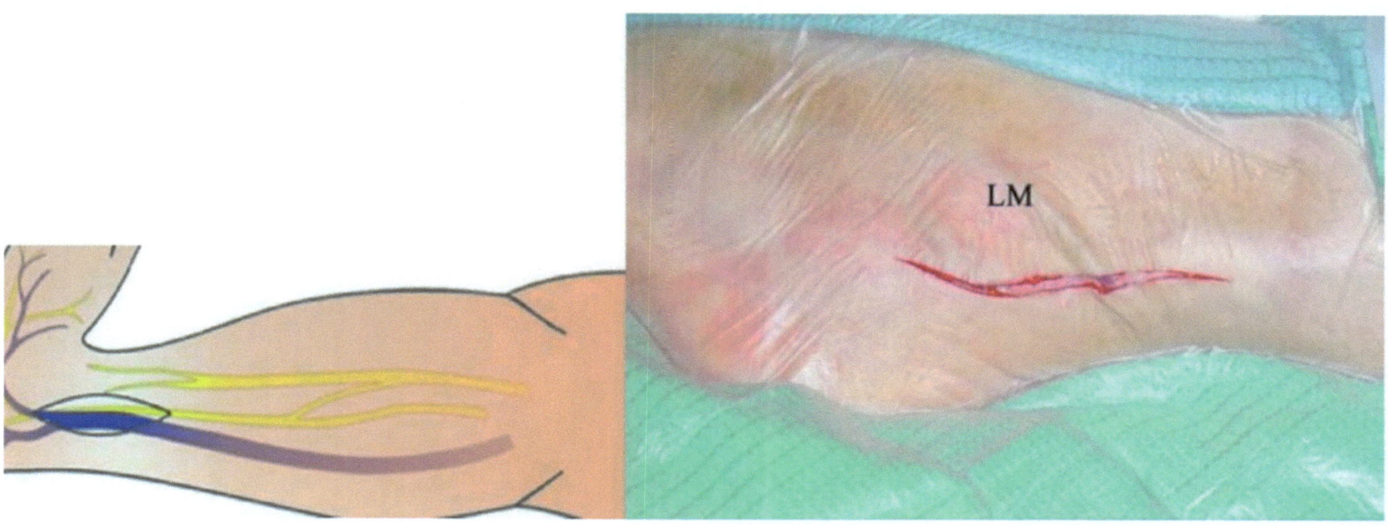

Fig. 9.188 Meanwhile, a sural nerve graft is harvested. An incision is made behind the inferior border of the lateral malleolus, extending superiorly for a distance of 8–9 cm.

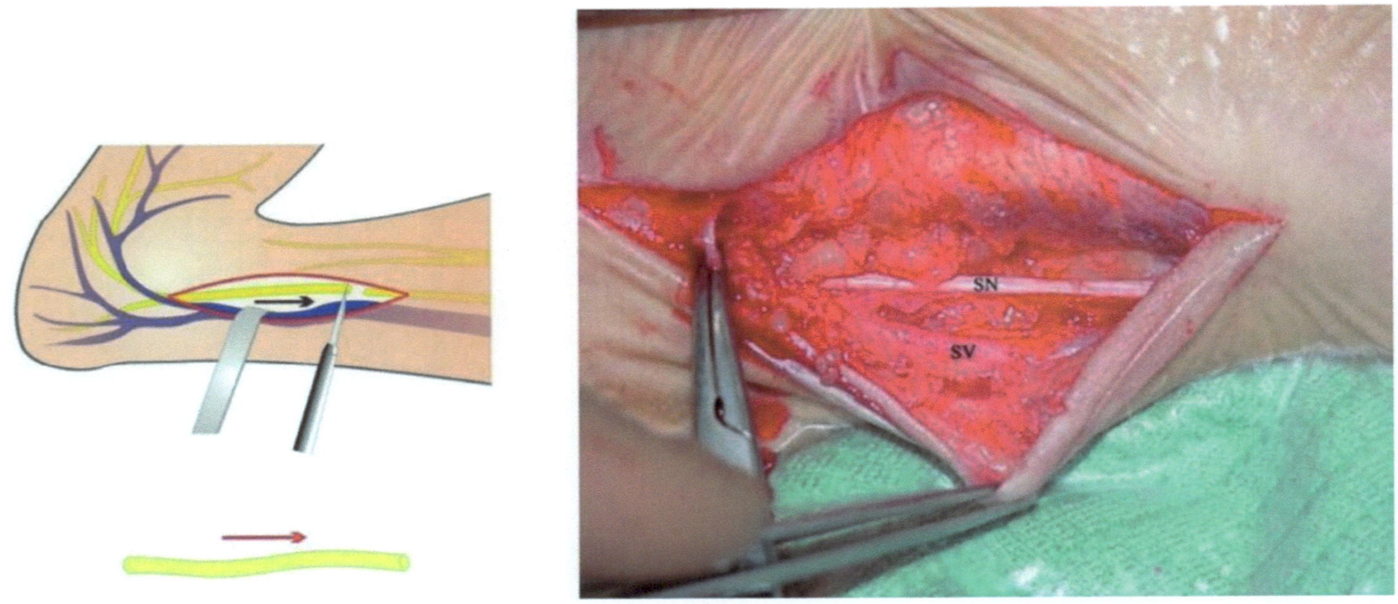

Fig. 9.189 The sural nerve (SN) is located anterior to the saphenous vein (SV).

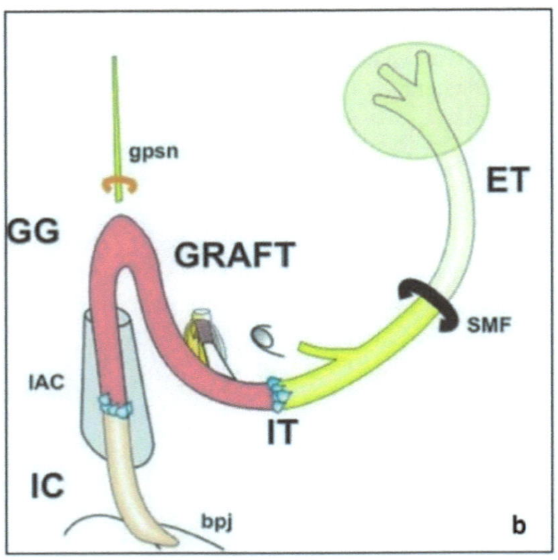

Fig. 9.190 Schematic drawing of the facial nerve anastomosis in this case. A sural nerve graft is placed between the intradural portion of the facial nerve and the mastoid segment. Stop the anastomosis to the mastoid segment of the facial nerve and to the intradural portion and put glue instead of suture. Make the anastomosis a little bit longer on the mastoid segment. The anastomosis is fixed with fibrin glue and temporalis fascia. *IAC*, internal auditory canal; *SMF*, stylomastoid foramen; *IC*, intracisternal; *bpj*, bulbopontine junction; *IT*, intratemporal; *GG*, geniculate ganglion; *ET*, extratemporal; *gspn*, greater superficial petrosal nerve.

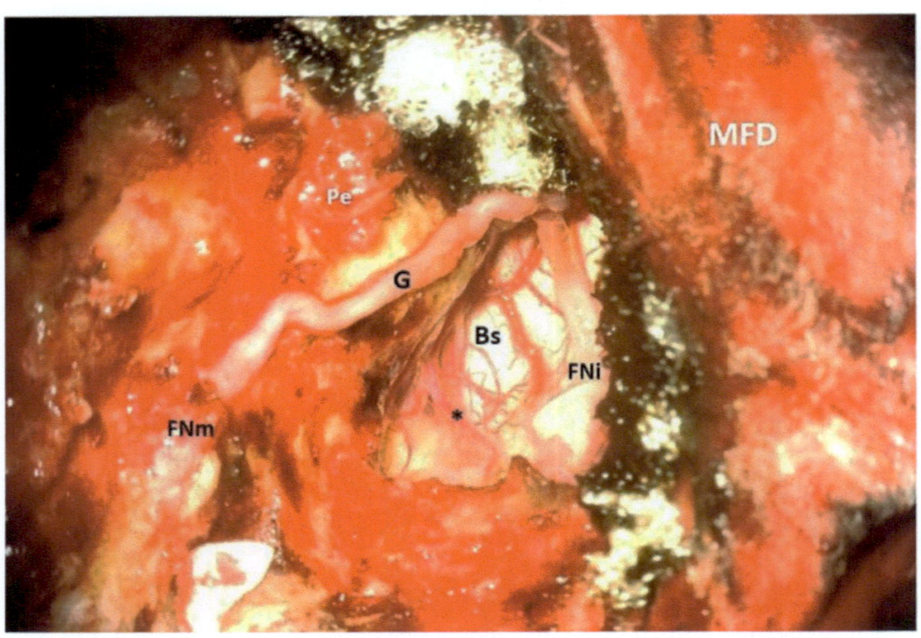

Fig. 9.191 Surgical view after completion of the anastomosis. The anastomosis is then fixed with fibrin glue and fascia. The cavity is then filled with abdominal fat. *MFD*, middle fossa dura; *G*, sural nerve graft; *FNm*, mastoid portion of the facial nerve; *FNi*, intradural portion of the facial nerve; *Bs*, brainstem; *, anterio-inferior cerebellar artery (AICA).

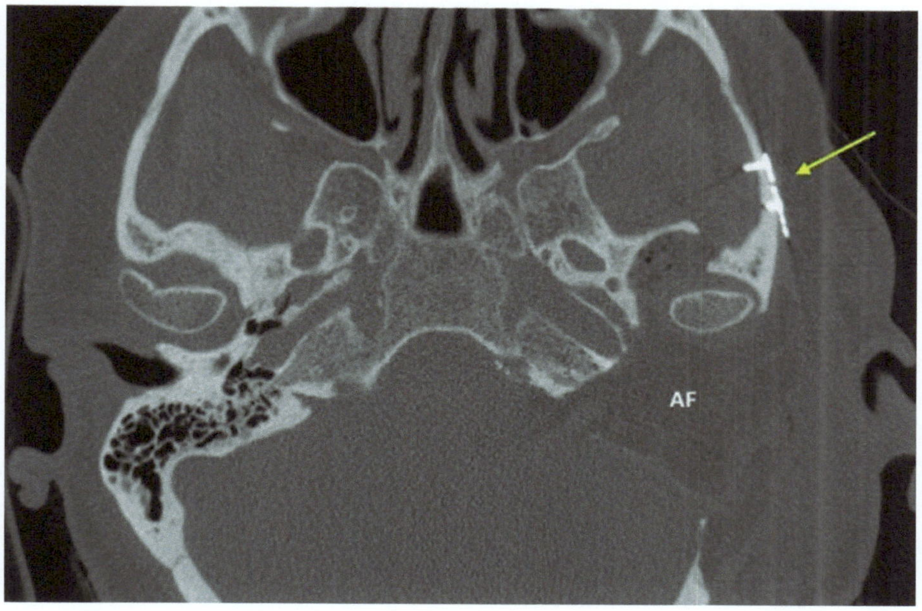

Fig. 9.192 Postoperative CT scan. The craniotomy plug has been replaced and fixed with titanium screws *(arrow)*. *AF*, abdominal fat.

Case 6: Lower Cranial Nerve Schwannoma Removed Through a TO Approach

See Figs. 9.193–9.213.

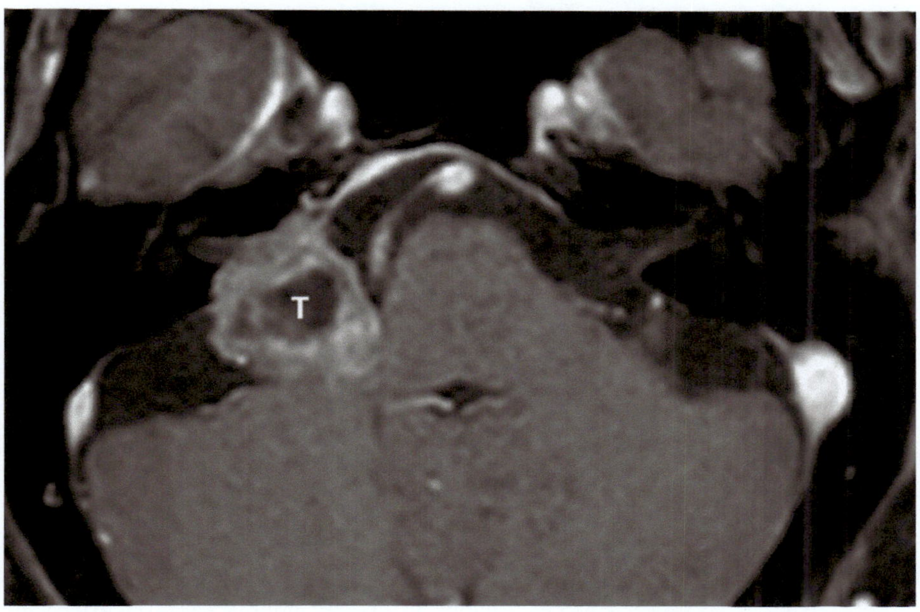

Fig. 9.193 T1W MRI of the brain+Gd enhancement, axial view. This 58-year-old male patient presented with paralysis of the right XII-IX-X cranial nerves and sensorineural hearing loss. The MRI showed a tumor (T) filling the cerebellopontine angle and part of the internal auditory canal, probably originating from the XII cranial nerve.

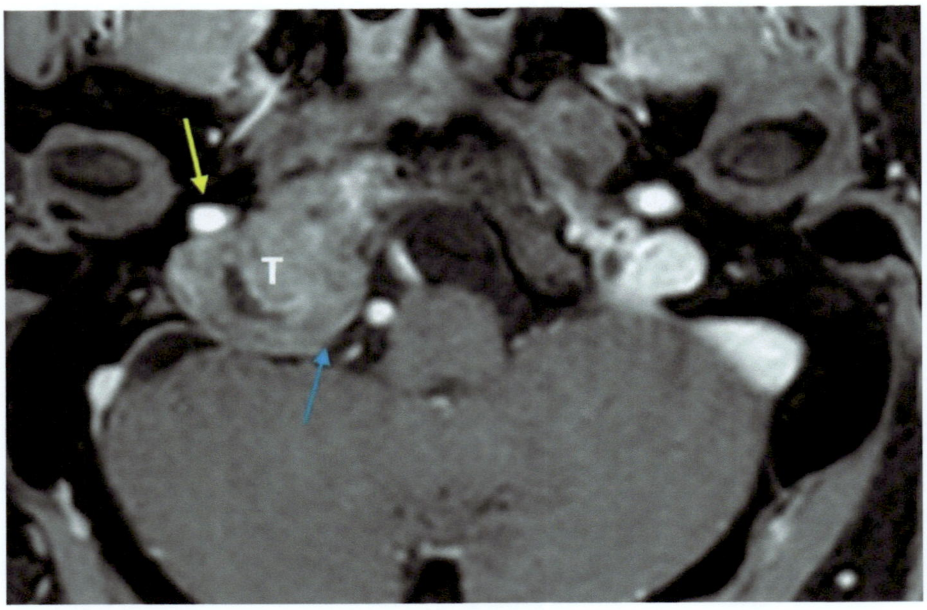

Fig. 9.194 T1W MRI of the brain+Gd enhancement, axial view. Note the extension of the tumor (T) toward the vertical portion of the petrous internal carotid artery *(yellow arrow)*. The jugular bulb is completely closed by the tumor *(light blue arrow)*.

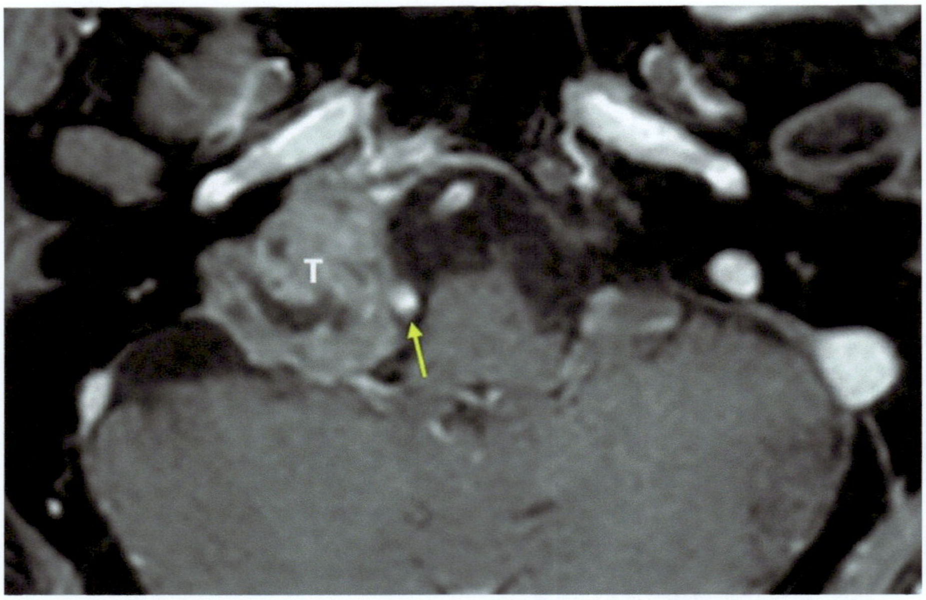

Fig. 9.195 T1W MRI of the brain+Gd enhancement, axial view. Note the extension of the tumor (T) toward the vertebral artery *(yellow arrow)*.

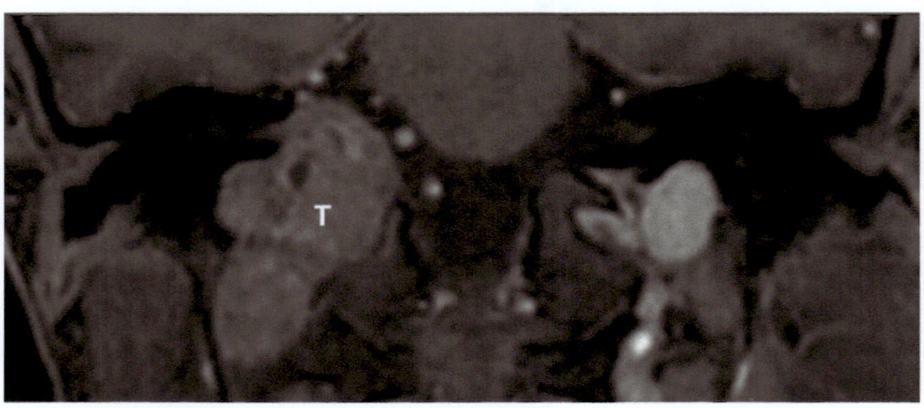

Fig. 9.196 T1W MRI of the brain+Gd enhancement, coronal view. The tumor extends into the neck with a coronal diameter of more than 5 cm. Given the neck extension, staged resection is recommended to avoid CSF leakage into the neck, increasing the risk of further meningitis. In this case, priority was given to the temporal and intradural tumor components to avoid the risk of further brainstem compression. Therefore, the cervical component was left for a second stage via a transcervical approach.

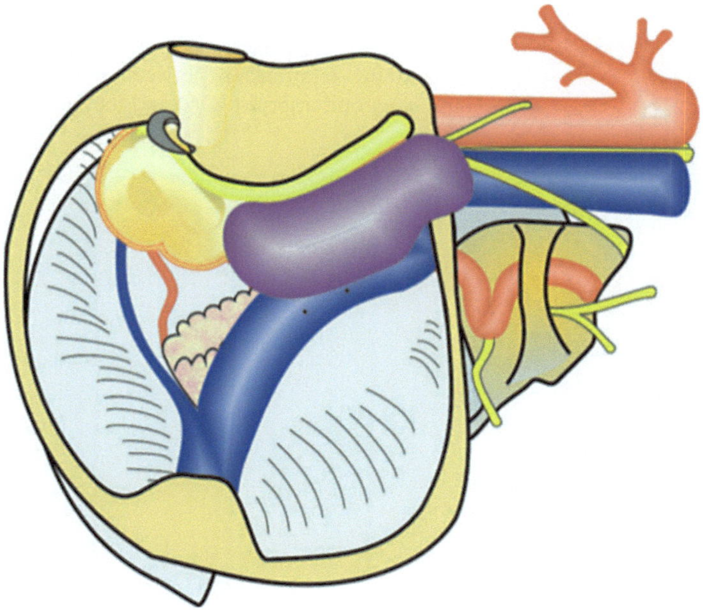

Fig. 9.197 According to our classification, this tumor is a type B lower cranial nerve schwannoma (intracranial) with intradural extension. Considering the extension to the petrous internal carotid artery, the petro-occipital-trans-sigmoid approach (POTS) is not sufficient. Therefore, a transotic approach was chosen.

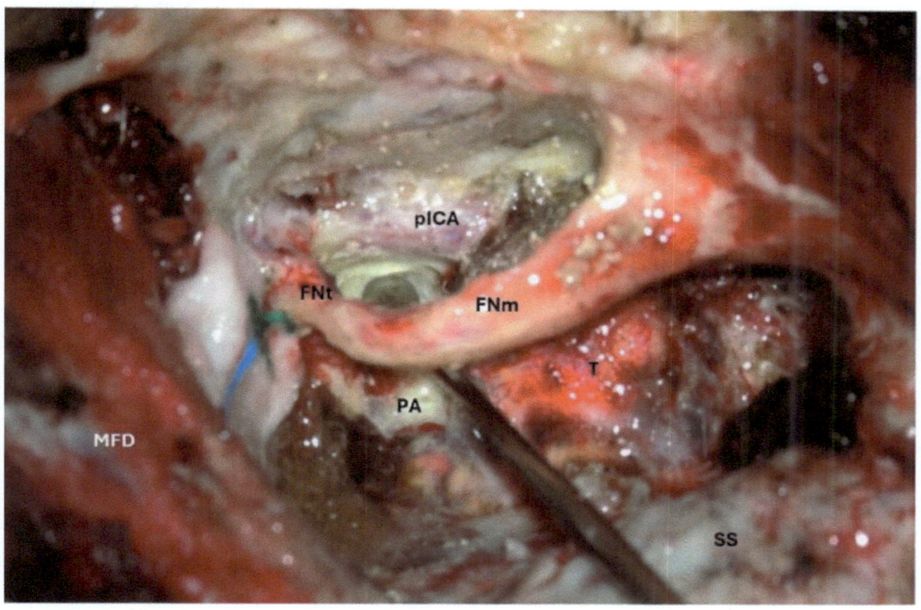

Fig. 9.198 The surgical approach is complete. Note the tumor filling the area of the jugular foramen. *FNm*, mastoid segment of the facial nerve; *FNt*, tympanic segment of the facial nerve; *pICA*, petrous portion of the internal carotid artery; *MFD*, middle fossa dura; *SS*, sigmoid sinus.

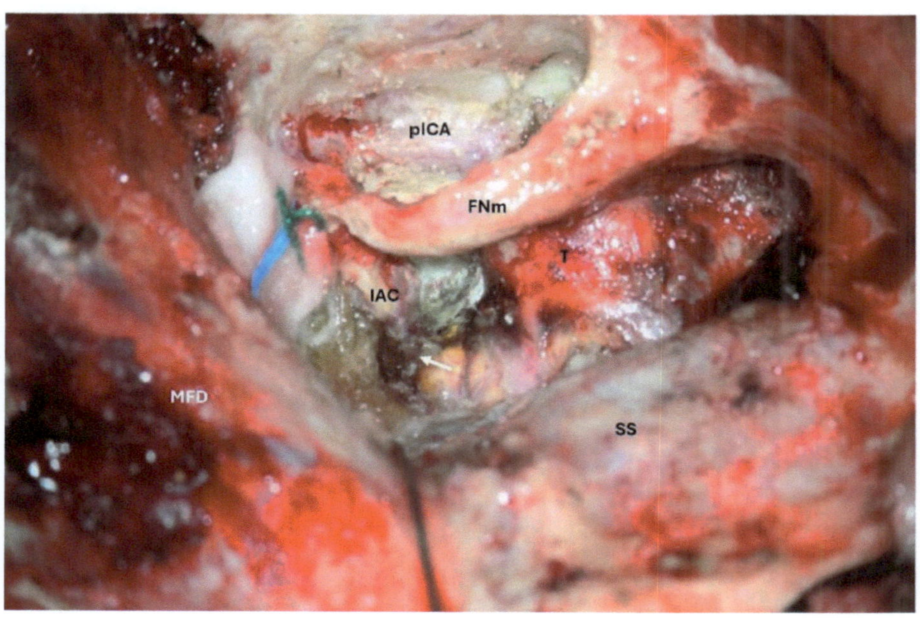

Fig. 9.199 The internal auditory canal (IAC) is decompressed. Note the tumor (T) compressing the inferior border of the IAC *(arrow)*. *FNm*, mastoid portion of the facial nerve; *SS*, sigmoid sinus; *MFD*, middle fossa dura; *pICA*, petrous portion of the internal carotid artery.

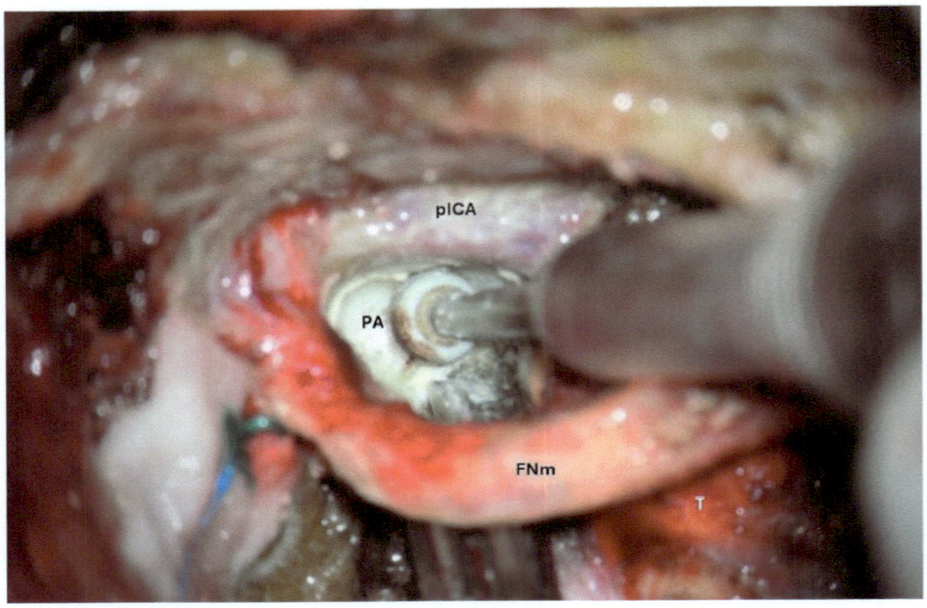

Fig. 9.200 The bone of the petrous apex (PA) is drilled to visualize the upper pole of the tumor (T) and the relationship between the petrous internal carotid artery (pICA) and the tumor.

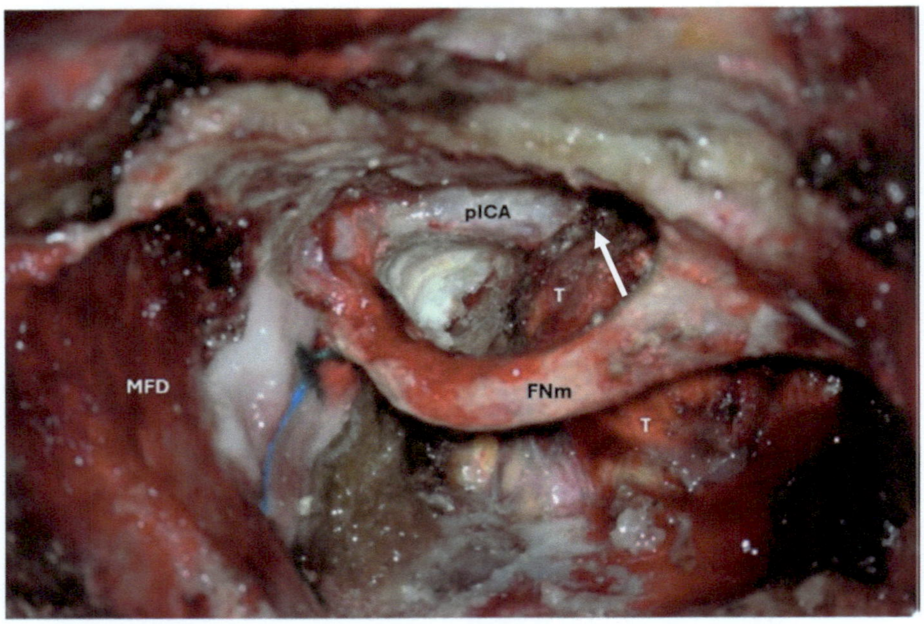

Fig. 9.201 Surgical view after further drilling of the petrous apex. Note the extension of the tumor (T) toward the entrance of the petrous internal carotid artery (pICA) into the skull base *(arrow)*. *MFD*, middle fossa dura; *FNm*, mastoid portion of the facial nerve.

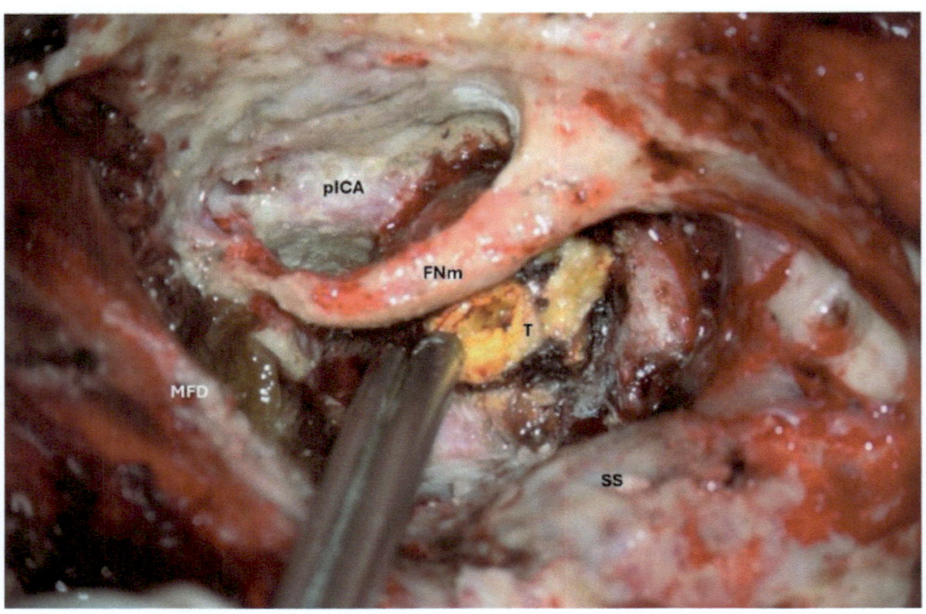

Fig. 9.202 Debulking of the tumor (T) begins. *MFD*, middle fossa dura; *FNm*, mastoid portion of the facial nerve; *SS*, sigmoid sinus; *FNm*, mastoid portion of the facial nerve.

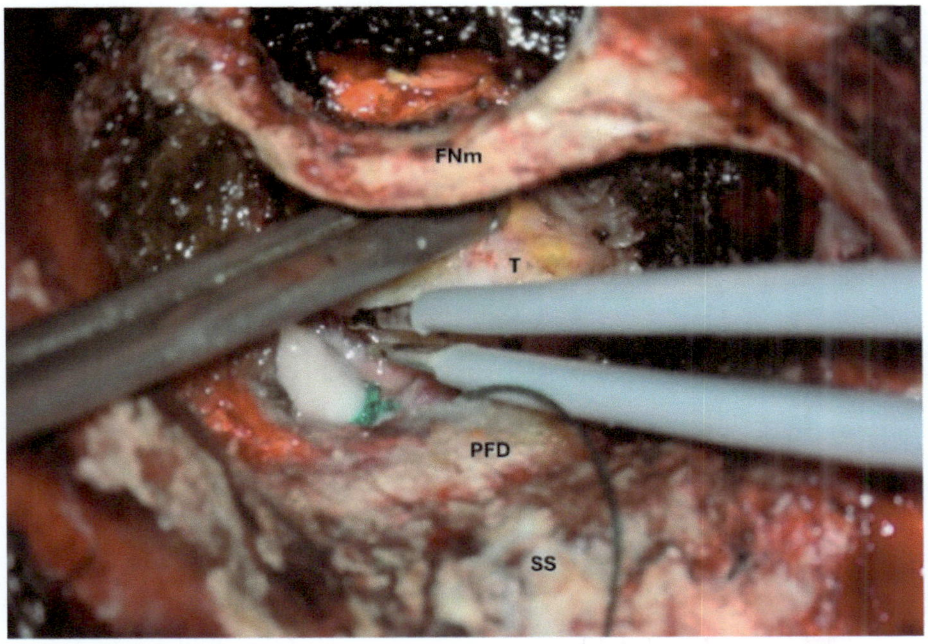

Fig. 9.203 Debulking of the tumor (T) begins. *MFD*, middle fossa dura; *FNm*, mastoid portion of the facial nerve; *SS*, sigmoid sinus; *FNm*, mastoid portion of the facial nerve.

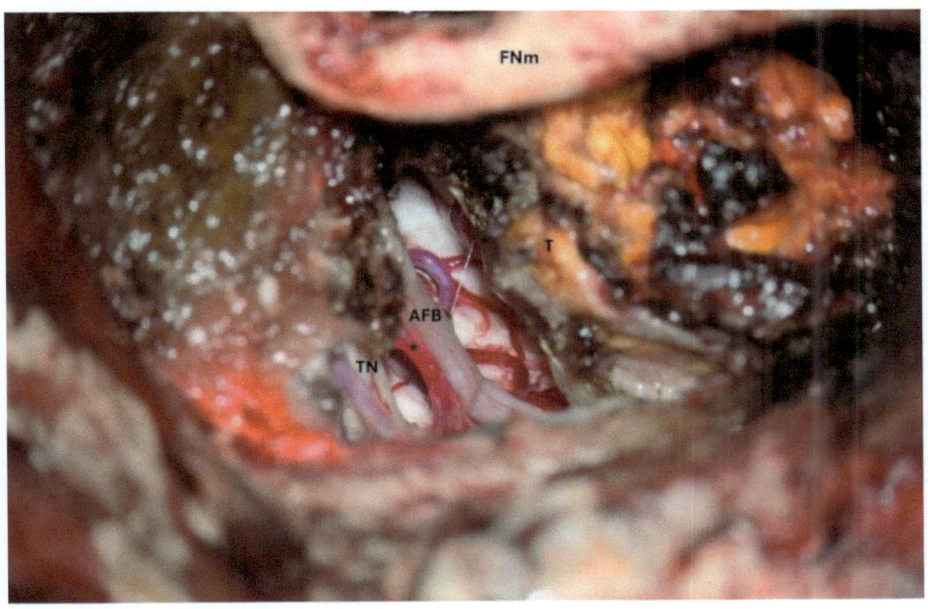

Fig. 9.204 The tumor (T) is separated from the acoustic-facial bundle (AFB). *, anterior-inferior cerebellar artery; *FNm*, mastoid portion of the facial nerve; *TN*, trigeminal nerve.

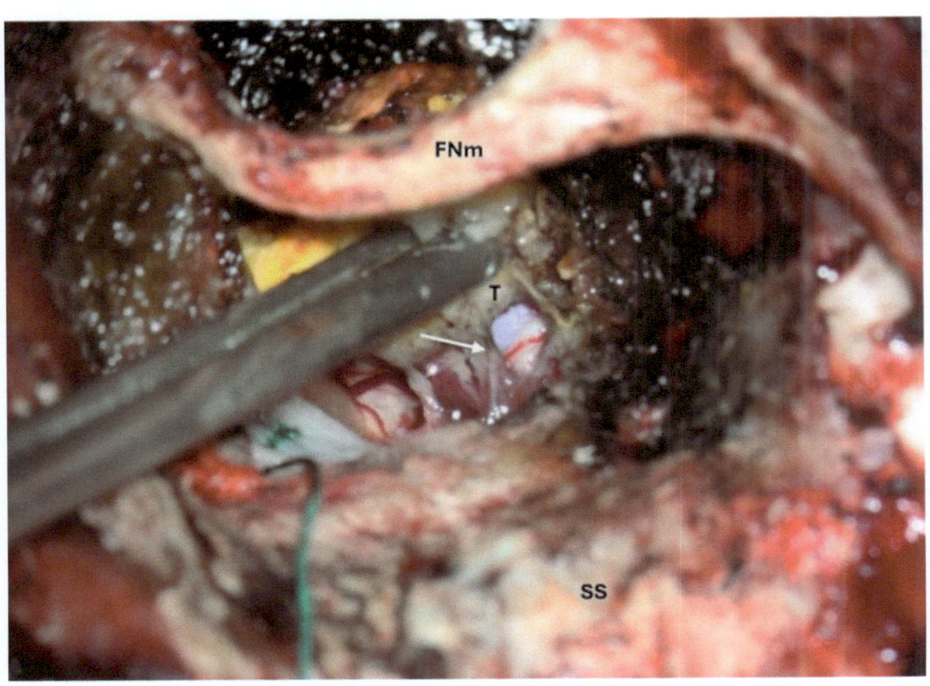

Fig. 9.205 The tumor (T) is dissected from the glossopharyngeal nerve *(arrow)*. *FNm*, mastoid portion of the facial nerve; *SS*, sigmoid sinus.

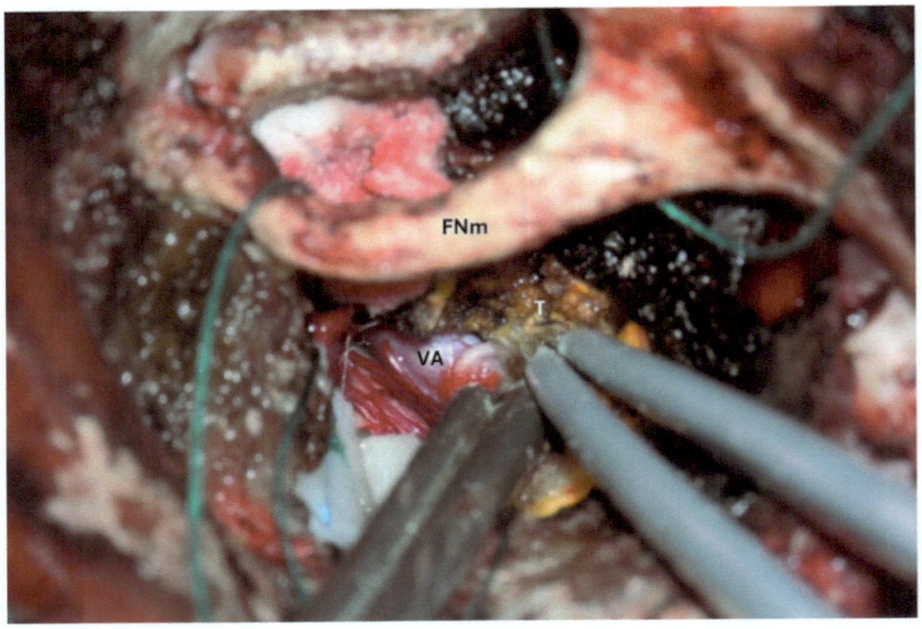

Fig. 9.206 The tumor (T) is then dissected from the vertebral artery (VA). *FNm*, mastoid portion of the facial nerve; *SS*, sigmoid sinus.

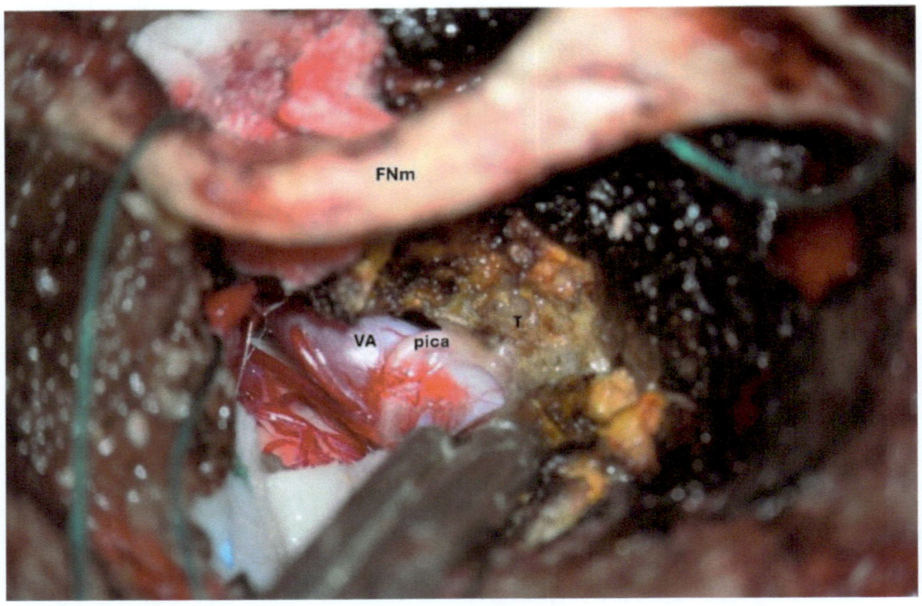

Fig. 9.207 Higher magnification surgical view. The vertebral artery (VA) is further freed from the tumor (T). Note the origin of the posterior inferior cerebellar artery (pica). *FNm*, mastoid portion of the facial nerve.

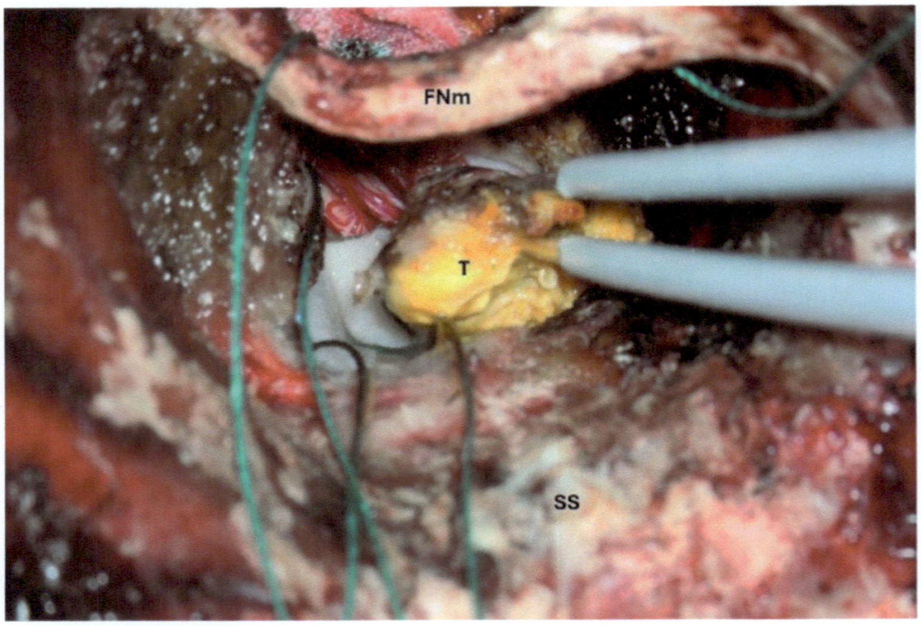

Fig. 9.208 The last piece of tumor (T) is removed. *SS*, sigmoid sinus; *FNm*, mastoid portion of the facial nerve.

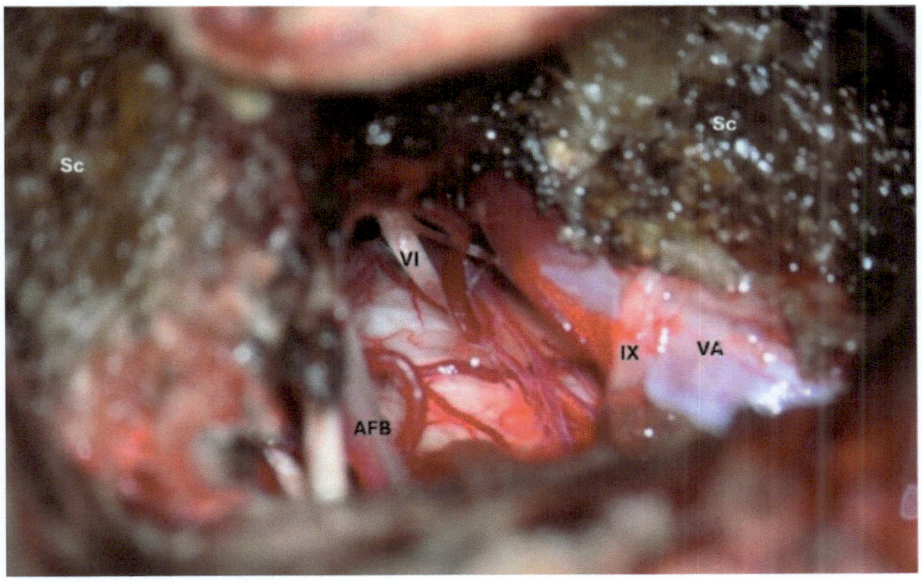

Fig. 9.209 Surgical view after tumor removal. *AFB*, acoustic-facial bundle; *VI*, abducens nerve; *IX*, glossopharyngeal nerve; *VA*, vertebral artery.

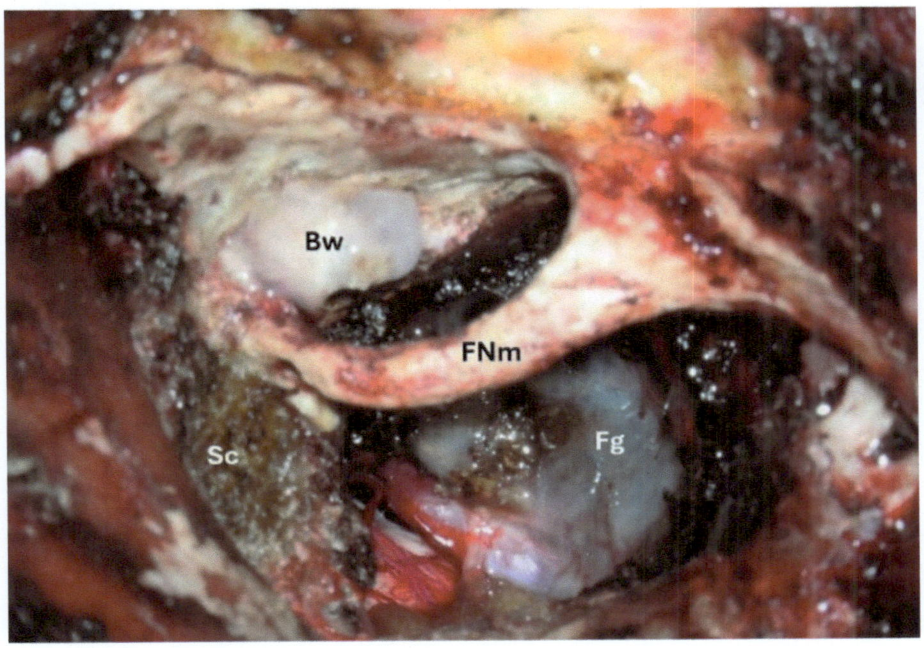

Fig. 9.210 Surgical view after tumor removal. The eustachian tube is closed with periosteum and bone wax (Bw). Bleeding is controlled with Surgicel (Sc) and fibrin glue (Fg).

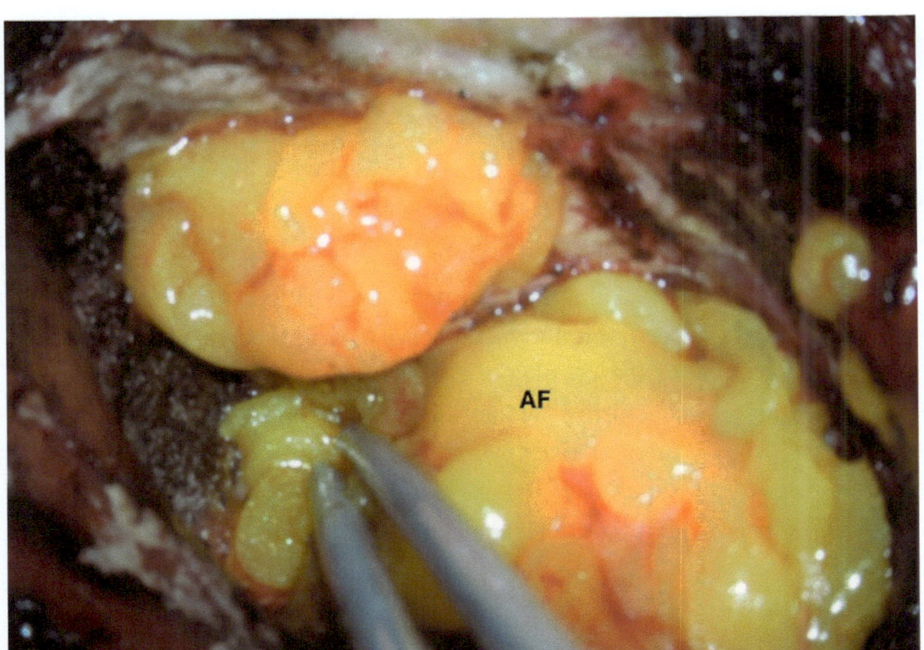

Fig. 9.211 The surgical cavity is filled with abdominal fat (AF).

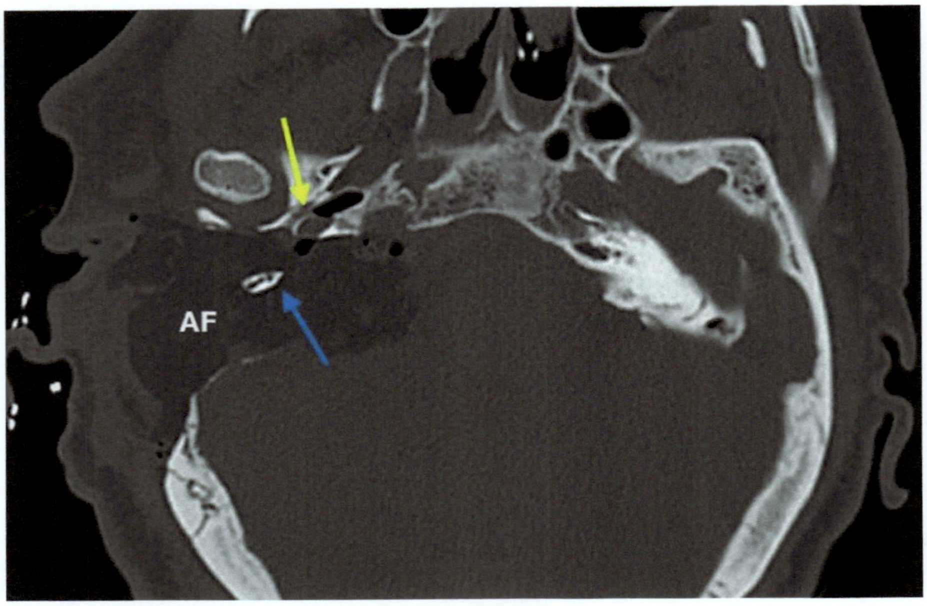

Fig. 9.212 Postoperative high-resolution CT scan. Note the abdominal fat (AF) filling the surgical cavity. *Light blue arrow*: mastoid portion of the facial nerve; *yellow arrow*: intrapetrous internal carotid artery (vertical portion).

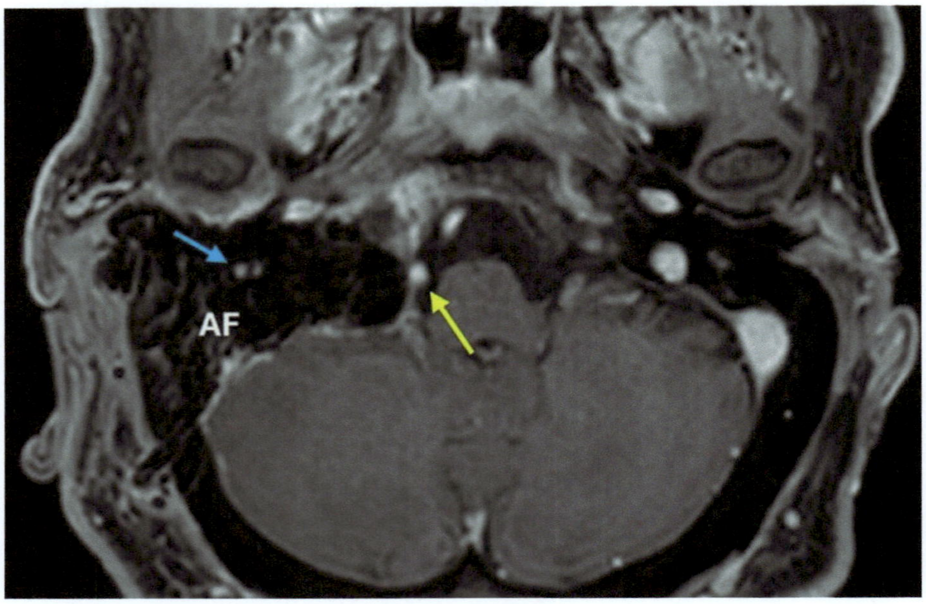

Fig. 9.213 Postoperative MRI scan. Note the abdominal fat (AF) filling the surgical cavity. All the tumors in the temporal bone and intradural compartment have been removed. *Light blue arrow*: mastoid portion of the facial nerve; *yellow arrow*: vertebral artery.

HINTS AND PITFALLS

- When dissecting the cartilage from the lateral part of the skin of the external auditory canal, care has to be taken to maintain the integrity of the skin. Any tear would interfere with a perfect cul-de-sac closure, with an attendant risk of postoperative CSF leakage.
- While drilling anterior to the facial nerve, care should be taken to avoid injuring the nerve with the rotating shaft of the burr. For this reason, the facial nerve should only be skeletonized, but not uncovered, especially in the mastoid segment. In addition, overzealous removal of bone under the fallopian canal should be avoided as it leads to loss of support of the nerve and renders it more liable to fracture during further manipulations.
- While drilling the cochlea, care should be taken to avoid opening the anterior wall of the internal auditory canal and injuring its contents.
- Once the cochlea has been drilled, the burr is passed medial to the facial nerve to drill the petrous apex medial to the cochlea. This should be done with care to avoid fracturing the laterally lying fallopian canal. Other structures at risk in this step are the internal auditory canal superiorly, the jugular bulb inferiorly, and the internal carotid artery anteriorly.
- At the end of the procedure, the facial nerve resembles a bridge in the middle of the approach. The utmost care should be taken during tumor removal so as not to exert

any pressure on the nerve with the instruments used, because this may lead to damage to the facial nerve.

- The internal carotid artery is immediately medial to the floor of the eustachian tube, and they are separated by a thin bony lamina. This bone may be absent in 2% of cases.
- The distance between the vertical portion of the internal carotid artery and the cochlea is variable, ranging from 1 to 5 mm.
- The distance between the facial nerve and the internal carotid artery is limited. The surgeon therefore has to work in a narrow space with an acute angle. Consequently, sufficient removal of the petrous apex is impeded.
- This approach should be combined with the enlarged translabyrinthine approach for the removal of the tumor extending to the cerebellopontine angle. With the TO approach alone it is difficult to remove the anterior wall of the internal auditory canal. For this reason, the transapical extension should be further added for removal of anterior-lying acoustic neurinoma, because the TO approach alone has a limited exposure for removing the anterior wall of the internal auditory canal.

REFERENCES

[1] Fisch U, Mattox D. Microsurgery of the skull base. Stuttgart: Georg Thieme Veriag; 1988. p. 74–135.

[2] Khrais TH, Falcioni M, Taibah A, et al. Cerebrospinal fluid leak prevention after translabyrinthine removal of vestibular schwannoma. Laryngoscope 2004;114:915–20. https://doi.org/10.1097/00005537-200406000-00011.

[3] Browne JD, Fisch U. Transotic approach to the cerebellopontine angle. Otolaryngol Clin North Am 1992;25:331–46.

[4] Chen JM, Fisch U. The transotic approach in acoustic neuroma surgery. J Otolaryngol 1993;22:331–6.

[5] Xia Y, Zhang W, Li Y, Ma X, Liu Q, Shi J. The transotic approach for vestibular schwannoma: indications and results. Eur Arch Otorhinolaryngol 2017;274(8):3041–7. https://doi.org/10.1007/s00405-017-4627-3.

[6] Day JD, Chen DA, Arriaga M. Translabyrinthine approach for acoustic neuroma. Neurosurgery 2004;54:391–6. https://doi.org/10.1227/01.NEU.0000103668.26590.5A.

[7] Van Leeuwen JP, Cremers CW, Theunissen EJ, et al. Translabyrinthine and transotic surgery for acoustic neuroma. Clin Otolaryngol Allied Sci 1994;19:491–5. https://doi.org/10.1111/j.1365-2273.1994.tb01275.x.

[8] Dandy W. Results of removal of acoustic tumors by the unilateral approach. Arch Surg 1941;42(6):926–33. https://doi.org/10.1001/archsurg.1941.01210120061006.

[9] Jenkins HA, Fisch U. The transotic approach to resection of difficult acoustic tumors of the cerebellopontine angle. Am J Otol 1980;2:70–6.

[10] House WF, Hitselberger WE. The transcochlear approach to the skull base. Arch Otolaryngol Head Neck Surg 1976;92:334–42.

[11] Gantz Bruce J, Fisch U. Modified transotic approach to the cerebellopontile angle. Arch Otolaryngol 1983;99:252–6. https://doi.org/10.1001/archotol.1983.00800180050010.

[12] Kuthubutheen J, Chen JM. Transotic approach. In: Brackmann DE, Shelton C, Arriaga MA, editors. Otologic surgery. 4th ed. Philadelphia: W.B. Saunders Co./Elsevier; 2016.

[13] Falcioni M, Taibah A, Di Trapani G, Khrais T, Sanna M. Inner ear extension of vestibular schwannomas. Laryngoscope 2003;113:1605Y8.

[14] Grayeli AB, Fond C, Kalamarides M, et al. Diagnosis and management of intracochlear schwannomas. Otol Neurotol 2007;28:951Y7.

[15] Kennedy RJ, Shelton C, Salzman KL, Davidson HC, Harnsberger HR. Intralabyrinthine schwannomas: diagnosis, management, and a new classification system. Otol Neurotol 2004;25:160Y7.

[16] Manoharan KS, Thakar A, Sharma SC. In reference to "The transotic approach for vestibular schwannoma: indications and results". Eur Arch Otorhinolaryngol 2017;274(12):4269–70. https://doi.org/10.1007/s00405-017-4692-7.

[17] Bittencourt AG, Alves RD, Ikari LS, Burke PR, Gebrim EM, Bento RF. Intracochlear schwannoma: diagnosis and management. Int Arch Otorhinolaryngol 2014;18(3):322–4. https://doi.org/10.1055/s-0033-1364170.

[18] Sanna M, Sunose H, Mancini F, Russo A, Taibah A. Microsurgical management of middle ear and petrous bone cholesteatoma. Thieme; 2019, ISBN:9783132000056.

[19] Sanna M, Khrais T, Mancini F, Russo A, Taibah A. The facial nerve in temporal bone and lateral skull-base microsurgery. Stuttgart-New York: Georg Thieme Verlag; 2006.

[20] Prasad SC, Piras G, Piccirillo E, Taibah A, Russo A, He J, Sanna M. Surgical strategy and facial nerve outcomes in petrous bone cholesteatoma. Audiol Neurootol 2016;21(5):275–85. https://doi.org/10.1159/000448584 [Epub 2016 Oct 7] PMID: 2779980.

[21] Yilala MH, Fancello G, Sanna M. Management of petrous bone cholesteatoma: the Gruppo Otologico experience. Eur Arch Otorhinolaryngol 2024;281(9):5211–24. https://doi.org/10.1007/s00405-024-08752-4 [Epub 2024 Jun 3].

[22] Yilala MH, Fancello G, Musumano LB, Lauda L, Sanna M. Long-term facial nerve outcome in surgically treated petrous bone cholesteatoma patients. Eur Arch Otorhinolaryngol 2024. https://doi.org/10.1007/s00405-024-09052-7.

[23] Sanna M, Pandya Y, Mancini F, Sequino G, Piccirillo E. Petrous bone cholesteatoma: classification, management and review of the literature. Audiol Neurootol 2011;16(2):124–36. https://doi.org/10.1159/000315900 [Epub 209 Jul 29].

CHAPTER 10

The Modified Transcochlear Approaches

The original transcochlear (TC) approach introduced by House and Hitselberger was designed for large petroclival lesions, cerebellopontine tumors with an anterior/medial extension, and medial intradural tumors. For a comprehensive management of these tumors, we have further classified the approach into four types denoted from A to D, according to the surgical extension. The transcochlear approach brought a significant improvement to the surgical access to the anterior CPA provided by the posterior transposition of the facial nerve. The procedure is accompanied by a temporary facial palsy that usually recovers partially in 6–12 months.

Introduction

House and Hitselberger designed the transcochlear approach to treat midline clival lesions and CPA tumors arising anterior to the IAC, without requiring the use of brain retractors [1–4]. These lesions could not have been reached through the traditional suboccipital and transoral approaches used at that time as they were limited by the cerebellum and the brainstem [4]. This approach has suffered various modifications and improvements, representing now a group of surgical techniques [5–10]. We have classified the approach into four types denoted from A to D described below. Type A represents the basic approach; types B, C, and D denote the anterior, superior, and inferior extensions, respectively. A schematic drawing of the classification of this approach is shown in Fig. 10.1.

The transcochlear approach is an extension of the translabyrinthine procedure combined with the posterior rerouting of the facial nerve, total exenteration of the otic capsule, and drilling of the petrous apex and lateral aspect of the clivus. A large access to the anterior CPA and clivus is thus attained, allowing visualization of cranial nerves V–XI, basilar and vertebral arteries, as well as contralateral cranial nerves and the opposite CPA [11–13]. Of particular importance, especially in the management of meningiomas [14,15], is that this wide exposure allows total tumor removal and also the closure of the arterial blood supply that comes from the internal carotid artery [4].

The main advantage of the approach is the large surgical field provided that results in optimal control of midline intradural, petroclival, and anterior CPA tumors, without the need for brain retraction. The disadvantages and associated morbidity are represented by hearing loss and facial nerve palsy. The high incidence of facial nerve paralysis associated with the TC approach is caused by the compromise of its vascular supply during posterior rerouting. Most patients recover the facial nerve function in the first year postsurgery but usually up to a grade III House-Brackmann. Other complications that may ensue include temporary vestibular symptoms, other cranial nerve dysfunctions depending on tumor extent, and cerebrovascular accidents [8,9,15–33].

Atlas of Microsurgery of the Lateral Skull Base. https://doi.org/10.1016/B978-0-443-19083-4.00010-5

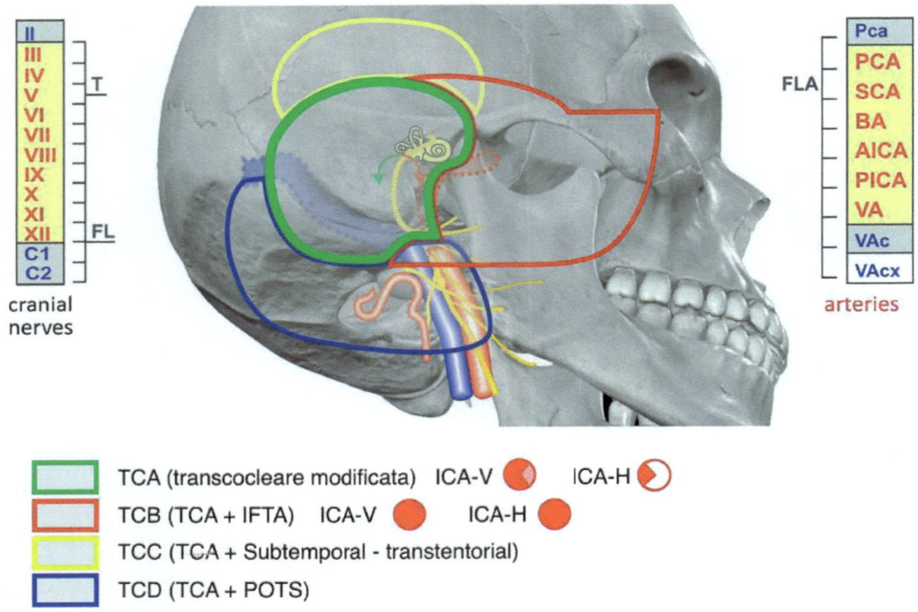

Fig. 10.1 A schematic drawing of the classification of this approach is shown. The extent of bone removal and the structures controlled with the transcochlear approach types A, B, C, and D.

Type A Modified Transcochlear Approach (The Basic Type)

Rationale

The original transcochlear approach described by House and Hitselberger in 1976 includes internal auditory canal identification, posterior rerouting of the facial nerve, removal of the cochlea and petrous apex with preservation of the middle ear and external auditory canal (Fig. 10.2). The modified transcochlear approach combines the removal of the external auditory canal (EAC) and middle ear with the posterior rerouting of the facial nerve (FN), thus removing the major impediment to the anterior extension of the approach. This allows better control of the vertical and horizontal intrapetrous internal carotid artery and facilitates total removal of the petrous apex. The extensive anterior bone removal provides excellent control of the ventral surface of the brainstem without cerebellar and brainstem retraction (Figs. 10.3 and 10.4).

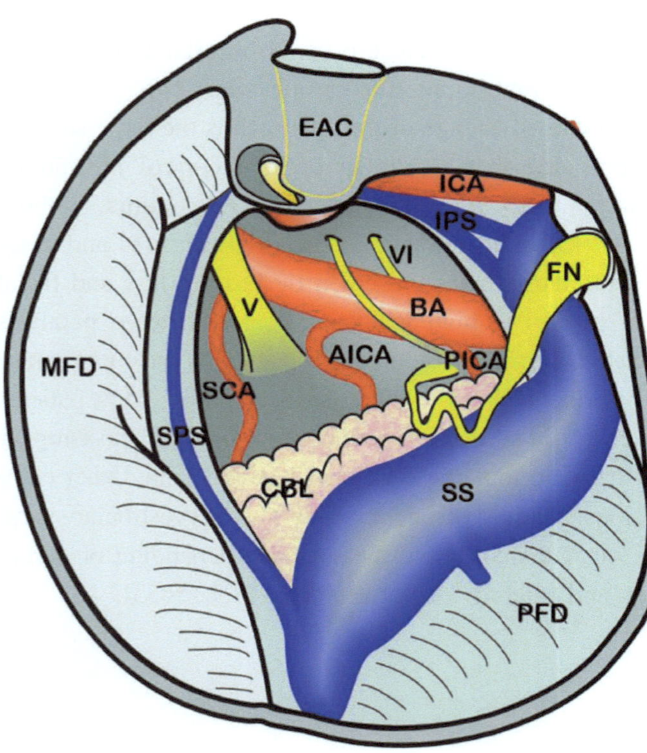

Fig. 10.2 The original transcochlear approach described by House and Hitselberger. *EAC*, external auditory canal; *AICA*, anterior inferior cerebellar artery; *BA*, basilar artery; *CBL*, cerebellum; *FN*, facial nerve; *ICA*, internal carotid artery; *SPS*, superior petrosal sinus; *IPS*, inferior petrosal sinus; *MFD*, middle fossa dura; *PFD*, posterior fossa dura; *PICA*, posterior inferior cerebellar artery; *SCA*, superior cerebellar artery; *SS*, sigmoid sinus; *V*, trigeminal nerve; *VI*, abducent nerve.

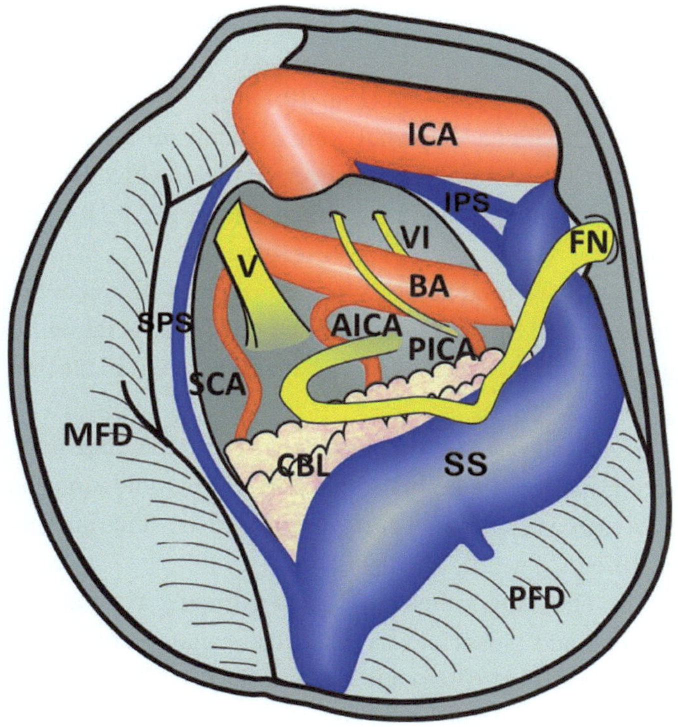

Fig. 10.3 Wide exposure obtained by the transcochlear approach type A. *AICA*, anterior inferior cerebellar artery; *BA*, basilar artery; *CBL*, cerebellum; *FN*, facial nerve; *ICA*, internal carotid artery; *SPS*, superior petrosal sinus; *IPS*, inferior petrosal sinus; *MFD*, middle fossa dura; *PFD*, posterior fossa dura; *PICA*, posterior inferior cerebellar artery; *SCA*, superior cerebellar artery; *SS*, sigmoid sinus; *V*, trigeminal nerve; *VI*, abducent nerve.

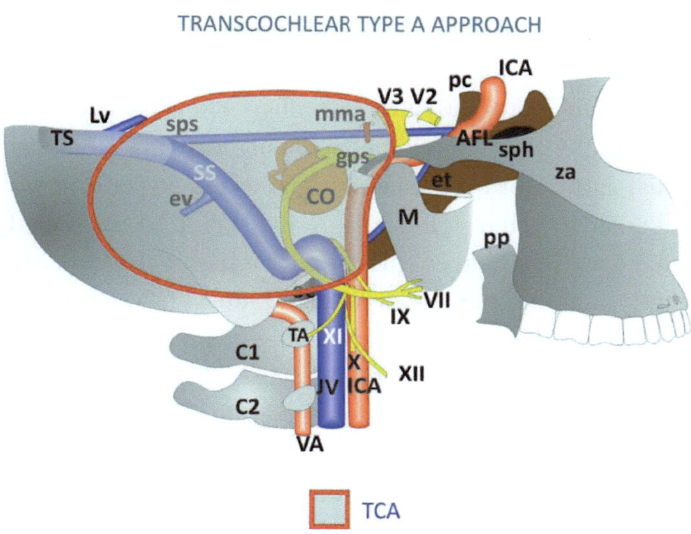

Fig. 10.4 The extent of bone removal and the structures controlled with the transcochlear approach type A. *VII*, facial nerve; *IX*, glossopharyngeal nerve; *X*, vagus nerve; *XI*, accessory nerve; *XII*, hypoglossal nerve; *AFL*, anterior foramen lacerum; *C1*, atlas; *CO*, cochlea; *et*, eustachian tube; *ev*, emissary vein; *gps*, greater superficial petrosal nerve; *ICA*, internal carotid artery; *JV*, jugular vein; *Lv*, Labbé's vein; *M*, mandible; *mma*, middle meningeal artery; *oc*, occipital condyle; *pc*, posterior clinoid; *pp*, pterygoid process; *sph*, sphenoid sinus; *sps*, superior petrosal sinus; *SS*, sigmoid sinus; *TA*, lateral process of the atlas; *TS*, transverse sinus; *V2*, second division of the trigeminal nerve; *V3*, third division of the trigeminal nerve; *za*, zygomatic arch.

Advantages

1. Wide exposure of the cerebellopontine angle and prepontine cistern without cerebellar and brainstem retraction.
2. Allows removal of invaded dura and bone.
3. Excellent control of the intrapetrous internal carotid artery.
4. The approach can be extended according to the extent of the disease.
5. In cases of meningioma, the blood supply of the tumor is controlled first before the dura is opened.

Disadvantages

1. Sacrifice of the ipsilateral hearing.
2. Facial nerve rerouting leads to postoperative paralysis, which usually recovers to grade III (rarely to grade II).
3. The procedure is lengthy. However, the wide exposure shortens greatly the time needed for tumor removal.

Indications

- **Extradural lesions**: Extensive petrous apex lesions with preoperative facial nerve and inner ear compromise, e.g.:
  - Petrous bone cholesteatoma of the massive, infralabyrinthine-apical, and (less commonly) supralabyrinthine types.
  - Recurrent acoustic neurinoma with petrous bone invasion.
  - Extensive facial nerve tumors.
  - Cholesterol granuloma (only with preoperative facial and hearing impairment).

- **Intradural lesions**:
  - Large clival and petroclival lesions lying ventral to the brainstem, e.g., petroclival meningiomas.
  - Previously irradiated petroclival meningiomas.
  - Residual or recurrent nonacoustic lesions of the posterior fossa with anterior extension into the prepontine cistern, particularly those with encasement of the vertebrobasilar artery or perforating arteries or both, e.g., huge posterior fossa epidermoids.
  - Recurrent acoustic neurinomas with facial nerve paralysis.
- Transdural lesions invading the petrous apex as en-plaque meningiomas, or primary clival or temporal bone lesions with secondary posterior fossa extension as chordomas, chondrosarcomas, and extensive glomus jugulare tumors.
- Some cases of epidermoids.

Contraindications

- Normal preoperative facial nerve function.
- Only hearing ear.
- Active middle ear infection.

Surgical Steps (Right Side)

(Figs. 10.5–10.59)

1. A wide C-shaped postauricular incision is made. The incision starts 3 cm above the auricle, curves posteriorly to about 4–5 cm posterior to the postauricular sulcus, and ends inferiorly at the level of the mastoid tip (see Fig. 10.5).

2. The skin and subcutaneous tissues are elevated, and the anterior skin flap is retracted using skin hooks (Fig. 10.6). The external auditory canal is transected at the bony cartilaginous junction (Fig. 10.7). The skin of the lateral part is dissected free from the cartilage for 1 cm (Fig. 10.8). Using artery forceps, the sutures are pulled through the external auditory canal meatus, thus everting the free skin edges. These are then closed with 3–0 silk sutures (Fig. 10.9). A second layer of closure is achieved by approximating the cartilage of the tragus and the concha (Fig. 10.10).

3. The inferiorly based musculoperiosteal layer is incised in a U-shaped pattern (Fig. 10.11). The musculoperiosteal layer is raised using a Lempert periosteal elevator and is kept retracted using 1/0 silk sutures (Fig. 10.12). We no longer use conventional retractors, which significantly limit the surgeon's access.

4. The tympanomeatal flap is detached from the bony canal wall (Fig. 10.13). The skin of the bony external auditory canal is removed, together with the tympanic membrane. An extended mastoidectomy is performed, with the removal of bone 2–3 cm posterior to the sigmoid sinus and over the middle fossa dura. The external auditory canal (posterior and superior walls) is drilled (Fig. 10.14), and the facial nerve is skeletonized (Fig. 10.15).

5. The incudostapedial joint is being disarticulated (Fig. 10.16), and the incus has been removed (Fig. 10.17). The tensor tympani tendon is being cut (Fig. 10.18), and the malleus has been removed (Fig. 10.19).

6. The retrofacial air cells are drilled to complete the skeletonization of the mastoid segment of the facial nerve (Fig. 10.20), and the jugular bulb has been identified (Fig. 10.21).

7. Using a large diamond burr, the remaining bone covering the middle cranial fossa and the sigmoid sinus as well as the posterior fossa dura posterior to the sinus are further drilled (Figs. 10.22–10.24).

8. The labyrinthectomy is performed. The lateral, posterior, and superior semicircular canals have been skeletonized (Fig. 10.25) and opened (Fig. 10.26).

9. The vestibule has been opened (Fig. 10.27). Care is taken to leave the anterior part of the lateral semicircular canal in order to protect the anteriorly lying facial nerve. Note the endolymphatic duct reaches the vestibule, where it bifurcates into the utricular duct and saccular duct (Fig. 10.28). The endolymphatic duct extends from the medial surface of the posterior canal to the posterior fossa dura anterior to the sigmoid sinus and should be transected with the use of a Beaver knife or microscissors (Fig. 10.29). Sharply cutting the endolymphatic duct facilitates the complete drilling of bone overlying the posterior fossa dura and the subsequent dural retraction (Fig. 10.30).

10. Identification of the lower edge of the internal auditory canal starts by drilling the bone between the jugular bulb and the expected position of the canal (Fig. 10.31). The cochlear aqueduct can be identified by drilling the bone between the jugular bulb and the expected position of the canal (Fig. 10.32). Drilling of the superior surface of the internal auditory canal is carried out (Fig. 10.33). The internal auditory canal is outlined with a thin shell of bone still covering it (Fig. 10.34). The remaining thin shell of bone is separated from the underlying dura of the internal auditory canal using a dissector and is then carefully removed (Fig. 10.35).

11. Using a diamond burr, the greater superficial petrosal nerve is identified by drilling anterior to the geniculate ganglion. Next, the middle fossa dura is gently retracted with the suction irrigator, and the labyrinthine segment of the nerve is identified. A small diamond burr is used at this crucial step. Drilling proceeds carefully parallel to the direction of the nerve, with the direction of rotation being away from the nerve. The mastoid, tympanic, and labyrinthine portions of the facial nerve are visible (Fig. 10.36).

12. The anterior wall of the external auditory canal and the tympanic bone are thinned out. Identification of the internal carotid artery is either started from the eustachian tube or at the anteroinferior limit of the promontory (Fig. 10.37). The tympanic bone has been partially drilled away, and the internal carotid artery is identified in front of the cochlea (Fig. 10.38).

13. After identification of the internal carotid artery, a diamond burr is used initially to open the cochlea (Fig. 10.39), and the basal turn, middle turn, and apical turn can be seen (Fig. 10.40).

14. The bone medial to the facial nerve is drilled to further skeletonize the internal carotid artery and jugular bulb (Fig. 10.41). The remaining part of the cochlea is removed. The internal auditory canal is drilled for 360 degrees of its circumference. The last shell of bone over the facial nerve is removed using a rasp. The facial nerve is now free over 270 degrees in its mastoid and tympanic parts and 180 degrees in the area of the geniculate ganglion and the labyrinthine segment (Fig. 10.42).

15. After identification, the greater superficial petrosal nerve and its accompanying vessels are now being cut. In live surgery, coagulation of this nerve should take place before it is cut (Fig. 10.43).

16. The superior ampullary nerve is dissected from its canal using a 90 degrees hook (Fig. 10.44). The superior and inferior vestibular nerves are dissected posteriorly. The anteriorly lying internal auditory canal segment of the facial nerve and cochlear nerve can be seen (Figs. 10.45 and 10.46).

17. Using an angled pick, the geniculate ganglion and the labyrinthine portion of the facial nerve are carefully separated from the underlying bone. The tympanic segment is next carefully freed. The mastoid segment is then freed using sharp dissection to cut the tough adhesions at this level. The nerve is rerouted posteriorly together with all the contents of the internal auditory canal. The superior vestibular, inferior vestibular, and cochlear nerves are avulsed at the fundus of the canal, and all the contents of the canal are rerouted posteriorly with the facial nerve. The nerve is fixed to the posterior fossa dura just anterior to the sigmoid sinus using fibrin glue (Fig. 10.47). An aluminum strip is used to protect the dura and facial nerve. The fallopian canal can now be removed (Fig. 10.48).

18. The fallopian canal is now removed with a rongeur (Fig. 10.49). An aluminum strip is used to protect the porus of the internal auditory canal while drilling the remaining part of the anterior canal wall (Fig. 10.50). The petrous apex is drilled to the level of the mid-clivus, exposing the dura on the posterior surface of the temporal bone (Fig. 10.51).

19. The incision starts superiorly 1–2 mm parallel to and inferior to the superior petrosal sinus, running from posterior to anterior using microsurgical scissors (Fig. 10.52). The inferior incision starts just in front of the distal part of the sigmoid sinus and follows the course of the sinus and jugular bulb to the inferior petrosal sinus (Fig. 10.53).

Surgical Anatomy After Opening the Dura

Figs. 10.54–10.59 show the different structures seen after opening the dura.

20. Closure at the end of the procedure, the eustachian tube is closed with a muscle plug, and the cavity is obliterated with a single piece of abdominal fat. The musculoperiosteal flap is repositioned and sutured with Vicryl 1/0, and the wound is closed in two layers. A tight dressing is applied, and no drains are inserted.

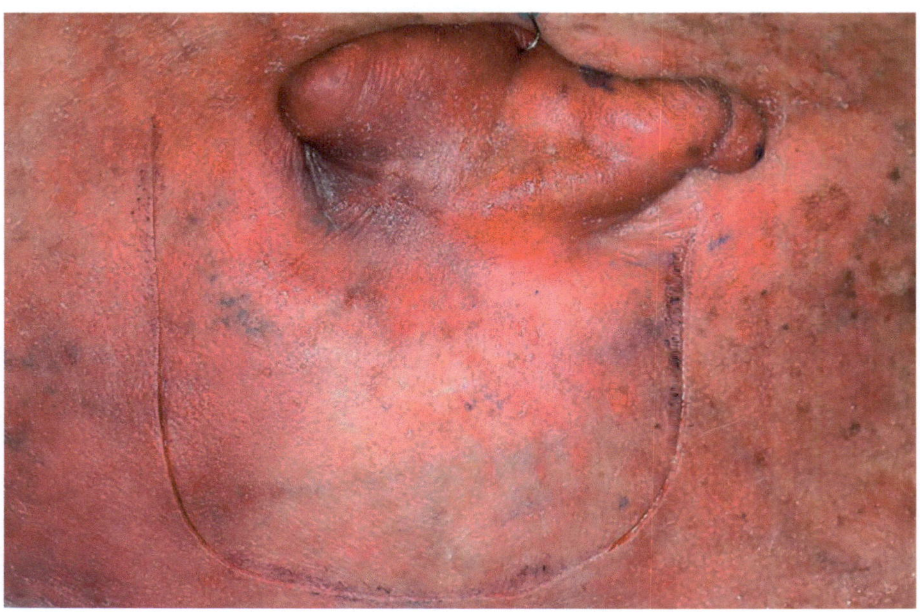

Fig. 10.5 A wide C-shaped postauricular incision is made. The incision starts 3 cm above the auricle, curves posteriorly to about 4–5 cm posterior to the postauricular sulcus, and ends inferiorly at the level of the mastoid tip.

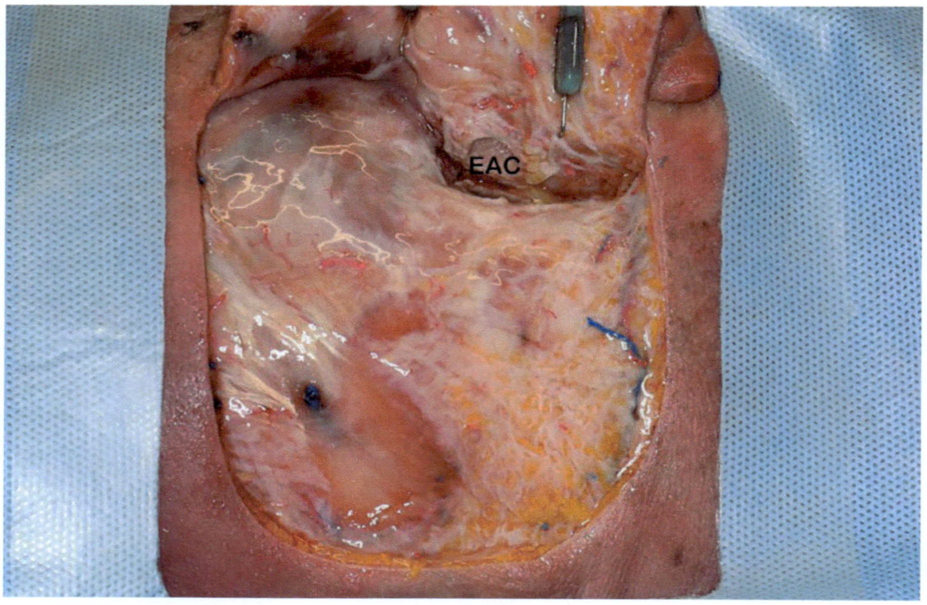

Fig. 10.6 Skin and subcutaneous tissues are elevated, and the external auditory canal is completely transected. *EAC*, external auditory canal.

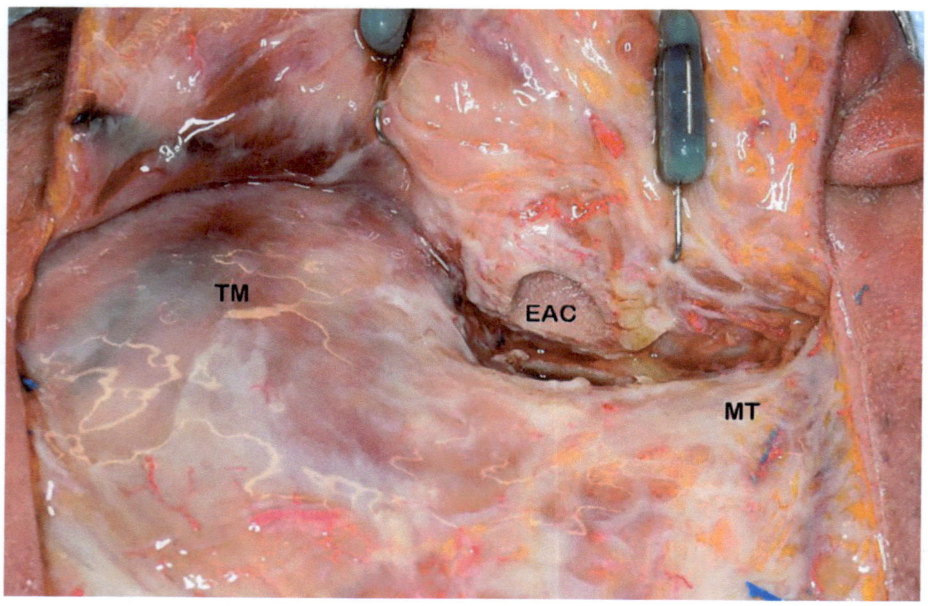

Fig. 10.7 At a higher magnification, the external auditory canal is transected at the bony cartilaginous junction. The external auditory canal is transected and closed as a cul-de-sac. "CuI de sac" closure of the external auditory canal is performed. *EAC*, external auditory canal; *MT*, mastoid tip; *TM*, temporalis muscle.

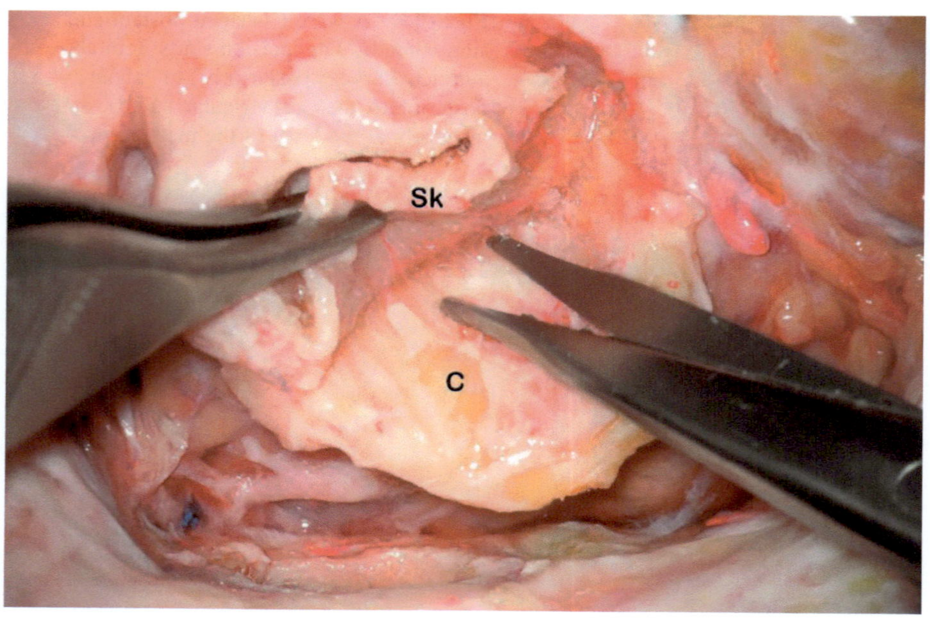

Fig. 10.8 The skin is then carefully separated from the cartilage with tympanoplasty scissors. *Sk*, skin; *C*, cartilage.

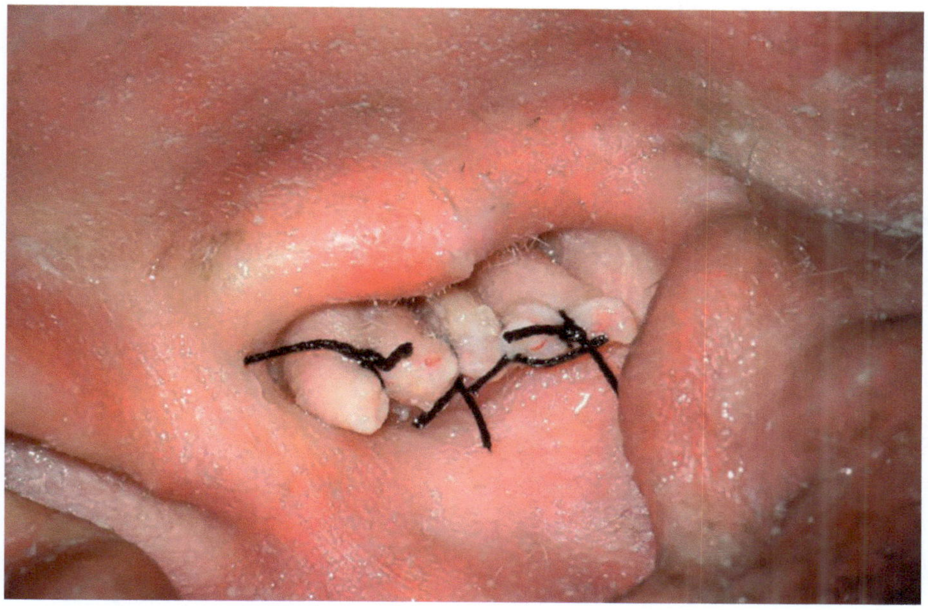

Fig. 10.9 The meatal skin is everted out through the canal and sutured with silk sutures.

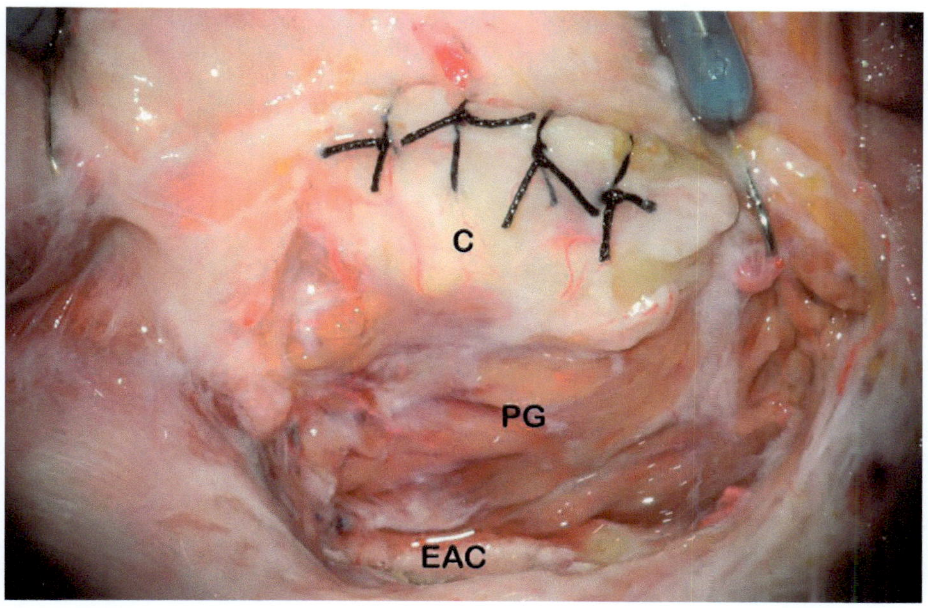

Fig. 10.10 A second layer of closure is achieved by approximating the cartilage of the tragus and the concha. *C*, cartilage; *PG*, parotid gland; *EAC*, external auditory canal.

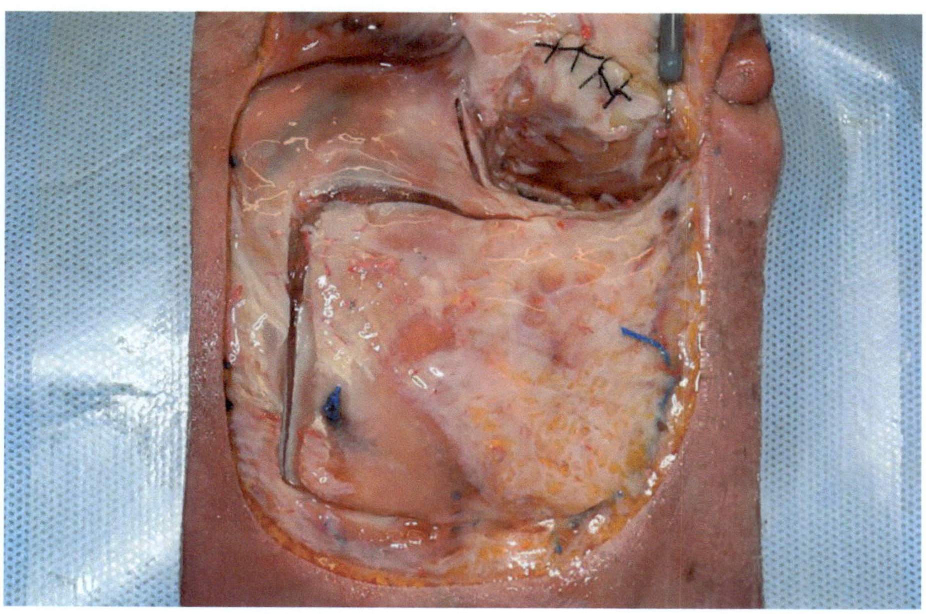

Fig. 10.11 A U-shaped, inferiorly based musculoperiosteal flap is outlined.

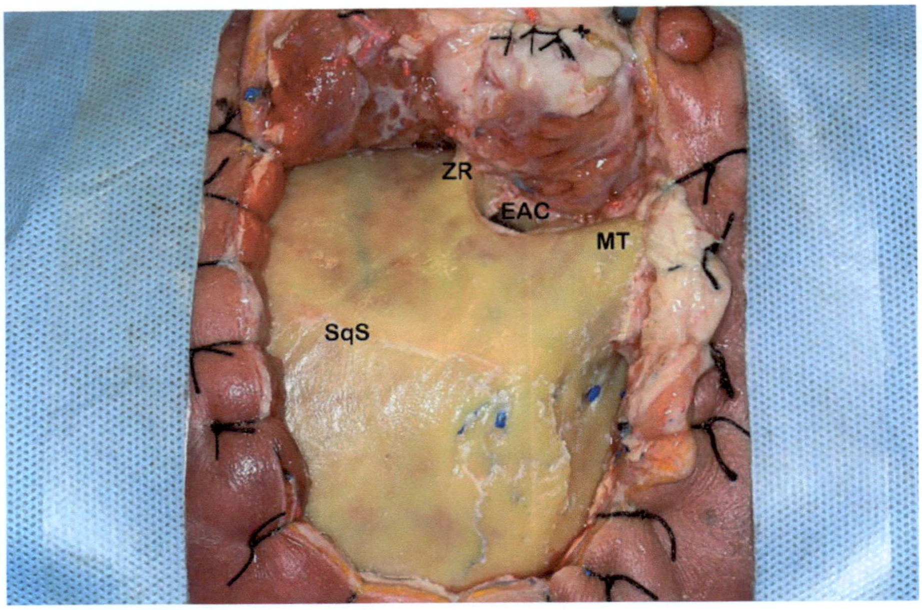

Fig. 10.12 The flap is elevated using a Lempert periosteal elevator. We use 1–0 silk sutures to stitch the soft tissue away from the surgical field, thus avoiding the use of retractors. *ZR*, zygomatic root; *EAC*, external auditory canal; *MT*, mastoid tip; *SqS*, squamous suture.

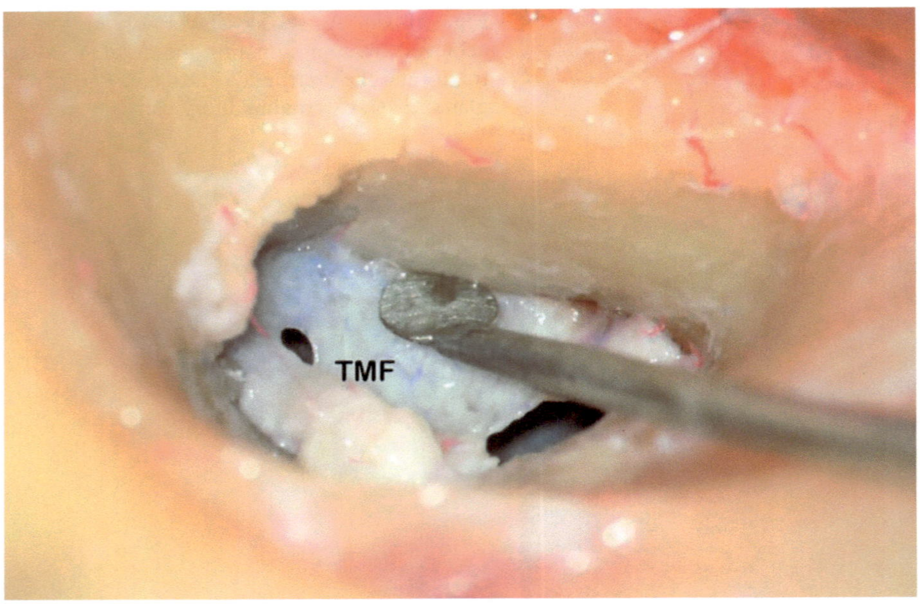

Fig. 10.13 The tympanomeatal flap is detached from the bony canal wall. The skin of the bony external auditory canal is removed, together with the tympanic membrane. *TMF*, tympanomeatal flap.

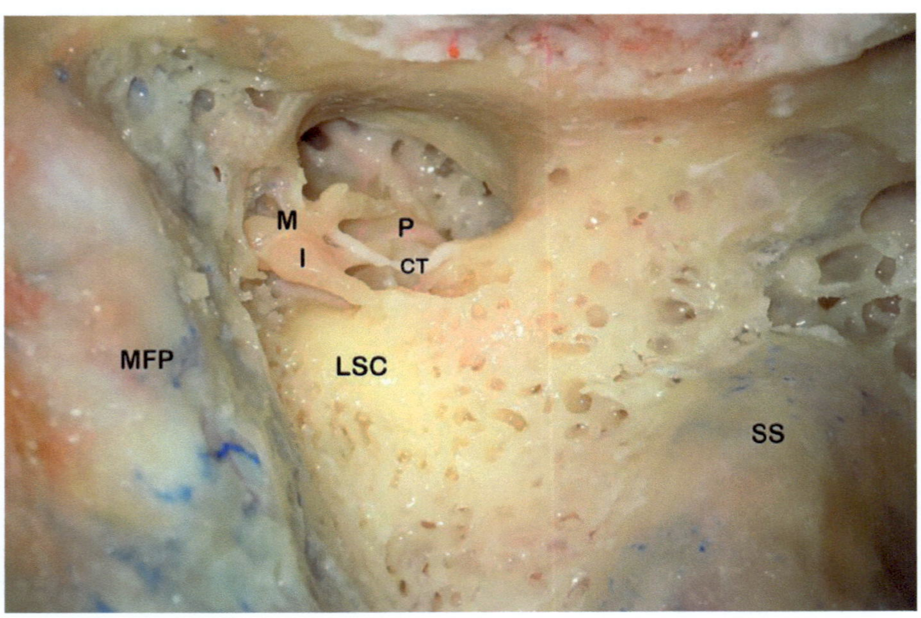

Fig. 10.14 An extended mastoidectomy is performed, with skeletonization of the middle fossa dura and sigmoid sinus. The posterior and superior bony external canal walls are drilled. *SS*, sigmoid sinus; *LSC*, lateral semicircular canal; *CT*, chorda tympani; *P*, promontory; *I*, incus; *M*, malleus; *MFP*, middle fossa plate.

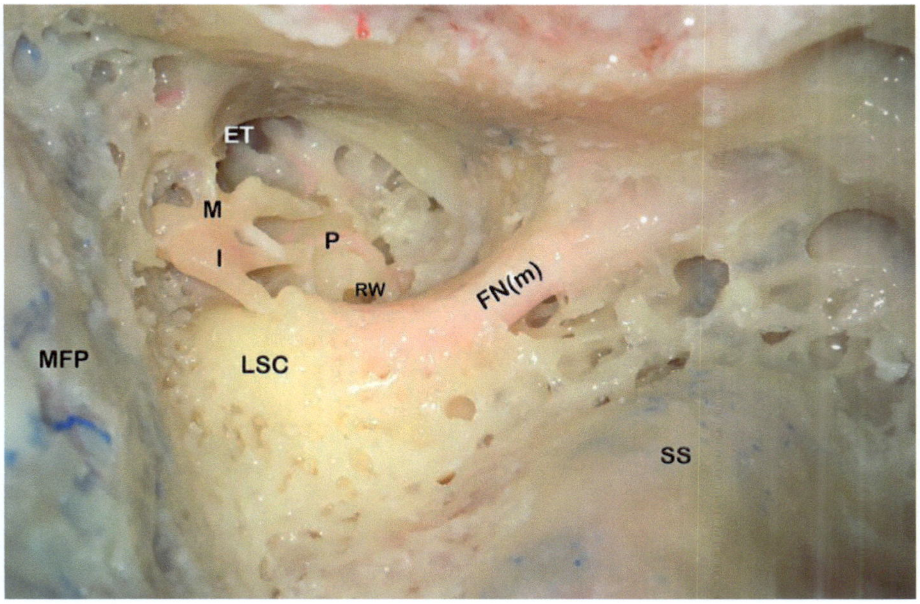

Fig. 10.15 The facial nerve is skeletonized. *SS*, sigmoid sinus; *RW*, round window; *ET*, eustachian tube; *M*, malleus; *I*, incus; *LSC*, lateral semicircular canal; *MFP*, middle fossa plate; *FN(m)*, mastoid segment of facial nerve.

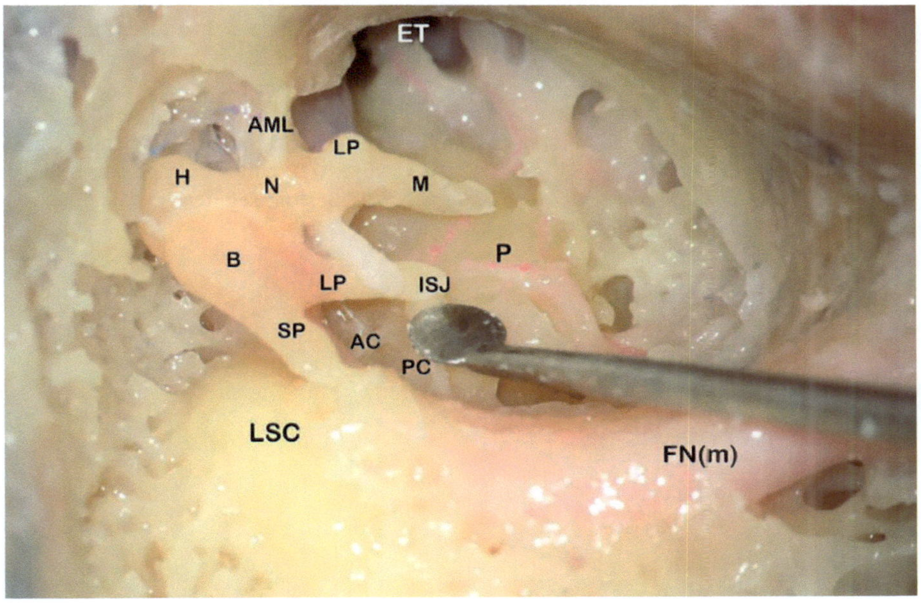

Fig. 10.16 The incudostapedial joint is being disarticulated. *ET*, eustachian tube; *M*, manubrium of the malleus; *LP*, lateral process of the malleus; *N*, neck of the malleus; *H*, head of the malleus; *B*, body of incus; *P*, promontory; *ISJ*, incudostapedial joint; *PC*, posterior crus; *AC*, anterior crus; *SP*, styloid process; *LSC*, lateral semicircular canal; *AML*, anterior malleal ligament; *FN(m)*, mastoid segment of facial nerve.

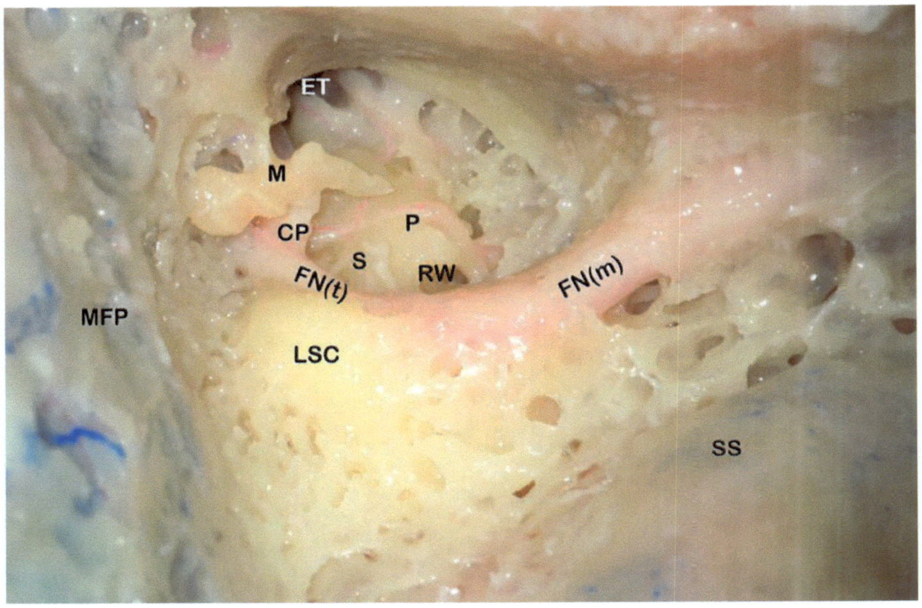

Fig. 10.17 The incus has been removed. *ET*, eustachian tube; *M*, malleus; *S*, stapes; *CP*, cochleariform process; *FN(t)*, tympanic segment of facial nerve; *FN(m)*, mastoid segment of facial nerve; *P*, promontory; *RW*, round window; *LSC*, lateral semicircular canal; *MFP*, middle fossa plate; *SS*, sigmoid sinus.

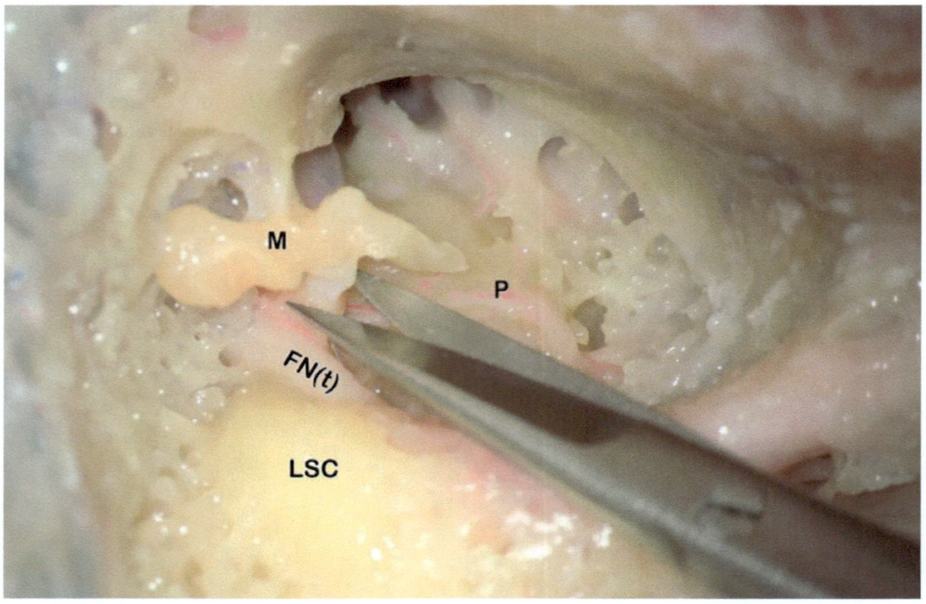

Fig. 10.18 The tensor tympani tendon is being cut. *M*, malleus; *P*, promontory; *FN(t)*, tympanic segment of facial nerve; *LSC*, lateral semicircular canal.

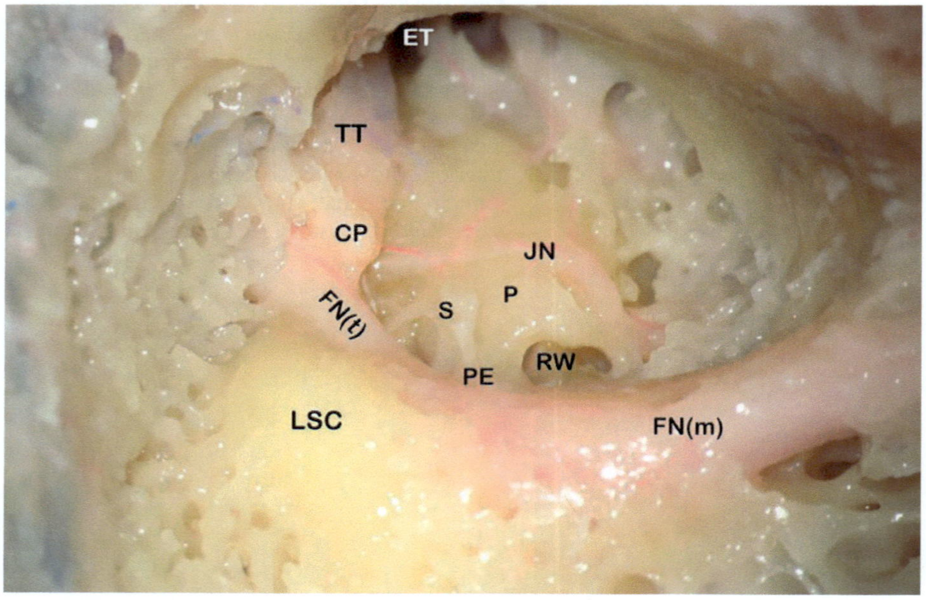

Fig. 10.19 The malleus has been removed. *ET*, eustachian tube; *TT*, tensor tympani muscle; *CP*, cochleariform process; *JN*, Jacobson's nerve; *P*, promontory; *RW*, round window; *PE*, pyramidal eminence; *S*, stapes; *LSC*, lateral semicircular canal; *FN(t)*, tympanic segment of facial nerve; *FN(m)*, mastoid segment of facial nerve.

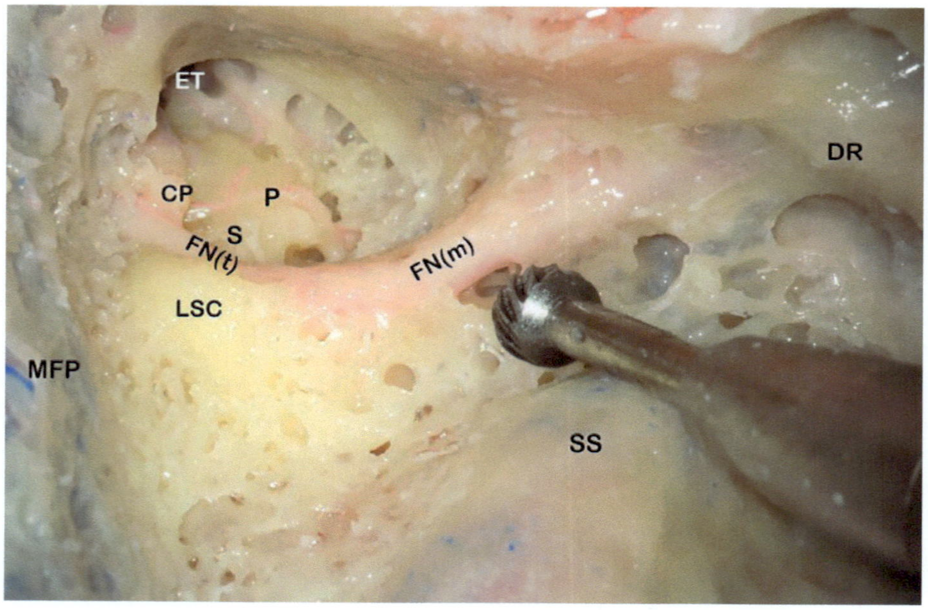

Fig. 10.20 The retrofacial air cells are drilled to complete the skeletonization of the facial nerve. *ET*, eustachian tube; *P*, promontory; *S*, stapes; *CP*, cochleariform process; *LSC*, lateral semicircular canal; *DR*, digastric ridge; *SS*, sigmoid sinus; *MFP*, middle fossa plate; *FN(t)*, tympanic segment of facial nerve; *FN(m)*, mastoid segment of facial nerve.

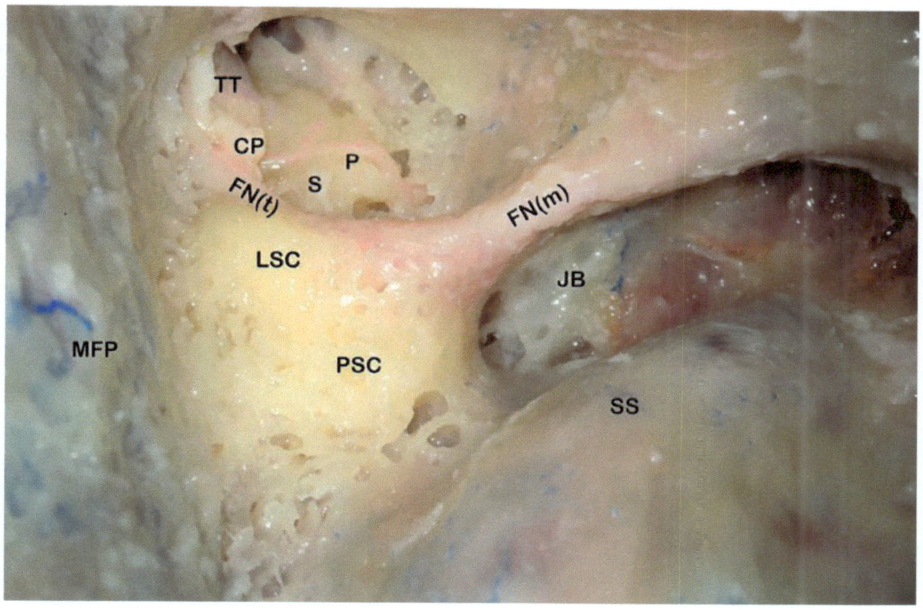

Fig. 10.21 The facial nerve has been well skeletonized, and the jugular bulb has been identified. *TT*, tensor tympani muscle; *FN(m)*, mastoid segment of facial nerve; *PSC*, posterior semicircular canal; *LSC*, lateral semicircular canal; *S*, stapes; *P*, promontory; *CP*, cochleariform process; *MFP*, middle fossa plate; *FN(t)*, tympanic segment of facial nerve; *SS*, sigmoid sinus; *JB*, jugular bulb.

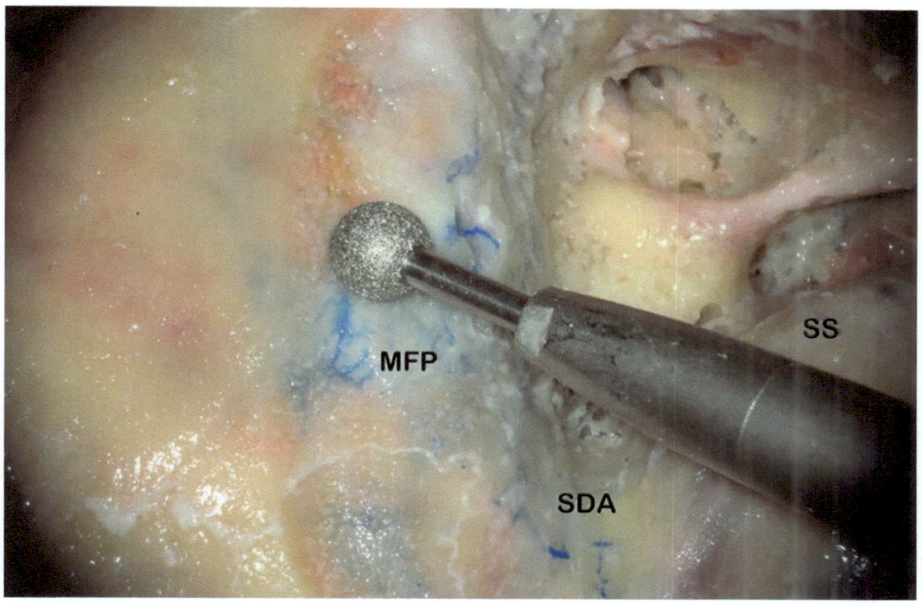

Fig. 10.22 A large diamond burr is used to remove the bone overlying the middle fossa dura. *SS*, sigmoid sinus; *MFP*, middle fossa plate; *SDA*, sinodural angle.

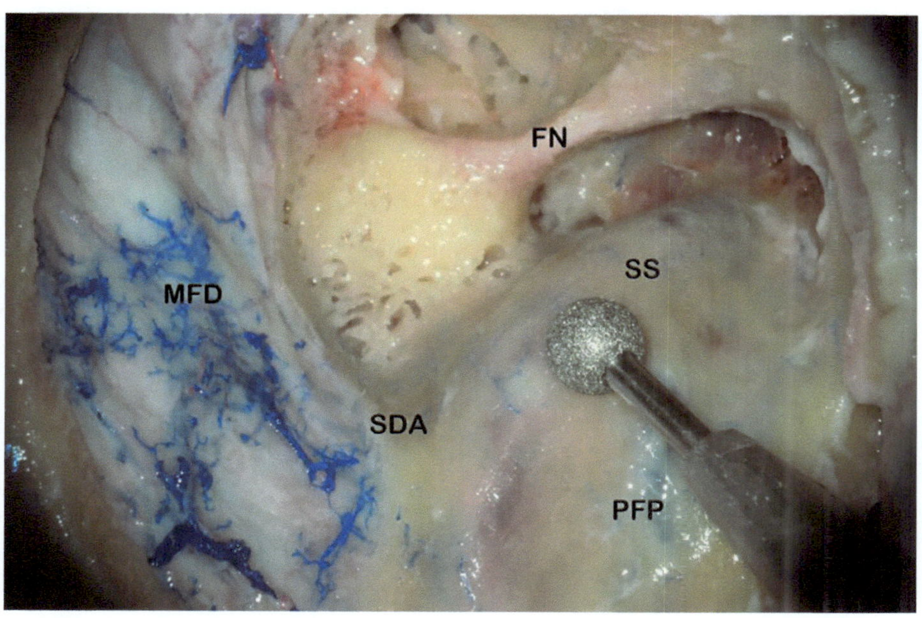

Fig. 10.23 A large diamond burr is used to remove the bone overlying the sigmoid sinus and the posterior fossa dura. *FN*, facial nerve; *SS*, sigmoid sinus; *PFP*, posterior fossa plate; *SDA*, sinodural angle; *MFD*, middle fossa dura.

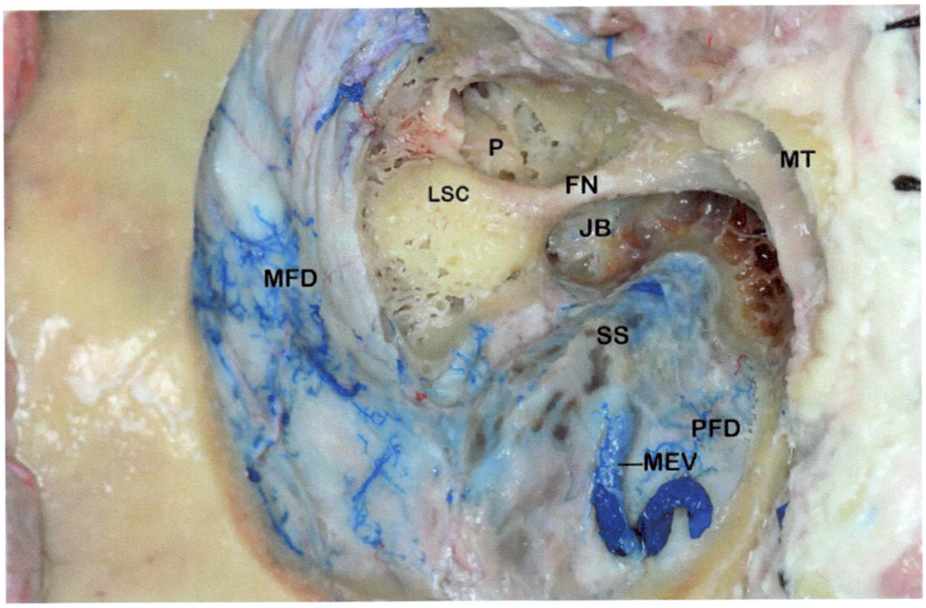

Fig. 10.24 The bone overlying the middle fossa dura, sigmoid sinus, and posterior fossa dura has already been uncovered and removed. *MT*, mastoid tip; *FN*, facial nerve; *JB*, jugular bulb; *P*, promontory; *LSC*, lateral semicircular canal; *MFD*, middle fossa dura; *PFD*, posterior fossa dura; *MEV*, mastoid emissary vein; *SS*, sigmoid sinus.

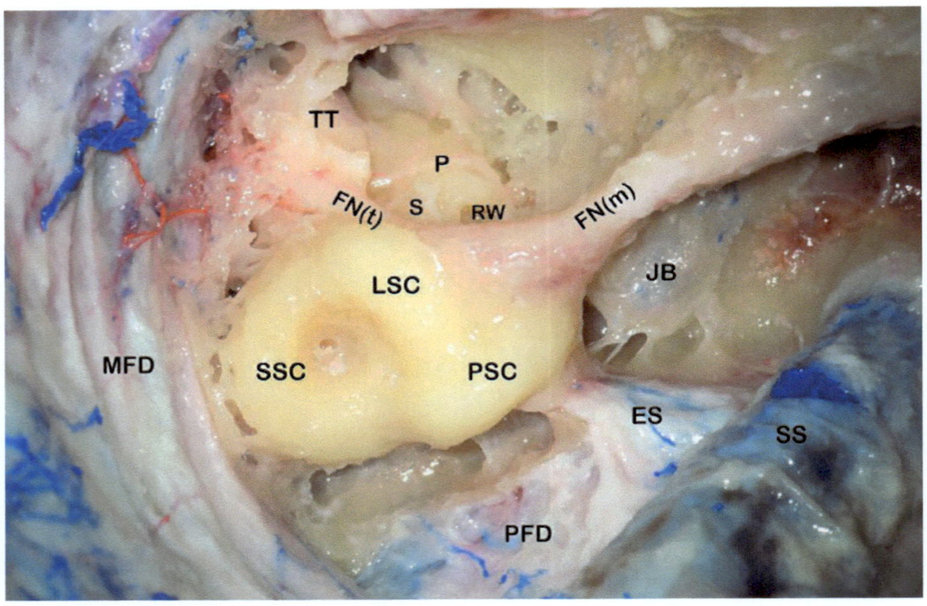

Fig. 10.25 The lateral, posterior, and superior semicircular canals have been skeletonized. *TT*, tensor tympani muscle; *P*, promontory; *S*, stapes; *RW*, round window; *FN(m)*, mastoid segment of facial nerve; *FN(t)*, tympanic segment of facial nerve; *SSC*, superior semicircular canal; *LSC*, lateral semicircular canal; *PSC*, posterior semicircular canal; *JB*, jugular bulb; *SS*, sigmoid sinus; *ES*, endolymphatic sac; *PFD*, posterior fossa dura; *MFD*, middle fossa dura.

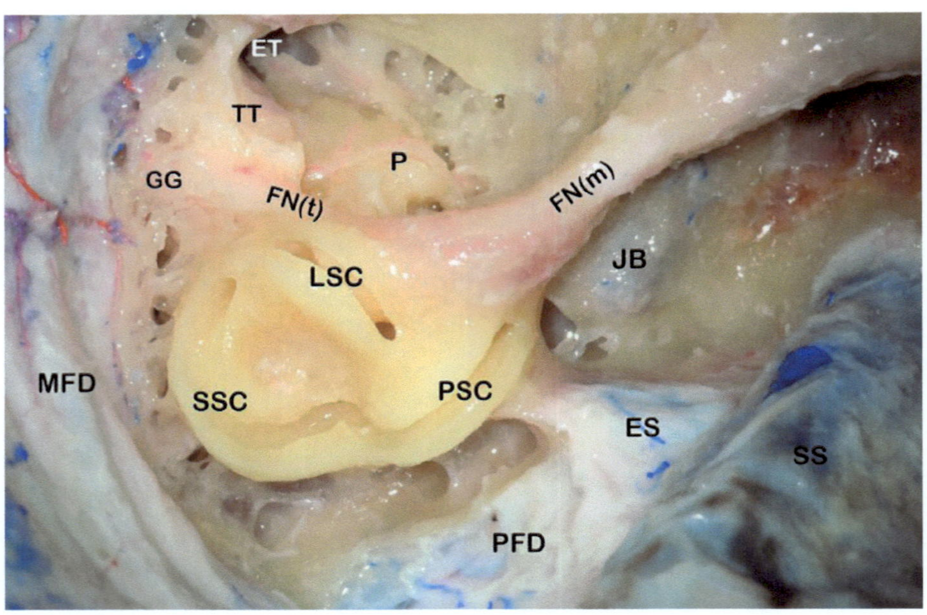

Fig. 10.26 The labyrinthectomy is performed. The lateral semicircular canal, posterior semicircular canal, and superior semicircular canal have been opened. *ET*, eustachian tube; *TT*, tensor tympani muscle; *P*, promontory; *GG*, geniculate ganglion; *FN(t)*, tympanic segment of facial nerve; *FN(m)*, mastoid segment of facial nerve; *SSC*, superior semicircular canal; *LSC*, lateral semicircular canal; *PSC*, posterior semicircular canal; *JB*, jugular bulb; *ES*, endolymphatic sac; *PFD*, posterior fossa dura; *MFD*, middle fossa dura; *SS*, sigmoid sinus.

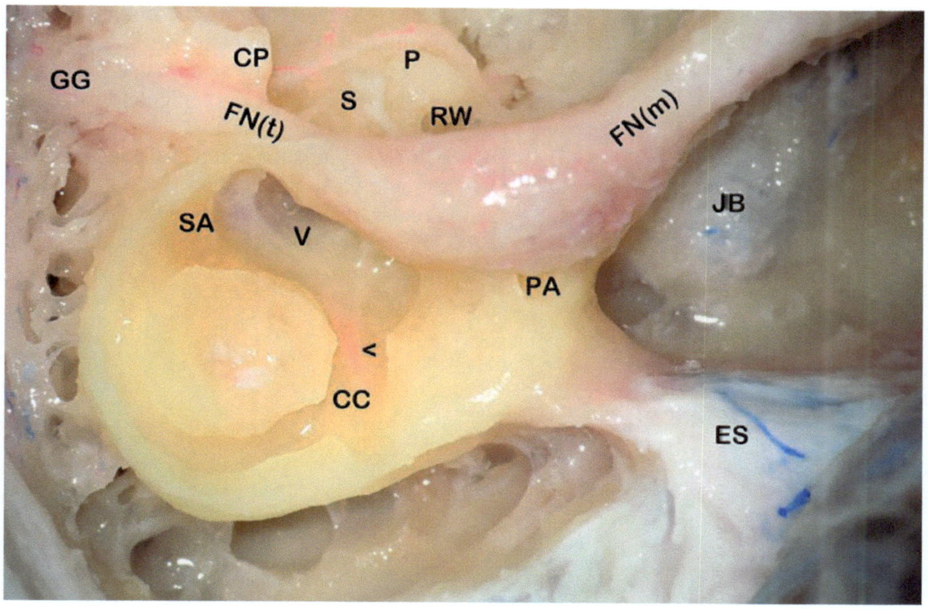

Fig. 10.27 The vestibule has been opened. Note the thin amount of bone separating the genu of the facial nerve from the vestibule. *JB*, jugular bulb; *ES*, endolymphatic sac; *PA*, ampullated end of posterior semicircular canal; *SA*, ampullated end of superior semicircular canal; *FN(m)*, mastoid segment of facial nerve; *FN(t)*, tympanic segment of facial nerve; *RW*, round window; *P*, promontory; *S*, stapes; *CP*, cochleariform process; *GG*, geniculate ganglion; *V*, vestibule; *CC*, common crus; <, bifurcation of endolymphatic duct.

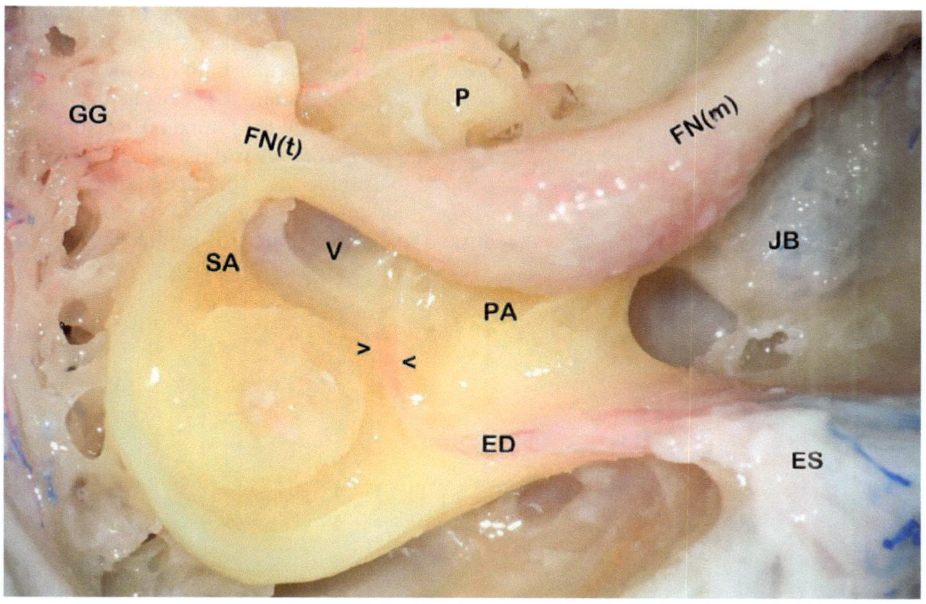

Fig. 10.28 The canals have been drilled away. Note that the endolymphatic duct reaches the vestibule, where it bifurcates into the utricular duct and saccular duct. *JB*, jugular bulb; *ES*, endolymphatic sac; *FN(m)*, mastoid segment of facial nerve; *P*, promontory; *PA*, ampullated end of PSC; *ED*, endolymphatic duct; *V*, vestibule; *SA*, ampullated end of superior semicircular canal; *FN(t)*, tympanic segment of facial nerve; *GG*, geniculate ganglion; <, saccular duct; >, utricular duct.

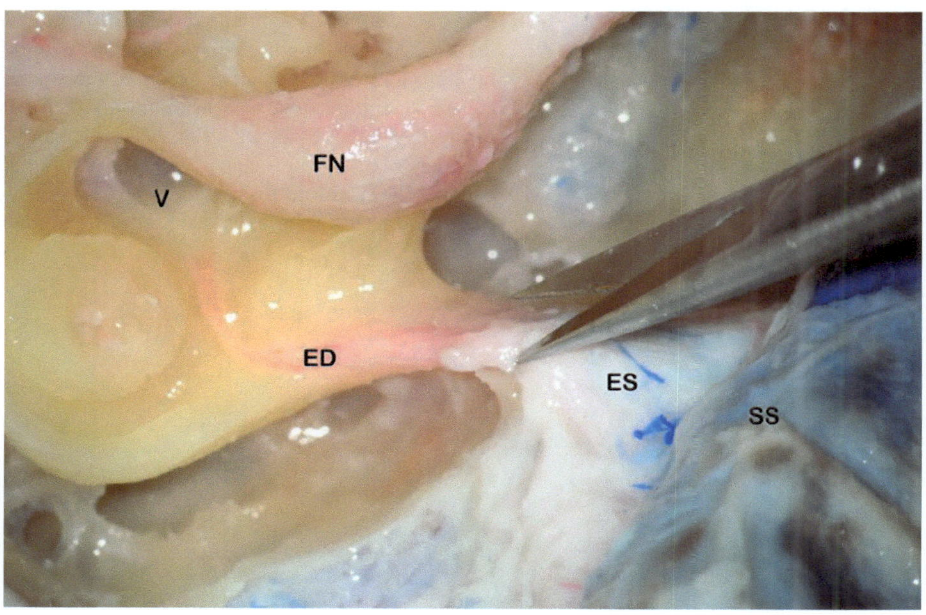

Fig. 10.29 The endolymphatic duct extends from the medial surface of the posterior canal to the posterior fossa dura anterior to the sigmoid sinus and should be transected with the use of a Beaver knife or microscissors. *FN*, facial nerve; *V*, vestibule; *ED*, endolymphatic duct; *ES*, endolymphatic sac; *SS*, sigmoid sinus.

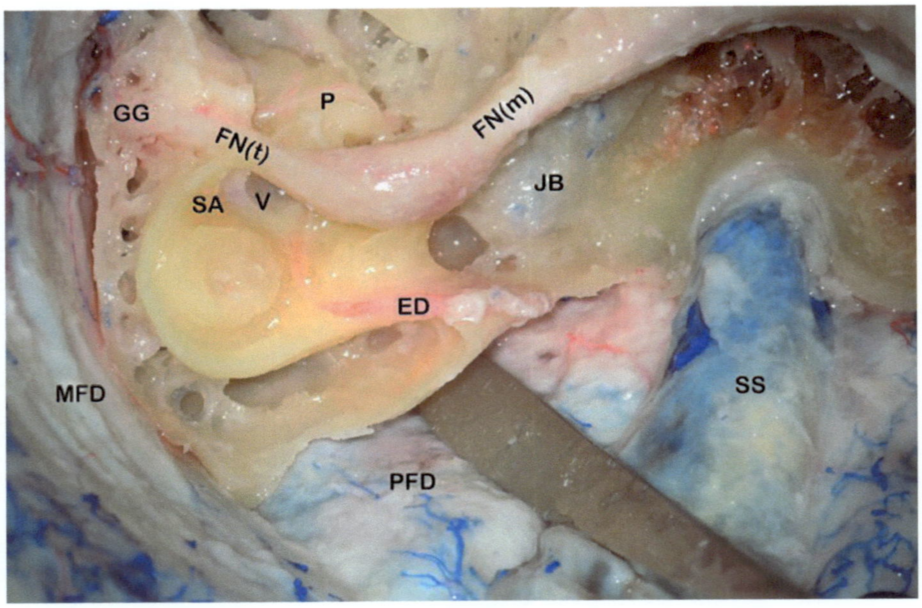

Fig. 10.30 Sharply cutting the endolymphatic duct facilitates the complete drilling of bone overlying the posterior fossa dura and the subsequent dural retraction. *GG*, geniculate ganglion; *P*, promontory; *FN(m)*, mastoid segment of facial nerve; *FN(t)*, tympanic segment of facial nerve; *V*, vestibule; *SA*, ampullated end of superior semicircular canal; *MFD*, middle fossa dura; *PFD*, posterior fossa dura; *SS*, sigmoid sinus; *JB*, jugular bulb.

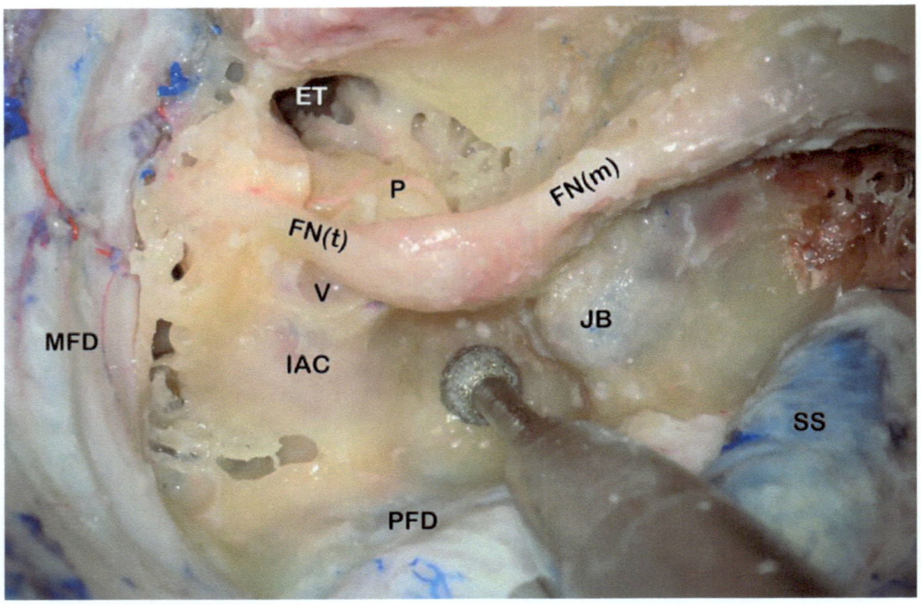

Fig. 10.31 Identification of the lower edge of the internal auditory canal starts by drilling the bone between the jugular bulb and the expected position of the canal. *ET*, eustachian tube; *P*, promontory; *FN(t)*, tympanic segment of facial nerve; *FN(m)*, mastoid segment of facial nerve; *V*, vestibule; *IAC*, internal auditory canal; *MFD*, middle fossa dura; *PFD*, posterior fossa dura; *SS*, sigmoid sinus; *JB*, jugular bulb.

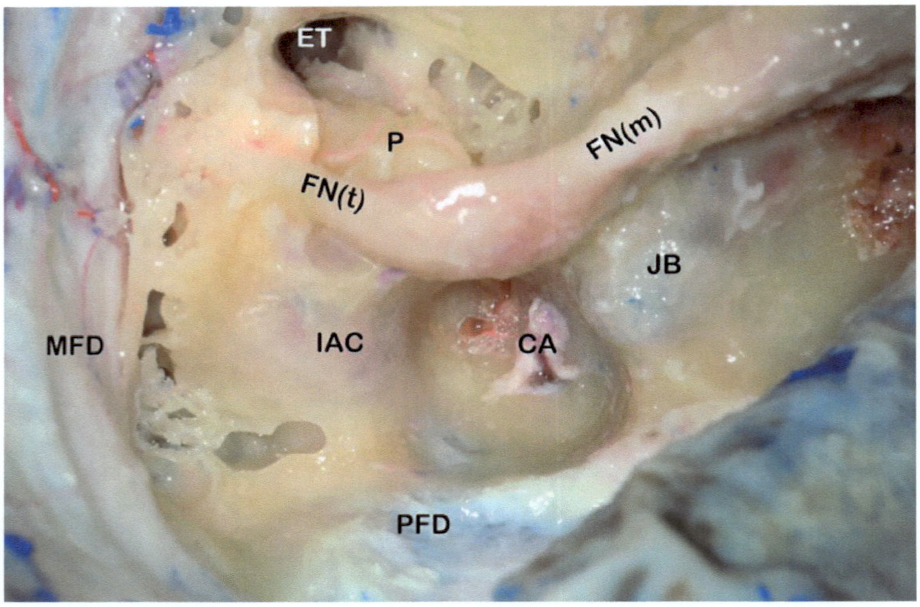

Fig. 10.32 The cochlear aqueduct can be identified by drilling the bone between the jugular bulb and the expected position of the canal. *ET*, eustachian tube; *P*, promontory; *FN(m)*, mastoid segment of facial nerve; *FN(t)*, tympanic segment of facial nerve; *IAC*, internal auditory canal; *MFD*, middle fossa dura; *PFD*, posterior fossa dura; *CA*, cochlear aqueduct; *JB*, jugular bulb.

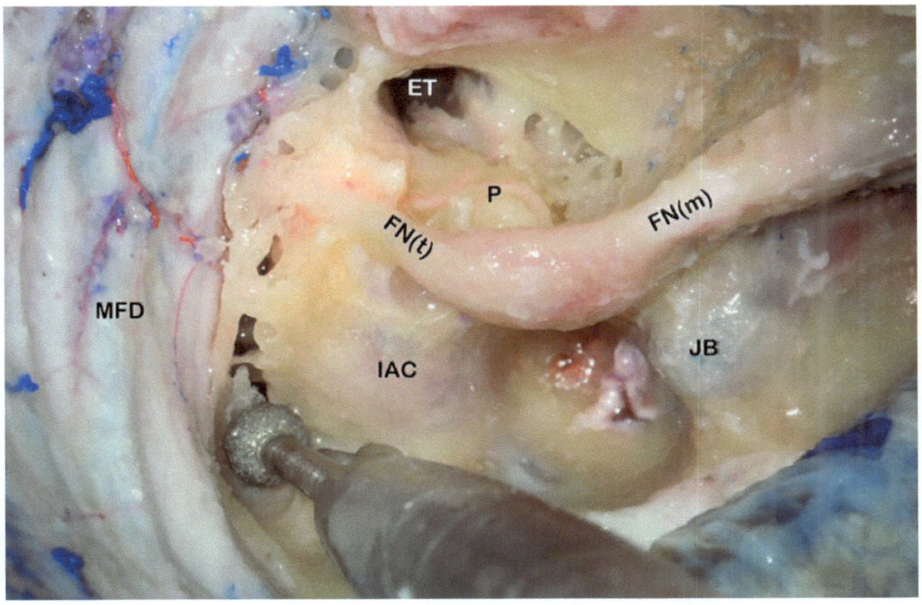

Fig. 10.33 Drilling of the superior surface of the internal auditory canal is carried out. *ET*, eustachian tube; *P*, promontory; *FN(m)*, mastoid segment of facial nerve; *FN(t)*, tympanic segment of facial nerve; *IAC*, internal auditory canal; *MFD*, middle fossa dura; *JB*, jugular bulb.

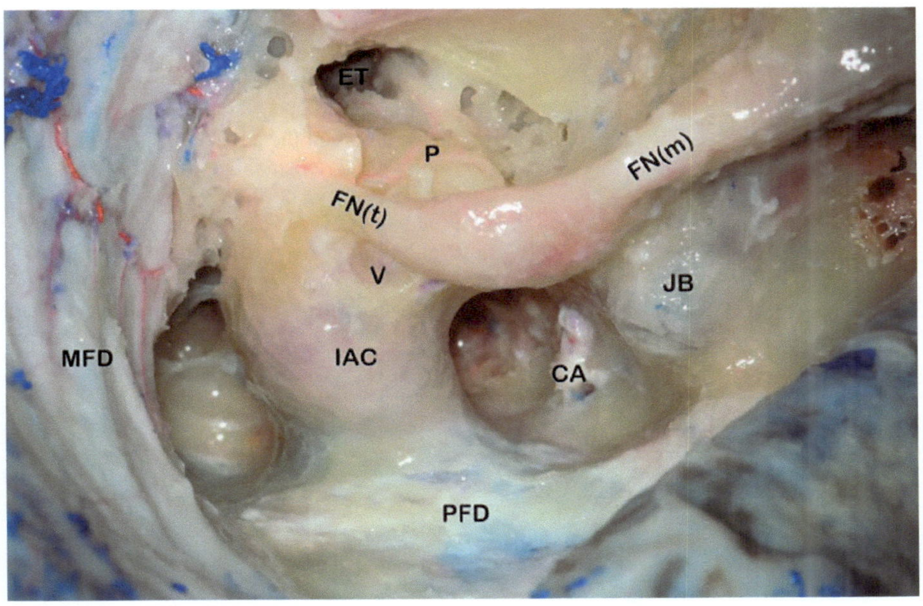

Fig. 10.34 The internal auditory canal is outlined with a thin shell of bone still covering it. *ET*, eustachian tube; *P*, promontory; *FN(t)*, tympanic segment of facial nerve; *FN(m)*, mastoid segment of facial nerve; *V*, vestibule; *IAC*, internal auditory canal; *CA*, cochlear aqueduct; *PFD*, posterior fossa dura; *MFD*, middle fossa dura; *JB*, jugular bulb.

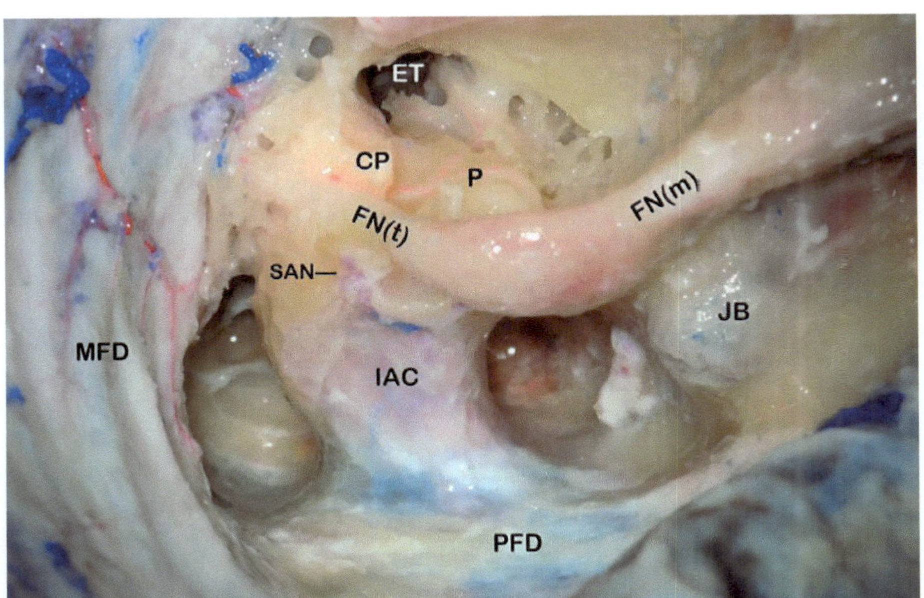

Fig. 10.35 The remaining thin shell of bone is separated from the underlying dura of the internal auditory canal using a dissector and is then carefully removed. *ET*, eustachian tube; *CP*, cochleariform process; *P*, promontory; *SAN*, superior ampullary nerve; *FN(t)*, tympanic segment of facial nerve; *FN(m)*, mastoid segment of facial nerve; *JB*, jugular bulb; *MFD*, middle fossa dura; *PFD*, posterior fossa dura; *IAC*, internal auditory canal.

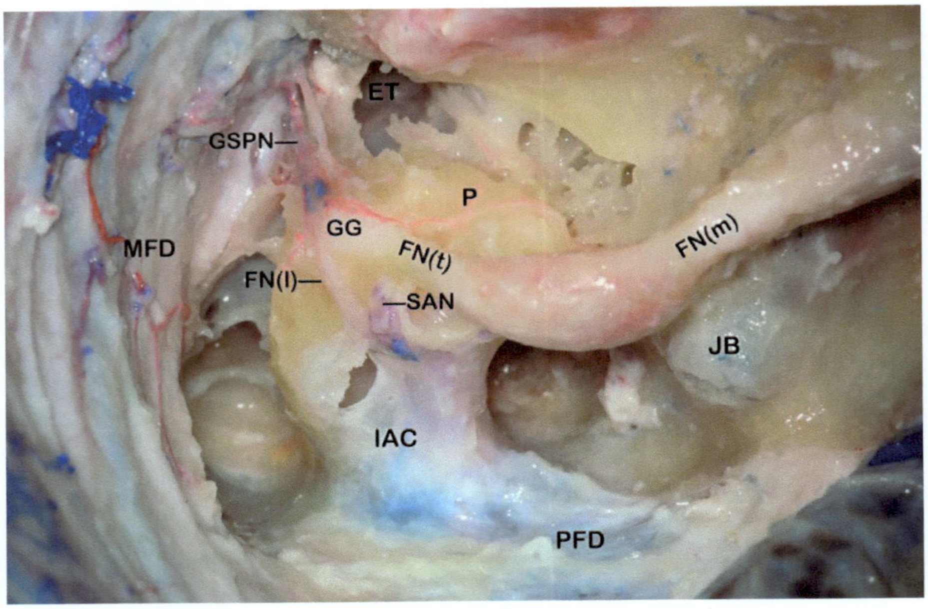

Fig. 10.36 The greater superficial petrosal nerve and geniculate ganglion are dissected. The mastoid, tympanic, and labyrinthine portions of the facial nerve are visible. *ET*, eustachian tube; *P*, promontory; *GSPN*, greater superficial petrosal nerve; *FN(t)*, tympanic segment of facial nerve; *FN(m)*, mastoid segment of facial nerve; *IAC*, internal auditory canal; *SAN*, superior ampullary nerve; *GG*, geniculate ganglion; *FN (l)*, labyrinthine segment of facial nerve; *MFD*, middle fossa dura; *PFD*, posterior fossa dura; *JB*, jugular bulb.

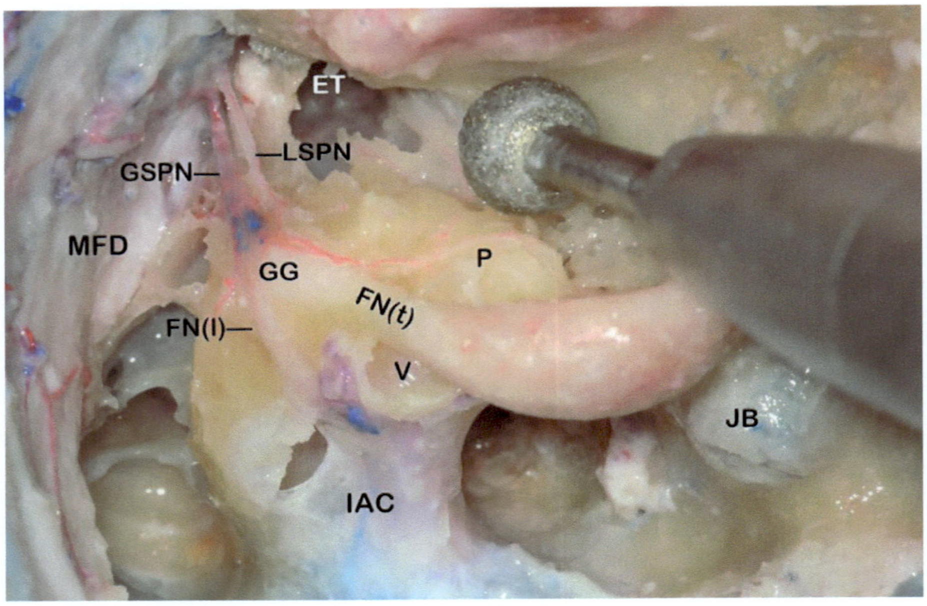

Fig. 10.37 The anterior wall of the external auditory canal and the tympanic bone are thinned out. Identification of the internal carotid artery is either started from the eustachian tube, where the burr is located or at the anteroinferior limit of the promontory. *ET*, eustachian tube; *LSPN*, lesser superficial petrosal nerve; *GSPN*, greater superficial petrosal nerve; *P*, promontory; *GG*, geniculate ganglion; *FN(l)*, labyrinthine segment of facial nerve; *FN(t)*, tympanic segment of facial nerve; *V*, vestibule; *IAC*, internal auditory canal; *MFD*, middle fossa dura; *JB*, jugular bulb.

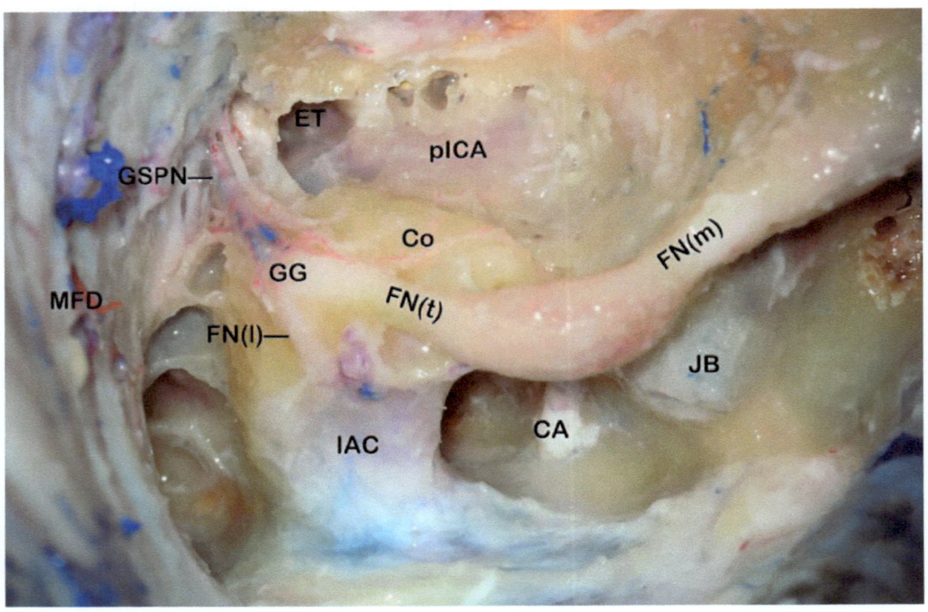

Fig. 10.38 The tympanic bone has been partially drilled away and the internal carotid artery is identified in front of the cochlea. *ET*, eustachian tube; *pICA*, petrous segment of internal carotid artery; *GSPN*, greater superficial petrosal nerve; *GG*, geniculate ganglion; *FN(l)*, labyrinthine segment of facial nerve; *FN(t)*, tympanic segment of facial nerve; *FN(m)*, mastoid segment of facial nerve; *CA*, cochlear aqueduct; *IAC*, internal auditory canal; *Co*, cochlea; *MFD*, middle fossa dura; *JB*, jugular bulb.

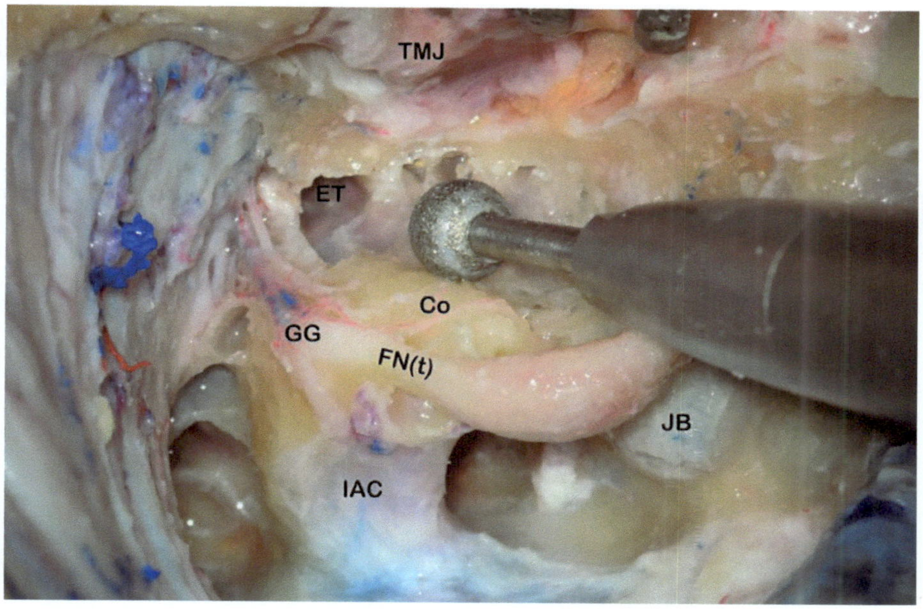

Fig. 10.39 After identification of the internal carotid artery, a diamond burr is used initially to open the cochlea. *TMJ*, temporomandibular joint; *ET*, eustachian tube; *Co*, cochlea; *GG*, geniculate ganglion; *FN(t)*, tympanic segment of facial nerve; *IAC*, internal auditory canal; *JB*, jugular bulb.

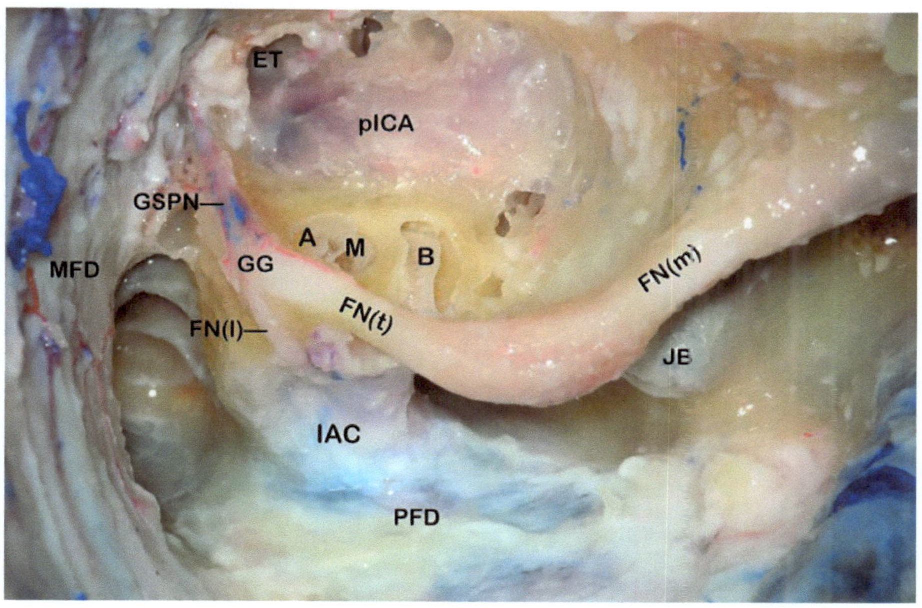

Fig. 10.40 The cochlea is partially drilled and the basal turn, middle turn, and apical turn can be seen. *ET*, eustachian tube; *pICA*, petrous segment of internal carotid artery; *A*, apical turn of cochlea; *M*, middle turn of cochlea; *B*, basal turn of cochlea; *GSPN*, greater superficial petrosal nerve; *GG*, geniculate ganglion; *FN(m)*, mastoid segment of facial nerve; *FN(t)*, tympanic segment of facial nerve; *FN(l)*, labyrinthine segment of facial nerve; *IAC*, internal auditory canal; *MFD*, middle fossa dura; *PFD*, posterior fossa dura; *JB*, jugular bulb.

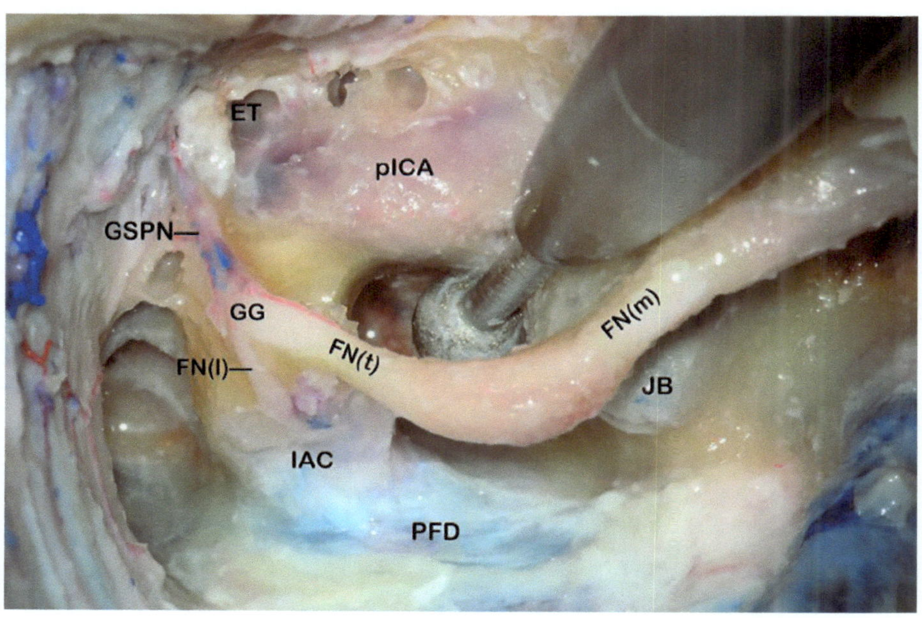

Fig. 10.41 The bone medial to the facial nerve is drilled to further skeletonize the internal carotid artery and jugular bulb. *ET*, eustachian tube; *pICA*, petrous segment of internal carotid artery; *GSPN*, greater superficial petrosal nerve; *GG*, geniculate ganglion; *FN(t)*, tympanic segment of facial nerve; *FN(l)*, labyrinthine segment of facial nerve; *FN(m)*, mastoid segment of facial nerve; *IAC*, internal auditory canal; *PFD*, posterior fossa dura; *JB*, jugular bulb.

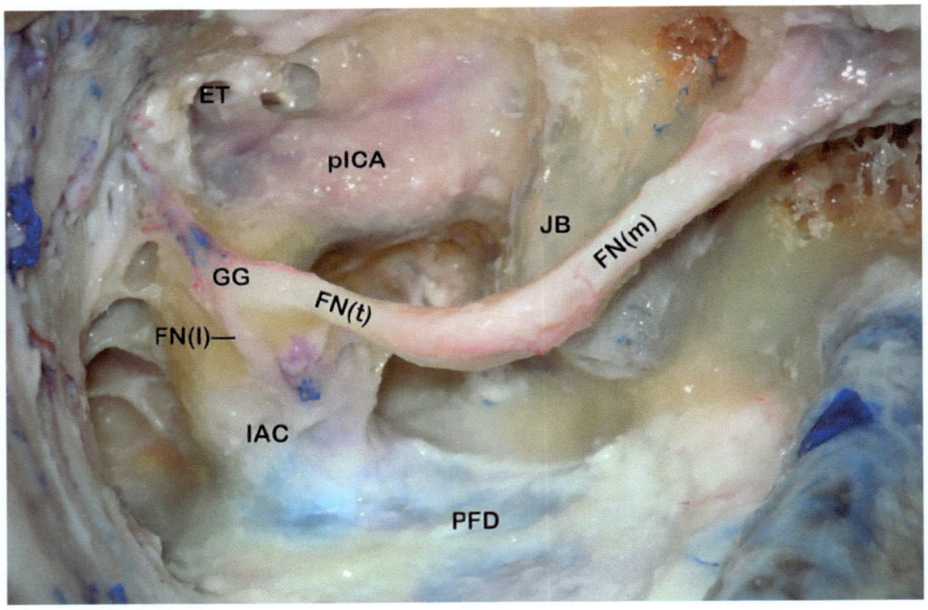

Fig. 10.42 The remaining part of the cochlea is removed. The internal auditory canal is drilled for 360 degrees of its circumference. The facial nerve has been decompressed, and the mastoid, tympanic, and labyrinthine portions are visible. *ET*, eustachian tube; *pICA*, petrous segment of internal carotid artery; *GG*, geniculate ganglion; *FN(t)*, tympanic segment of facial nerve; *FN(m)*, mastoid segment of facial nerve; *FN(l)*, labyrinthine segment of facial nerve; *IAC*, internal auditory canal; *PFD*, posterior fossa dura; *JB*, jugular bulb.

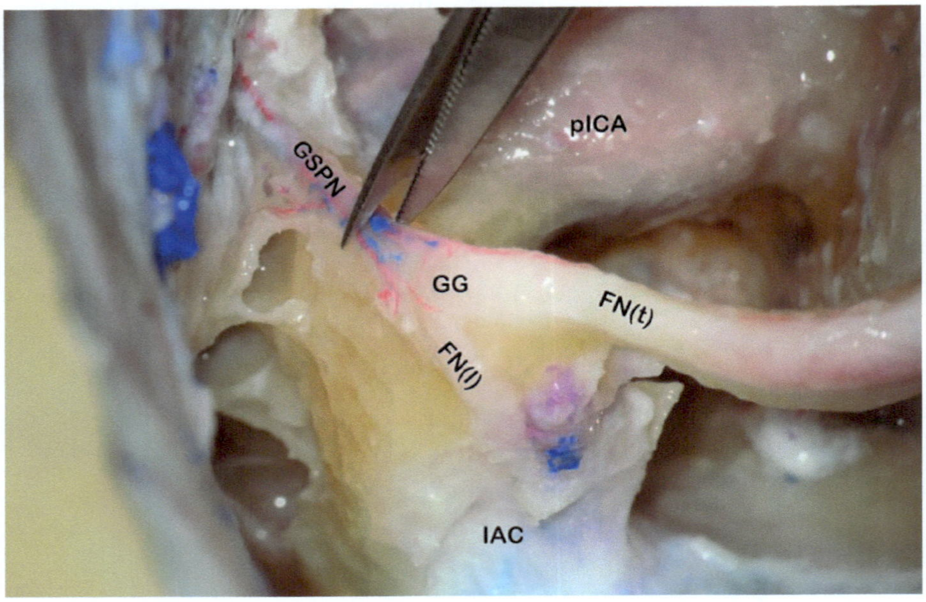

Fig. 10.43 After identification, the greater superficial petrosal nerve is now being cut. In live surgery, coagulation of this nerve should take place before it is cut. *GSPN*, greater superficial petrosal nerve; *pICA*, petrous segment of internal carotid artery; *FN(t)*, tympanic segment of facial nerve; *GG*, geniculate ganglion; *FN(l)*, labyrinthine segment of facial nerve; *IAC*, internal auditory canal.

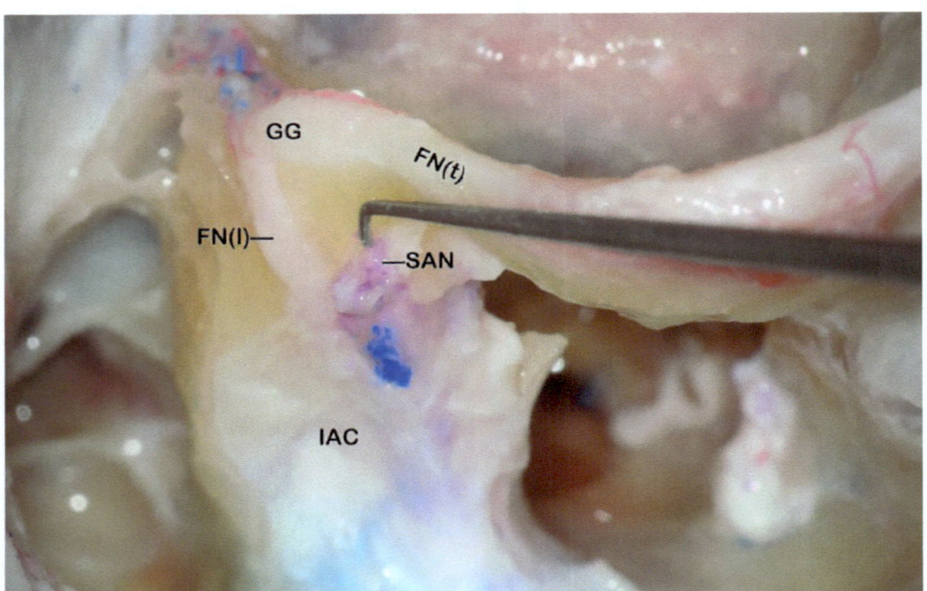

Fig. 10.44 The superior ampullary nerve is dissected from its canal using a 90degrees hook. *GG*, geniculate ganglion; *FN(t)*, tympanic segment of facial nerve; *FN(l)*, labyrinthine segment of facial nerve; *SAN*, superior ampullary nerve; *IAC*, internal auditory canal.

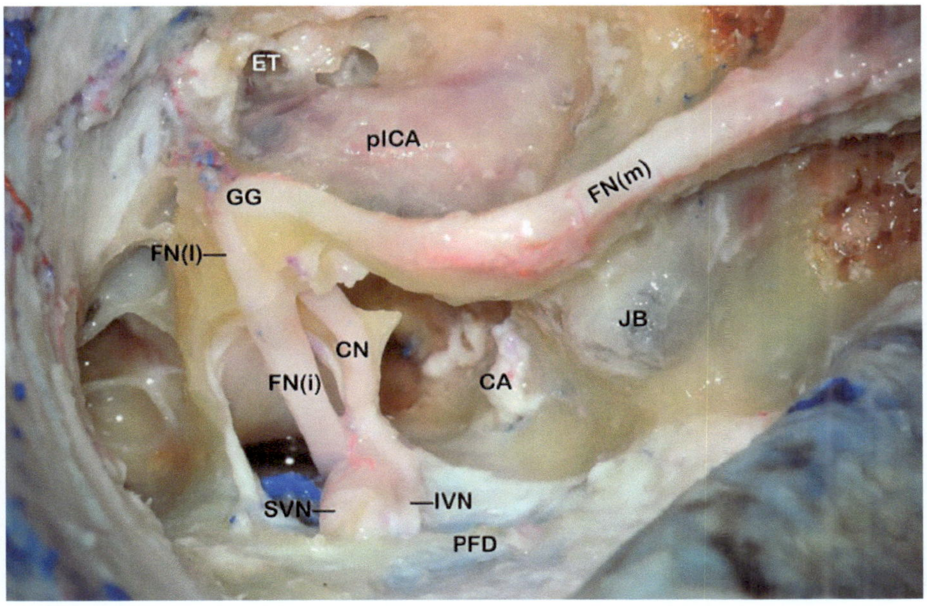

Fig. 10.45 The superior and inferior vestibular nerves are dissected posteriorly. The anteriorly lying internal auditory canal segment of the facial nerve and cochlear nerve can be seen. *ET*, eustachian tube; *pICA*, petrous segment of internal carotid artery; *FN(m)*, mastoid segment of facial nerve; *GG*, geniculate ganglion; *FN(l)*, labyrinthine segment of facial nerve; *FN(i)*, internal auditory canal segment of facial nerve; *CN*, cochlear nerve; *CA*, cochlear aqueduct; *SVN*, superior vestibular nerve; *IVN*, inferior vestibular nerve; *PFD*, posterior fossa dura; *JB*, jugular bulb.

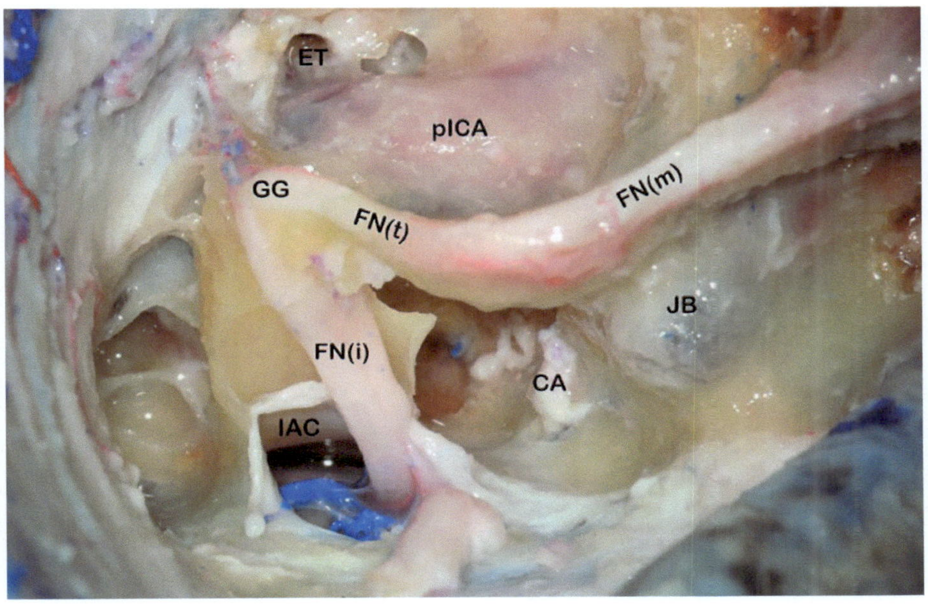

Fig. 10.46 The cochlear nerve is also dissected posteriorly. *ET*, eustachian tube; *pICA*, petrous segment of internal carotid artery; *FN(t)*, tympanic segment of facial nerve; *FN(m)*, mastoid segment of facial nerve; *GG*, geniculate ganglion; *FN(i)*, internal auditory canal segment of facial nerve; *IAC*, internal auditory canal; *CA*, cochlear aqueduct; *JB*, jugular bulb.

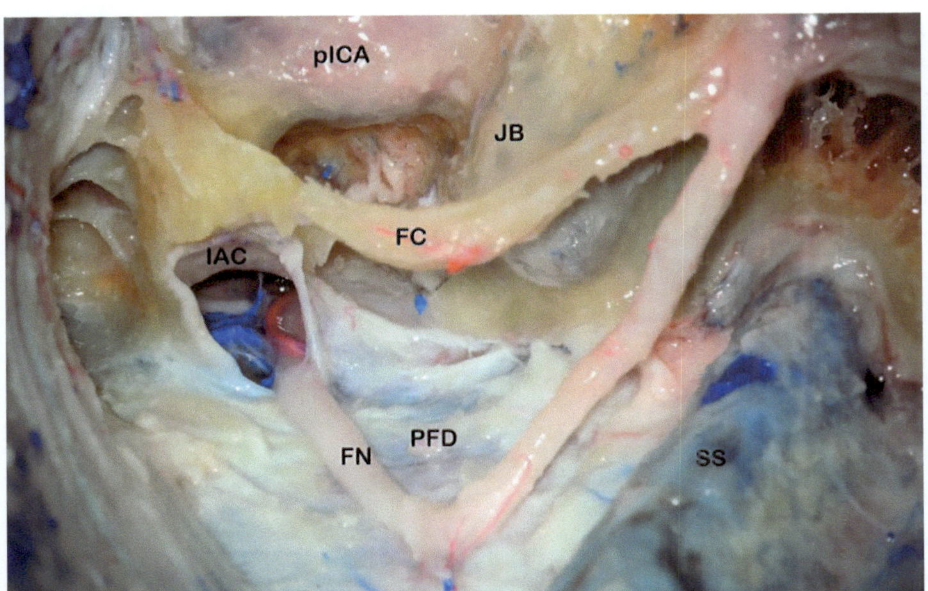

Fig. 10.47 The facial nerve has been posteriorly rerouted from the mastoid to the internal auditory canal segments. *pICA*, petrous segment of internal carotid artery; *FC*, fallopian canal; *JB*, jugular bulb; *IAC*, internal auditory canal; *FN*, facial nerve; *PFD*, posterior fossa dura; *SS*, sigmoid sinus.

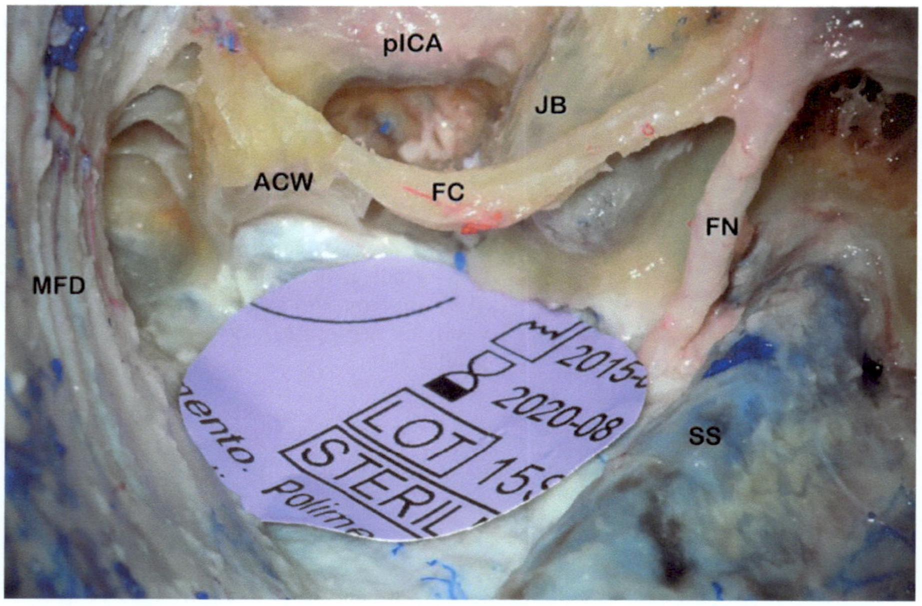

Fig. 10.48 The dura of the internal auditory canal is being dissected from the anterior wall of the canal. The whole contents of the internal auditory canal are transposed posteriorly with the facial nerve. An aluminum strip is used to protect the dura and facial nerve. The fallopian canal can now be removed. *pICA*, petrous segment of internal carotid artery; *FN*, facial nerve; *FC*, fallopian canal; *ACW*, anterior canal wall; *MFD*, middle fossa dura; *SS*, sigmoid sinus; *JB*, jugular bulb.

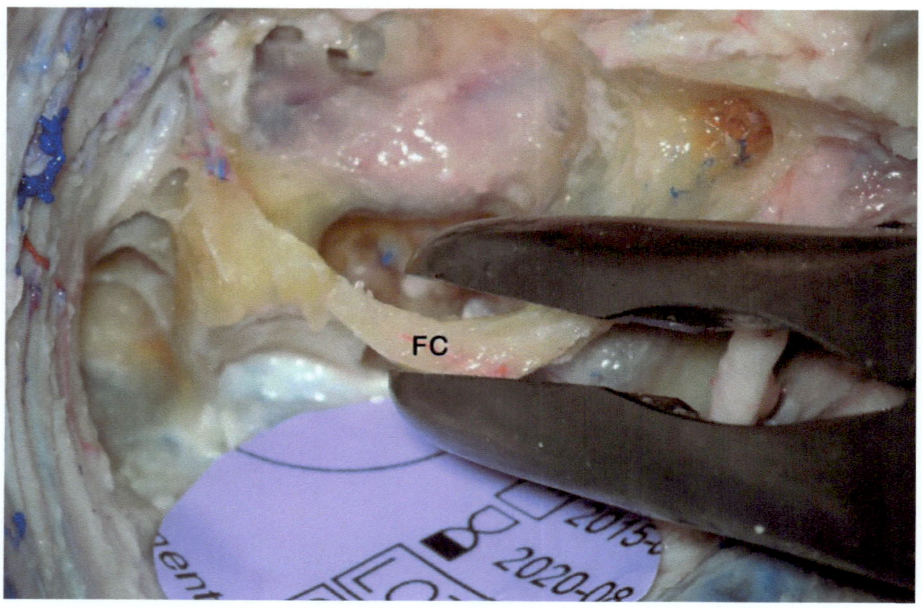

Fig. 10.49 Removal of the fallopian canal with a rongeur. *FC*, fallopian canal.

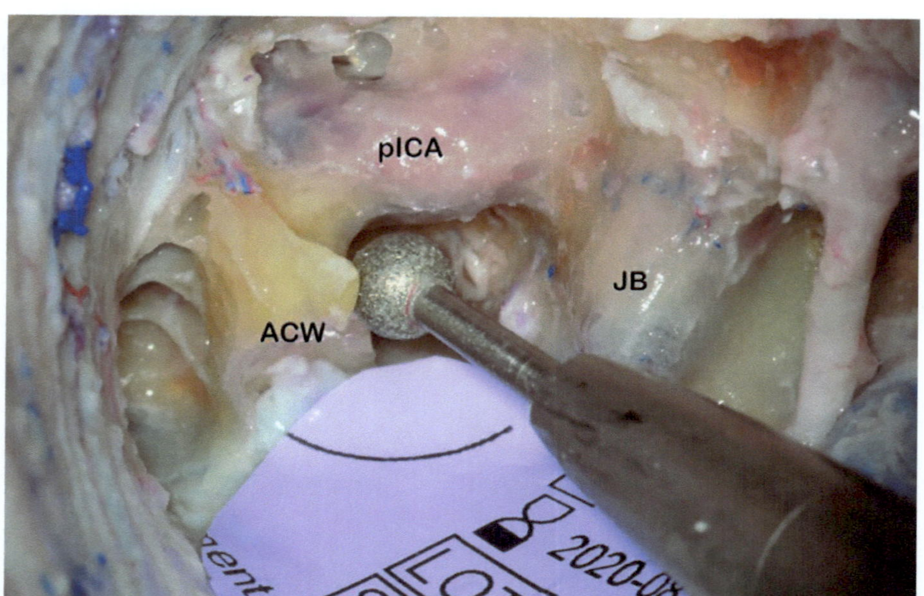

Fig. 10.50 Aluminum strip used to protect the porus of the internal auditory canal while drilling the anterior canal wall. *JB*, jugular bulb; *pICA*, petrous segment of internal carotid artery; *ACW*, anterior canal wall.

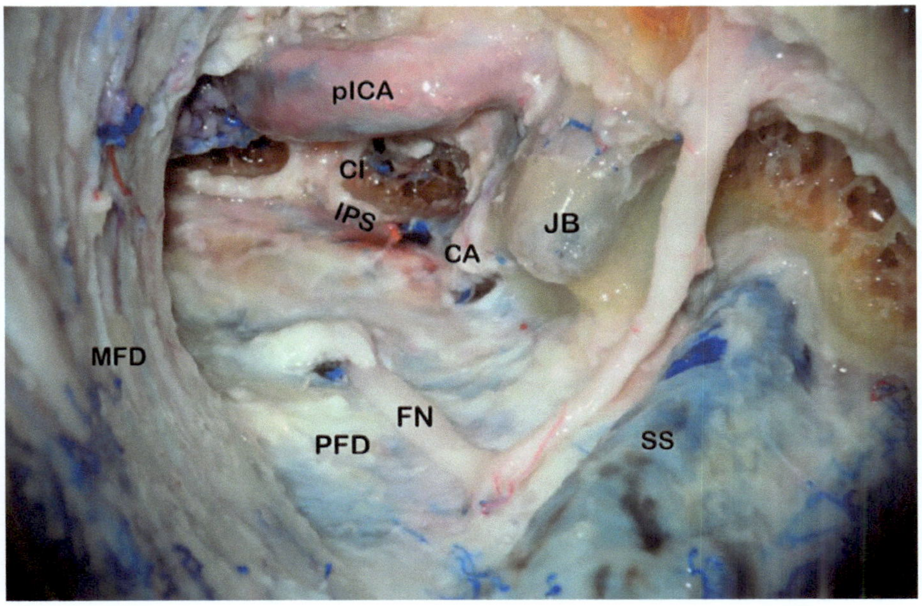

Fig. 10.51 View after the complete performance of this approach. *Cl*, clivus; *pICA*, petrous segment of internal carotid artery; *CA*, cochlear aqueduct; *FN*, facial nerve; *PFD*, posterior fossa dura; *IPS*, inferior petrosal sinus; *MFD*, middle fossa dura; *JB*, jugular bulb; *SS*, sigmoid sinus.

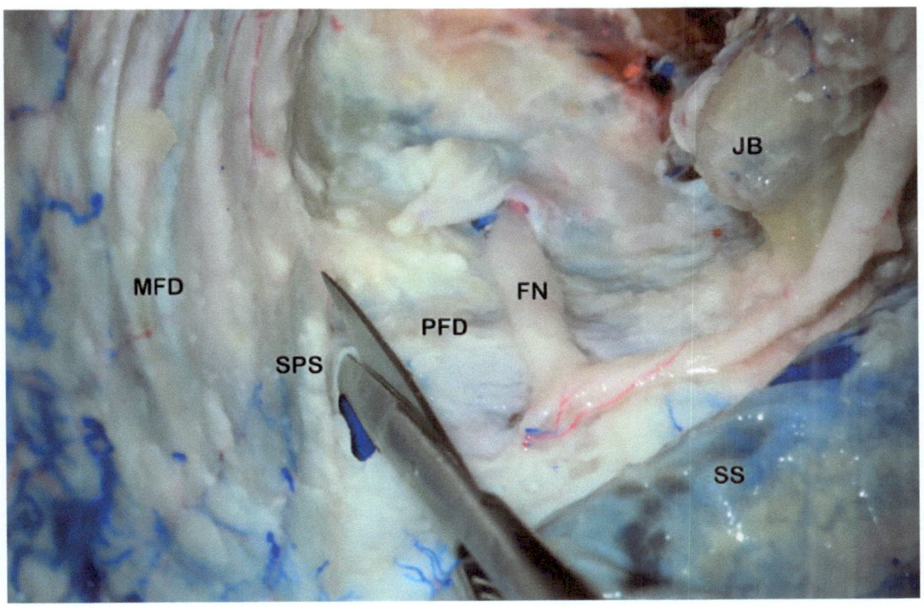

Fig. 10.52 The incision starts superiorly 1–2 mm parallel to and inferior to the superior petrosal sinus, running from posterior to anterior using microsurgical scissors. *JB*, jugular bulb; *FN*, facial nerve; *PFD*, posterior fossa dura; *SS*, sigmoid sinus; *SPS*, superior petrosal sinus; *MFD*, middle fossa dura.

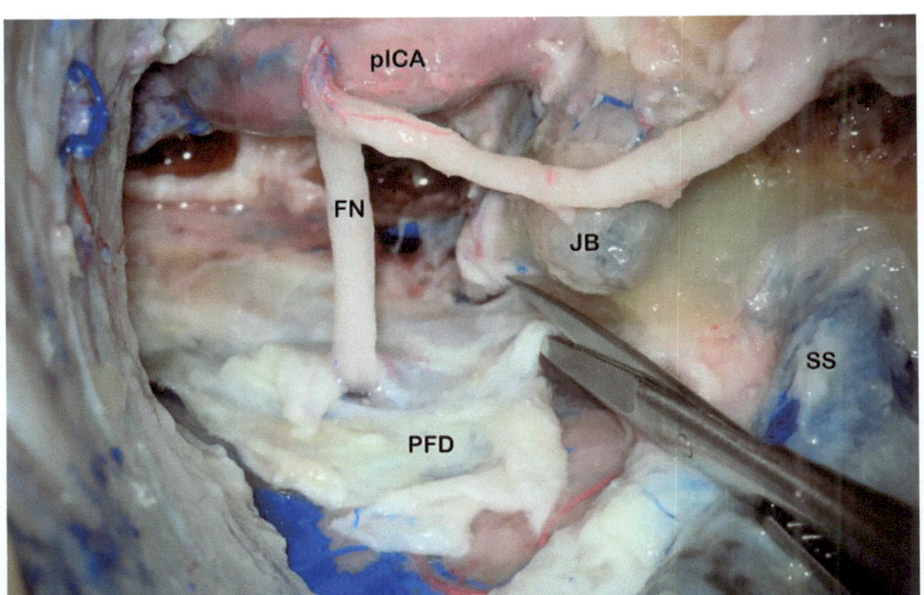

Fig. 10.53 The inferior incision starts just in front of the distal part of the sigmoid sinus and follows the course of the sinus and jugular bulb to the inferior petrosal sinus. *pICA*, petrous segment of internal carotid artery; *FN*, facial nerve; *PFD*, posterior fossa dura; *JB*, jugular bulb; *SS*, sigmoid sinus.

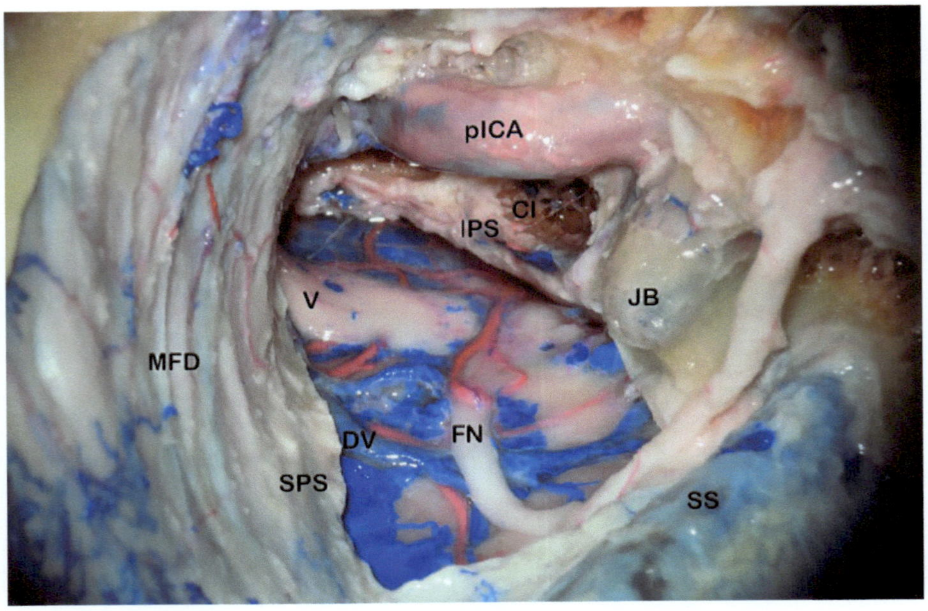

Fig. 10.54 View of the cerebellopontine angle and prepontine cistern as viewed after opening the dura. Excellent control of the prepontine cistern is obtained. *pICA*, petrous segment of internal carotid artery; *Cl*, clivus; *IPS*, inferior petrosal sinus; *FN*, facial nerve; *V*, trigeminal nerve; *DV*, Dandy's vein (superior petrosal vein); *SPS*, superior petrosal sinus; *MFD*, middle fossa dura; *JB*, jugular bulb; *SS*, sigmoid sinus.

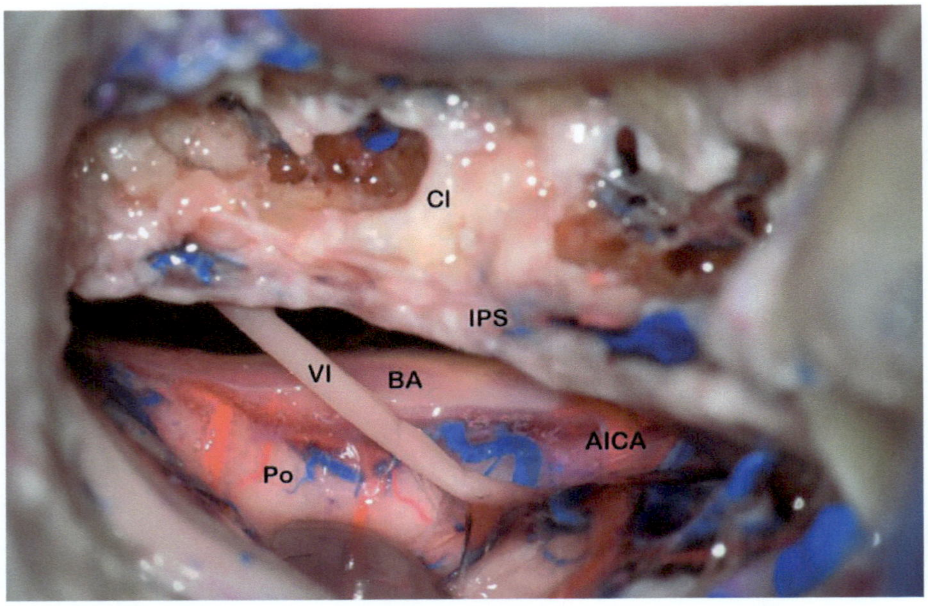

Fig. 10.55 A closer view, showing the basilar artery and abducent nerve in the prepontine cistern. The utmost care needs to be taken when incising the dura of the petrous apex so as not to injure the abducent nerve. *Cl*, clivus; *IPS*, inferior petrosal sinus; *AICA*, anteroinferior cerebellar artery; *BA*, basilar artery; *VI*, abducent nerve; *Po*, pons.

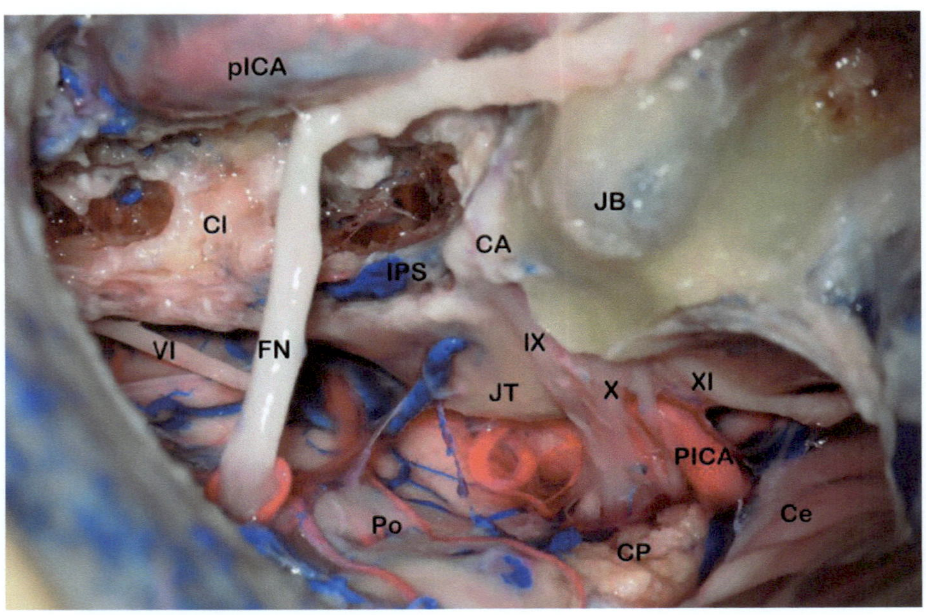

Fig. 10.56 Mild retraction of the cerebellum and flocculus provides a good view of the choroid plexus and lower cranial nerves. The relation between lower cranial nerves and the posterior inferior cerebellar artery is appreciated. *pICA*, petrous segment of internal carotid artery; *Cl*, clivus; *IPS*, inferior petrosal sinus; *XI*, accessory nerve; *X*, vagus nerve; *IX*, glossopharyngeal nerve; *JT*, jugular tubercle; *Po*, pons; *JB*, jugular bulb; *CA*, cochlear aqueduct; *FN*, facial nerve; *VI*, abducent nerve; *CP*, choroid plexus; *PICA*, posteroinferior cerebellar artery; *Ce*, cerebellum.

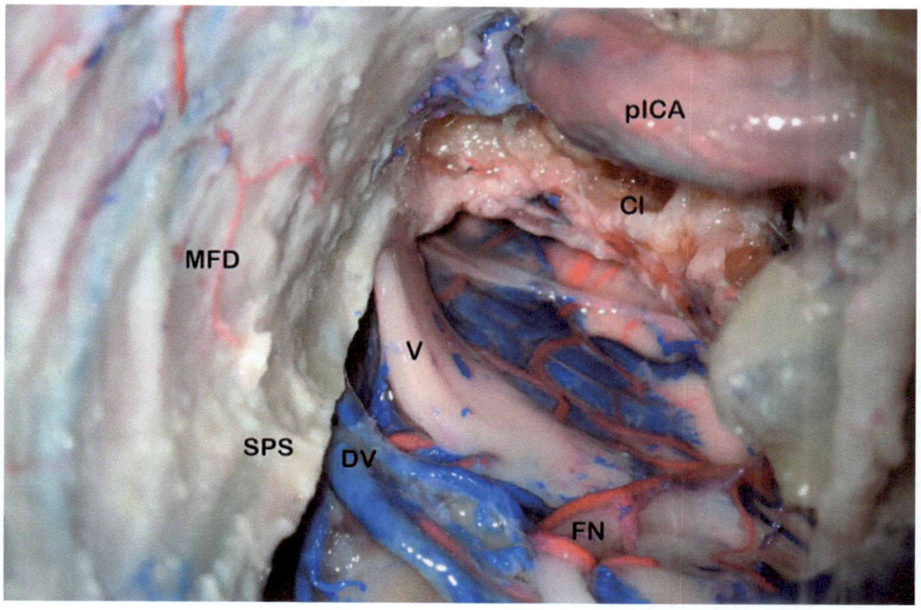

Fig. 10.57 Tilting the microscope upwards provides a good view of the Dandy's vein entering the superior petrosal sinus. Note the excellent exposure of the trigeminal nerve. *pICA*, petrous segment of internal carotid artery; *Cl*, clivus; *V*, trigeminal nerve; *MFD*, middle fossa dura; *SPS*, superior petrosal sinus; *DV*, Dandy's vein (superior petrosal vein); *FN*, facial nerve.

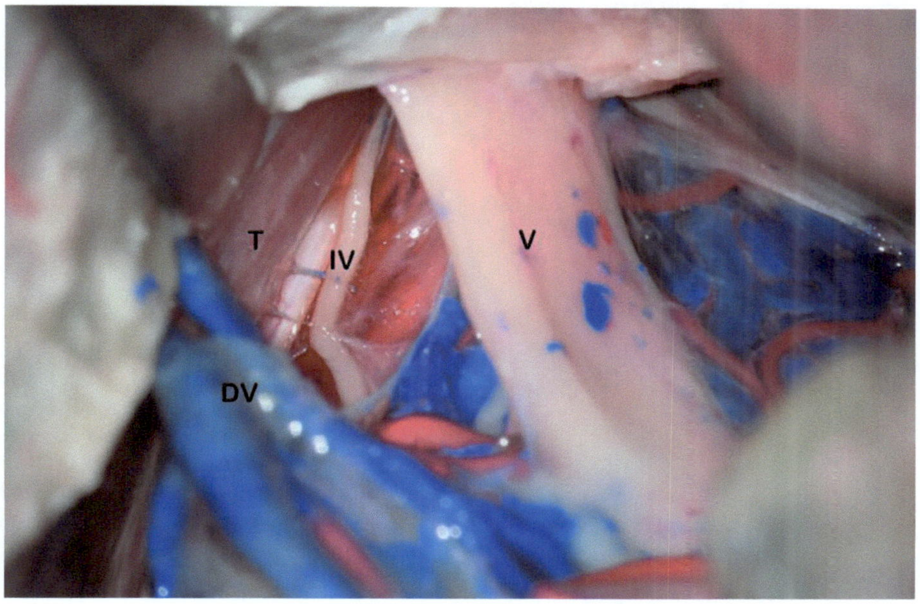

Fig. 10.58 Mild retraction of the tentorium provides a good view of the trochlear nerve. *V*, trigeminal nerve; *IV*, trochlear nerve; *T*, tentorium; *DV*, Dandy's vein (superior petrosal vein).

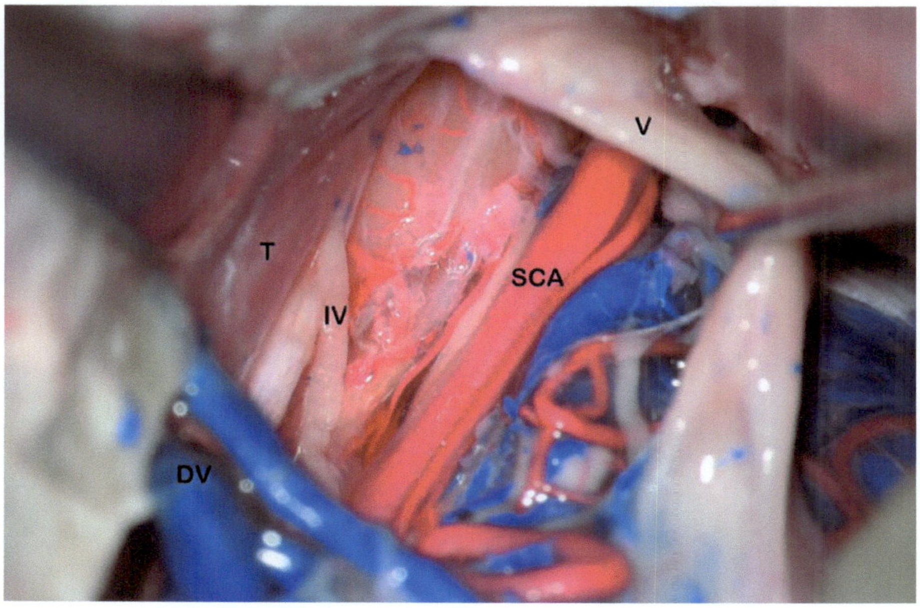

Fig. 10.59 Retraction of the tentorium upwards and trigeminal nerve downwards provides a good view of the trochlear nerve and its relation to the superior cerebellar artery. *V*, trigeminal nerve; *SCA*, superior cerebellar artery; *IV*, trochlear nerve; *T*, tentorium; *DV*, Dandy's vein (superior petrosal vein).

Case 1: Meningioma of the Cerebellopontine Angle Removed Through a Modified Transcochlear Approach Type A

See Figs. 10.60–10.89.

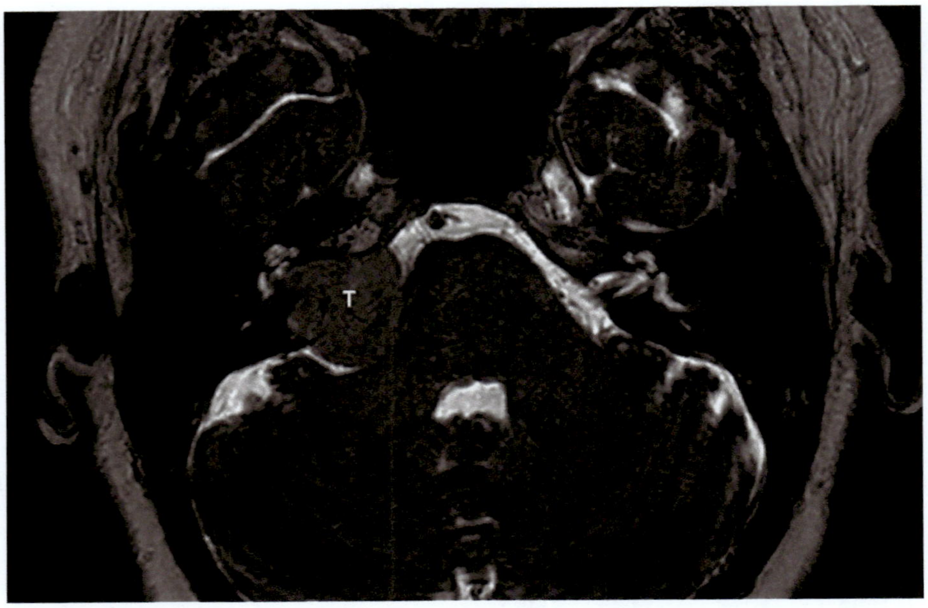

Fig. 10.60 T2W MRI of the brain shows a tumor of the right cerebellopontine angle in a 57-year-old patient. The tumor, histologically diagnosed as a meningioma, extended anteriorly, medially, and posteriorly in relation to the internal auditory canal. The patient had a preoperative House-Brackmann grade III facial nerve function. Therefore, a modified transcochlear type A approach was planned. *T*, tumor.

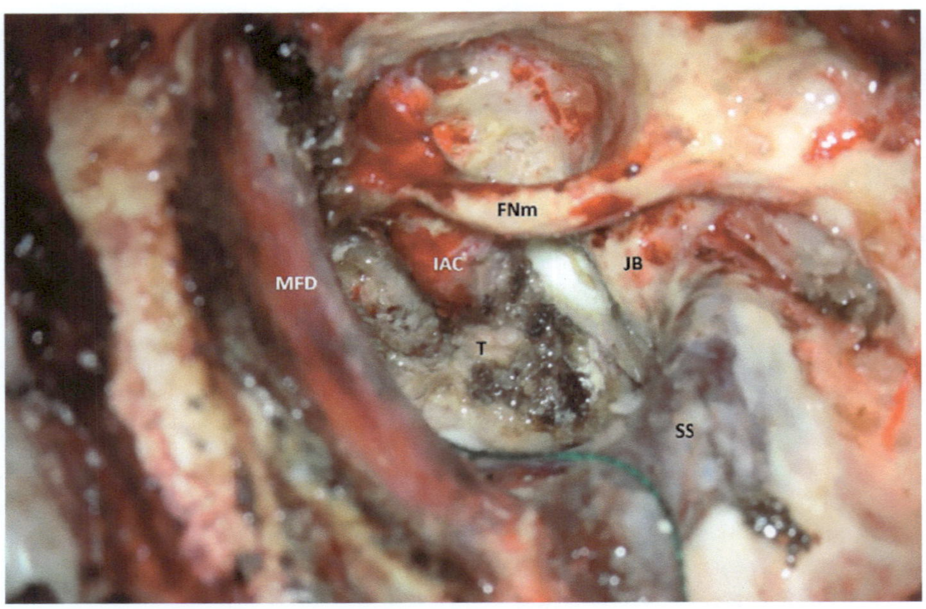

Fig. 10.61 The surgical approach is complete, and the posterior fossa is opened. The external auditory canal has been closed as a cul-de-sac, and the entire external and middle ear has been removed. The internal auditory canal (IAC) has been identified after the removal of the labyrinth. *T*, tumor; *MFD*, middle fossa dura; *SS*, sigmoid sinus; *JB*, jugular bulb.

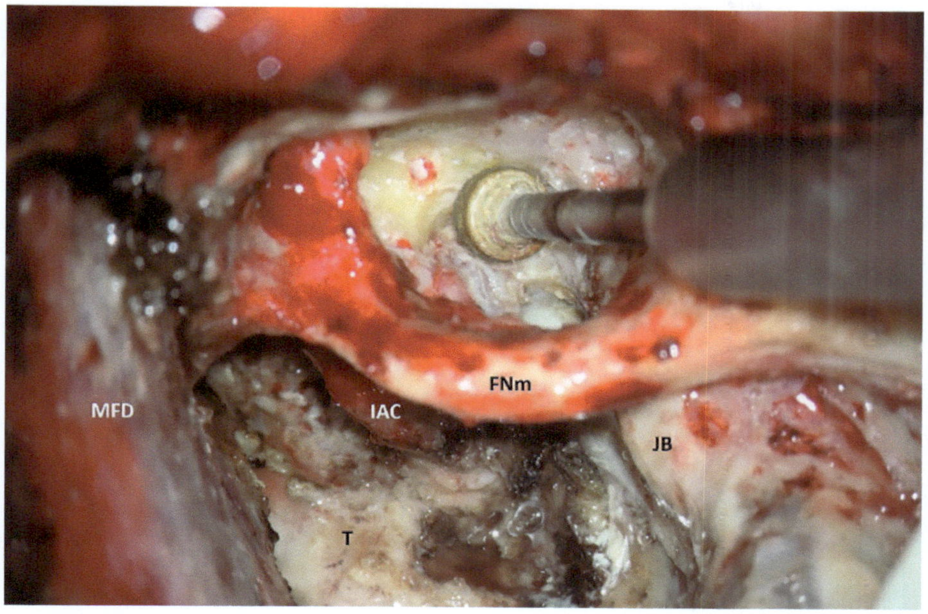

Fig. 10.62 The cochlea and anterior tympanic bone are drilled until the petrous portion of the internal carotid artery is identified. *T*, tumor; *MFD*, middle fossa dura; *IAC*, internal auditory canal; *JB*, jugular bulb; *FNm*, mastoid portion of the facial nerve.

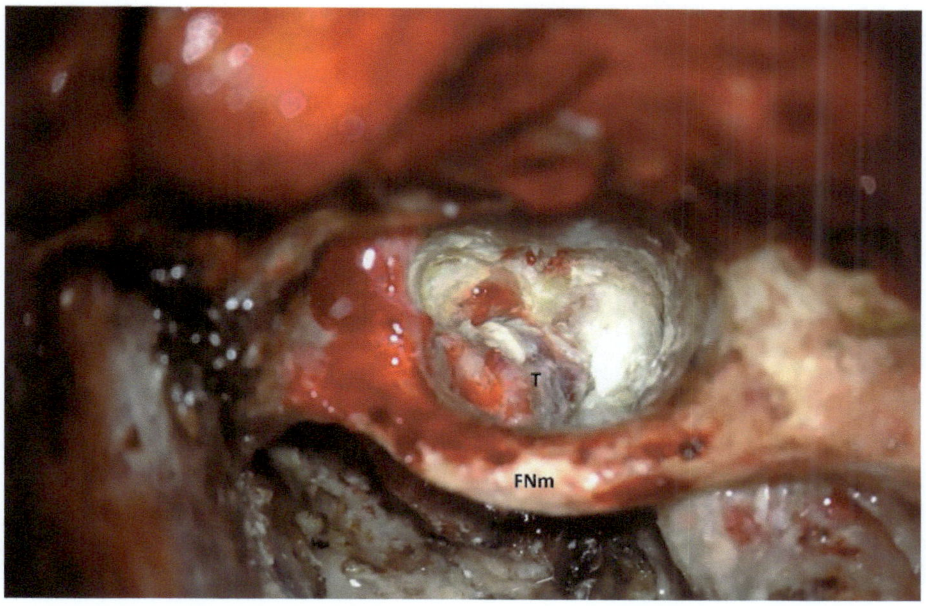

Fig. 10.63 Closer view. Note the extension of the tumor (T) anterior to the internal auditory canal. *FNm*, mastoid portion of the facial nerve.

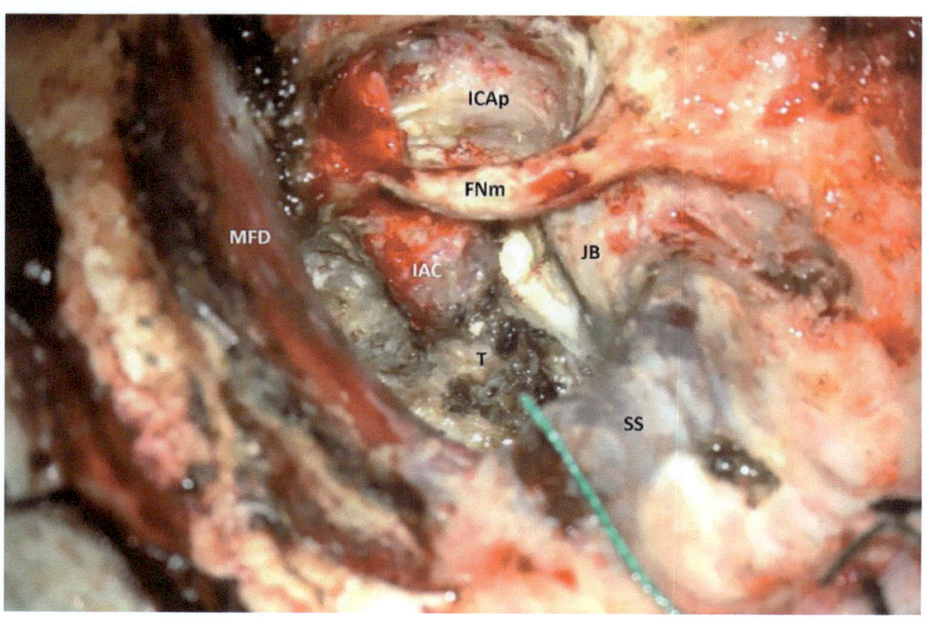

Fig. 10.64 After the removal of the anterior tympanic bone, the petrous internal carotid artery (ICAp) is visible. *T*, tumor; *MFD*, middle fossa dura; *IAC*, internal auditory canal; *JB*, jugular bulb; *FNm*, mastoid portion of the facial nerve; *SS*, sigmoid sinus.

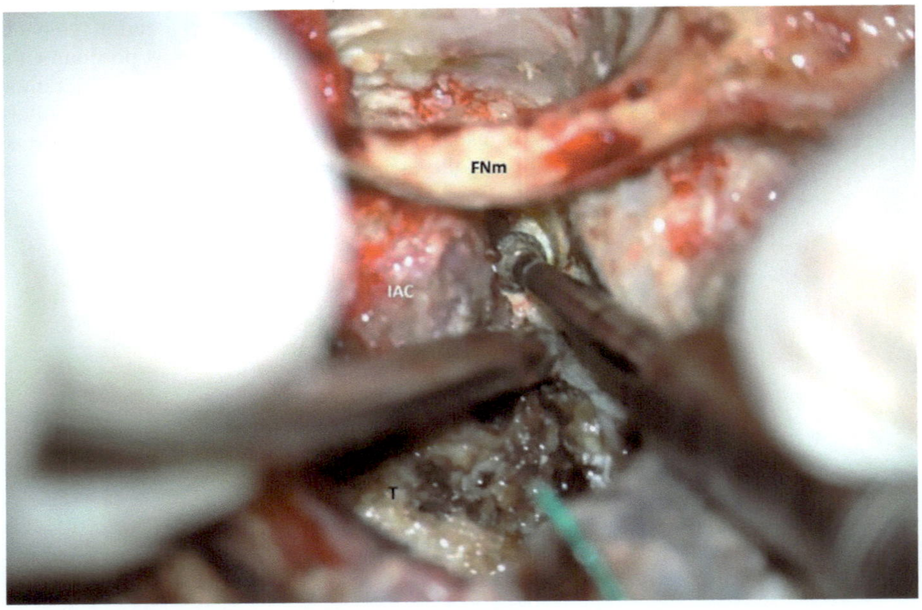

Fig. 10.65 Drilling the lower apical cells allows control of the lower part of the tumor. *FNm*, mastoid portion of the facial nerve; *IAC*, internal auditory canal.

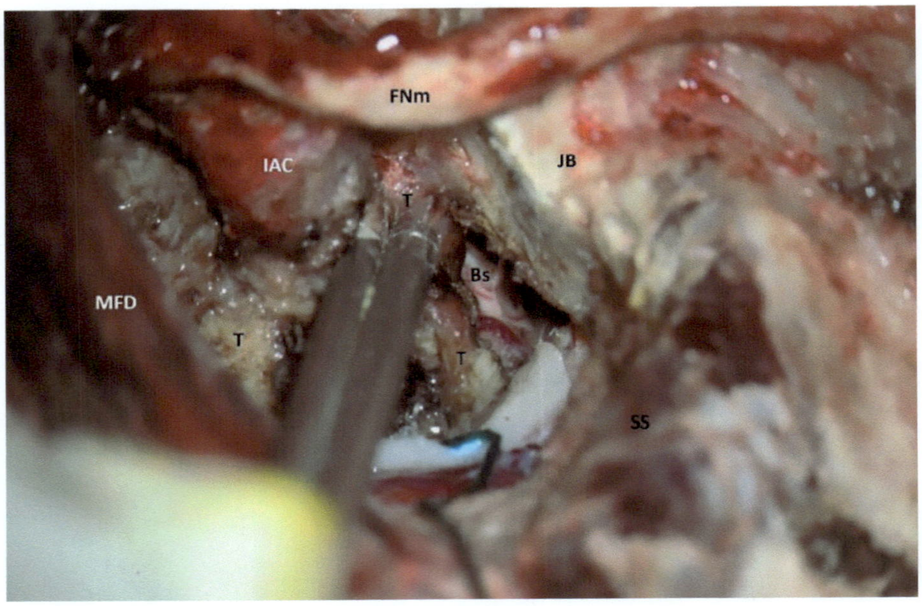

Fig. 10.66 The posterior fossa dura is opened corresponding to the lower part of the tumor. Previous bone removal allowed complete control of the lower cranial nerve area. *FNm*, mastoid portion of the facial nerve; *IAC*, internal auditory canal; *MFD*, middle fossa dura; *T*, tumor; *Bs*, brainstem; *SS*, sigmoid sinus; *JB*, jugular bulb.

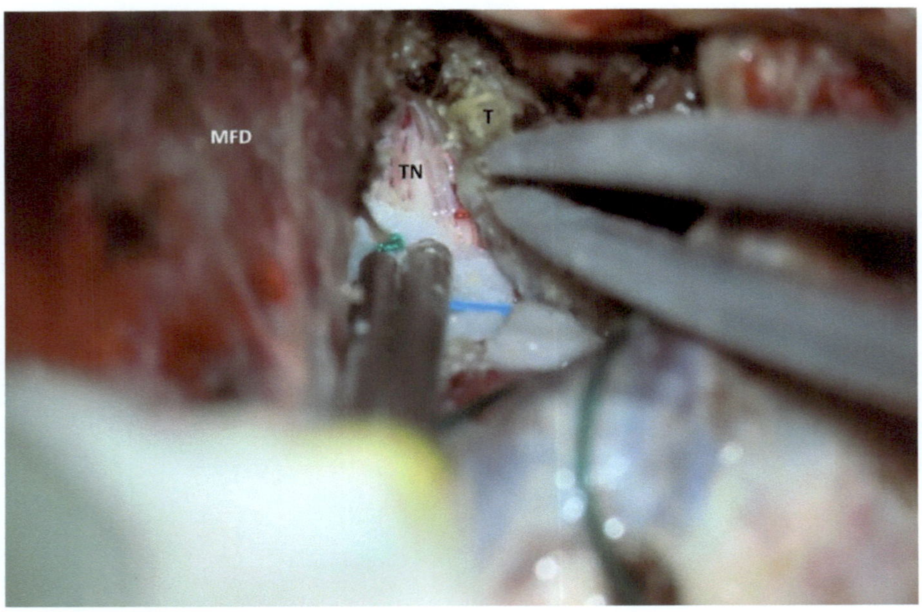

Fig. 10.67 The upper part of the tumor is dissected from the trigeminal nerve (TN). *MFD*, middle fossa dura; *T*, tumor.

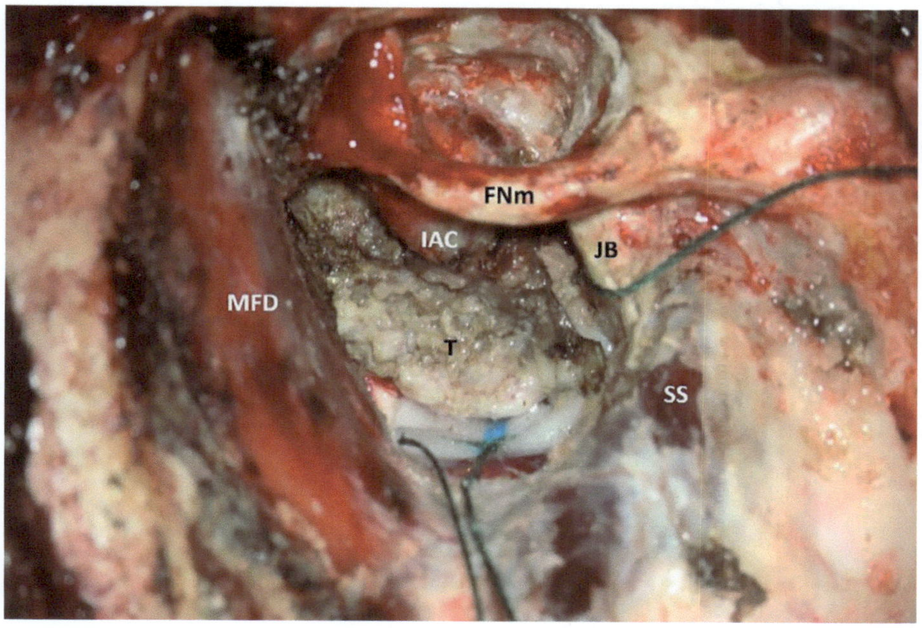

Fig. 10.68 The tumor (T) is debulked. Note that both the upper and lower parts of the lesion are isolated from the brainstem. *MFD*, middle fossa dura; *SS*, sigmoid sinus; *JB*, jugular bulb; *FNm*, mastoid portion of the facial nerve; *IAC*, internal auditory canal.

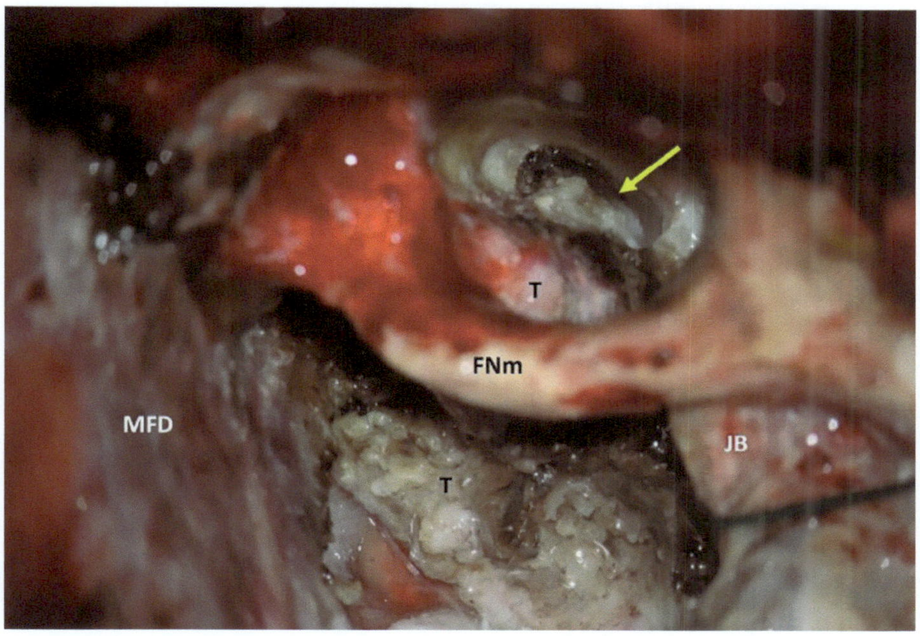

Fig. 10.69 Closer view. The bone anterior to the internal auditory canal is further drilled, showing tumor infiltration *(arrow)*. *MFD*, middle fossa dura; *JB*, jugular bulb; *FNm*, mastoid portion of the facial nerve; *T*, tumor.

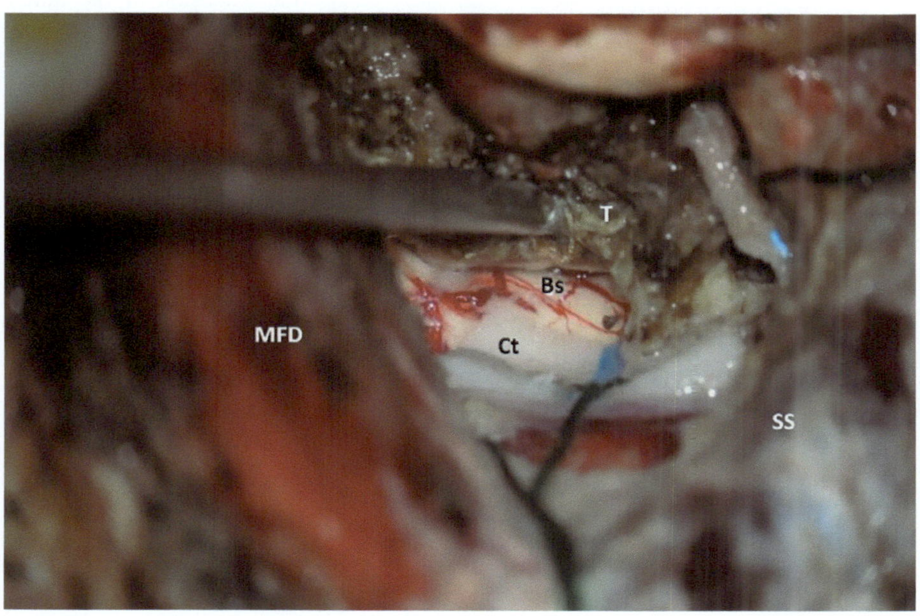

Fig. 10.70 The tumor (T) is dissected posteriorly. A cottonoid protects both the cerebellum and the brainstem (Bs). *MFD*, middle fossa dura; *SS*, sigmoid sinus.

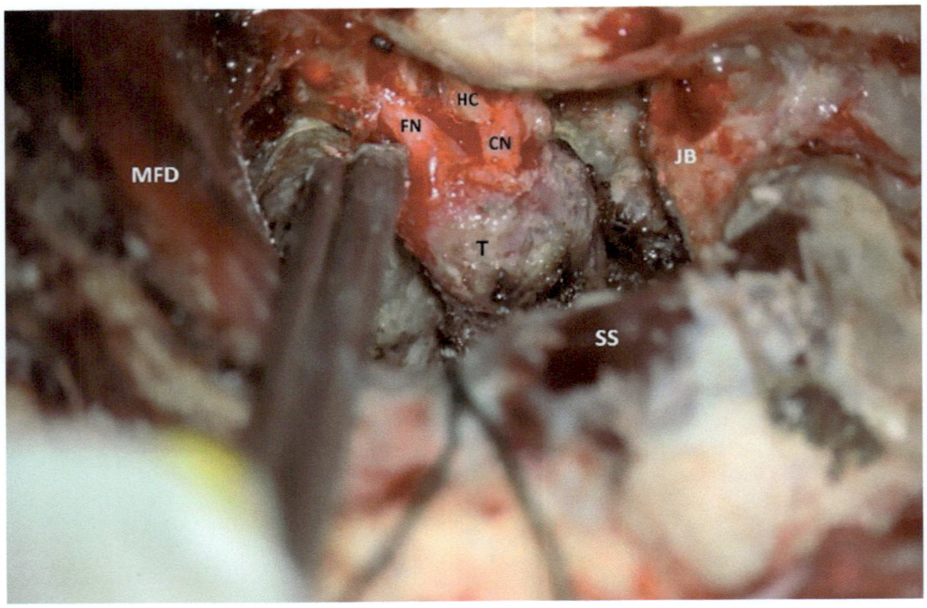

Fig. 10.71 Once the tumor (T) has been debulked and isolated from the brainstem, the internal auditory canal is opened. Note the infiltration of both the facial nerve (FN) and cochlear nerve (CN) at the level of the internal auditory canal. *MFD*, middle fossa dura; *SS*, sigmoid sinus; *JB*, jugular bulb; *HC*, horizontal crest.

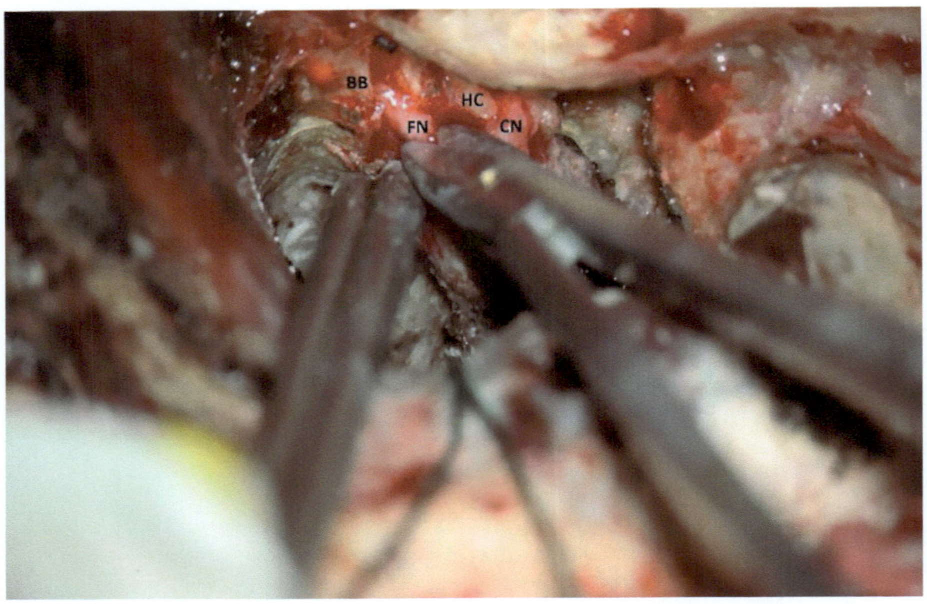

Fig. 10.72 As the tumor infiltrates both the anterior tympanic bone and the facial nerve (FN), the latter is transected. *CN*, cochlear nerve; *BB*, Bill's bar; *HC*, horizontal crest.

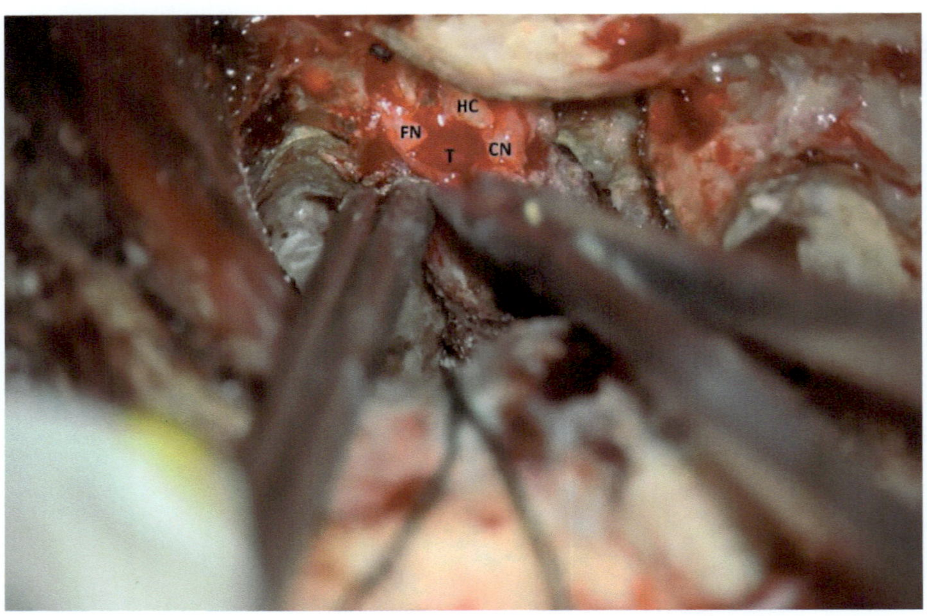

Fig. 10.73 Note the extension of the tumor (T) medial to the transected facial nerve (FN). *CN*, cochlear nerve; *HC*, horizontal crest.

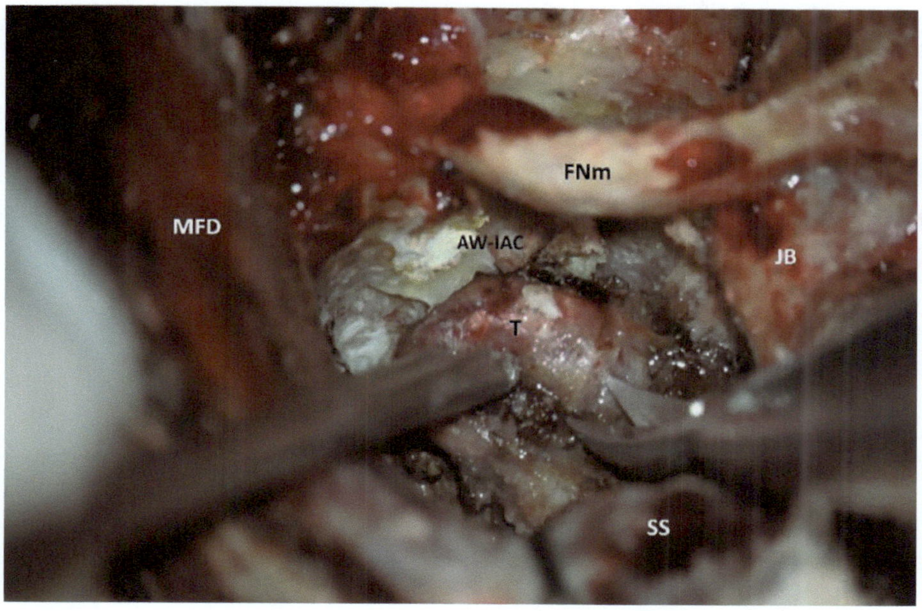

Fig. 10.74 The entire anterior wall of the internal auditory canal (AW-IAC) is infiltrated by the tumor (T), which is further debulked. *MFD*, middle fossa dura; *FNm*, mastoid portion of the facial nerve; *SS*, sigmoid sinus; *JB*, jugular bulb.

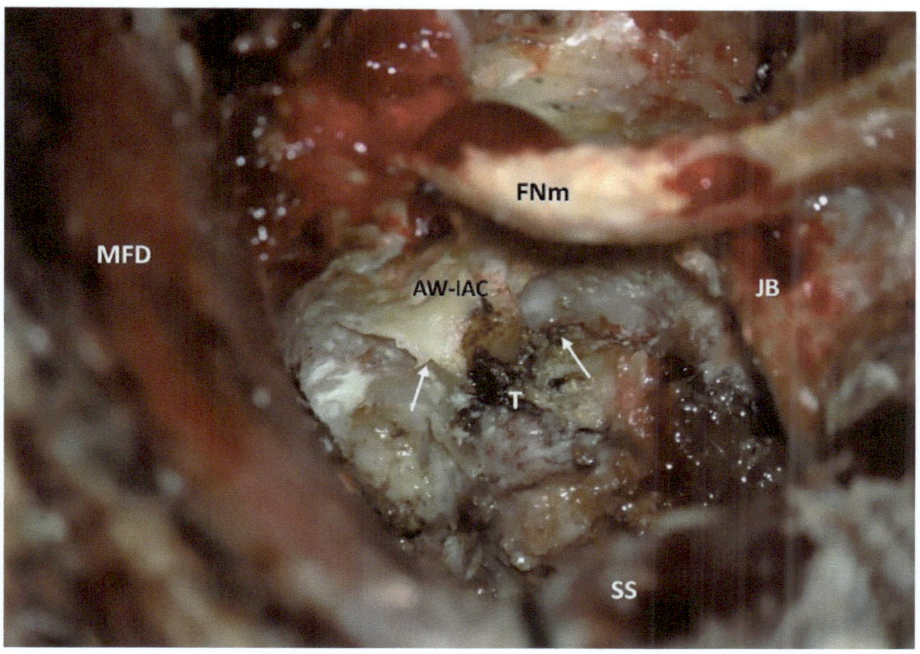

Fig. 10.75 Closer view. The tumor (T), histologically diagnosed as a meningioma, deeply infiltrates the bone *(arrows)* of the anterior wall of the internal auditory canal (AW-IAC) toward the area of the prepontine cistern. *MFD*, middle fossa dura; *FNm*, mastoid portion of the facial nerve; *SS*, sigmoid sinus; *JB*, jugular bulb.

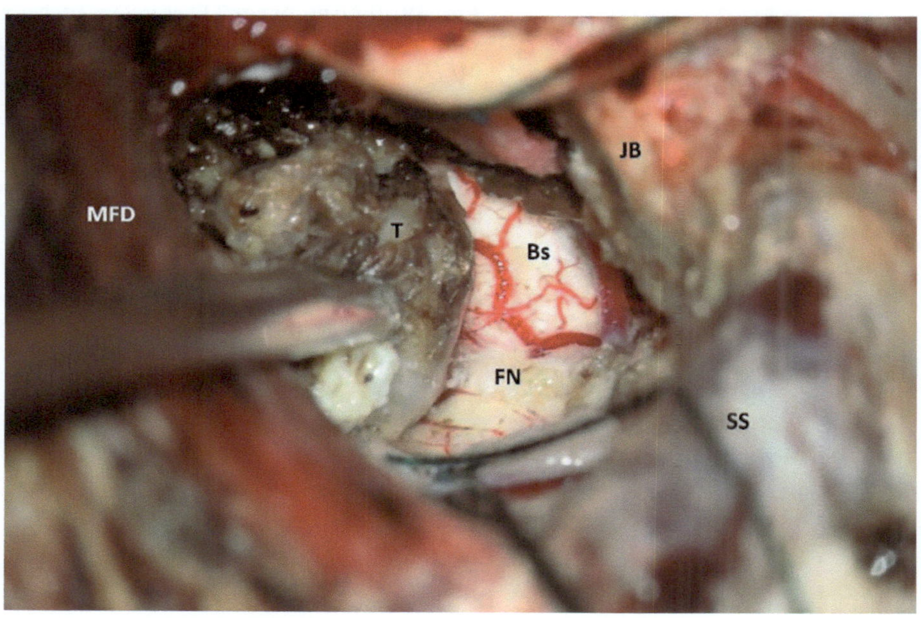

Fig. 10.76 The lower part of the tumor (T) is further dissected to reveal the proximal stump of the facial nerve (FN) at the level of the brainstem (Bs). Considering the tumor infiltration and the extreme thinning of this part of the FN, a sural nerve graft was not chosen for FN reconstruction. A masseteric nerve to FN anastomosis is then performed 3 months after this surgery. *MFD*, middle fossa dura; *SS*, sigmoid sinus; *JB*, jugular bulb.

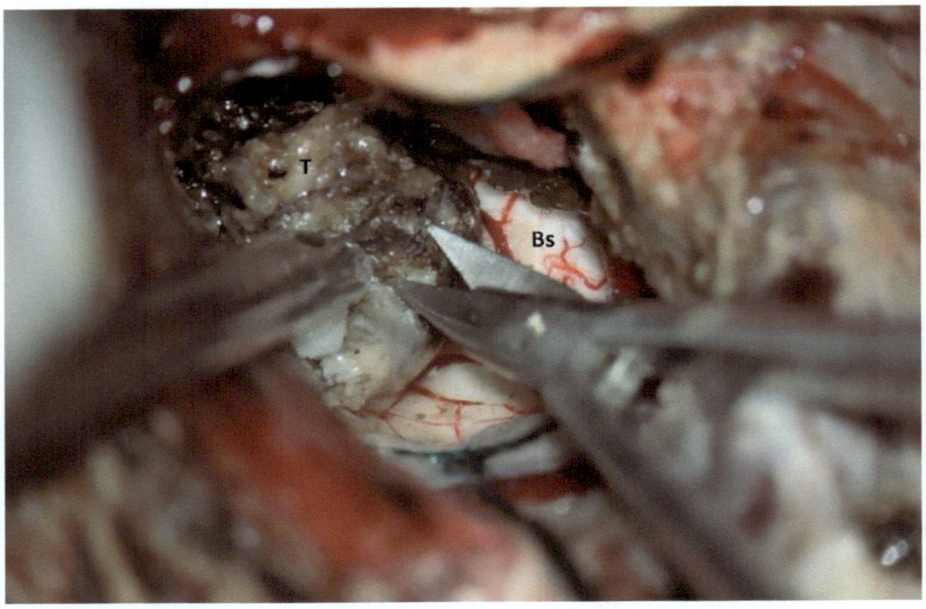

Fig. 10.77 Tumor debulking. *T*, tumor; *Bs*, brainstem.

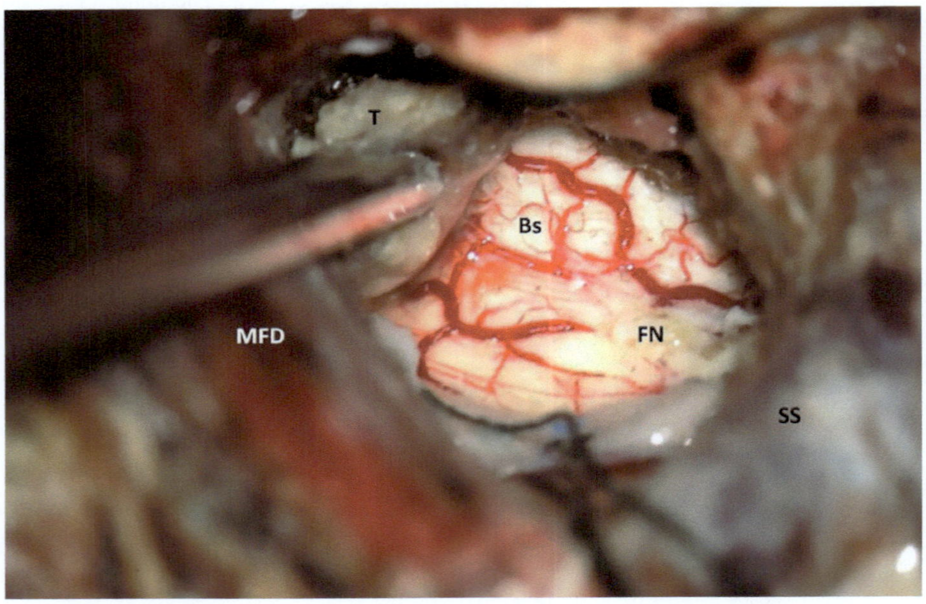

Fig. 10.78 The tumor (T) has been reduced and further separated from the brainstem (Bs). *FN*, facial nerve (proximal stump); *MFD*, middle fossa dura; *SS*, sigmoid sinus.

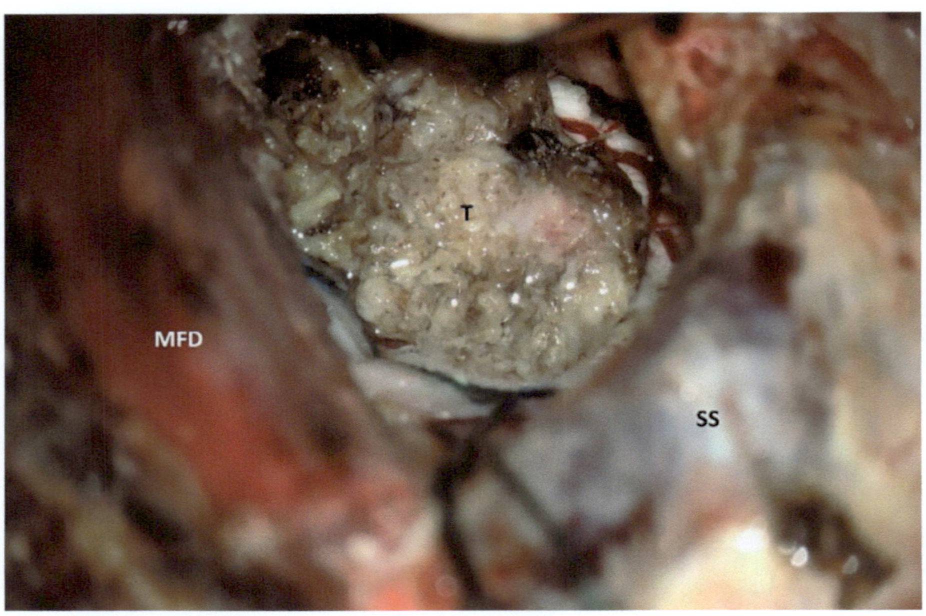

Fig. 10.79 The tumor (T) is reflected before further reduction of its upper and anterior parts. *MFD*, middle fossa dura; *SS*, sigmoid sinus.

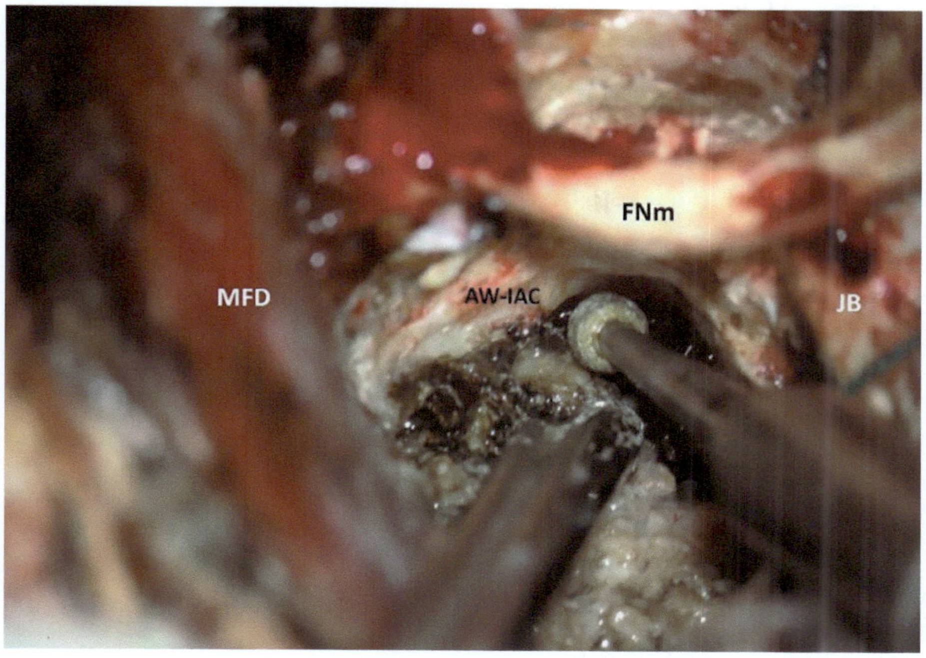

Fig. 10.80 The anterior wall of the internal auditory canal (AW-IAC) is drilled before opening the posterior fossa dura corresponding to the prepontine cistern. *MFD*, middle fossa dura; *FNm*, mastoid portion of the facial nerve; *JB*, jugular bulb.

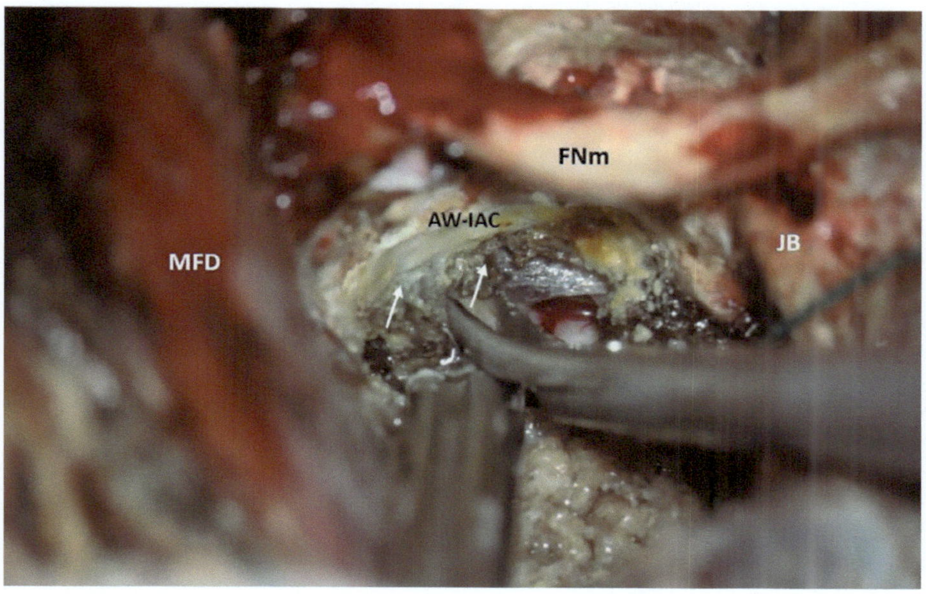

Fig. 10.81 Note the tumor infiltration *(arrows)* of the posterior fossa dura attached to the anterior wall of the internal auditory canal (AW-IAC). *MFD*, middle fossa dura; *FNm*, mastoid portion of the facial nerve; *JB*, jugular bulb.

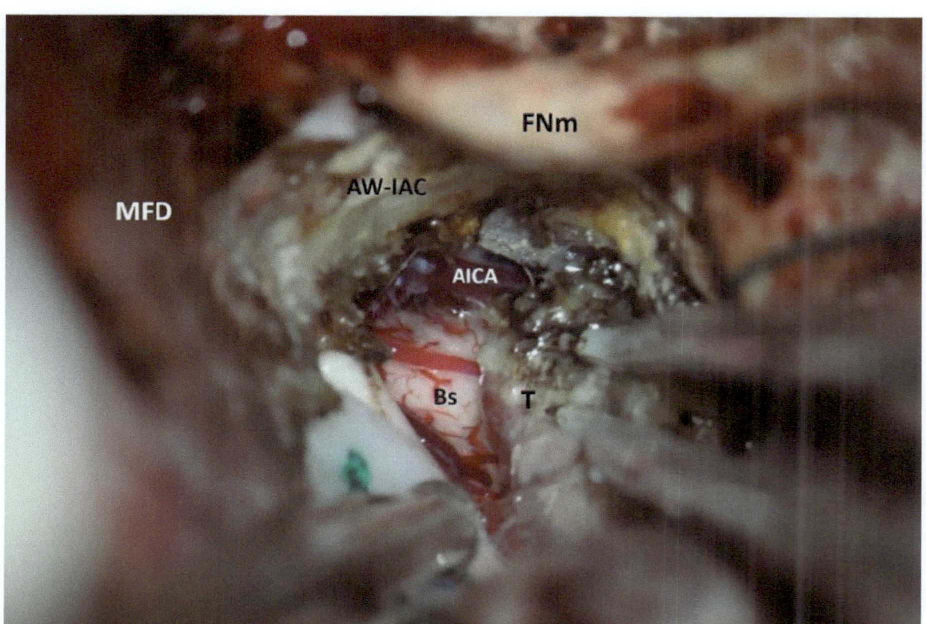

Fig. 10.82 The anterior-superior part of the tumor (T) is dissected, freeing the prepontine cistern. *MFD*, middle fossa dura; *FNm*, mastoid portion of the facial nerve; *AW-IAC*, anterior wall of the internal auditory canal; *Bs*, brainstem; *AICA*, anterior inferior cerebellar artery.

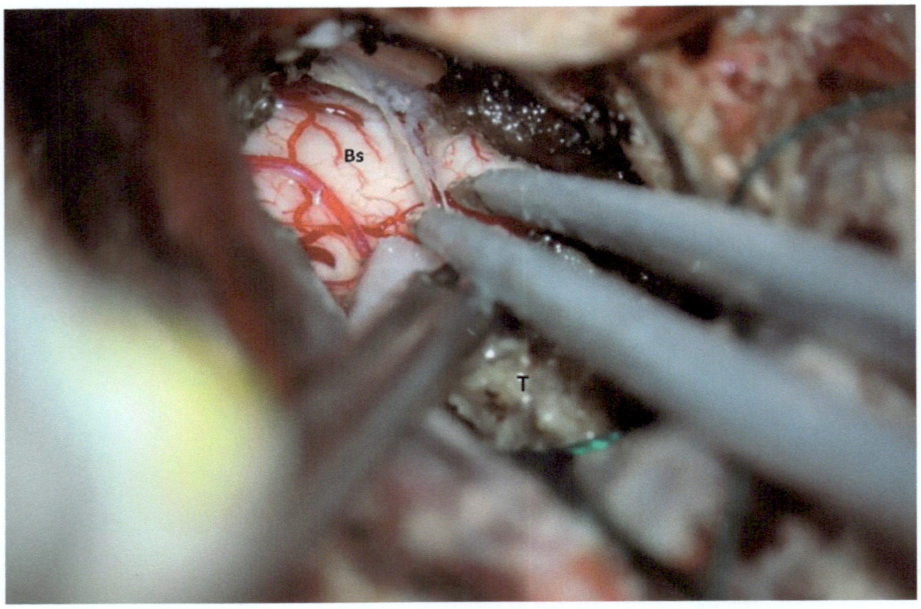

Fig. 10.83 The last anterior tumor (T) attachments are cut off. *Bs*, brainstem.

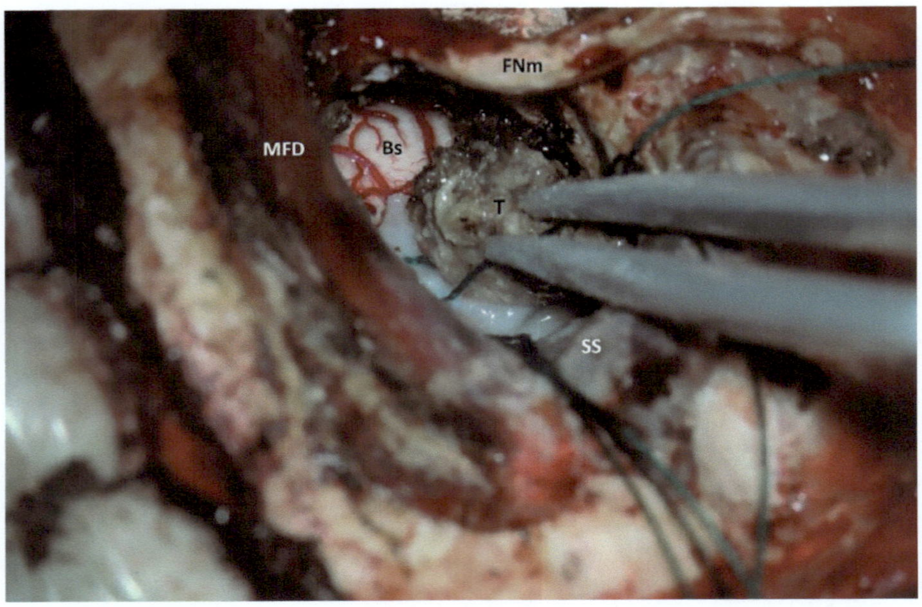

Fig. 10.84 The last part of the tumor is shrunk and removed. *Bs*, brainstem; *MFD*, middle fossa dura; *SS*, sigmoid sinus; *FNm*, mastoid portion of the facial nerve.

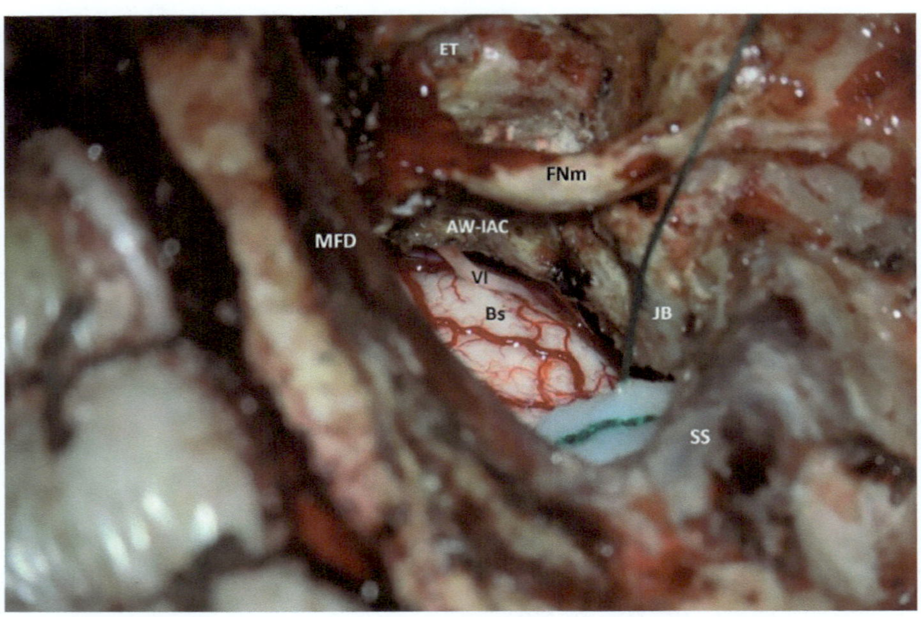

Fig. 10.85 Surgical view after total tumor removal. Note the control of the prepontine cistern. *Bs*, brainstem; *MFD*, middle fossa dura; *SS*, sigmoid sinus; *JB*, jugular bulb; *FNm*, mastoid portion of the facial nerve; *ET*, eustachian tube; *AW-IAC*, anterior wall of the internal auditory canal; *VI*, abducens nerve.

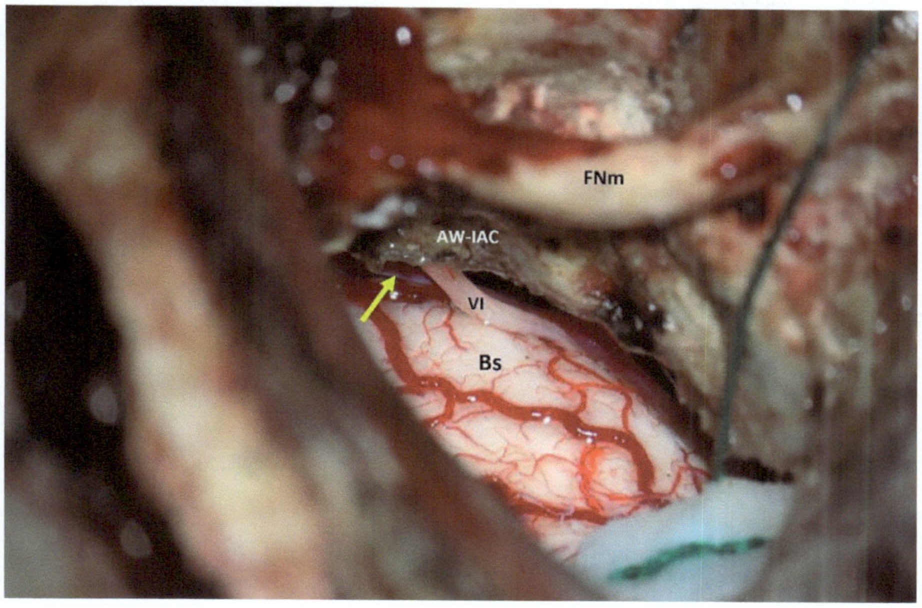

Fig. 10.86 Closer view. The approach allows complete control of the prepontine cistern, including the area of the basilar artery *(arrow)*. *Bs*, brainstem; *FNm*, mastoid portion of the facial nerve; *AW-IAC*, anterior wall of the internal auditory canal; *VI*, abducens nerve.

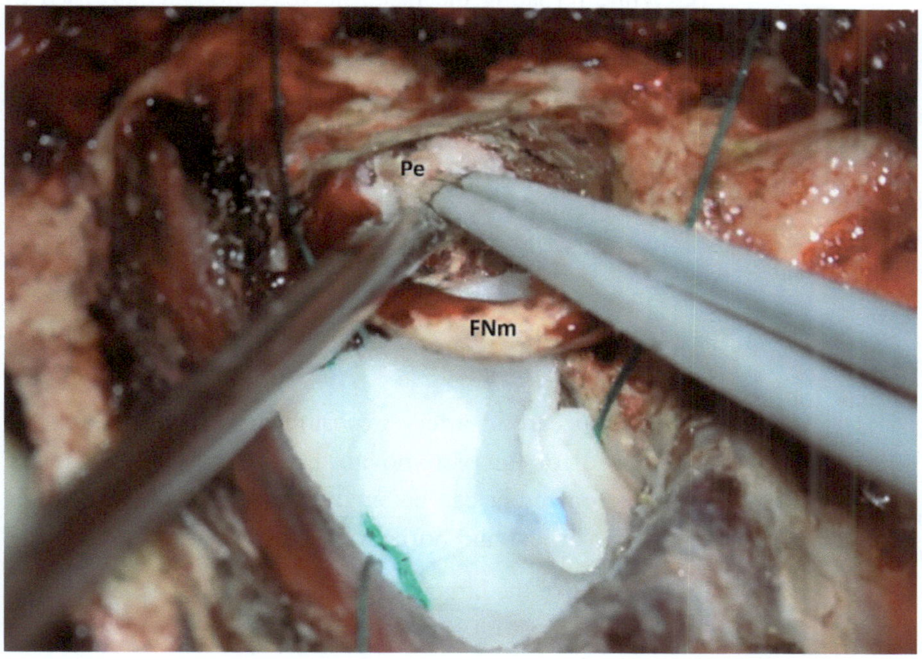

Fig. 10.87 The eustachian tube is closed by the periosteum (Pe). *FNm*, mastoid portion of the facial nerve.

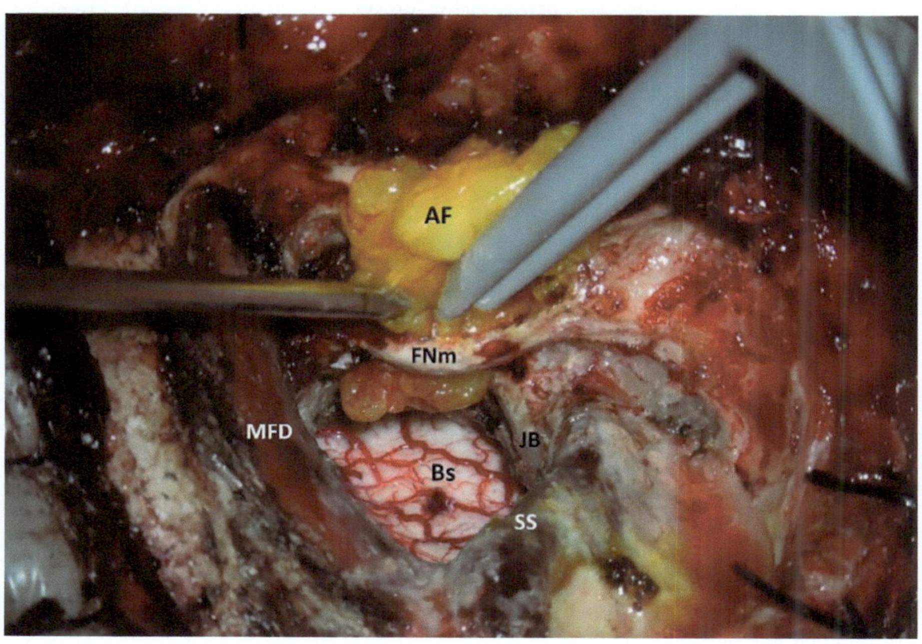

Fig. 10.88 The surgical cavity is filled with abdominal fat (AF). *FNm*, mastoid portion of the facial nerve; *MFD*, middle fossa dura; *SS*, sigmoid sinus; *JB*, jugular bulb.

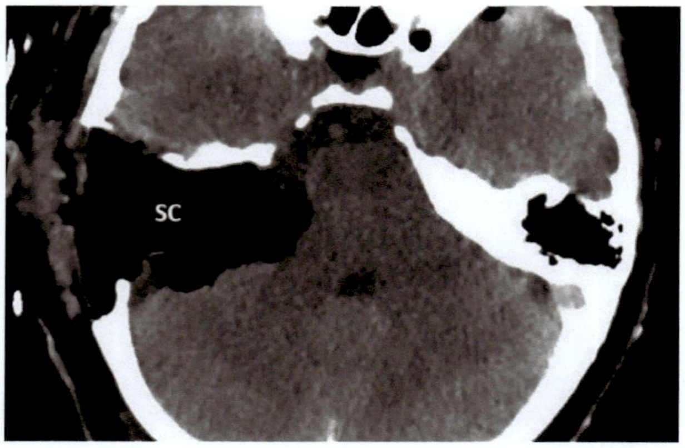

Fig. 10.89 Postoperative CT scan showing complete removal of the tumor and no postoperative complications. *SC*, surgical cavity.

HINTS AND PITFALLS

1. To avoid using retractors, sutures and skin hooks are used to hold the skin and subcutaneous tissue flaps. Retractors make the surgical field deeper and the angle of work more acute (Fig. 10.90).
2. Special care should be taken in the region of the geniculate ganglion. All bone should be removed here to avoid injury to the nerve during transposition (Fig. 10.91). The middle fossa dura is protected with an aluminum strip during this step. If the dura is prolapsing, the use of bipolar strokes from medial to lateral leads to retraction of the dura and gives more space to work (Fig. 10.92).
3. The vertical segment of the facial nerve has tough, fibrous attachments to the fallopian canal, and sharp dissection is therefore needed to avoid stretching the nerve at this level. The tympanic and labyrinthine segments, on the other hand, are delicate and require fine manipulation.
4. Failure to free the facial nerve up to the level of the stylomastoid foramen would leave the most distal part of its mastoid segment in situ. This part would hinder complete control of the jugular bulb and preclude visualization of the lower pole of the tumor (Fig. 10.93).
5. While rerouting the facial nerve, cottonoids are used to protect it against direct suction. Brackmann suction tips can be used instead. The whole contents of the internal auditory canal should be mobilized with the nerve to afford protection for this delicate part of the nerve (Fig. 10.94).
6. The jugular bulb has to be completely uncovered and, if necessary, lowered (as previously described).
7. An aluminum strip can be placed to protect the dura of the internal auditory canal while drilling the anterior wall of the canal (Fig. 10.95).
8. Internal carotid artery exposure depends on the extent needed for tumor removal. The artery may be merely skeletonized or can be exposed for 360 and 270 degrees in its vertical and horizontal segments, respectively.
9. Because of the large surgical defect, only one large piece of abdominal fat is used for obliteration, since the use of multiple smaller pieces might result in them slipping into the cerebellopontine angle.

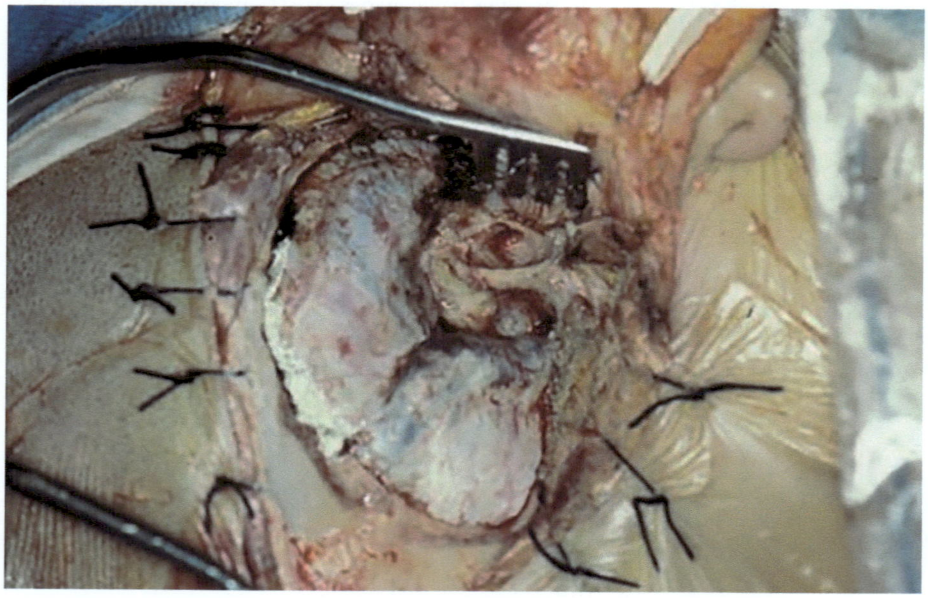

Fig. 10.90 Retractors make the surgical field deeper and the angle of work more acute.

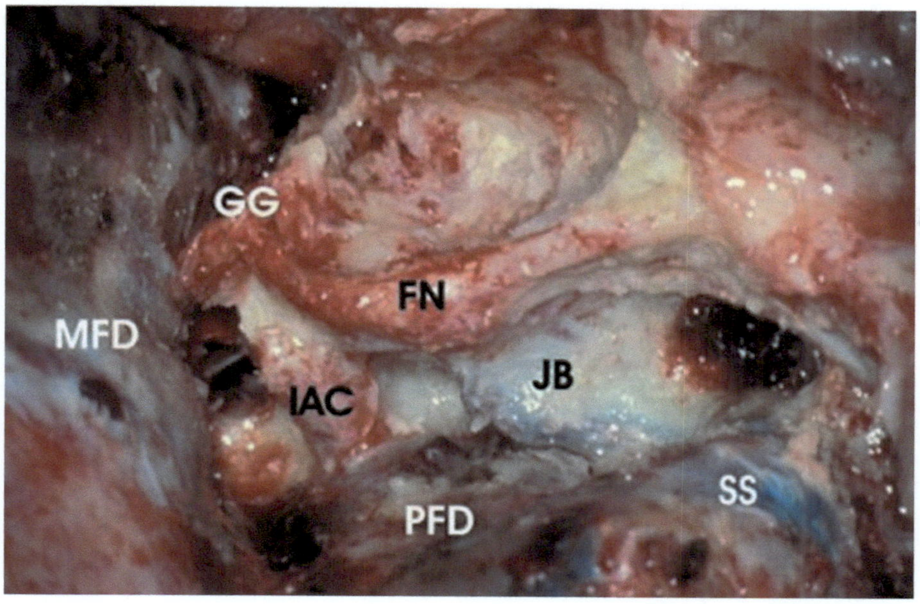

Fig. 10.91 Complete bone removal at the level of the geniculate ganglion (GG).

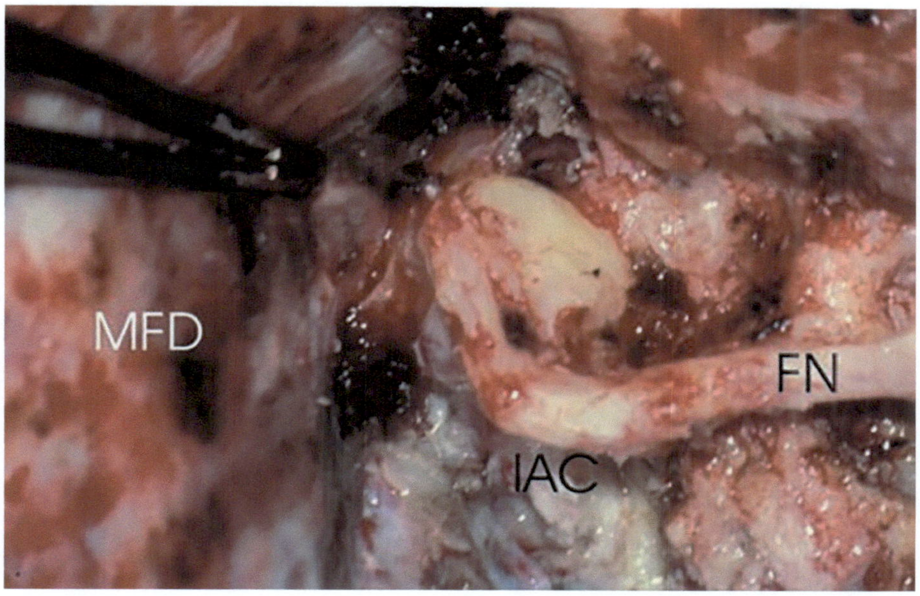

Fig. 10.92 Bipolar coagulation helps to retract the middle fossa dura (MFD).

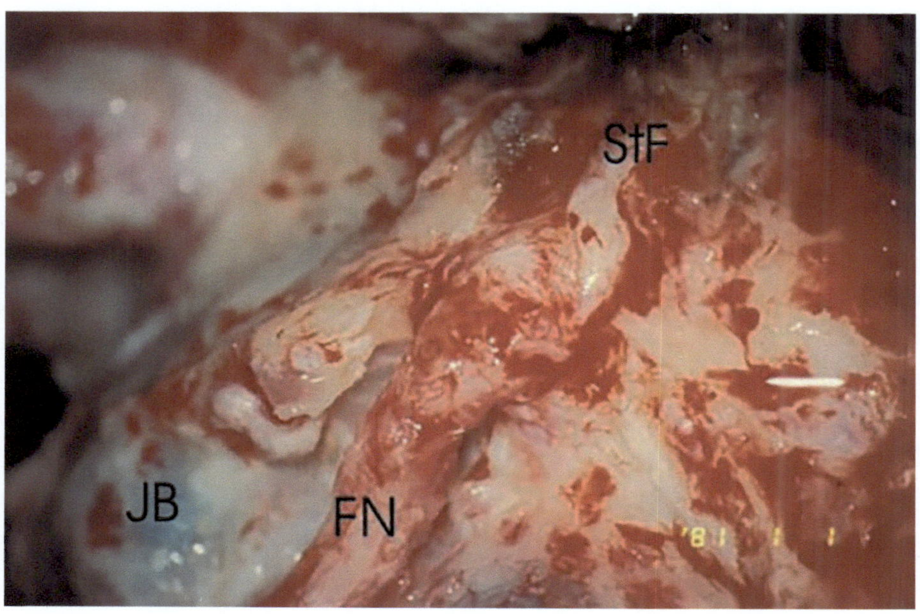

Fig. 10.93 Complete liberation of the facial nerve (FN) at the stylomastoid foramen (StF).

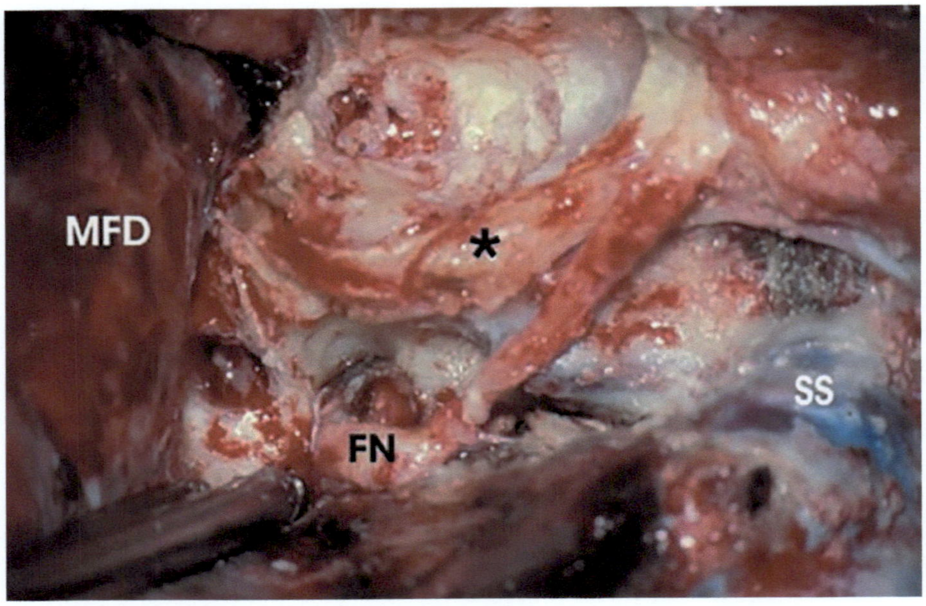

Fig. 10.94 The contents of the internal auditory canal afford protection to the transposed facial nerve (FN). The fallopian canal (*) can now be removed.

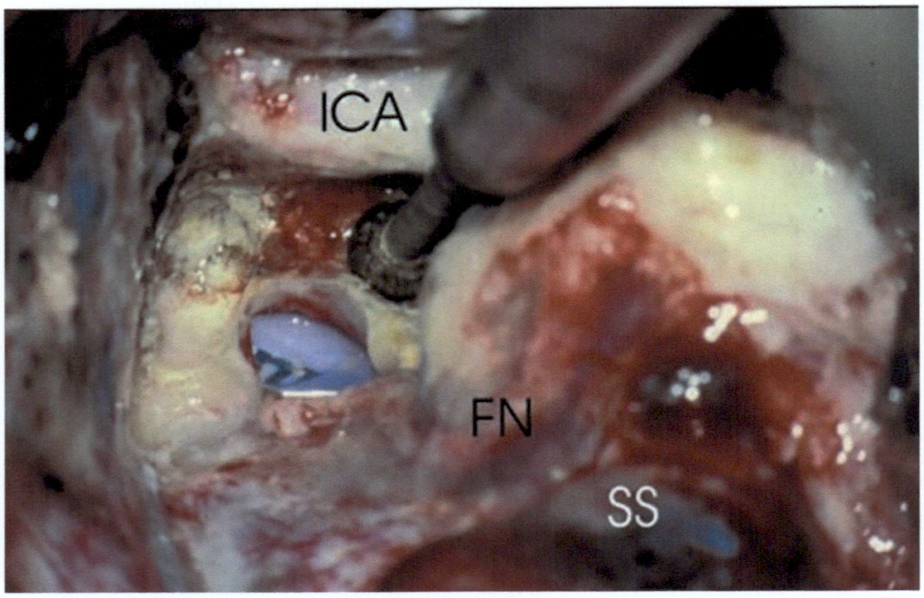

Fig. 10.95 Aluminum strip used to protect the porus of the internal auditory canal while drilling.

Type B Modified Transcochlear Approach

Rationale

Type B incorporates the Fisch infratemporal fossa approach of type B into the modified transcochlear approach of type A. In addition to the wide exposure of the infratemporal fossa, this type also affords additional access to anteromedial tumors lying ventral to the brainstem (Figs. 10.96 and 10.97).

Indications

- Extradural lesions with circumferential involvement of the internal carotid artery (e.g., clival chordomas and some cases of infralabyrinthine-apical and massive petrous bone cholesteatoma).
- Intradural lesions: petroclival lesions passing the midline, with limited contralateral extension (meningiomas, epidermoid cysts, etc.).
- Transdural lesions involving the parapharyngeal spaces (e.g., extensive clival chordomas or en-plaque meningiomas with extracranial extension). In these cases, the procedure is usually staged to avoid the risk of postoperative leakage of cerebrospinal fluid. The extradural part is removed first, followed by the removal of the intradural component in a second-stage procedure after 3–6 months.

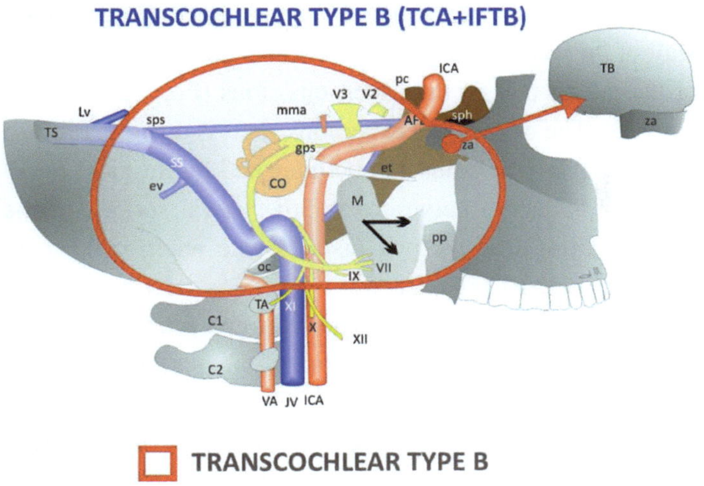

Fig. 10.96 Drawing showing the different structures controlled by this approach.

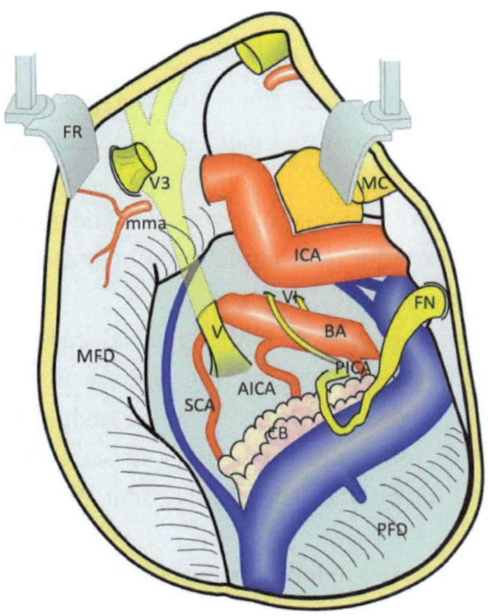

Fig. 10.97 Schematic drawing showing the limits of the transcochlear approach type B (TCB). Note the complete circumferential exposure of the vertical (ICA-V) and the horizontal (ICA-H) segments of the internal carotid artery.

Surgical Steps (Left Side)

(Figs. 10.98–10.109)

1. A wide postauricular incision is performed as in the type A approach. However, the incision extends more anteriorly up to the lateral margin of the orbit (Fig. 10.98).

2. The modified transcochlear approach type A is performed as previously described.

3. The temporalis muscle is detached from its bed to be further reflected anteriorly and inferiorly. The zygomatic arch is exposed, with care being taken not to injure the frontal branch of the facial nerve. The periosteum of the zygomatic arch is elevated (Fig. 10.99).

4. The capsule of the temporomandibular joint is detached and cut using strong scissors. The articular disk is removed, exposing the mandibular condyle (Fig. 10.100).

5. The arch is transected after two burr holes have been created for further refixation at the end of the procedure in live surgery. The zygomatic arch and part of the squamosal bone are osteotomy en block (Fig. 10.101).

6. The zygomatic arch and part of the squamosal bone are removed intact (Fig. 10.102). The Fisch infratemporal fossa retractor is applied with inferior displacement of the head of the mandible. The glenoid and the base of the middle fossa are drilled. The bony part of the eustachian tube is drilled as far as the isthmus. This provides complete control of the vertical segment of the internal carotid artery. The middle meningeal artery is identified (Fig. 10.103).

7. After bipolar coagulation, the middle meningeal artery is cut (Fig. 10.104). The mandibular nerve is identified (Fig. 10.105) and also transected after bipolar coagulation (Fig. 10.106).

8. Using a septal raspatory, the periosteum of the base of the pterygoid process is elevated together with the attachment of the lateral pterygoid muscle. The horizontal internal carotid artery is further exposed (Fig. 10.107). The cartilaginous eustachian tube is an important landmark and has a close spatial relation with the petrous segment of the internal carotid artery (Fig. 10.108).

9. By retracting the middle fossa dura, the horizontal segment of the internal carotid artery until it reaches the anterior foramen lacerum is seen. Note the exposure of the second branch of the trigeminal nerve. The base of the pterygoid process and the posterior opening of the vidian canal lie immediately in front of the foramen ovale (Fig. 10.109).

10. Further opening and retraction of the dura allow visualization of the cavernous segment of the internal carotid artery superior to the petrolingual ligament. The vidian nerve, which is formed by the junction of the greater superficial petrosal and the deep petrosal nerve, can be identified anterior to the foramen lacerum (Fig. 10.110).

11. The sphenoid sinus lies superomedial to the base of the pterygoid. The base of the pterygoid is now drilled. The lateral wall of the sphenoid sinus is opened, and the vidian nerve is seen just passing beneath the sinus. At a higher magnification, the relation between the vidian nerve and the petrous segment of the internal carotid artery is appreciated (Fig. 10.111).

The mucosa of the sphenoid sinus is now opened, and its relation to the different structures can be identified. The vidian canal is a reliable landmark to the petrous internal carotid artery. The canal opens anteriorly into the medial part of the pterygopalatine fossa and posteriorly at the upper part of the anterolateral edge of the foramen lacerum (Fig. 10.112).

12. The maxillary nerve running in the inferolateral wall of the cavernous sinus is sectioned. This allows control of the lateral part of the cavernous sinus (Fig. 10.113).

13. The dura of the cavernous sinus is opened. Profuse bleeding from the sinus can be managed by Surgicel packing. The abducent nerve crosses the lateral surface of the internal carotid artery just distal to the anterior foramen lacerum (Fig. 10.114).

14. Further opening of the dura and retraction of the temporal lobe allows visualization of the oculomotor nerve in the cavernous sinus (Fig. 10.115).

15. The eustachian tube can be followed inferomedially to its pharyngeal opening, cutting its tough attachments to the skull base. The medial pterygoid plate is drilled, the mucoperiosteum of the nasal cavity is opened at this level, and the nasopharynx is entered. The vomer can be seen (Fig. 10.116).

16. Further retraction of the dura and the position of the anterior clinoid process can be appreciated (Fig. 10.117).

17. An anterior clinoidectomy is performed. The paraclinoid area can be exposed after complete removal of the anterior clinoid process (Fig. 10.118).

18. The paraclinoid segment of the internal carotid artery lies between the proximal dural ring and the distal dural ring. The oculomotor nerve runs with the abducent nerve just below the proximal dural ring to exit the cavernous sinus and enter the superior orbital fissure (Fig. 10.119).

19. At the end of the procedure, the eustachian tube is sutured, and the temporalis muscle is used to obliterate the cavity.

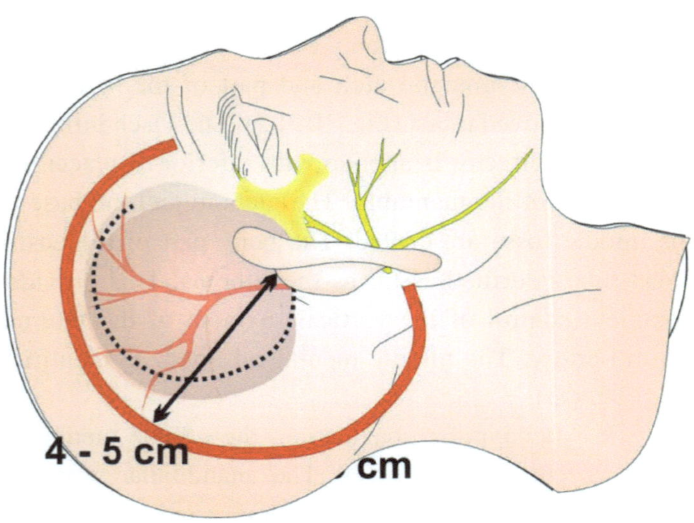

Fig. 10.98 Schematic drawing of the incision *(red line)* necessary to perform the trancochlear type B approach.

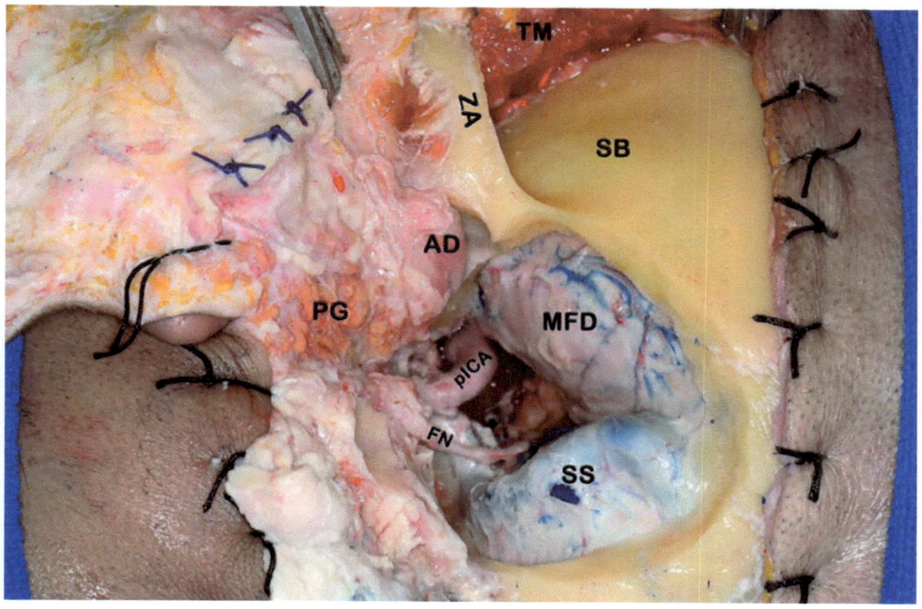

Fig. 10.99 The temporalis muscle is detached from its bed to be further reflected anteriorly and inferiorly. The zygomatic arch is exposed, with care being taken not to injure the frontal branch of the facial nerve. The periosteum of the zygomatic arch is elevated. *SB*, squamosal bone; *ZA*, zygomatic arch; *TM*, temporalis muscle (retracted); *AD*, articular disk; *MFD*, middle fossa dura; *pICA*, petrous segment of internal carotid artery; *SS*, sigmoid sinus; *FN*, facial nerve; *PG*, parotid gland.

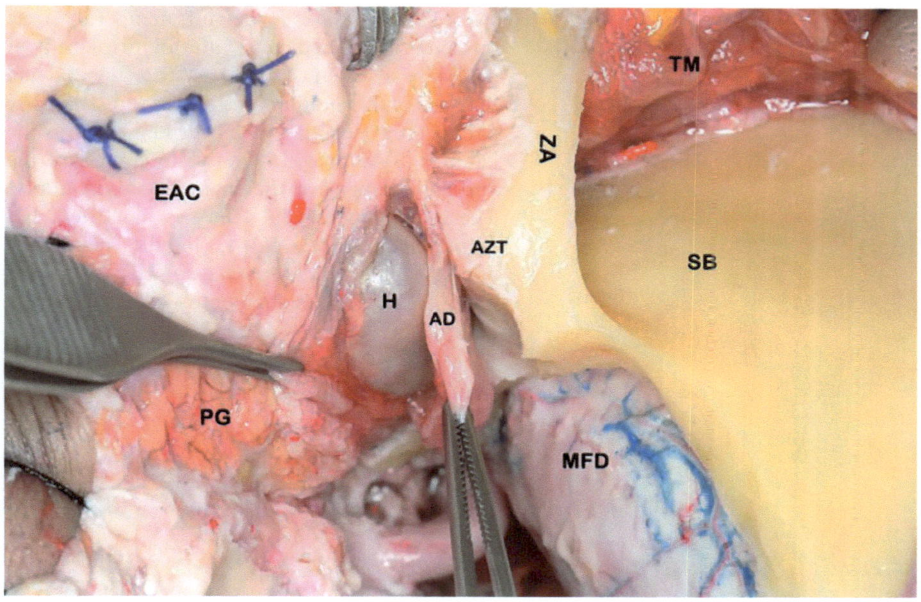

Fig. 10.100 The capsule of the temporomandibular joint is detached and cut using strong scissors. *TM*, temporalis muscle (retracted); *SB*, squamosal bone; *ZA*, zygomatic arch; *AZT*, anterior zygomatic tubercle; *AD*, articular disk; *H*, head of mandible; *EAC*, external auditory canal (closed); *PG*, parotid gland; *MFD*, middle fossa dura.

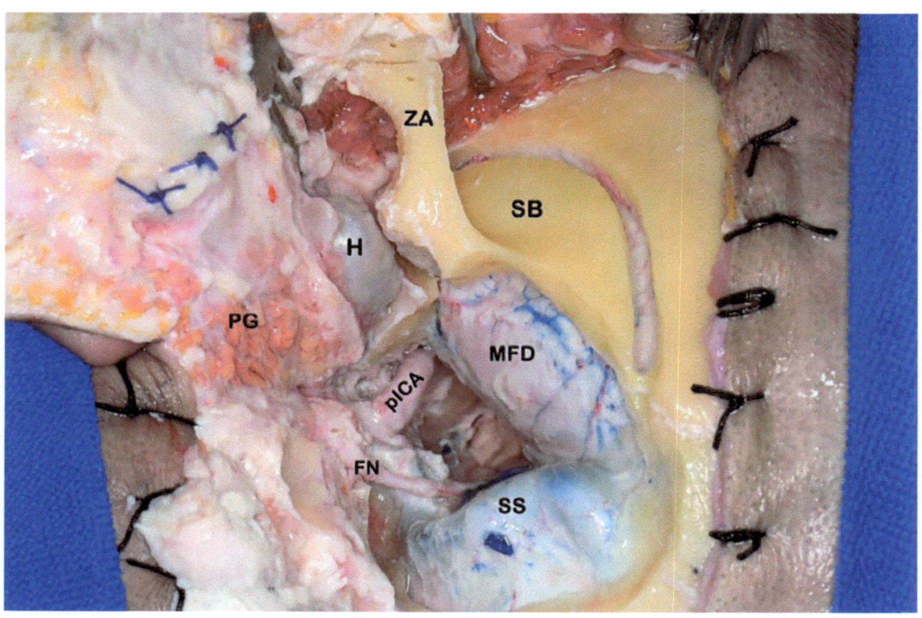

Fig. 10.101 The arch is transected after two burr holes have been created for further refixation at the end of the procedure in live surgery. The zygomatic arch and part of the squamosal bone are osteotomy en bloc. *ZA*, zygomatic arch; *SB*, squamosal bone; *H*, head of mandible; *MFD*, middle fossa dura; *SS*, sigmoid sinus; *pICA*, petrous segment of internal carotid artery; *FN*, facial nerve; *PG*, parotid gland.

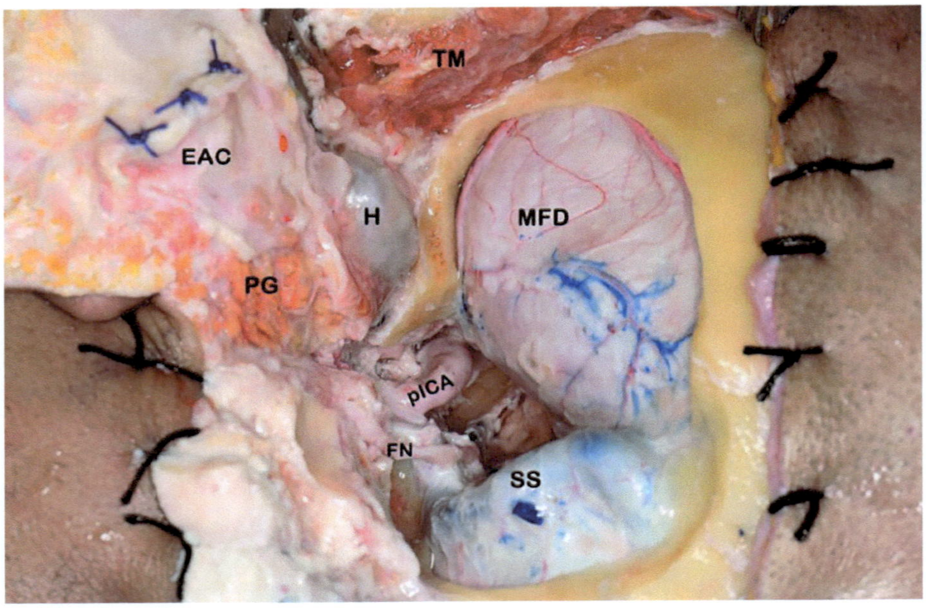

Fig. 10.102 The zygomatic arch and part of the squamosal bone are removed intact. *TM*, temporalis muscle (retracted); *EAC*, external auditory canal; *PG*, parotid gland; *H*, head of mandible; *MFD*, middle fossa dura; *SS*, sigmoid sinus; *pICA*, petrous segment of internal carotid artery; *FN*, facial nerve.

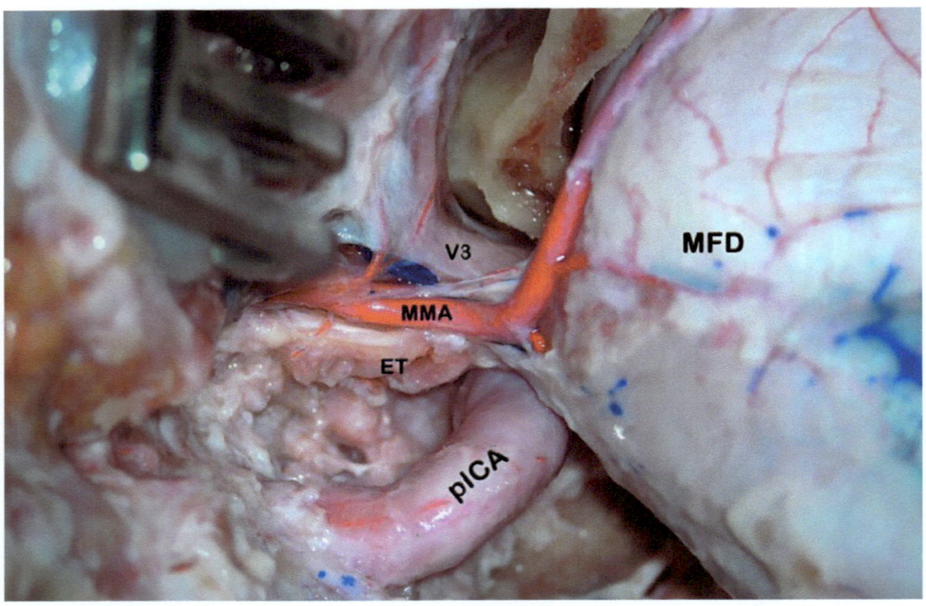

Fig. 10.103 The Fisch infratemporal fossa retractor is applied with inferior displacement of the head of the mandible. The glenoid and the base of the middle fossa are drilled. The bony part of the eustachian tube is drilled as far as the isthmus. This provides complete control of the vertical segment of the internal carotid artery. The middle meningeal artery is identified. *V3*, mandibular nerve; *MMA*, middle meningeal artery; *MFD*, middle fossa dura; *ET*, eustachian tube; *pICA*, petrous segment of internal carotid artery.

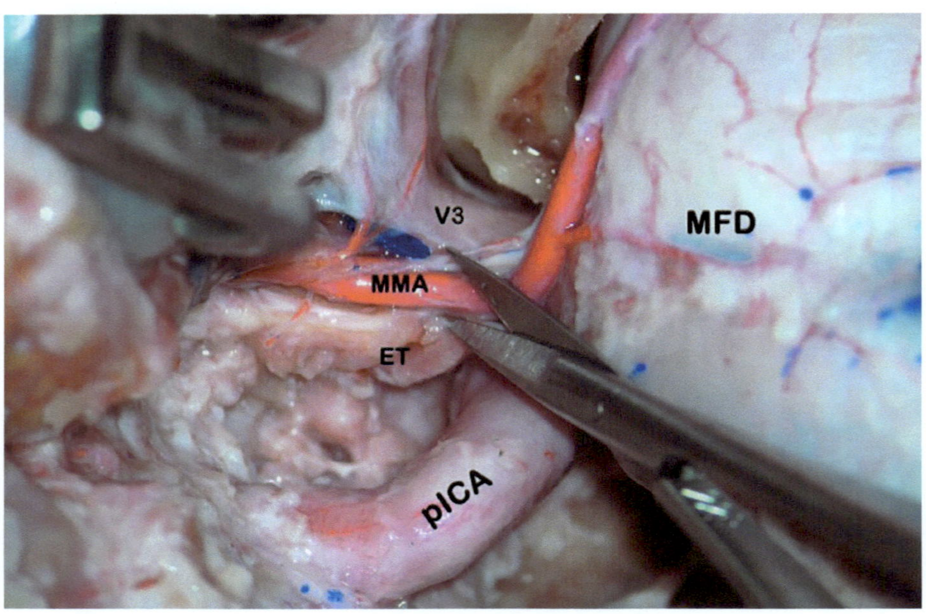

Fig. 10.104 After bipolar coagulation, the middle meningeal artery is cut. *MMA*, middle meningeal artery; *ET*, eustachian tube; *V3*, mandibular nerve; *MFD*, middle fossa dura; *pICA*, petrous segment of internal carotid artery.

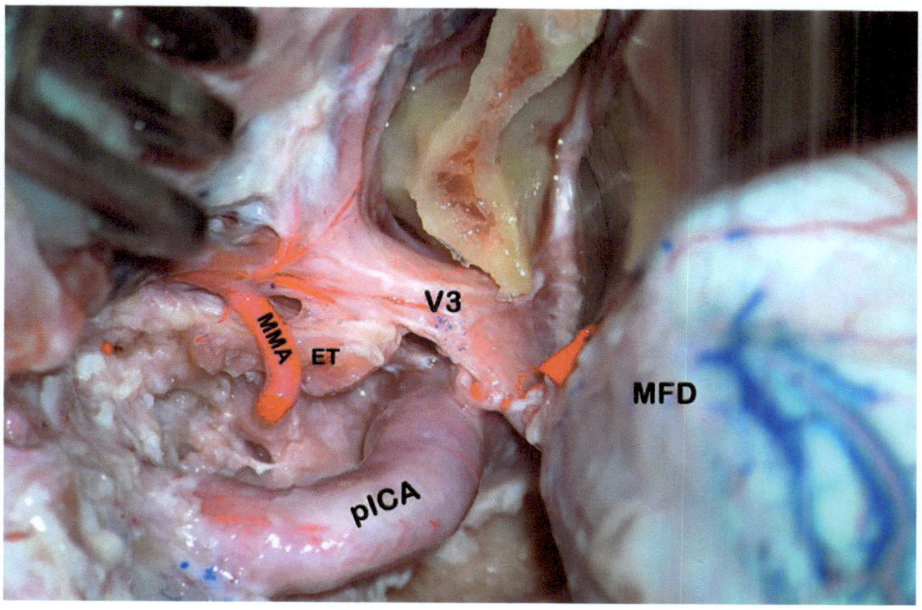

Fig. 10.105 The mandibular nerve is then identified. *MMA*, middle meningeal artery; *V3*, mandibular nerve; *ET*, eustachian tube; *MFD*, middle fossa dura; *pICA*, petrous segment of internal carotid artery.

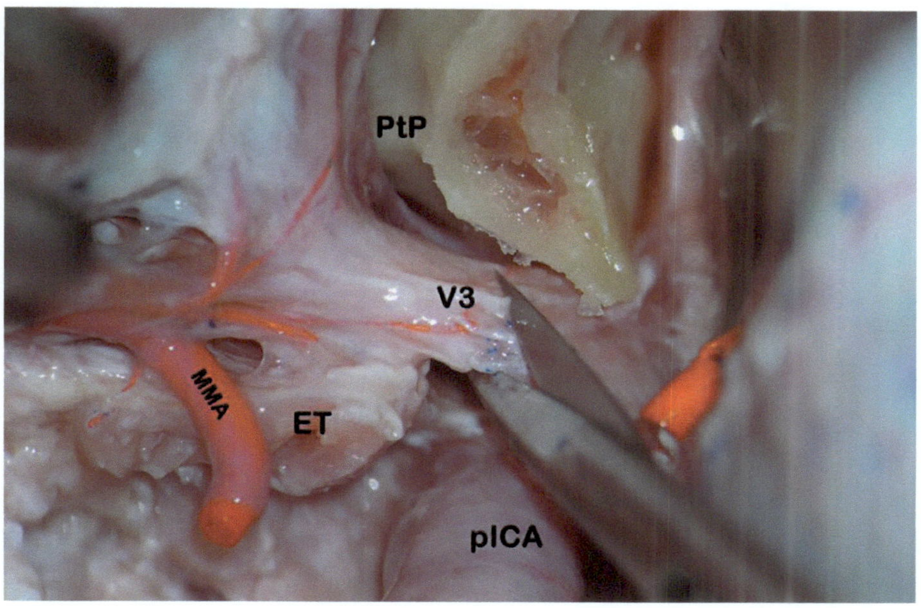

Fig. 10.106 The mandibular nerve is also transected after bipolar coagulation. *PtP*, pterygoid process; *V3*, mandibular nerve; *ET*, eustachian tube; *MMA*, middle meningeal artery; *pICA*, petrous segment of internal carotid artery.

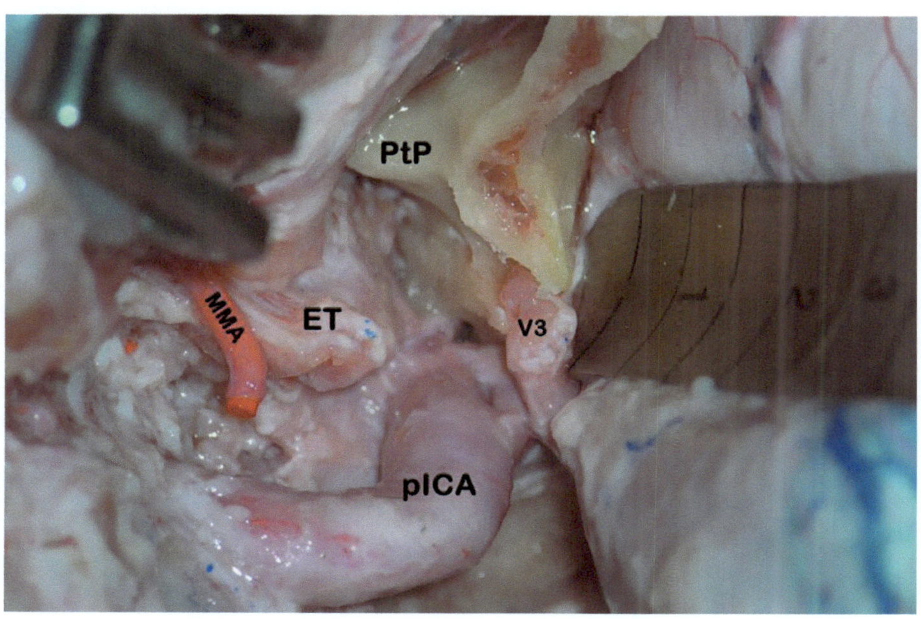

Fig. 10.107 Using a septal raspatory, the periosteum of the base of the pterygoid process is elevated together with the attachment of the lateral pterygoid muscle. *PtP*, pterygoid process; *V3*, mandibular nerve; *pICA*, petrous segment of internal carotid artery; *ET*, eustachian tube; *MMA*, middle meningeal artery.

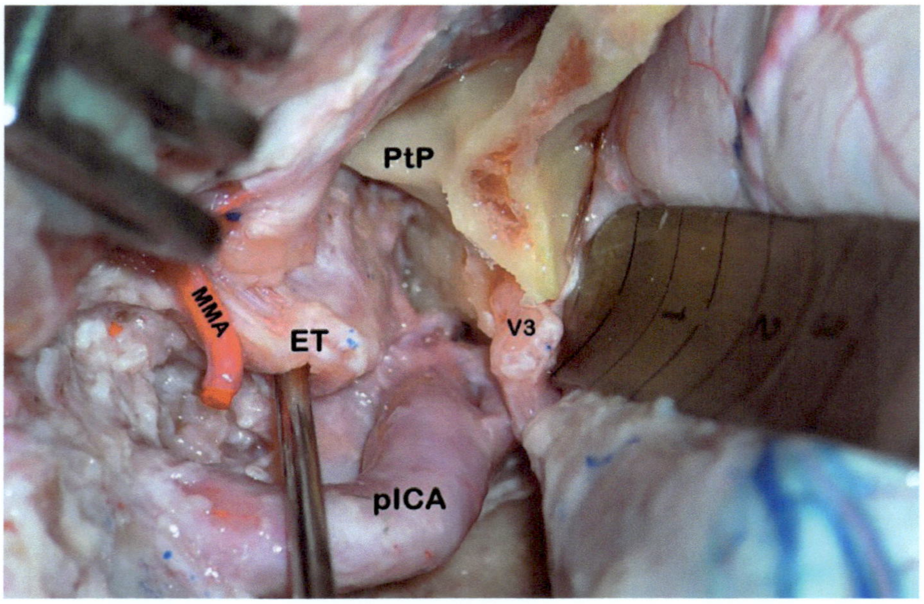

Fig. 10.108 A probe is inserted into the cartilaginous eustachian tube. The eustachian tube is an important landmark in skull base surgery and has a close spatial relation with the petrous segment of the internal carotid artery. *PtP*, pterygoid process; *V3*, mandibular nerve; *ET*, eustachian tube; *pICA*, petrous segment of internal carotid artery; *MMA*, middle meningeal artery.

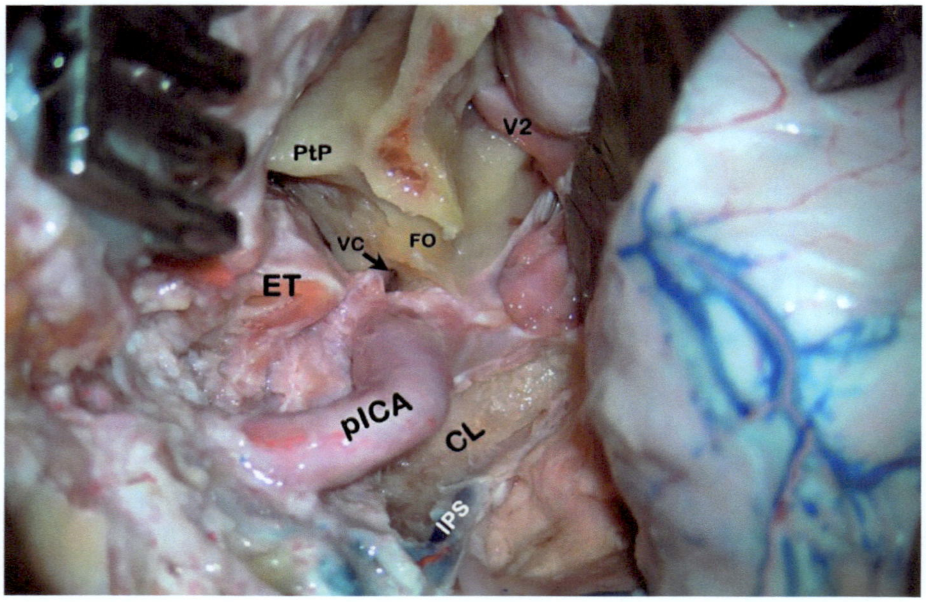

Fig. 10.109 By retracting the middle fossa dura, the horizontal segment of the internal carotid artery until it reaches the anterior foramen lacerum is seen. Note the exposure of the second branch of the trigeminal nerve. The base of the pterygoid process and the posterior opening of the vidian canal lie immediately in front of the foramen ovale. *Cl*, clivus; *IPS*, inferior petrosal sinus; *pICA*, petrous segment of internal carotid artery; *ET*, eustachian tube; *V2*, maxillary nerve; *FO*, foramen ovale; *VC*, vidian canal; *PtP*, pterygoid process.

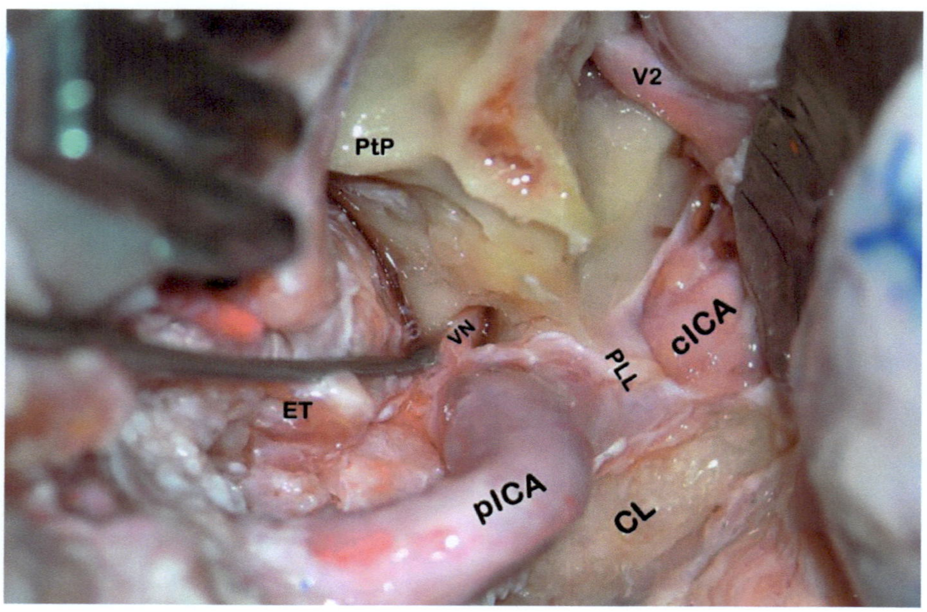

Fig. 10.110 Further opening and retraction of the dura allows visualization of the cavernous segment of the internal carotid artery superior to the petrolingual ligament. The vidian nerve, which is formed by the junction of the greater superficial petrosal and the deep petrosal nerve, can be identified anterior to the foramen lacerum. *V2*, maxillary nerve; *cICA*, cavernous segment of internal carotid artery; *PLL*, petrolingual ligament; *Cl*, clivus; *pICA*, petrous segment of internal carotid artery; *VN*, vidian nerve; *PtP*, pterygoid process; *ET*, eustachian tube.

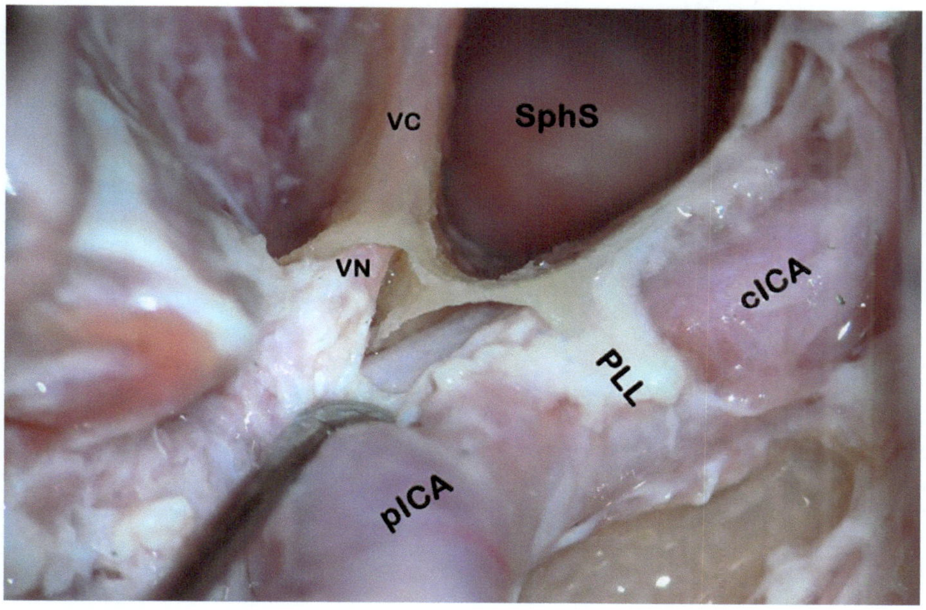

Fig. 10.111 The base of the pterygoid is now drilled. The lateral wall of the sphenoid sinus is opened, and the vidian nerve is seen just passing beneath the sinus. At a higher magnification, the relation between the vidian nerve and the petrous segment of the internal carotid artery is appreciated. *cICA*, cavernous segment of internal carotid artery; *PLL*, petrolingual ligament; *SphS*, sphenoid sinus; *VC*, vidian canal; *VN*, vidian nerve; *pICA*, petrous segment of internal carotid artery.

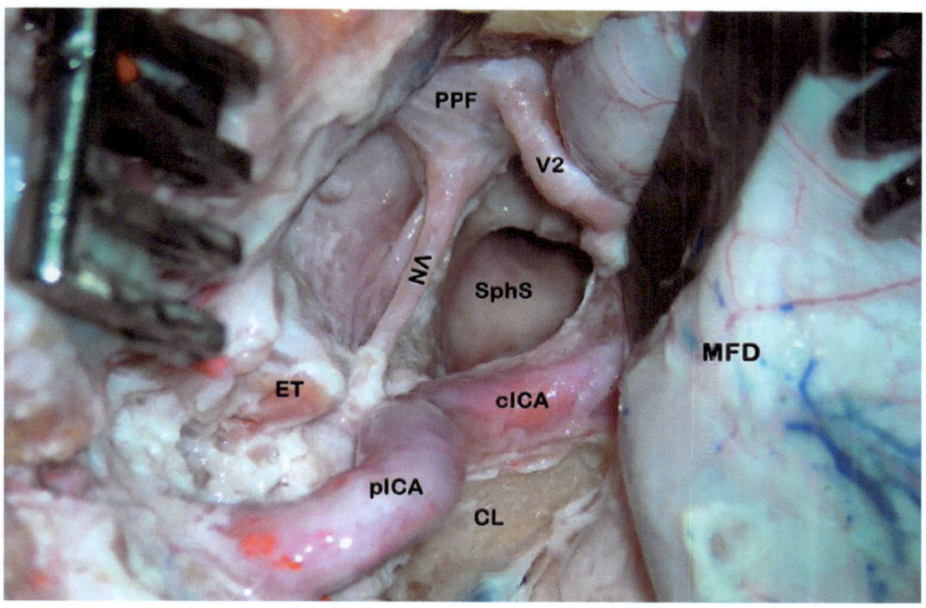

Fig. 10.112 The mucosa of the sphenoid sinus is now opened, and its relation to the different structures can be identified. The vidian canal is a reliable landmark to the petrous internal carotid artery. The canal opens anteriorly into the medial part of the pterygopalatine fossa and posteriorly at the upper part of the anterolateral edge of the foramen lacerum. *PPF*, pterygopalatine fossa; *VN*, vidian nerve; *SphS*, sphenoid sinus; *MFD*, middle fossa dura; *V2*, maxillary nerve; *cICA*, cavernous segment of internal carotid artery; *Cl*, clivus; *pICA*, petrous segment of internal carotid artery; *ET*, eustachian tube.

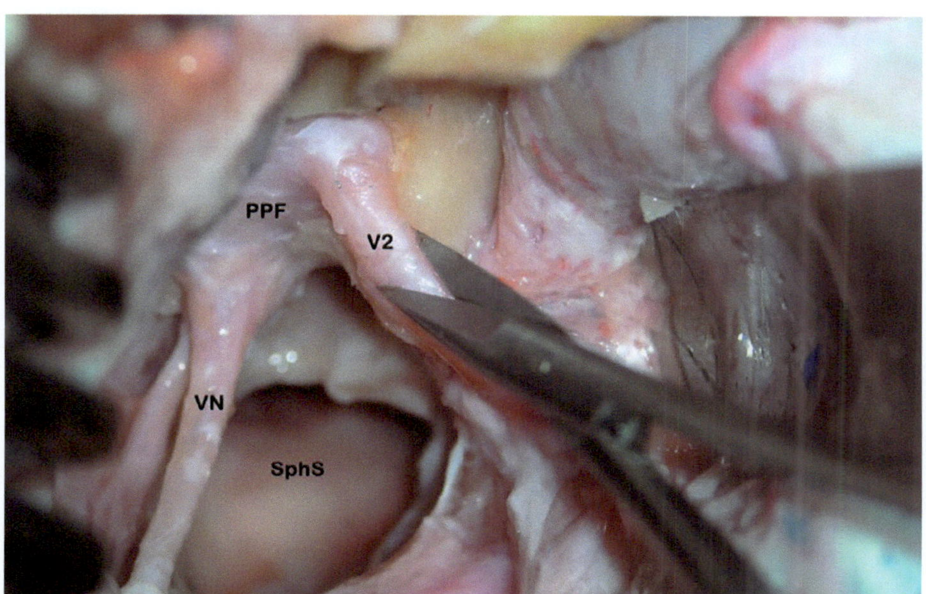

Fig. 10.113 The maxillary nerve running in the inferolateral wall of the cavernous sinus is sectioned. This allows control of the lateral part of the cavernous sinus. *V2*, maxillary nerve; *PPF*, pterygopalatine fossa; *VN*, vidian nerve; *SphS*, sphenoid sinus.

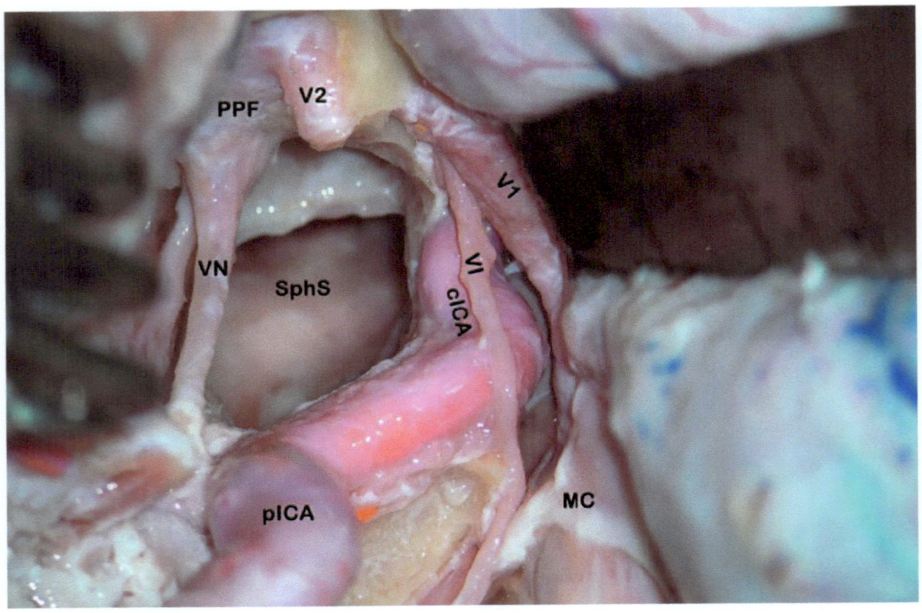

Fig. 10.114 The dura of the cavernous sinus is opened. Profuse bleeding from the sinus can be managed by Surgicel packing. The abducent nerve crosses the lateral surface of the internal carotid artery just distal to the anterior foramen lacerum. *V1*, ophthalmic nerve; *V2*, maxillary nerve; *PPF*, pterygopalatine fossa; *VN*, vidian nerve; *VI*, abducent nerve; *SphS*, sphenoid sinus; *cICA*, cavernous segment of internal carotid artery; *MC*, Meckel's cave; *pICA*, petrous segment of internal carotid artery.

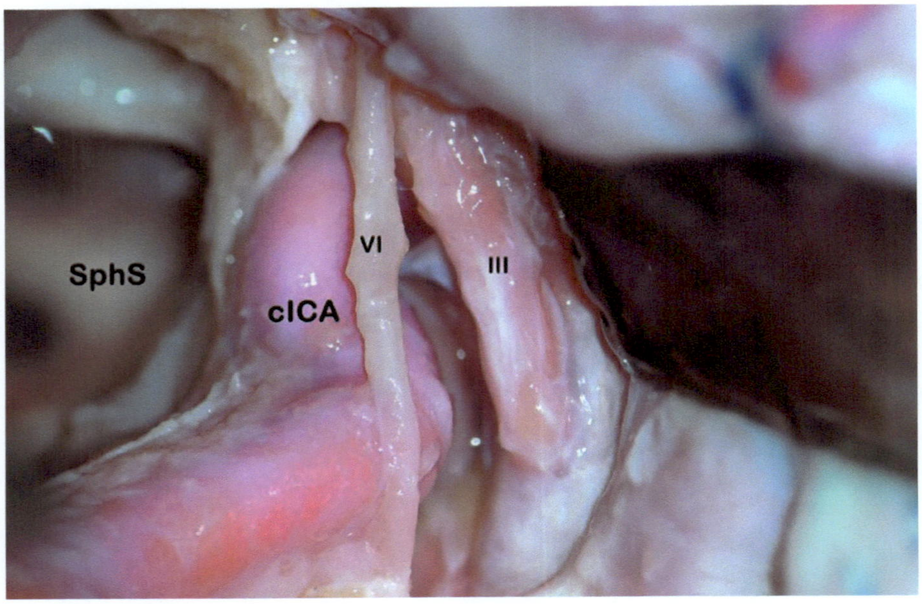

Fig. 10.115 Further opening of the dura and retraction of the temporal lobe allows visualization of the oculomotor nerve in the cavernous sinus. *III*, oculomotor nerve; *VI*, abducent nerve; *cICA*, cavernous segment of internal carotid artery; *SphS*, sphenoid sinus.

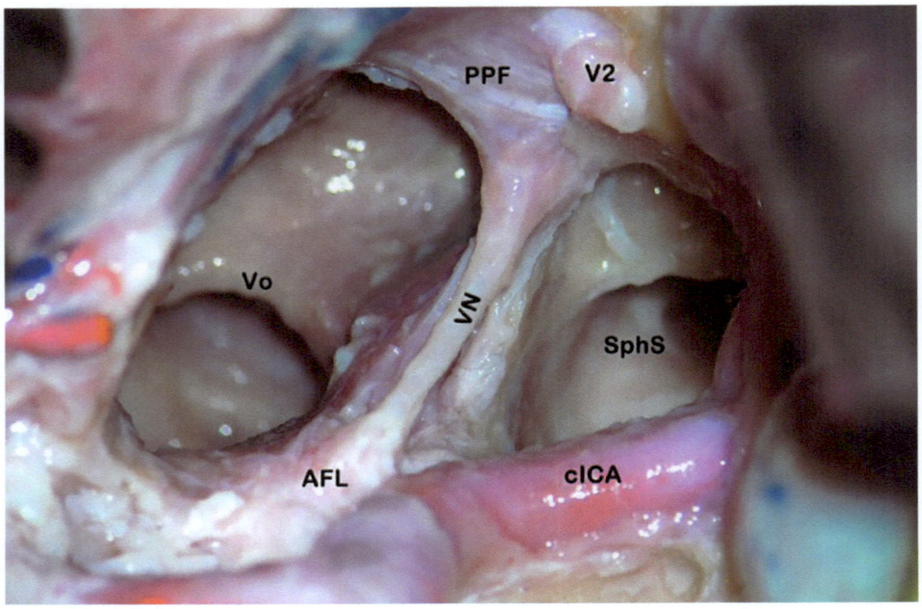

Fig. 10.116 The eustachian tube can be followed inferomedially to its pharyngeal opening, cutting its tough attachments to the skull base. The medial pterygoid plate is drilled, the mucoperiosteum of the nasal cavity is opened at this level, and the nasopharynx is entered. The vomer can be seen. *V2*, maxillary nerve; *PPF*, pterygopalatine fossa; *SphS*, sphenoid sinus; *VN*, vidian nerve; *AFL*, anterior foramen lacerum; *Vo*, vomer; *cICA*, cavernous segment of internal carotid artery.

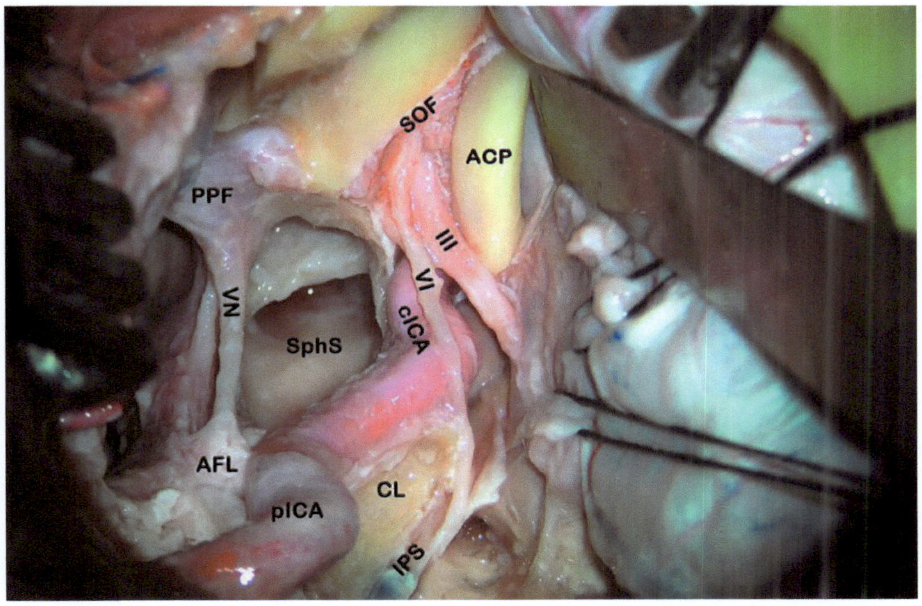

Fig. 10.117 Further retraction of the dura and the position of the anterior clinoid process can be appreciated. *ACP*, anterior clinoid process; *SOF*, superior orbital fissure; *III*, oculomotor nerve; *VI*, abducent nerve; *VN*, vidian nerve; *PPF*, pterygopalatine fossa; *SphS*, sphenoid sinus; *Cl*, clivus; *IPS*, inferior petrosal sinus; *cICA*, cavernous segment of internal carotid artery; *pICA*, petrous segment of internal carotid artery; *AFL*, anterior foramen lacerum.

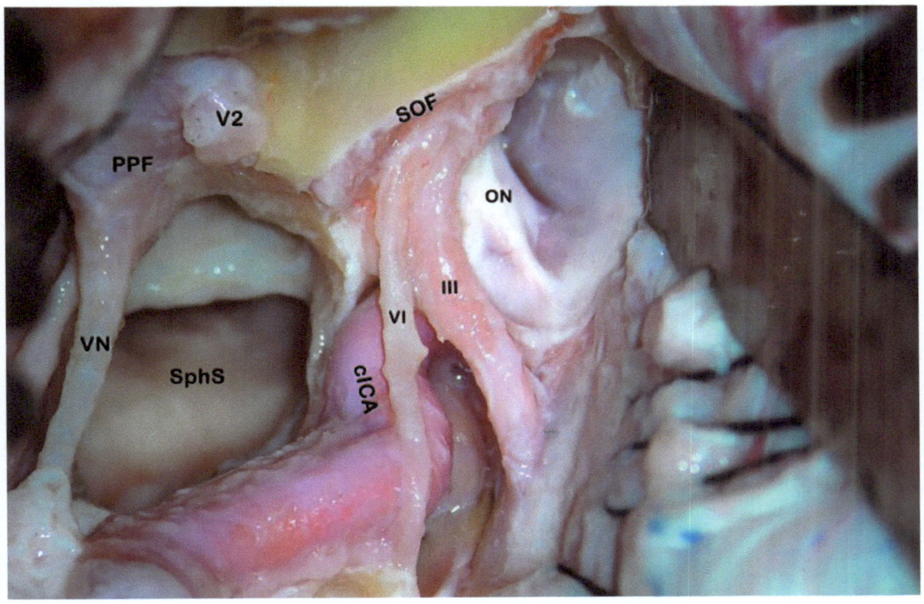

Fig. 10.118 Anterior clinoidectomy is performed. The paraclinoid area can be exposed after complete removal of the anterior clinoid process. *SOF*, superior orbital fissure; *ON*, optic nerve; *III*, oculomotor nerve; *VI*, abducent nerve; *cICA*, cavernous segment of internal carotid artery; *SphS*, sphenoid sinus; *VN*, vidian nerve; *PPF*, pterygopalatine fossa; *V2*, maxillary nerve.

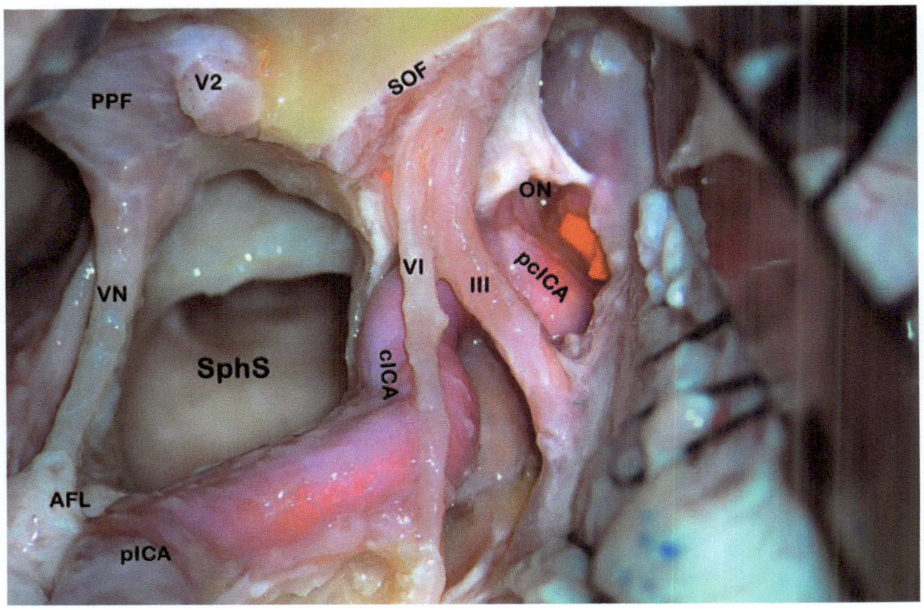

Fig. 10.119 The paraclinoid segment of the internal carotid artery lies between the proximal dural ring and the distal dural ring. The oculomotor nerve runs with the abducent nerve just below the proximal dural ring to exit the cavernous sinus and enter the superior orbital fissure. *ON*, optic nerve; *SOF*, superior orbital fissure; *pcICA*, paraclinoid segment of internal carotid artery; *III*, oculomotor nerve; *VI*, abducent nerve; *cICA*, cavernous segment of internal carotid artery; *SphS*, sphenoid sinus; *VN*, vidian nerve; *V2*, maxillary nerve; *PPF*, pterygopalatine fossa; *AFL*, anterior foramen lacerum; *pICA*, petrous segment of internal carotid artery.

Case 1: Massive Petrous Bone Cholesteatoma With Invasion of the Clivus

See Figs. 10.120–10.139.

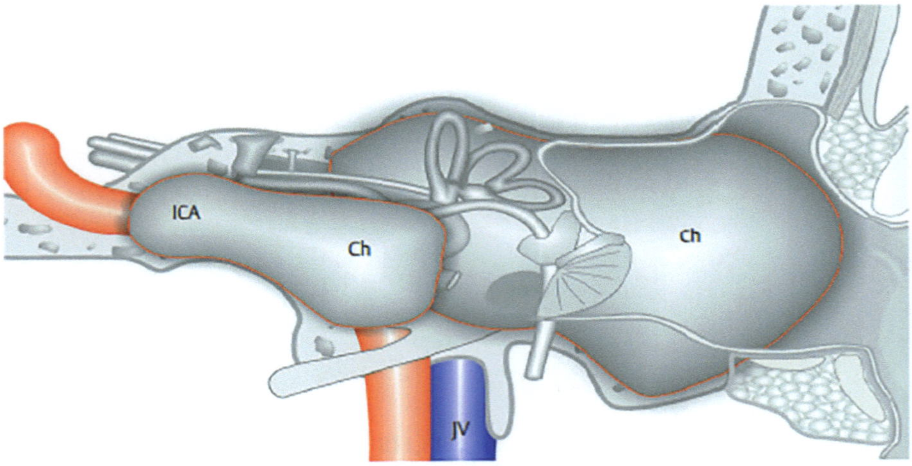

Fig. 10.120 Case of massive petrous bone cholesteatoma arose in the canal wall down cavity. The otic capsule is completely destroyed, and cholesteatoma infiltrates the area anteriorly to the horizontal segment of the internal carotid artery. *Ch*, cholesteatoma; *ICA*, internal carotid artery; *JV*, jugular vein.

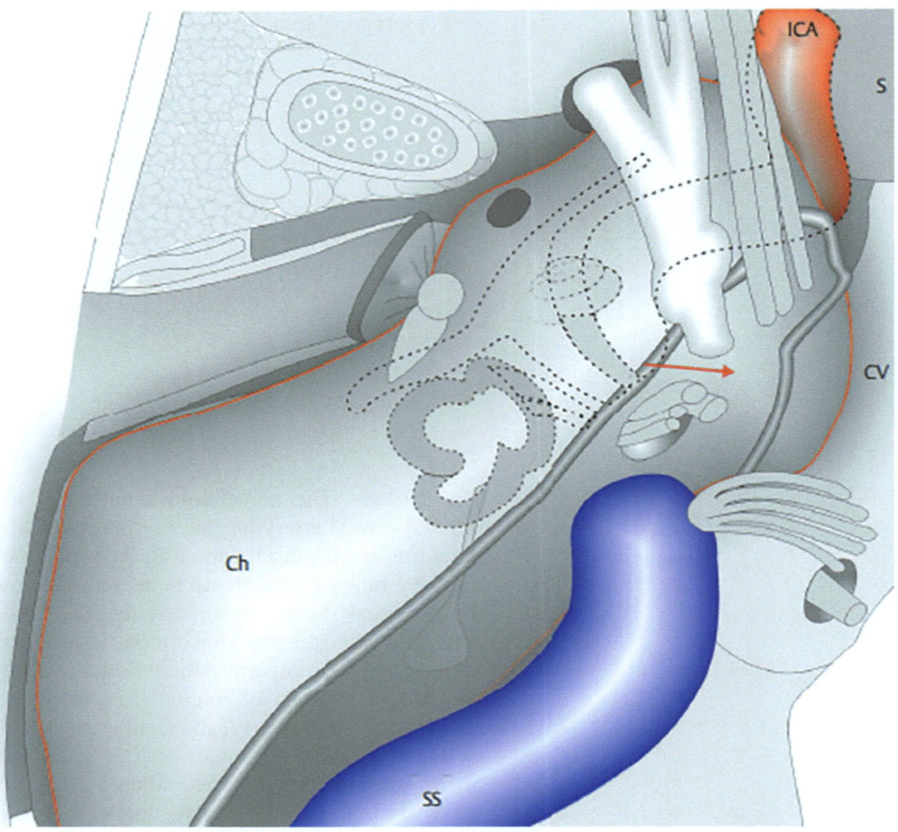

Fig. 10.121 The cholesteatoma went deep to erode the clivus *(arrow)*. To control the horizontal segment of the internal carotid artery, a modified transcochlear approach type B is indicated. *Ch*, cholesteatoma; *CV*, clivus; *ICA*, internal carotid artery; *S*, sphenoid sinus; *SS*, sigmoid sinus.

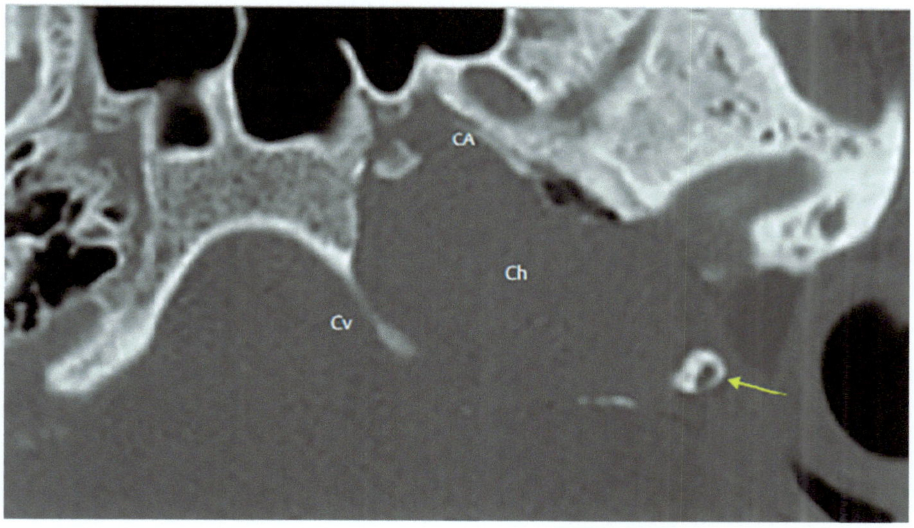

Fig. 10.122 The patient underwent surgery four times elsewhere, including a middle fossa approach with no radical removal of cholesteatoma. The axial CT demonstrates the presence of a massive round-shaped mass in the petrous apex. The majority of structures in the temporal bone have disappeared except the fallopian canal *(arrow)*. *CA*, carotid artery; *Ch*, cholesteatoma; *CV*, clivus.

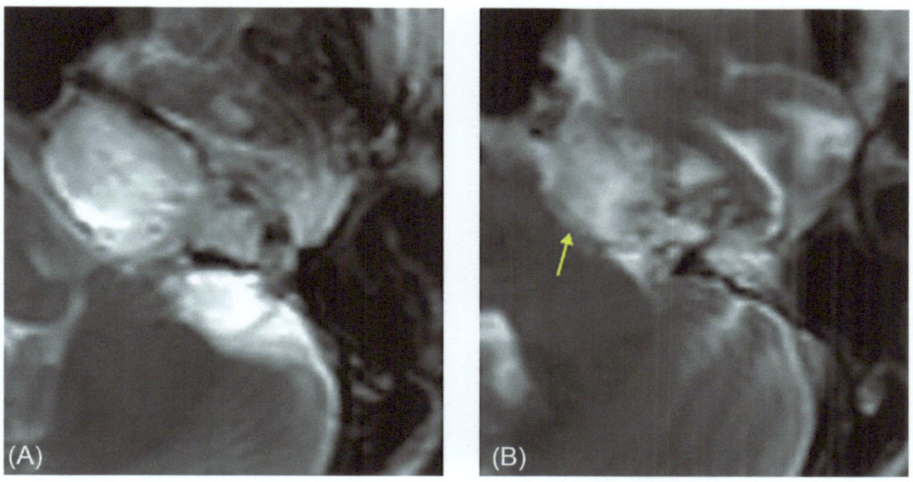

Fig. 10.123 The axial image of T2-weighted MRI (A) shows a massive high-signal lesion corresponding to cholesteatoma. The MRI, more cranial slice (B), shows that the cholesteatoma protrudes to the CP angle, starting to push the brainstem *(arrow)*.

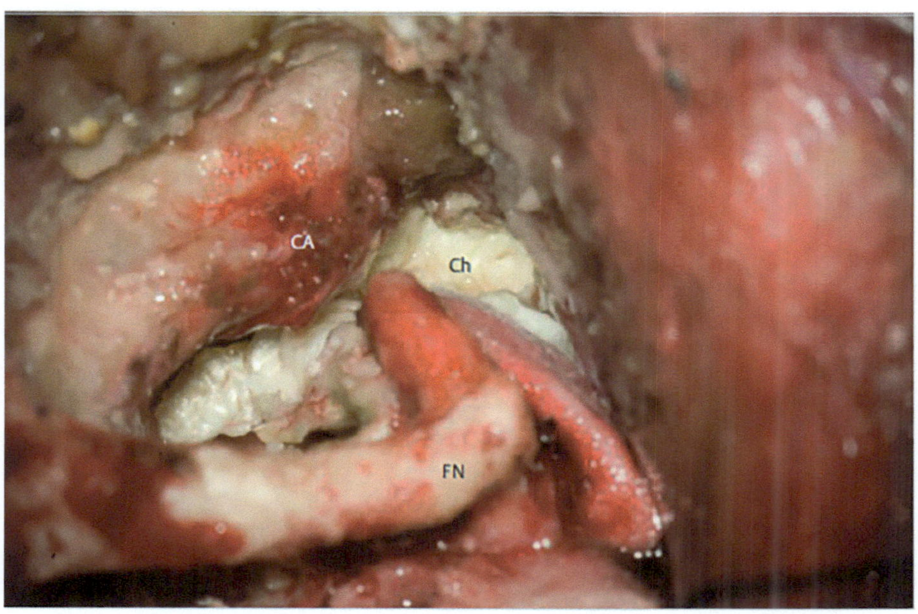

Fig. 10.124 The transotic approach, including labyrinthectomy and removal of the cochlea, has been completed. The facial nerve is well skeletonized. The greater superficial petrosal nerve is sacrificed. The carotid artery is also skeletonized. *CA*, internal carotid artery; *Ch*, cholesteatoma; *FN*, facial nerve.

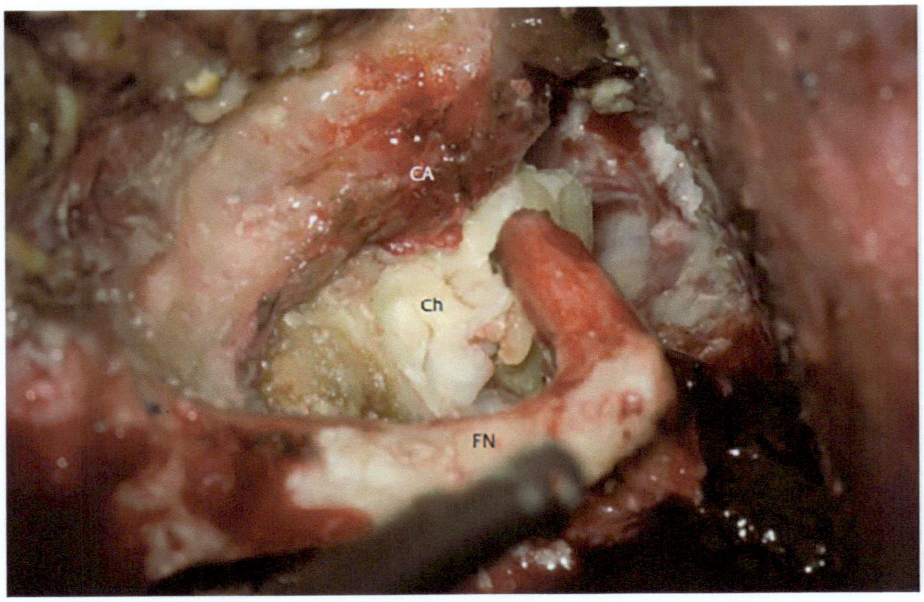

Fig. 10.125 Both the middle and posterior fossa dura are uncovered by the cholesteatoma. The matrix covering the middle fossa dura and the posterior fossa dura is dissected medially. *CA*, internal carotid artery; *Ch*, cholesteatoma; *FN*, facial nerve.

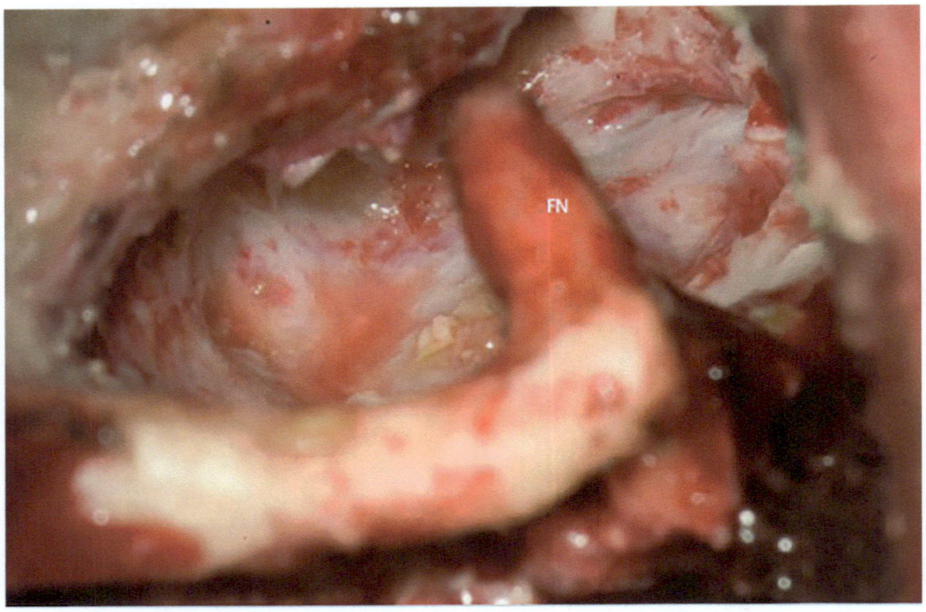

Fig. 10.126 The cholesteatoma is removed from the posterior fossa dura. *FN*, facial nerve.

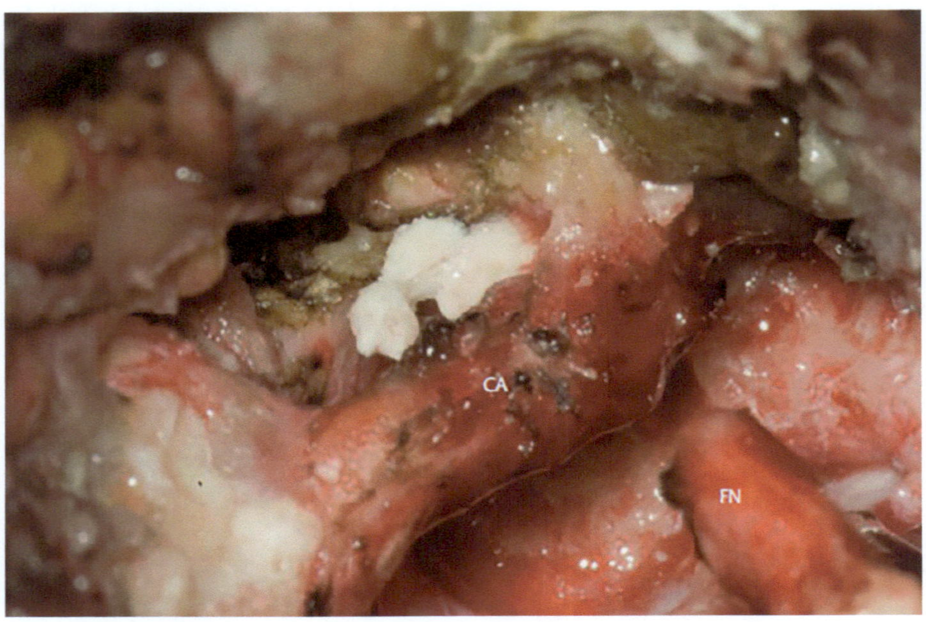

Fig. 10.127 Note the residual cholesteatoma at the level of the petrous internal carotid artery (CA). *FN*, facial nerve.

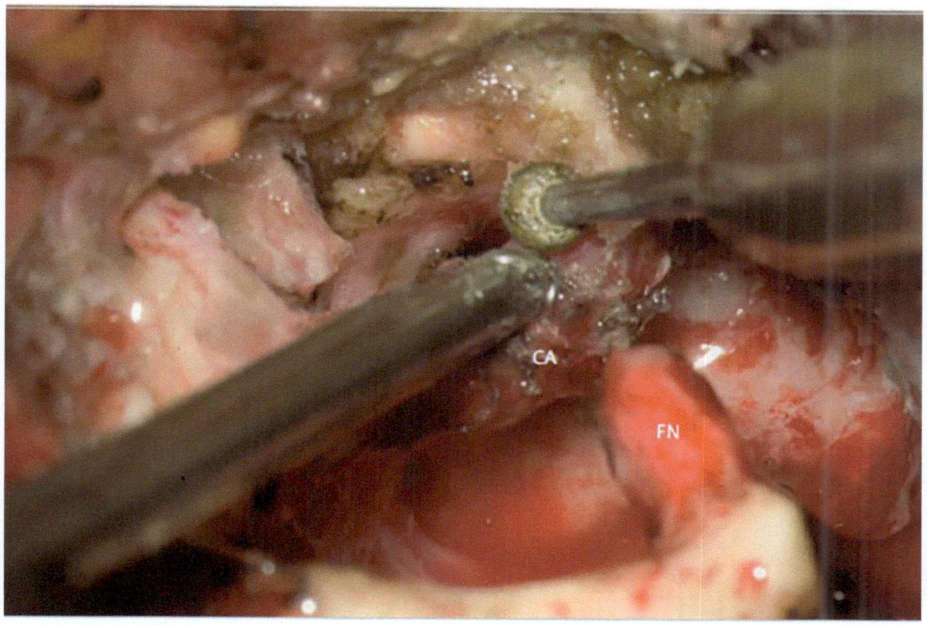

Fig. 10.128 The petrous internal carotid artery (CA) is decompressed to allow further matrix dissection. *FN*, facial nerve.

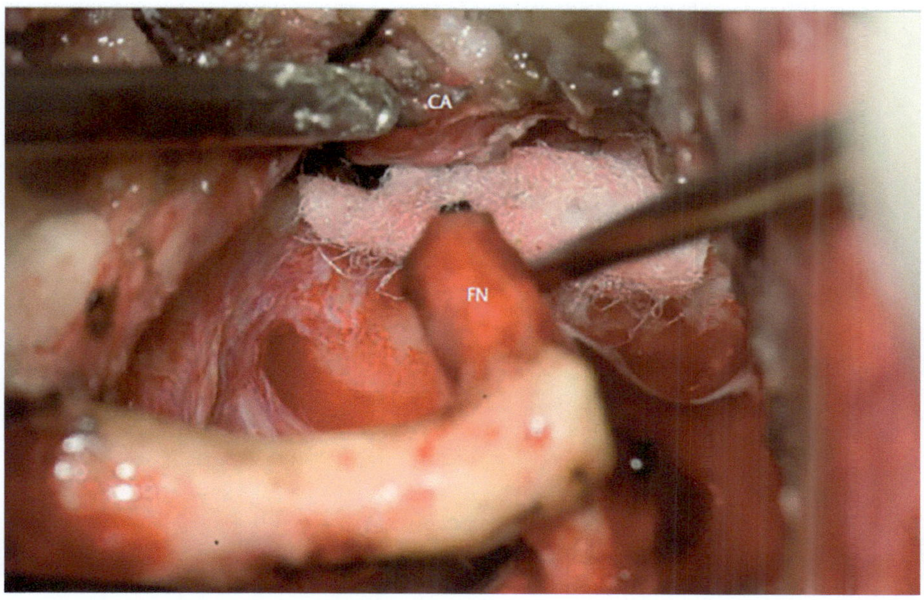

Fig. 10.129 To reduce the possibility of leaving cholesteatoma matrix on the bony wall, a piece of cottonoid is used to scrub the bony wall and invisible places such as the medial face of the carotid artery and the facial nerve and the anterior face of the internal auditory canal. *CA*, internal carotid artery; *FN*, facial nerve.

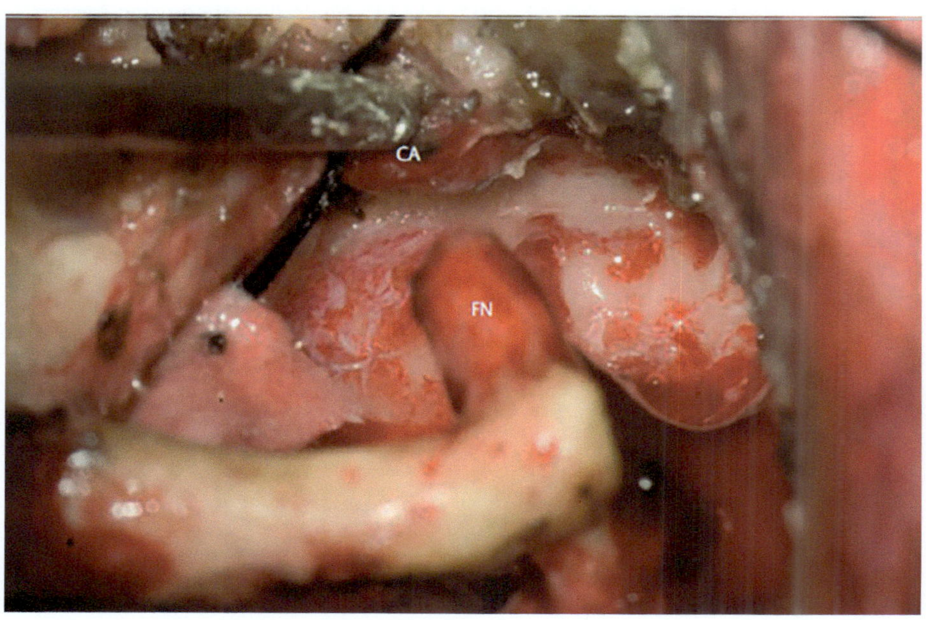

Fig. 10.130 The bony protrusion anterior to the facial nerve covering the carotid artery obstructs some view of the petrous apex. *CA*, internal carotid artery; *FN*, facial nerve

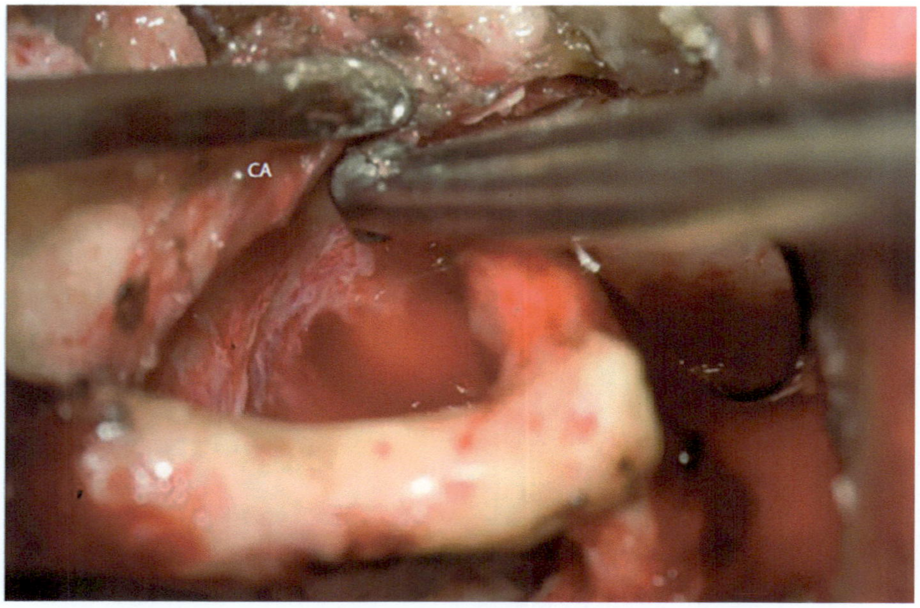

Fig. 10.131 The bone over the posterosuperior wall of the carotid artery can be removed safely with a curette. *CA*, internal carotid artery; *FN*, facial nerve.

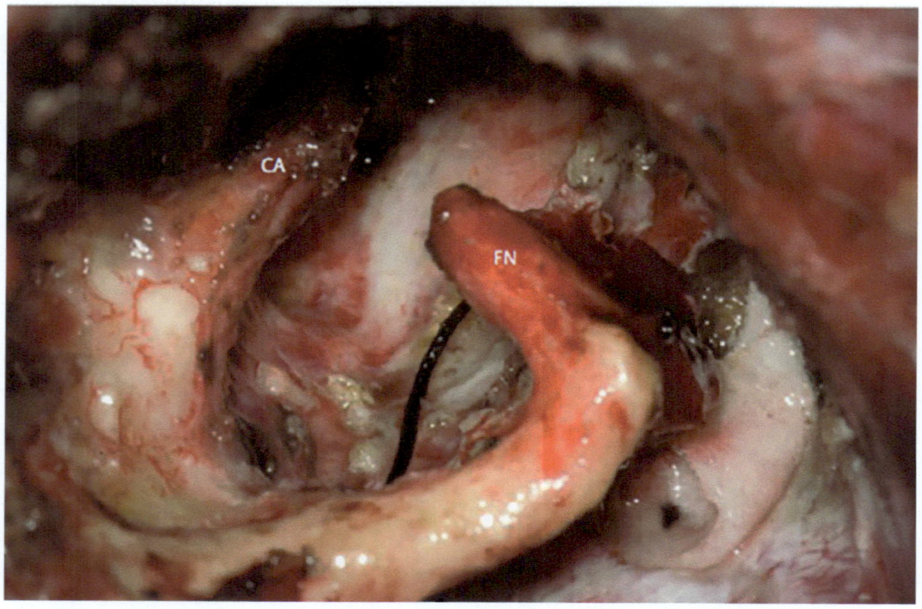

Fig. 10.132 The access to the bottom of the petrous apex becomes better. *CA*, internal carotid artery; *FN*, facial nerve.

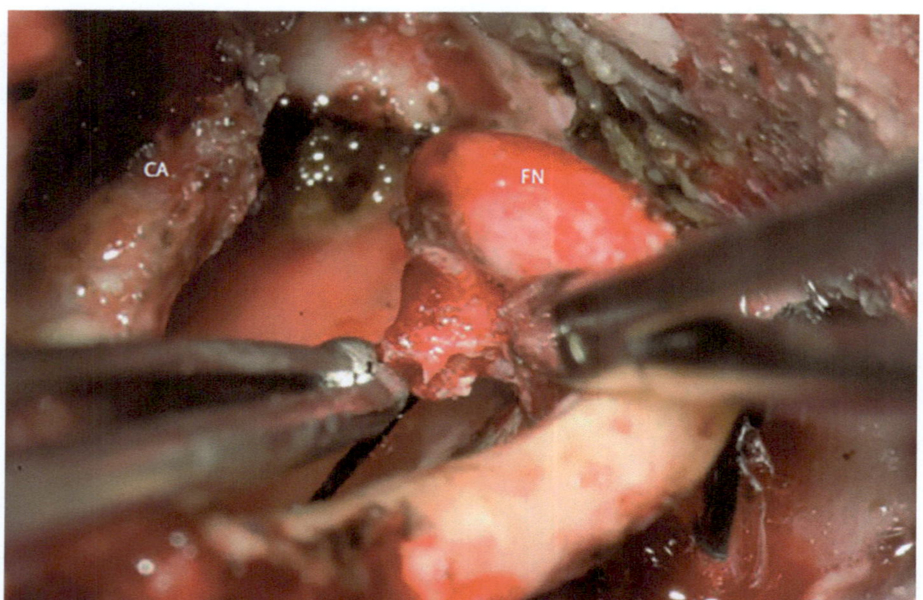

Fig. 10.133 Sharp dissection using scissors is carried out to remove the last piece of cholesteatoma adherent to the inferior face of the internal auditory canal. *CA*, internal carotid artery; *FN*, facial nerve.

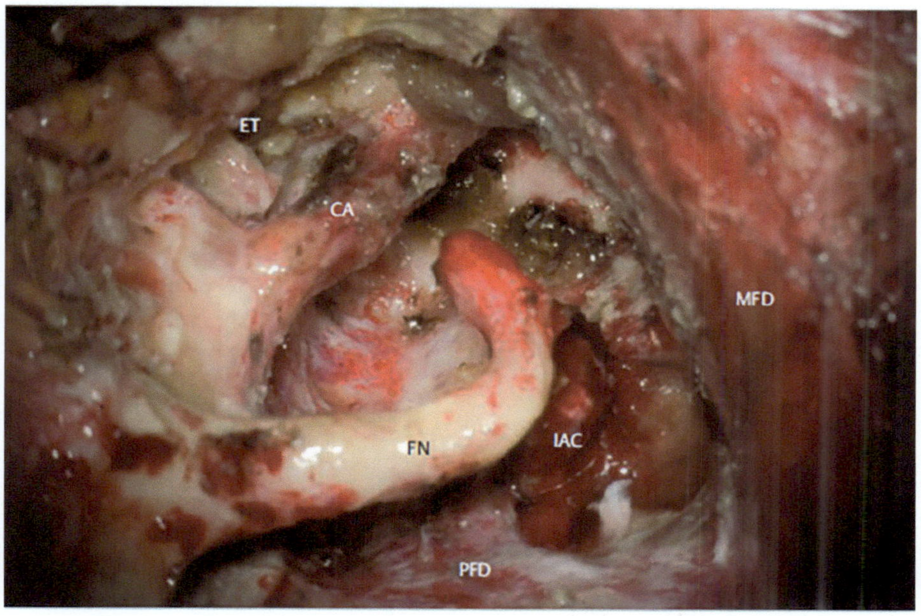

Fig. 10.134 The cholesteatoma is eradicated from the temporal bone. *CA*, carotid artery; *ET*, eustachian tube; *FN*, facial nerve; *IAC*, internal auditory canal; *MFD*, middle fossa dura; *PFD*, posterior fossa dura.

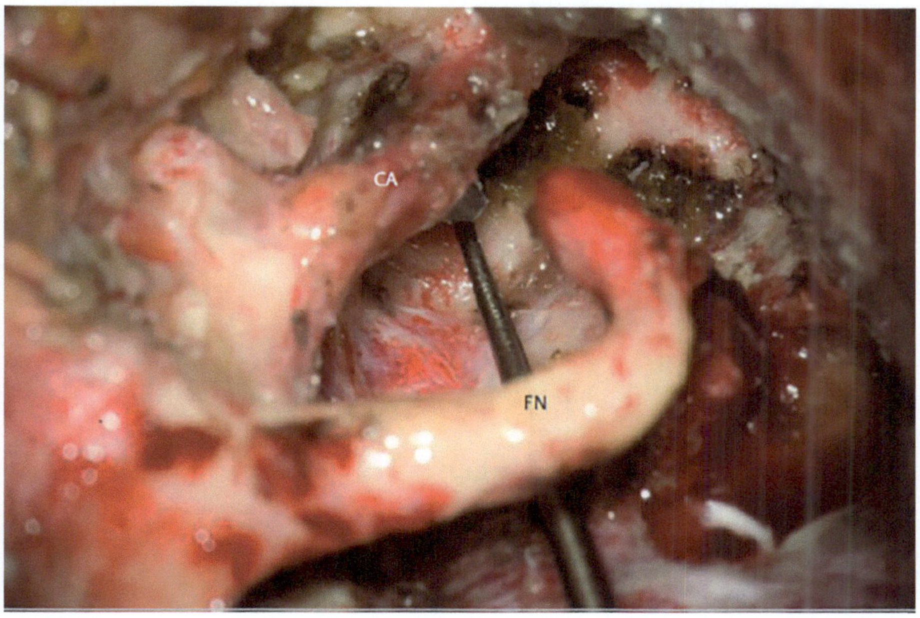

Fig. 10.135 Since the jugular bulb is low in this case, a large space was created under the facial nerve without difficulty. *CA*, carotid artery; *FN*, facial nerve.

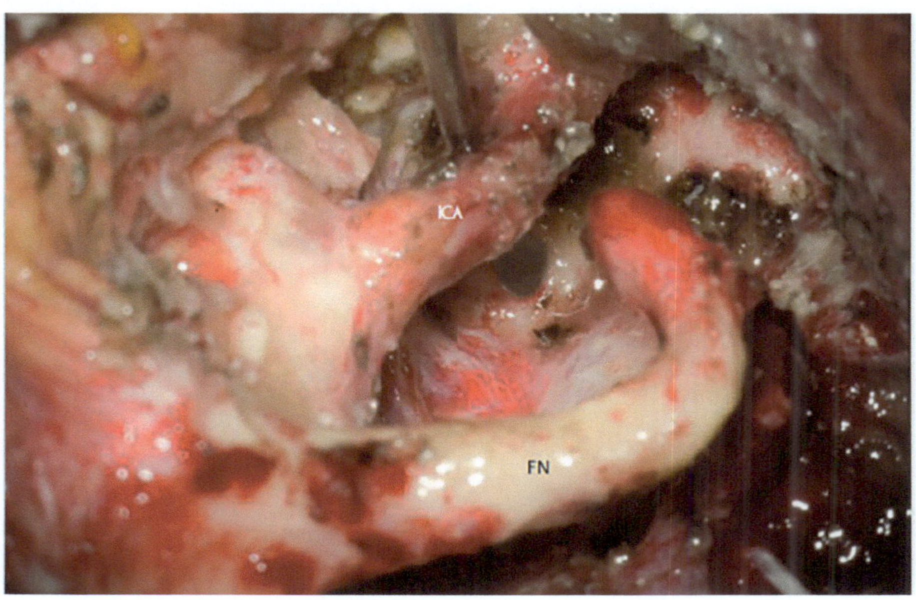

Fig. 10.136 The space created anteriorly to the carotid artery is shown. *CA*, carotid artery; *FN*, facial nerve.

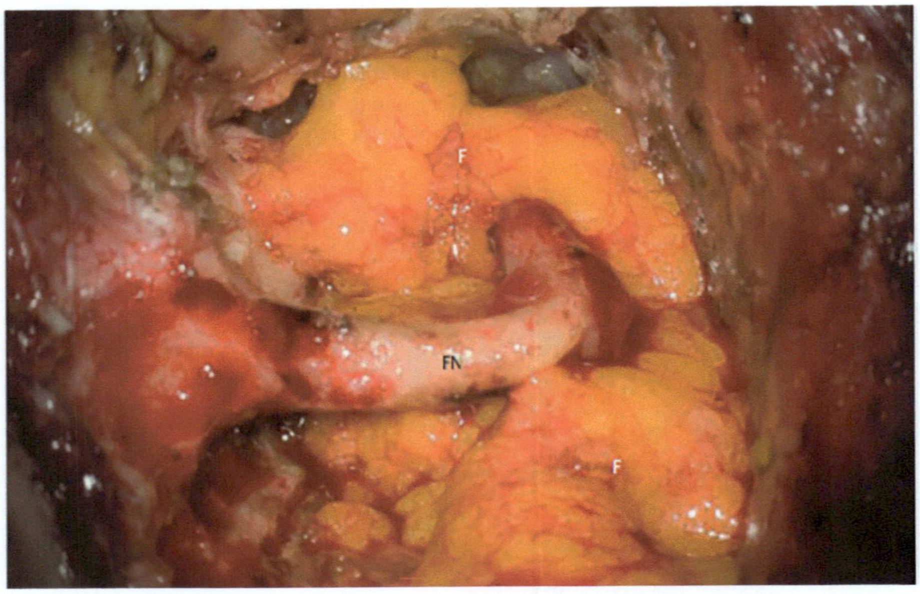

Fig. 10.137 After obliterating the eustachian tube with periosteum, the cavity is packed with abdominal fat. *F*, abdominal fat; *FN*, facial nerve.

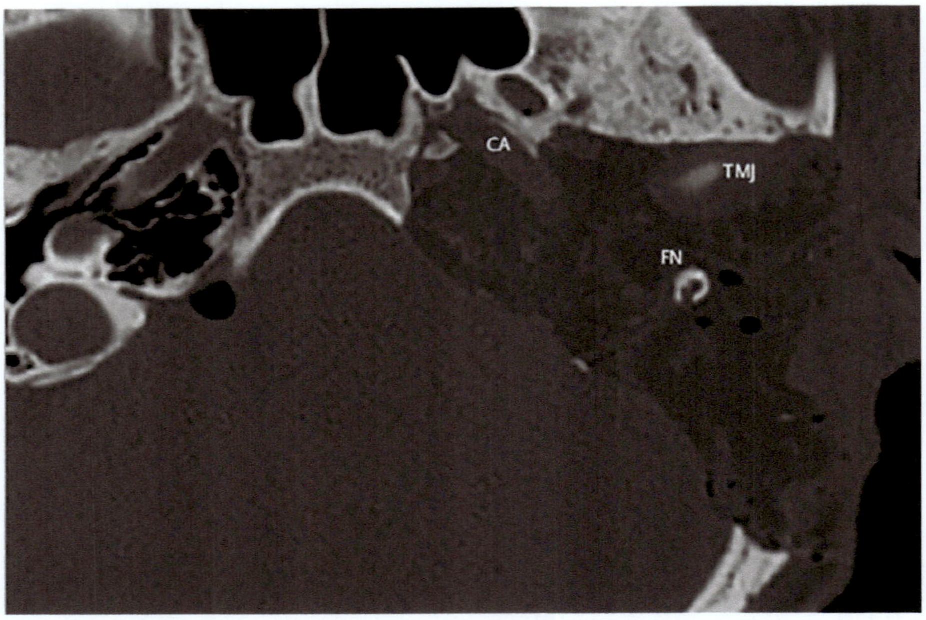

Fig. 10.138 The postoperative CT shows extensive removal of bone from the temporal bone. The axial section demonstrates the carotid artery, the temporomandibular joint, and the facial nerve in the fallopian canal in the abdominal fat. Note that the drilling reached the area close to the sphenoid sinus. *CA*, carotid artery; *FN*, facial nerve; *TMJ*, temporomandibular joint.

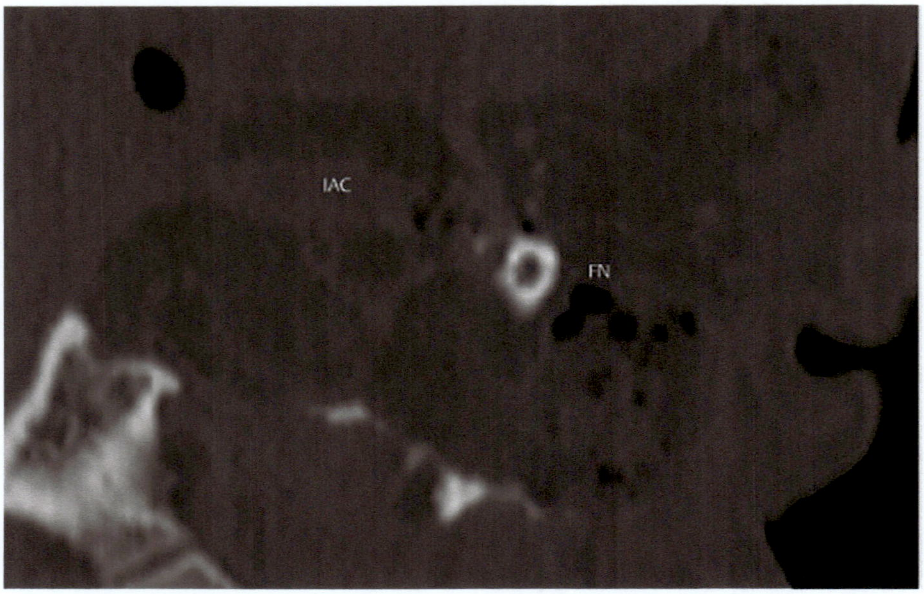

Fig. 10.139 In this coronal section, the exposed internal auditory canal is identified. The tympanic segment of the facial nerve runs in the fallopian canal. Note that the middle fossa dura and the posterior fossa dura are totally exposed to gain maximum access to the petrous apex. *IAC*, internal auditory canal; *FN*, facial nerve.

Case 2: Massive Petrous Bone Cholesteatoma With Intracranial and Sphenoid Invasion

See Figs. 10.140–10.162.

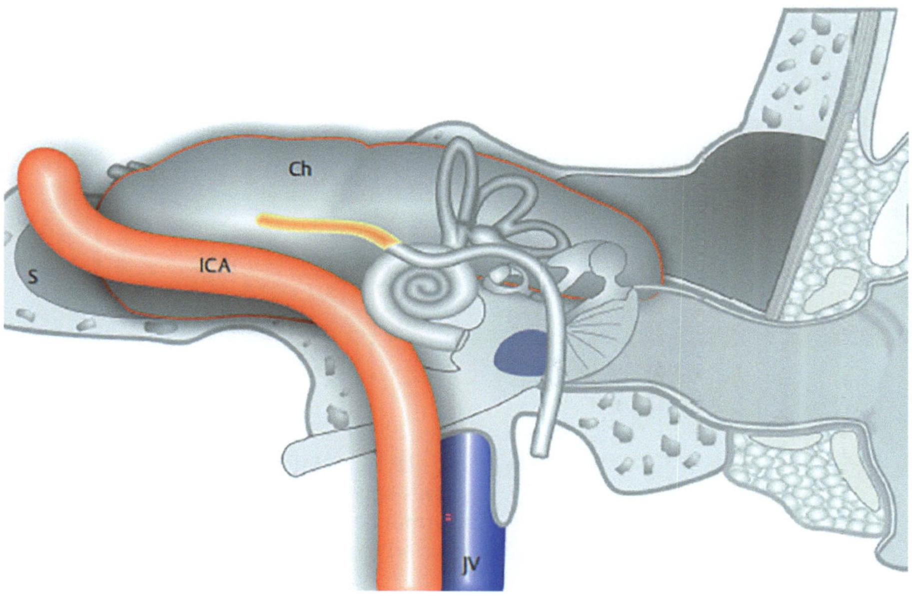

Fig. 10.140 A case of massive petrous bone cholesteatoma with intracranial and sphenoid extension. This 60-year-old patient underwent tympanoplasty elsewhere 10 years before. Even though no otorrhea was experienced, facial nerve palsy and hearing disturbance developed 5 years after the surgery. Physiologic examination revealed left anacusis together with complete left facial nerve palsy and well-compensated left vocal cord paralysis. Otoscopic examination revealed a recurrence of cholesteatoma in the pars flaccida. *Ch*, cholesteatoma; *ICA*, internal carotid artery; *JV*, jugular bulb; *S*, sphenoid sinus.

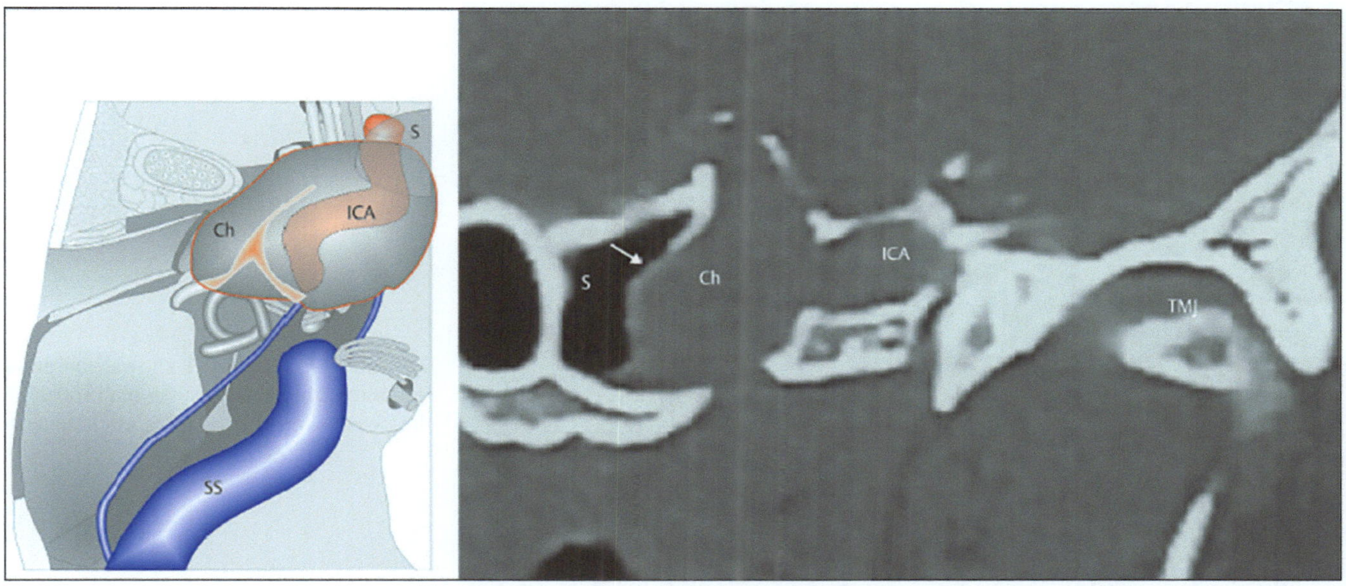

Fig. 10.141 The coronal CT shows the extension of cholesteatoma medially to the horizontal segment of the carotid artery, reaching the left sphenoid sinus *(arrow)*. *ICA*, internal carotid artery; *Ch*, cholesteatoma; *S*, sphenoid sinus; *SS*, sigmoid sinus; *TMJ*, temporomandibular joint. Physiologic examination revealed left anacusis together with complete left facial nerve palsy and well-compensated left vocal cord paralysis. Otoscopic examination revealed a recurrence of cholesteatoma in the pars flaccida. *Ch*, cholesteatoma; *ICA*, internal carotid artery; *JV*, jugular bulb; *S*, sphenoid sinus.

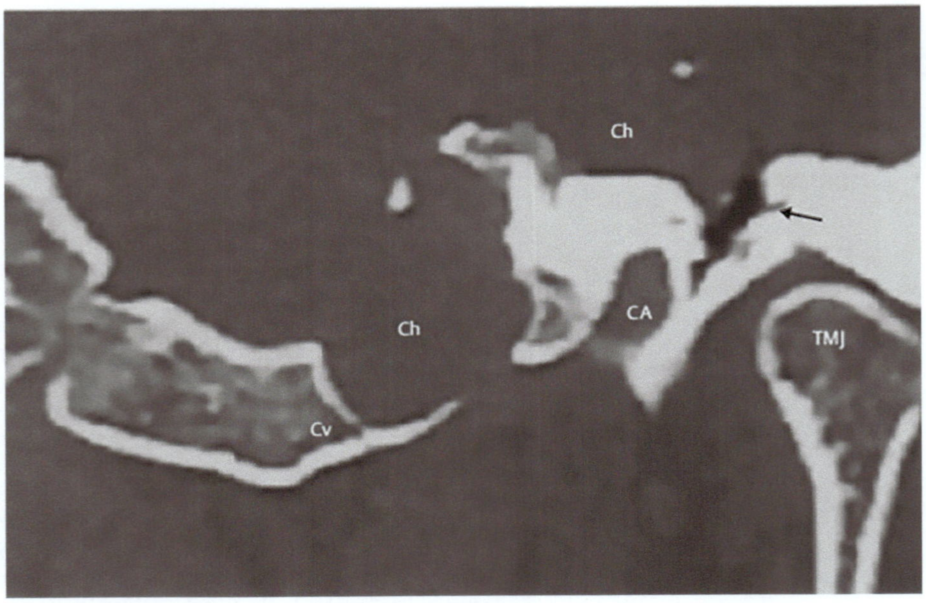

Fig. 10.142 More posterior sections at the level of the vertical segment of the carotid artery demonstrate the presence of cholesteatoma laterally to the protympanum *(arrow)* and far inferomedial extension eroding the clivus. *CA*, carotid artery; *Ch*, cholesteatoma; *Cv*, clivus; *TMJ*, temporomandibular joint.

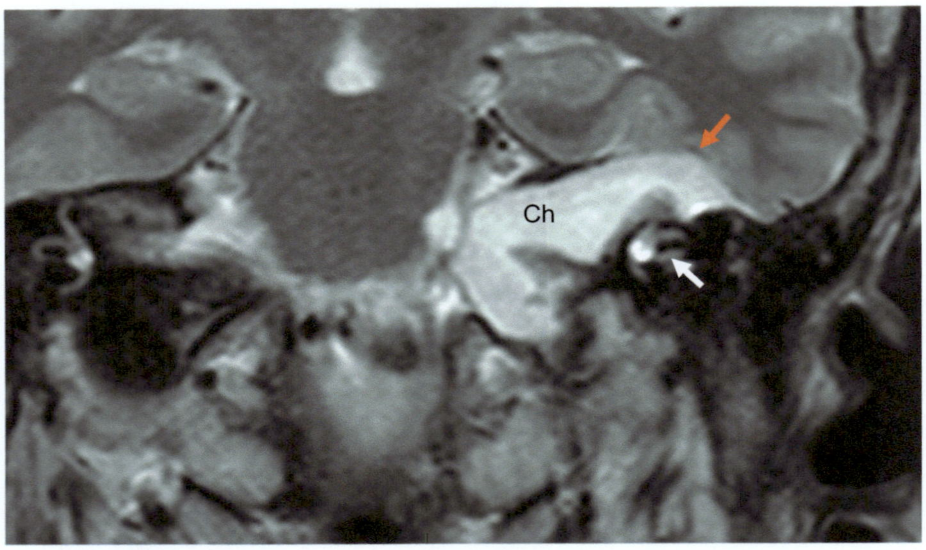

Fig. 10.143 MRI (coronal cut) shows the petrous bone cholesteatoma that, from the supralabyrinthine compartment *(white arrow)*, extends superiorly toward the temporal lobe *(red arrow)*. *Ch*, cholesteatoma.

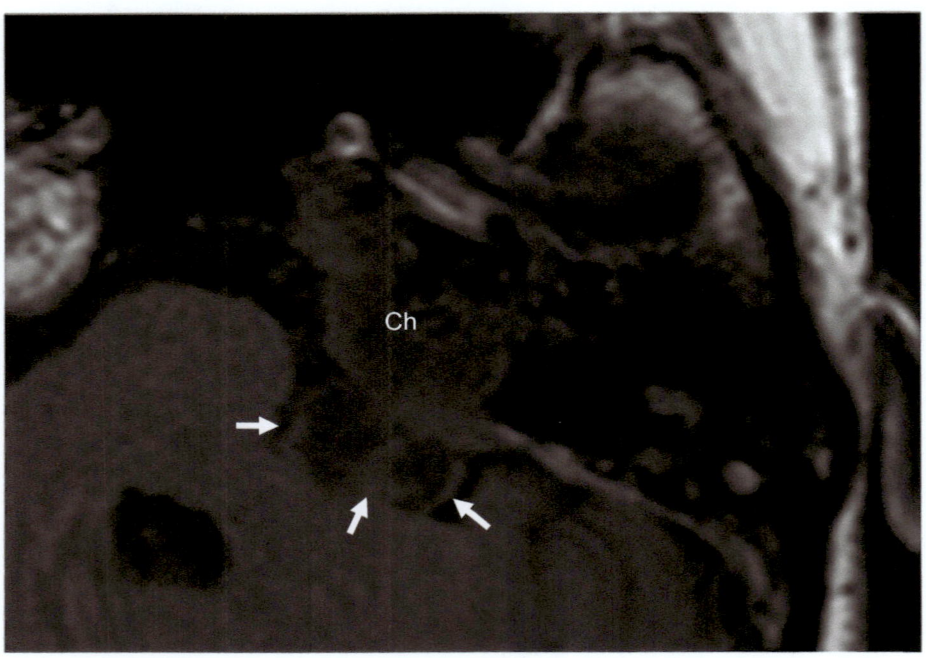

Fig. 10.144 The MRI shows compression of the brainstem by cholesteatoma protruding from the petrous apex *(arrows)*. *Ch*, cholesteatoma.

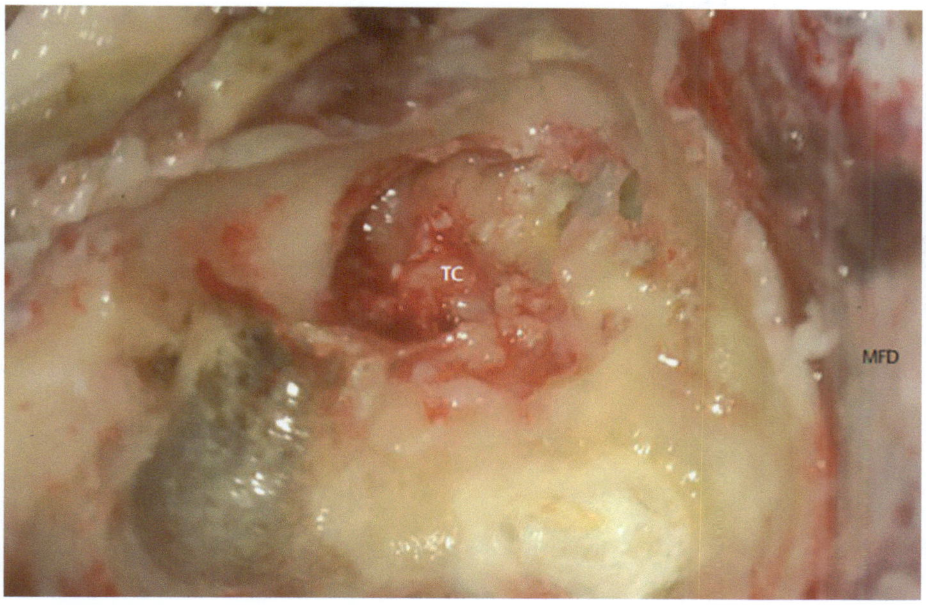

Fig. 10.145 The patient has a very long history of left deafness and facial nerve palsy. The radiologic examinations show extensive cholesteatoma involving the whole temporal bones with the extension of the cerebellopontine angle. Therefore, drilling of the inner ear and severing of the facial nerve without reconstruction are indicated in this particular case. Canal wall down mastoidectomy with exposure of the middle fossa dura is conducted. *MFD*, middle fossa dura; *TC*, tympanic cavity.

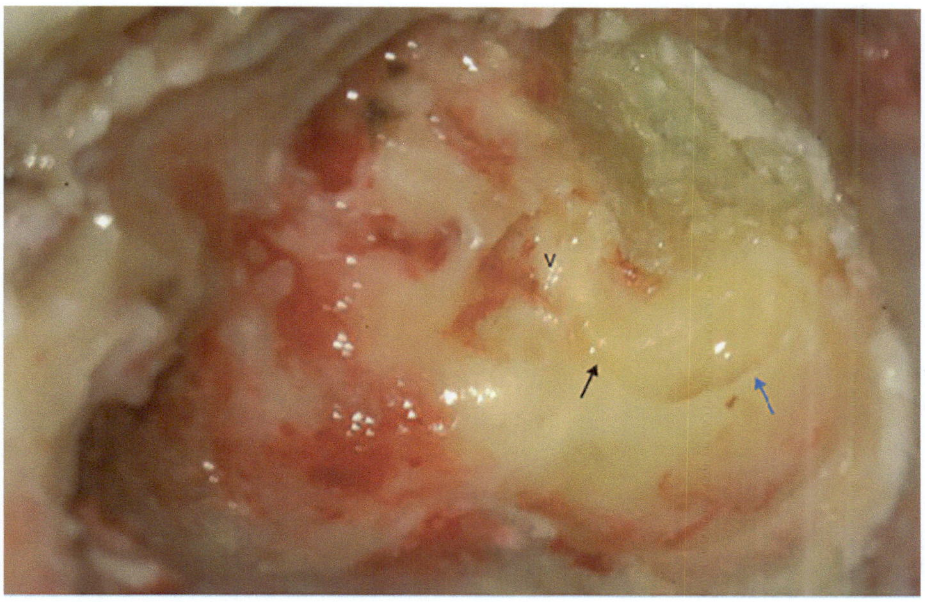

Fig. 10.146 The atrophic facial nerve is removed, and the lateral and posterior semicircular canals are drilled. The vestibule is opened from the lateral side of the posterior wall, which is lowered, and the attic is opened. The common crus that connects the superior *(blue arrow)* and the posterior semicircular canals to the vestibule can be identified in the medial wall *(black arrow)*. *V*, vestibule.

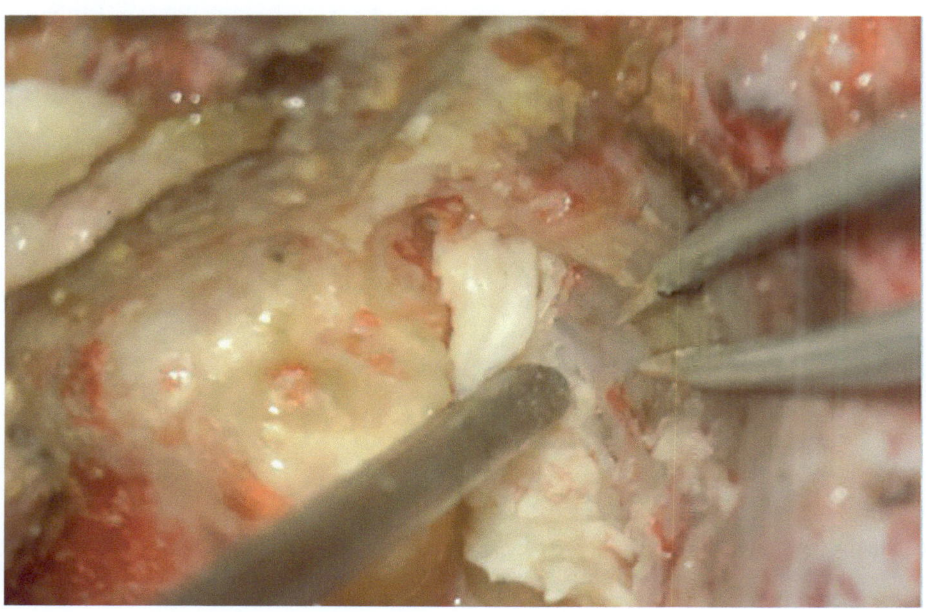

Fig. 10.147 The middle fossa dura is extensively covered by the very thin matrix of the cholesteatoma. The dura is coagulated with bipolar to reduce the possibility of leaving vital skin on it and to make room for dissection by shrinking the dura.

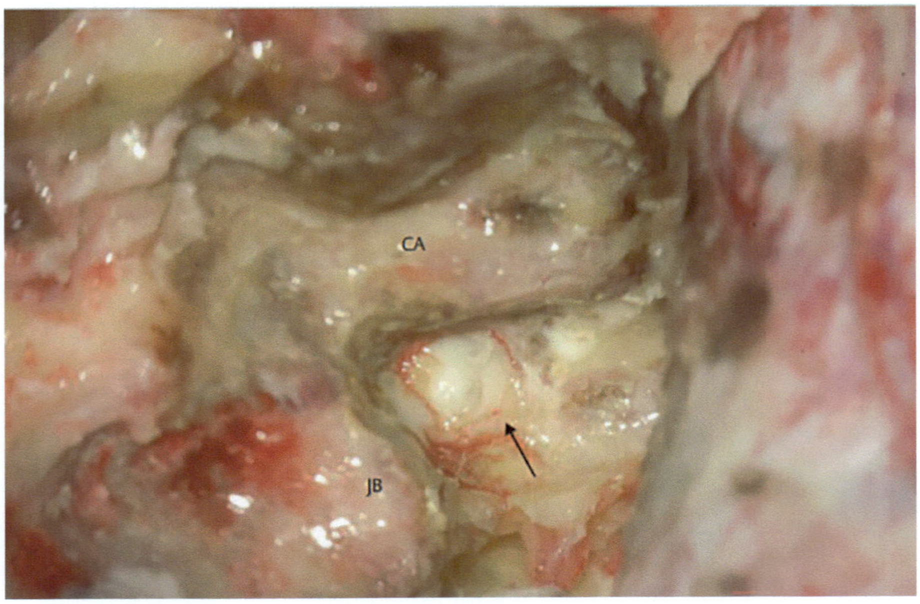

Fig. 10.148 The medial wall of the tympanic cavity, including the cochlea, is drilled, and the carotid artery and the jugular bulb are skeletonized. Medially to the carotid artery, the cholesteatoma occupying the petrous apex is starting to be seen *(arrow)*. *CA*, carotid artery; *JB*, jugular bulb.

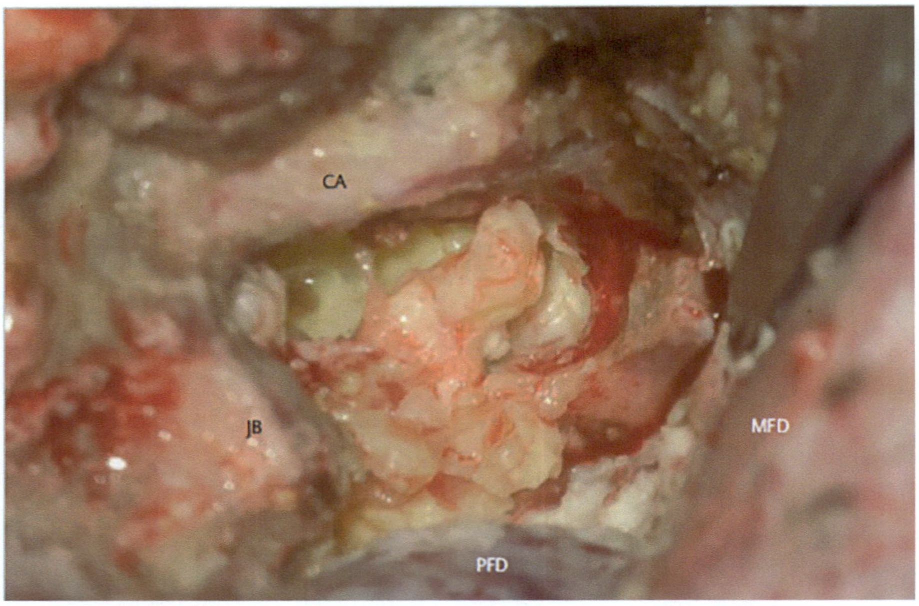

Fig. 10.149 Further drilling of the bone medially exposed massive cholesteatoma occupying the petrous apex. *CA*, carotid artery; *JB*, jugular bulb; *MFD*, middle fossa dura; *PFD*, posterior fossa dura.

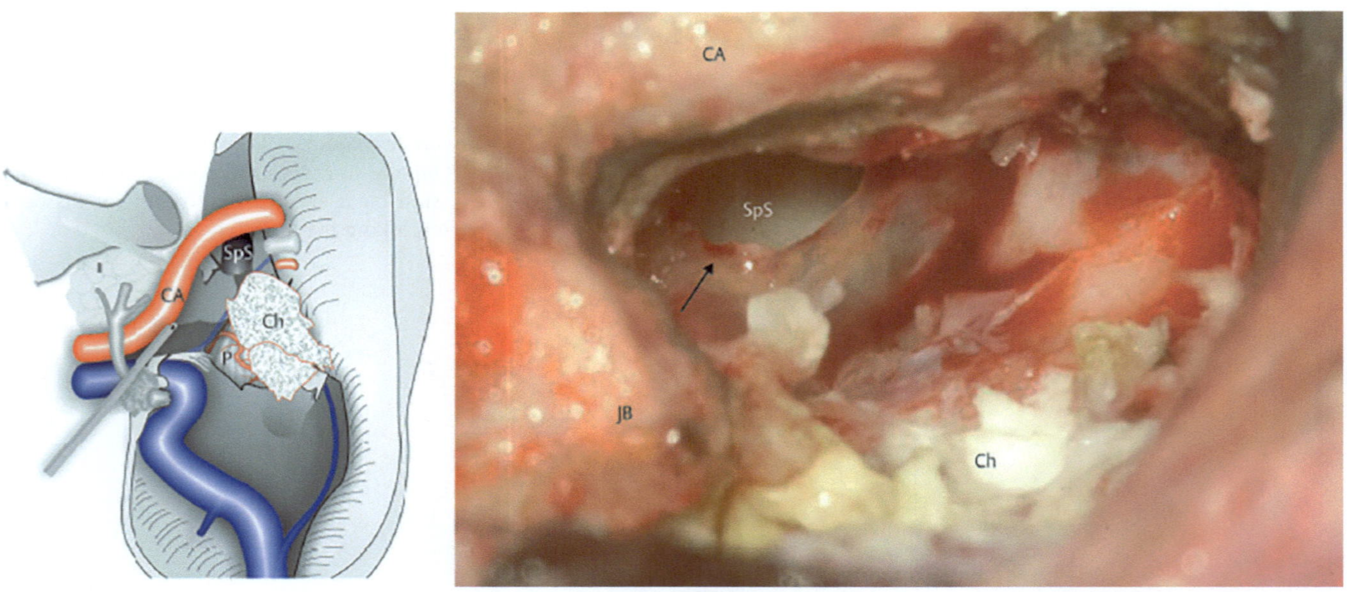

Fig. 10.150 An opening of the sphenoid sinus is seen in the petroclival area medially to the carotid artery *(arrow)*. The mucosa in the sphenoid is preserved and pushed medially, and the sinus is obliterated with a piece of muscle. *CA*, carotid artery; *Ch*, cholesteatoma; *JB*, jugular bulb; *SpS*, sphenoid sinus.

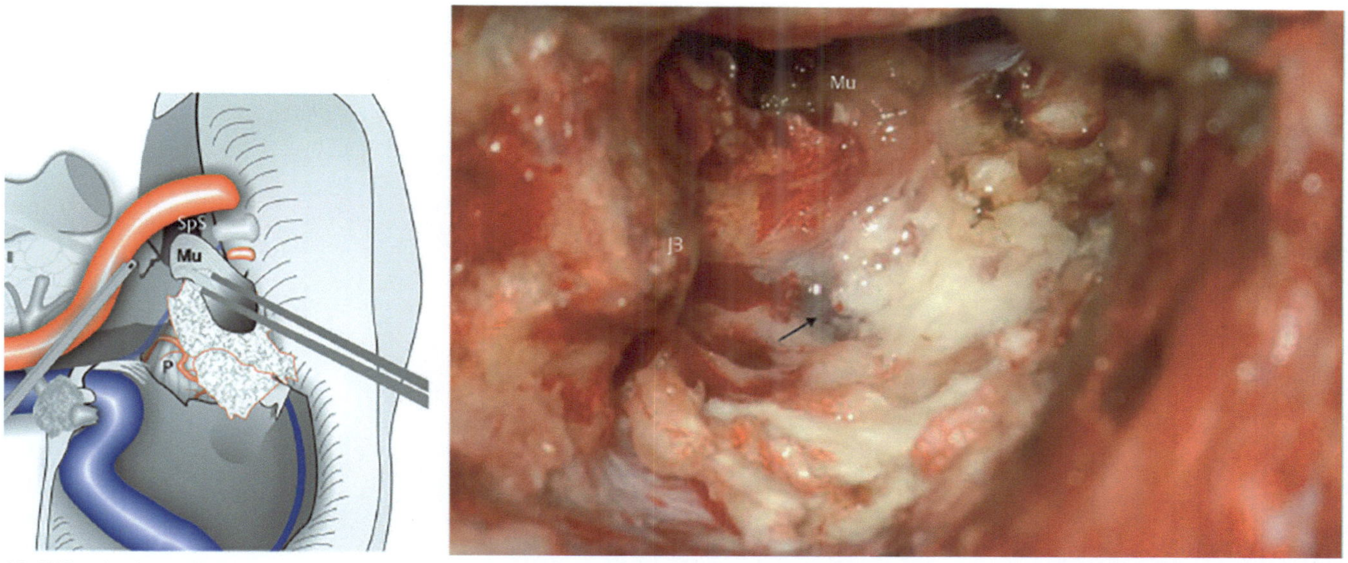

Fig. 10.151 A piece of muscle used to pack the sphenoid sinus is seen. The area of the internal auditory canal *(arrow)* is still covered with thick debris of cholesteatoma. *JB*, jugular bulb; *Mu*, muscle in the sphenoic sinus; *SpS*, sphenoid sinus.

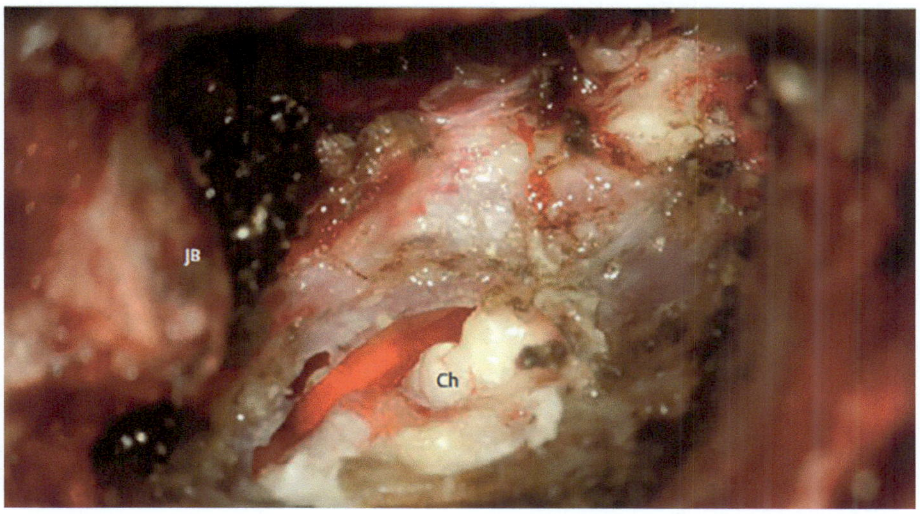

Fig. 10.152 Removal of the debris covering the posterior fossa dura exposed the other opening at the area of the internal auditory canal. From this opening, the cholesteatoma extending to the cerebellopontine angle is seen. *Ch*, cholesteatoma; *JB*, jugular bulb.

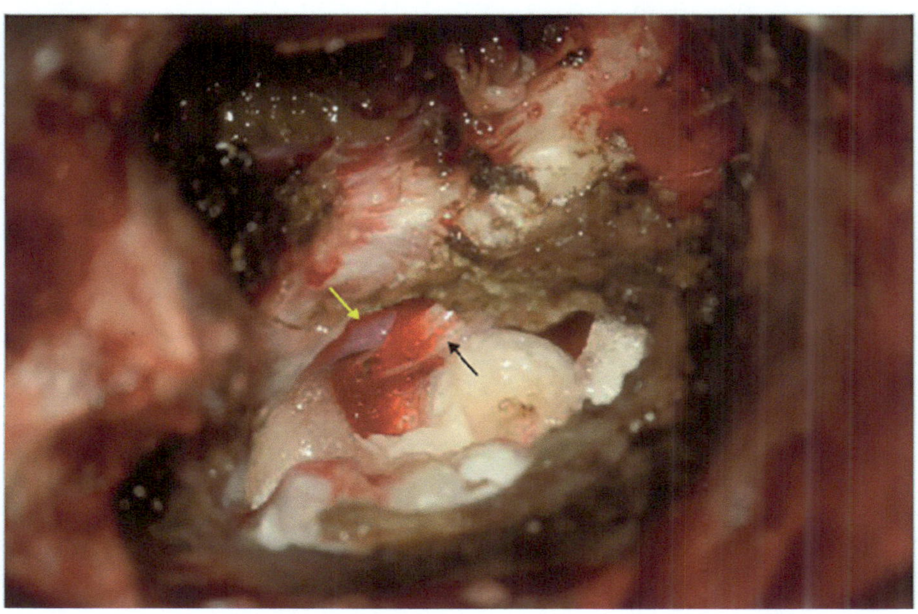

Fig. 10.153 The opening of the dura is enlarged to control the cerebellopontine angle better. The sixth cranial nerve (N. abducens; *black arrow*) running obliquely from the bulbopontine angle to the upper clivus is seen. The nerve has close relationships with the anterosuperior cerebellar artery (AICA; *yellow arrow*). Management of intracranial space requires different kinds of skill and anatomical knowledge from those of middle ear surgery. To manage an intracranial lesion, the surgeon should be prepared for such situations.

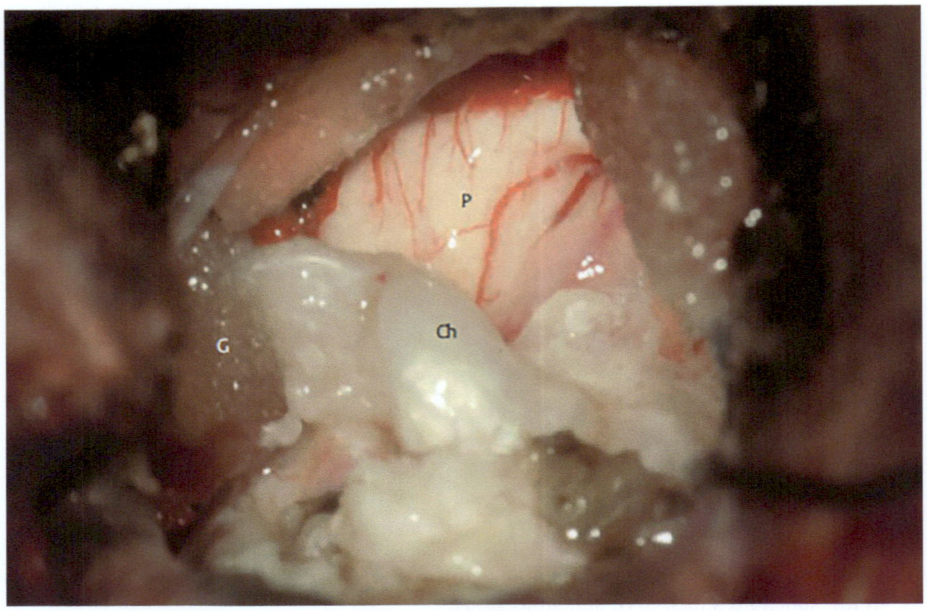

Fig. 10.154 The cholesteatoma attached to the pons is shown. *Ch*, cholesteatoma; *G*, Gelfoam; *P*, pons.

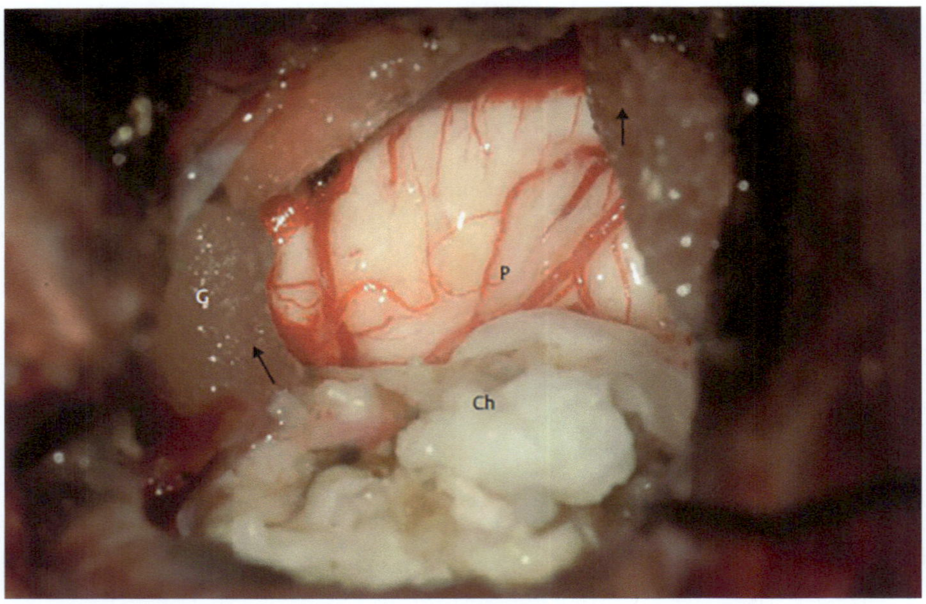

Fig. 10.155 The cholesteatoma is reduced, and the matrix is dissected carefully from the pons. Meticulous care should be taken in the cerebellopontine angle so as not to disrupt and coagulate major vessels, especially AICA, and perforating arteries on the pons. Lethal complications may follow. *Ch*, cholesteatoma; *G*, Gelfoam; *P*, pons.

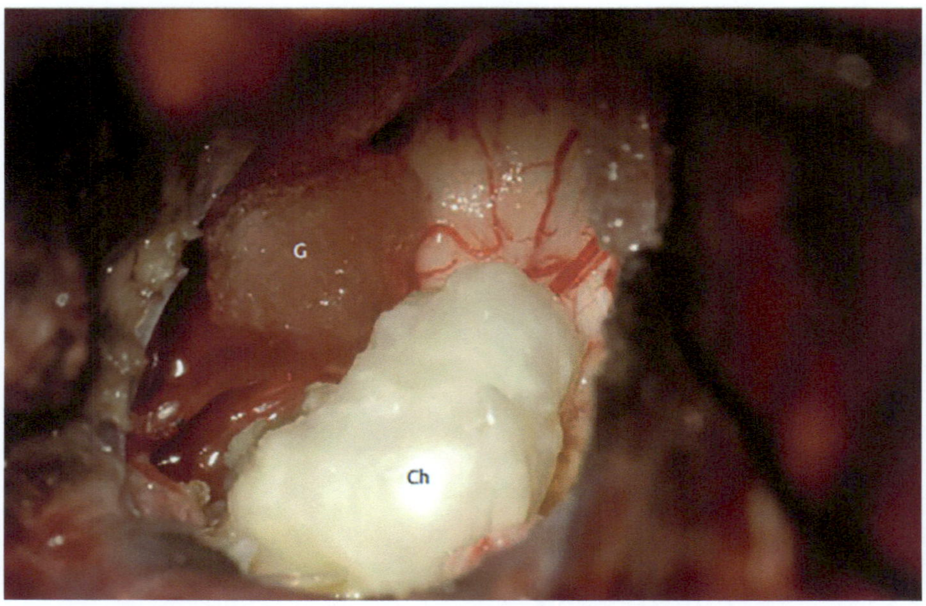

Fig. 10.156 The cholesteatoma is further dissected and reduced from the cerebellopontine angle, taking care not to leave the matrix behind. *Ch*, cholesteatoma; *G*, Gelfoam.

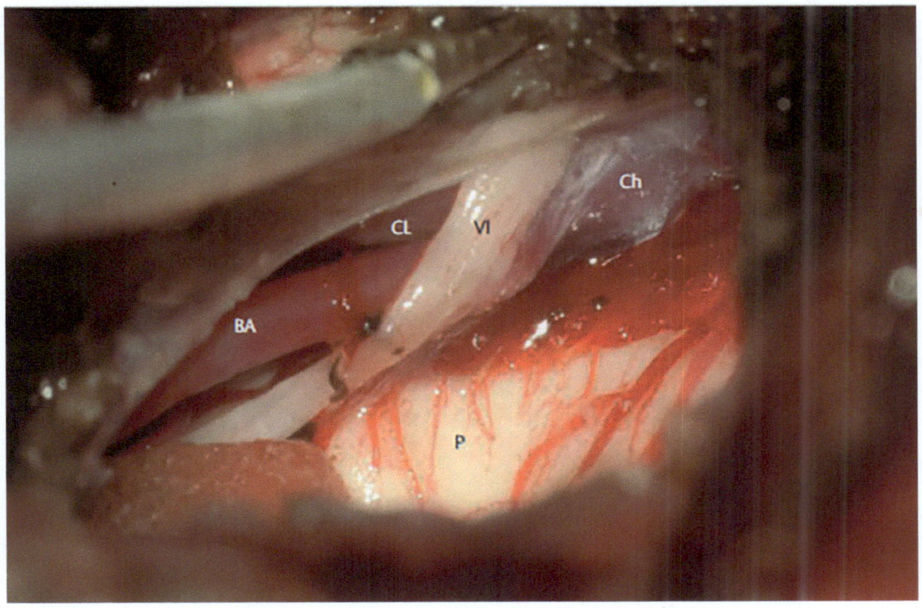

Fig. 10.157 The majority of the cholesteatoma is removed from the cerebellopontine angle, except a piece of matrix superomedially to the sixth cranial nerve. *BA*, basilar artery; *Ch*, cholesteatoma; *CL*, clivus; *P*, pons; *VI*, N. abducens.

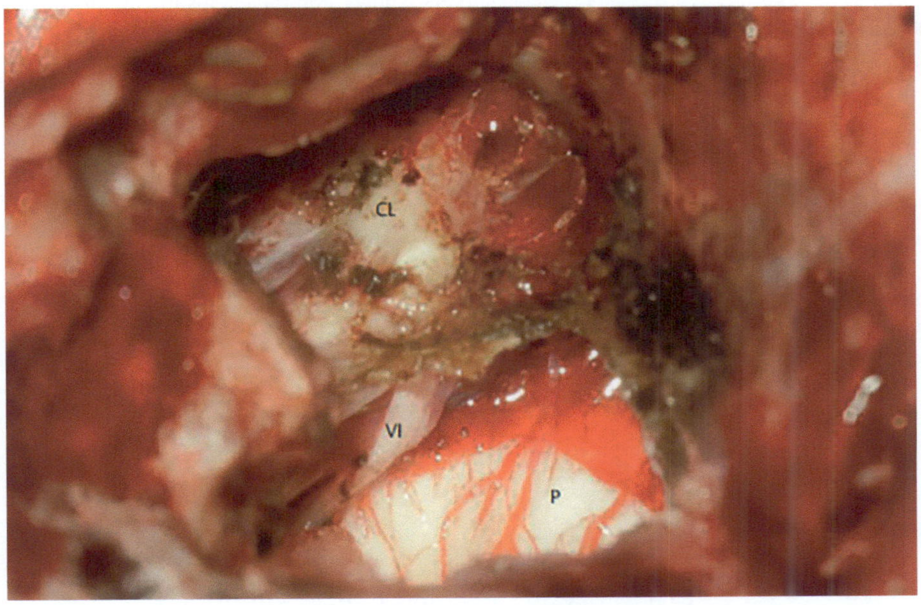

Fig. 10.158 The remnant of the cholesteatoma superiorly to the sixth cranial nerve is removed. *CL*, clivus; *P*, pons; *VI*, N. abducens.

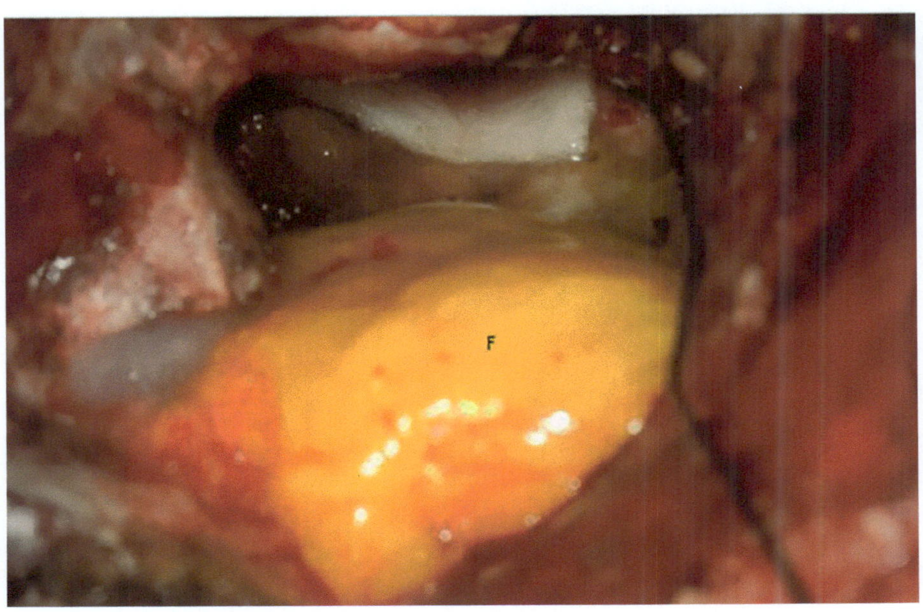

Fig. 10.159 The opening of the dura is closed with a small piece of abdominal fat. The cavity is further obliterated with abdominal fat. The cavity is further packed with the same material. *F*, abdominal fat.

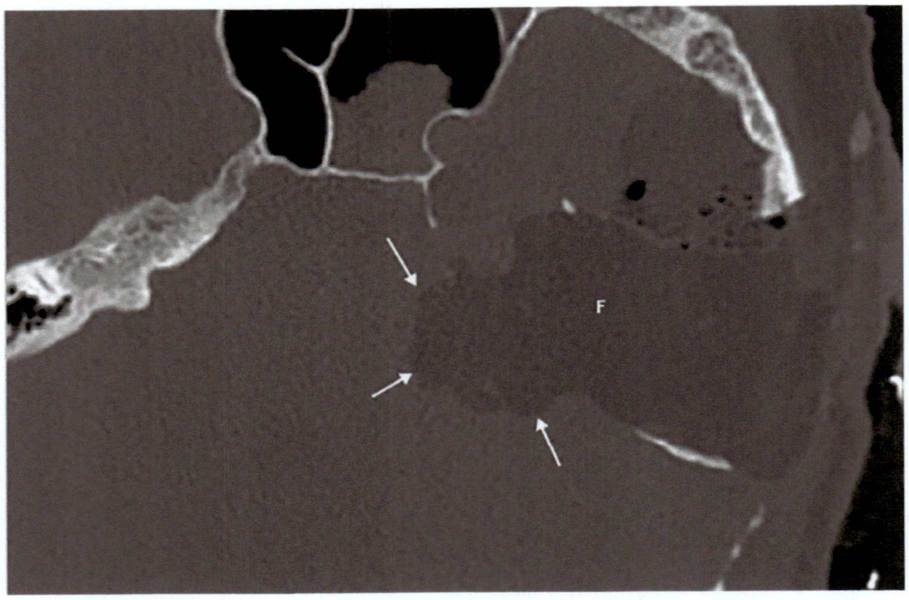

Fig. 10.160 The opening of the dura is closed with a small piece of abdominal fat. The cavity is further obliterated with abdominal fat. The cavity is further packed with the same material. *F*, abdominal.

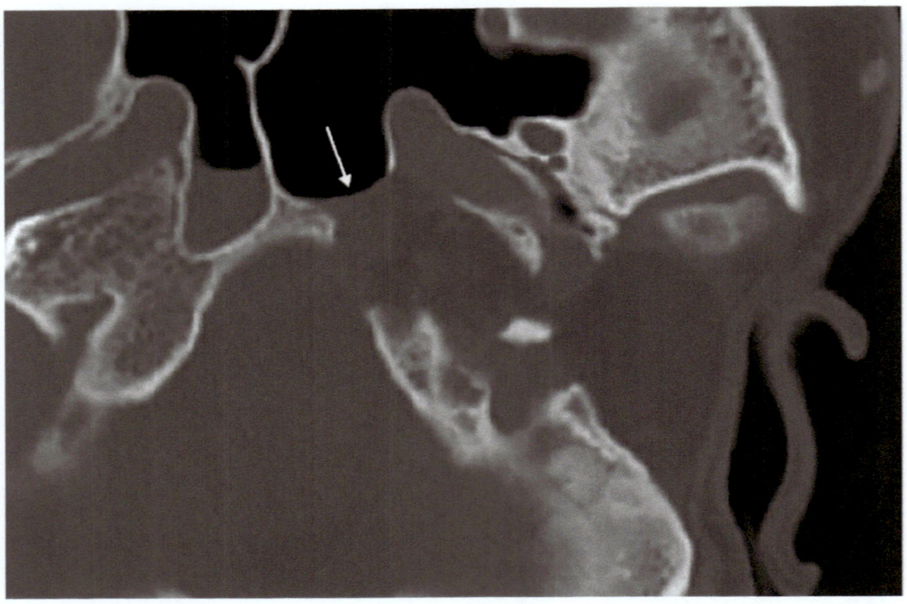

Fig. 10.161 Postoperative CT 4 years after the surgery shows complete closure of the opening in the petrous apex to the sphenoid sinus *(arrow)*.

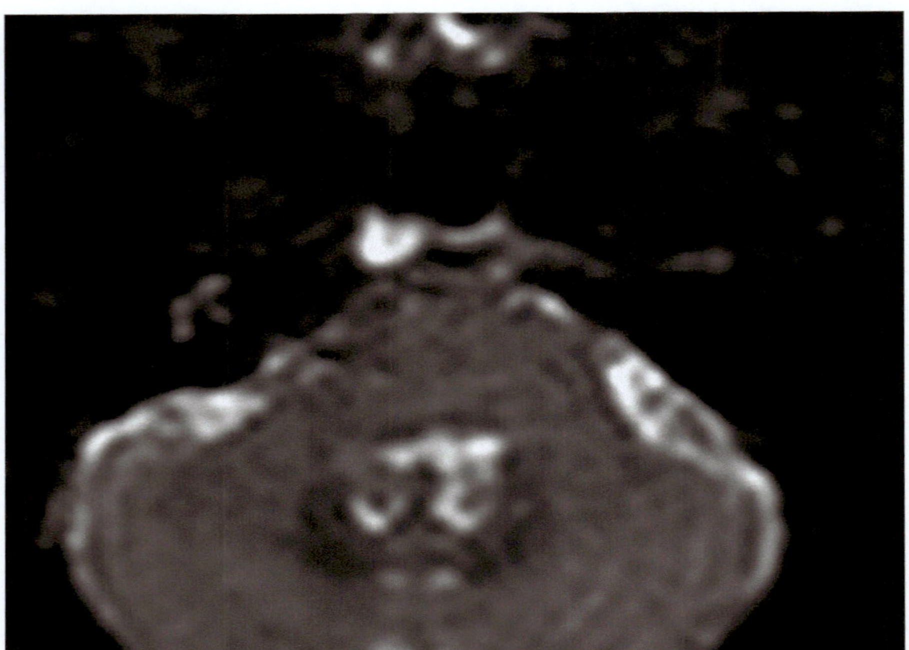

Fig. 10.162 Brain MRI-DWI shows no residual cholesteatoma 6 years after surgery.

1. If a larger space is needed, a transzygomatic extension is performed. In this case, the temporalis muscle is detached from the bone and reflected inferiorly; the zygomatic arch is transected after drilling two anterior holes at its anterior end to facilitate repositioning it at the end of the procedure.
2. In some cases, in which only a limited anterior extension is needed, the mandibular nerve is not sacrificed.

Type C Modified Transcochlear Approach

Rationale

Certain posterior fossa tumors, particularly petroclival meningiomas, extend to the middle fossa either by direct tentorial invasion or through Meckel's cave or the tentorial notch. The type C approach allows for control of both the infratentorial and the supratentorial parts of the tumor lying ventral to the pons and midbrain, as well as removal of the infiltrated tentorium with only minimal temporal lobe retraction. Fig. 10.163 indicates the extension of the craniotomy, while Fig. 10.164 shows the different structures controlled in this approach.

Indications

- Petroclival lesions with supratentorial extension (e.g., petroclival meningiomas).

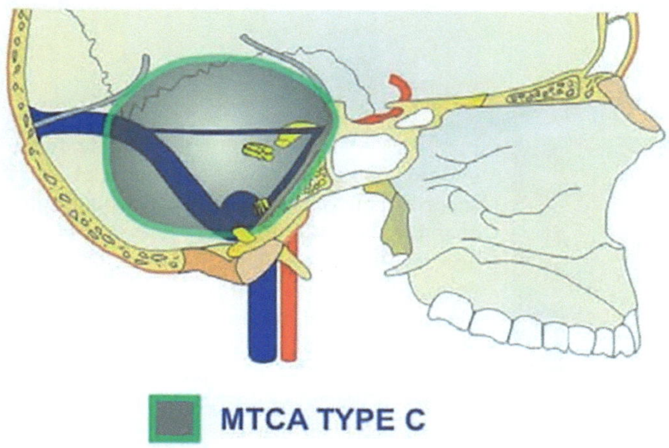

Fig. 10.163 Drawing showing the different structures controlled by this D approach.

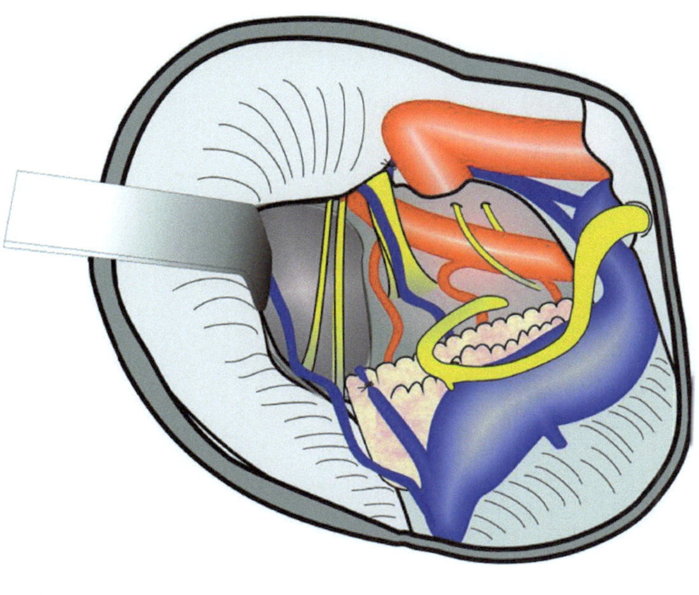

Fig. 10.164 Schematic drawing showing the limits of the transcochlear approach type C (TCC). Note the infratentorial and the supratentorial structures controlled in this approach.

Surgical Steps (Right Side) (Figs. 10.165–10.172)

1. The skin incision is performed as in the type B approach (Fig. 10.98). The approach is performed as previously described in type A. However, the middle fossa dura is widely uncovered. Bone removal at this level is much wider than in the type A approach.

2. The dura of the middle fossa is incised 3–4 mm above and parallel to the superior petrosal sinus (Fig. 10.165).

3. The superior petrosal sinus is now double-ligated and divided (Fig. 10.166). Special care should be taken not to injure the vein of Labbé. The vein is more vulnerable at its entrance into the transverse sinus (Fig. 10.167).

4. The tentorium is cut with microscissors, starting just in front of the junction between the superior petrosal sinus and the sigmoid sinus and coursing anteromedially. The tentorium should be incised vertical to the superior petrosal sinus. Cutting the tentorium should be carried out under direct vision to avoid injury to the trochlear nerve, which pierces the tentorium to reach the cavernous sinus (Fig. 10.168). If the tentorium is involved by a tumor, as in meningiomas, it is completely excised and removed.

5. After subtemporal dural opening and incision of the tentorium, the structures of the ambient cistern can be extensively approached. Structures exposed within the ambient cistern include the posterior cerebral artery, superior cerebellar artery, basal vein, and trochlear nerve. Note that Dandy's vein obstructs the control of the trigeminal nerve (Fig. 10.169). The Dandy's vein is removed, allowing free exposure of the trigeminal nerve (Fig. 10.170).

6. Mild retraction of the temporal lobe provides a good view of the oculomotor nerve and its relation to the superior cerebellar artery lying inferiorly and the posterior cerebral artery lying superiorly. The trochlear nerve is seen piercing the tentorium to reach the cavernous sinus. Note the excellent control of the trigeminal nerve; Meckel's cave can be opened when necessary (Fig. 10.171).

7. View after mild retraction of the pons, showing excellent control of the basilar artery and prepontine cistern. The abducens nerve passes behind the anterior inferior cerebellar artery and lateral to the basilar artery (Fig. 10.172).

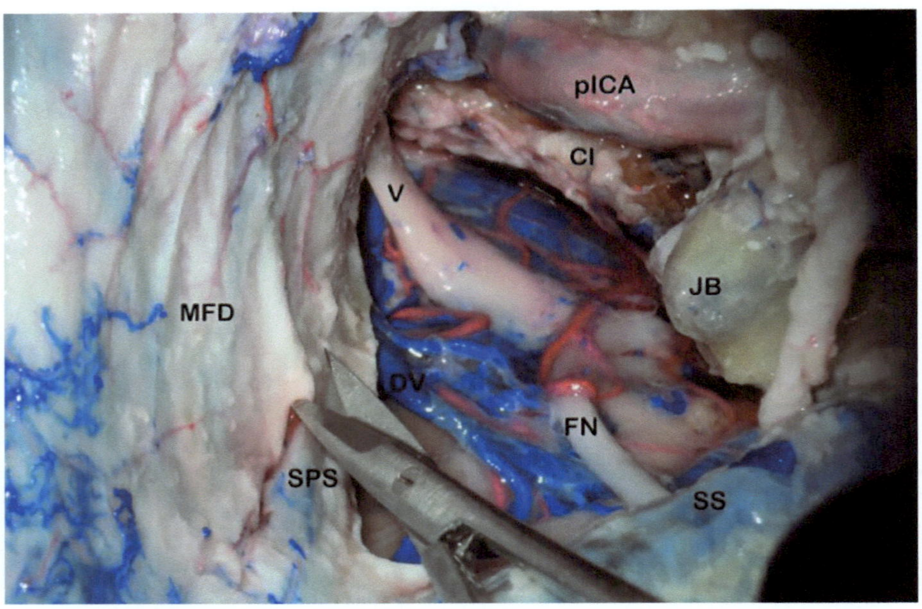

Fig. 10.165 The dura of the middle fossa is incised 3–4 mm above and parallel to the superior petrosal sinus. *pICA*, petrous segment of internal carotid artery; *Cl*, clivus; *FN*, facial nerve; *DV*, Dandy's vein (superior petrosal vein); *SPS*, superior petrosal sinus; *MFD*, middle fossa dura; *V*, trigeminal nerve; *JB*, jugular bulb; *SS*, sigmoid sinus.

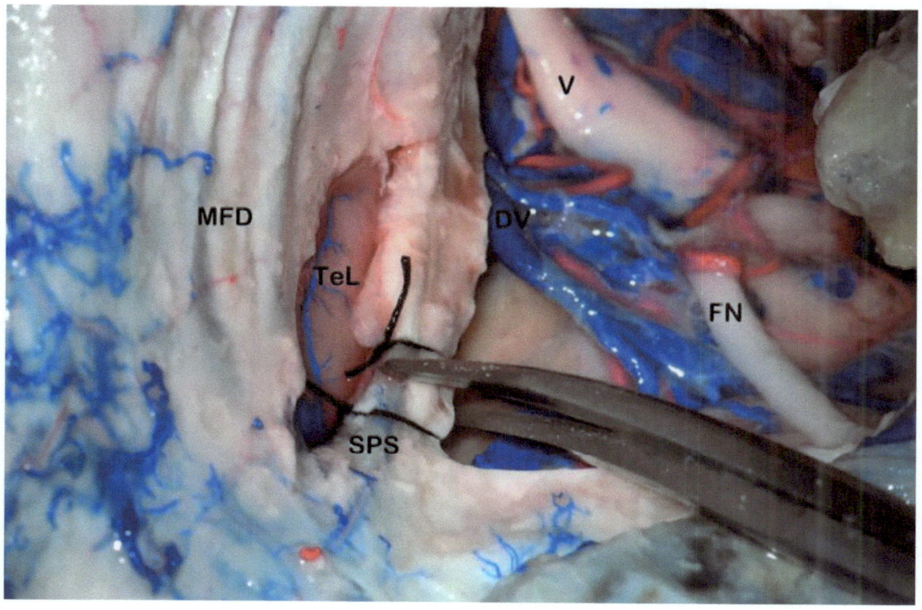

Fig. 10.166 The superior petrosal sinus is now double-ligated and divided. *V*, trigeminal nerve; *FN*, facial nerve; *DV*, Dandy's vein (superior petrosal vein); *TeL*, temporal lobe; *MFD*, middle fossa dura; *SPS*, superior petrosal sinus.

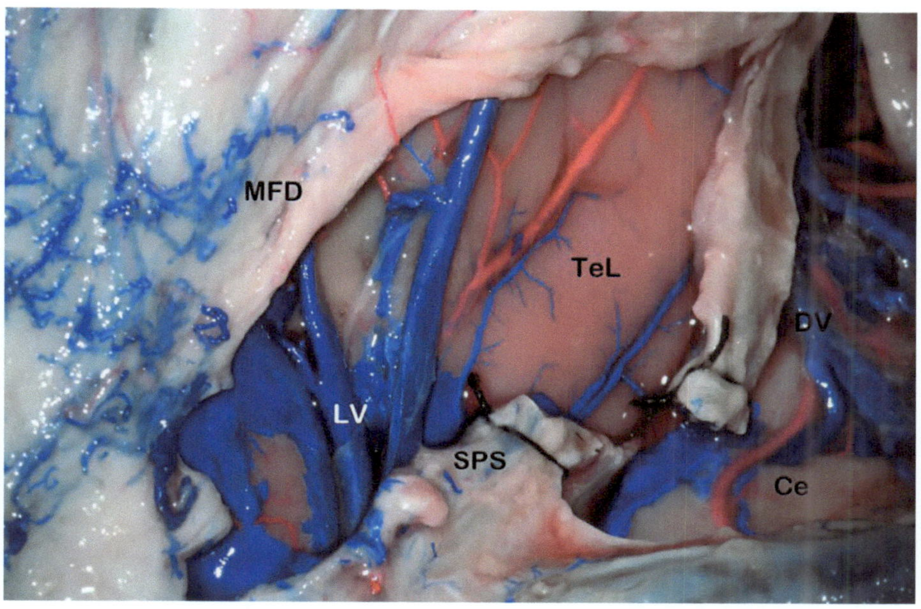

Fig. 10.167 Special care should be taken not to injure the vein of Labbé. The vein is more vulnerable at its entrance into the transverse sinus. *TeL*, temporal lobe; *DV*, Dandy's vein (superior petrosal vein); *Ce*, cerebellum; *SPS*, superior petrosal sinus; *LV*, Labbé's vein; *MFD*, middle fossa dura.

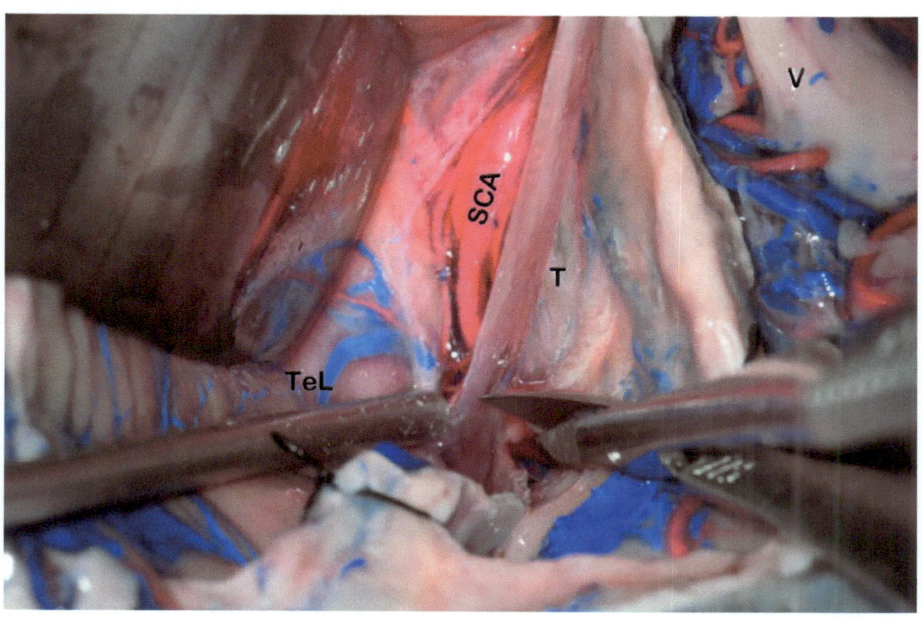

Fig. 10.168 The tentorium is cut with microscissors, starting just in front of the junction between the superior petrosal sinus and the sigmoid sinus and coursing anteromedially. The tentorium should be incised vertical to the superior petrosal sinus. Cutting the tentorium should be carried out under direct vision to avoid injury to the trochlear nerve, which pierces the tentorium to reach the cavernous sinus. *V*, trigeminal nerve; *T*, tentorium; *SCA*, superior cerebellar artery; *TeL*, temporal lobe.

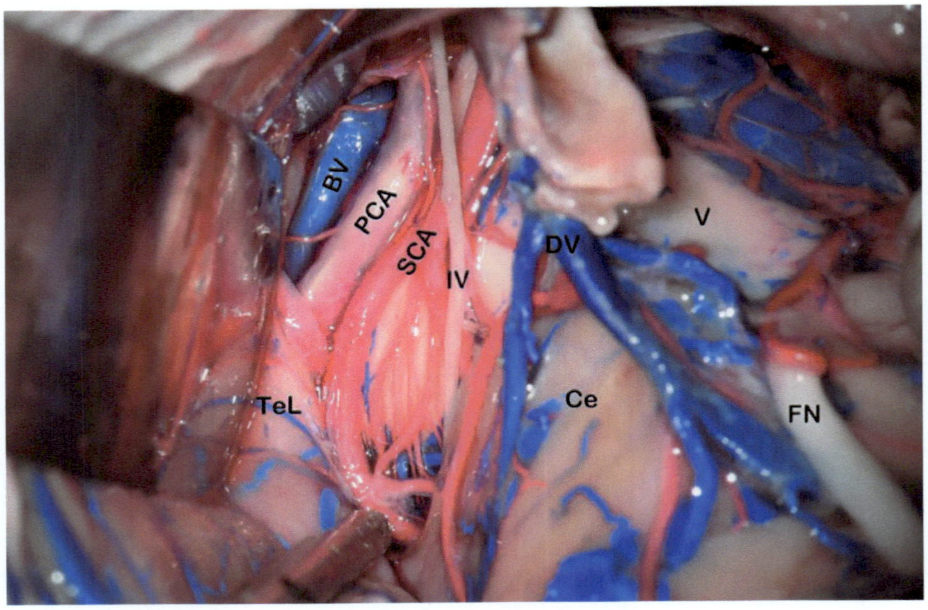

Fig. 10.169 After subtemporal dural opening and incision of the tentorium, the structures of the ambient cistern can be extensively approached. Structures exposed within the ambient cistern include the posterior cerebral artery, superior cerebellar artery, basal vein, and trochlear nerve. Note that Dandy's vein obstructs the control of the trigeminal nerve. *V*, trigeminal nerve; *FN*, facial nerve; *Ce*, cerebellum; *DV*, Dandy's vein (superior petrosal vein); *IV*, trochlear nerve; *SCA*, superior cerebellar artery; *PCA*, posterior cerebral artery; *BV*, basal vein; *TeL*, temporal lobe.

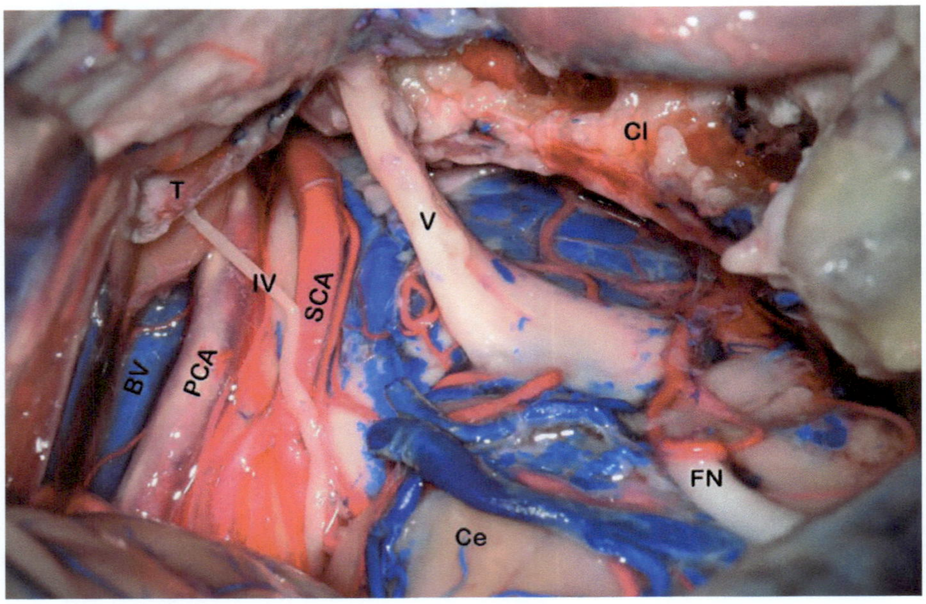

Fig. 10.170 The Dandy's vein is removed, allowing free exposure of the trigeminal nerve. *Cl*, clivus; *FN*, facial nerve; *V*, trigeminal nerve; *Ce*, cerebellum; *SCA*, superior cerebellar artery; *IV*, trochlear nerve; *T*, tegmen; *PCA*, posterior cerebral artery; *BV*, basal vein.

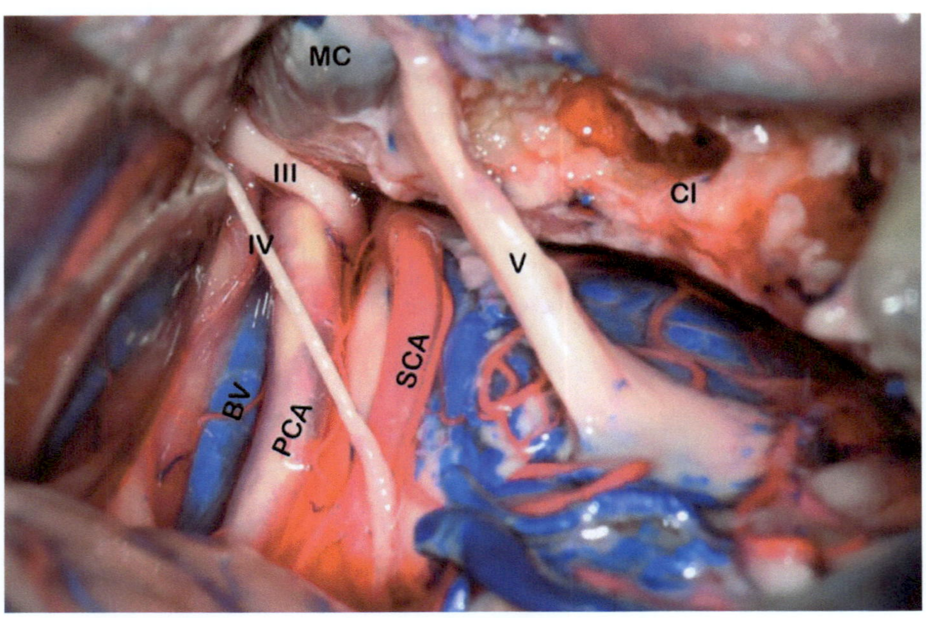

Fig. 10.171 Mild retraction of the temporal lobe provides a good view of the oculomotor nerve and its relation to the superior cerebellar artery lying inferiorly and the posterior cerebral artery lying superiorly. The trochlear nerve is seen piercing the tentorium to reach the cavernous sinus. Note the excellent control of the trigeminal nerve; Meckel's cave can be opened when necessary. *III*, oculomotor nerve; *IV*, trochlear nerve; *MC*, Meckel's cave; *Cl*, clivus; *V*, trigeminal nerve; *SCA*, superior cerebellar artery; *PCA*, posterior cerebral artery; *BV*, basal vein.

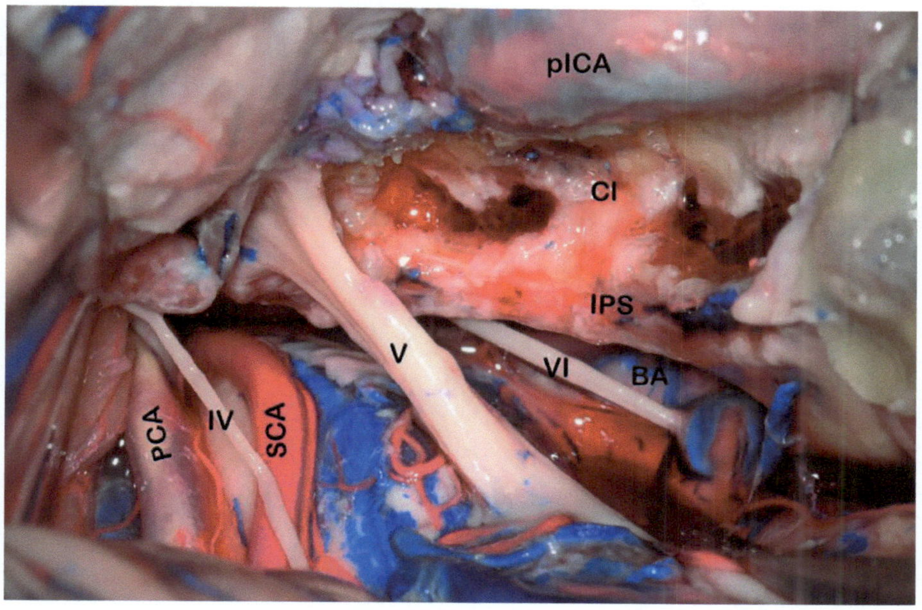

Fig. 10.172 View after mild retraction of the pons, showing excellent control of the basilar artery and prepontine cistern. The abducens nerve passes behind the anterior inferior cerebellar artery and lateral to the basilar artery. *pICA*, petrous segment of internal carotid artery; *Cl*, clivus; *IPS*, inferior petrosal sinus; *BA*, basilar artery; *VI*, abducent nerve; *V*, trigeminal nerve; *SCA*, superior cerebellar artery; *IV*, trochlear nerve; *PCA*, posterior cerebral artery.

Type D Modified Transcochlear Approach

Rationale

The major limitation of the type A approach is the limited access to tumors extending into the lower clivus and jugular foramen region. The type D approach is designed to overcome this limitation by combining the type A approach with the petro-occipital transsigmoid approach, thereby obtaining excellent control of these areas (Figs. 10.173 and 10.174).

Indications

- Lesions of the middle and lower clivus, e.g., chordomas with involvement of the lower clivus, some cases of petroclival meningiomas, and extensive lower cranial nerve schwannomas.

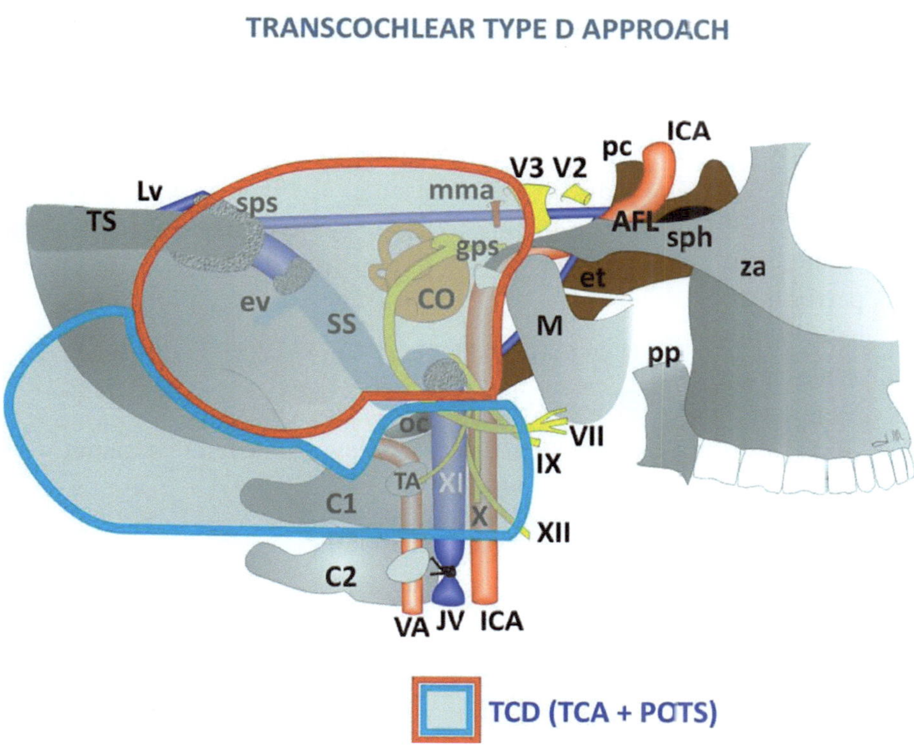

Fig. 10.173 Drawing showing the different structures controlled by this approach.

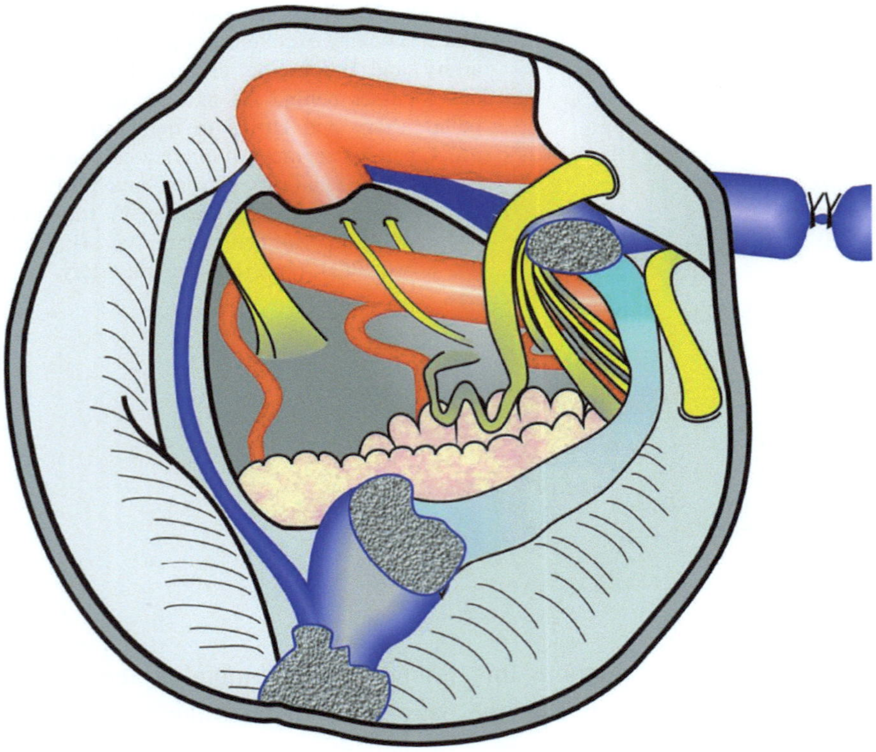

Fig. 10.174 The structures exposed through this approach. Note that the D sigmoid sinus has already been closed. The better exposure of the lower cranial nerves can be appreciated.

Surgical Steps (Right Side) (Figs. 10.175–10.181)

1. The incision is performed as in the type A approach. The lower limb of the incision slants inferiorly, extending 2 cm below the mastoid tip (Fig. 10.175).

2. An inferiorly based U-shaped musculoperiosteal flap is raised as described in the POTS approach. Anteriorly, a small Palva flap is created to form the second layer after the cul-de-sac closure of the external auditory canal.

3. The approach is performed as described for type A. In addition, a wider removal of the occipital bone posterior and inferior to the sigmoid sinus is performed.

4. The sigmoid sinus is closed extraluminally at the transverse sigmoid sinus, followed by opening the sinus and closing it intraluminally with large pieces of Surgicel.

5. The internal jugular vein is identified in the neck (Fig. 10.176) and is double-ligated. Fingertouch was used for the identification of the lateral process of C1. The lateral process of C1 is a dependable landmark for locating the internal jugular vein (Fig. 10.177).

6. The dura is further separated from the occipital bone and condyle. The occipital condyle is then partially drilled to expose the hypoglossal canal (Fig. 10.178).

7. The proximal part of the sigmoid sinus is compressed extraluminally with a piece of Surgicel placed between the sinus and the overlying bone at its junction with the lateral sinus. Next, the sinus is double-ligated and then opened with a Beaver knife. The dura is opened as before with the additional opening of the jugular bulb as described in the POTS approach (Fig. 10.179).

Surgical Anatomy After Opening the Dura

Figs. 10.180 and 10.181 show the different structures seen after opening the dura.

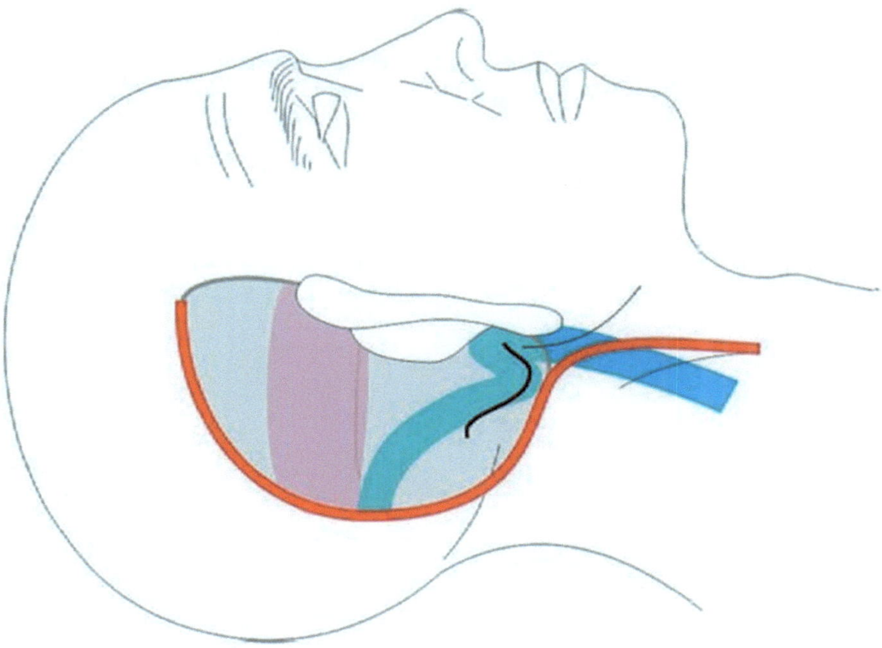

Fig. 10.175 The markings of the skin incision to be made.

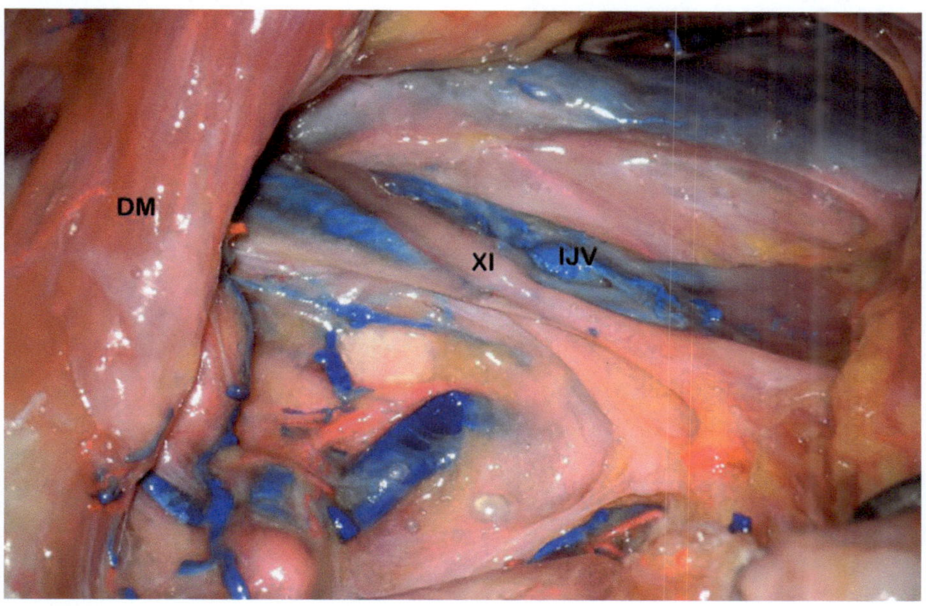

Fig. 10.176 The internal jugular vein is identified in the neck and then is double-ligated. *XI*, accessory nerve; *IJV*, internal jugular vein; *DM*, digastric muscle.

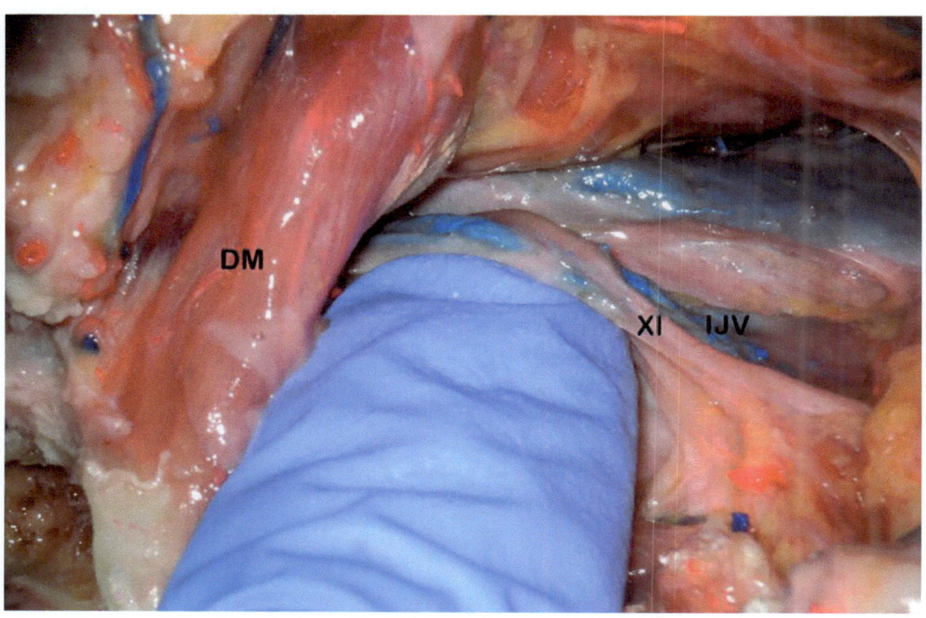

Fig. 10.177 Fingertouch was used for the identification of the lateral process of C1. The lateral process of C1 is a dependable landmark for locating the internal jugular vein. *XI*, accessory nerve; *IJV*, internal jugular vein; *DM*, digastric muscle.

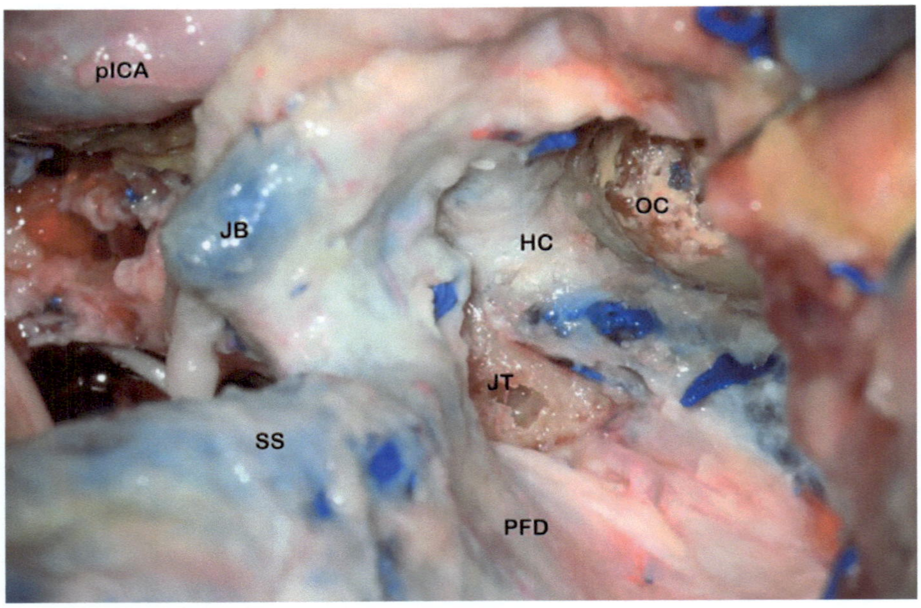

Fig. 10.178 The occipital condyle is then partially drilled to expose the hypoglossal canal. *pICA*, petrous segment of internal carotid artery; *OC*, occipital condyle; *HC*, hypoglossal canal; *JT*, jugular tubercle; *PFD*, posterior fossa dura; *SS*, sigmoid sinus; *JB*, jugular bulb.

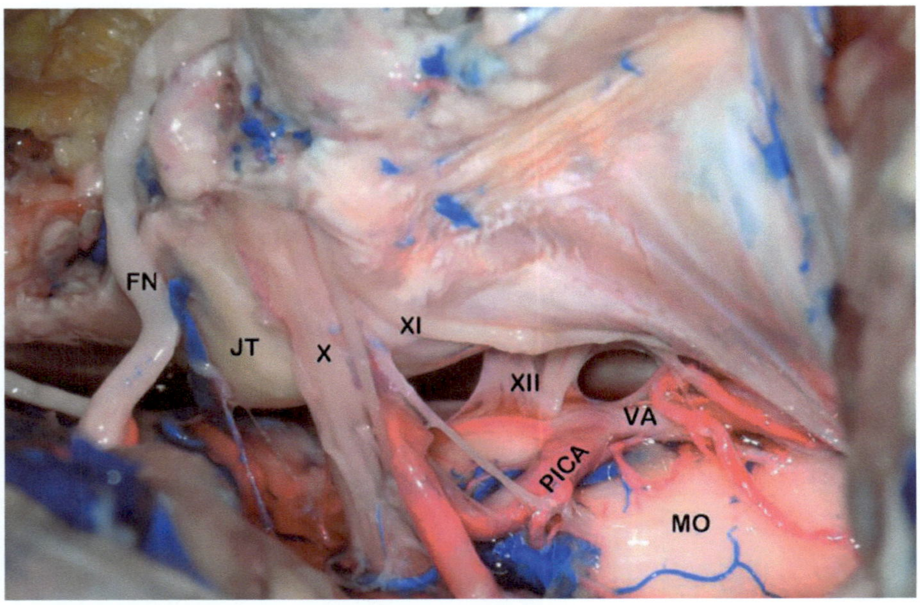

Fig. 10.179 The dura is opened, and the arachnoid is taken down from the neurovascular structures. Cranial nerves VII–XII can be visualized. *VA*, vertebral artery; *MO*, medulla oblongata; *PICA*, posteroinferior cerebellar artery; *XII*, hypoglossal nerve; *XI*, accessory nerve; *X*, vagus nerve; *JT*, jugular tubercle; *FN*, facial nerve.

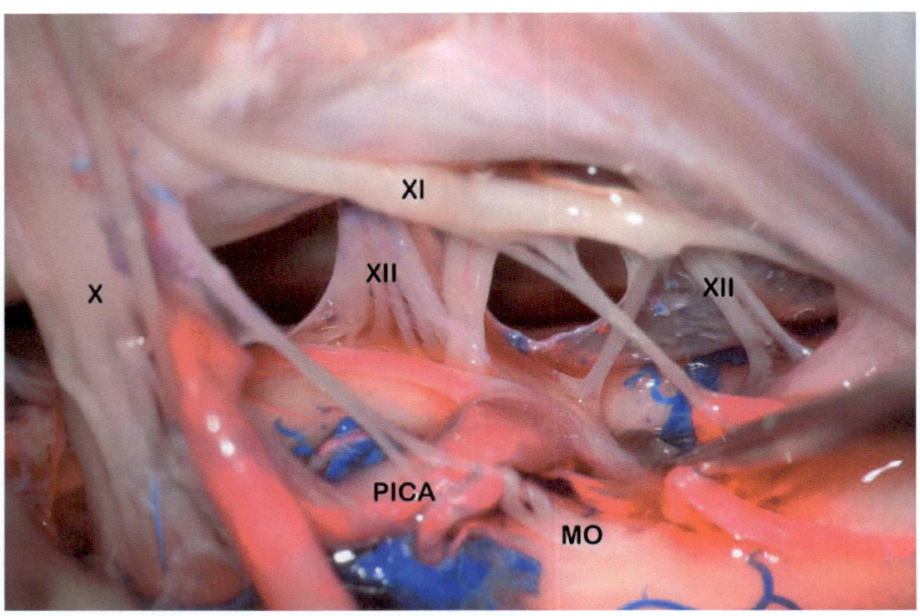

Fig. 10.180 At a higher magnification, the intimate relation between the hypoglossal nerve and the vertebral artery is well seen. *XII*, hypoglossal nerve; *XI*, accessory nerve; *X*, vagus nerve; *MO*, medulla oblongata; *PICA*, posteroinferior cerebellar artery.

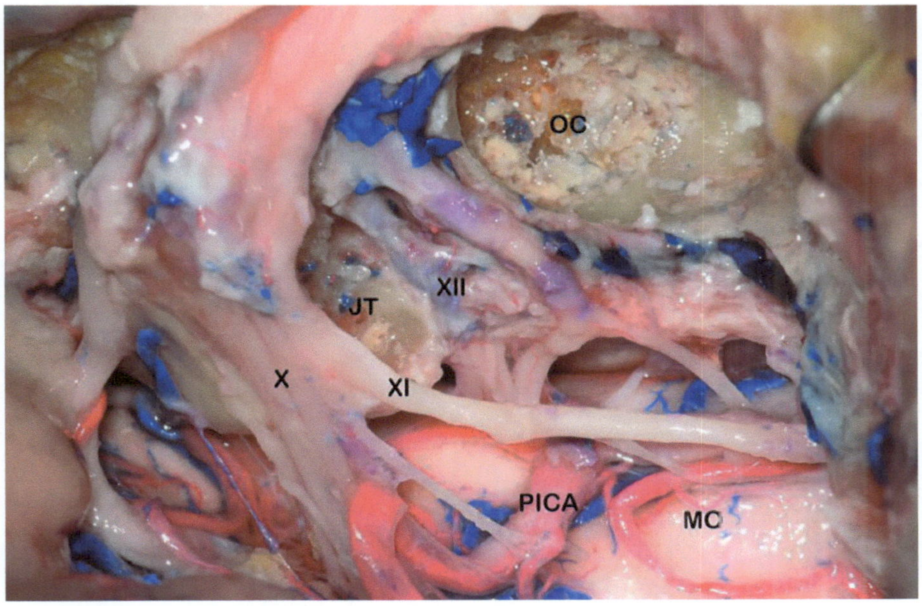

Fig. 10.181 The hypoglossal canal is opened. The two bundles forming the hypoglossal nerve unite to form a single nerve in the canal. *OC*, occipital condyle; *XII*, hypoglossal nerve; *XI*, accessory nerve; *X*, vagus nerve; *JT*, jugular tubercle; *MO*, medulla oblongata; *PICA*, posteroinferior cerebellar artery.

Clinical Application Transcochlear Type D Approach (Figs. 10.182–10.193)

Figs. 10.182–10.193 show the staged removal of the tumor using the transcochlear approach type D in the first stage and a transcervical approach for the second stage.

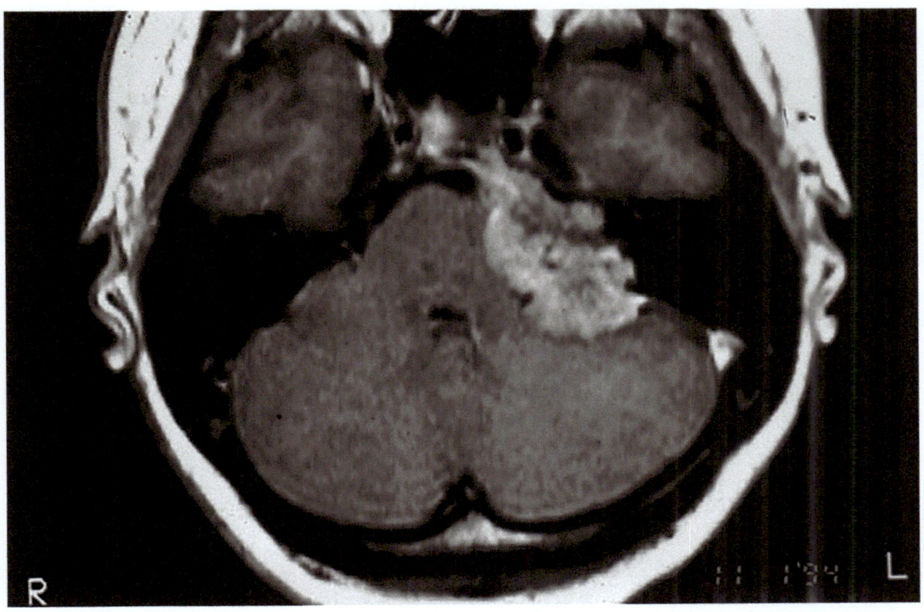

Fig. 10.182 Left jugular foramen En-plaque meningioma invading the temporal bone and extending into the neck.

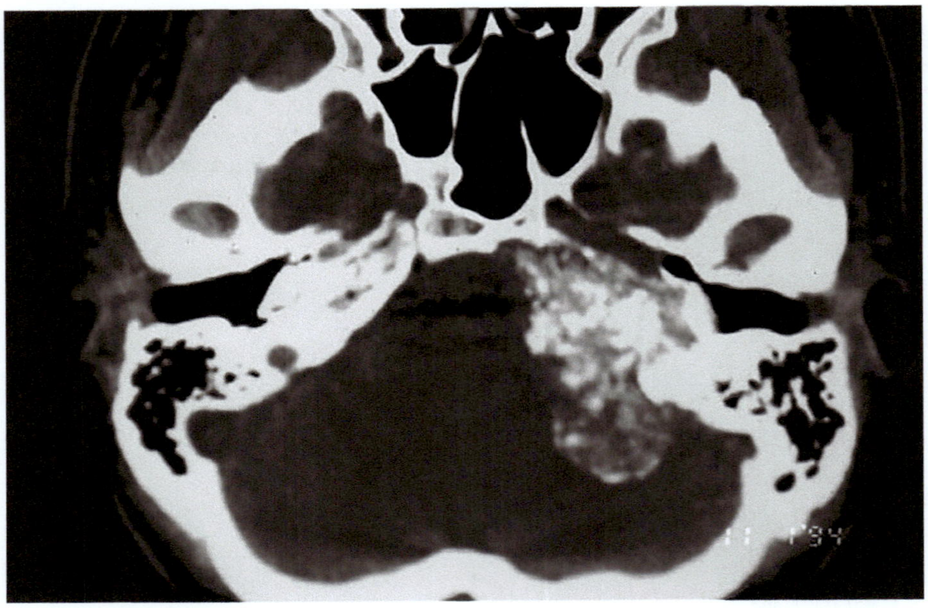

Fig. 10.183 High-resolution CT scan, bone window, showing marked invasion of the temporal bone.

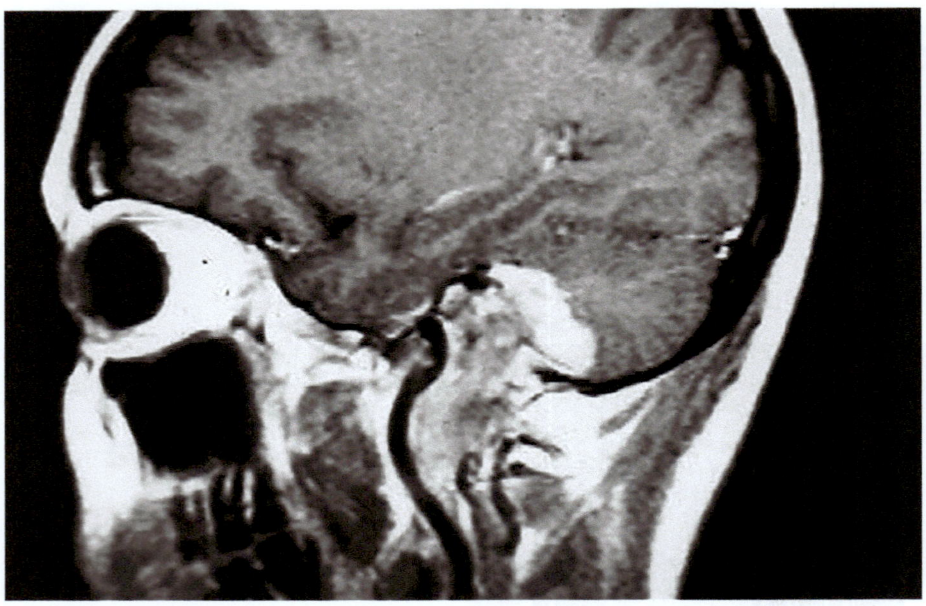

Fig. 10.184 MRI with gadolinium enhancement, sagittal view, showing the tumor extending into the neck.

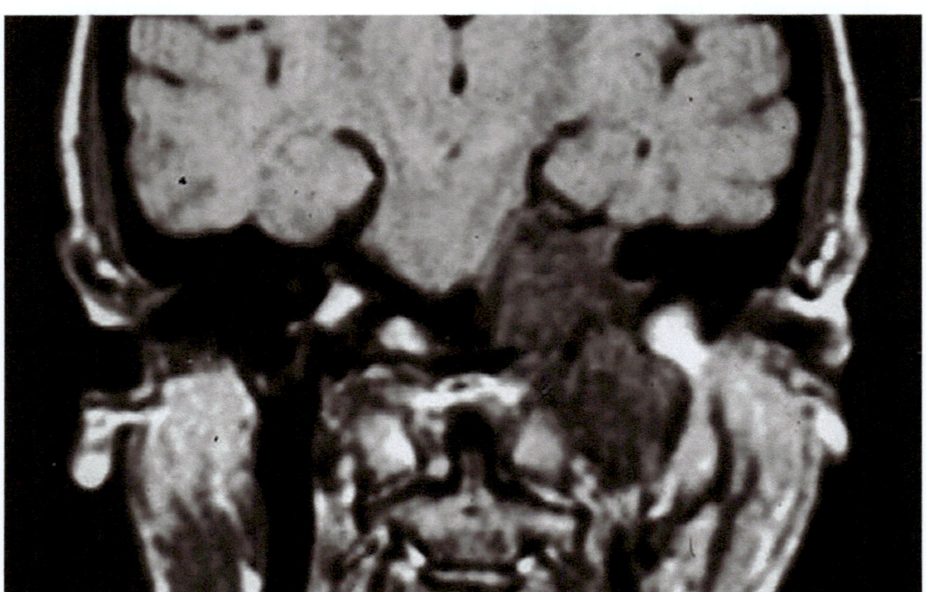

Fig. 10.185 MRI, coronal view, T1-weighted, showing neck extension.

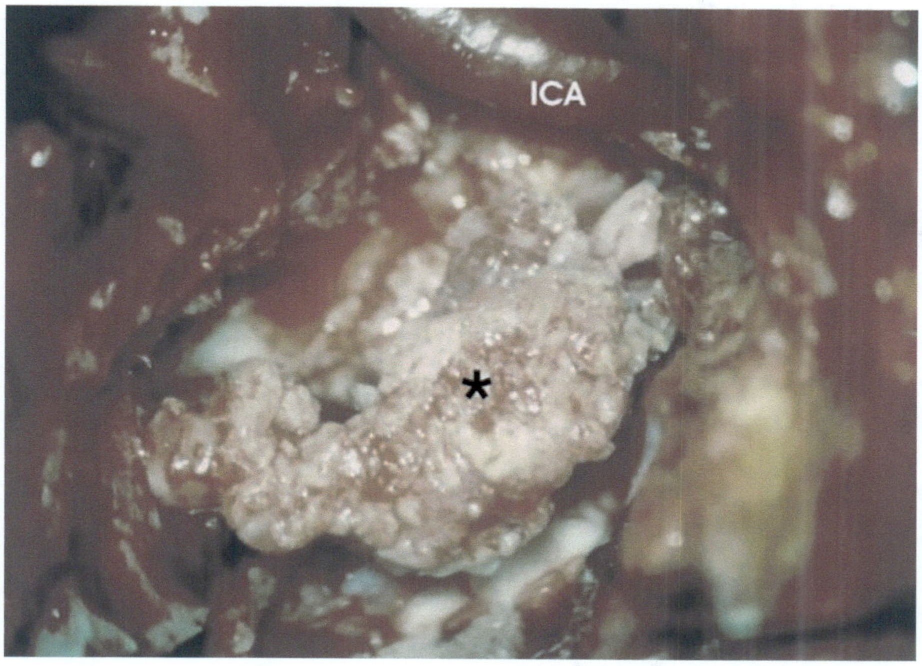

Fig. 10.186 The approach has been performed, and the internal carotid artery (ICA) has been dissected free from the tumor. The dura (infiltrated by the tumor) (*) is incised.

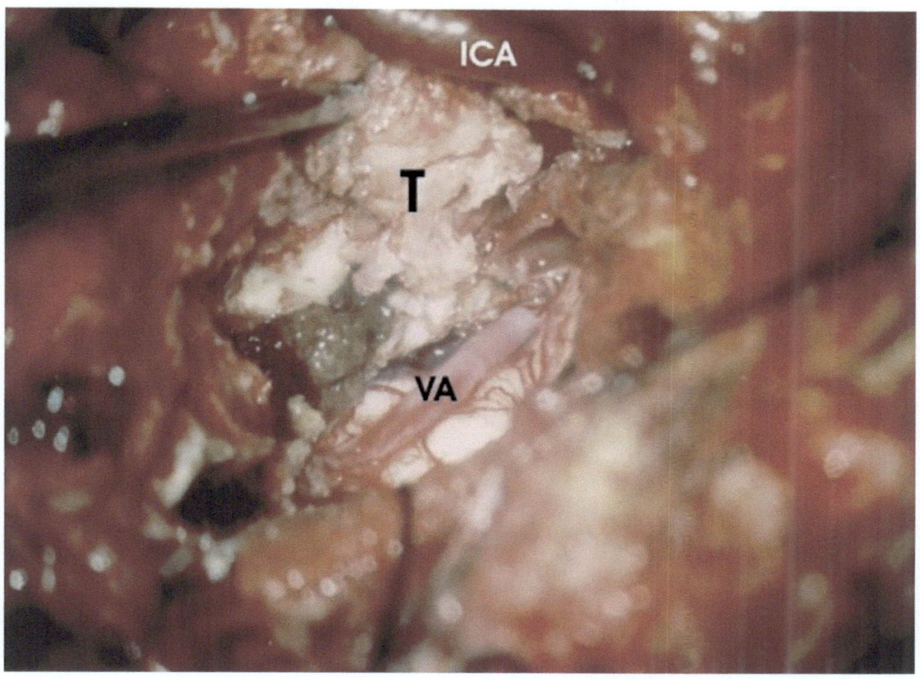

Fig. 10.187 With the opening of the dura, the tumor can be seen in close proximity to the vertebral artery (VA). *ICA*, internal carotid artery; *T*, tumor.

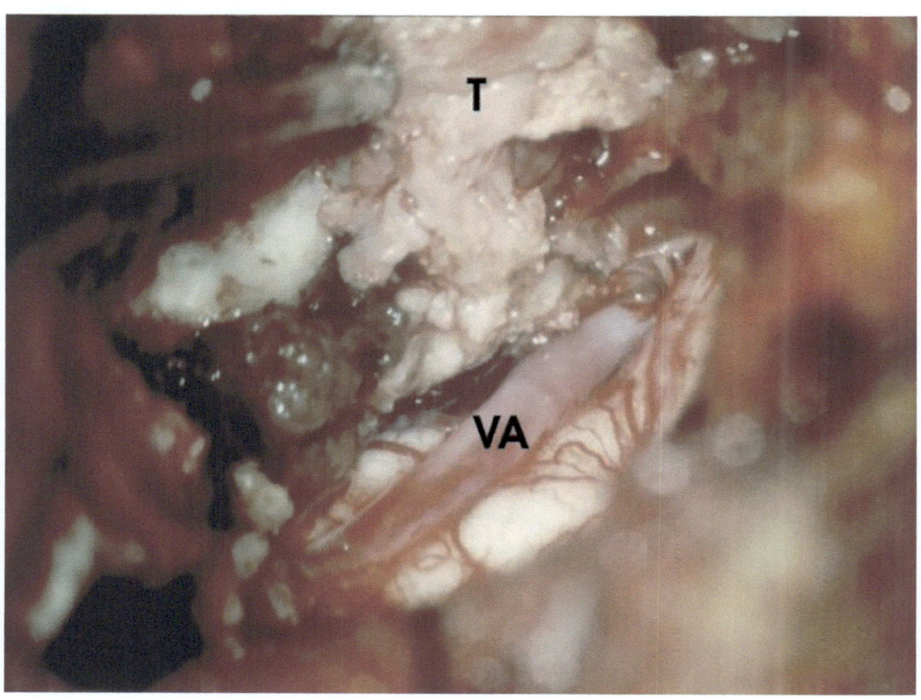

Fig. 10.188 At higher magnification, the vertebral artery (VA) and its relation to the tumor (T) can be well appreciated.

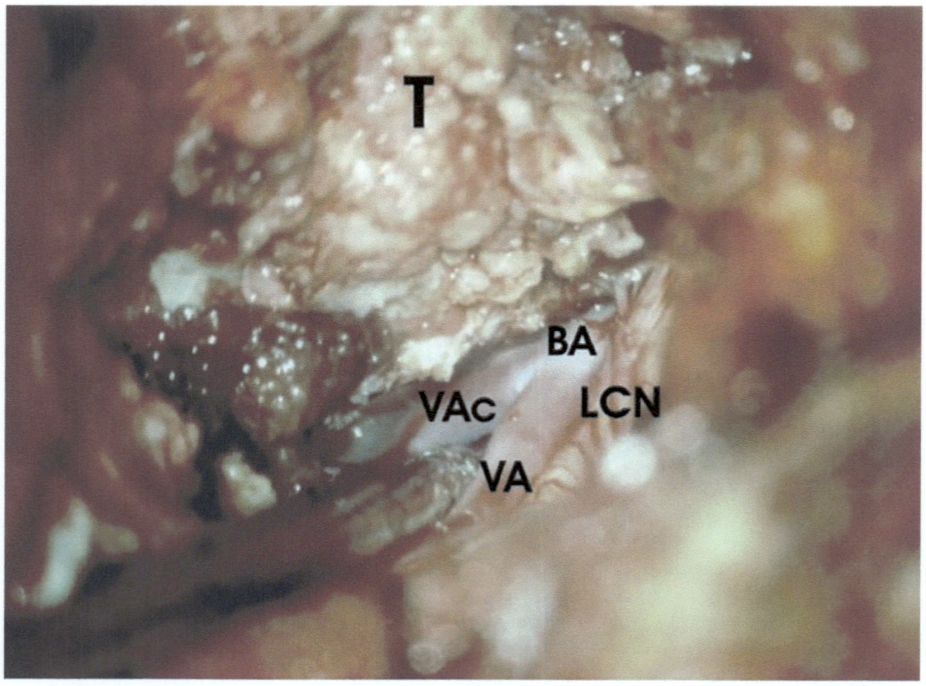

Fig. 10.189 With further dissection, the vertebrobasilar junction can be appreciated.

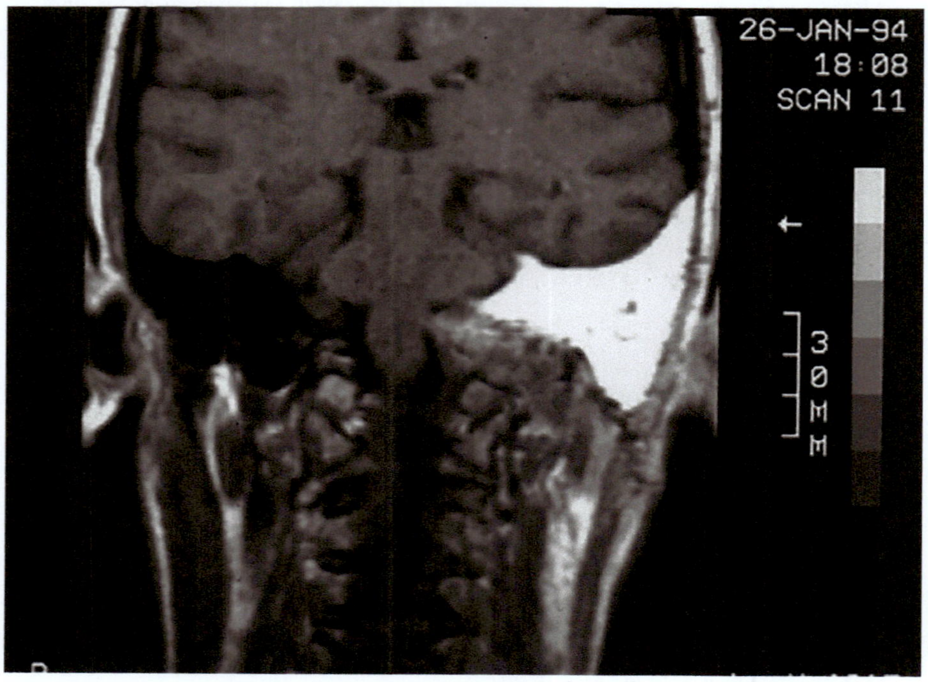

Fig. 10.190 Postoperative MRI after the first stage. The posterior fossa tumor has been removed. The residual tumor in the neck is seen.

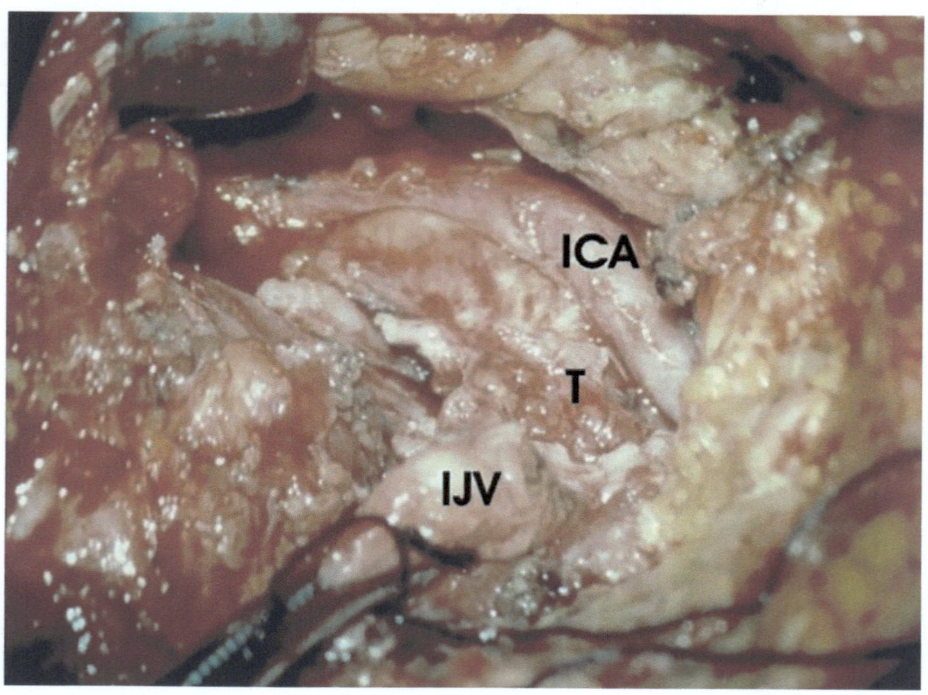

Fig. 10.191 Through a transcervical approach, the internal carotid artery (ICA) in the upper neck has been dissected free from the encasing tumor (T).

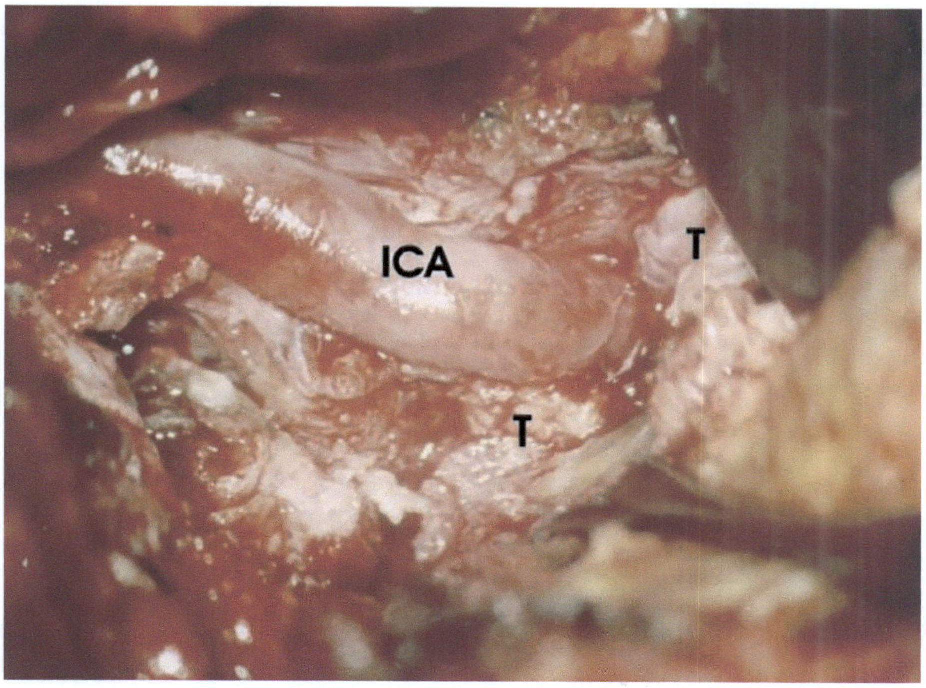

Fig. 10.192 At higher magnification, the relationship between the artery and the tumor can be appreciated. *ICA*, internal carotid artery; *T*, tumor.

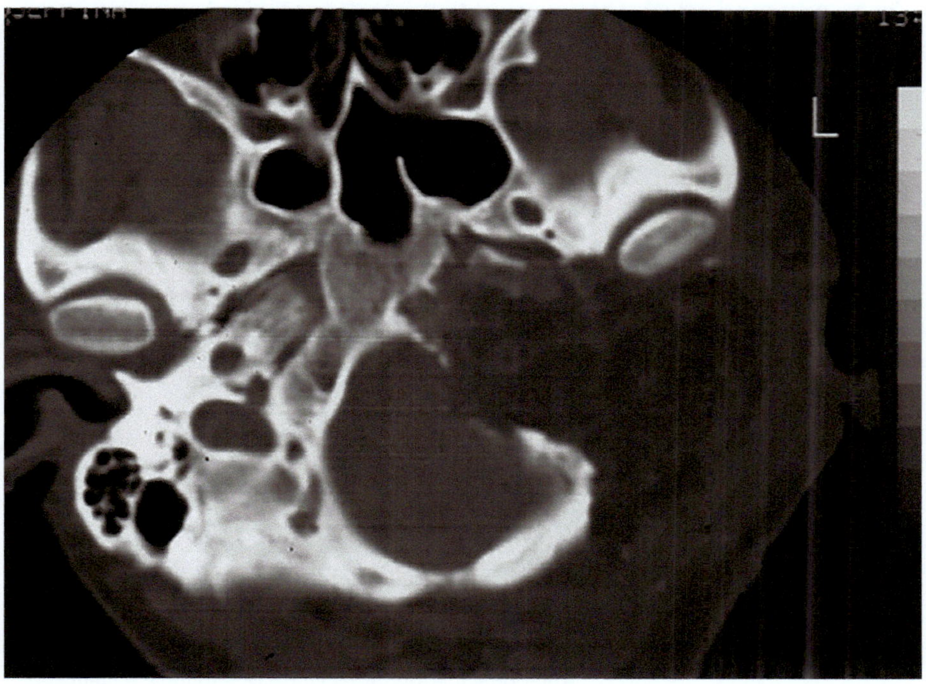

Fig. 10.193 Postoperative CT after total tumor excision.

REFERENCES

[1] House WF, Hitselberger WE. The transcochlear approach to the skull base. Arch Otolaryngol Head Neck Surg 1976; 102:334Y42.

[2] House WF, De la Cruz A. Transcochlear approach to the petrous apex and clivus. Trans Sect Otolaryngol Am Acad Ophthalmol Otolaryngol 1977;84:927Y31.

[3] De la Cruz A. The transcochlear approach to meningiomas and cholesteatomas of the cerebellopontine angle. In: Brackmann DE, editor. Neurological surgery of the ear and skull base. New York, NY: Raven Press; 1982. p. 353Y60.

[4] De la Cruz A, Teufert KB. Transcochlear approach to cerebellopontine angle and clivus lesions: indications, results, and complications. Otol Neurotol 2009;30(3):373–80. https://doi.org/10.1097/MAO.0b013e31819a892b.

[5] Sanna M, Mazzoni A, Saleh E, et al. The system of the modified transcochlear approach: a lateral avenue to the central skull base. Am J Otolaryngol 1998;19(1):88–97 [discussion 97–98].

[6] Tedeschi H, Rhoton Jr AL. Lateral approaches to the petroclival region. Surg Neurol 1994;41(3):180–216.

[7] Sanna M, Mazzoni A, Saleh EA, et al. Lateral approaches to the median skull base through the petrous bone: the system of the modified transcochlear approach. J Laryngol Otol 1994;108(12):1036–44.

[8] Angeli SI, De la Cruz A, Hitselberger W. The transcochlear approach revisited. Otol Neurotol 2001;22(5):690–5.

[9] Horn KL, Hankinson HL, Erasmus MD, Beauparalant PA. The modified transcochlear approach to the cerebellopontine angle. Otolaryngol Head Neck Surg 1991;104(1):37–41.

[10] Arriaga MA, Gorum M. Indications and variations of transcochlear exposure of the ventral brainstem. Laryngoscope 1996;106(5 Pt 1):639–44.

[11] Bernardo A, Stieg PE. Translabyrinthine and transcochlear petrosal approaches. In: Cappabianca P, Iaconetta G, Califano L, editors. Cranial, craniofacial and skull base surgery. Milano: Springer; 2010.

[12] Jackler RK, Sim DW, Gutin PH, Pitts LH. Systematic approach to intradural tumors ventral to the brain stem. Am J Otolaryngol 1995;16:39–51.

[13] Goddard JC, Theodore R, McRackan TB. Transcochlear approach to cerebellopontine angle lesions. In: Brackmann DE, Shelton C, Arriaga MA, editors. Otologic surgery. Philadelphia: W.B. Saunders Co./Elsevier; 2016. p. 4th ed.

[14] Arriaga M, Shelton C, Nassif P, Brackmann DE. Selection of surgical approaches for meningiomas affecting the temporal bone. Otolaryngol Head Neck Surg 1992;107:738–44.

[15] Thedinger BA, Glasscock III ME, Cueva RA. Transcochlear transtentorial approach for removal of large cerebellopontine angle meningiomas. Am J Otolaryngol 1992;13:408–15.

[16] Gonzalez LF, Alexander MJ, McDougall CG, Spetzler RF. Anteroinferior cerebellar artery aneurysms: surgical approaches and outcomes—a review of 34 cases. Neurosurgery 2004;55:1025–35.

[17] Siwanuwatn R, Deshmukh P, Figueiredo EG, et al. Quantitative analysis of the working area and angle of attack for the retrosigmoid, combined petrosal, and transcochlear approaches to the petroclival region. J Neurosurg 2006;104:137–42.

[18] Devèze A, Franco-Vidal V, Liguoro D, Guérin J, Darrouzet V. Transpetrosal approaches for meningiomas of the posterior aspect of the petrous bone: results in 43 consecutive patients. Clin Neurol Neurosurg 2007;109:578Y88.

[19] Bambakidis NC, Kakarla UK, Kim LJ, et al. Evolution of surgical approaches in the treatment of petroclival meningiomas: a retrospective review. Neurosurgery 2007;61:202Y9.

[20] Leonetti JP, Anderson DE, Marzo SJ, et al. Combined transtemporal access for large (>3 cm) meningiomas of the cerebellopontineangle. Otolaryngol Head Neck Surg 2006;134:949–52.

[21] Casazza GC, McCrary HC, Shelton C, Gurgel RK. Primary petrous apex epidermoids with skull base erosion. Otol Neurotol 2019;40(5):e556–61. https://doi.org/10.1097/MAO.0000000000002199.

[22] Peng KA, Chen BS, Lorenz MB, Lekovic GP, Schwartz MS, Slattery WH, Wilkinson EP. Revision surgery for vestibular schwannomas. J Neurol Surg B Skull Base 2018;79(6):528–32. https://doi.org/10.1055/s-0038-1635256 [Epub 2018 April 9].

[23] Prasad SC, Piras G, Piccirillo E, Taibah A, Russo A, He J, Sanna M. Surgical strategy and facial nerve outcomes in petrous bone cholesteatoma. Audiol Neurootol 2016;21(5):275–85. https://doi.org/10.1059/000448584 [Epub 2016 October 7].

[24] Kunimoto Y, Lauda L, Falcioni M, Taibah A, Hasegawa K, Sanna M. Staged resection for vestibular schwannoma. Acta Otolaryngol 2015;135(9):895–900. https://doi.org/10.3109/00016489.2015.1040170 [Epub 2015 May 8].

[25] Bernardo A, Evins AI, Visca A, Stieg PE. The intracranial facial nerve as seen through different surgical windows: an extensive anatomosurgical study. Neurosurgery 2013;72(2 Suppl. Operative). https://doi.org/10.1227/NEU.0b013e31827e5844. ons194-207 [discussion ons207].

[26] Macías-Reyes H, Ramos-Zúñiga R, Garcia-Estrada J, Jáuregui-Huerta F, Hidalgo-Mariscal ML. Combined approach for experimental oto-neurosurgical procedures. Surg Neurol Int 2012;3:68. https://doi.org/10.4103/2152-7806.97537 [Epub 2012 June 19].

[27] Graillon T, Fuentes S, Régis J, Metellus P, Brunel H, Roche PH, Dufour H. Multidisciplinary management of giant functional petrous bone paraganglioma. Acta Neurochir 2010;153(1):85–9. https://doi.org/10.1007/s00701-010-0818-z. Epub 2010 October 8 [discussion 89].

[28] Pandya Y, Piccirillo E, Mancini F, Sanna M. Management of complex cases of petrous bone cholesteatoma. Ann Otol Rhinol Laryngol 2010;109(8):514–25.

[29] Tummala RP, Coscarella E, Morcos JJ. Transpetrosal approaches to the posterior fossa. Neurosurg Focus 2005;19(2):E6.

[30] Kaylie DM, Warren 3rd FM, Haynes DS, Jackson CG. Neurotologic management of intracranial epidermoid tumors. Laryngoscope 2005;105(6):1082–6.

[31] Russo A, Piccirillo E, De Donato G, Agarwal M, Sanna M. Anterior and posterior facial nerve rerouting: a comparative study. Skull Base 2003;13(3):123–30.

[32] Shen T, Friedman RA, Brackmann DE, Slattery 3rd WH, Hitselberger WE, Schwartz MS, Fisher L. The evolution of surgical approaches for posterior fossa meningiomas. Otol Neurotol 2004;25(3):394–7.

[33] Mortini P, Mandelli C, Franzin A, Giugni E, Giovanelli M. Surgical excision of clival tumors via the enlarged transcochlear approach. Indications and results. J Neurosurg Sci 2001;45(3):127–39 [discussion 140].

CHAPTER 11

The Middle Fossa Approaches

The middle cranial fossa approach (MCF) is a well-established surgical technique that provides extradural access to the internal auditory canal (IAC) via a subtemporal corridor. It is usually indicated for the resection of small vestibular schwannomas (VS). The MCF includes a group of surgical techniques, classified into the following:

- Middle fossa approach to the IAC.
- The enlarged middle fossa approach.
- The middle fossa transpetrous approach, with removal of the petrous apex.

The major advantage of this technique is the ability to remove lesions involving the CPA while preserving the hearing function. The MCF can be combined with a transmastoid approach to achieve adequate exposure to the full length of the intratemporal facial nerve without compromising preoperatively preserved hearing.

Introduction

The MCF approach was first introduced by Hartley [1] in 1892, but the currently known microsurgical technique was promoted by House (1963) with the introduction of the operative microscope [2]. The MCF approach has gradually gained popularity due to the possibility of attaining optimal tumor control while preserving hearing and facial nerve function. With further advances in skull base surgery, the approach has undergone many variations and refinements to expand and to make it more reliable for tumor removal [3–12]. The procedure involves a temporal craniotomy, elevation of the temporal lobe, and exposure of the petrous bone and apex from above. Dissection of the IAC from this plane provides complete exposure of its contents and removal of local pathology with minimal damage to the acoustic or facial nerve. The extensions of the middle fossa approach allow access to the cerebellopontine angle (CPA), the petroclival area, the upper basilar artery, and the posterior cavernous sinus [13–19].

There are many applications to this technique, but the main indication remains the excision of small- to medium-sized VS (Grade 0 and 1) in patients with serviceable preoperative hearing [20–31]. The retrosigmoid approach is also indicated for the removal of small VS but is more appropriate for tumors located predominantly in the CPA with limited growth into the lateral aspect of the IAC. In the last decade, the use of the MCF approach has declined due to the increasing popularity of "wait and scan" management or the radiosurgical treatment. Nevertheless, microsurgery remains one of the primary options for the treatment of VS [23]. The morbidity associated with the technique is minimal, with CSF leak being the most commonly reported complication (2%–13%), and can be managed conservatively in most cases [22,23,32–36,37–43].

The advantages of the MCF approach are:

– Exposure of the whole length of the internal auditory canal: This allows total tumor removal, with no residual tumor left at the fundus, as is the case in the retrosigmoid approach.

– Possibility of hearing preservation: The overall long-term hearing preservation rate after VS removal reported is around 55% [30,44–46], but some report better results (up to 70%–80%) with proper selection of patients

Atlas of Microsurgery of the Lateral Skull Base. https://doi.org/10.1016/B978-0-443-19083-4.00011-8

[44,47–49] and improved preservation of low-tone PTA [30,50]. At the same time, tumor follow-up and radiosurgery are also associated with a high rate of hearing loss in time (45%–83%) [51–56], and neither the resolution of the disease.

– Minimum risk of CSF leakage (similar to the TLA and RS approaches in reported literature [36]).

– Very good postoperative facial nerve function: Initially, FN results were worse than for similar sizes of tumors dealt with using the enlarged TLA and RS approaches. The reason for this is that the facial nerve is the most superficial structure encountered in the enlarged MCF approach and is particularly at risk during tumor removal. Expertise and routine use of the facial nerve monitor have helped improve these results, with up to 95% of cases presenting a grade I or II HB postintervention [20–22,28,30,57–59].

Disadvantages:

- The need for significant temporal lobe retraction.
- The deep field and restricted access. Only limited exposure of the CPA is obtained through this approach. The lower cranial nerves are poorly controlled.
- Control of intraoperative bleeding is more difficult than the TLA and RS approaches. If postoperative hematoma of the CPA occurs, repeat surgery is more difficult than in the translabyrinthine approach.
- It still remains one of the most challenging surgical techniques because of the lack of strong surgical landmarks to identify the IAC.

Classification

The middle fossa approaches are classified into:

1. The middle fossa approach to the IAC (Fig. 11.1A).
2. The enlarged middle fossa approach, with exposure to the IAC and surrounding bone, increases the access to the CPA (Fig. 11.1B).
3. The middle fossa transpetrous approach, with removal of the petrous apex. This approach is designed basically for the anterior CPA, the ventral surface of the pons, and the upper clivus (Fig. 11.1C).

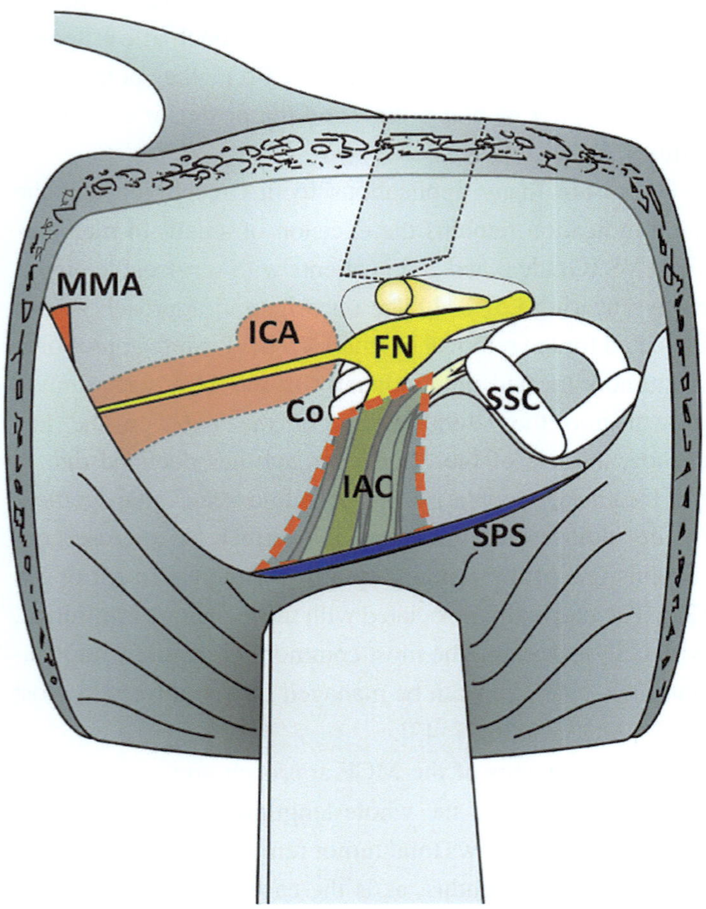

Fig. 11.1A The *gray* color represents the amount of bone removal using the classic middle cranial fossa approach. Note that the only bone overlying the internal auditory canal is removed. This technique is indicated in the case of facial nerve decompression or vestibular neurectomy. *Co*, cochlea; *FN*, facial nerve; *ICA*, internal carotid artery; *MMA*, middle meningeal artery; *SPS*, superior petrosal sinus; *SSC*, superior semicircular canal.

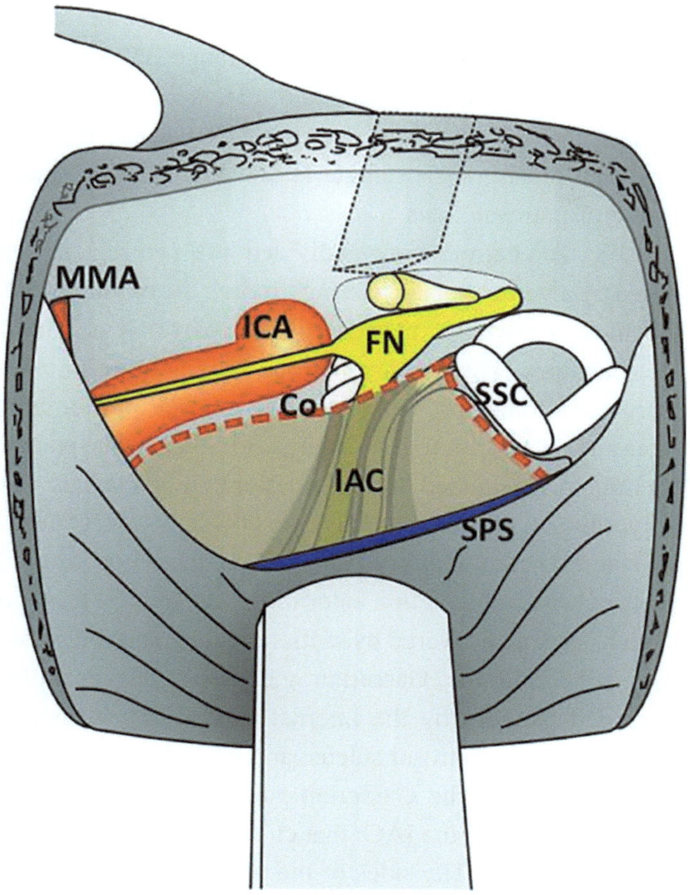

Fig. 11.1B The *gray* color represents the amount of bone removal using the enlarged middle cranial fossa approach. We would use this approach for the removal of small vestibular schwannomas reaching the fundus of the internal auditory canal with CPA extension of less than 0.5 cm when hearing preservation is aimed for. *Co*, cochlea; *FN*, facial nerve; *ICA*, internal carotid artery; *MMA*, middle meningeal artery; *SPS*, superior petrosal sinus; *SSC*, superior semicircular canal.

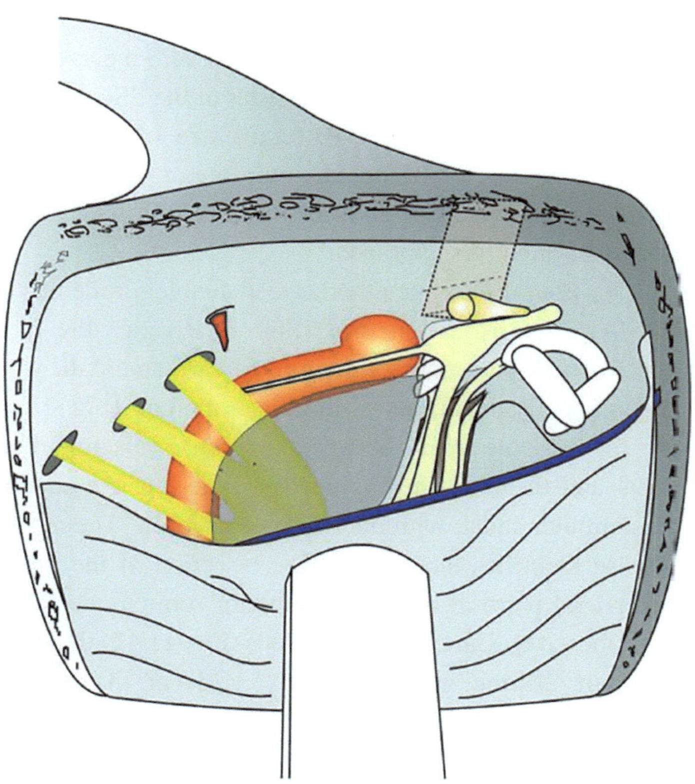

Fig. 11.1C The *gray* color represents the amount of bone removal using the middle fossa transpetrous approach. This approach is designed basically for the anterior cerebellopontine angle, the ventral surface of the pons, and the upper clivus.

Surgical Anatomy

1. The anatomy of the middle cranial fossa is very complex. Fig. 11.2 shows an anatomical dissection carried out through the middle fossa, illustrating the relationships between the various structures in this area. The IAC was exposed and opened. The angle between the superior semicircular canal and the IAC can be appreciated. The cochlea has been dissected to show its intimate proximity to the geniculate ganglion, the internal carotid artery, and the fundus of the internal auditory canal. The petrous apex has been partially drilled. The attic has been opened and the mastoid partially drilled to show the relative position of these structures when viewed through the middle fossa.

2. Fig. 11.3 shows the anatomy of the IAC seen from a MCF approach. The vertical crest (Bill's bar) separates the facial and superior vestibular nerves at the meatal fundus. The superior and inferior vestibular nerves are located posteriorly and the facial and cochlear nerves anteriorly in the meatus, with the cochlear nerve passing below the facial nerve to enter the modiolus. The labyrinthine segment of the facial nerve courses superolateral to the cochlea.

3. Fig. 11.4 shows a closer view of the lateral end of the internal auditory canal. Bill's bar separates the facial nerve anteriorly from the superior vestibular nerve posteriorly. The close proximity of the opened ampullae of the superior and posterior semicircular canals to the fundus of the IAC is clearly appreciated.

4. In Figs. 11.5–11.6, the greater superficial petrosal nerve, originating from the geniculate ganglion, is seen. The relationship between the fundus of the IAC and the ampullary ends of the lateral and superior semicircular canals, as well as the malleus, incus, and tympanic membrane, can be appreciated.

5. Despite the complexity of the anatomy of the middle fossa, the landmarks once the dura has been elevated are very few: namely, the middle meningeal artery, the greater superficial petrosal nerve, and the arcuate eminence, which is sometimes difficult to distinguish. In 48% of cases, the arcuate eminence does not correspond to the superior semicircular canal. Congenital blue lining of the superior canal is sometimes observed (Fig. 11.7).

6. The IAC has variable relations with the nearby structures (Fig. 11.8). It makes angles of 107 degrees (range 90–135 degrees) with the greater superficial petrosal nerve (1), 52 degrees (range 34–75 degrees) with the superior semicircular canal (2), 44 degrees (range 35–55 degrees) with the superior petrosal sinus (3), and 158.5 degrees (range 150–170 degrees) with the external auditory canal (4). The most constant angle is between the IAC and the superior petrosal sinus.

7. The petrous bone is divided into two regions separated by a vertical plane passing through the modiolus of the cochlea: a posterior perilabyrinthine part and an anterior petrous apex. The anterior petrous apex is more important surgically and is further divided into a posterior rhomboidal area and an anterior triangular area (Fig. 11.9). The posterior rhomboid is bounded by the superior petrosal sulcus medially, the IAC posteromedially, the cochlea posterolaterally, the internal carotid artery laterally, and the posterior border of the Gasserian ganglion anteriorly. The anterior triangular area has been uncovered by sectioning the mandibular nerve and reflecting the Gasserian ganglion. This triangle is bounded laterally by the internal carotid artery, medially by the superior petrosal sulcus, and posteriorly by the posterior margin of the Gasserian ganglion. A4 and A1 (the anterior border of the IAC) that closes the anterior quadrangular mute area. The sides of the posterior triangular mute area are also measured: the distance between the point where the superior vestibular nerve meets the ampulla of the SSC (P1) and the IAC point of the SPS with the imaginary line that passes through the posterior border of the (P2); the distance between P2 and the point where the SPS meets the line that passes through the anterior border of the SSC (P3); and the distance between P3 and P1 (at the level of the SPS) that closes the posterior triangular mute area. All these three areas are summated to give the value of the total area of access through this approach.

8. Four different methods are available for identifying the IAC. The House method (Fig. 11.10) puts the facial nerve at risk and requires the utmost care and skill, while the major disadvantages of the Fisch method (Fig. 11.11) are the variable angle between the superior semicircular canal and the IAC and the need to blue-line the superior semicircular canal, with the risk of opening it. The method we use identifies the canal at the bisection of the greater superficial petrosal nerve and superior semicircular canal, as proposed by Garcia-Ibanez (1980) (Fig. 11.12); however, we start drilling medially (Fig. 11.13) at the level of the porus.

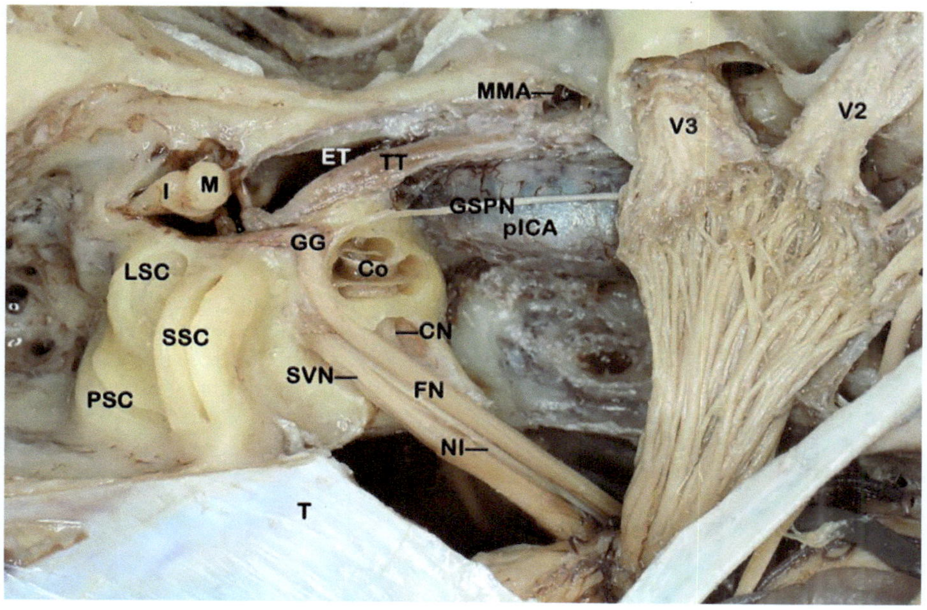

Fig. 11.2 An anatomical dissection carried out through the middle fossa, illustrating the relationships between the various structures in this area. *MMA*, middle meningeal artery; *M*, malleus; *I*, incus; *LSC*, lateral semicircular canal; *PSC*, posterior semicircular canal; *SSC*, superior semicircular canal; *GG*, geniculate ganglion; *Co*, cochlea; *FN*, facial nerve; *SVN*, superior vestibular nerve; *NI*, nervus intermedius; *pICA*, petrous segment of internal carotid artery; *GSPN*, greater superficial petrosal nerve; *TT*, tensor tympani muscle; *ET*, eustachian tube; *V3*, mandibular nerve; *V2*, maxillary nerve.

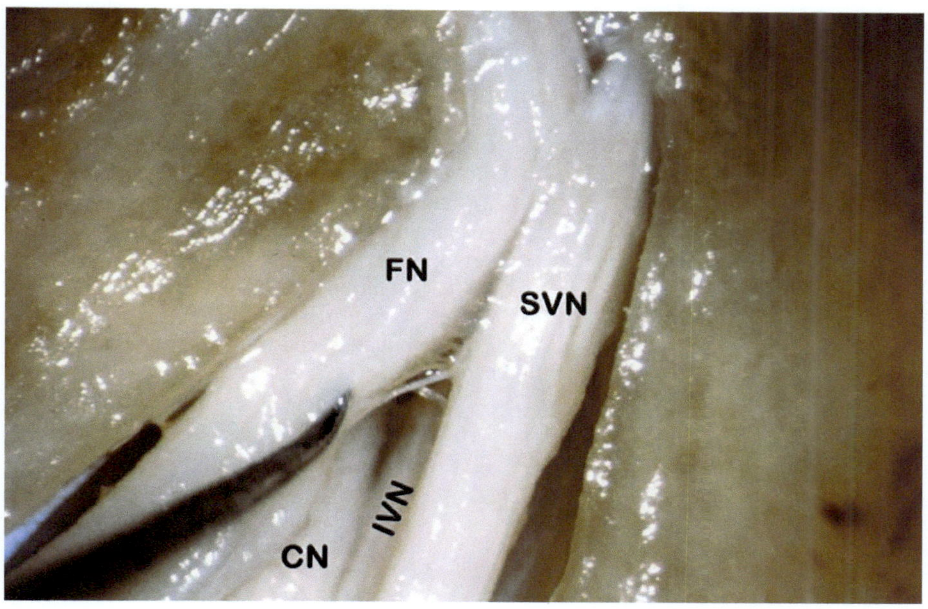

Fig. 11.3 Anatomy of the internal auditory canal seen from a middle cranial fossa approach. *FN*, facial nerve; *SVN*, superior vestibular nerve; *IVN*, inferior vestibular nerve; *CN*, cochlear nerve.

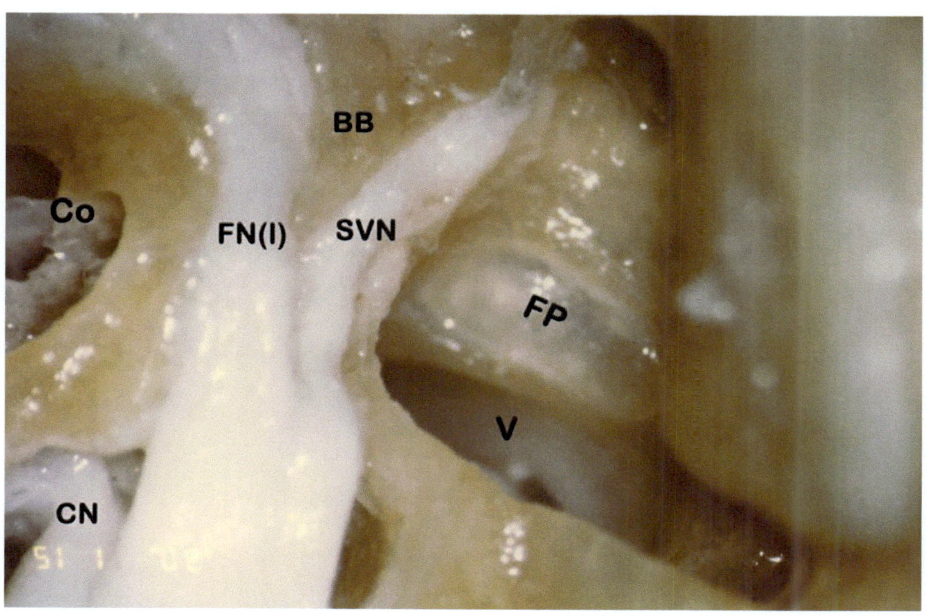

Fig. 11.4 Bill's bar separating the facial nerve from the superior vestibular nerve. *BB*, Bill's bar; *SVN*, superior vestibular nerve; *FP*, footplate of the stapes; *FN(I)*, labyrinthine segment of facial nerve; *Co*, cochlea; *V*, vestibule; *CN*, cochlear nerve.

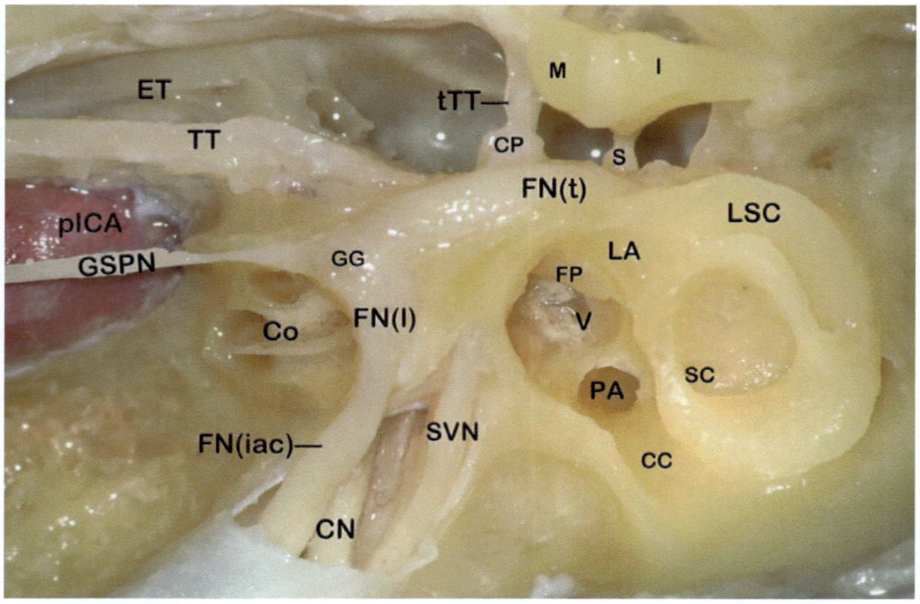

Fig. 11.5 *TT*, tensor tympani muscle; *ET*, eustachian tube; *GSPN*, greater superficial petrosal nerve; *pICA*, petrous segment of internal carotid artery; *Co*, cochlea; *GG*, geniculate ganglion; *M*, malleus; *I*, incus; *LSC*, lateral semicircular canal; *LA*, ampullated end of lateral semicircular canal; *PA*, ampullated end of posterior semicircular canal; *tTT*, tendon of tensor tympani muscle; *CP*, cochleariform process; *FP*, footplate of the stapes; *V*, vestibule; *CN*, cochlear nerve; *SVN*, superior vestibular nerve; *FN(iac)*, internal auditory canal segment of facial nerve; *FN(I)*, labyrinthine segment of facial nerve; *FN(t)*, tympanic segment of facial nerve; *CC*, common crus; *SC*, singular crus.

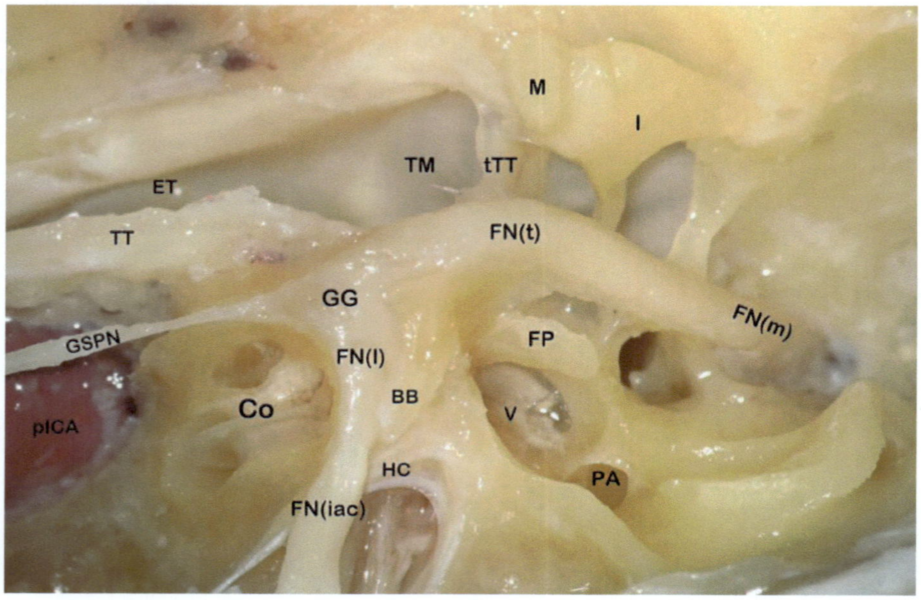

Fig. 11.6 *ET*, eustachian tube; *GSPN*, greater superficial petrosal nerve; *I*, incus; *M*, malleus; *TM*, tympanic membrane; *GG*, geniculate ganglion; *BB*, Bill's bar; *HC*, horizontal crest; *FP*, footplate of the stapes; *V*, vestibule; *Co*, cochlea; *pICA*, petrous segment of internal carotid artery; *TT*, tensor tympani muscle; *MMA*, middle meningeal artery; *FN(I)*, labyrinthine segment of facial nerve; *FN(t)*, tympanic segment of facial nerve; *FN(iac)*, internal auditory canal segment of facial nerve; *FN(m)*, mastoid segment of facial nerve; *PA*, ampullated end of posterior semicircular canal.

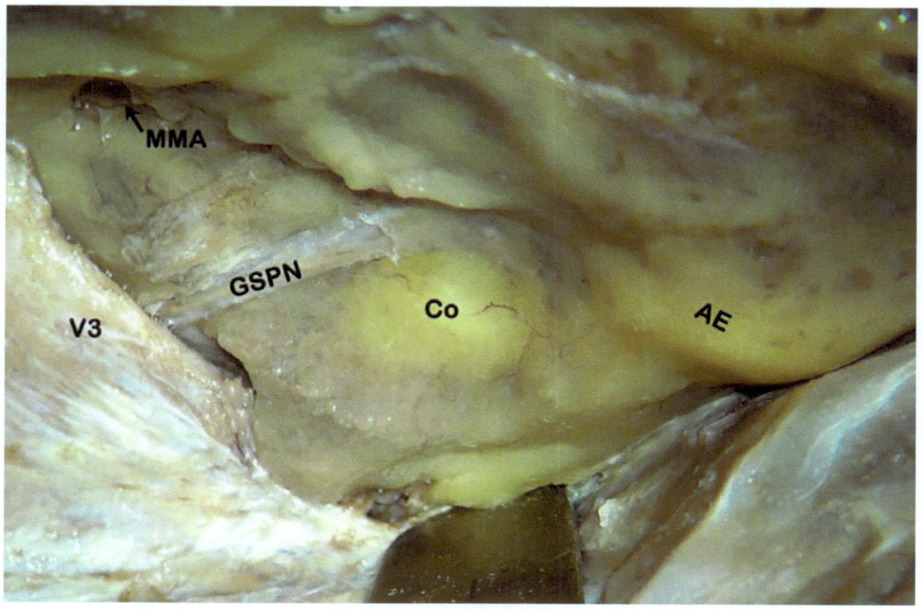

Fig. 11.7 Extradural right middle fossa exposure. The craniotomy is positioned above the root of the zygomatic arch and extends to the middle fossa floor. The dura has been elevated from the floor of the middle fossa to expose the arcuate eminence, greater petrosal nerve, and middle meningeal artery. *MMA*, middle meningeal artery; *V3*, mandibular nerve; *GSPN*, greater superficial petrosal nerve; *Co*, cochlea; *AE*, arcuate eminence.

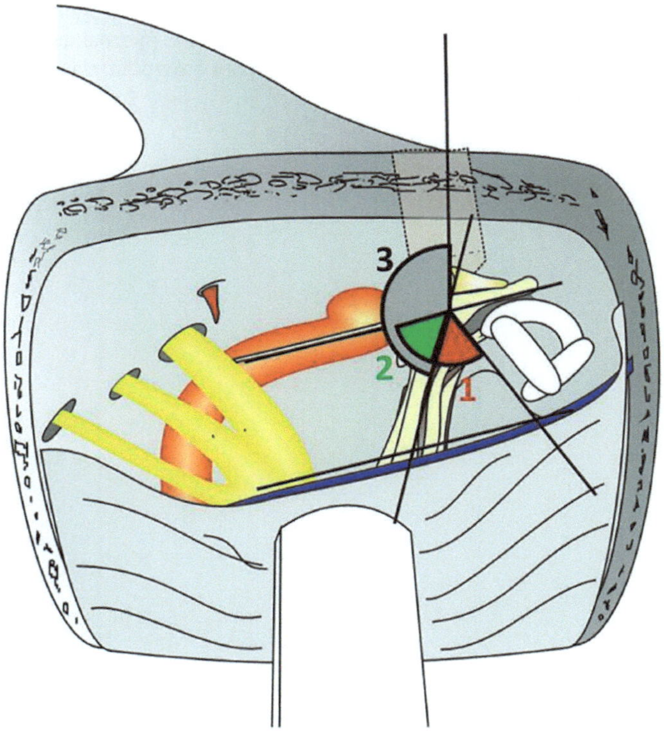

Fig. 11.8 The internal auditory canal has variable relations with the nearby structures (this figure). It makes angles of 107 degrees (range 90–135 degrees) with the greater superficial petrosal nerve (1), 52 degrees (range 34–75 degrees) with the superior semicircular canal (2), 44 degrees (range 35–55 degrees) with the superior petrosal sinus (3), and 158.5 degrees (range 150–170 degrees) with the external auditory canal (4) *EAC*, external auditory canal; *SSC*, superior semicircular canal; *IAC*, internal auditory canal; *SPS*, superior petrosal sinus; *GSPN*, greater superficial petrosal nerve.

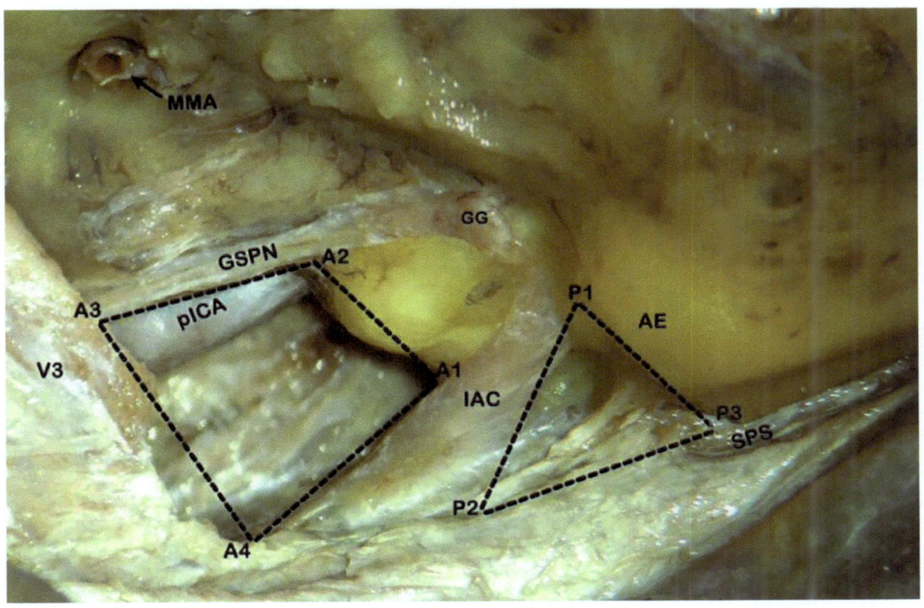

Fig. 11.9 The petrous bone is divided into two regions separated by a vertical plane passing through the modiolus of the cochlea: a posterior perilabyrinthine part and an anterior petrous apex. The anterior petrous apex is more important surgically and is further divided into a posterior rhomboidal area and an anterior triangular area Mute and IAC areas. A: anterior quadrangular area; P: posterior triangular area. For the description of the different points, see the text. *MMA*, middle meningeal artery; *GG*, geniculate ganglion; *AE*, arcuate eminence; *SPS*, superior petrosal sinus; *IAC*, internal auditory canal; *V3*, mandibular nerve; *GSPN*, greater superficial petrosal nerve; *pICA*, petrous segment of internal carotid artery.

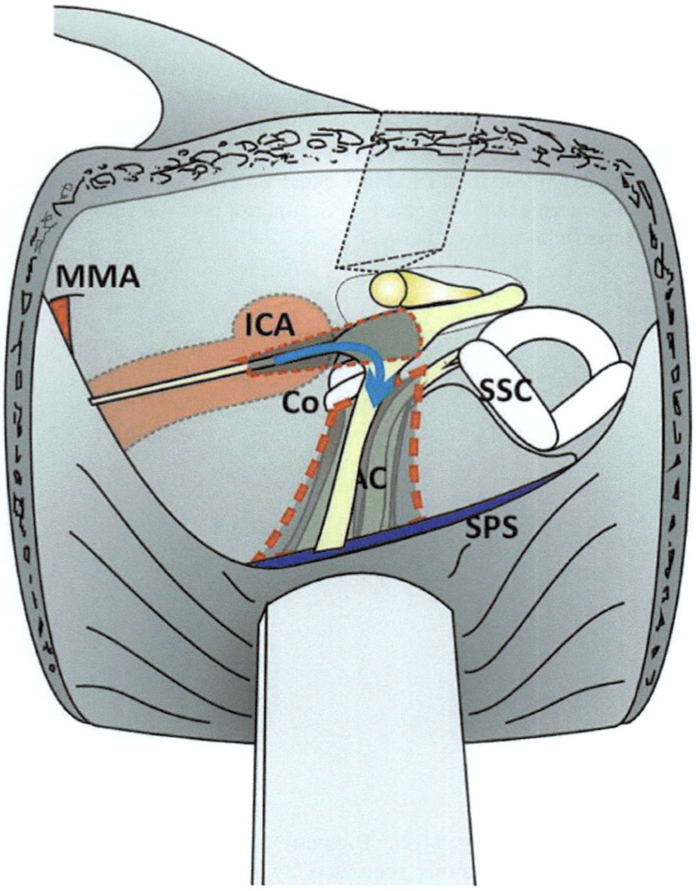

Fig. 11.10 *Right ear. House's method. MMA*, middle meningeal artery; *ICA*, internal carotid artery; *Co*, cochlea; *IAC*, internal auditory canal; *SPS*, superior petrosal sinus; *SSC*, superior semicircular canal.

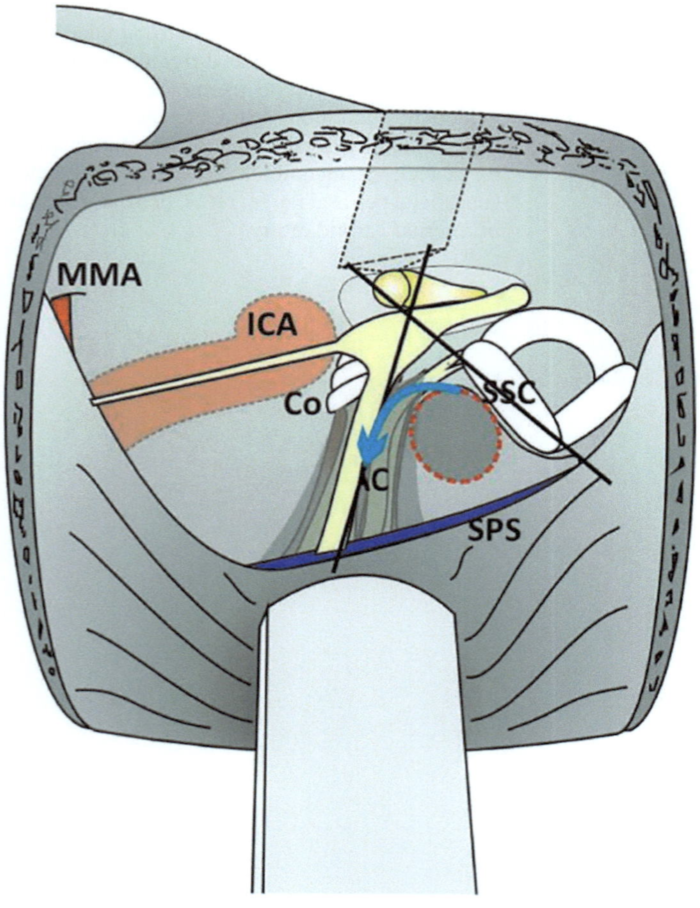

Fig. 11.11 *Right* ear. Fisch's method. *MMA*, middle meningeal artery; *ICA*, internal carotid artery; *Co*, cochlea; *IAC*, internal auditory canal; *SSC*, superior semicircular canal; *SPS*, superior petrosal sinus.

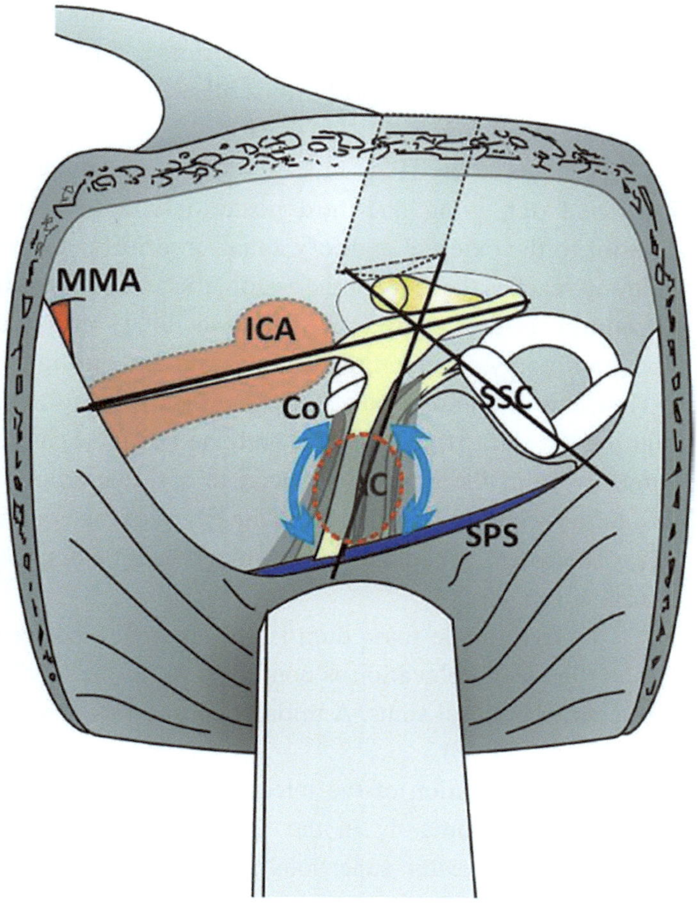

Fig. 11.12 *Right* ear. Garcia-Ibanez's method. *MMA*, middle meningeal artery; *ICA*, internal carotid artery; *Co*, cochlea; *SSC*, superior semicircular canal; *IAC*, internal auditory canal; *SPS*, superior petrosal sinus.

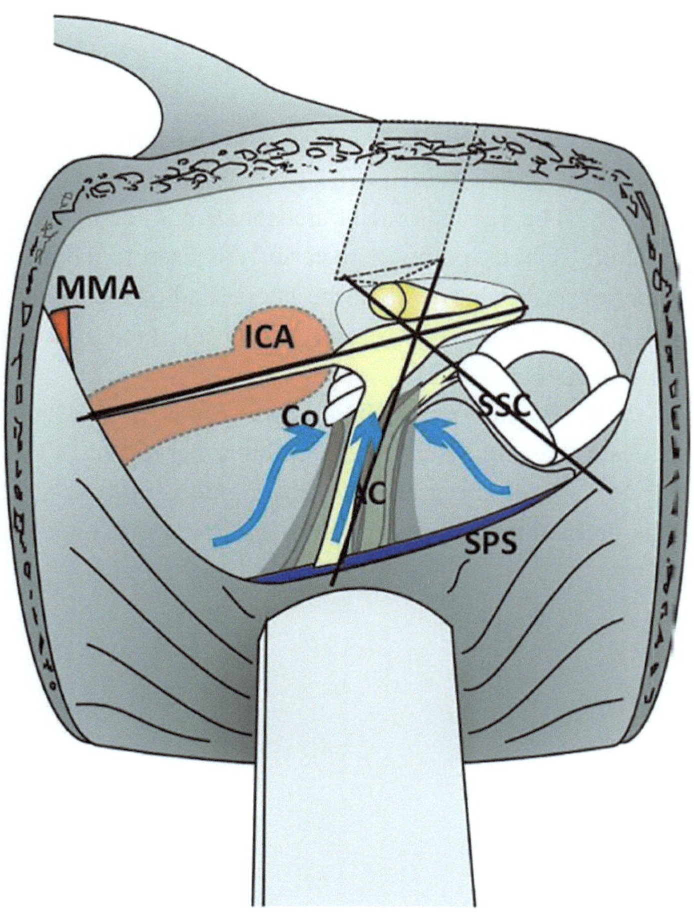

Fig. 11.13 *Right* ear. Sanna's method. *MMA*, middle meningeal artery; *ICA*, internal carotid artery; *Co*, cochlea; *SSC*, superior semicircular canal; *IAC*, internal auditory canal; *SPS*, superior petrosal sinus.

The Middle Fossa Approach to the Internal Auditory Canal

The standard MCF approach provides extradural exposure of the IAC and adjacent structures "from above," permitting preservation of hearing and middle ear function. Its utility is reduced in current practice, but it seems to be increasingly preferred in the management of CSF leak [60–63] and superior semicircular canal dehiscence. The approach exposes the entire middle fossa and is highly effective for the repair of tegmen mastoideum and tegmen tympani defects, particularly multiple noncontiguous and anteriorly located dehiscences. In cases of Bell's palsy with indication of surgical decompression or traumatic temporal bone fracture, the MCF approach provides the best access to the site of lesion.

Indications

- Vestibular neurectomy.
- Facial nerve decompression or repair following trauma.
- Geniculate ganglion tumors.
- CSF leak and meningoencephalocele.
- Repair of superior semicircular canal dehiscence.

Contraindications

- Nonserviceable hearing.
- This approach is relatively contraindicated in patients over 65 years of age. In the elderly, the dura is less resistant and may be torn during elevation from the temporal bone.

Surgical Steps (Right Side)

1. The patient is placed in the supine position, and the head is rotated to the contralateral side (see Figs. 11.14 and 11.15). The surgeon sits at the head of the table.
2. Incision (Fig. 11.16): An incision of 7–8 cm is made, starting just anterior to the helix base and extending superiorly, curving first posteriorly and then anteriorly. The temporalis muscle is then incised superiorly, curving first posteriorly and then anteriorly, thus allowing overlapping planes during closure.
3. Craniotomy (Fig. 11.17): A 3 × 4 cm craniotomy is carried out, lying one-third posterior and two-thirds anterior to the external auditory canal. Inferiorly, the craniotomy reaches the level of the zygomatic process in order to minimize the degree of temporal lobe retraction. The craniotomy is performed using a medium cutting burr at first and a diamond burr during the last part to avoid injuring the dura (fig. 4. 132). The bone flap is separated carefully from the dura and placed in a saline solution. The dura is further dissected from the edges of the craniotomy. The rough bone edges are trimmed using a diamond burr or a rongeur.
4. The middle fossa dura is elevated from posterior to anterior. Dural elevation is continued up to the level of the superior petrosal sinus. A middle fossa retractor is then applied (Fig. 11.18).
5. Identification of the internal auditory canal: The canal lies approximately in the line bisecting the angle formed by the greater superficial petrosal nerve and the arcuate eminence. Drilling along this line exposes the dura of the IAC (Fig. 11.19).
6. Further dissection continues laterally to identify Bill's bar, separating the facial nerve anteriorly from the superior vestibular nerve posteriorly (Fig. 11.20). The IAC dura is exposed over only 180 degrees of its circumference.
7. The dura of the canal is incised posteriorly over the superior vestibular nerve (Fig. 11.21).
8. The vestibulofacial anastomosis is identified and severed. A hook pointing posteriorly is used to avulse the superior vestibular nerve at the lateral end of the canal. At this point, the nerve is separated from the facial nerve by Bill's bar (Fig. 11.22).
9. The inferior vestibular nerve can now be identified beneath the avulsed superior vestibular nerve (Fig. 11.23).
10. A muscle graft with fibrin glue is used as a seal. The craniotomy flap is replaced, and the wound is closed in layers.

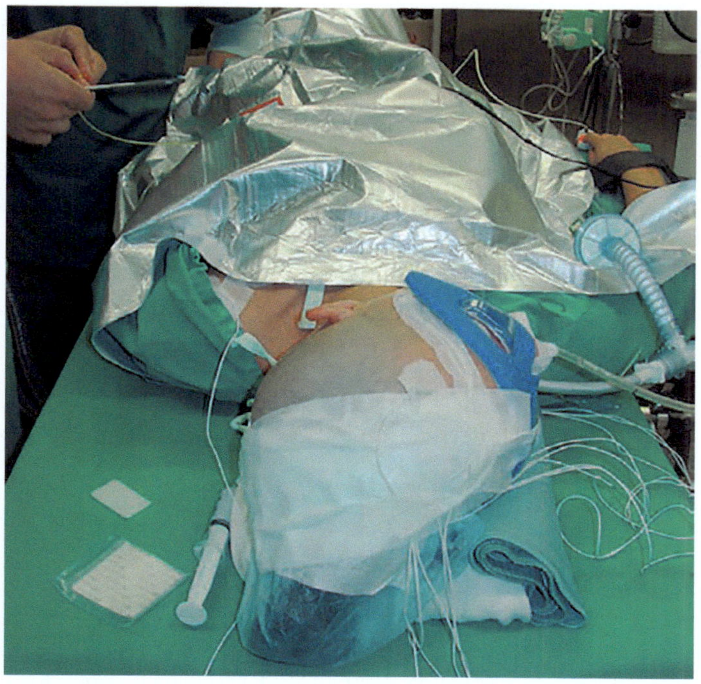

Fig. 11.14 The patient lies in the supine position on the bed. The patient's head is tilted to the contralateral side.

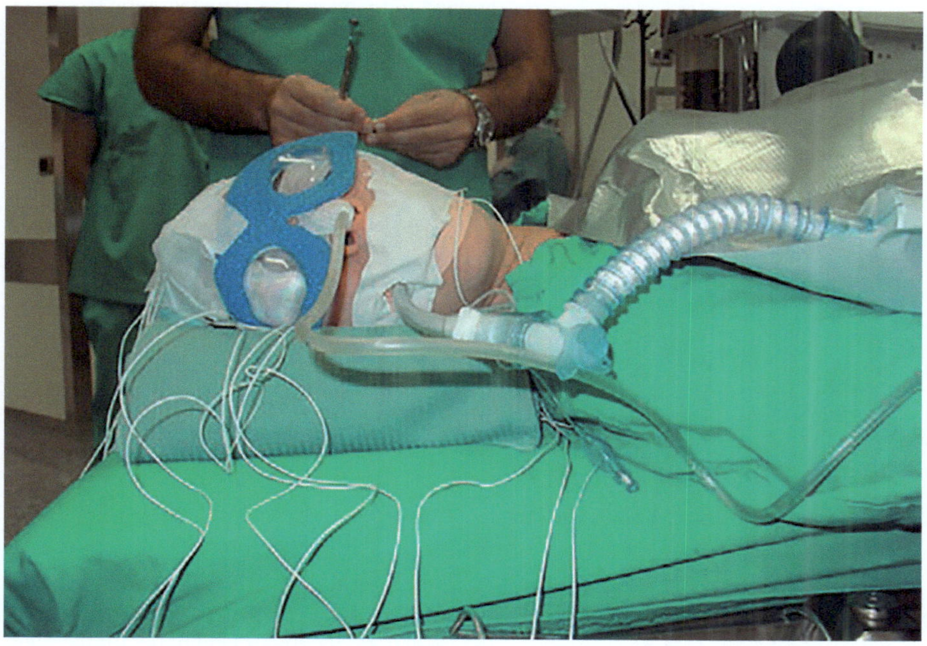

Fig. 11.15 Tilting angle of the head.

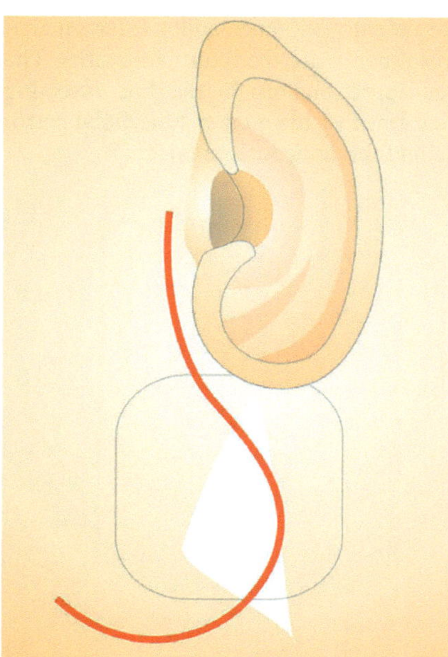

Fig. 11.16 A right-sided skin incision for the middle cranial fossa approach.

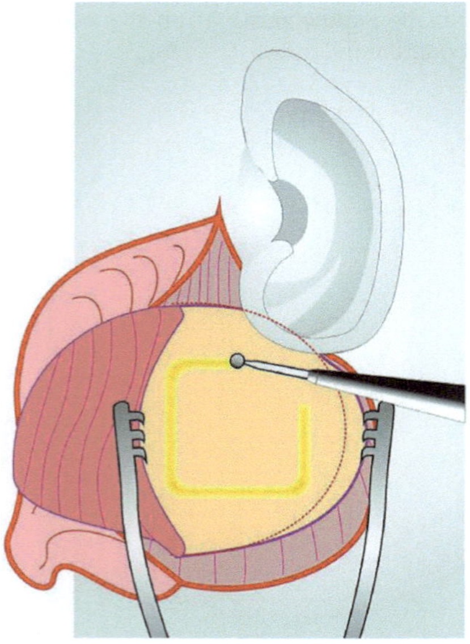

Fig. 11.17 The temporalis muscle is then incised superiorly, but curving first posteriorly and then anteriorly, thus allowing overlapping planes during closure. A large 4 × 5 cm craniotomy is performed.

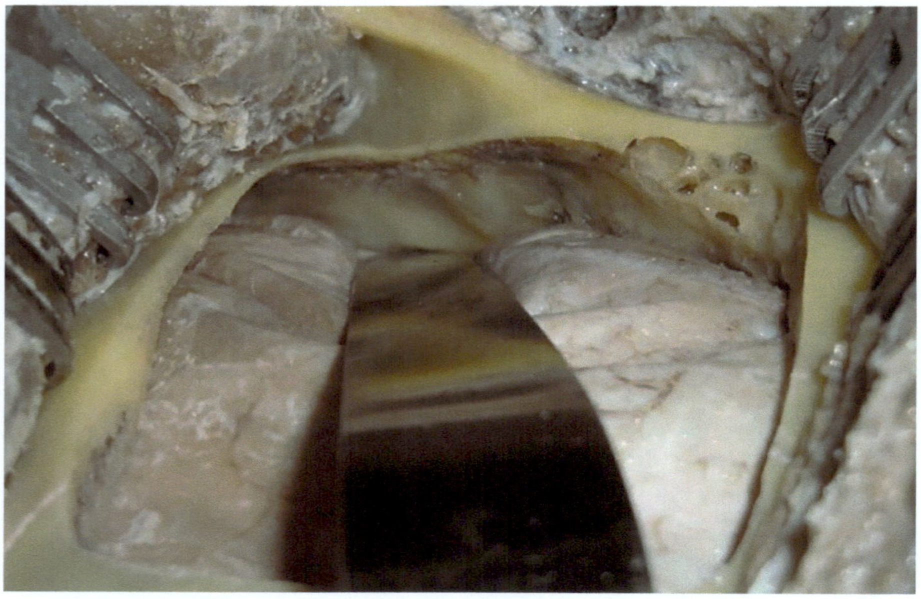

Fig. 11.18 The craniotomy flap has been elevated. The middle cranial fossa retractor has been applied to maintain the elevated dura.

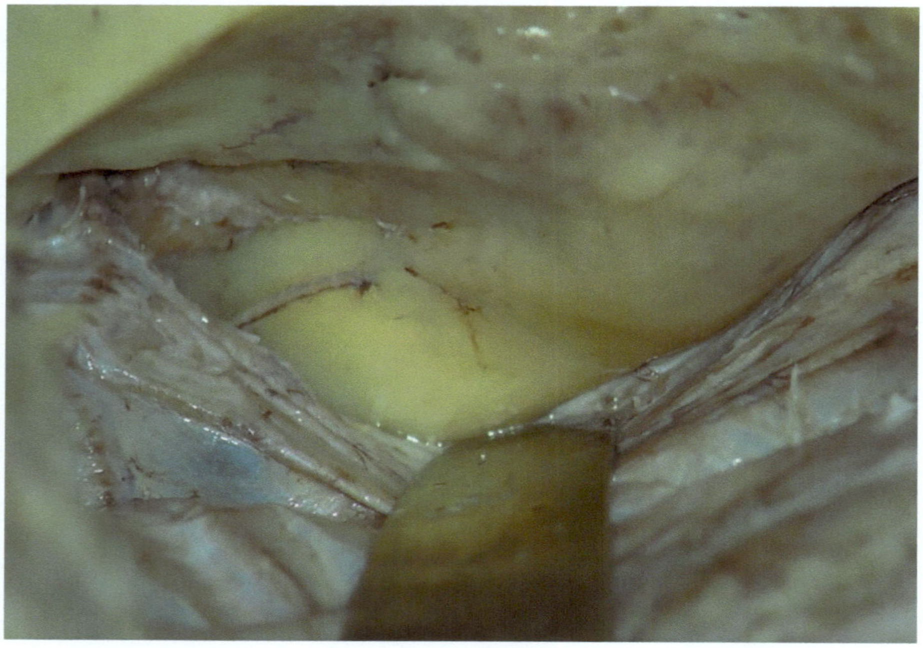

Fig. 11.19 After completely elevating the middle fossa dura, the major landmarks for the identification of the internal auditory canal can be seen. The arcuate eminence and greater superficial petrosal nerve are the main landmarks. The canal lies approximately in the line bisecting the angle formed by the greater superficial petrosal nerve and the arcuate eminence.

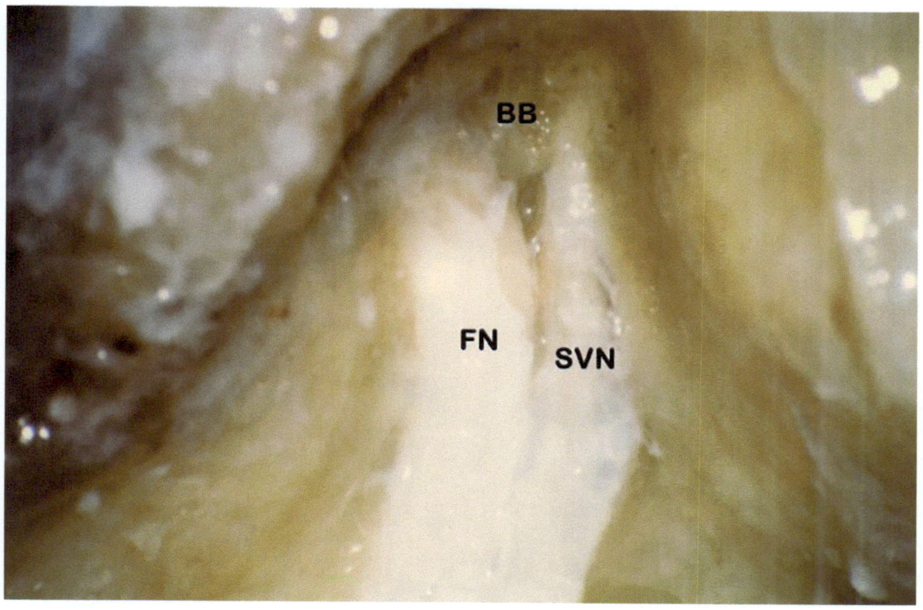

Fig. 11.20 Drilling along the line bisecting the angle exposes the dura of the internal auditory canal. Further dissection continues laterally to identify Bill's bar, separating the facial nerve anteriorly from the superior vestibular nerve posteriorly. The internal auditory canal dura is exposed over only 180 degrees of its circumference. *BB*, Bill's bar; *FN*, facial nerve; *SVN*, superior vestibular nerve.

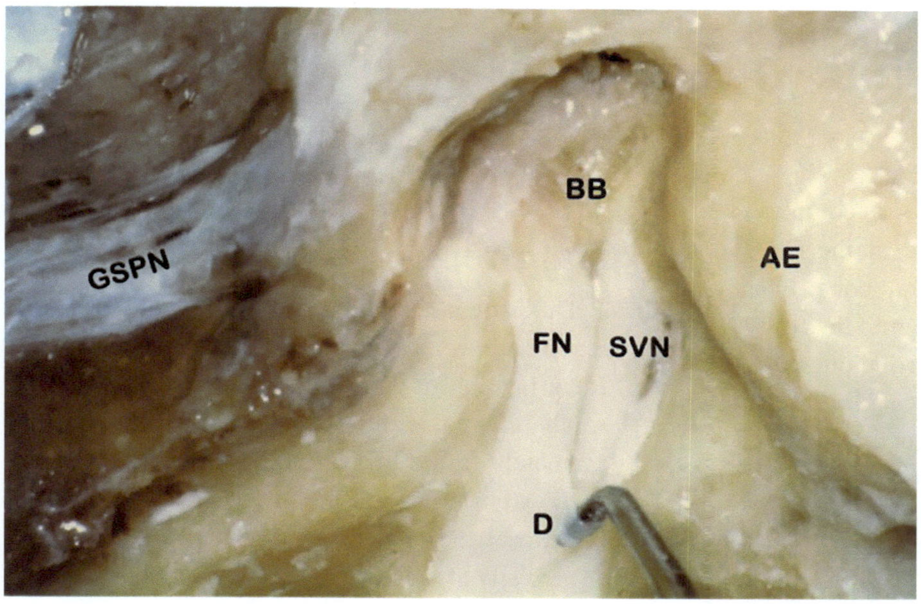

Fig. 11.21 The dura of the canal is incised posteriorly over the superior vestibular nerve. *BB*, Bill's bar; *AE*, arcuate eminence; *FN*, facial nerve; *SVN*, superior vestibular nerve; *D*, dura; *GSPN*, greater superficial petrosal nerve.

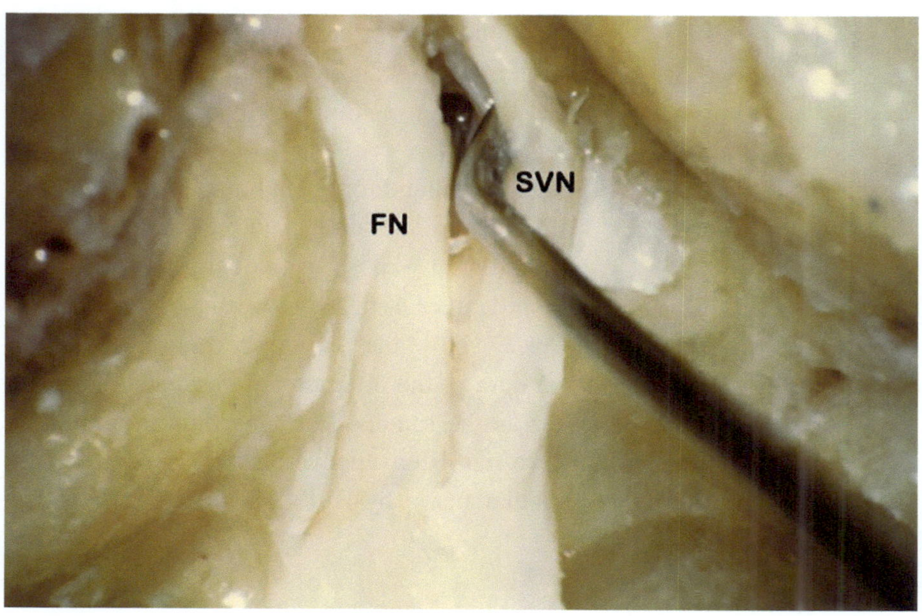

Fig. 11.22 A hook pointing posteriorly is used to avulse the superior vestibular nerve at the lateral end of the canal. At this point, the nerve is separated from the facial nerve by Bill's bar. *SVN*, superior vestibular nerve; *FN*, facial nerve.

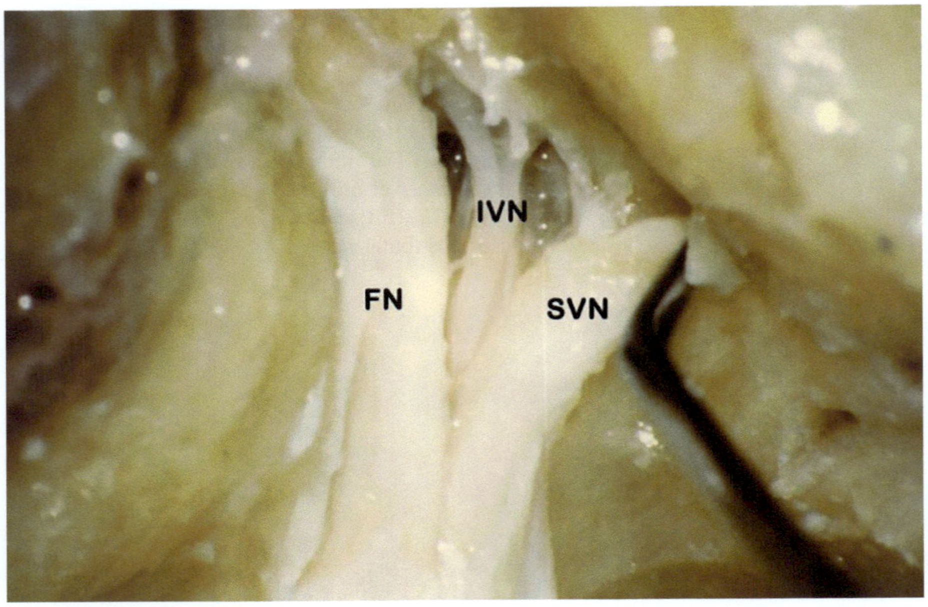

Fig. 11.23 The inferior vestibular nerve can now be identified beneath the avulsed superior vestibular nerve. *IVN*, inferior vestibular nerve; *FN*, facial nerve; *SVN*, superior vestibular nerve.

Enlarged Middle Cranial Fossa Approach

The enlarged middle fossa approach allows a complete exposure of the IAC but a limited exposure of the CPA through the superior surface of the temporal bone. The enlarged approach differs from the classic middle fossa approach, in which wider bone removal is carried out anteriorly and posteriorly to the internal auditory canal. As seen in Fig. 11.9, two areas devoid of important structures can be identified in relation to the internal auditory canal: a posterior triangular area, bounded by the superior semicircular canal, and an anterior rhomboidal area, bounded by the cochlea, internal carotid artery, and the posterior border of the Gasserian ganglion. This approach allows hearing preservation as well as a complete exposure of the full length of the internal auditory canal from the porus to the fundus.

Indications

- Acoustic neurinomas that are localized in the internal auditory canal or are less than 0.5 cm in the CPA when the aim is to preserve hearing.
- Facial nerve neuromas.
- Petrous bone cholesteatoma (supralabyrinthine or apical).
- Other tumors of the internal auditory canal.
- Cholesterol granuloma.

Contraindications

- Acoustic neurinomas larger than 0.5 cm in extrameatal diameter.
- Patients over 65 years of age.

Surgical Steps

1–4. The patient's position, incision, craniotomy, and dural elevation are as mentioned earlier.

5. Identification of the internal auditory canal:
    - The internal auditory canal corresponds approximately to the line bisecting the angle made by the arcuate eminence and the greater superficial petrosal nerve (Fig. 11.24). Drilling is started medially at the level of the superior petrosal sinus to identify the porus of the internal auditory canal. A wide area surrounding the canal is drilled.
    - Once the porus has been identified, three-quarters of the canal circumference is skeletonized, leaving a thin shell of bone over it (Fig. 11.25).
    - The internal auditory canal dura is followed laterally toward the fundus, with the amount of circumferential exposure gradually decreasing to reach degrees at the fundus (Fig. 11.26).

6. Fig. 11.27 shows the general view at the end of the approach. The dura of the internal auditory canal has been exposed as far as the fundus. The labyrinthine portion of the facial nerve has been uncovered, separated from the superior vestibular nerve by Bill's bar. The access to the

posterior fossa dura has been greatly enlarged by drilling the triangular and rhomboidal areas posterior and anterior to the internal auditory canal, respectively.

7. The posterior fossa dura is opened at the posterior edge of the canal, taking care not to injure a nearby anterior inferior cerebellar artery. This leads to the escape of cerebrospinal fluid, decreasing the intracranial tension (Fig. 11.28).

8. Fig. 11.29 shows the various structures seen after opening the posterior fossa dura. The anterior inferior cerebellar artery, looping lateral to the facial and vestibulocochlear nerves, is seen.

9. At a higher magnification, the relationship between the anterior inferior cerebellar artery and the facial and vestibulocochlear nerves is well observed (Fig. 11.30). The overall area and the view of the CPA afforded by the approach are limited in comparison with other approaches. Note also the poor control of the lower cranial nerves, which are concealed here by the flocculus.

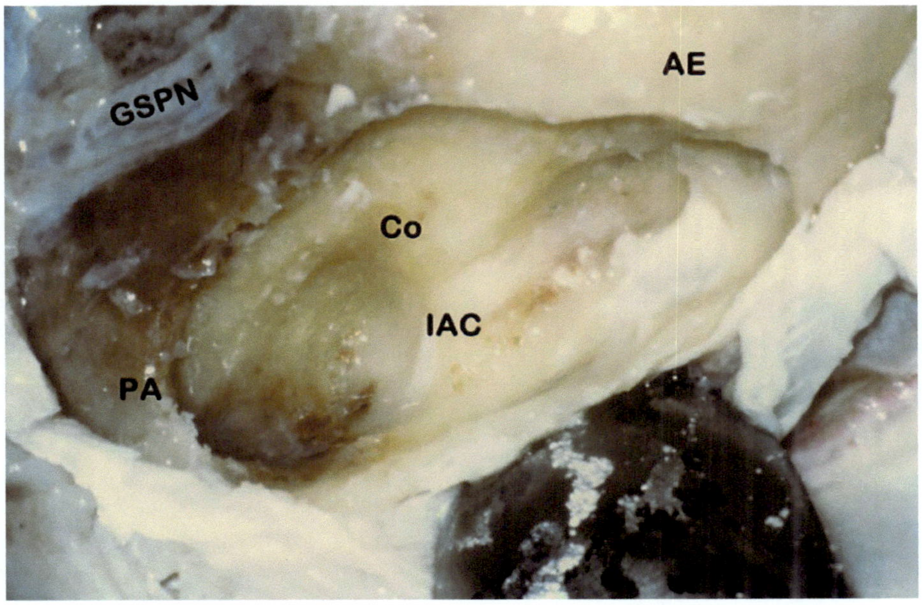

Fig. 11.24 Drilling is started medially at the level of the superior petrosal sinus to identify the porus of the internal auditory canal. A wide area surrounding the canal is drilled. *GSPN*, greater superficial petrosal nerve; *Co*, cochlea; *PA*, petrous apex; *IAC*, internal auditory canal; *AE*, arcuate eminence.

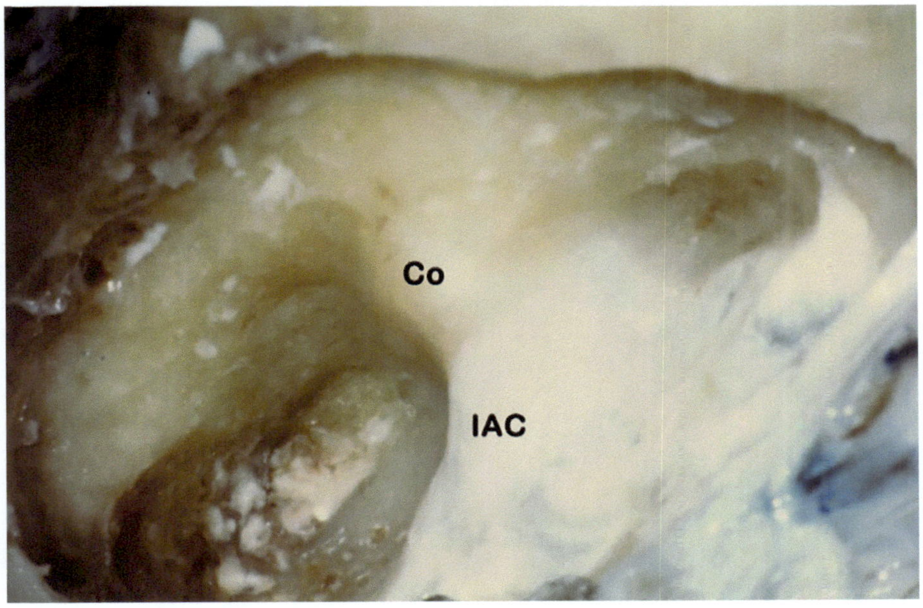

Fig. 11.25 Once the porus has been identified, three-quarters of the canal circumference is skeletonized, leaving a thin shell of bone over it. *Co*, cochlea; *IAC*, internal auditory canal.

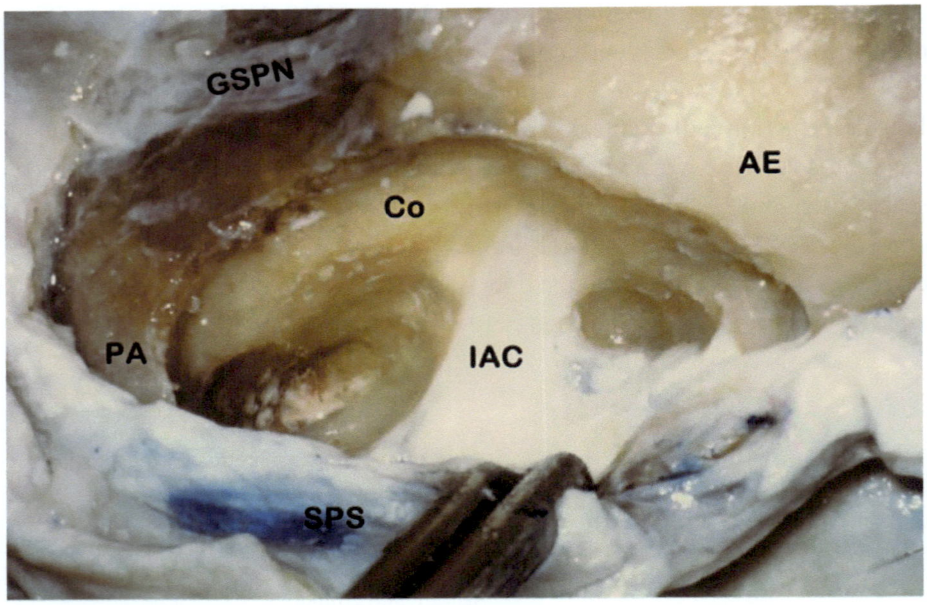

Fig. 11.26 The internal auditory canal dura is followed laterally toward the fundus. *GSPN*, greater superficial petrosal nerve; *Co*, cochlea; *PA*, petrous apex; *IAC*, internal auditory canal; *AE*, arcuate eminence; *SPS*, superior petrosal sinus.

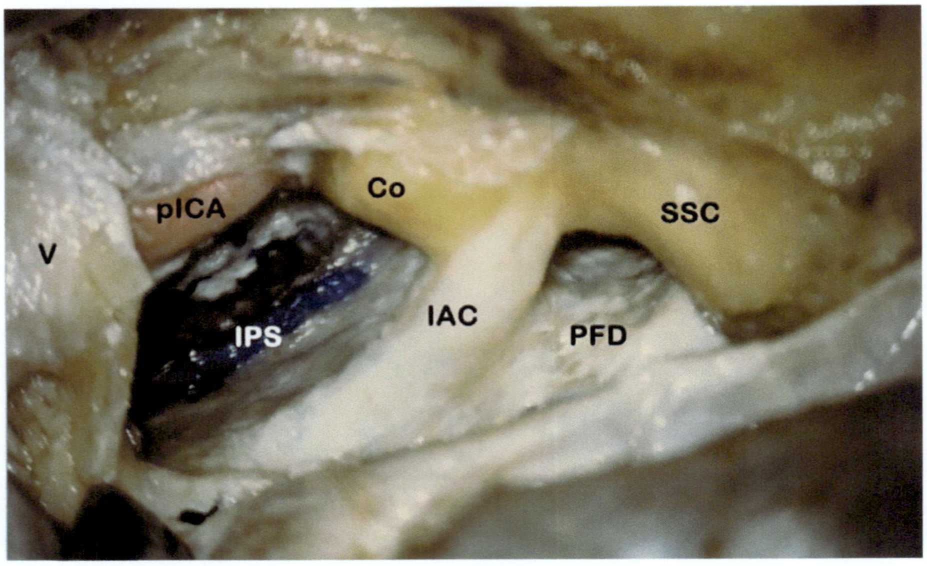

Fig. 11.27 The general view at the end of the approach. *V*, trigeminal nerve; *pICA*, petrous segment of internal carotid artery; *Co*, cochlea; *IAC*, internal auditory canal; *PFD*, posterior fossa dura; *SSC*, superior semicircular canal; *IPS*, inferior petrosal sinus.

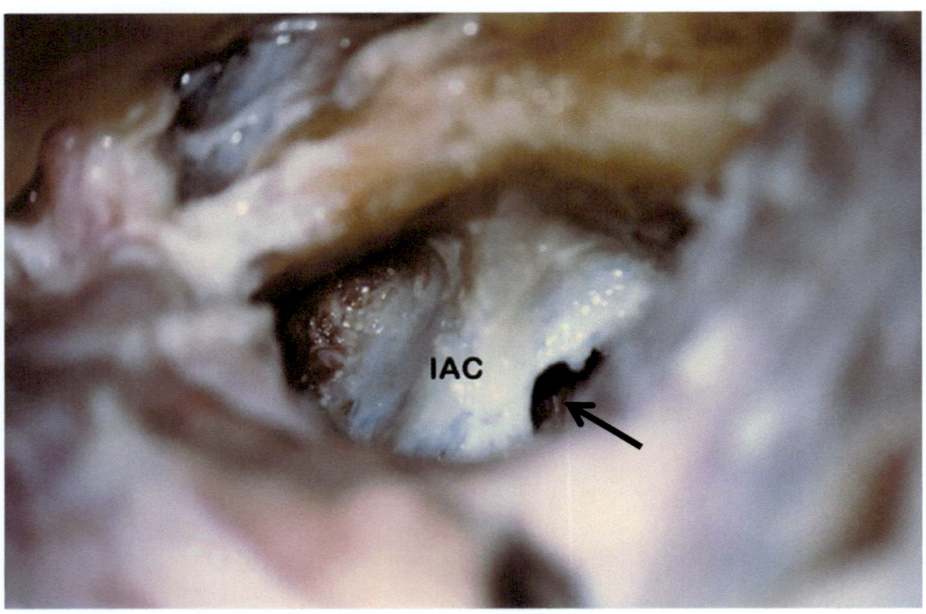

Fig. 11.28 The posterior fossa dura is opened at the posterior edge of the canal (*arrow*). This leads to the escape of cerebrospinal fluid, decreasing the intracranial tension. *IAC*, internal auditory canal.

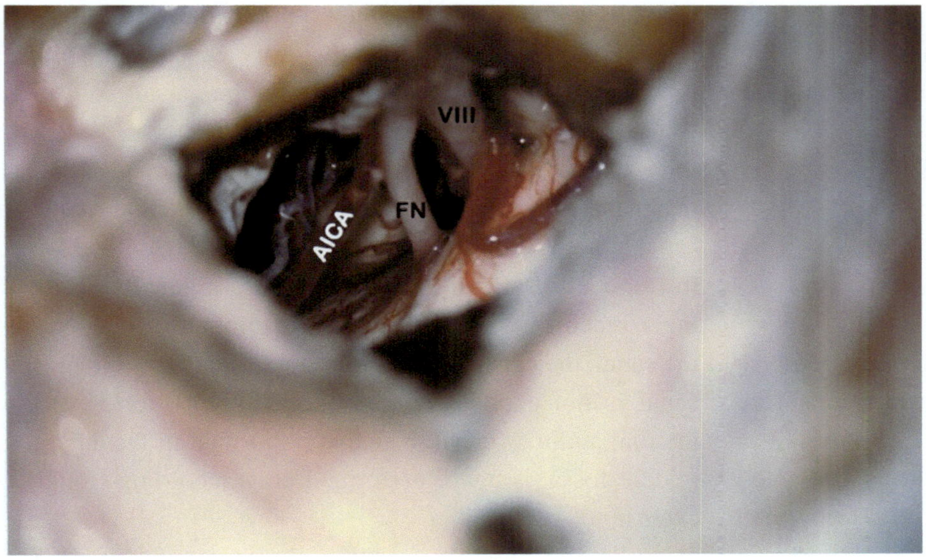

Fig. 11.29 The intracranial structures seen after opening the posterior fossa dura. The anterior inferior cerebellar artery, looping lateral to the facial and vestibulocochlear nerves, is seen. *AICA*, anterior inferior cerebellar artery; *FN*, facial nerve; *VIII*, vestibulocochlear nerve.

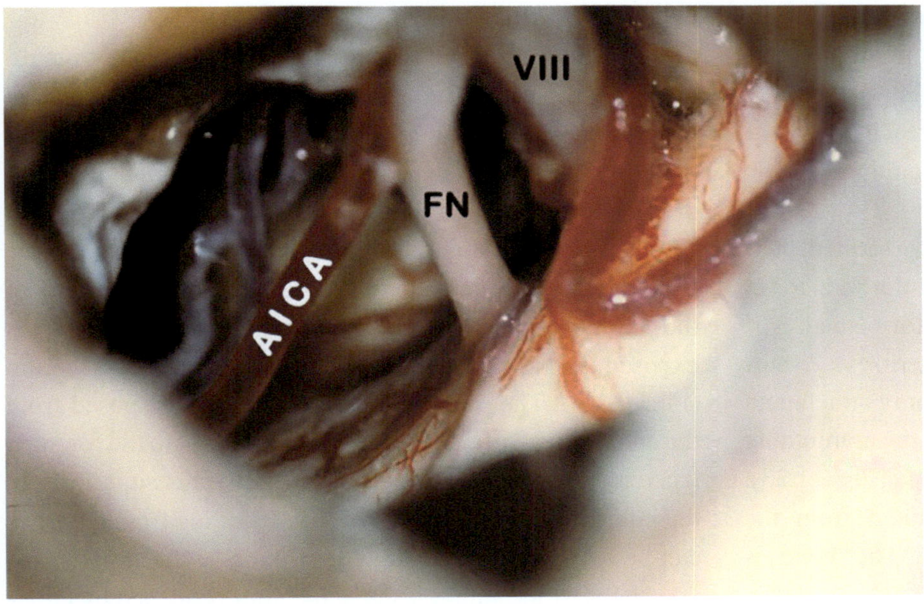

Fig. 11.30 At a higher magnification, the relationship between the anterior inferior cerebellar artery and the facial and vestibulocochlear nerves is well observed. *AICA*, anterior inferior cerebellar artery; *FN*, facial nerve; *VIII*, vestibulocochlear nerve.

The Middle Fossa Transpetrous Approach

This approach, also known as the extended middle cranial fossa approach (ECF), was initially described by Bochenek and Kukwa [4], and it was modified by Kawase [11]. He presented a modified approach that included dissection of the petrous apex initially with the purpose of treating aneurysms of the basilar artery. The indications of the technique widened later on to include the treatment of selected VS, petroclival meningiomas, and other tumors involving the petrous apex and clivus [64–67]. This is a valuable tool in the treatment of such lesions as it provides wide exposure of the CPA, prepontine cistern, anterior petrous apex, petrous carotid artery, Meckel's cave, cavernous sinus, mid and upper clivus [66,68–70]. Excision of larger tumors of the petrous apex and prepontine cistern can be achieved without the need for facial nerve rerouting as in the transcochlear approach [71]. The associated morbidity is low in experienced hands; the reported postoperative facial nerve function is excellent, and preservation of serviceable hearing can reach a 50% success rate [66]. Limitations to the technique are the need for temporal lobe retraction, lesions posterior to the IAC are difficult to access, requiring a combined petrosal approach [71], and because the approach is intended for larger lesions or those with challenging access, total tumor resection success rate varies between 50% and 90%. [66,70,72–74]

Indications

- Petrous bone cholesteatoma of the apical type, with serviceable hearing.

- Petrous apex or petroclival meningioma, with limited inferior extension and serviceable hearing.
- Some cases of upper clival and petrous apex chordomas, chondromas, and chondrosarcomas.
- Other selected tumors: trigeminal schwannoma, epidermoid cyst, glioma, and neurofibroma.

Surgical Steps

1. The patient is placed in the supine position, and the head is rotated to the contralateral side. The surgeon sits at the head of the table (see Figs. 11.14 and 11.15).

2. An incision of 7–8 cm is made, starting just anterior to the helix base and extending superiorly, curving first posteriorly and then anteriorly (Fig. 11.31).

3. The flap of skin and subcutaneous tissue is elevated above the level of the superficial fascia of the temporalis muscle. An incision is made in the temporalis muscle using monopolar cutting diathermy. Harvesting the temporalis fascia is useful. The incision should reach the bone. Using a Lempert periosteal elevator, the musculoperiosteum is elevated and retracted (Fig. 11.32).

4. A large 4 × 5 cm craniotomy is performed. The classical middle fossa craniotomy lies two-thirds anterior and one-third posterior to the external auditory canal. The middle fossa craniotomy extends more anteriorly. A burr hole is performed first, and then the craniotomy attachment of the Midas Rex is used. The lower edge of the craniotomy should be at the level of the base of the zygoma (Fig. 11.33).

5. The bone flap is carefully separated from the underlying dura using a septal dissector. The bone is removed and placed in saline solution.

6. The middle fossa dura is detached from the superior surface of the temporal bone. Dural elevation is carried from posterior to anterior. This avoids any traction on the greater superficial petrosal nerve and resultant damage to the facial nerve if a dehiscent geniculate ganglion is present. Dural elevation should reach medially to the level of the superior petrosal sinus. Anteriorly, the middle meningeal artery and the three branches of the trigeminal nerve can be identified extradurally. Posteriorly, good exposure of the arcuate eminence should be obtained. A middle fossa retractor is next applied (Fig. 11.34).

7. Identification of the internal auditory canal: The internal auditory canal corresponds approximately to the line bisecting the angle made by the arcuate eminence and greater superficial petrosal nerve. Drilling is started medially at the level of the superior petrosal sinus, to identify the porus of the internal auditory canal (Fig. 11.35). A wide area of bone surrounding the canal is drilled (Figs. 11.36–11.38). Once the porus has been identified, three-quarters of the canal circumference is skeletonized, leaving a thin shell of bone over it. The internal auditory canal dura is followed laterally toward the fundus, with the amount of circumferential exposure gradually decreasing to reach 180° at the fundus. If the tumor does not involve the canal, the canal is only skeletonized, and the dura of the canal is not exposed. Extensive drilling posterior to the internal auditory canal, which is performed in the enlarged middle fossa, is again performed only according to tumor extension.

8. The middle meningeal artery is identified (Fig. 11.39) and coagulated using bipolar coagulation and cut (Fig. 11.40).

9. Anteriorly, the mandibular branch of the trigeminal nerve and the greater superficial petrosal nerve are identified and cut (Figs. 11.41 and 11.42). The Gasserian ganglion is reflected, exposing the remaining triangular area of the petrous apex (Fig. 11.43) which is drilled next (Fig. 11.44). Bone removal extends as far inferiorly as the level of the inferior petrosal sinus. The whole length of the horizontal portion of the internal carotid artery is exposed as far as its exit through the foramen lacerum (Fig. 11.45). The posterior fossa dura is thus exposed from the anterior wall of the internal auditory canal to the foramen lacerum and from the superior petrosal sinus superiorly to the inferior petrosal sinus inferiorly. Bleeding from the latter can be controlled with Surgicel packing.

10. For intradural lesions, the dura is opened by creating an inferiorly based flap. The dashed lines in Fig. 11.46 show the dural incisions. If more exposure is needed, the superior petrosal sinus can also be transected after bipolar coagulation or clipping.

11. Surgical Anatomy after Opening of the Dura: Figs. 11.47–11.52 show the view of the CPA and prepontine cistern as viewed after opening the dura. For cases extending into the middle fossa, the middle fossa dura can be opened, and the approach can be further enhanced.

12. At the end of the approach, the surgical defect is obliterated by abdominal fat, and the craniotomy is replaced after effective hemostasis. Surgicel may be placed between the replaced bone and the middle fossa dura to ensure proper hemostasis. The temporalis muscle is sutured with Vicryl, and the skin is closed in layers.

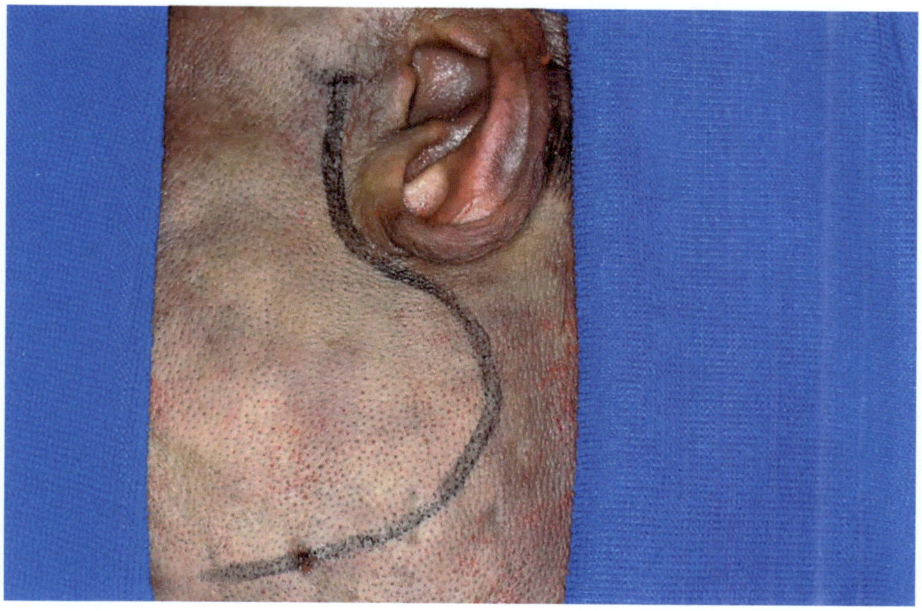

Fig. 11.31 An incision of 7–8 cm is made, starting just anterior to the helix base and extending superiorly, curving first posteriorly and then anteriorly.

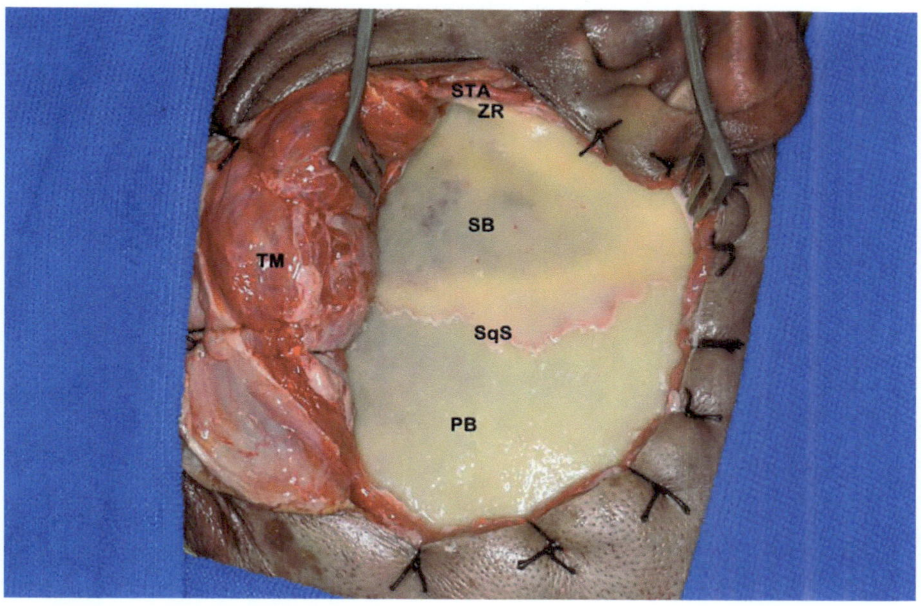

Fig. 11.32 The flap of skin and subcutaneous tissue is elevated above the level of the superficial fascia of the temporalis muscle. An incision is made in the temporalis muscle using monopolar cutting diathermy, and the musculoperiosteum is elevated and retracted anteriorly. *STA*, superficial temporal artery; *ZR*, zygomatic root; *SB*, squamosal part of the temporal bone; *TM*, temporalis muscle (retracted); *SqS*, squamous suture; *PB*, parietal bone.

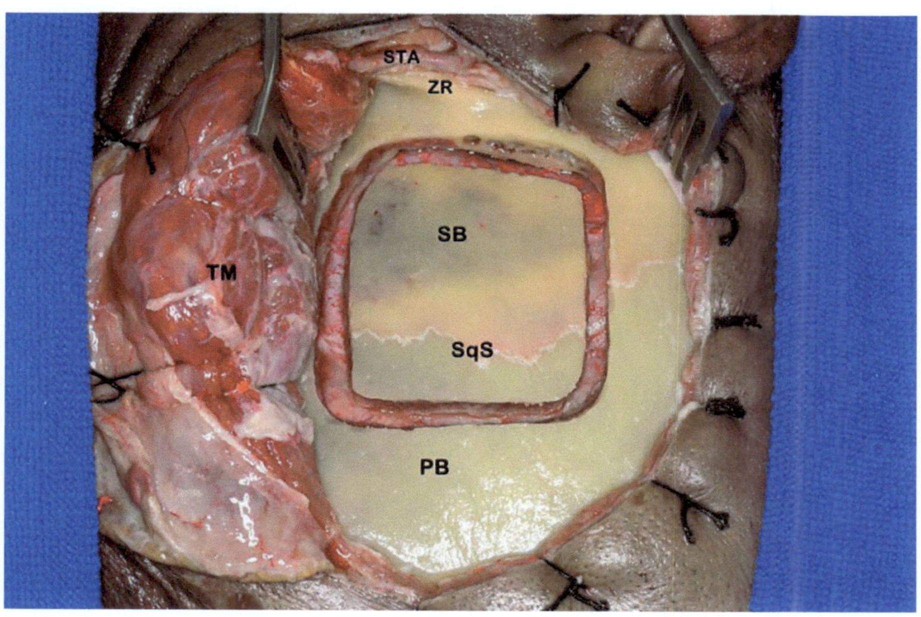

Fig. 11.33 A large 4 × 5 cm craniotomy is performed. *STA*, superficial temporal artery; *ZR*, zygomatic root; *SB*, squamosal part of the temporal bone; *SqS*, squamous suture; *PB*, parietal bone; *TM*, temporalis muscle(retracted).

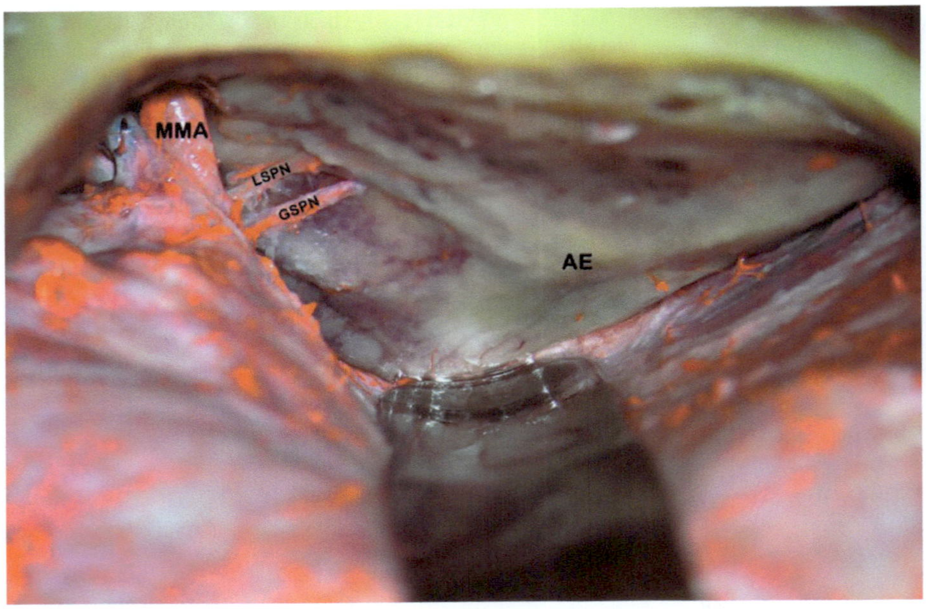

Fig. 11.34 The middle fossa dura is detached from the superior surface of the temporal bone. Dural elevation is carried from posterior to anterior. Dural elevation should reach medially to the level of the superior petrosal sinus. Anteriorly, the middle meningeal artery and the three branches of the trigeminal nerve can be identified extradurally. Posteriorly, good exposure of the arcuate eminence should be obtained. A middle fossa retractor is next applied. *MMA*, middle meningeal artery; *LSPN*, lesser superficial petrosal nerve; *GSPN*, greater superficial petrosal nerve; *AE*, arcuate eminence.

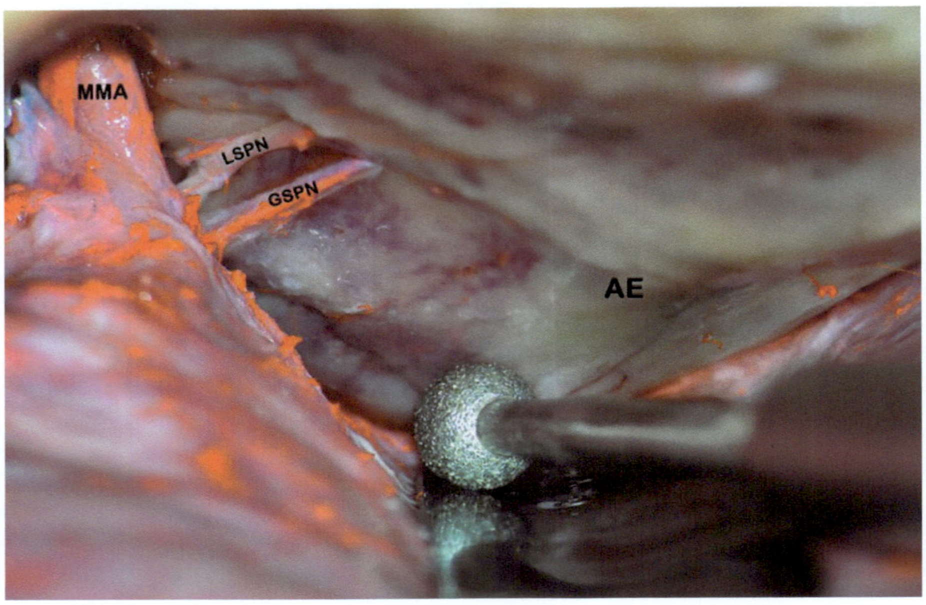

Fig. 11.35 Drilling is started medially at the level of the superior petrosal sinus to identify the porus of the internal auditory canal. *MMA*, middle meningeal artery; *LSPN*, lesser superficial petrosal nerve; *GSPN*, greater superficial petrosal nerve; *AE*, arcuate eminence.

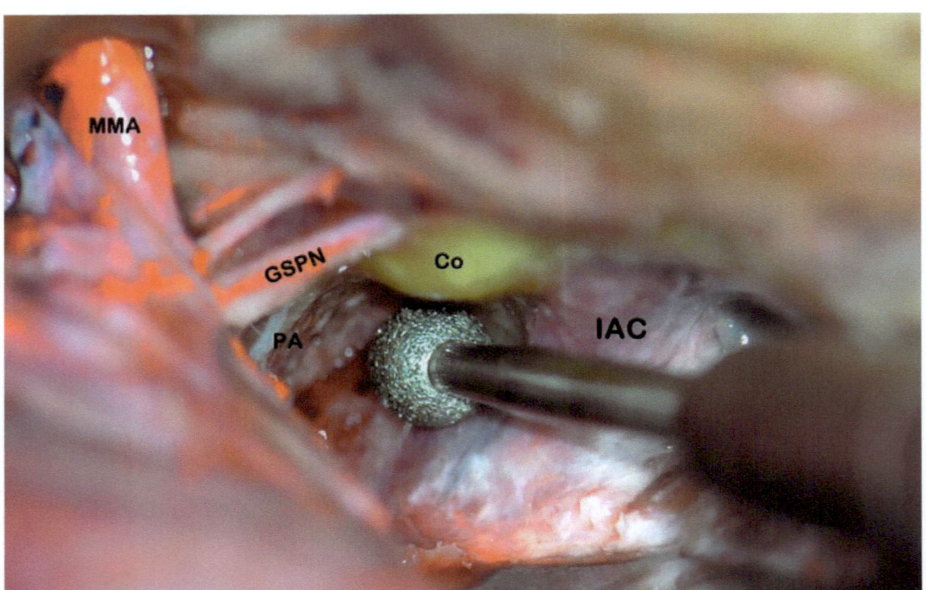

Fig. 11.36 A wide area of bone surrounding the canal is drilled. *MMA*, middle meningeal artery; *GSPN*, greater superficial petrosal nerve; *Co*, cochlea; *PA*, petrous apex; *IAC*, internal auditory canal.

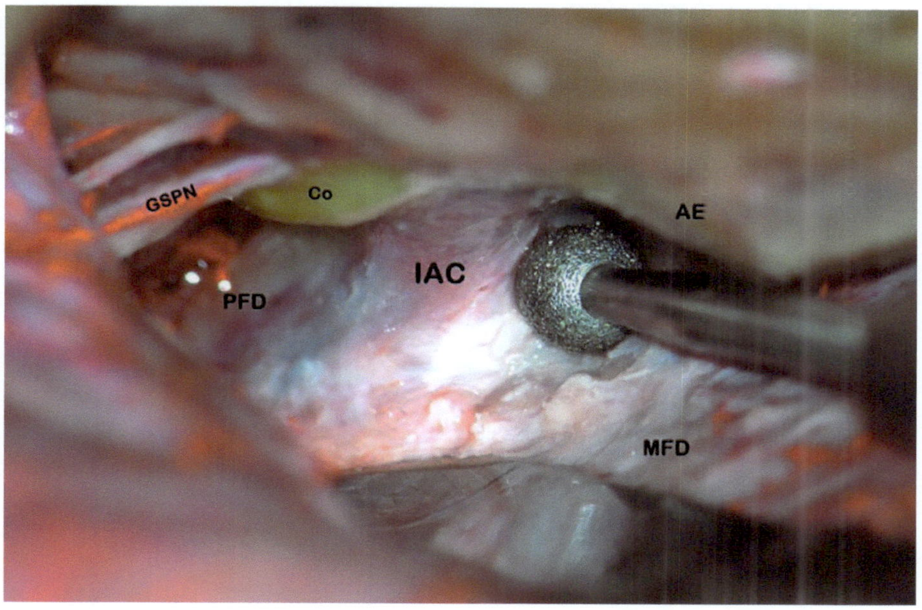

Fig. 11.37 The area posterior to the internal auditory canal is drilled next. *GSPN*, greater superficial petrosal nerve; *PFD*, posterior fossa dura; *Co*, cochlea; *IAC*, internal auditory canal; *AE*, arcuate eminence; *MFD*, middle fossa dura.

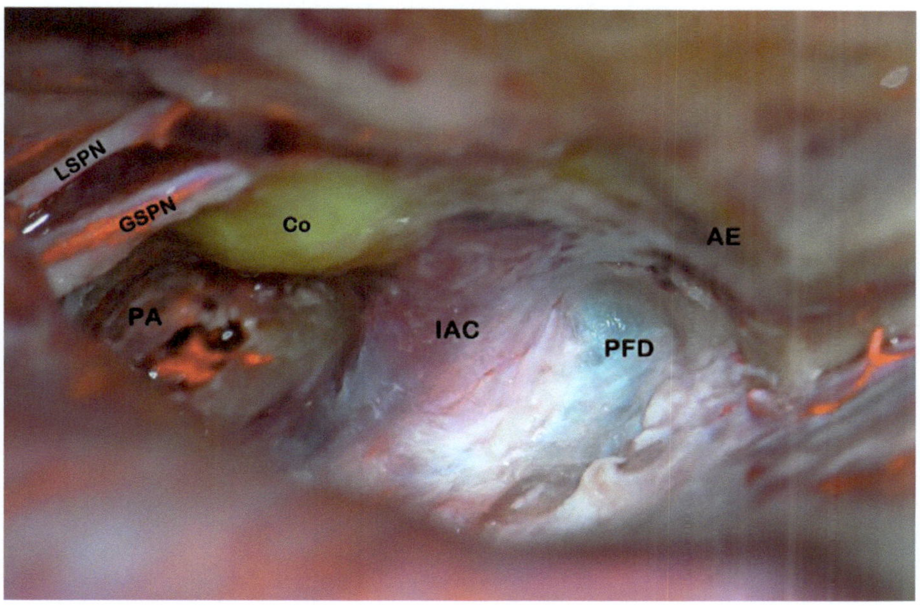

Fig. 11.38 Further drilling leads to visibility of the posterior fossa dura. *LSPN*, lesser superficial petrosal nerve; *GSPN*, greater superficial petrosal nerve; *Co*, cochlea; *PA*, petrous apex; *IAC*, internal auditory canal; *PFD*, posterior fossa dura; *AE*, arcuate eminence.

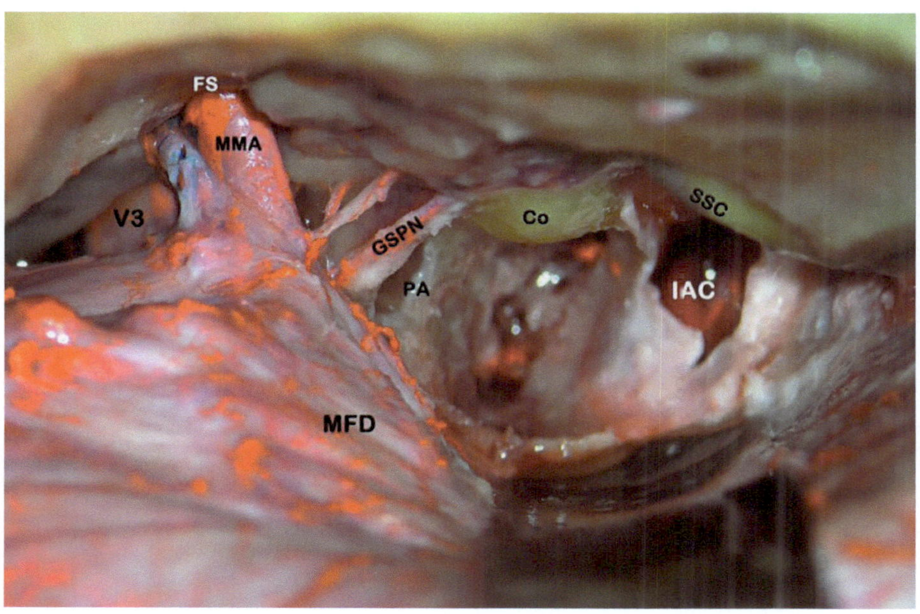

Fig. 11.39 The middle meningeal artery is identified. *FS*, foramen spinosum; *MMA*, middle meningeal artery; *V3*, mandibular nerve; *MFD*, middle fossa dura; *GSPN*, greater superficial petrosal nerve; *PA*, petrous apex; *Co*, cochlea; *SSC*, superior semicircular canal; *IAC*, internal auditory canal.

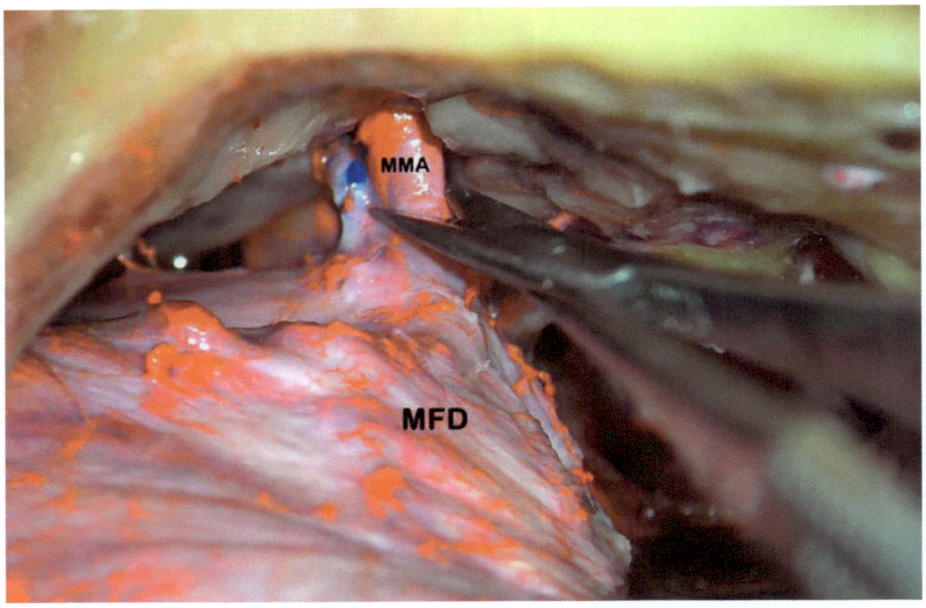

Fig. 11.40 The middle meningeal artery is coagulated using bipolar coagulation and cut. *MMA*, middle meningeal artery; *MFD*, middle fossa dura.

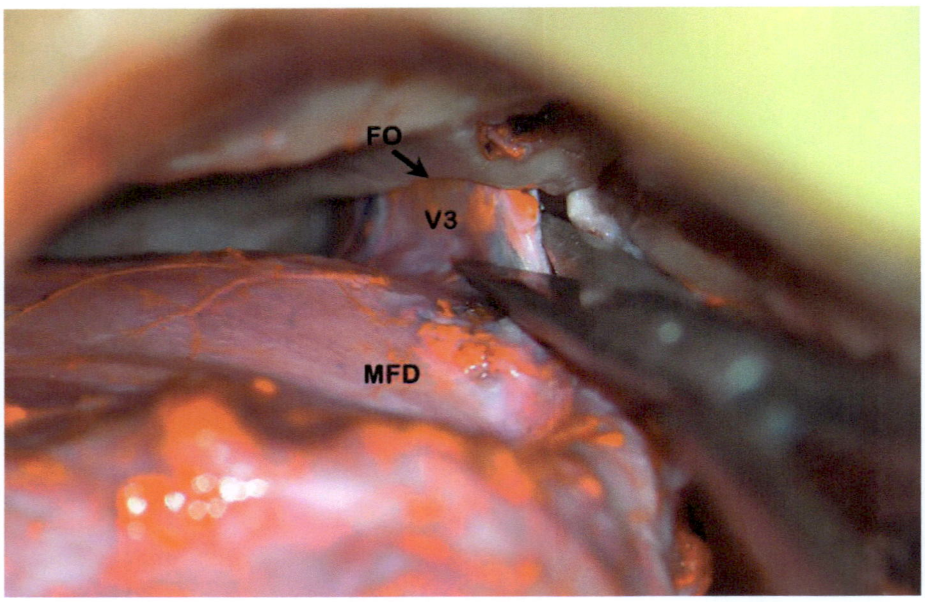

Fig. 11.41 Anteriorly, the mandibular branch of the trigeminal nerve is identified and cut. *FO*, foramen ovale; *V3*, mandibular nerve; *MFD*, middle fossa dura.

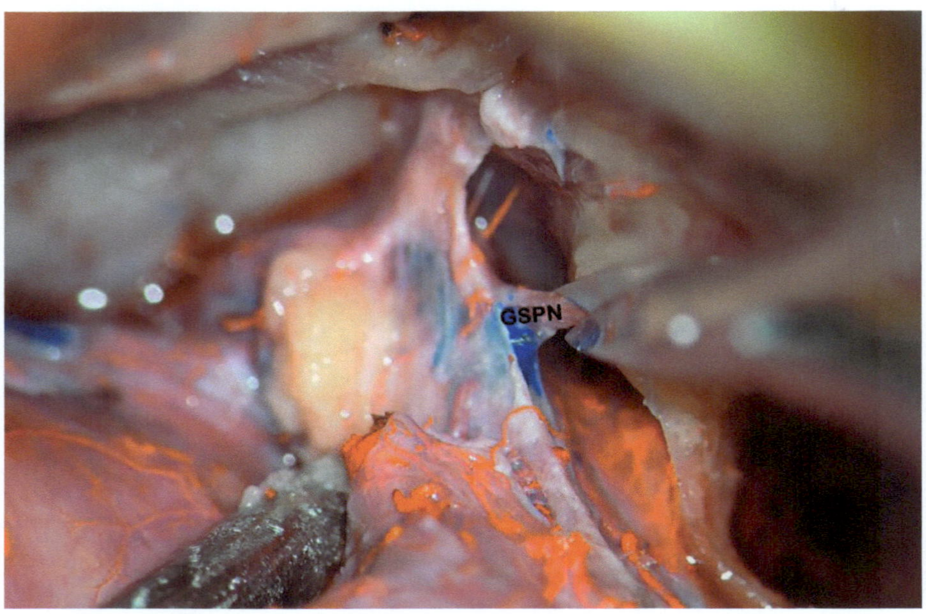

Fig. 11.42 The greater superficial petrosal nerve is cut. *GSPN*, greater superficial petrosal nerve.

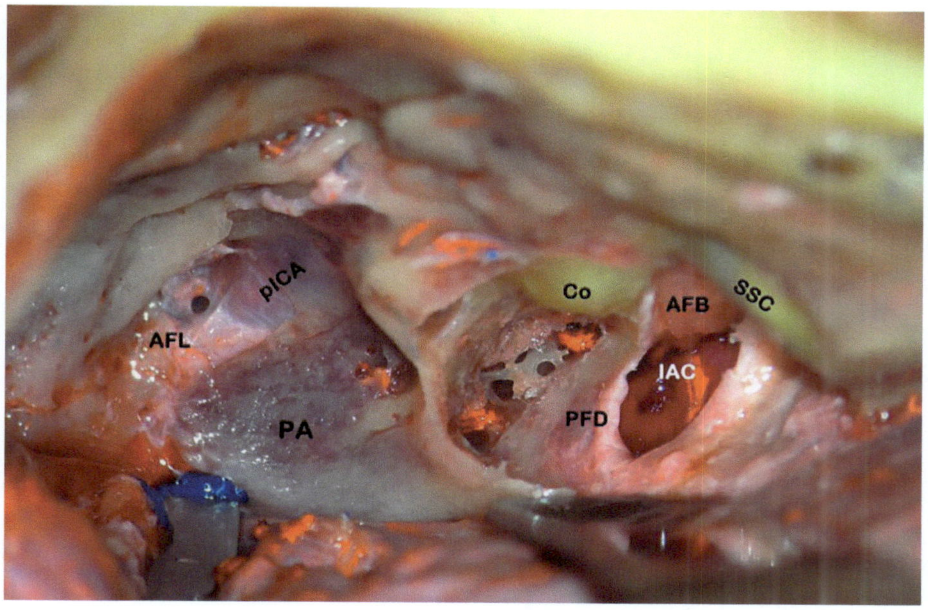

Fig. 11.43 The Gasserian ganglion is reflected, exposing the remaining triangular area of the petrous apex. SSC, superior semicircular canal; *AFB*, acoustic facial bundle; *IAC*, internal auditory canal; *PFD*, posterior fossa dura; *Co*, cochlea; *PA*, petrous apex; *pICA*, petrous segment of internal carotid artery; *AFL*, anterior foramen lacerum.

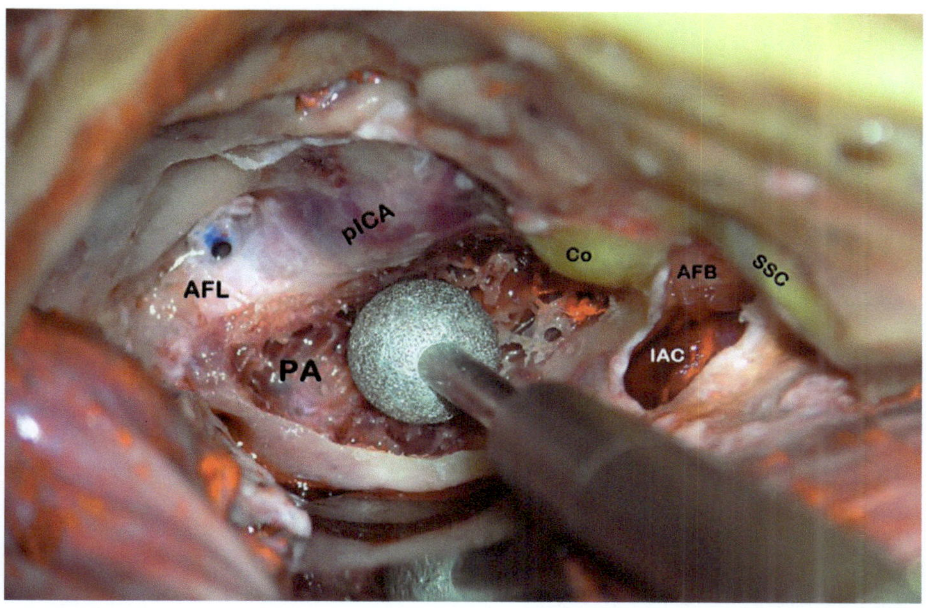

Fig. 11.44 The remaining triangular area of the petrous apex is drilled. *SSC*, superior semicircular canal; *AFB*, acoustic facial bundle; *IAC*, internal auditory canal; *Co*, cochlea; *pICA*, petrous segment of internal carotid artery; *AFL*, anterior foramen lacerum; *PA*, petrous apex.

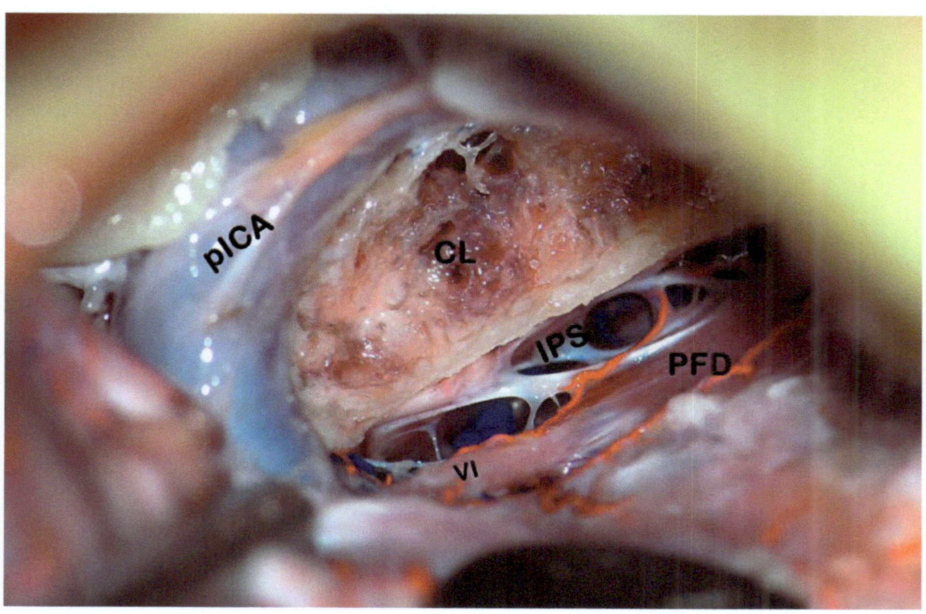

Fig. 11.45 Bone removal extends as far inferiorly as the level of the inferior petrosal sinus. The whole length of the horizontal portion of the internal carotid artery is exposed as far as its exit through the foramen lacerum. *pICA*, petrous segment of internal carotid artery; *Cl*, clivus; *IPS*, inferior petrosal sinus; *VI*, abducent nerve; *PFD*, posterior fossa dura.

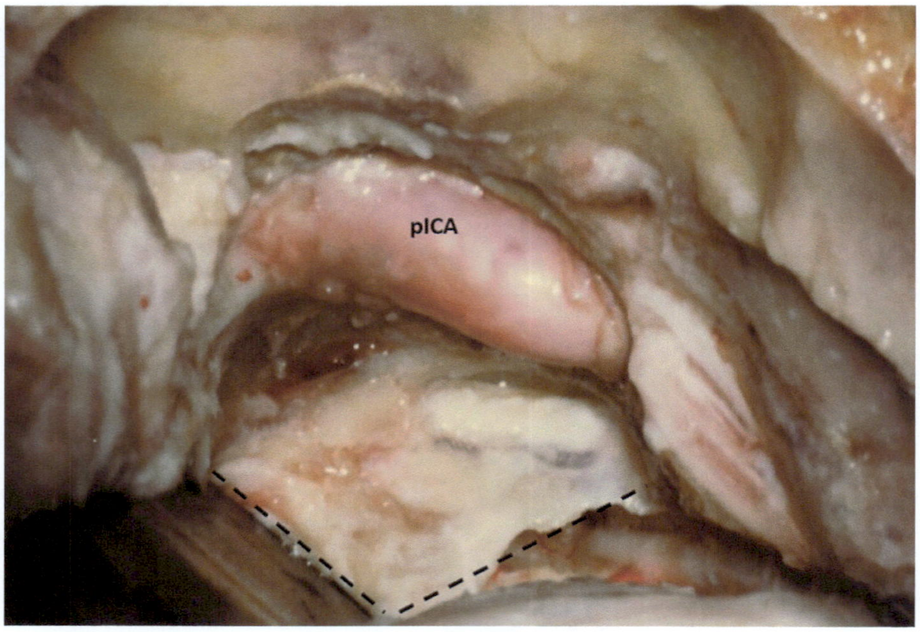

Fig. 11.46 The dura is opened by creating an inferiorly based flap. The *dashed lines* show the dural incisions. *pICA*, petrous segment of internal carotid artery.

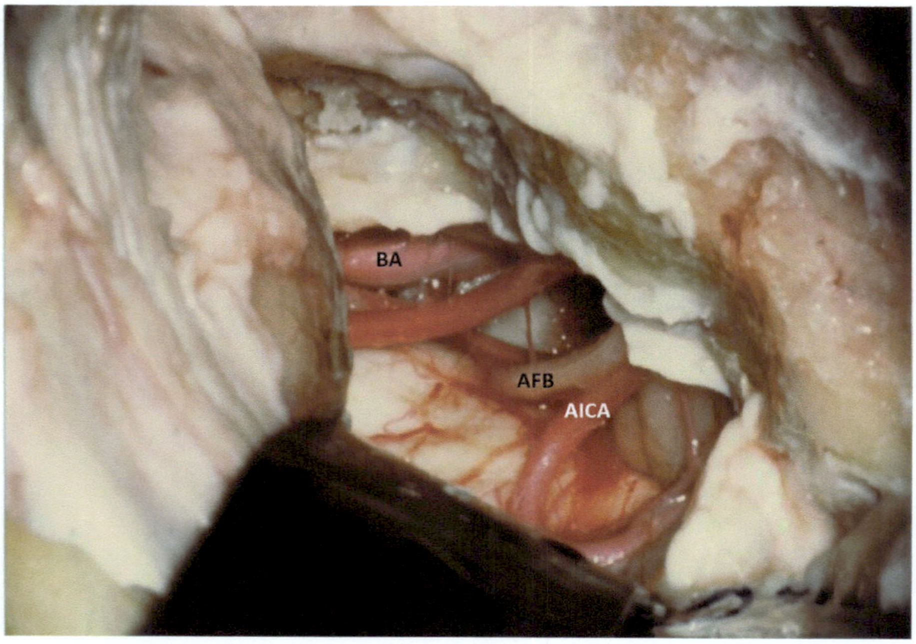

Fig. 11.47 The anterior inferior cerebellar artery (AICA) is seen looping around the acoustic facial bundle (AFB). A small segment of the basilar artery (BA) can be visualized.

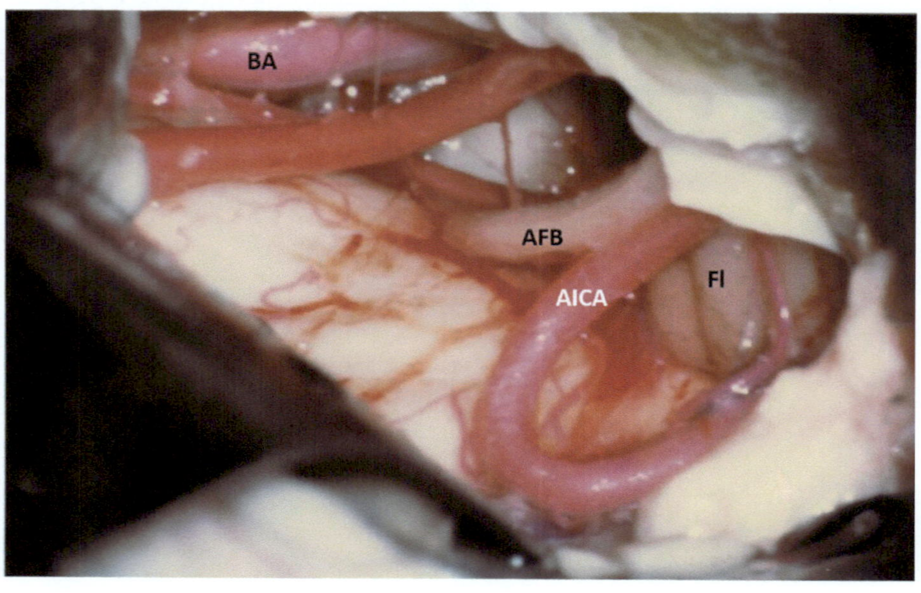

Fig. 11.48 At a higher magnification, a prominent flocculus (Fl) is observed. *BA*, basilar artery; *AFB*, acusticofacial bundle; *AICA*, anterior inferior cerebellar artery.

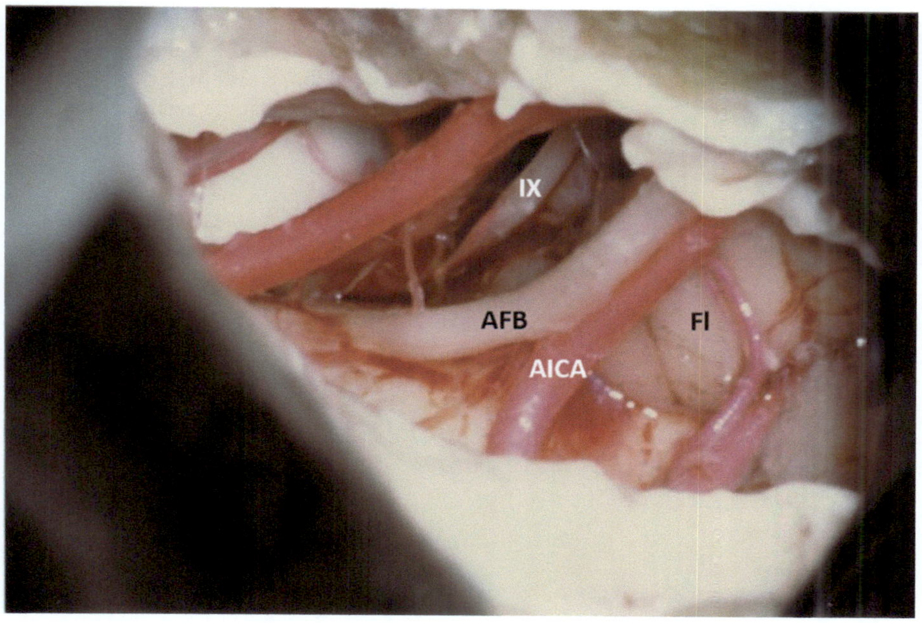

Fig. 11.49 Changing the tilt of the microscope, the glossopharyngeal nerve (IX) can be seen underneath the flocculus (Fl). Note the poor control of the lower cranial nerves provided by this approach. *AFB*, acusticofacial bundle; *AICA*, anterior inferior cerebellar artery.

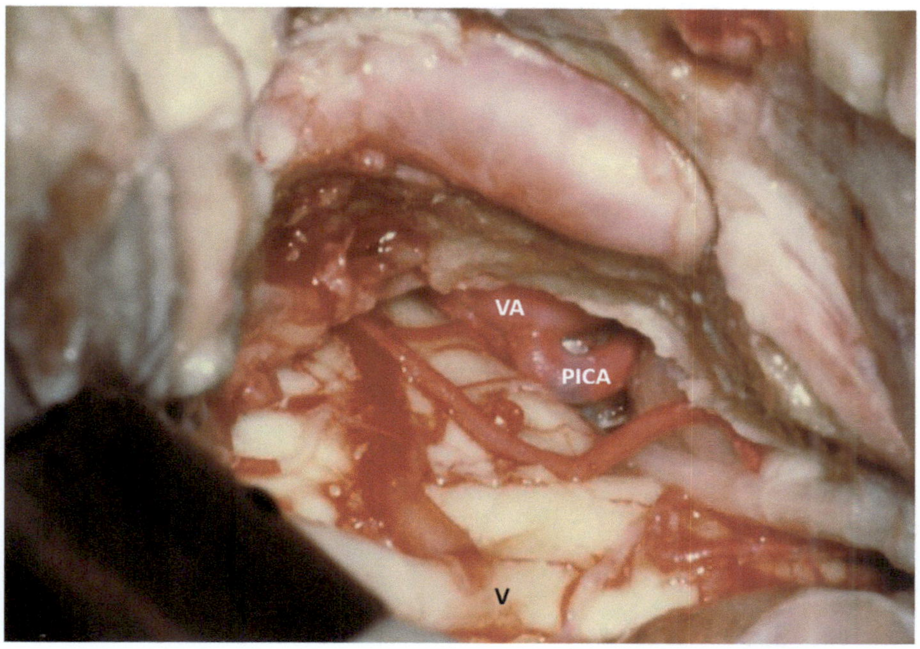

Fig. 11.50 In some cases, the distal part of the vertebral artery (VA) and the proximal segment of the posterior inferior cerebellar artery (PICA) can be seen. Note the partial control of the trigeminal nerve (V).

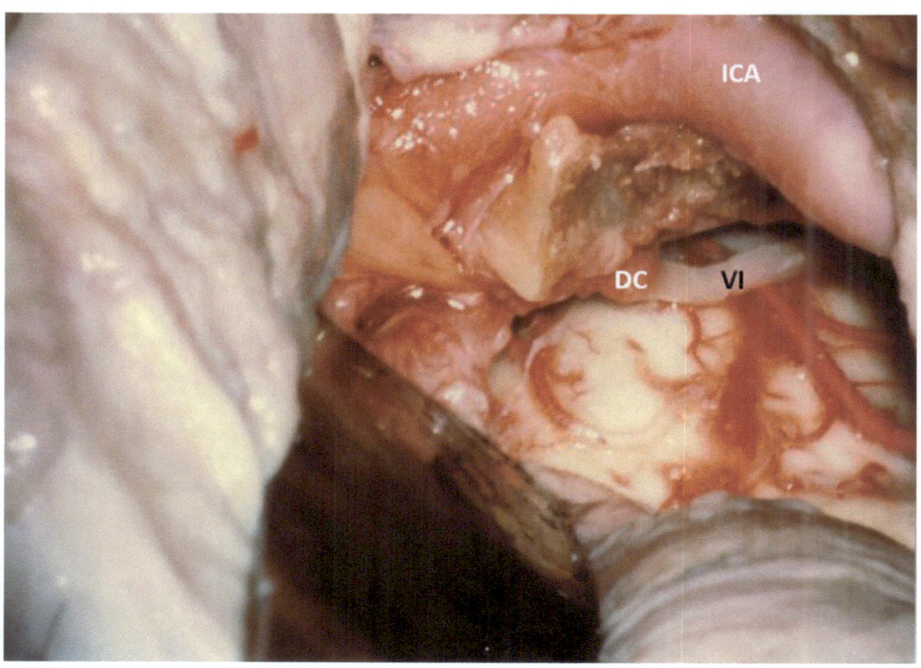

Fig. 11.51 The abducent nerve (VI) is seen coursing to enter into Dorello's canal (DC), after which it crosses the lateral aspect of the internal carotid artery (ICA).

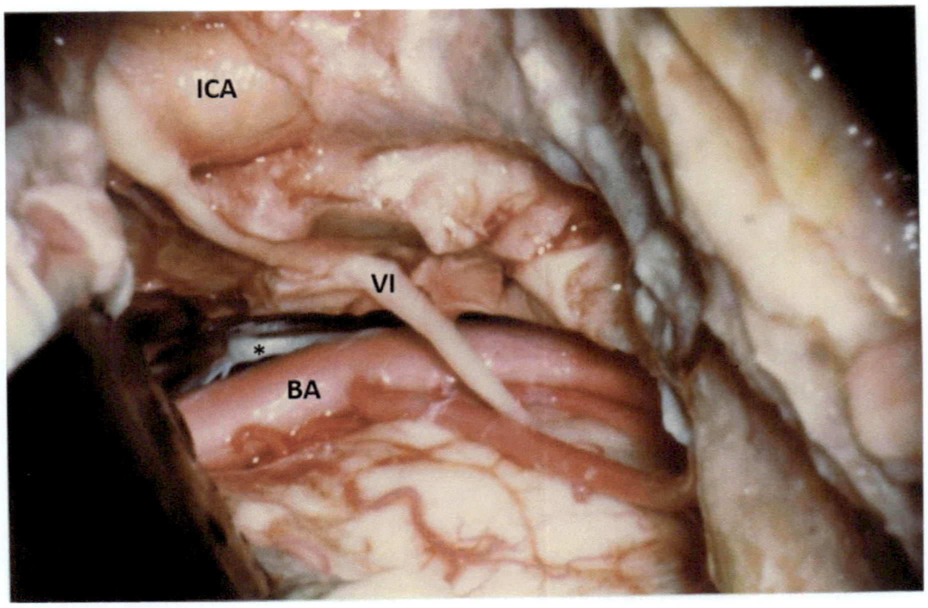

Fig. 11.52 Removing the remaining bone at the petrous apex allows control of the basilar artery (BA) in the prepontine cistern. Note the contralateral sixth nerve (*) can be seen. *VI*, abducent nerve; *ICA*, internal carotid artery.

Clinical Applications—Middle Fossa Approaches

Case 1: Intrameatal VS Removed Through an Enlarged Middle Cranial Fossa Approach

See Figs. 11.53–11.63.

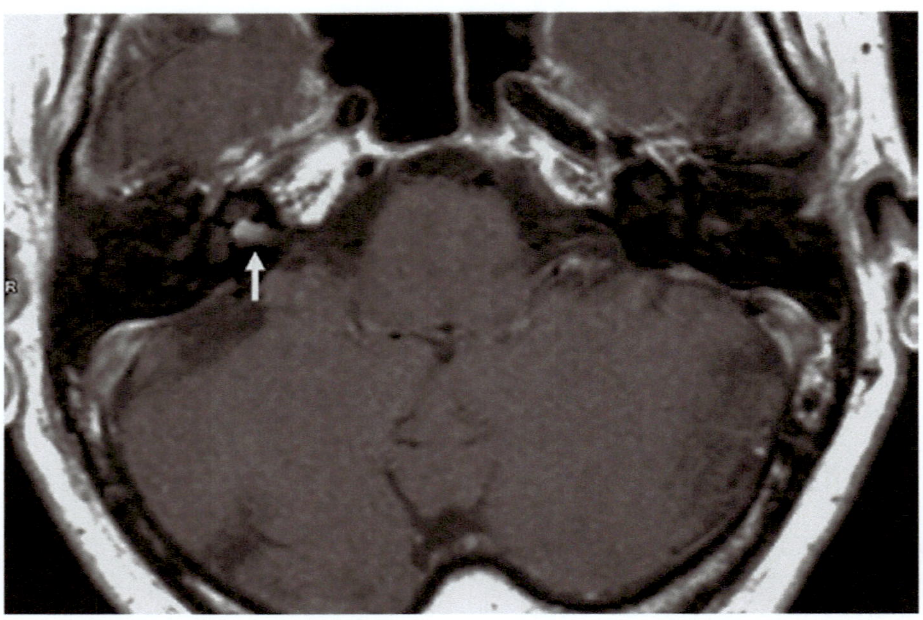

Fig. 11.53 Brain MRI (T1W+ Gd enhancement) showing a right-sided acoustic neurinoma localized in the internal auditory canal (*arrow*).

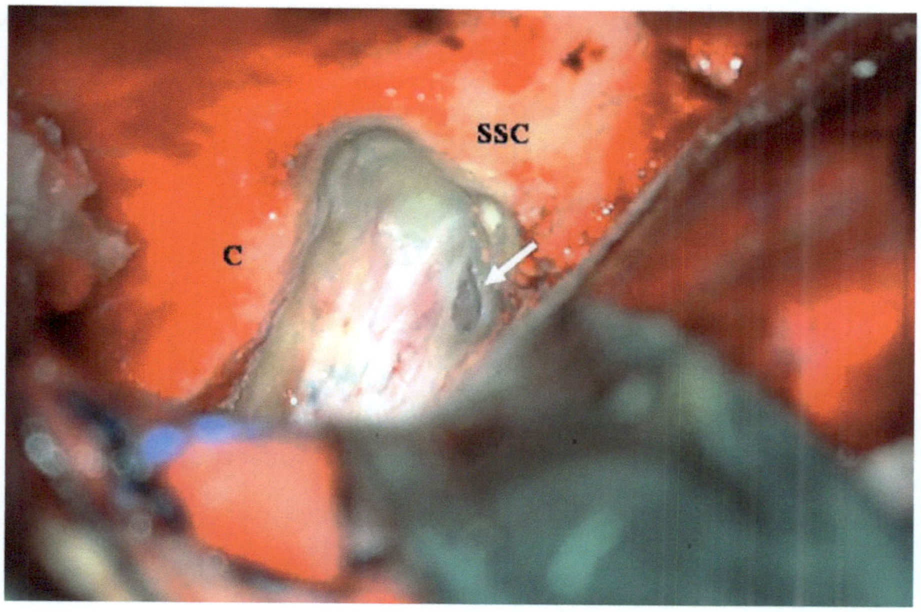

Fig. 11.54 The internal auditory canal is exposed up to near the fundus. A small opening (*arrow*) is made on the posterior fossa dura posterior to the internal auditory canal. *C*, cochlea; *SSC*, superior semicircular canal.

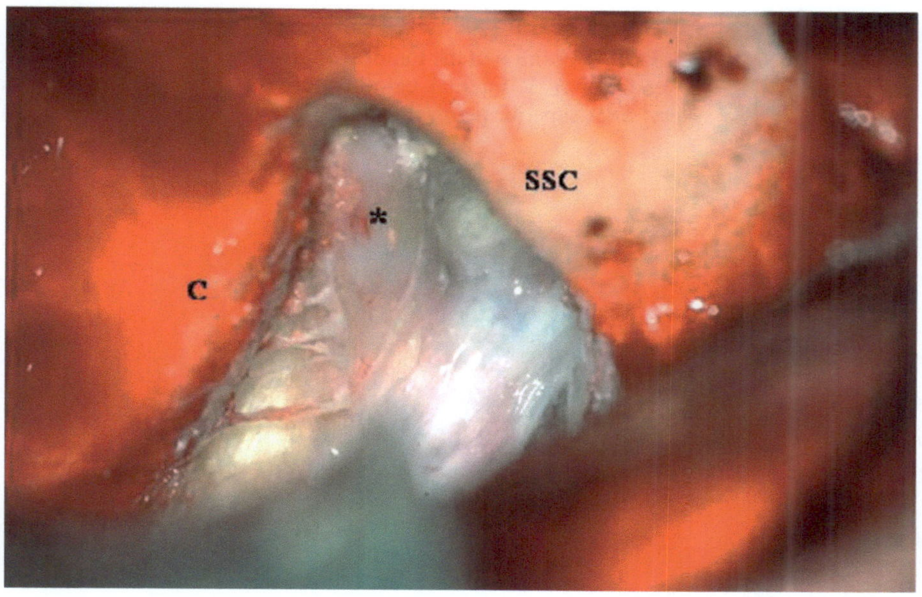

Fig. 11.55 After further drilling, the nerve lying superior to the tumor is observed (*). *C*, cochlea; *SSC*, superior semicircular canal.

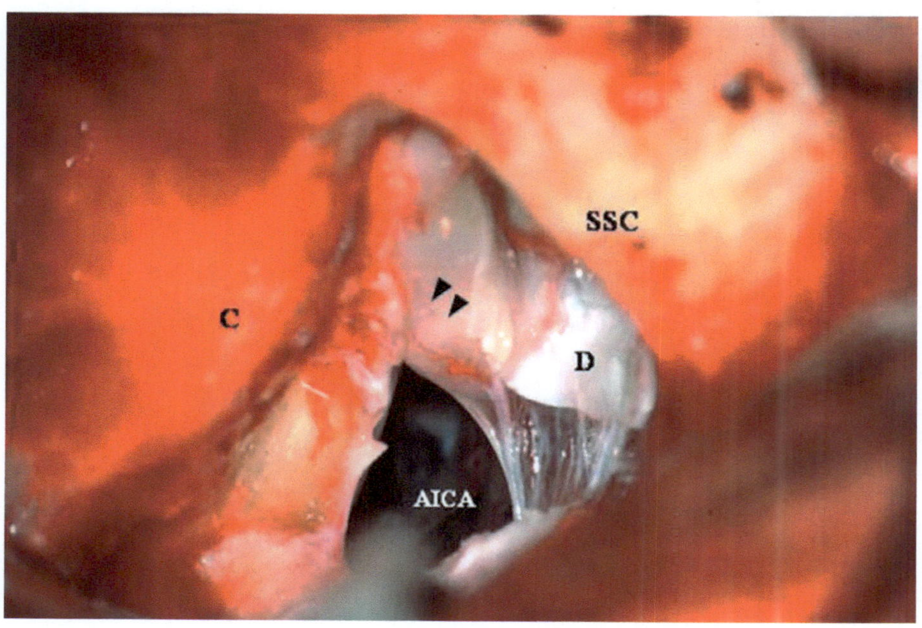

Fig. 11.56 The dura (D) is incised from the nerve posterior to the small opening previously made. In this case, the border between the facial nerve and the superior vestibular nerve can be appreciated (*arrowheads*). The anterior inferior cerebellar artery (AICA) is also seen. *C*, cochlea; *SSC*, superior semicircular canal.

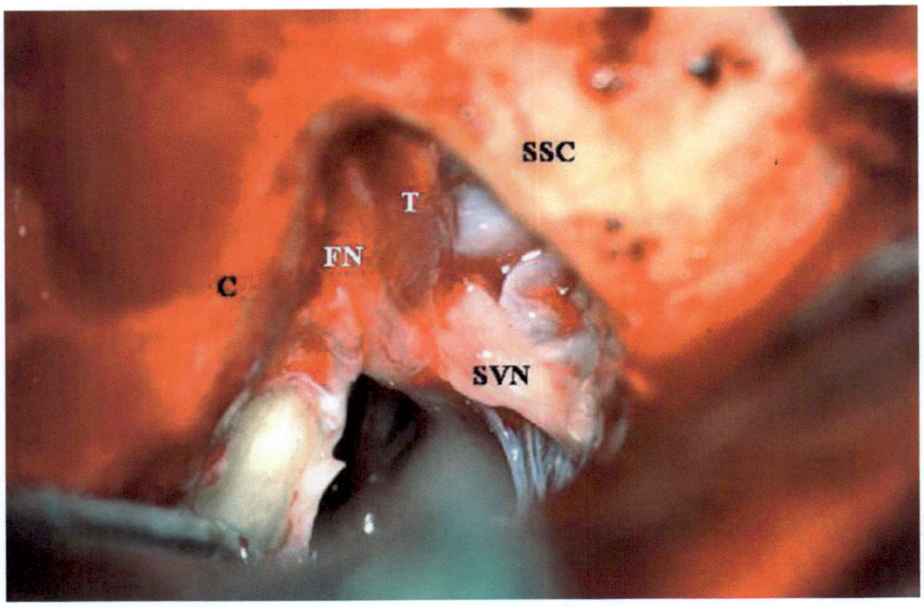

Fig. 11.57 The superior vestibular nerve (SVN) is transected and reflected. The tumor (T) is observed behind that nerve. *C*, cochlea; *FN*, facial nerve; *SSC*, superior semicircular canal.

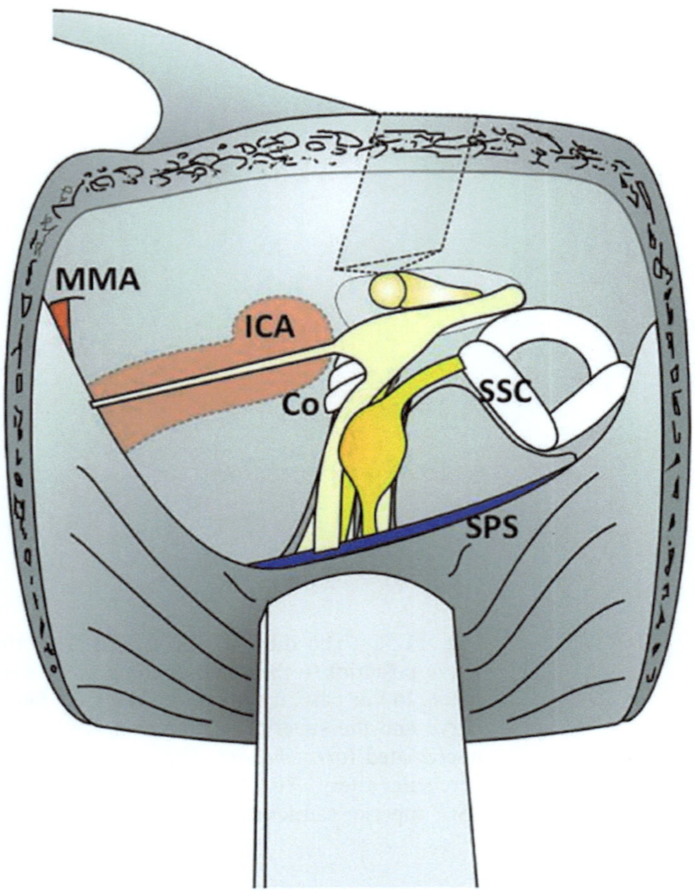

Fig. 11.58 Position of the tumor posterosuperiorly to the facial nerve. *Co* cochlea; *ICA*, internal carotid artery; *MMA*, middle meningeal artery; *SPS*, superior petrosal sinus; *SSC*, superior semicircular canal.

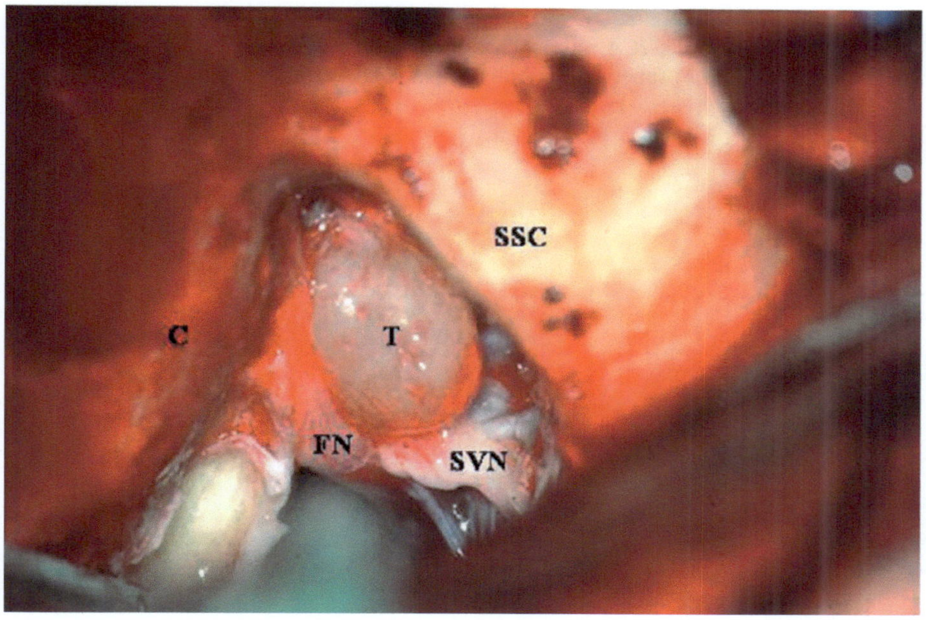

Fig. 11.59 Dissection of the tumor (T) from the facial nerve (FN). *C*, cochlea; *SSC*, superior semicircular canal; *SVN*, superior vestibular nerve.

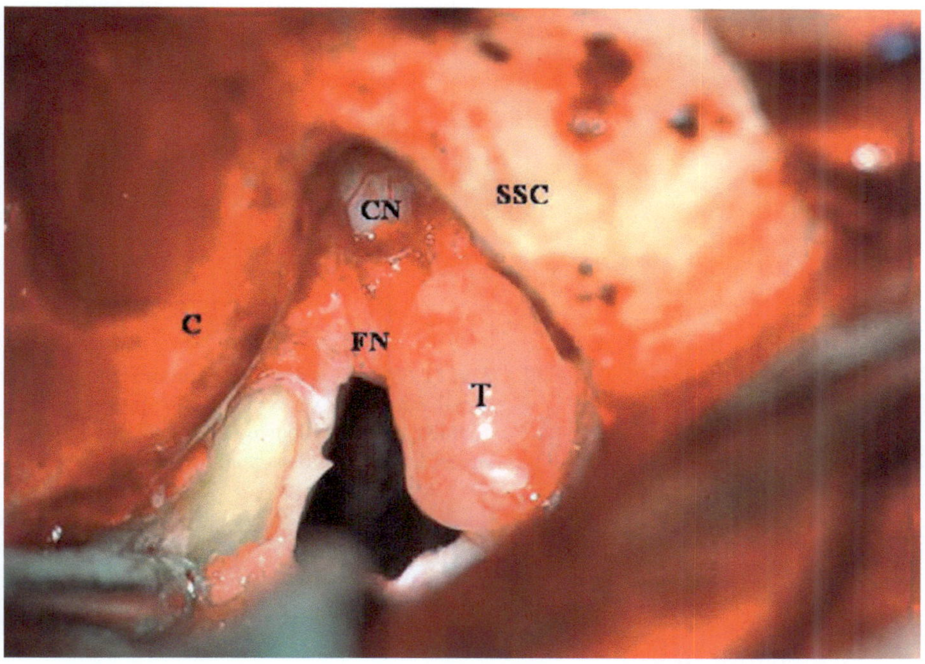

Fig. 11.60 Dissection of the tumor (T) from the facial nerve (FN). The cochlear nerve (CN) can be seen behind the tumor. *C*, cochlea; *SSC*, superior semicircular canal; *SVN*, superior vestibular nerve.

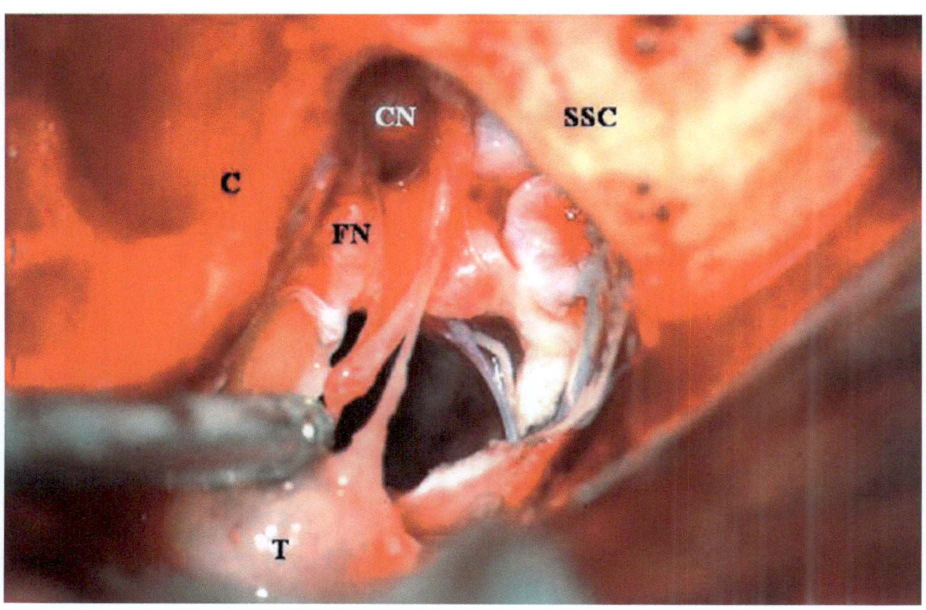

Fig. 11.61 Dissection of the tumor (T) from the facial nerve (FN). The adhesive band is sharply transected by the use of microscissors. *C*, cochlea; *SSC*, superior semicircular canal; *SVN*, superior vestibular nerve.

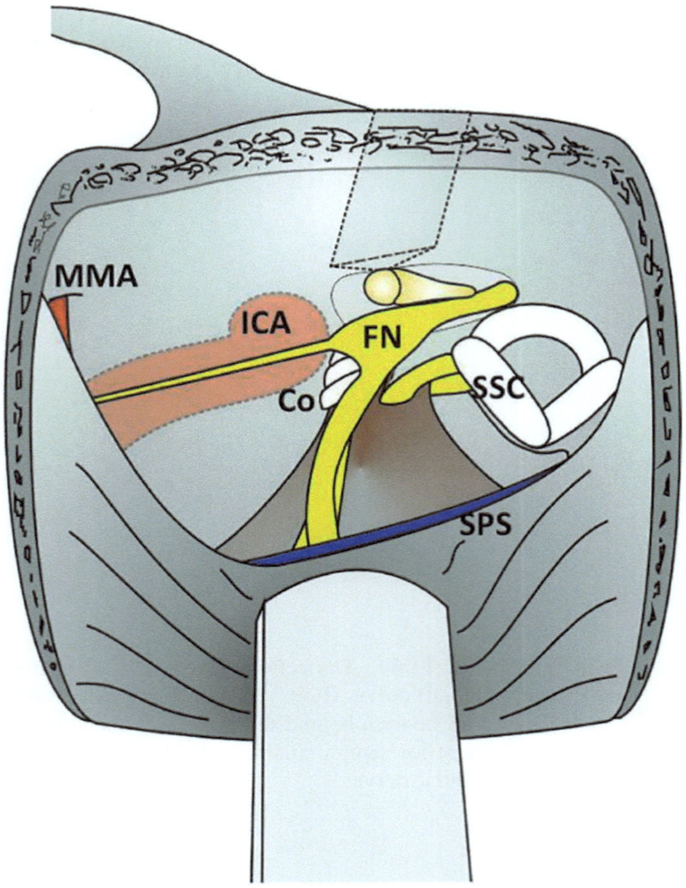

Fig. 11.62 Drawing after total tumor removal. *Co*, cochlea; *FN*, facial nerve; *ICA*, internal carotid artery; *MMA*, middle meningeal artery; *SPS*, superior petrosal sinus; *SSC*, superior semicircular canal.

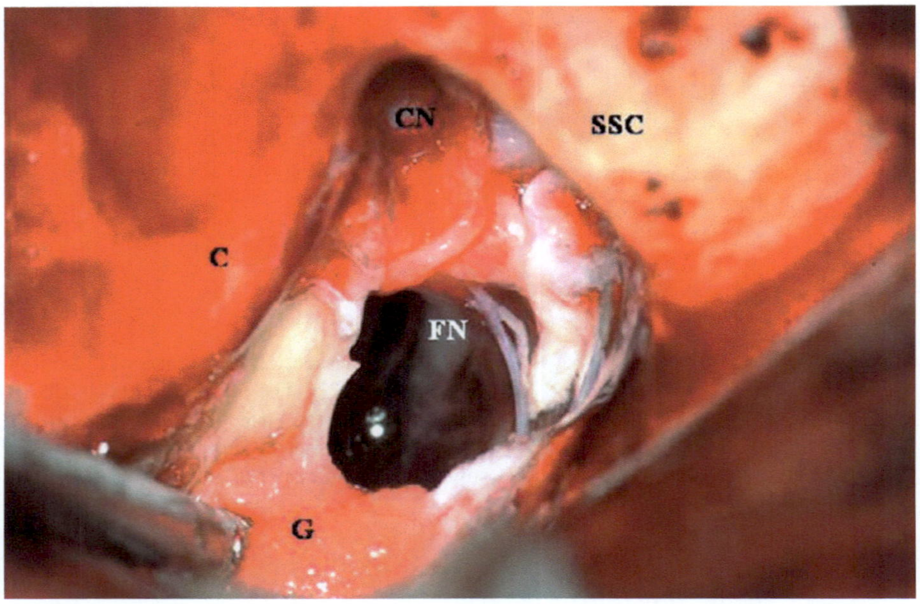

Fig. 11.63 View after total tumor removal. The facial nerve (FN) is perfectly preserved. *C*, cochlea; *CN*, cochlear nerve; *G*, Gelfoam; *SSC*, superior semicircular canal.

Case 2: Supralabyrinthine Petrous Bone Cholesteatoma in the Only Hearing Ear Removed Through a Combined Enlarged Middle Cranial Fossa and Transmastoid Approach

See Figs. 11.64–11.81.

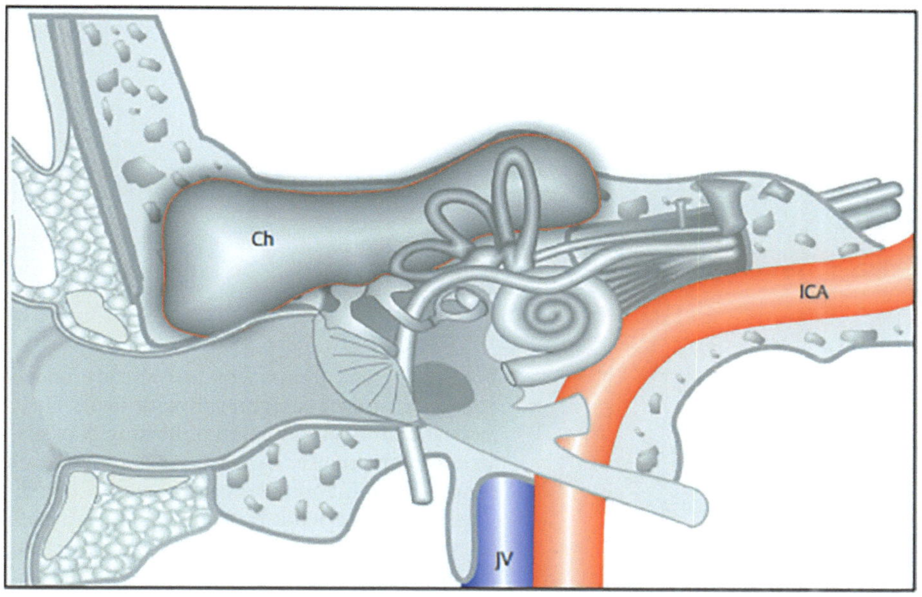

Fig. 11.64 This patient presented with a bilateral petrous bone cholesteatoma. The contralateral side underwent a transotic approach with intraoperative facial nerve reconstruction using a sural nerve graft. To preserve hearing in this ear, surgery was scheduled immediately after recovery from surgery on the contralateral side. *Ch*, cholesteatoma; *ICA*, internal carotid artery; *JV*, jugular vein.

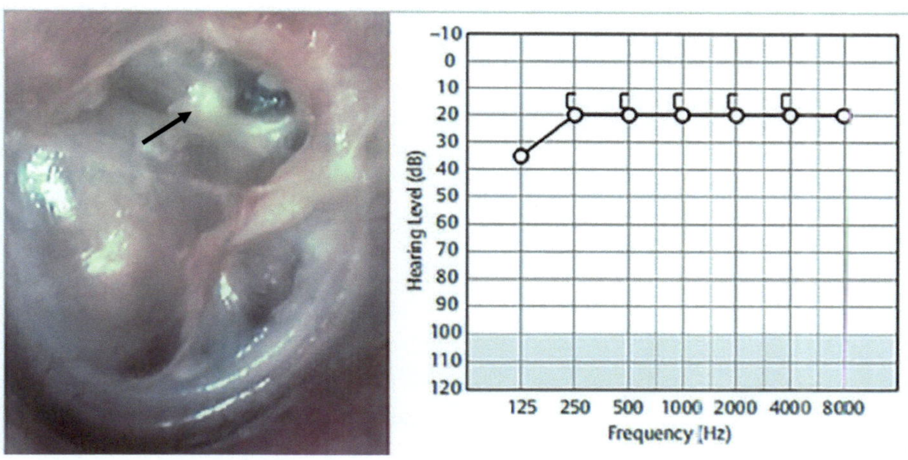

Fig. 11.65 The otoscopy of the right ear shows retraction of the posterior quadrants of the tympanic membrane. The posterosuperior quadrant goes deep toward the attic, touching the superstructure of the stapes (*arrow*). This spontaneous myringostapediopexy explains very good preoperative hearing.

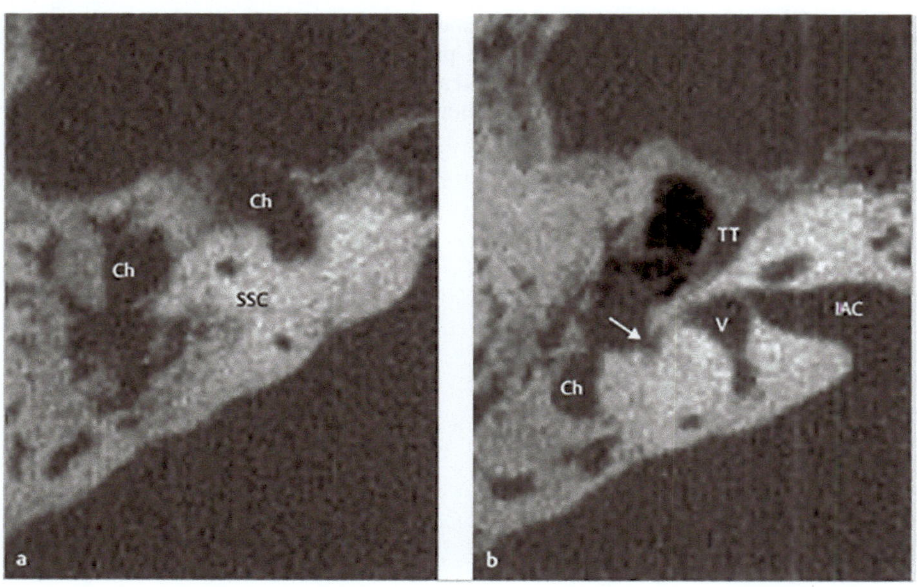

Fig. 11.66 The axial CT shows the medial extension of cholesteatoma surpassing the superior semicircular canal. Even though the tympanic cavity is pneumatized to some extent, a small labyrinthine fistula is identified in the lateral semicircular canal (*arrow*). The head of the malleus and the whole incus are missing in the attic. *Ch*, cholesteatoma; *IAC*, internal auditory canal; *SSC*, superior semicircular canal; *TT*, tensor tympani; *V*, vestibule.

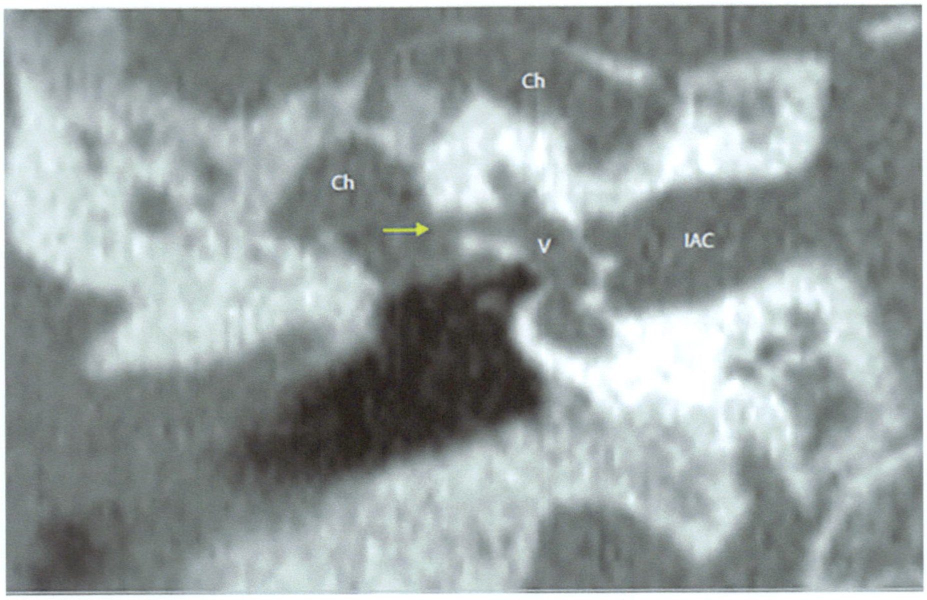

Fig. 11.67 In coronal view, a fistula in the lateral semicircular canal is clearly seen (*arrow*). The cholesteatoma infiltrates medially along the middle fossa plate to reach the petrous apex. *Ch*, cholesteatoma; *IAC*, internal auditory canal; *V*, vestibule.

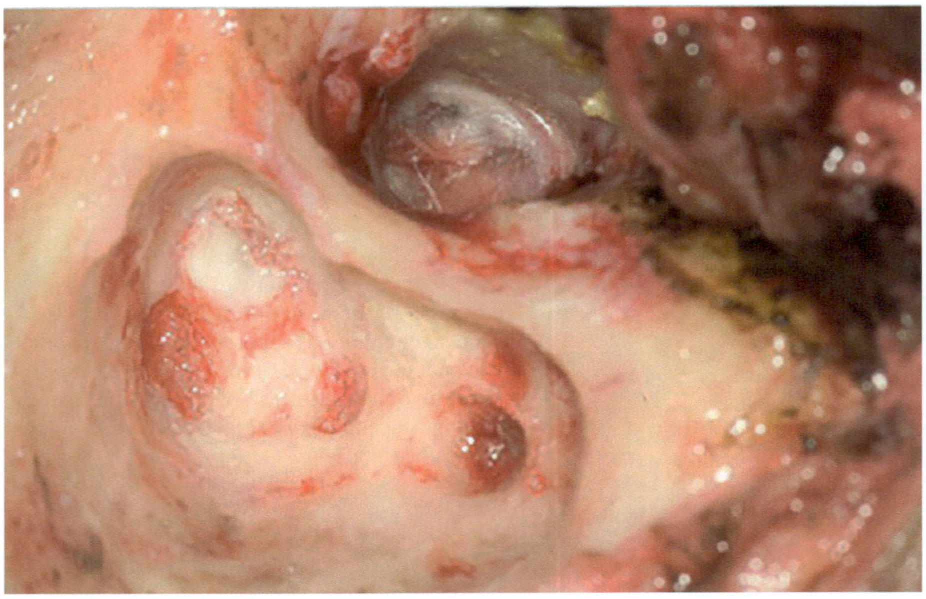

Fig. 11.68 As the patient had already undergone a left transotic approach with facial nerve reconstruction, preservation of hearing is of paramount importance. A transmastoid approach combined with a middle fossa approach is indicated. The bone work in the mastoid cavity has been completed and the cholesteatoma reaching the mastoid has already been removed.

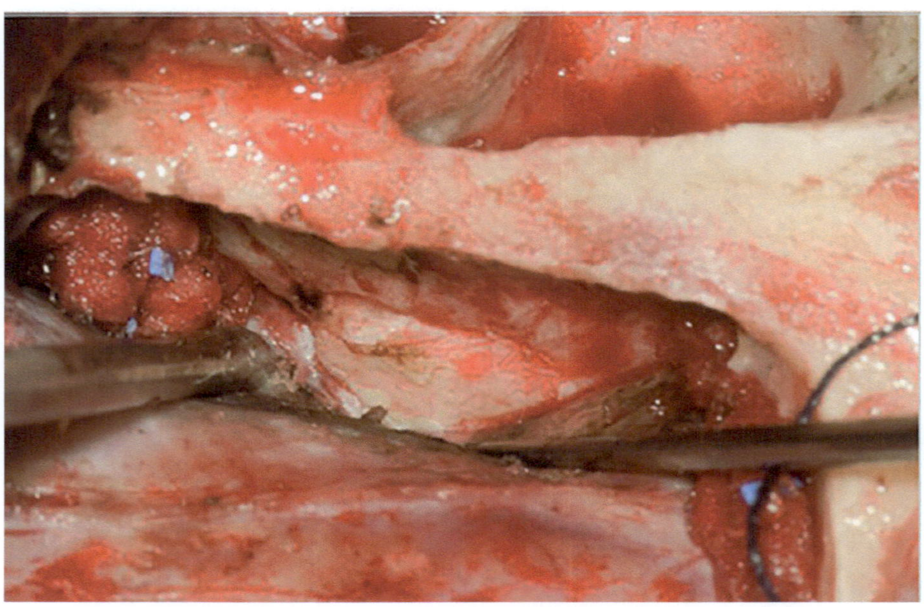

Fig. 11.69 Through the middle fossa approach, the superior face of the petrous bone is exposed from superior. The cholesteatoma eroding the tegmen of the mastoid is seen. The *left side* of the picture corresponds to the anterior of the patient, the top corresponds to caudal, and the bottom cephalad.

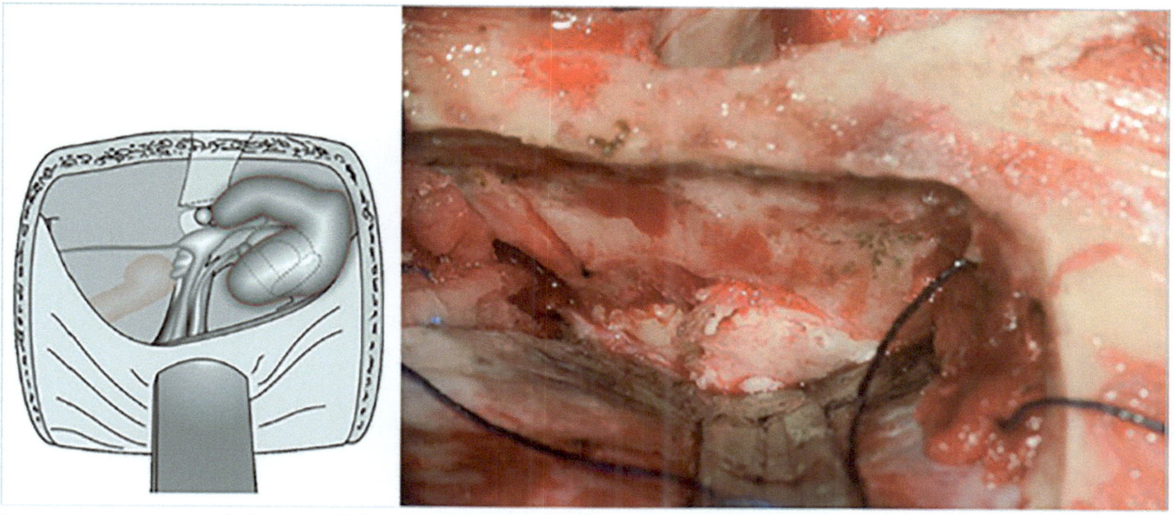

Fig. 11.70 A retractor is used to hold the middle fossa dura superiorly.

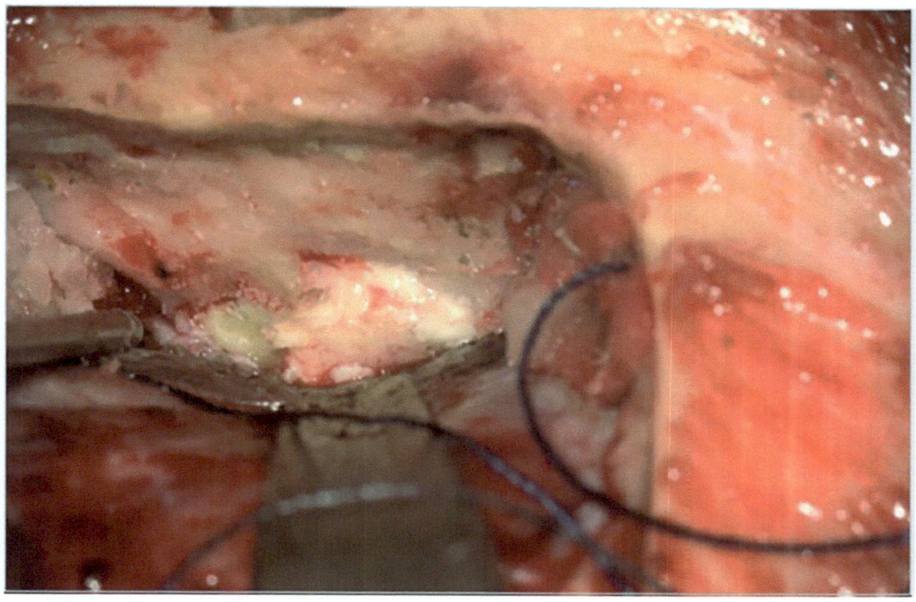

Fig. 11.71 The cholesteatoma is started to be dissected.

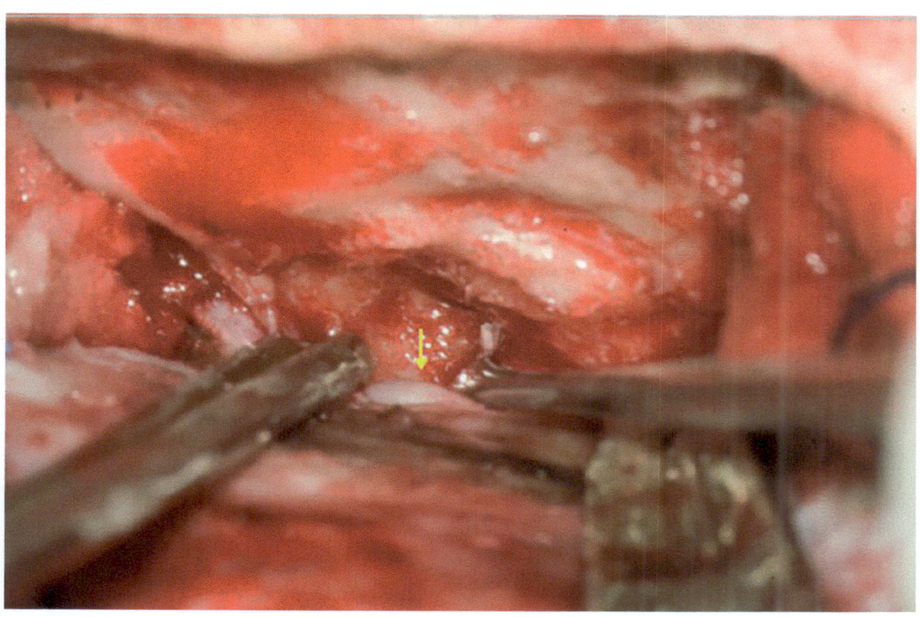

Fig. 11.72 There is still cholesteatoma. present (*arrow*) covering the dura.

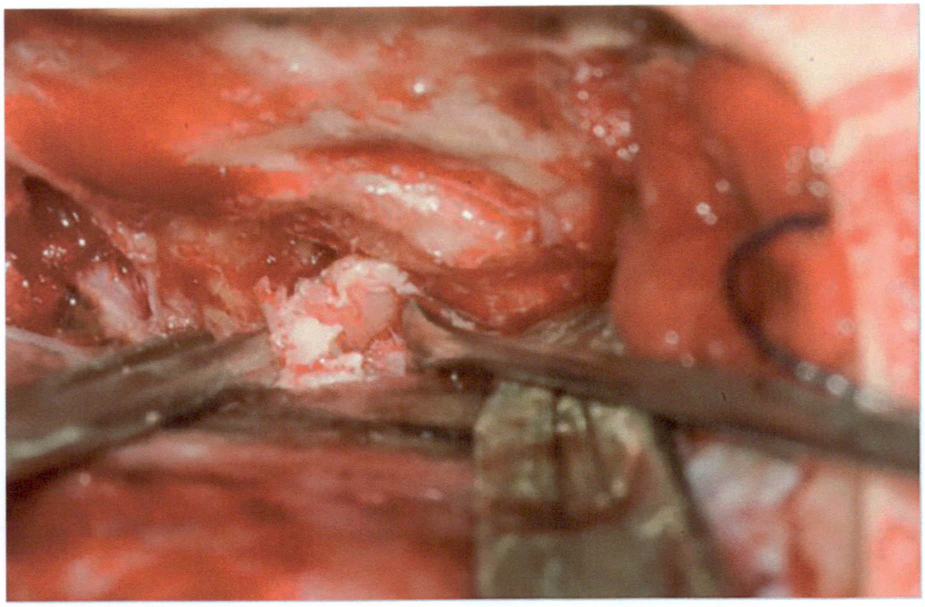

Fig. 11.73 The cholesteatoma is detached from the dura.

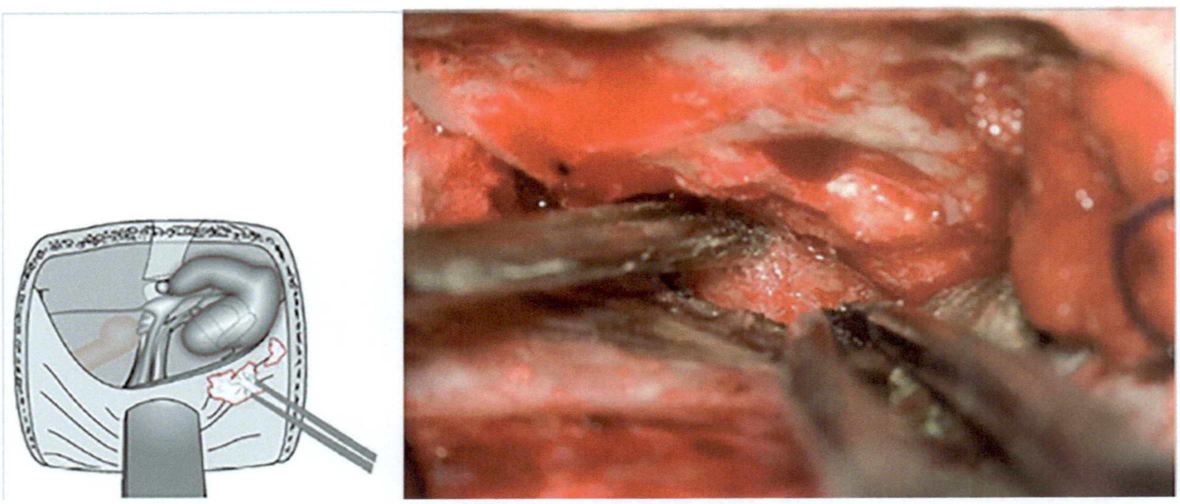

Fig. 11.74 The area of the dura suspicious of residual matrix should be coagulated with bipolar.

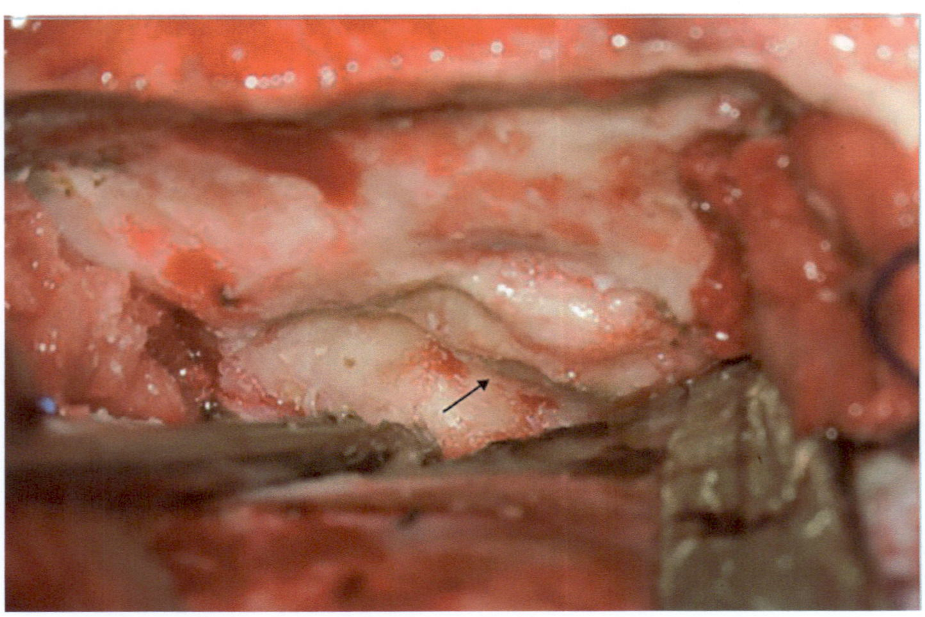

Fig. 11.75 Cholesteatoma infiltrating medially to the superior semicircular canal (*arrow*) is eradicated.

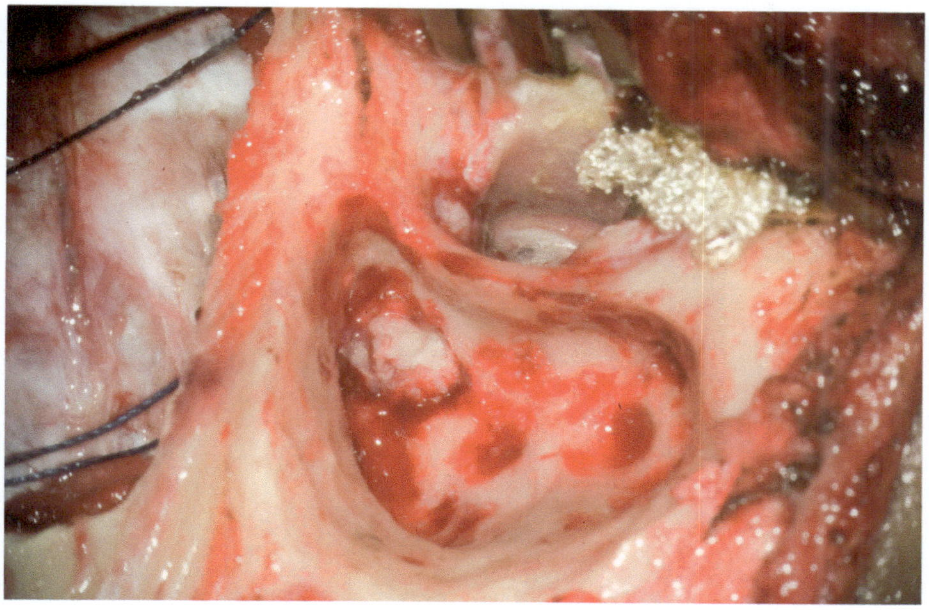

Fig. 11.76 The retractor is removed, and the surgery is conducted from the mastoid again. Cholesteatoma in the attic is shown.

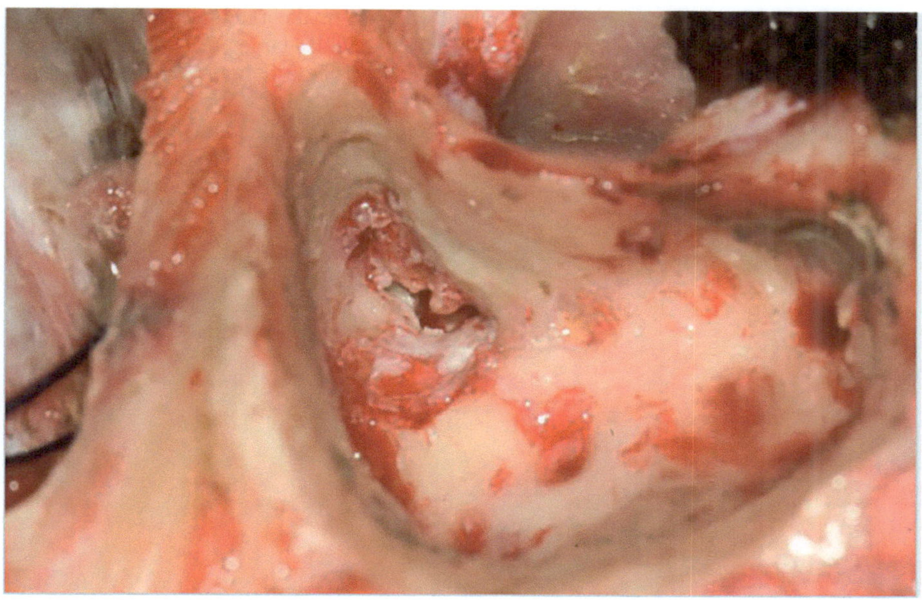

Fig. 11.77 Some debris has been evacuated from the mastoid. The cholesteatoma fills the entire attic.

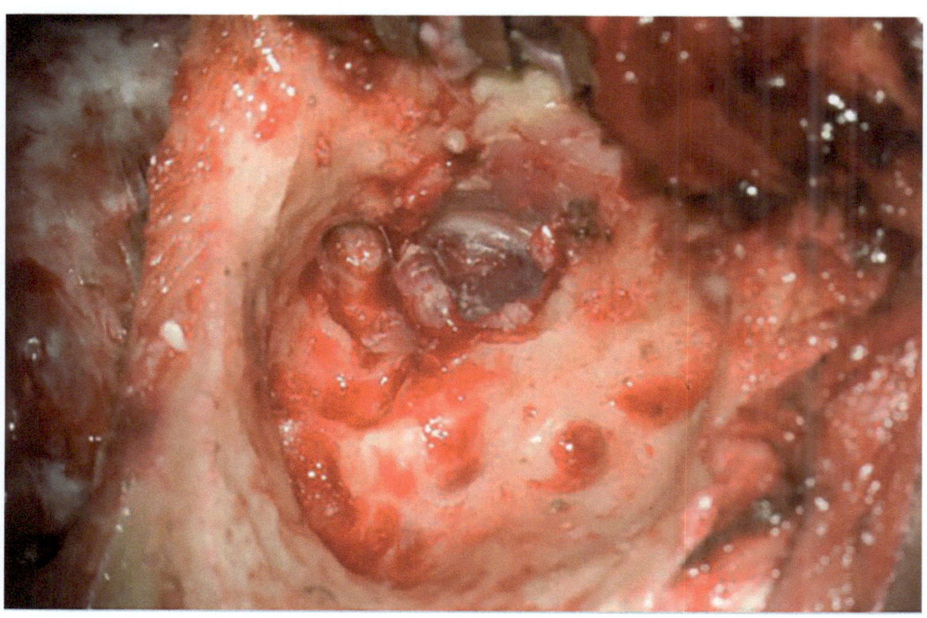

Fig. 11.78 The cholesteatoma is eradicated from the middle ear. Since this is the cholesteatoma surgery in only the hearing ear, the removal of the posterior wall to reduce the possibility of both recurrent and residual cholesteatoma is indicated. Since preoperative hearing is good, the tympanic membrane is left untouched.

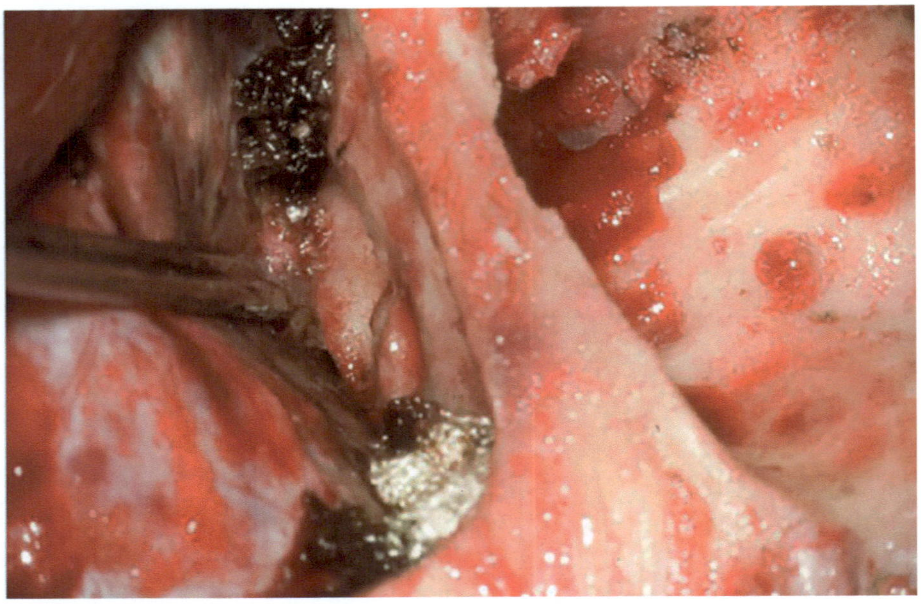

Fig. 11.79 View from the middle fossa. Absence of residual skin is verified.

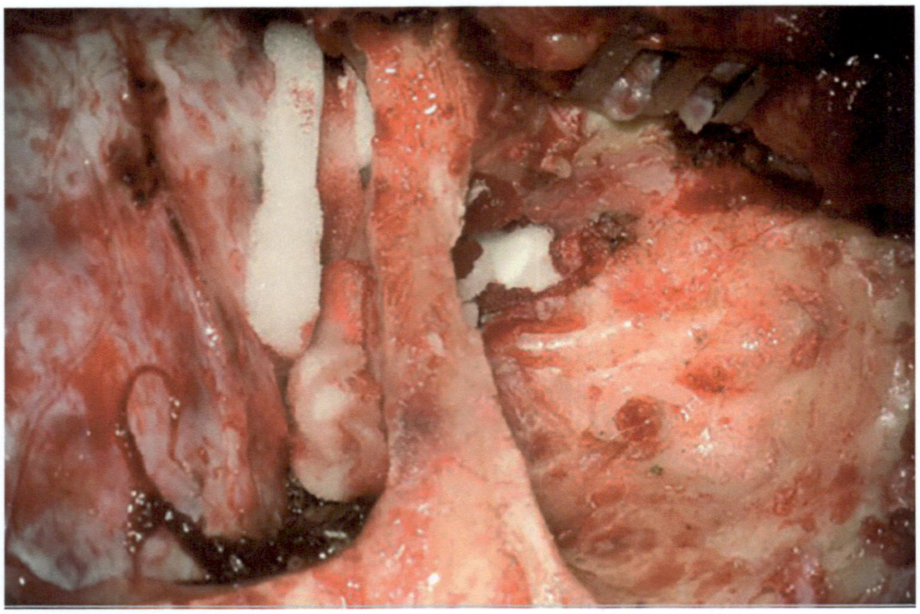

Fig. 11.80 From the middle fossa, the tegmen is reconstructed with a large piece of fascia and cartilage. The cartilage is held in place with Gelfoam. The bone flap is replaced to cover the lateral face of the middle cranial fossa. The mastoid cavity is covered with fascia, and the cavity is reconstructed in the same manner as a canal wall down tympanoplasty.

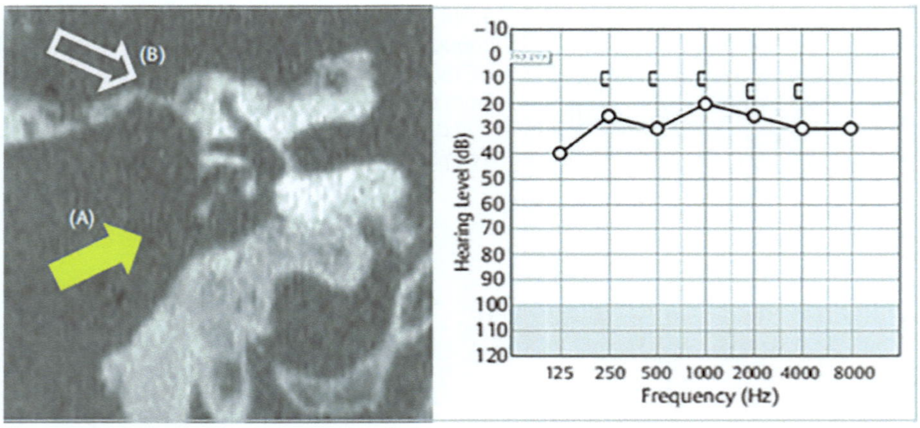

Fig. 11.81 The CT conducted immediately postoperatively clearly demonstrates two approaches: transmastoid (A) and middle fossa (B). The hearing stays near-normal postoperatively due to the preserved configuration of the tympanic cavity.

Case 3: Facial Nerve Neuroma Involving the Geniculate Ganglion Removed Through a Combined Enlarged Middle Cranial Fossa and Transmastoid Approach With Sural Nerve Grafting

See Figs. 11.82–11.107.

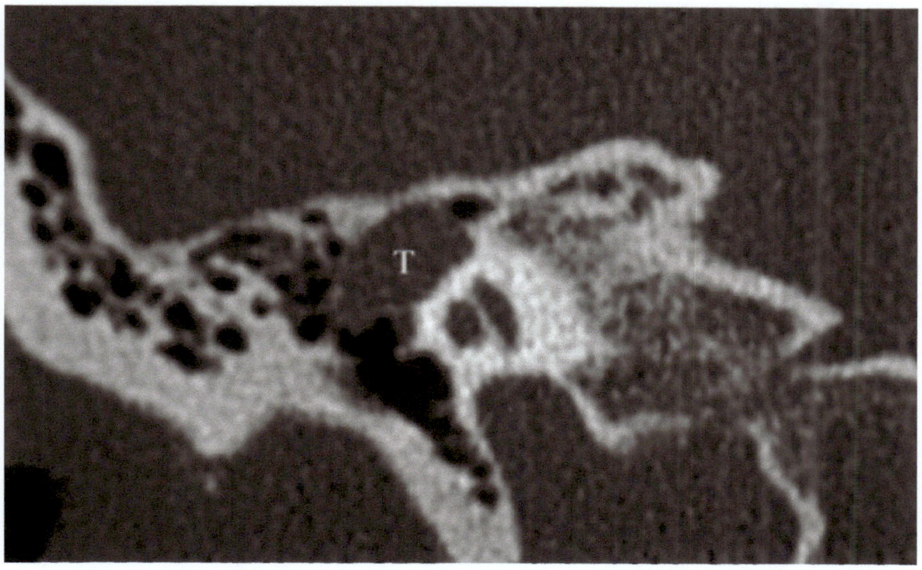

Fig. 11.82 This 36-year-old patient presented with progressive right facial nerve palsy (House Brackmann Scale grade V). The CT scan showed a tumor (T) involving the second portion of the facial nerve and the geniculate ganglion. Hearing was normal, so a combined enlarged middle fossa with a transmastoid approach was chosen.

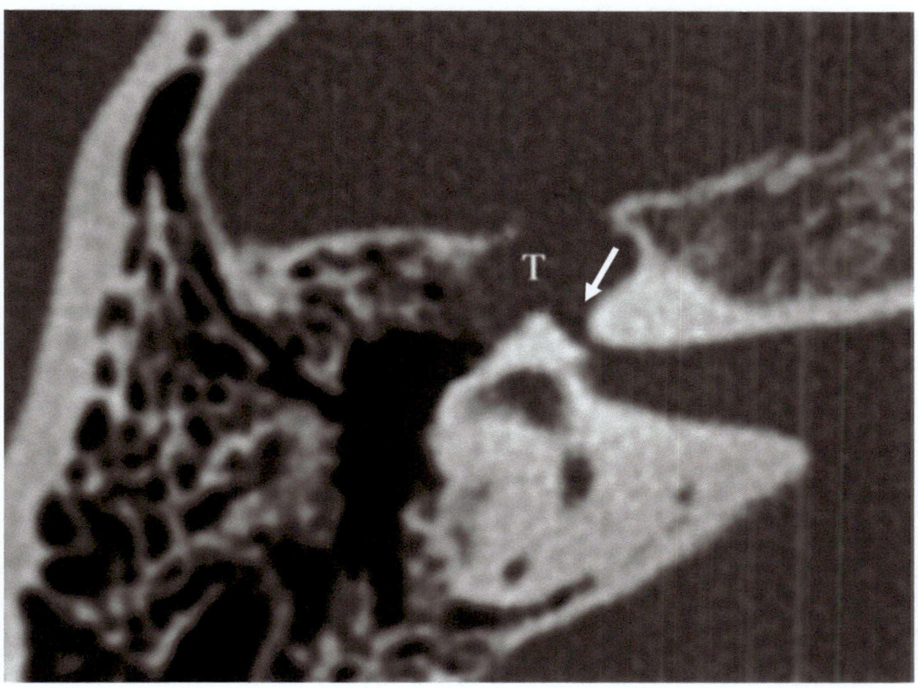

Fig. 11.83 The axial cut of the HRCT scan shows erosion of the geniculate ganglion and enlargement of the canal of the lateral part of the intralabyrinthine segment of the facial nerve (*arrow*). *T*, tumor.

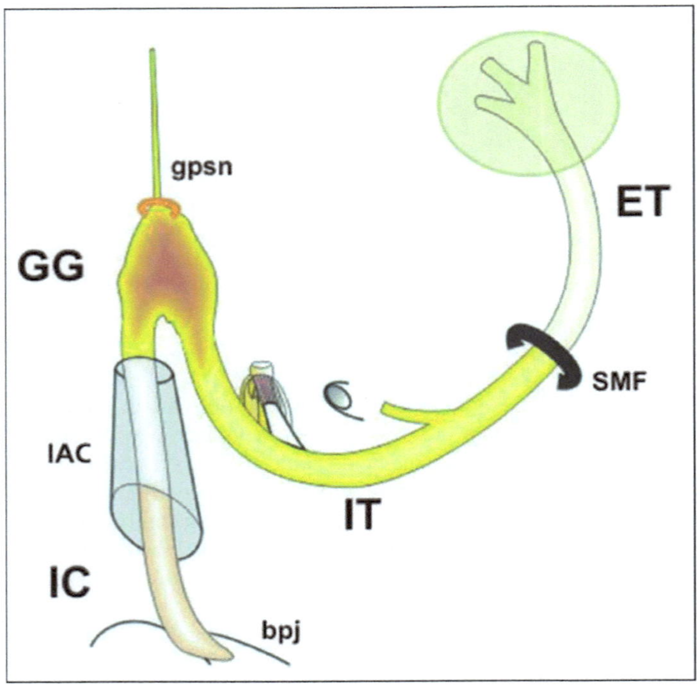

Fig. 11.84 Schematic representation of the tumor in relation to the entire course of the facial nerve. *IC*, intracisternal; *bpj*, bulbopontine junction; *IAC*, internal auditory canal; *GG*, geniculate ganglion; *gspn*, greater superficial petrosal nerve; *IT*, intratemporal; *SMF*, stylomastoid foramen; *ET*, extratemporal.

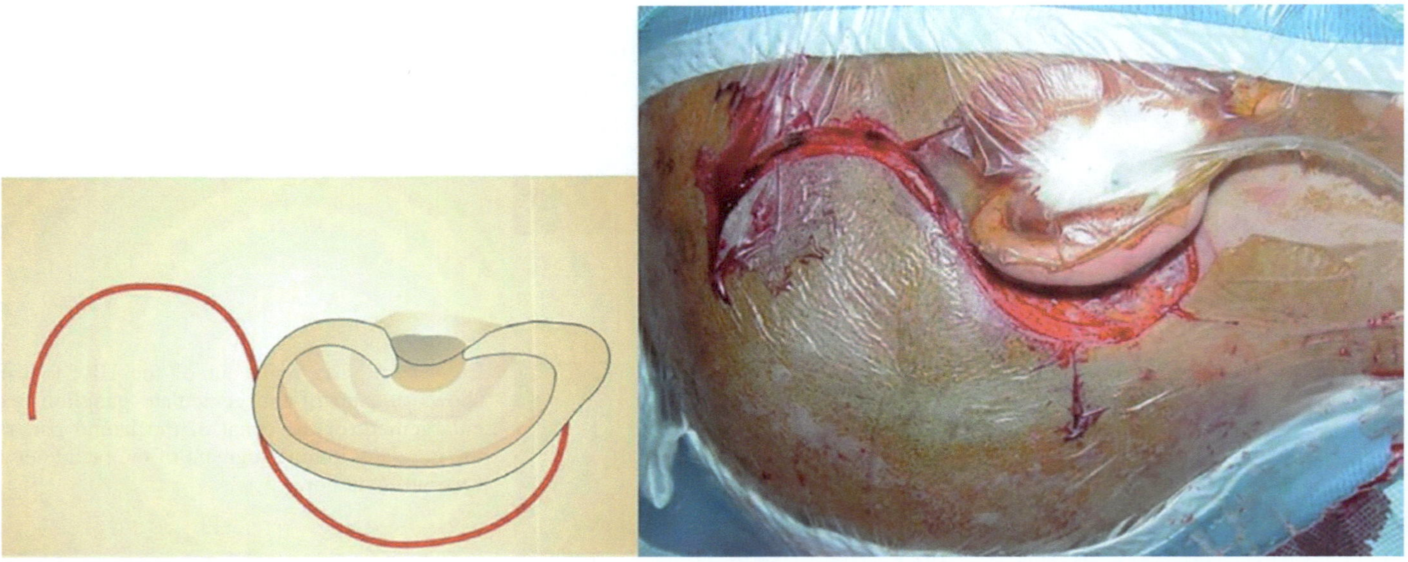

Fig. 11.85 Skin incision.

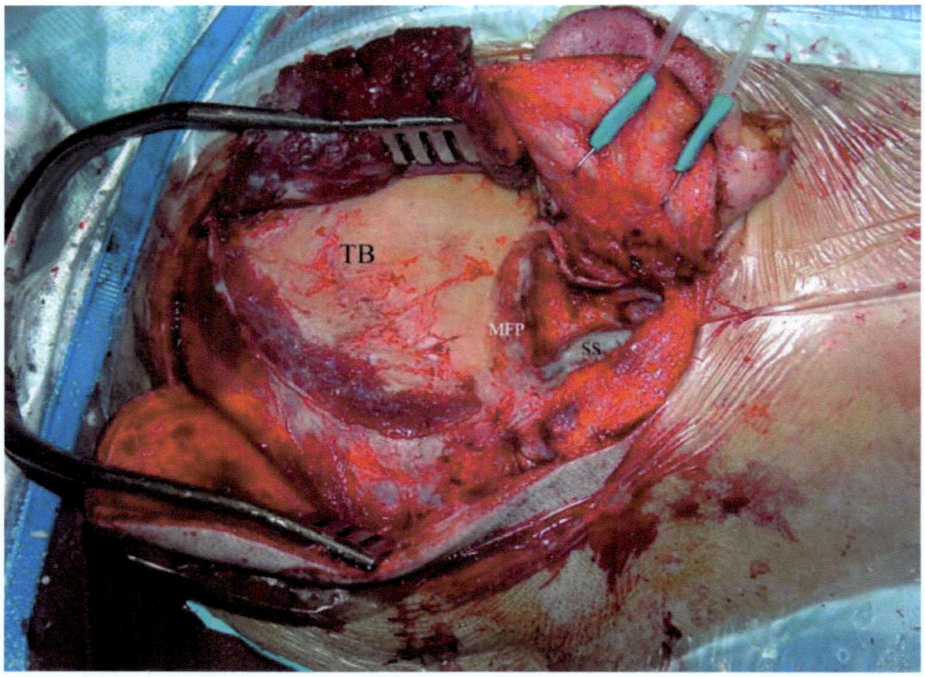

Fig. 11.86 The temporalis muscle together with the underlying periosteal tissues has elevated as an anterior-based flap. A mastoidectomy has already been performed. *TB*, temporal bone; *MFP*, middle fossa dura plate; *SS*, sigmoid sinus.

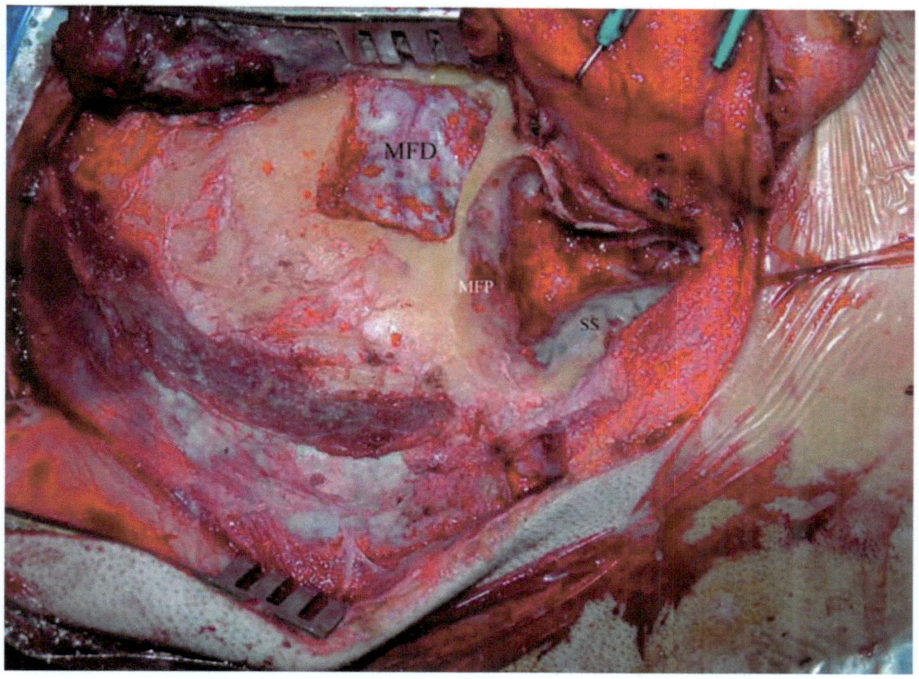

Fig. 11.87 The craniotomy has been done, and the middle fossa dura (MFD) is visible. The middle fossa dura plate is left intact to separate the two cavities. *SS*, sigmoid sinus.

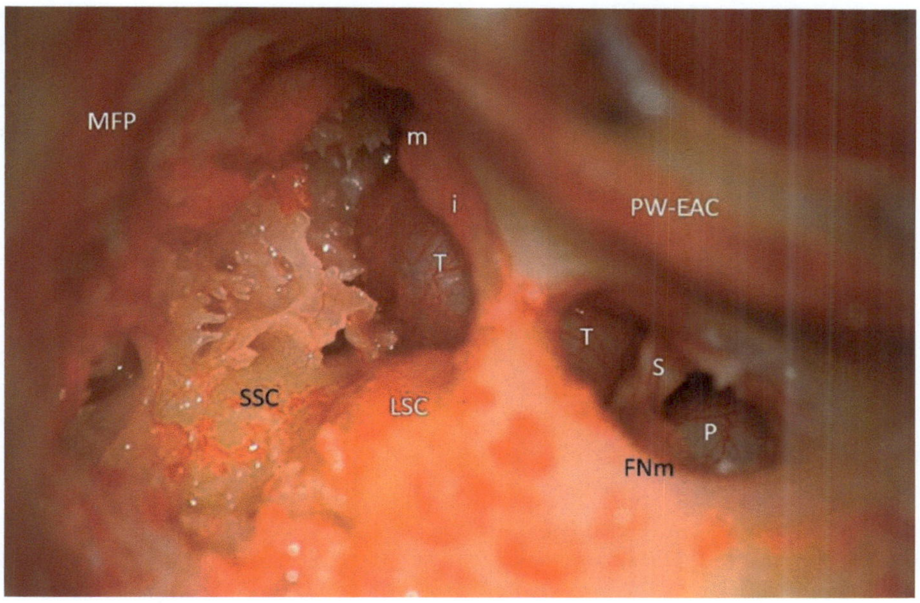

Fig. 11.88 Closer view. Note the tumor (T) within the middle ear cleft. A posterior tympanotomy has been performed, with visualization of tumor mass toward the stapes (S). *MFP*, middle fossa dura plate; *m*, malleus; *i*, incus; *P*, promontory; *LSC*, lateral semicircular canal; *SSC*, superior semicircular canal; *PW-EAC*, posterior wall of the external auditory canal.

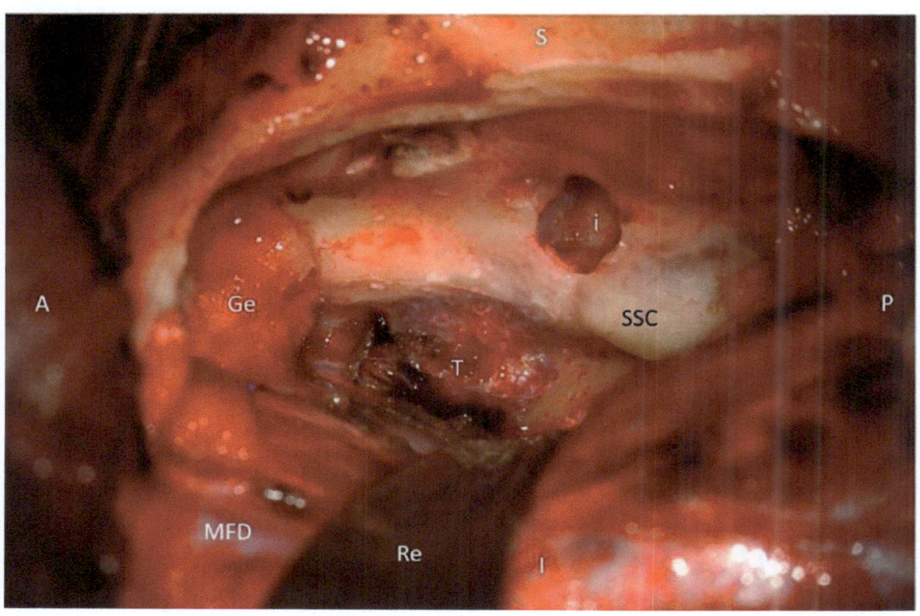

Fig. 11.89 View from the enlarged middle cranial fossa approach. The middle fossa dura (MFD) has been elevated with a retractor (Re). A natural dehiscence of the tegmen tympani allows visualization of the incus (i) from earlier. *Ge*, gelfoam; *SSC*, superior semicircular canal; *A*, anterior; *P*, posterior; *S*, superior; *I*, inferior.

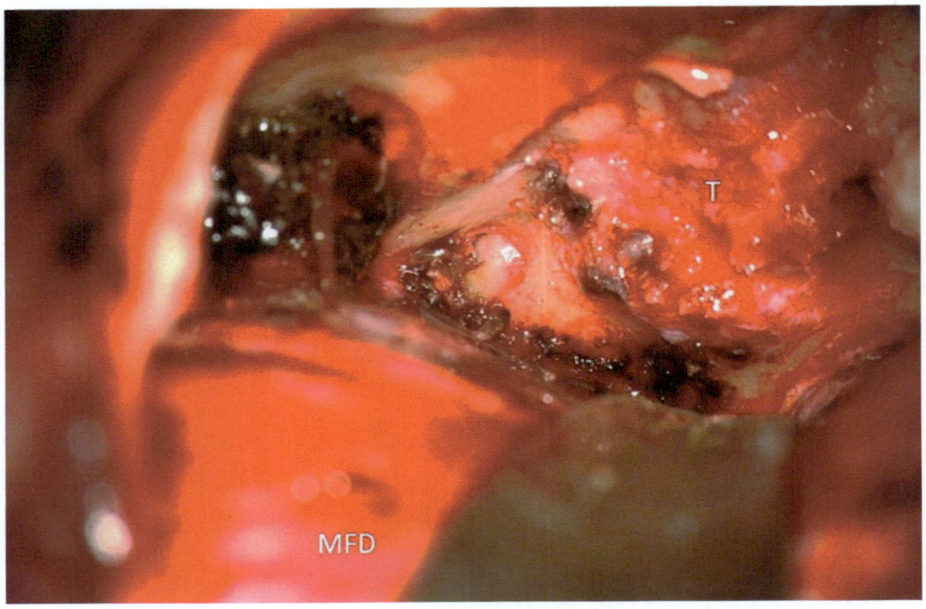

Fig. 11.90 Closer view of the tumor (T) involving the geniculate ganglion from the enlarged middle cranial fossa approach. *MFD*, middle fossa dura.

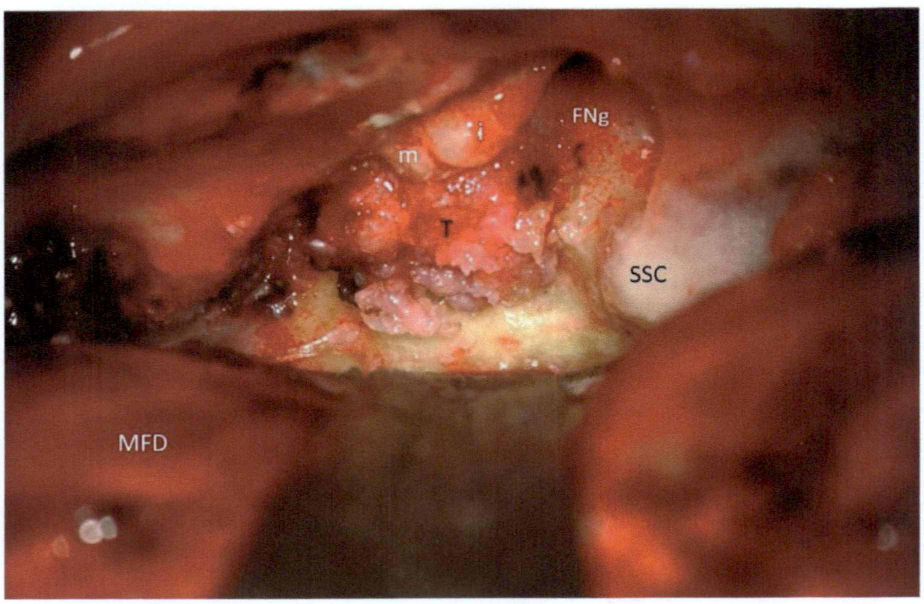

Fig. 11.91 The tumor on the geniculate ganglion has been completely exposed from the enlarged middle cranial fossa approach. Note the relationship of the tumor with the head of the malleus (m) and the incus (i). *SSC*, superior semicircular canal; *MFD*, middle fossa dura; *FNg*, genu of the facial nerve.

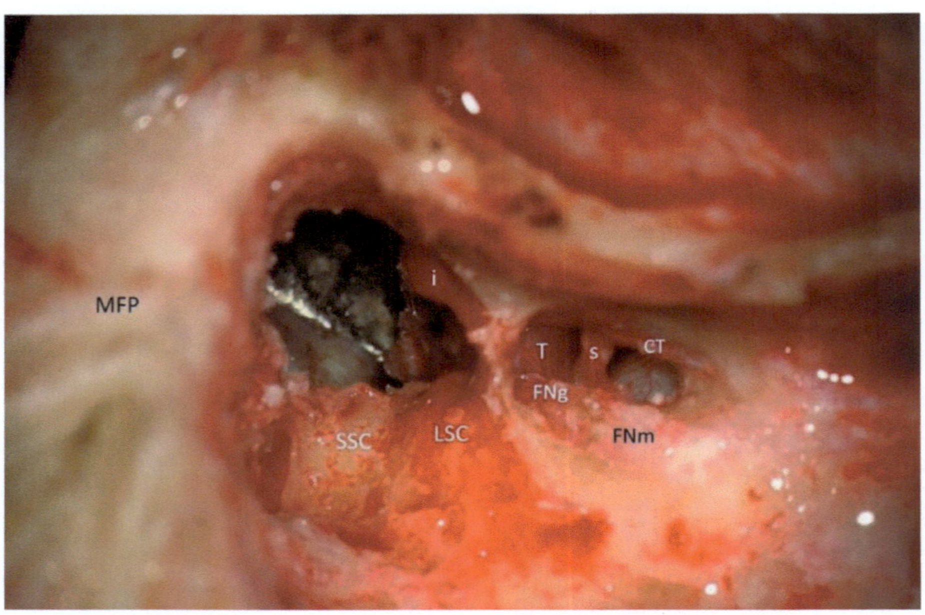

Fig. 11.92 The mastoid portion of the facial nerve (FNm) has been decompressed to the level of the second genu (FNg). The facial nerve is then cut at the level of the genu to provide a sufficient portion of healthy nerve for subsequent anastomosis. *MFP*, middle fossa dura plate; *i*, incus; *LSC*, lateral semicircular canal; *SSC*, superior semicircular canal; *S*, stapes; *CT*, chorda tympani; *T*, tumor.

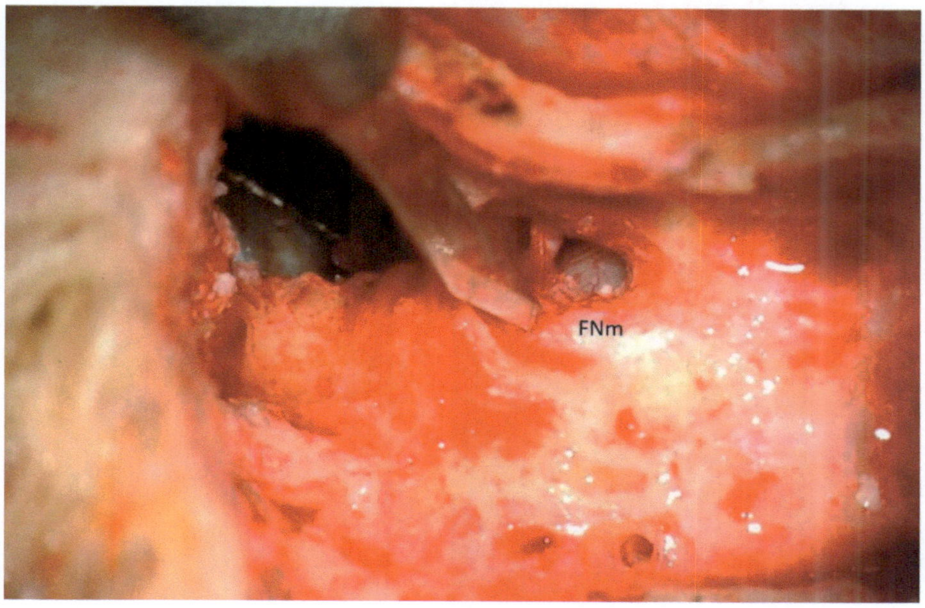

Fig. 11.93 The facial nerve is cut at the level of the second genu. *FNm*, mastoid portion of the facial nerve.

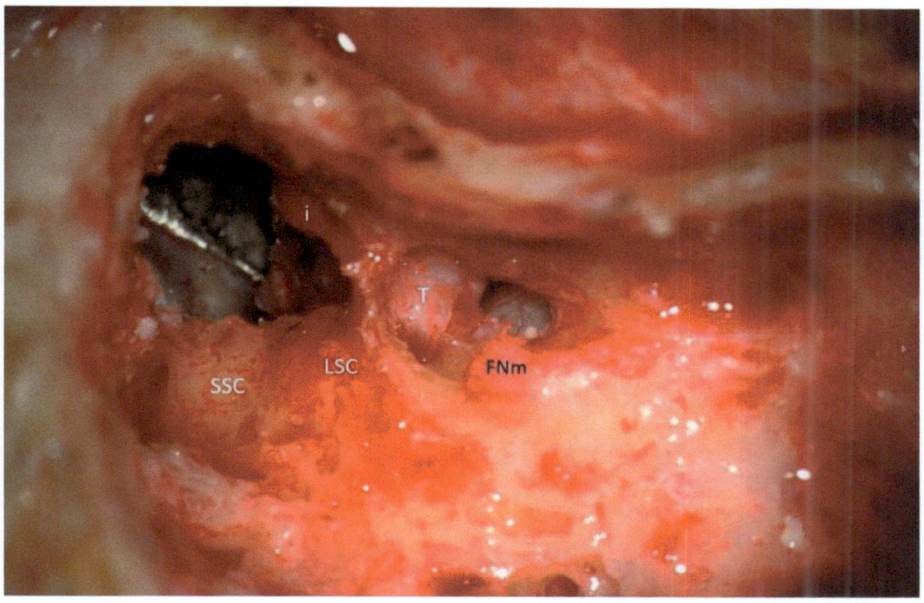

Fig. 11.94 The tumor (T) is then elevated. *FNm*, mastoid portion of the facial nerve; *i*, incus; *LSC*, lateral semicircular canal; *SSC*, superior semicircular canal.

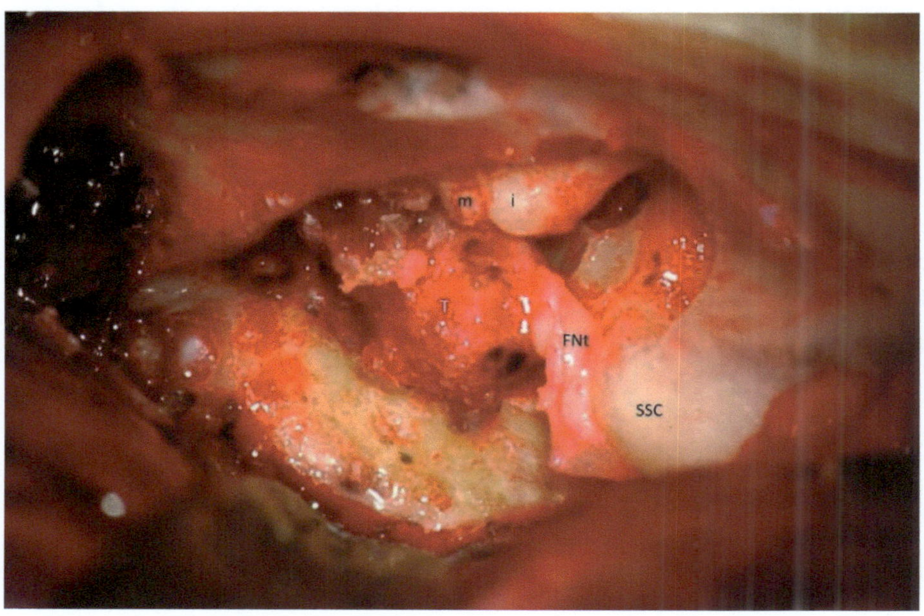

Fig. 11.95 The tympanic portion of the facial nerve (FNt), partially infiltrated by the tumor, is elevated, as can be seen from the enlarged middle cranial fossa approach. *SSC*, superior semicircular canal; *m*, malleus; *i*, incus; *T*, tumor.

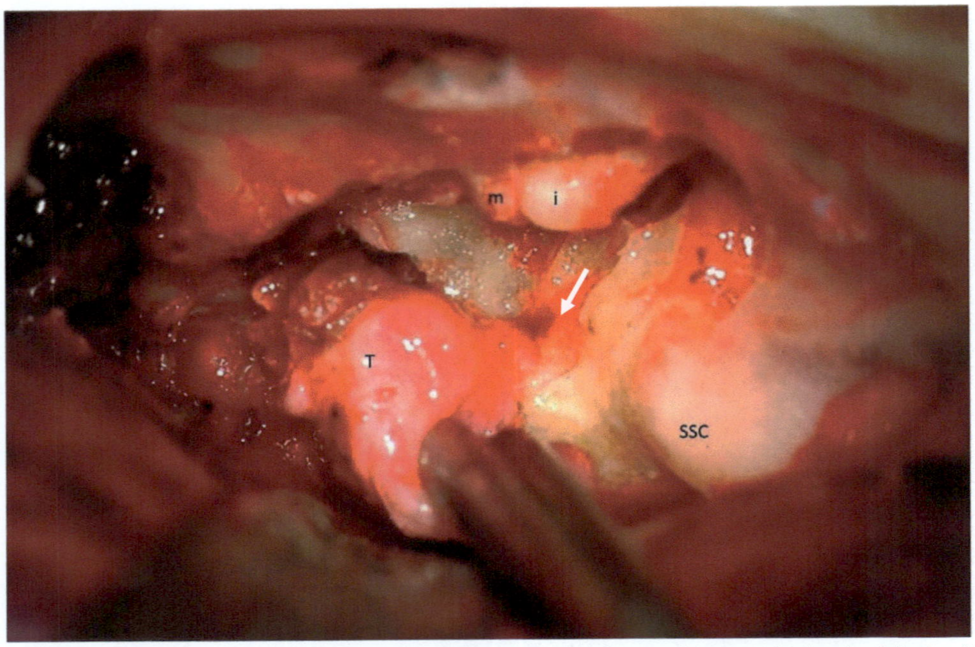

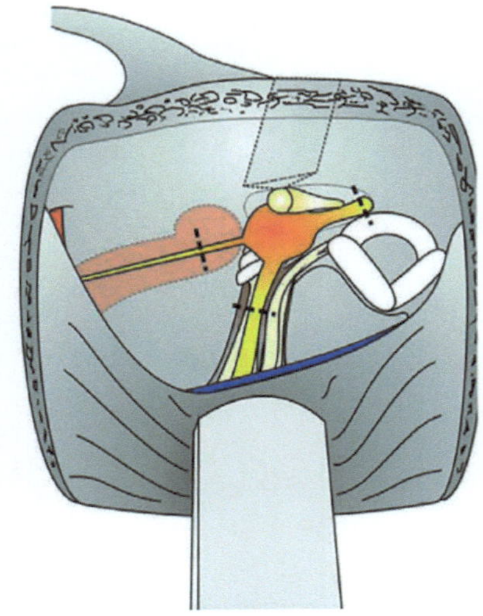

Fig. 11.96 The tumor is then totally removed, and the facial nerve is cut at the level of the healthy labyrinthine segment, as indicated in the drawing (*arrow*). *SSC*, superior semicircular canal; *m*, malleus; *i*, incus; *T*, tumor.

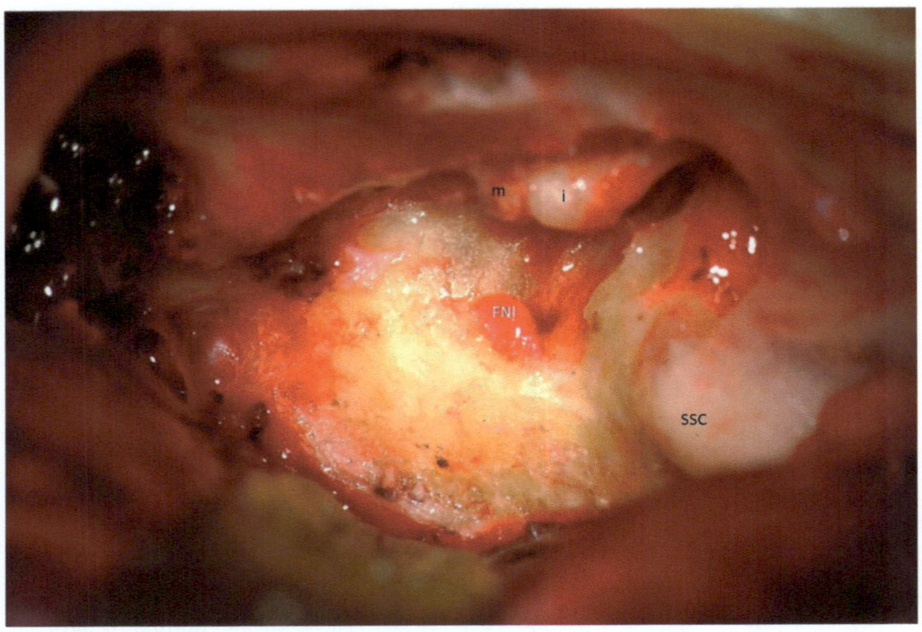

Fig. 11.97 The labyrinthine segment of the facial nerve (FNl) is visible. Complete removal of the tumor has been achieved. *SSC*, superior semicircular canal; *m*, malleus; *i*, incus.

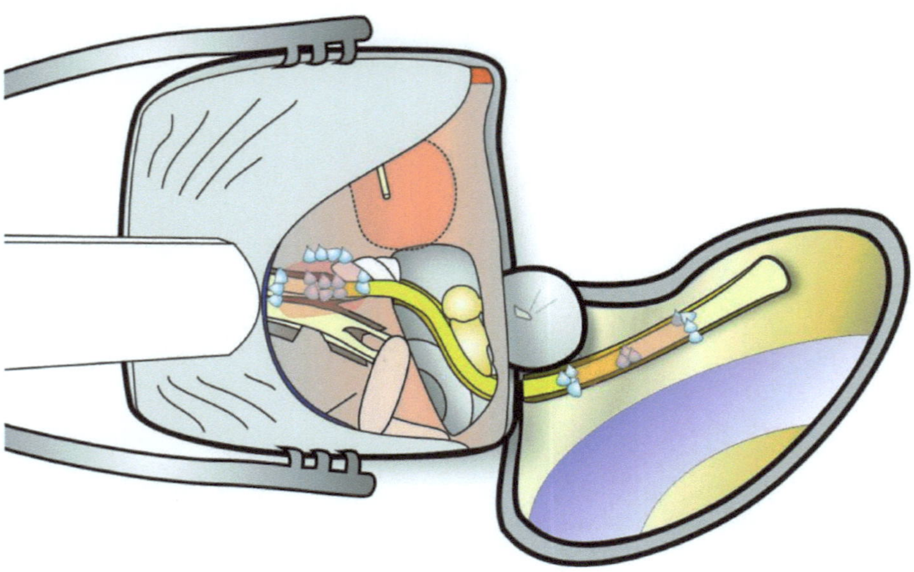

Fig. 11.98 Schematic drawing of facial nerve reconstruction after tumor removal. A sural nerve graft (*yellow line*) is placed between the labyrinthine segment of the facial nerve and the third segment of the facial nerve. The anastomosis is then fixed with fibrin glue.

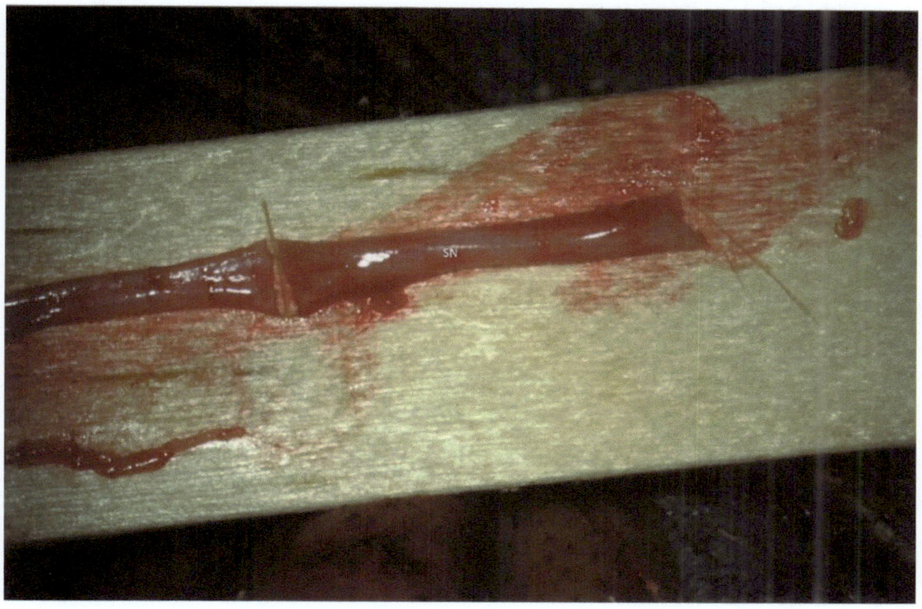

Fig. 11.99 The sural nerve (SN) graft is harvested.

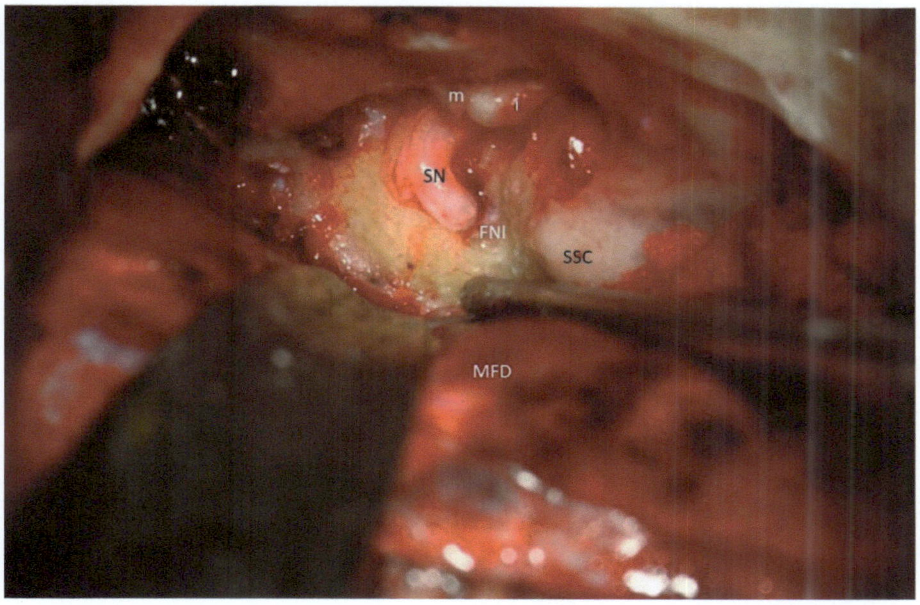

Fig. 11.100 The sural nerve (SN) graft is placed on the labyrinthine portion of the facial nerve (FNl) from the enlarged middle cranial fossa approach. *MFD*, middle fossa dura; *SSC*, superior semicircular canal; *m*, malleus; *i*, incus.

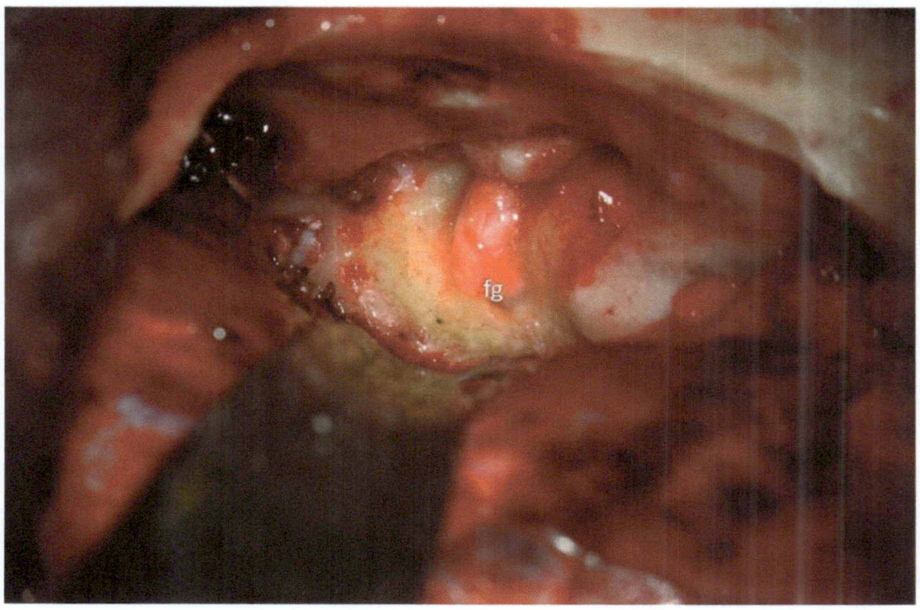

Fig. 11.101 The proximal end of the anastomosis is fixed with fibrin glue (fg).

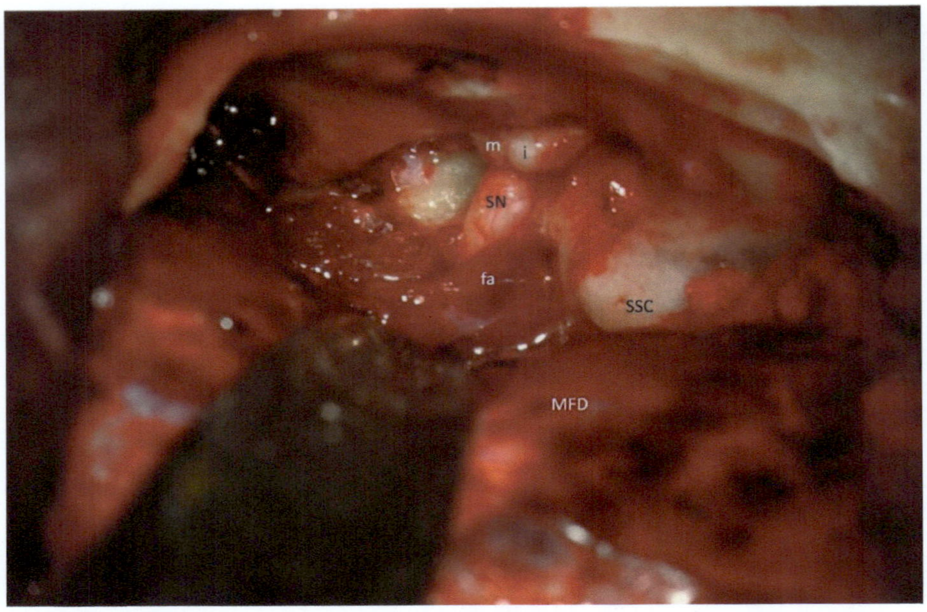

Fig. 11.102 A temporalis fascia (fa) is used to protect the proximal end of the anastomosis. *m*, malleus; *i*, incus; *SSC*, superior semicircular canal; *MFD*, middle fossa dura.

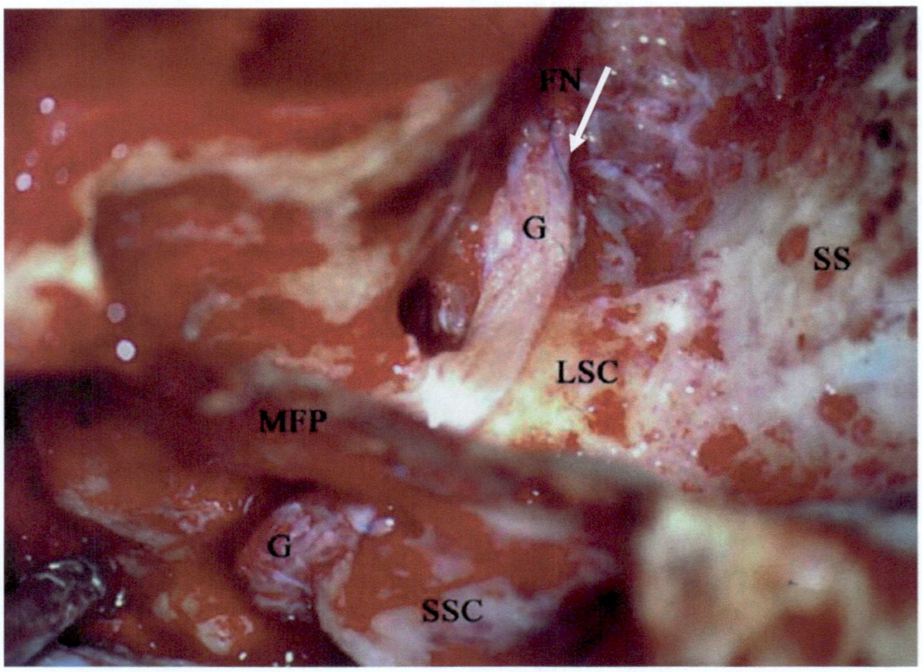

Fig. 11.103 The sural nerve graft (G) is then fixed distally at the level of the mastoid segment of the facial nerve (*arrow*). The fallopian canal has been used as the bed for the graft. *FN*, facial nerve; *SS*, sigmoid sinus; *LSC*, lateral semicircular canal; *SSC*, superior semicircular canal; *MFP*, middle fossa dura plate.

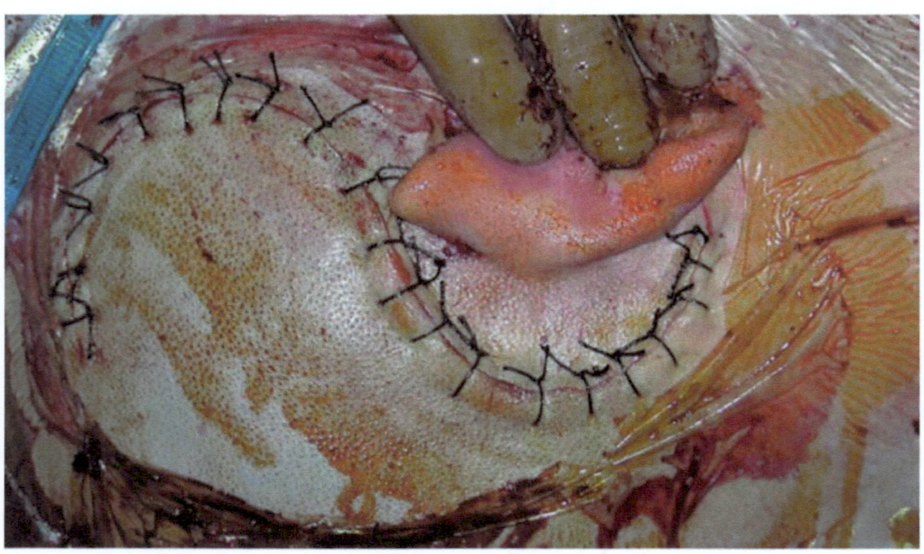

Fig. 11.104 The skin is closed with silk sutures.

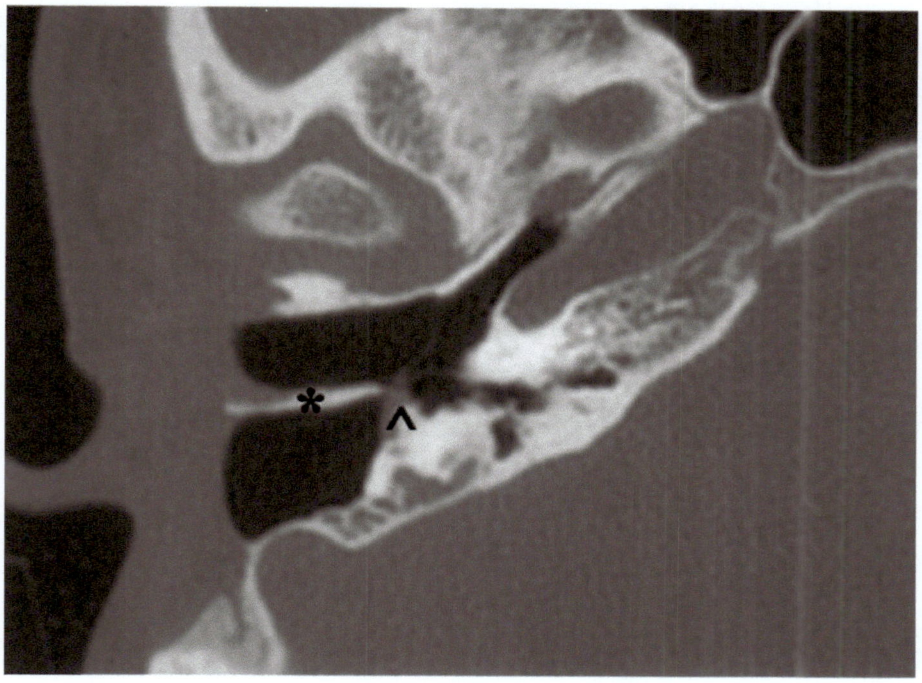

Fig. 11.105 The axial CT scan shows the mastoidectomy with the intact posterior canal wall (*). The soft tissue lying behind the canal wall (^) represents the facial nerve graft lying in the fallopian canal.

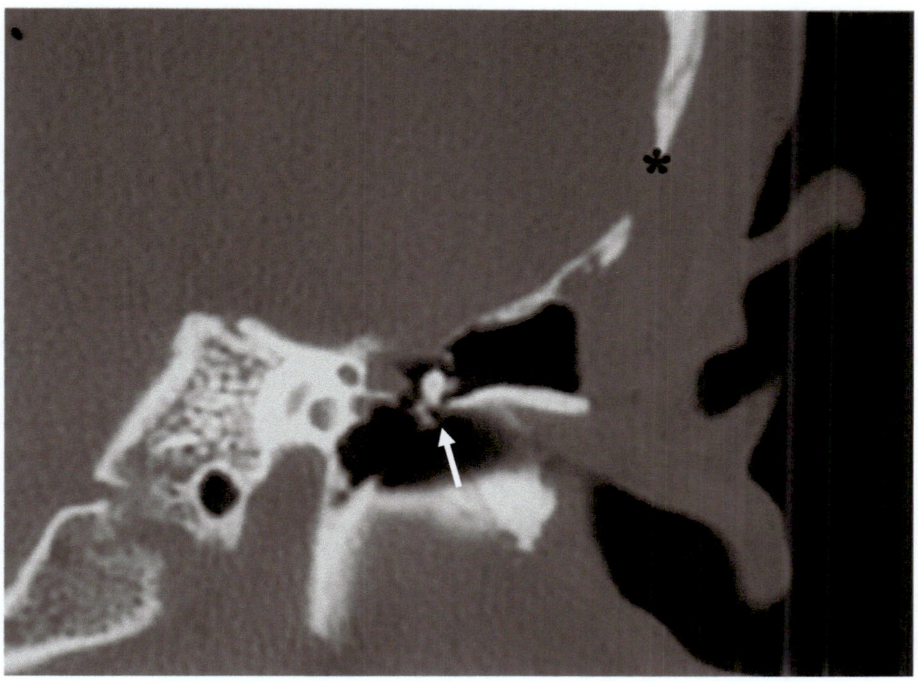

Fig. 11.106 The defect of the middle fossa craniotomy (*) can be seen. In the middle ear, the ossicles can be seen lying lateral to the facial nerve graft (*arrow*).

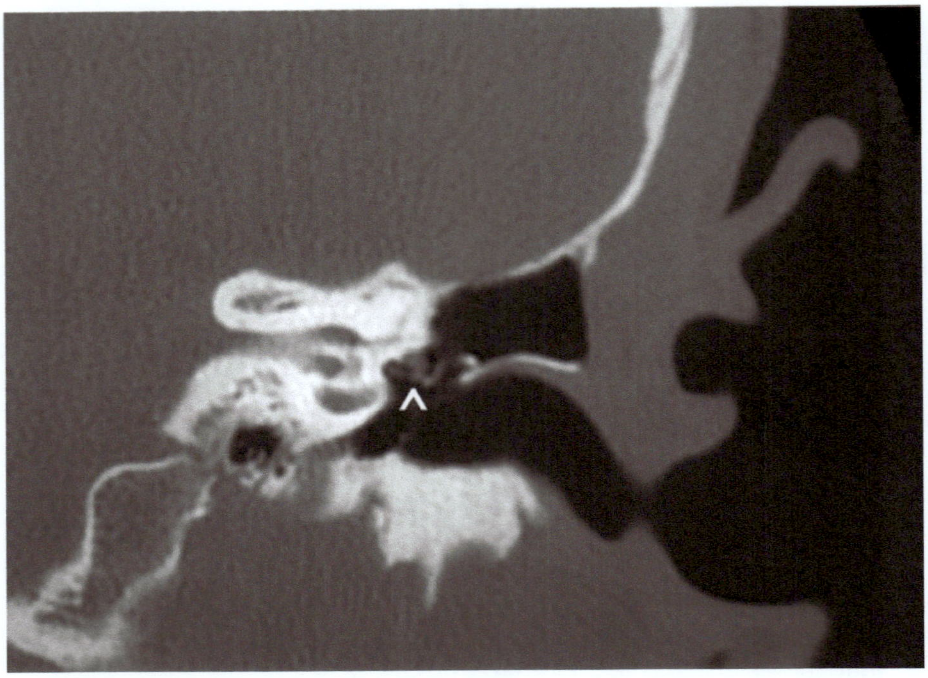

Fig. 11.107 In this section the facial nerve graft can be seen lying above the stapes (^) in the original fallopian canal.

HINTS AND PITFALLS

1. The craniotomy should be as low as possible, reaching the floor of the middle fossa to minimize the brain retraction.
2. The geniculate ganglion may be dehiscent in 16% of cases. Dural elevation should be done with care and from posterior to anterior to avoid any traction on the greater superficial petrosal nerve and subsequent injury to the dehiscent ganglion.
3. Dural elevation and temporal lobe retraction should be gentle and without rough manipulation. If the brain is tense, the surgeon should wait for dehydration measures to produce effective reduction in intracranial tension.
4. It is essential for dural elevation to reach the level of the superior petrosal sinus. Failure to achieve this would lead to more lateral drilling, with subsequent injury to vital structures.
5. Fukushima's Middle Fossa Retractor with Leyla's Arm offers better maneuverability than other retractors with less potential damage to the temporal lobe.
6. The blade of the retractor should be applied in such a way that it is centered on the porus. The blade should be advanced medially to be as near as possible to the superior petrosal sinus.
7. Various methods have been described to expose the internal auditory canal. Identification of the canal at the fundus as proposed by House and Fisch endangers the crowded and closely related important structures at this level. Fisch's method also has the disadvantage of using an in-constant angle between the superior semicircular canal and the internal auditory canal. Blue lining the superior semicircular canal also carries the risk of opening it. We use the method proposed by Garcia-Ibanez, but we start drilling medially at the level of the porus where no important structures are found (Figs. 11.10–11.13).
8. The internal carotid artery is intimately related to the apical and middle turns of the cochlea (Fig. 11.108). Keeping tangential to the internal surface of the internal carotid artery while carrying out the approach would put the cochlea at risk.
9. In rare cases with extensive supralabyrinthine pneumatization, air cells may be opened while drilling around the internal auditory canal. These cells should be closed with bone wax to avoid postoperative CSF leakage.
10. Identification of Bill's bar at the fundus of the internal auditory canal is an important step for positive identification of the facial nerve.
11. Opening of the posterior fossa dura is started posteriorly and continues anteriorly under vision to avoid injuring the abducent nerve (Figs. 11.109 and 11.110).

12. The facial nerve is the most superficial structure in the middle fossa approach and is therefore liable to trauma during tumor removal.
13. Tumor removal should be carried out in a medial-to-lateral direction whenever feasible. Dissection in the reverse direction might jeopardize the vascularization of the cochlear nerve and inner ear, with subsequent hearing loss.
14. In a few cases, the anterior inferior cerebellar artery loop enters the IAC and develops an intimate relationship with the canal contents, including the tumor. The utmost care should be taken in these cases to avoid injuring the artery while removing the tumor.
15. In cases of uncontrolled bleeding in the CPA, the superior petrosal sinus can be coagulated by use of the bipolar coagulator. The sinus is interrupted, and the tentorium is incised to provide enough space for better control.

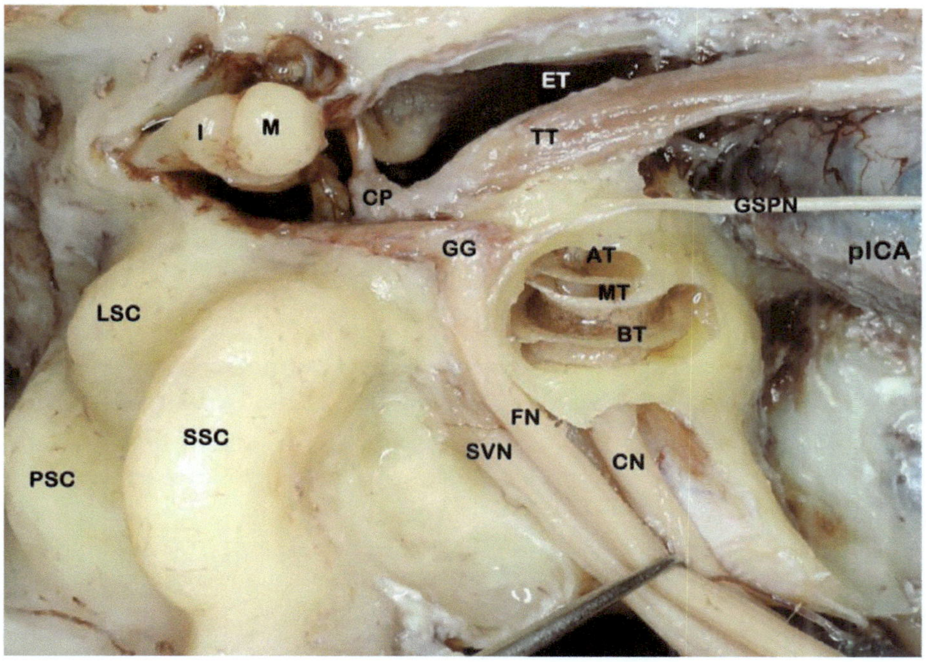

Fig. 11.108 The relationship of ICA to the turns of the cochlea (apical turn at, basal turn bt) can be appreciated. *M*, malleus; *I*, incus; *LSC*, lateral semicircular canal; *PSC*, posterior semicircular canal; *SSC*, superior semicircular canal; *GG*, geniculate ganglion; *AT*, apical turn of the cochlea; *BT*, basal turn of the cochlea; *MT*, middle turn of the cochlea; *FN*, facial nerve; *SVN*, superior vestibular nerve; *pICA*, petrous segment of internal carotid artery; *GSPN*, greater superficial petrosal nerve; *TT*, tensor tympani muscle; *ET*, eustachian tube; *CP*, cochleariform process; *CN*, cochlear nerve.

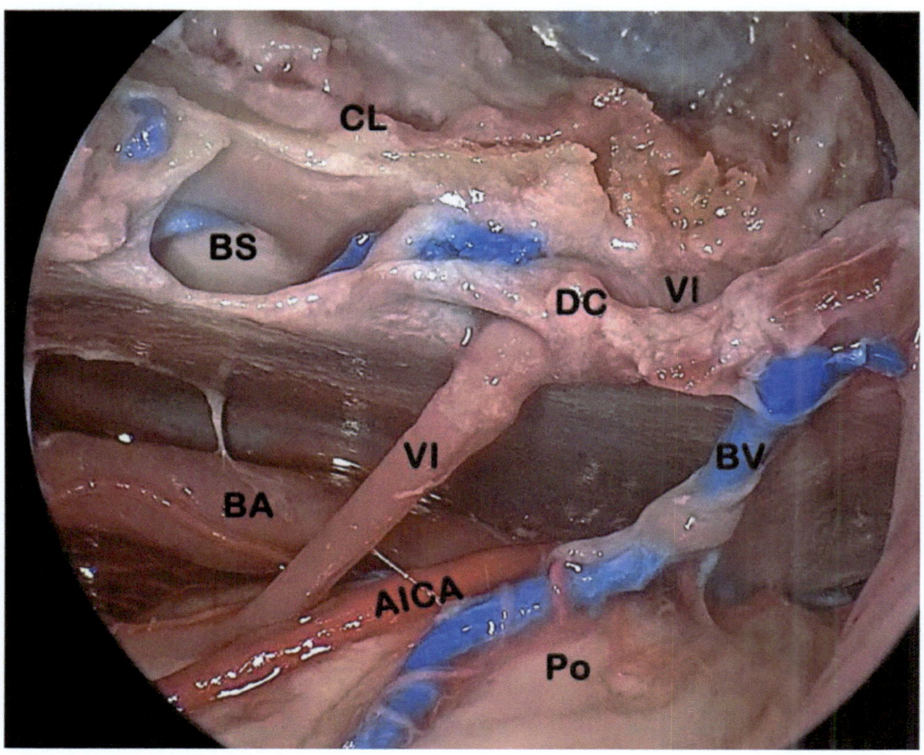

Fig. 11.109 The contents of the prepontine cistern. The abducens nerve passes above the anterior inferior cerebellar artery and through the inferior petrosal sinus, then passes through Dorello's canal (DC). *BA*, basilar artery; *AICA*, anterior inferior cerebellar artery; *Po*, pons; *VI*, abducens nerve; *BV*, bridge vein; *BS*, Basal sinus; *CL*, clivus.

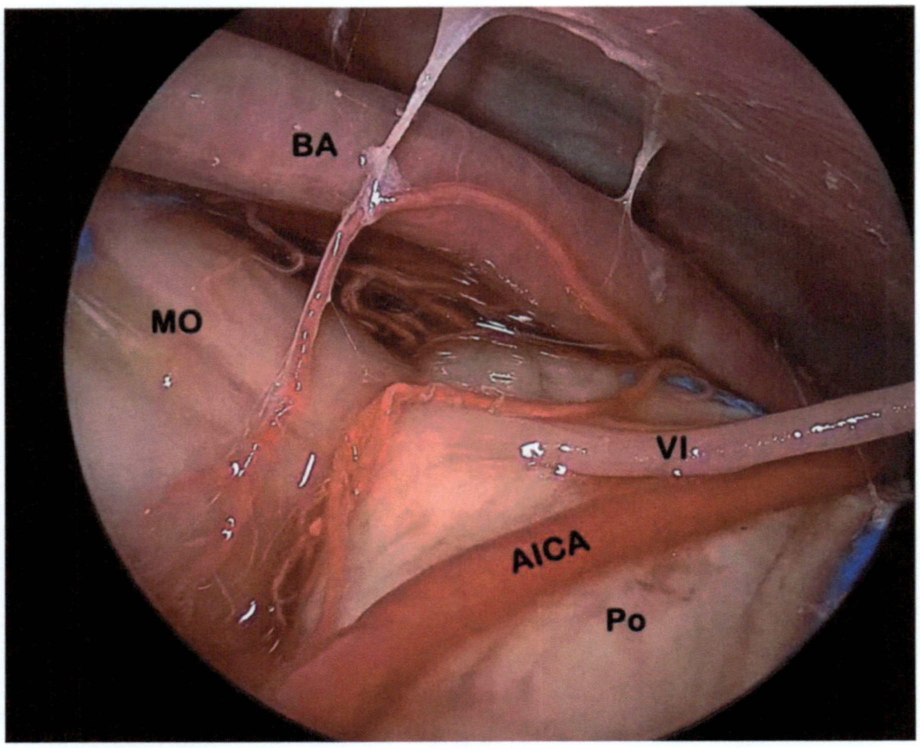

Fig. 11.110 The contents of the prepontine cistern. *BA*, basilar artery; *AICA*, anterior inferior cerebellar artery; *MO*, medulla oblongata; *Po*, pons;; *VI*, abducens nerve.

REFERENCES

[1] Hartley F. Intracranial neurectomy of the second and third divisions of the fifth nerve: a new method. N Y Med J 1892;55:317–9.

[2] House WF. Surgical exposure of the internal auditory canal and its contents through the middle, cranial fossa. Laryngoscope 1961;71:1363–85.

[3] Glasscock 3rd ME. Middle fossa approach to the temporal bone. An Otologic Frontier Arch Otolaryngol 1969;90:15–27.

[4] Bochenek Z, Kukwa A. An extended approach through the middle cranial fossa to the internal auditory meatus and the cerebello-pontine angle. Acta Otolaryngol 1975;80:410–4.

[5] Kanzaki J, Kawase T, Sano K, et al. A modified extended middle cranial fossa approach for acoustic tumors. Arch Otorhinolaryngol 1977;217:119–211.

[6] Morrison AW, King TT. Experiences with a translabyrinthine-transtentorial approach to the cerebellopontine angle. Technical note. J Neurosurg 1973;38:382–90.

[7] Kanzaki J, Shiobara R, Toya S. Classification of the extended middle cranial fossa approach. Acta Otolaryngol Suppl 1991;487:6–16.

[8] Wigand ME, Haid T, Berg M, Rettinger G. The enlarged transtemporal approach to the cerebello-pontine angle. Technique and indications. Acta Otorhinolaryngol Ital 1982;2:571–82.

[9] Wigand ME, Haid T, Berg M. The enlarged middle cranial fossa approach for surgery of the temporal bone and of the cerebellopontine angle. Arch Otorhinolaryngol 1989;246: 299–302.

[10] Hsu FP, Anderson GJ, Dogan A, et al. Extended middle fossa approach: quantitative analysis of petroclival exposure and surgical freedom as a function of successive temporal bone removal by using frameless stereotaxy. J Neurosurg 2004; 100:695–9.

[11] Kawase T, Toya S, Shiobara R, Mine T. Transpetrosal approach for aneurysms of the lower basilar artery. J Neurosurg 1985;63:857–61.

[12] House WF, Hitselberger WE, Horn KL. The middle fossa transpetrous approach to the anterior-superior cerebellopontine angle. Am J Otol 1986;7:1–4.

[13] Maina R, Ducati A, Lanzino G. The middle cranial fossa: morphometric study and surgical considerations. Skull Base 2007;17(6):395–403. https://doi.org/10.1055/s-2007-991117.

[14] Kawase T, Shiobara R, Toya S. Anterior transpetrosal-transtentorial approach for sphenopetroclival meningiomas: surgical method and results in 10 patients. Neurosurgery 1991;28:869–75.

[15] Kawase T, Shiobara R, Toya S. Middle fossa transpetrosal-transtentorial approaches for petroclival meningiomas: selective pyramid resection and radicality. Acta Neurochir 1994;119:113–210.

[16] Miller CG, Loveren HRv, Keller JT, et al. Transpetrosal approach: surgical anatomy and technique. Neurosurgery 1993;33:461–9.

[17] Slater PW, Welling DB, Goodman JH, Miner ME. Middle fossa transpetrosal approach for petroclival and brainstem tumors. Laryngoscope 1998;108:1408–11.

[18] Aziz KM, Loveren HRv, Tew JM, Jr Chicoine MR. The Kawase approach to retrosellar and upper clival basilar aneurysms. Neurosurgery 1999;44:1125–34.

[19] Kawase T, Bertalanffy H, Otani M, Shiobara R, Toya S. Surgical approaches for vertebro-basilar trunk aneurysms located in the midline. Acta Neurochir 1996; 138:402–10.

[20] Raheja A, Bowers CA, MacDonald JD, Shelton C, Gurgel RK, Brimley C, Couldwell WT. Middle Fossa Approach for vestibular schwannoma: good hearing and facial nerve outcomes with low morbidity. World Neurosurg 2016;92: 37–46. https://doi.org/10.1016/j.wneu.2016.04.085.

[21] Ahmed S, Arts HA, El-Kashlan H, Basura GJ, Thompson BG, Telian SA. Immediate and long-term hearing outcomes with the middle cranial Fossa Approach for vestibular schwannoma resection. Otol Neurotol 2018;39(1):92–8. https://doi.org/10.1097/MAO.0000000000001623.

[22] Scheich M, Ginzkey C, Ehrmann Müller D, Shehata Dieler W, Hagen R. Complications of the middle cranial Fossa Approach for acoustic neuroma removal. J Int Adv Otol 2017;13(2):186–90. https://doi.org/10.5152/iao.2017.3585.

[23] Scheich M, Ginzkey C, Ehrmann-Müller D, Shehata-Dieler W, Hagen R. Management of CSF leakage after microsurgery for vestibular schwannoma via the middle cranial fossa approach. Eur Arch Otorhinolaryngol 2016;273(10): 2975–81. https://doi.org/10.1007/s00405-015-3891-3.

[24] Alzhrani G, Shelton C, Couldwell WT. Middle fossa approach for resection of vestibular schwannoma. Acta Neurochir 2017;159(6):1023–6. https://doi.org/10.1007/s00701-017-3169-1 [Epub 2017 Apr 11].

[25] Scheich M, Ehrmann-Müller D, Shehata-Dieler W, Hagen R. Hearing results after middle fossa removal of small (T1/T2) vestibular schwannomas. HNO 2017;65(9):751–7. https://doi.org/10.1007/s00106-016-0228-5. German.

[26] Peng KA, Wilkinson EP. Optimal outcomes for hearing preservation in the management of small vestibular schwannomas. J Laryngol Otol 2016;130(7):606–10. https://doi.org/10.1017/S0022215116 007969 [Epub 2016 May 20].

[27] Quist TS, Givens DJ, Gurgel RK, Chamoun R, Shelton C. Hearing preservation after middle fossa vestibular schwannoma removal: are the results durable? Otolaryngol Head Neck Surg 2015;152(4):706–11. https://doi.org/10.1177/0194599814567874.

[28] Scheich M, Ginzkey C, Reuter E, Harnisch W, Ehrmann D, Hagen R. Quality of life after microsurgery for vestibular schwannoma via the middle cranial fossa approach. Eur Arch Otorhinolaryngol 2014;271(7):1909–16. https://doi.org/10.1007/s00405-013-2671-1.

[29] Peng KA, Lekovic GP, Wilkinson EP. Pearls for the middle fossa approach in acoustic neuroma surgery. Curr Opin Otolaryngol Head Neck Surg 2018;26(5):276–9. https://doi.org/10.1097/MOO.0000000000000479.

[30] Kosty JA, Stevens SM, Gozal YM, DiNapoli VA, Patel SK, Golub JS, et al. Middle Fossa Approach for resection of vestibular schwannomas: a decade of experience. Oper Neurosurg (Hagerstown) 2019;16(2):147–58. https://doi.org/10.1093/ons/opy116.

[31] Aihara N, Murakami S. Enlargement of the internal auditory canal and hearing preservation in the middle fossa approach for intracanalicular vestibular schwannomas. World Neurosurg 2015;84(6):1950–5. https://doi.org/10.1016/j.wneu.2015.08.041.

[32] Sameshima T, Fukushima T, McElveen Jr JT, Friedman AH. Critical assessment of operative approaches for hearing preservation in small acoustic neuroma surgery: retrosigmoid vs middle fossa approach. Neurosurgery 2010;67:640–4.

[33] Ansari SF, Terry C, Cohen-Gadol AA. Surgery for vestibular schwannomas: a systematic review of complications by approach. Neurosurg Focus 2011;33:E14.

[34] Arts HA, Telian SA, El-Kashlan H, Thompson BG. Hearing preservation and facial nerve outcomes in vestibular schwannoma surgery: results using the middle cranial fossa approach. Otol Neurotol 2006;27:234–41.

[35] Oghalai JS, Buxbaum JL, Pitts LH, Jackler RK. The effect of age on acoustic neuroma surgery outcomes. Otol Neurotol 2003;24:473–7.

[36] Lipschitz N, Kohlberg GD, Tawfik KO, Walters ZA, Breen JT, Zuccarello M, Andaluz N, Dinapoli VA, Pensak ML, Samy RN. Cerebrospinal fluid leak rate after vestibular schwannoma surgery via middle cranial fossa approach. J Neurol Surg B Skull Base 2019;80(4):437–40. https://doi.org/10.1055/s-0038-1675752.

[37] Ginzkey C, Scheich M, Harnisch W, Bonn V, Ehrmann-Müller D, Shehata-Dieler W, Mlynski R, Hagen R. Outcome on hearing and facial nerve function in microsurgical treatment of small vestibular schwannoma via the middle cranial fossa approach. Eur Arch Otorhinolaryngol 2013;270(4):1109–16. https://doi.org/10.1007/s00405-011-2074-8.

[38] Battelino S, Bošnjak R. Surgical approaches and anatomy of the lateral Skull Base. Encyclopedia Otolaryngol Head Neck Surg 2013;2644–63. https://doi.org/10.1007/978-3-642-23499-6_567.

[39] Chemtob RA, Noij KS, Qureshi AA, Klokker M, Nakajima HH, Lee DJ. Superior Canal dehiscence surgery outcomes following failed round window surgery. Otol Neurotol 2019;40(4): 535–42. https://doi.org/10.1097/MAO.0000000000002185.

[40] Johanis M, Yang I, Gopen Q. Incidence of intraoperative hearing loss during middle cranial fossa approach for repair of superior semicircular canal dehiscence. J Clin Neurosci 2018;54:109–12. https://doi.org/10.1016/j.jocn.2018.06.023.

[41] Schwartz SR, Almosnino G, Noonan KY, Banakis Hartl RM, Zeitler DM, Saunders JE, Cass SP. Comparison of transmastoid and middle fossa approaches for superior canal dehiscence repair: a multi-institutional study. Otolaryngol Head Neck Surg 2019;161(1):130–6. https://doi.org/10.1177/0194599819835173.

[42] Braca 3rd JA, Marzo S, Prabhu VC. Cerebrospinal fluid leakage from tegmen tympani defects repaired via the middle cranial fossa approach. J Neurol Surg B Skull Base 2013;74:103–7.

[43] Andresen NS, Sun DQ, Hansen MR. Facial nerve decompression. Curr Opin Otolaryngol Head Neck Surg 2018;1. https://doi.org/10.1097/moo.0000000000000478.

[44] Vincent C, Bonne NX, Guérin C, Lebreton JP, Devambez M, Dubrulle F, Haddad E, Schapira S, Lejeune JP, Vaneecloo FM. Middle fossa approach for resection of vestibular schwannoma: impact of cochlear fossa extension and auditory monitoring on hearing preservation. Otol Neurotol 2012;33(5): 849–52. https://doi.org/10.1097/MAO.0b013e318254ede3.

[45] Hilton CW, Haines SJ, Agrawal A, Levine SC. Late failure rate of hearing preservation after middle fossa approach for resection of vestibular schwannoma. Otol Neurotol 2011;32(1):132–5.

[46] Woodson EA, Dempewolf RD, Gubbels SP, et al. Long-term hearing preservation after microsurgical excision of vestibular schwannoma. Otol Neurotol 2010;31(7):1144–52.

[47] Quist TS, Givens DJ, Gurgel RK, et al. Hearing preservation after middle fossa vestibular schwannoma removal: are the results durable? Otolaryngol Head Neck Surg 2015;152: 706–11.

[48] Friedman RA, Kesser B, Brackmann DE, et al. Long-term hearing preservation after middle fossa removal of vestibular schwannoma. Otolaryngol Head Neck Surg 2003;119:660–5.

[49] Roche JP, Woodson EA, Hansen MR, Gantz BJ. Ultra long-term audiometric outcomes in the treatment of vestibular schwannoma with the middle cranial Fossa Approach. Otol Neurotol 2018;39(2):e151–7. https://doi.org/10.1097/MAO.0000000000001678.

[50] Gjuric M, Schneider W, Berg M. Audiologic classification of hearing retention after removal of acoustic neurinomas. HNO 1997;45(5):350–2.

[51] Hajioff D, Raut VV, Walsh RM, et al. Conservative management of vestibular schwannomas: third review of a 10-year prospective study. Clin Otolaryngol 2008;33(3):255–9.

[52] Kirchmann M, Karnov K, Hansen S, et al. Ten-year follow-up on tumor growth and hearing in patients observed with an intracanalicular vestibular schwannoma. Neurosurgery 2017;80(1):49–56.

[53] Coughlin AR, Willman TJ, Gubbels SP. Systematic review of hearing preservation after radiotherapy for vestibular schwannoma. Otol Neurotol 2018;39(03):273–83.

[54] Watanabe S, Yamamoto M, Kawabe T, et al. Stereotactic radiosurgery for vestibular schwannomas: average 10-year follow-up results focusing on long-term hearing preservation. J Neurosurg 2016;115:64–72.

[55] Carlson ML, Jacob JT, Pollock BE, et al. Long-term hearing outcomes following stereotactic radiosurgery for vestibular schwannoma: patterns of hearing loss and variables influencing audiometric decline. J Neurosurg 2013;118:579–87.

[56] Hasegawa T, Kida Y, Kato T, et al. Factors associated with hearingpreservation after Gamma Knife surgery for vestibular schwannomas in patients who retain serviceable hearing. J Neurosurg 2011;115:1078–86.

[57] Tolisano AM, Littlefield PD. Adverse events following vestibular schwannoma surgery: a comparison of surgical Approach. Otol Neurotol 2017;38(4):551–4. https://doi.org/10.1097/MAO.0 000000000001323.

[58] Kang WS, Kim SA, Yang CJ, Nam SH, Chung JW. Surgical outcomes of middle fossa approach in intracanalicular vestibular schwannoma. Acta Otolaryngol 2017;137(4):352–5. https://doi.org/10.1080/00016489.2016.1155992.

[59] Anaizi AN, DiNapoli VV, Pensak M, Theodosopoulos PV. Small vestibular schwannomas: does surgery remain a viable treatment option? J Neurol Surg B Skull Base 2016;77 (3):211–8. https://doi.org/10.1055/s-0035-1564591.

[60] Hoang S, Ortiz Torres MJ, Rivera AL, Litofsky NS. Middle cranial fossa approach to repair tegmen defects with autologous or alloplastic grafts. World Neurosurg 2018. https://doi.org/10.1016/j.wneu.2018.05.196.

[61] Tolisano AM, Kutz Jr JW. Middle fossa approach for spontaneous cerebrospinal fluid fistula and encephaloceles. Curr Opin Otolaryngol Head Neck Surg 2019;27(5):356–60. https://doi.org/10.1097/MOO.0000000000000560.

[62] Alwani M, Bandali E, Van Buren L, Yates CW, Nelson RF. Audiologic improvement following MCF Approach for spontaneous cerebrospinal fluid leaks. Otol Neurotol 2019;40(8): 1026–33. https://doi.org/10.1097/MAO.0000000000002302.

[63] Gupta A, Sikka K, Irugu DVK, Verma H, Bhalla AS, Thakar A. Temporal bone meningoencephalo-celes and cerebrospinal fluid leaks: experience in a tertiary care hospital. J Laryngol Otol 2019;133(3):192–200. https://doi.org/10.1017/S0022215119000203.

[64] Friedman RA, Pensak ML, Tauber M, et al. Anterior petrosectomy approach to infraclinoidal basilar artery aneurysms: the emerging role of the neuro-otologist in multidisciplinary management of basilar artery aneurysms. Laryngoscope 1997;107:977–83.

[65] Shen T, Friedman RA, Brackmann DE, et al. The evolution of surgical approaches for posterior fossa meningiomas. Otol Neurotol 2004;25:394–7.

[66] Lipschitz N, Kohlberg GD, Zuccarello M, Samy RN. Comprehensive review of the extended middle cranial fossa approach. Curr Opin Otolaryngol Head Neck Surg 2018;1. https://doi.org/10.1097/moo.0000000000000471.

[67] Liu JK. Extended middle fossa approach with anterior petrosectomy for resection of upper petroclival meningioma involving Meckel's cave: operative video and technical nuances. Neurosurg Focus 2017;43(VideoSuppl2):V8. https://doi.org/10.3171/2017.10.focusvid.17345.

[68] Natarajan SK, Sekhar LN, Schessel D, et al. Petroclival meningiomas: multimodality treatment and outcomes at long-term follow-up. Neurosurgery 2007;60:965–79 [Discussion 79–81].

[69] Abdel Aziz KM, Sanan A, van Loveren HR, et al. Petroclival meningiomas: predictive parameters for transpetrosal approaches. Neurosurgery 2000;47:139–50. Discussion 50–2.

[70] Roche JP, Goates AJ, Hasan DM, Howard MA, Menezes AH, Hansen MR, Gantz BJ. Treatment of lateral skull base and posterior cranial fossa lesions utilizing the extended middle cranial fossa approach. Otol Neurotol 2017;38(5): 742–50. https://doi.org/10.1097/mao.0000000000001356.

[71] Friedman RA, Ksenia AA. Extended middle cranial fossa approach. In: Brackmann DE, Shelton C, Arriaga MA, editors. Otologic surgery. 4th ed. Philadelphia: W.B. Saunders Co./Elsevier; 2016.

[72] Danner C, Cueva RA. Extended middle fossa approach to the petroclival junction and anterior cerebellopontine angle. Otol Neurotol 2004;25:762–8.

[73] Breuer T, Gjuric M, Wigand ME. Extended middle fossa surgery for meningiomas within or at the internal auditory canal. Am J Otol 2000;21:729–34.

[74] Gjuric M, Wigand ME, Wolf SR. Enlarged middle fossa vestibular schwannoma surgery: experience with 735 cases. Otol Neurotol 2001;22:223–30. discussion 30–31.

CHAPTER 12

The Infratemporal Fossa Approaches

Introduction

The infratemporal fossa approaches (IFT) developed by Ugo Fisch in the 1970s [1–3] are craniotemporocervical procedures that allow access to lesions occupying the jugular foramen, infratemporal fossa, clivus, cavernous sinus, and nasopharynx [4,5]. Surgery in these delicate regions is still fraught with danger, requiring the ability to accurately assess varied interrelated factors, their impact on the ultimate result, and their management. These approaches are classified into three postauricular procedures, types A, B, and C, and one preauricular procedure, the type D approach. The first three involve a subtotal petrosectomy causing an irreversible conductive hearing loss and possible facial nerve impairment, the type D approach does not impair hearing or nerve function, but it is usually limited to small, anteriorly located tumors (<2.5 cm) [6]. Although the IFT approach is associated with a degree of morbidity, it provides wide uninterrupted access to the jugular bulb and intrapetrous carotid artery [7]. The particularities of each surgical approach, indications, surgical technique, and limitations will be presented. The infratemporal fossa Type A is the most commonly used approach for the jugular foramen area; the type B is ideal for petrous apex lesions involving the internal carotid artery. With the advent of the transnasal endoscopic procedures, type C and D IFT approaches are rarely necessary in current skull base surgical practice.

Classification

- ITF type A (IFTA) is a complex approach that provides access to the infralabyrinthine and apical compartments as well as to the inferior surface of the petrous temporal bone. The technique entails performing an upper cervical dissection, subtotal petrosectomy, anterior transposition of the facial nerve, excision of the styloid process, and drilling skull base to expose the jugular foramen, vertical petrous carotid, and posterior infratemporal fossa.
- ITF type B involves inferior reflection of the temporal muscle, disarticulation of the temporo-madibular joint, and removal of the middle cranial fossa floor to provide access to the ITF and clivus.
- ITF type C has been developed for parasellar lesions and extended nasopharyngeal lesions invading the base of the skull along the Eustachian tube. The type C approach implies the resection of the pterygoid plates in addition to the ITFB technique.
- ITF type D is a preauricular variation of the ITF approach that provides access to the nasopharynx, parapharyngeal space, and pterygopalatine fossa without sacrificing the middle ear; it is performed by removing the posterior root of the zygomatic arch and drilling the bone at the base of MCF, anterior to the ICA toward the sphenoid sinus (Fig. 12.1).

Atlas of Microsurgery of the Lateral Skull Base. https://doi.org/10.1016/B978-0-443-19083-4.00012-X

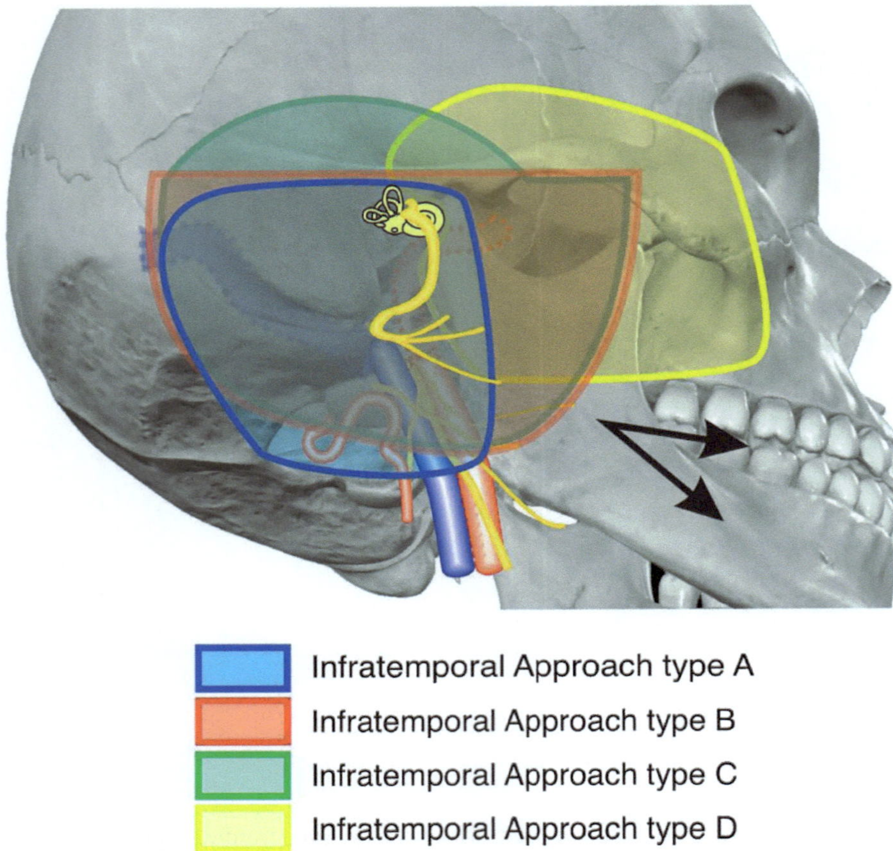

Fig. 12.1 A schematic drawing of the classification of the infratemporal fossa approach is shown. This approach is classified into three postauricular procedures, types A, B, and C, and one preauricular exposure, the type D approach.

Infratemporal Fossa Approach Type A

Rationale

This approach is designed to allow access to the jugular foramen area, the infralabyrinthine and apical compartments of the petrous bone, the vertical segment of the internal carotid artery, and the upper carotid space or poststyloid parapharyngeal space (Fig. 12.2A). The approach is designed primarily for extensive extradural lesions involving these areas. The key point in this approach is the anterior transposition of the facial nerve to provide optimal control of the targeted areas (Fig. 12.2B). The other structures that hinder lateral access to these areas are shown in Fig. 12.2C. As well as the facial nerve, they include the tympanic bone, the digastric muscle, and the styloid process. These structures are removed to allow unimpeded lateral access.

Circumferential dissection of the facial nerve in its vertical segment with preservation of a thin bony coverage is referred to as a fallopian bridge. This can be associated with either preservation or removal of the EAC and middle ear structures [8] and allows access to the jugular bulb on either side of the facial nerve [9]. This technique limits anterior control, and there remains a small risk of facial nerve injury due to possible fracture of the fallopian bridge. Facial nerve mobilization also allows the safe removal of the styloid process and the complete removal of the tympanic bone, which is invariably infiltrated in TJP. This would provide for optimal control of the upper parapharyngeal space, the ICA, and the lower cranial nerves (LCNs). The true results of facial function following rerouting must be analyzed in comparison with tumor recurrence rates and with associated morbidity when the facial nerve is not mobilized. We strongly feel that the relatively mild facial dysfunction associated with anterior rerouting from the geniculate ganglion is a small price to pay for improved access that provides definitive vascular control and ultimately reduces recurrence rates [10–21].

The morbidity associated with the classic ITFA includes:

- Conductive hearing loss.
- Facial nerve dysfunction: at least 80% of patients undergoing permanent anterior rerouting regain House-Brackmann (HB) II or III within 1 year postoperatively.
- LCN dysfunction, which is usually fully recovered or well compensated in time.
- Minor short-term masticatory difficulties.

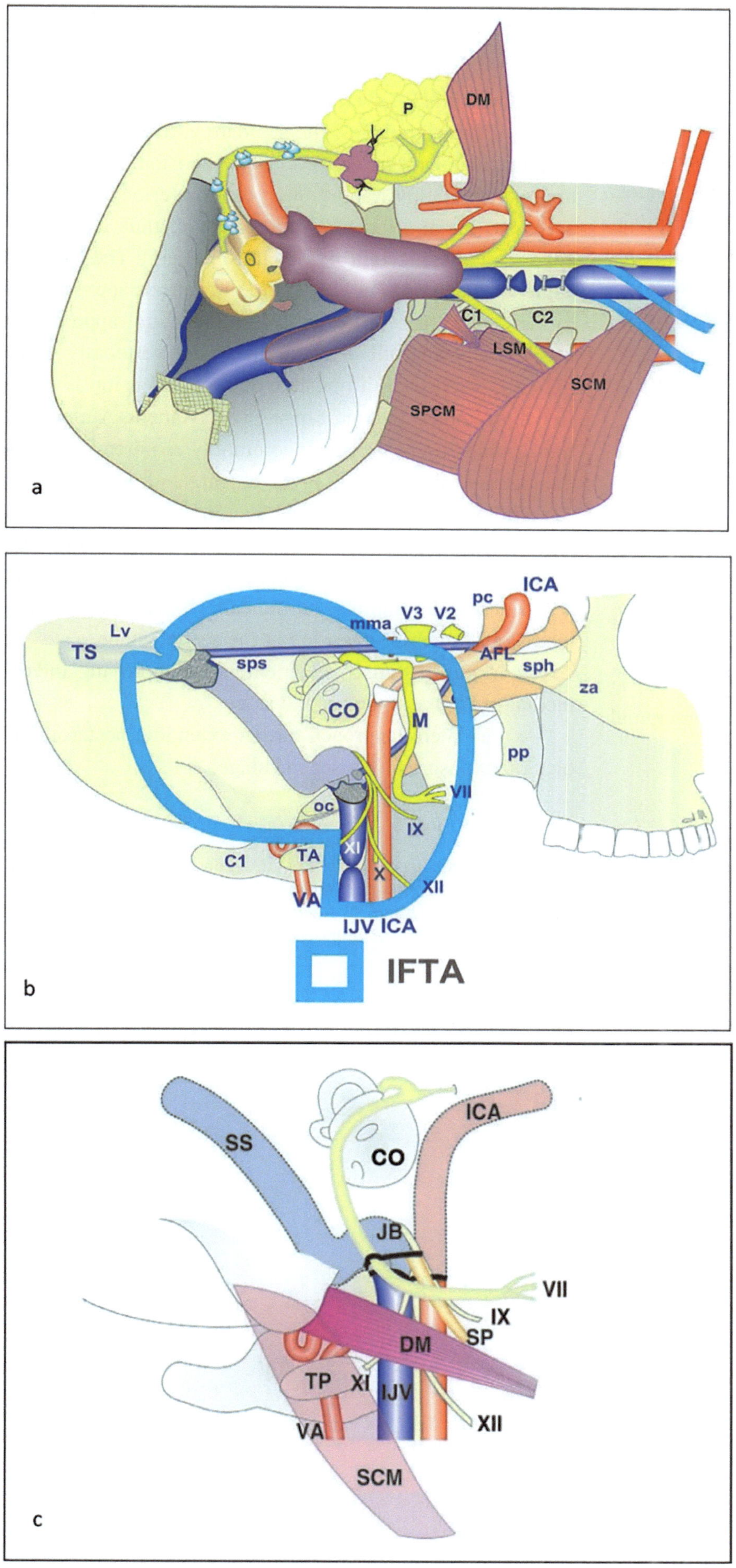

Fig. 12.2 (A–C) Illustrations for infratemporal fossa approach type A (ITFA). (A) An illustration of the surgical view in IFTA. (B) An illustration of the surgical limit in ITFA. (C) An illustration of obstacles to approaching the jugular bulb. *AFL*, anterior foramen lacerum; *C1*, atlas; *C2*, axis; *CO*, cochlea; *DM*, posterior belly of the digastric muscle; *ICA*, internal carotid artery; *IJV*, internal jugular vein; *JB*, jugular bulb; *LSM*, levator scapulae muscle; *Lv*, vein of Labbé; *mma*, middle meningeal artery; *M*, mandible; *OC*, occipital condyle; *P*, parotid gland; *pc*, clinoid process; *pp*, pterygoid plate; *SCM*, sternocleidomastoid muscle; *SP*, styloid process; *SPCM*, splenius capitis muscle; *sph*, sphenoid sinus; *sps*, superior petrosal sinus; *TP*, transverse process of the atlas; *TS*, transverse sinus; *V2*, maxillary branch of the trigeminal nerve; *V3*, mandibular branch of the trigeminal nerve; *za*, zygomatic arch; *VA*, vertebral artery; *VII*, facial nerve; *IX*, glossopharyngeal nerve; *XI*, spinal accessory nerve; *XII*, hypoglossal nerve.

Indications

- Lesions of the jugular foramen.
  - Types C and D tympanojugular paragangliomas. (This is the main indication for the approach.)
  - Lower cranial nerve schwannomas and meningiomas of the jugular foramen. However, in these cases, we can also resort to the petrooccipital transsigmoid approach with preservation of the middle ear function and without transposing the facial nerve.
- Lesions of the infralabyrinthine and apical compartments of the temporal bone.
  - Infralabyrinthine and some cases of apical cholesteatoma.
  - Lower clivus chordomas.
  - Some cases of extensive facial nerve neuromas.

Contraindications

- Severe contralateral hearing loss.
- Unique ipsilateral sigmoid sinus (anatomical or postsurgery).
- Contralateral vocal fold palsy.
- Inoperable disease due to extensive intracranial development.
- Advanced age >65 years.

Extensions of the Infratemporal Fossa Approach Type A

Based on the ITFA, various extensions are dictated by the extent of the pathology. The standard extension we use is a transcondylar, transtubercular extension for C2–C4 tumors. This allows additional posteroinferior and medial access to the JF, widening the exposure, thus facilitating venous and neural control (Fig. 12.3) [22–31]. The widened angle also affords better access to the petrous apex, medial to the carotid artery. Very rarely a far-lateral approach is employed with full exposure of the vertebral artery [32–36]. The use of a translabyrinthine extension is occasionally required for otic capsule involvement. A modified transcochlear approach is uncommonly required to access the petrous apex, clival involvement, and infratemporal fossa involvement:

- Transcondylar, transtubercular extension improving postero-infero-lateral and medial exposure at the first stage.
- Translabyrinthine or transotic extension allowing medial and superior exposure at the second stage.
- Modified transcochlear extension improving anteromedial exposure at the second stage.
- Far-lateral approach further extending posteroinferolateral exposure at the second stage.

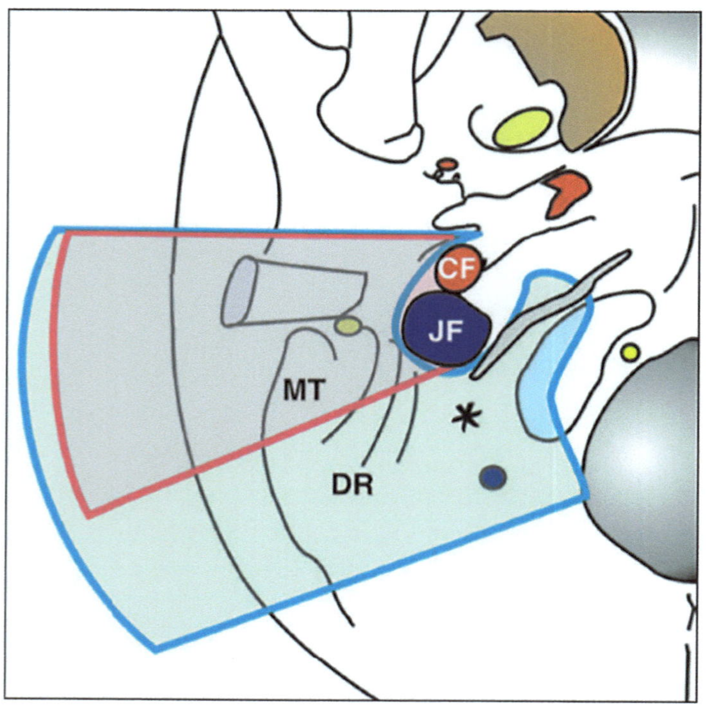

Fig. 12.3 Comparison of classic ITFA (zoned delimited by the *red line*) and ITFA with transcondylar-transtubercular extension (zone delimited by the *blue line*). *, Jugular process of the occipital condyle; *CF*, carotid foramen; *DR*, digastric ridge; *JF*, jugular foramen; *MT*, mastoid tip.

Surgical Steps (Left Side)

1. A postauricular skin incision that extends down to the neck is performed (Fig. 12.4).

2. The large postauricular flap is raised while keeping its adequate thickness. The skin flap is reflected anteriorly and maintained in position with two hooks (Fig. 12.5).

3. The musculoperiosteal layer is incised in a T-shaped fashion using a cutting diathermy (Fig. 12.6). The flap is elevated using a Lempert periosteal elevator. We use 1-0 silk sutures to stitch the soft tissue away from the surgical field, thus avoiding the use of retractors (Fig. 12.7). The inferior part of the musculoperiosteum is incised with diathermy and is then elevated, creating an inferiorly based flap continuous with the sternomastoid muscle. Sharp dissection and diathermy are needed to disinsert the sternomastoid muscle from the tip of the mastoid.

4. The external auditory canal is transected and closed in cul-de-sac (Figs. 12.8–12.13).

5. The sternocleidomastoid muscle has been dissected from the mastoid tip and sutured to the skin edge. Then the sternocleidomastoid muscle is retracted posteriorly to show the posterior belly of the digastric muscle. The digastric muscle can now be seen inserting into the digastric groove, medial to the mastoid tip (Fig. 12.14).

6. The lateral process of C1 is palpated as a landmark to identify the internal jugular vein and the point at which cranial nerve XI passes laterally to it (in 80%–85% of cases) (Fig. 12.15). The posterior belly of the digastric muscle is cut (Fig. 12.16) and retracted inferiorly. The underlying muscle that can be seen is the stylohyoid muscle inserting on the upper lateral segment of the styloid process. The internal jugular vein, occipital artery, and spinal accessory nerve can be seen (Fig. 12.17).

7. Further dissection with exposure of the internal carotid artery, vagus nerve, hypoglossal nerve, ansa cervicalis, and branches of external carotid artery (Fig. 12.18). Two vessel loops are placed in order to mark the internal carotid artery and the internal jugular vein (Fig. 12.19). At a higher magnification, the spinal accessory nerve, hypoglossal nerve, the internal carotid artery, external carotid artery, facial artery, ascending pharyngeal artery, and occipital artery are all well seen (Fig. 12.20).

8. The skin of the external auditory canal, the tympanic membrane, the malleus, and the incus are removed. A subtotal petrosectomy is performed (Figs. 12.21–12.26). The tendon of stapedius muscle is being cut using neurosurgical scissors (Fig. 12.27). The superstructure of the stapes is then removed using neurosurgical scissors. The crura of the stapes are transected to avoid trauma to the inner ear while dissecting the facial nerve during anterior rerouting. This is essential to avoid sensorineural hearing loss during anterior rerouting of the facial nerve due to footplate displacement (Figs. 12.28 and 12.29).

9. The vertical segment of the facial nerve is skeletonized. This is facilitated by drilling the retrofacial air cells and the tympanic bone anterior to the facial nerve. The retrofacial air cells are drilled to complete the skeletonization of the facial nerve (Fig. 12.30). The jugular bulb is identified after drilling the retrofacial air cells (Fig. 12.31).

10. Drilling of the tympanic bone between the internal carotid artery and the mastoid segment of the facial nerve is a very important step to get complete control of the jugular foramen area and the mastoid segment of the facial nerve (Fig. 12.32).

11. Decompression of the facial nerve is started. Drilling should be carried out parallel to the facial nerve (Fig. 12.33). The anterior wall of the external auditory canal is completely drilled. The facial nerve is skeletonized from the stylomastoid foramen to the geniculate ganglion (Fig. 12.34). The last shell of bone is removed with a double curved raspatory. Care must be taken to fully uncover the geniculate ganglion. Sharp bone left at this level might injure the anteriorly transposed nerve.

12. The mastoid tip is amputated. The tough attachments are sharply cut, taking care not to injure the facial nerve (Fig. 12.35).

13. The main trunk of extratemporal facial nerve is then identified (Fig. 12.36). The trunk is traced in the parotid until the proximal parts of the temporofacial and cervicofacial trunks are identified (Fig. 12.37). The facial nerve has to be liberated enough in the parotid gland. This provides an extra length of free nerve and prevents it from being stretched during the transposition.

14. Anterior rerouting of the facial nerve is starting. A new bony canal is drilled in the root of the zygoma superior to the Eustachian tube. The facial nerve is freed at the stylomastoid foramen using strong scissors. The soft tissue cuff surrounding the nerve at this level is not removed but rather kept intact. The mastoid segment is next separated using a Beaver knife to cut the fibrous attachments of the nerve to the fallopian canal. This tissue affords protection for the nerve during transposition. It also helps to maintain the vascularity of the nerve (Fig. 12.38).

15. Using a curved raspatory, the tympanic segment is carefully elevated. A nontoothed forceps are used to hold the soft tissue at the stylomastoid foramen and anterior transposition of the nerve is carried out. The nerve is fixed in its new

bony canal just superior to the Eustachian tube using tissue glue. A tunnel is created in the parotid gland where the transposed nerve is lodged and the tunnel is closed around the nerve using two sutures. Anterior rerouting of the facial nerve has been completed (Fig. 12.39). Displacement of the facial nerve represents the key point of the surgery because it allows unobstructed control of the jugular foramen and the vertical carotid canal.

16. The structures attached to the styloid process are severed. The process is then cut with strong scissors (Fig. 12.40). Removal of the fallopian canal with a rongeur (Fig. 12.41). The rest of the tympanic bone is drilled out, and the styloid process is removed to expose the petrous segment of internal carotid artery (Fig. 12.42). Note the relationship between the anterior wall of the jugular bulb and the tympanic bone. During the removal of the base of the styloid process, the utmost care has to be taken not to injure the ICA lying immediately deep in it. The infralabyrinthine air cells are drilled. These cells are usually infiltrated by the tumor and have to be completely removed. A diamond burr is used at this step to aid in hemostasis.

17. Drilling of the jugular process and occipital condyle is performed with a large diamond burr. Further bone has been removed posteroinferiorly to the sigmoid and jugular bulb, representing a partial removal of the posteromedial segment of the occipital condyle (Fig. 12.43).

18. Drilling of the occipital condyle is completed. Drilling out one-third of the lateral part of the occipital condyle permits complete visualization of the inferior aspect of the jugular bulb (Fig. 12.44).

19. The sigmoid sinus is completely uncovered (Fig. 12.45). The extraluminal compression of the sigmoid sinus is facilitated by the preservation of a bony shell over the proximal sigmoid sinus and prevents blocking of the transverse sinus (Fig. 12.46). This is the technique of extraluminal closure of the sigmoid sinus to avoid the risk of CSF leakage with the suture of the sigmoid sinus.

20. The internal jugular vein is then double ligated and transected. This is done following sigmoid closure to minimize venous congestion of the tumor (Fig. 12.47).

21. The lower cranial nerves can now be visualized after reflection of the internal jugular vein (Fig. 12.48). In cases in which the accessory nerve crosses the vein anterolaterally, the vein has to be pulled carefully under the nerve to prevent injury to it. The ICA is surrounded by a dense fibroperiosteal layer at its entrance in the skull base. The glossopharyngeal nerve lies deep in this layer and is slightly adherent to its undersurface. Care has to be taken not to injure the nerve while dissecting tumors at this level.

22. The lateral wall of the sigmoid sinus and jugular bulb is removed (Fig. 12.49). Bleeding usually occurs from the apertures of the inferior petrosal sinus and the condylar emissary vein. This is controlled with Surgicel packing. Preservation of the medial wall of the jugular bulb is necessary to protect lower cranial nerve function if there is no tumor infiltration.

23. The horizontal portion of the petrous internal carotid artery is under control after drilling the petrous bone lateral to the petrous carotid artery (Figs. 12.50 and 12.51).

24. The tip of the suction is used to displace the petrous carotid artery laterally (Fig. 12.52) while the petrous bone medial to the petrous carotid artery is drilled with a large diamond burr to reach the clivus (Fig. 12.53). This is especially important in the area lying between the basal turn of the cochlea and the ICA. The petrous apex was drilled until the inferior petrosal sinus and clivus were reached (Fig. 12.54).

25. The occipital condyle is starting to be drilled for the transcondylar-transtubercular extension of the approach (Fig. 12.55).

26. Drilling of the occipital condyle and jugular tubercle exposes the hypoglossal canal (Fig. 12.56).

27. View after finishing the transcondylar-transtubercular extension of the approach (Fig. 12.57). In addition to the removal of bone in classic ITFA of Fisch, drilling of the jugular process of the occipital bone and even some of the occipital condyle facilitates control of the area of the jugular bulb.

28. If there is intradural extension, the posterior fossa dura is opened (Fig. 12.58). View of the intradural space after opening of the posterior fossa dura. The glossopharyngeal and vagus nerves are well identified in the cerebellomedullary cistern before entering the jugular foramen.

29. Mild retraction of the cerebellum and flocculus provides a good view of the choroid plexus and lower cranial nerves (Fig. 12.59).

30. The facial and vestibulocochlear nerves and the anterior inferior cerebellar artery are visible. The anteroinferior cerebellar artery passes inferior to the abducent nerve and travels posteriorly to come into close relation to the facial nerve (Fig. 12.60).

31. View after finishing the intradural extension of the approach (Fig. 12.61).

32. The Eustachian tube is closed with a muscle plug. The cavity is obliterated with abdominal fat; the posterior belly of the digastric muscle is replaced and sutured. The musculoperiosteal flap is repositioned in place and sutured to the temporalis muscle. The skin is closed in layers.

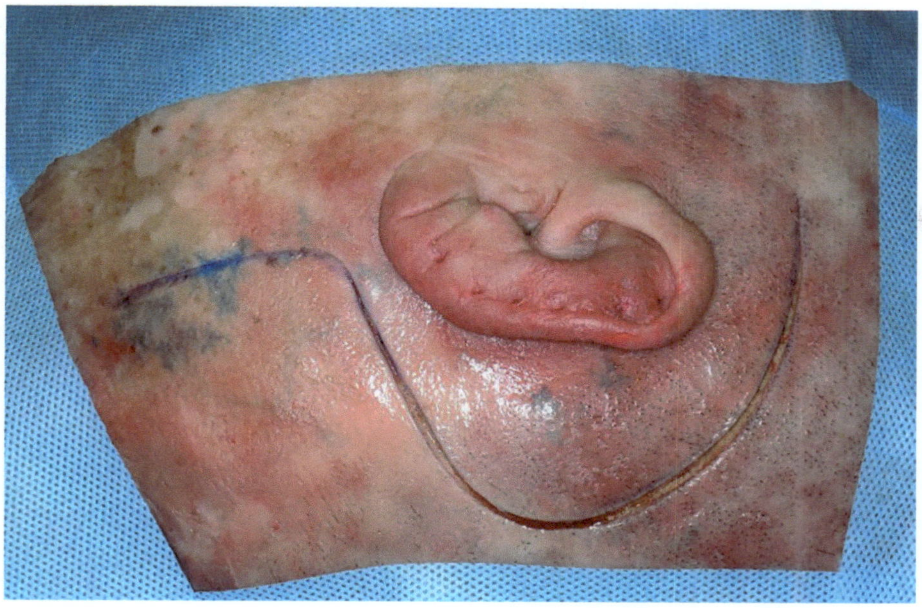

Fig. 12.4 A postauricular skin incision that extends down to the neck is performed.

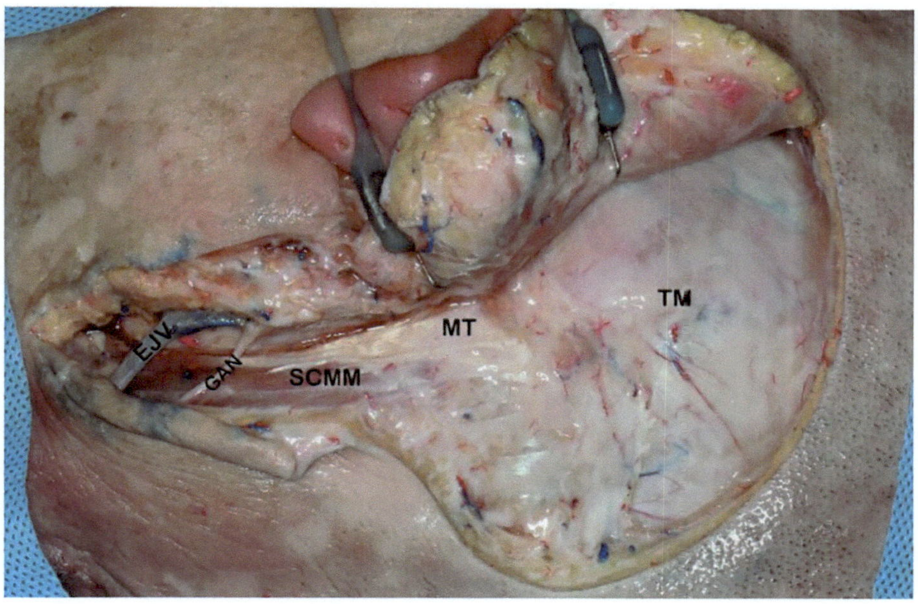

Fig. 12.5 The large postauricular flap is raised while keeping its adequate thickness. The skin flap is reflected anteriorly and maintained in position with two hooks. *TM*, temporalis muscle; *MT*, mastoid tip; *SCMM*, sternocleidomastoid muscle; *GAN*, greater auricular nerve; *EJV*, external jugular vein.

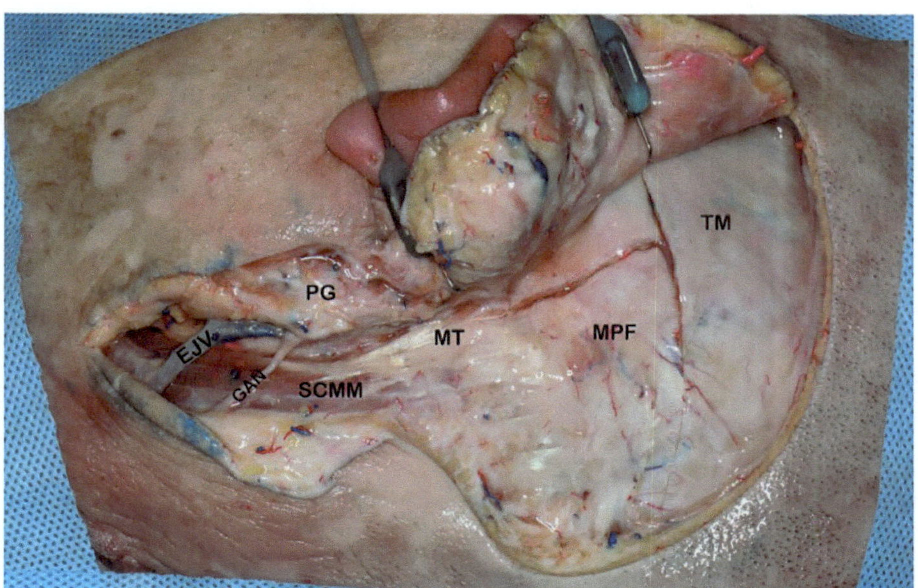

Fig. 12.6 The musculoperiosteal layer is incised in a T-shaped fashion using a cutting diathermy. *TM*, temporalis muscle; *MT*, mastoid tip; *MPF*, musculoperiosteal flap; *SCMM*, sternocleidomastoid muscle; *GAN*, greater auricular nerve; *EJV*, external jugular vein; *PG*, parotid gland.

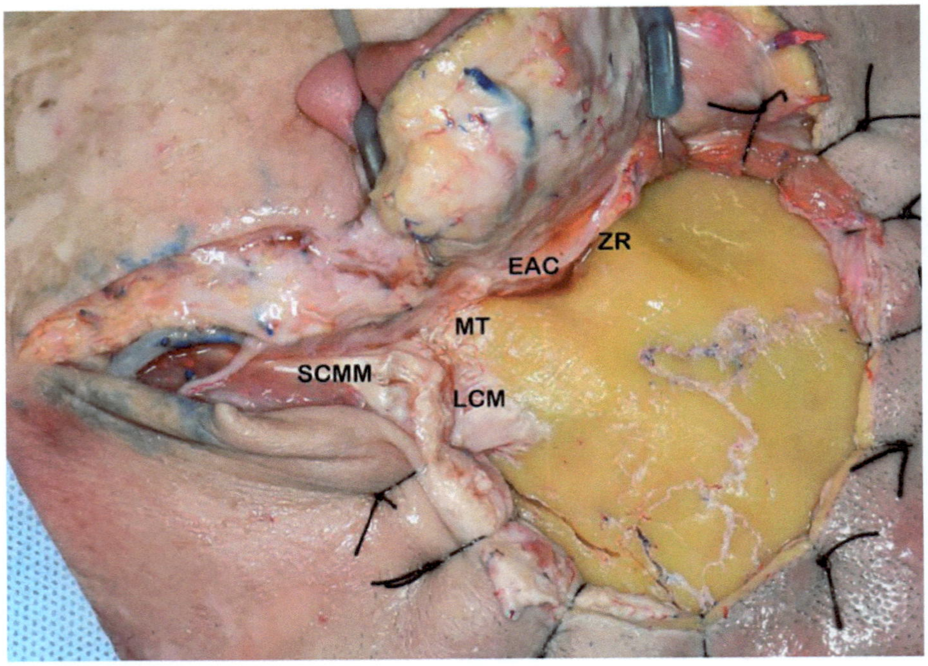

Fig. 12.7 The flap is elevated using a Lempert periosteal elevator. We use 1-0 silk sutures to stitch the soft tissue away from the surgical field, thus avoiding the use of retractors. The inferior part of the musculoperiosteum is incised with diathermy and is then elevated, creating an inferiorly based flap continuous with the sternomastoid muscle. Sharp dissection and diathermy are needed to disinsert the sternomastoid muscle from the tip of the mastoid. *ZR*, zygomatic root; *EAC*, external auditory canal; *MT*, mastoid tip; *SCMM*, sternocleidomastoid muscle; *LCM*, longissimus capitis muscle.

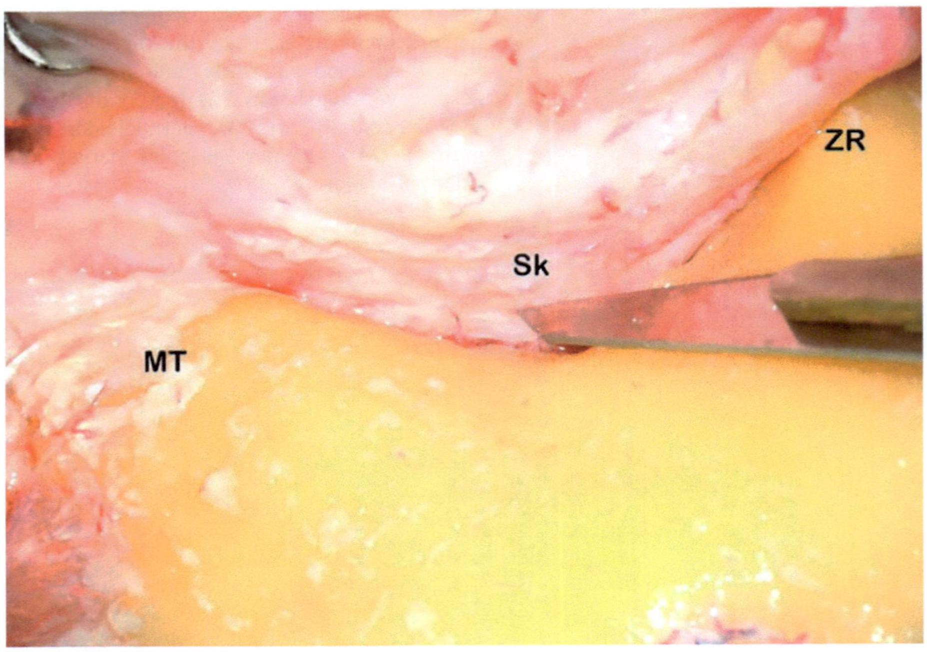

Fig. 12.8 The meatal skin is cut with a Beaver blade. *ZR*, zygomatic root; *MT*, mastoid tip; *Sk*, skin.

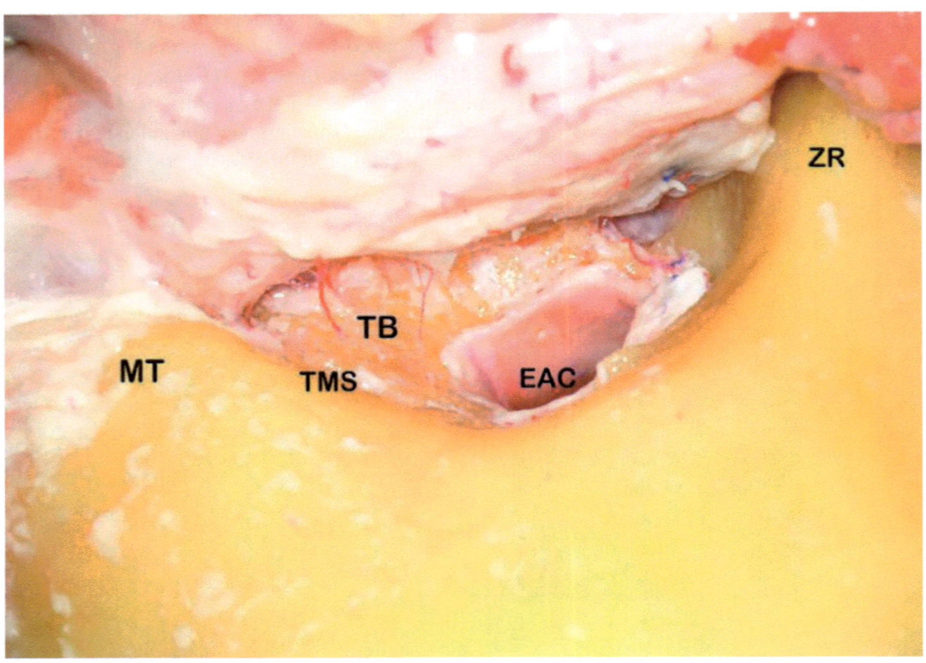

Fig. 12.9 The external auditory canal is transected for further blind sac closure. *ZR*, zygomatic root; *MT*, mastoid tip; *TB*, tympanic bone; *EAC*, external auditory canal; *TMS*, tympanomastoid suture.

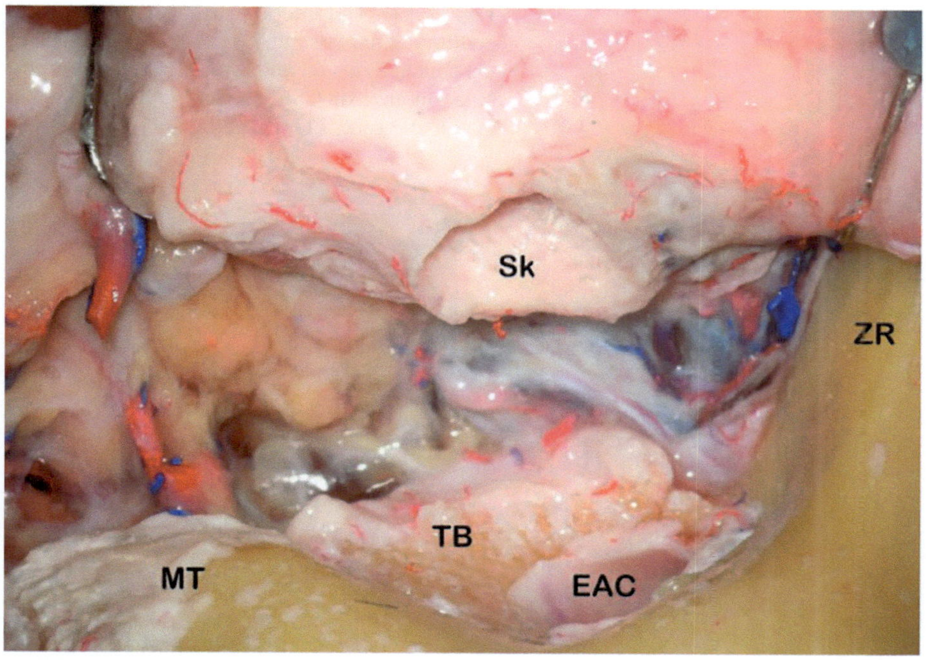

Fig. 12.10 The anterior skin flap is further retracted anteriorly and maintained in position with two skin hooks. *Sk*, skin; *ZR*, zygomatic root; *MT*, mastoid tip; *TB*, tympanic bone; *EAC*, external auditory canal.

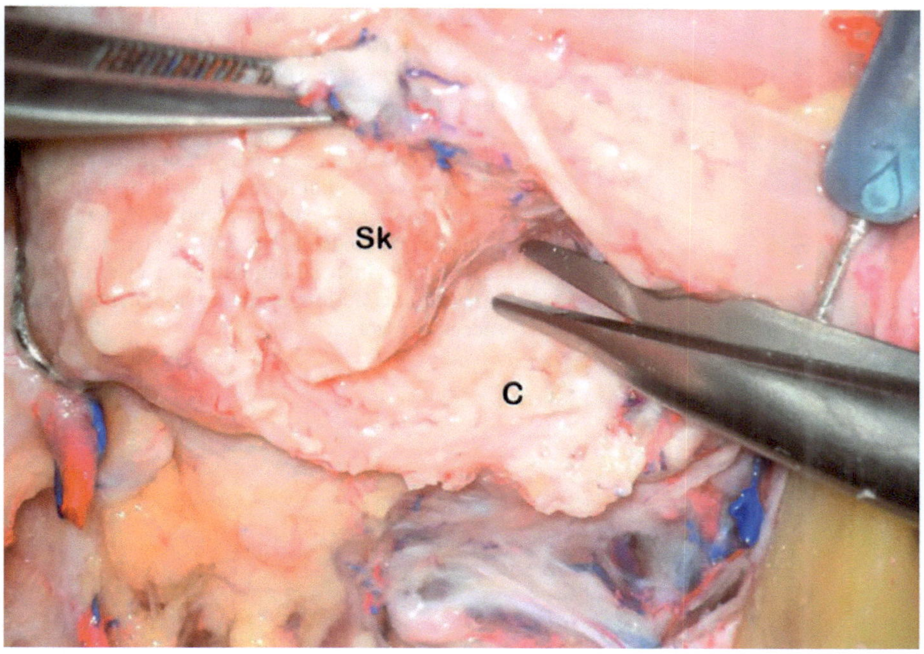

Fig. 12.11 Meticulous dissection of the skin from the anterior cartilage has to be performed. The skin of the cartilaginous canal is elevated for 1 cm from the cut margin toward the external meatus with tympanoplasty scissors to facilitate eversion of the canal skin. *Sk*, skin; *C*, cartilage.

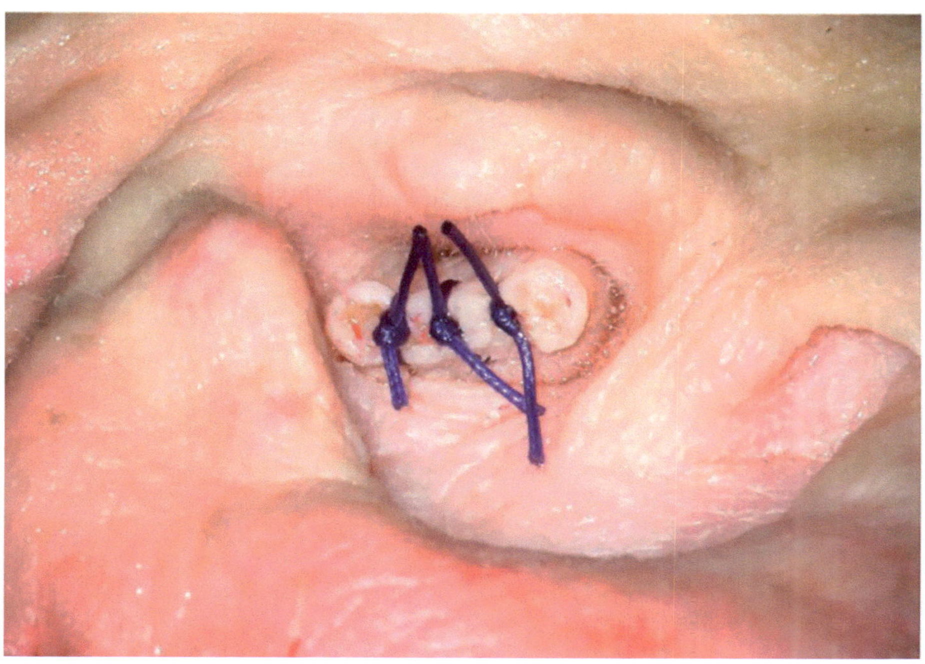

Fig. 12.12 The meatal skin is everted out through the canal and sutured with silk sutures.

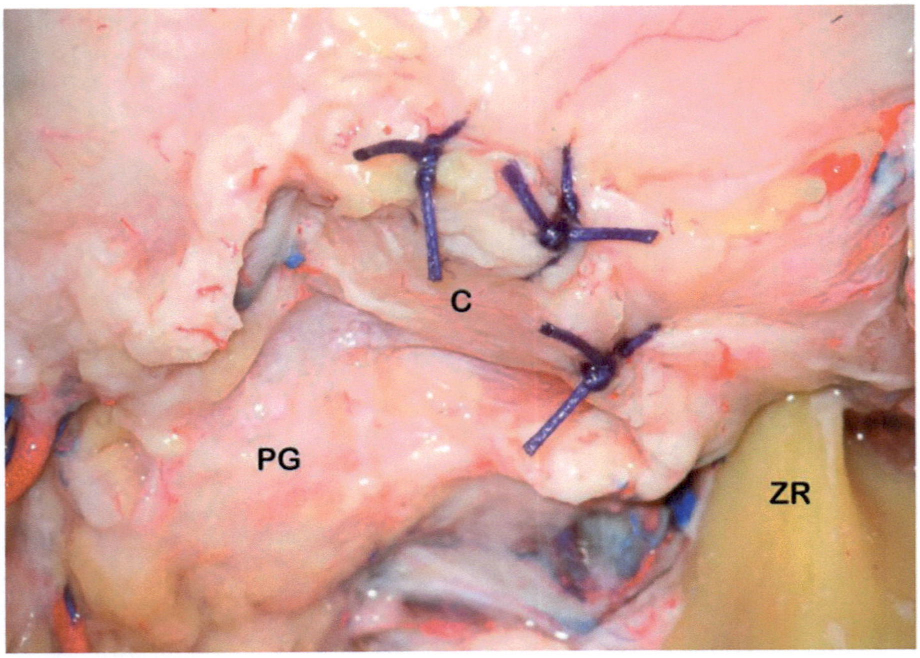

Fig. 12.13 The anterior cartilage of the external auditory canal is sutured with the soft tissues of the concha. This second layer is of utmost importance in case of CSF leakage. *C*, cartilage; *ZR*, zygomatic root; *PG*, parotid gland.

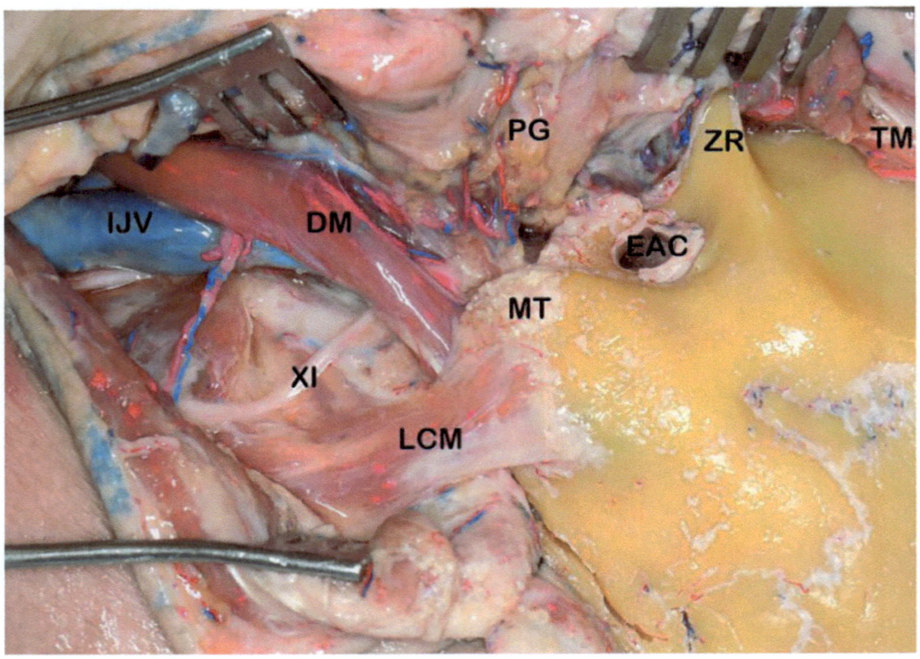

Fig. 12.14 The sternocleidomastoid muscle has been dissected from the mastoid tip and sutured to the skin edge. Then the sternocleidomastoid muscle is retracted posteriorly to show the posterior belly of the digastric muscle. The digastric muscle can now be seen inserting into the digastric groove, medial to the mastoid tip. *ZR*, zygomatic root; *PG*, parotid gland; *TM*, temporalis muscle; *MT*, mastoid tip; *EAC*, external auditory canal; *IJV*, internal jugular vein; *DM*, digastric muscle; *LCM*, longissimus capitis muscle; *XI*, accessory nerve.

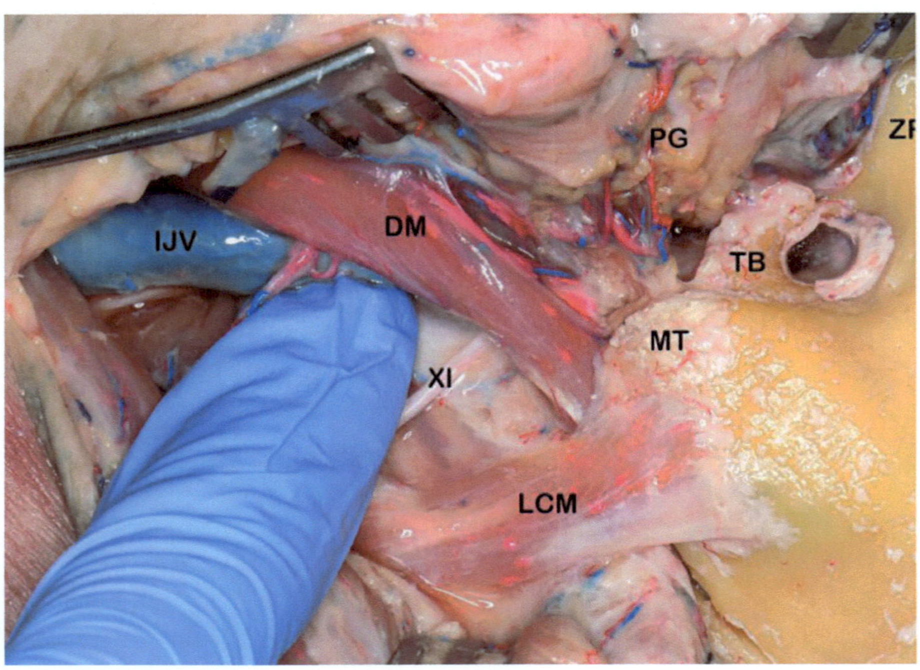

Fig. 12.15 The lateral process of C1 is palpated as a landmark to identify the internal jugular vein and the point at which cranial nerve XI passes laterally to it (in 80%–85% of cases). *ZR*, zygomatic root; *PG*, parotid gland; *TB*, tympanic bone; *MT*, mastoid tip; *DM*, digastric muscle; *IJV*, internal jugular vein; *LCM*, longissimus capitis muscle; *XI*, accessory nerve.

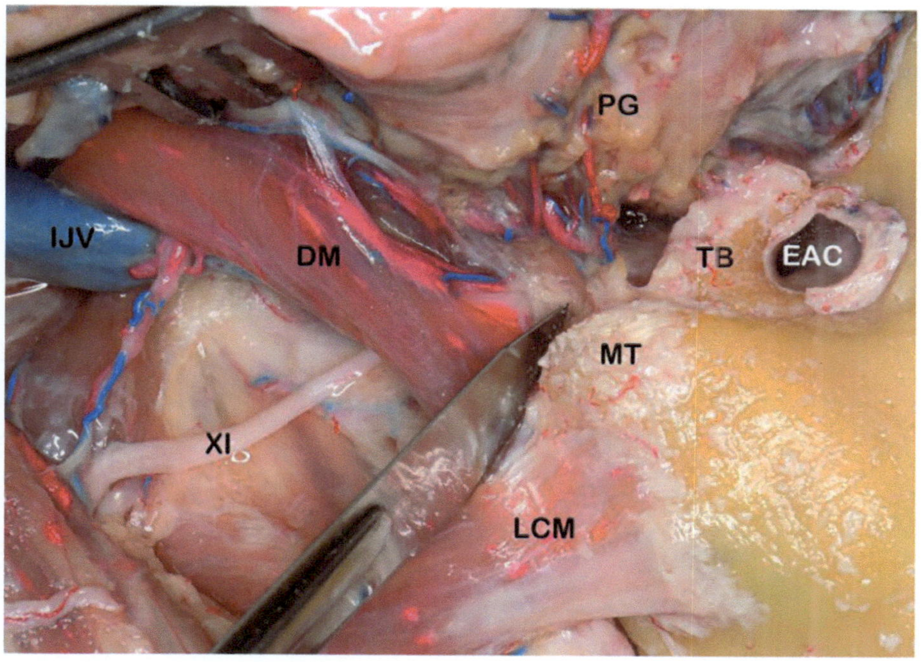

Fig. 12.16 The posterior belly of the digastric muscle is cut. *PG*, parotid gland; *TB*, tympanic bone; *EAC*, external auditory canal; *MT*, mastoid tip; *DM*, digastric muscle; *IJV*, internal jugular vein; *LCM*, longissimus capitis muscle; *XI*, accessory nerve.

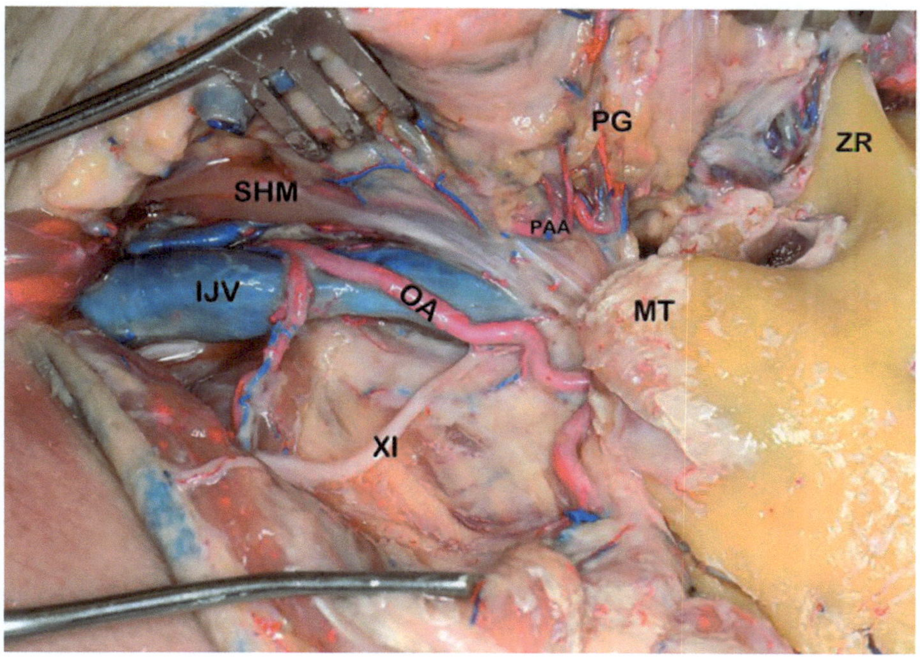

Fig. 12.17 The posterior belly of the digastric muscle has been cut and retracted inferiorly. The underlying muscle that can be seen is the stylohyoid muscle inserting on the upper lateral segment of the styloid process. The internal jugular vein, occipital artery, and spinal accessory nerve can be seen. *PG*, parotid gland; *ZR*, zygomatic root; *MT*, mastoid tip; *SHM*, stylohyoid muscle; *OA*, occipital artery; *IJV*, internal jugular vein; *PAA*, posterior auricular artery.

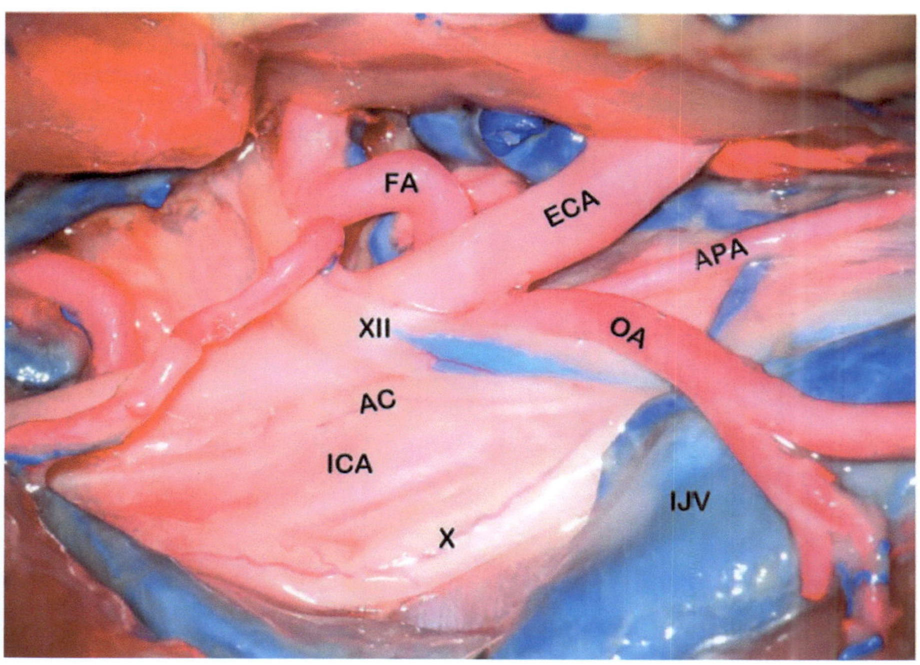

Fig. 12.18 Further dissection with exposure of the internal carotid artery, vagus nerve, hypoglossal nerve, ansa cervicalis, and branches of external carotid artery. *FA*, facial artery; *ECA*, external carotid artery; *APA*, ascending pharyngeal artery; *OA*, occipital artery; *IJV*, internal jugular vein; *XII*, hypoglossal nerve; *AC*, ansa cervicalis; *ICA*, internal carotid artery; *X*, vagus nerve.

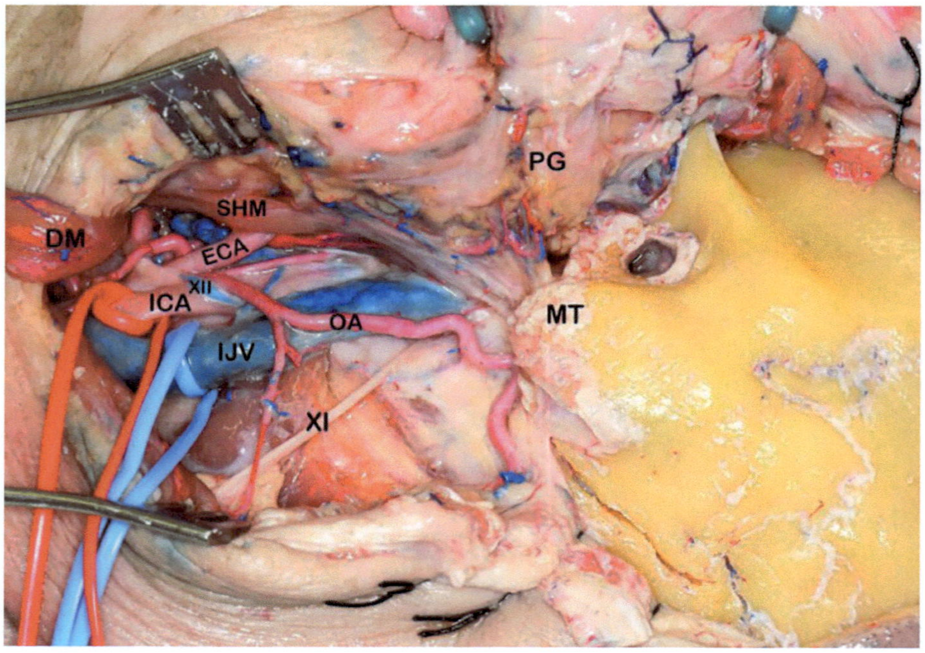

Fig. 12.19 Two vessel loops are placed in order to mark the internal carotid artery and the internal jugular vein. *PG*, parotid gland; *MT*, mastoid tip; *DM*, digastric muscle; *SHM*, stylohyoid muscle; *ECA*, external carotid artery; *XII*, hypoglossal nerve; *ICA*, internal carotid artery; *IJV*, internal jugular vein; *OA*, occipital artery; *XI*, accessory nerve.

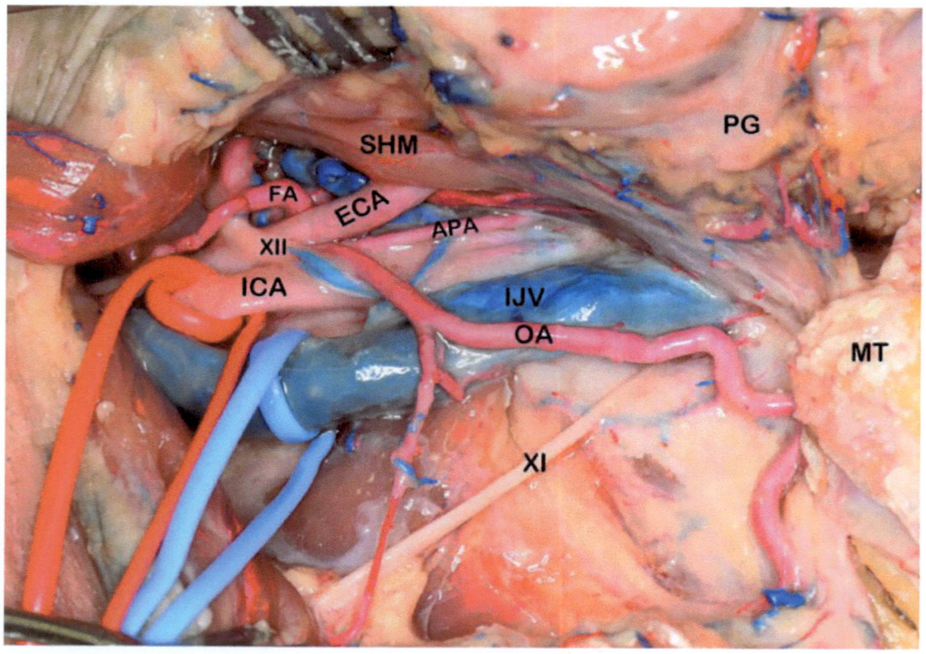

Fig. 12.20 At a higher magnification, the spinal accessory nerve, hypoglossal nerve, internal carotid artery, external carotid artery, facial artery, ascending pharyngeal artery, and occipital artery are all well seen. *PG*, parotid gland; *MT*, mastoid tip; *SHM*, stylohyoid muscle; *FA*, facial artery; *ECA*, external carotid artery; *APA*, ascending pharyngeal artery; *XII*, hypoglossal nerve; *ICA*, internal carotid artery; *IJV*, internal jugular vein; *OA*, occipital artery; *XI*, accessory nerve.

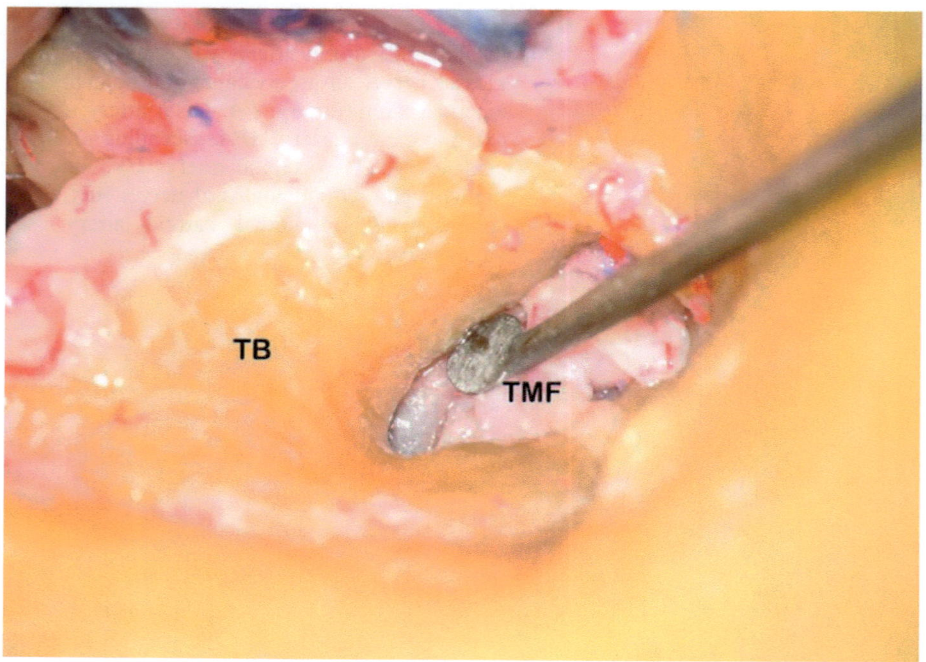

Fig. 12.21 Care to remove all the skin of the external auditory canal. All the skin of the external auditory canal, tympanic membrane, and ossicles should be removed. *TB*, tympanic bone; *TMF*, tympanomeatal flap.

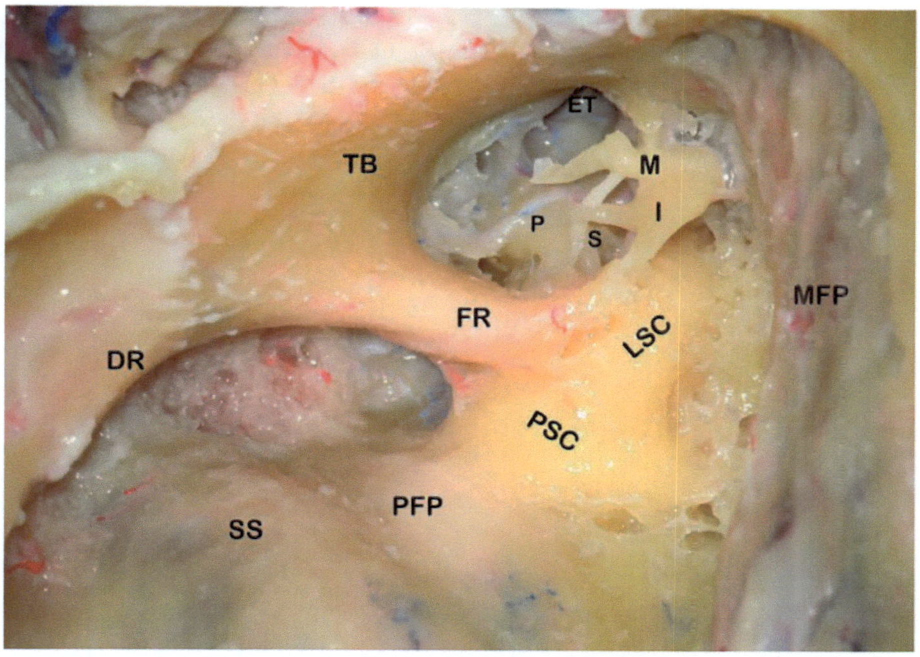

Fig. 12.22 A wide mastoidectomy is then performed. The posterior canal wall has been removed, exposing the ossicles and facial ridge. *ET*, Eustachian tube; *TB*, tympanic bone; *M*, malleus; *I*, incus; *S*, stapes; *P*, promontory; *FR*, facial ridge; *LSC*, lateral semicircular canal; *PSC*, posterior semicircular canal; *MFP*, middle fossa plate; *PFP*, posterior fossa plate; *SS*, sigmoid sinus; *DR*, digastric ridge.

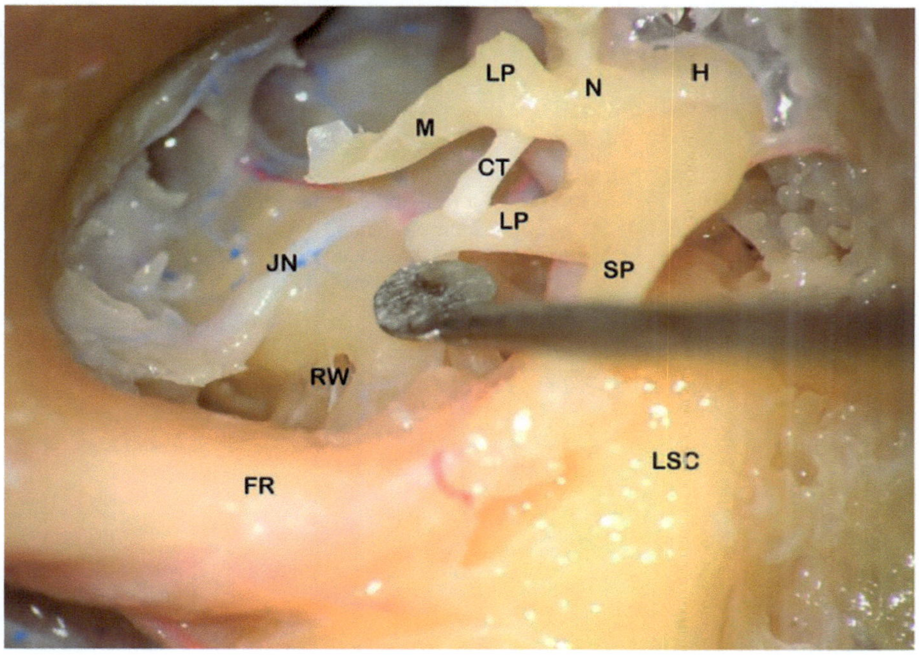

Fig. 12.23 The incudostapedial joint is being disarticulated. *M*, manubrium of the malleus; *LP*, lateral process of the malleus; *N*, neck of the malleus; *H*, head of the stapes; *CT*, chorda tympani; *LP*, long process of incus; *SP*, short process of incus; *JN*, Jackobson's nerve; *RW*, round window; *FR*, facial ridge; *LSC*, lateral semicircular canal.

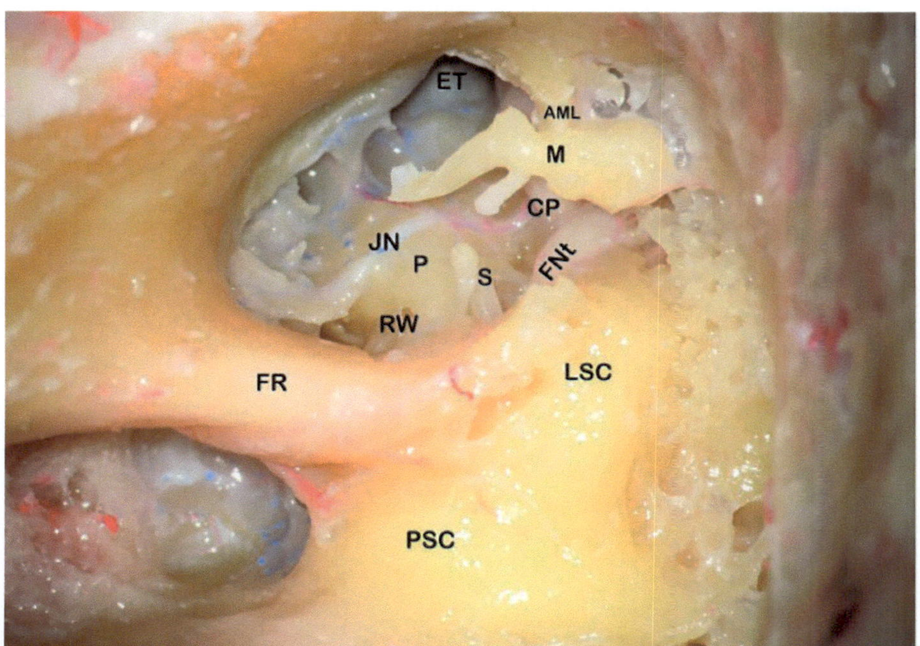

Fig. 12.24 The incus has been removed. *ET*, Eustachian tube; *AML*, anterior malleal ligament; *M*, malleus; *CP*, cochleariform process; *JN*, Jackobson's nerve; *P*, promontory; *S*, stapes; *FNt*, tympanic segment of facial nerve; *LSC*, lateral semicircular canal; *RW*, round window; *FR*, facial ridge; *PSC*, posterior semicircular canal.

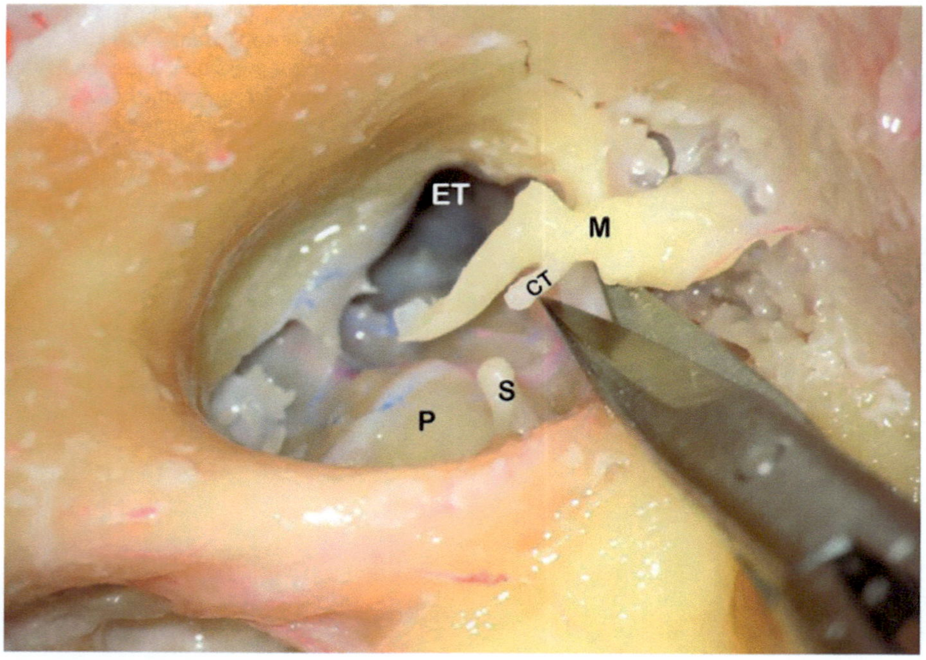

Fig. 12.25 The tensor tympani tendon is being cut. *ET*, Eustachian tube; *M*, malleus; *CT*, chorda tympani; *S*, stapes; *P*, promontory.

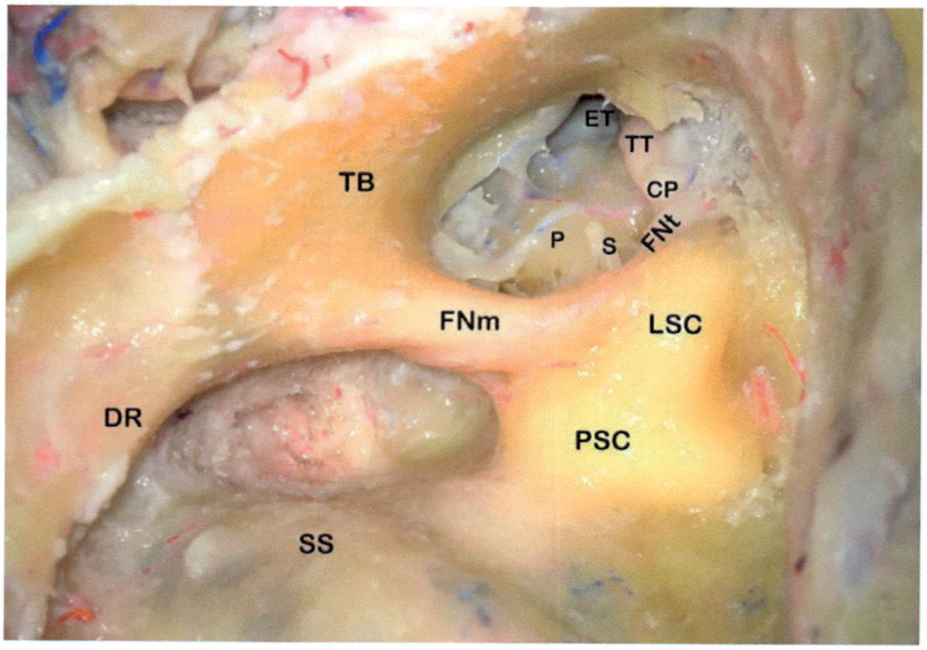

Fig. 12.26 The malleus has been removed. *ET*, Eustachian tube; *TT*, tensor tympani muscle; *CP*, cochleariform process; *FNt*, tympanic segment of facial nerve; *FNm*, mastoid segment of facial nerve; *P*, promontory; *S*, stapes; *TB*, tympanic bone; *LSC*, lateral semicircular canal; *PSC*, posterior semicircular canal; *DR*, digastric ridge; *SS*, sigmoid sinus.

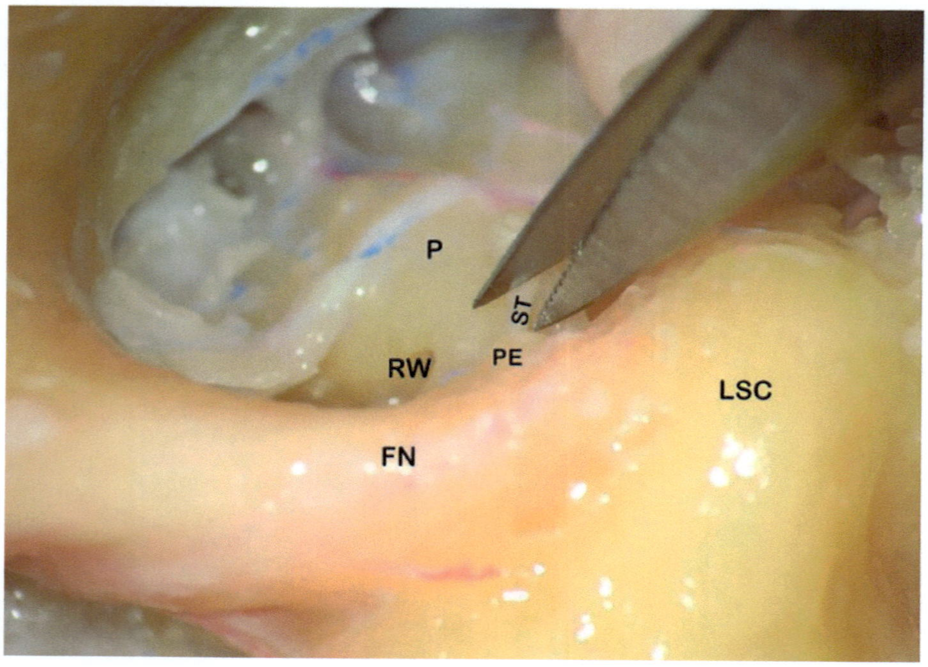

Fig. 12.27 The tendon of stapedius muscle is being cut using neurosurgical scissors. *P*, promontory; *ST*, tendon of stapedius muscle; *RW*, round window; *FN*, facial nerve; *PE*, pyramidal eminence; *LSC*, lateral semicircular canal.

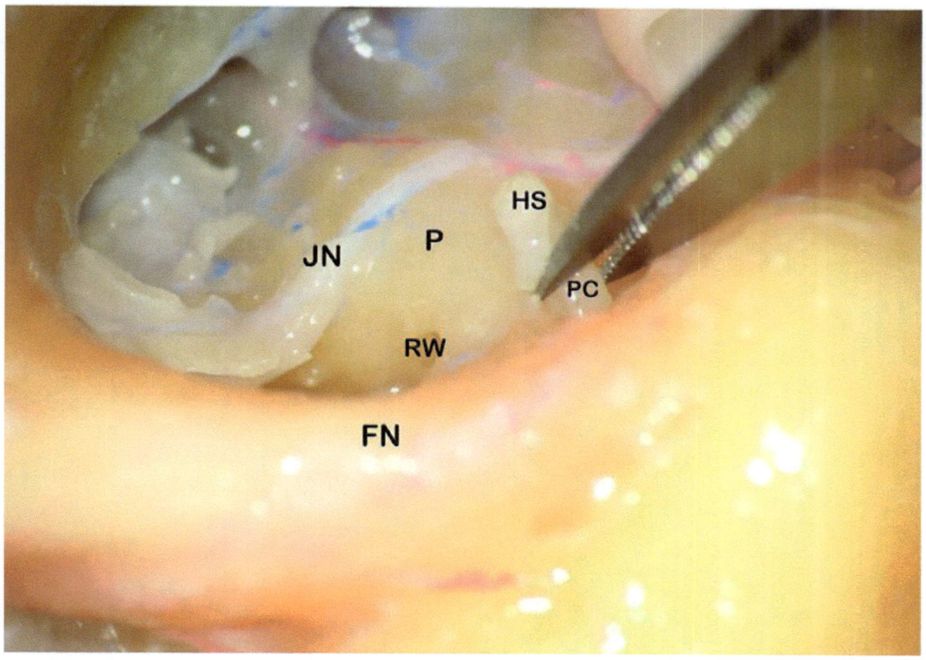

Fig. 12.28 The superstructure of the stapes is then removed using neurosurgical scissors. The crura of the stapes are transected to avoid trauma to the inner ear while dissecting the facial nerve during anterior rerouting. This is essential to avoid sensorineural hearing loss during anterior rerouting of the facial nerve due to footplate displacement. *HS*, head of stapes; *PC*, posterior crus; *P*, promontory; *JN*, Jackobson's nerve; *RW*, round window; *FN*, facial nerve.

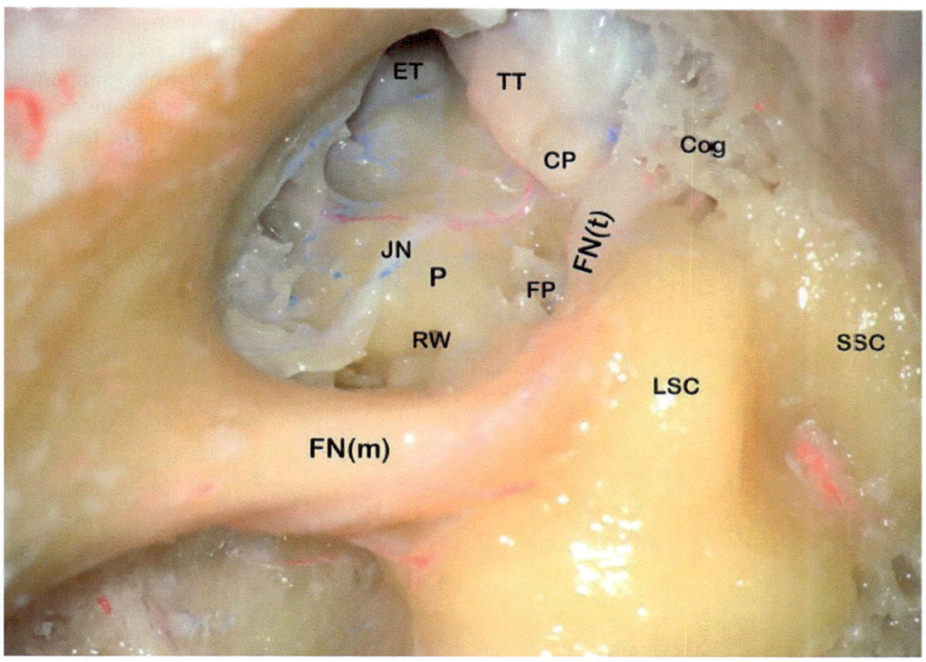

Fig. 12.29 The superstructure of the stapes has been removed. *ET*, Eustachian tube; *TT*, tensor tympani muscle; *CP*, cochleariform process; *FN(t)*, tympanic segment of facial nerve; *FN(m)*, mastoid segment of facial nerve; *FP*, footplate of the stapes; *P*, promontory; *JN*, Jackobson's nerve; *RW*, round window; *LSC*, lateral semicircular canal; *SSC*, superior semicircular canal.

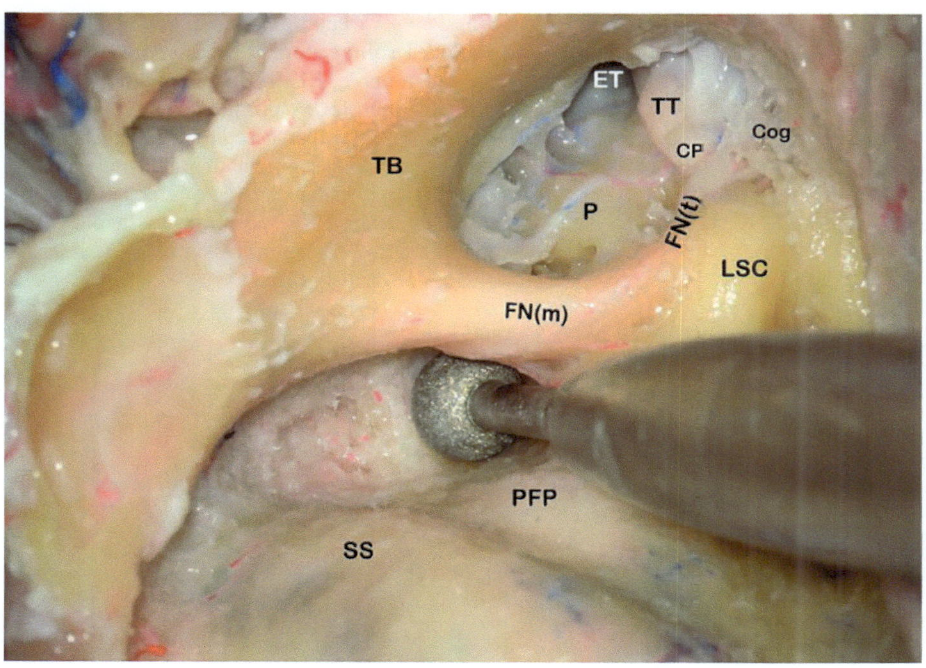

Fig. 12.30 The vertical segment of the facial nerve is skeletonized. This is facilitated by drilling the retrofacial air cells and the tympanic bone anterior to the facial nerve. The retrofacial air cells are drilled to complete the skeletonization of the facial nerve. *ET*, Eustachian tube; *TT*, tensor tympani muscle; *CP*, cochleariform process; *FN(m)*, mastoid segment of facial nerve; *FN(t)*, tympanic segment of facial nerve; *P*, promontory; *TB*, tympanic bone; *LSC*, lateral semicircular canal; *PFP*, posterior fossa plate; *SS*, sigmoid sinus.

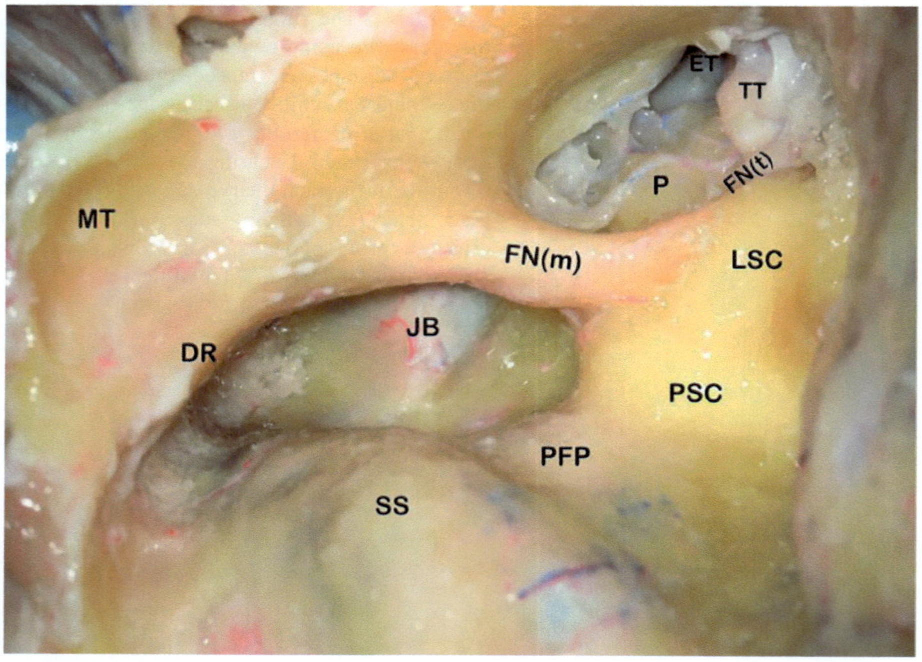

Fig. 12.31 The facial nerve has been well skeletonized, and the jugular bulb has been identified after drilling the retrofacial air cells. *ET*, Eustachian tube; *TT*, tensor tympani muscle; *P*, promontory; *FN(t)*, tympanic segment of facial nerve; *FN(m)*, mastoid segment of facial nerve; *LSC*, lateral semicircular canal; *PSC*, posterior semicircular canal; *PFP*, posterior fossa plate; *JB*, jugular bulb; *SS*, sigmoid sinus; *DR*, digastric ridge; *MT*, mastoid tip.

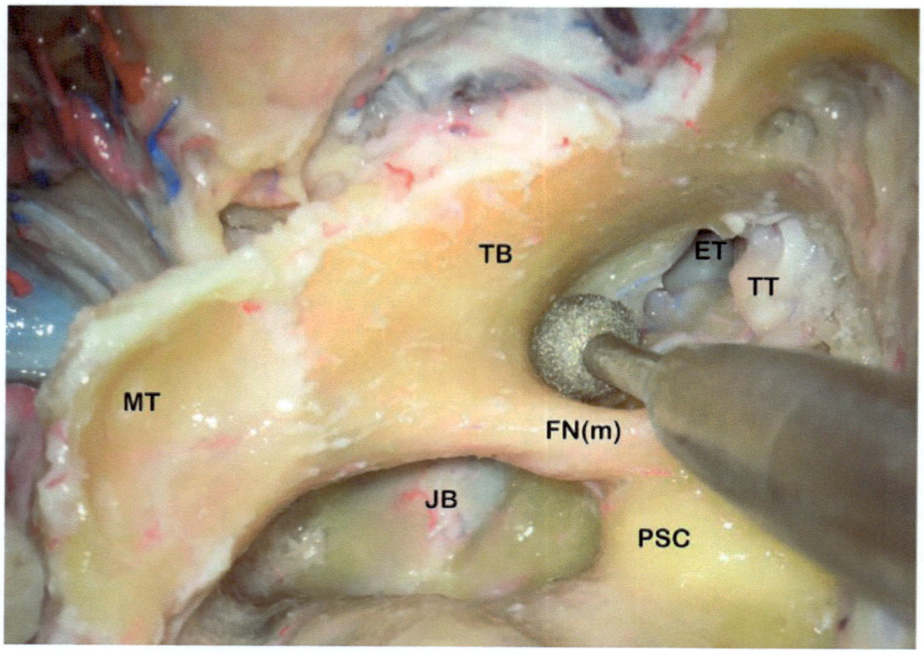

Fig. 12.32 Drilling of the tympanic bone between the internal carotid artery and the mastoid segment of the facial nerve is a very important step to get complete control of the jugular foramen area and the mastoid segment of the facial nerve. *ET*, Eustachian tube; *TT*, tensor tympani muscle; *TB*, tympanic bone; *FN(m)*, mastoid segment of facial nerve; *JB*, jugular bulb; *PSC*, posterior semicircular canal; *MT*, mastoid tip.

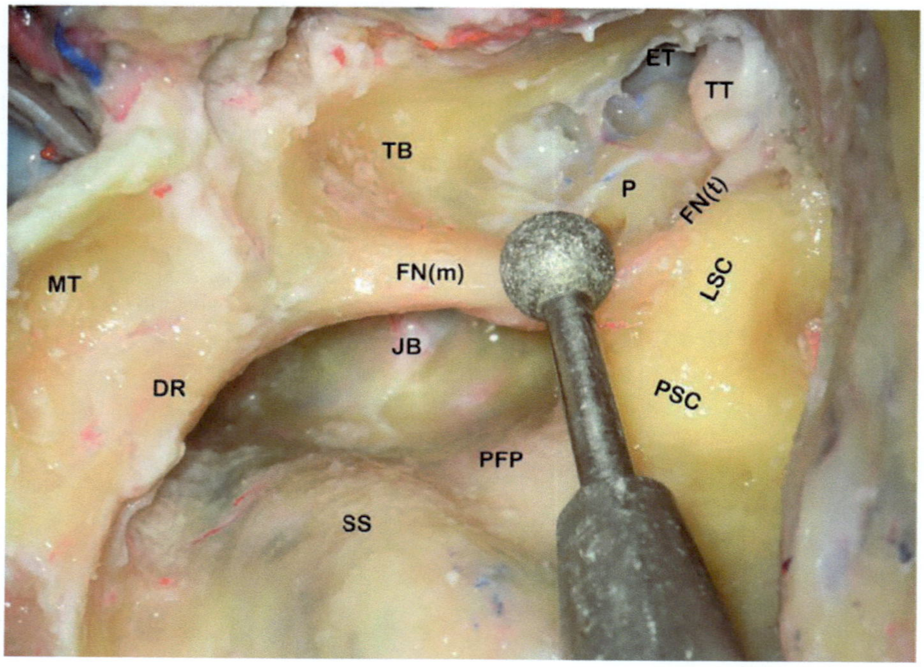

Fig. 12.33 Decompression of the facial nerve is started. Drilling should be carried out parallel to the facial nerve. *ET*, Eustachian tube; *TT*, tensor tympani muscle; *P*, promontory; *TB*, tympanic bone; *FN(t)*, tympanic segment of facial nerve; *FN(m)*, mastoid segment of facial nerve; *LSC*, lateral semicircular canal; *PSC*, posterior semicircular canal; *PFP*, posterior fossa plate; *JB*, jugular bulb; *SS*, sigmoid sinus; *DR*, digastric ridge; *MT*, mastoid tip.

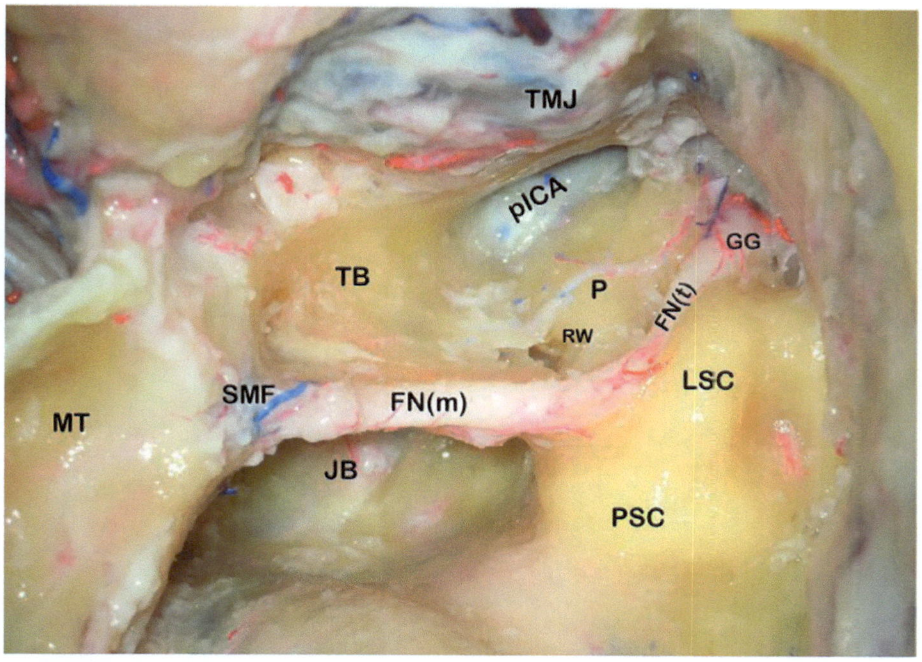

Fig. 12.34 The anterior wall of the external auditory canal is completely drilled. The facial nerve has been decompressed from the stylomastoid foramen to the geniculate ganglion. Care must be taken to fully uncover the geniculate ganglion. Sharp bone left at this level might injure the anteriorly transposed nerve. *TMJ*, temporomandibular joint; *pICA*, petrous segment of internal carotid artery; *P*, promontory; *GG*, geniculate ganglion; *FN(t)*, tympanic segment of facial nerve; *FN(m)*, mastoid segment of facial nerve; *RW*, round window; *LSC*, lateral semicircular canal; *PSC*, posterior semicircular canal; *JB*, jugular bulb; *TB*, tympanic bone; *SMF*, stylomastoid foramen; *MT*, mastoid tip.

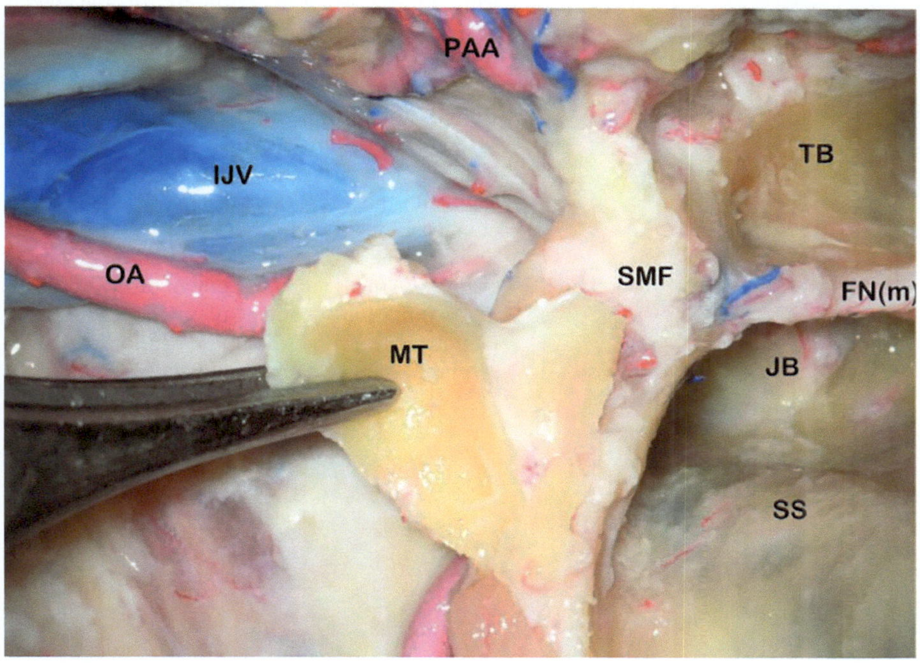

Fig. 12.35 The mastoid tip is amputated. The tough attachments are sharply cut, taking care not to injure the facial nerve. *PAA*, posterior auricular artery; *TB*, tympanic bone; *IJV*, internal jugular vein; *OA*, occipital artery; *MT*, mastoid tip; *SMF*, stylomastoid foramen; *FN(m)*, mastoid segment of facial nerve; *JB*, jugular bulb; *SS*, sigmoid sinus.

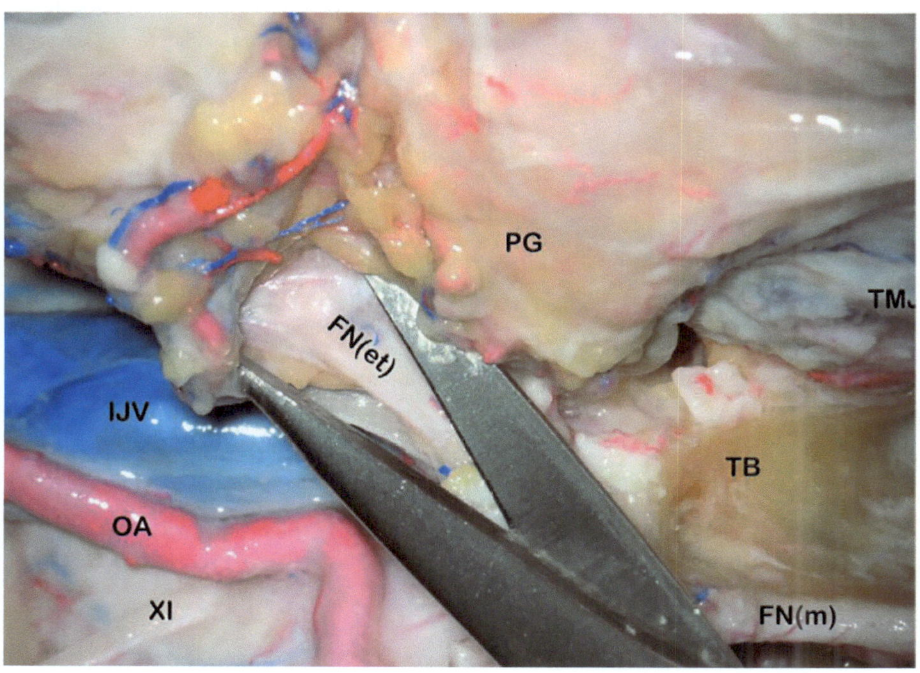

Fig. 12.36 The extratemporal facial nerve is then identified. *PG*, parotid gland; *TMJ*, temporomandibular joint; *TB*, tympanic bone; *FN(m)*, mastoid segment of facial nerve; *FN(et)*, extratemporal segment of facial nerve; *IJV*, internal jugular vein; *OA*, occipital artery; *XI*, accessory nerve.

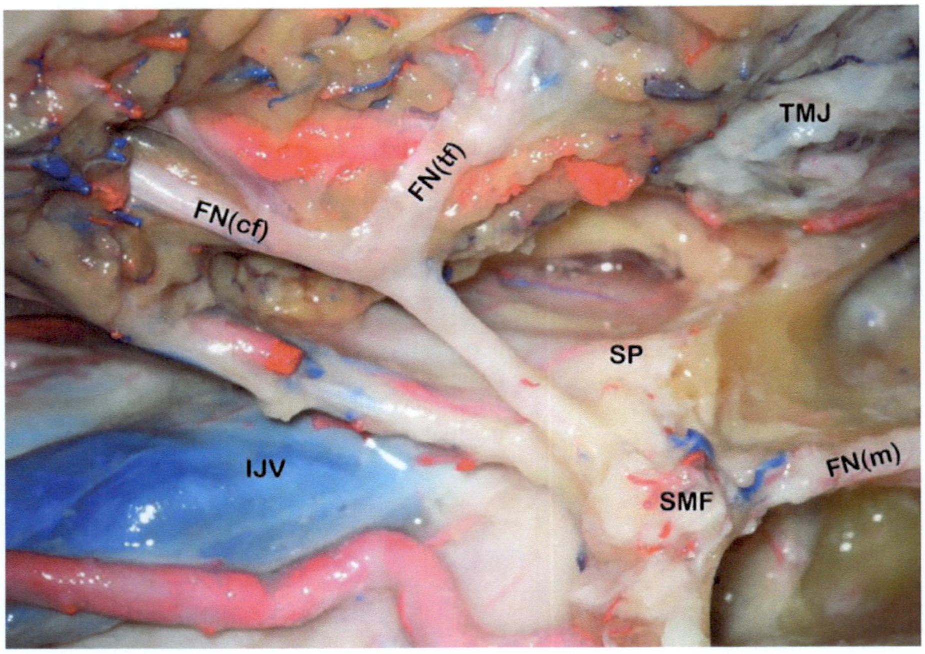

Fig. 12.37 The trunk is traced in the parotid until the proximal parts of the temporofacial and cervicofacial trunks are identified. The facial nerve has to be liberated enough in the parotid gland. This provides an extra length of free nerve and prevents it from being stretched during the transposition. *TMJ*, temporomandibular joint; *SP*, styloid process; *FN(tf)*, temporofacial trunk of facial nerve; *FN(cf)*, cervicofacial trunk of facial nerve; *IJV*, internal jugular vein; *SMF*, stylomastoid foramen; *FN(m)*, mastoid segment of facial nerve.

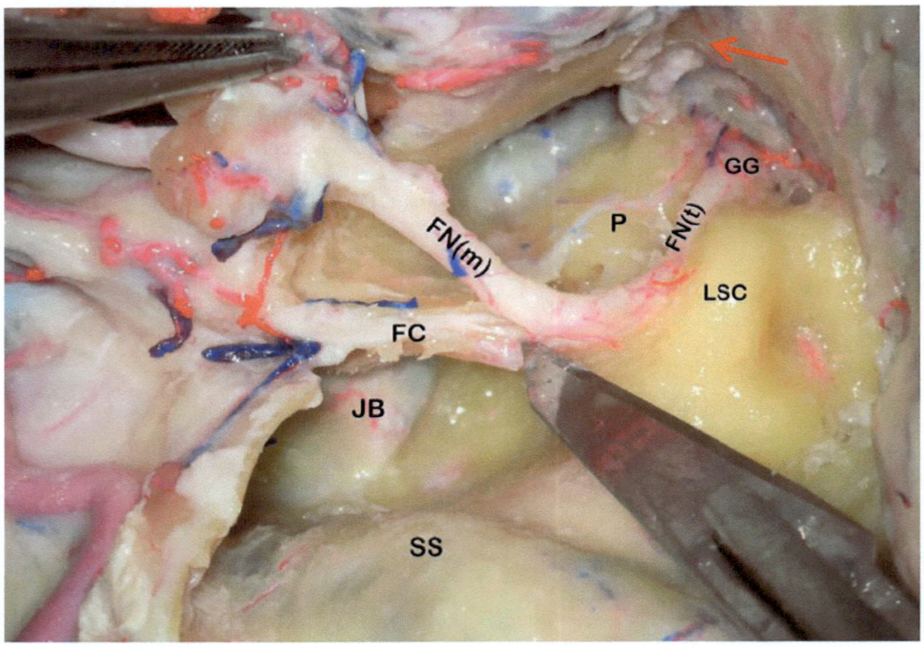

Fig. 12.38 Anterior rerouting of the facial nerve is starting. A new bony canal *(red arrow)* is drilled in the root of the zygoma superior to the Eustachian tube. The facial nerve is freed at the stylomastoid foramen using strong scissors. The soft tissue cuff surrounding the nerve at this level is not removed but rather kept intact. The mastoid segment is next separated using a Beaver knife to cut the fibrous attachments of the nerve to the fallopian canal. *GG*, geniculate ganglion; *P*, promontory; *FN(t)*, tympanic segment of facial nerve; *FN(m)*, mastoid segment of facial nerve; *LSC*, lateral semicircular canal; *FC*, fallopian canal; *JB*, jugular bulb; *SS*, sigmoid sinus.

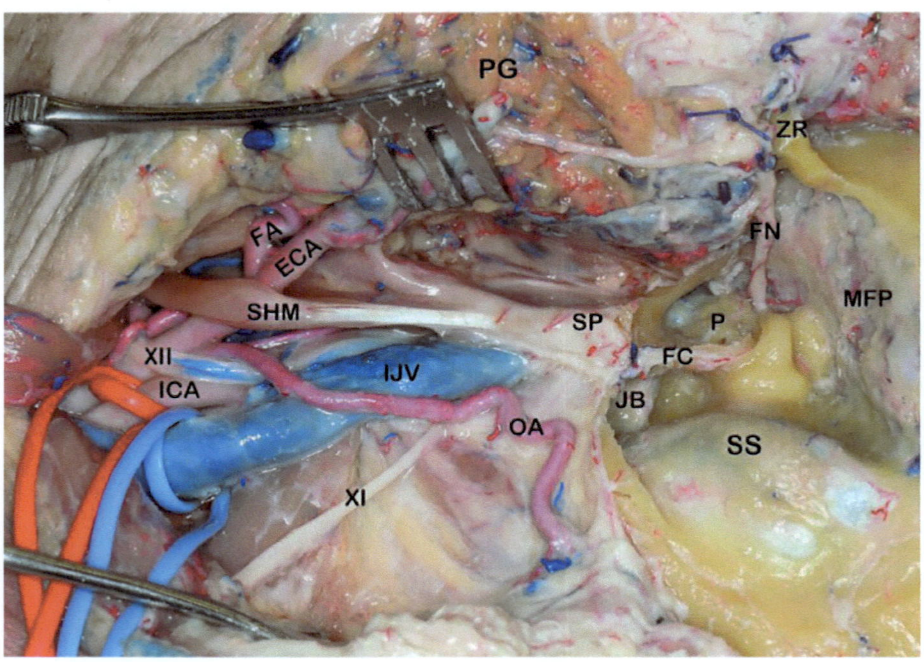

Fig. 12.39 Anterior rerouting of the facial nerve has been completed. The facial nerve is in its new bony canal, just superior to the Eustachian tube. In live surgery, the nerve is fixed to the new bony canal using fibrin glue. Displacement of the facial nerve represents the key point of the surgery because it allows unobstructed control of the jugular foramen and the vertical carotid canal. *PG*, parotid gland; *ZR*, zygomatic root; *FN*, rerouted part of the facial nerve; *MFP*, middle fossa plate; *P*, promontory; *SP*, styloid process; *FC*, fallopian canal; *JB*, jugular bulb; *SS*, sigmoid sinus; *OA*, occipital artery; *IJV*, internal jugular vein; *XI*, accessory nerve; *FA*, facial artery; *ECA*, external carotid artery; *SHM*, stylohyoid muscle; *XII*, hypoglossal nerve; *ICA*, internal carotid artery.

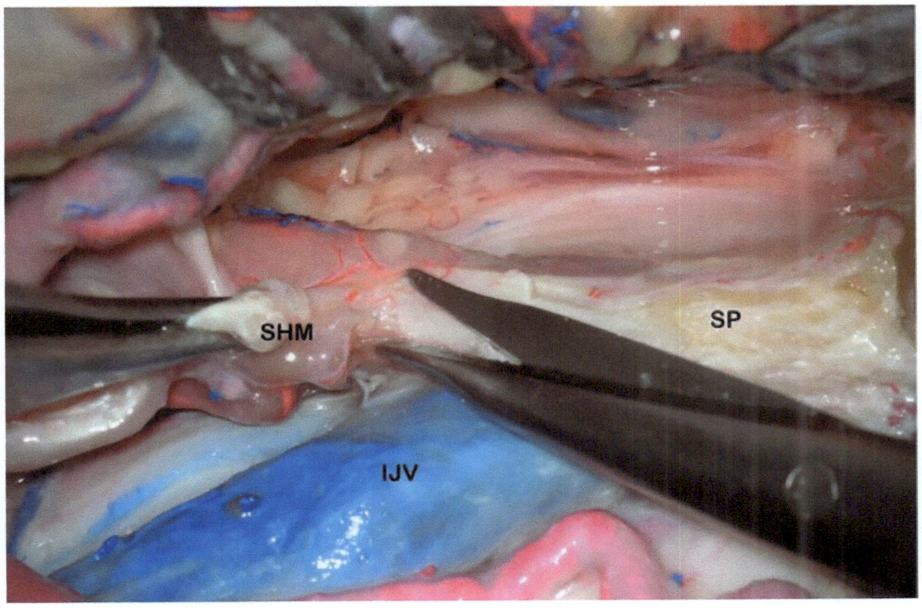

Fig. 12.40 The stylohyoid muscle is cut to free the styloid. *SP*, styloid process; *SHM*, stylohyoid muscle; *IJV*, internal jugular vein.

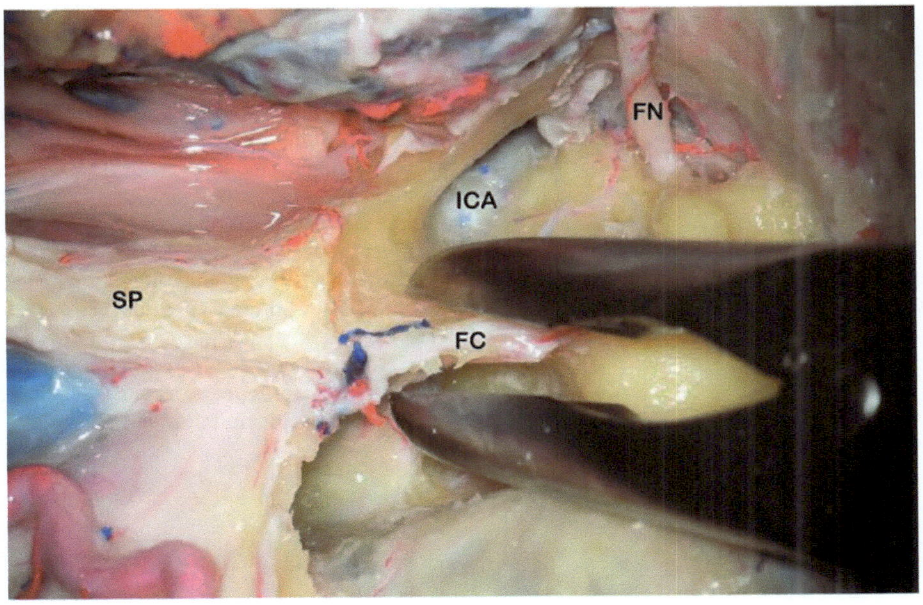

Fig. 12.41 Removal of the fallopian canal with a rongeur. *FN*, rerouted part of the facial nerve; *ICA*, internal carotid artery; *FC*, fallopian canal; *SP*, styloid process.

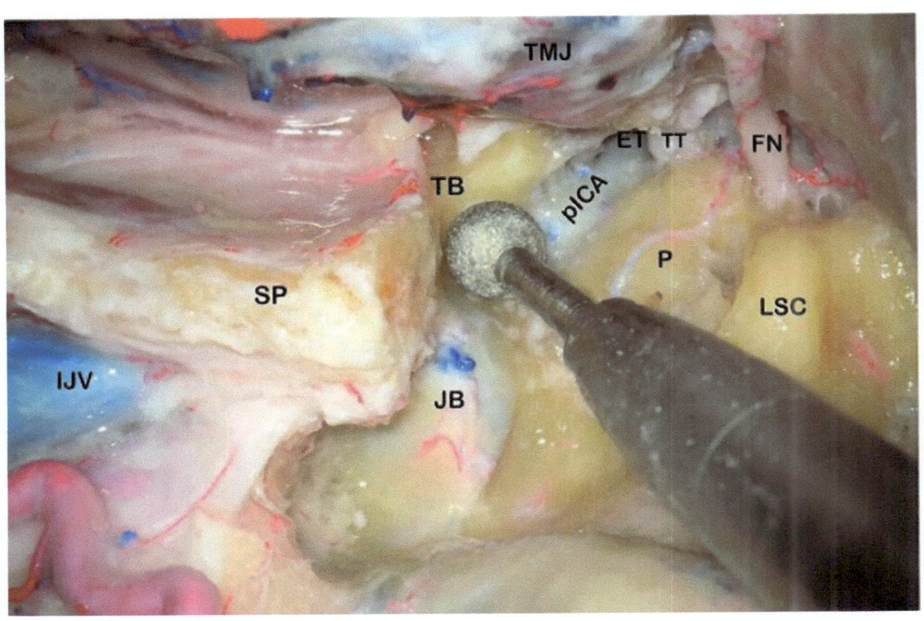

Fig. 12.42 The tympanic bone is drilled out, and the styloid process is removed to expose the petrous segment of internal carotid artery. Note the relationship between the anterior wall of the jugular bulb and the tympanic bone. During the removal of the base of the styloid process, the utmost care has to be taken not to injure the ICA lying immediately deep in it. *TMJ*, temporomandibular joint; *ET*, Eustachian tube; *TT*, tensor tympani muscle; *FN*, rerouted part of the facial nerve; *pICA*, petrous segment of internal carotid artery; *TB*, tympanic bone; *P*, promontory; *LSC*, lateral semicircular canal; *SP*, styloid process; *JB*, jugular bulb; *IJV*, internal jugular vein.

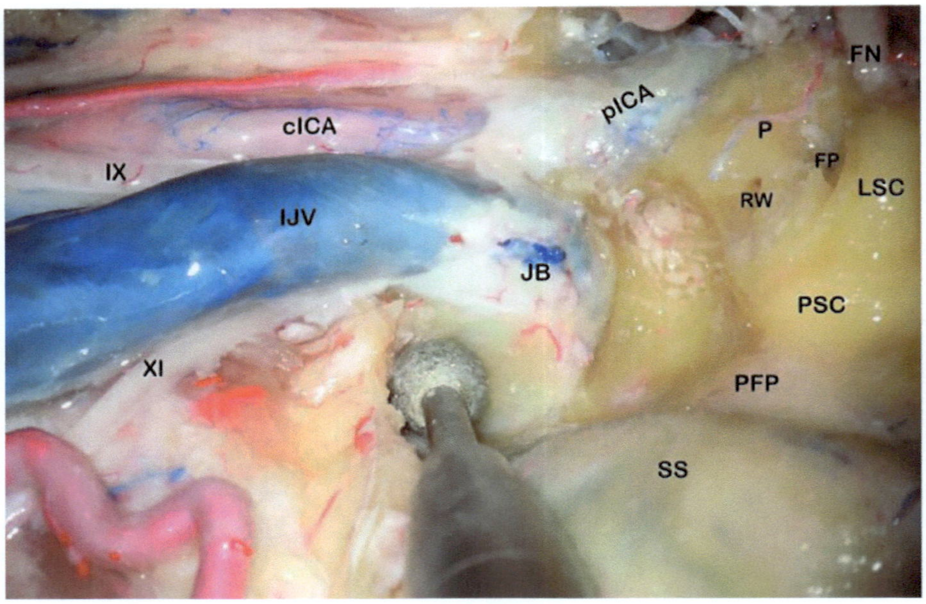

Fig. 12.43 Drilling of the jugular process and occipital condyle is performed with a large diamond burr. Further bone has been removed posteroinferiorly to the sigmoid and jugular bulb, representing a partial removal of the posteromedial segment of the occipital condyle. *FN*, rerouted part of the facial nerve; *P*, promontory; *FP*, footplate of the stapes; *RW*, round window; *LSC*, lateral semicircular canal; *PSC*, posterior semicircular canal; *pICA*, petrous segment of internal carotid artery; *PFP*, posterior fossa plate; *JB*, jugular bulb; *SS*, sigmoid sinus; *cICA*, cervical segment of internal carotid artery; *IJV*, internal jugular vein; *IX*, glossopharyngeal nerve; *XI*, accessory nerve.

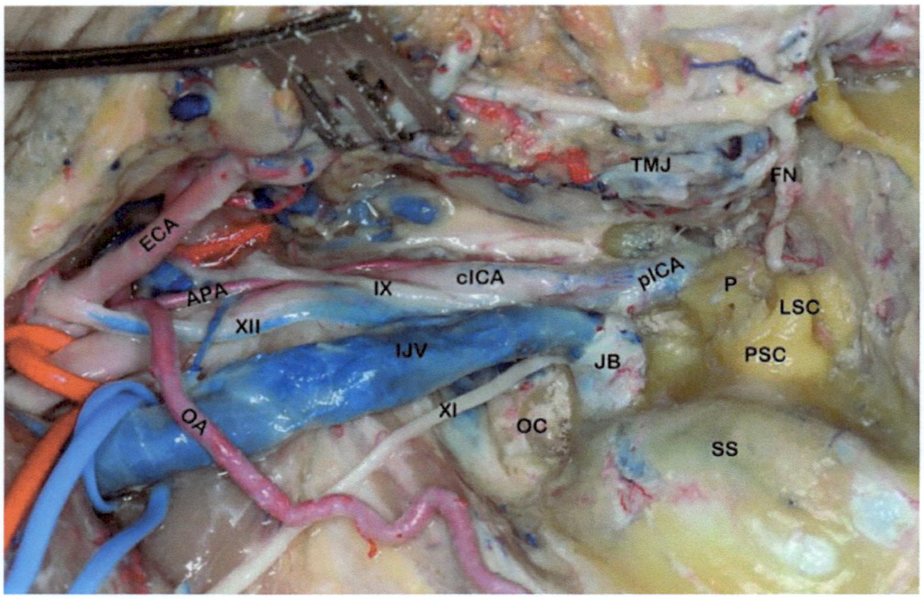

Fig. 12.44 Drilling of the occipital condyle is completed. Drilling out one-third of the lateral part of the occipital condyle permits complete visualization of the inferior aspect of the jugular bulb. *TMJ*, temporomandibular joint; *FN*, rerouted part of the facial nerve; *pICA*, petrous segment of internal carotid artery; *P*, promontory; *LSC*, lateral semicircular canal; *PSC*, posterior semicircular canal; *JB*, jugular bulb; *SS*, sigmoid sinus; *OC*, occipital condyle; *XI*, accessory nerve; *IJV*, internal jugular vein; *cICA*, cervical segment of internal carotid artery; *IX*, glossopharyngeal nerve; *XII*, hypoglossal nerve; *APA*, ascending pharyngeal artery; *ECA*, external carotid artery; *OA*, occipital artery.

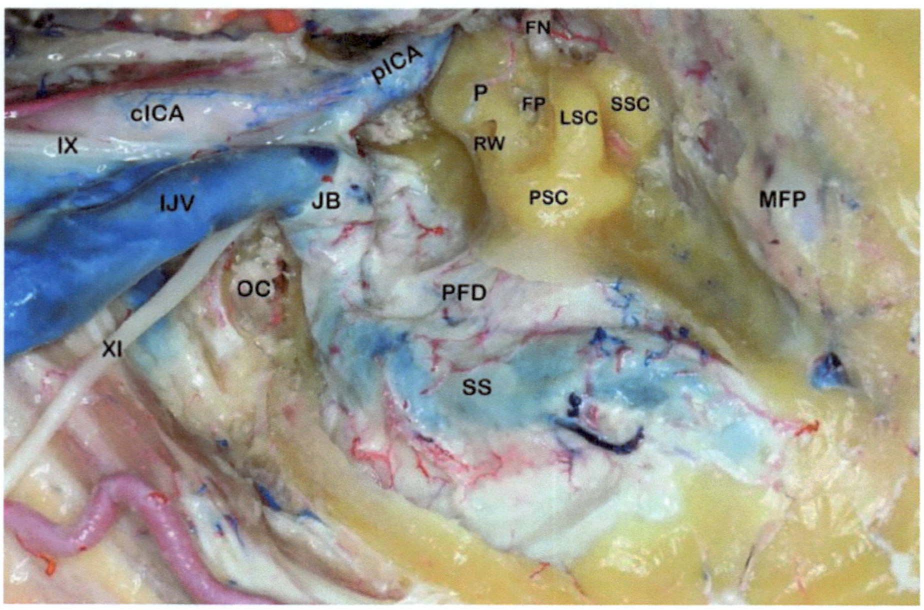

Fig. 12.45 The sigmoid sinus is completely uncovered. *MFP*, middle fossa plate; *FN*, rerouted part of the facial nerve; *P*, promontory; *FP*, footplate of the stapes; *RW*, round window; *SSC*, superior semicircular canal; *LSC*, lateral semicircular canal; *PSC*, posterior semicircular canal; *pICA*, petrous segment of internal carotid artery; *cICA*, cervical segment of internal carotid artery; *JB*, jugular bulb; *PFD*, posterior fossa dura; *SS*, sigmoid sinus; *OC*, occipital condyle; *IJV*, internal jugular vein; *IX*, glossopharyngeal nerve; *XI*, accessory nerve.

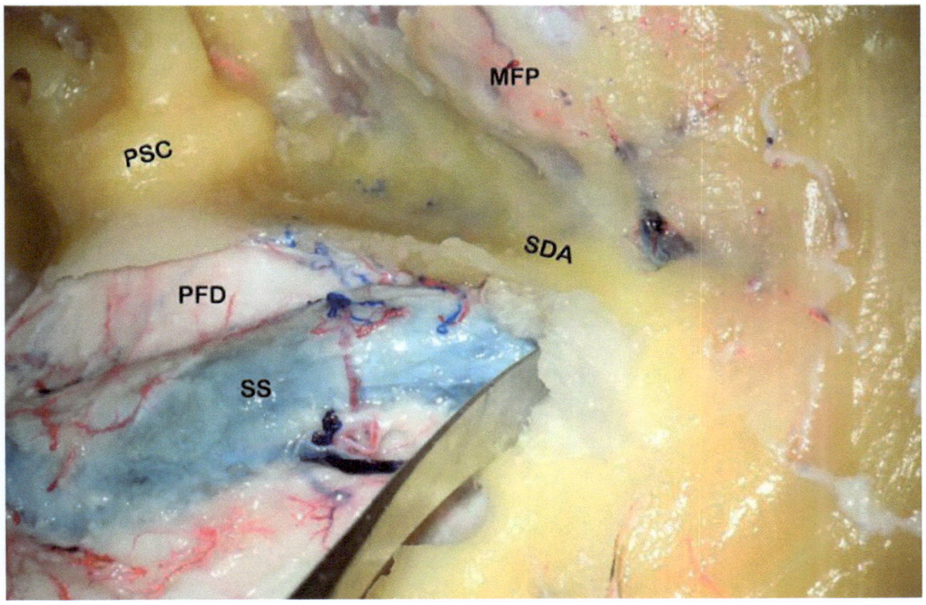

Fig. 12.46 The extraluminal compression of the sigmoid sinus is facilitated by the preservation of a bony shell over the proximal sigmoid sinus and prevents blocking of the transverse sinus. This is the technique of extraluminal closure of the sigmoid sinus to avoid the risk of CSF leakage with suture of the sigmoid sinus. *MFP*, middle fossa plate; *SDA*, sinodural angle; *PSC*, posterior semicircular canal; *PFD*, posterior fossa dura; *SS*, sigmoid sinus.

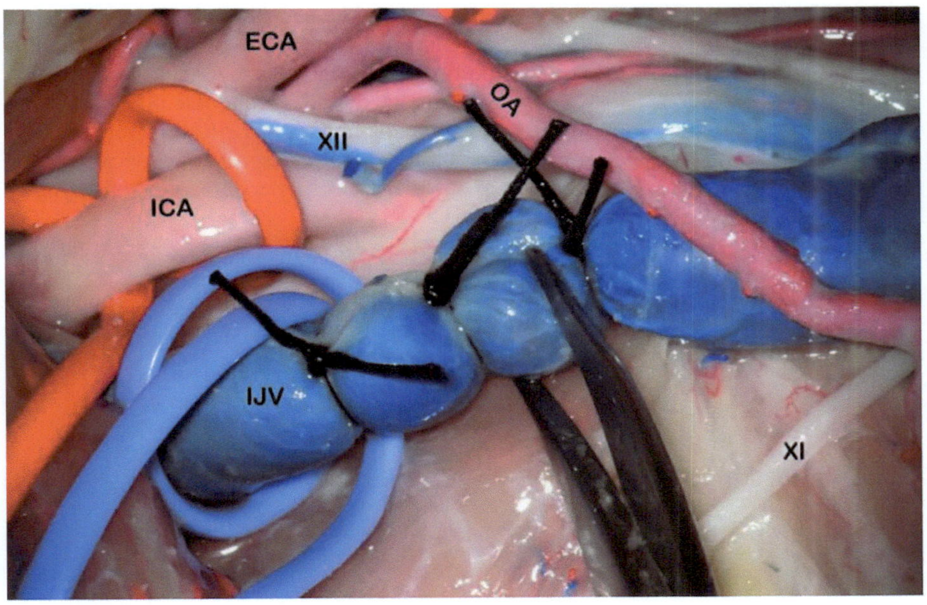

Fig. 12.47 The internal jugular vein is then double ligated and transected. This is done following sigmoid closure to minimize venous congestion of the tumor. *ECA*, external carotid artery; *OA*, occipital artery; *XII*, hypoglossal nerve; *ICA*, internal carotid artery; *IJV*, internal jugular vein; *XI*, accessory nerve.

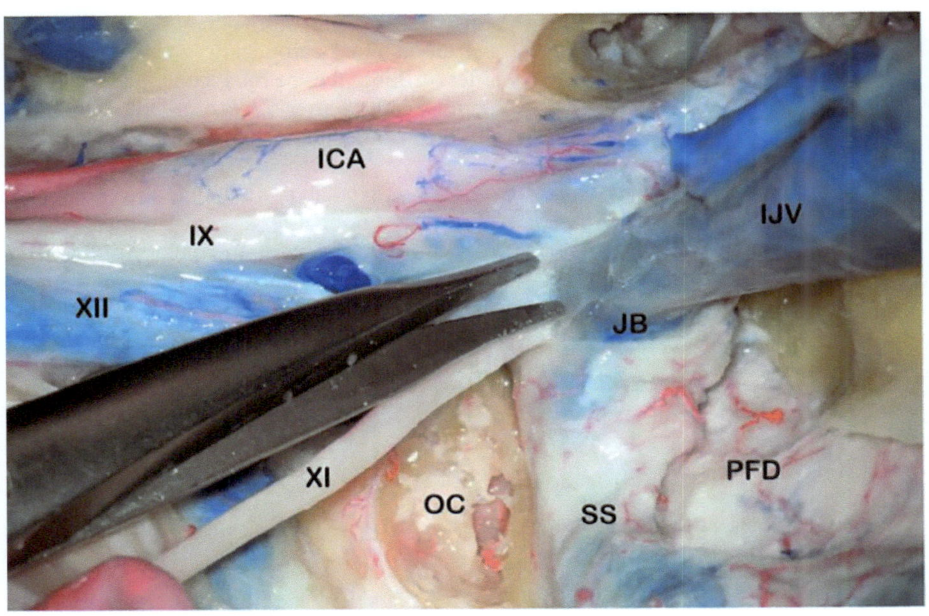

Fig. 12.48 The lower cranial nerves can now be visualized after reflection of the internal jugular vein. The ICA is surrounded by a dense fibroperiosteal layer at its entrance in the skull base. The glossopharyngeal nerve lies deep in this layer and is slightly adherent to its undersurface. Care has to be taken not to injure the nerve while dissecting tumors at this level. *ICA*, internal carotid artery; *IJV*, internal jugular vein; *JB*, jugular bulb; *PFD*, posterior fossa dura; *SS*, sigmoid sinus; *OC*, occipital condyle; *XI*, accessory nerve; *IX*, glossopharyngeal nerve; *XII*, hypoglossal nerve.

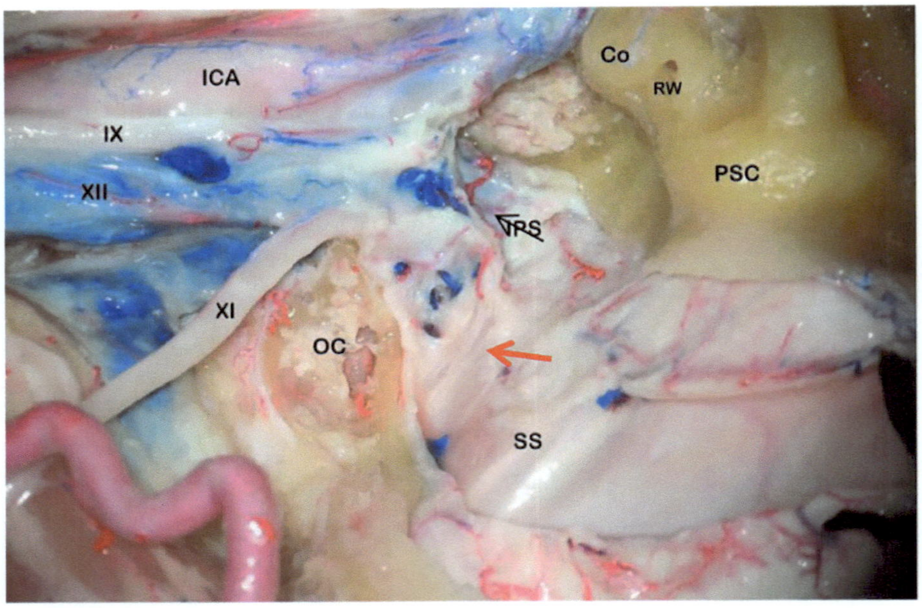

Fig. 12.49 The lateral wall of the sigmoid sinus and jugular bulb is removed. Bleeding usually occurs from the apertures of the inferior petrosal sinus and the condylar emissary vein *(red arrow)*. This is controlled with Surgicel packing. Preservation of the medial wall of the jugular bulb is necessary to protect lower cranial nerve function if there is no tumor infiltration. *Co*, cochlea; *RW*, round window; *PSC*, posterior semicircular canal; *IPS*, opening of inferior petrosal sinus; *SS*, sigmoid sinus; *OC*, occipital condyle; *XI*, accessory nerve; *ICA*, internal carotid artery; *IX*, glossopharyngeal nerve; *XII*, hypoglossal nerve.

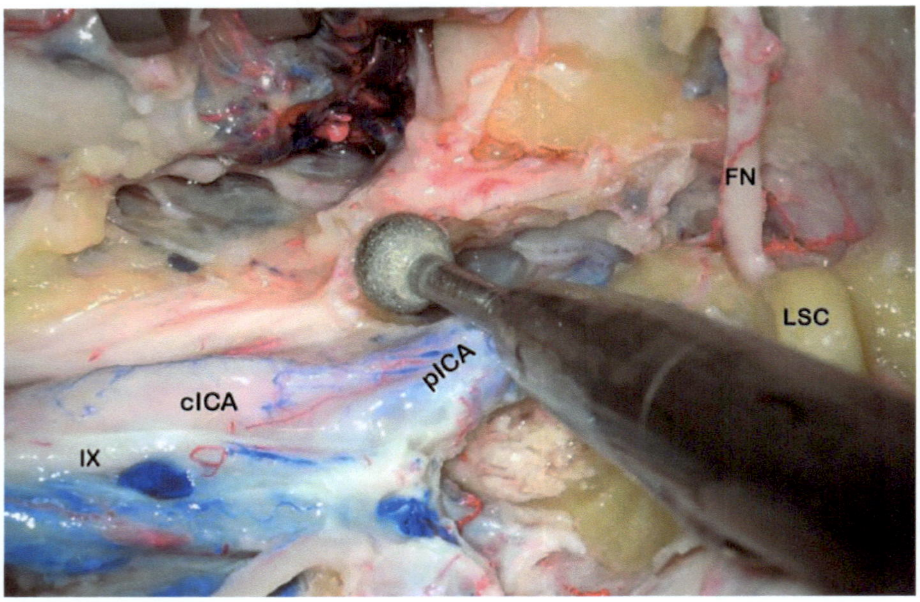

Fig. 12.50 The petrous bone in the area of the petrous apex is drilled to control the horizontal portion of the petrous internal carotid artery. *FN*, rerouted part of the facial nerve; *LSC*, lateral semicircular canal; *pICA*, petrous segment of internal carotid artery; *cICA*, cervical segment of internal carotid artery; *IX*, glossopharyngeal nerve.

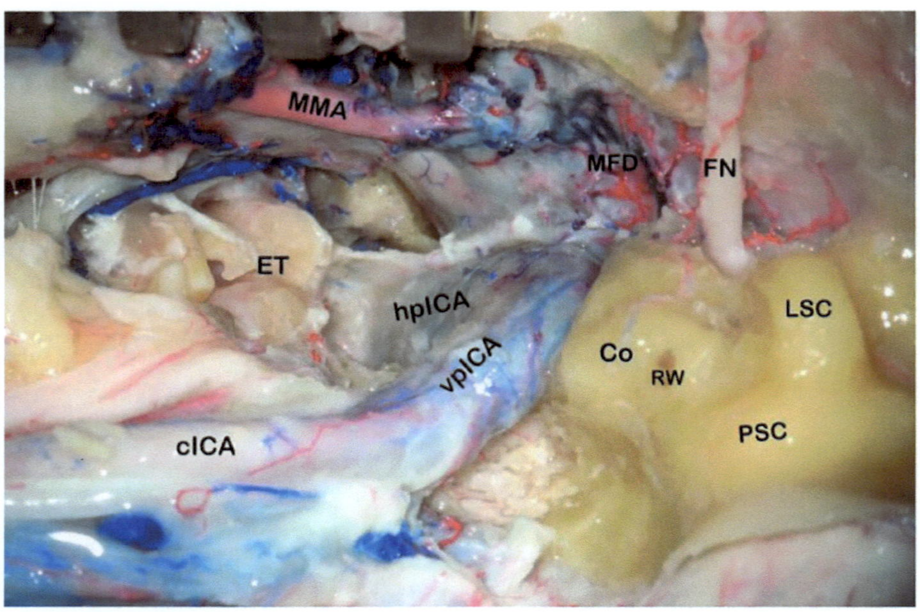

Fig. 12.51 The horizontal portion of the petrous internal carotid artery is under control after drilling the petrous bone lateral to the petrous carotid artery. *MMA*, middle meningeal artery; *MFD*, middle fossa dura; *FN*, rerouted part of the facial nerve; *ET*, Eustachian tube; *hpICA*, horizontal portion of petrous segment of internal carotid artery; *vpICA*, vertical portion of petrous segment of internal carotid artery; *cICA*, cervical segment of internal carotid artery; *Co*, cochlea; *RW*, round window; *LSC*, lateral semicircular canal; *PSC*, posterior semicircular canal.

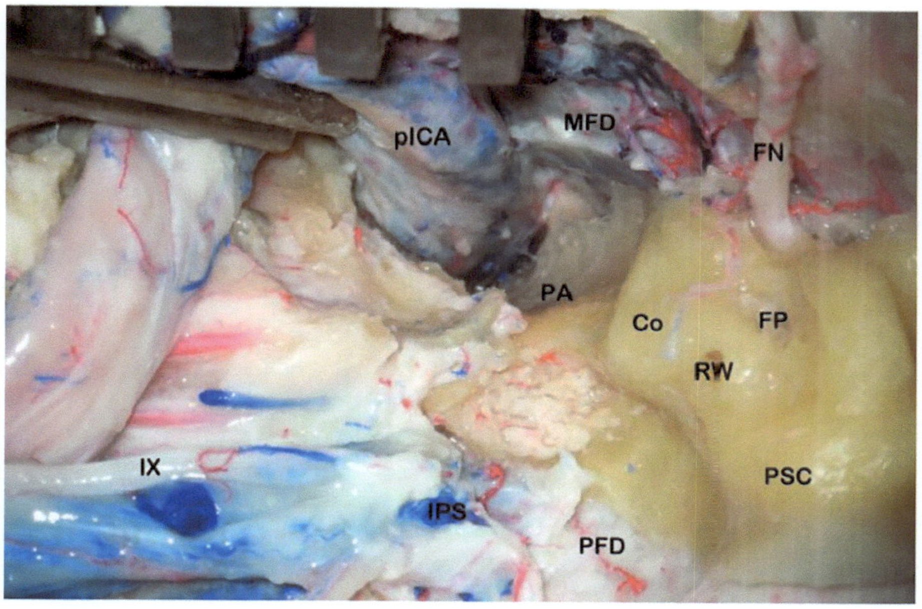

Fig. 12.52 The tip of the suction is used to displace the petrous carotid artery laterally while the petrous bone medial to the petrous carotid artery is drilled to reach the clivus. *pICA*, petrous segment of internal carotid artery; *MFD*, middle fossa dura; *FN*, rerouted part of the facial nerve; *PA*, petrous apex; *Co*, cochlea; *FP*, footplate of the stapes; *RW*, round window; *PSC*, posterior semicircular canal; *IPS*, opening of inferior petrosal sinus; *PFD*, posterior fossa dura; *IX*, glossopharyngeal nerve.

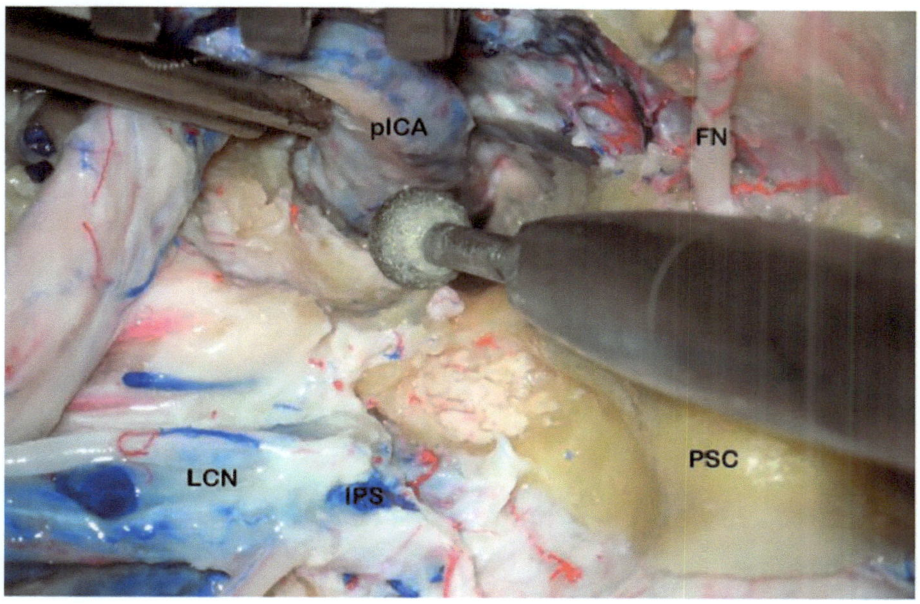

Fig. 12.53 Drilling of the petrous apex is started with a large diamond burr. This is especially important in the area lying between the basal turn of the cochlea and the ICA. *pICA*, internal carotid artery; *FN*, rerouted part of the facial nerve; *PSC*, posterior semicircular canal; *LCN*, lower cranial nerves; *IPS*, opening of inferior petrosal sinus.

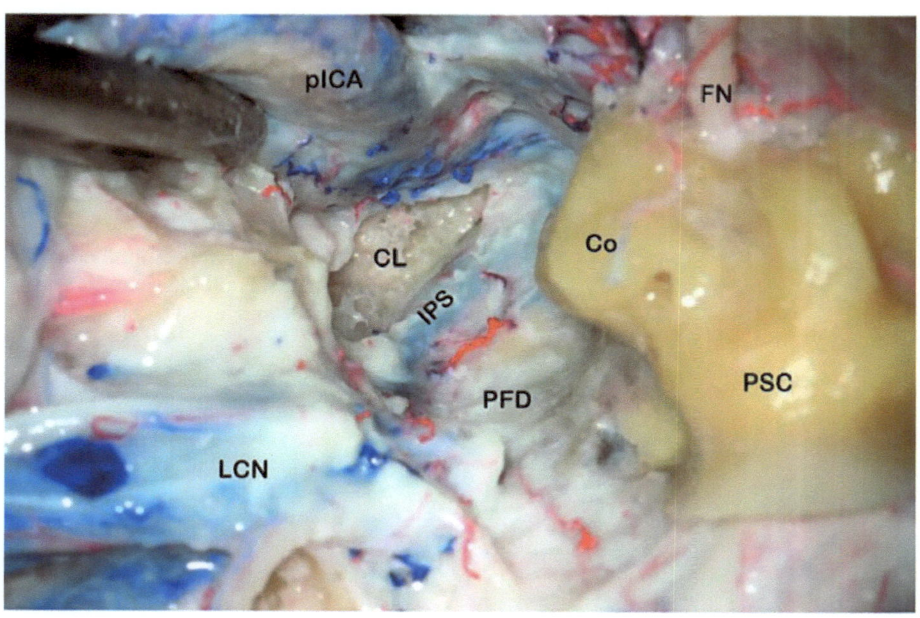

Fig. 12.54 The petrous apex was drilled until the inferior petrosal sinus and clivus were reached. *FN*, rerouted part of the facial nerve; *pICA*, petrous segment of internal carotid artery; *CL*, clivus; *Co*, cochlea; *IPS*, inferior petrosal sinus; *PFD*, posterior fossa dura; *PSC*, posterior semicircular canal; *LCN*, lower cranial nerves.

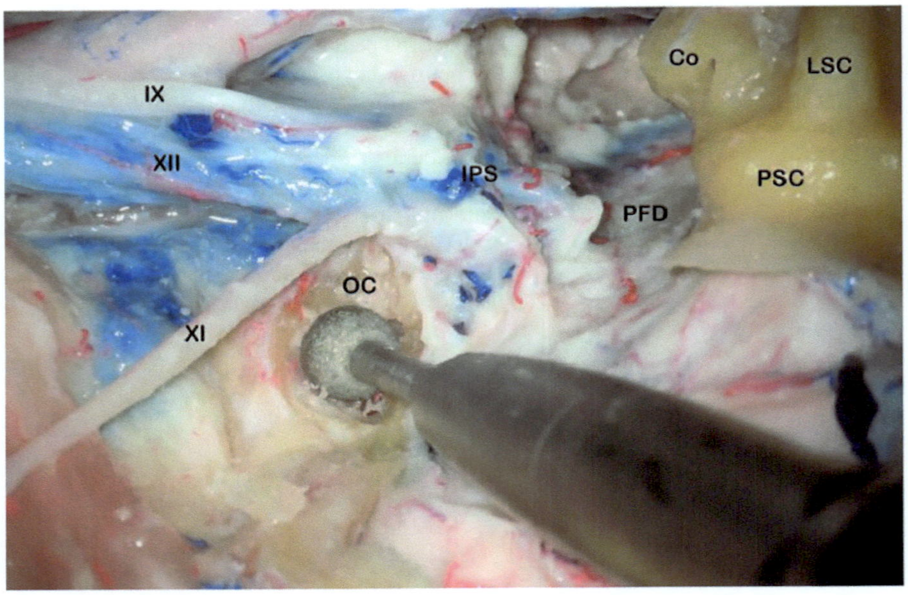

Fig. 12.55 The occipital condyle is starting to be drilled for the transcondylar-transtubercular extension of the approach. *Co*, cochlea; *LSC*, lateral semicircular canal; *PSC*, posterior semicircular canal; *PFD*, posterior fossa dura; *IPS*, opening of inferior petrosal sinus; *OC*, occipital condyle; *XI*, accessory nerve; *XII*, hypoglossal nerve; *IX*, glossopharyngeal nerve.

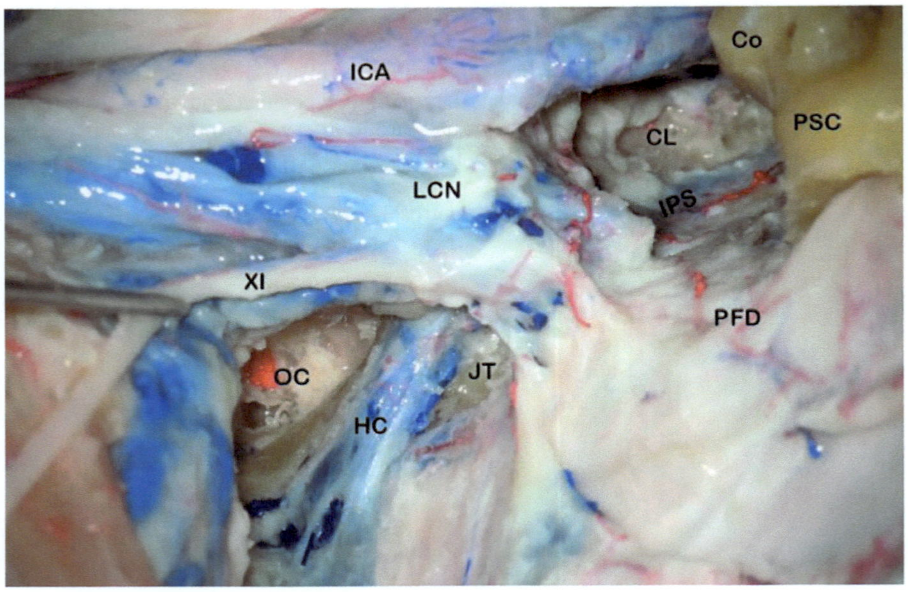

Fig. 12.56 Drilling of the occipital condyle and jugular tubercle exposes the hypoglossal canal. *Co*, cochlea; *PSC*, posterior semicircular canal; *CL*, clivus; *IPS*, inferior petrosal sinus; *PFD*, posterior fossa dura; *LCN*, lower cranial nerves; *ICA*, internal carotid artery; *XI*, accessory nerve; *OC*, occipital condyle; *HC*, hypoglossal canal; *JT*, jugular tubercle.

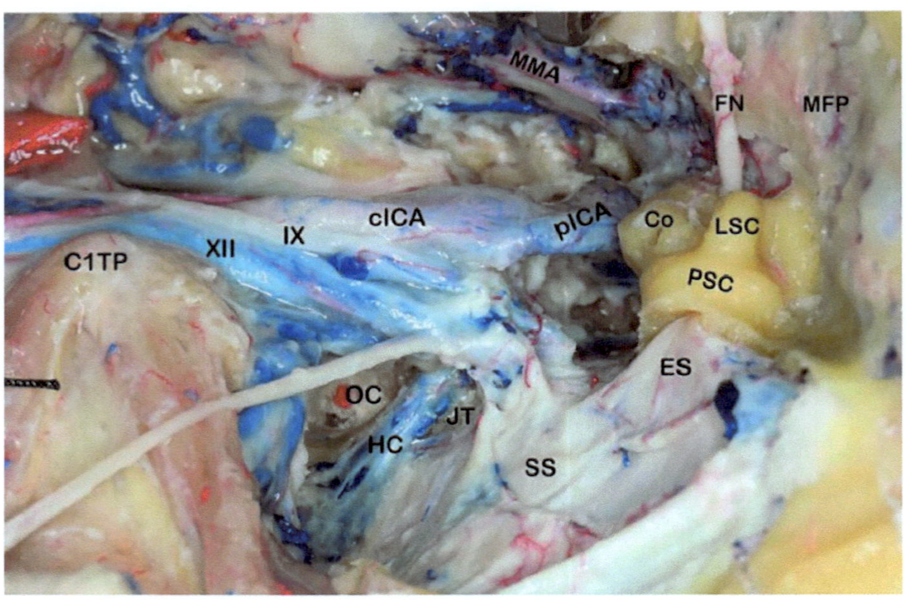

Fig. 12.57 View after finishing the transcondylar-transtubercular extension of the approach. *MMA*, middle meningeal artery; *FN*, rerouted part of the facial nerve; *MFP*, middle fossa plate; *pICA*, petrous segment of internal carotid artery; *Co*, cochlea; *LSC*, lateral semicircular canal; *PSC*, posterior semicircular canal; *ES*, endolymphatic sac; *SS*, sigmoid sinus; *JT*, jugular tubercle; *HC*, hypoglossal canal; *OC*, occipital condyle; *cICA*, cervical segment of internal carotid artery; *IX*, glossopharyngeal nerve; *XII*, hypoglossal nerve; *C1TP*, transverse process of C1.

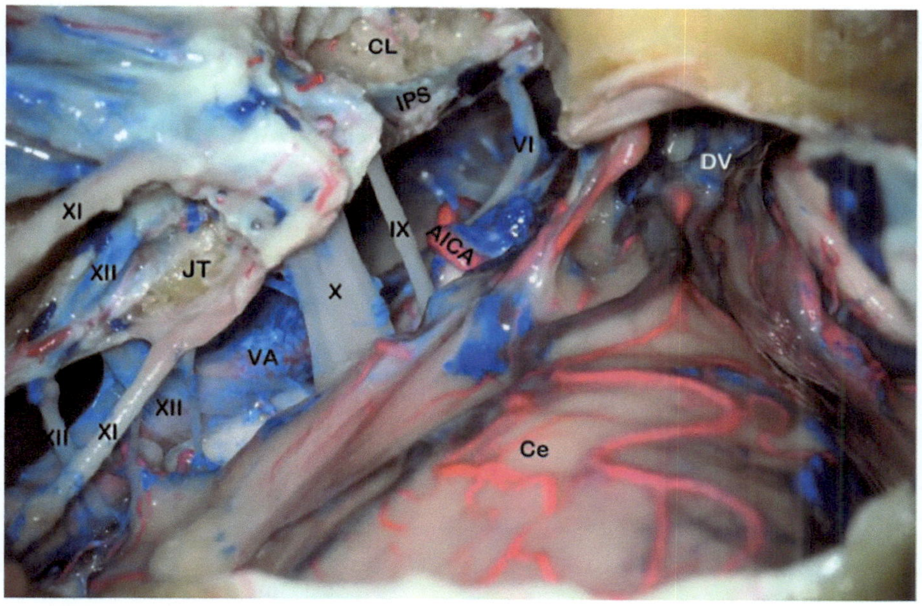

Fig. 12.58 If there is intradural extension, the posterior fossa dura is opened. View of the intradural space after opening of the posterior fossa dura. The glossopharyngeal and vagus nerves are well identified in the cerebellomedullary cistern before entering the jugular foramen. *DV*, Dandy's vein (superior petrosal vein); *VI*, abducent nerve; *IPS*, inferior petrosal sinus; *Cl*, clivus; *AICA*, anteroinferior cerebellar artery; *IX*, glossopharyngeal nerve; *X*, vagus nerve; *VA*, vertebral artery; *JT*, jugular tubercle; *XI*, accessory nerve; *XII*, hypoglossal nerve; *Ce*, cerebellum.

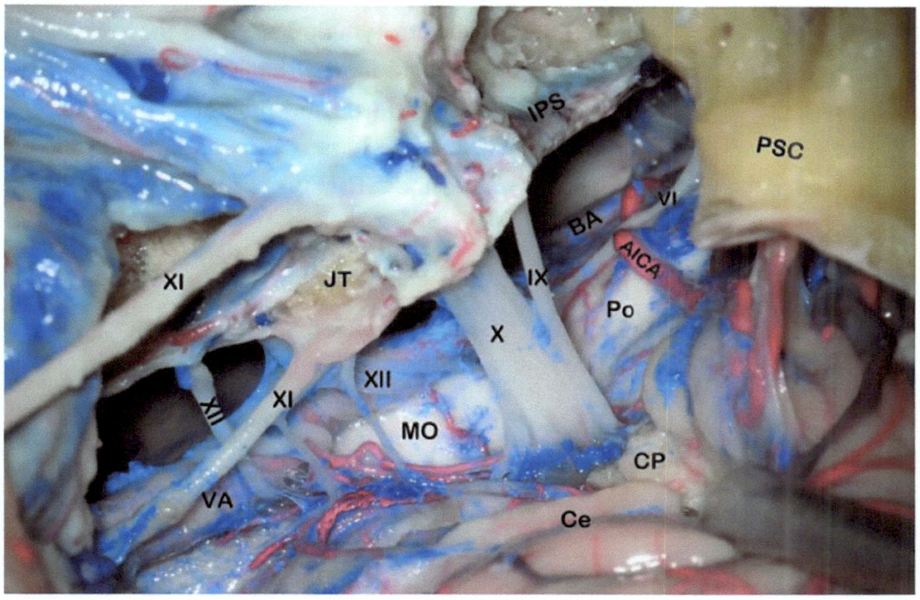

Fig. 12.59 Mild retraction of the cerebellum and flocculus provides a good view of the choroid plexus and lower cranial nerves. *PSC*, posterior semicircular canal; *IPS*, inferior petrosal sinus; *BA*, basilar artery; *AICA*, anteroinferior cerebellar artery; *VI*, abducent nerve; *CP*, choroid plexus; *Po*, pons; *Ce*, cerebellum; *MO*, medulla oblongata; *IX*, glossopharyngeal nerve; *X*, vagus nerve; *XI*, accessory nerve; *XII*, hypoglossal nerve; *JT*, jugular tubercle; *VA*, vertebral artery.

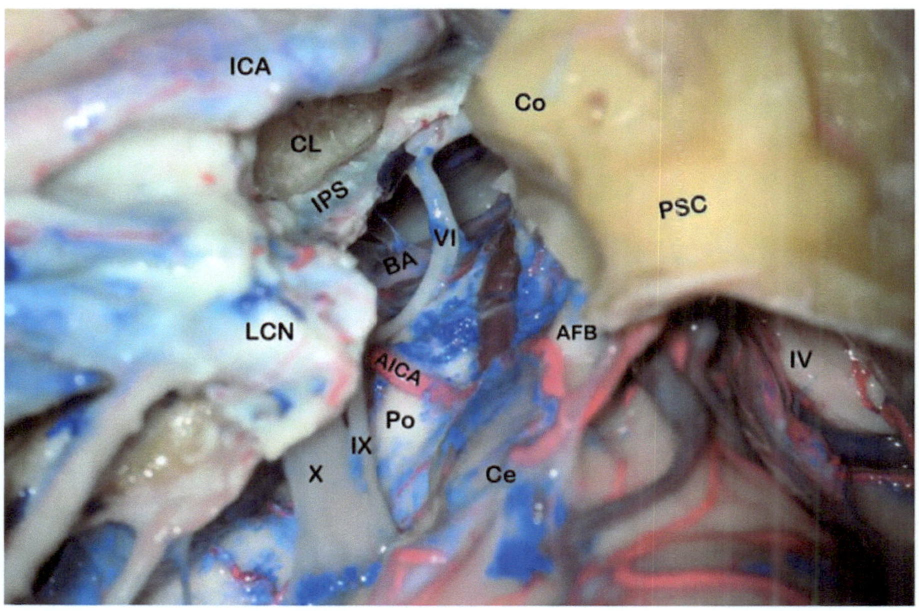

Fig. 12.60 The facial and vestibulocochlear nerves and the anterior inferior cerebellar artery are visible. The anteroinferior cerebellar artery passes inferior to the abducent nerve and travels posteriorly to come into close relation to the facial nerve. *ICA*, internal carotid artery; *Co*, cochlea; *CL*, clivus; *IPS*, inferior petrosal sinus; *BA*, basilar artery; *VI*, abducent nerve; *PSC*, posterior semicircular canal; *AFB*, acousticofacial bundle; *IV*, trochlear nerve; *AICA*, anteroinferior cerebellar artery; *LCN*, lower cranial nerves; *Po*, pons; *IX*, glossopharyngeal nerve; *X*, vagus nerve; *Ce*, cerebellum.

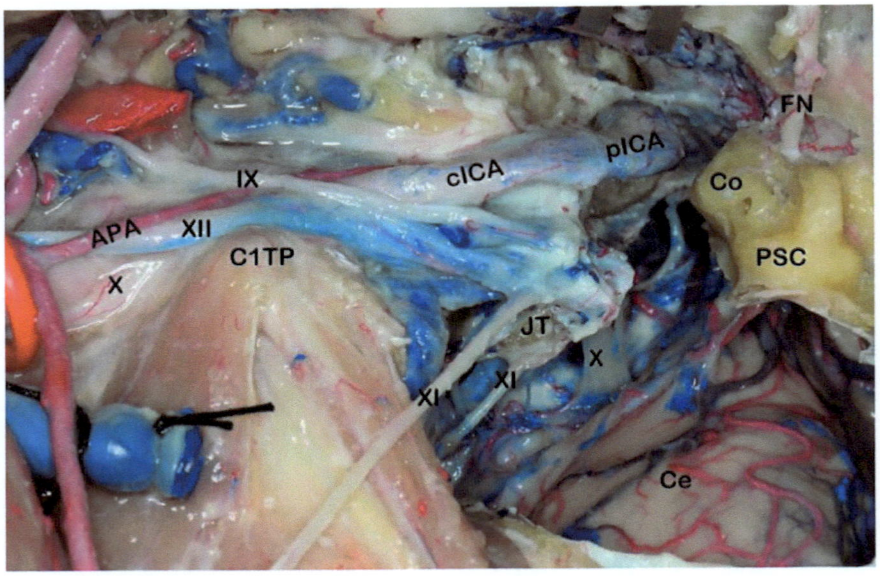

Fig. 12.61 View after finishing the intradural extension of the approach. *FN*, rerouted part of the facial nerve; *pICA*, petrous segment of internal carotid artery; *cICA*, cervical segment of internal carotid artery; *Co*, cochlea; *PSC*, posterior semicircular canal; *JT*, jugular tubercle; *X*, vagus nerve; *XI*, accessory nerve; *C1TP*, transverse process of C1; *XII*, hypoglossal nerve; *IX*, glossopharyngeal nerve; *APA*, ascending pharyngeal artery; *Ce*, cerebellum.

CASE 1: ITFA WITH TRANSCONDYLAR EXTENSION

A 33-year-old woman presented with dysphonia and left shoulder weakness. She had been prepared for surgery at another institute one year previously. During preoperative embolization, she suffered a cerebrovascular accident due to a connection between an unidentified branch of the ECA and the vertebral artery. She has no family history. On otoscopy, a clear retrotympanic reddish mass was noted. On audiography, she had normal hearing function. On cranial nerve examination, she had normal facial nerve function and complete palsy of cranial nerves IX and X and weakness of cranial nerves XI and XII (Figs. 12.62–12.98).

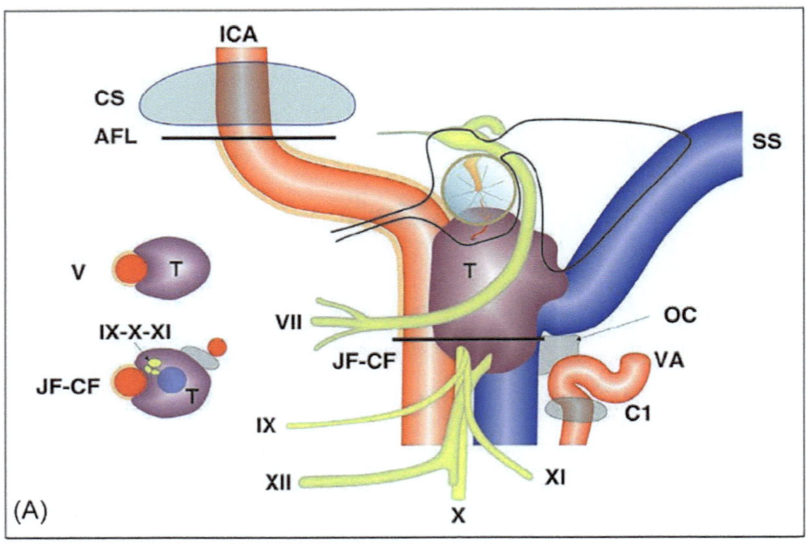

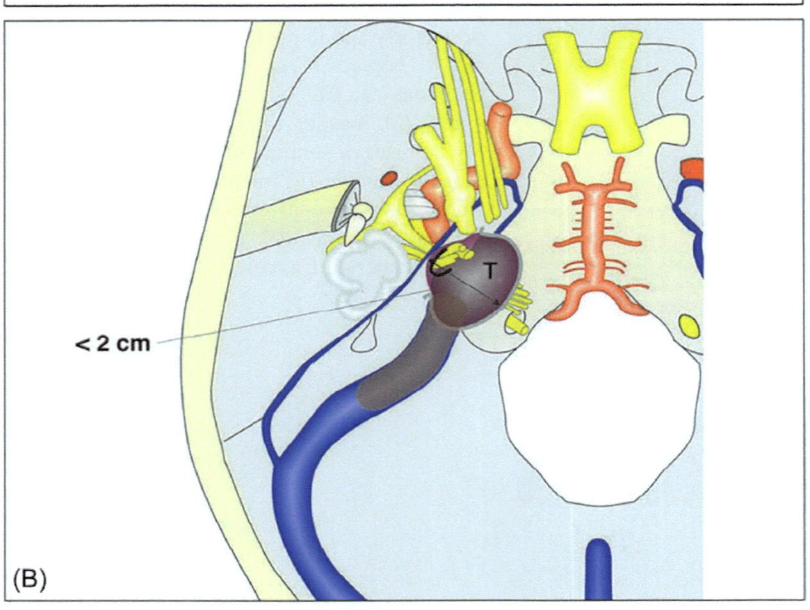

Fig. 12.62 (A, B) Class C2De1 (left side). *AFL*, anterior foramen lacerum; *C1*, atlas; *CS*, cavernous sinus; *ICA*, internal carotid artery; *JF-CF*, jugular foramen-carotid foramen; *OC*, occipital condyle; *SS*, sigmoid sinus; *T*, tumor; *V*, vertical portion of the internal carotid artery; *VA*, vertebral artery; *VII*, facial nerve; *IX*, glossopharyngeal nerve; *XI*, spinal accessory nerve; *XII*, hypoglossal nerve.

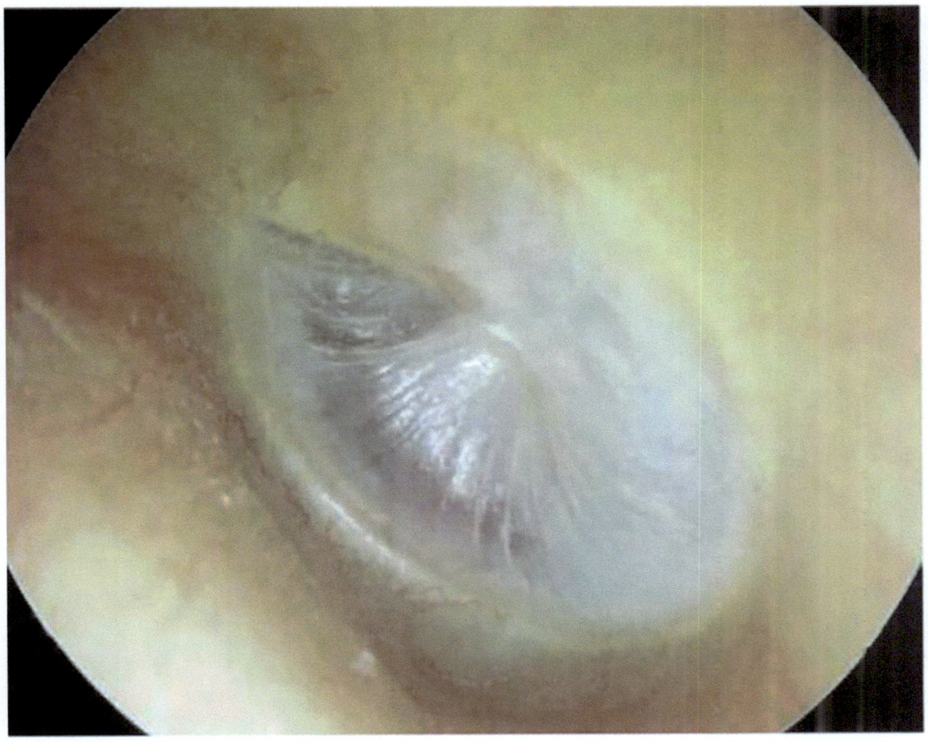

Fig. 12.63 Note the reddish retrotympanic mass in the anteroinferior portion of the tympanic membrane.

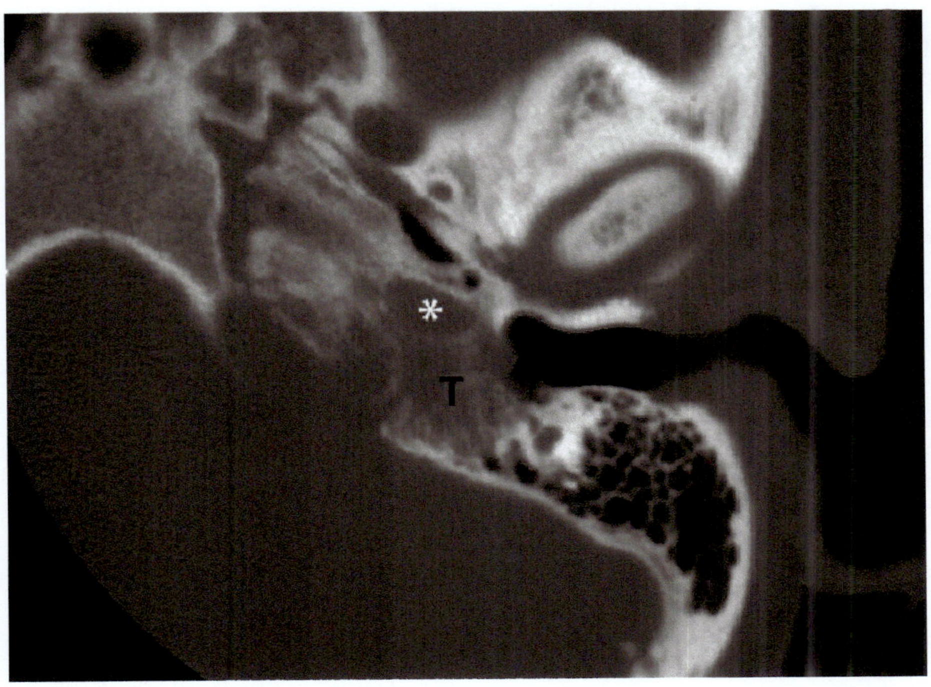

Fig. 12.64 CT, axial view. Infiltration of the bone surrounding the ICA (asterisk) is seen extending toward the genu. *T*, tumor.

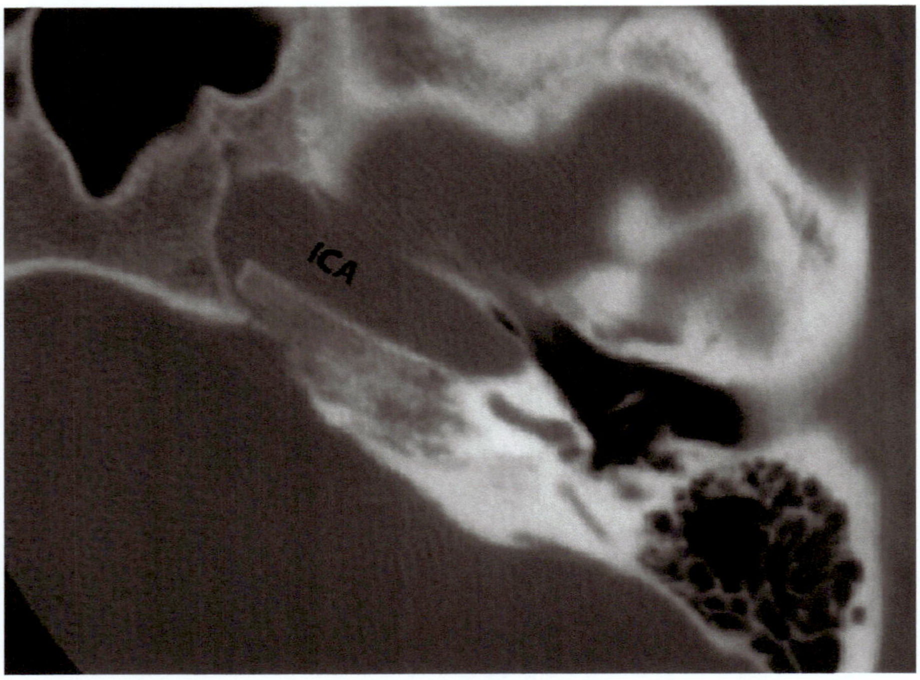

Fig. 12.65 The horizontal segment of the ICA appears free of tumor.

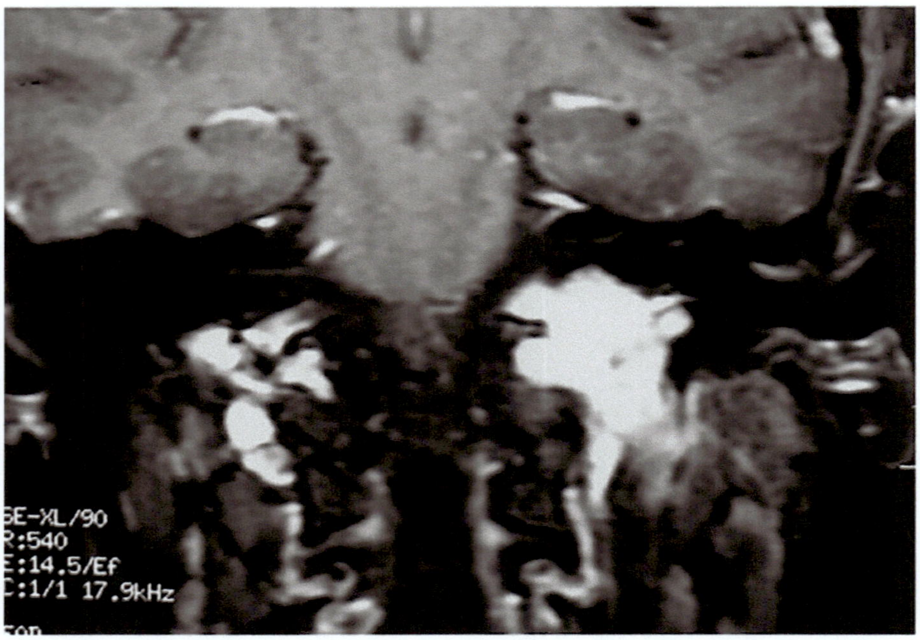

Fig. 12.66 Coronal T1W contrast-enhanced MRI showing tumor extending partly into the hypoglossal canal.

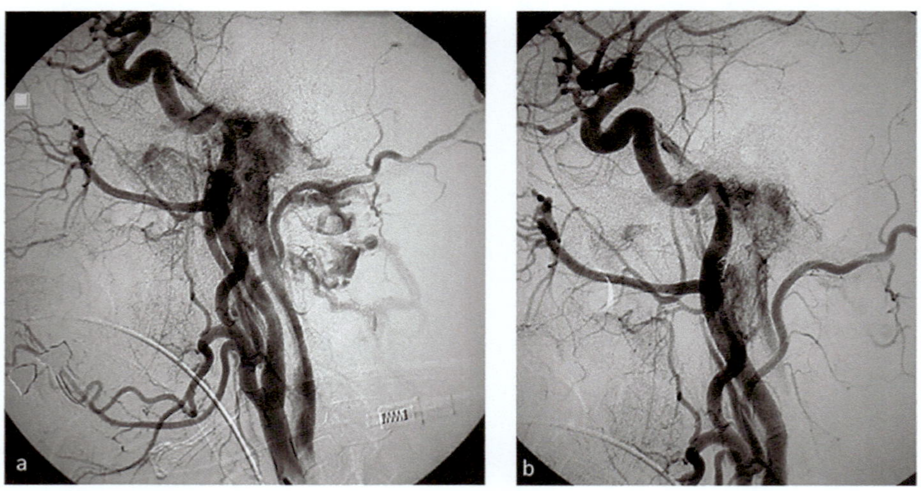

Fig. 12.67 (A, B) Angiography, revealing a characteristic tumor blush before (A) and after tumor embolization (B). Despite a significant reduction in tumor vascularity, tumor blush can still be seen adjacent to the distal vertical carotid. Platinum coils were used as the predominant material for embolization in this case due to the previous cerebrovascular accident. Devascularization by this technique is usually not as effective as that provided by microparticles.

Preoperative radiological assessment revealed a C2De1 TJP. A single-stage ITFA with preoperative embolization was planned.

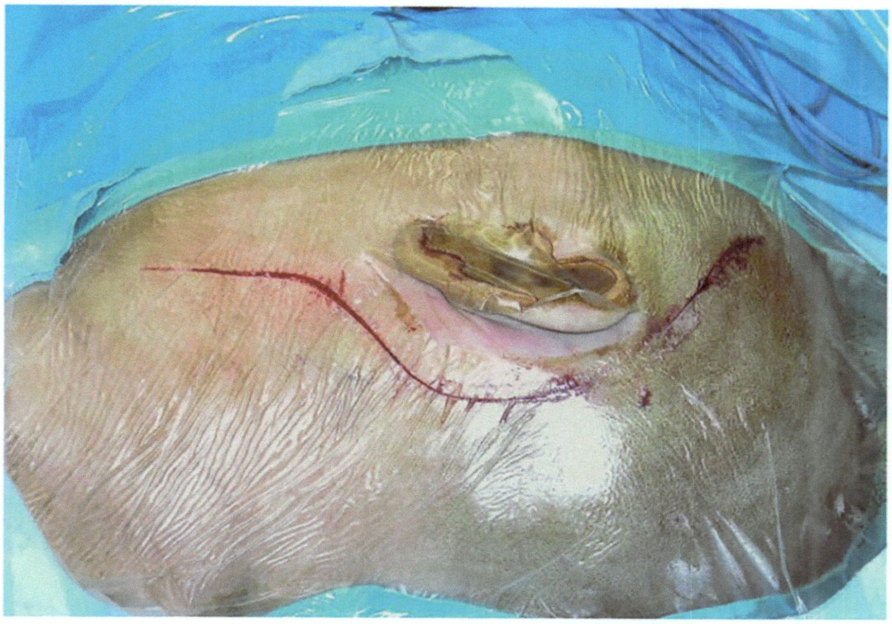

Fig. 12.68 A large C-shaped postauricular incision A wide postauricular incision is made with extension down to the neck.

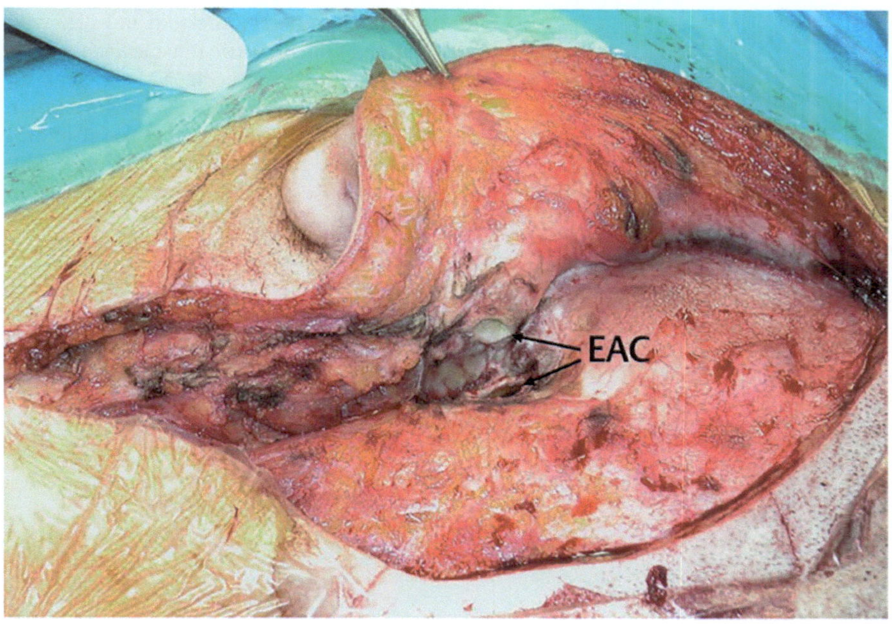

Fig. 12.69 The large postauricular flap is raised while keeping its adequate thickness. During flap elevation, the external auditory canal (EAC) is transected lateral to the bony cartilaginous junction.

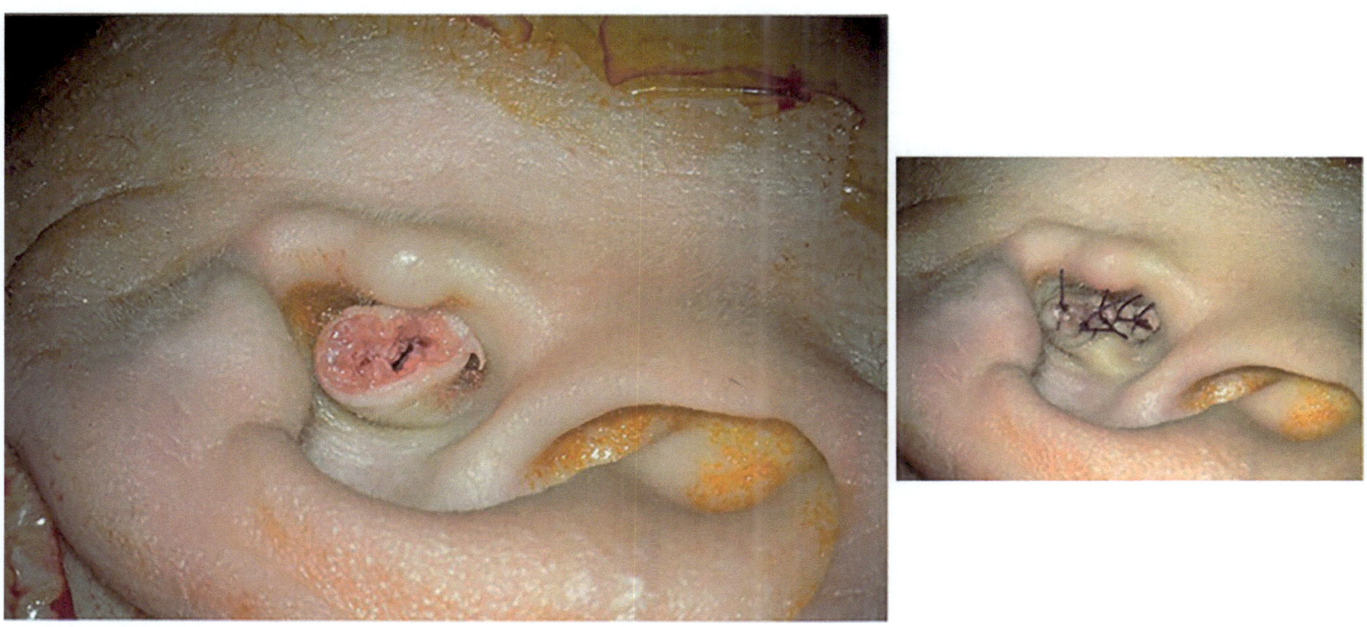

Fig. 12.70 The everted skin is tightly structured with vicryl 4-0.

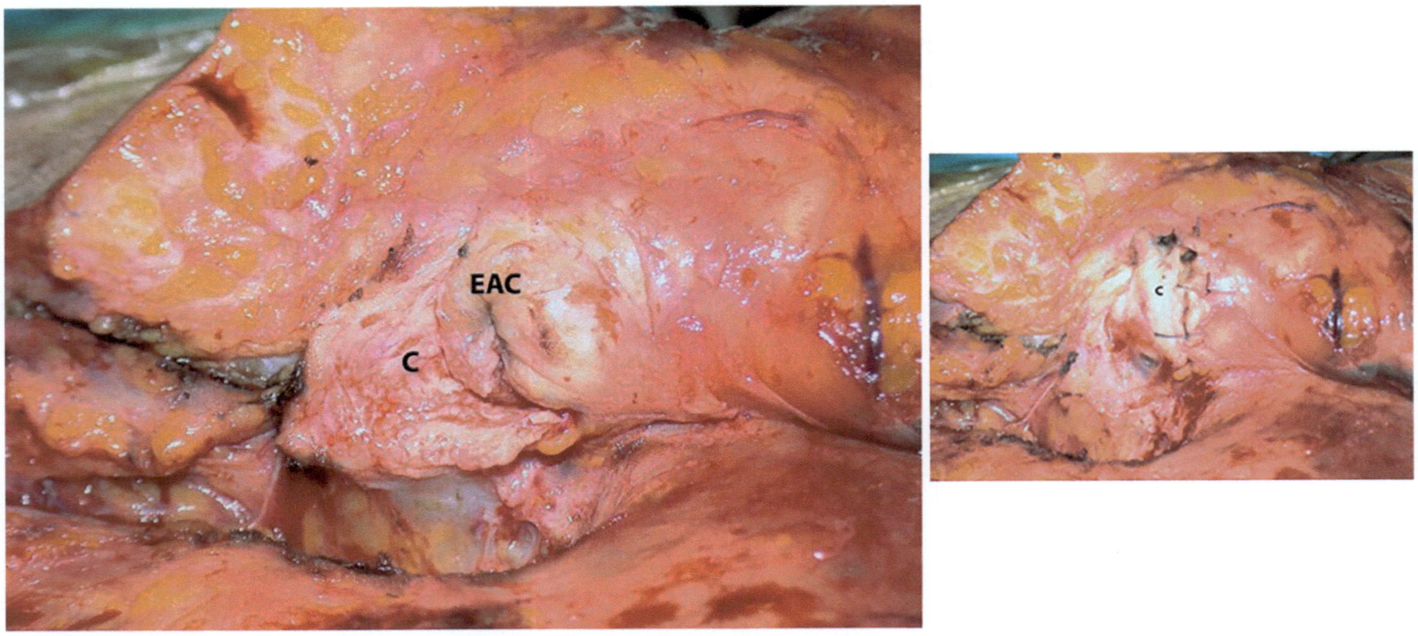

Fig. 12.71 The preserved cartilage then provides an ideal material with which to reinforce the blind sac closure. Note that the inferior angle of the cartilage represents the "tragal pointer." *C*, cartilage; *EAC*, everted external auditory canal.

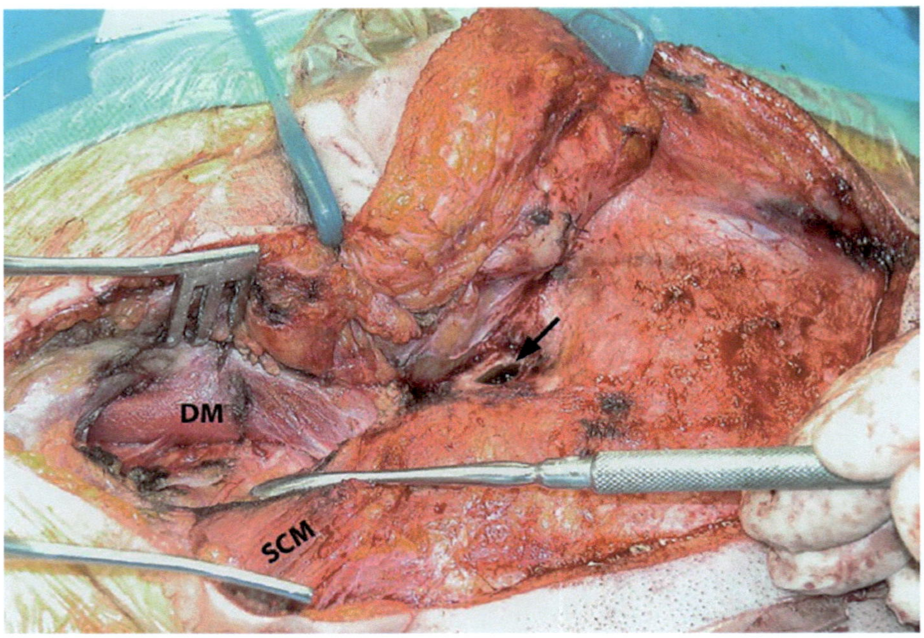

Fig. 12.72 An incision in the superficial layer of deep cervical fascia is made along the anterior border of the sternocleidomastoid muscle, with identification of the digastric muscle and the neurovascular bundle in the neck. Note the careful placement of the self-retaining retractor, which enhances visibility and allows the use of microscopic dissection. *arrow*, external auditory canal; *DM*, digastric muscle; *SCM*, sternocleidomastoid muscle.

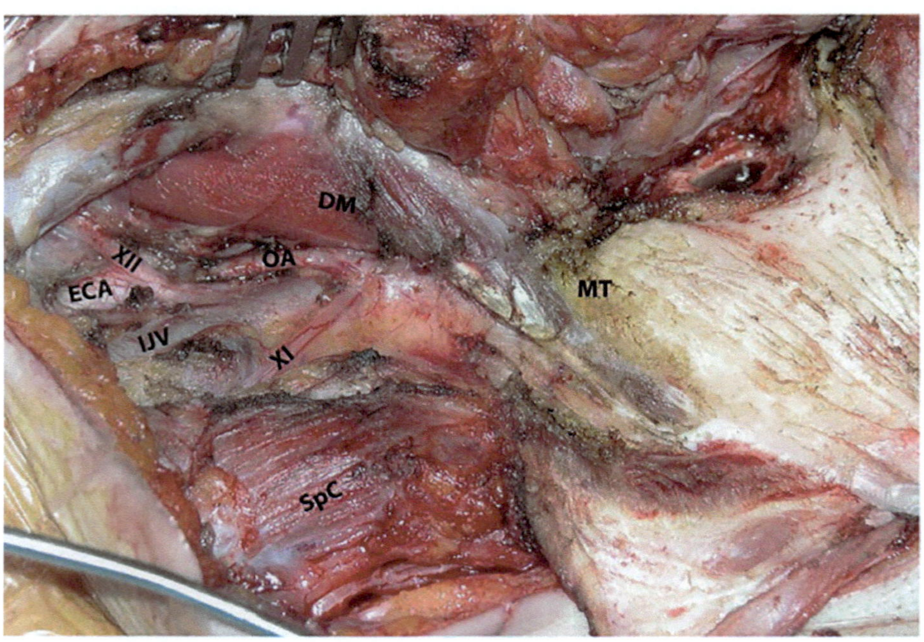

Fig. 12.73 The musculofascial incisions have been made: the superior limb along the zygomatic root, the inferior limb hugging the external auditory canal, and then continuous with the sternocleidomastoid muscle. The sternocleidomastoid muscle has been dissected from the mastoid tip and sutured to the skin edge. The digastric muscle can now be seen inserting into the digastric groove, medial to the mastoid tip. *DM*, digastric muscle; *ECA*, external carotid artery; *IJV*, internal jugular vein; *MT*, mastoid tip; *OA*, occipital artery; *SpC*, splenius capitis muscle; *XI*, spinal accessory nerve; *XII*, hypoglossal nerve.

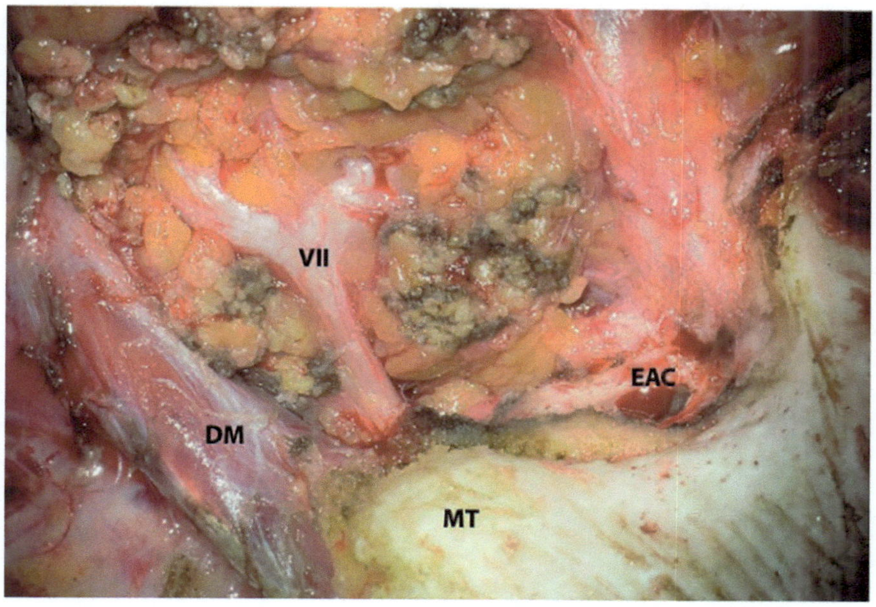

Fig. 12.74 The extratemporal facial nerve is then identified and exposed at the bifurcation. Note that the exit at the stylomastoid foramen is closely related to the digastric muscle. It is also 6–8 mm anteromedial to the inferior aspect of the tympanomastoid suture line. Importantly, with transection of the external auditory canal, and anterior retraction of the auricle, the facial nerve appears to be in a more superficial plane than when identifying the nerve via a preauricular incision. *EAC*, external auditory canal; *DM*, digastric muscle; *MT*, mastoid tip; *VII*, facial nerve.

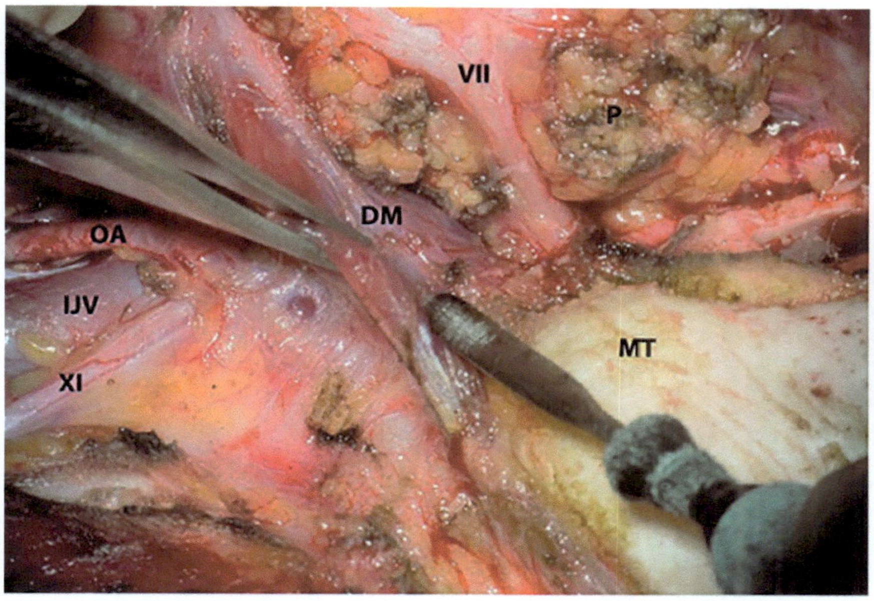

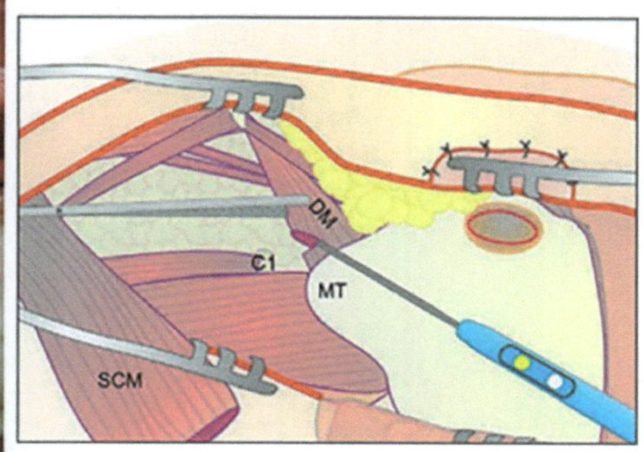

Fig. 12.75 The digastric muscle has been detached with monopolar diathermy. *C1*, atlas; *DM*, posterior belly of the digastric muscle; *IJV*, internal jugular vein; *MT*, mastoid tip; *OA*, occipital artery; *P*, parotid gland; *SCM*, sternocleidomastoid muscle; *VII*, facial nerve; *XI*, spinal accessory nerve.

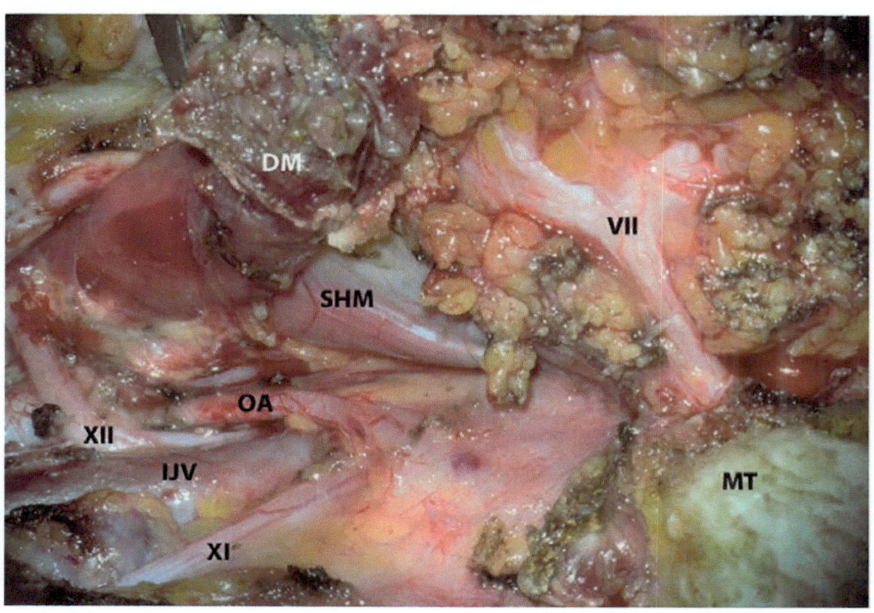

Fig. 12.76 Once the facial nerve has been definitively identified, the posterior belly of the digastric muscle is cut with monopolar diathermy. The transected muscle is retracted anteriorly, further enhancing exposure. The underlying muscle that can be seen is the stylohyoid muscle inserting on the upper lateral segment of the styloid process. Ligation of the facial vein is required to adequately access the carotid sheath. The occipital artery is then identified and ligated. The lateral process of C1 is palpated as a landmark to identify the internal jugular vein and the point at which cranial nerve XI passes laterally to it (in 80%–85% of cases). *DM*, posterior belly of the digastric muscle; *IJV*, internal jugular vein; *MT*, mastoid tip; *OA*, occipital artery; *SHM*, stylohyoid muscle; *VII*, facial nerve; *XI*, spinal accessory nerve; *XII*, hypoglossal nerve.

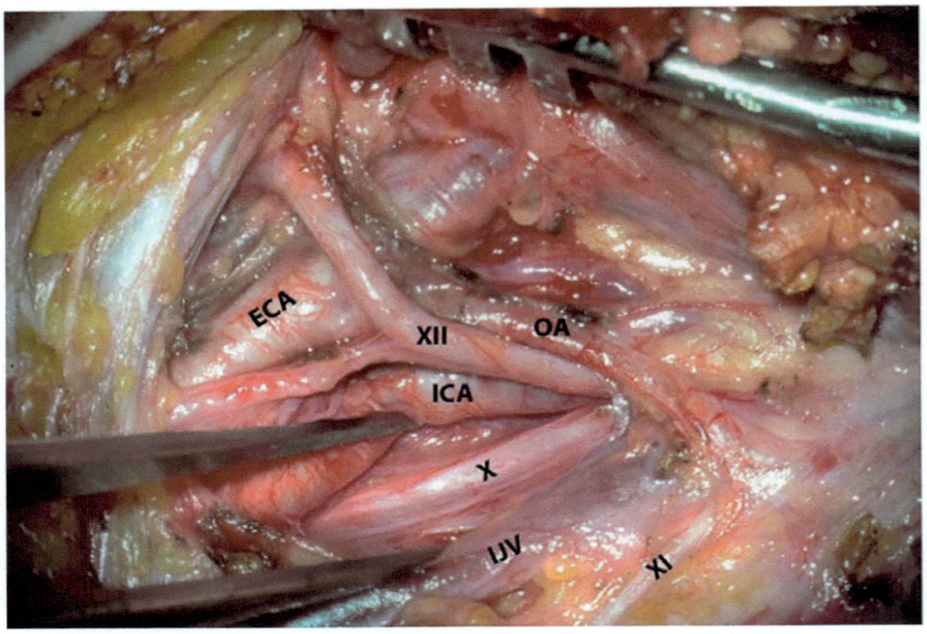

Fig. 12.77 The vagus, spinal accessory, and hypoglossal nerves, the ICA, ECA, and occipital artery are all well seen. *ECA*, external carotid artery; *ICA*, internal carotid artery; *IJV*, internal jugular vein; *OA*, occipital artery; *X*, vagus nerve; *XIs*, pinal accessory nerve; *XII*, hypoglossal nerve.

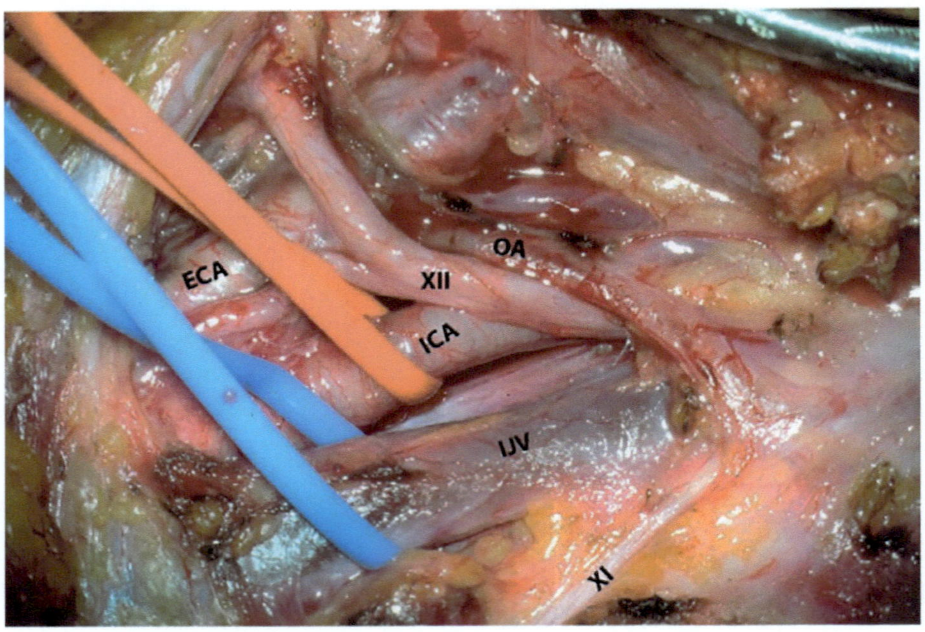

Fig. 12.78 Two vessel loops are placed in order to mark the internal carotid artery and the internal jugular vein. *ECA*, external carotid artery; *ICA*, internal carotid artery; *IJV*, internal jugular vein; *OA*, occipital artery; *XI*, spinal accessory nerve; *XII*, hypoglossal nerve.

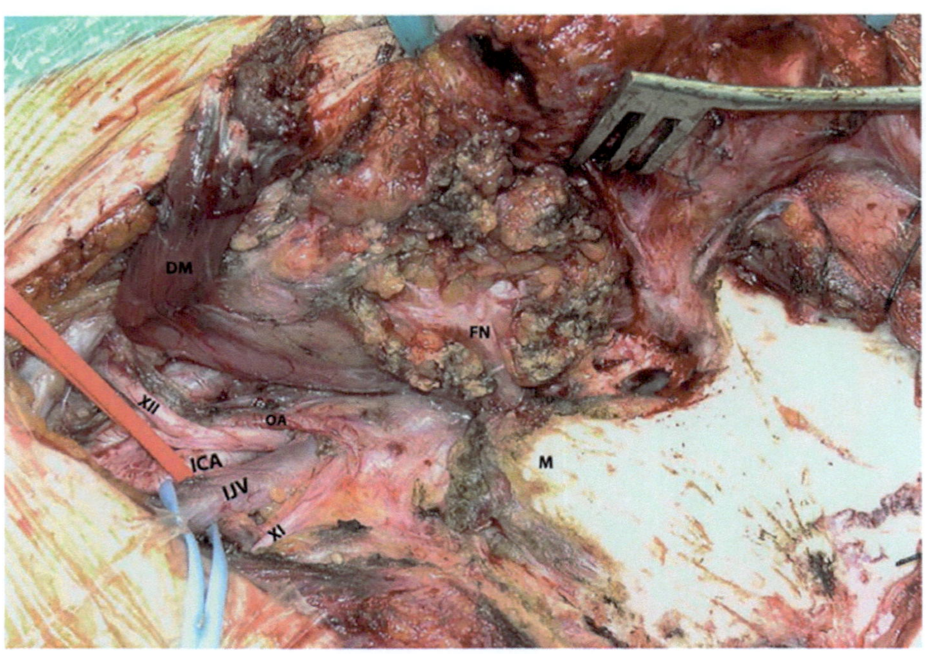

Fig. 12.79 A level II neck dissection is performed to sample the lymph nodes for the possibility of metastatic disease. The IJV is mobilized and marked with a blue rubber loop following definitive identification of the cranial nerves X and XII. A loop is placed around the ICA. The lateral process of the C1 is now clearly identified, with preservation of the overlying fascia. *DM*, digastric muscle; *FN*, facial nerve; *ICA*, internal carotid artery; *IJV*, internal jugular vein; *M*, mastoid bone; *OA*, occipital artery; *XI*, spinal accessory nerve; *XII*, hypoglossal nerve.

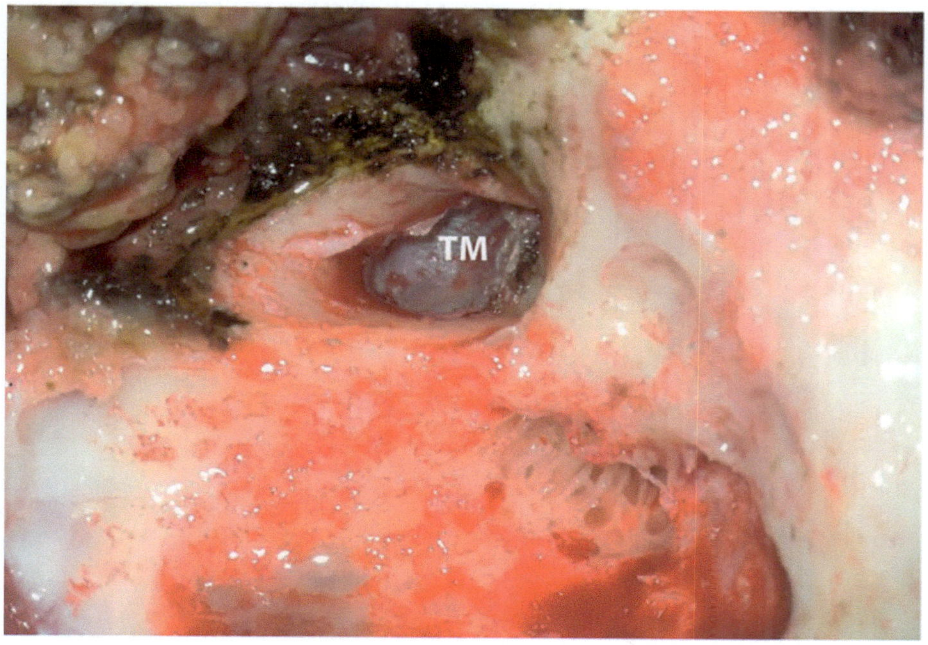

Fig. 12.80 A wide mastoidectomy is then performed, with care to remove all the skin of the external auditory canal. Note the extent of the tympanic bone once the soft tissues have been removed. *TM*, tympanic membrane.

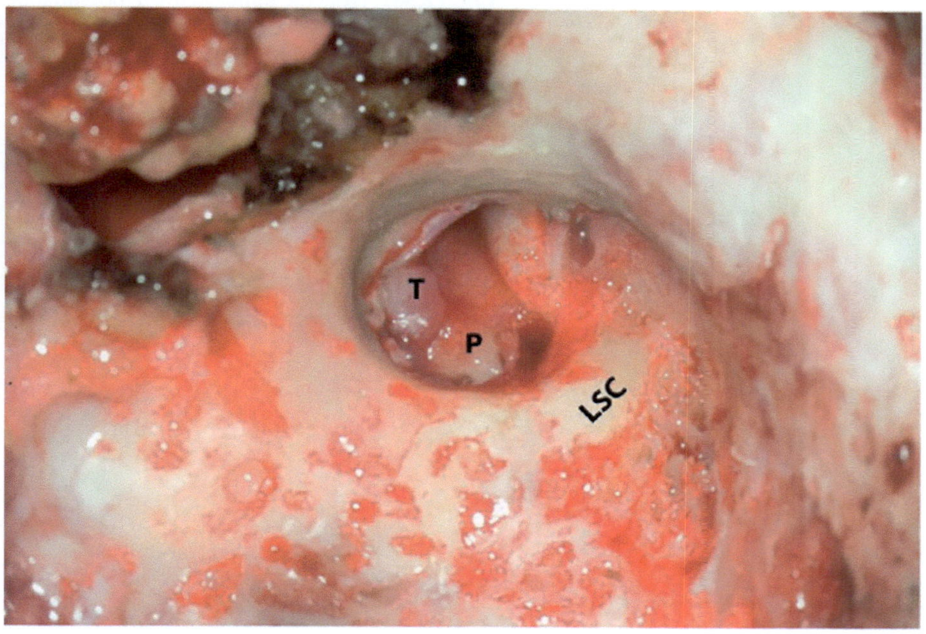

Fig. 12.81 The posterior canal wall, tympanic membrane, malleus, and incus have been removed. Note the complete removal of all squamous material. *LSC*, lateral semicircular canal; *P*, promontory; *T*, tumor.

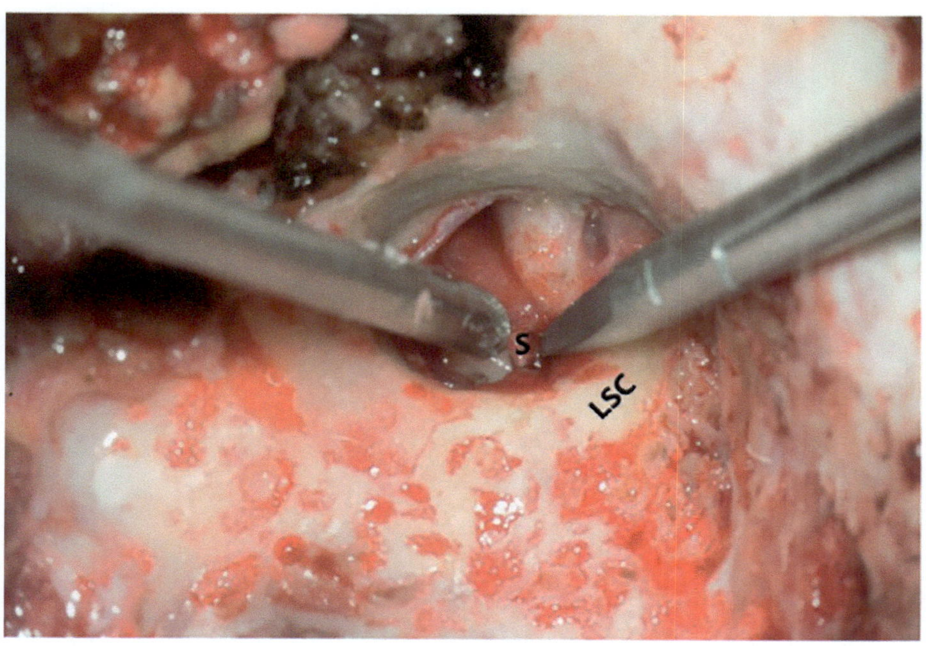

Fig. 12.82 The superstructure of the stapes is then removed using neurosurgical scissors. The crura of the stapes are transected to avoid trauma to the inner ear while dissecting the facial nerve during anterior rerouting. This is essential to avoid sensorineural hearing loss during anterior rerouting of the facial nerve due to footplate displacement. *LSC*, lateral semicircular canal; *S*, stapes.

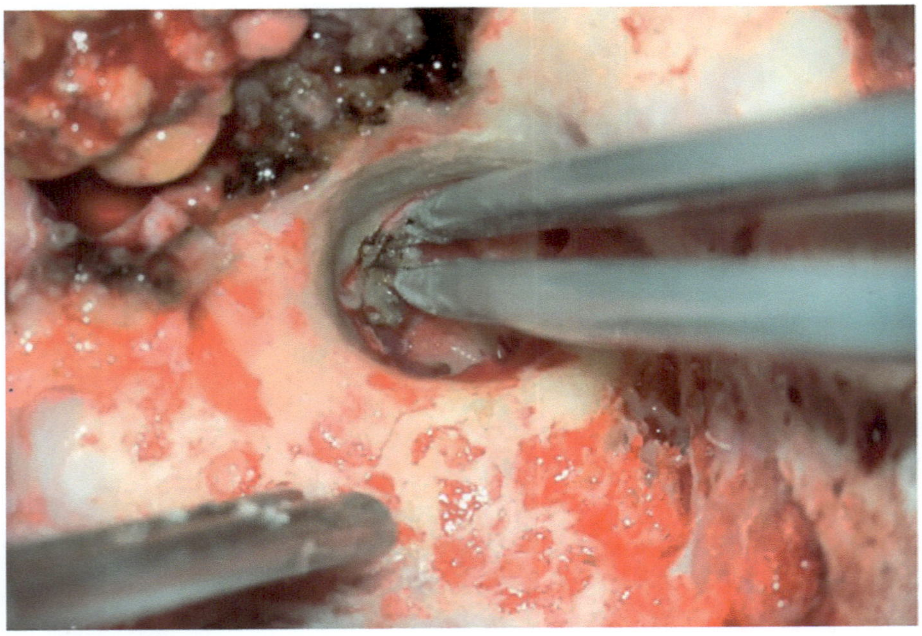

Fig. 12.83 Whether small or large, the middle ear component of the tumor is debulked with bipolar coagulation; Surgicel packing is often used at this stage to ensure that facial nerve dissection can proceed in a bloodless field. This is the only component of tumor that is dissected prior to wide exposure of all margins.

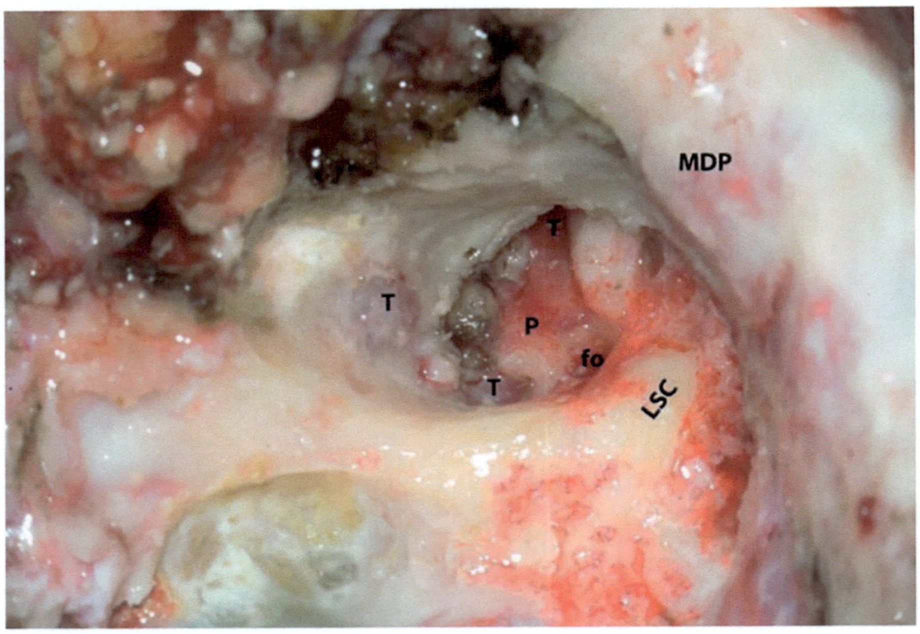

Fig. 12.84 Note the tumor in the hypotympanic area eroding the tympanic bone. *fo*, foramen ovale; *LSC*, lateral semicircular canal; *MDP*, middle cranial fossa dural plate; *P*, promontory; *T*, tumor.

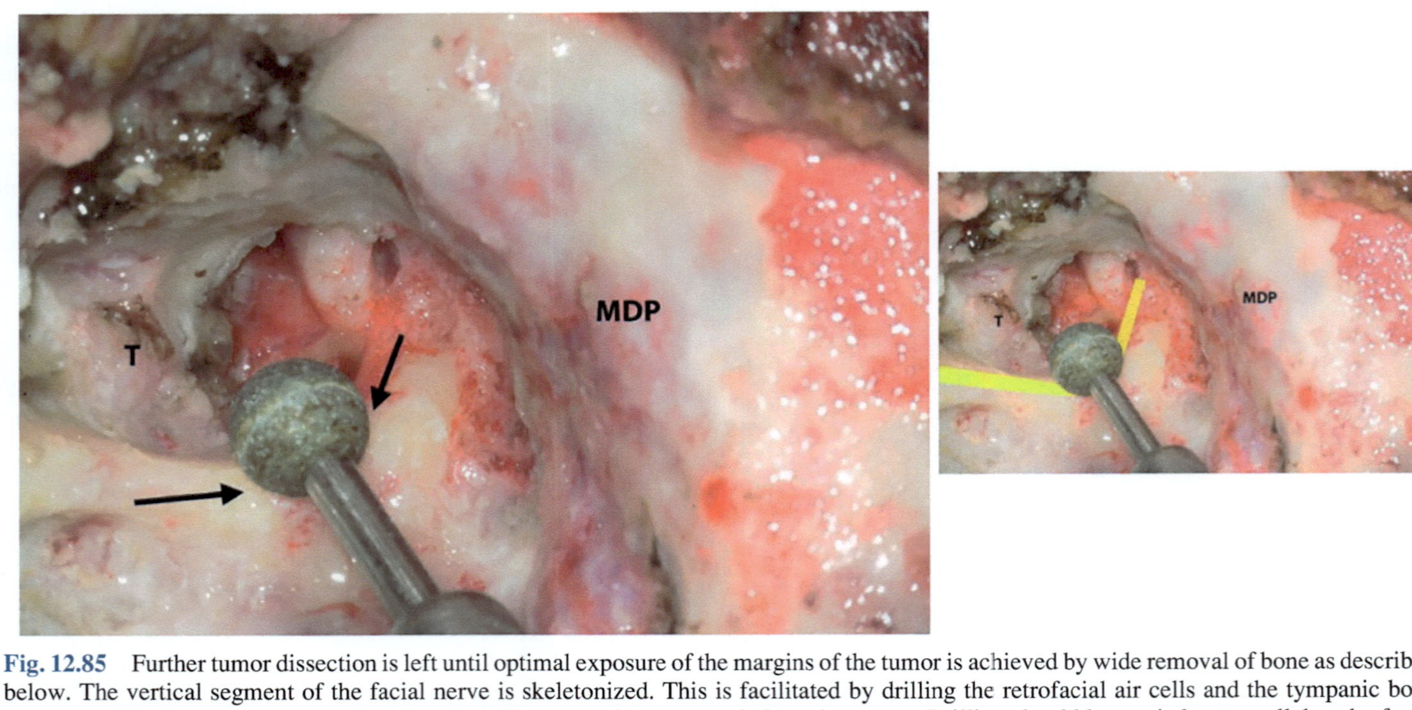

Fig. 12.85 Further tumor dissection is left until optimal exposure of the margins of the tumor is achieved by wide removal of bone as described below. The vertical segment of the facial nerve is skeletonized. This is facilitated by drilling the retrofacial air cells and the tympanic bone anterior to the facial nerve. Note the extensive infiltration of the tympanic bone by tumor. Drilling should be carried out parallel to the facial nerve (*arrows* in main figure, *yellow band* in indent). *MDP*, middle cranial fossa dural plate; *T*, tumor.

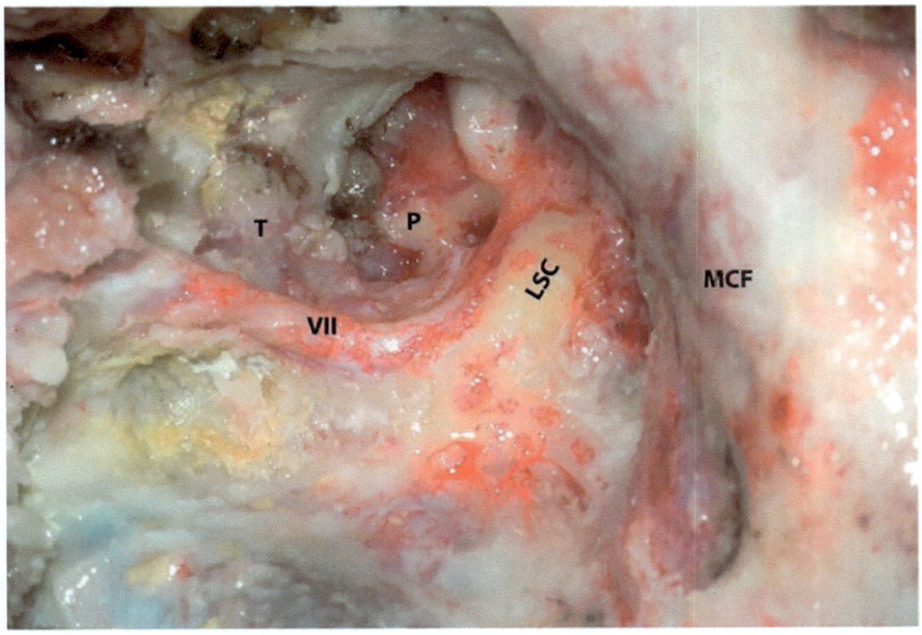

Fig. 12.86 Further dissection of the facial nerve has been done. *LSC*, lateral semicircular canal; *MCF*, middle cranial fossa; *P*, promontory; *T*, tumor; *VII*, facial nerve.

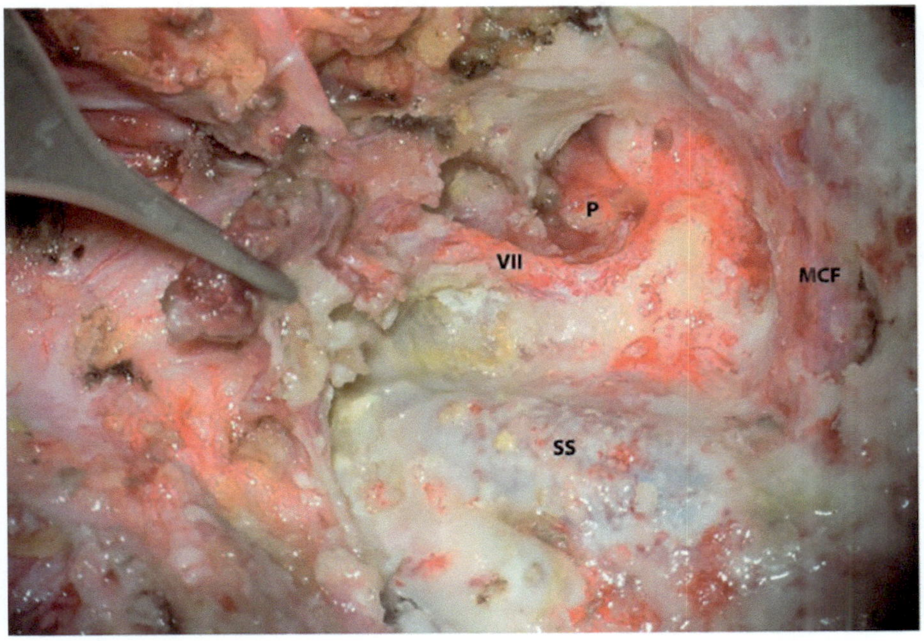

Fig. 12.87 The mastoid tip is removed. *MCF*, middle cranial fossa; *P*, promontory; *SS*, sigmoid sinus; *VII*, facial nerve.

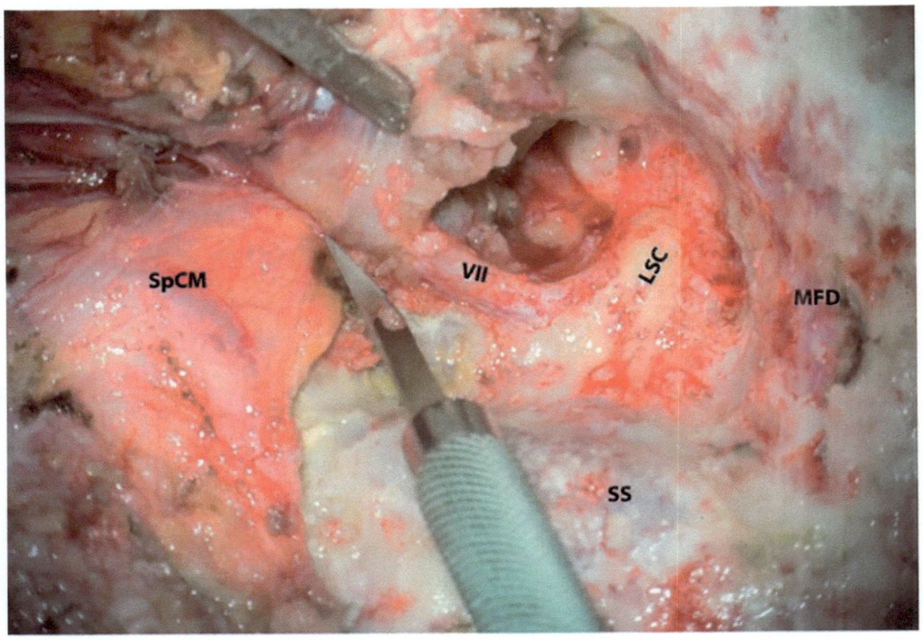

Fig. 12.88 The dense connective tissue at the level of stylomastoid foramen is freed. *LSC*, lateral semicircular canal; *MFD*, middle fossa dura; *SpCM*, splenius capitis muscle; *SS*, sigmoid sinus; *VII*, facial nerve.

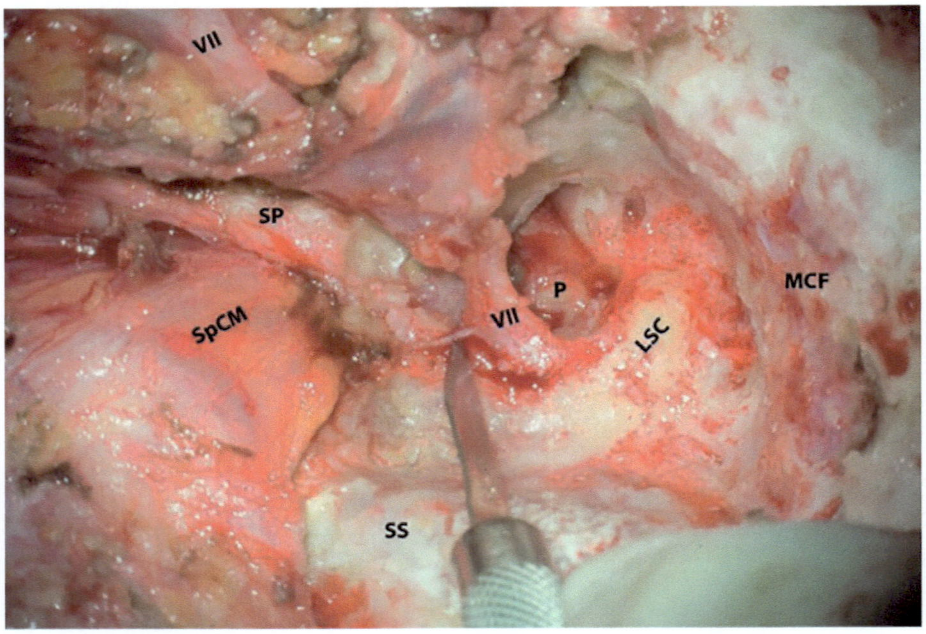

Fig. 12.89 The facial nerve is separated from the fallopian canal with a Beaver knife. *LSC*, lateral semicircular canal; *MCF*, middle cranial fossa; *P*, promontory; *SP*, styloid process; *SpCM*, splenius capitis muscle; *SS*, sigmoid sinus; *VII*, facial nerve.

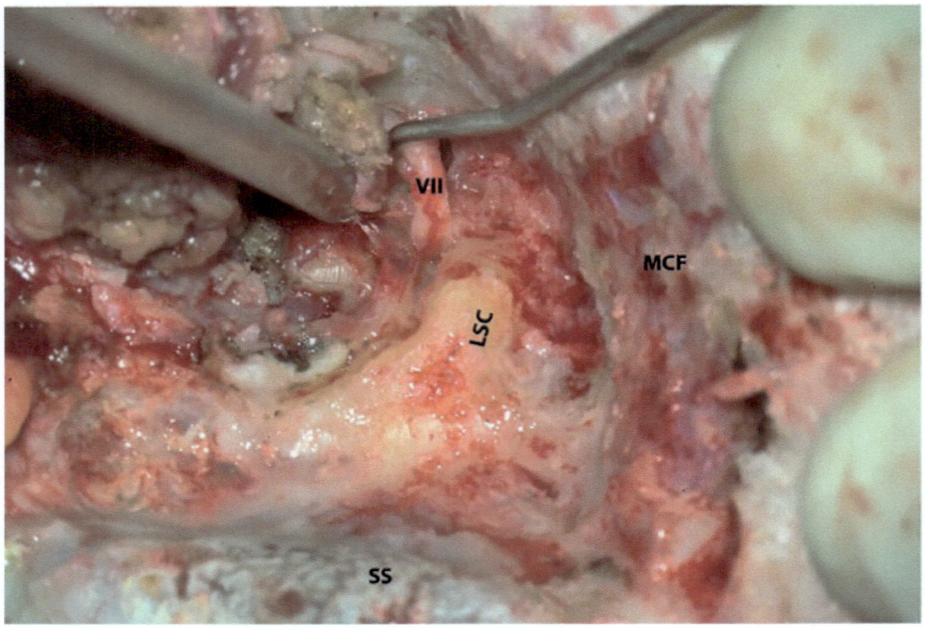

Fig. 12.90 Another groove for the terminal portion is created in the parotid gland. Displacement of the facial nerve represents the key point of the surgery because it allows unobstructed control of the jugular foramen and the vertical carotid canal. *LSC*, lateral semicircular canal; *MCF*, middle cranial fossa; *SS*, sigmoid sinus; *VII*, facial nerve.

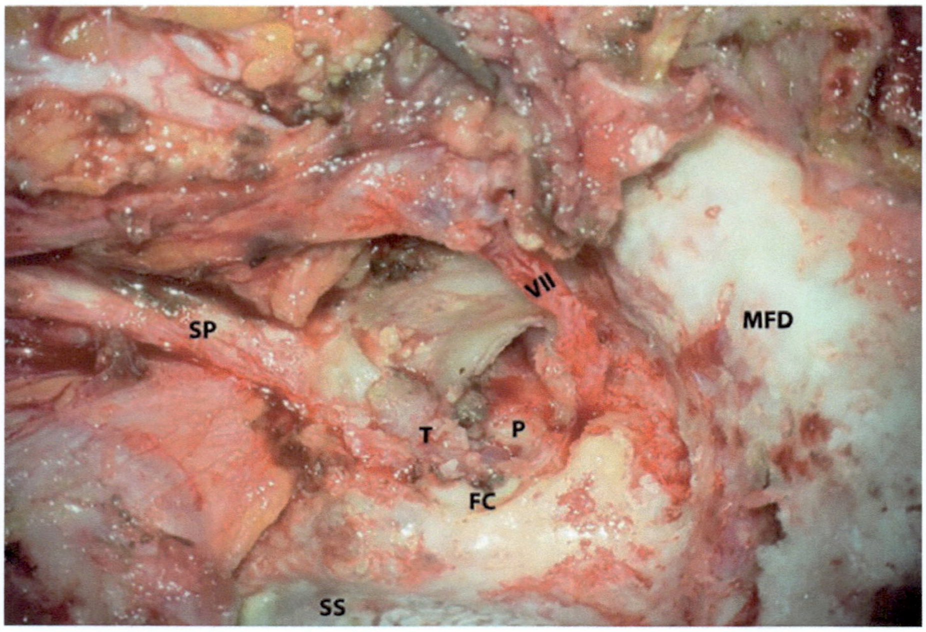

Fig. 12.91 Tips for correctly performing the anterior rerouting include complete bony decompression of the geniculate ganglion area and sharp dissection of the nerve from the fallopian canal. *FC*, fallopian canal; *MFD*, middle fossa dura; *P*, promontory; *SP*, styloid process; *SS*, sigmoid sinus; *T*, tumor; *VII*, facial nerve.

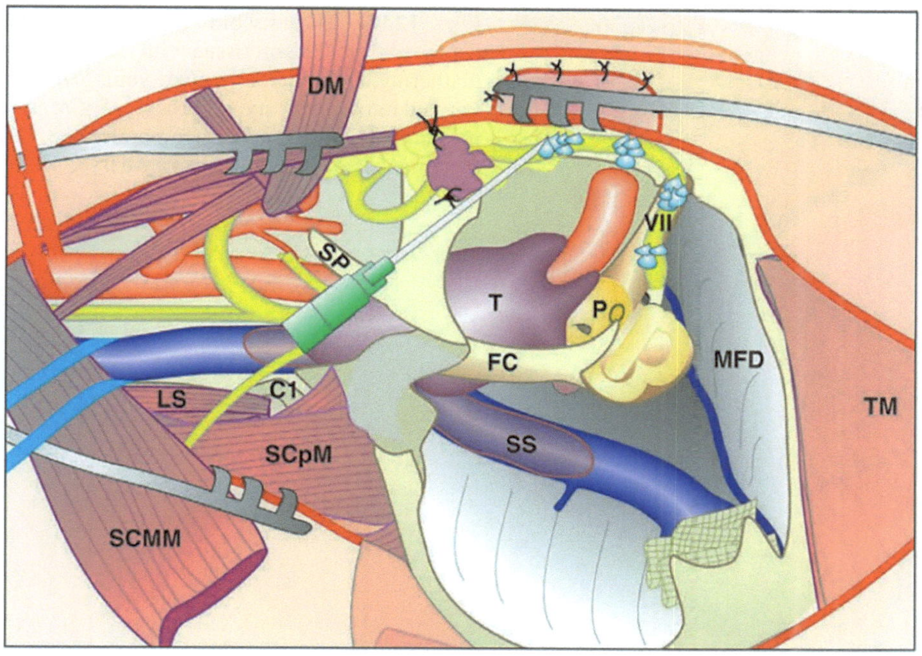

Fig. 12.92 Schematic illustration of anterior rerouting of the facial nerve. *C1*, atlas; *DM*, posterior belly of the digastric muscle; *FC*, fallopian canal; *LS*, levator scapulae muscle; *MFD*, middle fossa dura; *P*, promontory; *SCMM*, sternocleidomastoid muscle; *SCpM*, splenius capitis muscle; *SP*, styloid process; *SS*, sigmoid sinus; *T*, tumor; *TM*, temporalis muscle; *VII*, facial nerve.

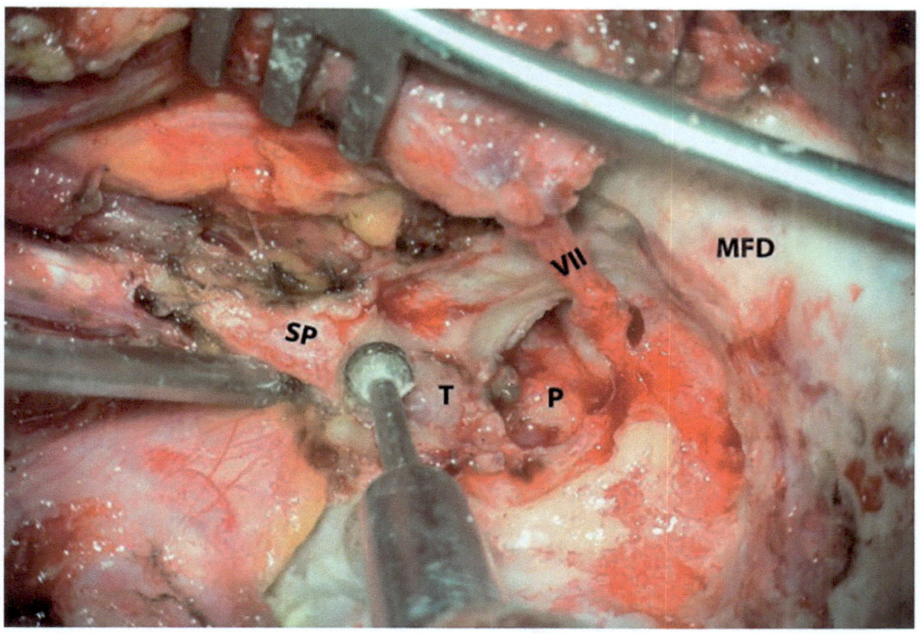

Fig. 12.93 The styloid process is drilled out to expose the internal carotid artery. *MFD*, middle cranial fossa dural plate; *P*, promontory; *SP*, styloid process; *T*, tumor; *VII*, facial nerve.

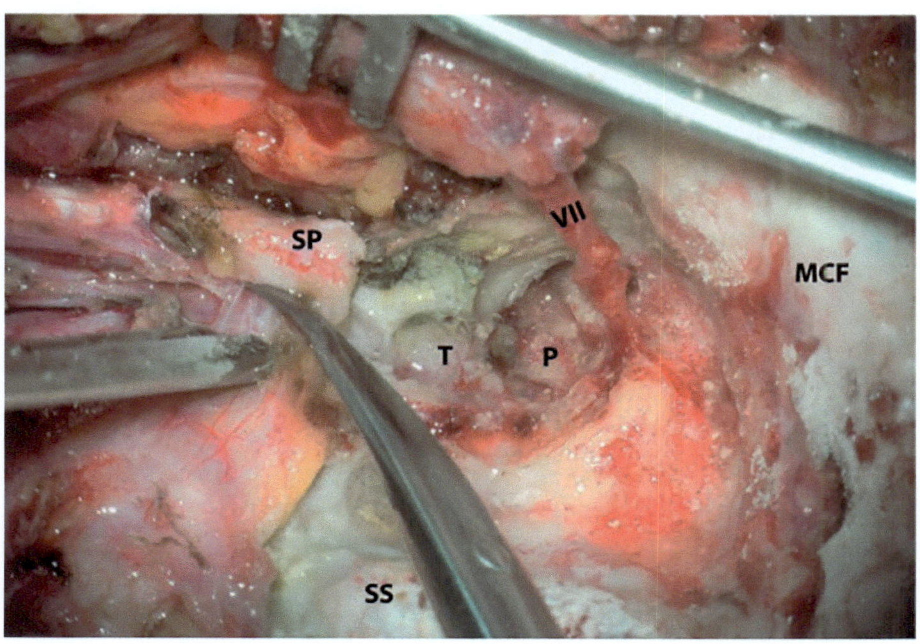

Fig. 12.94 The styloid process is detached from the soft tissue. *MCF*, middle cranial fossa; *P*, promontory; *SP*, styloid process; *SS*, sigmoid sinus; *T*, tumor; *VII*, facial nerve.

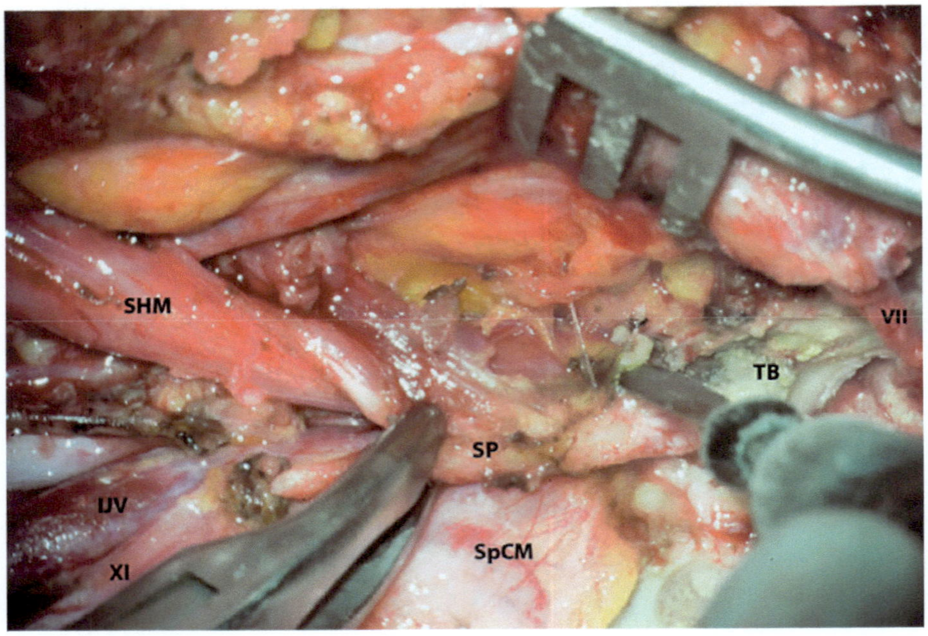

Fig. 12.95 The styloid process is further detached from the soft tissue with the monopolar diathermy. *IJV*, internal jugular vein; *SHM*, stylohyoid muscle; *SP*, styloid process; *SpCM*, splenius capitis muscle; *TB*, temporal bone; *VII*, facial nerve; *XI*, spinal accessory nerve.

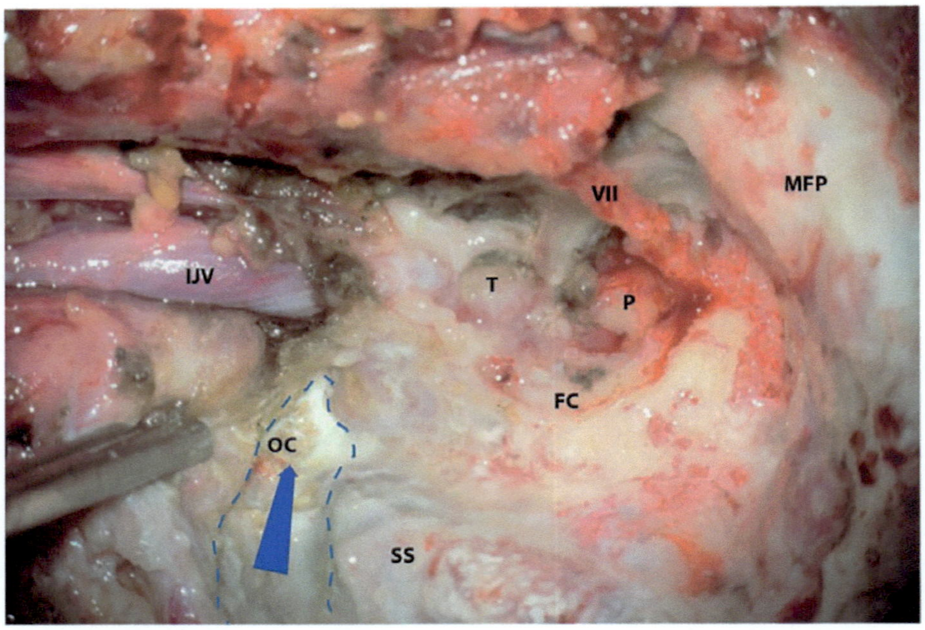

Fig. 12.96 Once the facial nerve has been anteriorly rerouted, the bone to be removed is posterior to the internal jugular vein. The direction of bone removal is indicated by the *blue arrow*. *FC*, fallopian canal; *IJV*, internal jugular vein; *MFP*, middle fossa plate; *OC*, occipital condyle; *P*, promontory; *SS*, sigmoid sinus; *T*, tumor; *VII*, facial nerve.

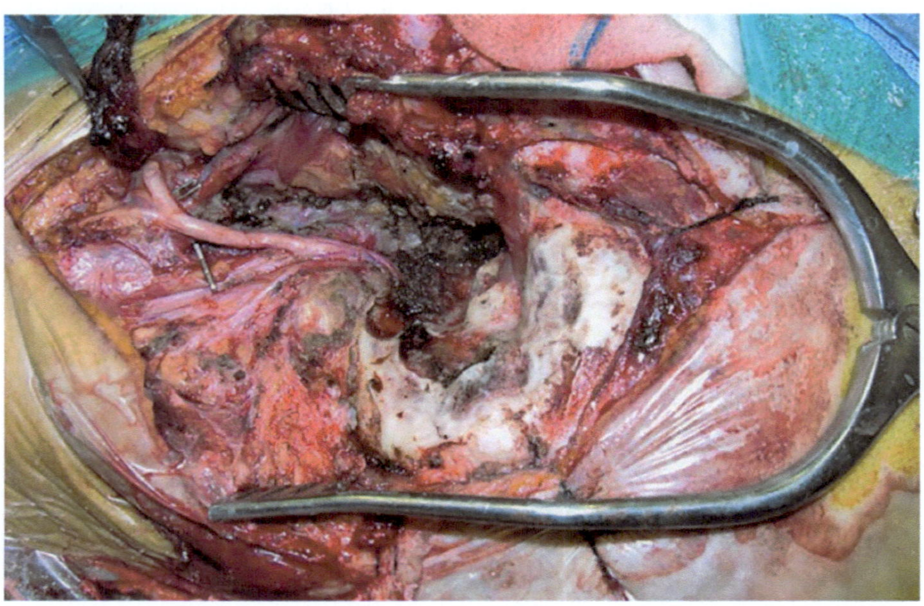

Fig. 12.97 Self-retaining retractor.

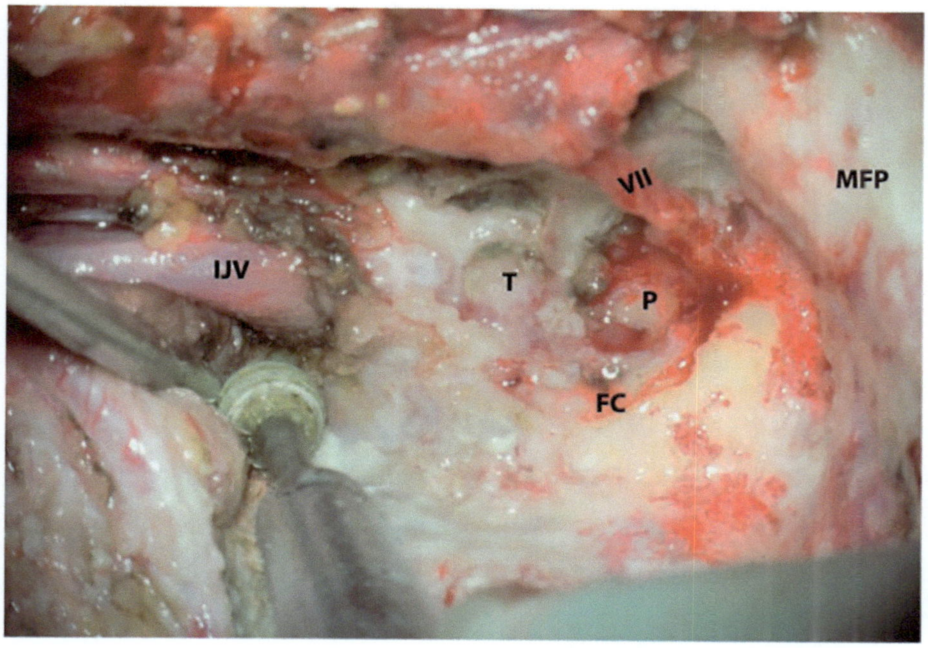

Fig. 12.98 Drilling of the occipital condyle is performed with a large diamond burr. *FC*, fallopian canal; *IJV*, internal jugular vein; *MFP*, middle fossa plate; *P*, promontory; *T*, tumor; *VII*, facial nerve.

Transcondylar Extension

(Figs. 12.99–12.124)

Further extradural bone removal extends anteromedially to the jugular bulb, superior to the hypoglossal canal, which represents the jugular tubercle. This is situated ~5 mm above the intracranial opening of the hypoglossal canal over which cranial nerves IX, X, and XI cross intradurally before entering the JF. This area of bone removal allows additional exposure to tumor margins from posterior, inferior, and medial directions. The bipolar is then used to devascularize the tumor, minimizing the degree of bleeding from the subsequent resection. Bony dissection can be made significantly more difficult by the presence of extensive bony infiltration, resulting in continuous bleeding. The use of bipolar diathermy, the diamond burr, bone wax, and judicious packing with Surgicel allow dissection to proceed.

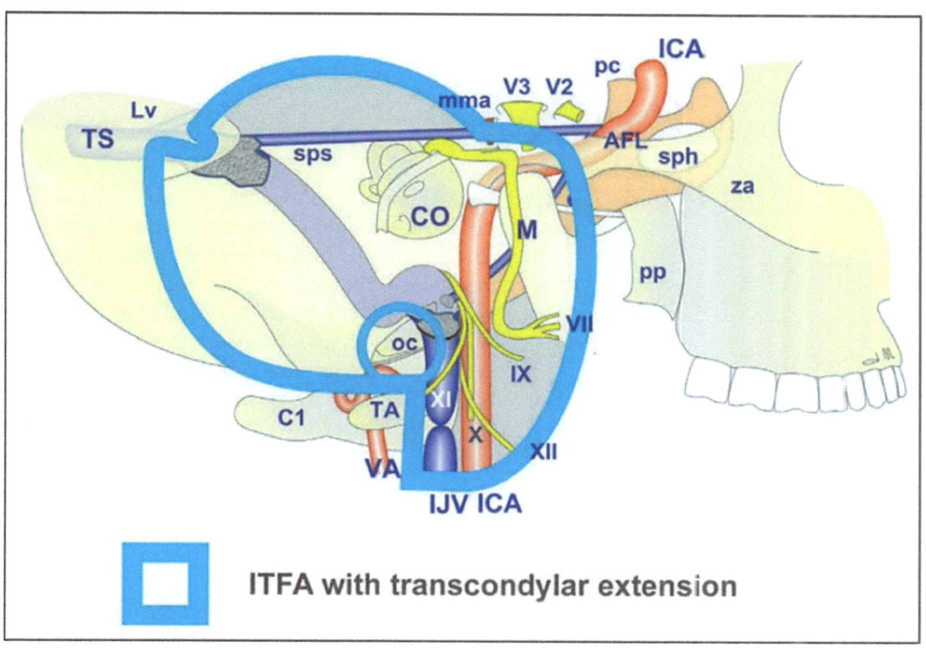

Fig. 12.99 Extent of ITFA with transcondylar extension. *AFL*, anterior foramen lacerum; *C1*, atlas; *CO*, cochlea; *ICA*, internal carotid artery; *IJV*, internal jugular vein; *Lv*, vein of Labbé; *M*, mandible; *mma*, middle meningeal artery; *OC*, occipital condyle; *pc*, clinoid process; *pp*, pterygoid plate; *sph*, sphenoid sinus; *sps*, superior petrosal sinus; *TA*, transverse process of the atlas; *TS*, transverse sinus; *VA*, vertebral artery; *VII*, facial nerve; *za*, zygomatic arch; *V2*, maxillary branch of the trigeminal nerve; *V3*, mandibular branch of the trigeminal nerve; *IX*, glossopharyngeal nerve; *X*, vagus nerve; *XI*, spinal accessory nerve; *XII*, hypoglossal nerve.

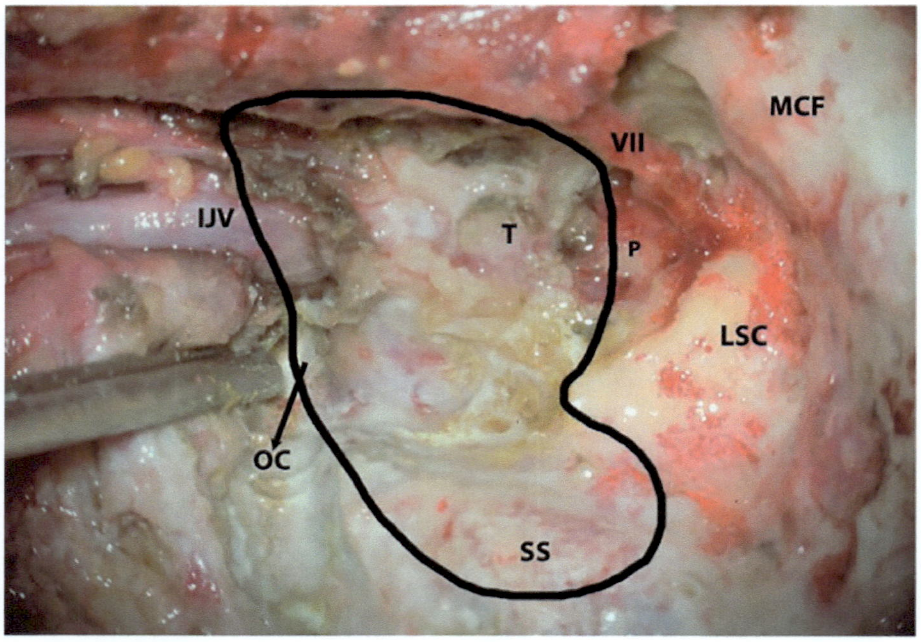

Fig. 12.100 Drilling of the occipital condyle is almost completed. *IJV*, internal jugular vein; *LSC*, lateral semicircular canal; *MCF*, middle cranial fossa; *OC*, occipital condyle; *P*, promontory; *SS*, sigmoid sinus; *T*, tumor; *VII*, facial nerve (*arrow* shows the direction to the occipital condyle).

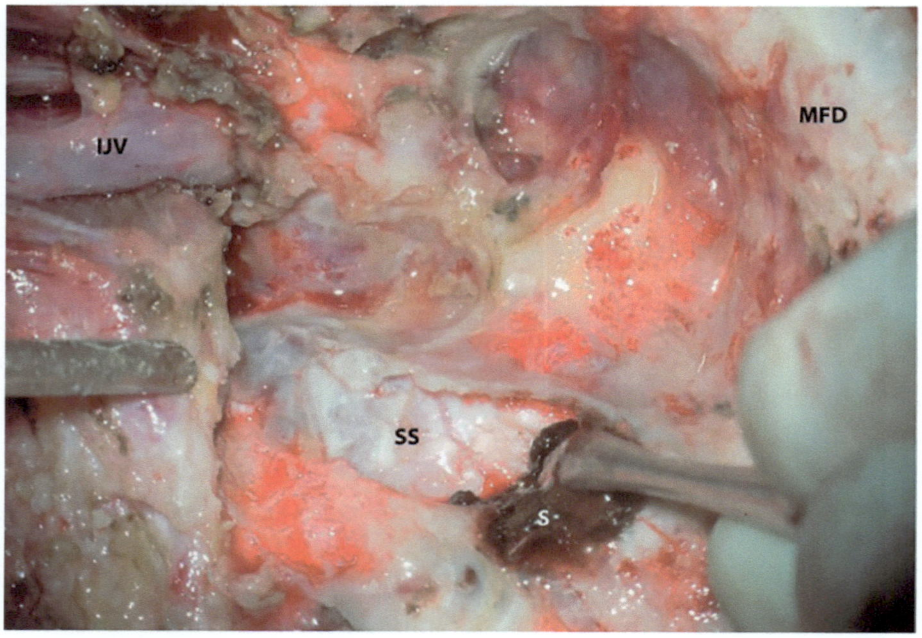

Fig. 12.101 The sigmoid sinus is occluded at the level just proximal to its transition from the transverse sinus by means of extraluminal packing. The extraluminal compression of the sigmoid sinus is facilitated by the preservation of a bony shell over the proximal sigmoid sinus and prevents blocking of the transverse sinus. *IJV*, internal jugular vein; *MCD*, middle cranial fossa dura; *SS*, sigmoid sinus.

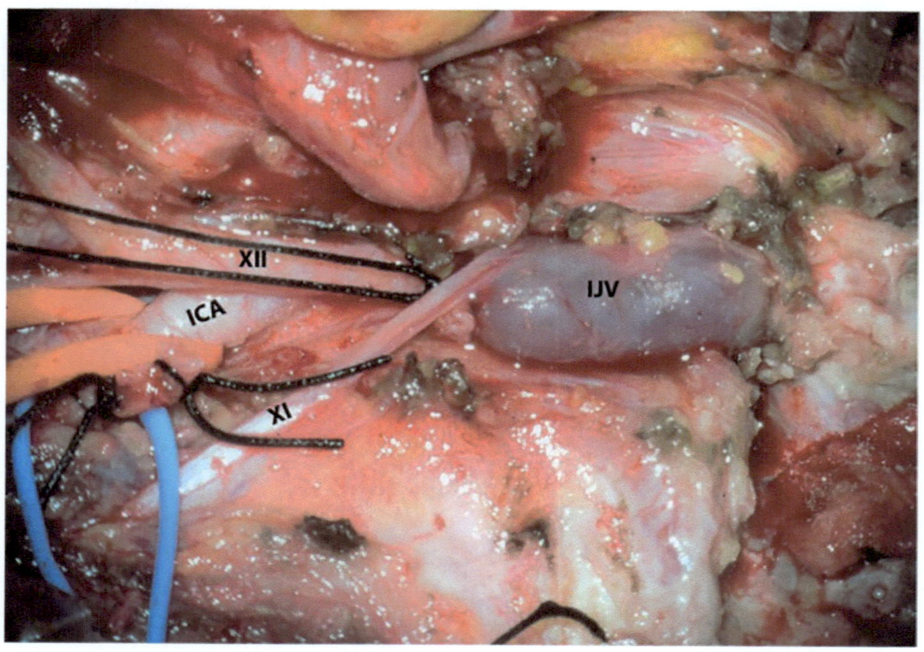

Fig. 12.102 The internal jugular vein is then double ligated and transected. This is done following sigmoid closure to minimize venous congestion of the tumor. The transected vein is passed medial to cranial nerve XI. The lower cranial nerves can now be visualized in an area free of tumor. *ICA*, internal carotid artery; *IJV*, internal jugular vein; *XI*, spinal accessory nerve; *XII*, hypoglossal nerve.

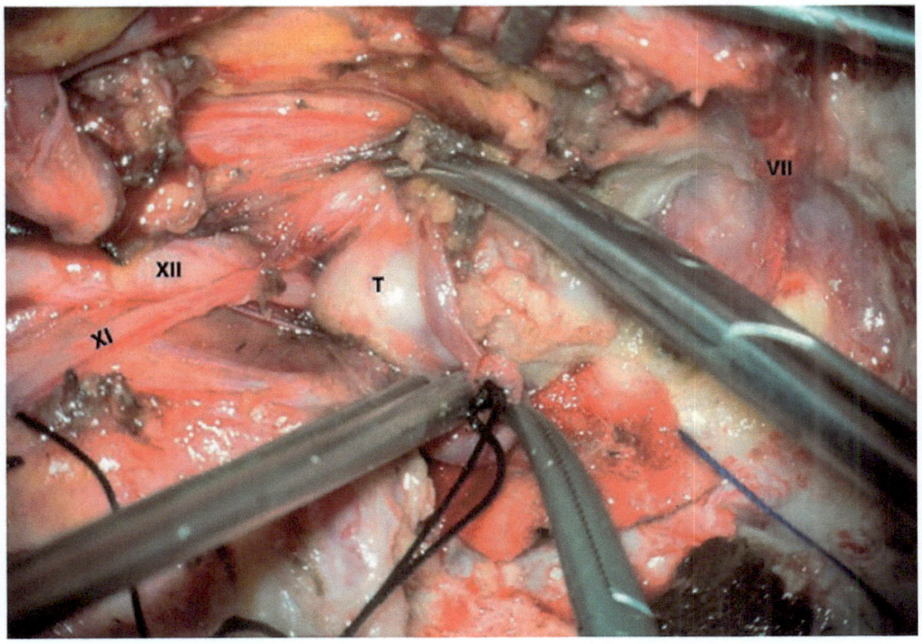

Fig. 12.103 After reflection of the internal jugular vein, the tumor is visible. *T*, tumor; *VII*, facial nerve; *XI*, spinal accessory nerve; *XII*, hypoglossal nerve.

The jugular bulb is opened, and tumor dissection performed in a piecemeal fashion using bipolar forceps or dissectors (Fig. 12.104). The critical point is represented by the invasion of the medial wall of the jugular bulb; in the presence of such infiltration, radical surgery requires the removal of the entire bulb, with inevitable sacrifice of the lower cranial nerves. In contrast, if the medial wall of the bulb can be preserved so too can neural function. This technique is also referred to as intrabulbar dissection. Prior identification of the lower cranial nerves in the uninvolved neck further aids in the preservation of these nerves. Care must be taken when packing branches of the inferior petrosal sinus. These can be multiple and usually enter among the glossopharyngeal, vagus, and spinal accessory nerves.

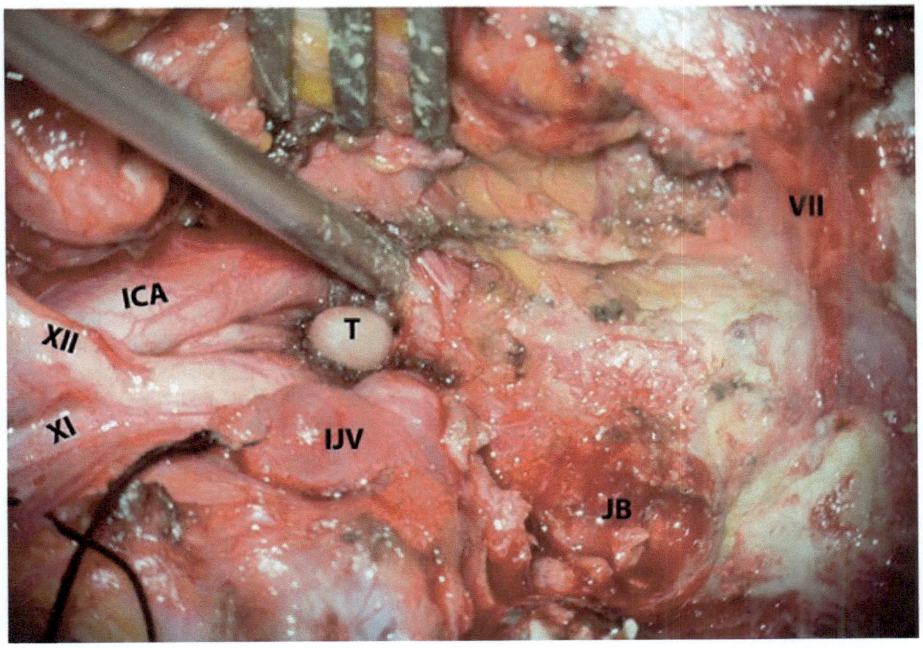

Fig. 12.104 After the jugular bulb (JB) is opened, a part of the tumor inside it is removed. Tumor involving the internal carotid artery (ICA) is visible. *ICA*, internal carotid artery; *IJV*, internal jugular vein; *T*, tumor; *VII*, facial nerve; *XI*, spinal accessory nerve; *XII*, hypoglossal nerve.

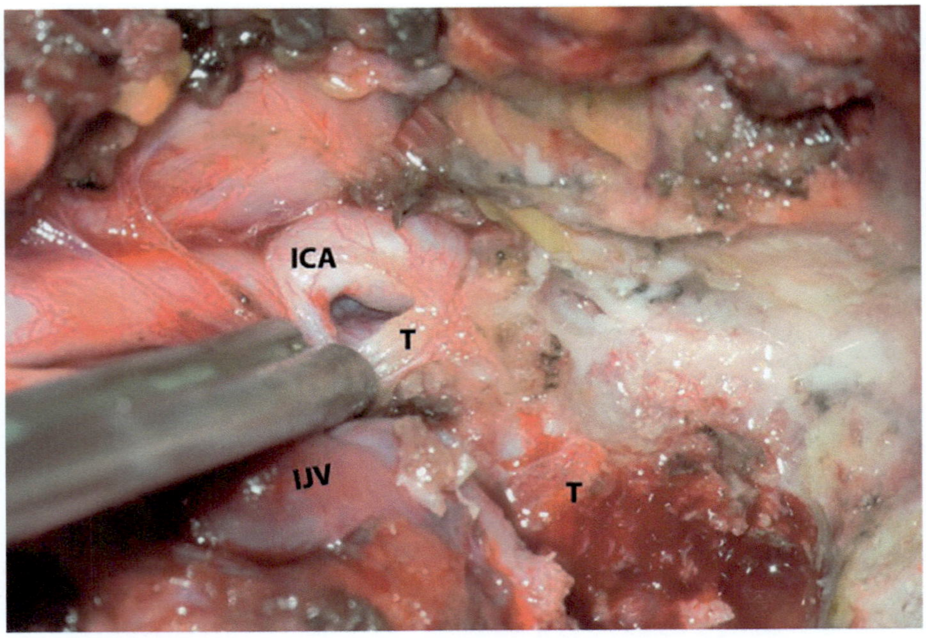

Fig. 12.105 The final stage of surgery involves the definitive management of carotid artery involvement. Identification of the correct plane of dissection around the ICA is best performed inferiorly. The robust periadventitial, periosteal tissue can be appreciated. *ICA*, internal carotid artery (adventitia); *IJV*, internal jugular vein; *T*, tumor.

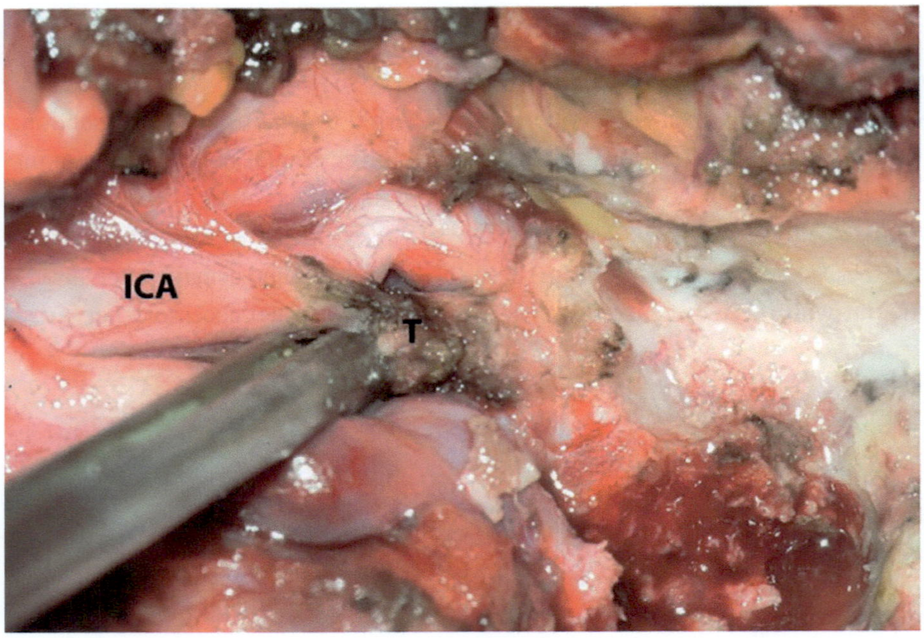

Fig. 12.106 Tumor adherent to this tissue is being dissected. There is no involvement of the ICA adventitia. *ICA*, internal carotid artery; *T*, tumor.

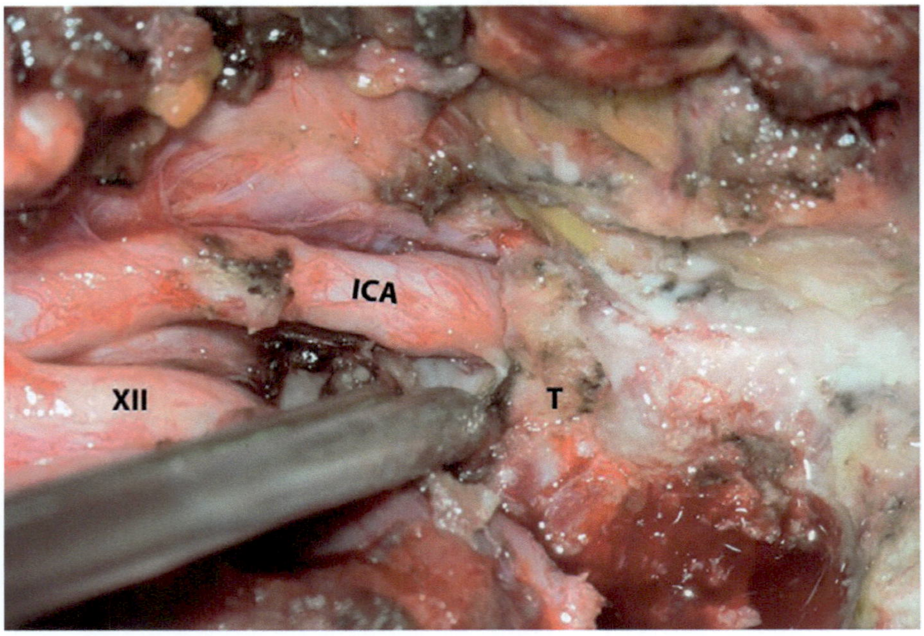

Fig. 12.107 The carotid artery has been freed up to its carotid canal. Note the periosteal layer retracted by the suction irrigator. Also, note that all of this dissection is performed under microscopic control. *ICA*, internal carotid artery; *T*, tumor; *XII*, hypoglossal nerve.

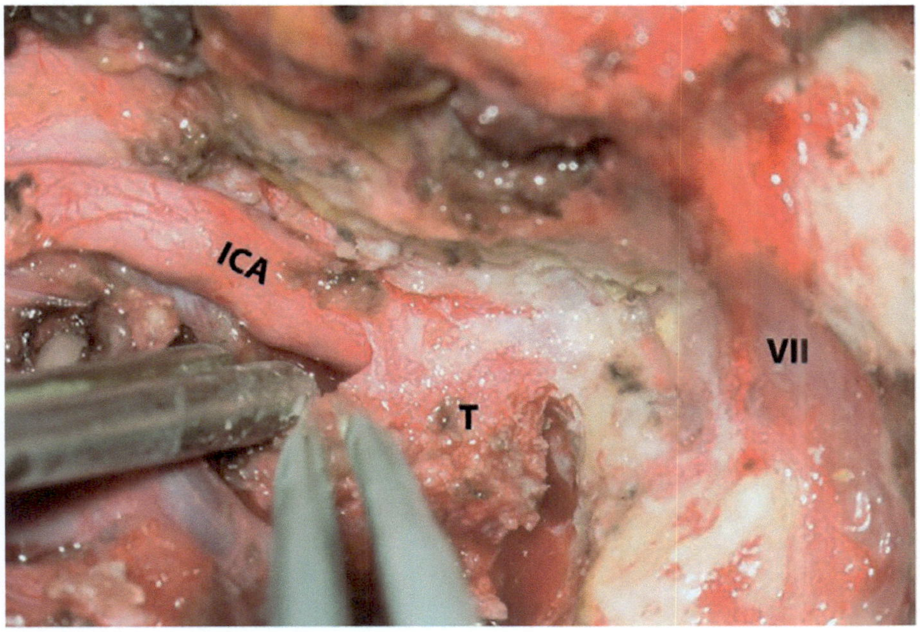

Fig. 12.108 Following further bone removal around the carotid artery, dissection is continued in a subperiosteal plane. *ICA*, internal carotid artery; *T*, tumor; *VII*, facial nerve.

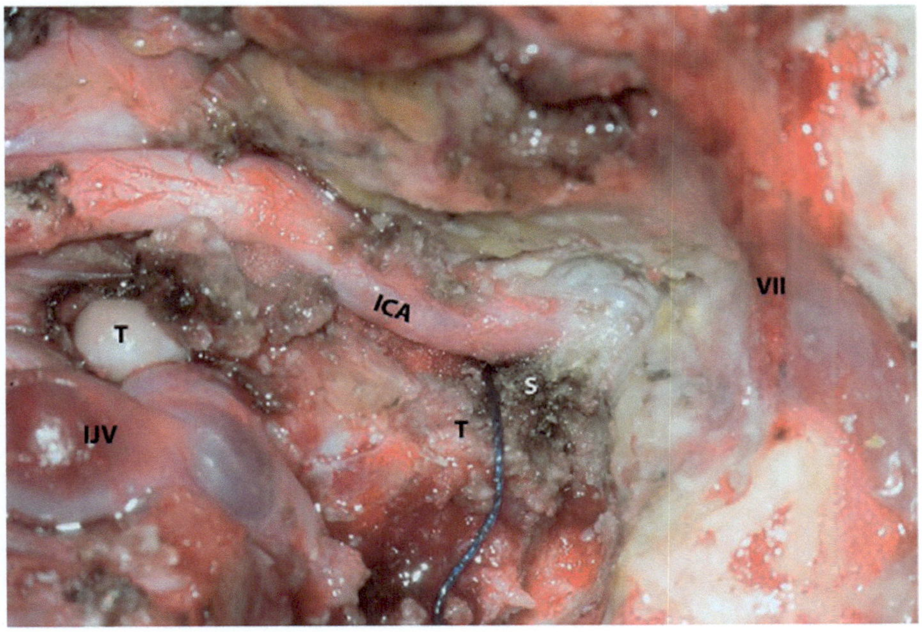

Fig. 12.109 Tumor can now be seen extending from the jugular foramen area anteromedially. A Merocel patty has been placed medial to the ICA. Bone lateral and medial to the ICA has been further drilled to the level of the genu, which will allow gentle mobilization of the artery. *ICA*, internal carotid artery; *IJV*, internal jugular vein; *S*, Surgicel; *T*, tumor; *VII*, facial nerve.

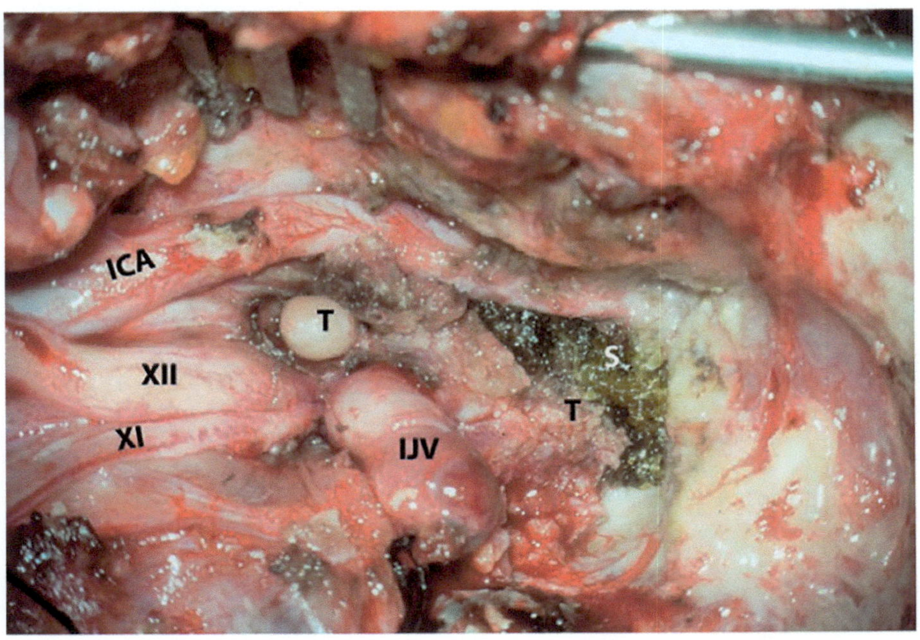

Fig. 12.110 The margins of the tumor have now been well exposed and controlled, allowing dissection of the main bulk of the tumor. *ICA*, internal carotid artery; *IJV*, internal jugular vein; *S*, Surgicel; *T*, tumor; *XI*, spinal accessory nerve; *XII*, hypoglossal nerve.

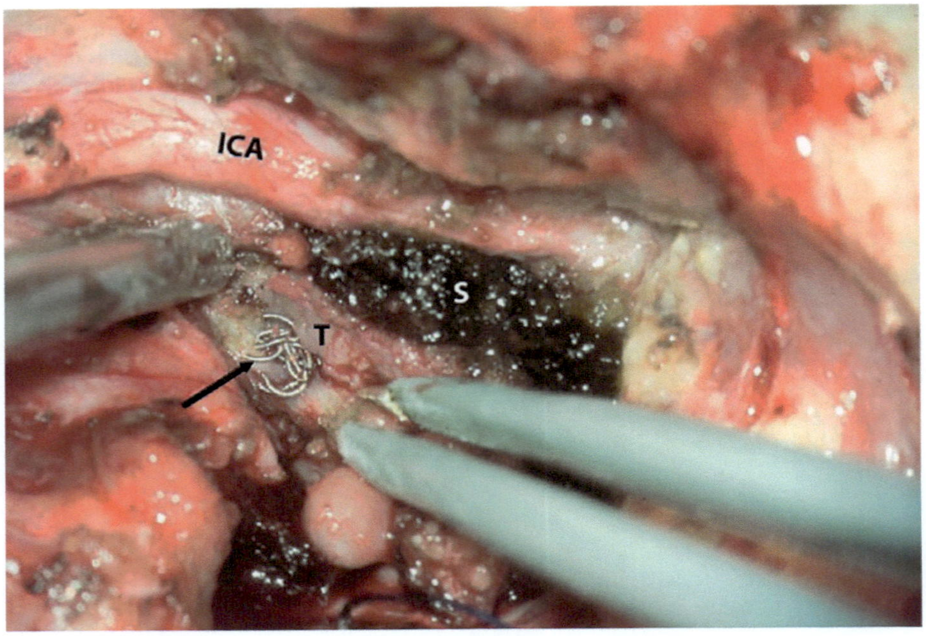

Fig. 12.111 The coils used to embolize the tumor can be seen *(arrow)*. Note the fibrous consistency of the tumor. *ICA*, internal carotid artery; *S*, Surgicel; *T*, tumor.

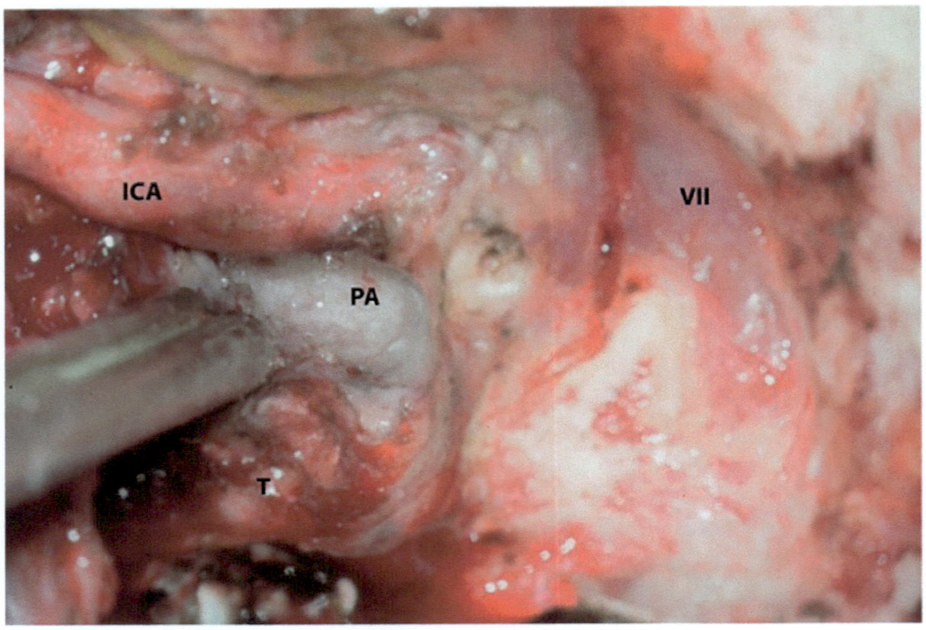

Fig. 12.112 Bone medial to the carotid genu is drilled to ensure removal of all infiltrated bone and to optimize superior control of the tumor margins. *ICA*, internal carotid artery; *PA*, petrous apex; *T*, tumor; *VII*, facial nerve.

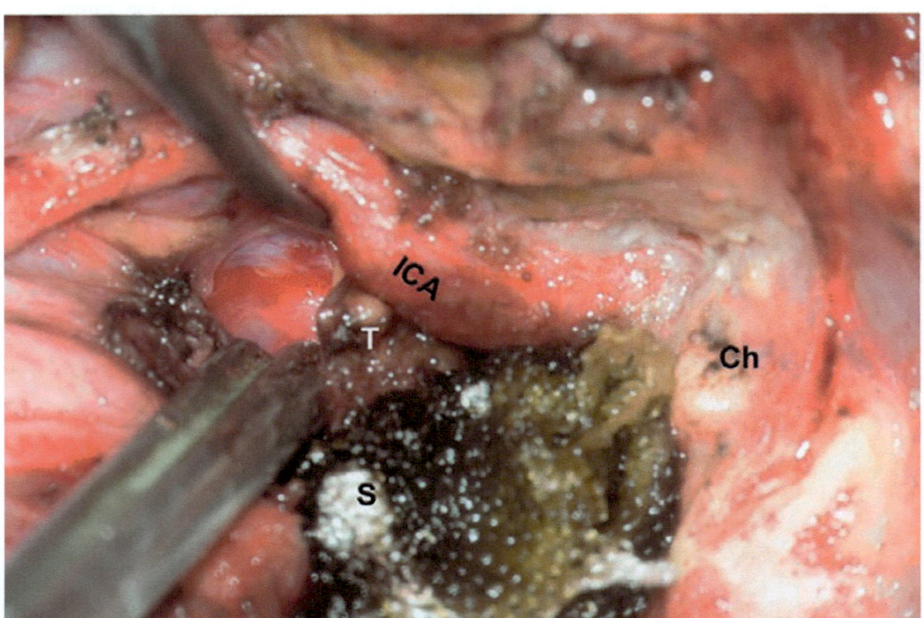

Fig. 12.113 The wide exposure of this approach allows further removal of bone until normal tissue is identified, to minimize recurrence. The tumor is controlled with temporary Surgicel packing, while the vertical carotid artery is gently mobilized anteriorly to access the tumor extending anteromedially. *Ch*, cochlea; *ICA*, internal carotid artery; *S*, Surgicel; *T*, tumor.

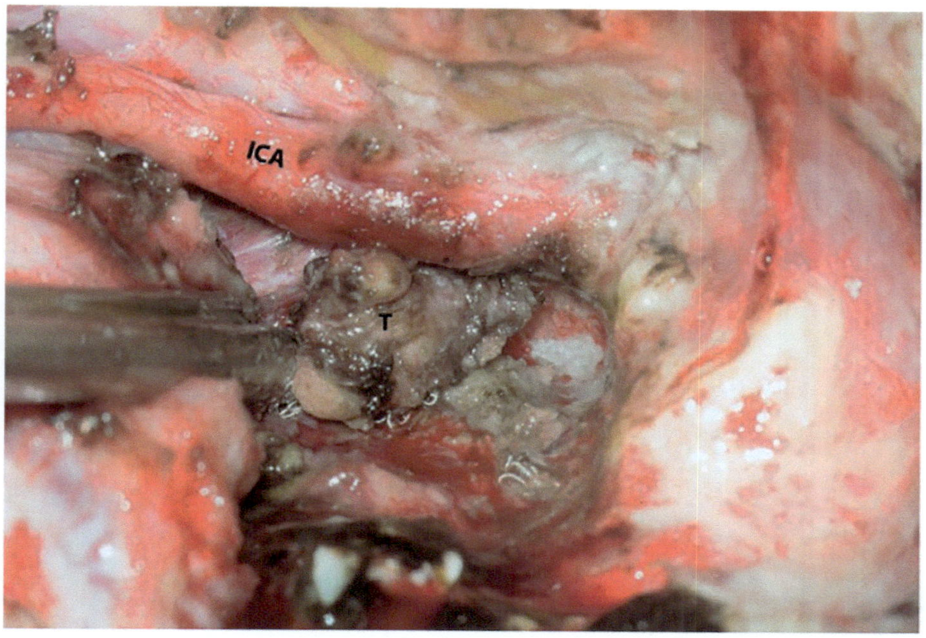

Fig. 12.114 The final anterior attachment of tumor can be seen in the area of the inferior petrosal sinus. *ICA*, internal carotid artery; *T*, tumor.

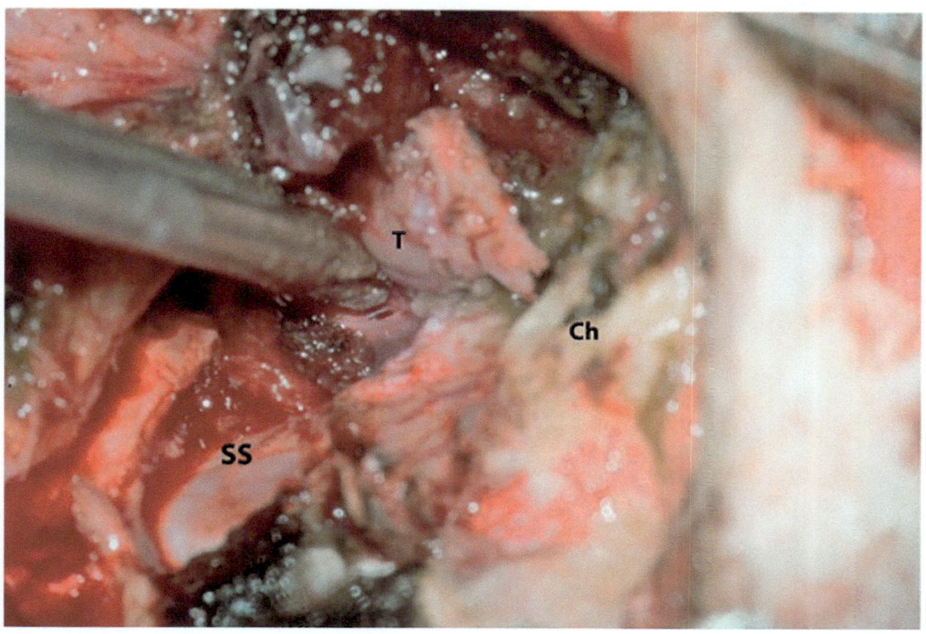

Fig. 12.115 The sigmoid has been opened, and the medial wall is seen. Two openings of the inferior petrosal sinus are identified. The inferior opening has been packed with Surgicel. Tumor is being peeled from the medial wall. *Ch*, cochlea; *SS*, sigmoid sinus (medial wall); *T*, tumor.

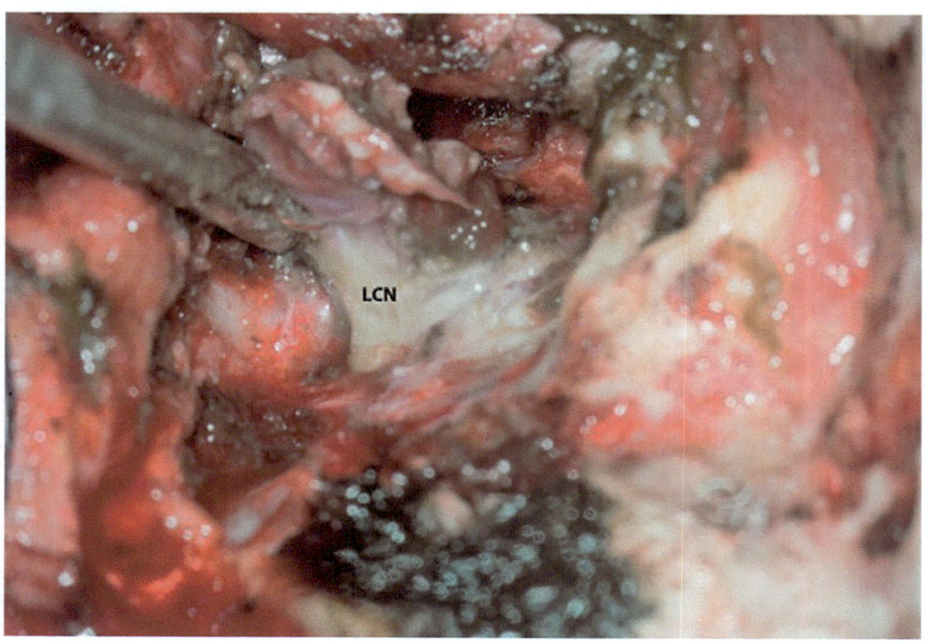

Fig. 12.116 The vagus and spinal accessory nerves can now be seen entering the jugular foramen. Tumor has infiltrated these nerves, explaining preoperative dysfunction. *LCN*, lower cranial nerves.

Identification and bipolar coagulation of an often enlarged caroticotympanic artery is an essential first step. With minimal involvement, this allows separation of the tumor and the ICA. The tumor usually invades only the periosteum rather than the carotid wall. A plane of cleavage is established between the tumor and the carotid artery by blunt dissection. The periosteum is then carefully separated from the artery (Figs. 12.117–12.124).

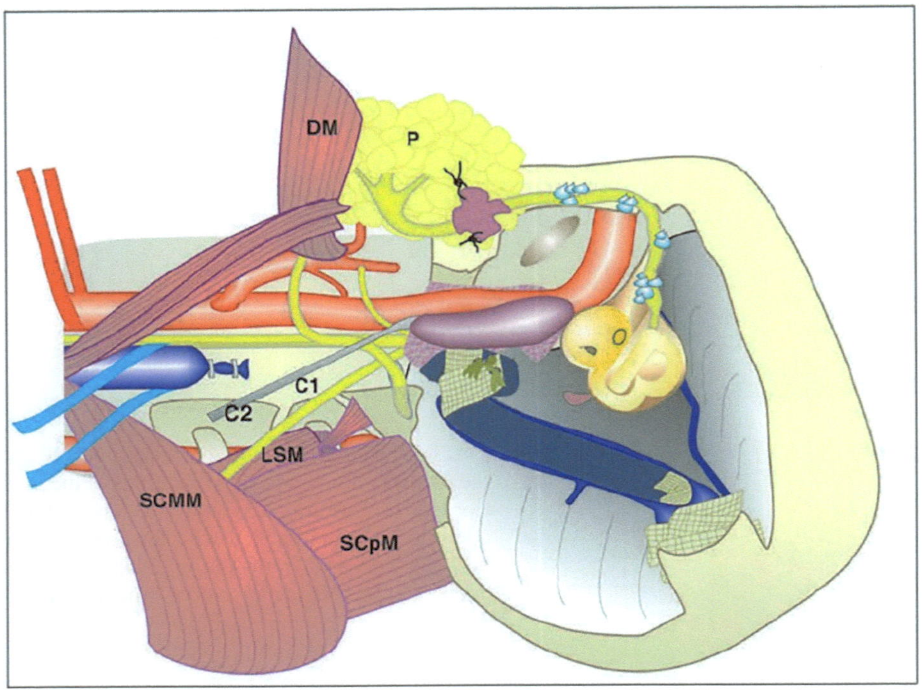

Fig. 12.117 A plane of cleavage is established between the tumor and the ICA. *C1*, atlas; *C2*, axis; *DM*, posterior belly of the digastric muscle; *LSM*, levator scapulae muscle; *P*, promontory; *SCMM*, sternocleidomastoid muscle; *SCpM*, splenius capitis muscle.

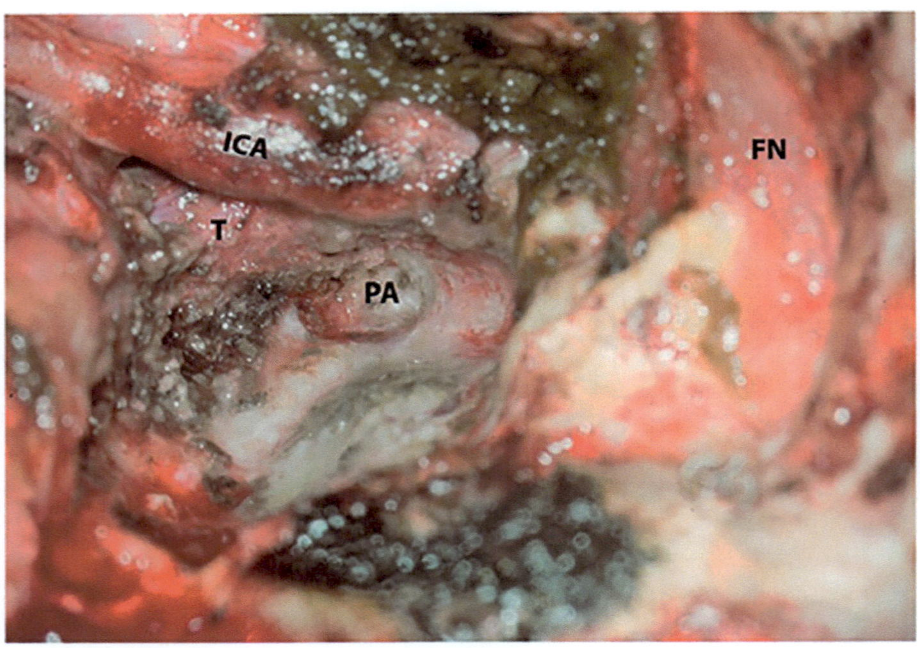

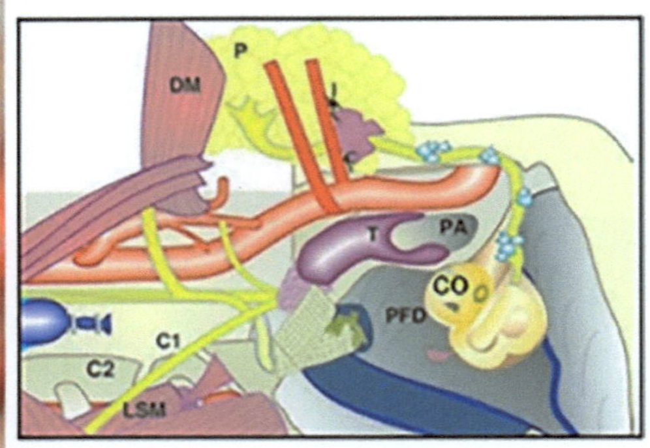

Fig. 12.118 It is important to reiterate that the degree of bony invasion is often difficult to assess even intraoperatively. *C1*, atlas; *C2*, axis; *CO*, cochlea; *DM*, posterior belly of the digastric muscle; *FN*, facial nerve; *ICA*, internal carotid artery; *LSM*, levator scapulae muscle; *P*, parotid gland; *PA*, petrous apex; *PFD*, posterior fossa dura; *T*, tumor.

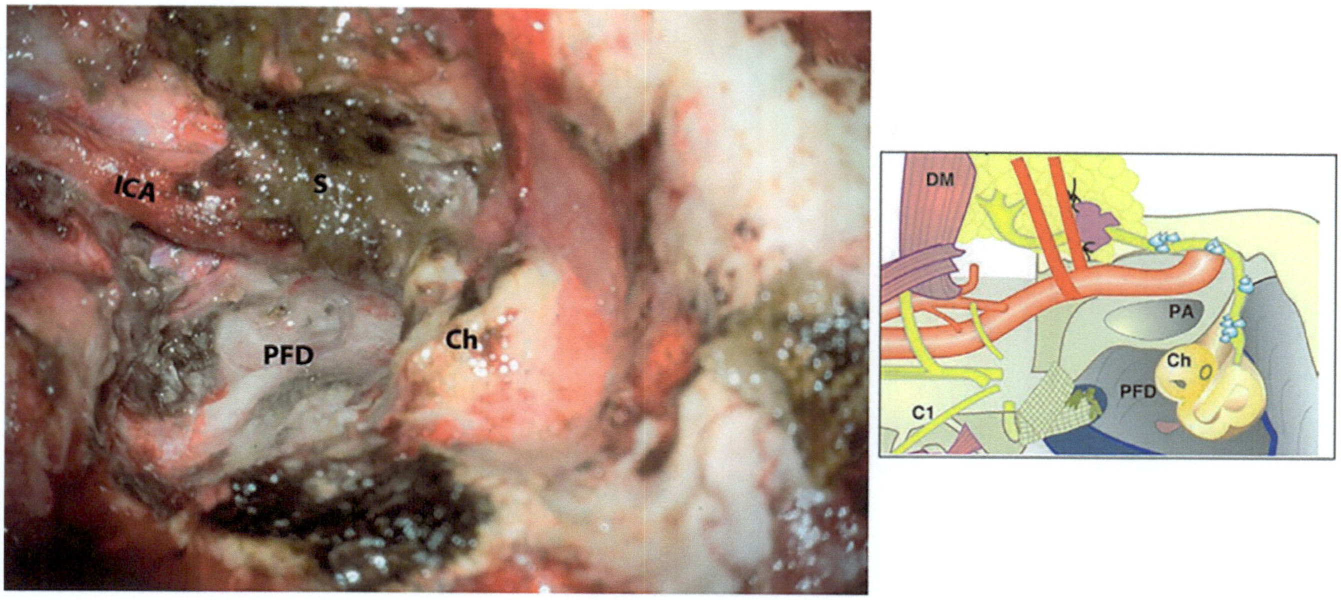

Fig. 12.119 The tumor has been removed, with a transection of the infiltrated nerves. Further bone removal of the lower clivus is performed to ensure complete removal of infiltrated bone. *C1*, atlas; *Ch*, cochlea; *DM*, posterior belly of the digastric muscle; *ICA*, internal carotid artery; *PA*, petrous apex; *PFD*, posterior fossa dura; *S*, Surgicel.

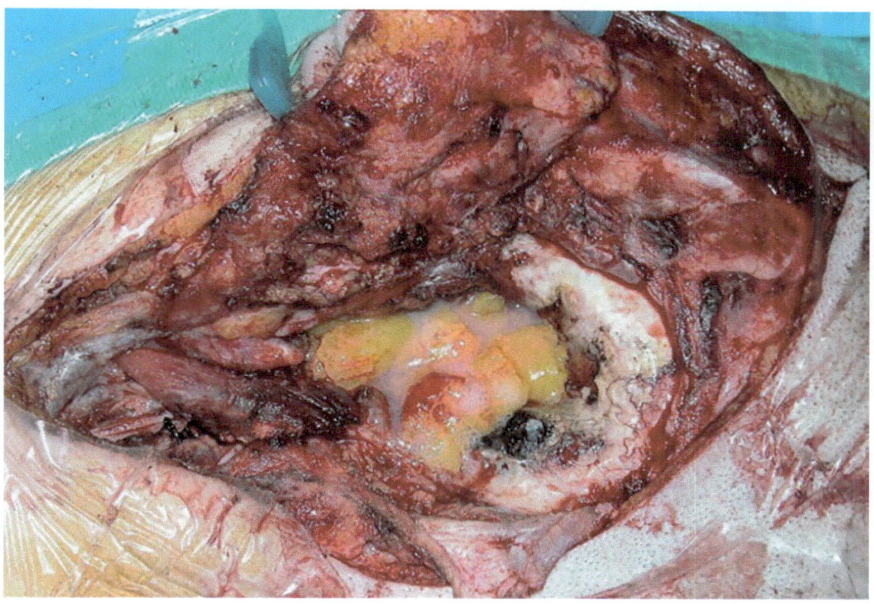

Fig. 12.120 Hemostasis is achieved using bipolar coagulation and Surgicel packing. The Eustachian tube is closed with a plug of periosteum. The cavity is obliterated with abdominal fat, and the dead space is filled with abdominal fat.

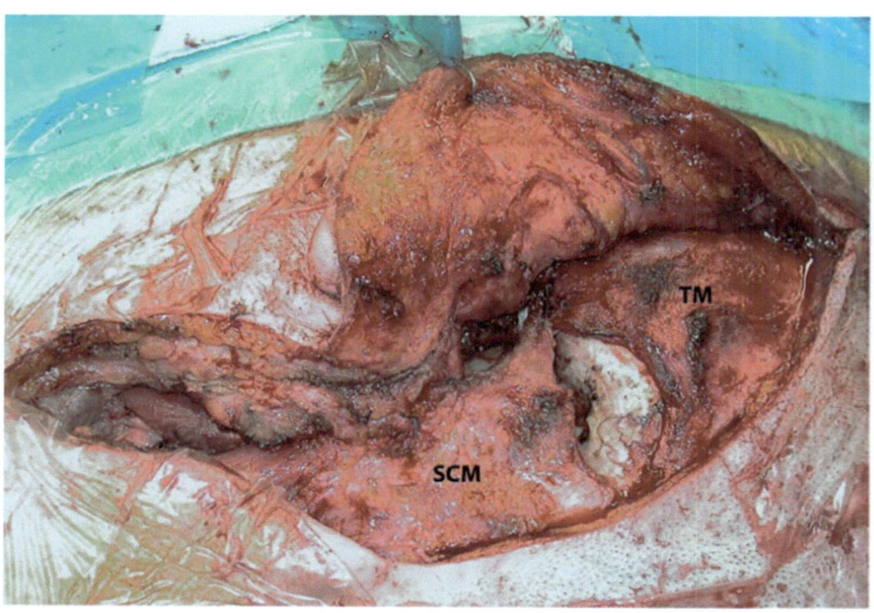

Fig. 12.121 We do not rotate the temporalis muscle as suggested by Fisch. The posterior belly of the digastric muscle is replaced and sutured. The musculoperiosteal flap is repositioned in place and sutured to the temporalis muscle. *TM*, temporalis muscle; *SCM*, sternocleidomastoid muscle.

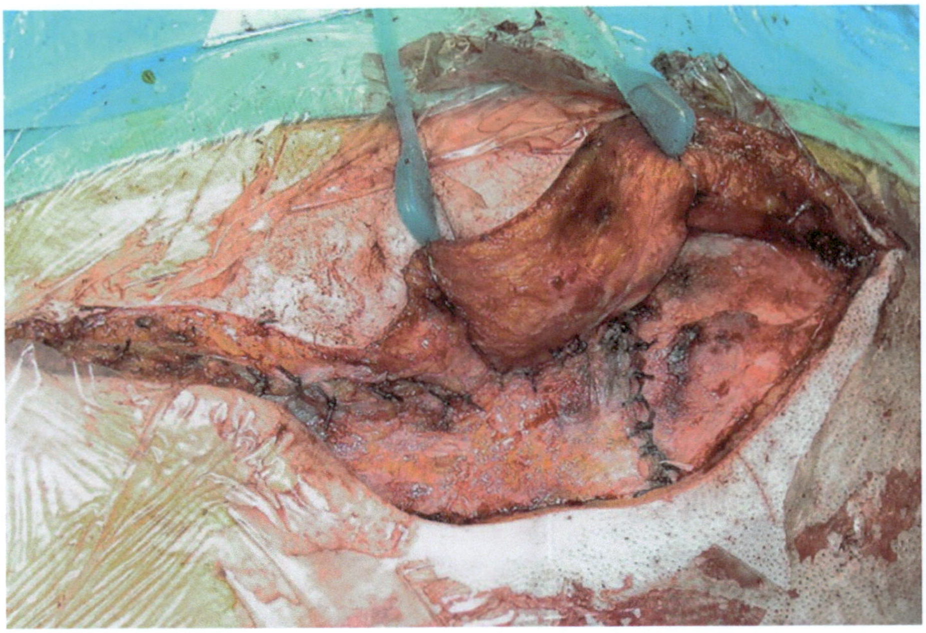

Fig. 12.122 Meticulous closure of the musculo-fascial flaps. The wound should be closed in watertight fashion.

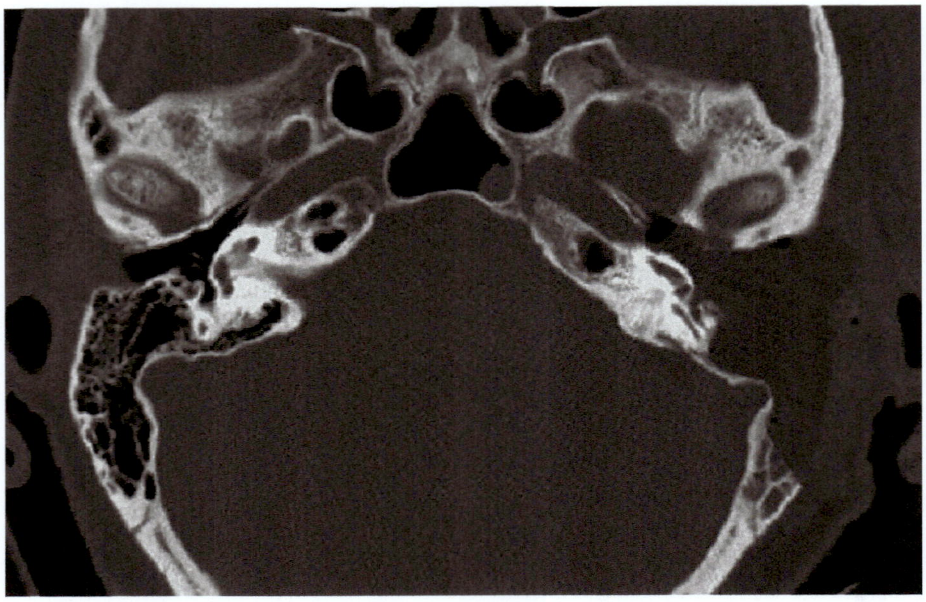

Fig. 12.123 Postoperative CT scan, axial view.

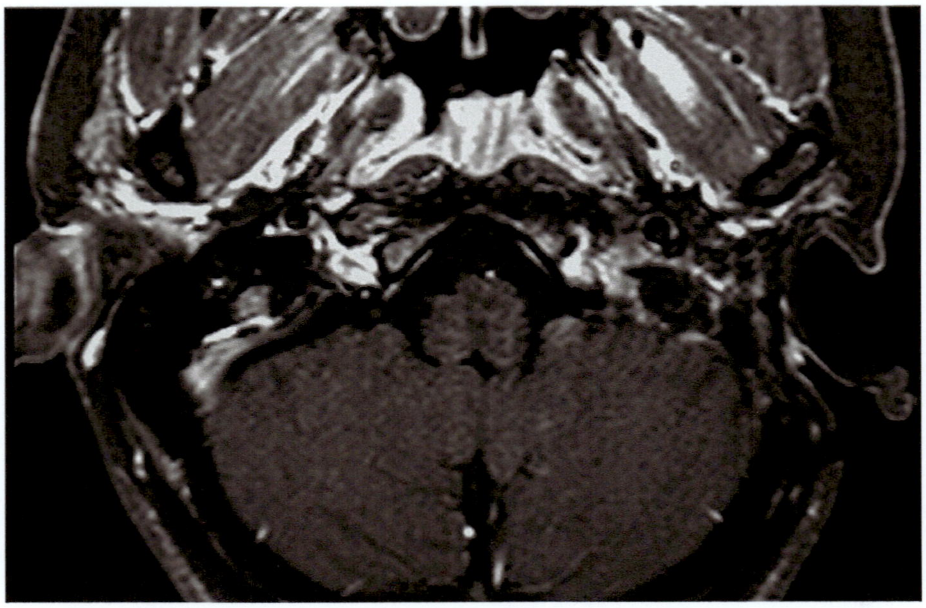

Fig. 12.124 Postoperative enhanced T1-weighted MRI, axial view. No residual tumor is visible.

Postoperatively, the patient showed grade III of facial nerve function. Despite lower cranial nerve palsy, she was well compensated with rehabilitation. Although the patient showed lower cranial nerve deficits, the contralateral carotid body tumor was surgically resected. Because the contralateral carotid body tumor was small, we expected that it might be removed without vagus nerve palsy. It was removed via a transcervical approach, and all lower cranial nerves were anatomically preserved: the patient showed no deficit of lower cranial nerves after surgery.

CASE 2: REMOVAL OF A CLASS C1 TYMPANOJUGULAR PARAGANGLIOMA WITH PRESERVATION OF THE MEDIAL WALL OF THE JUGULAR BULB

See Figs. 12.125–12.143.

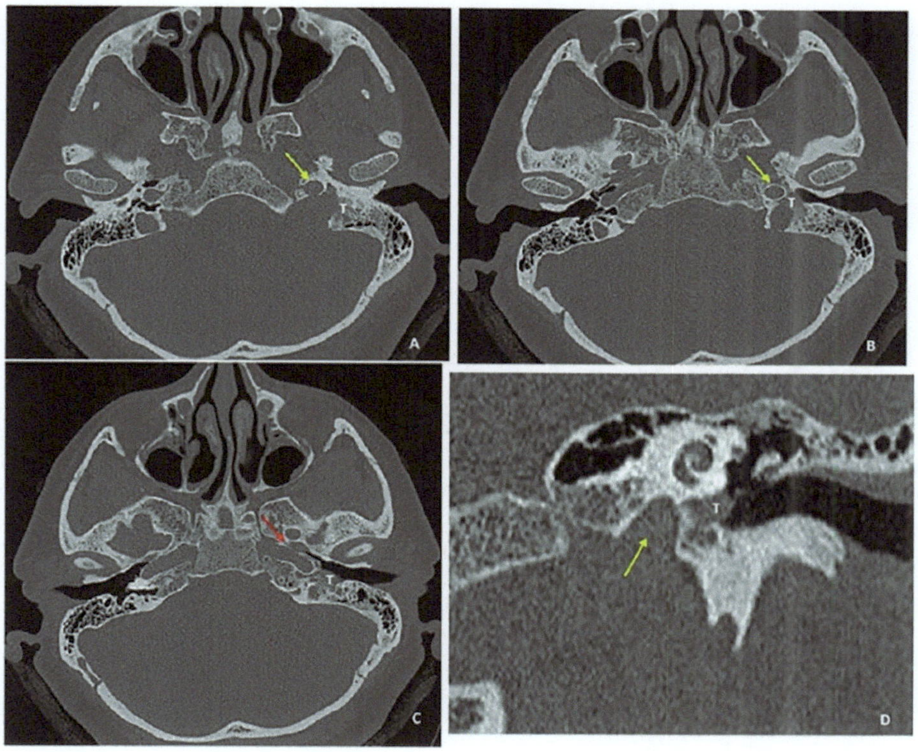

Fig. 12.125 This 45-year-old patient presented with left-sided pulsatile tinnitus and mixed hearing loss. The CT scan showed a tumor centered in the jugular foramen with initial erosion of the carotid canal (D). Neither the vertical *(yellow arrow)* nor the horizontal *(red arrow)* petrous internal carotid artery is infiltrated by the tumor, as can be seen in the axial slices (A–C).

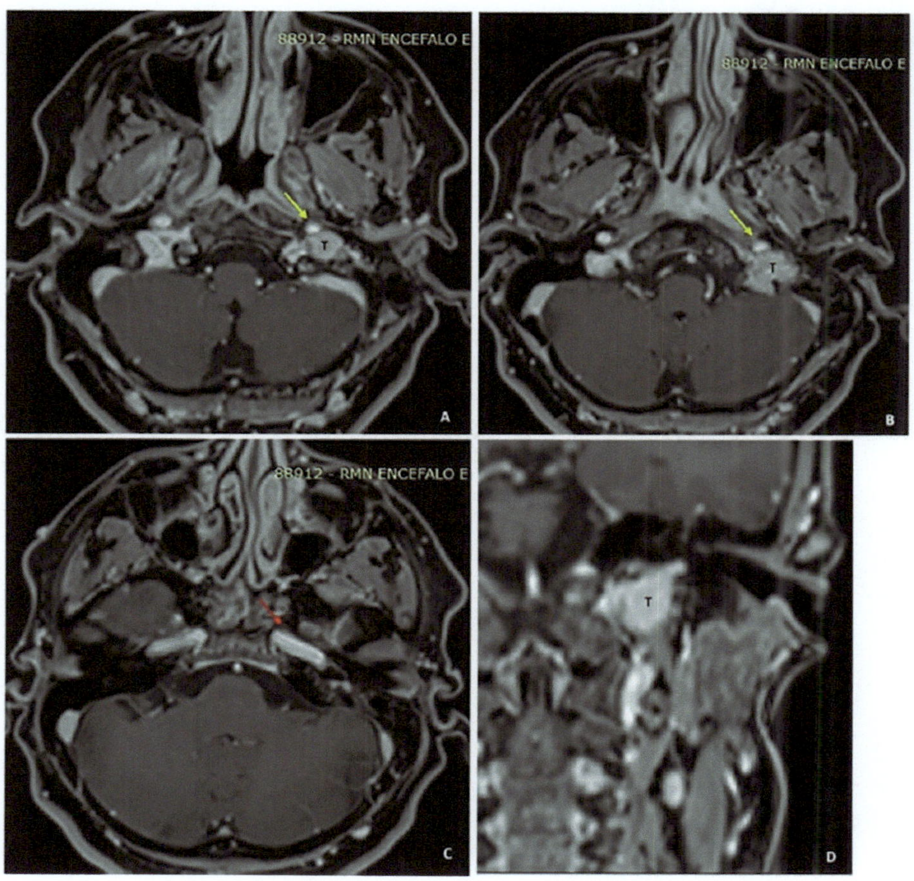

Fig. 12.126 T1W+Gd enhancement MRI of the brain. Note the tumor (T) invading the jugular foramen (A, B, D). The horizontal portion of the petrous internal carotid artery *(red arrow)* is free from the tumor (C). The presence of flow voids and a "salt and pepper" characteristic is suggestive of paraganglioma.

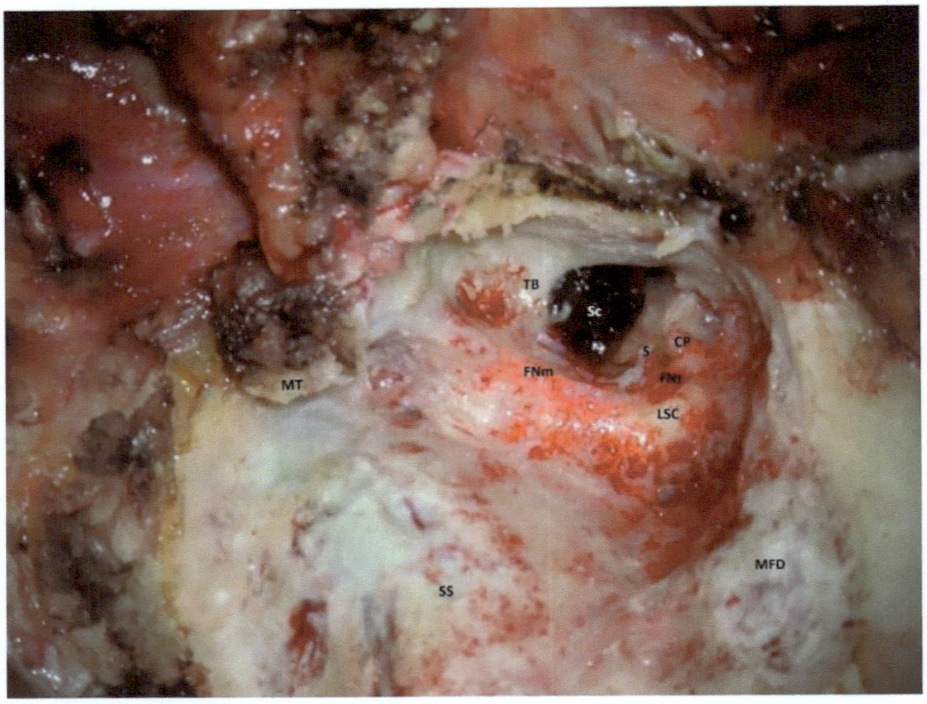

Fig. 12.127 Subtotal petrosectomy is completed. *MT*, mastoid tip; *FNm*, mastoid portion of the facial nerve; *FNt*, tympanic portion of the facial nerve; *LSC*, lateral semicircular canal; *MFD*, middle fossa dura; *S*, stapes; *TB*, tympanic bone; *CP*, cochleariform process; *Sc*, Surgicel; *SS*, sigmoid sinus.

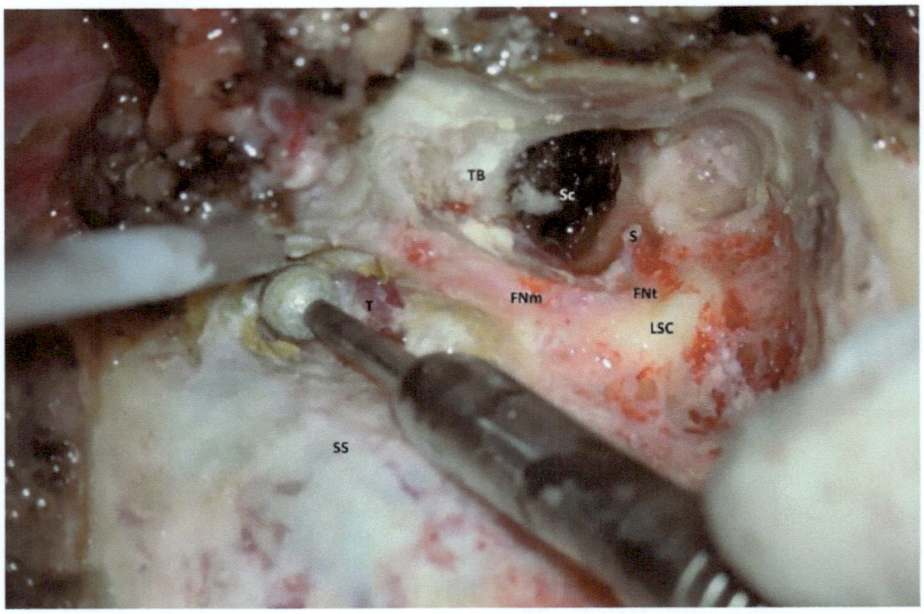

Fig. 12.128 The mastoid portion of the facial nerve (FNm) is decompressed and the bone at the mastoid tip is removed. Note the tumor (T) medial to the facial nerve. *FNt*, tympanic portion of the facial nerve; *LSC*, lateral semicircular canal; *S*, stapes; *SS*, sigmoid sinus; *Sc*, Surgicel; *TB*, tympanic bone.

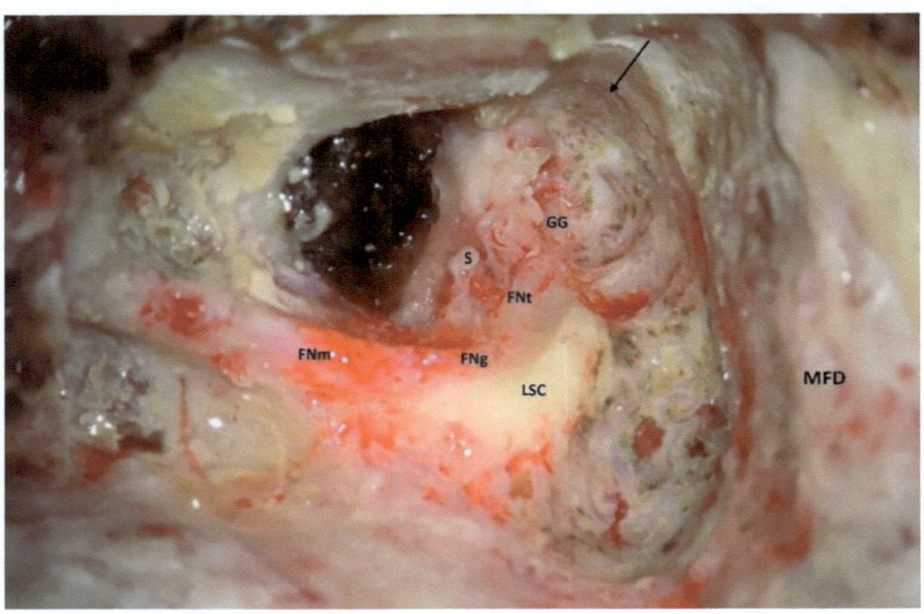

Fig. 12.129 The facial nerve is decompressed from the mastoid segment (FNm) to the geniculate ganglion (GG). A new bony canal for the rerouted facial nerve is prepared in the anterior attic *(arrow)*. *FNg*, second genu of the facial nerve; *LSC*, lateral semicircular canal; *FNt*, tympanic segment of the facial nerve; *MFD*, middle fossa dura; *S*, stapes.

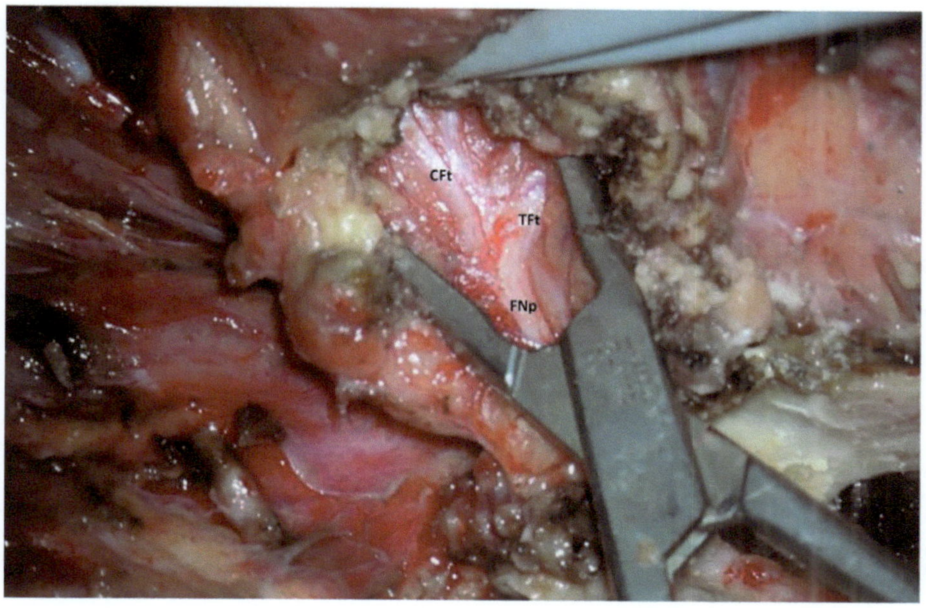

Fig. 12.130 The facial nerve is dissected at the level of the parotid gland. *FNp*, parotid portion of the facial nerve (main trunk); *CFt*, cervico-facial trunk; *TFt*, temporo-facial trunk.

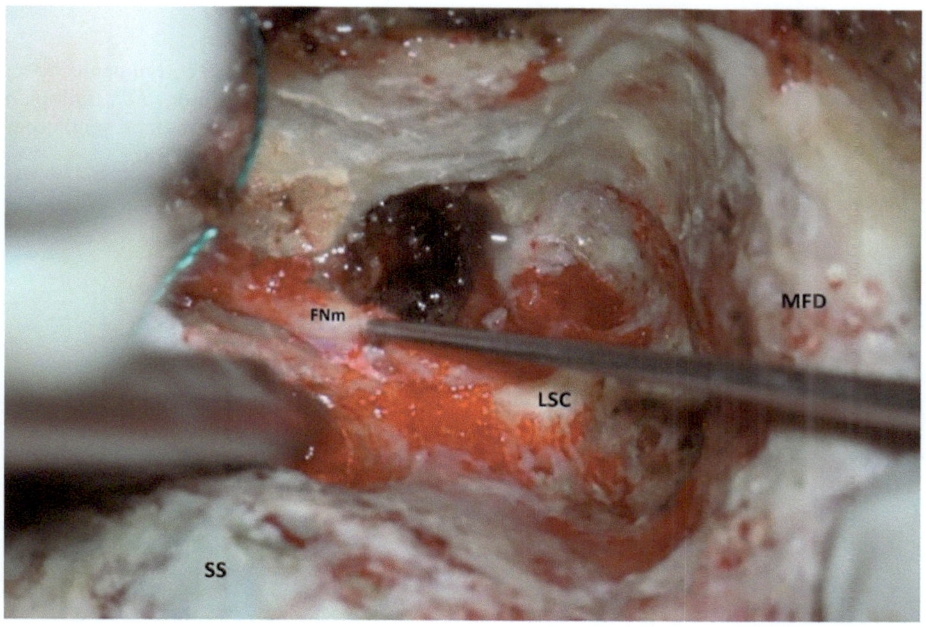

Fig. 12.131 The facial nerve begins to be rerouted from its mastoid segment (FNm). *LSC*, lateral semicircular canal; *MFD*, middle fossa dura; *SS*, sigmoid sinus.

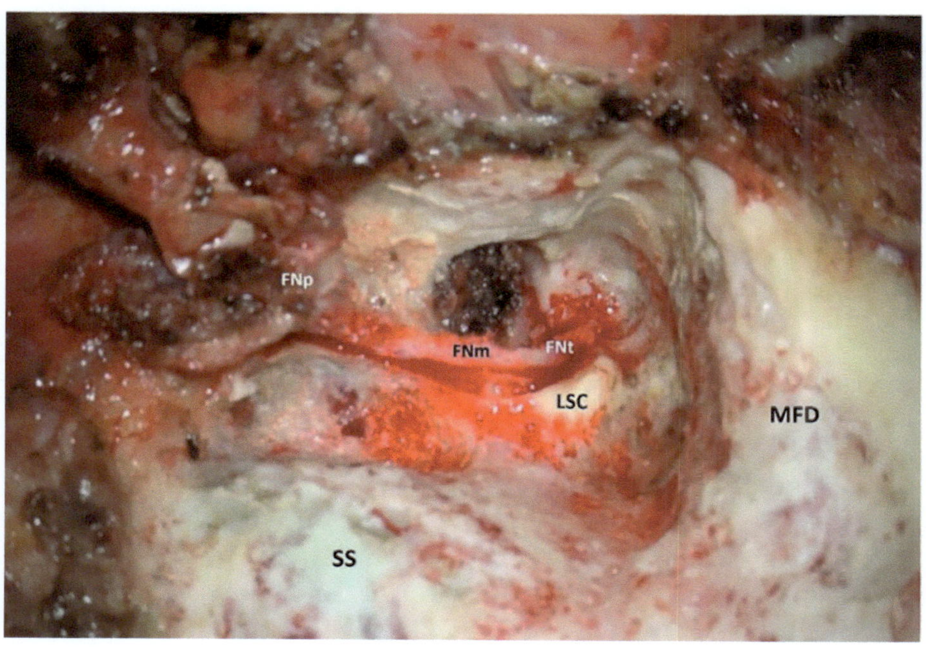

Fig. 12.132 The facial nerve is completely detached from the fallopian canal and ready for anterior rerouting. *FNp*, parotid portion of the facial nerve; *FNm*, mastoid portion of the facial nerve; *FNt*, tympanic portion of the facial nerve; *LSC*, lateral semicircular canal; *SS*, sigmoid sinus; *MFD*, middle fossa dura.

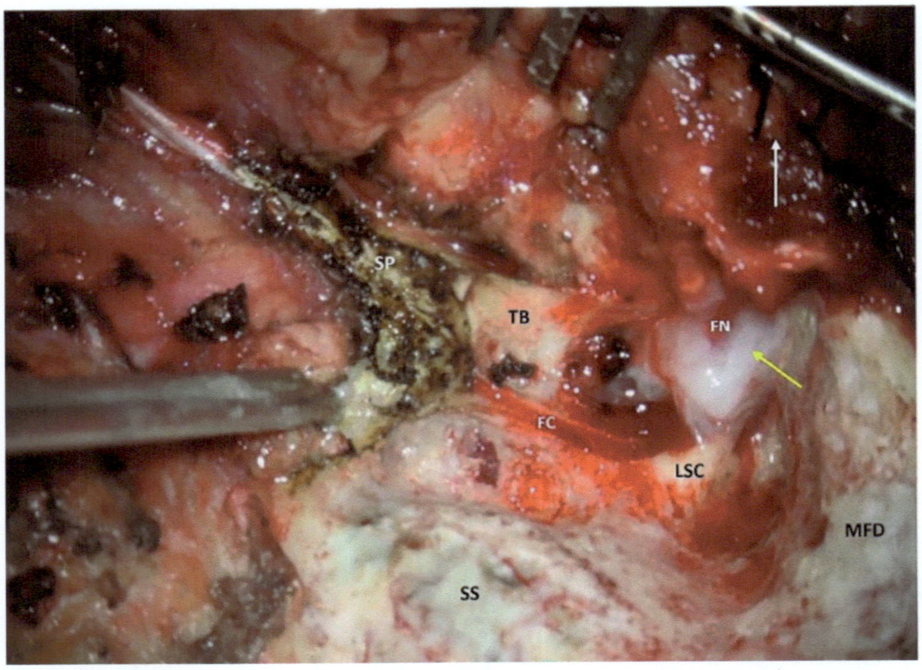

Fig. 12.133 The facial nerve (FN) has been anteriorly rerouted and fixed both in the parotid gland with a suture *(white arrow)* and in the supratubal canal with fibrin glue *(yellow arrow)*. The styloid process (SP) has been freed from the stylohyoid muscle and then removed together with the inferior tympanic bone (TB). This step allows complete control of the internal carotid artery at the entrance to the skull base. *FC*, fallopian canal; *SS*, sigmoid sinus; *LSC*, lateral semicircular canal; *MFD*, middle fossa dura.

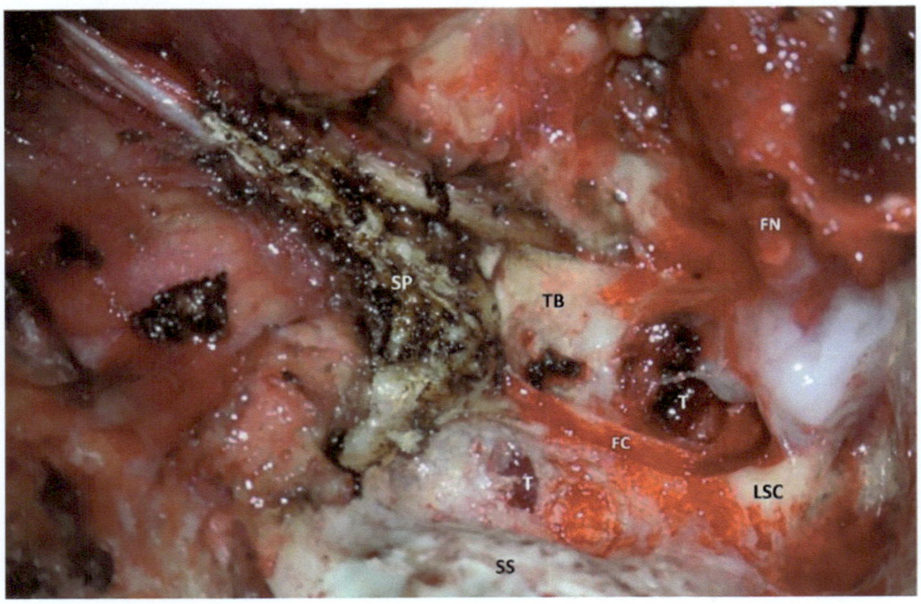

Fig. 12.134 Closer view. *TB*, tympanic bone; *SP*, styloid process; *T*, tumor; *FC*, fallopian canal; *SS*, sigmoid sinus; *LSC*, lateral semicircular canal; *MFD*, middle fossa dura.

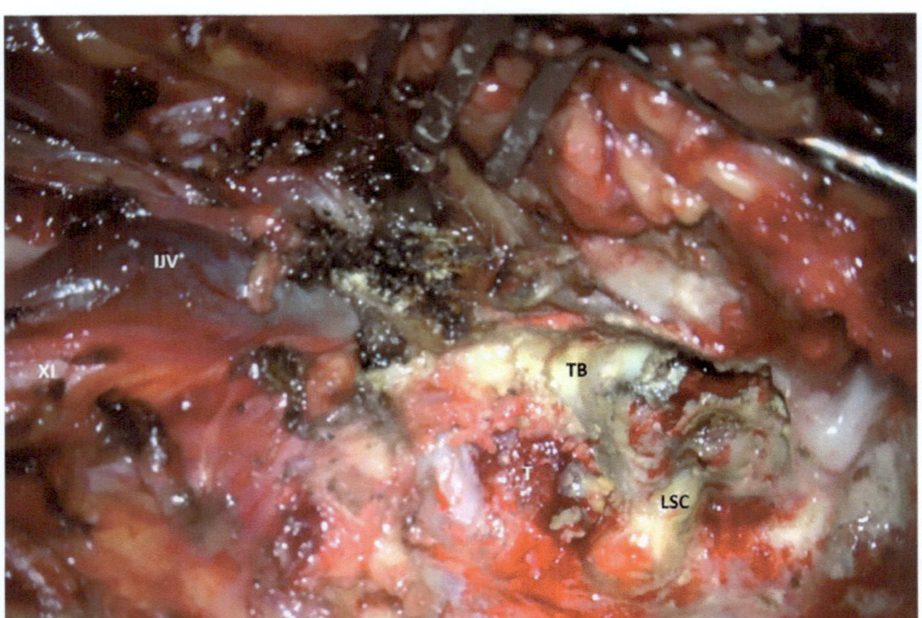

Fig. 12.135 The styloid process has been removed. The tympanic bone (TB) is drilled further. Note the tumor (T) invading the area of the jugular bulb. *IJV*, internal jugular vein; *LSC*, lateral semicircular canal.

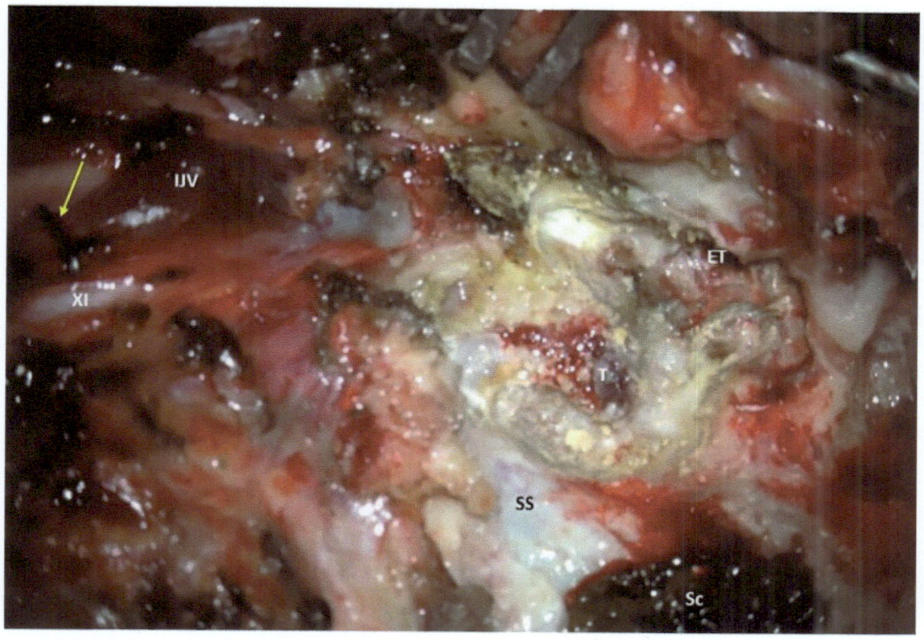

Fig. 12.136 The area of the jugular bulb invaded by the tumor (T) was progressively isolated. The inferior tympanic bone has also been drilled. The internal jugular vein (IJV) is closed with surgical clips *(yellow arrow)*, and the sigmoid sinus (SS) is packed extraluminally with Surgicel (Sc). *ET*, Eustachian tube; *XI*, spinal accessory nerve.

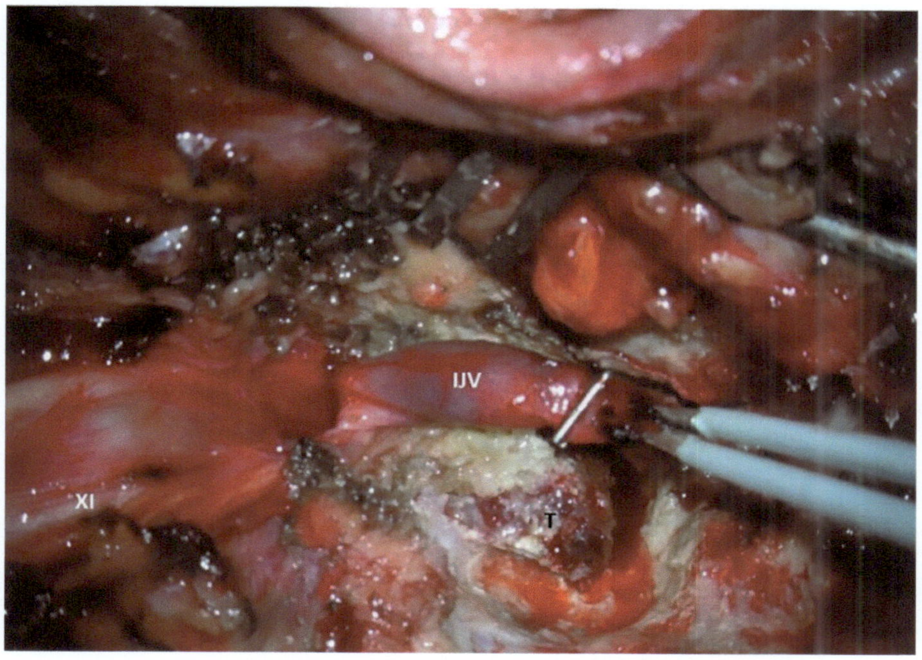

Fig. 12.137 Closer view. The internal jugular vein (IJV) is closed with surgical clips. *T*, tumor; *XI*, spinal accessory nerve.

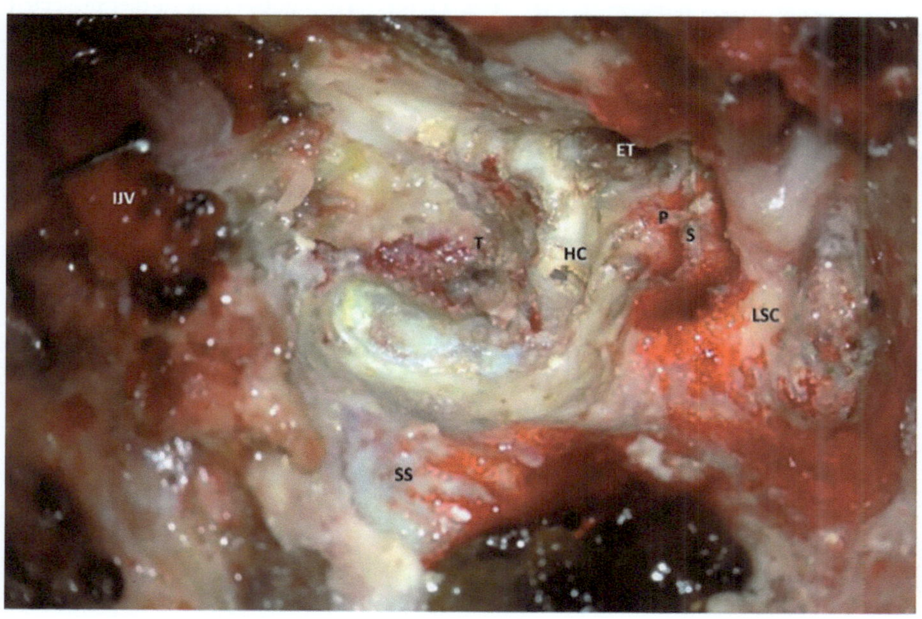

Fig. 12.138 The hypotympanic cells (HC) are drilled to isolate the tumor invading the jugular bulb. The pericarotid cells below the Eustachian tube (ET) are also drilled to assess invasion of the petrous internal carotid artery. *P*, promontory; *S*, stapes; *LSC*, lateral semicircular canal; *SS*, sigmoid sinus; *IJV*, internal jugular vein.

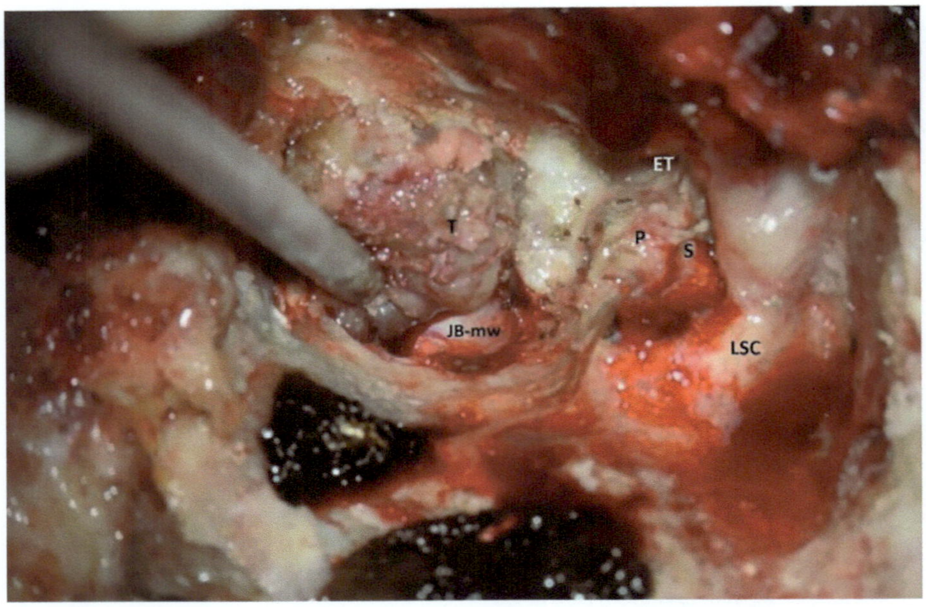

Fig. 12.139 The tumor (T) is dissected, leaving the medial wall of the jugular bulb (JB-mw) intact. *P*, promontory; *S*, stapes; *LSC*, lateral semicircular canal; *ET*, Eustachian tube.

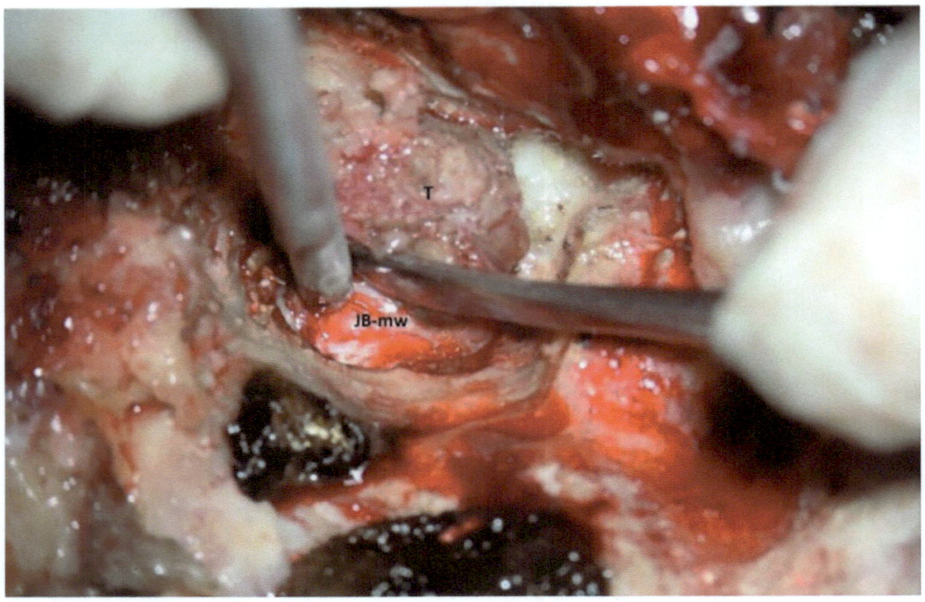

Fig. 12.140 Closer view. Preservation of the medial wall of the jugular bulb (JB-mw) allows anatomical and functional preservation of the lower cranial nerves. *T*, tumor.

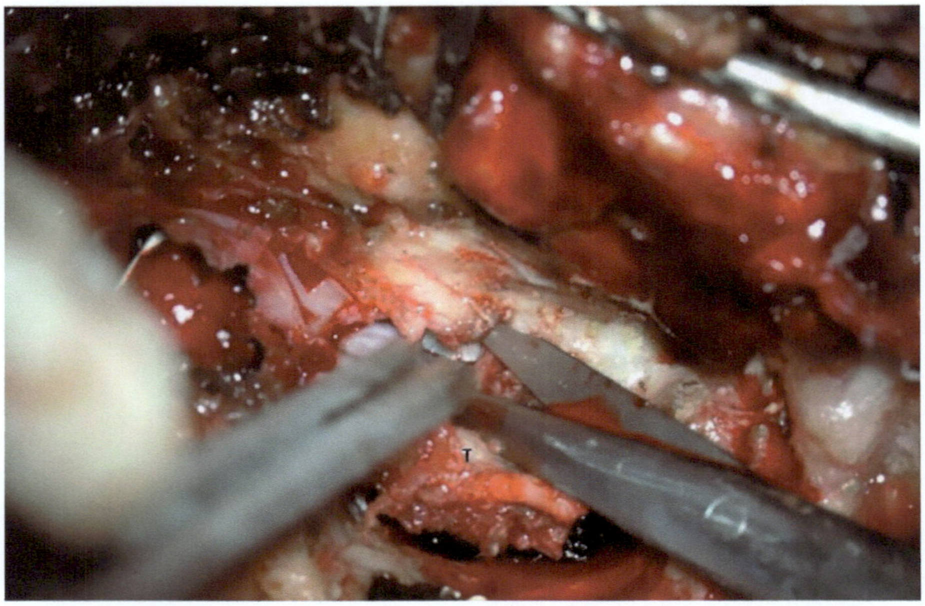

Fig. 12.141 The tumor (T) is further dissected from the jugular bulb and the internal carotid artery.

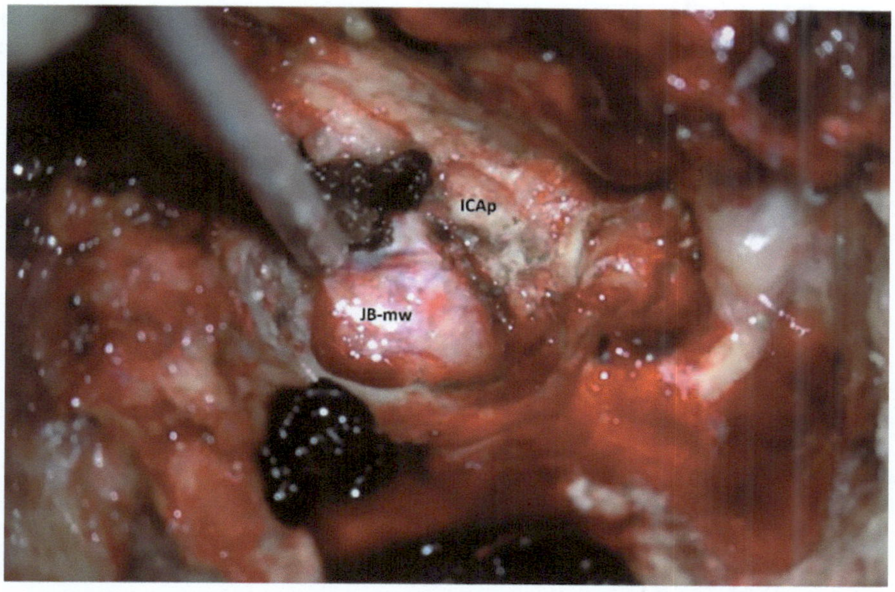

Fig. 12.142 The tumor is completely removed with complete preservation of the medial wall of the jugular bulb (JB-mw). *ICAp*, petrous portion of the internal carotid artery.

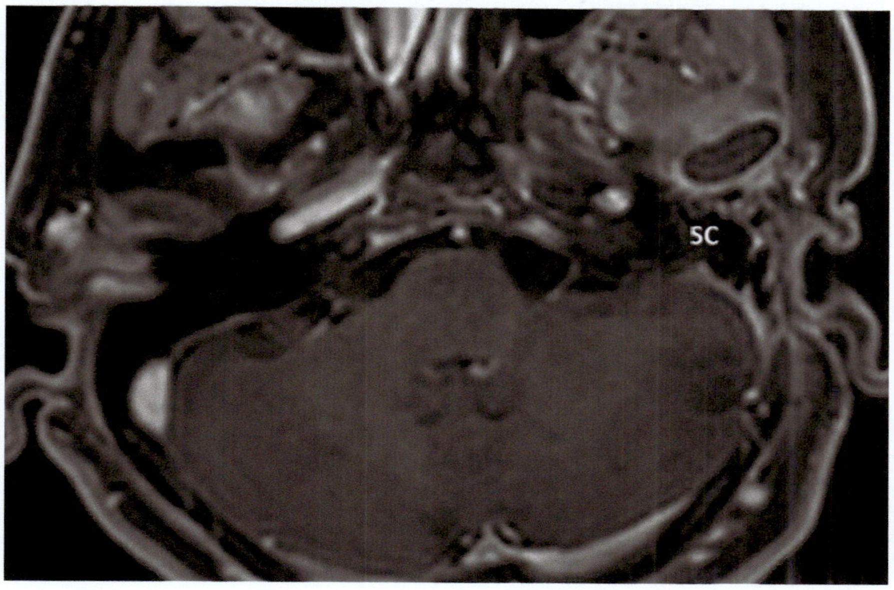

Fig. 12.143 Postoperative brain MRI with Gd enhancement showing no tumor recurrence at 2 years. *SC*, surgical cavity.

Special Considerations

- Removal of a left tumor with invasion of the inner ear (Figs. 12.144–12.147)

Not infrequently, glomus jugulare tumors invade the inner ear. In some cases, the cochlea might be invaded despite preoperative preserved hearing. To achieve total tumor removal and avoid recurrence, the cochlea has to be drilled as shown in Figs. 12.144–12.147.

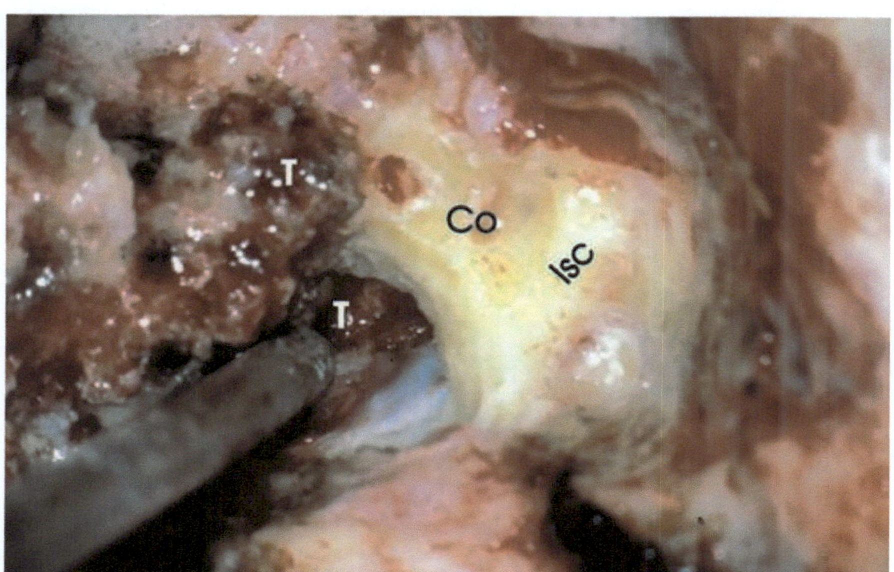

Fig. 12.144 The tumor (T) is seen extending along the posterior fossa dura medial to the cochlea (Co). The cochlea is also invaded by the tumor. *LSC*, lateral semicircular canal.

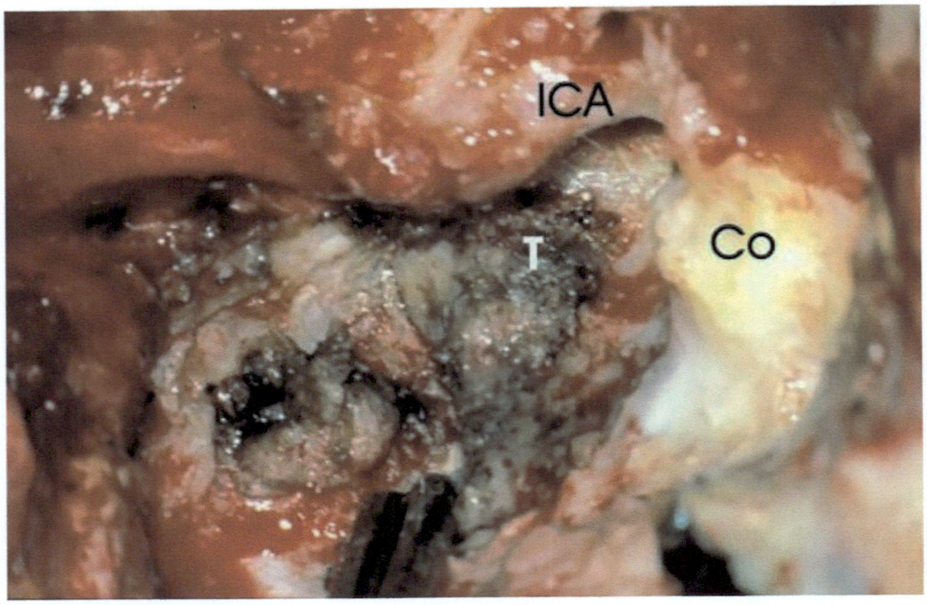

Fig. 12.145 The cochlea (Co) is drilled, and the tumor (T) medial to the cochlea can be appreciated. *ICA*, internal carotid artery.

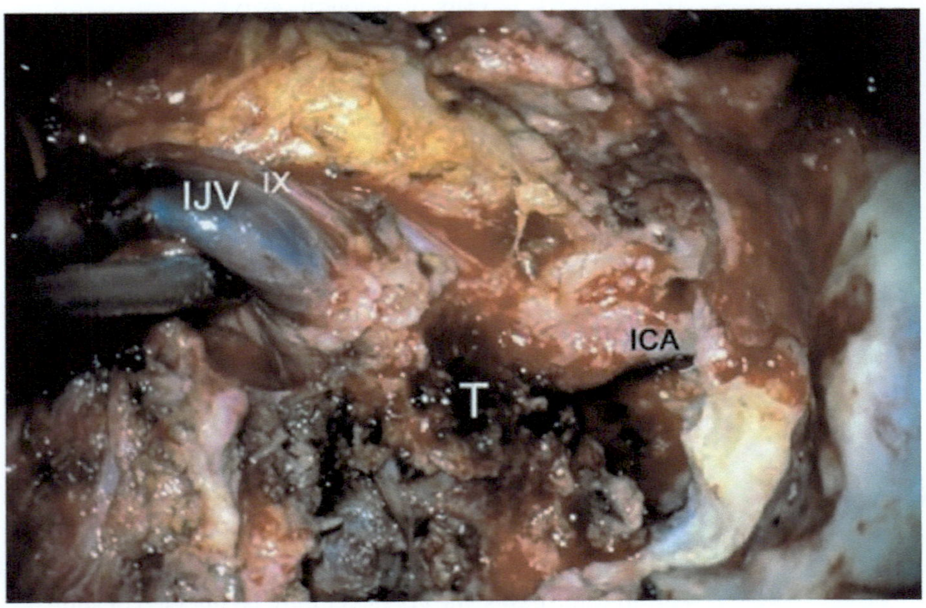

Fig. 12.146 The internal jugular vein (IJV) is ligated in the neck and is raised superiorly. Note that the glossopharyngeal nerve (IX) is invaded by the tumor (T). *ICA*, internal carotid artery.

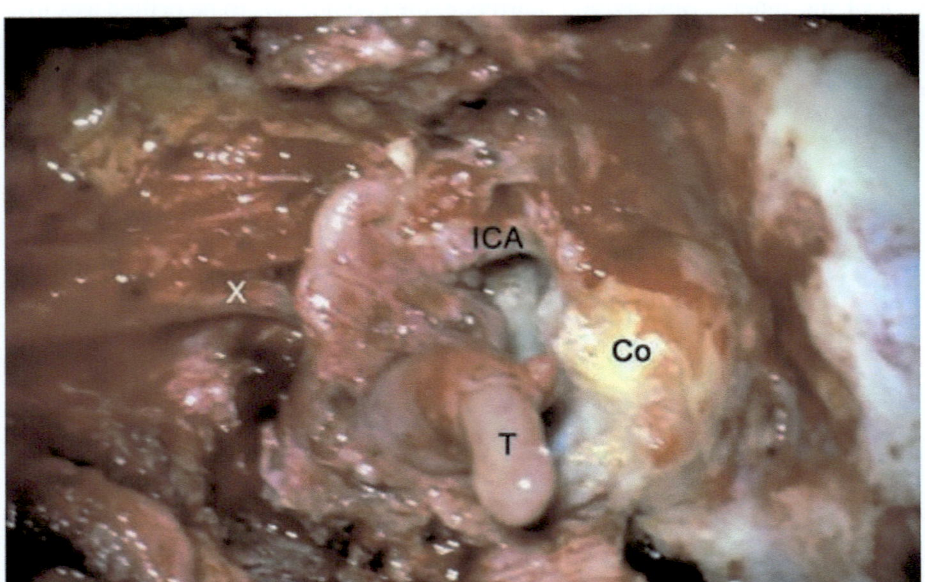

Fig. 12.147 The internal jugular vein (IJV) and the jugular bulb have been opened and most of the tumor has been removed. The tumor (T) is seen inside the opened jugular bulb. The vagus (X) nerve is seen. *ICA*, internal carotid artery; *Co*, cochlea.

- Sacrifice of the internal carotid artery

We emphasize that the degree of ICA involvement is the critical point in determining the surgical approach for TPJ paragangliomas [37–39]. The most accurate way in which to communicate the degree of ICA involvement is to stage TJP using the Fisch classification. It is our opinion that only selected cases of Class C1 tumors can be safely and adequately addressed without rerouting the facial nerve. We present the classic ITFA for Class C1 and certain Class C2 tumors and the ITFA with extensions as a simplified and unified paradigm in the approach to manage Class C2–C4 tumors.

The internal carotid artery can be sacrificed in the rare cases in which the artery is markedly encased by the tumor with subsequent stenosis or in cases with fragile wall of the artery due to previous surgery or irradiation. A preoperative balloon occlusion test is mandatory. If the test shows that the artery can be safely sacrificed, a permanent balloon is left to close the artery. In our early practice, carotid resection was performed more frequently; with time, we have adopted a less aggressive attitude for fear of long-term consequences.

- Removal of tumors with invasion of the left internal auditory canal (Figs. 12.148–12.150)

1. After performing the approach and transporting the facial nerve anteriorly, the tumor can be seen in the hypotympanum (Fig. 12.148).
2. Fig. 12.149 shows the tumor invading the internal auditory canal.
3. After dissection of the tumor from the canal, the lumen of the canal is occluded by a muscular plug (Fig. 12.150).

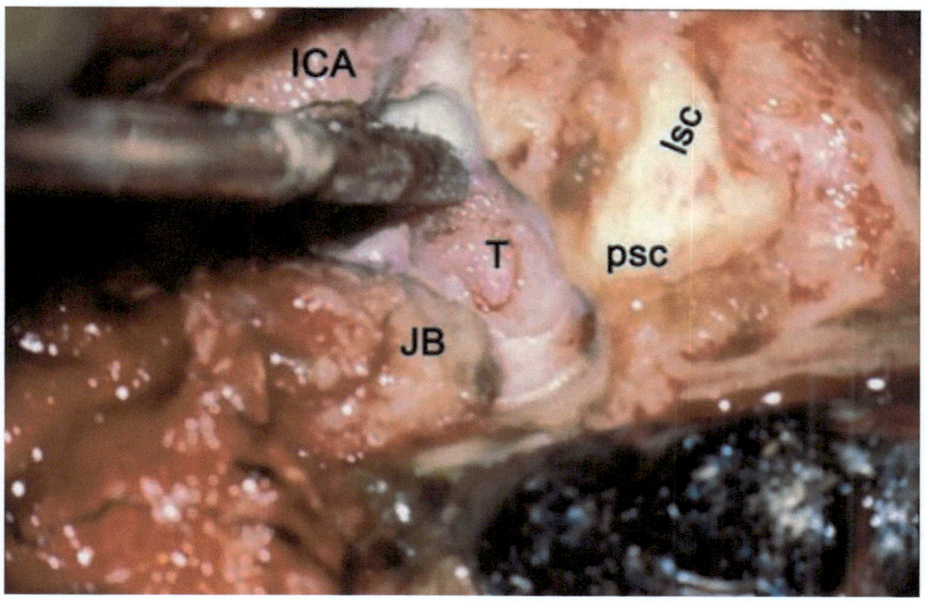

Fig. 12.148 The approach has been performed and the facial nerve has been transposed anteriorly. The tumor (T) is seen occupying the hypotympanum. *ICA*, internal carotid artery; *JB*, jugular bulb; *lsc*, lateral semicircular canal; *psc*, posterior semicircular canal.

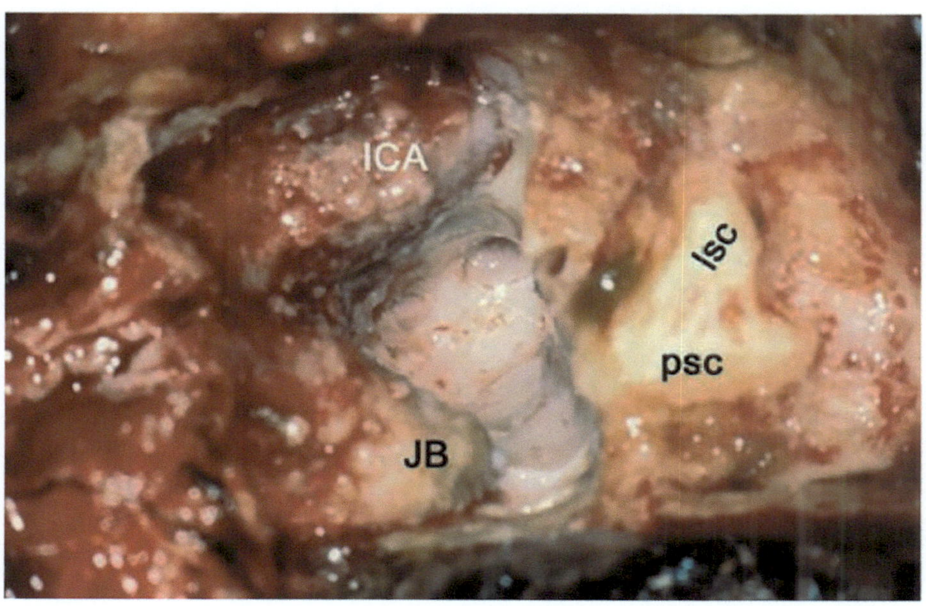

Fig. 12.149 The medial end of the internal auditory canal has been skeletonized. The tumor invades the canal at this level. *ICA*, internal carotid artery; *JB*, jugular bulb; *lsc*, lateral semicircular canal; *psc*, posterior semicircular canal.

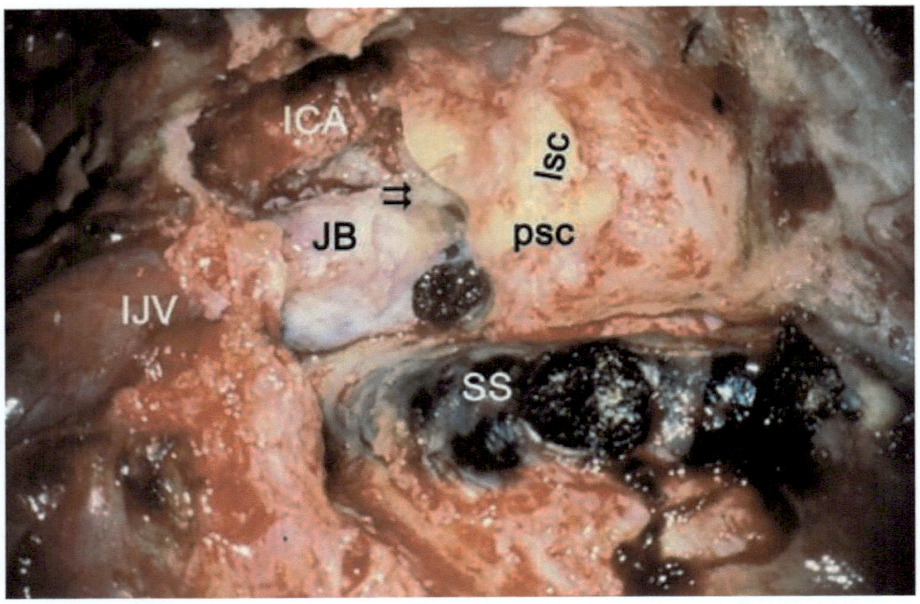

Fig. 12.150 The canal *(arrows)* has been opened and the tumor has been dissected from it. The canal is occluded by a muscle plug held in place by fibrin glue. Surgicel is placed to stop bleeding. *ICA*, internal carotid artery; *JB*, jugular bulb; *lsc*, lateral semicircular canal; *psc*, posterior semicircular canal; *IJV*, internal jugular vein; *SS*, sigmoid sinus.

- Facial nerve involvement (Figs. 12.151–12.160)

In some cases, the tumor invades the bone medial to the facial nerve at the level of the mastoid segment. This tumor can also infiltrate the nerve and can cause preoperative paralysis. In cases with invasion of the perineurium, the infiltrated segment is excised and an end-to-end anastomosis (or a sural graft if the segment is long) is performed. In rare cases, the facial nerve may be torn during anterior transposition. An end-to-end anastomosis is performed. A piece of abdominal fat is placed to support the repaired nerve.

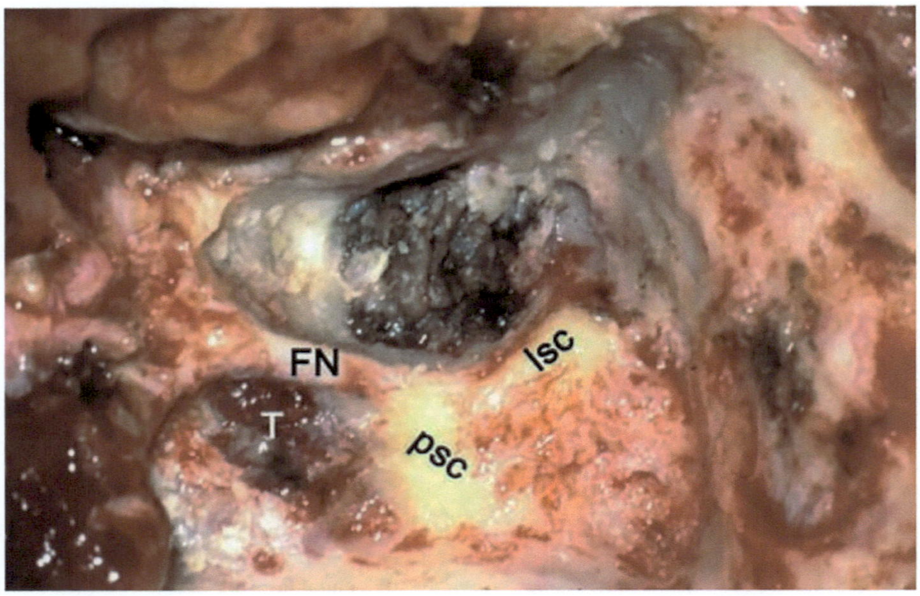

Fig. 12.151 The tumor (T) is seen infiltrating the fallopian canal immediately medial to the mastoid portion of the facial nerve (FN) just superior to the stylomastoid foramen. Note the Surgicel placed over the middle ear tumor to control bleeding. *lsc*, lateral semicircular canal; *psc*, posterior semicircular canal.

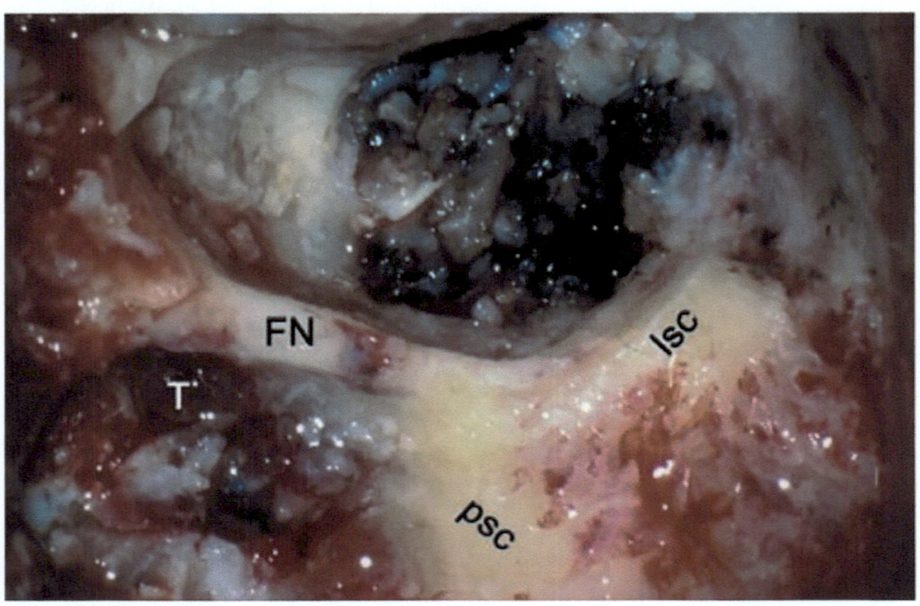

Fig. 12.152 At a higher magnification, the precise relation of the tumor (T) to the facial nerve (FN) can be appreciated. *lsc*, lateral semicircular canal; *psc*, posterior semicircular canal.

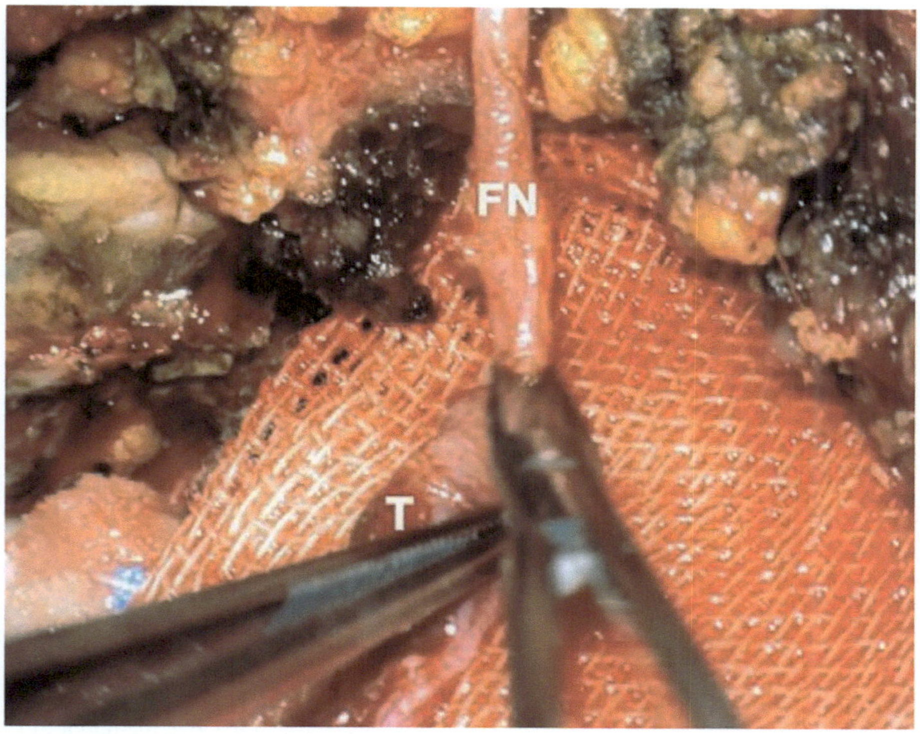

Fig. 12.153 The segment of the facial nerve (FN) infiltrated by the tumor (T) is excised. A sural graft is inserted.

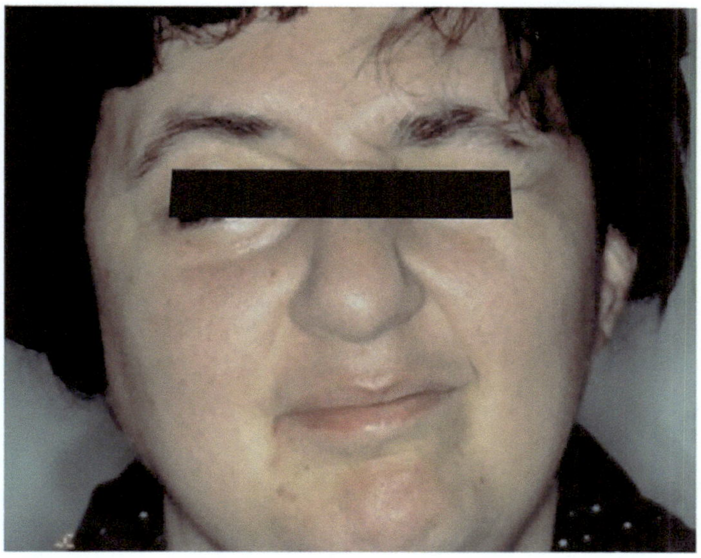

Fig. 12.154 One year postoperative view of the same case after sural nerve grafting. Note that the left facial nerve has recovered to grade III of the House-Brackmann system.

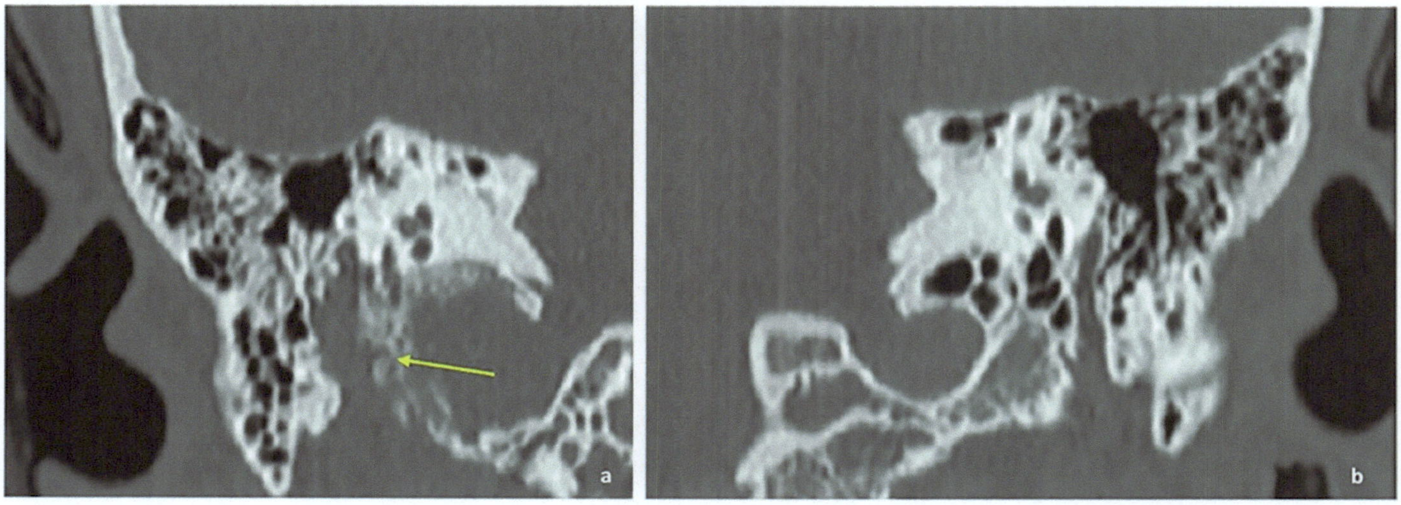

Fig. 12.155 HRCT scan, coronal view. Another case of tumor invasion of the fallopian canal. Note the enlargement and involvement of the fallopian canal on the side of the tumor (A, *yellow arrow*) compared to the healthy side (B).

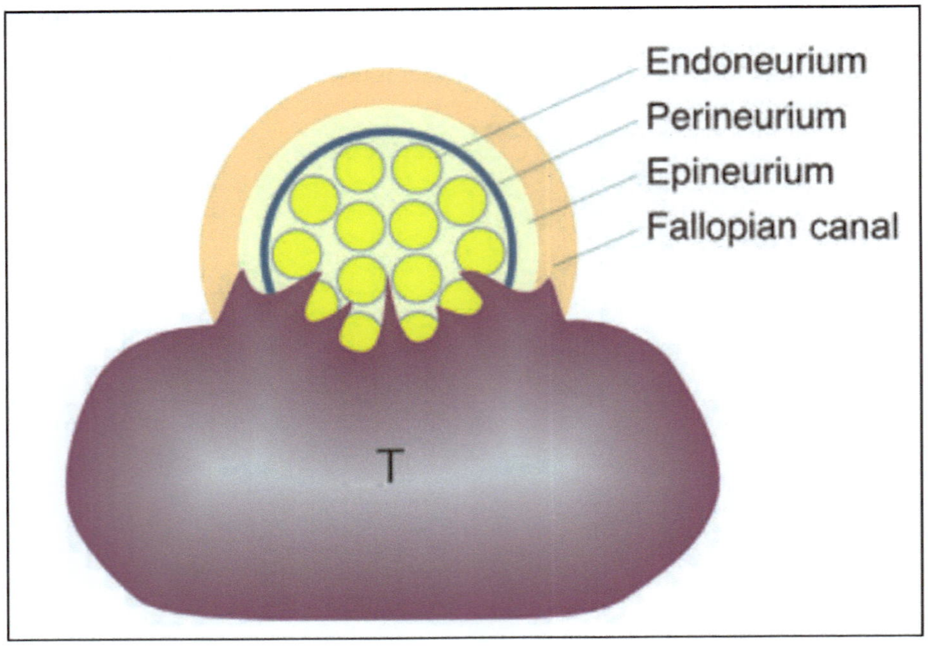

Fig. 12.156 Schematic drawing of fallopian canal invasion by the tumor. The nerve is invaded up to the endoneurium.

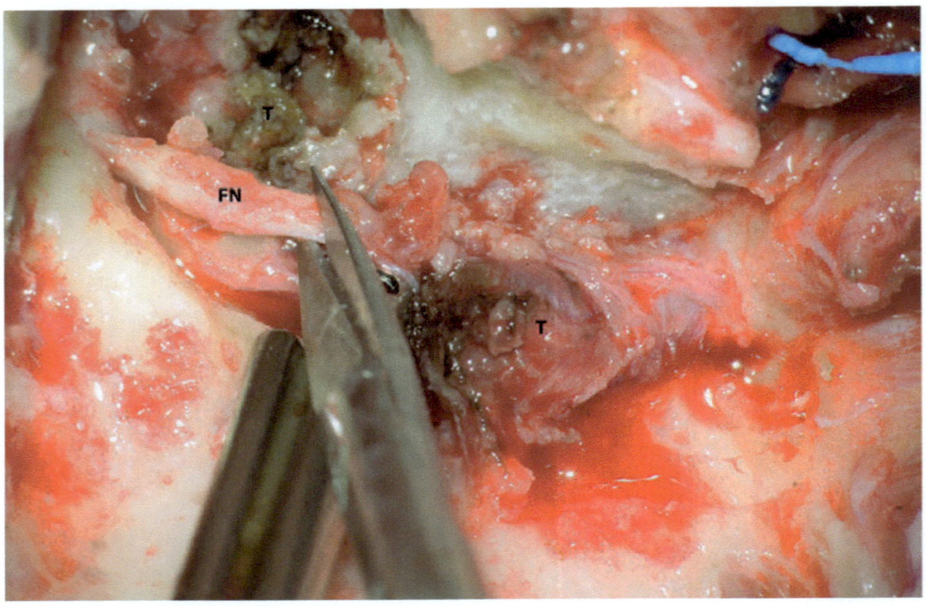

Fig. 12.157 The mastoid segment of the facial nerve (FN) is cut 1–2 mm above the tumor (T).

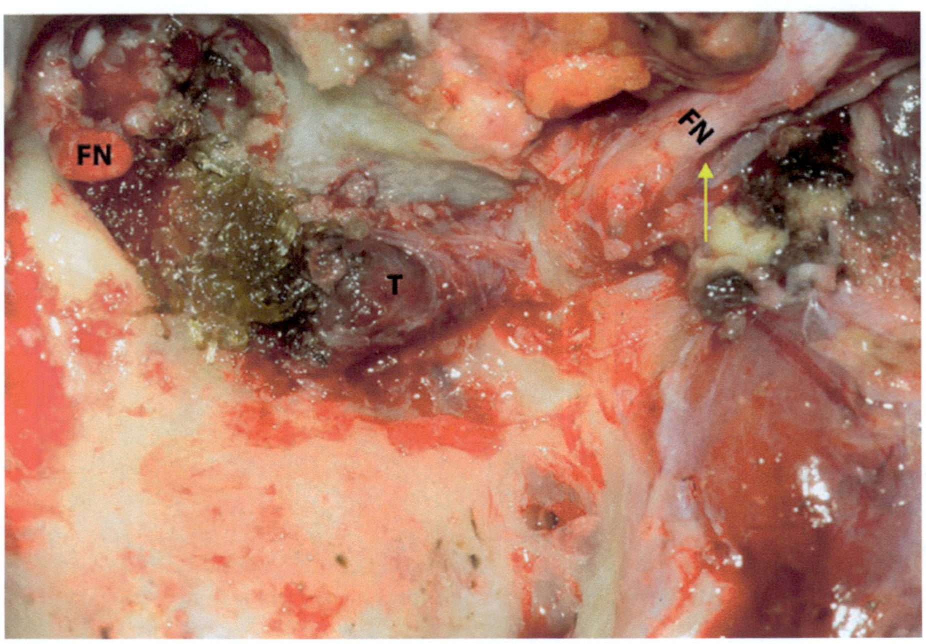

Fig. 12.158 The distal stump of the healthy facial nerve *(arrow)* is located at the level of the stylomastoid foramen. *T*, tumor; *FN*, facial nerve.

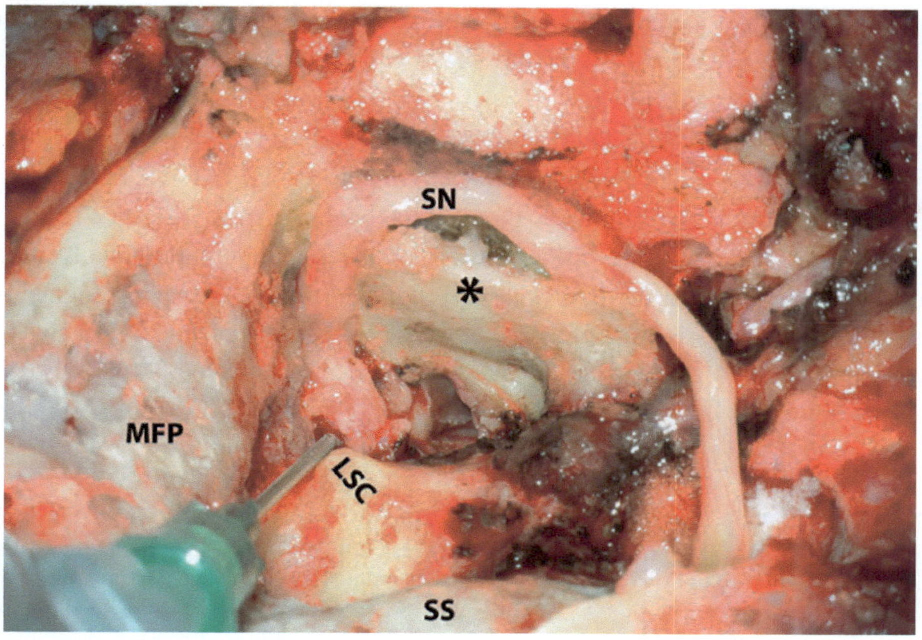

Fig. 12.159 The facial nerve is reconstructed with a sural nerve (SN) graft. The proximal stump is fixed with fibrin glue. *MFP*, middle fossa dura plate; *SS*, sigmoid sinus; *LSC*, lateral semicircular canal; *, bony canal for the new facial nerve.

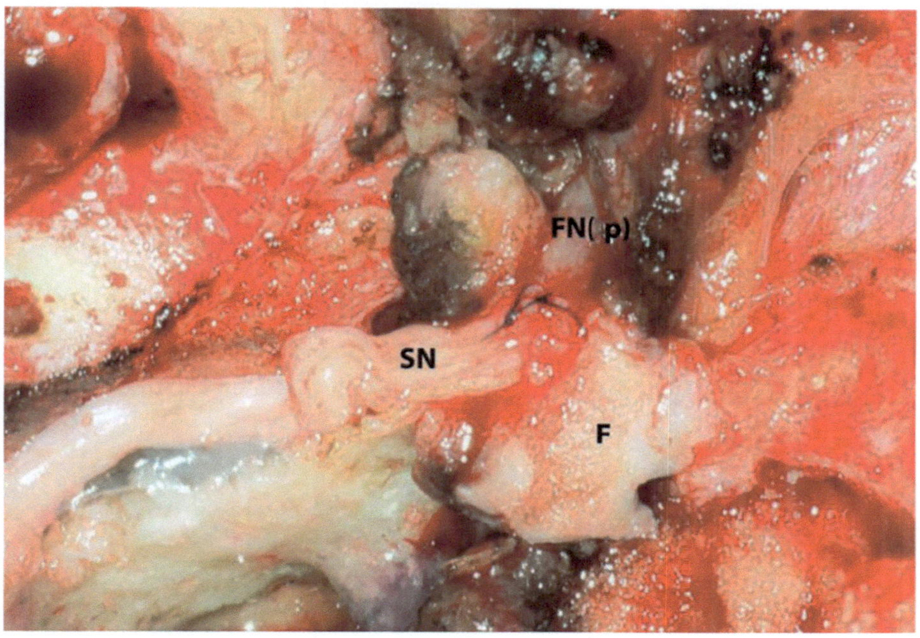

Fig. 12.160 The distal stump is sutured. A piece of temporalis fascia (F) and fibrin glue are then used to stabilize the reconstruction. *SN*, sural nerve; *FN(p)*, extratemporal portion of the facial nerve (parotid).

- Presence of ipsilateral carotid body tumor

Infrequently, a concomitant carotid body tumor can be found. If an ipsilateral tumor is present, it is removed at the same stage with the glomus jugulare tumor.

- Single-stage removal of tumor with small intradural extension (C3Di1)

Opening the dura and connecting it to the widely opened neck spaces has a high risk of CSF leakage. Since the tympanic bone and the tough fibrous tissue at the base of the skull are removed, there remains no rigid containment for the abdominal fat used for closure. Consequently, the fat slips and CSF leakage eventually occurs. Small intradural tumor extensions that need a small dural opening can be managed in a single stage.

- Staged removal of tumors with intradural extension

Glomus tumors with intradural extension are better managed in a staged operation. The extradural part is removed in the first stage, followed by a second stage for removal of the intradural residual tumor. This avoids the high risk of developing CSF leakage should a single-stage procedure be attempted. The choice of the approach for the second stage depends on the residual tumor location, its size, and the patient's hearing. Most cases can be dealt with using a POTS approach. In some cases, the modified transcochlear, the transotic, or the extreme lateral approaches may be necessary.

CASE 3: STAGED REMOVAL IN A COMPLEX CASE WITH A CLASS C3DI2 TYMPANOJUGULAR PARAGANGLIOMA, VAGAL PARAGANGLIOMA, CAROTID BODY TUMOR AND SACRIFICE OF THE INTERNAL CAROTID ARTERY (Figs. 12.161–12.200)

A 52-year-old woman presented with a 3- to 4-year history of dysphonia, progressive hearing loss, and pulsatile tinnitus. She had no family history. Otoscopy revealed a keratin-covered mass occupying the medial compartment of the external auditory canal. She showed paralysis of the vagus and hypoglossal nerves and had a dead ear.

Preoperative Imaging (Figs. 12.161–12.164)

In summary, the findings were of stage I large VP and a C3Di2 TJP. Because of the circumferential involvement of the ICA both in its cervical and intratemporal segments, the patient was assessed for permanent balloon occlusion prior to two-stage resection.

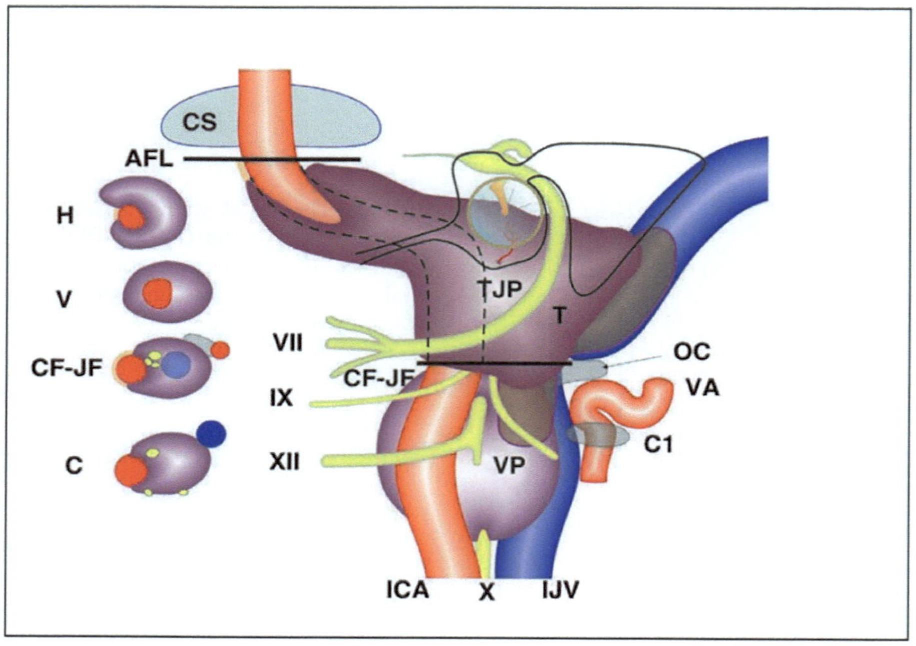

(A)

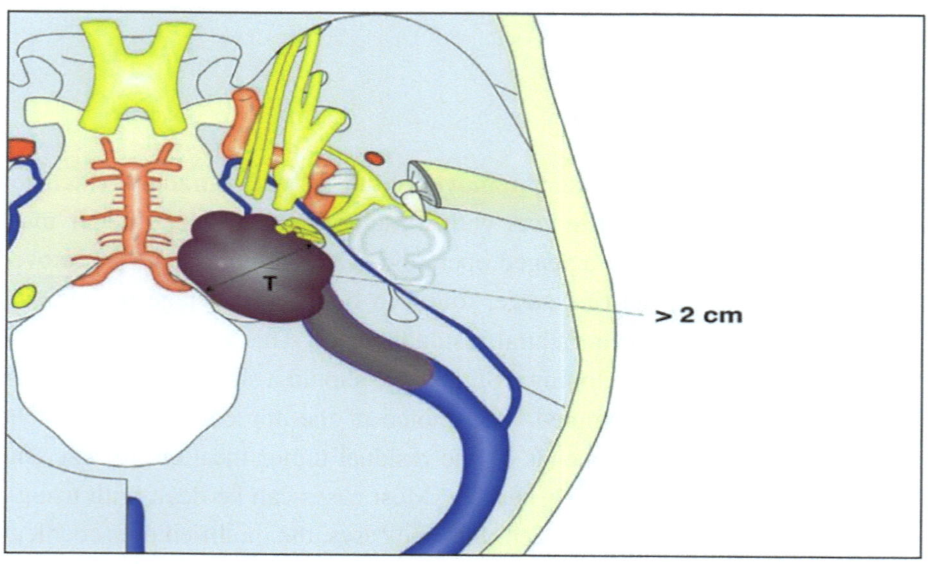

(B)

Fig. 12.161 (A, B) Class C3Di2 (left side). (A) Class C3. (B) Class Di2. *AFL*, anterior foramen lacerum; *C1*, atlas; *C*, extratemporal carotid level; *CS*, cavernous sinus; *H*, horizontal portion of the internal carotid artery; *ICA*, internal carotid artery; *IJV*, internal jugular vein; *CF-JF*, carotid foramen-jugular foramen; *OC*, occipital condyle; *T*, tumor; *TJP*, tympanojugular paraganglioma; *V*, vertical portion of the internal carotid artery; *VA*, vertebral artery; *VP*, vagal paraganglioma; *VII*, facial nerve; *IX*, glossopharyngeal nerve; *X*, vagus nerve; *XII*, hypoglossal nerve.

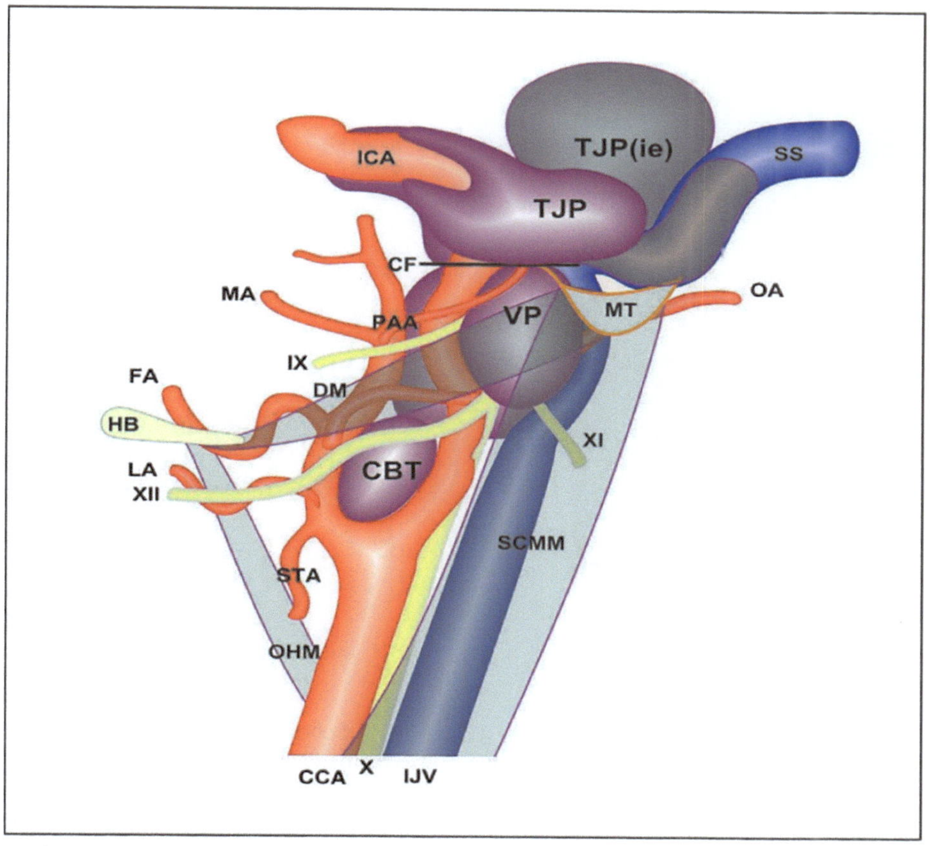

Fig. 12.162 Illustration for complex tympanojugular paragangliomas. *CBT*, carotid body tumor; *CCA*, common carotid artery; *CF*, carotid foramen; *DM*, posterior belly of the digastric muscle; *FA*, facial artery; *HB*, hyoid bone; *ICA*, internal carotid artery; *IJV*, internal jugular vein; *LA*, lingual artery; *MA*, maxillary artery; *MT*, mastoid tip; *OA*, occipital artery; *OHM*, omohyoid muscle; *PAA*, posterior auricular artery; *SCMM*, sternocleidomastoid muscle; *STA*, superior thyroid artery; *TJP*, tympanojugular paraganglioma; *TJP(ie)*, intradural or extradural TJP; *VP*, vagal paraganglioma; *IX*, glossopharyngeal nerve; *X*, vagus nerve; *XI*, spinal accessory nerve; *XII*, hypoglossal nerve.

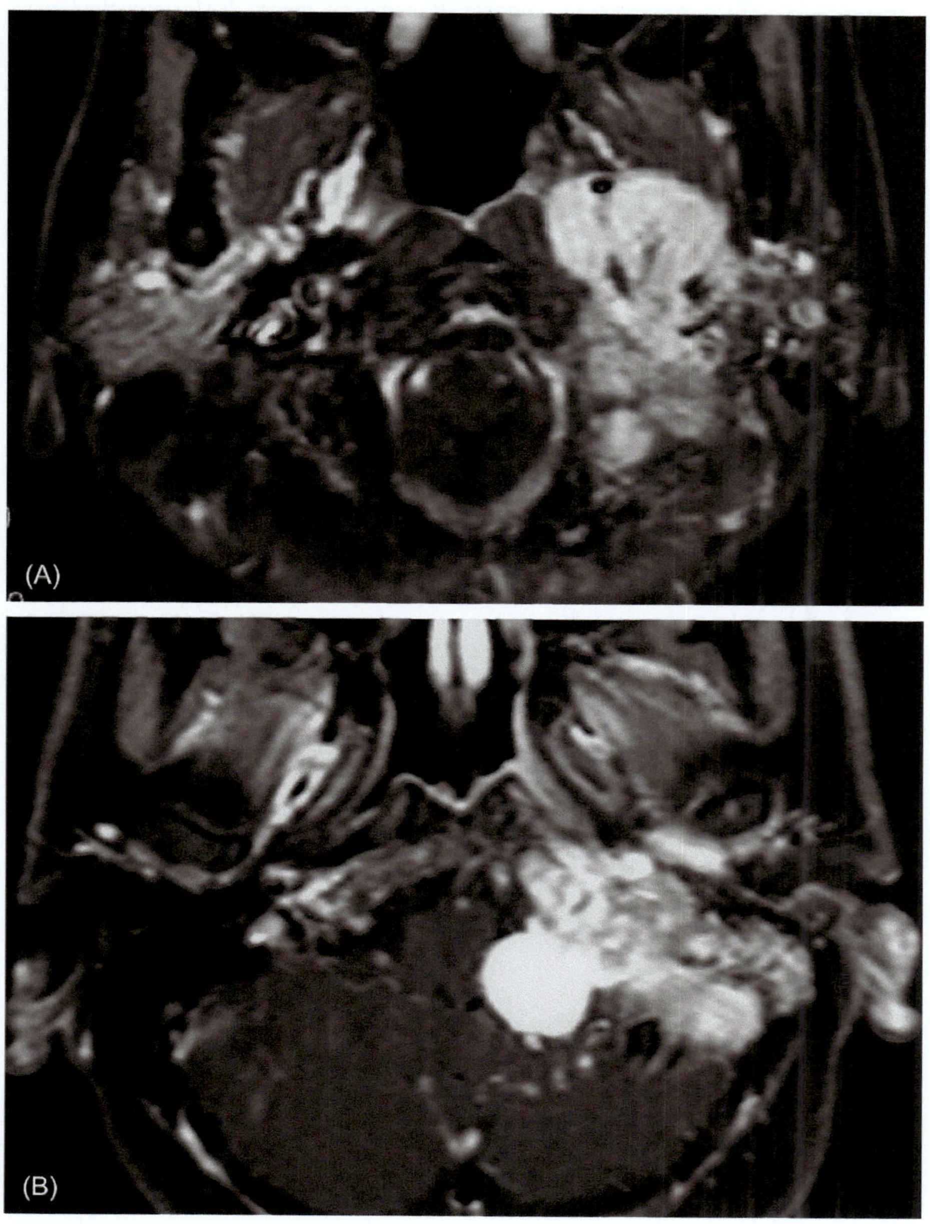

Fig. 12.163 (A, B) Axial T1W enhanced MRI scans revealing a large hypervascular mass occupying the poststyloid parapharyngeal space with complete encasement of the carotid artery. At a higher level, a large intradural component can be seen extending through the medial aspect of the jugular foramen. There is also dural infiltration along the entire posterior face of the petrous bone. These images are consistent with the coexistence of a tympanojugular and a vagal paraganglioma.

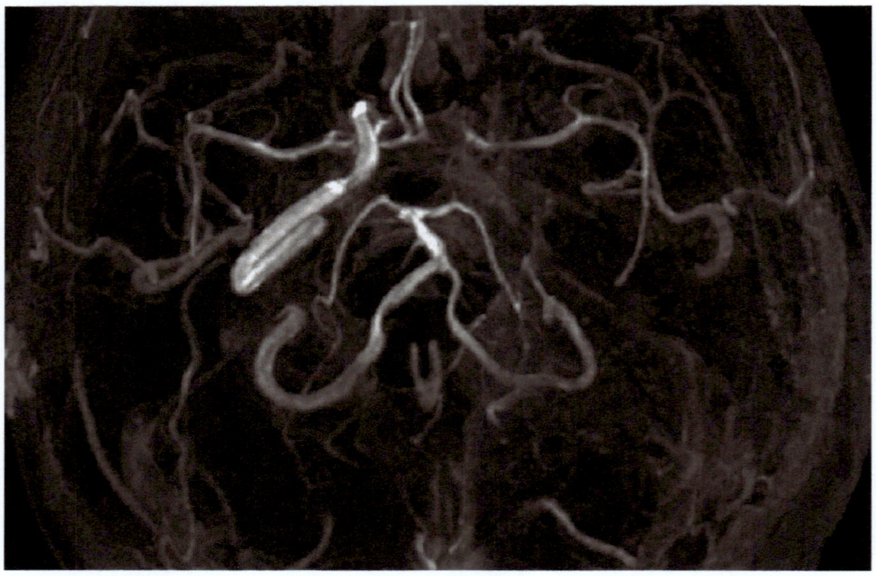

Fig. 12.164 Magnetic resonance angiography showing the absence of the left ICA but adequate collateral supply through the circle of Willis.

Surgical Resection

See Figs. 12.165–12.179.

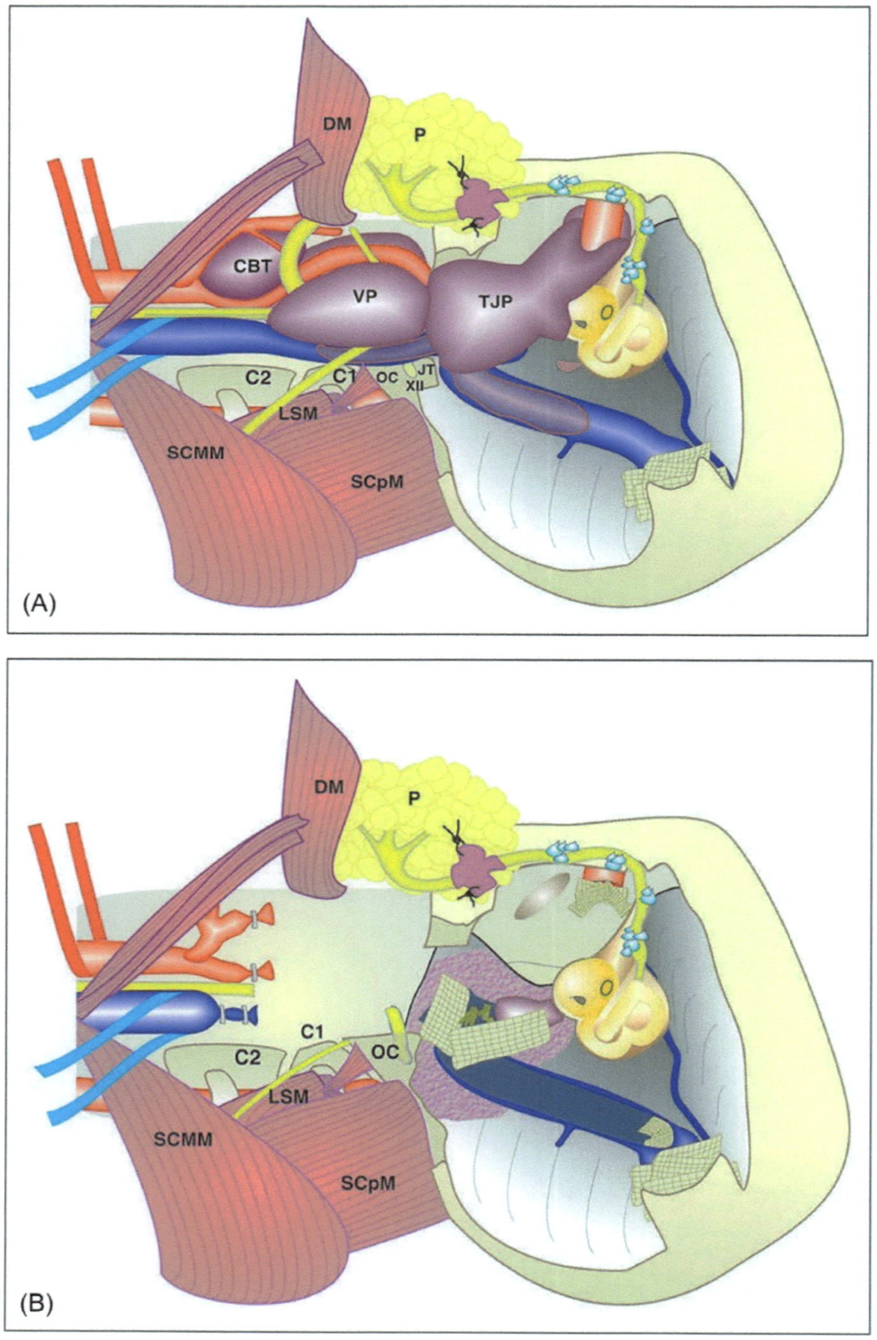

Fig. 12.165 (A) Schematic drawing for operative findings. (B) Schematic drawing after tumor removal. *C1*, atlas; *C2*, axis; *CBT*, carotid body tumor; *DM*, posterior belly of the digastric muscle; *JT*, jugular tubercle; *LSM*, levator scapulae muscle; *OC*, occipital condyle; *P*, parotid gland; *SCMM*, sternocleidomastoid muscle; *SCpM*, splenius capitis muscle; *TJP*, tympanojugular paraganglioma; *VP*, vagal paraganglioma; *XII*, hypoglossal nerve.

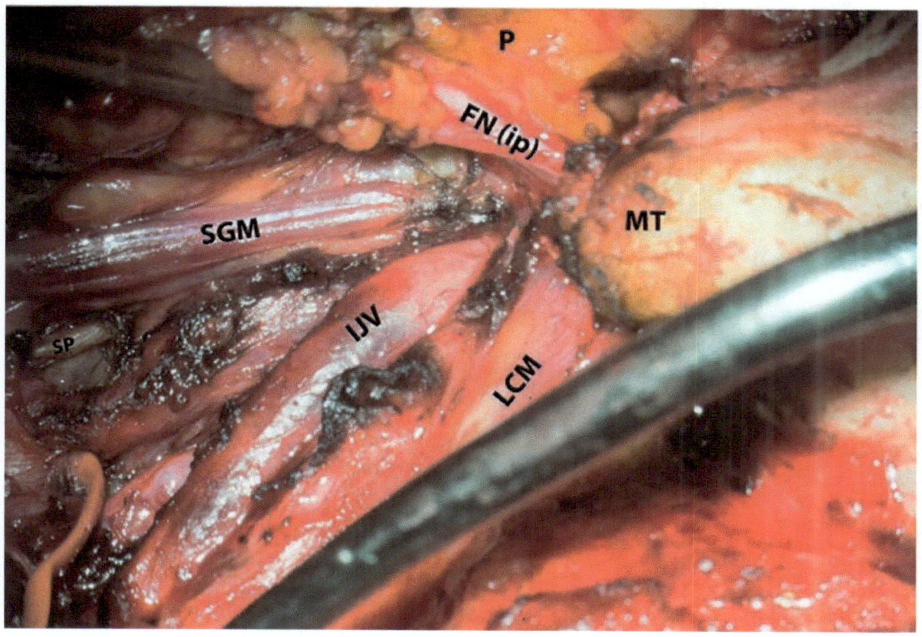

Fig. 12.166 Blind sac closure, neck dissection, and exposure of the extratemporal segment of the facial nerve have been completed. Note the relationship between the IJV, the styloid process, and the deep muscles in the neck. *FN(ip)*, intraparotid segment of the facial nerve; *LCM*, longissimus capitis muscle; *MT*, mastoid tip; *P*, parotid; *IJV*, internal jugular vein; *SGM*, styloglossus muscle; *SP*, styloid process.

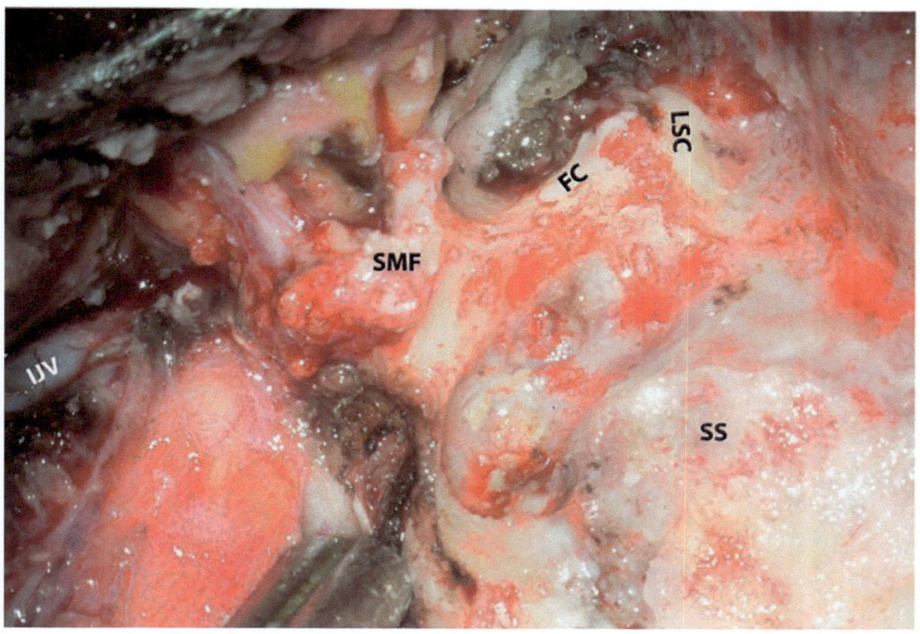

Fig. 12.167 Subtotal petrosectomy has been performed with removal of the mastoid tip. *FC*, fallopian canal; *IJV*, internal jugular vein; *LSC*, lateral semicircular canal; *SMF*, stylomastoid foramen; *SS*, sigmoid sinus.

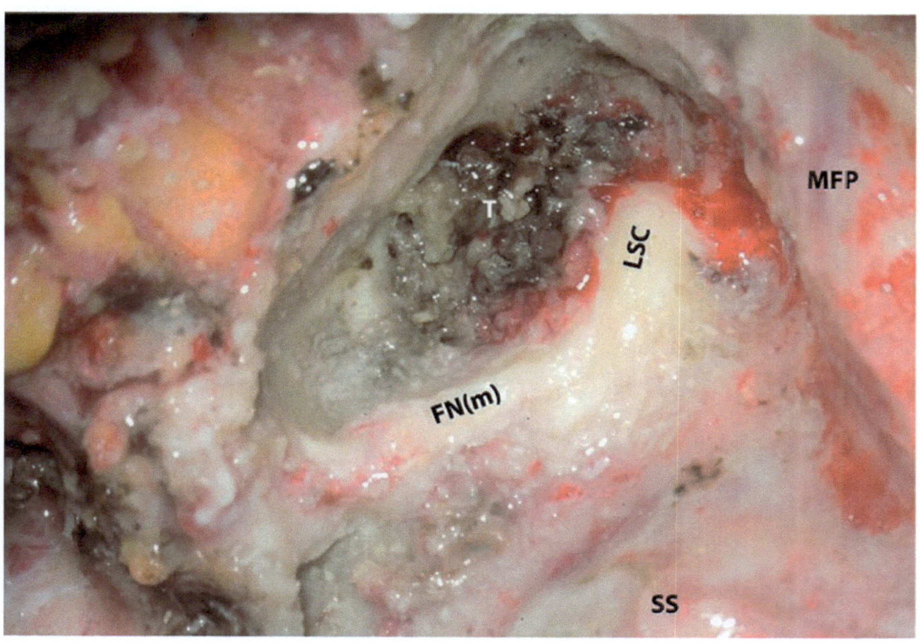

Fig. 12.168 Tumor in the middle ear has been extensively diathermized to control bleeding prior to rerouting of the facial nerve. This is the only segment of tumor approached prior to wide exposure of the tumor margins. Drilling the tympanic bone significantly improves exposure of the facial nerve. The tumor infiltrates the tympanic bone and the retrofacial air cells. *FN(m)*, mastoid segment of the facial nerve; *LSC*, lateral semicircular canal; *MFP*, middle cranial fossa dural plate; *SS*, sigmoid sinus; *T*, tumor.

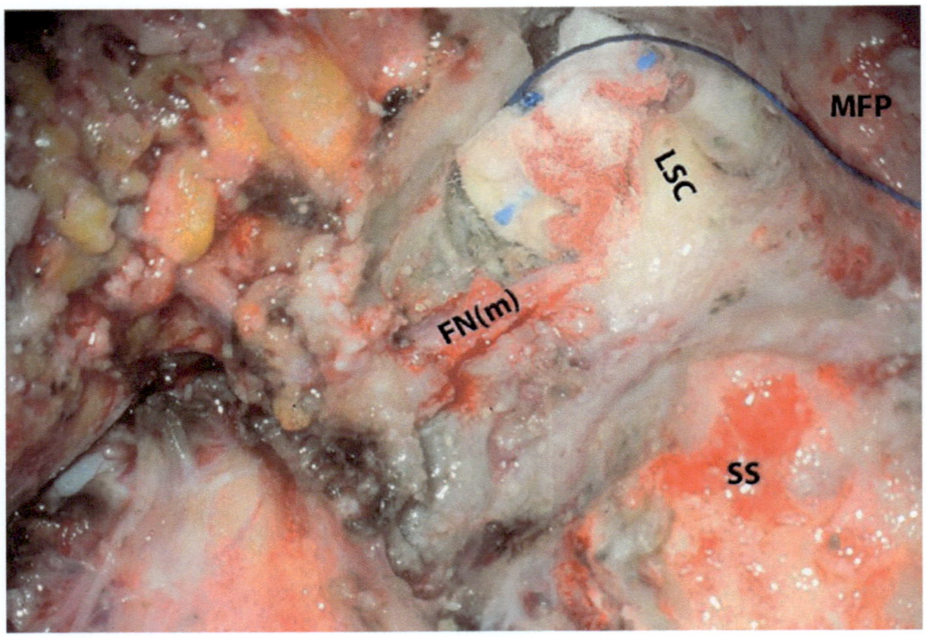

Fig. 12.169 Mobilization of the facial nerve has begun inferiorly. Further tumor bleeding is controlled with a Cottonoid. *FN(m)*, mastoid segment of the facial nerve; *LSC*, lateral semicircular canal; *MFP*, middle cranial fossa dural plate; *SS*, sigmoid sinus.

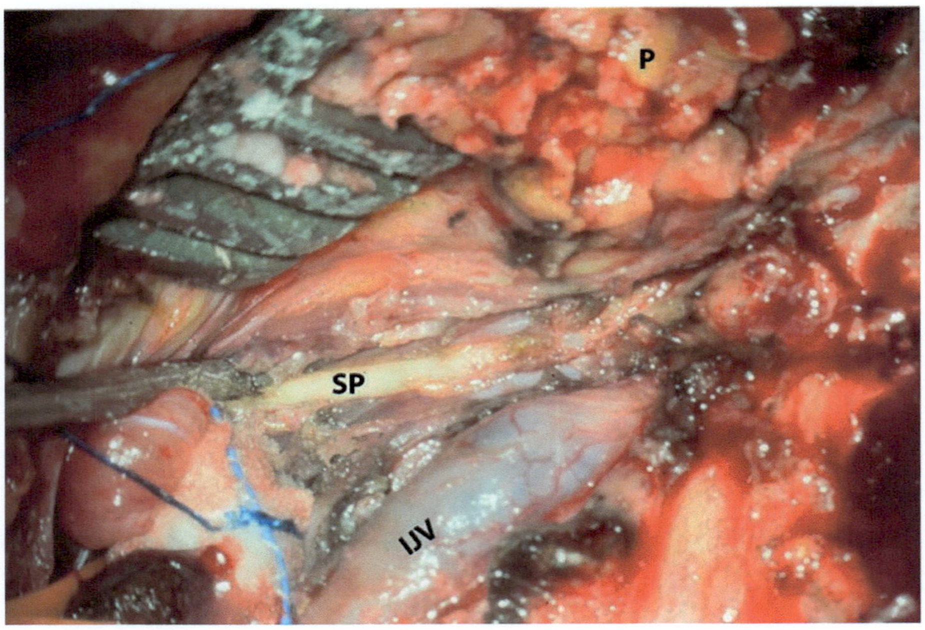

Fig. 12.170 Following completion of anterior facial nerve rerouting, the styloid process is freed from the attached muscles and ligament and removed to optimize ICA exposure. *IJV*, internal jugular vein; *SP*, styloid process; *P*, parotid.

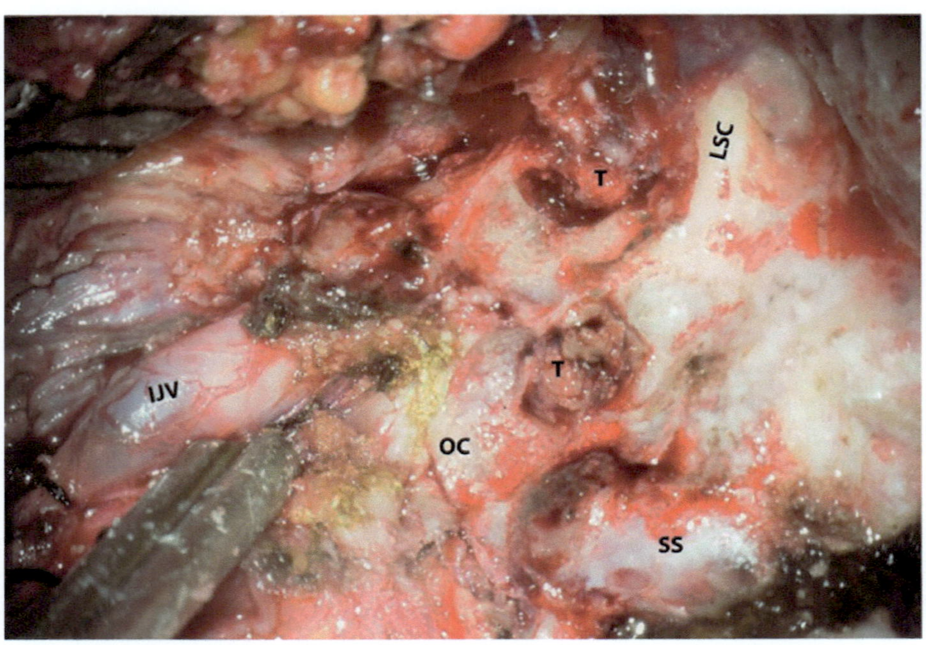

Fig. 12.171 The styloid process has been removed. The sigmoid sinus is extraluminally packed and the IJV is ligated in the neck. Further bone has been removed posteroinferiorly to the sigmoid and bulb, and tumor can be seen to be extending toward the occipital condyle. Note the wide exposure of the extended jugular foramen area due to anterior rerouting of the facial nerve and there being no styloid process. *IJV*, internal jugular vein (ligated); *LSC*, lateral semicircular canal; *OC*, occipital condyle; *SS*, sigmoid sinus; *T*, tumor.

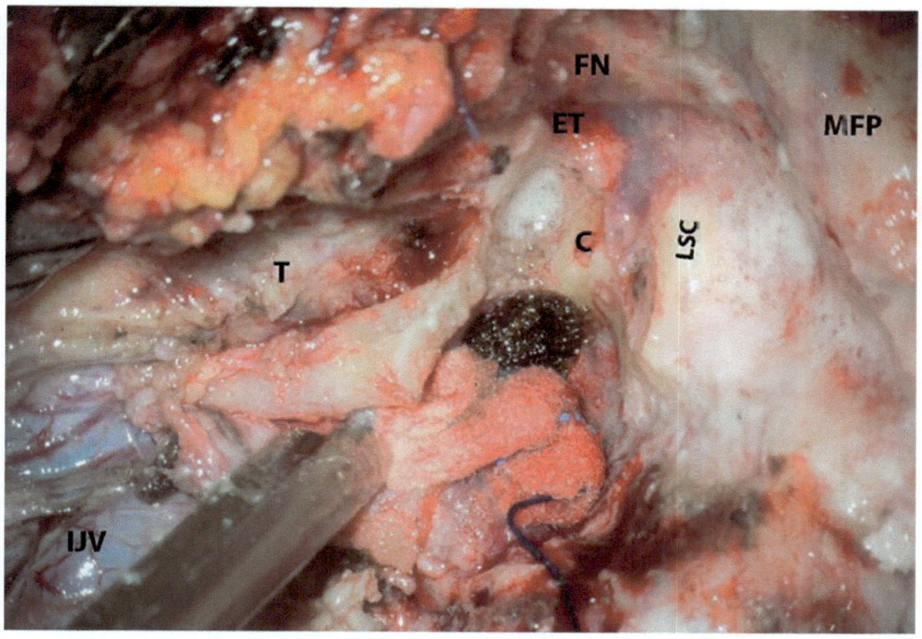

Fig. 12.172 The remainder of the tympanic bone has been removed. *C*, cochlea; *ET*, Eustachian tube; *FN*, facial nerve (rerouted); *IJV*, internal jugular vein; *LSC*, lateral semicircular canal; *MFP*, middle cranial fossa dural plate; *T*, tumor.

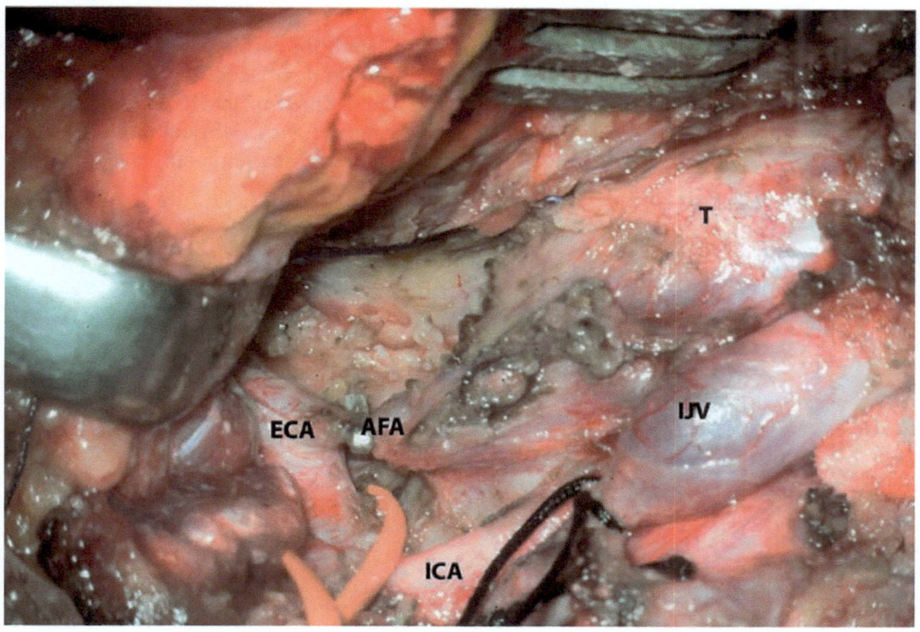

Fig. 12.173 A surgical loop has been placed around the external carotid artery. The ascending pharyngeal artery has been closed with a vascular clip. *AFA*, ascending pharyngeal artery; *ECA*, external carotid artery; *ICA*, internal carotid artery; *IJV*, internal jugular vein (ligated); *T*, tumor.

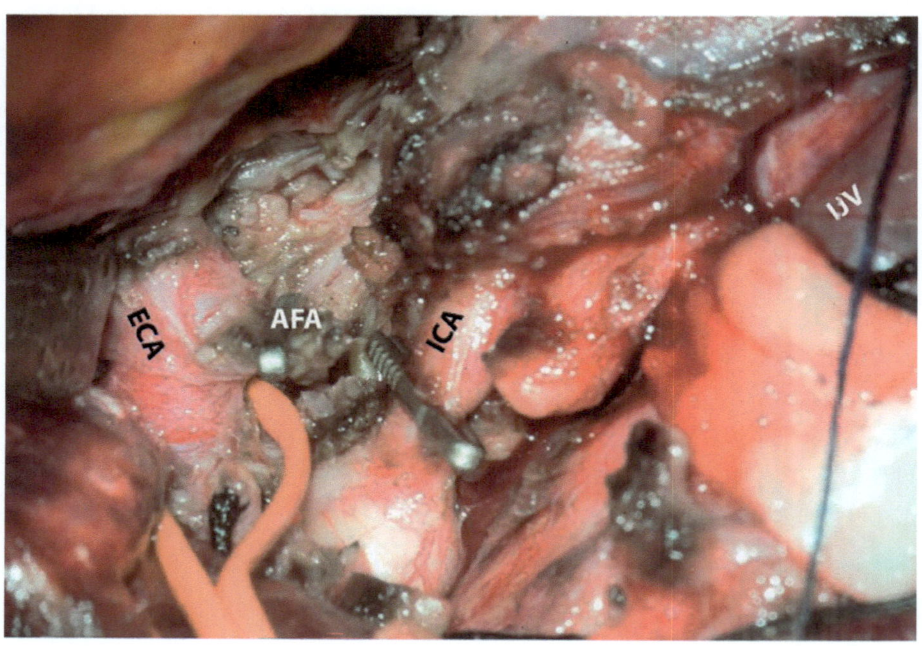

Fig. 12.174 A vascular clip has been used to close the balloon-occluded ICA. *AFA*, ascending pharyngeal artery; *ECA*, external carotid artery; *ICA*, internal carotid artery; *IJV*, internal jugular vein.

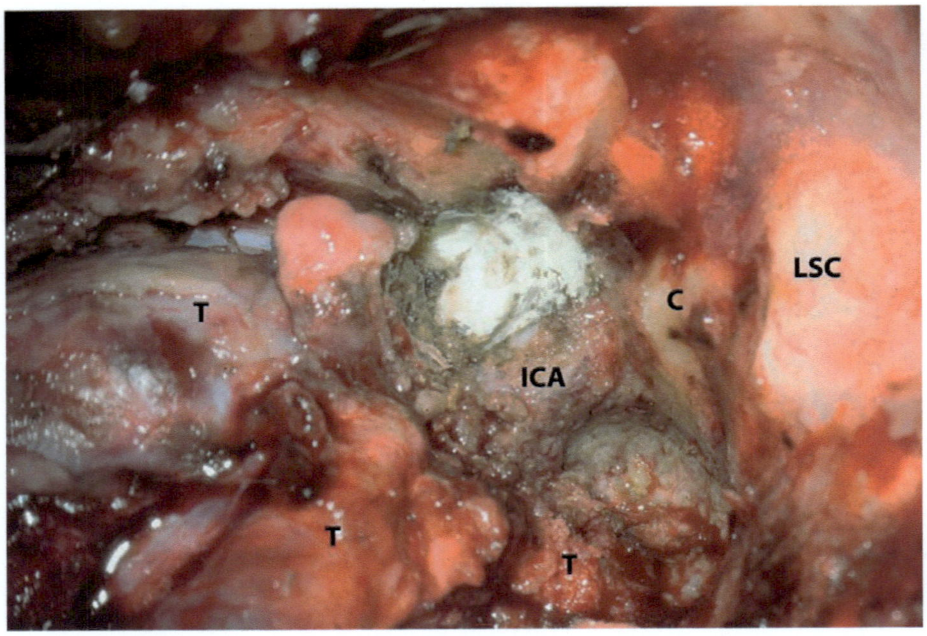

Fig. 12.175 The extent of the vagal paraganglioma anteromedial to the ICA can be appreciated. The periosteal sheath of the ICA is infiltrated by the tumor. *C*, cochlea; *ICA*, internal carotid artery; *LSC*, lateral semicircular canal; *T*, tumor.

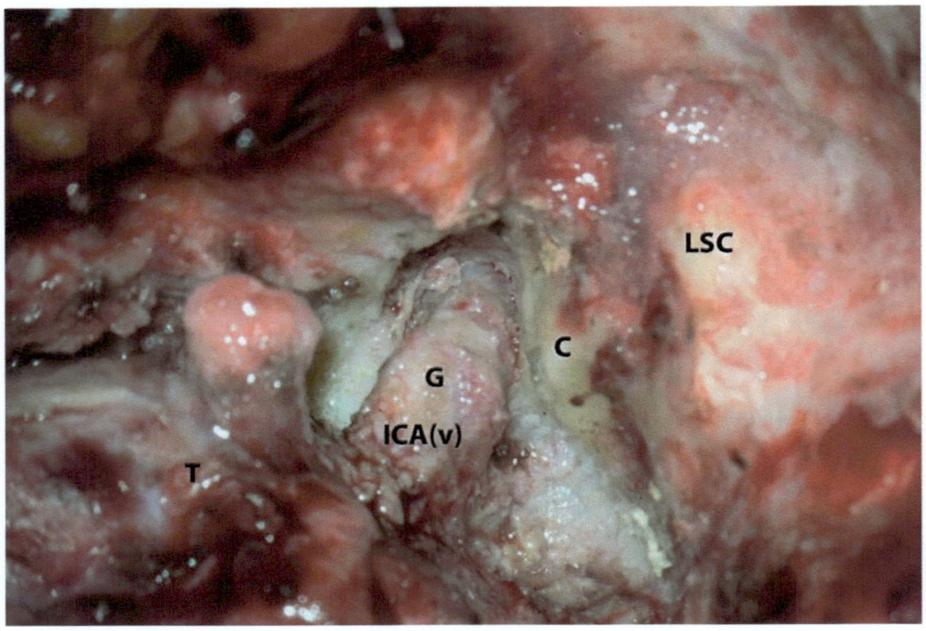

Fig. 12.176 Bone has been removed extensively around the horizontal segment of the ICA to the foramen lacerum. Note the relationship of the cochlea to the ICA. *C*, cochlea; *G*, carotid genu; *ICA(v)*, vertical segment of the ICA; *LSC*, lateral semicircular canal; *T*, tumor.

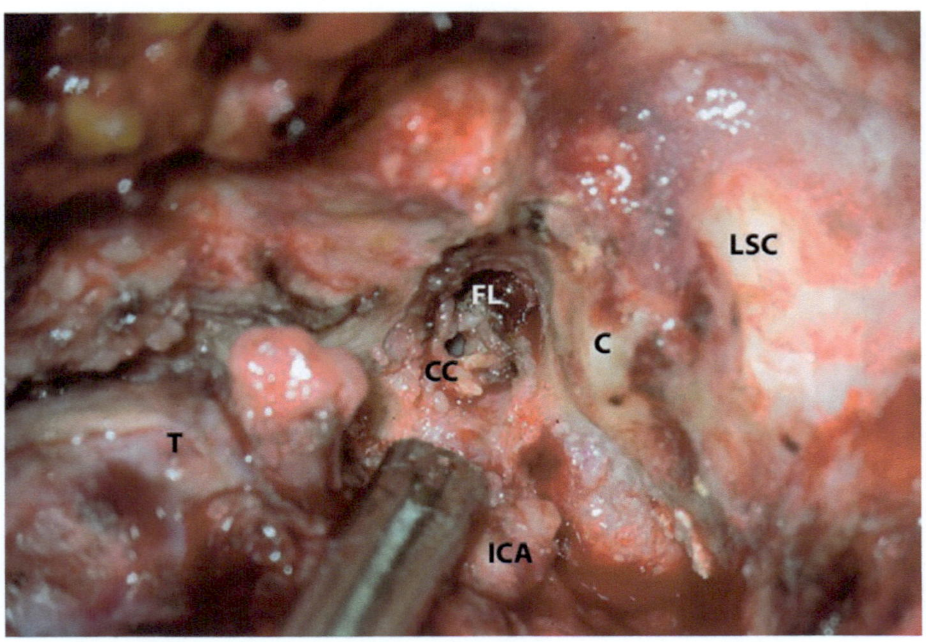

Fig. 12.177 The ICA has been transected at the level of the foramen lacerum, with the cell of the petrous apex now visible. The medial bony carotid canal is seen. The essentially bloodless dissection is a testament to the permanent balloon occlusion of the ICA. Superior control of the tumor has now been achieved. *C*, cochlea; *CC*, carotid canal; *FL*, foramen lacerum; *ICA*, internal carotid artery; *LSC*, lateral semicircular canal; *T*, tumor.

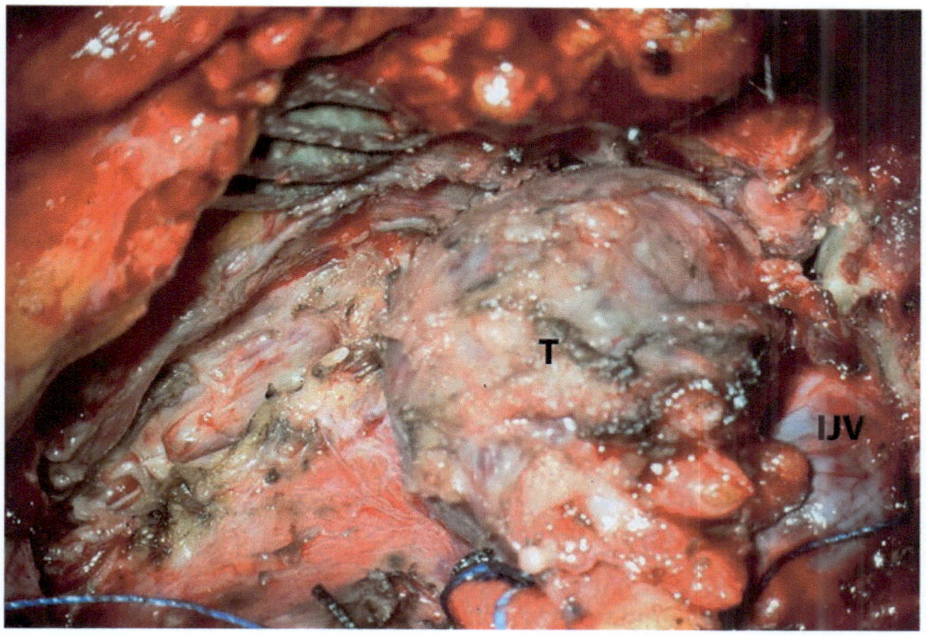

Fig. 12.178 The cervical component of the tumor (vagal paraganglioma) is now addressed. The ICA is within this large tumor mass. *IJV*, internal jugular vein; *T*, tumor.

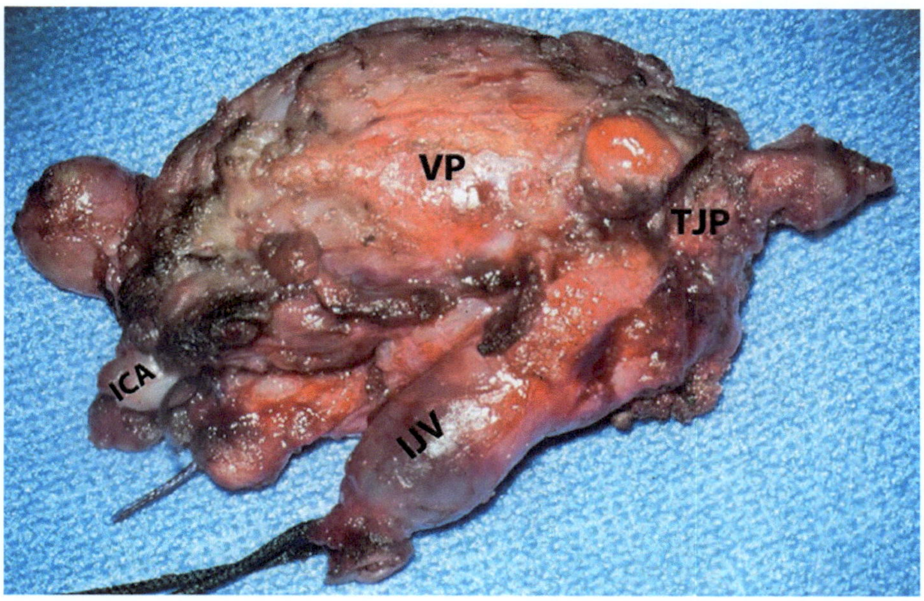

Fig. 12.179 The removed tumor, with IJV and ICA embedded. *ICA*, internal carotid artery; *IJV*, internal jugular vein; *VP*, vagal paraganglioma; *TJP*, tympanojugular paraganglioma.

Postoperative Imaging (Fig. 12.180)

There was additional spinal accessory nerve palsy, and facial cranial nerve palsy Grade III at 3 months postoperatively.

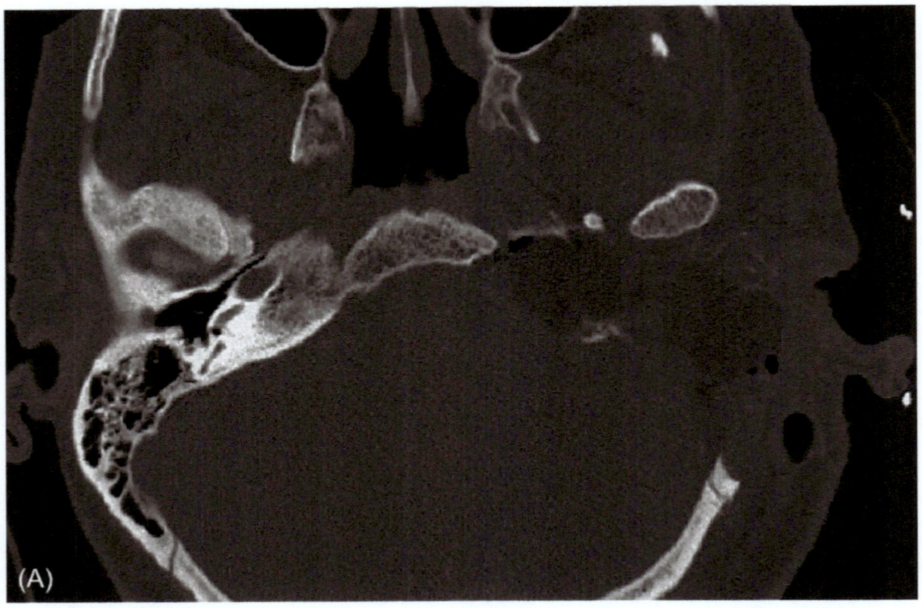

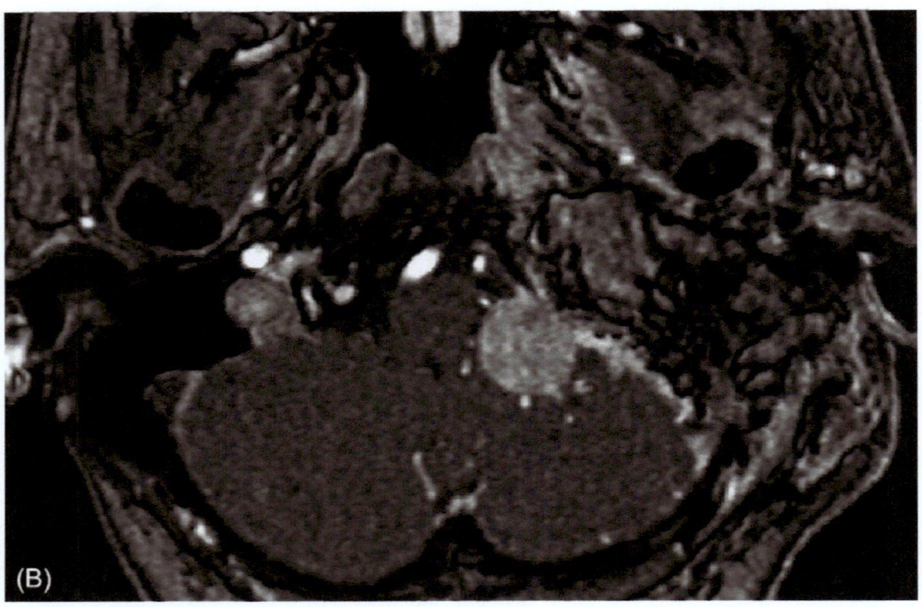

Fig. 12.180 (A) CT illustrating that the entire petrous apex surrounding the left ICA has been removed. (B) Enhanced axial T1W MRI revealing complete removal of the extradural component, with the residual intradural tumor and extensive dural enhancement seen.

Second Stage: Translabyrinthine and Transclival Approach

See Figs. 12.181–12.200.

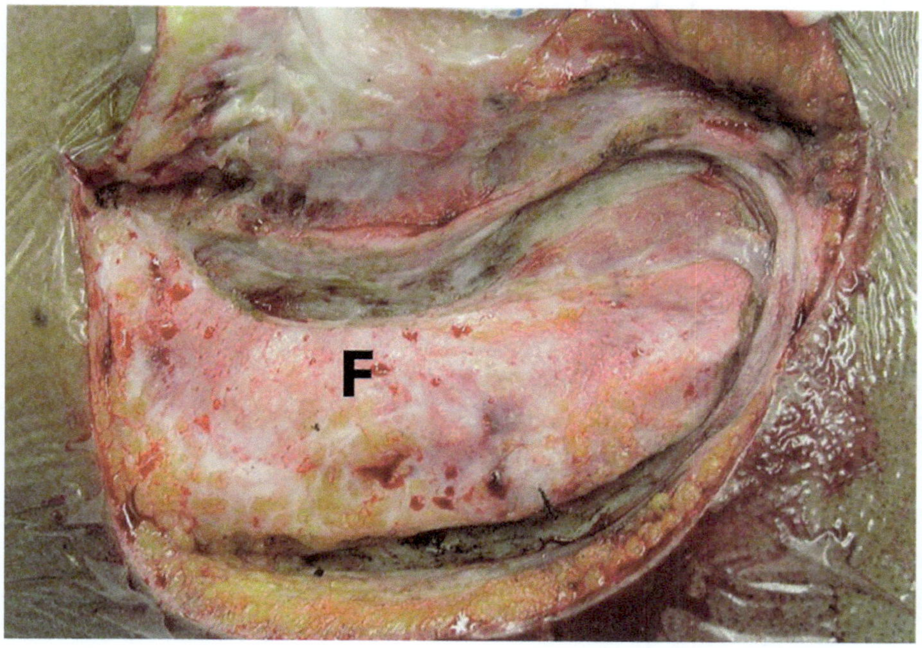

Fig. 12.181 A large postauricular C-shaped skin incision was made. An inferiorly based U-shaped musculoperiosteal flap (F) is used in this approach.

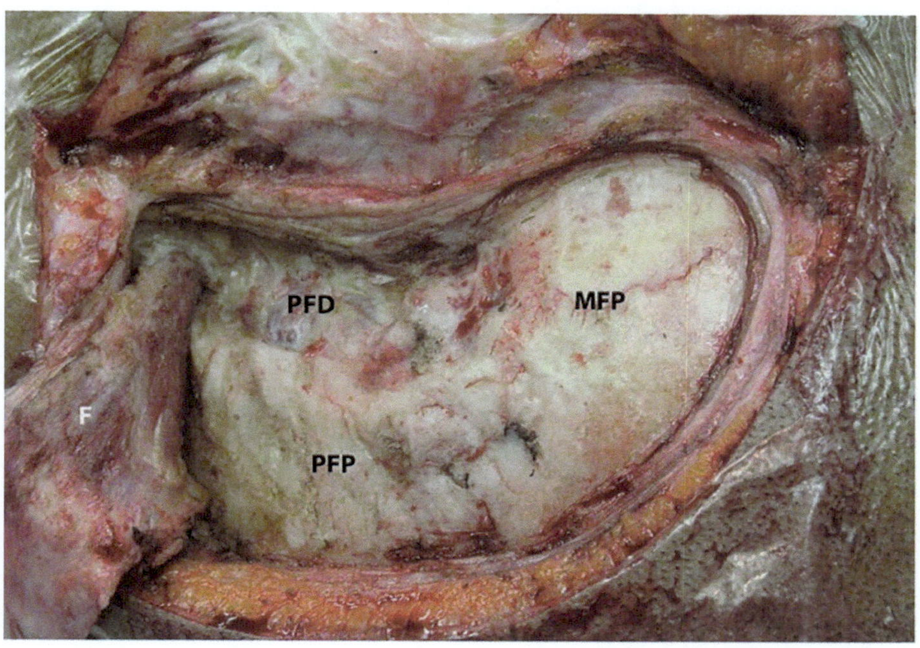

Fig. 12.182 The temporal bone has been exposed. The previously dissected mastoid cavity is seen. *MFP*, middle cranial fossa dural plate; *PFD*, posterior fossa dura; *PFP*, posterior fossa plate.

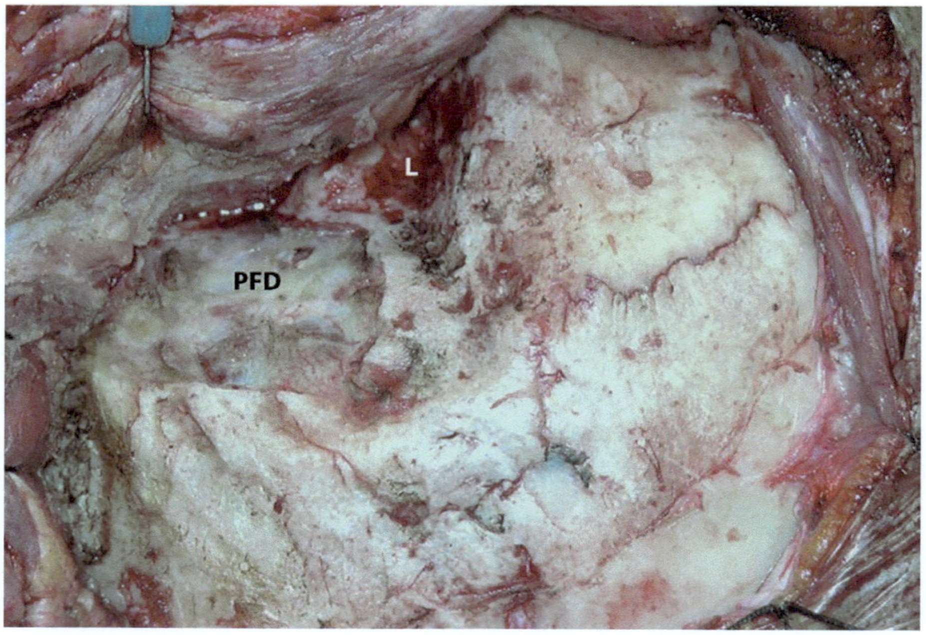

Fig. 12.183 Care must be taken to avoid injury to the anteriorly rerouted facial nerve during the soft tissue approach. The labyrinth is seen in the mastoid cavity, providing a ready landmark. *L*, labyrinth; *PFD*, posterior fossa dura.

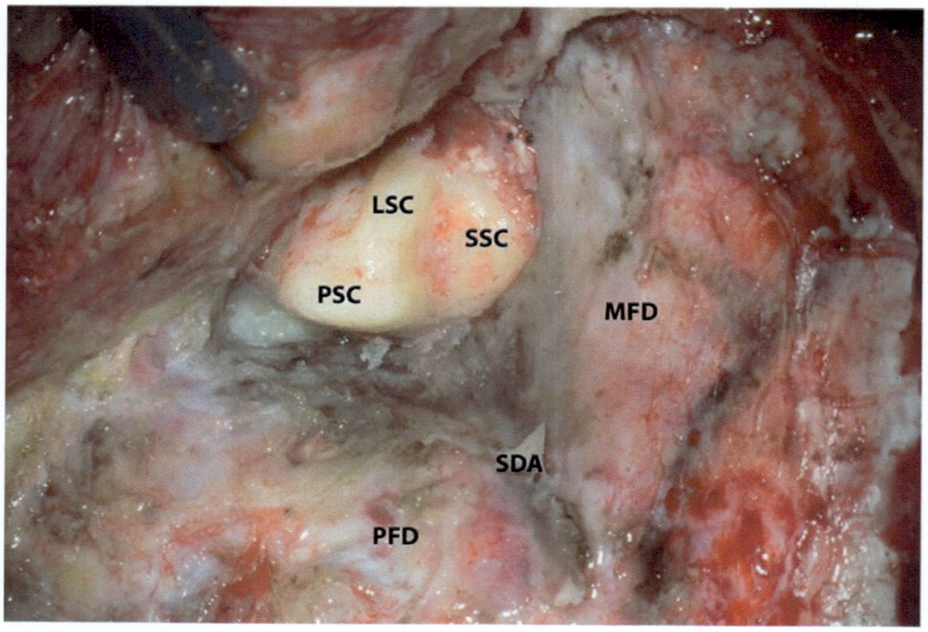

Fig. 12.184 Middle and posterior fossa dura have been uncovered, with the labyrinthine block better seen. Note that the sigmoid sinus was sacrificed at the previous surgery. *LSC*, lateral semicircular canal; *MFD*, middle fossa dura; *PFD*, posterior fossa dura; *PSC*, posterior semicircular canal; *SDA*, sinodural angle; *SSC*, superior semicircular canal.

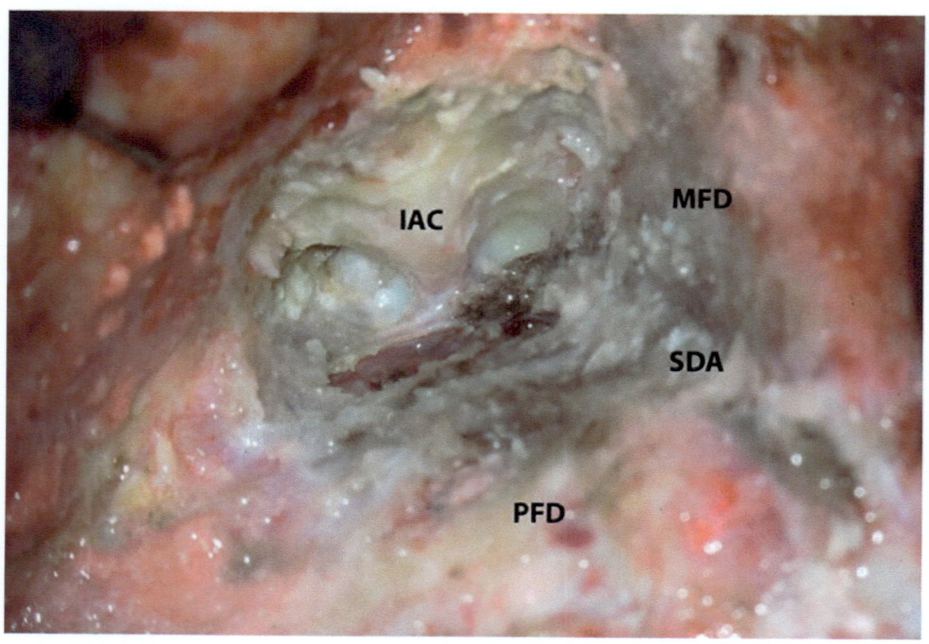

Fig. 12.185 The labyrinthectomy has been completed and the IAC skeletonized. *IAC*, internal auditory canal; *MFD*, middle fossa dura; *PFD*, posterior fossa dura; *SDA*, sinodural angle.

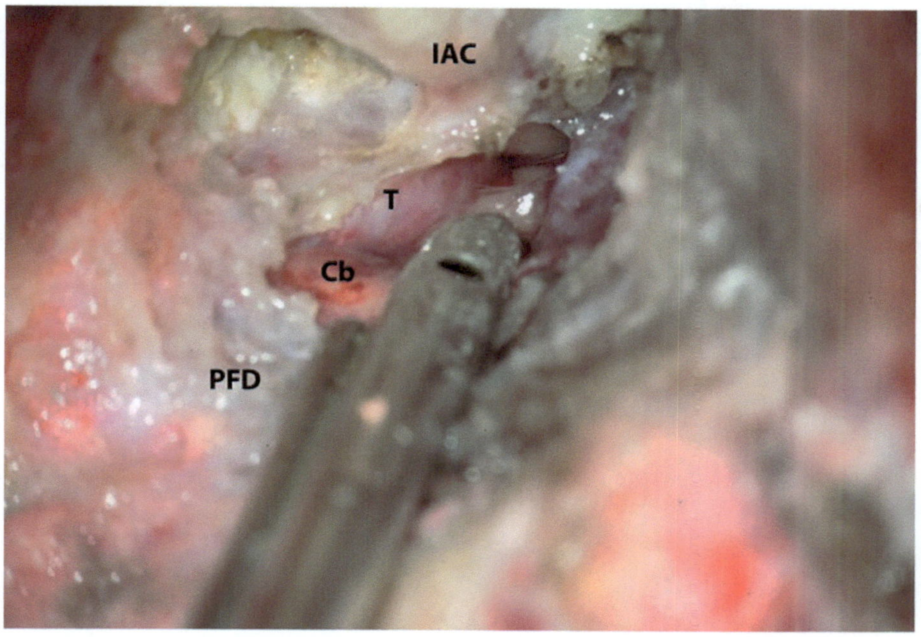

Fig. 12.186 The posterior fossa dura lateral to the porus acusticus was opened. The tumor was seen extending to the acousticofacial bundle. *Cb*, cerebellum; *IAC*, internal auditory canal; *PFD*, posterior fossa dura; *T*, tumor.

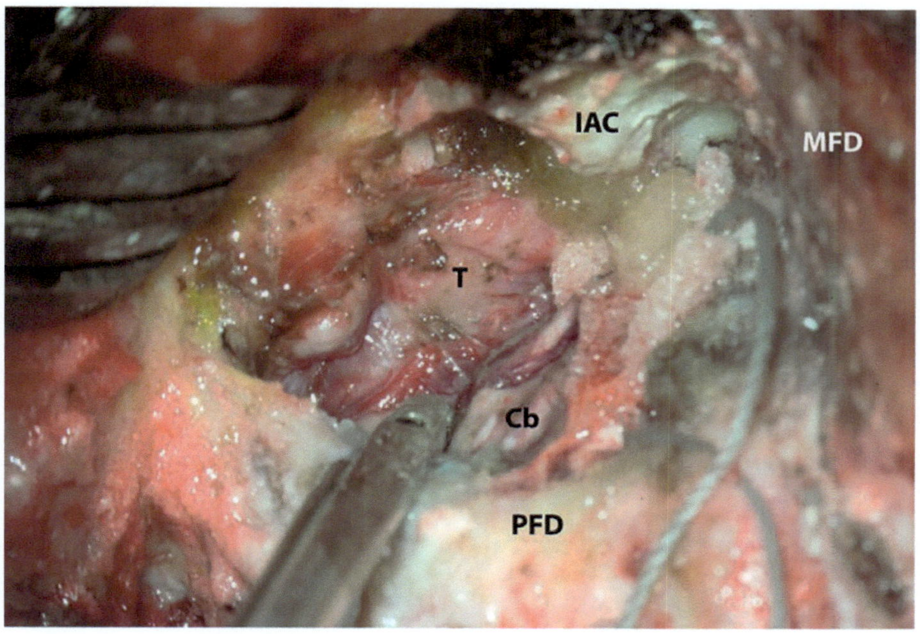

Fig. 12.187 The wide dural resection has been performed, with exposure to the large intradural tumor. *Cb*, cerebellum; *IAC*, internal auditory canal; *MFD*, middle fossa dura; *PFD*, posterior fossa dura; *T*, tumor.

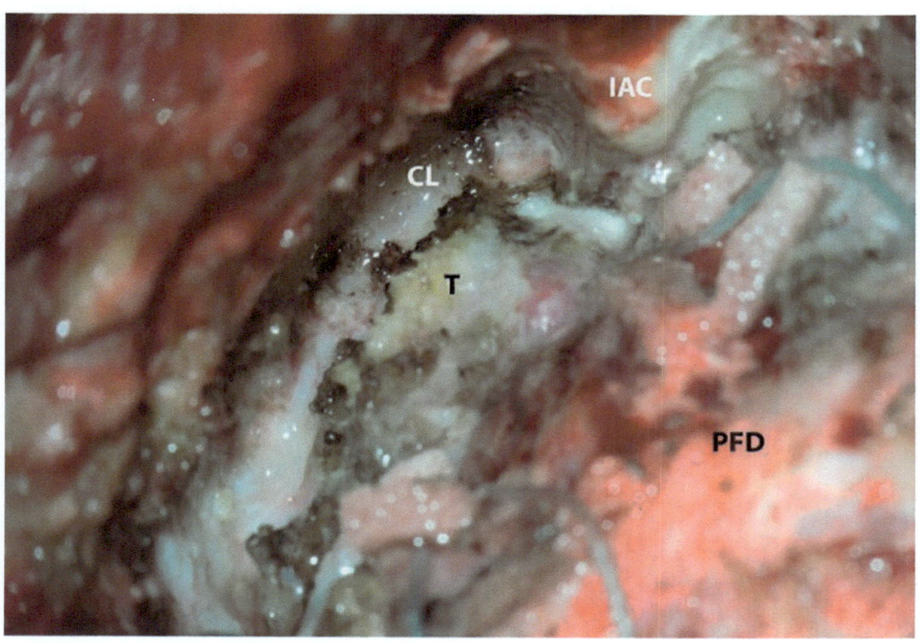

Fig. 12.188 The clivus anterior to the tumor was drilled until the normal cancellous bone had been exposed. Merocel patties were introduced between the tumor and the cerebellum. *CL*, clivus; *IAC*, internal auditory canal; *PFD*, posterior fossa dura; *T*, tumor.

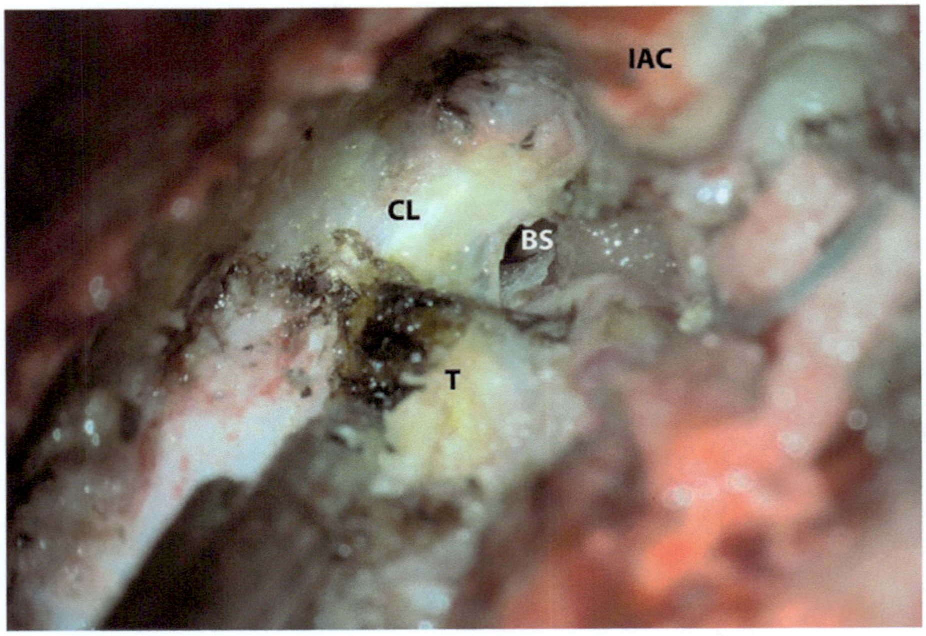

Fig. 12.189 The anterior margin of the tumor has been exposed and the brainstem is visible through the arachnoid membrane. *BS*, brainstem; *CL*, clivus; *IAC*, internal auditory canal; *T*, tumor.

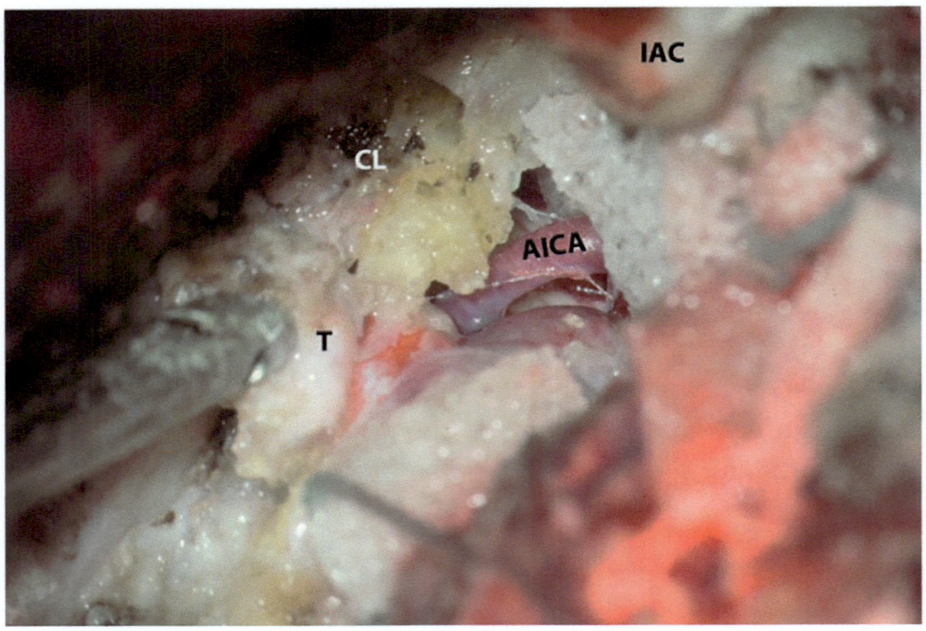

Fig. 12.190 During the debulking of the tumor, the anterior inferior cerebellar artery was exposed immediately beneath the tumor. Merocel superiorly is protecting the VII–VIII complex. *AICA*, anterior inferior cerebellar artery; *CL*, clivus; *IAC*, internal auditory canal; *T*, tumor.

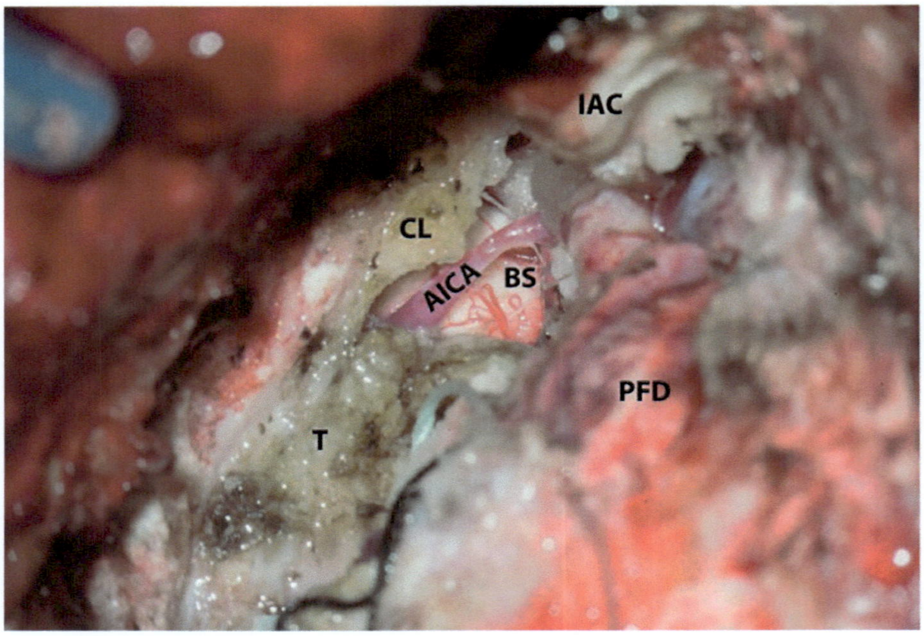

Fig. 12.191 Removal of tumor has been done from a superior direction. Note the gradual dissection of tumor from the anterior inferior cerebellar artery. *AICA*, anterior inferior cerebellar artery; *BS*, brainstem; *CL*, clivus; *IAC*, internal auditory canal; *PFD*, posterior fossa dura; *T*, tumor.

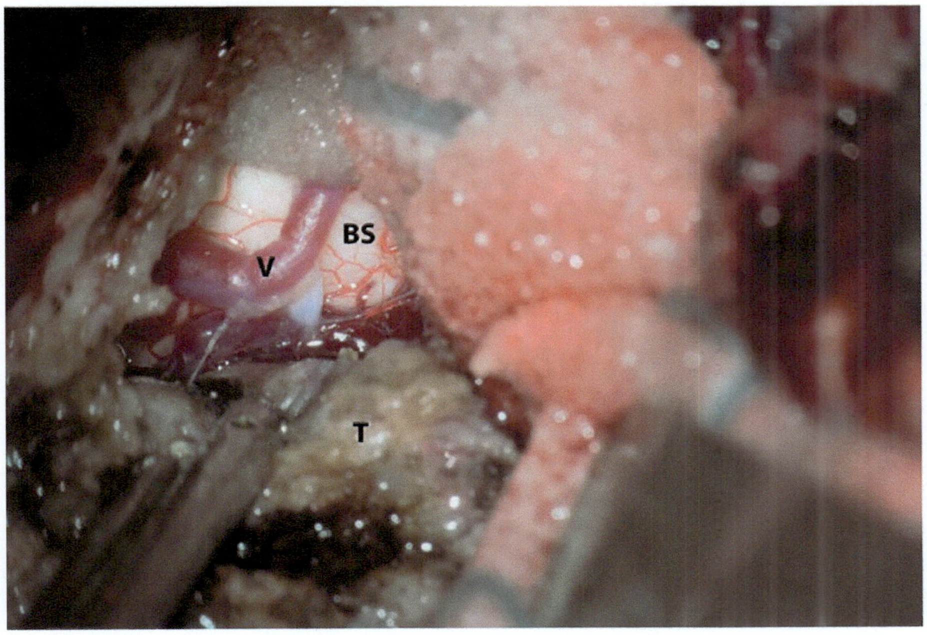

Fig. 12.192 The venous plexus on the brainstem is now seen. *BS*, brainstem; *T*, tumor; *V*, vein.

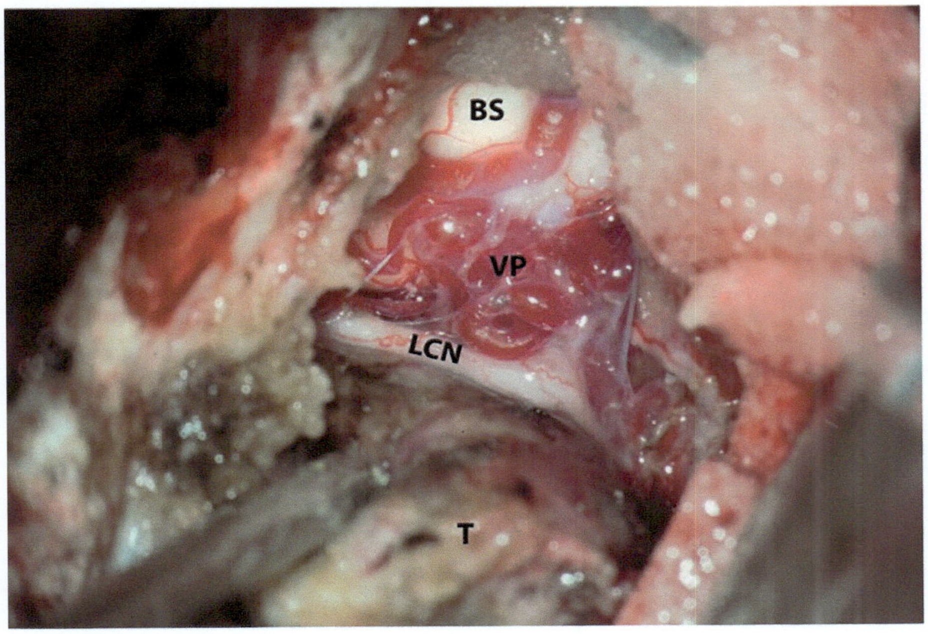

Fig. 12.193 The debulked tumor is gradually peeled from the brainstem and vasculature. The extensive vascularization can be appreciated. *BS*, brainstem; *LCN*, cranial nerve IX; *T*, tumor; *VP*, venous plexus of the tumor.

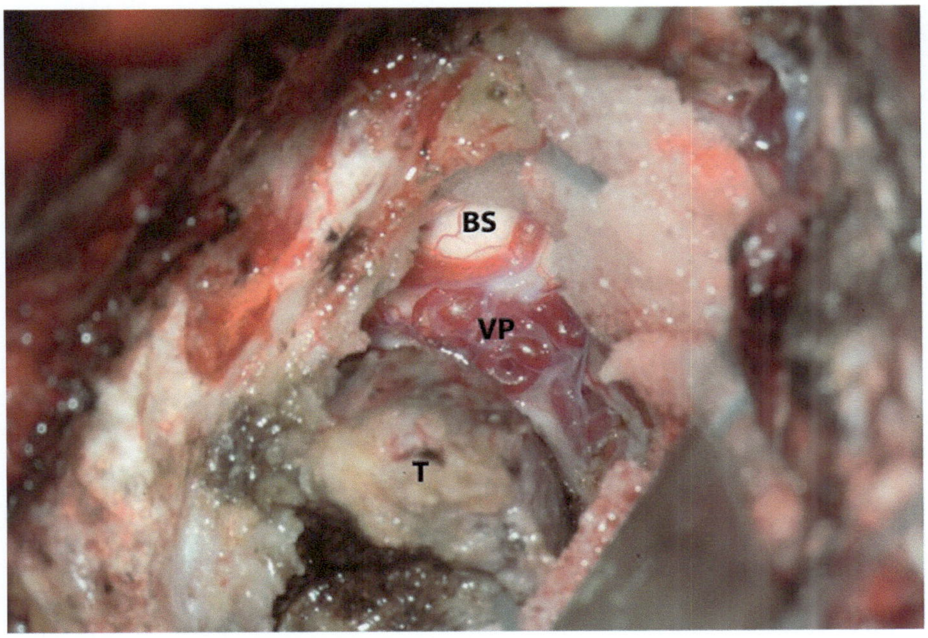

Fig. 12.194 A wider view of the surgical cavity. *BS*, brain stem; *T*, tumor; *VP*, venous plexus.

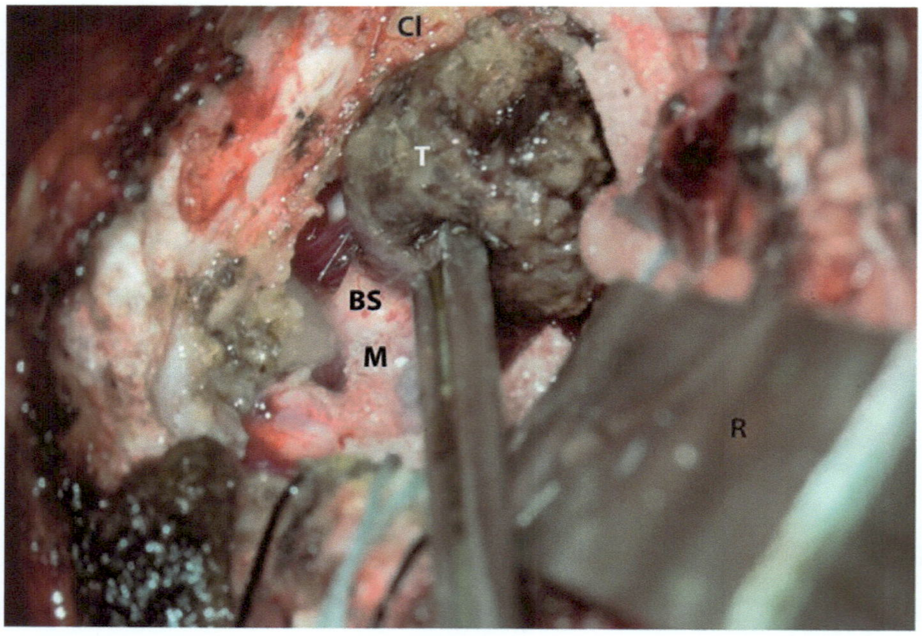

Fig. 12.195 The inferior pole of the tumor is now approached. Note the extensive devascularization of the tumor by surface bipolar diathermy. A Merocel patty is introduced between the tumor and the brainstem. *BS*, brainstem; *Cl*, clivus; *M*, Merocel; *R*, cerebellar retractor; *T*, tumor.

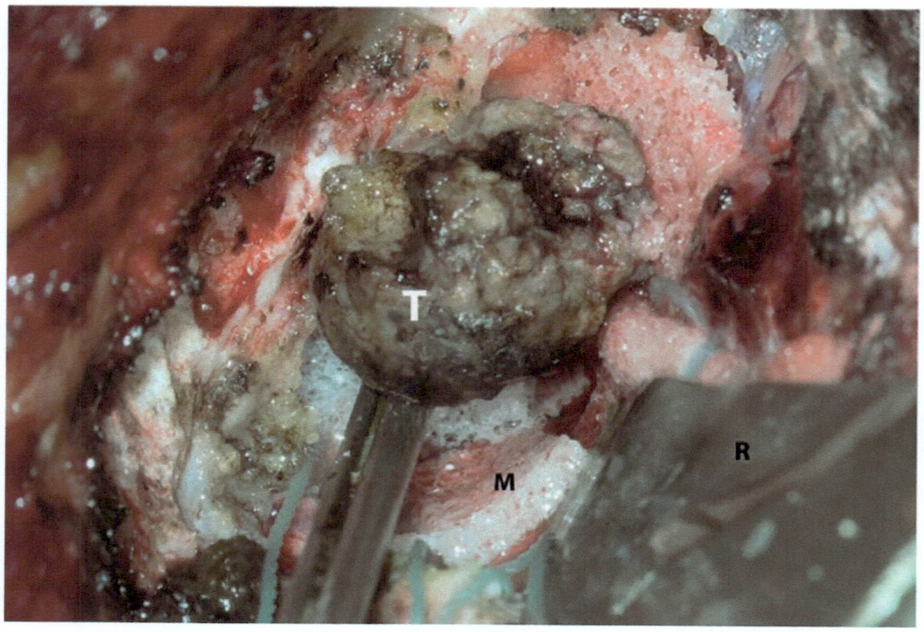

Fig. 12.196 The tumor (T) is now almost completely mobilized, with Merocel (M) seen covering the surface of the brainstem. *R*, cerebellar retractor; *M*, Merocel; *T*, tumor.

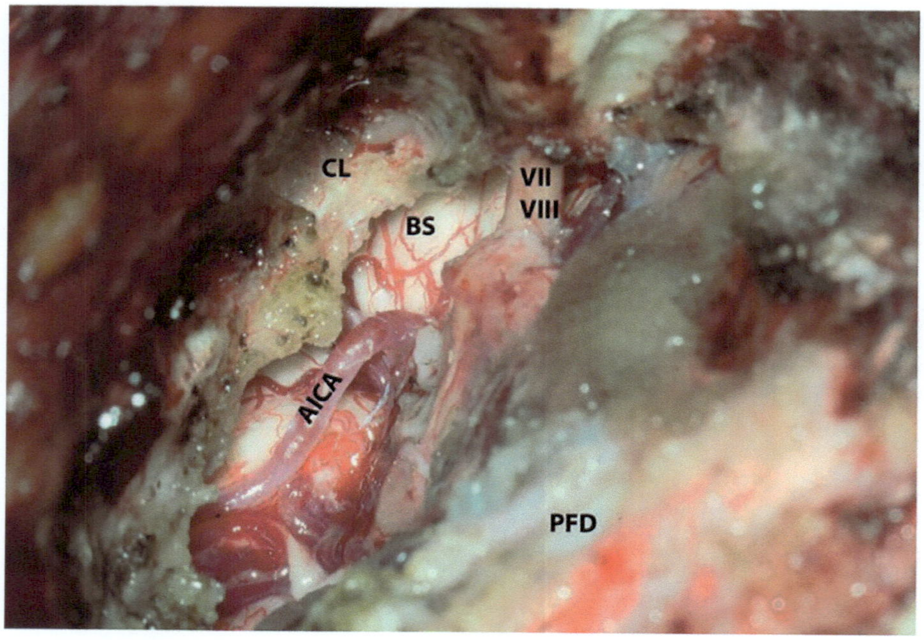

Fig. 12.197 The tumor has now been completely removed, with the VII–VIII complex seen superiorly. AICA anterior inferior cerebellar artery. *BS*, brainstem; *CL*, clivus; *PFD*, posterior fossa dura; *VII*, facial nerve; *VIII*, cochleovestibular nerve.

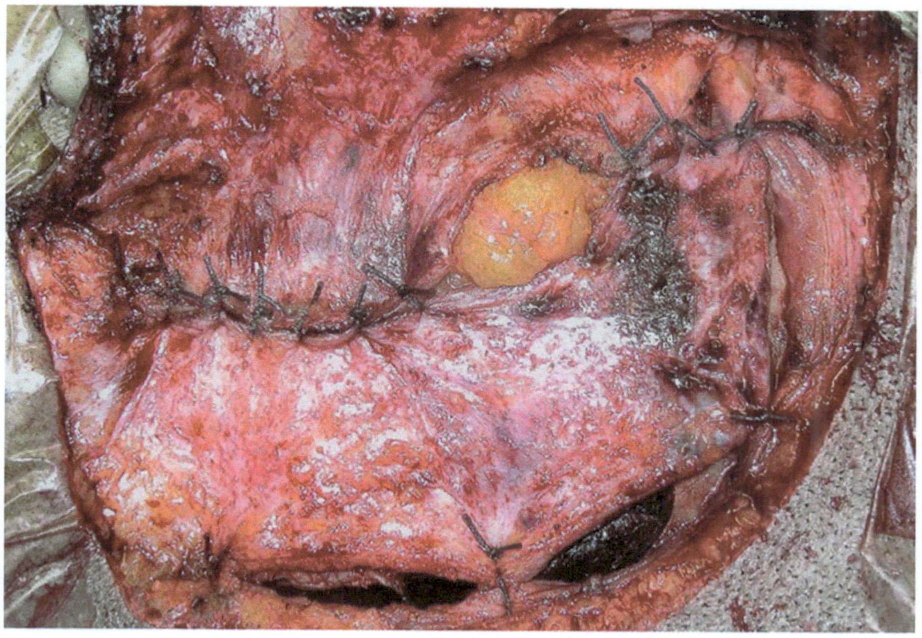

Fig. 12.198 The musculoperiostial flap is being closed, with abdominal fat having been placed through the dural defect into the intradural space. Complete watertight closure is ensured using this closure technique.

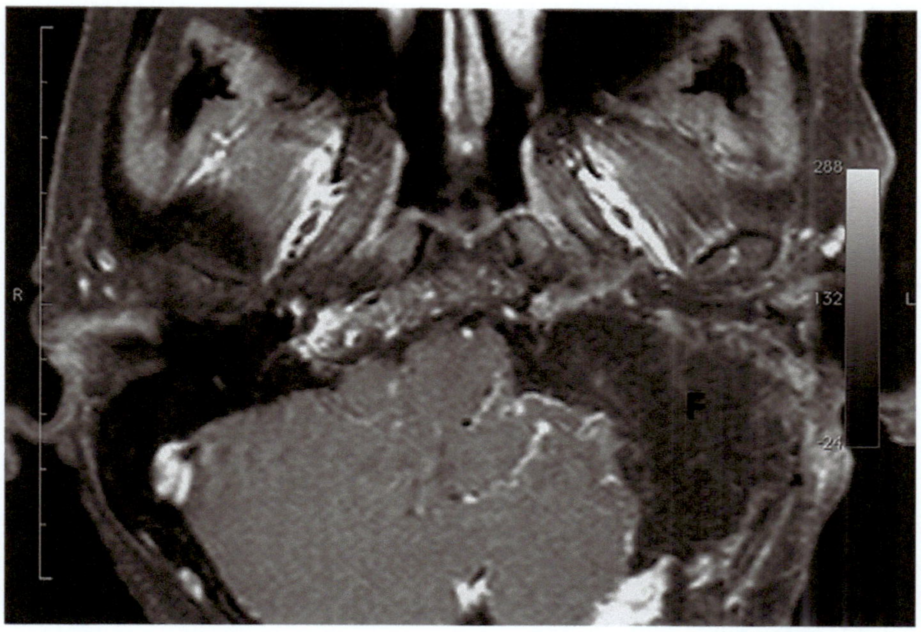

Fig. 12.199 Gd-enhanced MRI, axial view with fat suppression, after surgery shows that there is no residual tumor. *F*, abdominal fat.

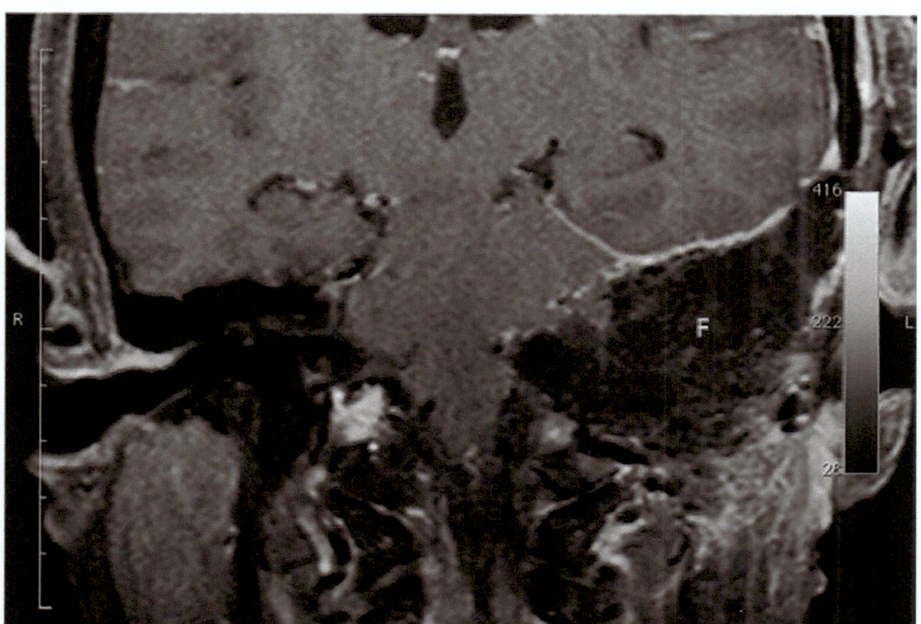

Fig. 12.200 MRI, coronal view. Total tumor removal is completed. *F*, abdominal fat.

CASE 4: REMOVAL OF A COMPLEX TYMPANOJUGULAR PARAGANGLIOMA CLASS C3Di2Ve

See Figs. 12.201–12.206.

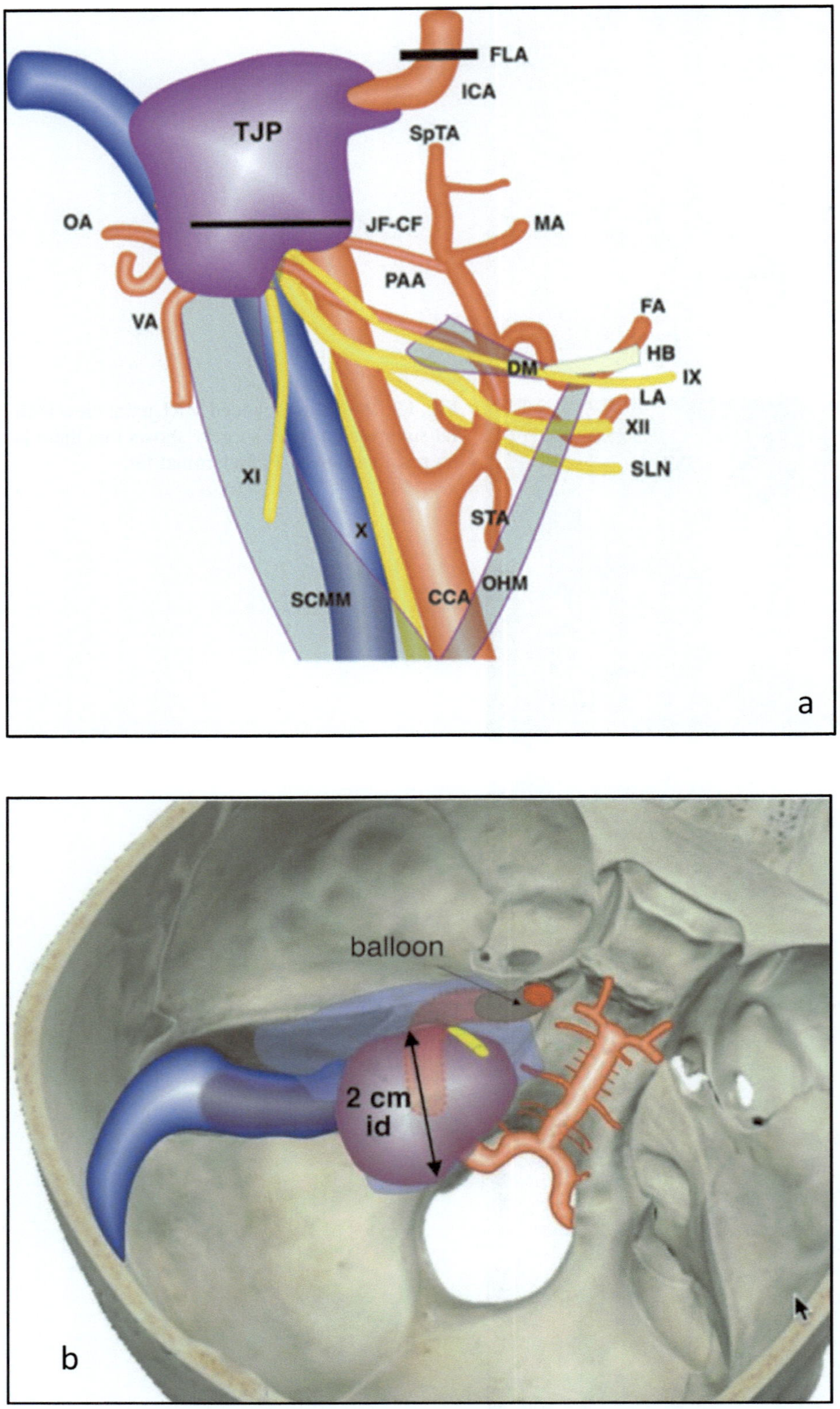

Fig. 12.201 This 53-year-old man presented with right total hearing loss and lower cranial nerve palsy. He had undergone three previous paraganglioma surgeries with partial resection. At the time of our consultation, he presented with a C3Di2Ve tympanojugular paraganglioma. Staged resection was planned. (A) Drawing showing the extension of the tumor toward the horizontal petrous internal carotid artery and the vertebral artery. (B) Drawing showing the extension of the tumor in the cerebellopontine angle.

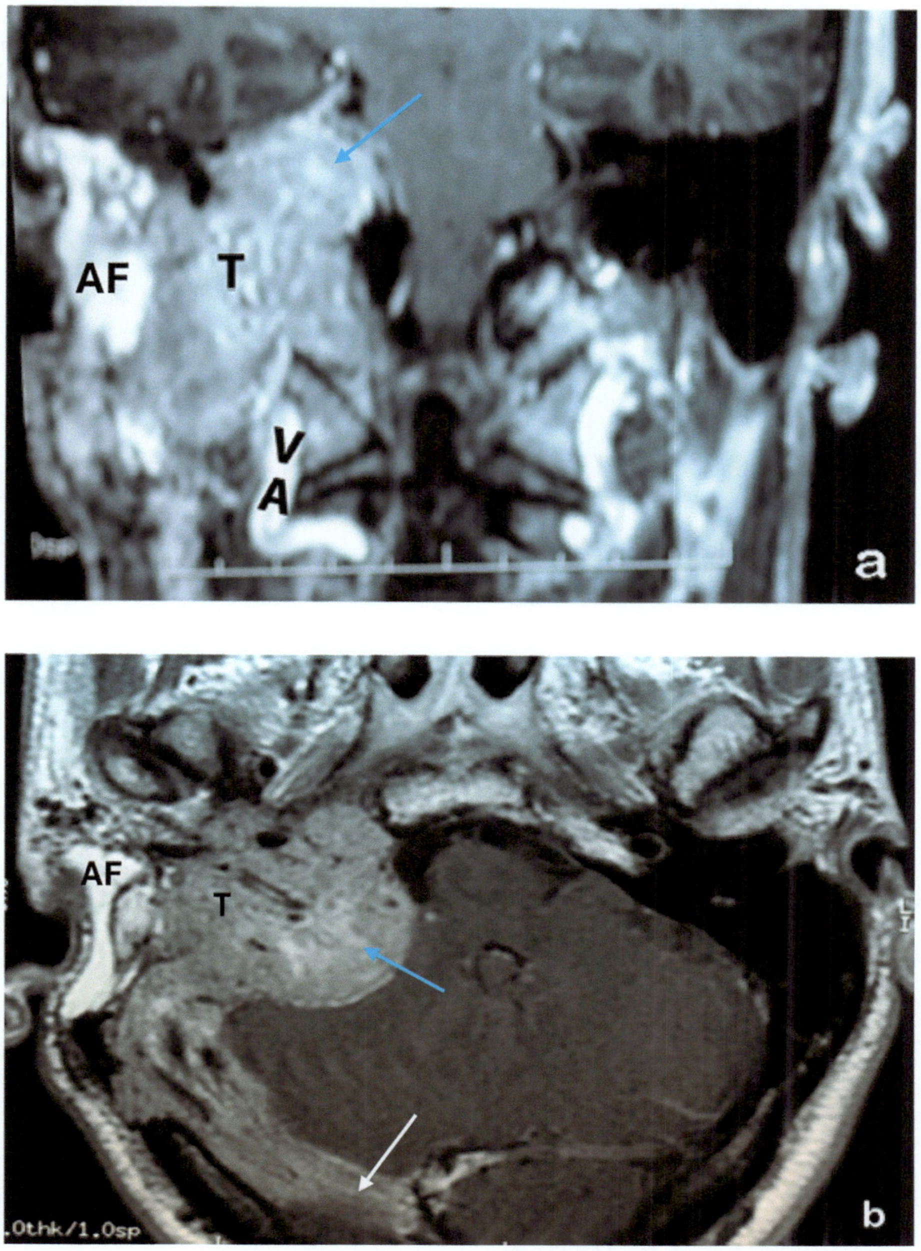

Fig. 12.202 Preoperative T1W+Gd-enhanced MRI. (A) Coronal view. Note the extension of the tumor (T) toward the vertebral artery (VA). (B) Axial view. Note the difference between the tumor (T) and the abdominal fat (AF) previously used to obliterate the surgical cavity. The tumor extends intradurally *(light blue arrow)* and involves the entire sigmoid sinus up to the torcular *(white arrow)*.

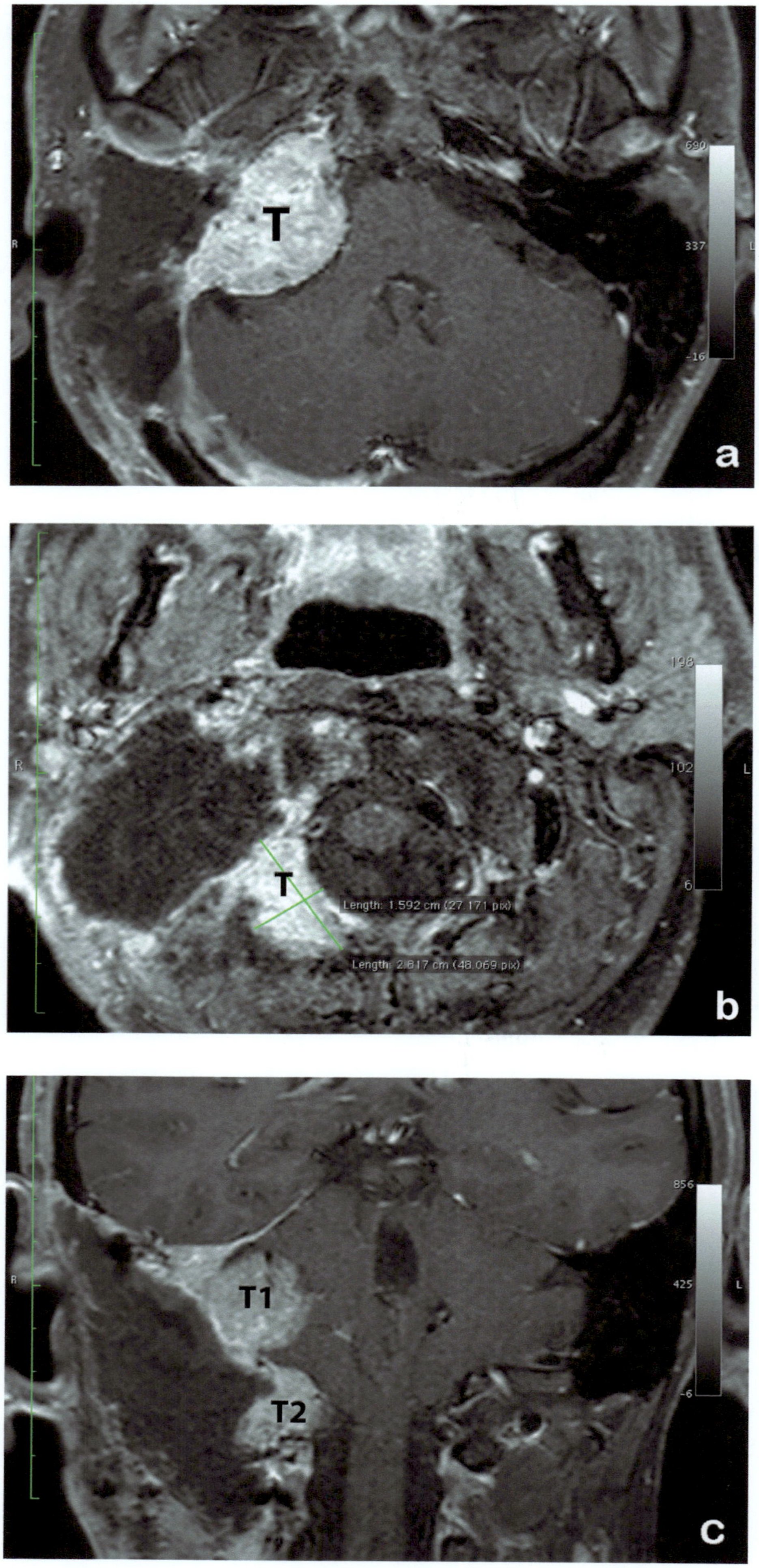

Fig. 12.203 Postoperative T1W+Gd-enhanced MRI after a first-stage IFTA approach. Both the internal carotid and vertebral arteries were balloon-occluded preoperatively. (A) Axial view. Note the tumor (T) left in the intradural compartment for the second stage. (B) Axial view. Note the tumor left in the area of the vertebral artery, close to the foramen magnum. (C) Coronal view. Note the two residual tumor components. *T1*, tumor in the cerebellopontine angle; *T2*, tumor in the vertebral artery.

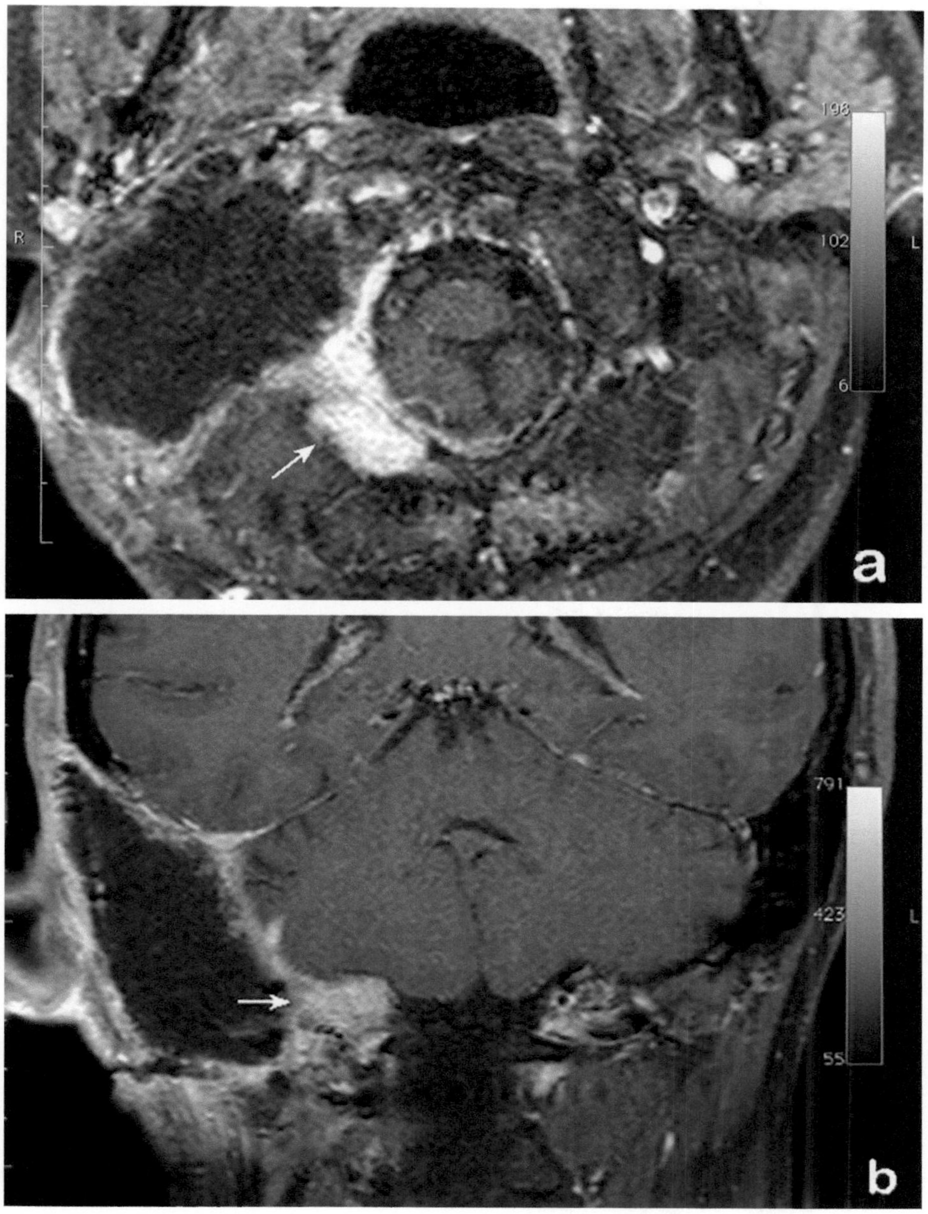

Fig. 12.204 Postoperative T1W + Gd-enhanced MRI after a second-stage IFTA approach. There is still some tumor in the area of the vertebral artery *(arrow)*. (A) Axial view. (B) Coronal view.

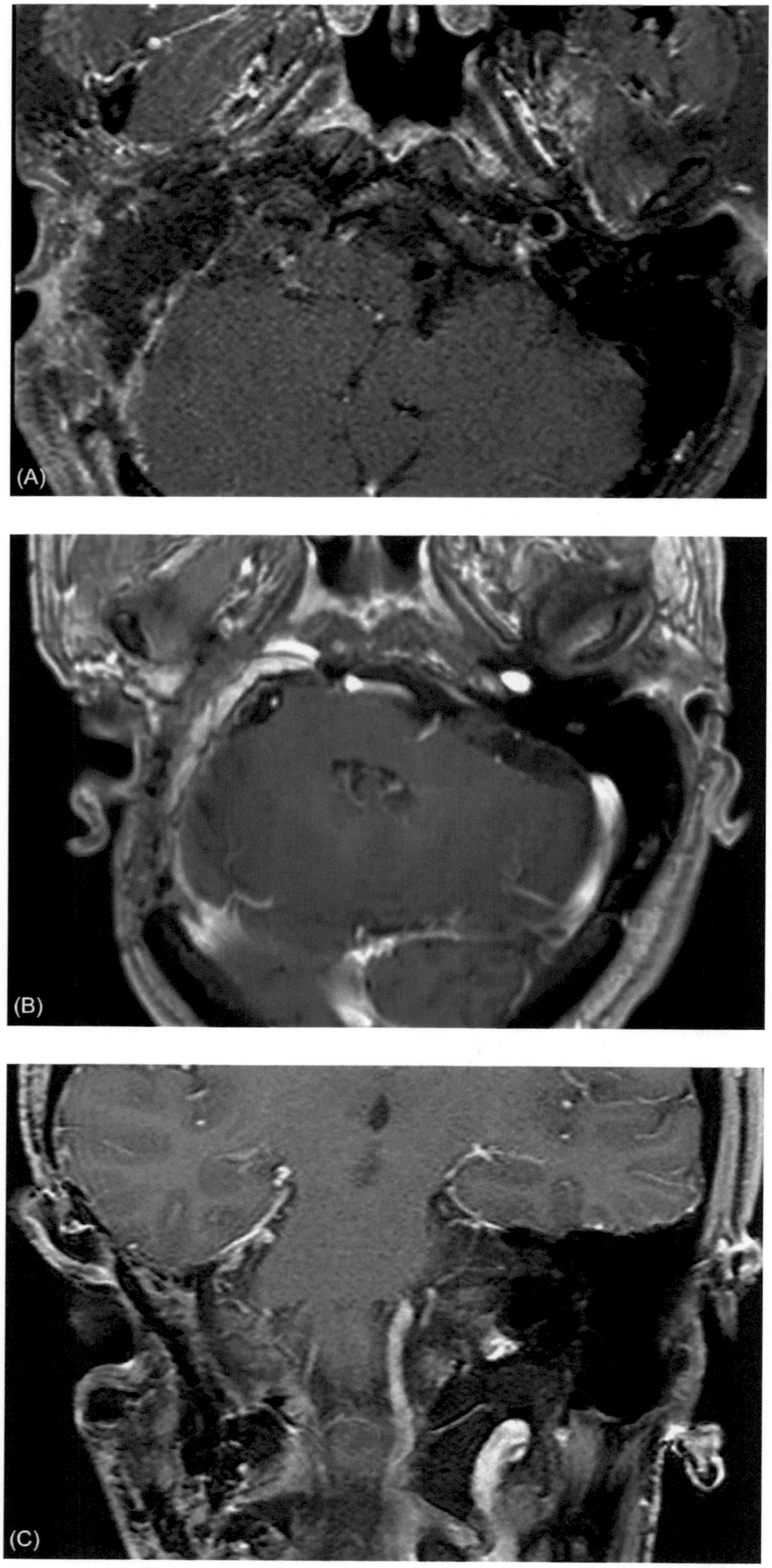

Fig. 12.205 Postoperative T1W+Gd-enhanced MRI after a third-stage surgery, performed via a far-lateral approach. Complete tumor removal was achieved. (A) Axial view (immediate postoperative period). (B) Axial view (after 10 years follow-up). Note the shrinkage of the abdominal fat filling the surgical cavity. (C) Coronal view (after 10 years follow-up).

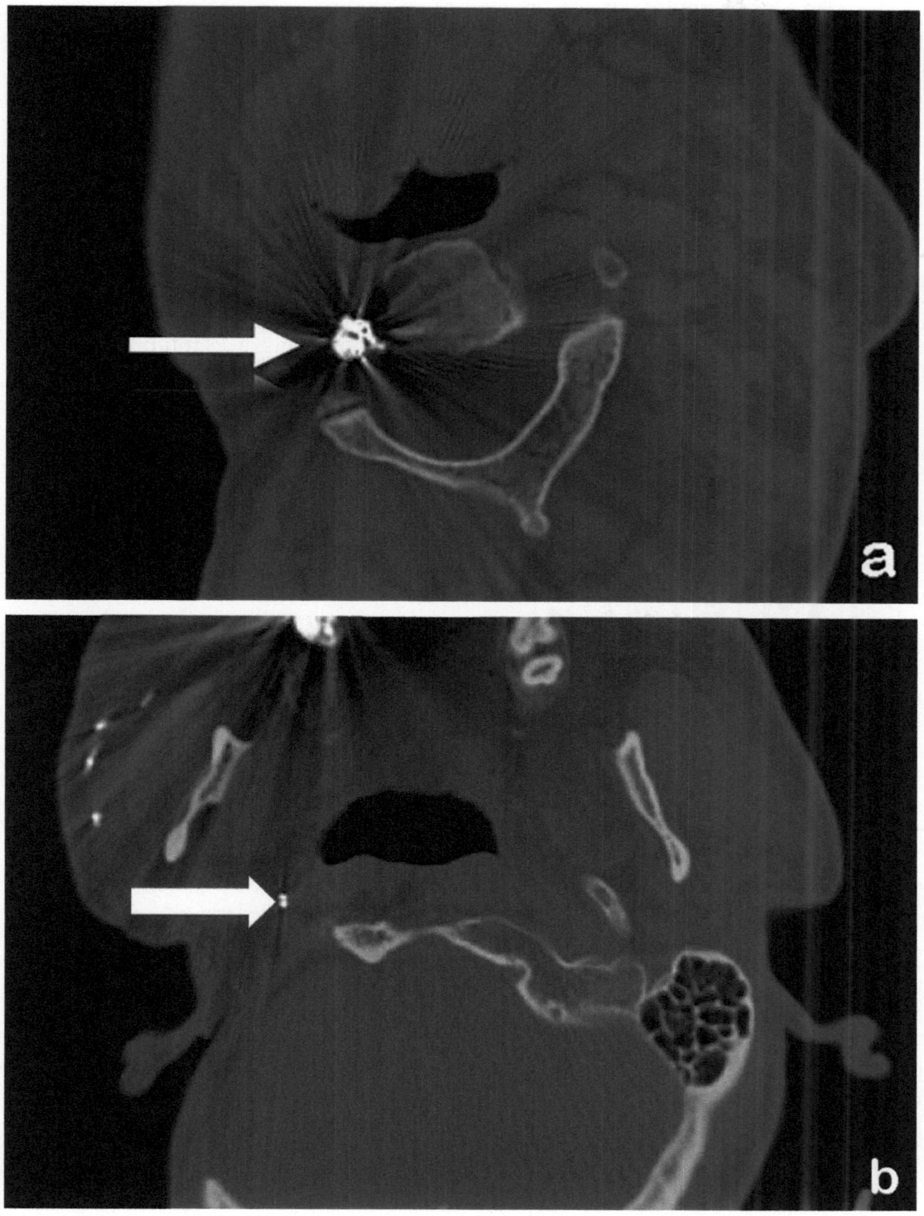

Fig. 12.206 Postoperative high-resolution CT scan, axial views. (A) Note the coils within the right vertebral artery *(arrow)*. (B) Note the coils within the right internal carotid artery *(arrow)*.

CASE 5: REMOVAL OF A COMPLEX TYMPANOJUGULAR PARAGANGLIOMA CLASS C3Di2Ve

See Figs. 12.207–12.210.

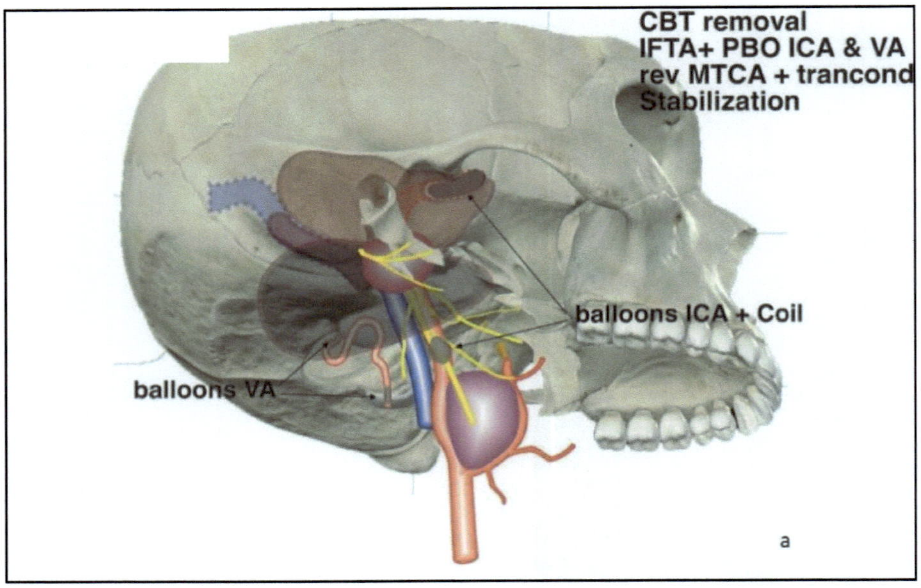

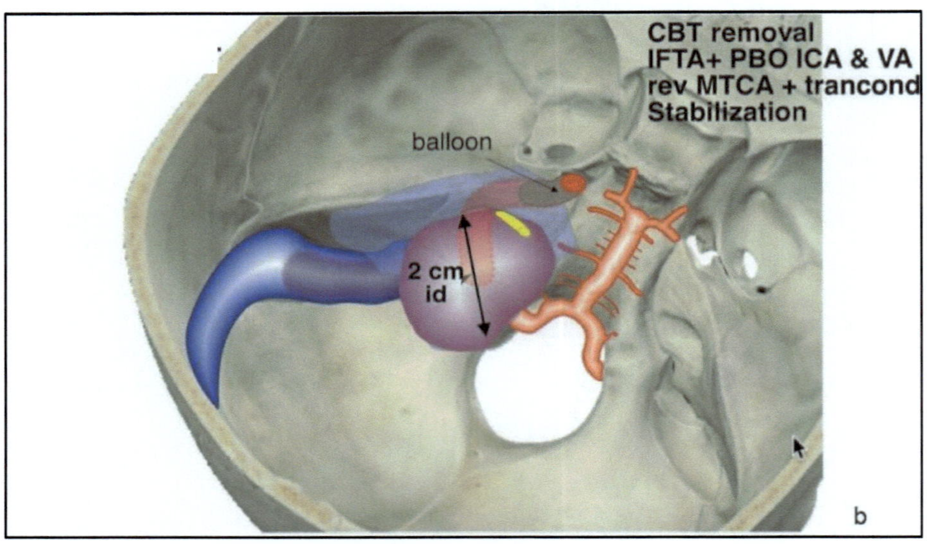

Fig. 12.207 This 51-year-old man presented with a history of radiosurgery and embolization for a right-sided C3Di1Ve tympanojugular paraganglioma (TJP) plus a carotid body tumor (CBT). Given the tumor growth after radiosurgery and the size, staged resection was planned. At the time of our consultation, the patient had paralysis of all the lower cranial nerves and anacusis on the tumor side. (A) Drawing showing the tumor extension (sagittal view). (B) Drawing showing intradural tumor extension (axial view).

(Continued)

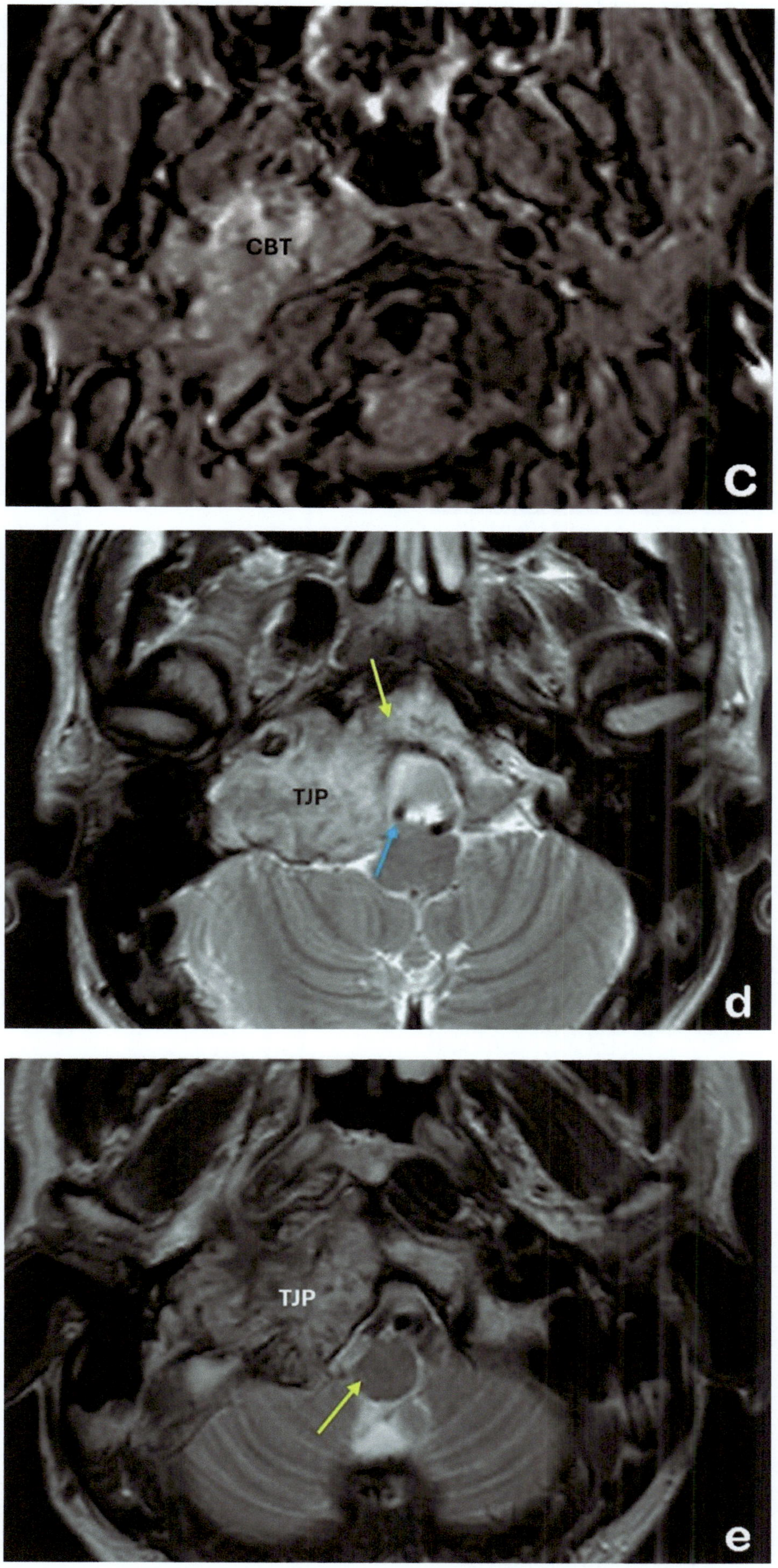

Fig. 12.207, cont'd (C) MRI scan (axial view) showing a class Shamblin 3 CBT extended into the parapharyngeal space. (D) Axial MRI scan showing a class C3 TJP extending into the clivus *(yellow arrow)* and vertebral artery *(light blue arrow)*. (E) Axial MRI scan showing extension of the TJP into the occipital condyle and toward the foramen magnum *(yellow arrow)*.

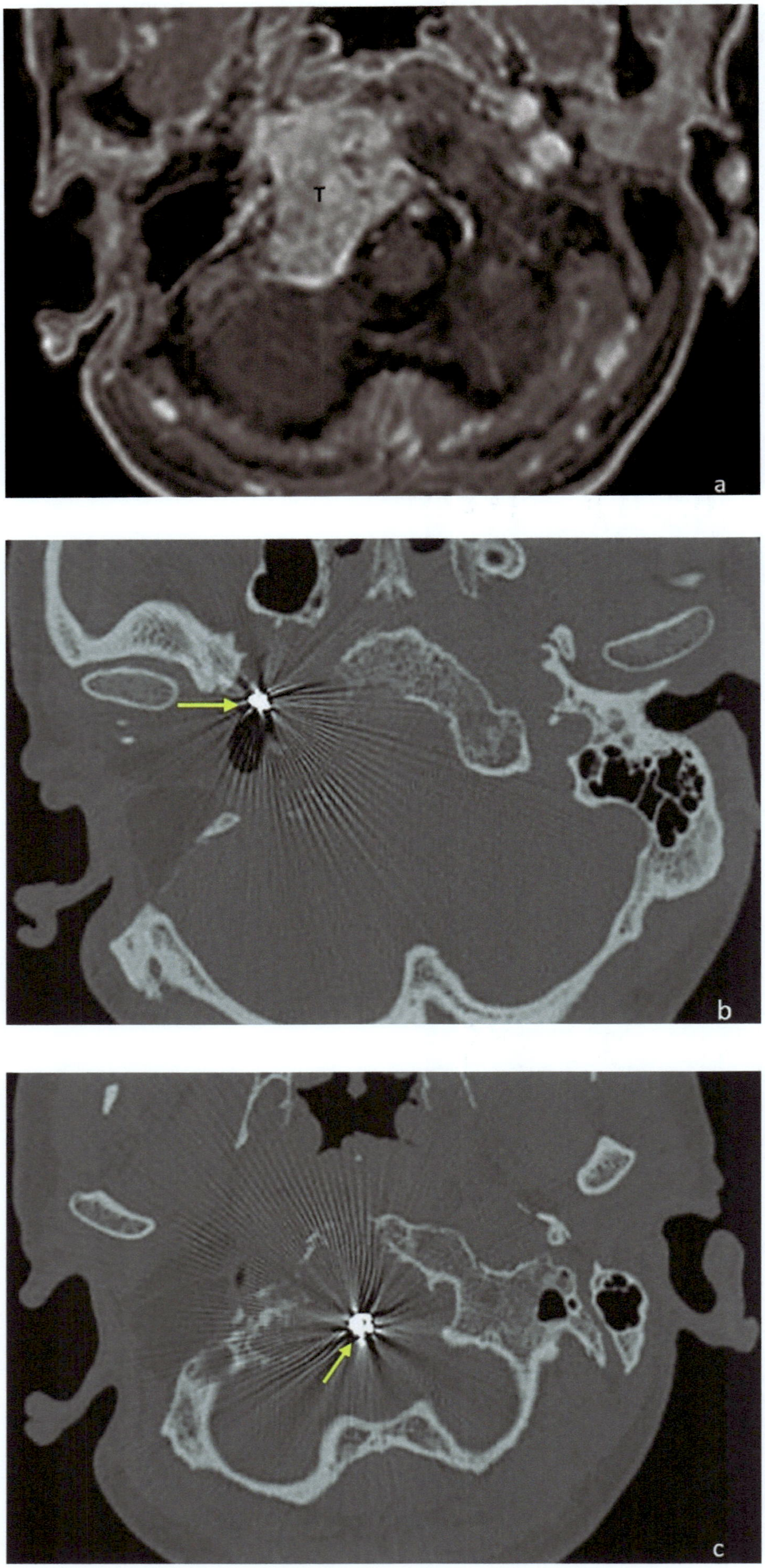

Fig. 12.208 The patient underwent a first-stage IFTA approach with complete resection of the CBT and the intratemporal TJP. Note the tumor remnant in the intradural and clival compartments (T). (A) Axial MRI showing the tumor. (B) Balloon and coil occlusion of the internal carotid artery was performed preoperatively *(arrow)*. (C) Balloon and coil occlusion of the vertebral artery was performed preoperatively *(arrow)*.

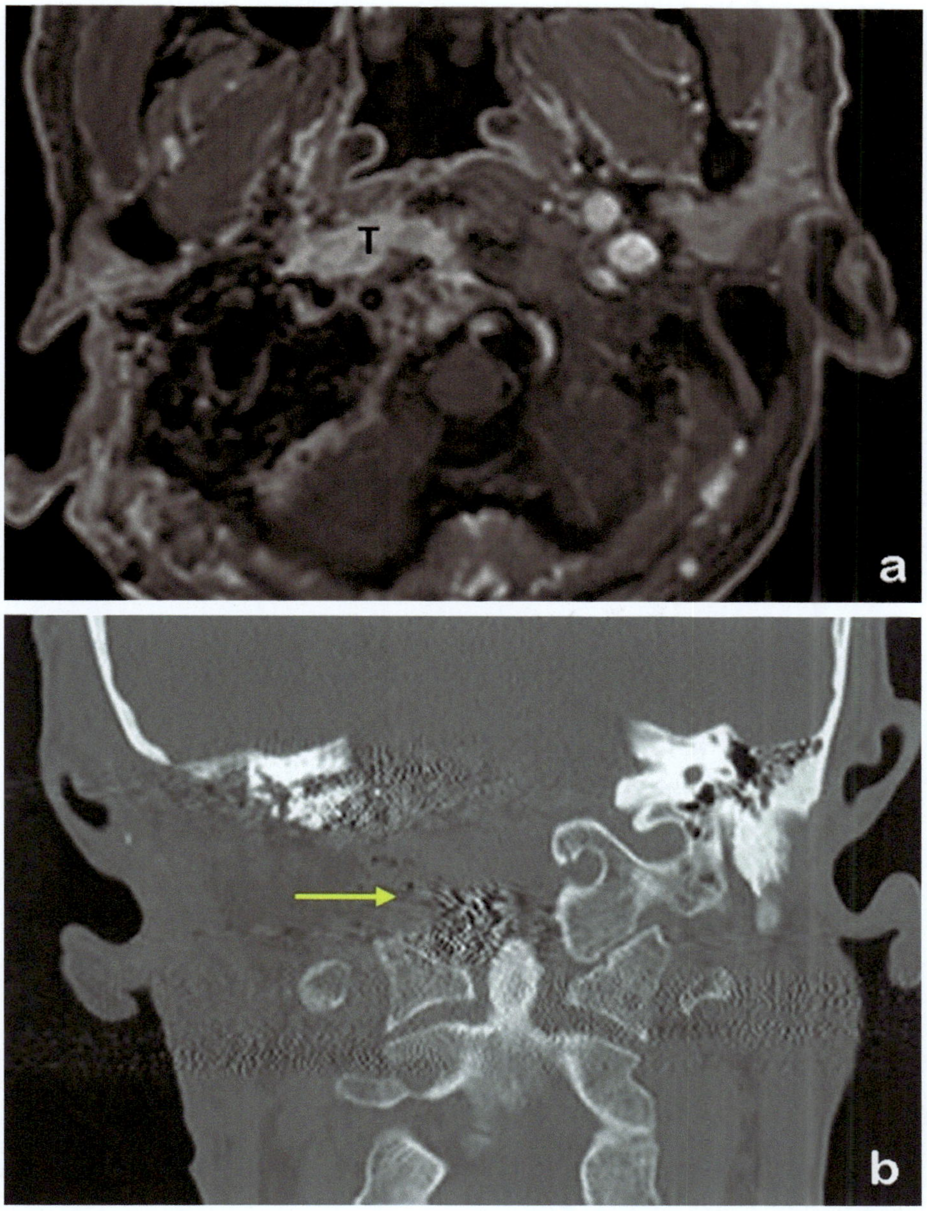

Fig. 12.209 The patient underwent a second-stage IFTA approach with complete removal of the tumor in the occipital condyle and foramen magnum area. Note the tumor remnant in the clivus (T). (A) Axial MRI showing the tumor remnant. (B) Coronal CT scan showing removal of the entire occipital condyle *(arrow)*.

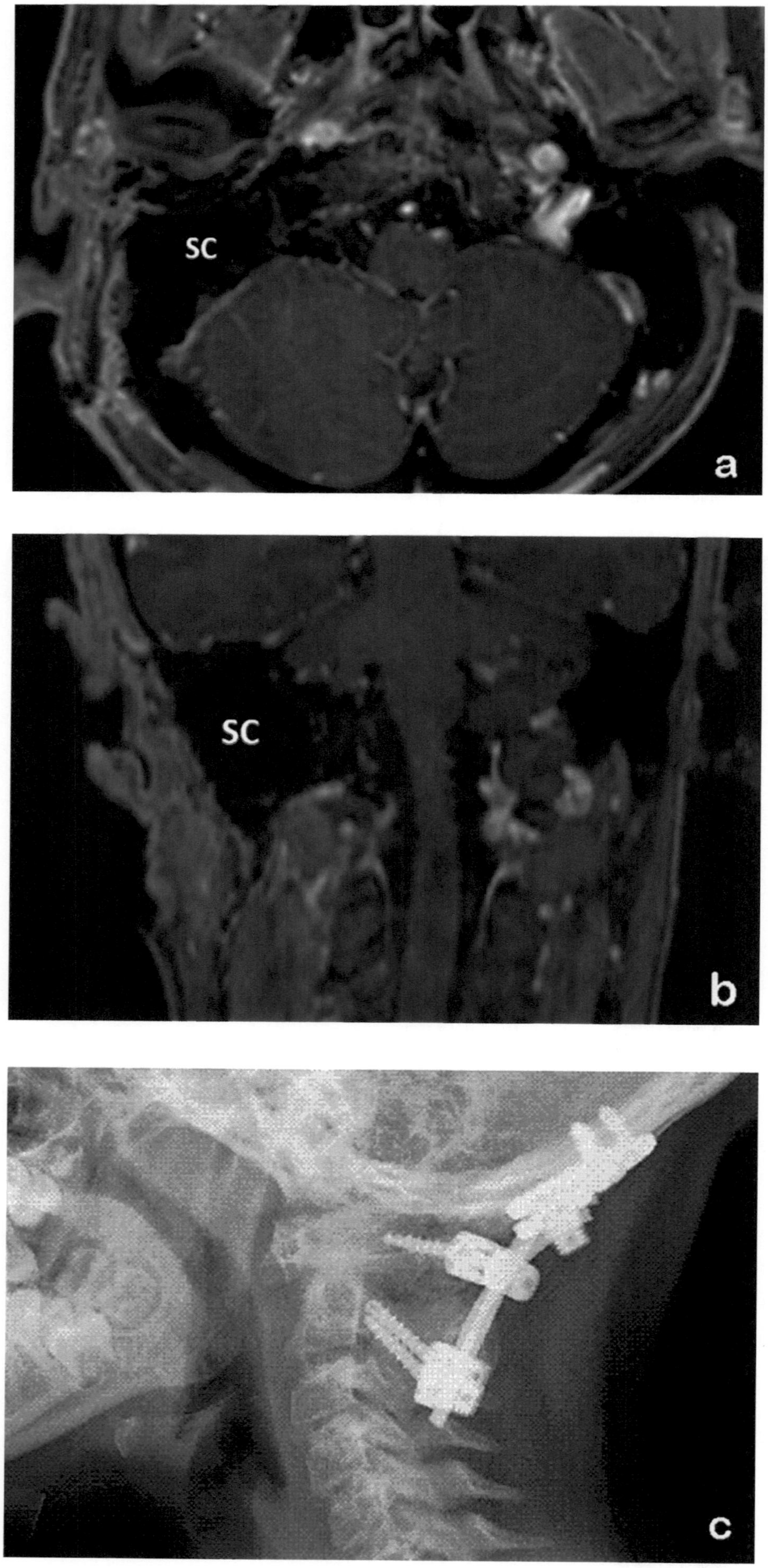

Fig. 12.210 The patient underwent a third-stage surgery with complete removal of the tumor in the clivus. Note the surgical cavity (SC). The patient also underwent an occipital cervical fixation due to the previous removal of the entire right occipital condyle. (A) Axial MRI showing no residual remnant. (B) Coronal MRI scan showing no residual tumor. (C) Cervical X-ray showing the occipital cervical fixation.

HINTS AND PITFALLS

- Care must be taken to fully uncover the geniculate ganglion. Sharp bone left at this level might injure the anteriorly transposed nerve.
- Before anterior transposition is carried out, the facial nerve has to be liberated in the parotid gland. This provides an extra length of free nerve and prevents it from being stretched during the transposition.
- A soft tissue cuff is left around the nerve at the level of the stylomastoid foramen. This tissue affords protection for the nerve during transposition. It also helps to maintain the vascularity of the nerve.
- During anterior rerouting of the facial nerve, a Beaver knife is used to cut the sharp attachments of the mastoid segment of the facial nerve.
- Direct suction over the facial nerve is to be avoided. A Brackmann suction tip is used instead. Cottonoids can also be used to avoid direct trauma from the suction tip.
- Care should be taken while applying the Fisch infratemporal retractor so as not to injure the facial nerve. We no longer use the Fisch retractor, using a self-retaining retractor instead.
- Complete removal of the tympanic bone is of paramount importance prior to any intervention on the tumor. Removal of the tympanic bone and the styloid allows the control of the internal carotid artery from the neck to its petrous portion (see Figs. 12.34, 12.42, 12.44, 12.93–12.98, 12.108). We therefore consider the tympanic bone to be the "gateway" to the lateral skull base.
- When removing the base of the styloid process, great care must be taken not to injure the ICA immediately below.
- The ICA is surrounded by a dense fibroperiosteal layer at its entrance in the skull base. The glossopharyngeal nerve lies deep in this layer and is slightly adherent to its undersurface. Care has to be taken not to injure the nerve while dissecting tumors at this level. In large TJPs, however, the nerve has usually been infiltrated, and it is therefore sacrificed.
- Due to excessive neck dissection and the connection between the soft tissues of the neck and the surgical cavity, large intradural extensions cannot be managed in one stage using this approach. Removing the large intradural part of the tumor would necessitate a larger dural opening, and therefore a higher risk of postoperative cerebrospinal fluid leakage. In such cases, a second-stage operation is needed to remove the intradural component of the tumor. When dural tears occur during tumor removal, or if the dura has been deliberately opened for a single-stage removal of a very small intradural tumor extension, the dura is closed with a muscle plug or with a long strip of abdominal fat as in the translabyrinthine approach for acoustic neuroma removal.
- With paragangliomas, all suspected cancellous bone should be removed. The Haversian system of this bone is usually invaded by the tumor and is a frequent cause of recurrence. Drilling should continue until sound bone is reached. This is especially important in the area lying between the basal turn of the cochlea and the ICA. Bone from the petrous apex lying medial to the ICA should also be drilled if suspect.
- In some cases, the occipital condyle is infiltrated by the tumor. The partial removal of the occipital condyle improves postero-inferolateral and medial exposure, reducing the possibility of recurrences. The condyle is partially drilled to remove the tumor completely. Bleeding from the condylar emissary vein at this step is controlled with Surgicel. Care is taken not to injure the hypoglossal nerve, which traverses the condyle at a more anterior level. In ~50% of cases, the accessory nerve passes anterolateral to the internal jugular vein in the neck. After ligating the vein, care is taken to not injure the nerve during elevation of the vein. We currently preserve the medial wall of the internal jugular vein and jugular bulb if not infiltrated, which affords protection for the closely related lower cranial nerves.
- TJPs usually infiltrate the jugular bulb and insinuate into the openings of the inferior petrosal sinus. Such finger-like tumor projections should be removed and bleeding should be controlled with Surgicel packing. However, overzealous packing should be avoided as it might cause excessive swelling, with resultant paralysis of the intimately related lower cranial nerves due to compression.
- The cochlea may be involved by a TJP despite normal preoperative hearing. In these cases, total tumor removal necessitates its removal.
- In Class C2 TJPs, the internal carotid artery should be freed from the adherent tumor. This is usually feasible, since the carotid sheath (periosteum), rather than the carotid wall itself, is the part involved in the tumor. The carotid sheath is cut in an area free of tumor and is held with a nontoothed forceps in one hand. The sheath with the infiltrating tumor is then dissected away from

the carotid using a pair of microneurosurgical scissors. Removal of the ICA may be required on rare occasions when there is marked stenosis in the artery, or its wall is too fragile as a result of previous irradiation or surgery. A preoperative balloon occlusion test is mandatory before removal of the carotid is attempted. In the early days of our experience, carotid resection was performed more frequently. We have adopted a less aggressive attitude nowadays, for fear of long-term consequences such as strokes, hemiplegia, and aneurysm of the contralateral ICA.

- With large paragangliomas (Class 3, involving the horizontal ICA; or Class 4, reaching the anterior foramen lacerum and extending to the cavernous sinus), the ITFA can still be utilized without resorting to an ITFB. This is achieved by the removal of the anterior wall of the external auditory canal and tympanic bone, anterior displacement of the mandible, and mobilization of the ICA.
- In cases of Class C3–C4 tumors with risk of carotid blowout or in previously irradiated cases, we can use definitive balloon occlusion instead of preoperative stenting of the ICA.
- Subtotal removal is indicated in tumor that extends to the cavernous sinus so as to avoid neurological deficits. In these cases, the patient receives radiotherapy if growth is demonstrated at follow-up.
- There is no need to use the transposition of the temporalis muscle for obliteration of dead space.
- Anterior displacement of the mandible is not required.
- Preservation of the branches of the external carotid artery would allow for the possibility of tumor embolization prior to any planned second-stage surgery.
- Mobilization or displacement of the vertebral artery is necessary only when the artery is embedded in the tumor.

Infratemporal Fossa Approach Type B (IFTB)

INTRODUCTION

This is an extralabyrinthine approach designed primarily for extradural lesions involving the petrous apex and midclivus (Fig. 12.211). The type B approach involves exposure of the ICA from the neck to the cavernous sinus posterior to the foramen ovale. To expose the horizontal segment of the ICA, the middle meningeal artery is divided and the Eustachian tube is sacrificed. By sacrificing the Eustachian tube at this point and preserving the middle ear fissure, hearing (bone conduction) can be preserved. Access to the clivus requires division of the mandibular nerve (V3) and complete removal of the bony Eustachian tube. Depending on the extent of the tumor, the technique can be modified and tumor resection can be achieved without transection of the middle meningeal artery and V3. The ITFB approach is an excellent technique that allows control of both the vertical and horizontal petrous ICA to the anterior foramen lacerum without significant brain retraction [31,40–44]. The morbidity associated with this procedure includes:

- Conductive hearing loss.
- Sacrifice of the mandibular nerve and disarticulation of the mandibular condyle. This usually results in no significant postoperative morbidity and is well tolerated by the patient [43–45].
- As the cochlea is preserved, tumors located between the medial aspect of the cochlea and the internal auditory canal are not well controlled. This may lead to tumor recurrence, particularly in chordomas.

ADVANTAGES

1. Preservation of the facial nerve and inner ear function, allowing for hearing rehabilitation.
2. Allows adequate control of the vertical and horizontal intrapetrous internal carotid artery.
3. No facial scarring.

INDICATIONS

– Petrous apex lesions such as petrous bone cholesteatoma of the apical and infralabyrinthine types and cholesterol granuloma [45,46].

– Clival lesions such as chordoma, chondrosarcoma, and extensive paragangliomas invading the clivus.

– Other rare lesions involving the infratemporal fossa such as choristoma of the Eustachian tube and giant cell tumors of the petrous bone.

CONTRAINDICATIONS

– Lesions extending far anteriorly to the sphenoid sinus, pterygopalatine fossa, or cavernous sinus. In such cases, a type C is performed.

– Transdural lesions.

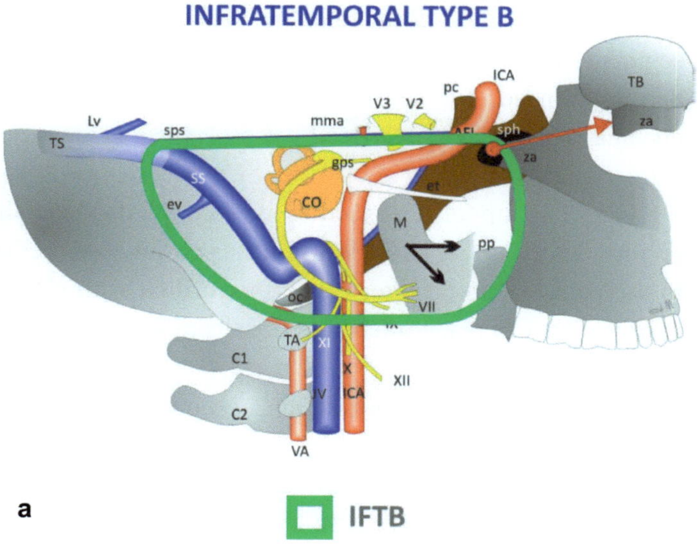

a

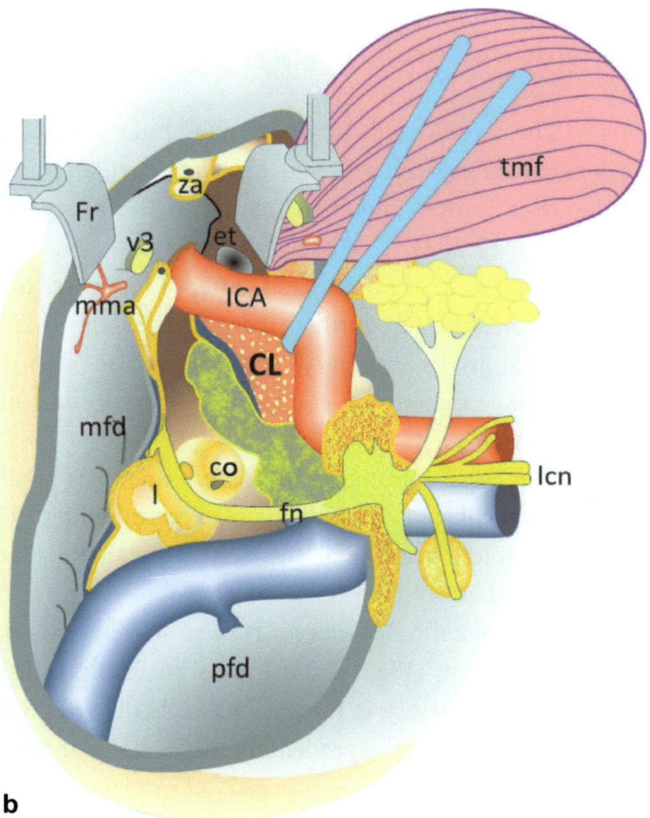

b

Fig. 12.211 (A, B) Drawings showing the hallmark of the infratemporal fossa approach type B (IFTB). *AFL*, anterior foramen lacerum; *C1*, first cervical vertebra; *CO*, cochlea; *et*, Eustachian tube; *ev*, emissary vein; *gps*, greater petrosal nerve; *ICA*, internal carotid artery; *JV*, jugular vein; *Lv*, vein of Labbé; *M*, mandible; *mma*, middle meningeal artery; *OC*, occipital condyle; *pc*, clinoid process; *pp*, pterygoid process; *sph*, sphenoid; *sps*, superior petrosal sinus; *TA*, transverse process of atlas; *TS*, transverse sinus; *za*, zygomatic arch; *V2*, maxillary nerve; *V3*, mandibular nerve; *VII*, facial nerve; *IX*, glossopharyngeal nerve; *XII*, hypoglossal nerve.

SURGICAL STEPS (RIGHT SIDE)

1. The skin incision is C-shaped, starting above and posterior to the lateral orbital margin, extending 3–4 cm behind the postauricular sulcus, and ending inferiorly at the mastoid tip (Fig. 12.212).

2. The skin and subcutaneous tissue flap is raised; the external auditory canal is transected (Fig. 12.213) and is closed as a cul-de-sac (Figs. 12.214 and 12.215).

3. The parotid gland and greater auricular nerve are identified (Fig. 12.216). The extratemporal facial nerve is exposed in the parotid (Fig. 12.217). The best landmark is that the main trunk of the nerve runs along the perpendicular bisection of a line joining the cartilaginous pointer and the mastoid tip. The temporal branches of the nerve are followed until it crosses the zygomatic arch. While identifying the extratemporal facial nerve, few veins are encountered at first before the nerve is found. Identifying the nerve is important to prevent it from being overstretched during the application of the infratemporal fossa retractor.

4. About 2–3 cm superior to the zygomatic arch, a semilunar incision is made in the superficial layer of the temporalis fascia, and the plane of dissection is changed to become deeper (beneath the superficial layer of the temporalis fascia), thereby affording protection for the temporal branches of facial nerve. This plane of dissection courses with the superficial temporal fat pad (Fig. 12.218).

The periosteum of the zygomatic arch is incised. The zygomatic arch is fully exposed taking care not to injure the frontal branch of the facial nerve after superficial layer of the temporalis fascia retracted anteriorly (Fig. 12.219).

5. The temporalis muscle is detached from its bed, to be further reflected anteriorly and inferiorly (Fig. 12.220).

6. The skin of the external auditory canal, the tympanic membrane, the incus, and the malleus are removed after disarticulation of the incudostapedial joint. A subtotal petrosectomy is then performed with preservation of the labyrinth. The facial nerve is skeletonized (Fig. 12.221).

7. The anterior wall of the external auditory canal is completely drilled to expose the capsule of the temporomandibular joint (Fig. 12.222).

8. The capsule of the temporomandibular joint is displaced anteriorly using a self-retaining retractor. This provides a wide approach to drilling the bone lateral to the petrous carotid artery (Fig. 12.223).

9. The glenoid fossa is drilled. The vertical portion of petrous carotid artery and middle meningeal artery are exposed (Fig. 12.224). The sphenoid spine serves as a landmark for the middle meningeal artery.

10. Drilling of the area lying between the basal turn of the cochlea and the vertical portion of petrous carotid artery with a large diamond burr (Fig. 12.225). However, further drilling of the petrous apex is limited by petrous carotid artery (Fig. 12.226). If more exposure is needed, the vertical internal carotid artery can be exposed the whole way around its circumference and mobilized using the suction (Fig. 12.227). The internal carotid artery is retracted anterolaterally, providing enough room to drill the ipsilateral petrous apex lying medial to the artery until the midclivus is reached anteriorly.

11. The horizontal and vertical portion of petrous carotid artery is now under control the whole way around its circumference (Fig. 12.228).

12. Note the close relation between the petrous segment of internal carotid artery and the cochlea. The petrous carotid has been exposed anteroinferior to the cochlea. The greater superficial petrosal nerve is identified (Fig. 12.229).

13. The tip of the suction is used to displace the petrous carotid artery laterally while the petrous apex medial to the petrous carotid artery is further drilled with a large diamond burr to reach inferior petrosal sinus and the clivus (Fig. 12.230).

14. After bipolar coagulation, the middle meningeal artery is cut (Fig. 12.231). The mandibular nerve is then identified. A probe is inserted into the cartilaginous Eustachian tube (Fig. 12.232). The Eustachian tube is an important landmark in skull base surgery and has a close spatial relation with the horizontal portion of petrous carotid artery.

15. The capsule of the temporomandibular joint is separated using strong scissors and bipolar coagulation. The articular disk is then removed, exposing the mandibular condyle (Fig. 12.233).

16. The arch is transected after two burr holes have been created for further refixation at the end of the procedure in live surgery. The zygomatic arch and part of the squamosal are osteotomized en bloc (Fig. 12.234).

17. The Fisch infratemporal fossa retractor is applied with inferior displacement of the head of the mandible. This provides complete control of the vertical segment of the internal carotid artery (Fig. 12.235).

18. The mandibular nerve is also transected after bipolar coagulation (Fig. 12.236). The horizontal portion of petrous carotid artery is further exposed (Fig. 12.237).

19. At the end of the procedure, the Eustachian tube is sutured, the zygomatic process is refixed in place, the temporalis muscle is used to obliterate the cavity, and the wound is closed in layers.

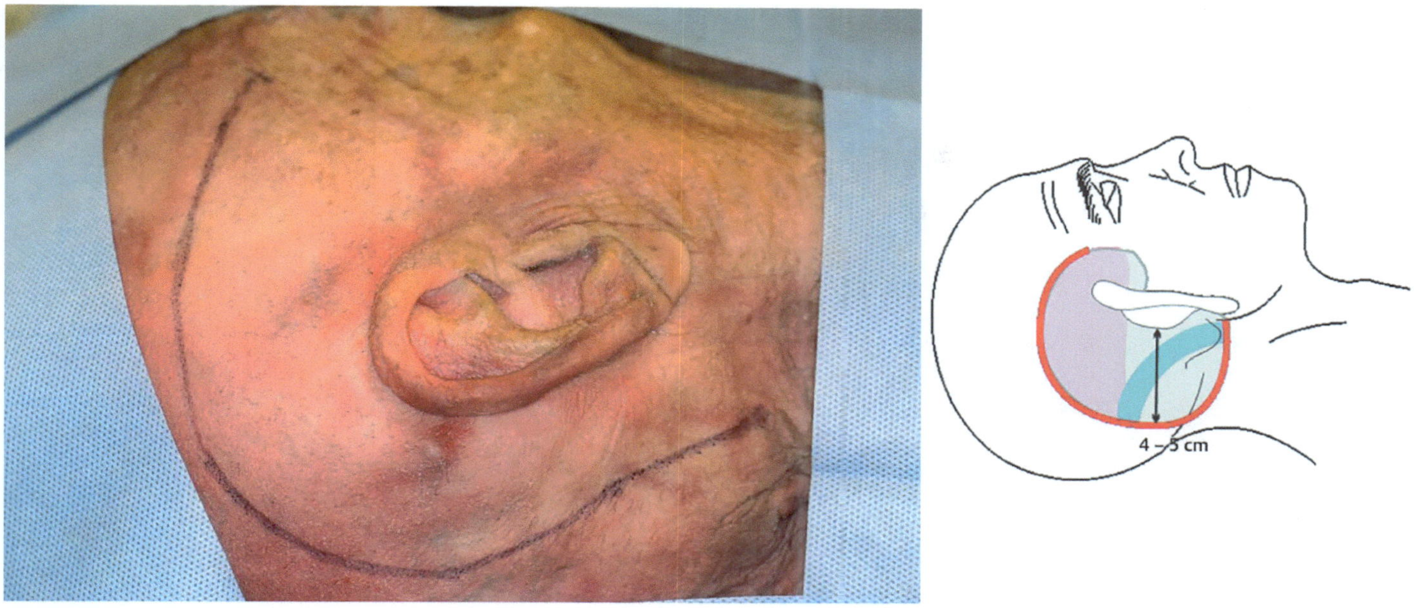

Fig. 12.212 The skin incision is C-shaped, starting above and posterior to the lateral orbital margin, extending 3–4 cm behind the postauricular sulcus, and ending inferiorly at the mastoid tip.

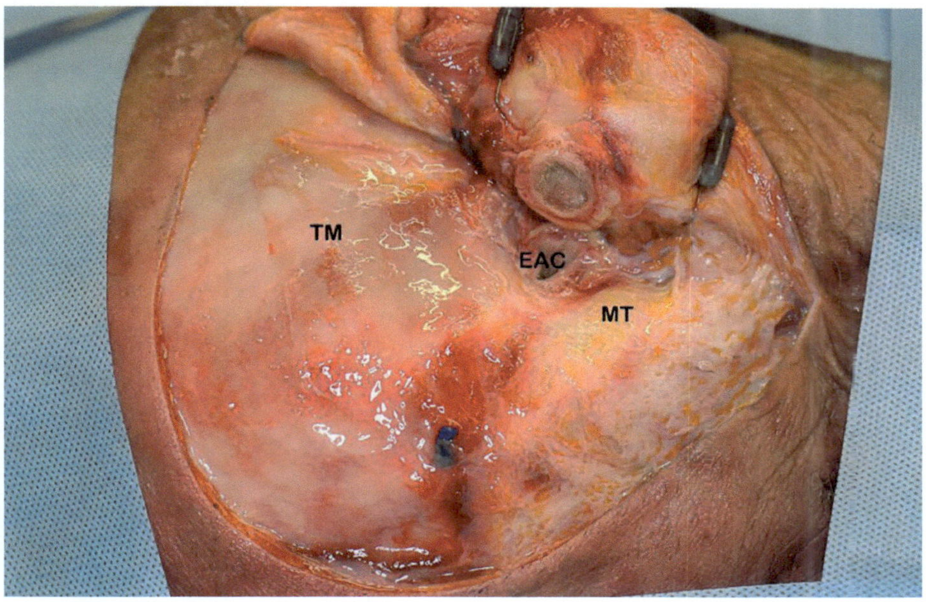

Fig. 12.213 The external auditory canal (EAC) is transected. *TM*, temporalis muscle; *MT*, mastoid tip.

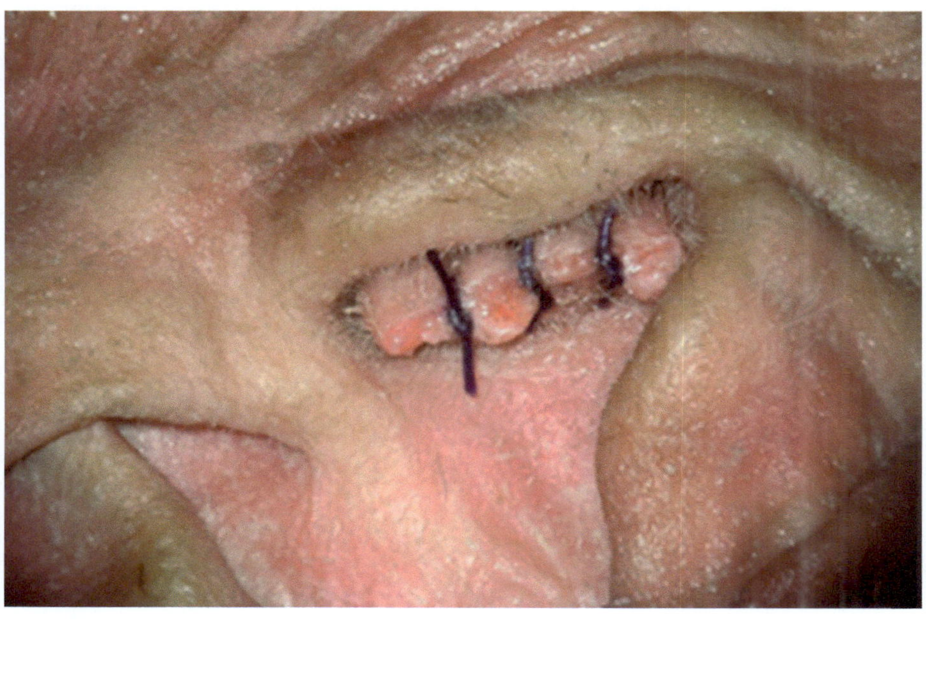

Fig. 12.214 Closure of the skin of the external auditory canal.

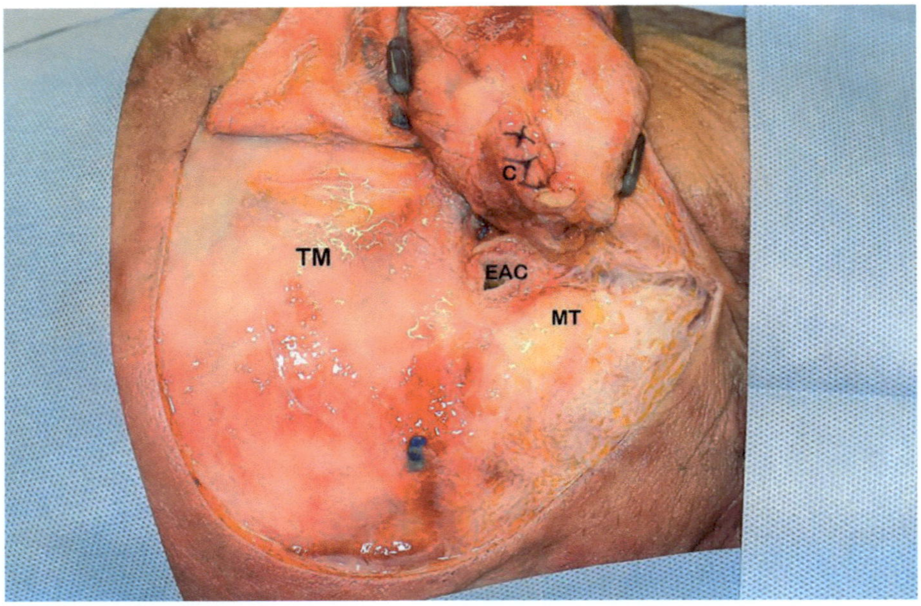

Fig. 12.215 Closure of the second layer of the blind sac closure of the external auditory canal (EAC). *C*, cartilage; *MT*, mastoid tip; *TM*, temporalis muscle.

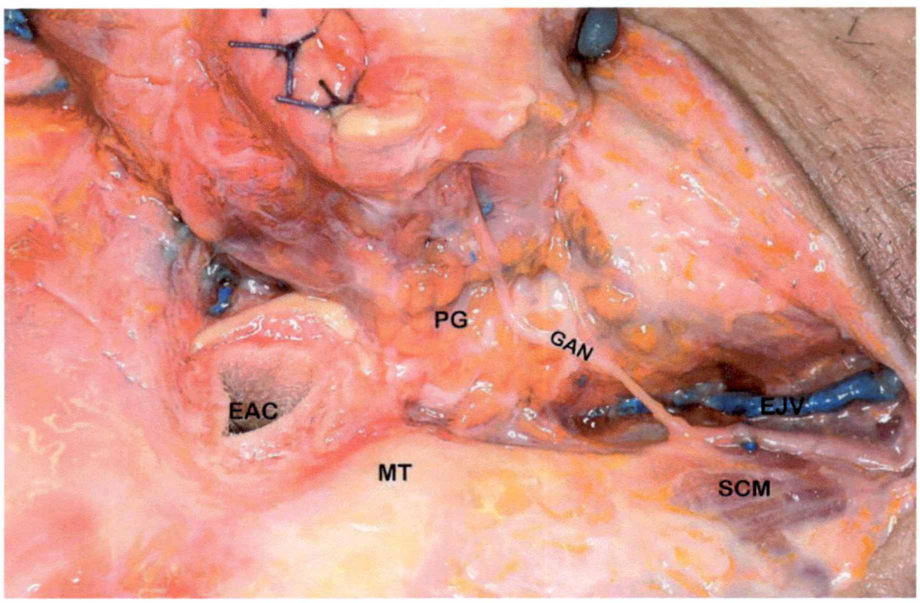

Fig. 12.216 The parotid gland and greater auricular nerve are identified. *PG*, parotid gland; *GAN*, greater auricular nerve; *EJV*, external jugular vein; *SCM*, sternocleidomastoid muscle; *MT*, mastoid tip; *EAC*, external auditory canal.

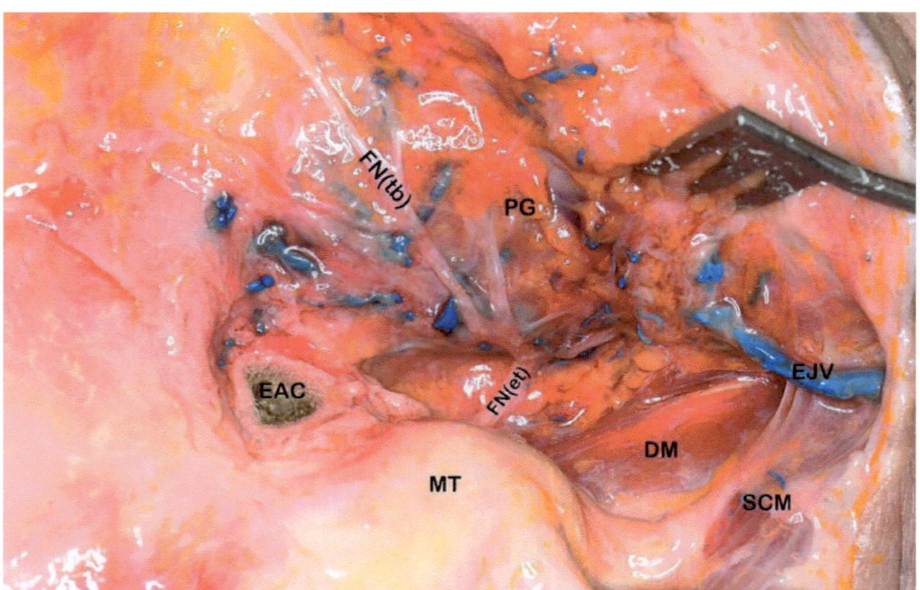

Fig. 12.217 The extratemporal facial nerve is exposed in the parotid. *FN(tb)*, temporal branches of facial nerve; *PG*, parotid gland; *FN (et)*, extratemporal segment of facial nerve; *EAC*, external auditory canal; *MT*, mastoid tip; *DM*, digastric muscle; *SCM*, sternocleidomastoid muscle; *EJV*, external jugular vein.

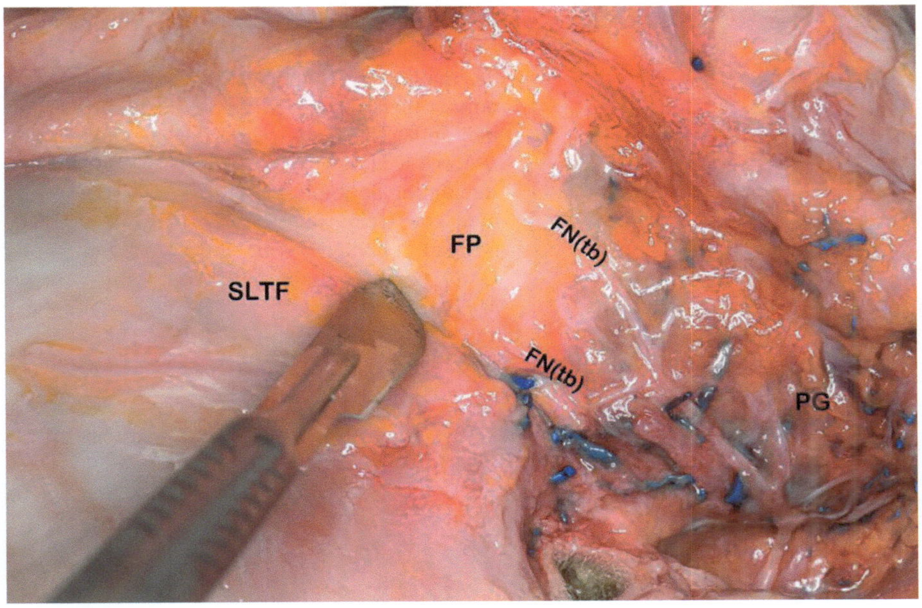

Fig. 12.218 About 2–3 cm superior to the zygomatic arch, a semilunar incision is made in the superficial layer of the temporalis fascia and the plane of dissection is changed to become deeper (beneath the superficial layer of the temporalis fascia), thereby affording protection for the temporal branches of facial nerve. This plane of dissection courses with the superficial temporal fat pad. *FN(tb)*, temporal branches of facial nerve; *PG*, parotid gland; *FP*, fat pad; *SLTF*, superior layer of the temporalis fascia.

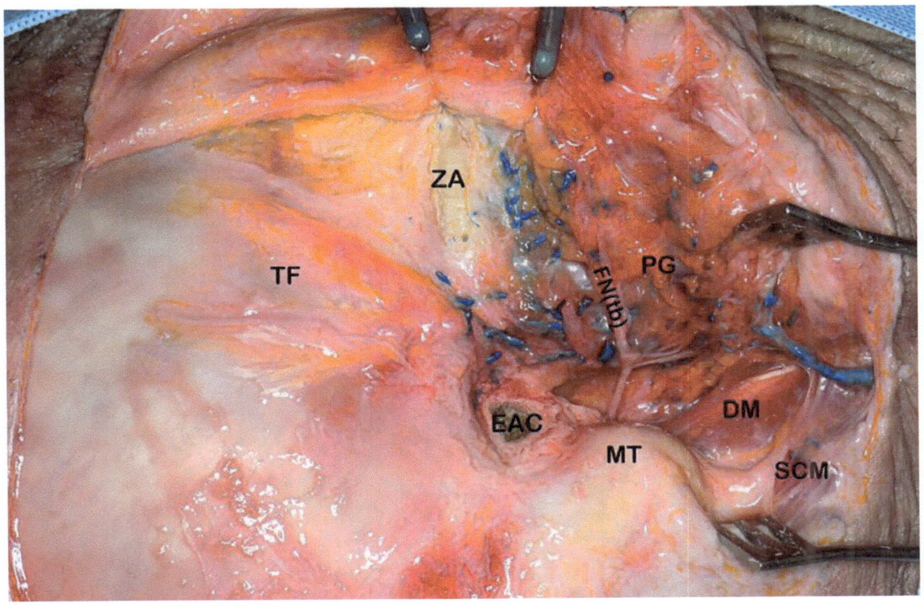

Fig. 12.219 The zygomatic arch is fully exposed after superficial layer of the temporalis fascia retracted anteriorly. *ZA*, zygomatic arch; *PG*, parotid gland; *FN(tb)*, temporal branches of facial nerve; *TF*, temporalis fascia; *DM*, digastric muscle; *SCM*, sternocleidomastoid muscle; *EAC*, external auditory canal; *MT*, mastoid tip.

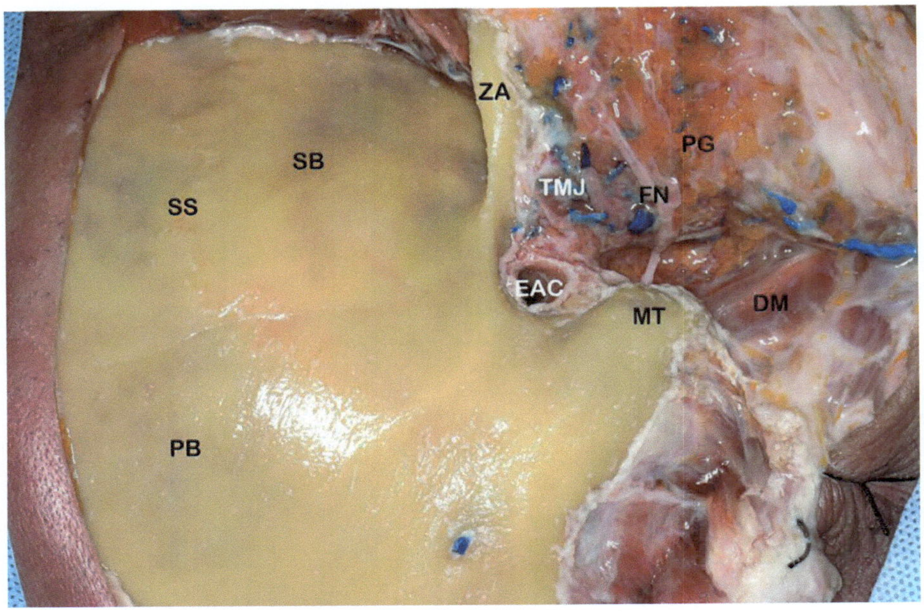

Fig. 12.220 The temporalis muscle is detached from its bed, to be further reflected anteriorly and inferiorly. *ZA*, zygomatic arch; *PG*, parotid gland; *FN*, facial nerve; *TMJ*, temporomandibular joint; *EAC*, external auditory canal; *MT*, mastoid tip; *DM*, digastric muscle; *SB*, squamosal bone; *SS*, squamous suture; *PB*, parietal bone.

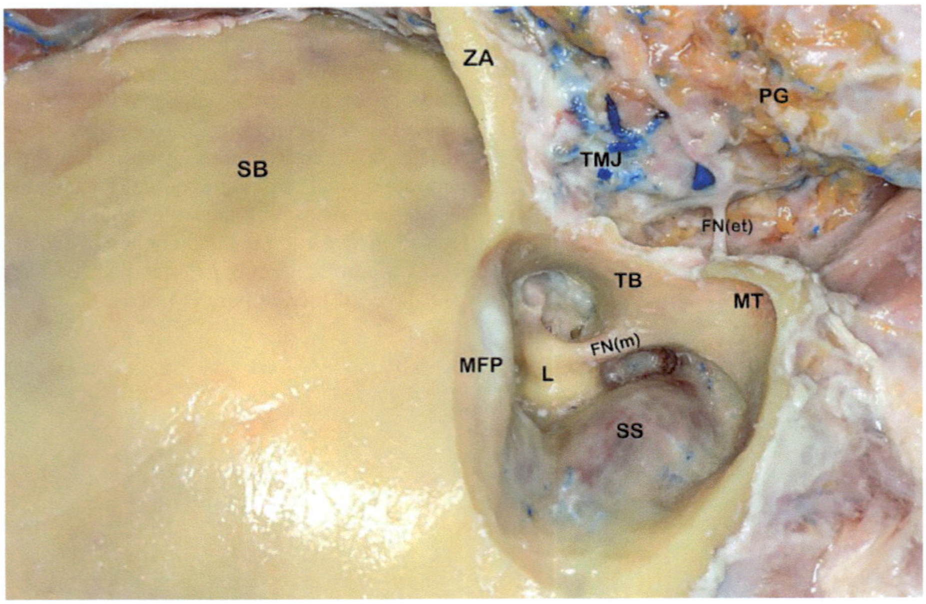

Fig. 12.221 The skin of the external auditory canal, the tympanic membrane, the incus, and the malleus are removed. A subtotal petrosectomy is then performed with preservation of the labyrinth. The facial nerve is skeletonized. *ZA*, zygomatic arch; *PG*, parotid gland; *TMJ*, temporomandibular joint; *FN(et)*, extratemporal segment of facial nerve; *MT*, mastoid tip; *TB*, tympanic bone; *FN(m)*, mastoid segment of facial nerve; *L*, labyrinth; *SS*, sigmoid sinus; *MFP*, middle fossa plate; *SB*, squamosal bone.

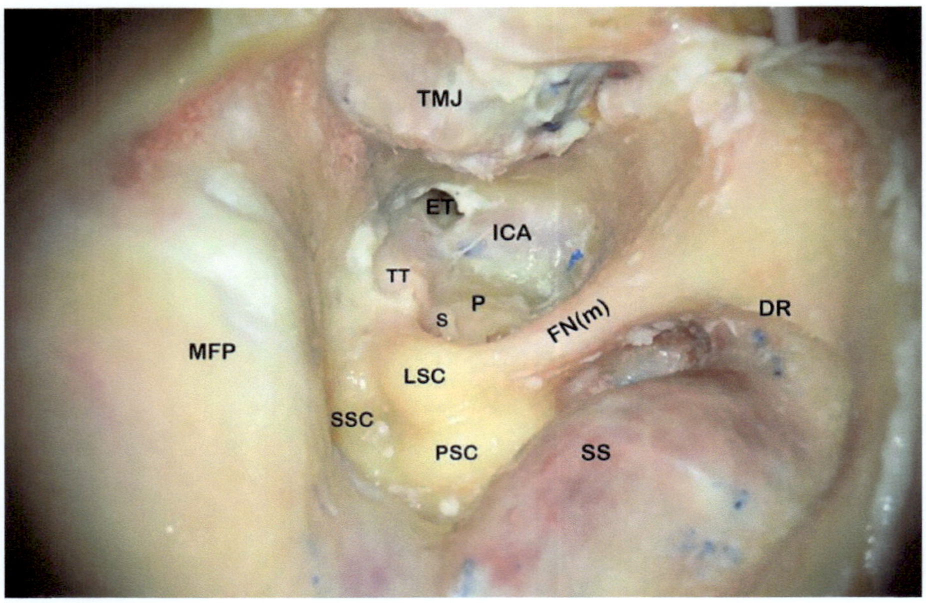

Fig. 12.222 The anterior wall of the external auditory canal is completely drilled to expose the capsule of the temporomandibular joint. *TMJ*, temporomandibular joint; *ET*, Eustachian tube; *ICA*, internal carotid artery; *TT*, tensor tympani muscle; *P*, promontory; *S*, stapes; *FN(m)*, mastoid segment of facial nerve; *DR*, digastric ridge; *LSC*, lateral semicircular canal; *PSC*, posterior semicircular canal; *SSC*, superior semicircular canal; *SS*, sigmoid sinus; *MFP*, middle fossa plate.

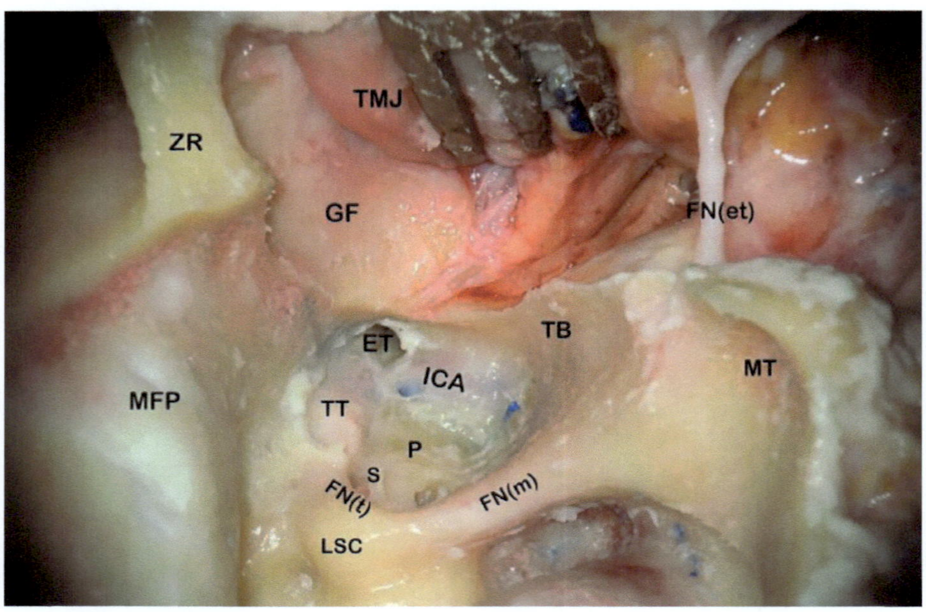

Fig. 12.223 The capsule of the temporomandibular joint is displaced anteriorly using a self-retaining retractor. This provides a wide approach to drilling the bone lateral to the petrous carotid artery. *TMJ*, temporomandibular joint; *FN(et)*, extratemporal segment of facial nerve; *MT*, mastoid tip; *TB*, tympanic bone; *GF*, glenoid fossa; *ZR*, zygomatic root; *ET*, Eustachian tube; *ICA*, internal carotid artery; *TT*, tensor tympani muscle; *P*, promontory; *S*, stapes; *FN(t)*, tympanic segment of facial nerve; *FN(m)*, mastoid segment of facial nerve; *LSC*, lateral semicircular canal; *MFP*, middle fossa plate.

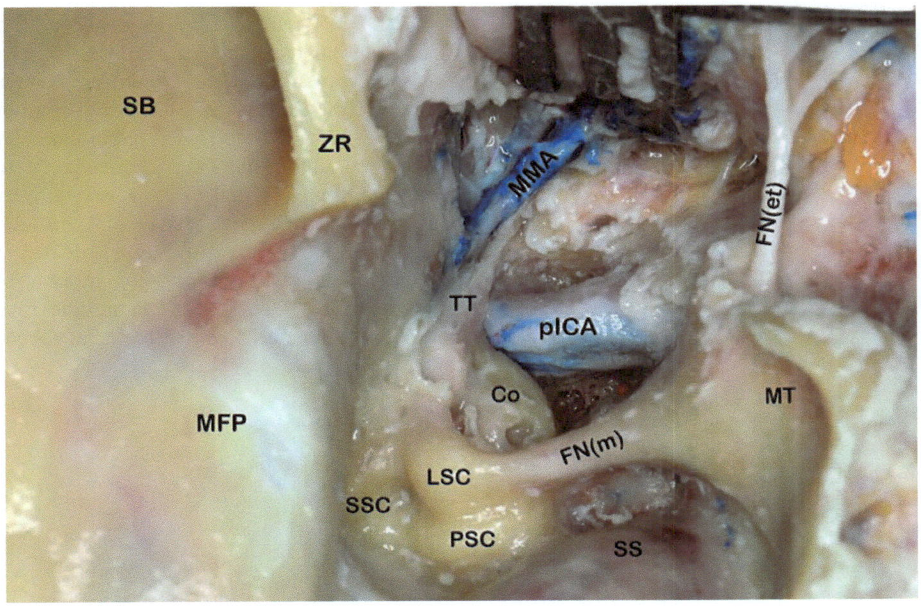

Fig. 12.224 The glenoid fossa is drilled. The vertical portion of petrous carotid artery and middle meningeal artery are exposed. The sphenoid spine serves as a landmark for the middle meningeal artery. *SB*, squamosal bone; *ZR*, zygomatic root; *MMA*, middle meningeal artery; *FN(et)*, extratemporal segment of facial nerve; *TT*, tensor tympani muscle; *pICA*, petrous segment of internal carotid artery; *Co*, cochlea; *FN(m)*, mastoid segment of facial nerve; *MT*, mastoid tip; *SS*, sigmoid sinus; *LSC*, lateral semicircular canal; *PSC*, posterior semicircular canal; *SSC*, superior semicircular canal; *MFP*, middle fossa plate.

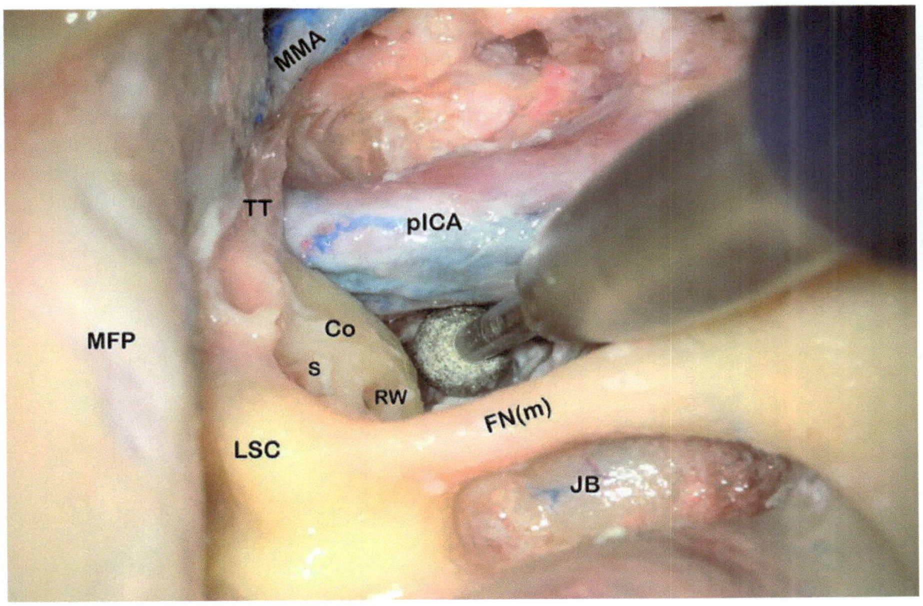

Fig. 12.225 Drilling of the area lying between the basal turn of the cochlea and the vertical portion of petrous carotid artery with a large diamond burr. *MMA*, middle meningeal artery; *TT*, tensor tympani muscle; *pICA*, petrous segment of internal carotid artery; *Co*, cochlea; *S*, stapes; *RW*, round window; *FN(m)*, mastoid segment of facial nerve; *JB*, jugular bulb; *LSC*, lateral semicircular canal; *MFP*, middle fossa plate.

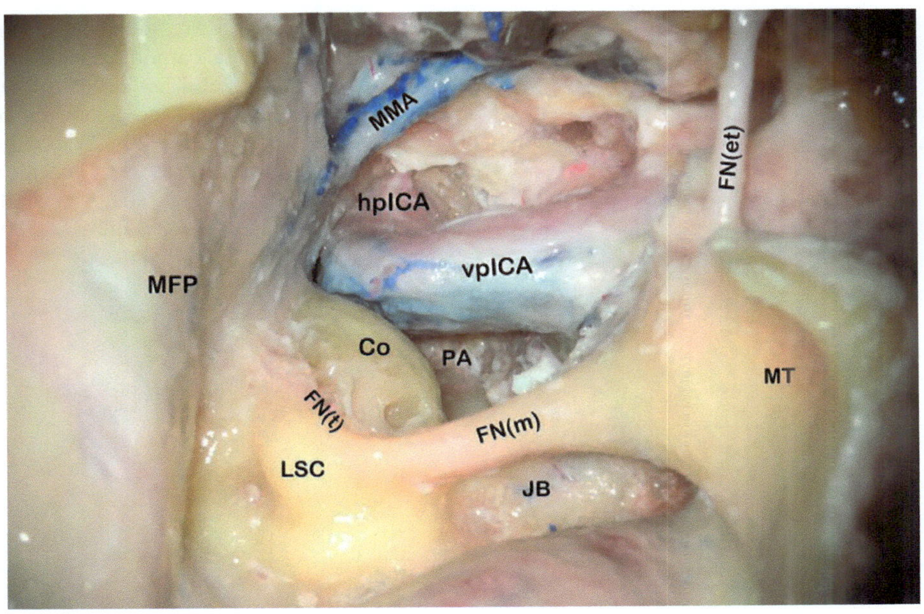

Fig. 12.226 Further drilling of the petrous apex is limited by petrous carotid artery. *MMA*, middle meningeal artery; *hpICA*, horizontal portion of petrous carotid artery; *vpICA*, vertical portion of petrous carotid artery; *FN(et)*, extratemporal segment of facial nerve; *JB*, jugular bulb; *FN (m)*, mastoid segment of facial nerve; *PA*, petrous apex; *Co*, cochlea; *FN(t)*, tympanic segment of facial nerve; *LSC*, lateral semicircular canal; *MFP*, middle fossa plate; *MT*, mastoid tip.

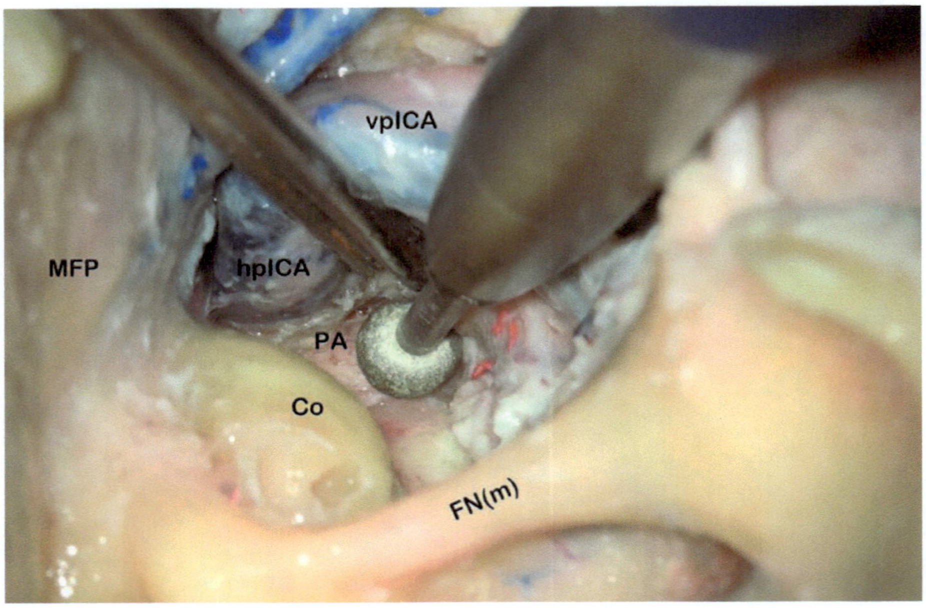

Fig. 12.227 If more exposure is needed, the vertical internal carotid artery can be exposed the whole way around its circumference and mobilized using the suction aspirator. The internal carotid artery is retracted anterolaterally, providing enough room to drill the ipsilateral petrous apex lying medial to the artery until the midclivus is reached anteriorly. *hpICA*, horizontal portion of petrous carotid artery; *vpICA*, vertical portion of petrous carotid artery; *PA*, petrous apex; *Co*, cochlea; *MFP*, middle fossa plate; *FN(m)*, mastoid segment of facial nerve.

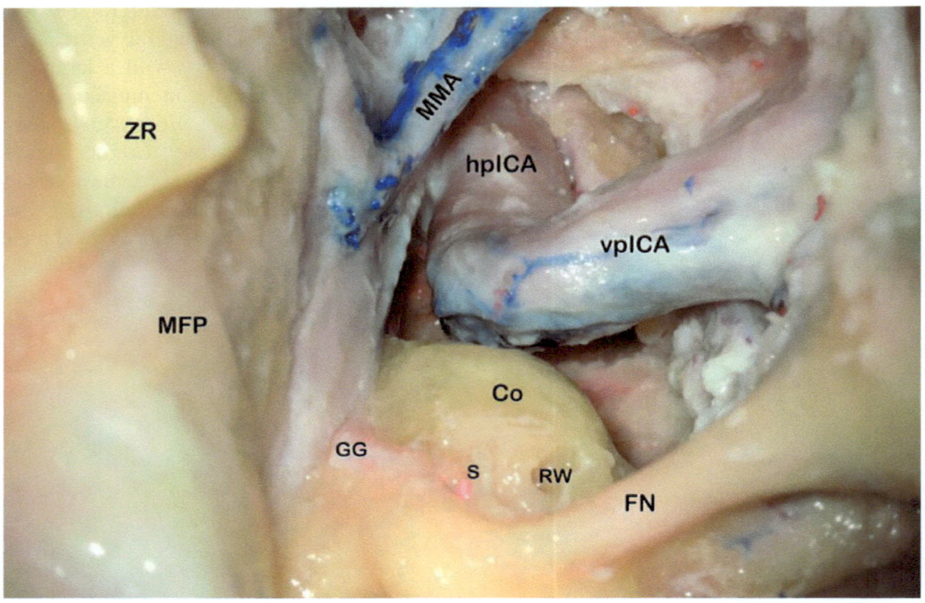

Fig. 12.228 The horizontal and vertical portion of petrous carotid artery is now under control the whole way around its circumference. *MMA*, middle meningeal artery; *hpICA*, horizontal portion of petrous carotid artery; *vpICA*, vertical portion of petrous carotid artery; *ZR*, zygomatic root; *MFP*, middle fossa plate; *GG*, geniculate ganglion; *Co*, cochlea; *S*, stapes; *RW*, round window; *FN*, facial nerve.

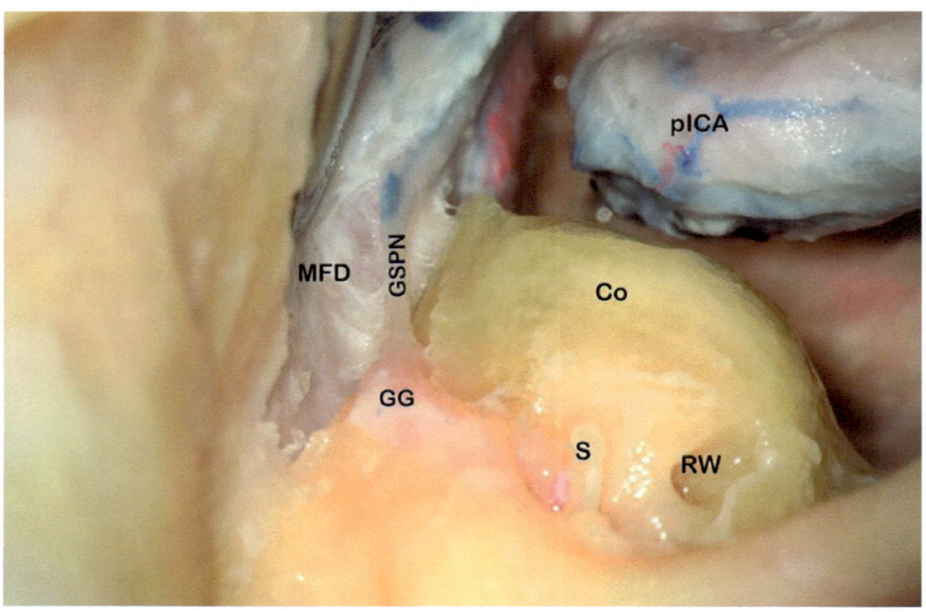

Fig. 12.229 Note the close relation between the petrous segment of internal carotid artery and the cochlea. The petrous carotid has been exposed anteroinferior to the cochlea. The greater superficial petrosal nerve is identified. *pICA*, petrous segment of internal carotid artery; *Co*, cochlea; *S*, stapes; *RW*, round window; *GG*, geniculate ganglion; *MFD*, middle fossa dura; *GSPN*, greater superficial petrosal nerve.

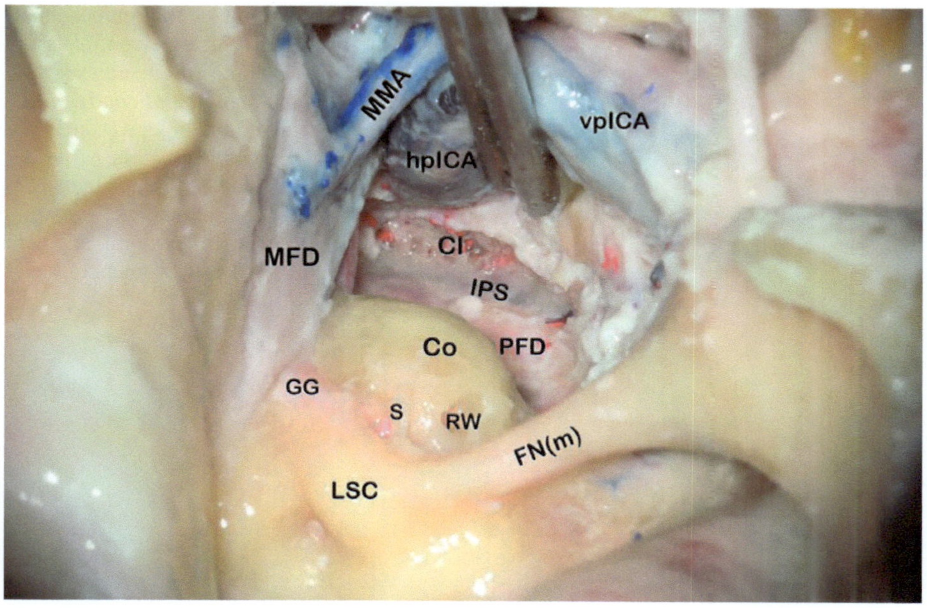

Fig. 12.230 The tip of the suction is used to displace the petrous carotid artery laterally while the petrous apex medial to the petrous carotid artery is further drilled with a large diamond burr to reach inferior petrosal sinus and the clivus. *hpICA*, horizontal portion of petrous carotid artery; *vpICA*, vertical portion of petrous carotid artery; *MMA*, middle meningeal artery; *MFD*, middle fossa dura; *CI*, clivus; *IPS*, inferior petrosal sinus; *PFD*, posterior fossa dura; *Co*, cochlea; *GG*, geniculate ganglion; *S*, stapes; *RW*, round window; *LSC*, lateral semicircular canal; *FN(m)*, mastoid segment of facial nerve.

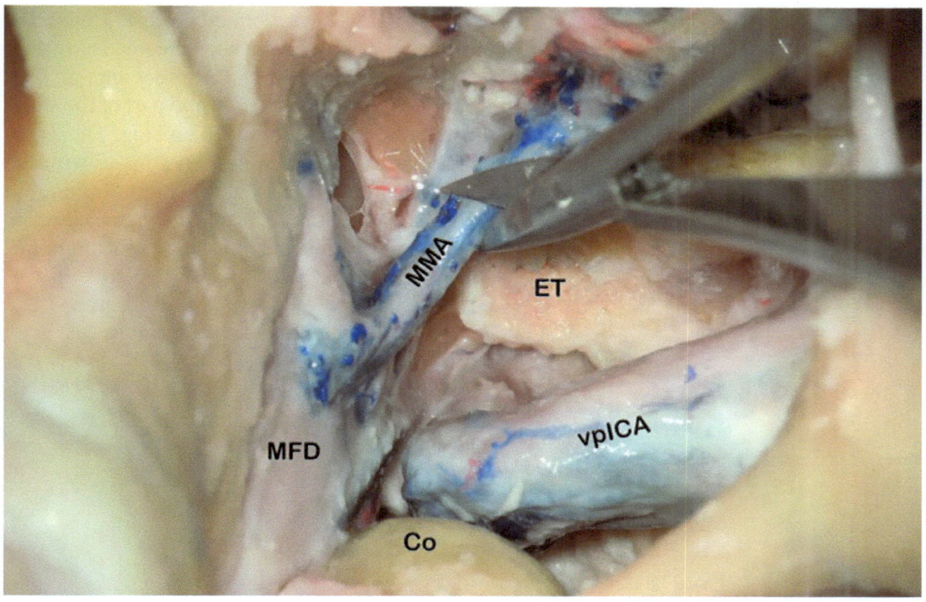

Fig. 12.231 After bipolar coagulation, the middle meningeal artery is cut. *MMA*, middle meningeal artery; *ET*, Eustachian tube; *vpICA*, vertical portion of petrous carotid artery; *MFD*, middle fossa dura; *Co*, cochlea.

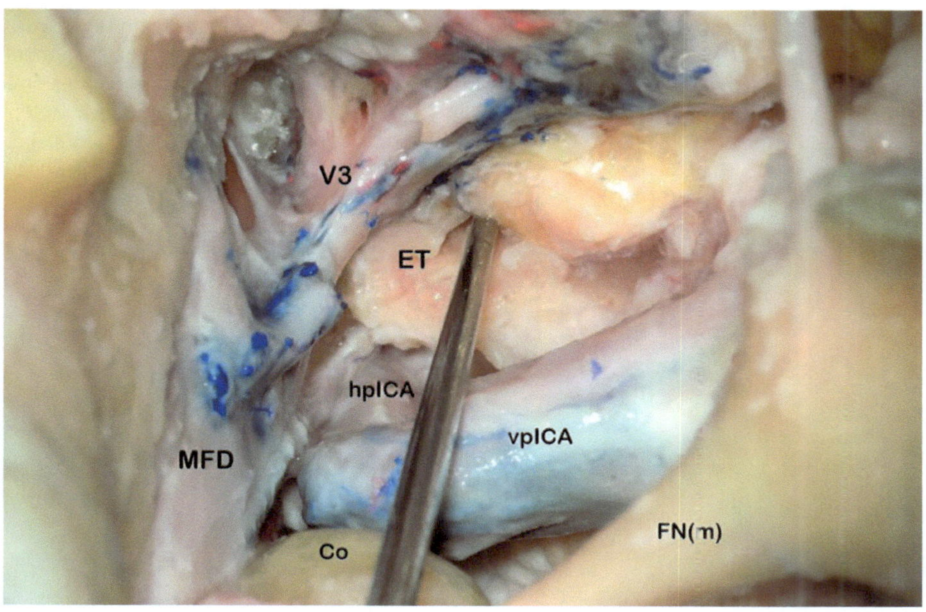

Fig. 12.232 The mandibular nerve is then identified. A probe is inserted into the cartilaginous Eustachian tube. The Eustachian tube is an important landmark in skull base surgery, and has a close spatial relation with the horizontal portion of petrous carotid artery. *hpICA*, horizontal portion of petrous carotid artery; *vpICA*, vertical portion of petrous carotid artery; *V3*, mandibular nerve; *ET*, Eustachian tube; *Co*, cochlea; *FN(m)*, mastoid segment of facial nerve; *MFD*, middle fossa dura.

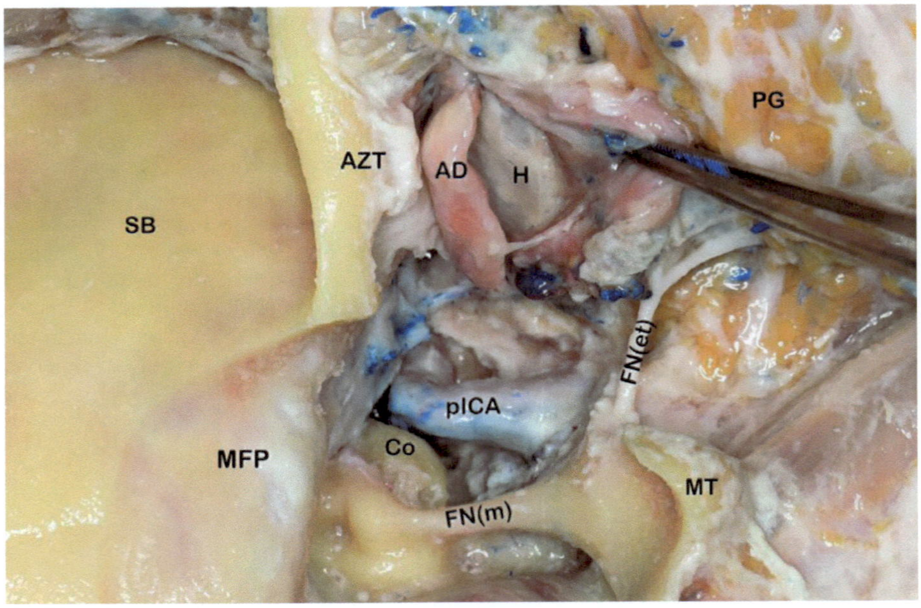

Fig. 12.233 The capsule of the temporomandibular joint is separated using strong scissors and bipolar coagulation. The articular disk is then removed, exposing the mandibular condyle. *PG*, parotid gland; *H*, head of mandible; *AD*, articular disk; *AZT*, anterior zygomatic tubercle; *SB*, squamosal bone; *MFP*, middle fossa plate; *Co*, cochlea; *pICA*, petrous segment of internal carotid artery; *FN(et)*, extratemporal segment of facial nerve; *FN(m)*, mastoid segment of facial nerve; *MT*, mastoid tip.

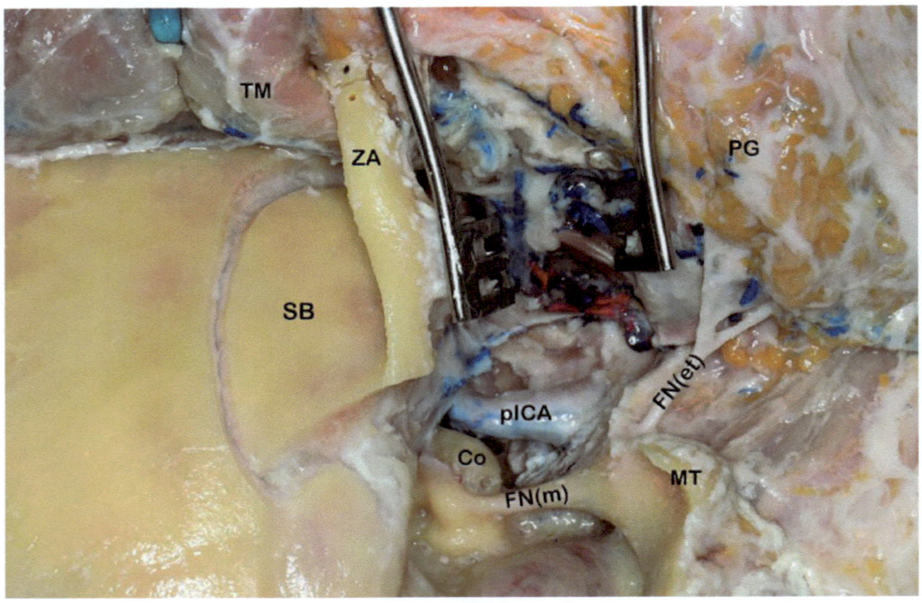

Fig. 12.234 The zygomatic arch is transected after two burr holes have been created for further refixation at the end of the procedure in live surgery. The zygomatic arch and part of the squamosal bone are osteotomized en bloc. The zygomatic process has been kept in contact with the vertical component of the temporal bone to achieve better cosmetic results on reconstruction. *TM*, temporalis muscle; *ZA*, zygomatic arch; *SB*, squamosal bone; *PG*, parotid gland; *pICA*, petrous segment of internal carotid artery; *Co*, cochlea; *FN(et)*, extratemporal segment of facial nerve; *FN(m)*, mastoid segment of facial nerve; *MT*, mastoid tip.

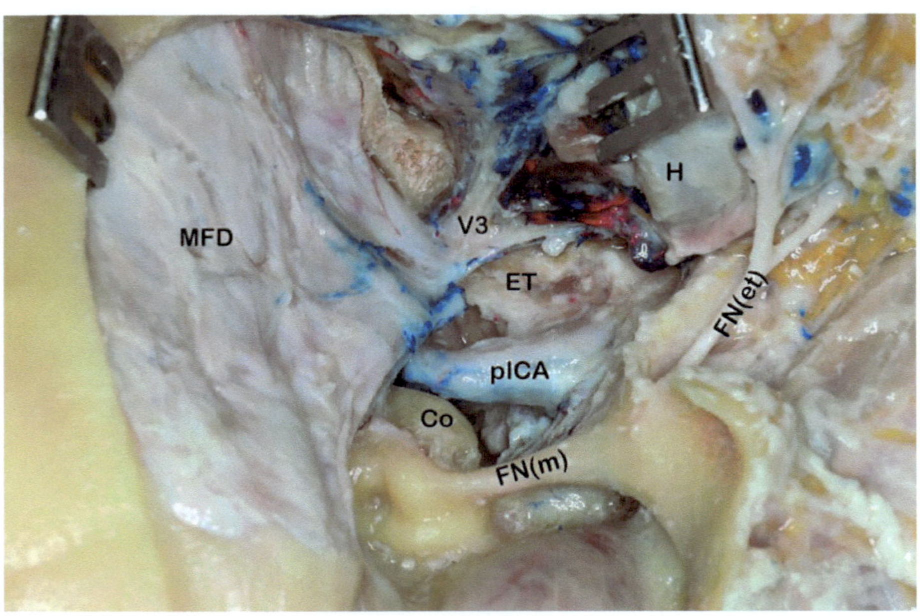

Fig. 12.235 The Fisch infratemporal fossa retractor is applied with inferior displacement of the head of the mandible. This provides complete control of the vertical segment of the internal carotid artery. *MFD*, middle fossa dura; *V3*, mandibular nerve; *ET*, Eustachian tube; *H*, head of mandible; *pICA*, petrous segment of internal carotid artery; *Co*, cochlea; *FN(et)*, extratemporal segment of facial nerve; *FN(m)*, mastoid segment of facial nerve.

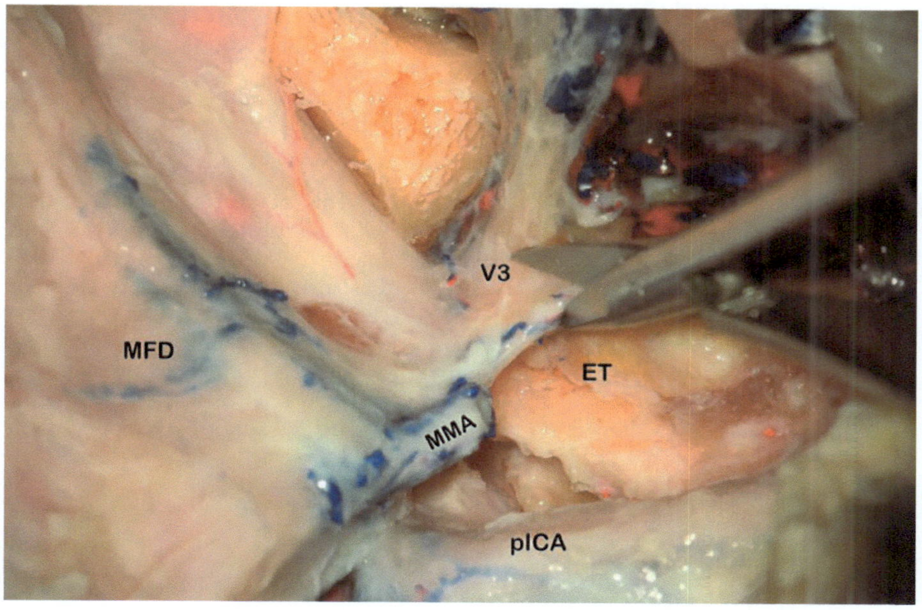

Fig. 12.236 The mandibular nerve is also transected after bipolar coagulation. *V3*, mandibular nerve; *ET*, Eustachian tube; *MMA*, middle meningeal artery; *pICA*, petrous segment of internal carotid artery; *MFD*, middle fossa dura.

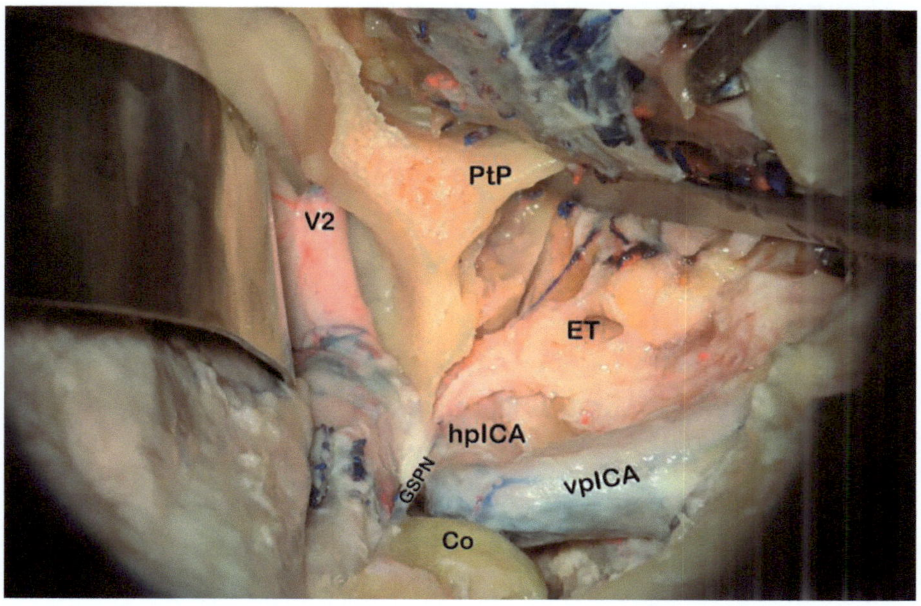

Fig. 12.237 The mandibular nerve is transected, and the horizontal portion of petrous carotid artery is further exposed. *PtP*, pterygoid process; *ET*, Eustachian tube; *V2*, maxillary nerve; *hpICA*, horizontal portion of petrous carotid artery; *vpICA*, vertical portion of petrous carotid artery; *GSPN*, greater superficial petrosal nerve; *Co*, cochlea.

Case 1: Petrous Apex Cholesterol Granuloma

See Figs. 12.238–12.264.

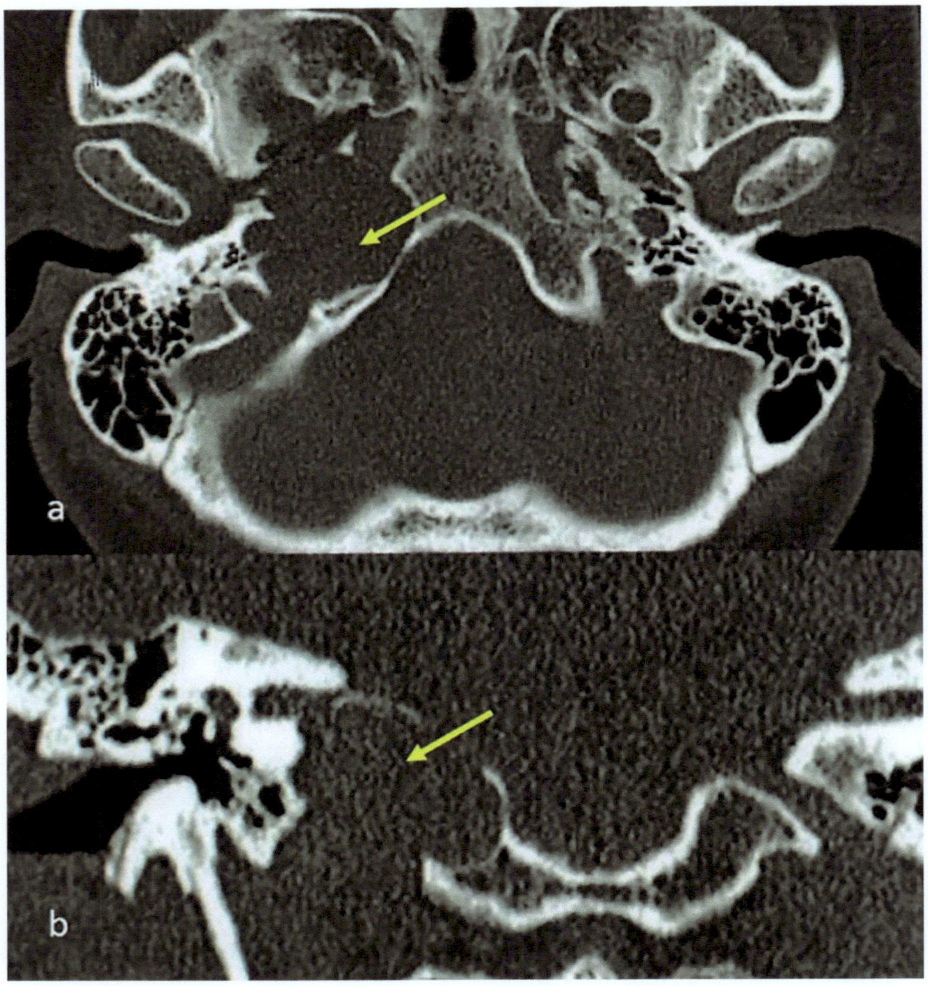

Fig. 12.238 This 58-year-old female patient presented with a growing right petrous apex lesion *(yellow arrow)*. The horizontal portion of the petrous internal carotid artery (ICA) is involved. The patient presented with moderate sensorineural hearing loss and diplopia. An infratemporal fossa type B approach was therefore chosen. (A) HRCT scan of the temporal bone, axial cut. (B) HRCT scan of the temporal bone, coronal cut.

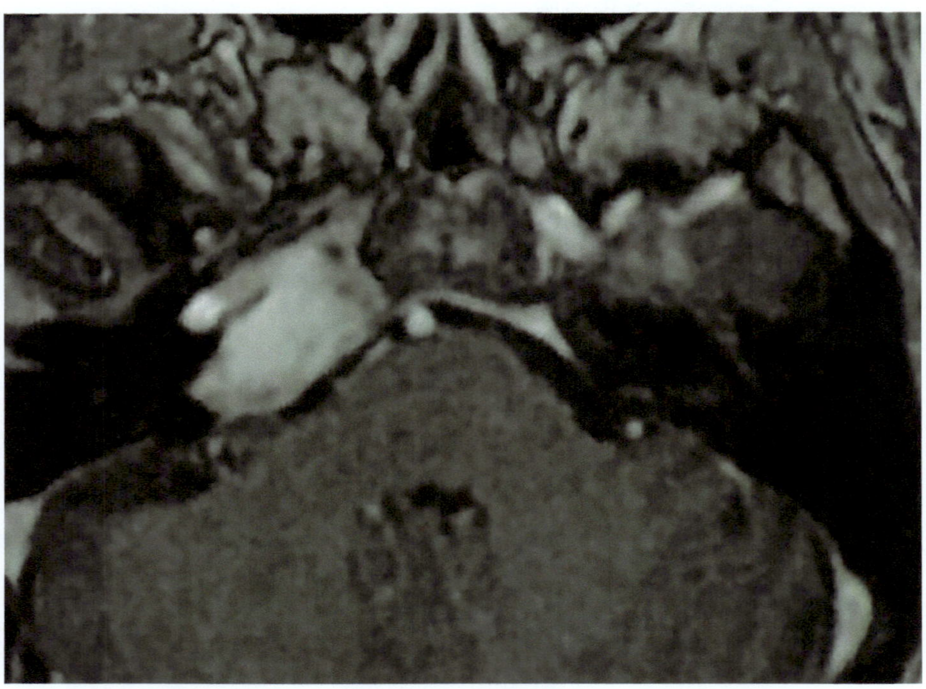

Fig. 12.239 T1W-MRI with Gd enhancement, axial cut. The lesion shows hyperintensity on T1 sequences. The presence of hyperintensity also on T2 suggests a cholesterol granuloma.

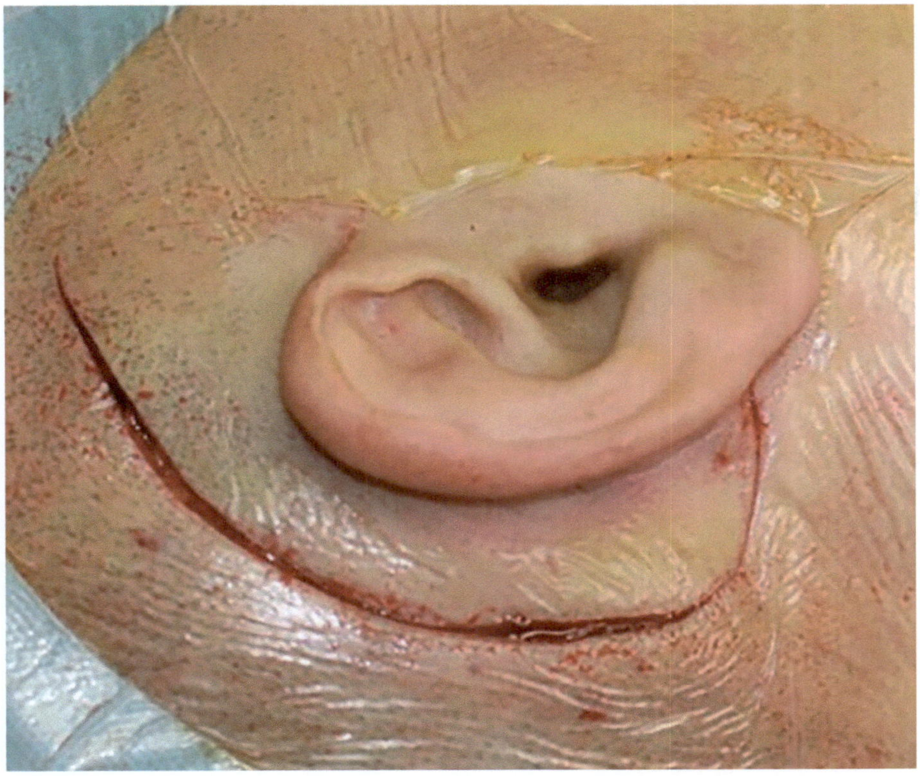

Fig. 12.240 C-shaped retroarticular skin incision.

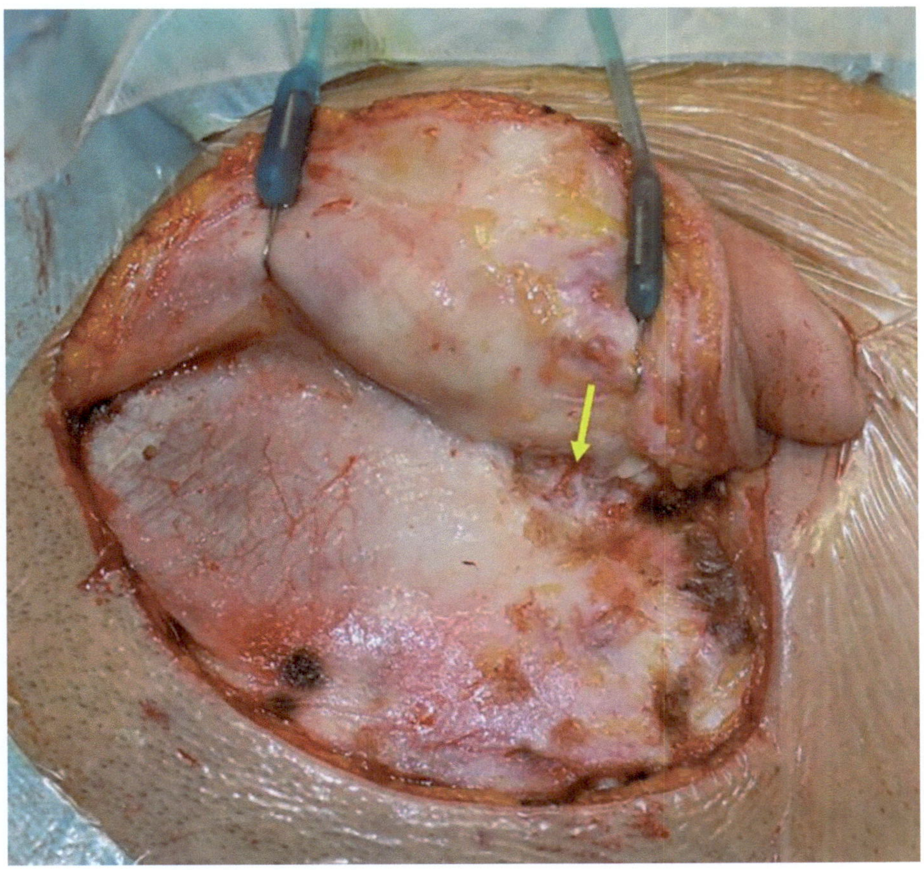

Fig. 12.241 The skin and subcutaneous tissue flap is raised. Note the level of the external auditory canal (EAC, *yellow arrow*).

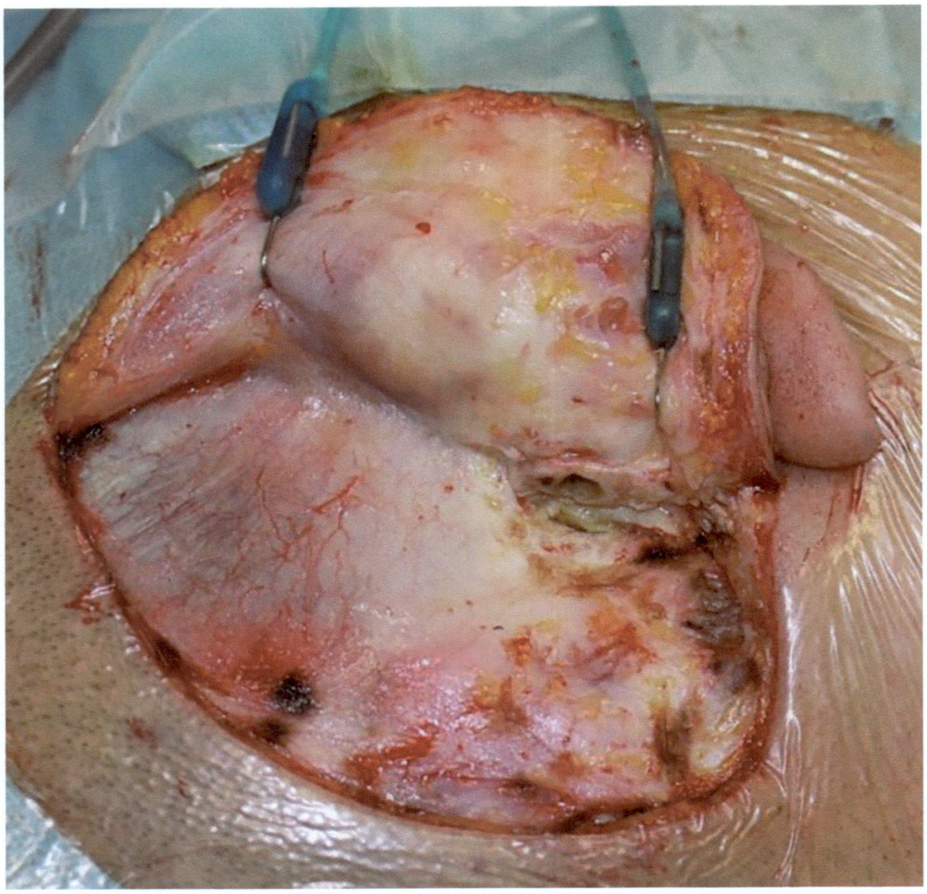

Fig. 12.242 The EAC is transected.

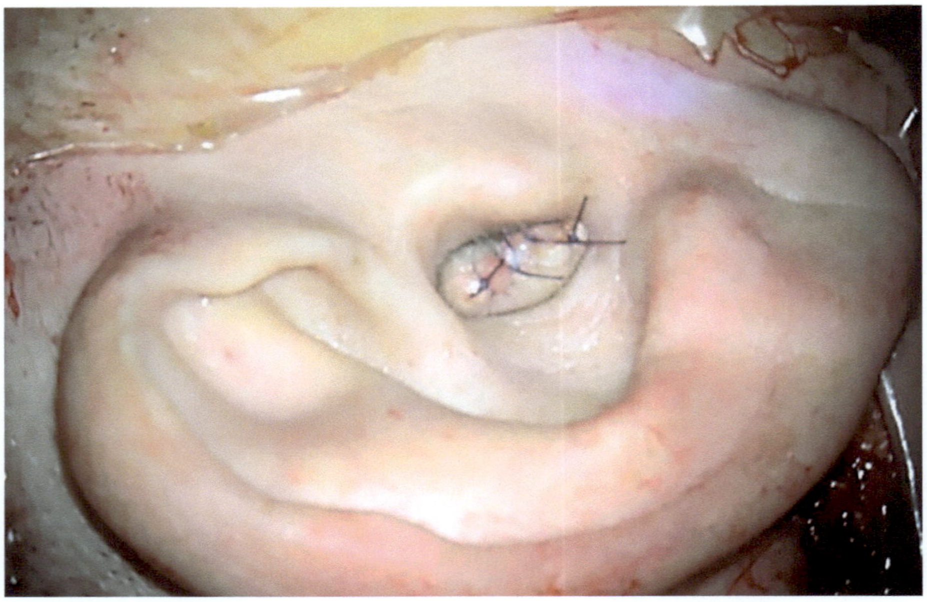

Fig. 12.243 Closure of the skin of the EAC (first layer).

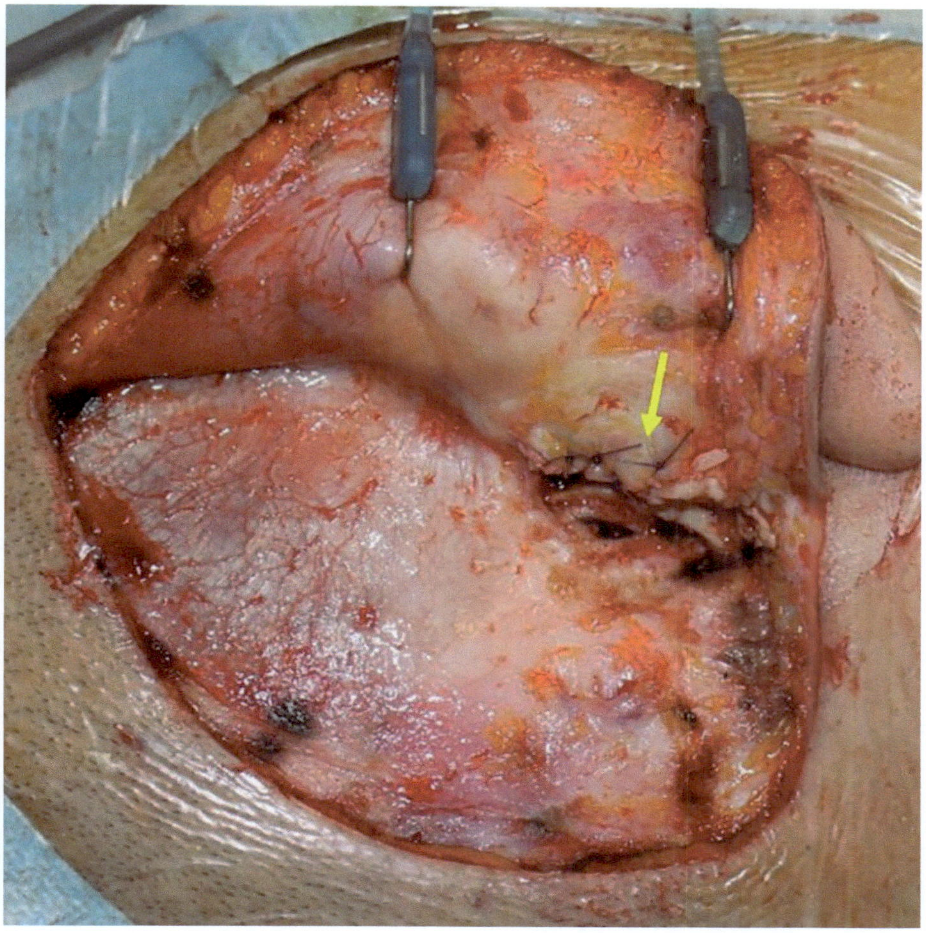

Fig. 12.244 Blind sac closure of the EAC (second layer), using fibrous-cartilaginous tissue *(yellow arrow)*.

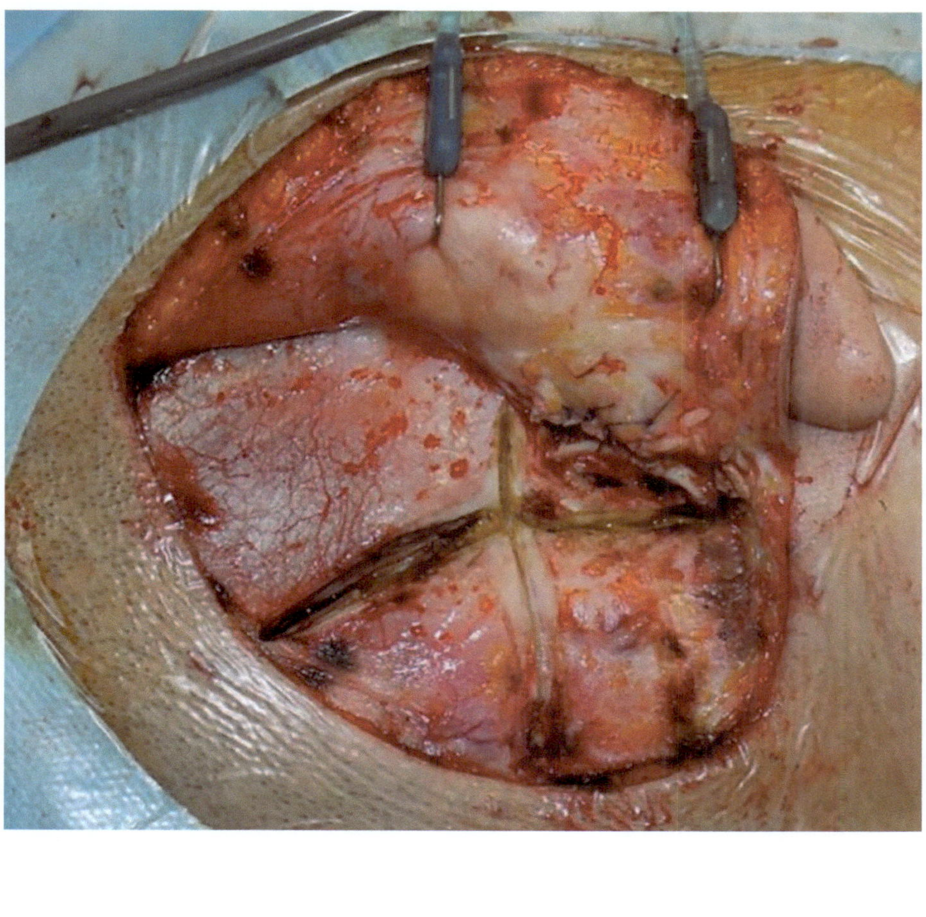

Fig. 12.245 Musculoperiosteal flap.

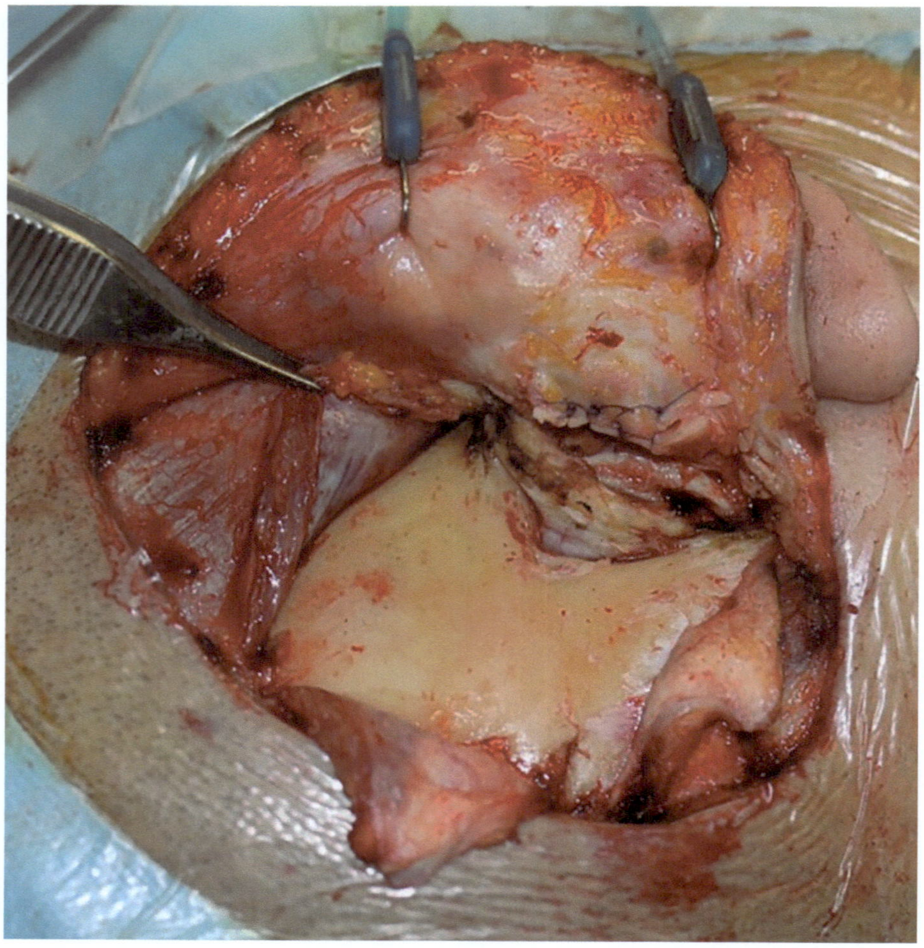

Fig. 12.246 Elevation of the musculoperiosteal flap to expose the mastoid cavity.

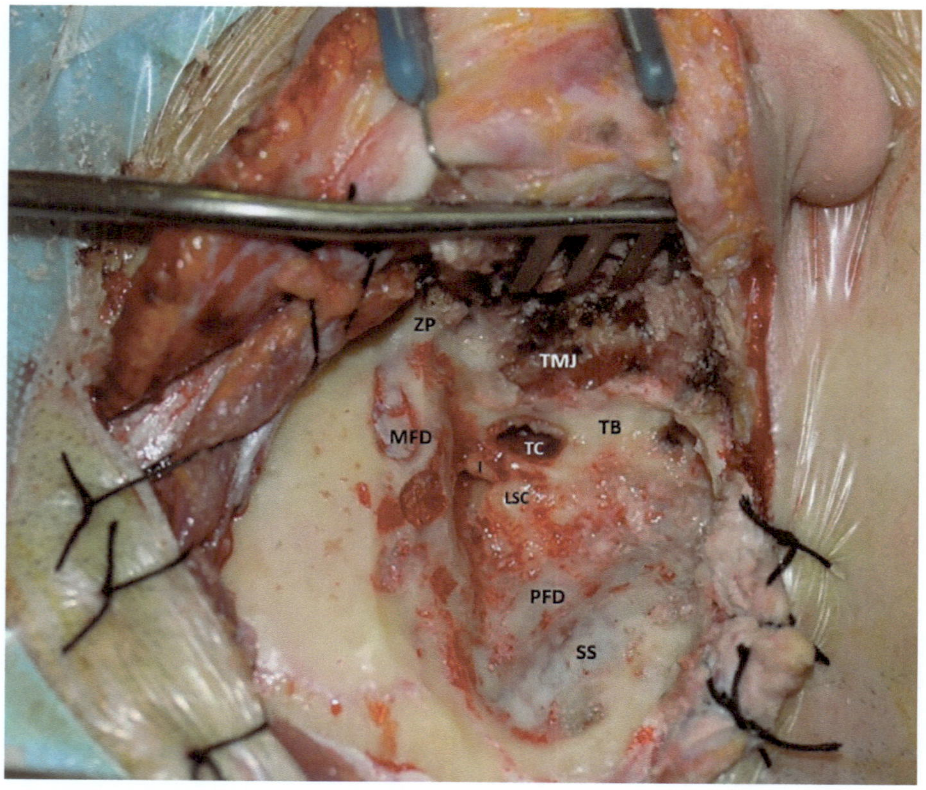

Fig. 12.247 Mastoidectomy has been performed. In this case, the zygomatic process (ZP) was not cut. To achieve control of the petrous internal carotid artery, the entire tympanic bone (TB) must be removed along with anterior displacement of the temporomandibular joint (TMJ). *MFD*, middle fossa dura; *TC*, tympanic cavity; *I*, incus; *LSC*, lateral semicircular canal; *PFD*, posterior fossa dura; *SS*, sigmoid sinus.

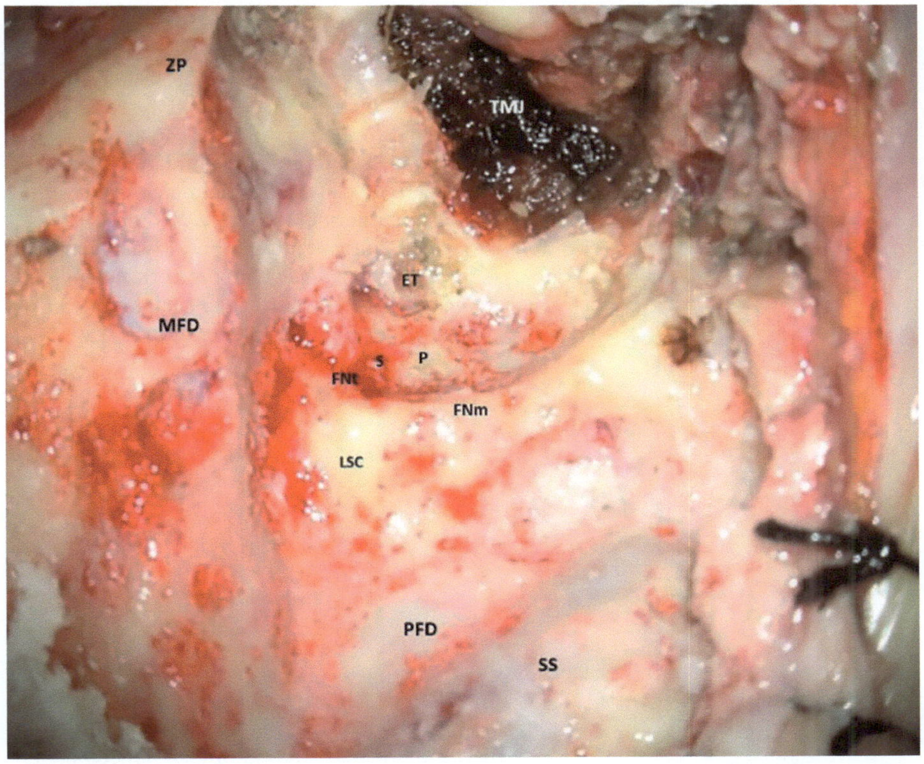

Fig. 12.248 Microscopic view. Subtotal petrosectomy was completed. *MFD*, middle fossa dura; *ZP*, zygomatic process; *LSC*, lateral semicircular canal; *PFD*, posterior fossa dura; *SS*, sigmoid sinus; *P*, promontory; *FNt*, tympanic segment of facial nerve; *FNm*, mastoid segment of facial nerve; *S*, stapes; *ET*, Eustachian tube; *TMJ*, temporomandibular joint (covered with Surgicel).

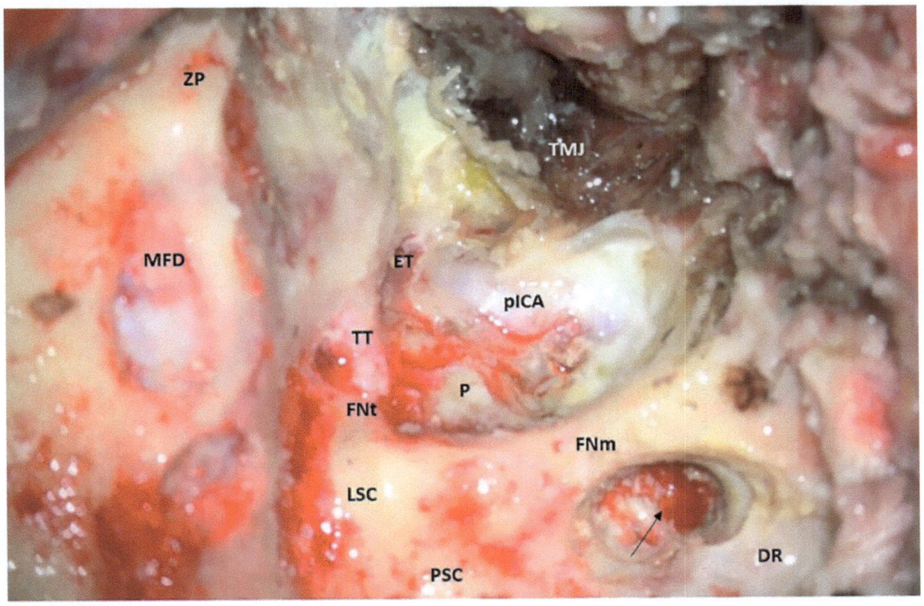

Fig. 12.249 The petrous internal carotid artery (pICA) is further decompressed, exposing its genu. The bone anterior to the pICA is drilled with initial exposure of the cholesterol granuloma (ChG) and the middle meningeal artery (mma). *MFD*, middle fossa dura; *TT*, tensor tympani muscle; *ET*, Eustachian tube; *P*, promontory.

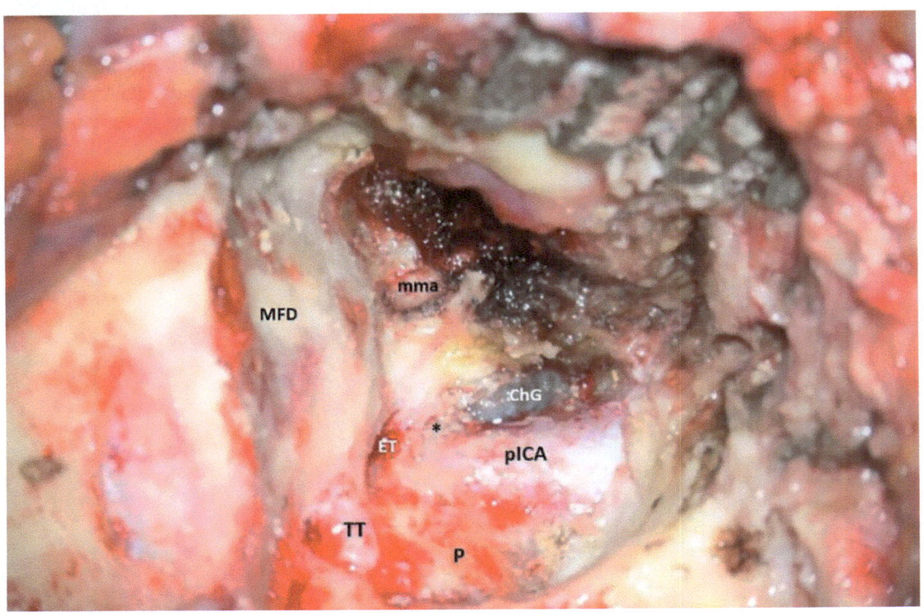

Fig. 12.250 The cholesterol granuloma has been drained from the petrous apex (PA), and its capsular component removed using cottonoids (Ct). Another remnant can be seen on the anterior surface of the petrous internal carotid artery (pICA). The middle meningeal artery was previously coagulated and cut. *, horizontal portion of the petrous internal carotid artery; *MFD*, middle fossa dura; *TT*, tensor tympani muscle; *LSC*, lateral semicircular canal; *S*, stapes; *P*, promontory; *JB*, jugular bulb; *FNm*, mastoid portion of the facial nerve; *FNt*, tympanic portion of the facial nerve.

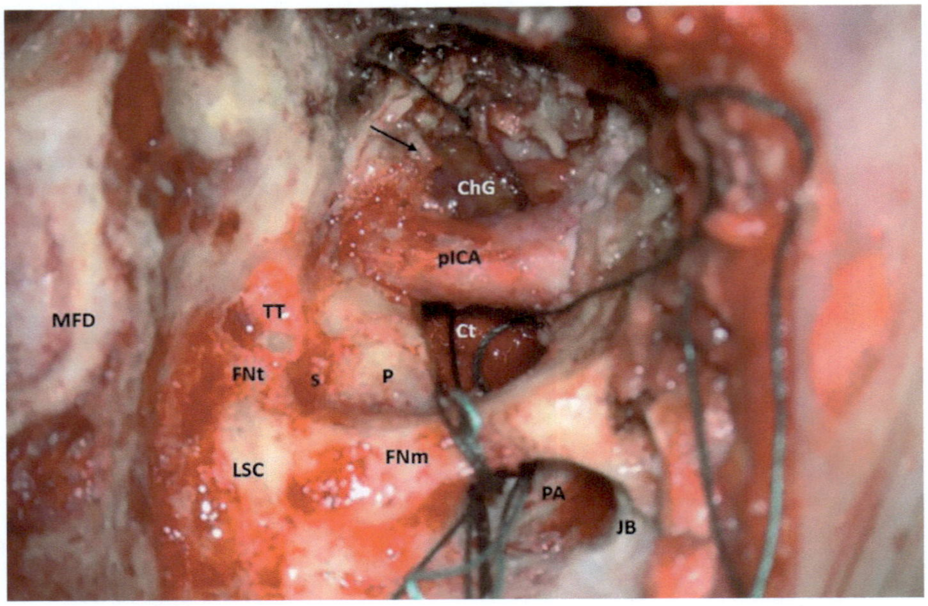

Fig. 12.251 Further drilling of the bone anterior to the promontory exposed the vertical portion of the petrous internal carotid artery (pICA). A retro- and subfacial tympanostomy is then performed to expose the petrous apex *(arrow)*. *MFD*, middle fossa dura; *ZP*, zygomatic process; *LSC*, lateral semicircular canal; *PSC*, posterior semicircular canal; *DR*, digastric ridge; *TT*, tensor tympani muscle; *ET*, Eustachian tube; *P*, promontory; *TMJ*, temporomandibular joint; *FNm*, mastoid portion of the facial nerve; *FNt*, tympanic portion of the facial nerve.

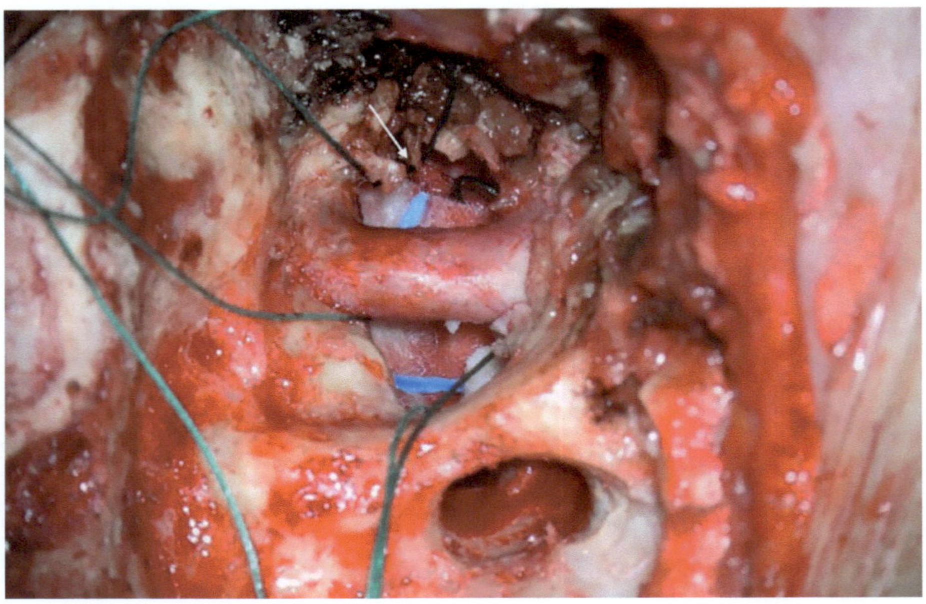

Fig. 12.252 The anterior remnant of the cholesterol granuloma has been removed *(arrow)*. Note that the internal carotid artery is completely decompressed (360 degrees) and cleaned.

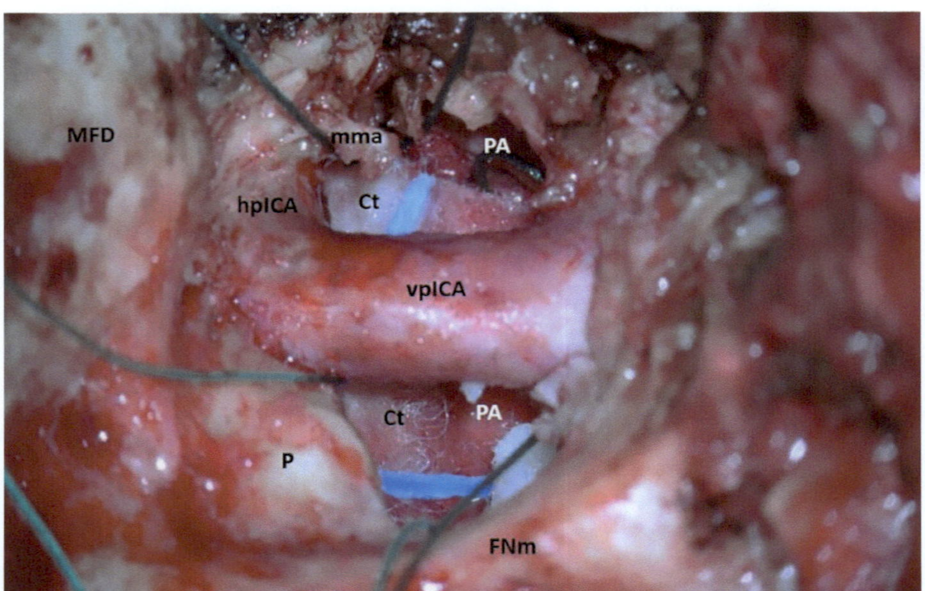

Fig. 12.253 Closer view. *MFD*, middle fossa dura; *mma*, middle meningeal artery (cut); *PA*, petrous apex; *P*, promontory; *Ct*, cottonoids; *FNm*, mastoid portion of the facial nerve; *vpICA*, vertical portion of the petrous ICA; *hpICA*, horizontal portion of the petrous ICA.

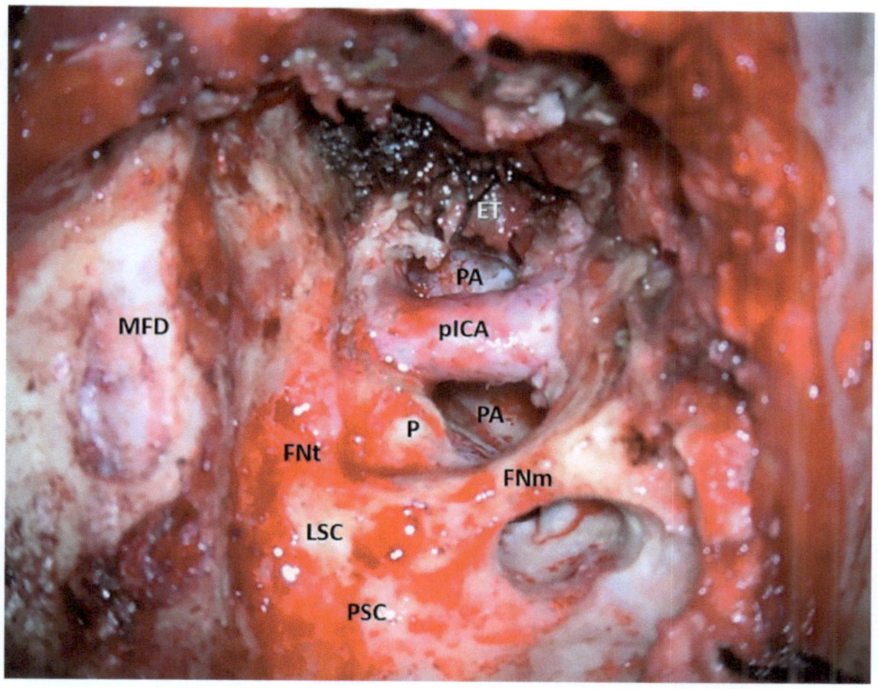

Fig. 12.254 Surgical view at the end of the procedure. The Eustachian tube (ET) has been sutured. *MFD*, middle fossa dura; *LSC*, lateral semicircular canal; *PSC*, posterior semicircular canal; *FNm*, mastoid portion of the facial nerve; *FNt*, tympanic portion of the facial nerve; *PA*, petrous apex; *pICA*, petrous portion of the internal carotid artery; *P*, promontory.

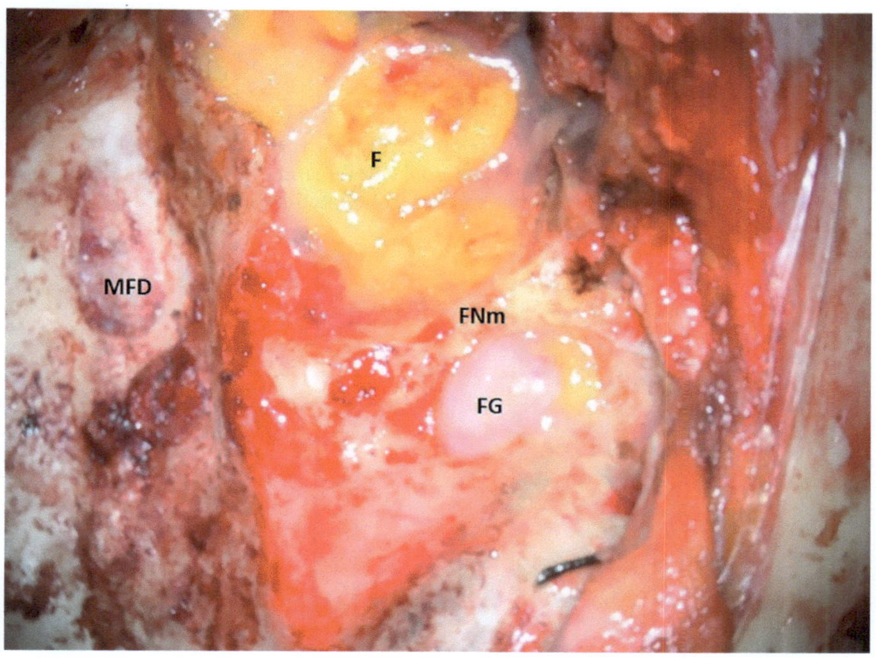

Fig. 12.255 Obliteration of the cavity with abdominal fat (F) and fibrin glue (FG). *MFD*, middle fossa dura; *FNm*, mastoid portion of the facial nerve.

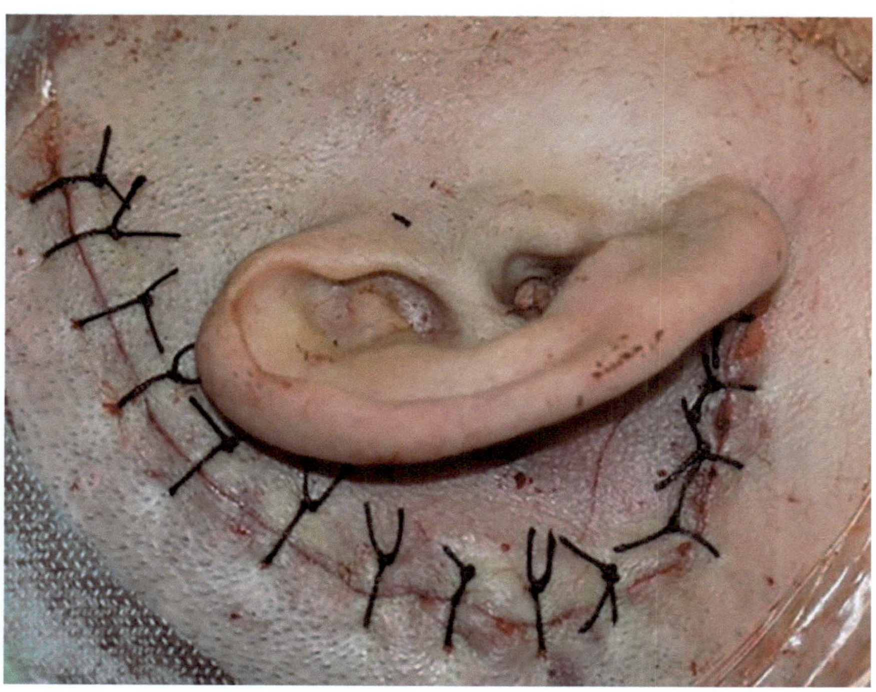

Fig. 12.256 Final view after skin closure.

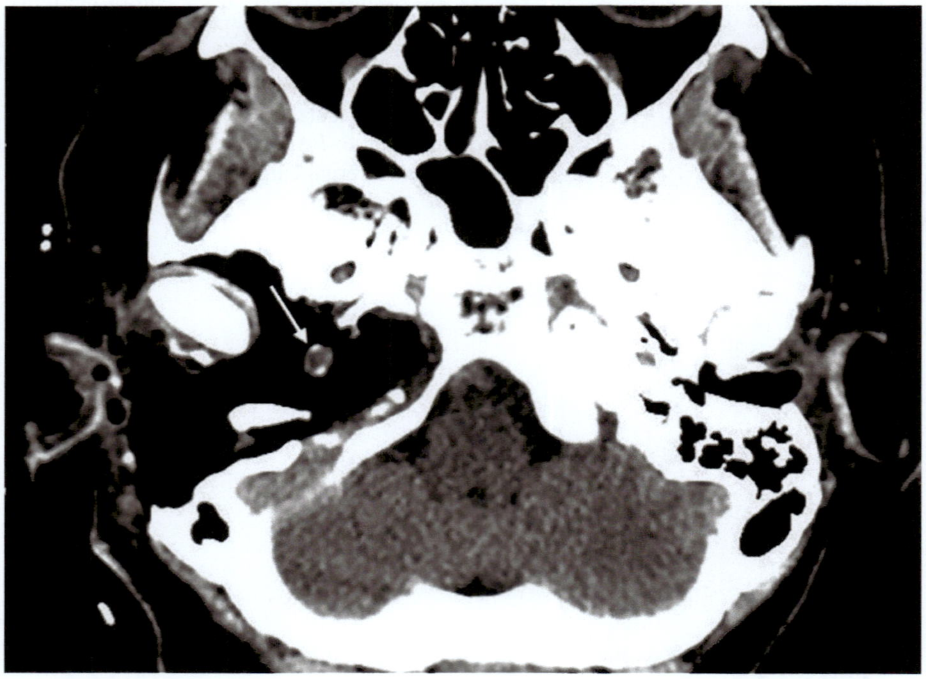

Fig. 12.257 Postoperative CT scan, axial view. Note the vertical portion of the petrous internal carotid artery *(arrow)* within the surgical cavity.

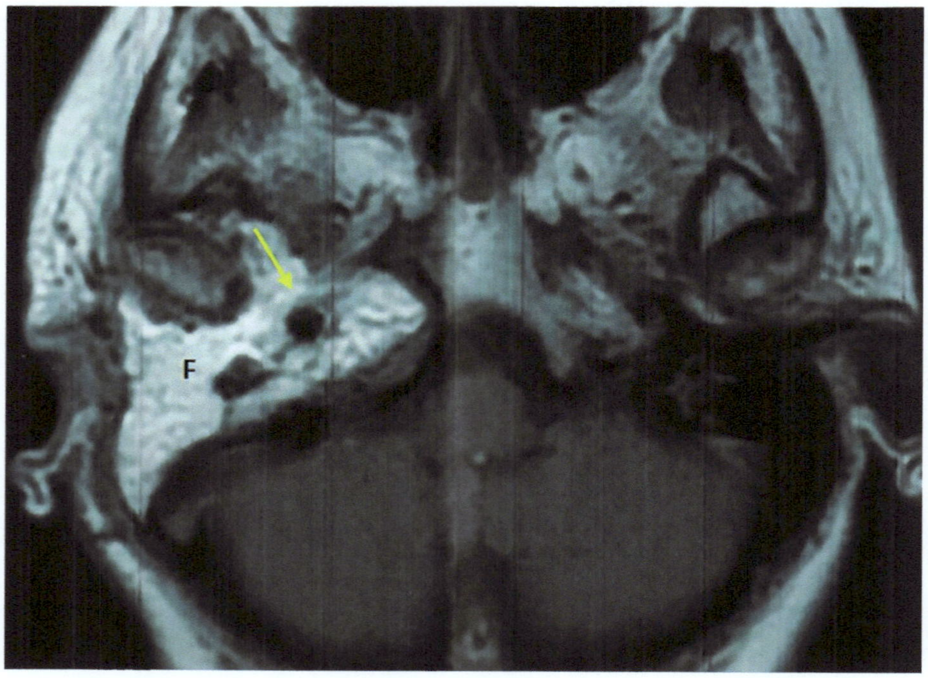

Fig. 12.258 Postoperative MRI scan, axial view. Note the vertical portion of the petrous internal carotid artery *(arrow)* within the surgical cavity, obliterated by abdominal fat (F).

Case 2: Chondrosarcoma of the Petrous Apex

See Figs. 12.259–12.264.

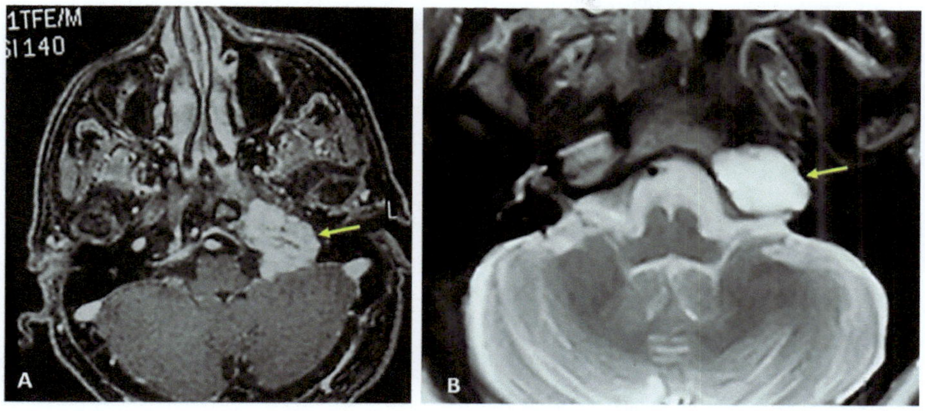

Fig. 12.259 This 57-year-old male patient presented with total left hearing loss and persistent diplopia. A brain MRI showed a left petrous apex lesion with Gd enhancement (Fig. A—*arrow*). The tumor also showed hyperintensity on T2W sequences (B). The patient underwent total tumor removal via a combined infratemporal fossa type B and transotic approach. The tumor was histologically diagnosed as a chondrosarcoma.

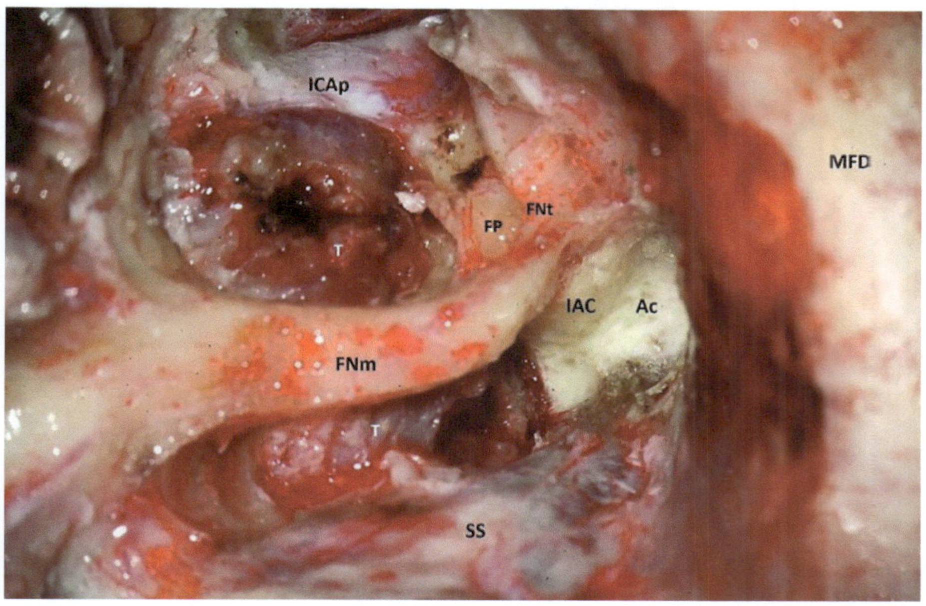

Fig. 12.260 The surgical approach is almost complete. Not that the tumor involves the infralabyrinthine apical cells and reaches the petrous internal carotid artery (ICAp). *FNm*, mastoid portion of the facial nerve; *IAC*, internal auditory canal; *Ac*, apical cells; *SS*, sigmoid sinus; *MFD*, middle fossa dura; *FNt*, tympanic portion of the facial nerve; *FP*, footplate; *T*, tumor.

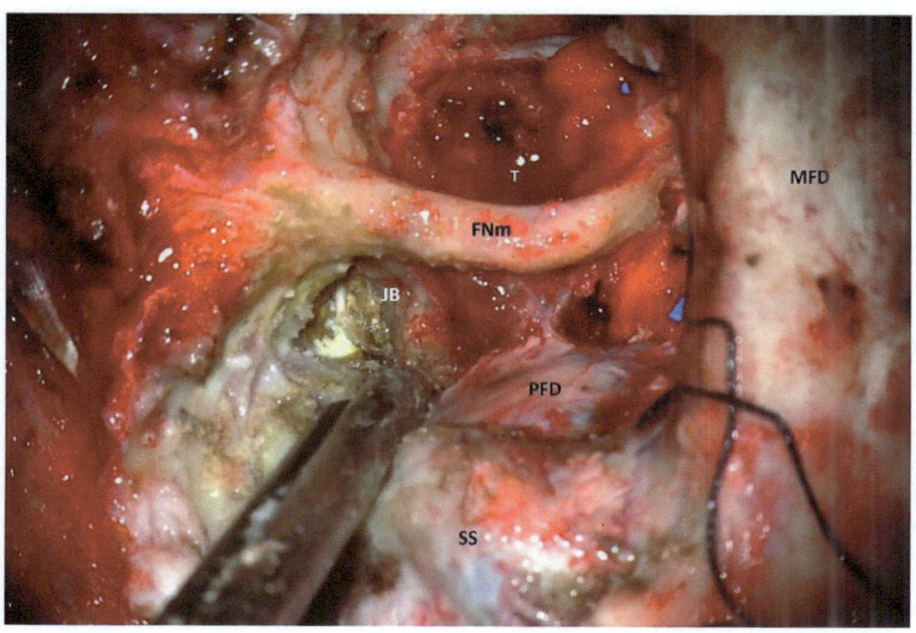

Fig. 12.261 The tumor was completely removed. The cochlea and labyrinth have been drilled to access all the infralabyrinthine apical cells. Note the healthy bone at the level of the petrous apex (PA) and the completely decompressed petrous internal carotid artery (ICAp). *FNm*, mastoid portion of the facial nerve; *MFD*, middle fossa dura; *JB*, jugular bulb; *PFD*, posterior fossa dura; *IAC*, internal auditory canal.

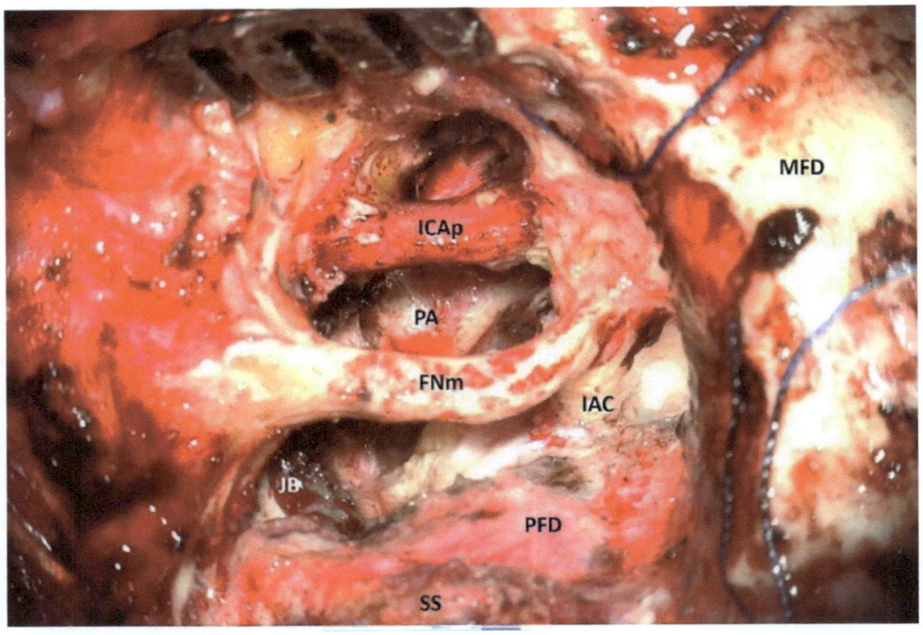

Fig. 12.262 Note the amount of space medial to the mastoid portion of the facial nerve (FNm) and the petrous internal carotid artery (ICAp). At the end of the procedure, the Eustachian tube is sutured, and the surgical cavity is filled with abdominal fat. *MFD*, middle fossa dura; *PFD*, posterior fossa dura; *IAC*, internal auditory canal.

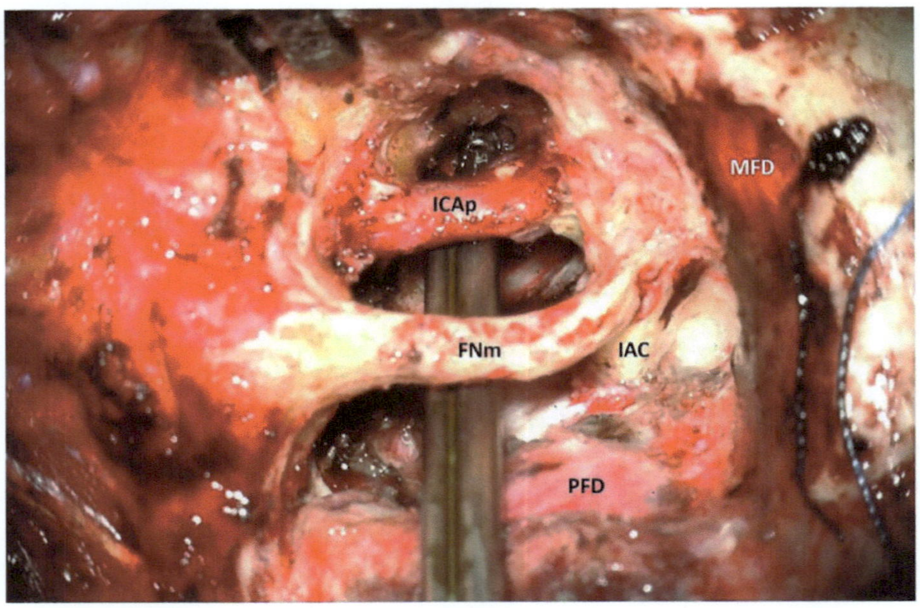

Fig. 12.263 The bleeding tumor (T) is further reduced. The posterior fossa dura (PFD) is decompressed, and the jugular bulb (JB) is skeletonized. *FNm*, mastoid portion of the facial nerve; *SS*, sigmoid sinus; *MFD*, middle fossa dura.

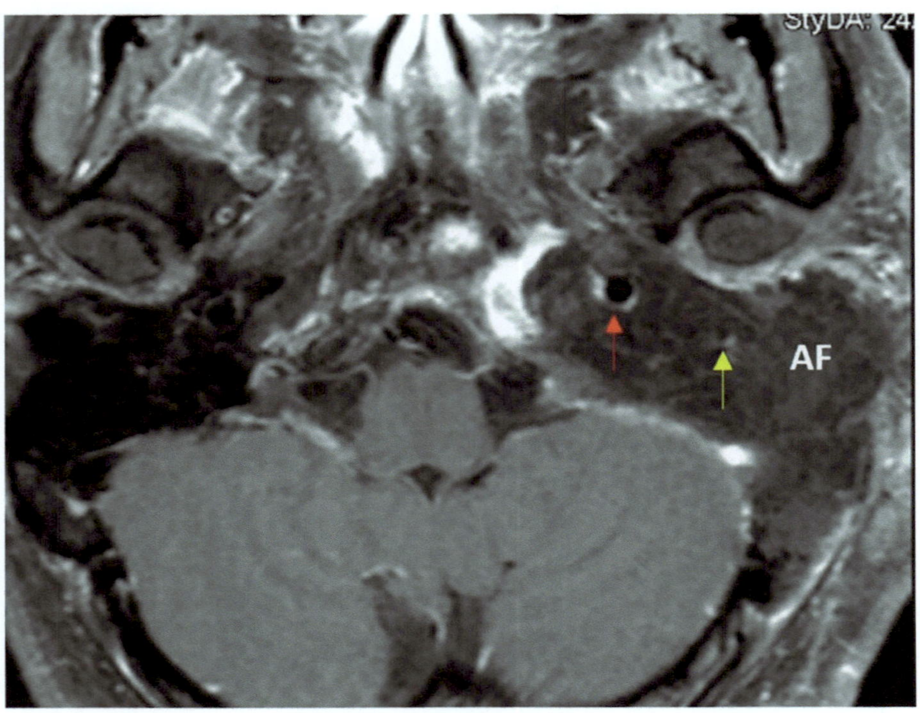

Fig. 12.264 Postoperative brain MRI showing the cavity filled with abdominal fat (AF). Note the petrous internal carotid artery *(red arrow)* and the mastoid portion of the facial nerve *(yellow arrow).*

HINTS AND PITFALLS

- Exposure of the extratemporal portion of the facial nerve is useful to prevent overstretching of the nerve during application of the infratemporal fossa retractor.
- The sphenoid spine serves as a landmark for the middle meningeal artery, which lies immediately anterior to it.
- Before the mandibular nerve is transected in live surgery, bipolar coagulation helps to reduce bleeding from the venous plexus surrounding the nerve.
- Drilling should always be parallel to the course of the internal carotid artery to reduce the risk of tearing the arterial wall.
- Bleeding from the veins inside the carotid sheath can be easily controlled with Surgicel packing.
- Bleeding from the inferior petrosal sinus can be controlled by intraluminal packing with Surgicel.

Infratemporal Fossa Approach Type C

Rationale

This is an anterior extension of the type B approach, in which the pterygoid process is drilled, providing control of the nasopharynx, the pterygopalatine fossa, and the sphenoid sinus (Fig. 12.265A and B). The approach is designed to give lateral access for the extirpation of extradural lesions involving the infratemporal fossa, the nasopharynx, the pterygopalatine fossa, the sphenoid sinus, and minimal extension into the cavernous sinus [47–54]. The pterygoid process with its medial and lateral laminae constitutes a key entity in the anatomy of this area. Fig. 12.266 illustrates the bony anatomy at the base of the skull. The hatched line represents the pharyngobasilar fascia. The nasopharynx can be entered by removing the lateral pharyngeal wall behind the medial pterygoid process. Exposure here gives visualization of the vomer, opposite inferior turbinate, and pharyngeal end of the opposite Eustachian tube. The pterygopalatine fossa is exposed by removing the pterygoid process. To expose the parasellar region, the zygoma and basal portion of the sphenoid are removed. The ipsilateral sphenoid and maxillary sinus are opened. For complete exposure of the cavernous sinus, the maxillary nerve is divided and the bone at the floor of the middle fossa is removed for extradural elevation of the temporal lobe. The most common complications reported are CSF leaks and wound difficulties, due to the wide communication between the nasopharynx and the operative field that makes closure more problematic [47,55–57]. Possible morbidity associated with the technique:

– Conductive hearing loss due to sacrifice of the middle ear.

– Sacrifice of the mandibular nerve. The resultant deficit, however, is of little significance to the patient.

– Drilling of the condylar fossa. However, this does not give a marked degree of temporomandibular dysfunction. Most patients suffer only a minimal degree of limited mobility and discomfort.

– A relatively lengthy procedure.

– Poor control of intranasal extension, particularly when the ethmoids are involved.

Advantages offered by the approach:

– Excellent control of the nasopharynx, pterygopalatine fossa, sphenoid sinus, and infratemporal fossa.

– Excellent control of the intrapetrous internal carotid artery reaching up to the precavernous segment.

– In cases of nasopharyngeal carcinoma, removal of the Eustachian tube, peritubal muscles, and surrounding soft tissue is performed en bloc with the tumor. Therefore, the lateral approaches are the only approaches that allow total tumor removal with adequate safety margin.

– No facial scars.

Indications

– Tumors of the infratemporal fossa and peritubal areas, such as cylindroma, adenoid cystic carcinoma, etc.

– Persistent or recurrent postirradiation nasopharyngeal carcinoma.

– Juvenile nasopharyngeal angiofibroma involving the pterygopalatine and infratemporal fossae.

– Tumors of the infratemporal fossa involving the parasellar region, such as extensive clival chordoma.

Contraindications

– Extensive involvement of the cavernous sinus.

– Tumor extension into the orbit.

– Nasopharyngeal carcinoma passing the midline.

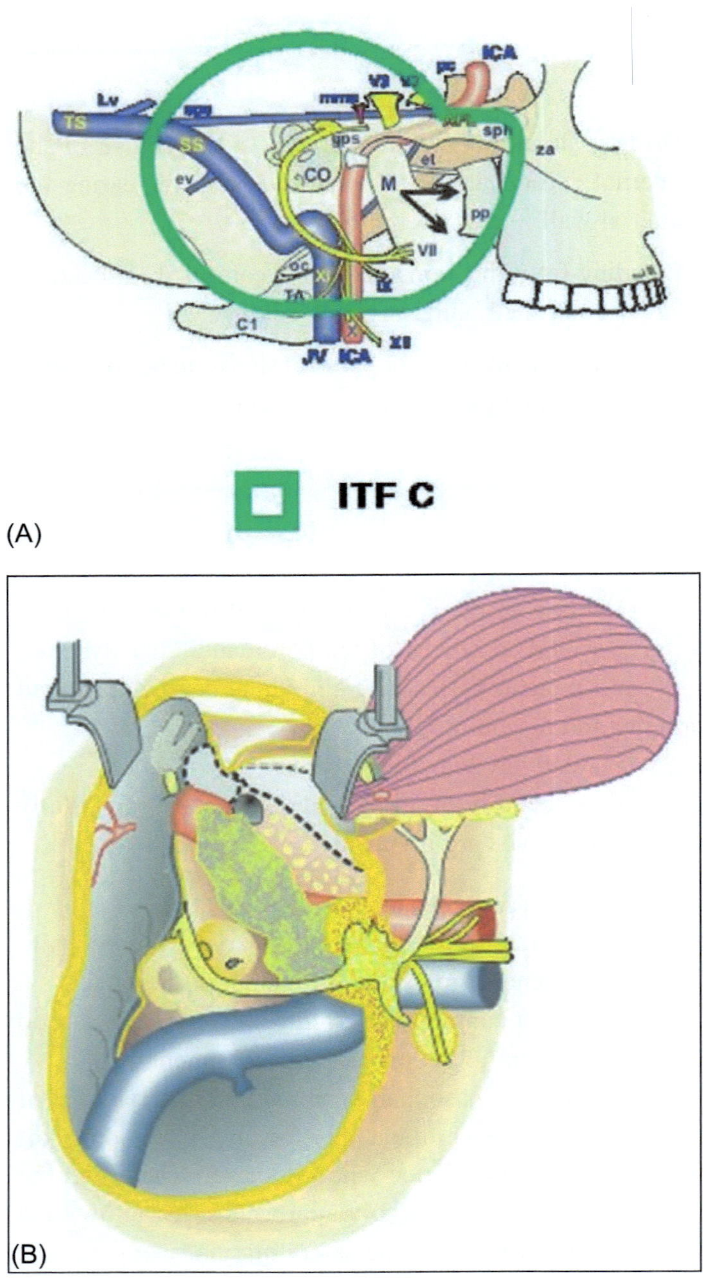

(A)

(B)

Fig. 12.265 (A, B) Drawings showing the hallmark of the infratemporal fossa approach type B (IFTB). *AFL*, anterior foramen lacerum; *C1*, first cervical vertebra; *CO*, cochlea; *et*, Eustachian tube; *ev*, emissary vein; *gps*, greater petrosal nerve; *ICA*, internal carotid artery; *JV*, jugular vein; *Lv*, vein of Labbé; *M*, mandible; *mma*, middle meningeal artery; *oc*, occipital condyle; *pc*, clinoid process; *pp*, pterygoid process; *sph*, sphenoid; *sps*, superior petrosal sinus; *TA*, transverse process of atlas; *TS*, transverse sinus; *za*, zygomatic arch; *V2*, maxillary nerve; *V3*, mandibular nerve; *VII*, facial nerve; *IX*, glossopharyngeal nerve; *XII*, hypoglossal nerve.

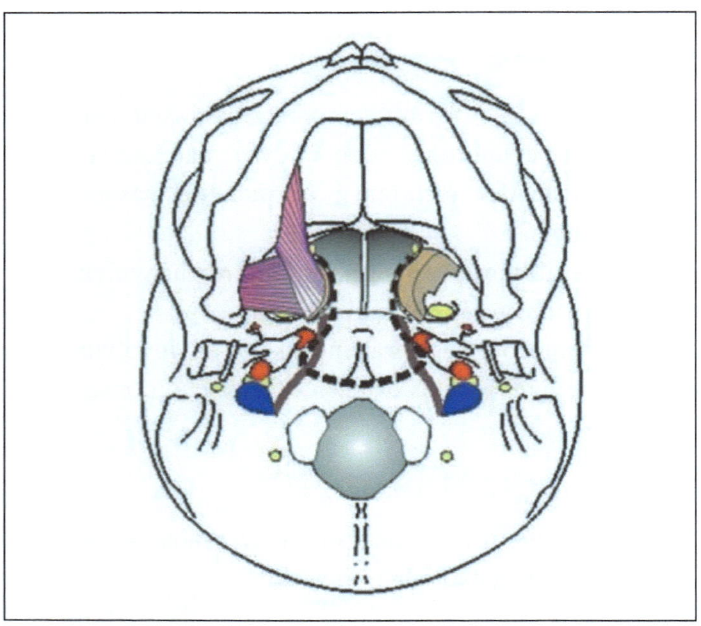

Fig. 12.266 The bony anatomy of the base of the skull. The hatched line represents the pharyngobasilar fascia.

Surgical Steps (Right Side)

1. The initial steps of the **infratemporal fossa B** are performed as before (Figs. 12.266–12.291).

2. Using a septal raspatory, the periosteum of the base of the pterygoid, the greater wing of the sphenoid, and the lateral lamina of the pterygoid process is elevated together with the attachment of the lateral pterygoid muscle (Fig. 12.267).

3. The base of the pterygoid is drilled (Fig. 12.268). According to the pathology and the site and extent of the lesion, the procedure varies as follows:

– **Tumors involving the sphenoid sinus and pterygopalatine fossa:**

1. The base of the greater wing of the sphenoid bone is drilled, uncovering the middle fossa dura. The sphenoid sinus can then be located. The vidian nerve usually passes just beneath the sinus, whereas the maxillary division of the trigeminal nerve is related to its roof (Fig. 12.269).
2. By retracting the middle fossa dura, the horizontal segment of the internal carotid artery until it reaches the anterior foramen lacerum is seen. The mucosa of the sinus is opened and its relation to the different structures can be appreciated.

– **In nasopharyngeal or peritubal tumors:**

1. The pterygoid muscles are cut near their insertions. The Eustachian tube is followed toward its pharyngeal opening, cutting its tough attachments to the skull base (Fig. 12.270). The periosteum of the petrous apex and clivus is sharply elevated using a septal raspatory. Scissors are used to cut the fibrous attachments at the level of the petro-occipital fissure (Fig. 9.23).
2. The medialpterygoid plate is drilled, the mucoperiosteum of the nasal cavity is opened at this level, and the nasopharynx is entered (Fig. 12.271). The tumor mass, the surrounding soft tissue, the Eustachian tube, and the pterygoid muscles are removed en bloc.

– Parasellar lesions:

1. The maxillary nerve running in the inferolateral wall of the cavernous sinus is sectioned (Fig. 12.272). This allows control of the lateral part of the sinus.
2. With further retraction of the middle fossa dura, the abducent nerve crossing the lateral surface of the internal carotid artery just distal to the anterior foramen lacerum can be seen (Fig. 12.273).
3. Further opening of the dura, and retraction of the temporal lobe allows visualization of the oculomotor nerve in the cavernous sinus (Fig. 12.274).
4. The tip of the suction is used to displace the petrous carotid artery laterally while the petrous apex medial to the petrous carotid artery is further drilled (Fig. 12.275).
5. The petrous apex is totally removed. The abducens nerve passes under the petrosphenoid ligament to enter the cavernous sinus (Fig. 12.276).
6. The abducens nerve pierces the dura forming the lower part of the posterior wall of the sinus at the upper border of the petrous apex and enters a dural cave, referred to as Dorello's canal, where it passes below the petrosphenoid ligament (Gruber's ligament) that extends from the lower part of the lateral edge of the dorsum sellae to the petrous apex, to enter the cavernous sinus (Fig. 12.277).
7. View after finishing the infratemporal fossa approach type C (Fig. 12.278).
8. After tumor removal, closure is performed as described in the infratemporal fossa type B approach.

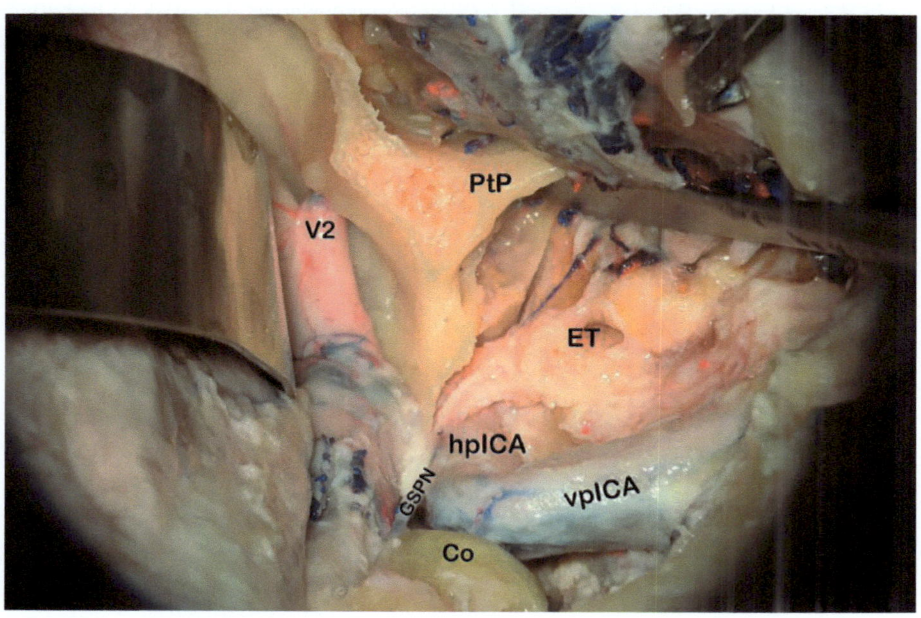

Fig. 12.267 Using a septal raspatory, the periosteum of the base of the pterygoid, the greater wing of the sphenoid, and the lateral lamina of the pterygoid process are elevated together with the attachment of the lateral pterygoid muscle. *PtP*, pterygoid process; *ET*, Eustachian tube; *V2*, maxillary nerve; *hpICA*, horizontal portion of petrous carotid artery; *vpICA*, vertical portion of petrous carotid artery; *GSPN*, greater superficial petrosal nerve; *Co*, cochlea.

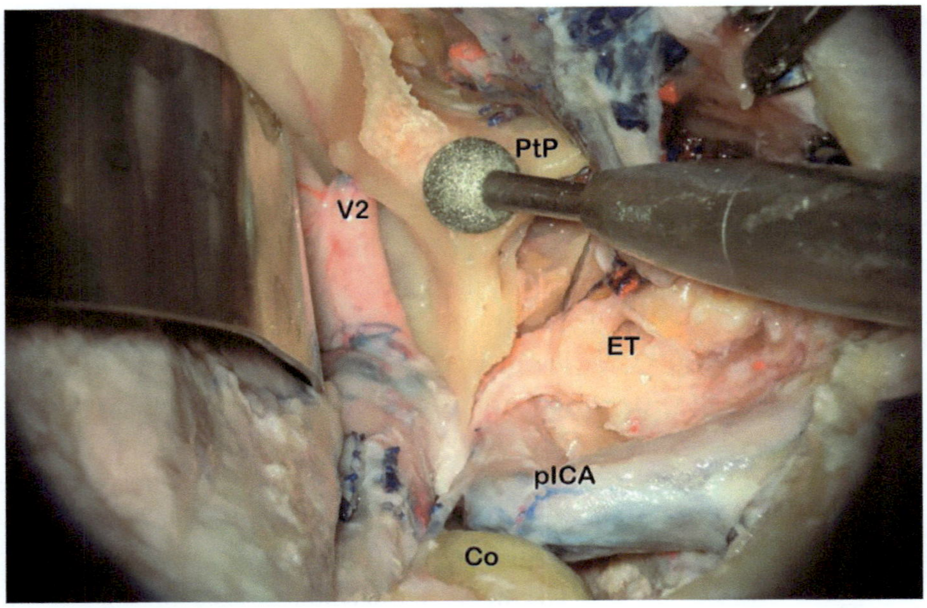

Fig. 12.268 The base of the pterygoid is now drilled with a large diamond burr. *PtP*, pterygoid process; *V2*, maxillary nerve; *ET*, Eustachian tube; *pICA*, petrous segment of internal carotid artery; *Co*, cochlea.

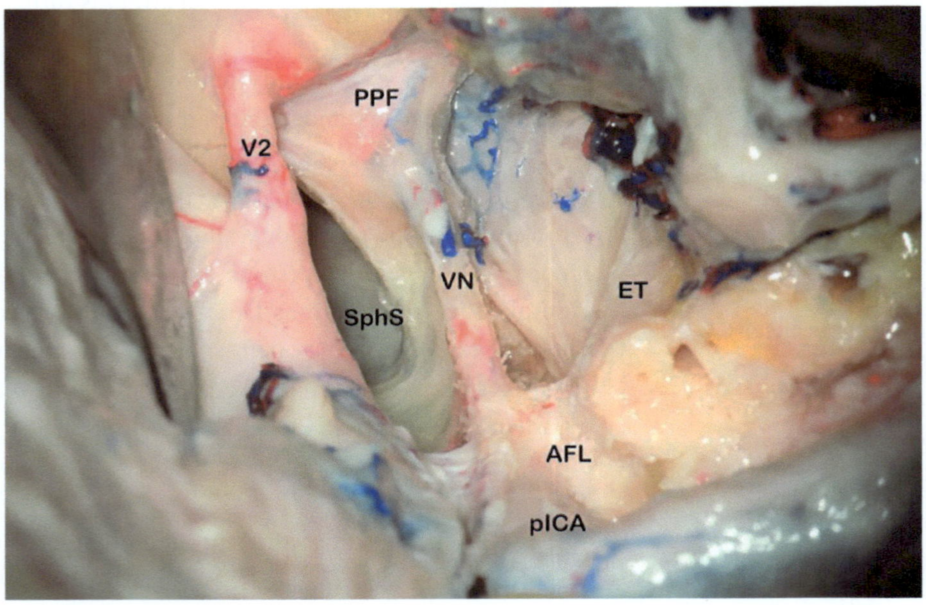

Fig. 12.269 The vidian nerve, which is formed by the junction of the greater superficial petrosal and the deep petrosal nerve, can be identified. Vidian nerve is a reliable landmark to the petrous internal carotid artery. The canal opens anteriorly into the medial part of the pterygopalatine fossa and posteriorly at the upper part of the anterolateral edge of the foramen lacerum. The sphenoid sinus mucosa is also seen. *PPF*, pterygopalatine fossa; *V2*, maxillary nerve; *VN*, vidian nerve; *ET*, Eustachian tube; *SphS*, sphenoid sinus; *AFL*, anterior foramen lacerum; *pICA*, petrous segment of internal carotid artery.

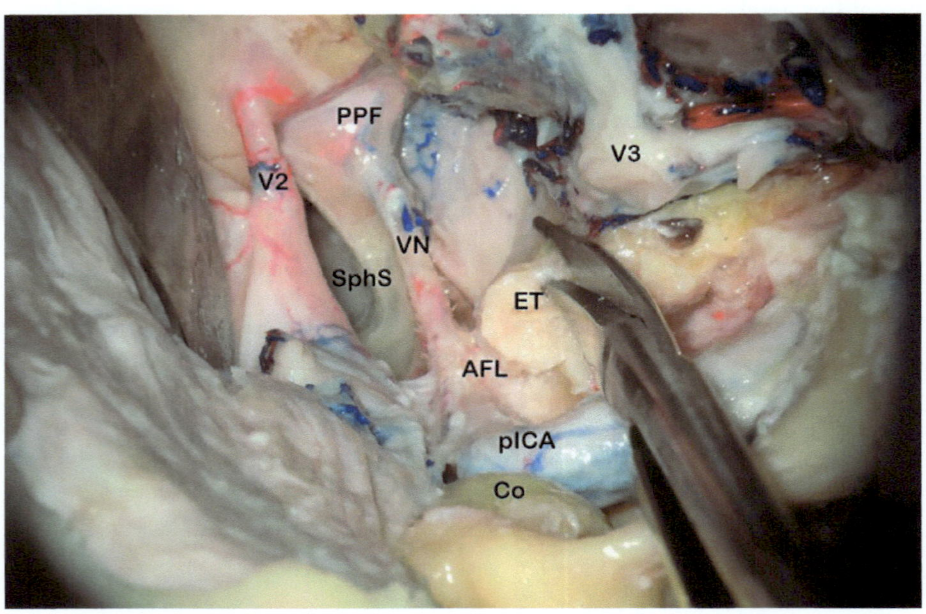

Fig. 12.270 The Eustachian tube can be followed inferomedially to its pharyngeal opening, cutting its tough attachments to the skull base. *PPF*, pterygopalatine fossa; *V3*, mandibular nerve; *V2*, maxillary nerve; *VN*, vidian nerve; *ET*, Eustachian tube; *SphS*, sphenoid sinus; *AFL*, anterior foramen lacerum; *pICA*, petrous segment of internal carotid artery; *Co*, cochlea.

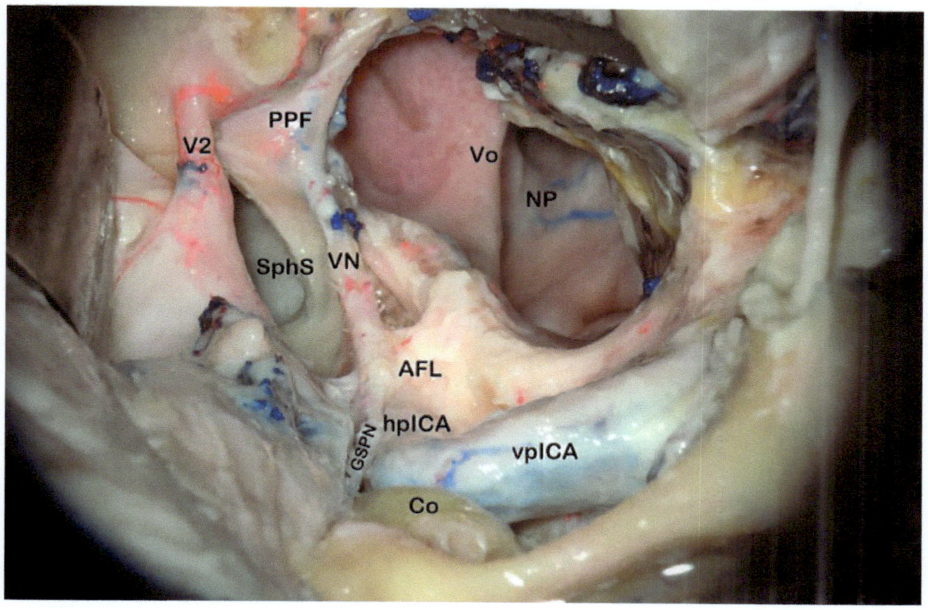

Fig. 12.271 The mucoperiosteum of the nasal cavity is opened at this level, and the nasopharynx is entered. The contralateral choana and the vomer can be seen. *V2*, maxillary nerve; *PPF*, pterygopalatine fossa; *SphS*, sphenoid sinus; *VN*, vidian nerve; *Vo*, vomer; *NP*, nasopharynx; *AFL*, anterior foramen lacerum; *hpICA*, horizontal portion of petrous carotid artery; *vpICA*, vertical portion of petrous carotid artery; *GSPN*, greater superficial petrosal nerve; *Co*, cochlea.

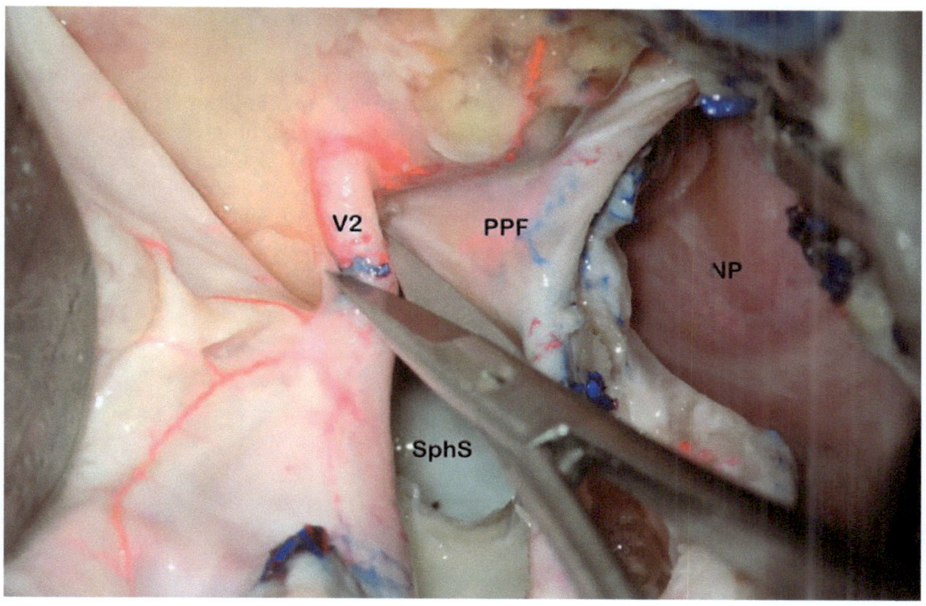

Fig. 12.272 The maxillary nerve running in the inferolateral wall of the cavernous sinus is sectioned. This allows control of the lateral part of the cavernous sinus. *V2*, maxillary nerve; *PPF*, pterygopalatine fossa; *NP*, nasopharynx; *SphS*, sphenoid sinus.

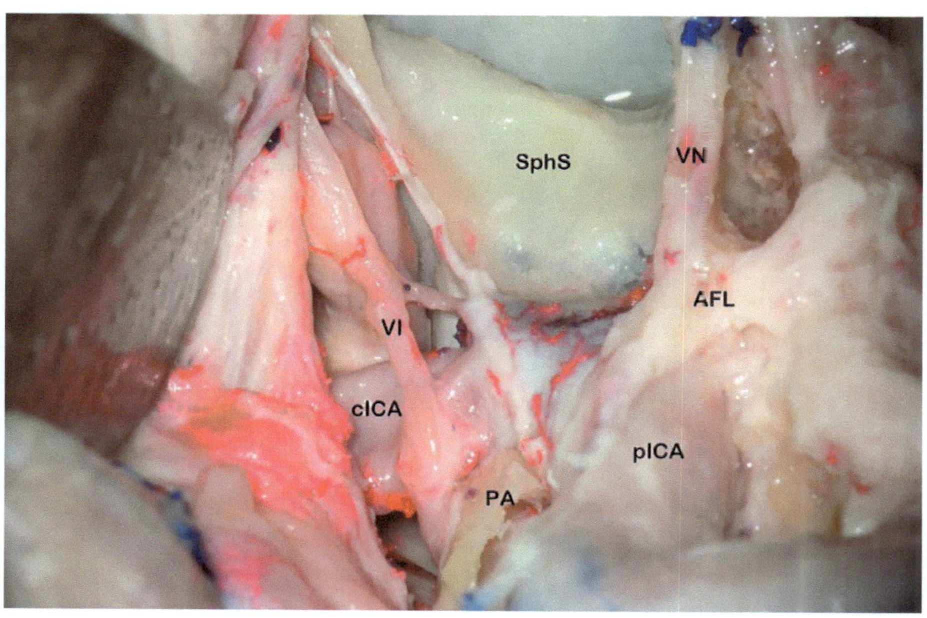

Fig. 12.273 The dura of the cavernous sinus is opened. Profuse bleeding from the sinus can be managed by Surgicel packing. The abducent nerve crosses the lateral surface of the cavernous segment of internal carotid artery just distal to the anterior foramen lacerum. *SphS*, sphenoid sinus; *VN*, vidian nerve; *AFL*, anterior foramen lacerum; *pICA*, petrous segment of internal carotid artery; *cICA*, cavernous segment of internal carotid artery; *VI*, abducent nerve; *PA*, petrous apex.

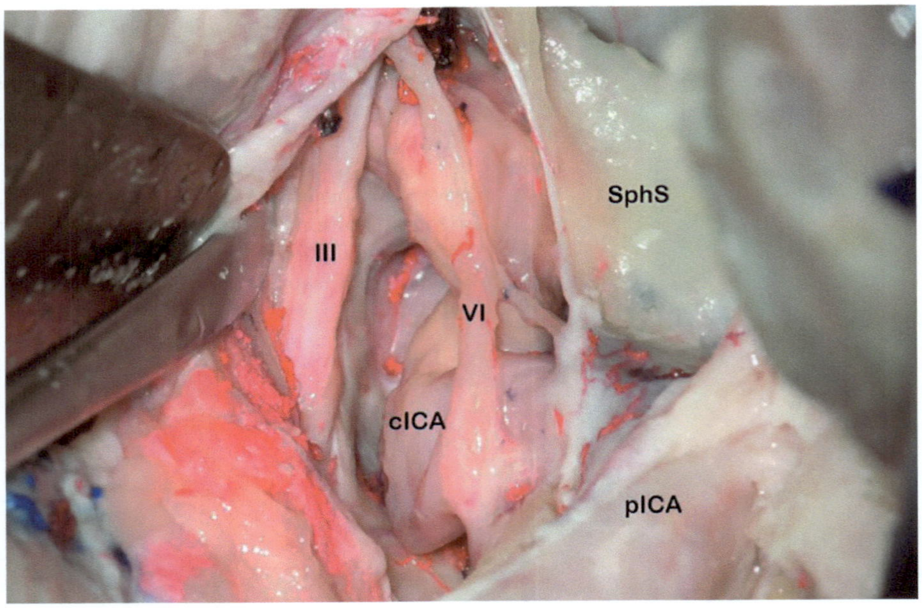

Fig. 12.274 Further opening of the dura and retraction of the temporal lobe allows visualization of the oculomotor nerve in the cavernous sinus. *SphS*, sphenoid sinus; *VI*, abducent nerve; *III*, oculomotor nerve; *cICA*, cavernous segment of internal carotid artery; *pICA*, petrous segment of internal carotid artery.

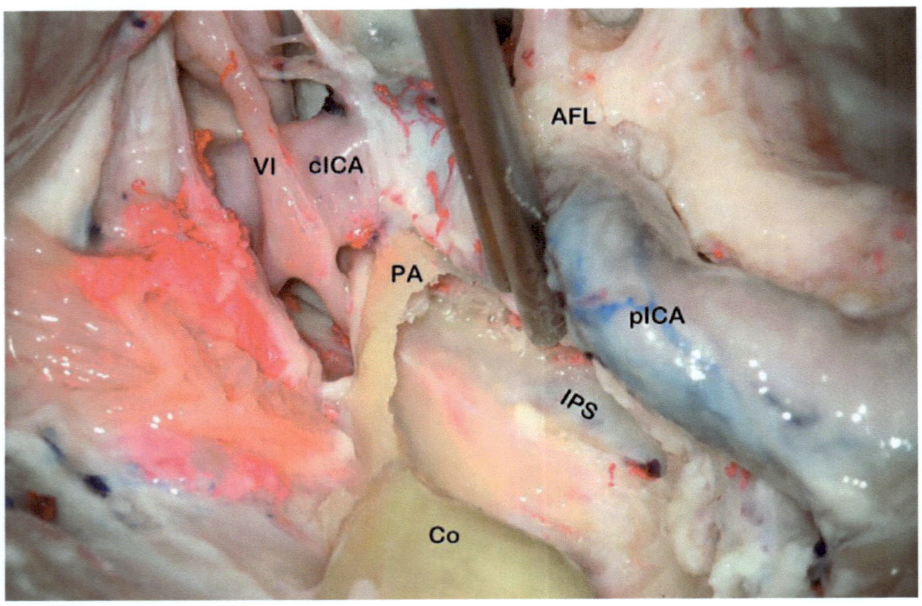

Fig. 12.275 The tip of the suction is used to displace the petrous carotid artery laterally while the petrous apex medial to the petrous carotid artery is further drilled. *AFL*, anterior foramen lacerum; *pICA*, petrous segment of internal carotid artery; *IPS*, inferior petrosal sinus; *Co*, cochlea; *PA*, petrous apex; *cICA*, cavernous segment of internal carotid artery; *VI*, abducent nerve.

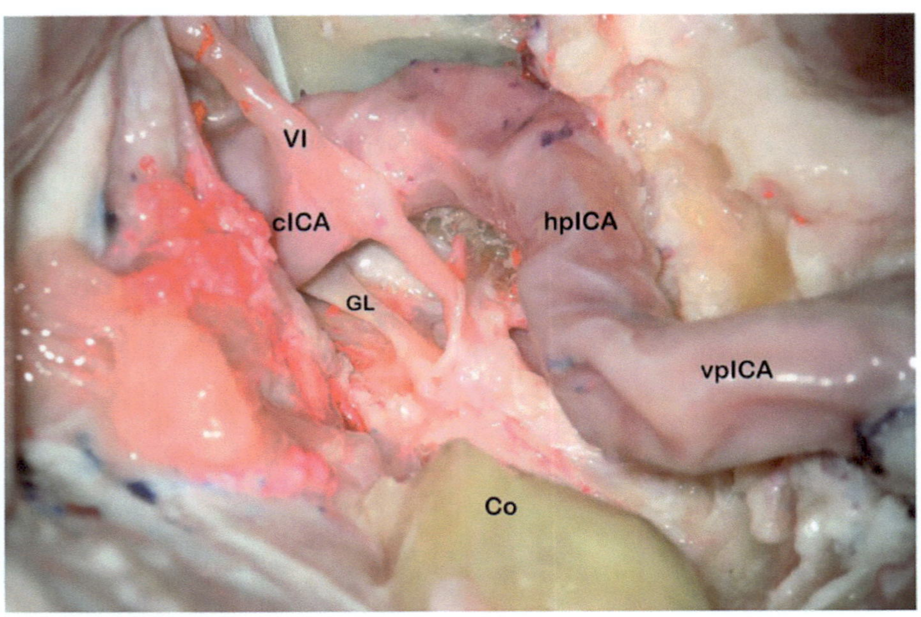

Fig. 12.276 The petrous apex is totally removed. The abducens nerve passes under the petrosphenoid ligament to enter the cavernous sinus. *VI*, abducent nerve; *cICA*, cavernous segment of internal carotid artery; *hpICA*, horizontal portion of petrous carotid artery; *vpICA*, vertical portion of petrous carotid artery; *GL*, Gruber's ligament; *Co*, cochlea.

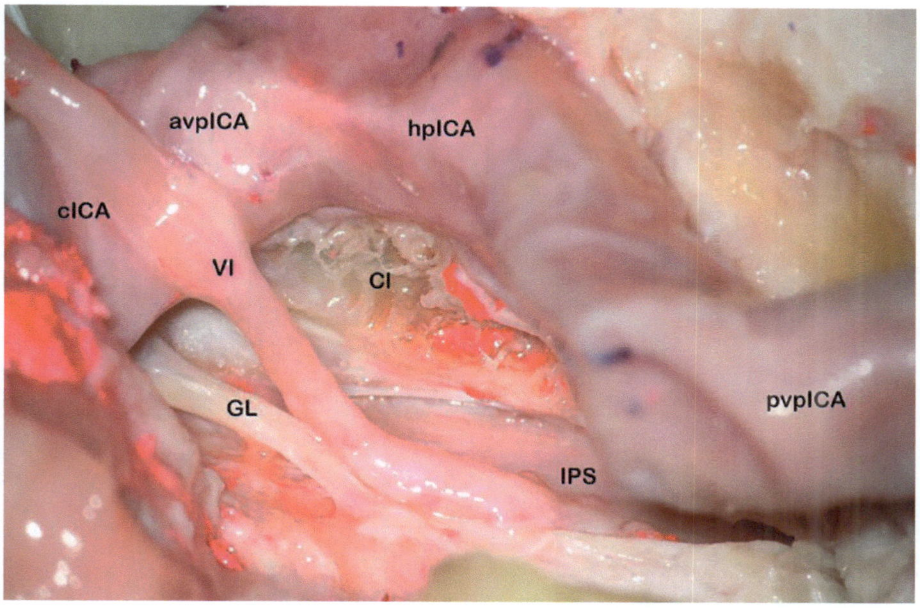

Fig. 12.277 The abducens nerve pierces the dura forming the lower part of the posterior wall of the sinus at the upper border of the petrous apex and enters a dural cave, referred to as Dorello's canal, where it passes below the petrosphenoid ligament (Gruber's ligament) that extends from the lower part of the lateral edge of the dorsum sellae to the petrous apex, to enter the cavernous sinus. *hpICA*, horizontal portion of petrous carotid artery; *avpICA*, anterior vertical portion of petrous carotid artery; *pvpICA*, posterior vertical portion of petrous carotid artery; *cICA*, cavernous segment of internal carotid artery; *VI*, abducent nerve; *CI*, clivus; *GL*, Gruber's ligament; *IPS*, inferior petrosal sinus.

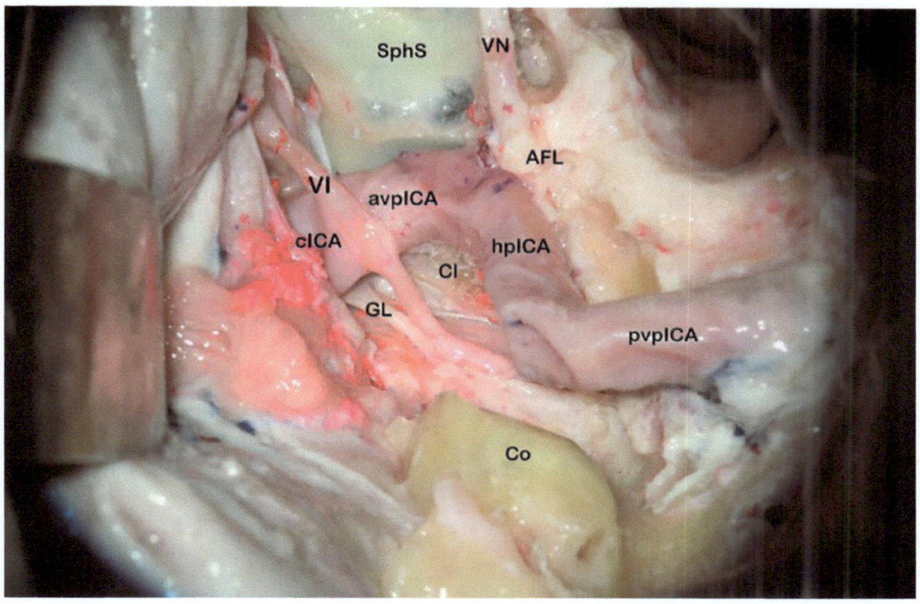

Fig. 12.278 View after finishing the infratemporal fossa approach type C. *hpICA*, horizontal portion of petrous carotid artery; *avpICA*, anterior vertical portion of petrous carotid artery; *pvpICA*, posterior vertical portion of petrous carotid artery; *cICA*, cavernous segment of internal carotid artery; *SphS*, sphenoid sinus; *VN*, vidian nerve; *AFL*, anterior foramen lacerum; *VI*, abducent nerve; *CI*, clivus; *GL*, Gruber's ligament; *Co*, cochlea.

Clinical Applications

CASE 1: LEFT POSTIRRADIATION RECURRENT NASOPHARYNGEAL CARCINOMA (Figs. 12.279–12.288)

MRI with gadolinium enhancement shows the tumor occupying the fossa of Rosenmüller (Fig. 12.279).

The approach has been performed as previously described and the infratemporal fossa retractor has been applied to retract the mandible inferiorly (Fig. 12.280). The middle meningeal artery has been cut after bipolar coagulation (Fig. 12.281). The mandibular nerve is next coagulated bipolarly and sacrificed (Fig. 12.282).

The pterygoid base and the lateral and medial laminae of the pterygoid process are drilled. The terminal part of the Eustachian tube and surrounding tissue can be seen. The tubal attachments are sharply elevated from the underlying tissues. The mucoperiosteum of the nasal cavity is opened after further drilling the medial lamina of the pterygoid process (Figs. 12.283 and 12.284).

The whole specimen including the tumor, the Eustachian tube, the pterygoid muscles, the surrounding soft tissue, and the mucoperiosteum of the ipsilateral side of the nasopharynx is removed en bloc (Fig. 12.285).

The widely opened choana is now appreciated after tumor removal (Figs. 12.286 and 12.287).

The specimen is shown in Fig. 12.288 after en bloc removal.

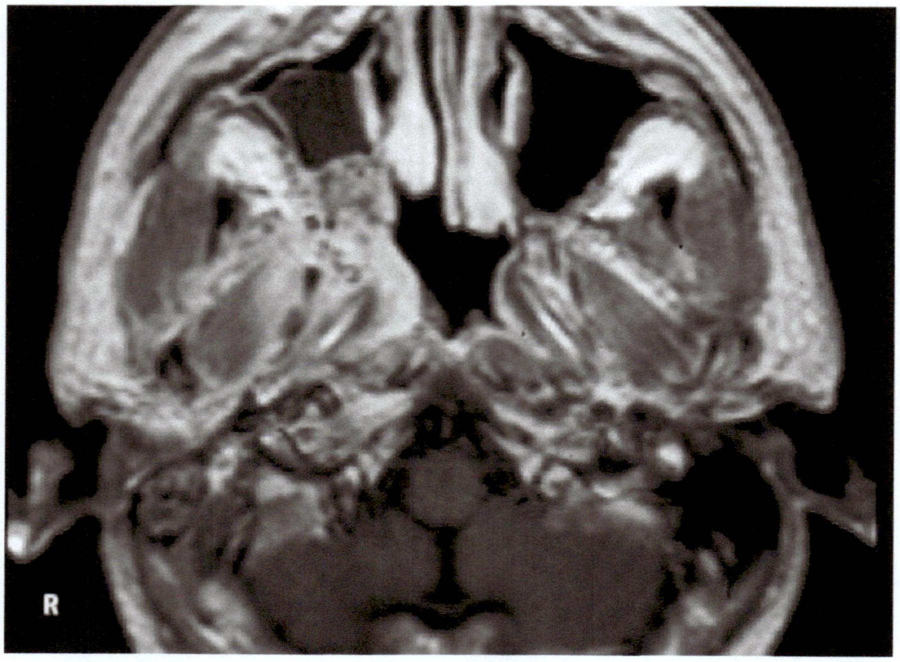

Fig. 12.279 MRI with gadolinium enhancement showing the enhancing tumor occupying the right fossa of Rosenmüller.

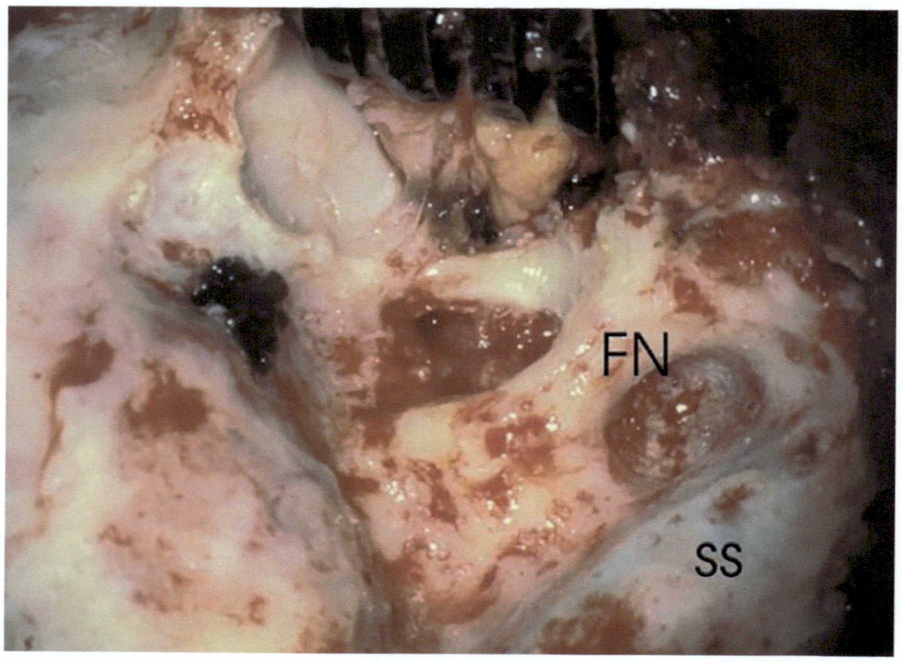

Fig. 12.280 The initial phase of the approach has been made.

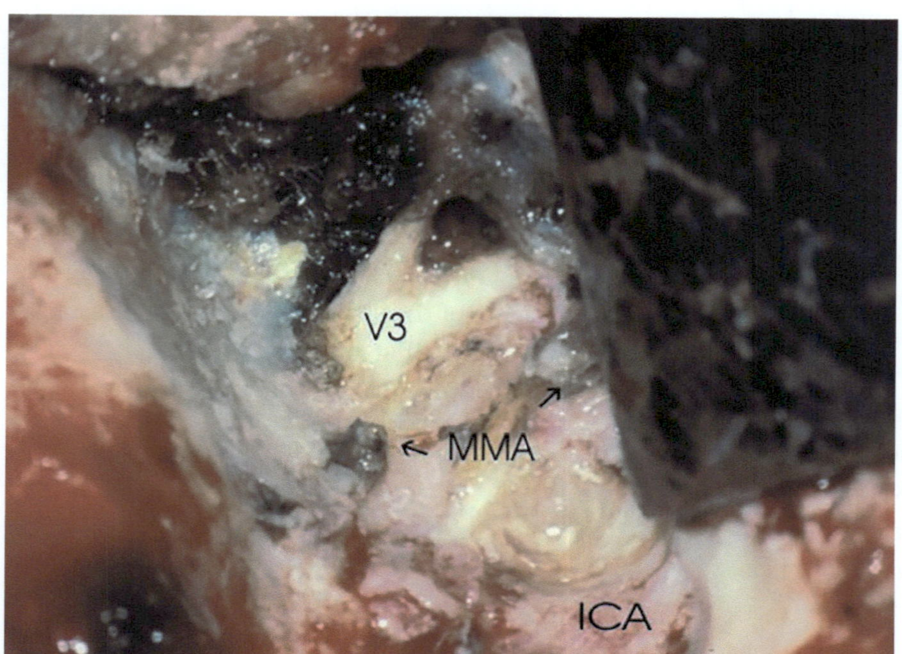

Fig. 12.281 The middle meningeal artery (MMA) has been cut.

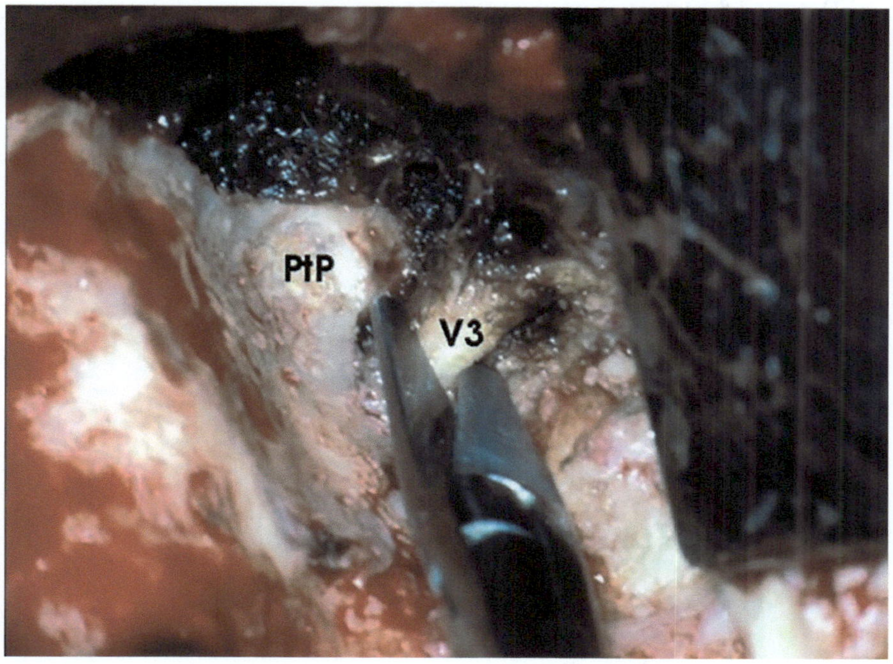

Fig. 12.282 Cutting the mandibular nerve (V3).

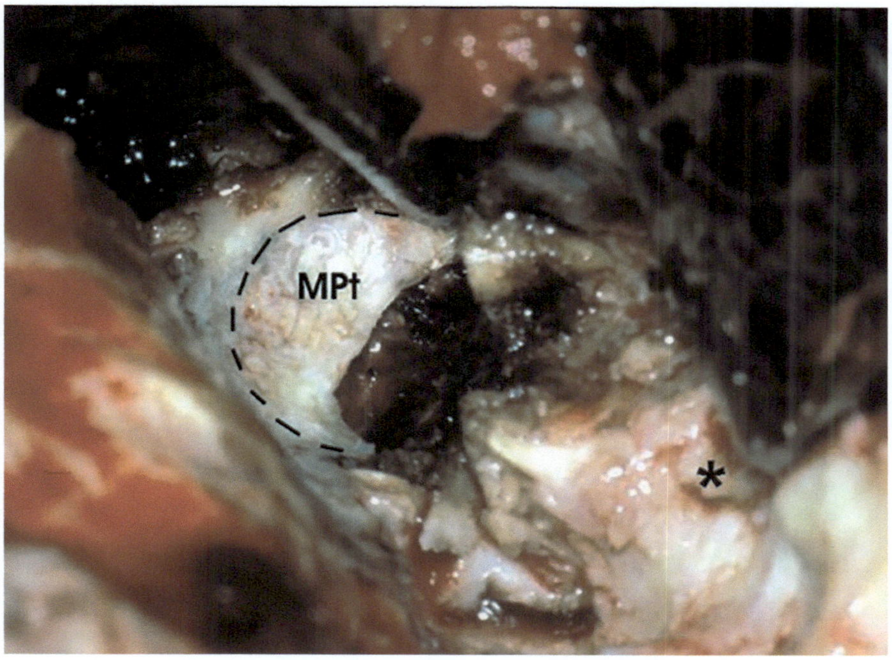

Fig. 12.283 The medial lamina of the pterygoid process (MPt) is to be drilled to the level of the hatched line. Note the terminal part of the Eustachian tube and the surrounding soft tissue (*) to be removed.

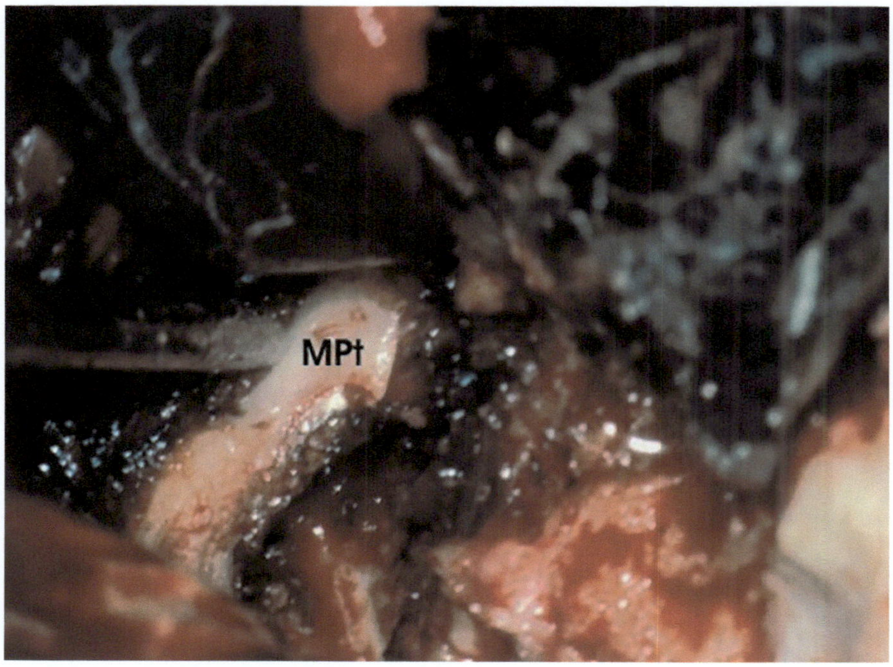

Fig. 12.284 The medial lamina of the pterygoid process (MPt) is now drilled. The nasopharynx is entered at a very anterior level.

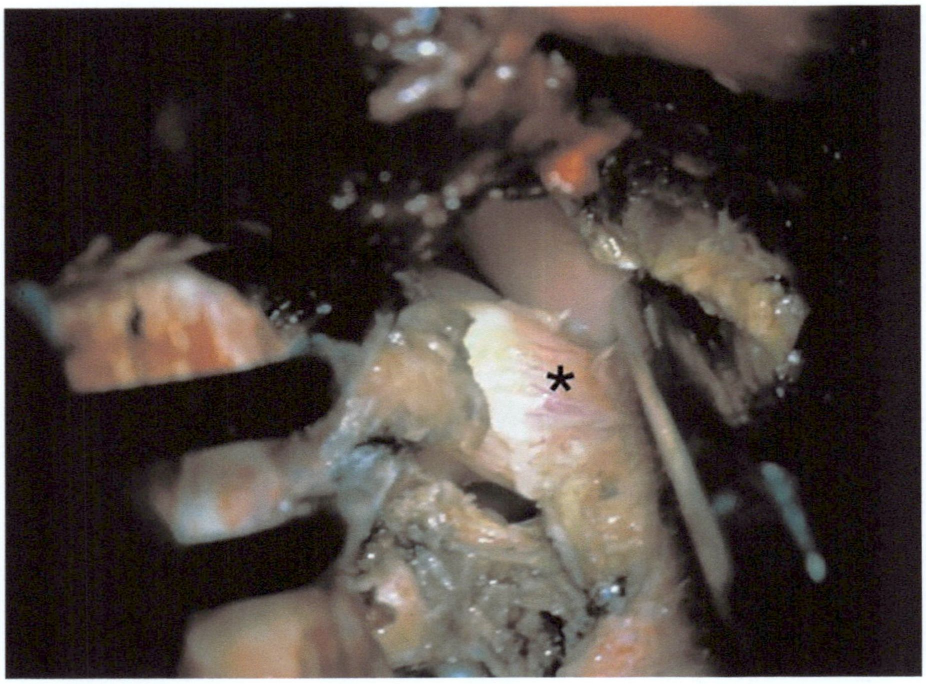

Fig. 12.285 Removal of the specimen (*) en bloc.

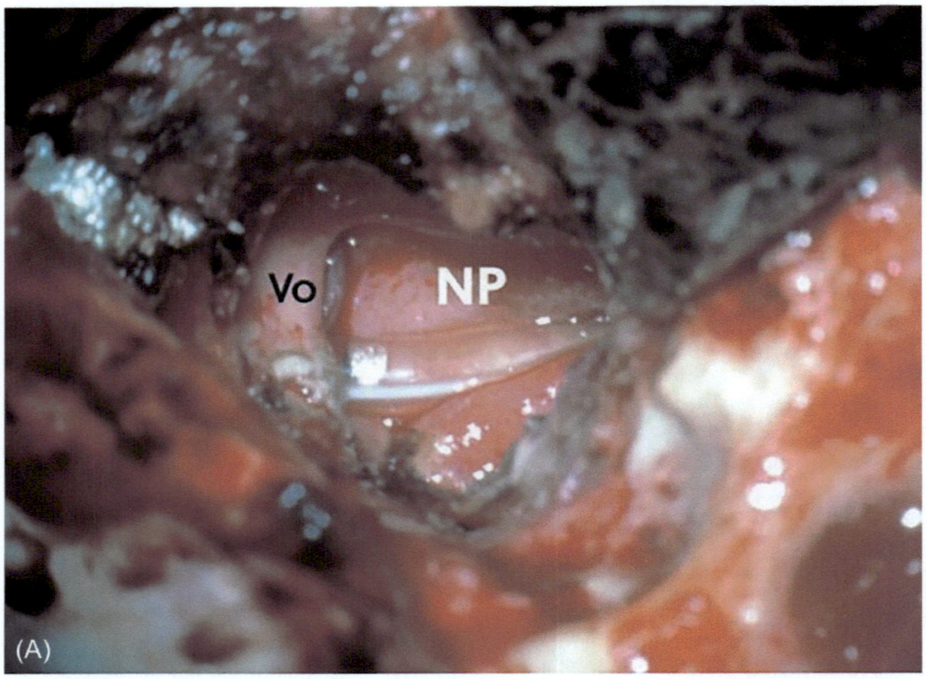

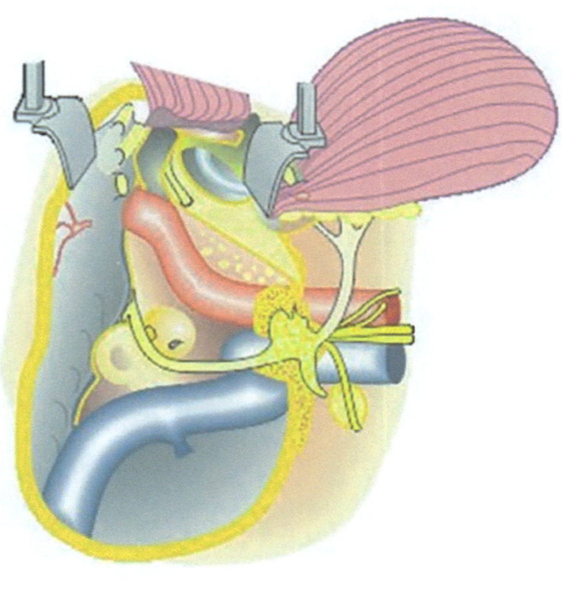

(B)

Fig. 12.286 (A, B) The widely opened choana is clearly seen after total tumor removal. *Vo*, vomer; *NP*, nasopharynx.

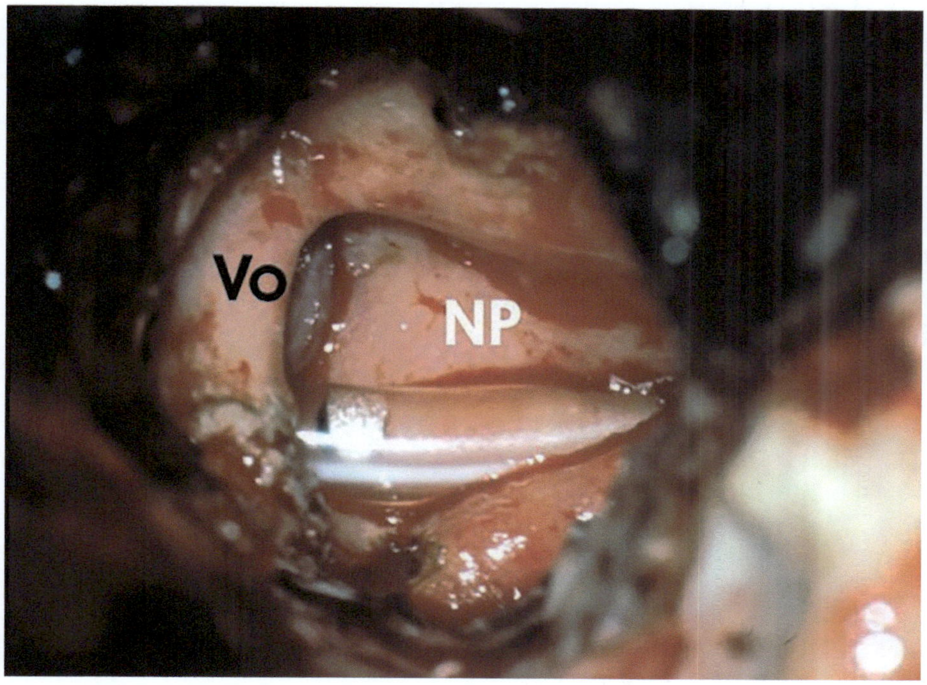

Fig. 12.287 At higher magnification, the nasotracheal tube (inserted from the contralateral side) is better seen.

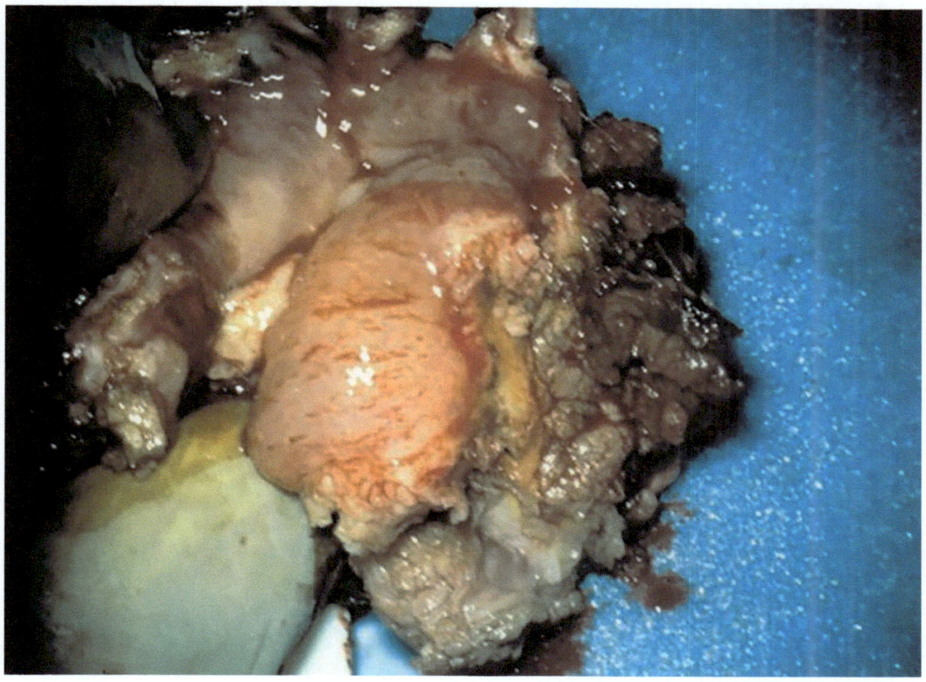

Fig. 12.288 The specimen after en-bloc removal.

CASE 2: LEFT NASOPHARYNGEAL ANGIOFIBROMA WITH INVOLVEMENT OF THE PTERYGOPALATINE AND INFRATEMPORAL FOSSAE

1. Preoperative CT scan of the case. The tumor is seen involving the left nasal cavity, maxillary sinus, pterygopalatine fossa, and infratemporal fossa. The sphenoid sinus is also involved (Figs. 12.289A–12.289C).

2. Preoperative angiography showing the blood supply of the highly vascular tumor. Preoperative immobilization was also done (Figs. 12.290A and 12.290B).

3. The approach has been performed and the infratemporal fossa retractor has been placed (Fig. 12.291). The middle meningeal artery and the mandibular branch of the trigeminal nerve are identified (Fig. 12.292).

4. The base of the pterygoid as well as the medial and lateral lamina have been drilled. The tumor occupying the infratemporal fossa is seen (Fig. 12.293).

5. With progressive removal, the tumor in the nasopharynx extending into the nasal cavity can be appreciated (Fig. 12.294).

6. At the end of the procedure, total tumor removal was achieved (Fig. 12.295). The cavity is closed in layers, and a Penrose drain is inserted.

7. A postoperative CT showing total tumor removal (Fig. 12.296).

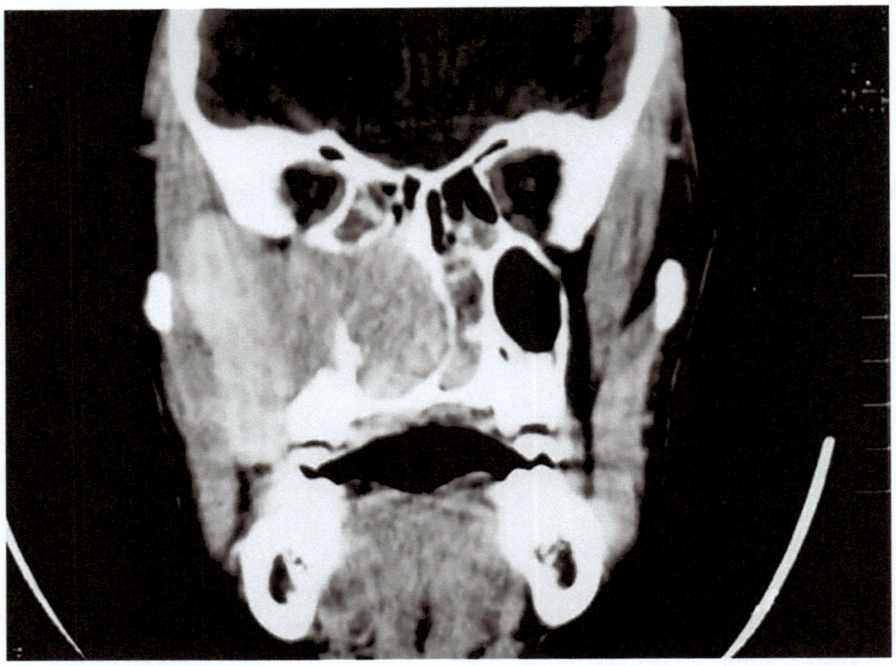

Fig. 12.289A Preoperative CT scan of the case. Coronal view with contrast. Tumor on the left side. Note the massive extension into the infratemporal fossa.

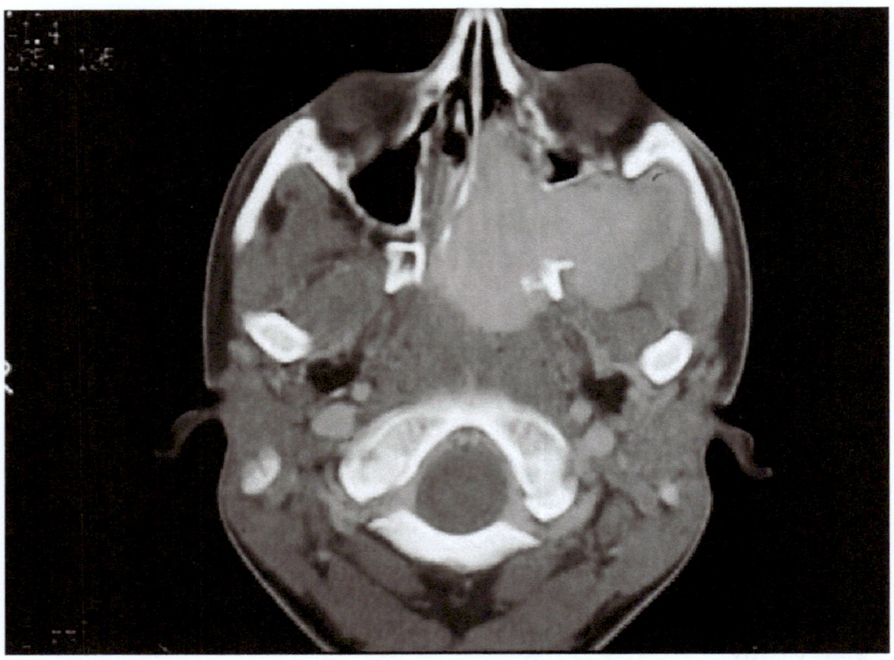

Fig. 12.289B Axial view. The pterygopalatine fossa is markedly enlarged by the tumor. The lateral extent into the infratemporal fossa is well appreciated.

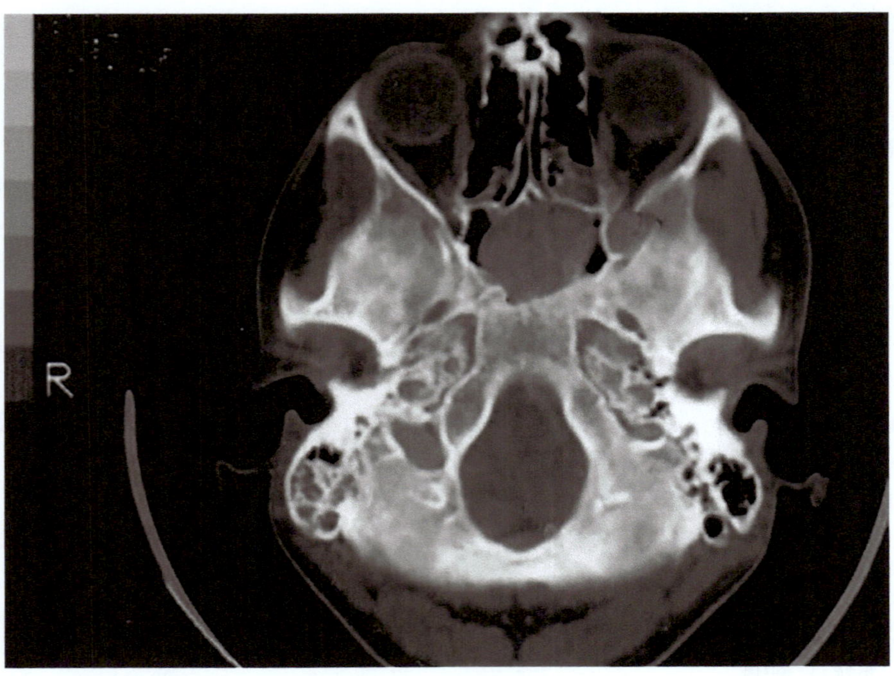

Fig. 12.289C Axial view at a higher level. Note the involvement of the sphenoid sinus.

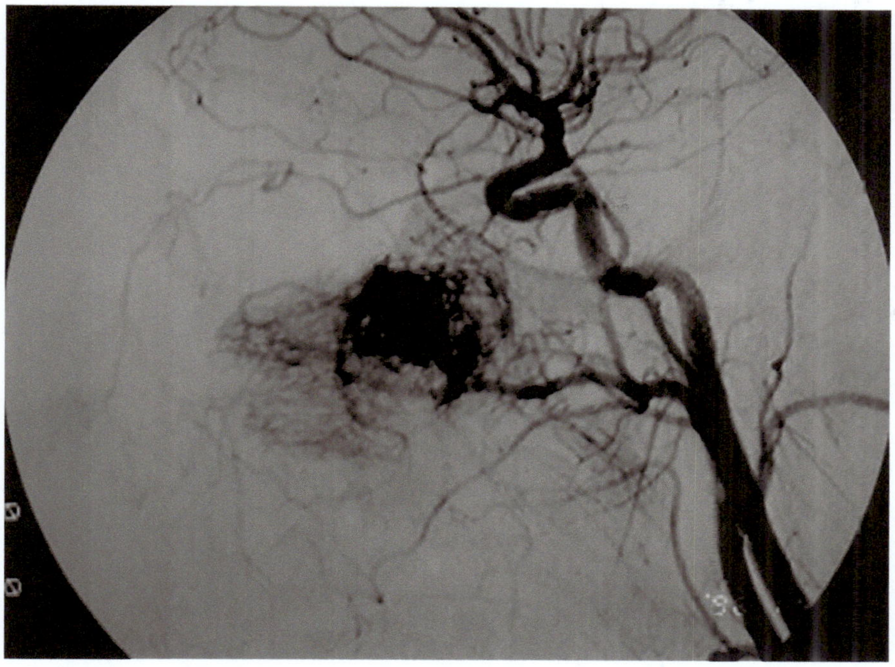

Fig. 12.290A Preoperative angiography showing an intense tumor blush with main blood supply from the internal maxillary artery.

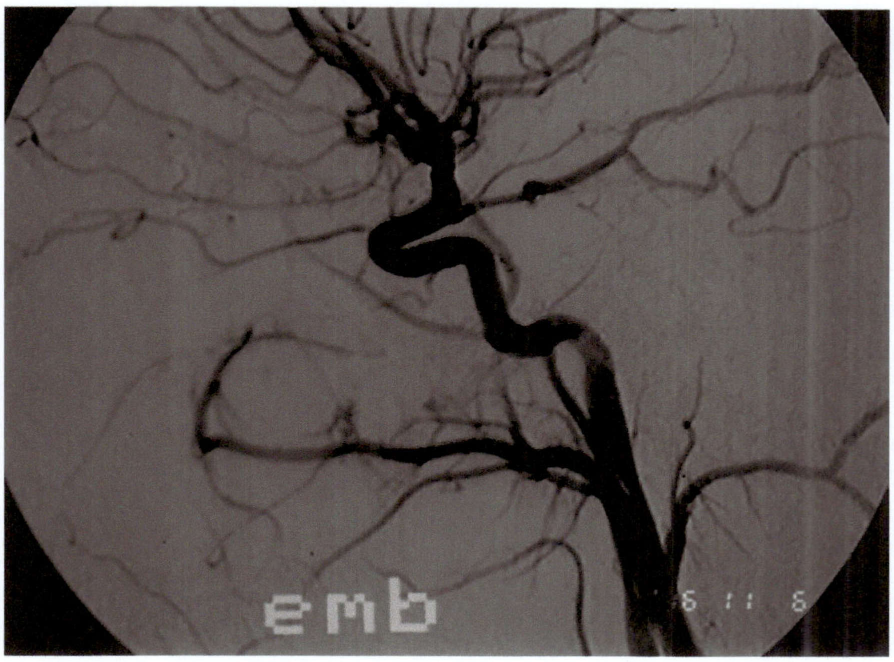

Fig. 12.290B View post embolization.

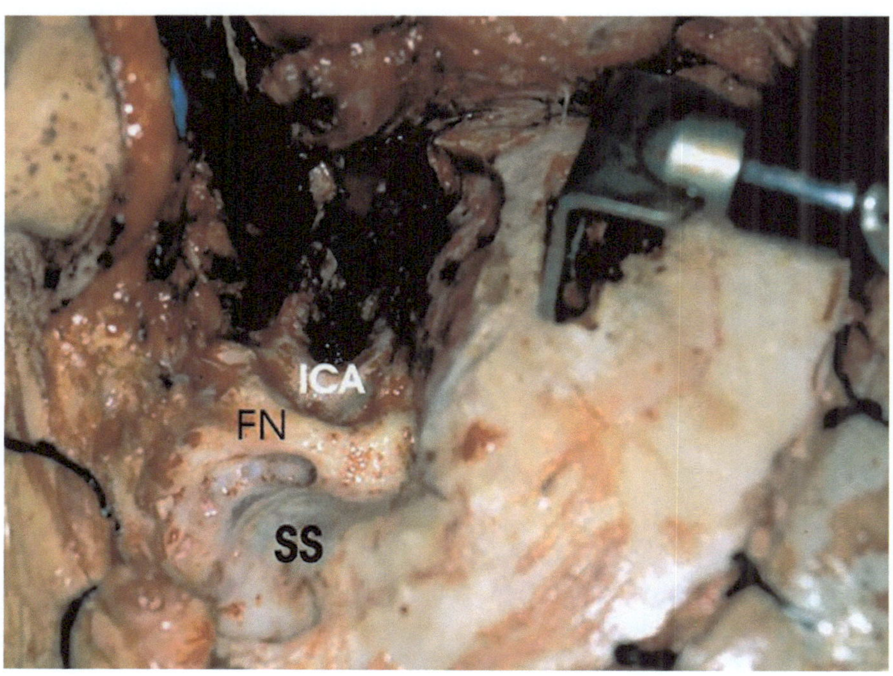

Fig. 12.291 The approach has been performed.

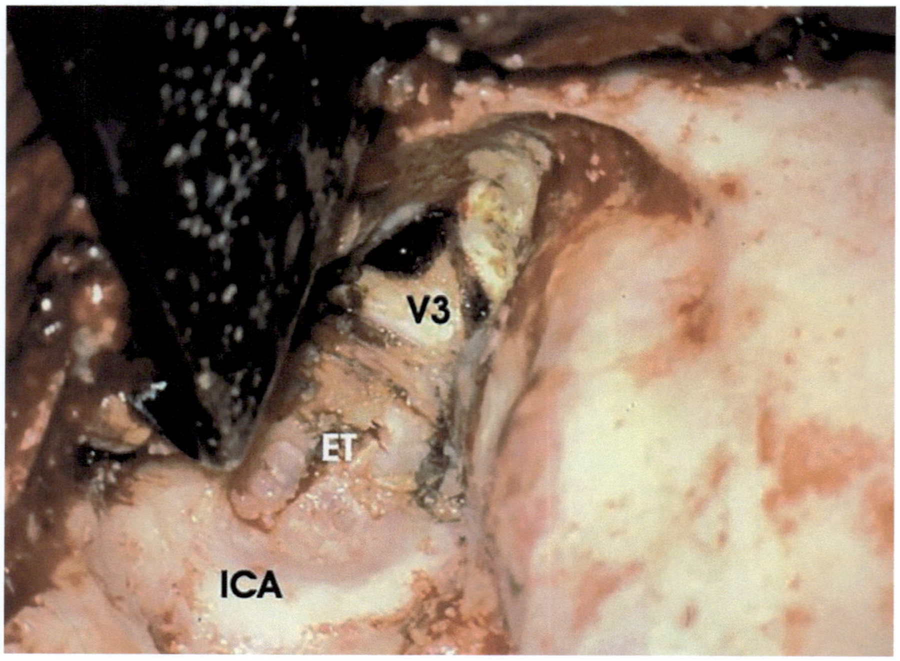

Fig. 12.292 Identification of the mandibular nerve (V3).

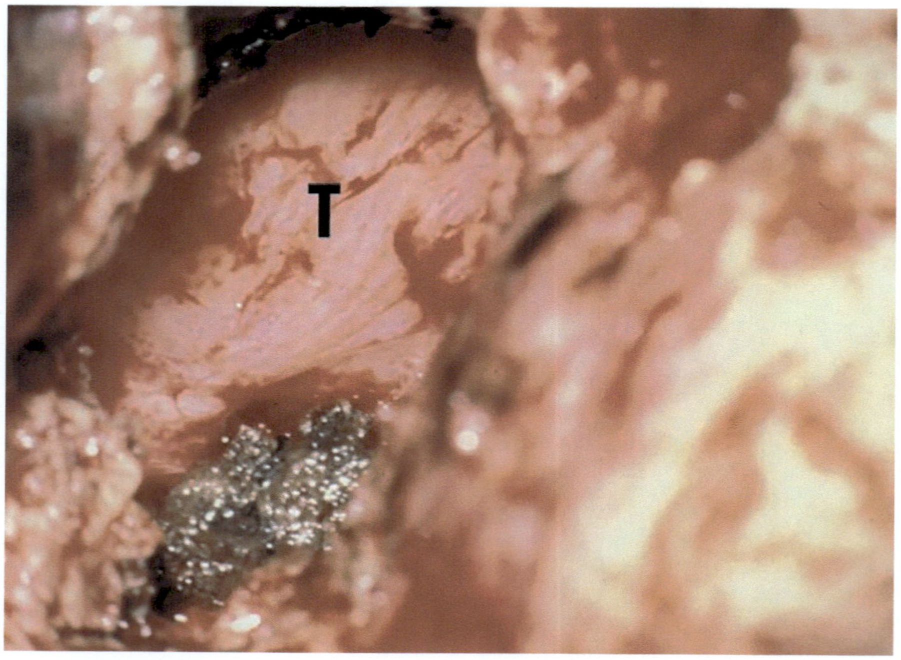

Fig. 12.293 The tumor (T) is seen after drilling the base of the pterygoid as well as the medial and lateral laminae.

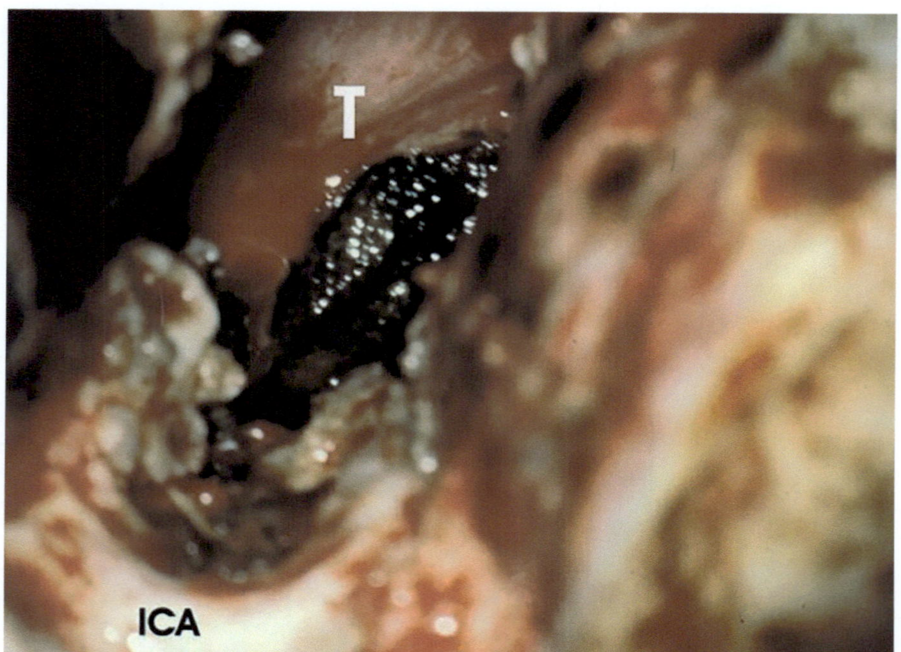

Fig. 12.294 Tumor (T) in the nasopharynx. *ICA*, internal carotid artery.

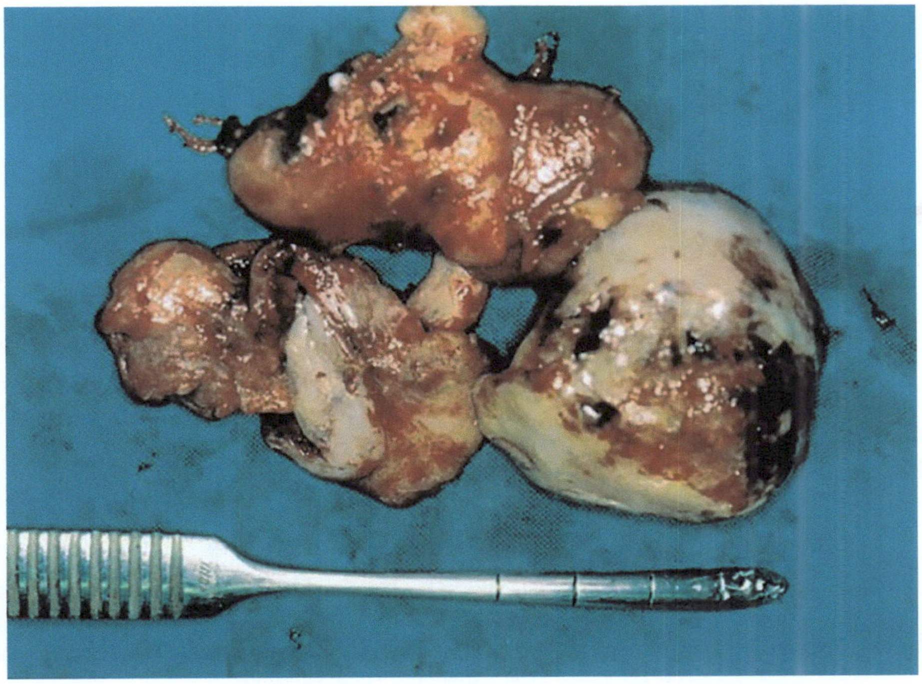

Fig. 12.295 The specimen after removal.

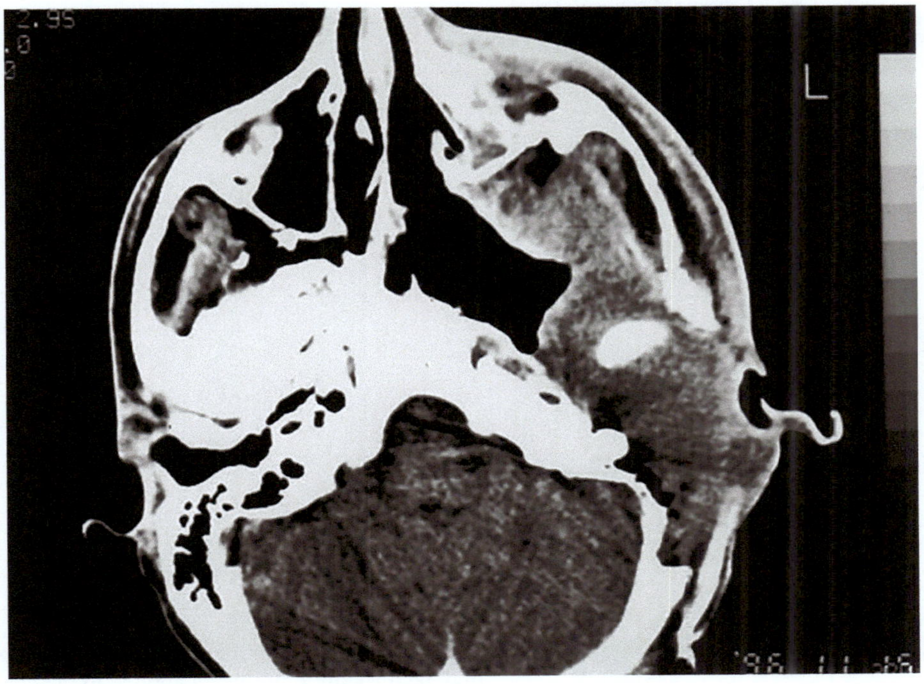

Fig. 12.296 Postoperative CT scan.

HINTS AND PITFALLS

1. The sphenoid spine serves as a useful landmark for the middle meningeal artery, which lies immediately anterior to it (Fig. 12.297).
2. The zygomatic arch provides an approximate indication of the level of the middle fossa dura, which lies slightly inferior to it.
3. When applied to the upper edge of the incision, the infratemporal fossa retractor sometimes does not give the necessary retraction needed to perform the approach (Fig. 12.298). A useful trick is to perform a minicraniotomy, which accommodates the upper blade of the retractor (Fig. 12.299).
4. Before transecting the mandibular nerve, bipolar coagulation is performed first to avoid bleeding from the venous plexus that accompanies the nerve in the foramen ovale (Fig. 12.300). Drilling in this approach should be parallel to the course of the internal carotid artery. The horizontal segment of the artery runs in an anteromedial

direction. This direction is approximately that of a line running from the spine of the sphenoid, foramen spinosum, and foramen ovale and ending at the pterygoid process.

5. In nasopharyngeal carcinoma, the nasopharynx is entered as far anteriorly as possible. This is done by drilling the medial lamina of the pterygoid plate (which is the lateral wall of the choana), and the nasal cavity is entered by incising the mucoperiosteum at this level (Fig. 12.301).
6. The vidian nerve is an important landmark for identifying the sphenoid sinus lying roughly on its floor (Fig. 12.302).
7. The pterygoid venous plexus can give rise to troublesome bleeding. This can be avoided by staying strictly in the sub-periosteal plane during the elevation of the soft tissue and muscles from the glenoid fossa and the base of the pterygoid, respectively.
8. This approach offers only a limited inferolateral access to the cavernous sinus. Infratemporal tumors extending to and abutting the lateral wall of the sinus can be successfully tackled through this approach. On the other hand, massive involvement or primary tumors of the cavernous sinus necessitate the performance of a larger craniotomy as in the frontotemporal orbitozygomatic approach.

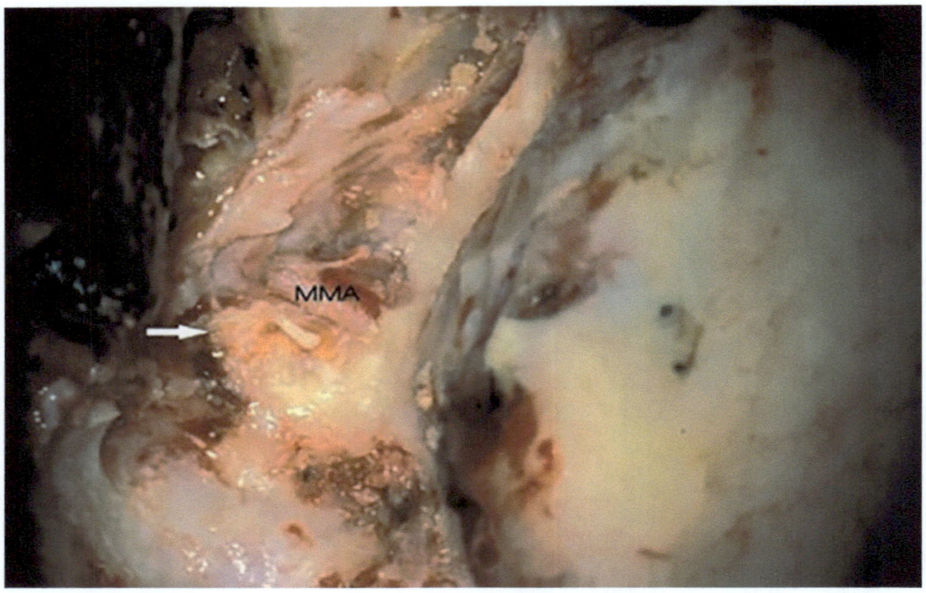

Fig. 12.297 *Left side*. The middle meningeal artery (MMA) lies immediately anterior to the sphenoid spine *(arrow)*.

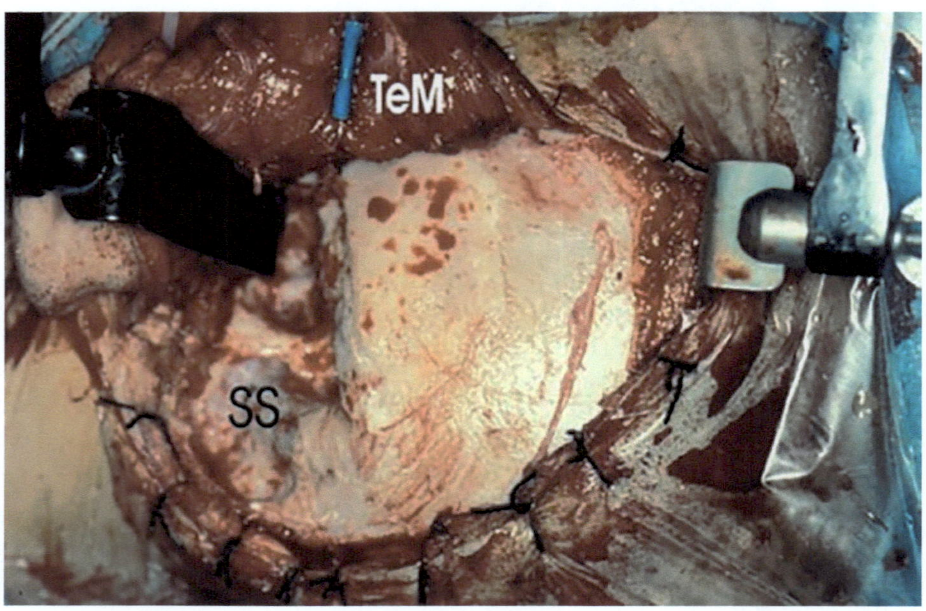

Fig. 12.298 The infratemporal fossa retractor may be difficult to use when applied to the edge of the incision.

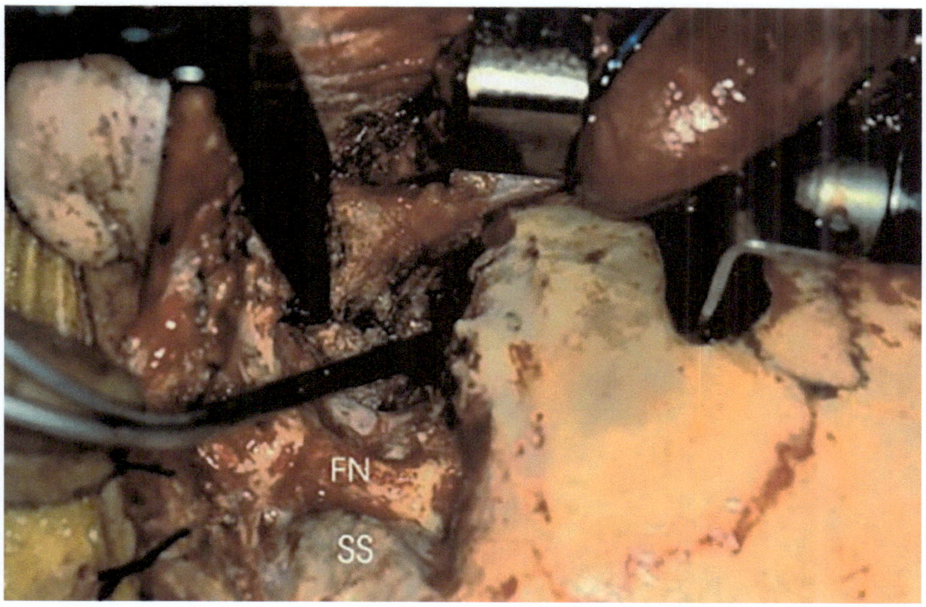

Fig. 12.299 A minicraniotomy used for better placement of the retractor.

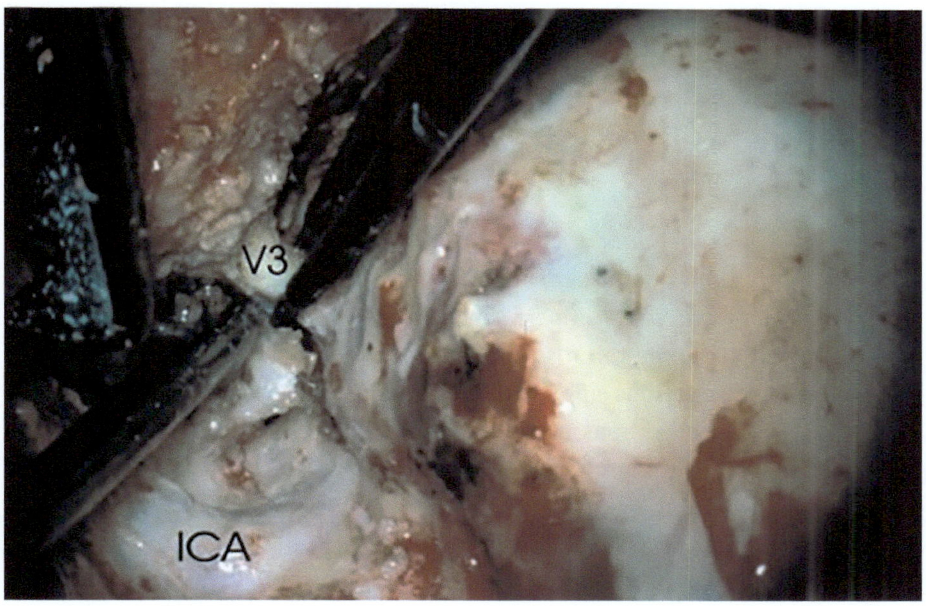

Fig. 12.300 Bipolar coagulation of the mandibular nerve (V3).

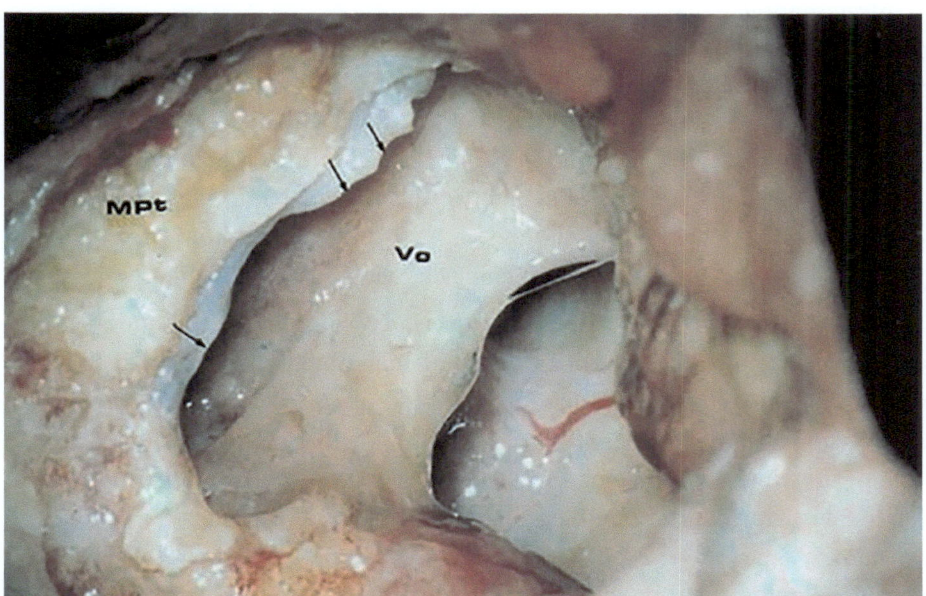

Fig. 12.301 To enter the nasopharynx in an area away from the tumor, the medial pterygoid (MPt) is drilled first. The nasopharynx is entered by incising the mucoperiosteum *(arrows)* at this level.

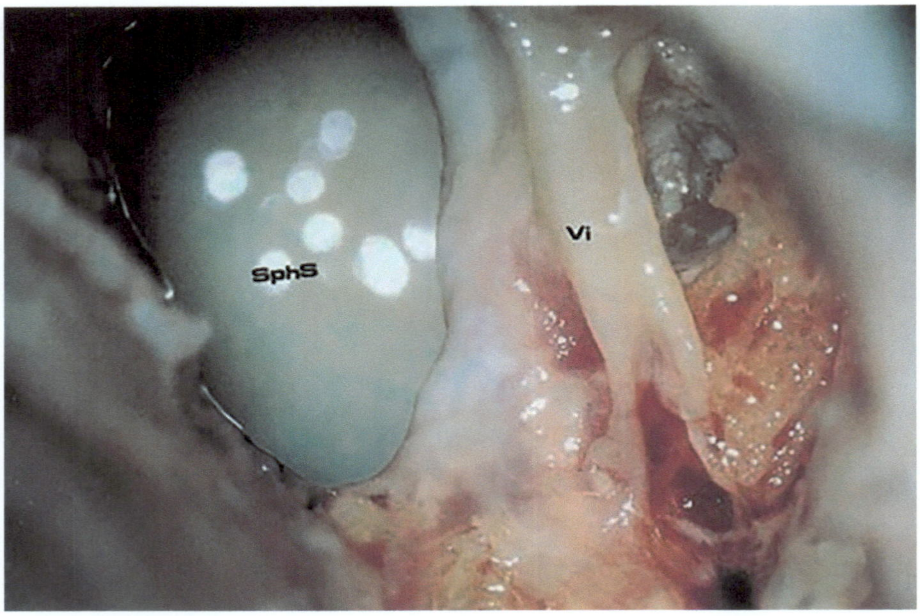

Fig. 12.302 The vidian nerve (Vi) roughly serves as a landmark for the floor of the sphenoid sinus (SphS).

Infratemporal Fossa Approach Type D

Rationale

The type D approach is similar to the type C but is achieved through a preauricular incision; drilling along the floor of the middle cranial fossa is performed anterior to the mastoid and the internal carotid artery, thus the middle ear is left intact. The Eustachian tube is not sacrificed. The approach is designed for infratemporal fossa lesions with or without extension to the pterygopalatine fossa, sphenoid sinus, and/or minimal cavernous sinus extension; for example, trigeminal neurinomas with predominant infratemporal fossa component and minimal middle fossa extension. The IFTD could be also used in the treatment of neurofibromas, teratomas, and tumors of the lateral and posterior parts of the orbit [58]. This type of intervention is, however, rarely needed in current practice due to the early diagnosis of pathology and progress achieved in the endoscopic surgery field.

The IFTD approach is part of the group of preauricular transzygomatic approaches. These approaches share primarily the same concept but with some variations in the technique and the extent of the craniotomy and the zygomatic bone removal. They collectively include the following subtypes. The approach will be discussed together with the more extensive preauricular approaches, namely, the preauricular infratemporal fossa and the preauricular frontotemporal fossa approaches.

Preauricular Infratemporal Transzygomatic Approach

The preauricular infratemporal approach is a variant of the infratemporal approach in which a large frontotemporal craniotomy is performed. This approach is suitable for infratemporal fossa lesions with or without extension to the petrous apex, sphenoid sinus, or upper clivus or minimal cavernous sinus extension [59,60]. It is suitable for both intradural and extradural lesions affecting the upper clivus and parasellar regions. The craniotomy allows tumor removal with minimal brain retraction.

Preauricular Frontotemporal Orbitozygomatic Approach

The orbitozygomatic approach is a variant in which the upper and lateral margins of the orbital rim and zygomatic arch are elevated with the bone flap. This affords additional exposure for suprasellar and parasellar lesions or infratemporal tumors extending into the orbit [61,62]. The orbitozygomatic craniotomy gives an enhanced view of the cavernous sinus.

Possible morbidity associated with these techniques:

– sacrifice of the mandibular nerve and drilling of the condylar fossa. The resultant deficit is, however, not significant.

– inadequate control of the vertical internal carotid artery and less exposure to the petrous apex.

– if it is not repaired properly, removal of the lateral orbital rim can lead to some esthetic disfigurement.

– a lengthy procedure.

Advantages offered by the approach:

1. Excellent control of the infratemporal fossa, pterygopalatine fossa, nasopharynx, sphenoid sinus, superolateral aspects of the orbit, and inferolateral cavernous sinus.
2. Excellent control of the horizontal intrapetrous internal carotid artery.
3. Preservation of the middle ear function.
4. No facial scars.

Indications

Lesions involving the infratemporal and middle fossae with or without extension into the petrous apex, sphenoid sinus, sphenopalatine fossa, and cavernous sinus (Figs. 12.303 and 12.304). Examples of these lesions include:

– Dumbbell-shaped trigeminal neurinoma. For those cases with massive extension into the posterior fossa, the approach is combined with a transpetrous approach (see later).

– Certain cases of meningiomas with parasellar involvement with or without extracranial extension into the infratemporal fossa.

– Extensive nasopharyngeal angiofibroma with cavernous sinus involvement.

– Other tumors involving the Meckel's cave area such as clival chordomas, epidermoids, etc.

– It should be noted that there are no rigid indications for the utilization of these variants. We usually tailor our approach according to each individual case.

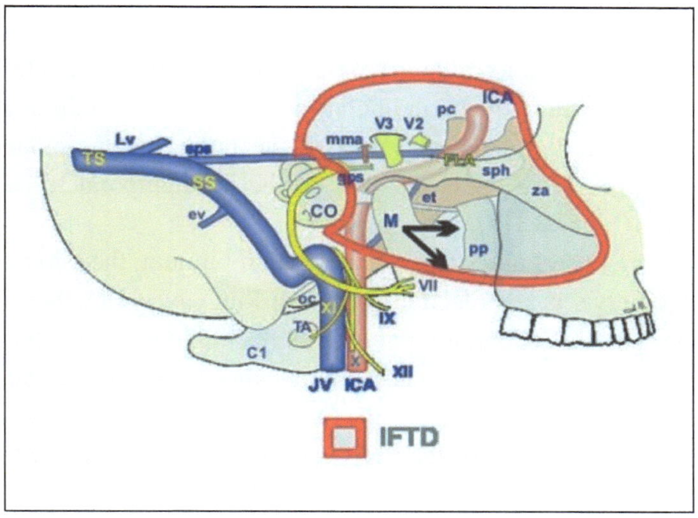

Fig. 12.303 A schematic presentation showing the structures exposed by the approach. Key as in Fig. 9.1.

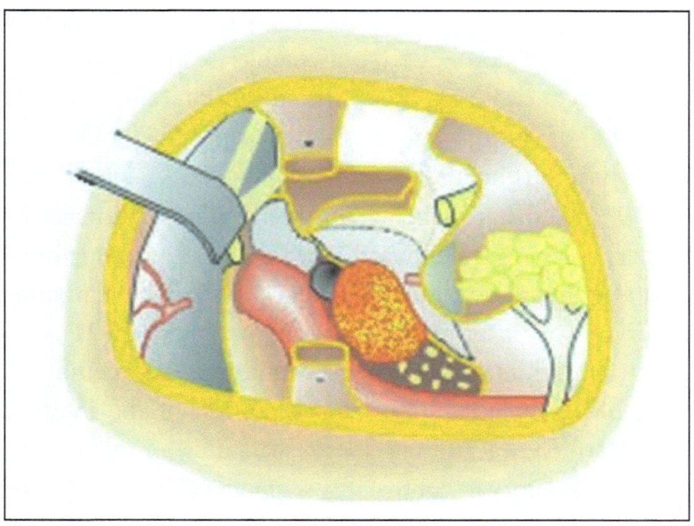

Fig. 12.304 A schematic presentation showing residual tumor lying in the petrous region anterior to the internal carotid artery.

Surgical Steps (Right Side)

1. A curvilinear preauricular incision is performed starting from in front of the tragus, coursing superiorly and then anteriorly in the frontal scalp to end superior to the supraorbital notch near the midline (Fig. 12.305).

2. The scalp is elevated, starting the dissection in the loose areolar layer of fascia (i.e., the subaponeurotic plane) (Fig. 12.306). About 2–3 cm superior to the zygomatic arch, a semilunar incision is made in the superficial layer of the deep temporal fascia and the plane of dissection is changed to become deeper (i.e., beneath the superficial layer of the deep temporal fascia), thereby affording protection for the facial nerve. This plane of dissection courses with the superficial temporal fat pad (Fig. 12.307).

3. The zygomatic arch and the lateral orbital rim are fully exposed. The periosteum of the zygomatic arch and orbital rim is elevated. More inferiorly, the dissection lies superficial to the masseter fascia, avoiding injury to the facial nerve branches at this level (Fig. 12.308). The temporalis muscle is incised in a line about 1 cm along its attachment. The temporalis muscle is then detached from the zygomatic arch using a large elevator (Fig. 12.309).

ZYGOMATIC AND ORBITOZYGOMATIC OSTEOTOMY (Fig. 12.310–12.315)

4. The zygomatic osteotomy is performed according to the extension of the lesions as follows:

(a) In limited cases, only the zygomatic arch is displaced inferiorly. The zygomatic bone and the lateral orbital rim are left intact. The zygomatic arch is transected after creating two burr holes for further refixation at the end of the procedure.

(b) For the majority of lesions only an "extended" zygomatic osteotomy is performed in which the zygomatic arch and part of the zygomatic bone are included in the zygomatic osteotomy. The orbitozygomatic osteotomy in more advanced cases in which the tumor extends into the orbital apex through the inferior or superior orbital fissures: The frontal and temporal dura are detached from the periorbita exposing the bony orbit. The following cuts are then performed: (1) A coronal cut at the level of the root of the zygoma, just in front of the condylar fossa. (2) A sagittal cut across the lateral orbital rim at the level of the supraorbital notch. This cut is carried deep until the lateral orbital wall is met and then carried downward across this wall until it meets the inferior orbital fissure. (3) From the orbital side, the saw is placed in the lateral aspect of the inferior orbital fissure, and a horizontal cut is carried from this point to the zygomatico-maxillary junction. The orbitozygomatic bone is removed and preserved in saline.

5. The temporalis muscle is reflected inferiorly. A large frontotemporal craniotomy is next performed. The dura has been elevated from the floor of the middle cranial fossa. The bone is then rongeured (or drilled) down to the base of the middle cranial fossa. The foramen ovale lies anterior and medial to the foramen spinosum. Shaving the bone superior to the temporomandibular joint (Fig. 12.316).

6. Bone removal extends to the floor of the middle fossa, uncovering the foramen spinosum, through which the middle meningeal artery passes, and the foramen ovale, through which the mandibular nerve passes. Maximum shaving of the anterior aspect of the temporomandibular joint provides a wide exposure of the infratemporal fossa space (Fig. 12.317).

7. After unroofing the foramen rotundum, the maxillary nerve is appreciated. After drilling away the bone between the maxillary nerve and mandibular nerve, the vidian nerve is seen just below the maxillary nerve (Fig. 12.318).

8. The base of the pterygoid process is now drilled. The maxillary nerve is followed distally into the pterygopalatine fossa, and the mandibular nerve is followed distally into the infratemporal fossa. Exposing the branches of the mandibular nerve without removing the lateral pterygoid muscle. The infraorbital artery, which arises from the maxillary artery, can be exposed. The infraorbital artery enters the orbital cavity through the posterior part of the inferior orbital fissure (Fig. 12.319).

9. The distal branches of the maxillary and mandibular nerves have been exposed. The middle meningeal artery is cut after bipolar coagulation (Fig. 12.320). The mandibular nerve is cut after bipolar coagulation (Fig. 12.321).

10. The mandibular nerve and middle meningeal artery lie lateral to the horizontal portion of the internal carotid artery. Sacrificing these structures or displacing them is mandatory to control this segment of the artery from the lateral aspect. By retracting the middle fossa dura, the greater petrosal nerve is exposed directly under the dura of the middle fossa. In this case, the terminal part of the petrous carotid artery is also exposed under the dura and

below the greater superficial petrosal nerve, as occurs in approximately 15% of temporal bones. The petrous carotid artery is usually covered by bone up to the lateral side of the third trigeminal division (Fig. 12.322).

11. The dura has been further elevated from the floor of the middle fossa to expose the greater petrosal nerve and petrous apex (Fig. 12.323).

12. The greater superficial petrosal nerve is cut (Fig. 12.324). The petrous apex bone medial to the internal carotid artery can now be drilled (Fig. 12.325).

13. The horizontal portion of petrous carotid artery is exposed by drilling anteromedial to the genu of the artery using a diamond burr. The carotid artery up to the level of the anterior foramen lacerum can be seen. Note the vidian nerve is a reliable landmark to the petrous internal carotid artery (Fig. 12.326).

14. The genu of the internal carotid artery is identified, and then the artery is followed down to identify its vertical part. The genu of the internal carotid artery is closely related to the cochlea. The distance between the two varies from 1 mm to 5mm (Fig. 12.327).

15. The ICA is surrounded by a dense fibroperiosteal layer at its entrance in the skull base. After removing the fibroperiosteal layer, the cervical portion of petrous carotid artery can be appreciated. The glossopharyngeal nerve lies deep in this layer and is slightly adherent to its undersurface. Note the relationship between the anterior wall of the jugular bulb and the carotid artery (Fig. 12.328).

16. Further drilling of the rest petrous apex is limited by petrous carotid artery (Fig. 12.329). A red vessel loop is placed in order to displace the petrous carotid artery laterally while the petrous apex medial to the petrous carotid artery is drilled with a large diamond burr to reach the clivus (Fig. 12.330).

17. The petrous apex was drilled until the inferior petrosal sinus and clivus were reached (Fig. 12.331).

18. The dura of the posterior fossa and cavernous sinus is opened. The abducens nerve courses in prepontine cistern and then pierces the dura and enters a dural cave, referred to as Dorello's canal to enter the cavernous sinus (Fig. 12.332).

19. After surgery, fat is used to obliterate the operative cavity. The craniotomy is fixed with miniplates. The temporalis muscle is sutured back. The zygomatic arch is fixed with a single miniplate placed anteriorly. The skin is closed in layers.

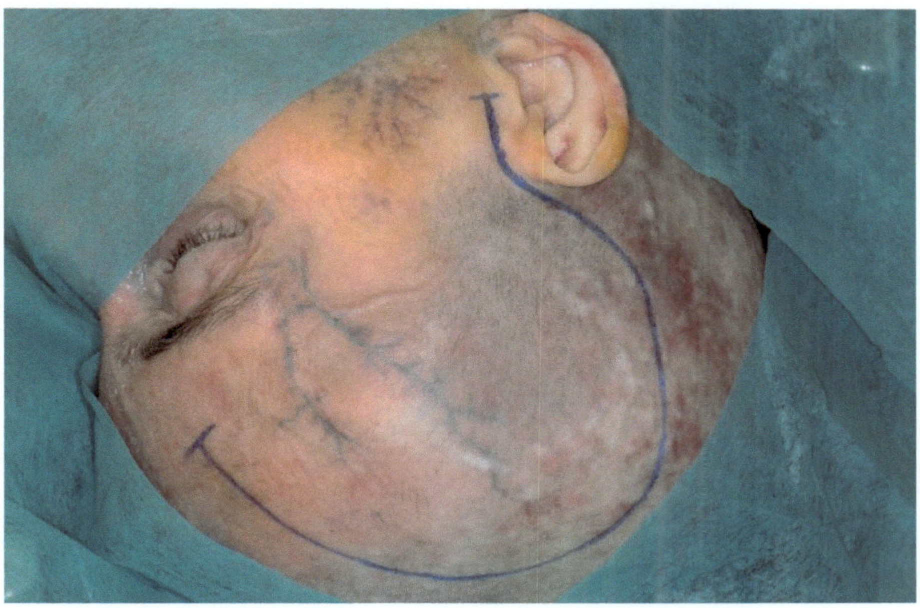

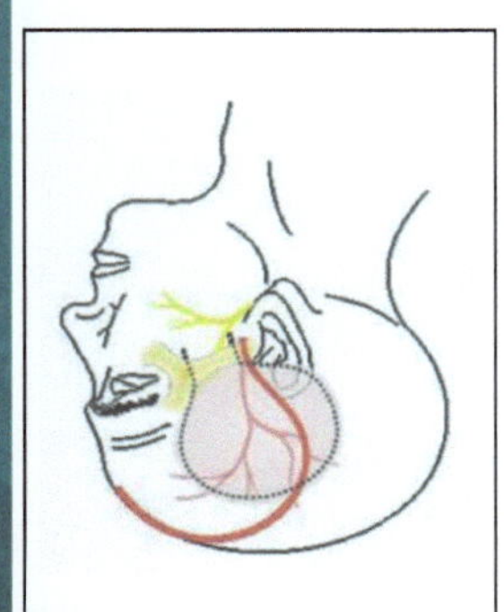

Fig. 12.305 A curvilinear preauricular incision is performed starting from in front of the tragus, coursing superiorly and then anteriorly in the frontal scalp to end superior to the supraorbital notch near the midline.

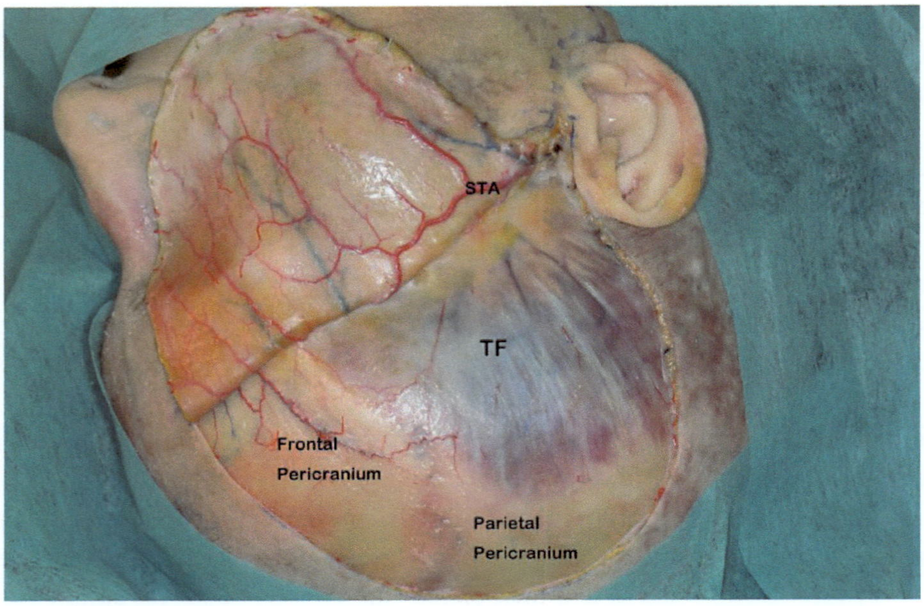

Fig. 12.306 The scalp is elevated, starting the dissection in the loose areolar layer of fascia. *STA*, superficial temporal artery; *TF*, temporalis fascia.

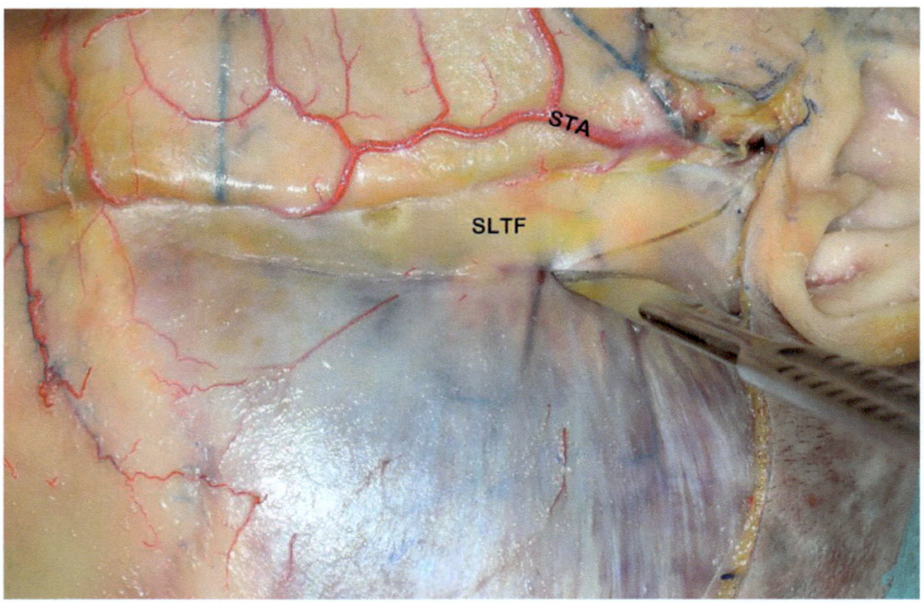

Fig. 12.307 About 2–3 cm superior to the zygomatic arch, a semilunar incision is made in the superficial layer of the deep temporal fascia, and the plane of dissection is changed to become deeper, thereby affording protection for the temporal branch of the facial nerve. This plane of dissection courses with the superficial temporal fat pad. *STA*, superficial temporal artery; *SLTF*, superior layer of the temporalis fascia.

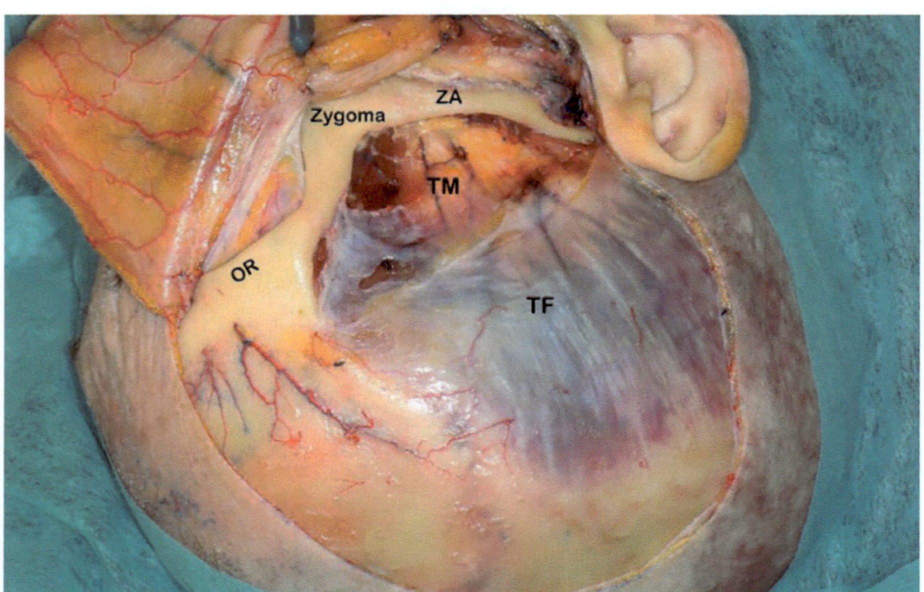

Fig. 12.308 The zygomatic arch and the lateral orbital rim are fully exposed. The periosteum of the zygomatic arch and orbital rim is elevated. More inferiorly, the dissection lies superficial to the masseter fascia, avoiding injury to the facial nerve branches at this level. *ZA*, zygomatic arch; *TF*, temporalis fascia; *TM*, temporalis muscle; *OR*, orbital roof.

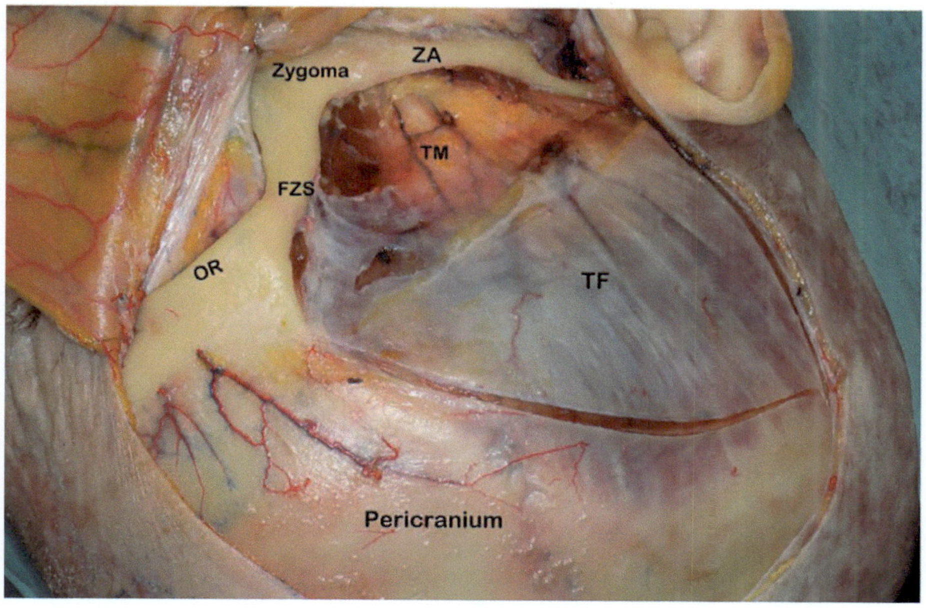

Fig. 12.309 The temporalis muscle is incised in a line about 1 cm along its attachment. The temporalis muscle is then detached from the zygomatic arch using a large elevator. *ZA*, zygomatic arch; *TF*, temporalis fascia; *TM*, temporalis muscle; *FZS*, frontozygomatic suture; *OR*, orbital roof.

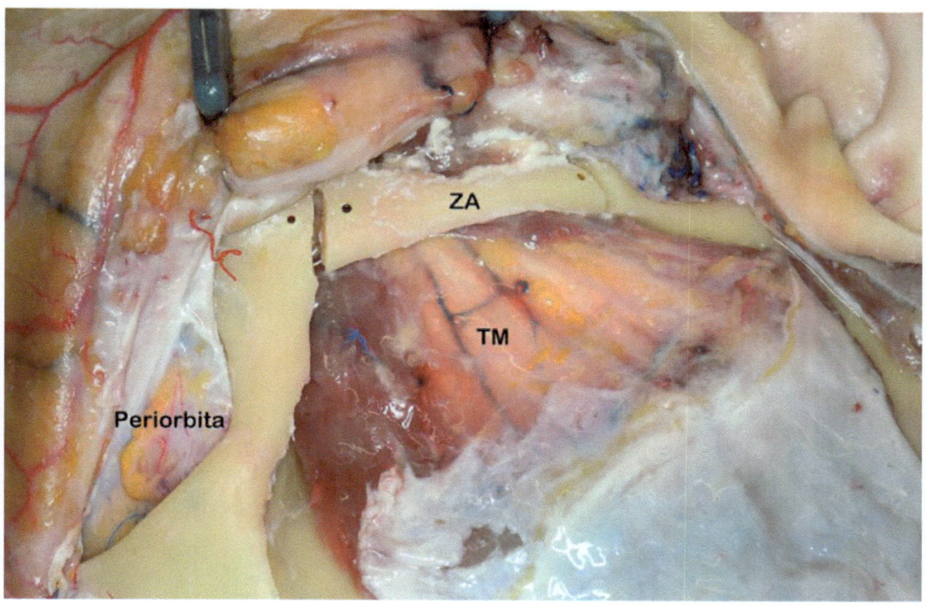

Fig. 12.310 The zygomatic arch is transected after creating two burr holes for further refixation at the end of the procedure. *ZA*, zygomatic arch; *TM*, temporalis muscle.

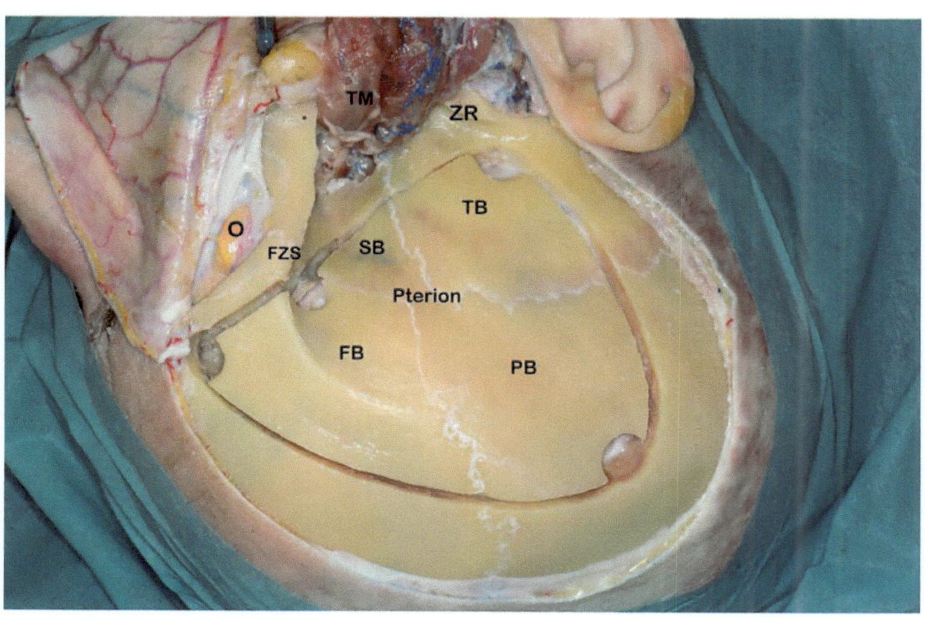

Fig. 12.311 The temporalis muscle is elevated from the attachment and reflected inferiorly. Next, a frontotemporal craniotomy is performed. The craniotomy extends anteriorly to just posterior to the lateral orbital rim; superiorly immediately superior to the pterion; and posteriorly behind the mandibular condyle. *ZR*, zygomatic root; *TB*, temporal bone; *PB*, parietal bone; *FB*, frontal bone; *SB*, sphenoid bone; *TM*, temporalis muscle; *O*, orbit; *FZS*, frontozygomatic suture.

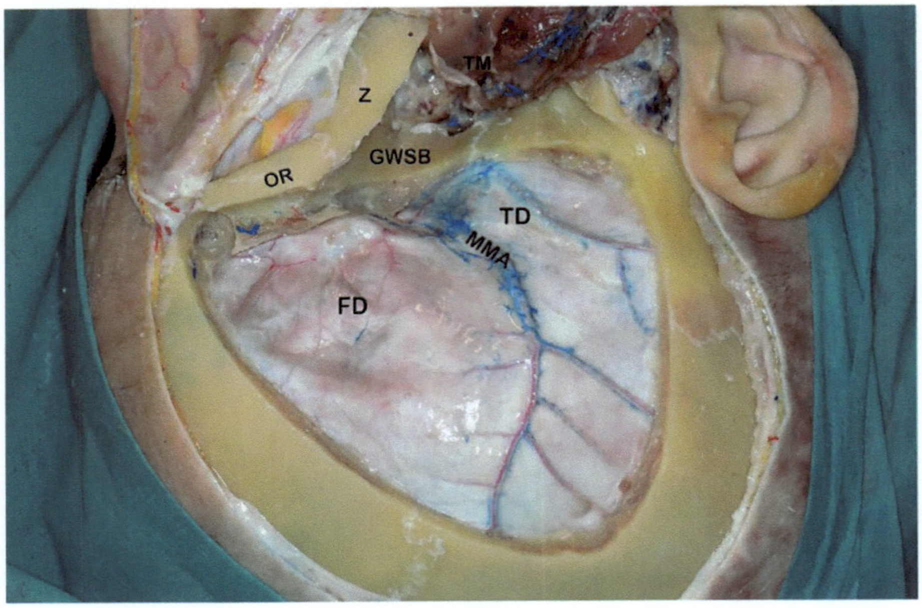

Fig. 12.312 The bone flap has been elevated to expose the frontal and temporal dura. *TD*, temporal dura; *FD*, frontal dura; *Z*, zygoma; *GWSB*, greater wing of sphenoid bone; *MMA*, middle meningeal artery; *OR*, orbital roof.

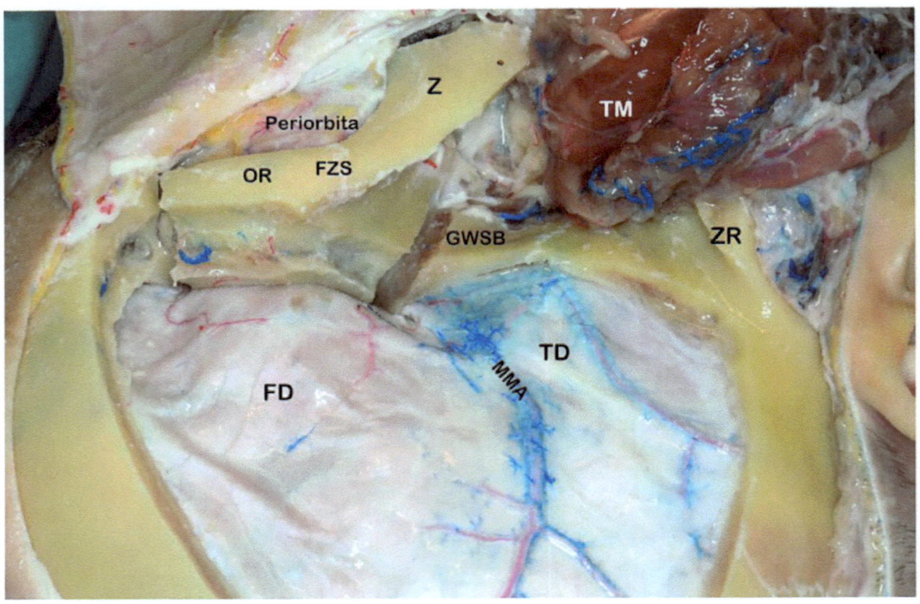

Fig. 12.313 If necessary, an OZ osteotomy can then be performed. *TD*, temporal dura; *FD*, frontal dura; *MMA*, middle meningeal artery; *FZS*, frontozygomatic suture; *OR*, orbital roof; *Z*, zygoma; *TM*, temporalis muscle; *ZR*, zygomatic root; *GWSB*, greater wing of sphenoid bone. The first cut is made with a thin-pointed drill at the level of the inferior aspect of the lateral orbital rim, just over the zygomatic root of the arch (which has been previously sectioned), directed toward the anterolateral border of the IOF.

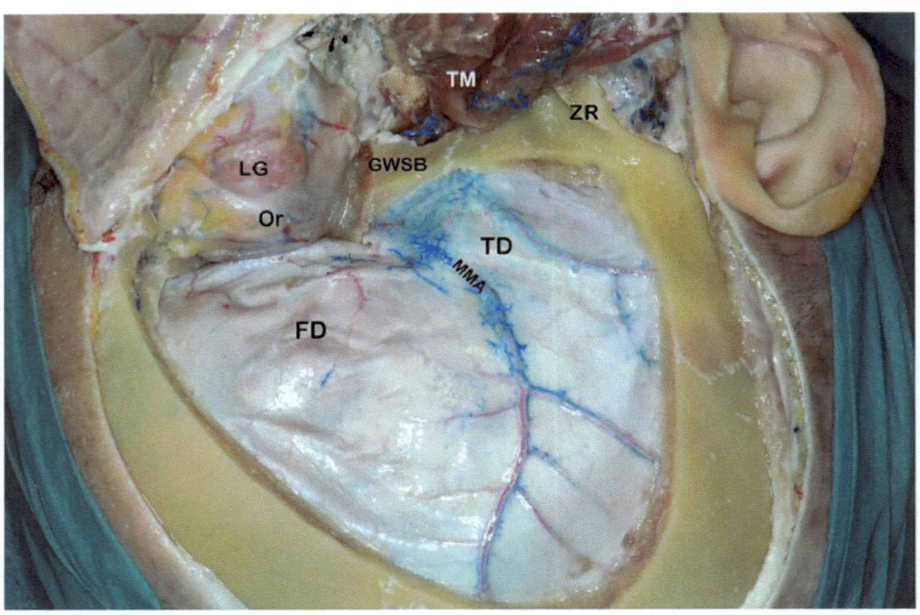

Fig. 12.314 After the lateral orbital rim is removed, the orbital contents are visualized. The approach provides excellent access to the superior and lateral aspects of the orbit. *TM*, temporalis muscle; *ZR*, zygomatic root; *MMA*, middle meningeal artery; *TD*, temporal dura; *FD*, frontal dura; *Or*, orbit; *LG*, lacrimal gland; *GWSB*, greater wing of sphenoid bone.

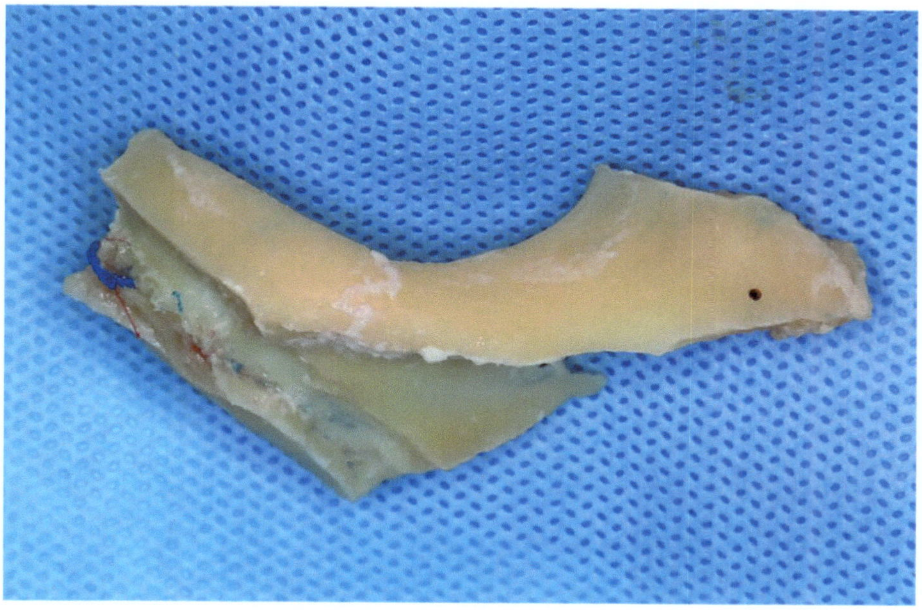

Fig. 12.315 Orbitozygomatic bone is removed and preserved in saline.

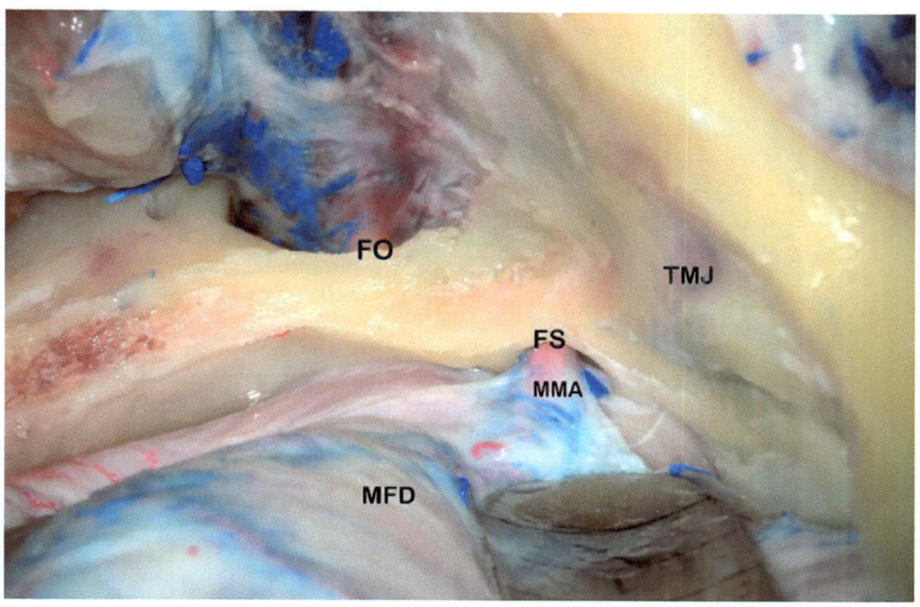

Fig. 12.316 The dura has been elevated from the floor of the middle cranial fossa. The bone is then rongeured (or drilled) down to the base of the middle cranial fossa. The foramen ovale lies anterior and medial to the foramen spinosum. Shaving the bone superior to the temporomandibular joint. *TMJ*, temporomandibular joint; *FS*, foramen spinosum; *MMA*, middle meningeal artery; *MFD*, middle fossa dura; *FO*, foramen ovale.

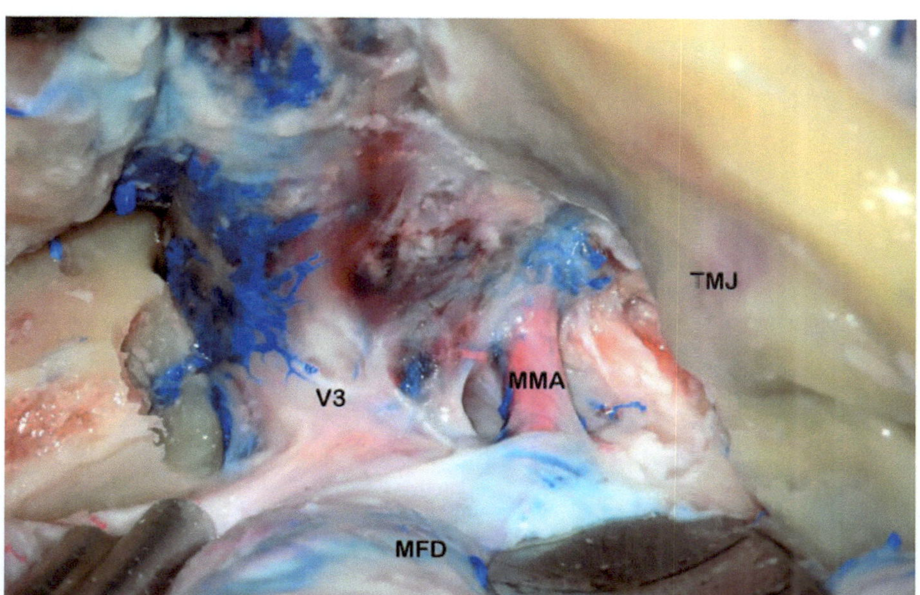

Fig. 12.317 Bone removal extends to the floor of the middle fossa, uncovering the foramen spinosum, through which the middle meningeal artery passes, and the foramen ovale, through which the mandibular nerve passes. Maximum shaving of the anterior aspect of the temporomandibular joint provides a wide exposure of the infratemporal fossa space. *TMJ*, temporomandibular joint; *MMA*, middle meningeal artery; *MFD*, middle fossa dura; *V3*, mandibular nerve.

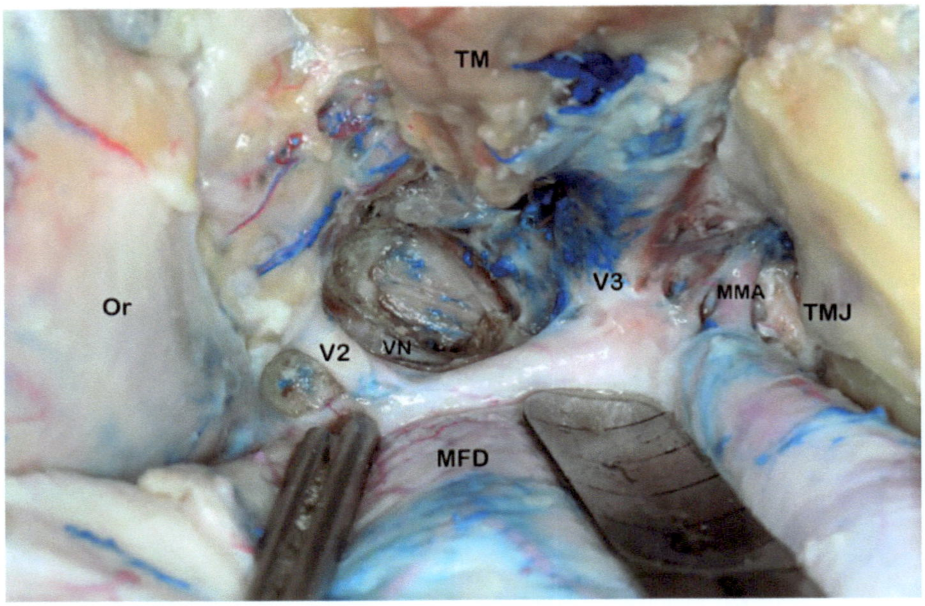

Fig. 12.318 After unroofing the foramen rotundum, the maxillary nerve is appreciated. After drilling away the bone between the maxillary nerve and mandibular nerve, the vidian nerve is seen just below the maxillary nerve. This specimen had a poorly developed sphenoid sinus. *TMJ*, temporomandibular joint; *MMA*, middle meningeal artery; *V3*, mandibular nerve; *V2*, maxillary nerve; *VN*, vidian nerve; *MFD*, middle fossa dura; *Or*, orbit; *TM*, temporalis muscle.

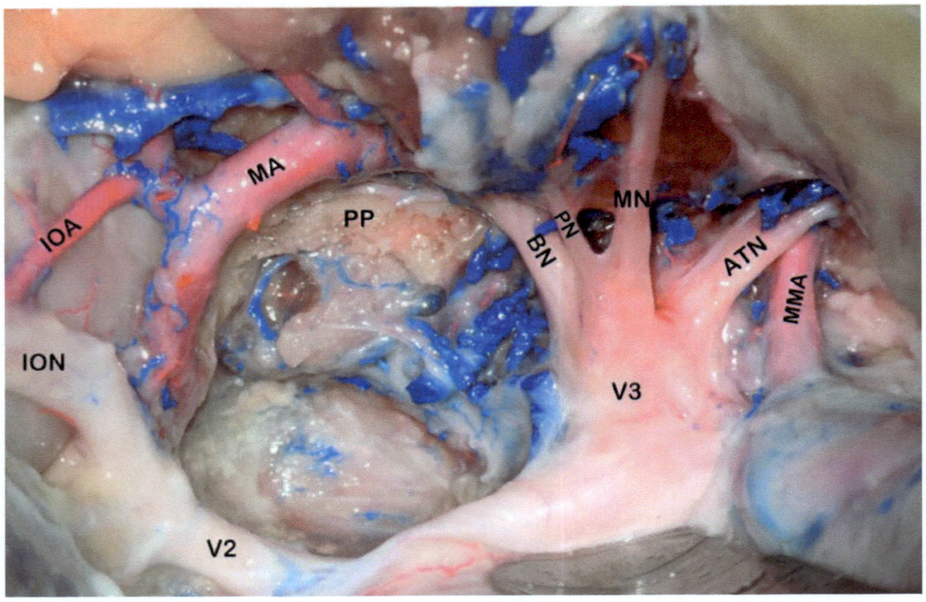

Fig. 12.319 The base of the pterygoid process is now drilled. The maxillary nerve is followed distally into the pterygopalatine fossa, and the mandibular nerve is followed distally into the infratemporal fossa. Exposing the branches of the mandibular nerve without removing the lateral pterygoid muscle. The infraorbital artery, which arises from the maxillary artery, can be exposed. The infraorbital artery enters the orbital cavity through the posterior part of the inferior orbital fissure. *MMA*, middle meningeal artery; *ATN*, auriculotemporal nerve; *MN*, masseteric nerve; *PN*, pterygoid nerve; *BN*, buccal nerve; *V2*, maxillary nerve; *V3*, mandibular nerve; *PP*, pterygoid process; *MA*, maxillary artery; *IOA*, infraorbital artery; *ION*, infraorbital nerve.

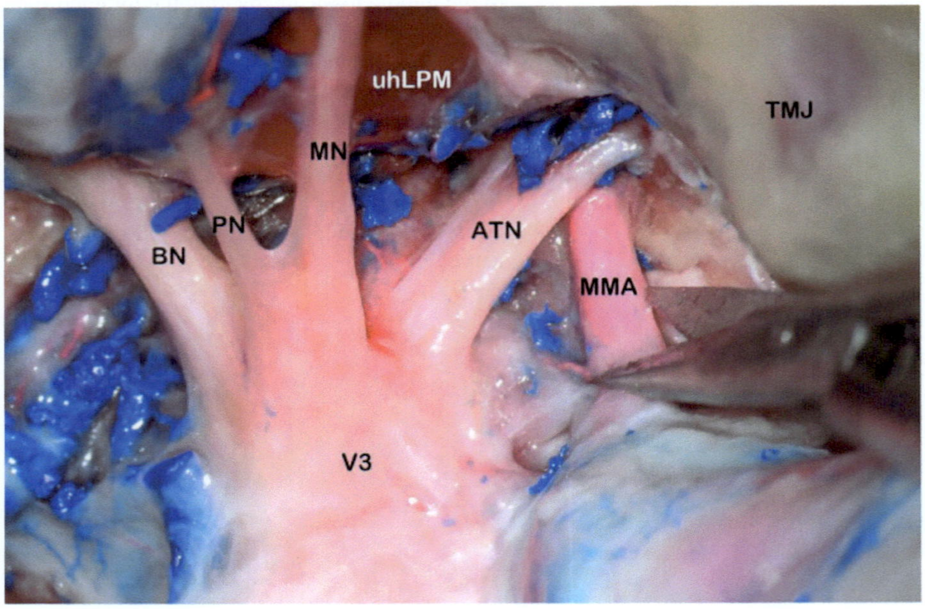

Fig. 12.320 The distal branches of the maxillary and mandibular nerves have been exposed. The middle meningeal artery is cut after bipolar coagulation. *TMJ*, temporomandibular joint; *MMA*, middle meningeal artery; *ATN*, auriculotemporal nerve; *V3*, mandibular nerve; *MN*, masseteric nerve; *PN*, pterygoid nerve; *BN*, buccal nerve; *uhLPM*, upper head of lateral pterygoid muscle.

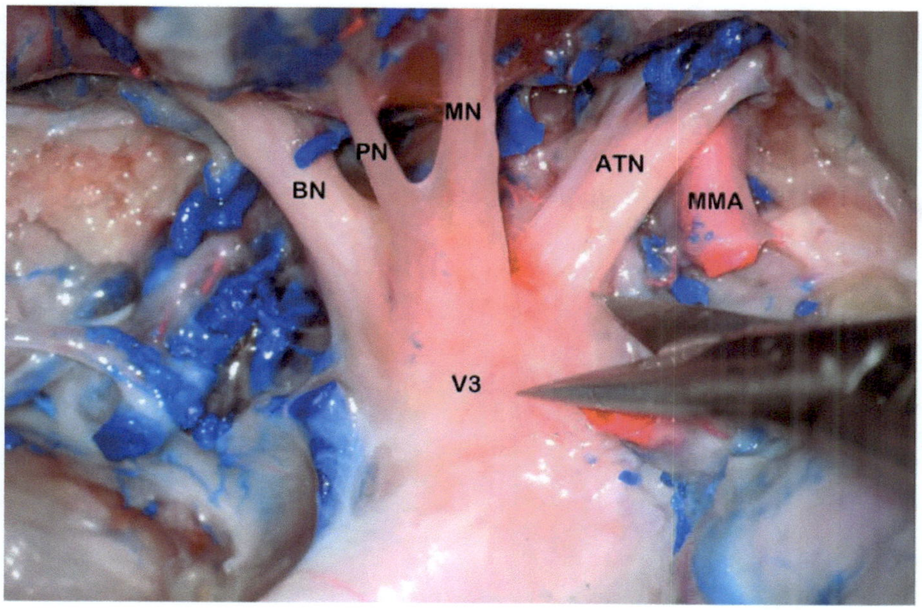

Fig. 12.321 The mandibular nerve is cut after bipolar coagulation. *MMA*, middle meningeal artery; *ATN*, auriculotemporal nerve; *MN*, masseteric nerve; *PN*, pterygoid nerve; *BN*, buccal nerve; *V3*, mandibular nerve.

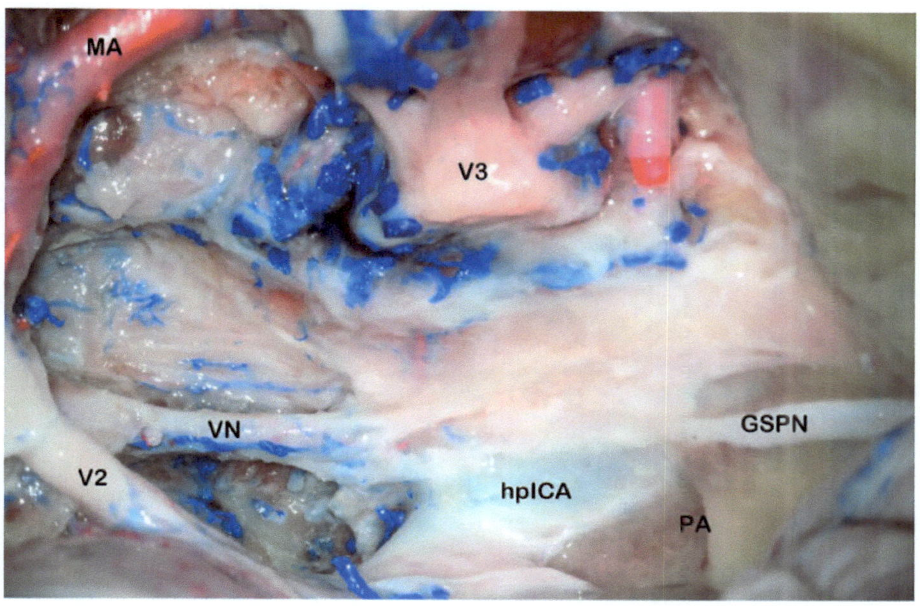

Fig. 12.322 By retracting the middle fossa dura, the greater petrosal nerve is exposed directly under the dura of the middle fossa. In this case, the terminal part of the petrous carotid artery is also exposed under the dura and below the greater superficial petrosal nerve, as occurs in approximately 15% of temporal bones. The petrous carotid artery is usually covered by bone up to the lateral side of the third trigeminal division. *VN*, vidian nerve; *V2*, maxillary nerve; *V3*, mandibular nerve; *MA*, maxillary artery; *GSPN*, greater superficial petrosal nerve; *PA*, petrous apex; *hpICA*, horizontal portion of petrous carotid artery.

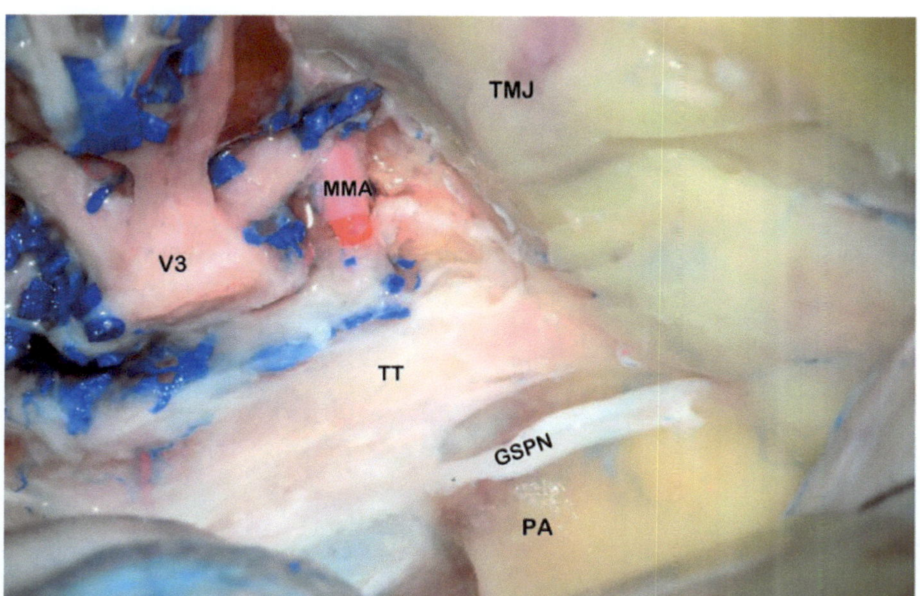

Fig. 12.323 The dura has been further elevated from the floor of the middle fossa to expose the greater petrosal nerve and petrous apex. *TMJ*, temporomandibular joint; *PA*, petrous apex; *GSPN*, greater superficial petrosal nerve; *TT*, tensor tympani muscle; *MMA*, middle meningeal artery; *V3*, mandibular nerve.

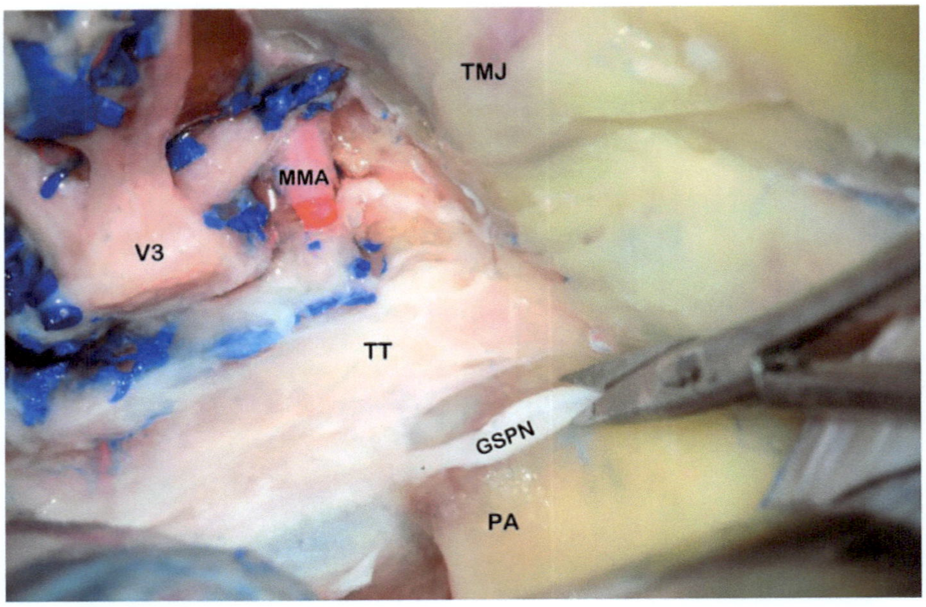

Fig. 12.324 The greater superficial petrosal nerve is cut. *TMJ*, temporomandibular joint; *TT*, tensor tympani muscle; *GSPN*, greater superficial petrosal nerve; *PA*, petrous apex; *V3*, mandibular nerve; *MMA*, middle meningeal artery.

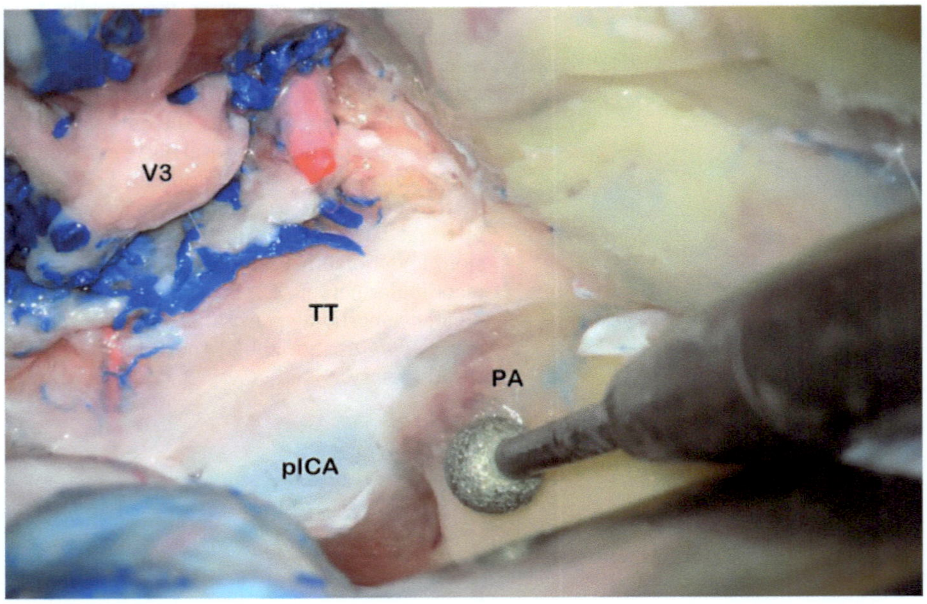

Fig. 12.325 The petrous apex bone medial to the internal carotid artery can now be drilled. *PA*, petrous apex; *TT*, tensor tympani muscle; *pICA*, petrous segment of internal carotid artery; *V3*, mandibular nerve.

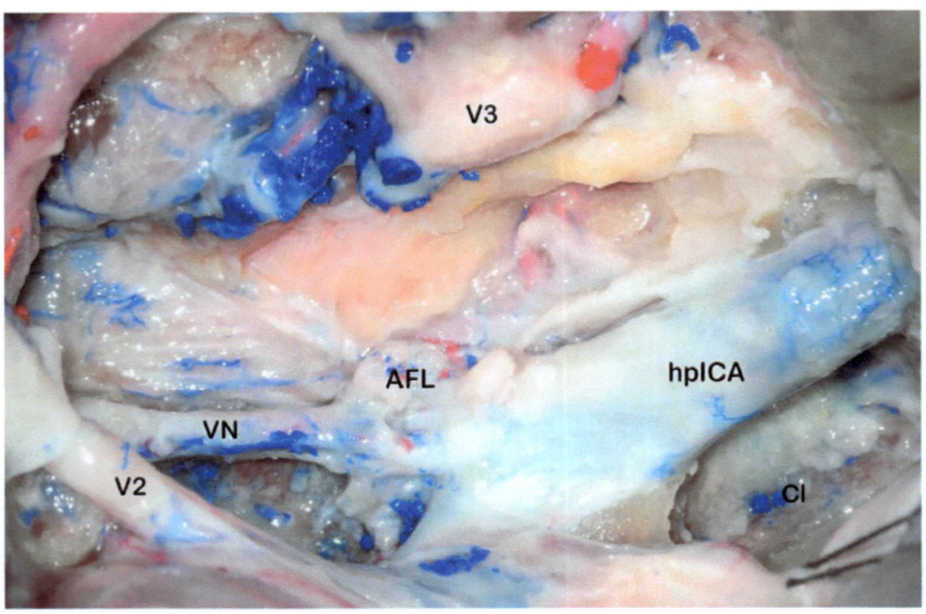

Fig. 12.326 The horizontal portion of petrous carotid artery is exposed by drilling anteromedial to the genu of the artery using a diamond burr. The carotid artery up to the level of the anterior foramen lacerum can be seen. Note the vidian nerve is a reliable landmark to the petrous internal carotid artery. *Cl*, clivus; *hpICA*, horizontal portion of petrous carotid artery; *AFL*, anterior foramen lacerum; *V3*, mandibular nerve; *VN*, vidian nerve; *V2*, maxillary nerve.

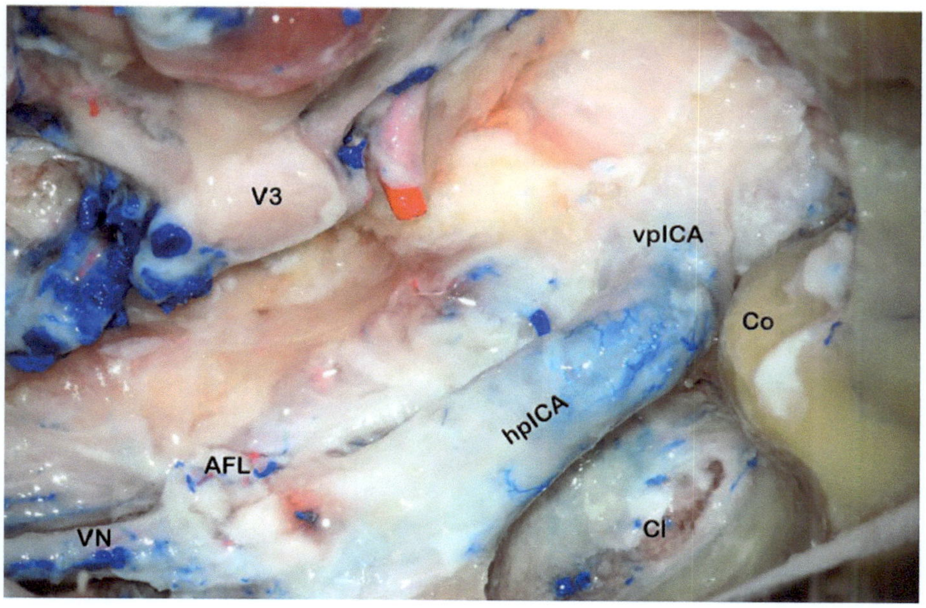

Fig. 12.327 The genu of the internal carotid artery is identified, and then the artery is followed down to identify its vertical part. The genu of the internal carotid artery is closely related to the cochlea. The distance between the two varies from 1 to 5 mm. *Co*, cochlea; *vpICA*, vertical portion of petrous carotid artery; *hpICA*, horizontal portion of petrous carotid artery; *Cl*, clivus; *AFL*, anterior foramen lacerum; *VN*, vidian nerve; *V3*, mandibular nerve.

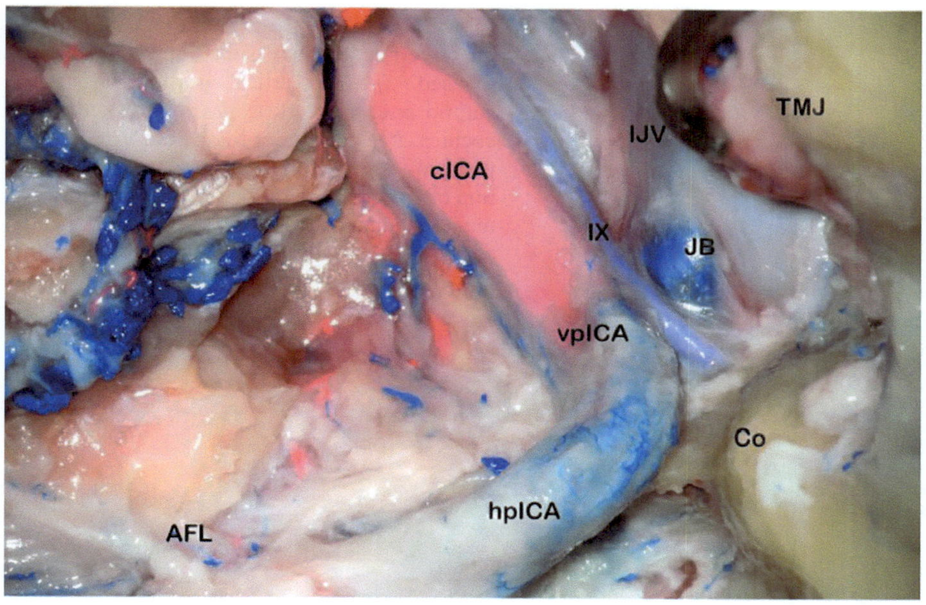

Fig. 12.328 The ICA is surrounded by a dense fibroperiosteal layer at its entrance in the skull base. After removed the fibroperiosteal layer, the cervical portion of petrous carotid artery can be appreciated. The glossopharyngeal nerve lies deep in this layer and is slightly adherent to its undersurface. Note the relationship between the anterior wall of the jugular bulb and the carotid artery. *TMJ*, temporomandibular joint; *JB*, jugular bulb; *Co*, cochlea; *vpICA*, vertical portion of petrous carotid artery; *IX*, glossopharyngeal nerve; *IJV*, internal jugular vein; *cICA*, cervical portion of petrous carotid artery; *hpICA*, horizontal portion of petrous carotid artery; *AFL*, anterior foramen lacerum.

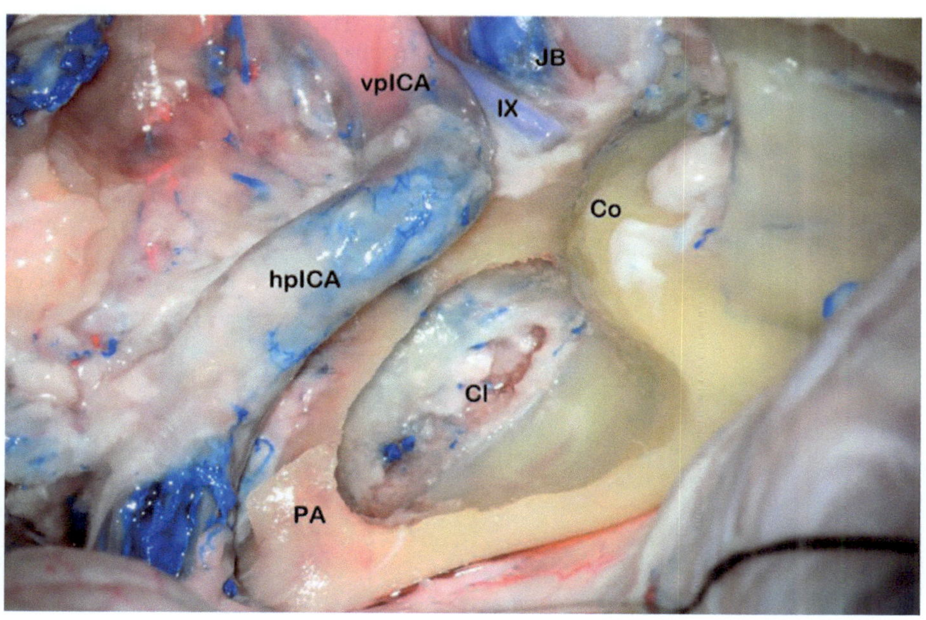

Fig. 12.329 Further drilling of the rest petrous apex is limited by petrous carotid artery. *Co*, cochlea; *Cl*, clivus; *IX*, glossopharyngeal nerve; *JB*, jugular bulb; *vpICA*, vertical portion of petrous carotid artery; *hpICA*, horizontal portion of petrous carotid artery; *PA*, petrous apex.

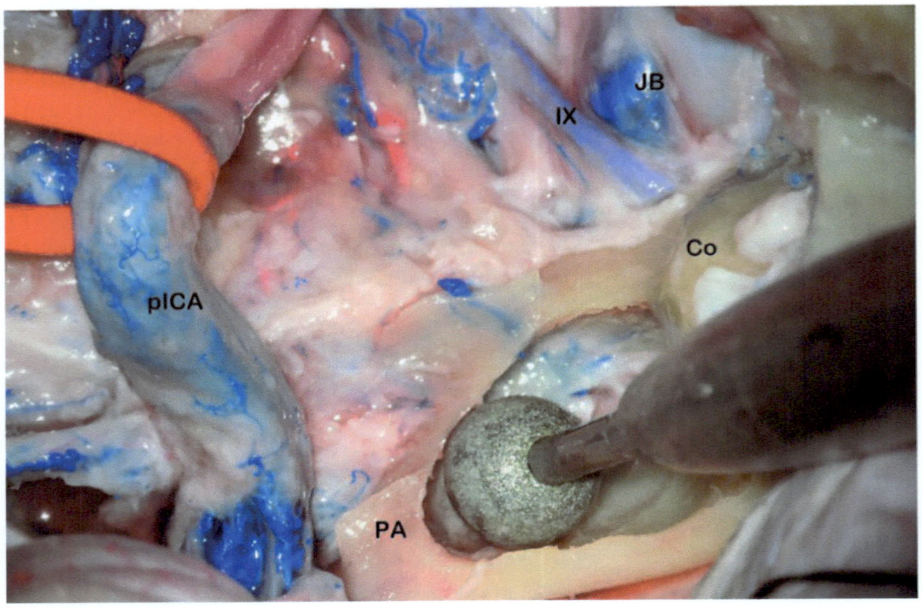

Fig. 12.330 A red vessel loop is placed in order to displace the petrous carotid artery laterally while the petrous apex medial to the petrous carotid artery is drilled with a large diamond burr to reach the clivus. *JB*, jugular bulb; *Co*, cochlea; *IX*, glossopharyngeal nerve; *PA*, petrous apex; *pICA*, petrous segment of internal carotid artery.

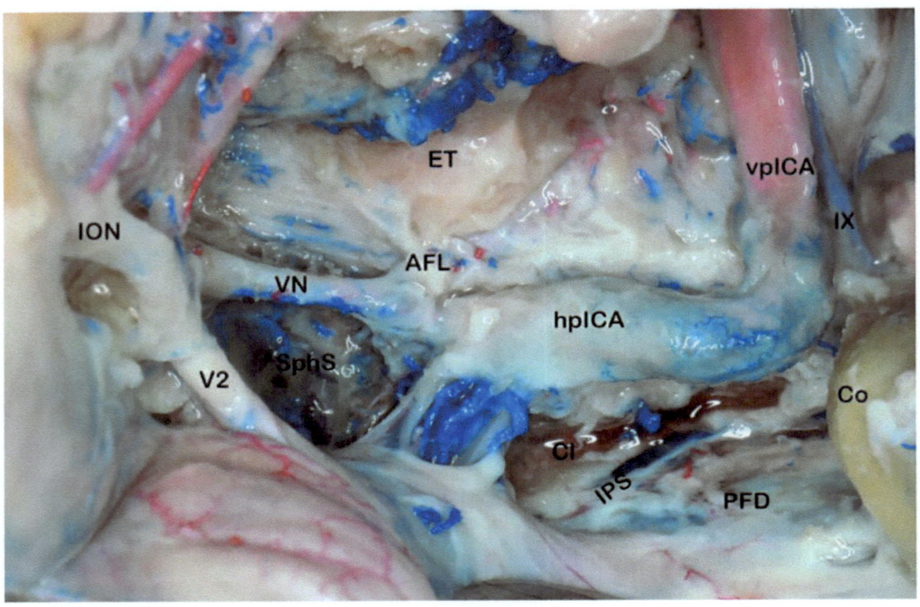

Fig. 12.331 The petrous apex was drilled until the inferior petrosal sinus and clivus were reached. *vpICA*, vertical portion of petrous carotid artery; *IX*, glossopharyngeal nerve; *Co*, cochlea; *PFD*, posterior fossa dura; *IPS*, inferior petrosal sinus; *Cl*, clivus; *hpICA*, horizontal portion of petrous carotid artery; *AFL*, anterior foramen lacerum; *VN*, vidian nerve; *SphS*, sphenoid sinus; *V2*, maxillary nerve; *ION*, infraorbital nerve; *ET*, Eustachian tube.

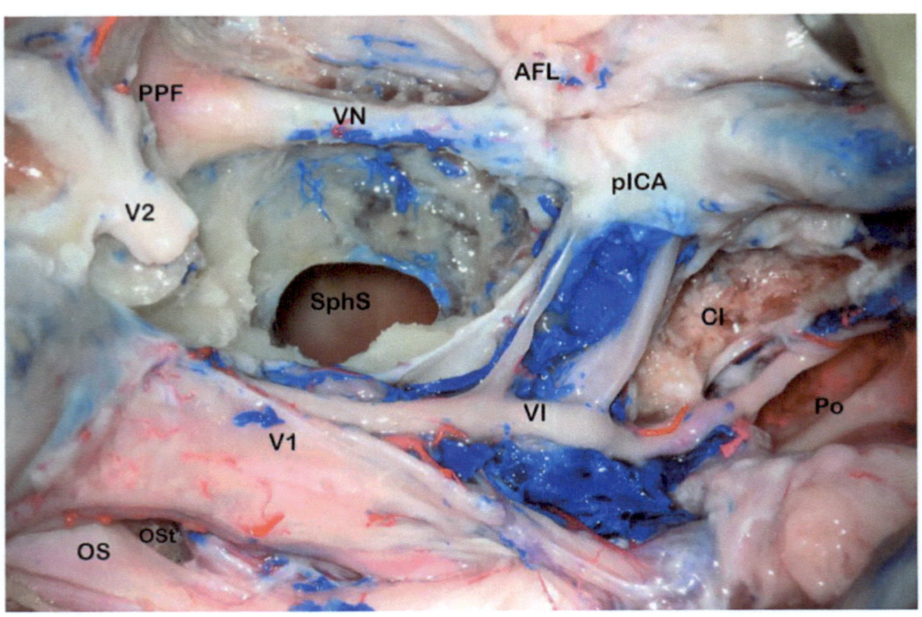

Fig. 12.332 The dura of the posterior fossa and cavernous sinus is opened. The abducens nerve courses in prepontine cistern and then pierces the dura and enters a dural cave, referred to as Dorello's canal to enter the cavernous sinus. *pICA*, petrous segment of internal carotid artery; *Cl*, clivus; *Po*, pons; *VI*, abducent nerve; *SphS*, sphenoid sinus; *V1*, ophthalmic nerve; *OSt*, optic strut; *OS*, optic sheath; *AFL*, anterior foramen lacerum; *VN*, vidian nerve; *V2*, maxillary nerve; *PPF*, pterygopalatine fossa.

Clinical Applications

CASE 1: RIGHT EXTRADURAL TRIGEMINAL NEURINOMA REMOVED THROUGH A PREAURICULAR INFRATEMPORAL ORBITOZYGOMATIC APPROACH (Figs. 12.333A–12.343)

1. Preoperative radiological imaging of the case. The tumor involves the floor of the middle fossa, the petrous apex with extension into the infratemporal fossa, the pterygopalatine fossa, the orbit, the sphenoid sinus, the maxillary sinus, and the nasal cavity (Figs. 12.333A–12.333D).

2. Nasal endoscopy showing the intranasal tumor (Fig. 12.334).

3. The incision has been performed, and the temporalis muscle has been reflected inferiorly. Note the limited exposure before doing the orbitozygomatic osteotomy (Fig. 12.335).

4. Using a small high-speed drill, the orbitozygomatic osteotomy is performed. This step is now better performed using the saw attachment of the Midas Rex (Figs. 12.336 and 12.337).

5. Once the orbitozygomatic bone has been removed, the tumor can be seen in close relation to the orbital contents (Fig. 12.338). A large frontotemporal craniotomy is performed. The temporomandibular joint is opened, and the mandibular condyle is displaced inferiorly using the infratemporal fossa retractor as previously described. The bony Eustachian tube is drilled, the middle meningeal artery is sacrificed after bipolar coagulation, and the mandibular branch is identified. The base of the pterygoid is drilled. Further drilling of the greater wing of the sphenoid bone opens the way to the foramen rotundum and the pterygopalatine fossa (Fig. 12.339).

6. The tumor is seen lying anterior to the mandibular branch of the trigeminal nerve. This latter is sacrificed after bipolar coagulation. This allows identification of the horizontal segment of the internal carotid artery. Removal of the tumor removal proceeds next (Fig. 12.340).

7. At the end of the procedure, the temporalis is used to obliterate the operative cavity. The osteotomy is replaced and fixed in place with miniplates (Fig. 12.341). The incision is sutured, and a drain is applied (Fig. 12.342).

8. Postoperative CT scan showing total tumor removal (Fig. 12.343).

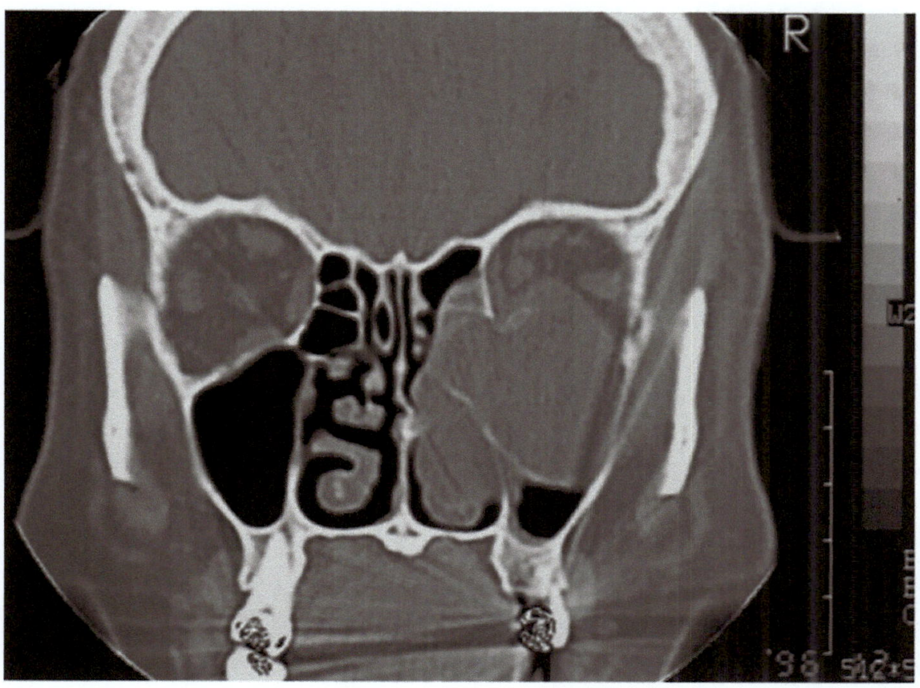

Fig. 12.333A Preoperative imaging of the case. A CT scan, coronal view, showing a right-sided tumor with massive involvement of the nasal cavity and maxillary sinus and extension into the orbit.

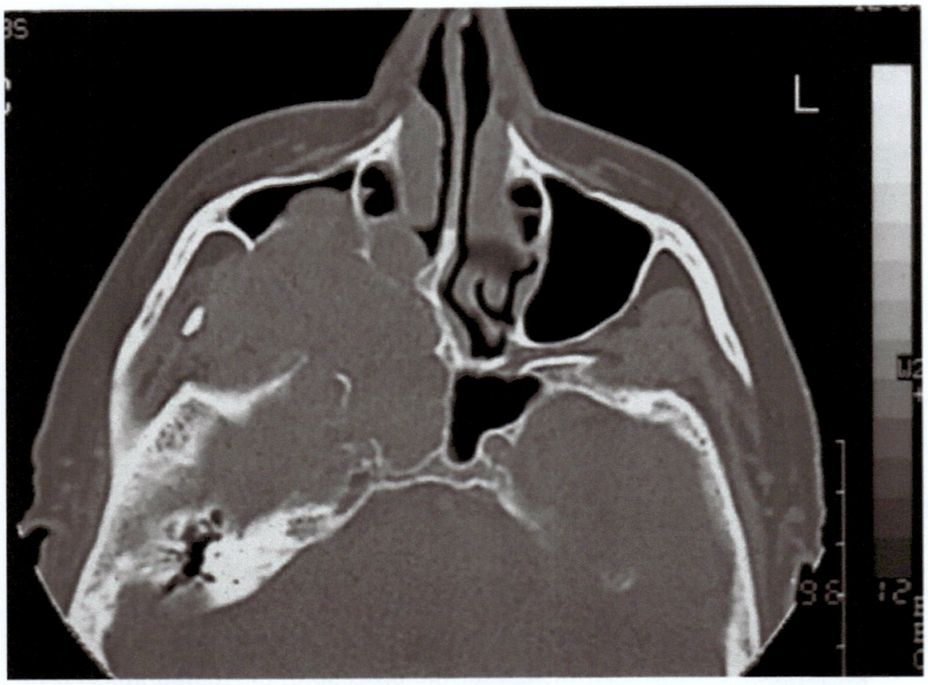

Fig. 12.333B CT scan, axial view. The tumor involves the petrous apex, sphenoid sinus, pterygopalatine fossa, infratemporal fossa, and maxillary sinus.

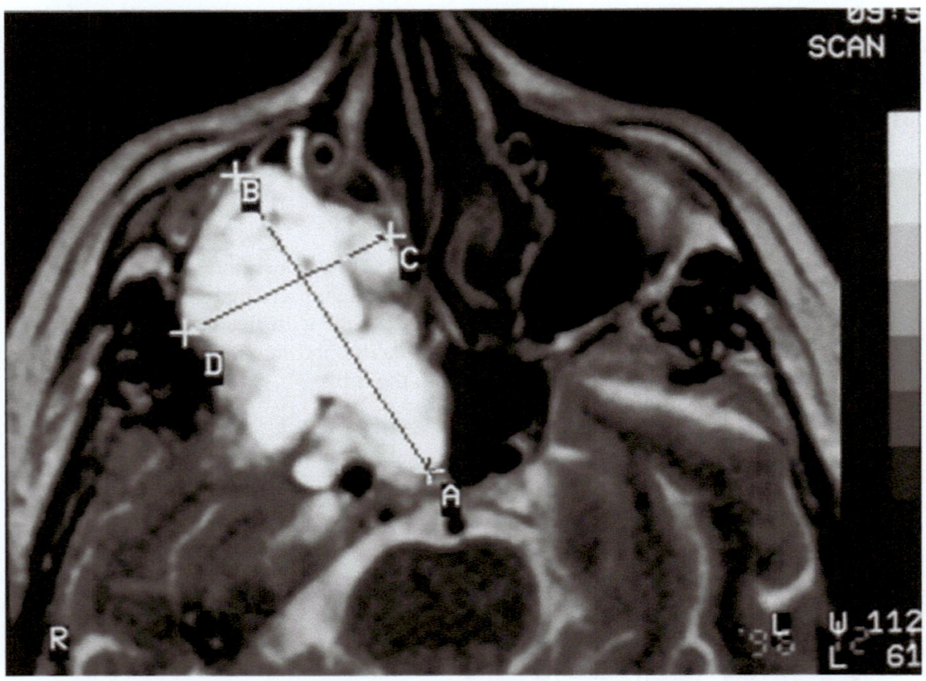

Fig. 12.333C MRI, T2-weighted image. Note the hyperintense tumor.

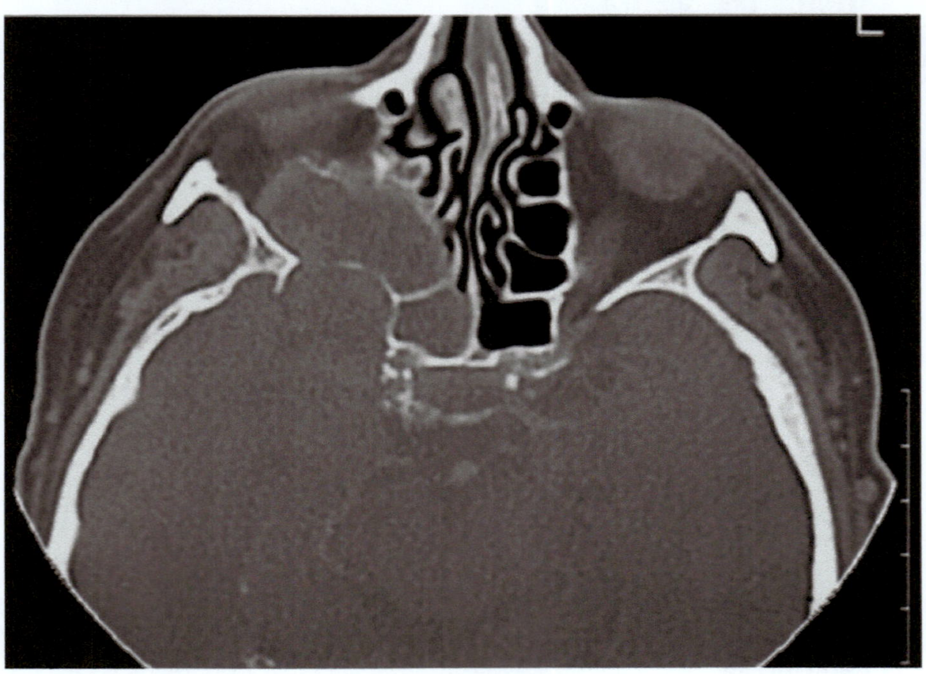

Fig. 12.333D CT scan, axial view at a higher level. Orbital extension can be seen.

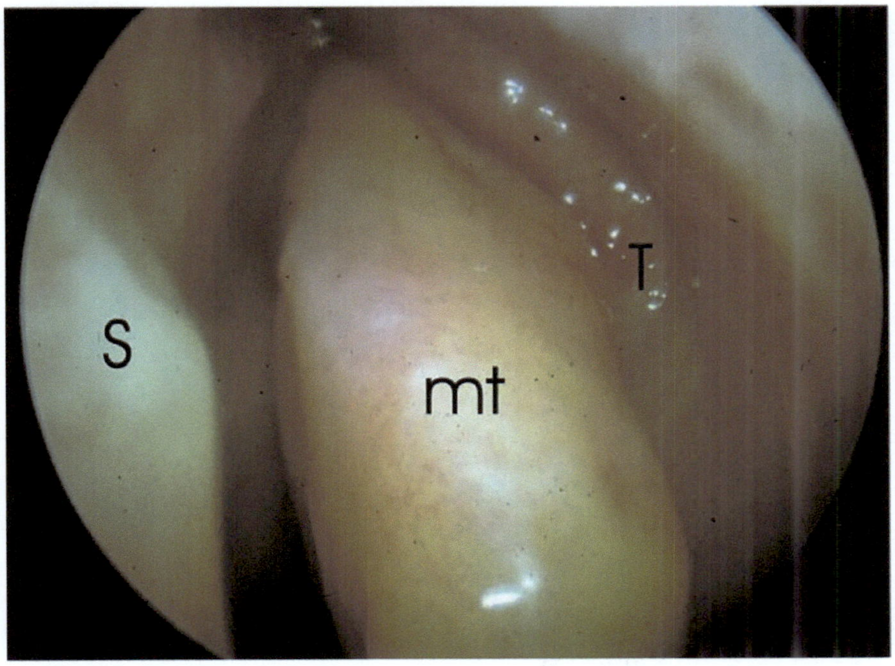

Fig. 12.334 Nasal endoscopy reveals the tumor (T) lateral to the middle turbinate (mt). *S*, nasal septum.

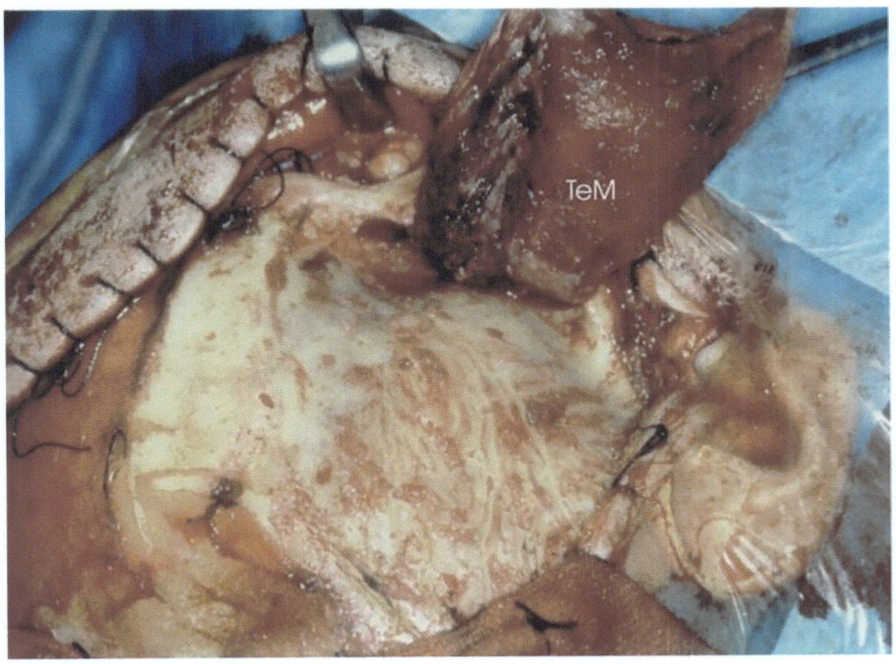

Fig. 12.335 The temporalis muscle (TeM) has been reflected anteriorly.

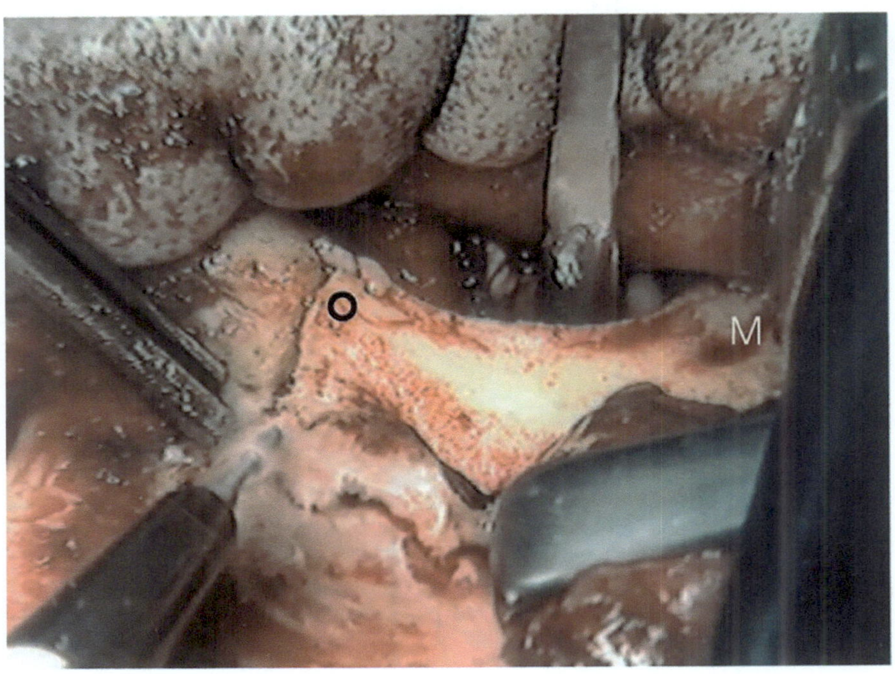

Fig. 12.336 The orbitozygomatic osteotomy is performed. The cut at the orbital process of the zygoma is nearly finished. Note the retractor protecting the orbital contents.

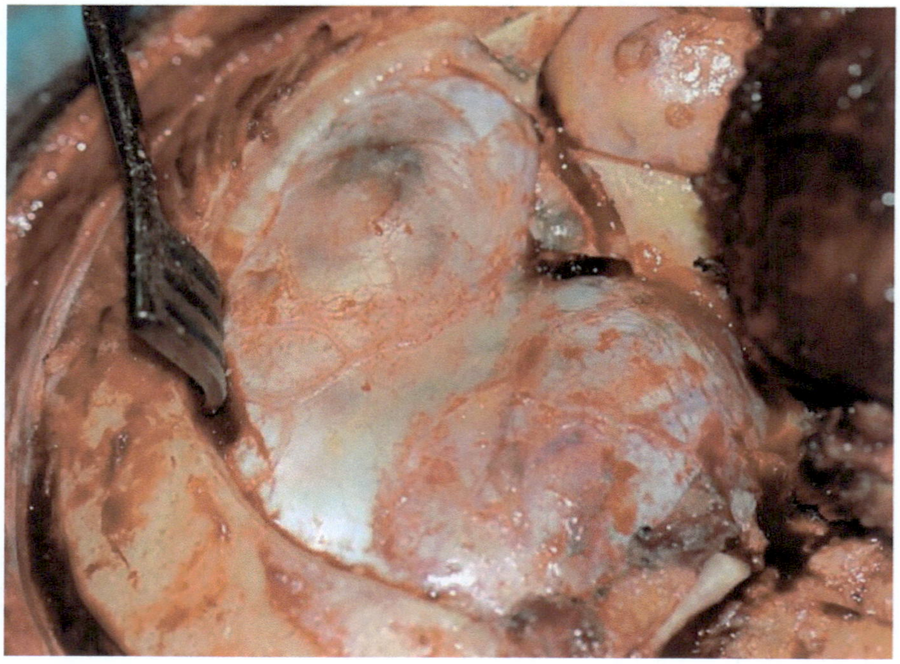

Fig. 12.337 Orbitozygomatic bone has been removed and orbital contents are visualized.

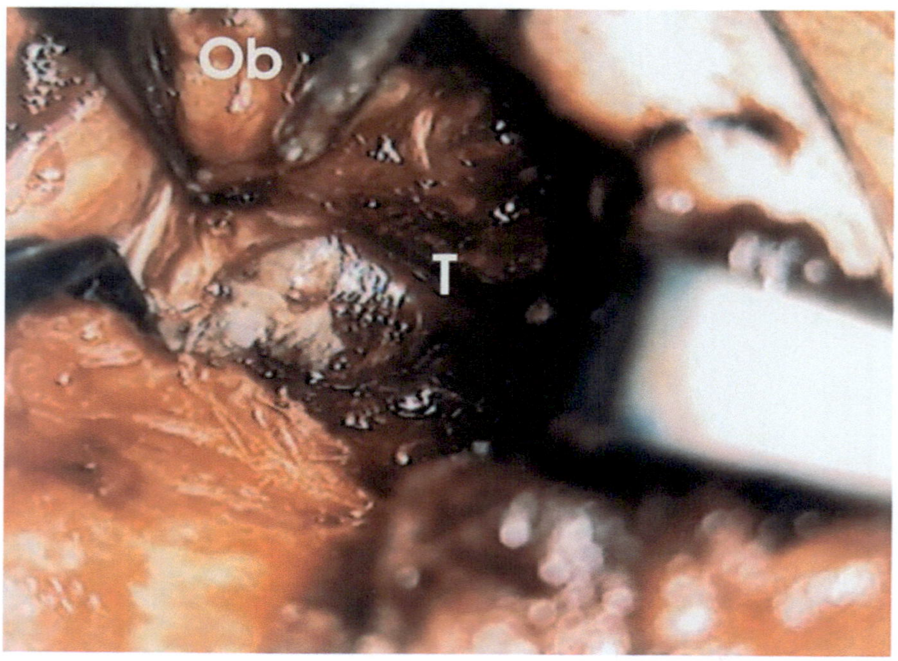

Fig. 12.338 The tumor (T) is seen in close relation to the orbital contents.

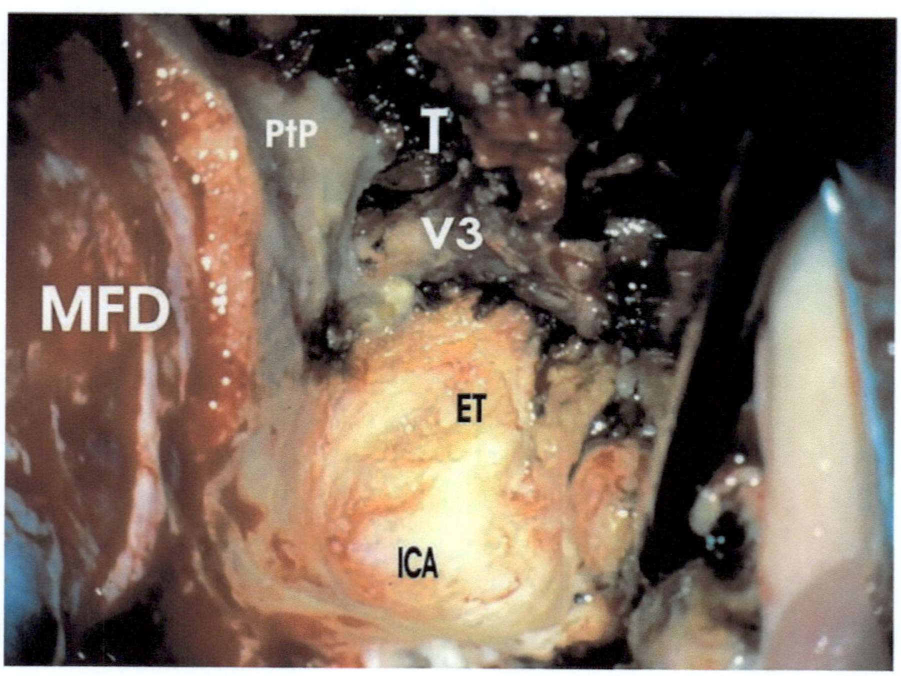

Fig. 12.339 The mandibular nerve (V3) is identified. The base of the pterygoid is still to be drilled.

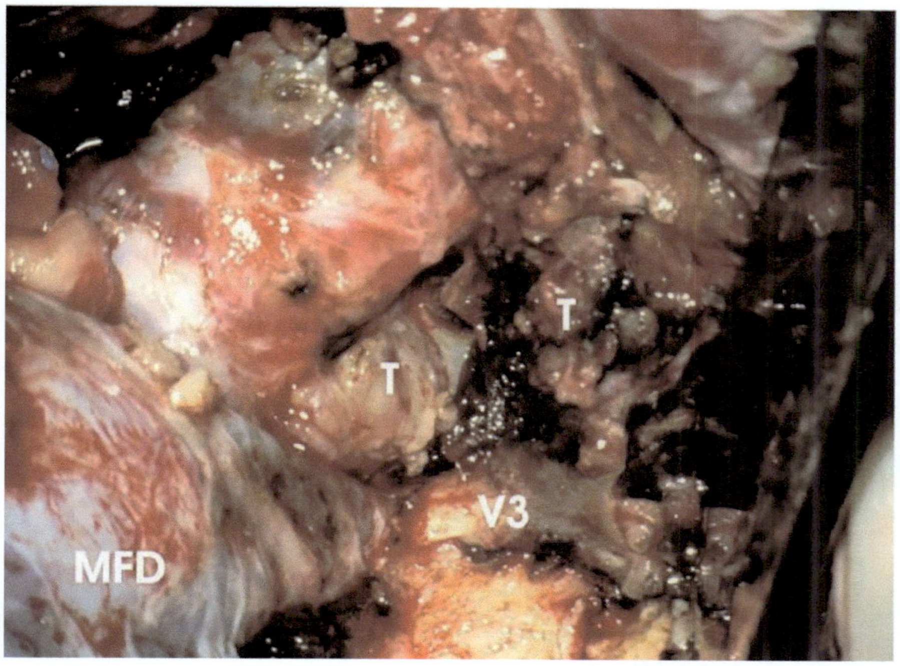

Fig. 12.340 After drilling the pterygoid process, the tumor (T) is seen anterior to the mandibular nerve (V3).

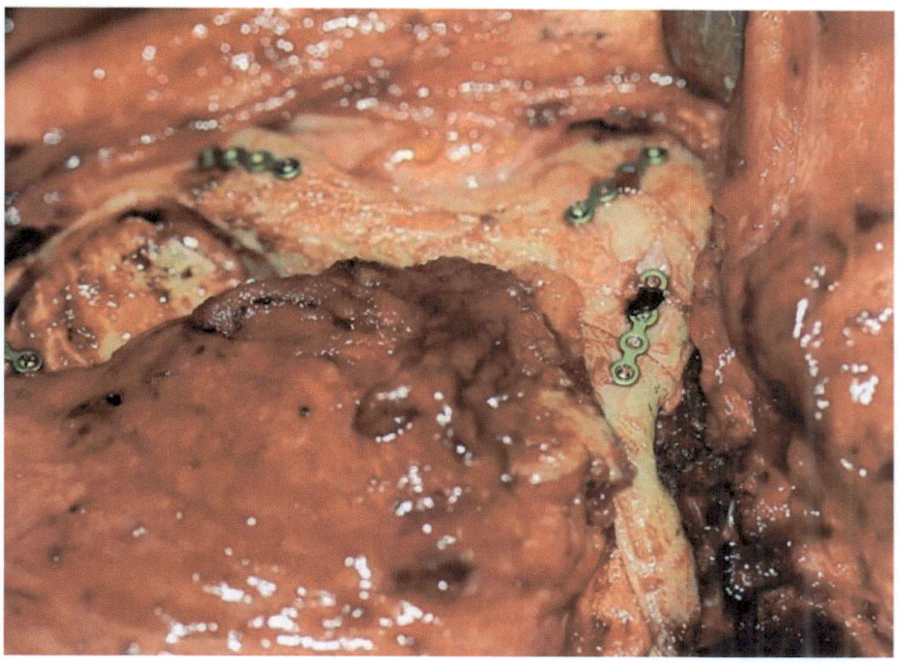

Fig. 12.341 Reconstruction is performed with plates and screws.

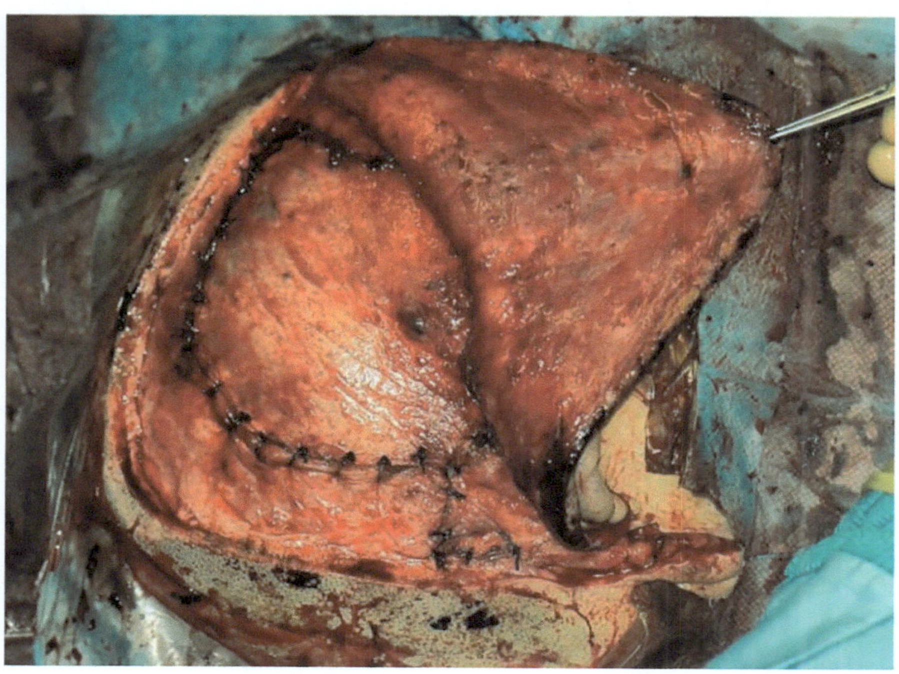

Fig. 12.342 Closure and drain insertion.

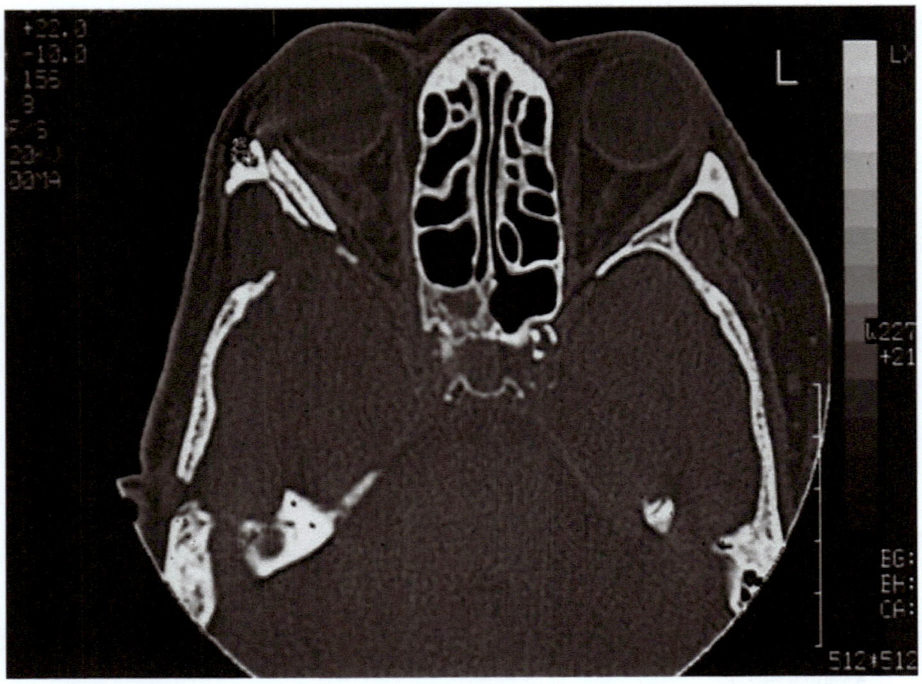

Fig. 12.343 Postoperative CT scan.

CASE 2: LEFT TRIGEMINAL SCHWANNOMA EXTENDING INTO THE INFRATEMPORAL SPACE MANAGED BY THE PREAURICULAR EXTENDED TRANSZYGOMATIC APPROACH WITH EXTENSION INTO THE NECK (Figs. 12.344A–12.355)

1. Preoperative MRI of the case. The extradural tumor involves Meckel's cave and extends extensively into the infratemporal fossa (Figs. 12.344A and 12.344B).

2. The incision is performed as previously described. In addition, the incision extends caudally below the ear, curving first posteriorly and then anteriorly to allow the better exposure of the inferior part of the tumor (Fig. 12.345).

3. The skin and subcutaneous tissues are elevated in the loose areolar layer of fascia. About 2–3 cm above the zygomatic arch a semilunar incision is made in the superficial layer of the deep temporal fascial and the plane of dissection becomes deeper to afford protection for the frontal branch of the facial nerve (Fig. 12.346). The zygomatic arch is exposed. The facial nerve is identified in the parotid and its branches are followed by blunt dissection. The tumor is already seen medial to the facial nerve branches anterior to the parotid (Figs. 12.347 and 12.348).

4. A frontotemporal craniotomy is performed. In this case, there was no need to perform an orbitozygomatic osteotomy as the tumor was mainly inferior rather than anterior (Figs. 12.349 and 12.350). The temporal lobe is retracted and the tumor component involving Meckel's cave is totally removed.

5. The cavity is obliterated with abdominal fat, and the craniotomy is replaced and fixed with miniplates. The temporalis is repositioned and sutured back, and finally, the zygomatic arch is returned and its anterior part is fixed with miniplates. The skin is closed in layers (Figs. 12.351–12.355).

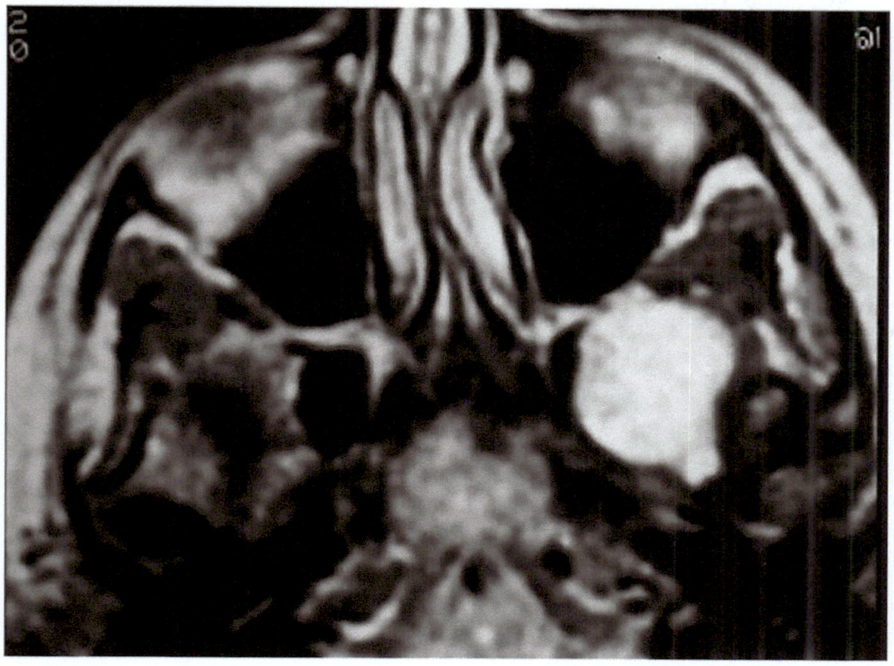

Fig. 12.344A Preoperative MRI with gadolinium enhancement. Axial view showing the tumor massively involving the infratemporal fossa.

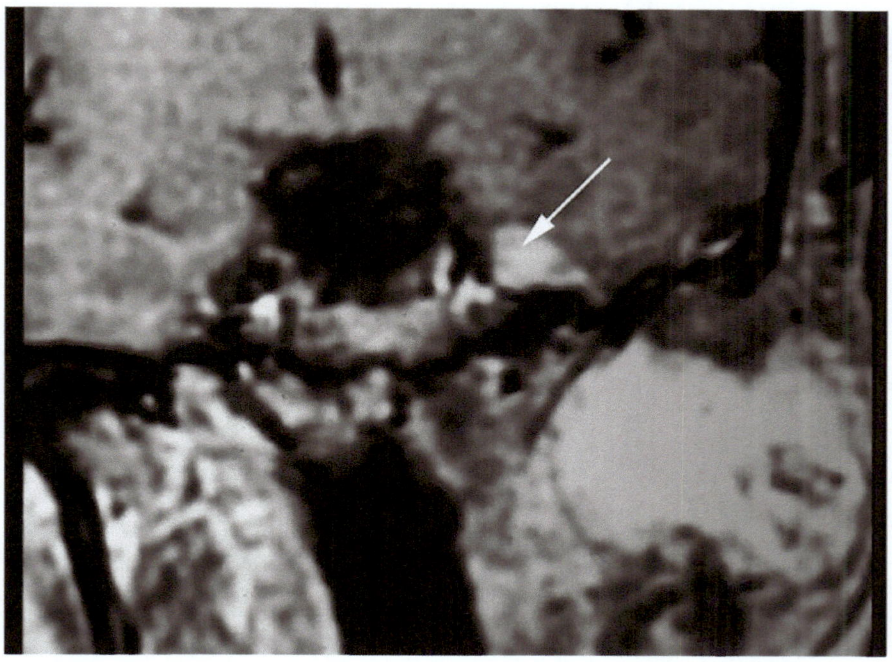

Fig. 12.344B Coronal view. The tumor involves Meckel's cave *(arrow)* and extends into the infratemporal fossa in a very low position.

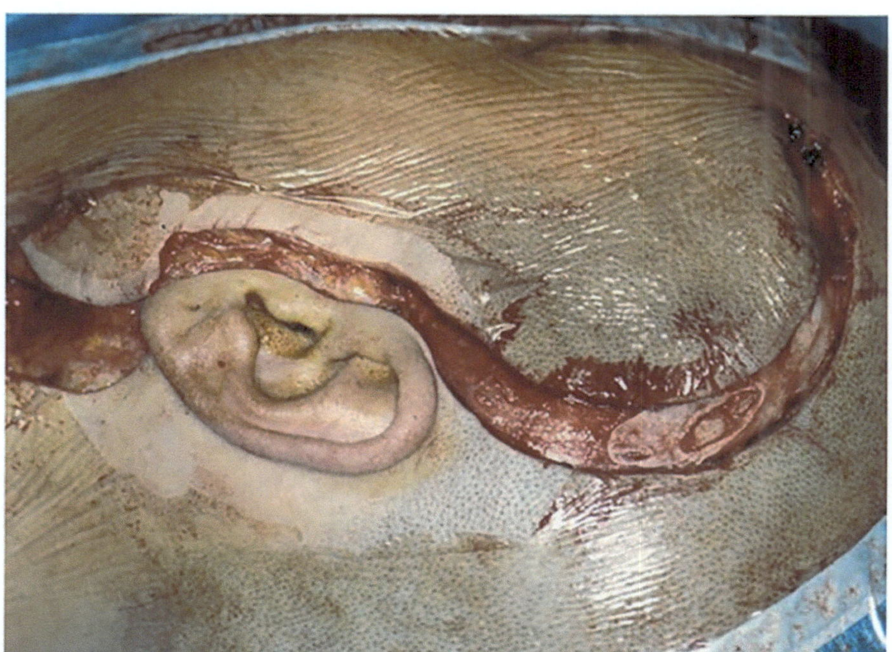

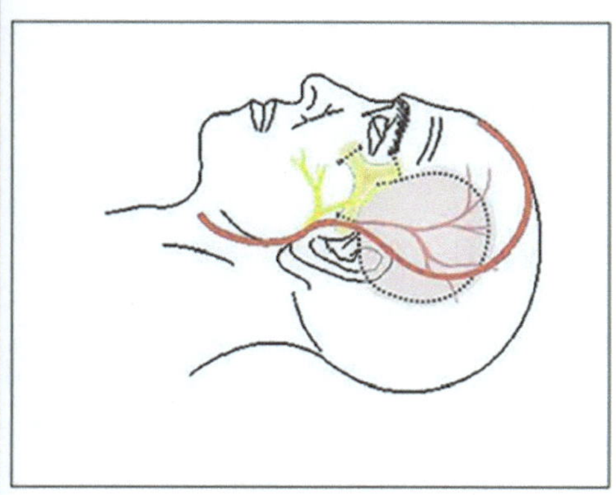

Fig. 12.345 The skin incision has been performed. Note the extension into the neck.

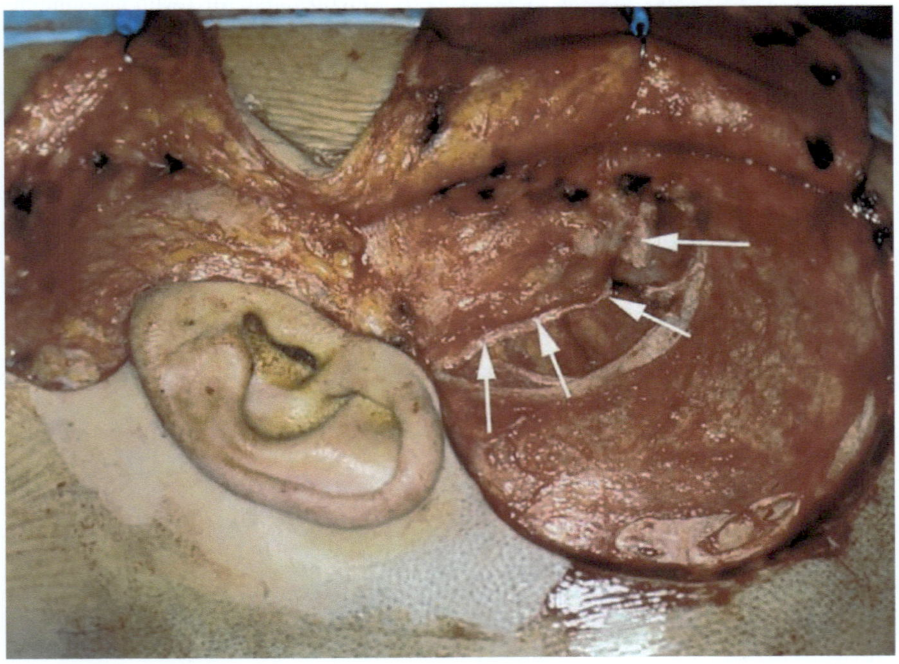

Fig. 12.346 The semilunar incision *(arrows)* is made above the zygomatic arch to change the plane of dissection to a deeper one.

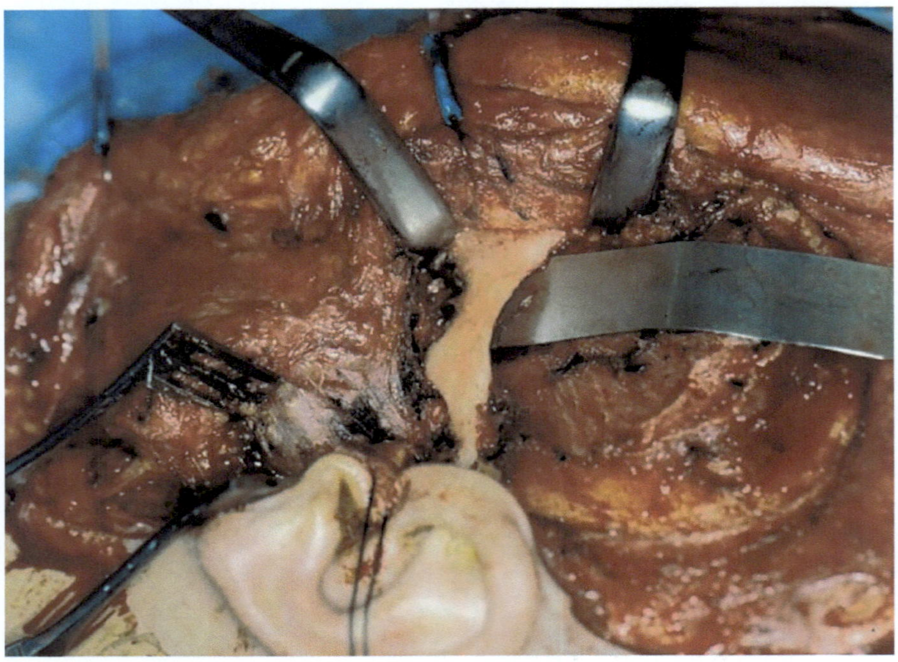

Fig. 12.347 Exposure of the zygomatic arch.

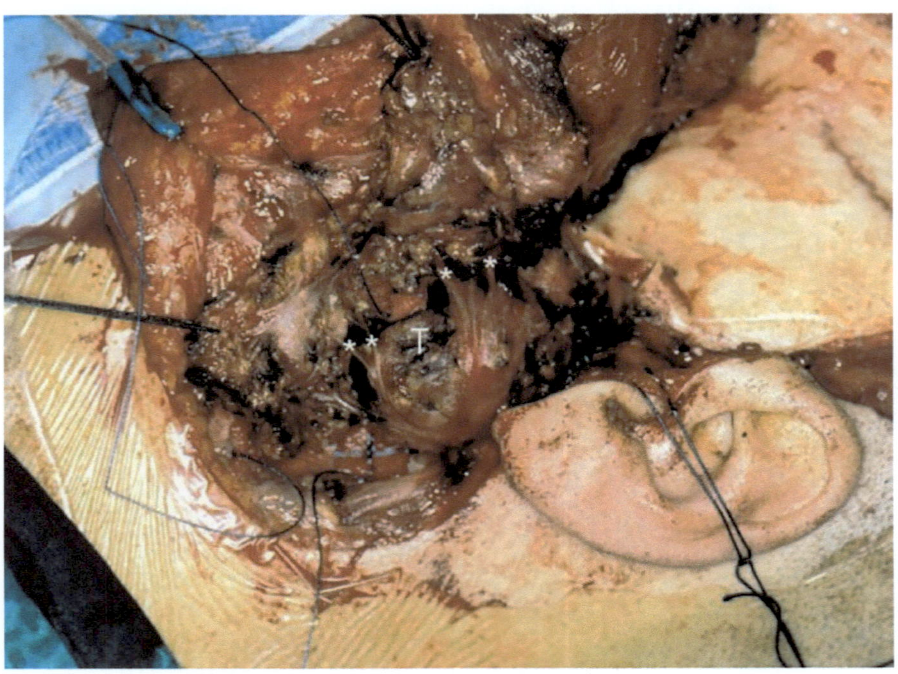

Fig. 12.348 The facial nerve branches (*) are already dissected. The tumor (T) is already seen involving the infratemporal fossa and lies anterior and inferior to the level of the parotid gland.

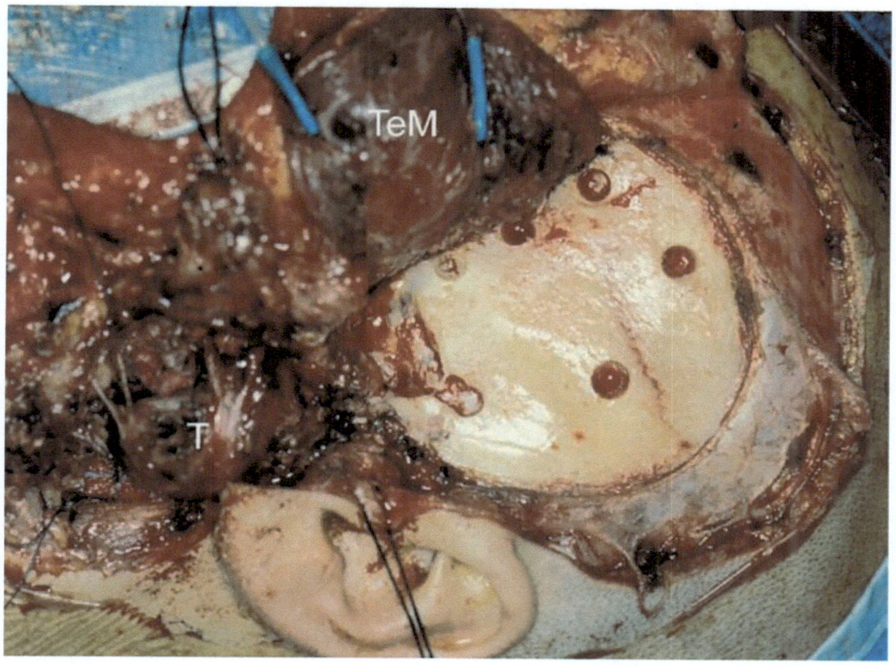

Fig. 12.349 Burr holes are drilled.

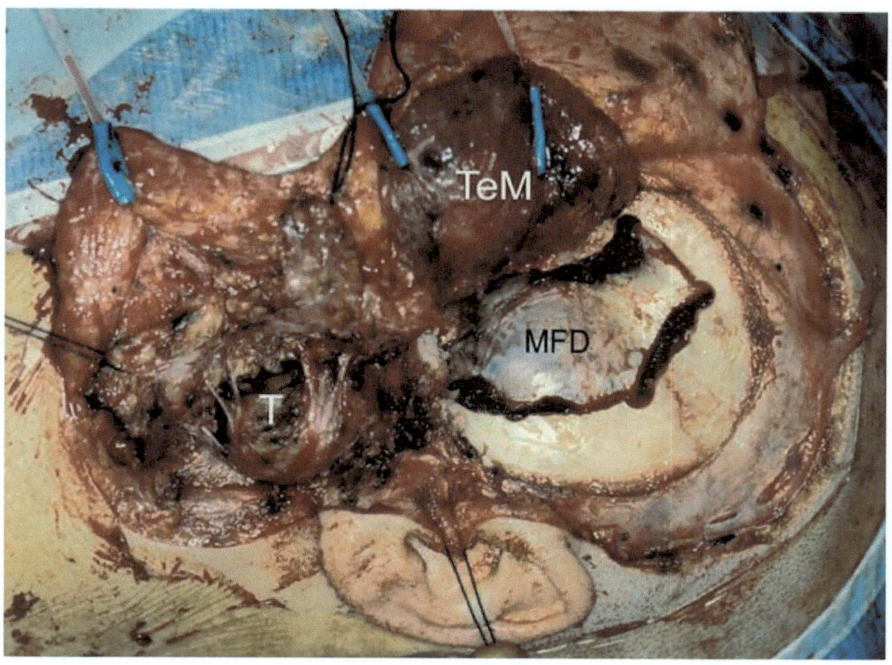

Fig. 12.350 A frontotemporal craniotomy is performed.

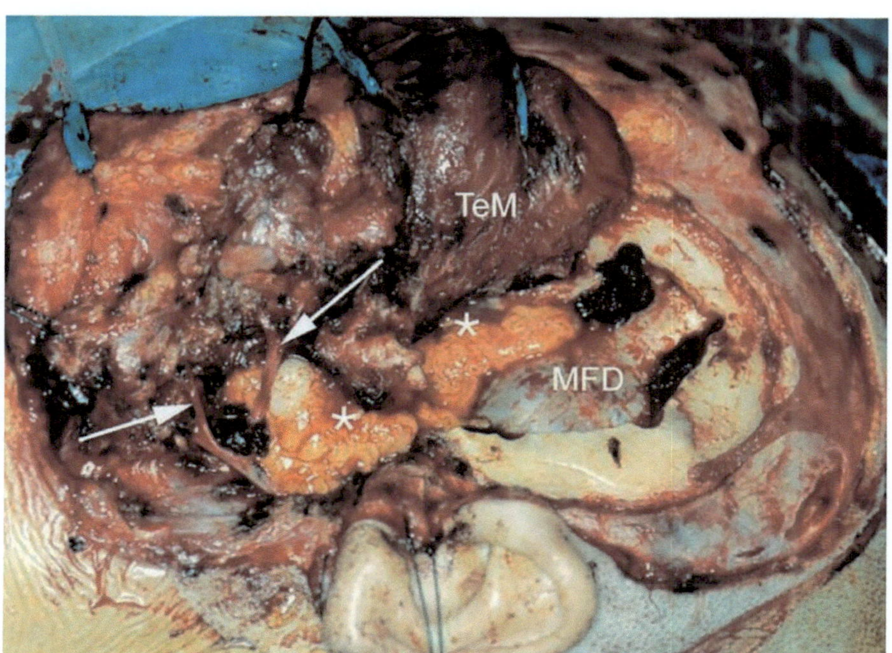

Fig. 12.351 Fat (*) is used to obliterate the operative cavity. The intact facial nerve branches *(arrows)* can be seen.

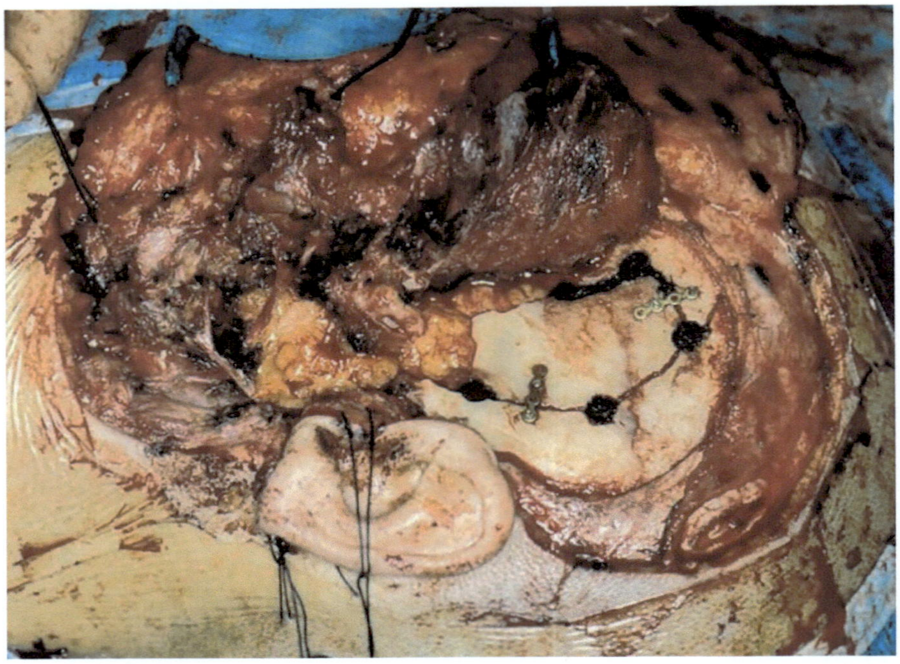

Fig. 12.352 The craniotomy is fixed with miniplates.

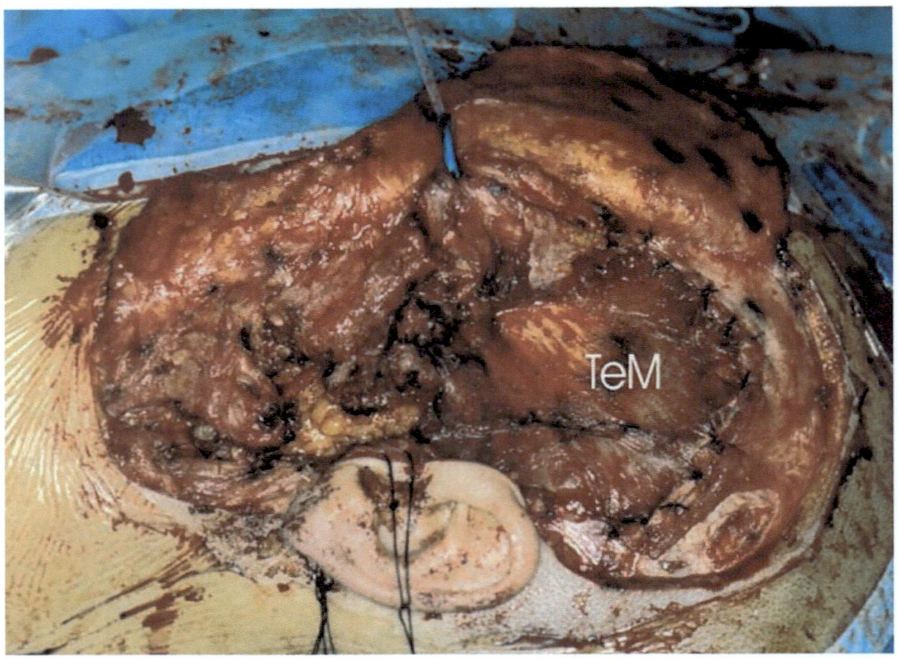

Fig. 12.353 The temporalis muscle (TeM) is sutured back.

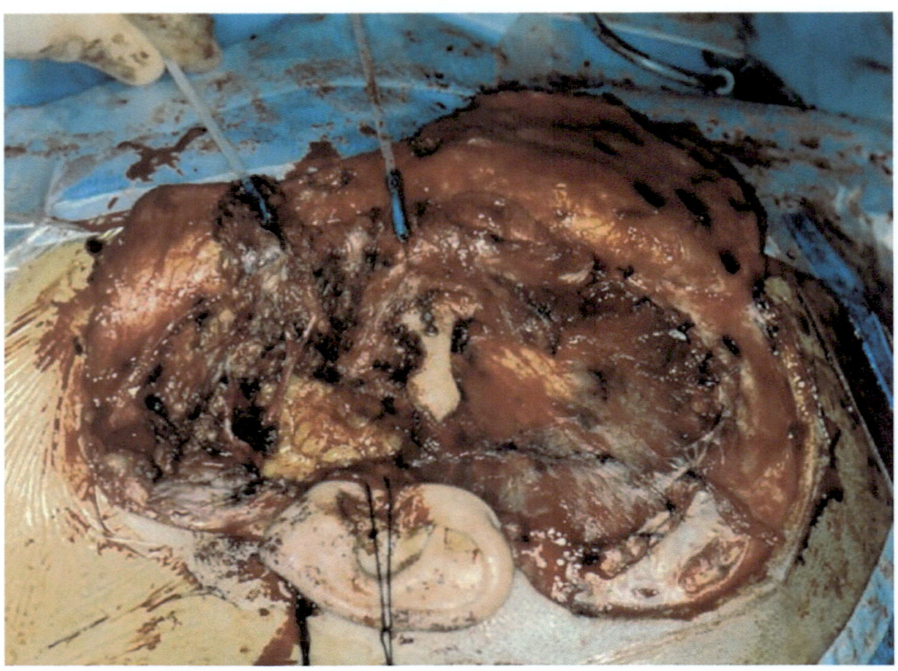

Fig. 12.354 The zygomatic arch is fixed with a single miniplate placed anteriorly.

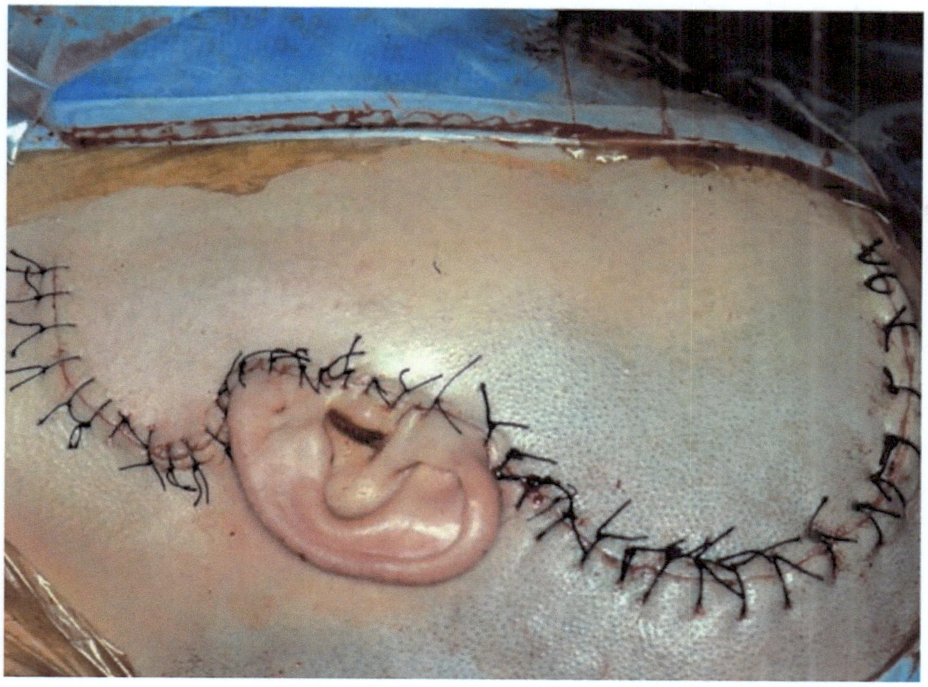

Fig. 12.355 Skin closure.

HINTS AND PITFALLS

1. The incision can be extended to the neck in cases where proximal control of the internal jugular vein and internal carotid artery is sought. Another reason is to allow the excision of certain extensive tumors with parapharyngeal space involvement (see Fig. 12.345).
2. Every effort should be made to preserve the blood supply of the temporalis muscle. Supply from the superficial temporal artery is sacrificed during the elevation of the muscle. However, blood supply from deep branches emerging from the internal maxillary artery has to be preserved.
3. The zygomatic osteotomy allows a lower frontotemporal craniotomy and hence less brain retraction and better control of the horizontal and precavernous internal carotid artery. The temporalis can be further displaced inferiorly when the zygoma is removed to allow for better exposure of the infratemporal fossa (Figs. 12.356 and 12.357).
4. Removal of the orbital rim provides a wider angle of attack, allowing better control of massive suprasellar involvement, and facilitates safer removal of orbital extension.
5. While the orbital osteotomy is being done, the contents of the orbit are protected using a large dissector (see Fig. 12.336). Every effort should be made to preserve the periorbita. If the orbital fat bulges out, it is cauterized using bipolar coagulation with consequent shrinking of the fat.
6. Sometimes it is difficult to have the frontotemporal craniotomy come out as one piece. The craniotomy can be made in two pieces for ease of accomplishment.
7. When a standard orbitozygomatic osteotomy is to be performed, the frontotemporal craniotomy is performed first. This allows brain retraction and facilitates the proper performance of the cuts at the orbital roof (Fig. 12.358).
8. During the closure, the frontotemporal craniotomy is closed first using miniplates. The temporalis muscle is next replaced, and then the zygomatic osteotomy is returned (Figs. 12.359 and 12.360). If the root of the zygoma has been drilled, it is usually enough to have good fixation by using miniplates only on the frontal and malar processes of the zygoma.
9. For transdural tumors involving the nasopharynx, nasal cavity, or maxillary sinus, the approach has to be staged with the removal of the extradural part first, followed by extirpation of the intradural component. A single-stage removal has a very high risk of CSF leakage and meningitis.
10. A wound drain is only inserted if the tumor is extradural and the dura has not been opened. If the approach is used for the treatment of intradural tumors, a wound drain is not inserted.

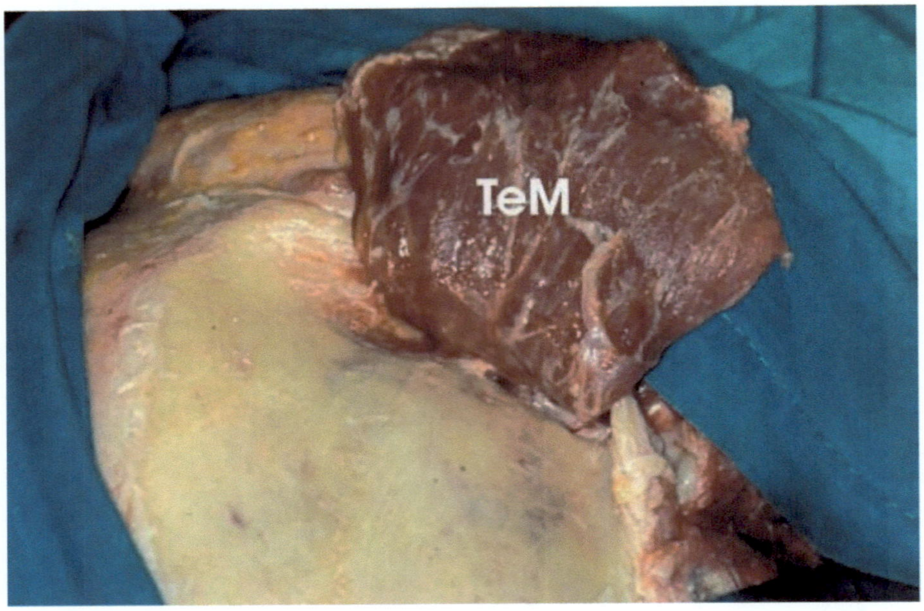

Fig. 12.356 The temporalis muscle (TeM) is reflected inferiorly. The zygomatic arch is in place. Note the limited exposure.

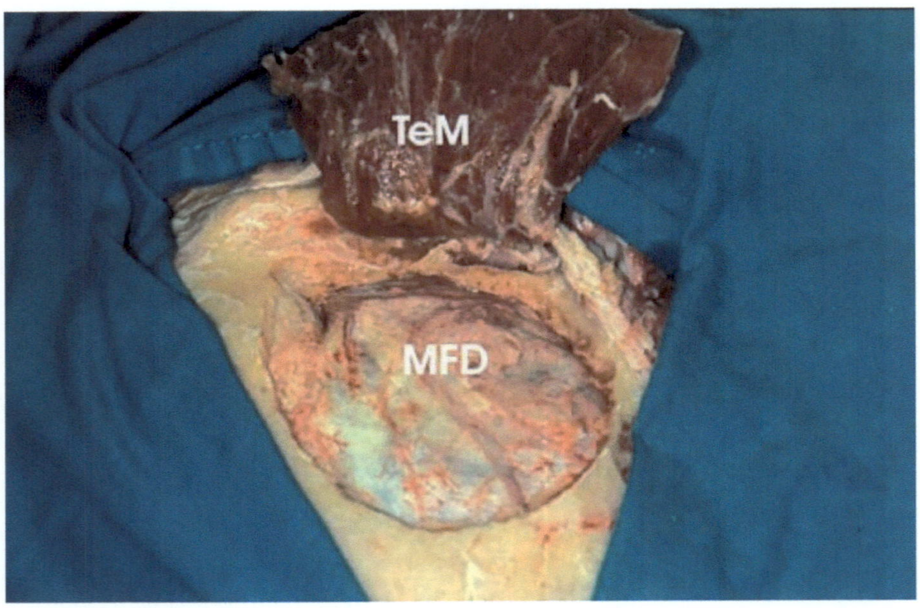

Fig. 12.357 The zygomatic arch is removed. Note the wider exposure gained by this step (cf. Fig. 12.356).

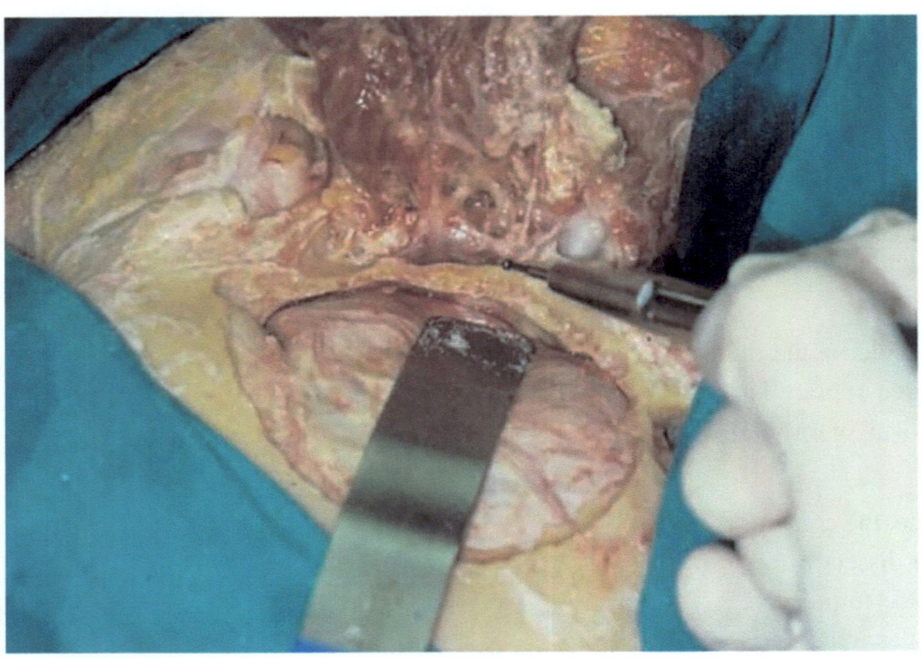

Fig. 12.358 The frontotemporal craniotomy performed first before the orbitozygomatic osteotomy.

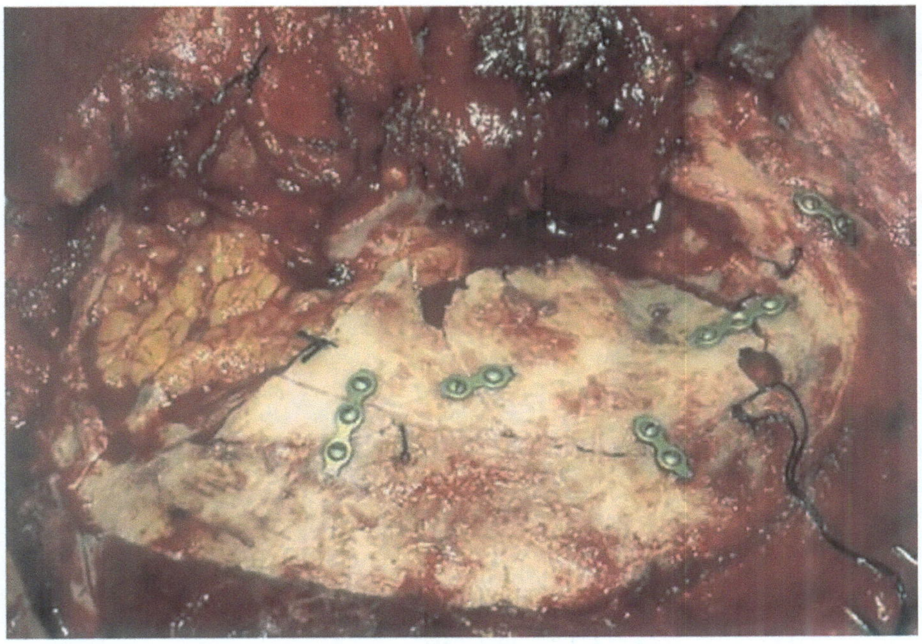

Fig. 12.359 The craniotomy pieces are fixed with miniplates. Note that in such a large craniotomy it is easier to separate it into two pieces rather than a single large piece of bone that may be difficult to elevate.

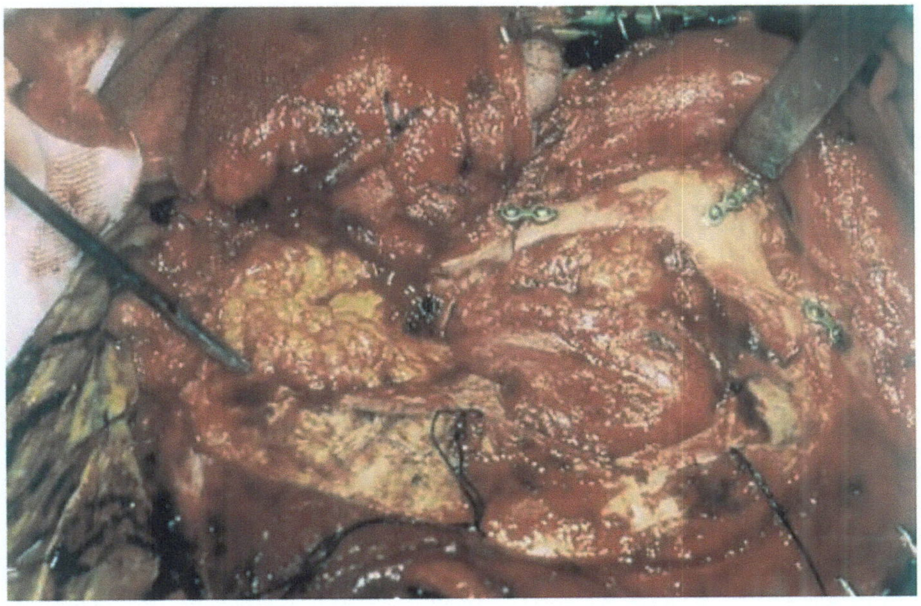

Fig. 12.360 The temporalis muscle is replaced first and then the zygomatic arch is fixed.

REFERENCES

[1] Fisch U. Infratemporal fossa approach for extensive tumors of the temporal bone and base of the skull. In: Silverstein H, Norrell H, editors. Neurological surgery of the ear. Birmingham: Aesculapius C; 1977. p. 34–53.

[2] Fisch U. Infratemporal fossa approach to tumors of the temporal bone and base of the skull. J Laryngol Otol 1978;92:949–67.

[3] Fisch U, Pillsbury HC. Infratemporal fossa approach to lesions in the temporal bone and base of the skull. Arch Otolaryngol 1979;105:99–107.

[4] Vrionis FD, Brodkey JA. Schmidek and Sweet Operative Neurosurgical Techniques: Indications, Methods, and Results. 6th edition, Elsevier, 2012.

[5] Leonetti JP, Benscoter BJ, Marzo SJ, Borrowdale RW, Pontikis GC, Malekpour M, Cohen-Gadol AA. Preauricular infratemporal fossa approach for advanced malignant parotid tumors. Laryngoscope 2012;122(9):1949–53. https://doi.org/10.1002/lary.23322 [Epub 2012 Jun 27].

[6] Shahinian HK, Suh RH, Jarrahy R. Combined infratemporal fossa and transfacial approach to excising massive tumors. Ear Nose Throat J 1999;78(5):350. 353-6.

[7] Fisch U, Mattox D. Microsurgery of the skull base. Stuttgart: Thieme; 1988.

[8] Fensak ML, Jackler RK. Removal of jugular foramen tumors: the fallopian bridge technique. Otolaryngol Head Neck Surg 1997;117(6):586–91.

[9] Aslan A, Falcioni M, Russo A, et al. Anatomical considerations of high jugular bulb in lateral skull base surgery. J Laryngol Otol 1997;111(4):333–6.
[10] Moe KS, Li D, Linder TE, Schmid S, Fisch U. An update on the surgical treatment of temporal bone paraganglioma. Skull Base Surg 1999;9(3):185–94.
[11] Sanna M, Jain Y, De Donato G, Rohit Lauda L, Taibah A. Management of jugular paragangliomas: the Gruppo Otologico experience. Otol Neurotol 2004;25:797–804.
[12] Sanna M, Piazza P, Shin SH, Flanagan S, Mancini F. Microsurgery of skull base paraganglioma microsurgery of skull base paraganglioma. Stuttgart: Thieme; 2012. p. 555–677.
[13] Sivalingam S, Konishi M, Shin SH, Lope Ahmed RA, Piazza P, Sanna M. Surgical management of tympanojugular paragangliomas with intradural extension, with a proposed revision of the Fisch classification. Audiol Neurootol 2012;17:243–55.
[14] Prasad SC, Mimoune HA, Khardaly M, Piazza P, Russo A, Sanna M. Strategies and long-term outcomes in the surgical management of tympanojugular paragangliomas. Head Neck 2016;38(6):871–85. https://doi.org/10.1002/hed.24177 [Epub 2015 Sep 7] PMID: 26343411.
[15] Sanna M, Al-Khateeb M, Yilala MH, Almashhadani M, Fancello G. Gruppo otologico's experience in managing the so-called inoperable tympanojugular paraganglioma. Brain Sci 2024;14(8):745. https://doi.org/10.3390/brainsci14080745. PMID: 39199440. PMCID: PMC11252639.
[16] Sanna M, Pacak K, Taïeb D, Mariani-Costantini R. A congress on head and neck paragangliomas: advancing clinical care. Nat Rev Endocrinol 2024;20(7):383–4. https://doi.org/10.1038/s41574-024-00987-9. PMID: 38693273.
[17] Grinblat G, Sanna M, Piccirillo E, Piras G, Guidi M, Shochat I, Munteanu SG. Comparison of lower cranial nerve function between tympanojugular paraganglioma class C1/C2 with and without intracranial extension: a four-decade experience. Otol Neurotol 2022;43(1):e122–30. https://doi.org/10.1097/MAO.0000000000003383. PMID: 34889847.
[18] Prasad SC, Laus M, Al-Ghamdi S, Vashishth A, Piazza P, Sanna M. Update in the classification and the role of intra-arterial stenting in the management of carotid body paragangliomas. Head Neck 2019;41(5):1279–86. https://doi.org/10.1002/hed.25567 [Epub 2019 Feb 16] PMID: 30771228.
[19] Bacciu A, Prasad SC, Sist N, Rossi G, Piazza P, Sanna M. Management of the cervico-petrous internal carotid artery in class C tympanojugular paragangliomas. Head Neck 2016;38(6):899–905. https://doi.org/10.1002/hed.24284 [Epub 2015 Nov 28] PMID: 26614405.
[20] Piazza P, Di Lella F, Bacciu A, Di Trapani G, Ait Mimoune H, Sanna M. Preoperative protective stenting of the internal carotid artery in the management of complex head and neck paragangliomas: long-term results. Audiol Neurootol 2012;18 (6):345–52. https://doi.org/10.1159/000354158 [Epub 2012 Oct 4] PMID: 24107406.
[21] Sanna M, Piazza P, Shin SH, Flanagan S, Mancini F. Microsurgery of skull base paragangliomas. 1st ed. Thieme; 2012. ISBN: 9783121486110.
[22] Sanna M, Shin SH, Piazza P, Pasanisi E, Vitullo F, Di Lella F, et al. Infratemporal fossa approach type a with transcondylar–transtubercular extension for Fisch type C2 to C4 tympanojugular paragangliomas. Head Neck 2012. https://doi.org/10.1002/hed.23480.
[23] Spektor S, Anderson GJ, McMenomey SO, Horgan MA, Kellogg JX, Delashaw Jr JB. Quantitative description of the far lateral transcondylar transtubercular approach to the foramen magnum and clivus. J Neurosurg 2000;92 (5):824–31.
[24] Cokkeser Y, Naguib MB, Kizilay A. Management of the vertebral artery at the craniocervical junction. Otolaryngol Head Neck Surg 2005;123(1):84–8.
[25] Matsushima T, Natori Y, Katsuta T, Ikezaki K, Fukui M, Rhoton AL. Microsurgical anatomy for lateral approaches to the foramen magnum with special reference to transcondylar fossa (supracondylar transjugular tubercle) approach. Skull Base Surg 1998;8(3):119–25. 4.
[26] Muthukumar N, Swaminathan R, Venkatesh G, Bhanumathy SP. A morphometric analysis of the foramen magnum region as it relates to the transcondylar approach. Acta Neurochir 2005;147(8):889–95.
[27] Seyfried DM, Rock JP. The transcondylar approach to the jugular foramen: a comparative anatomic study. Surg Neurol 1994;42(3):265–71.
[28] Liu JK, Sameshima T, Gottfried ON, Couldwell WT, Fukushima T. The combined transmastoid retro- and infralabyrinthine transjugular transcondylar transtubercular highc ervical approach for resection of glomus jugulare tumors. Neurosurgery 2006;59(1 Suppl 1):ONS115–125 [discussion ONS115–125].
[29] Sanna M, Piazza P, De Donato G, Menozzi R, Falcioni M. Combined endovascular–surgical management of the internal carotid artery in complex tympanojugular paragangliomas. Skull Base 2009;19:26–42.
[30] Sanna M, Flanagan S. The combined transmastoid retro- and infralabyrinthine transjugular transcondylar transtubercular high cervical approach for resection of glomus jugulare tumors. Neurosurgery 2007;61:E1240 [author reply E1240].
[31] Prasad SC, Piccirillo E, Chovanec M, La Melia C, De Donato G, Sanna M. Lateral skull base approaches in the management of benign parapharyngeal space tumors. Auris Nasus Larynx 2015;42(3):189–98. https://doi.org/10.1016/j.anl.2014.09.002.
[32] Wen HT, Rhoton Jr AL, Katsuta T, de Oliveira E. Microsurgical anatomy of the transcondylar, supracondylar, and paracondylar extensions of the far-lateral approach. J Neurosurg 1997;87(4):555–85.
[33] Rhoton Jr AL. The far-lateral approach and its transcondylar, supracondylar, and paracondylar extensions. Neurosurgery 2000;47(3 Suppl):S195–209.
[34] Banerji D, Behari S, Jain VK, Pandey T, Chhabra DK. Extreme lateral transcondylar approach to the skull base. Neurol India 1999;47(1):22–30.
[35] Babu RP, Sekhar LN, Wright DC. Extremelateral transcondylar approach: technical improvements and lessons learned. J Neurosurg 1994;81(1):49–59.
[36] Salas E, Sekhar LN, Ziyal IM, Caputy AJ, Wright DC. Variations of the extreme-lateral craniocervical approach: anatomical study and clinical analysis of 69 patients. J Neurosurg 1999;90(2 Suppl):206–19.
[37] Sanna M, Falcioni M, Flanagan S, Takata Y, DeDonato G. Tympano-jugular paragangliomas surgery. In: Kirtane MV, Brackmann DE, Borkar DM, de Souza C, editors. Comprehensive textbook of otology. Mumbai: Thomson Press (India); 2010. p. 521–31.

[38] Sanna M, Shin SH, DeDonato G, et al. Managementof complex tympanojugular paragangliomas including endovascular intervention. Laryngoscope 2011;121:1272–82.

[39] Sanna M, Khrais T, Menozi R, Piazza P. Surgical removal of jugular paragangliomas after stenting of the infratemporal internal carotid artery: a preliminary report. Laryngoscope 2006;116:742–6.

[40] Hatano M, De Donato G, Falcioni M, Sanna M. Chondroblastoma of the temporal bone. Acta Otolaryngol 2011;121(8):890–5. https://doi.org/10.3109/00016489.2011.566579.

[41] Pandya Y, Piccirillo E, Mancini F, Sanna M. Management of complex cases of petrous bone cholesteatoma. Ann Otol Rhinol Laryngol 2010;119(8):514–25. https://doi.org/10.1177/000348941011900803.

[42] Prasad SC, Piccirillo E, Nuseir A, Sequino G, De Donato G, Paties CT, Sanna M. Giant cell tumors of the skull base: case series and current concepts. Audiol Neurootol 2012;19(1):12–21. https://doi.org/10.1159/000355701.

[43] Sanna M, Dispenza F, Mathur N, De Stefano A, De Donato G. Otoneurological management of petrous apex cholesterol granuloma. Am J Otolaryngol 2009;30(6):407–14. https://doi.org/10.1016/j.amjoto.2008.08.007.

[44] Achilli V, Danesi G, Caverni L, Richichi M. Petrous apex arachnoid cyst: a case report and review of the literature. Acta Otorhinolaryngol Ital 2005;25(5):296–300.

[45] Grinblat G, Vashishth A, Galetti F, Caruso A, Sanna M. Petrous apex cholesterol granulomas: outcomes, complications, and hearing results from surgical and wait-and-scan management. Otol Neurotol 2017;38(10):e476–85. https://doi.org/10.1097/mao.0000000000001578. PMID: 28984806.

[46] Yilala MH, Fancello G, Sanna M. Management of petrous bone cholesteatoma: the Gruppo Otologico experience. Eur Arch Otorhinolaryngol 2024;281(10):5211–24. https://doi.org/10.1007/s00405-024-08752-4. [Epub 2024 Jun 3]. Erratum in: Eur Arch Otorhinolaryngol. 2024 Oct;281(10):5225–6. https://doi.org/10.1007/s00405-024-08856-x PMID: 38829556.

[47] Banuchi V, Kraus DH. The infratemporal fossa approach to the lateral skull base and parapharynx. Oper Tech Otolaryngol Head Neck Surg 2014;25(3):254–8. https://doi.org/10.1016/j.otot.2014.04.006.

[48] Kraus DH, Shah JP, Arbit E, Galicich JH, Strong EW. Complications of craniofacial resection for tumors involving the anterior skull base. Head Neck 1994;16:307–12.

[49] Mansour OI, Carrau RL, Snyderman CH, Kassam AB. Preauricular infra-temporal fossa surgical approach: modifications of the technique and surgical indications. Skull Base 2004;14:143–51.

[50] Guinto G, Abello J, Molina A, Gallegos F, Oviedo A, Nettel B, Lopez R. Zygomatic-transmandibular approach for giant tumors of the infratemporal fossa and parapharyngeal space. Neurosurgery 1999;45:1285–98.

[51] Richtsmeier WJ, Briggs RJ, Koch WM, Eisele DW, Loury MC, Price JC, Mattox DE, Carson BS. Complications and early outcome of anterior craniofacial resection. Arch Otolaryngol Head Neck Surg 1992;118:912–7.

[52] Sekhar LN, Schramm Jr VL, Jones NF. Subtemporal-preauricular infratemporal fossa approach to large lateral and posterior cranial base neoplasms. J Neurosurg 1987;67:488–99.

[53] Fisch U. The infratemporal fossa approach for nasopharyngeal tumors. Laryngoscope 1983;93:36–44.

[54] Danesi G, Zanoletti E, Mazzoni A. Salvage surgery in recurrent nasopharyngeal carcinoma. Skull Base 2007;17:173–80.

[55] Bigelow DC, Smith PG, Leonetti JP, Backer RL, Grubb RL, Kotapka MJ. Treatment of malignant neoplasms of the lateral cranial base with the combined frontotemporal-anterolateral approach: five-year follow-up. Otolaryngol Head Neck Surg 1999;120:17–24.

[56] Dos Santos LR, Cernea CR, Brandao LG, Siqueira MG, Vellutini EA, Velazco OP, Cruz OL, Morais-Besteiro J, Freitas CA. Results and prognostic factors in skull base surgery. Am J Surg 1994;168:481–4.

[57] Chan J, Carrau RL, Prevedello DM, Kassam AB, Arriagain MA. Anterior and subtemporal approaches to the infratemporal fossa. In: Brackmann DE, Shelton C, Arriaga MA, editors. Otologic surgery. 4th ed. Philadelphia: W.B. Saunders Co./Elsevier; 2016.

[58] Battelino S, Bošnjak R. Surgical approaches and anatomy of the lateral skull base. In: Encyclopedia of otolaryngology, head and neck surgery; 2012. p. 2644–63. https://doi.org/10.1007/978-3-642-23499-6_567.

[59] Isolan GR, Rowe R, Al-Mefty O. Microanatomy and surgical approaches to the infratemporal fossa: an anaglyphic three-dimensional stereoscopic printing study. Skull Base 2007;17:285–302.

[60] Sabit I, Schaefer SD, Couldwell WT. Modified infratemporal fossa approach via lateral transantral maxillotomy: a microsurgical model. Surg Neurol 2002;58:21–31.

[61] Guinto G, Abello J, Molina A, Gallegos F, Oviedo A, et al. Zygomatictransmandibular approach for giant tumors of the infratemporal fossa and parapharyngeal space. Neurosurgery 1999;45:1285–98.

[62] Kim SM, Paek SH, Lee JH. Infratemporal fossa approach: the modified zygomatico-transmandibular approach. Maxillofac Plast Reconstr Surg 2019;41(1). https://doi.org/10.1186/s40902-018-0185-x.

CHAPTER 13

The Petro-Occipital Transsigmoid (POTS) Approach

Rationale

The petro-occipital transsigmoid (POTS) approach was introduced by Mann [1] and later modified by Mazzoni and Sanna (1995) [2]. This approach was developed primarily for lesions of the jugular foramen with or without extension to the cerebellopontine angle and upper neck. Areas that can be controlled by this approach include the jugular foramen, the cerebellopontine angle, the occipital condyle, the ipsilateral clivus, the lower part of the petrous apex, the vertical part of the internal carotid artery below the eustachian tube, and the jugulocarotid space (Fig. 13.1A,B) [2–4]. It is most commonly used to remove lower cranial nerve (LCN) schwannomas and JF meningiomas [5–9] (Fig. 13.1C–F).

The rationale for this approach is to reach the jugular foramen while leaving the facial nerve in situ and preserving the external auditory canal and middle ear, in contrast to the IFTA approach. Exposure is achieved by moving the purely lateral approach posteriorly, making it a posterolateral approach. The approach involves a retro-labyrinthine petrosectomy combined with a retromastoid craniotomy and occlusion of the sigmoid sinus and jugular bulb. If necessary, the POTS approach can be easily combined with the translabyrinthine or transotic approaches, for example in cases of intradural extension or middle ear involvement and encasement of the vertical segment of the internal carotid artery [9,11]. Overall, the POTS approach to JF has been shown to be safe and effective [4,5,9,11–13].

Advantages of the approach:

- Direct access to the jugular foramen area.
- Preservation of external and middle ear function.
- The facial nerve is preserved.
- Single-stage removal of intradural tumors with extracranial extension.

Indications

- Primary jugular foramen tumors with or without intradural and/or extracranial extension, e.g.:
  - Lower cranial nerve neuromas with or without intradural and extracranial extension.
  - Meningiomas of the jugular foramen.
- Some cases of posteriorly located, C1 tympanojugular paragangliomas.
- Other rare tumors located at the jugular foramen area, e.g., infralabyrinthine petrous bone cholesteatoma with involvement of the jugular bulb, some cases of cholesterol granuloma, facial nerve neuromas

Contraindications

- Aggressive and vascular tumors of the jugular foramen with extensive involvement of petrous internal carotid artery, e.g., tympanojugular paragangliomas.
- Lack of adequate collateral venous circulation. In this case, closure of the sigmoid sinus would jeopardize venous drainage from the brain with disastrous consequences. However, in the majority of cases, the tumor would have already closed the jugular bulb preoperatively, creating an effective collateral circulation.
- Contralateral paralysis of the lower cranial nerves.

Atlas of Microsurgery of the Lateral Skull Base. https://doi.org/10.1016/B978-0-443-19083-4.00013-1

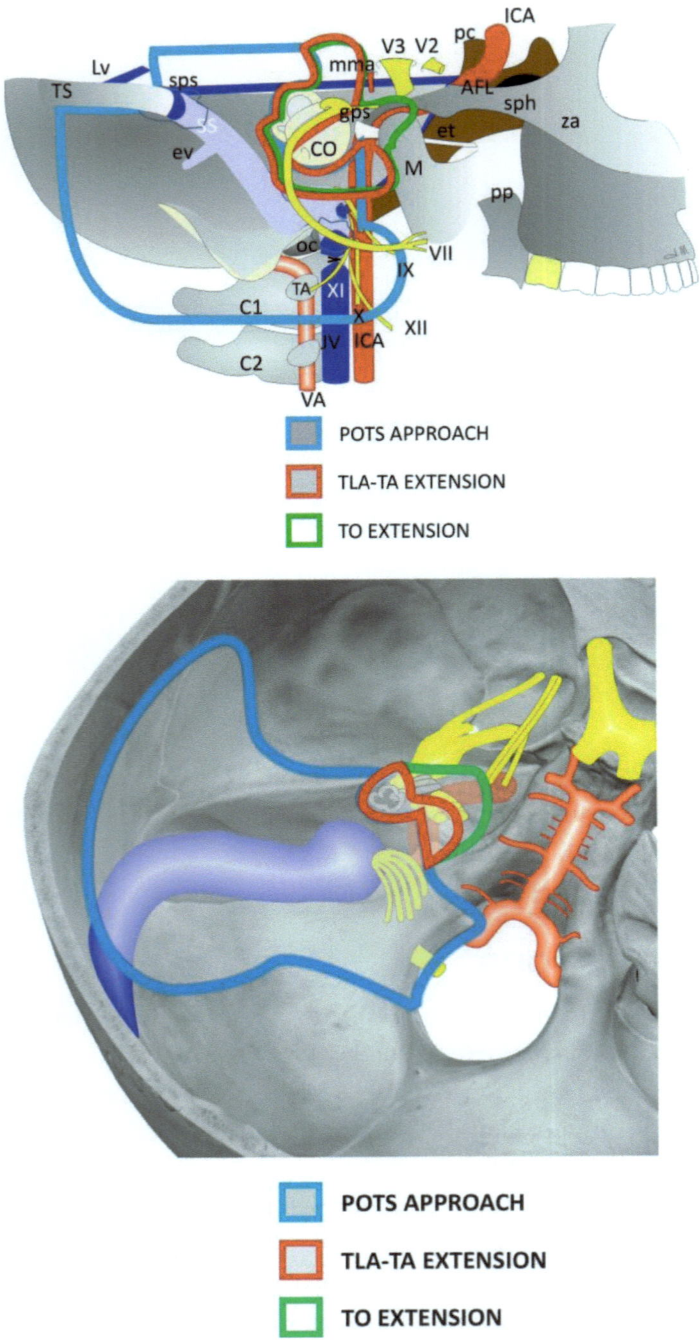

Fig. 13.1A,B Areas that can be controlled by POTS approach and its extension. *C1*, first cervical vertebra; *CO*, cochlea; *et*, eustachian tube; *ev*, emissary vein; *gps*, greater petrosal nerve; *ICA*, internal carotid artery; *JV*, jugular vein; *Lv*, vein of Labbé; *M*, mandible; *mma*, middle meningeal artery; *oc*, occipital condyle; *pc*, clinoid process; *pp*., pterygoid process; *sph*, sphenoid; *sps*, superior petrosal sinus; *TA*, transverse process of atlas; *TS*, transverse sinus; *za*, zygomatic arch; *V2*, maxillary nerve; *V3*, mandibular nerve; *VII*, facial nerve; *IX*, glossopharyngeal nerve; *X*, vagus nerve; *XI*, accessory nerve; *XII*, hypoglossal.

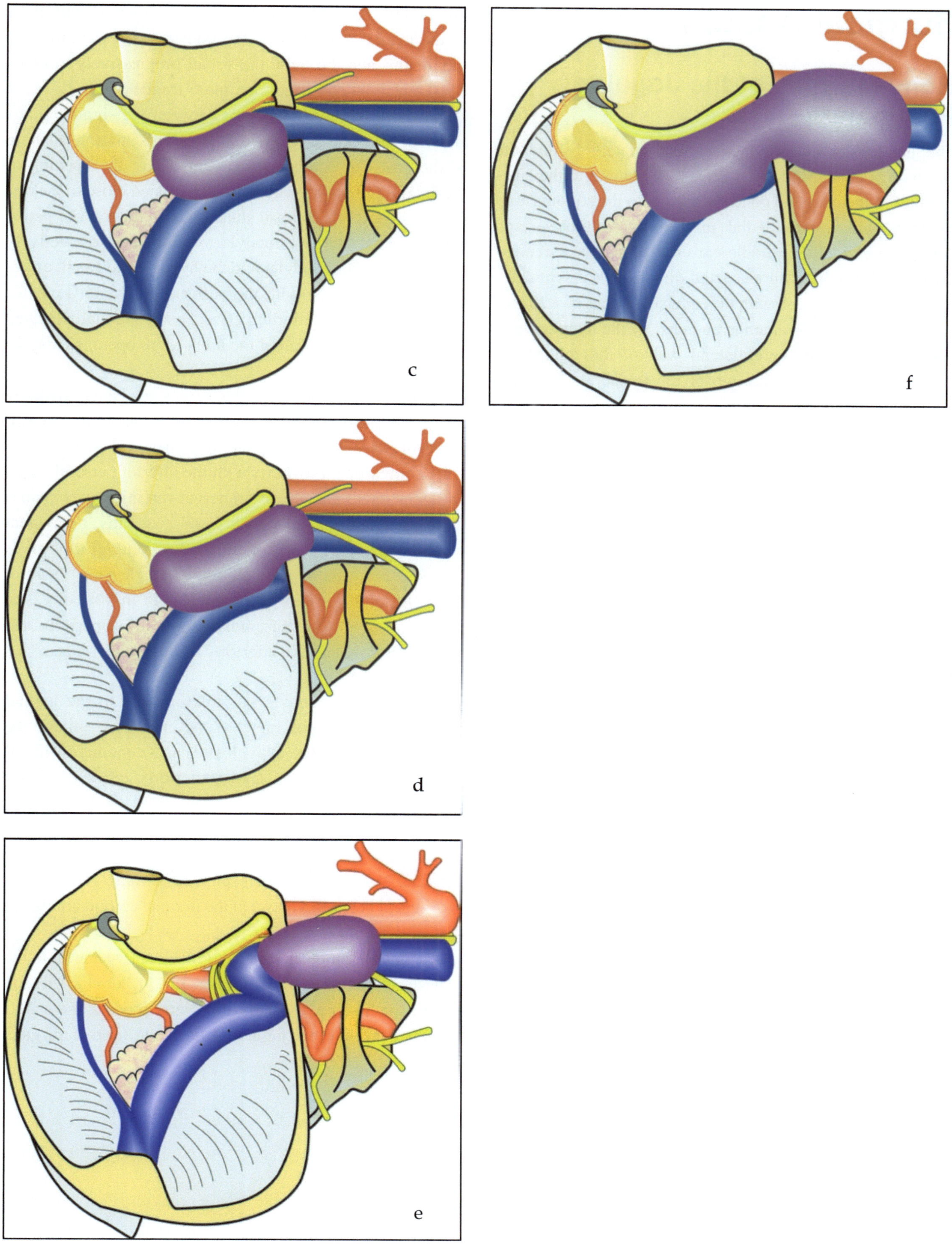

Fig. 13.1C–F Classification of JF schwannomas according to the classification system proposed by Kaye et al. [10] and modified by Pellet et al [10]. (C) Class A. Primary intracranial tumor with small bone extension. A POTS approach may be indicated in the presence of serviceable hearing. (D) Class B. Primary tumor in the JF with or without intracranial extension. A POTS approach may be indicated. (E) Class C. Primary extracranial tumor with extension into the JF or posterior cranial fossa. (F) Class D. Dumbbell-shaped tumor with intracranial and extracranial extension. In this case the POTS approach may be combined with a translabyrinthine and/or a transcervical approach. If the intradural portion is greater than 2 cm, a staged approach is indicated to avoid CSF leakage into the neck.

Surgical Anatomy

Bony Margins of the Jugular Foramen

The jugular foramen is located immediately behind the entrance of the carotid canal (Fig. 13.2), lateral to the anterior half of the occipital condyle, medial to the stylomastoid foramen, and posteromedial to the styloid process and posteroinferior to the middle ear cleft. It shows a tremendous variability in all three spatial dimensions. Asymmetry of the bulbs is frequent, with the right side larger in −70% of cases. The jugular foramen is bounded by the temporal and occiptal bones. The anterolateral wall of the foramen is formed by the petrous portion of the temporal bone, and the posteromedial wall is formed by the jugular process of the condylar part of the occipital bone. The petroclival fissure-the fissure between the lateral edge of the clival part of the occipital bone and the petrous part of the temporal bone-intersects the anteromedial edge of the foramen; and the occipitomastoid suture-the suture between the mastoid portion of the temporal bone and the condylar part of the occipital bone-intersects its posterolateral edge (Fig. 13.3).

Internal Anatomy

The jugular foramen can be divided into an anteromedial compartment, usually containing the glossopharyngeal nerve, inferior petrosal sinus, and meningeal branch of the ascending pharyngeal artery, and a posterolateral compartment containing the vagus and spinal accessory nerves with the jugular bulb. This arrangement shows a high degree of variability. The two compartments are divided by the jugular spine of the temporal and occipital bones, which are connected by a fibrous band that can occasionally ossify (Fig. 13.4). The endocranial opening is divided by a dural fold with the glossopharyngeal nerve usually running anterosuperomedially and the vagus and spinal accessory nerves posteroinferolaterally.

The sigmoid sinus joins the jugular foramen posteriorly. At this level it usually makes an almost 180° turn at the jugular bulb (Fig. 13.5), descending from there into the neck as the jugular vein. During the transition between the sigmoid sinus and the bulb, the vessel progressively loses its dural envelopment, becoming more fragile.

At the level of the junction between the sigmoid sinus and the bulb, the inferomedial wall presents the opening of the posterior condylar vein (Fig. 13.6); the opening of the anterior condylar vein is usually located close to the orifice of the inferior petrosal sinus if not in the inferior petrosal sinus itself. The two condylar veins then run respectively into the posterior condylar and the hypoglossal canal, establishing a connection with the perivertebral venous plexus. The opening of the inferior petrosal sinus is usually located in the anteromedial wall of the bulb. More openings may be present (Fig. 13.7).

In the jugular foramen, the nerves run anteromedially while the jugular bulb is located posterolaterally. In two thirds of cases, the glossopharyngeal nerve passes in a separate compartment together with the inferior petrosal sinus, while the vagus and accessory nerves run in the pars venosa together with the jugular bulb. Less frequently, no evident compartmentation can be distinguished, and all the nerves pass together with the jugular bulb in a single path. The inferior petrosal sinus joins the jugular bulb in 90% of cases, passing usually between cranial nerves IX superiorly and laterally and X and XI inferiorly and medially (Figs. 13.8 and 13.9).

The occipital condyle lies inferomedial to the jugular bulb and is crossed by the condylar emissary vein and the hypoglossal canal. The hypoglossal canal contains the hypoglossal nerve and a constant posteriorly located anterior condylar vein.

The vertical portion of the internal carotid artery is separated from the jugular bulb by a notch of triangular bone known as the caroticojugular spine, or "crutch." The glossopharyngeal nerve lies between the internal carotid artery and the jugular bulb (Fig. 13.10).

The internal jugular vein lies immediately anterior to the transverse process of the atlas (Fig. 13.11). And the surgeon can use the finger to touch the transverse process of the atlas to locate the internal jugular vein in the narrow surgical cavity (Fig.13.12).

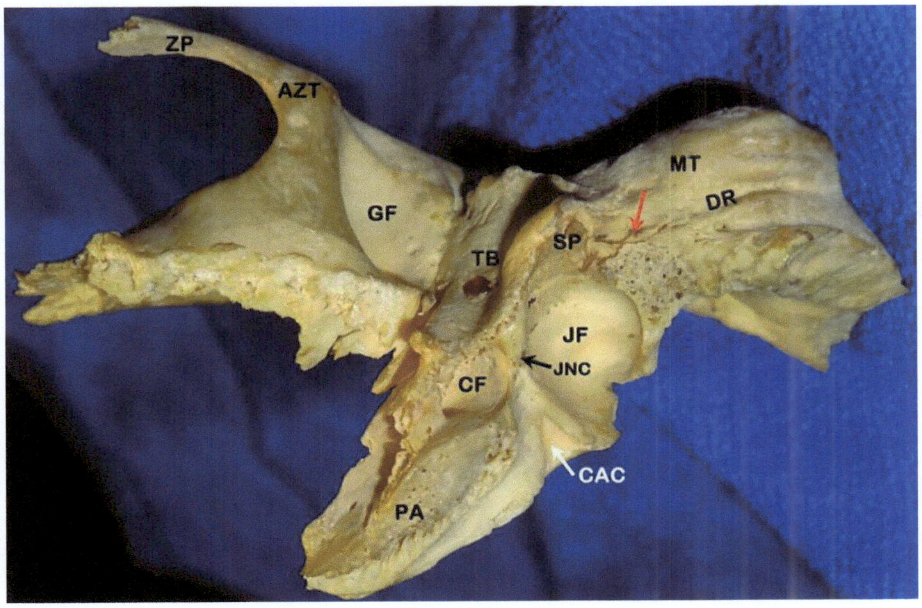

Fig. 13.2 An inferior view of a left temporal bone. *Red arrow*, stylomastoid foramen; *AZT*, anterior zygomatic tubercle; *CF*, carotid foramen; *DR*, digastric ridge; *GF*, glenoid fossa; *JF*, jugular foramen; *MT*, mastoid tip; *SP*, styloid process; *TB*, tympanic bone; *ZP*, zygomatic process; *PA*, petrous apex; *JNC*, Jacobson' s nerve canal; *CAC*, cochlear aqueduct canal.

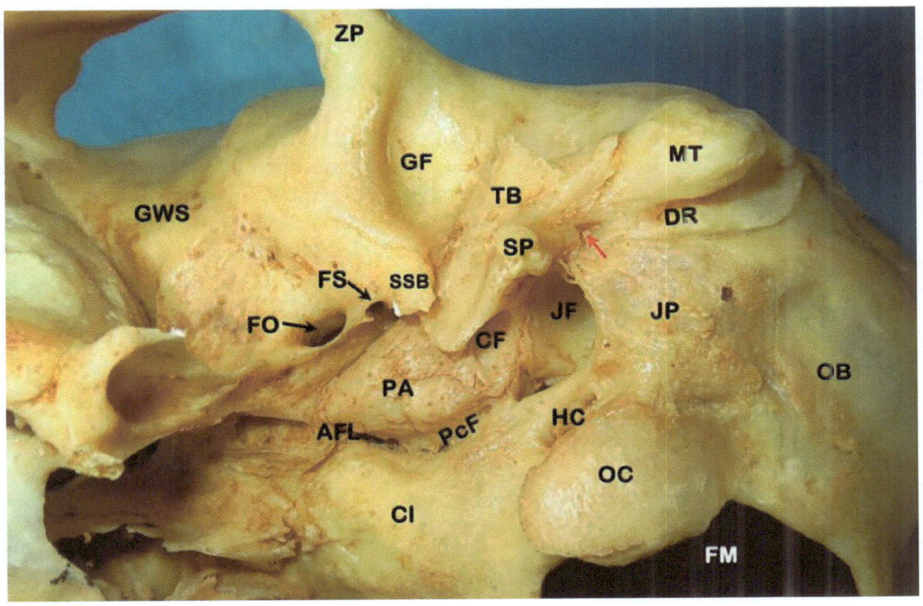

Fig. 13.3 An inferior view of an articulated temporal bone. *Red arrow*, stylomastoid foramen; *AFL*, anterior foramen lacerum; *FO*, foramen ovale; *OB*, occipital bone; *CF*, carotid foramen; *JF*, jugular foramen; *OC*, occipital condyle; *DR*, digastric ridge; *FM*, foramen magnum; *FS*, foramen spinosum; *GF*, glenoid fossa; *GWS*, greater wing of the sphenoid; *ZP*, zygomatic process; *MT*, mastoid tip; *TB*, tympanic bone; *PA*, petrous apex; *SSB*, spine of sphenoid bone; *PcF*, petroclival fissure; *JP*, jugular process of occipital bone.

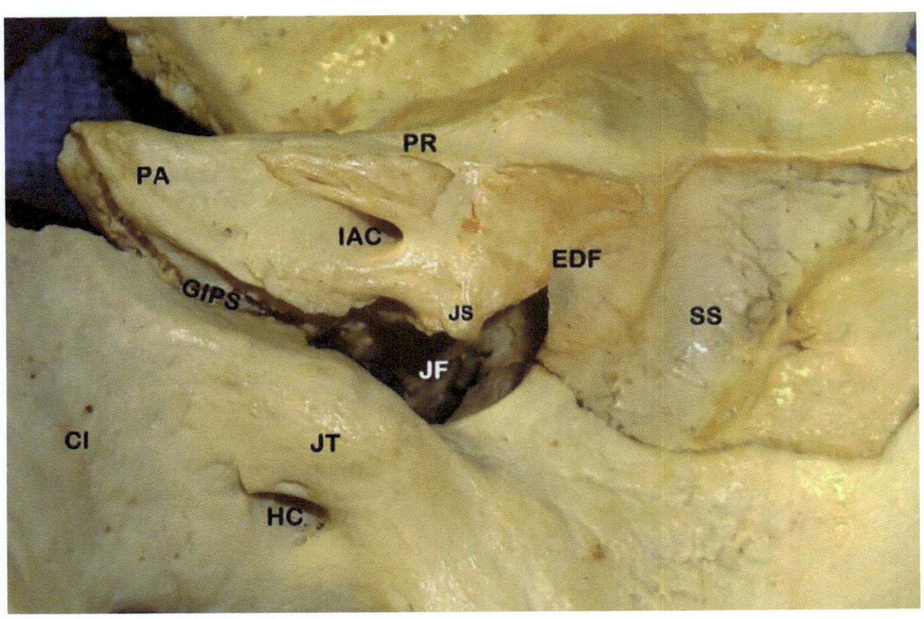

Fig. 13.4 An posterior view of an articulated temporal bone. *Cl*, clivus; *EDF*, endolymphatic duct foramen; *IAC*, internal auditory canal; *JF*, jugular foramen; *JT*, jugular tubercle; *PR*, petrous ridge; *SS*, sigmoid sinus groove; *PA*, petrous apex; *JS*, jugular spine; *GIPS*, groove of inferior petrosal sinus; *HC*, hypoglossal canal.

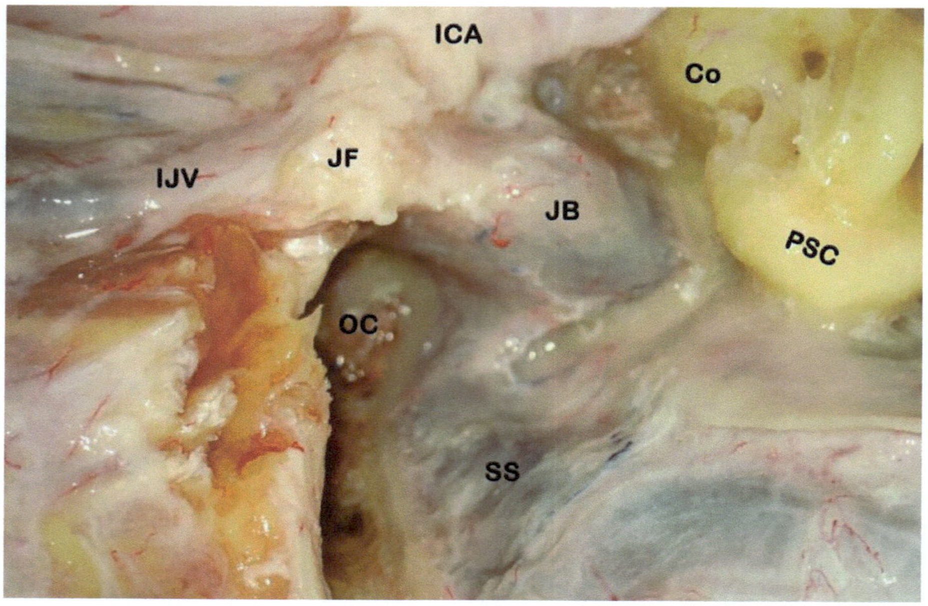

Fig. 13.5 The sigmoid sinus joins the jugular foramen posteriorly. At this level it usually makes an almost 180° turn at the jugular bulb. *ICA*, internal carotid artery; *IJV*, internal jugular vein; *Co*, cochlea; *PSC*, posterior semicircular canal; *JB*, jugular bulb; *JF*, jugular foramen; *OC*, occipital condyle; *SS*, sigmoid sinus.

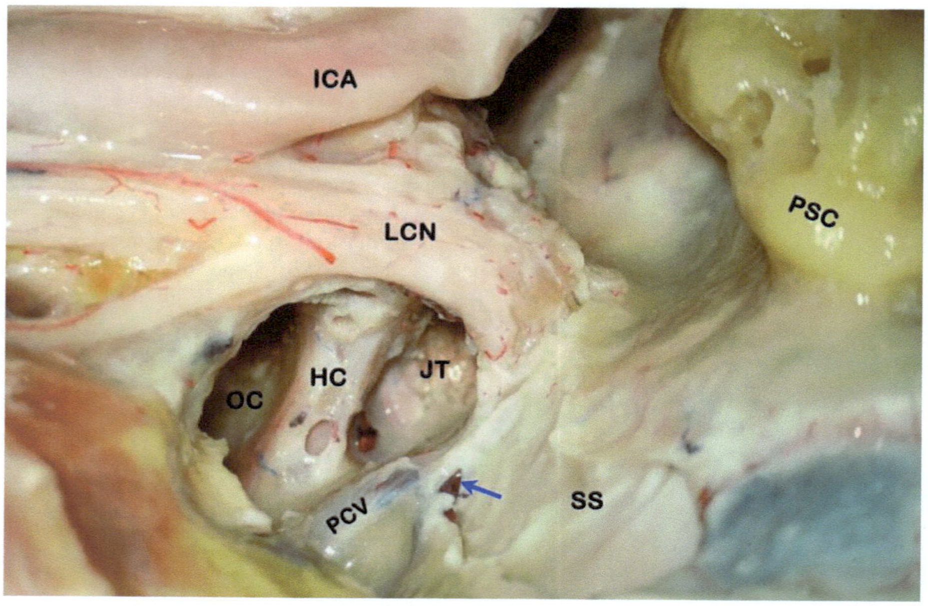

Fig. 13.6 At the level of the junction between the sigmoid sinus and the bulb, the inferomedial wall presents the opening of the posterior condylar vein. *Blue arrow*, opening of the posterior condylar vein; *ICA*, internal carotid artery; *PSC*, posterior semicircular canal; *OC*, occipital condyle; *SS*, sigmoid sinus; *JT*, jugular tubercle; *HC*, hypoglossal canal; *PCV*, posterior condylar vein; *LCN*, lower cranial nerve.

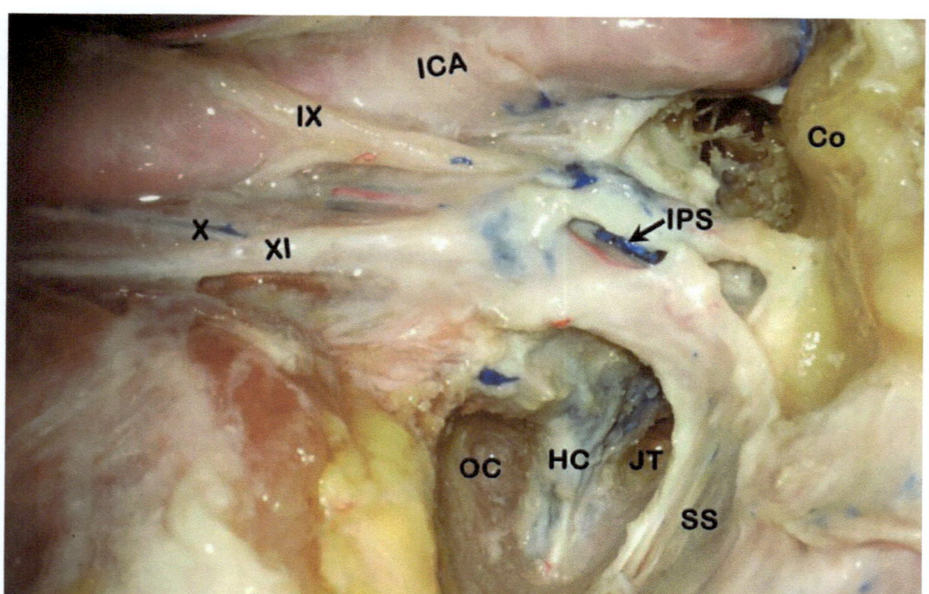

Fig. 13.7 The opening of the inferior petrosal sinus is usually located in the anteromedial wall of the bulb. More openings may be present. *ICA*, internal carotid artery; *Co*, cochlea; *OC*, occipital condyle; *SS*, sigmoid sinus; *JT*, jugular tubercle; *HC*, hypoglossal canal; *IPS*, opening of the inferior petrosal sinus; *IX*, glossopharyngeal nerve; *X*, vagus nerve; *XI*, accessory nerve.

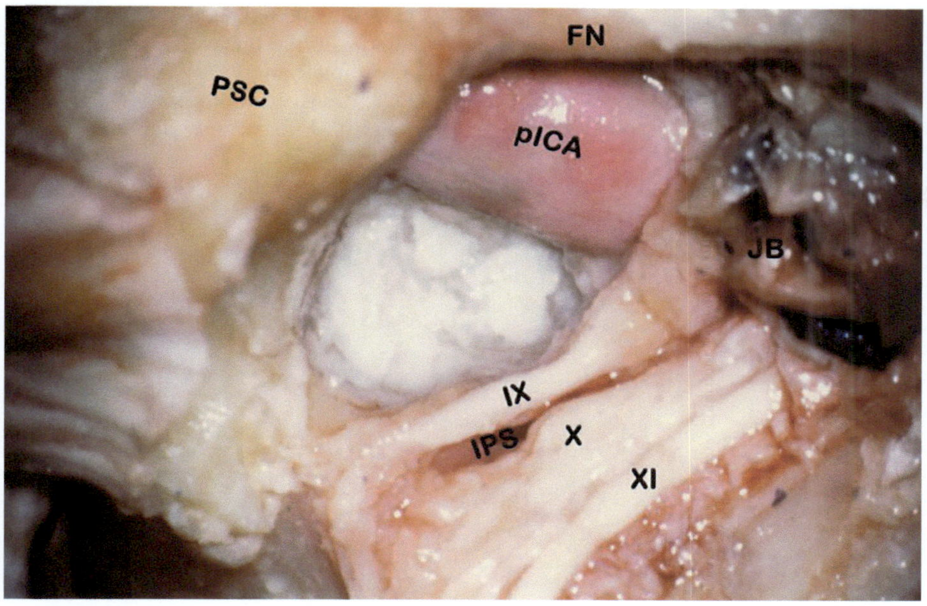

Fig. 13.8 The LCNs make a 90° rotation after entering the jugular foramen. *PSC*, posterior semicircular canal; *pICA*, petrous segment of internal carotid artery; *IPS*, opening of the inferior petrosal sinus; *FN*, facial nerve; *JB*, jugular bulb; *IX*, glossopharyngeal nerve; *X*, vagus nerve; *XI*, accessory nerve.

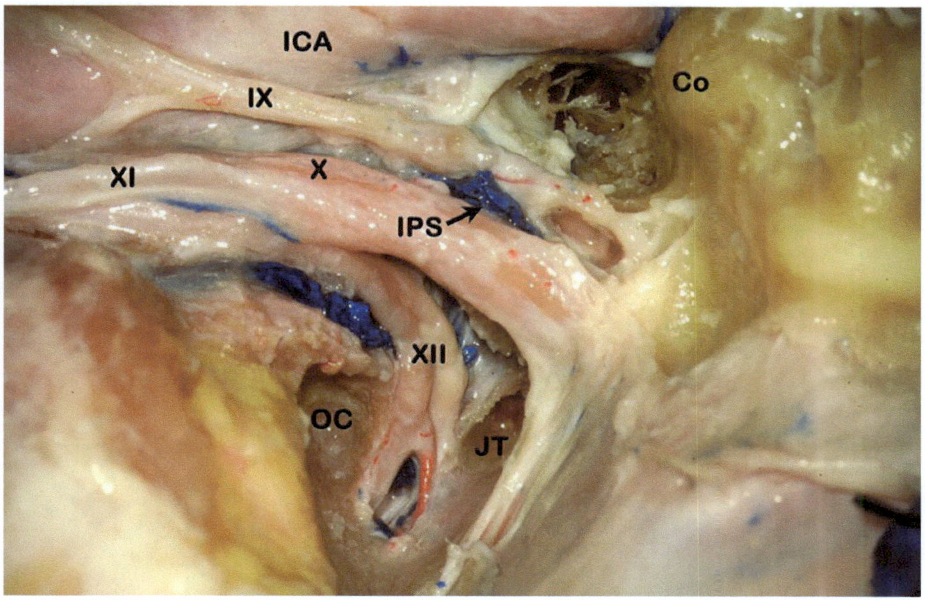

Fig. 13.9 The relationship between the hypoglossal canal and the lower cranial nerves is seen. *Co*, cochlea; *ICA*, internal carotid artery; *IPS*, opening of the inferior petrosal sinus; *OC*, occipital condyle; *JT*, jugular tubercle; *IX*, glossopharyngeal nerve; *X*, vagus nerve; *XI*, accessory nerve; *XII*, hypoglossal nerve.

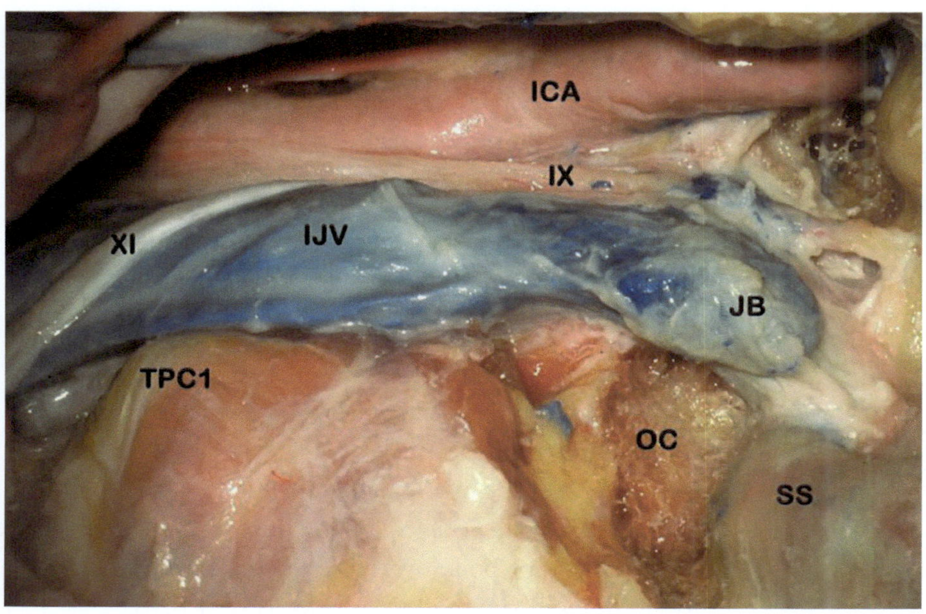

Fig. 13.10 The glossopharyngeal nerve lies between the internal carotid artery and the jugular bulb. *ICA*, internal carotid artery; *OC*, occipital condyle; *JB*, jugular bulb; *SS*, sigmoid sinus; *IJV*, internal jugular vein; *TPC1*, transverse process of the atlas; *IX*, glossopharyngeal nerve; *XI*, accessory nerve.

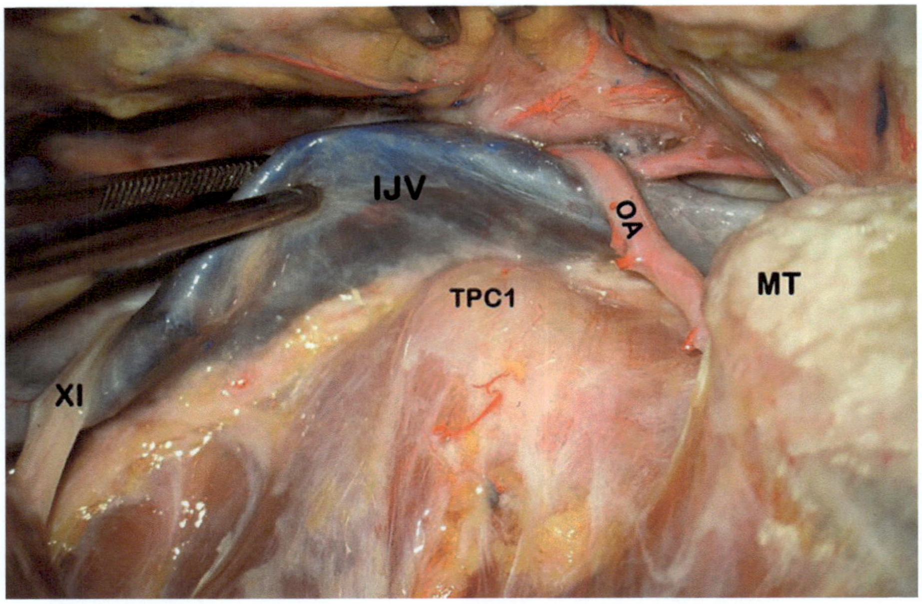

Fig. 13.11 The internal jugular vein lies immediately anterior to the transverse process of the atlas. *OA*, occipital artery; *IJV*, internal jugular vein; *TPC1*, transverse process of the atlas; *MT*, mastoid tip; *XI*, accessory nerve.

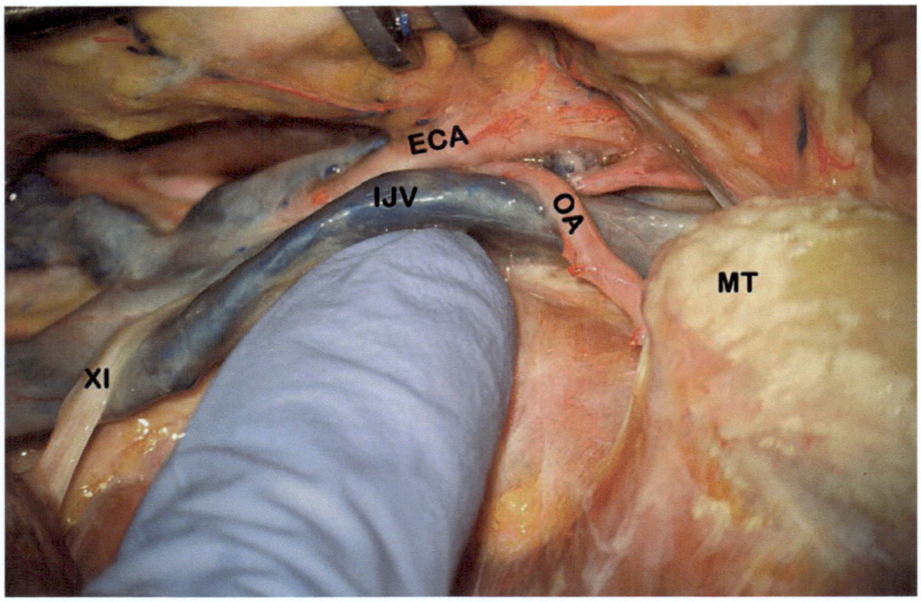

Fig. 13.12 The surgeon can use the finger to touch the transverse process of the atlas to locate the internal jugular vein. *OA*, occipital artery; *IJV*, internal jugular vein; *ECA*, external carotid artery; *MT*, mastoid tip; *XI*, accessory nerve.

Surgical Steps

1. A C-shaped postauricular skin incision is made that begins about 3 cm above the auricle, curves 4–5 cm behind the postauricular sulcus, and slants inferiorly to end 2–3 cm below the tip of the mastoid (Fig. 13.13).

2. The skin flap is raised and is retracted anteriorly using elastic-banded skin hooks (Fig. 13.14).

3. An inferiorly based U-shaped musculoperiosteal flap is outlined. The flap is raised using a Lempert periosteal elevator. It reaches down to the level of the insertion of the splenius capitis muscle (Fig. 13.15).

4. The sternomastoid muscle is retracted posteriorly. The lateral process of the atlas is felt by the finger. Blunt dissection anterior to this point reveals the internal jugular vein. After liberating the vein (Fig. 13.16), a suture is passed underneath it for later ligation.

5. An extended mastoidectomy is performed. The sigmoid sinus is skeletonized (Fig. 13.17). The digastric ridge is identified. The dura is separated from the underlying bone using a septal dissector. The retrofacial cells are next drilled. The mastoid segment of the facial nerve and the Jugular bulb is identified. The posterior semicircular canal is better identified but not blue-lined, with care being taken not to injure its ampulla while drilling the retrofacial cells superiorly (Fig. 13.18).

6. A 4 × 4 cm retrosigmoid craniotomy is performed. The craniotomy is limited anteriorly by the sigmoid sinus and superiorly by the transverse sinus. The bone flap is separated carefully from the underlying dura and is placed in saline solution (Fig. 13.19).

7. The bone overlying the skeletonized sigmoid sinus is removed. Bone at the sinodural angle is removed. However, a piece of bone overlying the genu from the lateral to the sigmoid sinus is left in place (Fig. 13.20).

8. With gentle pressure on the sigmoid sinus and the posterior fossa dura using the suction irrigator, the endolymphatic sac is identified passing medial to the posterior semicircular canal (Fig. 13.21). With further pressure, the dura is separated from the overlying bone to allow for safer exposure of the jugular foramen.

9. With further bone removal, the cochlear aqueduct is identified (Fig. 13.22). Next, the dura is further separated from the occipital bone and condyle. The occipital condyle is then partially drilled to expose the hypoglossal canal (Fig. 13.23).

10. The infratubal vertical segment of the internal carotid artery can be exposed by drilling the inferior tympanic bone while preserving the external auditory canal wall (Fig. 13.24).

11. The proximal part of the sigmoid sinus is compressed extraluminally with a piece of Surgicel placed between the sinus and the overlying bone at its junction with the lateral sinus. Next the sinus is opened with a Beaver knife and is packed proximally and distally with intraluminal Surgicel. The internal jugular vein in the neck is ligated. For intradural tumors, the posterior fossa dura is opened with a horizontal incision starting about 3 cm posterior to the sigmoid sinus and coursing anteriorly, traversing the medial wall of the sinus. The dural cut lies inferior to the endolymphatic sac. Bipolar coagulation of the dura is performed first before opening it in order to decrease bleeding.

12. The dura is held by two stay sutures, one for the superior dural flap and the other for the inferior one. The sutures are fixed to the skin. The intradural structures are seen covered by arachnoid (Fig. 13.25).

13. The arachnoid is taken down from the neurovascular structures. Cranial nerves IV–XI can be visualized. The superior cerebellar artery, anterior inferior cerebellar artery, and posterior inferior cerebellar artery can be observed (Fig. 13.26).

14. Tilting the microscope upwards identifies the terminal part of the basilar artery, seen through the tentorial notch, giving rise to the posterior cerebral artery. Note the branching of the superior cerebellar artery (Fig. 13.27)

15. The lateral wall of the jugular bulb can now be removed. The tumor would have already blocked the inferior petrosal sinus. However, if bleeding occurs, it can be controlled by Surgicel packing. If necessary, the medial wall of the jugular bulb can be carefully removed. This allows full exposure of the intracranial course of cranial nerves IX–XI (Fig. 13.28).

16. The relationship between the posterior inferior cerebellar artery and the lower cranial nerves is seen (Fig. 13.29).

17. Retraction of the vestibulocochlear nerve shows the anteriorly-lying facial nerve, with the anterior inferior cerebellar artery passing in between (Fig. 13.30).

18. At a higher magnification, the relationship between the anterior inferior cerebellar artery and the acoustic facial bundle is better appreciated (Fig. 13.31). The nervus intermedius arises on the anterior surface of the vestibulocochlear nerve, has a free segment in the cistern and/or meatus, and joins the facial nerve distally. The facial nerve is located anterior to the superior vestibular nerve and the cochlear nerve is anterior to the inferior vestibular nerve (Fig. 13.32).

19. With the patient turned forwards, the whole course of the anterior inferior cerebellar artery loop can be seen. Note the close proximity of the anterior inferior cerebellar artery to the internal auditory canal (Fig. 13.33).

20. Retraction of the acoustic facial bundle shows the basilar artery, abducens nerve and anterior inferior cerebellar artery (Fig. 13.34).

21. The relationship between the hypoglossal canal and the lower cranial nerves is seen (Fig. 13.35).

22. The dural of the drilled part of the condyle is excised. The hypoglossal canal is opened. Bleeding from the posteriorly lying venous plexus usually occurs during this step, and it is controlled by Surgicel packing. The two bundles forming the hypoglossal nerve unite to form a single nerve in the canal (Fig. 13.36).

23. At a higher magnification, the intimate relation between the hypoglossal nerve and the vertebral artery is well seen (Fig. 13.37). When needed, cranial nerves IX–XI can be gently retracted (or cut, if infiltrated by tumor), and drilling is continued to involve the ipsilateral lower clivus, down to the lateral border of the foramen magnum.

24. If more control of the infralabyrinthine compartment and the vertical portion of the internal carotid artery is needed, the mastoid segment of the facial nerve can be liberated as far as the stylomastoid foramen (Fig. 13.38).

25. The vertical segment of the nerve is then temporarily raised along the posterior canal wall (Fig. 13.39). If the pathology extends to the hypotympanum, an extended posterior tympanotomy is performed.

26. Fig. 13.40 shows the situation with the facial nerve in situ, demonstrating the exposure gained by partial rerouting of the nerve.

27. At the end of the procedure, the retrosigmoid dura is approximated with interrupted Vicryl 3/0 sutures. A small defect remains at the level of the jugular fossa. The defect and the operative cavity are closed with a single piece of abdominal fat.

28. The musculoperiosteal flap is closed and sutured with inverted Vicryl 1/0 sutures. The subcutaneous tissue is closed with catgut and the skin is closed with silk 3/0, and a tight dressing is applied.

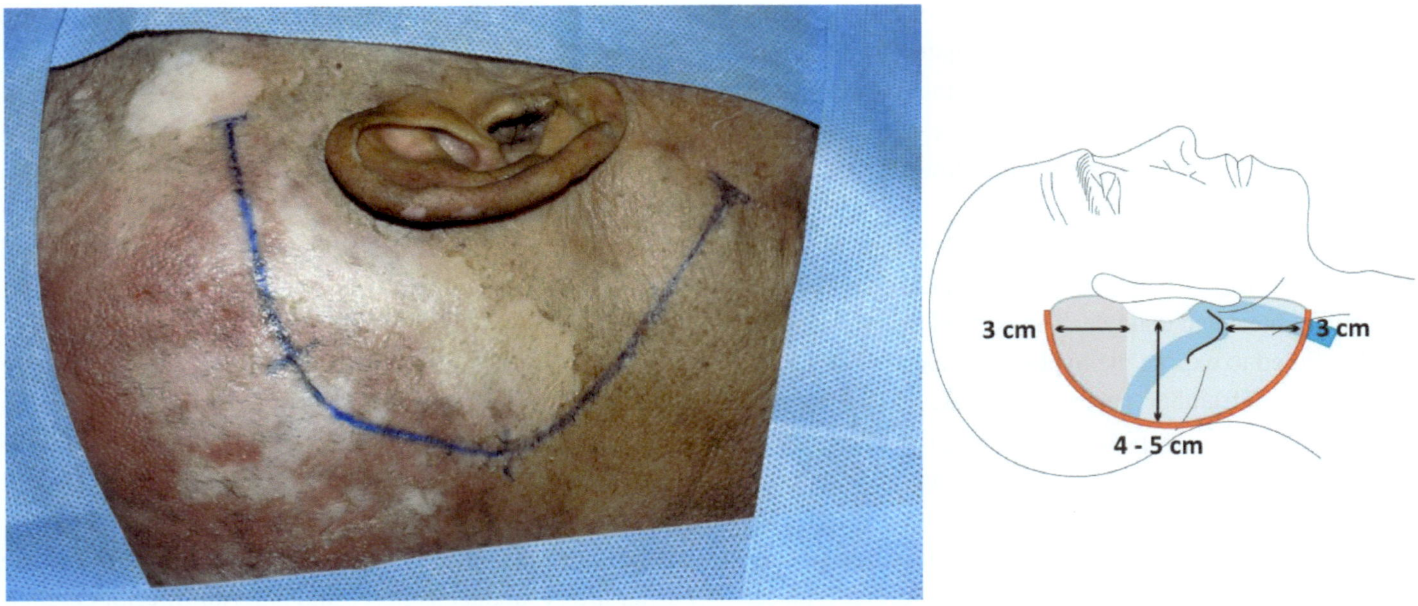

Fig. 13.13 A C-shaped postauricular skin incision is made that begins about 3 cm above the auricle, curves 4–5 cm behind the postauricular sulcus, and slants inferiorly to end 2–3 cm below the tip of the mastoid.

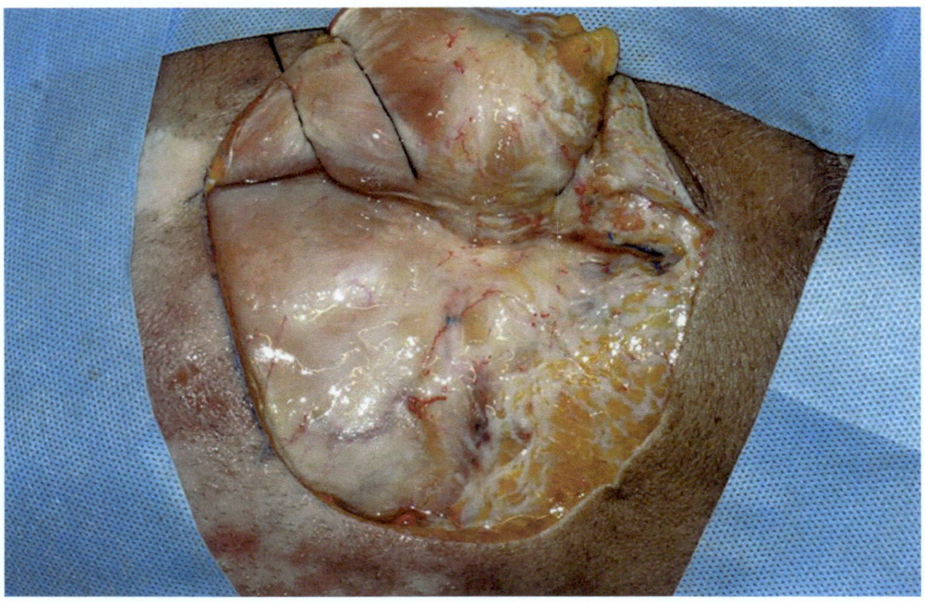

Fig. 13.14 The skin flap is raised and is retracted anteriorly using elastic-banded skin hooks.

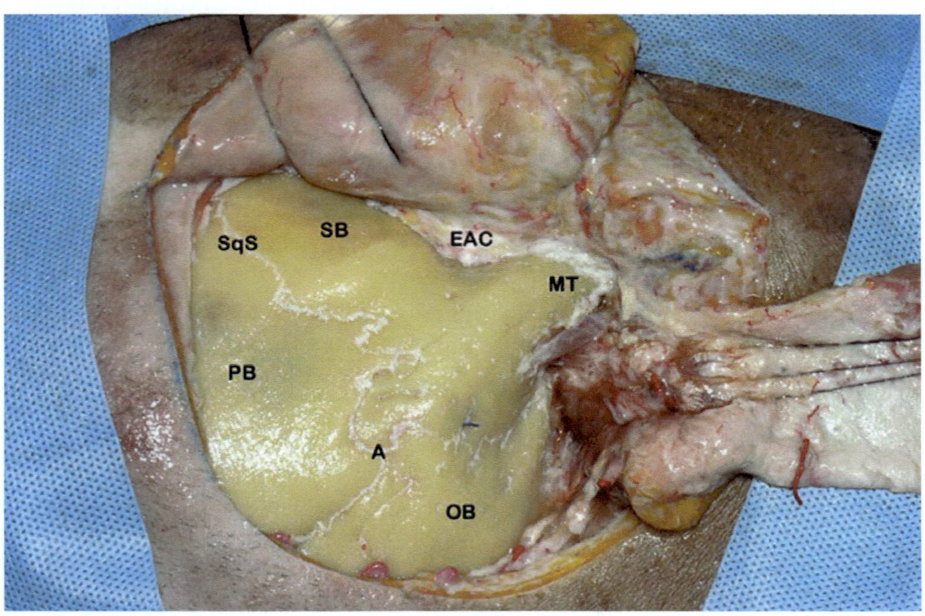

Fig. 13.15 An inferiorly based U-shaped musculoperiosteal flap is outlined. The flap is raised using a Lempert periosteal elevator. *EAC*, external auditory canal; *MT*, mastoid tip; *SB*, squamous bone; *SqS*, squamous suture; *PB*, parietal bone; *OB*, occipital bone; *A*, Asterion.

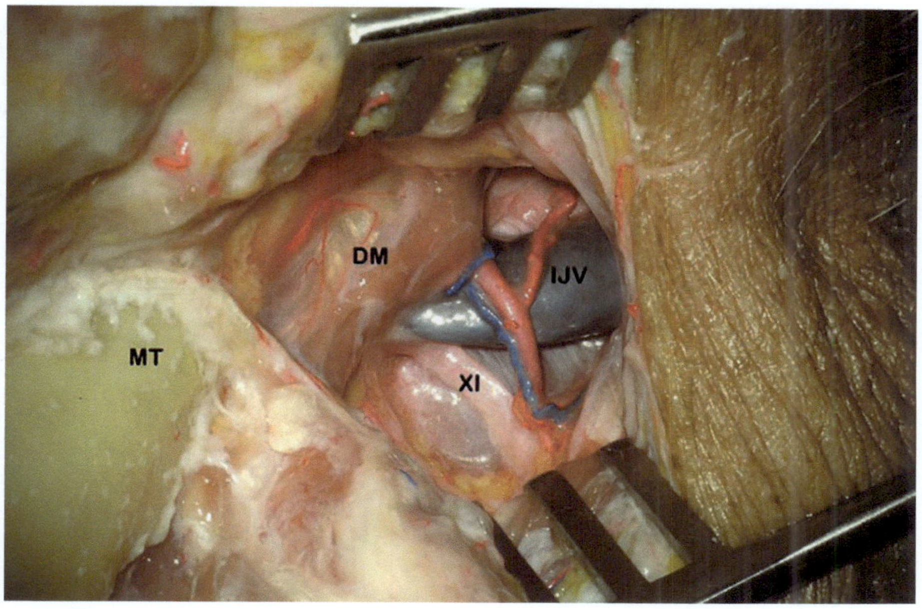

Fig. 13.16 Blunt dissection anterior to this point reveals the internal jugular vein. *MT*, mastoid tip; *DM*, digastric muscle; *IJV*, internal jugular vein; *XI*, accessory nerve.

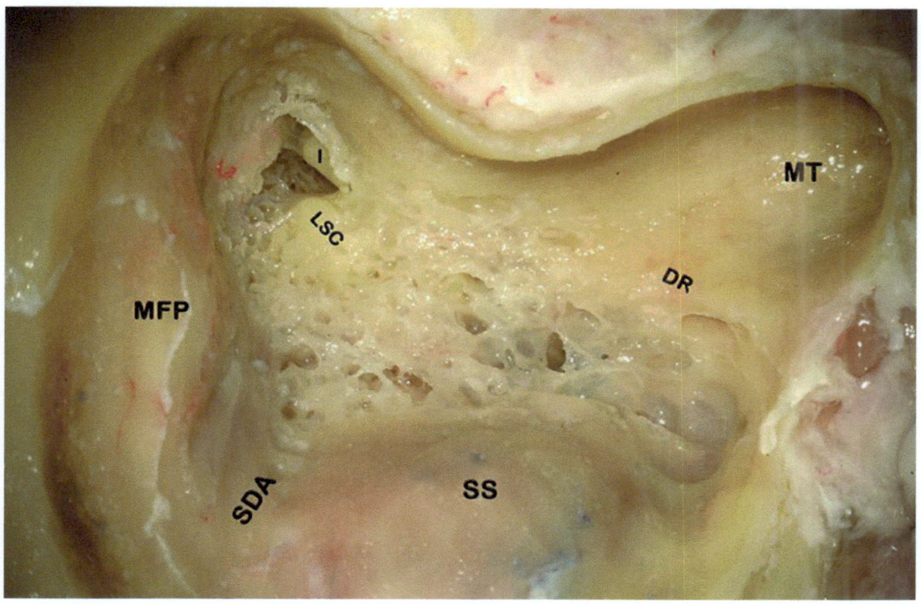

Fig. 13.17 An extended mastoidectomy is performed. *MT*, mastoid tip; *DR*, digastric ridge; *MFP*, middle fossa plate; *SS*, Sigmoid sinus; *SDA*, sinodural angle; *LSC*, lateral semicircular canal; *I*, incus.

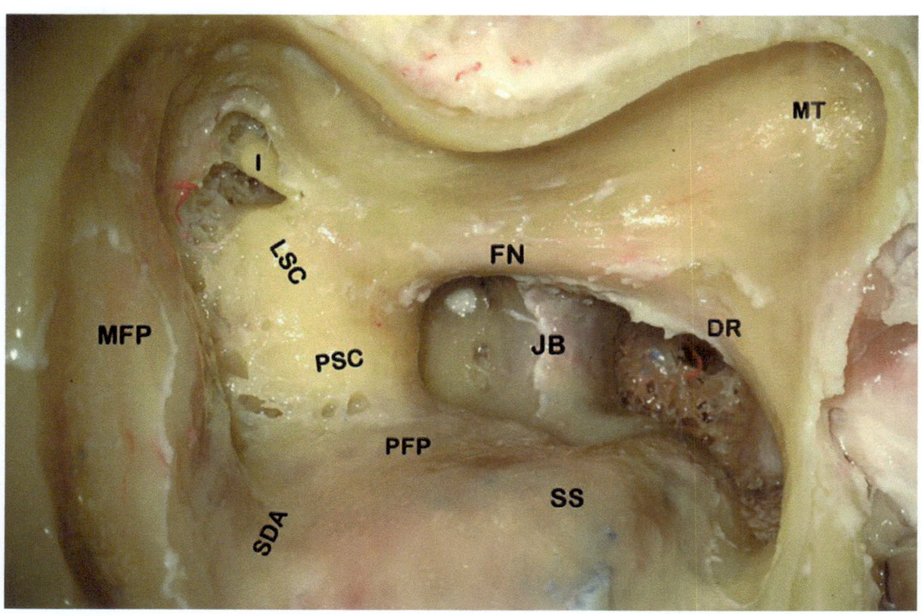

Fig. 13.18 The retrofacial cells are next drilled. The mastoid segment of the facial nerve and the Jugular bulb is identified. *MT*, mastoid tip; *DR*, digastric ridge; *MFP*, middle fossa plate; *PFP*, posterior fossa plate; *SS*, Sigmoid sinus; *SDA*, sinodural angle; *LSC*, lateral semicircular canal; *PSC*, posterior semicircular canal; *I*, incus; *JB*, jugular bulb.

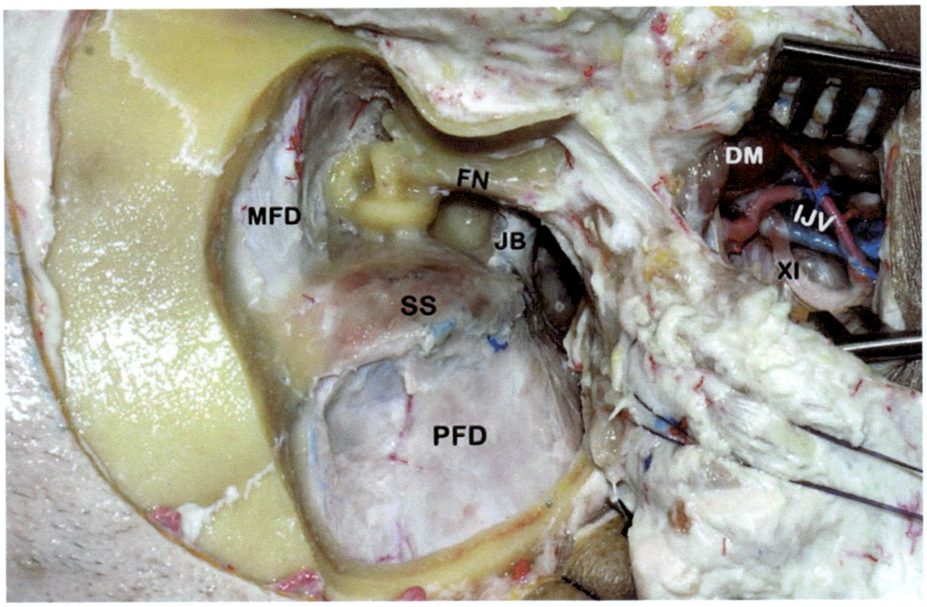

Fig. 13.19 A 4 × 4 cm retrosigmoid craniotomy is performed. The craniotomy is limited anteriorly by the sigmoid sinus and superiorly by the transverse sinus. *MFD*, middle fossa dura; *PFD*, posterior fossa dura; *SS*, Sigmoid sinus; *JB*, jugular bulb; *FN*, facial nerve; *DM*, digastric muscle; *IJV*, internal jugular vein; *XI*, accessory nerve.

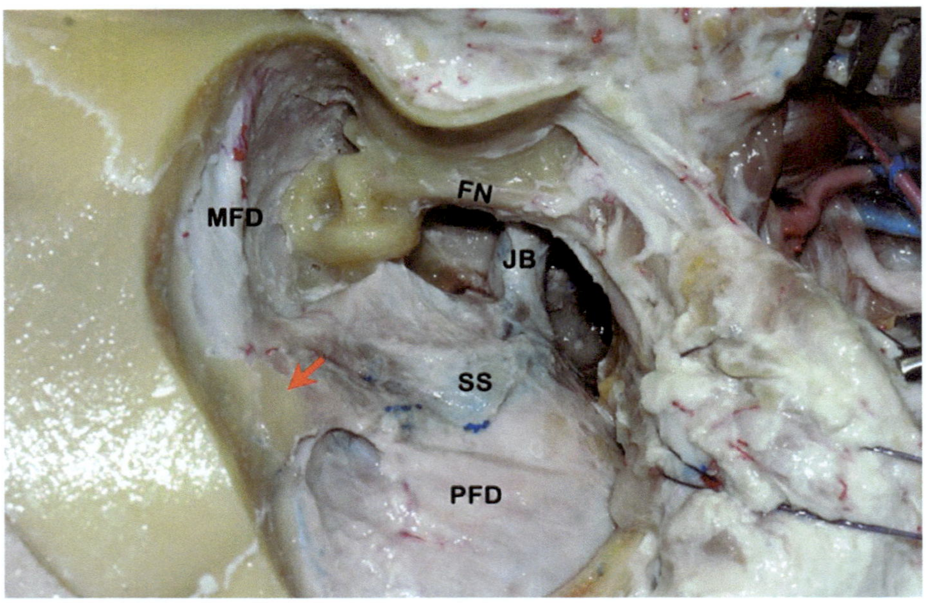

Fig. 13.20 The bone overlying the skeletonized sigmoid sinus is removed. However, a piece of bone overlying the genu from the lateral to the sigmoid sinus is *left* in place. *MFD*, middle fossa dura; *PFD*, posterior fossa dura; *SS*, Sigmoid sinus; *JB*, jugular bulb; *FN*, facial nerve.

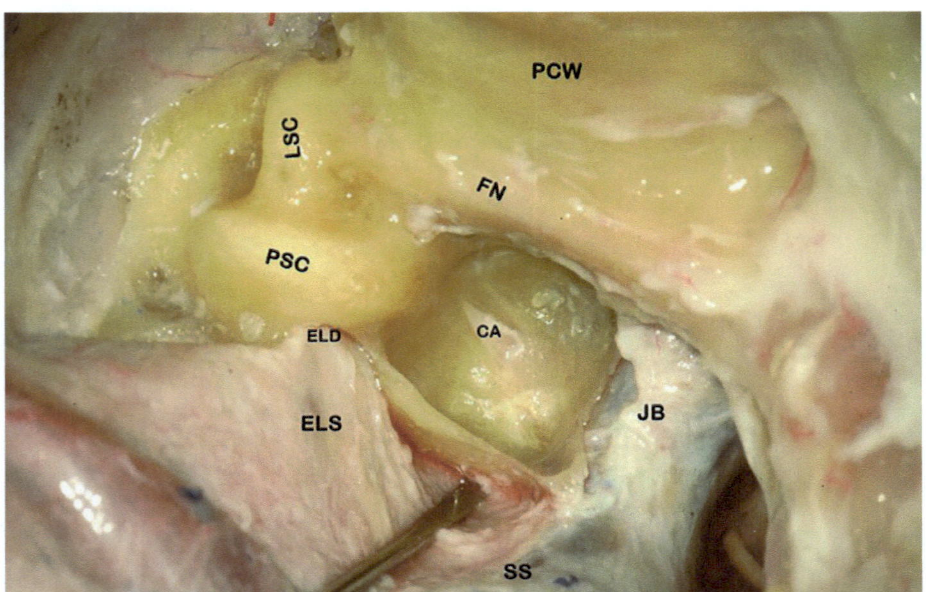

Fig. 13.21 With gentle pressure on the sigmoid sinus and the posterior fossa dura using the suction irrigator, the endolymphatic sac is identified passing medial to the posterior semicircular canal. *ELD*, endolymphatic duct; *ELS*, endolymphatic sac; *CA*, cochlear aqueduct; *SS*, sigmoid sinus; *JB*, jugular bulb; *FN*, facial nerve; *LSC*, lateral semicircular canal; *PSC*, posterior semicircular canal; *PCW*, posterior canal wall.

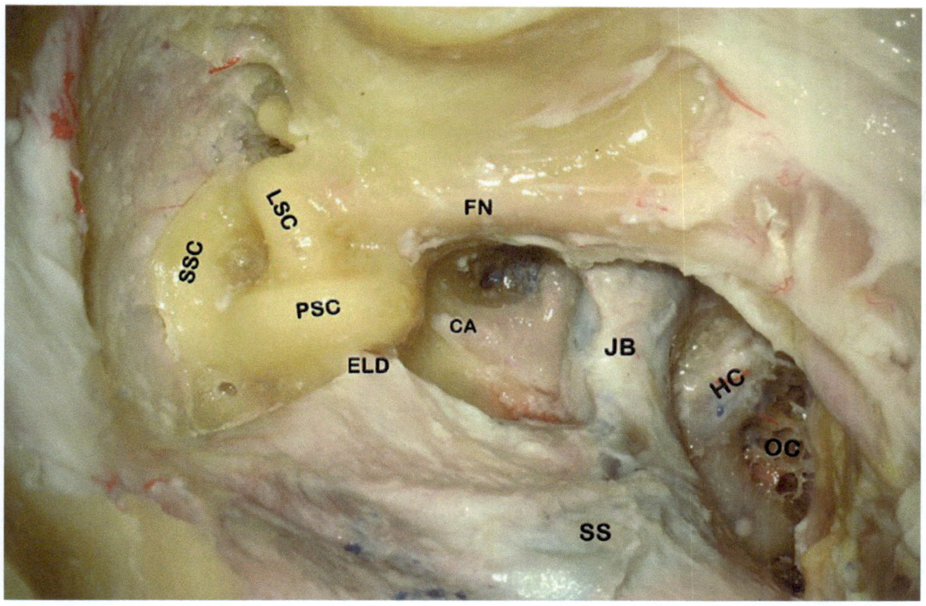

Fig. 13.22 With further bone removal, the cochlear aqueduct is identified. *ELD*, endolymphatic duct; *CA*, cochlear aqueduct; *SS*, sigmoid sinus; *JB*, jugular bulb; *FN*, facial nerve; *LSC*, lateral semicircular canal; *PSC*, posterior semicircular canal; *SSC*, superior semicircular canal; *OC*, occipital condyle; *HC*, hypoglossal canal.

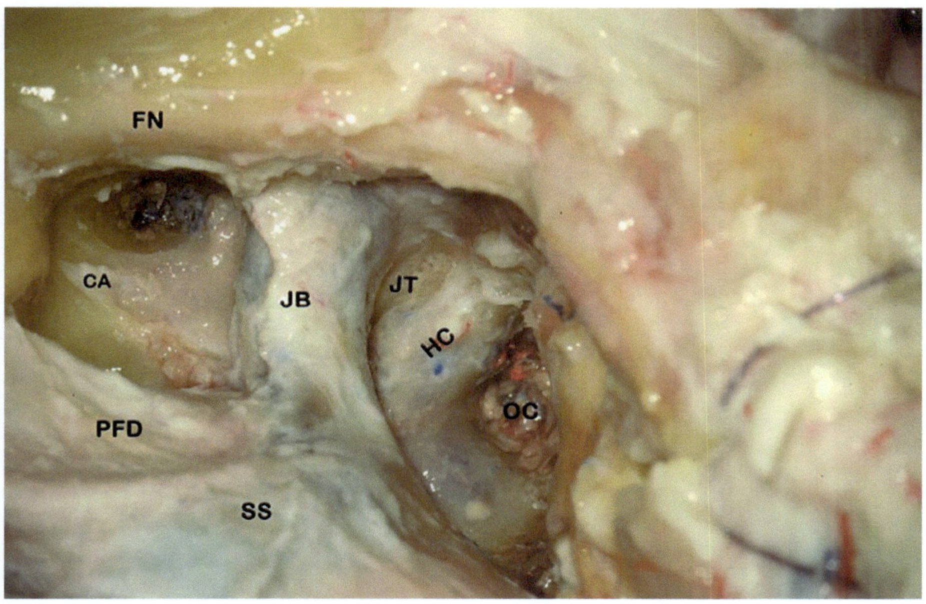

Fig. 13.23 The occipital condyle is partially drilled to expose the hypoglossal canal. *CA*, cochlear aqueduct; *SS*, sigmoid sinus; *JB*, jugular bulb; *FN*, facial nerve; *OC*, occipital condyle; *HC*, hypoglossal canal; *JT*, jugular tubercle; *PFD*, posterior fossa dura.

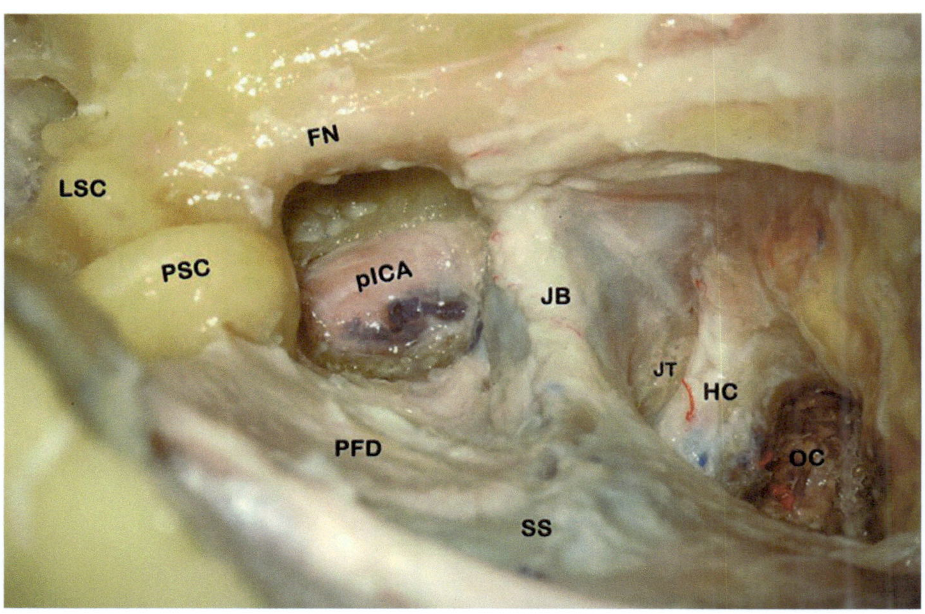

Fig. 13.24 The infratubal vertical segment of the internal carotid artery can be exposed by drilling the inferior tympanic bone. *SS*, sigmoid sinus; *JB*, jugular bulb; *FN*, facial nerve; *OC*, occipital condyle; *HC*, hypoglossal canal; *JT*, jugular tubercle; *PFD*, posterior fossa dura; *LSC*, lateral semicircular canal; *PSC*, posterior semicircular canal; *pICA*, petrous segment of internal carotid artery.

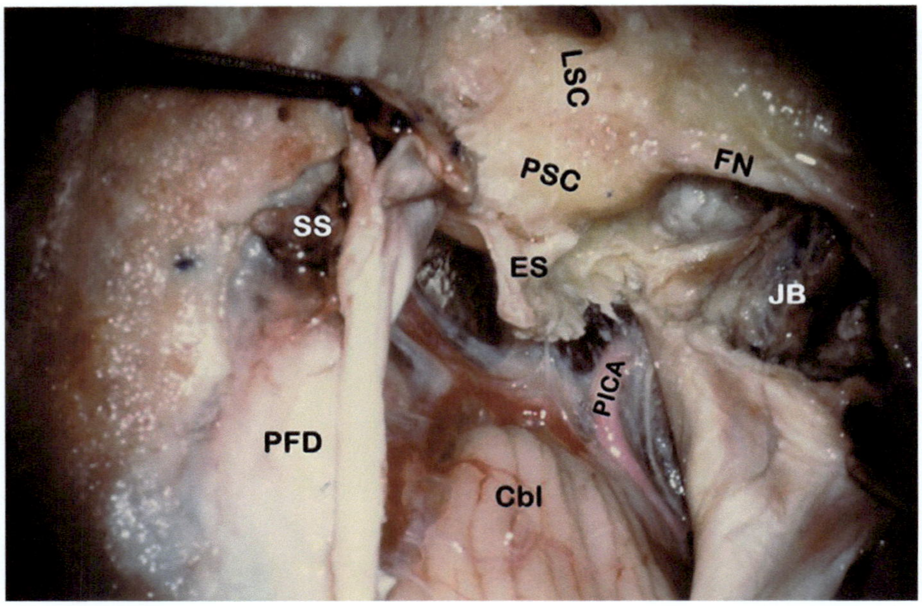

Fig. 13.25 The intradural structures are seen covered by arachnoid. *SS*, sigmoid sinus; *JB*, jugular bulb; *FN*, facial nerve; *ES*, endolymphatic sac; *PFD*, posterior fossa dura; *LSC*, lateral semicircular canal; *PSC*, posterior semicircular canal; *PICA*, posterior inferior cerebellar artery, *Cbl*, Cerebellum.

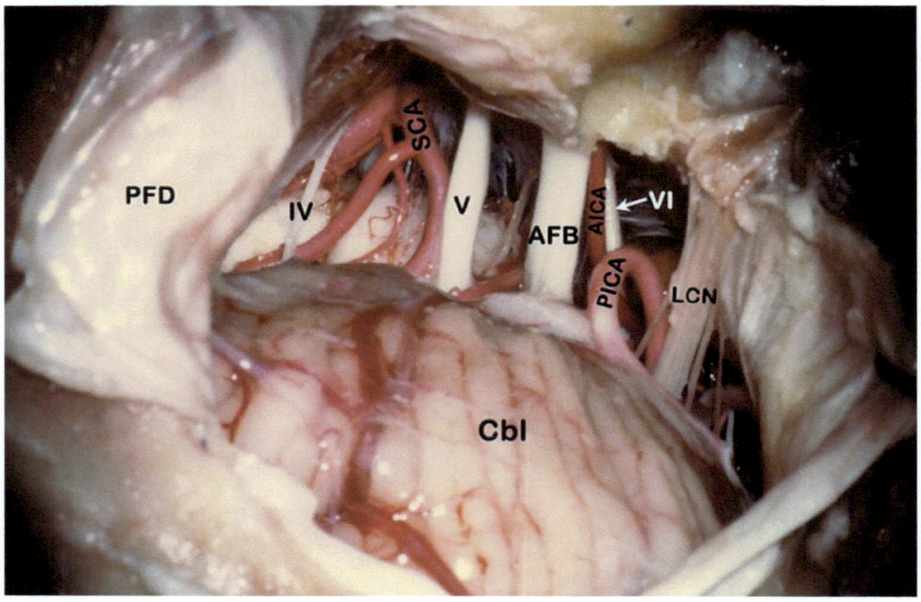

Fig. 13.26 The arachnoid is taken down from the neurovascular structures. Cranial nerves IV–XI can be visualized. *PFD*, posterior fossa dura; *SCA*, superior cerebellar artery; *AICA*, anterior inferior cerebellar artery; *PICA*, posterior inferior cerebellar artery; *Cbl*, Cerebellum; *AFB*, acoustic facial bundle; *IV*, trochlear nerve; *V*, trigeminal nerve; *VI*, abducent nerve; *LCN*, lower cranial nerves.

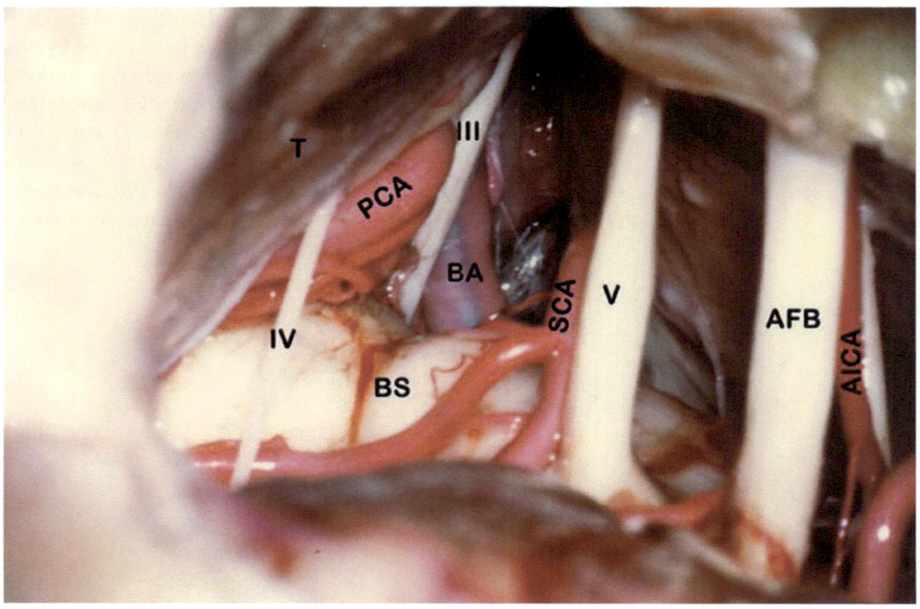

Fig. 13.27 Tilting the microscope upwards identifies the terminal part of the basilar artery, seen through the tentorial notch, giving rise to the posterior cerebral artery. *T*, tentorium; *BA*, basilar artery; *BS*, brainstem; *PCA*, posterior cerebral artery; *SCA*, superior cerebellar artery; *AICA*, anterior inferior cerebellar artery; *AFB*, acoustic facial bundle; *III*, oculomotor nerve; *IV*, trochlear nerve; *V*, trigeminal nerve.

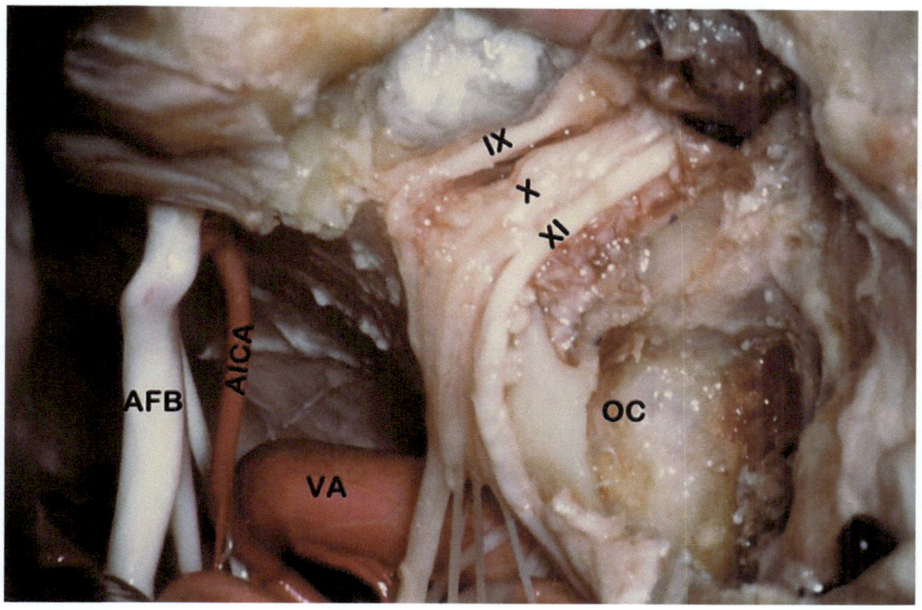

Fig. 13.28 The lateral wall of the jugular bulb can now be removed to exposure the intracranial course of cranial nerves IX–XI. *AICA*, anterior inferior cerebellar artery; *AFB*, acoustic facial bundle; *VA*, vertebral artery; *OC*, occipital condyle; *IX*, glossopharyngeal nerve; *X*, vagus nerve; *XI*, accessory nerve.

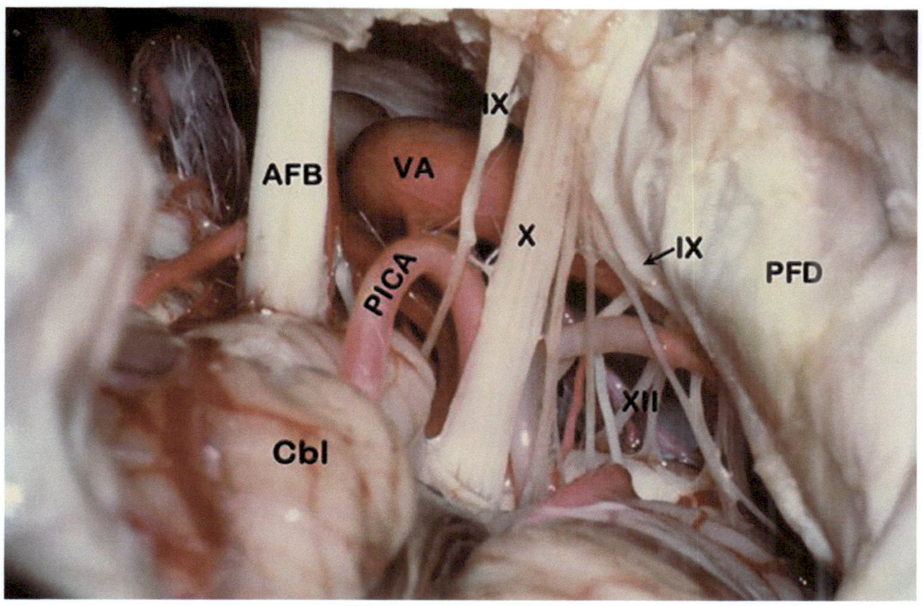

Fig. 13.29 The relationship between the posterior inferior cerebellar artery and the lower cranial nerves is seen. *PICA*, posterior inferior cerebellar artery; *AFB*, acoustic facial bundle; *VA*, vertebral artery; *Cbl*, Cerebellum; *PFD*, posterior fossa dura; *IX*, glossopharyngeal nerve; *X*, vagus nerve; *XI*, accessory nerve; *XII*, hypoglossal nerve.

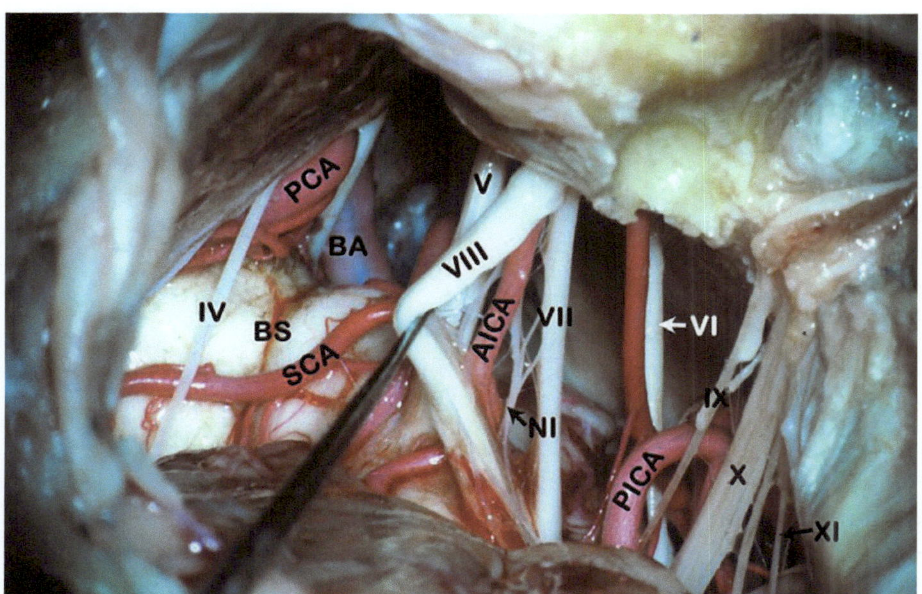

Fig. 13.30 Retraction of the vestibulocochlear nerve shows the anteriorly-lying facial nerve, with the anterior inferior cerebellar artery passing in between. *BA*, basilar artery; *BS*, brainstem; *PCA*, posterior cerebral artery; *SCA*, superior cerebellar artery; *AICA*, anterior inferior cerebellar artery; *PICA*, posterior inferior cerebellar artery; *IV*, trochlear nerve; *V*, trigeminal nerve; *VI*, abducent nerve; *VII*, facial nerve; *VIII*, vestibulocochlear nerve; *IX*, glossopharyngeal nerve; *X*, vagus nerve; *XI*, accessory nerve.

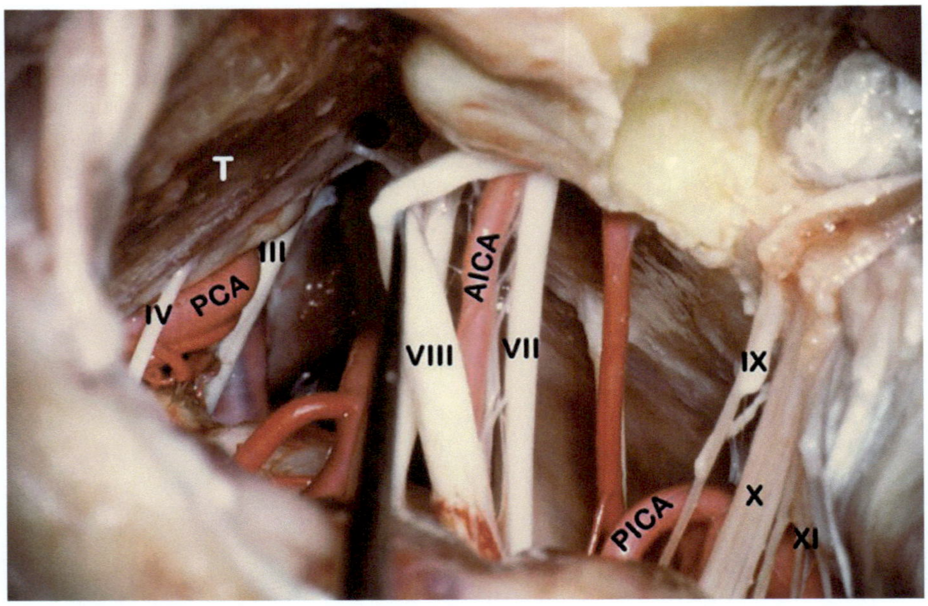

Fig. 13.31 At a higher magnification, the relationship between the anterior inferior cerebellar artery and the acoustic facial bundle is better appreciated. *T*, tentorium; *PCA*, posterior cerebral artery; *AICA*, anterior inferior cerebellar artery; *PICA*, posterior inferior cerebellar artery; *III*, oculomotor nerve; *IV*, trochlear nerve; *VII*, facial nerve; *VIII*, vestibulocochlear nerve; *IX*, glossopharyngeal nerve; *X*, vagus nerve; *XI*, accessory nerve.

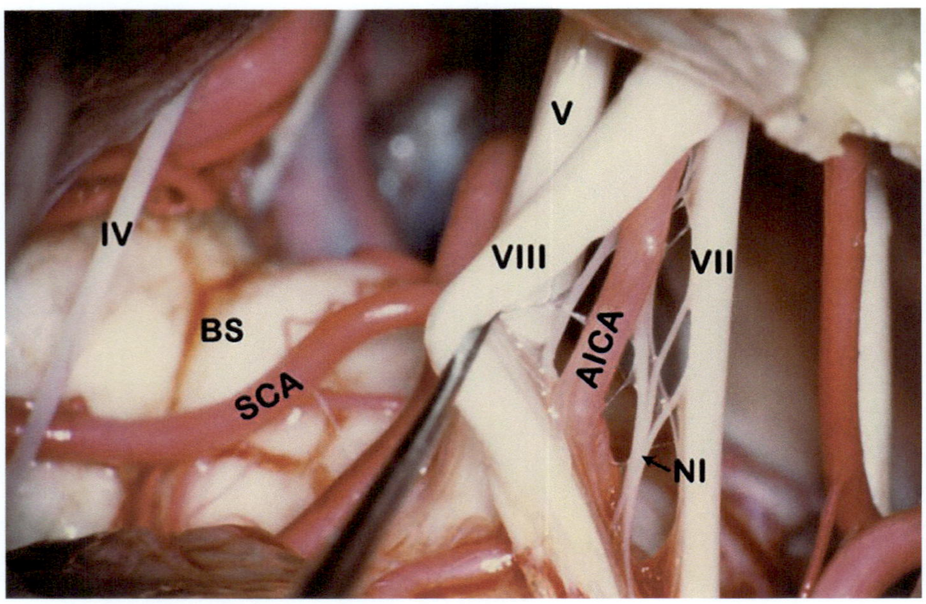

Fig. 13.32 The nervus intermedius arises on the anterior surface of the vestibulocochlear nerve, has a free segment in the cistern and/or meatus, and joins the facial nerve distally. *BS*, brainstem; *AICA*, anterior inferior cerebellar artery; *IV*, trochlear nerve; *V*, trigeminal nerve; *VII*, facial nerve; *VIII*, vestibulocochlear nerve; *NI*, nervus intermedius.

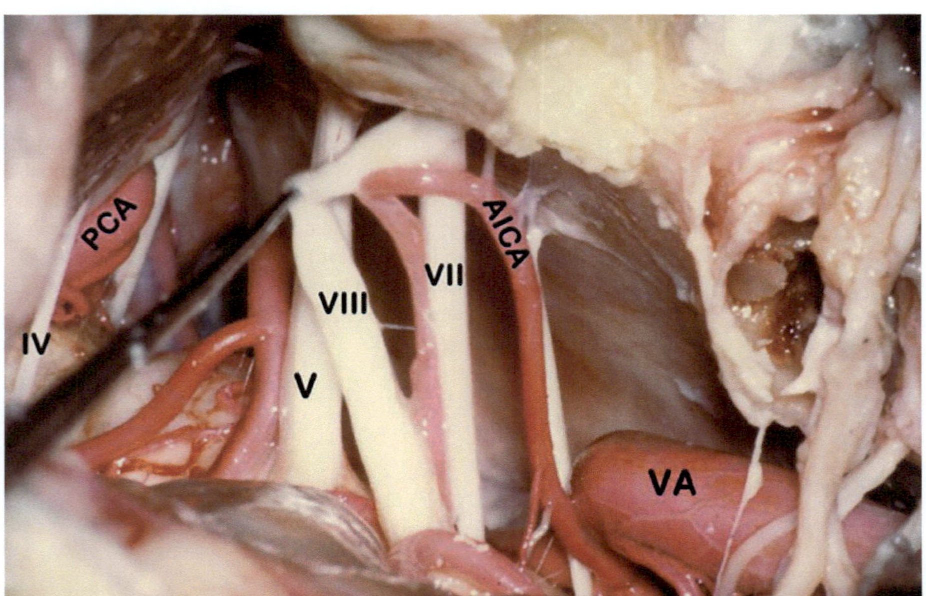

Fig. 13.33 With the patient turned forwards, the whole course of the anterior inferior cerebellar artery loop can be seen. *PCA*, posterior cerebral artery; *AICA*, anterior inferior cerebellar artery; *VA*, vertebral artery; *IV*, trochlear nerve; *V*, trigeminal nerve; *VII*, facial nerve; *VIII*, vestibulocochlear nerve.

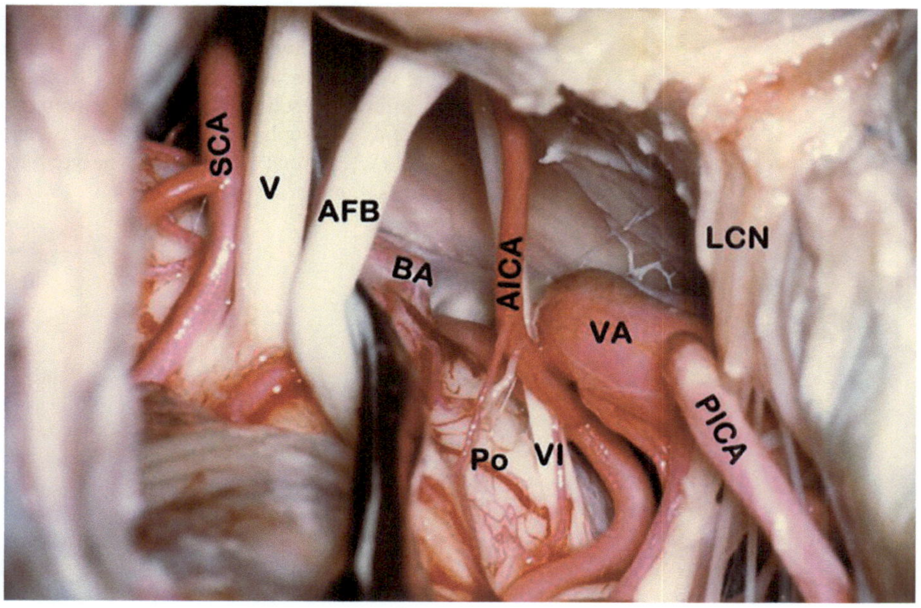

Fig. 13.34 Retraction of the acoustic facial bundle shows the basiler artery, abducens nerve and anterior inferior cerebellar artery. *BA*, basilar artery; *SCA*, superior cerebellar artery; *AICA*, anterior inferior cerebellar artery; *PICA*, posterior inferior cerebellar artery; *VA*, vertebral artery; *V*, trigeminal nerve; *VI*, abducent nerve; *AFB*, acoustic facial bundle; *Po*, pons; *LCN*, lower cranial nerves.

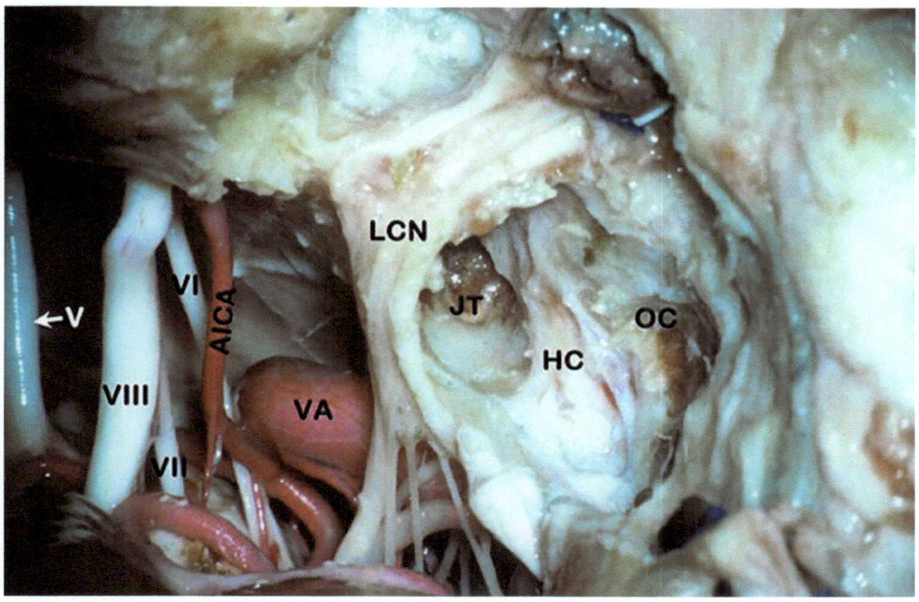

Fig. 13.35 The relationship between the hypoglossal canal and the lower cranial nerves is seen. *AICA*, anterior inferior cerebellar artery; *VA*, vertebral artery; *V*, trigeminal nerve; *VI*, abducent nerve; *VII*, facial nerve; *VIII*, vestibulocochlear nerve; *OC*, occipital condyle; *JT*, jugular tubercle; *HC*, hypoglossal canal; *LCN*, lower cranial nerves.

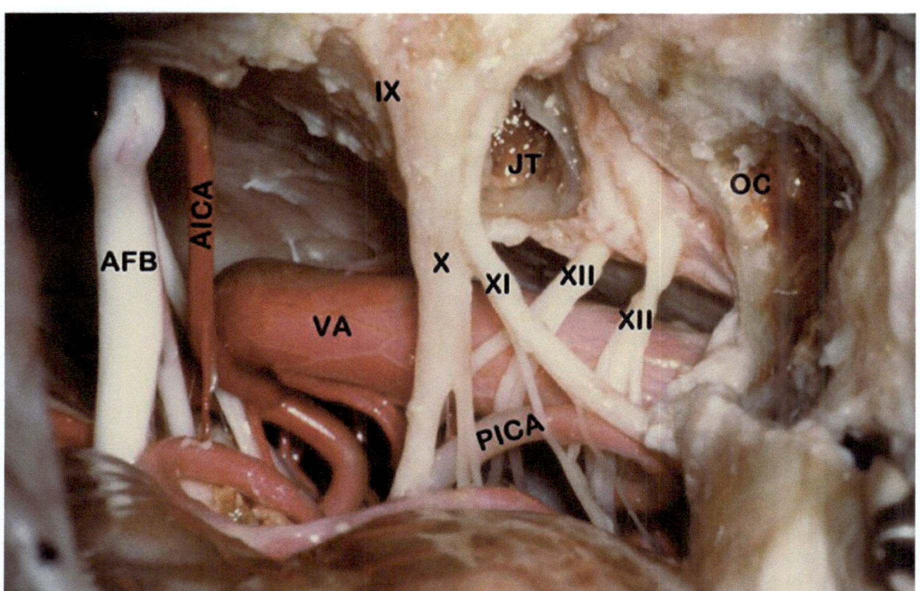

Fig. 13.36 The hypoglossal canal is opened. The two bundles forming the hypoglossal nerve unite to form a single nerve in the canal. *AICA*, anterior inferior cerebellar artery; *PICA*, posterior inferior cerebellar artery; *VA*, vertebral artery; *AFB*, acoustic facial bundle; *IX*, glossopharyngeal nerve; *X*, vagus nerve; *XI*, accessory nerve; *XII*, hypoglossal nerve; *OC*, occipital condyle; *JT*, jugular tubercle.

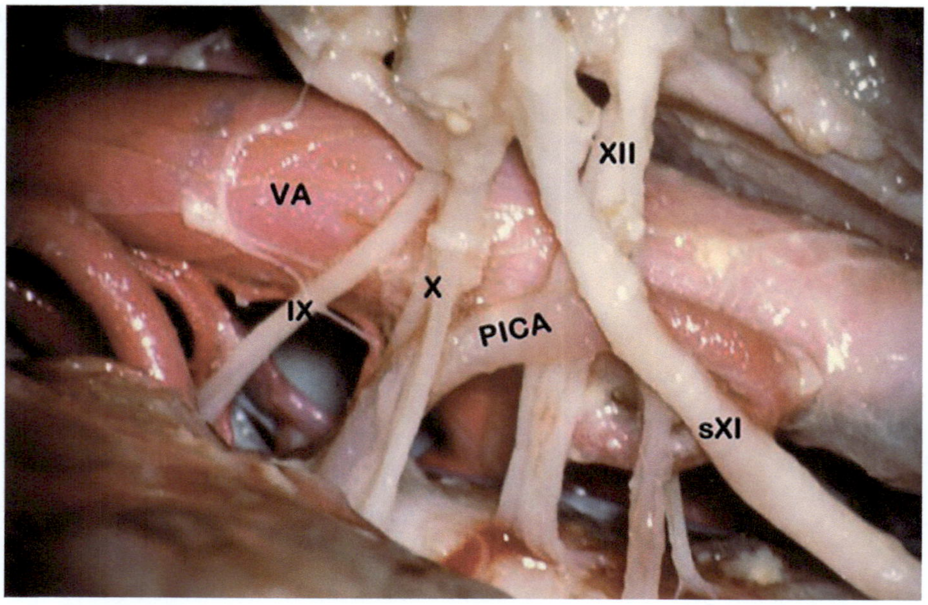

Fig. 13.37 At a higher magnification, the intimate relation between the hypoglossal nerve and the vertebral artery is well seen. *PICA*, posterior inferior cerebellar artery; *VA*, vertebral artery; *IX*, glossopharyngeal nerve; *X*, vagus nerve; *sXI*, spinal accessory nerve; *XII*, hypoglossal nerve.

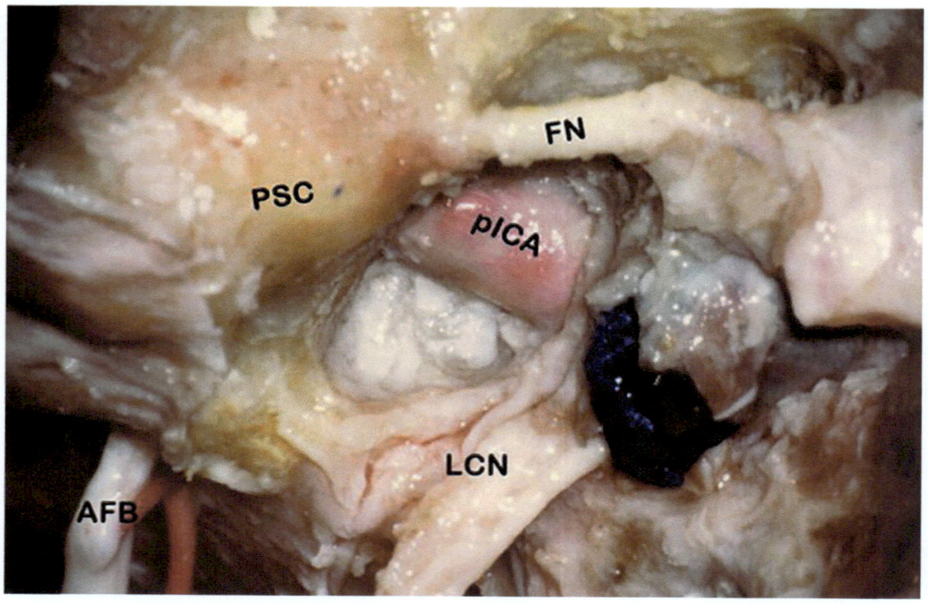

Fig. 13.38 The mastoid segment of the facial nerve can be liberated as far as the stylomastoid foramen to control the infralabyrinthine compartment and the vertical portion of the internal carotid artery. *PSC*, posterior semicircular canal; *pICA*, petrous segment of internal carotid artery; *FN*, facial nerve; *AFB*, acoustic facial bundle; *LCN*, lower cranial nerves.

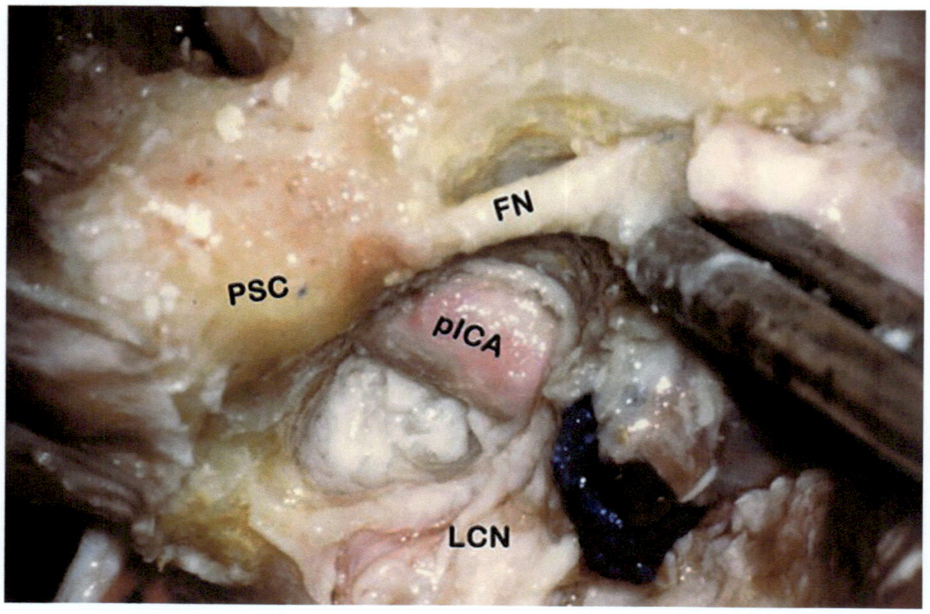

Fig. 13.39 The vertical segment of the nerve is raised along the posterior canal wall. *PSC*, posterior semicircular canal; *pICA*, petrous segment of internal carotid artery; *FN*, facial nerve; *LCN*, lower cranial nerves.

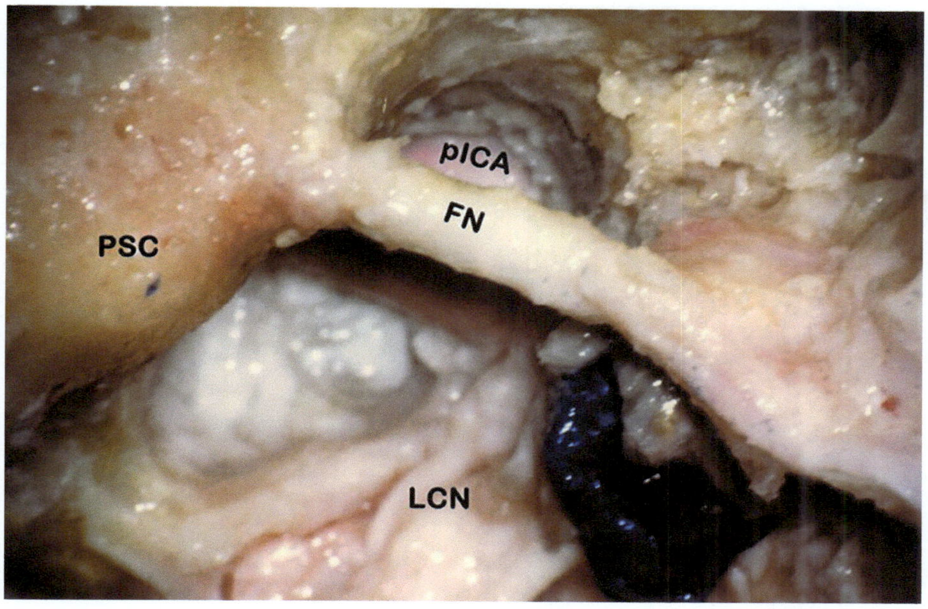

Fig. 13.40 The facial nerve was in situ. *PSC*, posterior semicircular canal; *pICA*, petrous segment of internal carotid artery; *FN*, facial nerve; *LCN*, lower cranial nerves.

Clinical Applications

Case 1: Left Lower Cranial Nerve Neuroma Size 2 × 2.5 cm With Minimal Involvement of the Jugular Foramen (Figs. 13.41–13.53)

1. Preoperative MRI with gadolinium enhancement at the level of the jugular foramen. Note the extent of the tumor inside the foramen (Fig. 13.41).

2. Preoperative audiogram showing ipsilateral hearing within normal except for high-tone sensorineural hearing loss (Fig. 13.42)

3. The approach has been performed. The posterior fossa dura has been opened as previously described, below the level of the endolymphatic sac. The cerebellum and the tumor can be identified (Fig. 13.43).

4. A plane of cleavage is established between the upper pole of the tumor and the acoustic facial bundle (Fig. 13.44)

5. Gelfoam is placed between the tumor and acoustic facial bundle to protect the latter. The tumor is further dissected from the glossopharyngeal nerve (Fig. 13.45).

6. The glossopharyngeal nerve is dissected free from the tumor at the level of the brainstem. More laterally, the nerve is still adherent to the tumor (Fig. 13.46).

7. At the end of the procedure, total tumor removal was accomplished. The acoustic facial bundle and the glossopharyngeal nerve can be identified. The vagus nerve, being the origin of the tumor, was sacrificed. Note the widely opened jugular foramen (Fig. 13.47).

8. A closer view at the end of the procedure. Note the preservation of the vascular supply to the vestibulocochlear nerve. The anterior inferior cerebellar artery is seen coursing anteromedial to the acoustic facial bundle (Fig. 13.48).

9. Tilting the microscope inferiorly, the vertebrobasilar junction is identified. The origin of the hypoglossal nerve from the medulla is seen (Fig. 13.49).

10. Postoperative CT scan with contrast, axial view. Total tumor removal can be appreciated (Fig. 13.50).

11. Postoperative CT scan, bone window. Note the preservation of the labyrinth. The posterior semicircular canal is the anterior limit of the approach (Fig. 13.51).

12. Postoperative audiogram. In comparison with the pre- operative audiogram (Fig. 13.42), slight hearing drop is seen with marked hearing loss only at high tones (Fig. 13.52).

13. One month postoperatively, showing grade I facial nerve function (Fig. 13.53)

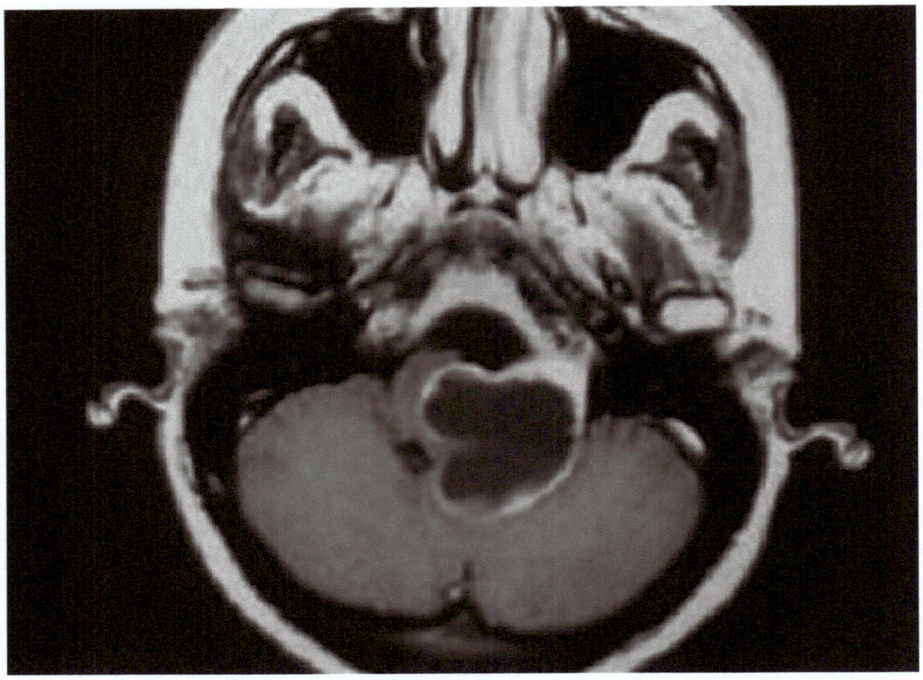

Fig. 13.41 Preoperative MRI of the case.

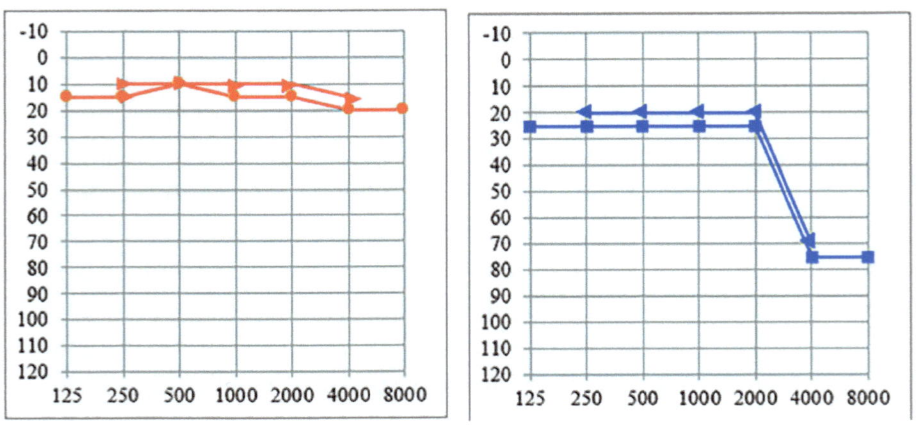

Fig. 13.42 Preoperative audiogram.

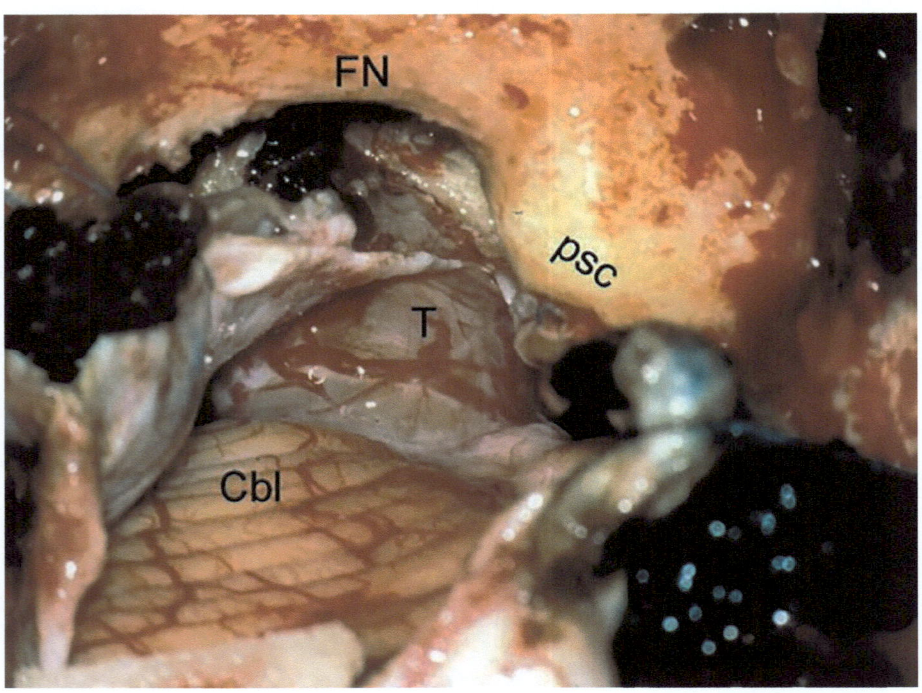

Fig. 13.43 Opening the dura.

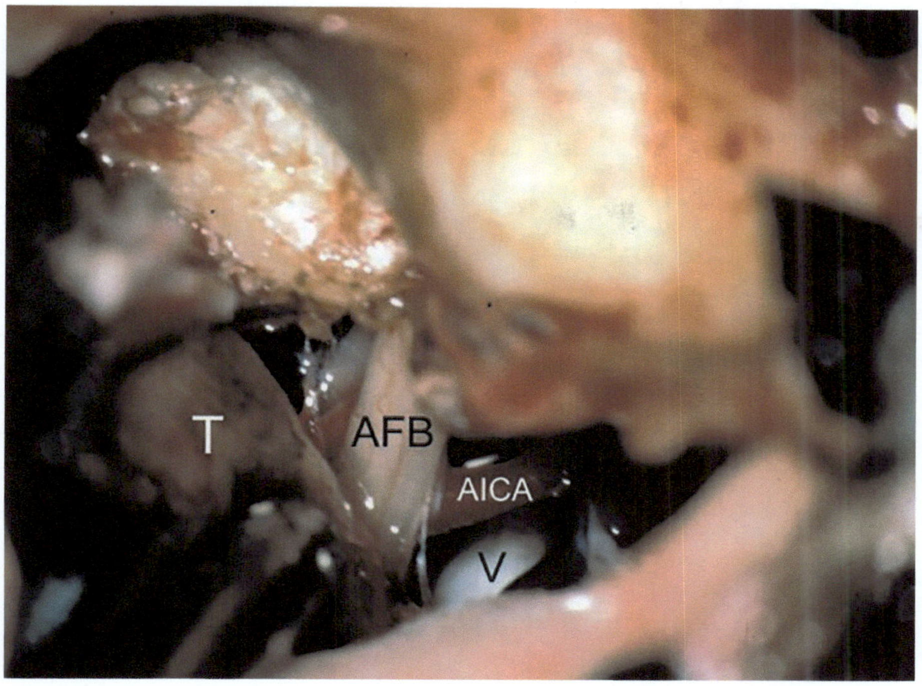

Fig. 13.44 Tumor separation from the acoustic facial bundle (AFB).

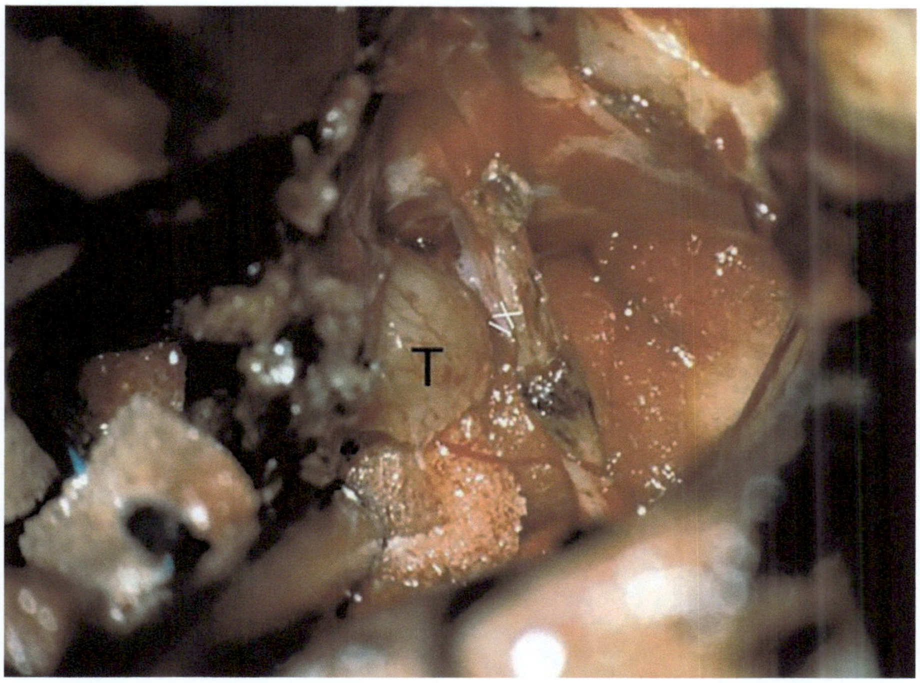

Fig. 13.45 Tumor (T) dissection from the glossopharyngeal nerve (IX).

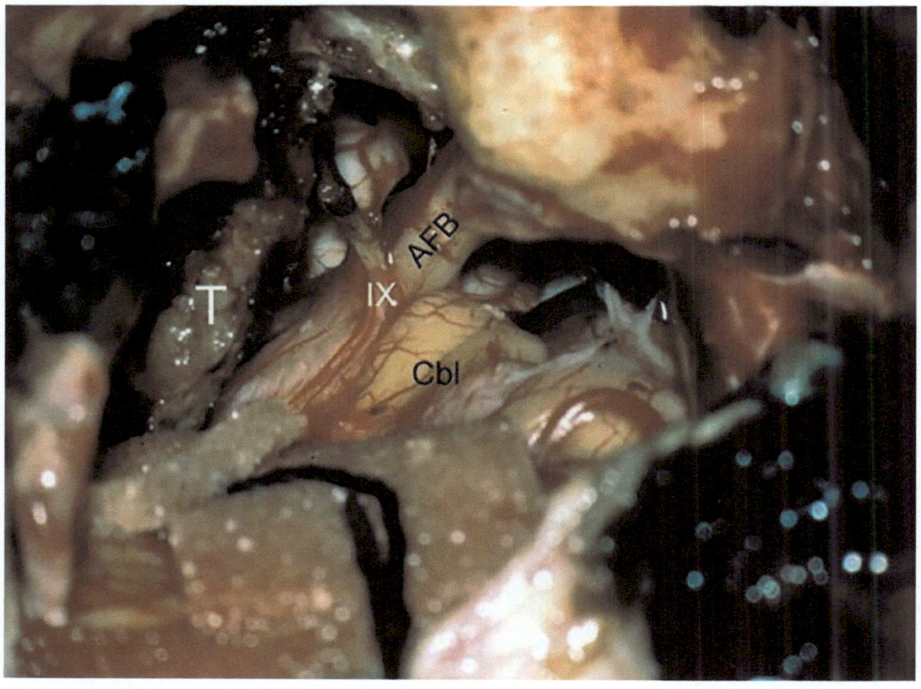

Fig. 13.46 The glossopharyngeal nerve (IX) has been freed from the tumor at the level of the brainstem.

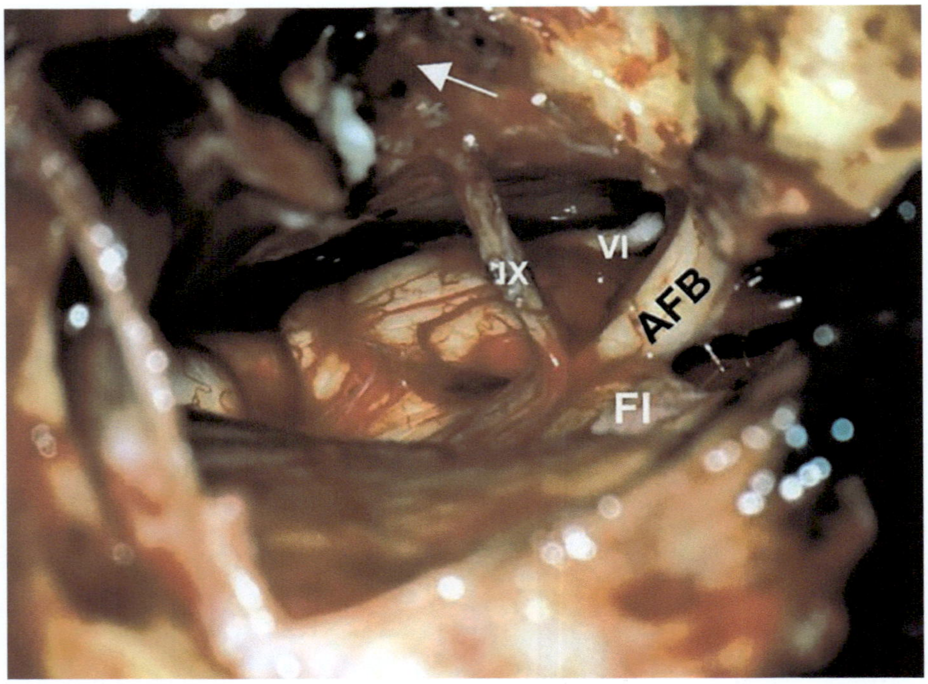

Fig. 13.47 View after total tumor removal. Arrow shows the opened jugular foramen.

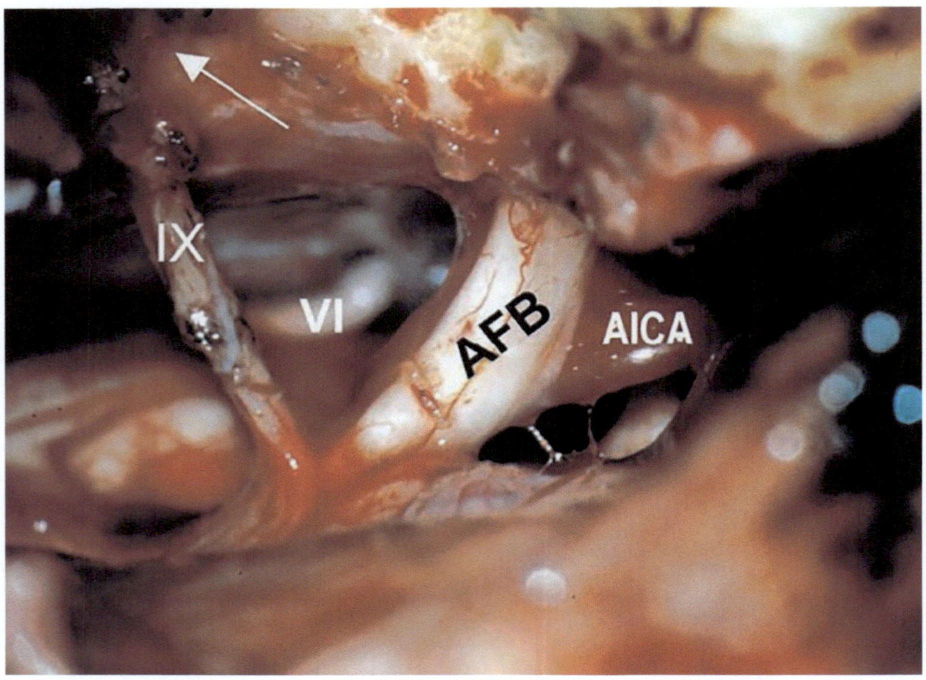

Fig. 13.48 Note the intact vasculature of the acoustic facial bundle (AFB) at higher magnification.

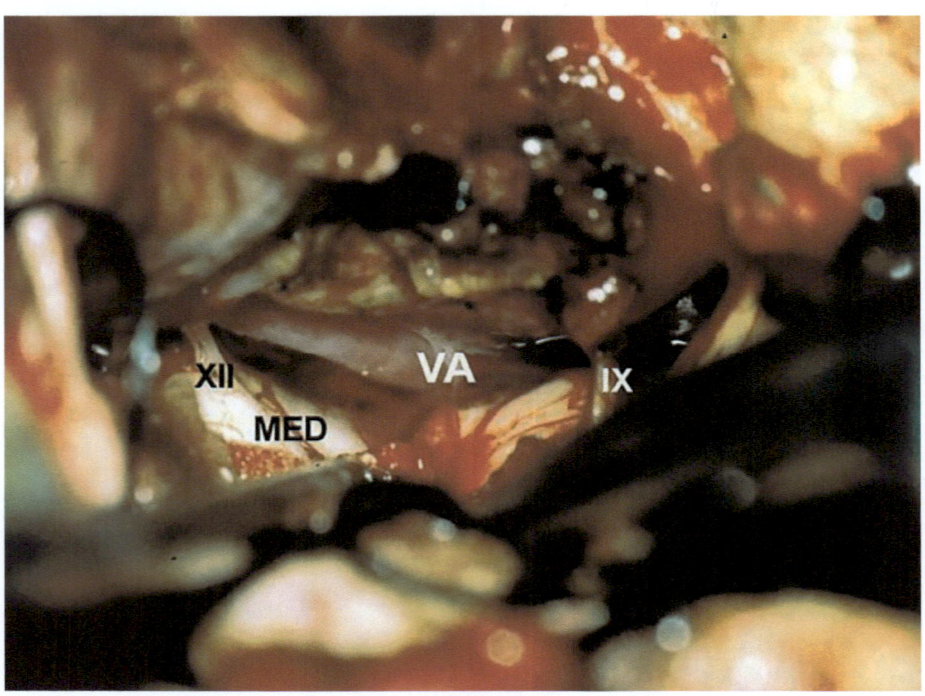

Fig. 13.49 View after total tumor removal. The hypoglossal nerve (XII), the vertebral artery (VA), the medulla (MED), and the ninth (IX) cranial nerve can be seen.

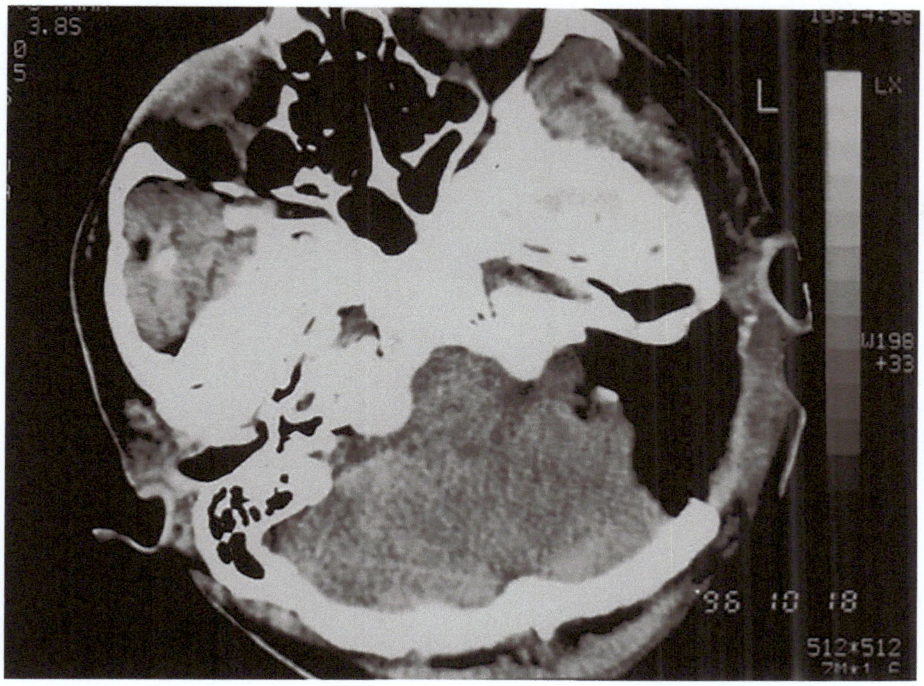

Fig. 13.50 Postoperative CT scan.

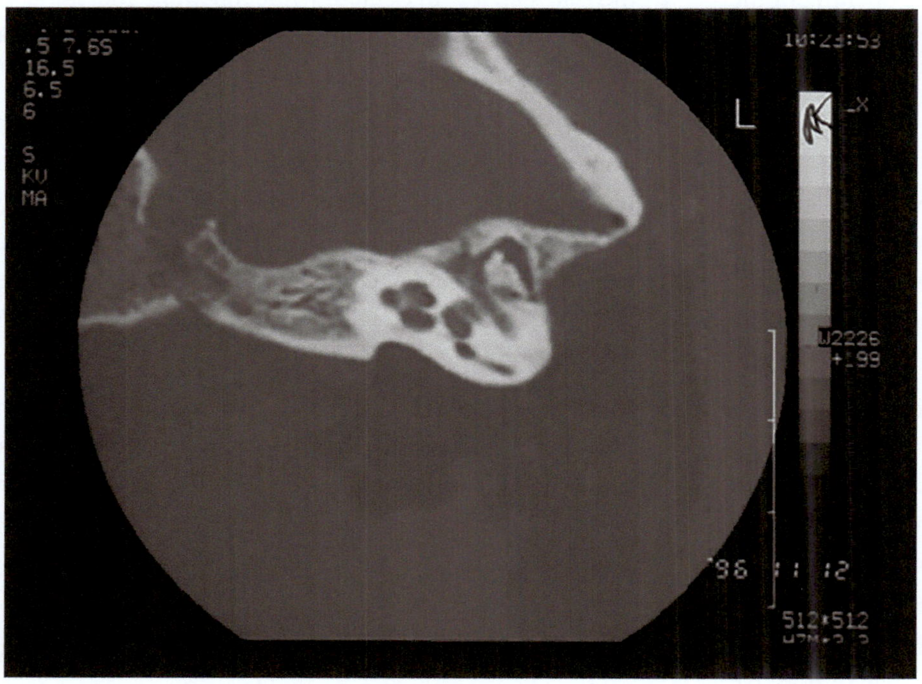

Fig. 13.51 Postoperative CT scan. Note the intact posterior semicircular canal.

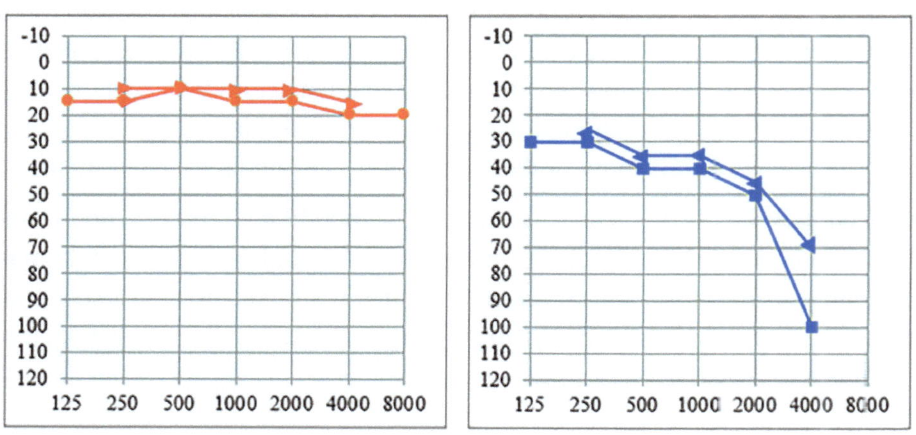

Fig. 13.52 Postoperative audiogram.

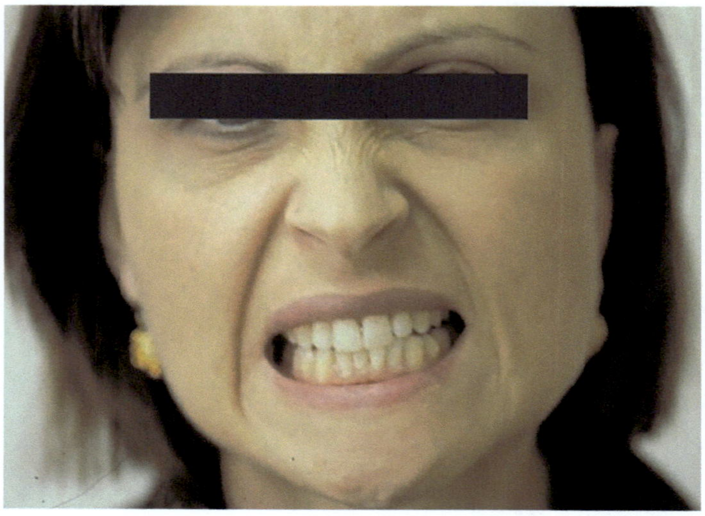

Fig. 13.53 Intact postoperative facial nerve function.

HINTS AND PITFALLS

1. Great care should be taken to avoid wide opening of the tissue spaces in the neck. Connecting wide open neck spaces to the subarachnoid space carries a high risk of postoperative CSF leakage
2. The transverse process of the atlas is first identified in the neck to facilitate identification of the internal jugular vein (Fig. 13.54). Before ligating the vein, great care must be taken not to injure the accessory nerve, which lies either medially or laterally to the vein (Fig. 13.55).
3. If necessary, the jugulocarotid crest is drilled to expose the vertical segment of the internal carotid artery. This allows exposure of the posterior and lateral aspects of the infratubal segment of the internal carotid artery (Fig. 13.56). If more control of the artery is desired, an extended posterior tympanotomy is performed and the mastoid segment of the facial nerve is temporarily transposed laterally (Fig. 13.57). In the case of massive involvement of the artery, an alternative approach should be used.
4. An adequate retrosigmoid craniotomy is an important part of the approach. The craniotomy should be extended caudally to allow wide access to the jugular foramen (Fig. 13.58).
5. The endolymphatic sac is an approximate landmark for the posterior semicircular canal. The sac can be identified with gentle depression of the sigmoid sinus using the suction irrigator (Fig. 13.59). The distal part of the sac can be dissected while opening the dura. In our experience, there is no risk to labyrinthine function as long as the proximal sac is not injured.
6. When performing a retrolabyrinthine petrosectomy, bone removal should extend to the posterior edge of the posterior semicircular canal and more superiorly to the level of the common crus. Failure to remove sufficient bone around the labyrinth will limit exposure. The posterior semicircular canal is skeletonised and not blue lined to avoid opening it.
7. Removal of the retrofacial air cells is a very important part of the procedure. The dura is progressively separated from the overlying bone to avoid injury to the dura (Fig. 13.60)
8. A piece of aluminum is placed between the separated dura and the bone to be drilled to protect the dura during drilling (Fig. 13.61). The limits of this area are the mastoid segment of the facial nerve laterally, the ampulla of the posterior semicircular canal superiorly and posteriorly, and the jugular bulb inferiorly and anteriorly. Further anterior drilling will locate the cochlea superiorly and laterally and the internal auditory canal more medially and anteriorly. Adequate knowledge of the anatomy of these structures is essential to avoid injury.
9. When packing the sigmoid sinus intraluminally with Surgicel, extreme care should be taken not to push the Surgicel beyond the sigmoid to transverse sinus junction (Fig. 13.62). Closure of the transverse sinus

would obstruct the venous drainage of the vein of Labbé with subsequent detrimental effects.

10. An alternative method of closing the sigmoid sinus is to use an aneurysm needle passed through two small dural incisions posterior and anterior to the sigmoid sinus (Fig. 13.63A,B). The sinus is closed by double ligation with a Vicryl suture.
11. In many cases the jugular foramen is completely occupied by the tumor, resulting in complete obliteration of the jugular bulb (Fig. 13.64).
12. Great care should be taken to preserve the vascular supply to the vestibulocochlear nerve when dissecting the upper pole of the tumor from the facial acoustic facial bundle. Injury to these vessels will result in hearing loss (Fig. 13.65).
13. Bleeding at the level of the jugular foramen or from the inferior petrosal sinus openings is controlled by the placement of Surgicel. Care should be taken to avoid excessive packing of the inferior petrosal sinus to avoid compression injury to the closely adjacent lower cranial nerves
14. Dural integrity should be maintained during drilling; the dural incision should be clean as the dura will be resutured at the end of the approach

Results

1. In our experience, single stage removal of lower cranial nerve neuroma and jugular foramen meningioma can be achieved in all cases using this approach. The majority of cases had normal facial nerve function postoperatively.

2. Hearing preservation is possible in the majority of lower cranial nerve neuromas. However, hearing results are less favorable in jugular foramen meningiomas. The main morbidity is the resulting lower cranial nerve palsy, which seems to be related to the pathology rather than the approach itself.

3. Despite the single stage removal of many intra-extracranial lesions using this approach, the incidence of CSF leakage was quite low. The reason for this is multifactorial. First, the approach avoids extensive contact between the cervical and subarachnoid spaces. The occipital dura and the medial wall of the sigmoid sinus are sutured and the remaining small defect is filled with abdominal fat. The last shell of the tympanic bone and the thick fibrous tissue at the base of the skull are left intact (Figs 13.66–13.67)). This, together with the specially designed myoaponeurotic flap, limits the size of the resulting surgical cavity and helps to ensure proper containment of the fat.

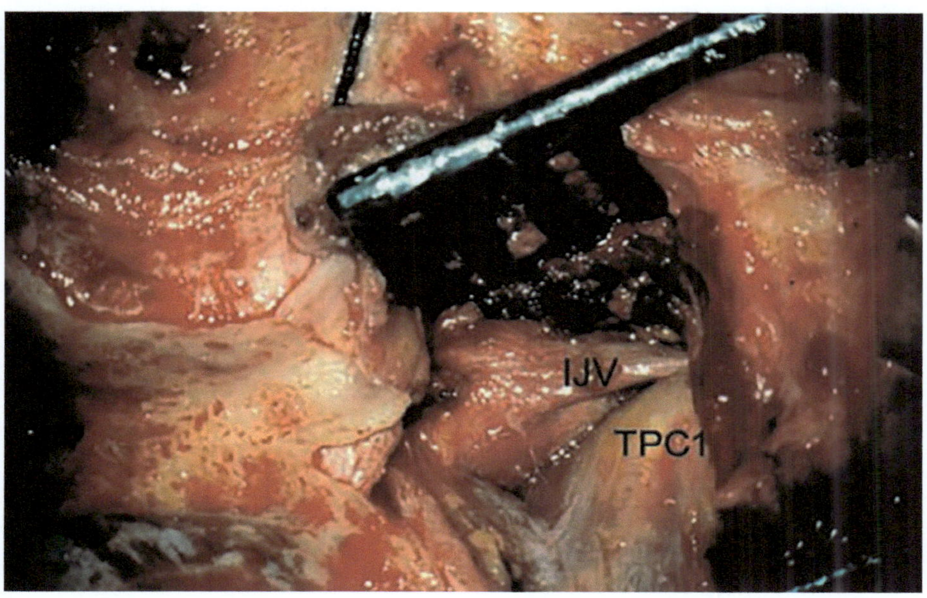

Fig. 13.54 Identification of the transverse process of the atlas (TPC1) (*right side*).

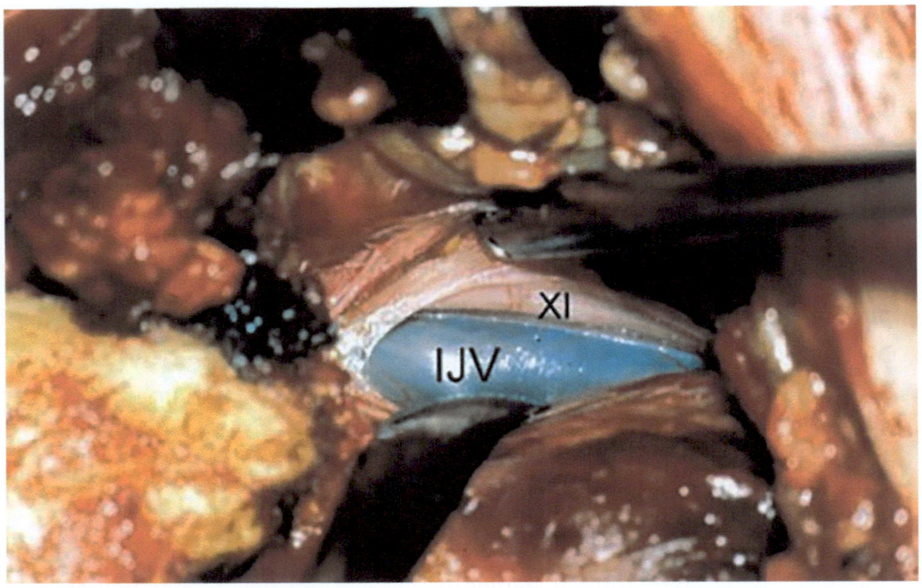

Fig. 13.55 *Right side*. The accessory nerve (XI) is immediately anterior to the internal jugular vein (IJV).

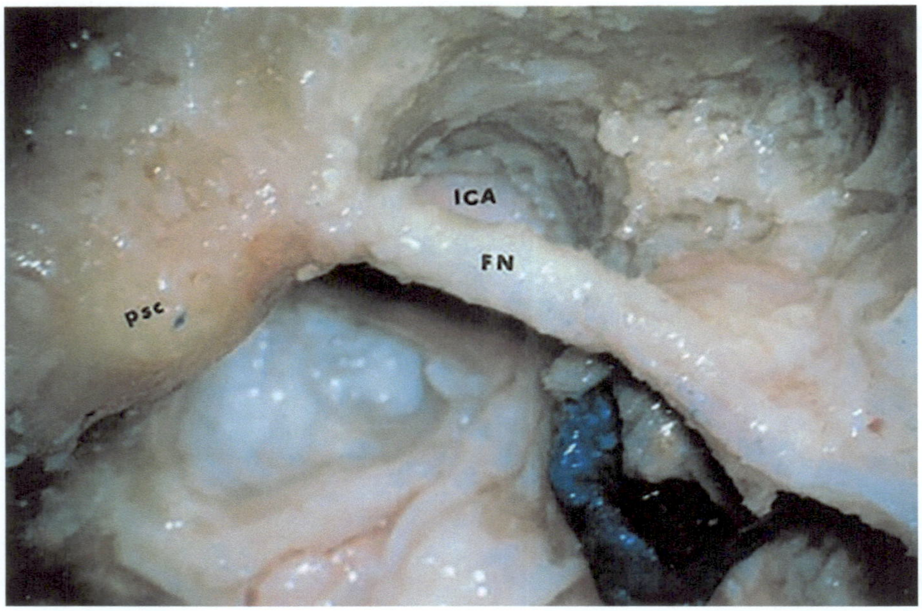

Fig. 13.56 Exposure of the infratubal internal carotid artery (ICA).

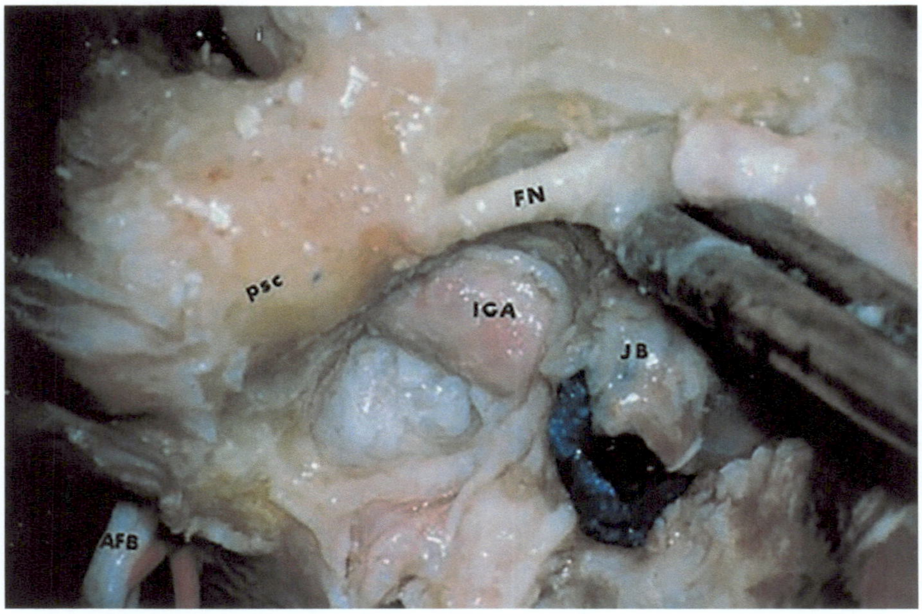

Fig. 13.57 Partial transposition of the facial nerve (FN) to afford more exposure of the IAC.

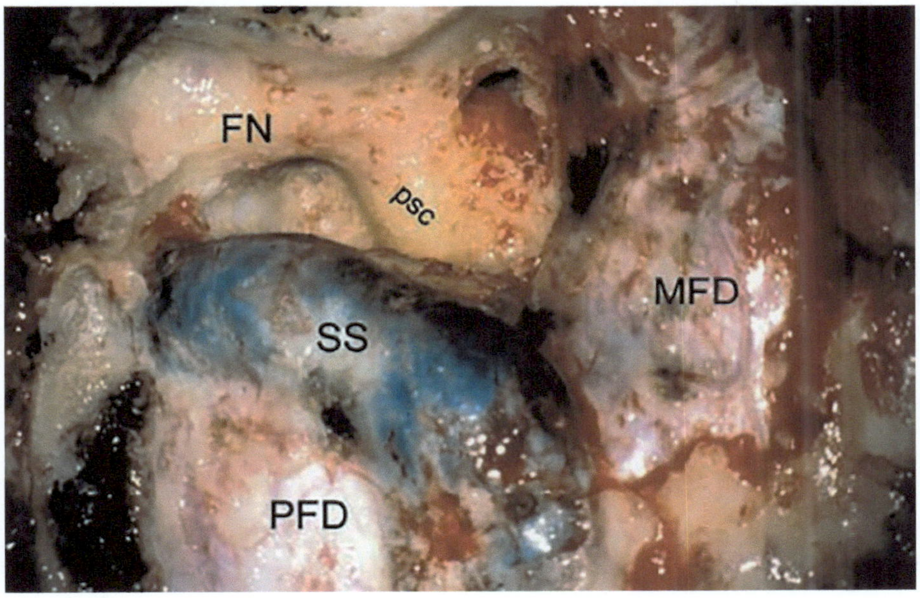

Fig. 13.58 Wide bone removal extending caudally.

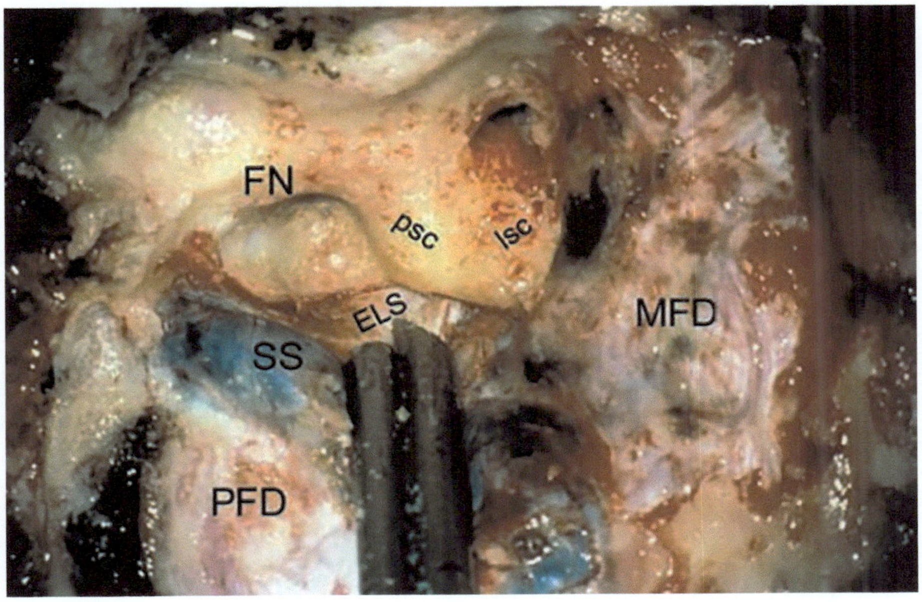

Fig. 13.59 The endolymphatic sac (ELS) is an approximate landmark for the posterior semicircular canal (PSC).

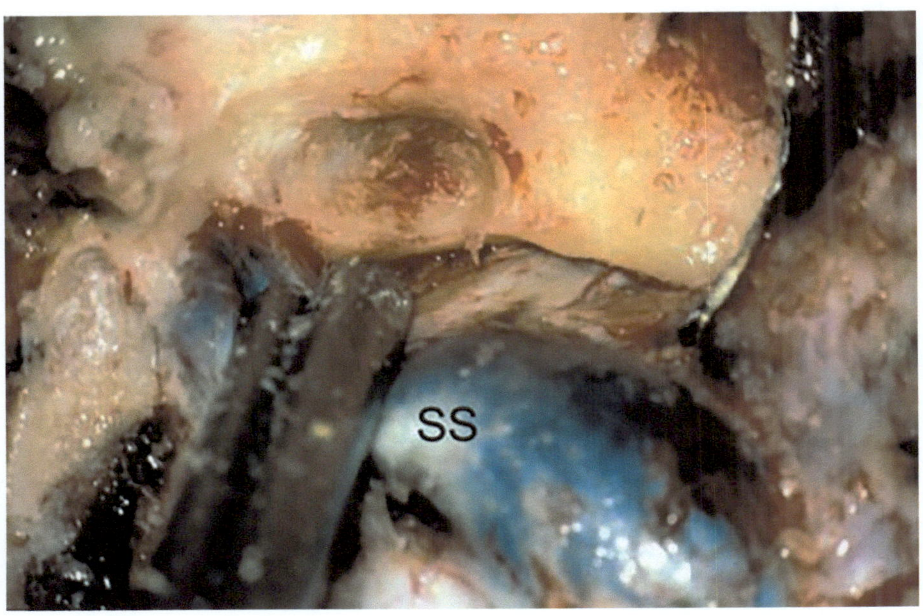

Fig. 13.60 Separation of the dura from the bone.

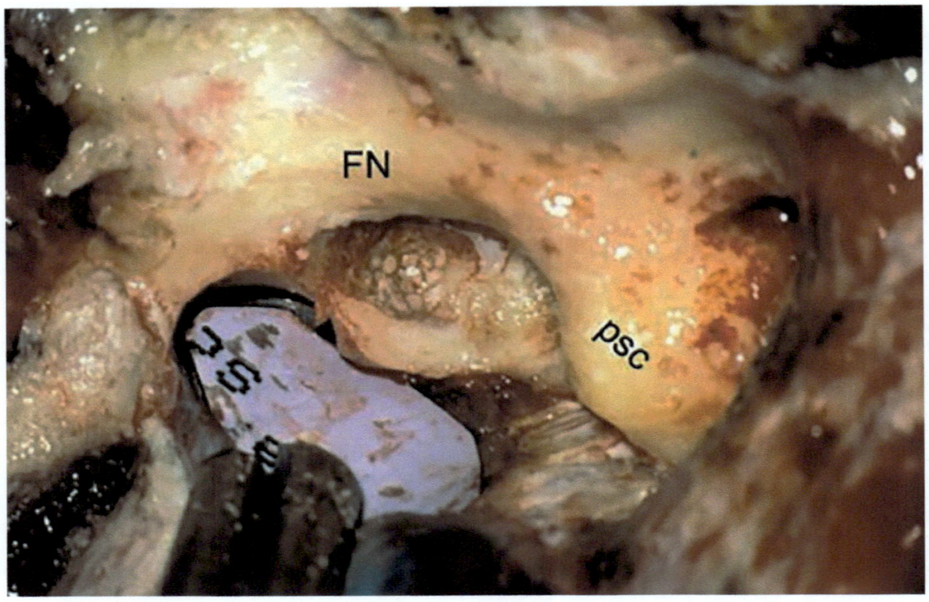

Fig. 13.61 Placement of aluminum to prevent dural injury.

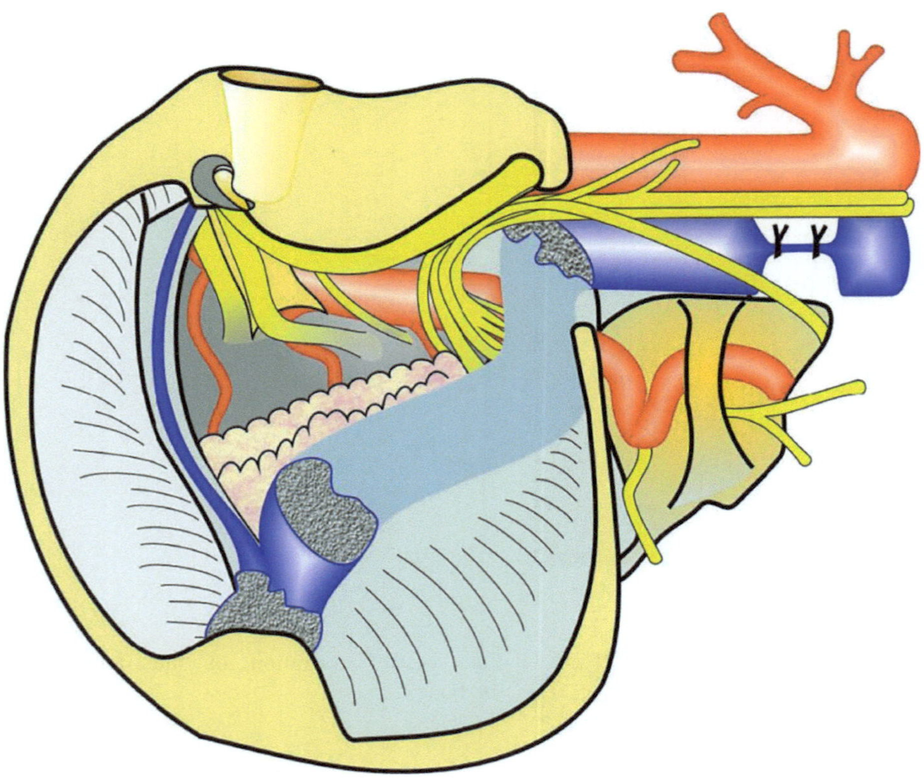

Fig. 13.62 The classic way of closing the sigmoid sinus by packing the sigmoid sinus intraluminally with Surgicel.

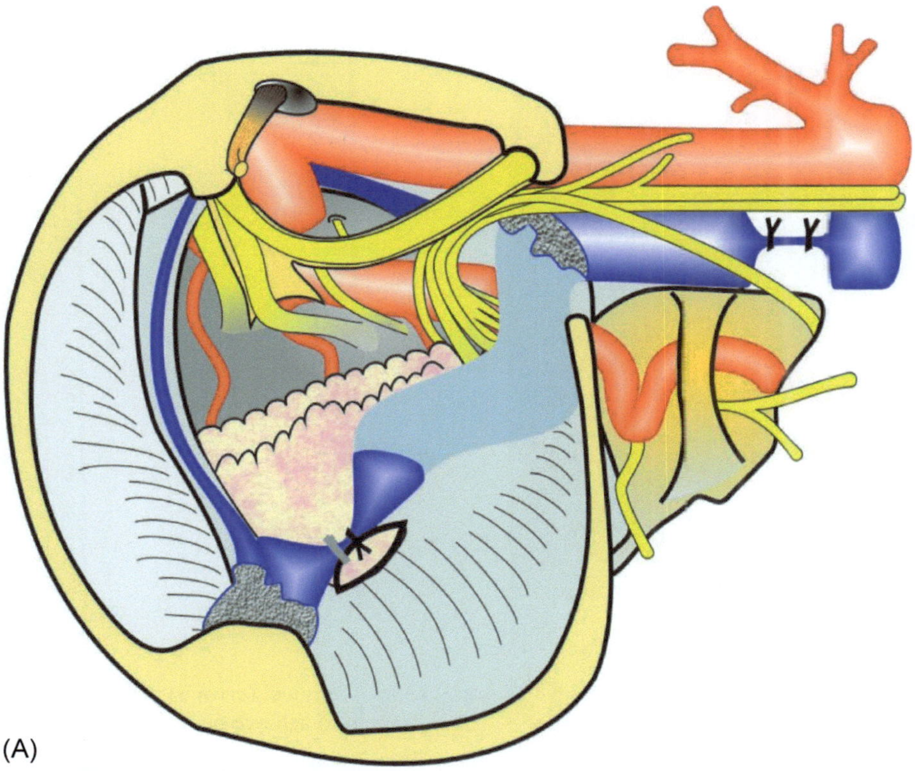

(A)

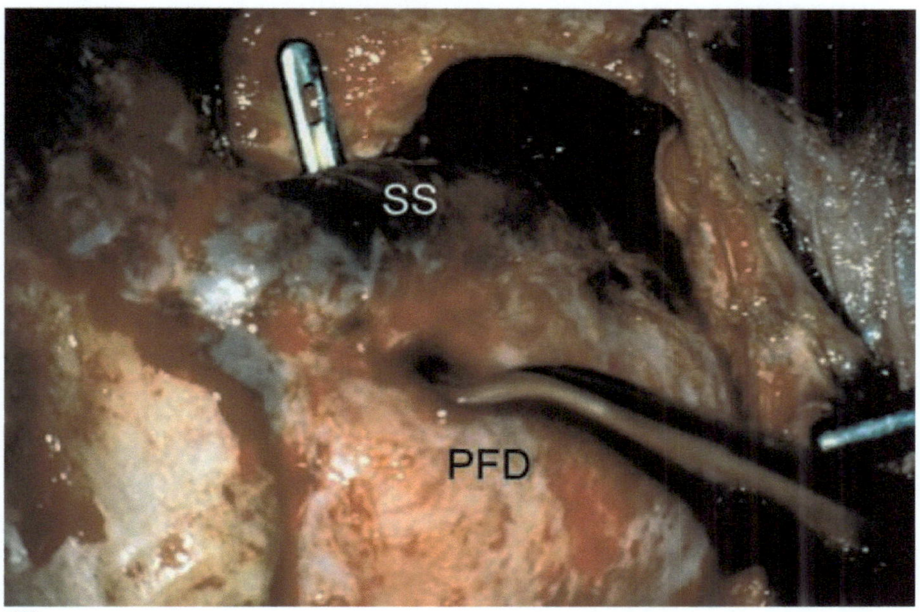

(B)

Fig. 13.63 (A,B) An alternative way of closing the sigmoid sinus.

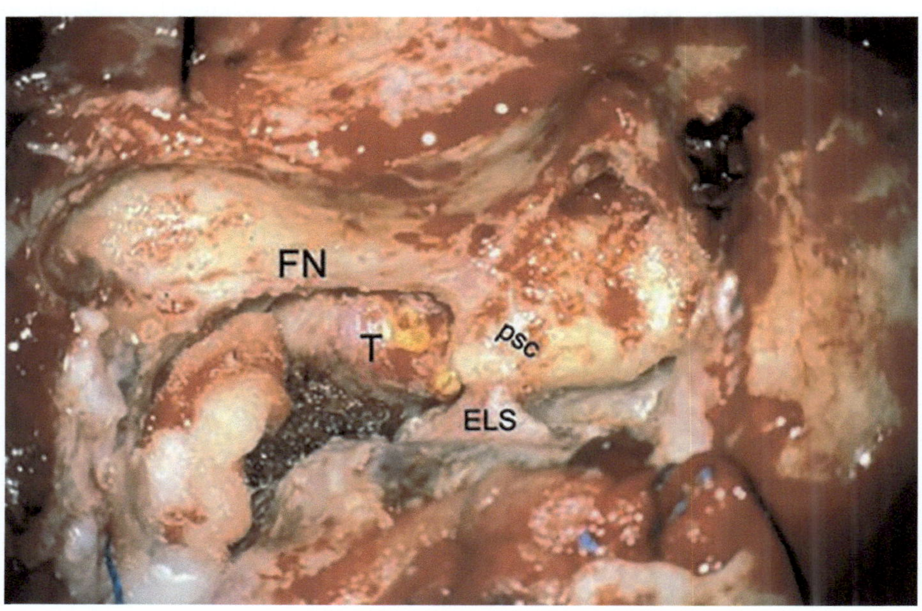

Fig. 13.64 The tumor (T) can be seen bulging into the *left* jugular foramen.

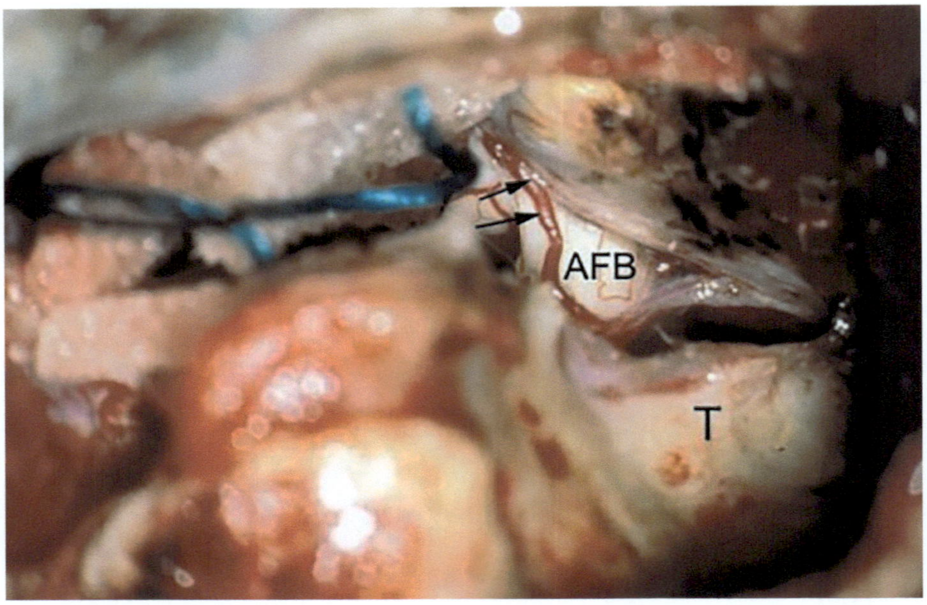

Fig. 13.65 Intact vasculature (*arrows*) over the acoustic facial bundle (AFB).

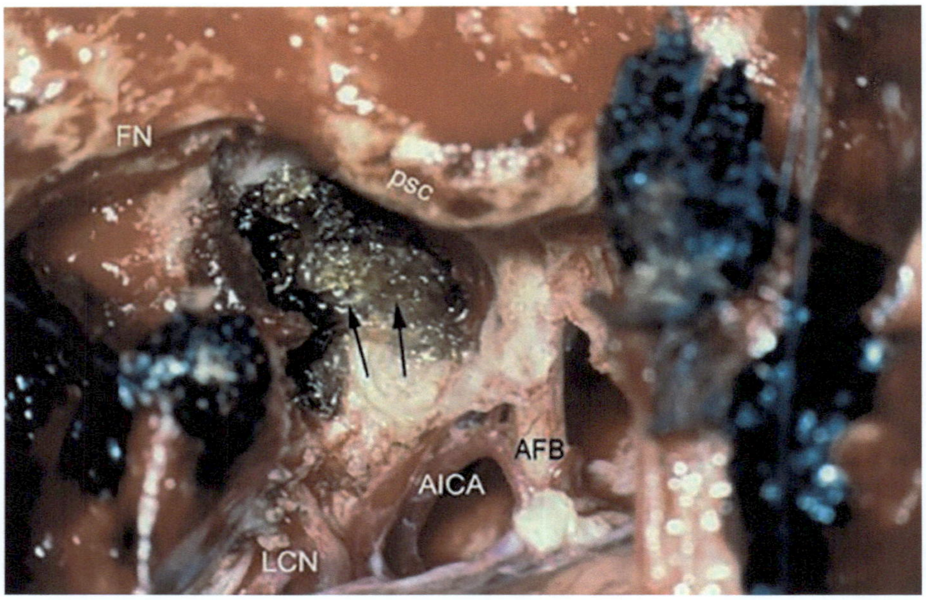

Fig. 13.66 Surgicel (*arrows*) placed in the jugular foramen to achieve hemostasis.

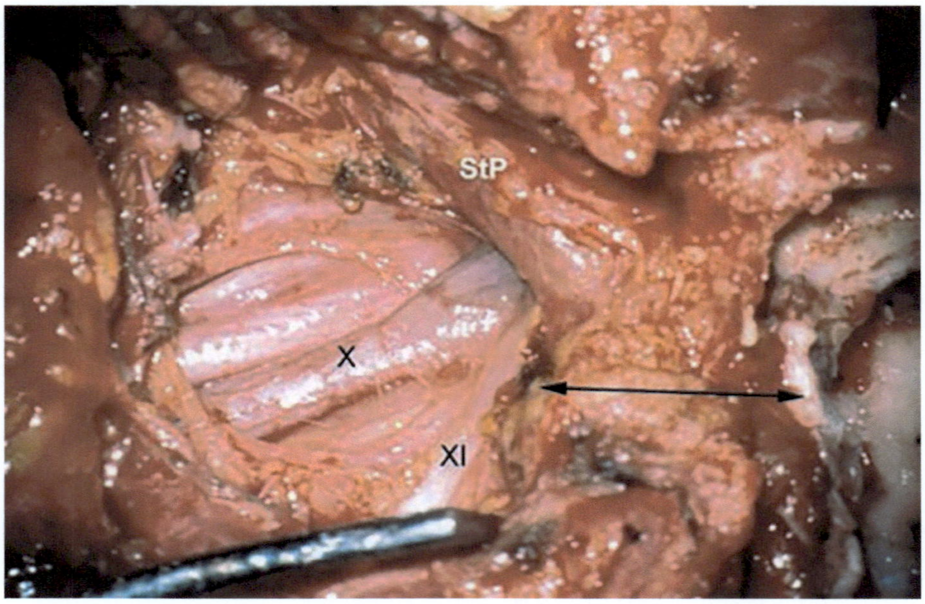

Fig. 13.67 *Left side* showing the styloid process (StP) and the intact cranial nerves in the neck. Note the bone and thick fibrous tissue separating the upper neck from the craniotomy site (double-headed line segment).

Combinations With the POTS Approach

The approach can be combined with other approaches, e.g.:

– The translabyrinthine approach in cases with preoperative ipsilateral sensorineural hearing loss (Fig. 13.68).

– The transotic approach in cases where the tumor shows massive involvement of the middle ear and internal carotid artery (Fig. 13.69). The following cases are an example of such a combination.

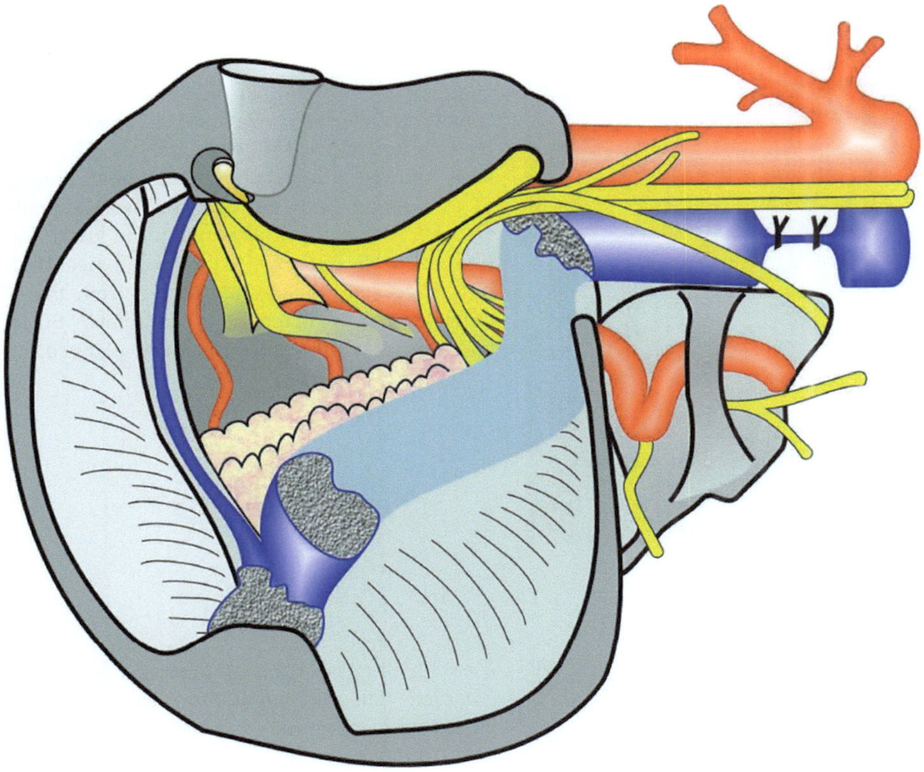

Fig. 13.68 Schematic drawing of a combined POTS and translabyrinthine approach.

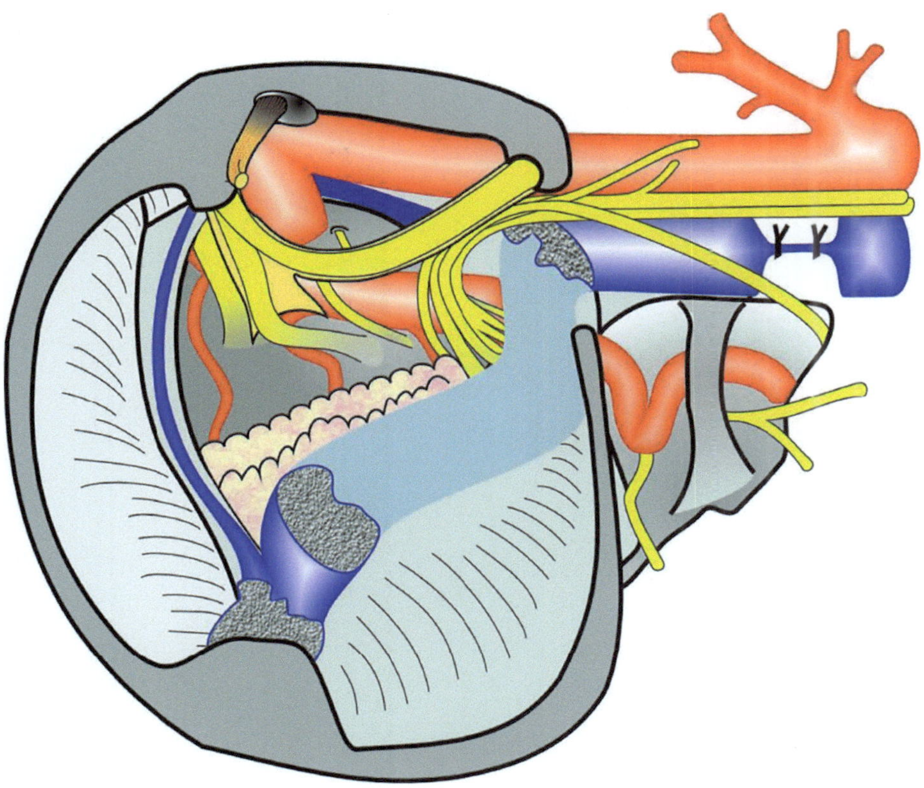

Fig. 13.69 Schematic drawing of a combined POTS and transotic approach.

Clinical Cases

Case 1: Lower Cranial Nerve Neuroma Removed With a Combined POTS+ Translabyrinthine Approach

See Figs. 13.70–13.82.

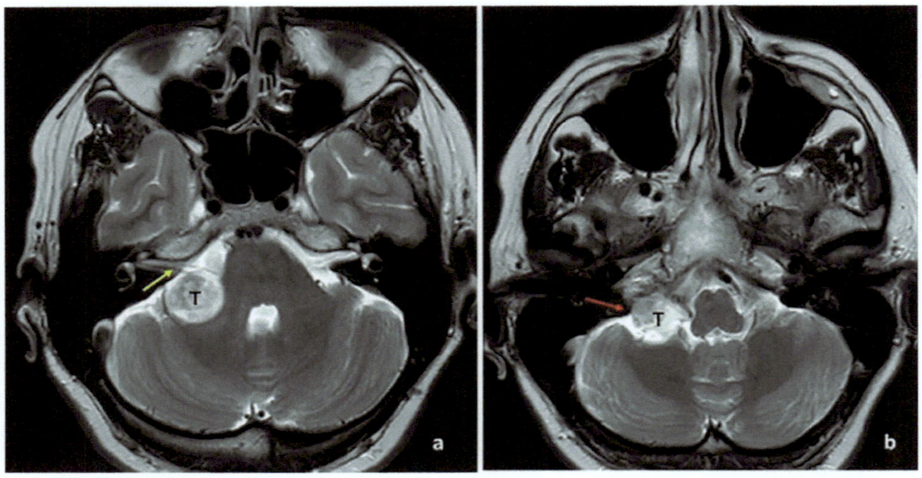

Fig. 13.70 This 50-year-old female patient presented with severe right-sided sensorineural hearing loss and vocal cord paralysis. The MRI scan showed a cerebellopontine angle tumor (T) involving both the acoustic facial bundle (A. *yellow arrow*) and the area of the lower cranial nerves (B. *red arrow*). The radiological findings suggested a neuroma arising from the X cranial nerve. Therefore, a combined POTS and translabyrinthine approach was planned.

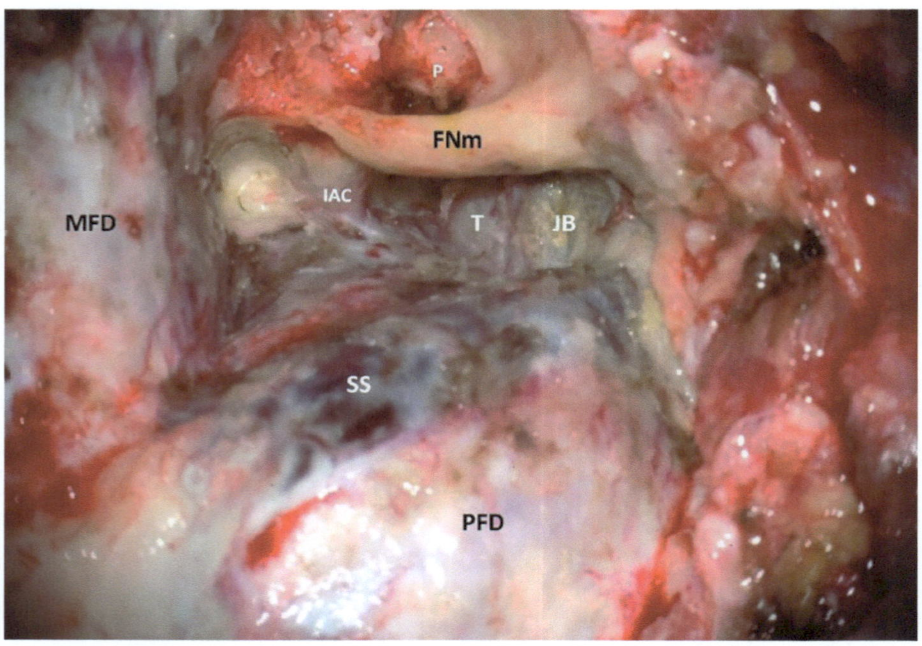

Fig. 13.71 The surgical approach is almost complete. The external auditory canal has been closed as a cul-de-sac and the middle ear has been removed. The internal auditory canal (IAC) has been identified and the middle fossa dura (MFD), posterior fossa dura (PFD) and sigmoid sinus (SS) have been decompressed. *FNm*, mastoid portion of the facial nerve; *T*, tumor; *P*, promontory.

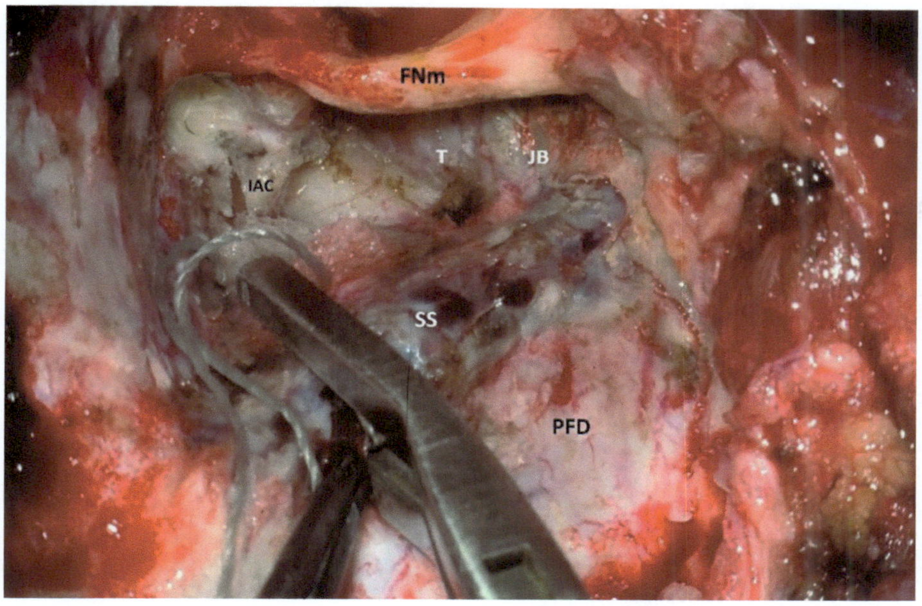

Fig. 13.72 After isolation of the sigmoid sinus (SS), the proximal part is closed with surgical clips. In this particular case, the internal jugular vein was not previously isolated in the neck to allow blood drainage from the emissary vein. *JB*, jugular bulb; *FNm*, mastoid portion of the facial nerve; *T*, tumor; *PFD*, posterior fossa dura.

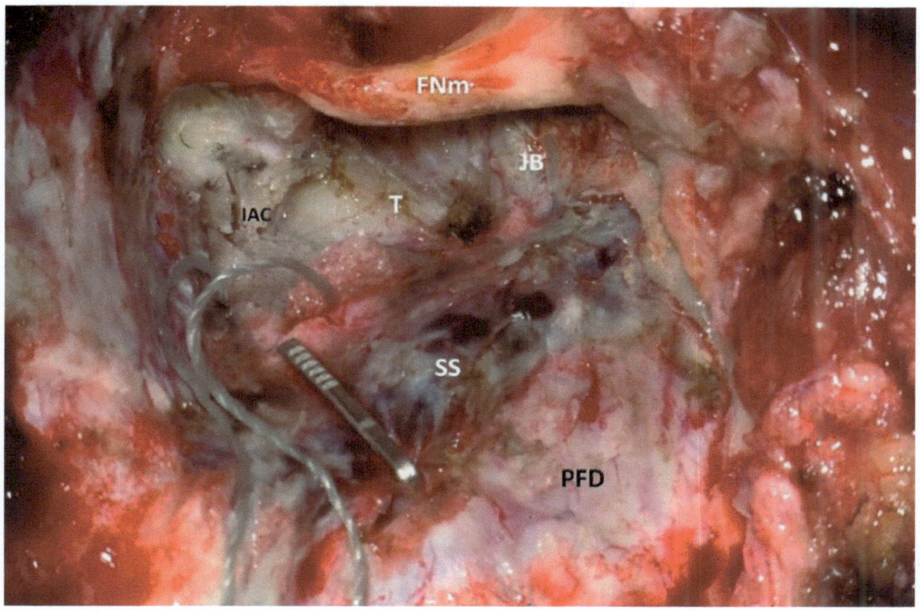

Fig. 13.73 Note the surgical clip. *JB*, jugular bulb; *FNm*, mastoid portion of the facial nerve; *T*, tumor; *PFD*, posterior fossa dura; *IAC*, internal auditory canal.

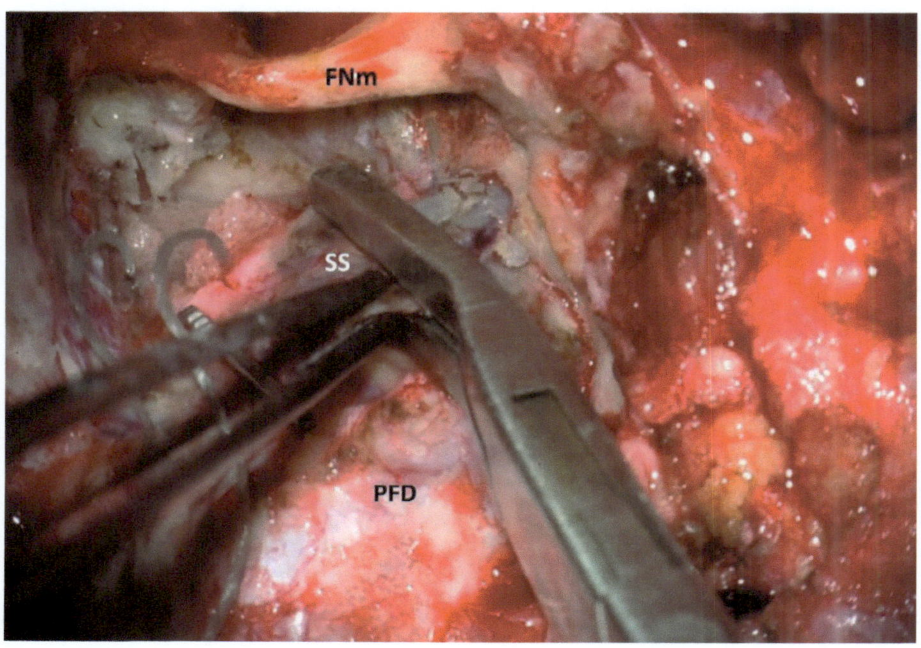

Fig. 13.74 Another surgical clip is placed inferiorly. *PFD*, posterior fossa dura; *FNm*, mastoid portion of the facial nerve; *SS*, sigmoid sinus.

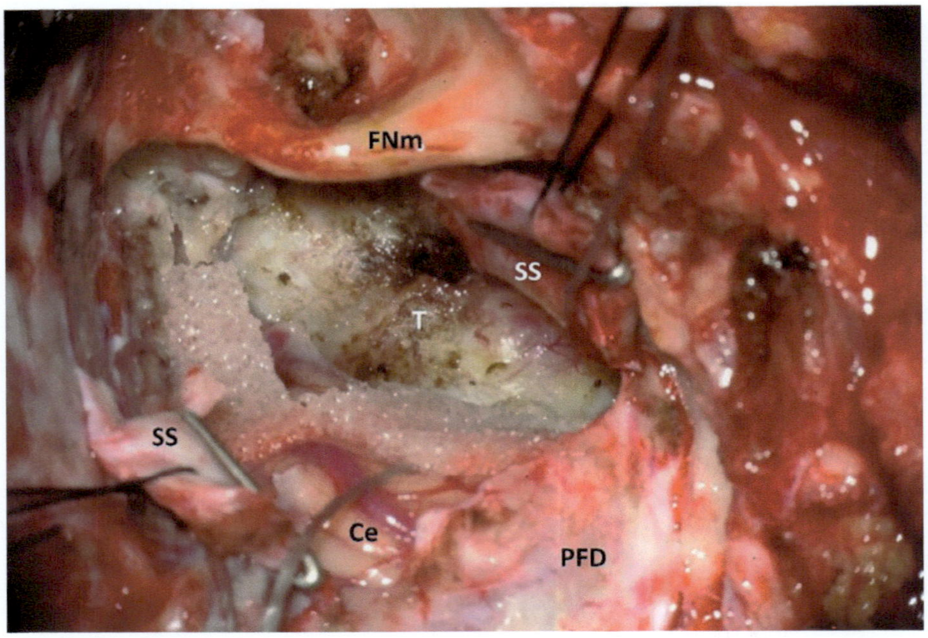

Fig. 13.75 The sigmoid sinus (SS) is cut between the two surgical clips and the ends are retracted. Note the tumor (T) occupying the cerebellopontine angle. *Ce*, cerebellum; *PFD*, posterior fossa dura; *FNm*, mastoid portion of the facial nerve.

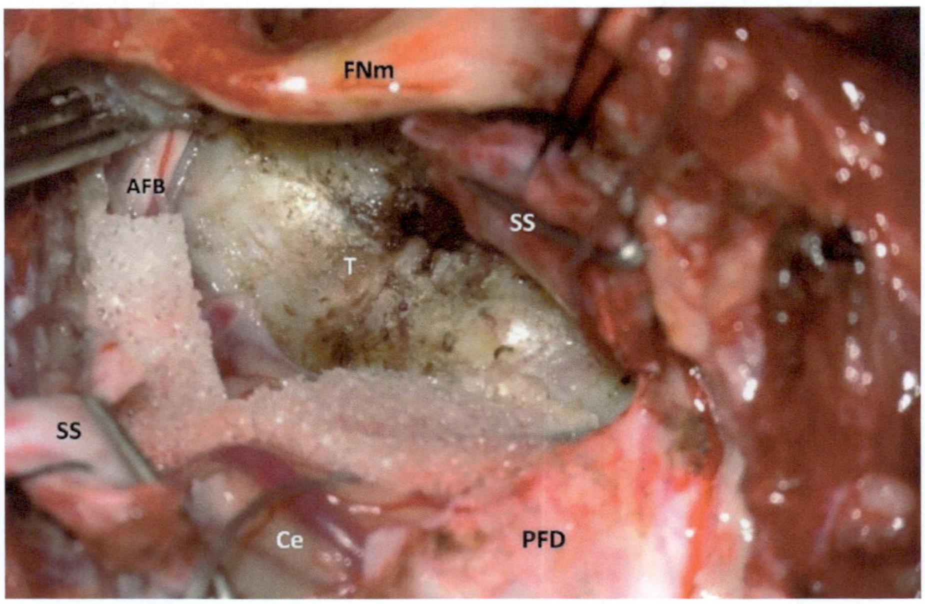

Fig. 13.76 The acoustic facial bundle (AFB) is isolated from the tumor (T). *Ce*, cerebellum; *PFD*, posterior fossa dura; *FNm*, mastoid portion of the facial nerve.

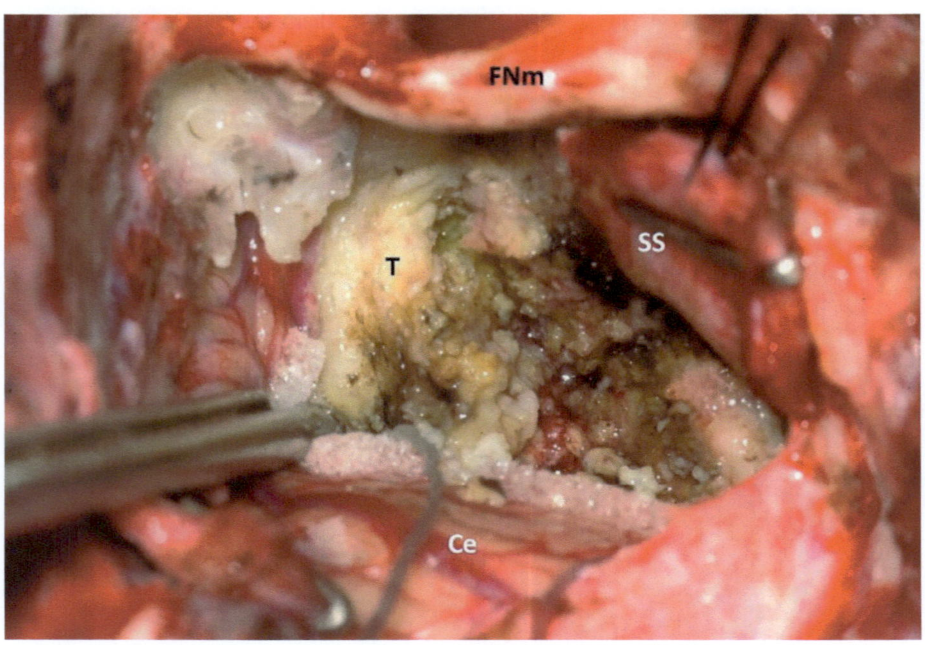

Fig. 13.77 Debulking of the tumor (T). *Ce*, cerebellum; *SS*, sigmoid sinus; *FNm*, mastoid portion of the facial nerve.

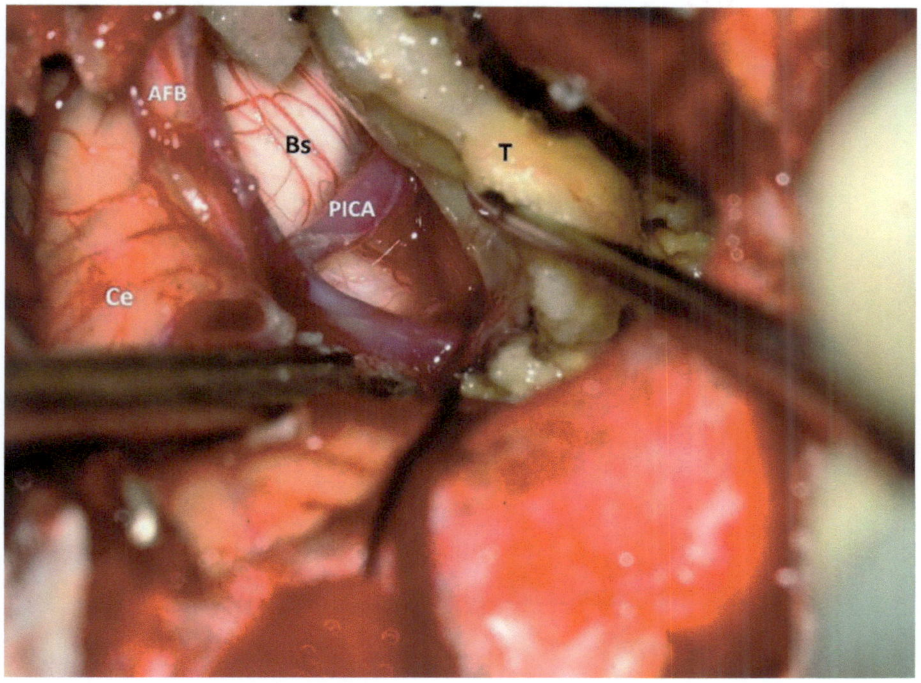

Fig. 13.78 After debulking, the tumor (T) is elevated from the brainstem (Bs). *Ce*, cerebellum; *AFB*, acoustic facial bundle; *PICA*, posterior inferior cerebellar artery.

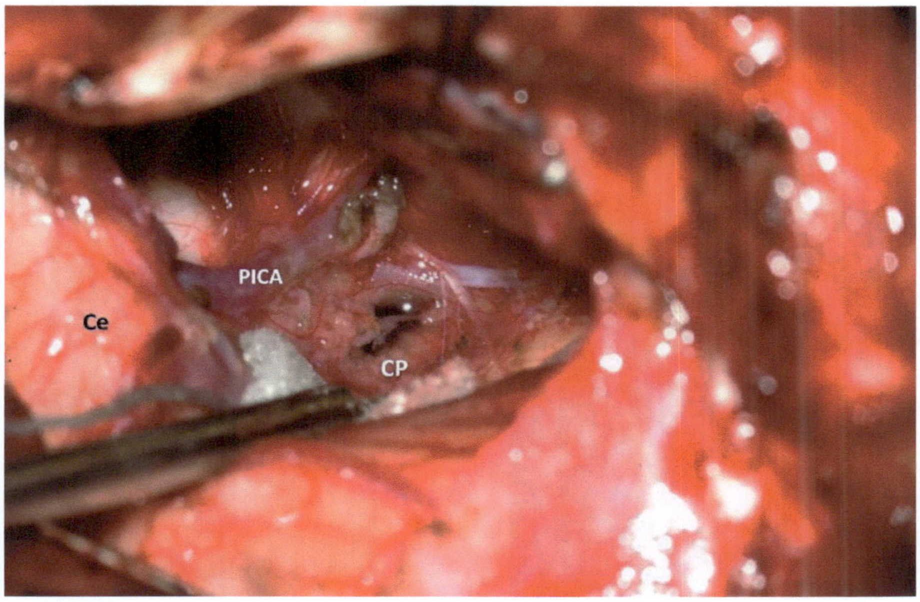

Fig. 13.79 The tumor (T) has been completely removed, exposing the choroid plexus (CP) and the entrance to the fourth ventricle. *Ce*, cerebellum; *PICA*, posterior inferior cerebellar artery.

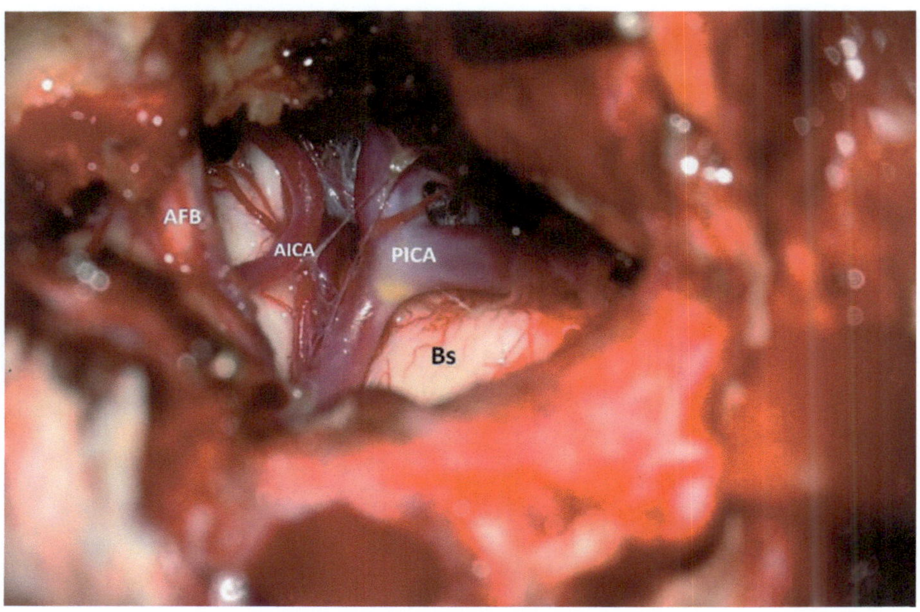

Fig. 13.80 Surgical view after complete tumor removal. *AFB*, acoustic facial bundle; *PICA*, posterior inferior cerebellar artery; *Bs*, brainstem; *AICA*, anterior inferior cerebellar artery.

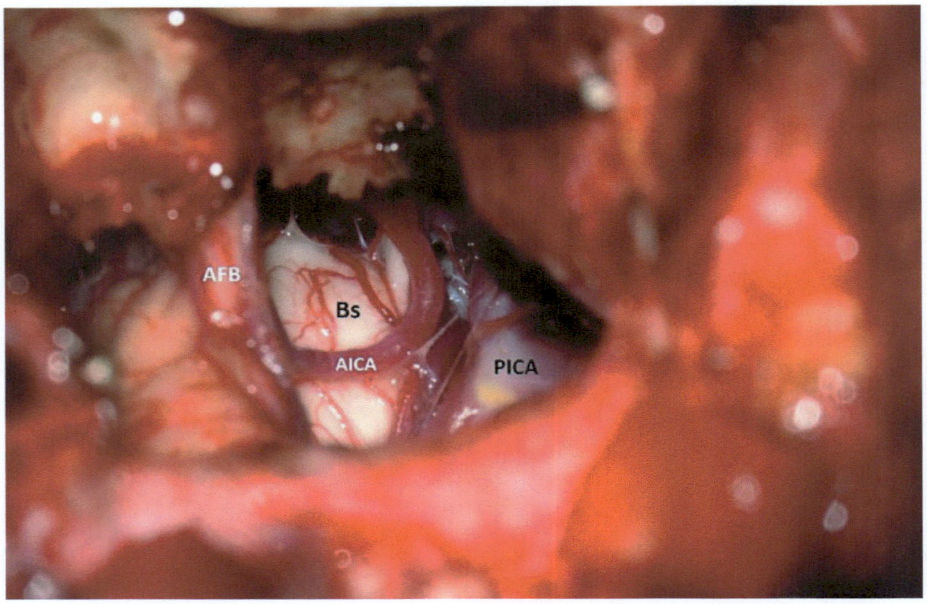

Fig. 13.81 Surgical view after complete tumor removal. Note the course of the anterior inferior cerebellar artery (AICA) originating from the basilar artery. *AFB*, acoustic facial bundle; *PICA*, posterior inferior cerebellar artery; *Bs*, brainstem.

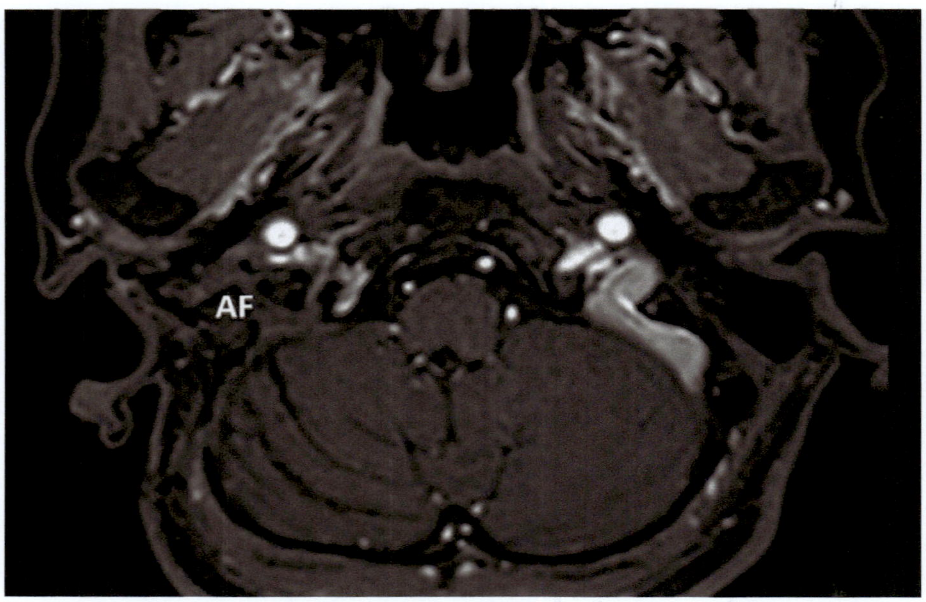

Fig. 13.82 Postoperative T1W+Gd enhancement MRI showing complete tumor removal. The cavity is filled with abdominal fat (AF).

Case 2: Lower Cranial Nerve Neuroma Removed With a Combined POTS+ Translabyrinthine Approach

See Figs. 13.83–13.95.

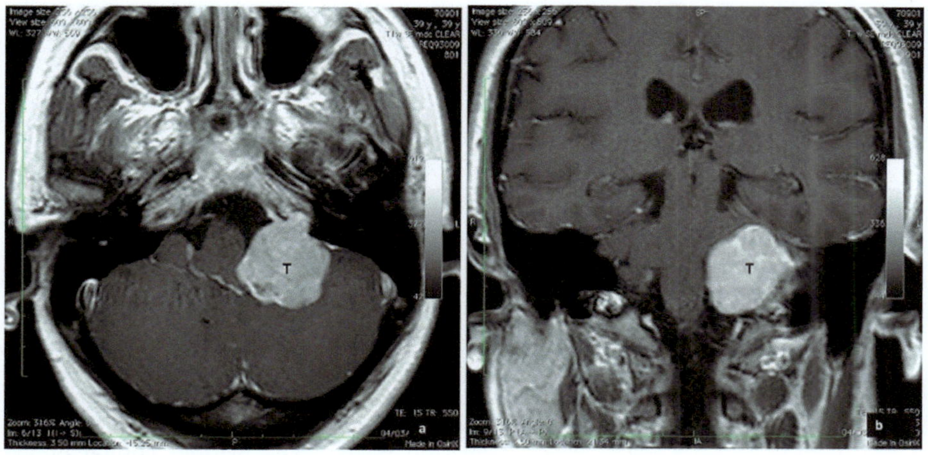

Fig. 13.83 Another case of a 45 year-old patient with a neuroma of the *left lower* cranial nerves. The patient presented with left IX, X, XII cranial nerve palsies and sensorineural hearing loss. *T*, tumor. (A) Brain T1W-MRI+Gd enhancement axial view, (B) Brain T1W-MRI+Gd enhancement coronal view.

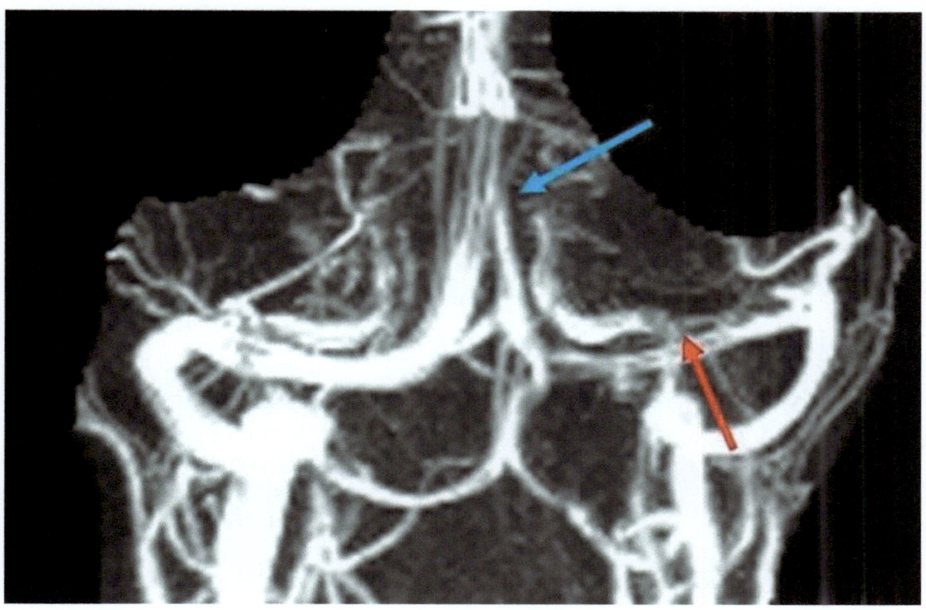

Fig. 13.84 MR Venogram showing reduced blood flow in the *left* sigmoid sinus (*red arrow*) with a weak connection at the level of the torcular erofili (*light blue arrow*).

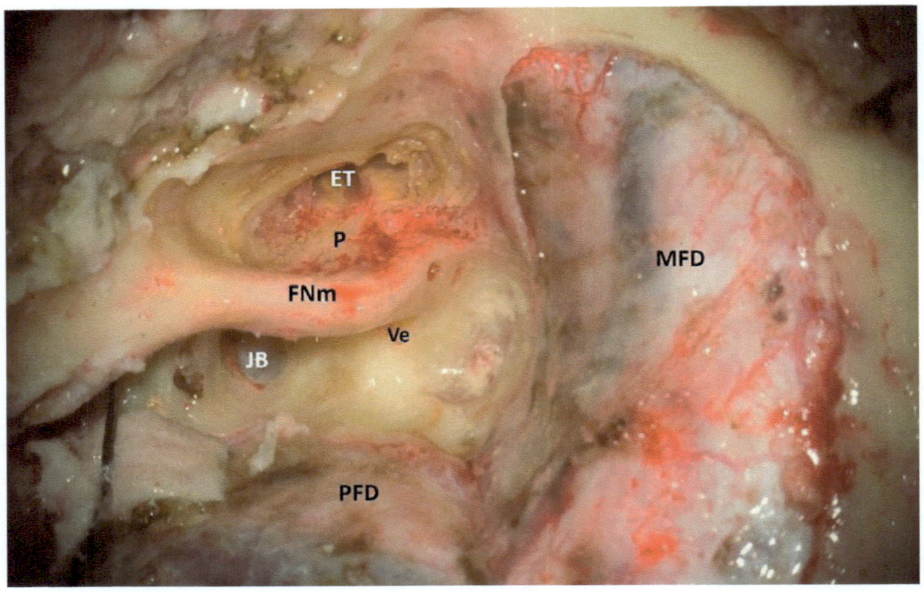

Fig. 13.85 A subtotal petrosectomy and labyrinthectomy have been performed. Note the extension of the middle and posterior fossa craniotomies. *ET*, eustachian tube; *P*, promontory; *FNm*, mastoid portion of the facial nerve; *JB*, jugular bulb; *Ve*, vestibule, *MFD*, middle fossa dura, *PFD*, posterior fossa dura.

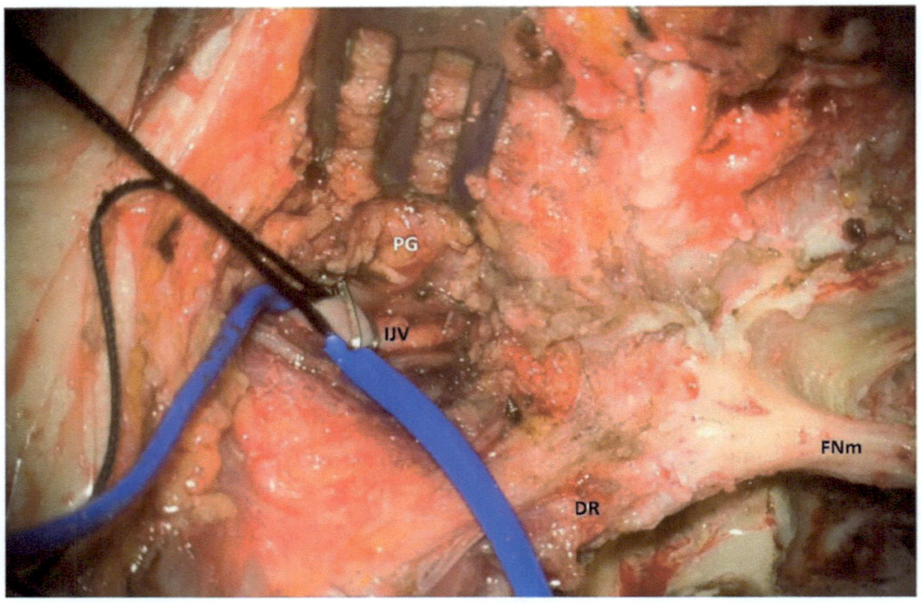

Fig. 13.86 The internal jugular vein (IJV) was been isolated and closed with surgical clips at the level of the upper neck. *PG*, parotid gland; *DR*, digastric ridge; *FNm*, mastoid portion of the facial nerve.

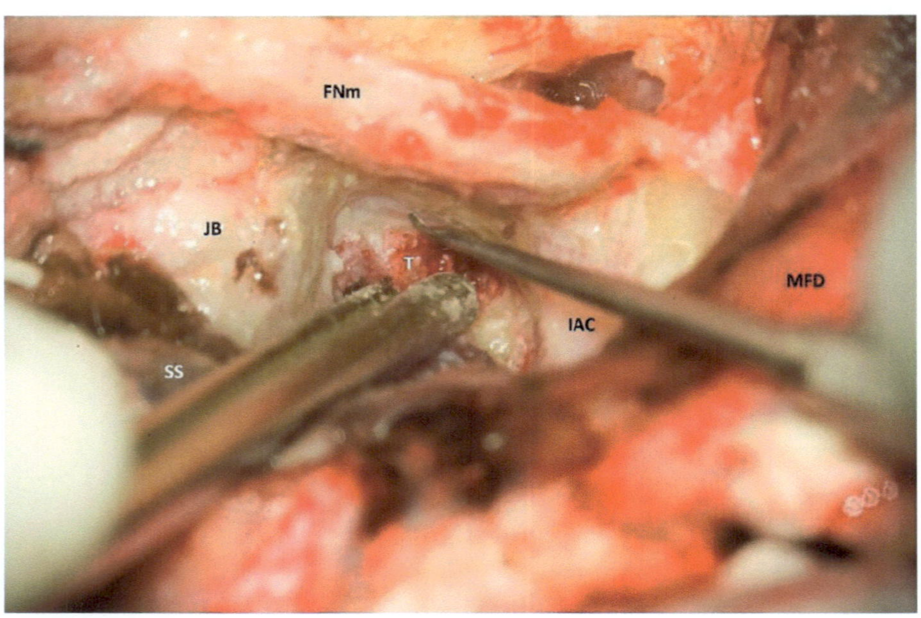

Fig. 13.87 The tumor (T) begins to appear medial to the mastoid portion of the facial nerve (FNm). *JB*, jugular bulb; *SS*, sigmoid sinus; *IAC*, internal auditory canal; *MFD*, middle fossa dura.

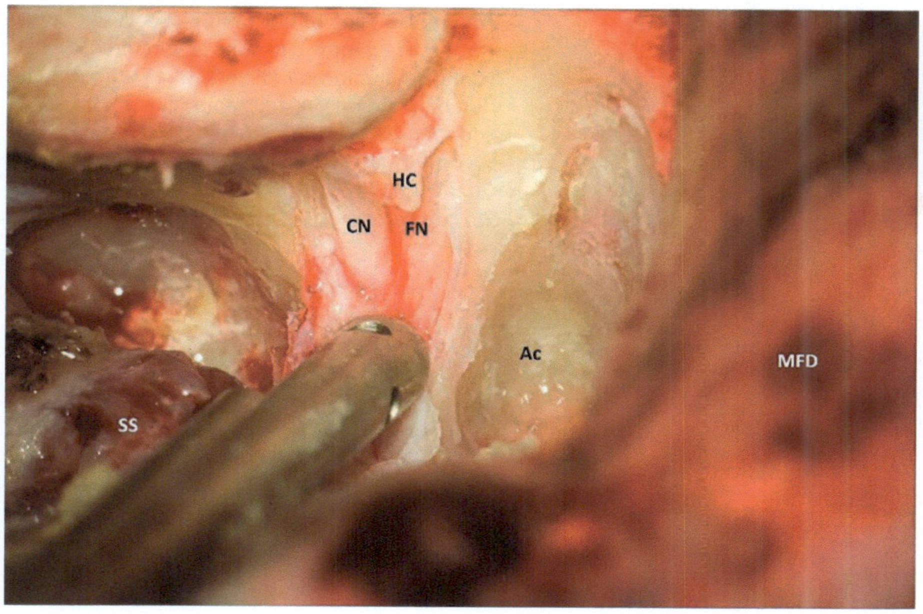

Fig. 13.88 The dura of the internal auditory canal is opened and the vestibular nerves are reflected inferiorly to expose the facial (FN) and the cochlear (CN) nerves. *AC*, apical cells; *SS*, sigmoid sinus; *MFD*, middle fossa dura; *HC*, horizontal crest.

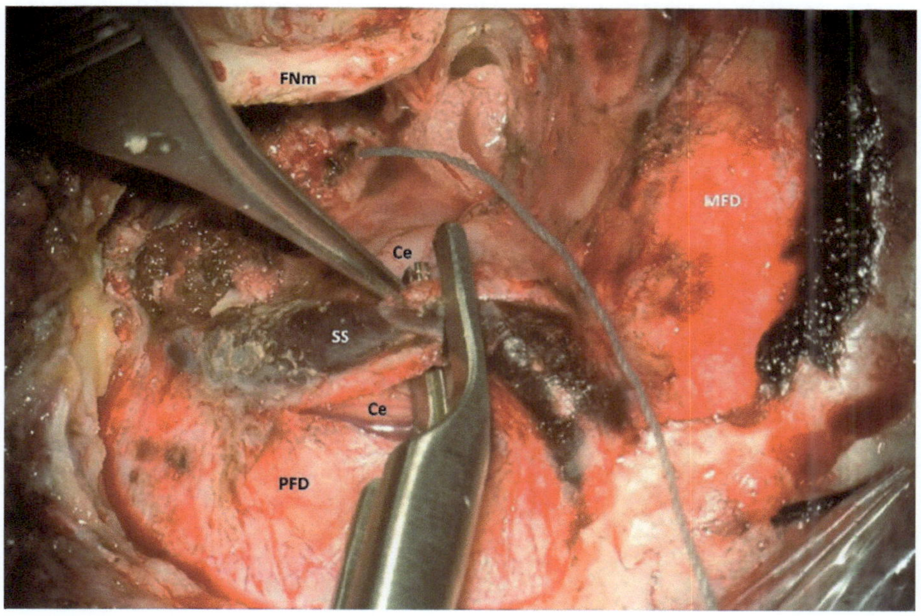

Fig. 13.89 The sigmoid sinus (SS) is isolated and closed with surgical clips. *MFD*, middle fossa dura; *PFD*, posterior fossa dura; *Ce*, cerebellum; *FNm*, mastoid portion of the facial nerve.

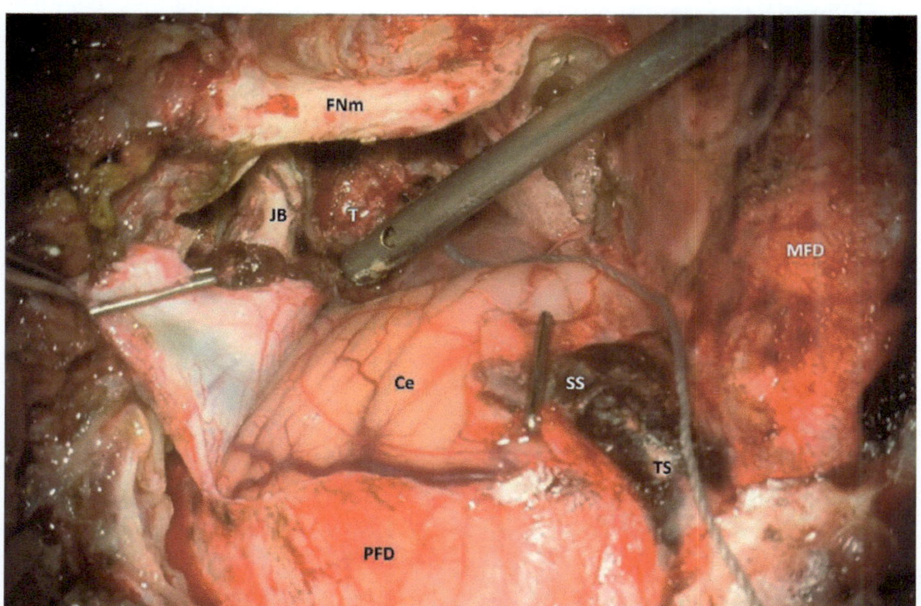

Fig. 13.90 The posterior fossa dura is widely opened after cutting the sigmoid sinus (SS). Note the extensive control of the cerebellum (Ce). *MFD*, middle fossa dura; *PFD*, posterior fossa dura; *TS*, transverse sinus; *JB*, jugular bulb; *FNm*, mastoid portion of the facial nerve.

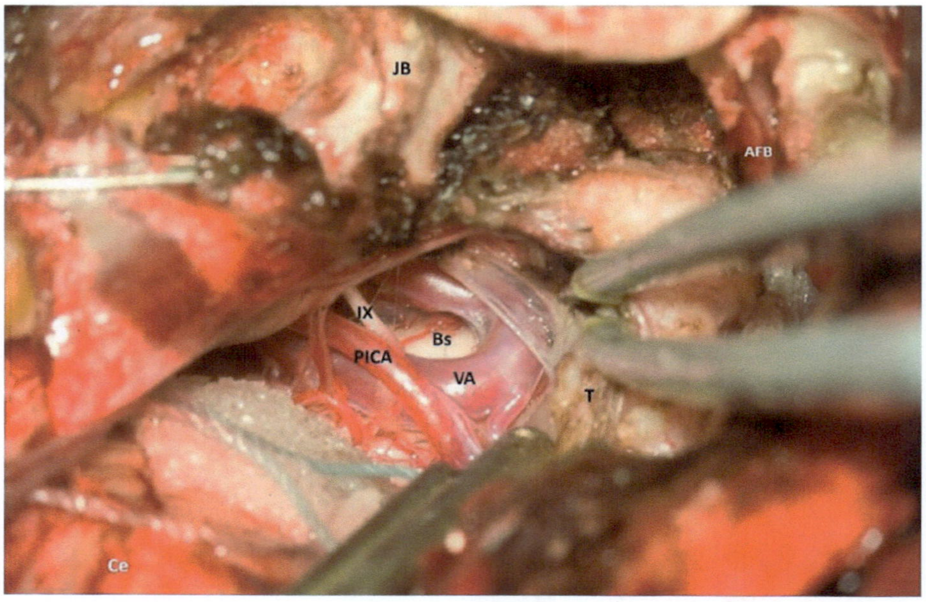

Fig. 13.91 The tumor (T) involving the posterior fossa is removed and the arachnoid dissected from the main intradural vessels. In this case, the vertebral artery (VA) was dominant on the tumor side. *Bs*, brainstem; *PICA*, posterior inferior cerebellar artery; *IX*, glossopharyngeal nerve; *JB*, jugular bulb; *AFB*, acoustic facial bundle; *Ce*, cerebellum.

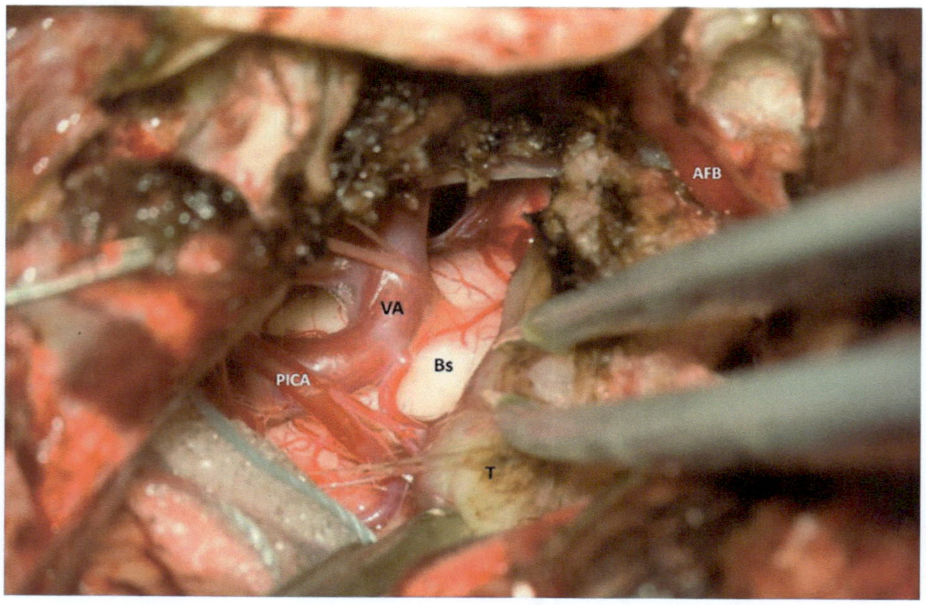

Fig. 13.92 The tumor (T) is further dissected. *VA*, vertebral artery; *PICA*, posterior inferior cerebellar artery; *AFB*, acoustic facial bundle; *Bs*, brainstem.

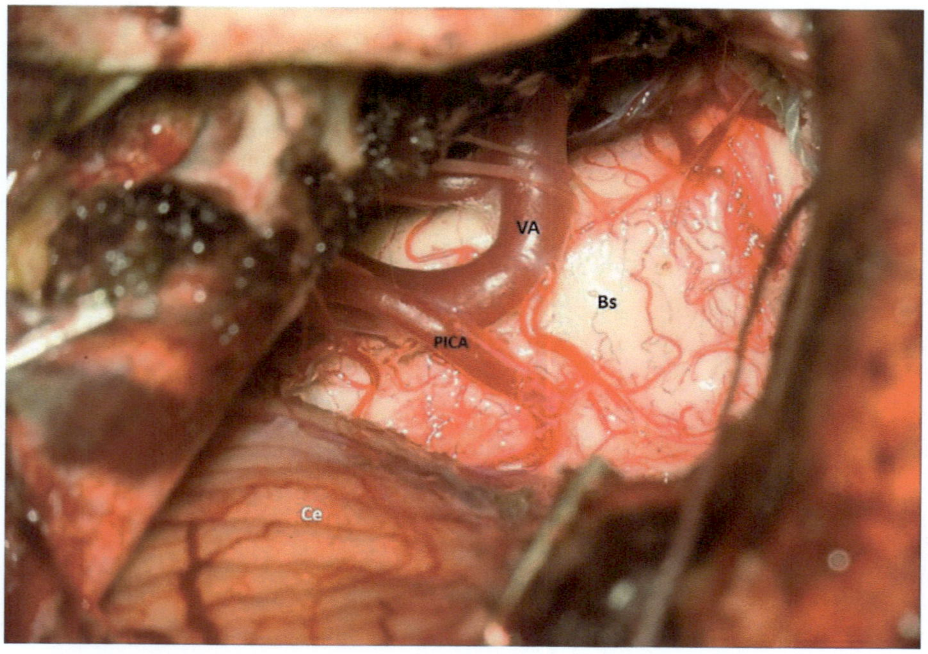

Fig. 13.93 The tumor is completely removed. *VA*, vertebral artery; *PICA*, posterior inferior cerebellar artery; *Bs*, brainstem; *Ce*, cerebellum.

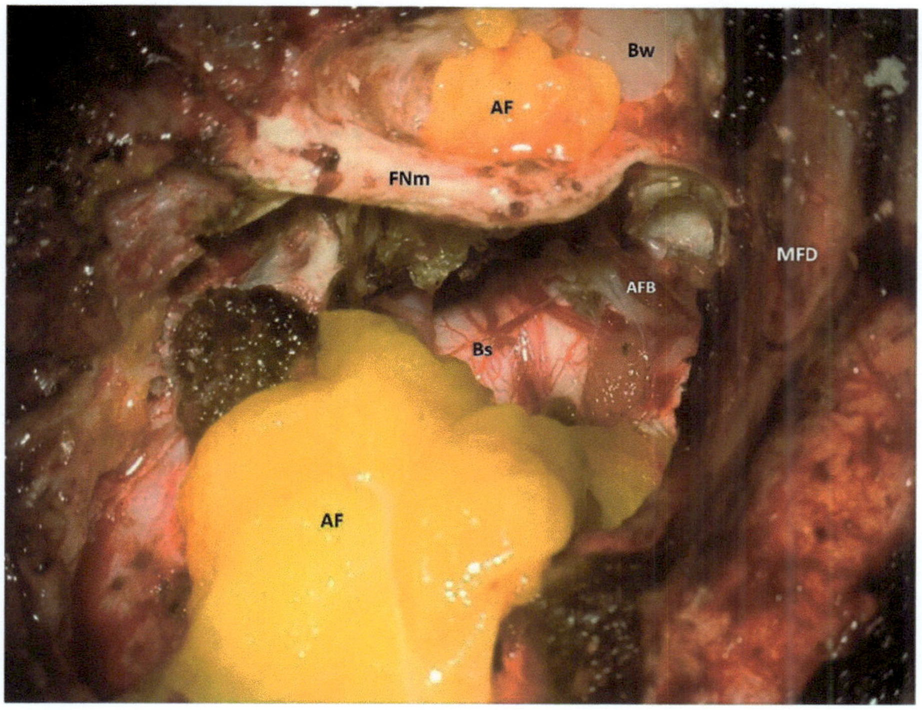

Fig. 13.94 At the end of the procedure the eustachian tube is closed with periosteum and bone wax (Bw). The surgical cavity is then filled with abdominal fat (AF). *MFD*, middle fossa dura; *Bs*, brainstem; *AFB*, acoustic facial bundle; *FNm*, mastoid portion of the facial nerve.

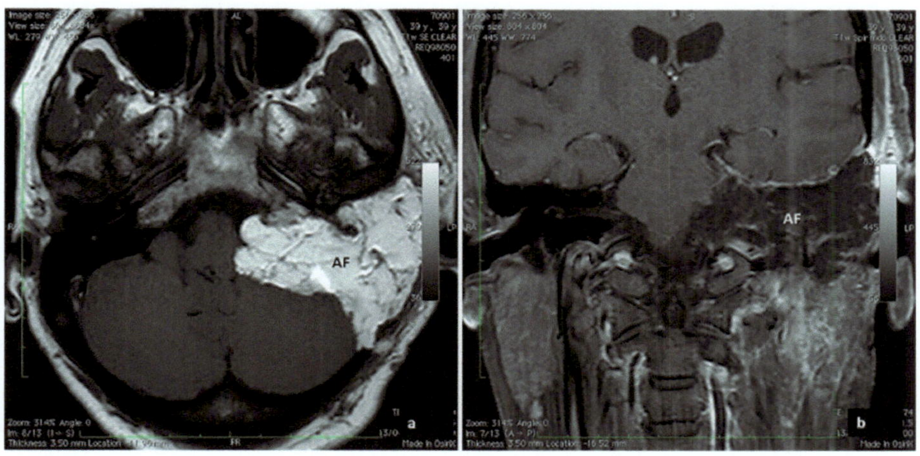

Fig. 13.95 Postoperative brain MRI. (A) T1W sequence without fat suppression, axial cut, showing the abdominal fat (AF) within the surgical cavity, (B) T1W+Gd enhancement sequence, coronal cut, showing no showing no residual or recurrent tumor.

REFERENCES

[1] Mann W, Amedee R, Gilsbach J, et al. Transsigmoid approach for tumors of the jugular foramen. Skull Base 1991;1(3):137–41.

[2] Mazzoni A, Sanna M. A posterolateral approach to the Skull Base: the petro-occipital transsigmoid approach. Skull Base 1995;5(03):157–67. https://doi.org/10.1055/s-2008-1058930.

[3] Anderson S, Panizza B. Petro-occipital transsigmoid approach. Oper Tech Otolaryngol Head Neck Surg 2013;24 (3):163–8. https://doi.org/10.1016/j.otot.2013.08.001.

[4] Sanna M, Bacciu A, Falconi M, et al. Surgical management of jugular foramen schwannomas with hearing and facial nerve function preservation: a series of 23 cases and review of the literature. Laryngoscope 2006;116:2191–204.

[5] Cokkeser Y, Wachter B, Fayad J. Conservative facial nerve management in jugular foramen schwannomas. Am J Otol 2000;21:270–4.

[6] Wilson M, Hillman T, Wiggins R, et al. Jugular foramen schwannomas: diagnosis, management and outcomes. Laryngoscope 2005;115:1386–92.

[7] Cokkeser Y, Wachter B, Fayad J. Conservative facial nerve management in jugular foramen schwannomas. Am J Otol 2000;21:270–4.

[8] Franklin D, Moore G, Fisch U. Jugular foramen peripheral nerve sheath tumors. Laryngoscope 1989;99:1081–7.

[9] Sanna M, Bacciu A, Falconi M, et al. Surgical management of the jugular foramen meningiomas: a series of 13 cases and review of the literature. Laryngoscope 2007;117:1710–9.

[10] Kaye AH, Hahn JF, Kinney SE, Hardy RW, Bay JW. Jugular foramen schwannomas. J Neurosurg 1984;60(5):1045–53.

[11] Lee JH. Meningiomas: diagnosis, treatment, and outcome. Am J Neuroradiol 2009;30(10):E157. https://doi.org/10.3174/ajnr.A1775.

[12] Zanoletti E, Faccioli C, Cazzador D, Mazzoni A, Martini A. Bilateral chondrosarcoma of the jugular foramen: literature review and personal experience. Eur Arch Otorhinolaryngol 2015;272(10):3071–5. https://doi.org/10.1007/s00405-015-3513-0 [Epub 2015 Feb 3].

[13] Zanoletti E, Cazzador D, Faccioli C, Martini A, Mazzoni A. Closure of the sigmoid sinus in lateral skull base surgery. Acta Otorhinolaryngol Ital 2013;34(3):184–8.

CHAPTER 14

The Extreme Lateral Approach (Far Lateral, Transcondylar)

Rationale

Access to craniovertebral junction lesions may be through a midline suboccipital approach, an anterior transoral approach, or the recently developed lateral approaches.

- *Access through traditional posterior approaches* requires significant brainstem retraction. Extradural control of the vertebral artery is not obtained, nor can an extradural extension of the lesion be removed through this route.
- *Access through the anterior transoral approaches* does not afford adequate control of the vertebral arteries. When used for intradural pathologies, it has the disadvantage of breaking the sterility by crossing a septic field.
- *Access through the lateral approaches* depends primarily on bone removal with or without mobilization of the vertebral artery, thereby obviating the need for brainstem retraction. It affords excellent control of the vertebral artery both extradurally and intradurally and permits the removal of intradural and extradural lesions in this area.

A schematic drawing of the approach and the extent of bone removal is shown in Fig. 14.1.

Indications

- Anterior and lateral lesions of the craniospinal junction, e.g., meningioma, chordoma, chondrosarcomas, some glomus jugulare tumors with vertebral artery involvement, and vertebrobasilar aneurysms.

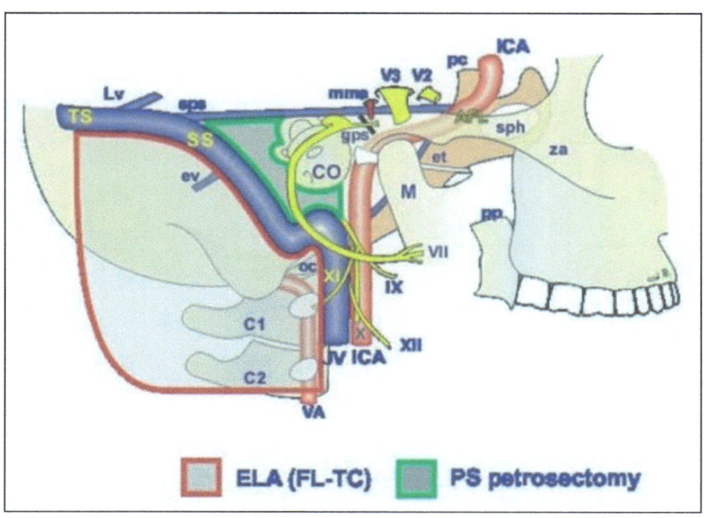

Fig. 14.1 Schematic drawing of the extreme lateral approach. *AFL*, anterior foramen lacerum; *C1*, first cervical vertebra; *CO*, cochlea; *et*, eustachian tube; *ev*, emissary vein; *gps*, greater petrosal nerve; *ICA*, internal carotid artery; *JV*, jugular vein; Lv, vein of Labbé; *M*, mandible; *mma*, middle meningeal artery; *oc*, occipital condyle; *pc*, clinoid process; *pp*, pterygoid process; *sph*, sphenoid; *sps*, superior petrosal sinus; *TA*, transverse process of the atlas; *TS*, transverse sinus; *za*, zygomatic arch; *V2*, maxillary nerve; *V3*, mandibular nerve; *VII*, facial nerve; *IX*, glossopharyngeal nerve.

Atlas of Microsurgery of the Lateral Skull Base. https://doi.org/10.1016/B978-0-443-19083-4.00014-3

Surgical Anatomy

1. The muscles in this region can be grouped into three levels. The sternocleidomastoid is the most superficial. More medially, and perpendicular to the sternocleidomastoid, lies the splenius capitis, and attached to its medial surface is the slender longissimus capitis. Anteriorly at the same level, the levator scapulae can be seen (Figs. 14.2 and 14.3).

2. The deepest layer is formed of the superior and inferior oblique muscles (Figs. 14.4 and 14.5).

3. The transverse process of the atlas is an important landmark. It gives attachment to the levator scapulae as well as the superior and inferior oblique muscles (Fig. 14.6). Just anterior lies the internal jugular vein and the accessory nerve. Removing the transverse process of the atlas reveals the vertebral artery running in the foramen transversarium.

4. After leaving the transverse process of the axis, the vertebral artery bends sideways before entering the foramen transversarium of the atlas. It then curves backward, indenting the posterior arch of the atlas (Fig. 14.7).

5. Landmarks for identification of the vertebral artery at this level include the oblique muscles and the transverse process of the atlas. The C2 nerve root is another important landmark since it constantly crosses the artery between C1 and C2 (Fig. 14.8).

6. To gain intradural access, the artery bends again anteromedially to pierce the dura posteromedial to the occipital condyle. At this point, the dura is firmly attached to the artery (Fig. 14.9).

7. At the point where the artery pierces the dura, the spinal accessory nerve is closely related to it (Fig. 14.10).

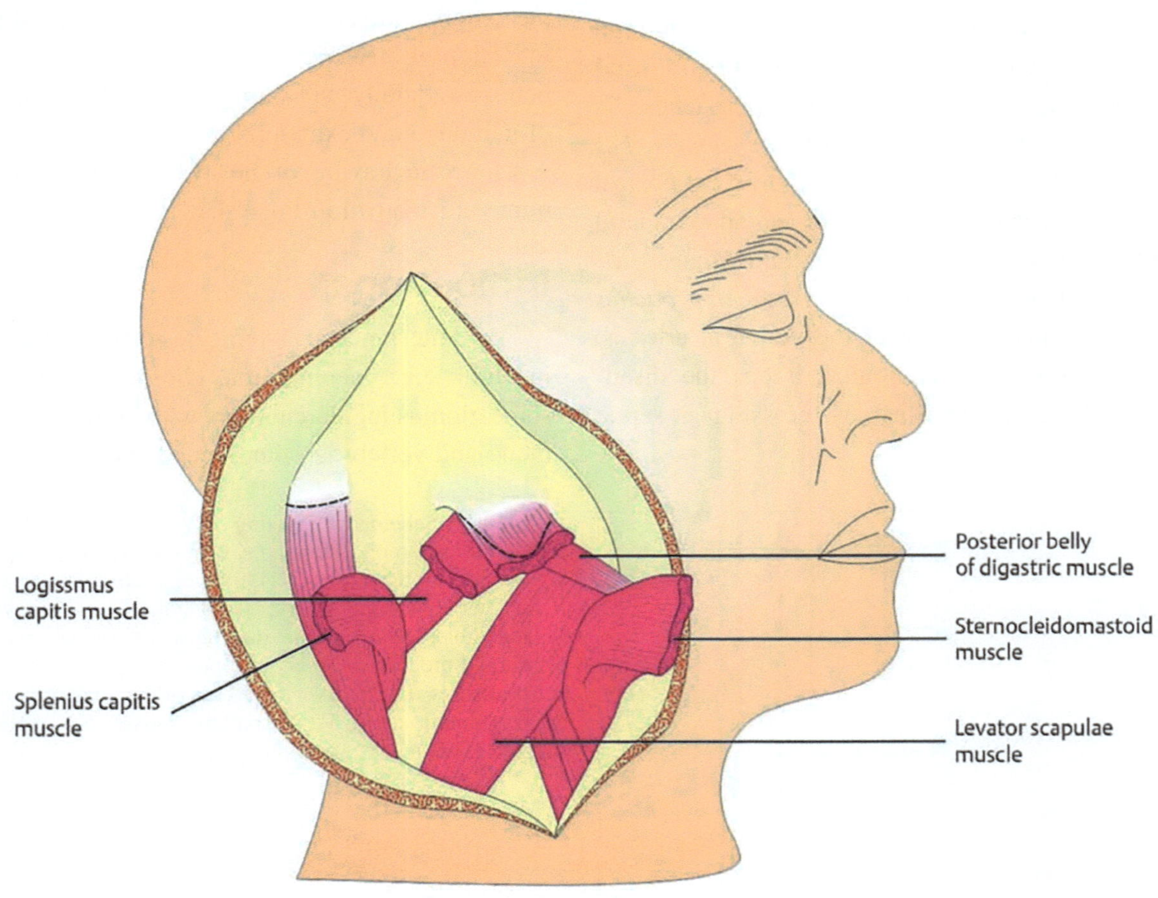

Fig. 14.2 Drawing of the muscles related to the vertebral artery at the suboccipital region. Reflection of the sternomastoid and the splenius capitis muscles reveals the deeper muscles in this area.

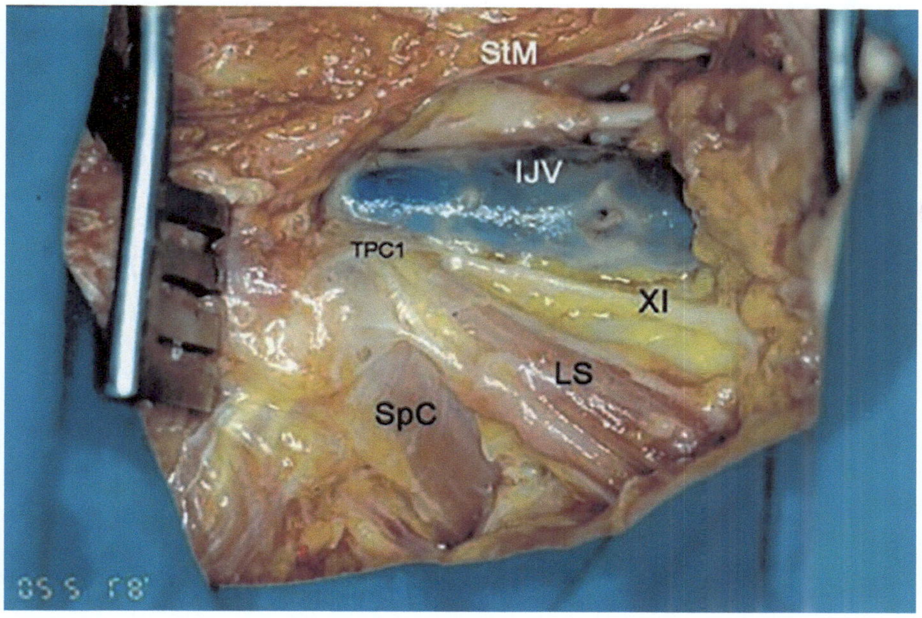

Fig. 14.3 Dissection of the right side. The sternomastoid muscle (StM) has been retracted anteriorly. The levator scapulae (LS) and the splenius capitis (SpC) muscles can be identified at a superficial level.

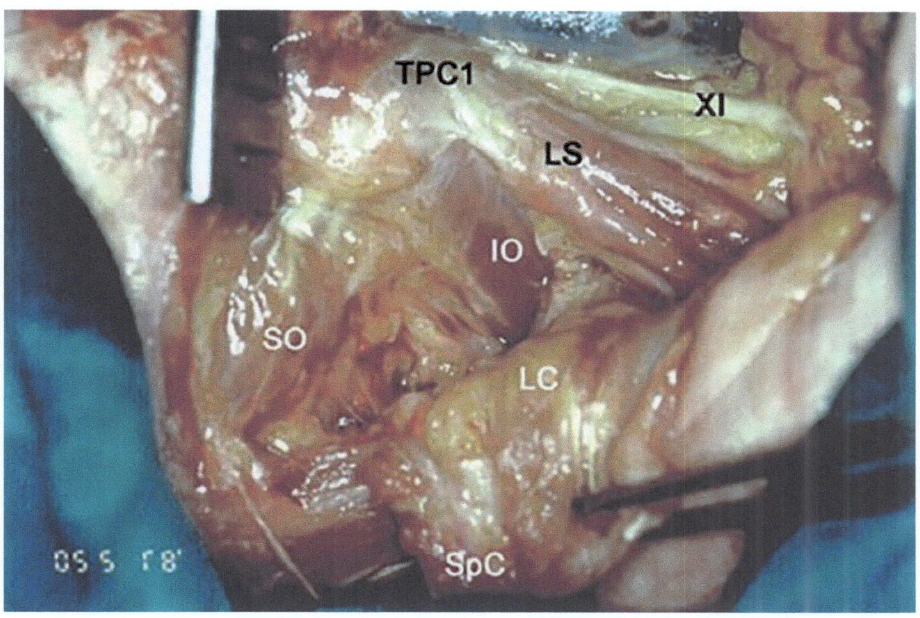

Fig. 14.4 Reflecting the splenius capitis (SpC) muscle together with the slender, deeply attached longissimus capitis (LC) muscle reveals the deep inferior (IO) and superior (SO) oblique muscles.

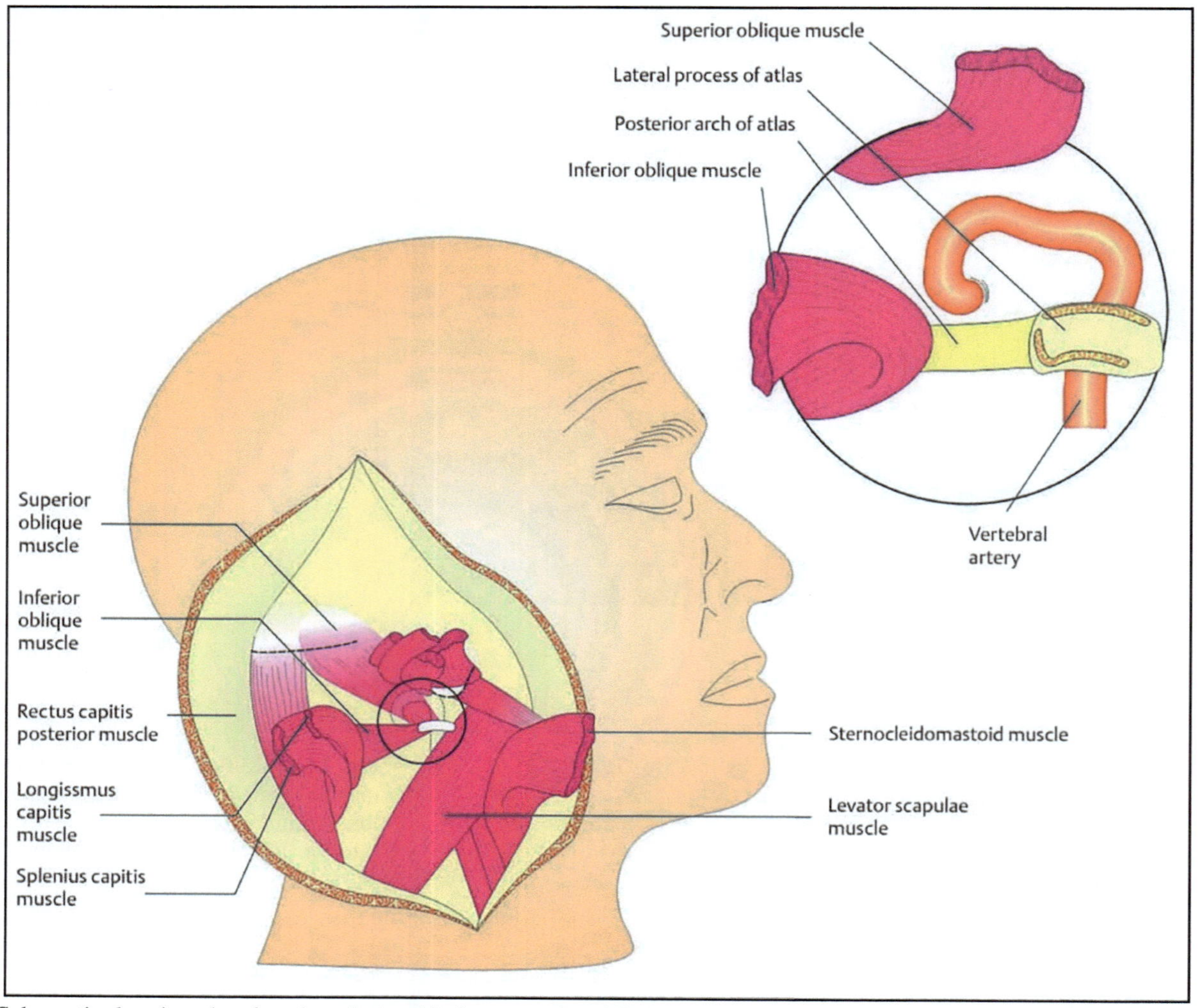

Fig. 14.5 Schematic drawing showing the relationship between the oblique muscles and the vertebral artery between C1 and C2.

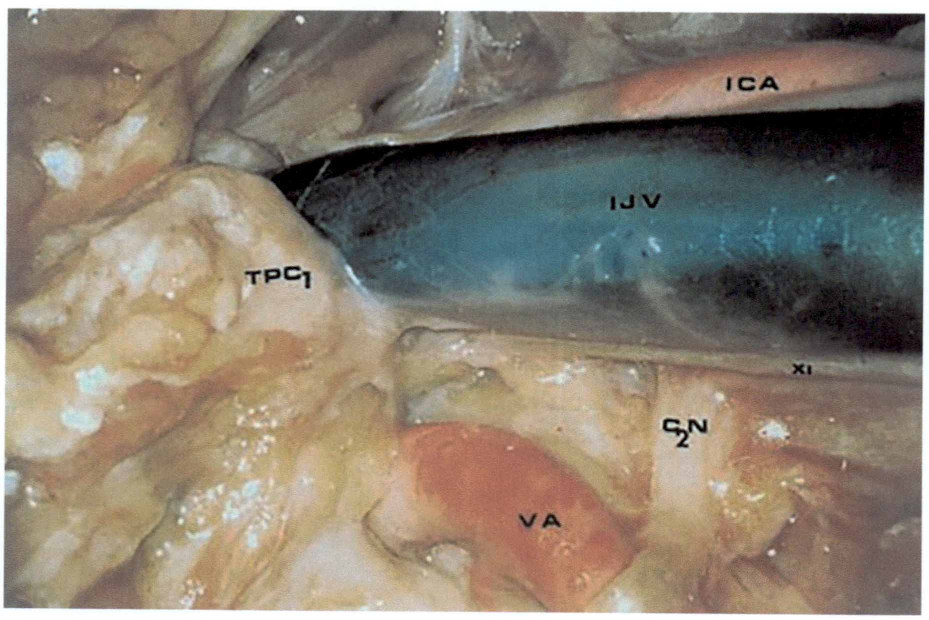

Fig. 14.6 The transverse process of the atlas (TPC1) forms an important landmark in this region.

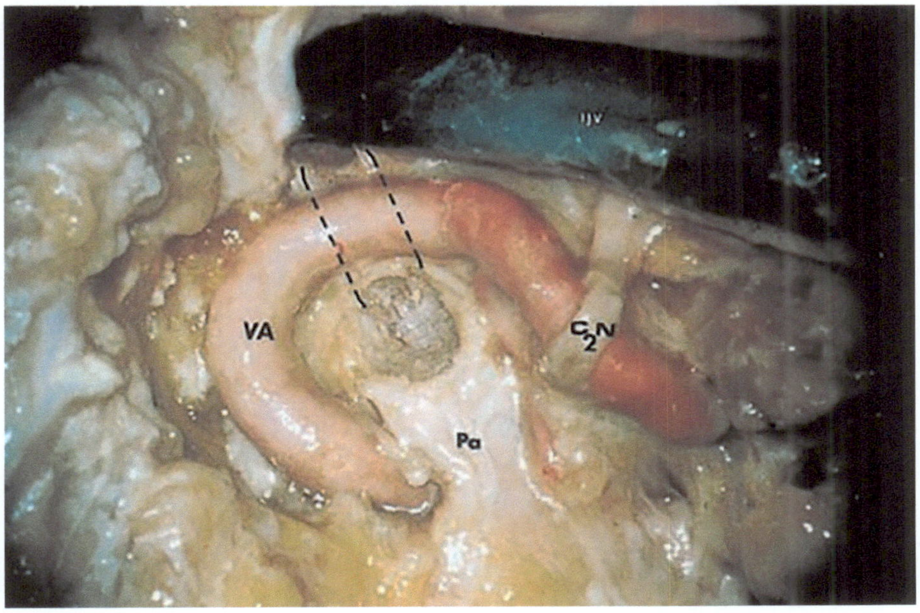

Fig. 14.7 Course of the vertebral artery (VA) after leaving the transverse process of the axis. The foramen transversarium of the atlas (hatched lines) has been opened. *Pa*, posterior arch of the atlas.

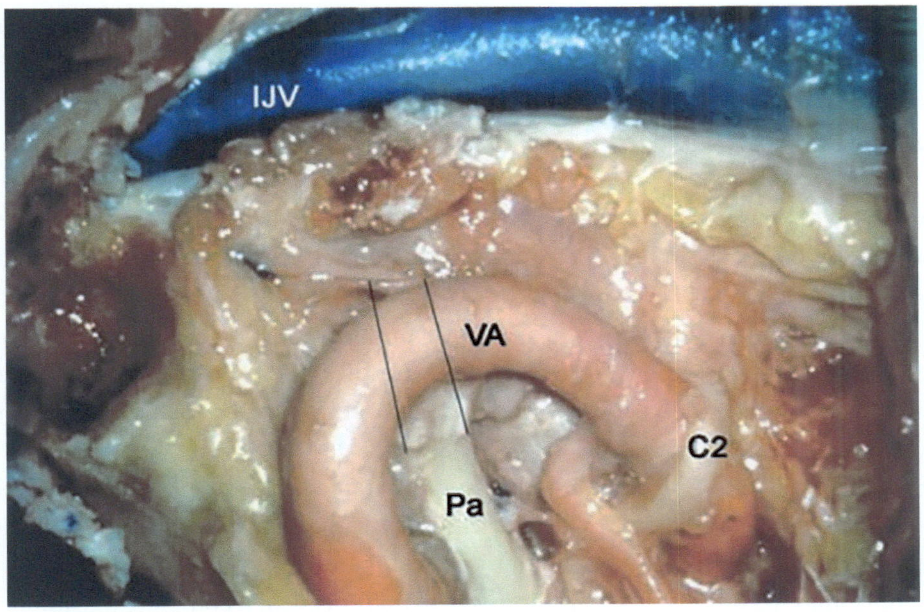

Fig. 14.8 At a higher magnification, the C2 nerve root is seen crossing the vertebral artery (VA).

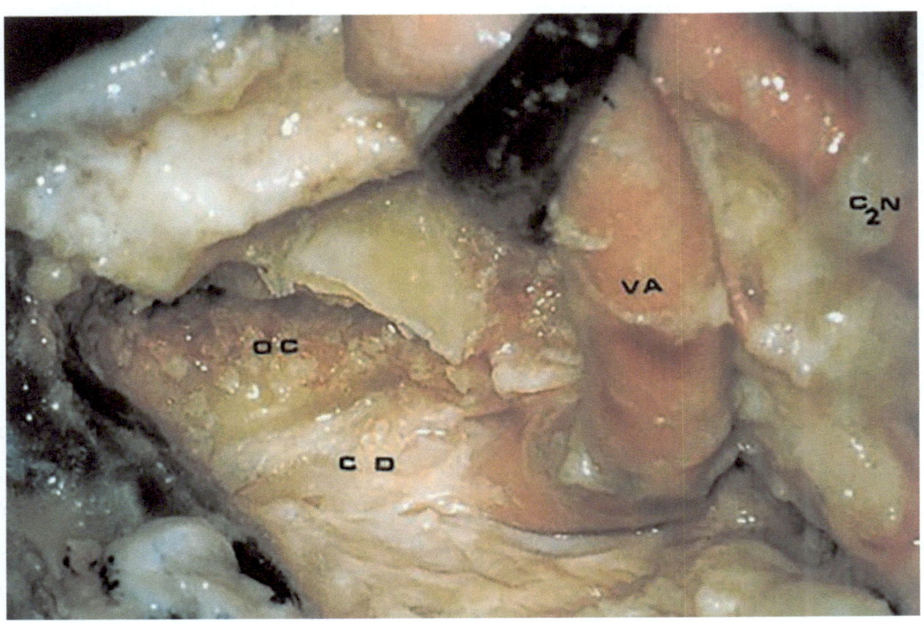

Fig. 14.9 The point where the vertebral artery (VA) pierces the dura.

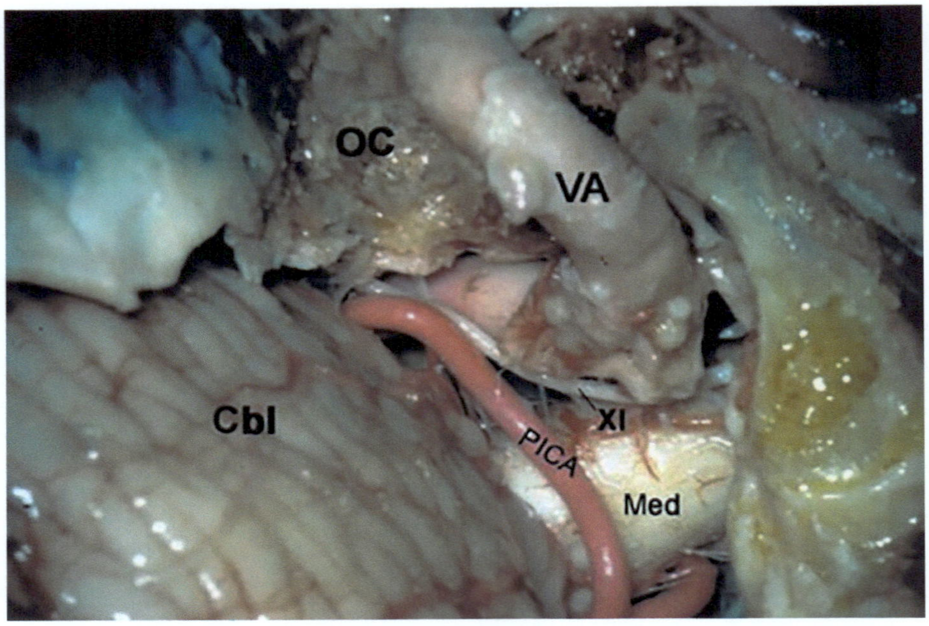

Fig. 14.10 The accessory nerve (XI) is closely related to the vertebral artery (VA) at the point of dural entrance. Note the dura attached to the artery.

Surgical Steps

1. Patient positioning: The patient is put in the lateral decubitus position. Care is taken to avoid rotation or flexion of the head.

2. The incision starts nearly 5 cm caudal to the mastoid tip, travels upward just posterior to the tip of the mastoid, and then curves posteriorly as an inverted "U" in the occipital region (Fig. 14.11).

3. The skin flap is elevated, taking care not to injure the accessory nerve. The sternomastoid muscle is detached and retracted inferiorly. A blue vessel loop is placed in order to mark the internal jugular vein (Fig. 14.12).

4. The splenius capitis is cut near its origin and retracted inferiorly, and then the semispinalis capitis and longissimus capitis are identified. The occipital artery may pass superficial or deep to the longissimus capitis (Fig. 14.13).

5. The underlying longissimus capitis and semispinalis capitis are also cut near their origin and retracted inferiorly to expose the suboccipital triangle located between the superior and inferior oblique muscles and the rectus capitis posterior major, in the depths of which the vertebral artery courses with a dense venous plexus. The levator scapulae is identified as anteriorly attached to the transverse process of the atlas (Fig. 14.14).

6. The muscles forming the margins of the suboccipital triangle have been removed to expose the floor of the suboccipital triangle, which is formed by the posterior atlanto-occipital membrane and the posterior arch of the atlas. The vertebral artery and the C1 nerve root, which are surrounded by the vertebral venous plexus, course along the upper surface of the posterior arch of the atlas (Fig. 14.15).

7. The surrounding venous plexus has been removed to expose the vertebral artery ascending through the C1 transverse process and behind the atlanto-occipital joint (Fig. 14.16).

8. After dissection of the suboccipital triangle, the extradural segment of the VA mobilization of this vessel may be performed. The C1 transverse foramen is unroofed. The vertebral artery is now exposed from the level of C2 up to its curve above the posterior arch of the atlas. The C2 nerve root crosses lateral to the artery between C1 and C2 and is another helpful landmark in identifying it (Fig. 14.17).

9. A retrosigmoid craniotomy reaching down to the lateral rim of the foramen magnum is carried out. The atlanto-occipital joint can be exposed (Fig. 14.18).

10. If more exposure is needed, a complete mastoidectomy is also performed. The sigmoid sinus is completely uncovered and followed down to the jugular bulb. This presigmoid petrosectomy allows anterior displacement of the sigmoid sinus, giving more exposure (Fig. 14.19).

11. The tip of the suction is used to displace the vertebral artery inferiorly to improve the exposure of the occipital condyle while the occipital condyle is drilled to reach the hypoglossal canal (Fig. 14.20). Drilling of the occipital condyle is the extension of this approach. Generally, the posterior and superior medial one-third of the condyle can be drilled away to expose the hypoglossal canal. However, aggressive removal of the occipital condyle may cause craniovertebral instability and necessitate the placement of hardware for stabilization.

12. The hypoglossal canal has been exposed by partially drilling the cancellous bone of the occipital condyle. The jugular tubercle is located above the hypoglossal canal

(Fig. 14.21). Drilling of the jugular tubercle is crucial to obtain a flat view to maximize surgical exposure without excessive cerebellum retraction. The accessory nerve is located over the jugular tubercle, so care must be taken to avoid injury to the nerve during the drilling of the jugular tubercle. Enlarged view. The cortical bone lining the hypoglossal canal has been opened to expose the hypoglossal nerve and the hypoglossal venous plexus in the canal (Fig. 14.22).

13. The dural incision begins behind the sigmoid sinus and extends behind the vertebral artery into the upper cervical area (Fig. 14.23). The upper extent of the dural opening depends on how much of the cerebellopontine angle is to be exposed.

14. A general view of the neurovascular structures controlled (Fig. 14.24). The relationship of the posteroinferior cerebellar artery and its relation to the lower cranial nerves is seen. Note the control of the ventrolateral control of the premedullary area.

15. Directing the microscope upward, the neurovascular structures of the CP angle can be appreciated (Fig. 14.25). This approach provides a better view of the CP angle in comparison with the retrosigmoid approach, since the additional presigmoid craniotomy allows anterior displacement of the sigmoid sinus, thereby permitting more anterolateral visualization.

16. The terminal part of the medulla and the beginning of the spinal cord can be seen (Fig. 14.26). In cases of pathology that extend further down, the spinal cord dura can be opened at the level of the axis, after dividing the C2 nerve root and mobilizing the artery. The first cervical nerve courses along the posteroinferior surface of the vertebral artery as it pierces the dura. The ventral root is located anterior to the dentate ligament. The rootlets forming the spinal portion of the accessory nerve, which arise from the cervical portion of the spinal cord midway between the dorsal and ventral rootlets as far caudally as C5, unite to form a trunk that ascends through the foramen magnum between the dentate ligament and the dorsal roots and enters the posterior fossa behind the vertebral artery.

17. At the craniocervical junction, the dentate ligament is located between the vertebral artery and ventral roots of C1 anteriorly and the branches of the posterior spinal artery and spinal accessory nerve posteriorly and is often incorporated into the dural cuff around the vertebral artery. The most rostral attachment of the dentate ligament is located at the level of the foramen magnum, where the vertebral artery pierces the dura and behind the accessory nerve (Fig. 14.27).

18. The hypoglossal nerve canal has been opened to expose the hypoglossal nerve (Fig. 14.28).

19. The jugular bulb is opened to expose the medial wall of the jugular bulb. The opening of inferior petrosal sinus can be appreciated (Fig. 14.29).

20. The medial wall of the jugular bulb is removed to expose the lower cranial nerve (Fig. 14.30).

21. At the end of the approach, the dura is sutured back. The occipital muscles are repositioned and sutured in place.

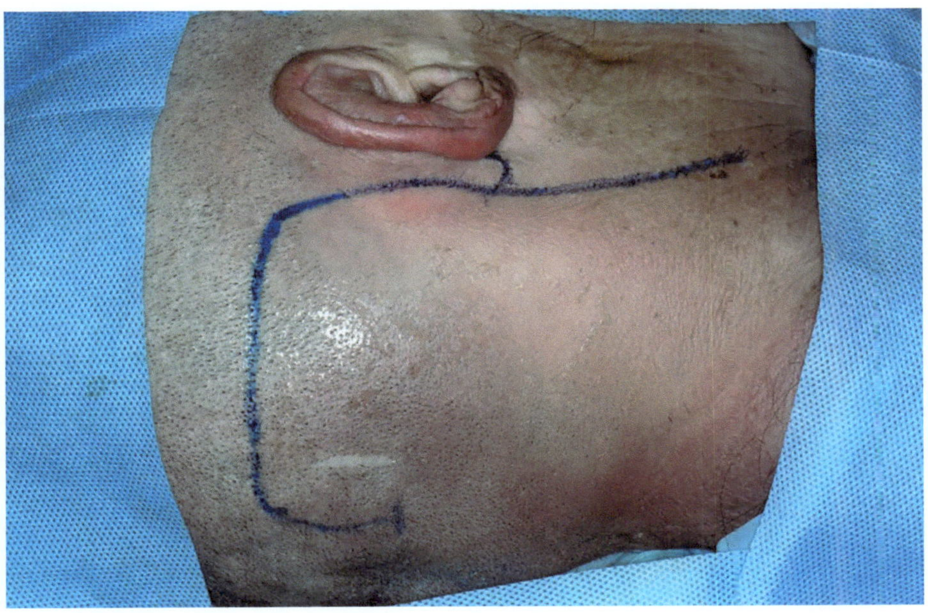

Fig. 14.11 The incision starts nearly 5 cm caudal to the mastoid tip, travels upward just posterior to the tip of the mastoid, and then curves posteriorly as an inverted "U" in the occipital region.

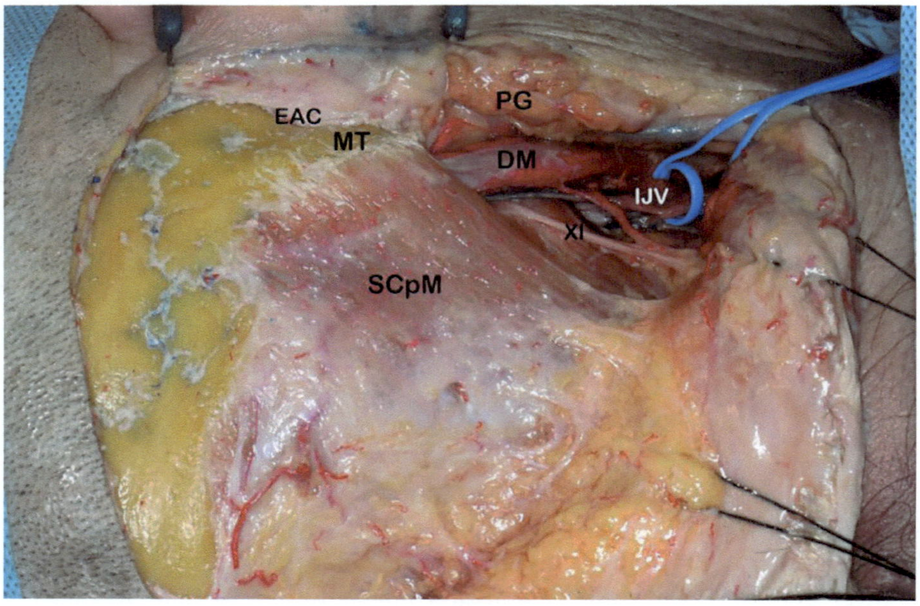

Fig. 14.12 The skin flap is elevated, taking care not to injure the accessory nerve. The sternomastoid muscle is detached and retracted inferiorly. A blue vessel loop is placed in order to mark the internal jugular vein. *PG*, parotid gland; *DM*, digastric muscle; *IJV*, internal jugular vein; *XI*, accessory nerve; *SCpM*, splenius capitis muscle; *MT*, mastoid tip; *EAC*, external auditory canal.

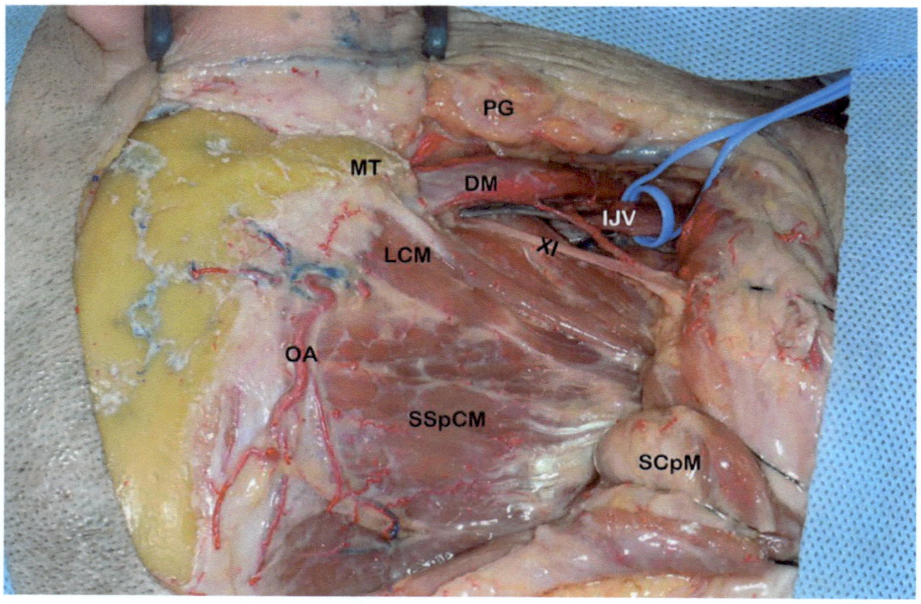

Fig. 14.13 The splenius capitis is cut near its origin and retracted inferiorly, and then the semispinalis capitis and longissimus capitis are identified. The occipital artery may pass superficial or deep to the longissimus capitis. *IJV*, internal jugular vein; *XI*, accessory nerve; *DM*, digastric muscle; *PG*, parotid gland; *MT*, mastoid tip; *LCM*, longissimus capitis muscle; *OA*, occipital artery; *SSPCM*, semispinalis capitis muscle; *SCpM*, splenius capitis muscle (retracted).

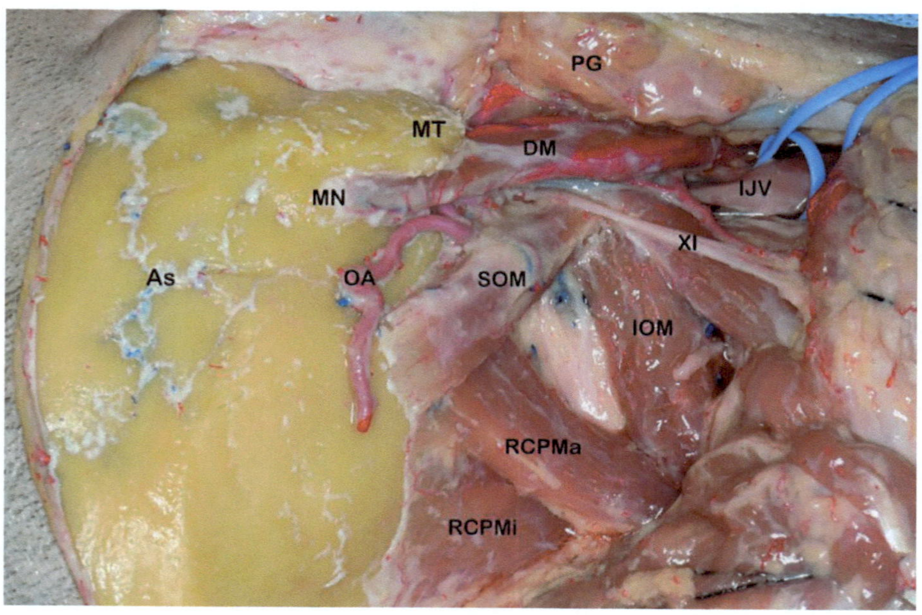

Fig. 14.14 The underlying longissimus capitis and semispinalis capitis are also cut near their origin and retracted inferiorly to expose the suboccipital triangle located between the superior and inferior oblique muscles and the rectus capitis posterior major, in the depths of which the vertebral artery courses with a dense venous plexus. *IJV*, internal jugular vein; *XI*, accessory nerve; *IOM*, inferior oblique muscle; *SOM*, superior oblique muscle; *DM*, digastric muscle; *PG*, parotid gland; *MT*, mastoid tip; *MN*, mastoid notch; *OA*, occipital artery; *As*, asterion; *RCPMa*, rectus capitis posterior major muscle; *RCPMi*, rectus capitis posterior minor muscle.

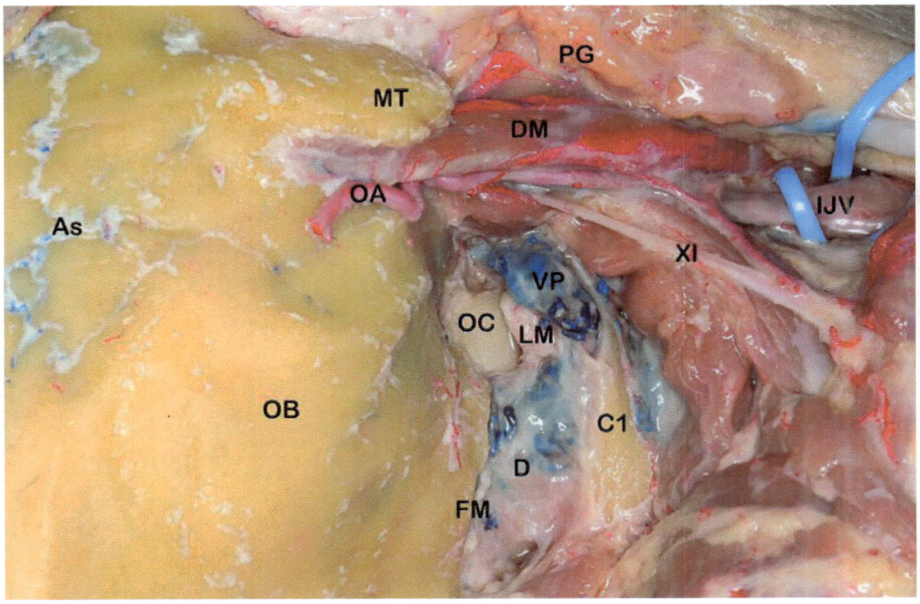

Fig. 14.15 The muscles forming the margins of the suboccipital triangle have been removed to expose the floor of the suboccipital triangle, which is formed by the posterior atlanto-occipital membrane and the posterior arch of the atlas. The vertebral artery and the C1 nerve root, which are surrounded by the vertebral venous plexus, course along the upper surface of the posterior arch of the atlas. *MT*, mastoid tip; *IJV*, internal jugular vein; *XI*, accessory nerve; *PG*, parotid gland; *DM*, digastric muscle; *OA*, occipital artery; *OC*, occipital condyle; *VP*, venous plexus; *LM*, lateral mass; *C1*, atlas; *D*, dura; *As*, asterion; *FM*, foramen magnum; *OB*, occipital bone.

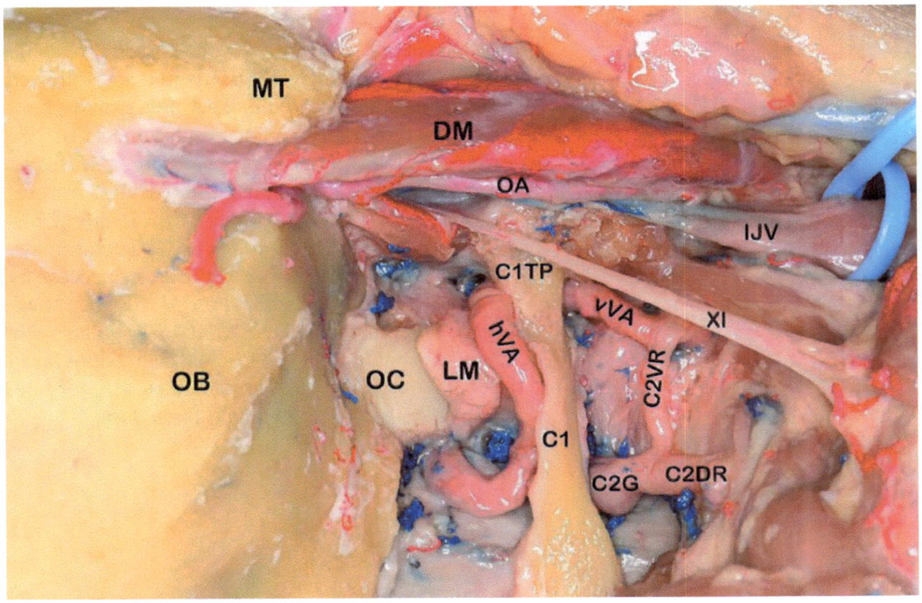

Fig. 14.16 The surrounding venous plexus has been removed to expose the vertebral artery ascending through the C1 transverse process and behind the atlanto-occipital joint. *IJV*, internal jugular vein; *XI*, accessory nerve; *vVA*, vertical segment of vertebral artery; *hVA*, horizontal segment of vertebral artery; *C2VR*, ventral ramus of C2; *C2DR*, dorsum ramus of C2; *C2G*, ganglion of C2; *C1*, atlas; *C1TP*, transverse process of atlas; *OA*, occipital artery; *DM*, digastric muscle; *MT*, mastoid tip; *OC*, occipital condyle; *LM*, lateral mass; *OB*, occipital bone.

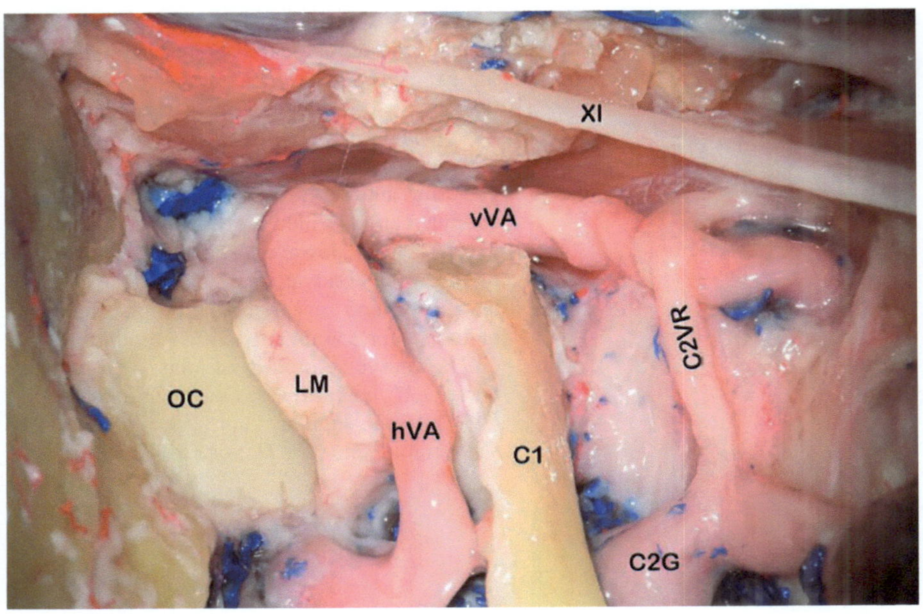

Fig. 14.17 After dissection of the suboccipital triangle, the extradural segment of the VA mobilization of this vessel may be performed. The C1 transverse foramen is unroofed. The vertebral artery is now exposed from the level of C2 up to its curve above the posterior arch of the atlas. The C2 nerve root crosses lateral to the artery between C1 and C2 and is another helpful landmark in identifying it. *XI*, accessory nerve; *vVA*, vertical segment of vertebral artery; *hVA*, horizontal segment of vertebral artery; *C2VR*, ventral ramus of C2; *C1*, atlas; *LM*, lateral mass; *OC*, occipital condyle; *C2G*, ganglion of C2.

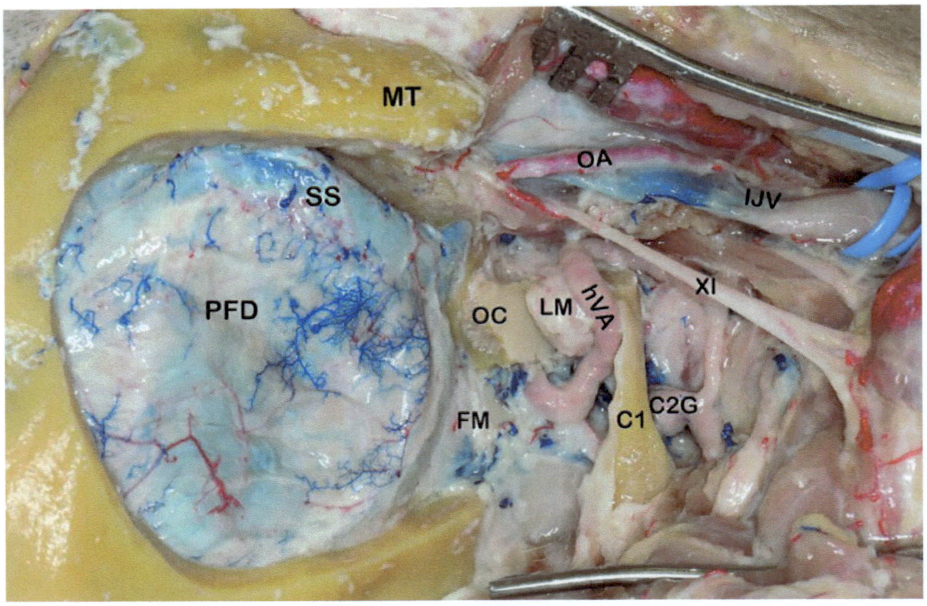

Fig. 14.18 A retrosigmoid craniotomy reaching down to the lateral rim of the foramen magnum is carried out. The atlanto-occipital joint can be exposed. *IJV*, internal jugular vein; *XI*, accessory nerve; *C1*, atlas; *C2G*, ganglion of C2; *hVA*, horizontal segment of vertebral artery; *OA*, occipital artery; *OC*, occipital condyle; *FM*, foramen magnum; *MT*, mastoid tip; *SS*, sigmoid sinus; *PFD*, posterior fossa dura; *LM*, lateral mass.

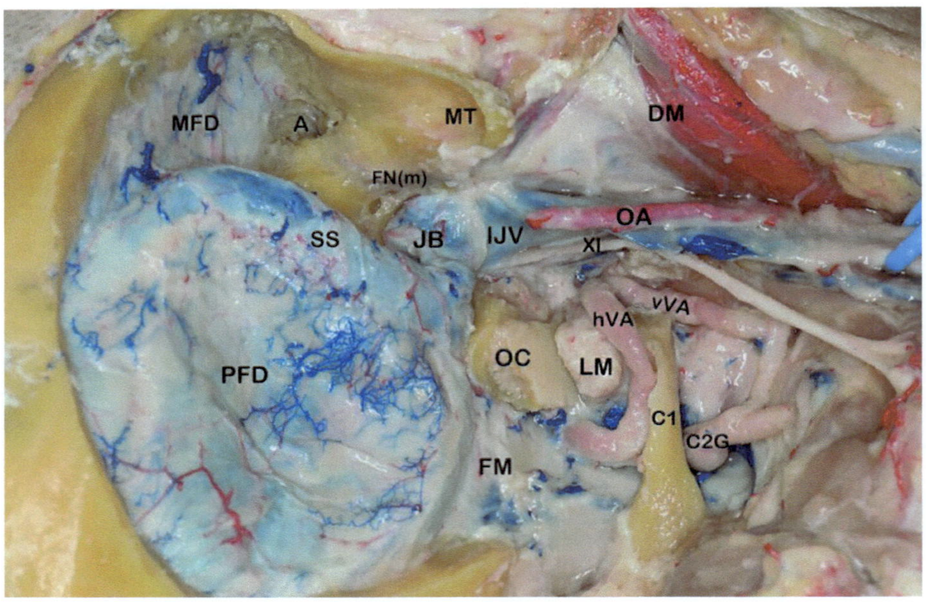

Fig. 14.19 If more exposure is needed, a complete mastoidectomy is also performed. The sigmoid sinus is completely uncovered and followed down to the jugular bulb. This presigmoid petrosectomy allows anterior displacement of the sigmoid sinus, giving more exposure. *DM*, digastric muscle; *OA*, occipital artery; *vVA*, vertical segment of vertebral artery; *hVA*, horizontal segment of vertebral artery; *LM*, lateral mass; *C1*, atlas; *C2G*, ganglion of C2; *XI*, accessory nerve; *OC*, occipital condyle; *FM*, foramen magnum; *IJV*, internal jugular vein; *JB*, jugular bulb; *MT*, mastoid tip; *FN(m)*, mastoid segment of facial nerve; *A*, antrum; *SS*, sigmoid sinus; *PFD*, posterior fossa dura; *MFD*, middle fossa dura.

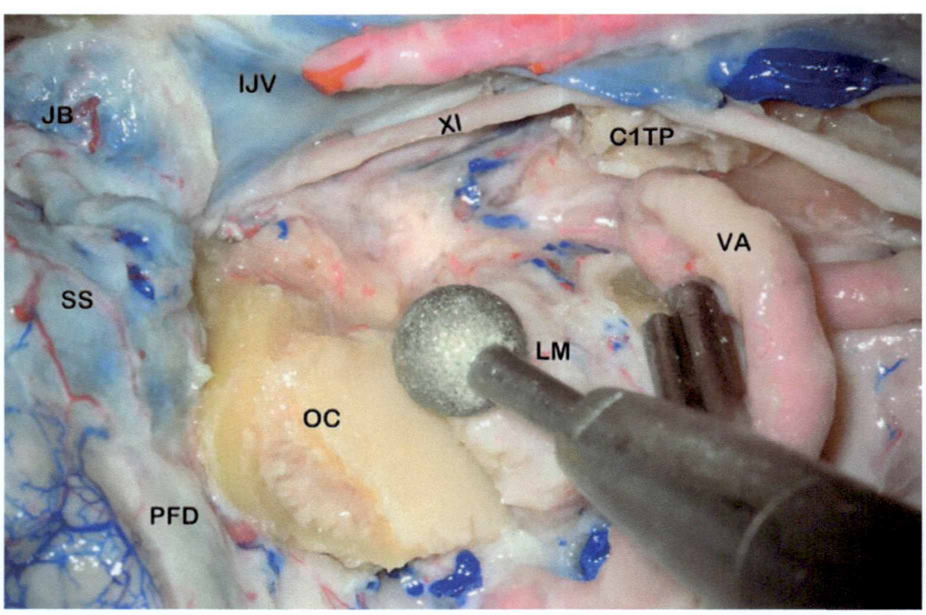

Fig. 14.20 The tip of the suction is used to displace the vertebral artery inferiorly to improve the exposure of the occipital condyle while the occipital condyle is drilled to reach the hypoglossal canal. Drilling of the occipital condyle is the extension of this approach. Generally, the posterior and superior medial one-third of the condyle can be drilled away to expose the hypoglossal canal. However, aggressive removal of the occipital condyle may cause craniovertebral instability and necessitate the placement of hardware for stabilization. *VA*, vertebral artery; *C1TP*, transverse process of atlas; *XI*, accessory nerve; *LM*, lateral mass; *OC*, occipital condyle; *IJV*, internal jugular vein; *JB*, jugular bulb; *SS*, sigmoid sinus; *PFD*, posterior fossa dura.

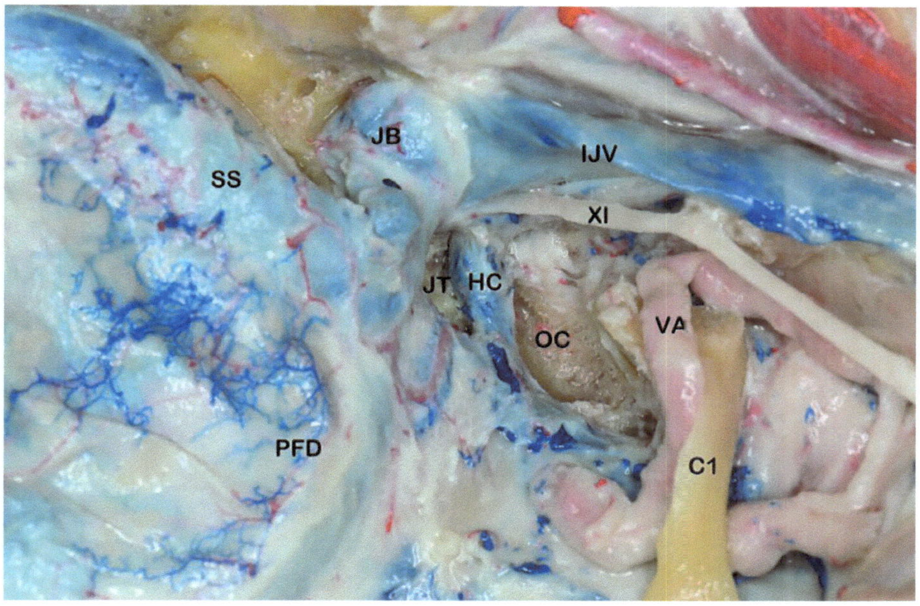

Fig. 14.21 The hypoglossal canal has been exposed by partially drilling the cancellous bone of the occipital condyle. The jugular tubercle is located above the hypoglossal canal. Drilling of the jugular tubercle is crucial to obtain a flat view to maximize surgical exposure without excessive cerebellum retraction. The accessory nerve is located over the jugular tubercle, so care must be taken to avoid injury to the nerve during the drilling of the jugular tubercle. *IJV*, internal jugular vein; *XI*, accessory nerve; *VA*, vertebral artery; *C1*, atlas; *OC*, occipital condyle; *HC*, hypoglossal canal; *JT*, jugular tubercle; *JB*, jugular bulb; *SS*, sigmoid sinus; *PFD*, posterior fossa dura.

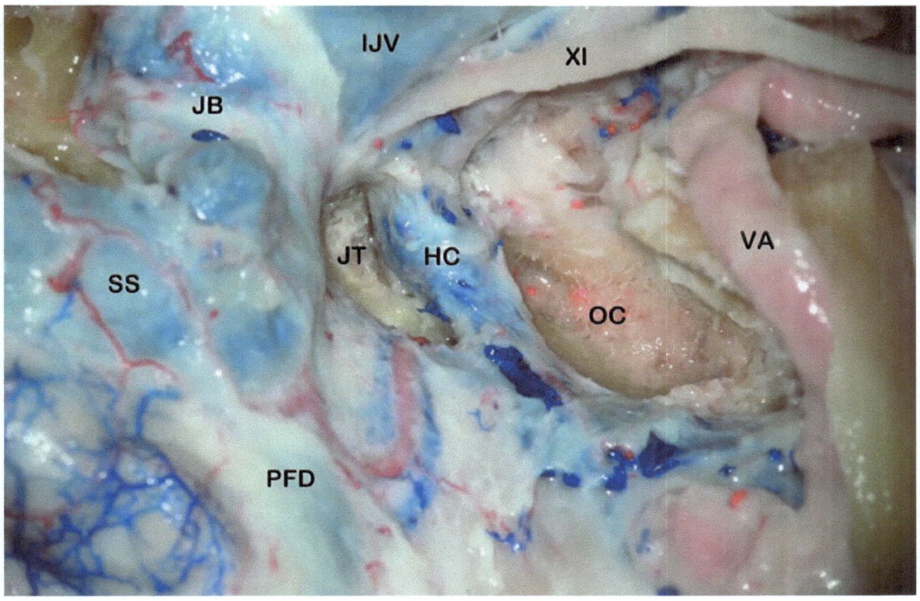

Fig. 14.22 Enlarged view. The cortical bone lining the hypoglossal canal has been opened to expose the hypoglossal nerve and the hypoglossal venous plexus in the canal. *VA*, vertebral artery; *OC*, occipital condyle; *XI*, accessory nerve; *HC*, hypoglossal canal; *JT*, jugular tubercle; *IJV*, internal jugular vein; *JB*, jugular bulb; *SS*, sigmoid sinus; *PFD*, posterior fossa dura.

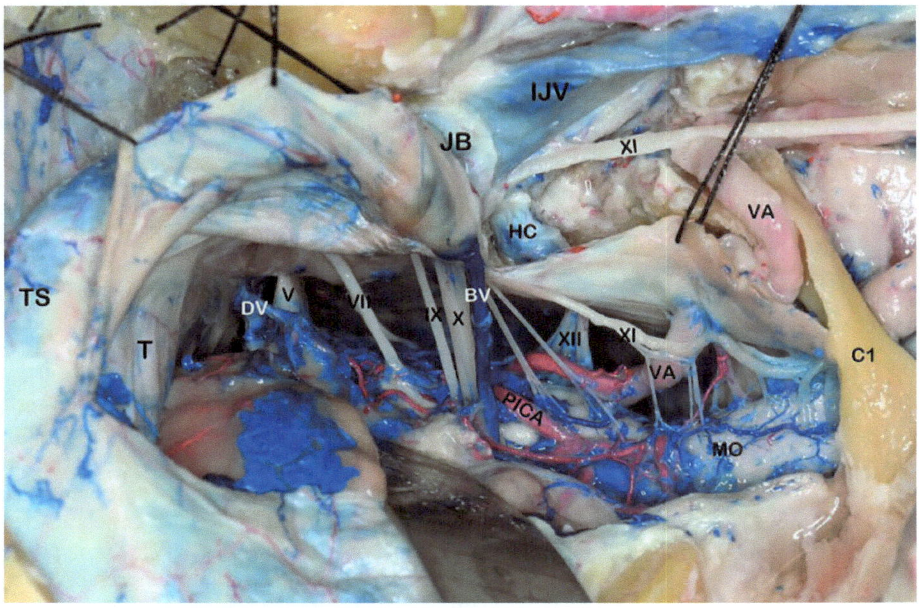

Fig. 14.23 The dural incision begins behind the sigmoid sinus and extends behind the vertebral artery into the upper cervical area. The upper extent of the dural opening depends on how much of the cerebellopontine angle is to be exposed. *VA*, vertebral artery; *C1*, atlas; *MO*, medulla oblongata; *V*, trigeminal nerve; *VII*, facial nerve; *IX*, glossopharyngeal nerve; *X*, vagus nerve; *XI*, accessory nerve; *XII*, hypoglossal nerve; *PICA*, posteroinferior cerebellar artery; *HC*, hypoglossal canal; *IJV*, internal jugular vein; *JB*, jugular bulb; *BV*, bridge vein; *T*, tentorium; *TS*, transverse sinus; *DV*, Dandy's vein (superior petrosal vein).

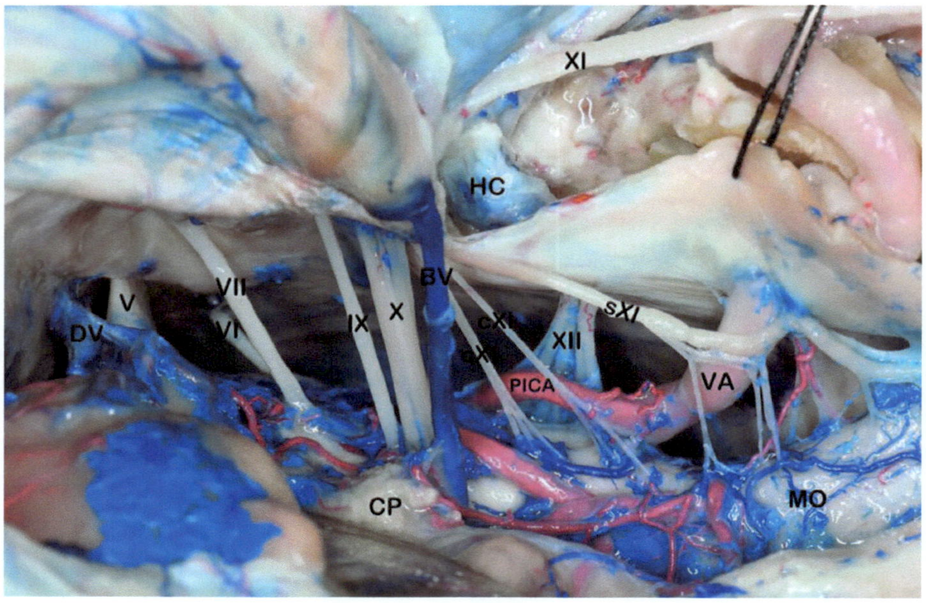

Fig. 14.24 A general view of the neurovascular structures controlled. The relationship of the posteroinferior cerebellar artery and its relation to the lower cranial nerves is seen. Note the control of the ventrolateral control of the premedullary area. *HC*, hypoglossal canal; *cXI*, cranial rootlets of accessory nerve; *sXI*, spinal root of accessory nerve; *XI*, accessory nerve; *XII*, hypoglossal nerve; *VA*, vertebral artery; *MO*, medulla oblongata; *PICA*, posteroinferior cerebellar artery; *BV*, bridge vein; *X*, vagus nerve; *IX*, glossopharyngeal nerve; *CP*, choroid plexus; *VI*, abducent nerve; *VII*, facial nerve; *V*, trigeminal nerve; *DV*, Dandy's vein (superior petrosal vein).

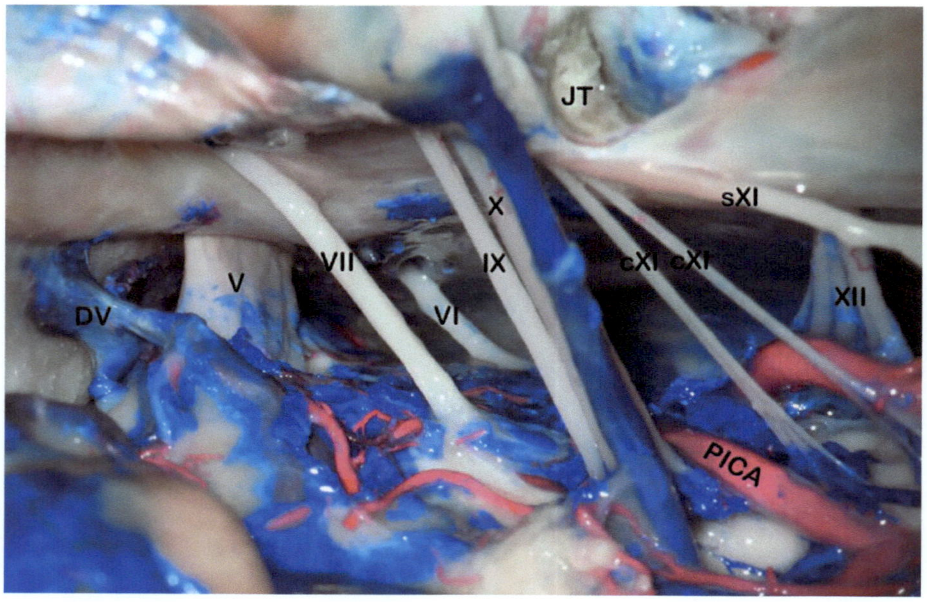

Fig. 14.25 Directing the microscope upward, the neurovascular structures of the CP angle can be appreciated. This approach provides a better view of the CP angle in comparison with the retrosigmoid approach, since the additional presigmoid craniotomy allows anterior displacement of the sigmoid sinus, thereby permitting more anterolateral visualization. *JT*, jugular tubercle; *sXI*, spinal root of accessory nerve; *cXI*, cranial rootlets of accessory nerve; *XII*, hypoglossal nerve; *PICA*, posteroinferior cerebellar artery; *V*, trigeminal nerve; *VI*, abducent nerve; *VII*, facial nerve; *X*, vagus nerve; *IX*, glossopharyngeal nerve; *DV*, Dandy's vein (superior petrosal vein).

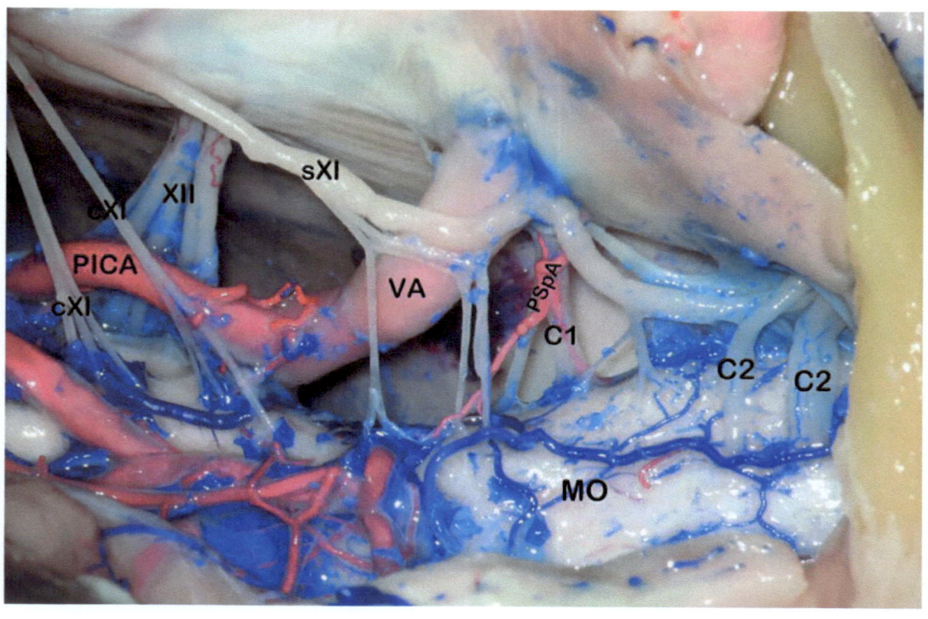

Fig. 14.26 The terminal part of the medulla and the beginning of the spinal cord can be seen. In cases of pathology that extend further down, the spinal cord dura can be opened at the level of the axis, after dividing the C2 nerve root and mobilizing the artery. The first cervical nerve courses along the posteroinferior surface of the vertebral artery as it pierces the dura. The ventral root is located anterior to the dentate ligament. The rootlets forming the spinal portion of the accessory nerve, which arise from the cervical portion of the spinal cord midway between the dorsal and ventral rootlets as far caudally as C5, unite to form a trunk that ascends through the foramen magnum between the dentate ligament and the dorsal roots and enters the posterior fossa behind the vertebral artery. *C2*, rootlets of C2; *C1*, rootlets of C1; *MO*, medulla oblongata; *PSpA*, posterior spinal artery; *VA*, vertebral artery; *sXI*, spinal root of accessory nerve; *XII*, hypoglossal nerve; *cXI*, cranial rootlets of accessory nerve; *PICA*, posteroinferior cerebellar artery.

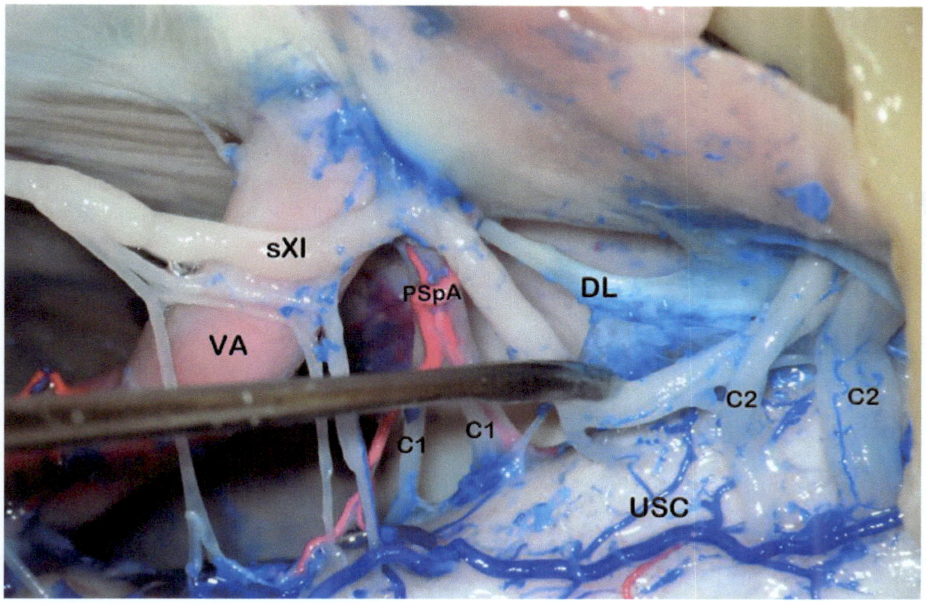

Fig. 14.27 At the craniocervical junction, the dentate ligament is located between the vertebral artery and ventral roots of C1 anteriorly and the branches of the posterior spinal artery and spinal accessory nerve posteriorly and is often incorporated into the dural cuff around the vertebral artery. The most rostral attachment of the dentate ligament is located at the level of the foramen magnum, where the vertebral artery pierces the dura and behind the accessory nerve. *C2*, rootlets of C2; *C1*, rootlets of C1; *USC*, upper spinal cord; *DL*, dentate ligament; *PSpA*, posterior spinal artery; *sXI*, spinal root of accessory nerve; *VA*, vertebral artery.

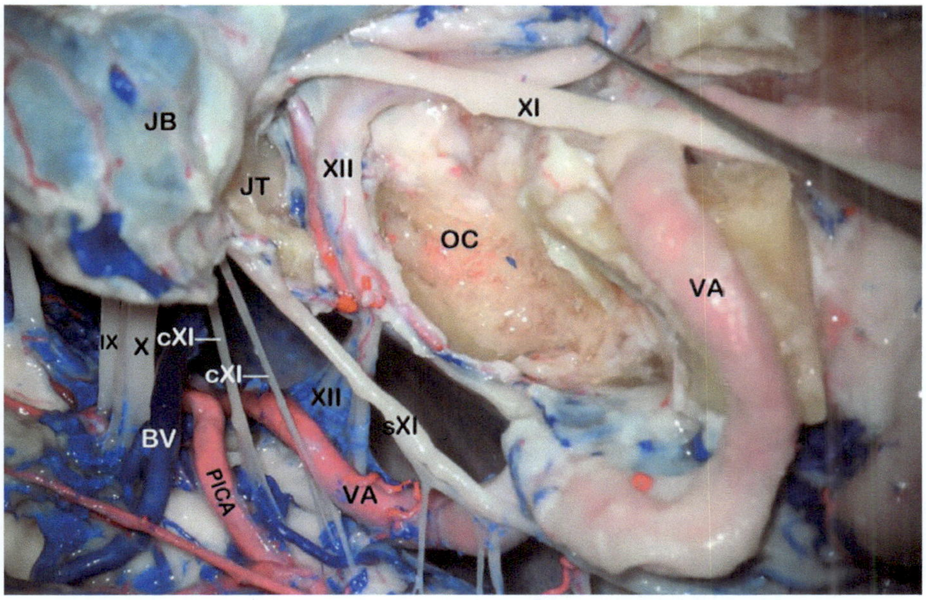

Fig. 14.28 The hypoglossal nerve canal has been opened to expose the hypoglossal nerve. *JB*, jugular bulb; *JT*, jugular tubercle; *OC*, occipital condyle; *XI*, accessory nerve; *sXI*, spinal root of accessory nerve; *cXI*, cranial rootlets of accessory nerve; *XII*, hypoglossal nerve; *PICA*, posteroinferior cerebellar artery; *X*, vagus nerve; *IX*, glossopharyngeal nerve; *VA*, vertebral artery; *BV*, bridge vein.

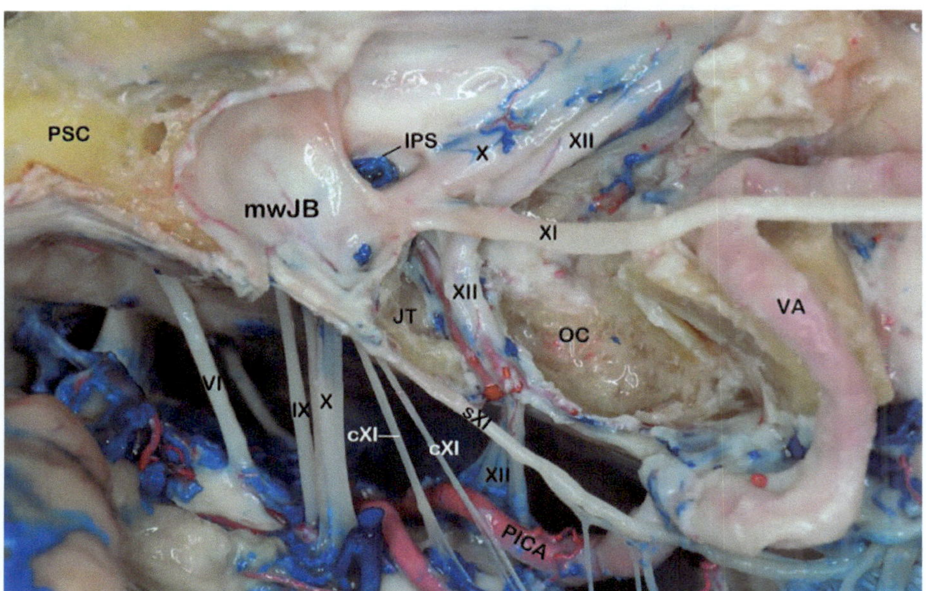

Fig. 14.29 The jugular bulb has been opened to expose the medial wall of the jugular bulb. The opening of the inferior petrosal sinus can be appreciated. *mwJB*, medial wall of the jugular bulb; *JT*, jugular tubercle; *OC*, occipital condyle; *XI*, accessory nerve; *sXI*, spinal root of accessory nerve; *cXI*, cranial rootlets of accessory nerve; *XII*, hypoglossal nerve; *PICA*, posteroinferior cerebellar artery; *X*, vagus nerve; *IX*, glossopharyngeal nerve; *VI*, abducent nerve; *VA*, vertebral artery; *PSC*, posterior semicircular canal.

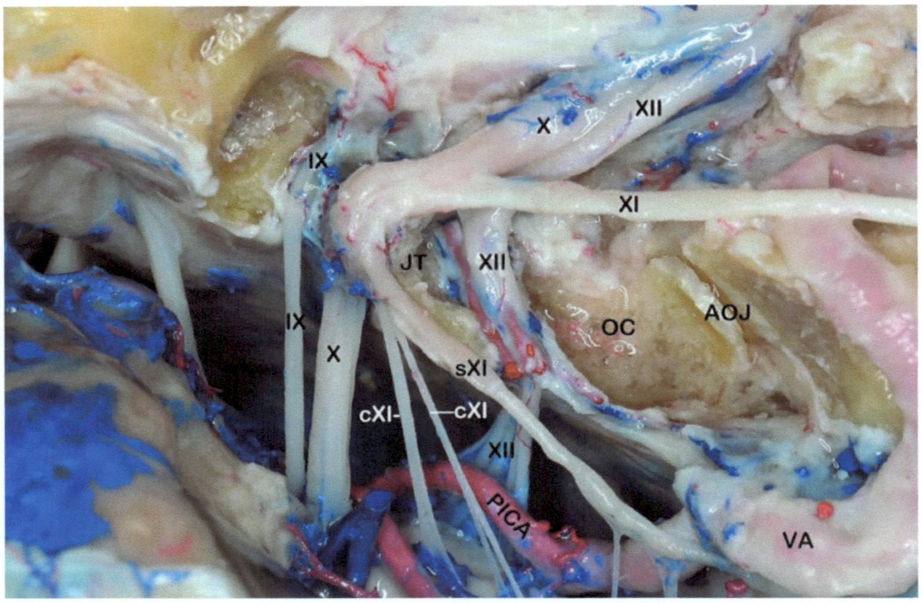

Fig. 14.30 The medial wall of the jugular bulb is removed to expose the lower cranial nerve. *JT*, jugular tubercle; *OC*, occipital condyle; *AOJ*, atlanto-occipital joint; *XI*, accessory nerve; *sXI*, spinal root of accessory nerve; *cXI*, cranial rootlets of accessory nerve; *XII*, hypoglossal nerve; *PICA*, posteroinferior cerebellar artery; *X*, vagus nerve; *IX*, glossopharyngeal nerve; *VA*, vertebral artery.

Clinical Applications

Case 1: Second-Stage Removal of Tympanojugular Paraganglioma Involving the Vertebral Artery

Preoperative balloon occlusion of the vertebral artery had been performed. The preoperative imaging and the steps for tumor removal are shown in Figs. 14.31–14.38.

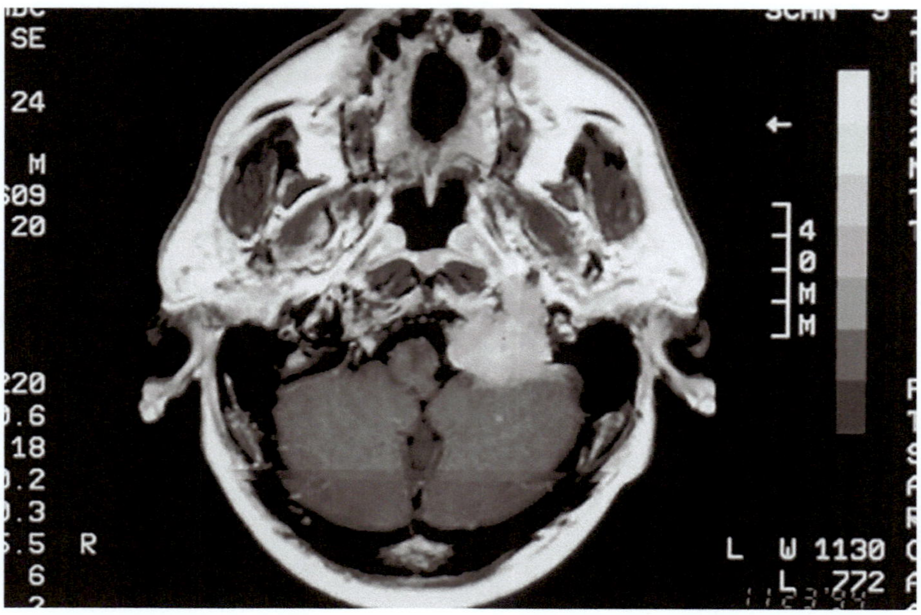

Fig. 14.31 Preoperative MRI of the case showing the residual tumor at the level of the lower clivus.

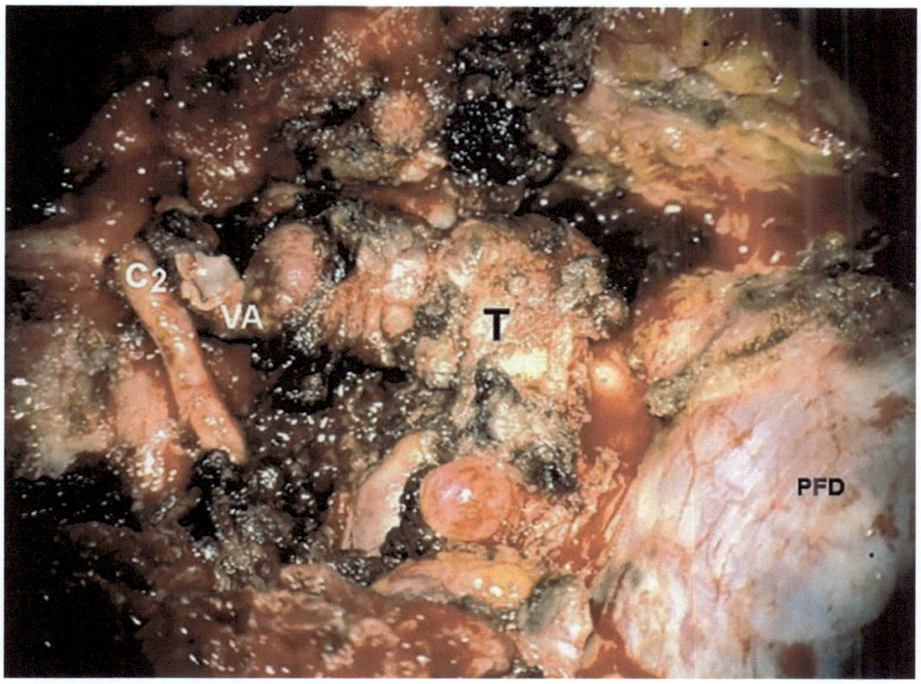

Fig. 14.32 The approach has been performed as previously described. The transverse process of C1 has been drilled. The C2 nerve root is seen crossing the vertebral artery (VA). The artery is double.

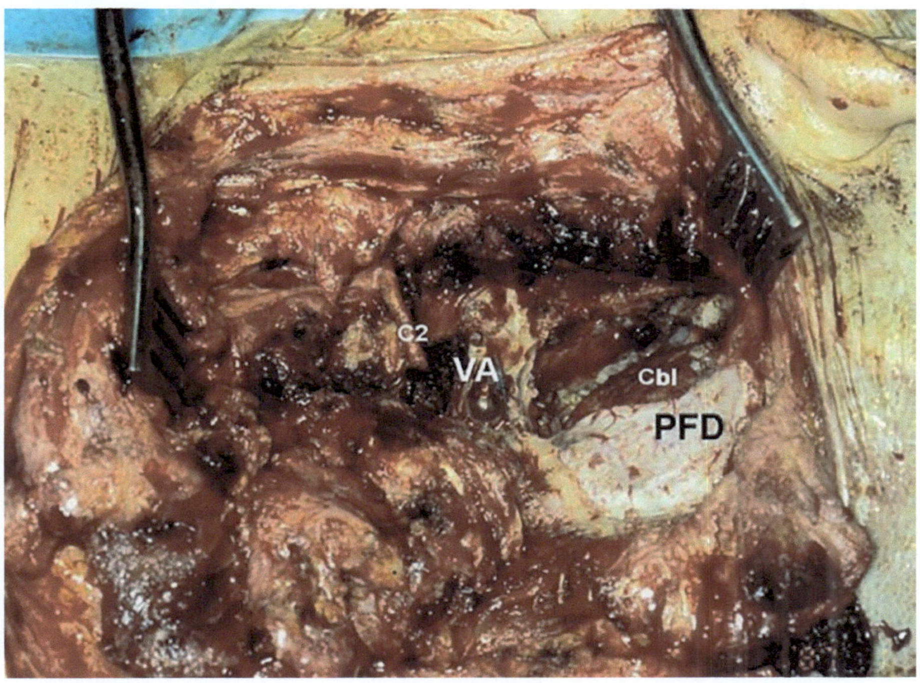

Fig. 14.33 The vertebral artery (VA) is seen just proximal to its intracranial entrance. The balloon can be seen from the open lumen of the ligated artery. The occipital condyle has been drilled. The retrosigmoid posterior fossa dura (PFD) has been opened.

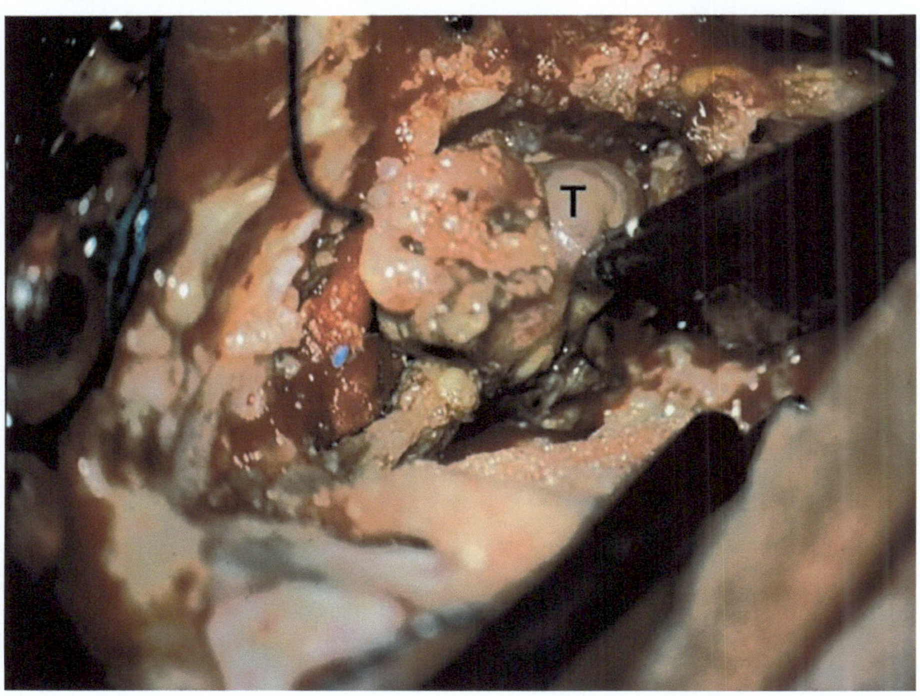

Fig. 14.34 The intradural tumor (T) can be seen.

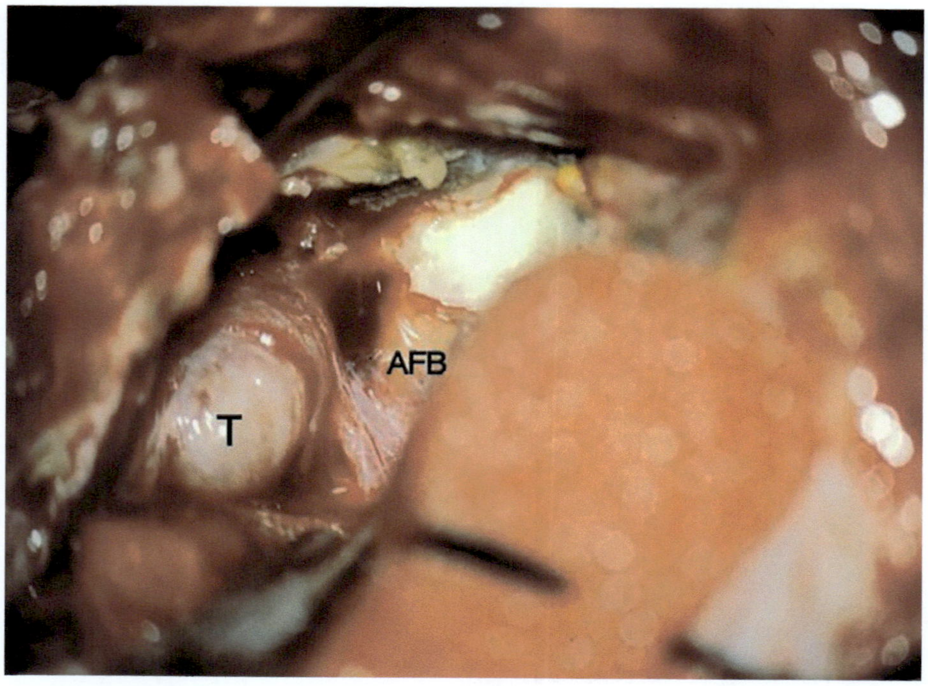

Fig. 14.35 The contact between the tumor and the acousticofacial bundle (AFB) can be appreciated.

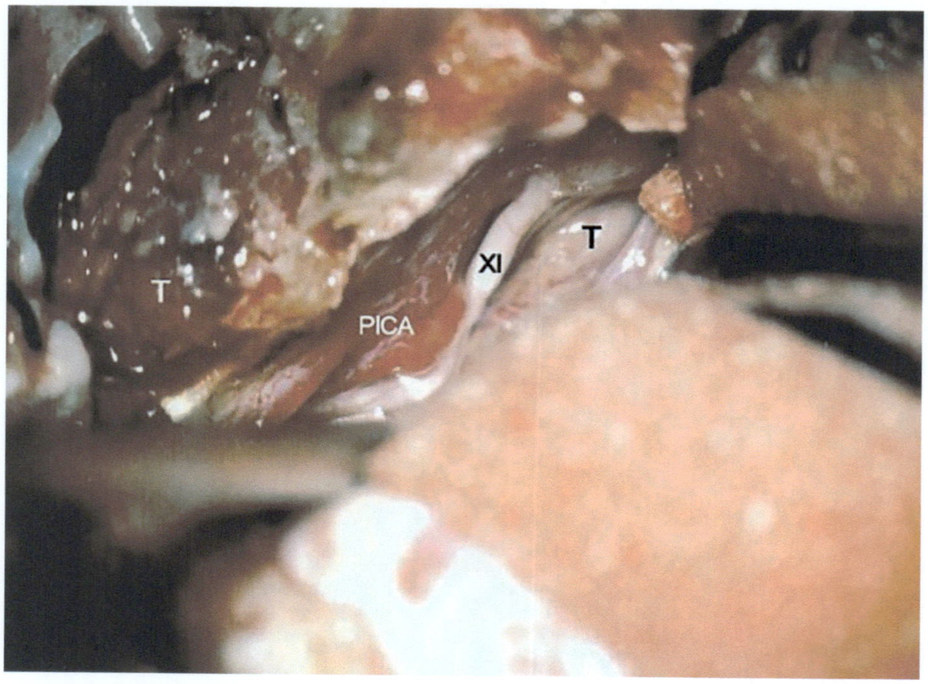

Fig. 14.36 After dissecting the tumor (T) from the acousticofacial bundle, the tumor is further separated from the lower cranial nerves and the posterior inferior cerebellar artery (PICA).

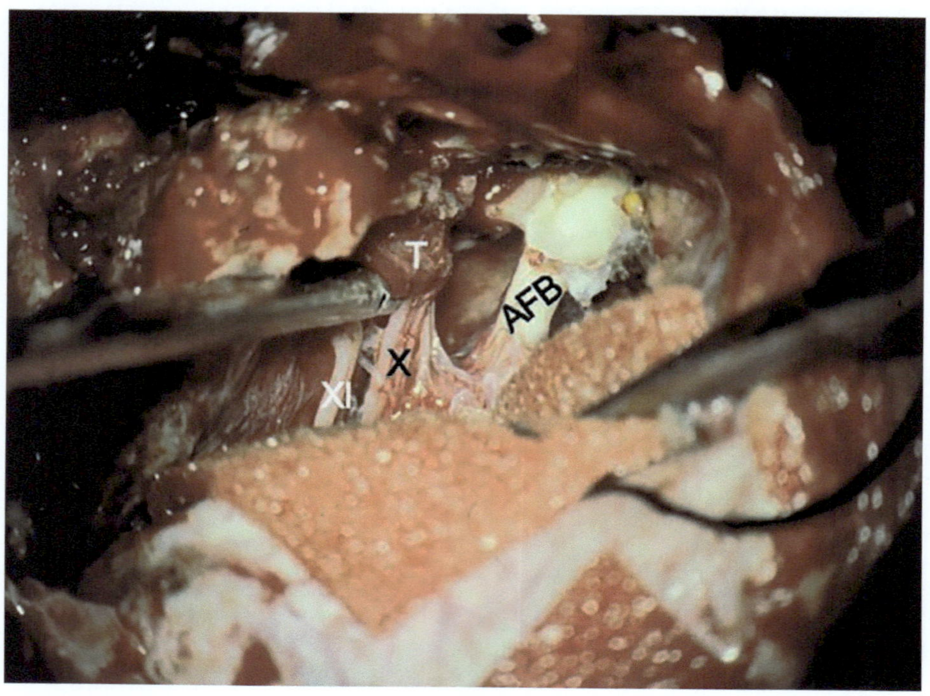

Fig. 14.37 General view showing the last piece of tumor related to the lower cranial nerves.

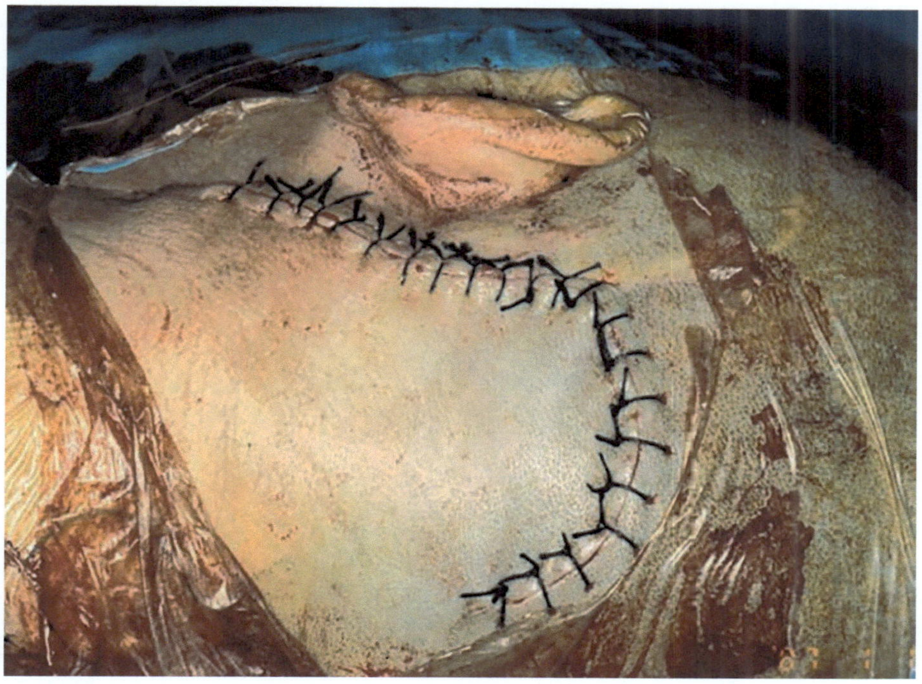

Fig. 14.38 After total tumor removal. The cavity is obliterated with abdominal fat. The dura is sutured. The muscles are sutured back. The skin and subcutaneous tissues are sutured in layers.

HINTS AND PITFALLS

1. The amount of bone removed depends on the extent of the tumor. In addition, vertebral artery management depends on the tumor site. For extradural tumors lying mainly anteriorly, the transverse process of the atlas is removed, and, if needed, the artery is mobilized posteriorly. On the other hand, for tumors lying intradurally and more posteriorly located, the posterior arch of the atlas is drilled. The transverse process of the atlas is left intact, and the artery can be mobilized anteriorly to have unhindered access to the tumor.
2. Care should be taken while drilling the transverse process of the atlas so as not to injure the internal jugular vein.
3. While mobilizing the vertebral artery, the periosteum of the posterior arch of the atlas is elevated and used to protect the artery and to avoid bleeding from the surrounding venous plexus.
4. As the dura is adherent to the vertebral artery at its intradural ingress, a cuff of dura is left around the artery at this point to avoid injuring it.
5. This approach provides a better view of the cerebellopontine angle in comparison with the retrosigmoid approach.

   The additional presigmoid craniotomy allows anterior displacement of the sigmoid sinus, thereby permitting more anterolateral visualization. Moreover, the caudal extension of the extreme lateral approach gives a far superior exposure of the foramen magnum.
6. More control of the vertebrobasilar junction and contralateral vertebral artery is obtained by occluding the jugular bulb and drilling the jugular tubercle, i.e., by combining the approach with the petro-occipital transsigmoid (POTS) approach.

CHAPTER 15

Combined Approach

Introduction

In a variety of complex skull base lesions, single approaches prove inadequate for total and safe tumor removal. A combination of more than one approach can be utilized for the management of such lesions. In general, combined approaches are indicated in:

- Extensive lesions involving more than one anatomical region, e.g., petroclival meningiomas with supratentorial extension or trigeminal schwannomas with marked posterior and middle fossa components. Combined approaches will provide access to the different anatomical compartments involved in the tumor.
- In order to provide enhanced exposure while preserving function, e.g., small petroclival meningiomas. A modified transcochlear approach offers very wide exposure but sacrifices hearing and results in impaired facial nerve function, whereas the traditional retromastoid approach is less safe and offers a narrow exposure. A combined retrolabyrinthine subtemporal approach offers relatively good access with no related functional deficit.

Retrolabyrinthine Subtemporal Transapical and Retrolabyrinthine Transtentorial Approaches

Rationale

The retrolabyrinthine approach entails opening the presigmoid posterior fossa dura (between the sigmoid sinus and the labyrinth). It offers only a narrow access to the posterior fossa. Exposure can be enhanced by drilling and opening the dura of the petrous apex, allowing both posterior (retrolabyrinthine) and superior (subtemporal) access to the tumor. Exposure can also be enhanced by cutting the tentorium after dividing the superior petrosal sinus. This releases the sigmoid sinus, which can be further depressed to enhance the exposure. In addition, cutting the tentorium allows the extirpation of the supratentorial tumor component (Figs. 15.1A, B and 15.2A, B).

Atlas of Microsurgery of the Lateral Skull Base. https://doi.org/10.1016/B978-0-443-19083-4.00015-5

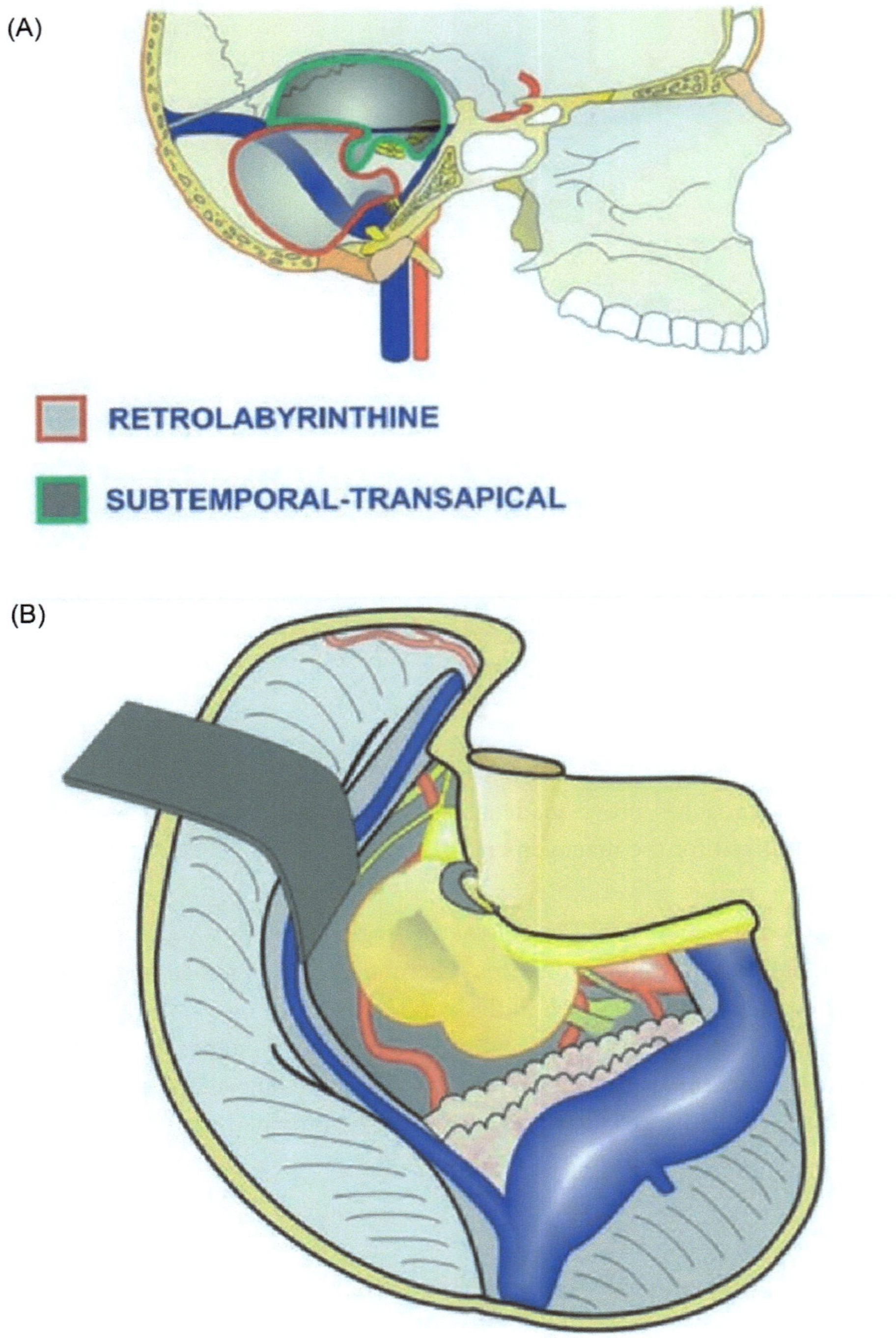

Fig. 15.1 (A) Schematic drawing showing the limits of the combined subtemporal transapical approach. (B) Drawing representing the structures seen after performing a combined retrolabyrinthine, subtemporal transapical approach. Note the wider and better exposure of the posterior fossa structures.

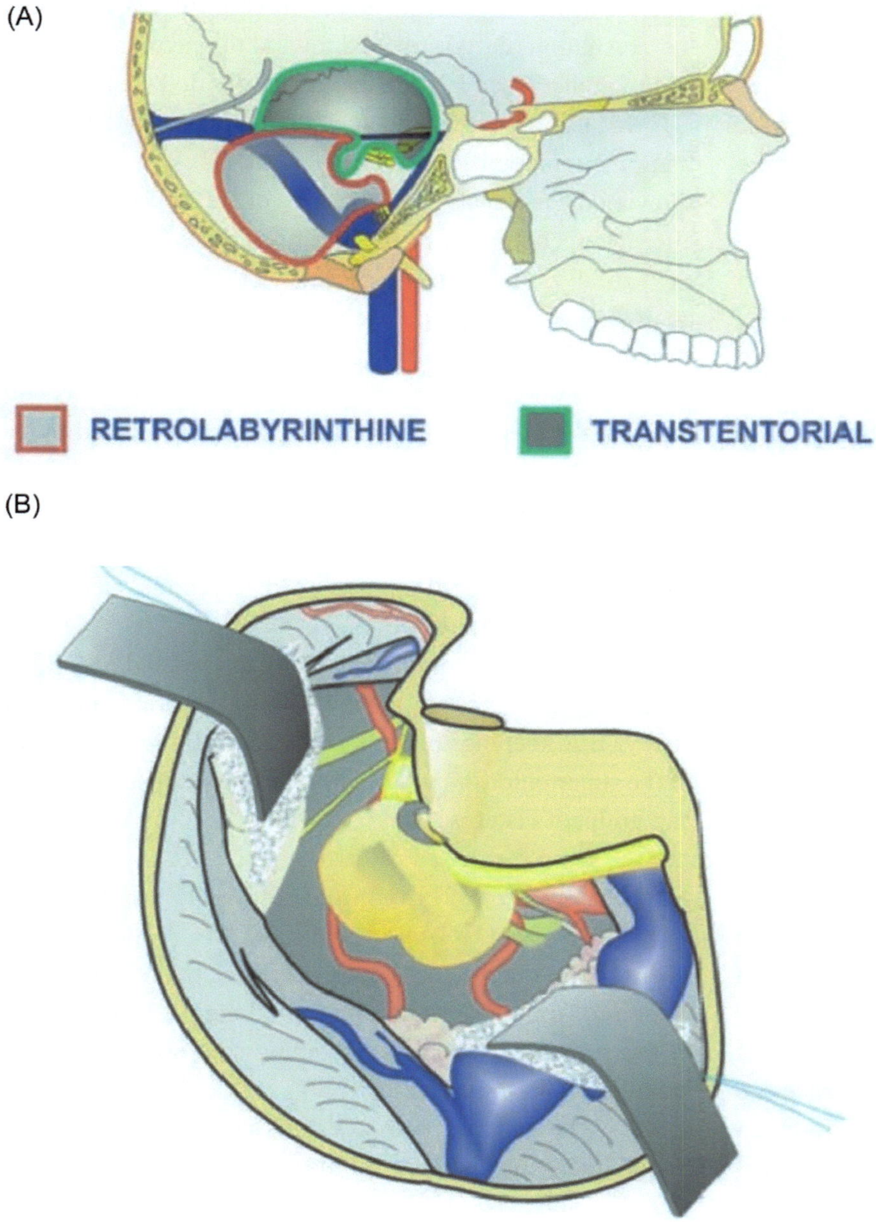

Fig. 15.2 (A) Schematic drawing showing the limits of the combined retrolabyrinthine transtentorial approach. (B) Drawing representing the structures seen after performing a combined retrolabyrinthine, transapical, and transtentorial approach.

Indications

1. Small lesions lying anterior and posterior to the internal auditory canal with intact preoperative hearing and facial nerve function (e.g., meningioma, epidermoids).
2. Small petroclival tumors not passing the midline with or without minimal supratentorial extension.

Surgical Steps

RETROLABYRINTHINE SUBTEMPORAL TRANSAPICAL (TRANSPETROUS APEX) APPROACH

1. A wide C-shaped incision is performed. The incision starts about superiorly behind the hairline for 6–7 cm, extends 4–5 cm posteriorly, and curves just inferior to the mastoid tip. It continues posteriorly parallel to the temporal line and turns inferiorly to pass through a point 2–2.5 cm dorsal to the asterion. The incision terminates just below the level of the mastoid tip (Fig. 15.3).
2. The skin and subcutaneous tissues are elevated and fixed anteriorly using skin hooks (Fig. 15.4).
3. An inferiorly based U-shaped musculoperiosteal flap is outlined and raised using the Lempert periosteal elevator as previously described. The temporalis muscle is detached from its bed and reflected anteriorly (Fig. 15.5).
4. The temporalis muscle is retracted anteriorly exposing the squamosal portion of the temporal bone, the root of the zygomatic arch, and the pterion. An inferiorly based musculoperiosteal flap is retracted inferiorly exposing the mastoid tip and lateral occipital bone. A monopolar coagulator should be used to separate the posterior neck muscles from the occipital bone (Fig. 15.6).
5. An extended canal wall-up mastoidectomy is carried out, uncovering the middle fossa plate and sigmoid

sinus as well as the dura anterior and 2–3 cm posterior to the sinus (Fig. 15.7).

6. The mastoid segment of the facial nerve is skeletonized. After drilling the retrofacial air cells, the jugular bulb has been identified medial to the mastoid segment of the facial nerve (Fig. 15.8). Identification of the facial nerve by facial nerve stimulator (Fig. 15.9).

7. The bone overlying the middle fossa dura, sigmoid sinus, and posterior fossa dura has been removed adequately. For maximal exposure in the retrolabyrinthine approach, the posterolateral portion of the bony labyrinth must be completely skeletonized by shaving the bony canal flat with 1–2 mm diamond burrs. The bone is often shaved until the *blue line* of the posterior and superior semicircular canal is seen (Fig. 15.10).

8. A retrolabyrinthine (presigmoid) craniotomy has been performed. Note the wide removal of bone posterior to the sigmoid sinus (Fig. 15.11).

9. A wide subtemporal craniotomy is performed (Fig. 15.12). After elevating the bone flap, the middle fossa dura can be exposed (Fig. 15.13).

10. The dura of the middle fossa is gradually detached from the superior surface of the temporal bone from posterior to anterior. The dura is retracted with a rigid retractor to expose the petrous apex. The most important landmark is the greater superficial petrosal nerve coming out of the facial hiatus. The superior semicircular canal is already skeletonized by the initial mastoidectomy, and the relation between the superior semicircular canal and the arcuate eminence can be appreciated. The middle meningeal artery and foramen spinosum are identified (Fig. 15.14). The middle meningeal artery is then coagulated using bipolar coagulation and cut (Fig. 15.15).

11. After transecting the middle meningeal artery, the middle fossa dura can be further retracted to expose the mandibular nerve (Fig. 15.16). The mandibular nerve is also coagulated using bipolar coagulation and cut (Fig. 15.17).

12. Overview after performing the craniotomy. The self-retaining retractors are used to support the dura, and the presigmoid and middle fossa dura can be well exposed (Fig. 15.18).

13. After transecting the mandibular nerve, the Gasserian ganglion can be mobilized to further expose the petrous apex area (Fig. 15.19). The rhomboid petrous apex area is now carefully drilled away to expose the petrous carotid artery, cochlea, internal auditory canal, posterior fossa dura, and petroclival junction (Fig. 15.20). Overview after performing the petrous apex removal (Fig. 15.21).

14. The incision starts 1–2 mm parallel to and inferior to the superior petrosal sinus, running from posterior to anterior using microsurgical scissors (Fig. 15.22). And then, the presigmoid dura is opened toward the jugular bulb. The endolymphatic sac is coagulated and divided (Fig. 15.23).

15. After opening the dura, the middle fossa dura overlying the temporal lobe is gently retracted. Panoramic view showing the neurovascular structures of the prepontine cistern after the opening of the posterior fossa dura (Fig. 15.24).

16. By tilting the microscope, the surgeon will see completely different perspectives of the posterior fossa structures (Fig. 15.25). At higher magnification, the anterior inferior cerebellar artery is seen stemming from the basilar artery at the prepontine cistern. The artery is crossed by the abducent nerve. Note the good control of the prepontine cistern through this approach.

17. Tilting the microscope downward as well as gently retracting the sigmoid sinus, the lower cranial nerves are well seen (Fig. 15.26).

18. Retraction of the tentorium upward and opening of Meckel's cave provide a view of the superior cerebellar artery. But keeping the tentorium intact limits the control of the supertentorium space as well as the exposure of crural and ambient cisterns (Fig. 15.27).

RETROLABYRINTHINE SUBTEMPORAL TRANSTENTORIAL APPROACH

1. The middle fossa dura overlying the temporal lobe is incised from posterior to anterior along the superior edge of the superior petrosal sinus, taking care not to injure the vein of Labbé (Fig. 15.28).

2. The superior petrosal sinus is now double-ligated and divided. Special care should be taken not to injure the vein of Labbé. The vein is more vulnerable at its entrance into the transverse sinus (Fig. 15.29).

3. The tentorium is cut with microscissors, starting just in front of the junction between the superior petrosal sinus and the sigmoid sinus and coursing anteromedially. The tentorium should be incised vertically to the superior petrosal sinus. Cutting the tentorium should be carried out under direct vision to avoid injury to the trochlear nerve, which pierces the tentorium to reach the cavernous sinus (Fig. 15.30).

4. After gently retracting the subtemporal dural and incision of the tentorium, the structures of the ambient cistern can be extensively approached. Structures exposed within the ambient cistern include the posterior cerebral artery, superior cerebellar artery, basal vein, and trochlear nerve. Note the excellent control of the trigeminal nerve. Meckel's cave can be opened when necessary (Fig. 15.31).

5. Mild retraction of the temporal lobe and tilting the microscope upward provide a good view of the oculomotor nerve and its relation to the posterior cerebral artery lying

superiorly. The trochlear nerve is seen piercing the tentorium to reach the cavernous sinus (Fig. 15.32).

6. With further retraction of the temporal lobe, the optic, ipsilateral oculomotor, and contralateral oculomotor nerves are seen. The internal carotid, anterior choroidal, and posterior communicating arteries and the perforators can be seen easily with minimal retraction of the temporal lobe (Fig. 15.33).

7. The tentorium is removed from the roof of Meckel's cave to expose the trigeminal nerve root and Gasserian ganglion. Bleeding from the superior petrosal sinus and veins within the roof of Meckel's cave is controlled with absorbable hemostatic fibers and sutures. Extradurally, all branches of the trigeminal nerve are under control (Fig. 15.34). Further extension into the cavernous sinus can be achieved by opening the dura of the sinus.

8. After total tumor removal and the accomplishment of hemostasis, the operative cavity is obliterated with abdominal fat that protrudes into the cerebellopontine angle. The musculoperiosteal flap is sutured back. Fat is also placed to obliterate the subtemporal aspect of the approach. The temporal craniotomy is replaced and is fixed in place using silk sutures. The temporalis is next repositioned and sutured using Vicryl 1/0 sutures.

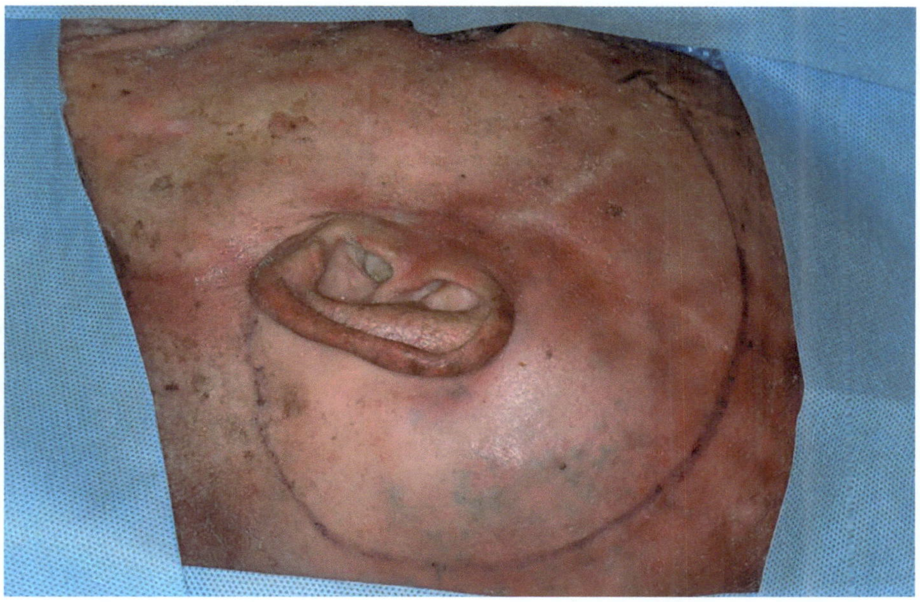

Fig. 15.3 A wide C-shaped incision is performed. The incision starts about superiorly behind the hairline for 6–7 cm, extends 4–5 cm posteriorly, and curves just inferior to the mastoid tip. It continues posteriorly parallel to the temporal line and turns inferiorly to pass through a point 2–2.5 cm dorsal to the asterion. The incision terminates just below the level of the mastoid tip.

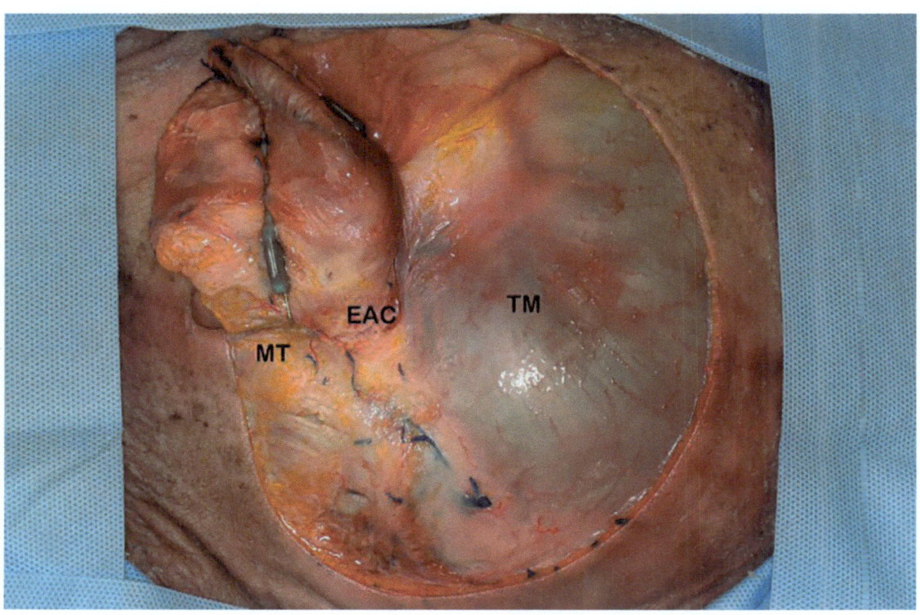

Fig. 15.4 The skin and subcutaneous tissues are elevated and fixed anteriorly using skin hooks. *TM*, temporalis muscle; *EAC*, external auditory canal; *MT*, mastoid tip.

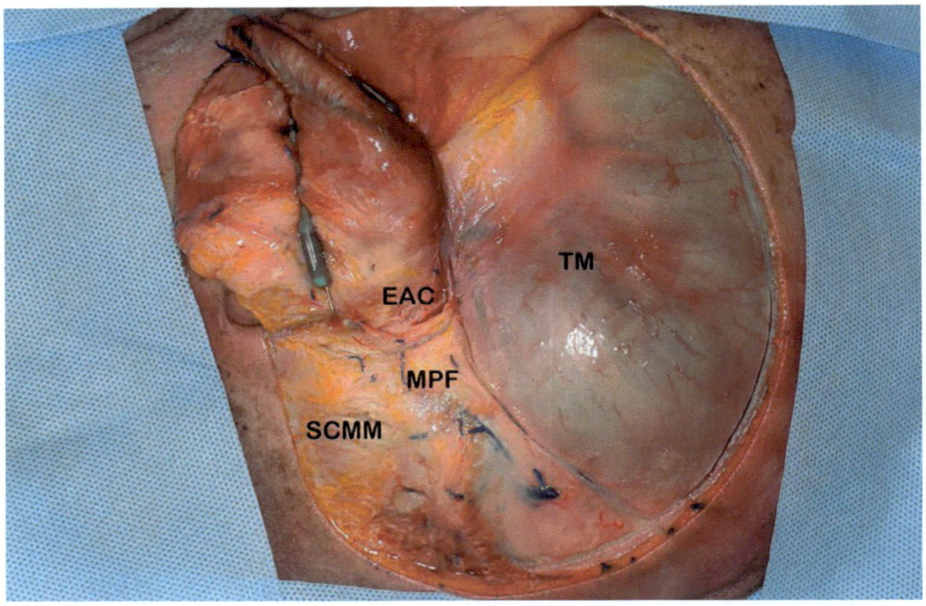

Fig. 15.5 An inferiorly based U-shaped musculoperiosteal flap is outlined and raised using the Lempert periosteal elevator as previously described. The temporalis muscle is detached from its bed and reflected anteriorly. *TM*, temporalis muscle; *EAC*, external auditory canal; *MPF*, musculoperiosteal flap; *SCMM*, sternocleidomastoid muscle.

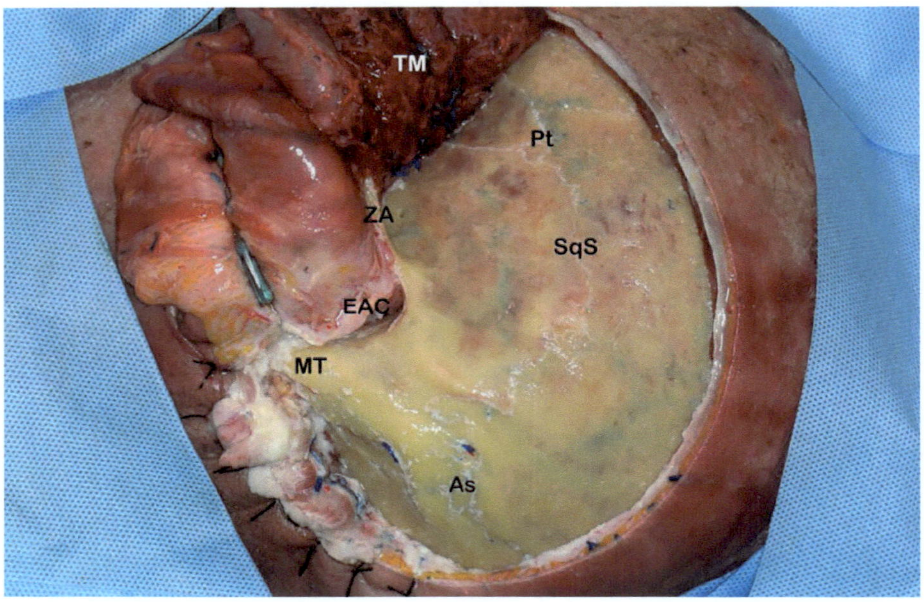

Fig. 15.6 The temporalis muscle is retracted anteriorly exposing the squamosal portion of the temporal bone, the root of the zygomatic arch, and the pterion. An inferiorly based musculoperiosteal flap is retracted inferiorly exposing the mastoid tip and lateral occipital bone. A monopolar coagulator should be used to separate the posterior neck muscles from the occipital bone. *Pt*, pterion; *SqS*, squamous suture; *TM*, temporalis muscle (retracted); *ZA*, zygomatic arch; *EAC*, external auditory canal; *MT*, mastoid tip; *As*, asterion.

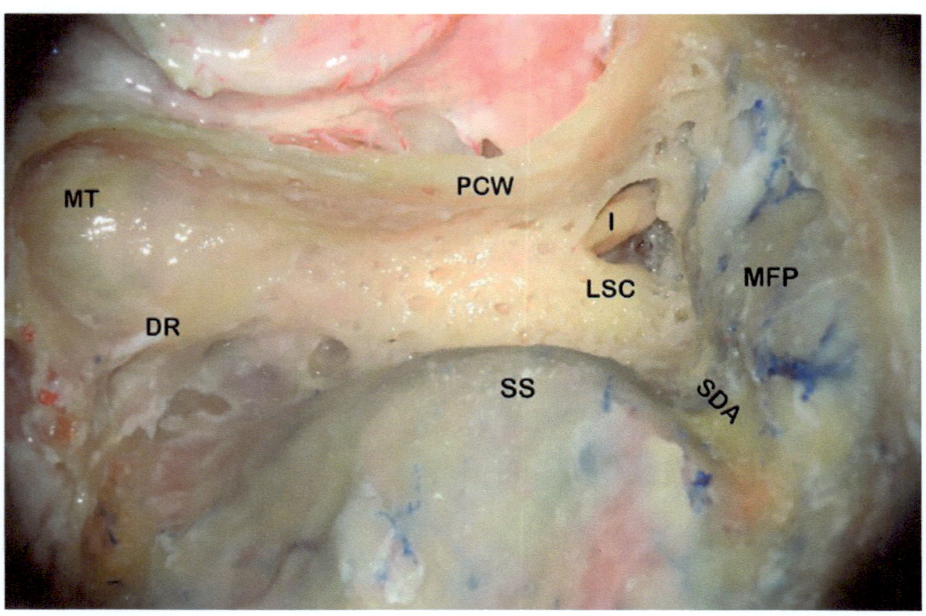

Fig. 15.7 An extended canal wall-up mastoidectomy is carried out, uncovering the middle fossa plate and sigmoid sinus as well as the dura anterior and 2–3 cm posterior to the sinus. *MFP*, middle fossa plate; *I*, incus; *LSC*, lateral semicircular canal; *SDA*, sinodural angle; *SS*, sigmoid sinus; *DR*, digastric ridge; *MT*, mastoid tip; *PCW*, posterior canal wall.

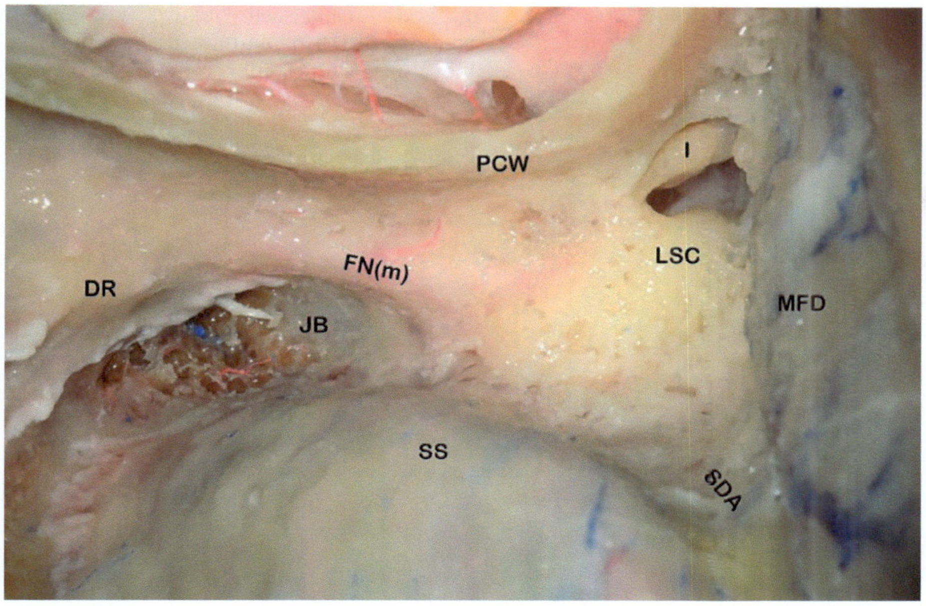

Fig. 15.8 The mastoid segment of the facial nerve is skeletonized. After drilling the retrofacial air cells, the jugular bulb has been identified medial to the mastoid segment of the facial nerve. *I*, incus; *LSC*, lateral semicircular canal; *MFD*, middle fossa dura; *SDA*, sinodural angle; *PCW*, posterior canal wall; *FN(m)*, mastoid segment of facial nerve; *JB*, jugular bulb; *SS*, sigmoid sinus; *DR*, digastric ridge.

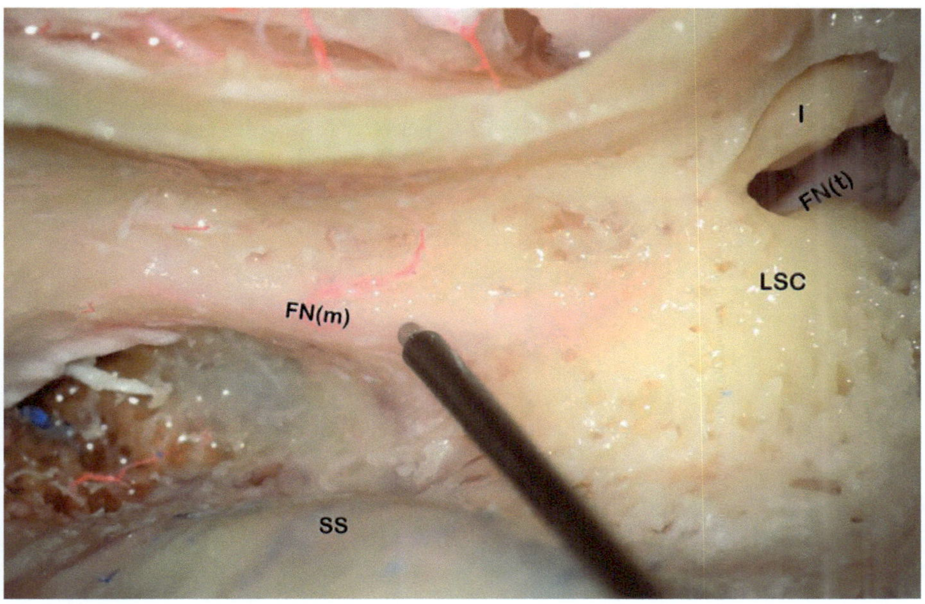

Fig. 15.9 Identification of the facial nerve by facial nerve stimulator. *I*, incus; *FN(t)*, tympanic segment of facial nerve; *FN(m)*, mastoid segment of facial nerve; *LSC*, lateral semicircular canal; *SS*, sigmoid sinus.

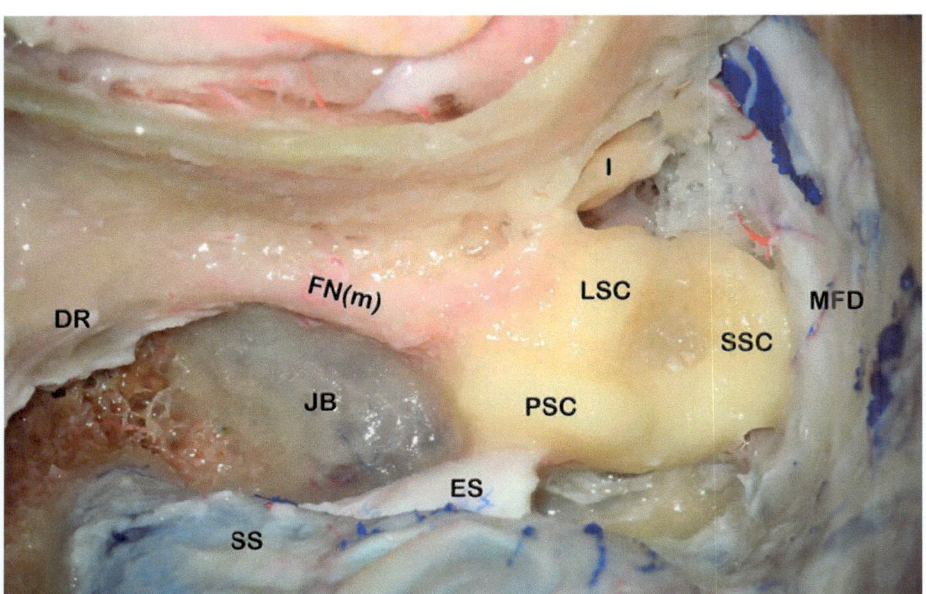

Fig. 15.10 The bone overlying the middle fossa dura, sigmoid sinus, and posterior fossa dura has been removed adequately. For maximal exposure in the retrolabyrinthine approach, the posterolateral portion of the bony labyrinth must be completely skeletonized by shaving the bony canal flat with 1–2 mm diamond burrs. The bone is often shaved until the *blue line* of the posterior and superior semicircular canals is seen. *MFD*, middle fossa dura; *I*, incus; *SSC*, superior semicircular canal; *LSC*, lateral semicircular canal; *PSC*, posterior semicircular canal; *ES*, endolymphatic sac; *JB*, jugular bulb; *FN(m)*, mastoid segment of facial nerve; *SS*, sigmoid sinus; *DR*, digastric ridge.

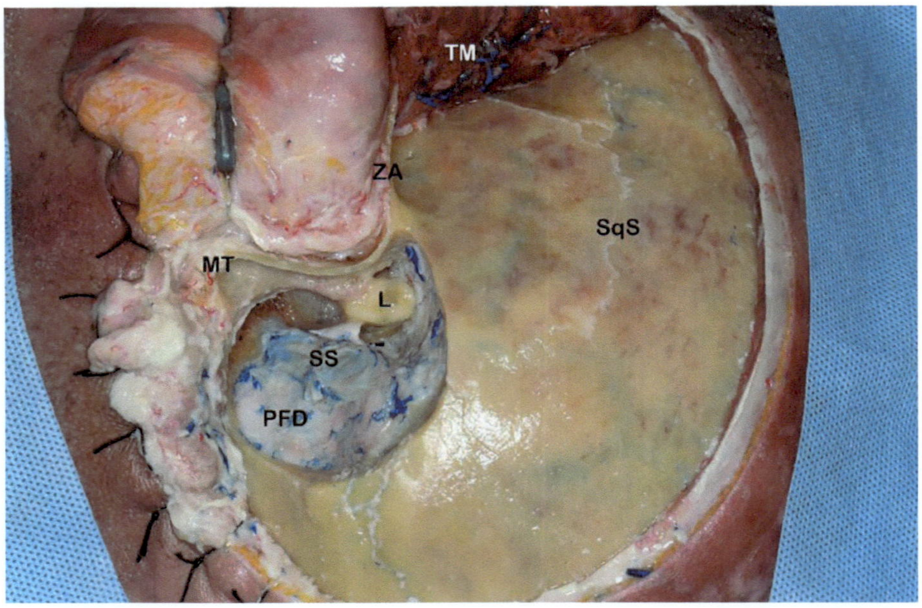

Fig. 15.11 A retrolabyrinthine (presigmoid) craniotomy has been performed. Note the wide removal of bone posterior to the sigmoid sinus. *TM*, temporalis muscle (retracted); *SqS*, squamous suture; *ZA*, zygomatic arch; *L*, labyrinth; *SS*, sigmoid sinus; *PFD*, posterior fossa dura; *MT*, mastoid tip.

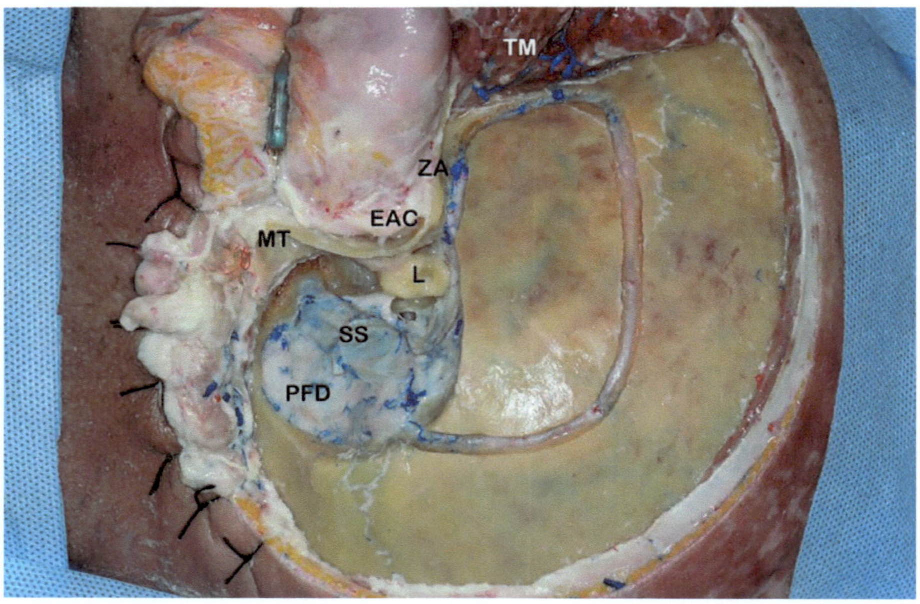

Fig. 15.12 A wide subtemporal craniotomy is performed. *TM*, temporalis muscle (retracted); *ZA*, zygomatic arch; *EAC*, external auditory canal; *L*, labyrinth; *SS*, sigmoid sinus; *MT*, mastoid tip; *PFD*, posterior fossa dura.

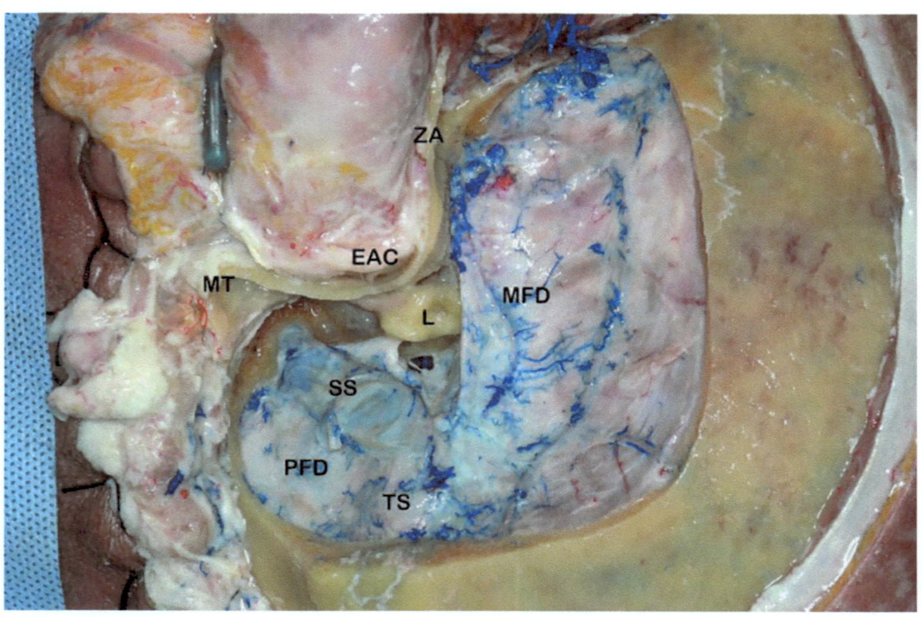

Fig. 15.13 Exposing the middle fossa dura after elevating the bone flap. *ZA*, zygomatic arch; *MFD*, middle fossa dura; *EAC*, external auditory canal; *L*, labyrinth; *SS*, sigmoid sinus; *PFD*, posterior fossa dura; *TS*, transverse sinus; *MT*, mastoid tip.

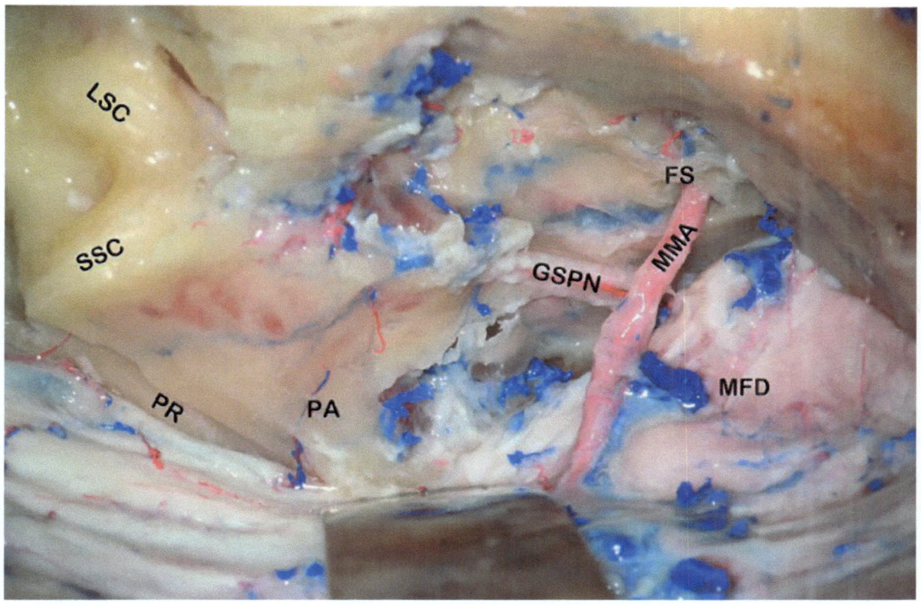

Fig. 15.14 The dura of the middle fossa is gradually detached from the superior surface of the temporal bone from posterior to anterior. The dura is retracted with a rigid retractor to expose the petrous apex. The middle meningeal artery and foramen spinosum are identified. The most important landmark is the greater superficial petrosal nerve coming out of the facial hiatus. The superior semicircular canal is already skeletonized by the initial mastoidectomy, and the relation between the superior semicircular canal and the arcuate eminence can be appreciated. *FS*, foramen spinosum; *MMA*, middle meningeal artery; *MFD*, middle fossa dura; *GSPN*, greater superficial petrosal nerve; *PA*, petrous apex; *PR*, petrous ridge; *SSC*, superior semicircular canal; *LSC*, lateral semicircular canal.

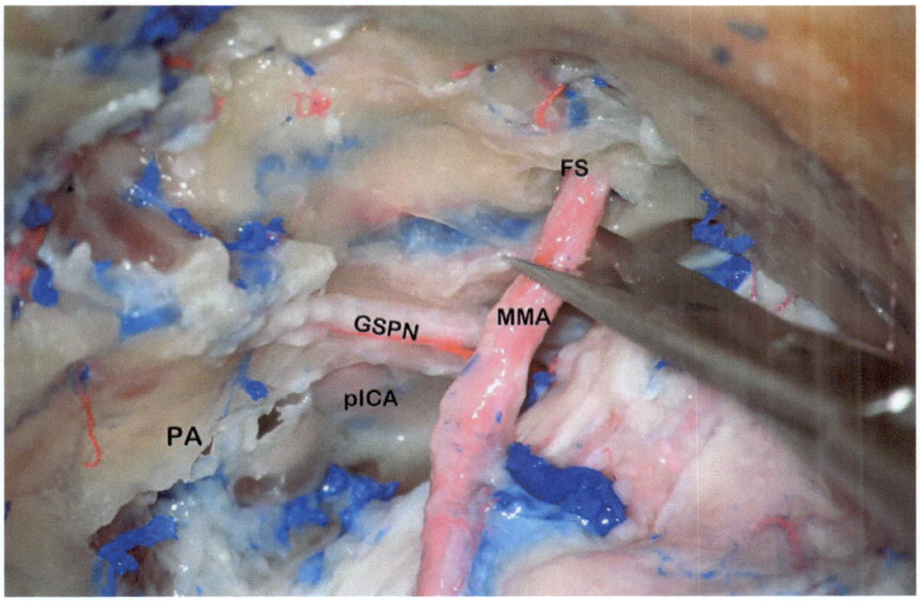

Fig. 15.15 The middle meningeal artery is coagulated using bipolar coagulation and cut. *FS*, foramen spinosum; *MMA*, middle meningeal artery; *GSPN*, greater superficial petrosal nerve; *pICA*, petrous segment of internal carotid artery; *PA*, petrous apex.

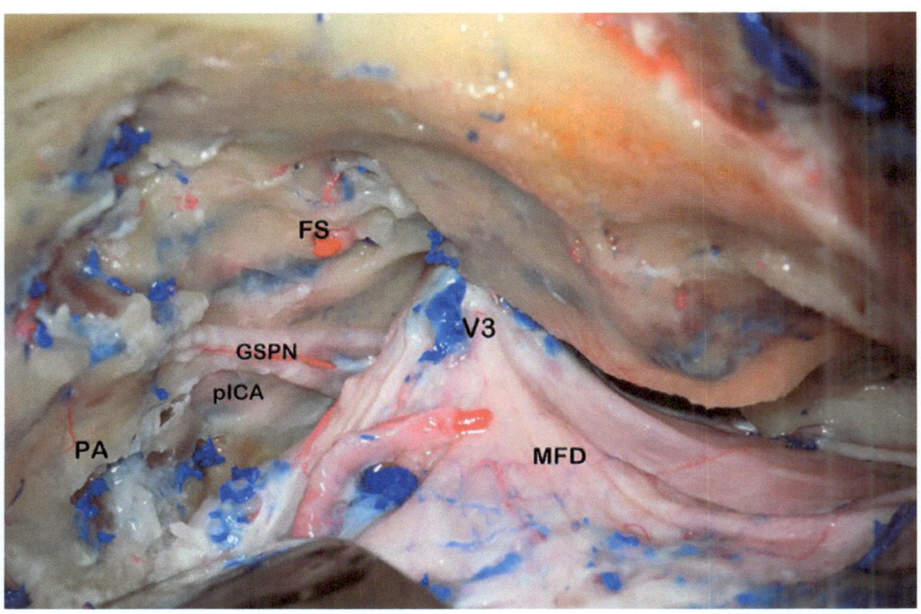

Fig. 15.16 After transecting the middle meningeal artery, the middle fossa dura can be further retracted to expose the mandibular nerve. *MFD*, middle fossa dura; *V3*, mandibular nerve; *FS*, foramen spinosum; *GSPN*, greater superficial petrosal nerve; *pICA*, petrous segment of internal carotid artery; *PA*, petrous apex.

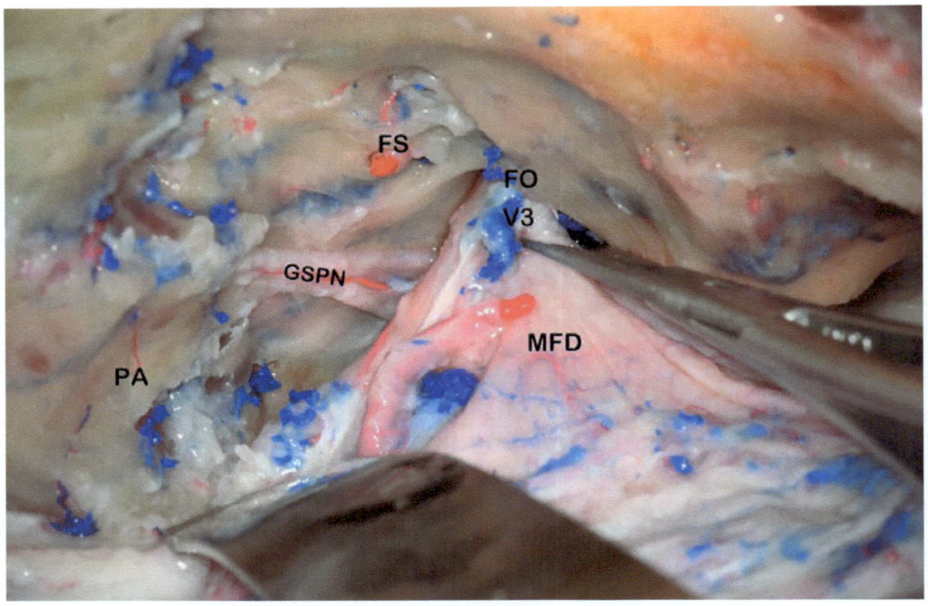

Fig. 15.17 The mandibular nerve is also coagulated using bipolar coagulation and cut. *FS*, foramen spinosum; *FO*, foramen ovale; *V3*, mandibular nerve; *MFD*, middle fossa dura; *GSPN*, greater superficial petrosal nerve; *PA*, petrous apex.

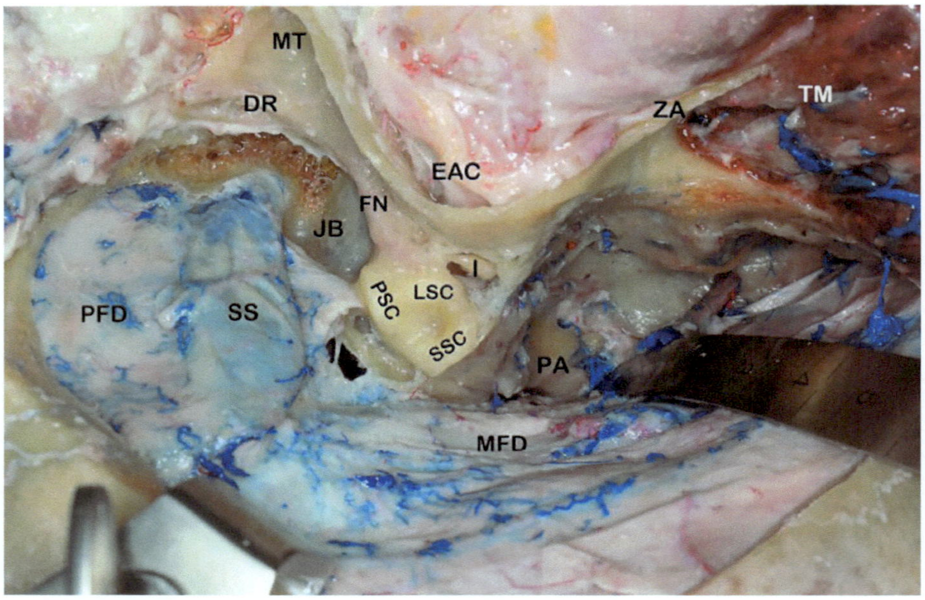

Fig. 15.18 Overview after performing the craniotomy. The self-retaining retractors are used to support the dura, and the presigmoid and middle fossa dura can be well exposed. *TM*, temporalis muscle (retracted); *ZA*, zygomatic arch; *EAC*, external auditory canal; *FN*, facial nerve; *I*, incus; *LSC*, lateral semicircular canal; *SSC*, superior semicircular canal; *PSC*, posterior semicircular canal; *PA*, petrous apex; *MFD*, middle fossa dura; *SS*, sigmoid sinus; *PFD*, posterior fossa dura; *JB*, jugular bulb; *DR*, digastric ridge; *MT*, mastoid tip.

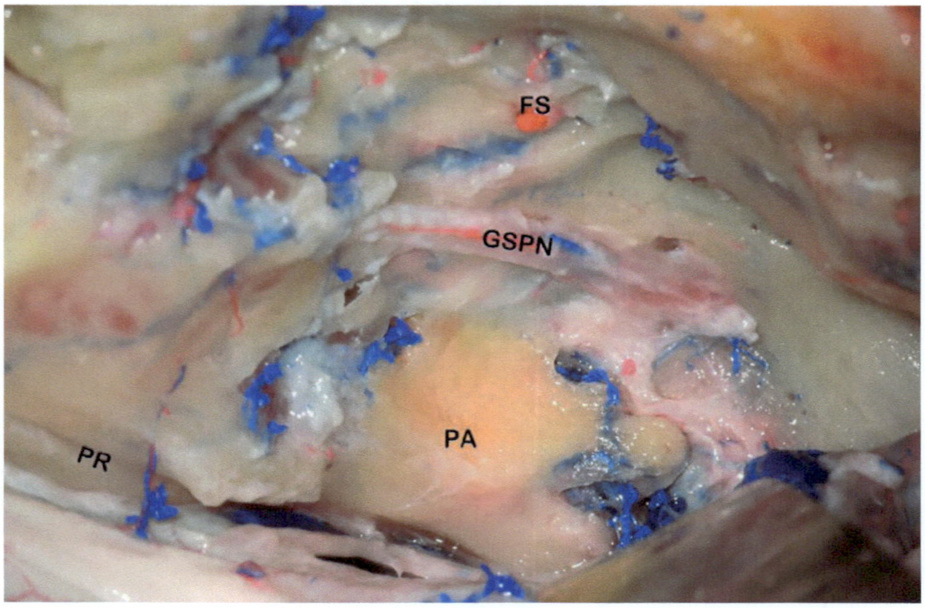

Fig. 15.19 After transecting the mandibular nerve, the Gasserian ganglion can be mobilized to further expose the petrous apex area. *FS*, foramen spinosum; *GSPN*, greater superficial petrosal nerve; *PA*, petrous apex; *PR*, petrous ridge.

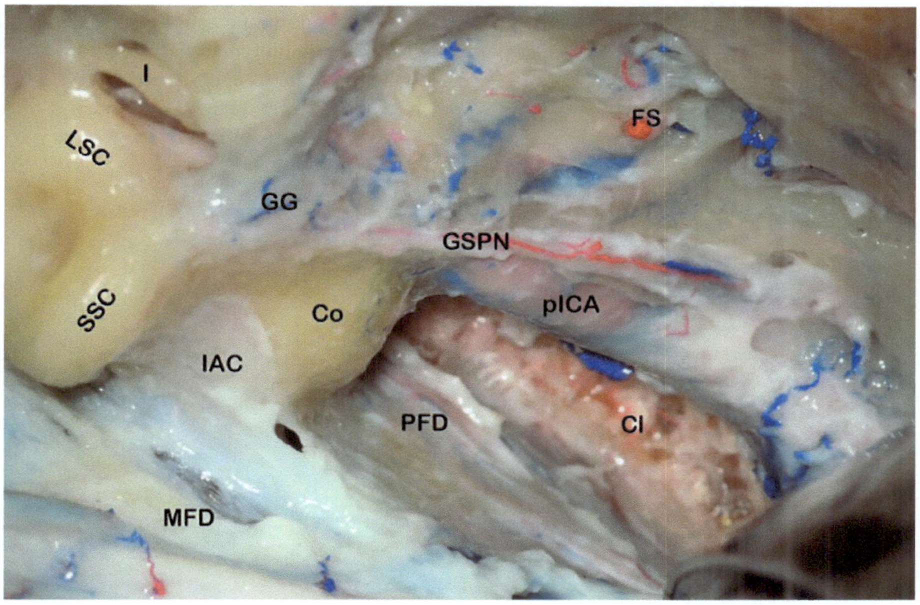

Fig. 15.20 The rhomboid petrous apex area is now carefully drilled away to expose the petrous carotid artery, cochlea, internal auditory canal, posterior fossa dura, and petroclival junction. *FS*, foramen spinosum; *GSPN*, greater superficial petrosal nerve; *pICA*, petrous segment of internal carotid artery; *Cl*, clivus; *PFD*, posterior fossa dura; *Co*, cochlea; *IAC*, internal auditory canal; *MFD*, middle fossa dura; *SSC*, superior semicircular canal; *GG*, geniculate ganglion; *I*, incus; *LSC*, lateral semicircular canal.

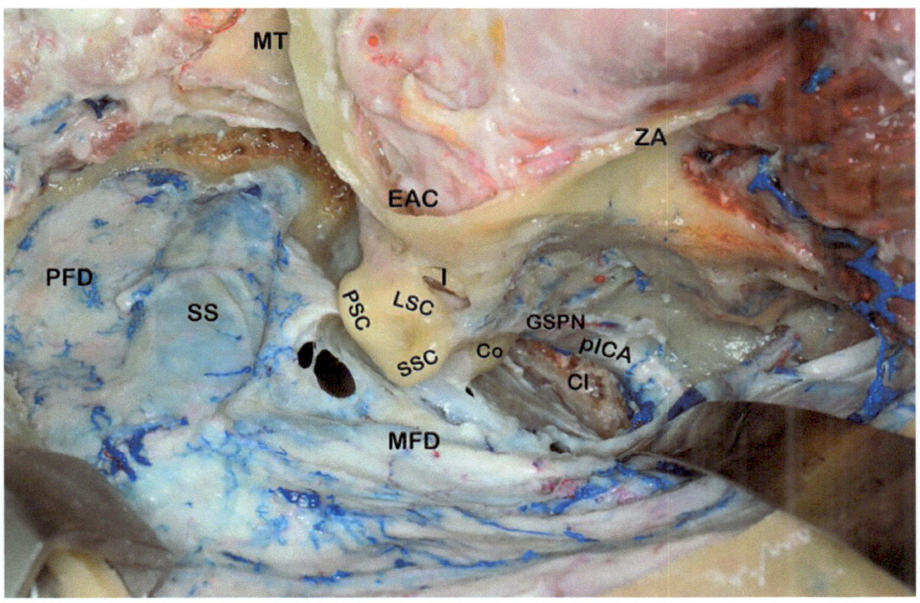

Fig. 15.21 Overview after performing the petrous apex removal. *ZA*, zygomatic arch; *Cl*, clivus; *pICA*, petrous segment of internal carotid artery; *GSPN*, greater superficial petrosal nerve; Co, cochlea; *MFD*, middle fossa dura; *SSC*, superior semicircular canal; *LSC*, lateral semicircular canal; *I*, incus; *PSC*, posterior semicircular canal; *EAC*, external auditory canal; *MT*, mastoid tip; *SS*, sigmoid sinus; *PFD*, posterior fossa dura.

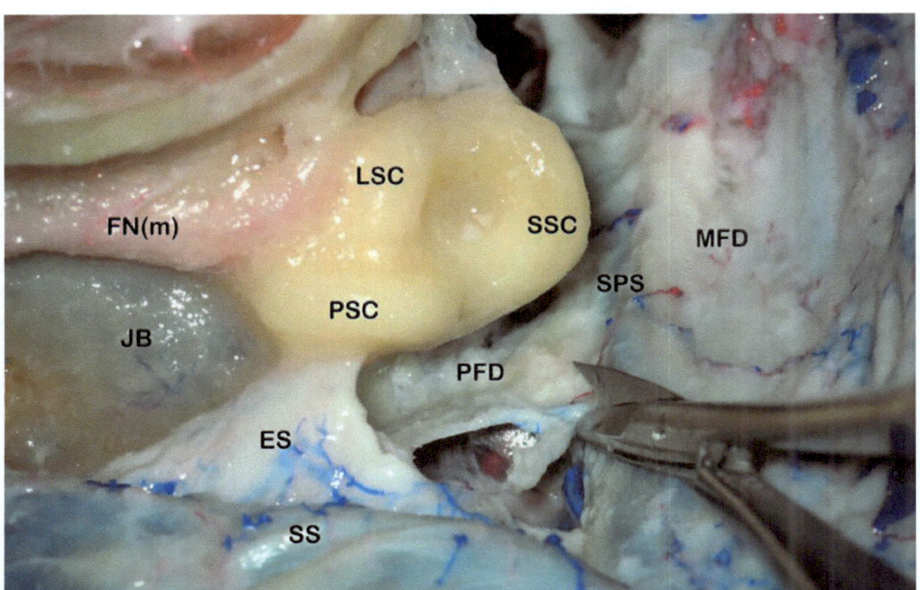

Fig. 15.22 The incision starts 1–2 mm parallel to and inferior to the superior petrosal sinus, running from posterior to anterior using microsurgical scissors. *MFD*, middle fossa dura; *SSC*, superior semicircular canal; *SPS*, superior petrosal sinus; *PFD*, posterior fossa dura; *PSC*, posterior semicircular canal; *LSC*, lateral semicircular canal; *FN(m)*, mastoid segment of facial nerve; *JB*, jugular bulb; *ES*, endolymphatic sac; *SS*, sigmoid sinus.

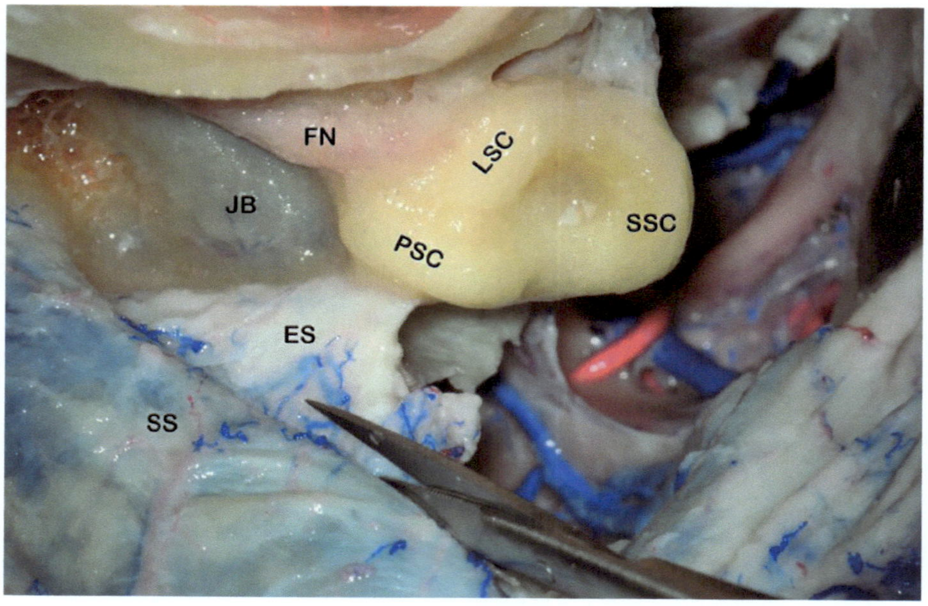

Fig. 15.23 And then, the presigmoid dura is opened toward the jugular bulb. The endolymphatic sac is coagulated and divided. *SSC*, superior semicircular canal; *LSC*, lateral semicircular canal; *PSC*, posterior semicircular canal; *FN*, facial nerve; *ES*, endolymphatic sac; *JB*, jugular bulb; *SS*, sigmoid sinus.

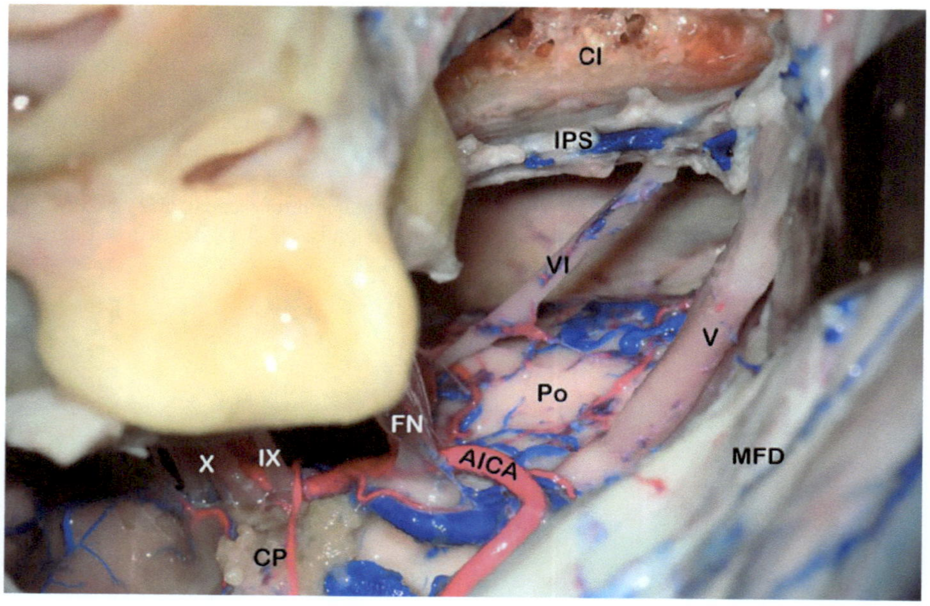

Fig. 15.24 After opening the dura, the middle fossa dura overlying the temporal lobe is gently retracted. Panoramic view showing the neurovascular structures of the prepontine cistern after the opening of the posterior fossa dura. *Cl*, clivus; *IPS*, inferior petrosal sinus; *VI*, abducent nerve; *V*, trigeminal nerve; *MFD*, middle fossa dura; *Po*, pons; *AICA*, anteroinferior cerebellar artery; *FN*, facial nerve; *IX*, glossopharyngeal nerve; *X*, vagus nerve; *CP*, choroid plexus.

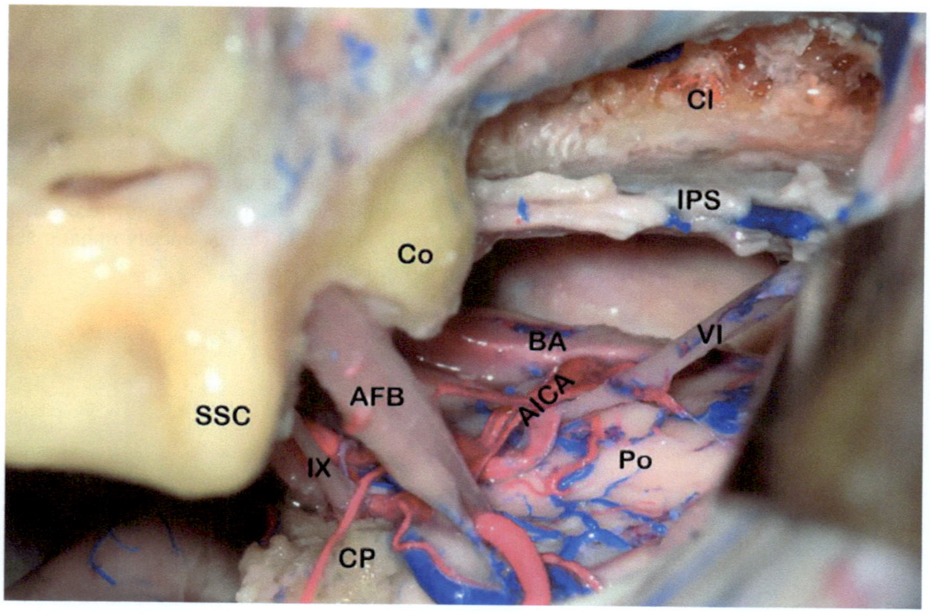

Fig. 15.25 By tilting the microscope, the surgeon will see completely different perspectives of the posterior fossa structures. At higher magnification, the anterior inferior cerebellar artery is seen stemming from the basilar artery at the prepontine cistern. The artery is crossed by the abducent nerve. Note the good control of the prepontine cistern through this approach. *Cl*, clivus; *IPS*, inferior petrosal sinus; *VI*, abducent nerve; *Po*, pons; *AICA*, anteroinferior cerebellar artery; *BA*, basilar artery; *Co*, cochlea; *AFB*, acousticofacial bundle; *IX*, glossopharyngeal nerve; *CP*, choroid plexus; *SSC*, superior semicircular canal.

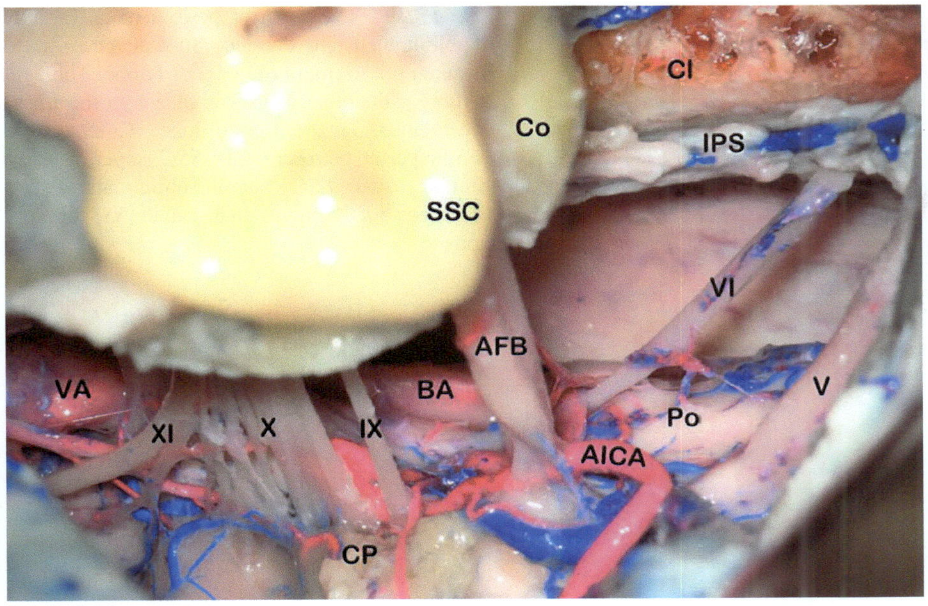

Fig. 15.26 Tilting the microscope downward as well as gently retracting the sigmoid sinus, the lower cranial nerves are well seen. *Cl*, clivus; *IPS*, inferior petrosal sinus; *VI*, abducent nerve; *V*, trigeminal nerve; *Po*, pons; *AICA*, anteroinferior cerebellar artery; *AFB*, acousticofacial bundle; *BA*, basilar artery; *SSC*, superior semicircular canal; *Co*, cochlea; *IX*, glossopharyngeal nerve; *CP*, cochleariform process; *X*, vagus nerve; *XI*, accessory nerve; *VA*, vertebral artery.

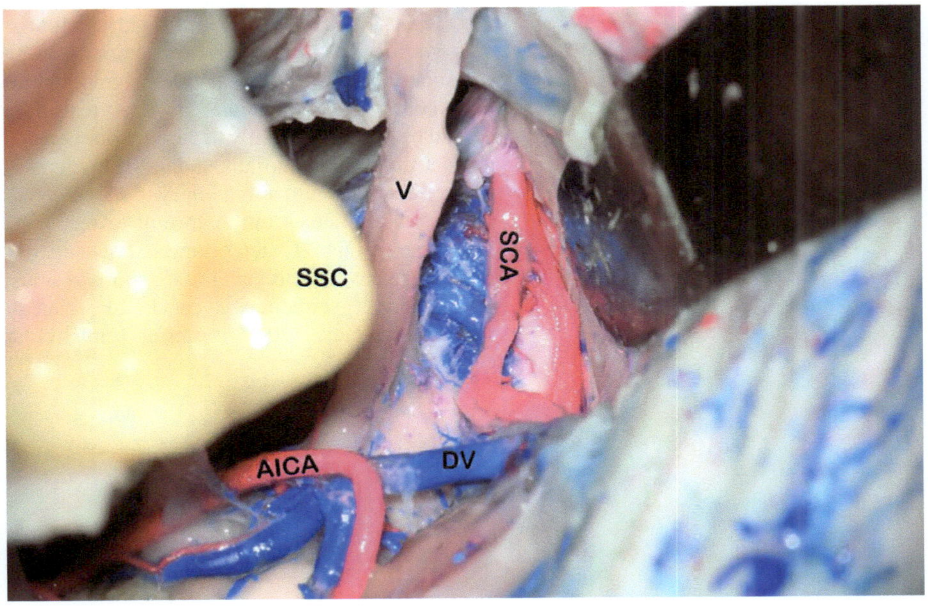

Fig. 15.27 Retraction of the tentorium upward and opening of Meckel's cave provide a view of the superior cerebellar artery. But keeping the tentorium intact limits the control of the supertentorium space as well as the exposure of crural and ambient cisterns. *SCA*, superior cerebellar artery; *DV*, Dandy's vein (superior petrosal vein); *SSC*, superior semicircular canal; *V*, trigeminal nerve; *AICA*, anteroinferior cerebellar artery.

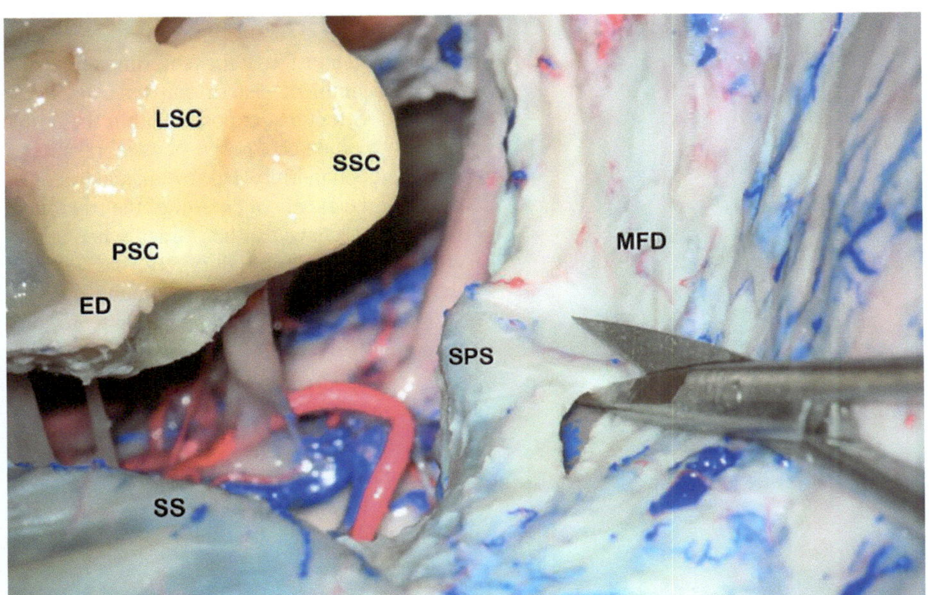

Fig. 15.28 Then, the subtemporal dura is opened posterior to anterior along the superior edge of the superior petrosal sinus. *MFD*, middle fossa dura; *SPS*, superior petrosal sinus; *SSC*, superior semicircular canal; *LSC*, lateral semicircular canal; *PSC*, posterior semicircular canal; *ED*, endolymphatic duct; *SS*, sigmoid sinus.

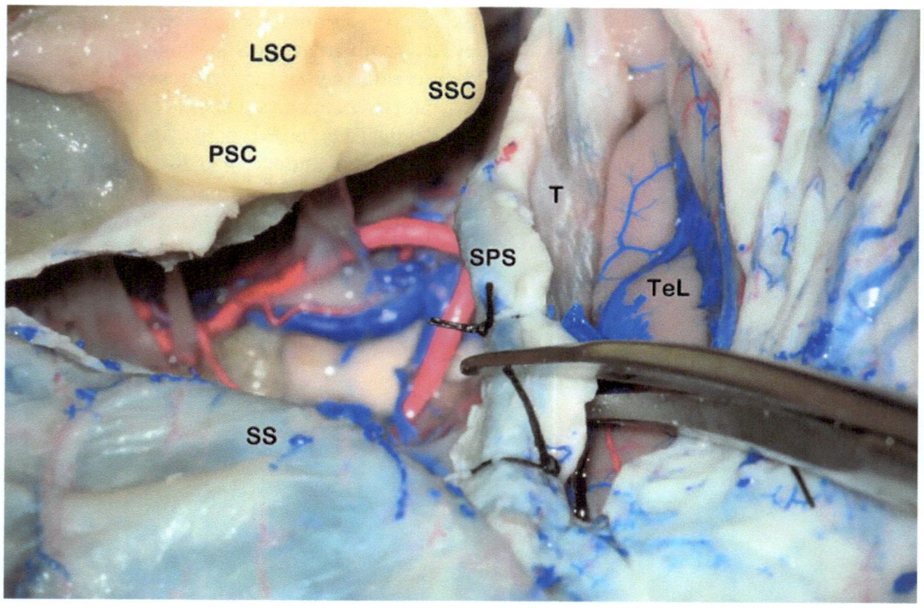

Fig. 15.29 The superior petrosal sinus is now double-ligated and divided. Special care should be taken not to injure the vein of Labbé. The vein is more vulnerable at its entrance into the transverse sinus. *TeL*, temporal lobe; *T*, tentorium; *SPS*, superior petrosal sinus; *SSC*, superior semicircular canal; *LSC*, lateral semicircular canal; *PSC*, posterior semicircular canal; *SS*, sigmoid sinus.

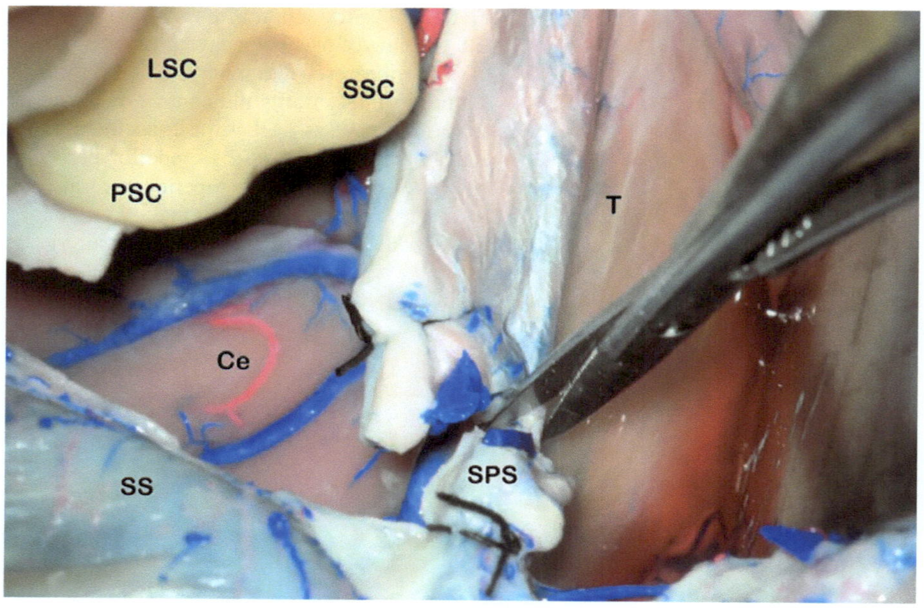

Fig. 15.30 The tentorium is cut with microscissors, starting just in front of the junction between the superior petrosal sinus and the sigmoid sinus and coursing anteromedially. The tentorium should be incised vertically to the superior petrosal sinus. Cutting the tentorium should be carried out under direct vision to avoid injury to the trochlear nerve, which pierces the tentorium to reach the cavernous sinus. *T*, tentorium; *SPS*, superior petrosal sinus; *Ce*, cerebellum; *SS*, sigmoid sinus; *SSC*, superior semicircular canal; *LSC*, lateral semicircular canal; *PSC*, posterior semicircular canal.

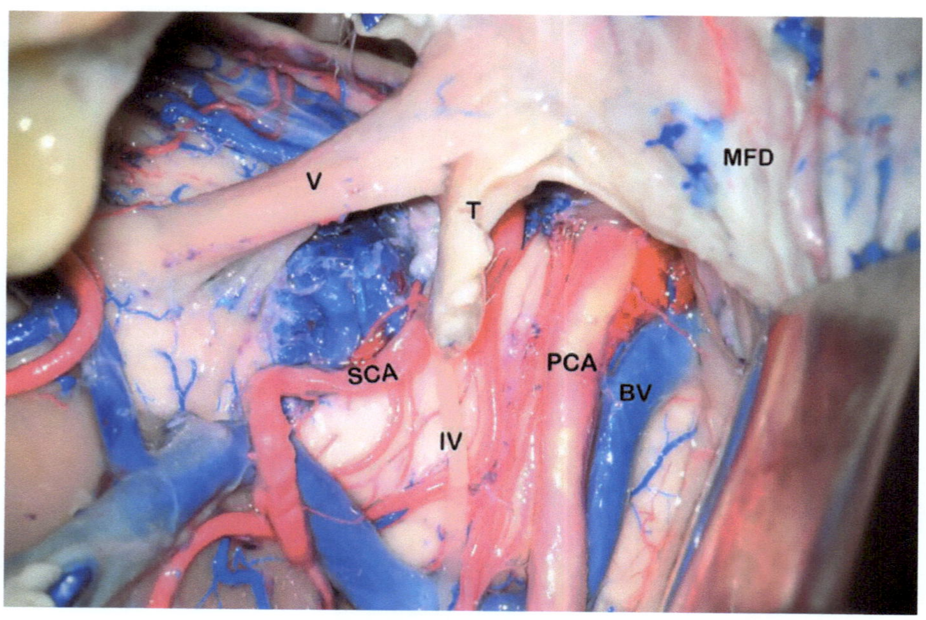

Fig. 15.31 After gently retracting the subtemporal dural and incision of the tentorium, the structures of the ambient cistern can be extensively approached. Structures exposed within the ambient cistern include the posterior cerebral artery, superior cerebellar artery, basal vein, and trochlear nerve. Note the excellent control of the trigeminal nerve. Meckel's cave can be opened when necessary. *MFD*, middle fossa dura; *BV*, basilar vein; *PCA*, posterior cerebral artery; *IV*, trochlear nerve; *SCA*, superior cerebellar artery; *T*, tentorium; *V*, trigeminal nerve.

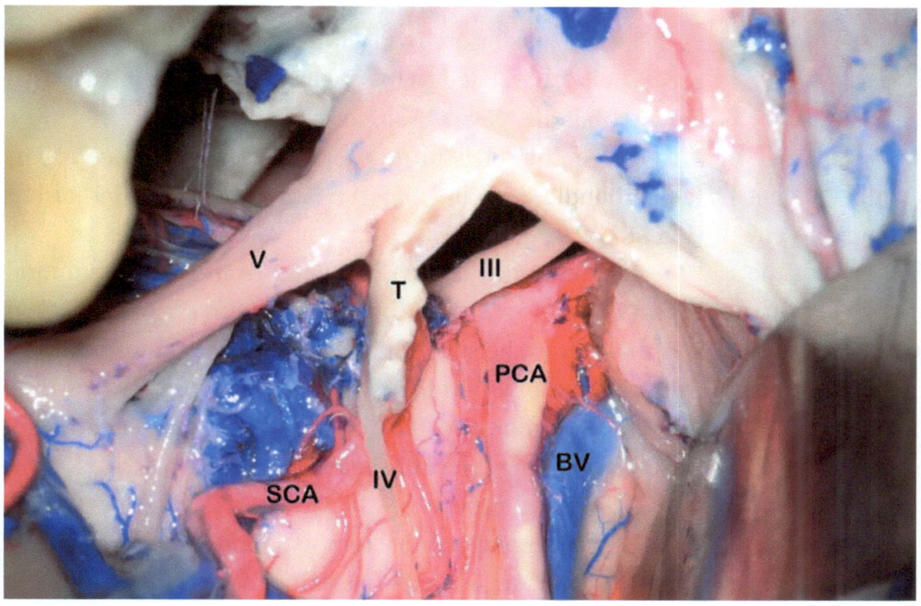

Fig. 15.32 Mild retraction of the temporal lobe and tilting the microscope upward provide a good view of the oculomotor nerve and its relation to the posterior cerebral artery lying superiorly. The trochlear nerve is seen piercing the tentorium to reach the cavernous sinus. *III*, oculomotor nerve; *PCA*, posterior cerebral artery; *BV*, basilar vein;*T*, tentorium; *IV*, trochlear nerve; *SCA*, superior cerebellar artery; *V*, trigeminal nerve.

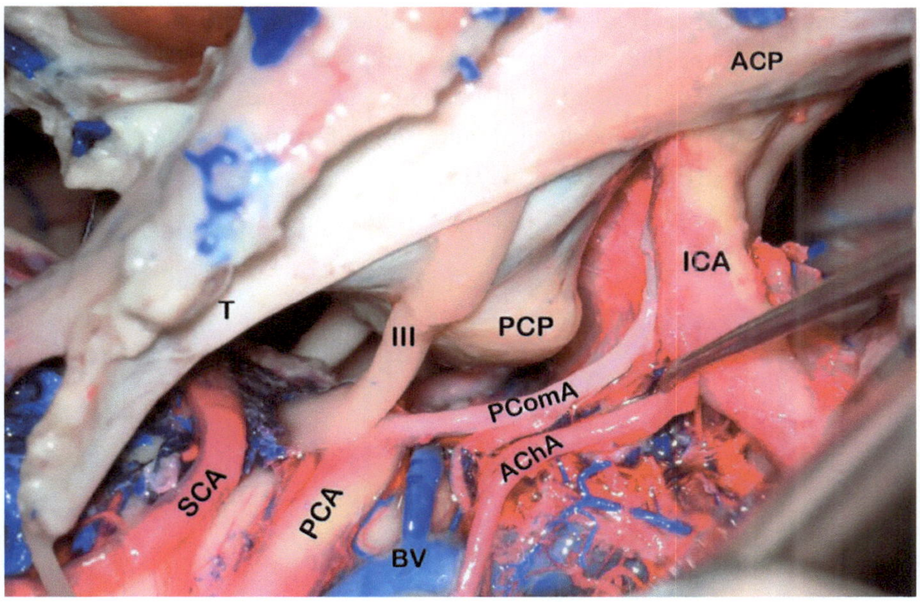

Fig. 15.33 With further retraction of the temporal lobe, the optic, ipsilateral oculomotor, and contralateral oculomotor nerves are seen. The internal carotid, anterior choroidal, and posterior communicating arteries and the perforators can be seen easily with minimal retraction of the temporal lobe. *ACP*, anterior clinoid process; *ICA*, internal carotid artery; *PCP*, posterior clinoid process; *PComA*, posterior communicating artery; *AChA*, anterior choroidal artery; *BV*, basilar vein; *PCA*, posterior cerebral artery; *SCA*, superior cerebellar artery; *T*, trigeminal nerve; *III*, oculomotor nerve.

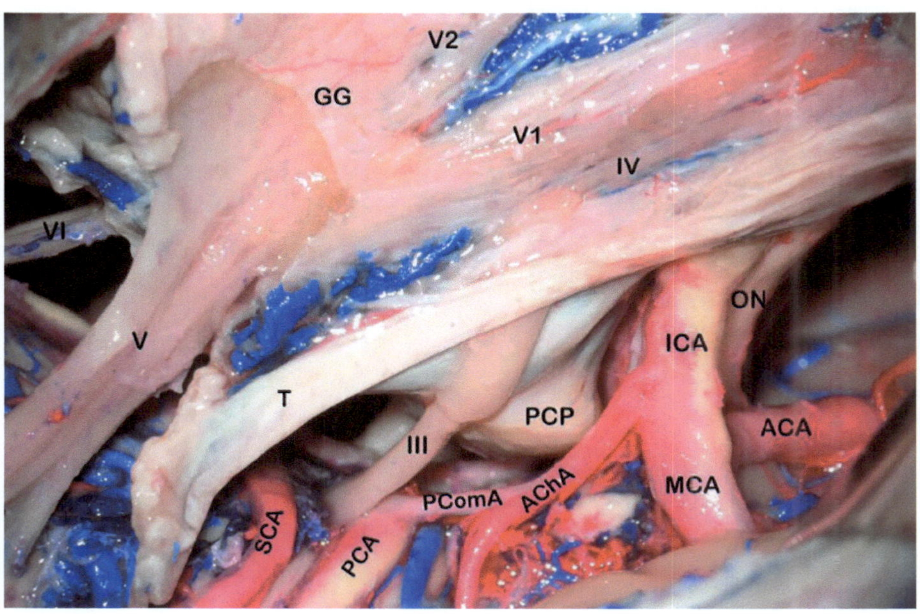

Fig. 15.34 The tentorium is removed from the roof of Meckel's cave to expose the trigeminal nerve root and Gasserian ganglion. Bleeding from the superior petrosal sinus and veins within the roof of Meckel's cave is controlled with absorbable hemostatic fibers and sutures. *ON*, optic nerve; *ICA*, internal carotid artery; *ACA*, anterior cerebral artery; *MCA*, middle cerebral artery; *PCP*, posterior clinoid process; *AChA*, anterior choroidal artery; *PComA*, posterior communicating artery; *III*, oculomotor nerve; *PCA*, posterior cerebral artery; *SCA*, superior cerebellar artery; *T*, tentorium; *V*, vestibule; *VI*, abducent nerve; *GG*, geniculate ganglion; *V2*, maxillary nerve; *IV*, trochlear nerve; *V1*, ophthalmic nerve.

Clinical Applications

CASE 1: MENINGIOMA OF THE RIGHT POSTERIOR SURFACE OF THE PETROUS BONE (Figs. 15.35–15.47)

1. Preoperative MRI with gadolinium enhancement showing a widely based posterior fossa meningioma centered on the internal auditory canal (Fig. 15.35). Note the skin incision that is required for this approach (Fig. 15.36).

2. A retrolabyrinthine (presigmoid) craniotomy has been performed. Note the wide removal of bone posterior to the sigmoid sinus (Fig. 15.37).

3. A wide middle fossa craniotomy is next performed. Surgicel is placed at the edges of the craniotomy to achieve hemostasis (Fig. 15.38).

4. The middle fossa dura is detached from the superior surface of the temporal bone using a large dissector and proceeding from posterior to anterior (Fig. 15.39). The middle meningeal artery and the mandibular branch of the trigeminal nerve are cut after bipolar coagulation. Using a diamond burr, the internal auditory canal is identified, starting from medial to lateral direction as previously described.

5. The posterior fossa dura is opened. The tumor is seen as closely related to the superior aspect of the acousticofacial bundle (Figs. 15.40 and 15.41).

6. The tumor is further separated from the acousticofacial bundle and the anterior inferior cerebellar artery (Fig. 15.42).

7. The tumor is next managed from the subtemporal aspect (Figs. 15.43 and 15.44). The internal auditory canal is opened from the subtemporal aspect. The tumor occupies the canal and involves the dura of the anterior border of the canal. The tumor and the infiltrated dura were removed. Tumor involving Meckel's cave was also removed from the subtemporal aspect.

8. After total tumor removal and the accomplishment of hemostasis, the operative cavity is obliterated with abdominal fat that protrudes into the cerebellopontine angle. The musculoperiosteal flap is sutured back. Fat is also placed to obliterate the subtemporal aspect of the approach. The temporal craniotomy is replaced and is fixed in place using silk sutures. The temporalis is next repositioned and sutured using Vicryl 1/0 sutures (Figs. 15.45–15.47).

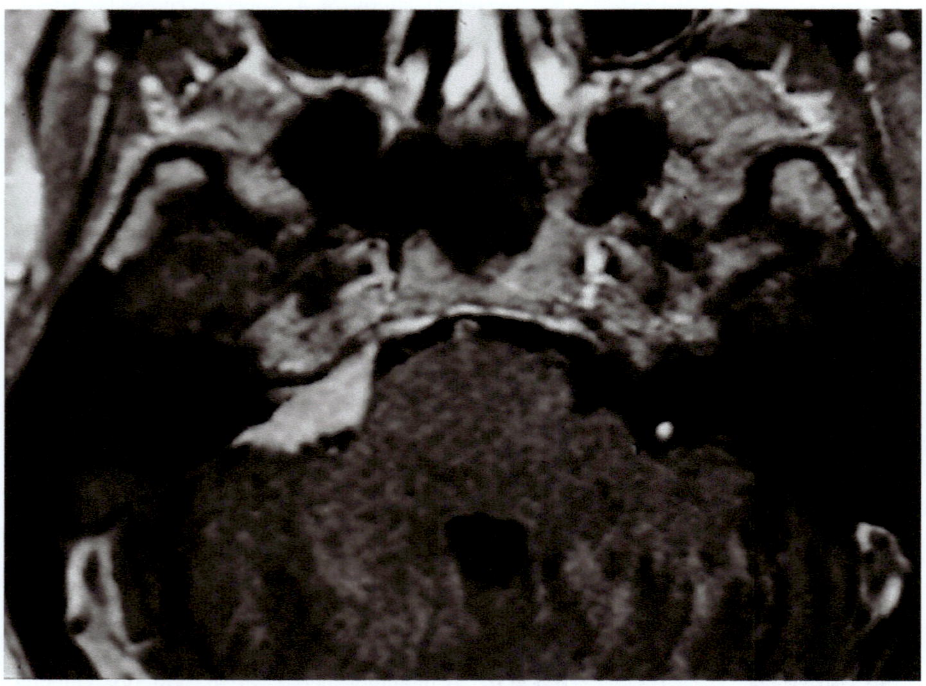

Fig. 15.35 Preoperative MRI with gadolinium enhancement showing the tumor.

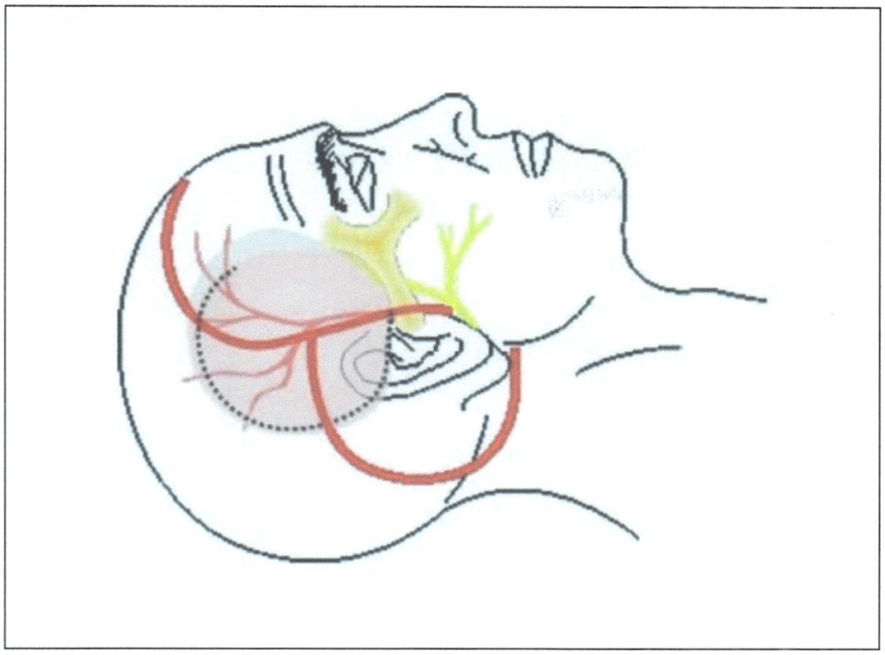

Fig. 15.36 Drawing showing the incision to be made.

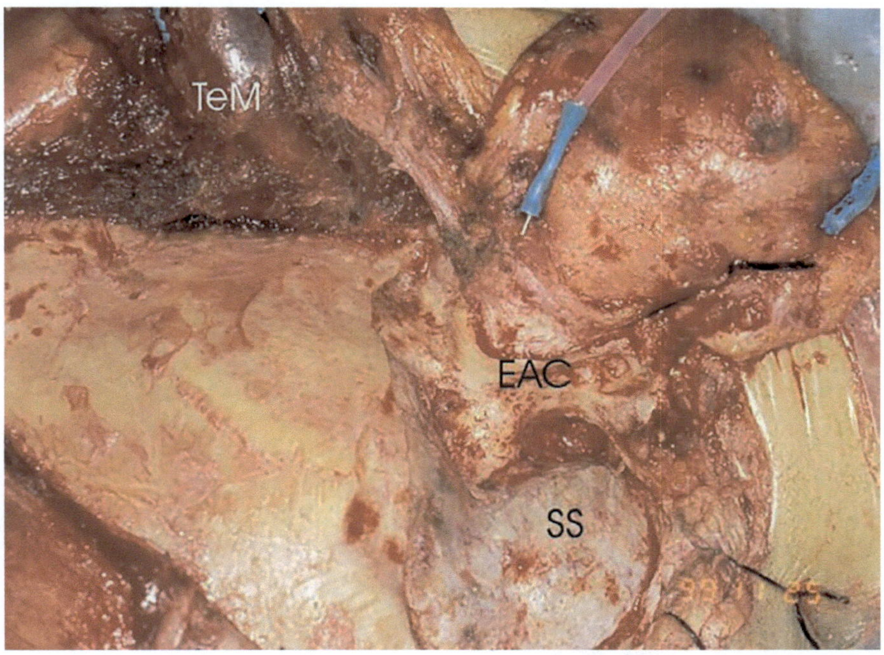

Fig. 15.37 Retrolabyrinthine craniotomy. The temporalis muscle (TeM) has been detached and retracted anteriorly.

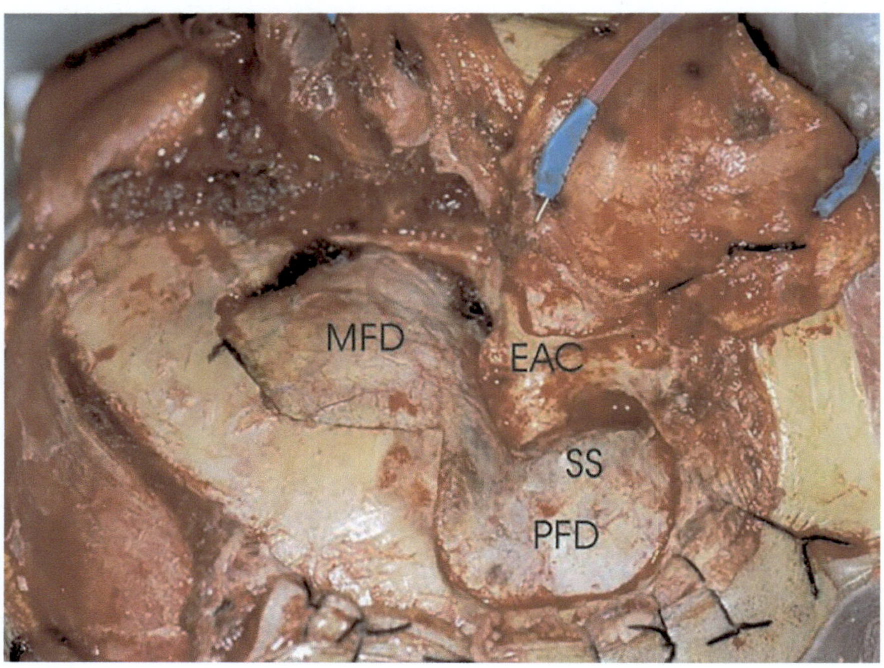

Fig. 15.38 Middle fossa craniotomy.

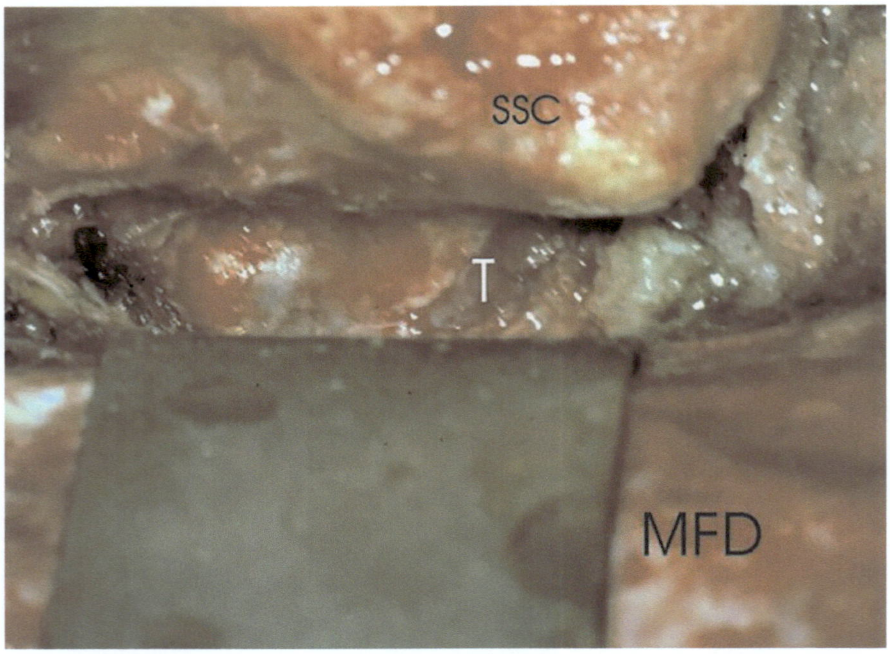

Fig. 15.39 Elevation of the middle fossa dura. The tumor (T) invading the petrous apex has already been identified.

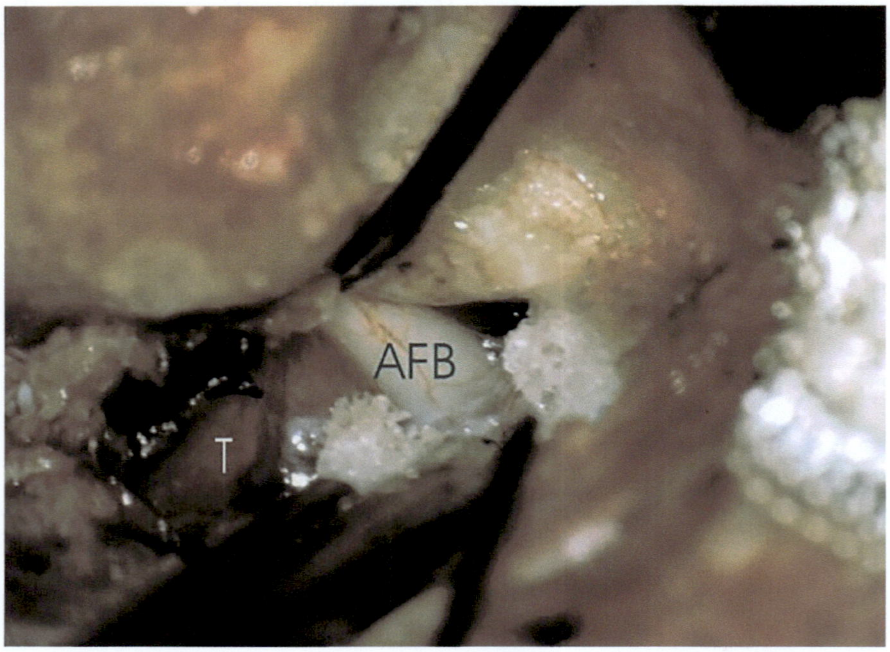

Fig. 15.40 The tumor (T) superior to the acoustico-facial bundle (AFB).

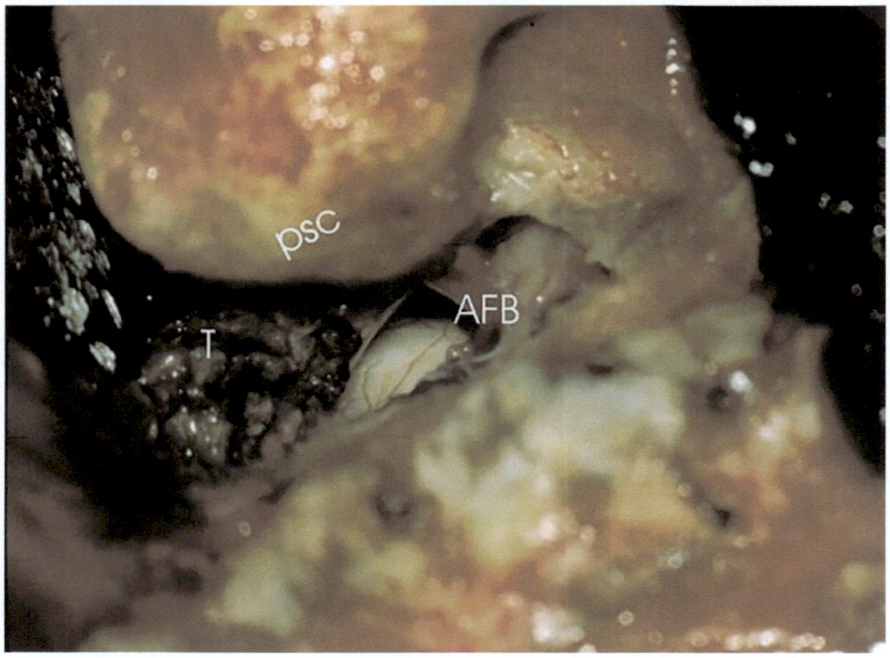

Fig. 15.41 At lower magnification, a panoramic view of the tumor in the posterior fossa is seen.

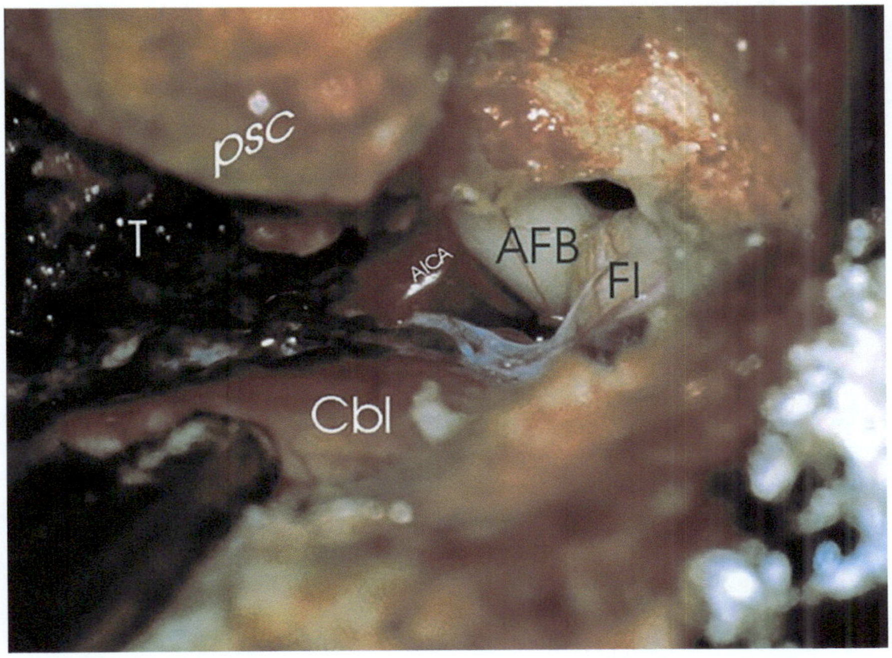

Fig. 15.42 The tumor (T) is separated from the acousticofacial bundle (AFB) and the anterior inferior cerebellar artery (AICA). *Cbl*, cerebellum; *Fl*, flocculus.

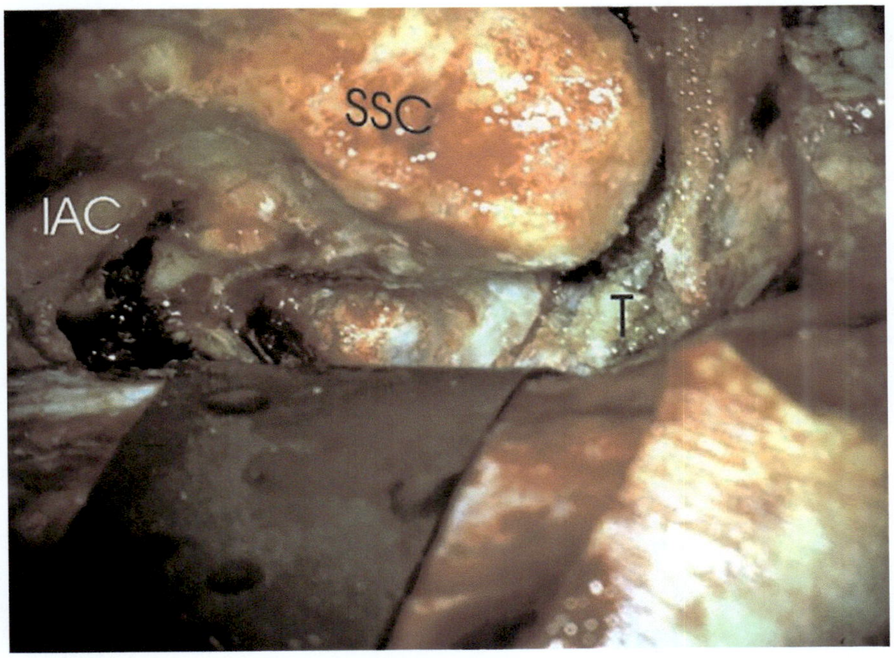

Fig. 15.43 View of the tumor (T) from the subtemporal aspect. The internal auditory canal (IAC) has been identified.

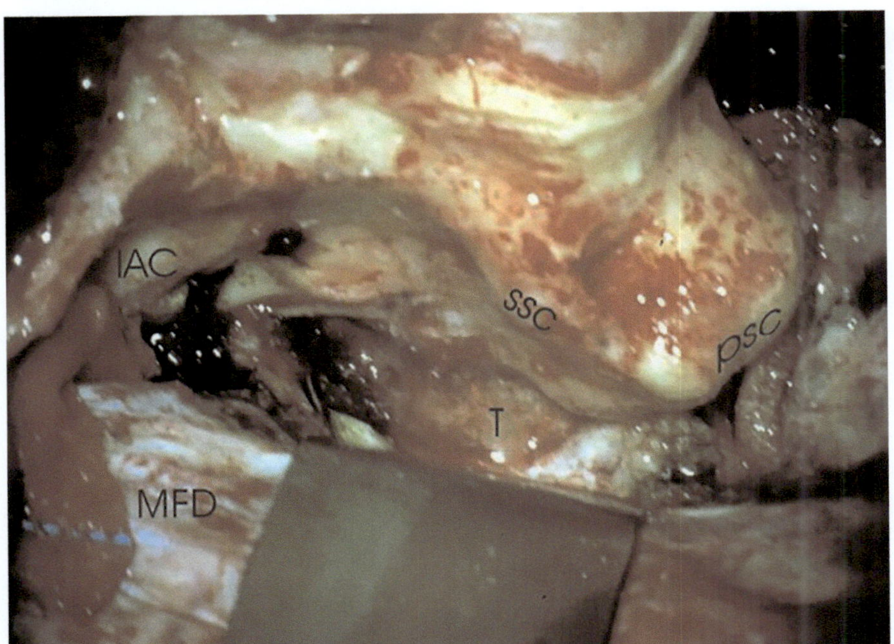

Fig. 15.44 The dura of the posterior fossa has been opened from the subtemporal aspect. The tumor can be well seen.

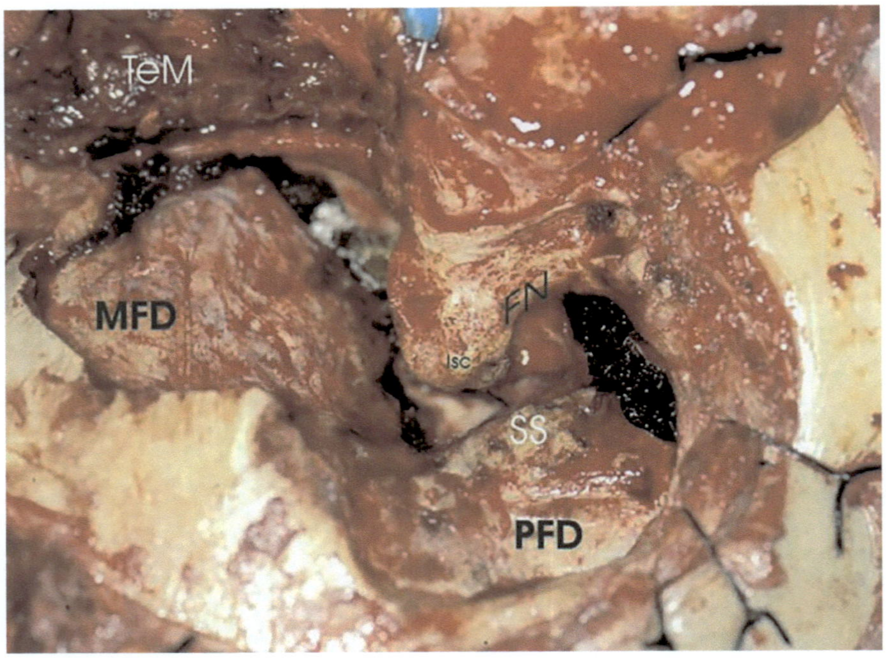

Fig. 15.45 View after total tumor removal.

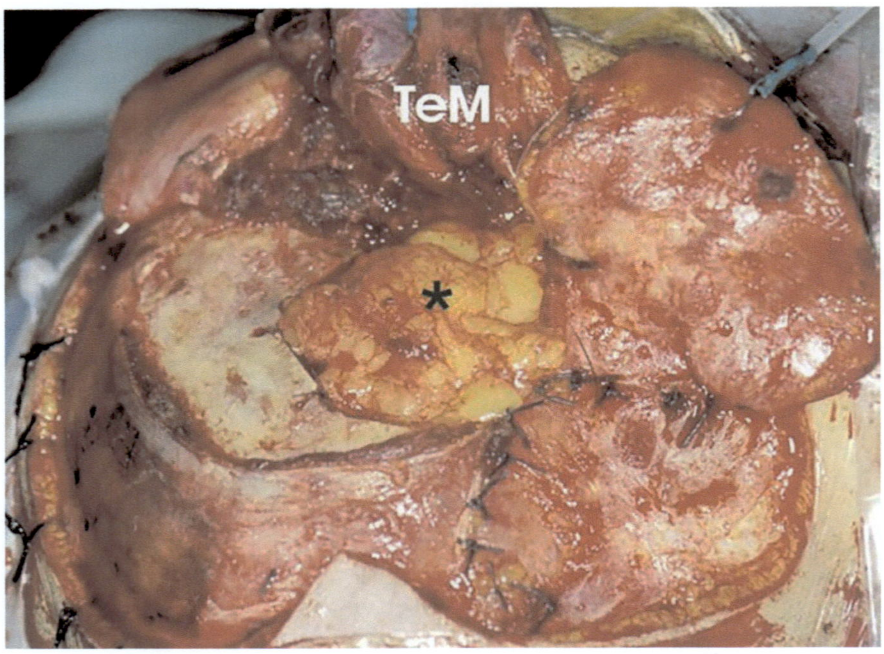

Fig. 15.46 The retrolabyrinthine aspect has been obliterated with abdominal fat, and the U-shaped musculoperiosteal flap has been repositioned and sutured. Abdominal fat (*) is also placed to obliterate the subtemporal opening.

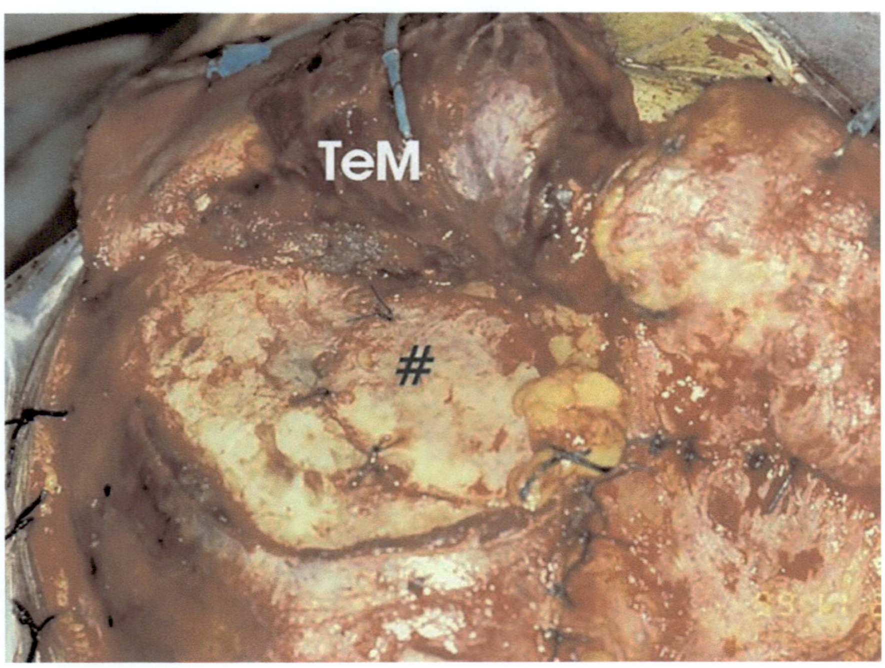

Fig. 15.47 The subtemporal craniotomy (#) is repositioned over the abdominal fat. The temporalis muscle (TeM) will next be sutured back.

CASE 2: LEFT CEREBELLOPONTINE ANGLE EPIDERMOID (Figs. 15.48–15.57)

1. MRI of the case showing the epidermoid lying anterior to the internal auditory canal (Fig. 15.48).

2. The approach has been performed with a wide retrolabyrinthine and subtemporal craniotomy (Fig. 15.49).

3. After opening the dura, cholesteatoma is seen in the posterior fossa with intimate relations to the neurovascular structures (Fig. 15.50).

4. Cholesteatoma is dissected from the anterior inferior cerebellar artery. The abducent nerve is seen in the prepontine cistern (Figs. 15.51–15.54).

5. Cholesteatoma is further removed from the acousticofacial bundle. A part of the cholesteatoma is still left anterior to the acousticofacial bundle and extends to the prepontine cistern (Fig. 15.55).

6. The remaining cholesteatoma is removed from the prepontine cistern and the basilar artery through the subtemporal transapical access (Fig. 15.56).

7. Postoperative CT scan showing total removal of tumor (Fig. 15.57).

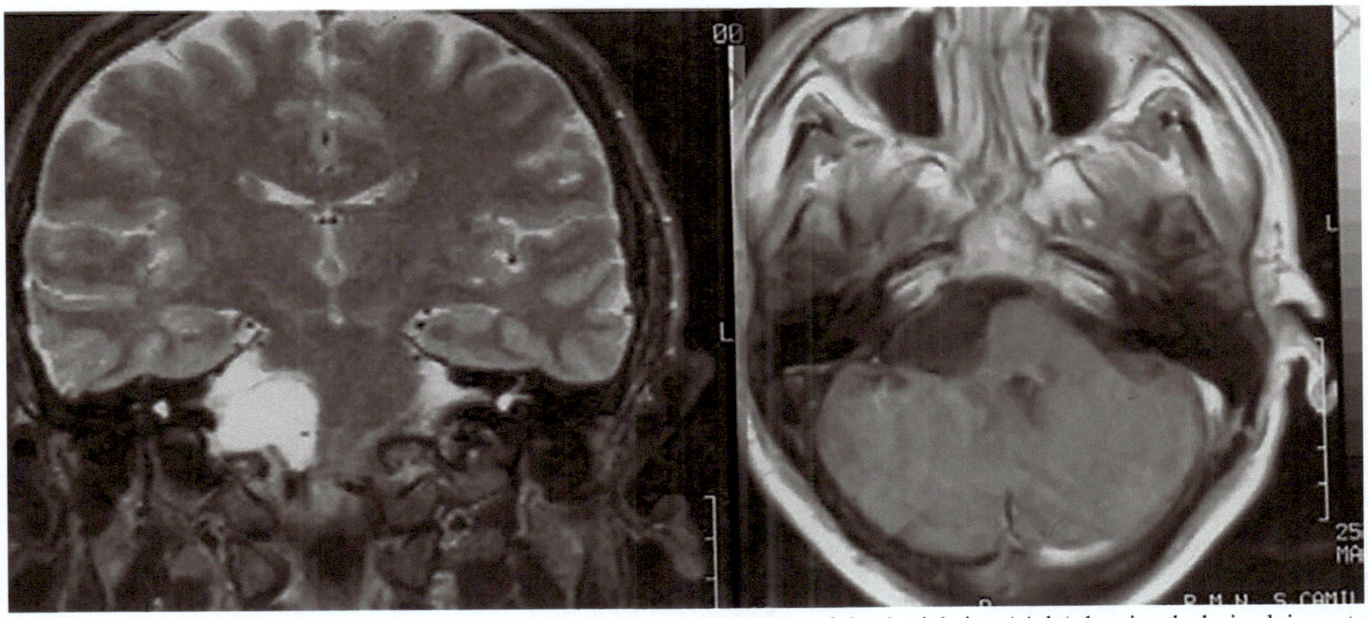

Fig. 15.48 Preoperative MRI of the case. T2-weighted coronal view *(left)* and T1-weighted axial view *(right)* showing the lesion lying anterior to the internal auditory canal. Encased vessels are seen in the T2-weighted image.

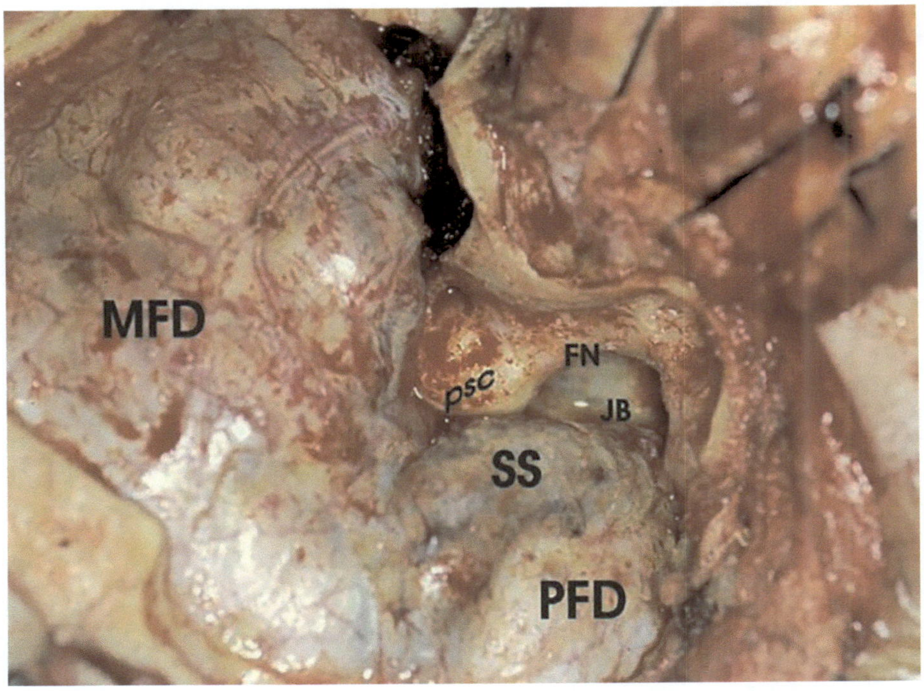

Fig. 15.49 The approach has been performed. Note the wide craniotomies.

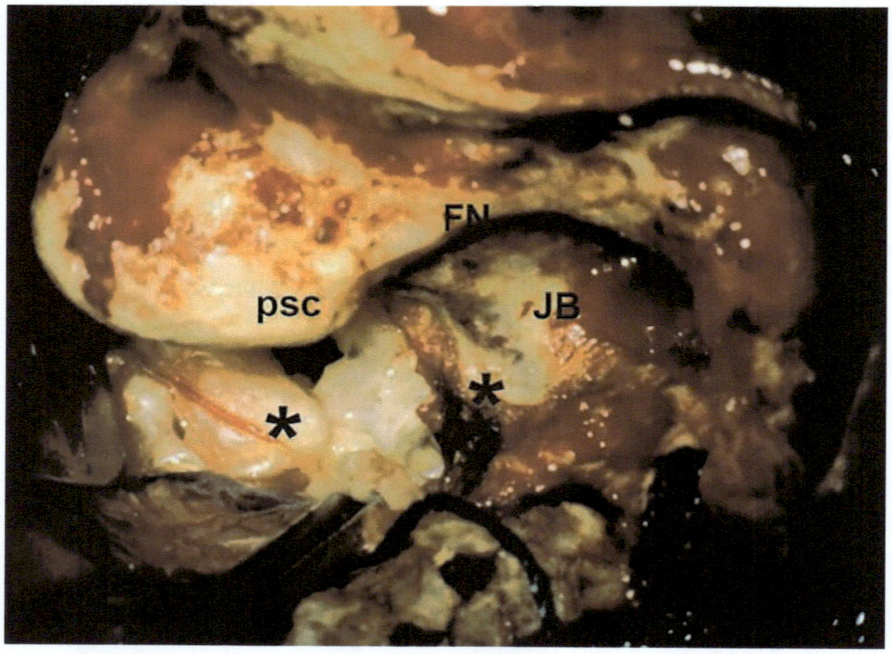

Fig. 15.50 Cholesteatoma (*) is seen in the posterior fossa.

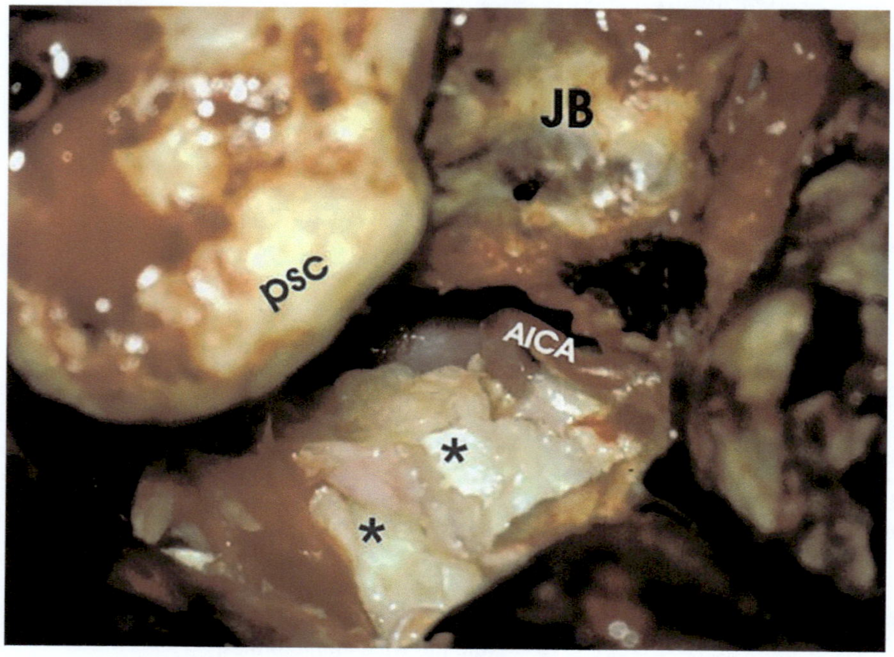

Fig. 15.51 Cholesteatoma (*) related to the anterior inferior cerebellar artery (AICA).

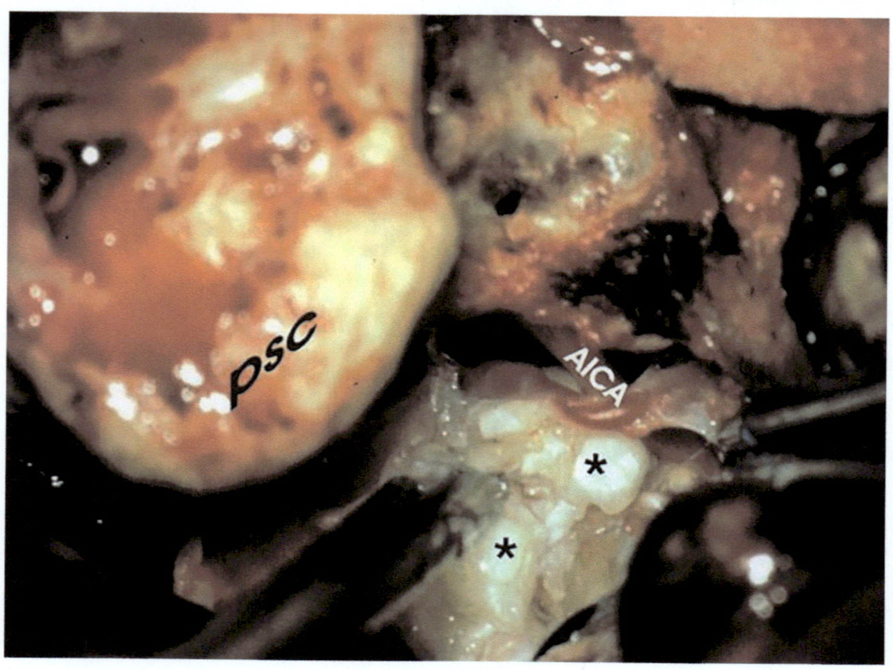

Fig. 15.52 At higher magnification, the relation between the cholesteatoma (*) and the anterior inferior cerebellar artery (AICA) is better appreciated.

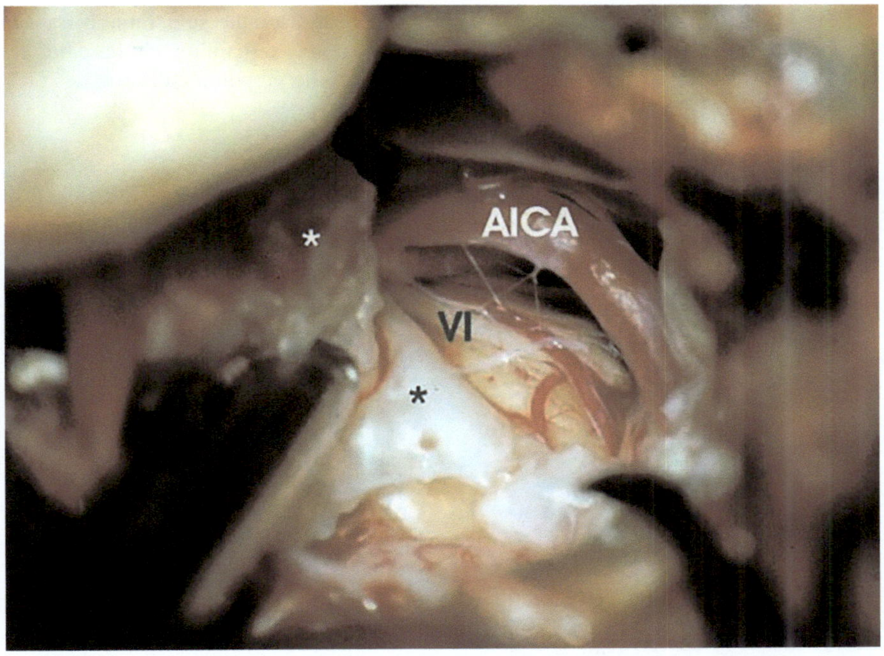

Fig. 15.53 Cholesteatoma (*) has been dissected from the anterior inferior cerebellar artery (AICA). The abducent nerve (VI) is clearly seen.

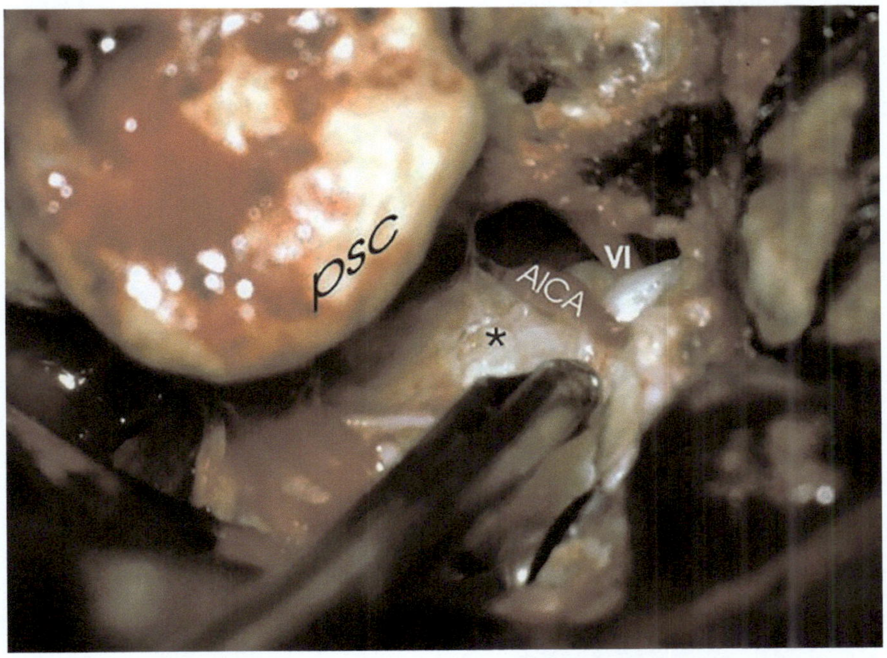

Fig. 15.54 The abducent nerve (VI) has been freed.

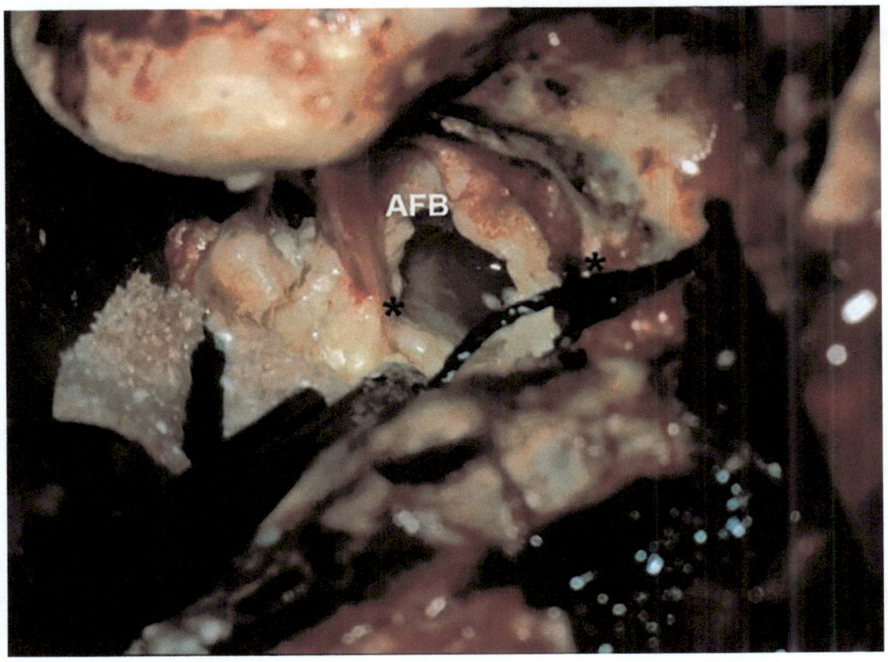

Fig. 15.55 Dissecting the cholesteatoma (*) from the acousticofacial bundle (AFB).

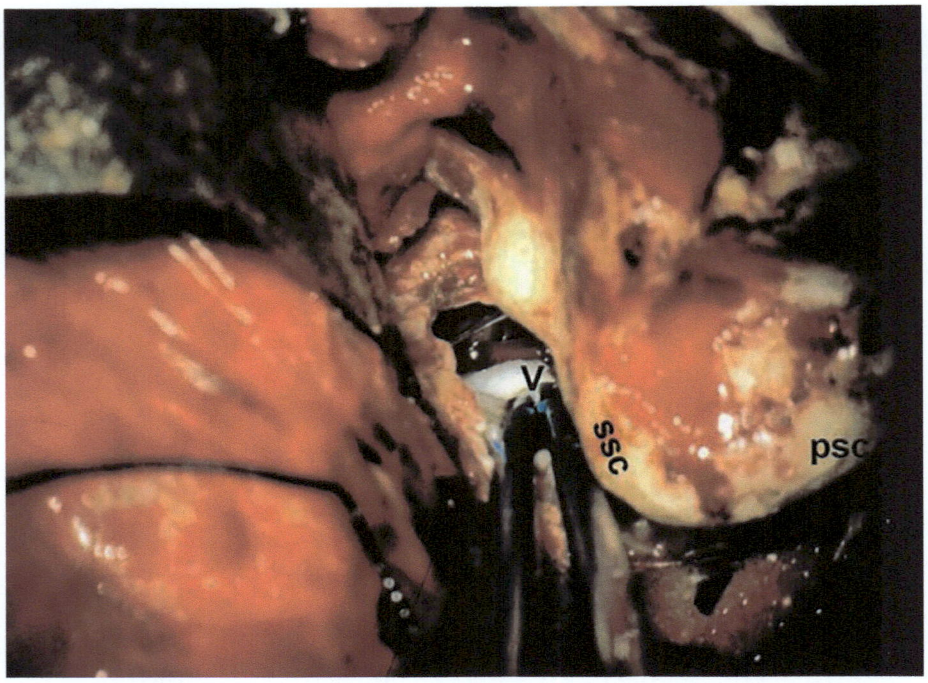

Fig. 15.56 Through the subtemporal transapical approach, the remaining cholesteatoma is removed from the prepontine cistern.

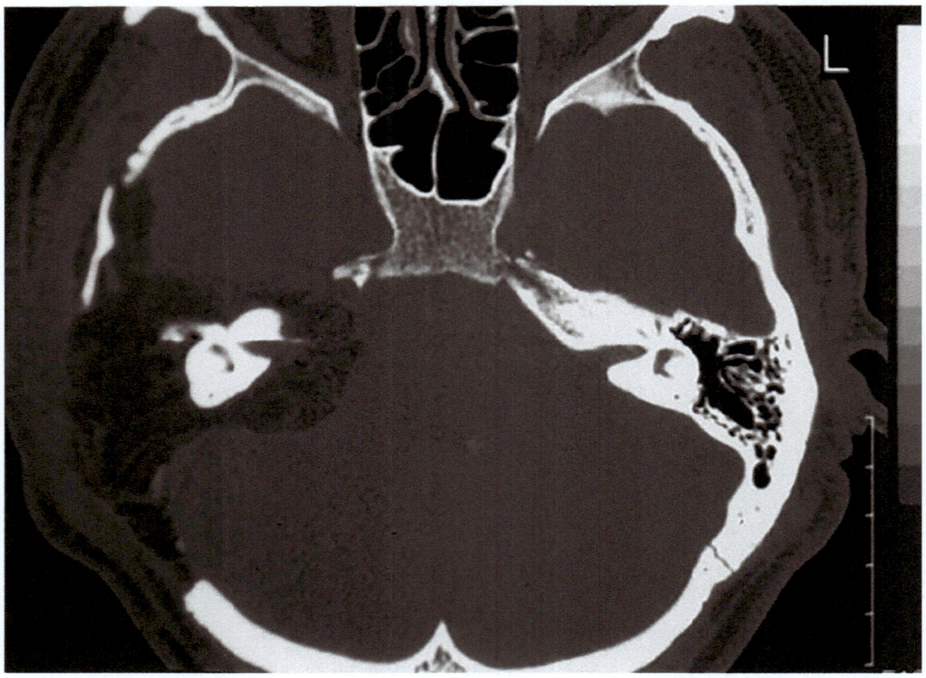

Fig. 15.57 Postoperative CT scan.

PEARLS AND PITFALLS

1. Wide bone removal posterior to the sigmoid sinus is mandatory. This allows gentle depression of the sinus during tumor removal and allows extra access and a better angle of work and vision (Fig. 15.58).
2. The endolymphatic sac can be cut as this releases the posterior fossa dura and allows for better depression of the sigmoid sinus (Fig. 15.59).
3. The retrofacial air cells should be drilled widely. This gives better exposure of the area of the lower cranial nerves (Fig. 15.60).
4. Bone removal should reach the posterior semicircular canal and the crus commune. Every millimeter of bone drilled at this step gives much better exposure. To avoid injuring the dura while removing bone around the semicircular canals, a septal dissector is used to detach the dura from the bone.
5. In some cases, a zygomatic osteotomy can be performed, and the zygomatic arch is displaced inferiorly. This decreases the amount of temporal lobe retraction and affords a better angle of vision through the subtemporal route (Figs. 15.61 and 15.62).

6. Care is taken not to injure the vein of Labbé while cutting the middle fossa dura. Injury to the vein can occur by direct interruption, coagulation, or even overstretching due to excessive temporal bone retraction. Injury to the vein of Labbé can cause seizures, aphasia (if the injury is on the dominant side), and less commonly massive brain edema (Fig. 15.63).
7. Cutting of the tentorium should be done under direct vision to avoid injury to the trochlear nerve, which pierces the tentorium traveling from the posterior fossa to the middle fossa (Fig. 15.64).
8. If involved by tumor, Meckel's cave can be opened through the subtemporal part of this approach (Figs. 15.65–15.67).

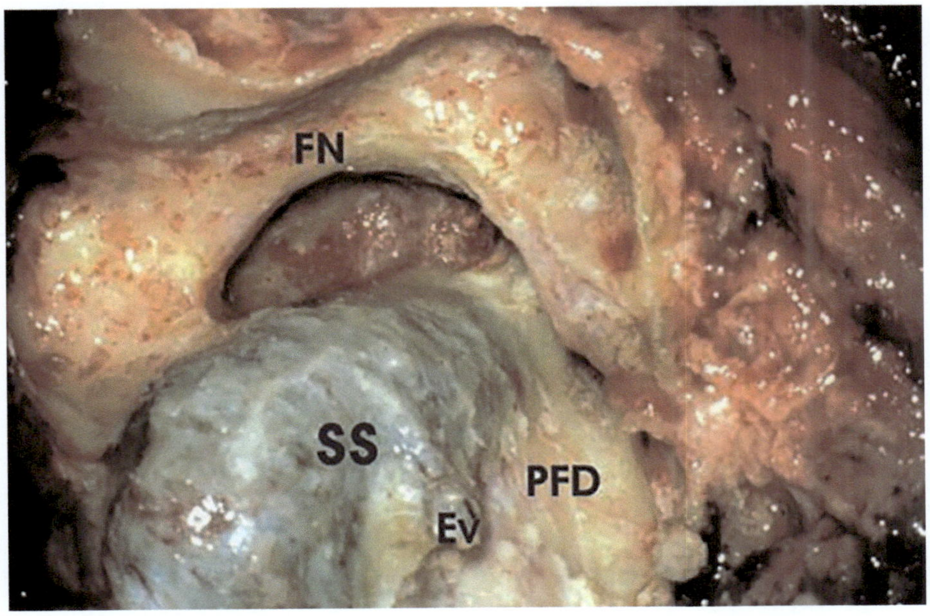

Fig. 15.58 Wide removal of bone posterior to the sigmoid sinus (SS).

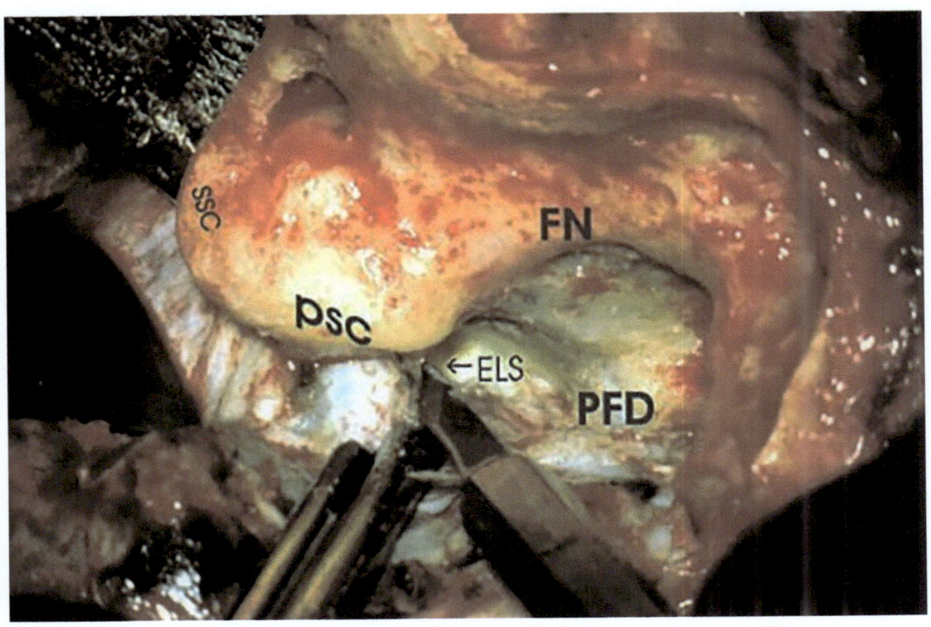

Fig. 15.59 Cutting the endolymphatic sac (ELS) to release the posterior fossa dura (PFD).

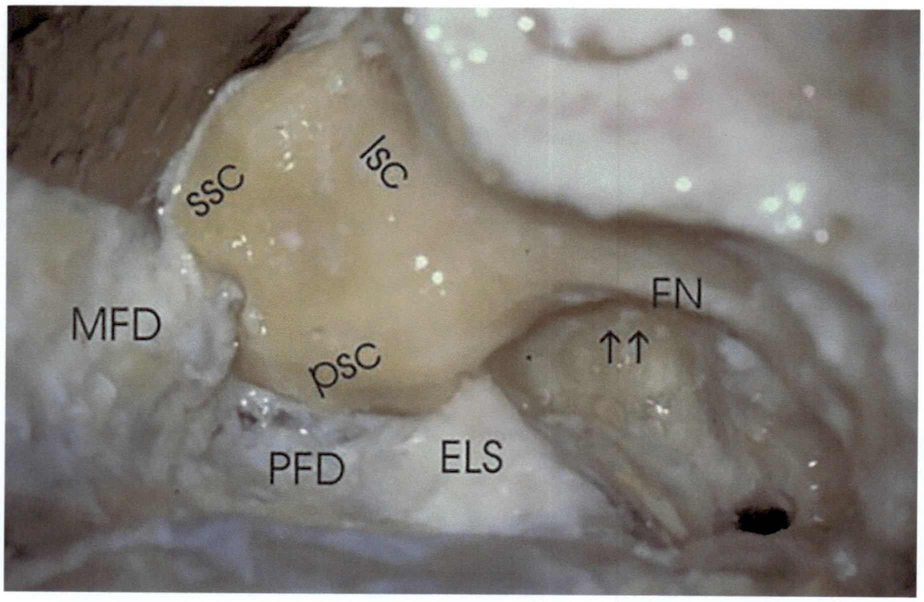

Fig. 15.60 The retrofacial air cells *(arrows)* should be drilled widely.

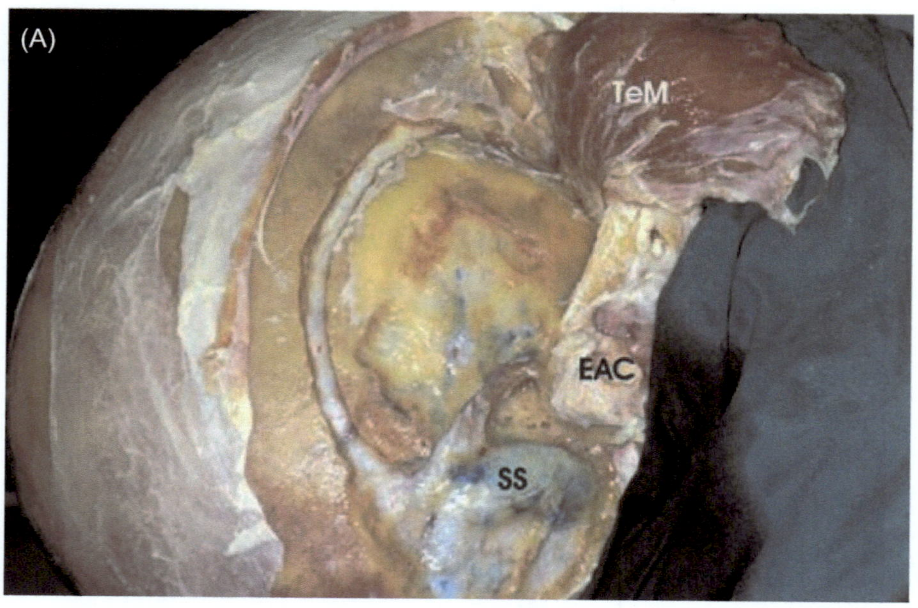

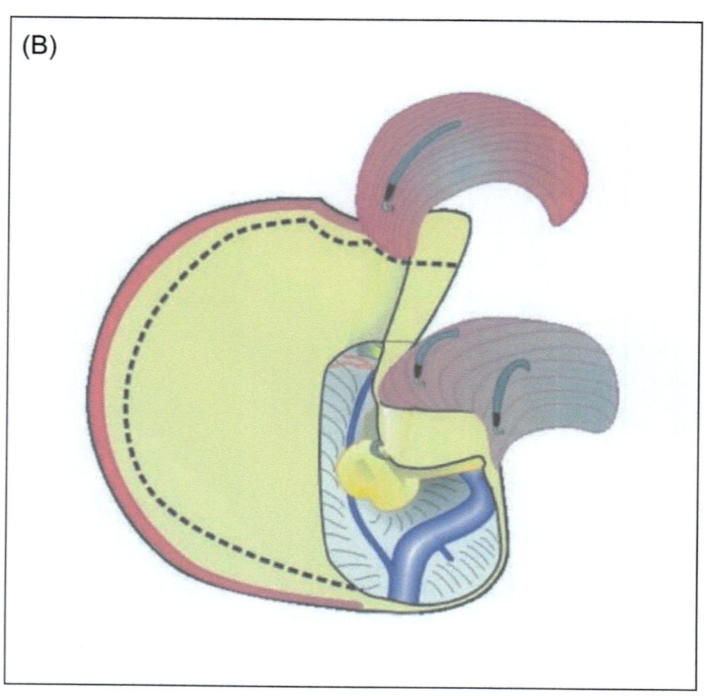

Fig. 15.61 (A, B) The craniotomy has been outlined, but the zygomatic arch is in situ.

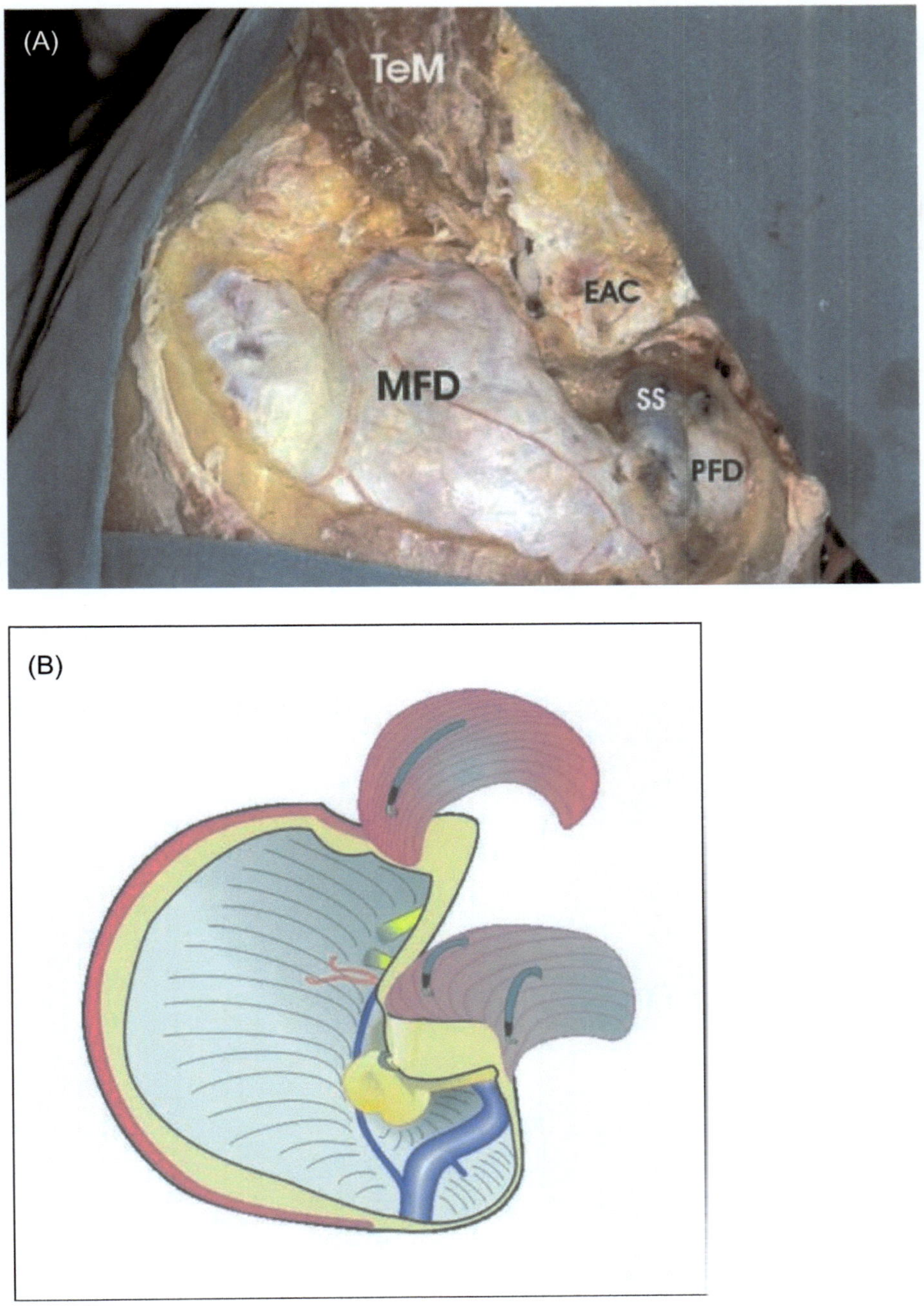

Fig. 15.62 (A, B) The craniotomy has been performed. The zygomatic arch is displaced inferiorly. Note how this can bring the craniotomy very low to reach the level of the base of the middle fossa. This reduces the amount of temporal lobe retraction needed to drill the petrous apex.

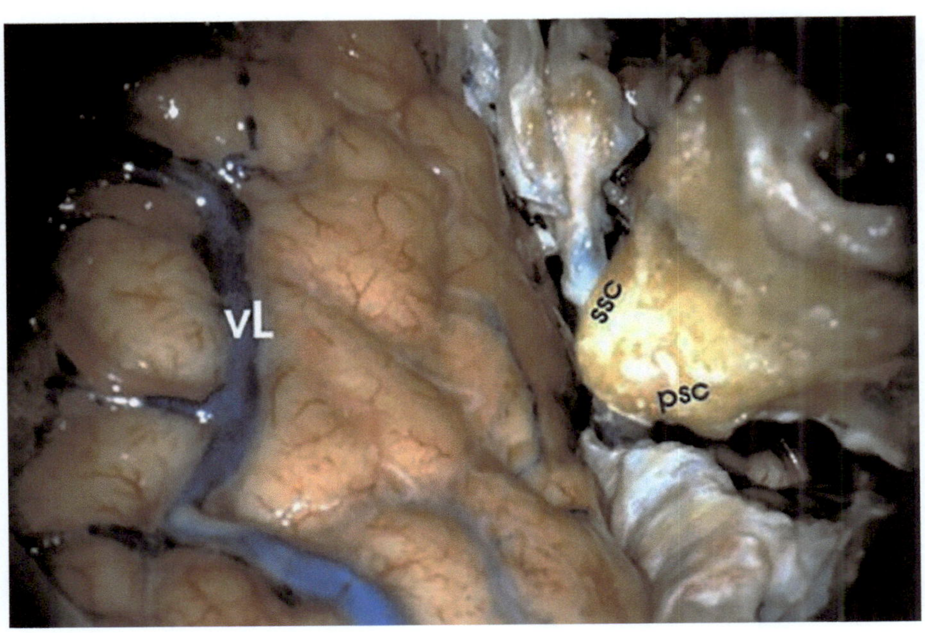

Fig. 15.63 The vein of Labbé (vL) is seen running on the temporal lobe.

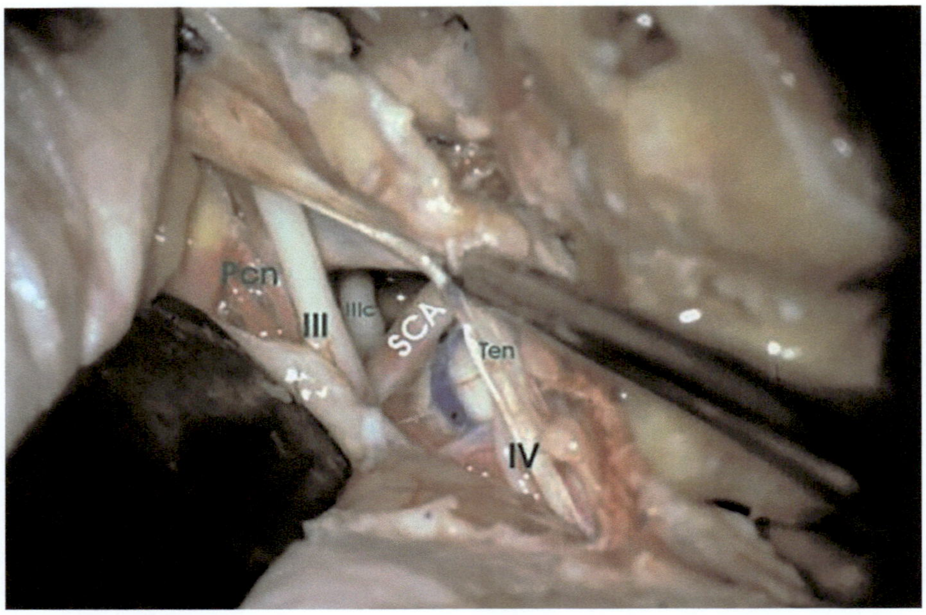

Fig. 15.64 Cutting the tentorium (Ten) after visualizing the trochlear nerve (IV) to avoid injuring the nerve as it pierces the tentorium.

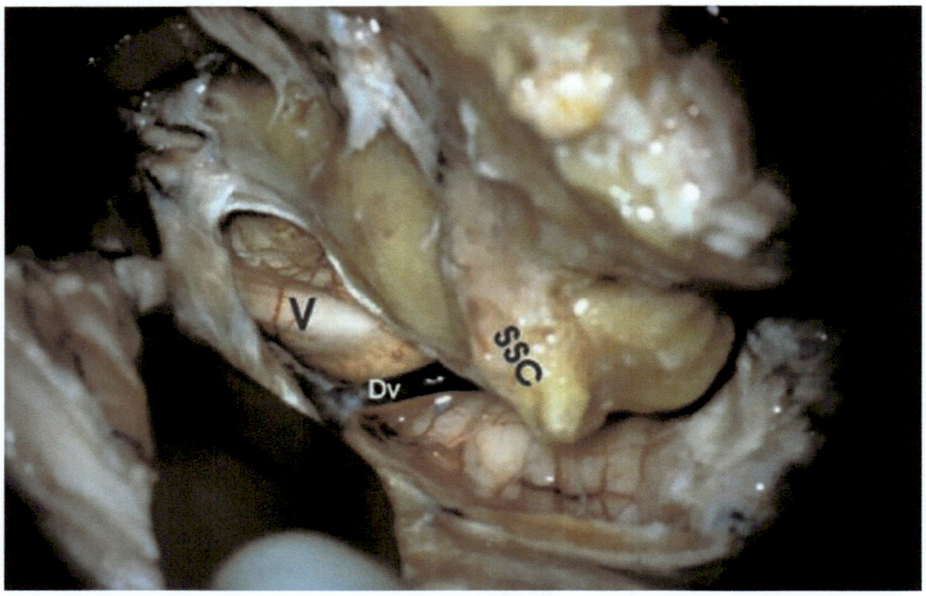

Fig. 15.65 The trigeminal nerve (V) is exposed.

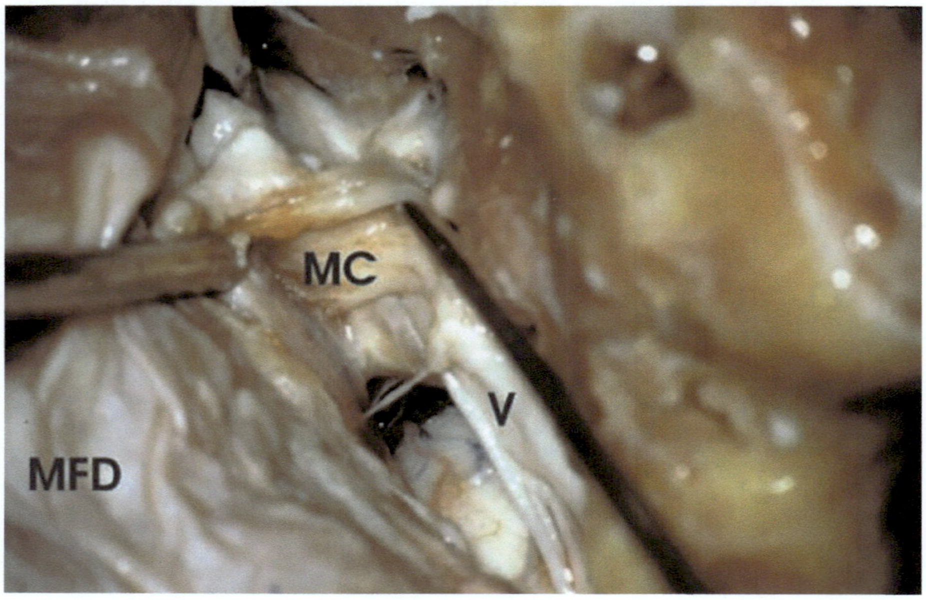

Fig. 15.66 Meckel's cave (MC) is opened through the subtemporal aspect.

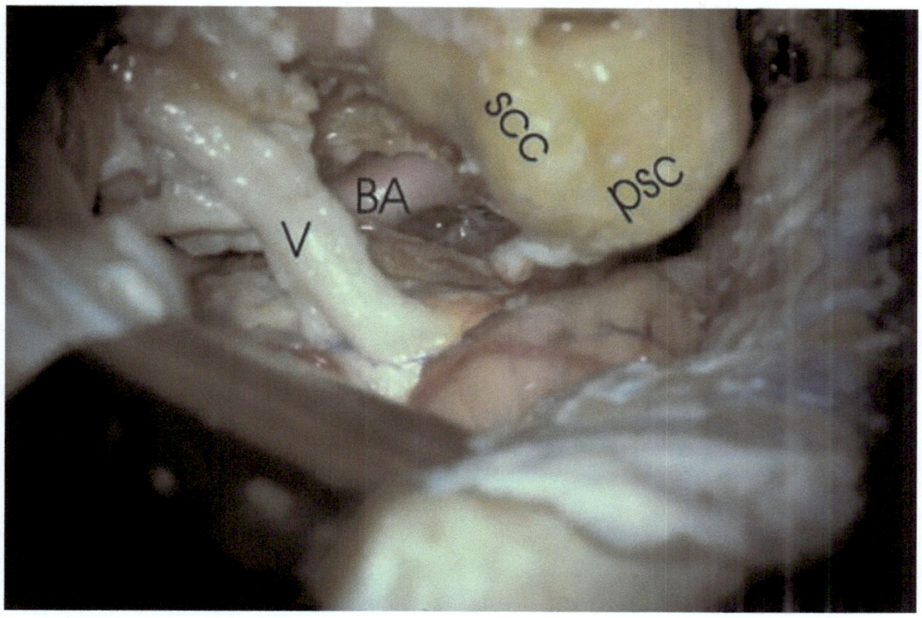

Fig. 15.67 Full exposure of the trigeminal nerve (V) from its origin at the brainstem to the opened Meckel's cave.

Combined Transpetrous Orbitozygomatic Approaches

Rationale

In addition to the posterior fossa component, certain lesions extend to the middle cranial fossa, cavernous sinus, or orbit, or even extracranially to the infratemporal or pterygopalatine fossae. These extensions can be managed safely through a wide skull base approach such as the orbitozygomatic-frontotemporal approach. The combination with a carefully chosen transpetrous approach allows the safe extirpation of the posterior fossa component and hence achieves total and safe tumor removal. The choice of the transpetrous approach depends on the site and size of the posterior fossa component, the preoperative ipsilateral facial function and hearing level, and the pathology to be dealt with.

Indications

1. Trigeminal neurinomas extending to both posterior and middle cranial fossae.
2. Other posterior fossa tumors with massive middle cranial fossa, cavernous sinus, or orbital extension, e.g., epidermoid tumors, chordomas, and petroclival meningiomas.

Approaches and Clinical Applications

COMBINED TRANSCOCHLEAR ORBITOZYGOMATIC APPROACH

The transcochlear component allows the safe removal of extensive posterior fossa tumors, particularly those that extend to the petroclival region and prepontine cistern. The orbitozygomatic component permits the control of the frequent middle fossa and/or cavernous sinus extension of these lesions. Fig. 15.68 shows a schematic presentation of this approach.

SURGICAL ANATOMY

The surgical anatomy of the combined transcochlear orbitozygomatic approach is seen in Figs. 15.69–15.73.

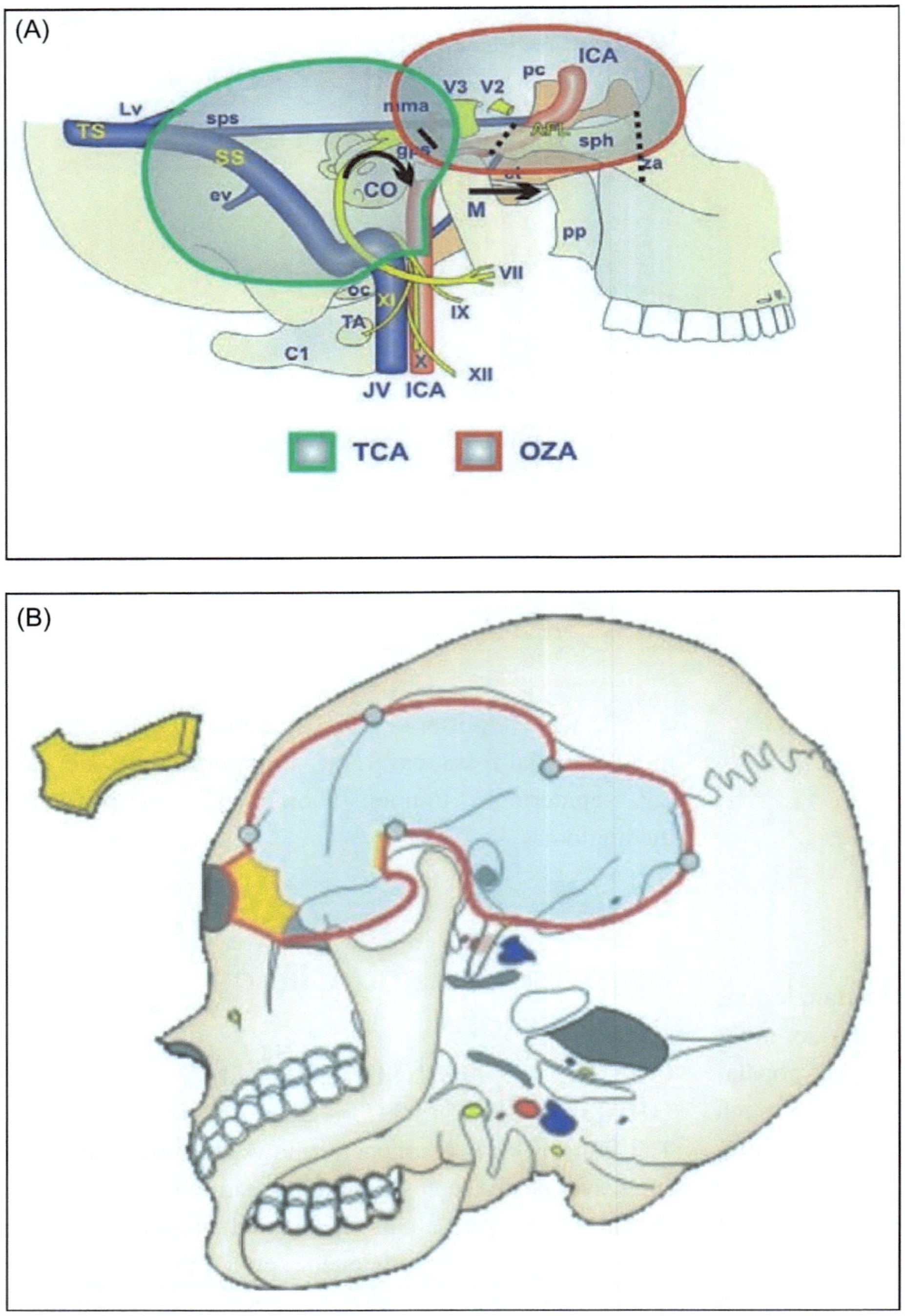

Fig. 15.68 (A, B) Schematic drawing representing the combined transcochlear (TCA) orbitozygomatic (OZA) approach. *AFL*, anterior foramen lacerum; *C1*, first cervical vertebra; *CO*, cochlea; *et*, eustachian tube; *ev*, emissary vein; *gps*, greater petrosal nerve; *ICA*, internal carotid artery; *JV*, jugular vein; *Lv*, vein of Labbé; *M*, mandible; *mma*, middle meningeal artery; *oc*, occipital condyle; *pc*, clinoid process; *pp*, pterygoid process; *sph*, sphenoid; *sps*, superior petrosal sinus; *TA*, transverse process of the atlas; *TS*, transverse sinus; *za*, zygomatic arch; *V2*, maxillary nerve; *V3*, mandibular nerve; *VII*, facial nerve; *IX*, glossopharyngeal nerve; *XII*, hypoglossal nerve.

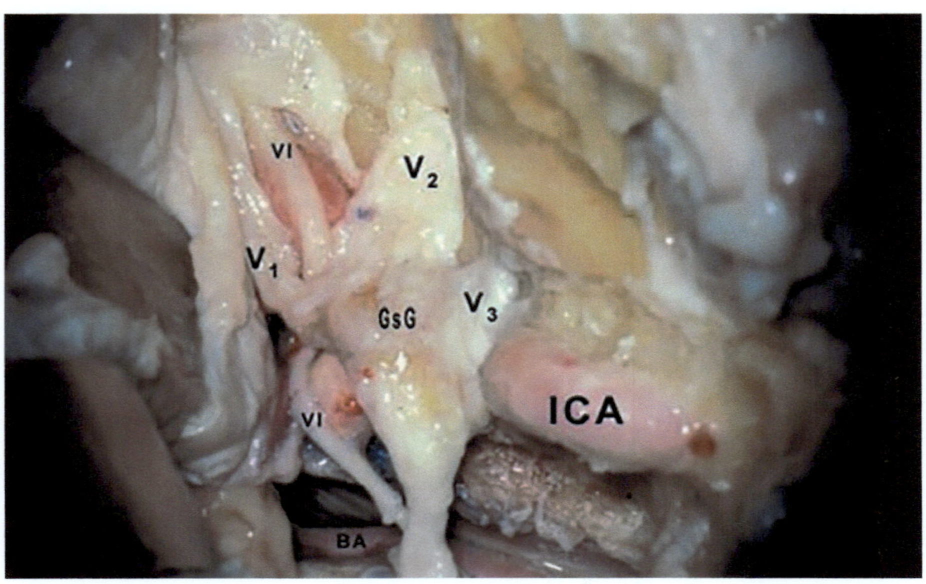

Fig. 15.69 *Right side*. The Gasserian ganglion (GsG) and the mandibular branch of the trigeminal nerve (V3) hinder the complete exposure of the horizontal transpetrous internal carotid artery (ICA).

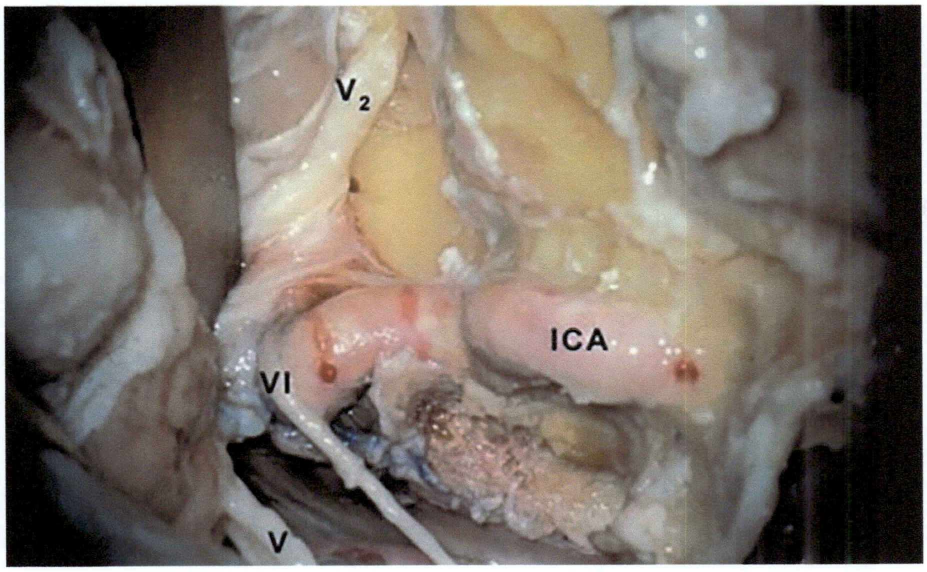

Fig. 15.70 The mandibular branch has been cut, and the Gasserian ganglion together with the maxillary branch of the trigeminal nerve (V2) has been retracted. The horizontal intrapetrous internal carotid artery (ICA) is better exposed.

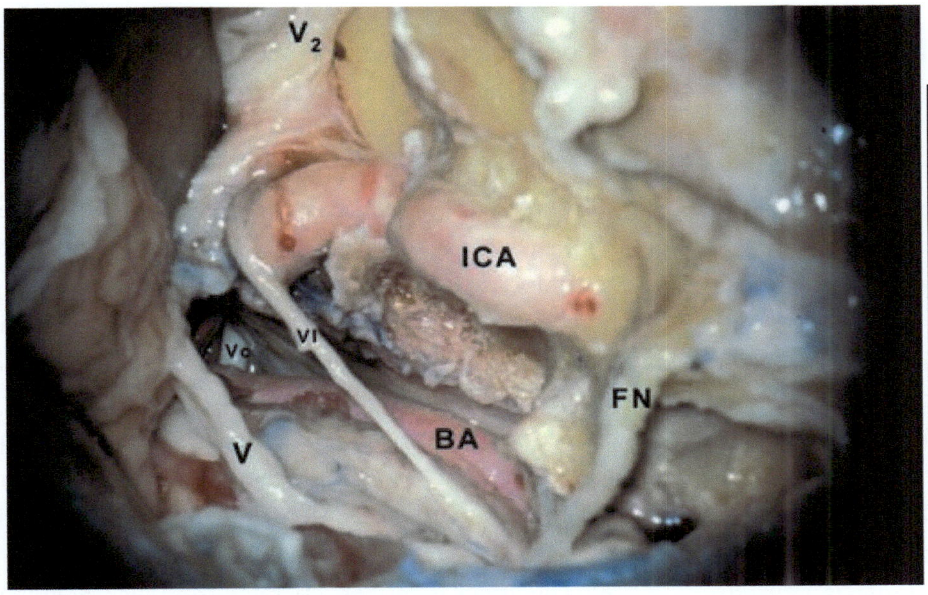

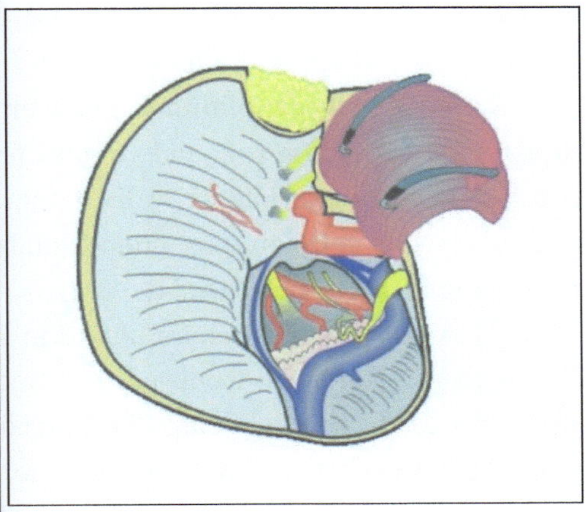

Fig. 15.71 A view of the posterior fossa through the modified transcochlear approach. The facial nerve (FN) has been rerouted. The whole intrapetrous internal carotid artery (ICA) is exposed. The contralateral trigeminal nerve (Vc) is also appreciated.

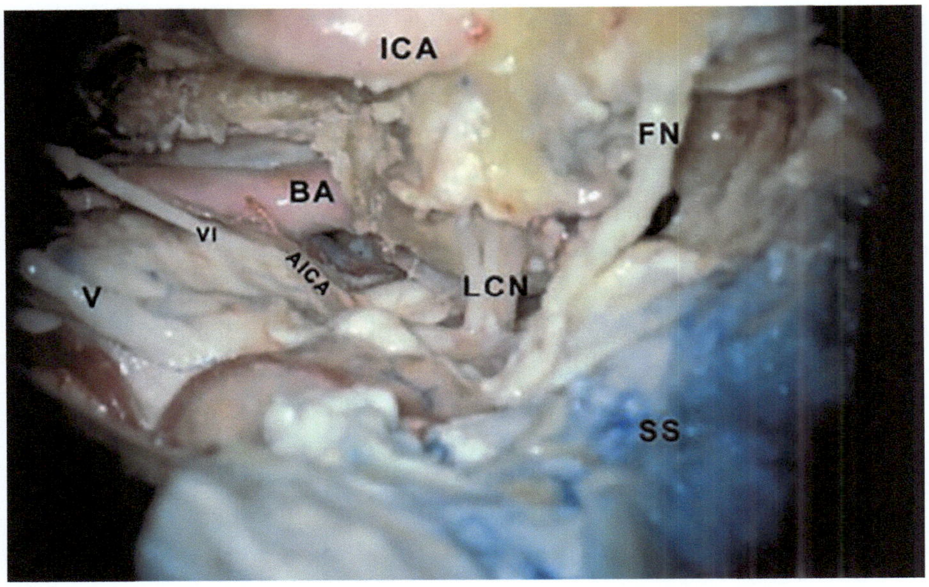

Fig. 15.72 Tilting the microscope downward shows the lower cranial nerves. A small anterior inferior cerebellar artery (AICA) is seen crossing the abducent nerve(VI).

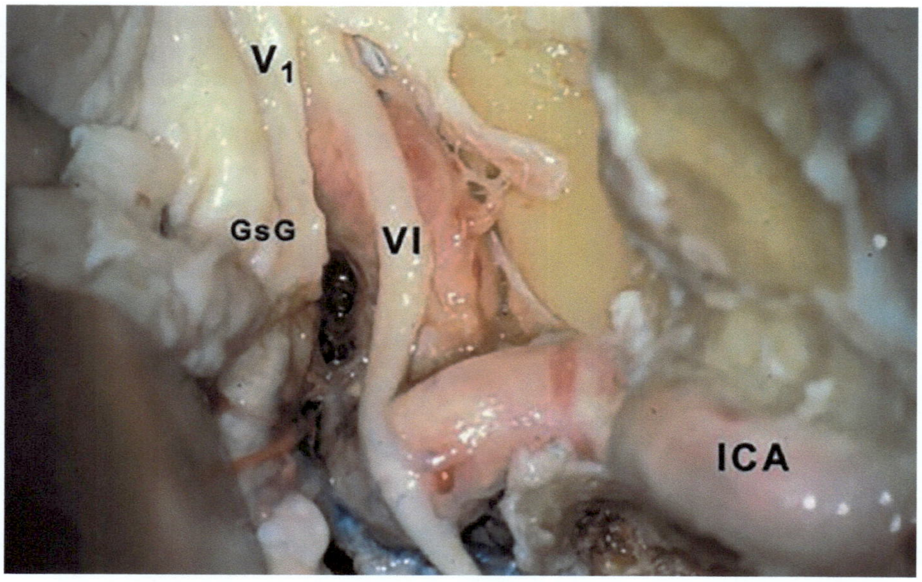

Fig. 15.73 The maxillary branch has been sacrificed, and the Gasserian ganglion (GsG) can be further retracted. The intracavernous internal carotid (ICA) artery starts to appear.

CASE 1: RIGHT TRIGEMINAL NEURINOMA WITH EXTENSIVE POSTERIOR AND MIDDLE FOSSA COMPONENTS (Figs. 15.74–15.89)

1. MRI with gadolinium enhancement showing the tumor with both posterior and middle fossa components. The tumor abuts the lateral wall of the cavernous sinus. It involves the petrous apex and extends partially to the infratemporal fossa (Fig. 15.74).

2. A curvilinear preauricular skin incision is made. The incision starts in front of the tragus and extends superiorly to the frontal scalp. The incision joins another wide C-shaped postauricular incision such as is performed for the modified transcochlear approach (Fig. 15.75).

3. The external auditory canal is transected and closed as a cul-de-sac. The temporalis muscle is incised and retracted inferiorly. The zygomatic arch and lateral orbital rim are removed using the reciprocating saw of the Midas Rex. A wide frontotemporal craniotomy is performed. The sigmoid sinus is exposed, and the posterior fossa dura anterior and posterior to the sinus is widely exposed (Fig. 15.76).

4. The labyrinth is drilled, the internal auditory canal is identified, and the facial nerve is decompressed from the level of the stylomastoid foramen to the fundus of the internal auditory canal. The nerve is rerouted posteriorly together with the contents of the canal as previously described. The cochlea and petrous apex are drilled to the level of the midclivus. The temporomandibular joint is opened, and the mandibular condyle is retracted inferiorly.

5. The middle meningeal artery and the mandibular branch of the trigeminal nerve are identified and cut after bipolar coagulation. With gentle retraction of the temporal lobe, the tumor is seen on the superior and medial aspects of the internal carotid artery (Figs. 15.77 and 15.78).

6. The posterior fossa dura is opened; the tumor is seen surrounding the abducent nerve (Fig. 15.79). After debulking the tumor, a plane of cleavage is established, and the tumor is dissected from the basilar artery. Using bipolar coagulation and gentle traction with the suction irrigator, the tumor gradually shrinks and becomes separated from the brainstem (Fig. 15.80).

7. More superiorly, the trigeminal nerve is seen completely surrounded by the tumor. With meticulous dissection, the tumor can be separated from the nerve (Figs. 15.81 and 15.82).

8. The tumor is further separated from the trigeminal nerve. The contralateral trigeminal and abducent nerves are clearly seen (Fig. 15.83).

9. With gentle retraction of the temporal lobe, the tumor is seen insinuating between the trigeminal and oculomotor nerves (Fig. 15.84). This last piece of tumor is removed. The posterior cerebral artery runs superior to the intact oculomotor nerve, whereas inferiorly, the superior cerebellar artery courses between the oculomotor and trigeminal nerves (Fig. 15.85).

10. At the end of the procedure, the cavity is obliterated with abdominal fat. The two craniotomies are replaced and secured using miniplates. The temporalis muscle is repositioned, and the wound is closed in layers. No drains are inserted (Figs. 15.86–15.88).

11. Postoperative CT scan, axial view showing total tumor removal (Fig. 15.89).

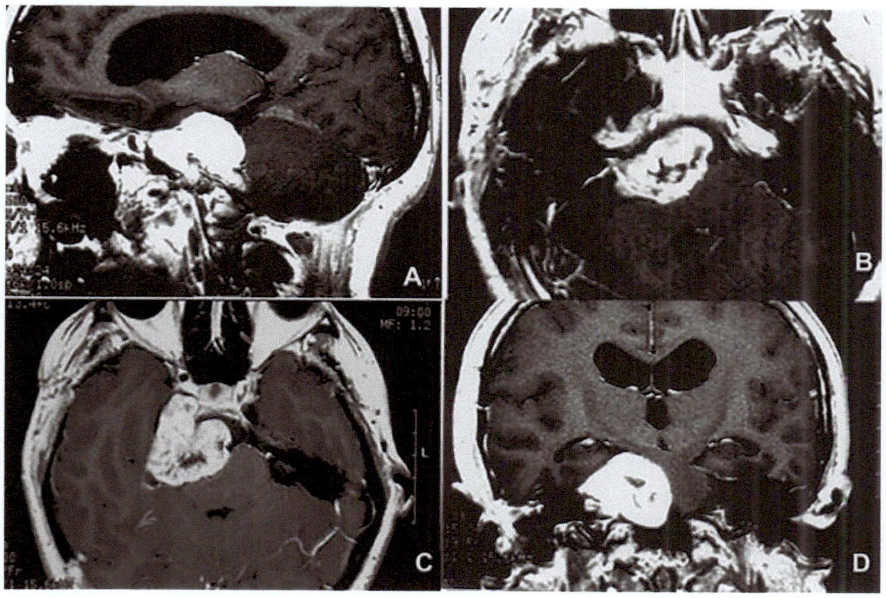

Fig. 15.74 MRI with gadolinium enhancement showing the tumor. (A) Sagittal view showing the tumor involving both the posterior and middle cranial fossa. (B) Axial view. The tumor is seen in the posterior fossa lying anterior to the internal auditory canal and involving the prepontine cistern. (C) Axial view at a higher level. The parasellar extension of the tumor is seen. (D) Coronal view. Note the marked displacement of the brainstem. Involvement in Meckel's cave is appreciated.

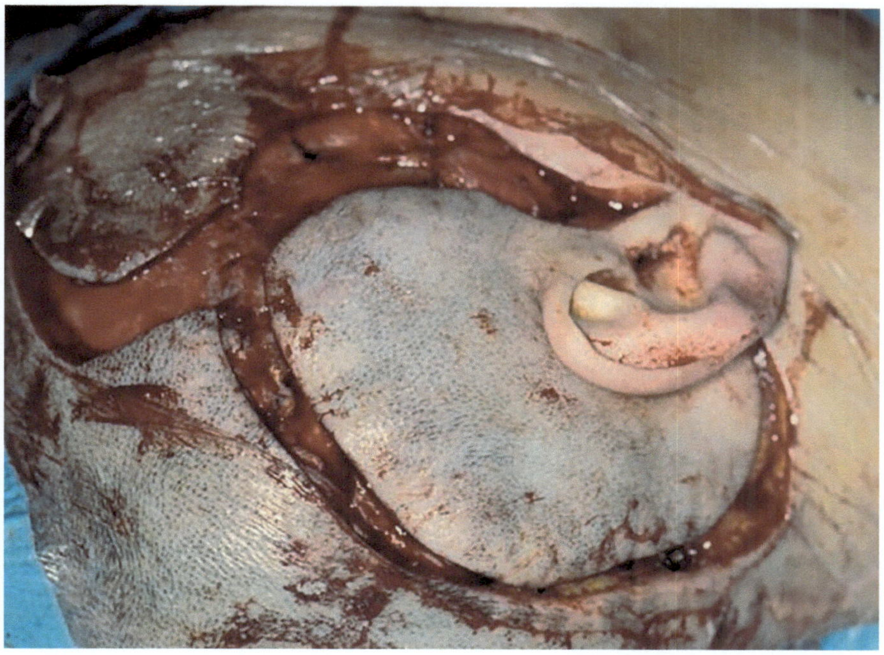

Fig. 15.75 Skin incision.

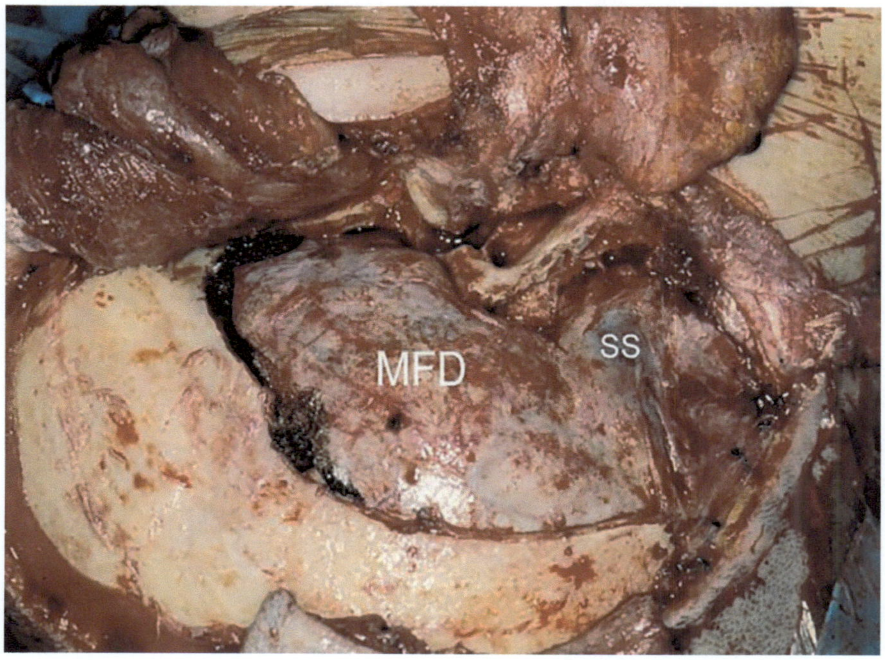

Fig. 15.76 The frontotemporal craniotomy. Note also the exposure of the posterior fossa dura.

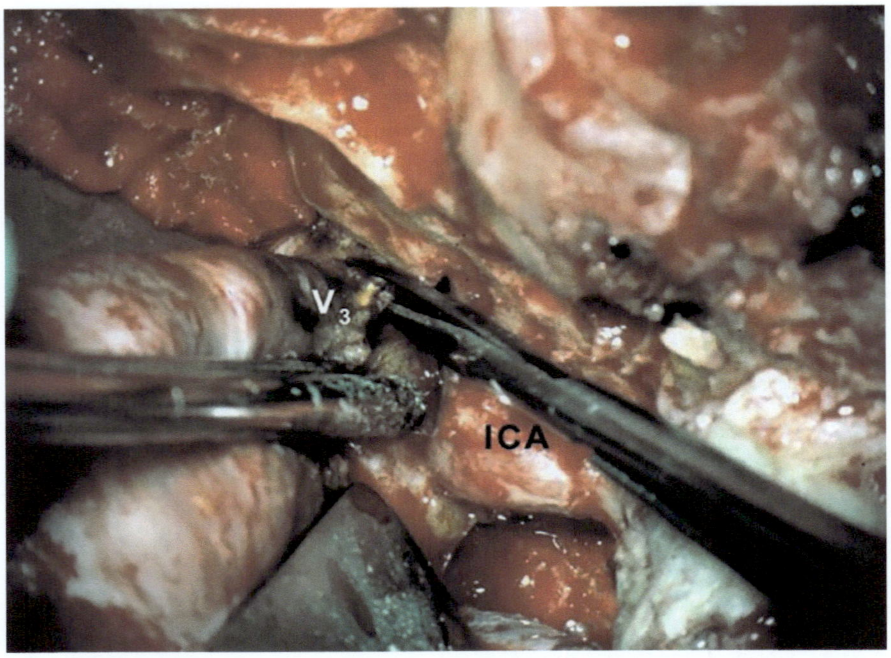

Fig. 15.77 Cutting the mandibular branch of the trigeminal nerve (V3).

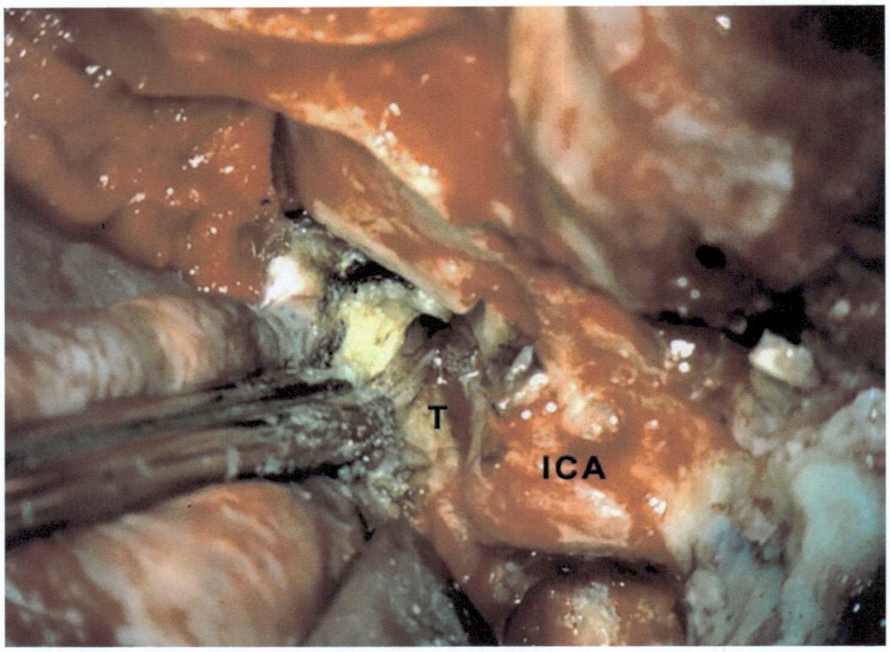

Fig. 15.78 With retraction of the temporal lobe, the relation of the tumor (T) to the internal carotid artery (ICA) is identified.

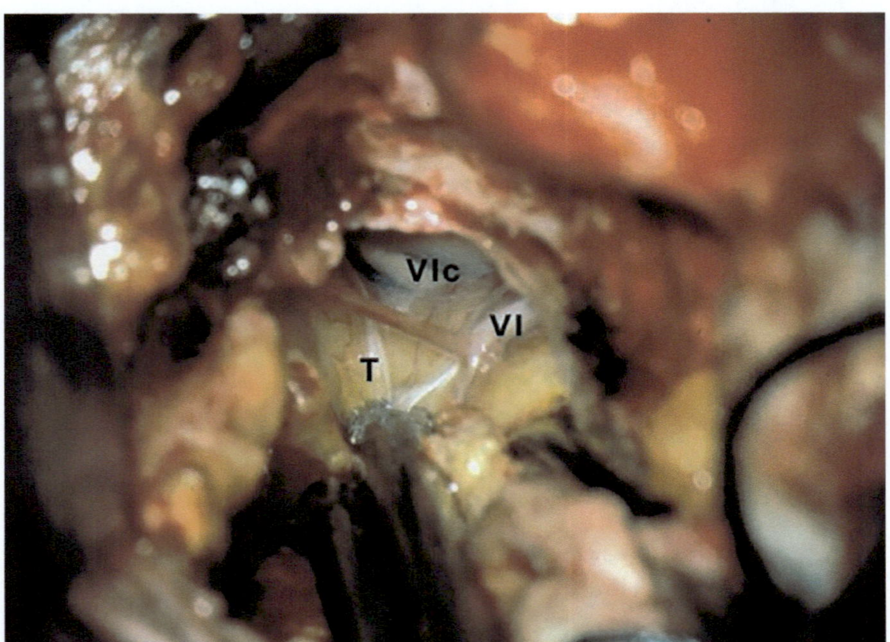

Fig. 15.79 The tumor (T) in the posterior fossa in relation to the abducent nerve (VI). The contralateral abducent (VIc) nerve is clearly seen.

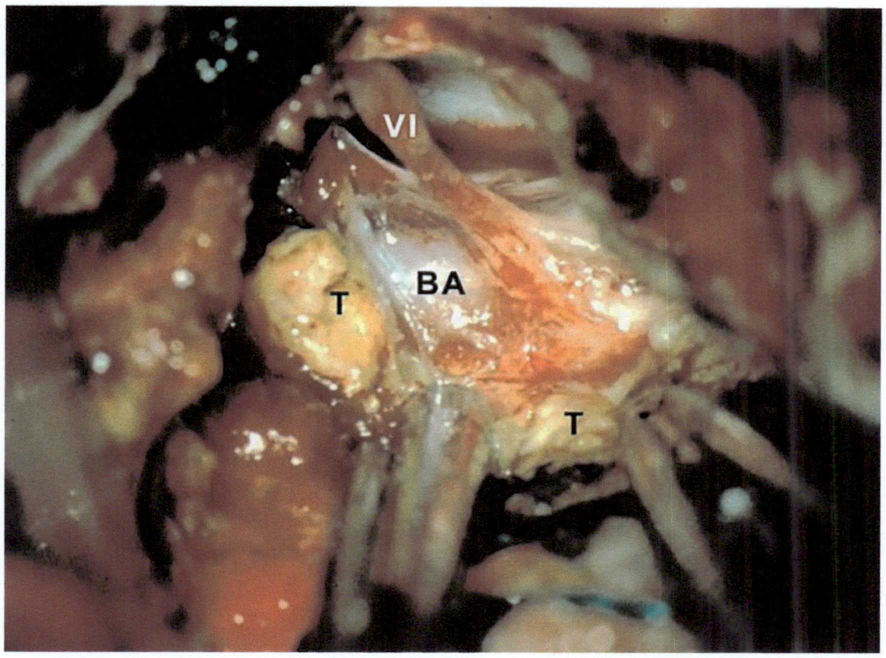

Fig. 15.80 The relation of the tumor (T) to the basilar artery (BA) in the prepontine cistern. Using bipolar coagulation and mild traction, the tumor is gently separated from the brainstem.

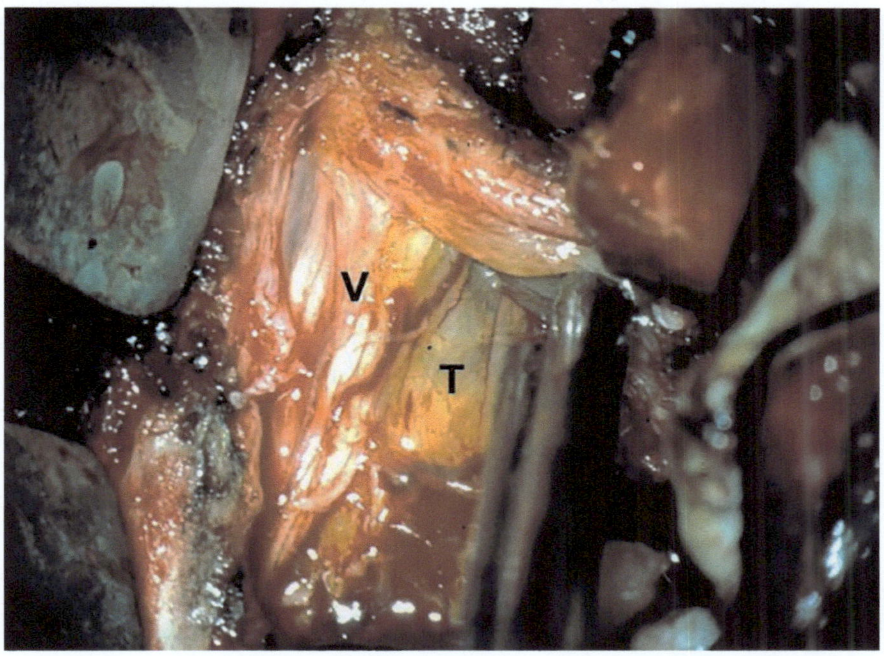

Fig. 15.81 Tumor (T) in Meckel's cave.

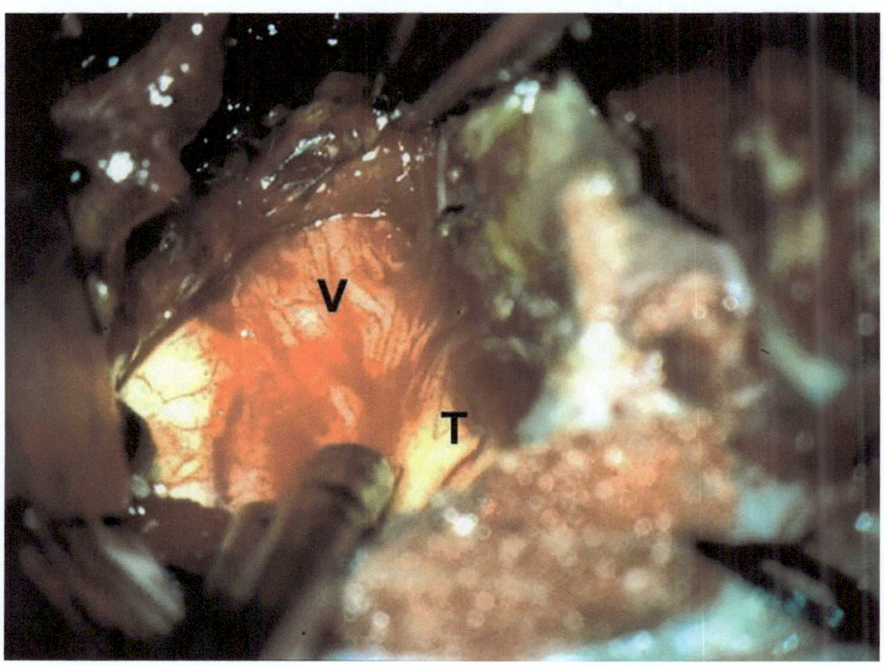

Fig. 15.82 Removal of tumor (T) from Meckel's cave.

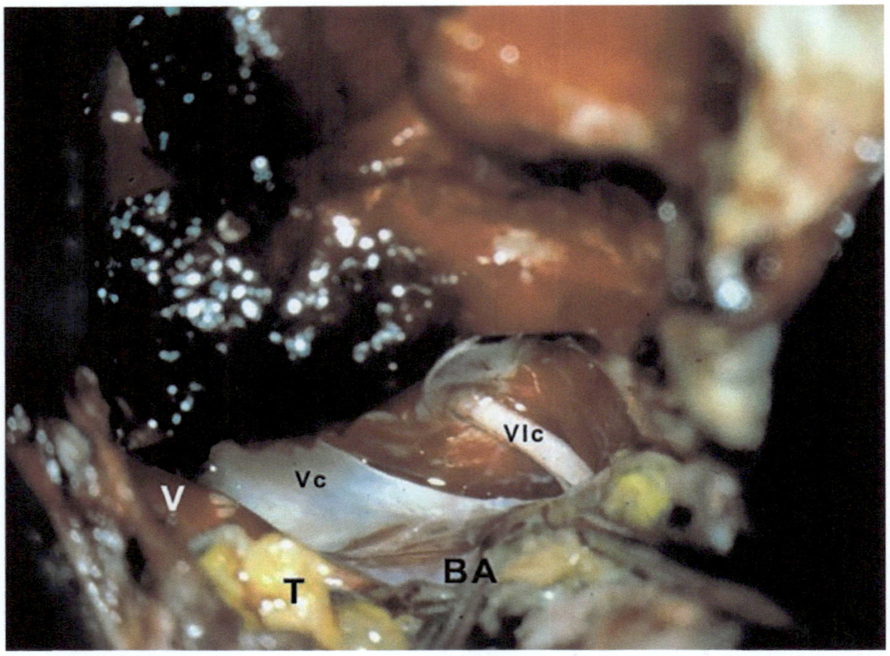

Fig. 15.83 View at the prepontine cistern. The contralateral trigeminal (Vc) and abducent (VIc) nerves are seen. The tumor still involves the basilar artery (BA) and the ipsilateral trigeminal nerve (V).

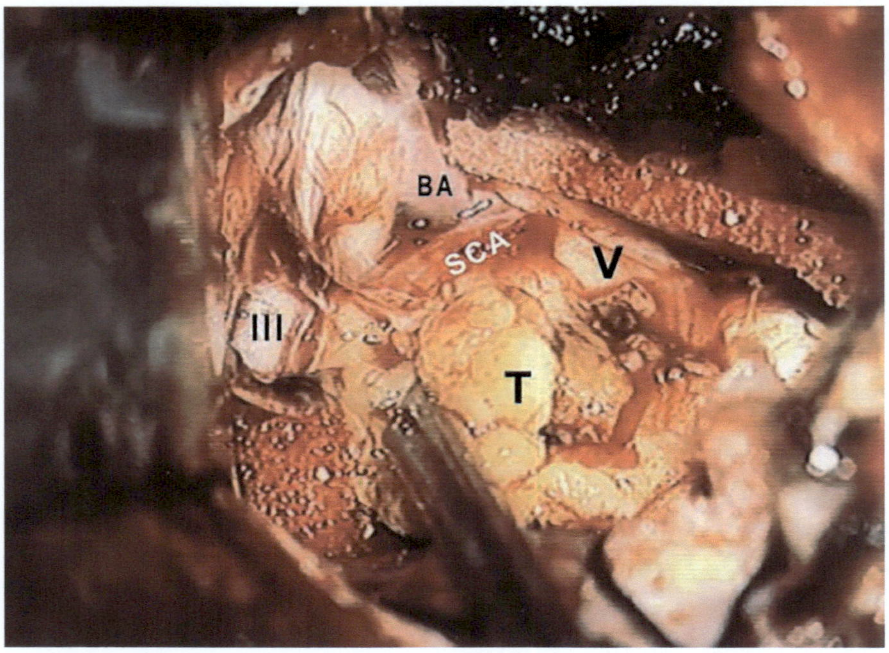

Fig. 15.84 The last piece of tumor (T) in the middle fossa is seen insinuating between the trigeminal (V) and the oculomotor (III) nerves. The terminal part of the basilar artery (BA) is identified.

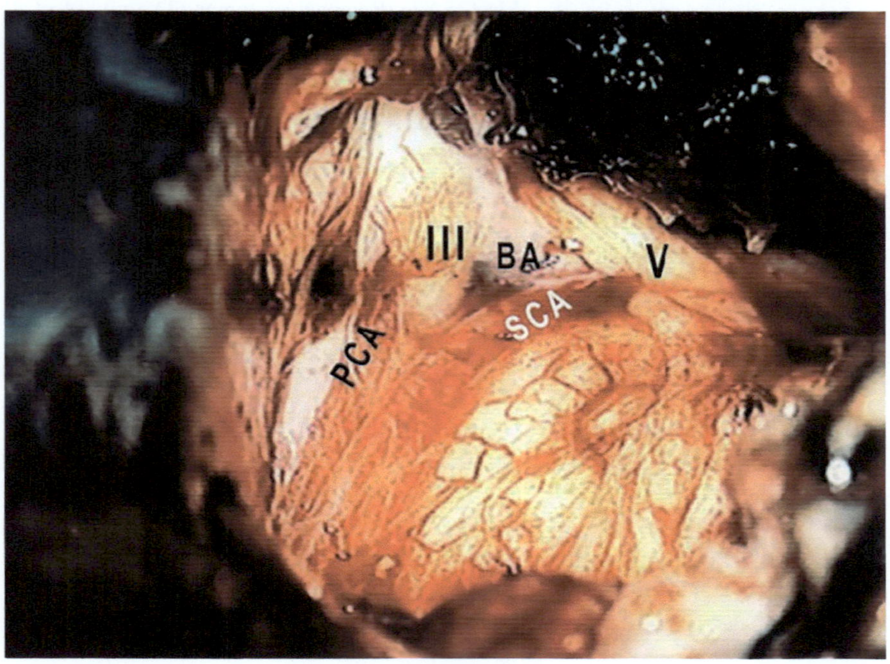

Fig. 15.85 View after total tumor removal. The oculomotor nerve (III) is sandwiched between the posterior cerebral artery (PCA) superiorly and the superior cerebellar artery (SCA) inferiorly.

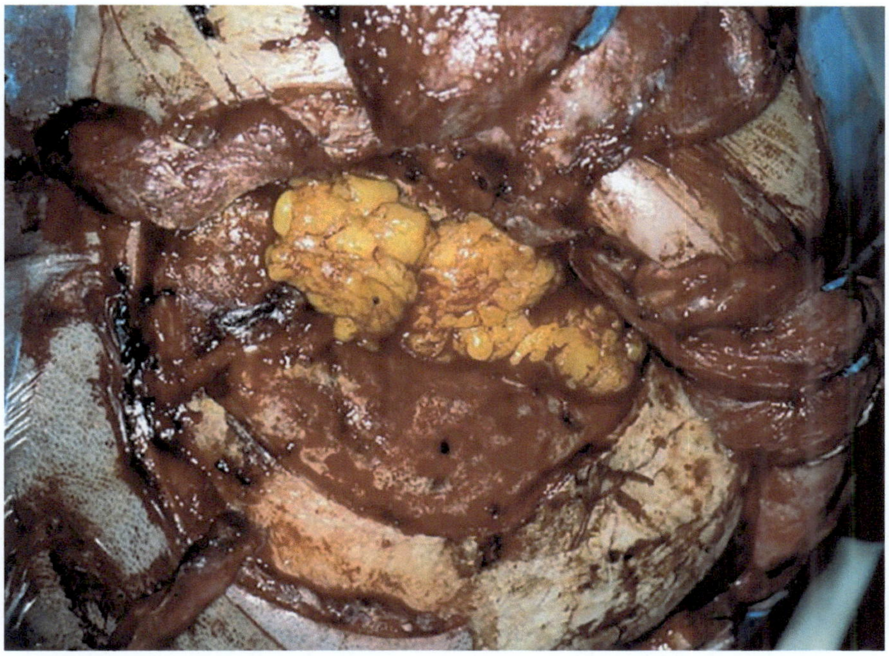

Fig. 15.86 The cavity is obliterated with abdominal fat.

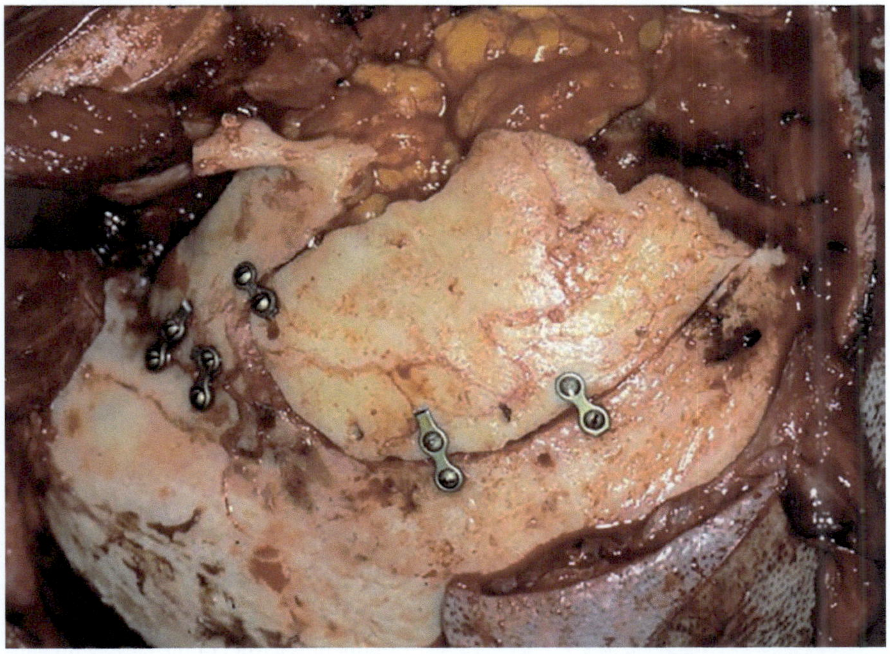

Fig. 15.87 The craniotomy bone is replaced and fixed with miniplates.

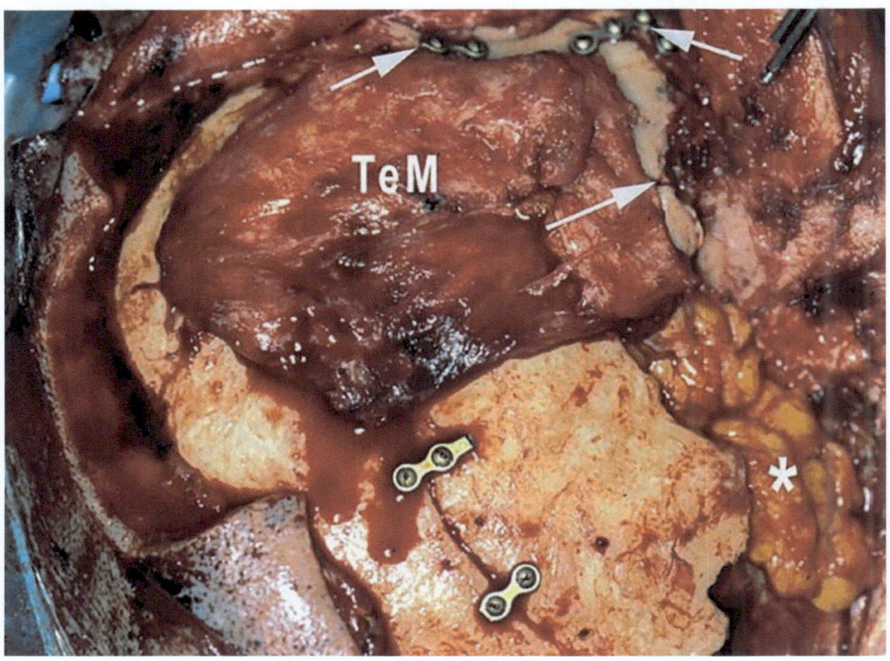

Fig. 15.88 The temporalis muscle (TeM) is repositioned. The orbitozygomatic craniotomy is returned and secured with miniplates. *Arrows* point to the three cuts of the orbitozygomatic bone.

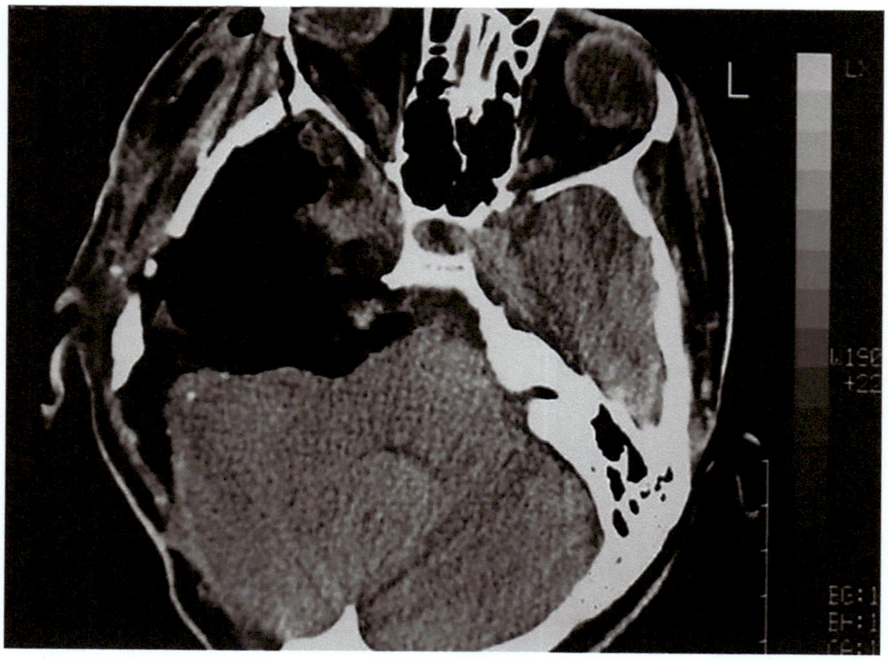

Fig. 15.89 Postoperative CT scan.

CASE 2: LEFT LARGE SPHENOPETROCLIVAL MENINGIOMA WITH MARKED MIDDLE FOSSA AND PARASELLAR EXTENSION (Figs. 15.90–15.98)

1. MRI of the case showing a large left sphenopetroclival meningioma (Fig. 15.90).
2. The incision is made (Fig. 15.91).
3. The temporalis muscle has been raised and reflected. The sigmoid sinus has been uncovered. The posterior fossa dura in front of and posterior to the sinus is widely exposed. Burr holes have been made in order to create the frontotemporal craniotomy (Fig. 15.92).
4. The transcochlear approach is performed as described, and the facial nerve is rerouted posteriorly (Fig. 15.93).
5. The posterior fossa dura is opened. The tumor is dissected free from the anterior inferior cerebellar artery, the superior cerebral artery, and the trigeminal and abducent nerves. The middle fossa dura is opened with a horizontal incision parallel to the superior petrosal sinus. The sinus is closed by bipolar coagulation. The temporal lobe is gently retracted. The tentorium is cut, and total tumor removal is accomplished (Figs. 15.94–15.97). The orbitozygomatic osteotomy allows better retraction of the anterior part of the temporal lobe and hence better control of the anterior and superior part of the tumor.
6. Postoperative CT scan shows total tumor removal (Fig. 15.98).

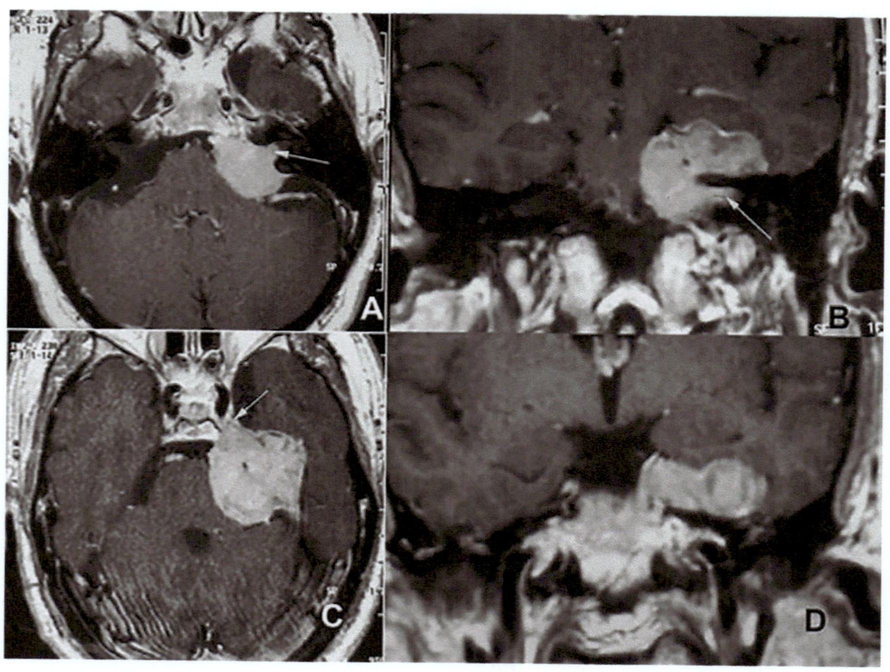

Fig. 15.90 Preoperative MRI of the case, with gadolinium enhancement. (A) Axial view at the level of the posterior fossa. The tumor lies mainly anterior to the internal auditory canal and grows into the prepontine cistern. Tumor extension into the internal auditory canal *(arrow)* is also seen. (B) Coronal view. The tumor involves both middle and posterior cranial fossae. Note the extent of the internal auditory canal *(arrow)*. (C) Axial view at a higher level. The middle fossa component is clearly seen. Note the parasellar component of the tumor *(arrow)*. (D) Coronal view. The tumor contacting the temporal lobe is appreciated.

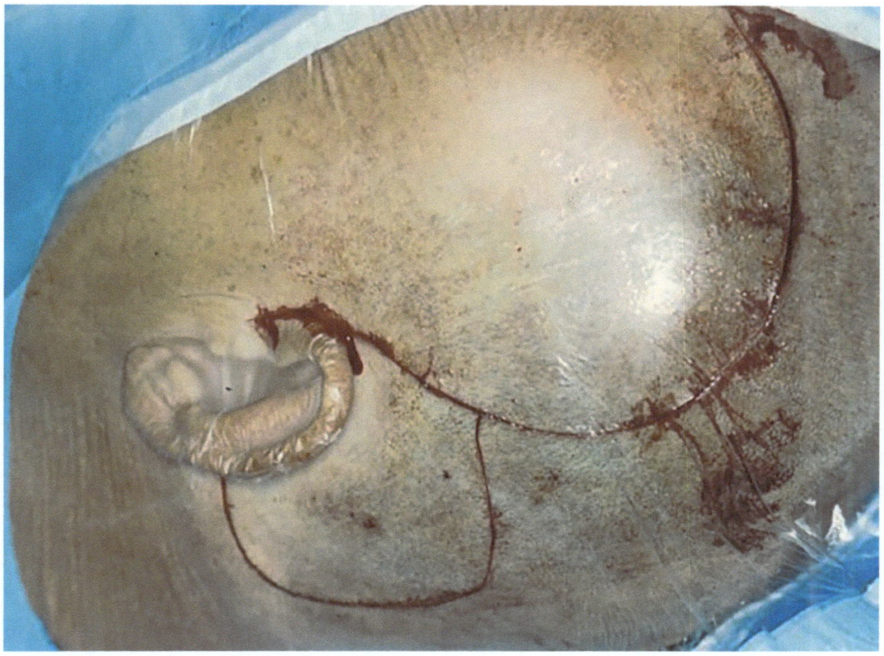

Fig. 15.91 Skin incision.

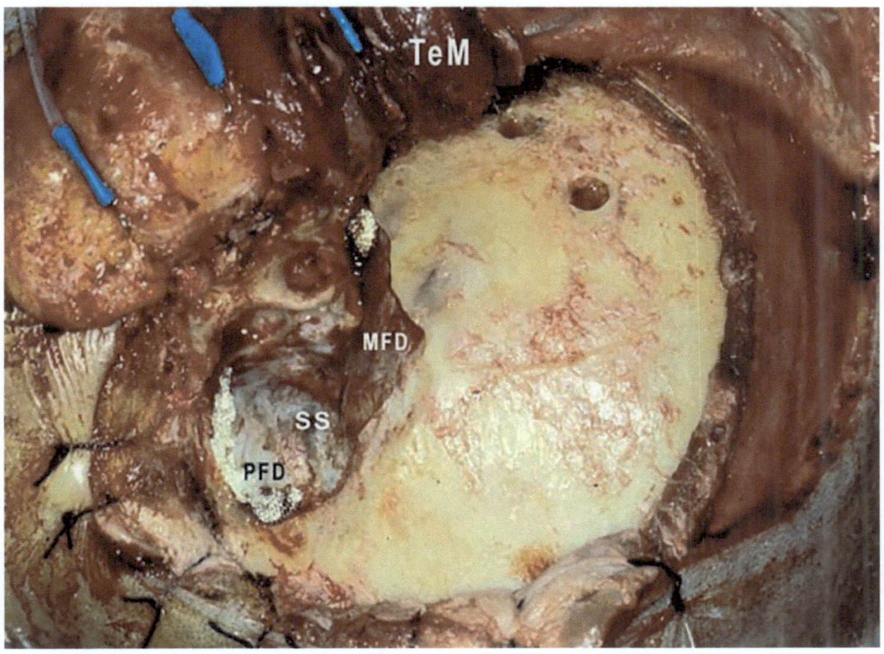

Fig. 15.92 The initial phase of the approach.

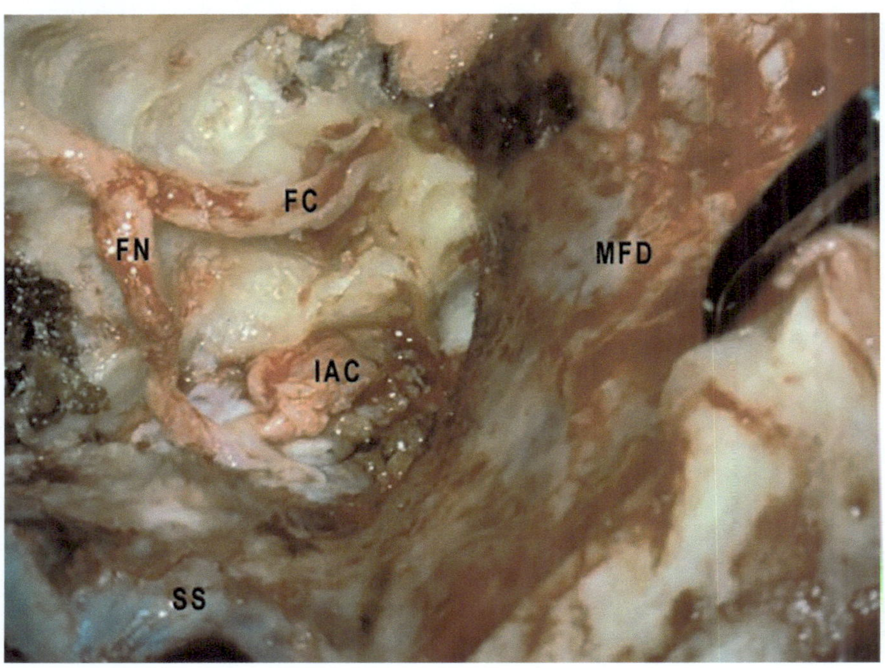

Fig. 15.93 Posterior rerouting of the facial nerve (FN).

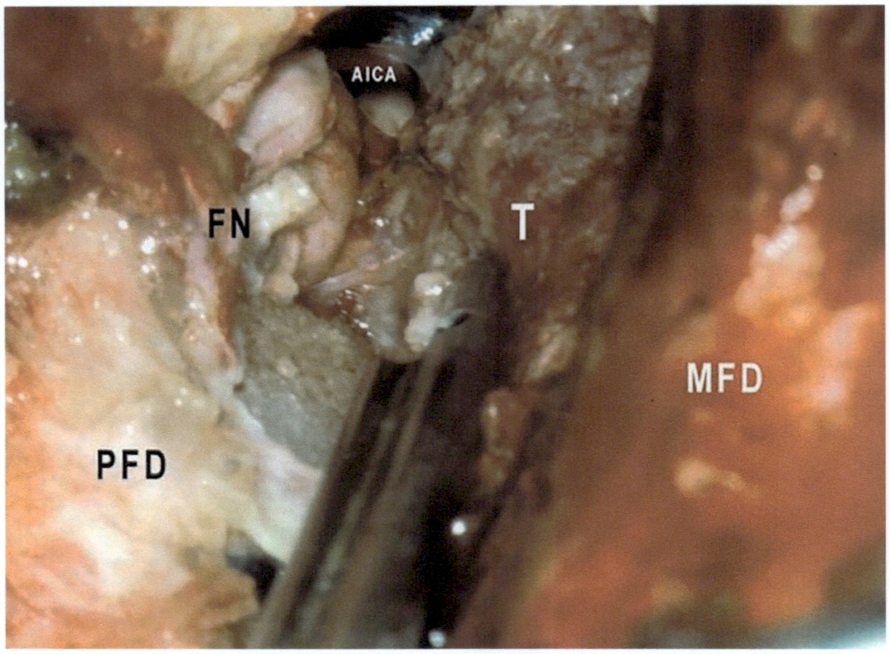

Fig. 15.94 Tumor (T) is dissected from the anterior inferior cerebellar artery (AICA).

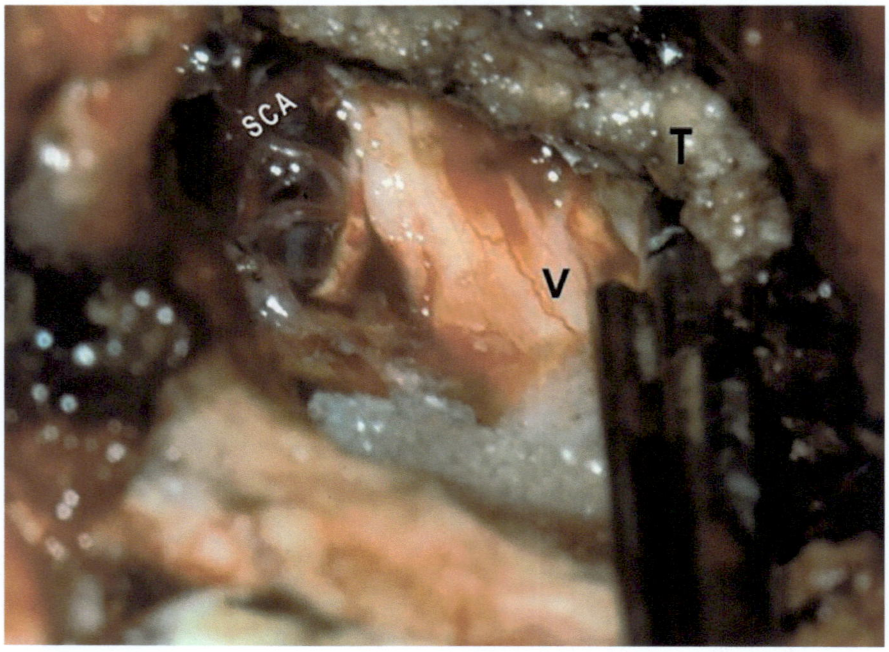

Fig. 15.95 The tumor (T) is dissected from the trigeminal nerve (V).

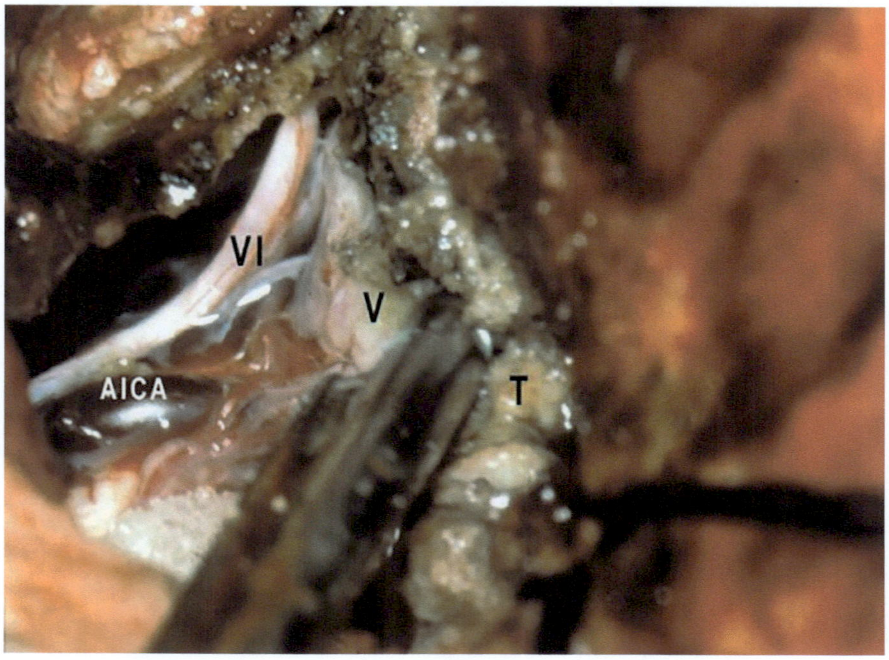

Fig. 15.96 With temporal lobe retraction, tumor (T) involving the abducent (VI) and trigeminal (V) nerves can be better controlled.

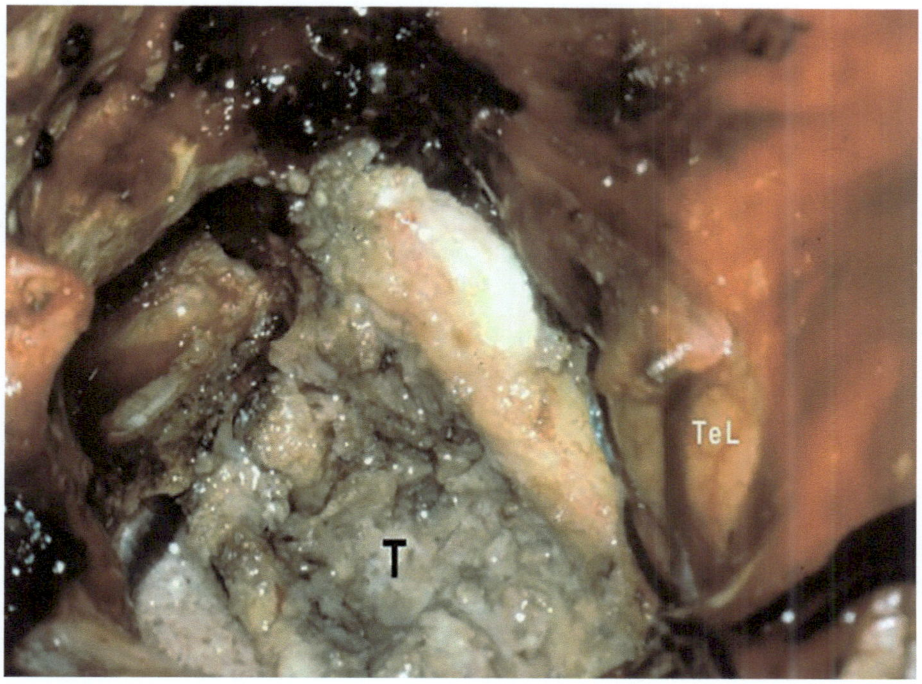

Fig. 15.97 The middle fossa dura has been incised. The temporal lobe (TeL) is already seen. The tumor (T) in the middle fossa and infiltrating the tumor is to be removed.

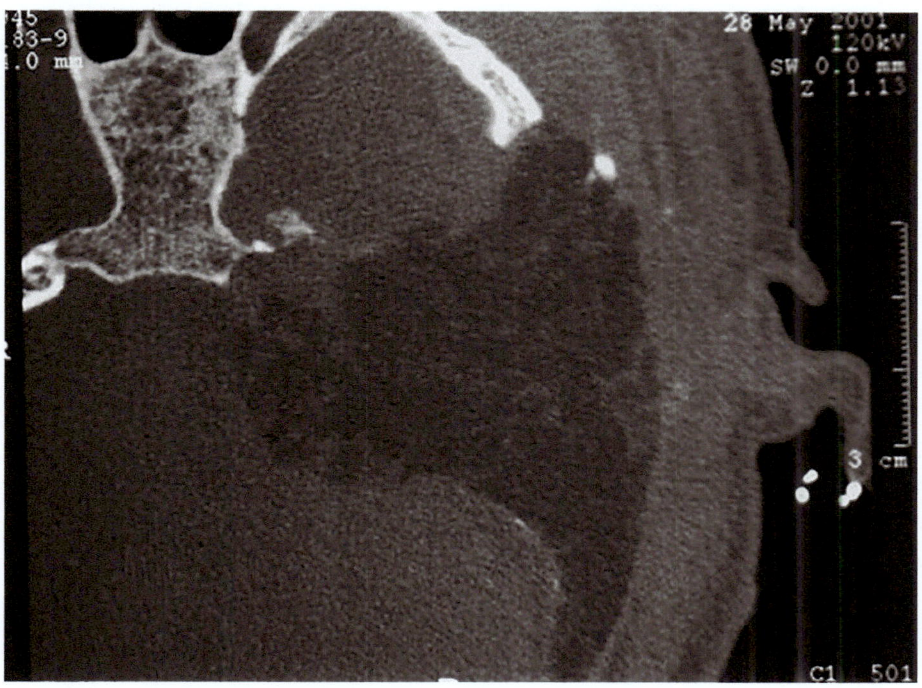

Fig. 15.98 Postoperative CT scan.

COMBINED RETROLABYRINTHINE SUBTEMPORAL ORBITOZYGOMATIC APPROACH

Fig. 15.99 shows a schematic presentation of this approach.

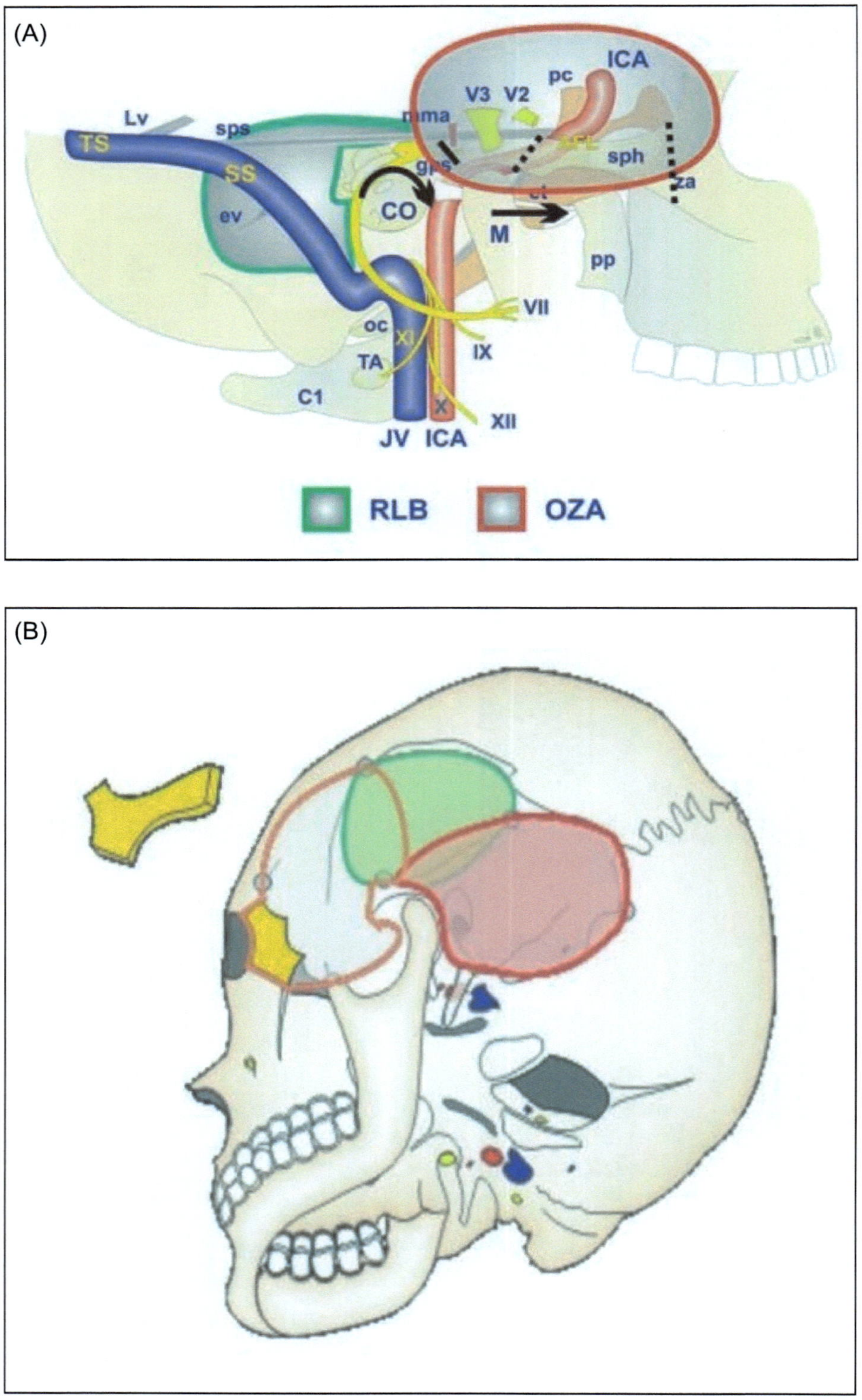

Fig. 15.99 (A, B) Schematic drawing of the combined retrolabyrinthine (RLB), orbitozygomatic (OZA) approach. Key as for Fig. 15.68.

CASE 3: RIGHT CLIVAL CHORDOMA WITH POSTERIOR AND MIDDLE FOSSA EXTENSIONS (Figs. 15.100–15.115)

1. Preoperative MRI with gadolinium enhancement showing the bilobed tumor with posterior and middle fossa components (Fig. 15.100).

2. The skin incision is a curvilinear preauricular incision starting in front of the tragus and extending superiorly to the frontal scalp. The incision joins another wide C-shaped postauricular incision (Fig. 15.101).

3. The skin and subcutaneous tissue are elevated, exposing the zygomatic arch and the lateral orbital rim (Fig. 15.102). The temporalis muscle is reflected inferiorly and the frontotemporal orbitozygomatic craniotomy is outlined and burr holes are made. The retrolabyrinthine craniotomy, on the other hand, is performed using a high-speed drill (Fig. 15.103).

4. The retrolabyrinthine approach is performed as previously described. Note the wide bone removal exposing the posterior fossa dura posterior to the sigmoid sinus. The frontotemporal orbitozygomatic osteotomy is also done (Figs. 15.104 and 15.105).

5. The middle fossa dura is elevated from the roof of the temporal bone. The mandibular nerve is sacrificed. The petrous apex lying anterior to the internal auditory canal is drilled, and the tumor in this area is removed. The horizontal segment of the internal carotid artery is clearly seen (Figs. 15.106 and 15.107).

6. The posterior fossa dura is opened. The abducent nerve is markedly displaced posteriorly. The tumor is seen pushing the acousticofacial bundle inferolaterally and the trigeminal nerve superolaterally. Debulking of the tumor follows (Figs. 15.108 and 15.109).

7. At the end of the procedure, total tumor removal is accomplished (Fig. 15.110). The different structures in the posterior fossa are seen (Fig. 15.111).

8. The operative cavity (at both the posterior and middle fossa levels) is obliterated with abdominal fat. The craniotomies are replaced and secured in place by miniplates, and the temporalis and musculoperiosteal flaps are repositioned and sutured (Figs. 15.112–15.114).

9. Postoperative CT scan showing total tumor removal. The areas of bone removal are clearly seen (Fig. 15.115).

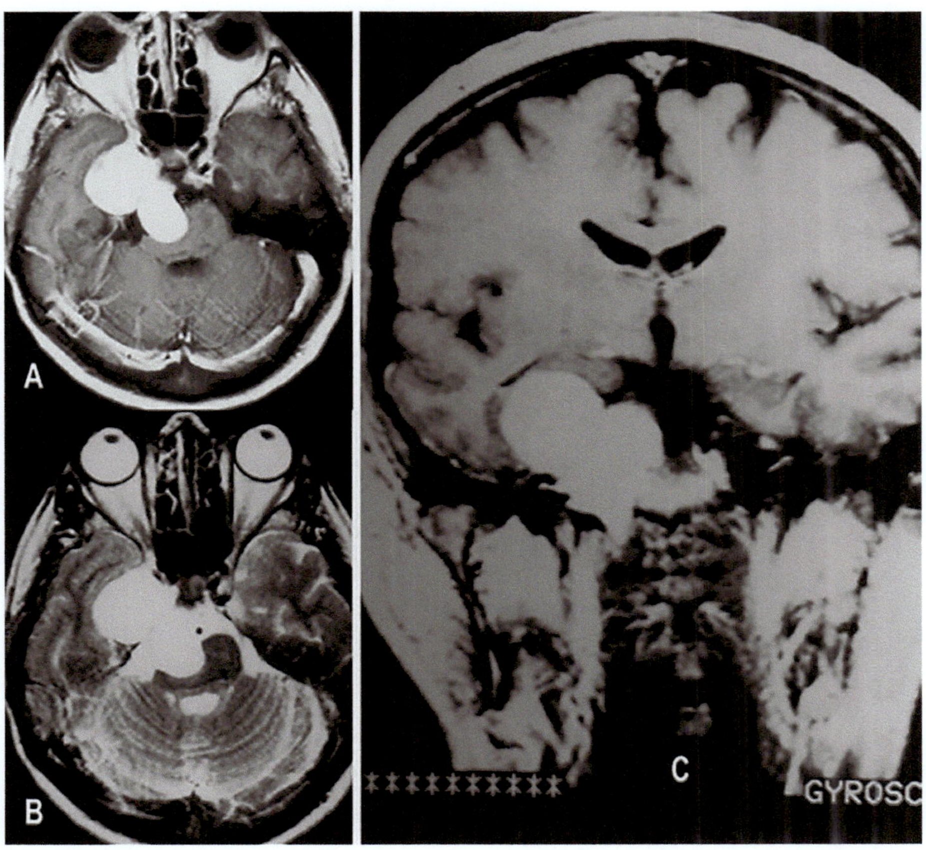

Fig. 15.100 (A–C) Preoperative MRI of the case. (A) MRI with gadolinium enhancement, axial view showing a dumbbell-shaped tumor occupying both the middle and posterior fossa. Not the displacement of the brainstem. (B) T2-weighted image, axial view showing the hyperintense mass. (C) Coronal view (with gadolinium enhancement). Temporal lobe displacement is evident.

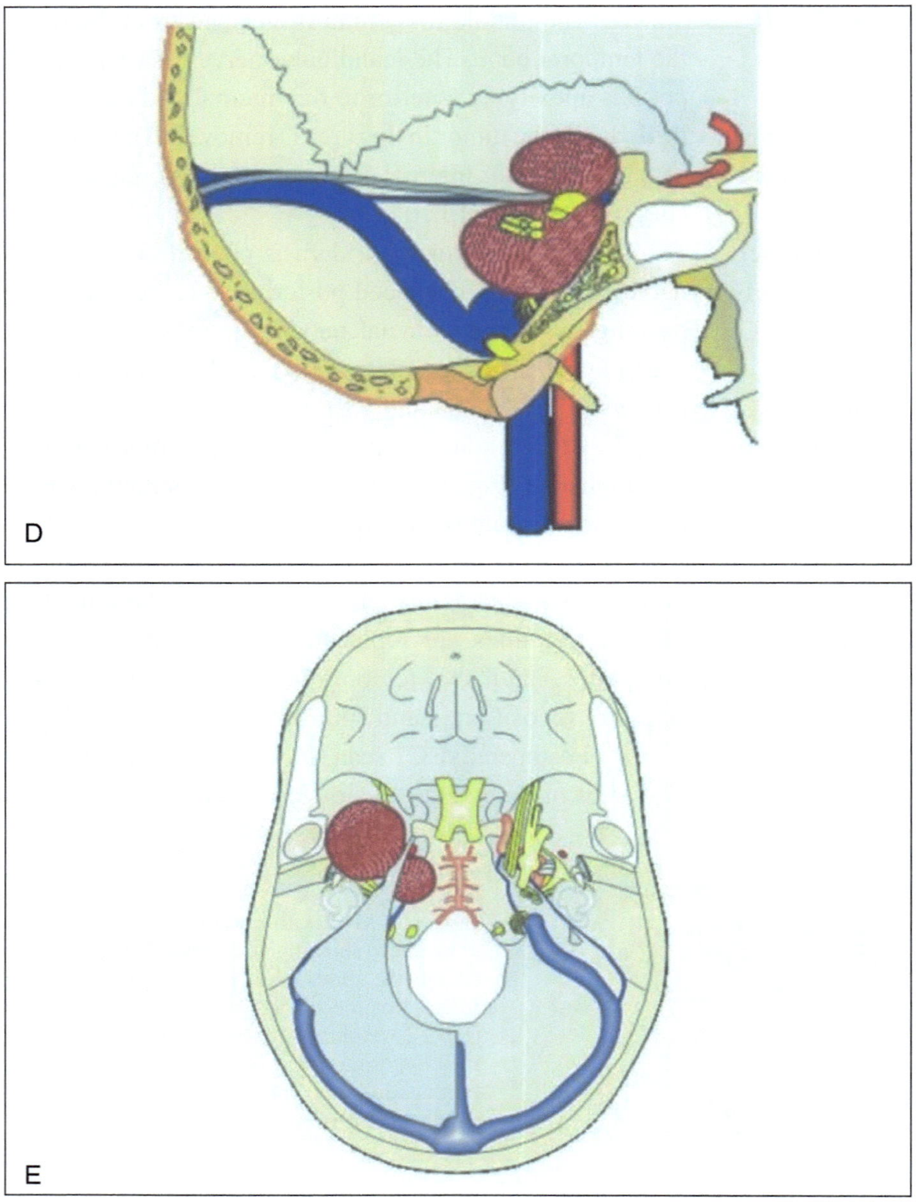

Fig. 15.100 (D,E) Illustrations showing the tumor shape.

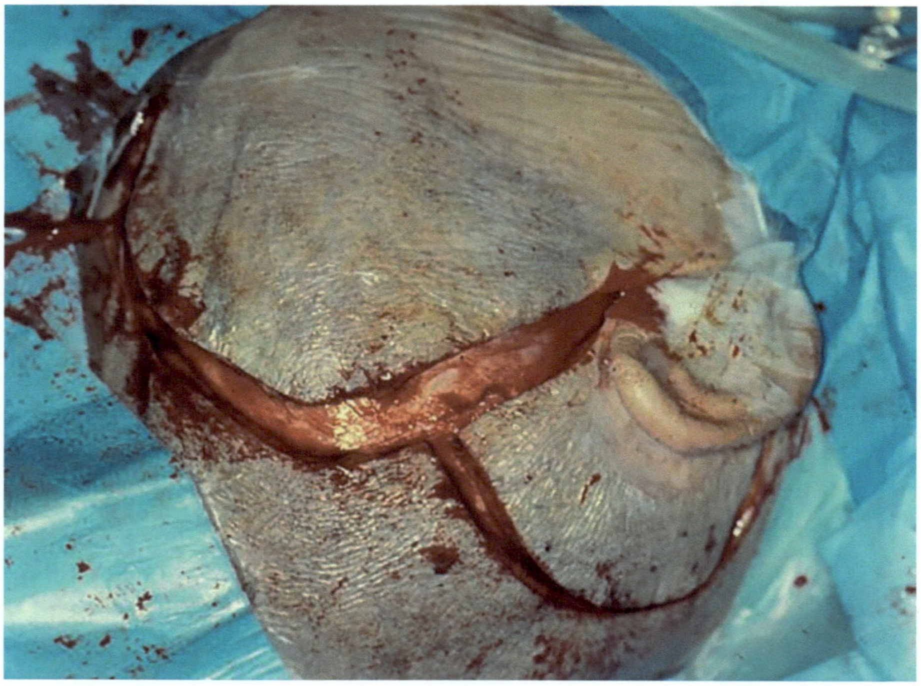

Fig. 15.101 Skin incision.

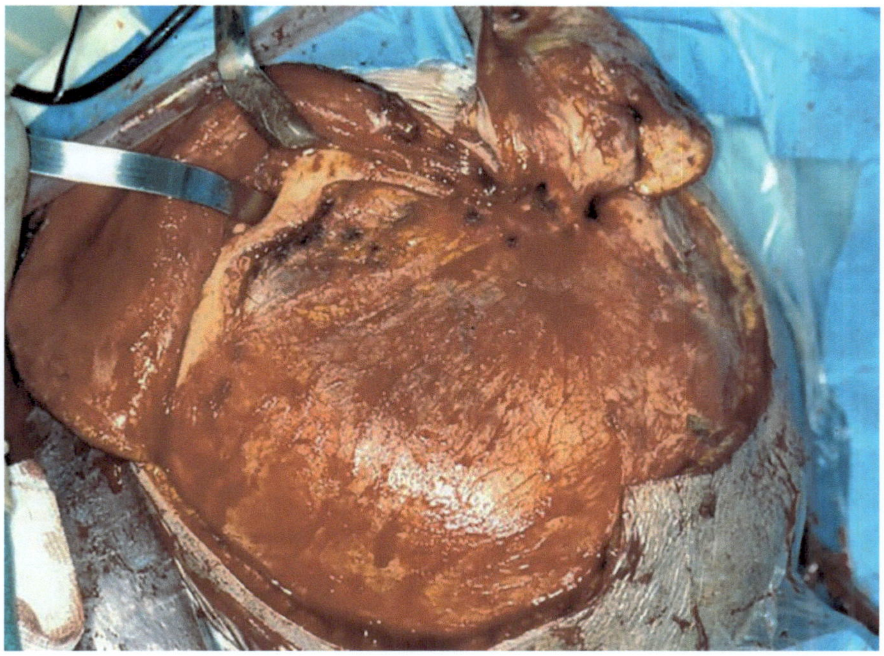

Fig. 15.102 Exposure of the orbital rim after elevation of the skin and subcutaneous tissue.

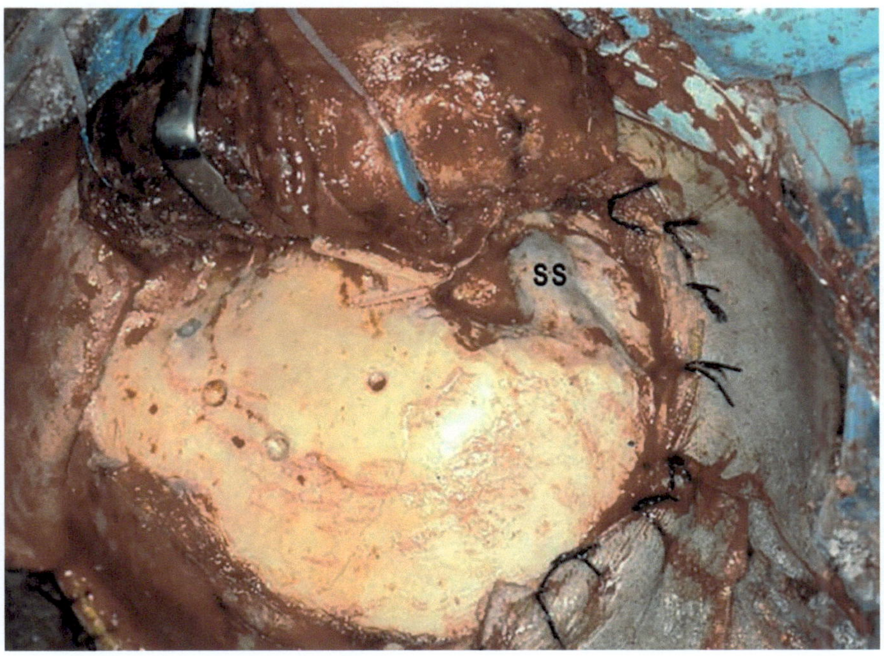

Fig. 15.103 Outlining the frontotemporal orbitozygomatic craniotomy.

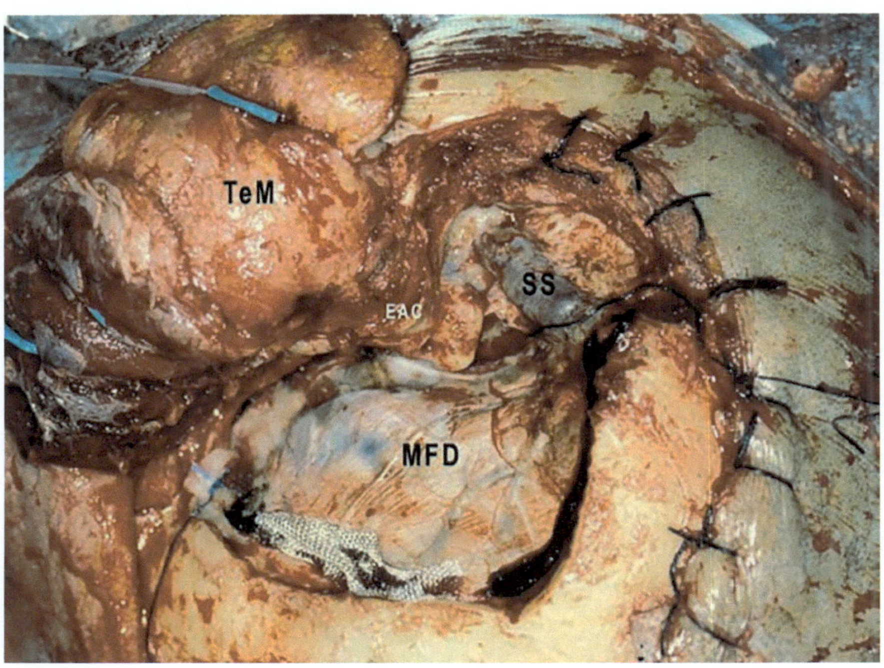

Fig. 15.104 The approach has been performed. Surgicel is used at the craniotomy edge for hemostasis.

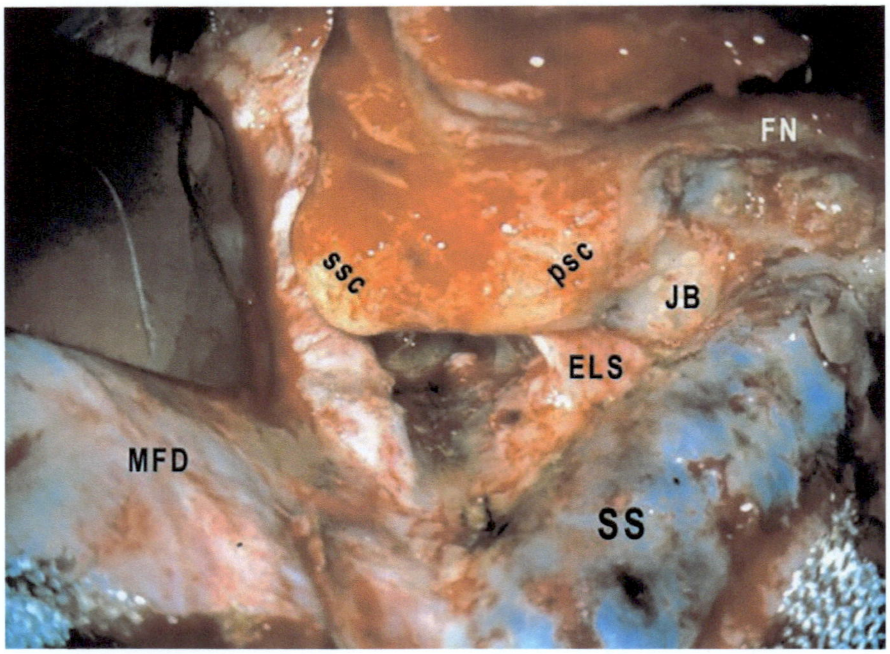

Fig. 15.105 A closer view showing the wide bone removal posterior to the sigmoid sinus (SS).

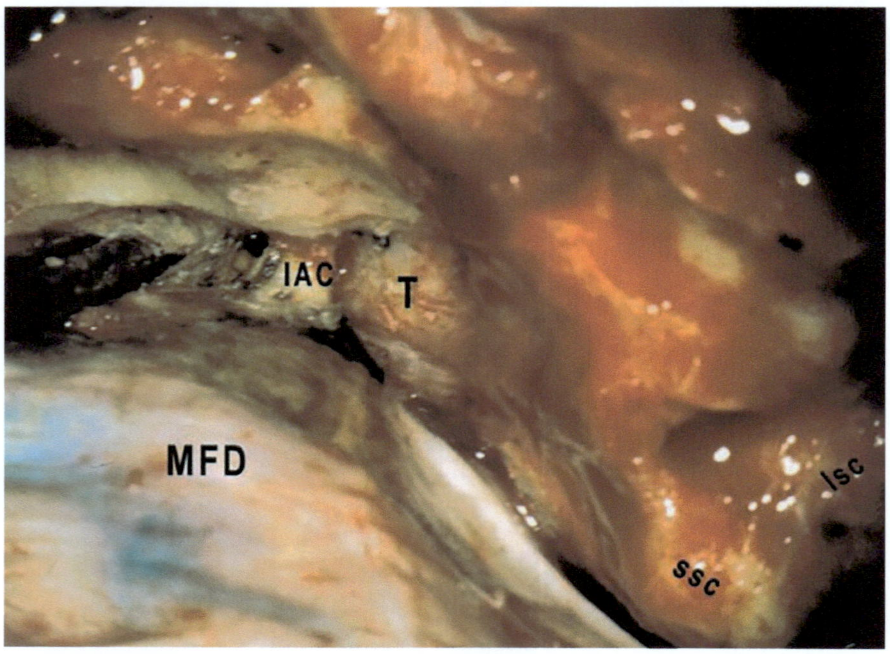

Fig. 15.106 The internal auditory canal (IAC) is identified. The tumor (T) is seen invading the petrous apex.

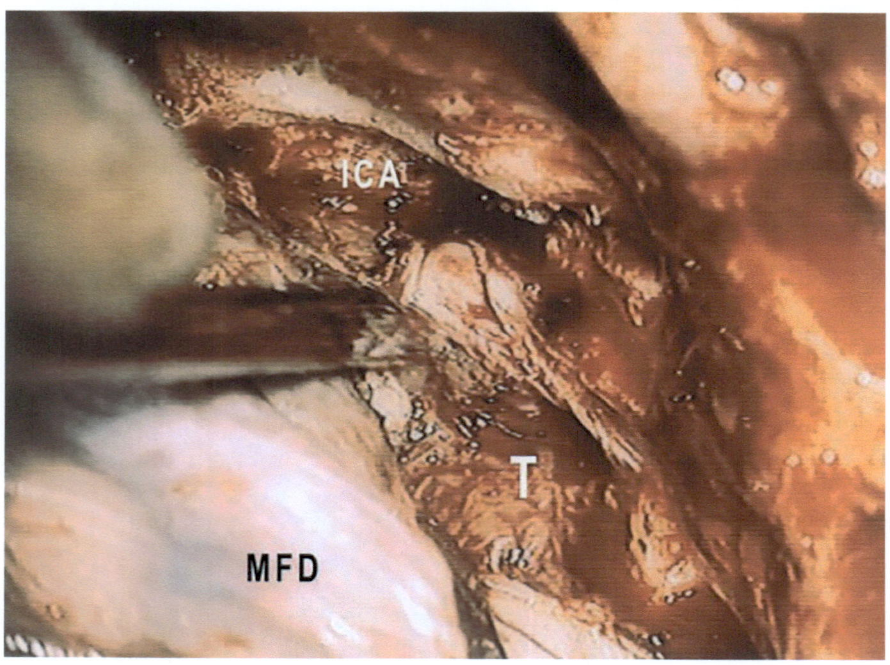

Fig. 15.107 With further drilling of the petrous apex, the internal carotid artery (ICA) is identified. Tumor at this level is removed.

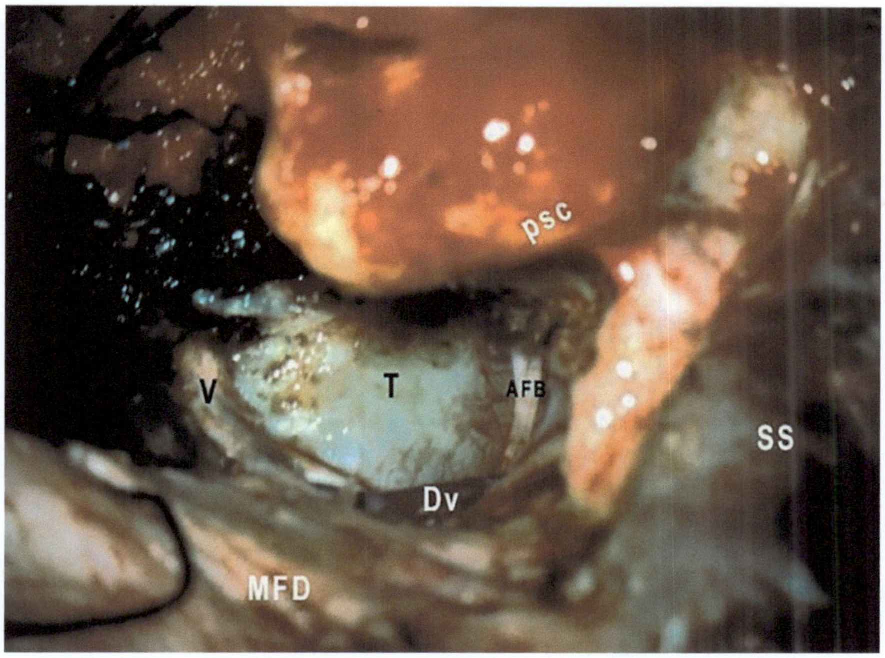

Fig. 15.108 The tumor in the posterior fossa markedly displaces the acousticofacial bundle (AFB) inferolaterally.

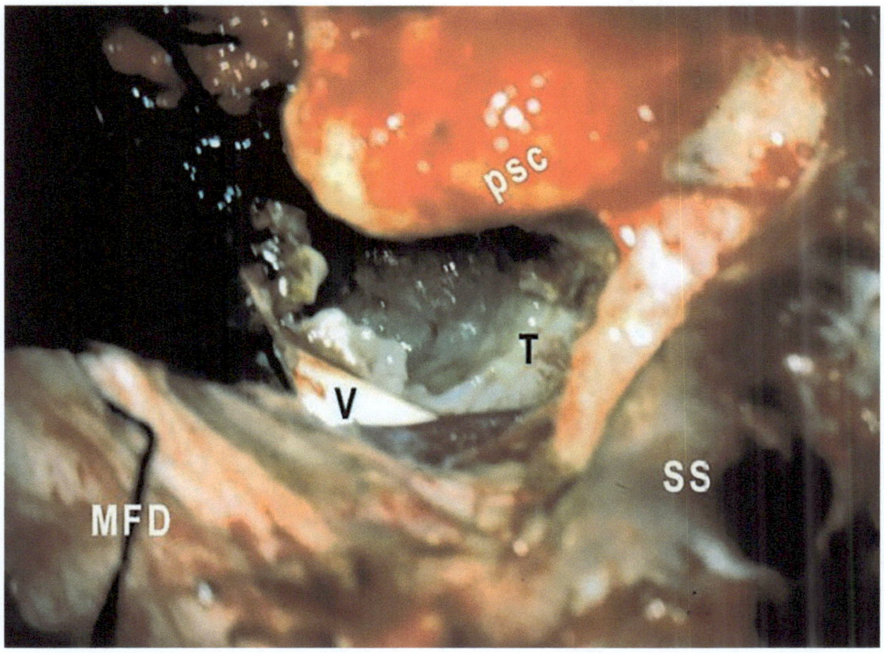

Fig. 15.109 Tumor debulking proceeds. Note the superiorly displaced trigeminal nerve (V).

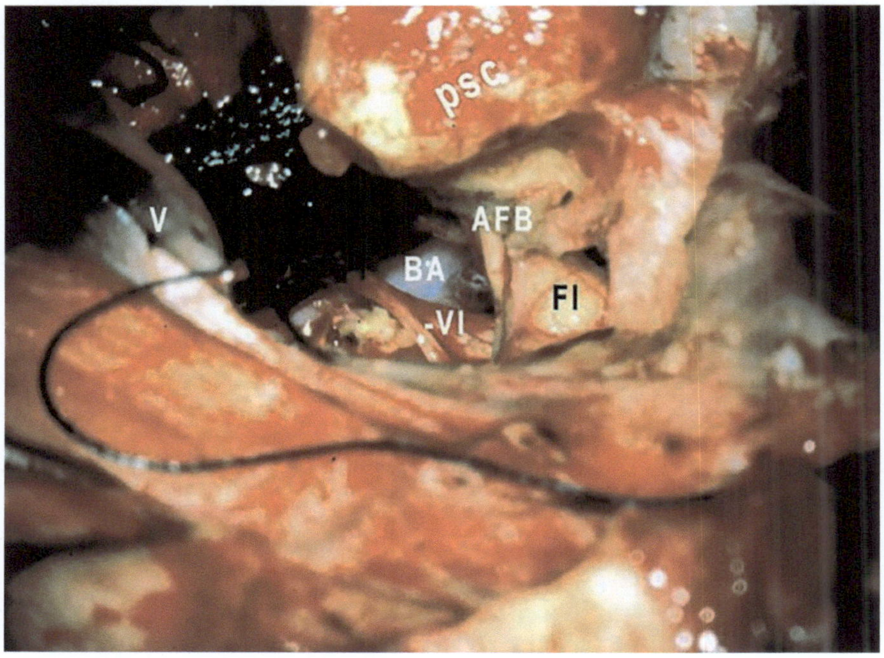

Fig. 15.110 View after total tumor removal.

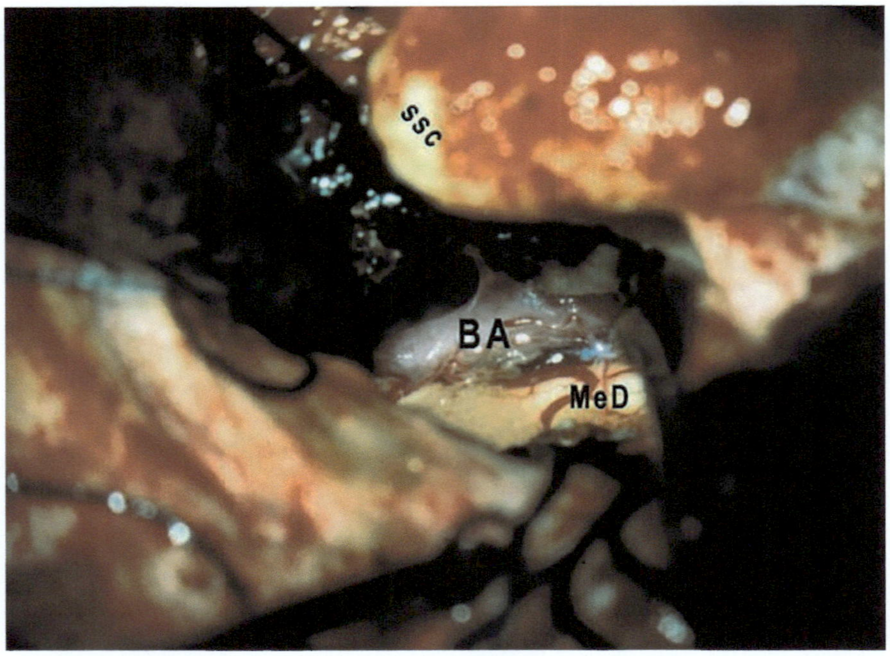

Fig. 15.111 At higher magnification, the basilar artery (BA) at the level of the vertebrobasilar junction is well seen.

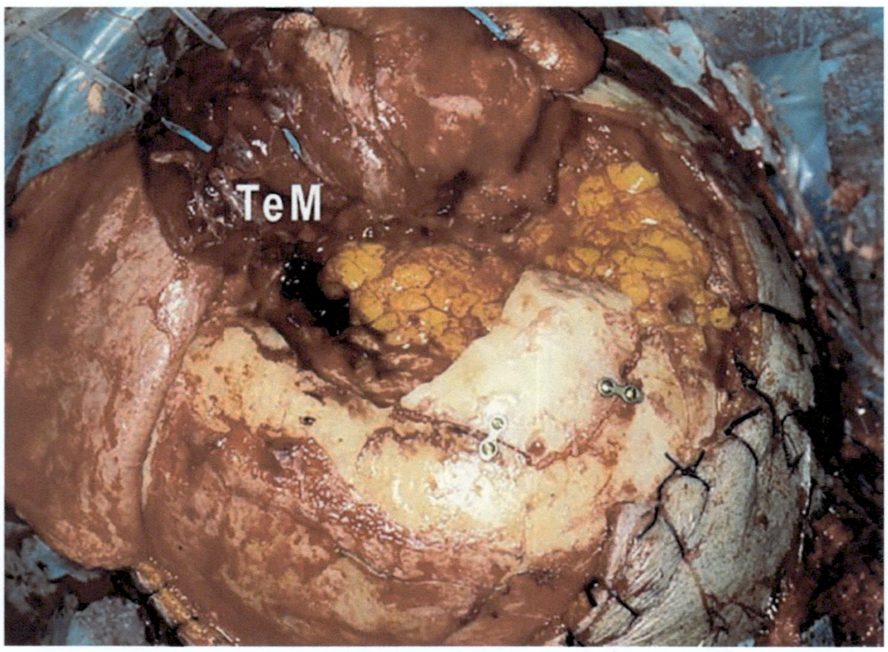

Fig. 15.112 Abdominal fat is used to obliterate both operative cavities.

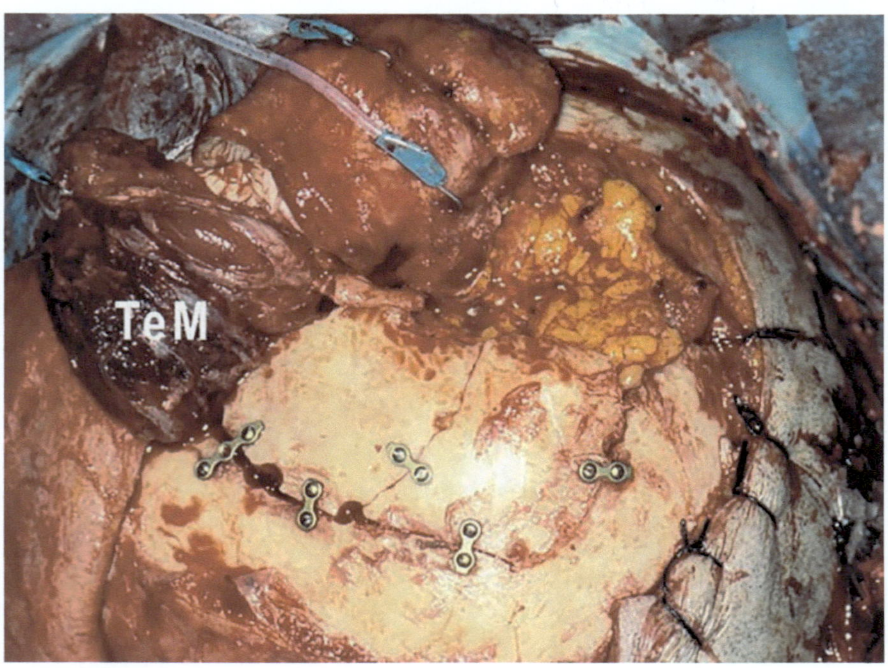

Fig. 15.113 Replacement of the craniotomy. Bone is secured using miniplates.

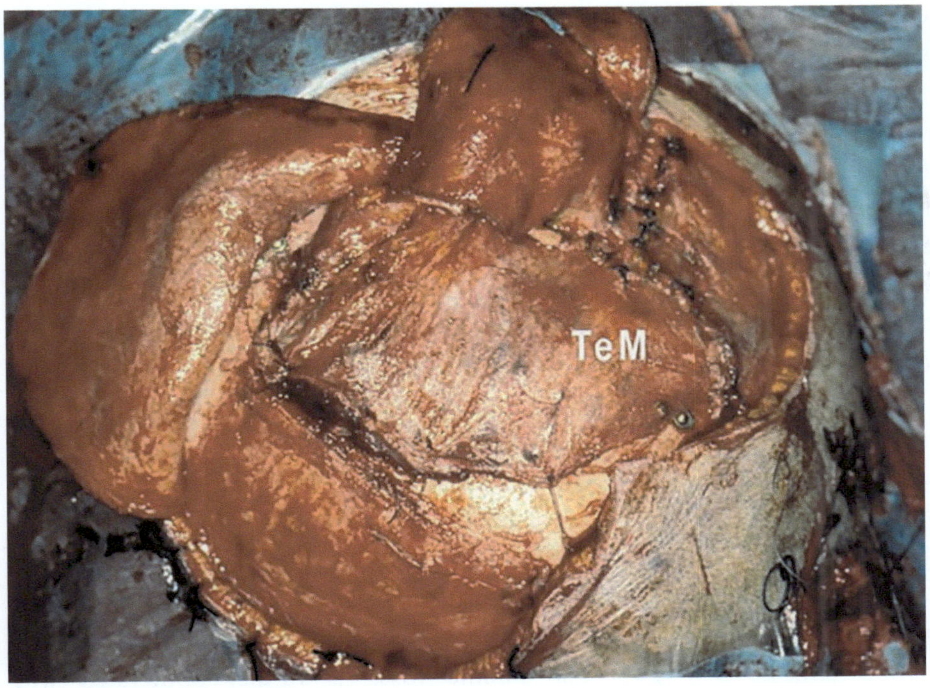

Fig. 15.114 The musculoperiosteal flap and the temporalis muscle (TeM) are repositioned and sutured.

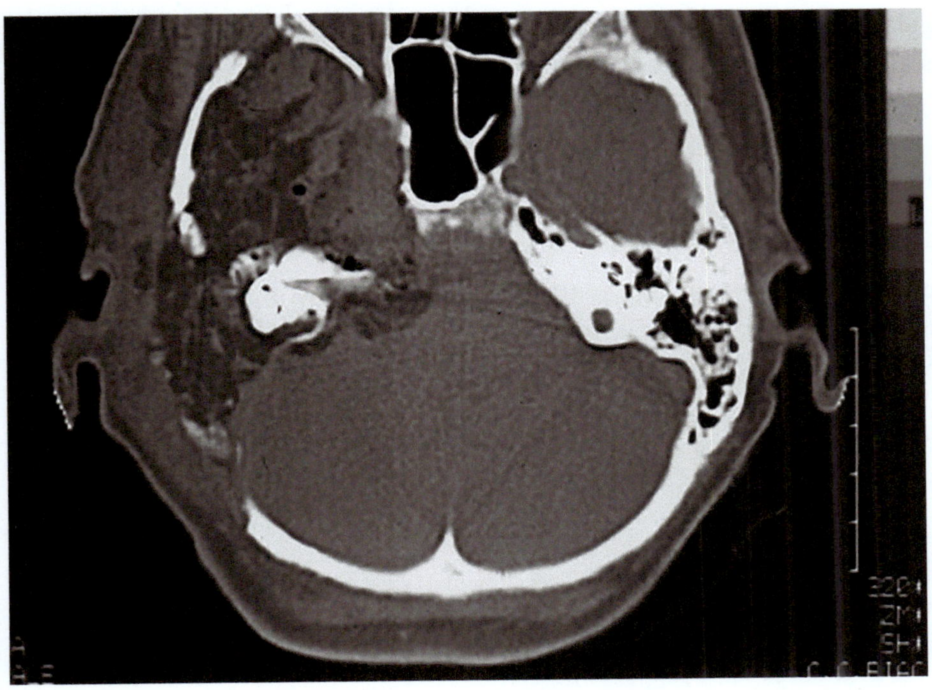

Fig. 15.115 Postoperative CT scan.

Combined Translabyrinthine Transapical Orbitozygomatic Approach

CASE 4: RIGHT POSTERIOR FOSSA DERMOID WITH EXTENSION INTO THE MIDDLE FOSSA AT THE AREA OF MECKEL'S CAVE (Figs. 15.116–15.133)

1. Preoperative MRI of the case. The tumor has a posterior fossa component and extends through the petrous apex and Meckel's cave to involve the middle fossa abutting the lateral surface of the cavernous sinus. In a three-dimensional reconstruction of the CT scan of the patient, the exact topographic tumor extension can be appreciated (Figs. 15.116 and 15.117).

2. The approach has been performed. The zygoma and the lateral orbital rim have been removed. A large frontotemporal craniotomy has been done (Fig. 15.118).

3. A retrolabyrinthine craniotomy has been performed as previously described. In addition, the posterior

and superior walls of the external auditory canal have already been removed. The external auditory canal is closed as a cul-de-sac. Labyrinthectomy then proceeds (Fig. 15.119).

4. The internal auditory canal is identified as previously described. Extensive bone removal superior and inferior to the internal auditory canal is performed (Fig. 15.120).

5. With more transapical bone removal, the tumor is seen invading the petrous apex anterior to the internal auditory canal. A dermoid cyst is seen superior to the canal, whereas inferiorly, an epidermoid can be appreciated (Fig. 15.121).

6. The dura is opened and the tumor is seen occupying the cerebellopontine angle (Fig. 15.122).

7. The tumor can be seen attached to and lying anterior to the acousticofacial bundle (Fig. 15.123). The acousticofacial bundle is dissected free from the tumor.

8. From the subtemporal aspect, the dura is retracted and dissected from the superior surface of the temporal bone. The middle meningeal artery is coagulated and cut. The mandibular branch of the trigeminal is also sacrificed after bipolar coagulation (Fig. 15.124).

9. The remnants of the petrous apex are drilled, and the tumor invading it is removed. Tumor removal continues. The two components of the tumor (epidermoid and dermoid) can be appreciated. The tumor is dissected free from the acousticofacial bundle (Fig. 15.125). After removal of the petrous apex bone and opening of the posterior fossa dura from the subtemporal aspect, the remaining superior pole of the tumor can be better controlled. The internal carotid artery is identified as its precavernous segment. The tumor extends up to the level of Meckel's cave (Fig. 15.126).

10. The remaining middle fossa component is tackled next. Tumor is freed from the brainstem. The last piece of tumor attached to the trigeminal nerve is finally removed (Figs. 15.127 and 15.128).

11. General view at the end of the procedure; total tumor removal has been achieved (Fig. 15.129).

12. The posterior fossa is obliterated with abdominal fat. The eustachian tube is closed with a piece of muscle (Fig. 15.130).

13. The middle fossa is further obliterated with another piece of abdominal fat (Fig. 15.131).

14. The craniotomy is replaced and fixed with miniplates. The incision is sutured without insertion of a drain (Fig. 15.132).

15. Postoperative CT scan showing the approach. Total removal can be seen (Fig. 15.133).

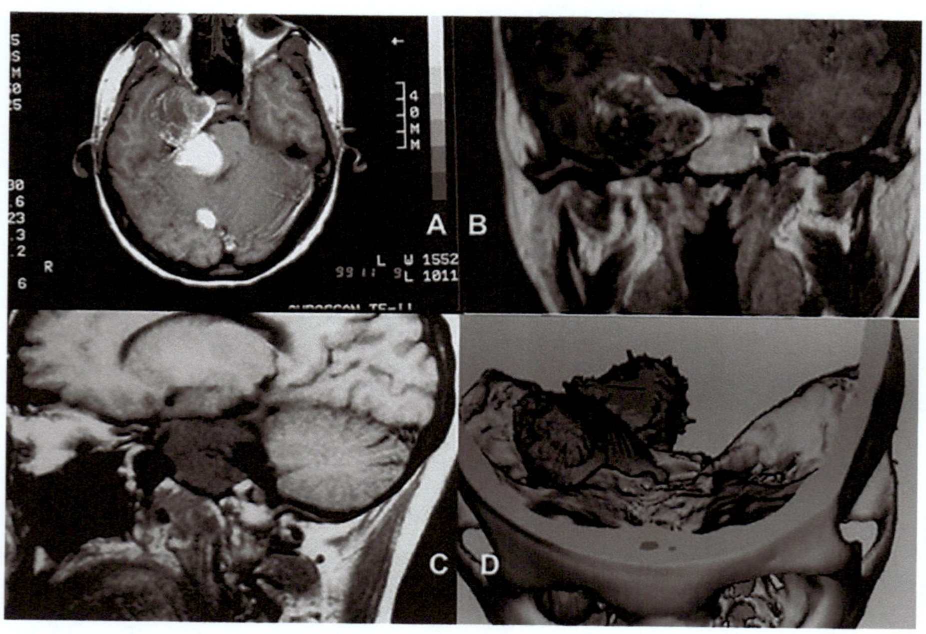

Fig. 15.116 (A–D) Preoperative Imaging of the case. (A) Preoperative MRI with gadolinium enhancement, axial view showing a homogeneously enhancing posterior fossa lesion displacing the brainstem and a heterogeneous middle fossa component at the area of Meckel's cave. (B) T1-weighted image with gadolinium enhancement, coronal view. The lesion displacing the temporal lobe at the level of Meckel's cave is appreciated. (C) T1-weighted image, sagittal view. The hypointense lesion and its anterosuperior extension are seen. (D) CT scan with 3D reconstruction. The exact three-dimensional relation of the lesion to the skull base is well demonstrated.

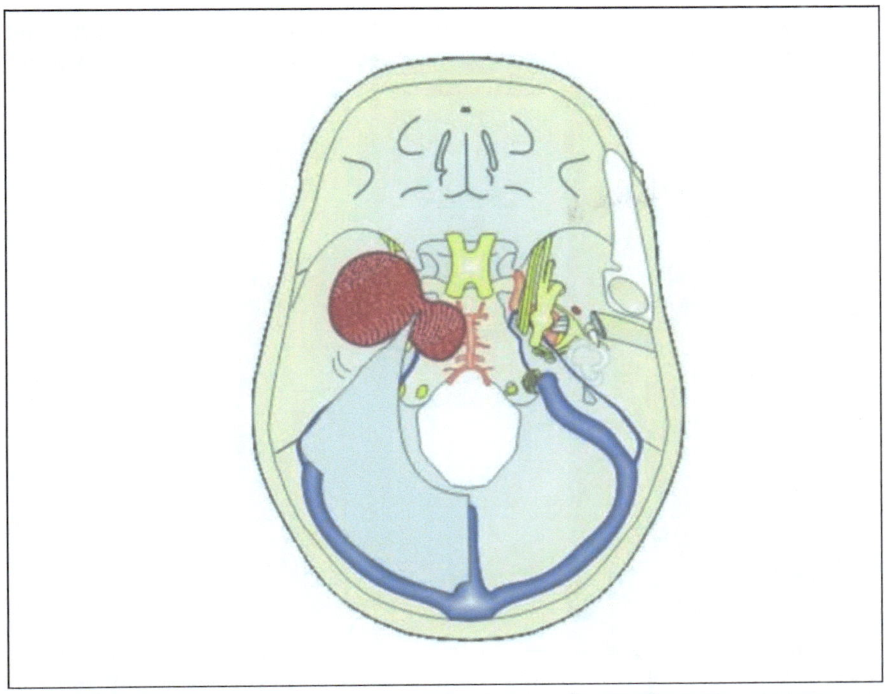

Fig. 15.117 An illustration showing the shape of the tumor.

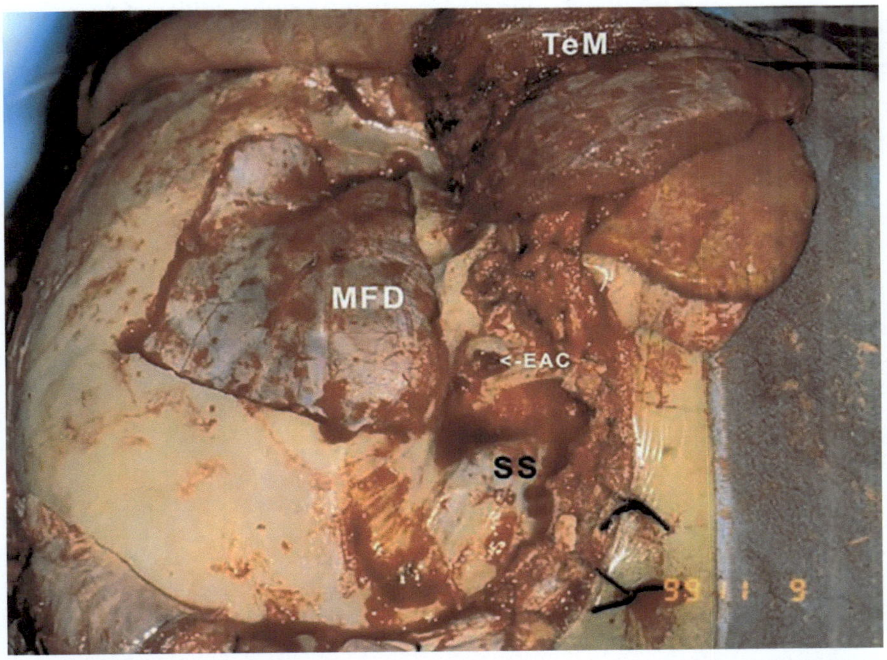

Fig. 15.118 A frontotemporal orbitozygomatic osteotomy has been performed.

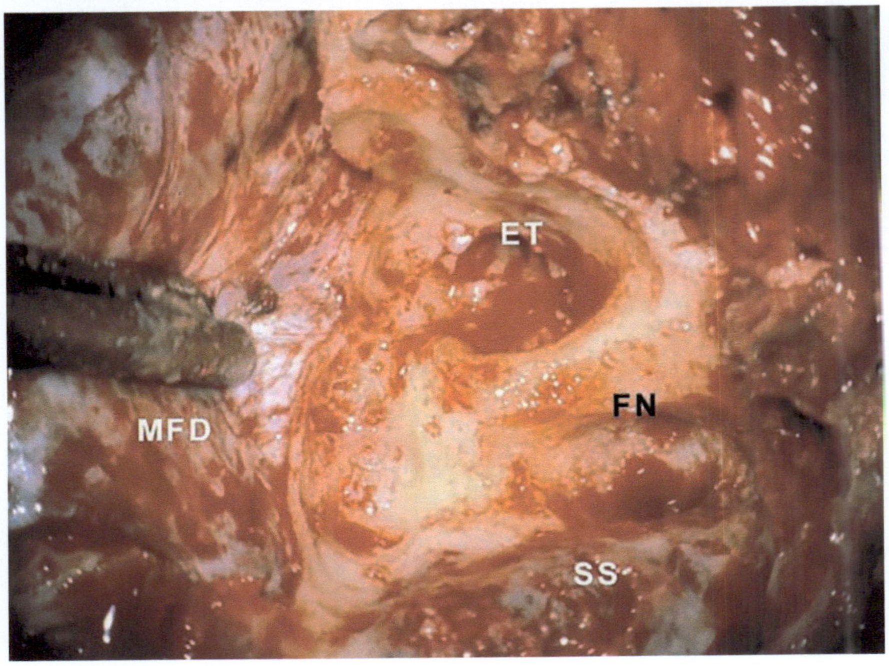

Fig. 15.119 Labyrinthectomy has been performed. The superior and posterior walls of the external auditory canal are also drilled.

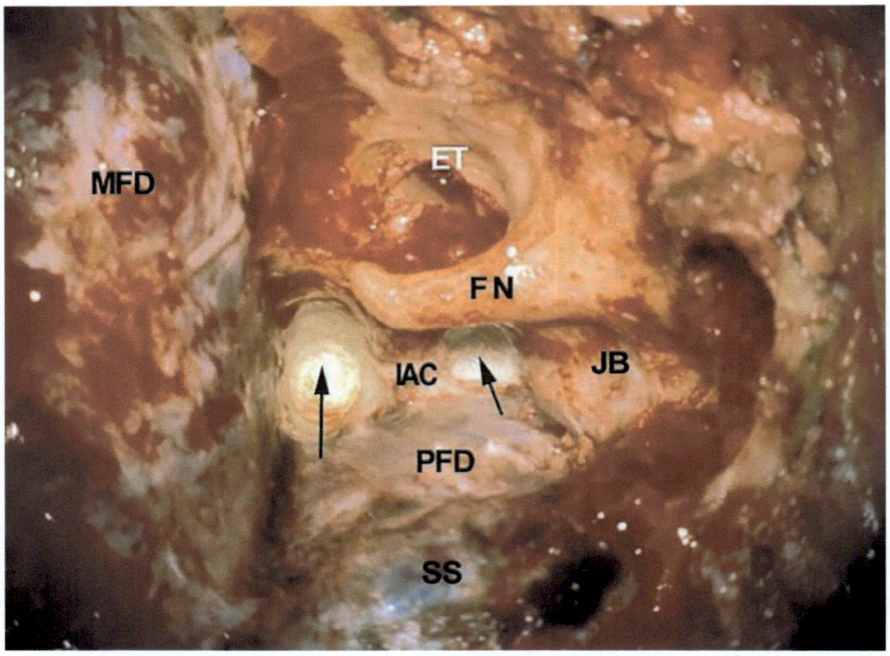

Fig. 15.120 Two deep troughs *(arrows)* superior and inferior to the internal auditory canal (IAC) have been created.

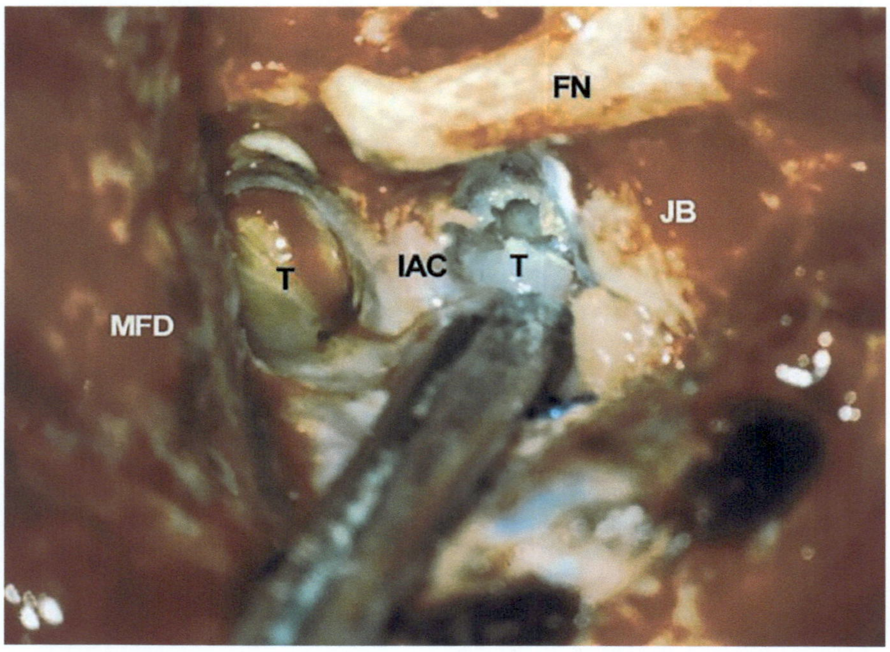

Fig. 15.121 With extensive drilling of the petrous apex, the two components (T) of the tumor (epidermoid inferiorly and dermoid superiorly) are identified.

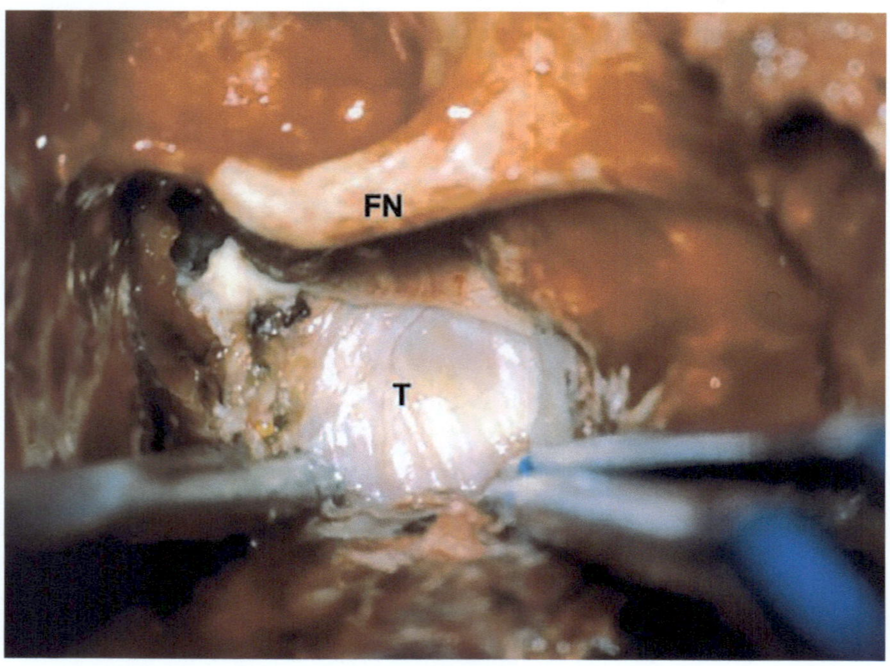

Fig. 15.122 The lesion in the cerebellopontine angle.

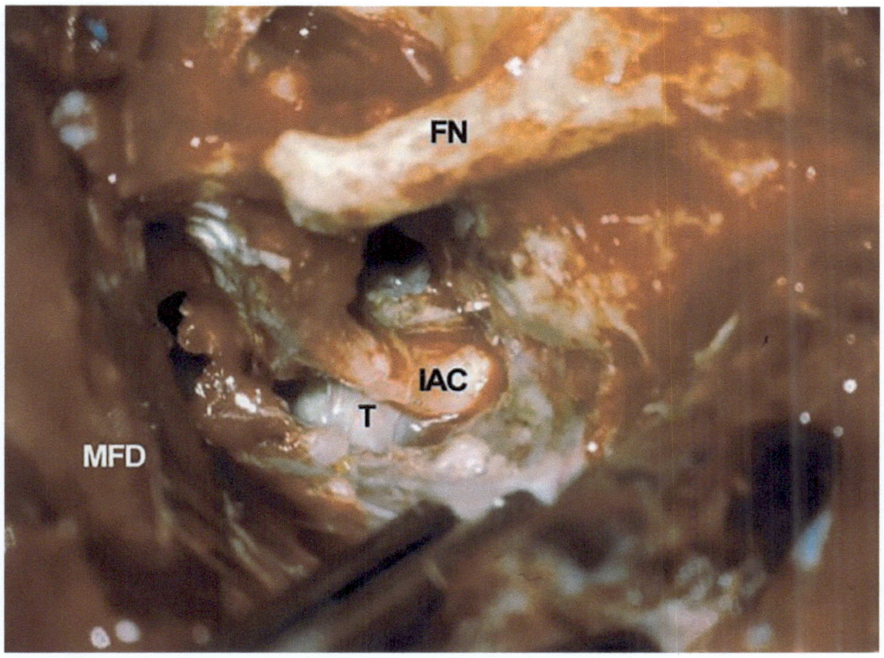

Fig. 15.123 Lesion lying anterior to the acousticofacial bundle.

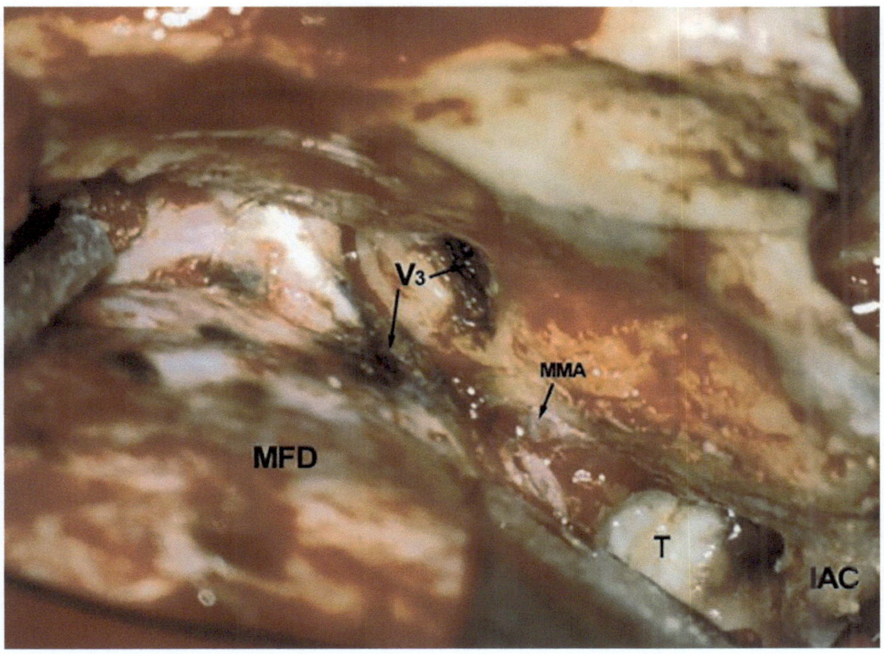

Fig. 15.124 View from the subtemporal access. The middle meningeal artery (MMA) and the mandibular branch of the trigeminal nerve (V3) have been cut.

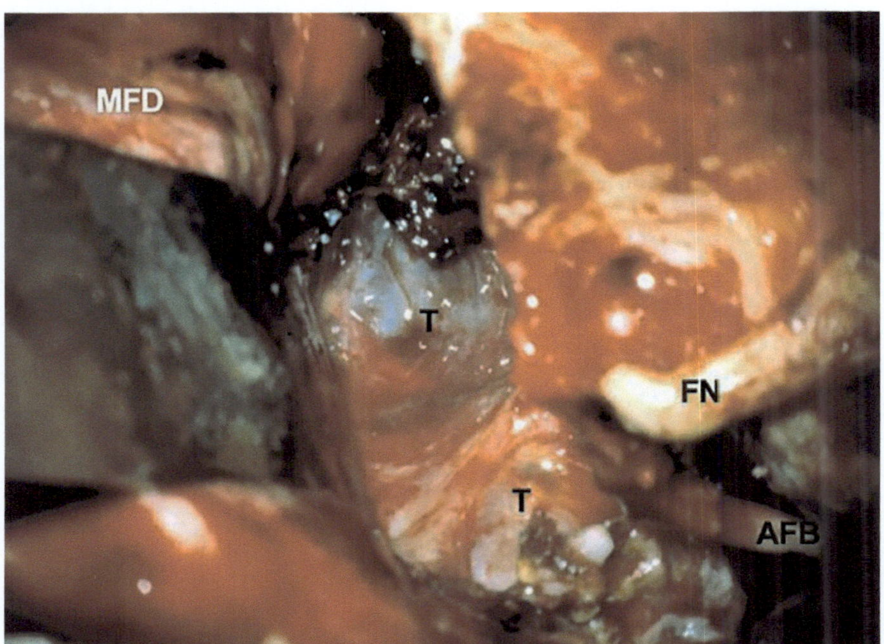

Fig. 15.125 Close view from the subtemporal access showing the dermoid component in the petrous apex. The epidermoid component is seen attached to the acousticofacial bundle (AFB).

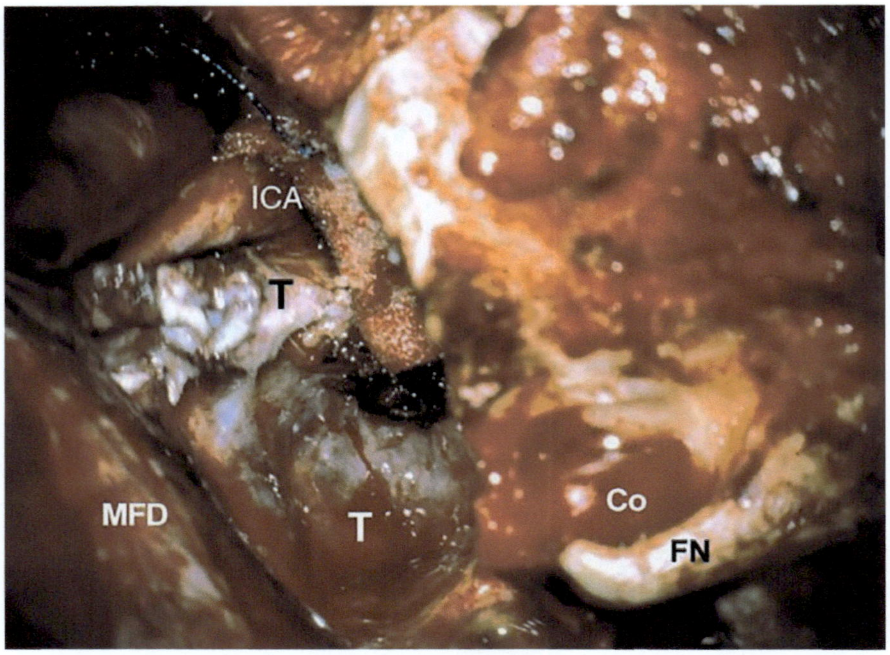

Fig. 15.126 The tumor occupying the whole petrous apex is seen. The precavernous carotid artery is identified.

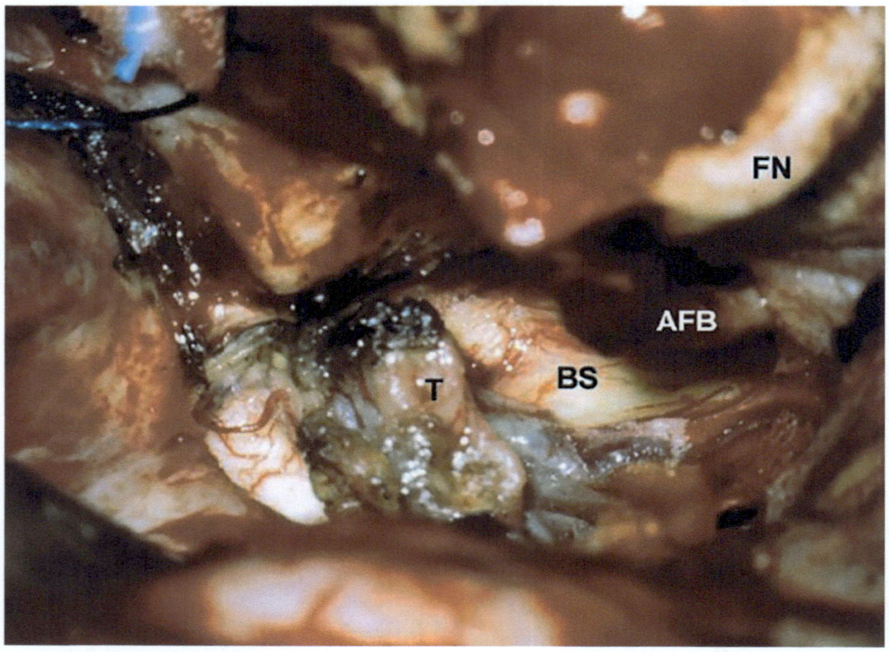

Fig. 15.127 Tumor is dissected from the brainstem.

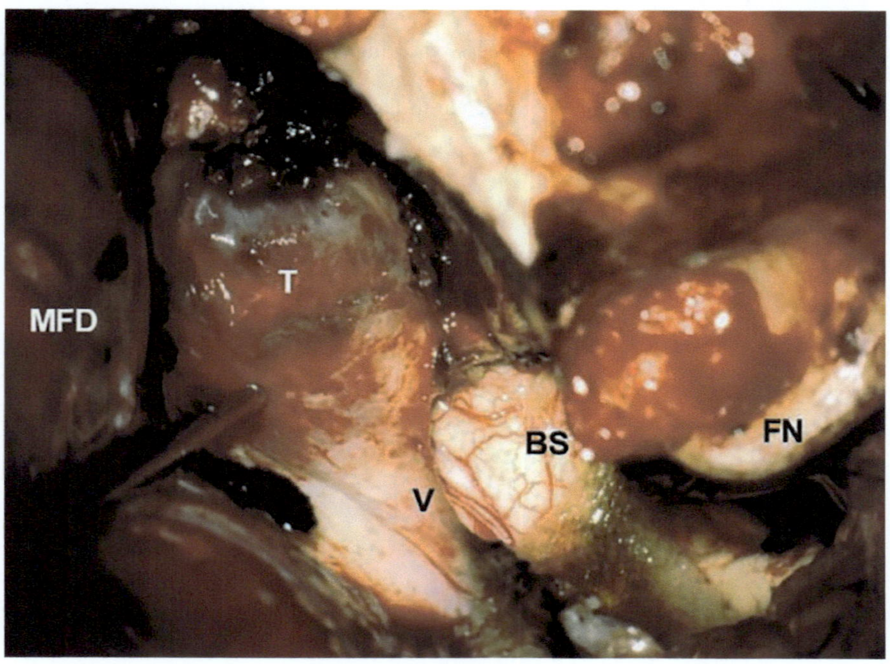

Fig. 15.128 The last piece of tumor is dissected from the trigeminal nerve (V).

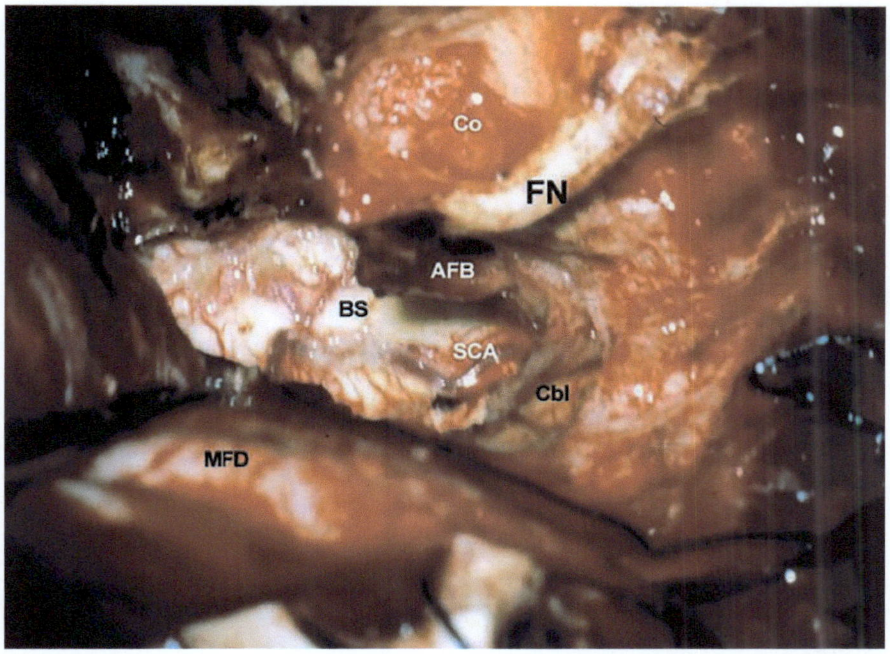

Fig. 15.129 General view after total removal.

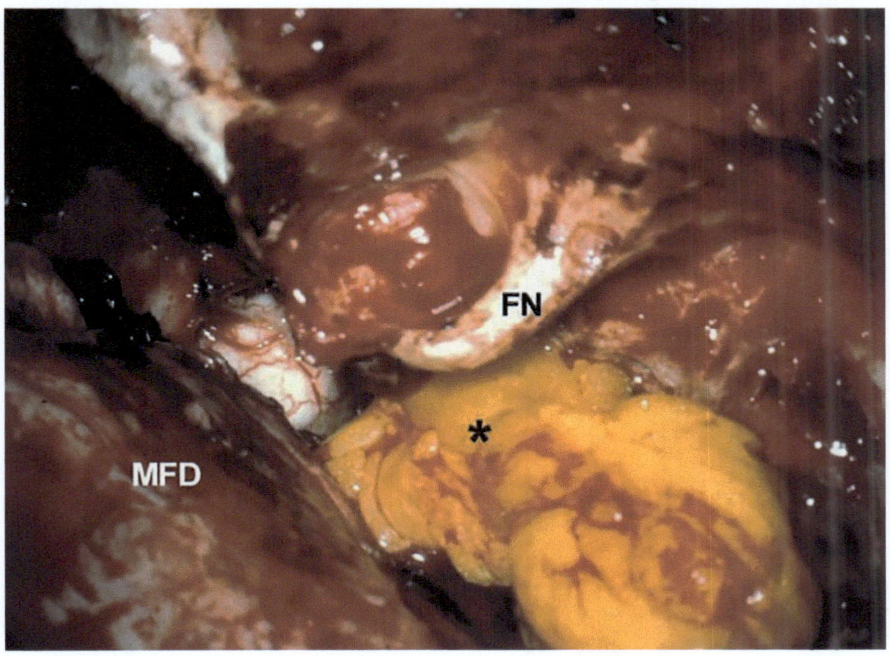

Fig. 15.130 The posterior fossa operative cavity is obliterated with abdominal fat (*).

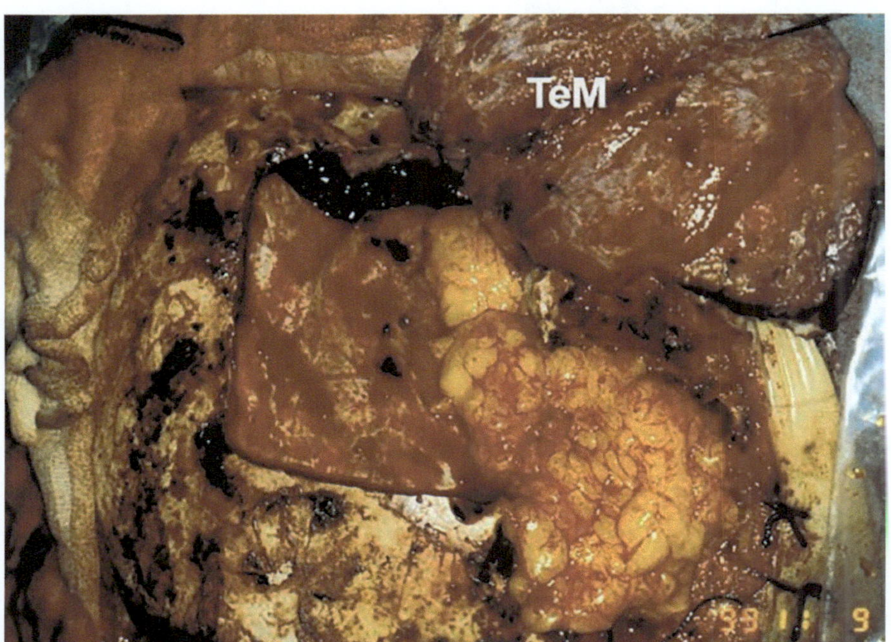

Fig. 15.131 The middle fossa cavity is further obliterated with abdominal fat.

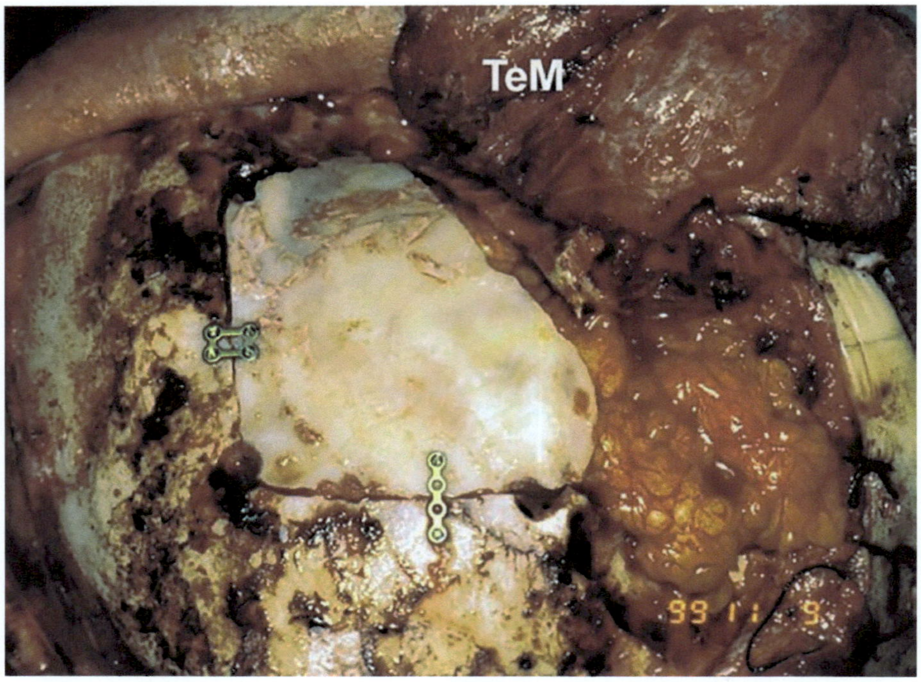

Fig. 15.132 The craniotomy is replaced and fixed with miniplates.

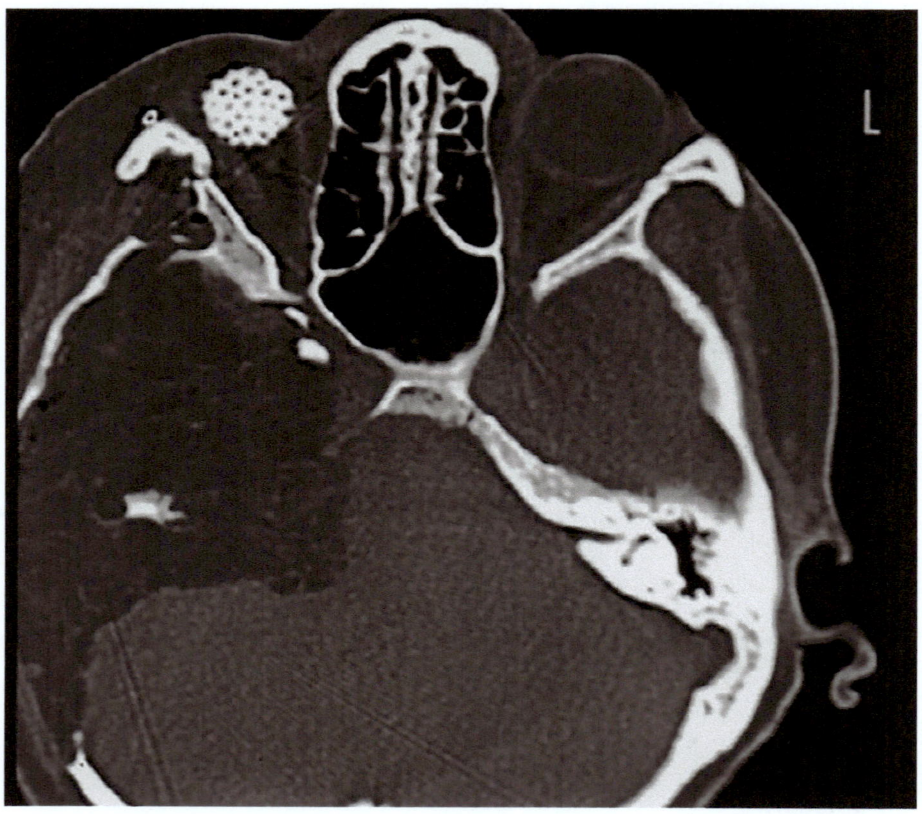

Fig. 15.133 Postoperative CT scan.

Combined Infratemporal Transzygomatic Approach

CASE 1: RIGHT INFRATEMPORAL SARCOMA REMOVED THROUGH A COMBINED INFRATEMPORAL B TRANSZYGOMATIC APPROACH (Figs. 15.134–15.155)

1. Preoperative imaging of the case showing the tumor involving the right infratemporal fossa and eroding its roof, lying in contact with the temporal lobe with no evidence of dural invasion. The tumor also erodes the tegmen tympani and involves the middle ear (Figs. 15.134–15.139).

2. The steps of surgery are shown in Figs. 15.140–15.155.

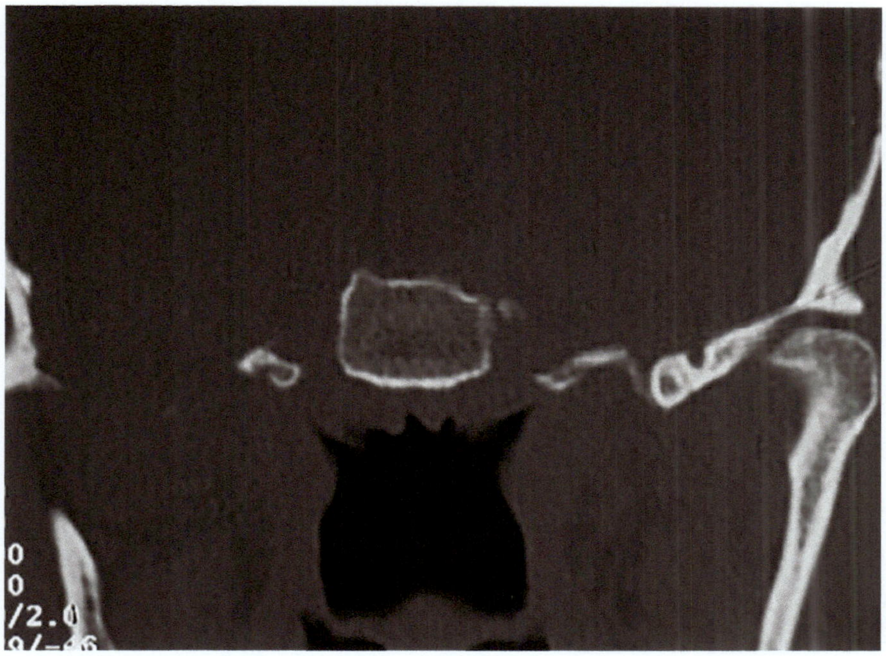

Fig. 15.134 Coronal CT scan showing erosion of the roof of the right infratemporal fossa.

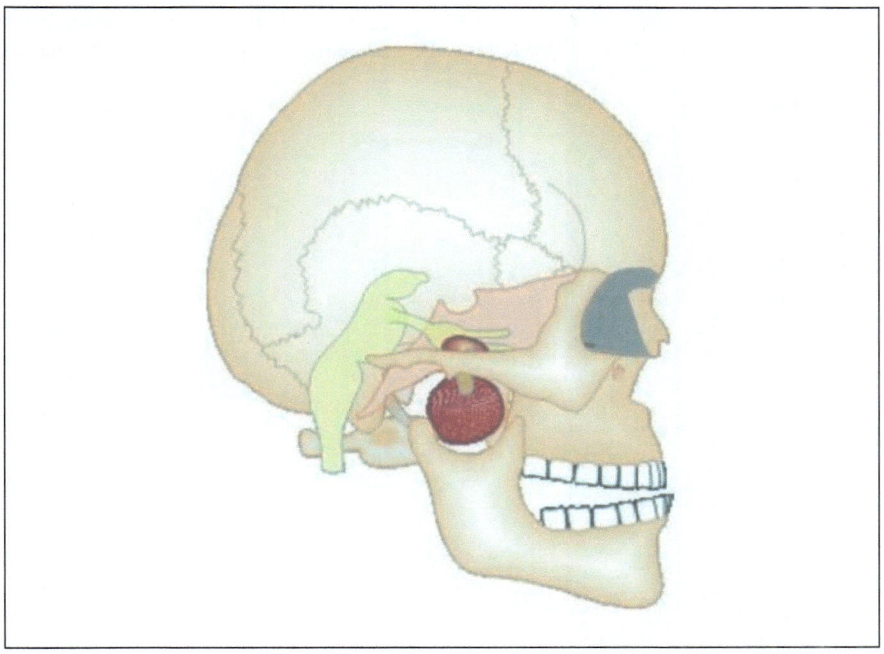

Fig. 15.135 Drawing showing the shape of the tumor.

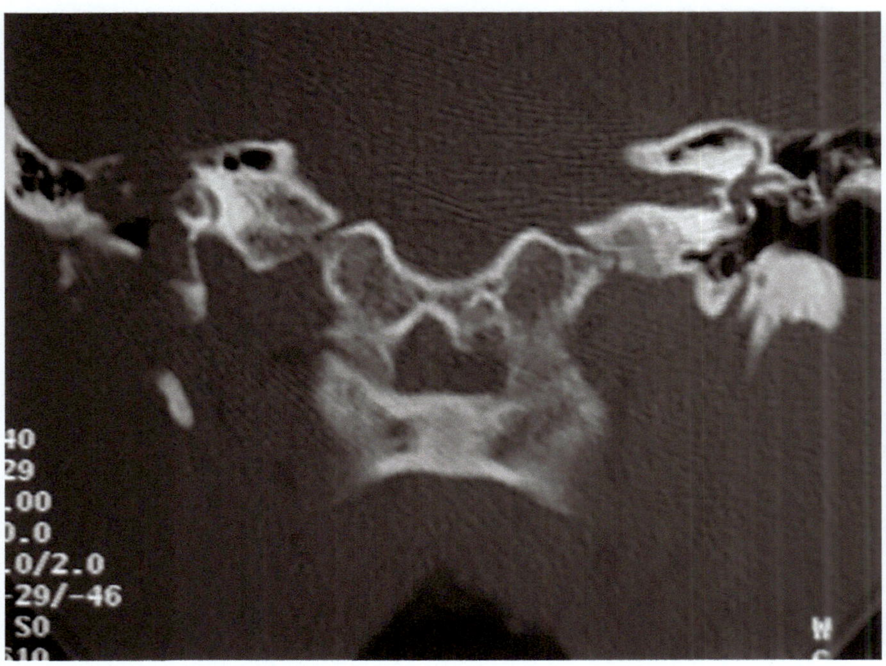

Fig. 15.136 A more posterior coronal view showing tumor extension into the middle ear through a tegmen erosion.

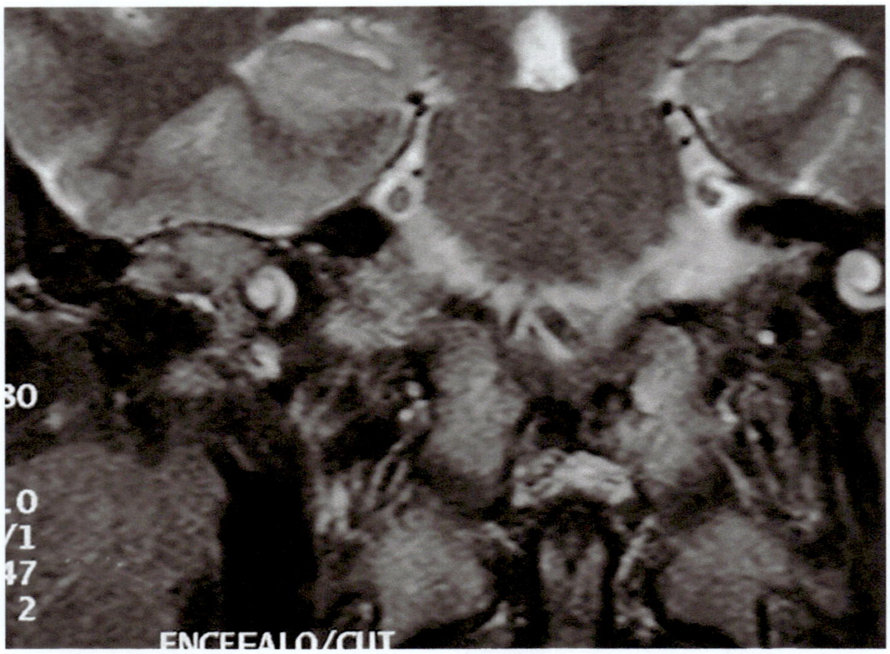

Fig. 15.137 Coronal MRI at the same level as Fig. 15.136. The mass involves the middle ear and erodes the tegmen tympani with no evidence of dural invasion.

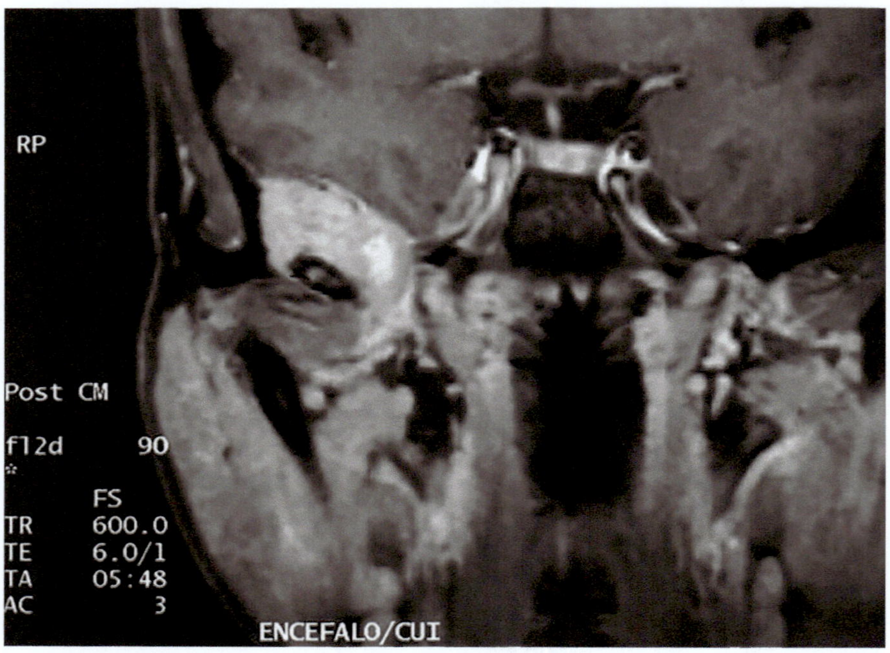

Fig. 15.138 Coronal MRI at a more anterior level showing the infratemporal component of the mass.

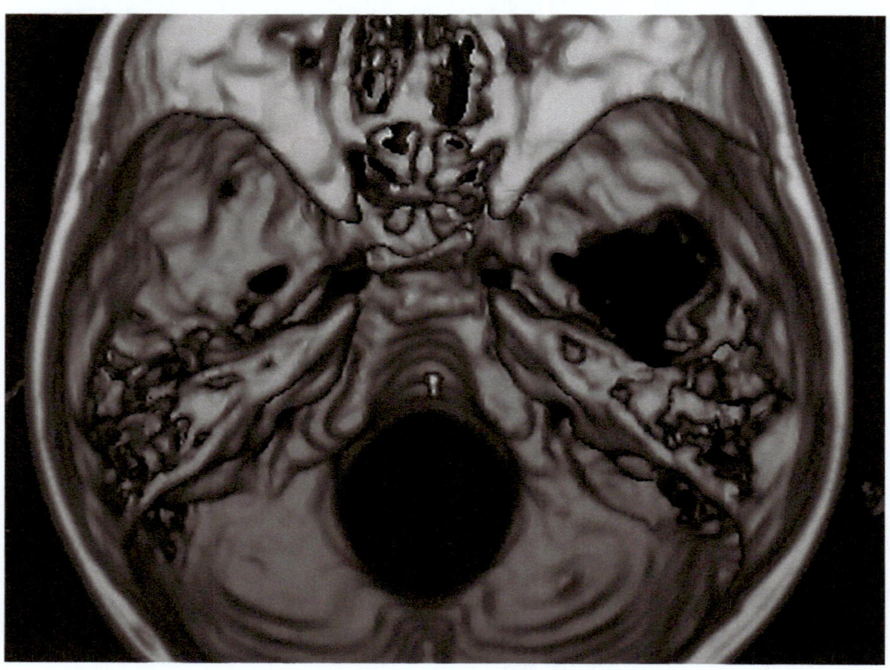

Fig. 15.139 A three-dimensional reconstructed CT image of the skull base showing the extent of erosion of the infratemporal fossa roof by the tumor. Note that the foramen ovale is partially eroded.

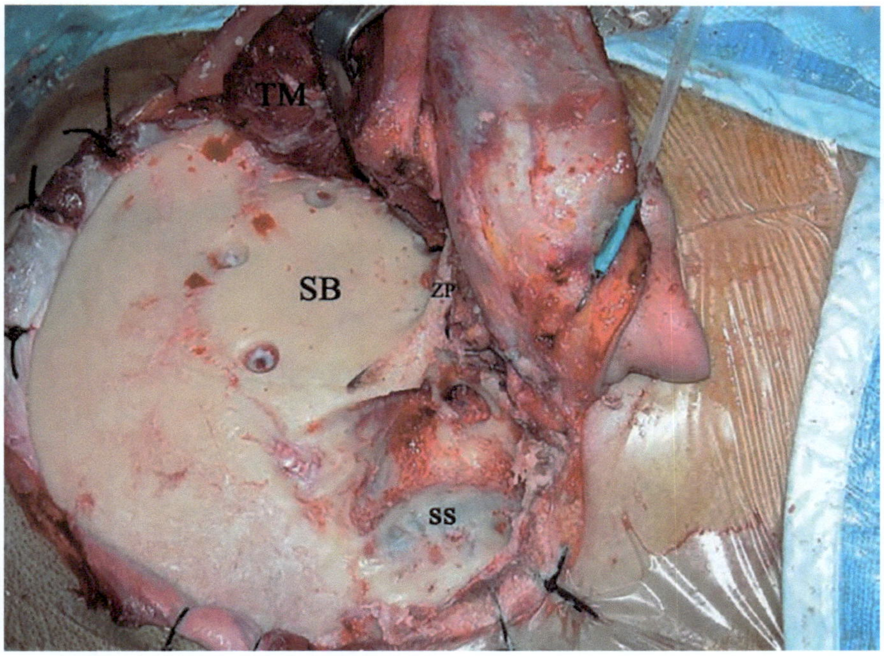

Fig. 15.140 A subtotal petrosectomy has been performed; the periosteum overlying the zygomatic process (ZP) has been dissected away intact to preserve the facial nerve. The temporalis muscle (TM) has been dissected from the squamous bone (SB), and several holes have been drilled to identify the margins of the craniotomy.

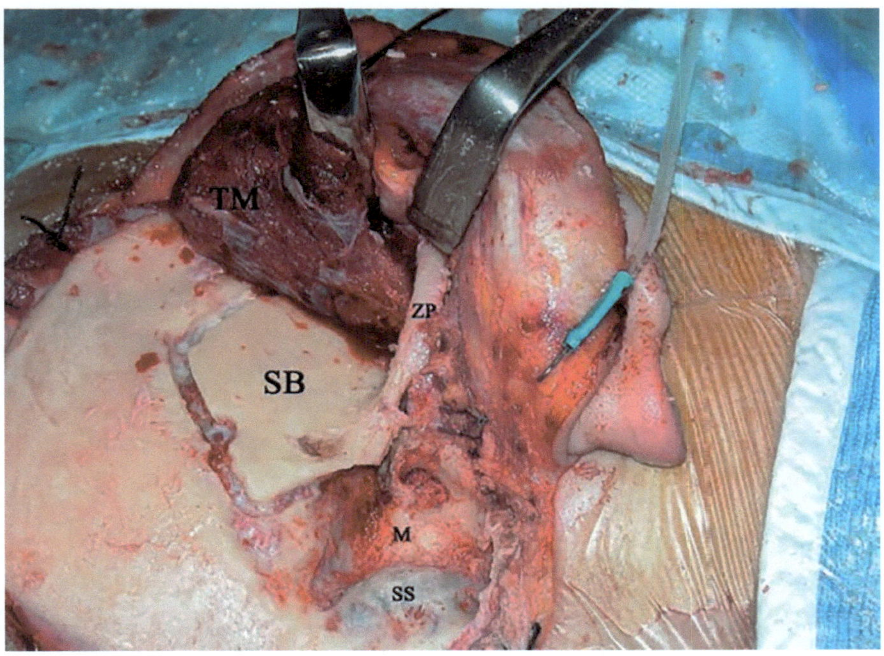

Fig. 15.141 The craniotomy has been performed.

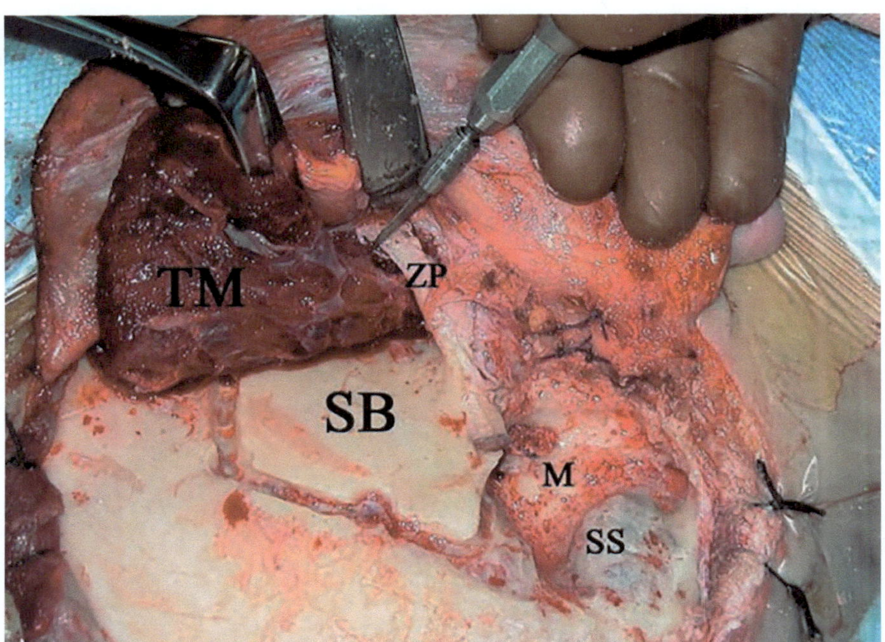

Fig. 15.142 A saw is being used to cut the zygomatic process (ZP) distally near its orbital component.

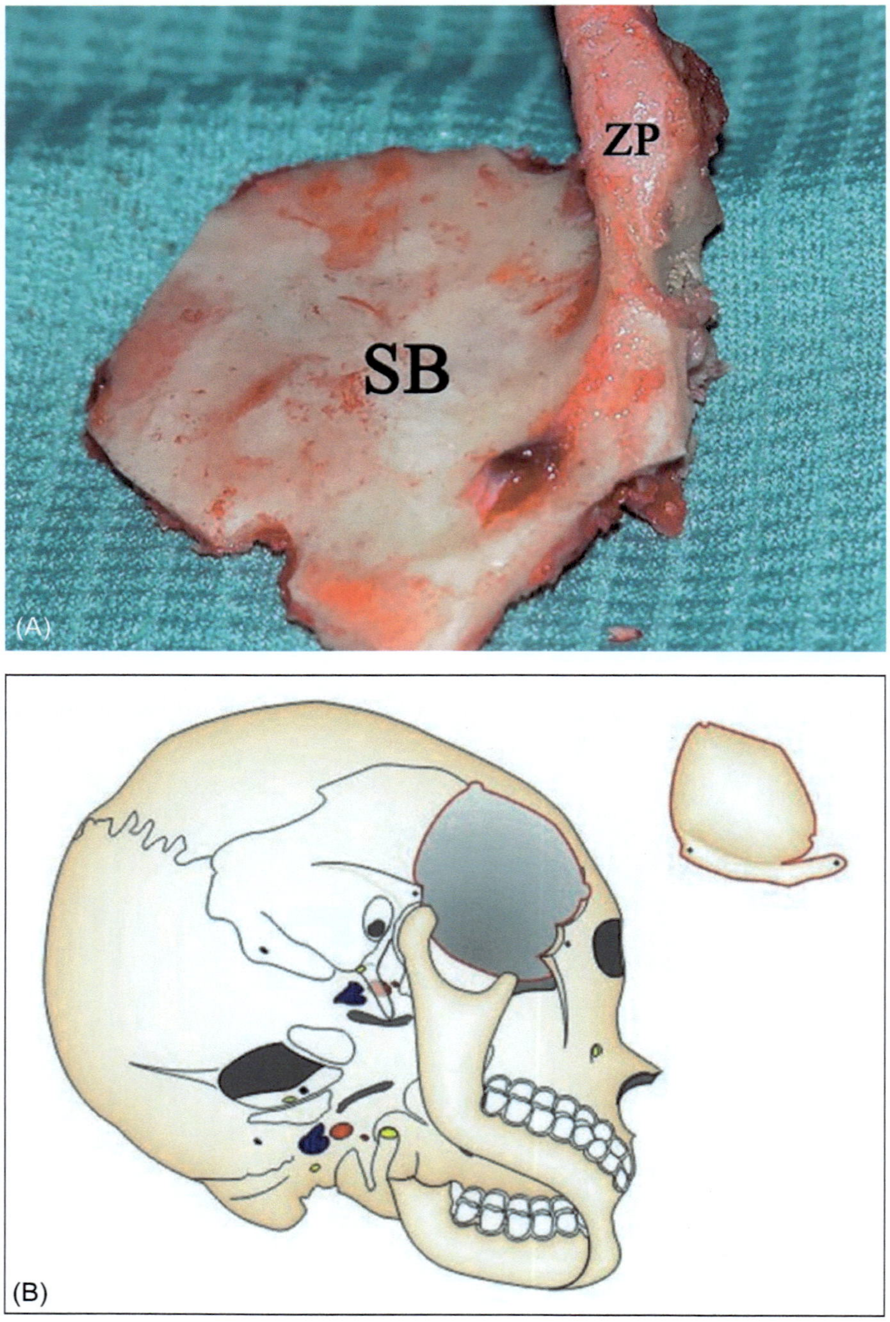

Fig. 15.143 (A) The transzygomatic craniotomy has been achieved. Notice that in this case the zygomatic process has been kept in contact with the vertical component of the temporal bone to achieve better cosmetic results on reconstruction. (B) Drawing representing the performed craniotomy.

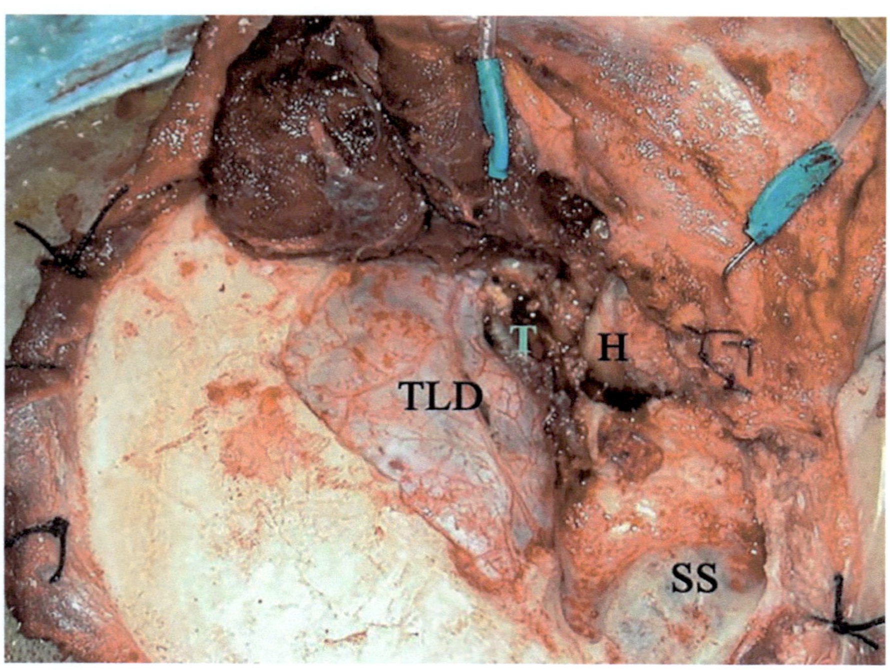

Fig. 15.144 The tumor (T) can be seen lying in the infratemporal fossa medial to the condyle of the mandible (H). TLD, temporal lobe dura.

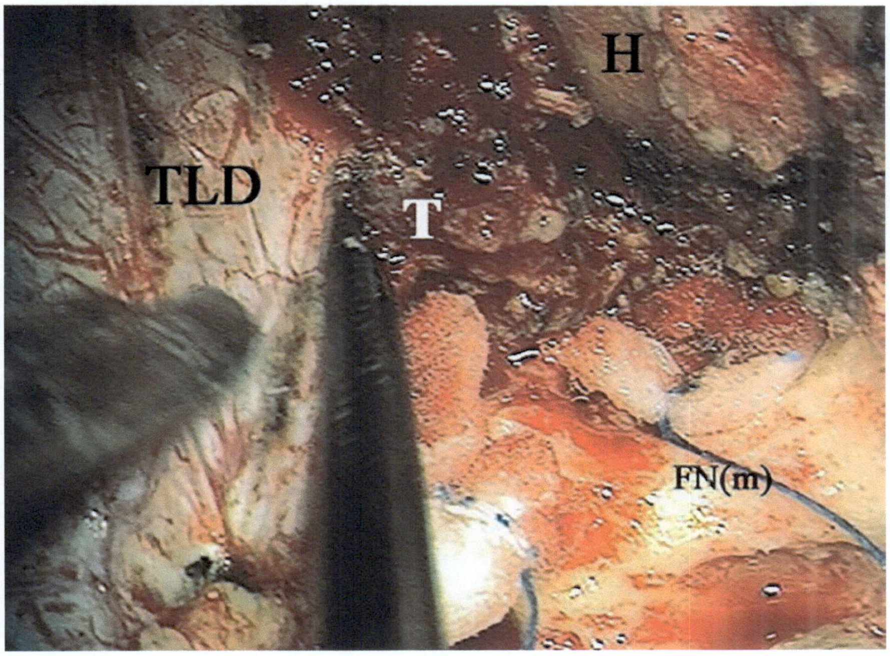

Fig. 15.145 Bipolar coagulation is used to achieve bloodless tumor removal.

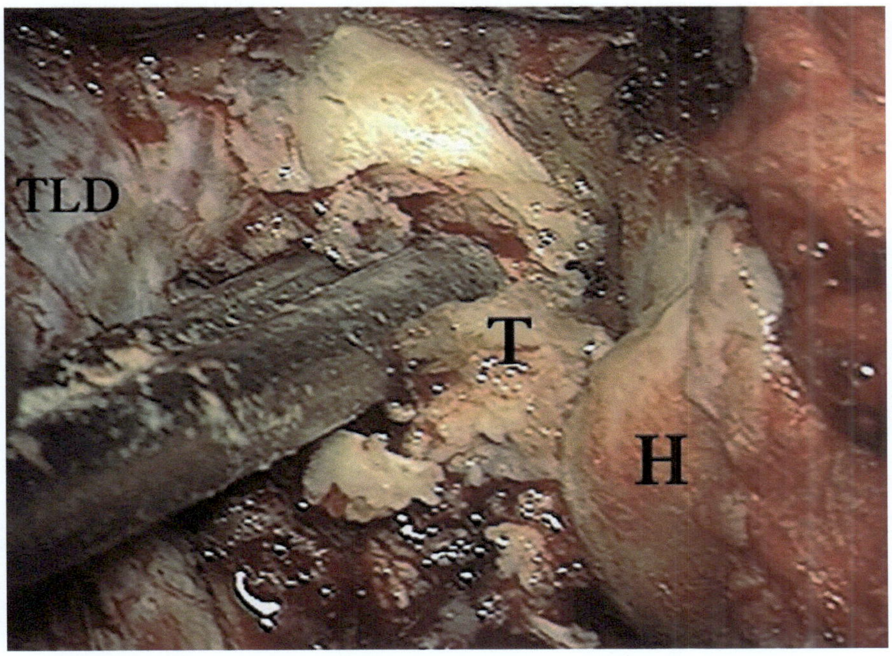

Fig. 15.146 The tumor (T) can be seen extending medial to the head of the mandible (H).

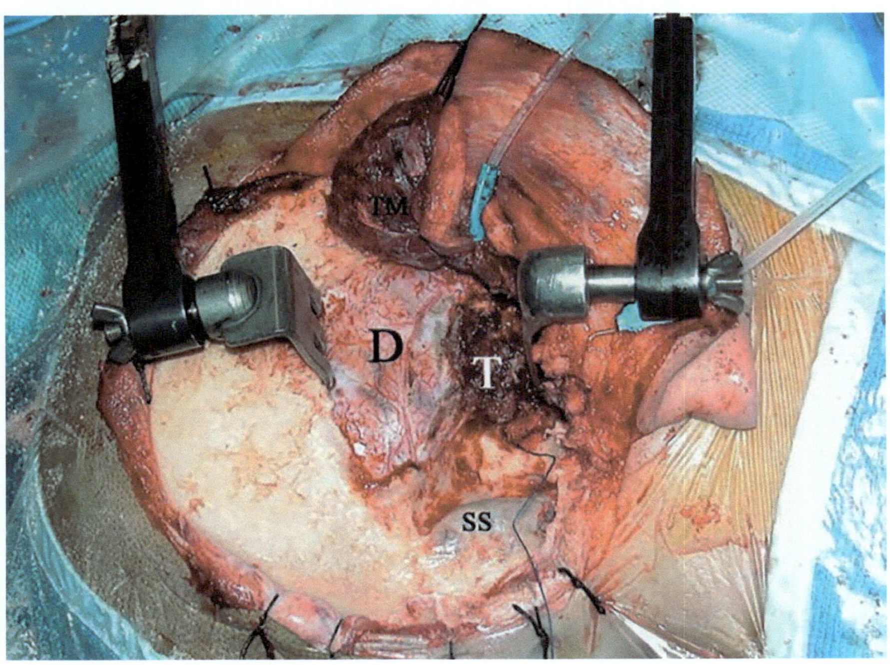

Fig. 15.147 The infratemporal fossa retractor is applied to achieve downward retraction of the head of the mandible in order to allow for tumor (T) removal. *D*, middle fossa dura.

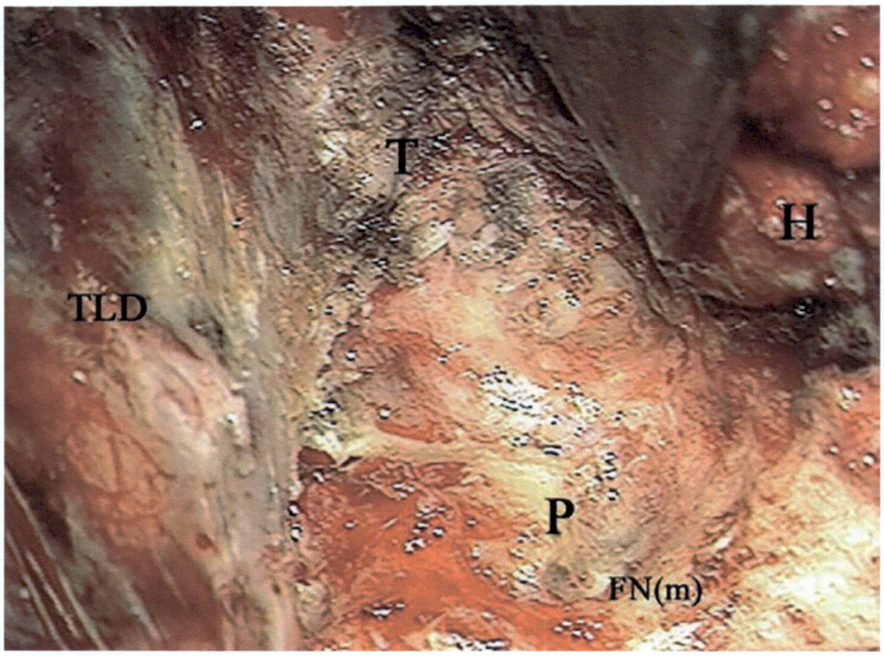

Fig. 15.148 Tumor remnants (T) can be seen extending anteriorly and have to be removed.

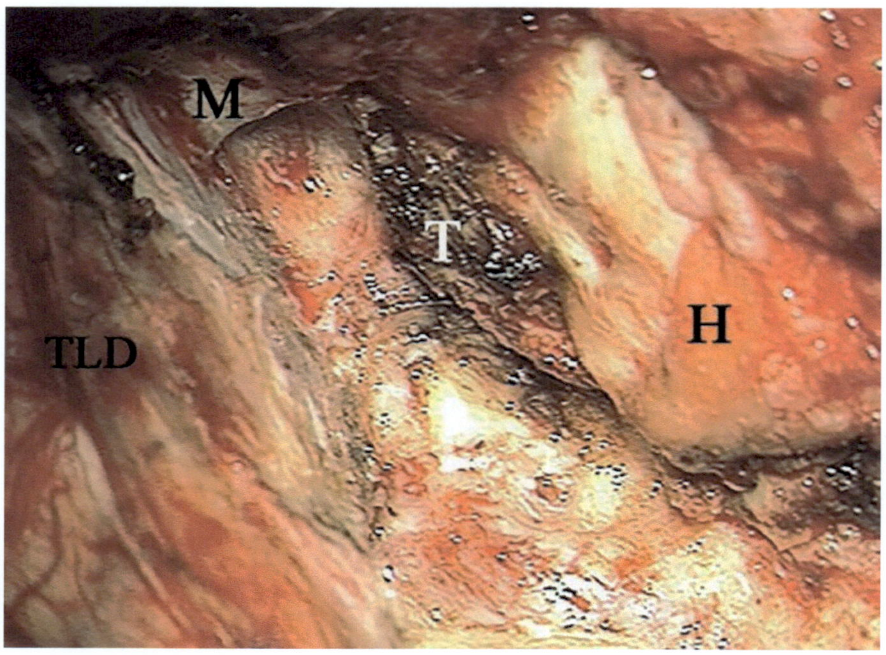

Fig. 15.149 The anterior component of the tumor has been completely removed, and the mandibular division of the trigeminal nerve has been kept intact. Attention is now directed to removing the part of the tumor (T) extending inferiorly within the infratemporal fossa medial to the head of the mandible (H).

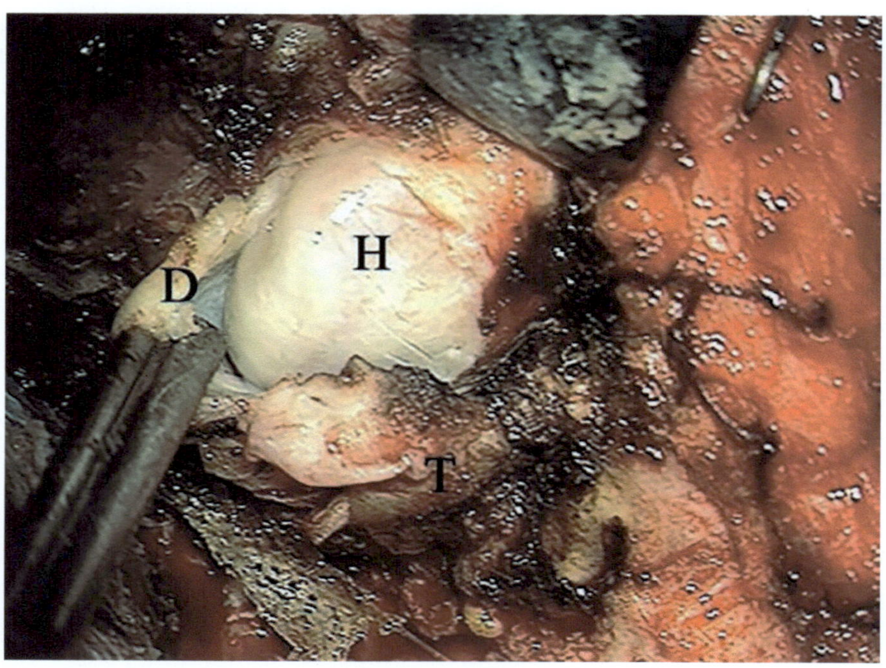

Fig. 15.150 The tumor was found to be infiltrating the articular disk (D) and the remnants of the capsule of the temporomandibular joint, and both are to be removed.

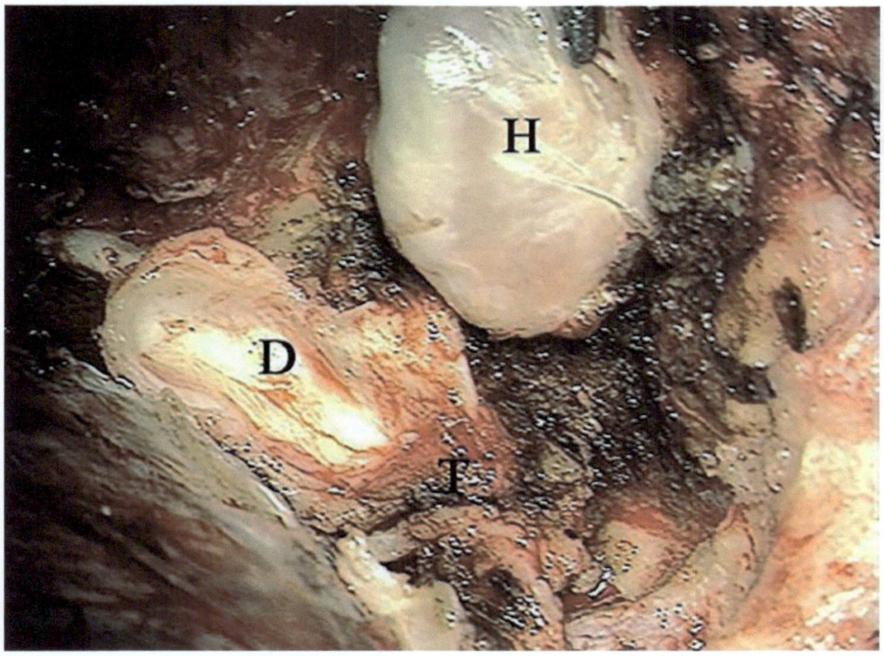

Fig. 15.151 Removal of the infiltrated disk (D) is being performed in order to achieve gross total tumor removal.

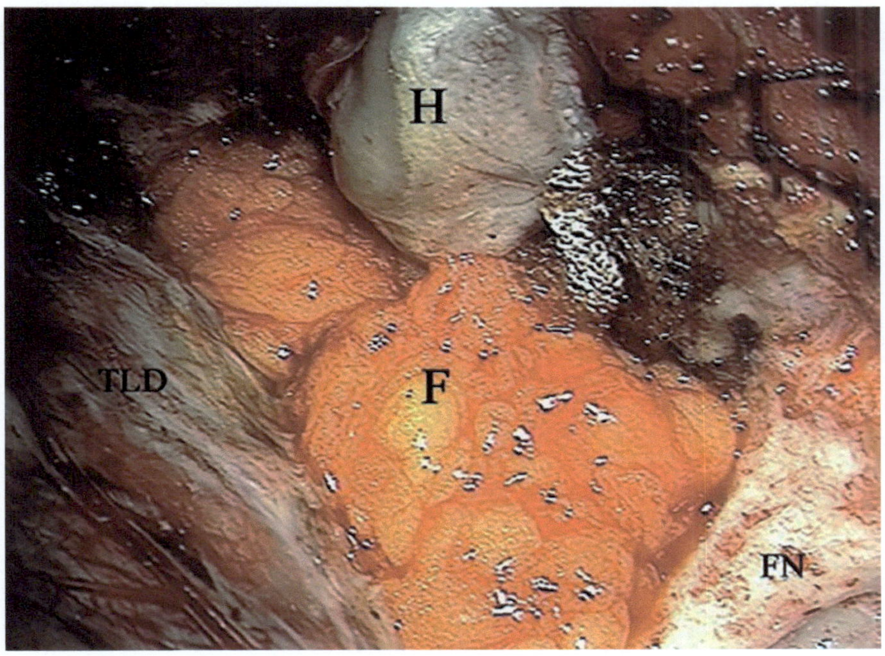

Fig. 15.152 Abdominal fat (F) is used to obliterate the surgical defect after blocking the eustachian tube with periosteal tissues.

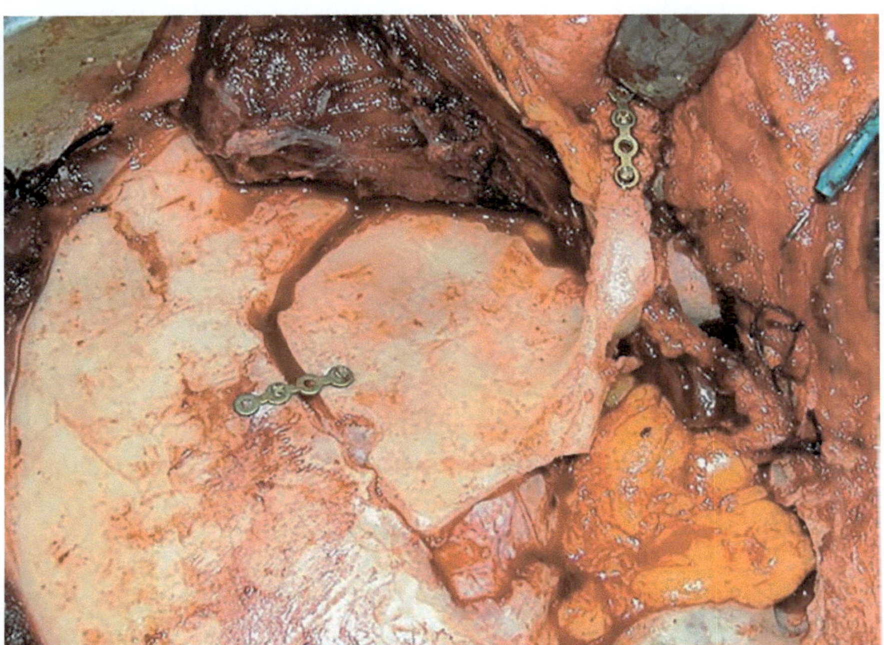

Fig. 15.153 The craniotomy has been fixed in its original place using surgical plates.

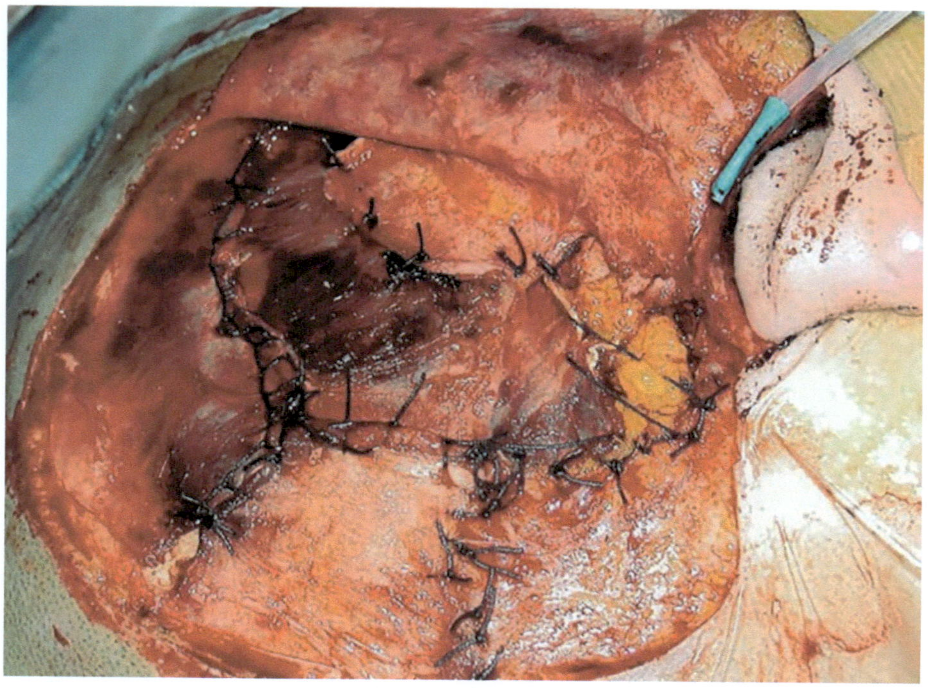

Fig. 15.154 The temporalis muscle has been repositioned and sutured together with the periosteal flaps.

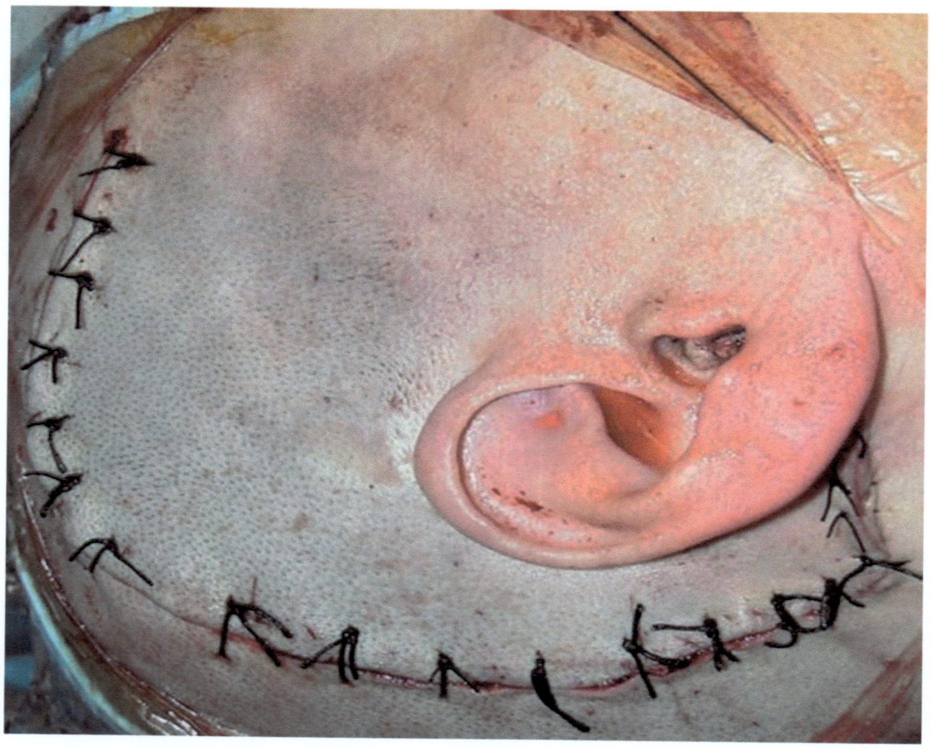

Fig. 15.155 Suturing of the skin flap has been achieved.

PEARLS AND PITFALLS FOR THE COMBINED TRANSPETROUS ORBITOZYGOMATIC APPROACHES

1. The utmost care should be taken to preserve the superficial temporal artery as it is the main blood supply for the skin flap.
2. The frontotemporal craniotomy can be made as two craniotomies rather than one large piece of bone. It is technically easier to separate two smaller bone flaps from the middle fossa dura than to separate a larger single flap.
3. The osteotomies are better performed with the Midas Rex saw attachment than with a small conical otological burr. This avoids bone loss and facilitates the replacement and coaptation of the bone during closure. It also gives better cosmetic results.
4. As in all the transpetrous approaches to the posterior fossa, the cavity is obliterated with abdominal fat.

Obliteration of the operative defect from the middle fossa aspect with a separate piece of abdominal fat is important to avoid any CSF leakage. Meticulous suturing of the temporalis muscle with the postauricular musculo-periosteal flaps helps to hold the fat in place and reduces any potential risk for postoperative leaks.

Advantages of the Combined Transpetrous Orbitozygomatic Approaches

1. Complete and single-stage removal of large tumors extending to both the posterior and middle fossae. The subtemporal orbitozygomatic approach offers excellent control of the middle and infratemporal fossa components but has poor control of the posterior fossa. On the other hand, the transpetrous approaches afford excellent exposure of the posterior fossa with minimal middle fossa control. The combination of both approaches provides excellent visibility of both fossae, particularly in the area of Meckel's cave where the tumor components usually interconnect.
2. The extensive bone removal allows maximal exposure while minimizing the amount of brain retraction needed.

Disadvantages

1. The main disadvantage is the long time required for the approach. However, the excellent control afforded shortens the time of tumor removal, and hence, the overall operative time is no longer in practice.
2. The extensive bone removal carries the theoretical risk of injury to the various neurovascular structures.
3. The orbitozygomatic osteotomy can produce cosmetic defects if not properly replaced.

CHAPTER 16

Decision-Making in Skull Base Surgery

Introduction

Decision-making for a given skull base lesion is not an easy task. The diversity and rarity of these lesions, the complexity of the area, as well as the multitude of approaches available, are all factors that further complicate the process of decision-making.

While the majority of skull base lesions are benign, they nevertheless differ significantly in their pathological characteristics. An understanding of the surgical pathology of the tumors in this area is mandatory before making the choice of management plan. The pathological behavior of the most common skull base lesions is shown in Table 16.1.

In general, management of a given skull base lesion follows three main lines: surgical removal, radiotherapy, or a follow-up policy.

TABLE 16.1 The pathological behavior of the most common skull base lesions.

Tumor	Characteristics
Meningioma	• Slowly growing • Aggressive behavior • Tends to invade the dura and bone • May engulf or encase the related vessels • Cranial nerve infiltration is common in certain locations (jugular foramen, cavernous sinus, etc.) • Some tumors are vascular • The plane of cleavage between the tumor and the brain is lost in some cases • High incidence of recurrence
Neurinoma	• Well circumscribed • Rarely erodes the bone • Relatively avascular • A plane of cleavage is found in most cases • Total removal is almost always feasible with no recurrence • NF2 cases are more aggressive
Paraganglioma	• Very aggressive • Very vascular • High tendency for bone and dural invasion • Can spread along the venous sinuses with fingerlike projections • Can invade the related nerves (facial and lower cranial nerves), leading to their dysfunction • Total removal is difficult in the involved bones due to invasion of the Haversian system • Posterior fossa extension is usually well encapsulated with a good plane of cleavage • Recurrence is very common

Continued

Atlas of Microsurgery of the Lateral Skull Base. https://doi.org/10.1016/B978-0-443-19083-4.00016-7

TABLE 16.1 The pathological behavior of the most common skull base lesions—cont'd

Tumor	Characteristics
Chordomas	• Locally malignant • Very aggressive • High tendency of bone and dural invasion • Nonvascular • Usually soft and nonencapsulated • Recurrence is very common and in a very short time
Intradural epidermoid	• Slowly growing and usually attaining a large size • Grow and expand along cleavage planes and occupy more than one intracranial compartment • Tend to envelop (but not invade) neurovascular structures • Easy to remove from surrounding tissues, but total removal is very difficult, as "tumor" can be left in hidden creases • Regrowth is common

Surgical Management

As the majority of skull base lesions are benign, surgery remains the mainstay of treatment. Total removal leads to complete cure in most cases. The complex anatomy and the close relation to vital structures mandate a wide exposure to facilitate tumor removal without endangering the related structures. Because of the deep-seated location, a marked degree of brain retraction is needed to gain access to the tumor in the traditional neurosurgical approach. The hallmark of skull base surgery is to remove the bone to provide the necessary access with minimal or even no brain retraction.

For a given lesion, an ideal skull base approach would achieve the following:

1. Wide access to the whole tumor, permitting total removal.
2. Minimal or no brain retraction.
3. Minimized morbidity and mortality.
4. No additional deficits related to the approach.
5. Access should be near the tumor so that the surgeon is working at a comfortable angle.
6. Ease and rapidity of performance.

Because of the diversity of skull base lesions and the variability in their presentation, treatment should be individualized in each case. Various factors determine the choice of surgical approach. These factors can be grouped under three main categories: tumor factors, patient factors, and surgeon factors.

1. Tumor factors include the pathology of the tumor, its site and extension, size, side, vascularity, and relationship to surrounding vital structures.
2. Patient factors include the patient's age, general condition, preoperative neurological and cranial nerve deficits, and the patient's expectations and preferences.
3. Surgeon factors include the surgeon's expertise and preference. The surgeon must have an array of approaches at hand in order to be able to choose the most appropriate approach for a given tumor. Surgeons who master only a few approaches tend to do the same procedure for different types of lesions, and this usually produces suboptimal results. The basic concepts of skull base surgery are summarized in Table 16.2.

TABLE 16.2 General concepts of skull base surgery.

- Remove the bone and leave the brain alone.
- Choose the most appropriate approach for each patient, i.e., adapt the approach to the patient and not the patient to the approach.

As the tumor type and site are the most important parameters that determine the management plan, tumors will be grouped under different categories according to the most difficult and challenging sites, and decision-making will be discussed accordingly.

Jugular Foramen Tumors

Because of their unique location, these tumors pose certain problems:

1. Injury to the lower cranial nerves.
2. The facial nerve is centered on and is closely related to the jugular bulb. In addition, most of these tumors have preoperative normal hearing; therefore, hearing preservation, if possible, is to be sought.
3. These tumors spread in three planes in three different "compartments": medial intradural, anterior intrapetrous, and inferior extracranial to the neck (Fig. 16.1).

The main lesions affecting the jugular foramen include paragangliomas, lower cranial nerve schwannomas, meningiomas, and others such as chondrosarcomas and petrous bone cholesteatoma. These tumors differ significantly in their vascularity, aggressiveness, pattern of spread, and invasion of nearby structures. Because of this diversity in tumor behavior and the unique pattern of extension, each case should be dealt with individually.

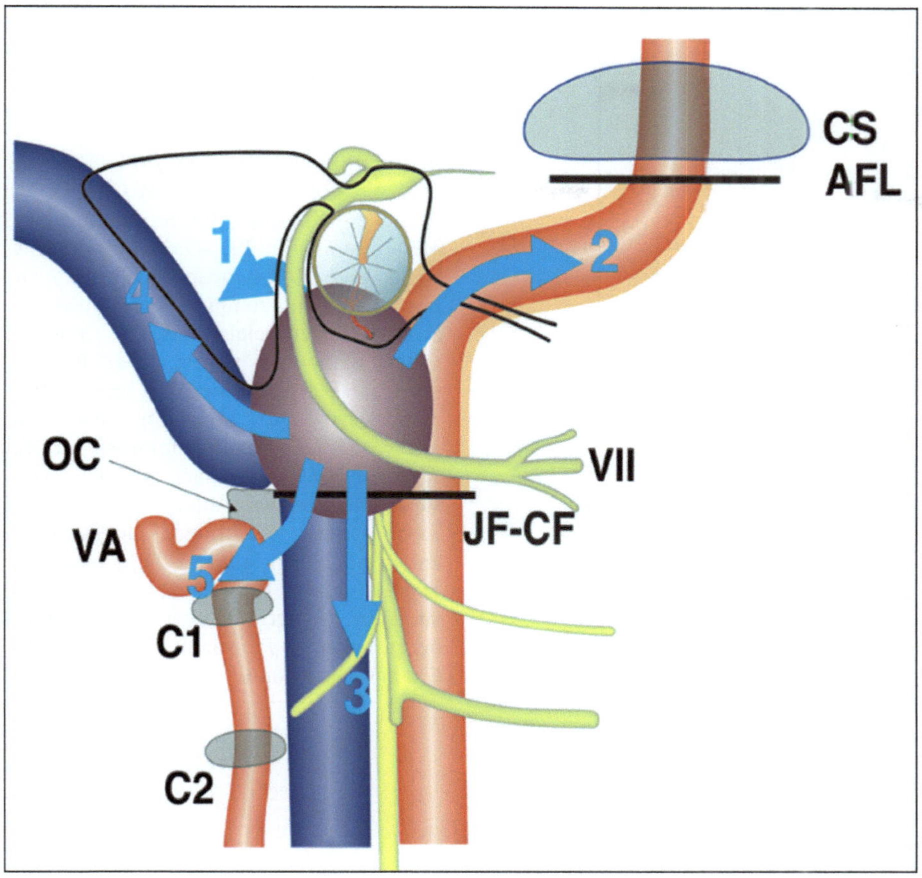

Fig. 16.1 Jugular foramen tumors can extend medially to become intradural, anteriorly to involve the petrous apex and internal carotid artery, and inferiorly into the neck.

Tympanojugular Paragangliomas

We follow the modified Fisch classification proposed by Gruppo Otologico (Table 16.3 and Fig. 16.2). Only class C and D tumors are discussed in this chapter. Management options include surgery, observation, and radiotherapy.

SURGERY

Surgery is the main line of treatment. The aim of surgery is total removal in most cases. Preoperative angiography should be performed. If the internal carotid artery (ICA) is compromised, a balloon occlusion test is done to determine the possibility of sacrificing the artery together with the invading tumor. When indicated, the tumor must be dissected from the arterial wall. This can be achieved by subperiosteal (or supra-adventitial) dissection of the ICA in the carotid canal (horizontal portion) or subadventitial dissection in the vertical portion. When the artery is completely surrounded by a tumor resulting in severe stenosis on arteriography, manipulation without proper endovascular management may lead to severe bleeding, incomplete removal, or a cerebral vascular accident. Permanent balloon occlusion (PBO) is performed when the ICA is infiltrated by the tumor and the collateral blood flow is sufficient. But in cases with insufficient collateral blood flow or class C3–C4 tumors, we routinely use ICA intraluminal stenting.

Preoperative embolization (when feasible) is performed in all class C and D cases. The management plan depends on the tumor class as follows:

- C1 and C2 tumors: Infratemporal fossa type A approach with anterior rerouting of the facial nerve.
- C3 tumors: Infratemporal fossa type A approach and ICA stenting/balloon occlusion.
- C4 tumors: Subtotal tumor removal with infratemporal fossa type A and B approach and ICA stenting/balloon occlusion postoperative radiotherapy is performed for the intracavernous part.

TABLE 16.3 Classification of paragangliomas according to the modified Fisch classification by Gruppo Otologico.

Class A	Tympanic
Class B	Tympanomastoid
Class C	Tympanojugular
• C1	• Carotid foramen
• C2	• Vertical ICA until genu
• C3	• Horizontal ICA
• C4	• ICA+foramen lacerum
Class D	Intracranial extension
• Di(1, 2)	• Intracranial intradural
Class V	Vertebral artery
• Ve	• Vertebral extradural
• Vi	• Vertebral Intradural

ICA, internal carotid artery.

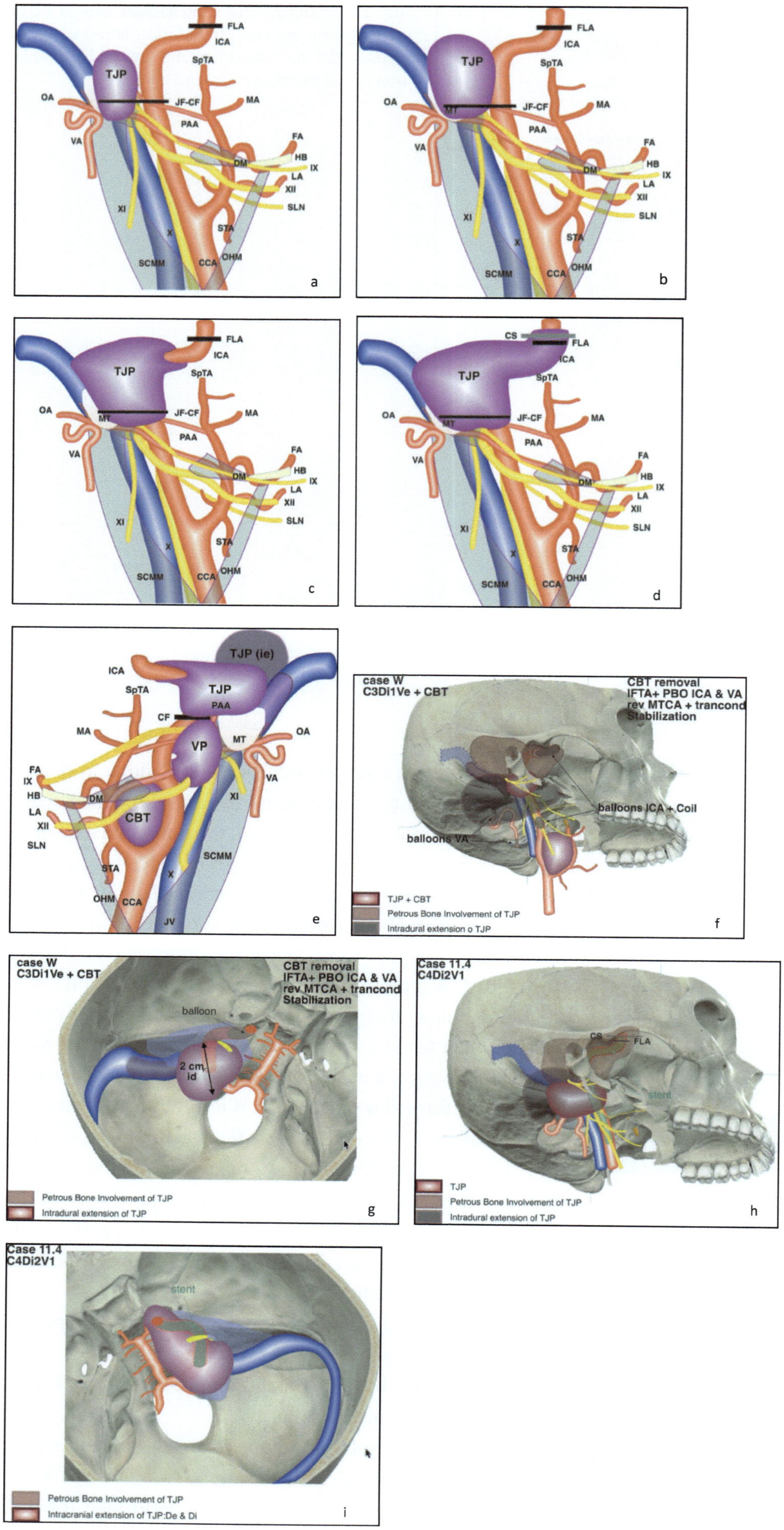

Fig. 16.2 Schematic drawings representing the different classes of tympanojugular paraganglioma according to the modified Fisch classification by Gruppo Otologico. (A) C1; (B) C2; (C) C3; (D) C4; (E) multiple tumors; (F, G) C3+Vi1+Ve+carotid body tumor; (H, I) C4 Di2 Vi. *OA*, occipital artery; *VA*, vertebral artery; *MT*, mastoid tip; *XI*, accessory nerve; *X*, vagal nerve; *TJP*, tympanojugular paraganglioma; *CCA*, common carotid artery; *SCMM*, sternocleidomastoid muscle; *OHM*, omohyoid muscle; *DM*, digastric muscle; *HB*, hyoid bone; *STA*, superior thyroid artery; *PAA*, ascending pharyngeal artery; *JF-CF*, jugular foramen carotid foramen; *MA*, maxillary artery; *SpTA*, superficial temporal artery; *ICA*, internal carotid artery; *LA*, lingual artery; *FLA*, anterior foramen lacerum; *IX*, glossopharyngeal nerve; *FA*, facial artery; *SLN*, superior laryngeal nerve; *CBT*, carotid body tumor; *VP*, vagal paraganglioma; *CS*, cavernous sinus.

INTRADURAL TUMORS

Because of their aggressive behavior, tympanojugular paragangliomas require wide opening of the neck spaces to achieve safe and complete removal. Removal of an intradural component at the same stage is not safe because connecting the subarachnoid space with the neck spaces during tumor removal would create a high risk of postoperative CSF leakage. Therefore, we use the following strategy in these cases:

- Di1 tumors: only small tumors (<1 cm) are removed at the same stage. The dura is closed with a muscle plug and a transfixion suture.
- Di2 tumors: staged removal is performed. The extradural part is removed first, followed by a second-stage POTS approach for the intradural part.

POSTOPERATIVE FOLLOW-UP

Regular follow-up is performed after 1, 3, 5, and 7 years. Follow-up is performed using magnetic resonance imaging (MRI), CT scan, and digital subtraction angiography (DSA).

OBSERVATION AND REGULAR FOLLOW-UP

This policy is indicated in:

- Small tumors in an elderly patient >65 years with intact lower cranial nerves or with poor general condition.
- Small tumors with insufficient collateral venous circulation (rare).

RADIOTHERAPY

Radiotherapy is indicated in:

- Refusal of surgery.
- Bad general condition.
- Preoperative intact lower cranial nerves in an elderly patient >65 years.
- Subtotal removal as in C4 tumors, carotid artery invasion with insufficient collaterals, etc.

Lower Cranial Nerve Schwannoma

Lower cranial nerve schwannomas differ from tympanojugular paragangliomas in being much less vascular, less aggressive, and less invasive of the nearby structures (the temporal bone and the ICA). The classification for these tumors is shown in Table 16.4 and Fig. 16.3.

SURGERY

Surgery is the mainstay of treatment. The POTS approach is appropriate in the majority of cases. It allows for single-stage total removal with preservation of facial nerve and middle ear function. The approach is suitable for the removal of tumors with or without cerebellopontine or upper neck extension, or both. A detailed flowchart of the surgical management is described in Fig. 16.4.

OBSERVATION

Observation is indicated in elderly patients with preoperative intact lower cranial nerve function.

TABLE 16.4 Classification of JF schwannomas according to the classification system proposed by Kaye et al. and modified by Pellet et al.

Class	Description
A	Primary intracranial tumor with small extension into the bone
B	Primary tumor in the jugular foramen with or without intracranial extension
C	Primary extracranial tumor with extension in the jugular foramen or posterior cranial fossa
D	Dumbbell-shaped tumor with intracranial and extracranial extension

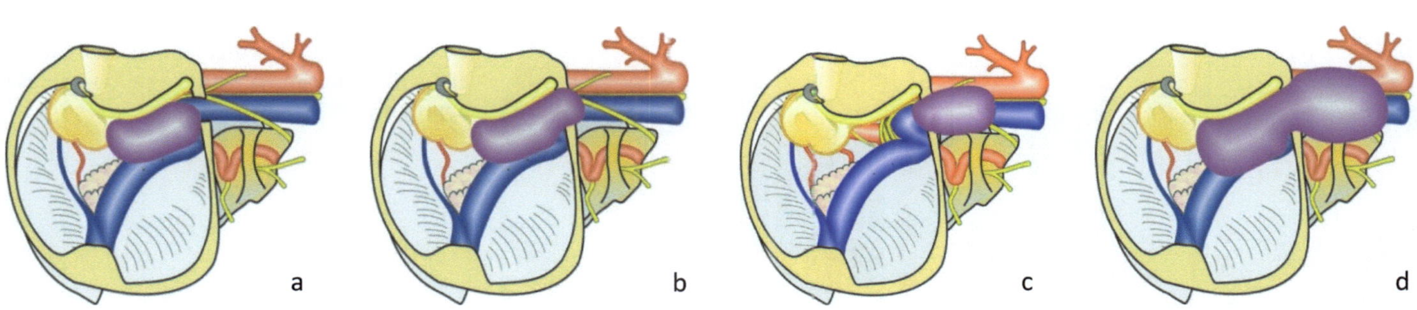

Fig. 16.3 Classification of jugular foramen schwannomas: (A) primary intracranial tumor, (B) primary tumor in the jugular foramen with or without intracranial extension, (C) primary extracranial tumor with extension in jugular foramen or posterior cranial fossa, (D) dumbbell-shaped tumor with intracranial and extracranial extension.

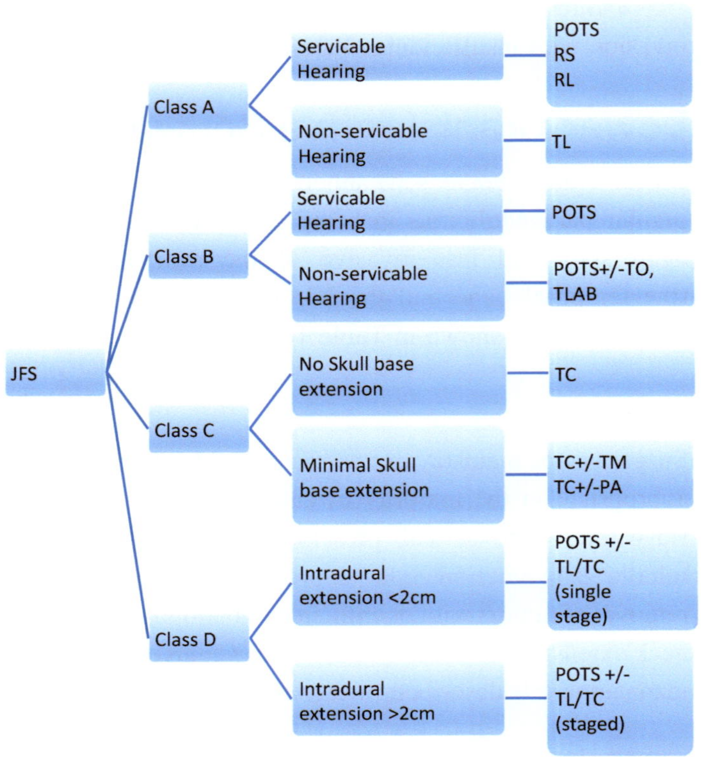

Fig. 16.4 Surgical approach algorithm for the treatment of jugular foramen schwannoma. *RS*, retrosigmoid; *RL*, retrolabyrinthine; *POTS*, petro-occipital trans sigmoid; *TL*, translabyrinthine; *TO*, transotic; *TC*, transcervical; *PA*, preauricular; *TM*, transmastoid.

Jugular Foramen Meningioma

These tumors are more aggressive than schwannomas. They very rarely extend to the neck. Management includes the following:

SURGERY

Surgery is the main line of treatment.

CHOICE OF SURGICAL APPROACH

- Tumors limited to the jugular fossa with intradural extension; the POTS approach.
- Tumors with intradural extension and limited neck extension; single-stage POTS approach (e.g., Fig. 16.5).
- Tumors with intradural extension and involvement of the middle ear and vertical ICA; POTS + transotic approach.
- Tumors with massive jugular foramen, intradural, and temporal bone involvement; modified type D transcochlear approach.
- En-plaque meningiomas invading the whole temporal bone with extension to the petroclival region and passing through the jugular foramen to the neck are rare variants.

In these tumors, we use a staged procedure in which a POTS or transcochlear approach is performed for intradural and intracranial parts in the first stage, and a transcervical or infratemporal type A approach is performed in the second stage for the extradural and extracranial parts. This strategy is essential to avoid inevitable postoperative CSF leakage. The subarachnoid space should not be connected to the widely opened neck spaces at the same stage (Figs. 16.6 and 16.7).

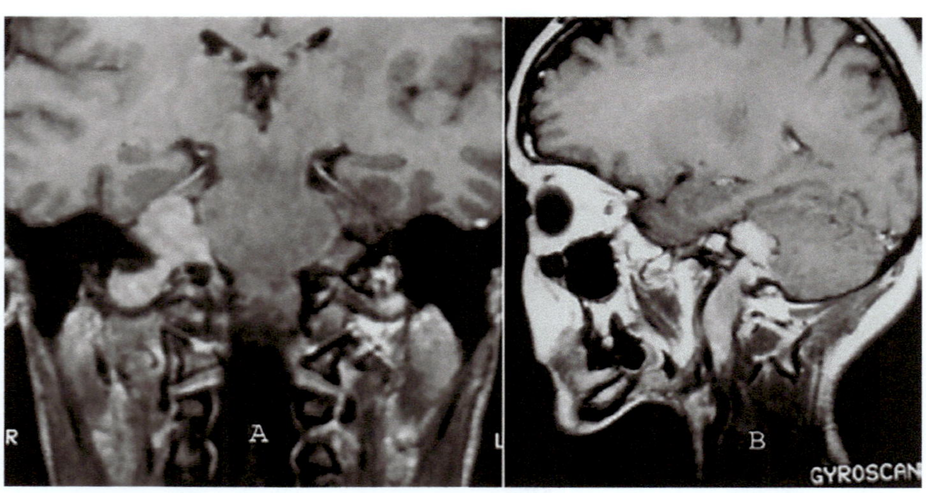

Fig. 16.5 Jugular foramen meningioma. (A: coronal, B: sagittal) with minimal neck extension can still be managed with a single-stage POTS approach.

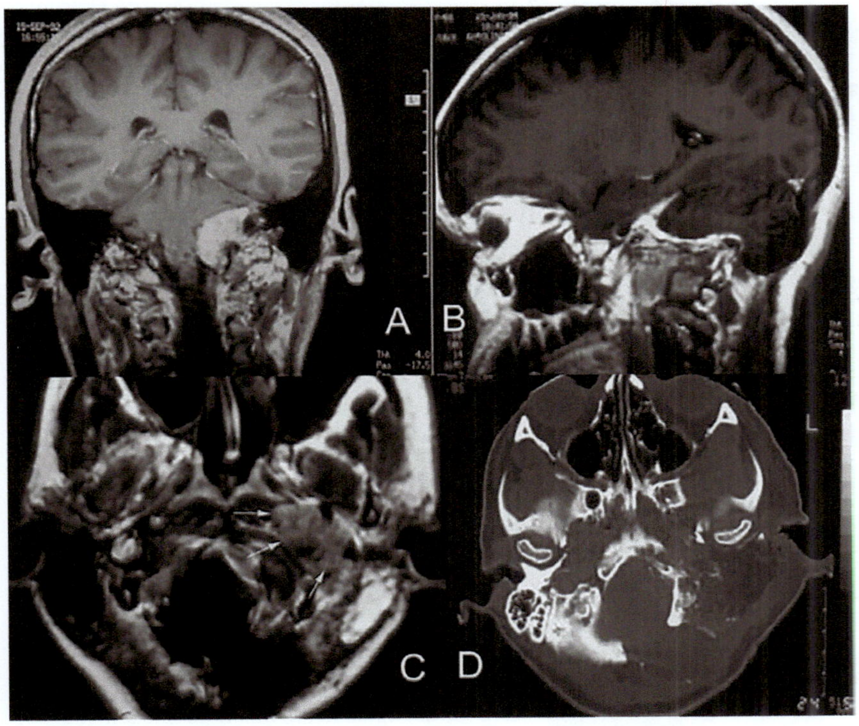

Fig. 16.6 En-plaque meningioma with massive extension into the neck. (A) Preoperative view. (B) Sagittal MRI after first-stage removal of the intradural part through a modified transcochlear approach type D. (C) Axial CT at the level of the neck before the second stage. *Arrows* point to the neck part of the tumor. Note the encasement of the internal carotid artery. (D) Postoperative CT scan after total tumor removal.

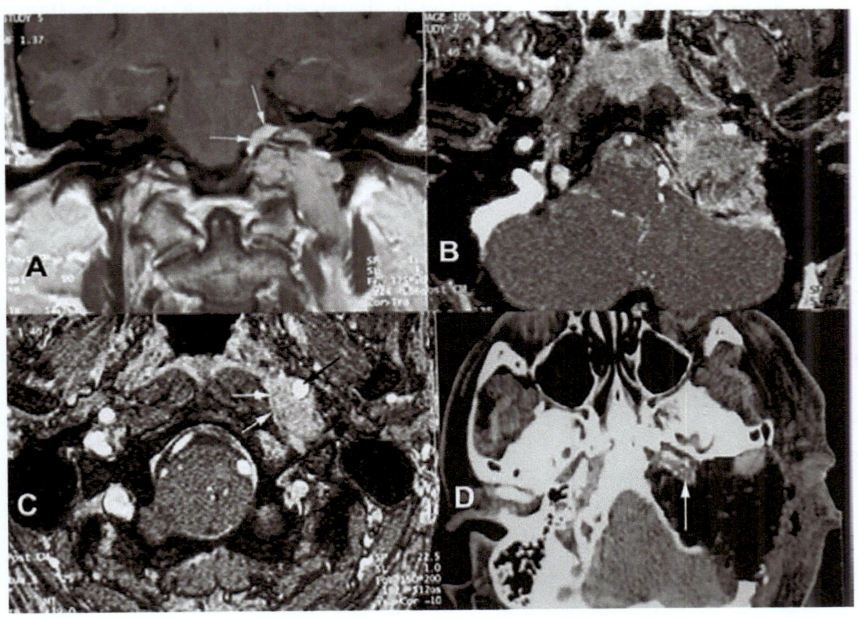

Fig. 16.7 MRI showing a left meningioma involving the whole temporal bone and extending into both the cerebellopontine angle and the neck. (A) Coronal view. The small tumor component in the cerebellopontine angle is seen *(arrows)*. (B) Axial view showing massive jugular foramen and petrous apex involvement. (C) Axial view at a lower level in the upper neck. Note the encasement of the internal carotid artery *(black arrow)* by the tumor *(white arrows)*. Such a massive involvement needs a wide opening of the neck spaces for complete and safe tumor removal. (D) Postoperative view after first-stage extradural modified transcochlear approach with extension into the neck. Note the freed horizontal internal carotid artery *(arrow)*. The dura was not opened at this stage for fear of CSF leakage. A second-stage modified transcochlear approach was later performed for removal of the residual intradural component.

TOTAL VERSUS SUBTOTAL REMOVAL

Total removal should be aimed at. However, because of tumor invasion of the lower cranial nerves observed in most cases, some surgeons advocate leaving a small piece of tumor around the infiltrated nerves, particularly in the elderly, to avoid significant postoperative morbidity (Fig. 16.8).

OBSERVATION

Old patients having small jugular foramen meningioma with intact lower cranial nerves can be managed by observation and follow-up.

Cerebellopontine Angle Tumors

The most common are vestibular schwannomas, followed by meningiomas, epidermoid, paragangliomas, etc.

VESTIBULAR SCHWANNOMA

The decision-making paradigm is summarized in Table 16.5.

TABLE 16.5 The management of vestibular schwannoma.

Tumor size	Good hearing	Poor hearing
<0.5 cm	MCF	TLA
0.5–1.5 cm	RL-RS	TLA
>1.5 cm	TLA	TLA

MCF, middle cranial fossa approach; *TLA*, translabyrinthine approach; *RL-RS*, retrolabyrinthine-retrosigmoid approach.

MENINGIOMA

Meningiomas can be classified as follows:

Posterior surface of the temporal bone meningioma:

- Lying anterior to the internal auditory canal (Fig. 16.9).
- Centered on the internal auditory canal (Fig. 16.10).
- Lying posterior to the internal auditory canal (Fig. 16.11).
- Very rarely, meningiomas are intracanalicular and are difficult to distinguish preoperatively from acoustic neurinomas.

Different combinations of the above are common. Some cases involve the whole surface of the temporal bone (Fig. 16.12).

Fig. 16.13A–C illustrates schematic drawings of different types of meningiomas related to the posterior surface of the temporal bone.

SURGERY

Surgery is the main line of treatment and is indicated in the majority of cases. Total removal is aimed at in most cases. Subtotal removal is performed in cases in which a total removal would carry a high risk of neurological deficit. The decision for subtotal removal can be taken preoperatively, as in giant lesions where the aim can be tumor debulking to relieve the tumor compression on the cerebellum and brainstem, particularly in elderly people, or intraoperatively, where tumor adherent to the brainstem or encasing important vessels is left.

The choice of surgical approach depends mainly on the site of the tumor. Other factors include the preoperative cranial nerve deficits, the side of the lesion, the patient's age, etc.

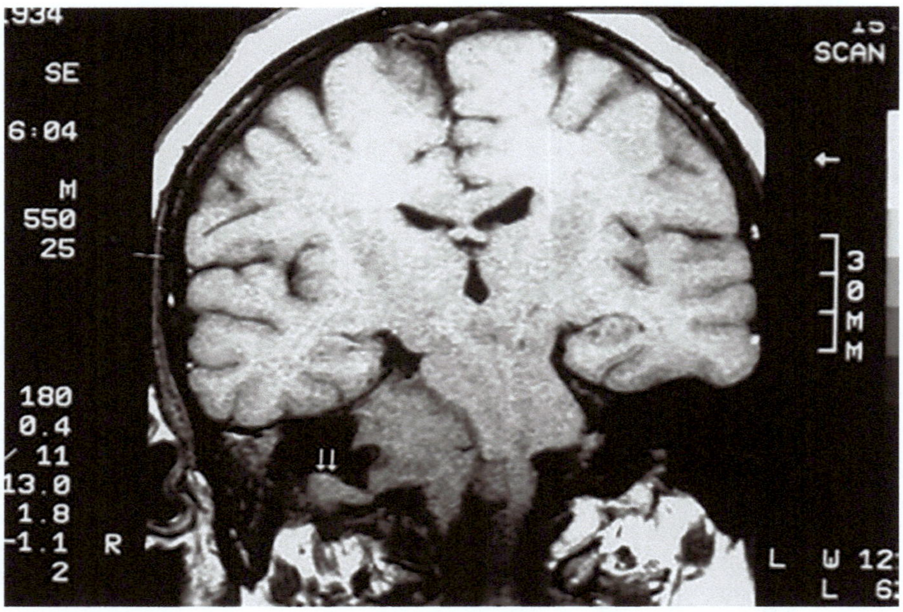

Fig. 16.8 MRI of a 65-year-old patient showing a right jugular foramen meningioma with preoperative lower cranial nerve function. The tumor extension into the jugular foramen *(arrows)* can be appreciated. Intraoperatively, a piece of tumor was left over the lower cranial nerves.

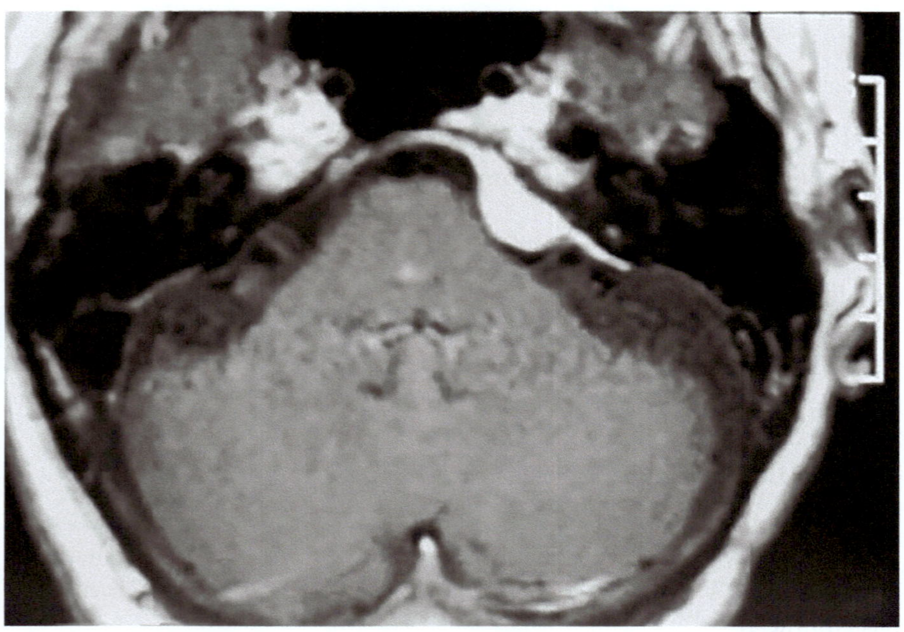

Fig. 16.9 Meningioma lying anterior to the internal auditory canal.

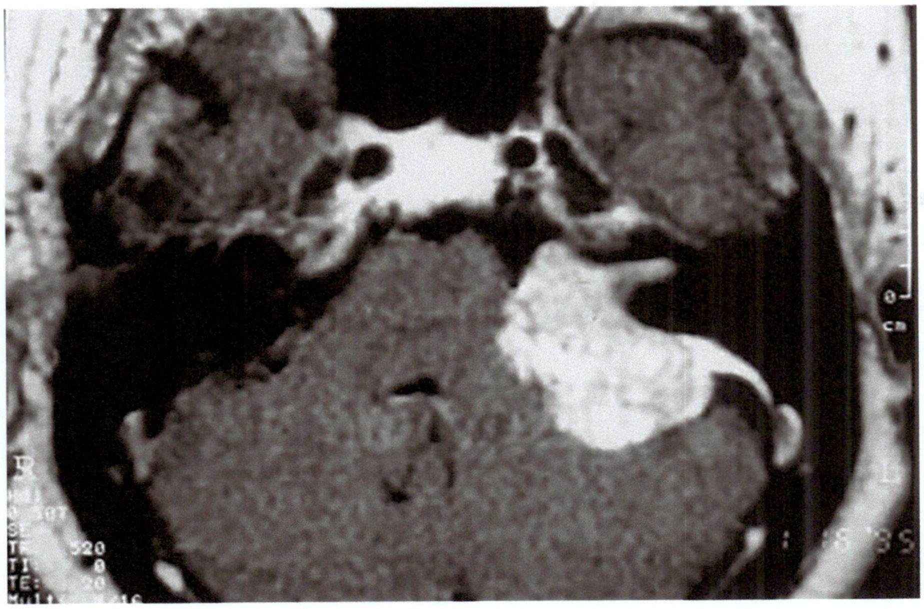

Fig. 16.10 Meningioma centered on the internal auditory canal.

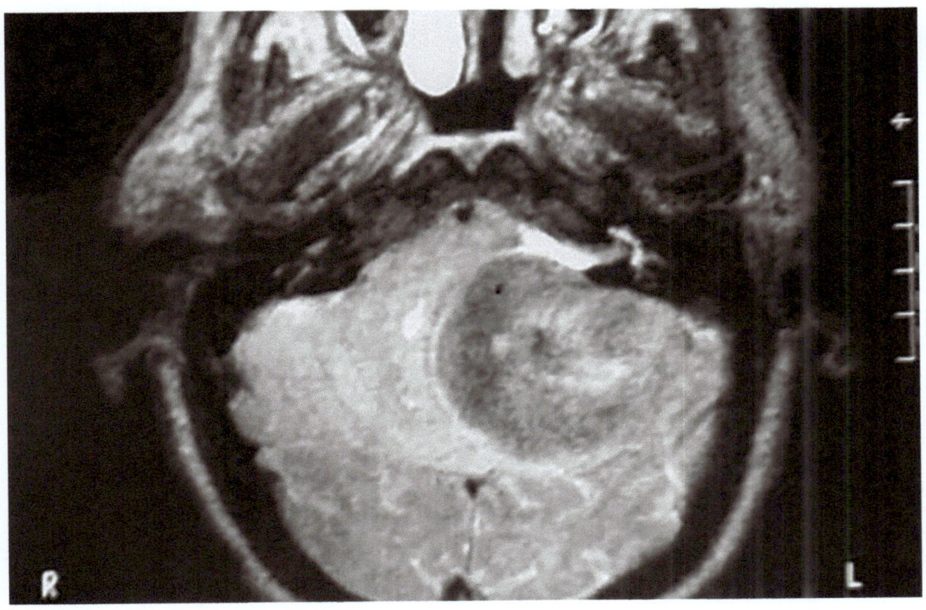

Fig. 16.11 Meningioma posterior to the internal auditory canal.

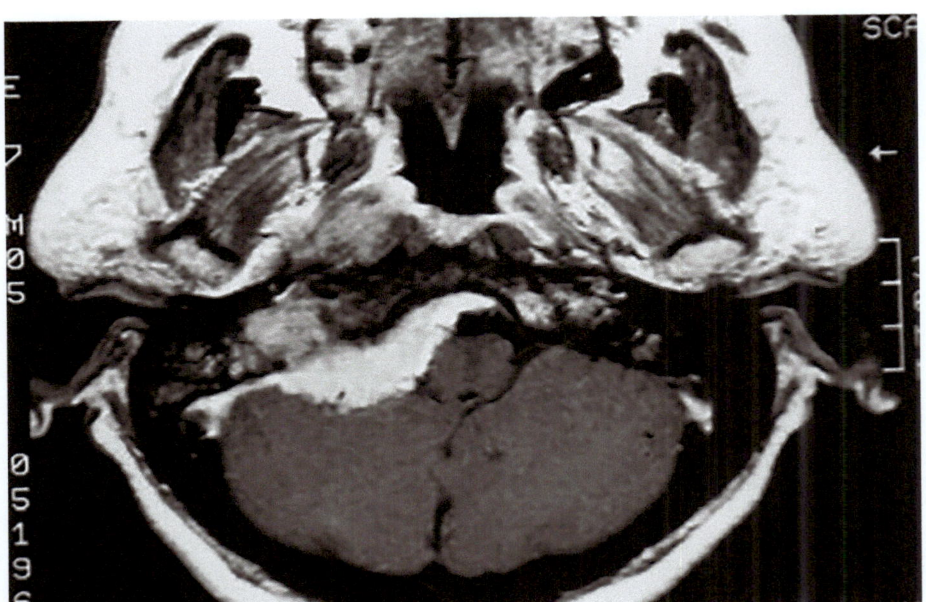

Fig. 16.12 Meningioma involving the whole posterior surface of the temporal bone.

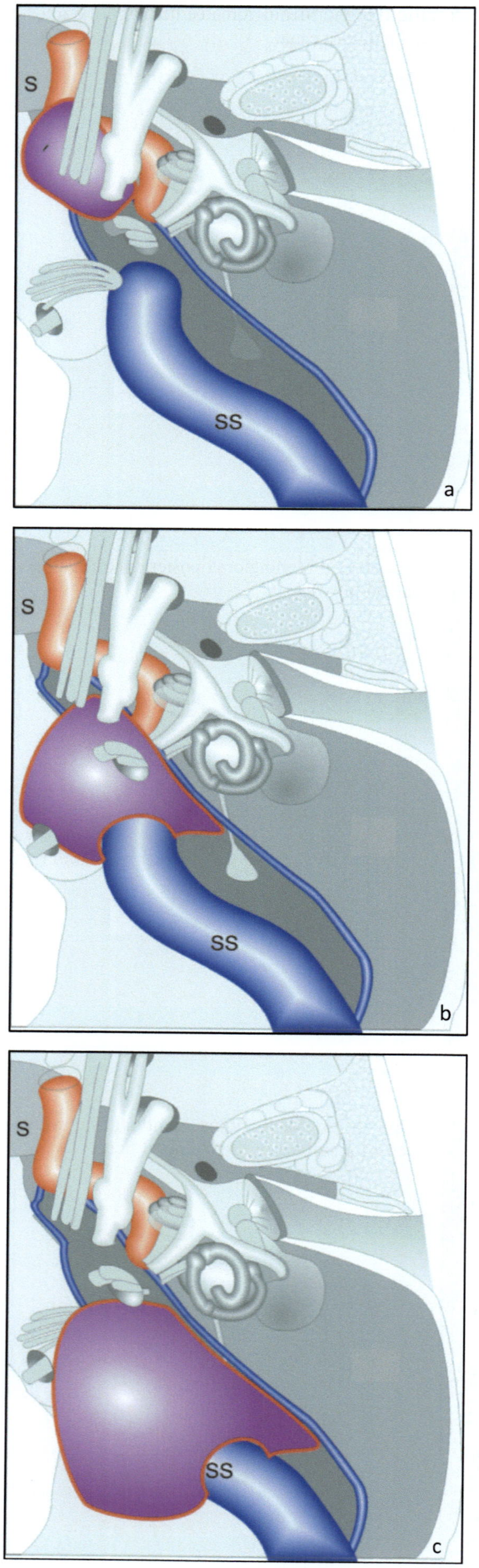

Fig. 16.13 (A–C) Schematic drawings showing the different types of meningioma in relation to the posterior surface of the temporal bone. (A) Anterior to the internal auditory canal, (B) centered on the internal auditory canal, (C) posterior to the internal auditory canal. *S*, sphenoid sinus; *SS*, sigmoid sinus.

CHOICE OF APPROACH

Meningiomas Lying Posterior to the Internal Auditory Canal

These are classically managed through the retrosigmoid approach. Cases with relative anterior extension can be managed with a combined retrosigmoid and retrolabyrinthine approach. In cases with poor preoperative hearing, the translabyrinthine approach can be used instead.

Meningiomas Centered on the Internal Auditory Canal

- Small lesions with good preoperative hearing are managed with a retrosigmoid approach, with or without a retrolabyrinthine approach (Fig. 16.14).
- Cases with anterior extension can be managed with a retrolabyrinthine subtemporal transapical or a retrolabyrinthine transtentorial approach (Fig. 16.15).
- Cases in which hearing preservation is not aimed at are best managed with the enlarged translabyrinthine approach. If significant anterior extension is present, a translabyrinthine transapical approach is adopted (Fig. 16.16).
- Intracanalicular meningiomas are managed with the enlarged middle fossa approach to preserve hearing, whereas the translabyrinthine approach is utilized in patients with preoperative unserviceable hearing.

Meningiomas Lying Anterior to the Internal Auditory Canal

These cases usually have good preoperative hearing. They are managed according to the following paradigm:

- Small lesions are managed by a middle fossa transpetrous approach (Fig. 16.17).
- Medium lesions not reaching the midline can be managed through a retrolabyrinthine subtemporal, transapical approach with or without a transtentorial extension.
- Large petroclival lesions extending into the prepontine cistern with or without superior (supratentorial) or inferior extension pose more difficulties and carry a high operative risk. These will be discussed in the section on clival and petroclival tumors.

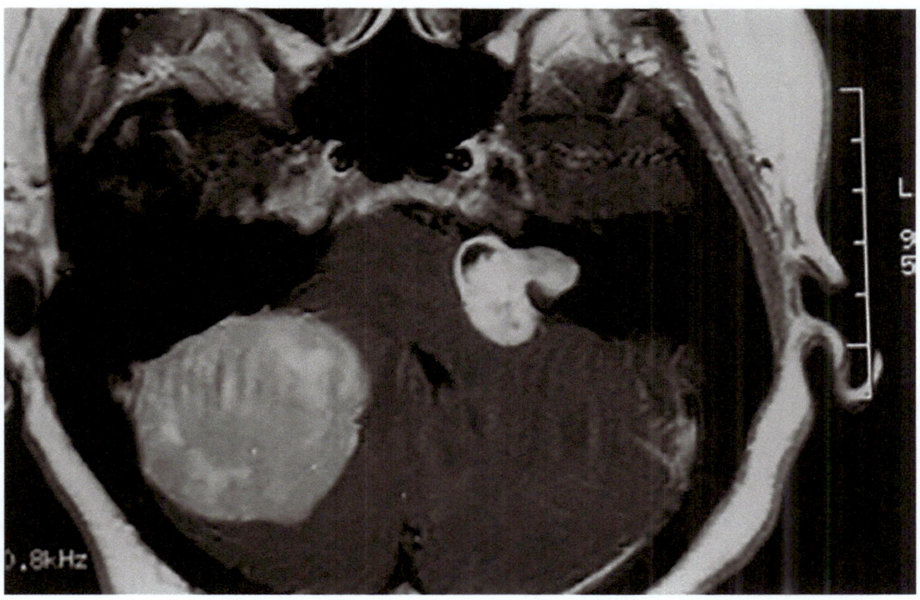

Fig. 16.14 MRI of an NF2 case showing a left acoustic neurinoma and a right meningioma lying mainly posterior to the internal canal with a rather anterior extension. The meningioma was managed through a combined retrosigmoid retrolabyrinthine approach.

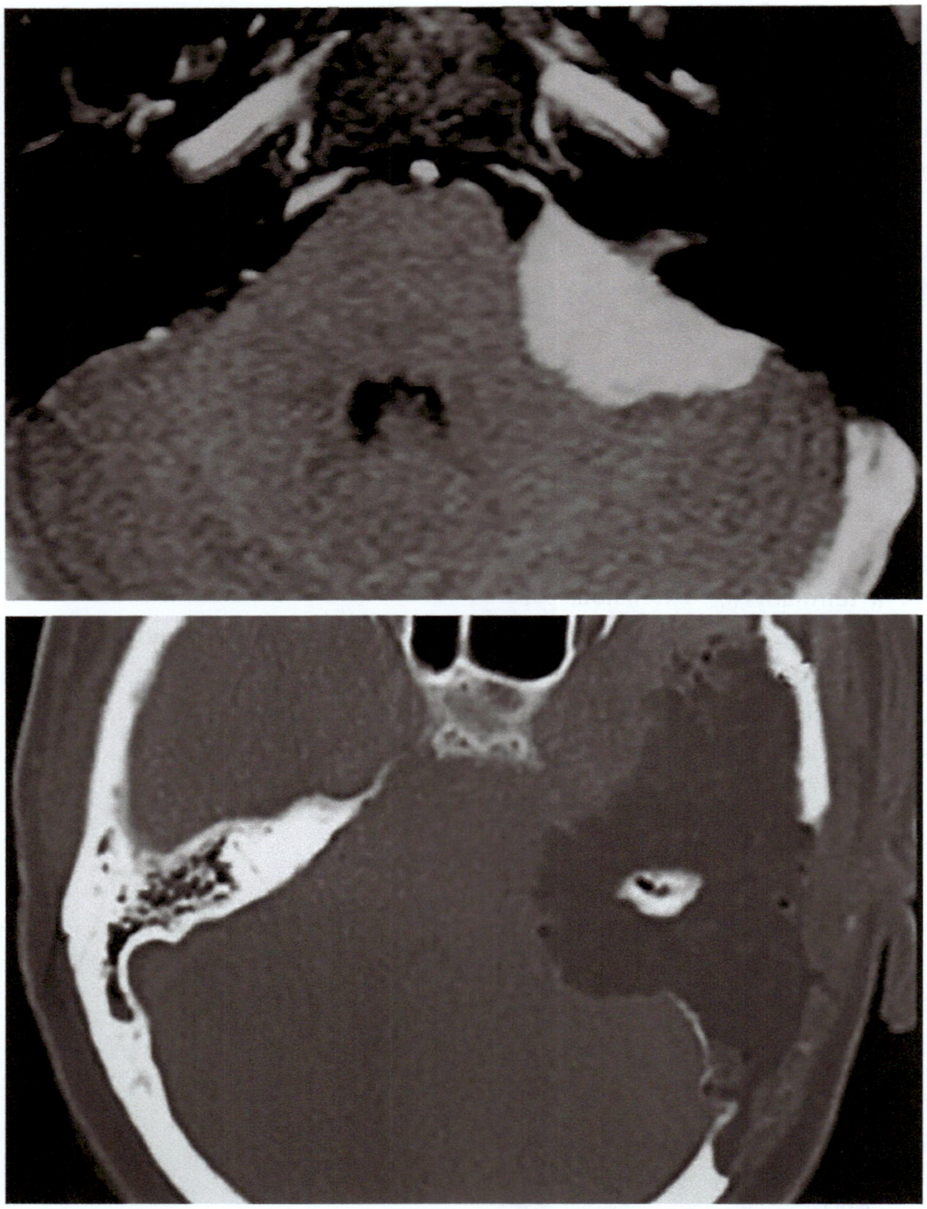

Fig. 16.15 (A) Left meningioma centered on the internal auditory canal with good preoperative hearing and definite anterior extension. This case was managed by a combined retrolabyrinthine, subtemporal transapical approach. (B) Postoperative CT scan showing total tumor removal.

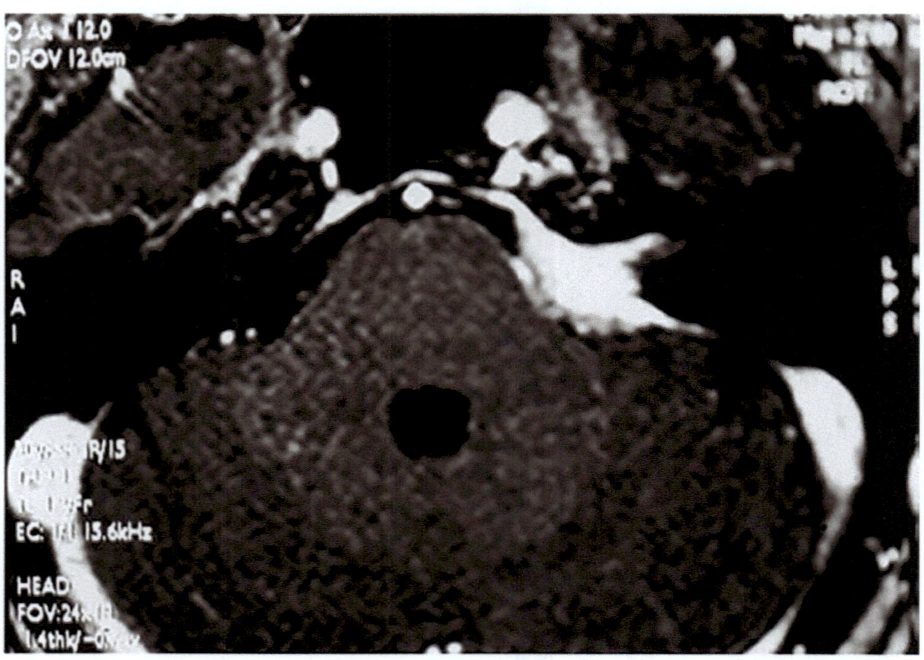

Fig. 16.16 A left meningioma centered on the internal auditory canal with poor preoperative hearing. This case is managed by a translabyrinthine transapical approach.

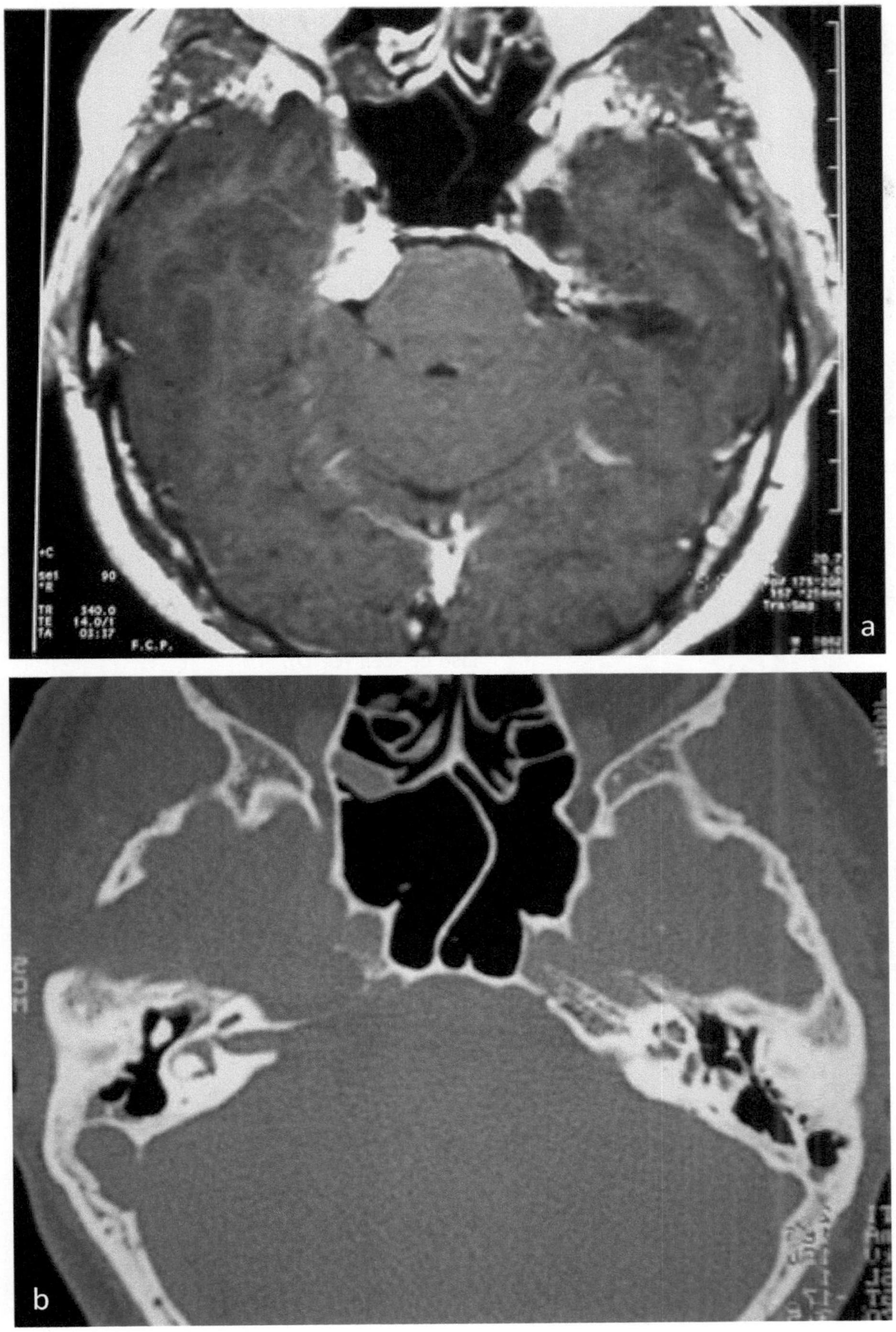

Fig. 16.17 (A) Right meningioma lying entirely anterior to the internal auditory canal with good preoperative hearing. (B) Postoperative CT scan after total removal through a middle fossa transpetrous approach.

Epidermoid

Epidermoid tends to involve more than one intracranial component and usually envelop vital neurovascular structures. Cerebellopontine angle epidermoid can be managed by a retrosigmoid approach. Moderately large lesions with more anterior extension to the prepontine cistern can be managed by the transotic or the translabyrinthine transapical approaches. Larger lesions passing the midline are best managed by the modified transcochlear approach (Fig. 16.18). In large lesions with massive supratentorial and parasellar involvement, a combined transpetrous orbitozygomatic approach is adopted (Fig. 16.19).

After removal of the mass, it is preferable to ascertain total removal using the endoscope (Fig. 16.20).

Fig. 16.21A–D illustrates the main types of posterior fossa epidermoid.

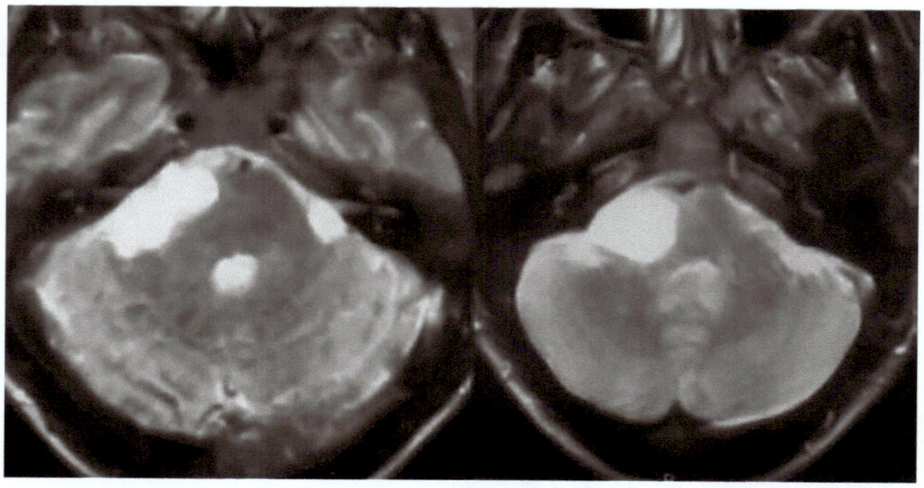

Fig. 16.18 A large right posterior fossa epidermoid reaching the prepontine cistern. This case is best managed by the modified transcochlear approach.

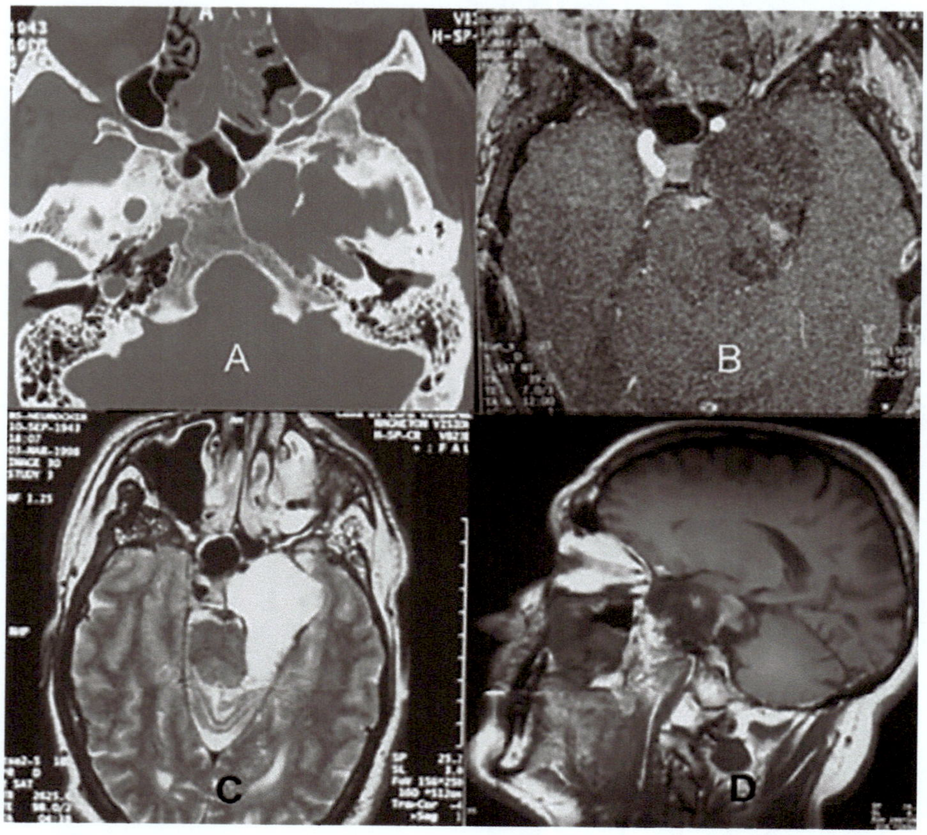

Fig. 16.19 A large left epidermoid that extends from the posterior to the middle fossa and involves the parasellar region. (A) The petrous apex and the base of the middle fossa are eroded by the tumor. (B) Note the involvement of the internal carotid artery at the level of the cavernous sinus. The posterior fossa component can also be seen (C, D).

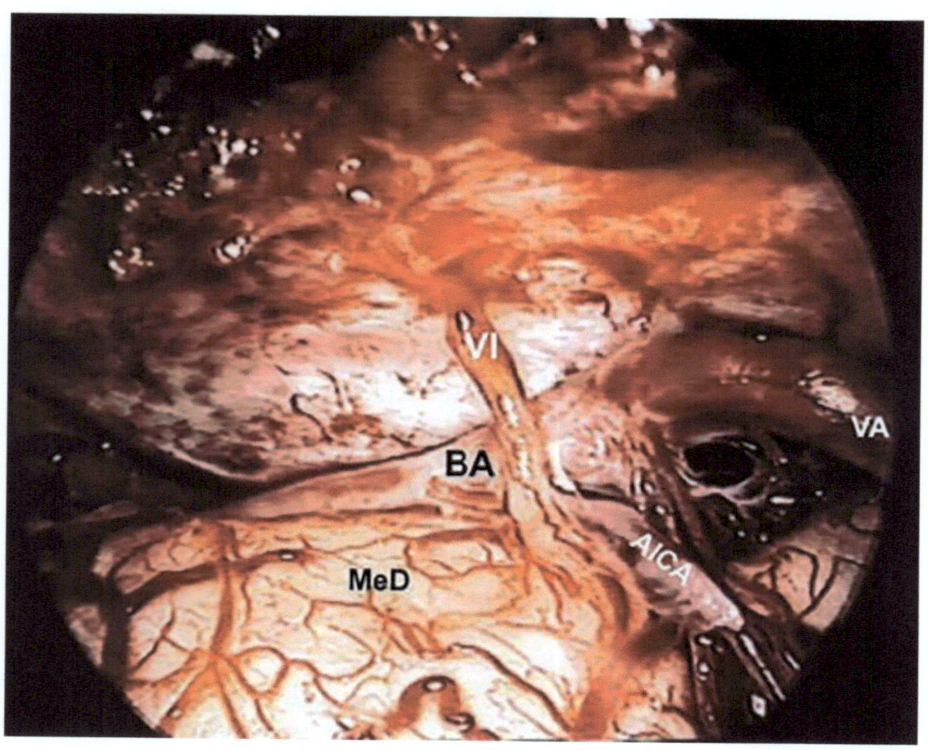

Fig. 16.20 An endoscopic view of the right cerebellopontine angle after total removal of an epidermoid. The endoscope allows viewing around angles and enables the detection and removal of any hidden residual.

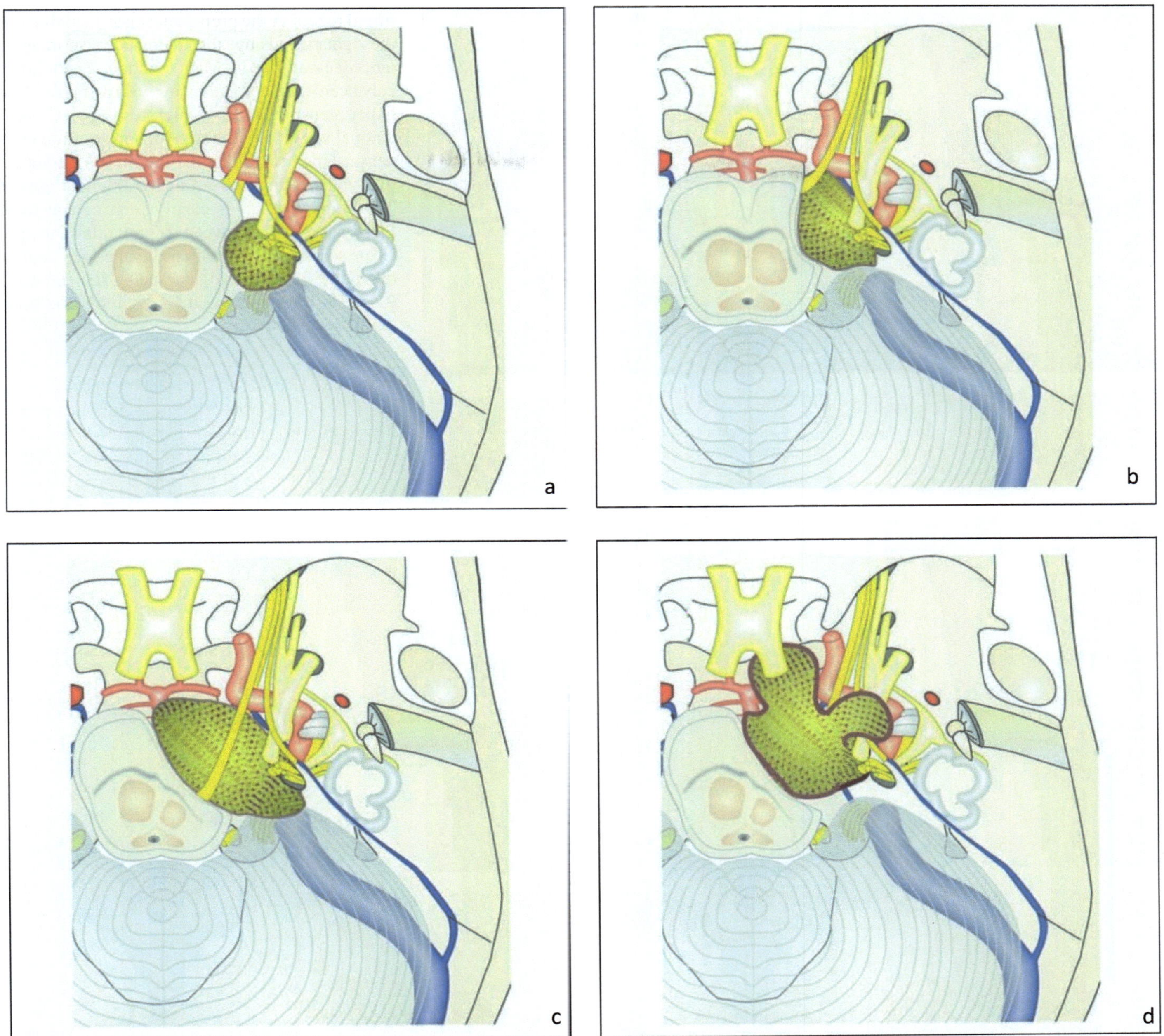

Fig. 16.21 Schematic drawings of the different types of epidermoid. (A) Small and centered on the internal auditory canal. This can be managed by a retrosigmoid approach. (B) A larger lesion that is mainly located anterior to the internal auditory canal. This is best managed by either a transotic or a translabyrinthine transapical approach or, in case of good hearing, a retrolabyrinthine transtentorial approach. (C) A lesion extensively involving the prepontine cistern is best managed by a modified transcochlear approach. (D) Posterior fossa lesion with middle fossa, Meckel's cave, and parasellar extension is best managed by a combined transpetrous orbitozygomatic approach.

Petrous Apex Lesions

Lesions of the petrous apex can be either primary or extension from other areas. The most common lesions are cholesterol granuloma, petrous bone cholesteatoma, chondroma, chondrosarcoma, chordoma, mucocele, petrous apex effusion, meningioma, nonacoustic cranial nerve neurinoma, and metastasis.

The choice of surgical approach depends mainly on the pathology of the lesion and its extent as well as the preoperative hearing and facial nerve function.

CHOLESTEROL GRANULOMA

As establishment of drainage is the main goal of surgery, a conservative approach can be attempted. These lesions can therefore be managed by an infracochlear or an infralabyrinthine approach (Fig. 16.22A–C). If the anatomy is unfavorable, as in cases with a high jugular bulb, a middle fossa approach can be used instead. In case of unfavorable anatomy together with a large lesion, an infratemporal fossa approach type B can be performed.

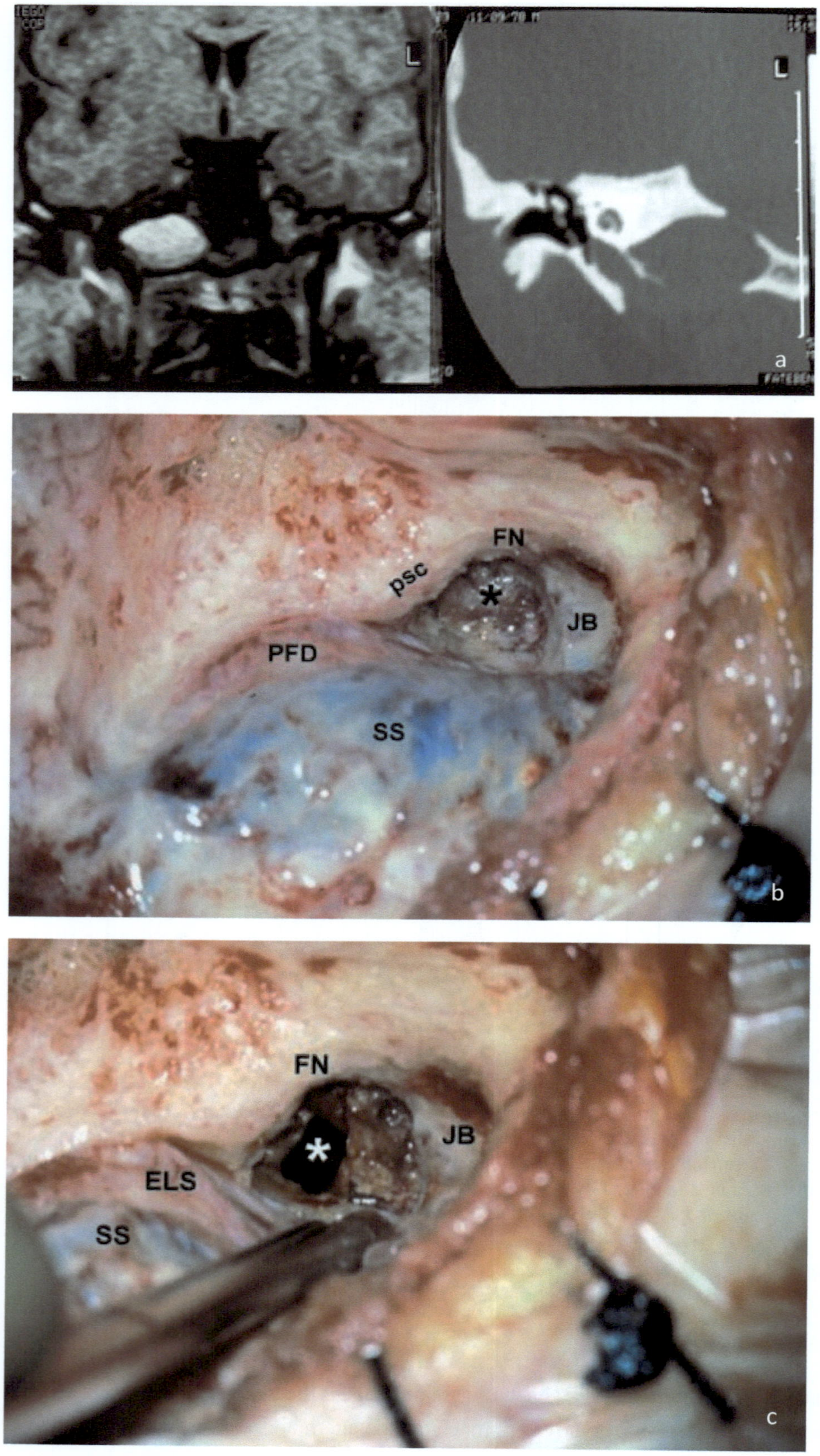

Fig. 16.22 A case of cholesterol granuloma of the right petrous apex. (A) Preoperative imaging: *left*, T1-weighted MRI without gadolinium enhancement, coronal view, showing a large hyperintense mass; *right*, high-resolution CT scan, coronal view. Note the smooth bone erosion of the petrous apex caused by the lesion. (B) Intraoperative view through an infralabyrinthine approach. The granuloma (*) is clearly seen medial to the facial nerve (FN) between the ampulla of the posterior semicircular canal (psc) superiorly and the relatively low-lying jugular bulb (JB) inferiorly. (C) View after opening the granuloma and draining its contents.

Small asymptomatic lesions that are accidentally discovered can be managed by clinical and radiological follow-up (Fig. 16.23).

CHOLESTEATOMA

Congenital cholesteatoma limited to the petrous apex with preoperative good hearing can be managed by the middle fossa approach. Acquired petrous bone cholesteatoma extending into the petrous apex usually needs a more aggressive approach to achieve total removal. These lesions usually cause extensive erosion of the otic capsule, and hearing cannot be preserved if such an erosion exists. A translabyrinthine transapical or a modified transcochlear approach (in more advanced cases) is adopted. Infralabyrinthine apical lesions without the involvement of the internal auditory canal can also be managed with an infratemporal fossa type B approach.

CHORDOMAS, CHONDROMAS, AND CHONDROSARCOMAS

Aggressive treatment is needed in these cases. Very small lesions with preoperative good hearing can be managed by the middle fossa transpetrous approach with extensive removal of the bone of the petrous apex (Fig. 16.24A and B). Larger lesions with involvement of the vertical or the inferolateral aspects of the horizontal ICA necessitate a wider approach, such as the infratemporal fossa type B approach. Extensive transdural cases will necessitate the addition of a modified transcochlear approach.

NONSURGICAL LESIONS

Petrous apex effusion and asymmetrical petrous apex pneumatization can be misdiagnosed as more aggressive lesions. These pathologies need no surgical interference (Fig. 16.25).

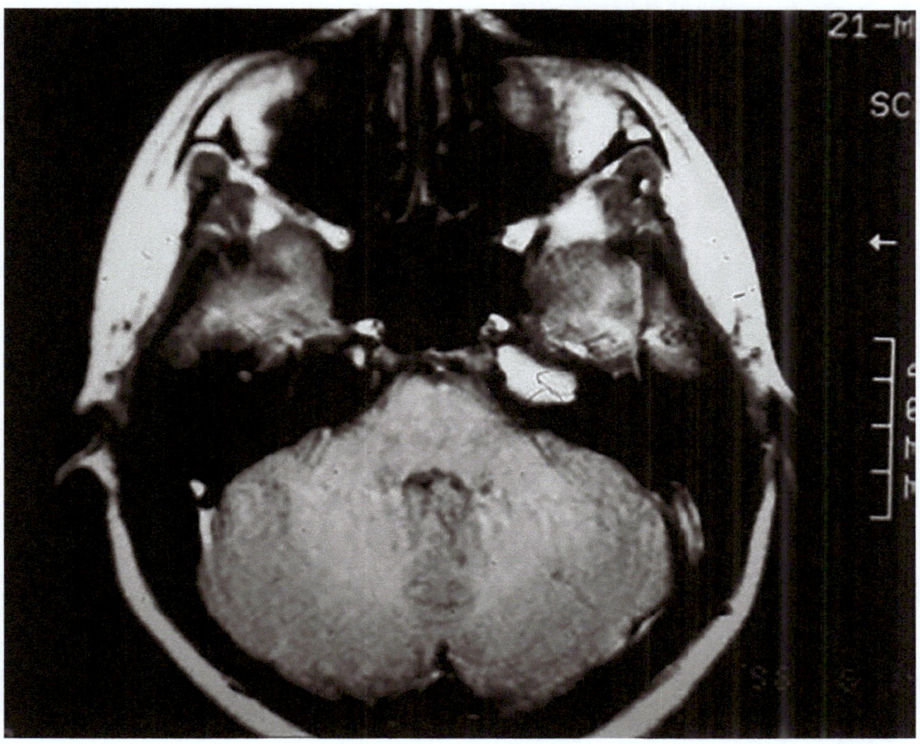

Fig. 16.23 MRI, T1-weighted without gadolinium enhancement, axial view, showing a small left asymptomatic petrous apex cholesterol granuloma. A follow-up policy was decided upon after discussing the treatment options with the patient.

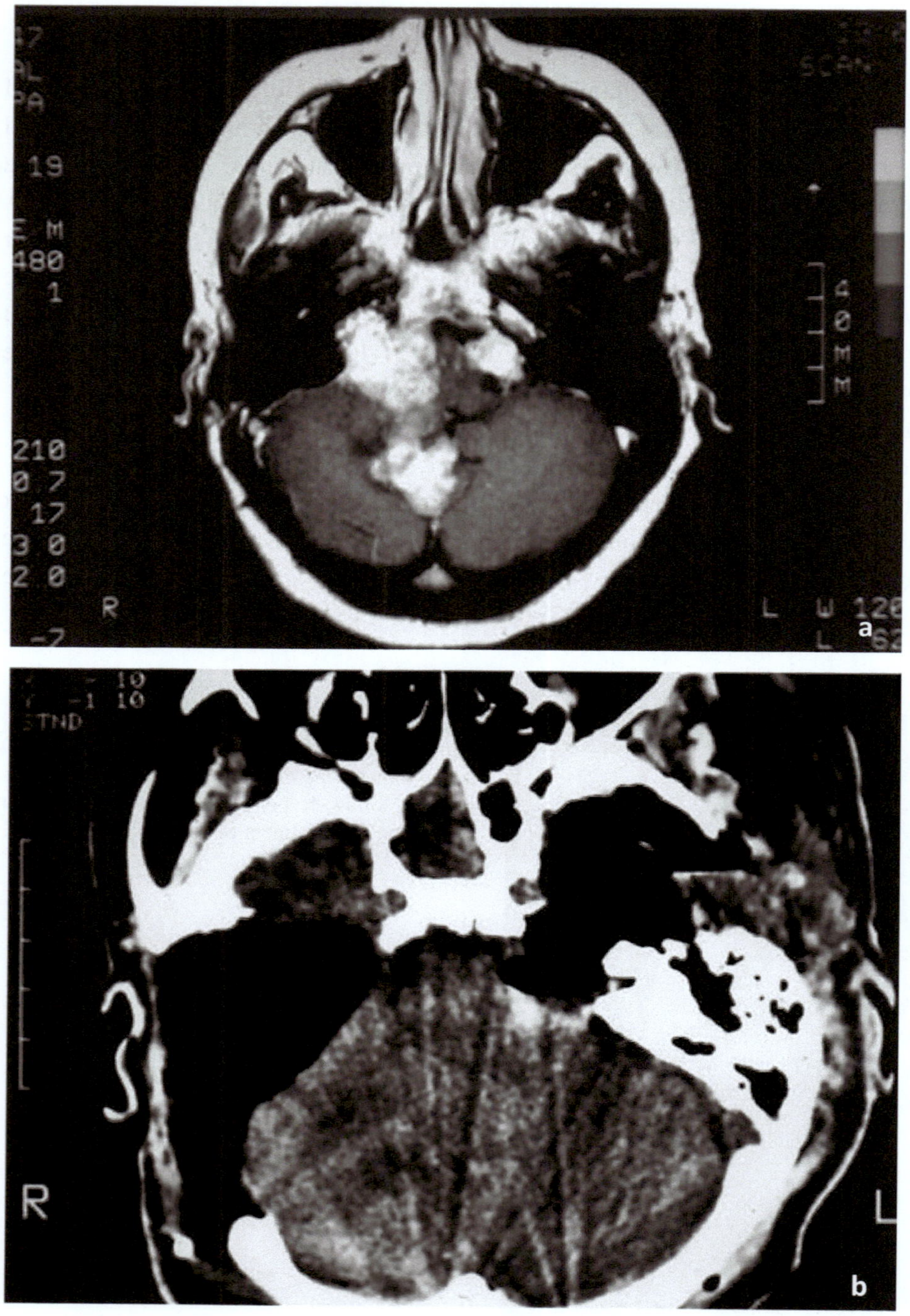

Fig. 16.24 (A) Preoperative MRI with gadolinium enhancement showing a multifocal recurrent clival chordoma. The tumor is located between the two cerebellar hemispheres, in the right petroclival region, and in the left petrous apex. (B) Postoperative view after staged total tumor removal. The transdural petrous apex tumor was removed through a middle fossa approach.

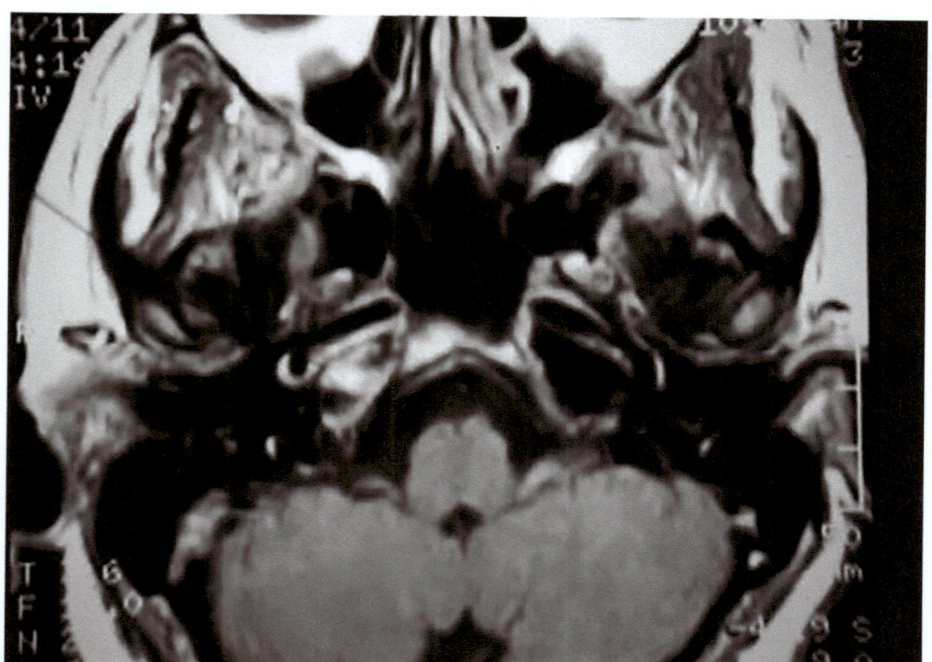

Fig. 16.25 MRI of a young woman showing an accidentally discovered right petrous apex effusion. The lesion enhances on T1-weighted imaging without gadolinium enhancement. The *fine dark strands* represent the intact bony septa of a well-aerated petrous apex. Note that there is no expansion of the petrous apex by the lesion (cf. the contralateral side). A simple petrous apex effusion is diagnosed.

Meckel's Cave Area

Lesions involving Meckel's cave include trigeminal schwannomas, meningiomas, chordomas, and epidermoid and dermoid cysts. Trigeminal schwannoma is a good representative of this group. Trigeminal schwannomas are classified according to their location into four types as follows:

- **Type I**: Predominantly middle fossa tumor (Fig. 16.26). The approach of choice is frontotemporal orbitozygomatic.
- **Type II**: Predominantly posterior fossa tumor with limited middle fossa involvement (Fig. 16.27). The choice of surgical approach varies as follows:
  - Large tumors with anterior extension to the prepontine cistern; a modified transcochlear approach type C is adopted.
  - Medium-sized lesions with preoperative good hearing; a combined retrolabyrinthine subtemporal, transtentorial approach can be performed.
  - Small tumors can be managed by the middle fossa transpetrous approach.
- **Type III**: Where there are both significant middle and posterior fossa components (Fig. 16.28), a combined frontotemporal orbitozygomatic transpetrous approach is chosen. The transpetrous combination varies according to the preoperative deficits as follows:
  - Preoperative facial paralysis and hearing loss; the modified transcochlear approach (Fig. 16.29A and B).
  - Preoperative hearing loss with intact facial nerve function; a translabyrinthine transapical approach is utilized.
  - Intact facial nerve and hearing; a retrolabyrinthine subtemporal approach is used.

A schematic drawing of the variable single or combined approaches for the management of types I–III lesions is seen in Fig. 16.30.

- **Type IV**: Where the tumor is exclusively extradural involving the infratemporal fossa and the surrounding structures (Fig. 16.31). The approach used is a preauricular infratemporal transzygomatic approach (type D infratemporal approach). If the vertical ICA is massively involved by the tumor, a type C infratemporal approach is preferred as it gives better control of the whole intrapetrous ICA (Figs. 16.32A–D and 16.33).

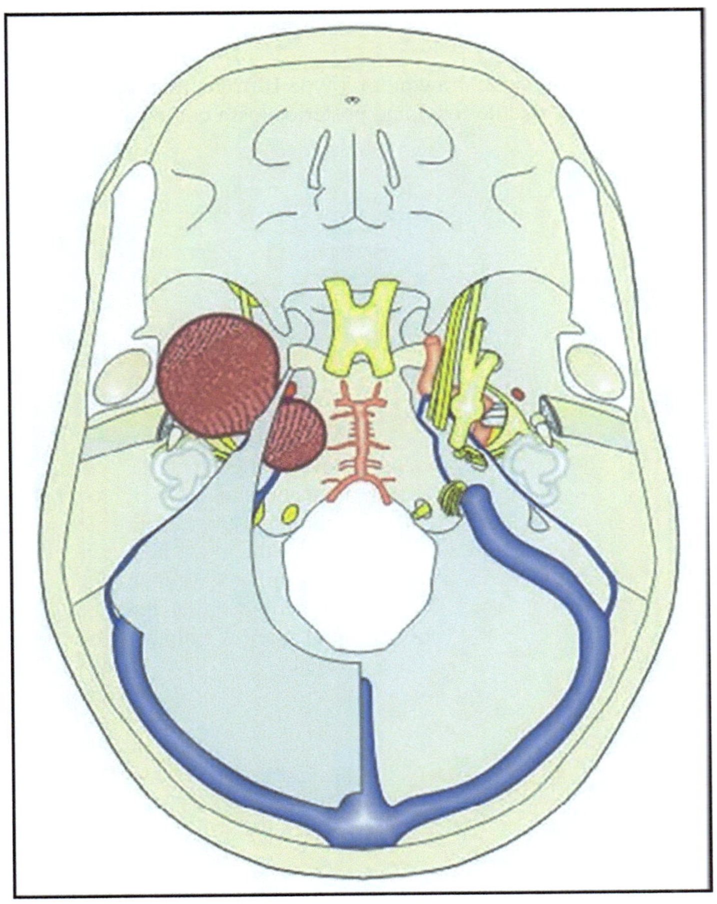

Fig. 16.26 Schematic drawing of a type I trigeminal schwannoma with a predominantly middle fossa component.

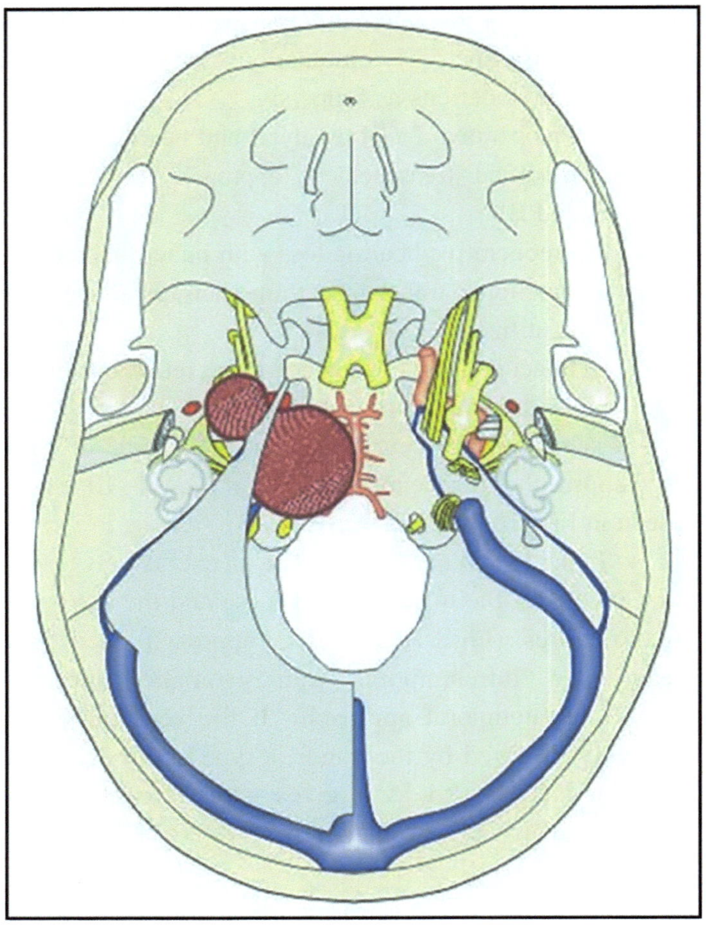

Fig. 16.27 Schematic drawing of a type II trigeminal schwannoma with a predominantly posterior fossa component.

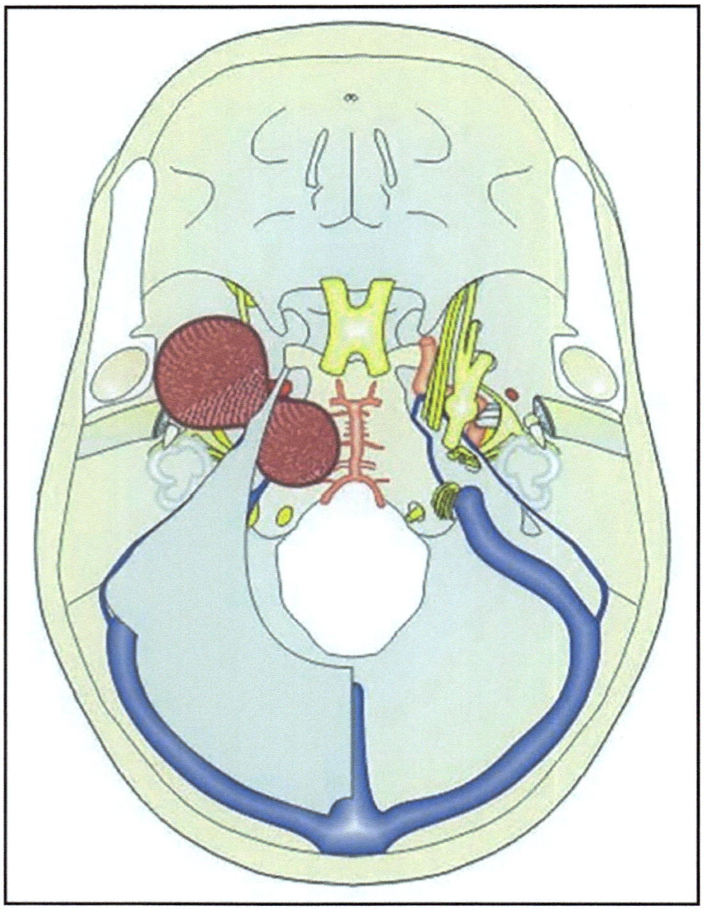

Fig. 16.28 Schematic drawing of a type III trigeminal schwannoma with significant middle fossa and posterior fossa components.

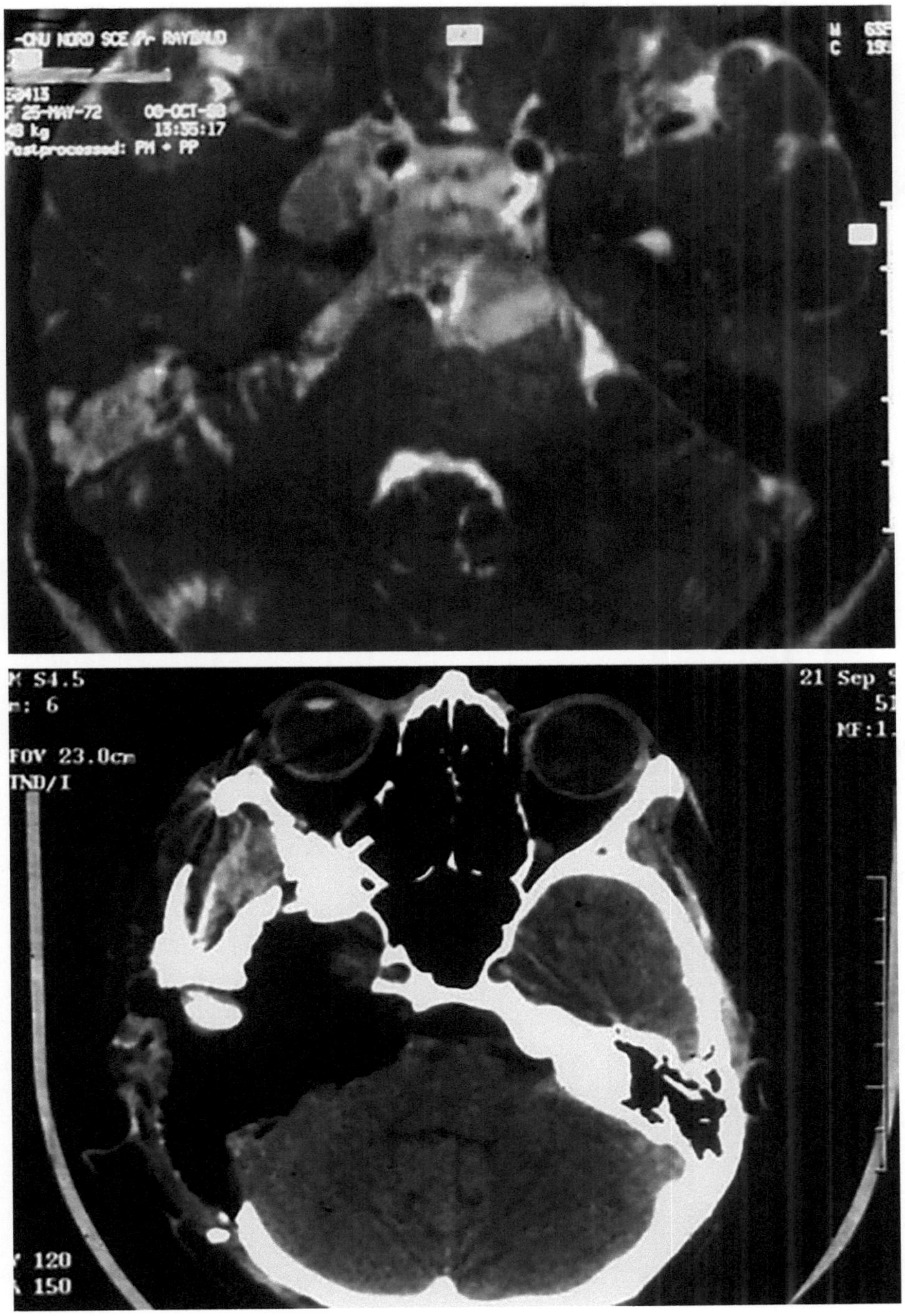

Fig. 16.29 (A) Preoperative MRI (T2-weighted image) of an extensive epidermoid of Meckel's cave area. The epidermoid is located in the prepontine cistern and extends into the middle fossa. (B) Postoperative CT scan following total excision via a combined frontotemporal orbito-zygomatic modified transcochlear approach.

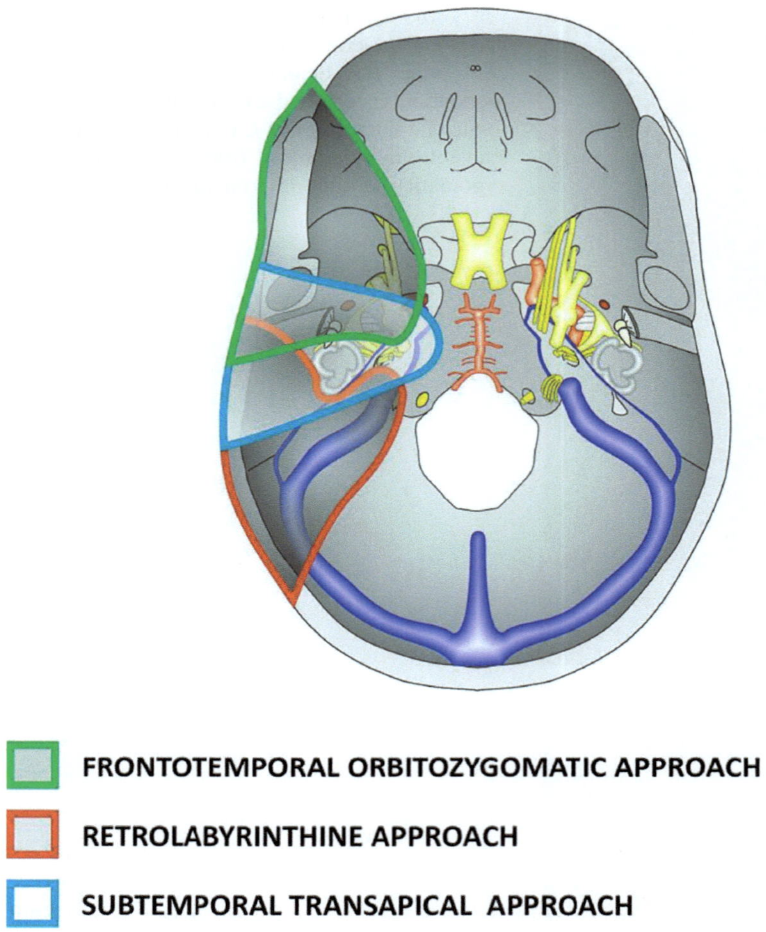

Fig. 16.30 Schematic drawing of the different craniotomies performed for the management of Meckel's cave lesions. The frontotemporal orbitozygomatic approach *(green outline)* is suitable for cases with significant middle fossa components, either alone or in combination with a suitable transpetrous approach, such as the retrolabyrinthine *(red outline)* for small posterior fossa components. Larger and anteriorly placed posterior fossa components require combination with other more extensive transpetrous approaches such as the translabyrinthine or the modified transcochlear approaches. For small middle fossa extensions, the subtemporal approach *(blue outline)* may be sufficient.

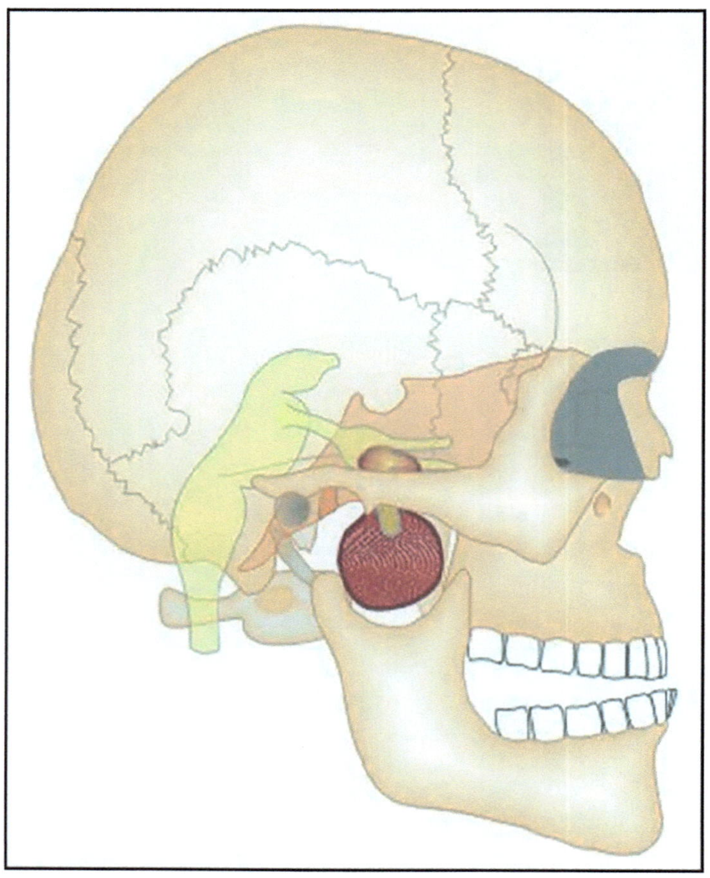

Fig. 16.31 Schematic drawing of a type IV, infratemporal extradural trigeminal schwannoma.

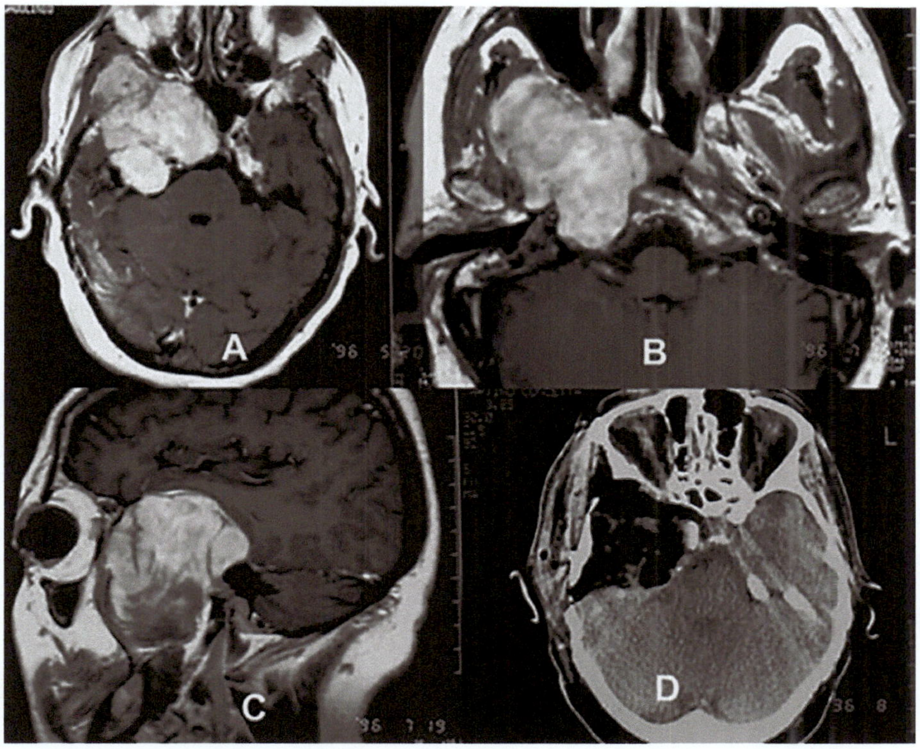

Fig. 16.32 Imaging of a case of a huge right trigeminal schwannoma. (A) MRI axial view showing the tumor extending into the sphenoid sinus and abutting the orbit. (B) Axial view at a lower level. Note the extensive involvement of the infratemporal fossa (type IV tumor). (C) Sagittal view. Note the impingement of the middle fossa by the tumor. (D) Postoperative CT scan after total removal via a combined infratemporal type C orbitozygomatic approach.

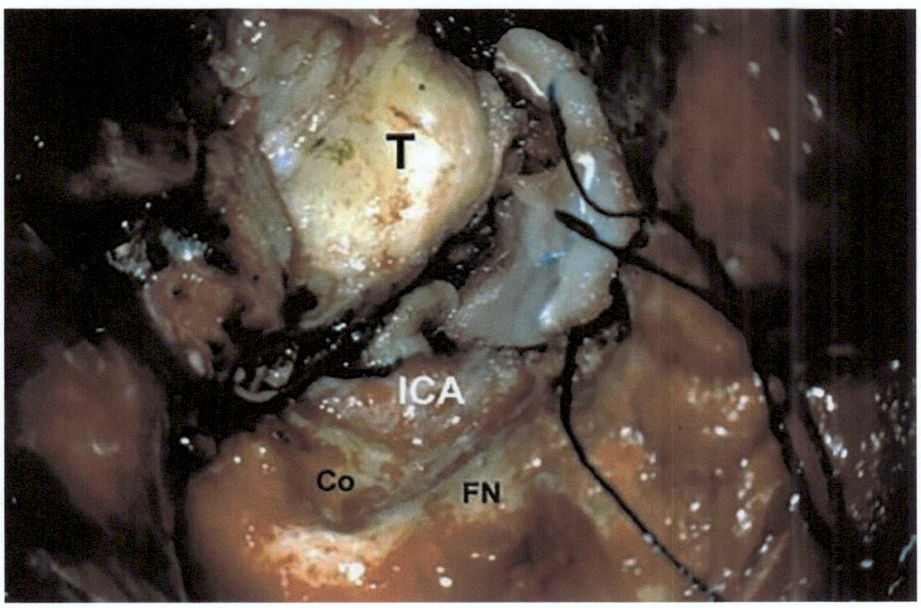

Fig. 16.33 Intraoperative view of the tumor in Fig. 16.32. The tumor (T) is seen in close relation to the vertical internal carotid artery (ICA). Such an extension is better controlled with the infratemporal type C approach, which gives an excellent control of the vertical internal carotid artery when compared with the preauricular infratemporal approach.

Special Considerations

STAGING

Single-stage tumor removal is aimed at in most cases. However, staging is adopted in the following situations:

- **Transdural tumors** with intradural and extracranial tumor components, e.g.:
  (i) Paragangliomas with intradural components.
  (ii) Intradural meningiomas with massive neck extension. The reason for staging in these two cases is to avoid the high possibility of postoperative CSF leakage should the widely opened neck spaces be connected with the subarachnoid space.
  (iii) Large clival chordomas involving the infratemporal fossa with intradural extension. Infratemporal fossa type B or C approach is performed in the first stage, followed by a modified transcochlear approach in the second stage. The reason for staging is to avoid connecting the subarachnoid space with a breached nasopharyngeal mucosa, leading to CSF leakage and meningitis (Fig. 16.34A–D).

Transdural tumors having an extradural intracranial component (which is still within the confines of the temporal bone) are managed in a single-stage procedure.

Intradural tumors extending into the neck that do not necessitate wide opening of the neck spaces for their removal can also be managed in a single-stage procedure (e.g., lower cranial nerve schwannomas).

- **To avoid a lengthy procedure**: e.g., huge petroclival tumors with large supratentorial components.

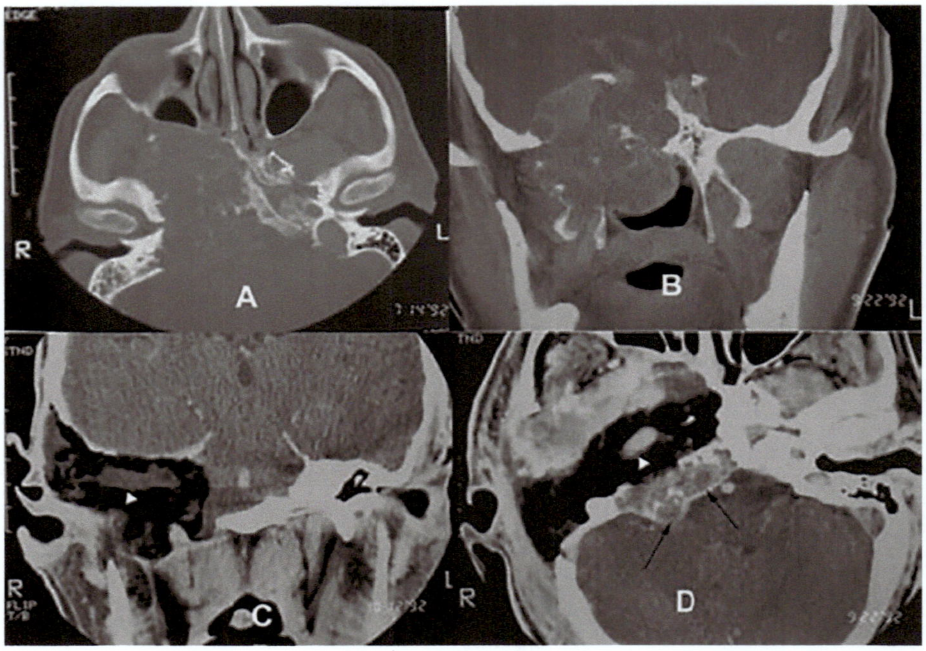

Fig. 16.34 (A) Preoperative CT scan of a large right transdural clival chordoma. Note the destruction of the clivus and the petrous apex. The tumor involves the ipsilateral nasal cavity and invades the right maxillary sinus. Note the extension into the posterior fossa. (B) CT scan, coronal view. The tumor involves the right nasopharynx, causing massive destruction of the clivus, the base of the pterygoid, and its processes. It also involves the infratemporal fossa. (C) CT scan, coronal view, following first stage removal of the extradural component via the infratemporal fossa approach. The liberated horizontal internal carotid artery *(arrowhead)* can be seen. (D) CT scan with contrast, axial view, following the first stage. Note the removal of the extradural component and the liberated internal carotid artery *(arrowhead)*. The residual intradural component *(black arrows)* is to be removed in a second-stage modified transcochlear approach.

TOTAL VS. SUBTOTAL REMOVAL

Total removal is aimed at in the majority of cases. Subtotal removal is performed in:

- Absence of a plane of cleavage between the tumor and brainstem, e.g., some petroclival meningiomas, particularly recurrent or irradiated tumors.
- Encasement of vital neurovascular structures such as the basilar artery or perforators.
- Massive involvement of the ICA in cases in which the artery cannot be sacrificed, as shown by the balloon occlusion test (Fig. 16.35A–C).
- In elderly patients with tumors adherent to preoperatively normal facial or lower cranial nerves, a small piece of tumor can be left over the nerves to avoid a postoperative deficit.

WHEN NOT TO DO SURGERY

In general, surgery is the mainstay of treatment of skull base lesions. Surgery is not performed in the following situations:

- *Patient factors:*
  - Poor general condition.
  - Contralateral loss of function:
  - Acoustic neurinoma with contralateral hearing loss.
  - Jugular foramen tumor with contralateral lower cranial nerve paralysis.
- *Tumor factors:*
  - Extensive, slowly growing tumors with little preoperative deficit.
  - Small asymptomatic tumors in old age (e.g., intracanalicular acoustic neurinoma, small asymptomatic cavernous sinus meningioma).

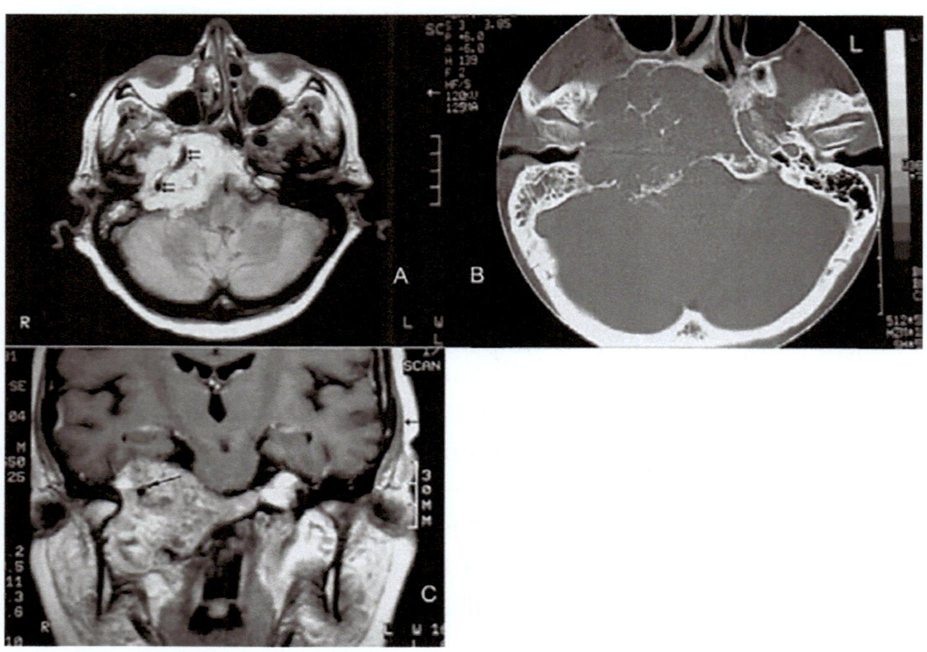

Fig. 16.35 (A) MRI with gadolinium enhancement, axial view of a case of right irradiated recurrent clival chordoma. Note the engulfment of the internal carotid artery *(arrows)* by the tumor. (B) CT scan, axial view, showing the massive extent of the tumor. (C) MRI with gadolinium enhancement, coronal view. The internal carotid artery *(arrow)* can be seen in the middle of the tumor.

General Principles of Embolization in Skull Base Tumors

Introduction

Embolization of skull base tumors is an endovascular neuroradiological technique that consists of selective obliteration of intratumoral vasculature through the superselective microcatheterization of the supplying arteries.

Tumor vascularization is explored using a transfemoral approach, nonionic contrast media, subtraction angiography, and coaxial systems of selective and superselective catheterization.

Embolization is generally performed under general anesthesia because the superselective injection of the branches of the external carotid artery is cumbersome. In addition, the prolonged procedure is not well tolerated by most patients, and any small movement could impair the quality of the images obtained, thus reducing the precision needed for the embolization in difficult situations such as in the anastomosis between the branches of the external and internal carotid arteries.

Embolization is generally performed preoperatively. Rarely, the aim is palliative reduction of tumor size. The objective of palliative reduction of tumor size is to reduce the mass effect of large tumors in patients who present with high surgical risk or in those who refuse surgical removal of the tumor. More frequently, however, the aim of preoperative embolization is to improve the surgical conditions during tumor removal.

Tumor devascularization produces a reduction of tumor size, tumor consistency, and blood loss during surgery. These effects improve the surgical conditions in terms of tumor exposure, manipulation and mobilization of the tumor, and the identification and preservation of important anatomical structures such as arteries and cranial nerves. All of this can help in the reduction of surgery time, increasing the chances of total tumor removal and reducing postoperative complications.

The main instruments utilized in embolization are guiding catheters, microcatheters, microguide wires, and embolizing agents. The 5–6 French caliber guiding catheters are positioned at the proximal portion of a major artery. This can be the external carotid, the internal carotid, or the vertebral arteries. At these sites, the guiding catheters are used to guide and support the microcatheters. The microcatheters are of 2–3 French caliber and are particularly flexible. Thanks to the fine micro guide wires with compressed calibers of 0.010 and 0.014 in., the micro catheters can be inserted within the subtle vessels that branch from major arteries.

Under fluoroscopic control, the embolizing agents are injected through the microcatheters positioned at the required point. The substance most commonly used for embolization in skull base tumors is polyvinyl alcohol. Small particles (50–400 μm) of this material are suspended in an iodinated contrast medium and injected under fluoroscopic control. In special situations, other agents can be used for the embolization. Examples include liquid acrylic substances and platinum micro coils.

From a technical-anatomical point of view, embolization of skull base tumors can be divided into the following classes:

- Embolization of tumors in the territory of the external carotid artery.
- Embolization of tumors in the territory of the ICA.
- Embolization of tumors in the territory of the vertebral artery.

Embolization of the branches of the external carotid artery is performed under free-flow conditions. Brief microcatheterization of the supplying arterioles has taken care of the pressure of injection to avoid reflux and arrest when there is stagnation of flow.

Embolization of the branches of the internal carotid and the vertebral arteries can be performed in two ways. The first involves microcatheterization of the required branches and embolizing them in free-flow conditions as for the external carotid artery. However, when this technique is not possible, the temporary balloon occlusion of the internal carotid or vertebral artery is employed. In this technique, a small nondetachable balloon mounted on a microcatheter is inserted through a 7 French guiding catheter and is used for temporary occlusion of the main artery. Another microcatheter is used for injection of the embolizing agent. After inflating the balloon, fluoroscopy is used to confirm the occlusion of the main artery and the presence of flow solely through the small branches of the tumor. At this point, the embolizing particles are injected under fluoroscopic control. The particles are carried directly to the tumor through the small branches originating from the main vessel. Once the flow to the tumor is reduced, embolization is stopped, 20–30 mL of blood is aspirated through the guiding catheter, and the balloon is deflated, thus opening the main artery.

The result of embolization can be assessed using various neuroradiological techniques. Experience has demonstrated that the immediate assessment of the result of embolization can give a false impression of complete devascularization of the tumor. Such a false impression can be caused by the spasm of the catheterized artery or the presence of collateral circulation that is not yet well-developed, especially in cases of proximal arterial occlusion. MRI with gadolinium contrast enhancement performed 24–48 h after embolization can better assess the degree of devascularization obtained. This can detect the degree of intratumoral necrosis and the reduction of gadolinium enhancement resulting from embolization. Avascular areas are detected as hypointense areas in T1-weighted images with contrast enhancement. CT scans with iodine contrast medium obtained 24 h after embolization can give similar results.

The ideal interval between a successful embolization and the surgical intervention in our experience is in the range of 48–72 h. Other authors have reported good results in terms of devascularization for periods as long as 7 days.

The potential complications of tumor embolization depend mainly on the arterial zone being embolized. At least theoretically, embolization of the internal carotid and vertebral arteries is associated with higher risk of major complications than embolization of the external carotid artery. The most serious complications are of ischemic type. These can manifest as ischemia of the central nervous system or as malfunction of cranial nerves. Generally, these incidents are transient. The incidence of complications is in the range of 0%–2% depending on the series. The risk of complications is related more to the angiographic technique employed and to the presence of atherosclerosis of the carotid trunks than to the embolization itself.

In the majority of cases, complications during embolization result from the application of inappropriate techniques or from lack of knowledge about the presence of collateral anastomosis between branches of the external carotid artery and the intracranial circulation—the so-called dangerous anastomosis.

The principal anastomotic circles found during embolization are as follows:

- Anastomosis between the neuromeningeal and musculospinal branches of the ascending pharyngeal and vertebral arteries.
- Anastomosis between the pharyngeal and neuromeningeal branches of the ascending pharyngeal and internal carotid arteries.
- Anastomosis between the internal maxillary and internal carotid arteries.
- Anastomosis between the medial meningeal and ophthalmic arteries.
- Anastomosis between the occipital and vertebral arteries.

These anastomotic circuits tend to be hidden during the initial phase of embolization when the blood flow to the tumor is still high. They tend to manifest as the embolization reduces the flow to the tumor. Alternatively, the vascularization of these circuits is sometimes dependent on the pressure of injection of the contrast medium. The presence of these dangerous collateral circuits is not an absolute contraindication for embolization, but it does require extra vigilance. Careful occlusion can be performed using platinum coils.

Complications related to the loss of cranial nerve function arise because the branches that supply the tumor are the same branches that supply the injured cranial nerve. The rate of such complications is in the range of 1%–3%, and they seem to affect the facial nerve more. To reduce the risk to cranial nerves to a minimum, we advise using particles larger than 150 μm.

There are some reports in the literature of cutaneous ischemia that has led to retardation of the healing of the surgical wound. In some cases, repair of the wound has required a vascular flap.

Finally, there are some reports of intratumoral or peritumoral hemorrhage. These are most probably due to the injection of the contrast medium at high pressure during the final phases of the procedure. In the majority of cases, these incidences are asymptomatic; in rare cases, however, they present with transient ischemic changes. Even more rarely, they present with permanent neurological deficits. Delicate injection of the contrast medium should avoid the majority of such occurrences (Fig. 16.36).

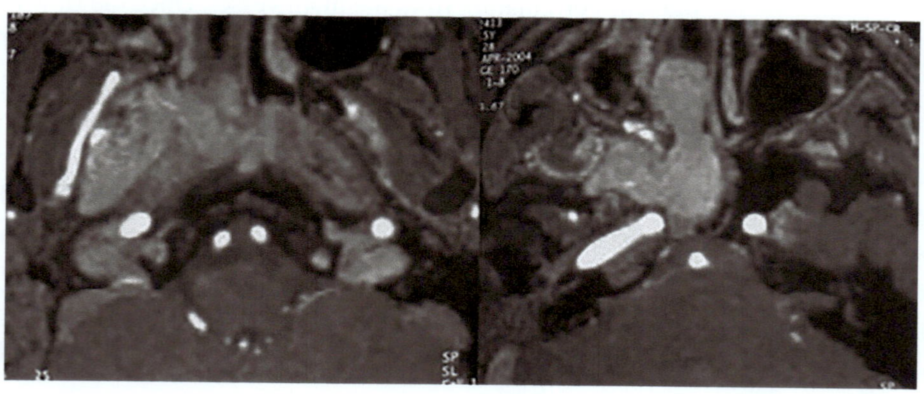

Fig. 16.36 Axial sections, time of flight (TOF) acquisition, show infiltration of the nasal cavity, pterygopalatine, and infratemporal fossae and the relationship of the tumoral mass to the internal carotid and internal maxillary arteries.

Embolization of Juvenile Nasopharyngeal Angiofibroma

Nasopharyngeal angiofibroma has a characteristic angiographic appearance, the special features of which are supplying arteries of normal or slightly enlarged diameters, an intense tumor blush that is slightly inhomogeneous, and the absence of arteriovenous shunts.

The supplying arteries (Figs. 16.37 and 16.38) are the ipsilateral internal maxillary artery and the superior pharyngeal branch of the ascending pharyngeal artery. Other possible supplying arteries include the ICA through the inferolateral trunk, the mandibular artery, and, in cases of orbital extension, the ophthalmic artery. When the tumor crosses the midline, it can derive blood supply from the contralateral external and sometimes the internal carotid arteries.

Fisch has classified these tumors according to their blood supply:

- **Class I**: The blood supply is derived from the internal maxillary artery, through the sphenopalatine branch for the nasal component and through the pterigovaginal branch for the nasopharyngeal component of the tumor.
- **Class II and class III**: Lateral extension to the pterygopalatine and infratemporal fossa derives its blood supply from the proximal internal maxillary artery mainly through the accessory meningeal artery. Further nasopharyngeal extension is supplied by the superior pharyngeal branch of the ascending pharyngeal artery. Orbital and ethmoid extensions are supplied by the ophthalmic artery, while parasellar intracranial extensions and extensions into the sphenoid sinus are supplied by the ICA through the inferolateral trunk.
- **Class IV**: Intracranial extradural, parasellar, and intracavernous extensions are supplied by the inferolateral trunk of the ICA and by the external carotid artery via the internal maxillary artery, the accessory meningeal artery, and the cavernous branch of the ascending pharyngeal artery. The rare intracranial intradural extensions are supplied by the cerebral branches of the middle cerebral artery. Although the arterial systems of the internal maxillary artery and the ICA are generally in hemodynamic equilibrium, one system can be dominant. These situations determine the

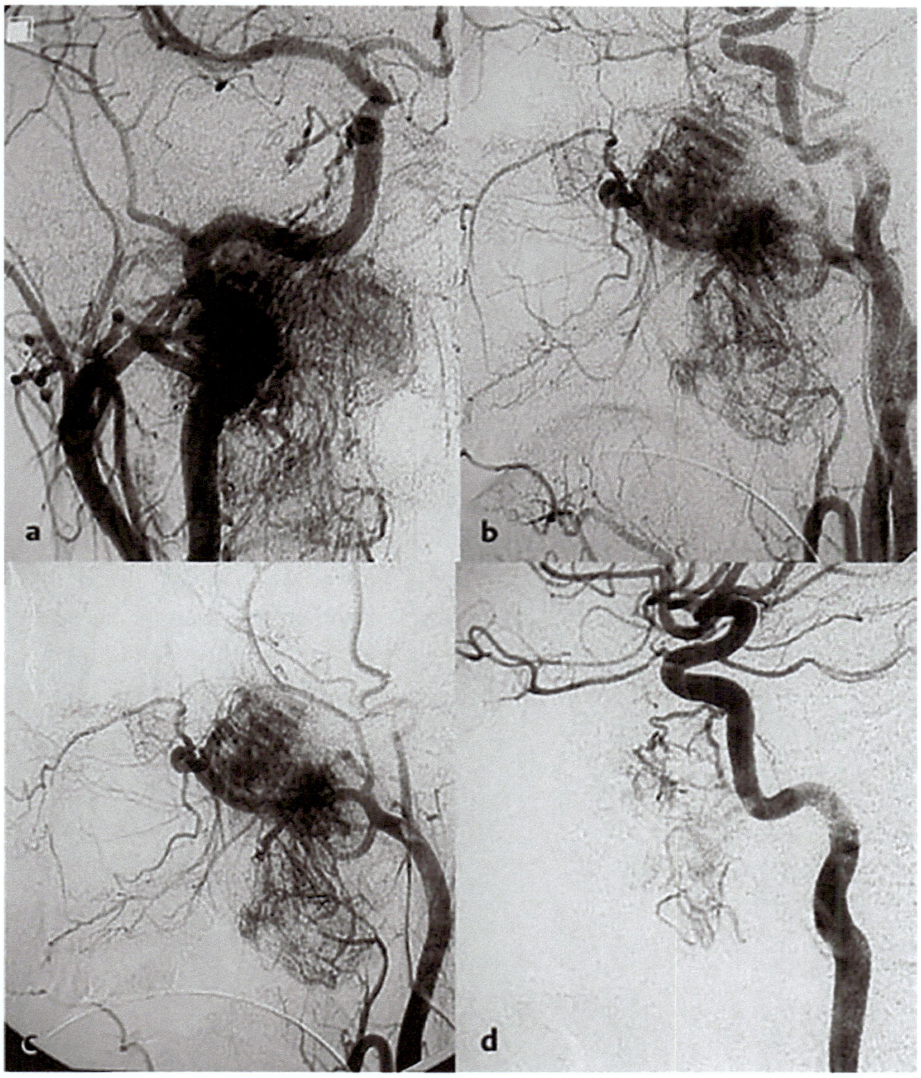

Fig. 16.37 (A–D) Common carotid injection in anteroposterior (AP) and laterolateral (LL) views; external and internal carotid injection in LL; a hypervascular mass with a dominant vascular supply from the external carotid and a small contribution from the inferolateral trunk of the internal carotid artery.

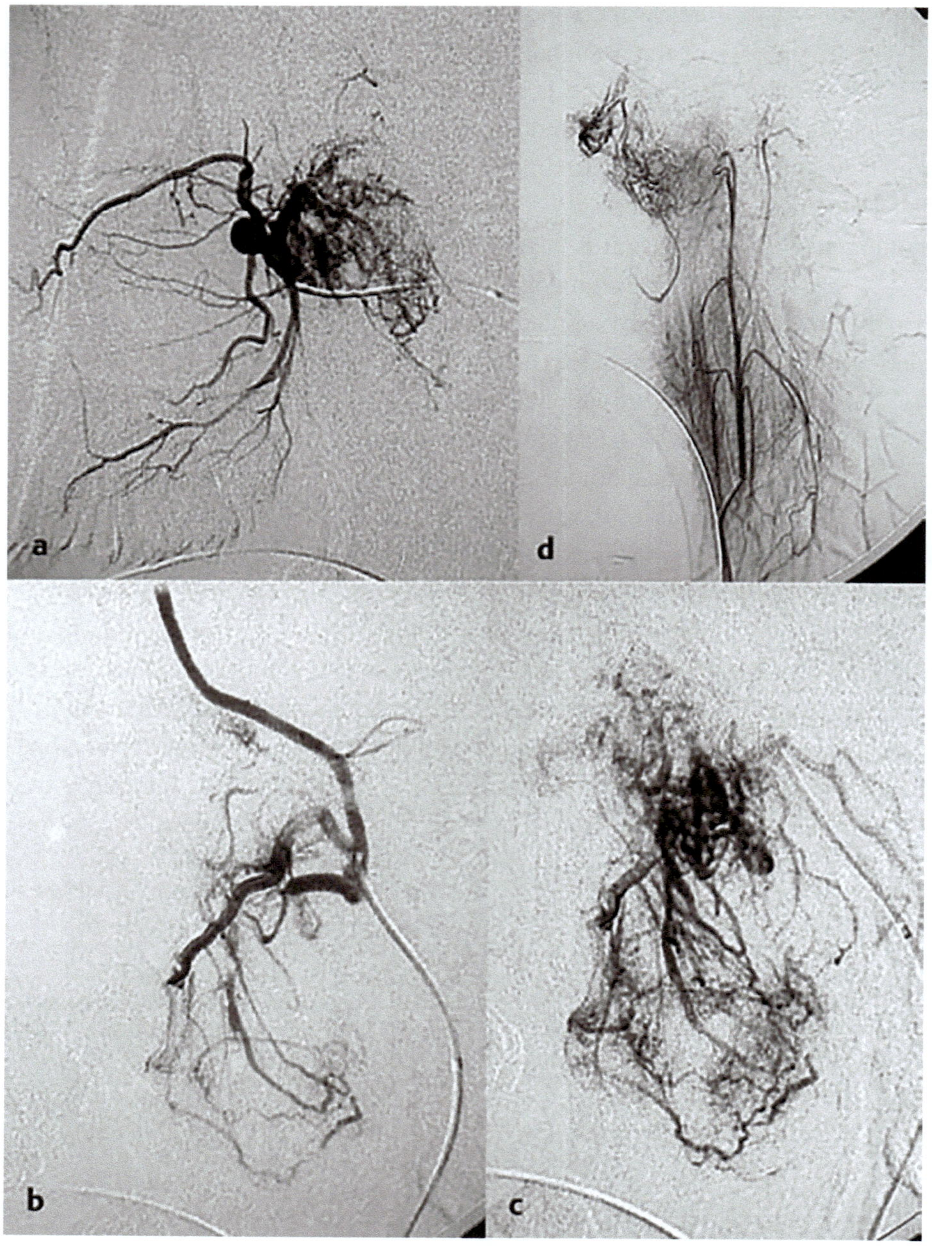

Fig. 16.38 Superselective injection of the internal maxillary (A) and accessory (B, C) meningeal arteries, which supply almost the entire mass; a small contribution from the ascending pharyngeal is demonstrated (D); there was no supply from the facial-ascending palatine artery (not shown).

strategy of embolization in class III and class IV tumors. Obviously, the embolization (Fig. 16.39) is less risky when the internal maxillary system is the dominant one. Otherwise, special endovascular tactics are required.

When the internal carotid system is dominant, the appropriate tactic is microcatheterization of the inferolateral trunk. If this is not feasible, an alternative technique is nonselective embolization of the inferolateral trunk accompanied by short temporary occlusion of the ICA. When equilibrium exists between the two systems, it is advisable to perform the embolization through the accessory meningeal artery while temporarily occluding the ICA with the aim of increasing the flow within the external carotid artery.

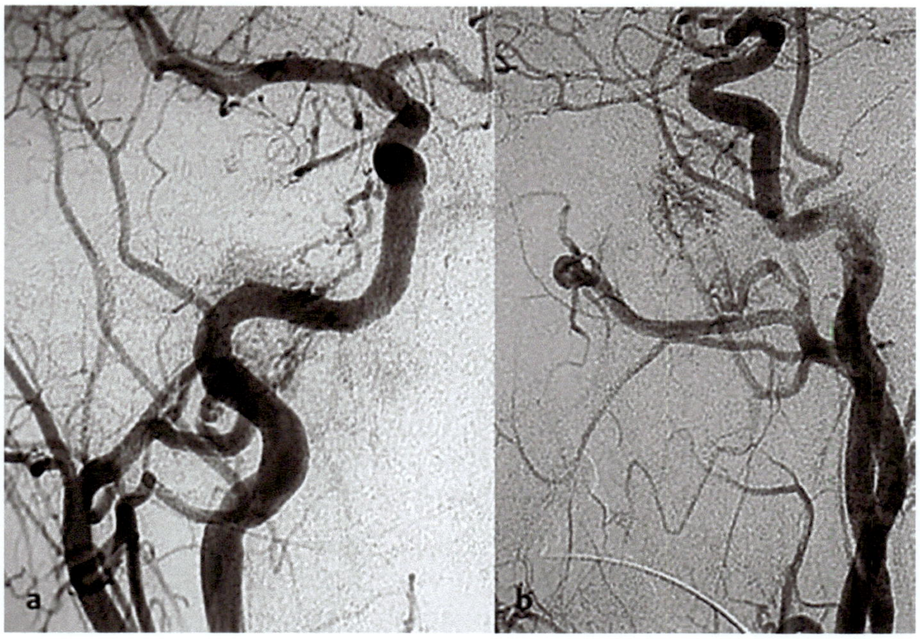

Fig. 16.39 Common carotid injection in AP (A) and LL (B) views at the end of the embolization, showing minimal residual tumoral filling from the internal carotid artery branches.

Embolization of Tympanojugular Paragangliomas

The angiographic appearance of paragangliomas is characterized by moderately dilated supplying arteries, intense and homogeneous tumor blush, the presence of arteriovenous shunts, and dilated venous drainage. The density of blood supply is directly related to the size and extension of the tumor and thus can be included in the tumor classification system (Figs. 16.40 and 16.41).

In relation to Fisch's classification, several types of vascularization can be distinguished:

- **Class A tumors**: The blood supply is mainly derived from the tympanic branches, the most important of which is the inferolateral tympanic branch of the ascending pharyngeal artery.
- **Class B tumors**: In addition to the inferolateral tympanic artery, the tumor derives blood supply from mastoid branches, the most important of these being the lateral stylomastoid artery originating from either the occipital or the posterior auricular arteries.
- **Class C tumors**: In addition to the supplying arteries already mentioned, the major blood supply is derived from the ascending pharyngeal artery. Depending on the degree of involvement of the internal carotid canal, additional blood supply from the ICA can also be present.
- **Class D tumors**: The intracranial extradural extensions can be supplied by the meningeal branches of the ascending pharyngeal, occipital, and middle meningeal arteries; the clival meningeal branches of the ICA; and the meningeal branches of the vertebral artery. Intradural extensions are supplied by the posteroinferior and anteroinferior cerebellar arteries.

Angiographically, it is possible to demonstrate multicompartmental structure in class C and D tumors. This structure was first pointed out by Morte and later confirmed by other authors. There are four compartments: inferomedial (Fig. 16.42), posterolateral (Figs. 16.43 and 16.44), anterior (Fig. 16.45), and superior.

The inferomedial compartment includes the hypotympanum, the jugular foramen with the jugular bulb, and the inferior petrosal sinus. There can also be an extradural or intradural intracranial extension. In the case of class C tumors, this compartment derives its blood supply from the ascending pharyngeal artery. For class D tumors, however, additional blood supply is provided by the meningeal branches of the internal carotid and vertebral arteries for the intradural extensions and by the meningeal branches of the occipital artery for extradural extensions.

The **posterolateral compartment** includes the posterior tympanic cavity, the mastoid, and the sigmoid sinus. There can also be an intracranial extradural extension. In class C tumors, this compartment is supplied by the stylomastoid branch of the occipital artery and by the posterior auricular artery. Intracranial extradural extensions of this compartment are supplied by the meningeal branches of the occipital artery.

The **anterior compartment** includes the anterior tympanic cavity, the internal carotid canal, and the cavernous sinus. In class C tumors, this compartment is supplied by the anterior tympanic branch of the internal maxillary artery and the caroticotympanic branch of the ICA. If the tumor extends to the foramen lacerum and the cavernous sinus, the cavernous branches of the ascending pharyngeal and internal carotid arteries provide additional blood supply. No class D has been observed for the anterior compartment.

The **superior compartment** includes the epitympanum and the supralabyrinthine region of the temporal bone. In class C tumors, this compartment is supplied by the superior tympanic branch of the middle meningeal artery and the accessory meningeal artery. The extradural intracranial extensions of this compartment are supplied by the middle meningeal artery and the inferolateral branch of the ICA.

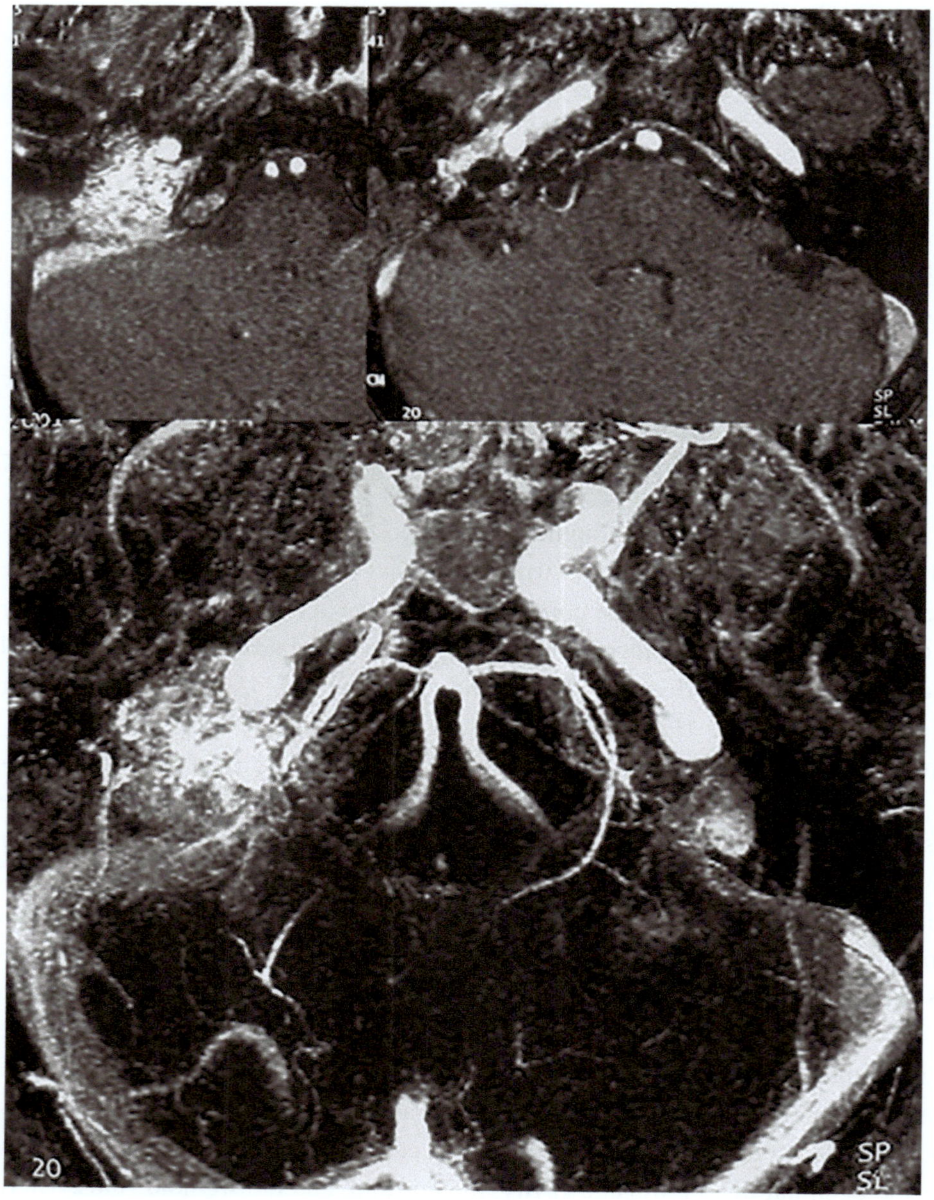

Fig. 16.40 Class C2 paraganglioma; axial partitions and MIP reformation of a TOF acquisition, showing involvement of the jugular foramen and the vertical portion of the carotid canal.

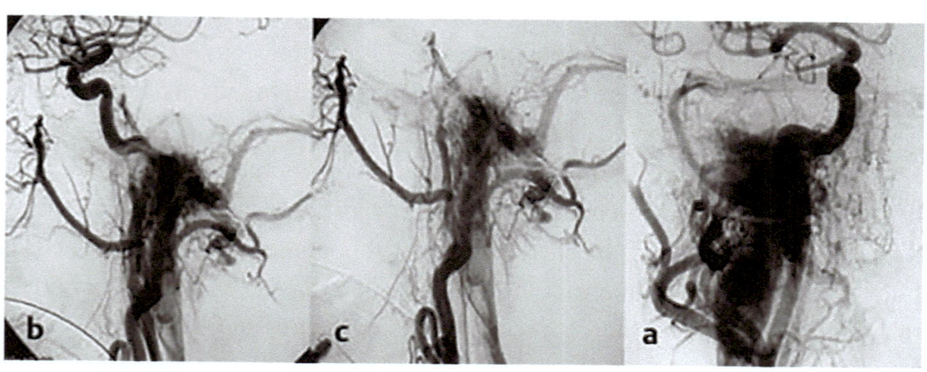

Fig. 16.41 (Same patient as in Fig. 16.40.) Common carotid injection in AP (A) and LL (B) views and external carotid injection in LL view (C), showing the tumoral blush.

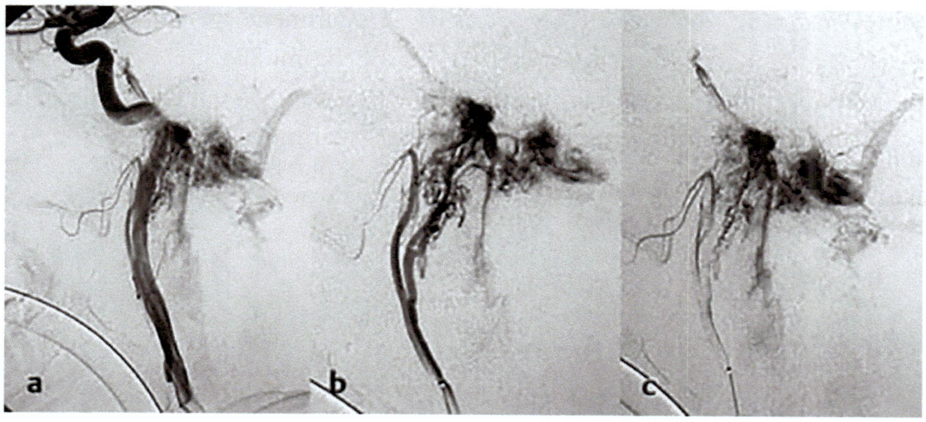

Fig. 16.42 Internal carotid artery injection showing anomalous origin (A) of the ascending pharyngeal artery; micro catheterization of the ascending pharyngeal branch showing the inferomedial compartment in early (B) and late (C) angiographic phases.

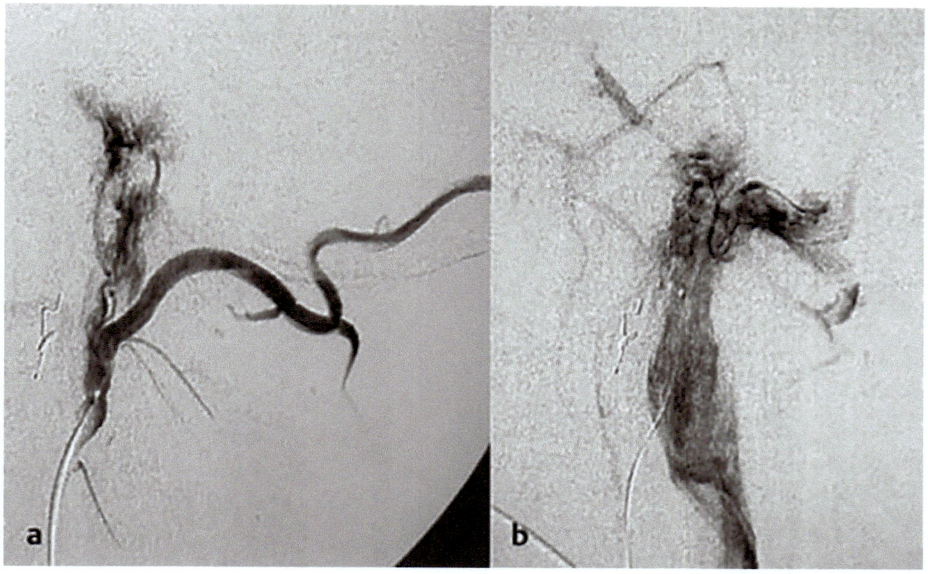

Fig. 16.43 Occipital artery injection, lateral views, showing the posterolateral compartment in early (A) and late (B) angiographic phases.

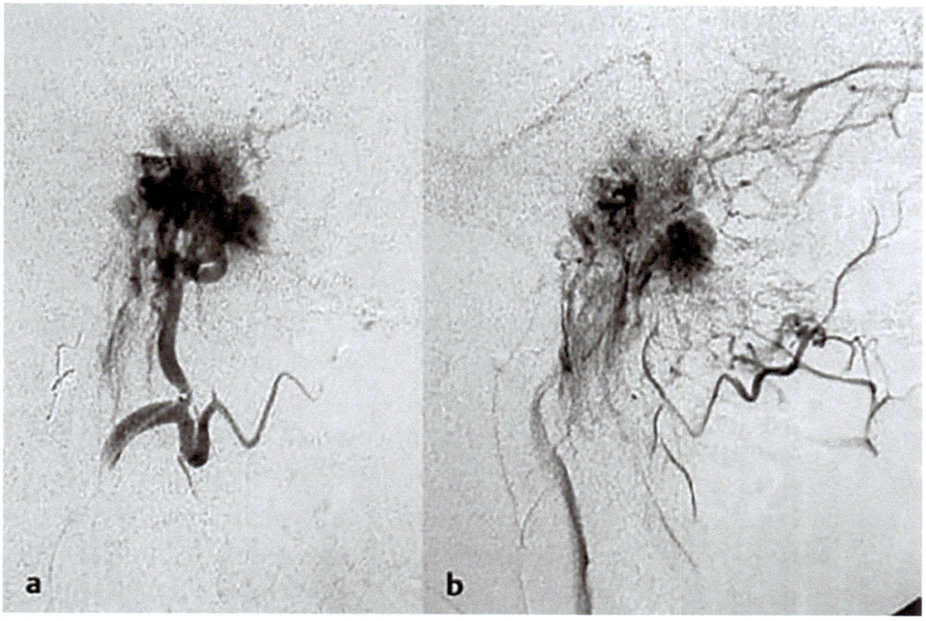

Fig. 16.44 Posterior auricular artery, lateral view, showing the posterolateral compartment in early (A) and late (B) angiographic phases.

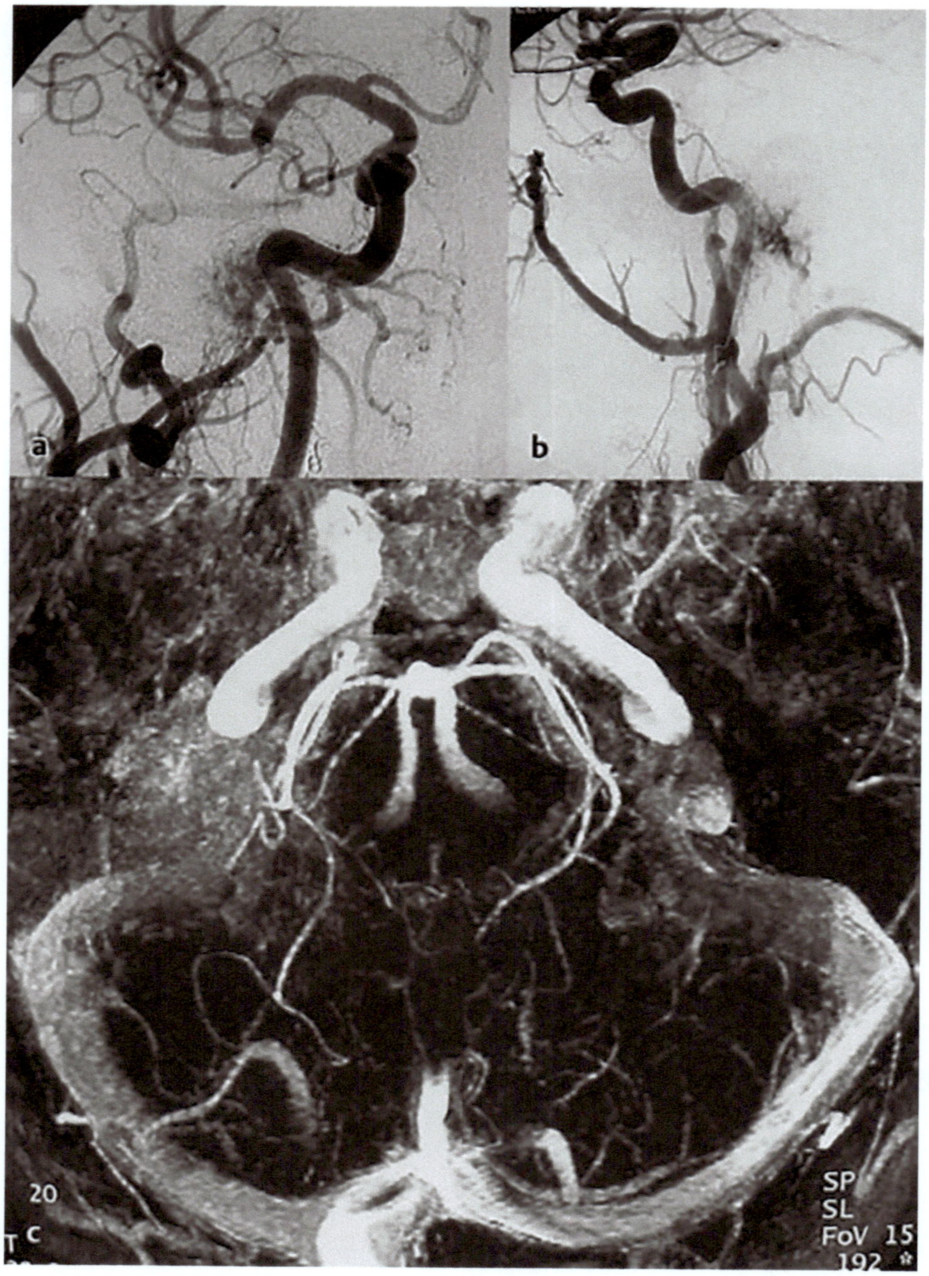

Fig. 16.45 Common carotid injection after embolization, with persistence of a small anterior compartment (A, B) and MR angiography (C) showing marked reduction of the vascularization.

Embolization of Meningiomas of the Skull Base

The angiographic appearance of meningiomas is characterized by a mass with different grades of hypervascularization, well delineated during all the angiographic phases and homogeneously and persistently injected in the venous phase. Rarely, arteriovenous shunting can be seen.

Both the arterial supply and the venous drainage can be dural or mixed dural and pial. When mixed, they can be balanced or one system can predominate. In smaller meningiomas, the blood supply is exclusively dural. As meningiomas grow in size, they tend to pick up pial vessels. Embolization is much more effective when the dural vascular component is predominant.

The number and type of supplying blood vessels are related to the site of origin and the size of the tumor. However, there is a wide individual variation between tumors apparently having the same site of origin. This variation is related to the hemodynamic balance between the meningeal branches of the external carotid artery on one side and the meningeal branches of the internal carotid and the vertebral arteries on the other. In some cases, these two systems can be equally balanced; in other cases, one of the systems dominates.

The meningiomas that should be considered when dealing with the approaches of the lateral skull base are those of the parasellar region, including clinoid and cavernous sinus meningiomas; posterior cranial fossa meningiomas, including meningiomas of the foramen magnum (Figs. 16.46–16.53); meningiomas of the posterior surface of the temporal bone; and petroclival meningiomas (Figs. 16.54–16.57).

The vascularization of cavernous and sphenoidal meningiomas is mostly complex, and an equilibrium between three vascular groups can be found with dense anastomosis between them. The three groups are the basal meningeal branches of the middle meningeal artery, the cavernous meningeal branches of the ICA, and the recurrent meningeal branches of the intraorbital ophthalmic system.

The middle meningeal artery usually originates from the internal maxillary artery. After entering the skull through the foramen spinosum, it gives off a cavernous branch that anastomoses with the ascending pharyngeal, the accessory meningeal arteries, and the inferolateral trunk. In addition, it gives off a petrosal branch for the geniculate ganglion and the dura mater of the cerebellopontine angle.

The intracavernous portion of the ICA gives off the meningohypophyseal trunk and the inferolateral trunk. From the meningohypophyseal trunk, there originate numerous meningeal branches, the posterior inferior hypophyseal artery, the lateral artery of the clivus, the recurrent artery to the foramen lacerum, and the lateral artery of the trigeminal ganglion. From the inferolateral trunk, there originates a superior branch for the tentorium cerebellar, an anteromedial branch behind the orbit and the superior orbital fissure, an anterolateral branch for the foramen rotundum, and a posterior branch for the foramen ovale.

The ophthalmic system affords recurrent neuromeningeal branches that originate from the intraorbital portion of the ophthalmic artery.

The vascularization of posterior fossa meningiomas is generally in equilibrium between the external carotid, vertebral, and cerebellar arteries.

The external carotid artery participates in the vascularization through the ascending pharyngeal and the occipital arteries. The ascending pharyngeal artery affords a neuromeningeal branch that sometimes divides into hypoglossal and jugular branches behind the corresponding foramina. The occipital artery has two branches: the mastoid and the falx cerebelli arteries.

The vertebral artery contributes only to the dura of the posterior cranial fossa through its posterior and anterior meningeal branches.

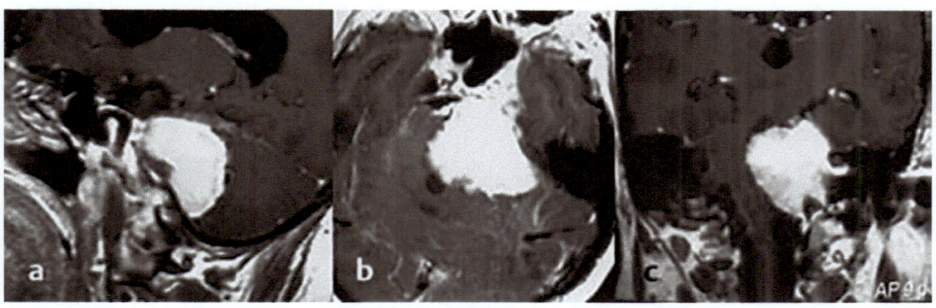

Fig. 16.46 Meningioma of the posterior surface of the temporal bone. T1-weighted MRI, gadolinium-enhanced, sagittal (A), axial (B), and coronal (C) sections of a petroclival meningioma with extension into the internal auditory canal and Meckel's cave.

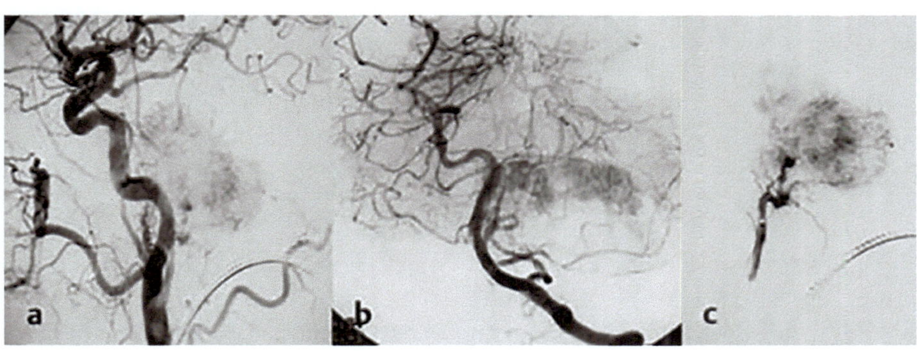

Fig. 16.47 (Same patient as in Fig. 16.46.) Angiography of the common (A) carotid and vertebral (B) arteries; superselective injection of the ascending pharyngeal artery (C).

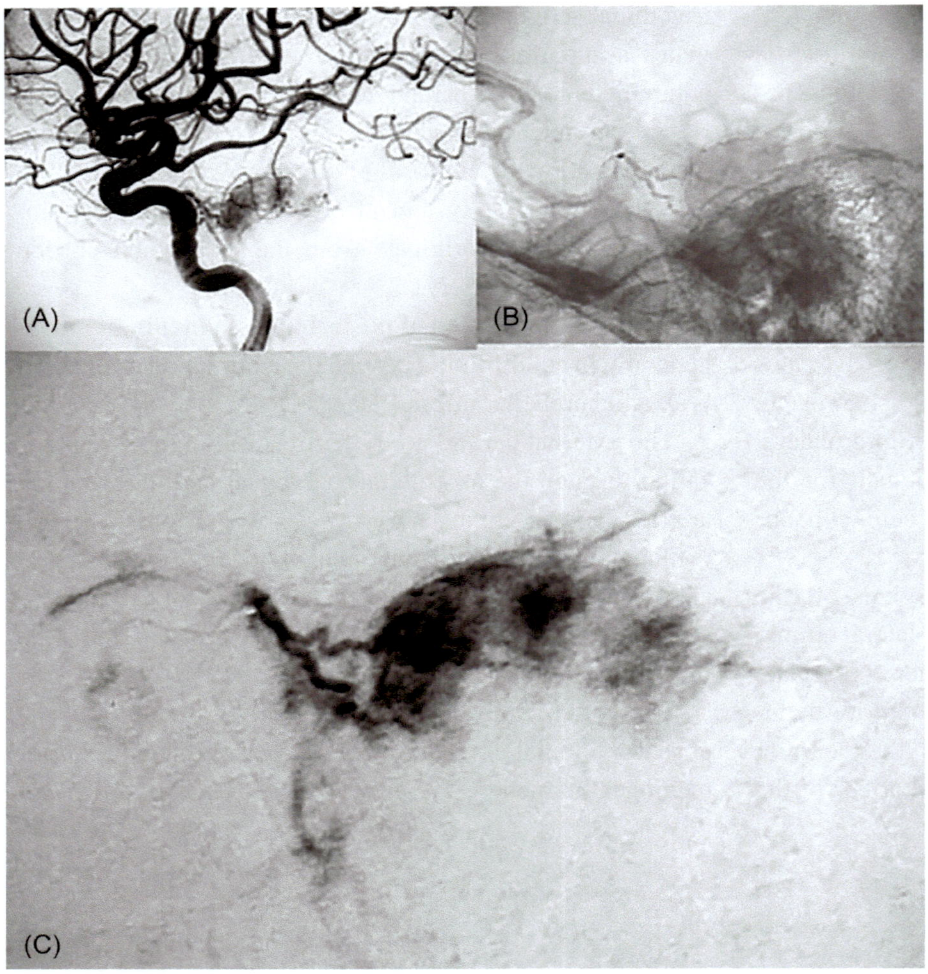

Fig. 16.48 (Same patient as in Fig. 16.46.) Angiography of the internal (A) carotid artery and superselective injection of the meningohypophyseal trunk (B, C).

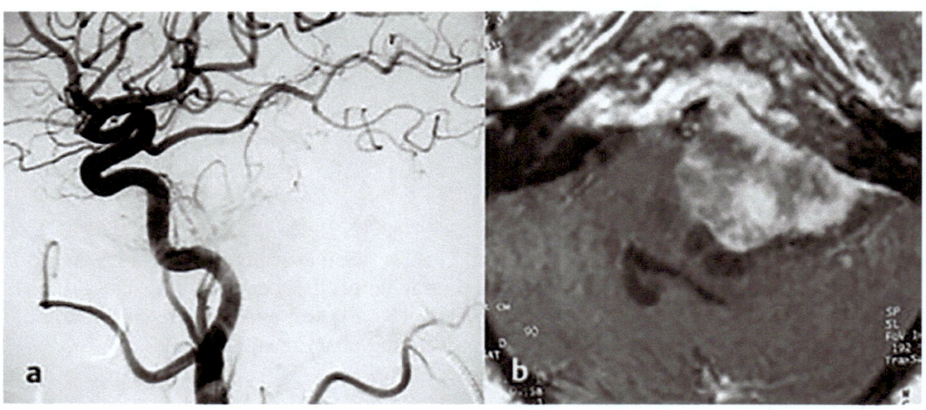

Fig. 16.49 (Same patient as in Fig. 16.46.) Common carotid angiography at the end (A) and axial contrast-enhanced MRI 24 h after embolization (B).

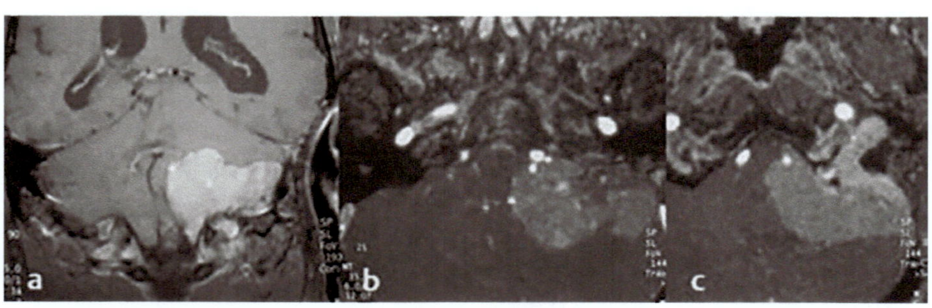

Fig. 16.50 Meningioma of the foramen magnum. T1-weighted coronal MRI (A) of a foramen magnum meningioma and coronal TOF partitions (B, C) showing encasement of the left vertebral artery and posterior inferior cerebellar artery.

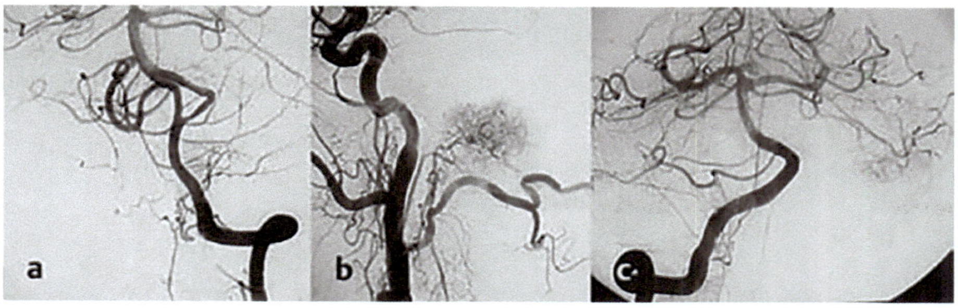

Fig. 16.51 (Same patient as in Fig. 16.50.) Left vertebral artery showing supply from the anterior meningeal artery (A); common carotid artery showing supply from a branch of the external carotid (B); right dominant vertebral artery showing supply from the superior cerebellar artery (C).

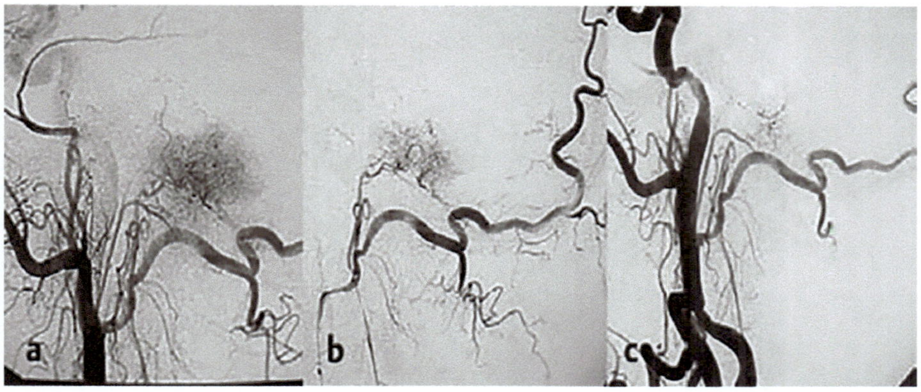

Fig. 16.52 (Same patient as in Fig. 16.50.) External carotid artery (A) and occipital artery (B) injection showing supply from the mastoid meningeal branch; common carotid artery (C) after embolization.

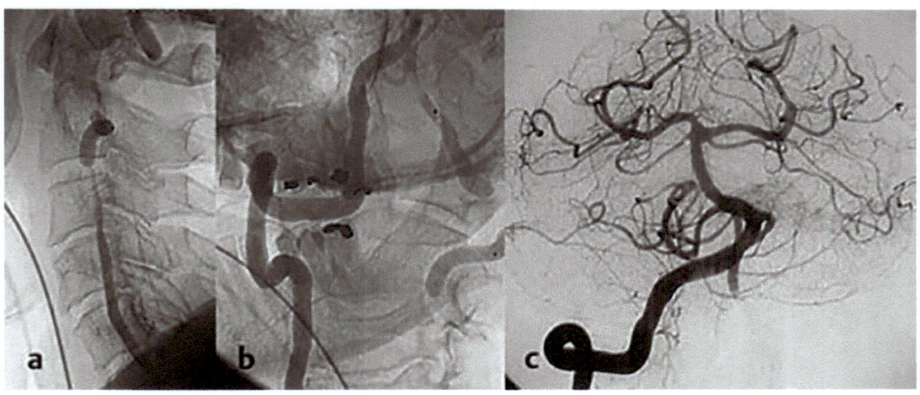

Fig. 16.53 (Same patient as in Fig. 16.50.) Permanent balloon occlusion of the left vertebral artery with two latex balloons detached in V3 and in V4 segments, before the origin of the posterior inferior cerebellar artery (A, B); right vertebral artery injection showing compensation of the left posterior inferior cerebellar artery (C).

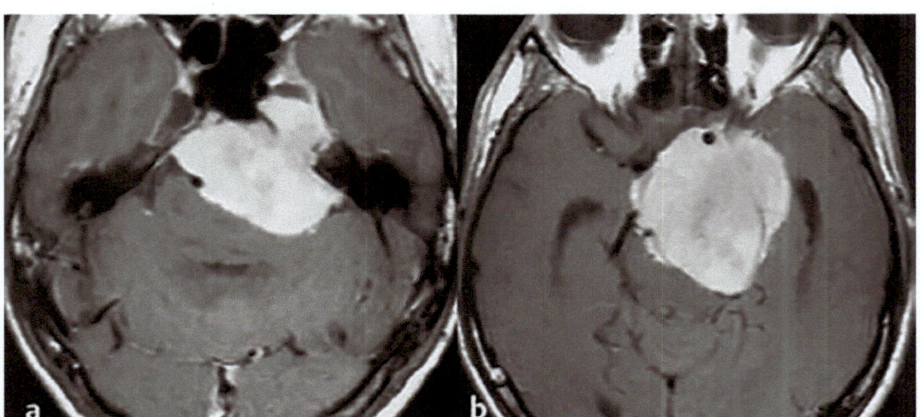

Fig. 16.54 Petroclival meningioma. Axial T1-weighted MRI of petroclival meningioma showing encasement of the basilar (A) and internal carotid (B) arteries.

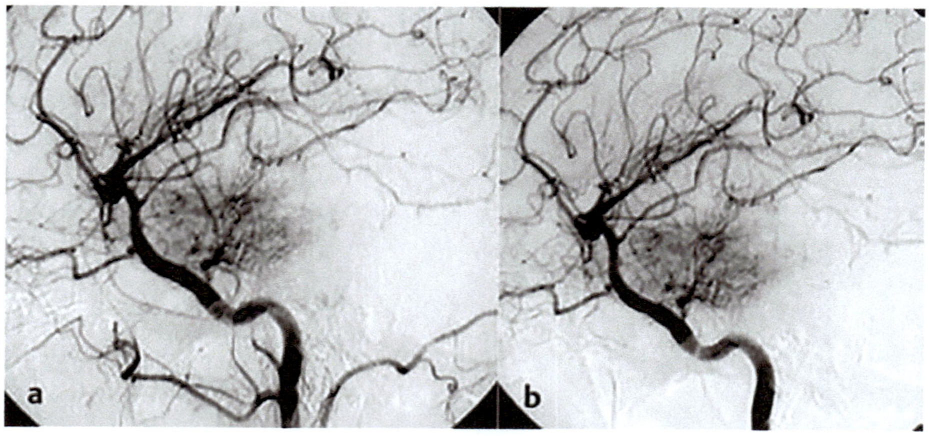

Fig. 16.55 (Same patient as in Fig. 16.54.) Common (A) and internal (B) carotid artery injection showing supply from the meningohypophyseal trunk.

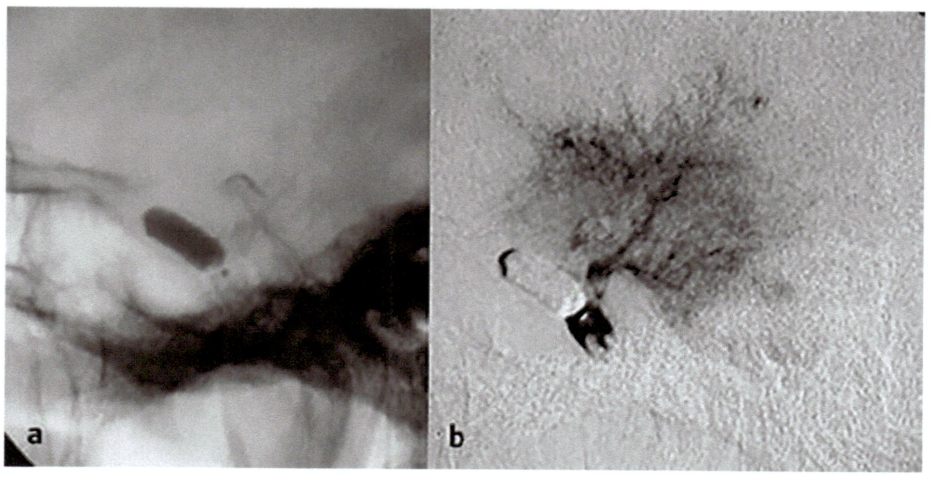

Fig. 16.56 (Same patient as in Fig. 16.54.) Temporary balloon occlusion of the internal carotid artery (A) and embolization of the meningohypophyseal trunk with a microcatheter inserted into the internal carotid lumen (B).

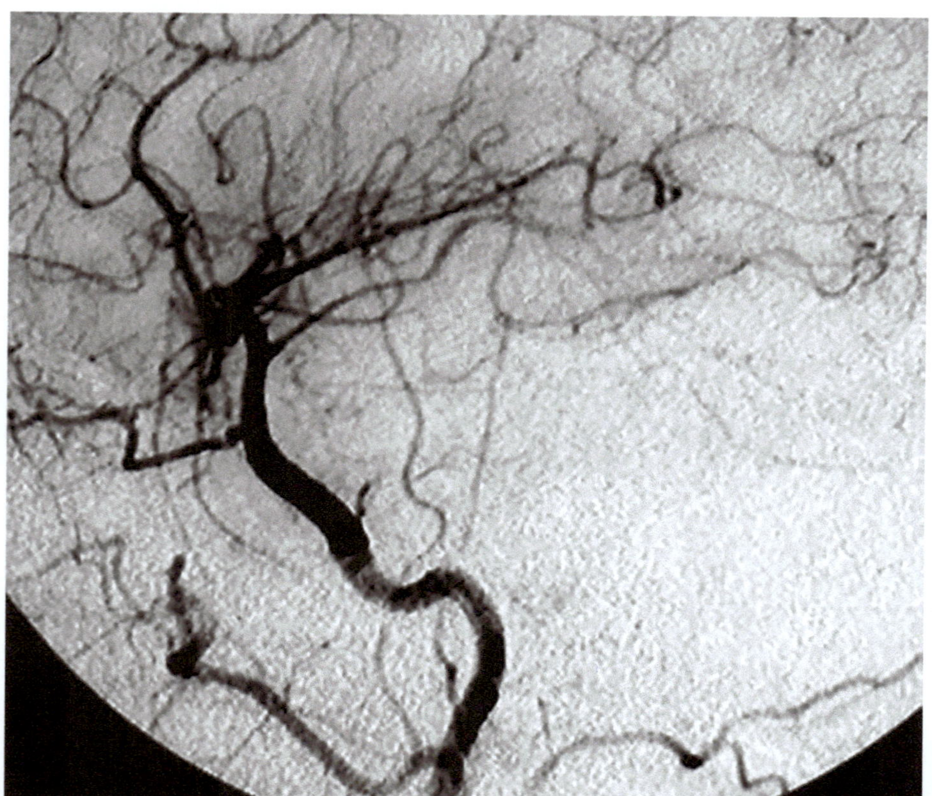

Fig. 16.57 (Same patient as in Fig. 16.54.) Common carotid artery injection after embolization.

Management of the ICA in Skull Base Surgery

Introduction

In the recent past, many skull base lesions were considered inoperable because of their extensive nature and deep location. Anatomical boundaries were among the important factors that hindered access to tumors or prevented radical removal.

The most important of these were the facial nerve and the ICA. With advances in skull base surgery, the facial nerve could be tackled and successfully mobilized both anteriorly (infratemporal fossa approach) and posteriorly (transcochlear approach), thus opening the way for access to tumors. The ICA, however, posed more difficulty. Thanks to advances in imaging modalities and recent techniques of interventional neuroradiology, management of tumors involving the ICA has become safer and more radical. This has led to the removal of a large variety of tumors that were previously considered inoperable.

Surgical Anatomy

1. The styloid process separates the external carotid (located laterally) from the cervical ICA (located medially). Control of the distal part of the ICA at the level of the upper neck, before its ingress into the carotid canal, is hampered by three structures: the styloid process and attached structures, the posterior belly of the digastric muscle, and the extratemporal facial nerve. At its ingress into the skull base, the ICA curves slightly medially. The glossopharyngeal nerve passes lateral to the artery; it lies deep and adherent to tough fibrous tissue surrounding the artery at this level.

2. The petrous vertical segment of the ICA is surrounded by a venous plexus and a thick fibrous tissue layer. It is separated from the jugular bulb by the jugulocarotid spine (Fig. 16.58).

3. The artery ascends for about 14 mm. It runs anteromedial to the cochlea and then runs to form the petrous horizontal segment, which runs in an anteromedial direction. The genu lies medial to the eustachian tube and is also related to the geniculate ganglion (Fig. 16.59).

4. The petrous horizontal segment of the ICA has a thin fibrous layer. It is crossed laterally by the eustachian tube, the middle meningeal artery, and the mandibular branch of the trigeminal nerve (Fig. 16.60).

5. The greater superficial petrosal nerve is related to the superior aspect of the artery and runs parallel to its course. The Gasserian ganglion covers the superior aspect of the anteriormost segment of the artery. In many cases, the petrous horizontal segment of the carotid canal of the ICA is dehiscent, with no bone intervening between it and the Gasserian ganglion (Fig. 16.61).

6. At the level of the foramen lacerum, the ICA turns 90 degrees anterosuperiorly to assume a vertical course (lacerum segment). At this point, the artery enters the cavernous sinus. The abducent nerve enters Dorello's canal, after which it crosses the lateral aspect of the ICA at this step (Fig. 16.62).

7. The cavernous segment of the ICA can be divided into different subsegments. The posterior vertical segment is short and turns forward to form the horizontal segment. At the anterior part of the cavernous sinus, the artery makes a smooth curve, running medially and superiorly to form the anterior vertical segment, which leaves the cavernous sinus to enter the subarachnoid space.

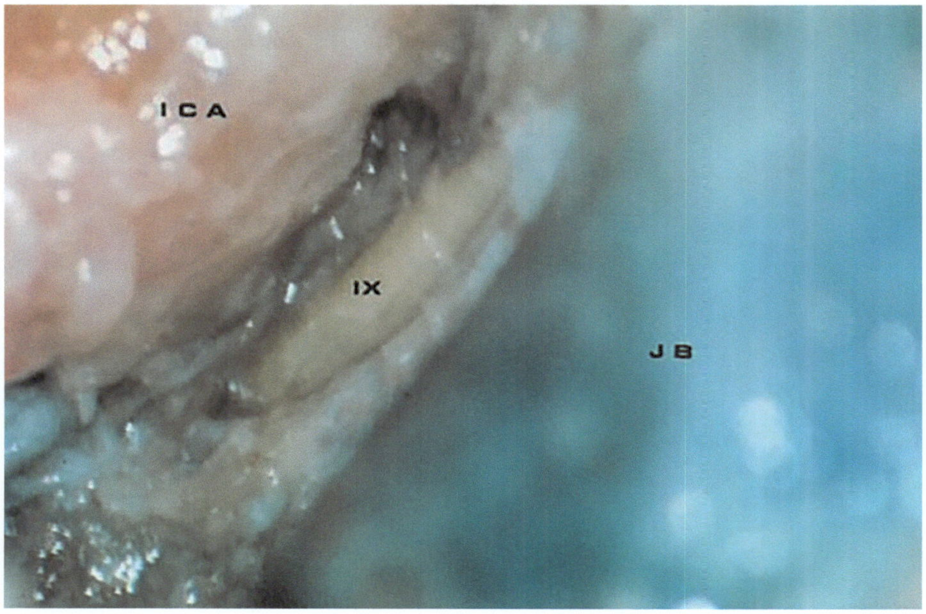

Fig. 16.58 The jugulocarotid spine, which separates the internal carotid artery (ICA) from the internal jugular vein, can be seen. Partial drilling of the spine reveals the glossopharyngeal nerve (IX) before its exit from the skull base.

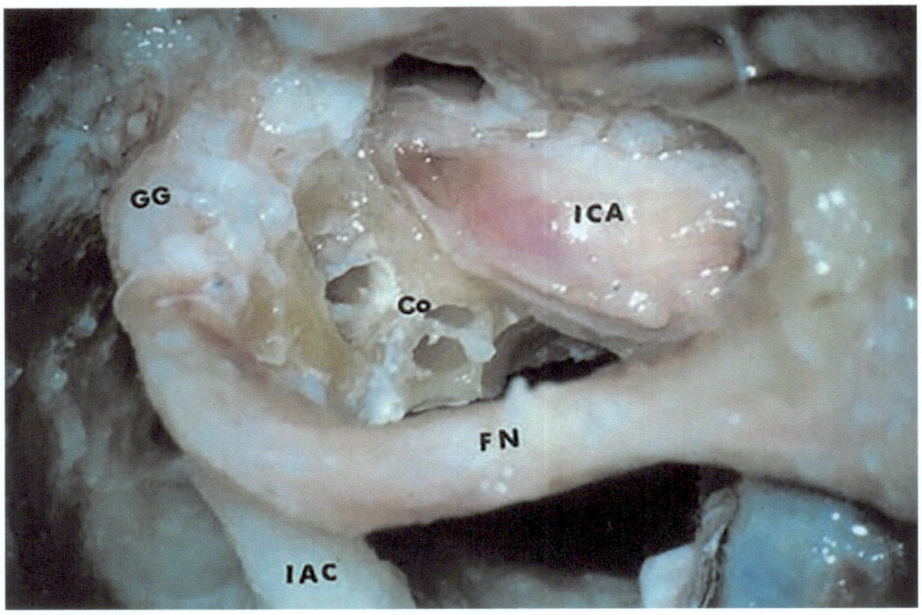

Fig. 16.59 The cochlea has been partially drilled to show its relation to the internal carotid artery (ICA). Note the close relation between the geniculate ganglion (GG) and the genu of the artery. The orifice of the eustachian tube lies lateral to the genu.

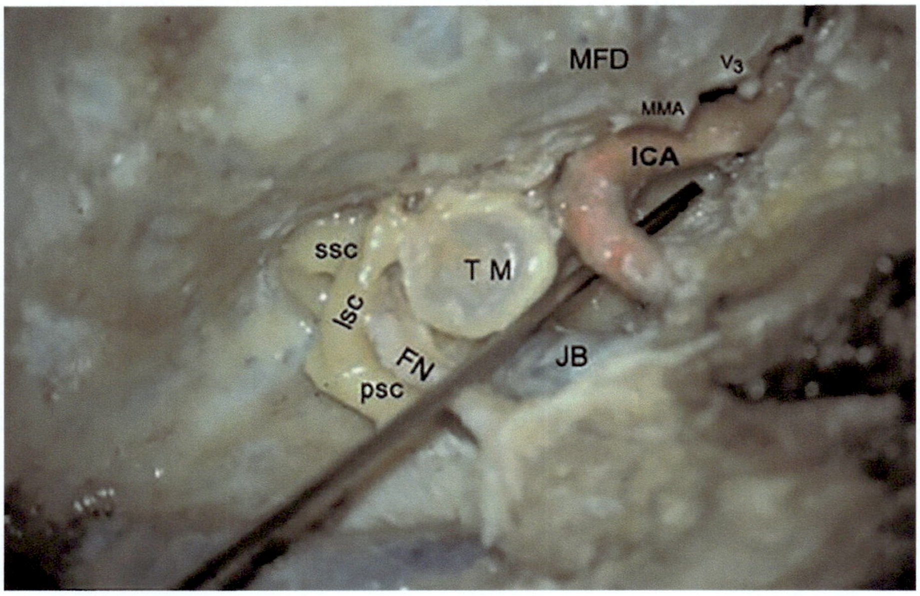

Fig. 16.60 A dissection of the skull base showing the vertical and horizontal segments of the internal carotid artery (ICA) in relation to the tympanic membrane (TM), the facial nerve (FN), and the jugular bulb (JB). The horizontal segment of the artery has been exposed after sacrificing the eustachian tube, the middle meningeal artery (MMA), and the mandibular branch of the trigeminal nerve (V3). An instrument has been directed medial to the artery toward the petrous apex.

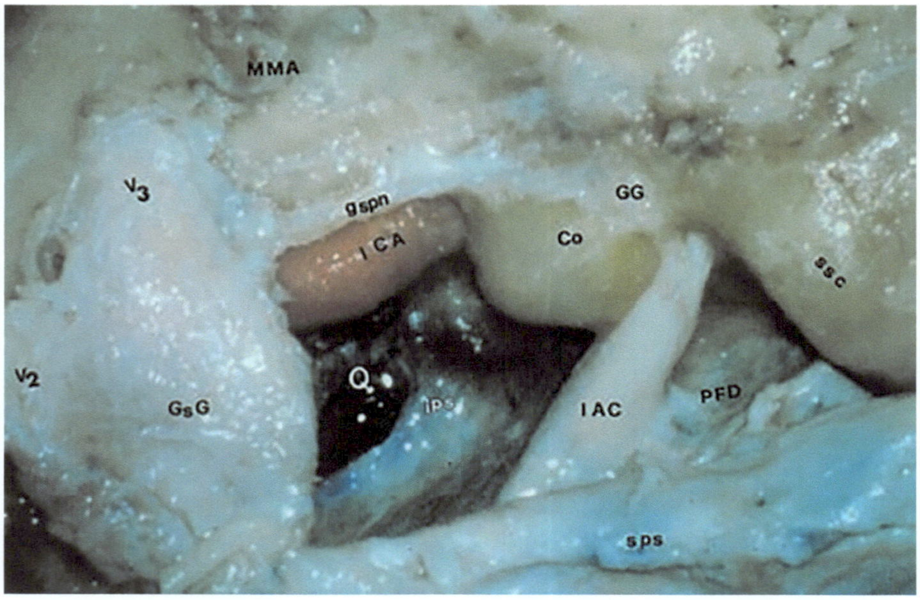

Fig. 16.61 A dissection from the superior aspect of the temporal bone. The horizontal portion of the internal carotid artery (ICA) is identified, and its anterior medial part is covered by the Gasserian ganglion (GsG). The greater superficial petrosal (gspn) nerve runs on the superolateral aspect of the artery.

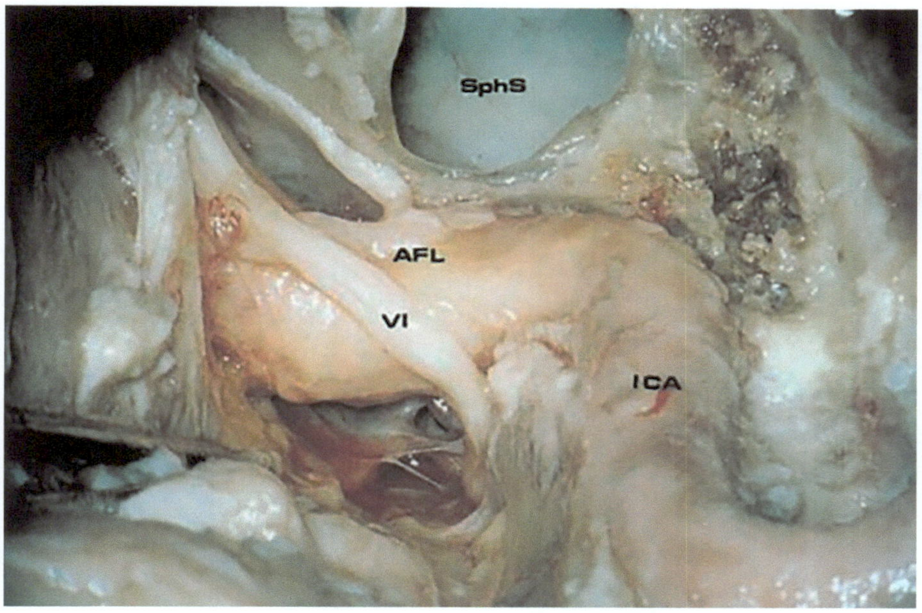

Fig. 16.62 The horizontal segment of the internal carotid artery (ICA) is completely uncovered. Note the vertical turn of the artery at the foramen lacerum. The lateral aspect of the artery is crossed by the abducent nerve (VI) at the entrance of both structures into the cavernous sinus. The relation of the artery to the sphenoid sinus (SphS) can be appreciated.

Pathologies

Most skull base tumors can involve the ICA. The most common pathologies in this context are as follows:

1. Type C2–C4 tympanojugular paragangliomas (Fig. 16.63)
2. Type III–IV juvenile nasopharyngeal angiofibroma
3. Petrous bone cholesteatoma
4. Certain types of meningiomas, particularly sphenopetroclival, en-plaque meningiomas involving the petrous bone, and extensive jugular foramen meningiomas (Fig. 16.64)
5. Clival chordomas (Fig. 16.65A and B)
6. Carcinomas of the temporal bone
7. Trigeminal schwannomas (Fig. 16.66)

Of these, paragangliomas, meningiomas, juvenile nasopharyngeal angiofibromas, and chordomas are aggressive tumors that may be difficult to dissect from the artery. Trigeminal schwannomas and petrous bone cholesteatomas are less aggressive tumors that very rarely may be difficult to dissect from the ICA. Carcinomas of the temporal bone involving the ICA usually require sacrifice of the artery when sound oncological tumor resection is sought.

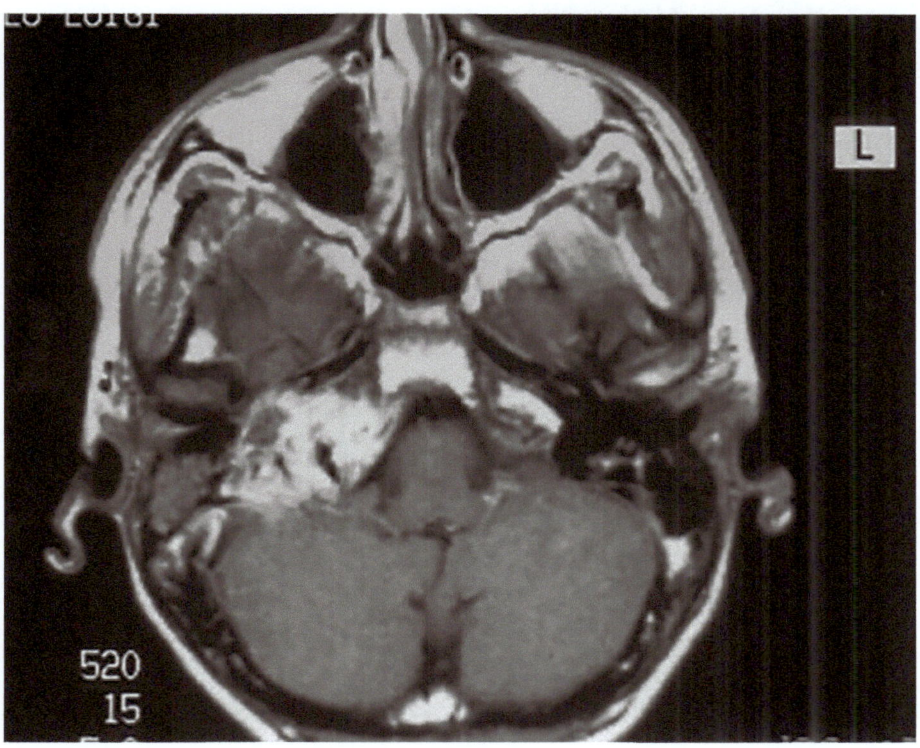

Fig. 16.63 MRI with gadolinium enhancement (axial view) of a case of right C3 tympanojugular paraganglioma.

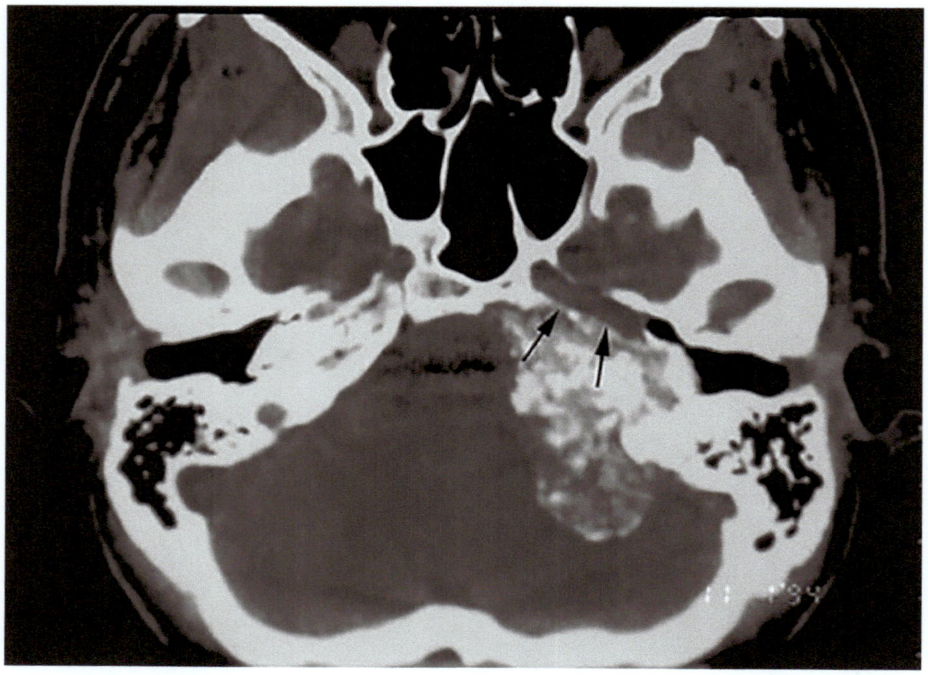

Fig. 16.64 CT scan with contrast (axial view) of a left en-plaque meningioma. Note the massive involvement of the horizontal segment of the internal carotid artery *(arrows)*.

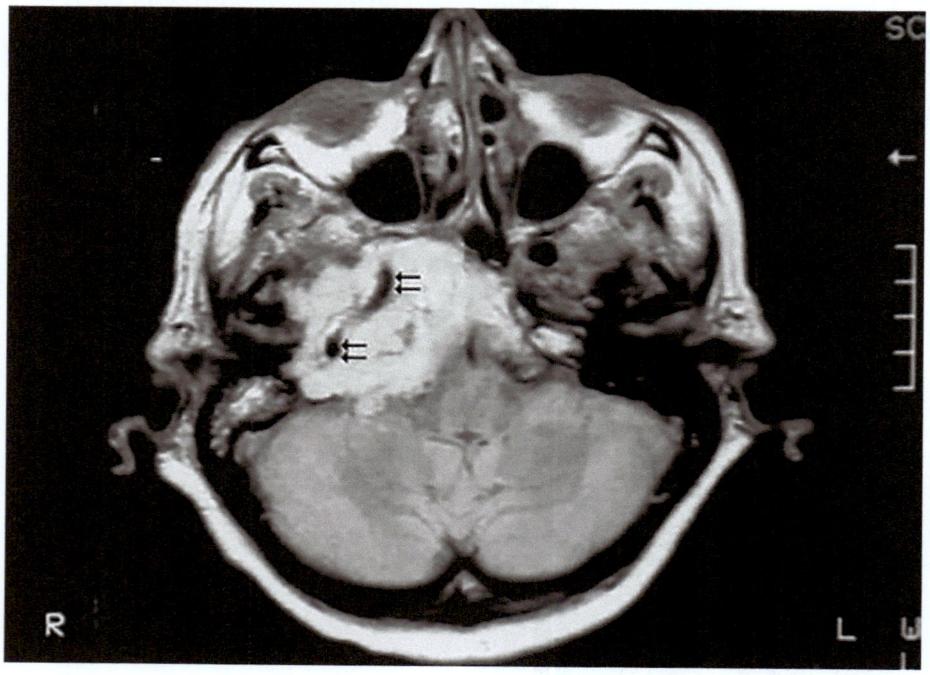

Fig. 16.65 (A) MRI with gadolinium enhancement (axial view) of a right recurrent, postirradiation clival chordoma. The artery *(arrows)* is encased by the tumor.

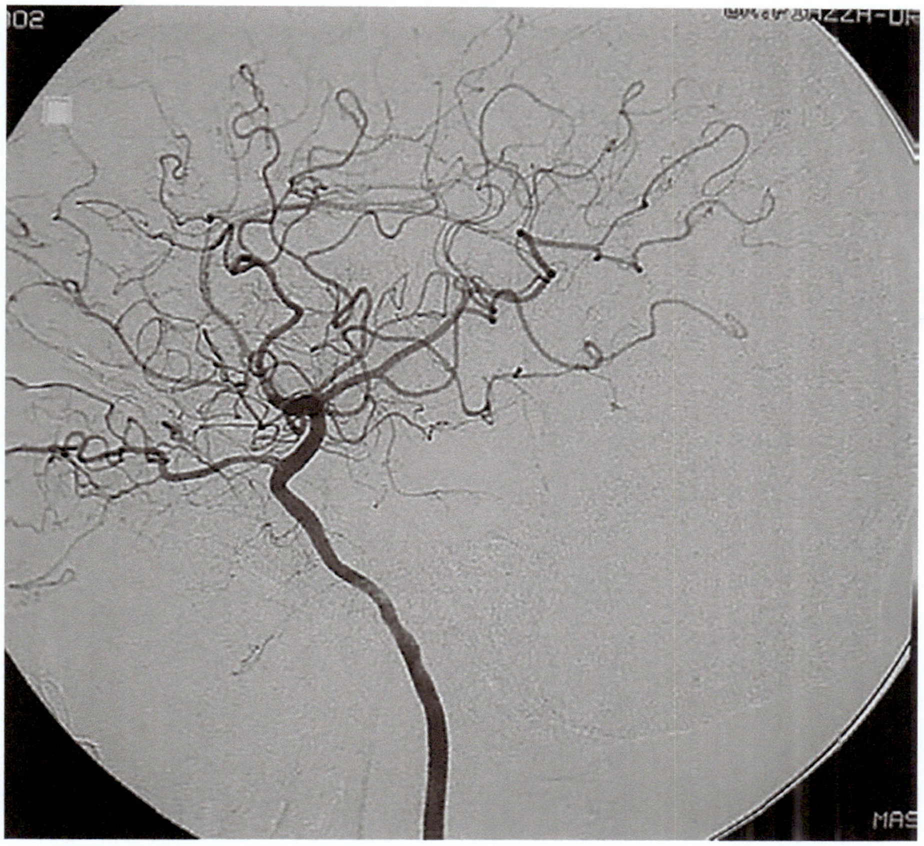

Fig. 16.65 (B) Angiogram of a case of clival chordoma (lateral view). The right internal carotid artery shows evidence of stenosis and displacement.

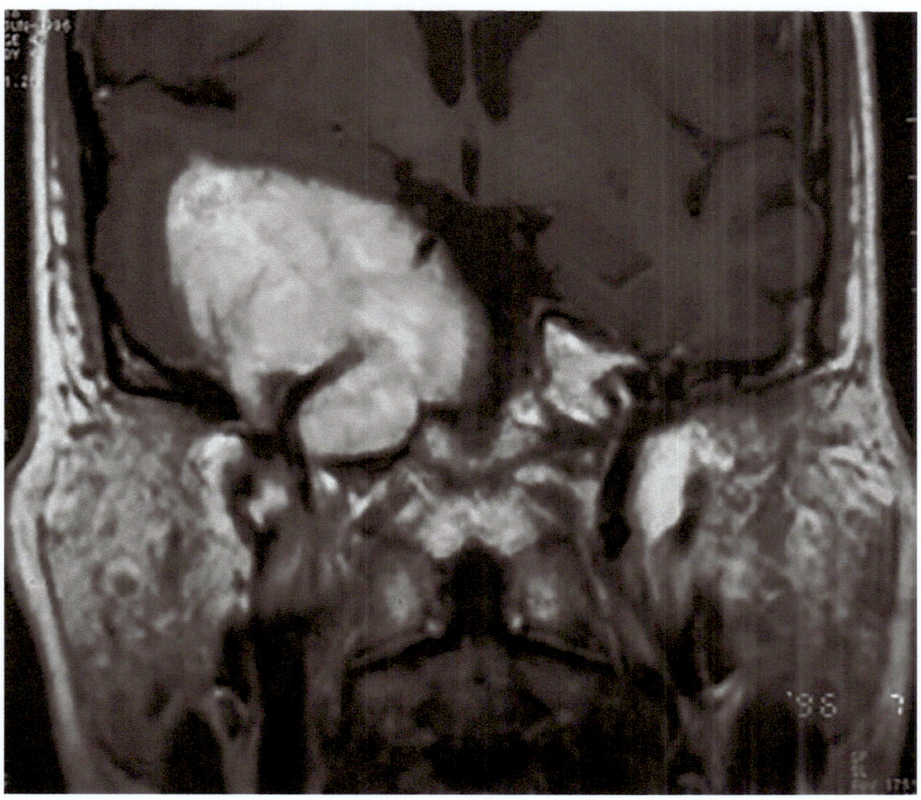

Fig. 16.66 MRI with gadolinium enhancement (coronal view) of a huge right trigeminal schwannoma. Note the complete encasement of the internal carotid artery by the tumor.

Normal Angiographic Anatomy of the ICA and the Circle of Willis

THE INTERNAL CAROTID ARTERY

The classification proposed by Bouthillier et al. identifies ICA segments according to their adjacent structures and the compartments they traverse. Seven anatomically distinct segments (Fig. 16.67A–C) are described, numbered in the direction of normal blood flow:

- Cervical segment (C1)
- Petrous segment (C2)
- Lacerum segment (C3)
- Cavernous segment (C4)
- Clinoid segment (C5)
- Ophthalmic segment (C6)
- Communicating segment (C7)

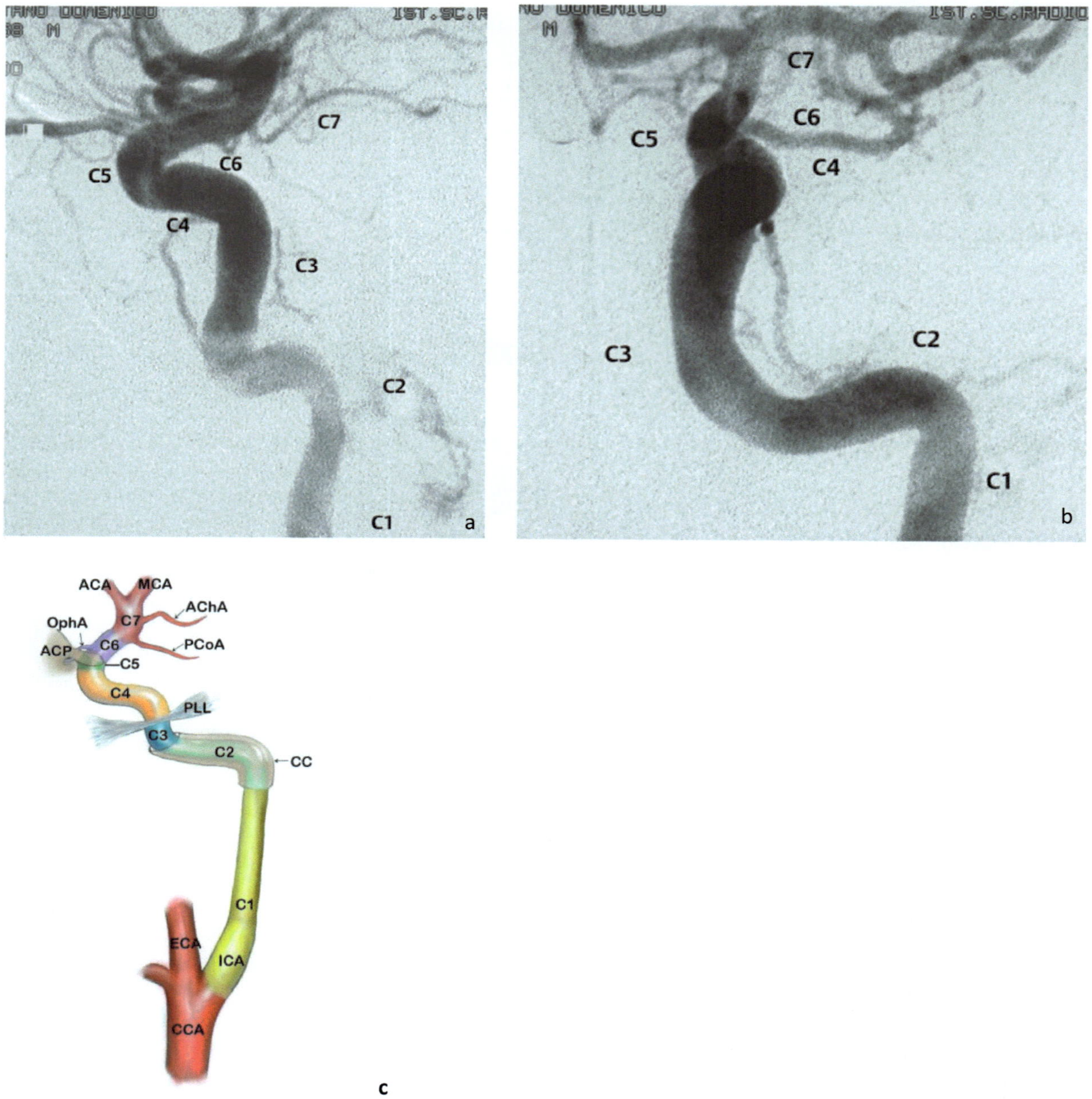

Fig. 16.67 ICA anatomy in lateral (A) and anteroposterior (B) views, showing the seven segments according to Bouthillier's classification system. (C) schematic drawing. *PLL*, petrolingual ligament; *ACP*, anterior clinoid process; *OphA*, ophthalmic artery; *PcoA*, posterior communicating artery; *AchA*, anterior choroidal artery; *ACA*, anterior cerebral artery; *MCA*, middle cerebral artery.

The C1 segment of the ICA has two parts, represented by the carotid bulb and the ascending cervical segment. The carotid bulb is the most proximal aspect of the cervical ICA and forms a significant focal dilatation; the diameter of the normal bulb is approximately 8–9 mm, compared with diameters of 7 mm for the common carotid artery and 5 mm for the ICA distal to the bulb. The ascending segment courses cephalad within the carotid space; tortuosity or kinking of the cervical ICA is relatively common.

No named branches arise from the carotid bulb or cervical ICA segment; however, a number of anomalous vessels may originate from the extracranial ICA. Anomalous branches reported as arising from the extracranial ICA include vessels that normally originate from the external carotid artery (e.g., the ascending pharyngeal and occipital arteries), from other ICA segments (vidian artery), or the vertebrobasilar circulation (e.g., cerebellar and posterior meningeal arteries).

The C2 segment of the ICA enters the skull base at the carotid canal and is contained throughout its course by the petrous temporal bone. The petrous ICA has two distinct segments: a vertical or ascending segment and a horizontal segment; the junction between the two segments forms a knee. The vertical segment averages approximately 10 mm in length. From the genu, the ICA turns anteromedially toward the apex of the petrous temporal bone. The horizontal segment of the petrous ICA is nearly twice the length of its vertical segment. The intrapetrous ICA branches are represented by the vidian and caroticotympanic arteries. The vidian artery, also termed the artery of the pterygoid canal, usually arises from the external carotid artery (however, it may originate from the horizontal petrous ICA segment), passes through the foramen lacerum, and anastomoses with branches of the external carotid artery. The caroticotympanic artery arises near the genu; it anastomoses with the inferior tympanic branch of the ascending pharyngeal artery.

The C3 segment of the ICA begins where the petrous carotid canal ends; here the ICA runs above the foramen lacerum, surrounded by the periosteum and fibrocartilaginous tissue. Then the ICA ascends in a posterior extension of the carotid sulcus, part of the basisphenoid. The lacerum segment ends at the petrolingual ligament, 1 cm below the posterior genu of the cavernous segment; it usually has no branches.

The C4 segment of the ICA begins at the superior margin of the petrolingual ligament; it has three subsegments: posterior or ascending vertical portion, horizontal portion, and anterior vertical portion. At their junction with the horizontal portion, the two vertical subsegments form gently rounded curves; these bends are called the posterior genu and the anterior genu. The C4 segment exits the cavernous sinus through a dural ring in the superior wall.

Several branches arise from the C4 ICA segment. The posterior trunk, also called the meningohypophyseal artery, arises from the posterior genu; it has hypophyseal and dural tentorial and clival branches. The lateral trunk, also called the inferolateral trunk, originates from the horizontal segment; it has dural branches for the cavernous sinus and Gasserian ganglion and for cranial nerves III, IV, and VI. It anastomoses with the internal maxillary artery.

The C5 segment of the ICA is the shortest of all the ICA segments; it begins at the proximal dural ring, just above the curve of the anterior genu, and ends at the distal dural ring, where the ICA enters the subarachnoid space; it is thus an interdural structure. This part of the ICA usually has no discernible branches.

The C6 segment of the ICA begins at the distal dural ring and terminates just proximal to the posterior communicating artery; it represents the most proximal intradural portion of the supraclinoid ICA. Two important branches arise from the C6 segment: the ophthalmic artery and the superior hypophyseal artery.

The C7 segment of the ICA begins just proximal to the origin of the posterior communicating artery and ends as the ICA bifurcates into its two terminal branches, the anterior and middle cerebral arteries. Two major branches, the posterior communicating artery and the anterior choroidal artery, arise from the communicating segment.

THE CIRCLE OF WILLIS

The great anastomotic ring that connects both halves of the anterior circulation with each other and with the vertebral-basilar system is called the circle of Willis.

Because the circle of Willis provides the best potential source of collateral blood flow, awareness of its normal anatomy as well as its common variations is essential and represents an integral part of preoperative planning for skull base surgery.

The circle of Willis is an arterial polygon; a complete circle of Willis in which no part is hypoplastic or absent has 10 components:

- The two terminal C7 segments of the ICA.
- The two proximal segments of the anterior cerebral arteries, also called the precommunicating or A1 segments.
- The anterior communicating artery, a small segment that connects the two A1 segments.
- The two posterior communicating arteries arise from the C7 segment of the ICAs; they provide the major anastomotic link between the anterior and posterior circulations.
- The bifurcation of the basilar artery.

- The two proximal terminal branches of the basilar artery, also termed the precommunicating or P1 segments of the posterior cerebral arteries, which at a variable distance from their origin are joined by the ipsilateral posterior communicating arteries, close the circle of Willis.

The anterior part of the circle is formed by the ICAs, the A1s, and the anterior communicating artery (Fig. 16.68). The posterior part of the circle is formed by the posterior communicating arteries, the P1s, and the basilar artery (Fig. 16.69A and B).

Noninvasive imaging studies, such as spiral computed tomography (CT) angiography, high-resolution MR angiography, and transcranial color Doppler flow imaging, can provide an overview of the major arteries at the base of the brain; however, detailed examination of the circle of Willis is still largely the domain of high-resolution DSA.

The entire circle of Willis is rarely seen on a single cerebral angiogram, and its components are usually visualized sequentially; temporary cross-compression of the contralateral common carotid artery during injection of one ICA is sometimes necessary to demonstrate the anterior communicating artery.

Transient reflux of contrast into the posterior communicating artery during injection of either the ICA or a vertebral artery often occurs; temporary cross-compression of the common carotid artery during injection into the vertebral artery may improve reflux into the posterior communicating artery.

Anatomical variations in the circle of Willis are the rule, not the exception. A so-called complete circle of Willis, with well-developed symmetric components, is present in less than half of all cases; in at least 60% of anatomical specimens, one or more vessels is sufficiently narrowed to reduce its role as a potential collateral route. Common circle of Willis variations are as follows:

- Hypoplastic or absent A1 segment (10%)
- Absent anterior communicating artery (5%)
- Hypoplastic or absent posterior communicating artery (25%–30%)
- Hypoplastic or absent P1 segment (15%–20%)

Circle of Willis variations are especially common in the posterior part.

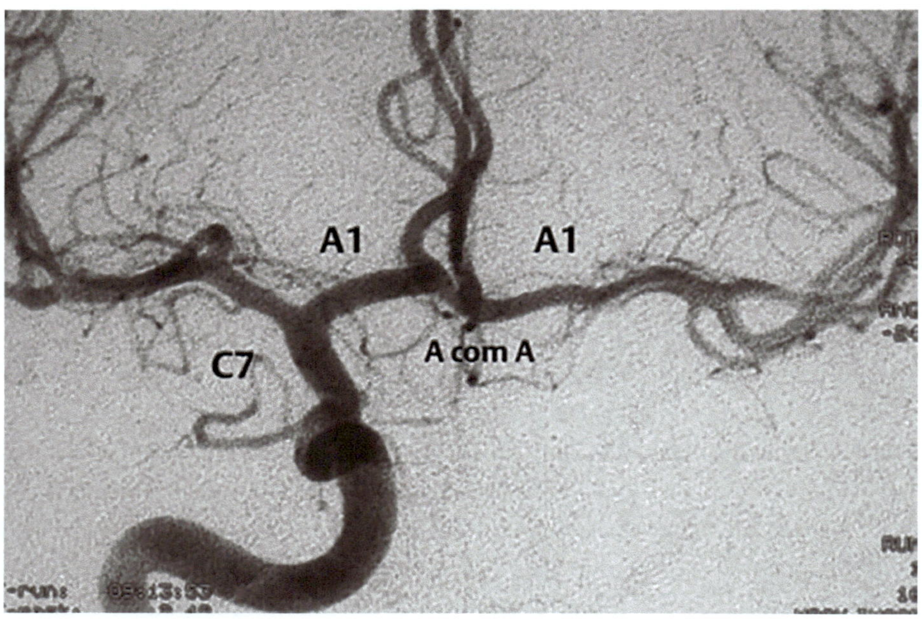

Fig. 16.68 Anatomy of the circle of Willis; notice the anterior portion with the right C7 segment of the ICA, right A1 segment, anterior communicating artery, and left A1 segment.

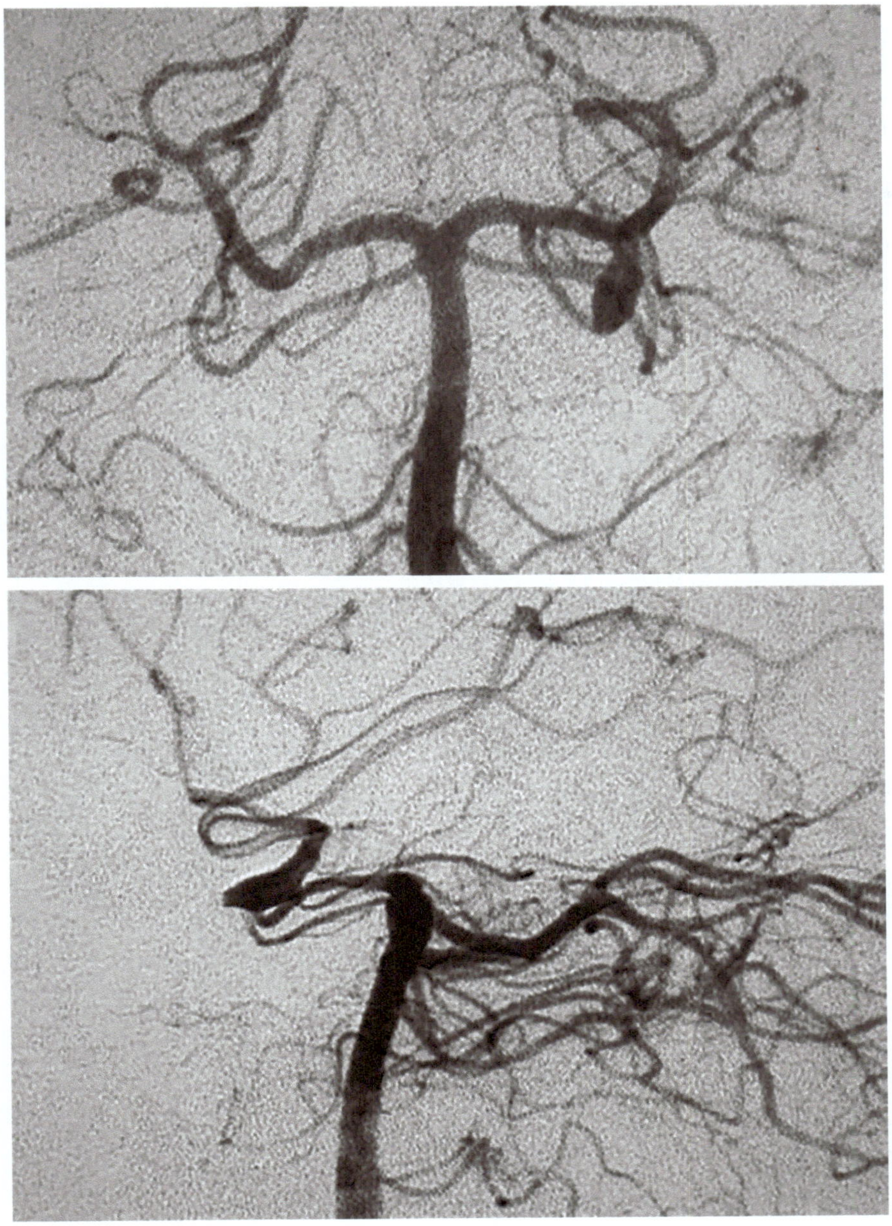

Fig. 16.69 (A, B) Anatomy of the posterior circle of Willis in anterior-posterior (A) and lateral views (B) and C7 segment of left ICA.

Preoperative Neuroradiological Assessment of the ICA

The aims of preoperative neuroradiological assessment of the ICA are:

- To determine the site, the degree, and the extent of involvement of the ICA by the tumor. The investigations used for this purpose include high-resolution CT, MRI, magnetic resonance angiography, and intraarterial DSA. The CT images are evaluated mainly for bone modifications, especially erosion, of the vertical and horizontal segments of the petrous carotid canal. The MR images, with the technique's multiplanar capabilities, are reviewed in order to ascertain the presence of tumor along the distal cervical, petrous, lacerum, and cavernous segments of the ICA and to establish the degree of encasement of each arterial segment. The DSA images are evaluated for the presence of angiographic signs of infiltration of the arterial wall, such as irregularities and stenosis of the arterial lumen, and for the presence of vascular supply from clival and cavernous branches of the ICA.
- To determine the efficacy of collateral circulation in maintaining the perfusion of the areas that would be affected by the manipulation or sacrifice of the artery. A preliminary evaluation of the collateral circulation through the circle of Willis is done at our center with angiography of the contralateral carotid artery (Matas test) and of the dominant vertebral artery (Allcock test) during manual compression of the carotid artery to be tested.

The Matas test is performed by injecting the contralateral ICA during manual compression of the common carotid artery to be tested; rapid and complete filling of the anterior and middle cerebral arteries of the occluded side indicates a patent anterior communicating artery system; downward filling of the ICA, stimulated by the sump effect of the external carotid artery, is another good indicator of collateral circulation (Fig. 16.70A).

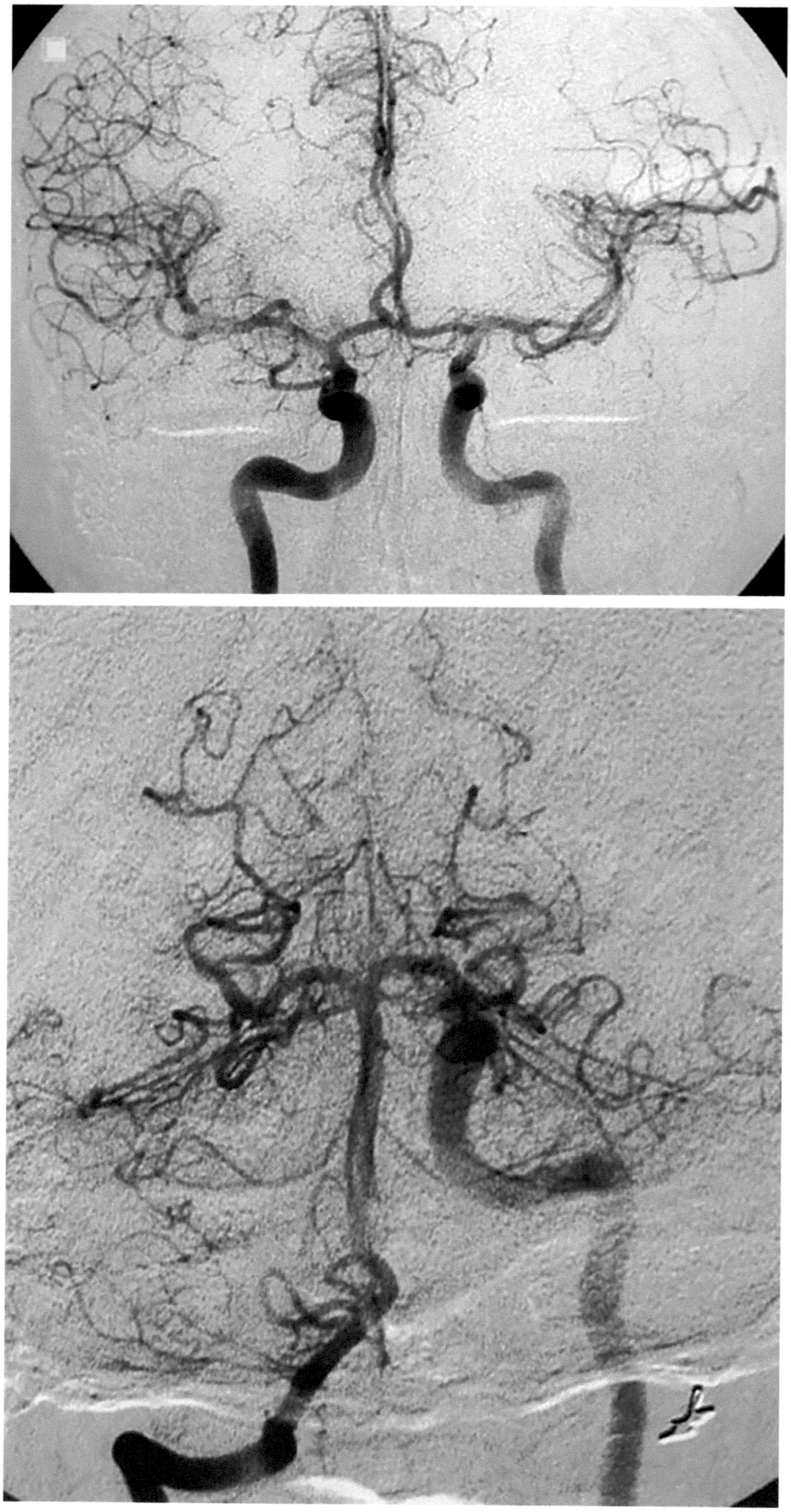

Fig. 16.70 (A) Matas test and (B) Allcock test.

The Allcock test is performed by injecting the dominant vertebral artery during manual compression of the common carotid artery on the occlusion side; rapid and complete filling of the middle cerebral artery of the occluded side and/or downward filling of the ICA indicates a patent posterior communicating artery system (Fig. 16.70B).

Although symmetry of the arterial, capillary, and venous angiographic phases is better evaluated during the balloon occlusion test, a clear asymmetry between right and left hemispheres during the Matas test and between supra- and infratentorial compartments during the Allcock test should alert the physician.

Manual compression of the common carotid artery should be firm in order to stop the flow completely; the technique consists of applying pressure some 2–3 s before starting the injection and maintaining it during angiography until the arterial phase is completed.

Only in the presence of a good cross-filling from at least one of the two communicating systems is the patient scheduled for a balloon occlusion test with clinical and angiographic evaluation, generally followed by PBO of the ICA. In our experience, when cross-filling of the middle cerebral artery is absent, the patient does not clinically tolerate temporary balloon occlusion even of very short duration.

Interventional Neuroradiological Management of the ICA

The main form of interventional neuroradiological management of the ICA in skull base surgery is represented by PBO of the ICA. More recently, the insertion of stents to reinforce the arterial wall of the ICA has been introduced into clinical practice.

PERMANENT BALLOON OCCLUSION OF THE INTERNAL CAROTID ARTERY

The rationale of PBO of the ICA is to allow easy surgical mobilization of the artery and of the pericarotid portion of the tumor without the risk of an uncontrollable hemorrhage from laceration of the arterial wall; moreover, PBO devascularizes the pericarotid portion of the tumor and facilitates radical tumor removal. The main indications for PBO (Fig. 16.71) of the ICA have traditionally been as follows:

- Invasion of the wall of the ICA, shown at angiography, as moderate to marked stenosis of the arterial lumen
- Extensive blood supply to the tumor coming from branches of the ICA
- Circumferential encasement, of 270 degrees or more, of the distal cervical and vertical petrous segments of the ICA
- Extension of the tumor along the horizontal petrous, precavernous, and cavernous segments of the ICA
- Aggressive tumors in previously operated or irradiated patients

Permanent occlusion and removal of the ICA should, however, be carefully evaluated because of the potential morbidity related to the procedure. Transient or permanent neurological deficits can occur due to dissection of the ICA wall during the procedure itself, in the order of 0.5%–1%. Periprocedural complications such as thromboembolic and hemodynamic infarction and even long-term complications such as the development of saccular aneurysm have been reported by some authors.

Distal insufficiency is one of the two major mechanisms that may account for delayed cerebral ischemia after PBO of the ICA. The distal insufficiency concept for ischemia implies decreased vascular perfusion of those areas of the brain located at the greatest distance from the site of occlusion. As a consequence, stagnation thrombus may develop from local circulatory failure at those sites and infarction follows. The areas at risk include the most distal segments of the cortical branches of the middle cerebral artery, in particular the superior frontal, the superior parietal, and the posterior temporo-occipital areas, which represent the border zone region between major blood vessels; these areas have been labeled as distal fields, comparable to the far zones of an agricultural irrigation system. However, we must not forget that simple global hypoperfusion can also cause true territorial infarcts, not watershed infarcts, probably unrelated to embolization.

When stenosis or occlusion of the ICA progresses slowly, secondary collateral pathways other than the circle of Willis may be recruited; these may progressively enlarge and contribute to cerebral perfusion preservation. Such important pathways include retrograde flow in the ophthalmic artery from the ipsilateral external carotid artery and the leptomeningeal collateral vessels; however, these connections do not contribute significant cerebral collateral blood flow under circumstances of acute occlusion of the ICA such as PBO. Thus the circle of Willis is the principal collateral pathway that is able to maintain cerebral perfusion during PBO.

Abrupt occlusion of the ICA without balloon test occlusion resulted in 25% stroke rate and 10% mortality rate. A more careful selection of patients, based on the clinical tolerance balloon occlusion test, has greatly reduced the stroke rate to 5%. Nevertheless, in order to further diminish periprocedural ischemic complications, a battery of adjunctive tests has been introduced in clinical practice. These adjunctive functional tests are represented by xenon CT perfusion scan, xenon-133 scintillation scan, single-photon

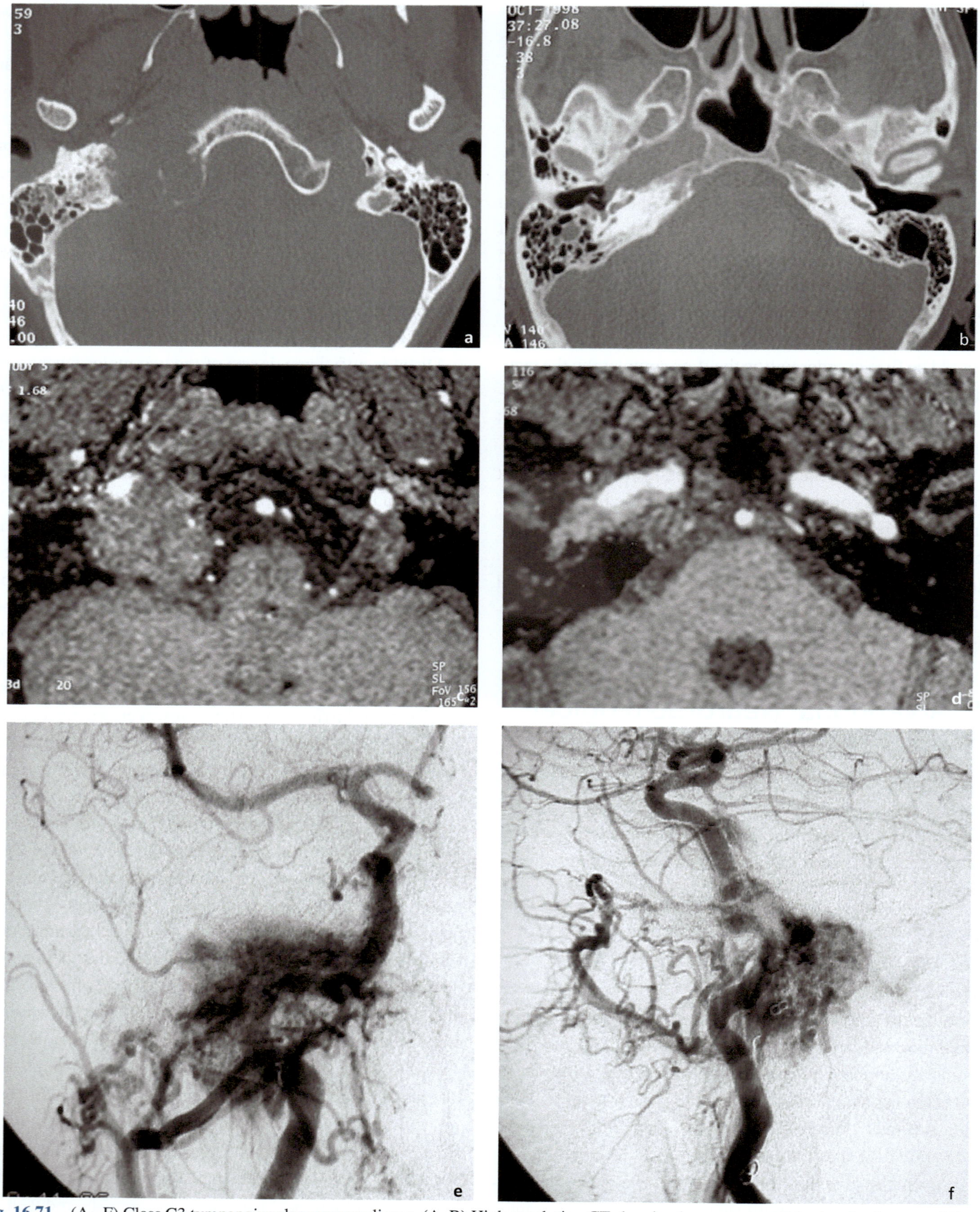

Fig. 16.71 (A–F) Class C3 tympanojugular paraganglioma. (A, B) High-resolution CT showing bone erosion of the jugular foramen (A) and of the horizontal portion of the petrous carotid canal (B). (C, D) MRI: axial TOF partition showing tumor mass growing along the cervical (C) and petrous segments (D) of the ICA. (E, F) Angiography: common carotid injection in anteroposterior (E) and lateral (F) projections, showing tumoral blush coming from both the external and internal carotid arteries.

emission CT scan, transcranial Doppler monitoring, positron emission tomography scan, stump pressure monitoring, electroencephalographic monitoring, perfusion MR imaging, noninvasive infrared cerebral oxygen saturation monitoring, hypotensive challenge test, and angiography. Nowadays these are available for a better selection of those patients who are able to tolerate a temporary occlusion for the short period of the test but after 24–48 h show a delayed neurological deficit, sometimes associated with episodes of profound hypotension. Introduction of these functional tests has further reduced but not completely eliminated the number of delayed ischemic deficits; moreover, these tests cannot predict the occurrence of embolic stroke. At our center, the patient's functional test of tolerance to permanent occlusion of the ICA is based on angiography. PBO of the ICA is generally performed under local anesthesia, mild sedation, and systemic heparinization. The technique requires a double femoral approach; in one of the femoral arteries, usually the right, an 8 F guiding catheter is inserted and subsequently positioned in the ICA to be occluded, while the contralateral femoral artery is used for the angiographic functional evaluation.

To perform the balloon occlusion test of the ICA, the GVB 16 balloon (Minyvasis), mounted on its delivery catheter, is employed; this is the same balloon that will be detached to permanently occlude the ICA. Such a balloon is usually placed into the cavernous segment of the ICA, just proximal to or at the origin of the ophthalmic artery, and is inflated with isotonic contrast material. At this point, occlusion of the ICA is confirmed through injection of the contrast medium into the 8 F guiding catheter.

Then the collateral circulation through the circle of Willis is rechecked to confirm the data obtained with the manual compression test and especially to add new data representing the symmetry of the angiographic phases, which is difficult to evaluate during the manual compression test because the site of occlusion is not the internal but the common carotid artery. This is an important point because the balloon occlusion test must be done with the balloon in the same position as the permanent occlusion, in order to create the same hemodynamic conditions. A more proximal occlusion, in the cervical ICA, can lead to false-positive and false-negative results of the test due to the presence of an arterial steal from the tumor or a collateral circulation that will subsequently be sacrificed. Another reason to position the first balloon in the cavernous segment of the ICA and not at the cervical level is tc reduce the length of thrombotic mass that develops distal to the occlusion site, which could act as a source of embolic stroke.

Soon after occlusion is achieved in each patient, we start to evaluate the clinical and angiographic conditions for a period of 15–20 min. The clinical evaluation is done by an assistant, who checks the main neurological functions. The angiographic evaluation consists of checking the symmetry of the arterial, capillary, and venous phases of the two cerebral hemispheres in the frontal view during injection in the contralateral carotid artery (Fig. 16.72A and B); then the angiographic symmetry of the infra- and supratentorial compartments, in both frontal and lateral views, during injection of the dominant vertebral artery is evaluated (Fig. 16.72C–F).

The rationale for utilizing the symmetry of the angiographic phases as an indicator of tolerance to PBO of the ICA is based upon the principles of cerebral hemodynamics assessed by angiography in the work of Greitz et al. The average supratentorial circulation time is between 3 and 4 s; it is ordinarily defined as the time interval between the maximal filling of the cavernous segment of the ICA and the maximal filling of the parietal veins. The arterial phase lasts approximately 1.0–2.5 s, the capillary phase 0.25–1 s, and the venous phase 1.5–4.5 s. The infratentorial circulation time is the time from maximum opacification of the basilar artery to maximum opacification of the superficial cerebellar and petrosal veins. The posterior supratentorial circulation time is the time from maximum opacification of the basilar artery to maximum opacification of the occipital veins. The circulation times in vertebral angiograms are comparable to those encountered in the carotid system, and the occipital circulation time is essentially the same as that in the infratentorial region.

Familiarity with the sequence of arterial and venous filling in the cerebral angiogram is also essential for correct interpretation of the angiographic functional test. The arterial circulation time is slightly less in the frontal region than in the parietal and occipital regions; the regional variation is related to the different lengths of the arteries supplying these areas from the carotid siphon. The sequence of venous filling in both the superficial and deep venous systems is relatively constant: the superficial venous system fills sequentially from anterior to posterior, with the frontal veins filling always before the parietal veins. The deep venous system usually begins to opacify later than the superficial system and continues to be opacified longer because flow is slower in the white matter, which the deep system largely drains; the thalamostriate veins, the internal cerebral veins, and the basal veins usually fill at the same time as the parietal veins.

A good correlation between the circulation time and regional blood flow measurement has been demonstrated

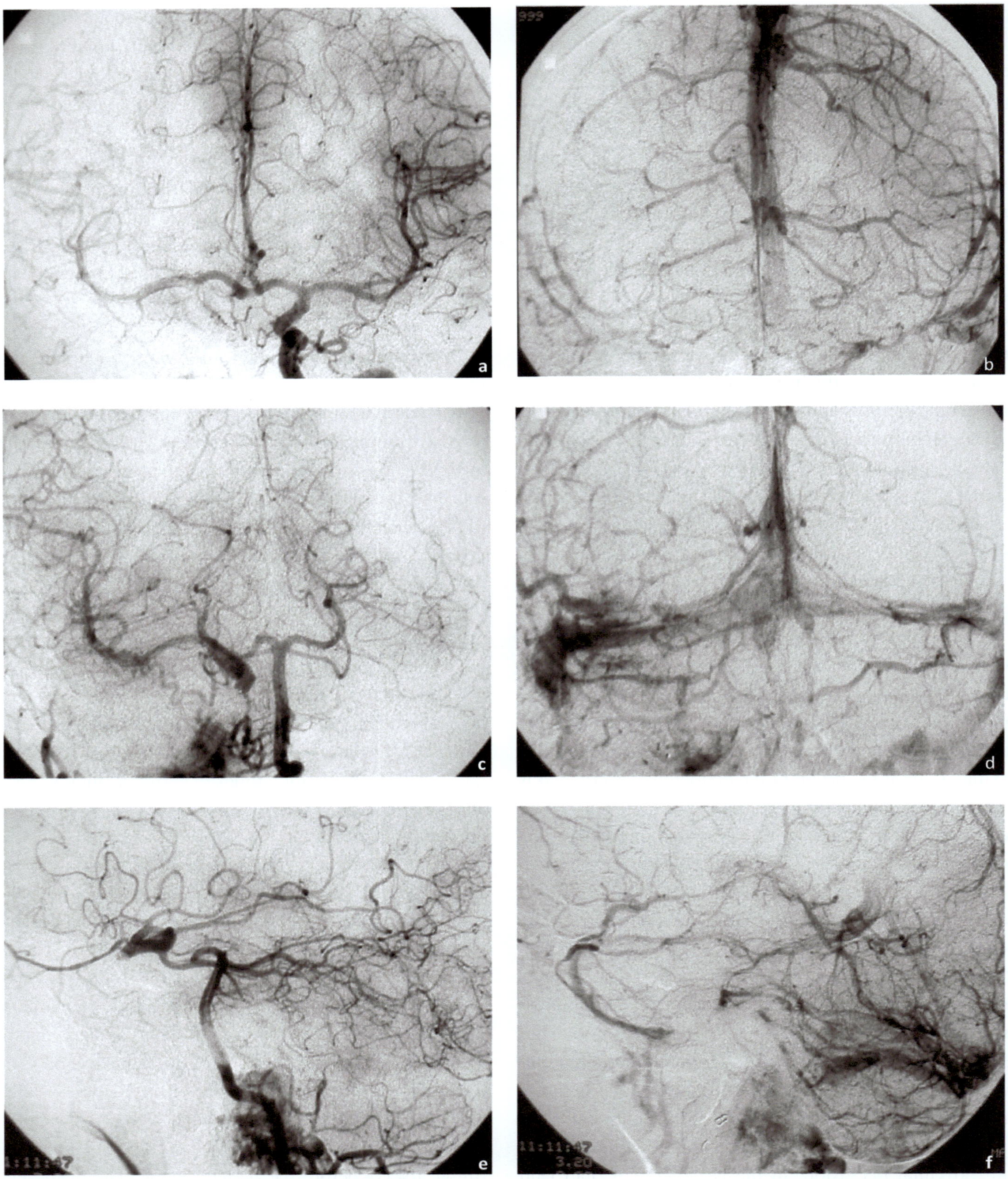

Fig. 16.72 (A–F) Class C3 tympanojugular paraganglioma; same case as shown in Fig. 16.71A–F. Permanent balloon occlusion of the internal carotid artery. (A, B) Injection of the left carotid artery showing collateral circulation from anterior communicating artery in the arterial (A) and venous (B) phases, after balloon occlusion of the right carotid artery; the slight asymmetry between the two hemispheres is due to unopacified blood entering the C7 segment of the ICA from the posterior circulation. (C, D) Injection of the left vertebral artery in anterior-posterior view showing collateral circulation from the posterior communicating artery. In the arterial (C) and venous (D) phases, after balloon occlusion of the right carotid artery. Notice the symmetry of the angiographic phases in the infratentorial and supratentorial compartments. Note also blood supply from a branch of the vertebral artery and occlusion of the sigmoid sinus. (E, F) Injection of the left vertebral artery in lateral view showing collateral circulation from the posterior communicating artery in the arterial (E) and venous (F) phases, after balloon occlusion of the right carotid artery. Note the symmetry of the angiographic phases. Note also blood supply from branches of the vertebral artery.

using the xenon washout technique. The crucial pressure that determines the adequacy of cerebral blood flow is the cerebral perfusion pressure, represented by the difference in pressure between the arteries entering the cranial subarachnoid space and the veins leaving it. In our experience, we have found a strict correlation between patency of communicating arteries, clinical tolerance to balloon occlusion test, and symmetry of the angiographic phases.

In case of clinical tolerance and symmetry of the angiographic phases, or slight asymmetry of less than 1 s, the balloon is detached in the cavernous segment of the ICA, and permanent occlusion is achieved by inserting and detaching two other balloons in the cervical-petrous and proximal cervical segments, respectively, of the ICA (Fig. 16.73A and B).

A few days before surgery, MR imaging can confirm the presence of thrombus in the petrous segment of the ICA (Fig. 16.73C). Finally, preoperative embolization of the external carotid artery branches is performed (Fig. 16.74A and B).

In case of clinical tolerance and slight asymmetry of the angiographic phases, between 1 and 2 s, we perform a second angiographic evaluation after a few minutes in order to allow the regulatory mechanisms of the cerebral blood flow to adapt. If a prolongation of the circulation time on the occlusion side persists, the balloon is deflated and the patient is scheduled for bypass; if there is improvement of the circulation time, we proceed to PBO.

In case of clinical intolerance, which usually appears in the very first minutes after temporary occlusion of the ICA, the balloon is immediately deflated. In our experience of balloon occlusion test of the ICA in skull base tumors, we do not usually have cases of clinical intolerance to the

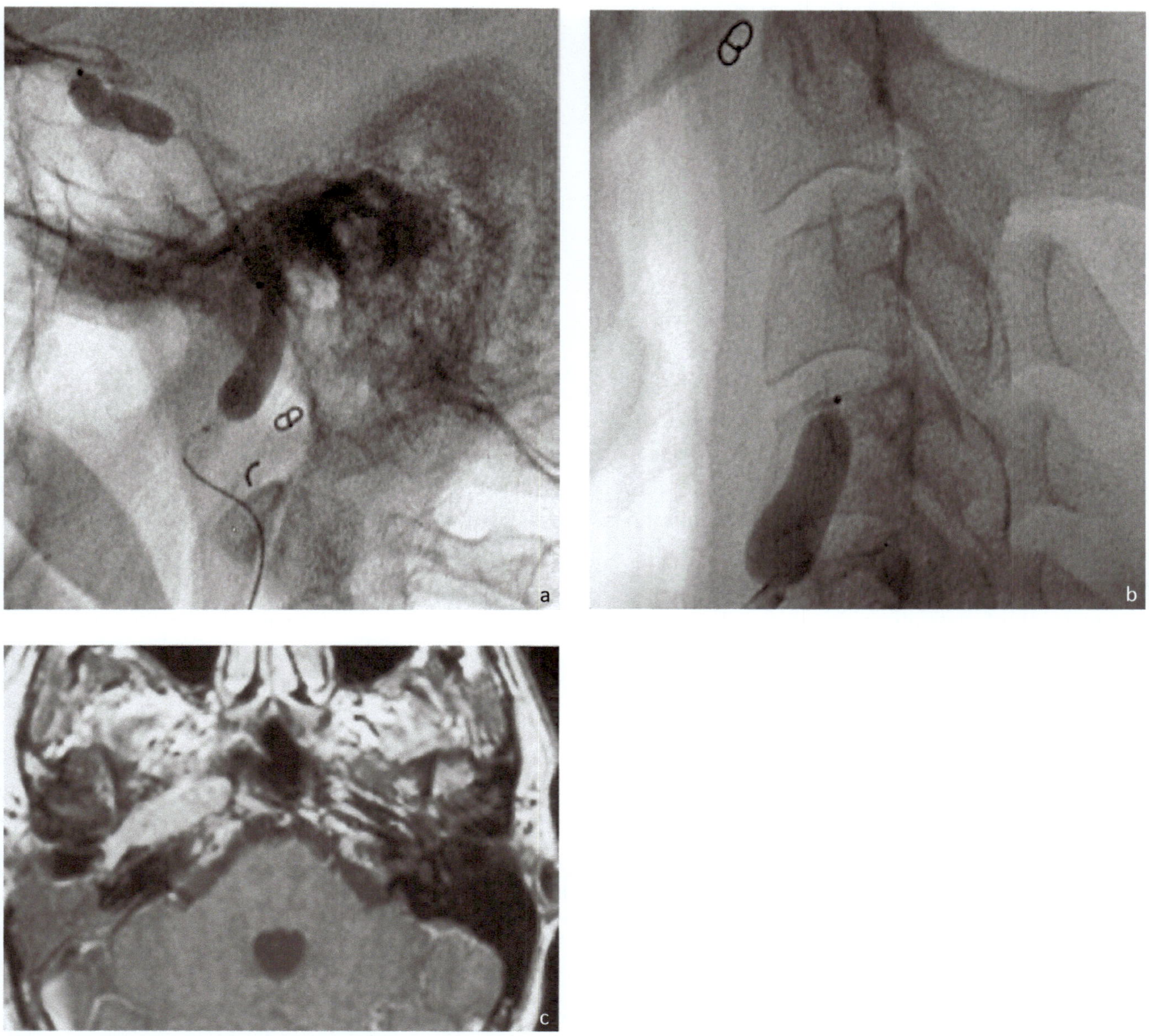

Fig. 16.73 (A–C) Class C3 tympanojugular paraganglioma. Permanent balloon occlusion of the internal carotid artery. (A) Lateral plain film showing balloons inside the cavernous and cervico-petrosal segments of the ICA. (B) A third balloon is visible in the proximal cervical segment of the ICA. (C) T1-weighted MRI after occlusion of the internal carotid artery showing hyperintense thrombus in the petrous segment of the ICA.

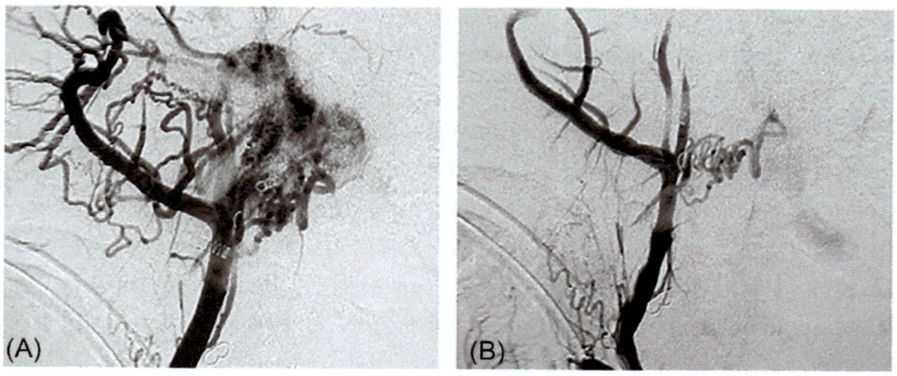

Fig. 16.74 (A, B) Class C3 tympanojugular paraganglioma. Permanent balloon occlusion of the internal carotid artery. Common carotid injection after balloon occlusion of the internal carotid artery, showing vascular supply from external carotid artery branches, (A) before and (B) after embolization. The right vertebral artery is slightly opacified from reflux (B).

temporary balloon occlusion test because the patients were previously selected on the basis of Matas and Allcock tests. Moreover, for further support to the functional test based on angiography, PBO of the ICA has been performed under general anesthesia in some uncooperative patients, without any problem, only relying upon the angiographic criterion. After PBO of the ICA has been accomplished, the patient is monitored for 24 h in an intensive care unit in order to avoid any sudden drop in systemic blood pressure; if any symptom should develop, prompt elevation of blood pressure and volume expansion should be undertaken. With adequate management, intracranial circulation and collaterals usually adapt, and the patient's condition improves.

The timing of PBO of the ICA in relation to surgery also plays an important role in determining the final outcome. An interval of 3–4 weeks between the occlusion and surgery is usually advocated because surgery under general anesthesia causes hypotension with possible hemodynamic insufficiency. An adequate period of time for adaptation of the cerebral vasculature to the new hemodynamic situation after occlusion of the ICA will reduce the chances of ischemic complications.

PREOPERATIVE STENTING OF THE ICA

The management of patients with advanced skull base tumors involving the ICA is controversial. Although complete tumor removal intuitively represents the optimal management strategy, the potential complications of carotid sacrifice with or without reconstruction have tempered enthusiasm for an aggressive approach in this patient population.

Recent introduction of preoperative stenting of the internal carotid has represented a new option for treating those patients who are at risk for intraoperative management of the ICA. The rationale of stenting of the ICA is, as for PBO, to allow an easier mobilization of the artery and of the pericarotid portion of the tumor without the risk of uncontrollable hemorrhage from laceration of the arterial wall. In contrast to PBO, stenting does not devascularize the pericarotid portion of the tumor, and some concerns exist in terms of radical tumor removal.

The main indications for stenting of the ICA in skull base tumors (Fig. 16.75) are similar to indications for PBO but with a fundamental difference represented by the patients who do not tolerate PBO, either clinically or/and angiographically (Fig. 16.76). This is a major point of improvement in the treatment of such patients, who, before the stent era, were generally only partially treated. Nevertheless, there are a few conditions in which we think the stent is contraindicated; such situations are mainly represented by:

- Marked stenosis of the arterial lumen of the ICA shown at angiography
- Extensive blood supply from cavernous branches of the ICA not amenable to embolization
- Previously irradiated patients or previously operated patients with extensive manipulation of the ICA
- Kinking and/or marked tortuosity of the cervical C1 segment of the ICA
- Contraindication to antiplatelet therapy

In such conditions, PBO of the ICA, preceded by high-flow bypass in case of intolerance, is usually adopted.

In order to reduce the possibility of stent thrombosis, pre- and poststenting double antiplatelet therapy, with ticlopidine or clopidogrel and salicylates, must be introduced, starting from one week before stent insertion, and has to be continued for at least 30 days. Then salicylate alone is continued indefinitely, with the only exception of the perioperative period, during which it is replaced by low-molecular-weight heparin.

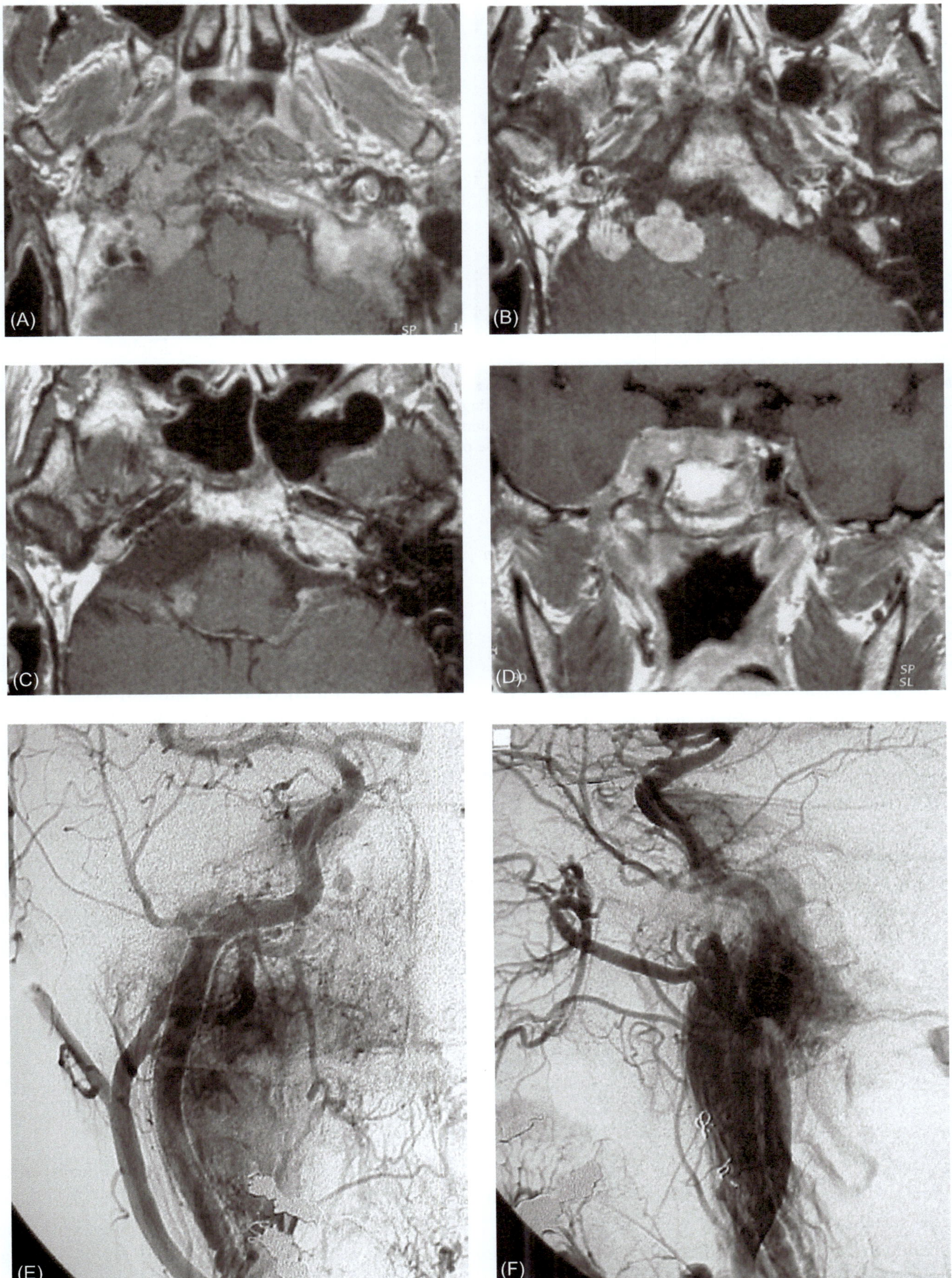

Fig. 16.75 (A–F) Class C4 tympanojugular paraganglioma. Stenting of the internal carotid artery. (A, B) T1-weighted MRI showing encasement of the mid (A) and distal (B) cervical segments of the carotid artery, which is surrounded by the tumor at 360 degrees. (C, D) Axial (C) and coronal (D) T1-weighted MRI showing, respectively, encasement of the petrous and cavernous portions of the internal carotid artery. (E, F) Angiography in anteroposterior (E) and lateral (F) projections. Common carotid artery injection showing the tumoral blush, which came mainly from branches of the external carotid artery.

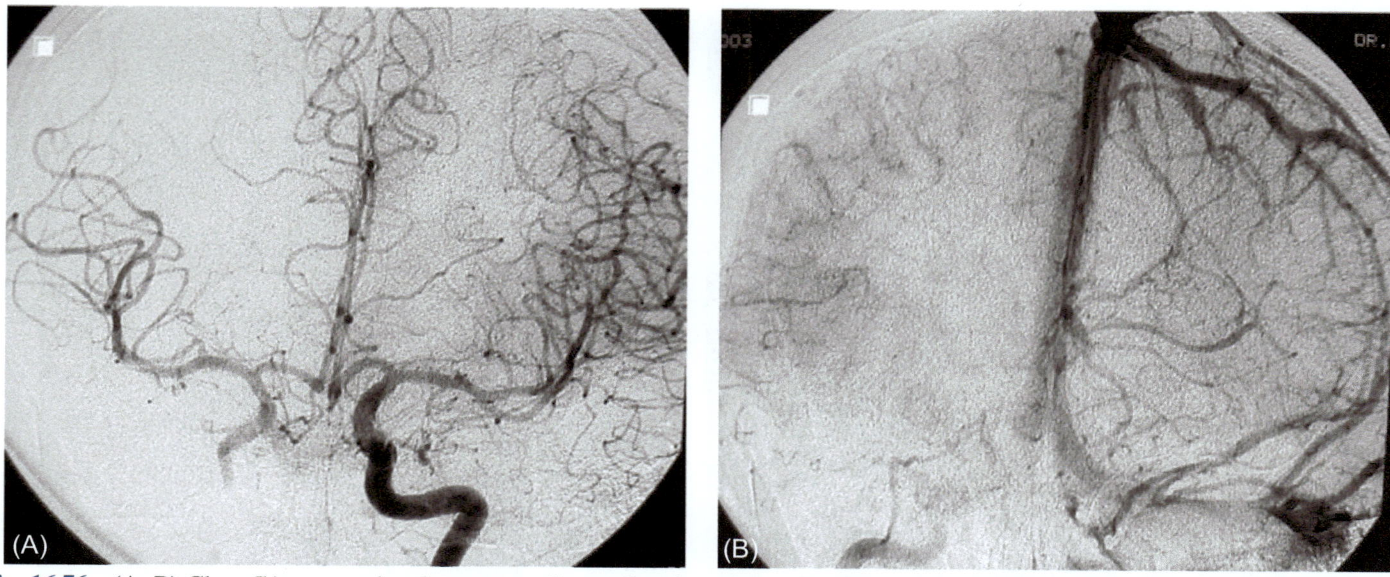

Fig. 16.76 (A, B) Class C4 tympanojugular paraganglioma. Stenting of the ICA. Same case as shown in Fig. 16.75A–F. Matas test, with injection of the left carotid artery during manual cross-compression of the right carotid artery, showing marked asymmetry of arterial (A) and venous (B) phases with delay in the right hemisphere because of hyperplasia of the anterior communicating artery; compensation from posterior circulation (not shown) was absent.

Self-expandable Nitinol stents are the most suitable for reinforcement of both the cervical and intratemporal portions of the ICA (Fig. 16.77A); to facilitate embedding of the stent into the arterial wall, a little oversizing by some 15%–20% of the diameter of the ICA must be done; this is usually accomplished with 5–6 mm stents for the C1 cervical segment and with a 4–5 mm stent for the C2 temporal segment of the ICA. In order to reduce the possibility of injuring the ICA at its junction with the tumor, we believe that at least 10 mm of tumor-free vessel wall should be reinforced with stent, both proximally and distally. To achieve this it is sometimes necessary to insert more than one stent; the intratemporal stent is usually inserted first, followed by insertion of the cervical stent, with at least 5 mm of overlap (Fig. 16.77B and C).

Difficulties encountered during stent positioning are usually related to the occurrence of transitory vasospasm of the C1 cervical segment (Fig. 16.77C) or to the impossibility of navigating the stent through kinking of the C1 cervical segment or through the genu of the C2 temporal segment of the ICA. In such situations, a softer and more flexible stent must be chosen from the ones available, even to the detriment of reinforcement of the arterial wall, without forcing the progression of the more rigid stent, which could cause an intimal dissection.

Completion angiography confirms the good result at the end of the procedure (Fig. 16.77D). High-resolution CT imaging shows the stent inside the ICA (Fig. 16.78A–C).

The timing of reinforcement with stent(s) also plays an important role; an interval of at least 4–6 weeks has been advocated between stenting and surgery in order to allow the formation of a stabilized neointimal lining on the luminal surface of the stent.

A very special situation is represented by the presence of a significant blood supply to the tumor from branches of the ICA; in such a condition, a bare self-expandable Nitinol stent is unable to consistently reduce the vascular supply, and insertion of a covered stent represents an attractive alternative.

There are a few case reports of insertion of balloon-expandable and self-expandable covered stents, both in tumors and in pseudoaneurysms; however, at present, covered stents have several at least theoretical disadvantages, such as increased thrombogenicity, rigidity, and subsequent difficulty in positioning at arterial angles, in comparison with bare self-expandable Nitinol stents.

The assessment of long-term patency of the stented ICA is easily accomplished with ultrasound; although in our preliminary series, there were no cases of carotid occlusion due to in-stent thrombosis or in-stent stenosis due to intimal hyperplasia, a longer period of follow-up is necessary for definitive evaluation.

Some concerns still exist in terms of radicality of surgery after reinforcement with stent; while gross total removal has been achieved in the majority of our cases, only long-term evaluation will say the last word. CT and plain radiography a few days before surgery can confirm the stability of the result and the relationship of the stents to the bony structures. Embolization is performed as usual a few days before surgery (Fig. 16.79).

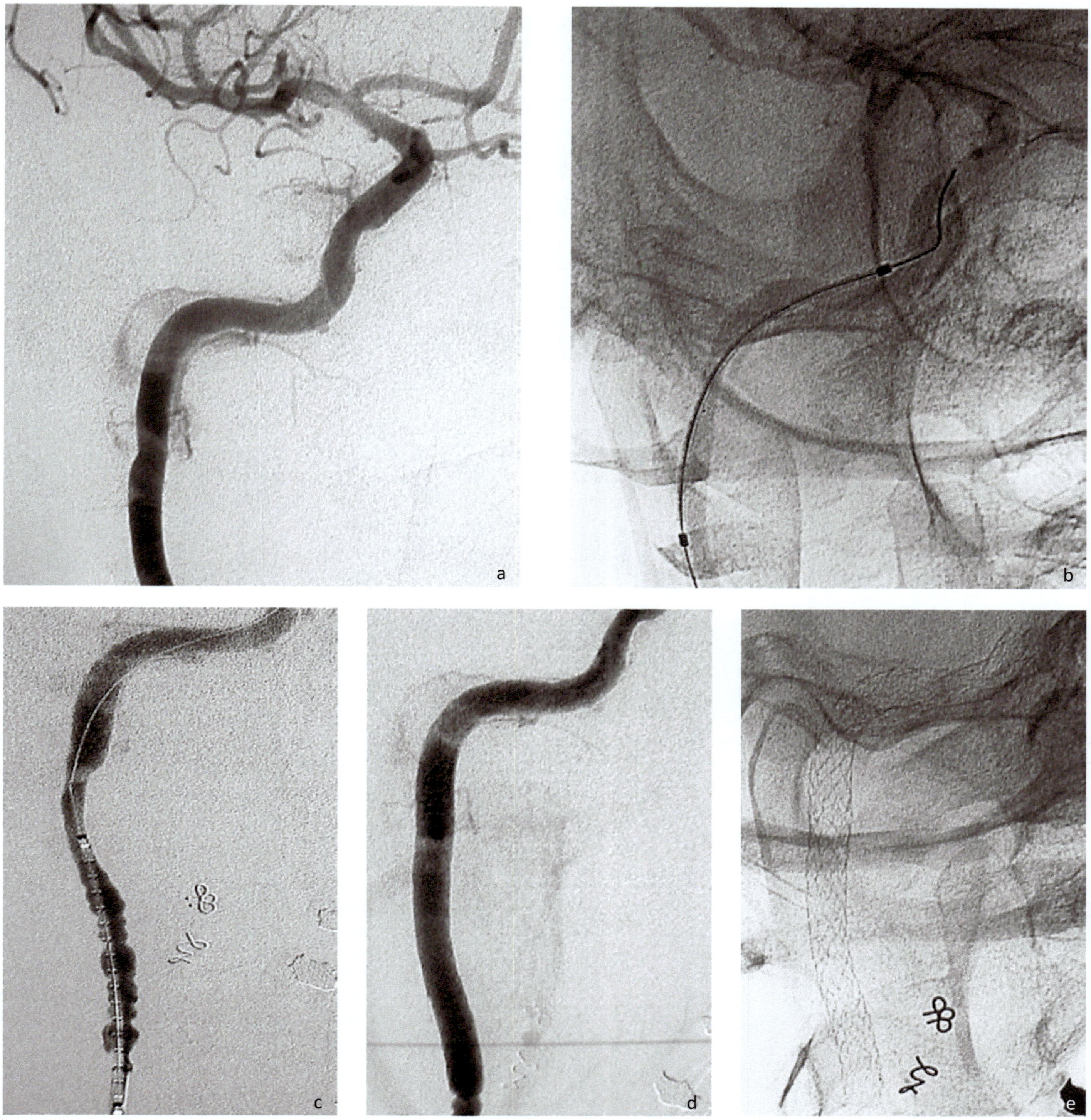

Fig. 16.77 (A–E) Class C4 tympanojugular paraganglioma. Stenting of the internal carotid artery. Same case as shown in Fig. 16.85A–F. (A, B) Internal carotid injection (A) showing the slight tumoral blush, a favorable condition for insertion of a stent; (B) unsubtracted view during ICA injection, showing the position of the stent before deployment. (C–E) Internal carotid injection showing the insertion of the second stent and the vasospasm generated by the endovascular maneuver (C); internal carotid artery injection at the end of the procedure (D), with resolution of the vasospasm. Unsubtracted view (E) showing the stent covering the C1, C2, and C3 segments of the internal carotid artery.

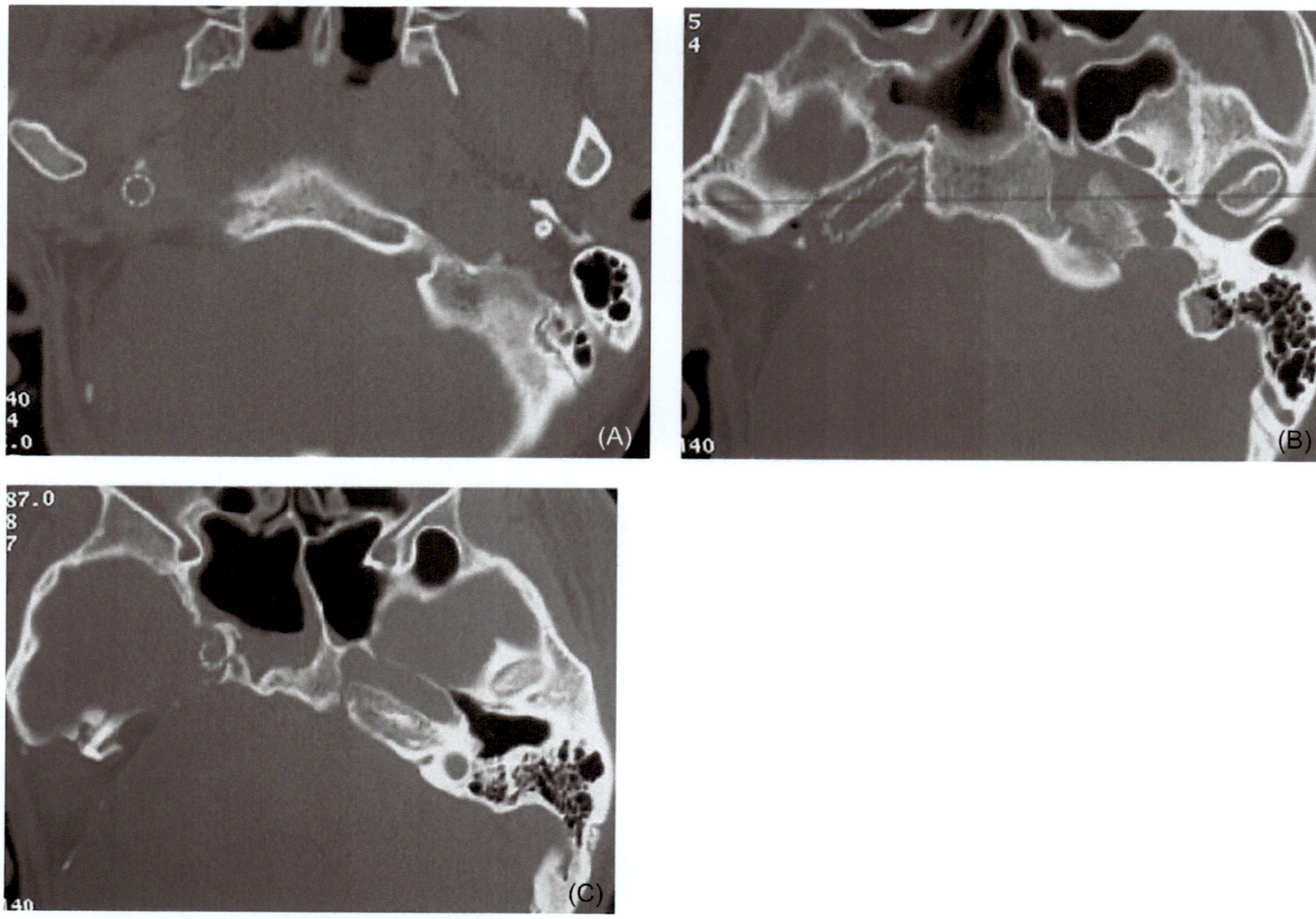

Fig. 16.78 (A–C) Class C4 tympanojugular paraganglioma tumor. Stenting of the internal carotid artery. Same case as shown in Fig. 16.85A–F. High-resolution CT showing the stent inside the C1 (A), C2 (B), and C3 (C) segments of the internal carotid artery.

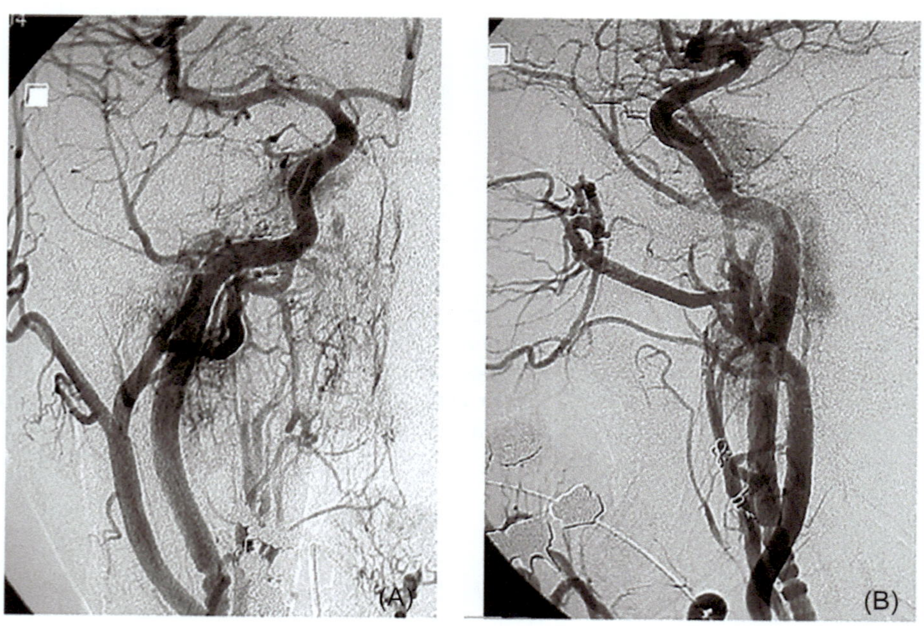

Fig. 16.79 (A, B) Class C4 tympanojugular paraganglioma. Stenting of the internal carotid artery. Same case as shown in Fig. 16.85A–F. Common carotid injection, in anteroposterior (A) and lateral (B) views after embolization, showing marked reduction of the tumoral blush.

Surgical Management Modalities of the ICA

Modalities of surgical management of the ICA include:

- Skeletonization
- Displacement
- Subperiosteal/subadventitial dissection
- Dissection and resection after PBO
- Subadventitial dissection after reinforcement with stent

SKELETONIZATION

This is done in tumors reaching but not adhering to the artery. The most common lesions are represented by petrous bone cholesteatomas and type C1 tympanojugular paragangliomas (Fig. 16.80). The artery can be exposed in certain approaches to provide proximal control, e.g., the infratemporal fossa approach or the modified transcochlear approach type A.

In the middle fossa transpetrous approach, the artery is one of the anatomical boundaries that are skeletonized to avoid injury while drilling the petrous apex. Skeletonization carries little risk in experienced hands.

An exhaustive knowledge of anatomy is mandatory; a large diamond burr parallel to the course of the artery is used to remove the last shell of bone covering the artery.

DISPLACEMENT

Displacement is used to gain access, e.g., during an infratemporal fossa type B approach to the petrous apex. Displacement should be done gently, and complete liberation of the artery is needed first (Fig. 16.81).

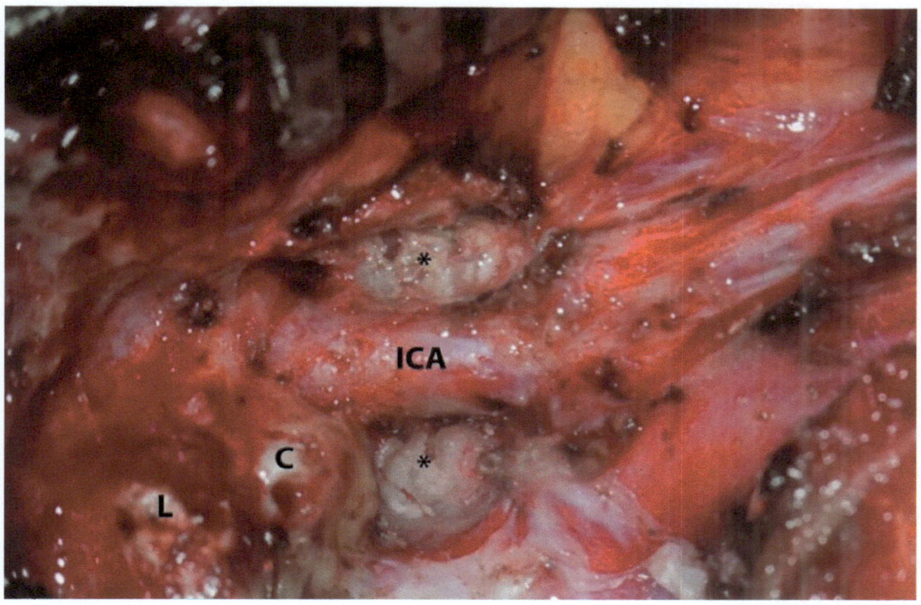

Fig. 16.80 Skeletonization of the internal carotid artery (ICA) in a case of C1 tympanojugular paraganglioma. *C*, cochlea; *L*, labyrinth; *, infracarotid space.

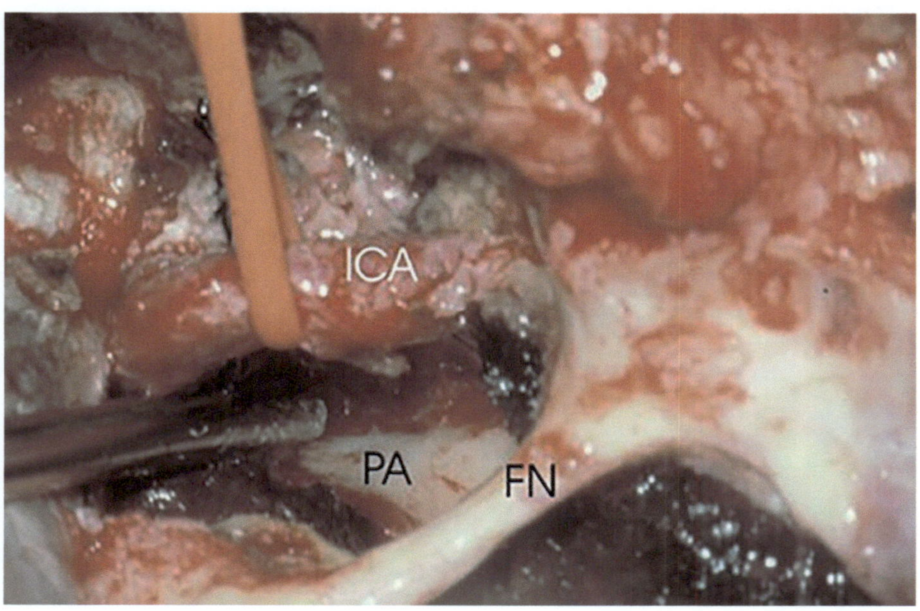

Fig. 16.81 A case of right clival chordoma. The vertical internal carotid artery (ICA) is gently displaced to allow proper control of the petrous apex (PA) lying medial to the artery.

SUBPERIOSTEAL/SUBADVENTITIAL DISSECTION

Subperiosteal/subadventitial dissection is accomplished for tumors that involve the ICA to a greater extent, such as C2 tympanojugular paragangliomas and meningiomas (Fig. 16.82A–F). In general, dissection of the tumor from the artery is relatively easier and safer in the vertical intrapetrous segment, which is thicker and more accessible than the horizontal intrapetrous segment.

A plane of cleavage between the tumor and the artery should be found first. In most cases, the tumor is attached to the periosteum surrounding the artery. Dissection is better

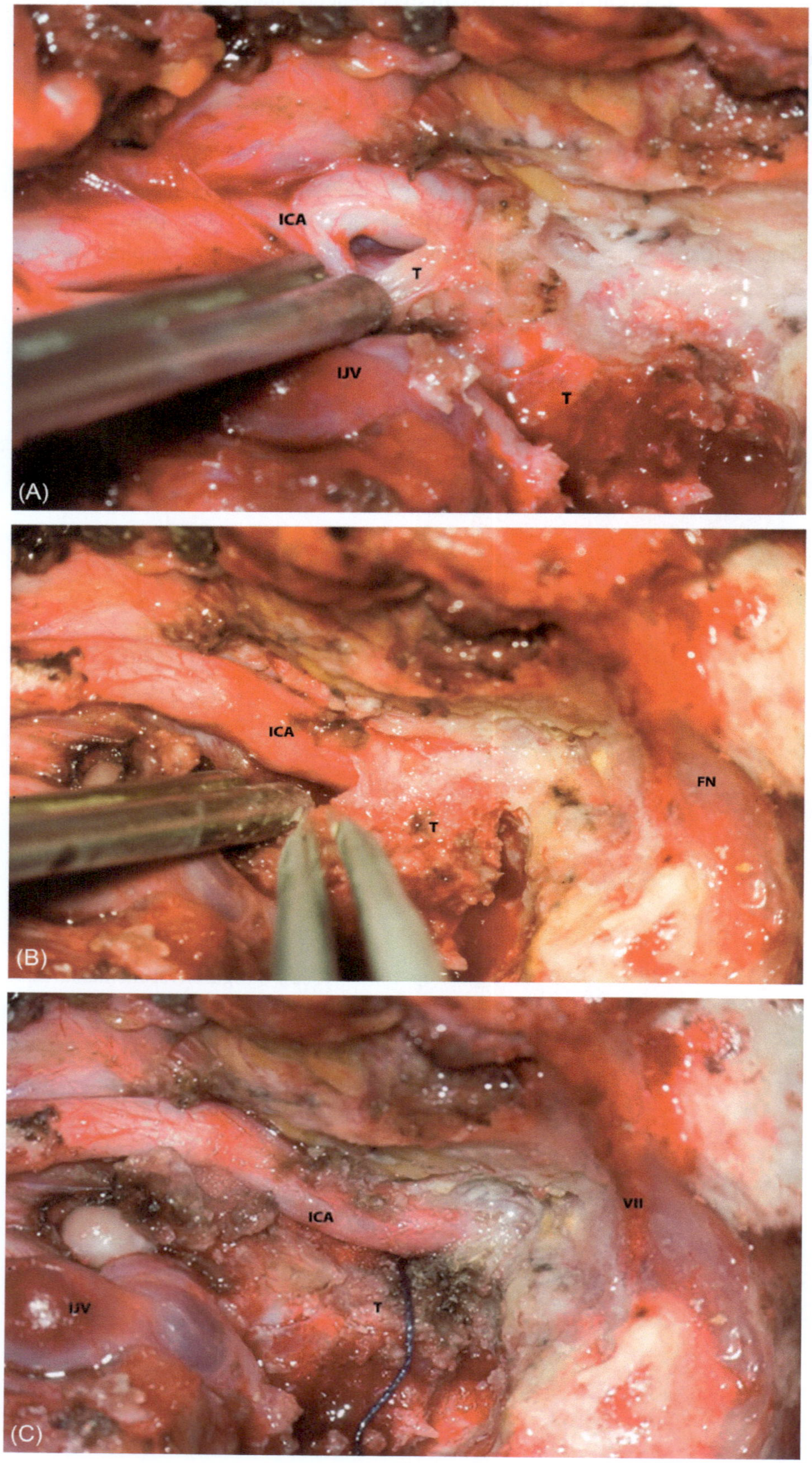

Fig. 16.82 (A–F). Subadventitial dissection and mobilization of the internal carotid artery (ICA) in a case of C2 tympanojugular paraganglioma. Note the coils used for the preoperative embolization (D). The ICA is completely free of tumor at the end of the procedure (F). *T*, tumor; *PA*, petrous apex; *FN*, facial nerve; *IJV*, internal jugular vein.

(Continued)

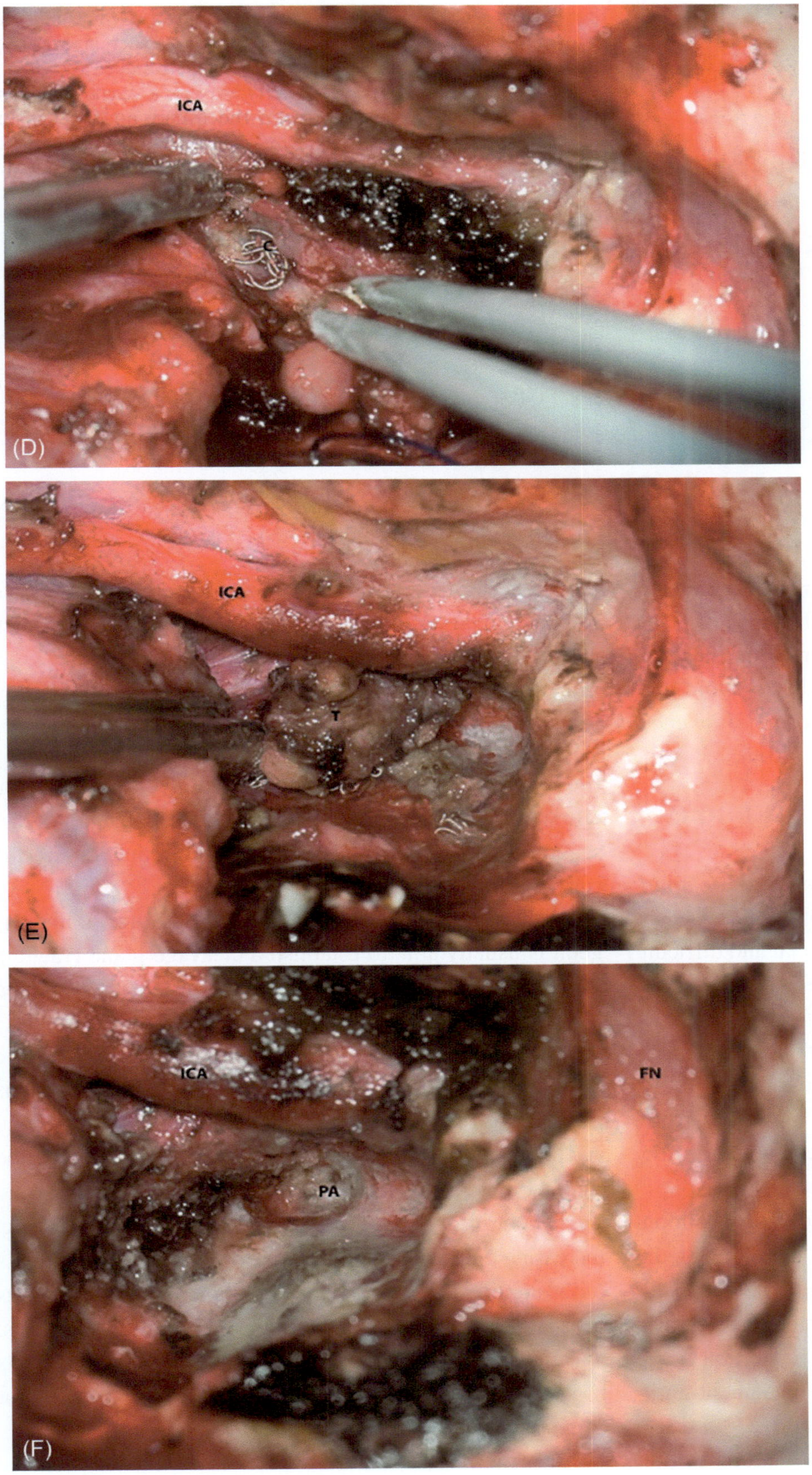

Fig. 16.82, cont'd

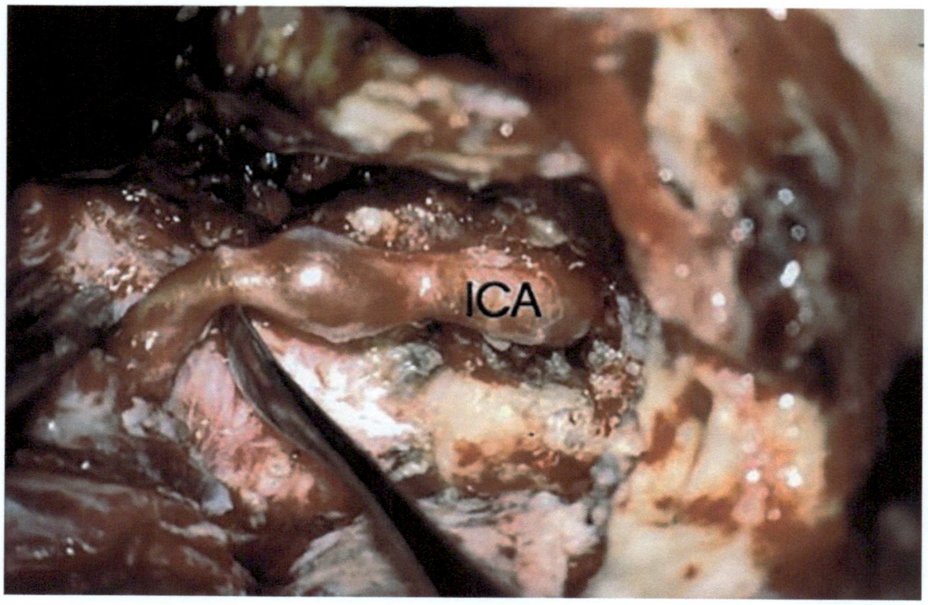

Fig. 16.83 A case of left tympanojugular paraganglioma in our early experience. Subadventitial dissection was performed; after tumor removal the artery was so weakened with the formation of a pseudoaneurism. Although the patient had no relevant complications postoperatively, such excessive manipulation is better avoided, and permanent balloon occlusion or stenting is preferably tried preoperatively.

started in an area immediately free of tumor. Aggressive tumors may, however, extend even to the adventitia of the artery and subadventitial dissection may be needed. This should be done very carefully in order to avoid any tear to the arterial wall, which can become weakened (Fig. 16.83), with the risk of subsequent blowout.

DISSECTION AND RESECTION AFTER PBO OF THE ICA

In patients in whom the tumor is adjacent to the carotid artery, the preoperative examination to determine whether the carotid artery has to be resected is a crucial and difficult task that can lead to false-negative and false-positive results. Carotid artery invasion is difficult to assess even at operation: often the tumor obscures a portion of the carotid artery or completely surrounds it; malignant tumors, by their infiltrative nature, do not allow for their separation from the ICA; manipulating vascular tumors can increase the difficulty, as bleeding impairs visualization.

When the carotid artery has been controlled by balloon occlusion, these problems are lessened but not eliminated.

Dissection of the occluded ICA is started at the cervical level; after isolation, the artery is ligated immediately after the proximal balloon; then dissection and separation from the tumor proceed from the vertical petrosal segment until the junction between horizontal petrosal and lacerum segments; finally, thanks to the presence of a balloon in the cavernous segment, the petrous portion of the artery is resected, possibly with a portion of tumor adherent to it, and sent for pathological examination (Fig. 16.84). Indeed, only after serial sections of the suspected arterial segment are made can a determination be made about whether there has been vascular invasion.

Despite the lack of carotid wall infiltration, removal of these tumors and of the adventitia can significantly weaken the carotid wall and lead to blow out; therefore, PBO of the ICA should be strongly considered in skull base tumors with massive radiological involvement of the ICA.

SUBADVENTITIAL DISSECTION AFTER REINFORCEMENT WITH STENTING

From a surgical point of view, preoperative stent insertion allows the skull base surgeon to perform subadventitial dissection of the ICA with a significant reduction of the surgical risk (Figs.16.85–16.94).

In the presence of an intraluminal stent, in fact, the surgeon is usually able to establish a cleavage plane reaching the external surface of the stent, so removing all the involved portions of the arterial wall.

At the same time, the presence of the metallic net of the stent represents protection against accidental rupture; this is particularly true when working at the level of the carotid genu and the horizontal segment of the petrous ICA. In this area, the surgical room and the mobility of the artery are reduced, and direct control of the medial wall is particularly demanding, increasing the difficulty and the risk of surgery. The thickness of the struts of the stent, which determines its rigidity and its resistance to crushing, can offer different surgical sensations: although surgical dissection in the presence of thicker stents has seemed more comfortable, it has been possible without surgical problems even in the presence of softer stents.

Dissection usually starts at the cervical level, away from the tumor, where it is easier to find the correct cleavage plane and proceed distally; the anteromedial wall of the artery is considered the most difficult to manipulate because direct visualization requires bony decompression and anterior displacement of the intrapetrous segment of the ICA. The unsolved surgical problem remains the medial wall of the ICA at the level of the anterior foramen lacerum, until now unreachable with the available surgical approaches.

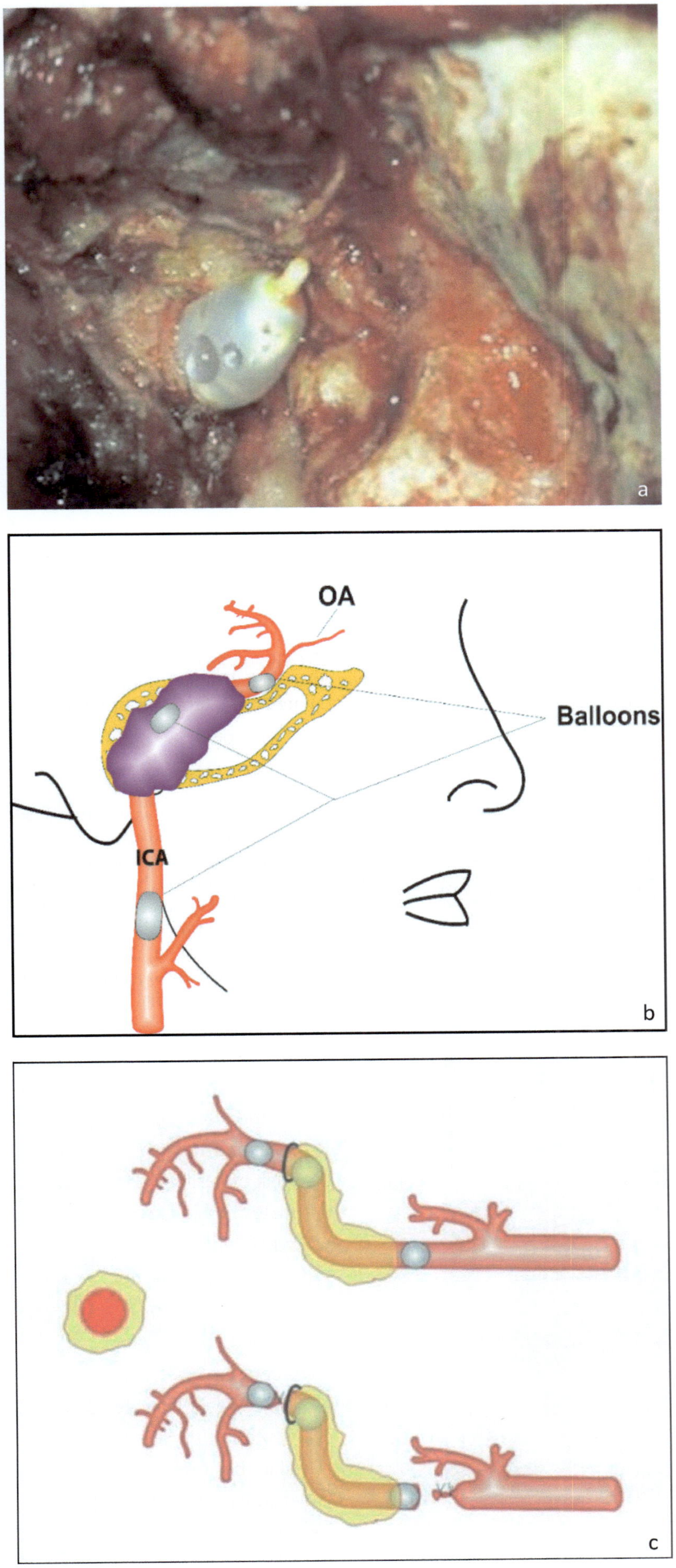

Fig. 16.84 (A) Intraoperative view of the balloon used to permanently occlude the internal carotid artery. (B, C) Schematic drawings showing the permanent balloon occlusion of the internal carotid artery. *MCA*, middle cerebral artery; *BA*, basilar artery; *ACA*, anterior cerebral artery; *OA*, ophthalmic artery.

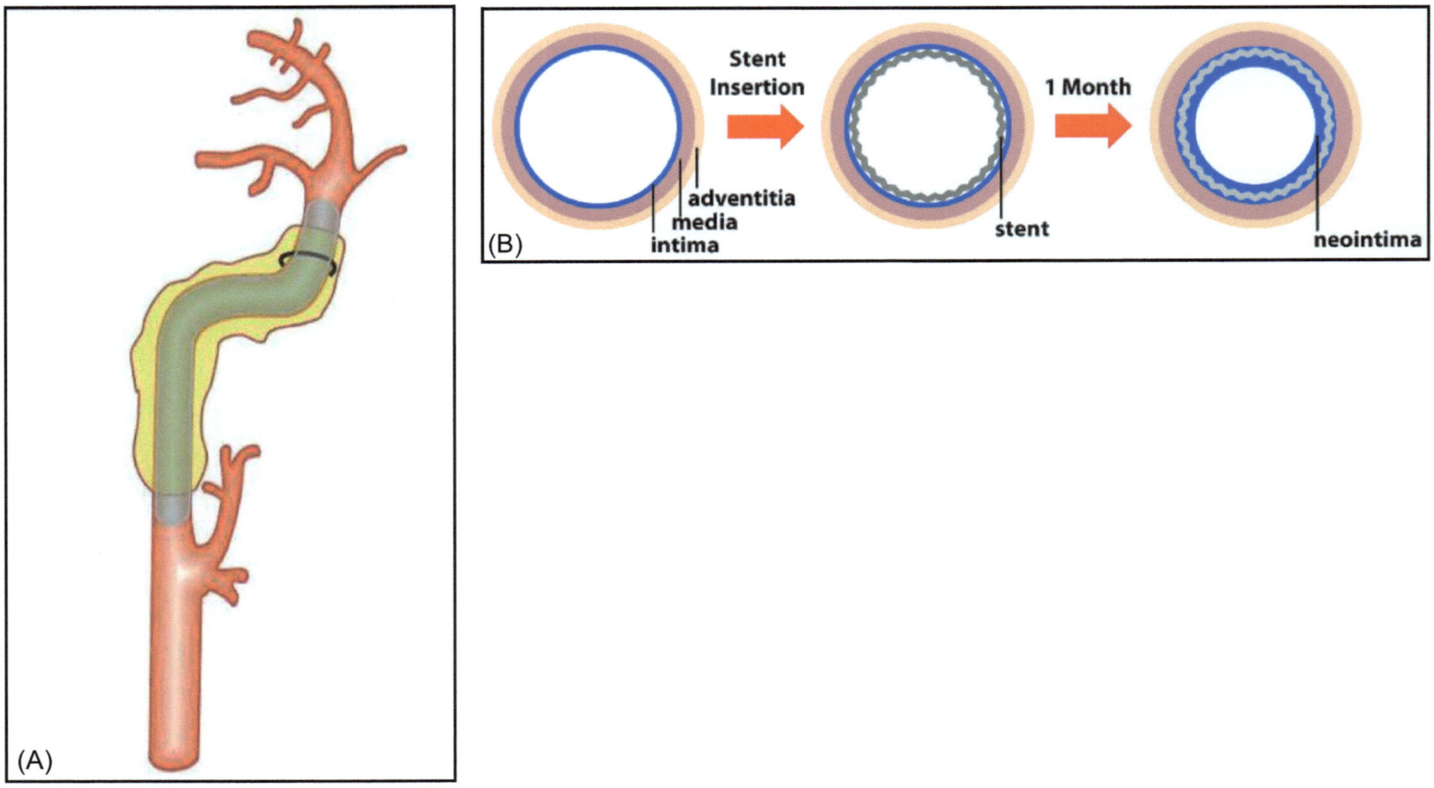

Fig. 16.85 (A, B) Internal carotid artery stenting. One to three overlapping stents are placed in the carotid lumen to prevent carotid blowout during tumor dissection. To reduce the possibility of injuring the ICA at the stent-tumor border, at least 10 mm of tumor-free vessel wall must be reinforced with the stent, both proximally and distally (A). The result of intraarterial stenting is the formation of a stabilized neo-intima 4–6 weeks after placement (B).

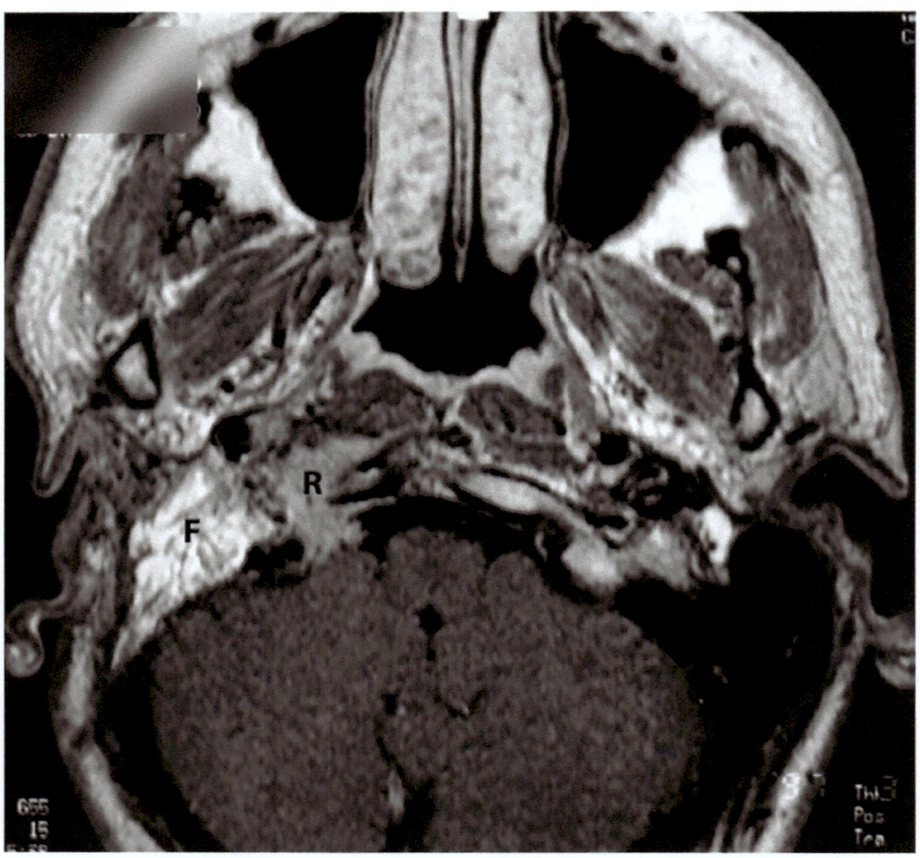

Fig. 16.86 This patient underwent a first-stage partial resection of a right-sided class C3 tympanojugular paraganglioma due to inadequate contralateral carotid compensation. Note the residual tumor (R) on the T1W + Gd-enhanced MRI of the brain. After the introduction of stenting of the ICA into surgical practice in 2003, the case was evaluated for surgical revision and complete tumor removal. *F*, abdominal fat.

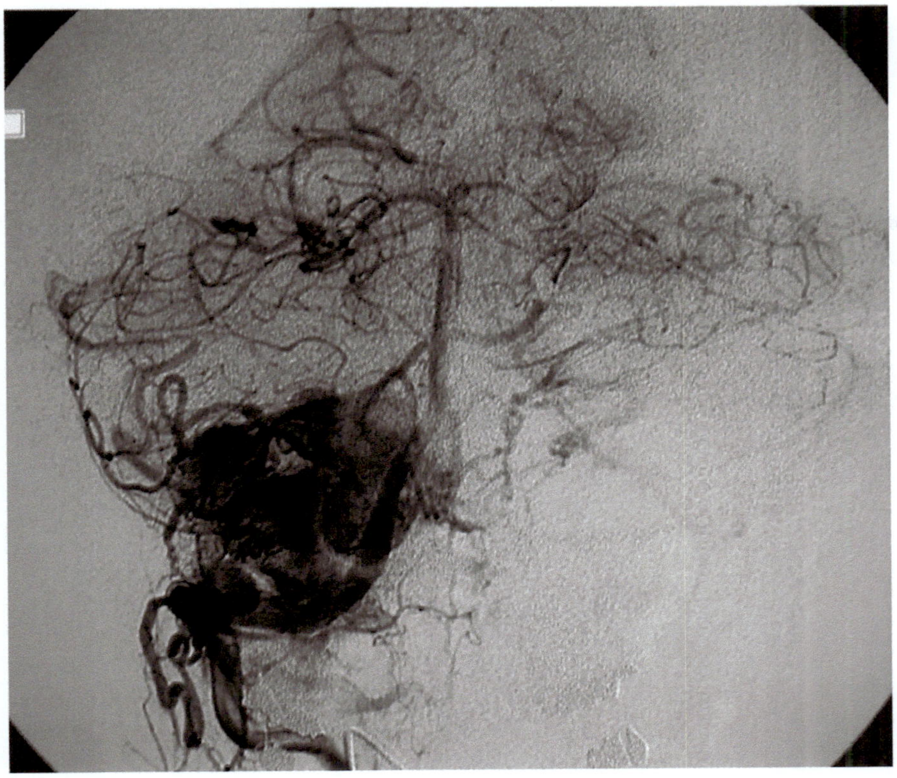

Fig. 16.87 Angiography showed high tumor vascularization and infiltration of the horizontal petrous internal carotid artery.

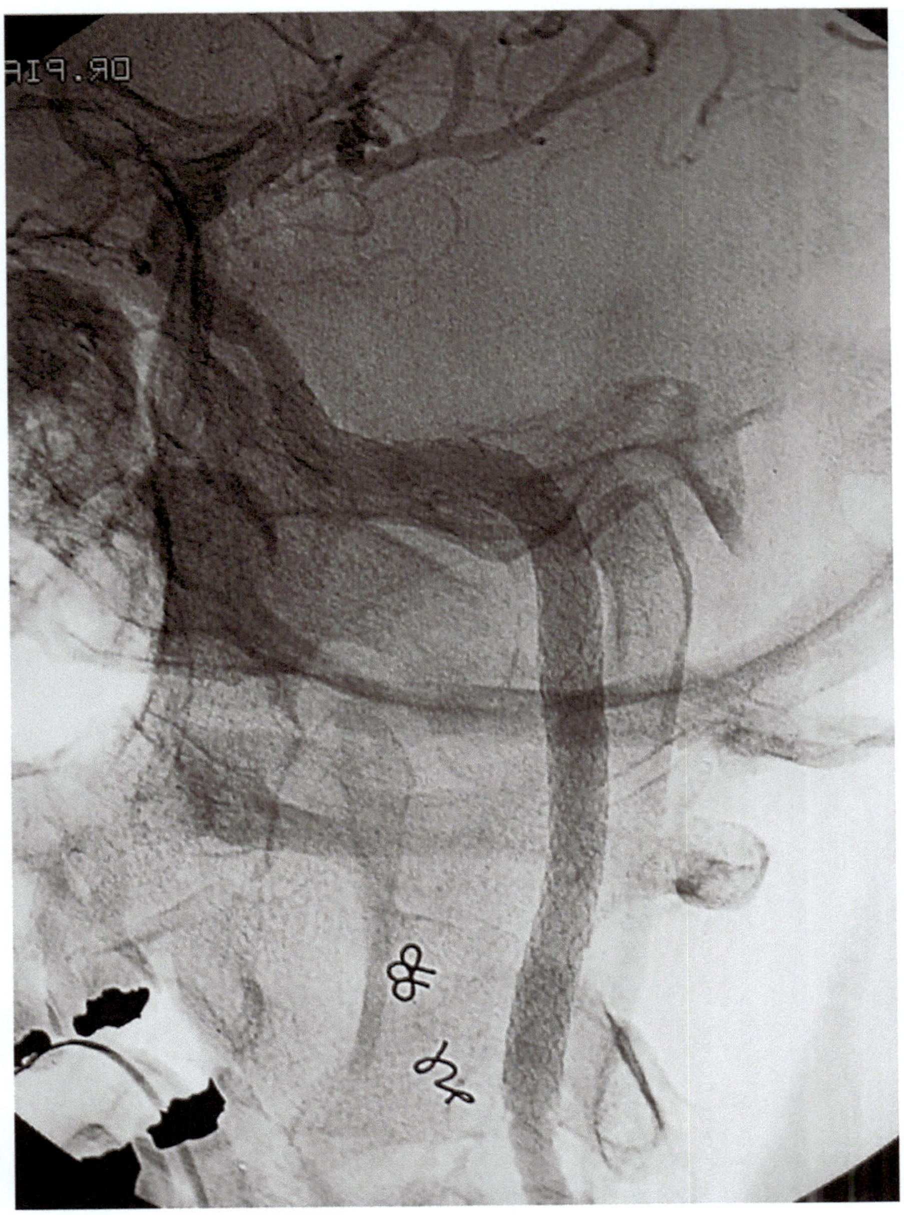

Fig. 16.88 An intraarterial stent was inserted from the cervical internal carotid artery to the anterior foramen lacerum segment.

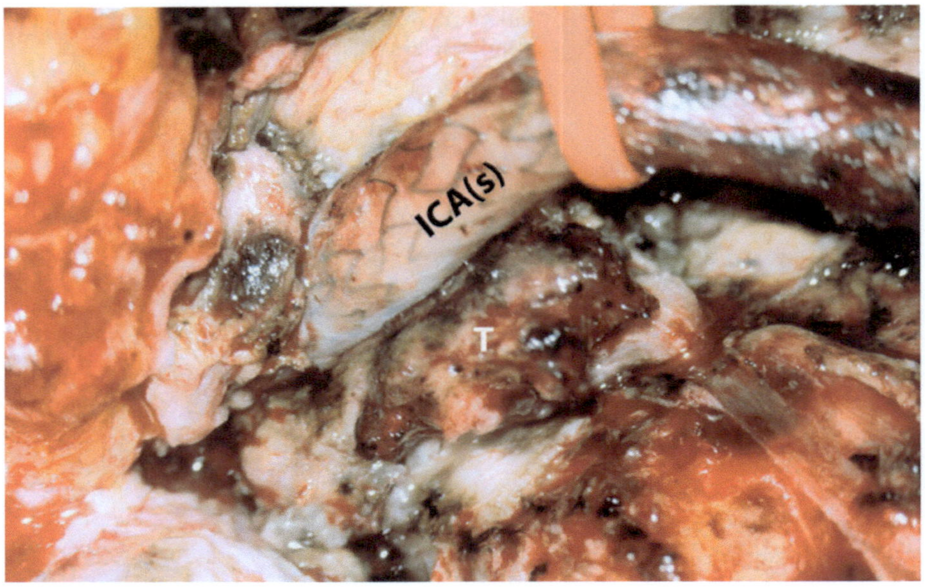

Fig. 16.89 The stented internal carotid artery (ICA (s)) is displaced. Note the residual tumor (T) medial to the artery.

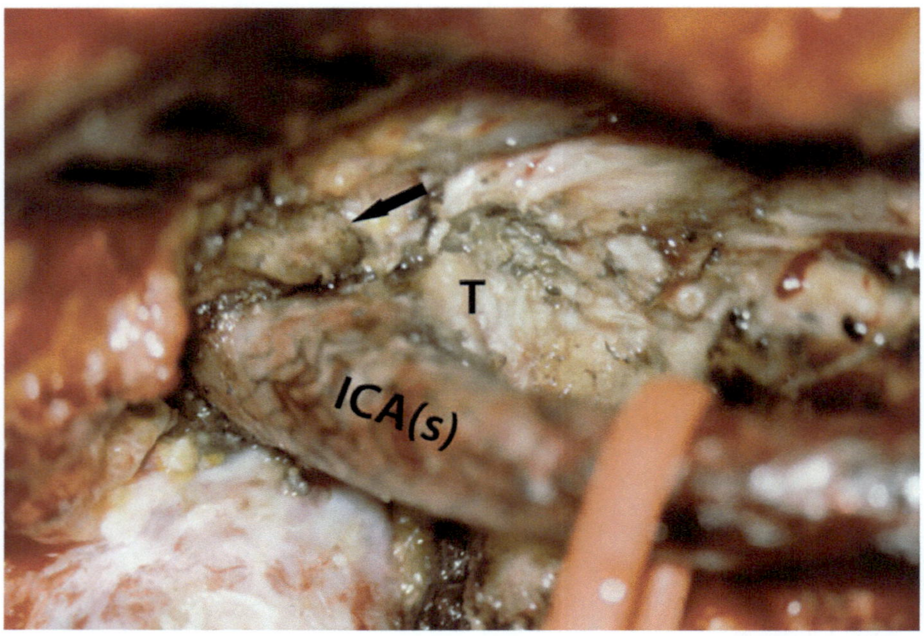

Fig. 16.90 The tumor (T) involved the horizontal portion of the petrous ICA *(arrow)*. *ICA(s)*, stented internal carotid artery.

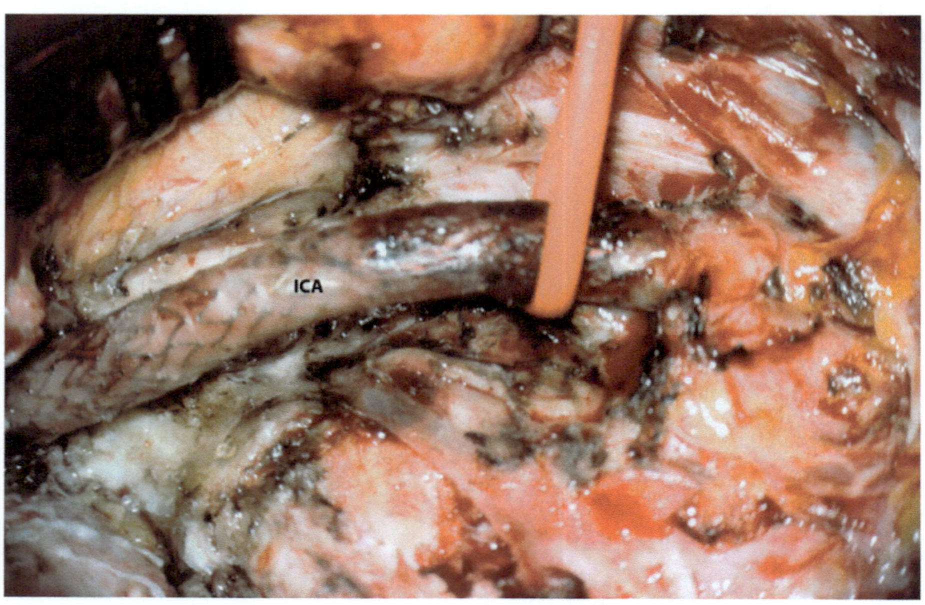

Fig. 16.91 33 Surgical view at the end of the procedure. *ICA*, internal carotid artery.

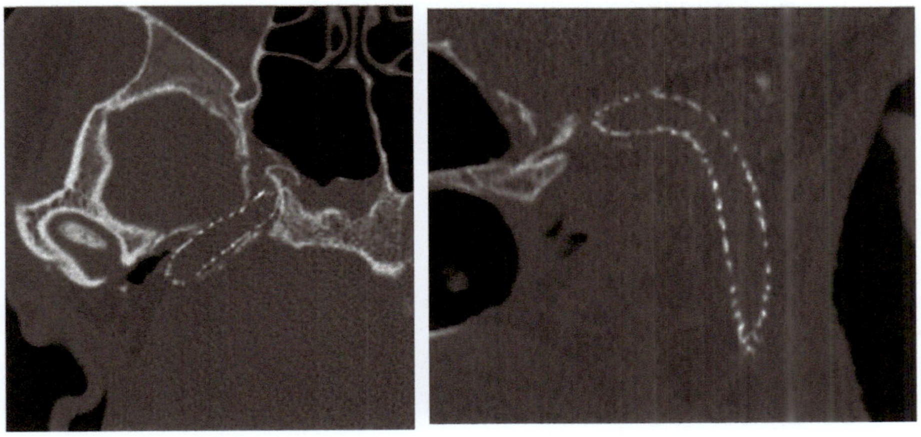

Fig. 16.92 Postoperative HRCT scans showing the intraarterial carotid stent from the cervical internal carotid artery to the anterior foramen lacerum.

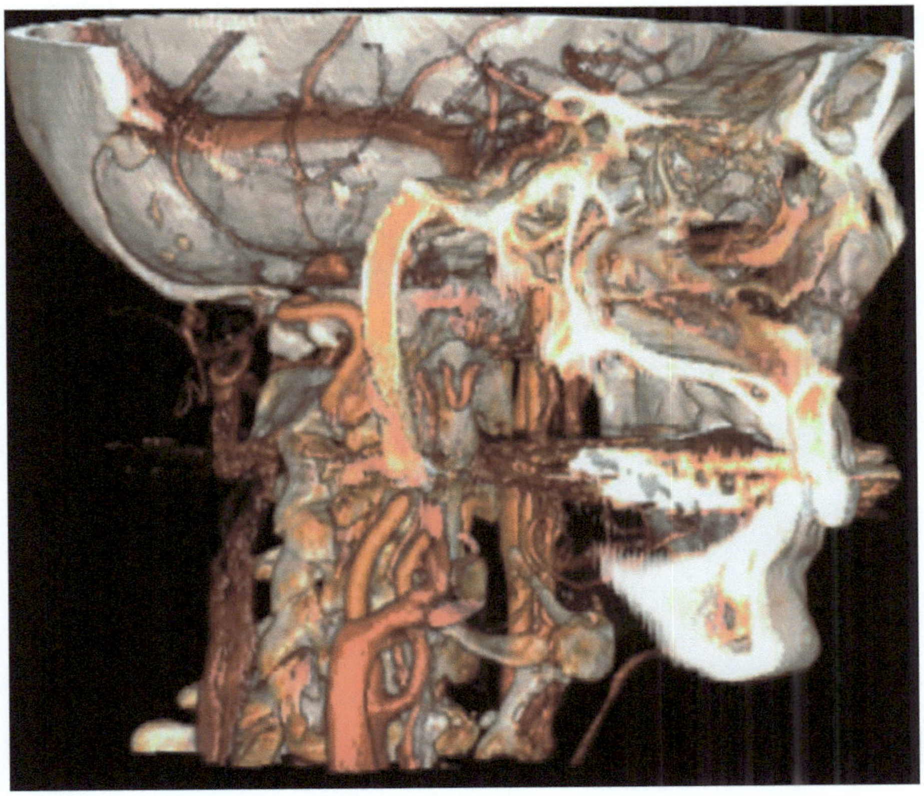

Fig. 16.93 Color 3D angio CT showing the stent and the bone removal.

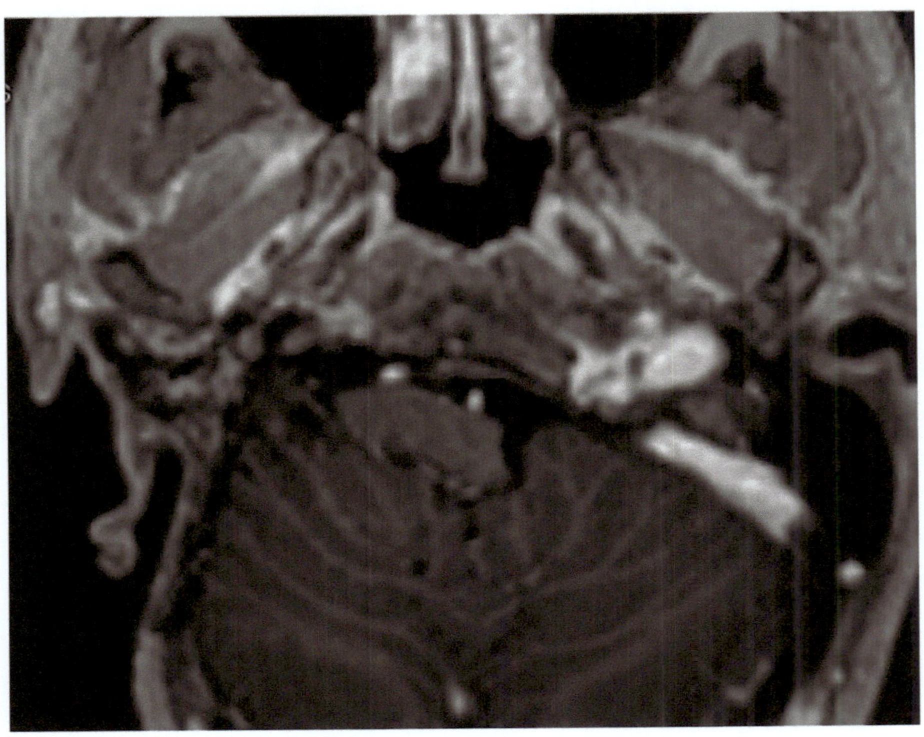

Fig. 16.94 Brain MRI shows no tumor recurrence after 20 years.

INDEX

Note: Page numbers followed by *f* indicate figures and *t* indicate tables.

F

G

H

I

J

N

O

P

R

S

T